To Julie

GW01459451

A Biographical Dictionary of Civil Engineers in Great Britain and Ireland

Volume 1: 1500–1830

Menai Bridge, opened 1826

A Biographical Dictionary of Civil Engineers in Great Britain and Ireland

Volume 1: 1500–1830

Edited by

A. W. Skempton
M. M. Chrimes
R. C. Cox
P. S. M. Cross-Rudkin
R. W. Rennison
E. C. Ruddock

Thomas Telford

THE INSTITUTION OF
CIVIL ENGINEERS

Published by Thomas Telford Publishing on behalf of The Institution of Civil Engineers, Thomas Telford Ltd, 1 Heron Quay, London E14 4JD. URL: http://www.thomastelford.com

Distributors for Thomas Telford books are
USA: ASCE Press, 1801 Alexander Bell Drive, Reston, VA 20191-4400
Japan: Maruzen Co. Ltd, Book Department, 3–10 Nihonbashi 2-chome, Chuo-ku, Tokyo 103
Australia: DA Books and Journals, 648 Whitehorse Road, Mitcham 3132, Victoria

First published 2002

A catalogue record for this book is available from the British Library

ISBN: 0 7277 2939 X

© ICE and Thomas Telford Limited 2002

All rights, including translation, reserved. Except as permitted by the Copyright, Designs and Patents Act 1988, no part of this publication may be reproduced, stored in a retrieval system or transmitted in any form or by any means, electronic, mechanical, photocopying or otherwise, without the prior written permission of the Publishing Director, Thomas Telford Publishing, Thomas Telford Ltd, 1 Heron Quay, London E14 4JD

This book is published on the understanding that the authors are solely responsible for the statements made and opinions expressed in it and that its publication does not necessarily imply that such statements and/or opinions are or reflect the views or opinions of the publishers. While every effort has been made to ensure that the statements made and the opinions expressed in this publication provide a safe and accurate guide, no liability or responsibility can be accepted in this respect by the authors or publishers

ISBN 0-7277-2939-X

9 780727 729392

Typeset by Helius, Brighton and Rochester
Printed and bound in Great Britain by MPG Books, Bodmin

Contents

Dedication

Professor Sir Alec Skempton FRS (1914–2001)

The death of Professor Skempton—'Skem' to his closer friends—occurred on 9 August 2001, tragically before the publication of this book, on which he latterly spent so much of his time.

Few authors have done as much to further the knowledge and appreciation of the history of civil engineering than has Professor Skempton and none has brought to the subject such scholarship as he has done in his many books and papers on the subject. As a research engineer in reinforced concrete, a pioneer in soil mechanics, an academic and a consultant, he was actively involved in the civil engineering profession for some sixty-five years and his publications on the history of the subject extend over a period of fifty-five years.

Skem was firmly of the opinion that history is about people and that is the emphasis he placed on his published material. His works have always been informative and guaranteed to stimulate further research. Many were derived from the published engineering reports and maps which he collected avidly. His writings made constant use of primary sources.

By many of his friends Skem will be remembered for his membership and Presidency of the Newcomen Society, by others for his chairmanship of the Panel for Historical Engineering Works of the Institution of Civil Engineers. To both of these bodies he brought his considerable scholarship and personality, the two combining to lead to a marked increase in the professionalism of research and writing. His Fellowship of the Royal Society was well deserved and the award of his knighthood, belatedly in January 2000, brought much satisfaction to all his colleagues and friends.

By the members of the Editorial Board responsible for this volume he will be sorely missed; his knowledge of the history of civil engineering was always a source of wonder and admiration and his guidance and encouragement were much appreciated. We owe a great debt to him.

M. M. CHRIMES
R. C. COX
P. S. M. CROSS-RUDKIN
R. W. RENNISON
E. C. RUDDOCK

Preface

In 1995 the Archives Panel of the Institution of Civil Engineers began to investigate the practicability of compiling a biographical dictionary of civil engineers. Such a work was seen as answering a need within the profession and filling a gap in the published history of Great Britain and Ireland. Biographical works dealing with related professions are well established with Sir Howard Colvin's *Biographical Dictionary of British Architects, 1600 to 1840* (3rd edn., 1995) and Sarah Bendall's *Dictionary of Land Surveyors of Great Britain and Ireland 1530–1850* (1997) developed from the earlier work edited by Peter Eden (1975–1979).

The Council of the Institution agreed to support the enterprise. An Editorial Board, reporting to the Archives Panel, was appointed in May 1996 consisting of:

Chairman: Professor Sir Alec Skempton, Imperial College
Secretary: Mr. M. M. Chrimes, Head Librarian, Institution of Civil Engineers
Members: Dr. Ron Cox, Trinity College, Dublin; Mr. P. S. M. Cross-Rudkin, Coventry University; Dr. R. W. Rennison, Newcastle; Mr. Ted Ruddock, MBE, Edinburgh

Acknowledgement is made to the contributors whose names are appended to the articles they have written. Julie Gorman gave valuable service as copy editor. Susan Hots, Robert Thomas and Carol Morgan assisted in archival research. Valerie Lawless and Paul Parkes provided administrative support.

From the outset it was realised that more than one volume would be necessary, owing to the large number of entries and the desirability of describing works before the mid-eighteenth century in some detail, since such information, with few exceptions, is not readily available elsewhere. The present volume contains biographies of men engaged in civil engineering in Great Britain and Ireland between 1500 and 1830. Engineers of British birth working overseas are included.

The years around 1830 provide a convenient division between the classic period represented by Smeaton, Brindley, Jessop, Rennie and Telford, and a new era: the railway age, dominated initially by Robert Stephenson, I. K. Brunel and Joseph Locke. As for the starting date, civil engineering works obviously were carried out long before the term 'civil engineer' was introduced and the profession began to be recognised in the 1760s; on fen drainage schemes of the seventeenth century, for example, and on Dover harbour in the reigns of Henry VIII and Elizabeth I. The biographies therefore include anyone, whether amateur, tradesman or professional, who played a significant role in civil engineering from the early years of the sixteenth century. Biographies of the master masons, carpenters and designers of cathedrals in earlier times can be found in John Harvey's dictionary of *English Mediaeval Architects* (2nd edn., 1987).

Works covered in this volume of the Dictionary are principally: harbours and lighthouses, docks, water supply to towns, fen drainage and land reclamation, river navigations, canals, roads, bridges and early railways, and associated research. A chronological table listing major works and their engineers is set out in Appendix II, preceding the Index of Places.

There were, of course, individuals active in the period who would not now be considered civil engineers but who were connected with the development of the profession. There are, for example, architects responsible for structural innovations, military engineers involved with major construction projects, and land surveyors particularly associated with civil engineering schemes. A representative selection of such persons has naturally been included. There is, too, a problem of changing definitions since, when the term 'civil engineer' was introduced in the mid-eighteenth century, it was taken to embrace all branches of engineering outside the military sphere. This is exemplified by the career of John Smeaton, who devoted much attention to the design of millwork and steam engines in addition to his great public works, and several distinguished early members of the Institution would in later terminology be classed as mechanical or mining engineers.* For this reason the Dictionary also includes brief articles on engineers in those disciplines.

Research is ongoing and the Editors would be pleased to receive additional information or material on new subjects. This can be forwarded to the Librarian at ICE for inclusion in the Newsletter of the Panel for Historical Engineering Works, available from the library and on the ICE website (http://www.ice.org.uk).

* The earliest use of the term 'mechanical engineer' so far discovered dates from 1838. The Institution of Mechanical Engineers was founded in 1847, and the North of England Institute of Mining Engineers in 1852.

Notes

Dates
In England until 1751 in the majority of manuscript and printed documents, and for all official purposes, the Old Style (O.S.) calendar was used, in which the new year begins on 25 March. By an Act of Parliament the day following 31 December 1751 became 1 January 1752 (not 1751). To avoid confusion, earlier dates are here given in the New Style (N.S.). A further change was made: 2 September 1752 was followed by 14 September, but this correction (the 'lost' eleven days) is not applied retrospectively. Thus:

- 20 May 1660 in the original is given as 20 May 1660
- 12 February 1660 in the original is given as 12 February 1661.

In a quotation the N.S. date is added in square brackets, e.g. in Mepal church is a memorial to 'Samuel Fortrey Esq. of Byal Fen in this parish ob. 10 February 1688 [1689 N.S.] aet.38'.

Place names
Names of the traditional British counties have been retained, as they represent a continuity which late twentieth century local government changes were only to confuse. Except in quotations and titles of reports, etc., place names are given in their modern form, although this is not always possible for places in the Indian sub-continent. Following the American revolution British North America comprised four colonies: Nova Scotia, Newfoundland, Lower Canada (now Quebec) and Upper Canada (now Ontario). The Dominion of Canada was not created until 1867, and thus in the present volume the colonial names are used.

Conversion of units
Weights and measures are given in their original (Imperial system) units. Metric equivalents are:

Length
1 inch (in.) = 25.4 millimetres
1 foot (ft.) = 12 in. = 0.305 metre
1 yard (yd.) = 3 ft. = 0.914 metre
1 mile (1760 yd.) = 1.609 kilometres

Area
1 square inch (sq. in.) = 645 square millimetres
1 square foot (sq. ft.) = 0.0930 square metre
1 acre (4840 sq. yd.) = 0.405 hectare
1 square mile (640 acres) = 259 hectares

Volume
1 gallon = 4.55 litres
1 cu. ft. = 0.0283 cubic metre
1 cu. yd. = 0.765 cubic metre

Mass
1 pound (lb.) = 4.45 N = 0.454 kilogram
1 ton (2240 lb.) = 9.97 N = 1.016 tonnes

Pressure
1 pound per sq. in. = 6.90 kN/m^2 = 0.070 kg/m^2
1 ton per sq. ft. = 107 kN/m^2 = 1.09 kg/cm^2

Power
1 horsepower (hp) = 0.746 kilowatt

Currency
Up to 1971 the English pound sterling was divided into shillings and pence:

1 shilling (s) = 12 pence (d)
1 pound (£) = 20 shillings (s)
1 guinea = 21 shillings = £1.05

Standard abbreviations

Bendall	Bendall, S., *Dictionary of Land Surveyors of Great Britain and Ireland, 1530–1850*
BL	British Library
Colvin (1/2/3)	Colvin, H. M., *A Biographical Dictionary of British Architects 1600–1840* (edition)
DBB	*Dictionary of Business Biography*
DNB	*Dictionary of National Biography*
DSB	*Dictionary of Scientific Biography*
ICE	Institution of Civil Engineers
Kings Works (1)	Colvin, H. M., and others, *History of the Kings Works* (volume)
Min Proc ICE	*Minutes of Proceedings of the Institution of Civil Engineers*
NLS	National Library of Scotland
NMRS	National Monuments Record for Scotland
PRO	Public Records Office
RO	Records Office
Rep PI	Repository of Patent Inventions
R S Cat	Royal Society (London) Catalogue of Scientific Papers
Singer (1)	Singer, C., and others, *History of Technology* (volume)
Skempton	Skempton, A. W., *British Civil Engineering 1640–1840, a Bibliography of Contemporary Printed Reports, Plans and Books*
SRO	Scottish Records Office, now National Archives of Scotland (NAS)

List of illustrations*

*Unless otherwise credited, from the collections of the Institution of Civil Engineers.

The practice of civil engineering 1500–1830

From medieval times to the mid-seventeenth century in England the term 'engineer' applied to military engineering. Thereafter, it was increasingly also used in a civil (non-military) context. The term 'civil engineer' appears in 1763, in a London directory,[1] and the Society of Civil Engineers was founded in 1771.

Later known as the Smeatonian Society, it still exists essentially in its original form as a club for senior members of the profession and some of their friends (48 engineers and 12 gentlemen). The need for a more broadly based organisation was identified when, in 1818, eight young engineers led by Henry Robinson Palmer founded the Institution of Civil Engineers, primarily as a learned society for the presentation of papers and discussion. Thomas Telford, the head of the profession at that time, accepted a request to become President in 1820. Under his aegis a Royal Charter was granted in 1828 to the Institution, defined (in the words of Thomas Tredgold) as 'A Society for the general advancement of Mechanical Science and more particularly for promoting the acquisition of that species of knowledge which constitutes the profession of a Civil Engineer, being the art of directing the Great Sources of Power in Nature for the use and convenience of man'.

While the Institution continued to play its part as a learned society, it very soon developed also as a professional association. Membership was limited to those 'practising in the profession of a civil engineer', and the proposal form for admission required the signature of at least three Members who could certify, from personal knowledge of the candidate, his qualification and suitability for membership. The class of Associate was open to those whose pursuits were related to engineering but were not engineers by profession (surveyors, instrument makers, millwrights, etc.). Honorary Members were eminent in science; they had to be proposed by at least five Members. In 1823 Members paid an annual subscription of 3 guineas and Associates 2.5 guineas. New members paid an admission fee of 1 guinea. By 1824 the total membership stood at 87, and by 1829 at 156. The designation MInstCE was authorised in 1831.

Minute Books of the Smeatonian Society survive from 1771, as do those of the Institution from 1818. The Society was reorganised on a more formal basis in 1793. Its principal contribution to learning was the publication in 1812 of a complete set of Smeaton's reports. The early papers and discussions at the Institution are kept in manuscript as 'Original Communications' and 'Minutes of Conversations'. The regular publication of papers began with the *Transactions* in 1836.

Before that date publications on civil engineering and related topics, besides a relatively small number of books, consisted mainly of printed reports. These were circulated to the clients and to fellow engineers[2] and are a distinctive feature of civil engineering practice in the eighteenth and early nineteenth centuries.

Table 1. The five periods of civil engineering covered by this volume

Period	Major works started		Engineers named
	Total	Per decade	
1500–1599	11	1	12
1600–1689	25	3	22
1690–1759	59	8	52
1760–1789	85	28	60
1790–1830	155	39	87

In viewing the development of civil engineering over the three centuries covered in this volume it is possible to recognise five periods, each with its own characteristics. These periods are listed in Table 1 together with certain statistics extracted from the chronological table given in Appendix II.

Sixteenth century

During this period there was evidently not enough work to provide employment for 'professional' British engineers; the sole known exception is John Trew, engineer of the Exeter Canal, and his background was in mining. Otherwise the usual procedure appears to have been to employ engineers from The Netherlands or Germany on specific projects. For instance, Matthew Hake for the sluice on the Witham at Boston (1500), Joas Johnson for piers at Yarmouth Harbour (1566) and Humphrey Bradley from Brabant on the first investigations for draining the Great Level of the Fens (1589), although the key surveying work here was by two English land surveyors, John Hexham and Ralph Agas. However, when it became necessary to extend the short pier already built on the south side of Dover Harbour, and the town petitioned Henry VIII for a grant, it was a local cleric, John Thompson, who took the petition to London along with his plans for the new pier. He received a favourable response. Commissioners were sent to Dover, they approved the plans and work began in 1535. Thompson was appointed Surveyor of the Works with four 'Overseers', chosen from seamen at Dover, a paymaster and comptroller and a workforce of masons, labourers and men transporting rocks from the shore by sea to form the subaqueous foundations of the pier.

This is not the only example of a cleric being actively involved in civil engineering in the late mediaeval period. John Clerk, Thompson's predecessor at Dover, seems to have had responsibility for building the first pier there, completed *c.* 1495, and was brought in to assist over arrangements for building the sluice at Boston in 1500; John Morton, Bishop of Ely (1479–1486), is credited with planning and supervising a new 12 mile channel for the river Nene in the Fens (known as Morton's Leam), and the Reverend Alexander Galloway probably designed and certainly directed construction of the Bridge of Dee at Aberdeen, built 1520–1527 by the Scottish master mason Thomas Franche.

Thompson's Dover pier, completed in 1540, provided shelter from south-westerly storms but failed in its main purpose of preventing the harbour mouth from being choked by shingle driven along the coast. On the King's orders it was therefore extended further out to sea, the work then being directed by the military engineer, Sir Richard Cavendish, and others. Completed at great expense in 1551 it too failed, against reasonable expectation. Shingle driven around its head formed an offshore bar stretching more than half a mile across the bay to the cliffs below Dover Castle. In 1576, with the outer part of the King's pier now in ruins, the Privy Council consulted the mariner William Borough. He proposed a brilliant solution: to build a sea wall along the bar and a cross wall back to the shore, enclosing a large basin of water which, admitted at high tide, could be released at low tide through a sluice in the cross wall to scour the harbour entrance. The cost appeared too great. After a delay, work began (by John Trew) on rebuilding the pier. In a report of 1581 Borough pointed out the fallacy of this. The eminent mathematical practitioner Thomas Digges supported Borough's analysis and made an accurate survey of the whole site. Sir Thomas Scott, a local landowner, suggested an economical method of building the long sea wall, which greatly reduced the estimated costs, and the works were carried out in 1583–1586 with renowned success. Digges acted as 'Surveyor General', reporting to the Principal Secretary of State, assisted by an administrator. A large workforce was recruited. Scott and the Lieutenant of Dover Castle directed operations on site and two master shipwrights built the sluice.

1600–1689

The next major contribution came from John Hunt, a fenland landowner. Bradley had already deduced from the surveys that drainage of the Great Level was possible by gravitational flow, provided the course of the winding rivers could be shortened, to increase the gradients. But he made only a brief suggestion as to how this might be accomplished. By contrast, Hunt in 1605 made a series of bold and practical proposals, the principal one being a 21-mile cut from Earith to Denver on the river Great Ouse. His plans were approved and work began in a small way. But the scheme was too ambitious for its time. When the very large necessary funding eventually became available, through efforts of the fourth Earl of Bedford, the works (1631–1636) followed Hunt's plans closely. Details of the organisation are not available, although several names of those involved are known, including Andrewes Burrell and two surveyors, Thomas Wright and William Hayward (who had produced a fine map of the Great Level at the time of Hunt's scheme in 1605).

Meanwhile the largest bridge of the seventeenth century was built in 1611–1624 at Berwick-on-Tweed. As a King's work it is well documented. The designer was James Burrell, trained as a mason and formerly Deputy Surveyor of Works of the Berwick fortifications. Appointed 'Surveyor', he directed construction assisted by a clerk and a staff consisting of a leading mason and carpenter and a paymaster. The workforce included thirty-one masons, twelve carpenters, twenty-two quarrymen and seventy-six labourers.

Berwick Bridge can be considered as a fully 'professional' work. Most others in the early seventeenth century were (like the Elizabethan Dover Harbour) 'amateur' undertakings, but nonetheless successful. The outstanding example is the New River, a 40 mile long channel bringing water from springs in Hertfordshire to London on an average gradient of 5 in. per mile. Edmund Colthurst, a landowner at Bath and ex-army Captain, conceived the idea and started work, but with inadequate funds. With support from the City and his own considerable wealth, the goldsmith and banker Sir Hugh Myddleton then took full managerial responsibility for the scheme, and under his active direction the work was completed in 1609–1614. He employed a clerk. Colthurst was appointed 'Overseer' (resident engineer), and Edward Wright carried out the precise levelling needed in the early stage of the work. Excavators worked in gangs at piecework rates, bricklayers and carpenters were paid on a daily basis; the only contracts were for particular timber structures.

Myddleton is the first of the civil engineering 'gentleman' entrepreneurs. Next is the Bedfordshire landowner Arnold Spencer, who undertook to make the Great Ouse navigable from St. Ives to St. Neots (15 miles with six pound locks in 1618–1621). He was followed by William Sandys who spent a small fortune on the Avon river navigation, 43 miles from Tewkesbury to Stratford in 1636–1639, and notably by the agriculturist and landowner Sir Richard Weston, who personally planned and directed the Wey Navigation from Weybridge on the Thames to Guildford in 1651–1653 (15 miles, with 9 miles in new cuts and ten pound locks).

In none of the seventeenth-century works mentioned so far is there any evidence of foreign engineers. Nevertheless, the dominant figure of the period is the Dutch engineer Sir Cornelius Vermuyden. He was invited to England in 1621 by James I with a view to draining the Great Level. Nothing came of this at the time, but Vermuyden stayed on in England and in 1626 was granted powers by Charles I to drain a large area of low ground in Hatfield Chase (owned by the Crown) in return for an agreed area of the drained land. Vermuyden quickly raised the necessary capital, in Holland, and brought over a Dutch workforce. The works, including the diversion of two rivers, were completed by 1631. Further work was required, however, to provide a better outfall for the diverted river Don. This involved a new channel, later known as the Dutch River, built under the direction of Vermuyden's nephew John Liens, who had worked as his assistant on the Chase.

Liens went on to be Director of Works on the drainage of Lindsey Level, south of Boston in Lincolnshire (1636–1639), at a salary of £200 per annum. On his staff were 'Mr. Wright' the 'general Overseer' (probably Thomas Wright, previously on the Great Level), a land surveyor, paymaster, book-keeper and assistant overseers.

In the meantime Vermuyden had been consulted (again) on the Earl of Bedford's Great Level scheme in 1630–1631 but, contrary to the usual assumption, he was not engaged on the works. Instead, he went into partnership in what proved to be a highly profitable lead mine in Derbyshire.

The Earl of Bedford's works on the Great Level resulted in the Fens being dry from April or May every year; 'summer grounds' had been achieved. The next step was to prevent winter flooding and thereby secure 'winter grounds' fit for arable farming This was beyond the Earl's resources, and the King declared himself willing to be undertaker. Vermuyden submitted a 'Discourse' to the King in 1639 (published in 1642) on the method to be adopted. The second campaign began under his direction in 1640, but came almost to a halt by the end of 1641 and was completely abandoned on the outbreak of the Civil War the next year.

In 1649 the fifth Earl of Bedford and his associates formed themselves into a Company of Adventurers determined to complete the undertaking. Their Minute Books survive. Vermuyden was appointed Director of Works in January 1650. He would receive 4,000 acres of drained land plus £1,000 (paid £300 down and the balance at £20 per month). Anthony Hammond, one of the Adventurers, acted as 'Superintendent' or project director (£20 per month); as administrators were the Expenditor, Comptroller and Clerk to the Company. Vermuyden's senior staff comprised the mathematical practitioner Jonas (later Sir Jonas) Moore as 'Surveyor General' (£200 per annum) and Robert Burton as Principal Overseer (£100 per annum).[3] Under them were three overseers and two, later increased to four, land surveyors (£52 per annum).

Vermuyden's duties were to plan the scheme (he considerably revised his 1639 proposals), to decide what works should be done, where and when, and issue directions accordingly to the Surveyor and Overseer. Moore had to 'sett out the works and see that they bee done', in Vermuyden's words, and with his assistants to divide the land after drainage. Burton 'let out' contracts. He and his assistants had to oversee construction and 'take up' or measure quantities for payment of work done. The works were completed in 1656 on a scale and at an expenditure (about

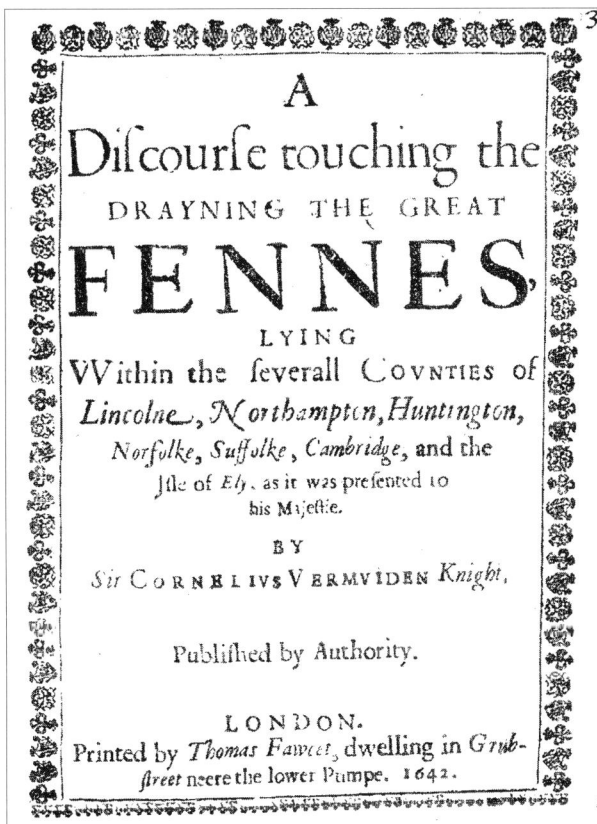

Title page of Vermuyden's report on draining the Great Level of the Fens, 1642

£65 million in modern costs, see Appendix I) rarely exceeded in England until the nineteenth century. At its peak, in the summer of 1652, the workforce numbered 10,000 men.[4] In 1658 Moore published a large-scale map of remarkable accuracy and detail of the entire Great Level (some 300,000 acres, known as Bedford Level from 1663).

For purposes of administration the area was divided into three 'Levels', each under the direction of a 'Surveyor' responsible for maintenance of the drains, river banks, roads and sluices, and any new works required from time to time. In this way a distinct class came into being: engineers on long-term appointments. Mark Le Pla, for example, held the post as Surveyor of the North Level, 1665–1696. It is typical of such appointments that he was able, on leave of absence, to undertake some outside consulting work in Romney Marsh in 1689.

The Fens also gave rise to the first printed reports (Vermuyden 1642, Andrewes Burrell 1642, William Dodson 1665). Sir William Dugdale's great *History of Im-banking and Drayning* appeared in 1662.

Several works illustrate points of interest in the next thirty years: the 1.5-acre wet dock in the famous Blackwall shipyard on the Thames 1659–1661, possibly the first in England; the Fleet Canal 1671–1674, planned by Sir Christopher Wren as part of the rebuilding of London after the Great Fire and constructed by Thomas Fitch, a rare early example of a 'general contract' involving the whole work in contrast to the usual procedure of letting separate small contracts to various tradesmen or carrying out the work by direct labour; an extension of the Exeter Canal 1676–1677, by Richard Hurd, one of the earliest to be called engineer (spelt 'Ingineere') in a civil context; and the rebuilding of Denver Sluice 1682–1684, by Ralph Peirson, Surveyor of the South Level from 1671 to 1685 (at £100 per annum) before becoming land agent to the Earl of Bedford's estate at Thorney.

A work of exceptional magnitude was the pier or 'Mole' of the naval harbour at Tangier built by Sir Hugh Cholmley of Whitby, as contractor from 1663 and as Surveyor General 1669–1676, with a staff of military officers comprising a Deputy Surveyor, Comptroller and Principal Overseer. Work continued under the military engineer Henry Sheres, but was abandoned in 1683 for political reasons.

1690–1760

A new era opens in the 1690s, running on to the 1750s; a period of increasing trade and prosperity, political stability and technological advance symbolised by Thomas Newcomen's steam engines (from 1712). The number of new civil engineering works proved sufficient to provide employment for 'professionals', in the sense that they could earn a living as engineers for all or the greater part of their career, and an increasing number of long-term salaried appointments were made. The leading practitioners in this period, drawn from a range of different backgrounds, are as follows:

John Hadley   Described by Sir Godfrey Copley as 'a man of Mathemat and Bookish', he built pumped water supply schemes at Worcester and Chester in the 1690s, and was engineer of the Aire & Calder Navigation, giving evidence in Parliament on the scheme in 1698 and 1699.

London Bridge waterworks, 1701

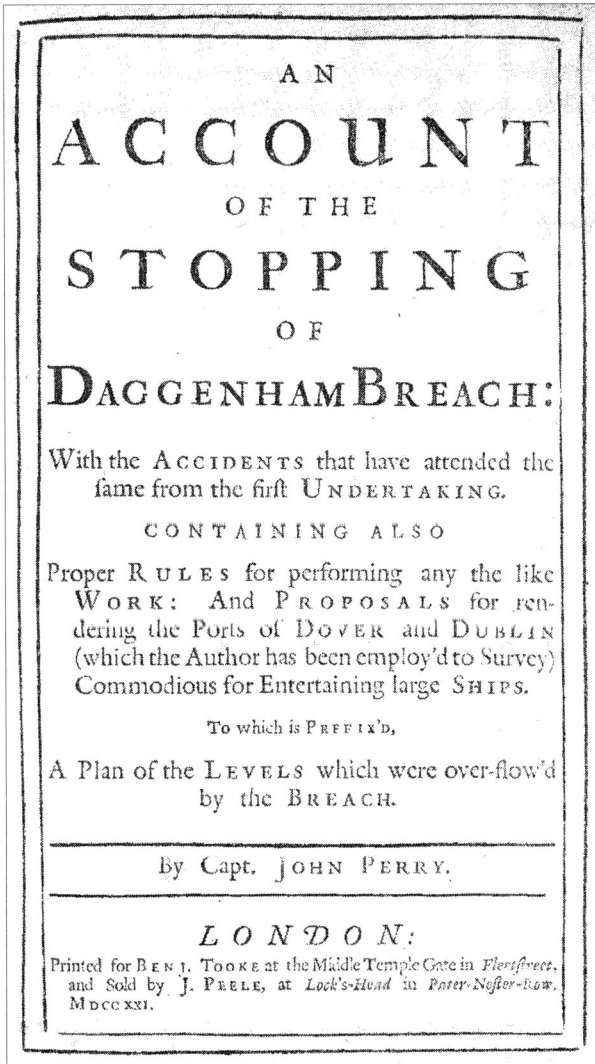

Title page of Perry's account of stopping Dagenham Breach, 1721

George Sorocold   Referred to by a contemporary (in 1708) as 'that great English engineer', he built waterworks (similar to Hadley's) at Derby, Leeds and several other cities and the much admired waterwheels for London Bridge waterworks (1701). He was one of the first consulting engineers, reporting on river navigations and other projects in Derbyshire, Yorkshire, Liverpool and Scotland.

Edmund Dummer   Surveyor to the Navy, planned and built dockyards at Plymouth and Portsmouth in the 1690s, which included two dry docks and three wet docks of advanced design.

John Perry   An ex-naval Captain, he worked on canals and docks in Russia before returning to England in 1712. He became famous for his work in stopping a serious breach in the Thames embankment at Dagenham, on a 'design and build' contract, and published an account of this undertaking in 1721. He went on to plan and direct works at Rye New Harbour and Deeping Fen. His printed report on improving drainage of the North Level of the Fens (1727) is a model of its kind.

Thomas Steers Engineer of the first dock at Liverpool, opened 1715, and of the second dock there, and of the Newry Canal in Ireland (1734–1741). He drew up plans for the Mersey & Irwell and the Calder & Hebble Navigations, and several other schemes.

John Hore Probably began as a millwright. He was engineer of the Kennet Navigation, constructed to his plan 1718–1723 (18 miles, 11 miles in cut with twenty locks), and then of the Bristol Avon Navigation 1725–1727, on which he charged the high fees of 9 guineas per week (to include expenses). He acted as consultant on what later became the Stroudwater Canal, and from 1734 to 1761 was Surveyor on the Kennet.

Nathaniel Kinderley A fenland landowner, he proposed in 1721 the major improvement in drainage of the South Level carried out much later as the Eau Brink Cut. He directed work on a new outfall channel of the river Nene, which was destroyed by rioters but subsequently restored as 'Kinderley's Cut'. As an entrepreneur he undertook construction of the ship canal in the Dee estuary below Chester and the associated land-reclamation scheme.

John Reynolds Trained as a carpenter, he worked under Perry on the Dagenham job, undertook several small harbour works, on design and build contracts, and became engineer on the later stages of the Dee estuary works 1737–1743.

William Lellam After four years' work at Bridlington harbour he went to Sunderland as harbour engineer in 1722, a post he held until moving to Scarborough in 1731. At Sunderland he built the 1,000 ft. long south pier.

Humphry Smith In his early career was land agent to the Duke of Bedford. He carried out works in the Fens, including 'Smith's Leam' on the Nene upstream of Wisbech, and was appointed Surveyor (engineer) of Deeping Fen in 1738.

John Grundy Sr. A land surveyor and teacher of mathematics, he became deeply interested in fen drainage and worked with (and then succeeded) Humphry Smith on Deeping Fen. On leave of absence he reported on improvement of the river Witham in 1744.

Charles Labelye Born in Switzerland of French refugee parents, he came to London *c*. 1725, worked as assistant to the scientist J. T. Desaguliers FRS, taught mathematics to the Navy, planned a new harbour at Sandwich, and in 1738 was appointed engineer for Westminster Bridge. He worked practically full-time on the bridge until its completion in 1750, assisted by Richard Graham, but was able to do consulting work on occasional leave of absence. He produced printed reports on Fen drainage and on harbours at Sunderland and Yarmouth, and advised on the (second) rebuilding of Denver Sluice, which was carried out in 1746–1750 by William Cole, Surveyor of the South Level, and his assistant John Leaford.

William Vincent Engineer of Scarborough Harbour on a long-term appointment from 1734, building 'Vincent's Pier' and starting the great East Pier before moving to Sunderland Harbour in 1752; he died in office two years later. He and John Grundy Sr. were among the subscribers to Desaguliers's two-volume *Experimental Philosophy* (1734 and 1744).

The works in this period, although numerous, were mostly on a relatively small scale. Consequently the organisation was simple and the salaries modest. In 1724 on Rye Harbour, for instance, Perry had a salary of £170 per annum for himself and a clerk, and was expected to devote three-quarters of his time, while his full-time 'foreman' or resident engineer, Edward Rubie, received 25s per week (£65 per annum).[5] A very different pattern obtained on Westminster Bridge, by far the largest

Table 2. Reports and engraved plans published in the period 1700–1830

Name	Period	Books	Printed reports[a]	Engraved plans
Thomas Steers	1712–1741			2
John Perry	1721–1727	2	5	
Nathaniel Kinderley	1721–1740		2	1
Charles Labelye	1737–1750	2	4	4
John Grundy	1753–1774		12	7
Thomas Yeoman	1754–1775		7	4
John Smeaton	1757–1791	2	21	14
James Brindley	1760–1772	1[b]	3	10
Robert Mylne	1770–1802		12	1
Robert Whitworth	1770–1797		9	13
William Jessop	1772–1807		28	16
William Chapman	1785–1832	2	38	3
John Rennie	1790–1821		36	26
Thomas Telford	1800–1831	1[c]	27	11
Robert Stevenson[d]	1814–1829		15	2

[a] Excluding reports printed in Parliamentary Papers.
[b] Attributed to Brindley.
[c] Published posthumously.
[d] Reports by Stevenson jointly with his sons Alan and David appear from 1832 to 1839.

project of the early eighteenth century, where expenditure on the works amounted to £198,000 (roughly £100 million in modern costs). The Bridge Commissioners employed two administrative officers, the Treasurer and Clerk (Secretary), both on £200 per annum. A 'Surveyor and Agent', also on £200 per annum, dealt with the buying and disposal of property in relation to the bridge approaches. For the bridge itself Labelye, as 'Engineer', was responsible for design and direction of construction, with a salary of £100 per annum plus 10s per day for attendance on site, and a gratuity of £2,000 on completion. Richard Graham's duties as 'Surveyor and Comptroller of Works', at £300 per annum, involved constant supervision of progress, letting contracts, measurement of work done and keeping accounts.

The majority of contractors in the period, and indeed throughout most of the eighteenth century, were local craftsmen of whom little is known apart from their name and connection with a particular work. Three exceptions are: William Ogbourne, a prominent London master carpenter, who built the large (timber-walled) Howland Dock at Rotherhithe, completed in 1700; the builder Thomas Phillips, who constructed the timber bridge over the Thames between Fulham and Putney in 1729; and Andrews Jelfe, a master mason with an extensive practice, who with his partner Samuel Tufnell executed the masonry of Westminster Bridge in fourteen successive contracts totalling £155,000 in value.

1760–1790

The state of civil engineering changed decisively in the 1760s with a dramatic increase in the number of works commissioned: at least thirty-five in that decade and fifty in the next two decades to 1790. Engineers forming a small but distinguished group were now fully employed in consulting, designing, giving evidence in Parliament and directing works, often performing two or more of these roles simultaneously. A clearly defined management structure became essential, and printed reports and engraved plans appear in increasing numbers (Table 2), and engineers worked together on site, met in Parliament, and kept in touch by correspondence. To mention a few examples from the 1760s: Smeaton worked with James Brindley on site investigation for the Bridgewater and the Trent & Mersey

Canals; Smeaton and John Grundy Jr. collaborated in planning the Holderness Drainage scheme; they and Langley Edwards reported jointly on the River Witham improvements; Yeoman and Smeaton worked together on the River Lea Navigation; and Brindley recommended Golborne as 'the properest person' to improve the Clyde. All this means that the engineers were not only working as professionals, but also partook in a degree of association characteristic of a profession. Its existence is implicit in the founding of the Society of Civil Engineers in 1771.

At the first meeting in March 1771 it was 'Agreed that the Civil Engineers of this Kingdom do form themselves into a Society consisting of a President, Vice President, Treasurer and Secretary and other Members who shall meet once a fortnight on Saturday evenings at seven o'clock from Christmas or so soon as the country members come to Town … to the end of sitting of the Parliament, at the Kings Head in Holborn'. Thomas Yeoman took the chair as President; others present were Smeaton, Grundy, Robert Mylne and Joseph Nickalls. Brindley was not a member, but his principal assistants Hugh Henshall and Robert Whitworth joined within the next month, as did John Golborne. Within the next two years another six engineers had been elected, including Langley Edwards, John Gott and William Jessop, Smeaton's former pupil and assistant.

The leading civil engineers of the 1760s are as follows:

John Smeaton FRS   Regarded as the founder of the civil engineering profession. He completed the Eddystone Lighthouse to national acclaim in 1759, and in that year was awarded the Copley Medal by the Royal Society for research on the natural powers of water and wind (begun in 1752). He proceeded to engineer the Calder & Hebble Navigation, to design two handsome bridges at Coldstream and Perth, and in 1768 began work on the Forth & Clyde Canal. These were followed by a series of works of remarkable diversity and distinction: harbours, fen drainage, river navigations, steam engines and industrial water mills. He kept manuscript copies of all his reports after 1760; twenty-one of them were issued in printed form during his lifetime. His drawings, over 1,000 in number, are in the Royal Society library.

John Grundy Jr.   Trained by his father, whom he succeeded in 1748 as engineer of Deeping Fen. He produced several reports in the 1750s and was principal consultant on the Witham. His works include two small canals and three fen drainage schemes, and he designed the first dock at Hull. Up to the early 1760s his professional fees were 1 guinea per day plus expenses, the standard rate for civil engineers since 1700. In 1767, at or near the peak of his career, he earned £591 in fees, a figure indicating a rate of 2.5 guineas per day. His (manuscript) Report Books (1741–1779) survive: sixteen quarto volumes in Leeds University Library and one in the Institution of Civil Engineers.

Thomas Yeoman FRS   Began his career as a millwright, and from 1763 was a member of the Society of Arts, holding the position as joint chairman of the Committee on Mechanics for fifteen years. He was consulted on the proposed River Nene Navigation in the 1750s and on several other river navigations of the 1760s, was engineer in charge of the river Lea improvements and planned the Stroudwater Canal.

James Brindley   He had established a high reputation as a millwright and made a preliminary survey for what was to become the Trent & Mersey Canal when called in as engineer on the proposed Bridgewater Canal in 1759. He immediately revised its route to Manchester, and in 1761 planned the extension towards Runcorn. Between 1766 and 1771 he was appointed engineer for the Trent & Mersey and six other canals, totalling 300 miles in length. To carry out this enormous task he recruited and trained a staff of assistants and resident engineers. By 1768–1769 his combined salary from the canals then under construction amounted to at least £760 per annum

Eddystone Lighthouse, completed 1759

Hugh Henshall   Surveyed the Trent & Mersey Canal under Brindley's direction in 1765 and was appointed its resident engineer. He completed this canal and the Chesterfield Canal as chief engineer after Brindley died in 1772.

Robert Whitworth   'Levelled for that great Genius Mr. Brindley' on four of his canal surveys. He was later appointed engineer for that part of the Thames Navigation administered by the City of London; he planned the Thames and Seven Canal and in the 1780s and 1790s carried out notable works on the Forth & Clyde and Leeds & Liverpool Canals. An obituary notice refers to 'Mr. Robert Whitworth, one of the most able engineers in England'.

Robert Mylne FRS   Coming from a long line of Scottish master masons, and fresh from his studies in Italy, he won the competition for Blackfriars Bridge with a design of exceptional quality. He personally directed construction in 1760–1769 with one or two assistants. He practised with distinction as an architect, held the post of engineer to the New River Company for forty years, and was much in demand as a consulting engineer.

Langley Edwards   Land agent on the Duke of Bedford's Thorney estate and Commissioner of the North Level of the Fens before becoming in the late 1750s consultant and engineer of several river navigations, on one of which he worked with Thomas Yeoman. He reported on the Witham scheme in 1760 and again in 1761, this time jointly with Grundy and Smeaton, and was appointed engineer in charge

Blackfriars Bridge under construction, 1766

of construction in 1762. The works were completed in 1767, including the Grand Sluice at Boston built to Edwards's design. He was also engineer of the Black Sluice Drainage, and in 1771 presented a decisive report on improvement of the Nene Outfall.

John Golborne   Engineer of the River Dee Company from 1754, continuing the works begun by Kinderley and Reynolds. He is best remembered for his successful deepening of the Clyde for 14 miles below Glasgow, and he published two influential reports on drainage of the Fens.

By 1790 some 850 miles of canal had been opened, 'the arteries of the Industrial Revolution'. Major improvements were made in harbours at Aberdeen and Ramsgate (both by Smeaton), two new docks were built at Liverpool (Henry Berry) and the largest dock of the eighteenth century at Hull, 10 acres in area. Roads were being improved (the General Turnpike Act dates from 1773 and the mail coach system from 1784). An outburst of activity took place in the building of county bridges, especially in the 1770s. Several counties now had Surveyors on long-term (but part-time) appointments. John Gott, for example, was engineer of the Aire & Calder Navigation from 1760 and Surveyor of Bridges for the West Riding of Yorkshire from 1777, holding both posts until a year before he died in 1793. Besides the large bridge at Blackfriars, many others were built including those at Richmond and Kew (both to designs by the architect James Paine) and at Newcastle (Mylne and John Wooler) and also the first iron bridge, at Coalbrookdale (by the iron master Abraham Darby), opened in 1780.

The organisation of a large project in this period is illustrated by the Forth & Clyde Canal. In 1768, when work started, Smeaton was appointed 'Engineer in Chief' with a salary of £500 per annum (to include all expenses). Robert Mackell became

'Resident Engineer' (a term introduced by Smeaton on this occasion) at £315 per annum, and the lawyer Alexander Hope was appointed 'Agent and Clerk' to the Company with a salary of £200 per annum. Shortly afterwards Alexander Stephen joined the team as 'Clerk and Overseer' under Mackell at £1 per week (later raised to £70 per annum) and in 1770 Archibald Hope came in as 'Comptroller and Auditor' at £60 per annum. In addition, there were Superintendents of masonry and carpentry, and the land surveyor John Laurie was employed as and when required at 0.5 guineas per day.

Smeaton had planned the scheme, and produced typical working drawings. He visited the site quite often in the early stages but resigned in 1773, being happy to leave completion (achieved in 1777) in the hands of Mackell, although he kept actively in touch by correspondence. Mackell was in charge of day-to-day operations throughout, letting contracts, making detailed drawings and even proposing an improved line of part of the canal. Reporting on this, Smeaton said 'In the whole of this Business Mr. Mackell has shown a great deal of Judgement and attention … and I have the further pleasure to acquaint the Company that, having carefully inspected the Works now going on, they are in general and in things the most material, exceedingly well executed and much to my satisfaction'.

More typically in this period resident engineers were paid £100 per annum (John Gwyn on Ramsgate Harbour 1788) to £150 per annum (Hugh Henshall on the Trent & Mersey Canal 1766) and £200 per annum (Benjamin Outram on the Cromford Canal 1789). Salaries of engineers on long-term appointments covered a similar range: worth about £25,000 to £35,000 in the 1990s (see Appendix I).

The usual practice of letting small contracts continued, but with the significant development that in 1767 John Dyson and James Pinkerton set up as earthwork contractors and before long were operating on a national scale. At first in partnership and then separately, they established family firms which continued into the early decades of the nineteenth century. Moreover, in 1788 John Pinkerton took on a contract for constructing the Basingstoke Canal as a whole. This marks the beginning of the 'general contract' system.

1790–1830

This period witnessed a further increase in works commissioned and particularly in the number of very large works, several of which were Government funded. The dominant figures are William Jessop, John Rennie[6] and Thomas Telford. Not far behind in reputation are William Chapman and Robert Stevenson. A glance at the chronological table (Appendix 2) will show many others, while among those whose career began before, but lasted after, 1830 are notably James Walker (second President of the Institution), Sir Marc Isambard Brunel, George Stephenson and Rennie's son Sir John. Printed reports now appear in considerable number (see Table 2).

By 1830 practically the entire canal network had been completed, totalling over 2,000 miles in length. Major works include the Lancaster (Rennie), Grand Junction (Jessop and James Barnes), Rochdale (Jessop and William Crosley), Ellesmere (Jessop and Telford), Kennet & Avon (Rennie), Huddersfield (Benjamin Outram), Union (Hugh Baird) and the Birmingham & Liverpool Junction Canal (Telford and William Cubitt). In addition, Telford built the Government-funded Caledonian (ship) Canal, with Jessop as consultant in the early years of construction.

Docks of unprecedented size were constructed in Dublin, 25 acres (Jessop), and in the Port of London—West India, 54 acres (Jessop), London, 20 acres (Rennie) and East India, 26 acres (Rennie and Ralph Walker)—all three built between 1800 and 1806. Jessop's work at Brisol, bypassing the River Avon and converting its original course into a 70-acre 'floating harbour', is in a class of its own. Meanwhile an additional dock had been provided at Liverpool and at Hull, followed by another at each port and in London by the Commercial Docks (James Walker) and St. Katharine's (Telford).

Among maritime works the Bell Rock Lighthouse (Rennie and Stevenson) and Plymouth Breakwater (Rennie and Whidbey) are impressive achievements, as are the harbours at Aberdeen (Telford), Kingstown (Rennie), Scarborough and Seaham (both by Chapman) and Donaghadee (Sir John Rennie). Sheerness naval dockyard, by Rennie, is the largest work in terms of expenditure before 1830: over £200 million in modern costs.

Two other vast Government projects were carried out by Telford, for the Commissioners of Highland Roads and Bridges (and harbours) in Scotland and the London–Holyhead Road. He also constructed the Chester–Bangor and the Glasgow–Carlisle roads. At the same time J. L. McAdam and his sons were improving hundreds of miles of road. County bridges continued to be built in ever increasing numbers, and Rennie's Waterloo Bridge ranks with Mylne's Blackfriars Bridge as a masterpiece of architectural and engineering design.

A characteristic feature of the period was the development of iron arch bridges from the pioneer Sunderland Bridge (Rowland Burdon and Thomas Wilson) to the mature bridges of Telford and William Hazledine at Craigellachie and Tewkesbury and James Walker's Vauxhall Bridge in London, and the equally remarkable development of suspension bridges from the Union Bridge of Sir Samuel Brown to Telford's 580-ft. span Menai Bridge on the Holyhead Road.

The success of the Menai Bridge depended upon testing of materials and mathematical analysis (by Davies Gilbert FRS) as well as Telford's skill and imagination. Also of importance, in leading to the construction of multi-storey buildings with internal iron framing and fire-proof floors, was the work of two 'amateurs', the cotton manufacturer William Strutt FRS and his friend Charles Bage of Shrewsbury. They erected the first buildings of this type in the years around 1800, based on full-scale structural tests and a theory on the strength of cast iron beams.

St. Katharine's Docks, opening, October 1828

Bristol Harbour, opened 1810

Another feature of the period was the introduction of public railways. Private lines at collieries and ironworks had long been in use, first on timber and from *c*. 1790 on iron rails. The Surrey Iron Railway by Jessop, completed in 1803, was the first open to the public. The next twenty years saw a rapid expansion of (horse-drawn) railways. Jessop's younger son William, at the Butterley Company, produced an improved cast iron rail used by his elder brother Josias on the Mansfield & Pinxton and the Cromford & High Peak Railways. Meanwhile George Stephenson had been developing steam locomotives on colliery lines in the Tyneside and Durham area, and John Birkinshaw introduced wrought iron rails in 1820. Stephenson used such rails as well as locomotives on his Stockton & Darlington Railway, opened in 1825 as the world's first public steam railway. The opening of his Liverpool & Manchester Railway in 1830, with its tunnel, cuttings, embankments and viaducts, and regular passenger service operated by eight locomotives, including the *Rocket*, signals the beginning of the railway age.

Work also continued in the more traditional field of fen drainage, with two beautifully designed schemes in east Yorkshire by Chapman and another by Rennie in Lincolnshire: the East, West and Wildmore Fens. The Eau Brink Cut was eventually made (by Rennie) with immense benefit to the South Level, and Telford's last completed work was a major improvement in drainage of the North Level.

In terms of practice, the number and scale of works demanded more than ever on a highly organised management structure. On Rennie's Lancaster Canal, for example, in 1797 the staff comprised two resident engineers (each on £400 per annum), Archibald Millar on the north and Henry Eastburn on the south divisions of the 55 miles of canal then under construction, with an assistant engineer William Cartwright (£250), two Superintendents of Works (£100), a land surveyor (£100) and five Superintendents of earthwork (£54 to £78) together with a Solicitor (£300) and the Clerk or Secretary (£260). Masons were paid 18s per week (say £45 per

Sheerness Dockyard under construction, 1824

annum) and labourers 13s. The reference given for Millar on his appointment in 1793 says 'that he is a good Engineer, can carry an Accurate Level, and has a perfect Knowledge of Cutting, Banking, etc. and that he is a compleat Mason'. Eastburn had recently completed the Basingstoke Canal as RE under Jessop. The sum of £400 a year, equivalent to £500 per annum in the 1820s, allowing for inflation, is a typical salary for resident engineers on major works in the post-1790 period. It is worth about £55,000 per annum in modern terms.

When appointed Principal Engineer in 1792 Rennie's salary was £600 for five months' residence per year at Lancaster and on other occasions as requested. By 1797, with work well under way, less of his time was required and he received 250 guineas per annum. Jessop and Mylne charged 5 guineas per day at this period, as did the fashionable architect James Wyatt; Rennie increased his fees from 3 to 5 guineas per day in 1800.

On the Holyhead Road, under the administrative leadership of Sir Henry Parnell, the senior staff comprised: Alexander Milne, Secretary to the Commissioners at £200 per annum; Telford, appointed Chief Engineer in 1815 at 5 guineas per day; William Provis, Assistant Engineer on the Welsh section (Shrewsbury to Holyhead) at £200 per annum, although while acting as RE on Menai Bridge[7] he received £300 per annum plus an allowance of £50 per annum for a Clerk to assist in keeping the accounts; and John Easton, Assistant Engineer on the London–Shrewsbury section, also on £200 per annum. In addition, John Sinclair, one of at least four Inspectors, superintended construction of the iron arch bridge at Bettws-y-Coed, and Thomas Rhodes supervised the iron work on Menai Bridge. Payments for contract work were made by the Secretary upon monthly certificates, from the engineers, of the works performed. From Telford's total fees from 1815 to 1830 (£6,295) it appears that on average he devoted 80 days a year to the project.

On a few works chief engineers were engaged practically full-time; for instance Sir Marc Brunel on the Thames Tunnel from 1824 and George Stephenson on the

Liverpool & Manchester Railway from 1827, both at £1,000 per annum. With inflation this is worth about the same as Whitworth's £630 a year on the Leeds and Liverpool Canal in 1790 and equivalent to roughly £100,000 per annum today.

Engineers on long-term appointments continued to play an important role, and many counties now had their own Surveyor. The career of James Green, County Surveyor of Devon, is well documented. Stephen Leach, who held his post on the Thames Navigation for thirty years, engineered the six great locks and weirs between Teddington and Staines.

While much of the work was still being carried out on numerous relatively small contracts, with varying degrees of success, the period witnessed the rise of the general contractor. The two outstanding firms are those of Hugh McIntosh and Sir Edward Banks.[8] Both began in the 1790s on small canal contracts, but by 1810–1820 were capable of handling several large works at the same time nationwide with a site 'agent', counterpart of the resident engineer, on each contract.

Finally, it was in this period that text books became readily available. They include: Charles Hutton on the Theory of (arch) Bridges (1801), Olinthus Gregory on Mechanics (1806, later editions 1807, 1815 and 1826), Smeaton's Reports (1812; 2nd edn., 1837), John Loudon McAdam on Road Making (1816 and eight further editions to 1829), Peter Barlow on the Strength of Timber and Iron (1817, 1824 and 1826), Thomas Tredgold on Carpentry (1820, 1828 and 1840) and on Iron (1822, 1824 and 1831) and Henry Moseley on Hydrostatics and Hydrodynamics (1830).

Pupillage and training

In 1759 Smeaton took William Jessop as a pupil, and in 1768 Henry Easburn. Both served seven years and went on to work as his assistant. James Golborne (who was to become Superintendent General of Bedford Level) served a seven-year apprenticeship with his uncle, John Golborne, from 1760. These are the first formal pupillages known in civil engineering, although there is no doubt that John Grundy was trained by his father from the mid-1730s and there are several later instances of such (probably informal) father and son arrangements. More generally, in the eighteenth century young engineers learnt by experience on the job. This was the case with John Reynolds working under John Perry on the Dagenham Breach 1716–1720, and Henry Berry worked as Steers's assistant before taking over from him as Liverpool's dock engineer in 1750. Later examples abound.

From about 1800 the pupillage system was adopted to an increasing extent. James Walker, after five years general study at Glasgow University, became a pupil of his uncle Ralph Walker in 1800, and in 1811 took Alfred Burges as a pupil (who joined him as partner in 1829). Rennie's son John trained in his father's workshop and drawing office, learnt surveying as a pupil of Francis Giles, and then gained practical experience from 1813 as assistant to James Hollinsworth, the RE on Waterloo Bridge. William Provis, having spent three years with his father, joined Telford in 1808 almost certainly as his first pupil, and later as assistant. H. R. Palmer, another of Telford's assistants (from 1816) had been a pupil of Bryan Donkin. Telford himself took two pupils from 1821: Alexander Gibb and Joseph Mitchell. Chapman's pupil Thomas Elliot Harrison in due course became President of the Institution.

In the earliest years of the Institution a pupil could attend meetings only as a guest of a Member, but this soon changed when the class of Associates could include young persons studying for the profession. I. K. Brunel, admitted as an Associate in 1824, was one of those. The separate class of 'Graduate' for such persons was authorised in 1831.

A. W. SKEMPTON

Notes

1. Mortimer's *Universal Director* (1763), in which Smeaton and Yeoman are both listed as 'Surveyor and Civil Engineer'.
2. From John Grundy's catalogue of his library in 1774 it is known that he had twenty-five printed reports by other engineers.
3. For comparison, around 1650 an army Colonel received £218 per annum and naval Captains between £273 and £91 depending on the rating of their ship.
4. Comparable numbers appear on the railways: 10,000 on the London–Birmingham in 1836 and 6,250 on the Great Western in 1837.
5. Masons and carpenters at this time were paid about 12s per week.
6. A complete set of Rennie's manuscript Report Books is in the ICE Library, nineteen large folio volumes 1790–1821.
7. His book on the Menai Bridge (1828) ranks with Smeaton's on the Eddystone Lighthouse (1791) as a classic in civil engineering literature.
8. In the firm of Jolliffe and Banks from 1807.

Signatures

Signatures of leading engineers and contractors in the Dictionary are given on the following pages.

Sixteenth to early eighteenth century engineers

John Trew (fl. 1564–1588)

Hugh Myddleton (1560–1631)

Cornelius Vermuyden (1590–1679)

John Liens (fl. 1627–1641)

Jonas Moore (1617–1679)

Andrew Yarranton (1619–1684)

Samuel Fortrey (1622–1682)

George Sorocold (1668–c. 1716)

John Perry (1669–1733)

Thomas Steers (1672–1750)

William Lellam (fl. 1716–d.1733)

John Hore (1680–1763)

Humphry Smith (1671–1743)

William Vincent (fl. 1734–d. 1754)

Charles Labelye (c. 1705–1762)

Mid-eighteenth to early nineteenth century engineers

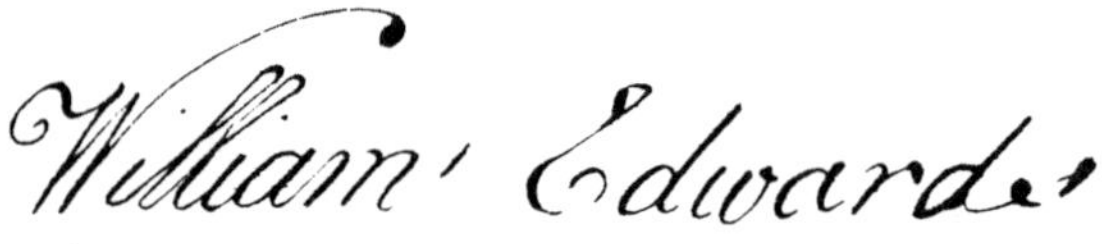

William Edwards (1719–1789)

John Smeaton (1724–1792)

James Brindley (1716–1772)

John Grundy (1719–1783)

Robert Mylne (1734–1811)

Hugh Henshall (1734–1816)

Robert Whitworth (1734–1799)

John Golborne (1724–1783)

William Jessop (1745–1814)

Joseph Huddart (1741–1816)

James Watt (1736–1819)

John Rennie (1761–1821)

Thomas Telford (1757–1834)

William Chapman (1749–1832)

Ralph Walker (1749–1824)

Joseph Whidbey (1755–1833)

Hugh Baird (1770–1827)

Alexander Nimmo (1783–1832)

Contractors

William Adam (1689–1748)

John Pinkerton (fl. 1770–d. 1813)

Thomas Dadford (1730–1809)

Alexander Stevens (1730–1796)

Hugh McIntosh (1768–1840)

Edward Banks (1770–1835)

John Wilson (1773–1831)

John Cargill (1772–1848)

William Hazledine (1763–1840)

Thomas Hughes (fl. 1790–1846)

Thomas Townshend (1771–1846)

Daniel Pritchard (1777–1843)

George Burge (1795–1874)

Biographical dictionary

A

ABERCROMBIE, Charles (1750–1817), eminent Scottish road surveyor and engineer, was born in 1750 at St. Cyrus, Kincardineshire, where his father, James, was gardener or factor to the Laird of Morphie. Abercrombie is thought to have been educated at Montrose burgh school but nothing is known of his early life and work. His earliest extant plan is of the Mans [sic] of Kinnalty (1771). Improvement of Scotland's main roads began about 1750 and gathered momentum from the late 1780s, by which time he was playing a leading role in their planning and construction. Unusually for the time, his road layouts were effectively engineered to lines and levels which facilitated horse traction, similar to the practice later institutionalised by Thomas Telford (q.v.) which formed the basis of modern design.

Abercrombie adopted a robust flexible form of road construction; for example, in 1813 for part of the Glasgow to Carlisle Road, he proposed an 18 ft. wide carriageway consisting of a 16-in. layer of broken stone, with the hardest (whinstone) in the top 8 in., and side drainage. This was a stronger form of construction than that subsequently adopted by John Loudon McAdam (q.v.) in his widely promoted and implemented system of road-making from 1816.

Although Abercrombie planned much of the present-day Scottish road network his achievements are less well-known than those of Telford and McAdam, probably because he generally worked on his own. Telford was particularly appreciative of his work and mentioned it in several reports. For example, in one of 26 March 1824 relating to the Maybole to Girvan eastern line of the Glasgow to Portpatrick road, he commented that the road had been executed 'according to the surveys of that excellent road surveyor, the late Mr. Abercrombie'.

An indication of the immense scale of Abercrombie's work is provided by Sir John Sinclair's comment in his *General Report on the Agricultural State ... of Scotland* (1814), that the 'art of reducing [road] ascents and descents by a well-directed line has been brought to its utmost perfection by Mr. Charles Abercrombie, who has lined out above 10,000 [sic] miles of road in Scotland, besides what he has executed in Ireland, with uniform success, even in the most difficult districts'.

Abercrombie's early work included the roads radiating from Perth. The prominent Perthshire road trustee, Sir Patrick Murray, stated in 1814 that he 'never knew an instance where any material deviation took place from lines of road laid down by Abercrombie, when such deviations were not found to be disadvantageous and when they were not afterwards regretted or required to be altered'.

Further north, Abercrombie surveyed and engineered many roads in and around Dundee and Aberdeen. At Aberdeen in 1799 he was described in minutes as an 'engineer of great eminence' and from 1800 to 1807 his work included planning and constructing the main approach from the south and the city's principal thoroughfare, Union Street. Contemporaneously he planned a succession of road improvements in Ayrshire (1802–1807) centred on Girvan. In 1807 he surveyed an alternative line of road from Tarbet (Loch Lomond) to Cairndow (Loch Fyne). In the central belt of Scotland Abercrombie surveyed roads in the Edinburgh area; for example, the Newmills Bridge approaches at Dalkeith on the Edinburgh to Newcastle road in 1814 and much of the east coast main road from Edinburgh to Berwick-on-Tweed. He worked in and around Glasgow and on the Glasgow to Carlisle road as early as 1798 and on another section of this road in 1813. He also put forward proposals for improving some 'military roads', for example, in 1809, over the 'Rest and be Thankful' hill at the head of Glencroe on the Dumbarton to Inveraray road. Abercrombie also prepared a report with plans for at least two harbours: Montrose, in June 1806, estimating the cost of its improvement at £8,223, and Dunure in 1811.

Abercrombie eventually retired to Doonfoot, on the coast near Ayr, where he built the house that was known for over a century as 'Abercrombie House'. Prior to its completion he lived in a herring-boat with a wooden house built on its deck, which he used to sail to locations on the west coast where surveying was required. He died on 13 August 1817 and was buried in Ayr Parish Churchyard where a stone monument still stands in his memory. He had a brother, James, who became a surveyor/engineer in England on harbour and bridge construction, and worked for Rennie on Waterloo Bridge, London (1810–1817), and afterwards as a surveyor at Southwark.

R. A. PAXTON

[Various Scottish Turnpike Trust Minutes; Plans (5001, 658/4, 22477, 6694, 6782, 21854, 21856, 21857, 2028, 2739), SRO; Business records of Robert Stevenson & Sons. MS. Acc. 10706, 128, 363–369, NLS; Ms. S. A. Robertson (n.d.) *The Roads and Bridges of Scotland*, chap. 8, Charles Abercrombie (1750–1817) (now in the papers of the late Prof. Percy Johnson-Marshall, Edinburgh University); T. Telford (1824) Report on roads from Glasgow to Port Patrick; H. Hamilton (1963) *An Economic History of Scotland in the Eighteenth Century* 27; R. A. Paxton (1998) *Road and Bridge Making on Main Routes in and around Dalkeith 1750–1850*]

Publications
1813. Report of Mr. Charles Abercrombie. In: *Reports Respecting a New Line of Road from Hamilton to Elvanfoot*

Works

(Note: Abercrombie surveyed more than 1,000 miles of roads, and surveys survive for Aberdeenshire, Argyll, Angus, Ayr, Berwick, Caithness, Fife, Lanarkshire, Midlothian, Perthshire, Renfrewshire and Sutherland; a selection of works carried out to his recommendations follows)

1787–1794. Perth to Crieff (30 ft. broad (between fences?), maximum gradient 1 in 50, replacing 1 in 10 over the brae of Foulis], on to Comrie, 1812 (Abercrombie's proposed maximum gradients of 1 in 30 were, according to Trust chairman Sir Patrick Murray not implemented in places with very injurious effects)
1789–1808. Perth to Forfar (in four sections)
1790–1791. Perth to Invergowrie
1793–1796. Perth to Dunblane, and Blairgowrie
1798 and 1813. Glasgow–Carlisle Road (parts)
1800–1807. Aberdeen, Union Street and southern approaches
1800–1811. Perth (towards Stirling)–Auchterarder–Greenloaning
1802–1804. New Inn (from Burntisland) to Newport via Rathillet
1802–1807. Roads in Ayrshire in Girvan area
1803–1820. Milnathort via Auchtermuchty to Cupar and St. Andrews
1805. Cranstoun Bridge, Midlothian
1808–1818. Bridge of Earn to Glenfarg, Milnathort to Cowdenbeath, and towards N. Queensferry (at Glenfarg the gradient was improved from 1 in 10 to 1 in 55)
1814. Dalkeith area, Edinburgh–Newcastle Road
1815. Glendevon to Gleneagles (nearly 600 ft. of rises and falls were eliminated and the gradient reduced to one seldom exceeding 1 in 40)

ABERNETHY, George (fl. 1819–1853), mechanical engineer, is first known as manager of a small works in Aberdeen. In 1823 he went to Wales to manage the Dowlais ironworks for John Guest and in 1827 he took over an iron foundry in Southwark. While living at 10 Hermitage Street, London, he was elected a Member of the Institution of Civil Engineers in 1828, and remained a member until 1853 when he resigned. His proposal form was signed by Thomas Rhodes (q.v.) and others involved in the construction of St. Katharine's Dock and he is believed to have become involved in dock work at that time. He then acted as assistant engineer with Rhodes on the timber pier designed by Telford at Herne Bay in 1831. He must have displayed some talent as in 1833 Thomas Telford (q.v.) sought his advice in relation to an opinion he had been asked to give on Seaham Harbour. Abernethy then acted as resident engineer under Henry Robinson Palmer (q.v.) at London Docks.

He had two sons, George and James, the latter becoming President of the Institution of Civil Engineers in 1880.

MIKE CHRIMES

[T. Telford (1833) *Diary*. NLS; *Membership Records*. ICE]

ABRAHAM, Robert (1774–1850). The son of a builder, Abraham trained and worked as a London building surveyor until in middle age he established a successful architectural practice (Colvin lists works between 1819 and 1842). His contribution to engineering was the concrete raft foundation for the buildings of the Westminster New Bridewell prison in Tothill Fields close to the Thames, and especially the process of groundwater lowering introduced by him (in 1830) on this job. The raft covered an area of rather more than 5 acres. It was 7 ft. thick, increased to 9 ft. under the boundary walls at a depth of 12 ft. below high tide level, and contained 60,000 cu. yd. of concrete.

Concrete raft foundations had been used earlier, in similar situations on Thames alluvium, by Sir Robert Smirke (q.v.) at Millbank Penitentiary (1817) and the Custom House (1825). Abraham's innovation was to lower the groundwater before excavation to a level below the base of the raft and keep it lowered during construction until the concrete had hardened. By this procedure, described by Pasley (1838) as 'perfectly original', excavation and concreting could be carried out in the dry.

Abraham's specification required the contractor to sink a 6 ft. diameter well near the middle of the area, some 300 ft. from the boundary walls, to a depth of 20 ft. below high tide level (into gravel) from which pumps operated by an 8 hp steam engine would lower groundwater to a depth of a few feet below the base of the raft. Excavation could then proceed and, when completed, concrete was pitched in from a height of 9 ft. and immediately trodden down in 9-in. to 12-in. layers. The concrete specification, an advance on previous practice, called for 1 part of Dorking or Merstham lime, 2 parts of sand and 6 of gravel (by volume). These to be well mixed, then 4 cu. ft. of water added for every 3 cu. ft. of lime, and then mixed again. Dorking lime contains about 8% of clay and when burnt produces hydraulic lime capable of setting under water.

Pumping, excavation and concreting were carried out during 1830–1831 with complete success at a cost of £30,000. The buildings were completed in 1834.

Abraham passed on his extensive knowledge of building practice to a number of pupils. They presented a gold box to him on retirement in 1842. He died on 11 December 1850 and was buried in Hampstead churchyard.

A. W. SKEMPTON

[George Godwin (1836) On the nature and properties of concrete, *Trans Inst British Architects*, **1**, 1–37; Plan of Westminster New Bridewell, RIBA Library; Charles Pasley (1838) *Observations on Limes, Calcareous Cements and Concrete*; Colvin]

Publications
1836 Letter to Thomas Donaldson on the concrete used at Westminster New Bridewell, *Trans Inst British Architects*, **1**, 38–39

ACONCIO (ACONTIUS), Jacopo (1492–1566).
Jacopo (or Giacomo) Aconcio was a scientist, historian and famed theologian, advocating religious tolerance. It is typical of Elizabethan pragmatism that he should have been invited to work in England as a reserve military engineer who 'was not to be compared to (Giovanni) Portinari (q.v.) for deep judgement, but of service in matters of fortification' yet it was five years before he was actually asked to be of service by reporting on the defences of Berwick. In the meantime he earned a living as a civil engineer in areas of water power and land reclamation and became an English subject on 8 October 1561.

Aconcio's year of birth, in Ossano in the Tyrol, now in northern Italy, is sometimes given as 1492, but an alternative post-1500 date seems more likely as there is a paucity of sightings before he is found as a notary in Trent in 1548. From 1549 to 1556 he was either at the Court of Charles V in Vienna or in the field in Piedmont and Lombardy; in 1557 he was in Milan as Secretary to Cardinal Madruzzo. During this period he managed to learn the Italian method of fortifications from two of the Emperor's most famous engineers: Count Francesco Landriano, and Giovanni Maria Olgiati who as well as designing the fortifications of Milan also worked on the fabric of Milan Cathedral.

Aconcio's Protestantism forced him to flee from Italy in 1557 and he was subsequently invited to England, arriving in 1559, then being given an annuity of £60 per year. Two months after his arrival he petitioned Queen Elizabeth for a patent for 'new designs of machines of all sorts that use waterwheels and a new design for building furnaces for dyers and those who make beer with a great saving of fuel', adding that 'those who at trouble and expense provide inventions that serve the commonwealth should receive the privilege of monopoly'. This is arguably the first English description of the purpose of a patent.

The patent when granted in 1565 was for 'machines for grinding cutting and crushing wood by the use of water powered machinery'. In some cases power was to be supplied by the wind, so that whatever the wind direction the mill would operate while the rest of the structure remained unmoved, apparently a variation of the vertical-axle windmill used in Italian engineering. A further patent, for a perpetual motion syphon, predated Zonca's concept by four decades.

In 1563, by Act of Parliament, he gained the right to attempt to reclaim 2,000 acres in Plumstead Marshes which had once been good pasture land but had been devastated by floods that had also silted up the Thames. A Commission reported in 1566 that 'six hundred acres thereof were then won and inned with [cross] walls and banked [to protect] from the water and flood of the river Thames'. The first attempt was flooded, but land was soon regained and the partial success of this scheme may well have influenced the many drainage schemes which arose in the next decades.

Aconcio was appointed to the Commission to consult on the defences of Berwick in 1564, the other Commissioners being Sir Richard Lee (q.v.), William Pelham and Portinari, whose 'deep judgement' meant he was paid £27 13s 4d to Aconcio's £20. At £130,000 Berwick fortifications were the most expensive building work of the Elizabethan period, costing the Crown ten times more than any other military structure. Unsurprisingly therefore the Commissioners were to report on the cost in general as well as on the specific design of Brass Bastion. Equally unsurprisingly there was no agreement among any of their numerous reports. Aconcio's views were that the existing scheme was inadequate for protecting the bridge over the river Tweed. The walls were too high and inadequately sloped, and therefore vulnerable to artillery, and the flankers and casemates were too small. More controversially, he disapproved of reinforced earthworks for permanent structures, considering that masonry was cheaper in the long run, as initial savings in cost would be eaten up by the expense of moving the earth to replace rotting beams and fascine work, and that a masonry wall resting on earth backing was liable to far less damage by cannon. In vindication of his opinions, the Flankers were enlarged to 34 ft. and 50 ft., with tunnels 9 ft. wide; the batter was increased to 26° and it was agreed that the river wall should be completely rebuilt.

Aconcio's last known action was the turning over of his financial share in the drainage scheme to his partners in 1566 to reimburse their original £5,000 investment. It is believed that he died in either 1566 or 1567.

SUSAN HOTS

[PRO: Calendar of State papers foreign, Elizabeth 1558–1559, 1564–1565, ; Calendar of patent rolls Elizabeth I, 1560–1563, 1563–1566; Calendar State papers domestic, Elizabeth 1601–1603, addenda 1547–1565; Cottonian ms TITUS F.XIII, fols. 232–234, BL; Sir William Dugdale (1662) *History of Imbanking and Draining of Divers Fens and Marshes*; I. Maclvor (1965) The Elizabethan fortifications of Berwick on Tweed, *Antiquaries Journal*, **65**, 64–96; L. White Jr. (1967) Jacopo Aconcio as an engineer, *American Historical Review*, **3**(72), 425–444; E. R. Briggs (1976) An apostle of the incomplete reformation, *Proc Huguenot Society*, **22**(6), 481–495; J. R. Hale (1977) Renaissance fortifications: art or engineering, 35; *King's Works*, **4**, 409, 657–658]

Publications
Publications include:

1558. *De Methodo, hoc est, de Recta Investigandarum Scientarium Ratione*
1585. *Ars Muniendrum Oppidarum* [now lost]

Works

Drainage 600 acres Plumstead marshes, *c.* £5,000

ADAIR, John, FRS (1647–1718), mathematician, geographer and surveyor, made the first competent survey for a canal from the Forth to the Clyde. He was licensed in 1681 to make maps of the counties of Scotland for a fee of £100 spread over two years, and he began surveying in the counties of the south-east. He soon moved into hydrographic surveys of the eastern coastline and from 1686 onwards a tonnage was levied on ships trading in and to Scotland to support his work. He continued to work on the coastal surveys and less constantly on county maps throughout his life, although suffering from gout from about 1710. He tried, sometimes by advertisement to local residents, to collect information on the antiquities, geology, botany, wild life and general curiosities of the landscape and include it in his surveys; before he died he had surveyed many of the counties and the whole mainland coastline and also visited, though not measured, most if not all of the northern and western isles. It was very expensive work, needing, at its peak, a chartered vessel and twenty men. He had great difficulty in obtaining payment of the receipts of the tonnage dues and he lived in constant debt.

Adair worked partly in concert and sometimes in conflict with Sir Robert Sibbald (1641–1722) who had proposed at one time to publish descriptions of all the counties of Scotland, and with John Slezer (d. 1717) who was engaged in making 'prospects' (i.e. landscape views) of the towns, major buildings and some natural phenomena of Scotland. Slezer's drawings were engraved and published in 1693 as *Theatrum Scotiae,* the verbal part of the book being extracted from Sibbald's previous publication *Scotia Antiqua et Moderna.* Slezer's work was also entitled to funds from the tonnage levy, which caused more than one fierce argument between him and Adair.

From the beginning of his career Adair had also undertaken surveys on the estates of the nobility. This brought him into contact with influential landowners, notably Sir John Clerk of Penicuik, and leading architects who worked for them, including Sir William Bruce at the outset of Adair's career and, later, James Smith (q.v.) and Alexander McGill. His proposal for a canal from the Forth to the Clyde is known mostly from several letters of Sir John Clerk which suggest that his survey was made about 1703, but the date may be too early. A petition on his behalf was made to James Duke of Queensberry, Clerk's father-in-law, in 1707 by William Paterson, founder of the Bank of England, recommending Adair's ideas to Her Majesty Queen Anne. Adair was suggesting the making of wet and dry docks in the 'firth of Edinburgh' and 'a passage, or canal, through which boats and barges might, by means of sluces [i.e. locks], be conveyed from the Forth to the Clide [*sic*]'. He had considered two possible routes, which were basically similar to the routes considered by Smeaton when the canal project came alive in the 1760s: one from the Forth above Stirling to Loch Lomond and from the Loch to the Clyde, and another from the river Carron through the central lowlands to a point on the Clyde 5 or 6 miles downstream from Glasgow.

The positive outcome of Paterson's petition was a commission issued by the Queen in 1710 to a group of men including Adair to form a substantial port at Bo'ness, with sub-ports (presumably landing places subject to Bo'ness for customs control) at Culross, Queensferry, Inverkeithing and Alloa; Bo'ness was expected to supersede Leith as the main port in the Forth. Adair appears to have taken the lead and recommended that there be a dry dock for cleaning ships and a pier but no wet dock because there were adequate anchorages in the firth. He proposed that stone for construction should be brought from a quarry on the north shore owned by Alexander McGill. How much development took place is not clear, but work was still under way in 1715.

His canal proposals were revived when he 'travelled twice over the ground' in company with McGill and James Smith (q.v.), who were partners as architect/builders and together had interests in quarrying and mining; also with the party was George Sorocold (q.v.) who was brought to Scotland more than once in the years 1710–1715, initially to advise the Earl of Mar about drainage of his mines near Alloa. They judged that adequate water supply was available, planned necessary locks and estimated the cost at £30,000, although saying that twice that sum would probably be justified by the potential profits. Sorocold's presence probably shows that this survey took place after 1710.

Adair, Smith and McGill also co-operated in a survey of the Bridge of Dee which was reported in a letter-book of Aberdeen Town Council in July 1713 (Adair had already been made a burgess of Aberdeen in 1706). In August and September 1716 he was in Perth reporting on floods in the Tay and drawing river plans.

John Adair bore the title of 'the King's Geographer in Scotland' in 1688 and was later known as 'Hydrographer for Scotland'. He visited Holland in 1687 and was elected a Fellow of the Royal Society of London the following year. He spent considerable sums on surveying instruments and paid £70 to bring the engraver, Moxon, from Holland to Scotland about 1689 to engrave some of his maps. He married Jean Oliphant about 1687 and had several sons and daughters. He died on 15 May 1718 leaving debts and claims for money due which occupied his widow and family for several years. The widow eventually received a pension from the government of £40 per annum. A number of his maps and drawings lay in manuscript form at the Exchequer Office and were consulted by others in subsequent years, but have since disappeared.

TED RUDDOCK

[DNB; H. R. G. Inglis and others (1934) *The Early Maps of Scotland*; P. G. Vasey (1992) The Forth–Clyde Canal: John Adair, progenitor of early schemes, *Journal of the Railway and Canal History Society*, **30**(7, No. 150), 373–377, and ms. sources therein; P. G. Vasey (comp.) and I. Hill (ed.) (1998) John Adair, Geographer and Surveyor: a source list, unpublished dossier, SRO (made available by Dr. Tristram Clarke)]

Publications
Many maps and charts including:

1683. Engraved maps included in Sir Robert Sibbald's *Scottish Atlas:* The East part of Fife; East Lothian; Mid Lothian; West Lothian; Strathern, Stormont and the Carse of Gowrie; The Country about Stirling; Clackmannan-shire with the Turnings of the Forth; West of Scotland, containing Clidsdale, Nithsdale, Renfrew and the Shires of Air and Galloway

c. 1683. *River Forth Estuary*, engraved by David Lindsay

1689. *The South of England with the Coast of France and Flanders*, engraved by Moxon

1703. *Description of the Sea Coasts and Islands of Scotland.* Includes *A True and Exact Hydrographic Description of the Sea Coast and Islands of Scotland ...* [1688] and *The Toune and Water of Montross* [*sic*] [1693]

Many of his County surveys were published posthumously by the engraver Richard Cooper

ADAM, John (1721–1792), architect and builder, was the first son of William Adam (q.v.) and Mary Robertson. He was born (or baptised?) on 5 March 1721 at Abbotshall, Fife, was later sent to Dalkeith Grammar School but afterwards withdrawn 'from the pursuit of his studies at the College' (Edinburgh University) to assist with his father's extensive business. John worked first at masonry and architecture, then took progressively increasing responsibility for supervision and management and, on William's death in 1748, full control of the business.

He also inherited his father's lucrative appointment as Mason to the Board of Ordnance in North Britain and he quickly made his brother Robert (q.v.) a partner in this. They became contractors for the construction of Fort George on Ardersier Point, much the largest military fortress yet constructed in Scotland and designed by Colonel William Skinner (q.v.). Construction of the main bastions and curtains extended over ten summer seasons and involved huge earthworks, masonry retaining walls, casemates and bastions, to be followed over a further ten years by the construction of the buildings within.

John seems to have left architectural design work largely to Robert until 1754 when Robert left for a Grand Tour on the Continent. From then on John was the everyday Adam partner in Scotland, designing, building and managing projects for private clients as well as for the Ordnance Board.

His architectural works have been listed by Colvin but he also engaged, like his father, in the manufacture and supply of bricks and tiles, and probably in coal-mining. He took two leases of ground to quarry near Aberdeen in 1766–1767, noting that he already had a business of stone supply there, and he shipped paving stones to London (possibly meaning granite setts) for at least ten years. During the 1760s he became a partner in the famous Carron Iron Company and took an active part in its affairs.

He became an acknowledged expert on bridge design and construction. As an adjunct to the task of supervising the building of Inveraray Castle for the Duke of Argyll, a task laid on William in 1744 and inherited by John, he supervised the construction in 1747–1749 of Garron Bridge at the head of Loch Shira. It was a bridge on a military road but for which the Duke's architect, Roger Morris, had provided a special design of architectural quality. In about 1756, when another military bridge was required much nearer to the Duke's castle to take the road into the new village of Inveraray, Adam made the design, a three-arch bridge over the river Aray with Gothic detail and tall castellated turrets at piers and abutments, all matching the architecture of the castle. It was destroyed by a flood in 1772 but another bridge within the estate, called the Garden Bridge, which he designed in 1759 still stands; with a single arch of 60 ft. span it also has thick turrets at the abutments but no Gothic details. Both of these bridges were constructed by local builders, but in 1760–1762 Adam was both designer and builder of another, rather larger, estate bridge of three arches at Dumfries House, by Cumnock in Ayrshire, and which also still stands, faithfully restored in 1970 after serious mining subsidence. It bears four obelisks on the parapets over the middle arch, a feature obviously derived from William Adam's famous bridge at Aberfeldy. A notable feature of the Inveraray Garden and Dumfries House bridges is their elliptical arches, almost the first bridge arches of that form to be built in Britain.

In the 1760s Adam was consulted repeatedly by the leaders of the Town Council of Edinburgh about designs submitted by others for the city's North Bridge, about details of the contract made with William Mylne (q.v.) in 1765, about the quality of materials during construction, and later together with John Smeaton and John Baxter (qq.v.) to report on its partial collapse in 1769. When plans for a bridge over the North Esk just north of Montrose were under discussion in 1769–1770 the bridge committee requested a design from Adam, which he declined to give them, but introduced Baxter as a possible architect and contractor while Adam himself became unpaid advisor to the committee. He visited the site at least three times, drafted the contract and, after the committee had decided on Andrew Barrie and Patrick Brown as contractors, made the final choice between the designs of Baxter and Barrie in favour of the latter. In 1783–1784, when a committee was preparing for a bridge (now called Bridge of Dun) across the South Esk 2–3 miles from Montrose, Adam again visited, fixed the site for the bridge, sent a sketch design with recommendations for the dimensions of arches, foundations and other details, commented on the draft contract and design(s) submitted and, after the completion of the bridge by Alexander Stevens (q.v.) in 1786, engaged on behalf of the committee an Edinburgh architect, William Christie, to make a final inspection and report. Again Christie was paid for his visit but Adam apparently not.

In 1764 he made a brief return to practical bridge work when in financial difficulties following the collapse of Fairholme's bank. In response to a request from Smeaton he agreed to direct the site work for the new bridge of Perth (Smeaton (1812) *Reports*, I, 181) He seems, in the records, to have limited himself to the masonry while John Gwyn (q.v.) directed the wright-work; he employed two overseers, initially William and James Mitchell, one at the quarry and one at the bridge. Gwyn increasingly took over the whole project and Adam is not mentioned after 1767.

The financial disaster for Adam foreseen in 1764 was averted, but the four brothers set up at that time a joint firm called William Adam & Co. which Robert and James later used for various enterprises which failed, including the Adelphi development in London, and John's capital was greatly depleted by their adventures, leaving him with most of his property mortgaged and insufficient money to improve his estate of Blair Adam as he wished.

He married Jean Ramsay of Kirkcaldy on 8 July 1750. His eldest son, William, became a prominent advocate first at the Scottish Bar and later at the English Bar; another son, John, died while at Eton. John Adam died in Edinburgh on 25 June 1792, three months after the death in London of his brother, Robert.

Two portraits of him exist, a youthful likeness painted *c*. 1750 by Francis Coates which is at Blair Adam and is reproduced by Fleming as plate 21, and a paste medallion by James Tassie in 1791 held at the Scottish National Portrait Gallery and reproduced by Ruddock as Fig. 114.

TED RUDDOCK

[Perth Bridge Commissioners Papers, PE23, and Minutes 1765–1879 and account books, etc., PE1/7/1–6, at Perth and Kinross Council Archive, the A. K. Bell Library, Perth; Minutes of Northesk Bridge Commissioners 1765–1880 (2 vols); J. Smeaton (1812) *Reports*; Montrose Town Council Minutes and Bridge of Dun Papers, in Angus Archives at Montrose Library; William Adam (1812) *Vitruvius Scoticus* 1980 edn by J. Simpson with biographical note on John Adam; T. C. Barker (1958) Aberdeen Quarries, *The Architectural Review*, Feb.; R. H. Campbell (1961) *Carron Company*; J. Fleming (1962) *Robert Adam and his*

Circle; I. G. Lindsay and M. Cosh (1973) *Inveraray and the Dukes of Argyll*; A. J. Rowan (1974) After the Adelphi: forgotten years in the Adam brothers practice, *Journal of the Society of Arts*, **123**, 659–710; E. C. Ruddock (1974) The building of North Bridge, Edinburgh 1763–1775, *Transactions of the Newcomen Society*, **47**, 9–33; I. McIvor (1976) Fort George, *Country Life*, 12 & 19 August; Ted Ruddock (1979) *Arch Bridges and their Builders 1735–1835*; Colvin (3)]

Publications

1812. Report of Messrs. Smeaton, Adam and Baxter concerning the Bridge of Edinburgh [dated 22 August 1769], in J. Smeaton, *Reports ...* **3**, 218–222, and in *The Scots Magazine*, **31**, 464–467.

Works

1747–1749. Garron Bridge, near Inveraray, supervisor

1748–1759. Fort George, Arderseir Point, Invernesshire, for the Board of Ordnance, contractor for ramparts, buildings, etc.

1754. Jetty at Inveraray, 100 ft. long, design (constructed by others 1758–1762, extended *c*. 1805)

1757–1759. Aray Bridge, Inveraray, three arches, middle arch 50 ft., design (collapsed 1772)

1758–1761. Garden Bridge, Inveraray, one elliptical arch 60 ft. span, cost £752, design

1760–1762. Bridge at Dumfries House, three elliptical arches, middle arch 50 ft., cost £543, design and contractor

1764–1767. Perth Bridge, directed early quarrying and masonry

ADAM, Robert, FRS (1728–1792), architect, was the second son of William Adam (q.v.) and Mary Robertson, born on 3 July 1728 in Kirkcaldy but brought up in Edinburgh. He attended the High School to learn Latin and at fifteen matriculated into the College, now the University of Edinburgh. He left in 1745 and after a very serious illness joined his father's practice in 1746. When William died two years later Robert was taken into partnership in his brother John's (q.v.) appointment as Mason to the Board of Ordnance in North Britain, which is said to have earned him £5,000 in the next six years. During the construction of Fort George he spent several full summers at the site, supervising huge earthworks and masonry construction. In 1754 he departed on a well-funded Grand Tour of Europe, basing himself in Rome and travelling, with assistants whom he employed, to several parts of Italy and, as his special contribution to the archaeology of the classical world, to Spalatro (now Split) on the Dalmatian Coast to record the ruins of the Emperor Diocletian's Palace. He was back in London at the beginning of 1758 ready to establish an architectural practice on the friendships he had made in Italy with English nobles and gentlemen.

After the summer seasons he had spent at Fort George between 1749 and 1753, Adam's activity was never primarily engineering. His mind was focused on architectural design (his works listed by Colvin), but with an ambition to shape whole cities and landscapes. He even canvassed busily while in Rome for a commission to plan the reconstruction of Lisbon after the great earthquake of 1755, but without success. His surviving drawings therefore show very few structural or constructional details, although his designs seem to be informed by a good understanding of construction—an assertion which cannot be made of the work of his brother, James (1734–1794), who joined him in the London practice in 1763.

An early design attributed to Robert was that of the mausoleum to William Adam in Greyfriars Kirkyard, Edinburgh (c. 1753), an arched and domed stone structure in which the masonry was extensively tied both vertically and horizontally by embedded wrought iron bars, an early use of such tying which may be related to his father's experience in the use of iron and/or his brother, John's, interests in ironwork.

Bridges were of great interest to Robert. While on the Continent he made many sketches of bridges and design esquisses for bridges, borrowing designs or motifs from Palladio, Piranesi ('bridges of magnificence') and other sources, and conceiving some bridges simply as features in landscape. Throughout his life he painted landscapes, both real and imaginary, in which bridges were a very common feature, most of them arched, some ruinous, but always drawn realistically with practical details showing.

These studies were carried through to his architectural practice. About ten bridges were built on the estates where he designed or altered houses but they are only a fraction of the bridge designs he drew. Every bridge was conceived as a feature of the landscape and structural considerations were secondary, but nevertheless rational. He therefore adopted without inhibition segmental arches (contrary to the views of architectural purists who favoured the Roman semicircle) and a wide range of masonry finishes and decorative motifs. A few of his unbuilt designs were for bridge ruins in the landscapes, and the winding 'antique' viaduct (1780) approaching Culzean Castle is a deliberate landscape ornament, although it also formed the main access to the Castle and wisely incorporated a clay layer under the roadway which excluded rainwater from the structure until it was breached during installation of the first electricity supply cable to the Castle in the 1950s, leading to extensive decay. Good examples of Adam's other estate bridges are at Audley End, Essex (1764), Kedleston Hall, Derbyshire (1770–1771), where the balusters are of cast iron painted to match the stone, and Dalkeith House, Midlothian (c. 1792).

Adam made a number of drawings for a new three-arch public bridge at Ayr during the 1780s and was paid later for his design, but a different bridge was built (see: STEVENS, Alexander). A similar design, but of five arches, was proposed for a bridge at Saltmarket Street in Glasgow but as the drawing is dated December 1793 it was presumably made by James. It also was not adopted.

Several of Adam's grandiose plans for developments in cities involved bridges. The first was drawn as a long elevation of two-storey buildings apparently fronting a viaduct stretching across the valley between the Old and New Towns of Edinburgh, probably drawn about 1752 and unsigned but attributed to Robert Adam (see MacInnes (1993)). The second produced Pulteney Bridge in Bath in 1768–1774, and was intended to give access to a large town development for which he made plans in 1777 and 1782 but which were not pursued. The three-arch bridge was designed as a level street lined with elegant shops on both sides, a unique creation in its time. After about twenty years the bridge required underpinning, which was directed by Thomas Telford (q.v.), and the shops have been much altered, but all survive. In the case of the South Bridge in Edinburgh (c. 1784), a much higher viaduct, eventually of nineteen arches, was required, flanked on both sides with buildings built up from the ground. Adam opted to design a very grand street with unified classical facades but his ideas were not accepted and he had to go to law to obtain his fees for the designs.

The business speculations of Robert and James within the family firm William Adam & Co. (see: ADAM, John) included the acquisition of the patents for two different stucco materials and the creation of an enormous building organisation to construct their Adelphi development in London; although achieving prominence for them, these activities in sum brought all the family to the edge of bankruptcy. Throughout his life Adam possessed an enormous capacity for work, as well as his brilliant design talent and unquenchable ambition. He never married and, perhaps more surprisingly, retained the love of his sisters through all his escapades.

Robert Adam died suddenly in London on 3 March 1792 and was buried in Westminster Abbey. He had been a member of the Society of Arts and Fellow of the Royal Society, as well as having been Member of Parliament for Kinross-shire from 1768 to 1774. A portrait of him in oils, attributed to David Martin, is in the National Portrait Gallery, London; the Adam family retain another likeness on an ivory plaque; and the Scottish National Portrait Gallery hold three paste-relief medallions bearing his head by James Tassie, but at least two of them are said by Colvin to have been made posthumously.

TED RUDDOCK

[MS drawings of most of the bridge designs of Robert and James Adam, Sir John Soane's Museum, London; other design drawings in the Adam collection at Victoria and Albert Museum, London; T. Telford (1830) Bridge, *The Edinburgh Encyclopaedia*, **4**, 524–5 and pl. 96; A. T. Bolton (1922) *The Architecture of Robert and James*

Adam, 2 vols.; John Fleming (1962) *Robert Adam and his Circle*; Ted Ruddock (1979) *Arch Bridges and their Builders 1735–1835*, 116–120; David King (1991) *The Complete Works of Robert and James Adam*; Margaret H. B. Sanderson (1992) *Robert Adam and Scotland. Portrait of an Architect*; *Architectural Heritage*, **4** (1993), 9 papers on Robert Adam; R. MacInnes (1993) Robert Adam's public buildings, *Architectural Heritage*, **4**, 10–22; Jean Manco (1995) Pulteney Bridge, *Architectural History*, **38**, 129–145; Colvin (3)]

Publications

1764. *The Ruins of the Emperor Diocletian at Spalatro*
1773–1779. *The Works in Architecture of Robert and James Adam*, 2 vols

Works

(Architectural works without specific engineering interest are not listed)

1750s. Fort George, Invernesshire, rampart almost 1 mile long and barracks, joint contractor
c. 1753. Mausoleum to William Adam, Greyfriars Kirkyard, Edinburgh
1764. Audley End, Essex, three-arch bridge in estate, middle arch 25 ft. span
1764. Kedleston Hall, Derbyshire, three-arch estate bridge over a weir, with cast iron balusters in the parapet (built 1770–1771) *c.* 1768. Osterley Park, Middlesex, single-arch estate bridge, now ruinous
1769–1774. Pulteney Bridge, Bath, three equal segmental arches carrying level public road and lines of shops on both sides (shops much altered)
1773. The Lion Bridge, Alnwick, public bridge built for the Alnwick Estate, three arches 40, 48 and 40 ft. span carrying level road, castellated parapet bearing iron sculpture of a lion
1780. Viaduct for access to Culzean Castle, Ayrshire, six small arches, irregular in size and spacing
1782–1783. Audley End, Essex, tea-house bridge, one arch carrying tea house
1780s? Mistley Hall, Essex, small single-arch bridge, since mostly rebuilt
1788. Kirkdale House estate, Dumfries and Galloway, three-arch bridge over a small river
1792–1793. Montagu Bridge, Dalkeith House estate, Midlothian, one semicircular arch 70 ft. span

ADAM, William (1689–1748), architect, builder and industrialist in many fields, was the son of John Adam (*c.* 1650–1715), master mason in Linktown of Abbotshall, and Helen Cranstoun. He was baptised on 24 October 1689, educated at Kirkcaldy High School, served his apprenticeship as a mason and then continued to work in his father's firm. He travelled at an early age to the Low Countries where he studied construction methods and industrial machinery, including the production of bricks and tiles, and inland navigation structures. In 1714 he entered into partnership with William Robertson of Gladney to manufacture bricks and pantiles, the latter made in Scotland for the first time. He also engaged, for most of his life, in coal-mining and salt-panning, and in the import of timber, marble, tiles, glass and other goods to his store at Leith. His activities in Kirkcaldy in 1728 were listed by a visitor, Sir John Clerk of Penicuik, as 'barley mills, timber mills, coal works, salt pans, marble works, highways, farms, houses of his own a-building and houses belonging to others not a few'.

In 1716 he married his partner's daughter, Mary Robertson, and took up residence in Gladney House, Linktown, which he had built a few years earlier. His business was run from there until 1728 when he moved his family and the centre of his business to Edinburgh, having become Scotland's most esteemed and busiest architect (his works listed by Colvin). In 1731 he purchased an estate in Kinross-shire which, with several later purchases of adjacent land, became a country seat of 3,000 acres; he called it Blair Adam and it remains in the family's ownership today, with a house of modest size, first built by him.

Of his four sons, three, named John (q.v.), Robert (q.v.) and James, all became architects like their father. There were also five daughters, some of whom became associated with their father's and brothers' work, keeping records, making drawings and/or house-keeping for them.

Adam contracted for the building of most of the large houses and public buildings which he designed, and also worked his mines and saltpans, mills, etc., all of which involved what is now called structural engineering. There are, for instance, major wooden roof structures in the wings and stables of Hopetoun House, Midlothian, where his work extended from 1723 to his death and after, and at Duff House, Banffshire, built 1735–1739, while the new Royal Infirmary in Edinburgh (1738–1748) had a very large plan with four-storey walls of masonry. When he reported on the roof of the Palace of Holyroodhouse, Edinburgh, in 1733, he detailed the repairs required with an estimated total cost of £4,000. Also during the 1730s, he inserted a new floor of span 29 ft. and structural depth of only 10.75 in. into Yester House, East Lothian, involving a sophisticated system of timber beams trussed with wrought iron.

Adam designed and built stone arch bridges of modest size at Dalkeith House, Midlothian (1741–1742), at Broxmouth, East Lothian, and the large West Bridge at Cullen House, Banffshire (1743). After he was appointed Clerk and Storekeeper of the Works in 1728 and Mason to the Board of Ordnance in North Britain in 1730, his largest bridge, of five arches and 400 ft. long, was built to the order of General Wade (q.v.) to span the river Tay at Aberfeldy. A mixed workforce was assembled from Scotland and the northern counties of England, and the foundations required twelve-hundred piles to be driven. Adam provided a pretentious design, but the most prominent features, four tall obelisks on the parapet over the middle arch, were added by late request of the General (see letter from Adam to Mungo Greene, 18 October 1734, GD220/5/1318/3 at SRO). Adam's unique profile of the parapet at Aberfeldy was widely copied by other designers, including his sons. In 1744 he submitted a proposal for an even larger bridge over the Tay, at Perth and of nine arches. It was estimated to cost £12,529, but was not built (MS estimate at Perth Museum).

Subsequent works on military forts in the Highlands involved considerable earthworks, but much the largest ramparts, those of the new Fort George at Ardersier Point, were constructed by William's sons after his death.

In 1726 William Adam made an end-to-end survey of at least one route for a canal from the Forth to the Clyde, accompanying the antiquary Alexander Gordon (1692–1754) whose antiquarian interests had brought him into contact with Sir John Clerk of Penicuik. Clerk was patron of John Adair (q.v.) who made the first careful survey for the canal, and held some of Adair's manuscripts as well as having access to others by his office of Baron of the Exchequer, which had financed Adair. Gordon wished to revive the canal project and Sir John persuaded him to take Adam, his esteemed friend, on the survey. They followed the route from the Carron's mouth on the Forth to the Kelvin's on the Clyde. Adam found the project feasible, proposing 'sluices' (meaning locks) of the type he had seen raising ships from sea-level into the Ostend–Bruges canal on his early visit to the Low Countries. In a letter of 1741 he said 'I am as much master of the form and disposition of [sluices or locks] as of the building of a house' and that he was ready to make another survey. He believed the canal could be made for £200,000 at most (letter to Sir John Clerk of Penicuik, 6 October 1741, GD18/4736 at SRO).

William Adam died on 24 June 1748 and was buried in Greyfriars Kirkyard, Edinburgh. His sons erected an impressive classical mausoleum over his grave about five years later. It contains a marble bust, as well as reliefs in marble of two of his buildings. A portrait of Adam in 1727 by William Aikman is held by the family and is reproduced by Fleming, Gifford and Ruddock.

TED RUDDOCK

[Letter to Sir John Clerk of Penicuik, 6 October 1741, GD18/4736 at SRO, J. Fleming (1962). *Robert Adam and his Circle*; Ted Ruddock (1979) *Arch Bridges and their Builders 1735–1835*; J. Simpson (1980) Introduction and notes to facsimile edition of William Adam (1812) *Vitruvius Scoticus*; M. Stancliffe (1986) The repair of the saloon floor at Yester House, Gifford, *Association for Studies in the Conservation of Historic Buildings, Transactions*, **11**, 3–10; J. Gifford (1989) *William Adam 1689–1748*; J. Simpson (1989) *William Adam 1689–1748. A Tercentenary Exhibition; Architectural Heritage I* (1990), nine conference papers on William Adam, including W. R. M. Kay, 'What's his Line? Some recent research discoveries', with sources quoted; P. G. Vasey (1992) The Forth–Clyde Canal: John Adair, progenitor of early schemes, *Journal of the Railway and Canal History Society*, **30**, 373–377; Colvin (3)]

Publications
1812. *Vitruvius Scoticus*. Drawings, most made by himself and engraved in his lifetime, stored and eventually bound and published in 1812. (Facsimile in 1980, ed. James Simpson)

Works
1726. Survey for a canal from the Forth to the Clyde
1730s. Works (unrecorded?) on buildings and fortifications in the Highlands of Scotland for the Board of Ordnance
1730s. Yester House, Gifford. Trussed floor of saloon, 29 ft. span
1733–1735. Tay Bridge, Aberfeldy, designed and directed construction, five arches, middle arch 60 ft., cost £4,095
1741–1742. Masonry bridge on the estate of Dalkeith House, Midlothian
n.d. Bridge at Broxmouth, East Lothian
1743. West Bridge at Cullen House, Banffshire, one large arch, c. 82 ft. span

AGAS, Ralph (c. 1545–1621), land surveyor, lived at Stoke-by-Nyland in Suffolk and, starting his career c. 1566, developed an extensive practice. He was one of the earliest in England to use the plane-table and to adopt the theodolite after its description by Thomas Digges (q.v.) in 1571. From a letter written in 1592 it appears that his own theodolite had a 20-in. diameter horizontal circle with a 12-in. vertical semicircle. For distance measurements he used a steel wire 2 poles (33 ft.) long linked at 1 ft. intervals, the precursor of Edmund Gunter's surveying chain of 1620.

In March 1589 the Privy Council recommended the Brabant engineer Humphrey Bradley (q.v.) and the surveyors John Hexham (q.v.) and Agas as being well qualified to 'view and platt' the fens and marshes in the counties of Cambridge, Northampton, Huntingdon and Lincoln and 'to observe the true dyssentes of waters and qualyties of the soile' in order to investigate the feasibility of draining these lands. Agas and Hexham had already, in 1588, surveyed South Holland in relation to a scheme by George Carleton (q.v.) for improving the drainage of this part of south Lincolnshire. In response to the question of general draining of the Fens, Hexham surveyed between the Witham and the old Nene; his map still exists in Lord Burghley's collection now at Hatfield House. Agas probably surveyed the area south of the Nene, although no record seems to have survived. In a letter he wrote to Burghley in 1597, on the late payment of his fees, Agas reminded him that he had 'some time ago platted those grounds and well considered the quantities of their waters, gauged their ascents and noted their slow passages'. In other words, he had taken levels along the rivers.

Bradley reported on the investigations in December 1589. His report included the crucial observations that the greater part of the silt lands in South Holland and Norfolk Marshland lay 6 or 7 ft. lower than the peat fens further inland and that these fens—today lowered by three hundred years of drainage—were above high sea-level. Thus they could be drained by gravity flow provided the courses of the rivers wandering across the fens were shortened by new channels leading to the outfalls. Nothing came from these investigations at the time, although they provided a basis for the scheme put forward by John Hunt (q.v.) in 1605 and acted upon by the Earl of Bedford and his followers from 1631 to 1656.

In 1596 Agas published *A Preparative to Platting of Landes ... Shewing the Diversitie of Sundrie Instruments Applied Thereto*. He mentions his thirty years' experience as a surveyor and the fact that for twenty-five years he had used the plane-table, but recent experience of surveying large areas (in the Fens?) had shown the advantage of the theodolite, here named for the first time instead of the term 'topographical instrument' used by Digges.

Agas was still active in 1606, now with forty years' experience, but may have been nearing retirement. He died on 26 November 1621.

A. W. SKEMPTON

[Andrewes Burrell (1642) *A Briefe Relation Discovering Plainely the True Causes why the great Levell of the Fenns ... have been Drowned ... and How they may be Drained*; William Dodson (1665) *The Design for the Perfect Draining of the Great Level of the Fens*; DNB; E. G. R. Taylor (1954) *The Mathematical Practitioners of Tudor & Stuart England*; L. E. Harris (1961) *The Two Netherlanders, Humphrey Bradley and Cornelius Drebbel*; A. W. Richardson (1966) *English Land Measuring to 1800: Instruments and Practice*; Bendall]

AHER, David (1780–1841), civil engineer, was born in 1780. He rapidly developed a proficiency in the physical sciences and commenced his studies as a civil engineer at the early age of fifteen. In 1803 the Grand Canal Company leased some collieries at Doonane on the borders of Counties Laois and Kilkenny near Athy. It was proposed to construct a branch canal from Athy to the collieries to act both for drainage of the mines and for transportation. Aher was charged with designing a tunnel to take the canal to the coal face, but the scheme was abandoned in 1805 and Aher resigned to become manager of Lady Ormond's colliery at Castlecomer, an occupation for which he was well suited from his knowledge of geology. By his judicious direction of borings, discoveries of coal seams were made which proved commercially valuable.

His inventions and improvements in mining and boring machinery displayed remarkable ingenuity, and between 1810 and 1812, whilst engaged in the direction of the collieries, he undertook experiments and compiled reports for the government Bog Commissioners (see also A. Nimmo, The Edgeworths, W. Bald). In 210 days, at 2 guineas a day, he surveyed 50,000 acres of bog. Commencing in 1807, he laid out nearly all the new lines of road through County Kilkenny, including the main roads from Castlecomer to Carlow (12 miles), Castlecomer to Athy (17 miles) and Castlecomer to Kilkenny (12 miles). Between 1812 and 1820 he laid out roads for the Post-Master General and was later responsible for surveying a route for the Great Leinster and Munster Railway.

He was elected a Member of the Institution of Civil Engineers in 1836. In 1840, he suffered depression due to a number of major disappointments and died the following year.

RON COX

[Commissioners ... into ... Bogs in Ireland (1810–1814) *Reports* 1–4; Obit. of David Aher (1844) *Min Proc ICE*, **3**, 14; M. B. Mullins (1859) Presidential address, *Trans ICEI*, **6**, 1859–1861, 111; R. Delany (1973) *The Grand Canal of Ireland*, 146–147; J. Andrews (1990) David Aher and Hill Clements, maps of county Kilkenny, 1812–1824, in W. Nolan and K. Whelan (eds.), *Kilkenny History and Society*, 437–463]

Publications
1814. Report on the district of Bog No. 8, laying between Roscrea and Kilkenard (1811), *Commissioners on the Bogs in Ireland, Reports*, **3**, 61–82
1814. Report on the District No. 12, in the Queen's County, *Commissioners on the Bogs in Ireland, Reports*, **3**, 83–96

Works
1807–1810. Main roads in counties Kilkenny and Carlow over 40 miles, engineer
c. 1812–1820. Roads in Ireland for Post-Master General, engineer

AIKIN, Arthur (1773–1854), was an early member of the Institution of Civil Engineers. Born on 19 May 1773 at Warrington, Lancashire, he was the son of Dr. John Aikin, MD, a popular author, whose *Letters to his Son* were especially admired. Arthur Aikin was educated at Palgrave School, where his aunt, the famous Mrs. Barbauld, taught, and he later attended Hackney College. He became pastor of a Presbyterian congregation in Shrewsbury, but soon decided he was unsuited to the ministry. He then made extensive tours, resulting in *A Journal of a Tour in North Wales &c.*, and articles for the Geological Society.

After settling in London, for six years beginning in 1802 he edited the *Annual Review* for Messrs. Longman and Co. He lived with his brother, Charles Rochemont Aikin, MRCS, and together they produced the *Chemical Dictionary* from 1807 to 1814.

He became chemical lecturer at Guy's Hospital, a post he held for thirty-six years, and was active in many scientific organizations, including the formation of the Geological Society in 1807, the foundation of the Zoo, the Linnaean Society and the Chemical Society. From 1817 to 1839 he was Secretary of the Society of Arts, Manufactures and Commerce.

He became Honorary Secretary of the Institution of Civil Engineers in 1823 after the resignation of John Jones (q.v.), but there is no evidence that he performed active duties. He had been elected to the Institution in 1820 and he remained on the membership lists until 1841. He was involved in the design of improved chemical manufactures and in making chemical analyses for patentees and public companies. Possibly this work led to him being consulted about patent law reform in 1829. He died unmarried on 15 April 1854 in his house at 7 Bloomsbury Square, London. Portraits of him were published in the *European Magazine* in 1819 and in *Profiles of Warrington Worthies* in 1854.

His younger brother, Edmund (1789–1820), settled in Liverpool in 1814 after being articled to a builder or surveyor practising in London. He assisted General Sir Samuel Bentham (q.v.) on Sheerness and Portsmouth dockyard works and in 1812 on Vauxhall Bridge; in 1813, with Bentham, he published designs for a bridge over the Swale.

TESS CANFIELD

[Membership records, ICE archives; Biography (1819) *European Magazine*, **75**, 398 (with portrait); House of Commons, *Select Committee on the Law Relating to Patents for Inventions* (1829) *Minutes of Evidence*, 40–46; Obituary (1854–1855) *Min Proc ICE*, **14**, 120–123]

Publications

1797. *Journal of a Tour through North Wales and Part of Shropshire, with Observations on Mineralogy and other Branches of Natural History*
1800. *Proposal for a Mineralogical Survey of the County of Salop*
1802–1808. *Annual Review*, editor
1803. *Travels in Upper and Lower Egypt*, D. V. Denon (translator)
1807–1814. *Dictionary of Chemistry and Mineralogy with an Account of the Processes Employed in Many of the most Important Chemical Manufactures* (with Aitkin, C. R.)
1807. Botanical chapters in Pinkerton's *Geography*
1811. Observations on the Wrekin and on the great coal field of Shropshire, *Trans Geological Soc*
1814–1815. *Manual of Mineralogy* (with others)
1816. Some observations of a bed of trap occurring in the Colliery of Birch Hill near Walsall in Staffordshire, *Trans Geological Soc*
1817. Address to the Society of Arts, 27 May 1817
1818. *Observations on the Valleys and Watercourses of Shropshire, and of Parts of the Adjacent Counties*
1832. On pottery …, *Trans Society of Arts*
1841. *Illustrations of Arts and Manufactures* (with others)

1848. *Report upon the Vulcanized India-rubber Joints for Water and Gas Socket-pipes* (with T. Wicksteed)
1850. *Preliminary Report upon the Sewerage, Drainage and Supply of Water for … Leicester: and a Report upon the Analysis of the Sewage-water and the Water of the Streams* (with T. Wicksteed and A. S. Taylor)

AINSLIE, John (1745–1828), surveyor and cartographer, was born in Jedburgh on 22 April 1745, the younger child of John Ainslie, druggist. After attending Jedburgh Grammar School and, one assumes, some training as a surveyor, from 1765 he was employed by Thomas Jeffery, geographer to the King, in a survey of Bedfordshire, Buckinghamshire and Yorkshire. After Jeffery's death in 1771 he returned to Jedburgh and embarked on a large scale programme of mapping central Scotland and the Borders, which was largely completed by 1797. Having removed to Edinburgh he also began producing charts of the Scottish coasts in the 1780s and in January 1788 began to publish his map of Scotland, the first map of its coastline which had any accuracy.

Ainslie's great cartographic achievements were supplemented by private surveying work, which included surveys for leading engineers, no doubt drawn to him by the quality of his published maps. He worked for Robert Whitworth (q.v.) and John Rennie (q.v.) on various canals and harbour surveys in Scotland. His treatise on land surveying was one of the leading texts of the time; it is copiously illustrated and details canal levelling and coastal surveying techniques.

Ainslie married Christian Caverhill of Jedburgh on 27 October 1776 and he died at his home, 53 Nicholson Street, Edinburgh, on 29 February 1828, leaving nearly £9,000. Among his pupils was William Bald (q.v.), civil engineer.

MIKE CHRIMES

[DNB; Bendall; Skempton]

Publications

1785. *Plan of Forth and Clyde Canal* (with R. Whitworth)
1794. *Report on Proposed Canal betwixt Edinburgh and Glasgow* (with R. Whitworth, Jr.)
1798. *Plan of Different Lines Proposed for a Canal between Edinburgh and Glasgow* (for J. Rennie)
1803. *Plan of Proposed Canal between Glasgow and Saltcoats* (for J. Rennie)
1809. *Plan of Harbour at Burntisland* (for J. Rennie)
1812. *Comprehensive Treatise on Land Surveying* (repr. 1842)

AIRD, John (c. 1760–1832), civil engineer, was born in Scotland about 1760. He was first employed by John Rennie (q.v.) about the year 1794, almost certainly on the Crinan Canal, after an initial approach had been turned down because Aird was asking too high a salary. He remained with Rennie and his son (Sir) John (q.v.) for the rest of his life. He was at Greenock in 1807–1808 and reported on the state of the works at the New Harbour in November 1807.

In 1809, John Rennie appointed Aird as acting resident engineer at Howth Harbour to the north of Dublin Bay, and he lived in Ireland for the rest of his life, taking on increasing responsibilities. His report dated 18 May 1809 describes the manner in which he proposed to carry out the construction of Howth harbour. In November 1812, Rennie proposed to send Aird to Ramsgate Harbour to gain experience in the use of the diving bell, Rennie himself using the bell for laying the foundations of the east pier at Howth. Work was substantially completed at Howth by 1813, although a lighthouse was added in 1818.

Rennie had by then been consulted about other works in the area, including the Custom House Dock, where work began on an entrance lock and basin in 1815. Aird assisted here, although he remained at Howth until his appointment in 1816 as resident engineer under John Rennie, and later under Sir John Rennie, at Kingstown (now Dun Laoghaire) Harbour. Sir John, when describing Kingstown Harbour in his treatise on harbours, referred to Aird as 'the able resident engineer'.

Following the death of John Rennie in 1821 responsibility for his various works in the Dublin area changed. Although Sir John Rennie was retained as engineer at Kingstown, in February 1822 Thomas Telford (q.v.) was appointed engineer for the Custom House Dock scheme. The interest of the Holyhead Road Commisioners in communications with Dublin led to the transfer of responsibility for Howth Harbour to their Engineer, also Telford, and they invested in road improvements between Howth and Dublin. Continuity was in all cases provided by Aird. He continued as resident engineer at Kingstown, and Dublin, where the lock gates had yet to be installed and the warehouses completed; he was retained at Howth, where he was paid 2 guineas a day for each visit, plus half a guinea travelling expenses, and supervised the roadwork. This workload could only be achieved with the assistance of other staff. James Sowter was the salaried engineer at Howth, where £16,000 was spent, mostly to repair gale damage 1823–1830. Elsewhere he was assisted by William Wilkie and John L. Rodgers; a millwright was employed at Dublin Docks who may have been the Thomas Kearney who was briefly a member of Institution of Civil Engineers (1825).

The whole of Aird's career was spent on harbour construction and his opinion was sought in 1818 in connection with works proposed at Balbriggan Harbour, and at Wexford in 1822.

John Aird died at Dun Laoghaire in 1832.

RON COX

[Telford mss, ICE Library; Rennie mss, NLS; J. Rennie (1807–1812) Reports, vols. 5–7. ICE archives; Letters from John Aird to John Rennie (1817) [on Howth Harbour], in *Fourth Report from the Select Committee on the Roads from Holyhead to London*. HC; Sir J. Rennie (1854) *The Theory, Formation and Construction of British and Foreign Harbours*, **1**, 197]

Publications
1817. *Design for the New Proposed Asylum Harbour at Dunleary* (with J. Rennie)

Works
1794–1800? Crinan canal, assistant engineer
1801–1806. East Dock, Leith, assistant engineer
1807–1808? Greenock harbour, assistant engineer
1809–1816. Howth Harbour, resident engineer, £347,000
1815–1824. Dublin Custom House Docks, resident engineer
1816–1832. Kingstown Harbour, resident engineer, c. £500,000
1823–1824. Howth–Dublin Road, resident engineer, £10,000

AITCHISON, George (1792–1861), architect, was the son of a carpenter and builder at Leyton, Essex, whose apprentice he became. From 1813 he was a pupil of Henry Blake Seward, a district surveyor in Westminster, before working for Thomas Hardwick and Thomas Lee.

He was Clerk of Works for Philip Hardwick (q.v.) on the construction of the architectural work at St. Katharine's Docks from 1827 in succession to Richard Douthwaite who had previously assisted Daniel Alexander (q.v.) in the later stages of work at London Docks before joining Hardwick in 1825. When work was complete at St. Katharine's in December 1830, Hardwick left but Aitchison remained in charge of the architects department.

Aitchison joined the Institution of Civil Engineers as an Associate in 1828, contributing to its discussions, and was recognised by his contemporaries as a proficient civil engineer and surveyor as well as an architect. His architectural work included the design of banks, such as Union Bank, Fleet Street (1856–1857), and private houses.

Aitchison's work for the St. Katharine's Dock Company included the designs of E, H and I warehouses, as well as a new engine and accommodation house. Through this work he came into contact with some of the directors of the London–Birmingham Railway, and he devised a system of book-keeping for the construction works there, as well as acting as architect for the intermediate stations. He also, according to his ICE obituary, gave Robert Stephenson 'the aid of his experience in carrying out the structural works on the line'. He held a variety of appointments as District Surveyor or architect to local boards, and designed many warehouses and wharves along the Thames. His Irongate Wharf of 1848 was probably the first 'fire-proof' building in London.

Aitchison died after a stroke on 12 June 1861 and is buried at the City of London Cemetery, Ilford. His son, also named George, became President of the Royal Institute of British Architects.

MIKE CHRIMES

[PRO RAIL 384/291; ICE membership records; On the formation of concrete masses (OC 60); A general historical account of the Old London Bridge (OC 80), Minutes of conversations 115, 116, 118 (1832), ICE archives; *Min Proc ICE*, **xxi** (1861–1862), 569–571; A. W. Skempton (1981–1982), Engineering in the Port of London 1808–1834, *Trans Newc Soc*, **53**, 82–87; Colvin (3)]

Works

1827–. St. Katharine's Docks, architectural work
1834–1838. London–Birmingham Railway, architect for stations, in charge of bookkeeping, £150 p.a.

ALEXANDER, Daniel Asher (1768–1846), architect, engineer and surveyor, was born in London and educated at St. Paul's School. In 1780 he began training as a pupil of Samuel Robinson and attended the Royal Academy School from 1782. Alexander is best known for his work as surveyor to the London Docks Company. He had already designed warehouses at Bankside and Mark Lane when he began working for the prospective company in 1794 following the proposals of William Vaughan who called for enclosed 'wet' docks as a means of improving the Port of London.

Alexander planned two linked docks with a total area of 35 acres, an entrance basin at Wapping and a second entrance at Shadwell or Blackwall. City merchants soon raised £800,000 and petitioned Parliament for an Act. This first came before Parliament in May 1796 with Alexander and John Rennie (q.v.) giving evidence in support. Following alternative proposals, chiefly involving a rival West India Dock scheme, a modified version of Alexander's proposals finally obtained its Act in June 1800 and in the following month the company appointed Alexander surveyor at a salary of £1,200 per annum.

Although Rennie, along with Robert Mylne (q.v.) and Joseph Huddart (q.v.), were consulted in March 1801, Rennie was not appointed engineer until May 1801, with James Murray (q.v.) as resident engineer. By this time Alexander had drawn up detailed plans for the dock and warehouse, the former following Rennie's recommendations. Rennie seems to have been responsible for the detailed design of the docks, while Alexander was responsible for that of the warehouses which were all completed by December 1806. The next phase involved the Tobacco Dock and its warehouses, where again Rennie acted as engineer and Alexander as architect, with Richard Douthwaite as his assistant. The Tobacco Warehouse, a single-storey structure with vaults, featured an unusual form of branched cast iron columnar supports; it is unclear who detailed this design. With the completion of the second,

Hermitage, entrance in 1821 Alexander left in December that year. He was generally well known for his utilitarian architecture, which was also seen in his prisons at Dartmoor (1806–1809) and Maidstone (1811–1819), although at the latter he failed to detect fraud among the tradesmen and his reputation suffered.

Alexander's engineering work also encompassed bridges and lighthouses and he was responsible for repairs to Rochester Bridge, Lower Bridge, Tonbridge and a survey of the Old London Bridge. He was appointed as the first salaried architect and engineer for the medieval bridge trust at Rochester in 1792, having successfully won a competition for improvements, and he remained involved until 1818. His plan for a new bridge was abandoned in favour of John Rennie's.

He was appointed surveyor to Trinity House and designed a number of lighthouses for them. Generally they displayed some innovation compared with what had gone before, exemplified in the use of cavity walls. Access at South Stack was provided by a rope suspension bridge and the work was supervised by Captain Hugh Evans, harbourmaster at Holyhead. His architectural practice, incorporating major houses and religious buildings, is detailed by Colvin.

Alexander apparently retired to Yarmouth in the 1820s, and then to Exeter. He is commemorated by a portrait held by the Port of London Authority and was buried at Yarmouth, Isle of Wight, where in 1829 he had designed the church tower as a memorial to his third son, Henry. His eldest son Daniel was trained by him, but never practised.

MIKE CHRIMES

[London Docks Company, Minutes and Drawings, Museum of London in Docklands; House of Commons Committee on the Port of London (1796) *Report*, 227–230; House of Commons (1799) *Minutes of Evidence Committee on Wet Docks*; Second Report of the Select Committee on the Port of London; House of Commons (1803) *Reports from Committees*, vol. 14; Tidal Harbours Commission (1845) *1st Report*, 190–192; *Select Committee on Rochester Bridge* (1820); Anon. (1831) [South Stack lighthouse] *Mech Mag*, **13**, 457–458; D. A. Stevenson (1959) *The World's Lighthouses before 1820*; D. B. Hague and R. Christie (1975) *Lighthouses*; A. W. Skempton (1978–1982) Engineering in the Port of London, *Trans Newc Soc*, **50**, 87–108; **53**, 73–96; R. Thorne (ed.) (1990) *Iron Revolution*; N. Yates and J. M. Gibson (1994) *Traffic and Politics—Rochester Bridge*; Colvin (3)]

Publications

1794. *The London Docks* (repr. 1796–1803)
1796. *The London Docks* (repr. 1799)
1796. *Plan of the Proposed London Docks* (various editions 1796–1803)
1802. *A Correct Plan of the London Docks*
1813. *Report Relative to Rye Harbour*

1814. *Report on London Bridge* (with Dance and Chapman)
1826. *Rye Harbour Plan of Works*

Works
1792–1818. Rochester Bridge improvements, £20,000
1793–1821. London Docks, planning and warehouses, £1.4 million
1808–1809. South Stack lighthouse, Anglesey, masonry
1810–1811. Farnes lighthouse
1811. Heligoland lighthouse, with brick cavity walls
1812. Hurst lighthouse
1814. Lower Bridge, Tonbridge
1816–1822. Harwich lighthouses, brick towers
1818. Lundy lighthouse, with granite cavity walls

ALEXANDER, George (fl. 1810–1823), architect and builder, was based at Golspie in Sutherland. In addition to various buildings in the Highlands he built Helmsdale harbour; begun in 1818 it comprised a quay on the north bank of the river Helmsdale with a small rectangular basin. John Rennie (q.v.) proposed breakwaters in 1820 to shelter the river mouth, but this work was not carried out, although Alexander undertook extensions in 1823.

MIKE CHRIMES

[J. Gifford (1992) *Buildings of Scotland: Highland and Islands*; Colvin (3)]

Works
1818–1823. Helmsdale harbour

ALLAM, Samuel (fl. *c.* 1767–1785) of Spalding, carpenter, was brought in by John Grundy (q.v.) as 'Surveyor to attend the Execution of the Works' at the start of construction on the Driffield Canal, east Yorkshire, in October 1767 at an annual salary of 52 guineas. He held this post until the opening of the canal in 1770. In that year he acted as surveyor for work on another Grundy scheme, the Cherry Cobb embankment in Yorkshire. By 1771, if not earlier, he moved to Hedon, near Hull. In June 1772 he was appointed Surveyor of Works on the Market Weighton Drainage and Navigation scheme, at the same salary. Working under Grundy's direction, he supervised construction of the outfall sluice (his name is recorded on a plaque) and excavation of the main drain; the work was completed in 1775 and cost about £20,000. He then took full charge of the next stage of operations, completing the work in 1780 at an additional cost of £14,000.
While the Weighton works were in progress, in November and December 1774, he prepared estimates in connection with Grundy's proposed improvements of Hedon Haven. Finally, in 1785, Allam took levels and made plans and estimates for the Beverley and Skidby Drainage scheme, and acted as engineer for the first year's work.

A. W. SKEMPTON

[Minute Books of the various authorities mentioned, East Riding RO; Reports, etc., on Weighton drainage (DRA & DDJ series), Hull University Library]

ALLNUTT, Zachary (fl. 1801–1856) of Henley, son of Henry Allnutt, General Clerk to the Thames Navigation Commission from 1771 to 1820, was appointed Surveyor of the Lower Districts of the navigation (Staines to Mapledurham) in February 1804 in place of the recently deceased John Clarke (q.v.). At the same time John Treacher, Jr. (q.v.) became Surveyor of the Upper Districts (Whitchurch to Lechlade). However, for seventeen years thereafter the Commission employed no General Surveyor; Treacher and Allnutt shared the responsibility (at no extra payment) while looking after the interests of their own Districts. Throughout the period of his appointment Allnutt received a salary of 2 guineas per week.
He had already carried out some work for the Commission on a freelance basis before 1804, including a survey in 1802 of a proposed new cut at Culham to bypass the very awkward arrangement at Sutton Courtenay mill. In 1805 he published on his own initiative a scheme for improving navigation in the river from Richmond up to Staines within the jurisdiction of the City of London. This involved building locks near Teddington, Sunbury and Chertsey, each with weirs that could be completely opened in times of flood. Alternative schemes were proposed, but eventually, in 1810 and following recommendations of Stephen Leach (q.v.), engineer to the City's Navigation Committee, an Act was obtained and four locks built between 1810 and 1813 at or near these three places, and another at Shepperton, with solid weirs having very long overfalls or 'tumbling bays' capable of passing floods.
Meanwhile in 1809 Allnutt produced a definitive survey for the Culham lock and 1400-yd. cut, the works being carried out 1809–1810 under Treacher's supervision. In 1815 Allnutt completed a survey of all the locks of the Thames Commission from Romney up to Lechlade. Each of the 27 locks was briefly described, with its date and style of construction, in a list by Allnutt and Treacher appended to the plans.
From about 1816 determined efforts were made to improve the navigation between Staines and Maidenhead. Allnutt and Treacher drew up plans for two locks (additional to the one already built at Romney): Bell lock, built 1812–1818 and Old Windsor lock with a 1000-yd. cut built 1821–1822 (plans dated 1820). Much dredging was also carried out, sometimes with as many as twelve dredger boats at work.
In September 1821 Treacher was appointed General Surveyor, and Allnutt became General Receiver of the Tolls; a post he held for many years, returning from time to time to update the lock plans and add new ones.

A. W. SKEMPTON

[Minute Books of the Thames Navigation Commission (1801–1856), Berkshire RO; Zachary Allnutt (1815) Plans of the pound locks and lands belonging to the Commissioners of the Thames Navigation (MS), Berkshire R.; A. W. Skempton (1984) Engineering on the Thames Navigation, *Trans Newc Soc*, **55**, 153–176]

Publications

1805. *Considerations on the Best Mode of Improving the Present Imperfect State of the Navigation of the River Thames from Richmond to Staines*

ANDERSON, Adam (*c*. 1780–1846), scientist, the third of four sons of Andrew Anderson and Anne Moon, was born in Tulliallan parish on 29 June 1780. Anderson's father was a Kincardine-on-Forth shipmaster, and Adam attended St. Andrews University from 1797, graduating MA. His main course of study was theology, but he also studied natural philosophy. After some private teaching work he became Rector of Perth Academy from 1809–1837 and subsequently Professor of Natural Philosophy at St. Andrews.

While in Perth he was involved in various schemes for municipal improvement, the most notable of which was the new waterworks. John Rennie (q.v.) had been consulted in 1806 on the early polluted supply, but nothing had transpired and in 1819 Anderson was approached by the Council, subsequently carrying out a chemical analysis of possible supplies. He identified Clathymore spring as a pure source but the expense of carrying the water 6 miles precluded further action. In 1823 James Jardine (q.v.) suggested that Perth was ideally suited for a pumped water supply using a steam engine, but again nothing was done. In 1827 the water supply system failed for a fortnight, and Anderson, having studied and improved upon a similar establishment on the Clyde for Glasgow, then devised a scheme for taking water from the river Tay, establishing filtration beds on Moncrieffe Island, taking water in cast iron pipes beneath the river to a reservoir, and building a pumping station. He proposed that the 146,000-gallon reservoir be housed in a Doric-style building covered by an iron dome; Jardine supported his recommendations.

Anderson was appointed engineer with John Turnbull acting as Inspector of Works and construction began in 1830, the Dundee Foundry supplying the steam engine and ironwork. Having overcome problems in sinking foundations 4 ft. below the level of the river bed, the works were largely completed by 1834, although Anderson remained in charge until 1837. By this time he was totally disillusioned over the issue of his fees, but eventually settled for £850, although considering himself entitled to more.

While in Perth Anderson was also involved with coal-mining investigations near the Bridge of Earn and from 1822 with the Perth Gas Light

Company. In 1823 he visited gasworks at Glasgow, Edinburgh and Berwick, subsequently designing and building a new plant, brought into production in 1824; for his work he was presented with 300 guineas worth of silver plate.

Anderson's knowledge of gas supply led to him being consulted about works in Arbroath and Montrose; he became involved in the local standardisation of weights and measures; in the 1830s he was associated with a railway; and in 1830 he helped to establish Perth Mechanics Institute.

He was a distinguished writer on all aspects of science and, in addition to numerous articles in scientific journals, he contributed articles on subjects such as dyeing, gas lighting, navigation and hygrometry to the *Edinburgh Encyclopaedia* and the *Encyclopaedia Britannica* (7th edn.).

In 1837 he was appointed to the chair of natural philosophy at St. Andrews, where he was involved in broadening the scientific content of the curriculum, and in 1838 he suggested the inclusion of civil engineering; unfortunately he fell out with the Principal, David Brewster (q.v.). Anderson died at St. Leonard's Bank, Perth, on 5 December 1846.

MIKE CHRIMES

[Tidal Harbours Commission (1847) *2nd Report. App. C*, 176; K. J. Cameron (1988) *The Schoolmaster Engineer*; Colvin (3)]

Works

1822–1824. Perth gas supply
1828–1837. Perth Water supply, £13,609 11s 11½d

ANDERSON, James, FRSE (1793–1861), land surveyor and civil engineer, was born on 22 January 1793 in the old town of Edinburgh, the son of James Anderson, a flax dresser, and Elisabeth Wittit. His early education and career are uncertain but it is known that he was a pupil of James Jardine (q.v.).

In 1818 Anderson published his proposals for a wrought iron suspension bridge over the Firth of Forth at Queensferry, near Edinburgh. He outlined two designs, both involving timber-decked spans of up to 2000 ft., suspended from either rod-stays or catenary bar-cables supported on cast-iron towers on top of masonry piers. He based his design calculations on the results of experiments carried out by himself and by Thomas Telford (q.v.), but his ambitious plans were an order of magnitude beyond the capabilities of early nineteenth century bridge engineering.

He deserves credit, however, for visualising what the suspension principle would eventually achieve, and also for incorporating in his designs some provision against oscillation of the deck. Anderson later (in 1830) successfully renewed the seaward timber abutment of Samuel Brown's (q.v.) Trinity Chain Pier, erected in 1821 at Newhaven near Edinburgh, and built up a harbour engineering practice in Edinburgh. He was elected a Fellow of the Royal Society of Edinburgh in 1836.

He died in Edinburgh at 29 St. Bernard's Crescent on 19 February 1861 and was buried in Old Calton Cemetery.

RON BIRSE

[Skempton; R. Paxton (ed.) (1990) *100 Yrs of the Forth Bridge*, 20–24; W. T. Johnston (1998) *Scottish Engineers and Shipbuilders* (publ. as computer database)]

Publications

1818. *Report Relative to a Design for a Chain Bridge, Proposed to be Thrown over the Frith of Forth at Queensferry* (drawn under the direction of James Jardine, Civil Engineer, by James Anderson)

1818. *Plan of Callender Park situate in the County of Stirling ... shewing the Deviation through the Park from the Parliamentary Line of the Edinburgh and Glasgow Union Canal, proposed by the Canal Company, November 1818*

1834. *Report on the present state of Leith Harbour, and the Practicability of Rendering available its Wet Dock, by means of a Deep-Water Entrance at Newhaven, and a Communicating Dock or Ship-Canal.* Repr. in Tidal Harbours Commission (1847), *2nd Report*, appendix C, 48–54

ANDERSON, William (d. 1844), civil engineer, worked for John Rennie from around 1800 onwards and was the first engineer of the Grand Junction Waterworks scheme. The original proposal stemmed from a provision in the 1798 Grand Junction Canal Act to enable the Canal Company to supply water to the Paddington area. The Company was unable to make this provision, and in 1810 it was decided to transfer powers to a separate company; John Rennie (q.v.) was appointed (consulting) engineer.

It was originally believed that the water in the canal, derived from the Brent and Colne, was superior to the Thames and that the level of the canal would be sufficient to supply water by gravity; neither expectation was realised. An alternative supply from a reservoir at Ruislip proved no more successful and a Boulton and Watt pumping engine was installed at Paddington in 1812. In 1820 it was decided to derive the water supply from the Thames in Chelsea, from where it was pumped by two steam engines to three reservoirs at Paddington where it was distributed to some 8,000 homes. Rennie was the engineer for all these proposals, and a shareholder in the Stone Pipe Company, which was awarded some of the contracts for the supply of piping for the works, a decision which the Company was to regret bitterly, as the pipes proved an expensive failure.

Anderson's role in the early stages of the Grand Junction Waterworks Company was effectively as resident engineer. As such he is known to have made extensive observations on the failure of the stone and wooden pipes used by the Company, and in 1813 was seeking an effective means of testing cast iron pipes. The 42 in. cast-iron mains used by the Company were an early large scale application of cast iron for water supply. Following Rennie's death in 1821 it can be assumed that Anderson played a more significant role in the Company's affairs, then suffering financial problems partly as a result of the stone pipe debacle, and also as a result of problems in supply; by the mid-1820s it had an unenviable reputation for the poor quality of its water. Anderson investigated the quality of the water in the Thames, its flow, and the impact of the removal of London Bridge on the influence of the tide up-river. Eventually he fixed upon a site at Kew Bridge for a new intake, and in 1835 a new Act authorised extraction there. The existing pumping stations were closed down and the engines removed from Chelsea to Kew.

In the meantime Anderson was appointed consulting engineer to the Southwark Waterworks Company, formed in 1822 as heir to the Borough Waterworks and London Bridge Waterworks. The original works had their intake near London Bridge, but Anderson recommended an intake near Battersea and the construction of a reservoir there. An Act was obtained in 1834 but the Company was not fully incorporated until 1839, acquiring 18 acres of land in Battersea by 1841. Anderson had, in 1828, recommended Teddington as an intake source for the Metropolitan water supply.

In addition to his knowledge of the Metropolitan water supply Anderson was involved in other schemes, and in 1823 he submitted, in association with the architect Charles Tyrrell (q.v.), a proposal for a New London Bridge. He was engineer for South Bridge, Northampton (1815–1816) and acted as consultant to the Grand Junction Canal in the 1830s.

Anderson married in *c.* 1806 and became a member of the Institution of Civil Engineers in 1821. He died in 1844.

His son, William Daniel Anderson (q.v.), was a pupil of Telford before working for his father for a short time in the early 1830s.

MIKE CHRIMES

[Grand Junction Waterworks Company, Minutes (Greater London Record Office) 1811–1844; Rennie Reports, ICE Library; *House of Commons Select Committee on Metropolitan Water Supply* (1821); *The Dolphin; or Grand Junction Nuisance* (1827); *Select Committee on Metropolis Water* (1828); *Select Committee on Metropolis Water* (1834); (HLRO c1,d3,5, 1835); F. S. Peppercorne (1840) *Supply of Water to the Metropolis*; H. W. Dickinson (1954); *Water Supply of Greater London*; A. H. Faulkner (1972) *The Grand Junction Canal*; *Northamptonshire Archaeology*, **15** (1980), 150–151; G. M. Saul (1987–1988) John Rennie: one of his contributions to waterworks technology, *Trans Newc Soc*, **59**, 3–14]

Publications

1823. *London Bridge* (with C. Tyrrell)

Works

1811–1844. Grand Junction Waterworks, £300,000 by 1828

1815–1816. South Bridge, Northampton

c. 1833–1844. Southwark Waterworks

1836–1839. Grand Junction Canal Company, Tringford pumping station and feeders, c. £5,000

ANDERSON, William Daniel (1808–1842), civil engineer, was baptised on 22 May 1808 at Southwark, the son of William Anderson (q.v.) and his wife, Elizabeth. Anderson's career began with his father and he then worked under Thomas Telford (q.v.) in whose house he lived. Under Telford he prepared plans for St. Katharines Dock and was also conversant with the Mythe and Over Bridge works and the Gloucester and Berkeley and the Liverpool and Birmingham canals, together with other major works. He accompanied Telford on inspections of works in progress, among them several road projects, the Crinan, the Birmingham and the Caledonian canals, Glasgow bridge and the Clyde Navigation. Sponsored by Telford he became a member of the Institution of Civil Engineers in 1825.

In 1829–1830 Anderson visited France, Switzerland and Italy, part of the journey at least in the company of his father, who had been consulted by the French concerning waterworks. As a result of his stay in Florence he prepared drawings of the sixteenth century Ponte Sante Trinita, and on his return to England presented them to the Institution of Civil Engineers.

In 1831 he left Telford to practise at Cannon Row, Westminster, and was there engaged with his father on works for the Grand Junction Waterworks, involving the use of cofferdams and a diving bell. He was appointed engineer for the Exeter Waterworks in 1832 and executed new works there, planning to abstract water from the river Exe for the supply of the town. His proposals 'provided for the erection of water-wheels, pumps, reservoirs and steam-engines at the mill site at Pynes'. Again with his father, he carried out work for the Southwark Water Works.

Telford's death in 1834 brought to Anderson a legacy of £400 (possibly £1000). He was engaged under William Provis (q.v.) on the surveys for the proposed South Eastern Railway and its Brighton and Maidstone branches c. 1836. In January 1838 he was appointed river engineer to Newcastle Corporation, for which position there were 20 applicants; his salary was £300 per year. In his application he stated that he had designed and constructed piers and/or wharves at Queenhithe, Dyers Hall, Westminster and Hungerford.

His first major task at Newcastle was the construction of the first section of a 1000 ft. long quay wall to the town's river frontage. The quay had originally been designed by John Dobson, the Newcastle architect, and had been reported upon by William Cubitt (q.v.), whose comments were, in turn, the subject of reports by John Blackmore (q.v.), John Murray and James Milne (q.v.). The contract was completed in August 1840, after a change of contractor and some acrimony involving a member of the Corporation who considered that the River Committee had not managed the work well. Anderson was responsible for the partial removal of Bill Point, a major hazard to navigation on the river downstream of Newcastle. He also ordered barges for use with dredging plant and it was under him that river improvements were first placed on a systematic footing. He submitted reports on the Newcastle and Carlisle Railway Company's bridge over the river Tyne at Scotswood and on the proposals for a wet dock at Jarrow for the Stanhope and Tyne Railway.

In Newcastle, Anderson lived in the newly developed Rye Hill area of the town. His stay there was short (although he did become a member of the Literary and Philosophical Society) and, in July 1841, writing from Bow, Middlesex, when on leave of absence on health grounds, he resigned his appointment, having been advised by his physician that he must not return to the north-east. His resignation was accepted, the Corporation writing to him that he could be assured that he would be remembered with 'the kindest and most grateful feelings (for) the very satisfactory manner in which (he had) in every respect' conducted himself.

In 1830, Anderson married Isabella English in London and three years later he married Sarah English, perhaps her sister. He died in London on 21 March 1842, survived by her and at least one child, a daughter.

R. W. RENNISON

[Gibb Papers, ICE archives; *River Tyne*, **3**, 1824–39. Newcastle Central Library: L942.8; *Corporation of Newcastle upon Tyne: River Committee Minute Book* (1836–41). Tyne and Wear Archives: 589/455; *Proceedings: Newcastle Town Council* (1838–42); *Newcastle Courant*, 25 March 1842; *Min Proc ICE*, (1843), **2**, 13–14; W. Minchinton (1987), *Life to the City*, 38–39]

Drawings

Mythe Bridge, Tewkesbury, 1824

Plan and Elevation of a Bridge to be Erected near Swansea, 1824

Design for a Bridge over the River Dee (at) Aberdeen, 1827

Plan for supplying the City of Exeter and the Vicinity with Water from the River Exe, 1832

Works

1831. Grand Junction and Southwark waterworks (with Anderson Sr.)

1832–1837. Exeter water supply

1838–1841. River Engineer, Newcastle upon Tyne: construction of Newcastle quay, removal of Bill Point, systematic dredging and improvement of river Tyne

ANSTICE, Robert (1757–1845), civil engineer and Somerset's first County Surveyor, was baptised on 9 December 1757 at St. Mary's

Bridgwater, Somerset, the son of William (d. 1783) and Mary Anstice. He eloped to Gretna Green c. 1778 with a Quaker girl, Susannah Ball (c. 1761–1816), who was promptly expelled from the Quakers for marrying outside the Order; Robert remained a devout member of the Established Church all his life. In 1789 Susannah's younger sister, Hannah, married William Reynolds (q.v.), a Shropshire ironmaster and, after Reynolds' death, one of Robert and Susannah's eight children, William, took over the management of Madeley Wood, or Bedlam, Ironworks, near Coalbrookdale.

Robert Anstice followed his father into business as a merchant and shipowner and after Reynolds died took up his share in the Coalport China Works. His house was close to Bridgwater Town Bridge, facing the River Parrett. He was Custom House Comptroller by 1794, served as a councillor from 1799 until 1834, was Mayor in 1804, 1817 and 1825, and an Alderman from 1827.

A self-educated man, he was a keen amateur geologist, ornithologist and student of advanced mathematics. During the 1790s he published two books on mathematical topics and he was elected a Fellow of the Geological Society of London in 1818. He was an active anti-slavery reformer.

Plans for improving the drainage of King's Sedgemoor and lowlands adjoining the rivers Brue and Axe, in the Somerset Levels, were put forward by John Billingsley, a farmer from Sutton Mallet and author of *A General View of the Agriculture of Somerset*, and the surveyor William White (q.v.). Work on King's Sedgemoor began under an Act of May 1791, involving a new main drain 8 miles long passing through a 2 mile cut in the coastal clay belt to a new sluice, Dunball Clyse, on the Parrett estuary. Details of the organisation are missing, but it is known that contractors wishing to tender for two masonry bridges over the drain in January 1794 could see plans, elevations and sections 'on applying to Mr Robert Anstice in Bridgwater', and since he acted later as resident engineer on the Brue Drainage he probably held a similar post on King's Sedgemoor. The scheme was completed in 1795 (total cost of £32,000; engineering works £14,000, area drained 11,000 acres).

On the Brue Drainage work began in 1801 with William Jessop (q.v.) as engineer, White as surveyor and Anstice as 'Superintendent of Works' at 4 guineas per week. Lessons had evidently been learnt from mistakes on the Sedgemoor job and, although not free from problems, construction of the outfall sluice at Highbridge and cutting the new South Drain were completed successfully in 1806 (total cost £53,000, engineering works £34,000, area drained 22,000 acres).

Operations on the Axe Drainage started in 1803 with new cuts on the river. Jessop sent in drawings of the outfall sluice at Hobb's Boat, near Bleadon, in 1804 and a contract was let later that year. Anstice came in to supervise construction. In May 1805 he reported completion of the sluice

pit, 20 ft. deep, but a boring 40 ft. deep from ground level showed 15 ft. of soft clay beneath foundation level followed by 5 ft. of water-bearing sand. Bearing piles would therefore have to be adopted and the sluice floor planked throughout, although Anstice also suggested, as an alternative, preloading with 'an adequate temporary weight to consolidate the substrata': an interesting and novel idea, not appropriate in the circumstances, but used with great success by Thomas Telford (q.v.) and Jessop in building the Clachnaharry sea-lock on the Caledonian Canal (1809–1810) to consolidate a 50 ft. depth of clayey silt. Anstice completed the sluice and other works on Axe Drainage in 1810 (area drained 16,000 acres, total cost £41,000, engineering works £25,000). Of the latter sum, about £10,000 can be attributed to works carried out from 1805 after Jessop left direction to Anstice, who received £1,080 in fees.

It is probable that he designed and supervised reconstruction of Huntspill sea wall embankment in 1799 after its breach in September and November 1798. Here he would have been acting for the Commission of Sewers (the land drainage authority), of which he was a member. He also designed the reconstruction of Blue Anchor sea wall in 1817.

From 1810 he designed and supervised the repair, improvement and reconstruction of bridges and other public buildings for the County Justices of Somerset, but he was not formally appointed their first County Surveyor until 1817. He resigned this post the following year, probably because it compromised his position as a Commissioner of Sewers and interfered with his business interests, which he had maintained throughout this period.

There is no record of his active involvement in civil engineering after 1818, or of his contact with contemporary engineers, apart from William Jessop and ambiguous references suggesting his possible acquaintance c. 1826–1830 with Henry Jessop (1791–1834), fifth son of William. However, he was numbered among the friends of the influential Poole family, as were Robert Southey and John Rickman, both closely associated with Thomas Telford. In view of this and his connection with the Reynolds family, it is possible that he had some indirect contact.

He was said to be a man of commanding personal appearance 'with good means and much originality and force of character'. However, his grandson's widow wrote of him, 'Like many of the Anstices, he lacked the impulse which pushes men to the front. Shyness and a sort of apparent indolence prevented their very uncommon talents from making the mark they ought to have done'.

His memorial in St. Mary's, Bridgwater, reads: 'Sacred to the memory of Robert Anstice, Gentleman, who departed this life April 30th 1845, in the 88th year of his age'.

DAVID GREENFIELD

[Bridgwater Borough minutes and papers, Somerset River Authority papers (Commission of Sewers, King's Sedgemoor, Brue and Axe Drainage), Luttrell papers (Blue Anchor Sea Wall), Quarter Sessions bridge papers, Somerset RO; advertisement for tenders for bridge in King's Sedgemoor (January 1794) *The Western Flying Post*; E. Sandford (1888) *Thomas Poole and his Friends*; Michael Williams (1970) *The Draining of the Somerset Levels*; H. S. Torrens (1982) The Reynolds–Anstice Shropshire geological collection, *Archives of National History*, **10**, 429–441; J. B. Bentley (1985) More on Blue Anchor sea defences, *Somerset Archaeolog Soc Bull*, **39**, 3–5; J. B. Bentley (1967) Robert Anstice, Somerset's first county surveyor, *ibid.*, **46**, 10–15]

Publications

1790. *Remarks on the Comparative Advantages of Wheel—Carriages of Different Structure*
1794. *An Enquiry on the Laws of Falling Bodies &c.*
1814. *Report on the River Parrett Outfall* (with J. Easton)
1816. *Report on the State of Dunball Clyse, the Main Drain and Main Subsidiary Drains* (of King's Sedgemoor)

Works

1791–1795. King's Sedgemoor Drainage, resident engineer (attrib.), £15,000
1799. Huntspill sea-wall reconstruction (attrib.)
1801–1805. River Brue Drainage, resident engineer, £34,000
1805–1810. Hobb's Boat Clyse, Bleadon, and other works on River Axe drainage, *c.* £10,000
c. 1810. Ivelchester Gaol, alterations and improvements
1810–1818. County bridges in Somerset, repairs and improvements
c. 1813. St. Mary's Church, Bridgwater, repairs to spire
1817. Blue Anchor sea-wall reconstruction, designer

ARMSTRONG, John, FRS (1674–1742), Chief Engineer of England, 1714–1742. Despite an impressive collection of military titles—Colonel of the Royal Regiment of Foot of Ireland, Quartermaster General and Major General, and Surveyor General of his Majesty's Ordnance—Armstrong is actually better known as a civil than a military engineer through his *Report with Proposals for Draining the Fens and Amending the Harbour of King's Lynn*, produced in 1724.

Armstrong was born in Ballyard, King's County on 31 March 1674, the eldest son of Robert Armstrong, and served with the Duke of Marlborough from 1702 until 1712. Marlborough described him as an 'intelligent good officer' who 'on all occasions distinguished himself during the war in this country'. This high regard is presumably the reason that Sarah, Duchess of Marlborough, later employed him at Blenheim.

John Armstrong FRS

One of the conditions of the peace treaty of Utrecht was the complete demolition of Vauban's Great Fortress at Dunkirk and Armstrong and Thomas Lascelles (his eventual successor as Chief Engineer) were appointed to superintend the operation which lasted from July 1713 to September 1714 for which period he was paid £20 per day, double the usual allowance. The harbours and canals were also to be filled in and the sluice gates demolished so that (it was hoped) the port might never again be used against British shipping.

An ongoing debate with the citizens of Dunkirk ensued as to which were repairs needed solely to restore navigation and which abrogated the treaty. As late as 1730 both Armstrong and Lascelles, separately, had to return to view and advise on these repairs, resulting in the destruction of the jetties and quays on both sides of the canal

In 1714 Armstrong was appointed Chief Engineer (i.e. military engineer) of England in succession to Michael Richards (q.v.) and became a member of the Board of Ordnance; presumably it was in this capacity that he has been credited with the establishment of the foundry at Woolwich in 1716. He also worked at Portsmouth and on the cistern and battery at Sheerness and became Surveyor General of Ordnance in 1722 after the death of Richards.

In 1714 Armstrong was consulted over the arguments which had arisen over solutions to the Dagenham Breach although quoted as then saying that 'the undertaking of the Breach was out of his depth and foreign to his profession'; a decade later he was considered sufficiently expert in the field to be thrown into the even more acrimonious conflict between the Bedford Level Corporation, the South Level proprietors and the Corporation of (King's) Lynn. He was engaged by Robert Walpole to 'view the port and harbour of Lyn ... and also ... the river Ouse and all the navigable rivers that branch out of it ... and act as consistently as navigation and drainage shall appear inseparable', an unlikely occurrence as the views of the two sides had been totally incompatible for nearly a century since Cornelius Vermuyden (q.v.) created the New Bedford River and sluices at Denver and Salters Lode. The immediate problem was the wrecking of the Denver Sluice in 1713, its destruction diverting the floodwater upstream and exacerbating both flooding and silting of the harbour at Lynn.

Armstrong viewed both the Fens and the harbour and had maps and surveys made using both authentic records and current observations. This work was done by Thomas Badeslade (q.v.). Published in 1724, Armstrong's report suggested that the solution to both the navigation and drainage problems was to find ways of reopening the river Ouse to its former depth and breadth, the excavated material to be used to build embankments to prevent winter flooding. Impediments to the river's flow were to be removed to enable floods to carry away silt and sand, easing passage to the sea; the controversial sluice at Denver was either to be removed or a new cut made close by to widen the river from 80 ft. to its former 150 ft. The main engineering proposal was to re-site the Hermitage sluice on the New Bedford river to prevent the waters from the Ouse flooding inland towards Cambridge instead of the sea. The only action taken in the next half century was the rebuilding of the Denver sluice in 1746–1750.

At the same time as Armstrong was unsuccessfully advocating natural solutions in the Fens, he was involved with the construction of a canal and basin, in the rigidly geometrical fashion of the time, at Blenheim where the Duchess of Marlborough had hired Armstrong to remedy arrangements which displeased her. His original design was for a canal 1,840 ft. long and 100 ft. wide, passing under the main span of the bridge to a basin 300 ft. in diameter, with surplus water carried in small channels through the side arches, which also supplied Aldersea's pumping engine which raised water 120 ft. to the cistern above the palace kitchens. Instructions for the canal are included in a contract drawn up for the masons William Townsend and Bartholomew Peisley on 24 April 1722, and the canal and basin were built in 1722, the small lake with a twelve-step cascade in 1723 on the other side of the bridge and a further extension southward, including three more cascades, in 1724.

Armstrong's reputation in the scientific community was confirmed by his being elected a Fellow of the Royal Society in 1723, but he undertook no further civil works in almost twenty years although as a member of the Board of Ordnance he was involved with the organisation of engineering works at Gibraltar, also acquired as part of the treaty of Utrecht. His last action as a member of the Board of Ordnance in 1741 was to establish officially the Royal Military Academy at Woolwich which he had planned with Richards thirty years earlier. His will was drawn up in May of that year. He died on the 15 April 1742 at the Tower of London and was buried in the graveyard. A portrait exists by James MacArdell.

From his marriage to Priscilla Burrough (d. 1725) he is thought to have had five daughters. There is certainly a family connection with John Armstrong (Jr.) (fl. 1733–1758), who was ordered home from Minorca in 1742 to take care of affairs relative to the death of General Armstrong. He had a distinguished career as a military engineer in England, America, Nova Scotia and Minorca; in 1752 he published *The History of Minorca*.

SUSAN HOTS

[Blenheim muniments [F.1.49], Correspondence and papers, c. 1705–1730, Kenneth Spencer Research Library, University of Kansas; W. Boswell (1717) *Impartial Account of the Frauds and Abuses at Dagenham Breach and of the Hardships Sustained by Mr. W. Boswell Late Undertaker of the Works there*; T. Badeslade (1725) *History of the Ancient and Present State of the Port of King's Lynn and of Cambridge*; W. Porter (1889) *History of the Corps of Engineers*, **1**; S. R. Taylor (1966) *Mathematical Practitioners of Hanoverian England*; A. Crossley (1990), VCH, Oxford, **12**; J. Bond and K. Tiller (1997) *Blenheim, Landscape for a Palace*, 2nd edn.; DNB; Bendall]

Publications
1724. *Report with Proposals for Draining the Fens and Amending the Harbour of Lynn*

Works
1713–1714. Demolition of fortress and port facilities at Dunkirk
1722–1724. Canal, circular basin and cascades at Blenheim

ARMSTRONG, John (1775–1854), civil engineer, was born at Ingram, Northumberland on 13 October 1775. From an agricultural background he had little education until he was apprenticed to a local millwright. He worked for Thomas Dodgin, a Newcastle millwright, on the White Lead Works, Bill Quay, where his brother was foreman. Around 1800 he moved to Bath where he worked on the repairs to the foundations of Pulteney Bridge, where John Simpson (q.v.) was the contractor; Thomas Telford (q.v.) was the engineer for the repairs. From 1804 he worked on the construction of Bristol docks under William Jessop (q.v.), being responsible for the installation of the lock machinery and swing bridges. For a number of years he worked on similar projects in the Gloucestershire area, before acting as Clerk of Works for Sir Robert Smirke (q.v.) on Gloucester Bridge. With this experience he was employed in 1821 as resident engineer on the repairs to Rochester Bridge under John Rennie (q.v.) and Thomas Telford at a salary of £50 per year plus a house. His civil engineering career continued with work on the Grosvenor Canal in London, involving a tidal lock and basin.

Armstrong's most famous post was as resident engineer for Marc Isambard Brunel (q.v.) on the Thames Tunnel, but he held this position only for a short period as the strain of the work began to undermine his health, and in August 1826 he resigned and was eventually replaced by Isambard Kingdom Brunel. After a period with Bramah and Robinson (see: BRAMAH, Joseph), building the lock gates and installing equipment for St. Katharine's Dock—and managing their property development at Calverley Park, Tunbridge Wells—he applied in 1831 for the post of City Surveyor to Bristol, an appointment he held until his death on 17 March 1854.

Armstrong was elected an Associate of the Institution of Civil Engineers in 1828, while engaged at St. Katharine's Docks. William Armstrong (d. 1858) was probably his son; he entered the Clifton Bridge competition in 1829, designed a number of religious buildings in Bristol and neighbouring counties, and was a building surveyor in Bristol from 1836 until 1858.

MIKE CHRIMES

[Thames Tunnel mss, ICE Library; Membership records, ICE; G. T. Clark (1850) *Report to the General Board of Health ... Bristol*; Obituary (1855) *Min Proc ICE*, 147–148; R. Beamish (1862) *Memoir of the Life of Sir Marc Isambard Brunel*, 214–230; R. Sharp (1991) *Bramah and Robinson Drawings*, inv. 1991–349, Science Museum; G. Body (1976) *Clifton Suspension Bridge*, 20; N. Yates and J. M. Gibson (1994) *Traffic and Politics*, 217–218; Colvin (3)]

Works
1800–1802. Pulteney Bridge, Bath, repairs
1804–1810. Bristol docks, lock machinery, etc., assistant engineer/millwright
1813. Congresbury drainage, Gloucester, engineer/millwright
1813. Lydney lock gates, engineer/millwright
1814–1817. Westgate Bridge, Gloucester, resident engineer
c. 1821–1824. Rochester Bridge, repairs, resident engineer
1824. Grosvenor Canal, resident engineer
1825–1826. Thames Tunnel, resident engineer
1826–1829. St. Katharine's Dock, lock gates, resident engineer/millwright

ARROWSMITH, Aaron (1750–1823), map-maker and geographer, was born on 14 July 1750 in Winston, County Durham. He came to London in about 1770 and was there employed by Carey, the eminent map seller. By 1790 he was established in Castle Street, Long Acre, where he began producing maps and charts. His reputation was established by his *Large Map of the World on Mercator's Projection*, published in 1790. From 1802 to 1808 he was at Rathbone Place where in 1807 he produced his *Map of Scotland ...*, engraved on four sheets to a scale of 4 miles to the inch, and the subsequent *Memoir* of 1808. In 1814 he moved to Soho Square, London, where in 1822 he produced his *Atlas of Southern India*. He published over 130 maps and was for several years Hydrographer to the King.

Arrowsmith became an Associate of the Institution of Civil Engineers in 1820, recommended by Thomas Telford (q.v.), on account of 'the vast services he has rendered to the profession of Civil Engineer by his indefatigable exertions for the attainment of accurate geographical delineations and the great success which has followed those exertions'.

He died on 23 April 1823, leaving two sons, Aaron (d. 1854) and Samuel (d. 1839), who continued their father's business until the death of Samuel. Aaron, Jr., was educated at Oxford and wrote several geography textbooks, three of which were published by his brother in 1832. Both sons seem to have maintained an association with Telford after the death of their father, of whom an engraved portrait exists. The *Atlas* to the *Life of Telford*, published in 1838, contains ten plates for which payment was made to 'Arrowsmith'. These plates are mostly based on maps and appear anonymously in the *Atlas*, or are plates credited to Edmund Turrell (q.v.). The Telford Bequest to the Institution of Civil Engineers included many of Arrowsmith's maps.

TESS CANFIELD

[*Gibb Papers*, ICE; *British Library Catalogue of Maps Charts and Plans; Map Catalogue of the Royal Geographical Society*; Skempton; DNB]

Cartographic works
Maps and plans engraved for Thomas Telford by father and sons include:

1806. *General Map of the Intended Caledonian Canal*
1807. *Plan of the Caledonian Canal* (updated 1809 and 1810)
1807. *Map of Scotland*
1819. *Map of that Part of the Menai Strait which includes the Site of the New Bridge ...*
1819. *Map of the ... Mail Roads between Holyhead & Bangor Ferry...*
1820. *Holyhead Road ...*
1821. *Caledonian Canal ...*
1821. *Map and Sections of a Proposed Mail Road between London and Edinburgh ...*
1822. *Holyhead Road ...*
1823. *Plan of the Crinan Canal ....*
1824. *Section of a Part of the Old Holyhead Road Improved ...*
1824. *Map of the Roads between Girvan and Stranraer ...*

1824. *Map of the South West Parts of England & Wales ...*

1825. *Map of the present Mail Road between Edinburgh and Morpeth ...*

1826. *Plan of the Proposed Newport Branch of the Birmingham and Liverpool Junction Canal*

1828. *Map of the Country between the Towns of Liverpool and Talk on the Hill ...*

1828. *Sections of Sundry Distinct Lines of Mail Road between Liverpool and Talk on the Hill*

1834. Three plans to accompany *Metropolis Water Supply, Report of Thomas Telford*

1838. Ten plates in J. Rickman, *Life of Thomas Telford*

Publications

Other publications include:

1794. *A Companion to the Map of the World*

1809. *Memoir Relative to Construction of the Map of Scotland*

1825. *Geometrical Projection of Maps*

ASHLEY, Henry (fl. 1674–d. 1710), engineer-undertaker of the Great Ouse navigation. Between 1618 and 1621, the 15 mile length of the river Ouse was made navigable from St. Ives to St. Neots by Arnold Spencer (q.v.) and by 1638 he had extended the navigation 8 miles up to Great Barford. Trading declined during the Civil War, before Spencer had recovered his expenditure, and he died in 1658 without realising his original intention of extending the navigation to Bedford. After the War little more was done other than essential maintenance of the sluices below St. Neots, while the river upstream was allowed to deteriorate.

Eventually, in 1674, the navigation, now held in trust for Spencer's grand-daughters Anne and Elizabeth and their husbands the brothers John and Nathaniel Jemmatt, was leased to 'Henry Ashley of Eynesbury, tanner' for twenty-one years at a rent of £160 per year. Eynesbury is close to St. Neots. In 1688 Ashley said he took the lease 'having for some years before lived near the river and sluices, and being skilful in such works'. Joseph Hunt (q.v.), who in all probability constructed some of the locks and stanches in the 1680s, referred to Ashley as being 'skilled in works of navigation'. How he gained his skill is not known but he proceeded to improve and extend the navigation with considerable success.

The first task was to 'cleanse' the river between Eaton Mills and Great Barford, a stretch of 7 miles with no watermills but many shallows which required repeated scouring. Moreover, shallows had developed downstream of St. Ives due to a lowering of river level by the cutting of New Bedford River between 1651 and 1653 as part of the work by Sir Cornelius Vermuyden (q.v.) in draining the Great Level of the Fens. To overcome this obstacle Ashley built a stanch and weir half a mile below St. Ives in 1678 at his own expense (£200), even though it lay outside the boundaries of his lease.

At about the same time he rebuilt the lock at Houghton Mill. Next, in 1680, he built a new lock at Eaton Mills, and then provided a permanent solution to some of the worst shallows upstream by a lock in a new cut to Roxton with a stanch in the river (1681–1682), a stanch at Tempsford and another near Barford, both built about 1686. The cost of these five works according to Ashley's accounts came to the surprisingly small total of £960.

The way was now open to extend the navigation 7 miles up to Bedford. In March 1687 Commissioners acting for the Crown appointed 'Henry Ashley the elder of Eynesbury, gentleman' to be undertaker for making the river navigable from Eaton Mills to Bedford. He proceeded to build five pound locks above Great Barford and the navigation was opened in September 1689.

Meanwhile his son Henry Ashley Jr. (c. 1654–1729) purchased in 1680 from Nathaniel and Elizabeth Jemmatt their half share in the navigation between St. Ives and Barford. Ashley senior's appointment as undertaker in 1687 obviously affected the rights of John and Anne Jemmatt in the river above St. Neots to Barford and in 1688 they sought protection for those rights in the Court of Chancery on the grounds that Ashley's works in this stretch had been made while he held a lease; the dispute dragged on for nine years.

Henry Ashley junior appears as a well-to-do man with an address in Gray's Inn in 1690, the year in which he purchased the manor of Eaton Socon (including Eaton Mills). In the later stages of the dispute he acted as his father's counsel. Under an Act of 1701 he became undertaker of the river Lark navigations from Mildenhall to Bury St. Edmunds, and he rebuilt St. Ives stanch in 1719–1720. Henry Ashley senior is last heard of in the affairs of the navigation in May 1696 and he died at Eaton Socon in January 1710.

A. W. SKEMPTON

[T. S. Willan (1946) The navigation of the Great Ouse between St. Ives and Bedford in the seventeenth century, *Bedfordshire Hist Record Soc*, **24**; D. Summers (1973) *The Great Ouse, the History of a River Navigation*; J. Boyes & R. Russell (1977) *Canals of Eastern England*]

Works on the Great Ouse

1678. Stanch and weir downstream of St. Ives, £200

1680–1686. Two locks and three stanches between St. Neots and Great Barford, 8 miles, £1,000

1687–1689 Five locks between Great Barford and Bedford, 7 miles

ASPDIN, Joseph (1778–1855), cement manufacturer, was born in Hunslet, Leeds, the eldest of six children of Thomas and Mary Aspdin. Joseph followed his father's trade of bricklayer and builder at Briggate, Leeds. Around 1800 there was increasing interest in cement manufacture following the publication of Smeaton's researches into the properties of cements and James Parker's 1796 patent for Roman cement, which encouraged imitators and competitors to experiment. Aspdin's experiments with cement making resulted in him taking out a patent (5022), *An Improvement in the Modes of Producing an Artificial Stone*, granted in June 1824.

There were a number of other manufacturers experimenting at the time, notably James Frost (q.v.), and the strength and characteristics of their products may have been of higher quality than Aspdin's were at that time, but Aspdin chose the name 'Portland cement' for his patent material, doubtless inspired by the high reputation of Portland stone. To have produced the kind of strength associated with modern 'Portland cements' would have required high kiln temperatures in addition to the right blend of correctly ground materials, and not enough was known about his process in the 1820s. It has been suggested that he used a kiln like the furnaces used in the glassworks at Hunslet and Wakefield. Initially the impact was local, although there is an obscure reference in an early (unpublished) Institution of Civil Engineers discussion on cements, to a Yorkshire manufacturer using stones gathered from roads as his source of cement.

Aspdin had set up a partnership in the mid 1820s as 'patent Portland cement manufacturers', and operated from Leeds and Wakefield, later expanding to an agency at Liverpool. In 1838 he moved to a new site in Wakefield, with his two sons, James and William, and took James into partnership three years later. One assumes that the new kiln reflected the experience gained over fifteen years and was capable of producing stronger cements. The products of this works, and the one set up by William with J. M. Maude in Rotherhithe *c.* 1841, were those which established Aspdin's reputation as the founder of the modern cement industry, although other manufacturers, notably Parker, Frost, and Francis and White, were of greater importance in the period.

When Joseph retired in 1842, James continued the business, while William was involved in ventures in Gateshead and Hamburg and promoted his father's reputation.

Aspdin married Mary Fotherby at Hunslet, Leeds, in 1811. They had five daughters, two of whom died in infancy, as well as their two sons. Aspdin died on 20 March 1855 and was buried in St. John's churchyard, Wakefield.

MIKE CHRIMES

[P. E. Halstead (1961–1962) The early history of Portland cement, *Trans Newc Soc*, **xxxiv**, 1961–1962; A. W. Skempton (1962–1963) Portland cements, 1843–1887, *Trans Newc Soc*, **xxxv**, 117–152; B. L. Hurst (1996) Concrete and the structural use of cements in England before 1890, *ICE Proc, Struct Build*, **116**, 283–294; DNB]

ASHWELL, James (1799–1881), engineer, was the youngest of the founders of the Institution of Civil Engineers present at the inaugural meeting on Friday, 2 January 1818 at the Kendal Coffee House in Fleet Street, London. He was born in Nottingham, the son of an iron-master and colliery owner. He attended a grammar school in Nottingham and then graduated in Classics from Edinburgh University. From 1817 to 1823 he was a pupil of Bryan Donkin (q.v.).

Interested in ironworks, steam mills and machinery, he followed a career it what would now be considered mechanical engineering. He worked for a Thames Street firm in their works in Derbyshire and Scotland. Then from 1836 until 1841 he was managing director of the Blaenavon Iron and Coal Works. He inherited property and, wanting his sons to attend Cambridge, he moved there, also entering himself as a Fellow-Commoner at Jesus College. He kept the regular terms, passed the examinations and took BA and MA degrees.

When it was recommended that he travel after the breakdown of his health in 1845, he visited the continent, along the way making a report to the Directors of the Great Luxembourg Railway Company, which operated canals, ironworks and coal-mines as well as railways. He was offered the post of Managing Director and Engineer in Chief, based in Brussels, where he remained from 1847 to 1852. In a lawsuit involving the Belgian government and the railway company, Ashwell was named as respondent and, although the company was acquitted, Ashwell suffered great financial loss. His health deteriorated and he returned to England and took up the study of Hebrew and harmony.

Ashwell had married when young and he died on 2 July 1881 at his home, Mildmay Lodge, Weston Super Mare.

TESS CANFIELD

[ICE membership records, ICE archives; Obituary (1881) *Min Proc ICE*, **66**, 372–375; J. G. Watson (1988) *The Civils*]

ATHERTON, Charles (1805–1875), civil engineer, was born at Calne, Wiltshire, the son of Nathan Atherton, a solicitor. At the age of nineteen he went to Cambridge, graduating after four years as a Wrangler with a firm grasp of scientific principles and determined on an engineering career. He began working for Thomas Telford (q.v.) on St. Katharine's Dock, assisting Thomas Rhodes (q.v.), and joined the Institution of Civil Engineers in 1828. On the completion of the dock in 1829, he moved to Scotland to act as Resident Engineer on Telford's Dean Bridge. This bridge, and other late examples of Telford's designs at Pathhead and Stonebyres, is of some interest, as Telford deliberately attempted to lighten the visual appearance by placing thin projections on the faces of the piers and springing subsidiary arches from these well above the springings of the main arches. It was Atherton rather than

Telford who described this device, in the *Encyclopaedia Britannica*, explaining that it was done to counter criticism of the heavy appearance of tall viaducts. Telford referred to this article in his autobiography.

In early 1832 Atherton went to Glasgow to act as resident engineer on the new bridge (Broomielaw) there. He was responsible for the site investigation which governed the foundation design and shortly afterwards he was appointed resident engineer, on the recommendation of James Walker, to the River Clyde Trustees. For them he prepared plans for the improvement of the Broomielaw Quay.

In late 1834 Atherton resigned his position on the Clyde, and took over the management of Claud Girdwood and Company, well established iron founders. He remained with them until 1835, producing high quality marine engines. Those to the designs of the well-known marine engineer, John Bourne, for the *Don Juan* included several innovations, including the marine governor.

After reporting on navigation and harbour improvements at Malines with another Telford protégé, Alexander Gordon (q.v.), in 1837, Atherton worked for two years in Canada on improvements to the St. Lawrence. After working in the USA Atherton returned to England in 1845, and in June the following year became an Assistant Engineer in Woolwich Dockyard, becoming Chief Engineer in April 1847. In that capacity he issued a succession of reports on marine engineering and contributed to the Institution of Civil Engineers *Minutes of Proceedings*. He was transferred to Devonport from 1848 until 1851, but returned to Woolwich, where he remained until 1861.

On his retirement from the Admiralty he practised as a consultant in Whitehall until 1870 when he retired to Sandown, Isle of Wight. Atherton was unusual for his formal scientific training at a time when most of his contemporaries had been self-taught. He carried through his interests in retirement as an amateur astronomer. He died on 24 May 1875, aged 70.

MIKE CHRIMES

[Membership records, ICE archives; Drawings of Broomielaw Bridge, and Port Glasgow Wet Dock, 1832–1834 ICE archives; J. Rickman (1838) *Life of Telford*, 191–201; Memoir (1875) *Min Proc ICE*, **42**, 252–255; Sir A. Gibb (1935) *The Story of Telford*, 243, 260; T. Ruddock (1979) *Arch Bridges and their Builders*]

Publications

1837. *Rapport Preliminaire sur les Moyens d'Ameliorer la Navigation de la Dyle et de Construire un Port a Malines* (with A. Gordon)

[1842]. Article on Dean Bridge, *Encyclopaedia Britannica*, 7th edn.

1855. *The Capability of Steam Ships, Based on the Mutual Relations of Displacement, Power and Speed*, 2nd edn.

Works

1829–1831. Dean Bridge, Edinburgh, resident engineer, £18,556

1832–1834. Broomielaw Bridge, Glasgow, resident engineer, £34,427

c. 1833–1834. River Clyde Commission, resident engineer

ATKINSON, George (1663–1728), was baptised on 8 March 1663, a son of Richard and Elizabeth Atkinson of Roxby, Lincolnshire. About 1696 he moved to Thorne, and by that time had become a Quaker. In 1702, with James Mitchell (q.v.), he contracted to build locks and weirs at Beal and Haddlesey on the river Aire. The work was completed about a year later at a cost of rather more than £1,600. Atkinson continued to carry out various works on the Aire and Calder navigation at least to 1718. In December 1722 he bought timber to a value of £1,200 with an eye to building locks on the proposed river Don navigation, though in the event work did not start until several years later. He owned land in the neighbourhood of Thorne, mentioned in his will. He died on 20 March 1728.

A. W. SKEMPTON

ATKINSON, Joseph (1693–1760), surveyor and carpenter, elder son of George and Milcah Atkinson of Roxby, was born in March 1693 and taken with his parents when they settled in Thorne. He married Sarah Johnson at the Friends' Meeting House, Thorne, in 1718.

In 1722 he was engaged by the Corporation of Doncaster to survey the river Don with a view to improving its navigation. At the same time William Palmer (q.v.) was employed by the Company of Cutlers of Hallamshire to survey the river up to Sheffield. They brought in Joshua Mitchell (q.v.) to help. The survey was finished in November after four weeks; Mitchell and Palmer took the levels or 'falls', Atkinson measured the distances. The map by 'Will. Palmer and Partners' was published, with tables of the falls and distances, in 1723.

Palmer and Atkinson attended a meeting in November 1722 to explain their proposals. Soon afterwards one of the promoters of the scheme wrote to a friend saying that Atkinson was very willing to serve as surveyor (i.e. engineer) and 'how can we have a better man?' He had meanwhile estimated the cost of the works for Doncaster at £6,000. After lengthy debate, an Act was passed in May 1726 for the navigation from Doncaster up to Tinsley (3 miles north east of Sheffield), Palmer giving evidence. An Act for the navigation below Doncaster followed in April 1727, Atkinson giving evidence as engineer supported by Richard Ellison.

Work began on the locks above Doncaster in June 1726 with Palmer in charge. In the absence of records it may be assumed that Atkinson directed work for the Doncaster proprietors from 1727. By 1731 six locks and several cuts had been made, Sheffield having spent about £8,500 and Doncaster about £3,500. The two sets of proprietors were amalgamated in August 1731 to form a single company with John Smith (q.v.) as their engineer.

Atkinson had become one of the lessees of the Aire and Calder Navigation in 1723 and he continued as such until about 1756, having 'amassed a large property'. The will of Joseph Atkinson, merchant of Rawcliffe, where he had lived from about 1730, is dated June 1757. He returned to Thorne, where he died on 21 February 1760 leaving property in Thorne, Rawcliffe and Snaith to his widow and several children.

A. W. SKEMPTON

[Will, Palmer and Partners (1723) *A Survey of the River Dunn*; H. W. Atkinson (1933) *The Family of Atkinson of Roxby and Thorne*; T .S. Willan (1956) *The Early History of the Don Navigation*; R.W. Unwin (1962) The Aire and Calder Navigation, *Bradford Antiquary*, **43**, 151–186; C. Hadfield (1972) *Canals of Yorkshire and North East England*]

ATKYNS, Richard (fl. 1596–1618). In the Preface of his book on the history of the port of King's Lynn and the Great Level of the Fens, Thomas Badeslade (q.v.) refers to a 'valuable manuscript of Mr. Richard Atkins of Outwell (near Wisbech) who flourished in the Reign of King James I. He was an eminent Commissioner of Sewers, a man of great Learning and Experience and, as Sir William Dugdale says, a notable Observer of the Fens. He was consulted by all the Undertakers of his Time, and was Surveyor of the Works of Sewers'.

The Commissioners of Sewers were the land drainage authorities, organised on a county basis, responsible for maintaining the drains, rivers, banks and sluices. From time to time new works were carried out, on a relatively modest scale, but in 1589 and again in 1604 and 1618 Commissions drawn from all six counties—Norfolk, Cambridge and the Isle of Ely, Northampton, Huntington and Lincoln—considered the 'general draining' of the Fens.

In the 1604 proceedings John Hunt (q.v.) acted as the 'artist' or engineer, Atkyns presented reports on the nature of the fens and the river Nene outfall, while William Hayward (q.v.) was the land surveyor who produced a map of the whole region and, in July 1605, a detailed list of all the fens with the area of each, totalling 307,242 acres. In April 1605, as Dugdale relates, 'Mr. Richard Atkins of Outwell (a person whose observations on these ferry grounds were very notable) made search with an auger 11 ft. long on the skirts of the New [Morton's] Leam from Guyhirn to Stanground, to find the nature of the soil'.

The results at thirty-six places along this 12 mile length are given, together with similar investigations at a dozen sites north of the river

Ouse from Earith to Denver. The soils are described as moor (peat), silt, clay and gravel. By June, Hunt had decided on the principal works required and in August he and Atkyns set out the first of them: a cut 5.5 miles long from the old Nene river to Well Creek at Nordelph. The cut, subsequently known as Popham's Eau, was the only work executed at the time though Hunt's other proposals were eventually carried out, notably the 21 mile long channel for Earith to Salters Lode, called the Bedford River, made 1632–1635. Meanwhile in June 1605 Atkyns reported on the channels, salt marshes and creeks in the Nene estuary seaward of a point about 4 miles below Wisbech, at a time before the reclamation of the marshes lying outside the old, so-called 'Roman sea bank'. Wells refers to 'Mr. Atkyns and Mr Hunt, two of the most able and scientific fen men of that time', a judgement that seems fully justified.

Atkyns was doubtless engaged on other matters following the 1604–1605 project and one such is on record: an accurate survey of the bank along the north side of Well Creek protecting the Marshland townships from floods caused by the river Ouse on fens to the south. But the next (and last) time he comes into prominence is in connection with investigations in 1618. Resulting from a petition from the Commissioners of the six counties, the Privy Council ordered their Clerk, Sir Clement Edmondes, to report on the rivers Ouse, Nene and Welland, which he did in September 1618, and a little later that year Atkyns presented a more general report on the Fens at the request of the Bishop of Ely. On his seven-day field visit Edmondes was accompanied by three Commissioners from each county and 'my Lord Russell—who gave great assistance in the business'. To quote Wells again, the 'very able report by Mr. Atkyns' drew on long experience and keen observation over a period of twenty-two years. Together, the reports provide the classic account of the Fens preceding the general draining effected between 1631 and 1656. They existed as manuscript copies, and were still being quoted in the eighteenth century; both are printed in Wells' *History* of 1828.

A. W. SKEMPTON

[*Relatio R. A. de Mariscis & Eorum Statu. Anno 1604. Mense Januarii & Februarii* (1605 N.S.); Sir William Dugdale (1662) *The History of Imbanking and Drayning*; Thomas Badeslade (1725) *The History … of the Port of King's Lyn … and the Draining of the Great Level of the Fens*; Samuel Wells (1830) *The History of the Drainage of … Bedford Level*, and vol. 2 (1828); H. C. Darby (1940) *The Draining of the Fens*; L. E. Harris (1953) *Vermuyden and the Fens*]

ATWOOD, George, FRS (1746–1807), mathematician, was educated at Westminster School and Trinity College, Cambridge, where he became a Fellow and a distinguished lecturer. Elected an

FRS in 1776, he left Cambridge in 1784 and obtained a government sinecure as a Patent Searcher for the Customs, with an income £500 per annum, thanks to the influence of William Pitt. Until his health broke down he was used by successive administrations to carry out the calculations connected with government finances.

Atwood's prestige as a scientist led Thomas Telford (q.v.) to write to him in April 1801 regarding his proposal for a cast-iron arch bridge of 600 ft. span to cross the Thames at London. Atwood had written a number of textbooks on natural philosophy, no doubt based on his lectures, and his *Dissertation on the Construction and Properties of Arches* was published soon after Telford's approach. It was one of the first books presented to the Institution of Civil Engineers for its Library and was given by the Reverend H. T. Ellicombe, who had worked at Chatham Dockyard before taking Holy Orders.

Atwood produced a supplement to his book in 1804 describing his experiments with model arches of brass. To investigate the properties of voussoirs and the equilibrium of arches he used an apparatus incorporating weights and pulleys to measure the pressure on the voussoir surfaces. As a result of earlier work of Charles Hutton (q.v.), an understanding of arch theory had become relatively common among the scientists and better-read engineers and, although Atwood's work was well known at the time, it was generally forgotten within a generation as the work of Henry Moseley and French engineering scientists became familiar.

Atwood also made contributions to the theory of the stability of ships and other mechanical subjects. He died at his home in Westminster in July 1807 and is buried locally at St. Margaret's.

MIKE CHRIMES

[DNB; Evidence (1800–1801) *House of Commons Select Committee Upon the Improvement of the Port of London*, 3rd report and appendix, 48–52 (*Report*, **xiv** (1803), 621–623); Ted Ruddock (1979) *Arch Bridges and their Builders*, 156–159; A. W. Skempton (1980) *Telford and the Design for a New London Bridge*, in A. Penfold (ed.), *Thomas Telford, Engineer*, 62–83; J. Heyman (1982) *The Masonry Arch*]

Publications

Publications include:

1784. *An Analysis of a Course of Lectures on the Principles of Natural Philosophy*
1794. Investigations founded on the theory of motion for determining the times of vibration of watch balances, *Phil Trans*, **84**, 119–168
1796. The construction and analysis of geometrical proportions determining the positions assumed by homogeneal bodies which float freely, and at rest, on a fluid's surface, *Phil Trans*, **86**, 46–130
1798. Disquisition on the stability of ships, *Phil Trans*, **88**, 201–310

1801. *A Dissertation on the Construction and Properties of Arches*
1804. *A Supplement to a Tract, entitled Treatise on the Construction and Properties of Arches*

AYDON (AYTON), Samuel (fl. 1790–1820), iron-master, of Flanshaw, near Wakefield, is best known for his involvement in the Shelf ironworks. In 1794 he purchased a share in Fall Ings ironworks from John Elwell (q.v.) and was subsequently granted a 25% interest in the Shelf works. The firm built steam engines for the local textile industry including that for the Ossett Mill Company; by 1810 Aydon and Elwell's connection with Fall Ings was over. The firm established a reputation for the production of elements for cast iron bridges, particularly moveable bridges for docks.

In May 1796 Aydon's brother (or son) Isaac, married Maria, the daughter of John Banks (q.v.), the lecturer in natural philosophy, whose work *On the Power of Machines* (1803) contains reference to tests carried out in 1795 on the strength of cast iron produced at the Shelf blast furnace. This is among the earliest works containing experimental research on the strength of cast iron. Isaac appears to have continued with the Wakefield foundries and was responsible for some of the tramway from Barnby Basin—on the Barnsley canal—to Silkstone Cross, along with the Low Moor works.

Samuel married Mrs Hollas of Halifax in October 1814. His partnership with Elwell apparently was bankrupted in 1810 but was subsequently refinanced as many bridges were erected using their iron prior to 1820. However, in late 1821 Samuel Aydon, also of Halifax, was reported bankrupt and in 1824 the Shelf works were taken over by the Low Moor Company.

Other than for dock works, the significance of Aydon and Elwell's contribution to the development of iron bridges was minimal, possibly because they lacked a connection with a distinguished designer.

MIKE CHRIMES

[East India Docks minute books, etc., Museum of London in Docklands; Rennie mss, National Library of Scotland; J. G. James collection, ICE Library; J. Banks (1803) *On the Power of Machines*; *Monthly Mag*, **20** (1805), 278; *Monthly Mag*, **30** (1810), 263; *Monthly Mag*, **38** (1814), 382; *European Mag*, December 1821; W. Cudworth (1891) *Histories of Bolton and Bowling*, 205–213; J. Goodchild (1968) The Ossett Mill Company, *Textile History*, 46–61; W. N. Slatcher (1968) The Barnsley canal, *Transport History*, **1**(1), 48–66; W. L. Norman (1969) Fall Ings, Wakefield, *Ind Archaeol*, **6**(1), 74–79; M. J. T. Lewis (1970) *Early Wooden Railways*; J. Goodchild (1977) *The Lake Lock Rail Road*; J. G. James (1987–1988) Evolution of early iron arched bridge designs, *Trans Newc Soc*, **59**, 153–186]

Works
(see: *ELWELL, John*)

B

BACON, Anthony (b. 1772, fl. 1826–1828), ironmaster, of Elcott Park, Newbury, Berkshire, was elected an Associate of the Institution of Civil Engineers in 1826. He was an illegitimate son and heir of Anthony Bacon MP (d. 1786) a shipowner and merchant who, with his partner William Brownrigg, acquired the mineral rights to 4,000 acres near Merthyr Tydfil in 1765. In 1766–1767 the first blast furnaces were erected for him there, the origins of the Cyfartha ironworks. In 1773 Bacon Sr. obtained a government contract for cannon and, with a new partner, Richard Crawshay (q.v.), from 1777 began to produce them at Cyfartha; in 1794 Bacon Jr. negotiated the sale of the works. He maintained an interest in scientific and technical matters and as a member of the ICE described in some detail his method of heating hothouses etc. on his estate near Reading. Bacon's mother was Mary Bushby, and he only assumed his name in 1792. He probably died in 1828.

MIKE CHRIMES

[Membership records and minutes of conversations, ICE archives; L. Namier and J. Brooke (1964) *The House of Commons 1754–1790*]

BADESLADE, Thomas (fl. 1712–1745), land surveyor, topographical draughtsman and author of a book on fen drainage. In the early 1720s the Corporation of King's Lynn represented to their MP, Sir Robert Walpole, their 'dreadful Apprehension' concerning the decay of Lynn harbour and navigation on the river Great Ouse. In response, Sir Robert consulted Colonel John Armstrong (q.v.) who—then in the area—ordered that maps and surveys of the ancient and present state of the Fens should be copied, authentic records of drainage and the harbour collected and tidal observations made. By his direction and at the request of Lynn Corporation Badeslade carried out these tasks. After three years research he published the results in 1725 in a 148-page folio volume dedicated to Walpole and illustrated with hand-coloured engraved plates including a copy of the map by William Hayward (q.v.) and Badeslade's own map of 1723. Also included are several seventeenth century documents, printed here for the first time, and a copy of Armstrong's report which had been separately published in 1724. Badeslade's book is a valuable source for the history of Fen drainage, even if his (and Armstrong's) interpretation of the facts are in some respects mistaken.

Armstrong claimed to have discovered the main cause of the present bad state of the Fens and Lynn harbour: namely that for a considerable time past, since the works by Sir Cornelius Vermuyden (q.v.) in the 1650s, there had not been a sufficient indraught for the tides to flow up the Great Ouse as they formerly did for many miles. Consequently as a remedy the sluices at Denver and Earith should be removed, the river widened and deepened where necessary and its banks raised. These controversial proposals were almost immediately opposed in a report by Charles Bridgeman, commissioned by the Earl of Lincoln. He adopted a suggestion by Nathaniel Kinderley (q.v.) of making a new cut to bypass a long and shallow bend in the river above King's Lynn but with a navigable sluice at its lower end to keep out the tides. Bridgeman's report was answered paragraph by paragraph in a pamphlet, unsigned but certainly written by Badeslade on behalf of King's Lynn, and bound in with many copies of his book.

A succession of reports followed, by John Perry and Humphry Smith (qq.v.) and by Kinderley, and in 1729 Badeslade returned to the attack. He reported Armstrong's view of the Ouse problem but now provided additional evidence from the decay of Rye harbour, attributed to a reduction of tidal indraught by inclosure of salt marshes and sluices on the river Rother. Badeslade was not against sluices in their proper places, for example for scouring the harbours at Dunkirk and Dover, and in his 1729 report he mentions that he himself had lately cleansed the harbour of Bideford by such means.

Before this, in about 1724, he had taken levels for a proposed new cut to improve the outfall of Shire Drain, the main drain of the North Level of the Fens. The cut was to run for 3 or 4 miles from Hills Sluice to the coast. It is shown on a map of the North Level by Badeslade published in 1726, but was not made, probably on grounds of cost.

In 1730 Badeslade was carrying out some 'waterworks' for the Earl of Gainsborough at Exton Park, near Stamford. Hearing that John Perry was short of workmen on the Welland at Spalding, Badeslade sent a gang to help out, and in the summer of 1731 spent several days with Perry at the opening of the great sluice on the river there. As he expected, the scouring effect was limited to a rather short distance downstream.

Badeslade's concept of the importance of tidal indraught formed his criticism of Kinderley's scheme for a ship canal from Chester through the salt marshes on the south side of the Dee estuary to a channel in the sands about 7 miles below the city. This was being made to improve navigation and also, when completed, to allow reclamation of a large area of land from the sea in the upper estuary. He visited the works in 1735 on his way to Ireland and was so concerned by the possible consequences that he published a critical report on the subject. The loss of tidal waters would in his opinion lead to destruction of the navigation downstream and drowning of low land adjacent to the river. He repeated the now familiar arguments regarding the Ouse and Rother and added a criticism of the drainage works on Deeping Fen, near Spalding.

As it happens, John Grundy (q.v.) had visited the Chester works earlier that year. He, too, was critical but more with regard to the canal itself. He also disagreed with what Badeslade had to say about Deeping Fen. He therefore published 'Remarks' on Badeslade's report and this, in turn, prompted a lengthy reply to which Grundy responded with an equally long dissertation, touching on flow in open channels, levelling, and his own work in the Spalding district.

This dispute with Grundy seems to mark the end of Badeslade's involvement in civil engineering problems. With John Rocque he published the fourth volume of *Vitruvius Britannicus* in 1739 containing general views and plans of houses and their gardens: Kensington Palace, Hampton Court and so on. In 1741 he completed his *Chorographia Britanniae*, a set of maps of all the counties of England and Wales (published 1742, 2nd edn., 1745). His *Views of Noblemen and Gentlemen's Seats in Kent, All Designed Upon the Spot by the Late T. Badeslade* is tentatively dated 1750. The date of his death has not been found.

A. W. SKEMPTON

[John Grundy (1736) *An Examination and Refutation of Mr Badeslade's New Cut Canal*; Richard Gough (1780) *British Topography*; Bendall]

Publications

1725. *The History of the ... Port of King's Lyn ... and of Draining the [Great Level] of the Fens*
1726. *A Map of the North Level and of the Marshes*
1729. *A Scheme for Draining the Great Level of the Fens, Called Beford Level, and of Improving the Navigation of Lyn-Regis*
1735. *Reasons ... Shewing How the Works now Executing ... to Recover and Preserve the Navigation of the River Dee will Destroy the Navigation*
1736. *The New Cut Canal Intended for Improving the Navigation of the City of Chester ... Compared with the Welland ... and Deeping Fens*

BAGE, Charles Woolley (*c*. 1752–1822), a pioneer of the use of iron for the internal framing of industrial buildings, was the eldest son of Robert and Elizabeth Bage. He was born at Darley-Abbey, near Derby, where for several generations the family had worked a paper mill on the river Derwent.

His youth was spent at Elford, between Lichfield and Tamworth, where his father had bought a paper mill on the river Tame in 1754. In 1766 Robert Bage sold his Elford mill to raise capital for a new iron rolling and slitting mill at nearby Wychnor. A wide cut made from the site to the river Trent provided waterpower for the ironworks and soon afterwards became part of the Trent and Mersey Canal. Partners with Bage were John Barker and Dr. Erasmus Darwin, the great eighteenth century polymath, both of Lichfield, and Samuel Garbett, the Birmingham industrialist and co-founder of the famous Carron ironworks in Scotland. The Wychnor works proved to be unprofitable and were sold in about 1783, after the death of Barker, who had been manager. Through his father's business interests, Charles Bage would have been exposed to engineering works at an early age, including mills and their machinery and perhaps, more significantly, the manufacture and properties of iron.

Although he worked the Elford mill as a tenant until he died in 1801, Robert Bage had lost money in the ironworks venture and, as a consolation, he took up his pen and wrote the novels that were to make his reputation. Bage's father was a cultivated man with both literary and scientific interests, on the edge of the circle of Midland's philosophers that had Darwin at its centre. In 1760 he had decided to study mathematics and had weekly lessons with a well-known teacher, Thomas Hanson of Birmingham. Thirty years later his friend from childhood, William Hutton, was full of praise for his abilities as a philosopher and mathematician.

In a family with a Quaker background, the formal education of Charles and his brothers would have been handled with care. His younger brother, Edward, became a surgeon and apothecary in Tamworth, before moving to Shrewsbury where he died in 1812. Charles himself married Margaret Botevyle, the daughter of a Shrewsbury apothecary, in 1781, set up in business as a wine merchant and gradually became established as a public figure. From 1784 to 1787 he was a Director of the House of Industry, the Shrewsbury workhouse, with some responsibility for the fabric of the buildings and in 1789 he was admitted a burgess of the corporation, in 1807 serving as Mayor.

Charles Bage's most important engineering work began through his association with the Benyon brothers, Shrewsbury wool merchants. The Benyons had entered into partnership in December 1793 with one of their suppliers, draper turned flax spinner John Marshall of Leeds. With the injection of capital from the Benyons, a second steam-powered mill was completed in Leeds by September 1795 but early the following year this building was destroyed by what the *Leeds Intelligencer* reported as 'the most dreadful fire we ever remember to have happened in this country'. This had been the fate of other large industrial buildings with internal structures of combustible timber in which highly flammable materials were processed. By the end of the following year the mill was rebuilt and back in production, but when Benjamin Benyon pushed for expansion of the business to his home town and into thread manufacture, Bage was on hand to propose a new and safer form for the building.

In September 1796, a site was acquired at Castle Foregate in Shrewsbury. Bage gave up the wine trade, joined Marshall and the Benyons as a partner with a one eighth share of profits, and took on responsibility for the design of the mill. He was a keen student of new engineering methods and had gleaned useful information from William Strutt (q.v.), who had designed a 'fireproof' cotton mill in Derby, completed in 1793. Strutt's floor construction comprised shallow arches of brick spanning between timber beams, the exposed lower parts of which were protected against fire by thin sheet iron. Cruciform section cast iron columns provided intermediate support to the beams between the external load-bearing walls. Bage's innovation was to use cast iron beams instead of timber beams as Strutt had used at Derby, thus providing a complete internal frame of iron. The Castle Foregate mill was complete and in production in late 1797 and was extended two years later.

Bage continued to experiment with enthusiasm and set out his method for sizing iron beams in a letter in reply to a request from Strutt in August 1803, at the same time reporting the results of load tests on (rather crude) cast iron roof trusses. His later ideas failed to match his initial innovation but modern concerns about progressive collapse were anticipated, with thought given to the question of 'how far the failure of one pillar would affect the rest'. There is no evidence that he put into practice his idea to use 'flat pillars' of iron embedded in the brickwork 'to take the weight of the walls and enable us to attain greater accuracy in the workmanship', although this was a further step towards the modern fully framed building.

In 1804 the original partnership agreement expired and Bage stayed with the Benyons. A short time earlier, the Benyons had built two fireproof mills to Bage's design, one at Meadow Lane in Leeds and another at the Canal Terminus in Shrewsbury. Part of the latter building was converted to housing in the 1850s. In these buildings the iron beam section and support details were an improvement on the details of the earlier Castle Foregate mill. He provided Thomas Telford (q.v.) with the results of full-scale tests on a form of beam intended for these later mills and contributed to the Select Committee report on Telford's proposed 600 ft. cast iron arch over the Thames. Unlike Telford, he never became well known outside Shropshire.

Bage's first wife died in 1806 and three years later he married 16 year old Ann Harding in Derby, with whom he had six children. His only known non-industrial building was the Lancasterian School in Shrewsbury, a building of traditional construction completed in 1812, with a factory like appearance. With the Benyons he had several small weaving factories in the Shrewsbury area and he remained interested in power-loom weaving after he withdrew from the partnership in 1815. The judgement of his erstwhile partner, John Marshall, had been that 'Mr Bage was possessed of talent and understanding but was not a man of business'. Bage died on 30 December

1822, aged 70, and his remaining linen factory failed four years later, leaving his wife bankrupt.

Charles Bage is buried at the new St. Chad's Church in Shrewsbury, which boasts a fine memorial to William Hazledine (q.v.), the Shrewsbury iron founder who had provided the ironwork for the Castle Foregate mill. This church, in which iron columns support the upper galleries, was built in 1790–1802 by John Simpson (q.v.).

T. SWAILES

[Charles Bage (1801) Letter to Thomas Telford, ICE archives; Select Committee upon Improvement of the Port of London, *4th Report* (1801); Charles Bage (*c.* 1797–) Letters to William Strutt., Shrewsbury RO; J. L. Hobbs (1950). This Shropshire building is the oldest Surviving iron framed one in Britain, *Shropshire Mag*, 43–44; F. R. Gameson (1954) The story of Charles Bage, wine merchant, who tried to put Shrewsbury on the industrial map, *Shropshire Mag*, 27–28; A. W. Skempton (1956) The origin of iron beams, *Actes du 8e Congr. Int. d'Historie des Science*, 1029–1039; W. G. Rimmer (1960) *Marshalls of Leeds Flax Spinners*; A. W. Skempton and H. R. Johnson (1962) The first iron frames with description of Castle Foregate and Meadow Lane mills, *Architectural Rev*, **131**, 175–186; H. A. Green (1976) An early fireproof cast iron framed building (Castle Fields mill), *Foundry Trade Journal*, 891–896; P. Faulkner (1979) *Robert Bage*; J. Gould (1981) The Lichfield Canal and the Wychnor Ironworks, *Trans South Saffs Arch Hist Soc*, **23**, 109–117; R. Fitzgerald (1988) The development of the cast iron frame in textile mills, *Ind Arch Rev*, **10**, 127–145]

Works

1796–1797. Castle Foregate Mill, Shrewsbury
1802–1803. Meadow Lane Mill, Leeds
1803–1804. Castle Fields Mill, Shrewsbury
1812. The Lancasterian School, Shrewsbury

BAILDON, John (1772–1846), metallurgical engineer, was born at Larbert, Stirlingshire, on 11 January 1772, the first son of William Baildon, employed at the Carron Ironworks. He was responsible for the first coke-fired iron furnace, the first successful cylinder boring equipment and the first cast iron bridge in Prussia.

Following the acquisition of Silesia by Prussia in 1740, Frederick the Great encouraged the development of the area. Friedrich Wilhelm von Reden, who was appointed Oberbergrat in 1778, with responsibility for the mining and metalworking industries in the area, asked John Smeaton (q.v.) to recommend a British engineer to oversee the development of these industries and Baildon was suggested. He arrived in Silesia in the Spring of 1793.

In 1794–1796 he designed and built, in Gliwice, Prussia's first blast-furnace specifically designed to use coke; to supply the works with coal he built a canal with two inclined planes. Two further furnaces were built at Königshütte in 1797–1801, increasing to four by 1820. The works cast the first iron bridge in Prussia, erected in 1796 at Lassan, close to Breslau, and also the Long Bridge over the Havel in Potsdam.

Steam engines were also produced in Gliwice, with Baildon responsible for the first cylinder boring machine on the mainland of Europe. The first steam engine to be built in Prussia—for which he received the gold medal of the Prussian Academy of Science—was made by Baildon for the Porcellan factory in Berlin in 1800. He also built the first Boulton and Watt type steam engine to be used in Moravia in 1814.

Baildon was builder and part-owner of a new iron works at Bytkow (Welnowiec), opened in 1828 and also had shares in the zinc works there. He developed metal industries in Moravia and in Czech areas, rebuilding the blast furnace in Frydland. With his brother William (1781–1833) he built works at Stepanovo, Benezovo, Polnieska and Zeladna.

In 1804 Baildon married Helena Antonia Jozefa Galli (1784–1859) and they had five sons and two daughters. William (1805–1833), Maria (1820–1850) and Artur (1822–1909) survived into adulthood, Artur taking over his father's business at the time of his death in Silesia on 7 August 1846. The family had houses in Belk, Gliwice, Kopienice, Zubie and Pogrzebieo.

MIKE CLARKE

[Anon. (1825) Etablissement d'un pont en fer sur la Havel, *Revue encyclopedique*, **27**, 585; W. Palley (1912) *Baildon and the Baildons, a History of a Yorkshire Manor and Family*; G. Reitböck (1920) John Baildon und Baildonhütte in Oberschlesien, *Die Bergstadt*, issue 7; C. Matschoss (1925, 1985) *Männer der Technik*; W. O. Henderson (1958) *The State and the Industrial Revolution in Prussia 1740–1870*; H. Christoph (1996) *John Baildon zarys biografii*]

Works
1794–1796. Gliwice Canal and inclined planes
1796. Lassan Bridge, cast iron arch
1824. Havel Bridge, cast iron arch

BAIRD, Charles (1766–1843), ironmaster, was born at Westertown, Bothkennar, near Falkirk, on 22 December 1766, the second son of Nicol Baird (*c.* 1734–1807) and Christian Pringle (*c.* 1734–1823). His father was initially a toll collector and from 11 November 1779 until his death surveyor or superintendent of works on the Forth and Clyde Canal. Charles had three sisters and six brothers, several of whom became involved in engineering, and Hugh (q.v.) succeeded his father. In 1782 Charles joined the Carron Company and by the age of nineteen he was manager of the gun department. At this time the company was approached by the Russian Government about improving cannon production in Russia and the works manager, Charles Gascogine (q.v.) agreed to go there; unsurprisingly given his expertise, he

took Charles and his younger brother, James, with him. Having helped Gascogine establish new blast furnaces at Koutherserky, a gun foundry at Alexandrovsk (1786–1789), and a cannon ball foundry at Kronstadt (1789), Baird went into business with Francis Morgan in 1792. Morgan had a small factory in St. Petersburg which under Baird's ingenious management became one of the largest engineering works in Russia, producing large numbers of steam engines (from 1811), steamships (from 1815) all kinds of equipment for the Russian armed forces from buttons to trumpets, and all manner of architectural and structural ironwork, including the framing for the dome of St. Isaac's Cathedral and several of St. Petersburg's earliest suspension bridges. As a consequence of building the first Russian steamship, the Elizabeth, in 1815, he obtained a ten year monopoly on its route. He supplied work machinery to the government, particularly the Admiralty and the Mint, as well as the Imperial Glassworks.

While the foundations of Baird's success lay in the knowledge he brought with him from Carron, he built on this by assiduously following the latest engineering developments in Britain, taking and exploiting patent rights in Russia for new inventions whenever he saw the potential for making money. In this he was helped by the continuing involvement of his family in Scottish industry and mill engineering. His brother, Hugh, went on to become engineer of the Union Canal and brothers John and James were also involved in Shotts Ironworks.

He sent his son Francis (1802–1864) to Edinburgh University, and returned himself to Scotland on regular visits. On one of these he persuaded his nephew William Handyside (q.v.) to return with him in 1810 and it was Handyside and Francis, who joined the firm at the age of seventeen, who were responsible for many of the firm's important achievements; because of their shared involvement it is difficult to ascribe precise responsibility for these works. Under Charles the works grew steadily, employing 400 men in 1812, 500 in 1820 and 1000 by 1832.

The earliest work of civil engineering interest was the first cast iron arch bridge to be erected in Russia, crossing the canal adjacent to the works (1805); it was in the style of Thomas Wilson (q.v.). No further bridge works are known until the 1820s when Baird supplied the chains for the suspension bridges designed by Wilhelm Traitteur in St. Petersburg. The works also developed a hydraulic chain testing machine, apparently designed by Augustin de Betancourt, for proving the bridge chains in 1824, slightly earlier than the large machine built by Murray for Alexander Wilson (q.v.).

This was followed by a long period of collaboration with the architect Montferrand, in which William Handyside took the lead, in the construction of St. Isaac's Cathedral and the erection and decoration of the Alexander I column. In both cases Bairds were not only involved in supplying structural cast and wrought iron work and copper cladding, but also decorative castings for bas reliefs, with Handyside even developing methods for gilding the decorative surface; Bairds also developed methods for transporting and erecting the large masonry monoliths used in the construction. Special foundry work was involved in the castings, generally ascribed to Handyside, although the later castings were made after his departure from Russia and must have been supervised by Francis Baird. To get some idea of the scale of the works the alto reliefs for the Alexander column weighed upwards of 20 tons and those on St. Isaac's pediments were 105 ft. × 14 ft.; the columns were 8 ft. in diameter and 56 ft. high while the Alexander column was 84 ft. high and 12 ft. in diameter. Montferrand had originally planned a traditional dome and only determined on the use of iron framing in the 1830s. According to him it was Baird who recommended the alteration in the number of ribs used from 32 to 24.

Baird died on 10 December 1843. His reputation was high among fellow engineers and Sir John Rennie (q.v.) remarked that he was 'a shrewd, intelligent, clever, active, indefatigable person, wholly devoted to making money'. He had married the daughter of his partner Francis Morgan in 1794 and although he had three sons, two of them—Nicol (c. 1800–1830) and James (d. 1837)—predeceased him, so Francis took over the works. Born on 16 February 1802, he had married Dorothea Halliday (d. 1880) in 1828 and had ten children, three of whom survived him and maintained the family business after his death on 25 March 1864. The works continued to prosper under Francis, employing 1,200–1,500 in 1860, and producing 500,000 roubles-worth of goods per annum; output included 800 hp marine engines. Both Bairds were showered with Russian Imperial honours and Charles was elected a member of the Institution of Civil Engineers in 1841, presumably being in Britain on a visit, but Francis was one of the earliest members, elected in February 1823.

MIKE CHRIMES

[Membership records, ICE archives; J. G. James collection, ICE archives; G. Lamé (1825) Sur les ponts de chaines, *Annales des Mines*, **10**, 311–330; W. Traitteur (1825) Description des ponts en chains executees à St Petersburg; Henry (1826–1827) Mémoire sur l'emploi fer dans les ponts suspendus, *Journal des Voies de Communications*, **5**, 19–43, **9**, 29–55 (also *Journal du Genie Civil*, **1** (1828), 219–236; G. Lamé (1826) Mémoire sur les ponts suspendus, *Journal des Voies de Communications*, **3**, 49–71 (also *Journal du Genie Civil*, **1** (1828), 245–260); T. Tower (1867) Memoir of the late Charles Baird ...; Memoir (of Francis Baird) (1870) *Min Procs ICE*, **30**, 461–465; Sir John Rennie (1875) Autobiography, 253–255; J. G. James (1983) The Application of iron to bridges and other structures in Russia to about 1850; R. Christian (1990) *Butterley Brick*; S. G. Fedorov (1992) Der Badische Ingenieur Wilhelm von Traitteur, *Inst Baugeschichte, Univ Karlsruhe, Materialen zu Bauforschung und Baugeschichte*, **4**; S. G. Fedorov (1995) Early iron domed roofs in Russian church architecture: 1800–1840, *Construction History*, **12**, 41–66]

Works

Selected works:

1805–1806. Chugunny cast iron arch bridge, Tallow Wharf, St. Petersburg, 63 ft. span
c. 1824–1858. St. Isaac's Cathedral, column bases, dome, bas reliefs, etc.
1824. Post office (suspension) footbridge, St. Petersburg, 114 ft., 11 in. span
1824. Pantaleymon (suspension) road bridge, St. Petersburg, 129 ft., 6 in. span
1824. Hydraulic chain testing machine
1826. Egipetsky (suspension) road bridge, St. Petersburg, 180 ft. span
1826. Griffin (suspension) foot bridge, St. Petersburg, 77 ft. span
1826. Lions (suspension/footbridge) St. Petersburg, 77 ft. span
1834. Alexander I column, St. Petersburg
1842–1850. Nikolaevsky Bridge, St. Petersburg, decorative railings

BAIRD, Hugh (1770–1827), surveyor and civil engineer, was born on 10 September 1770 at Westerton in Bothkennar Parish, Stirlingshire, the son of Nicol Baird (1735–1807) and Christian Pringle. In 1779 Nicol Baird, who had been for six years the toll-collector at the sea-lock linking the Forth and Clyde Canal to the river Carron, was appointed surveyor on the death of Robert Mackell (q.v.). The canal had been opened to shipping from the Firth of Forth as far as Kirkintilloch in 1773, reaching Stockingfield on the northern outskirts of Glasgow two years later. Owing to lack of funds work had to be suspended for the next nine years but it resumed in 1784 and the main line of the canal, 39 miles in length from Grangemouth on the Forth to Bowling on the Clyde, was completed by 1790 under Robert Whitworth (q.v.) as chief engineer.

Hugh Baird's early life, education and career arc unknown but in 1796 he is noted as having inspected the sites of two proposed reservoirs for the canal. He was 37 when he succeeded his father as surveyor on the latter's death in 1807, and five years later he was appointed resident engineer to the Forth & Clyde Canal Company at a salary of £250 per annum plus £50 for 'all contingencies'. The port of Grangemouth, which had not existed fifty years before, was now expanding rapidly and in 1810 Baird submitted plans for extending the harbour. He put forward a new proposal in 1814 for a £17,629 wet-dock to be entered from the Forth, but preference was given to a more ambitious scheme suggested by John

Rennie (q.v.) which was to cost £125,000. Neither of these proposals was acted on at the time.

As early as 1791–1792 proposals had been put forward for a canal linking Edinburgh with Glasgow and the Clyde and in 1797 Rennie was asked to comment on four alternative lines surveyed by Ainslie and Whitworth (qq.v.). Strong opposition greeted all of these proposals. In 1813, nothing having been done in the meantime, Baird was commissioned to draw up a new plan for a canal linking Edinburgh to the Forth & Clyde Canal at Falkirk, but again the objectors were about as numerous as the supporters. In 1815, however, Telford came out in favour of Baird's plan and the necessary Act was passed in June 1817.

Baird was appointed engineer at a salary of £500 pa and work on the Union Canal began in March 1818. The line finally selected ran from the centre of Edinburgh via Linlithgow to Lock No 16 on the Forth & Clyde Canal. There was only one flight of eleven locks, at the Falkirk end of the canal, but the work included the only canal tunnel in Scotland, 690 yd. in length, and three major aqueducts over the Water of Leith and the rivers Almond and Avon. They were designed by Baird but modelled on Telford's aqueduct at Chirk on the Ellesmere Canal; Telford's advice was sought both before they were built and during their construction. The canal was opened in May and fully completed by the end of 1822.

Hugh Baird died at Kelvinhead, Stirlingshire on 24 September 1827 and was buried in Kilsyth Old Cemetery. He left estate valued at £592 7s 1d.

RON BIRSE

[Jean Lindsay (1968) *The Canals of Scotland*; Skempton; W. T. Johnston (1998) *Scottish Engineers and Shipbuilders* (publ. as computer database)]

Publications
1813. *Report on the Proposed Edinburgh and Glasgow Union Canal* (2 edns.)
1814. *Report of Mr Hugh Baird on the Proposed Plan for making the Dike or Track-Path to Carron Water Mouth*
1814. *Subsidiary Report ... Edinburgh and Glasgow Union Canal* (with John Rennie Sr. (q.v.))
1814. *Second Subsidiary Report ... Edinburgh and Glasgow Union Canal*
1814. *Report to the Committee of subscribers ... Edinburgh and Glasgow Union Canal*
1814. *Additional Report to the Committee of Subscribers ... Edinburgh and Glasgow Union Canal*
1814. *Remarks, etc. ... Edinburgh and Glasgow Union Canal*
1824. *Reports on the Improvements of the River Leven and Loch Lomond*
1824. *Report on the Proposed Railway, from the Union Canal at Ryal, to Whitburn, Polkemmet, and Benhar: or, the West Lothian Railway*

Works
c. 1793–1796. Illverston Canal, assistant engineer
1796–1827. Forth & Clyde Canal (completed 1790), surveyor 1807–1812, resident engineer 1812–1827
1817–1827. Edinburgh & Glasgow Union Canal (constructed 1818–1822), engineer 1817–1827

BAKER, Samuel (fl. 1787?–1836), contractor, probably the son of Samuel Baker of Rochester, operated principally as a timber merchant, carpenter and building contractor, in association with Thomas and George Baker. A Samuel Baker was elected a Member of Rochester Common Council in 1787 and served as Mayor in 1797 and 1802. A namesake acted as Warden of Rochester Bridge between 1826 and 1836, his brother, Captain Thomas Baker, joining him in 1830–1831; they formed part of the Tory group on the Council. A further son/brother, George, became based in London.

Samuel Baker and Son began contracting in the Napoleonic period, undertaking government contracts and being particularly associated with Robert Smirke (q.v.), Samuel marrying Smirke's sister, Sarah, in 1815. The families shared political allegiances. The earliest major contract undertaken by Samuel Baker, in partnership with a Mr. Nicholson, was the Brompton Artillery Barracks, Chatham (1804–1806). Nicholson went on to work at Sheerness Dockyard with Jolliffe and Banks (q.v.), whilst Baker worked with his family. In the history of civil engineering the particular interest of their work lies in their involvement in the use of concrete and iron.

In the autumn of 1816 Smirke was appointed architect for the Millbank Penitentiary, under construction since 1813 on ground condemned by Samuel Bentham (q.v.) as unfit for building. Approximately one third of the prison had been completed but was showing considerable evidence of distress due to subsidence when Smirke was brought in. He replaced the existing contractors by Baker who, following a joint report by John Rennie (q.v.) and Smirke, began to rebuild to Smirke's specification. The existing foundations were underpinned with a layer of ballast covered with a thick layer or raft of 'grouted gravel' or concrete comprising one part Dorking lime and 5–7 parts of gravel, built up in courses of 6 in., 'puddled' by the labourers. In places the raft was 18 ft. deep.

While work proceeded at Millbank, Smirke was planning the development of the Savoy Precinct, laying out a general plan with standard designs which Baker, and another speculative building contractor, Henry Peto, could follow. Concrete was not yet established in Smirke's mind and the Savoy Wharf, which Baker built to Smirke's designs, was founded on a timber grid and brick cells filled with 'rubbish'. Lancaster Place on the other hand, which Smirke also directly supervised, had foundations 8 ft. wide and 8 ft. deep made up of rammed rubble on which was a layer of concrete, and then brick footings enclosing timber. Baker also had the contract for the concrete foundations of Robert Peel's House (1822–1823), again designed by Smirke, but their most famous collaboration, which established concrete as a foundation material, was at the Custom House (1825–1827).

The Custom House was a contemporary cause celebre. Work began in 1813 under David Lang (q.v.) and the project rapidly gained a reputation for excessive expenditure and poor execution. Subsidence began in 1820 and was followed in 1824–1825 by collapse. Smirke was appointed architect for the reconstruction and brought in Baker, who underpinned parts of the existing structure with 'grouted gravel'. Existing walls were then underpinned in 10 ft. sections by a layer of concrete capped with Yorkshire landings 5 ft. x 5 ft., 6 in. thick, and with courses of brickwork alongside existing brickwork.

Success at the Custom House was followed by a succession of contracts in the 1830s where concrete was regularly employed in the foundations and, after a period of experimentation, this practice became standardised. Thus whilst Baker's first contract for the East Wing of the British Museum (1823–1827) used brickwork for the foundations, all his later contracts there used concrete varying in thickness from 2 ft. 6 in. to 6 ft. according to the superstructure. In these early days not all went well and at the General Post Office Baker allegedly lost money when he misjudged the quantities required. The last mention of Samuel Baker dates from 1836.

While Baker is generally referred to as being based in Rochester, his son, George, was based at 2 Montague Place, London and in Lambeth. George joined the Institution of Civil Engineers as an Associate in 1836, resigning in 1840. In the 1840s the firm of George Baker and Sons maintained Samuel's tradition of involvement in state-of-the-art building techniques through their involvement with building work in the Royal Dockyards. At the close of the Napoleonic Wars there had been a move to cover the shipbuilding slipways in the dockyards with roofs of timber. In a new wave of ship construction from the late 1830s there was a transition from timber to iron-framed roofs, in which George Baker played the leading contracting role, installing eleven roofs in comparison to the five of his more famous competitor, Fox Henderson. Moreover, the later roofs his firm supplied show a more advanced design than Fox Henderson's with greater spans between the frames which were of braced arch rather than trussed form. Unfortunately, the designer employed by Baker is unknown.

Although the family carried out some railway work, for example on stations on the London to Greenwich Railway, they are best known for their continuing involvement in government contracts, often on a grand scale. They continued as building contractors into the second half of the nineteenth century as (George) Baker and Son of Stangate, Lambeth. At that time H. Fielder seems to have been the leading partner. In 1870, with the end of a building and maintenance contract at the British Museum Library, they announced that they would cease trading; their work was to be taken over by Markwick and Thurgood. In 1872, however, George Dashwood Baker had joined this firm, and the name of George Baker and Son was continued and they built a new graving dock in the eastern half of Blackwall Yard (1876). They were bankrupted (1880–1881) over their work on the Natural History Museum (1872–1880), a contract originally tendered for at £352,100, but where their claims totalled over £250,000 for alterations (c. £142,000) and omissions (c. £121,000).

This did not mark the end of the connection with the construction industry, as Arthur Baker (1842–1897) and his nephew, Sir Herbert Baker (1862–1946), both practised as architects. Sir Herbert was Thomas Baker's grandson.

MIKE CHRIMES

[On concrete (1830) Minutes of conversation, ICE archives; C. W. Pasley (1838) *Observations on Limes, Calcareous Cements ...*; C. W. Pasley (1862) *Outline of a Course of Practical Architecture* (reproduction of 1826 edition); J. M. Crook (1965–1966) Sir Robert Smirke: a pioneer of concrete construction, *Newc Soc Trans*, **38**, 5–22; H. M. Colvin *et al.* (1982) *History of the King's Works*, **6**; A. S. Gray (1985) *Edwardian Architecture*, 97–99; R. J. M. Sutherland (1989) Shipbuilding and the long span roof, *Trans Newcomen Soc*, **60**, 107–126; RIBA (1993) *Directory of British Architects 1834–1900*; N. Yates and J. M. Gibson (eds.) (1994) *Traffic and Politics: the Construction and Management of Rochester Bridge*; M. H. Port (1995) *Imperial London: Civil Government Building in London 1850–1915*; S. Porter (ed.) (1995) Poplar, Blackwall and the Isle of Dogs, *Survey of London*, **xliv**, 572; J. Douet (1998) *British Barracks 1600–1914*, 86–87]

Works

As contractor:

1804–1806. Brompton Barracks, Chatham
1816–1819. Millbank Penitentiary, reconstruction, c. £320,000
1817–1823. Savoy Precinct development (including 1820–1823, Lancaster Place)
1819. Rochester, temporary bridge erection during repairs to Rochester Bridge
1822–1823. 4 Whitehall Gardens (Sir Robert Peel's House)
1823–. British Museum (1st contract East Wing, brickwork, joinery, etc. (total for wing £130,000); work continued into 1850s)
1823–1830. Lincoln County Court
1824–1829. General Post Office
1825–1827. Custom House, London, reconstruction
1825–1830. Temple Church, London
1827. Kensington Gardens, five lodges
1831. Clarence House, private entrance, £383

1835. Banqueting Hall, Whitehall, restoration, c. £2,000
1835. Houses of Parliament, temporary accommodation
1836–(1845). Fitzwilliam Museum, Cambridge, £36,000
1837. Juvenile Prison, Isle of Wight, £20,000
1839. Surrey Lunatic Asylum, £45,000
1839–1840. Greenwich Station, London and Greenwich Railway
1841. British Museum façade
1841. Deptford Station, London and Greenwich Railway
1842–1843. Buckingham Palace Chapel
1844–1847. Iron roofs for Royal Dockyard ships at Portsmouth (three), Pembroke (two), Deptford (three) and Chatham (three)
1845–1853. Keyham Dockyard, Devonport, £1 million
1847. New Custom House, Southampton Docks, £13,000
1854–1857. British Museum Dome and Reading Room
1862–1865. Bovisland Fort, Plymouth
1865–1869. Crown Hill Fort, Plymouth
1873–1880. Natural History Museum, c. £500,000
1876. New Graving Dock, Blackwall Yard

BALD, Robert, FRSE (c. 1775–1861), mining engineer, was born in Culross, the eldest son among the six known children of Alexander Bald (c. 1743–1823) of Alloa. His father was colliery agent for the 7th Earl of Mar for 49 years. Bald was trained as a colliery manager by his father and is first mentioned on his own account in the Alloa area in 1799–1802. He was responsible for a large number of mineral surveys and reports, as well as some estate and road surveys, with an extensive practice in Scotland, although he also carried out surveys in the South Wales coalfield. Inevitably this led to his involvement in canal and railway schemes serving the coalfields. His reputation was sufficient in his mid-thirties to be asked to accompany Thomas Telford (q.v.) on his visit of 1808 to Sweden to report on the Gotha Canal project. He concerned himself with the living and working conditions of the miners and tried to encourage more sober behaviour on the coalfield. His *General View of the Coal Trade of Scotland* is a most authoritative account, full of historical detail, which, from extant correspondence, was thoroughly researched.

He belonged to a large number of scientific societies, including the Geological Society of London, the Highland and Agricultural Society, the Wernerian Natural History Society, and was a Fellow of the Royal Society of Edinburgh. He was elected a Corresponding Member of the Institution of Civil Engineers in 1824; although he resigned due to ill health in 1841, he was reinstated in 1843. For the bulk of his career he had offices in Edinburgh. He died at his home in Bedford Place, Alloa, on 28 December 1861, aged 86.

MIKE CHRIMES

[Membership records, ICE archives; Reports, 1808–1825, SRO; *Royal Society Catalogue of Scientific Papers*; I. G. Grant (1982) *An Enterprising Family: the Balds of Alloa*; B. F. Duckham (1969) The emergence of the professional manager in the Scottish coal industry 1760–1815, *Business History*, 21–38; Gulley (2000) in lit; Skempton; Bendall]

Publications
1808. On the coal-formation of Clackmannanshire, *Edinb Mem Wern Soc*, **I**, 479–503
1814. *Report of a Mineral Survey of the Country Through which the North or Level Line of Canal Passes, as Projected by John Rennie, Esq. Anno 1798*
1819. Additional observations on the coal field of Clackmannanshire, and a description of the absolute shape or form of the coal fields in Great Britain, *Edinb Mem Wern Soc*, **III**, 123–155
1819. On the temperature of air and of water in the coal mines of Great Britain, *Edinb Phil Journ*, **I**, 134–137
1819. Notice respecting the discovery of the skeleton of a whale on the estate of Airthrey, near Stirling, *Edinb Phil Journ*, **I**, 393–396
1820. On the steam and sediment of water found in the boilers of steam-engines, *Edinb Phil Journ*, **II**, 340–341
1821. Notices regarding the fossil elephant of Scotland, *Edinb Mem Wern Soc*, **IV**, 58–66
1825. *General View of the Coal Trade of Scotland*
1828. Observations on the coal field and accompanying strata in the vicinity of Dalkeith, Mid-Lothian, *Edinb New Phil Journ*, **IV**, 115–122
1828. On the fires that take place in Collieries. *Edinb New Phil Journ*, **V**, 101–121
1829. Observations on the spontaneous emissions of inflammable gas, particularly of carburetted hydrogen, *Edinb Journ Sci*, **I**, 71–75
1832. *Notice to Colliers, Workmen and Others, Employed and Residing at the Alloa, Collyland and New Sauchie Collieries*
1833. On the manufacture of cast iron. *Trans Royal Soc Sci*, 18 February.
1850. An account of the mineral field between Airdrie and Bathgate, and from Bathgate to Edinburgh and Leith, *Edinb New Phil Journ*, **XLVIII**, 173–180

Works
1811–1813. Kincardine on Forth shipping pier, c. 100 ft.

BALD, William, FRSE, MRIA (1789–1857), civil engineer, surveyor and cartographer, was probably born in Burntisland in Fife around 1789. He was educated locally and in Edinburgh, where he became apprenticed in 1803 to John Ainslie (q.v.), surveyor and cartographer. Bald surveyed and prepared maps for some 300,000 acres of private property in Scotland, especially in the Hebrides and the Western Isles. These surveys extended General Roy's military surveys.

Bald married in June 1807 and had a large family, his eldest sons, Charles and William, later assisting him with his professional work.

In 1809, he moved to Ireland to direct the Trigonometrical Survey of County Mayo, ordered by the Grand Jury, and spent the next 7 years carrying out a theodolite survey and completing a map of the county at a scale of two inches to one Irish mile (1:42,240); this is still considered a work of outstanding value by geographers. His reports for the Bog Commissioners (1811–1813, published 1814) covered 20,000 acres in the County.

Bald was elected a Fellow of the Geological Society of London in 1816 and a Member of the Royal Irish Academy in 1822. Following a report on Cork Harbour of 1821, little is known of his activities during the period 1825 to 1830 except that he was occupied as a 'man of science' and that he was based in Paris. He was elected a Member of the Société de Géographie de Paris in 1828.

On his return to Ireland in or about 1830, Bald developed his career as a civil engineer. He planned and supervised the construction of the Antrim coast road for the Board of Public Works and was responsible for several hundred miles of roads and associated bridges in Ireland. He undertook a number of harbour and river improvement schemes, including the river Boyne and port of Drogheda on which he first reported 1831–1832. He also reported on proposals for Londonderry and Belfast. Although he surveyed routes for railways, such as that between Limerick and Carrick-on-Suir, there is no evidence that he supervised any actual railway construction.

Bald's interests in Ireland continued to 1845 but he had, in 1839, been appointed engineer to the Clyde River Trust and placed in charge of the deepening, widening and improving of the river. While working in Scotland he drew up a scheme for improvements to Troon Harbour, including a wet dock, which was completed to his designs in 1846. He reported on the state of the Tay for the burgh of Perth in 1843.

In 1845, he returned to France for a few years, before settling in London in 1851. In that year Bald was called upon to head the Admiralty Inquiry into the *Tees Conservancy Bill*, proposed with the intention of transferring the conservatorship of the river Tees from the Tees Navigation Company to a Commission. He considered that the river was capable of improvement, stating that the aim should be 'to hold fast to that which exists in being good and useful, and to assist Nature by Works of Art'. So comprehensive was his report that the Admiralty recommended that it be laid before Parliament in its entirety and although the Bill then failed, it was reintroduced the following year and was then passed.

William Bald died in London on 26 March 1857 and is buried in Highgate cemetery.

RON COX

[Public Works Commission of Ireland (1832–) *Reports*; House of Commons (1835) *1st & 2nd Reports on Public Works in Ireland*; House of Commons (1835) *Report of Select Committee on Connaught Lakes, Minutes of Evidence of W Bald* [HC 354]; Tidal Harbours Commission (1845–1847) *1st Report, Minutes of Evidence*, 1–13; 2nd report, appendix B 103–105, 119–120, 171–172, appendix C 418–422, 464–472; W. Bald (1851) *Report to the Lords Commissioners of the Admiralty* (March); J. E. Portlock (1858) Obituary of William Bald. *Q J Geol Soc Lond*, **14**, 42–44; F. Boase (1892–1921) *Modern English Biography*; M. C. Storrie (1969) William Bald, FRSE, *c.* 1789–1857, surveyor, cartographer and civil engineer. *Trans Inst Modern Brit Geographers*, **47**, 205–231]

Publications

1813–1814. Reports on three districts in the County of Mayo, *Bog Commissioners*, vol. VI, appendices 7–9.

1821. An account of the trigonometrical survey of Mayo, one of the maritime counties of Ireland, *Trans Roy Ir Acad*, **14**, 51–61.

1821. Report … made to Cork Harbour Commissioners, *Tidal Harbour Commissioners Reports*

1836. River Suir improvements, *Tidal Harbour Commissioners Reports*

1837. Report on the formation of a harbour in Lady Cove, Tramore Bay, *Tidal Harbour Commissioners Reports*

1839. General report on a part of the River Clyde, *Tidal Harbour Commissioners Reports*

1840. Observations on the tides in the harbour of Glasgow, *British Association Reports*, 49–50

1844. Report (with others) on the propriety of laying rails on the quays of Glasgow, for mineral traffic, *Tidal Harbour Commissioners Reports*

1844. *Report on Proposed New Works for Improving the Harbour of Glasgow*

1844. *Report on the Improvements of the Foyle Navigation*

Chronological list of extant maps, plans and sections compiled by William Bald, or prepared under his supervision, between 1805 and 1825, is given in M. C. Storrie, *op. cit.*, 1969, 226–229.

Works

1803–1809. Surveying and mapping in Scotland

1809–1824. Surveying and mapping in Ireland

1831–. Antrim Coast Road, 33 miles, engineer, including Corratevey Bridge, *c.* £13,000

1832. Aghivey timber bridge, engineer

1833–1845. Improvement of Drogheda Harbour and Boyne navigation, *c.* £10,000

1833–1840. Kenmare-Bantry Bay Road, including Kenmare suspension bridge, £21,000

1834. Antrim—Coleraine Road, engineer

1836–1840. River Suir improvements

1839–1845. Clyde navigation improvements, engineer

BANKS, Sir Edward (1770–1835), civil engineering contractor, was born in the hamlet of Hutton Hang near Middleham, North Yorkshire, on 4 January 1770. After a brief period as a merchant seaman, Banks began the type of work to which he devoted his life through experience of sea banking in the Holderness district of East Yorkshire. His early experience may have been under the Pinkerton family (q.v.). Between 1791 and 1800 he is believed to have worked on the Leeds & Liverpool, Lancaster, Ulverston (under Pinkerton), Huddersfield, Peak Forest, Ashton-under-Lyne and Nottingham canals. The scale of his business operations at this time can be gauged from the fact that in return for various services for the Cromford Canal between 1802 and 1807 he received total payments of £1,315.

On moving south in the early 1800s, Banks assisted in the development of the Merstham terminus of the Croydon, Merstham & Godstone Iron Railway in order to provide a technically advanced system of plateways for the exploitation of the stone and chalk of the area. These arrangements led to his first meeting with William John Jolliffe (1770–1835) who, although he had a radically different upbringing and social position from that of Banks, was destined to be his partner to the last year of their lives. It is difficult to be precise about the contribution which Jolliffe made to their joint success but it may well have been confined chiefly to providing the necessary finance. Colonel Hylton Jolliffe, William's brother, acted as a surety for several of their projects, including New London Bridge.

The work of the partnership commenced in 1807 with the construction of a Court House in Croydon but the civil engineering projects of Jolliffe and Banks, which are equalled in aggregate, variety and magnitude in the early nineteenth century only by the achievements of Hugh McIntosh, are so numerous as to preclude reference to all but the most important. Much of their work took place under the engineering direction of John Rennie (q.v.) and his son, Sir John (q.v.).

In 1812 Jolliffe and Banks worked on the entrance lodge for Millbank Penitentiary, the most imposing prison constructed in Britain to that time; it stood on the site of today's Tate Gallery. The lodge, based on a rubble and pile foundation, soon began to show signs of instability and in 1816, when Sir Robert Smirke (q.v.) took over as the project's engineer, they were dismissed as contractors and the lodge eventually replaced. They are also said to have contributed in 1809–1810 to the construction of Dartmoor Prison and to another of the new-style prisons of the period, Surrey House of Correction at Brixton, opened in 1820.

Jolliffe and Banks made their reputation and attained their highest level of proficiency when working on projects involving the control of water, whether in a canal, along a river bank or around a cofferdam. Between 1819 and 1821, under the direction of both Thomas Telford (q.v.) and John Rennie, they made Eau Brink Cut, a waterway to the south of King's Lynn which was designed to improve the outfall of the Bedford Level

Sir Edward Banks

and it is reputed that the partnership carried out work valued at more than £40,000 before receiving any payment from the commissioners for the scheme. To the west of this, and of even greater importance, from 1827 to 1831 they were engaged in cutting a more direct course for the River Nene over a distance of 5–6 miles between Wisbech and The Wash, together with a new bridge at Sutton along Nene Outfall Cut.

These projects ameliorated the traditional problems associated with drainage in the Fenlands but they were eclipsed by the schemes which Jolliffe and Banks carried out for the Undertakers of one of the country's oldest waterways, the Aire & Calder Navigation. Commencing in 1822, under the direction of George Leather Jr. (q.v.), they created in Goole a basin with locks at the entrance to the river Ouse, and so to the Humber, and a barge dock and ship dock. A canal almost 19 miles in length was cut westwards from Goole to Ferrybridge on the River Aire. Banks augmented the assets of the new port by finding £4,000 for the erection of a 'large and commodious Inn', the Banks Arms Hotel. Approximately 45% of the expenditure on this scheme (£361,484) went to Jolliffe and Banks.

An immense programme of works for the Royal Navy with a final contract price of c. £1.5 million began in 1817 at Sheerness Dockyard and was not completed until 1830. Under contracts made in 1817, 1819 and 1825, Jolliffe and Banks created a basin, mast pond, several slips, a tunnel, a sea wall, boat and frigate docks and three dry docks. Although they were not amongst the leading contractors in the initial expansion of London's dock system, this partnership did provide new entrances for West India Docks (Limehouse Lock, 1811–1812) and London Docks (the Hermitage, 1818–1821), and some years later carried out the same operation and built walls for HM Dockyard, Deptford.

Although the majority of Jolliffe and Banks' work, especially in their later years, took place in and around London, it should be mentioned that further afield they were responsible, *inter alia*, for a lighthouse on Heligoland and embankments in Cardiff Bay and the nearby Leckwith.

In 1812 Jolliffe and Banks began work on Waterloo Bridge, the first of the three major Thames crossings which set the seal on their reputation. The Strand Bridge Company had already put in the foundations of the south abutment and established the cofferdams—rather than use caissons—for two of the piers. It proved difficult to obtain granite in the required quantities and therefore Rennie permitted the use of Craigleith stone for the intrados of the arches; the work proceeded without incident until in 1815 four of the eight ribs supporting the fifth arch fell when the centering was being removed. Nevertheless, later in that year, Rennie had no hesitation in reporting 'that these Works are of an extent far exceeding anything of the kind ever executed in the British Dominions ... (and) the last Arch was closed within the period mentioned in the Contract ...'; the bridge was opened in June 1817.

In 1814 and 1816 Jolliffe and Banks signed contracts for the masonry work and approaches of Southwark Bridge, the iron arches being the responsibility of Joshua Walker & Co. of Rotherham, and purchased shares to the value of £70,000 in the Southwark Bridge Company. Although their operations with cofferdams were 'much retarded by the unruly and vicious conduct of the Bargemen navigating the River who frequently injured the Moorings and Booms' they completed their work in 1819.

Waterloo and Southwark were but the prelude to the project for the improvement of the capital which will always be associated with this partnership, the construction of New London Bridge. In eight contracts signed between March 1824 and June 1830, they undertook to build the new bridge and its approaches and, in order to keep a crossing in existence at this point, maintain the old bridge. John Rennie Jr., the engineer for this prestigious work, preferred their tender to those of Hugh McIntosh and Henry Peto because of their experience in bridge building and expertise with cofferdams. A prominent feature of this bridge was the extensive use of granite in its construction, chiefly from Aberdeenshire and Haytor, near Ashburton, Devon. The specification called for granite carefully matched for colour and free of pyrites. The bridge, which was opened in August 1831, had been constructed entirely to Rennie's satisfaction and it is therefore regrettable that the final payments to Jolliffe and Banks were

withheld by the Bridge Committee for over two years at a time when they were 'much in want of money'. The sequence of crossings over the Thames was completed with the opening by William IV in 1832 of Staines bridge, comprising three segmental arches.

The feature of Jolliffe and Banks' career which distinguished it from that of Hugh McIntosh, their contemporary and equal as a contractor in the field of civil engineering, was the extension of their interests into forms of business outside that of construction. In this diversity they were the forerunners in business practice of many of the contractors of Victorian Britain. A close working relationship developed between this partnership and Butterley Iron Works through the former's purchases of ironwork for bridges and pumping and ships' engines. It had its origins at least 10 years before Benjamin Outram's (q.v.) death in 1805, when Banks had begun working on the Huddersfield Canal, an Outram project, and Banks subsequent move to the East Midlands involved work for Outram and Butterley. On Outram's death Banks took a 14 year lease on the 'Butterley' quarrying and limeworks operations with two other partners. At the same time he had a fleet of thirty narrow boats for coal carrying. The major programme of development carried out at Sheerness Dockyard led to the purchase by Banks of land in the area with the aim of turning Sheerness into a popular watering-place, an unsuccessful venture. In 1816 Jolliffe and Banks made their first investment in a steam packet service recently established by George Dodd on the Thames; this ran vessels to Richmond, Gravesend and Margate. The partnership's interest in sea transport strengthened in 1824 when they joined the promoters of the General Steam Navigation Company which, capitalised in the sum of £200,000, by the end of that year had a fleet of thirteen steamers and twenty-six by 1826.

Banks was knighted in June 1822 in recognition of his role in the construction of Waterloo and Southwark bridges, an honour rarely awarded to civil as opposed to military engineers. Many in the profession in the nineteenth century, such as John Rennie Sr., eschewed such distinctions. Banks, whose London residential address in the 1820s was Adelphi Terrace, Strand, ran his business affairs from Beaufort Wharf, Strand; here, a long-serving employee, John Plews (q.v.), kept his records since, as he told a Parliamentary committee in later years, 'Sir Edward Banks was no accountant, nor Mr. Jolliffe ...'. Banks purchased the seats of Oxney Court, near Dover, and Sheppey Court in Kent and left assets valued at approximately £250,000. His first wife, by whom he had six children, died in 1815; his second marriage was to Amelia Pytches, the sister of Jolliffe's wife. Banks died on 5 July 1835 at Tilgate Lodge, Sussex, a home of the Jolliffe family, and was buried in an elaborate tomb depicting his achievements in Chipstead churchyard, Surrey.

DAVID BROOKE

[Aire & Calder Navigation, PRO: RAIL 800: General Assembly & Directors' Minutes, 1799–1820 (/9), General Assembly of the Undertakers, 1820–1841 (/10), Cash Books, 1775–1827 (/169–171), Reports & Accounts, 1822–1829, RAIL 1113/7–14; Ashton-under-Lyne Canal, RAIL 804: Committee Proceedings (no records before 1798), 1798–1815 (/1); Cromford Canal, RAIL 819: General Assemblies & Committee; 1789–1799 (/1), Cash Book, 1802–1818 (/4); Huddersfield Canal, RAIL 838: General Assemblies, 1794–1845 (/1), Canal Committee, 1794–1801 (/2); Lancaster Canal, RAIL 844: Committee Minutes, 1792–1813 (/230, 231), Reports to the Canal Committee by Millar, Engineer, 1793–1797 (/240), Reports to the Canal Committee by Gregson and Eastburn, 1794–1798 (/241), Copies of Articles of Agreements, 1793–1795 (/247). Copies of Contracts & Estimates, 1801–1803 (/248); Leeds & Liverpool Canal, RAIL 846: Minutes of the General Assembly & Leeds Committee, 1790–1793 (/4), Minutes of the General Assembly & Liverpool Committee, 1788–1793 (/44); Peak Forest Canal, RAIL 856: Committee Minutes, 1794–1800 (/1), Ulverston Canal, RAIL 880: Committee Meetings (no records before 1797), 1797–1818 (/1); Corporation of London Records Office: two Contracts for the Building of Southwark Bridge & Approaches, 1814–1816, GBWT/18/2–3; eight Contracts for the Building of New London Bridge & Approaches, 1824–1830, Misc. MSS 200.8; Journals of the New London Bridge Committee, 1823–1835; ICE: Volumes of Reports of John Rennie and Sir John Rennie, 1807–1831; an Index of all the Works executed by Jolliffe & Banks from the Commencement of the Firm to … and also the mount received for them, Somerset RO: Hylton Estate Paper DD/HY 23; S. Wells (1830) *The History of the Drainage of the Bedford Level*; Evidence of N. Plews, *House of Commons' Committee, London & Brighton Railway Bill*, 28 April 1836; J. Mordaunt Crook (1965) Sir Robert Smirke: a Pioneer of Concrete Construction, *Newc Soc Tr* **38**, 5–22; A. W. Skempton (1978–1979) Engineering in the Port of London, 1789–1808, *Newc Soc Tr*, **50**, 87–108; (1981–1982) Engineering in the Port of London, 1808–1834, *Newc Soc Tr*, **53**, 73–96: H. W. Dickson (1931) Jolliffee and Banks, contractors, *Newc Soc Tr*, **11–12**, 1–8; P. W. Sowan (1984) Jolliffe & Banks, Civil Engineers of Merstham, 1807–1832, *The Bourne Society Local History Records*, **23**, 36–44; J. Greenwood, *Jolliffe & Banks—Transportation Entrepreneurs in Kent* (typescript); DNB]

Works
Canals and waterways (*attributed involvement):

1791. Leeds & Liverpool*
1793. Lancaster*
1793–1794. Ulverston*
1795. Huddersfield*
1795–1797. Peak Forest*
1797. Ashton-under-Lyne*
1800. Nottingham*

1802–1807. Cromford [& Pinxton Railway] construction and improvements, payment £1,315
1808. Cardiff and Leckwith Embankments, contractors
1819–1824. Eau Brink Cut, construction, 1819–1821, contractor, payments £312,610; navigation improvements, c. 1824, £33,000
1822–1826. Aire & Calder Navigation: excavation and masonry for canal and port, Goole, contractors, payment £169,815; canal work, Knottingley, contractors, contract £56,000
1827–1831. Nene Outfall Cut, contractor, payments £207,113
1828. River Witham Outfall, contractors
1828. Level of Ancholme, improvements in drainage and navigation, contractors, payment £50,188

Bridges:

1812–1817. Waterloo Bridge, construction with approaches, contractors, payment £481,742
1814–1819. Southwark Bridge, masonry and approaches, contractors, payment £302,361
1824–1831. London Bridge, construction and approaches, contractors, payment £671,122
1827. Serpentine, construction, contractors, payments £58,513
1828–1832. Staines, construction, contractors, estimate £44,938

Docks:

1811–1812. West India Docks Entrance, Limehouse Lock, contractors, payments £78,881
1817–1830. H.M. Dockyard Sheerness, contractors, payment c. £1,500,000
1814–1817. HM Deptford Dockyard Entrance Basin & Victualling Yard Wall, contractors, payment £66,877
1818–1821. London Docks Entrance, contractors, payments £30,456
1817–1819. London Customs House Quay Wall and Billingsgate Dock Wall, contractors, payment £72,759
n.d. Howth Harbour, £13,746

Prisons:

1809–1810. Dartmoor, additions, contractors, payment £4,386
1812–1816. Millbank, entrance, contractors, payment c. £5,000
c. 1818–1820. Brixton, completed 1820, £26,677
n.d. Chelmsford gaol, £12,886

Miscellaneous construction and interests:

1802–1807. Alfreton–Ripley–Derby Turnpike Rd, £8,000
1807–1810. Croydon, Merstham & Godstone Railway Terminus, construction, contractors
1807. Croydon Court House, construction, contractors, payment £4,968
1807. Heligoland Lighthouse, construction, contractors, payment £7,771
1816–. Thames Steam Ship Service, partners

1824–. General Steam Navigation Company, partners
n.d. Croydon and Reigate Turnpike, contractors, £3,981

BARKAS, George (1639–*c.* 1710), 'Surveyor of Highways', was the son of George Barkas and was baptised at Ryton, six miles upstream of Newcastle, on 28 April 1639. It is possible that he was married twice, first to Marjory Robinson in 1664 and subsequently to Anna Snowden of Heddon-on-the-Wall in 1675.

Nothing is known of his early life but *c.* 1684 he was appointed as Surveyor to Northumberland County, possibly the first such appointment in England; his salary was initially £20 p.a. but was later increased to £30. One of his first instructions was to make 'a particular survey yearly of all the Highways within the County'. In 1692 he was paid £20 'for the repaire of Corbridge Bridge ... for the cutting of land-beds, making of wears and putting in of credles belonging to the said bridge'.

Barkas was also responsible for the upkeep of the Moot Hall in Newcastle, the administrative centre of the County, and when a new gaol was proposed at Alnwick, he suggested that the old one at Morpeth be restored instead. He was paid £15 in 1705 for his 'very great trouble and pains ... in and about the building of the Gaol' which had cost £1,338.

Barkas retired at Easter 1709, 'being old and of late infirm and unable to officiate in his Office and Employment as he formerly did', having served the County for more than 25 years. As a token of their appreciation the County Justices awarded him a pension of £10 p.a.; he died within eighteen months.

R. W. RENNISON

[Northumberland Quarter Sessions: Orders (1687–1710), Northumberland RO: QSO/2–4; J. Hodgson (1832) *A History of Morpeth*, 72, 73]

BARLOW, Peter, FRS (1776–1862), mathematician and scientist, was born in Norwich, Norfolk in October 1776. After local schooling, and whilst still young, he worked in a commercial establishment in Norwich during which time he established a junior scientific society to discuss the mathematical and physical sciences. Subsequently, Barlow acquired an impressive scientific knowledge and turning his attention to teaching, became a schoolmaster. He became a regular contributor to *The Ladies Diary* then managed by Dr. Hutton (q.v.), Professor of Mathematics at Woolwich Royal Military Academy. Through Hutton's agency Barlow became a member of staff at the Academy in 1801 and remained there until he retired on full pay in 1847 at the age of 71. His interests and consequent publications cover a very wide range of subjects including mathematics, magnetism, strength of materials, locomotive performance, and he was frequently appointed to Government Commissions. All of

Peter Barlow FRS

which brought him into regular contact with the civil engineering profession. In 1817 Barlow published *An Essay on the Strength and Stress of Timber*—the first major work on the strength of materials in Britain, which, in successively enlarged editions remained a standard textbook for half a century. The book was the result of experiments made on a testing machine of his own design in Woolwich Dockyard, and attracted the attention of Thomas Telford. Telford made available the results of his tests on the strength of iron and subsequently Barlow assisted Telford by making experiments and calculations for the proposed Menai suspension bridge. Barlow's materials experiments were the subject of discussion at the first meeting of the Institution of Civil Engineers in February 1818. In 1819 he studied the variation of navigational compass needles and made experiments and devised a means of correcting the errors. Barlow was elected FRS in May 1823 and received the Society's Copley Medal. He was awarded the Government's prize for useful discoveries by the contemporary Board of Longitude. In January 1825 he presented a paper to the Institution of Civil Engineers on 'The Force exerted by Hydraulic Pressure in a Bramah Press' and produced design rules for such presses. Barlow was also an early experimenter in electromagnetism and was one of the first to make an electric signal deflect a needle. The experiment was made in the garden of his home in Woolwich using a mile of un-insulated copper wire carried on posts, the energy being provided by a battery. Pressure of other work led him to discontinue the experiments which might have led to the earlier use of the electric telegraph in Britain.

Professor Peter Barlow was an important figure in that scientific and engineering community in Woolwich based on The Royal Military Academy, The Royal Arsenal, and the Royal Naval Dockyard. Barlow had a European-wide reputation for his mathematical, scientific and engineering work when he died in Woolwich on 1 March 1862. His sons, William Henry, and Peter William (q.v.) went on to eminent careers in civil Engineering.

DENIS SMITH

[Obituary (1863) *Min Proc ICE*, 615–618; *Royal Society Catalogue of Scientific Publications*; Professor Peter Barlow Engraved by Samuel Cousins after William Boxall (*ICE Cat. of Prints & Drawings* (1978), No. 508); DNB]

Publications

Publications include:

1808. Articles for Rees's *Cyclopaedia, of Universal Dictionary of Arts and Sciences*
1809. On the motion of floating bodies in running water, *Phil. Mag.*, **XXXIII**, 300–302
1811. *An Elementary Investigation of the Theory of Numbers*
1814. *A New Mathematical and Philosophical Dictionary*
1817. *Essay on the Strength and Stress of Timber and Other Materials* (six editions in English between 1817 and 1867, and a French edition in 1828)
1818. Calculation of the stress and strength of the projected Iron Hanging Bridge over the Menai Strait, *Papers Relating to the Building a Bridge Over the Menai Strait, Near Bangor Ferry. House of Commons, 1819,* 13–14
1821. On the effects produced in the rates of Chronometers by the proximity of masses of iron, *Phil Trans*, 361–390
1822. On the anomalous magnetic action of hot-iron between the white and blood-red heat, *Phil Trans*, 117–126
1823. *An Essay on Magnetic Attractions*
1823. Account of a new series of electro-magnetic experiments; with observations on the mathematical laws of electro-magnetism, *Edin Phil Journ*, **VIII**, 368–382
1823. Observations and experiments on the daily variation of the horizontal and dipping needles under a reduced directive power, *Phil Trans*, 326–341
1824. *An Account of the Experiments Made on Board H.M's. Ships*, Leven, Conway, *and* Griper, *for Correcting the Local Attractions of Those Vessels*
1825. On the force exerted by hydraulic pressure in a Bramah press, *Edin Journ Sci*, **II**, 293–296; *Trans Instn Civ Engrs*, **1** (1836), 133–140
1827. Improvement of achromatic object-glasses, *Phil Trans*, 231–267
1827. Some particulars relative to the tides in the upper part of the river Thames, and of the obstructions caused by the present (i.e., old) London Bridge, *Edinb New Phil Journ*, **II**, 49–59
1827. On the secondary deflections produced in a magnetized needle by an iron shell, in consequence of an unequal distribution of magnetism in its two branches. First noticed by Captain J. P. Wilson, HEICS, 'Hythe', *Phil Trans*, 276–285
1828. An account of a series of experiments made with a view to the construction of an achromatic telescope with a fluid concave lens, instead of the usual lens of flint glass, *Phil Trans*, 105–112
1828. Experiments relative to the effect of temperature on the refractive index and dispersive power of expansible fluids, and on the influence of these changes in a telescope with a fluid lens, *Phil, Trans*, 313–318
1828. An account of the preliminary experiments and ultimate construction of a refracting telescope of 7.8 in. aperture, with a fluid concave lens, *Phil Trans*, 33–46
1830. On the performance of fluid refracting telescopes, and on the applicability of this principle of construction to very large instruments, *Phil Trans*, 9–16
1831. On the probable electric origin of all the phenomena of terrestrial magnetism, *Phil Trans*, 99–108
1831. On the errors in the course of vessels, occasioned by local attraction; with some remarks on the recent loss of HMS 'Thetis', *Phil Trans*, 215–222
1833. Report on the present state of our knowledge respecting the strength of materials, *Brit Assoc Rep*, 93–104
1833. An account of the construction of a fluid lens refracting telescope of 8 in. aperture and 8¾ ft. long made for the Royal Society by George Dolland, *Phil Trans*, 1–14
1833. On the present situation of the magnetic lines of equal variation, and their changes on the terrestrial surface, *Phil. Trans*, 667–674
1834. On the principle of construction and general application of the negative achromatic lens to telescopes and eye-pieces of every description, *Phil Trans*, 205–208
1835. *Experiments on the Transverse Strength and Other Properties of Malleable Iron, with Reference to its Uses for Railway Bars; and a Report Founded on the Same, Addressed to the Directors of the London and Birmingham Railway Company*
1835. *Second Report Addressed to the Directors and Proprietors of the London and Birmingham Railway, Founded on an Inspection of, and Experiments Made on the Liverpool and Manchester Railway* (repr. 1837)
1836. Letter to the Rev. D. Lardner, on the theory of gradients in railways, *Phil Mag*, VIII, 97–100
1836. Remarks on Lecount's treatise on iron rails, *Phil Mag*, **VIII**, 291–293
1836. Observations on the valuation of the mechanical effect of gradients on a line of railroad, *Roy Soc Proc*, **III**, 390–391

1837. On the electro-magnetic conducting power of wires of different qualities and dimensions, *Phil Mag*, **XI**, 1–11

1837–1838. *Two Reports of the Commissioners on Railways in Ireland* (with others)

1838. D. Mahan, *Course of Civil Engineering*, edited by Barlow (repr. 1848)

1840. T. Tredgold, *Principles of Carpentry*, edited by Barlow

1842. An investigation into the power of locomotive engines, and the effect produced by that power at different velocities, *ICE Trans*, **III**, 183–196

1848. *The Encyclopaedia of Arts, Manufacturers and Machinery* (in *Encyclopaedia Metropolitana*, 1836 etc.)

BARLOW, Peter William (1809–1885) was born at Woolwich on 1 February 1809, the eldest son of Peter Barlow (q.v.), Professor at the Royal Military Academy. Educated at private schools, Barlow, perhaps through his father's contacts with Thomas Telford (q.v.) and other leading engineers, resolved on a career in civil engineering and became the pupil of Henry Robinson Palmer (q.v.), founder of the Institution of Civil Engineers, in 1824.

Palmer, after several years working for Telford, was just beginning an independent career and Barlow assisted him with survey drawings for the Liverpool–Manchester Railway, against which Palmer gave evidence on behalf of the Mersey–Irwell navigation in Telford's absence.

Palmer proposed Barlow as an Associate of the Institution of Civil Engineers in 1826 and later that year he was appointed (Resident) Engineer at London Docks. Barlow gained experience both there and on various surveys with which Palmer was involved for early railways, etc. From 1832 Palmer was heavily involved in proposals for what was to become the South Eastern Railway and Barlow assisted him, becoming resident engineer on the Tonbridge district under William Cubitt (q.v.) following Palmer's resignation due to poor health. Barlow subsequently became Engineer for the railway.

Barlow followed his engineering career in tandem with scientific pursuits and contributed several papers to the *Philosophical Transactions* and other scientific journals.

He died at 56 Lansdowne Road, Notting Hill, London on 19 May 1885.

MIKE CHRIMES

[ICE membership records, ICE Archives; Telford mss, ICE Archives; DNB]

BARNES, James (c. 1739–1819), brewer and civil engineer, ran two careers in parallel. The details of his birth are unknown, though it was probably not in Banbury, where he later became a leading citizen. In 1778, his brewery in North End, Banbury, was sufficiently prosperous to allow him to lend the Oxford Canal Company £80, at

4½% interest. Construction of the canal, which was then open from Coventry to Banbury, rested for eight years until, in 1786, a concerted move was made to establish a through navigation and work resumed towards Oxford. The plan was by Robert Whitworth (q.v.) and Barnes, who apparently had no previous engineering experience, was appointed on 26 May 1786 as Surveyor of the Works of the Canal from Banbury to Oxford at £50 p.a. (Thomas Carpenter, the manager of the length open for trade, received £80 p.a.) The canal was constructed by a consortium of six contractors, of whom Samuel Simcock and Samuel Weston (qq.v.) appear to have been the active members. After a few visits during the early months, Whitworth does not appear to have taken an active part and the canal committee relied on Barnes or Simcock, or sometimes both, for engineering advice. In 1788, as part of the continual quest to deal with the lack of water at locks, Barnes proposed side ponds and was authorised to make a trial—possibly the first such in Britain, though the technique had been known in France for some time. The canal was completed in 1790 and Barnes must have given satisfaction to his employers as his salary was raised progressively to £200 p.a. He maintained his interest in his brewery, though, and by December 1791 the canal committee was complaining of his lack of attention to their interests.

In 1789 Barnes, Simcock and Weston made the initial survey for a Western Canal, but the time was not ripe and it was a revised route by John Rennie (q.v.) which was authorised in 1794 as the Kennet & Avon Canal. Meanwhile the commercial success of the Oxford and other canals led to a mania of canal speculation and in 1792 Barnes made two surveys to improve the route from the Midlands to London. His proposed shortening of the windings in the northern part of the Oxford Canal was not pursued until 1829, to the plans of Charles Vignoles (q.v.). However, his plan for a canal from Braunston, in Northamptonshire, to the river Thames was approved by the promoters on 20 July 1792 and with minor changes by William Jessop (q.v.) it became the Grand Junction Canal. On 3 June 1793 Barnes was appointed 'Engineer to execute the canal, he engaging to devote his whole time to the undertaking and to consider himself responsible for the works and people employed ...' at a salary of two guineas a day with half a guinea for expenses; Jessop became Chief Engineer two days later.

Work proceeded rapidly at first but by 1795 there was trouble on two fronts. Jessop's estimate of £400,000 was clearly quite inadequate and the canal committee was having to allocate available funds. Meanwhile poor workmanship and financial losses by the contractors for the tunnels at Braunston and Blisworth had led the company to take over the works themselves. On 1 June 'the Committee took into consideration the works at the tunnels and observe the necessity of Mr. Barnes' attention to them, but as many inconveniences might arise from limiting him to specific portions of duty there Ordered that Mr. Barnes do devote as much time as he possibly can to the same, and that he do allow nothing but the most pressing circumstances to call him away'. Braunston tunnel was completed in June 1796 but ingress of water at Blisworth had caused the virtual abandonment of the works there. Jessop advised substituting locks over the hill but Barnes considered a tunnel on a slightly different line feasible, and preferable for operational reasons. Whitworth and John Rennie (qq.v.), who were consulted, agreed and the canal committee resolved 'that the tunnel at Blisworth be executed in the line originally proposed by Mr. Barnes ... and that Mr. Barnes shall have the entire direction of it'. Barnes's expertise now was recognised when the committee permitted him to spend three days inspecting the problems at Saddington tunnel on the Leicestershire & Northamptonshire Union Canal; his report to that company was presented in August. In 1797 the Grand Junction referred to Barnes as 'our Chief Engineer' and Jessop was politely removed from office on the pretext of saving money.

The company's financial constraints prevented Barnes doing more at Blisworth than driving a drainage heading and the question of an early connection between the completed sections to the north and south became more pressing. Barnes, on instructions, built a road directly above the line of the tunnel but a committee member, Joseph Wilkes, pressed for a joint inspection of the works by Barnes and Benjamin Outram (q.v.) of the Butterley Company. Wilkes was a colliery owner who had had a railway built at Brinsley in Nottinghamshire by Outram in 1797; he was also a major shareholder in the Ashby Canal in Leicestershire and was active at this time in persuading them to employ Outram there. His motion was negatived but he returned to the attack early in 1799, requesting a special meeting to consider the state of the works. The committee agreed to go over the matter with Barnes and another engineer and Jessop, Outram's partner at Butterley, was desired to attend; two weeks later Jessop proposed a cast iron railway over Blisworth Hill. After visiting Wilkes's line the committee entered into a contract with Outram for the railway but Wilkes continued his war with Barnes and the committee felt compelled to minute its continuing confidence in its Engineer.

Barnes himself, though, must have had his doubts. In June 1799 he had been given permission by the committee to make the survey for what became the Grand Union Canal (later, made by Benjamin Bevan (q.v.)), which was expected to bring additional traffic to the Grand Junction. It appears that he was unauthorised to engage in several works which he now undertook as consulting engineer or contractor. In July he contracted with the Leicester Navigation to repair their Blackbrook reservoir, which had failed. Using agents, Barnes rebuilt the damaged works, though not entirely successfully as there was another partial failure in March 1800, whether due to sabotage or bad workmanship was not established with certainty. In September 1800 he branched out further afield, becoming contractor for the section of the Penydarren Tramroad from Plymouth Furnace to Navigation House in Glamorgan. Two months later he was involved in a preliminary survey for the Sirhowy Tramroad in Monmouthshire and in 1801, with a colleague named Morris, he projected and laid out the Carmarthenshire Railway, which gained its Act in 1802. Part of this sizeable tramroad, which was sixteen miles long and had significant earthworks, has been claimed to be the first public railway in Britain; the Surrey Iron Railway was authorised earlier but opened later.

Meanwhile, completion of Outram's railway had opened a through route of sorts on the Grand Junction Canal from Braunston to Brentford. Work continued slowly on Blisworth tunnel, a contractors' joint venture being chosen this time by public tender. Once again the contractors had underpriced the work and when they were obliged to give up in 1804, Barnes resumed control. The tunnel was completed by March 1805 and canal navigation established throughout. Despite heavy commitments on other canals, Jessop provided design consultancy for much of the twelve years of construction but it was Barnes who bore the direct burden of driving the work forward. Jessop's estimate of the cost of 'this stupendous and most useful line of navigation' had been £400,000 but in the event it cost more than four times that amount, making it easily the most expensive canal built in Britain before the Manchester Ship Canal.

In August 1809, Barnes contracted for the Parkend to Lydney section of the Severn and Wye Railway but by July the following year it was alleged that great defects existed in his work and he was ordered to replace some of it. In February 1811, he stopped work and Roger Hopkins (q.v.), the company's engineer, arranged for others to complete it. The dispute was settled in favour of the company, Thomas Sheasby Jr. (q.v.) being the arbitrator.

Barnes took an active role in the public life of Banbury. He was elected a Capital Burgess of Banbury Corporation on 8 March 1799, an Alderman from c. 1806 and was Mayor in 1801 and 1809. He was also Bridgemaster of Banbury Charities from 1812 to 1815 and was a member of the Supervising Committee of Banbury Savings Bank when it was established in March 1818.

Barnes and his wife, Mary, had a daughter, also called Mary, who was baptised at Banbury in 1764. In 1803 she married Richard Austin, who became a partner with his father-in-law in the brewing business. Barnes's wife died in 1807 and when he himself died on 18 January 1819 he was described in *Jackson's Oxford Journal* simply as 'an eminent brewer'. He was buried at Bodicote, just south of Banbury, where his monument describes him as Principal Engineer of the Grand Junction Canal.

P. S. M. CROSS-RUDKIN

[PRO: RAIL 830/39–41 and 855/3–5; *Rennie Reports*, **1**, 346 (1796) at ICE; Dowlais letters, Glamorgan RO; Carmarthenshire RO; H. W. Paar (1963) *The Severn & Wye Railway*; *Priestley's Navigable Rivers and Canals* (repr. 1969); Alan H. Faulkner (1972) *The Grand Junction Canal*; Hugh Compton (1994) *Two Canal Entrepeneurs from Banbury*, in Cake and Cockhorse, 12/9, 230–238]

Publications

1792. *A Plan of the Proposed Canal from the Oxford Canal at Braunston … to Join the River Thames at New Brentford … to be Called the Grand Junction Canal; With the Collateral Cuts or Branches*
1801. *The Report of James Barnes, Engineer, as to the Intended Rail Way or Tram Road, From the Flats, Near Llanelly, to the Lime Rocks, Near Castle-y-Garreg*
1802. *A Report and Estimate … the Leicestershire and Northamptonshire Union Canal*

Works

1786–1791. Oxford Canal, Banbury–Oxford, resident engineer
1793–1805. Grand Junction Canal, resident engineer
1793. Newport Pagnell Canal, survey; not built then
1793. Grand Junction Canal branches, survey
1796. Grand Junction Canal, Northampton branch, survey
1799, 1802. Leicestershire & Northamptonshire Union Canal, survey for connection to Braunston; later built as Grand Union Canal by Benjamin Bevan (q.v.)
1799–1801. Blackbrook Reservoir, Leicestershire, repairs
1800. Grand Junction Canal, Old Stratford branch, supervised construction
1800. Penydarren Tramroad, Plymouth Furnace-Navigation House, contractor
1801–1805. Carmarthenshire Railway, engineer
1808. Grand Union Canal, survey
1809–1811. Severn & Wye Railway, Parkend–Lydney, contractor

BARNES, John (1798–1852), marine engineer, was born at Walker Colliery, Newcastle-upon-Tyne, on 12 August 1798, the son of Thomas Barnes (q.v.). His father died in 1801 and his mother later encouraged him to study works such as Desaguliers's (q.v.) *Natural Philosophy*. He was educated by Rev. W. Rawes of Houghton-le-Spring before, at the age of 15, he was taken on by Boulton and Watt; James Watt (q.v.) was his godfather. From 1815–1817 he attended Edinburgh University and was then instructed by Francis Giles (q.v.) in civil engineering surveying.

In 1818 he was commissioned by Gordon and Murphy to design and install a Cornish engine which he subsequently ordered from Butterley, where Joseph Miller, whom he had met at Boulton and Watt, was now working. They decided to set up a business in London building marine engines for steam ships and in 1822 the firm of Barnes and Miller was established.

From the start they used steam expansively in their engines, a practice then uncommon. At that time steamships were in their infancy and the quality of their work helped establish their viability. In 1835 the business was wound up, and Barnes worked independently, designing engines which he had built at Horseley Ironworks. At this time he introduced an improved, 'feathered', paddle wheel, several being supplied to the French shipbuilder, M. Normand, beginning with *Le Rotterdam* in 1837 and culminating with *Le Napoleon* in 1841, at the time of its construction the fastest paddle steamer in the world.

After a period as a consulting engineer, in 1845, on Robert Stephenson's recommendation, Barnes was appointed manager of Louis Benet et Cie's works at La Ciotat near Marseilles and was responsible for 14 important vessels including the *Charlemagne*, a screw steam frigate, for which he received the Legion d'Honeur. Barnes was interested in the classics, antiquities, geology and archaeology.

He was elected a Member of the Institution of Civil Engineers in 1823 and frequently attended meetings. He had a devoted wife, from Staffordshire; he died at La Ciotat on 24 September 1852 and was buried alongside his parents at Longbenton. He left a collection of his father's correspondence with John Smeaton and Boulton and Watt—no longer extant—to the Institution of Civil Engineers.

MIKE CHRIMES

[Membership records, ICE Library; Discussion: The Great Britain (1845) *Min Procs ICE*, **4**, 165–185; *Min Procs ICE* (1853), **12**, 140–148 (History of Ships)]

BARNES, Thomas (1765–1801), a colliery viewer, was the son of John Barnes (b. 1732) and his wife Elizabeth Hunter. His father was also a colliery viewer as was his grandfather, Amos (b. *c.* 1708) and his father before him. He was the fourth of six children and was baptised in Sunderland on 4 March 1765.

Barnes was apprenticed to his father and, following the death of the latter, reported on many of the collieries in the area. In 1797 he married Ann Forster (1767–1813) in Newcastle and the marriage produced two sons. Barnes came to be viewer at Walker colliery and although highly thought of by his contemporaries 'in the application of Engineering Science' he was blamed for the flooding of the Felling pit and an accident at the Hope pit, Sheriff Hill, for both of which collieries he was responsible. His scientific interests were confirmed by his membership of Newcastle's Literary and Philosophical Society and by the fact that he subscribed to Howard's book on spherical geometry in 1798.

When viewer at Walker, he named the Byker 'A' pit Lawson's Main and 'boasted that that pit would be the best pit in the Coal Trade both for serving Newcastle and also the shipping'. He enlarged the shaft from seven to twelve feet, opening a quarry to line with stone the upper 30 fathoms of its 133 fathom depth. He also installed a Boulton and Watt steam engine for both sinking and drawing the coal.

Barnes laid a railway from the pit to join the Byker waggonway, leading to the river Tyne, using cast-iron fish bellied edge rails three feet long laid on transverse stone sleepers, stated by Nicholas Wood (q.v.) to have been the first such application. The rails were of T section, varying in depth from 3¾ ins at the ends to 5 ins at midspan; the horizontal flange was 3 ins wide. Laid to a gauge of some 4 ft. 3 ins the line was brought into operation in 1799/1800. A major part of its length comprised rails cast by Boulton and Watt at their Soho Foundry, Birmingham; the iron shoes were cast locally.

Barnes was considered to be one of the leading viewers in the North East but Matthias Dunn (q.v.), noting his eminence, commented that although he had been well taught, he 'was not always right, but he was so daring that none durst tell him that he was doing wrong'. Barnes died at Walker on 19 April 1801, the *Newcastle Courant* likening his death to 'a public calamity'. He was buried at Longbenton three days later, the same day as his second son, Thomas, was baptised, and his estate of some £600 was left to his widow.

R. W. RENNISON

[Forster, Buddle, Easton, Watson and Johnson Collections, Northumberland RO: Acc. No. 3410; *Newcastle Courant*, 25 April 1801; Matthias Dunn (1811) *History of the Viewers*, Northumberland RO: EAST/3b; N. Wood (1825) *A Practical Treatise on Rail-Roads*, 46; A. W. Skempton and A. Andrews (1976–1977) Cast iron edge rails at Walker Colliery 1798, *Newc Soc Tr*, **48**, 119–122]

BAXTER, John (fl. 1767–1798), architect and bridge designer, was known as John Baxter junior, being the son of John Baxter (fl. 1722–1769) who is thought to have died in 1770. John Baxter Sr. was a mason who worked for all his active life for the second Sir John Clerk, Bart. (1676–1755) of Penicuik, some ten miles south of Edinburgh, and his son and successor to the title, Sir James Clerk. Sir John was a leading figure among Scottish antiquarians, naturalists and men of letters of his day, and chief patron of the architect William Adam (q.v.). For Sir John, John Baxter built in the late 1720s Mavisbank, a house of unique style and quality to whose design Sir John had contributed nearly as much as Adam; for Sir James, Baxter built in the 1760s a new

Penicuik House, for which Sir James made all the initial drawings, Baxter himself copying them more expertly for use in the construction. Having approached so near to the status of architect, Baxter sent both his sons, John and Alexander, to Rome in 1761 for study which would enable them to make their way as architects. John spent six years there and the experience he acquired had the desired result.

While John Jr. was away his father found himself helping Sir James Clerk in a controversy concerning designs for the first North Bridge in Edinburgh (see: *MYLNE, William*). Sir James believed that the bridge should be built 'in the style of a Roman aqueduct' and offered to the city a sketch design with Baxter Sr., advising him on its construction and probable cost. He also wrote to Baxter Jr. to send him a design from Rome. When the North Bridge as designed and built by Mylne suffered a partial collapse in 1769, one of the John Baxters joined with John Smeaton and John Adam (qq.v.) in a report to the city's committee on the accident. As the elder John Baxter was said—in a letter of February 1769—to be in such poor health that 'he had devolved his business on his son' it must be assumed that John Baxter Jr. was the reporter on the North Bridge. His practice as architect to the fourth Duke of Gordon resulted from a recommendation in the same letter, and work for other noblemen and public bodies followed.

He had already designed and built a bridge of modest size for Sir James in the Penicuik estate and from December 1769 to May 1770 acted with John Adam as consultant in the choice of site and designs for a major bridge over the river North Esk just north of Montrose, though his own design, of ten arches, and estimate of £5,584 were not accepted. That project was aided by a grant from the Commissioners for the Forfeited Estates and Baxter designed and built at least three smaller bridges for which they were the chief funding agency. The most prominent was Kenmore Bridge at the mouth of Loch Tay, of three arches with a circular arched cylinder through each of the spandrels. Architecturally the spandrel arches and other masonry details may have been copied from Smeaton's larger bridges at Coldstream and Perth, the latter of which also received some funds from the Commissioners, though Smeaton's bridges never had open cylinders, only arched rings on the spandrels used as an aesthetic device. In 1776 Baxter charged the Commissioners £1 for making a drawing for the large inscription stone which adorns the Perth Bridge.

Baxter's practice as architect and builder included a considerable variety of work, from country houses and churches to a layout plan for the new town of Fochabers and the town house (with steeple) of Peterhead. All his architectural works are listed by Colvin.

TED RUDDOCK

['Report of Messrs Smeaton, Adam and Baxter ...', see below; Minutes of the Northesk Bridge Commissioners 1765–1880, ref. 1/6/5–6 in Royal Burgh of Montrose Records, Angus Archives at Montrose Public Library; accounts of John Baxter Jr. with the Commissioners for the Annexed Estates 1771–1779, SRO: ref. E732/18; SRO, Register House Plans 971/1 and 1944; A. Rowan (1968) Penicuik House, Midlothian, *Country Life*, **150**, 386–7; Anne M. Simpson (1971) The architectural work of the Baxter family in Scotland 1722–1798, unpublished thesis, Edinburgh University (copy at National Monuments Record of Scotland); Ted Ruddock (1979) *Arch Bridges and their Builders* 1735–1835, 50, 87, 122; Colvin (1995) 111–113]

Publications
1812. Report of Messrs Smeaton, Adam and Baxter concerning the Bridge of Edinburgh, dated 22 August 1769, in J. Smeaton (1812) *Reports ...*, **3**, 218–222, and *The Scots Magazine*, **31**, 464–467

Works
1768. Bridge on the Penicuik estate, design and builder, one arch 40 ft. span
1772. Bridge over Balveg Water, near Strathyre, Perths., design and builder, for Commissioners of Annexed Estates
1772. Bridge over Gartchonzie Water, Perths., design and builder, for Commissioners of Annexed Estates
1774. Kenmore Bridge over River Tay, designer and builder, for Commissioners of Annexed Estates, three arches, middle arch *c.* 55 ft. span, commissioners' grant £1,000

BAXTER, William (d. 1841), architect, of Penrhyn Castle, Bangor, North Wales, became a member of the Institution of Civil Engineers in 1829 and remained on the membership lists until 1838. Although distant from London, he contributed to the Institution by adding to its library and to its collection of models. It seems likely that Baxter was involved in the rebuilding of Penrhyn Castle, undertaken by Thomas Hopper (1771–1856) for G. H. Dawkins Pennant during the period 1822–1837.

TESS CANFIELD

[Membership records, ICE; Obituary, *Min Procs ICE* (1842), 12–13]

BAYLIS, Thomas (fl. 1812–1829), contractor, was probably born in Birmingham and, by his own evidence, he started to work as a civil engineering contractor about 1812. In October 1817 he and his younger brother, Benjamin (1796–1859), won the contract to extend the Gloucester & Berkeley Canal from Hardwicke, where it had rested uncompleted since 1799, to a junction with the Stroudwater Canal at Saul. A significant term of the contract was that they agreed not to apply for any interim payments until £8,000 worth of work had been done. By December they

had cut the ground along the whole of the line and completed a large part of it; their first payment was also due. The Canal Company now invited tenders for the remaining length to the river at Sharpness, clearly expecting Messrs Baylis to win it and so to provide continuity for their workforce. In this they were successful and by May 1818 they had received £26,000 for progress on the works. Thomas Telford (q.v.), who was Engineer, described them approvingly as 'in the prime of life (and devoting) their whole attention to the works'. John Woodhouse (q.v.), however, who had been an unsuccessful tenderer for the work, was appointed Resident Engineer and he noticed that Messrs Baylis were using stone at Sharpness other than that specified. Thomas Baylis asserted that he would lose at least £20,000 if the terms of the specification were enforced and in January 1819, on Woodhouse's recommendation, the Baylises were allowed to resign their contract. Ironically, Woodhouse himself was dismissed by the Company in 1820 for accepting at Sharpness inferior stone which had been supplied by his son. To the Company, the cost of completion was several times more than the outstanding value of Baylis's contract and the £20,000 taken together. Telford continued to show his faith in Baylis, employing him on a project for sea defences in Norfolk.

In 1820 Baylis published a plan for a 16-mile tramroad from Stratford-upon-Avon to Moreton-in-the-Marsh and followed it up in 1821 with an offer to construct the whole line for £33,456 16s 8d. This was not accepted and the line was built, with only minor variations in the route, in 1823–1826 under John Urpeth Rastrick (q.v.). Benjamin Baylis operated the railway from 1829 under a repairing lease, which led to some contention with the Company in 1832 when it was found that he was not maintaining the line to the agreed standard.

In 1821 Thomas Baylis tendered successfully for five contracts on the English length of the Holyhead road, on which a comprehensive programme of upgrading under Telford had just started. Over the next eight years he added a further fourteen, as far apart as Barnet in Hertfordshire and Shifnal in Shropshire. He was resident in Stratford-upon-Avon in 1824 when he undertook to rebuild a bridge for the County on the Warwickshire length, quoting Telford as a referee. In 1829, however, when he was engaged in seven contracts worth £17,808, he failed. Sir Henry Parnell, the Chairman of the Holyhead Road Commissioners said that Baylis had 'executed the greater part of the works (in England), and executed them exceedingly well, and against whom no kind of complaint could be made'. The client had been withholding 10% retention on the monthly progress payments and, when other contractors had been brought in to complete the works and had been paid from the retention monies, there was a balance in Baylis's favour of £29 5s 9d. Why a man who had been able twelve years earlier to finance work in progress to the value of £8,000 and who, in his own words, would 'nearly' be ruined by a loss of £20,000, should have been unable to deal with this situation remains to be explained.

By April 1830 Thomas Baylis had left his house in Greenhill Street, Stratford-upon-Avon, and was living in Small Heath, Birmingham. Benjamin Baylis remained, first at Mulberry Terrace and later at Great William Street, pursuing a career as a national railway contractor, his most visible work being the Ouse Viaduct on the London & Brighton Railway.

A Thomas Baylis 'at Nantwich, Cheshire, Civil Engineer' was cited as a potential surety for about £800 in an unsuccessful tender for a bridge in Warwickshire in 1835. If this is the same man, which is quite probable, it shows an interesting change of fortune.

P. S. M. CROSS-RUDKIN

[PRO: RAIL 829/4 and /5 and 875/3; Main Papers 9 May 1827, HL RO; Holyhead Road Commissioners' Report; Ironbridge Gorge Museum Library; Shakespeare Birthplace Trust; *Warwickshire Advertiser* 7 Jan. 1860; Warwickshire RO; Worcestershire RO]

Publications

1821. *Plan of the Stratford upon Avon and Moreton in the Marsh Rail Way or Tram Road showing its communication with the Stratford upon Avon Canal*

Works

1817–1819. Gloucester & Berkeley Canal, contractor; three contracts worth altogether £86,200
1819? Burnham Marshes Embankment, Norfolk contractor
1821. Stratford upon Avon Canal, Earlswood reservoir, contractor, £2,660+
1821–1829. Holyhead Road, contractor for nineteen sections, in various locations from Hertfordshire to Shropshire, aggregate value £60,489:
1821. at Cuttle Mill £2,200; at Braunston Hill, £3,170; at Meriden Hill, £4,300; at Hockliffe, £5,700; at Little Brickhill, £1,272
1822. at Summerhouse Hill, £1,125; at Blackwell Hill, £200
1823. at Stowe Hills, £1,920; at St. Albans, £4,020
1825. at Summerhouse Hill, £2,250
1826. at Llewllyn, £1,900; at Wednesbury, £3,400; from Barnet to South Mimms, £894; at Fosters Booth, £2,582
1827. at Coventry, £10,329
1828. at Gullet Hills, £1,952; at Shifnal, £1,037; at Knowles Bank, Shifnal, £2,110
1829. at Wolverhampton, £2,080
1824–1826. Stone Bridge, Warwickshire, contractor; was paid £20,115

BEAMISH, Richard, FRS (1798–1873), engineer, was the fourth son of William Beamish of Beaumont, County Cork. At ten he was sent to school at Clifton, Bristol, and spent his holidays at Doddershall Park, Buckinghamshire, with Colonel

Pigott, an arrangement which Beamish found very satisfactory. He went to the Royal Military Academy at Marlow, and was successful in mathematics, drawing, and other subjects, qualifying for a commission without purchase; his father, however, bought a commission in the Coldstream Guards for him. He was in France and Belgium in 1815 and 1816, and as a young guardsman, conducted himself in a manner which he later came to regard as dissipated. In 1818 he accepted an offer from Sir Stamford Raffles to go east with him as aide-de-camp, but was prevented by his parents. While on half-pay from the Guards, Beamish took up music, elocution and languages. He returned to Ireland in about 1819. After being stopped from marrying by his parents, Beamish decided to become independent of them, and in 1824 settled on the profession of civil engineer.

After a period of concentrated study, Beamish moved to London in 1826 where he visited Alexander Nimmo (q.v.) who introduced him to Thomas Telford (q.v.). Beamish was able to study the progress of the new London Bridge, extensions to the London Docks, and other important works and in June 1826 he was taken on at the works of the Thames Tunnel for a month's trial. In August he was appointed as an assistant to the Resident Engineer, Isambard Kingdom Brunel (1806–1859), where he remained until the closing of the works in 1828; his experiences form the basis of his *Memoir of Sir M. I. Brunel*. His meticulous journals record the progress and costs of the work. At the behest of the Brunels, he twice visited Liverpool to investigate the possibility of a Mersey Tunnel, and in November 1826 visited the works of he Liverpool and Manchester Railway, where he met George Stephenson.

Beamish returned to Ireland in 1828, having by then inherited considerable wealth from his father. For nearly six years he was Engineer of Cork and surrounding counties, carrying out largely routine public works. In late 1834, Beamish accepted the Brunels' invitation to become Resident Engineer for the Thames Tunnel but was forced to retire in August 1836 by failing health. During this period he married, and lived at Rotherhithe, virtually at the entrance to the works. He became a Fellow of the Royal Society in 1836, and lived a fairly retired life thereafter, taking a few pupils for basic instruction to qualify them to enter engineering apprenticeships. In 1846 at the request of I. K. Brunel, he prepared parliamentary plans for the Cork and Waterford Railway. Having suffered serious financial reverses, Beamish was appointed Resident Engineer by Brunel for the construction of the Gloucester and Forest of Dean Railway. The work was completed in 1850, effectively ending Beamish's engineering career. He spent two years in Hanover with his family and his wife's German relatives, educating his children and in 1854 he settled in Cheltenham, where he organised the 1856 meeting of the British Association. From 1865 to 1872 he lived in retirement in the Isle of Wight and in Woolston, near Southampton. He died at

Bournemouth on 20 November 1873. A portrait by Henry Wyatt exhibited at the Royal Academy in 1836 was sold at auction in 1996.

For all his scientific thoroughness, Beamish embraced the irrational. He believed he had an intuitive power of reading the character of men from their faces and was a follower of D'Arpentigny's theory of the form of the hand as an indication of mental characteristics. Beamish's book on the subject includes I. K. Brunel's hand, a full, strong example. He was variously a follower of Spurzheim, the German phrenologist, and of George Combe. He believed in mesmeric second sight, and through mediums, had contacted the spirit of I. K. Brunel.

TESS CANFIELD

[Diary, April–December 1827 and January 1835–May 1836 concerning the construction of the Thames Tunnel, 2 vols., and R. Beamish and T. Page (1835–1844) Resident engineers' special reports, ICE archives; Obituary, *Min Procs ICE*, **40** (1874–1875), pt. 2, 246–251]

Publications

1837. Notice concerning the Thames Tunnel, *Min Procs ICE*, **1**, 32–33
1839. L. A. J. Quetelet, *Popular Instructions on the Calculations of Probabilities*, with notes by R. Beamish
1843. *Approximate Rationale of the Cold Water Cure*
1854. *Treatise on Elocution*
1856. *Statistical Account of the Town and Parish of Cheltenham*
1862. *Memoir of the Life of Sir Marc Isambard Brunel*
1865. *The Psychonomy of the Hand; or, the Hand as an Index of Mental Development ...* (2nd edn., 1879)

BEATSON, Alexander, Lieutenant-General

(1759–1830). Although most of his thirty year career in India (1776–1808) was spent as either an engineer or a surveyor Alexander Beatson never held an official position in any of the Indian Corps of Engineers. He was born in 1759, the son of Robert Beatson and originally went to Madras as an ensign in the infantry. However, heavy losses among military engineers at the siege of Pondicherry caused him to be sent to Masulipatam as superintending engineer. He held the post from 1778–1782, but piqued by the refusal of a commission in the engineers became a Captain in the Corps of Guides and 'During the interval of peace ... was indefatigable in surveying and exploring the whole face of the Carnatic'. These surveys included Palnud, on a scale of half inch to the geographical mile, the Godavary River (6 in. to the mile) and the roads of Masulipatam (Bandar). This knowledge was incorporated into a map of the Coromandel coast in 1789. It was described as 'a monument of great industry skill and minute accuracy' and included corrections to existing information on the Kistna

River which, allied with his engineering background made him the most qualified person to report on irrigation schemes for the Kistna and Godavari Rivers in 1792.

Beatson proposed two possible dam designs to solve the problem, both to be built at Bezwada where the river Kistna was confined by two mountains to a width of 1,100 yd., expanding within a short distance to over 2½ miles. A large and thick stone dam could divert the entire river to Masulipiatam, but he considered that a smaller dam allied to a series of aqueducts could easily divert sufficient quantities of water at less expense and he recommended 'a series of levels should be taken before any final decision was taken'. This was entrusted to Michael Topping (q.v.) with the assistance of James Caldwell (q.v.). Beatson was on sick leave in England from 1795–1797, but after Topping's death he asked to replace him and was ordered to complete work on the surveys by any means. But he had barely begun work in the Circars when the political situation deteriorated and he was summoned to report to the Governor General Lord Mornington who 'received great satisfaction from his knowledge' and made him his ADC. When war broke out in 1798 he was made surveyor general to the 'grand army' where his successful plan for a breach in the north-west angle of the wall at Seringapatam was adopted over strenuous objections from the expedition's official engineers. For this he was promoted to Colonel in 1801, but never returned to work on the Kistna, which project languished until 1855 when the Bezwada Anicut was finally built by C. A. Orr.

Beatson did, however, maintain an interest in agricultural problems both during his period as commander-in-chief of St. Helena where he introduced a better cultivation system and during his retirement in England. He died on the 15 October 1830. A portrait is reproduced in the *Gentleman's Magazine*, **111** (1900), 394.

SUSAN HOTS

[E. Buckle (1839) Wulloorpullum sluice, *Prof Papers, Madras Engineers*, **1**; D. Sim (1839) Coleroon annicuts, *Prof Papers, Madras Engineers*, **1**; E. Buckle (1846) Report on the resources of the districts bordering the Kistna, *Prof Papers, Madras Engineers*, **2**, 1–13; R. B. Smith (1856) The Cauvery, Kistna and Godavary, 1–47; G. T. Walch (1899) *The Engineering Works of the Kistna Delta*, **1**, 9; F. W. C. Sandes (1933) *The Military Engineer in India*, **I**, 173–175; R. H. Phillimore (1945) *Historical Records of the Survey of India*, vol. 1]

Publications (including maps)

1788–1789. *The Countries between Pennar and Godavari rivers compiled for Sir Archibald Campbell* (scale 6 in. to 1 mile)
c. 1790. *The roads to Madras from Masulipatam*
1790–1792. *Geographical Observations in Mysore and Baramahal*
1800. *A view of the Origin and Conduct of the War with Tippoo Sultan*

1820. *A New System of Cultivation without Lime or Dung or Summer Fallowing*

BEAUMONT, Huntingdon (*c.* 1560–1624), entrepreneur, and the builder of Britain's first horse-drawn waggonways, was a younger son of the Beaumonts of Coleorton in the Leicestershire coalfield. He gained his first experience in coal-mining management at Bedworth in Warwickshire, before moving to Nottingham where he leased coal pits at Wollaton in 1601 and at Strelley in 1602. A new lease on these pits in 1604 allowed him to carry coals from Strelley through Wollaton 'alonge the passage now laide with Railes and with suche of the lyke Carriages as are now in use for that purpose', the first known horse-drawn waggonway. By *c.* 1605, the waggonway system had also come into use in Shropshire, and Beaumont had taken it to the north east of England. According to Gray, the first historian of Newcastle, in his *Chorographia* (1649):

> Master Beaumont, a gentleman of great ingenuity, and rare parts, adventured into our Mines, with his thirty thousand pounds; who brought with him many rare engines, not known then in these parts; as the art to boore with iron rodds to try the deepnesse and thickness of the coale; rare engines to draw water out of the pits; waggons with one Horse to carry down coales from the pits, to the stathes, to the river, & c. Within few yeares, he consumed all his money, and rode home upon his light horse.

£30,000 is probably an exaggeration, as is the figure of £20,000 to which it has been amended in Gray's own copy of his book, but Beaumont's venture was clearly ill-fated, and it is clear that he was quite unable to break into Newcastle's monopoly of the trade in coal to London.

In fact Beaumont's activities were not on Tyneside, but in the Bedlington/Blyth area of Northumberland, where he took mine leases in 1605. The assignments for these leases seem not to have survived, but since Beaumont had built his Strelley waggonway in *c.* 1604, it can be assumed that his leases in Northumberland allowed him to build a waggonway, and that he laid it promptly. This waggonway served coal pits and salt pans at Cowpen and Bebside, and a bishopric lease of 1608 specified carriage 'upon Rayles'. Beaumont was bought out by his partners in 1614, but his successors were unsuccessful, deserting the business in 1616 and leaving behind them three boats and 'many diverse Rayles, Utensils and implements'.

This, then, was the first waggonway in the north east of England. It was to be followed by the Whickham Grand Lease waggonway to the Tyne (*c.* 1621), the Benwell Way (*c.* 1627), the Winlaton or Brockwell Way (pre 1633?), and the Stella Grand Lease Way (pre 1635?). Many more waggonways were built to the coal trading rivers of the North East after 1650.

Beaumont was to die an undischarged bankrupt in Nottingham Gaol, but he had started a revolution in inland transport which was eventually to engage most of the great engineers of the nineteenth century.

STAFFORD M. LINSLEY

[W. Gray (1649, repr. 1790) *Chorographia: or a Survey of Newcastle upon Tyne*, 20; Richard S. Smith (1957) Huntingdon Beaumont, adventurer in coal mines, *Renaissance and Modern Studies*, **I**, 115–153; Richard S. Smith (1960) England's first rails, a reconsideration, *Renaissance and Modern Studies*, **IV**, 119–134; M. J. T. Lewis (1970) *Early Wooden Railways*; G. Bennett, E. Clavering and A. Rounding (1990) *A Fighting Trade: Rail Transport in Tyne coal 1600–1800*]

BECKMAN, Sir Martin (d. 1702), Chief Engineer of England 1685–1702, was a Swedish Captain of Artillery employed in England as a reward for the services of his late brother Diderich ('our incomparable engineer') to the Royalists in the Civil War. Despite forty years employment by the Board of Ordnance, Beckman's career is mainly of interest in two areas, his involvement in Tangier, the largest engineering project of the period, and his employment of the Fitch brothers (q.v.) as contractors at Portsmouth and Hull.

He was the first engineer to visit Tangier in 1662, producing a plan of the town and its proposed fortifications, estimated at £200,000. He was engineer in charge there periodically for almost the entire period of English rule from 1662–1684, including the period of the siege, 1679–1680, when resources for the mole were diverted to the defences, working with Sir Bernard de Gomme (q.v.) and Sir Henry Sheres (q.v.). He was a member of the committee which reported that £4,798,561 would be needed to make repairs, although Beckman had already drawn up a detailed memorandum for the demolition of Tangiers before he even left England.

In England, in 1678, he was Commissioner for the Defences of both Portsmouth, with de Gomme and Sir Jonas Moore (q.v.) and Hull. In 1699 he reported that £15,000 (Portsmouth) and £150,000 (Hull) were still needed to be spent, not least because of the working methods of the Fitches of which he remarked 'If Mr Fitch doth finish this year's work in 10 days I will finish the whole citadel in 5 days, the one as possible considering the season and the want of stores and masons as the other'.

In 1685 he was appointed as Chief Engineer with a salary of £300 pa, was knighted and, three years later, was made 'Comptroller of the King's Fireworks as well for War (he was in charge of the naval bombardment of French and Belgian ports) as for Triumph and all firemasters, bombardiers and petardiers', and first head of the Royal Laboratory, Woolwich. He was naturalised in

1691. He took soundings in the Humber and the Tweed and Portsmouth harbour and as well as his 'fine map of Tangier' (Pepys) and a watercolour of Kingston upon Hull (BL MS 33233), produced charts and plans of Berwick, Cliffords Tower on the Tyne, and the Channel Islands, Milford Haven and the River Avon.

He died in London on 24 June 1702.

SUSAN HOTS

[PRO: MPH1; PRO WO 46.1/2, WO 55/1788; Dartmouth papers, app. vol. 15; House of Lords manuscripts, vol. 111 (new series) 1965; Sir James Halkett (1680) *A Short and True Account of the Most Remarkable Things that Passed During the Late Wars with the Moors at Tangier in the Year 1680*, first published with introduction by H. MaCance (1922), *Journal, Society of Army Historical Research*, **1**, Dec. 1922; DNB; W. Porter (1889) *History of the Corps of Royal Engineers*, I; E. M. S. Routh (1912) *Tangier: England's Lost Atlantic Outpost, 1661–1684*; Edwin Chappell (ed.) (1935) *The Tangier Papers of Samuel Pepys*; A. H. W. Robinson (1962) *Marine Cartography in Britain*; H. Tomlinson (1973) *The Ordnance Office and the King's Forts 1660–1714, Architectural History*, vol. 16, 5–25; G. H. Williams (1979) *Western Defences of Portsmouth Harbour 1400–1800*, Portsmouth papers, No. 30; *King's Works*, vol. iv (2) (1982), 475/664; M. Foreman (1989) *The defences of Hull, Fortress*, 2, 36–45; A. Saunders (1989) *Fortress Britain*]

BEIGHTON, Henry, FRS (1687–1743), land surveyor and engineer, was baptised on 20 August 1687, the son of a yeoman family long settled in Chilvers Coton near Nuneaton, Warwickshire. For most of his life he lived in the neighbouring village of Griff, working from 1707 as a land surveyor and, a little later, as a topographical draughtsman.

In 1714 Thomas Newcomen (q.v.) began erecting his second steam pumping engine, at a colliery near Griff. Beighton studied this 'Fire Engine' and published a detailed engraving of it in 1717, the first illustration of a Newcomen engine. From an analysis of the work done by the Griff engine Beighton calculated that the effective pressure on the piston could be taken as 8 lb per sq. in. and he was therefore able to publish in 1717 a table giving the amount of water pumped from a given depth by engines of various cylinder sizes, assuming a 6 ft. stroke at 16 strokes per minute.

Beighton was then employed in the Durham coalfield, where he lived for three years near Gateshead. He built an engine in 1718 at Oxclose Colliery in Washington Fell with an improved valve gear and from experiments on this engine and at Griff he measured the volume of steam at a pressure of 1 lb per sq. in. evaporated for a given volume of water.

Returning to Griff he designed and made a new plane-table in 1721 and from 1722 to 1725 he surveyed and mapped the entire county of Warwickshire. Engraved and published in 1728, this was the second county map on a scale of 1 in. to a mile and the first to be based on trigonometrical survey. It was reissued with more detail in 1750 by his widow, Elizabeth. Early in 1734 it was to Beighton that John Grundy (q.v.) chose to demonstrate his new precise levelling instrument, made for him by Jonathan Sisson.

Beighton was elected a Fellow of the Royal Society in November 1720 and contributed four papers to the *Philosophical Transactions*. These include a description of his plane-table, meteorological observations 1735–1736 and in 1731 a description, illustration and analysis of the waterwheel and pumps of the London Bridge waterworks built by George Sorocold (q.v.). He calculated the power of the 20 ft. diameter wheel from the work done by the pumps and also calculated the velocity of flow between the piers of the bridge, deriving what is perhaps the earliest numerical application of the basic flow equation in hydraulic engineering.

It was through a shared interest in the Griff engine that he first met the distinguished experimental philosopher John Theophilos Desaguliers (q.v.) who later refers to 'my ingenius and very good friend Mr Henry Beighton'. Over a period of years Beighton contributed descriptions and drawings of several hydraulic machines for inclusion in the second volume of Desaguliers's *Course of Experimental Philosophy* published after long delays in 1744.

Beighton died on 9 October 1743 and was buried at Chilvers Caton where there is a mural tablet in his memory.

A. W. SKEMPTON

[J. T. Desaguliers (1744) *A Course of Experimental Philosophy*, vol. 2; Gordon Goodwin (1885) Henry Beighton, DNB, **4**, 132; Rhys Jenkins (1936) *Collected Papers*; P. D. A. Harvey and H. Thorpe (1959) *The Printed Maps of Warwickshire*; E. G. R. Taylor (1966) *Mathematical Practitioners of Hanoverian England*; Elizabeth M. Rodger (1972) *The Large Scale County Maps of the British Isles*; L. T. C. Rolt and J. S. Allen (1977) *The Steam Engine of Thomas Newcomen*]

Publications

1717. *The Engine for Raising Water (with a Power Made) by Fire*.

1728. *A Map of Warwickshire* (four sheets on scale of 1 in. to a mile), reissued by Elizabeth Beighton 1750

1731. A description of the water-works at London-Bridge, *Phil Trans Roy Soc*, **37**, 5–12

BELL, Henry (1767–1830), engineer, architect, and pioneer of steam navigation was born 7 April 1767, at Torphichen Mill, near Linlithgow, West Lothian, the fifth son of a millwright, Patrick Bell (1727–1793) and Margaret Easton. Bell claimed a long history of family involvement in harbours, canals, bridges and mechanical improvements, and relatives who were inspectors and resident engineers under Thomas Telford (q.v.). After a rudimentary education at a parish school in Falkirk, he was trained as a mason with a near relation from 1780. In 1783 he was bound in the millwright trade for three years with his uncle, another Henry Bell. Bell then joined Messrs. Shaw & Hart at Bo'ness for a year to learn ship building. He claimed to be the Scotsman, 'name of Bell', who about 1785 at Mosney near Preston, made one of the great improvements in cloth manufacture, the invention of cylinder printing. In 1787 he became a pupil of James Inglis, engineer, of Glasgow and subsequently went to London to work for John Rennie (q.v.) for about eighteen months.

In 1790 he settled in Glasgow, forming in 1791 the firm of Bell and Patterson, architects, builders, and cabinet makers, with James Patterson. His partnership with Patterson lasted about seven years, until in 1798 Bell turned to entrepreneurship and steam navigation. In 1794 he married Margaret Young (1770–1856) of Kilbride, who was to be a great support throughout his difficult career.

Bell's claim that he twice, in 1800 and 1803, appealed unsuccessfully to the Admiralty for support for his steamboat experiments is unsubstantiated. He had observed the steam boat built by Symington at the Carron Works on the Forth and Clyde Canal, successfully demonstrated in 1801. Bell's interest was in a steam ship for the carriage of passengers whereas Symington's had been intended as a towing vessel. It seems possible that in 1804, Bell and Robert Fulton (q.v.) saw Symington's *Charlotte Dundas* (the second) laid up on the canal. Bell was apparently in regular correspondence with Fulton and claimed to provide information used by Fulton in his successful 1807 steamship.

In about 1806 Bell built an extensive range of hot and cold baths and the Baths Hotel in Helensburgh, a salubrious small town on the bank of the Clyde opposite Greenock. A steam engine pumped water and heated the baths and a conservatory. Although Helensburgh had been granted burghal status in 1802, it was not implemented until 1807, when Bell became its first Provost. Visitors to the town and the Baths had to choose between a 22 mile road journey from Glasgow and an unpredictable voyage on a small sail boat. In 1812, with the success of the *Comet*, a 42 ft. double paddle wheel steam boat, Bell established the first passenger steam ship service in Europe, between Glasgow, Greenock and Helensburgh. The boiler and engine were built by David Napier and the hull by John Wood at Port Glasgow, to Bell's design. As he had difficulty in paying for the highly expensive undertaking, Bell applied to the Admiralty for support in 1813. In 1819, with the encouragement of William Thompson, engineer, the service was extended to Fort William via the Crinan Canal.

The *Comet*, with only a four horsepower engine, was overwhelmed in a gale and sank on the Point of Craignish in 1820; all those on board, including Bell himself, escaped injury. Shares were sold, and a new *Comet* was launched in 1821 but it sank after

colliding with the Ayr steamer on a voyage from Inverness to Glasgow in 1825, with the loss of more than seventy lives. By 1826 a steamer named *Henry Bell* was plying the Glasgow-Liverpool run. Although steam navigation expanded rapidly, Bell seems to have lacked the capital to take full advantage of his great success.

Bell was also interested in canals and water supply. In 1806 Thomas Telford (q.v.) was commissioned to prepare plans for extracting water from the Clyde at Dalmarnock. Bell published a broadsheet proposing an alternative scheme, bringing water from the Falls of Clyde to Glasgow by a canal, providing clean water by gravity. He advised strongly against the use of steam engines to raise water, on the grounds that the engines would damage land values, create a nuisance, and raise the price of coal, and suggested that the proposed site of extraction would be subject to industrial pollution. However, on Telford's advice, steam power was adopted and by 1829 Glasgow had the largest engine-powered waterworks in Britain.

Bell's superior scheme was vindicated when in the mid nineteenth century, Glasgow implemented a similar scheme based on Loch Katrine. In 1810, while Bell was Provost of Helensburgh, he proposed a scheme to supply the town with water from a new reservoir in Glen Fruin. In 1820 Bell proposed to the Trustees of Clyde Navigation the construction of a small canal from the Broomielaw eastward to proposed docks at Glasgow Green, leaving Broomielaw for passenger vessels. Glasgow Green was a popular park, and local protest stopped the scheme. His last project was a plan for a sea level canal from East to West Tarbert, cutting across the long Kintyre peninsula, to allow improved access from the Clyde to the West Coast; it came to nothing.

In 1829 the city of Glasgow raised a subscription for Bell, by then impoverished and in poor health. The appeal was endorsed by Telford, Brunel, and other notables and Canning, as Prime Minister, awarded Bell £200. In his last years Bell was dependent on a pension of £50 per year from the Trustees of the River Clyde, raised to £100 in 1829, which continued to benefit his widow.

Bell died on 14 November 1830 at Helensburgh and is buried in Rhu (Row) parish churchyard where Napier erected a statue to mark his grave in 1853. His estate was valued at £707 15s 6d. There is an obelisk dedicated to his memory on the esplanade at Helensburgh and a monument to him at the old castle of Dunglass on the River Clyde near Bowling.

TESS CANFIELD

[Gibb Papers, ICE Library; E. Baines (1835) *History of Cotton Manufacture in Great Britain*; J. W. Gibb (1838) *Appendix 2, Steam Navigation*, in Rickman, J. (ed.) *Life of Telford* (London); B. Boyman (1840) *Steam Navigation ... and Symington's Inventions* (London); Morris (1844) *The Life of Henry Bell*; DNB 1885; D. Bell (1912) *Henry Bell* and *Henry Bell and the Comet*, in *David Napier Engineer*; A. Bowman

(1981) *Symington and the Charlotte Dundas*; Colvin (3); B. D. Osborne (1995) *The Ingenious Mr. Bell*; L. Day and I. Mcneil (1996) *Biographical Dictionary of the History of Technology*]

Publications

1806. *To the Committee … Interested in Supplying the City of Glasgow with Water* Glasgow
1813. *Observations on the Utility of Applying Steam Engines to Vessels …*
1820 … *Plan for the Improvement of the Harbour, Construction of Wet and Graving Docks, and Extension of the Navigation to Clyde Iron Works and Blantyre Mills*

Works

c. 1795. Clayslap Mill, Glasgow, Partick
c. 1795. Melville Place, Glasgow
1799. Carluke Church, Lanarkshire
c. 1806. Baths and The Bath Hotel, Helensburgh, Dunbartonshire
1811–1812. The steamship *Comet*, Glasgow
1821. The steamship *Comet*, Glasgow

BENNET(T), William (fl. 1790–1826), civil engineer, apparently came from the Lancashire area. He surveyed extensions of the Manchester, Bolton and Bury canal from Bolton to the proposed Leeds-Liverpool canal at Red Moss, and also an alternative route to the Rochdale canal, the Bury and Sladen canal (1792). He surveyed the Haslingden canal from Bury to Accrington which obtained its Act in 1794 but was never built.

His work brought him into contact with Robert Whitworth (q.v.), who employed him and recommended him in September 1793 to the Dorset and Somerset Canal Committee for detailed surveys of their scheme. Whitworth cited pressure of work as the reason he could not act in any capacity other than a consultant, but Bennet himself was busy on other schemes, notably the surveys of the proposed Ivelchester and Langport navigation. Consequently his report on the Dorset scheme was not completed until June 1795. In July his proposals were discussed and his suggestion of joining the Kennet and Avon at the same point as the recently authorised Somerset Coal canal was rejected. He then carried out an additional survey and the final route of the canal, from Rode Common via Beckington to Widbrook on the Kennet and Avon, was approved on 13 August 1795, it included a branch from Frome to collieries around Coleford.

The canal obtained its Act in 1796, Bennet presenting the engineering evidence and estimating the costs at £146,008. Major engineering works were a 1,009 yd. long Brewham tunnel, together with locks, intended to be designed to Robert Weldon's (q.v.) patent for 'caisson' locks. By September 1796 work was under way, although progress on the locks was delayed awaiting the verdict on trials on the Somersetshire Coal canal, for which Bennet was also by then engineer.

The original engineer for the Somersetshire Coal canal had been John Rennie, with William Smith (qq.v.) undertaking surveys; Acts were obtained in 1794 and 1796. By December 1794 John Sutcliffe (q.v.) had become the engineer and Bennet's involvement seems to have begun around April 1795, when he was consulted about proposed deviations. In November he was appointed engineer. By this time the Canal Company had become interested in various alternatives to traditional locks and decided to make a trial of Weldon's caisson lock. It was estimated that, if the trials were successful, the caisson locks would save the Canal Company £10,000; preparations took some time and it was not until February 1798 they could be begun. Practical problems dogged the caisson trials, culminating in bulging brickwork in the lock walls. Although Bennet was consulted, in the end an outside engineer, Benjamin Outram (q.v.), was brought in and in February 1800 he recommended the use of inclined planes. In April, Bennet reported on both Outram's proposal and alternative schemes of James Fussell and Norton and Whitmore (q.v.) for a balanced lock, or boat lift. Bennet preferred trials of the lock schemes and a further opinion was sought of Sutcliffe, who also criticised Outram's scheme. An inclined plane was finally adopted in June 1800. In the event this was no more successful and a further Act was obtained in 1802 to enable further work to proceed, with provision for conventional locks rather than other mechanical contrivances. Bennet underestimated the cost of these works by a third and by 1806, when this had become clear, his services were dispensed with.

Bennet's work on the Dorset and Somerset line had proved even less successful. Shareholders were loath to pay for calls on their shares as little progress had been made. The lack of success of the Somersetshire Coal canal's trials of Weldon's caisson lift compounded these problems and in April 1800 Bennet recommended a trial of Fussell's system, which proved so successful that half a dozen boat lifts were ordered. This was the last note of optimism in the story of the canal as two years later all funds had run out. All that had been built was a stretch of the Frome branch. There were attempts to revive the scheme and they culminated in a railway proposal in 1825, on which Bennet reported with Thomas Tredgold (q.v.).

Little else is known of Bennet's later career other than the fact that he was based in Eccles Green, Lancashire, between 1821 and 1825. He carried out some work for the Kennet and Avon Canal in 1808 in connection with an inclined plane from Bathampton quarries to the canal and in the same year he and Robert Anstice (q.v.) were consulted about the Axe drainage.

MIKE CHRIMES

[Rennie reports, ICE archives; E. Baines (1824–1825) *History, Directory and Gazeteer of the County of Lancaster*; C. Hadfield and G. Biddle (1970) *The Canals of North-west England II*, 245–261; K. R. Clew (1970) *The Somersetshire Coal Canal and Railways*; K. R. Clew (1971) *The Dorset and Somerset Canal*]

Works

1795–1803. Dorset and Somerset Canal, engineer, 8 miles, c. £30,000

1795–1806. Somersetshire Coal Canal, engineer, 11 miles, £110,000

BENTHAM, Sir Samuel (1757–1831), was born on 11 January 1757 into a well-connected family; he was the youngest son of Jeremiah Bentham, an attorney, and brother of Jeremy Bentham, the utilitarian philosopher, and step-brother to Charles Abbot, Speaker of the House of Commons. His mother died in 1766 and his father then married the widow of Rev John Abbott. Early in life Bentham showed an aptitude for mechanical contrivances and a desire to learn about shipbuilding. At the age of fourteen he was apprenticed to Mr. Gray, the Master Shipwright at Woolwich; from there he moved to Chatham before completing his seven year apprenticeship at the Naval Academy at Portsmouth. In 1778 he gained sea experience in Lord Keppel's fleet before leaving with letters of introduction from Lord Howe, First Lord of the Admiralty, to study naval architecture in northern Europe and Russia. He arrived in St. Petersburg in March 1780 and gained military experience in the Empress Catherine II's service against the Turks, rising to the rank of Brigadier General, travelling to the Chinese frontier and commanding a Russian flotilla against the Turks. While in Russia he observed their practice and developed a number of woodworking machines; these he patented in England when he was given temporary leave of absence in 1791 to settle his father's affairs (Patent Nos. 1838 (1791), 1947 (1793), 1951 (1793)). During this period, in collaboration with Jeremy, he devised a panopticon prison based on a concept he had developed in workshops at Kritchev; the structure made extensive use of iron. In 1795 he was due to return to Russia but instead offered to assist the Admiralty in the modernisation and mechanisation of the dockyards. This was a propitious moment. In Parliament there was disquiet at naval inefficiencies, while at the Admiralty, Lord Spencer was First Lord and Charles Middleton was the senior Naval Lord; both were keen on reforming the dockyards and improving output. As a result, Bentham was asked to fill a new post of Inspector General of Naval Works. Although an Admiralty appointment, it trespassed firmly into preserves normally the responsibility of the Navy Board, a situation which was to lead to great friction. Bentham's instructions were to look at all aspects relating to 'the improvement of the building, fitting out and arming of … ships and vessels as well as what may conduce to the better navigating and

victualling of them; the construction of docks, slips, basins, jetties and other works subservient to the construction of ships and vessels; together with the choice, preservation and economical employment of the several stores and provisions made use of in the navy'.

Bentham threw himself into this extraordinarily wide-ranging task with enthusiasm. He recruited as Chemist James Sadler (1753–1828), who designed the dockyards' first steam engine and should be credited with the invention of the table engine, while Bentham fulfilled the twin roles of architect and engineer. In 1796 Samuel Bunce was appointed to his staff as the navy's first salaried architect; he had already worked with Bentham on the panopticon project. He remained in post until his death in October 1802 when he was succeeded by Edward Holl (q.v.). In 1799, Simon Goodrich (q.v.) was upgraded from draughtsman and given the title of Mechanist. These men formed the core of Bentham's department.

Bentham's influence on warship design and construction was probably slight, although he apparently introduced more efficient working methods for ships' repair and maintenance and made use of diagonal bracing and a new system of planking. His chief work was within the dockyards themselves and at Portsmouth Bentham roofed over with two tiers of brick vaults the former North Basin, in use since the 1770s as a giant reservoir into which all the dry-docks drained by gravity. Here he installed the navy's first steam engine late in 1798; this powered his woodworking machines by day and emptied the reservoir by night. Few details are known of the woodworking machines, but the vaulted reservoir still remains in use.

To make efficient use of steam engines Bentham had to find simple repetitive tasks capable of being mechanised. He achieved his greatest success in collaboration with Marc Isambard Brunel (q.v.) when he saw the potential for the latter's invention of block-making machinery. Henry Maudslay (q.v.) had the skills to translate the ideas into reality and the first sets of machinery began working at Portsmouth in 1803. The Block Mills, which incorporate Bentham's woodmills, still stand and have been described as the world's first use of machine tools for mass-production; by 1808, 130,000 blocks were being produced annually. In 1811 Brunel and Bentham again collaborated in the design and construction of the navy's first steam sawmill at Chatham. This was to be completed in 1814, two year's after Bentham's departure from office. The sawmill, which still exists, used batteries of reciprocating saws; timber was transported to it using a remarkable sequence of canal tunnel, elevator and overhead railway. In collaboration with W. E. Sheffield, Bentham installed the first metal smelting and rolling mill at Portsmouth to reprocess copper sheathing from warships as well as to manufacture other copper articles.

One of Bentham's most successful developments was his harnessing of steam power to dredging. He began this work in 1800 and in 1803 the first bucket-ladder steam dredger started operating with great success in Portsmouth harbour. Others followed, based on the Thames and Medway, helping solve the hitherto largely intractable problems of shoaling in the latter river and by so doing helping to give Chatham a more secure future. In 1805 Bentham sailed to Russia to see if warships for the Royal Navy could be built there. Although this aspect of the mission was not a great success, he designed a panopticon at Ochta, using timber rather than iron. On his way back in 1807 he visited Karlskrona to inspect the covered slip and it is probably no coincidence that such structures began to appear in the British dockyards within a few years. Bentham recommended such a system in November 1811.

Bentham's absence had given the Navy Board an opportunity to bring his department under its direct control. This change took place in 1808, Bentham being given a seat on the Navy Board and the revised title of Civil Architect and Engineer. His remaining years in office were clouded by quarrels with the Board, inevitably leading to his dismissal in November 1812. His ideas concerning the development of Sheerness dockyard brought him into conflict with John Rennie (q.v.), who had produced a highly critical report on the Navy's dock facilities during Bentham's absence. He produced designs for a cast iron piled river wall at Sheerness, and in 1811 built an experimental wall using hollow brick cylinder caissons (Patent Nos. 3429, 1811, and 3544, 1812). At much the same time he became involved with Vauxhall bridge, and work began on foundations to his brick caisson design in 1812. Bentham hoped to get Goodrich to help supervise the work, but he wisely held back. The second pier failed and work was halted; James Walker (q.v.) was brought in as engineer and all Bentham's work was removed.

In 1814 he emigrated to France, a country he had visited in his youth to study the language and his eldest daughter, Mary Louisa, married the Marquis de Chesnel in 1819. Bentham seems to have spent much of the following years in sorting out his papers for (re)publication, as part of a process of self-justification, and this makes it difficult to be definitive about his personal responsibility for works carried out under his charge. He returned to England in 1827 and died on 31 May 1831. His wife, Mary Sophia, nee Fordyce, lived until the age of 93, having spent much time in enhancing her husband's reputation; she died on 18 May 1858. Their son George (1800–1884) became a distinguished botanist.

Bentham had an enquiring mind, an ability to recognise potentially rewarding inventions and a determination to succeed in his mission to reform the dockyards. The list of his own inventions, many of which were concerned with woodworking, but ranged from egg-shaped drains to pile-drivers, caisson lock gates and amphibious baggage wagons, is a remarkable testimony to the range of his interests. His work for a while brought the royal dockyards into the van of technical progress, albeit with the aid of Marc Brunel. More importantly, his small department was the seed from which ultimately grew the great civil engineering and architecture departments of Admiralty.

JONATHAN COAD

[M. S. Bentham (1846) Paper on the first introduction of steam engines into naval arsenals, enumerating of the principal inventions of … Bentham, etc, *Quarterly Papers on Engineering*, **6**, xi–lll,, xii–I, III; M. S. Bentham (1862) *Life of Brigadier-General Sir Samuel Bentham—By His Widow*; K. R. Gilbert (1965) *The Portsmouth Blocking Machinery*; A. W. Skempton (1975) A history of the steam dredger, *Trans Newc Soc*, **42**, 97–115; R. A. Morris (1981) Samuel Bentham and the management of the Royal Dockyards, 1796–1807, *Bull Inst Hist Res*, **54**, 225–240; R. J. M. Sutherland (1986) Shipbuilding and the long span roof, *Trans Newc Soc*, **60**, 108–111; R. [A.] Morris (1983) *The Royal Dockyards During the Revolutionary and Napoleonic Wars*; J. G. Coad (1989) *The Royal Dockyards 1690–1850*; DNB]

Publications (including patents)

1793. *Repertory of Arts and Manufactures*, **1**, 10, 22, 293, 367

1795. *Report to the Admiralty Submitting Proposals of Enlarging the Basin, and the Forming of Jetties and Other Accomodation Suited to the Largest Classes of Ships*

1796. Specification for his invention of a new method of planing wood [1791], *Repertory of Arts and Manufactures*, **5**, 293–317

1797. Specification for his new method of performing and facilitating the business of divers manufacturing and economical processes [1795], *Repertory of Arts and Manufactures*, **7**, 145–164

1797–1813. *Letters on Certain Experimental Vessels*

1800. *Answers to the Comptroller's Objections*

[1811]. Breakwaters, *Mechs Mag*, 13 July 1844

1812. *Minute on the Subject of the Breakwater in Plymouth Sound*

1812. Specification … laying the foundation of works of stone, brick …, *Repertory of Arts and Manufactures*, **20**, 129

1812. Specification … excluding the water of the sea … during the erection of masonry under water, *Repertory of Arts and Manufactures*, **21**, 1

1812. *Desiderata in a Naval Arsenal … Outline of a Plan for the Improvement of the Naval Arsenal at Sheerness* (repr. 1814)

1814. *Letter … Respecting the Real Causes of the Defeat of the English Flotilla on Lake Erie*

1814. *Representations on the Causes of Decay in Ships of War* [1812]

1827. *Naval Papers and Documents* [1795–]

1827. *A Statement of Services Rendered in the Civil Department of the Navy* [1813 etc.]

1828. *Letters and Papers Relative to the Arming Vessels of War* [1796–1804]

1830. *Financial Reform Scrutinized*

Works

1798–1805. Portsmouth Dockyard, improvements, c. £270,000

1806. Ochta Panopticon

1811. Sheerness River Wall

1811. Chatham Sawmill (with M. I. Brunel)

1812–1813. Vauxhall Bridge, London, experimental foundations

BERRY, Henry (1719–1812), dock and canal engineer, was probably born in 1719, the son of John Berry, yeoman, of Parr, near St. Helens, Lancashire. His origins are obscure, as are his beginnings as an engineer although it is probable that he had gained experience by working under Thomas Steers (q.v.); indeed there are not many other places he could have done. The first recorded mention of him is in the Liverpool *Town Books*, 7 November 1750: 'Whereas Mr Alderman Steers is lately dead, it is ordered that Mr Henry Berey [*sic.*], lately clerk to him, be continued to oversee the works until further notice'. Further notice was duly given the following year when he was appointed successor to Steers and granted the status of Freeman.

At the time of his appointment he was heavily engaged in completing the long-planned South Dock (later known as Salthouse), opened in 1753. Soon afterwards he became involved in the proposal to construct a navigation to connect the St. Helens coalfield with the Mersey. In 1754, with William Taylor, he made the initial survey for the Sankey navigation at a fee of £66. The political heavyweights behind this scheme had interests both in Liverpool and St. Helens and were able to persuade the Dock Trustees to release Berry for two days a week. Either they or Berry contrived an ingenious deception: the Sankey Navigation is thought to be unique among British canals in that nobody petitioned against its Bill in Parliament (1755). This was probably because it was inaccurately but successfully portrayed as an improvement to an existing river navigation rather than what it really became, namely a true dead water canal. The Act permitted such cuts as were necessary and, on Berry's advice, the whole navigation was taken as a cut, 8 miles long, with of fall of 78 ft. in nine locks. Be that as it may, it opened successfully in 1757. Two branches were added by 1761, making a total length of 10 miles. The only weakness of the Sankey was its poor river access and with a later extension downstream, on the Mersey, it worked successfully for two centuries.

Work on the Sankey was followed by designs for improvements to the Weaver in January 1758. Construction problems in 1759, culminating in a foundation failure in a lock at Pickerings, led to Berry being replaced by Pownall, with John Golborne (qq.v.) acting as consultant.

In the Liverpool docks, Berry was more succesful. He far exceeded the output of his respected predecessor, with South (Salthouse) Dock, three graving docks and the George's (1771) and Kings (1788) Docks. Work had also begun on Queen's Dock when he retired in 1789. Kings Dock had settlement problems with its river wall but the others were sound, which represents a good success rate for the time. Small though they were these were the docks which saw Liverpool first become a major oceanic port.

Not until 1840, when Jesse Hartley (q.v.) was forbidden to undertake outside work, did Liverpool have sufficient work to keep a top-flight dock engineer gainfully occupied full-time and, in addition to minor works such as street improvements and bridge repairs around Liverpool, Berry was involved in various works for other port authorities; there is evidence that he had contact with Hull as early as 1756. What is certain is that in 1775 he worked for the Hull Dock Company, on the then largest dock in the country (designed by John Grundy Jr. (q.v.) and opened 1778), though it appears that he had a less-than-ideal relationship with his clients. His limited time on site meant that his overall influence must have been small and the work was largely that of Luke Holt (q.v.). He also did a small amount of work for the Lancaster Port Commissioners in the period 1776–1783. His successor in Liverpool, Thomas Morris (q.v.) seems to have taken over from him there, presumably because Berry's commitment to the completion of Kings Dock was pressing.

Almost nothing is known of Berry's personal life. His maximum salary, a modest £100 p.a., was trifling compared with Steers's income, yet he died worth £12,000 in personal property. His dissenting background placed him outside the parish record system and no records have yet been found to trace his family with any accuracy. We cannot, therefore, establish whether this substantial sum was inherited, accumulated from his own earnings (unlikely) or (perhaps more likely) from property speculations. His income in retirement is said to have been about £1,000 p.a., suggesting a significant property portfolio. His own house, on the corner of Berry Street and Duke Street—then a very fashionable part of town—was substantial. When he died, his body was carried back to St. Helens for burial in the Dissenters' Chapel.

During Berry's long lifetime, dock and canal engineering began to emerge as an identifiable specialism in Britain. Although eclipsed by such Smilesian heroes as James Brindley and John Rennie (qq.v.), he played his part in that process. He consolidated and passed on the achievements of Steers, with worthwhile contributions of his own. He more than quintupled the area of wet docks in Liverpool, and the dues revenue of the docks, which helped prevent damage to the river wall by skippers 'casting out their anchors thereon' in order to careen their vessels on the falling tide. Two of his three graving docks survive, albeit modernised-but most docks of that age have disappeared completely.

ADRIAN JARVIS

[Minutes of the Liverpool Dock Trustees, Merseyside Maritime Museum; S. A. Harris (1937) Henry Berry (1720–1812): Liverpool's Second Dock Engineer, *Trans Historic Soc Lancashire Cheshire*, **89**, 91–111; S. A. Harris (1939) Further information concerning Henry Berry, *Trans Historic Soc Lancashire Cheshire*, **90**, 197–198; T. C. Barker (1949) The Sankey Navigation, *Trans Historic Soc Lancashire Cheshire*, **100**, 121–155; A. W. Skempton (1953) Engineers of the English river navigations, *Trans Newc Soc.*, **29**, 25–54; S. A. Harris and T. C. Barker (1960). Henry Berry (1719–1812): an inventory of his professional papers, *Trans Historic Soc Lancashire Cheshire*, **112**, 57–63. (NB: only the inventory, not the papers themselves, survives); G. Jackson (1972) *Hull in the Eighteenth Century*; M. W. Baldwin (1973) The engineering history of Hull's earliest docks, *Trans Newc Soc*, **46**, 1–12; N. Ritchie-Noakes (1984) *Liverpool's Historic Waterfront*]

Works

1738–1753. Salthouse Dock, Liverpool, completion. 4.5 acres, £21,000
1755–1757. Sankey Navigation, c. £18,600
1756–1765. Liverpool Graving Docks, Nos. 1–3
1767–1771. George's Dock, Liverpool, 5.1 acres, £21,000 (estimate)
1774–1778. Hull [old] Dock, consultant, 10 acres
1774–1787. Various works for Lancaster Port Commissioners, including design for Glasson Dock
1785–1788. Kings Dock, Liverpool, 5.3 acres, £25,000

BETTS, William (1790–1867), contractor, was born at Charing, Kent, on 30 September 1790, the third of four children of William Betts and Anne Baker. His early career and training is obscure, but he is assumed to have been trained as an engineer. In 1813 he married Elizabeth Haywood Ladd (1797–1844) in Deal. Their eldest child, (Frederick) Edward Ladd Betts was born at Bucklands near Dover on 5 January 1815 and was to become an eminent railway contractor. Of their eleven known younger children William (1820–186?), the second son, also worked for a short time with his father.

At some stage in his career Betts became an important agent of Hugh McIntosh but documentary evidence is lacking, so preventing a definite date to be established. Betts was living near Dover in 1815 when work was being carried out under James Walker, and this may mark the start of their relationship. By 1820 he had moved to Weymouth where his second and third sons, William and George, were born. At that time the town bridge was under construction (1820–1824) and a 'Mr. Betts' was also harbour master. It is known that William Betts reported on the bridge in 1833 and can be assumed to have been involved in its original construction.

It is possible that Betts then worked in Sweden as in 1825 von Platen (q.v.) and his colleagues on the Gotha Canal were looking for additional staff to complete the complement of workers at the Motala foundry; on 28 September a William Betts was invited to join the staff as a founder for a salary of £200 per annum. Betts became ill and in June 1826 he returned to England, von Platen remarking he 'has been of immense use by the absolute new proceedings in this kind of trade he was introduced tho ill and sickly almost from the beginning he came into the country'.

In the late 1820s and early 1830s Betts was based near Gosport and can be assumed to have been involved with McIntosh's contracts at the Portsmouth Dockyard. Following this activity he moved on to contracts in the Southampton area. In 1831 an Act of Parliament was passed for the construction of a new wooden pier at Southampton and Betts, on behalf of Hugh McIntosh, was awarded the contract for this work, opened in July 1833. Shortly afterwards, in August 1833, a Company was formed to building a bridge across the Itchen near the site of an ancient ferry and a subsequent Act of Parliament provided for a floating bridge designed by James Meadows Rendel (q.v.). Betts, as McIntosh's local foreman, began work which was halted by the Admiralty. In July 1835 a revised Act was passed and in February 1836 the Company agreed to Betts proposals to improve the approach road. The Company was in poor financial condition with McIntosh as chief creditor and although its affairs were settled in McIntosh's favour, in 1849 it was bankrupt again, and in the refloated Company of 1851 Betts, now retired, was a director. In 1854 he visited Torpoint and Saltash ferries and recommended a new iron 'bridge', which was in use 1854–1896.

In the meantime Betts was acting for McIntosh elsewhere. He was agent for the works at Dover harbour (1833–1836) and then became involved in railway work. His son Edward Ladd Betts had been appointed agent for the Dutton viaduct on the Grand Junction Railway (1834–1837) and in 1835 father or son was obliged to withdraw a tender they had made on the Grand Junction without Hugh McIntosh's approval. In 1837 William Betts was acting for David McIntosh on the London and Southampton Railway. At that time David McIntosh obtained a large contract on the Midland Counties Railway and Betts moved with his younger children to Leicester to act as agent, describing himself as railway contractor of Southfields House at Regent Street, Leicester. Betts was held responsible when Crow Mills Viaduct collapsed on 24 November 1840.

In 1840 Betts was still tendering on behalf of David McIntosh on the Manchester-Birmingham Railway. However, following Hugh McIntosh's death in 1840 David withdrew from contracting and Betts, perhaps encouraged by Edward's success as contractor's agent, began trading as William Betts and Sons. They took contracts on the South Eastern Railway and the Chester and Holyhead Railway. The latter resulted in part from William's involvement with William Mackenzie (q.v.) and others in the attempt to revive Brymbo Ironworks and establish rail communications between Brymbo and the Dee via the North Wales Mineral Railway.

In 1844, perhaps prompted by his wife's death, Betts retired and the firm's contracts were completed by Edward. William purchased a large property at Bevois Mount, Southampton, and in July 1845 was remarried, to Elizabeth Bailey Arnett. Having sold part of the property for housing and made extensive alterations to the house, he sold it in 1854, allegedly at something of a loss as his financial affairs were in a poor state. This may have been due to the activities of his son William, with whom he broke off at some stage.

Betts retired to Sandown, Kent and died at Minster, Isle of Thanet, Kent on 14 August 1867. He left an estate of around £30,000, considerably more than his son Edward who had just been declared bankrupt.

MIKE CHRIMES

[F. Smith archive, ICE; PRO RAIL: Minutes of: Grand Junction Railway; London-Southampton (South Western) Railway; Manchester and Birmingham Railway; Midlands Counties Railway; North Wales Minerals Line; Weymouth Library; Flintshire RO; Southampton Local History Collection; *London Gazette*, 1845, 6725; 1847, 4485; B. C. Jones, Crossing the Itchen, *Southampton Papers*, 1; M. M. Chrimes (1994–1995) Hugh McIntosh (1768–1840) national contractor, *Newc Soc Trans*, **64**, 175–192]

Works

c. 1826–1831. Portsmouth Dockyard, agent
1830. Weymouth Harbour Report
1831–1833. Southampton Pier, agent
1833–1836. Dover harbour, agent
1834–1836. Southampton Floating Bridge, agent
1837–1838. Southampton Pier, agent
1837–. London–South Western Railway, agent
1837–1841. Midland Counties Railway (Rugby–Leicester), agent
1837–1841. Manchester and Salford Junction Canal, shares profits with Mackenzie
1841–. Brymbo Ironworks and Mines, shareholder
1841–. South Eastern Railway (Marsden-Ashford), contractor
1843–1844. South Eastern Railway (Maidstone Branch), contractor
1843. South Eastern Railway (Saltwood Tunnel), contractor, £18,700
1844–. North Wales Mineral Railway/Chester-Holyhead Railway contract 1, contractor

BEVAN, Benjamin (1773–1833), canal engineer, was born on 26 December 1773, the eldest of the four children of Joseph Bevan, a yeoman of Ridgmont, Bedfordshire, and his wife, Mary Ravens. On his father's death in 1782 he inherited a farm in Ridgmont and Husborne Crawley. In 1799 he married Mary Allen at Bedford and they had five children. Bevan worked from his home at Leighton Buzzard but when he died, on 2 July 1833, he was buried at Ridgmont.

In 1826, giving expert evidence on the Bill for the Norwich and Lowestoft Navigation, Bevan claimed to have been a surveyor and civil engineer for thirty years. However, the first record of his employment as a civil engineer comes from 1802 when he was requested to provide a design for rebuilding Lake Bridge at Leighton Buzzard. In early 1804 he was employed on the Grand Junction Canal Wendover branch where he had to remedy a problem of leakage of water and at Wolverton, with Henry Provis (q.v.), to deal with difficulties with the aqueduct under construction over the Ouse. Whilst at Wendover he invented a new method of closing swing bridges but it seems not to have been adopted generally. His work must have proved satisfactory for he was one of three engineers—John Woodhouse (q.v.) and Henry Provis being the others—who succeeded James Barnes when the latter retired on completion of the through route. In 1808 the aqueduct at Wolverton collapsed and, although it was in Woodhouse's area, Bevan was called upon to supervise its replacement. He suggested a trough of cast iron, then still a relatively new medium, and after inspecting Pontcysyllte, his proposal was agreed. This work was carried out under contract by William Reynolds (q.v.) and Anstice of Ketley.

His work on the Grand Junction allowed him to be involved in proposals for other canals and in 1807 he surveyed the river Ivel from Biggleswade to Shefford with a view to making it navigable. Nothing was done at the time and the work was carried out by Francis Giles (q.v.) in 1822–1823. In 1808–1809 he was involved in the complicated gestation of the final link in the direct route from the Thames to the Trent, which became the original Grand Union Canal. His initial route, which was based on earlier work by James Barnes, would have linked Foxton with Braunston, the junction of the Oxford and Grand Junction Canals. Thomas Telford (q.v.) proposed another route further east and Bevan modified his own route slightly to a junction at Norton instead of Braunston. The tunnel at Crick, which this new line entailed, proved impossible to construct when quicksand was encountered and a new tunnel on a different alignment had to be built. Less productive were his surveys in 1809–1811 for a connection between the canals at Market Harborough and Stamford. This would have given the Midlands a direct link to the east coast ports and was promoted in rivalry with a scheme from Oakham to Stamford. Bevan made a further survey in 1814, but neither scheme came to fruition. Bevan was still employed by the Grand Junction Canal and built their Northampton branch some 22 years after it had been authorised in the original Act. He was also Engineer to the Newport Pagnell Canal, effectively another branch, though an independent company.

He had already worked outside the East Midlands, reporting in 1811 on the best means of bringing the Worcester and Birmingham Canal down the Lickey Hills, where his erstwhile colleague John Woodhouse had built a canal lift. Also in 1811 he declined an invitation to become the Engineer of the Wey & Arun Canal in Sussex, but in 1812 he did submit reports on the navigation of the river (Great) Ouse and on the navigation and drainage of the river Welland, as well as reporting on a possible reservoir for the Stratford upon Avon Canal. He was one of several engineers who were consulted about possible completion of the Gloucester and Berkeley Canal, which had languished uncompleted for sixteen years. His proposal to alter the southern terminus to Sharpness was taken up by Telford, who completed the canal with funds obtained under the Poor Employment Act of 1817. Another canal which Bevan proposed to complete was the Grand Western, which John Rennie (q.v.) had built for 11 miles east from Tiverton but which had also run out of funds. Bevan proposed a narrow canal in 1818 to make the link to Taunton but it was not until 1831 that work resumed, under James Green (q.v.). He had inspected the privately owned canals and reservoirs at Arbury near Nuneaton in 1815 and 1817 and in 1819 he supported the owner against claims about excess water flowing from them into the Coventry Communication canal, surely an unusual complaint. In 1823 he followed up his work of 1812 by reporting on the navigation of the river Nene and the drainage of Deeping Fen in Lincolnshire; under the resulting Act he was responsible for the installation of two steam engines to drain 30,000 acres, one of the earlier uses of steam power for this purpose.

As the pace of canal construction slowed, and possibly through personal choice, Bevan seems to have wound down this aspect of his career. He continued to produce occasional reports on the navigation of the river Great Ouse, whose maintenance was hindered by an archaic toll income, and later he replaced some sluices on another outdated river navigation, the Warwickshire Avon, by pound locks. His client here was William James (q.v.), whose son W. H. James was articled to Bevan and himself became a civil engineer. Bevan also undertook surveys of towns and estates in the area round his base at Leighton Buzzard but the emphasis now was on mathematics, astronomy and the science of materials. He prepared articles which were published in the periodical magazines of the day: on the properties of cast iron, wood, bone, ice and glue as well as more general matters such as the weather, the sun's declination and the correct method of measuring the heights of places. He had his own experimental press made for use in his experiments and developed a rain gauge which was considerably more accurate than others then in use.

In 1823, Bevan was a member of committee which promoted the London Mechanics' Institution, and it was he who introduced Dr. Birkbeck to the chair of the public meeting which led to its establishment. Writing in support of it, he asserted that 'nothing is accomplished without labour, and nothing is denied to application and perseverance', sentiments which are more usually associated with the subsequent works of Samuel Smiles.

Bevan was a keen, self-taught amateur astronomer and died, probably of a heart attack, while watching an eclipse of the moon.

Bevan's career fell towards the end of the main canal age and his completed works are less than they might have been otherwise. His development of a system for closing swing bridges without the necessity of leaving the canal boat was considered worthy of detailed description in Rees's Cyclopaedia and he has been credited with installing the first side ponds at canal locks in Britain, at Berkhamsted, though this is dubious.

Bevan's eldest son, Benjamin Jr., followed his father as a civil engineer and surveyor and was based at Wellingborough, Northamptonshire. He succeeded his father on the Grand Union Canal and was involved with canal schemes in the east of England.

J. B. POWELL and P. S. M. CROSS-RUDKIN

[*Mech Mag*, **1**, 146, 177, 199, **9**, 140, 256, **11**, 3, **16**, 116, 191, **18**, 31, 168, **19**, 103, 256; *Monthly Mag*, **14** (1802), **22** (1806)—Farey; *Phil Mag*, **1** (1827), 14, **2**, 74, 291, **3** (1828), 29, 153; Canal, in *Rees' Cyclopaedia*, vol. 6; *Royal Society Catalogue*, **1**; Bedfordshire RO; Cambridgeshire RO; Kent AO; House of Lords RO; PRO; Warwickshire RO; W. H. Wheeler (1896) *A History of the Fens of South Lincolnshire*]

Publications

1808–9. *Plan of the Proposed Grand Union Canal*
1810. *A Map of the Intended Line of Canal from Market Harborough to Stamford*
1820. *Report Respecting the Mills on the Line of (the Leicester) Navigation*
1820. *A Practical Treatise on the Sliding Rule*, two parts (repr. 1838)
1823. *The Report on the Bridge and Causeway Near St Neots (Huntingdonshire) and on the Effect of the Same on the Height of the Floods*
1824. *Report … on the Stour Navigation and Sandwich Harbour*
1832. *A Guide to the Carpenter's Rule*

Works

1804–1817. Grand Junction Canal, engineer of Middle District (Leighton Buzzard to Hunton Bridge)
1808–1811. Ouse Aqueduct, Wolverton, designed and supervised construction of replacement for failed structure

1810–1814. Grand Union Canal, engineer, £500 p.a. plus £200 expenses; consulting engineer, 1814–1833, £150 p.a. (estimate £219,000, cost £292,000)

1815–1817. Newport Pagnell Canal, plan and estimate, 1813; constructed, but replaced by a railway in 1864

1812–1815. Grand Junction Canal Northampton branch, constructed

1823. Great Bridge, Cambridge, supervised rebuilding in cast iron for County (Arthur Browne of Norwich, architect)

1823–1825. Deeping Fen, civil engineer; installed two steam engines of 60 hp and 80 hp at Pode Hole in 1824–1825, cost c. £17,000

1824. Stour (Kent) River, Engineer to Commissioners of Sewers

1827–1828. Upper Avon Navigation, Warwickshire, substituted locks for weirs, £5,000

BIDDER, George Parker (1806–1878), was born on 13 or 14 June 1806 at Moretonhampstead near Dartmoor, the third son of William Bidder, a stonemason, and Elizabeth (*née* Parker). As an infant Bidder was sent to a village school where he learnt little, but when he was about six years of age his elder brother, a mason, introduced him to figures, which he taught him up to a hundred. By personal perseverance Bidder developed methods to manipulate these numbers mentally and subsequently was able to perform much more complex calculations with larger numbers, an ability which he retained until his death. This remarkable talent led his father to promote his prodigy around local fairs from around 1813.

Bidder's fame as the 'Calculating Boy' spread nationally and in 1816 his father took him to Cambridge where he met several mathematicians. Subsequent to this, but not necessarily a direct consequence of it, John Herschel and Rev. Thomas Jephson persuaded Bidder's parents to let him attend Camberwell Grammar School. So lucrative had Bidder's calculating ability been that after a year he was removed from the school by his father and resumed his career as a showboy. In the summer of 1819 he was in Edinburgh and was rescued once more from his peripatetic existence by (Sir) Henry Jardine, the King's Remembrancer, and other Edinburgh gentlemen who were determined he should receive an education befitting his talents; they offered his family money in compensation. Bidder was placed with a private tutor and then, in October 1820, entered the university, being taught mathematics by William Wallace (q.v.), for which subject he was awarded a prize in 1822. Bidder attended the class in natural philosophy by Leslie and that in geology by Jameson. It was in Leslie's class that Bidder met Robert Stephenson for the first time.

In May 1824, with Jardine's support, Bidder obtained a post as a trainee surveyor with Robert Dawson, of the Ordnance Survey, then based in Cardiff. In early July he met Henry Robinson Palmer and Peter William Barlow (qq.v.) and got on well. In August he was directed by Major Colby, head of the survey, to remove to London to assist in the 'Computation of the Trigonometric part of the Survey'.

Bidder evidently conveyed an aura of competence far in excess of his engineering and surveying experience as in March 1825 Palmer invited him to join him in business, Bidder beginning work the following month. Initially he worked on surveys for the Kentish Railway and the Liverpool and Birmingham line. He gave evidence in support of the Mersey & Irwell navigation in opposition to the second, successful, Liverpool–Manchester Railway Bill, and immediately distinguished himself as a Parliamentary witness, made formidable by his calculating ability. Bidder continued to work for Palmer on the early stages of extension works at London Docks when Palmer was appointed engineer in 1826 but was clearly dissatisfied with his pay. In mid-1828 he began work on a job at Dymchurch Sea Wall, remaining there until early 1829 when he began work for James Walker (q.v.).

Bidder gained a variety of civil engineering experience under several of the leading engineers until 1834 when he began working on the London–Birmingham Railway with Robert Stephenson, a position which blossomed into a lifelong friendship.

Bidder joined the Institution of Civil Engineers in 1825 and was elected President in 1860–1861. He died at his country home in Dartmouth on 28 September 1878.

MIKE CHRIMES

[ICE membership records, ICE archives; Bidder collection, Science Museum; *Min Procs ICE*, **57** (1878–1879), 294–309; E. F. Clark (1983) *George Parker Bidder*; E. F. Clark Bidder at London Docks, *Trans Newc Soc*, **55**]

BIRKINSHAW, John (fl. 1811–1844), was agent or technical manager at the Bedlington ironworks, Northumberland, in the early nineteenth century. It is possible that his family was involved with the Birkinshaw foundry near Bradford which was flourishing in the 1780s.

The Bedlington ironworks originated around 1736 and were one of the few ironworks operating in the north eastern coalfield. In 1782 the works were taken over by William Hawks and Thomas Longridge of Gateshead and production was broadened to encompass rods, hoop iron and heavy forgings. In 1809 they were bought by Gordon and Biddulph. It seems likely that Birkinshaw became involved at this time as his son was born there. Shortly afterwards Michael Longridge, Thomas's nephew, came to manage the works.

In 1819 an agreement was made with Thomas Mason of the Engine Pit, Choppington, to build an iron wagonway from the colliery to the quay at Bedlington. In return, the works would obtain a supply of cheaper coal. One assumes it was as a result of this experience that in October 1820 Birkinshaw took out his patent (No. 4503) for 'certain improvements in the manufacturing and construction of a wrought or malleable iron railroad or way'. Experimental lengths of wrought iron rail had already been laid elsewhere and Robert Stevenson (q.v.) had noted that some success had been achieved on the Tindale Fell wagonway, near Brampton. He had sent his report on this to George Stephenson (q.v.), who drew it to the attention of Longridge. Birkinshaw determined the wedge-shaped profile of the rail to ensure a similar bearing surface to a cast iron rail, and greater depth for added strength without unnecessary weight. It is possible that Stephenson laid out the wagonway using Birkinshaw's rails as he became financially involved in the colliery at the time. Certainly by the autumn of 1821 both Stevenson and William James (q.v.) had seen and been impressed by the wrought iron railway at Bedlington. Although cast iron rails were also used on the Stockton and Darlington Railway, the successful application of Birkinshaw's patent rails there, and Stephenson's support, ensured the success of wrought iron. Birkinshaw is assumed to have remained at Bedlington until the 1840s as he built Hollymount Hall near there in 1844. He subsequently moved south, selling the Hall to Longridge.

His son, John Cass (1811–1867), became involved in civil engineering at an early age and was articled to Robert Stephenson (q.v.) when the latter returned from South America, and he worked on early railways such as the Leicester and Swannington, the Canterbury and Whitstable and the London end of the London and Birmingham Railway. As engineer to the Birmingham-Derby Railway from 1837 he may have been one of the first civil engineers to specify the use of cement rather than lime mortar for arch bridges, probably following practice on the London & Birmingham. He died in poverty in 1867 after several years of illness.

MIKE CHRIMES

[Memoir of J. C. Birkinshaw (1871) *Min Procs ICE*, **31**, 202–207; R. Stevenson (1819) *Report of a Proposed Railway from the Coalfield of Mid-Lothian to the City of Edinburgh*; N. Wood (1825) *Practical Treatise on Railroads*; W. W. Tomlinson (1914) *The North Eastern Railway*, 14–16, M. J. T. Lewis (1970) *Early Wooden Railways*, 294; W. O. Skeat (1973) *George Stephenson: The Engineer and His Letters*]

BLACKETT, Thomas Oswald (1790–1847), surveyor, was the son of John Blackett, a Newcastle slater, and his wife, Judith Oswald. One of at least three children he was baptised on 31 October 1790 at St. Andrews's church, Newcastle. In 1818 he married Alice Rowell at St. Mary's church, Heworth.

He described himself as a 'land, mining and engineering surveyor' and it is likely that he received his training as a land surveyor. His earliest known association with engineering surveying was the work he carried out for William Chapman (q.v.) at Scarborough in 1821–1822.

Blackett is best known for his railway surveys, particularly on the Liverpool and Manchester Railway where George Stephenson (q.v.) engaged him following his appointment as engineer in place of William James (q.v.) in May 1824. Stephenson's plans of November 1824 were prepared by Blackett who may well have had previous experience of surveying wagonways.

Blackett was next employed on the survey of the eastern end of the Newcastle and Carlisle Railway, undertaking this work under the direction of Benjamin Thompson (q.v.) in 1828; details of it are reproduced in his *Essay on the Use of the Spirit Level* (1838). In the House of Commons Committee hearing, Blackett stated that he had undertaken the work from Haltwhistle to Newcastle and had also taken borings for the Scotswood and Warden bridges, taken down to 32 ft. Errors had been made in the deposited plans, principally the omission of details of the bridge at Scotswood, and Blackett admitted his mistake, attributing it to a clerical error.

In 1830 he was appointed as an assistant to Francis Giles (q.v.) on the Newcastle and Carlisle Railway and the following year, with Giles, he surveyed a proposed South Shields and Monkwearmouth Railway. In 1836 he was involved in surveying an alternative to the Great North of England Railway, then under consideration, and he gave evidence during the Act's passage in Parliament. Blackett's practice encompassed estate surveying as well as engineering work but his professional publications seem to have concentrated on the latter.

Blackett met an untimely end, dying from multiple injuries sustained when run down by a train on the Newcastle to Carlisle line. Having suffered the amputation of a hand he died in the Newcastle Infirmary on 19 December 1847 and was buried in the Newcastle General Cemetery, Jesmond, four days later.

MIKE CHRIMES and R. W. RENNISON

[Evidence given regarding the Great North of England Railway (1836/7) (DCRO: D/HH. 2/16/48 and 49); *Newcastle and Carlisle Railway: House of Commons Committee* (1829); *Newcastle Journal*, 24 December 1847; W. W. Tomlinson (1914) *The North Eastern Railway*; J. S. Maclean (1948) *The Newcastle and Carlisle Railway*; R. H. G. Thomas (1980) *The Liverpool and Manchester Railway*; Skempton; Bendall]

Publications
1821. *Plan of the Harbour With Part of the Town of Scarborough*
1824. *Plan and Section of an Intended Railway or Tramroad from Liverpool to Manchester*
1828. *Plan and Section of an Intended Railway or Tramroad from … Newcastle … to Carlisle …*

Surveyed Under the Direction of Benjamin Thompson by T. O. Blackett and I. Studholme
1831. *South Shields to Monkwearmouth Railway, F. Giles and T. O. Blackett, Engineers*
1838. *An Essay on the Use of the Spirit Level*
1847. *A Treatise for the More Readily Finding By Inspection the Cubical Contents of Earthwork*

BLACKMORE, John (1801/2–1844), civil engineer, spent much of his career on railways in north east England and was in part responsible for the complete length of railway between the Irish Sea and the river Tyne. Blackmore, his origins not ascertained, was noted in 1828 as being chief assistant to Francis Giles (q.v.) and it is possible that he carried out surveying work for the extension of the Ivel Navigation at Shefford and perhaps surveys for roads and bridges in Bedfordshire.

Giles was appointed as engineer to the Newcastle and Carlisle Railway on its formation in 1830 and Blackmore continued to serve him following his new appointment; it was under his day-to-day control that construction of the line took place. In 1833 the company dispensed with the services of Giles—who had been responsible for the design of the principal works on the western section of the line—and Blackmore was appointed as 'operative engineer' acting under a committee of management comprising Nicholas Wood, George Johnson, and Benjamin Thompson (qq.v.). In 1835, while construction work was in progress, he became a member of the Institution of Civil Engineers, his main proposer being John Buddle (q.v.).

Under the direction of Blackmore, the 60 mile long line was completed in sections and opened throughout in 1838, the major works for which he was responsible being the crossings of the river Tyne at Scotswood and at Warden. Both of the bridges were of timber construction, at Scotswood comprising eleven spans of 60 ft. and at Warden five spans of 50 ft. In 1838 Blackmore submitted a report on a 96 mile long extension of the N&CR into Scotland by means of a branch northwards from Hexham, an idea which found little support; it was commented that it was in fact almost the same proposal as that put forward by Joshua Richardson two years before. His proposal was later revived in a slightly different form by Wood and Johnson but, again, no progress was made.

The Blaydon, Gateshead and Hebburn Railway was established in 1834, leaving the Newcastle and Carlisle Railway at Blaydon and running eastwards along the river Tyne on its south bank in order to provide an outlet for coal mined in the Tanfield area. Blackmore became engineer to the company but, as a result of events involving the Newcastle and Carlisle Railway and the Brandling Junction Railway, only two miles of the 11 miles sanctioned came to be built by the Blaydon, Gateshead and Hebburn Railway; it was completed in 1836.

The completion of these railways led to Blackmore becoming involved in other work and in 1838 he reported on the design of a new quay wall in Newcastle, designed first by John Dobson and amended by William Cubitt (q.v.). He also designed a bridge over the river Tweed at Norham; completed in 1841 it comprised a laminated timber two-span arch structure with spans of 190 ft. and a piled central pier. In 1839 he was involved in plans for the provision of a new supply of water to the town of Newcastle by the Newcastle and Gateshead Union Joint Stock Water Company. The Subscription Water Company had, until then, provided a supply from the river Tyne at Elswick but it had been the cause of much complaint. Blackmore suggested that water be taken from the river Pont and from Prestwick Carr, eight miles from the town, but work did not proceed in spite of the fact that an Act of Parliament was obtained.

In 1840 Blackmore carried out a valuation of the railway from Rainton Bridge to Seaham Harbour, the port completed by the Marquess of Londonderry in 1832, and also in 1840 he was appointed as engineer to the Maryport and Carlisle Railway, originally set out by George Stephenson (q.v.); as engineer, he reported progress half-yearly to the directors and shareholders only until 1842 when he left the company. The line was opened over its full length in 1845.

At this period of his career he maintained an office in the Arcade, Newcastle, while living in the west end of the town, in Greenfield Place. Aged 42, he died there on 15 March 1844 'having scalded himself in a steam bath the week before' and was buried in Jesmond Old Cemetery. He left some £5,000 to his widow, Ann, nee Nixon, and two young daughters.

R. W. RENNISON

[*Newcastle and Carlisle Railway: Minute Book* (1829–1844) PRO: RAIL 509; *Senhouse Collection* (1839–1844) Cumbria RO: D/Sen.M&CR.19; *Forster Collection*, Northumberland RO: FOR/3/42; *Collections Relating to the River Tyne*, 4: Newcastle Central Library, L942.8; Hann, Hughes and Hosking (1839) *Bridges in Theory, Practice and Architecture*; W. W. Tomlinson (1914), *North Eastern Railway*; R. W. Rennison (1979) *Water to Tyneside*; Skempton]

Publications

1833. *Plan and Sections of an Intended Railway or Tramroad from Blaydon in the County of Durham to Gateshead and Hebburn and Ending at the River Tyne at Hebburn Quay*
1834. *Railway from Blaydon to Newcastle Spittal*
1836. *Views on the Newcastle and Carlisle Railway, From Original Drawings* (with J. W. Carmichael)
1838. *Great North Junction Railway, or Inland Line of Railway from Newcastle-upon-Tyne to Edinburgh.* [Report] *To the Directors of the Newcastle-upon-Tyne and Carlisle Railway*

1839. *Newcastle and Gateshead Union Joint Stock Water Company*

Works

1830–1833. Newcastle and Carlisle Railway, 63 miles long, assistant engineer
1833–1844. Newcastle and Carlisle Railway, engineer
1834–1836. Blaydon, Gateshead and Hebburn Railway, engineer, 2 miles constructed of 11 miles planned
1839–1841. Ladykirk and Norham Bridge, two spans of 190 ft., cost *c*. £7,000
1840–1842. Maryport and Carlisle Railway, 28 miles long, engineer

BLACKWELL, John (*c*. 1775–1840), civil engineer, was for many years resident engineer on the Kennet and Avon Canal between Bath and Newbury. Work had begun on the construction of the canal, designed by John Rennie (q.v.), in 1794. At 57 miles it was one of the longest canals in the country and inevitably required a large amount of capital. This reason alone meant that construction could only proceed in sections with partial opening. Blackwell joined the canal staff in about 1806 and was probably employed in supervising the construction of the flight of 29 locks at Devizes, completed in 1810, which he regarded as his greatest achievement; two years later he was appointed resident engineer.

The canal as a whole was opened on 28 December 1810 and from 1814 until the early 1830s enjoyed increasing prosperity, with the proprietors expanding their interests in neighbouring navigations on the Kennet, Avon, and elsewhere and also introducing a fast boat service which covered the distance between London and Bristol in five days. Blackwell's role was therefore one of maintenance and incremental improvements such as the introduction of gas lighting on the Devizes locks in 1829 and the reconstruction of turf-sided and timber-framed locks on the Kennet from 1833, as well as an additional low lock at Ufton.

As early as 1812 Blackwell surveyed a feeder railway from Coalpit Heath collieries to the Avon. The estimated cost of £20,000 proved prohibitive, although a similar scheme, the Avon and Gloucester Railway, was carried out in 1828 with H. Cotterell as engineer. In 1824 Blackwell visited the North-East to report on railways and locomotives on behalf of the Kennet and Avon, who were concerned at possible competition from a Bristol-Bath railway proposal.

Blackwell died in Hungerford, where he had lived for many years, on 28 September 1840; he had been predeceased by his wife, Fanny, by six months. A memorial to him, his wife, and his son, Thomas Edward (*c*. 1820–1863), who succeeded him on the canal before going to work in Canada, remains in Hungerford churchyard. Blackwell was elected a Member of the Institution of Civil Engineers in 1833.

MIKE CHRIMES

[ICE membership records, ICE archives; K. R. Clew (1968) *The Kennet and Avon Canal*; J. Russell (1996) in lit.]

Works

1807–1810. Devizes locks, Kennet and Avon Canal, resident engineer
1810. Oakhill locks (two), Kennet and Avon Canal, engineer
1812. Horse towpath, Avon navigation, engineer
1830–1835. Lock improvements, Kennet navigation, engineer

BLANE, George Rodney, Captain (1791–1821), Bengal Engineers, Superintendent of Canals, Delhi Territory, was born on 7 January 1791, the third son of Sir Gilbert Blane, Physician to King George III, and Elizabeth, his wife.

After education at Charterhouse, 1802–1806, he entered the East India Company's service as a cadet, joining the Bengal Engineers in 1808, attaining the rank of Captain in 1818.

Blane received some early experience as an assistant surveyor under Captain Frederick Sackville in Cuttack. Between 1811 and 1813 he carried out survey work at Sagar Island near the mouth of the Hugli at Calcutta, extending this survey the following year to the east near Diamond Harbour. Late in 1814 he was appointed to the station at Ludhiana in the Punjab, but was called to the Nepal War of 1814–1815, serving with the 2nd division Dehra; he was wounded at Kalanga.

On his return from the war he was based at Ludhiana and became interested in the restoration of the old Mughal canal on the west bank of the Jumna (Yamuna) (later to become part of the Western Jumna system) which passed through Karnal to Delhi. This was a 185 mile long irrigation canal designed by Ali Mardan Khan in the early seventeenth century, which had fallen into disuse since *c*. 1750.

Working from the report of an earlier survey by Lieutenant John Macartney (1781–1811) made in 1809 and 1810, Blane's proposal for the restoration led to his appointment to this task in 1817. He successfully established a head for the canal and was able to bring water to Delhi in 1820. His achievement was rewarded by his appointment as the first Superintendent of Canals, Delhi Territory in 1820, an post he enjoyed only too briefly as he died of malaria at Ludhiana on 18 May 1821.

To Blane must be given the credit for the first success in canal engineering by the British in India in this period in which there was little knowledge and certainly no training in the hydraulics of flow in open channels, which was not to come until very much later in the century. He was succeeded by (Lieutenant General) Richard Tickell (q.v.)

JOYCE BROWN

[W Baker (1849) *Memoranda on the Western Jumna Canals, in the North Western Provinces of the Bengal Presidency*; E. W. C. Sandes (1935) *The*

Military Engineer in India, II, 2–4; E. W. C. Sandes (1948) *The Indian Sappers and Miners*, 55–56; J. Brown (1978) Sir Proby Cautley (1802–1871), a pioneer of Indian irrigation, *History of Technology*, **3**, 44–45]

Works

1817–1821. Western Jumna Canal restoration (Delhi Canal), 185 miles

BLENKINSOP, John (1783–1831), coal viewer, was born near Leeds in 1783. He was a cousin of Thomas Barnes (q.v.) and was initially apprenticed to Barnes before becoming, from about 1808, principal agent for John Charles Brandling at his Middleton colliery near Leeds. In 1811 he patented (No. 3431) a rack railway system as a 'tramway for conveying coals', intended to be powered by a steam locomotive. It was agreed to utilise the system on a line from Middleton to Leeds. The details of the locomotive were worked out by Matthew Murray (q.v.) at the works of Fenton Murray and Wood, and around 20 June 1812 the first journey was made. In March 1814 Blenkinsop wrote to *Monthly Magazine* describing the success of his system so far, referring to the use of it at Orrell colliery, Wigan, and at the Kenton and Coxlodge collieries near Newcastle-upon-Tyne, as well as Middleton. From correspondence with John Watson, the coal viewer at Kenton and Coxlodge, it appears that Blenkinsop had prepared figures demonstrating the favourable performance of his locomotive over horse traction in August 1812. Whilst the Newcastle locomotives were built again by Murray, those at Orrell were apparently all built by Robert Daglish (q.v.) of the Haigh Foundry. The design was also copied by the Royal Iron Foundry at Berlin in 1816 following the visit of Frederick Krigar and Eckardt in 1815. It would appear that a boiler explosion in 1818 caused a temporary suspension of the locomotive service but it was running again in 1834.

Blenkinsop died in Leeds on 22 January 1831 at the age of 47 and is buried at the parish church of Rothwell.

MIKE CHRIMES

[Andrieux (1815) *Bulletin Societe d'Encouragement de l'Industrie Nationale*; Mr Blenkinsop's patent steam carriage and railway, *Repertory of Arts & Manufactures* (1818), 2, 33, 19; L. G. G. Gallois (1818) Des chemins de fer en Angleterre, *Annales Des Mines*, III, 129–144; T. Gray (1821, etc.) *Observations on a General Iron Railway*; N. Wood (1825) *Treatise on Railways*; Anon. (1825) Locomotive engine upon the cogwheel principle, *London mechanics register*, 1, 225–230; Anon. (1910) Links in the history of the locomotive, *Engineer*, 109, 29 April, 432–433; C. F. D. Marshall (1953) *A history of railways down to the end of 1831*, 28–54; M. J. T. Lewis (1970) *Early wooden railways*, 115, 131, 296–297; C. E. Lee (ed) (1971) *Railways in England, 1826–1827*, 20; DNB]

Publications

1814. Steam carriage, *Monthly Magazine*, **37**, 394–395

Works

1812. Middleton (rack) railway

BONOMI, Ignatius, RNF (1787–1870), an architect, was the son of Giuseppe Bonomi—born in Rome in 1739—and his wife Rosa Florini; they married in 1775. Bonomi was the fourth child, and the eldest son, of a family of ten and was born in London in 1787.

Bonomi's training was with his father, also an architect, and through the latter's involvement at Lambton Hall, he came to work in the north east of England, where he built up an extensive practice, mainly in County Durham. Colvin lists 30 churches, both Anglican and Catholic, designed or restored by him in addition to his many secular commissions.

In addition to his architectural work, Bonomi was appointed as Durham County Bridge surveyor in 1813 and, between this time and 1827, he was responsible for building several masonry bridges, the first at Frosterley, a three-span structure. At Shincliffe his bridge comprised two spans with semi-elliptical arches of 60 ft. and others were built at Waskerley, Dipton, Fernhall and Chester-le-Street. One of his works, a private commission, was the Lamb Bridge at Lambton castle, a masonry structure with a single span of 82 ft.; it was built in 1819. He remained bridge surveyor until 1850.

Bonomi was also the designer of the country's first major bridge for steam railway locomotives, the Skerne bridge on the Stockton and Darlington Railway. There were two significant bridges on the line, the first over the river Gaunless, near West Auckland, and the second over the river Skerne, at Darlington. For the former, an iron bridge was designed under the direction of George Stephenson (q.v.) whereas the latter became the responsibility of Bonomi. Stephenson was advised by the Stockton and Darlington Railway directors that he should discuss the construction of the bridge—originally intended to be of iron also—with Bonomi and in June 1824 it was resolved that the company's secretary 'do write to Mr. Bonomi and request his professional assistance'. Bonomi made alterations to Stephenson's design and they were approved on 2 July 1824, the foundation stone being laid four days later, with completion the following year.

Bonomi entered into a partnership with John Augustus Cory in 1824 and in 1855–1856 he left Durham to live at Wimbledon, to be near his younger brother Joseph, to whom he had given financial assistance. The Durham practice continued after his retirement; when in Durham, Bonomi had maintained an office in Old Elvet and later, with Cory, in North Bailey. He lived in both North Bailey and at Elvet Hill.

In 1837 Bonomi had married Charlotte Ann Fielding (1800–1860), the daughter of Rev. Israel Fielding of Barnard Castle and, as neither wished to convert, two marriage ceremonies took place, the first Anglican at Lanchester and the second Roman Catholic, at St. Cuthbert's church, Durham, designed by him ten years earlier.

Bonomi—of whom a portrait is extant—died at his home, The Camels, in Wimbledon in January 1870, his wife, Charlotte, having predeceased him. He was said to have been hard-working, kindly and generous, especially so far as his family was concerned. He died childless.

R. W. RENNISON

[*Local Biography*, 47, 121 (Newcastle Central Library); *Penny Magazine* (1843), 501; W. W. Tomlinson (1914) *North Eastern Railway*, **94**, 5; Colvin (1978), 122–123; J. H. Crosbie (1987) *Ignatius Bonomi of Durham*; Information from Environment Department, Durham County Council; N. Pevsner (ed.) *Buildings of England*]

Works
1813. Frosterley Bridge, spans 52, 63 and 52 ft.
1817. Waskerley Bridge, 55 ft. span
1819. Floaters Mill Bridge, Chester-le-Street, 17 ft. span
1819. Lamb Bridge, 82 ft. span
1820. Dipton Bridge
1824–1825. Skerne Bridge, Darlington, spans 8, 40 and 8 ft.
1824–1825. Shincliffe Bridge, two spans each 60 ft.
1827. Fernhall Bridge, 25 ft. span

BOORER, John (fl. 1820–1852), architect, obtained second prize in the London Bridge competition of 1823 and the following year, then of 3 Western Street, Pentonville, was nominated as Associate membership of the Institution of Civil Engineers, remaining a member until 1851 when he 'retired'. At that time he was resident at Stowe House, Buckingham.

MIKE CHRIMES

[ICE membership records; Colvin (2)]

BOROUGH, William (c. 1537–1598), mariner and consultant on Dover harbour, was born at Northam, Devon, the younger son of John A'Borough, mariner, who was consulted on Dover harbour in 1541 with Richard Cavendish (q.v.). As a boy of sixteen, William Borough sailed with his elder brother, Stephen, on Richard Chancellor's ship on the expedition of 1553 in search of the North-East Passage. During the next twenty-five years he made many voyages of discovery and prepared charts, including one in 1576 covering the north Atlantic for use on Frobisher's expedition. By 1579 he was residing at Limehouse, already in the service of the Crown, and in 1581 published an important work on the compass (reprinted several times).

Meanwhile he had been commissioned by the Privy Council to report on Winchelsea, Rye and Dover harbours in 1576 and three years later he was brought in to advise a commission appointed to consider what could be done to improve the harbour at Dover. As a result a plan emerged for an entirely new scheme which formed the basis for the most successful civil engineering work of Elizabeth's reign.

Cessation of work in 1551 on the King's Pier, south of the old harbour, marked the end of a long-sustained effort started in 1535 by John Thompson (q.v.) to provide protection from south-westerly storms and from shingle driven by littoral drift. In the latter function it failed. Shingle banked up against the pier and moved around its head, forming an off-shore bar which extended more than half a mile north to the shore beneath Dover castle. Behind the bar the small river Dour flowed through a tidal lagoon to the harbour mouth but the combined energy of the river and the ebb tide was insufficient to maintain an adequate depth of water for ships. Moreover, by the 1570s the pier, built of timber framework infilled with large stones, was in ruins; only its foundation of massive rocks, known as the 'Mole', remained intact.

Dover was not the only harbour on this part of the coast to be in decline. In 1576 Francis Walsingham (Principal Secretary of State) appointed a four-man commission under Borough to report to the Privy Council on Winchelsea, Rye and Dover. Their report, in May, recommended Dover as the most promising for improvement and suggested a scheme by which improvement could be achieved. This brilliant concept was to build a long wall on the shingle bar and a cross wall back to the shore forming a basin in the northern part of the lagoon, later known as the Pent. The basin would fill at high tide and the water thus stored could be released at low tide through a sluice in the cross wall to scour the new harbour south of the Pent.

There can be no doubt that the proposal is due to Borough. An estimate, probably prepared by Matthew Baker, royal master shipwright and the other technical man on the commission, amounted to £30,000. This was far too high a cost for the Council to approve so nothing was done. But in February 1579 a violent storm destroyed what was left of the King's Pier. Action could no longer be delayed. The Council established a commission under Lord Cobham (Warden of the Cinque Ports). They called in Matthew Rickworth, sluice master of Dunkirk, who recommended two new groynes as a first-aid measure and Flemish workmen were procured to execute this work. In August the commission turned to the wider issue of a reformed harbour, bringing to their conference William Borough and Peter Pett, master shipwright of Deptford; Rickworth would have been present. A 'platt' or plan 'very handsomely set out' emerged and was sent to the Council, to be explained by Borough. The 1579 platt appears to have been lost but is almost certainly represented in a drawing made for Lord

Cobham *c.* 1581 and now in the collection of maps at Hatfield House, tentatively but erroneously attributed to Thomas Digges. It shows two jetties at a new harbour mouth; from the inner end of the north jetty a wall runs north for about 3000 ft. on the bar to the shore just east of the Dour outfall into the lagoon; another wall extends some 400 ft. out from the shore, at a point about two-thirds of the way down the lagoon, and then runs south-east to the jetty; in the 400 ft. portion there is a lifting-gate sluice and nearby a pair of mitre gates pointing upstream to allow the passage of (small) vessels into the Pent at the slackwater period of high tide. The sluice is shown discharging water at low tide. The cross wall is timber-framed. Its extension down to the jetty was presumably intended to form a north wall of the harbour. No details of the long wall are shown. The estimated cost was now (1579) about £21,000 though the Flemings seem to have been willing to undertake the scheme for £18,200.

To the Council these figures still appeared to be excessive, so for the time being the scheme lay in abeyance. The next move came from John Trew (q.v.) who proposed to rebuild the King's Pier on its existing foundations at a cost between £8,000 and £10,000. The pier would have two masonry walls with a chalk infill, thus providing a more solid and permanent structure than the original one.

In August 1580 the Council, 'having of late persued two several plats drawn for the repairing of Dover haven, the one by William Borough with the advice of some skilful persons of Dunkirk, the other by [John] Trew [for] building the pier unto the molehead', recommended the latter to the commissioners. But with the proviso that a 33 ft. trial section should first be made to assess the unit cost more accurately, a wise decision since the whole pier would be over 1000 ft. in length.

By December 1580 Trew, who received the high salary of £3 10s a week, had forty hewers (at 6s a week) and sixty labourers (at 4s a week) working in Folkestone quarries. At the same time he required two boats of 40 tons burden to transport the stone and ordered a hundred labourers to begin digging chalk at Dover.

Funding for the harbour works was secured early in 1581 by an Act laying tax on vessels over 20 tons entering English ports during a seven year period. But all was not well. Trew's work showed signs of costing more than the estimates and, more fundamentally, Borough expressed serious doubts on the likelihood of its success. This he did in an 8-page folio report entitled *Notes on Dover Haven*. He discusses the phenomenon of littoral drift and describes the nature of beach growth at Rye. For the better understanding of his discourse, which 'may seem to be intricate and obscure', he refers the reader to 'platts that I have made for that purpose'. One of these is his map of the south-east coast with detailed notes on the

tides. Proceeding from the general to the particular, he examines the history of Dover harbour showing, contrary to the opinion held in certain quarters, that destruction of the pier merely altered the off-shore bar and was not the cause of its formation. Therefore rebuilding the pier would in this respect be of little or no benefit.

Shortly afterwards the same point was made by Thomas Digges (q.v.) the famous mathematical practitioner whose name first appears in the Dover records in July 1581. The Council now had to admit that Trew's plan 'is not likely to take that good effect we looked for' and informed the commissioners that he would have to be discharged.

Attention then returned to Borough's scheme. The drawing made for Lord Cobham may well date from this time. Before the end of the year an accurate survey was made by Digges of the existing harbour and the off-shore bar. In October 1581 Fernando Poyntz, an engineer–entrepreneur, who had recently completed repairs to the Thames bank of Plumstead Marshes, offered to undertake the scheme. Nothing came of this although he was later employed on several relatively minor works at Dover.

In December preparations were under way to obtain 1000 tons of timber and an 8 cwt iron ram for the pile driver, clearly for the timber-framed walls. Early in 1582 Trew was paid off, after spending about £1,300, and Poyntz came up with a revised estimate of £14,580 having decided to use a less costly form of construction for the wall: banks made of shingle mixed with partially dried mud. The Council sent Borough and Digges to Dover for a conference. At about that time Digges presented to the Queen his 'Brieff Discourse … [on] the making of Dover Haven' with a new scheme estimated at £13,365. This has the sluice and lock gate close to the inner end of the jetties, so producing a much larger basin which becomes the harbour (or, strictly, a dock) to be entered at high tide. His long wall was to be a bank made of shingle and chalk with a central core of mud.

In March new commissioners were appointed, again under Lord Cobham but now including Digges and Sir Thomas Scott (q.v.). On 11 April 1582 they wrote to the Council saying that 'by the good advice of Sir William Wynter, Mr Digges and Mr Borrowe we have finally decided on such a plat as we judge most convenient'. Wynter at that time was Surveyor and Master of Ordnance of the Navy, brought in, like Borough, as a consultant.

The revised plan was a sensible simplification of Borough's 1579 scheme: in the first stage to construct the Pent with a wall 2000 ft. long on the off-shore bar and a cross wall running about 600 ft. back to the shore with a sluice (but no mitre gates) near its landward end, and later to extend the long wall to the harbour mouth and to build two jetties at the new entrance, leaving the mole (or the outer part of it) as a breakwater.

Many details had yet to be decided but after April 1582 progress was rapid. Digges became

General Surveyor, Scott played a very active role and Borough continued as an independent consultant until 1584. Work on the first stage envisaged in the 1582 plan was completed in 1586 at a cost of £12,000, adopting for the walls the economical and well-tried method used for building and repairing the sea wall of Romney Marsh. The long wall was extended to the harbour mouth in 1592–1593 and entrance jetties were built in 1594–1595.

In 1585 Borough was appointed Master of Trinity House. He took part in the 1587 expedition to Cadiz under Francis Drake, with whom he had a furious dispute, and commanded a ship in the Armada. At that time (1588) he became Comptroller of the Navy. He remained active in his official duties until a year before he died. His will, dated July 1598, was proved in December of that year.

A. W. SKEMPTON

[T. W. Wrighte (1794) Transcript of Thomas Digges 'A briefe discourse ... [on] the making of Dover Haven', *Archaeologia*, **11**, 212–254; S. Robertson (1876) Medieval Folkestone, *Archaeologia Cantiana*, **10**, 113–119; DNB; A. H. W. Robinson (1962) *Marine Cartography in Britain*; R. A. Skelton (1971) *A Description of Maps ... in the Collection made by [Lord] Burghley now at Hatfield House*; John Summerson (1982) Dover harbour, *King's Works*, **4**, 755–764; Stephen Johnson (1994) Mathematical practice ... in Elizabeth England (unpublished Ph.D. Thesis, Cambridge University, contains important new material on Dover harbour)]

Works

1576–1582. Dover harbour, planning

BOUGH, family of contractors (fl. 1770–1830). **James Bough** (fl. 1770–d. 1796), stonemason and superintendent of works, came from the West Midlands and became involved in the first phase of canal construction in the late 1760s. In March 1771 he was asked to supply stone for locks on the Birmingham Canal beyond Wolverhampton. Work on this canal was largely complete in 1772, and a namesake is next found carrying out masonry work on the locks and bridges of the Chesterfield Canal from 'Worksop River to Retford River' in 1774. It seems probable that Bough found work on other canals before being recommended to the Stroudwater navigation Company in 1776.

The Stroudwater linked Stroud to the Severn Estuary via Framilode, and had its origins in a scheme designed by John Hore (q.v.) as early as 1729. The idea was revisited several times over the next 40 years, and engineers such as Thomas Yeoman, Thomas Dadford Jr. and John Priddey (qq.v.), became involved. The latter two had both worked in the West Midlands and may have recommended Bough. Priddey succeeded Samuel Jones (q.v.) as resident engineer in February 1775, and was still there in April 1776 when

Bough was contracted to bring four or six hands to assist with the masonry. In July he undertook all the stonework on the navigation, at prices 20% below those of his competitors. In the face of the general inexperience of the Committee Bough's role was rather more than that of mason; he advised on sources of stone, Aberthaw pebbles for cement, and clay for bricks, and assisted Edward Lingard, the resident engineer, on survey work. So regularly was his assistance required that in November 1777 he was allowed 5s a day when Company business kept him away from his contracts. He built bridges at Bristol Road, Stonehouse farm, Bridgend Road, Reddall's swing bridge, Hayward's field, and Wallbridge, culverts at Ryeford, Lodgemore and Wallbridge, locks at Bristol Road, Westfield, Count Orchard, Eastington, Lower and Upper Nassfield, as well as supervision elsewhere.

Work on the Stroudwater was largely complete in 1779, and by then Bough had resumed work on the Birmingham Canal, replacing a swivel bridge at Mr. Carver's farm in September 1778 with a brick arch. In May 1780 he built a bridge at Randall's Lane Tipton, and henceforth regularly appears in the Company Minutes, working with Samuel Bull (q.v.), and repairing and building bridges. In August 1783 he was in attendance with John Pinkerton (q.v.) to tender for the new work on the Fazeley extension however shortly afterwards, in January 1784 he became a full time employee of the Birmingham Canal Company at £120 p.a. His inexperience of surveying and possibly his hard drinking may have led to an acrimonious dispute with Pinkertons (q.v.), the contractors. In 1788 he and Bull condemned their work at the Dunton Tunnel. Bough continued to work for the Company as surveyor of works until late in 1796 when he was replaced by Robert Hood and died shortly afterwards.

It seems likely that James was the father of **Thomas** (d. 1802) and **William** (c. 1770–1813) **Bough**, best known for their work in London's docks. A Thomas Bough was involved on the Stroudwater. William Buff [sic] was variously agent for John Holmes on the Kennet and Avon Canal at its western end and a canal cutter both there and on the Somersetshire Coal Canal. It may be of relevance that Holmes had previously worked on the Royal Canal in Ireland and it would certainly appear John Rennie (q.v.), the engineer for these works, must have been familiar with his work throughout.

William Bough of Church Row, Limehouse, and Thomas (d. 1802) began work on West India Docks in 1800. Thomas died in an accident there on 22 July 1802. They must have been reasonably successful as William Bough was given charge of the construction of the 'chalk ridge communication' between Guildford and Rochester, a military road following the Pilgrim's Way built in two months in late 1804, and probably achieved by removing the top soil to expose a harder wearing surface. This defence work was occasioned by the Napoleonic threat, and was followed by a contract with Dyson and Hollinsworth (qq.v.) on the Royal Military canal; it was taken out of their hands within six months.

On other contracts, which included work at Dartmoor Prison, they had as their partner John Hughes (q.v.). The dredging of Greenland Dock, however, begun in 1807, was completed by Bough alone. In addition to their early work at West India Docks, Bough laid out railways for Butterley & Co. at London Dock (1801–1802). In a later contract (1809) at the Limehouse basin of West India Docks the wing walls had to be reconstructed by Jolliffe and Banks as Bough had not driven the sheet piles deeply enough and water burst through in 1810. In 1811 he corresponded with Rennie regarding Highgate Archway and supplied labour for the work. In 1811–1813 Bough worked on the excavation of the Tobacco Dock at London Docks and followed this by excavation work at the Hermitage entrance. Bough's final contract, completed by his successors, Bough & Co., was for the excavation of a new basin at East India Docks (1815–1816). William Bough died on 23 May 1815 at the age of forty-five. Another **William Bough**, possibly his son, civil engineer of 15 Dalby Terrace, City Road, London, unsuccessfully tendered with his partner Smith for Marlow suspension bridge in 1829. He died on 6 January 1831, leaving a widow, Ann, and a daughter, Anne, who married William Pierson Gordon, banker, of Bridgnorth.

MIKE CHRIMES

[Probate records, 1874; Q/AB/43/767, Bucks RO; Butterley archive, Derbyshire CRO; PLA archive, Museum of London in Docklands; Kennet and Avon Canal, Western Committee Minutes, PRO: RAIL 842/2; Birmingham Canal Co. Minutes, PRO: RAIL 810/2–7; *Monthly Magazine* (1802), 14, 181; *London Chronicle*, 29 May 1815; *Architect, Engineer and Surveyor* 4 (1843), 1, 6; K. R. Clew (1970) *The Somersetshire Coal Canal and Railways*, 33; A. W. Skempton (1982) Engineering in the Port of London, *Trans Newc Soc*, **53**, 75–79; C. Richardson (1996) Minutes of the Chesterfield Canal Company 1771–80, *Derbyshire Record Society*, **xxiv**, 83]

Works
By James Bough:

–1772. Birmingham Canal, masonry work on locks, etc.
1774. Chesterfield Canal, masonry work on locks, bridges, etc.
1776–1778. Stroudwater Canal, masonry work, seven bridges, six locks, *c*. £5,000
1778–1783. Birmingham Canal, masonry work
1784–1796. Birmingham canal, Birmingham Fazeley extension, surveyor of works, etc.

By William Bough:

1795. Kennet & Avon Canal, canal cutter

1795–1798. Somersetshire Coal Canal, contractor, £4,900
1800–1802. West India dock extension, contractor
1801–1802. London dock railways, contractor
1804. Chalk Ridge Road, contractor
1804. Dartmoor Prison, contractor
1807. Greenland dock dredging works
1809. Limehouse basin, West India docks, contractor
1811–1813. Tobacco dock excavation, London docks, contractor
1815–(1816). East India dock basin excavation, contractor

BOWER, Anthony (fl. 1781–1821), of Lincoln, surveyor and resident engineer. He surveyed Holderness Drainage in 1781 and eleven years later returned there for his first known employment as Resident Engineer, for William Jessop (q.v.). It was while in this area that he married Mary Tayler, at Hull, on 6 May 1784. With James Murray he was employed by John Rennie (qq.v.) in 1800 to survey the East, West and Wildmore Fens near Boston in Lincolnshire. As a result a comprehensive scheme to drain 59,196 acres at an estimated cost of £188,552 was authorised by an Act of 1801, amended by another in 1803. Bower was appointed Resident Engineer and supervised the construction by contractors, of whom John Pinkerton (q.v.) had the largest part of the work. Bower appears to have been very competent technically but to have been less fortunate in his dealings with the other people involved. As was often the case with a scheme of fen improvement, there was the potential of violent opposition, and Bower was warned not to go out at night without a guard. On one such occasion he was shot at but, he noted laconically, was missed. Pinkerton did not proceed with the works as fast as intended, with some justification as ground conditions were worse than expected, but when the Commissioners appeared inclined to admit a claim for this, the amount increased from £4,300 to £7,200 and sympathy disappeared. But worst of all was the open warfare which existed between Bower and some of the Commissioners. Civil engineering was still a young profession and many shareholders in construction schemes considered themselves equally capable as their Engineer. Bower was particularly troubled by a Mr. Whitelock. Sir Joseph Banks reported to Rennie in 1804 that 'poor Bower is like a tennis ball between Special and General Committees ... . I fear he is too pert in his criticisms of the projects of Mr. Whitelock who cannot be suffered to understand the extensive business of our drainage which Bower is so perfectly master of'. Charles Wedge (q.v.) four years later noted the deep rooted enmity between Bower and Whitelock but also that 'Bower does not seem on good terms with anybody'. Later that year though Bower had the satisfaction of reporting to Rennie 'We went through the East Fen with a post chaise—the first that ever passed through since the world began'.

Bower continued to undertake surveys for Rennie. In 1806 he submitted a report and estimate on the River Witham; after a further report by Rennie raised the amount required from Bower's £92,736 to £106,720, an Act was obtained in 1808. Money was not forthcoming but an Act of 1812 increased the funds required still further to £120,000 and it was only in 1813–1816 that the works were carried out to Rennie's revised plan. Bower reported on Minsmere Level in Suffolk in 1808–1809 and gave evidence to Parliament the next year during the passage of the Act for its drainage. Hew was appointed Engineer but failed to provide the drainage required for the works, which he had submitted to Rennie for approval but been unable to obtain due to the latter's absence in Ireland. He also failed to attend meetings of the Commissioners because he was required to attend Assizes at Lincoln and Ely and Sessions at Meaford, and was also employed at Downham, Norfolk, in some capacity on the Bedford Level. There is also a hint in extant correspondence of strong antipathy from one of the Commissioners, and the almost inevitable outcome was his dismissal in March 1811. He had to wait three years for his bill of £363 to be paid. In 1810 Bower surveyed the sea banks from Wainfleet to the River Glen but on this occasion little was done. He was still settling the fen accounts in 1817 and in 1821 Rennie recommended him to survey Boston Haven. It was Bower's son Samuel (q.v.) who replied from Boston to Rennie and who was clearly in partnership with his father. Rennie died two months later and his son employed Francis Giles (q.v.) instead. There is no further mention of Anthony Bower and the record of his death has not been found.

P. S. M. CROSS-RUDKIN

[Rennie reports, ICE; Rennie collection at NLS; W. H. Wheeler (1896) *A History of the Fens of South Lincolnshire*]

Publications

1781. *A Plan of the Holderness Drainage in the County of York*
1786. *A Plan of the Town & Harbour of Kingston upon Hull*
1804. *A Plan of the River Foss Navigation …*
1806. *Mr. Bower's Report Respecting the Improvements of the River Witham …*
1811. *Plan of the East, West and Wildmore Fens, near Boston …*
1814. *Mr. Bower's Statement as to the Drainage and Levels of the Fens North of Boston, and Comparison with the Levels and Drainage of the Low Lands of South Holland and the Bedford Level*

Works

1802–1808. East, West and Wildmore Fens, resident engineer
1810. Boston East Enclosure Commissioner
1813–. Wainfleet Enclosure and Embankment Commissioner

BOWER, Samuel (d. 1847), civil engineer, son of Anthony Bower, noticed above, appears in commercial directories at Bridge Street, Boston, in 1830 and 1835. He was active in estate and inclosure work and was admitted to membership of the Institution of Civil Engineers in 1825; he appears in membership lists for 1827 and 1830. He is best known for drainage and navigation suveys in the Boston area.

TESS CANFIELD

[Membership records, ICE; Lincolnshire RO; PRO; Bendall]

BRADLEY, Humphrey (fl. 1584–1625), civil engineer, came from the Brabant town of Bergen op Zoom and was probably a son of the merchant John Bradley who settled there and in 1553 married Anna van der Dilft. Though nothing is known of his early career he must have gained some reputation as an engineer since in June 1584 he was called in as a consultant on Dover harbour works. As part of a scheme for restoring and improving the harbour a large sluicing basin (the Pent) had recently been completed by building a bank along a shingle ridge and a cross wall, with a sluice, back to the shore line, a plan devised chiefly by William Borough (q.v.). Problems still existed, however, in relation to leakage in the sluice, protecting the beach in front of the bank, and Bradley's presence there almost certainly resulted from efforts by Sir Francis Walsingham, on behalf of the Privy Council, to obtain assistance from experts in the Low Counties; it was to Walsingham that he reported on 31 October 1584.

Whether Bradley remained in England, or more probably returned to the Netherlands, is uncertain. He is next heard of in March 1589 in a reply from the Privy Council to a request from a Commission of Sewers (the land drainage authorities) that 'some special good service might be performed for the drowned and decaied grounds in the counties of Northampton, Lycolne, Cambridge and Huntingdon'. The reply, dated 21 March, from the Council stated that they had 'received information by men of good credyt, skill and judgement that Humfrey Bradley of Bergen op Zome in Brabant, John Hexhame in Huntingdon and Ralfe Agasse of Suffolk, be known sufficient and great experience in their faculties, were able to make viewe and platt for the severall fennes, marshes and decaied grounds in the said fower shires, and to observe the true dysentes of waters and quallyties of the soile, through which the waters should be carryed, with other needful circumstances. [And the Council] commends these three persons unto [the Commissioners] desiring and requiringe them to authoryse the said persons … speedylie to repair downe into the sayd fower countyes, there to make ready … according to their best skill all matters and circumstances to that service belonging for the making of the said platt'.

This marks the first attempt to tackle the general draining of the Fens. Ralph Agas (q.v.) was a leading land surveyor of his time, John Hexham (q.v.) had surveyed fens in south Lincolnshire, and both of them had recently been involved in an enquiry relating to a scheme by George Carleton (q.v.) for draining parts of South Holland. Agas, writing later to Lord Burghley (Elizabeth's Secretary State and Lord High Treasurer) about the late payment of his fees, reminds him that he had 'some time ago platted those grounds and well considered the quantities of their waters, gauged their ascents and noted their slow passages' while Hexham's contribution was a map of the Fens between the Welland and the Old Nene. No complete map of the area seems to have been produced, since this was first accomplished so far as is known by William Hayward (q.v.) in 1604–1605. But evidently the two surveyors put together a good deal of topographic information, and Bradley presented an engineering report to Burghley on 3 December 1589.

The main points were that the cause of the inundation lay in the failure of the drainage of the water that comes into the Fens from the rivers Granta, Ouse, Nene, Welland, Glen and Witham and was chiefly a result of the shallow beds of the rivers. He considered it would be possible to drain these lands, since practically the entire surface was above high sea-level, the sea rising and falling 16 or 17 ft. from high to low tide. He also stated that the greater part of South Holland and Marshland [the silt lands], lying lower than the inundated [peat] lands by 6 or 7 ft., remained dry, ensuring that the higher ground could be drained. He concluded by stating that the only way to redeem the lands was to draw off the waters by directing them along the shortest tracks to the greatest outfalls, in canals dug of such width and depth as would serve to make the water run out into the sea. In Bradley's time the peat fens lay above high tide level, whereas today their surface is at about mean sea-level (Ordnance Datum), the result of 300 years of drainage.

Moreover, substantial local improvements in drainage had already been brought about by a new cut for the river Nene made by John Morton, Bishop of Ely 1479–1485. The cut, known as the New or Morton's Leam, ran in a more less straight line for 12 miles greatly shortening the course of the river from Stanguard near Peterborough to Guyhirn, and at the same time the river was deepened from Guyhirn to its outfall below Wisbech. Nothing on this scale was attempted during the next century, and Bradley's report is reticent on specific proposals. Perhaps for this reason no action followed but in 1593, realising that a more positive plan was required, he wrote to Burghley saying that he was ready to declare his scheme, and a few days later on 3 April sent his *Project for the Draining of the Fennes*. This sets out the need for a central authority, in view of the various tenures and ownerships prevailing, repeats the general principle of draining the Fens (and

contrasts the conditions there with those in Holland) and outlines his main proposal: for a new cut 4 miles in length from Welney to the river Ouse which he thought could be made in six months with 700 or 800 men at a cost of £5,000. This was no more than a useful first step towards general draining, a fact of which Bradley was probably well aware though he claimed that large benefits would result. Nevertheless no action seems to have been taken, and this is the last connection of Bradley with the Fens. However, during 1589 he had been busily engaged on several projects.

In May he and Hexham were appointed to design the course and outfall of a new drain in South Holland originally proposed by Carleton, which was not constructed, and in August they were again consulted, this time on the north bank of the Witham near Boston. Meanwhile Bradley was working on the great sluice on the Nene at the Horseshoe about a mile below Wisbech, possibly with Hexham as clerk of works. This was the second or third attempt to build a sluice on the tidal river there and, like its predecessors, it failed soon after construction. Also in 1589, on 27 June, Bradley married Anna Sermartens, and two of their children were baptised at the Dutch Reformed Church, Austin Friars, in November 1592 and January 1594.

After that he presumably left England and worked in the Netherlands, for in June 1596, responding to a request from Henry IV of France for 'qualified individuals experienced in the art of dyking' to work in his service, the States General sent Bradley with a dyker from Zeeland and another from Holland; the outcome proved successful. In 1599 he was appointed Maitre des digues du royaume, granting him a practical monopoly of draining low-lying and drowned land in the Kingdom of France. There he remained at least until 1625, but the date of his death is not known.

A. W. SKEMPTON

[M. Champion (1858–1864) *Les inondations en France*; H. C. Darby (1940) *The Draining of the Fens*; L. E. Harris (1961) *The Two Netherlands, Humprey Bradley and Cornelis Drebbel*; R. A. Skelton and John Summerson (1971) *A Description of Maps and Architectural Drawings in the Collection Made by [Lord] Burghley now at Hatfield House*]

Publications

Manuscript reports:

Advys of Humfre Bradley touching Dover works for the beche, sluys, harbour or pente, and banck. Mr. L. E. Harris, who discovered this document in State Papers Domestic, describes it as a very able report but gives no details; nor is it more than mentioned subsequently (and even then without the title) by historians of the harbour

A Treatise Concerning the State of the Marshes or Inundated Lands (Commonly Called the Fens) in

the Counties of Norfolk, Huntingdon, Cambridge, Northampton and Lincoln, drawn up by Humphrey Bradley, a Brabanter, on 3rd December 1589 Project for the Draining of the Fennes (B. M. Lansdowne MS 74/65)

BRADLEY, Thomas (1753–1833), civil engineer of Halifax, was trained as a joiner, but was clearly of some ability as by the age of 22 he had his designs accepted for Halifax Piece Hall, erected 1775–1779 under the supervision of Samuel and John Hope. His only other known architectural work is Crownest Hall, Lightcliffe, Yorkshire (1788) and soon after he became involved in civil engineering, which profession he followed the remainder of his life.

In early 1792 he was appointed 'Superintendent Surveyor' or (Resident) Engineer of the Calder and Hebble Navigation at a salary of £105 per annum at a time when a link with the Rochdale Canal was obtaining Parliamentary approval, and work was required to provide wharves and warehousing at Sowerby Bridge. Henceforth William Jessop's (q.v.) services as consultant to the navigation were largely dispensed with. The navigation prospered in the 1790s; in 1798 the first section of the Rochdale Canal was opened from Sowerby Bridge to Todmorden.

By 1803, in recognition of his services Bradley was earning £400 per annum, increased to £500 per annum in 1811 when he received an ex-gratia payment of £500.

In 1812 a new cut and pair of locks were completed at Fall Ing, to improve the link to the Aire and Calder. This was one of a number of improvements carried out under Bradley's stewardship. In 1816 he compiled a report on dredgers he had seen at work on the Thames and Trent. In 1825 Parliamentary sanction was obtained for a 1¾ mile branch from Salterhebble to Halifax; it was opened on 28 March 1828 having cost of £57,800.

By 1832 Bradley's health was failing and, on Bryan Donkin's (q.v.) recommendation William Gravatt (q.v.) was appointed engineer, nominally under Bradley. It was an unsatisfactory appointment and in 1833 William Bull (q.v.) was appointed engineer.

Bradley joined the Institution of Civil Engineers as a Corresponding Member in 1820 and he died in 1833 and was buried in the Square Independent Chapel graveyard, Halifax.

MIKE CHRIMES

[ICE membership records, ICE archives: C. Hadfield (1972) *The Canals of Yorkshire and North East England*, passim; Colvin (3)]

Works

1792–1833. Calder and Hebble Navigation, engineer

BRADSHAIGH, Sir Roger (1628–1684), MP, baronet and coal owner, the eldest surviving son of James Bradshaigh (d. 1631) and Anne *née* Norreys, was born at Haigh, Lancashire on 14 January 1628. His family, on both sides, were important Lancashire gentry, and the Bradshaighs had been mining cannel coal at Haigh since before 1540.

Consideration had already been given to the drainage of the Haigh mines and around 1652 he resolved upon driving the 'Great Sough', a drainage channel more than two thirds of a mile long, 6 ft. wide and 4 ft. high. Although in the Tudor period Wollaton Colliery had a sough nearly a mile long, the Haigh Sough was unusual in the amount of tunnelling involved. There were ten 9 ft. diameter airshafts, up to 147 ft. deep, used for extracting the rock during construction, and the sough was not completed until 1670, after 17 years labour; work continued thereafter on maintenance and extensions.

Bradshaigh became Deputy Lord Lieutenant, and later Sheriff, and MP from the Restoration until 1679, when he was created a baronet. He married Elizabeth Pennington (d. 1695) in 1647, and had seven children of which a son, Sir Roger (1649–1686) and daughter, Elizabeth, survived into adulthood.

He died on 31 March 1684 and is buried at Wigan. Much of his time and money was sunk into his 'Great Sough' a remarkable tunnelling achievement.

MIKE CHRIMES

[A. J. Hawes (1945) *Sir Roger Bradshaigh*; J. Hatcher (1993) *History of the British Coal Industry*, vol. 1]

Works

1652–1670. Grand Sough tunnel, *c.* 1,200 yd.

BRAITHWAITE, John (1797–1870), engineer, was the third of seven sons of John Braithwaite (1760–1818), engineer, and his wife Elizabeth. A John Braithwaite had inherited a foundry or forge in St. Albans in 1695, and the family operated it until 1770 when John (Sr.)'s father William (*c.* 1732–1800) moved to Soho, then a centre of London's metal working trade. The family's business was mostly concerned with well sinking, and manufacture of pumps. This was developed into diving bell design, and John Sr. and his elder brother William (*c.* 1757–1802) developed a profitable business in raising wrecks, being at sea fairly continuously between 1783 and 1792. On his voyage to the Cape Verde Islands he met his wife Eliza Doyle, and on their return to London they married on 3 June 1794. The Braithwaites built a terrace of houses 1–4 Bath Place, New (Euston) Road behind which they built workshops. John was born here on 19 March 1797. Although educated at Dr. Lord's School, Tooting, he was brought up in the family business, and accompanied his father in 1806–1807 securing £130,000 of coins as well as the general cargo of the East India man *Earl of Abergavenny*. From the proceeds, his father was able to purchase the Old

Manor House, Westbourne Green. In 1818 John senior was killed as the result of a highway robbery, leaving £30,000 to his children, of whom Francis (d. 1823) and John inherited the engineering business, although the other brothers, notably the eldest, William, referred to themselves as engineers.

John clearly had an aptitude for mechanical inventiveness. He expanded the business of well-sinking and supplying London brewers and waterworks with pumps and steam engines to include Scotland, Ireland and the West Indies and developed high pressure steam engines. He was called upon by the 'Committee of Engineers' to report on the Norwich steam boat explosion to the House of Commons and in 1820 he installed air pumps to ventilate the House of Commons. In 1827 he cast the statue of the Duke of Kent and at the same time he became interested in locomotive engines, meeting George and Robert Stephenson (qq.v.) and Captain John Ericsson. Together he and Ericcson developed the *Novelty* engine for the Rainhill trials, the first steam engine to run above 60 m.p.h. He also developed steam fire engines which were used in several prominent fires at the English Opera House, the Houses of Parliament, and Bishop's Brewery. Opposition of the firemen led to its abandonment, although four engines were built, for Berlin, Liverpool and St. Petersburg, to where John's eldest brother William emigrated in 1844 at the Tsar's invitation.

From 1834 John ceased to be actively involved in manufacturing, management of which was taken over by his younger brother Frederick (1798–1865), and although he continued to take an interest in steam and locomotive engines—for road and rail—he became a consulting civil engineer. His first report, with Charles Blacker Vignoles (q.v.) was in 1834 for the Eastern Counties Railway and in 1836, following the passage of its Act, he was appointed Engineer for the line. Braithwaite adopted a five foot gauge, in the erroneous assumption that the line would not be connected to the rest of the rail network. In 1843 he left the company and the following year the standard gauge was adopted. In 1837 he co-founded the *Railway Times*, of which he continued as proprietor until 1845. Although he carried out further railway surveys on the continent in the 1840s, few of his projects were executed.

He was elected a Member of the Institution of Civil Engineers in 1838 and died suddenly on 25 September 1870. He is buried at Kensal Green Cemetery.

MIKE CHRIMES

[J. Braithwaite (mss), Sketches on engineering subjects, ICE Archives; Evidence (1817) *Select Committee on Steam Boats*, etc., 26–27 (HC 422); F. Braithwaite (1843) Statement of the results of some experiments with various description of fuel with an without a blowing apparatus, and

with various widths of fire-bars and airspace, OC 612, ICE Archives; Memoir of Frederick Braithwaite, *ICE Minutes of Proceedings*, **26** (1866–1867), 560–561; Memoir of John Braith- waite, *ICE Minutes of Proceedings*, **31**, 207–210; W. Braith-waite (1992) A short history of the Braithwaites, *Family Tree Magazine*, **8**, **11**, 8–9]

Publications

(Note: His brother Frederick was a very active contributor to ICE meetings.)

1835. *Eastern Counties Railway from London to Ipswich, Norwich and Yarmouth*
1836. *Eastern Counties Railroad*
1850. *Danchells Patent Steam Pressure Gauge*

Works

1834–1843. Eastern Counties Railway

BRAMAH (BRAMMA), Joseph (1749–1814), mechanical engineer and inventor, was born on 2 April 1749 at Stainborough, Barnsley, the second of five children of Joseph Bramah (1713–1800) and Mary Denton. His father had worked as coachman for Lord Stafford before becoming a tenant farmer in 1746. As a child he became interested in woodwork and at the age of sixteen was apprenticed to Thomas Allott, a local carpenter, for seven years.

On completion of his apprenticeship Bramah determined to walk to London to make his fortune. He obtained work as a joiner or cabinet maker, and for a time worked for a 'Mr. Allen' who was involved in installing water closets to the horologist Alexander Cumming's 1775 patent (1105). Allen himself improved this device, but it was Bramah who obtained a patent (1177) in January 1778, giving his address as Cross Court, Carnaby Market. This was the first of eighteen patents Bramah took out down to the time of his death, of which the most important are those for the hydraulic press (No. 2045, 1795) and locks (No. 1478, 1784 and No. 2232, 1798).

In 1783 Bramah joined the Society of Arts at a time when there was much interest in 'pick-proof' locks. Bramah came up with a solution in 1784, at which time he moved to more fashionable premises at 124 Piccadilly and married Mary Lawton, of Mapplewell, Yorkshire. Bramah had problems in the manufacture of the locks which were not satisfactorily solved until after 1789 when he was joined by Henry Maudslay (q.v.) the great machine tool designer. Maudslay remained with Bramah until 1797, developing (*c*. 1794) the slide tool with slide rest.

Around 1785 Bramah began to describe himself as an engineer, using the description in his patent (1478) for a 'Hydrostatical machine and bodies, propelling vessels, carriages, etc'. This was indicative in a change of emphasis of his inventiveness from the domestic to industrial scale and it was followed by patents for rotary engines (with Thomas Dickinson, No. 1720, 1790) and fire engines (No. 1948, 1793). Bramah inevitably took an interest in steam engine manufacture and became involved in the controversy over the Watt patents, giving evidence against Watt and in support of Edward Bull and the Hornblowers in the celebrated patent infringement cases of the early 1790s. The somewhat heated nature of some of his evidence suggests he was a difficult man to work with.

This view is reinforced by the story of his involvement with Norwich waterworks, his only civil engineering scheme. In 1788 Robert Mylne (q.v.) was asked to report on the waterworks, which had originally been designed by Peter Morris (q.v.) in the sixteenth century. Mylne's report, published in January 1789 recommended that the existing use of the town mills should be restricted to water supply and grinding corn, and a total overhaul of the water supply system including new reservoirs. Following Parliamentary approval in 1790 progress was slow, and the new leasees of the mill, White and Crane, brought in Bramah in 1795 to build the new waterworks. Bramah was critical of the lack of detail in Mylne's report, but it was not until September 1796 that Bramah's designs were ready. Mylne, perhaps understandably, was concerned at these develop ments and in 1797 brought in Boulton and Watt to comment, no doubt suspecting patent infringement in Bramah's steam engine design. Mylne reported critically on Bramah's scheme and in 1798 Bramah issued a vituperative pamphlet against Mylne It would appear that despite Mylne's views Bramah carried out the work.

Bramah is probably best-known today for his development of the hydraulic press. The principles behind this had been identified by Stevin and Pascal around 1600, but Bramah was the first to bring them to practical fruition, possibly aided by Maudslay. The take up of his 1795 patent was initially affected by luddite views among some sections of industry, but use became widespread. Bramah himself saw the potential of hydraulic power and his patent No. 3611 (1812) included proposals for using water mains to distribute hydraulic power as well as potable water. Bramah's other inventions included beer engines and brewing (No. 2196, 1793) a planing machine (No. 2652, 1802) and improvements to paper manufacture and printing (No. 2840, 1805; No. 2977, 1806), pens (No. 3260, 1809), and carriages (No. 3270, 1809; No. 3616, 1812) His patent of 1806 (No. 2957) was applied to the printing and automatic numbering of Bank of England notes.

Bramah's applications of hydraulic power included hydraulic cranes, which he had in his works from around 1802, and pile driving and extraction, applied at Waterloo Bridge. His demonstration of the use of hydraulic power to extract trees by the roots, at Alice Holt Forest, led to him catching a fatal chill and he died surrounded by his family on 9 December 1814 and was buried in Paddington churchyard.

Bramah had four sons and a daughter, Hannah. One son John (d. 1825) trained as an architect before deciding to become a clergyman, but the others, Timothy (1784–1838), Francis (1786–1840) and Edward (1788–1854) all joined his firm, which was known as J. Bramah and Son from 1807, and Bramah and Sons from 1817. Bramah had obtained a factory site in Pimlico around 1806.

Timothy specialized in developing hot water heating which was most famously applied for the hot house at Windsor Castle in 1828. Installations were supplied for St. George's, St. Thomas's and Westminster hospitals and the National Gallery. There survives extensive correspondence on the subject with Thomas Tredgold (q.v.) and Timothy revised the third edition of Tredgold's book.

The firm, latterly known as Bramah and Robinson, supplied all kinds of steam engines and machinery, including testing machinery for the dockyards, ironwork and equipment for St. Katharine's Docks, and ironwork for a number of bridges, such as the suspension bridge at Haslar Lake. Structural ironwork, including trussed girders as detailed by Amhurst Hawkes Renton (q.v.), was supplied for St. Katharine's Docks, Buckingham Palace, Windsor Castle, Hungerford Market, and Millbank Penitentiary. In the face of considerable concern over the strength of cast iron beams, Francis carried out tests in 1834 to verify the principles enunciated by Tredgold in his work on the subject. Such materials testing was clearly a feature of the works.

Some confusion exists between the firm of Bramah and Sons and that established by Joseph's nephew, John Joseph Bramah (1798–1846), which was also located in Pimlico. This business, which was also trading in the 1820s, expanded greatly in the 1830s as suppliers of structural ironwork to the early railway companies, and was eventually to become Fox Henderson.

MIKE CHRIMES

[Memoir (1814) *Gentleman's Magazine*, **84**, 2, 613; C. Brown (1815) Memoir, *New Monthly Magazine*; T. Bramah (1817) *Evidence*, Select Committee on Steamboat Explosions; A. Rees (1820) *Cyclopedia*; T. Bramah (1824) *Evidence, Select Committee on Artisans and Machinery*, **1**, 28–42; Bramah–Tredgold correspondence, ICE archives; Select Committee on Windsor Castle and Buckingham Palace (1831) *2nd report*; T. Bramah (1834) *Minutes of Evidence*, Select Committee on Metropolitan Sewers, HC584; T. Tredgold (1836) *The Principles of Warming and Ventilating Public Buildings*, 3rd edn.; W. Freeman (1836) Experiments on the force required to fracture and crush stones, *ICE Trans*, **2**, 234–235; F. Bramah (1838) A series of experiments on the strength of cast iron, *ICE Trans*, **2**, 113–135; S. Smiles (1863) *Industrial Biography*, 183–197; H. W. Dickinson (1941–1942) Joseph Bramah and his inventions, *Trans Newc Soc*, **22**, 169–186; I. McNeil (1968) *Joseph Bramah: a Century of Invention 1749–*

1851; R. Sharp (comp.) (1991) *The Bramah and Robinson Drawings*]

Publications
1798. *Strictures on a Plan Drawn by Robert Mylne for the Improvement of the Mills and Water Works in the City of Norwich*
1812. *Specification of Patent for Improvements in the Method of Constructing … the Main and Other Pipes for Conveyance of Water for the Supply of the Metropolis*

Works
1795–1798. Norwich waterworks

BRAMHAM, James, Major-General (fl. 1740–1786), military engineer, was Chief Engineer of Great Britain between 1781 and 1786. In 1744 Bramham became the first Tower of London trained draftsman to receive a Warrant from the Corps of Engineers

His forty year career encompassed the Netherlands (1748) the West African survey with Justly Watson (q.v.) (1756), Chief Engineer at the siege of Belle Isle (1761) and Lieutenant Governor of the Royal Military Academy from 1776 until 1781. Of particular interest are his work on the King's Road at Blairgowrie (1749) and three years in Jersey where he modernised Elizabeth Castle originally designed by Paul Ive (q.v.). While Chief Engineer at Woolwich he interested himself in wooden piers, a 'boring engine' and a wooden wharf.

Bramham died at Greenwich on 11 November 1786.

SUSAN HOTS

[Royal Engineers Library: Conolly papers; Porter, History of the Royal Engineers, vol. 2, 389–391; Bendall, 59; D. W. Marshall (1976) British military engineers 1741–1783 (unpub. Ph.D. diss. University of Michigan, 117/118; F. J. H. Hebbert (1990) Belle Isle and the Siege of 1761, *Fort*, **18**, 69–81]

Works
1749. King's Road from Blairgowrie to the Spittal of Glenshee

Publications (reports)
1748. Survey of the town of Breda fortifications and environs
1756. Report and estimates of the condition of several works erected for the defence of Jersey
1757. Report and estimates for Jersey. Also for Elizabeth Castle
1757. Bays of archirondelle St. Catherines and Belval, showing proposed work
1758. Bays of Beaufort, St. Brelade, Portelet and St. Helier with the town Elizabeth Castle and vicinity
1759. Report and estimates for Jersey. Master gunners house in Elizabeth Castle
1761. State of the Citadel of the Palais (Belle Isle) (BL additional manuscript, 36995)

BREMNER, James (1784–1856), shipbuilder, marine salvage expert, and engineer was born on 25 September 1784 at Stain near the village of Keiss in the county of Caithness. He was the youngest child of James (1744–1819) and Janet Bremner who had a family of nine children, six of whom died in infancy. Bremner received a rudimentary education at the village school and entered into an apprenticeship with Robert Steele & Sons, shipbuilders of Greenock.

On completion of his apprenticeship he made two voyages to Nova Scotia where he toyed with the idea of setting up as a shipbuilder in Pictou. However, he was persuaded to settle in Caithness by the very favourable terms being offered by the British Fisheries Society at their new settlement of Pulteneytown and here Bremner took a life lease of a site beside the proposed harbour and commenced ship-building. After starting modestly it is said that during his lifetime his receipts from shipbuilding alone totalled almost £170,000. This steady income enabled him to branch out into salvage work and harbour building.

In the days of sail, the coasts around the Pentland Firth saw many wrecks and groundings and, as the local ship-builder, Bremner was frequently consulted on the feasibility of repairing and refloating stranded ships. His reputation as a wreck-raiser quickly grew and over his lifetime he refloated 236 vessels. According to his contemporary, Joseph Mitchell, Bremner earned £600 a year from these 'God-sends' from wrecks. In 1841 the *Uncertain* of Sunderland sank in Broad Bay, Lewis, in 11 fathoms of water; she was only six months old and carried a valuable cargo of 600 tons of coal. During the next two years, three unsuccessful attempts to lift her were made by other salvage engineers but eventually Bremner, using the combined buoyancy from five small vessels and twelve large casks, eased the wreck into shallow water, where the coal was unloaded and the hull patched. She was then towed into Stornoway and fully recommissioned.

The publicity which followed the recovery of this heavily laden vessel from such a depth brought Bremner to the notice of a wider public and almost certainly led to his involvement with refloating the largest ship of all, Brunel's SS *Great Britain*, which had run aground in Dundrum Bay in Ireland in 1845. In his memoirs, Joseph Mitchell, after describing the grounding and the unsuccessful attempts to refloat the *Great Britain*, continued, 'At last Bremner was thought of and sent for, and by his ingenuity, practical knowledge, and talent triumphantly succeeded', adding that 'Bremner was rewarded with a sum of £1,000 for his six months services, and all his expenses'.

George Burn (q.v.), the contractor engaged by the British Fisheries Society to build the first harbour at Pulteneytown, found difficulty in laying the foundation stones of the pier and consulted Bremner, who provided a crane barge of sufficient capacity to allow construction to continue. The crane, designed to operate

James Bremner

weights on a rising tide, at times using the hulls of boats tied together like a catamaran to lift stones weighing up to forty tons from a quarry on the foreshore. In a patent for harbour building granted to Bremner in 1844 he proposed to lift and transport to their final location sections of harbour wall which had been prefabricated in sheltered conditions on the foreshore, using a purpose-built vessel but it is unlikely that the system was ever used.

As a prominent businessman, Bremner was well respected by his fellow townsmen who recorded their appreciation as early as 1833 by presenting him with his portrait. The accompanying citation notes 'his superior talent and genius in Harbour building and engineering, and all under water works, and above all, on account of the generous intrepidity with which he has so often, to the eminent danger of his own life, saved that of the shipwrecked and perishing mariner'. His bravery is confirmed by Mitchell who described Bremner as 'a man of Herculean frame and undaunted courage' who did not comprehend what fear was.

In politics he was a Liberal and the only blemish on an otherwise impeccable record was a spurious charge of mobbing and rioting during the Parliamentary election of 1826. When the case appeared before the Circuit Court at Inverness on 24 April 1827 the charges were abandoned.

Telford sponsored Bremner's membership of the Institution of Civil Engineers and in 1833 he was formally elected a Corresponding Member. In 1836 he presented a paper on a dredge which he had invented to recover a cargo of iron which had sunk into the sands of Sinclairs Bay in 1829 and in 1845 he was the recipient of a Telford silver medal for papers on Pulteney Town Harbour, Sarclet Harbour, a new Piling Engine and an Apparatus for floating large Stones for Harbour Works. In the same year he wrote a short Treatise on Harbour Building which encapsulated his main achievements up to that time.

Eventually Bremner's iron constitution weakened and he died on 12 August 1856 and was buried in Wick Parish churchyard, beside his wife, Christina Sinclair, who had been interred just three months earlier; they had been married for forty-five years. Of their eleven children only three daughters and two sons survived their parents. His sons and grandsons carried on the engineering tradition of their forebear. In 1903 a granite obelisk was erected to Bremner's memory and his portrait was returned to the township of Wick by his great-great-grandson, James Bremner; it now hangs in the Town Hall.

GEORGE WATSON

simultaneously on both sides of the barge, was one of the first of Bremner's inventions to be mentioned. By 1807, with the harbour walls partially complete, Burn, whose previous experience lay in road and bridge building, persuaded Bremner to take over the submarine building of the breakwater terminations. On completion in about 1811, the work was inspected and approved by the consulting engineer for the British Fisheries Society, Thomas Telford (q.v.).

During his lifetime Bremner built or carried out major repairs to a total of 18 harbours and surveyed and planned several more which were not built due to a downturn in the fishing industry. His first harbour, which he started in 1818, was at his native village of Keiss. The stones of the outer walls were laid with their long axis nearly vertical, a technique which became his trade-mark in subsequent harbours. In 1822 with the project almost finished, construction had to be suspended and it was not until 1834 that the work was completed.

During the hiatus at Keiss, several other harbours were completed, among them Pulteneytown, Sarclet and Lossiemouth, where, in 1836, when faced with excavating a large area of submerged rock within the harbour, he sealed the walls and dammed the entrance to the basin which was then pumped dry using a 12 hp pump. This is one of the few portable engines mentioned by Bremner who generally used manual equipment.

Throughout his working life he constantly improved his technique of lifting and floating heavy

[Memoir of the Late Mr James Bremner, CE, *The Northern Ensign*, 21 August 1856; Memoir to Mr James Bremner (1856–1807), *Minutes of Proceedings of ICE*, 16; H. Miller (1883) *Rambles of a Geologist*, chap. X, 382; J. Mitchell (1883) *Reminiscences of My Life in the Highlands*; P. F. Anson

(1974) *Fishing Boats and Fisher Folk on the East Coast of Scotland*; D. Macdonald (1978) *Lewis*, 98; J. Mowat (1980) *James Bremner Wreck Raiser*; A. Graham and J. Gordon (1987) Old Harbours in Northern and Western Scotland, *PSAS* , **117**, 265; R. P. Gunn (1998) *Inventors and Engineers of Caithness*; Various contemporary entries in the local newspapers, *The Northern Ensign, The John O' Groat Journal*, and the *Inverness Courier*]

Harbour works

1811–1834. Pulteneytown, Caithness (ND 369 505), pier heads 1811; outer basin south break-water plus an extension to it 1826, later contracts completed the outer basin and converted the old harbour to an inner basin, completed 1834, cost £20,900

1818–1834. Keiss, Caithness, commenced in 1818, suspended in 1821, completed 1833–1834 with Government grant £5,000

1825. Castlehill, Caithness, £1,500

1830. Pitullie and Sandhaven, Buchan and Banff, £4,200

c. 1830. Ham, Caithness, £1,000

1834. Banff repairs, £3,000

1835. Sandside, Caithness, £3,000

1834–1836. Sarclet, Caithness

1835. Cairnish, Lewis, £1,810

1835. Callicott, Lewis (NB 540 638), £1,645

1836. Lossiemouth, Moray and Nairn, £20,000

1842. Helmsdale, Sutherland Harbour repairs, jointly supervised construction

1845. Hopeman, Moray and Nairn (NJ 145 699), surveyed, planned and built

1847. Occumster Caithness, c. 1847, joint engineer

1849. Lybster Caithness, joint engineer

1851. Skullomie, Sutherland, £3,500

c. 1851. Ferry-boat piers, Kyle of Tongue and Loch Eriboll

BREWSTER, Sir David, FRS, FRSE (1781–1868), natural philosopher, was born on 11 December 1781 at Jedburgh, the third child of James Brewster and Margaret Key. Brewster entered Edinburgh University in 1794 where he received an MA in 1800 and was licensed to preach in the Church of Scotland in 1804. Brewster was an evangelical and was one of the leaders of the Free Church of Scotland which was organised following the 'Disruption' in 1843. Until 1838 he earned most of his income from writing and thereafter he was Principal of St. Andrews University until 1859 when he was elected Principal of Edinburgh University. He was knighted in 1832 at the suggestion of the Lord Chancellor, Henry Brougham, who had been a fellow student at Edinburgh.

Brewster joined the Institution of Civil Engineers in 1820. His main practical contribution to civil engineering was in lighthouse technology and, following some of his optical theories, he developed a polyzonal lens in 1823 and also proposed, to the Northern Lighthouse Board in 1831, the use oxygen instead of air in lighthouse lamps.

For both of these pieces of work, Brewster had to defend his priority (the former against the supporters of Fresnel) which he did vigorously. His interest in a lens based system of lighthouse illumination brought him into conflict with Robert Stevenson (q.v.) who had improved the efficiency of the catoptric system using reflectors in the lighthouses of the Northern Lighthouse Board. Stevenson resisted the introduction of lenses on grounds of cost, and Brewster became very critical of the Board for what he saw as a backward policy. Eventually, following a visit by Robert's son Alan Stevenson to France in 1834, when he met Fresnel, the Board introduced its first dioptric light, in 1835. It is unlikely Brewster derived much personal satisfaction as it was designed by Alan and based on French practice.

He was active in promoting science and science education, for example in playing a role in founding organisations such as the British Association and the Royal Scottish Society of Arts. His research interests centred on optics, where he invented the kaleidoscope and argued strongly against the theory proposed by Fresnel and supported by most other British natural philosophers that light was undulatory in nature; Brewster argued that it was particulate in nature. He also wrote on the history of science, including a two volume biography of Isaac Newton (1855).

His promotion of the multi-volume *Edinburgh Encyclopaedia* brought him into contact with Thomas Telford (q.v.). who was commissioned to write articles on Bridges with Alexander Nimmo (q.v.), Civil architecture, and Navigation, Inland. It has been suggested Telford may have written other articles notably that on William Jessop, and Telford recommended the work. These articles were important sources on best practice at the time. It was a great source of frustration to Telford that the completion of the work was delayed for almost twenty years until 1830, by which time much of the detail on bridges had lost its currency -suspension bridges not being covered at all. Brewster himself contributed entries on Rennie among others of interest to civil engineers

Brewster died at Allerby on 10 February 1868, following an attack of pneumonia and bronchitis. He was survived by his second wife and children from both marriages. An engraved portrait is held by the Scottish National Portrait Gallery

FRANK A. J. L. JAMES

[Membership records, ICE archives; Telford mss, ICE archives; Faraday collection, IEE; D. Stevenson (1859) *Reply to Sir David Brewster's memorial to the … Treasury, on the New System of Dioptric Lights* (and appendix); M. M. Gordon (1870) *The Home Life of Sir David Brewster*; A. D. Morrision-Low and J. R. R. Christie (eds.) (1984) *'Martyr of Science': Sir David Brewster 1781–1868* (this contains an excellent bibliography of Brewster's writings)]

Publications

(Note: Brewster wrote extensively, and a large number of papers were published by the Royal Society of Edinburgh on light, etc.)

1806. *Ferguson's Lectures on Select Subjects in Mechanics ...* 2nd edn.

1808–1830. *Edinburgh Encyclopaedia* (editor)

1812. Demonstration of the fundamental property of the lever, *Trans Roy Soc Ed*

1818. The article 'Steam and steam engines' written for the *Encyclopedia Britannica* (by John Robison), ed. D. Brewster

1824. *Robison's Mechanical Philosophy* (editor)

1859. *Memorial on the New System of Dioptric Lights*

1867. *The History of the Invention of Dioptric Lights*

BRINDLEY, James (1716–1772) was born in 1716 in a then relatively isolated part of Derbyshire at Tunstead, between Tideswell and Buxton. His parents were a yeoman family of modest means with apparently little evidence of a tradition of mechanical aptitude. In 1725 they moved to Low Hall near Leek, Staffordshire, where his father bought a farm and James is supposed to have followed the usual pattern of casual farm labouring until 1733 when, aged seventeen, he was apprenticed to Abraham Bennett, mill-wright and wheelwright, at Sutton near Macclesfield. He had a basic ability to read and write though, as his notebooks illustrate, his spelling of place names and unusual words remained phonetic throughout his life.

During his apprenticeship he was engaged in the construction of corn mills and their machinery and proved his ability by mastering his conception of mechanical principles and his comprehension and application of hydraulics. He demonstrated his innate theoretical and practical competence during the erection of machinery at a new paper mill at Wildboarclough on the river Dane. Bennett was experiencing great difficulty in making and fitting the drives and machinery, and on Brindley hearing about it, though not engaged in the project, he walked to Manchester and back, 25 miles each way, without telling Bennett, in order to examine a similar plant. He then modified Bennett's plan and successfully completed the mill. This and other work established his reputation in the local community before the end of his apprenticeship.

After Bennett's death in 1742 Brindley moved from Sutton to Leek and began business on his own account at a workshop in Hill Street. Here he undertook work in the immediate neighbourhood on agricultural machinery improvements and repairs, including watermills, but he also began to seek other outlets for his skill. In 1750, noting the industrial development in the Stoke area, he rented a workshop from the Wedgwood family in High Street, then Rotton Row, in Burslem.

James Brindley

In 1752 Brindley was engaged by John Heathcote to advise on draining the flooded parts of Wet Earth Colliery at Clifton, Salford. This was successfully accomplished by constructing a weir, Ragley Weir, across the river Irwell and leading a leat from intakes on the east bank via a tunnel and an inverted syphon under the river to an underground water wheel which in turn operated pumps at the colliery.

During the 1750s Brindley's interest was aroused in other forms of technical development. It had been found that powdered flint, which had previously been calcined, added to the pottery body resulted in a very acceptable white ware. In the process the flint pebbles were calcined in a kiln and then crushed under stamps. Tragically the flint powder being virtually pure silica, when inhaled, led rapidly to silicosis, known as 'Potter's Rot'. To overcome this the practice of grinding it wet under edge runner mills had been developed. The idea had been patented by Benson in 1726 but Brindley modified and improved the process when he was commissioned to construct a number of water-powered flint mills. On 26 February 1756 he records in his diary that 'he began to work on the engine for Mr John Baddeley's flint mill', located in the Moddershall valley, near Stone and again in 1757 there are several entries in his diary relating to work on Mr. Baddeley's mill. In 1758 he was engaged by Josiah Wedgwood to build a windmill at Jenkins, Burslem, for the same purpose. Although this failed on the first day of operation due to a high wind blowing off the sails Brindley repaired it and it worked successfully for many years.

Meanwhile, also in 1756, he received a commission from Thomas Broade of Little Fenton for a steam engine. In accordance with his principle of obtaining maximum information before finalising his designs he visited existing engines in Coventry and Birmingham and then arranged for a boiler to his design to be built at Coalbrookdale. A second engine for Miss C. M. Broad followed in 1757 and in 1758 he took out a Patent (No. 730) for a boiler made on the arched principle of brick and stone with an internal firebox. During this time he continued constructing and repairing flint mills.

Although fully occupied in millwrighting which at that stage was his main income, and which included the construction of mill leats and other water channels, it was inevitable that he and his associates would realise that on a larger scale these channels could carry traffic more efficiently than the wretched eighteenth century roads. The fulfilment of such an idea led ultimately to millwrighting becoming subordinate to his canal engineering practice.

His first major commission relating to a canal was his engagement in 1758 to survey the route of a proposed canal from Stoke-on-Trent to Wilden Ferry on the Trent on behalf of Earl Gower and others which resulted in the submission of *A Plan for a Navigation Chiefly by Canal from Longbridge near Burslem ... . to Newcastle, Lichfield and Tamworth and to Wilden in the County of Derby*, approved by John Smeaton (q.v.). The branches to Lichfield and Tamworth were not made. Meanwhile the Duke of Bridgewater and his estate agent, John Gilbert, had produced a plan for which they had received an Act in 1759 for a canal linking the Duke's mine at Worsley with Manchester in order to reduce transport costs arising from road haulage by horse and cart and also to increase the quantity of coal available for distribution in Manchester. Brindley was called in to provide the technical knowledge necessary for the plan to succeed. The Duke's original plan was to take the canal north of the Irwell into Salford but Brindley, with his involvement with the Trent and Mersey project, wisely foresaw the possibility of a canal network linking it with London, Birmingham, Liverpool and Bristol.

Following his six day visit to Worsley in 1759 he suggested taking the proposed canal across the Irwell and via Stretford into Manchester. This modification was accepted by the Duke and Gilbert as they realised that this held greater potential than their own scheme. Brindley then resurveyed the route and included an aqueduct across the Irwell at Barton, spending forty-six days at Worsley preparing the plans. With the Duke's full confidence he was sent in January 1760 to London to give evidence before the Parliamentary Committee. Clearly his presentation was adequately articulate and technically convincing in that the Act for the construction from Worsley to Manchester across the Irwell was obtained without major opposition; this Act superseded the earlier one. Work commenced immediately though not without ridicule from contemporary engineers on the feasibility of the aqueduct. Water was successfully run into the stone arched aqueduct twelve months later on 17 July 1761.

Brindley's great contribution to the Duke's or Bridgewater Canal was in engineering the work on a single contour level from Worsley to Manchester without locks, which further demonstrates his competency as a practical surveyor. His hydraulic ingenuity was also displayed in the system adopted for operating the Castlefield, Manchester, hoists for raising coal, and later warehouse goods, from the boats to street level.

The Castlefield terminus of the Bridgewater Canal's first phase was not completed until 1765, and by then Brindley had surveyed, in September 1761, and helped obtain Parliamentary approval for an act, in 1762, for an extension from Longford bridge to the Hempstones, on the Mersey, above Runcorn. The scheme incurred the wrath of the Mersey and Irwell Navigation, whose line it avoided, and did not really have a great economic viability until the Trent and Mersey obtained its Act (see below). Work began in 1762, but was not completed down to the Mersey until 1773, shortly after Brindley's death. The most notable engineering features were an embankment a mile long across the Mersey valley, incorporating a single arch aqueduct, and the ten lock staircase down to the Mersey near Runcorn. The embankment was across boggy ground and Brindley solved the problem by using timber supports for an earth embankment lined with puddle clay. Under Brindley, puddle linings became the standard waterproofing for canals until the twentieth century.

As a result of his success with the Bridgewater Canal his ability and integrity became famous and he began establishing a reliable staff of assistants. Among those with whom he became closely associated was John Henshall, whose son, Hugh (q.v.), became his chief assistant and successor, and whose daughter, Anne, he married in 1765 at Wolstanton parish church; Brindley was then forty-nine and Anne nineteen. After the marriage they settled at Turnhurst Hall, Newchapel, where they had two daughters, Susannah and Anne. There is a possibility that Brindley had been proprietor or leaseholder at Turnhurst since 1760.

At this period Brindley's business interests were expanding and he had entered into partnership with his brother John in a pottery and he also had a quarter share in Golden Hall Colliery near the line of the Trent and Mersey Canal (or the Grand Trunk as it was also known), a colliery which was later linked via underground waterways to the Harecastle Tunnel.

During the early 1760s the case of the Trent and Mersey was under active discussion and Brindley again gave evidence before Parliament. On 14 May 1766 the Act received the Royal

Assent. The executive committee formed under the Act included Brindley as Surveyor-general at a salary of £200 p.a. while Hugh Henshall was appointed Clerk of works, or resident engineer. Brindley was present when the first sod was cut on 26 July 1766. The most important engineering feature on the canal and the one which caused Brindley the most trouble was the Harecastle Tunnel.

Harecastle Tunnel was the pioneer canal tunnel in the British Isles and Brindley had to work from first principles. He ranged the line of the tunnel over the hill and from this sank 15 vertical shafts from which headings were driven on the canal line. Unfortunately the ground varied from soft earth to Millstone Grit, added to which large quantities of water flooded parts of the workings, a problem which he overcame by building steam engines to operate the pumps. The excess water later proved a blessing as it provided a constant supply at the summit level. Another problem overcome was that of ventilation. This was effected by installing stoves at the bottom of upcast pipes. Owing to the difficulties encountered the tunnel was not finished until after Brindley's death. He also arranged for the excavation of side tunnels to allow for coal from the colliery workings to be brought direct into the tunnel.

With his maturing canal engineering experience Brindley was realising the value of standardised design for features such as locks and bridges which could be constructed in either brick or stone according to the locality and the local materials. He worked to standard dimensions based on a 7 ft. beam narrow boat with locks 74 ft. 9 in. long and this became the norm for Midland canals.

His masterly presentation of evidence before Parliament so impressed those who were becoming interested in canal promotion that his advice was regarded as an essential element to their success. Thus after 1762 his time was occupied in consultancy on a national scale, despite the fact that he was later officially the engineer to the Trent and Mersey and coping with the unique and continuous problems of the Harecastle Tunnel.

As examples of his consultancy in 1762 were his surveys for the Chester Canal; for a branch canal from the Bridgewater at Sale Moor to Stockport, though this was not built; and for navigations from Chester to Shrewsbury and from Rotherham to Doncaster. In 1763 he spent ten days on the Lower Avon Navigation advising George Perrott on improvements which enabled 40 ton barges to reach Evesham.

In 1764 the Common Council of Liverpool sent for him to advise on an effective way of cleaning out their docks. Another sequel to the high regard in which he was held was that parties opposed to others realised that Brindley's evidence might be used to counter surveys of other engineers. On 31 January 1765 a newly appointed committee on the Calder and Hebble Navigation dismissed John Smeaton who had completed the work from Wakefield to Brighouse, and appointed Brindley in his place despite the fact that Smeaton was regarded as the pre-eminent civil engineer of his day. Brindley stayed only until 1766 but the reason for his departure is not recorded. In 1765 he was approached to oppose Thomas Yeoman's proposals for a Chelmer and Blackwater Canal in Essex but he refused the invitation on the grounds that he was too busy but possibly also because of his marriage that year.

On the same day that the Trent and Mersey Act was passed, 14 May 1766, the Royal Assent was given to the Staffordshire and Worcestershire Act which provided Brindley's southern link with the Severn. James Perry, who promoted it, became Treasurer, but Brindley was appointed Surveyor and it was built to his standards. Work started in 1766 and the canal was opened in 1772, within his lifetime.

Following the first proposals for a Leeds and Liverpool Canal, a meeting of Rochdale merchants was held to promote a trans-Pennine canal to link West Yorkshire with Manchester and provide a shorter link between Hull and Liverpool and so benefit their trade. This was the forerunner of the Rochdale Canal and Brindley was engaged to survey the line. He produced two surveys, one of which included Bury and the other served Todmorden and although there was a delay of over twenty years it was basically Brindley's latter line which was constructed.

In 1767 he was requested to survey a possible canal between Stockton and Darlington to convey coal from the South Durham Coalfield to the Tees at Stockton. Although this was never made his surveyed line was that taken fifty years later by the Stockton and Darlington Railway. Brindley's national reputation was heightened the following year when he was commissioned by promoters to survey routes for the Birmingham Canals, the Coventry Canal and the shorter Droitwich Canal.

With this amount of work in hand, it was clear that Brindley was no longer able to exercise day to day supervision over individual construction sites and therefore was fundamentally that of Consultant with trusted assistants such as Henshall, Samuel Simcock, Robert Whitworth (qq.v.) and others dealing with routine practical problems. His influential position and relationship with the proprietors of a canal was becoming more fully recognised and how much the committee accepted this relationship was obvious when on 26 December 1771 the Clerk to the Chesterfield Canal was ordered 'to write to Mr. Brindley to know when he will next attend the Works that this Committee may stand adjourned until such time'. Nevertheless when he was appointed Engineer and Surveyor to the Coventry Canal he undertook 'to give at least two month's attendance in the whole in every year upon the design'.

As a consultant his presence could be demanded at very short notice and the Chesterfield request suggests this degree of tolerance as well as recognition that travel by horse on appalling roads could not be accomplished in minimal time. The Coventry Committee was less tolerant of his failure to attend the site in accordance with the terms of his appointment and uniquely dismissed him. On the other hand, when he was criticised by the Oxford Committee he tendered his resignation, so disconcerting the Committee that they sent a letter of apology and hoped he would reconsider his decision.

Despite the pressure of work he continued to accept commissions in various parts of the country and in 1769 he was concerned in surveys outside his home territory for a canal from the Bristol Channel to Exeter, from Langport to Exmouth, and for improvements to the Thames Navigation with a canal from Boulter's Lock, Maidenhead, to Mortlake. None of these were built or even started, but they were a reflection of the canal mania which was sweeping the country and the pre-eminent position Brindley held in the hierarchy of engineers.

It is evident that Brindley preferred to use the natural contours of the ground to determine the course of the waterway and so limit the amount of lockage required and minimise, except where absolutely essential, substantial engineering features such as tunnels and aqueducts. For this reason, too, he tried to avoid the inclusion of natural rivers in his proposals. This was particularly noticeable when the City of London asked him in 1770 how the navigation of the Lower Thames could best be improved and his reply was to build a canal between Monkey Island, Maidenhead, and Isleworth, Middlesex, so as to bypass the great and tortuous river loop to the south. The City was unhappy about this and he was requested to re-examine the Thames river. His view in his reply in December was unequivocal. 'It (the Thames) hath been found by long experience to be impassable for barges in time of flood ... and is out of the power of Art to remedy. The problem of drought could be overcome by building about twelve locks and weirs but the expense of improving so large a river in this way will be so great I suppose it will not be put into practice'. The canal was not built.

In 1771 he was back again in the North Midlands. The preliminary surveys and final submission to Parliament had been made incorporating Brindley's advice and evidence for the Chesterfield Canal, and on 18 April 1771 the first General Assembly of the proprietors was held in Worksop. At this meeting it was deferentially 'Ordered that Mr Brindley be waited upon and desired to attend the Committee Meeting at Worksop on 25 June'. At this meeting his salary was authorised and at the following meeting he was instructed to 'immediately begin to make the Navigation at a place called Norwood' with John Varley as Clerk of Works.

Meanwhile Brindley saw the culmination of one of his schemes with the opening in 1771 of the Droitwich Canal linking Droitwich with the river Severn. His role in its construction was almost certainly only supervisory but it is unusual for Brindley's work in that it was built to a 64 ft. × 14 ft. 6 in. standard and not to his normal 74 ft. × 7 ft. Towards the end of the year he was approached for a survey for a Lancaster Canal but again he was too busy and passed the commission to Robert Whitworth.

In June 1772 he again met the Chesterfield Committee and outlined the forthcoming stages of the construction programme, which included his second great tunnel at Norwood and also the impressive engineering of the two flights of locks at each end of the tunnel at Norwood and Thorpe; he was back again in September but this was his last visit. He returned to Staffordshire to carry out a survey for a branch canal from Etruria, on the Trent and Mersey, to Froghall for the Caldon Canal. While engaged on this he was soaked in torrential rain and caught a chill. He was put into a damp bed in the Inn at Ipstones and developed pneumonia which, in turn, so aggravated his long-standing diabetes that he died on 27 September 1772.

Brindley is buried in the churchyard at Newchapel in a grave marked by a stone inscribed: 'James Brindley of Turnhurst Engineer was interred September 30 1772'. He is commemorated by a portrait which now hangs in the Institution of Civil Engineers.

JOHN H. BOYES

[DNB; Brindley's diaries (1754–1763), ICE archives; PRO Rail/various; *Wentworth–Wolley Papers*, Brotherton Library, Leeds; *Weales Quarterly Papers*; Plans of canals and rivers, Manchester Reference Library (FF 386 (9)); S. Smiles (1861) *Lives of the engineers* 1; A. R. L. Saul (1939) *James Brindley and His Staffordshire Associates*; H. Malet (1961) *The Canal Duke*; C. T. G. Boucher (1968) *James Brindley Engineer 1716–1772*; R. G. Banks and R. B. Schofield (1968) *Brindley at Wet Earth Colliery*; C. Hadfield (1969) *The Canals of the West Midlands*, 2nd edn.; C. Hadfield (1970) *The Canals of the East Midlands*, 2nd edn.; C. Hadfield and G. Biddle (1970) *The Canals of North West England 1*; A. Burton (1972) *The Canal Builders*; S. R. Broadbridge (1974) *The Birmingham Canal Navigations*; H. J. Compton (1976) *The Oxford Canal*; J. Lindsay (1979) *The Trent and Mersey Canal*; C. Richardson (1996) *Minutes of the Chesterfield Canal Company*, Derbyshire Records Society, 24; B. Elliott (1998) Pop, crisps, boats and corn: rediscovering Wolley Dam and Mill, *Aspects of Wakefield*, 163–182]

Publications

1760. *A Plan for Navigation Chiefly by a Canal from Longbridge near Burslem in the County of Stafford to Newcastle, Lichfield and Tamworth, and to Wilden in the County of Derby* (taken by Hugh Oldham)

James Brindley: canal works

Year	Canal	Length (miles)	Cost (£)	Survey assistant	Resident engineers (engineers, etc., post-Brindley)
1760–1763	Bridgewater I	7	7,500		
1762–1773	Bridgewater II	26	70,000	Hugh Oldham, 1761	Thomas Morris?, 1766–1773
1766–1777	Trent & Mersey	93	300,000	Hugh Henshall, 1765	Hugh Henshall (Hugh Henshall, 1772–1777) (R. E. Josiah Clowes?)
1766–1772	Staffs & Worcs	46	100,000	Hugh Henshall and Samuel Simcock, 1768	John Baker[a] with Thomas Dadford as Superintendent
1768–1771	Droitwich	7	23,500	Robert Whitworth, 1767	John Priddey
1768–1771	Coventry	16	50,000	Robert Whitworth, 1767	Joseph Parker, 1768–1769 (Edmund Lingard, 1769–1771) (R. E. Samuel Bull?)
1768–1772	Birmingham	22	112,000	Robert Whitworth, 1767	George Holloway[a]
1769–1778	Oxford	64	200,000	Robert Whitworth, 1768	James King, 1769–1772[a] (Samuel Simcock, 1772–1778) (R. E. John Priddey)
1771–1777	Chesterfield	46	150,000	John Varley, 1769	John Varley, 1771–1777 (Hugh Henshall, 1773–1777)

1761. *A Plan of the Rivers Irwell and Mersey from Manchester to Runcorn Gap and from Longford Bridge to the Hempstones in the County of Chester* (Hugh Henshall)

1765. *A Plan of the Great Navigable Canal, Intended to be Made from the Trent to the Mersey; and also of the Duke of Bridgewaters Navigations, Proposed to Communicate Therewith* (Hugh Henshall del.)

1767. [*Report Relating to the Petition of the Proprietors of the London-Bridge Water-works*]

1767. *A Plan of the Intended Navigation from Birmingham, in the County of Warwick, to the Canal at Aldersley near Wolverhampton in the County of Stafford* (2 edns.) (Robert Whitworth del.)

1767. *A Plan of the Intended Navigable Canal from the City of Coventry to the Canal on Friday Heath, in the County of Stafford* (Robert Whitworth del.)

1767. *A Plan of the River Salwarp, and of the Intended Navigable Canal, from Droitwich to the River Severn in the County of Worcester* (Robert Whitworth del.)

1768. *A Plan of the Intended Navigable Canal, from the Coventry Canal near the City of Coventry, to the City of Oxford* (Robert Whitworth del.)

1769. *A Plan of the Navigable Canals now Making in the Inland Parts of the Kingdom, for Opening a Communication to the Ports of London, Bristol, Liverpool and Hull, with Adjacent Towns and Rivers* (drawn by Robert Whitworth)

1768. *The Report of James Brindley, Upon his Being Requested to Give his Opinion of the Best Plan of Making a Navigable Communication between the Friths of Forth and Clyde.*

1769. *A Plan of the Intended Navigable Canal, from Chesterfield to the River Trent Near Stockwith* (John Varley del.)

1770. *A Plan of the Intended Canal in Berkshire from Sunning to Monkey Island* (Robert Whitworth del.)

1770. *Queries proposed by the Committee of the Common-Council of the City of London, about the Intended Canal from Monkey Island to Isleworth, Answered.*

1770. *A Plan of the River Thames from the Kennets Mouth ... to Isleworth* [and] *A Plan of the Intended Canal in Berkshire from Reading to Monkey Island* (rev. edn.)

1770. *The Report of Mess. Brindley and Whitworth, Engineers, Concerning the Practicability and Expense of Making a Navigable Canal, from Stockton by Darlington to Winston, in the County of Durham.*

1770. *Report of James Brindley, Engineer, for Improving the Navigation and Drainage of Wisbeach*

1771. [Report] *To the Committee of the Common Council of the City of London* [and] *A Description of the Profile or Section of the River Thames, from Boulter's Lock to Mortlake* (attributed to Brindley)

1766. *The History of Inland Navigations. Particularly those of the Duke of Bridgwater, in Lancashire and Cheshire; and the Intended one Promoted by Earl Gower and other Persons of Distinction in Staffordshire, Cheshire, and Derbyshire* (3 edns.)

Works (excluding millwork)

1752–1756. Wet Earth Colliery Drainage, 800 yd. tunnel, inverted siphon

1765–1766. Calder and Hebble Navigation, Sowerby Bridge extension, and Brookesmouth—Salterhebble Bridge extension, engineer

See the table for canal works

BRISCOE, Henry, Captain (?–1837), was a military engineer who, after graduating from the Royal Military Academy at Woolwich, was commissioned a Second Lieutenant in the Royal Engineers in 1813.

It is not clear when Briscoe was posted to Lower Canada (now Quebec) but in 1822 he prepared a survey of lands around the fortifications at Amherstburg on the Detroit River with notes on the condition of bridges and buildings. In 1826, and again in 1827, he was sent up the Ottawa River to survey a water route to Lake Simcoe and to Georgian Bay on Lake Huron. This route, which ultimately came to be known as the Georgian Bay Ship Canal, would have provided an 'all Canadian' shipping route through the interior to the upper Great Lakes. Briscoe's extensive report is the first documented survey of this canal which, although never built, was long considered an important transportation project and was nearly approved in lieu of the now famous St. Lawrence Seaway. Briscoe was later appointed to the Rideau Canal project under John By (q.v.) and assigned to the Jones Falls division where he might have been responsible, among other things, for the supervision of the construction of the impressive Jones Falls Dam. This dam is a cut-stone, masonry (in vertical courses) arch dam 61 ft. high with a crest 350 ft. long; it was the largest dam then constructed in North America and the most significant structure on the Rideau Canal.

Promoted to Second Captain in 1831, Briscoe was given a permanent appointment to the canal staff in 1832 after the canal was completed but apparently soon after returned to England. He died on 17 August 1837 at Demerera, British West Indies.

MARK ANDREWS

[T. W. J. Conolly (1898) *Roll of Officers of the Corps of Royal Engineers*; R. F., Legget (1955) *Rideau Waterway*; E. F., Bush (1976) *The Builders of the Rideau Canal*, Parks Canada, MSS Report No. 185; J. Winearls (1991) *Mapping Upper Canada 1780–1867; An Annotated Bibliography of Manuscript and Printed Maps*]

Publications

1826, *Report of a Survey to Examine the Water Communication between Lake Simcoe and the River Ottawa*, Rideau Canal Office (now Ottawa) (updated and reissued 1827)

BROCKEDON, William (1787–1854), painter, writer and inventor, was born at Totnes on 13 October 1787, the son of a watchmaker. He was educated at a private school and learnt his mechanical skills from his father. Following his father's death in 1802, Brockedon spent six months in London with a watch manufacturer before returning to Totnes where in ran the family business for the next five years. During this period his talent for drawing became apparent and he found the patronage to study painting, especially in Continental Europe, and as a consequence of this he also wrote a number of travel books.

Despite his painting and literary activities, Brockedon also found the time to make a number of inventions. The most important of these was his method of drawing very fine wires through small holes in gemstones and his method of making pencils using finely ground graphite compressed *in vacuo*. He patented both these (and other) inventions and Mordans manufactured the pencils. His pencils were very popular with his fellow artists since they were free of grit.

Brockedon's interest in invention doubtless led to his being elected an Associate of the Institution of Civil Engineers in 1824 but he ceased paying his subscriptions in 1831. He does seem, however, to have retained some link with the Institution since in 1834 he drew a pencil sketch of Thomas Telford (q.v.). He was a friend of Michael Faraday (whom he also drew) and delivered eleven Friday Evening Discourses at the Royal Institution on various aspects of his work. Brockedon's only son became a pupil of Isambard Kingdom Brunel, but died 1849. Brockedon never recovered from the shock of his son's death; his health deteriorated and he died on 29 August 1854 and was buried in the burial ground of St. George, Brunswick Square.

FRANK A. J. L. JAMES

[Membership records, ICE archives; F. James (in progress) *The Correspondence of Michael Faraday*; DNB]

BROCKET, Andrew (fl. 1787–1812), of Glasgow, contractor, may well have been the man who married Janet Brown in 1783, the ceremony being recorded twice, at Lesmahagow in Lanarkshire and Stewarton in Ayrshire. Originally a mason in Stewarton, he worked on the locks of the Crinan Canal for at least six years, receiving over £14,000 in payments in that time. There were difficulties in recruiting and keeping skilled labour there. His workmanship and energy won the praise of the Resident Engineer, John Paterson (q.v.) who noted also that Brocket and some of the Company's staff 'made trials of their gifts in prayer before such judges as could be found to decide between them'.

He is first noted in the Glasgow directories in 1803 and became much involved in building the newer parts of the city, opening quarries for building stone and laying out streets and squares there. He also erected an obelisk 144 ft. high on Glasgow Green, the first public monument in Britain to Nelson's victories, to the designs of the architect David Hamilton.

Later, Brocket left his contract at Greenock Harbour amidst some recrimination, although it may be noted that the client also had trouble with his successor, John Murray (q.v.). From 1810, while building Inchinnan Bridge at Renfrew, his address in Glasgow was in the same street as Robertson Buchanan (q.v.), the Engineer for the bridge. He moved back to the Anderston district in 1820 and was still noted there in 1835.

He employed William Stuart and Alexander Easton (qq.v.) as young men and was described in the obituary of the former as 'a contractor of considerable eminence in the West of Scotland'. His only son John, matriculated at Glasgow University in 1797. Any relationship with G. H. Brockett whose firm of Brocket and Mitchell worked for John Rennie (q.v.) at Plymouth Dockyard in 1816 has not been established.

P. S. M. CROSS-RUDKIN

[Rennie reports, ICE archives; Rennie mss, NLS; A2/1/107, Mitchell Library, Glasgow; Obituary of William Stuart (1854) *Mtn Procs ICE*, **14**, 138]

Works

1787–1789. Forth & Clyde Canal, contractor (with William Ross) for locks 28, 29, 30, 38 and 39 (sea lock)

1793–1799. Crinan Canal, contractor for locks (signed 20 November 1793), £14,000

c. 1800–1806. Firth of Clyde Navigation, contractor, £2,700 p.a.

1806. Nelson's Monument, Glasgow, contractor

1807–1809. Greenock Harbour, contractor

1809–1812. Inchinnan Bridge, contractor, cost, £19,130

BROTHERHOOD, William (1778–1839), contractor, the eldest son of Henry Brotherhood, in the early 1800s developed a contracting business in estate improvement in the area to the west of London including Uxbridge, Slough, Windsor and Staines. In 1811 Brotherhood married Charlotte Wilder (1790–1859) and their son Rowland (1812–1883) was born at Whitton Deane on 5 October 1812. He began working for his father in December 1821. By the 1830s they had considerable experience in earthworking contracts, including the excavations for the Manchester Zoological Gardens (*c.* 1835) and, unsurprisingly, became involved in the construction of the Great Western Railway.

Their first contract was for the foundations of the Wharncliffe Viaduct, and Rowland acted as agent. William resolved to concentrate on this type of work and further contracts followed on an associated new road to Southall, embankments at Maidenhead, and the line between West Drayton and Langley. With the opening of the line to Maidenhead in 1838 they took on maintenance contracts on that section of the line following the decision of the Great Western Railway to relieve William Ranger (q.v.) of his contract at Sonning Cutting; in September 1838 the Brotherhoods took the contract on. Unfortunately William died in a fall following a visit to Rowland on 26 January 1839. His other sons returned to estate work, but Rowland persevered and became an important contractor, particularly for the Great Western Railway, as well as establishing a bridge fabrication and locomotive works at Chippenham.

MIKE CHRIMES

[F. Smith archive, ICE archives; ICE membership records; S. A. Leleux (1965) *Brotherhoods, Engineers*]

BROWN(E), James (fl. 1802–1840), civil engineer, from Scotland, worked under Thomas Townshend (q.v.) on the Royal Canal of Ireland, when John Rennie (q.v.) was Engineer. He was also employed by Thomas Townshend as one of his assistants, along with Patrick Leahy and John Thomas (d. 1844) (q.v.), on work for the Irish Bog Commissioners. He carried out 124 days of surveying 1809–1810, and 64 days 1810–1811, on district 6. Townshend then removed to survey work in Ulster, and there seems little doubt he was able to recommend Brown to John Rennie when he was looking for a resident engineer for the Admiralty Pier at Holyhead Harbour when he was appointed engineer in 1810. This was a major civil engineering project. Rennie died in 1821 and his works were largely completed in 1824. Some minor work was carried out by Brown under Thomas Telford acting as engineer to the Holyhead Road Commissioners, including a dry dock. Under the Commissioners his salary was £320 per annum.

With completion of works at Holyhead Sir John Rennie (q.v.) recommended Brown as resident engineer for small improvements to the pier at Douglas, Isle of Man (1832), and, shortly afterwards as resident engineer at Hartlepool. Here radical proposals for harbour and railway improvements had been authorised for the Hartlepool Dock and Railway Company, with Sir John Rennie acting as consulting engineer for the dock. James Milne (q.v.) another former Rennie associate, was initially appointed engineer for the dock, but his proposals were at variance with Rennie's, and this led to his dismissal in 1833. There were construction difficulties arising from the ground conditions, particularly ingress of water through fractured limestone, but work on the main basin was completed under Brown by the end of 1835. Work then continued on what became known as the Victoria Dock, which was opened in 1840. Brown's date of death is unknown, but he had died by 1851 when Sir John Rennie noted him in his great work on harbours. He is referred to as James Browne in the reports of the Holyhead Road Commissioners.

A namesake worked for Boulton and Watt as a marine engineer for many years; his notebook is preserved in the Institution of Civil Engineers archives.

MIKE CHRIMES and P. S. M. CROSS-RUDKIN

[Commissioners ... into ... Bogs in Ireland (1810–1814) *1st–4th reports*; Holyhead Road Commissioners (1823–1830) *Reports*; J. Rennie (1875) *Autobiography of Sir John Rennie*; R. W. Rennison (1987) The development of the North Sea coal ports (Ph.D. thesis, University of Newcastle-upon-Tyne)]

Works
1810–1828. Holyhead harbour, resident engineer, £80,000+
1832–1833. Douglas pier, Isle of Man, resident engineer
1833–1835. Hartlepool harbour, new basin, resident engineer, £160,000
1835–1840. Hartlepool, Victoria Dock, resident engineer, £80,000

BROWN, James, John and **William** (fl. 1800–1820), mill owners of Dundee, were pioneers in the application of iron and steam power to textile mills in the Dundee area. In the late eighteenth and early nineteenth century Dundee was a centre of the linen industry, but had limited water power for mechanical flax spinning which was being introduced in Scotland from the late 1780s. East Mill, Dundee, was converted from a tannery to a flax mill by 1799 by G. Wilkie and powered by a Boulton and Watt steam engine. The Napoleonic wars caused a crisis in the flax trade and in 1809 the mill was acquired by the Browns who carried out a series of improvements, installing a cast iron beam on the engine. The mill has an unusual (single) row of 'Y' columns supporting two longitudinal timber beams. The columns are cylindrical, but the arms of the Y are 'T' in section. The Y was designed to accommodate the horizontal shaft driven directly by the flywheel. Originally there was one line shaft but in 1823 Brown installed a second.

In 1806–1807 James Brown built the Bell Mill, Dundee, chiefly modelled on Bage's Meadow Lane Mill, Leeds, with cruciform columns, cast iron beams and jack arches. The roof was of timber supported on slender octagonal cast iron columns, with cast iron fire plates below. The steam engine was by Matthew Murray (q.v.).

The J. & W. Brown's South Mill of 1825 was a more conservative two-storey structure with cast iron columns supporting timber floors, and John Brown's Column Mill had no columns at all.

William Brown made important advances in flax processing, including the spinning of twine.

MIKE CHRIMES

[M. Watson (1990) *Jute and Flax Mills in Dundee*]

Works
1806–1807. Bell Mill, Dundee
c. 1809. East Mill, Dundee
c. 1825. Old Mill, South Mills, Dundee
1828. Column's Mill, Dundee
c. 1846. Arch Mill, Dundee

BROWN, John Stevenson (1782–1852), ironmonger, smith and bridge-builder, was born at Hallyne, Peeblesshire, on 16 October 1782. His father, William Brown, was 'a highly respected millwright' at Hallyne and his mother, Margaret Stevenson, was daughter of a farmer, tenant of Fulford Farm south of Edinburgh. He had a brother, William, four years younger, and four or more sisters, three of them older than him.

He was apprenticed in May 1798 to Thomas Edmondston, ironmonger in the West Bow, Edinburgh, where John Redpath, ten years his senior and his subsequent business partner, was already working. The partnership began when Brown's indentures were completed in May 1803 and the business, which comprised the sale of ironmongery, some but not all being of their own manufacture, prospered. Contract works were also undertaken both in Edinburgh, where Brown was made a burgess on 15 September 1807, and elsewhere in Scotland, including Peebles, four miles from his father's home, where he received his burgess ticket in 1817. It was near Peebles in that year that the partners undertook their first recorded civil engineering contract, to build a suspension footbridge across the river Tweed on the King's Meadows estate of Sir John Hay. The spanning structure was all of iron wire, the second bridge of wire to have been built in Scotland and possibly in Europe; some supporting elements were of wrought iron bar and cast iron, and timber was used for the deck and foundations. Brown was the designer, with Redpath and Brown contracting to construct the bridge. Suspension was achieved by stays from the heads of the end portals to the deck.

For the firm's second bridge the initial design, presumed to include both ironwork and masonry, was made by John Smith (q.v.) of Darnick in 1825, Redpath and Brown contracting to build the ironwork for £600. It was the footbridge from Melrose to Gattonside, of 300 ft. span, with stone pylons and catenary cables of wrought iron eyebars, and at a celebration dinner after its completion in October 1826, Brown received praise and an engraved silver punch-bowl from the assembled trustees. It has been rumoured that his partner was unhappy at Brown's undertaking of these bridge contracts.

In 1821 Brown married Maria Badger, of Walsall, who was related to a family called James involved in the iron trade there and they had two daughters, Catherine and Maria, born in 1822 and 1823. Brown's wife died in 1827 and he never remarried. He was resident between 1815 and 1825 at 5 Graham Street in Edinburgh but by 1831 he lived at 17 George Square and later bought no. 18 as well. Having retired from the business in 1841, he moved out of the city in 1847 and lived at several addresses to the south and south-east, dying at Smeaton Park, Inveresk, a mansion-house he had built, on 29 November 1852. He was buried in St. John's Episcopal churchyard, Princes Street, Edinburgh, where there is a memorial to him and his wife. He owned property at Inveresk and Bridge of Allan, and some shops on George IV Bridge, Edinburgh, as well as the two houses in George Square which he bequeathed to his two daughters.

The firm he founded with John Redpath continued trading under their names until the 1960s and became a major stockholder of steel sections and ships plates and also contractor for the design and erection of steel structures.

TED RUDDOCK

[A. W. Turner (1996) *Redpath Brown: 200 Constructive Years* (privately circulated, copy in NLS) and sources quoted; Ted Ruddock (1999) Blacksmith Bridges in Scotland and Ireland 1816–1834, *Proceedings of Historic Bridges Conference, Wheeling, West Virginia*, October 1999, and sources quoted]

Works

1817. King's Meadows footbridge, near Peebles (designer and contractor), span 110 ft., cost £160, destroyed by flood 1954

1825–1826. Gattonside footbridge, by Melrose (contractor for ironwork), span 300 ft., contract price £600. Refurbished insensitively with much loss of the ironwork in 1991

BROWN, Lancelot (1716–1783), the celebrated landscape gardener universally known as 'Capability' Brown. For thirty years he held almost a monopoly in designing and executing 'improved' landscapes in the grounds of great country houses. Many of his schemes involved the creation of ornamental lakes. These usually required the construction of earth embankment dams. Details of six of them are available.

Brown was not a pioneer in this field; earlier examples are dams for the Serpentine lake in Hyde Park (1730–1731) by James Horne (q.v.) and the Great Water at Grimsthorpe, Lincolnshire (1746–1748) by John Grundy (q.v.). But Brown's dams are among the largest of the mid-eighteenth century and some pre-date the dams for canal reservoirs, one of the earliest being Townshead reservoir for the Forth & Clyde Canal (1770–1773) by John Smeaton (q.v.).

Born in Northumberland, Brown attended a village school and from the age of 16 was employed as a gardener by Sir William Loraine of Kirkharle, where he was later entrusted with laying out an addition to the grounds. After small commissions at other places in the county he moved south in 1739 and in 1741 became head gardener to Lord Cobham at Stowe. He was soon acting also as Clerk of Works for the various alterations in the grounds and for buildings designed by the architect James Gibbs. It seems that Cobham lent the services of his gardener to the Duke of Grafton whose estate at Wakefield Lodge in Northamptonshire lay five miles from Stowe, and by tradition it was Brown who created the great lake there, which laid the foundations for his fame and fortune. Dating from *c.* 1748 the lake is impounded by a dam 700 ft. long with a maximum height of 25 ft., probably built of boulder clay in a 'homogeneous' section without the need for a special watertight element.

Brown left Stowe in 1751 to set up an independent practice. His first big commission came in 1752 to landscape the grounds of Petworth House, Surrey, for the Earl of Egremont. The scheme included a lake with an earth dam 540 ft. long, about 23 ft. high and a 35 ft. wide crest carrying an estate road. The upper half of the lake is on Atherfield Clay while the lower part of the lake and the dam itself lie on the sandy Hythe Beds. This part of the lake was therefore lined with a clay blanket which continues up the inner slope of the dam, protected by stone pitching near the crest and by sand and rough stone lower down. The main bank consists of random fill. Beneath the bank is a brick culvert controlled by a wooden sluice gate operated from the crest through a vertical shaft, the shaft and culvert being surrounded by clay.

A contract of June 1755 refers to digging all such parts as are not deep enough (for the lake) and making the 'clay walls'. By July 1756 the lake had been filled but there were signs of leakage. Whatever remedial measures were taken they must have been successful as no further trouble is recorded.

The work at Blenheim for the Duke of Marlborough constitutes Brown's most notable achievement. The Great Lake, 125 acres in area, is impounded by a dam known as the Grand Cascade three-quarters of a mile below the bridge and extends nearly half a mile upstream in the valley of the river Glyme. It completely submerged the relatively small 'canal' created by John Armstrong (q.v.) 1722–1724. Brown's plan of the lake is dated 1765 and his sketch of the Grand Cascade a year earlier. Work was completed in 1770. Below the cascade the river Glyme was widened, deepened and in places re-channelled with a half-mile long embankment, known as Lince dam, before falling over a cascade into the river Evenlode a mile downstream of the lake. These works were started in 1767 and also finished in 1770. The whole site is underlain by Great Oolite limestone.

Lince dam embankment, 13 ft. high, is built of rubble limestone with a clay blanket on the inner face, as at Petworth. The Grand Cascade dam has a length of 440 ft. and a maximum height of 29 ft. with limestone facing on the upstream slope and a central clay core. Between 1764 and 1774 Brown was paid a total of £21,500 for work at Blenheim, but this includes reshaping the ground around the lake and new tree plantations in the estate.

Brown's next work of this kind was at Harewood House in Yorkshire. His 'plan for the intended water' dates from 1772. Earth moving began in 1774 and the lake was ready for filling in 1777. The dam is 380 ft. long, 25 ft. high with a 20 ft. crest width and slopes of 2.5:1 upstream and 3:1 downstream. It is almost certainly of the 'homogeneous' type, built of boulder clay. The overflow is at one end of the dam in natural ground. Cost about £900.

In 1775 work began on the lake at Burghley House, near Stamford, for the Earl of Exeter. The 23 acre lake was filled in 1777. The dam has a clay blanket on the upstream slope under stone pitching, and outlet arrangements similar to those at Petworth. The lake, on Upper Lias Clay, needed no lining.

Lastly, Brown signed a contract in 1775 for general landscaping at Sherborne Castle, Dorset. The work was carried out during the next four years at a cost of £1,250 the main item being a 49 acre lake impounded by a dam 23 ft. high and 500 ft. long, built mainly of clay founded on clay of the Fullers Earth formation, with side slopes of about 2.5:1.

In his landscaping work Brown acted as designer and contractor. He visited the sites fairly frequently but day-to-day supervision was the responsibility of 'foremen'. Some of their names are recorded: Benjamin Read at Blenheim, Cornelius Dickinson at Harewood and Sher- borne, and William Ireland at Burghley. On his staff he had two surveyors, John Spyers and Samuel Lapidge, both of whom were fully employed from 1765 or earlier until Brown's death in 1783, making preliminary surveys, preparing plans and as draughtsmen.

Brown was appointed Master Gardener to the royal parks at Hampton Court and Richmond (including Kew) in 1764. For this he received £2,000 per annum, but had to pay the undergardeners

and meet the cost of replacing trees and plants and gravel. Even so it ranked as a highly paid post, especially as it entitled him to an official residence at Hampton and allowed him to continue his private practice. The foremen, George Lowe at Hampton and Michael Milliken at Kew, were paid £150 per annum. Milliken had previously worked for Brown at Chatsworth, Derbyshire, where a large amount of earth moving was carried out 1760–1763 (costing well over £2,000) to 'improve' the landscaping, and a loop in the River Derwent was straightened.

Brown's activities included architecture. His architectural works, including some bridges in country house parks, are listed by Colvin. Several of the buildings were erected by Henry Holland, a master builder of Fulham. In 1771 Brown took Holland's son Henry into an informal partnership, gradually handing over to him the architectural side of his practice. Henry Holland junior (1745–1806) became one of the leading English architects of the late eighteenth century.

Brown married Bridget Wayet, from a 'very respectable county family' of Lincolnshire, in 1744. The wedding took place at Stowe. Their eldest son Lance went to Eton in 1761, their daughter Bridget married Henry Holland junior in 1773. Having lived since 1751 in Hammersmith the family moved to Hampton Court in 1765. Brown died on 6 February 1783 leaving a fortune of well over £20,000 to his wife, three sons and two daughters.

A. W. SKEMPTON

[Geological Surveys maps, Sheets 157, 170, 202, 236, 312; Dorothy Stroud (1975) *Capability Brown*; M. F. Kennard (1975) Eighteenth century dams in Britain, *Symposium on Inspection, Operation and Improvement of Earth Dams*, paper 5.1.1; Thomas Hinde (1986) *Capability Brown, the Story of a Master Gardener*; G. M. Binnie (1987) *Early Dam Builders in Britain*; A. Crowley (1993) Blenheim Park, *VCH Oxfordshire*, **12**, 460–468; H. Maggridge (1997) Capability Brown at Blenheim, in J. Bond and K. Tiller, *Blenheim, Landscape for a Palace*, 90–114; Colvin; Tom Wilkinson (1998) *Chatsworth, the Park and Gardens*]

Works (lakes with earth dams)
These include:

c. 1748. Wakefield Lodge, Northamptonshire
1755–1757. Petworth, Sussex
1765–1770. Blenheim, Oxfordshire, total cost of works £21,500
1774–1777. Harewood, Yorkshire, £900
1775–1777. Burghley, Northamptonshire
1775–1778. Sherborne Castle, Dorset, £1,250

BROWN, Nicholas (fl. 1793–1838), civil engineer, of Saddleworth, and later Wakefield, is first mentioned in connection with civil engineering work as surveyor of a proposed Huddersfield Canal in 1793. The canal, which was to form part of a Manchester–Hull link, was promoted by shareholders of the Ashton Canal and obtained its Act the following year with Benjamin Outram (q.v.) as engineer. Construction involved two short tunnels and two large flights of thirty-two and forty-two locks up to and down from the 3 miles long summit tunnel at Standedge.

Although Brown's survey had been crude and of little value for engineering purposes, nevertheless he was appointed to act as agent and superintendent under Outram at a salary of £315 per annum. While progress on the bulk of the canal was satisfactory the costs and difficulty of the tunnel delayed its completion and in early 1799, with the completion of the Stalybridge-Saddleworth section, Brown was given notice, although the canal was not complete. His lack of experience had resulted in much criticism of his supervision.

He is next heard of in connection with a survey of a proposed High Peak Junction Canal in 1810, intended to link the Cromford Canal to the Peak Forest and on to Manchester; this was not proceeded with. In 1814 he was taken up, once more, by the Huddersfield Canal Company as their engineer, the tunnel having been completed and the canal opened in 1811; John Rooth had supervised the final stage of the work. Brown appears to have done little to improve the canal, perhaps stymied by the lack of funds, and in 1819 he was succeeded by John Raistrick [sic], who set about improvements, continuing in office until 1843.

In the 1820s Brown was involved in the creation of the Gorton Reservoirs for the Manchester and Salford Waterworks Company. The lower reservoir was created by lining the upstream face of the embankment of the Ashton Canal branch across the Gore Brook with puddle and building an outlet shaft at the upstream end of the culvert. With its neighbour it enclosed 35 million cu. ft. of water with a total surface area of 56.5 acres. Based on canal reservoir practice it was one of the earliest impounding reservoirs for public water supply. Associated works included a 3 mile cast iron pipe main of 18 in. diameter to convey the water to service reservoirs and settling ponds at Beswick. Once completed Brown determined the works were supplying Manchester with 943,250 gallons a day. The whole scheme was overhauled by James Simpson who became engineer to the company and reported on the condition of the supply in 1842.

Brown also carried out surveys for the Mersey and Irwell Navigation, Rochdale Canal, and reported on water supply for the proposed Macclesfield Canal. He was probably responsible for the initial design of the Todd Brook Reservoir on the Peak Forest Canal, submitting reports in 1835–1836, following an initial report by William Crosley (q.v.). Once work had begun on the embankment by the contractor William Collinge (q.v.), William Mackenzie (q.v.) acted as the consultant.

Brown was elected a Member of the Institution of Civil Engineers in 1826, when resident in Wakefield, and is last heard of in 1838. As an ICE member he contributed to its early discussions, and referred to the cement being manufactured by Aspdin (q.v.). He carried out estate surveys and acted for Enclosure Commissioners in the West Riding of Yorkshire. His relationship with Thomas Brown (q.v.) is uncertain.

MIKE CHRIMES

[Membership records, ICE library; Macclesfield Canal, Minutes of General meetings (PRO: RAIL 850/1); J. F. La Trobe Bateman (1884) *History and Description of Manchester Waterworks*, 18–25; C. Hadfield and G. Biddle (1970) *The Canals of North West England 2*; R. B. Schofield (1981) The construction of the Huddersfield narrow canal 1794–1811, *Trans Newc Soc*, **53**, 17–38; R. W. Rennison (1996) *Civil Engineering Heritage: Northern England*]

Publications

1793. *Plan of the Canal between Huddersfield in the County of York and Ashton-under-Lyme in the County of Lancashire* (Outram, engineer)
1827. *Plan of the River Mersey from Runcorn to Woolston Weir Showing the Works of the Mersey and Irwell Navigation Company at Runcorn*

Works

1793–1799. Huddersfield canal, surveyor and superintendent
1814–1819. Huddersfield canal, engineer
1823–1825. Lower Gorton Reservoir, engineer

BROWN, Sir Samuel, FRSE, Captain (1774–1852), born in London on 10 January 1774, was the eldest son of William Brown of Borland, co. Galloway, and Charlotte, third daughter of the Reverend Robert Hogg of Roxburgh. In August 1822 Brown married Mary, daughter of John Home, of Edinburgh. There was no issue to the marriage. Brown entered the Royal Navy in 1795 as an able seaman, rose through the ranks, was promoted Lieutenant in 1801, was advanced to the rank of Commander in 1811 and in 1842 accepted the rank of retired Captain. He was appointed a Knight of the Hanoverian Guelphic Order in January 1835 and in 1838 was made a Knight Bachelor. Brown saw action in a number of ships during the Napoleonic Wars, and was present as first lieutenant of the frigate *Phoenix* (Captain Thomas Baker) at the capture, after a furious single ship action, of the larger French frigate *La Didon*. Subsequently he had a number of other appointments, and served for a short time under Captain Lord Cochrane, later 10th Earl of Dundonald.

While still on active service Brown took the first steps towards his later involvement with the promotion and construction of suspension bridges and chain piers. He did not make the transition from naval officer to engineer/entrepreneur in a single step, but began by

Captain Sir Samuel Brown FRSE

experimenting with the use of wrought iron for standing rigging in ships and for mooring vessels at anchor. In 1808 he registered a patent (No. 3107) for the use of iron for standing rigging and subsequently tested his proposals on a sea-trial to the West Indies of the *Penelope*, a vessel of 400 tons. Whilst Brown's proposals for iron rigging were not successful, the use of iron anchor chains was. Chains for mooring were given further trials on four naval vessels, each of which was equipped with 100 fathoms of chain cable. The trials proved successful and chain cables were subsequently taken into use by the Royal Navy.

At this time Brown established two small chain works near the Pool of London and in 1812, when these works became too cramped, he moved to new premises at Mill Wall, Poplar. In 1816 Brown, in conjunction with Philip Thomas, registered a patent (No. 4090) for a new method of manufacturing chains which reduced the number of heating operations required to form links. The finance needed to establish the chain works at Poplar came from his cousin Samuel Lenox and in 1818 the two men opened another works for the manufacture of chains at Newbridge, South Wales; from 1823 this firm traded as Brown Lenox and Company. The demand for chain cables was sufficient for Brown to grant licences to other firms to use his specification for the manufacture of chains and his success as a chain maker provided the finance for him to experiment with the construction of bridges using suspension cables manufactured from bar-links, and for him to become a leading promoter of the genre. In 1813

Brown constructed a prototype suspension bridge using the bar-link design adapted from his proposals for standing rigging. The bridge, 105 ft. (31.98 m) span, was erected at his Millwall works and was inspected by Thomas Telford (q.v.) and by John Rennie (q.v.) who reported favourably on it. Telford later used the principles developed by Brown for the construction of the Menai suspension bridge.

The construction of the prototype bridge led, in 1817, to the registration of Brown's most important patent (No. 4137) *Construction of a Bridge by the Formation and Uniting of its Component Parts.* This patent for suspension bridges registered the method developed by Brown for the construction of the main suspension chains from multi-bar links of either round or rectangular cross-section, and it gave details of the connections to the short links from which were hung the vertical hangers that supported the deck beams. Brown standardised his proposals so that transverse deck beams were spaced at five feet centres and were suspended from alternate chains when more than a single pair of cables was used to support the deck of a bridge. The design for the wrought-iron chains was based on a tensile stress of 10 tons per square inch ($150 N/mm^2$), resulting in the main structural elements of the bridges being very slender.

In about 1817 Brown made his first proposal for the construction of a suspension bridge. The proposed bridge of 1000 ft. (306 m) span was to be built over the river Mersey at Runcorn but neither his nor Telford's proposal was built. In 1820 the Union bridge spanning the river Tweed, the first suspension bridge in the United Kingdom capable of sustaining vehicular traffic, was completed to a design by Brown. This bridge, the first of a number of bridges and piers designed by Brown, established the veracity of the suspension principle that he had earlier promoted.

He used both iron and masonry for the towers and pillars supporting main suspension cables. When arched masonry towers were used for bridges, Brown, having little expertise in the design of masonry, collaborated with local architects and only contracted for the iron and timber elements of bridges. In the case of the Union bridge over the river Tweed, John Rennie designed the masonry towers and advised on the integrity of Brown's design for the ironwork. Generally the masonry design of the towers was simple and utilitarian, but there was a rare exception when castellated gothic details were used for the masonry towers of the bridge over the river Findhorn at Forres, Morayshire.

Brown acted as sub-contractor to supply the ironwork for the suspension bridge at Hammersmith, designed by William Tierney Clark (q.v.), and this resulted in his adoption of Clark's specification for drilling and marking groups of bar-links to ensure their correct location in suspension chains. With the exception of the chain pier at Newhaven, Leith, chains on bridges and piers designed by Brown were constant cross-section throughout their length, but for this pier he graduated the cross-section of the bar-links within the length of the chains to correspond with the load induced in the chains. In the early decades of the nineteenth century Brown was amongst the most eager of those promoting the use of suspension bridges and, because of their lightness and in an age when wind induced oscillations and the problems associated with un-stiffened bridge decks were not understood, damage was sustained by some of his bridges and piers. The chain pier at Brighton was twice, in 1833 and 1836, partially destroyed by wind-induced oscillations of the deck which caused the vertical hangers to fracture. Similar wind-excited oscillations induced in 'hurricane' conditions caused the much more substantial bridge at Montrose to fail in a like manner. Brown's bridge at Stockton was the first suspension bridge built to carry railway traffic and although the bridge had lattice timber parapets, the deck was unstiffened and it was unable to support the weight of a locomotive without undue deflection. During the short working life of the bridge it was necessary to prop it at mid-span and to restrict the load on the deck of the bridge to wagons alone

Brown was involved with a succession of proposals for bridges and piers about which little detail survives including Runcorn, 1811, Saltash, over the Tamar, 1823, St. Katharine's, London, 1824, Shields, 1824–1825, Clifton, 1830, Deal pier, 1824, Wye, 1825, Lockerbie, 1825, Whorlton, 1830, Dollar, 1830, Dalkeith, 1830, Sluie, 1832, Gainslaw, 1832, Lambeth, c. 1835, Greenwich pier, c. 1836, Metropolitan (London), 1836, and Norham, 1837–1838. It is unclear to what extent these were fully worked out designs.

Brown registered patents for a wide range of interests that included, beside those for chain-making and the construction of suspension bridges, proposals for the propulsion of vessels and railway locomotives, improvements to breakwaters, the construction of iron lighthouses, and refinements to the mariner's compass. He also prepared reports for the improvement of the harbour at Leith (1828) and on the state of Westminster bridge (1836).

Brown was a member of the Society of Arts (1820), and was elected a member of the Royal Society of Edinburgh in 1831. He died at Vanbrugh Lodge, Blackheath on 15 March 1852.

TOM DAY

[Telford mss, ICE Library; Strathern and Blair W. S. (1825–1832) GD 314, SRO; papers regarding Tamar suspensionbridge, Duchy of Cornwall archives; QS Records, Isle of Ely RO; DD 749, Gwent CRO; DNB; Patents associated with civil engineering; No. 4090, 1816; No. 4137, 1817; No. 9630, 1843. *Records of the Patent Office*; Anon. (1814) On chain cables or moorings, *Phil Mag*, **44**, 294–303; Anon. (1820) Facts respecting the comparative strength of chain cables on the construction of Capt. Samuel Brown …, *Phil Mag*, **55**, 456–463; S. Brown, 'On iron bars and cables, made at the patent cable manufactory of Captain S. Brown …', in P. Barlow (1817) *An Essay on the Strength and Stress of Timber*, 232–237; R. Stevenson (1821) Description of bridges of suspension, *Ed (New) Phil J*, **5**, 237–256; B. Hall (1825) Some remarks respecting the utility of chain cables, *Ed (New) Phil J*, **13**, 316–325; J. Marshall (1833) *Royal Naval Biography*, 4.1, 20–6; W. O'Bryne (1849) *A Naval Biographical Dictionary*, 183; *Gent's Mag*, **i** (1852), 519; E. L. Kemp (1977) Samuel Brown: Britain's pioneer suspension bridge builder, *Hist Tech*, 1–37; R. A. Paxton (1977–1978) Menai Bridge (1816–1826) and its influence on suspension bridge development, *Newc Soc Trans*, **49**, 87–110; T. Day, An examination of the contribution made by Samuel Brown to the architecture and design of suspension bridges (M.Litt. thesis, University of Aberdeen, 1982); T. Day (1983) Samuel Brown: his influence on the design of suspension bridges, *Hist Tech*, 154–70; T. Day (1985) Samuel Brown in North-East Scotland, *Industrial Archaeology Review*, 7.2, 154–70; D. Smith (1991–1992) The Works of William Tierney Clark (1783–1852), *Newc Soc Trans*, **63**, 186–188]

Publications

1816. *Reports and Observations on the Patent Iron Cables*

1822. *Report and Broadsheet to the Brighton Pier Committee*

1822. Description of the Trinity Pier of Suspension at Newhaven near Edinburgh, *Edinburgh (New) Philosophical Journal*, **6** (1822), 22–28

1824. *Prospectus and Report of the Proposed Plan of Erecting a Patent Wrought Iron Bridge of Suspension, over the Thames, near Irongate and Horslydown* (with J. Walker)

c. 1825. *Plan and Report of the Patent Wrought iron Suspension Bridge … from North to South Shields* (n.d.)

1825. *Broadsheet and Prospectus, 'North and South Shields … Suspension Bridge designed by Captain Samuel Brown'*

1828. *Prospectus for Reconstruction of Leith Harbour*

1836. *Private Report on the Present State of Westminster Bridge*

Works

1819–1820. Union Bridge, Norham, Northumberland/Berwickshire

1821. Trinity Pier, Newhaven, Leith

1822–1823. Brighton Chain Pier

1825. Netherbyres Bridge, Eyemouth, Berwickshire

1825–1826. Warden Bridge, Northumberland

1826. Welney Bridge, Cambridgeshire, £2,710

1828–1829. Montrose Brridge

1829. Chain Bridge, Kemeys Commander, Monmouthshire

1829. Llandovery Bridge

1829–1830. Stockton Railway Bridge
1830–1831. Wellington Bridge, Aberdeen
1830. Boharm Bridge, Banffshire
1830. Bath Chain Bridge
1832. Findhorn Bridge, Forres
c. 1835. Kalemouth Bridge, Kelso
c. 1840. Kenmare Bridge, Co. Kerry
1843–1844. Earith Bridge, Cambridgeshire, c. £1,500

BROWN, Thomas (1772–1850), engineer, was born at Mowhole, Disley where his father owned some 11 acres in small plots of land and property. He is believed to have been apprenticed to Benjamin Outram (q.v.), for whom he worked on a survey for the Ashton Canal in 1791 and was subsequently supervising surveyor or resident engineer on the Peak Forest Canal 1793–1801. He acted as Company Engineer 1801–1804 and was subsequently consulting engineer for the company 1804–1846, representing them in Parliament. Both he and Outram were paid half yearly salaries, and Brown was increasingly relied upon, becoming a leading shareholder. He was responsible for the impressive flight of locks at Marple.

The Peak Forest Company, with its Derbyshire shareholders, supported the idea of a Macclesfield Canal link with the Trent and Mersey which Outram had surveyed as early as 1796. When the idea of a Macclesfield Canal was revived in the mid 1820s the Peak Forest may well have recommended Thomas Brown as engineer/surveyor. Thomas Telford (q.v.) drew up the parliamentary proposal but Thomas Brown was on the Committee of Management until 1827 and was thanked for his work by the Company. Actual construction was the work of William Crosley (q.v.).

Brown was still being consulted about the Peak Forest in the 1830s when Nicholas Brown, William Crosley and William Mackenzie (qq.v.) were brought in about the reservoir at Tods Head. He had an extensive practice as a mining surveyor in the Hyde area of Cheshire and married Elizabeth Hancock, the daughter of his partner in coal and lime business. They had three children, William, Elizabeth and Richard. He died at 16 Ardwick Green, Manchester, on 24 January 1850.
MIKE CHRIMES

[C. Hadfield and G. Biddle (1970) *The Canals of North West England 2*; B. Lamb (1999) in lit.]

Works
1793–1846. Peak Forest Canal, (assistant) engineer

BROWN, William (d. 1782), colliery viewer and engineer, was probably born at Throckley, near Newcastle upon Tyne, and first came to notice in the 1740s, when he was an overman for one of Lord Carlisle's collieries. It is said, perhaps apocryphally, that his career really took off after a chance discussion with Matthew Bell, a well-to-do draper in Newcastle, from whom Brown was buying some flannel for his pit clothes. On mentioning to Bell that he thought it a pity that Heddon pit in Northumberland should be laid in, Bell took over collieries at Throckley and appointed Brown as his viewer. Matthias Dunn (q.v.), however, suggested that he was recommended to the Throckley position by another viewer, William Newton. Whichever, if either, story is true, Brown became the most important viewer of his day, with responsibilities for the Throckley, Heworth, Shiremor, Walbottle, Washington Oxclose, Beamish, Hartley and Ford collieries.

By 1751, Brown had laid a new waggonway from Throckley to the Tyne, and waggonway building became another of his specialities. Thus in 1754, he advised on underground waggonways at Bo'ness colliery, on the south side of the Firth of Forth, he surveyed a waggonway route for the Lambtons in 1770, and c. 1779 advised on a waggonway for Inverkeithing as part of his plan for improving the harbour there to the benefit of the coal trade in Scotland. Between 1749 and 1765 he also regularly corresponded with Carlisle Spedding (q.v.) on matters of waggonways, waggonway gauge and gradients, iron waggon wheels, waggon brakes, laying ashes on rails, etc., as well as on colliery matters such as the driving of drainage levels, Brown's preference for circular shafts rather than Spedding's oval shafts, horse gins for winding, the use of fire baskets to aid ventilation, Newcomen engines and their cylinders, the use of steel mills for underground lighting, etc.

After constructing a Newcomen pumping engine for one of the Throckley pits in 1756, Brown went on to become pre-eminent in the building of such engines, at least twenty being built in the North East, and three in Scotland, before 1776; most of them are included in his 1769 list of those then working in the North and Scotland. In 1759, in his capacity as viewer to John Hussey Delaval's collieries in Northumber- land at Hartley and Ford, he visited Staffordshire to see one of James Brindley's (q.v.) Newcomen engines in action, and particularly to study Brindley's new kind of boiler, with which he was not too impressed. In the same capacity he despaired of ever being able to increase the sale of Hartley coals, while the natural harbour remained unimproved. In 1761, Brown advertised in local newspapers for contractors to cut the new harbour but, as the person nominally in charge of developing it, he felt ill-suited to his task, suggesting in vain to Delaval that 'one Robson of Sunderland' should be called in for a consultation; this was undoubtedly Joseph Robson (q.v.), then engineer to the River Wear Commissioners. The new harbour opened in March 1764 but there is, however, no indication that Brown was directly involved with its construction after December 1761.

Brown became much in demand for his advice, not just in the North East, and not just on coal mining. He visited Scotland in 1768, to take borings in, among other places, Cromarty, to advise the landowner on the prospects for working coal there. He reported on the poor performance of a water wheel used for working the bellows at the Blaydon on Tyne lead refinery. He was also asked to make an estimate of the cost of building 'six cynder ovens' at an unspecified location in c. 1774.

In 1772 he joined in partnership with Matthew Bell and William Gibson, an attorney at Law and town clerk of Newcastle, to take leases on mineral royalties at Willington on Tyne and to work the coals there. By 28 October 1775, the Willington Engine shaft had reached the 6 ft. 4 in. thick main coal at a depth of 100 fathoms, the event being celebrated with a dinner for 262 people, and an additional 150 people partaking of liquor—a waggon load of punch and a large quantity of ale. Although Brown provided two Newcomen engines for the shaft, he also drove the 'Tyne Level Drift' some 2,030 yd. towards the Tyne to assist drainage. By May 1776, the colliery was equipped with a powerful 'water gin' after the style of Joseph Oxley (q.v.). Brown erected the first coal screens at this time and also experimented with a form of robot miner which became known as 'Willy Brown's Iron Man'. The partnership later became well known throughout the whole of the coal trade as Bells & Brown.

Brown died at his house at Willington, on 14 February 1782, and the *Newcastle Courant* noted that he had been 'a considerable coal-owner and principal viewer of the collieries in this country, a gentleman greatly respected for his skill and integrity'. He had two sons, Richard and William, whose names appear in various mining reports. William Jr. was born in 1743 and died in 1812.

STAFFORD M. LINSLEY

[Brown, William, *Letterbook*, Northumberland County RO, NRO 3410; *Delaval Papers*, Northumberland County RO; Matthias Dunn (1811) *The History of the Viewers*, mss in the Northumberland County RO, NEIMME collection East/3B; John Holland (1841) *Fossil Fuel, the Collieries and Coal Trade* (repr. Frank Cass, 1968); R. L. Galloway (1898) *Annals of Coal Mining and the Coal Trade*, vol. 2 (repr. David and Charles, 1971); William Richardson (1923) *History of the Parish of Wallsend* (Newcastle upon Tyne); *Northumberland County History* **13** (1930); T. V. Simpson (1930–1931) Old Mining records and plans, *Trans Inst Mining Engineers*, **LXXXI**, 75–108, 292–94]

BROWNRIGG, John (c. 1749–1838), surveyor and canal engineer, was possibly the son of Henry Brownrigg, a military engineer. It is thought that he may have had two sons, James (b. 1770) and Henry. John was a surveyor of estates, roads and canals. Together with Thomas Sherrard, Brownrigg served an apprenticeship with Peter Bernard Scalé, who was agent to Lord Downshire at Edenderry in county Offaly. At the Society of Artists in Ireland in 1770, Brownrigg exhibited a survey drawing of Passage, county Waterford and, in 1771, a survey drawing of the demesne of Wentworth Thewles at Harristown, county Kildare. His address at this time is given as 'at Mr Jones's cabinet maker in Charles Street' (Dublin). In about 1773, Scalé took Sherrard and Brownrigg into partnership at 123 Lower Abbey Street and later Sherrard and Brownrigg formed a partnership of their own, which was dissolved in 1778 when Brownrigg set up independently in Grafton Street in Dublin.

From the 1780s, Brownrigg worked with the Grand Canal company. He supervised the laying out of the Barrow Branch of the Grand Canal and, in 1788, prepared a map and section of 'that part of the Grand Canal now perfected', which he published the following year. Also in 1788, he proposed connecting the Grand Canal with the river Liffey by a flight of locks at James's Street in Dublin. Two years later he was appointed assistant engineer to Richard Evans (q.v.) on the Upper Boyne Navigation. He was also at this time Director of the Barrow Navigation and employed Tarrant to survey the navigation.

In 1789, Brownrigg was engaged by the Directors of the Royal Canal company to make a survey of the proposed route for the Royal Canal. He prepared another survey of the Royal Canal in 1801, including the proposed extension to Coolnahay in county Westmeath. He also appears to have designed the company's hotel at Moyvalley in 1805–1806. In or about 1803, Brownrigg was, along with John Killaly (q.v.), appointed as an engineer to the Directors General of Inland Navigation, a position he held until their dissolution in 1831 on the establishment of the Irish Board of Works. He applied for a pension in 1833 at the age of eighty-four. He worked on the Shannon navigation between 1803 and 1809 and, with Killaly, inspected the state of the navigation in July 1810. In 1803 he produced an estimate for £13,157 for work required on the canal from Limerick to Killaloe. He was employed on the Tyrone and Newry navigations until 1812.

In 1830, Brownrigg made a proposal for a possible ship canal between Dun Laoghaire and the Grand Canal Docks in Dublin. In this he agreed with Killaly and Rennie, but the railway was opened in 1834 and passage across the Bar to Dublin Port improved, thereby removing the need for any canal link.

Brownrigg was admitted a Freeman of the City of Dublin in 1794 and from about 1799 until about 1806 held the position of Surveyor of the Dublin Paving Board, a post which had previously been held by Thomas Brownrigg. His plan for a road linking Kevin Street with Portobello is mentioned in the Minutes of the Wide Streets Commissioners for 13 June 1806. In the 1820s he was the government's examiner of surveys of Crown lands.

John Brownrigg died in 1838.

RON COX

[M. B. Mullins (1859) Presidential address *Trans ICEI*, **6**, 1859–1861; J. H. Andrews (1985) *Plantation Acres*; Bendall]

Publications

1801. *Report to the Directors of Inland Navigation*

Works

1780s. Barrow line of Grand Canal, supervising engineer

1790s. Boyne navigation, assistant engineer

1803–1809. Shannon navigation improvements, Limerick–Killaloe

1810–1812. Tyrone and Newry navigation, engineer

BRUNEL, Sir Marc Isambard, FRS (1769–1849), civil engineer—always known as Isambard [or Isambart]—was born on 25 April 1769 at Hacqueville, Normandy, where his family had farmed for more than three hundred years. He was the second son of Jean Charles Brunel, farmer and hereditary Maître des Postes, by his second wife (of four), Marie Victoire (Lefèvre). His parents intended him for the church and at the age of eleven he was sent to the seminary of St. Nicaise at Rouen whose principal, accepting that Brunel's talents lay elsewhere, helped his transfer to pupillage with François Carpentier, a retired sea captain, for training in hydrography (under M. Dutagne) and draughtsmanship, in preparation for service in the French Navy. In 1786 he was commissioned as 'Voluntaire d'honneur' (an honour only previously conferred on M. de Bougainville), and sailed for the West Indies aboard the corvette *Le Maréchal de Castries*.

Paris in 1792 was hostile to his royalist sympathies, so he returned to stay with the Carpentier family, where he first met his future wife Sophia, orphan daughter (the youngest of 16 children), of William Kingdom, contracting agent for the Navy and the Army. From Rouen, he managed to obtain a passport for America, sailing for New York on 7 July 1793. He set out initially with two fellow French refugees to undertake a large land survey near Lake Ontario and the Black River. Through a meeting with an American, Thurman, he was next commissioned to survey the line of a canal to link the Hudson river with Lake Champlain; at this point he decided to become an engineer in preference to rejoining the French Navy. His first major success was the winning design for a new Congress Building in Washington, which was never built on grounds of cost. In modified form, resembling the Halle aux Grains in Paris, it was built as the Park Theatre in the Bowery, New York, but was destroyed by fire in 1821.

At the age of twenty-seven he was appointed Chief Engineer of New York, in which capacity his achievements included a design for a new cannon factory and advice on the military defences for the navigation channel between Long Island and Staten Island. In 1796, Brunel was made a citizen of USA—District of New York.

Sir Marc Isambard Brunel FRS

Through Major General Hamilton, then serving as a British diplomat in Washington, Brunel was introduced to Earl Spencer of Althorp, First Lord of the Admiralty. He sailed from New York on 20 January 1799 and on arrival in Britain used this influence to prepare novel designs of ships' block-making machinery for the Royal Navy, manufactured by Henry Maudslay (q.v.). After initial rebuffs, support by Sir Samuel Bentham, Inspector-General of Naval Works, led to the government adopting his proposals in 1803. By 1806, forty three machines for executing the several processes in making blocks were in use at Portsmouth dockyard, with the claim that six men could thereby do the previous work of sixty. This was the first of many initiatives by Brunel to improve current manufacturing techniques which ranged from pantographs for copying drawings to textiles, metal sheet printing and timber working machinery. Too often Brunel failed to establish rights to his inventions although he obtained 18 patents between 1799 and 1825.

From his work in designing wood-working machinery (1805–1812) for government mills at Woolwich, Brunel invested in his own sawmills at Battersea. His observations of the wretched footwear of soldiers returning from Corunna set him to persuade the Government to commission him in supplying machine-stitched boots. There was no formal contract, however, and peace in 1815 caused the accumulation of large surplus stocks at Brunel's expense (The boots were subsequently offered to the Prussian army). Fire in 1814 had destroyed his sawmills, the loss compounded by failure of his partner to control the finance of the undertaking. These circumstances contributed to his arrest for debt in 1821 and confinement, in the company of Sophia, in the King's Bench Prison until patrons and friends, including the Duke of Wellington, to secure his release, obtained an honorarium of £5,000 from the government in recognition of past unpaid services.

In 1810 Brunel designed an air engine, developed to a high pressure gas engine following Faraday's experiment in 1823 on the liquefaction of gases. In 1812 he undertook experiments in steam navigation on the Thames using a boat powered by a double-acting piston engine with paddle-driven propulsion. He demonstrated this capability to the Navy which in 1814 accepted his proposals for steam-driven tugs. In 1816, however, conservatism prevailed, and the Navy Board described his scheme as 'too chimerical to be seriously entertained'; Brunel was left to bear the cost of the experiments.

Brunel's resumption of civil engineering included designs for a sawmill in Chatham dockyard, where surface transport incorporated a wide gauge (7 ft.) rail track, together with sawmills for Trinidad and Berbice, British Guyana. His designs in 1820 for a composite low timber bridge across the Seine to La Croix Island and for a timber arch bridge of 850 ft. span across the Neva at St Petersburg were not built but his design for a 122 ft. long chain suspension bridge with lateral arched stays for Bourbon, French West Indies, to withstand hurricane winds was manufactured and pre-erected in Britain, then shipped to the West Indies in 1823. In 1822 Brunel designed a swing-bridge to Liverpool docks—it may have been built—and prepared an elegant design for a new bridge over the Serpentine, in Hyde Park, London. In 1826, he designed a tidal landing stage for Liverpool.

During a visit to Chatham in 1816, Brunel had observed the method of boring through ship's timbers used by the teredo which inspired his inventions of the tunnelling shield, patented in 1818, in two different circular forms. The first projected application was in 1820 for a 22-foot diameter carriage tunnel beneath the Neva at St Petersburg. By 1823, Brunel had become a principal advocate for a road crossing beneath the Thames primarily to connect the docks currently under construction on each bank of the river Thames. He prepared plans for a tunnel between Rotherhithe and Wapping, explored the availability of land and prepared descriptive plans and models. He modified the design of the shield to a rectangular form, principally to standardise iron castings for the structure and to provide suitable working cells for the miners. His design comprised 12 frames, each of three cells set one above the other; at each side and over the top of the shield were fitted sliding staves, advanced independently of the frames. Each frame was supported by a massive iron foot, advanced separately from the frame and the working face was supported by poling boards, held against the face by screwjacks attached to the frame. At the rear, the shield thrust against completed brickwork lining by screw jacks. The shield was designed to progress by 4½ in. advances, applied to alternate frames, in order to ensure continuous support of the face at all times.

Shaft sinking at Rotherhithe began in 1825, tunnelling the following year. Immediately great problems were encountered, boreholes having erroneously suggested that the tunnel would be in clay throughout; in reality it ran also in sands and silts of the Woolwich and Reading beds. Major river interruptions occurred, the most serious in 1828 leading to nearly seven years stoppage while funds were raised to complete the project. Tunnelling was completed in 1841 and the project officially opened in 1843 as a pedestrian tunnel between the two working shafts. The original intention of extension to 200 ft. access shafts with helical roadways was never undertaken.

While the tunnel was stopped, Brunel was engaged on re-designing the shield, in planning the mid-river shield chamber—in itself a major undertaking—and in fund-raising. He also undertook surveys for the Grand Junction Canal and the Oxford Canal, for docks at Woolwich and for a bridge over the Vistula at Warsaw. Additionally, he worked on details of the suspension links for Isambard Kingdom Brunel's Clifton Bridge. He undertook experiments within the tunnel of cantilevers of brickwork in roman cement mortar reinforced with tendons of vegetable fibre, timber laths and steel reinforcement.

Brunel investigated schemes for bridges at Kingston-on-Thames and across the Tamar, an aqueduct from Hampstead to Hammersmith, two suspension bridges for the Huddersfield Canal Company, improvements to the Bermondsey Docks, for new docks in South London, a rubble masonry and a brick arch bridge across the Dee at Chester, a bridge at Totnes and a canal, partly in tunnel, from Fowey to Padstow. He was consulted by Richard Beamish (q.v.) regarding a bridge in Cork.

In 1814 Brunel was elected Fellow of the Royal Society and in 1841 he was knighted as Sir Isambard. He received several recognitions from France, as a corresponding member of the Institut de France, as member of several organisations local to Rouen and as a recipient of the Légion d'Honneur in 1829. He was elected a member of the Royal Academy of Sciences of Stockholm in 1818 and in 1823 he became a Member of the Institution of Civil Engineers, taking an active part in its meetings; he received the Telford silver medal in 1839. Brunel married Sophia Kingdom in 1799, the marriage resulting in two daughters, Sophia and Emma, and a son, Isambard Kingdom. Brunel Sr. suffered strokes in 1842 and 1845 and he died on 12 December 1849. There is a memorial to him in Kensal Green Cemetery, a development of the scheme prepared by Brunel with Pugin for a Necropolis of London, and at least three portraits of him exist.

SIR ALAN MUIR WOOD

[Thames tunnel diaries, journals, etc., ICE Archives; The Brunel Collection, University of Bristol Library; R. Beamish (1862) *Memoir of the Life of Sir Marc Isambard Brunel*; *Nouvelle Biographe Universelle* (1852–1865); Cynthia Gladwyn (1971) The Isambard Brunels, *Proc ICE*, **50**, 1–14; Jonathan G. Coad (1989) *The Royal Dockyards 1690–1850*; A. M. Muir Wood (1994)—The Thames Tunnel—where shield tunnelling began, *Proc ICE*, **102**, 130–139]

Works (civil engineering)

1812–1814. Chatham Dockyard Sawmills design
1821–1823. Bourbon Suspension Bridge, design and procurement.
1826. Liverpool Landing stage, design.
1825–1843. Thames Tunnel, London, design and development, £454,000

BRUNTON, William, Sr. (1777–1857), civil engineer, was the eldest son of a watch and clock maker at Dalkeith in Ayrshire and studied mechanics in his father's shop and engineering from his grandfather who was a viewer at a local colliery. William's brothers John (fl. 1790–1835) and Robert (1796–1852) also became engineers. In 1790 William began work in the fitting shops at New Lanark Mills and in 1796 moved to Soho Works where he worked closely with James Watt (q.v.) and became superintendent of the engine manufactory in 1802 with a salary rising to £100 p.a. At some stage his brother, John, joined him at Soho; although he was generally to be associated with the gas industry, they were to work together on occasion for the rest of their lives.

In 1807 he moved to manage the newly enlarged engine manufactory at Jessop and Outram's (qq.v.) Butterley Ironworks. This pioneered early marine steam engines. In 1810 for example he installed a steam engine in a dredger for the Bristol Dock Company. He made a well-publicised intervention in steam locomotive design, doubting the adhesion of railway locomotives, he experimented with engines propelled by legs (Patent No. 3700, 1813) in 1813–1815 at Crich and at Newbottle Colliery, Newcastle, the latter exploding with five fatalities. Despite this calamity Brunton continued to develop inventions throughout his career and took out eight further patents, mostly related to furnaces and ore dressing (Patents: No. 4387, 1819; No. 4449, 1820; No. 4685, 1822; No. 5621, 1828; No. 5722, 1828; No. 6500, 1833; No. 9135, 1841; No. 9351, 1842).

In 1815 he moved to Regent Street, Birmingham, as a partner and as the mechanical engineer of the Eagle Foundry which built marine engines for early steamers on the Humber, Mersey and Trent, including the *Lady Stanley*, based at Runcorn.

Around 1824 Brunton set up as a London-based consulting engineer. Among his clients was the Harvey's foundry in Hayle, Cornwall; N. O. Harvey had been his pupil in Birmingham. In 1824–1826 Brunton engineered the 10 mile long Redruth and Chacewater Railway in Cornwall. He joined the Instiution of Civil Engineers in 1824 and the following year moved to Canonbury Square, London, and entered independent practice as a civil engineer. He became a regular contributor to the early meetings of the Institution on topics such such as iron manufacture, steam engines, locomotives and railways. He patented an arsenic calciner that held the ore on a rotating iron table and was engaged to carry out a series of improvements to the Ynysgedwyn Ironworks in the Swansea Valley in 1828. The latter included a river weir driving a blowing-machine and a short colliery tramroad with traction provided by a two-way differential winding-drum. He assisted in the surveys for the London–Birmingham Railway, 1830–1832. In 1832–1835 he supervised his sons in building the nearby 5 mile long Claypon's Tramroad which used large earthworks rather than the masonry causeways used earlier and he published works advocating the use of steam-driven excavators on railway excavations (Patent No. 6500, 1833). He was also involved in the 1830s with an alternative route to Brunel's Great Western Railway to Bristol.

In 1835 he became managing-director of the Cwmavon Tinplate and Copperworks and developed its infrastructure until 1838 when he left and lost a substantial sum by investing in the Vale of Neath Brewery. In 1849 he was living in Newport (Monmouthshire) and built an early steam-driven colliery fan at Gelligaer Colliery: he published the design and exhibited a model at the Great Exhibition.

He retired in 1850 and died at Camborne, Cornwall, on 5 October 1851. He married A. E. Button on 30 October 1810. They had two daughters and six sons of whom John (1812–1899), Robert, William (1817–1881) and George (1823–1900) all became engineers; he is not known to have been related to Thomas Brunton, whose cable manufactory was the location of Thomas Telford's (q.v.) famous Runcorn Bridge tests.

STEPHEN HUGHES

[Membership records and Minutes of conversations, ICE archives; Evidence (1817) *Select Committee on Steam Boats*, etc., 34–37; (Experiments on iron) (1843) *Min Procs ICE*, **2**, 126–127; *Min Procs ICE*, **10** (1851), 55 (On the ventilation of mines); 1854 *Min Procs ICE*, **13**, 390 (Description of smoke-consuming fire grates); Obituaries of William Brunton Senior and Junior, *Min Procs ICE*, **xi** (1851–1852), 95–98, **lxvii** (1881–1882), 395–396; William Brunton Senior, *Dictionary of National Biography*, VII, 148f; H. W. Dickinson and R. Jenkins (1927) *James Watt and the Steam Engine*, 280–297; J. Marshall (1978) *A Biographical Dictionary of Railway Engineers*, 43–4.; C. C. Short (1989) The William Bruntons and their buddle, *Industrial Archaeology Review*, **XII**, I, 50–54; S. R. Hughes (1990) *The Archaeology of an Early Railway System: The Brecon Forest*

Tramroads, 127–131; A. Guy and J. Rees, eds. (2001) *Early Railways*, 125–128]

Publications

1811. Improved pump for raising the water whilst wells are sinking or making, *Soc Arts Trans*, **29**, 98–105 (repr. *Phil Mag*, **39**, 362–367)
1813. *Repertory of Arts and Manufactures*, **2**, 29, 65
1831. 'Mr. Brunton's Report' and 'Supplementary Report From Mr. Brunton', in *Reports on the Formation of a Floating Harbour at Swansea with Reference to Plans Submitted to the Trustees*
1833. *Plan of the Intended Basing and Bath Railway* (with F. Giles)
1834. *Observations on Railways*
1836. *Description of a Practical and Economical Method of Excavating Ground and Forming Embankments for Railways &c.; with Practical Observations on the Construction of Railways*
1849. *On the Ventilation of Coal Mines*

BUCHANAN, George, FRSE (1790?–1852), civil engineer, was born about 1790, the third son of David Buchanan, a printer and publisher at Montrose (1745–1812). His father was an accomplished classical scholar who published numerous editions of the Latin classics. Buchanan was educated at Edinburgh University, where he was a favourite pupil of Sir John Leslie. About 1812 he began business as a land surveyor, but his strong scientific interests led him to devote himself to the profession of a civil engineer. In this capacity his work in the construction of harbours and bridges made a considerable local reputation. In 1822, on the invitation of the directors of the Edinburgh School of Arts, he delivered a course of lectures on mechanical philosophy in the Freemasons' Hall, remarkable for the originality of his experiments. Although he subsequently gave some courses of lectures on natural philosophy, his increasing business as an engineer interfered with any further educational work.

In 1827 he drew up a report on the South Esk estuary at Montrose in relation to a dispute concerning salmon fishing. This report attracted the attention and gained the marked commendation of Lord-justice-clerk Hope, then solicitor-general, who afterwards, as long as he remained at the bar, always gave the advice in any case involving scientific evidence to 'secure Buchanan'. Subsequently in all the important salmon-fishing questions which arose, and which embraced nearly every estuary in Scotland, Buchanan's services were enlisted, the point being generally to determine where the river ended and the sea began.

When the tunnel of the Edinburgh and Granton Railway was being constructed under the new town, and the adjacent buildings were considered in imminent danger, Buchanan was commissioned by the sheriff of Edinburgh to supervise the works on behalf of the city. In 1848 he began the work of erecting the huge chimney, nearly

400 ft. in height, of the Edinburgh Gasworks and he carried out an exhaustive series of experiments to assure its stability.

Buchanan was the author of several scientific works, as well as contributing to a large number of scientific journals, including the *Courant* newspaper and the *Transactions of the Royal Scottish Society of Arts*. He also contributed the article on 'Furnaces' to the eighth edition of the *Encyclopaedia Britannica*. He was a fellow of the Royal Society of Edinburgh and was elected president of the Royal Scottish Society of Arts for the session 1847–1848. He died of lung disease on 30 October 1852. David Buchanan (1779–1848) and William Buchanan (1781–1863) were Buchanan's elder brothers. His son George Buchanan (1834–1902) was also a distinguished civil engineer.

RON BIRSE

[NLS (Acc. 10706, 545); Minute Book 1837–1857, Moray Gardens Committee, Edinburgh; *Scotsman*, November 1852; *Courant*, 19 June 1851; *Proc Roy Scot Soc Arts*; DNB; *ICE Min Procs*, **150** (1902), 425–427; Skempton]

Publications

1824. *Report Relative to the Theory and Application of Professor Leslie's Photometer, Edinburgh*
1824. *Report on the Present State of the Wooden Bridge at Montrose, and the Practicability of Erecting a Suspended Bridge of Iron in its Stead*
1824. *Plans of the Mouth of the River South Esk Shewing the Various States of the Tide*
1827. *Reduced Plan of the Lower Division of the River South Esk and of the Annat Bank and Stake Nets Thereon with Part of the Bay of Montrose from a Survey by George Buchanan, Civil Engineer*
1827. *Report on the Erecting of Low-water Landing-places on the Coast between Kinghorn and Pettycur*
1829. *Tables for converting the Weights and Measures ... in Use in Great Britain into Those of the Imperial Standards*
1830. *Plan of the Estate of Ingliston Belonging to James Gibson Craig, Esq. Drawn by George Buchanan* (n.d.)
1832. *Views of the opening of the Glasgow and Garnkirk Railway. By D. O Hill Esq ... Also, an Account of That and Other Railways in Lanarkshire. Drawn up by George Buchanan*
1835. *Report on the Erecting of Low-water Landing-places, on the Coast between Kinghorn and Pettycur*
1849. *Abstract of Expositions on the Strength of Materials, Read before the Royal Scottish Society of Arts*. Parts 1 and 2. Edinburgh, 1848. Part 3. Edinburgh, 1849

Works

1837–1840. Remedial works by masonry counterforts and arches after ground slip behind Ainslie Place and Moray Place, Edinburgh, adjacent to Dean Bridge, for Gardens Committee of the Moray Estate development (this involved a large bulk of masonry; other parties were represented by James Jardine, Alexander Stevenson, Thomas Grainger and John Miller)

BUCHANAN, Robertson (1769–1816), engineer, was born on 14 July 1769 in Glasgow. His family was well-connected; his uncle Neil Buchanan was the MP for Glasgow in 1768, and his mother, who died in childbirth, was the daughter of Arthur Robertson, Chamberlain of the City of Glasgow. Buchanan's father died when he was about fifteen years of age and he was placed with a carpenter, and subsequently received some training as a millwright.

After spending a short time working in London Buchanan returned to Glasgow and set up in business as a millwright. In 1791 he was appointed manager of the Rothesay cotton mill on the Isle of Bute. Although this was not ultimately financially successful, and contributed to a breakdown in his health, his work provided him with a wealth of experience which he later exploited in his writings which brought him national eminence.

In 1796 notice of his inventiveness was revealed in articles in the *Repertory of Arts and Manufactures*. That year he patented a ship's pump, whose novel feature was the discharge of water below the piston to avoid clogging with solids. By 1800 he was actively promoting its sales, with agents in London, Liverpool, Glasgow and Greenock. In the following year he moved to London and received an attentive hearing from the Navy Board and was actively considering setting up a manufactory there, but it seems little came of it. He did however meet Count Rumford and Professor Pictet, who encouraged his interest in heating buildings. This stimulated his career as a technical author. He had already begun work on *An Essay on the Teeth of Wheels* but there were problems with the publisher, and in 1807 he published his first work on heating. By then he had returned to Glasgow and in 1809 he was resident at York Street Foundry in the Port Dundas district. For the remainder of his life he produced a succession of technical publications.

As a civil engineer Buchanan had a varied practice in the south-west of Scotland, but is remembered almost entirely for one completed work. He was the designer of the unusual double bridge over the White and Black Cart Waters immediately above their confluence at Inchinnan, near Renfrew. The bridge of 1759 had crossed the combined river a little lower down, but was swept away in 1809. Rebuilding on the new site necessitated an abrupt turn in the road where the two bridges abutted. The arches are four-centred with a pronounced curvature towards the springing and there are apertures above the piers, unique in Britain. Following on, he was appointed Engineer for the construction of Cree Bridge at Newton Stewart, which had been designed by John Rennie (q.v.). As John Hall (q.v.) was resident inspector, Buchanan's duties

involved a visit to site every three months, and sufficient progress had been made by October 1813 for his services to be dispensed with, six months before completion.

He was employed to resurvey the Nith Navigation after an Act had been obtained in 1811, based on a plan by James Hollinsworth (q.v.), but the work was carried out under the supervision of the local cabinetmaker/architect/engineer, Walter Newall. In 1812 Buchanan was consulted about the Clyde Navigation, and prepared a report on Port William in Wigtownshire; in 1816 he reported on the navigation of the river Urr in Kirkcudbrightshire. It is not clear whether action ensued from any of these.

In 1811 he published a plan for a 30 mile railway from Sanquhar down Nithsdale to Dumfries, which was notable on several counts. It showed that he was familiar with practice on the Sirhowy Railway in South Wales, where passengers had been carried for four years, and although the Sanquhar line was projected to carry coal, he was one of the first to make provision in his estimates for newly generated classes of traffic. He also envisaged the line as forming part of a through route from Carlisle to Glasgow, very much along the course taken more than thirty years later.

He last appears in the Glasgow directories in 1816, when he had a house in Nile Street as well as the foundry at Port Dundas. He died at Creech St. Michael, Somerset on the 22 July of that year. Described by Thomas Tredgold (q.v.) as 'a man of amiable character, with a strong sense of religious and moral duty … His practical knowledge in mechanics was extensive … a close and accurate observer …', his published work on millwork is the most detailed available to that time, with a wealth of information on strength of materials. It would appear he was unable to translate his ability into financial success.

P. S. M. CROSS-RUDKIN

[GD171/692 and 734 at Falkirk Museum; Kirkcudbrightshire Commissioners of Supply minutes at Ewart Library, Dumfries; Glasgow directories, Mitchell Library, Glasgow; *New Statistical Account* **vii**; *Tidal Harbours Commission* (1846); C. J. A. Robertson (1983) *The Origins of the Scottish Railway System, 1722–1844*]

Publications
1807. *An Essay on the Warming of Mills and other Buildings by Steam*
1808. *General Account of the Cranston Hill Waterworks*
1808. *An Essay on the Teeth of Wheels* (revised by P. Nicholson (q.v.))
1810. *Practical and Descriptive Essays on the Economy of Fuel and Management of Heat*
1811. *Report Relative to the Proposed Rail-Way from Dumfries to Sanquhar*
1814. *Practical Essays on Millwork and other Machinery* (later edns. revised by Tredgold and G. Rennie (qq.v.))

1816. *A Practical Treatise on Propelling Vessels by Steam*
Articles on 'Cotton spinning' and 'Sir Richard Arkwright' in *Edinburgh Encyclopaedia* (*c.* 1810)

Works
1809–1812. Inchinnan Bridge, engineer, cost £19,100
1812–1813. Cree Bridge, Newton Stewart, engineer (was paid £2 2s per day plus travelling expenses)

BUCK, George Watson (1789–1854), civil engineer, was born at Stoke Holy Cross, Norfolk, on 9 April 1789, the son of Quakers. He and his two brothers attended the Quaker School at Ackworth, Yorkshire, and subsequently he studied Latin and French for a year with a clergyman. His father wanted him to become involved in trade and he was placed with a wholesaler at Tower Hill. Buck was unhappy there and he was eventually able to obtain a position at Old Ford pumping station then under construction for the East London Water Works to designs by Ralph Walker (q.v.), *c.* 1807. He then worked as resident engineer, under Walker, on the Farlington water supply scheme near Portsmouth, which obtained its Act in 1809. The original engineer was Barodall Dodd (q.v.). The scheme was intended to supply both Portsmouth and neighbouring towns such as Gosport. A rival scheme was engineered by William Nicholson (q.v.). Both experienced problems with continuity of supply.

In 1819 Buck was appointed engineer to the Eastern branch of the Montgomeryshire Canal, a waterway designed to run from Porth-y-waen, north of Llanymynech, to Newtown. In the first phase of construction (1794–1797) it was, in fact, only built between Carreghofa on the Ellesmere canal and Garthmyl, which became known as the Eastern branch, and the work on the remaining (western) branch did not recommence until 1815 due to lack of funds, with Josias Jessop (q.v.) as consultant. The eastern branch was the work of the relatively inexperienced John Dadford assisted by his elder brother Thomas Dadford Jr. (qq.v.), with John Simpson and William Hazledine (qq.v.) acting as contractors for major structures like the Vyrnwy aqueduct. This, and other aqueducts on the line were showing signs of distress when Buck became engineer and he immediately began a major series of reconstruction, rebuilding Luggy aqueduct, Brithdir, as a cast iron trough (1819) and introducing cast iron lock gates, and probably cast iron, fish-bellied, beam bridges. In 1823 work began on the bulging distorted masonry of the Vyrnwy aqueduct using wrought-iron tie rods and cast-iron facing plates, together with cast-iron beams on the actual arch face. Buck also used cast iron for railings on the aqueducts and probably introduced a new form of paddle gearing for the lock sluices in 1831. In

1821 he designed a waterwheel to pump water to the top pond on the western branch at Newtown.

In December 1832 Buck was also made engineer to the western branch, an appointment which was short-lived as in December 1833 he left to become one of Robert Stephenson's (q.v.) senior assistant engineers on the London-Birmingham Railway in charge of the section from Tring to Camden Town. He had already shown interest in railway affairs, visiting the Stockton and Darlington Railway in 1828 and attending the Rainhill trials in 1829. His salary—£600 per annum—on the London and Birmingham Railway reflects his experience and ability. He was a candidate for the post of resident engineer when the railway opened in 1837 but turned down the conditions offered. He then became Engineer-in-Chief on the Manchester and Birmingham Railway where the Dane and Stockport viaducts can be seen as his masterpieces. In 1840 he briefly visited Germany to work on the Altona–Kiel Railroad, but ill-health forced his early return. His work on oblique bridges (1839), possibly written in response to the demands of junior engineers working under him, became a standard textbook and was last reprinted in 1895. He became a close personal friend of Robert Stephenson who supported him in later life.

Buck was elected a Member of the Institution of Civil Engineers in 1821 and contributed to the early Minutes of Proceedings. Pressure of work in the railway mania brought on a crisis in his health and as a consequence he became deaf and retired to Ramsay, Isle of Man, and became increasingly engrossed in a study of the Scriptures. His death was brought on by an attack of Scarlatina and he died on 9 March 1854. He was buried in Manghold Churchyard, alongside his wife and a daughter who died within a fortnight, also from Scarlatina. He is apparently described by F. R. Conder as 'of the stamp of the old county surveyor, or the engineers of the school of Telford, than of a mechanical turn. His shrewd grey eye, half inquisitive, half deficient, twinkled with apparent love of fun …'.

MIKE CHRIMES

[F. R. Conder (1868) *Personal Recollections of English Engineers*, see 1983 reprint edn., J. Simmonds, p. 11; D. H. Thomson (1957) *A History of the Portsmouth and Gosport Water Supply*; M. Hallett (1971) Portsmouth's water supply, 1800–1860, *Portsmouth Papers*, **12**, 3–25; S. Hughes (1988) *The Archaeology of the Montgomeryshire Canal*, 4th edn.]

Publications
1838. (contributions to) F. W. Simms, *Public Works of Great Britain*
1839. *A Practical and Theoretical Essay on Oblique Bridges*

Works
c. 1807–1809. East London Waterworks, assistant engineer

1809–1813?. Portsmouth (Farlington) Waterworks, resident engineer

1819–1833. Montgomeryshire Canal, (resident) engineer

1833–1837. London and Birmingham Railroad, resident engineer

1838–1842. Manchester and Birmingham Railroad, engineer

1840. Altona–Kiel railroad, engineer

c. 1840. Tralee and Blennenville Ship Canal, engineer

BUDDLE, John, Jr. (1773–1843), colliery viewer, coal owner, the 'King of the Coal Trade', was born in 1773 at Bushblades, Kyo, in County Durham, where his father, also John, began his working life as a miner, before becoming a schoolmaster, and then a viewer at Greenside colliery, near Ryton on Tyne. In 1792, John Sr. was viewer at Wallsend colliery, where he was one of the first, if not the first, to introduce cast-iron tubbing into mine shafts; he died in 1806. John Jr. (henceforth 'Buddle') had only one year of formal education, although he obviously received a good deal of instruction from his father, whom he succeeded as viewer at Wallsend; he was soon to be regarded as 'head and shoulders' above all the other viewers in the early nineteenth century.

As a viewer, one of Buddle's early personal concerns was the large number of accidents caused by poor ventilation underground, hence his interest in Carlisle Spedding's (q.v.) system of air coursing. During 1807–1810, he experimented with several alternatives to furnace ventilation, for there had been incidents, or suspicion of incidents, where explosions had occurred when the mine gases passed through or over the shaft-bottom furnace. Amongst the methods tried by Buddle was a steam-jet ventilator at Hebburn colliery, but he had greater success with his improvements to Spedding's system when he split the ventilation air into a number of different currents, each of which only ventilated a part of the mine. Thus each current was cooler than it would have been if it had coursed the whole mine, and also carried less gas. The upcast shaft was sectioned for each current flow, each upcast section having its own furnace, and by reducing the quantity of air, and therefore the amount of explosive gas, passing over any individual furnace, there was a reduced likelihood of explosions at the shaft bottom. 'Compounding' the ventilation, as this system became known, was sometimes taken to great lengths, limited only by the diameter of the upcast shaft and the number of sections which could be fitted into it.

Buddle also modified and greatly improved his father's system of cast-iron tubbing, and he was passionate about the need to keep and preserve mining records as a precaution against certain forms of mining accident, his efforts resulting in the creation of the Mining Record Office. He entertained Sir Humphrey Davy at Wallsend colliery

in August 1815, explained the problem of explosions in mines and the need for a safe light, and subsequently assisted Davy by supplying samples of gases and testing the lamps which Davy devised. Buddle was initially dubious about the possibility of a safe lamp but after he had successfully tried the Davy, he noted that: 'If it had been a monster destroyed, I could not have felt more exultation than I did ... [and I was] overwhelmed with feelings of gratitude to that great genius which had produced it'.

He continued to use Davy lamps in his mines for over thirty years. The South Shields Committee was later to criticise the lamps, but Buddle was quick to their defence because, although explosions did still occur underground when Davy lamps were being used, Buddle claimed, rightly or wrongly, that these were usually caused by carelessness on the part of certain miners, who would open the lamp in order to re-light it, or to obtain more light from it. His position with regard to safety lamps was actually more complex than that, for he and colleagues like Nicholas Wood (q.v.), constantly argued that their use was not compatible with the use of gunpowder underground. Moreover, he argued, the use of candles in certain parts of a mine offered the men better protection than the Davy lamp, because the quantity of air passing could be better determined by a candle. It is difficult to avoid the suggestion that Buddle was in fact arguing in favour of greater profits for the coal-owners. As he informed an 1835 Select Committee, it would be detrimental to the mining interests of the North of England were the coal-owners to be deprived of the use of candles, for: 'It would be almost as bad as depriving them of the use of the Davy lamp; because if you take away the candles you take away the gunpowder, and the produce of small coal, and waste would be so great that it would not be worth their while to go on'.

Buddle became viewer for many collieries. He was, along with Thomas Sopwith, a Commissioner under the Forest of Dean Mining Act, and he was a witness before many Select Committees of inquiry into the coal trade, as well as that into 'Child Employment and Mines' Safety', to which Buddle gave seventy-one pages of evidence. He was often called upon to advise on non-mining projects and, following his appointment in 1819 as Agent to the third Marquis of Londonderry, he became deeply involved—with William Chapman (q.v.)—in the development of the new town and harbour of Seaham Harbour in County Durham. The harbour scheme brought him into contact with Thomas Telford (q.v.) whose advice he first sought in 1822. The objective was to develop Londonderry's collieries in the Rainton area and construct a railway to Seaham (then Dalden) which Londonderry acquired in 1821. Problems in financing the project, particularly the new harbour, meant work did not begin at Seaham until 1828. Work on the railroad, built by Benjamin Thompson (q.v.), started the following year. Buddle had long been involved in colliery railroads, and with Chapman whose locomotive patent he had paid for. This involvement included the trials of Chapman's first locomotive at Heaton, ordering the second for Lampton Colliery and rebuilding that railway. This was followed by a third locomomotive he ordered from Hawks (q.v.) in late 1814 for Wallsend Colliery, subsequently worked on the Washington line. On his advice Chapman also received an order from Whitehaven for a locomotive. In 1820 he had the Lampton engine rebuilt for service on a new line at Heaton. In 1822 a further engine was ordered from Joseph Smith, a Heaton enginewright, for use on the Seaham line, although first trialed at Runton. Such co-operation with Chapman continued in locomotive cranes for use at Seaham. Further railway promotion included setting out a proposed Durham to South Shields line in 1832. In time he became a landowner, and a coal-owner in his own right with the Stella Coal Co., and he also became involved in shipping. He and Telford advised the Albion Mines, Nova Scotia, Buddle recommending the use of three locomotives.

Buddle's stature within his profession in the North East was without compare. When the Duke of Wellington visited the North East in 1827, amongst the many functions he attended was a 'cold collation' at Buddle's house at Penshaw. But he was, apparently, also well liked by the pitmen, and made regular visits underground, even into his sixties. He was never slow to descend a pit after an accident, and tried, without much success, to provide funds for widows and orphans, and men injured in the pits. His feelings towards his workforce come across clearly in his reports, that of the 1835 Wallsend explosion making fairly harrowing reading. This is in contrast to Matthias Dunn's (q.v.) reaction after the 1812 explosion at Felling, where he seemed surprised that the families of the ninety-two whose bodies were still underground, wanted to recover them for proper burial.

Buddle was an amateur violinist before pressures of work caused him to abandon it, but later took to the double bass; he was a member of many learned societies, and a Unitarian. He died at his home in Wallsend on 10 October 1843, aged seventy years, but his body was taken to Benwell, where he had latterly lived, for burial in ground which he had donated for a cemetery. The attendance at the funeral showed not only that he was admired and respected by his influential friends of the coal-owning class, but was also held in high regard by the men who had worked under him. There were 'sixty gentlemen on Horse back, while seventy carriages and a vast multitude of miners afoot followed the hearse'; the cortege was said to be a mile long. He had been a bachelor all his life, and left most of his £150,000 to nephews and other relations.

STAFFORD M. LINSLEY

[T. Telford (1833) Diary, NLS; Obituary, *Min Procs ICE*, **3** (1844) 12 13; R. Welford (1895) *Men of Mark 'twixt Tyne and Tweed*; T. V. Simpson (1930–1931) Old mining records and plans, *Trans Inst Mining Engineers*, **LXXXI**, 75–108, 292–94; F. S. Hewitt (1961) Papers of John Buddle (M.A. thesis, Durham University); F. S. Hewitt (1962) An assessment of the value of the papers of John Buddle, *Mining Engineer*, 25–36; C. E. Hiskey (1978) John Buddle, Agent and Entrepreneur (M.Litt, Durham University); D. Bowman (1998) The Rainton to Seaham Railway, 1820–1840, *Trans Newc. Soc.*, **69**, 2, 249–269; A. Guy and J. Rees, eds. (2001) *Early Railways*]

Publications
1831. Notice of a Whin dyke lately discovered in the Fenham Division of Benwell Colliery, *Trans. Natural History Society of Northumberland, Durham, and Newcastle upon Tyne*, **1** (1831), 9–16
1831. An account of the explosion which took place in Jarrow Colliery, on the 3rd of August last, *Trans. Natural History Society of Northumberland, Durham, and Newcastle upon Tyne*, **1** (1831), 184–205
1838. On making the Society a place of deposit for the mining records of the district, *Trans. Natural History Society of Northumberland, Durham, and Newcastle upon Tyne*, **2** (1838), 309–336
1838. Narration of the explosion which occurred at Wallsend Colliery, on the 18th of June, 1835, *Trans. Natural History Society of Northumberland, Durham, and Newcastle upon Tyne*, **2** (1838), 346–383

Works
1828–1835. Seaham harbour, William Chapman engineer
1829–1831. Rainton–Seaham Railway, *c.* 6 miles, £45,600

BUFFERY, David (fl. 1767–1772) was appointed Surveyor of the Works (resident engineer) on the Adlingfleet Drainage scheme in Yorkshire, under the direction of John Grundy (q.v.), 'he residing on the place and giving constant attendance' at an annual salary of £50. He remained in post to the completion of the works in April 1769. The total cost was about £7,000.

In February 1769, while Grundy was preparing his plans for Laneham Drainage, Buffery came over from Adlingfleet and spent a week on site in Nottinghamshire, checking levels. After the Act had passed he was appointed Surveyor of Works in May. The scheme involved 7 miles of embankment along the river Trent, a 12 ft. outfall sluice, main and branch drains and a catchwater drain along the edge of the higher ground to the west. The works were completed in 1772 at a cost of about £15,000.

A. W. SKEMPTON

[Adlingfleet & Whitgift Commission (1767–1772) *Adlingfleet Drainage Minute Book*, Crowle; John Grundy (1768–1772) *Report Books on Laneham Drainage*, vols. 10 and 11, Leeds University Library]

BULL, Samuel (fl. 1768–1798), civil engineer, was first employed by the Coventry Canal as Assistant Clerk of Works for four years at £70 p.a. His lack of experience is reflected in the need to send him to the Staffordshire and Worcestershire Canal for three weeks instruction in lock building. With some experience on the Oxford Canal, he moved on to the Birmingham Canal Company, by whom he was employed, probably, until his death or retirement. The initial phase of construction, designed by James Brindley, was engineered by his assistants Simcock and Whitworth (qq.v.) and completed by 1771. It must already have been evident to the company that the commitments of experienced engineers elsewhere meant they would have to employ their own staff. Bull may have already worked on the canal when, on 6 September 1771 it was agreed he should be 'retained in the Company's service to look after the whole course of the canal between Birmingham and Bilson [sic], and Birmingham and Wednesbury till Lady Day next' at a salary of £70 p.a. This was modest by the standards of the engineering profession of the time.

Thereafter Bull's name crops up regularly in company affairs, carrying out surveys and valuations, measuring water levels, and supervising the locks. His contract was renewed in 1773 and 1774; with his assistants he was in charge of the summit locks, maintenance of the reservoirs and feeders, and towpaths, at a salary of £2 10s a week. Other duties included looking after the company's engines and property. He complained about his workload in January 1780.

By the early 1780s Bull was a trusted employee of the Birmingham Canal Co, carrying out more responsible engineering work, including surveys (1782–1783) with John Snape (q.v.), including the new line to Fazeley. As work progressed on new works he was joined by James Bough (q.v.). Bull was involved in surveying a junction between the Dudley Canal and Birmingham Canal in 1785 with Snape, Thomas Dadford senior acting as consultant. On the Birmingham Canal Bull was involved in resurveying the levels between Fazeley and Birmingham in 1786–1787, an issue which was the cause of a dispute between the contractors, the Pinkertons (q.v.), and Bough, an argument he found in favour of the Pinkertons. The following year he condemned their work on the Dunton tunnel.

In the 1790s Bull was loaned to other canal companies. In 1791 he assisted John Snape in marking out the Worcester and Birmingham Canal, returning to the Birmingham in February 1792, declaring he wanted to stay with them 'so long as he lives'. In 1796 Bough was replaced by Robert Hood, but Bull continued to appear in the

Company minutes until 1798. In March 1797 he requested an additional £50 p.a. for superintending work on the Walsall Canal. In 1790 he and Bough lowered the summit on the Birmingham Canal.

It is possible Bull was related to Edward Bull, the engine erector employed by Boulton and Watt, who subsequently became involved in the notorious patent infringement case in the 1790s. The Birmingham Canal Company were great customers of Boulton & Watt, although Samuel Bull had a very sceptical view of the value of steam engines.

MIKE CHRIMES

[Birmingham Canal Co Minutes, PRO: RAIL 810/2–7; Coventry Canal Minutes, PRO: RAIL 818/1; Worcester & Birmingham Canal Minutes, PRO: Rail 886/4; H. W. Dickinson and R. Jenkins (1927) *James Watt and the Steam Engine*; J. Tann (1981) *The Selected Papers of Boulton & Watt*; C. Hadfield (1969) *The Canals of the West Midlands*, 2nd edn.; S. R. Broadbridge (1974) *The Birmingham Canal Navigations*]

Works

1768–1771. Coventry Canal, assistant to the clerk of works
1771–1798. Birmingham Canal, superintendent

BULL, William (fl. 1824–1846), civil engineer, was appointed engineer to the Calder and Hebble Navigation in succession to William Gravatt (q.v.) in May 1833. Under Thomas Bradley (q.v.) the navigation had been steadily improved—as had his salary which increased from £105 to £500 per annum—most recently with a branch to Halifax, opened in March 1828. On his appointment Bull recommended further works, estimated at £60,000, a response to contemporary improvements in the Aire and Calder; in 1834 an Act was passed permitting improvements up to Mirfield. Henry Robinson Palmer (q.v.) recommended further cuts but his ideas were rejected as too ambitious and in December 1834 it was decided that the improved cuts should be 7 ft. deep and 50 ft. wide, with a 26 ft. width waterway beneath bridges, and locks 70 ft. × 18 ft. 6 in., as on the Aire and Calder.

Under Bull's stewardship a succession of cuts and improvements were carried out, several of which were reported to the Institution of Civil Engineers. In March 1843 the navigation was leased to the Manchester and Leeds Railway, so ending its independent existence. Bull gave evidence on its behalf in 1846.

Bull was probably originally trained as a surveyor and described himself as such when he joined the Institution of Civil Engineers in 1824. He carried out surveys for the Fairbanks, famous railway surveyors, on the Gainsborough–Sheffield–Doncaster line in 1836. In 1824 he was living at 83 Great Tichfield Street in London, but in 1833 was in Halifax and in 1841 in Wakefield. He left the ICE membership in 1844.

MIKE CHRIMES

[ICE membership records, ICE archives; [On Kyanization of timber] *Min Procs ICE*, **2** (1842), 86; [On stone pitching on the Calder] *Min Procs ICE*, **2** (1842), 130–131; C. Hadfield (1972) *The Canals of Yorkshire and North East England*, **1**, 189–206]

Publications

1837. Drawing and description of a wooden bridge erected over the River Calder, at Mirfield, *Min Proc ICE*, **1** (1837), 27; *Trans*, **2**, 87–88, plate viii

Works

1834–1838. Calder and Hebble navigation improvements (Horbury cut, Broad cut locks, Thornes lock)
1835. Mirfield Bridge, timber arch, 147 ft. 6 in., £743 8s 5d

BULMER, Sir Bevis (1536–1615), a mining engineer, was so successful that much of his career can be traced despite its relatively early date. He was born in difficult circumstances to Sir John Bulmer and Elizabeth Elmedon of Clevedon, Yorks., in 1536 perhaps explaining the latter's delayed arrival in the Tower of London for participating in the Pilgrimage of Grace rising. Neither parent acknowledged him as their heir before their execution early in 1537, nor was Bevis immediately made a Ward of Court. However, Bevis probably took his name from the nearby Brotherhood of Papey or the Threescore Priests of Bevis Marks which survived the dissolution of monasteries until 1548. Catholic priests may therefore be responsible for his early education as his strongly Catholic parents' lands were forfeit to the Crown, albeit some were later granted to Sir George Bowes. Under this educational influence and the lifelong guidance of Sir George Bowes (1527–1580) and his sons, all investors in the Company of Mines Royal, Bevis maintained a long interest in that company from 1568 to 1603.

Bevis's professional career began with an interest in deep mining and smelting enterprise on some of the old Bulmer family estates near Wilton in North Yorkshire. Bevis Bulmer took a youthful interest in Sir John Manner's iron smelter beside Rievaulx Abbey, again returning there in 1577 while a big new smelter was being commissioned. His later surviving water and drainage works betray experience of this old monastic site's water supply, and ideas illustrated in Georg Agricola's *De Re Metallica* (Basle, 1556).

About 1562 Bulmer made his name as a mining engineer by founding the Mendip calamine and lead mines of Worle Hill, Greenacre and Rew Pilot, and also the Rowpits mine near Chewton Minery, Somerset. The calamine ores and galena of those deep Mendip mines were to prove crucial to the success which the Saxon émigré engineer Christopher (alias Jonas) Schutz enjoyed, operating the new Tintern smelter of the Company of Mineral and Battery Works from 1565 to 1586, for they were readily taken to the

Purton ferry for transhipment over the Severn to Lydney. Thereafter Bulmer remained on the fringe of the technical contacts of Michael Lok and Jonas Schutz during their disastrous involvement in Martin Frobisher's voyages to the Artic Mines around Baffin Bay in 1576 to 1578.

As a mining engineer, Bevis Bulmer grasped the implications of the interest which Schutz and Lok had taken in the Irish silver and lead mines of Artully in Desmond in 1580 and their abortive proposal of that year for the re-use of the short-lived Dartford smelter built in 1577–1578 to Schutz's design. About 1581 Bulmer visited Artully and the complex of silver mines and smelters at Bannow Bay and Clonmines in Wexford because Richard Hayward, a major London investor in the Company of Mines Royal, was asked to prepare a geographical survey concerning the rights to prospect for, mine and smelt lead ores in England and Ireland.

At that time William Humfrey was acting against John Manners, later Earl of Rutland, and Mr. Folijambe to enforce an injunction stopping them from smelting lead and calamine ores at Rievaulx. Lord Burghley was concerned for the next three years about the geographical limits of the original twenty-one year charter accorded to Humfrey and Schutz. Bulmer's Irish visit was probably the direct result of this initiative from the Privy Council, which preceded his writing a well-reasoned petition during 1584 with help from Lok's son-in-law, Sir Julius Adelmare Caesar, for a patent to build lighthouses to be operated with Trinity House help. The experience of petitioning the Privy Council intelligently stood Bulmer in good stead for later in 1584 Bulmer was granted a perpetual interest in ores and the prospect of mixing of calamine and lead ores with other ores to facilitate smelting. His reputation established, in February 1585 he was one of three asked by an Admiralty Court Commission to assay the Spanish bullion on the *Volante* at Bristol

Although the Dartford smelter's waterwheels were put to other use in milling by 1582, Bulmer, Dee and Schutz all remained interested in the potential of coastal blast furnaces including the Cadoxton furnaces built in 1582 near Swansea for the Company of Mines Royal and then an older seaside complex of silver mines at Coombe Martin. On 13 March 1583 John Dee signed a lease to work any ores found under Knap Down (680 ft.) near Coombe Martin. Itinerant miners hold the key to the usurping of Dee's early interest in Coombe Martin in 1587 by John Poppler and Adrian Gilbert. They took advantage of Dee's self-imposed exile in central Europe to formally encourage Bulmer and about two hundred expert miners to dig and drain the appropriate workings. They offered him half of the output of future precious metal ores.

Bulmer developed Faye's Mine 32 fathoms under Knap Down and eventually his works ran seawards for about the same distance. Bulmer solved the drainage and storage problems of grading ores near the mine's new adits by opening a small coastal smelter there, and shipping Gilbert's share straight to Bere Ferrers for smelting until 1593. On 19th December 1589, on his return from exile, Dee was offered generous compensations for his Coombe Martin mining lease by Gilbert, his former pupil. Gilbert (now acting to protect Bevis Bulmer too) could well afford to pay off Dee's debts because during 1587 and 1588 the Coombe Martin mines had yielded Bulmer and Gilbert about £10,000 each in silver and lead sales with output reaching a total value of £44,000 by 1593 against costs and levies which never reached £20,000.

The huge wealth found in this mine was recalled by three long poetic inscriptions which Bevis Bulmer had engraved on three fine silver cups made from the mine's final year of production in 1593. The idea for them must have come from Schutz's tales of the Rappolsteiner Pokal cup. Of these only the Earl of Bath's pot survives; the City's cup bore an engraved portrait of Bulmer and included a scene of a miner at work. It also recorded Bulmer's role on the Corporation of London's estate at Broken Wharf.

In 1593 Bulmer had gained a patent for the raising of river (Thames) water at Broken Wharf by means of a pumping engine and the City lent him £1000 towards it, as well as use of the Leadenhall to assemble pumps. His riverside works built between 1593 and 1595 would supply the areas around west Cheapside and St. Paul's and later Fleet Street. The system was modelled on the system of lead pipes and cisterns installed in 1583 in the neighbouring Ward of St. Nicholas Colechurch. Bulmer built a pumping tower of 120 ft. high just inside the gatehouse of the City Brewhouse. Its tower features in seventeenth century images of London. A team of horses operated it as a large chain powered water pump.

During the 1590s the same technology did not prove quite adequate to pump out the Treworthy tin mines in conjunction with Robert Denham. In 1596 Bevis Bulmer was employed with Master Dimocke in assaying the South American ores of El Dorado brought back to London with unfounded hope from Sir Walter Raleigh's Guyana expedition.

The opportunities of the 1590s were the inverse of the legal position which Schultz and his co-patentee William Humfrey had enforced through the Company of Mineral and Battery Works in the 1580s. In 1584, despite all his Court connections, Sir John Blagrave was unsuccessful in his petition for a twenty-one year patent for improving the performance of blast furnaces with water powered bellows and an allegedly new way of building a water powered crushing mill. Bulmer, however, succeeded in the 1590s in exploiting rather similar ore crushing means in a Scottish stamp house modelled on the pioneering one built at Grasmere in 1570–1571.

In 1588 Bevis Bulmer was granted a patent for slitting iron bars, a process vital to comb manufacture, suggesting his continued association with Schutz in the financially successful Tintern and in the Beauchief (Sheffield) ironworks both run as profitable parts of the Company of Mineral and Battery Works. These works specialised successfully in steel and brass comb manufacture for a growing English market in large wool carding mills giving the Company a value added product. This differed greatly from the simple copper ingots which characterised the works of the Mines Royal Company and the wealth which Bevis Bulmer would earn from his mining and smelting of gold and silver in Scotland where any related lead ingot sales were very welcome. The Registers of Privy Council of Scotland for 1597 show Bulmer cautioned for diverting lead to such sales.

Once Humfrey Cole had located supplies of sulphurous sea coal from Newcastle-upon-Tyne which Schutz found he could effectively use among the furnace additives at the new Dartford smelter in 1578–1579, interest grew quickly in using sea coals in the furnaces of London until a Parliamentary bill passed during 1581 curtailed the location of new iron mills near London. Nonetheless, there was much money to be made by bringing sea coal for domestic fires from the Tyne to London. In 1591 Bevis Bulmer bid unsuccessfully for the farm of the dues on the 'Sea Coals from Durham' a concession which he won in 1599 but abandoned afterwards on payment of £1,000 by Elizabeth I towards his expenses.

Further northern opportunities opened to Bevis Bulmer in 1592 when the abolition of tacksmens' privileges and the death of Schutz, opened the way for Bulmer's rapid advance. The opportunity was confirmed by Act of the Scottish Parliament in 1593 whereupon Bulmer took up Schutz's former office of 'Master of the [metal] works for ores from Cathay and the North West Parts', soon gaining a reputation for effective silver extraction and the operation of his water pumping engines.

From 1594 Bulmer was involved with a partner called Foullis, an Edinburgh goldsmith and the jeweller to Queen Anne of Denmark, James VI's wife, in some Lanarkshire lead mines the 'Leadhills. In searching for lead galena they also found small quantities of gold allegedly in deposits of 'Sapper stone'. In 1598 Bulmer improved the water supply to domestic buildings near the palace complex at Linlithgow before turning his attention the following year to making improvements to Edinburgh's water supply by bringing in supplies from the South Loch to new conduits built at Libberton Wynd, Market Cross and Blackfriars.

In 1599 Bulmer tendered £10,000 a year for the right of pre-emption of the whole output of tin from Cornwall's mines for the equivalent of a figure of £26 3s 4d per 1,000 lbs produced. Ultimately, such pre-emptive bids operated first by Bulmer and Sir Walter Raleigh and from 1601 by Brigham and Wennes were all thwarted in 1603 due to thorough-going opposition from two established tin dealers, who acted with the Levant Company and Michael Lok working from Venice to secure most of Cornwall's tin output for export.

By 1603 several more mines had been assigned to Bevis Bulmer to work in Scotland as twenty four gentlemen were invited by King James to contribute £500 each for shares—a venture which uniquely carried the right to the title of 'A Knight of Golden Mines'. Sir John Claypool subscribed his £500 readily before his commercially earned title was duly awarded to Bulmer in 1604 along with an initial advance of £2,000 from King James I towards the costs of further mining development. However, opposition from Robert Cecil to such an order of knighthood put an end to this fundraising idea originated by Bulmer.

From November 1603 Bulmer was considering the possibilities of the Crawford Moor mines and their separate processing facilities at Short-clough water, Long Clough Head, and Fryers Moor and other similarly isolated facilities at Glengonnar in the Forest of Attrick. Three possible Scottish minesties were considered by prospecting parties sent at the instance of James I and working under Sir George Bowes (Jr.). Sir William Godolphin and Bulmer were due to meet him at Carlisle for an inspection starting on 12th November. Soon afterwards Bulmer learned of the death of his mentor Sir George Bowes.

Bevis Bulmer and Sir Thomas Hamilton obtained on 4th April 1607 a lease to exploit all the precious minerals in lands around Ballinkrief, Baithgate (Bathgate), Drumcrose, Tortrevin and Hilderston. The Hilderston mine was the result of the discovery of a metallic seam in a rock sample on Sir Thomas's land which was referred to Bulmer who was then seeking alluvial gold deposits near Wanlockhead but realised its value at once. On 14th April next Bulmer was paid £5,000 by the King of Scotland presumably because he had found at least one workable site. In 1608 Sir Bevis was made Master and Surveyor of the recently discovered gold and silver mines at Hilderston in Scotland. He took legal action and petitioned the Scottish Privy Council to enforce a penalty of over £1,000 on the Williamson family for deliberate destruction of one of his processing mills that year. It was rebuilt as a modern stamping mill at Killeith, near Linlithgow. About 1610 Bulmer's senior assaying assistant, Stephen Atkinson, took some samples of the Hilderston ore to London to show to King James I but they was appropriated this led to his partners' disillusion, and Bulmer claiming the enterprise was 'more fitting for Kings than subjects'.

In 1613 the goldsmith Walter Busbie briefly took over some of Bevis Bulmer's mining interests in North Yorkshire. Slightly later in 1613 Bulmer once more became involved in partnership with Foullis and the Portuguese financier, Paulo Pinto, in an attempt to further develop the Hilderston and Wanlockhead mines by building a new dam and at least two gold and silver scouring sluices and associated settlement beds at Upper Straight Clough water and Long Clough Bra[yn]e. Meanwhile Bulmer had become famous for the generosity of his hospitality at Wanlockhead as well as in his own large house called Whitfield beside the growing lead mining complex he had opened at Alston Moor in Northumberland during 1615. Bulmer's mining verses above Whitfield's front door survived until about 1920, as did a similar example at Wanlockhead. Atkinson records that Bulmer died at Whitfield near Alston during 1615.

Bulmer's involvement in several Irish mining schemes, together with the lead mines re-opened speculatively on Alston Moor in Northumberland, led to his owing Stephen Atkinson £300 by the time of his death. After his death Atkinson copied parts of several manuscript texts left to him by Bulmer, thereby creating a curious amalgam of biblical, metallurgical and mining insights covering his own working life and Bevis Bulmer's from the London assays of Frobisher's ores in 1577–1579 to the Hilderston venture of 1621. From these one can appreciate the significance of Bulmer's obvious engineering and mining skills to the English and Scottish Crowns

Atkinson called the resultant manuscript colla tion, finished about 1621, *The Discoverie and Historie of the Gold Mines in Scotland*. By then he notes Bulmer's initial sluice system at South Clough Water was almost derelict. Further subsequent reference to Bulmer's wealth appears in Ben Jonson's play *The Staple of News* (1631) where his wealth is lampooned under the guise of Sir Bevis Bullion

Bulmer's daughter Elizabeth Bulmer married Oliver Moon on 4th May 1633 at St. Martin-in-the-Fields, London. She was born at Berwick-on-Tweed on 30th November 1600 presumably during one of his many crossings of that border. His Catholicity may account for no other obvious extant record of his children. His will was administered by his son John. He also had at least two other daughters, Prudence and her much younger sister, Elizabeth. His generosity over the patent of 1588 may indicate a familial interest, or an illegitimate son, because in 1606 Bevis Bulmer's patent for slitting iron bars was renewed and transferred to Clement Dawbeney, who was again to renew it for twenty-one years in 1618.

ROBERT BALDWIN

[For early drafts of Stephen Atkinson's manuscript on mining and Bulmer's works see BL Harleian Ms. 4621 and, slightly differing, the Sibbald Mss in the National Library of Scotland; for Dee's Diary with its entries about Coombe Martin's mine see Bodleian Library Ms. Ashmole 488; for Bulmer's mining investments up to 1599 see SP12/168/9, SP12/251/71 (100, 129, SP12/252/49, SP12/253/69, SP12/254/58 and *Calendar of the Salisbury Mss at Hatfield House*, vols. v, ix and xii (1910); M. A. B. Hobson (*c.* 1937) *The Bulmer Family Chronicles, 1050–1936*, BL 9907 t.16; W. Maitland (1754) *The*

History and Survey of London, vol. ii, 1030–1031; T. Thomson and C. Innes (1814) *The Acts of Parliaments of Scotland, 1567–1592*, vol. iii, 555; S. Atkinson (1825) in G. L. Meason (ed.), *The Discoverie and Historie of the Gold Mines of Scotland*, 1–77; R. Cochran Patrick (1878) *Early Records of Mining in Scotland*, xviii–xxii, 4–9; S. H. Barton (1953) *Walks in North Devon*, 69–70 (full texts of the two inscriptions as originally cited in Prince's, *Worthies of Devon* (1810)), 115; G. W. Green (1958) The central Mendip Lead/zinc orefield, *Bulletin of the Geological Survey of Britain*, **14**, 70–90; M. B. Donald (1961) *Elizabethan Monopolies, The History of the Company of Mineral and Battery Works*; D. Stuckey (1965) *Adventurers' Slopes: The Story of the Silver and Other Mines of Coombe Martin in Devon*; J. McDonnell (1965) *A History of Helmsley, Rievaulx and District*; J. W. Gough (1967) *Mines of Mendip*; T. C. Barker and J. Hatcher (1974) *A History of British Pewter*, 235; P. Swinbank (1976) Wanlockhead: the maps, the documents, the relics and the confusion, *Scottish Archaeological Forum*, **viii**, 23–36; P. Hunting (1988) *St Paul's Vista, A History Commissioned by Lep Group to Mark Redevelopment of the Sunlight Wharf Site*, 49–52; E. Masters (1978) The history of the civic plate, 1567–1731, *Collectanea Londinensia*, LAMAS, 312; B. Weinreb and C. Hibbert (1983) *The London Encyclopedia*, 95, 928; P. F. Clayton (1989) *Out of the World and into Coombe Martin*, Coombe Martin Local History Group, 33–37; R. C. D. Baldwin (1998) Speculative ambitions and the reputations of Frobisher's metallurgists, *in* T. H. B. Symons (ed.), *Meta Incognita, A Discourse of Discovery, Martin Frobisher's Arctic Voyages, 1576–1578*, 401–476; for Bulmer's grant on Lighthouses, 1584, see BL Add Ms. 12497 fols. 395–404]

Works

1593–1595. Broken wharf waterworks
1598. Linlithgow water supply
1599. Edinburgh water supply

BURDON, Rowland (1756–1838), gentleman, the only child of Rowland Burdon (1724–1786) and his wife, Elizabeth Smith, was born on 28 December 1756 and was baptised in All Saints church, Newcastle, two days later. The family hailed from Stockton but the Castle Eden estate, some seven miles north-west of Hartlepool, was bought in 1758 by Burdon Sr., a proprietor of the Exchange Bank and, it was written, 'famous for having twice won the first prize in the State Lottery'. Following the purchase, he rebuilt the adjoining church in 1764 and completed the building of the hall by 1786.

Burdon was educated at the Royal Free Grammar School in Newcastle and later studied law at University College, Oxford. For a period of twelve months in Italy, he studied architecture as a pupil of (Sir) John Soane. He married Margaret, the daughter of Charles Brandling of Gosforth in 1780 but she and their daughter died in 1791.

Three years later he married Cotsford (1773–1860), the posthumous daughter, and only child, of General Richard Matthews, 'horribly murdered in India by Tippoo Sahib'.

Burdon put in hand improvements to his estate, building a cotton factory, a bleachery and a foundry, as well as forming a road from the house to the sea. He represented Durham County in the House of Commons from 1790 to 1806 and was Mayor of Stockton in 1793 and 1794. When in Parliament, Burdon was responsible for the sanctioning of a road from Sunderland to Stockton and its later extension to Thirsk. It was as a continuation of these roads that he came to be involved in the building of the Wearmouth bridge at Sunderland, only the second large iron bridge in Britain.

In 1792, Burdon secured an Act of Parliament for the construction of the bridge, its intention the easing of communication between Sunderland and Newcastle. He sought the views of Soane, John Nash, Robert Mylne, Charles Hutton, John Smeaton, Joshua Walker (qq.v.) and others prior to finalising his design for the 236 ft. span arch. The structure comprised six ribs each made up of 105 individual castings five feet deep; in form they replicated masonry. The bridge deck was 100 ft. above river level and the spandrels were filled with cast circles of iron, diminishing in radius towards the centre of the span.

Burdon patented his invention in 1795, defining the use of cast blocks with 'wrought- or cast-iron braces being affixed to their sides and passing horizontally between the ribs'. Under the direction of Thomas Wilson (q.v.), the ribs were erected on a timber scaffold and the bridge was opened on 9 August 1796 with great Masonic ceremony. Following the bridge's completion, Burdon—with Wilson—was granted a further patent in 1802. It covered modifications to the method of connecting the 'metallic patent blocks' used earlier and envisaged, too, the provision of greater rigidity.

Without Burdon's direct involvement, Wilson built further bridges in accordance with the second patent but they generally proved unstable and most soon came to be demolished; similarly, the Wearmouth bridge was found to lack rigidity and was strengthened in 1805 by John Grimshaw (q.v.), rebuilt by Robert Stephenson (q.v.) in 1859 and replaced in 1929. Burdon's knowledge was however brought to bear on the Parliamentary Committee considering the various proposals, including Wilson's, for a new London Bridge in 1803.

In 1806, Burdon was declared bankrupt as the result of the failure of the bank of Surtees & Co., in which a large amount of his assets had been invested. His financial failure—in which he had played no part—enforced his resignation from Parliament and it led, in 1816, to the disposal of the Wearmouth bridge by lottery, Burdon having financed its construction largely from his own funds. When, in 1836, it was suggested that a testimonial be established to reward his work for the town and county, Burdon refused it, writing that the money should instead be used to free the bridge from tolls.

Burdon was a member of the Literary and Philosophical Society in Newcastle and was described as 'a stout, stalwart, independent Englishman (who) as a practical man of business ... stood second to none, and as a commercial man he was known and respected by the wealthiest merchants of the Tyne, Wear and Tees'. He was not a great orator, having an impediment which caused him 'to speak with a slowness and deliberation often painful to witness' but nevertheless he earned the respect and esteem of those with whom he came into contact. It was written that, following Burdon's bankruptcy and the necessity for him to part with his house and estate, his friends purchased them with the intention of making them over to him as a gift. He refused to accept it as such but was prepared to receive it as a loan, which he was successful in repaying before his death.

Burdon died on 17 September 1838 at Castle Eden, his widow and six of their eight children—born between 1800 and 1815—surviving him. His estate was valued at some £10,000 but this can not be taken as a true indication of the family's wealth.

R. W. RENNISON and J. G. JAMES

[*Particulars of Mr. Burdon's Interest in the Property of the Late General Matthews* (1809) Literary and Philosophical Society, Newcastle, 8vo Tr, 434; *Gent Mag* (1816), **86**, 2; *Local Biography*, 2, Newcastle Central Library, L920; J. G. James (1986) *The Cast Iron Bridge at Sunderland (1796); Clayton & Gibson Collection*, Durham RO: DCG/5/49–104]

Publications

1795. *Making, Uniting and Applying Cast-iron Blocks in Lieu of Keystones, in the Construction of Arches*, patent 2066, 18 September
1802. *Uniting, Combining and Connecting the 'Metallic Patent Blocks' for the Construction of Arches* (with Thomas Wilson), patent 2635, 23 July

Works

1792–1796. Wearmouth Bridge, Sunderland, 236 ft. span, cost £41,500

BURGE, George (1795–1874), was baptised on 1 November 1795 at Spa Fields, Clerkenwell, the son of John and Elizabeth Burge but is best known as 'contractor of Herne Bay', which he helped develop. He rose to prominence as the main contractor for St. Katharine's Docks in London in the late 1820s; the scale of these works is well illustrated in W. Ranwell's famous watercolour, engraved by J. Phelps. 2,500 men were occupied in the excavations.

His success at St. Katharine's must have helped secure him further contracts, on Telford schemes,

for a timber pier at Herne Bay, and for tunnel works at Dover harbour. Burge was the original contractor for the Western Dock at Southampton, but his best-known work, aside from St. Katharine's, was as principal contractor for Box Tunnel on the Great Western Railway.

The original contractors for the shafts, Horton, ran into problems and the work had to be completed by the railway company. This gave rise to concerns about ground conditions and the practicability of the tunnel and, in this situation, Burge, who was clearly a contractor of some resources, and indeed had an unknown contract in Germany at this time, was awarded contracts 4–6, and 6 extension, on the Tunnel in June 1838. Nevertheless, in the face of the difficult ground conditions, Burge fell behind on the work. The work, due to be completed June 1841, had as late as March 1840, 650 yd. outstanding, and it was not until April 1841 that Burge was on target. The cost was high; 100 men died. The bulk of the work was carried out by manual labour; between 1,100 and 1,200 men, rising to 4,000 at the end, and over 100 horses worked day and night. An estimated thirty million bricks were used for the lining in three rings.

The Tunnel was not Burge's only railway contract, and a John Burge, presumed to be his son—as George was his surety—worked on the London–Birmingham and Eastern Counties railways.

Burge was contractor for the foundations of the Szechenyi chain bridge designed by William Tierney Clark (q.v.) in Budapest, and was still active in 1860 when he was involved with the London, Chatham and Dover Railway scheme. Burge died of gangrene and congestion of the lungs at 32 Hawley Square, Margate, on 2 May 1874 at the age of seventy-eight.

MIKE CHRIMES

[Marlow Bridge papers, Bucks RO; F. Smith collection, ICE; Contract for Szechenyi (Budapest) bridge foundations, ICE; GWR; London–Birmingham; Eastern Counties, London–Chatham and Dover; South Eastern Railway; London and South Western Railway; Manchester and Birmingham Railway, minutes, etc., PRO RAIL; *Civil Engineer and Architects Journal* **59** (1837), 172; *Railway Times* (1838), 3; J. Rickman (ed.) (1838) *Life of Thomas Telford*, 652; Jeaffreasson (1864) *Life of Robert Stephenson*, 192; Sir J. Rennie (1875) *Autobiography, passim*; T. Gale (1884) *a brief Account of the Making and the Working of the Great Box Tunnel*; A. W. Skempton (1983) *Engineering in the Port of London*; *Newc Soc Trans*, **53**, 82–87]

Works (as contractor)

1826–1830. St. Katharine's Dock, 11.5 acres, £190,000
1830–1832. Herne Bay pier, 3,613 ft. timber
1834. Dover harbour tunnels and associated works, *c.* £23,000
1837. Western dock, Southampton

1838–1841. Box tunnel, three contracts, 9,680 ft., 414,000 cu. yd. of excavation
South Eastern railway–Rochester–Canterbury
1841–1842. Foundations for Szechenyi bridge, Budapest
1852–. Royal Swedish Railways, 50 miles, c. £500,000
1858–1860. London–Chatham and Dover railway (with Crampton); Stroud–Canterbury, £94,000; Faversham–Herne Bay, c. 10 miles

BURN, George (1759–c. 1820), civil engineer and contractor, son of George Burn and Helen Pearson, and a younger brother of James Burn (q.v.), was born at Yester, East Lothian. Throughout his working life Burn practised as an architect and he prepared designs for a number of buildings in northern Scotland but he is better remembered for his work as an engineer and contractor. Burn commenced his career as an engineer and bridge builder in 1798 when he designed and constructed a modest two-span bridge at Bilsdean, East Lothian. This was followed the next year by an unexecuted proposal for a bridge at Ballater. Between 1800 and 1804 he constructed a number of bridges over rivers that drained the Moray coastal plain and he subsequently moved further north to build bridges and harbours in the northernmost counties of Scotland. By 1806 Burn had become domiciled at Wick, Caithness, and from there, until his death circa 1820, he continued his professional activities.

Besides the bridge at Bilsdean, he designed and constructed bridges at Thurso and Nairn, the Ballindalloch Bridge, Banffshire, and the impressive bridge of four arches at Fochabers that spanned the river Spey. The last-named bridge, completed in 1804, was built as a joint contract with his brother James. It remained in use until 1829 when the western river pier and the adjacent arches were destroyed during flood conditions. The two fallen spans were replaced by a single timber arch designed by Archibald Simpson of Aberdeen and constructed by William Minto (q.v.) and William Leslie.

Despite success as an engineer, the fragility of Burn's finances, a problem that dogged him for many years, showed itself when his property was sequestered during the construction of the bridge at Fochabers. His financial problems may have influenced the direction of his career for his subsequent involvement with civil engineering was as a masonry contractor employed by the Commissioners for Roads and Bridges in the Highlands of Scotland. Between 1806 and 1817 he constructed bridges, to designs by Thomas Telford (q.v.), at Wick, Helmsdale, Beauly (Lovat bridge), and Fairn-ness, beside which he built the bridges, including that at Greystones, on the line of the Commission's Thurso road. When constructing the bridges at Beauly and Fairn-ness, Burn suffered a prolonged spell of ill health and the successful completion of these contracts was due largely to the efforts of his foreman and the vigilance of Telford's inspectors. Burn's enforced absence from the work did nothing to remedy his fragile financial situation.

In 1806, although he had no experience of marine work, Burn's ability as an engineer and contractor was sufficiently well established for Telford to recommend him to the British Fisheries Society who were then developing their fishing station at Pulteneytown, Wick. Here Burn, with practical assistance from James Bremner (q.v.), built the two piers and the breastwork that formed the harbour. The successful completion of this contract later led to his appointment as a sub-contractor for the construction of the harbour piers at Kirkwall, Orkney and Portmahomack, Ross-shire.

TOM DAY

[NRA(S) 61, Farquharson of Invercauld Archives, Specification and Estimate for Bridge at Ballater; British Fisheries Society (1806–1813), Records, GD9, SRO; RHP 2008/8 and 5504, SRO; Reports of the Commissioners for Roads and Bridges in the Highlands of Scotland (1815, 1817, 1821); A. Graham (1965) Archaeology on a Great Post Road, *Proceedings of the Society of Antiquaries of Scotland*, **96**, 338; A. Gibb (1935) *The Story of Telford*, London, 302–303, 308, 310–311; A. R. B. Haldane (1962) *New Ways through the Glens*, Edinburgh, 129; J. Dunlop (1978) *The British Fisheries Society, 1786–1893*, Edinburgh, 158–67; T. Day (1997) Did Telford rely, in Northern Scotland, on vigilant inspectors or competent contractors?, *Construction History*, **13**, 5–7; T. Day (1987) The Old Spey Bridge, Fochabers, *Industrial Archaeology Review*, **10.1**, 71–5; H. Colvin, *A Biographical Dictionary of British Architects, 1600–1840*, 3rd edn., London, 180–81]

Works
1798. Bilsdean Bridge, East Lothian, £105
1800. Thurso Bridge, possibly designed by John Rennie Sr.
1800. Bridge of Avon, Ballindalloch, Banffshire
1801–1804. Spey Bridge, Fochabers
1803–1804. Nairn Bridge
1806–1809. Wick Bridge
1807–1811. Wick Harbour
1809–1811. Helmsdale Bridge, Sutherland
1809–1813. Kirkwall Harbour
1811–1814. Lovat Bridge, Beauly
1813–1816. Portmahomack Pier, Ross and Cromarty
1814–1816. Fairn-Ness Bridge, Nairnshire
c. 1817. Greystones Bridge, Caithness

BURN, James (1748–1816), builder and architect, son of George Burn and Helen Pearson, and elder brother of George Burn (q.v.), was born in Yester, East Lothian. Burn, a wright by trade, became a successful builder in Haddington, established an architectural practice, and was responsible for the design and construction of domestic and public buildings in East Lothian and in Aberdeen. Colvin details his architectural works.

Burn chiefly practised as an architect but he designed and built seven bridges. Four of them had masonry arches and three had timber arches. He was also joint contractor, with his brother George, for the construction of the Spey bridge at Fochabers, Morayshire. The masonry bridges were built at Dunglass and Saltoun, East Lothian and at Inverbervie and Uras, Kincardineshire. The first and second to be built, at Dunglass and Inverbervie, spanned steep-sided valleys, and had arches of large dimensions.

Burn's background as a wright may have prompted his interest in the design of timber arch bridges and he made a significant contribution to the design process. His first timber bridge (1798), a triple span structure, was built at Brechin Castle, Angus. The second (1803) and third (c. 1810) both spanned the river Don in Aberdeenshire. The second bridge, the Bridge of Dyce, had a single span of 109 ft.

The constructional method developed by Burn used short braced timber frames laid in the plane of the arch ring like elongated voussoir blocks. The deck beams of the roadway were carried by vertical posts that stood on the timber frames of the arch ring. The use of short lengths of timber to carry loads parallel to the longitudinal grain of the main elements of the arch ring aided the dimensional stability of the arches and Telford acknowledged the 'very ingenious mode' of construction used by Burn and stated that there were 'no wooden bridges in Britain so judiciously constructed'. Although Burn's design proposals received favourable comment from Telford, he did not utilise them, and it was subsequently left to Joseph Mitchell to use Burn's principles for the construction of bridges in the Highlands of Scotland.

TOM DAY

[Minute Book, Commissioners of Supply for Kincardineshire, 1762–1804, 268–70 (Aberdeenshire Archives); Minute Book, Trustees of the Stonehaven and Bervie Turnpike Road, 1796–1812, 20, 64–76 (Aberdeenshire Archives); Minute Book, Trustees of the Old Meldrum Turnpike Road, 1802–1821, 6–19 and 25–30 (Aberdeenshire Archives); Minute Book, Haddingtonshire Road Trustees, 1783–1800, 56–65, 280–82 (SRO); T. Telford (1830) Bridge, *Edinburgh Encyclopaedia*, 488, 537–38, plate 88; Ted Ruddock (1979) *Arch Bridges and their Builders, 1735–1835*, 169–72; Colvin (3), 180–181]

Works
1797–1800. Dunglass New Bridge, East Lothian, £1,000
1797–1799. Inverbervie Bridge, Kincardineshire
1798. Brechin Castle Bridge
1798. Burn of Uras, Kincardineshire
1801–1804. Spey Bridge, Fochabers (joint contractor)

1802–1803. Bridge of Don, Dyce, Aberdeenshire
1805. Milton Ford Bridge, Saltoun, East Lothian
c. 1810. Grandholm Bridge, Aberdeen

BURNETT, Thomas (fl. 1812–d. 1824) civil engineer, worked on several projects associated with John Rennie (q.v.) before leaving Britain for Canada. He worked at Greenock harbour for three years, and in the autumn of 1815 carried out some surveys for Rennie at Ardglass harbour. He was then temporarily employed by Major Taylor in Dublin. In 1817 he asked Rennie for a reference as Engineer to the River Wear Commissioners. This post was awarded to Thomas Milton (q.v.), and Burnett went instead to work on harbours in Canada. In this capacity he was elected a Corresponding Member of the Institution of Civil Engineers in 1820, on Telford's recommendation. He was the first overseas member.

His most famous work was as engineer for the Lachine Canal. An attempt to bypass the Lachine rapids and link Lake St. Louis to Montreal was made by the 'Messieurs de Saint Sulpice' 1689–1701, but was not completed due to lack of funds, although part was used to power mills. The idea may have been revived under Sir Frederick Haldimand (1778–1786) but details are lacking, and the next proposal was made in 1796 by John Richardson, a Montreal merchant.

The 1812 naval war with the USA emphasised the need for a proper navigation and in 1821 the Government of Lower Canada organised the necessary finance. Work began that November with Burnett as engineer. The Canal, 8.5 miles in length had seven stone locks, 100 ft. long and 20 ft. wide. It was 40 ft. wide at water level, and 28 ft. at the bed, much larger than the traditional English narrow boat canal. Burnett died on 9 November 1824, leaving the canal to be completed the following year by his son John. Burnett's original estimates were for £78,000, but costs rose to £107,000. Some of the contractors went on to work on the Rideau Canal, designed by John By (q.v.).

The masonry of the locks, the work of the Scottish mason John Redpath, was described by Lord Dalhousie as 'the finest masonry I ever saw'.

MIKE CHRIMES

[ICE membership records, ICE archives; Rennie mss, NLS; R. F. Leggett (1976) *Canals of Canada*, 144–149; H. Kalman (1994) *History of Canadian Architecture*, 206; P. Desjardins (1995, 1999) in lit.]

Works
1821–1824. Lachine canal, £107,000

BURRELL, Andrewes (fl. 1631–1664), landowner, played an active role in the first campaign (1631–1636) of draining the Fens. He was the son of William Burrell, a London merchant who had acquired some 3000 acres in Waldersey, south of Wisbech, and in 1629 became one of the Commissioners of Sewers (the land drainage authority) of Cambridgeshire; also in that year he and Sir Thomas Crooke submitted a proposal for Fens drainage. Nothing came of this, but in January 1631 Andrewes Burrell appears among the members of a Commission passing the so-called Lynn Law which granted Francis Russell, fourth Earl of Bedford, powers to undertake the draining of the Great Level, an area of rather more than 300,000 acres. Burrell joined forces with the Earl as one of the fourteen Adventurers in the undertaking, holding one of the twenty lots into which 80,000 acres of the drained land would be divided among them, though he sold his share in 1633. At the same period, 1631–1636, the Earl of Bedford with another group of Adventurers undertook the drainage of Deeping Fen, an area of 30,000 acres adjacent to the Great Level north of the Welland.

Works in the Great Level, beginning in July 1631, essentially followed the scheme designed by John Hunt (q.v.) and promoted by Sir John Popham in 1605, the main features being improvements of the rivers Welland and Nene and cutting a new 21 mile long channel for the Ouse. Other works may have been adopted from a plan put forward by Sir Cornelius Vermuyden (q.v.) in September 1630 of which unfortunately no details are known.

Records of the Proceedings under the Earl of Bedford no longer exist; they were probably lost in the Great Fire of 1666 which destroyed the Fen Office in London. But they were available to Sir William Dugdale (q.v.) when writing his *History of Imbanking and Drayning* in 1662. He lists the following principal works, where those marked with an asterisk were broadly as proposed by Hunt; notes are added from other sources: (1)* River Welland improved from Peakirk, at the edge of the Fens, past Crowland and Spalding to its outfall at Fosdyke, with a new bank from Peakirk to Crowland and the other (existing) banks heightened. Except for the new bank the Welland improvements were carried out in connection with Deeping Fen drainage. Visitors in August 1634 saw 600 men engaged on the work and record that the river was being widened to 50 ft. and deepened to 6 ft. (Darby, 1940). (2) New South Eau, 7 miles long from near Crowland to Clows Cross. (3)* Shire Drain improved from Clows Cross to its outfall at the old sea bank near Tydd St. Giles with a new outfall sluice. This is Hill's Sluice built 1632 (Wheeler, 1896). (4) Peakirk Drain, 10 miles long and 17 ft. wide, to the Nene at Guyhirn. (5)* Morton's Leam improved. This is the 12 mile long channel of the Nene from Stanground near Peterborough to Guyhirn made in the late 15th century. Burrell says it was widened to 50 ft. (6) The great stone-built sluice on the Nene at the Horseshoe about a mile below Wisbech. Burrell records its cost as £6,000 and adds that the river was in 1635 deepened in this locality by removing up to 8 ft. of silt. The same visitors in the summer of 1634 saw 'a little army of artificers' working in this locality,

presumably on the sluice. (7) Bevill's Leam, 10 miles long and 40 ft. wide, from Whittlesey to Guyhirn. (8)* Bedford River, a new cut 70 ft. wide and 21 miles long from Earith to Salters Lode, embanked and with sluices at each end. Burrell says the banks were originally 7 ft. high and gives the cost of the whole work as £26,000. (9)* Hermitage navigable sluice on the Ouse, a short distance below Earith, to direct water down Bedford River. (10)* Popham's Eau restored. Made in 1605 this runs 5½ miles from the Old Nene to Nordelph on Well Creek. (11)* Well Creek improved from Nordelph to its outfall into the Ouse at Salters Lode, where a navigable sluice was built. (12) Sam's Cut, 6 miles long and 20 ft. wide, from Feltwell to the Ouse.

Where the dimensions differed from those proposed by Hunt they were always smaller. The most conspicuous example is Bedford River, with a 70 ft. width and banks not more than 200 ft. apart, whereas in 1605 they were to be not less than about 500 ft. apart and Hunt had proposed a 120 ft. wide waterway for the channel. The results in this case were disastrous. The first great flood to be sent down the River breached its banks. Vermuyden later made a New Bedford River (1651–1653) with 100 ft. width and banks up to half a mile apart to provide a washland.

Even with these economies the cost of the Great Level works came to the enormous sum, for those days, of £97,000 mostly spent during 1632–1635 at an average rate of £20,000 a year (Wells 1830). Work on the Welland would have been largely charged to Deeping Fen drainage, the total cost of which is given by Burrell as £23,000.

Little is known of the organisation of the undertaking. However, by analogy with operations under the fifth Earl of Bedford in the third campaign (1650–1656) in the Great Level, of which details are available, it seems possible that the enterprise was directed by a small group of Adventurers acting as administrative officers and supervisors in the field. On the technical side there would have been land surveyors and 'overseers' or resident engineers. Thomas Wright (q.v.) was one of the surveyors, Benjamin Hare was another. The surveyors would set out the works and divide the land after drainage. William Hayward (q.v.) may have held a post equivalent to Surveyor General; he certainly carried out a detailed survey of the Level in 1635–1636. The overseers would have been responsible for letting contracts, conducting the works and 'taking up' or measuring quantities for assessment of payments to the workforce. From his own account it is probable that William Dodson (q.v.) was an overseer. To Simon Hill (q.v.) may be attributed the Shire Drain outfall sluice apparently named after him, while Burrell in 1641 says he was 'imployed in those workes … from the first day to the last of the late Earle's undertaking' and in particular had charge of building Horseshoe Sluice and deepening the river.

Two, if not three, previous attempts to build a sluice at the Horseshoe had been made, the last by the Netherlands engineer Humphrey Bradley (q.v.) in 1589; all failed within a short time, probably due to inadequate foundations. Burrell's sluice proved successful. Nevertheless it was bypassed by Vermuyden in 1650, at the outset of work in the third Fens campaign, and demolished about 1656. The wisdom of providing sluices on the main rivers was, and for many years remained, a subject of controversy.

The final 'award' of land was made by the Commissioners in October 1637. For the most part the Fens were now dry by April or May each year; in other words 'summer grounds' had been achieved. To keep the land free from winter flooding was the next objective, requiring a fresh input of funds. King Charles in July 1638 declared himself willing to undertake works needed to create 'winter grounds' in the Great Level and Deeping Fen. He sought advice from 'gentlemen expert in such adventures'. Burrell submitted proposals in October. These included a sluice with an adjacent lock on the Welland near Spalding, and others probably resembling those in his submission of 1642 (see below). Vermuyden presented a scheme in January 1639. This embraced three major propositions: diversion of the Welland into the Nene; provision of large 'receptacles' or washlands for the temporary storage of flood water; and a cut-off drain along the eastern edge of the Fens leading into the lower reaches of the Ouse.

After debate, Vermuyden's scheme was provisionally approved and he became the King's 'agent' in September 1639. Work under his direction began soon after August 1640, following some revisions, but was slowing up for lack of funds by October 1641. Operations involved opening a new channel from the Horseshoe for 2½ miles towards the sea; deepening the river from Wisbech up to Guyhirn; building along part of this length a new south bank which, together with the existing north bank on the south side of Morton's Leam, would form another, far larger, washland of about 3000 acres; and a lock on the Old Nene at Stanground. The cost of these works is given by Burrell as £23,500. Diversion of the Welland was not carried out then or at any later time. The eastern cut-off, however, eventually formed a basis for the successful flood-relief scheme of 1954–1964.

The Waldersey Receptacle covered nearly 500 acres of land belonging to Burrell, for which he received inadequate compensation. Incensed by this and by Vermuyden's scornful view of works in the first campaign, Burrell issued late in 1641 a pamphlet entitled *An Explanation of the Drayning Workes ... Lately Made for the King's Majesty*. Critical of what he considered to be Vermuyden's waste of money in inept construction of banks and the timber-framed Stanground lock, he adds the final condemnation: 'I have often heard Sir Cornelius sleight all the workes

which were made by the late Earle, as if they were rather hurtful than beneficial to the Country'. He refers to the fourth Earl who died in May 1641.

In February 1642 a Parliamentary Committee, set up to consider the whole matter of the Great Level, ordered Vermuyden's *Discourse* to be printed 'that all men whom it may concerne ... may thereby inform themselves, and may make their exceptions against it, and likewise may offer any other designe, and they shall have notice to be heard'. Burrell responded with his *Briefe Relation Discovering ... how* [the Fens] *May be Drained*. After some preliminaries he gives an analysis of reasons why the Fens are 'drowned' and states the general principles of preventing floods: to embank and enlarge the main rivers, of which the recently made Bedford River is one; to place sluices at their outfalls 'to repell the Sea floods'; and to make cut-off drains along the borders of the Fens. He then goes on to specify in considerable detail the works required with estimates of their cost, the total being £180,000.

Most of his main proposals were improvements on what had already been done. For example, to double the width of Morton's Leam to 100 ft. and deepen it to 6 ft. throughout its 12 mile length with a bank on the north side (£12,000) and to widen Bedford River from 70 to 80 ft. and make it 3 ft. deeper, using the excavated material to heighten the banks (£18,000). As new works, he proposed banks 6 ft. high along the Ouse (£9,000) and its tributaries the Cam, Lark, Little Ouse and Wissey (£16,000), all of which were later carried out, and he followed Vermuyden as regards the eastern cut-off which he estimated at £16,800. Mistakenly he never approved the principle of washlands.

Burrell made several other, relatively minor, proposals such as a new sluice at an extended sea outfall of Shire Drain with a 20 ft. waterway (£2,400) and a navigation lock (Sasse) 24 ft. wide on the Ouse at Littleport (£2,200). He also allowed £4,500 for horsemills (for draining excavations), barrows and other equipment as well as houses for the workmen, and £4,000 for salaries of officers over four years. But his boldest suggestion was a great sluice with an adjacent lock on the Ouse at Magdalen, 5 miles south of King's Lynn. With a waterway of 160 ft. this was to be much larger that the Horseshoe Sluice and he estimated its total cost at £21,000. There would be a pair of pointing sea doors and a single land door in each arch. He realised that the sluice might be opposed by navigation interests of Lynn and Cambridge but emphasised its value in keeping out the sea floods and scouring the river downstream. The Magdalen sluice was not built but the idea was repeated, at a slightly different location, in the scheme proposed by William Dodson (q.v.) in 1649 and both to some extent foreshadow Vermuyden's lock and sluice at Denver (1651–1656).

Shortly after Vermuyden's *Disclosure* appeared in print Burrell published his *Exceptions* against it. The pamphlet adds several points of interests and makes some valid criticisms but its ill-tempered tone had damaged his reputation at the expense of his *Briefe Relation*, a serious engineering report and one of the earliest to be printed.

Before anything more could be done the Civil War broke out. When peace had temporarily been restored, Parliament again took up the matter of the Great Level. In June 1646 William, fifth Earl of Bedford, gave evidence concerning expenditure on his father's work. In January 1647 Vermuyden declared that draining the Fens to give winter grounds was entirely feasible, and in evidence a month later 'Andrewes Burrell of Wisbech, gent, sayeth that he conceives it very feasible and sayeth that after the (late) Earl of Bedford had made his workes, which were not finished, the main body of the Fennes were not drowned in summer time for seven years together'.

Meanwhile, during the Civil War Burrell held an appointment under the Navy Board and in 1646 published another of his pamphlets, this time on the *Reformation of England's Navie*. It met with a cool response but nevertheless he later became Master Shipwright at Woolwich.

Burrell left the public service in 1651 or 1652. He played no part in the third and final campaign of draining the Fens. His will, dated 2 March 1664, suggests a man in reduced circumstances. The date of his death is not known.

A. W. SKEMPTON

[Sir Cornelius Vermuyden (1642) *A Discourse touching the Drayning of the Great Fennes*; Sir William Dugdale (1662) *The History of Imbanking and Drayning of divers Fenns and Marshes*; C. N. Cole (1784) *Extracts from the Report of a view of the South Level*; Samuel Wells (1830) *The History of the Draining of the Great Level of the Fens, called Bedford Level*; H. C. Darby (1940) *The Draining of the Fens*; L. E. Harris (1953) *Vermuyden and the Fens*; G. E. Aylmer (1973) *The State's Servants*; Margaret Knittl (in lit., 1996) and Frances Willmoth (1997), further information]

Publications

1641. *An Explanation of the Drayning Workes which have Lately been Made for the Kings Majesty*

1642. *A Briefe Relation Discovering Plainely the True Causes why the Great Levell of Fenns ... have been Drowned ... and as Briefly how they may be Drained, and Preserved from Inundation*

1642. *Exceptions against Sir Cornelius Virmudens Discourse ... wherein His Majesty was mis-informed ... and the Great and Advantageous Workes Made by the Late Earle of Bedford, Slighted*

Works

1634–1635. Sluice on the river Nene at the Horseshoe, near Wisbech, £6,000

BURRELL, James (fl. 1604–d. 1631) was the designer and overseer of Berwick Bridge. On the accession of James VI of Scotland to the English throne, the garrison and fortifications of Berwick became redundant. Nevertheless the post of Surveyor of Works survived, nominally, and from 1604 was held by the former Deputy Surveyor, James Burrell. His main task, however, was maintenance of the old timber bridge over the river Tweed which had been rebuilt several times, the last occasion being in 1570–1573 by Rowland Johnson (q.v.). By the latter part of the century it needed frequent repairs and finally in February 1608 ice coming down the river in flood carried away ten of the (timber) piers. Burrell wrote to Lord Salisbury, Secretary of State, that the only real solution was to build a new bridge in stone; meanwhile he organised a ferry service.

Arrangements were made in May 1608 for collecting funds but by 1611 only some £3,300 had been received and a small amount of preliminary work achieved. Dissatisfied with this state of affairs, Sir William Bowyer, Captain of Berwick, pressed for more urgent action, enclosing an estimate by Burrell for a bridge of seven stone arches over the deepest part of the river, the remainder in timber. A few days later, in March 1611, Burrell and some others left for London to attend the Privy Council, but on the evening of their departure two more of the old bridge piers collapsed. A further estimate for a thirteen-arch stone bridge was then submitted, amounting to £8,462, over and above the money already collected, signed by Burrell as 'Bridge Master'.

After deliberation the King directed the Lord Treasurer on 16 May to issue a warrant to pay £8,000 in instalments for building the bridge. No time was wasted and on 27 May George Nicholson was appointed Paymaster and Burrell (described here as master mason) became Overseer of the works at 2s 6d per day (15s a week) with a clerk. Work began on 19 June 1611. By September the workforce numbered one hundred and seventy men, including thity-one masons, twelve carpenters, twenty-two quarrymen and seventy-six labourers. The leading mason (Lancelot Bramston) and carpenter (Roger Richardson) each received 1s 8d per day, the labourers 6d a day. Each pier was founded on piles within a starling built up to a little above low water level (the tidal range being about 15 ft.). The piles, 12–18 in. square and iron shod, were between 6 and 23 ft. in length. In all, for the foundations and perimeter walls of the starlings, 873 oak trees were used, nearly all from Chopwell forest and shipped from Newcastle. Sandstone came up the river in lighters from a quarry at Tweedmouth. Burrell devised a new type of pile driver hoisting a ram of 9 or 10 cwt and operated by five men instead of the normal fifteen or twenty. The piers had triangular cut-waters which extended to the parapet as refuges.

Work proceeded steadily but slowly. A further sum of £4,000 was granted in 1618 but by 1620, this money having been spent, the Council became worried and they directed the Bishop of Durham to act as agent in forwarding the business. He drew up a contract with Burrrell (now named as Surveyor) and Bramston to complete the work for £1,750, and installed the Newcastle bridge master John Johnson as supervisor.

At some early stage in the proceedings it had been decided to add two arches. The bridge, with fifteen arches, was completed by September 1621, apart from the parapet and paving, but in October an unprecedented flood destroyed the most recently built masonry and the centering of some of the arches still in place. No one could be blamed for the accident. A further grant of £3,000 was sanctioned. Work re-started in March 1622 and was complete, or very nearly so, in 1624 at a total cost of about £18,000.

The bridge, the largest built in Britain during the seventeenth century, carried the main road from London to Edinburgh for three hundred years. It was superceded by a new bridge in 1928 but still exists in good condition, carrying a minor road. It has a length of 1164 ft. and is 17 ft. wide between parapets; the arches are segmental (slightly pointed) with spans ranging from 24 to 75 ft.

Burrell, besides designing the bridge and directing construction from first to last, served as an alderman for Berwick and held the office of mayor in 1606 and 1611. He died in January 1631.

A. W. SKEMPTON

[E. Jervoise (1931) *The Ancient Bridges of the North of England*; J. M. Johnson and C. W. Scott-Giles (1933) *British Bridges*; John Summerson (1982) *King's Works*, **4**, 769- 788; Ted Ruddock (1984) Bridges and roads in Scotland 1400–1750, in A. Fenton and G. Stell (eds.), *Loads and Roads in Scotland*, 67–91]

Works

1611–1624. Berwick Bridge, fifteen arches, max. span 75 ft., £18,000

BURTON, Robert (fl. 1649–d. 1653) of Spalding, was Overseer (resident engineer) in the draining of the Great Level of the Fens. This vast project proceeded in three campaigns: 1631–1636, 1640–1641 and 1650–1656 with Sir Cornelius Vermuyden (q.v.) as Director of Works in the second and third. Records of the first two campaigns are largely missing but the Proceedings of the 'Company of Adventurers' in the third still exist. They show that, following the enabling Act of 29 May 1649, the Company began negotiations with Vermuyden but for various reasons his appointment as Director was delayed. Meanwhile, in the hope of getting something started, Sir Miles Sandys and two other Adventurers, assisted by William Dodson (q.v.), submitted on 22 June 1649 'a Designe and estimate of the workes … thought fitt to be done this summer'. After due consideration their plan was put into action with Dodson as acting Director and Burton as Overseer (from 23 August).

Nothing has yet come to light regarding Burton's earlier career, though his appointment to a senior post at the start of operations, on the recommendation of the Expenditor, Robert Castell, suggests that he was well known, possibly with previous experience (? in the first campaign). Unfortunately, work had to stop for lack of funds towards the end of September. Further negotiations ensued leading finally to approval of Vermuyden's plan and his appointment on 25 January 1650. Overall control resided in a small group of Adventurers including the Treasurer, the Expenditor and the Superintendent, Anthony Hammond, who acted in effect as project director.

Burton's position as Overseer was confirmed on 27 February. Having received 'directions' from Vermuyden he went to Wisbech to organise work in that locality with instructions from the Company, issued on 8 March, to 'let out' contracts and engage workmen. A week later Vermuyden joined him 'to stake out the worke' a task normally done by the Surveyor but Jonas Moore (q.v.) did not arrive on site, as Surveyor General, until September. Burton already had an assistant, another was added to the staff towards the end of March.

The Surveyor's duties, as defined by Vermuyden, were 'to sett out the works and see that they bee done'. Day-to-day conduct of the works, however, was carried out by the overseers. They also had to 'take up' or measure quantities for assessment of payments to the workforce. Hammond took a view of the operations early in May by which time construction was under way. In October he and Burton were involved in a survey of the river outfalls.

Works in the first twelve months were confined to the North and Middle Levels, an area totalling about 170,000 acres. They included a 12 mile bank on the north side of Morton's Leam (the main channel of the river Nene from Peterborough to Guyhirn, near Wisbech); the Forty Foot Drain running 10 miles from Ramsey to the 'old' Bedford River at Welches Dam; the Sixteen Foot Drain 9 miles in length draining fens near Chatteris; and the restoration of works neglected in the Civil War period. Hammond and three other members of the management group prepared the case for adjudication by Commissioners who in March 1651 examined witnesses, including Moore and Burton, and declared the drainage of this area to be complete.

Vermuyden then produced his plan for the South Level, some 140,000 acres. This involved cutting the Hundred Foot or New Bedford River from Earith to Salters Lode and building barrier banks along the whole 21 mile length, typically half a mile apart, to form a large washland or 'receptacle' for the temporary storage of flood water; building a dam and navigable sluice on the Ouse at Denver; cutting a slaker channel 114 ft.

wide and 4 miles long from Denver to Stow, with outfall sluices providing a 72 ft. waterway; and embanking the four tributaries of the Ouse (Cam, Lark, Little Ouse and Wissey).

These works were completed within two years, employing (at a maximum) 10,000 men. Commissioners adjudged the drainage of the South Level complete in March 1653. Burton's position as Principal Overseer, which in effect he held from the start, was formally recognised on 26 March 1651. From July he was empowered to let contracts at his own discretion in Vermuyden's absence. During 1651 another overseer and a land surveyor joined the staff. Even so, pressure on site and administrative personnel would have been severe, with a rate of work not rivalled in intensity until the canal age more than a hundred years later.

Faced by problems of manpower, Burton at times had to agree higher rates of pay than the Company was prepared to sanction. Whether for this reason or on account of failing health, he was replaced in March 1652 by Edmund Welch (q.v.). Burton's salary (probably £100 a year) was reduced in May and by August he left. At that time there were four overseers, including Welch; all four were empowered to let out certain contracts and 'act as Burton would have done'.

The will of Robert Burton of Spalding, yeoman, is dated 19 August 1652. He died before 26 March next year, leaving his house and land to his wife, Margaret, a married daughter and two sons. The will was proved on 20 April 1653.

FRANCES WILLMOTH and A. W. SKEMPTON

[*Proceedings of the Company of Adventurers (1649–1662)*, Cambridgeshire RO; Samuel Wells (1830) *The History of the Draining of the Great Level of the Fens*; Frances Willmoth (1993) *Sir Jonas Moore, Practical Mathematics and Restoration Science*]

BUSBY, Charles Augustin (1786–1834), architect, was born in London on 27 June 1786, one of seven children, and the eldest son to survive childhood, of Dr. Thomas Busby (1755–1838) and Miss Angier. He was educated by his parents and his father's scientific interests were reciprocated by his son. Around 1802, at the age of sixteen, Busby commenced his pupillage with Daniel Asher Alexander (q.v.), then active in the London Docks scheme, exposing Busby to the structural and civil engineering developments in London's docks. On Alexander's recommendation Busby entered the Royal Academy Schools in 1803.

Busby's early career was that of a promising aspirant architect. In 1806 he helped form the London Architectural Society, acting as its Secretary, and began to win commissions such as the Commercial Coffee Rooms, Bristol (1809). His interest in engineering matters was revealed in November 1812 when he submitted an innovative lock system for use on the Regent's Canal, with a view to saving water. Although the Company did

not adopt this proposal, Busby patented (No. 3683) his system in 1813. He also developed an iron roof truss of unconventional, and almost certainly unstable, design, and the failure of one of his roofs, presumably to this design, precipitated his emigration to the United States of America in 1817. He was accompanied by his wife Louisa Mary Williams, whom he had married in 1811, and their children, Charles Stanhope Burke, born 1814, and Ellen Mary, born in 1817.

Busby made his home in New York, where he was immediately fascinated by the steam ferry boats. While he sought architectural commissions he investigated steam boat design and developed upright paddles which were tried out unsuccessfully on the York ferry. In 1818 he developed his ideas further with a mechanically driven boat 80 ft. long. He also became involved in the activities of the New York Institution.

In many ways Busby's most enduring contributions to the development of civil engineering were his measured drawings of major timber bridges: the Trenton Bridge, New Jersey, and the Schuylkill and Upper Schuylkill bridges, Philadelphia. These were subsequently engraved and published by Taylor following Busby's return to London. Despite travelling extensively Busby was largely unsuccessful in obtaining work and in 1819 returned to England, where his subsequent work frequently showed American influences.

In England he began working for Francis Goodwin (1784–1835), who was then engaged in designing churches for the Church Commissioners and he attempted, unsuccessfully, to get them to accept two designs incorporating iron roof trusses. As this controversy continued in 1822 he won the Gold Vulcan Medal of the Royal Society of Arts for a hydraulic powered orrery, a result of his investigations of the resistance of solid bodies in their passage through fluids. In the following year he came third in the competition to design a new London Bridge.

In terms of his later career the decisive movement came when in May 1823 he entered an architectural partnership with the Brighton architect, Amos Henry Wilds (1784–1857), a move made at the suggestion of Thomas Read Kemp, who was anxious to develop his property in Brighton. Busby then became a key player in the architectural development of Brighton, not only as architect but also investing heavily. Ultimately he overstretched himself financially and in 1833 became bankrupt. In the meantime he had maintained his interest in engineering.

In 1824 he had advocated hot running water for his houses in Brunswick Town and in 1832 took out a patent for a form of central heating (6268) which was installed at Bow Street Police Station. In 1829 he advocated using the moat of the Tower of London, then abandoned, as a dock for steam boats.

He developed further his ideas on steamboat propulsion and in May 1832 used his ideas to propel a small vessel along the coast near

Brighton. He reported on Ericsson's steam engine in the *Mechanics Magazine*.

Busby died at his home in Stanhope Place, Brighton, on 18 September 1834 and is buried at St. Andrews, Hove. He left an estate of under £200.

MIKE CHRIMES

[Busby drawings, RIBA drawings collection, PRO: RAIL 860/5; J. Taylor (*c.* 1823) *Bridges of Britain and North America*; W. Baddeley (1830) Busby's hydraulic orrery, *Mechanics Magazine*, **14**, 241–242, **15**, 57–58; *Gentleman's Magazine*, **104** (1834), 2, 446; N. Bingham (1991) *C. A. Busby: The Regency architect of Brighton and Hove*]

Publications

1818. *An Essay on the Propulsion of Navigable Bodies*

1823. Hydraulic orrery, *Trans RSA*, **40**, 98–104

1832. Mr Busby's volcanic steam engine, *Mechanics Magazine*, **16**, 329–330

1832. Ericsson's steam engine and waterwheel, *Mechanics Magazine*, **17**, 420–421; ibid. **18**, 5, 20–22, 39, 72

1832. Mr Hancock's steam coach at Brighton, *Mechanics Magazine*, **18**, 84–85

1833. Extraordinary phosphorescent meter, *Mechanics Magazine*, **20**, 5

1833. Destruction of Brighton chain pier, *Mechanics Magazine*, **20**, 53–54

1834. Rotation of the earth, *Mechanics Magazine*, **21**, 104–105

BUTLER, Thomas (fl. 1824–1827) was appointed Assistant Secretary of the Institution of Civil Engineers in April 1823. Not being a Member, a Special General Meeting was required to permit his appointment. His salary was 20 guineas a session, then approximating to a Parliamentary session. He resigned the following January on the grounds that the salary was inadequate for the responsibilities and was succeeded by Nathaniel J. Larkin (q.v.). Butler was elected an Associate of the Institution of Civil Engineers in 1825 and remained a Member through 1827, resident in Bridge Street, Westminster. No more is known of him.

MIKE CHRIMES

[Membership records, ICE archives; J. G. Watson (1988) *The Civils*, 199–200]

BY, John, Lieutenant Colonel (1783–1836), military engineer, was born into an upper middle-class home in Lambeth in 1783, the second son of John and Mary By. His grandfather and father operated a shipping interest on the river Thames in London and were successive Chief Searchers in the London Customhouse (as was, eventually, his elder brother, George). He received an early education in a local grammar school and then attended the Royal Military Academy at Woolwich. Upon graduation in 1799, he was gazetted a Second lieutenant in the Royal Artillery and later

Lieutenant-Colonel John By

he transferred to the Royal Engineers. By was first posted to the fortifications at Plymouth where he likely worked on upgrading and repairing the main batteries in the Royal Citadel.

In 1802 he was posted to Quebec City in Lower Canada (now Quebec). Initially By was assigned duties in Quebec but by 1803 or 1804 was sent up the St. Lawrence River to work on improvements to the existing small canals at Coteau du Lac and at Cascades. In 1805 he was placed in charge of the reconstruction of the timber lock at Cascade as a masonry one and remained there for two years until it was completed. This lock was first constructed under William Twiss (q.v.) in 1779 and is generally regarded as being the first lock constructed in North America. By returned to Quebec where he carried out various duties for the next five years. His most important work at this time was the creation of a large model of the Quebec Fortifications which was intended to help garner approval for the expansion proposed by the Board of Ordnance. By worked closely with several other Engineer Department staff members, particularly one of the best regarded draughtsman, Jean-Baptiste Duberger, in crafting the 5.5 m square model. By returned to England in 1810 and took the model with him for presentation to the Board of Ordnance. It is reported that the model remained on display in the Model Room at Woolwich for several decades; it is now on display in Quebec City.

Upon his return to England, By was assigned to Wellington's forces in the Iberian Peninsula. In 1811 he was present at the battle of Almeida and was awarded prize money for his involvement in the first two, unsuccessful, sieges of the fortified town of Badajoz on the Spanish–Portuguese frontier. By was also in action at Pardaleras and San Christoval and was invalided home in July 1811. Early in 1812 By was assigned to the gunpowder manufactory at Waltham Abbey to reconstruct works destroyed in a recent explosion and was eventually appointed Commanding Royal Engineer for the Royal Gunpowder Mills. At this time he developed new, safer gunpowder and charcoal presses, the new machines incorporating hydraulic rams which had been invented a few years earlier by Joseph Bramah (q.v.) and with whom By worked closely. Also at this time By was responsible for the design and construction of a new Small Arms Manufactory at Enfield Lock and he collaborated with John Rennie (q.v.), who had previous experience designing improvements to the river Lea at Enfield, in the design and specification of hydraulic machinery. Apparently By kept up interest in bridge engineering and in 1816 submitted a design and model for a 1,000 ft. wood truss bridge to Board of Ordnance. This design may have been based on the short lived Terrebonne Bridge which was erected across Riviere des Mille-Illes, the north branch of the St. Lawrence River north-east of Montreal, at the time of By's first posting to Canada. In 1821 By was placed on half-pay and took up residence at Shernfold Park, his estate located near Frant. At this time he became active in local affairs and was appointed 'Surveyor for Frant Green Division'. He was also active on the parish council and was involved in the construction of the new St. Alban's church in Frant. He was promoted to Lieutenant-Colonel in 1824.

By was recalled to service in 1826 to undertake what would become his masterwork, the Rideau Canal in Upper Canada (now Ontario). The War of 1812 proved the vulnerability of Upper Canada—and for that matter all of British North America—to an invasion by American troops. By 1815 preliminary surveys were underway to establish an inland water route that would bypass the transportation route along the St. Lawrence River. The Americans had failed to cut this vital British supply link during the war and it was feared that should hostilities recur the river would be the first focus of attack. In 1825 the Duke of Wellington, Master-General of the Board of Ordnance and soon to be prime minister, appointed Sir Carmichael-Smyth, RE, to head up a three-person reconnaissance team to review defence options for British North America. In his final report, Carmichael-Smyth recommended, among other things, the preferred alignment of the proposed Rideau Canal that would join the Ottawa River and Lake Ontario at Kingston. Approval to commence construction of the canal, along with initial financing, was obtained later that year. The canal was to be constructed by the Board of Ordnance under the direction of the Royal Engineers and was to be financed by the Colonial Office.

By was appointed Commanding Royal Engineer for the canal project. He travelled to Canada and early in 1826 established his base of operations at the Ottawa River (eastern) terminus of the canal. The semi-permanent construction camp that developed around the Engineer's establishment soon came to be known as 'Bytown' in honour of the commanding officer and was changed to 'Ottawa' in 1854 when it was chosen as the seat of government of Canada.

The proposed 124 mile waterway involved the construction of forty-seven masonry locks, fifty-two dams and numerous weirs and spillways. The eight 'staircase' entrance locks at Ottawa, which lift the canal some 79 ft. above the Ottawa River, present one of the most impressive vistas on the route. The most notable single structure, however, is the dam at Jones Falls. This 60 ft. high, masonry (in unusual vertical courses) arched structure was by far the highest dam in North America at the time. Given the wilderness condition of the working area and the limited availability of resources along the route, By recognised that the Rideau Canal could not be completed according to the ambitious schedule using the traditional approach of starting at one end and systematically working along the route. For the Rideau Canal, By commenced work simultaneously in five distinct divisions. Royal Engineer officers were placed in charge of each division and all resources, including contractors, were allocated to each work site. In this way each division was completed independently and the entire project was completed in a remarkably short five working seasons. Among all of By's achievements, his orchestration of this great engineering work, which even by today's standards is impressive, is no doubt the most important. The Rideau Canal was the first significant project undertaken by the Royal Engineers using private contractors. Although this novel approach on such a large undertaking proved troublesome from the point of view of Parliamentary spending control, it was one of the more critical factors in allowing By to schedule the work to a speedy conclusion.

By was repeatedly charged with overspending his budget and with spending without formal Parliamentary approval. A Select Parliamentary Committee was convened in London in March 1831, and again in June 1832, to investigate cost overruns on the project. Numerous witnesses were called but By never appeared before the Committee. The Committee criticised the way the Board of Ordnance conducted the financial affairs of the Rideau Canal project and the way in which it appeared intentionally to delay reports to the Treasury and Parliament. However, the Board exonerated By and credited him with completing the canal under trying conditions. None-the-less, By was later recalled from Canada to face further charges of overspending but there is no evidence that he was ever called before any committee or tribunal or that he was ever charged with any wrongdoing. Although exonerated, By never actually received any formal commendation for his achievements on the Rideau Canal and to this day the shadow of charges clouds his career. On another front however, By did face a tribunal, in Bytown, over charges of misappropriation of funds which were brought against him during construction by Henry Burgess, a disgruntled Assistant Clerk of Works. An inquiry was held by the Royal Engineers in Bytown late in 1831 which found that By was fully innocent of these charges.

The Rideau Canal was substantially complete by the fall of 1831 but disputes over water rights prevented formal opening until early 1832. By traveled, with his wife and two children, through the eastern United States and Nova Scotia in the summer of 1832, then returned to England in the fall and lived at his home, Shernfold Park, where he started to expand his land holdings.

John By died suddenly in February 1836. Although he had suffered, perhaps throughout most of his life, from what is thought to have been malaria he apparently succumbed to some other 'sudden attack' and died within two days. He had married Elizabeth Johnson Baines at Madron on 12 November 1801 but she died at Waltham Cross in December 1814 without surviving children. By was remarried on 14 March 1818, to Esther March—the wealthy heiress of John March, a printer—at Cheshunt. With Esther, By had two daughters, Esther March and Harriet Martha. Harriet never married and Esther married Percey Ashburnham in 1838, by whom she had two daughters who died young. Several contemporary authors describe By as being a jovial, good natured man with a 'showy hospitable mode of living [which] made him universally popular and beloved'. McNab, one of the contractors on the Rideau Canal described him 'as a very handsome man, of fine presence, well built and the 'beau ideal' of the British officer'. He has been characterised as nearly six feet tall, with dark hair and a florid complexion, stoutly built, with boundless energy and determination. At the time of his death he owned nearly a thousand acres of land, mostly around Frant, but also near Bytown, including his estate and farms. He also held substantial cash and investments in both England and Canada.

MARK ANDREWS

[E. F. Bush (1976) *The Builders of the Rideau Canal*, Parks Canada, MSS Report 185; *Dict Can Biog*, **VII**, 128, 1988; R. F. Legget (1955) *Rideau Waterway*; R. F. Legget (1982) *John By: Builder of the Rideau Canal, Founder of Ottawa*; M. E. Andrews (1998) *For King and Country, Lieutenant-Colonel John By, R. E. Indefatigable Civil-Military*]

Works

1803–1805. Coteau du Lac and Cascades Canal improvements
1826–1831. Rideau Canal, 124 miles, engineer

C

CALDWELL, Sir James Lilliman, General (1770–1863), military engineer, the son of Major Arthur Caldwell (d. 1786) of the Bengal Engineers and Elizabeth *née* Weed was born on 22 November 1770, and entered the service of the East India Company as a cadet in 1788. He was appointed Ensign in the Madras Engineers on 22 July 1789 and was successively promoted to General, a rank achieved in 1854. His distinguished military career led to him being created CB in 1815, KCB in 1837 and GCB in 1848.

Shortly after his arrival in India, Caldwell saw active service in the Third Mysore War against Tipu (1791–1792), and again in the Fourth Mysore War of 1799, when he was wounded at the second siege of Seringapatam; in 1810 he was mentioned in despatches following the capture of Mauritius.

On his return to Madras in 1811 he was appointed Engineer in Charge of the centre division of the Madras Army, and the following year rebuilt the fortifications at Seringapatam. In 1813, he was made Special Surveyor of Fortresses and became Chief Engineer of Madras in 1816; he was Commissioner for the restoration of French settlements on the Malabar and Coromandel coasts. In 1824, he became Lieutenant Colonel Commandant of the Madras Engineers.

Despite his active military service, Caldwell spent a considerable amount of his early career on public works between and after the Mysore Wars. In the 1790s he served as Assistant Surveyor to Michael Topping (q.v.) on various surveys around the Kistna (Krishna) and Godavari rivers. In 1792 Alexander Beatson recommended a dam, the Bezwada anicut, on the Kistna as a means of providing irrigation water for the area and Caldwell participated in the ensuing surveys, taking levels. Topping's 1794 recommendation was for cuts on the Kistna, thus avoiding the expense of a dam. Between 1798 and 1803 Caldwell surveyed the area between Bezwada and Tallapoody on the Godavari and in 1800 he reported on Beatson's dam proposal. The anicut was later constructed 1852–1855 by C. A. Orr, of the Madras Engineers. He also considered restoring the Cavilashapooram tank (reservoir) but felt it would be too costly, although the local official, Oakes, subsequently carried out the work with the support of the local inhabitants.

Caldwell was himself directly involved in the execution of some of the earliest British irrigation schemes in Madras. On Topping's death in 1796, Caldwell succeeded him as Superintendent of the Department of Tank Repairs, probably the first department of public works to be established in British India. Between 1796 and 1797 he completed Topping's drainage of Masulipatam (Machilipatnam) near the Kistna.

Moving further south, in 1804 he was sent to report on the state of the Cauvery River, as a means of providing irrigation for the district. The river divides about 18 miles from the head of the delta and Caldwell found that the bulk of the water passed down the Coleroon whilst the irrigation water coming from the Cauvery was often reduced to a trickle. The initial solution was to raise the Grand Anicut (weir) on the Coleroon, but it began to silt dangerously and the problem was not solved until 1836 when (Sir) Arthur Cotton built a permanent masonry weir 20 miles upstream of the Grand Anicut.

Another irrigation scheme, which Caldwell examined in 1808, involved the diversion of the Periyan River into Madurai via a cutting in excess of 100 ft. in depth, but the scale of the work meant that Caldwell felt that he could not recommend it. It was subsequently carried out in the 1880s.

Caldwell's most durable civil legacy was St. George's Cathedral, Madras, which was completed to his designs by Thomas Fiott De Havilland (q.v.) in 1815.

Caldwell married Jeanne Baptiste Johnston, the widow of Captain Charles Johnston of the Madras Army, in India in 1796. They had two children, (Major) Arthur James (1799–1843) and Elizabeth Maria (1797–1870). He retired to Paris in 1837 after fifty years service and after his wife's death he bought Beechlands, Ryde, Isle of Wight, where he died on 28 June 1863. His portrait appears as the frontispiece to Vibart.

JOYCE BROWN

[DNB; E. Buckle (1839) Wulloorpullum sluice, *Prof Papers, Madras Engineers*, **I**, 34; D. Sim (1839) Coleroon anicuts, *Prof Papers, Madras Engineers*, **1**, 131–133; E. Buckle (1846) Report on the resources of the districts bordering the Kistra, *Prof Papers, Madras Engineers*, **2**, 1–33; R. B. Smith (1856) *The Cauvery, Kistna and Godavery*, 1–47; H. M. Vibart (1881) *The Military History of the Madras Engineers*, **I**, 213–273, 280–281, 288–331, 364–368, 417, 436–437; H. M. Vibart (1883) *The Military History of the Madras Engineers*, **II**, iii–vi; G. T. Walch (1899) *The Engineering Works of the Kistna Delta*, **I**, 9; E. W. C. Sandes (1933) *The Military Engineer in India*, **I**, 163, 166, 173, 177, 192–193, 232; E. W. C. Sandes (1935) *The Military Engineer in India*, **II**, 20, 24, 28, 89; R. H. Phillimore (1945) *Historical Records of the Survey of India*, **I**, 105–109, 321, 345–347]

Works

1796–1797. Masulipatam (Machilipatnam), drainage completion
1804. Cauvery irrigation works
1815. St. George's Cathedral, Madras, architect

CAMPBELL, Dougal (fl. 1725–d. 1757), military engineer, had a versatile career on the cusp of the military and civil engineering and architectural professions working on roads, water supply and barracks, plans and reports of many of which

survive in the Public Records Office and the National Library of Scotland.

From 1725 to 1729 he was Deputy Store Master at Edinburgh and also Cadet Gunner from 1728. In 1734 he joined the civil branch of the Ordnance as Sub-Engineer, was promoted to Clerk of Ordnance in 1739 (salary £60 p.a., doubled five years later) and finally to Sub-Director, 1751–1755. He joined the Corps of Engineers as Engineer Extraordinary in 1739, eventually rising to Major and Chief Engineer in North America in 1757.

In 1743 he created a reservoir at Berwick to provide water for the barracks and gave evidence in the lawsuit of William Adam (q.v.) vs. Lord Braco. He had been working with Adam at Edinburgh Castle on the modernisation of the walls and designing a new Governor's House and was probably present throughout the Jacobite siege in 1745, being one of only three engineers in Scotland at the time. He was then sent to Aberdeen where he converted a house and garden for poor boys into an emergency fort and was present at the Battle of Culloden along with John Romer (q.v.) in 1746. Later that year he drew up a plan of the town and castle of Stirling, the river Forth and country adjacent. He also rebuilt and extended the barracks at Fort William, where his original work in 1744 had suffered during the rising, to accommodate one-hundred and sixty men as well as building houses for the Governor and Storekeeper.

Also while in Scotland in 1746 the Duke of Cumberland ordered William Skinner (q.v.) to build Campbell's design for a new Fort George on the site of the old Cromwellian fort at Inverness, but construction was prevented by local magistrates demanding compensation for the loss of the use of their harbour. There are also drawings in the Duke of Argyl's archives for a polygonal fort at Inverary; presumably the changed political climate after 1746 rendered it unnecessary.

In 1746–1747 he was Chief Engineer in Flanders and 1748 saw him back at Berwick working on the magazine and counterscarp and surveying the road thence to Carlisle, where he also provided a plan of the city and vicinity with proposals to modernise the fortifications. In 1749 he continued his surveying eastward, providing a report, map and estimates for the road from Carlisle to Newcastle-upon-Tyne, where he was assisted by Hugh Debbieg (q.v.). The road was completed in 1752 at a cost of £22,450. Its fourteen bridges (overall estimate £4,500) included one over the River Irthing of three arches, one of 40 ft. span and two of 35 ft. and one at New Burn Dean of three arches of 53 ft. each, the latter bridge estimated at £1,500. Campbell was given leave to go abroad in June 1750 and the care and direction of his works were committed to Leonard Smelt who later married his sister, Janet.

Campbell's final barracks work was at the Tower of London where he reported on the state of the Irish barracks in 1752. Both repairs and new works were completed by 1755 when he was promoted to Chief Engineer of the Medway. In 1757 he was sent to Halifax as Chief Engineer in North America. However, the planned expedition to Louisberg was cancelled due to a change of ministry and he died at sea on his way from Halifax to New York.

SUSAN HOTS

[PRO 31/129–131; WO 47/34; NLS Z2/4A, 6A, 28, 30A; PROBATE 11/834, F35; Zan.MS.M.5, Northumberland CRO; Royal Engineers Library, Conolly papers (ms); Colvin (3); Bendall; *Notes and Queries*, 2nd series, 7, 19 February 1859; W. Porter (1889) *History of the Corps of Royal Engineers*, 1; W. Lawson (1966) The origin of the military road from Newcastle to Carlisle, *Arch Ael*, **4**, 44; D. W. Marshall (1976) The British Military Engineers 1741–1783 (unpublished Ph.D. dissertation, University of Michigan); J. Gifford (1992) *Buildings of Scotland, Highlands and Islands*; C. Tabraham and D. Grove (1995) *Fortress Scotland and the Jacobites*]

Surveys
Include:

1749. Report and estimates of the road between Newcastle and Carlisle showing present and proposed roads with the course of the Roman wall [original report missing, but quoted by Conolly]
1750. Map [of above], engraved by Nathaniel Hill [BM K. Top 5.103]

Works
Berwick Barracks Reservoir
Fort William Barracks
1749–1750. Newcastle Carlisle Military Road, £22,000 excluding land costs, on completion in 1752

CARGILL, John (1772–1848), contractor, was baptised on 27 September 1772 in Newcastle-upon-Tyne, the eldest of five children of James Cargill, a house carpenter and Freeman of Newcastle, and Elizabeth Carr. The youngest brother, Daniel Lawrence (*c*. 1783–1821) became a doctor and one can assume that John had a reasonable education. His family were possibly Quakers.

It is unclear where Cargill obtained his early experience, but in 1804 he was appointed foreman for the masonry and lockworks at the eastern end of the Caledonian Canal under John Simpson (q.v.). In Scotland, Simpson and Cargill continued to act as contractors together until Simpson's death in 1815, after which Cargill continued as a contractor on the Caledonian Canal.

Following its completion, Cargill was contractor for one of Thomas Telford's (q.v.) most famous masonry bridges, the Over Bridge at Gloucester. He then acted as resident engineer on Telford's Morpeth Bridge, completed in 1831 by Thomas King and William Beldon, contractors.

Cargill had two children, Donald and Margaret, with his wife Sarah (*née* Brough) and he died in Rouen on 27 January 1848.

MIKE CHRIMES

[Gibb mss, ICE Library; J. G. James collection, ICE Library; *Caledonian Canal Commissioners Reports, Reports of Commissioners for Highland Roads and Bridges*; A. Gibb (1935) *The Story of Telford*; T. Day (1995) Telford's Aberdeenshire bridges, *Industrial Archaeology Review*, **16**, 193–207; T. Day (1996) Gentlemen contractors: the Farquharsons of Monaltrie and construction of Ballater's bridges 1775–1812, *Northern Scotland*, **16**, 87–106]

Works
With J. Simpson until 1815:

1804–1822. Caledonian Canal (eastern end), £56,500 excluding excavation
1807–1810. Ballater Bridge, five arches, *c*. £3,800
1811–1812. Bonar Bridge masonry work, £13,971
1811–1813. Creech Road, 2 miles 68 yd., £808 6d
1812–1814. Craigellachie Bridge masonry work, £8,200
1826–1828. Over Bridge, Gloucester, 150 ft. span arch, £43,500
1829–1831. Morpeth Bridge, three spans, resident engineer

CARLETON, George (fl. 1578–d. 1590), landowner and pioneer of fen drainage, held the manor of Coldham, an area of 420 acres south of Wisbech, probably from 1578, and also owned or leased 1000 acres in South Holland between Whaplode and Holbeach rivers. In 1581 he was appointed to the Commission of Sewers (the land drainage authority) for Cambridgeshire. A year or so earlier Carleton and Humphrey Michell (possibly a previous holder of Coldham manor) took over the 1578 Patent of Peter Morris (q.v.) for draining 'fennes and lowe groundes … by certain engines [unspecified] and devices never known or used before'. In June 1580 they were experiencing delays in the field 'for want of workmen and labourers', but nevertheless the work was completed in due course, a hollow-post windmill being erected 3 miles south of Wisbech at the north end of Coldham Fen close to Friday Bridge. The windmill, discharging water into Elm Leam, is the earliest drainage mill in England for which positive evidence exists. Its location is accurately shown on the large-scale map of the Fens by Jonas Moore (q.v.) published in 1658. It is not known what mechanism was adopted for raising the water. Possibly it consisted of twin reciprocating pumps, as used by Morris himself at London Bridge waterworks in 1582. More probably, as a 'watermill with sails', the Coldham mill operated a paddle wheel; this would be better suited for lifting large quantities of water through a small height than pumps, which deliver a small quantity through a large height.

When peat shrinkage began to be a serious problem in the 1680s, following the general draining of the Fens between 1631 and 1656, and it became necessary to raise water several feet into the rivers and channels, the typical drainage engine employed was a smock mill operating a scoop-wheel, a design used in The Netherlands since the sixteenth century and probably transferred to England. But the less costly post mill with paddle wheel was still in use in the eighteenth century where a small lift, or even no more than an accelerated flow, sufficed.

Carleton next turned his attention to South Holland, with a scheme for draining land south of Ravens Dyke in Whaplode, Holbeach and Fleet to an engine and outfall sluice at Sutton sea bank. This met with opposition from the Lincolnshire Commission of Sewers, based chiefly upon a dislike of 'innovations'. The dispute dragged on for years. Meanwhile, perhaps in despair of a resolution, he erected an engine below the sea bank at the outfall of Holbeach river north of the town.

This, too, met with opposition, but of a violent kind. The mill was maliciously destroyed four days after completion, the probable date being December 1587. In the trial of the culprit it emerged that 'a beam in the water work that bore the most part of the frame standing upon it containing about 7 tons of timber was cut underneath'; it also emerged that threats had been made against another of Carleton's proposals, for an engine 3 miles south of Holbeach at Saturday Bridge on Ravens Dyke (apparently never built), and finally it was pointed out that the 'Ingen' at the sea bank posed no threat nor any charge to Holbeach since this 'was given by Mr. Carleton at his own cost', and the like was still on offer in his scheme for an engine at Sutton. And had he better neighbours, like those 'about Wisbech [i.e. at Coldham], the Holland Ingyns might have prospered'.

The dispute over the South Holland drain rumbled on until July 1588 when Lord Burghley desired on behalf of the Privy Council that 'a perfect platt for South Holland and of all the banks, draynes and divisions be committed to Mr. Hexham ... for the better deciding of this present controversy'. Hexham, joined by Ralph Agas (q.v.), completed the survey before April 1589. A committee set up to consider the matter reported next month that, contrary to previous opinion, the low grounds in question could best be drained by a new drain to the sea, thus endorsing Carleton's scheme in principle. Four special commissioners, of whom Carleton was one, acting with professional advice from Hexham and the engineer Humprey Bradley (q.v.), were then appointed to design the as yet unagreed course and outfall of the new drain. Investigations by this group began in July but no evidence exists of any material results, though it may be noted that the South Holland Drain was eventually constructed in 1793–1796.

There is, however, an interesting sequel. In response to an awakened interest in the possibility of a general draining of the Fens the Privy Council in March 1589 appointed Bradley, Hexham and Agas to survey the whole area and report on the feasibility of achieving this objective. Hexham's map was made as a result, and Bradley reported to Lord Burghley in December 1589. This first attempt to tackle the general drainage problem thus in part grew out of Carleton's scheme. He died in January 1590. His interests were purchased by John Hunt (q.v.) who continued where Bradley had left off, producing in 1605 the first detailed engineering proposals which formed the basis for much of the work eventually carried out under the Earl of Bedford and others from 1631 onwards in the Great Level of the Fens.

A. W. SKEMPTON

[Jonas Moore (1658) *A Mapp of the Great Level of Fenns*; H. C. Darby (1940) *The Draining of the Fens*; L. E. Harris (1957) Land drainage and reclamation, *A History of Technology* (C. Singer, ed.), vol. 3, 300–323; Richard. L. Hills (1967) *Machines, Mills and Uncountable Costly Necessities; Victoria County History of Cambridgeshire*, **4**, 180–186; R. A. Skelton and John Summerson (1971) *A Description of Maps and Architectural Drawings in the Collection made by [Lord] Burghley now at Hatfield House*; Richard L. Hills (1994) *Power from Wind*; J. S. P. Buckland (in lit. 1997), further information; British Library (Add. MS 71,126)]

CARLINE, John II (*c.* 1758–1834), architect and stonemason, was the second of three generations to bear the same name and vocation. Both of his parents came originally from Lincoln. His father, John Carline I (1730–1793), married Ann Hayward, sister of William Hayward (q.v.), and was brought by the latter to be foreman mason during the rebuilding of English Bridge at Shrewsbury in 1769–1774. He subsequently had a lease of the well-known stone quarries at Grinshill in Shropshire and worked the stone at his yard at Carline's Fields, Shrewsbury. He built a bridge over the Rea Brook in 1771 to improve access to it and about the same time supplied stone worth more than £5000 for John Gwynne's Atcham Bridge.

John Carline II followed his father from Lincoln to Shrewsbury in September 1780 and worked with him until the latter returned to Lincoln. He then removed his yard to a site at the north-east corner of English Bridge, Shrewsbury; his house was demolished in 1932 and Wakeman School now occupies the site. In 1788 he was unsuccessful in his tender to design and build a replacement for the mediaeval Meole Brace Bridge outside Shrewsbury as only widening works were carried out then; when it was eventually replaced in cast iron in 1811 he had the satisfaction of providing the temporary timber bridge. Among the contracts he won for masonry was that on 28 November 1789 for stonemasons' and bricklayers' work at new St. Chad's Church, Shrewsbury, at a schedule of rates. In this he was associated with John Tilley (*c.* 1749–1795), stonemason, and Jonathan Scoltock, bricklayer; Carline signed the contract but Tilley put his mark to it. Carline and Tilley were the successful contractors for Thomas Telford's first bridge in Shropshire, at Montford, and when, in December 1792 they submitted the winning design to replace the old Welsh Bridge in Shrewsbury, its details were based closely on Montford. Carline and Tilley now traded as architects and on 8 May 1794 they were awarded the contract to design and rebuild St. Alkmond's church, Shrewsbury, for £2,490. It was to be complete by Michaelmas 1795 but the firm still had the capacity to rebuild Carline's father's bridge over the Rea, which had been swept away in the great flood of February that year. Tilley died in December and Carline does not appear to have undertaken many significant commissions in the years following. Nevertheless, he enjoyed a sufficient reputation to be asked by the magistrates of Staffordshire in 1799 to arbitrate on their claim for damage to Blackbrook Bridge caused by the breaking of the Wyrley and Essington Canal Company's Sneyd Reservoir. He was in partnership with Tilley's stepson, Henry Linell, in 1801–1803 and later with his own eldest son, John, as J. & J. Carline. In 1814 he was employed by the Bridgnorth Bridge Commissioners to inspect the works carried out there by John Simpson (q.v.) and to certify the monies due.

Carline married Mary Cotton of Little Ness on 10 November 1789 in Shrewsbury and they had nine children. The third child and eldest son, John III, followed his father and was for a time in partnership with him, as noted above. At the time that the rebuilding of the Welsh Bridge was under consideration there had been criticism of its siting, at an angle to the flow of the river. Problems with scour under the piers did indeed occur. Telford and James Trubshaw (q.v.) submitted proposals which the Corporation did not feel inclined to accept and when John Carline III, who had by now taken over his father's business, submitted a cheaper solution, it was accepted. He carried out the work in 1833 and it has proved satisfactory since then. In the 1840s he became financially embarrassed and returned to Lincoln.

The sixth child, Thomas, was the only one of John II's five sons to marry. He continued the family tradition as a sculptor but was unable to sustain his initial success. He later became County Surveyor of the North Riding of Yorkshire.

John Carline II died on 8 December 1834.

P. S. M. CROSS-RUDKIN

[Shropshire RO; Staffordshire RO; A. W. Ward (1935) *The Bridges of Shrewsbury*; J. L. Hobbs (1960) Famous Shropshire architects, *Shropshire Magazine*, Jan.–Apr.; Anthony Blackwall (1985) *Historic Bridges of Shropshire*; Colvin (3)]

Works

1790–1792. Montford Bridge, contractor (with John Tilley) for construction
1793–1795. Welsh Bridge, Shrewsbury, contractor (with John Tilley), to design and build replacement for mediaeval bridge, for £6,600
1795. Longden Coleham Bridge, Shrewsbury, contractor (with John Tilley), £450

1815–1816. Overton Bridge, supervised rebuilding by Thomas Penson Jr. (q.v.), after failure of Thomas Penson Sr.'s structure during construction
1818. Cound Bridge, stonework £790 plus old materials (ironwork by William Hazledine)

CARR, John (1723–1807), architect and bridge designer, was baptised on 15 May 1723 at Horbury in the West Riding of Yorkshire, the eldest of six sons of Robert Carr Sr. (q.v.). He was educated at the village school and trained as a stone mason in his father's quarries at Horbury. The experience thus gained on masonry building contracts, together with the use of the standard treatises of the day, seem to have comprised his training in architecture. By the age of twenty-five, however, he was able to design the simple yet elegant Huthwaite Hall at Thurgoland. Designs for a well house in York and for a grandstand at the racecourse in the 1750s led to the freedom of that city in 1757 and to commissions for private and public buildings throughout the north of England particularly, but also as far afield as Portugal. Now firmly established in York, he was one of the city chamberlains in 1766, was excused from serving as a sheriff in 1769 but was elected Mayor in 1770; he served again in 1785 when the incumbent died in office. His monument in Horbury church asserts that in civic life he attained the highest place on account of his singular hard work, agreeable manners, outstanding generosity and integrity of life, a claim which it is not hard to believe. His architectural work is dealt with in Colvin.

In 1752 John Carr provided many of the neat measured drawings for the West Riding's *Book of Bridges*, made under the direction of his father and John Watson Sr., the Bridgemasters. In April 1761, four months after his father's death, he joined John Watson Jr., who had succeeded his own father, as Joint Bridgemaster; they had already been working together on widening the important Lady's Bridge in the centre of Sheffield. They also supervised the reconstruction in 1764–1765 of the substantial bridge at Ferrybridge, to a design by the contractor, John Gott (q.v.). This was considered to be outside their normal duties as Bridgemasters and they received extra recompense for their trouble.

On 16 July 1772 Carr was appointed Surveyor of Bridges to the North Riding at a salary of £100 p.a. His standing as an architect was recognised by the appointment also of Henry King as deputy, at £50 p.a., with particular responsibility for supervision of construction, an unusual post for the time. Soon though a more intractable call on Carr's time became clear. Having made his usual survey of West Riding bridges for the Michaelmas Quarter Sessions at Leeds that year, he was obliged to attend at Thirsk also to let some bridge contracts. His colleague, John Billington, presented the report instead and Carr resigned his position with the West Riding in early 1773.

Over the next forty years Carr provided designs for the widening or rebuilding of 39 bridges in the North Riding. Unlike some of his contemporaries, he appears to have avoided any attempt at a standard architectural treatment. Probably his most ornate design is that at Greta Bridge, near Barnard Castle, which has been the subject of paintings by Cotman, Girtin and others. His largest span, one of 82 ft., is over the same river at Rutherford Bridge, but here in the moors the treatment is elegant in its simplicity.

By far the most expensive of Carr's bridge designs saw him return to Ferrybridge. At first it was intended to widen the bridge whose construction he had supervised in 1765, but the contract was cancelled after work had started and a bridge on a new alignment built instead. The contractor was Bernard Hartley (q.v.). Later in life Carr was one of the engineers named in the Selby Bridge Act of 1791.

Carr married Sarah Hinchcliffe in 1746, but they had no children. The profits of his architectural practice and building business were substantial and he rebuilt Horbury parish church to his own design and at his own expense in 1791–1793. He was reputed to have left property worth £130,000 to his relations at his death, which occurred on 22 March 1807.

P. S. M CROSS-RUDKIN

[Book of bridges, North Yorkshire CRO; *Gentleman's Magazine* (1807); DNB; R. B. Wragg (1971) The bridges of John Carr, *Trans Hunterian Society*, **10**, 315–334; York Georgian Society (1973) *The Works in Architecture of John Carr*; Colvin (3)]

Works
Include:

1760. Lady's Bridge, Sheffield, with John Watson Jr., widened
1763. Coniston Cold Bridge, with John Watson Jr.
1764. Went Bridge, Wentbridge, with John Watson Jr., widened
1765. Horton Bridge, with John Watson Jr., widened
1766. Marle's Bridge, Darfield, with John Watson Jr.
1768. Rotherham Bridge, with John Watson Jr., widened
1772–1773. Reeth Bridge, rebuilt after floods, £1,300
1773. Downholme Bridge, Marske, £1,200
1773. Rutherford Bridge, near Bowes
1773. Greta Bridge, rebuilt after floods, £850
1780. Whitby Bridge, repairs; rebuilt 1911
1781. Skipton-on-Swale Bridge
1785. Boroughbridge Bridge, widened, £420, widened further since
1785. Crambeck Bridge, Welburn
1786. Topcliffe Bridge, widened, £994
1788. Aysgarth Bridge, widened, £420
1789. Richmond Bridge, £899
1791. Tadcaster Bridge, widened

1792. Catterick Bridge, widened, £2,980
1793. Yore Bridge, Bainbridge, £884
1797–1804. Ferrybridge Bridge, £29,678 15s 6¾d
1797. Grinton Bridge, widened
1798. Strensall Bridge, £1,363
1803. Morton Bridge, Morton-on-Swale
1806–1810. Yarm Bridge, widened

York Georgian Society (1973) details further minor bridgeworks.

CARR, Robert, Sr. (1697–1760), was a mason of Horbury in the West Riding of Yorkshire. He was the grandson of Robert Carr (1644–1689) and son of John Carr (1668–1736), who had also been masons and quarry owners in Horbury, and by his wife, Rose Lascelles, he had sons John and Robert Jr. (qq.v.).

In April 1743 he was appointed County Bridgemaster of the West Riding jointly with John Watson of Hobroyd, to be paid £15 p.a. each. In 1740 the maintenance of the 110 Riding bridges had been contracted out to John Westerman, John Gott Sr. (q.v.), William Gott and John Gott Jr., for £300 p.a., but in their initial survey Carr and Watson were very critical of the standard of workmanship, claiming that at twenty-eight bridges nothing at all had been done. The contractors submitted accounts demonstrating an expenditure of only £200 but the matter must have been resolved satisfactorily as their contract was renewed in 1748 and was still being continued in the 1760s.

It seems that the duties of the Bridgemasters were confined to surveying and reporting on the state of the bridge stock. When bridges were to be rebuilt in the 1750s, tenders would be sought from contractors based on their own plans and specifications, which would then be incorporated into the contract. The largest was Cattal Bridge, near Wetherby, for which the contract price in 1751 was £400. The Bridgemasters were not prevented from contracting in this way, as they had rebuilt Turn Bridge at East Cowick for £135 in 1743–1745. The largest bridge designed by Carr lay just over the border in the North Riding at Masham, built by local contractors in 1754.

In 1752–1753, Carr, his son, John, and John Watson Sr., prepared the *Book of Bridges*, which contained elevations and plans of all of the bridges for whose maintenance the Riding was then responsible. Although not the first of its kind—Richard Goodman, late Surveyor of Bridges, had been paid by Cumberland Justices in 1730 for drawing plans of twenty bridges—nor unique, it remains an early and invaluable source of information about one county's bridge stock. Carr remained joint Bridgemaster until his death in 1760, his colleague, Watson, having died in 1757 and been replaced by John Watson Jr.

P. S. M. CROSS-RUDKIN

[Robert Carr and John Watson (1752) *The Book of Bridges belonging in Whole or Part to the West Riding of the County of York*, West Yorkshire RO; Colvin]

CARR, Robert, Jr. (1724–1777), was baptised on 26 December 1724. He was a mason of Horbury like his father, Robert Carr Sr. (q.v.). He appears to have undertaken contract work in the West Riding, sometimes with Luke Holt (q.v.) when mason's and carpenter's work were required together. For one such contract, at the House of Correction in Wakefield, where his elder brother, John, was one of the architects, he produced his only known design; it was for the Infirmary.

In 1769, as part of the reorganisation of the Calder & Hebble Navigation, he and Luke Holt were appointed joint Resident Engineers. The main work was repair of the extensive damage which had been caused by the floods of October 1767 and February 1768 and this was substantially complete by September 1770. Improvements to the canal and its water supply continued and Carr carried on alone after Holt left in 1772 to work on the Huddersfield Broad Canal.

Additionally, in 1773 he was appointed County Bridgemaster to the West Riding in the place of his brother, John, who had succeeded their father in that position but resigned in April 1773 to take up the more valuable equivalent in the North Riding. Robert shared the office with John Billington of Foulby who had been in post since April 1771, and who was replaced by Jonathan Sykes of Oulton in July 1774. During his four-year tenure at least sixteen of the Riding bridges were widened or rebuilt, usually to a plan by either John or William Gott.

Carr was buried at Horbury on 11 March 1777.

P. S .M. CROSS-RUDKIN

[West Yorkshire RO; RAIL 815 at PRO; *Trans Hunterian Soc*, **10**]

Works

1766. Shipley Bridge, contractor (with Luke Holt), to repair, for £89 2s 5d
1766–1768. House Of Correction, Wakefield, contractor (with Luke Holt), to build to design of John Carr and John Watson, for £2650
1769–1777. Calder & Hebble Navigation, Resident Engineer (with Luke Holt until 1772)
1773–1777. West Riding Of Yorkshire, joint Bridgemaster

CARTWRIGHT, Thomas (fl. 1790–d. 1810), civil engineer, brother of William Cartwright (q.v.) is best known as engineer for the Worcester and Birmingham Canal from its inception until 1809, when he retired and was succeeded by John Woodhouse (q.v.). The Act authorising the canal had been passed in 1791 with Josiah Clowes (q.v.) giving evidence. In the autumn of that year Samuel Bull (q.v.) and John Snape surveyed and marked out the line. In January 1792 Cartwright applied for a post on the canal as superintendent of lock and bridge building at a salary of 1½ guineas a week. He came with a reference to his 'very good character from John Bough, clerk of the Birmingham Canal, suggesting he had carried out related work for that Canal Company previously. His post was regularly reviewed and In early 1794 he used an offer of a post on the Horncastle navigation to persuade the Worcester Company to raise his salary to £180 p.a., further increased to £200 at midsummer 1794. This was not the end of such dealings as in 1799 an attempt was made, doubtless prompted by a financial crisis to cut his salary to £100 p.a., and in contrast in March 1809 he was offered 400 guineas p.a.

Clowes's other responsibilities and early demise in December 1794 meant that Cartwright was effectively the engineer, Bull having returned to Birmingham Canal duties in February 1792. He was assisted by Samuel Hodgkinson (q.v.) from 1794 to 1800. Perhaps reflecting Cartwright's limited experience other engineers were consulted—Benjamin Outram (1798) and John Rennie (1808) (qq.v.). Progress was slow and further Acts were required in 1798, 1804 and 1805.

The Canal was opened to Selly Oak in October 1795, and to the junction with the Stratford navigation at King's Norton in May 1796. Although the West Hill tunnel was opened in March 1797, Cartwright almost lost his life in a flood there the following January. Possibly as a consequence of the financial problems facing the company from 1802 he seems to have carried out contract work. In July 1804 he drew up a specification for completing the canal from Hopwood Wharf to Tardebigge tunnel, and the reservoirs, and contracted for the work 'giving bonds himself in £950 and a surety in £500 for the execution of the work'. By July 1808 his claims for extras amounted to £1,630 and were referred to Rennie.

While these works proceeded Cartwright was consulted about various other schemes in the Midlands and South Wales. These included two surveys, in late 1800 and 1802, for tramroad routes connecting the Neath Canal with Aberdare. The latter scheme, largely financed by the Tappendens of the Abernant works, linked the works to Hirwaun Ponds and was engineered by Evan Hopkin between 1803 and 1805.

Cartwright also surveyed a tramroad between Brecon and Hay-on-Wye in 1805. This was not acted upon until 1810 when William Crosley Jr. (q.v.) was asked to survey an alternative line, John Hodgkinson (q.v.) acting as engineer to the final scheme. Earlier he had been asked to surveyed an extension from Gilwern to Llanfoist, on the Brecknock and Abergavenny Canal in 1802. The construction of the canal from Brecon to Gilwern, with Thomas Dadford as engineer, was then complete. The extension was completed in 1805.

Cartwright was also responsible for superintending the construction of the Stratford Canal's tunnel at King's Norton. He was brought in by Josiah Clowes.

Cartwright resigned his position on the Worcester and Birmingham due to ill health in May 1809, and in July his outstanding account for £1,817 was settled. He died in 1810 at Longbridge.

MIKE CHRIMES

[Worcester & Birmingham Canal Minutes, RAIL 886/4, PRO; C. Hadfield and J. Norris (1962) *Waterways to Stratford*, 80–82; C. Hadfield (1967) *The Canals of South Wales and the Border*, *passim*; C. Hadfield (1969) *The Canals of the West Midlands*, 138–140; W. M. Hunt (1979) The Horncastle navigation engineers, *RCHS J*, **25**, 1, 2–11]

Works
1794. King's Norton Tunnel, Stratford Canal, 352 yd., resident engineer
1791–1794. Worcester and Birmingham Canal, resident engineer
1795–1809. Worcester and Birmingham Canal, engineer and contractor
1805–1807. Worcester and Birmingham Canal, Hopwood Wharf–Tardebigge, contractor

CARTWRIGHT, William (*c*. 1765–1804), civil engineer, the brother of Thomas Cartwright (q.v.) was on the Basingstoke Canal under Henry Eastburn (q.v.). As work on this canal was winding down in late 1793 both he and Eastburn were looking for alternative positions. Although he surveyed a proposed junction between the Basingstoke and Bristol and Bath canals at this time nothing came of this. There was a general dearth of suitable qualified engineers in a period of canal mania and John Rennie discussed positions on the Lancaster Canal, Horncastle Navigation, Medway, and Crinan Canal. In the end his offer of 31 December 1793 of a post of resident engineer on the Lancaster Canal at £250 persuaded him. At this salary he was expected to stay for five years. Cartwright's salary and position were beneath that of his colleagues Archibald Millar and Eastburn, both paid £400. In canal documentation he is described variously as assistant or resident engineer.

He was appointed assistant engineer in January 1794 with specific responsibilities for supervising the installation of the foundations of the Lune Aqueduct. He was clearly competent as following the completion of the piers in October 1795 he was presented with a silver cup 'as a reward for his extra care and attention in superintending the Foundation of the Lune Aqueduct'. By 1799 Cartwright was resident engineer for the whole canal, of which much work was now complete. He recommended a tramroad from Clayton Green to Preston, a distance of 5 miles; it included the first powered incline plane in Britain. A new Act was obtained in 1800 for both this work and an extension of the canal from Spital Moss to Preston. There was some discussion about the proposal and Jessop and Rennie (qq.v.) were asked to report, both men suggesting that an aqueduct over the Ribble be considered, but supporting Cartwright's tramroad; Cartwright drafted a design for the tramroad bridge and work began in 1802, although completion was delayed until June 1803 due to problems in driving the Whittle Hills tunnel. Cartwright's other works for the

canal included water supply schemes from the river Keer and a tunnel from the canal at Preston down to the Ribble—completed by William Miller after his death—together with a proposed tramroad from Tewitfield to Kellet Seeds quarries.

No doubt as a result of his reputation on the canal, in August 1800 Cartwright was asked to prepare plans and estimates for the Crofton drainage scheme on the Douglas, Lancashire. He set out the works and by a contract of 18 March 1802 was responsible for carrying out approximately £6,000 of drainage works.

William must have kept in regular contact with his brother as Thomas was apparently consulted about tramroad surveys in South Wales at more or less the same time as William was building his. Cartwright died after a short illness, probably brought on by overwork, on 19 January 1804, in Preston, at the age of thirty-nine. He left a substantial Georgian town house, only completed by himself in 1802, and an estate of £10,000 to his wife Mary. His obituary in the Lancaster Gazette speaks of his 'happy cheeriness of temper, which gained him the esteem of all who had the pleasure of knowing him, and made the workmen employed under him, execute his commands with a promptitude few could excite ... What seldom falls to the lot of one man, was united in him—a strong theoretical genius, accompanied by practical knowledge in every branch, and a persevering spirit, that overcame all difficulties'.

MIKE CHRIMES

[Lancaster Canal minutes, PRO RAIL; Crofton Drainage Commision, Minutes, Lancashire CRO; Lancaster Canal specifications, etc., ICE Library; Rennie reports, ICE Library; *Lancaster Gazette*, 21 January 1804; G. Biddle, The Lancaster Canal tramroad, *RCHS J*, **IX**, 5, 88; C. Hadfield and G. Biddle (1970) *The Canals of North West England*, **1**, 187–191; R. Philpotts (1993) *Building the Lancaster Canal*]

Publications
1800. *A Plan of the Intended Reservoirs and Feeders for Supplying the Lancaster Canal Navigation with Water from the River Keer*

Works
1794–1804. Lancaster Canal, assistant engineer
1800–1804. Croft.on drainage, Lancashire, engineer and contractor, £6,000

CASEBOURNE, Thomas (1797–1864), civil engineer, was born in Hemel Hempstead, Hertfordshire, in 1797. He was trained as a 'civil engineer', almost certainly by his uncle Thomas Townshend (q.v.), under whom he is known to have worked on the Eau Brink Cut (1818–1821), Townshend being Resident Engineer. The original Cut was found to be inadequate in times of flood and, following reports by Thomas Telford and John Rennie (Jr.) (qq.v.), it was widened in 1826. Casebourne monitored the effects of the Cut on the river Ouse, reporting on them to the Institution of Civil Engineers in 1833. Casebourne's involvement appears to have ceased by 1828, by which time he was working for Thomas Telford.

Casebourne is known to have surveyed two sections of the London–Liverpool Road for Telford between 1826 and 1828 and assisted him with the Metropolitan (London) Water Supply Inquiry between 1828 and 1833. Telford had already been consulted about the Ulster Canal when Casebourne was appointed (Resident) Engineer in 1833 and he remained in charge of the construction of the canal for twelve years under (Sir) William Cubitt (q.v.) who, on the death of Telford, succeeded him as consultant.

In 1845 he became Resident Engineer for the construction of the port of West Hartlepool, principally the formation of the protective breakwaters designed initially by James Simpson (q.v.). He became actively involved in the development of this new town, acting as a Commissioner for the town from 1854 until his death.

Casebourne joined the Institution as an Associate in 1828 and was transferred to the class of Member in 1837. Some idea of the regard with which his survey work was held can be gauged from the fact that William Mackenzie (q.v.) contemplated asking him to survey the Marne–Rhone canal when he was considering taking out a contract on it in 1843. He was also entrusted with preparing five of the plates for the *Atlas* published to accompany the *Life of Telford*.

Casebourne died on 2 January 1864.

MIKE CHRIMES

[T. Telford (1833) Diary, NLS; T. Casebourne, Notebook (1833–1842) and correspondence, ICE archives; Some account of the effects produced on the river Ouse between Lynn and Denver Sluice, after the opening of the Eau Cut [ICE OC 159, 1833]; ICE membership records; [it should be noted that the drawings presented to ICE regarding the Ouse; Ulster Canal, etc., are no longer there, and may be in the local records offices or with the present owners]; *Report from the Select Committee on Metropolitan Water* (1833) (HC 571); Obituary notice (1864–1865) *Min Proc ICE*, **24**, 527–528]

Publications

1829. *Mr Telford's Reports, Estimates and Plans for Improving the Road from London to Liverpool*, Map of the country between the village of Weedon and the City of Lichfield; three maps of the county between the towns of Liverpool and Talk-on-the-Hill (surveyed 1826–1828)
1842. Description of a portion of the works of the Ulster Canal, *Min Proc ICE*, **2**, 52–53;
1851. Description of a raft., or float, used for submarine blasting, on the works of West Hartlepool harbour and docks, *Min Proc ICE*, 1850–1851, **10**, 293–295

Works

1833–1844. Ulster Canal (Resident), Engineer
1845–1864. West Hartlepool Docks, Engineer

CASTLE (CASSELS), Richard (1695–1751), engineer and architect, was probably a Huguenot refugee, born in Kassel, Germany in 1695. It is likely that he was a student of Paul du Rey at the Collegium Carolinium in Kassel. He became an officer in a regiment of engineers in about 1715 and studied fortifications and canals in Germany, France and the Low Countries. By 1725 he was studying waterworks and architecture in England. He was brought by Sir Gustavius Hume to Ireland to design his house at Castle Hume, County Fermanagh in about 1727–1728. He then took up a post as architectural assistant in the office of Edward Lovatt Pearce in Dublin 1728.

In 1729 the parliament in Dublin had established the 'Commissioners of Inland Navigation for Ireland' in an attempt to improve the country's communications. Pearce was appointed Surveyor General in 1730 and it seems likely that this was the link between Castle and the commissioners, since Pearce's job involved a good deal of work for the commissioners. Pearce was asked to superintend their first navigation scheme, the Newry Canal, the first summit level canal in these islands, but he delegated the work to Castle.

In 1730, Castle published *An Essay on Artificial Navigation*. His knowledge of continental canal engineering is apparent from his references to examples in Holland and France. The significance of this essay in transport history is its early date, preceding the period of canal building in England, Scotland and Wales, which began effectively only after the middle of the eighteenth century. The original 'Castle manuscript' is in the National Library of Ireland, Dublin. In his essay, Castle considered some of the main problems in canal building, such as water supply, excavation methods, and the provision of adequate foundations for locks, and proposed solutions based on best continental practice.

Castle was also interested in other engineering issues and published *An Essay towards Supplying the City of Dublin with Water* in 1735. In the same year, Gabriel Stokes published an essay on the same topic and also published observations on Castle's essay.

When Pearce died in 1733, Castle took complete charge of the canal work and supervised the building of the first canal lock ever built in Ireland. However, he was later dismissed from the Newry canal works and replaced by Thomas Steers (q.v.), who had come to Ireland to advise on Ballycastle harbour. A local overseer, William Gilbert, completed the canal in 1741.

It was primarily as an architect that Castle made his name in Ireland. In 1728 Pearce commended him to the Irish parliament and this seems to have borne ample fruit, For over twenty years, until his death in 1751, he was the leading designer of country seats and Dublin houses for the Protestant establishment. He has been described as a man of strictest integrity, though by temperament convivial, eccentric and improvident. Intemperance and late hours were said to have caused the gout from which he suffered. He married Jane Truffet (d. 1744) of Lisburn, County Antrim, at the Huguenot church in Dublin on 28 June 1733, but they had no children. He resided for many years in Dublin, first in Sussex Street, then in Proud's Lane, but his death, from a sudden fit or stroke, occurred at Carton, county Kildare on 19 February 1751, and he was buried in the churchyard at Maynooth. In his will he left his property to his sister-in-law, Anne Truffet, and his brothers in Saxony, Samuel, Daniel and Benjamin (Castles) de Richardi.

RON COX

[G. Stokes (1735) *An Essay on Supplying the City of Dublin with Water*, Royal Irish Academy, Haliday Collection; G. Stokes (1735) *Observations on a Late Essay of Mr. Richard Castle*, Royal Irish Academy, Haliday Collection; Obituary (1751) *Dublin Gazette*, February; T. U. Sadleir (1911) Richard Castle Architect, *J R Soc Antiq Ir*, **XLI**, 31, 241–245; Knight of Glin (1964) Richard Castle architect: his biography and works, a synopsis, *Irish Georgian Society*, **VII**, 1, 31–38; Irish Architectural Archive (in prep.) *Index of Irish Architects*]

Publications
1735. *An Essay on Artificial Navigation*, edited by J. H. Harrington; reprinted in *Transport History*, **5**, 1972; original manuscript in National Library of Ireland, Dublin, MS 2737
1735. *An Essay Towards supplying the City of Dublin with Water*, Dublin, Sylvester Pepyat

Works
1730–. Newry Canal, resident engineer

CAULFEILD, William (d. 1767), of Raheenduff in Queen's County, soldier and road-maker, was the only surviving son of Toby Caulfeild of Clone, County Kilkenny, himself the third son of the fifth Baron, first Viscount Charlemont.

Road building by the army under General Wade (q.v.) in the Highlands of Scotland had started in 1725 and Caulfeild had joined the work at least by 1729. In 1732 he was one of six junior officers working under Major Scipio Duroure; based at Fort Augustus. He was promoted to Major and appointed Baggage Master and Inspector of Roads in North Britain. As such he was entrusted with the money to procure the materials for the Bridge of Tay at Aberfeldy. The first great campaign of road construction under General Wade was now almost over and Caulfeild's work was mainly confined to maintenance and improvement. A survey for a new route from Ruthven in Badenoch to Invercauld on Deeside, for which Caulfeild was giving instructions in 1735, came to nought and it remains unbuilt to this day. Then, in 1741, under Wade's successor, General Clayton, he began the extension of the network. As an army officer, he was also required to maintain surveillance for signs of renewed Jacobite insurrection but he seems to

have treated much of it as 'flying rumour'. When the landing did take place in 1745 he was working on the Dumbarton to Inveraray road and withdrew the road parties immediately to Glasgow; he himself went to Edinburgh to act as Quartermaster to the unfortunate General Cope.

Almost immediately after the final defeat at Culloden, work on the roads resumed. Their value had been amply demonstrated during the war and over the next twenty years Caulfeild extended and infilled the Highland system, as well as building a new route from Carlisle to Portpatrick. For 1748, General Churchill ordered 1,350 men to be deployed on the three roads then under construction, as well as eighty men on maintenance of existing roads, whose increasing importance was recognised that year by a larger annual allowance of £500, up from the £200 which it had been for fifteen years. Bridges were built under contract.

Resistance to higher expenditure was growing, however, and when Caulfeild's estimate of £6,796 1s 4d for 1749 was presented to the Commander-in-Chief, the Duke of Cumberland, it was roundly condemned as a 'very bill' and it own expenses were peremptorily cut. A second estimate for £7,117 16s 6½d submitted three months later, when much of the work of the season had been done, included a calculation to show that the increase in cost amounted to £24 2s 3½d only. At almost the same time, Caulfeild received a stinging reprimand from General Bland, who suggested that work paid for on the Inveraray road had been done badly or not at all. Bland seems to have been determined to rein in Caulfeild, having rebuked him earlier for failing to keep him informed. On this occasion the matter must have been resolved satisfactorily but the following year, when the road was supposed to have been completed but yet more work was still required, Caulfeild was making arrangements for local people to be employed during the winter months and the costs charged to him, so that he could place them in some other account in 1751.

The success of the military road-building programme brought requests from local dignitaries for assistance. In 1748, General Bland took it upon himself to accede to a request from the Provost of Stirling, and similar petitions from Midlothian and Moray were also received favourably. In January 1754, though, he asked Caulfeild, who was then in London, to find out if the Duke agreed to a proposal for 46 miles of road in Aberdeenshire. In this case a sergeant and twelve men were to work at the army's expense to instruct the country labourers and this model was repeated elsewhere. In 1756 the Duke of Atholl wanted work to start on two roads in Perthshire but was asked to choose one, as the army could not afford to assist with both.

In 1732, under General Wade, the road building season lasted from April to October. In 1746–1747 it seems that some work continued throughout the winter, but by 1749 the main campaign was limited to ninety-two days; this could be seriously disrupted by bad weather and by other factors too. In 1754 the start was delayed until after the middle of June because of lack of provisions for the men; the previous year the men near Fort George had subsisted for the first three weeks on bread. In the same year Caulfeild was advised that he might also be delayed by the failure of the Treasury to issue money soon enough. The army itself could create problems. For obvious reasons, the issue of gunpowder was tightly controlled and the miners removing rock obstructions sometimes ran out before the season was over; new supplies had to come from Stirling or Edinburgh. In 1748 General Bland himself had to apologise to the Board of Ordnance for taking a few of their tools without informing them.

Caulfeild made a tour of the whole extent of the roads each year, usually quite early in the working season. He also personally reconnoitred any proposed new route, though sometimes the initiative came from above. In the early days, his assistants were officers seconded from the regular army regiments and in 1747 Caulfeild complained to General Churchill that one had been recalled by his commanding officer after only a few weeks. He was especially upset because the man was local and was therefore particularly useful, but pointed out that if officers were changed frequently like this, he would be unable to train any of them properly. The officers were responsible for certifying that the work had been done and the problem with General Bland in 1749 may have arisen from this. One, Ensign Archibald McCorkell, did remain with Caulfeild from at least 1745 to 1755, based at Dumbarton Castle and responsible for the roads north and west from there, and Caulfeild seems to have put a lot of trust in him. From 1749, however, he had three engineers on his staff, able to deal equally with the professional and administrative duties of their respective districts.

The road from Gretna to Portpatrick was surveyed at the request of the surrounding counties in 1757 but the matter rested there until 1763. As a direct route for the military to Ireland, the government assumed responsibility and substantial sums were spent over the next twenty years on improvements and maintenance. The construction was carried out under Caulfeild's overall control by Major Rickson, Deputy Quarter Master-General in Scotland.

Caulfeild had been appointed Deputy Governor of Fort George—at first Inverness Castle, before the construction of a new Fort George at Ardersier—and made his home in Inverness. In 1751 he was promoted to Lieutenant-Colonel, but seems always to have been described as Major or Governor, even in military orders.

He died on 14 August 1767 and the obituary notice in the Gentleman's Magazine refers to him as 'late of New York'.

P. S. M. CROSS-RUDKIN

[Field Marshal Wade's Orders, General Bland's Orders, General Churchill's Letters and Saltoun Collection, at NLS; *Gentleman's Magazine* (1767) 430; J. B. Salmond (1934) *Wade in Scotland*; W. Taylor (1976) *The Military Roads in Scotland*]

Works
1741–1742. Stirling–Crieff road, 20 miles
1744–1750. Dumbarton–Inveraray road, 44 miles
1748–1753. Stirling–Fort William road, 93 miles
1748–1757. Coupar Angus–Fort George road, 100 miles
1752–1754. Tarbet–Crianlarich road, 16 miles
1752–1754. Dalmally–Bonawe road, 16 miles
1754–1756. Corgarff–Aberdeen road, 46 miles
1755–1763. Fort Augustus–Bernera road, 43 miles
1757–1761. Inveraray–Tyndrum road 22 miles
1750s. Stonehaven–Aberdeen–Fochabers road, 84 miles
1750s. Huntly–Portsoy road, 20 miles
1761. Dunkeld–Amulree road, 9 miles
1761. Fettercairn–Huntly road, 48 miles
1761–1763. Contin–Poolewe road, 52 miles (part built)
1763–1765. Gretna–Portpatrick road, 105 miles
1760s. Coupar Angus–Dunkeld road, 15 miles
Undated. Stirling–Dumbarton road
Undated. Coupar Angus–Amulree road
Undated. Contin–Loch Maree road
Undated. Sluggan Bridge–Dulnain Bridge road
Undated. Grantown–Duthill road
Undated. Dulsie Bridge–Aviemore road
Undated. Grantown–Forres road
Undated. Fort George–Inverness road
Undated. Fettercairn–Fochabers road

CAVENDISH (CAUNDISH), Sir Richard (c. 1490–1554), military engineer and Comptroller of Dover harbour works, was born c. 1490, the elder son of Sir Richard Cavendish (d. 1515) and his wife Elizabeth. The family had lived at Grimston Hall in Trimley St. Martin, Suffolk, for at least five generations. As a young officer he gained experience in construction at Tournai (1516), where he installed four horse-mills operating suction pumps to de-water an excavation, and Berwick (1522–1524); he also acquired proficiency in surveying. He continued in the border country for several years, but occasionally came south as is apparent from a payment he received in 1525 for a survey of south-east Suffolk; it was later used as a basis for a chart of the coast which is usually attributed to him. Additionally, a chart of the Thames estuary of c. 1533 carries his signature. In the late 1520s he married Beatrice Golde; they had two sons and two daughters. After a short spell in Denmark he returned to England and in the summer of 1537 was posted to Dover where construction of the south pier was proceeding under the direction of John Thompson (q.v.) as Surveyor of the works.

The nature of Cavendish's initial appointment at Dover is uncertain, but in April 1539 he took over as Comptroller from Thomas Wingfield who left. to work on the castles in the Downs: Deal, Sandown and Walmer. Cavendish played an active role in building the defensive bulwarks of the harbour as well as construction of the pier, especially in the later part of 1540 while Thompson was ill.

By then it had become clear that the pier would have to be extended further than originally planned since shingle, driven northward by littoral drift, was moving around the pier head and forming an off-shore bar which blocked the harbour mouth. In January 1541 Thompson, Cavendish, Auchar, Bartlet and A'Borough were commissioned to 'oversee' the works and procure materials and workmen. Antony Auchar was Paymaster at Dover; John Bartlet and John A'Borough were experienced mariners and navigators.

The King (Henry VIII), who had taken a keen interest in the harbour works from the start, came to Dover in March (his fifth visit). As a result a magnificent plan emerged, shown on a drawing signed by Cavendish, Bartlet, A'Borough and Auchar; it is perhaps the earliest surviving plan of a major civil engineering work in England. The south pier is to be extended 1017 ft. into water 21 ft. at low tide, founded (like Thompson's pier) on a mound of massive rocks brought above low-water level, and a new north pier is to be built 1985 ft. in length.

Work on the south pier extension, known as the King's Pier, started early in 1541. Thompson seems to have retired at about that time, and Cavendish probably directed operations until he left. Dover in 1544, to be succeeded as Comptroller by Auchar. Whether A'Borough was employed at Dover in this period is not known. Certainly he reported, jointly with John Rogers (q.v.) and Auchar, on the state of the pier in 1548. Work continued under increasingly difficult conditions until 1551 when the pier had reached almost to its planned length. But although the mound foundations remained stable, the superstructure of timber framing filled with large stones was already much damaged by storms and, against reasonable expectation, was failing in its main purpose of preventing northward movement of shingle across the harbour mouth. The proposed north pier was never built. Instead, after a period of thirty years in which little was done, work began in 1582 on an entirely new plan largely devised by William Borough (q.v.) and carried into effect by Thomas Digges (q.v.) and others.

On leaving Dover Cavendish was knighted in September 1544 and went to Guines as Master of Ordnance. In 1546 he took charge of works at Blackness (Cap Gris-Nez) and next year became Comptroller at Boulogne under John Rogers. There he would have been involved in construction of the Mole built in the harbour between 1547 and 1549.

Little is known of his career after 1549. He died on 12 March 1554. His widow held the manor of Grimston Hall jointly with her elder son William Cavendish (c. 1530–1572) who married Mary, daughter of Lord Wentworth, and was elected MP for Suffolk. Their son, Thomas Cavendish (1560–1592), a friend of Sir Walter Raleigh, was the second Englishman to sail round the world. Sir Richard's younger son Richard (d. 1601) was for many years a member of Corpus Christ College, Cambridge, also an MP and author of a translation of Euclid into English.

A. W. SKEMPTON

[A. H. W. Robinson (1962) *Marine Cartography in Britain*; H. M. Colvin (1975) Tournai and Boulogne, *King's Works*, **3**, 375–393; Martin Biddle (1982) Dover harbour, *King's Works*, **4**, 729–755; M. R. Pickering and N. M. Fuidge (1982) *The House of Commons 1558–1603*, **1**, 567–568; Joan Corder (1984) The Visitation of Suffolk, 1561, *Harleian Society*, **3**, 207–211; B. L. Cotton, MS, Aug. I.i.53; drawing (B. L. Cotton, MS, Aug. I.i.26) reproduced by Robinson (1962)]

CHAPMAN, William, MRIA (1749–1832), was born into a Quaker family on 7 March 1749 in Whitby, the son of Captain William Chapman (1713–1793) and his second wife, Hannah Baynes (d. 1786). Three daughters had been born to his first wife, and of the ten children of the second marriage William Jr. was the eldest. In 1765 Chapman Sr. left Whitby for the north-east, living first at Barnes, near Sunderland, and then moving to Newcastle where he was instrumental in establishing a ropeworks at Willington in 1789.

Chapman Jr. became a mariner at the age of eighteen, sailing for some three years from the Tyne to the Continent and becoming a member of the Guild of Master Mariners in Newcastle in 1769; he also became a freeman of the town. With his brother John (1750–1814) he went into business as a merchant and coal fitter and the two men leased the St. Anthonys and Wallsend collieries in 1778. In that year, as a result of Chapman's correspondence with James Watt (q.v.), a Boulton & Watt steam engine was installed for de-watering the mine at St. Anthonys but, after an initial successful beginning, the project failed as a result of financial problems with the sinking of the new shaft. at Wallsend. The mortgagor foreclosed and the Chapmans were declared bankrupt in 1782, it being said at the time that they had been the victims of some sharp practice.

The failure of the mining venture led Chapman to establish himself as an engineer. He had already developed an improved method of coal-winding and had corresponded with Watt on steam engines other than that installed at Byker. He had also won second prize for measures to improve an engine at Rotterdam and it was natural that he should accompany Matthew Boulton (q.v.) on a visit to Ireland in 1783. The visit was to lead to a stay there of some ten years, during which time he acted initially as agent for Boulton & Watt, among other work supervising

William Chapman MRIA

the erection of a colliery engine in county Tyrone.

From this time Chapman's activities turned towards civil engineering, being given practical help by Watt who supplied him with drawings and specifications of work undertaken by John Pinkerton (q.v.) on the Birmingham canal. The information supplied enabled him in 1785 to produce a report—unsolicited—proposing a canal around the southern part of Dublin. At that time the Grand Canal terminated in the western part of the town, without a connection to the river Liffey. Although a short link to the river had been proposed by the canal company, Chapman suggested that a new cut, 4 miles long with eight locks, would be preferable, entering the Liffey through a dock either at Ringsend or opposite the new Customs House. Built later with some modifications by William Jessop (q.v.), the canal proposal brought Chapman to the notice of the public and led to his first engineering appointment, as engineer to the Kildare Canal, authorised by Act of Parliament in 1786.

The Kildare works comprised a canal 2½ miles long with five locks giving a rise of 44 ft.; the locks were 14 ft. wide. A basin was to be provided at Naas and there were to be three bridges over the waterway. It was here between 1787 and 1789, at Osberstown, Oldtown and Naas, that Chapman designed and built the first oblique arch bridges, later describing them in Rees's *Cyclopaedia*. Additionally, he undertook work on the Grand Canal of Ireland between 1789 and 1794, acting as assistant to Jessop. Work

comprised a survey of the canal's 53 mile length, the rebuilding of sixteen of its eighteen locks and the design of several bridges. While engaged on this work Chapman, at the request of the Duke of Leinster, designed a five span masonry bridge over the river Liffey.

His work in Ireland continued with the construction of the Ringsend Docks, Dublin, for the Grand Canal Company. He had reported on these works in 1785, construction was authorised by an Act of Parliament in 1791, and when construction began the following year he was appointed as Jessop's assistant during the early stages of design and construction; his brother, Edward Walton Chapman (1762–1847) was involved as resident engineer. Undertaken at first by contract, the work comprised the formation of two wet docks totalling 26 acres, three dry docks, two ship locks and a barge lock; the docks were the largest built in the U.K. before 1800 and their formal opening took place in April 1796.

Further work in Ireland followed. In 1790 he was appointed engineer to the Barrow Navigation Company, producing an extensive report on the company's strategy and subsequently carrying out the rebuilding of four of the eight original locks and constructing ten new ones. Chapman remained as engineer until 1795 by which time the first steps had been taken to enable the canal to accommodate boats of 80 tons, rather than the former limit of 15 tons. The last of his ventures in Ireland was the rebuilding of several locks on the Cloonlara cut of the Limerick–Killaloe Canal, the setting out of a 2 mile Moya cut and the provision of inverted arch foundations to the Cusane lock. Chapman was elected a Member of the Royal Irish Academy in 1794 and soon afterwards became a member of the Smeatonian Society of Civil Engineers. Proposed by Jessop and supported by John Rennie (q.v.) and Robert Mylne (q.v.) he attended meetings of the Society over a period of some thirty years.

Following the death of his father, Chapman returned to Newcastle in 1794 in order to take over the running of the Willington ropeworks and also to undertake a commission to report on a projected canal from the Tyne to the Solway. At the ropeworks he became actively involved and an 8 hp Boulton & Watt engine was installed in 1799, the second to be used in a rope manufactory. That he was so involved is confirmed by the fact that over a period of ten years beginning in 1797 he took out seven patents relating to the manufacture of ropes. The works were managed by his brother Edward but Chapman is thought to have maintained a financial interest in the works until about 1816, by which time his engineering practice had grown.

During 1795 and 1796 Chapman submitted several reports on the feasibility of building a canal from the river Tyne to Carlisle, some 60 miles, and extending it to Maryport. Chapman's proposals were subjected to the scrutiny of other engineers, including Jessop, John Sutcliffe (q.v.) and Ralph Dodd (q.v.) and eventually the promoters of the scheme deposited plans in Parliament in November 1796, later calling in Robert Whitworth (q.v.) to give an opinion on the line of canal passing along the north bank of the Tyne. The positions of the several engineers polarised in 1797 with Chapman and Jessop supporting the canal on the north side of the river and Sutcliffe and Whitworth favouring one on the south. The ensuing Parliamentary contest led to the withdrawal of the Bill and it was not until 1818 that Chapman produced a further report on a canal, but only from Carlisle to the Solway.

The ending of the consideration of a canal between the east and west coasts led to Chapman being consulted in 1796 on three large land drainage and flood protection works in Yorkshire. The first of them was the Beverley and Barmston Drainage, an Act for the work being obtained in June 1798, two years after Chapman's report. As engineer, Chapman formed the two mile long Barmston cut, 23 miles of drainage cuts, 20 miles of river embankments and an outfall sluice at Hull. In addition there were eleven tunnels carrying drainage cuts under waterways, including the Driffield Canal, and twenty-seven road bridges, together with numerous culverts and occupation bridges. The scheme cost £115,000 and had the effect of draining and protecting from floods some 12,600 acres of land to the west of the river Hull. Work was completed in 1810.

The second of these undertakings was the Muston and Yedingham Drainage—its Act of Parliament dated 28 July 1800—comprising a sea cut 3 miles long, the Everley weir, the New Hertford channel 7 miles in length, the 4 mile long Derwent channel, 22 miles of drainage channels, eight masonry bridges and the Yedingham sluice. By this work 10,500 acres of land south west of Scarborough were drained and protected from flooding. Having cost a total of £42,000 the scheme, begun in 1800, was completed in 1808 under Chapman's direction.

The third of the major schemes was the Keyingham Level Drainage, on the north bank of the Humber estuary, for which an Act was obtained in 1802. In concept similar to that for the Muston and Yedingham works, this undertaking comprised a new cut 1½ miles long to Hedon Haven, new cuts across Keyingham Marsh and Cherry Cob Sands, 2 miles and 1½ miles long respectively, two outfall sluices and the improvement and deepening of the existing drainage channels. Begun in 1803 and completed in 1807 the works drained 5,500 acres and cost £30,000. Between 1808 and 1811 the Ottrington Drainage, an extension of the Keyingham Drainage, was built; by it a further 2000 acres were improved by a new cut 1½ miles long, together with an outfall sluice at Stone Creek.

In July 1801 an Act of Parliament was obtained for the construction of the Driffield Canal and work began the following year with Chapman as engineer. It involved the formation of a new channel ¾ mile long bypassing a loop in the River Hull, the building of a lock and weir at the south end of the channel near Struncheon Hill and the deepening of the river from the entrance of the Driffield Canal to Frodingham Bridge. The Beverley and Barmston drain was carried under the new river channel by a tunnel upstream of the lock. Having cost £6,000 the work was completed in 1805.

In 1804 Chapman, while working in Yorkshire, and for a time living there, married Elizabeth Knowsley at Lowthorpe, near Driffield; it was his second marriage, the first having taken place in 1786 when he married Martha Rae, daughter of Robert Rae of Ross, Wexford. The second marriage resulted in the birth of a daughter, baptised in York the following year.

Chapman's work in Yorkshire continued with his employment on the formation of a dock at Hull. As a joint venture between the Hull Dock Company and Hull Corporation, Chapman acted as engineer for the Corporation and Rennie for the dock company; Chapman oversaw the planning and Rennie the structural design. Work comprised the construction of a wet dock of 7¼ acres having an entrance lock 42 ft. wide and 158 ft. long, its inverted arch construction based on piled foundations, together with a 2¾ acre entrance basin protected by projecting jetties. The works, capable of accommodating seventy vessels of 180 tons burden, were authorised by an Act of Parliament in 1802 and were completed in 1809, having cost £230,000.

The involvement of Chapman at Hull ceased in 1806 at which time his association with Scarborough Harbour became more time-consuming, a result of the Commissioners obtaining an Act of Parliament in that year for the execution of major works. He had first reported on the harbour in 1800, the works then comprising a pier dating from the sixteenth century with another dating from the seventeenth and eighteenth centuries. Under Chapman, major modifications were put in hand: the East Pier was extended by 500 ft. between 1800 and 1826; a West Pier 480 ft. long was built between 1818 and 1820 and was widened between 1827 and 1829; and the end of the original Vincent's Pier was rebuilt between 1823 and 1826 together with a spur, added in 1829–1830. These works were all built in massive masonry and came to form inner and outer harbours of 8½ acres and 5½ acres respectively. The cost of the works between 1801 and 1831 was some £65,000.

Chapman had reported twice on the River Orwell Navigation, in 1797 and 1803, before the Commissioners obtained their Act for further work in 1805. His appointment as engineer led to a new 550 yd. cut from Ipswich shipyard to Cliff Reach, the widening and deepening of the river channel to Ipswich Quays and new cuts at Upper Hearth Point and Round Ooze. He was also instrumental in designing a dredger for use by the commissioners, the vessel built by William Cordingley, its 6 hp engine obtained from Boulton & Watt and the bucket-ladder machinery from the Butterley Company; its cost was £3,000.

The river Tees had been reported upon by Jonathan Pickernell Sr. (q.v.) in 1791 when a dramatic shortening of the course of the river at Mandale was suggested. Nothing further transpired until 1804 when a further report was submitted by Chapman as a result of a decision to form the Tees Navigation Company to undertake the work first proposed. An Act was obtained in 1808 and the projected work was put in hand immediately, the 220 yd. long Mandale Cut reducing the river's sinuous course by approximately 2 miles. The 250 ft. wide cut, giving 16 ft. depth of water, was opened in September 1810 and led to a doubling in the tonnage of shipping handled by the upriver port of Stockton; it had cost £9,300.

Between 1806 and 1816 there would seem to have been little new involvement in civil engineering on the part of Chapman although he continued to oversee those works in progress. There was, however, a resumption in his mining interests, taking two forms: the leasing of another colliery and the acquisition of several patents, all related to mining. In 1806 he took a share with Benjamin Blaydon, Edward Middleton, James Tindall, Robert Knowsley and George Knowsley—the two last-named relatives of his second wife—in the East Kenton and Coxlodge collieries, north of Newcastle. The previous year, with James Tindall, he had negotiated a wayleave for the carriage of coal over ground at Kenton and in 1808 further agreements were made so as to permit the construction of the Kenton and Coxlodge Railway to proceed. With a gauge of 4 ft. 7½ in., its 4 mile length of iron track—an early example—was completed in 1809; rails were obtained from the Butterley Company. Plans for winning a new pit at Coxlodge were prepared by Chapman in 1807 and approved by John Buddle (q.v.), an eminent colliery viewer. The sinking resulted in the purchase of a 24 hp engine and although the new pit, the Jubilee Pit, proved a success Chapman and Knowsley sold their shares to Tindall in 1811.

While actively engaged in mining, Chapman in 1807 obtained a patent for a coal 'drop' whereby loaded wagons could be lowered vertically from the extremity of the staith directly into the ships' holds, avoiding breakage of the coal and reducing handling costs. His patent was first used at the Benwell Colliery staiths on the Tyne, they were built at a cost of £270. No other applications appear to have followed but in 1812 a variant of the design was introduced by Benjamin Thompson (q.v.), his design adopting a swinging motion whereby wagons were lowered in an arc to the deck of the ship. It was Thompson's design which came to be almost universally used on the Tyne although Chapman's vertical lowering system was used at other ports in the region.

In 1813 Chapman became concerned with the design of locomotives, the first based on his patent of the previous year, financed by Buddle and taken out in conjunction with his brother, Edward, for a chain-haulage engine with a pivoted bogie. It was built for him by the Butterley Company and incorporated a cast iron boiler 7 ft. long and 3 ft. 6 in. diameter with safety valves designed to operate at 60 psi. Power to the wheels was provided by two vertical cylinders with a stroke of 2 ft., the piston rods connected to side levers working the crankshaft. which, through gearing, drove the band wheel. The engine was mounted on a timber frame carried on a pair of leading wheels with a fixed axle, and also a four wheel bogie, the first such design built and providing a reduction and partial equalisation of wheel load on the rails. The bogie frame was pivoted so as to allow rotation about both a vertical axis and a horizontal traverse axis. The engine cost £514 and, arriving at Heaton colliery in 1813, was tried out during the following year, following which Buddle wrote that it could 'haul itself and its load up hill, as has been effectually proved under very disadvantageous circumstances on the wooden railway from Heaton colliery to the river Tyne'. After serious flooding of the colliery it was used for pumping.

A second engine designed by Chapman was built in 1814 by Phineas Crowther of the Ouseburn Foundry, Newcastle, at a cost of £483. Trials of this engine took place at the end of 1814 and, weighing six tons, it was mounted on eight wheels seemingly mounted on two bogies; it then drew eighteen loaded coal wagons up an incline of 1 in 115 at a rate of 4 m.p.h. The wagonway, with cast iron rails and a gauge of 4 ft. 2 in., ran from the Lambton collieries to the river Wear at Penshaw. It is thought that the locomotive used chain haulage for steep gradients but reverted to normal haulage for less severe inclines. A major advantage was the fact that the bogies reduced wheel loadings to what would have been experienced by horsedrawn wagons, so eliminating the need for replacing wooden rails with iron. Further locomotives built with Buddle and Chapman's involvement included that at Whitehaven (1816).

The defeat of the French in 1815 brought about a loosening of financial constraints, one result of which was an increase in the volume of civil engineering work obtained by Chapman. The first such work actually predated the cessation of hostilities, a report on the Sheffield Canal having been prepared by him in 1814, followed by an enabling Act in June 1815. He was appointed as engineer and construction of the canal began in 1816; running parallel with the river Don, its line began at Sheffield and terminated at the head of the River Don Navigation at Tinsley. Works comprised 4½ miles of canal cut with twelve locks 64 ft. by 16 ft. giving a rise of 70 ft. Additionally there was a further cutting at Athercliffe, an aqueduct over a road and seven road bridges. A basin 600 ft. by 100 ft. was built at Sheffield with a warehouse and wharves. Steam pumping plant supplied water to the canal from disused mine workings. Opened in February 1819 the canal had cost £76,000.

In 1760 an attempt had been made to open a new entrance to Shoreham Harbour opposite Kingston and also dam off the eastern arm of the harbour. This proposal failed and in 1800 William Jessop propounded the view that little could be achieved at the harbour other than by dredging. In 1810 a very expensive wet dock was suggested by Robert Vazie (q.v.) and John Rennie but their scheme was considered inappropriate and, as a result, Chapman—at the instigation of Trinity House—was called upon to submit a report. It was accepted by the Shoreham Harbour Commissioners and improvement work in accordance with his recommendations began in 1817 with him as engineer. Built by contract the work on the harbour cost £58,000, of which £3,400 was for the supply of a dredger. The principal work was the formation of a 250 ft. wide entrance through the shingle bank opposite Kingston protected by chalk-filled timber-framed piers, groynes extending 200 ft. beyond the ends of the piers and the formation of a 700 ft. long middle pier forming the Light House Point. Dredging was included, as were two sets of sluice gates for scouring the eastern arm of the harbour; the sluices were only partially successful and their use was abandoned in 1828. Chapman was instrumental in securing for the Commissioners a £15,000 loan but in spite of this his fees were not paid in full, leaving a balance of £180 unpaid.

Chapman's interest in canal communication between the Tyne and the Solway was revived in 1818 when he reported upon a canal westwards from Carlisle to the Solway near Bowness. An Act for the formation of the Carlisle Canal Company was obtained in 1819 and Chapman was appointed as engineer for the construction of a canal 11¼ miles long with a surface width of 54 ft. and a depth of 8½ ft. A basin 250 ft. by 80 ft.—with a sea lock and timber jetty—was formed at Fishers Cross, on the Solway, and another at Carlisle, 450 ft. by 100 ft., with a warehouse and wharves. The rise of 46 ft. from the sea to Carlisle was accommodated by seven locks 78 ft. by 18½ ft. and a reservoir was formed at Mill Beck, near Grinsdale. The opening in March 1823 was marred by the fact that disputes between Chapman and the directors had led to his dismissal three months prior to the opening; the resident engineer, Richard Buck, whose part Chapman had taken left two months after the canal was brought into operation. The canal designed by Chapman and built at a cost of some £80,000 proved a success, especially after the completion of the Newcastle and Carlisle Railway in 1839. The canal was forced to close in 1853 and its line was converted to a railway.

Chapman's interest in transport in this area did not cease with the completion of the Carlisle canal and in 1824 he wrote on the feasibility of adopting his 1795 line for a canal between Newcastle and Carlisle for railway use, noting that a deviation of the line would be needed; a railway would be both cheaper to build and less costly to operate than a canal. Further details were supplied by him and he then estimated the cost of a canal at £800,000 and a railway £252,000. Thomas Telford (q.v.) and Josias Jessop (q.v.) were then asked for their views and the comments of the latter resulted in the estimate for the railway being increased to £300,000. It was thereupon decided that the railway proposals should go ahead under the direction of Chapman and Thompson, a director of the new company. The version of the line to be adopted differed from Chapman's original proposal but the Parliamentary plans prepared in 1828 showed a line which differed little from Chapman's original route for the section west of the summit. Construction began in 1830 and was completed in 1839 although the engineer, at least initially, was Francis Giles (q.v.) rather than Chapman.

Immediately following the completion of the Carlisle Canal, Chapman became involved in 1824 in the planning of the East London Dock where the first London Dock, at Wapping, had been built by Rennie between 1801 and 1806. It had been Rennie's intention to provide a second dock to the east and connecting to the first. Work on it had not been carried out and Rennie's death in 1821 led to the Dock Company commissioning Chapman to draw up the plans necessary. Discussions between Daniel Alexander (q.v.), the company's surveyor, and Chapman led to the submission of estimates of £138,000 for the engineering works and £114,000 for storage vaults and warehouses; contingencies brought the total to £280,000. Alexander was appointed as executive engineer and Chapman, in addition to his fees, was paid £110 'for his Zeal and active exertions relative to the projected New Works'. All work, comprising a seven acre dock capable of accepting vessels of 700 tons together with iron swivel bridges, a warehouse for 11,000 tons of merchandise and vaults for 10,000 casks of wine or spirits was completed in 1828, having cost £302,000.

The penultimate of Chapman's major works was at Leith, the port of Edinburgh, where docks had been built to Rennie's design between 1801 and 1817. An Act of Parliament was obtained in July 1825 for the construction of protection works, a 1500 ft. long eastern pier, a 1000 ft. long gravel extension to it and a 1200 ft. western breakwater; dredging of the harbour was included. Under Chapman's direction as engineer, work began in 1826 and was completed in 1831 having cost £40,000, exclusive of dredging. As a result of this work a large outer harbour with a fairway 19 ft. deep at HWST was provided, giving a safe approach to the earlier harbour.

In 1821, Buddle—acting for the third Marquess of Londonderry—sought the views of Chapman concerning the transhipment of coal from small capacity keels (or barges) into seagoing ships. He designed an upright counterbalanced frame rotating about a horizontal axis whereby a container could be lifted from a barge, swung across the floating crane vessel and lowered onto the ship. Known as the 'Tansferrer', several such floating cranes came to be used on the river Wear, offloading barges each carrying eight containers of some 2½ tons capacity. The device was patented in 1822.

The two men again worked together when Seaham Harbour came to be built for Londonderry. A harbour had first been proposed at Daldon Ness by Sir Ralph Millbanke, owner of the Seaham Estate, he then envisaging that it could become an outlet for the coal raised at Hetton colliery, then about to begin production. Chapman had prepared plans for a small harbour in 1820 but the decision of the Hetton Company in 1821 to build its own line to Sunderland led to Millbanke abandoning his ideas of building a harbour and instead he sold the estate to Londonderry the following year. Chapman was retained by the estate's new owner to prepare plans for the construction of an enlarged port and Buddle sought an opinion on Chapman's proposals from both Telford and Rennie; both considered that the costs would exceed those estimated by Chapman.

Construction began in September 1828 and the first stage was completed by July 1831. Work carried out by that time comprised a 560 ft. long North Pier which with a 450 ft. section of the South Pier formed a harbour 3 acres in extent having an entrance 110 ft. wide. A 2½ acre inner harbour—its depth maintained by a gate—was cut from the headland, the excavated material used as fill for the masonry piers or, if limestone, burnt in kilns on the site and sold. The second stage of the works, completed after Chapman's death, consisted of the completion of the South Pier to a length of 1000 ft., the forming of the 3 acre South Harbour, the completion of the gated 2½ acre South Dock, and additional loading berths. These works were completed by July 1835. Built principally by direct labour under Chapman's direction, the harbour works cost £165,000 and completion enabled 230,000 tons of coal to be shipped in 1832. By 1845 the annual tonnage had reached 700,000. Chapman, justifiably proud of what was to be his last work, wrote in 1832 that there was 'no instance of a private Harbour being constructed with such rapidity and to such extent as the present Harbour has been ... and I, alone, should have been to blame, had the scheme failed of success, as the whole was left to my decision and opinion'.

In addition to the major works noted, Chapman was concerned in many of a minor nature. Among those in the North East were: comments on and estimates for the widening of the Tyne bridge at Newcastle in 1801 in conjunction with David Stephenson (q.v.); proposals for protective work on the harbour at Seaton Sluice in 1826; a report on the siting of bridges over the river Tyne relative to the extension westwards of a turnpike

from Newcastle in 1828; a report on the foundations of a proposed bridge over the river Wansbeck at Morpeth in 1830; advice given in the same year on a replacement skew bridge on the Great North Road some 3 miles north of Newcastle; and a report regarding the propping of the Tees suspension bridge of the Stockton and Darlington Railway after its having been found inadequate for railway use in 1831. Chapman was also involved with the proposal to erect a suspension bridge of 760 ft. span over the river Tyne between North and South Shields in 1825; although a provisional committee was formed, construction did not proceed.

Following the removal of the family from Whitby to Newcastle, Chapman had lived in Ireland from 1783 until 1794, near York in 1804–1805 and at Morton, in County Durham, in 1811–1812. Much of the time, however, had been spent in Newcastle where he had an office at 10 Saville Row—perhaps the former home of his parents—and a residence at 26 Ridley Place, nearby. He was a member of the Literary and Philosophical Society in Newcastle from its inception in 1793 and several of his writings were recorded in the Society's *Minutes*. Chapman remained active until shortly before his death having 'retained the full enjoyment of his faculties, and followed the active pursuits of his profession, till within a very short period of his decease'.

He died on 29 May 1832, the *Newcastle Courant* reporting his death 'at an advanced age, the much deservedly lamented William Chapman'. He was survived by his widow, Elizabeth, and his daughter Elizabeth Hannah who three years later married Frederick William Tappenden. By his Will, made in 1818, he left his business interests and property—valued at less than £8,000—to his wife, his daughter, his sister Hannah and his brother Edward Walton. The two last-named, administrators of the will, affirmed as Quakers but it would seem that Chapman had possibly deserted the Society of Friends, perhaps at the time of his marriage in 1804, or even earlier. His library of 535 volumes was sold by auction in 1833, but his printed reports were donated by his widow to the Institution of Civil Engineers in 1837. He was buried at St. Andrew's church in Newcastle and is commemorated there by a plaque in the north wall of the chancel; the headstone formerly at his grave was in 1895 moved to St. Andrew's Cemetery.

R. W. RENNISON

[*Report Relative to Making the Catenary Bridge [at Stockton] Capable of Carrying 125 Tons* (1832?) Durham CRO; Londonderry Papers, Durham County RO; Will of William Chapman (1818) Northumberland RO, ZMD 168/2; report on the Foundations of Morpeth Bridge (1830) Northumberland RO, 2/39/1–3; Northumberland Quarter Sessions, Orders (1830) Northumberland RO, QSO; correspondence re. widening of Tyne Bridge, Newcastle (1801) Tyne & Wear AS,

612/345, 616/345; Local Directories; Watson, Buddle, Forster and Bell Collections, Northumberland RO; *Report on the Comparative Eligibility of Bridges over the Tyne at Newburn, Stella and Scotswood* ... (1827) *Newc L&P Tracts*, **191**, No. 10; E. Mackenzie (1827) *Descriptive and Historical Account ... of Newcastle*, 469–471; J. Sykes (1833) Local Records; *Report of the Literary and Philosophical Society of Newcastle upon Tyne*, No. 40, 6; J. Weale (ed.) (1843) Memoir of William Chapman, *Quarterly Papers on Engineering*, **1**, Pt. 1; *Inventions, Improvements and Practice of Benjamin Thompson* (1847); L. G. Charlton (1974) *The First Locomotive Engineers*; A. W. Skempton (1974) William Chapman (1749–1832), Civil Engineer. *TNS*, **46**, 45–82, and sources cited therein; Biography Database, CD-ROM. University of Newcastle upon Tyne; A. Gray and J. Rees (2001) *Early Railways*]

Publications
Reports:

1789. *Observations on the Advantages of bringing the Grand Canal round by the Circular Road into the River Liffey*
1789. *Report on the Means of Perfecting the Navigation of the River Barrow from St. Mullin's to Athy*
1789. *Estimates of the Expences of completing the Navigation of the River Barrow, from St. Mullin's to Athy*
1791. *Report on the River Shannon, from Lough Allen to Killaloe, with Estimates*
1791. *Remarks on the Proposed Plan of an Inland Cut from Burn's Island to Leighlin*
1792. *Report on the Improvement of the Harbour of Arklow, and the Practicability of a Navigation from thence by the Vales of the ... Ovoca*
1793. *Report ... on the means of making Woodford River Navigable, from Lough-Erne to Woodford-Lough*
1795. [Letter] *to the Subscribers of the Limerick Navigation*
1795. *Report on the Measures to be attended to in the Survey of a Line of Navigation, from Newcastle upon Tyne to the Irish Channel ...*
1795. *Report on the Proposed Navigation between the East and West Seas ... from Newcastle to Haydon Bridge, with Observations on the Separate Advantages of the North and South Sides of the River Tyne*
1795. *Second Part of a Report on the Proposed Navigation between the East and West Seas: viz. from Haydon-Bridge to Maryport*
1795. *Third and Last Part of a Report on the Proposed Navigation between the East and West Seas ...*
1795. *Report on the Proposed Line of Navigation, between Newcastle and Maryport ... And also that from Stella to Hexham* (with William Jessop)
1795. *Postscript to Mr. Jessop's Report*
1796. *Report on Measures to be Attended to in the Survey of a Line of Navigation from Newcastle-upon-Tyne to the Irish Channel*

1796. *The Report ... respecting the Drainage of the Low Ground lying below the Wolds, on the West side of the River Hull, and in Frodingham Carrs, and at Lisset ...*
1797. *Observations on Mr. John Sutcliffe's Report on a Proposed Line of Canal from Stella to Hexham ...*
1797. *Report ... on the Drainage and Navigation of Keyingham Level, in Holderness*
1800. *Report ... on the Means of Draining the Low Grounds in the Vales of Derwent and Hertford ... in the County of York*
1800. *Sundry Papers and Reports, relative to the Defence of the Estate of Cherry Cobb Sands, against the Humber*
1800. *Report on the Proposed Navigation to Knaresbro'*
1800. *Report on Mr. Golbourne's Plan for Draining Keyingham Level*
1800. *To the Proprietors of Lands within the Keyingham Level Drainage*
1800. *Report on the Proposed Branch Navigations from the River Hull ...*
1800. *Report on the Harbour of Scarborough ...*
1800, 1801. *Observations on the Improvement of Boston Haven ...*
1801. *Drainage of Keyingham Marshes; and ... the Fore-Shore, opposite Foul Holme Sands*
1801. *Report to the Committee of Proprietors of the Beverley and Barmston Drainage*
1802. *Report on the proposed navigation to Knaresbro'*
1802. *Reports of the Committee Appointed to Procure the Levels, Surveys and Estimates of the Proposed Navigation to Knaresbro' ...*
1803. *General Principles of Public Drains: Addressed to the Directors of the Muston Drainage*
1804. *To Subscribers for Improving the Port of Stockton*
1803, 1804, 1806. *Report(s) on the Beverley and Barmston Drainage*
1806. *Report on the last Act for the Improvement of Scarborough Piers*
1807. *Report on the Means of Obtaining a Safe and Commodious Communication from Carlisle to the Sea*
1808. *Further Report ... on Mr. Telford's Report respecting the Cumberland Canal*
1809. *Report on the state of the Beverley and Barmston Drainage*
1810. *Report on the Drainage of the Marshes, from Ancroft Fen to Wainfleet Haven ...*
1811. *Report on the Measures Necessary to the Final Completion of Scarborough Harbour ...*
1813. *First Report on the Laneham Drainage*
1813. *Second Report and Supplement on the Laneham Drainage*
1813. *Report ... on Various Projected Lines of Navigation from Sheffield*
1814. *Report of Dance, Chapman and others respecting the Enlargement and Improvement of the Waterway under London Bridge*
1814. *Report ... on the Proposed Canal, from Castle Orchards, Sheffield, to the River Dun below Tinsley*

1815. *Report on the Harbour of New Shoreham*
1815. *Supplementary Report on the Improvement of the Harbour of New Shoreham*
1815. *Report on a projected improvement of Scarborough Harbour*
1816. *Report to the Commissioners of Scarborough Piers*
1818. *Report on the Proposed Canal Navigation between Carlisle and Solway Frith*
1818. *Particulars and Specification of the Proposed new West Pier of Scarborough*
1818. *Report on a Line of Canal from Lemmington towards Carlisle*
1818, 1819, 1820, 1821, 1825, 1826, 1827, 1828, 1829, 1830. [Reports] *to the Commissioners of Scarborough Pier*
1821. *Case of Scarborough Harbour*
1821. *Report to the Chairman and Commissioners of Shoreham Harbour*
1821. *Report ... on ... Improving and Enlarging the Harbour of Whitehaven*
1821. *Case of Scarborough Harbour*
1822. *Observations on Shoreham Harbour*
1823. *Address to the Subscribers to the Canal from Carlisle to Fisher's Cross*
1823. *Reply to the Report of Messrs. Telford and Rennie (on the improvements in the tideway of* the River Mersey at Runcorn)
1823. *Address to the Commissioners of the Harbour of New Shoreham*
1823, 1824. [Reports] *on the works of Scarborough Harbour*
1824. *Observations on the Most Advisable Measures to be adopted in Forming a Communication ... from Newcastle [to] Carlisle*
1824. *A Report on the Cost and Separate Advantages of a Ship Canal and of a Rail-Way, from Newcastle to Carlisle*
1824. *Report relative to the Improvement of the Harbour of Leith*
1824. *Report on the Works, Improvement, Revenue and Expenditure of Scarborough Harbour*
1824. *Supplementary Report* [on] *the Harbour of Leith*
1825. *Manchester and Dee Ship Canal*
1825. *Report on the projected patent wrought iron Suspension Bridge, across the River Tyne, at North and South Shields*
1828. *Report on the Improbability of the Formation of a Useful Harbour at Lake Lothing, and on the Means of Improving the Navigation from Norwich to Yarmouth ...*
1829. *Report on ... the Harbour of Scarborough, and on the Measures Requisite for its Final Improvements*
1829. *Supplementary Report on Scarborough Harbour*
1829. *Additional Supplement ... on the ... Harbour of Scarborough*
1830. *A Description of the Port of Seaham ...*
1832. *Report of the Rise, Progress, Present State, and Projected Extension of the Harbour of Seaham*

Other publications:

1797. *Observations on the Various Systems of Canal Navigation ... [vis-à-vis] Mr. Fulton's Plan of Wheel-Boats, and the Utility ... of Small Canals are particularly investigated ...*
1798. An Essay on a Universal Standard of Weights and Measures, 10 April, in: *Papers Read at the Monthly Meetings: Literary and Philosophical Society of Newcstle upon Tyne*, vol. 2, 1794–1814, MS
1800. *Sundry Papers and Reports, relative to ... Cherry Cobb Sands; ... the Facts and Remarks relative to the Witham and Welland ... and ... the Means of Improving the Channel of the Witham and the Port of Boston.*
1808. *A Treatise on the Progressive Endeavours to improve the Manufacture and Duration of Cordage*
1811. A Description of a Stone found during the combustion of a large quantity of small Coal at Kenton Colliery, in: *Papers Read at the Monthly Meetings: Literary and Philosophical Society of Newcstle upon Tyne*, vol. 2, 1794–1814, MS
1815. Instructions for easily correcting a spirit level, in: *Views and Estimates*, vol. 5, 1802–1814, Northumberland RO, JOHN/5/99, MS
1815. *Hints for Establishing an Office in Newcastle for Collecting and Recording ... Information relative to the Strata of the Collieries in its Neighbourhood*, Literary and Philosophical Society, Newcastle
1815. *Observations on the Necessity of adopting Legislative Measures to Diminish the Probability of the Recurrence of Fatal Accidents in Collieries*, Literary and Philosophical Society, Newcastle
1817. *A Treatise containing the Results of numerous Experiments on the Preservation of Timber from Premature Decay*
c. 1818. Consumption of coals in England and Wales, estimated by Mr. William Chapman, *Colliery Memoranda*, 1814–1825, Northumberland RO, BUD/30/90
1818. *Observations on the Effects that would be produced by the Proposed Corn Laws*
1819. Observations on mineral coal and fossil trees, in: *Papers Read at the Monthly Meetings: Literary and Philosophical Society of Newcstle upon Tyne*, vol. 3, 1814–1819, MS
1832. An account of some antiquities presented to the [Literary and Philosophical] Society [of Newcastle upon Tyne], *Archaeologia Aeliana*, **2**, 115–119

Patents:

Patent 2191, 13 September 1797. *Making Ropes of any Number of Yarns and Strands, Tarred or Untarred*
Patent 2219, 6 March 1798. *Making Ropes of any Number of Yarns and Strands, Tarred or Untarred; Coiling up the Same while Making*
Patent 2265, 8 November 1798. *Making Ropes of any Number of Yarns and Strands, Tarred or Untarred; Coiling up the Same*

Patent 2326, 16 July 1799. *Making Cord, Ropes, and Cordage, Tarred or Untarred, from the Spinning of the Yarn to the Finishing of the Rope or Cordage* (with Edward Walton Chapman)

Patent 2513, 5 June 1801. *Preserving Cordage by the Application of certain Substances, Separately or Combined*

Patent 3026, 8 April 1807. *Reducing the Wear and prolonging the Duration of Ropes for Drawing Coals or other Minerals from Pits or Mines*

Patent 3030, 11 April 1807. *Putting Coals on board Ships, Lighters, and other Vessels, so as to prevent Breakage*

Patent 3078, 30 October 1807. *Making Belts or Flat Bands for Drawing Coals and other Minerals out of Pits and Mines, and for Raising Weights* (with William Edward Chapman)

Patent 3126, 27 April 1808. *Conveying Coals and other Minerals in the Working of Mines; Returning the Empty Carriages*

Patent 3335, 9 May 1810. *Wheels for Mechanical Movements, which Wheels are to be Worked by Water, Steam, or other suitable Fluids or Gases*

Patent 3632, 30 December 1812. *Facilitating and Reducing the Expense of Carriage on Railways and other Roads* (with Edward Walton Chapman)

Patent 4550, 12 April 1821. *Transferring the Ladings of Lighters and Barges, into Ships or Vessels, and vice versa*

Patent 5330, 7 February 1826. *Machinery for Loading or Unloading Ships, Vessels, or Craft*

Patent 5540, 14 August 1827. *Construction of Waggons for Railways or Tramways*

Drawings:

1785. *A Plan for Communicating the Grand Canal with the Liffey and Improving the City of Dublin*

1795. *Plan of the Proposed Canal between Newcastle and Maryport and of the Adjacent County*

1796. *Plan of the Proposed Navigation from Newcastle upon Tyne, to Haydon Bridge*

1797. *Plan of the Drainage of the Levels of Keyingham, Burstwick etc. with the Improvements Projected*

1799. *Plan of Proposed situation of Graving Docks below St. Peters Quay* [Newcastle]

1800. *Plan of the Rivers Derwent and Hertford, shewing the Alterations in their Course through the flooded Grounds*

1800. *Plan of Cherry Cobb Marsh*

1800. *A Plan of Scarborough Harbour*

1801. *A Plan of Cherry Cobb Sands: Paul Holme and Foul Holme Sands, and Keyingham Marshes*

1804. *Plan of the River Tees between Stockton and Portrack*

1804. *Plan of the Proposed New Cut* [at Stockton]

1811. *A Plan of Scarborough Harbour*

1814. *A Plan of the Intended Canal from … Sheffield into the Township of Tinsley … in the West Riding of the County of York …*

1815. *A Plan of the Harbour of New Shoreham*

1818. *Carlisle Canal*

1821. *A Plan of the Harbour with Part of the Town of Scarborough*

1821. *Chart of the Entrance to Leith Harbour and Docks …*

1826. *Stone Pier on the North Side of the Entrance to the New Cut* [at Seaton Sluice]

1830. *Plan of Seaham Harbour as Projected … by William Chapman …*

1831. *Plan of Scarborough Harbour*

1832. *Plan of the Harbour of Seaham and of its Future Extension*

Works

1787–1789. Kildare Canal, cost £10,000

1790–1795. River Barrow Navigation, cost £58,000, resident engineer Humphrey Mitchell

1791–1794. Limerick–Killaloe Canal, cost £22,000, resident engineer Mr. Lascelles

1799–1710. Beverley and Barmston Drainage, cost £115,000, resident engineer William Settle

1800–1808. Muston and Yedingham Drainage, cost £42,000, resident engineer Robert Wilson

1801–1831. Scarborough Harbour, cost £65,000, resident engineers Matthew Shout (1801–1804), Roger Nixon (1804–1822), William Barry (1822–1831)

1802–1805. Driffield Canal improvements, cost £6,000, resident engineer William Settle

1803–1807. Keyingham Level Drainage, cost £30,000

1803–1809. Humber Dock (with John Rennie), cost £230,000, resident engineer John Harrap

1806–1808. River Orwell Navigation, cost £10,000, resident engineer William Shaw

1809–1810. River Tees Navigation, cost £9,000

1816–1819. Sheffield Canal, cost £76,000, resident engineer Henry Buck

1817–1821. Shoreham Harbour, cost £58,000, William Clegram

1820–1823. Carlisle Canal, cost £80,000, Richard Buck

1826–1831. Leith Harbour, cost £50,000, James Leslie

1826–1832. Seaham Harbour (completed 1835 by John Buddle), cost £165,000, resident engineer Thomas Nicholson

CHAPPLE, James (1771–1846), bridge surveyor, was appointed in 1806 as Surveyor of Bridges for the Eastern Division of Cornwall and was then described as 'James Chapple the younger … of Bodmin, builder'. Although the responsibility of the County Surveyor under Quarter Sessions was essentially one of maintenance, two of the new bridges built during his term of office deserve notice.

The new Polston Bridge was built over the river Tamar near Launceston in 1833, in conjunction with James Green (q.v.) who was at that time County Surveyor of Devon. Green had reported unfavourably following his initial inspection in 1809 but it was not until 1830 that the two Surveyors met 'for the purpose of fixing the most eligible site for a new bridge'; their consultations led to the construction of a three span bridge with an iron middle span of 50 ft.

The first Glynn Bridge, over the river Fowey, was built in 1836–1837 by the Bodmin Turnpike, for which the surveyor was William Chapple. The new and improved route for the turnpike through the Glynn valley was opened on 29 December 1837 but in January 1839 Chapple reported considerable damage to the bridge, a consequence of the late floods. The County's magistrates were not amused, but by August 1839 he was permitted to advertise for 'Plans, Estimates and Tenders for the building of a County Bridge'. The new bridge, built by the contractor William Burt to the design of Robert Coade was eventually reported as finished in 1841.

Chapple died in Bodmin on 18 December 1846, aged seventy-five. He had held the post of Bridge Surveyor for forty years.

R. P. TRUSCOTT

[Quarter Session records, Cornwall CRO, Truro]

Works
1830–1833. Polston Bridge
1836–1837. Glynn Bridge (1)
1839–1841. Glynn Bridge (2)

CHOLMLEY, Sir Hugh (1632–1689), contractor and later Surveyor General of the Mole at Tangier, and Member of Parliament, was born on 21 July 1632 the second surviving son of Sir Hugh Cholmley of Whitby, Bt. (1600–1657) and his wife Elizabeth Twysden. He attended St. Paul's School from the age of ten but, his father having been a royalist in the Civil War, the family left England for France in 1645. Four years later they returned and in February 1650 regained their estate. Hugh afterwards spent a year or two at Cambridge. He enrolled as student at the Inner Temple in 1656 and in 1662 was given a post in the Queen's household. Before this he gained practical engineering experience at Whitby harbour. The west pier there had been rebuilt in the 1630s under the initiative of his father. Exactly what was done then and some twenty years later is uncertain. The pier, which was about 230 yd. long, is said to have been much improved and protected by a breakwater in the form of a palisade of timber piles.

Whatever the nature of this Whitby experience it was a sufficient recommendation for his appointment to the Tangier Committee in December 1662. The Earl of Sandwich led the expedition that took control of Tangier in January 1662, installing a garrison and Governor. Its perceived importance was such that it came under supervision of the King in Council and a harbour wall was regarded as essential from the start. Sandwich viewed the Bay and picked a suitable site for the wall or 'Mole'. The naval commander, Sir John Lawson, produced a plan. The first Governor was told to 'speedily consider' the making of a Mole but complained of lack of artificers and materials. The committee therefore decided to let a contract for its construction and did so on 26 March 1663, reaching agreement with Andrew Rutherford, the new Governor (soon to become Lord Teviot), Lawson and Cholmley. The contract was for building 'a Mole or Pier' of stone from the shore running out to sea 400 yd. east-north-east and thence 200 yd. east-south-east, and 30 yd. broad at its foundation. An advance of £2,000 would be made for tools and materials, to be deducted from future payments. From later evidence it appears that the cost was estimated at £10,000 with a four-year period of construction; in the event both proved exceedingly optimistic.

In June 1663 an expedition set forth, including Jonas Moore (q.v.) and Cholmley. The latter brought with him 'about forty masons, miners and other proper artists and workmen'. Moore, formerly engaged with Sir Cornelius Vermuyden (q.v.) in draining the Great Level of the Fens, was involved in 'setting out the Bounds of the Mole' and taking soundings. He stayed for two months.

Work began in August under Cholmley's direction. The wall was to be founded on a flat-topped mound of very large stones deposited in the sea up to low-water level, a technique—later known as *pierres perdues*—used at Dover harbour in the 1530s by John Thompson (q.v.) and elsewhere in deep water conditions. The wall itself, 90 ft. wide and rising 18 ft. above low-water, would be built in solid masonry cemented in lime-tarras (hydraulic) mortar and cramped with lead and iron. The range of spring tides was 9 ft. Quarries lay close to the shore west of Tangier. The stone was blasted out by gunpowder and transported to the site by boats and by carts. A little town called 'Whitby' by the Yorkshire workmen, some of whom had been employed on the west pier, was built nearby with quarters for the men and their families and stabling for ninety horses.

Cholmley went home in 1664 following the death of his elder brother. Some delays ensued but by January 1665, soon after his return to Tangier, more than 10,000 cu. yd. of stone had been deposited for the foundations. About two-hundred soldiers were at work, in addition to the miners and other skilled men.

Cholmley's nephew, from whom he inherited the title, died later that year. Sir Hugh, as he had now become, again went back to England and while there married Lady Anne Compton, daughter of the Earl of Northampton, in February 1666. In his absence he left Major Samuel Taylor in charge. By August 1668 about 350 yd. of the Mole had been completed, a notable achievement in four years considering the scale of the operations and unavoidable delays caused by winter storms.

Much of the work was now being supervised by Henry Sheres (q.v.), a young and afterwards well-known military engineer. But even so Sir Hugh felt he could no longer continue as sole contractor; Lord Teviot had died in 1664 and Lawson was killed in action against the Dutch in 1665. Once again he returned to England, this time with a plan for the establishment of an 'Office for the Mole'. This was examined and approved by Jonas Moore, recently appointed Surveyor General of Ordnance, and Sir Christopher Wren, and accepted by the Privy Council in August 1669. The principal officers were: Surveyor General, Sir Hugh Cholmley with the high salary of £1,500 a year; Deputy Surveyor, Major Taylor at £500; a Comtroller and the Clerk Examiner, Henry Sheres, both at £250 p.a. Other officers included the principal Overseer, Captain Bolland, at £100 and five overseers at £50 to £80, and a resident surgeon at £60 p.a.

Exceptional storms in the winter of 1669–1670 caused some damage to the Mole and to Cholmley's reputation. Moreover a body of opinion, led by Sheres, favoured a different method of construction along the lines of that used at Genoa since 1638. This involved building the wall (concrete faced with masonry) in great chests or caissons sunk on the *pierre perdues* foundation. Some Genoese engineers had been invited to Tangier in 1663 to give their advice. Cholmley opposed the system as inappropriate in the stormy, tidal waters of Tangier Bay; these conditions differed from the calmer and almost non-tidal waters at Genoa where divers were able to level the foundations for the chests and it was possible to float them accurately into position.

His arguments prevailed in 1663 but opinion was now against him. With some misgivings he consented to a scheme drawn up by Sheres in March 1670. The Mole had now reached a length of 380 yd. On 18 September the first chest was placed, though not without difficulty and even then not very firmly. With experience the technique of placing chests improved. They were founded at a depth of 6 ft. below low-water. By February 1673 a further 60 yd. had been built. Several ships of the Mediterranean squadron sheltered in the harbour in the winter of 1673–1674. But storms next winter caused further damage, raising old doubts about the project as a whole and provoking an official enquiry in 1675. This revealed that the enormous sum of £240,000 had been spent. Proposals for continuing the work at a lower unit cost were submitted by Cholmley, and by a London syndicate, and also by Sheres. The latter won the day and succeeded as Surveyor General in June 1676. The mole then extended 457 yd. into a depth of 20 ft. below low-water.

Work proceeded, but slowly, the chief hindrance now being attacks from the Moors, which ultimately stopped the undertaking completely. Outright war with Morocco began in 1678 and culminated in a lengthy siege in 1680. Temporary peace was restored, but fears that the permanent retention of Tangier would be too expensive led to the decision in 1683 to abandon it and demolish the Mole, then 480 yd. in length built at a total cost of £340,000, rather than let it fall into other hands. The difficulty of this operation demonstrated the solidity of the structure, in the pre-

1670 portion as much if not more than in the later part.

Cholmley was elected MP for Northampton in 1679. A year or so later he published *A Short Account of the Progress of the Mole at Tangier*, and in 1685 re-entered Parliament as member for Thirsk. Towards the end of his life he wrote a longer account of the Mole which, along with some of his speeches as an MP, was privately published a century later, in 1787. He died on 9 January 1689 and was buried at Whitby, leaving his widow and a daughter Mary. She married a distant cousin and their eldest son inherited the estate. Samuel Pepys, who served with Cholmley on the Tangier Committee, described him as 'a fine, worthy and well-disposed gentleman'.

A. W. SKEMPTON and MIKE CHRIMES

[G. Young (1817) *A History of Whitby*; E. M. G. Routh (1912) *Tangier, England's Lost Atlantic Outpost 1661–1684*; Ada Russell (1923) in: *Victoria County History of Yorkshire North Riding* **2**, 510–511, 517; M. Biddle (1982) Dover Harbour, *King's Works*, **4**, 729–755; P. A. Bolton and Paula Watson (1983) Sir Hugh Cholmley, *The House of Commons 1660–1690*, **2**, 62–63; F. Willmoth (1993) *Sir Jonas Moore*; L. Franco (1996) History of coastal engineering in Italy, in: *A History and Heritage of Coastal Engineering*, ed. N. C. Kraus, 275–335]

Publications

c. 1680. *A Short Account of the Progress of the Mole at Tangier, from the First beginning of that Work*

1787. *An Account of Tangier by Sir Hugh Cholmley*, North Riding CRO

Works

1663–1676. The Mole, Tangier; 457 yd. long, £240,000 (final cost £340,000)

CLARK, Dudley (c. 1750–1823), civil engineer, was in treaty with John Rennie (q.v.) for employment on improvements to the river Witham in Lincolnshire in December 1794 when James Watt (q.v.) supplied the following reference:

After 20 years ... he was then young perhaps 24 or 25 showed some ability in mechanical matters, and had abilities as a bookkeeper. It was in the latter capacity only he was employed by me; though not entrusted with money himself he contrived to borrow it from the canal clerk, who was a silly rascal, and betwixt them lost the money, as far as I remember ... I always found him good humoured, accurate and intelligent in what I had to do with him, and I certainly wish him well ...

If, as seems probable, this relates to the Monkland Canal with which Watt was closely associated, it is difficult to date this episode precisely. It is known that Clark was working on the Forth and Clyde Canal in 1775, and it is not likely he would have been employed there if he had been dismissed by the neighbouring concern. His job was to take care of the stores and oversee the la bourers, with a salary of 15s a week, but as part of the retrenchment that year, when the company ran out of capital, he was asked to agree to a reduction to 7s 6d or be dismissed. He evidently preferred the latter. Nicol Baird, the Surveyor of the Forth & Clyde, also referred to him as 'clever at what he professed, such as erecting fire engines ...'

In 1794 Clark still had a house at Carronshore at the east end of the canal, but had been working in Dundee for some time. The appointment to the Witham fell through but doubtless Rennie gained a favourable impression of him, as he was appointed Resident Engineer of the Western District of the Kennet and Avon Canal between Bath and Seend (between Melksham and Devizes) at £300 p.a., plus expenses in March 1795. Work had already begun on the canal in October 1794 but progress was restricted by lack of funds, financial problems and shortages of skilled labour. In 1798 Clark produced plans for the canal through Sydney Gardens, Bath, which became the subject of lengthy negotiations with residents. In 1800 he was admonished by the Canal Company for not following Rennie's specifications, and thus incurring extra expense, and, failing to give satisfaction in July 1800 Clark was dismissed.

Rennie himself cannot have been too dissatisfied as in 1801 they were associated again, in the promotion of the Croydon Canal, for which Rennie was consultant. Following the passage of the Act, Clark supervised construction until 1809, and remained associated with the Company until at least 1823.

In 1802, Clark was involved in a survey for Rennie for an abortive scheme to extend the Croydon Canal to the River Arun and on to Portsmouth. Clark is subsequently found being consulted by the River Dee Company in November 1805, with Charles Wedge, another of Rennie's preferred surveyors.

Clark's fate after 1823 is unknown.

MIKE CHRIMES and P. S. M. CROSS-RUDKIN

[Chester RO; Rennie mss, National Library of Scotland; Rennie reports, ICE Library; Kennet and Avon minutes, RAIL 842, PRO; BR/FCN, SRO; C. T. G. Boucher (1963) *John Rennie*, 72–73; K. R. Clew (1968) *The Kennet and Avon Canal*; C. Hadfield (1969) *The Canals of South and South East England*]

Works

1795–1800. Kennet & Avon Canal, Resident engineer, Western district
1801–1823. Croydon Canal, engineer

CLARK, Matthew (1776–1846). Matvej Yegorovich Clark, the son of George (Yegor) Clark, mining engineer (d. 1804), was born in 1776. In 1786 his father accompanied Charles Gascoigne (q.v.) to Russia, taking his family, and worked as a mining surveyor for the State Mining Department at the Olonetz ironworks in Kardia. Both Matthew and his brother Vasilij (d. 1842) worked for the Mining Department. Matthew's first position, in 1792, was at the Konstadt ironworks under Gascoigne. In 1795, described as a casting, locksmith and pattern maker, Clark was transferred to the Alexandrovsky cannon works near Petrozarodsk, in 1801 he was appointed 'locksmith, machinery and blacksmith master' at the St. Petersburg State Ironworks and in 1802 he was moved again, to the Izhorskij Admiralty factory in St. Petersburg, run by Gascoigne and, subsequently, by Alexander Wilson (q.v.); he remained there until 1811. While there his ability was recognised with the award of a prize of 1,000 roubles by the Tsar in 1808. In 1811 he returned to the St. Petersburg State Ironworks where he was steadily promoted, reaching the rank of Oberhültenverwalter of the eighth rank in 1819.

Whatever were Clark's achievements in the ironworks during the early part of his career, his modern significance stems from his contribution to the structural use of iron in Russian architecture in the first third of the nineteenth century. This seems to have commenced in 1817 when he collaborated with Stasov on a cast iron gateway weighing over 100 tons and incorporating eight 6 m high columns in Ekaterinskiy Park at Tsarskoe Selo. Further collaboration with Stasov followed at Baron Arakcheev's Palace at Gruzino where he built a cast iron temple (1820–1821) and spire (1824–1825). Of greater significance structurally was his collaboration with the architect Carlo Rossi.

In 1819 work began on the 'Main Headquarters' Building in Palace Square, St. Petersburg where a structural frame in three tiers of cast iron columns and beams with spiral rings, cast iron safes and wrought iron shelves made up the interior of the 'archives' building. Clark supplied the ironwork from the St. Petersburg State Ironworks and was awarded the Order of St. Vladimir (fourth degree) for his skill and diligence. The headquarters building was crowned with a group of iron framed copper sculptured horses built by Clark.

Clark's next collaboration with Rossi was the cast iron portico with six columns for the Nicholas Gate in 1826. The ironwork for the Aleksandrinskij Theatre roof (1828–1830) was the product of the Alexandrovsky ironworks built by Clark (1825–1827) on the left bank of the Neva. Clark was made its director in December 1826 and was awarded the Diamond Order of St. Ekaterina in February 1828 for his work. Clark's other works by this time had included involvement in oil gas lighting in 1820, organising the design of equipment for the St. Petersburg Arsenal in 1824, building a new steam engine for the St. Petersburg Mint in 1825, and work in the Okhta gun powder factory.

The Aleksandrinskij Theatre had a 29.16 m span iron arch roof, and 10.4–10.6 m trusses supported on cast iron cantilevers. Such a wide span structure aroused intense discussion concerning its design in engineering circles and Bazaine and Rossi argued about its safety. As a consequence the arches were load tested with 47 tons.

The use of iron arches without ties displayed startling structural innovation, which was seen again in the roofing for the halls of the Winter Palace following a disastrous fire in December 1837. For spans up to 15 m thin walled, riveted, elliptical beams of wrought iron were employed; for larger spans over the Concert Hall and Emblem Hall, parallel chord trusses of 20.5 m span and 1.8 m depth were used, braced with an arch and chain. Because of the complexity of the system all the girders were load tested. For the St. George's Hall roof I-beams were initially used, but one of these failed and a catenary structure was used between upper and lower chords, 21.3 m span and 2.9 m high. The roofs were supported on triangulated trusses, generally of the Polonceau type.

The Winter Palace roofs represented the peak of Clark's structural engineering achievements. In the late 1820s and 1830s he had continued to work with Stasov but his design for a wrought iron dome at the Ishmajlovskij Regiment's Trinity Cathedral (1828–1834) proved too light and proved susceptible to torsion in a storm and was replaced by P. D. Bazaine's design. Again with Stasov, Clark supplied decorative copper plates and sculpted details for the Narva Gates. These Moscow gates, of copper and cast iron, contained seven hollow cast iron columns each weighing 37.5 tons and the structure was regarded by the Russians as the largest cast iron structure in the world at that time.

In addition to his architectural work, Clark's factories produced steam engines and other machinery and, along with those of Alexander Wilson (q.v.), were the leading state engineering enterprises. Despite his prestige and undoubted skills Clark was summarily dismissed in May 1842 as a consequence of defective work at the Marinskij Palace, St. Isaac's Square, for which the architect, Stakenschneider, had ordered three-hundred and eighty-four beams from Clark. While Clark claimed he lacked all the necessary drawings, he failed to complete the order by the June 1841 deadline. This coincided with failures at St. George's Hall and a commission looked into the beams supplied at Marinskij Palace and found some were crooked and lacked bracing. Rossi was called in and despite Clark's protestations that the beams were structurally adequate, he discovered that old iron had been reused rather than the new ironwork specified. Clark was ordered to make good the defects at his own expense. In 1846 he was dismissed from all state employment and although awarded a state pension of 1,254 roubles a year, he died shortly afterwards on 24 September 1846 in St. Petersburg.

MIKE CHRIMES

[J. G. James collection, ICE archives; C. Eck (1841) *Traité de l'Application de Fer, de la Fonte et de la Tôle*; J. G. James (1983) The application of iron to bridges and other structures in Russia to about 1850, abridged in *Trans Newc Soc*, **54**, 79–104; A. L. Punin (1990) *Arkhitektura Peterburga seredin" xix veka*, 97–99; S. G. Federov (1992) Matthew Clark and the origins of Russian structural engineering, 1810–1840s: an introductory biography, *Construction History*, 69–88; S. G. Federov (1996) Early iron domed roofs in Russian church architecture, *Construction History*, 12, 41–66]

Works
Major structures:

1817. Comrades in Arms Gate, Tsarskoe Selo, cast iron
1819–1823. Main Headquarters Building archives, fireproof construction
1820–1821. Cast iron chapel, Grozino
1821. Cadetsgate, Tsarskoe Selo
1825. St. Petersburg Mint, steam engine
1825–1826. Griffons and Lions Suspension Bridges, St. Petersburg, decorative castings
1826. Nicholas Gates, Pavlovsk
1826–1827. Alexandrovsky ironworks
1827. General Staff Building, St. Petersburg
1827–1828. Large Stables Bridge, St. Petersburg
1828–1830. First Engineers Bridge, St. Petersburg
1828–1830. Maly Konyushenuy Bridge, St. Petersburg
1828–1830. Treatralny Bridge, St. Petersburg
1828–1832. Aleksandrinskij Theatre, iron roof structure
1831–1834. Narva Gates, brickwork and copper plate
1834–1836. Demidov Bridge, St. Petersburg
1834–1838. Moscow Gates, cast iron and copper
1835–1840. Nikolsky Bridge, St. Petersburg
1838–1842. Winter Palace reconstruction, iron roof structures
1839–1840. Pevchesky Bridge, St. Petersburg
1840–1841. Marinskij Palace roof beams

CLARK, William Tierney, FRS (1783–1852), civil engineer, was born at Bristol on 23 August 1783, the son of Thomas Clark. The early death of his father deprived William of a formal education and resulted in William's early apprenticeship to a Bristol millwright. Having served his time he went to the Coalbrookdale Ironworks in Shropshire. John Rennie visited Coalbrookdale in 1808 where he met Tierney Clark and, recognising his abilities, offered him a post in London at his Albion Works at Blackfriars. In January 1811, on Rennie's recommendation, Clark was appointed Engineer to the West Middlesex Waterworks at a salary of £200 p.a. The company provided him with a house at Hammersmith, 'with coals and candles free', where he lived for the rest of his life and which was also to become the base for his extensive consulting engineering practice. In his early years with the West Middlesex Company Clark spent much of his time in procuring cast iron pipes and in laying down the water mains. The Company already had had reservoirs at Hammersmith and Campden Hill in Kensington built before Clark became Engineer. In 1820 the Company acquired 3 acres of land on Barrow Hill (Primrose Hill) on which Clark built a reservoir of 4.5 million gallons capacity and installed a 105 hp steam engine at Hammersmith to pump to Primrose Hill. By 1818 the West Middlesex works were sufficiently complete for Clark to accept additional work elsewhere, with the Company's agreement.

His first job as an independent consulting engineer was on the Thames and Medway Canal which included the 4,012 yd. long tunnel from Higham to Frindsbury where the canal terminated in a basin with a lock into the River Medway. The tunnel was begun in April 1819 and on this project Clark pioneered the use of the transit instrument in setting out the line of the tunnel. The canal was completed in 1824. In 1825 Clark was consulted about a project to build the Alford Canal in Lincolnshire, but this was not built; considerable doubt was expressed by Thomas Tredgold (q.v.) about the practicability of the design. In 1830, during the cessation of work in building Sir Marc Brunel's Thames Tunnel from Rotherhithe to Wapping, Clark was consulted and recommended continuing the work to Brunel's design. The following year he reported with Thomas Telford and James Walker (qq.v.) on the state of the new London Bridge, designed by Sir John Rennie.

Clark became one of the most successful British designers of suspension bridges. He designed the Hammersmith Bridge, the first suspension bridge over the Thames, which was opened on 6 October 1827. His river piers for this bridge were used for the present structure opened in 1887. Clark's second suspension bridge was that at Marlow over the Thames. An Act was obtained in 1829 and the bridge was opened in 1832. Clark replaced John Millington the original engineer because of concerns over the design and poor supervision. In 1830 the Duke of Norfolk undertook to build a bridge across the river Adur at New Shoreham in Sussex, and Clark designed a handsome suspension bridge with masonry towers. The contract was let to William Ranger (q.v.) in October 1830 and the bridge was completed in 1833.

The major part of Clark's consulting engineering work, including the prestigious suspension bridge over the Danube at Budapest (1839–1849), was undertaken after 1830. He was approached by Counts Szechenyi and Andrassy in 1832, on the recommendation of William Yates (q.v.). He visited in 1834 and in 1837 produced a definitive report. Work began in 1839, with Adam Clark (no relation) as resident engineer; Clark's cousin Bland William Croker supervised the ironwork. The contractor for the cofferdam foundations was George Burge (q.v.), and ironwork for the chains was supplied by Howard and Ravenhill, who developed their patent for rolling the eye links, while working on the bridge.

Perhaps Clark's principal contribution to suspension bridge design was his distinctive use of diagonal braced wrought iron balustrades to stiffen the decking. This was first used on Hammersmith Bridge and was justified by tests on a large model exposed to the wind.

Clark's experience in ironwork at Coalbrookdale and with Rennie in London equipped him to be a first-rate designer in cast and wrought iron and, in addition, his superb detailing of masonry work, displayed in his suspension bridges and North Parade Bridge, Bath, crosses the notional boundary between engineering and architecture. His mastery of iron was confirmed in his design for Gravesend Town pier, probably the first pier to use cast iron piles in its substructure. Constructed in 1834, Clark used an ingenious template to position the piles.

While his work as a bridge engineer was gaining him international recognition, he continued to develop West Middlesex Waterworks. In 1838 work began on settling tanks and reservoirs at Barn Elms, a site acquired in 1829. As a consultant on water supply he was planning a scheme in Amsterdam at the time of his death.

Clark was elected a Member of the Institution of Civil Engineers in 1823 and a Fellow of the Royal Society in 1837. William Tierney Clark died in the waterworks house in Hammersmith on 22 September 1852 and is buried in St. Paul's Church, Hammersmith where his memorial plaque on the north wall features a suspension bridge.

DENIS SMITH

[Minute Books West Middlesex Waterworks; Hammersmith and Fulham Archives; ICE archives; F. W. Simms (1838) *The Public Works of Great Britain*; E. Cresy (1847) *Encyclopaedia of Civil Engineering*, J. Priestley (1830) *Navigable Rivers and Canals*; D. Smith (1991–1992) The works of William Tierney Clark (1783–1852), *Trans Newc Soc*, **63**, 181–208]

Publications

1825. *Report on the Practicability and Advantage of a Navigable Canal, from the Sea at Anderby Haven to the Town of Alford*
1831. *Report (and Second Report) of William Tierney Clark Esq., relative to New London Bridge*
1835. *Plan of a Proposed Canal from the River Thames, to the Town of Dartford, Kent, with a Branch to Crayford Creek: also a Ferry from the entrance to the Canal to the opposite Shore at Purfleet in Essex*
1838. Account of the Gravesend pier, *Trans ICE*, **3**, 245–256
1853. Account of the Buda Pesth suspension bridge, in: *Supplement* to *Weale's Bridges*

Works

1819–1824. Thames and Medway Canal, completion, engineer

1820–1822. Barrow Hill (Primrose Hill) Reservoir (West Middlesex Waterworks)
1825–1827. Hammersmith Suspension Bridge, Engineer, £43,341
1829–1832. Marlow Suspension Bridge, Engineer, £22,000
1830–1833. Shoreham suspension bridge, Engineer

CLARKE, John (1744–1804), baptised in March 1744, the son of John and Elizabeth Clarke of Marlow, was appointed Surveyor (of works) in the Lower Districts (Staines to Mapledurham) of the Thames Navigation in August 1773 at a salary of 1 guinea per week. Eight timber-framed locks had recently been built between Maidenhead and Sonning under the direction of Joseph Nickalls (q.v.). The locks were 113 ft. in length between the gates, 18 ft. wide, with 3.5 ft. of water on the sills, accommodating barges up to about 120 tons burden. Another two locks, completed 1777–1778, were made at Caversham and Mapledurham with Thomas Yeoman (q.v.) as consultant. Clarke would have superintended their construction. To save as much money as possible all these locks were built in fir or 'Memel' timber but this proved a false economy as after seven or eight years they had to be rebuilt in oak.

In July 1780 Clarke reported the first (1773) set of locks to be in a bad state and estimated that about £4,000 would have to be spent on rebuilding over the next ten years. Work started a month later on the lowest lock (Boulters) and proceeded upstream until by 1788 all eight locks had been rebuilt, seven in oak and one with brick abutments. The works were carried out by contractors under Clarke's supervision, except in one case where Clarke himself acted as contractor (his bid being the lowest of five) under supervision by one of the Thames Commissioners.

Meanwhile in August 1786 work began on the first of a new set of locks in the Upper Districts, at Whitchurch. William Jessop (q.v.) had been consulted. He agreed with the proposed location and the lock, timber-framed in oak, was built by direct labour to Clarke's plan and under his supervision at a cost of about £1,500 including some modifications to the (existing) weir and dredging of the approach channels.

In January and February 1787 Clarke had been preparing plans and estimates for the next two locks, at Goring and Cleeve, and was still working on the rebuilding programme. Clearly he could not continue to superintend the Lower Districts while undertaking new works many miles upstream, so in March 1787 John Treacher (q.v.) was appointed to take charge of construction at Goring and Cleeve and to become Surveyor of the Upper Districts at a salary of 2 guineas per week 'for his constant superintendence'. Clarke still had plenty to do in the Lower Districts, rebuilding the remaining old locks, improving several of the weirs and so on. In 1790 the Commissioners decided 'he is justly intitled to an increase' in his salary to 2 guineas per week.

In 1793 he gave evidence to a House of Commons committee on the Thames, saying that navigation from Maidenhead upwards was good, but downstream to Staines it was necessary to build three locks: at Romney, Boveney and near Staines, and these would require weirs across the river. After long discussions it was agreed to build Romney lock and weir. Treacher, since 1791 General Surveyor but retaining responsibility for the Upper Districts, drew up the plans in April 1796. Next month Joseph Kimber, millwright, signed the contract and work began under Clarke's supervision. He reported completion of the lock in December, dredging was finished in April 1797 and the weir in October 1798; the total cost was about £2,200. Romney lock and weir proved to be very successful. For reasons that are not clear, no other locks were built between Maidenhead and Staines until the period around 1820.

Clarke is last mentioned in the Thames Commissioners Minute Book in October 1803. He died at Marlow in January 1804 and Zachary Allnutt (q.v.) was appointed his successor a month later.

A. W. SKEMPTON

[Minutes Books of the Thames Navigation Commission (1773–1804) Berkshire RO; *Report from the Committee on the Navigation of the Thames* (1793) House of Commons; A. W. Skempton (1984) Engineering on the Thames Navigation, *Trans Newc Soc*, **55**, 153–176]

Works

1780–1788. Rebuilding locks on Thames, £4,000
1786–1787. Whitchruch lock, £1,500
1796–1798. Romney lock and weir, £2,200

CLEGG, Samuel, Sr. (1781–1861), was born in Manchester on 2 March 1781 the son of Wheatley Clegg, a businessman. He was educated under John Dalton (1766–1844), attending New College Manchester 1794–1797. After a spell in his uncle's counting house in Manchester he was apprenticed to Boulton and Watt (q.v.) where, from 1802, he was concerned in the gas-lighting experiments of William Murdoch (q.v.). He left the firm in 1805. At the end of the year Clegg was employed to introduce gas lighting at Henry Lodge's cotton mills at Sowerby Bridge, Yorkshire, probably completed ahead of Murdoch's work at Phillips and Lee. He had a small works off Deansgate in Manchester, and began to equip textile works and mill-owners' houses with gas lighting. His early experience included gas installation work at Knight's factory, Longsight which Charles Bage reported on to William Strutt (qq.v.) in 1808. His experience led to his initial development of the lime purifier and its use at the Harris factory at Coventry. In 1808 he was awarded a silver medal for this work by the Society of Arts, and moved to larger premises in Major Street, Manchester. Work continued with a factory at Dolphin Holme (1810–1811), which had a complicated flue arrangement and two cast iron retorts, and more or less simultaneously he introduced the wet lime purifier at Stonyhurst Catholic College. In 1811 he introduced hydraulic mains for the Greenaway works. The following year he installed gas lighting for Ashton Bros works in Hyde, with improved cylindrical retorts.

It was during this period that attempts were made to produce gas on a commercial scale. Frederick Albert Winsor (1763–1830) first introduced gas street lighting in London in Pall Mall in 1807, and, in 1809 he applied to Parliament to establish the The New Patriotic Imperial and National Heat and Light Company but was unsuccessful due to the opposition of Boulton and Watt; the company was succeeded in the following year by the (Chartered) Gas Light and Coke Company.

Meanwhile Clegg was called to London and in 1812 illuminated the showroom of Ackerman's, the print-sellers. Late that year he was asked to become engineer to the Chartered Gas Company. Having closed down his Manchester works, he took up his appointment in January 1813 at a salary of £500 p.a. He set about improving the plant, with the assistance of John Farey Jr. and Aaron Manby (qq.v.). In 1813 the Company first lit Westminster Bridge and then the parish of St. Margaret's Westminster; by December 1816 London had 26 miles of gas mains, and by 1823 this had been extended to more than 200 miles, much Clegg's work.

In 1813 he married Ursula, seven years his senior, they had a son and daughter, born in 1814 and 1817 respectively. At that time they were living in Lambeth.

While working for the Chartered Company Clegg took on consultancies elsewhere, including other London companies; this led to a disagreement and he left the Company in 1817 and was engaged in constructing gas works for a number of years. That year he installed gas at the Royal Mint. By 1823 he had fitted up over thirty works, including those of the Imperial Gas Co. in London and the City of London Works. He then turned to marine engineering and from 1824–1828 was involved in fitting steam engines in ships. After a brief spell establishing the Imperial Company's works at Fulham he effectively retired from gas works construction in 1829. Clegg was elected a Member of the Institution of Civil Engineers in that year.

Although his name is normally associated with gas equipment, as engineer to a large number of works, he was responsible for a lot of the 'structural' engineering of the early gas works. The early gasholders were limited in their design by concern over explosions. Clegg allegedly dramatically disproved this by puncturing a holder in the view of onlookers and then setting fire to the stream of leaking gas. The holder at Stonyhurst College had a capacity of 1,000 cu. ft., and was enclosed in a building, but those at Chester and Birmingham were the first not to be so confined.

Unsurprisingly the early gasworks sites were on poor ground, that at Great Peter Street was described by Clegg as being 'on a swamp', and as a consequence care had to be taken over the foundations. Clegg seems to have been an early employer of concrete in foundations, most spectacularly for the chimney at the Fulham works, which was founded on quicksand, with a concrete base 22 ft. wide and 8 ft. deep.

Clegg's involvement with an engineering works at Liverpool 1829–1833 resulted in the loss of all his money when it failed. Afterwards he spent some time (1834–1838) employed as an engineer in Portugal, where he reconstructed the mint at Lisbon and undertook work for the government there. On his return to England he joined with Jacob Samuda, the marine engineer, in the development of the atmospheric railway system which spectacularly failed on the South Devon Railway. Their 1838 patent (No. 7920) addressed the problem of leakage but otherwise their system was based on that of Henry Pinkus. This plagiarism was one reason that the editor of *Railway Magazine*, John Herapeth, criticised the system from the start, although many senior engineers expressed approval at first.

Clegg patented eight inventions before 1852, mostly concerning gas-making, of which the most important was his wet gas meter of 1818. Towards the end of his life Clegg worked as a government surveyor, investigating the feasibility of new Parliamentary gas bills (1847–1856), and he helped his son, Samuel, in writing his book on gas lighting. His son's death on 25 July 1856 was a bitter blow to him. Following it he moved in with his (married) daughter in Putney Heath. His daughter was the wife of the gas engineer and patentee, John Malam, who in 1819, improved upon Clegg's gas meter.

In 1860 he moved to Haverstock Hill, and died of bronchitis at his residence, Fairfield House, Adelaide Road on 6 January 1861. His wife died on 3 October 1865, aged ninety-one.

His son, Samuel Clegg Jr. (1814–1856), in 1836 worked as a surveyor in Portugal and on his return to Britain was employed on a number of railway projects. In 1849 he was appointed Professor of Civil Engineering and Architecture at Putney College, and also lectured on civil engineering to Royal Engineers at Chatham. He died at Putney at the age of forty-two.

A. P. WOOLRICH

[Membership records, ICE archives; Institution of Gas Engineers; F. Accum (1819) *Description of the Process of Manufacturing Coal Gas*; *DNB*; Rees's *Cyclopaedia* (1802–1819) articles on gas lighting; *House of Commons, Select Committee on Gas-light establishments* (1823) Minutes of evidence, 51–63, 64–69; S. Clegg Jr. (1841–1859) *A Practical Treatise on the Manufacture and Distribution of Coal Gas*, 3 edns.; J. Samuda (1844) The atmospheric railway, *Min Proc ICE*, 256–283; *House of Commons, Select Committee on Atmospheric*

Railways (1845) Report and minutes of evidence; T. Newbigging and W. T. Fewtrell (1878) *King's Treatise on the Science and Practice of the Manufacture and Distribution of Coal Gas*; C. Hadfield (1967) *Atmospheric railways*; S. Bennett (1997) *Samuel Clegg—his Contribution to the Development of the Gas Industry*, IGasE Manchester Assoc.]

Publications

1808. Improved apparatus for extracting carbonated hydrogen gas from pit coal, *Soc Arts Trans*, **26**, 204–210, ill.
1815. Patent 3968: an improved gas apparatus, *RAM*, **2**, xxx, 1–10, 65–75, 129–137, ill.
1818. Patent 4283: an improved gazometer, *RAM*, **2**, xxxvii, 193
1820. Description of an apparatus by which 25,000 cubic feet of gas are obtained from each cauldron of coal (abridged in: *London J Arts*, 226)
1830. Patent 6020: improved gas meter, *Register of Arts & Sciences*, **vi**, ns, 70; *Mechs Mag*, **xv**, 321–322
1838. Patent 7674: description of … dry gas-meter
1839. Patent 7920: description of patent atmospheric railway
1840. Atmospheric railway (with J. Samuda)
1858. New hydraulic gas meter

Works

All gas lighting installation, unless otherwise described:

1805. Lodge's Mill gas lighting installation
1806. Willow Hall, Lodge's private house, gas lighting installation
1806. Potter's Mill, nr. Bury, gas lighting installation
1806. Barton's Dye House, Lower Ardwick, gas lighting installation
1806. Thornton's Drawing Academy, Manchester, gas lighting installation
1806. Gallymore's print works, Bury, gas lighting installation
1806. King Street Police Station, Manchester, gas street light
1807. Knight's Mill, Longsight, Manchester, gas lighting installation
1809. Harris's Works, Coventry, gas lighting installation
1810–1811. Dolphin Holme Worsted Mill, Lancaster, gas lighting installation
1810–1811. Stonyhurst College, Clitheroe, gas lighting installation, £400
1811. Greenway's Mill, Manchester, gas lighting installation
1812. Ashton & Bros. Mill, Hyde, gas lighting installation
1812. Ackermann's print shop, London
1813. Chartered Gas Light and Coke Co., Works, Peter Street/Artillery Row, Westminster
1813–1814. Chartered Gas Light and Coke Co., Gas Works, Norton Folgate, London
Gas Mains, St. Margaret's Parish, Westminster.
Chartered Gas Light & Coke Co., distribution gas holder

c. 1816. Bristol Gas Works
1817. Royal Mint, gas lighting installation.
1817. London Gas Light & Coke Co., Whitechapel works
c. 1817. Birmingham Gas Works
c. 1818. Dudley Gas Works
c. 1819. Warwick Gas Works
c. 1819. Chester Gas Works
c. 1820. Worcester Gas Works
c. 1821. Kidderminster Gas Works, iron work by J. U. Rastrick (q.v.)
1821–1823. Imperial Gas Co., Shoreditch & St. Pancras Works
1823. City of London Gas Light & Coke Co., Works
1828–1829. Imperial Gas Co., Fulham Gas Works

CLEGRAM, William (1784–1863), sea captain and civil engineer, was born at Ryhope, County Durham, on 20 May 1784. He went to sea as a boy in a ship belonging to his father, hailing from Sunderland. His mathematical knowledge and a love of engineering he gained from Thomas Wilson (q.v.), schoolmaster and engineer of the iron bridge at Sunderland. By 1809 he was master of a ship at Shoreham, on the south coast near Brighton.

At that time Shoreham Harbour was in a deplorable state. In 1760–1762 an entrance had been made through the long offshore shingle bar at a site proposed by J. P. Desmaretz (q.v.) about 1.5 miles east of the town. But this was soon destroyed by the sea and new natural openings developed successively at various places further eastward. Several engineers were consulted including William Jessop in 1800, who proposed too little, and John Rennie in 1810, who proposed too elaborate a scheme. Finally, in December 1814 the merchants and shipowners of Shoreham submitted a memorial to the Harbour Commissioners stating that the remedy was to reopen the 1760 entrance with proper engineering works. The memorial was signed by Clegram among others. In February 1815 he submitted a report and a plan. On the recommendation of Trinity House, William Chapman (q.v.) was asked to report. He did so in July, endorsing Clegram's proposals and estimating the cost. A Bill went to Parliament and passed on 1 July 1816. Chapman was appointed Engineer. Clegram received £226 for 'taking and drawing Plans of the Harbour and acting as Secretary to the Committee of Subscribers' and in October was appointed Surveyor of the Works at a salary of £150. He also became Harbour Master, with £100 p.a.

The main works included two timber framed chalk-filled piers each about 320 ft. long forming an entrance 250 ft. wide through the shingle bank, with groynes extending 200 ft. seaward beyond the pier heads. Work began in April 1817. The first contractor gave up after a few months and Hugh McIntosh (q.v.) took over. By September 1820 the new entrance was open to shipping. Clegram received £300 'for his claims and compensation for his great attention and valuable services'. Ancillary work continued and in June 1821 Clegram signed an affidavit that the harbour had been completed agreeable to the Plan referred to in the Act. Chapman made nine site visits during the period of construction and kept up a very active correspondence with Clegram.

Heavy storms in November breached the groyne at the head of the east pier and drove shingle round the west pier head. Clegram drew up plans for restoring and improving the entrance. His report was approved by Thomas Telford (q.v.) in April 1822 and work began soon afterwards. Meanwhile, Samuel Brown (q.v.) had designed the Chain Pier at Brighton, the Act for its construction passed in July 1822. McIntosh was to be contractor and Clegram the Superintending Engineer, on leave of absence from Shoreham. Work began in September and the official opening took place in November 1823.

Back at Shoreham, three new groynes built in 1825 on the shore of the shingle bank east of the east pier completed the works, the total cost for which amounted to about £62,000, including £3,500 for a steam dredger.

Presumably again on leave of absence, Clegram then supervised construction of Brown's 315 ft. span suspension bridge over the river South Tyne near Hexham (also known as Warden Bridge) built in 1825–1826.

In October 1826 he resigned his job at Shoreham, with a gratuity of £100 and thanks from the Harbour Commissioners, to take up a post as Harbour Master, Engineer and General Superintendent of the Gloucester & Berkeley Canal at a salary of £370, with a house. This ship canal, with docks at Gloucester and an entrance lock and basin at Sharpness on the Severn Estuary, was nearing completion under Telford's direction. Thomas Fletcher (q.v.), resident engineer since 1820, continued until the opening of the canal in April 1827. Thereafter Clegram remained in charge until 1861 when he was succeeded as Engineer and Superintendent by his son, William Brown Clegram. He became Clerk to the canal company in 1829, a post which left time for him to assist his father in designing new basins, quays and docks at Gloucester.

Clegram Sr. was elected a Member of the Institution of Civil Engineers in 1832 on Telford's nomination. In his later years he was a keen amateur astronomer. He died on 20 October 1863, aged seventy-nine.

A. W. SKEMPTON

[Minute books of Shoreham Harbour Commissioners (1760–1826), Memorial from the Merchants and Shipowners (1814), Letter Books (1816–1825), all in West Sussex RO; William Chapman (1815) *Report on the Harbour of New Shoreham*; Memoir of William Clegram (1864) *MPICE*, **23**, 485–486; Burton Green (1877) New Shoreham, *Sussex Archaeol Coll*, **27**, 69–109; Obituary of W. B. Clegram (1889) *MPICE*, **98**,

388–390; J. G. Bishop (1897) *The Brighton Chain Pier*; H. C. Brookfield (1949) A critical period in the history of Shoreham harbour 1760–1810, *Sussex Archaeol Coll*, **82**, 42–50; Charles Hadfield (1969) *Canals of South and South East England*; A. W. Skempton (1974) William Chapman 1749–1832, civil engineer, *Trans Newcomen Soc*, **46**, 45–82; E. L. Kemp (1977) Samuel Brown: Britain's pioneer suspension bridge builder, *Hist Technol*, **2**, 1–37]

Publications

1815. *A Plan of the Harbour of New Shoreham*
1815. *Report on a Survey of New Shoreham Harbour*
1822. *Report and Plan for the Further Security and Improvement of New Shoreham Harbour*

Works

1817–1825. Shoreham Harbour, resident engineer, £62,000
1823. Brighton Chain Pier, superintending engineer, £30,000
1825–1826. Hexham Suspension Bridge, superintending engineer, £50,000

CLELAND, James (1770–1840), civil engineer and statistician, was trained as a cabinet maker in Glasgow before moving to London where he worked as a Clerk of Works. In 1813 he became involved in proposals for Glasgow water supply and improvements, and in 1819 was employed by Glasgow Corporation to take a census, the first such taken in the British Isles. He established a reputation in this field, investigating mortality rates in Glasgow and other large towns. Cleland knew Thomas Telford (q.v.), presenting him with a copy of his *Enumeration of the Inhabitants of the City of Glasgow* and doubtless this contact led to him being elected a Corresponding Member of the Institution of Civil Engineers in 1825. He married Mary Stewart and had two sons, Alexander Stewart and Henry Wilson, and also a daughter, Jane.

He died in Glasgow on 14 October 1840 and left an estate of £5,955 6s 5d.

MIKE CHRIMES

[Membership records, ICE; ICE Library catalogue; steam vessels plying on the Clyde in 1832–1833 (OC 90), ICE archives; DNB; Skempton; W. Johnstone (1997) in lit.]

Publications

Selected printed works:

1813. *A Scheme for Raising and Distributing Water, and Erecting Public Baths in the City of Glasgow*
1820. *Enumeration of the Inhabitants of the City of Glasgow*
1829. *Report Relative to the Proposed Road to Connect the Inchbelly Bridge Road with Garscube Road*
1831. *A Description of the Manner of Improving the Green of Glasgow, of Raising Water—for the Supply of Public Buildings of that City*

CLERK, John (fl. 1484–d. 1535) was a cleric and builder of the first harbour pier at Dover. Reginald Scot, Surveyor of Romney Marsh and closely acquainted with Dover harbour in the 1580s, recorded that the first pier there was built about 1500 by John Clerk, Master of the Maison Dieu. No contemporary documents exist to prove (or disprove) this statement but there can be very little doubt that it is correct. Clerk was certainly Master of the Maison Dieu, a religious house or 'hospital' providing support for the poor and infirm of Dover; he was elected to the post in 1484. There is some evidence that the pier was completed c. 1495 and a defensive tower at its outer end might well date from c. 1500. The earliest surviving pier accounts, of 1510, record masons working on a structure already in existence; later accounts of 1513–1515 show both masonry and timber work taking place. Clerk is not mentioned in these but in 1518 there is a reference to the pier 'made by the seid master of mesonduwe'. Clerk himself kept accounts of maintenance work from 1518 to 1529 and in 1521 he was paid £40 13s. for his 'oversight' of the pier for the past four years.

Moreover, it has recently been discovered that by the late 1490s he had gained recognition as an expert in engineering matters, since in March 1500 he was brought in to assist with arrangements for the building of a large sluice on the river Witham at Boston by the Flemish engineer Matthew Hake (q.v.). Clerk was recommended for his job in Lincolnshire by John Morton who, when Bishop of Ely between 1479 and 1486, had planned and supervised construction of a new 12 mile channel for the river Nene in the Fens (still known as Morton's Leam) and in 1488 was consulted by the men of New Romney on a new cut to prevent silting up of their harbour. Dover is less than 20 miles from New Romney, but in any case Morton, now Archbishop of Canterbury and Lord Chancellor, would have been aware of what was going on at Dover in the 1490s.

Clerk's pier, later called the Stone Jetty, ran some 500 ft. out to low-water mark on the south side of the harbour. It provided protection from south-westerly storms and acted as a barrier against shingle driven northwards along the coast. By 1523, however, shingle which had been accumulating behind the pier began encroaching around its end and by 1532 the harbour became completely choked. The only sensible, though expensive, solution to this problem would be to extend the pier into deeper water. Clerk was now an old man in ill health and he died in January 1535. The extension was planned in 1533 by John Thompson (q.v.), who succeeded Clerk as Master of the Maison Dieu, and work began under Thompson's direction in July 1535.

A. W. SKEMPTON

[W. A. J. Archbold (1894) John Morton *DNB*, **39**, 151–153; R. C. Fowler (1926) The hospital of St. Mary or the Maison Dieu, Dover, *VCH Kent*, **2**, 217–219; Martin Biddle (1982) Dover harbour, *King's Works*, **4**, 729–755; H. C. Darby (1983) *The Changing Fenland*; Michael K. Jones (1986) Lady Margaret Beaufort ... and an early Fenland drainage scheme, *Lincolnshire History and Archaeology*, **21**, 11–15]

Works

c. 1490–1495. Dover harbour pier, c. 500 ft. long

CLOWES, Josiah (1735–1794), engineer, surveyor, contractor and canal trader, was perhaps the son of William Clowes and Maria Whitlock. He was born in 1735 in North Staffordshire. He was the sixth and youngest child of the family, one brother being William (1728–1782). William Clowes probably owned coalfields at Whitfield, Norton and Sneyd Green and Josiah came to have interests in the mining concern of Samuel Perry and Co. at Sneyd Green where he met Charles Bagnall (1747–1814), a partner in the enterprise. In 1762 Josiah Clowes was established well enough to contemplate matrimony and he then married by special licence Charles Bagnall's sister, Elizabeth, from Wolstanton; she died at Chell Heath only seven weeks and five days after the marriage and was buried at Norton churchyard on 19 February 1763. A witness of the marriage had been Hugh Henshall (q.v.), a clear indication of the friendship already existing between the two men. They were to become related by marriage.

The Act for the construction of the Trent and Mersey Canal was passed in 1766 and excavations began at Harecastle in July of the same year. Clowes undoubtedly had first hand experience of the work particularly following James Brindley's death in 1772 when Henshall became Engineer. Clowes appears to have been acting in the role of contractor during these years as in June 1775 he was advertising from Middlewich for labourers in order to get the project completed. As early lock construction involved carpenter's skills in the laying down of a wooden framework in the excavated area prior to brickwork and masonry construction proceeding, Clowes may have been involved in building locks as well as driving the Harecastle tunnel on the Trent and Mersey canal; during his later work on the Thames and Severn Canal he was described as resident engineer and carpenter.

Once the Trent and Mersey Canal had been completed Clowes's interest focused on trade and business and, despite approaches from the Chester Canal Company in 1776, he went into partnership with Hugh Henshall in the latter's canal carrying company, the official carrier for the Trent and Mersey Canal. He worked from Middlewich in Chester, now his home and the proposed exchange port for broad to narrow barges working on the new canal between Preston Brook and Stoke. It is at this time that he may have married Margaret, his second wife, but the date of this marriage is unknown.

Clowes invested in the expanding canal system by buying shares in the Staffordshire and Worcester Canal and in 1778 he returned to engineering, agreeing in that year to assist the Chester Canal Company and replacing Thomas Morris as 'general surveyor and overseer of the works'; his annual salary was £200. In June and July, he was engaged in repairing locks between Chester and Beeston Brook but the following month the company wrote to him complaining that he had been absenting himself from their works and threatened a stoppage of pay if this continued. In 1778 Clowes travelled to Gloucestershire where he assisted another of Brindley's apprentice engineers, Thomas Dadford, in building some of the locks on the Stroudwater Navigation. This work—plus the pressure of his trading business in Middlewich—resulted in a rate of attendance on site that was unacceptable to the Chester Canal Company and it ordered that he 'be discharged for not attending the works agreeable to contract.' For the next five years Clowes concentrated on his trading and business interests on the Trent and Mersey, in 1781 occupying property in Dog Lane, Middlewich.

In 1783 Clowes was appointed by the Thames and Severn Canal Company as 'surveyor, engineer and head carpenter' to its ambitious east–west canal link, his duties to assist Robert Whitworth (q.v.), the surveyor, in setting out the navigation; he became the resident engineer and was paid £300 p.a. The canal was to link the rivers Thames and Severn, piercing the Cotswolds with the world's then longest canal tunnel, 3,817 yd., at Sapperton. Whitworth's initial survey did little other than mark out the line of the 9 mile summit level which included the tunnel at Sapperton and the project was then left in the hands of Clowes and James Perry, the salaried superintendent of the company, with no previous canal experience. The Canal company made clear the part that Whitworth played when they stated that he 'was not at any time the only surveyor and engineer for the construction of the canal'. Clowes, now in his late forties, took up residence at Castle Street in Cirencester, near the works at Sapperton, and began what became an extremely difficult and onerous task.

Leakage was the major area of concern with the tunnel during its construction and during its subsequent use. In spite of Clowes's attempts to seal the tunnel of which he remarked to the Committee 'I have done all that I could to discover any errors in dangers of my life many times'. The real problem related to the properties of Fullers earth which swelled when wet, the consequent movement of soil causing significant leakage in the invert. There were problems with underground springs finally resolved by Robert Mylne (q.v.) in 1790. He also had major problems with one of his contractors, Charles Jones (q.v.), and in 1789 he wrote plaintively to the proprietors that 'I wish you would send me a man that understands canal navigations and then he may inform the

gentlemen (the committee) his opinions'. The work, including the tunnel, was eventually completed in five and a half years. It appears he developed a 'driving frame' during the construction of the tunnel, possibly moveable centering.

Despite the many difficulties recorded in the building of the canal and particularly the tunnel at Sapperton, Clowes left the work of the Thames & Severn with his reputation enhanced, gaining good reports from Whitworth and John Smeaton (q.v.) and the French engineer Dupin. In the remaining five years of his life he became much sought after. In 1789, Clowes left the Thames and Severn to complete the Dudley tunnel, 3,172 yd. long, and to remedy the errors in its alignment. This canal and its tunnel were built to facilitate the easy transportation of coal from Lord Dudley's mines at Tipton to the Thames and Severn Canal, via the Stourbridge and the Staffordshire and Worcestershire canals. Clowes received from the Dudley Canal Company 1½ guineas a day for his advice and all his expenses were paid. He was now becoming a wealthy man and to underline the point he bought property in Marsh Green, in Biddulph, and in Lewin Street, Middlewich. After two years the tunnel was opened throughout and the reservoir at Gads Green finished.

In 1791 Clowes gave evidence to the Commons on the Leominster Canal, the Worcester–Birmingham Canal and the Upper Thames Navigation; he was also involved in surveying and engineering the Hereford and Gloucester Canal. On Whitworth and Henshall's advice a tunnel 2192 yd. long was built at Oxenhall, and the gin wheels from Sapperton were transferred for this work. All these canals were designed to link to the Thames and Severn Canal and stimulate trade on this, Clowes's first canal. In the same year Clowes was invited to inspect the building of the Leeds and Liverpool Company's summit tunnel at Foulridge, the company suspecting that Whitworth had committed errors in its construction. Clowes indicated that a tunnel shorter by 600 or 700 yd. could have been dug but concluded that it would be more sensible to complete Whitworth's line as planned. In the same year, he had to turn down, due to the demands of other work, a request from the Worcester–Birmingham Canal to survey its line.

In January 1792 the blandishments of the Worcester–Birmingham Canal Company eventually resulted in Clowes accepting the position of consultant to the company for 29 guineas a day. He also came to be employed to survey and engineer the Stratford and Dudley Number 2 Canals. The Dudley was tied to the development of a more direct route from Birmingham and the Black Country coalfields to the Severn at Worcester, needed to circumvent the heavy tolls of the BCN. On the Stratford Canal Clowes engineered the Brandwood Tunnel and on the Dudley Number 2 Canal, tunnels again dominated. The entire Dudley Number 2 line was only 13 miles long but over 5 miles of it were in tunnel, at Gosty Hill and Lappal; at the latter, a 3,795 yd. tunnel was constructed under the California coal mines. Clowes did not live to see the line completed and, with Thomas Green, the aptly named William Underhill (q.v.) took over the enterprise after Clowes's death.

Between 1792 and 1794 Clowes was working for six canal companies. In 1792 he surveyed the line of the Gloucester–Berkeley Canal and estimated its building cost at £102,108; later built according to the survey by Robert Mylne from 1793 it was a lateral canal from Sharpness to Gloucester designed to obviate the problems in navigating the treacherous waters of the Severn. Clowes continued to work as surveyor and engineer to the Hereford and Gloucester Canal whilst undertaking the survey and engineering of the Shrewsbury (tub boat) Canal. On this canal he tunnelled again at Berwick, 970 yd., the first tunnel of any length to be built with a wooden cantilevered towpath, an addition suggested by William Reynolds (q.v.), the iron master. His masonry aqueduct at London on Tern was destroyed by floods in 1795 and replaced by Telford's famous design in cast iron. In 1794 Clowes returned again to the Stroudwater canal to report on improvements to be made to the canal's entrance with the Severn at Framilode.

Clowes died at Middlewich during late December 1794 and he was interred at Norton-le-Moors on 1 January 1795. A significant canal engineer he was arguably the first great tunnel engineer, responsible for three of the four longest tunnels in the country, and probably consulted on the longest, that at Standedge. He died a relatively wealthy man with shares in excess of a thousand pounds in many of the companies he was working for. He also owned several properties in Cheshire and Staffordshire. These he left principally to his second wife, Margaret, who survived him by only a few months; as there were no children his nephew, William Clowes, inherited his estate, and developed his Newhall pottery business, founding a pottery at Longport with Henshall and Robert Williamson, the husband of Anne Brindley, James's widow. Clowes's great niece Anne married their son Hugh Henshall Williamson, linking the Henshall, Brindley and Clowes families who were brought together by the world of pottery and canals.

It is not known whether there was any family relationship with Samuel Clowes and his sons John and James who worked as contractors in North America in the 1820s, initially on the Erie Canal and subsequently on the Welland Canals.

CHRISTOPHER LEWIS

[Norton-in-the-Moors, Parish Records. *Pub Staffs Parish Record Soc (SPRS) 1574–1751*, vol. 1 (1924), 1754–1837, vol. 2 (1942–1943) (these transcripts are incomplete); Norton-in-the-Moors Parish Records for record of deaths after 1794 (the registers are in the church, marriages are in a separate volume); Burslem Parish Records, *Pub SPRS 1578–1812*, vols. 1–3, 1913; Stoke-upon-Trent Parish Records, *Pub SPRS 1629/1812*, vols. 1–3, 1914–1927; Wolstanton Parish Records, *Pub SRPS 1624–1812*, 2 vols., 1914–1920; Josiah Clowes' will, Cheshire RO, proved 1795, ref. WS 1795; William Clowes' will, Lichfield Joint Record Office, proved 1823, with codicil; Land Tax Assessment, Middlewich. Cheshire RO, ref. QDV 2/288; *Aris Birmingham Gazette*, 12 June 1775, advertisement for labourers on the Trent and Mersey Canal; document concerned with the transfer of property by 'William Clowes of Porthill in the parish of Wolstanton, potter, nephew and devisee, named and appointed in the will of Josiah Clowes, late of Middlewich in the County of Chester, deceased', The Adams Collection, City Central Library, Bethesda Street, Hanley, Stoke, doc. ref. EMT8/796; report of Clowes's death, *Gloucester Journal*, 9 February 1795; minute books of the Chester Canal Company, 1776–1780, PRO RAIL; minute books of the Huddersfield Canal Company, entries from 26 June 1794 until 24 June 1813, no mention of Josiah Clowes; State of Accounts of the Herefordshire and Gloucestershire Canal Company, 7 November 1796, shows Clowes held four shares, £520 fully paid up, report of Company to General Assembly of Hereford and Gloucester, 10 November 1796, shows Clowes' report and estimate was not considered altogether satisfactory; C. Weaver, *Dudley Canal: Old Tunnel* (notes prepared by Railway and Canal Historical Society); RAIL PRO, letter by Smeaton, 22 July 1791, re. Worcester and Birmingham Canal; reports and correspondence between Josiah Clowes and the Thames and Severn Company, June to July, 1789, ref. TS 194/4, Shire Hall, Gloucester; rate books, Castle Ward, Cirencester, 1788/9; Shire Hall, Gloucester; Thames Navigation Minute books, 1 July 1789–14 July 1790, Berkshire CRO; Stroudwater Canal minute book, April/May 1794; C. Hadfield, notes on individual engineers, London School of Economics; R. D. Field (n.d.) *The Grand Scheme ...*, printed by W. Langsbury, Cheltenham; J. Priestley (1830) *Historical Accounts of the Navigable Rivers, Canals and Railways throughout Great Britain*; S. Smiles (1861) *Lives of the Engineers*, **1**; L. T. C. Rolt (1958) *The Life of Thomas Telford*; C. Hadfield and Norris (1962) *Canals to Stratford*; C. Hadfield (1966) *Canals of the West Midlands*, Greenslade and Jenkins (eds.) (1967) *Victoria History of the County of Stafford*, **2**; C. Hadfield (1967) *Canals of South Wales and the Border*; W. E. Duncan Young (1968–1969) *The Thames and Severn Canal*; H. Household (1969) *The Thames and Severn Canal*; L. T. C. Rolt (1969) *Navigable Waterways*; C. Hadfield and G. Biddle (1970) *Canals of North West England*; D. Holgate (1971) *Newhall and its Imitators*; R. Russell (1971) *The Lost Canals of England*; A. W. Skempton and H. C. Wright (1971–1972) Early Members of the Smeatonian Society, *Trans Newcomen Soc*, **XLIV**; H. Malet (1977) *Bridgewater, the Canal Duke*

1736/1803; C. G. Lewis (1978) Josiah Clowes, 2 parts, *Waterways News*, April–May, 79–80; C. G. Lewis (1978) Josiah Clowes, *Trans Newcomen Soc*, **50**, 1978–1979; C. G. Lewis, (1980) Summit problem. Sapperton Tunnel, *Waterways News*, May, 100]

Publications

n.d. *A Plan of the Canals to be made …*

1791. *Plan of the Intended Navigable Canal from Hereford to Gloucester*

Works

1777. Trent and Mersey Canal, contractor/other work not known, total cost £300,000

1778. Chester Canal Locks at Beeston, Clowes's annual salary £200

1778. Stroudwater Canal, individual lock repairs

1783–1789. Thames and Severn Canal, Engineer, £220,000

1789–1790. Building of locks on Upper Thames Navigation

1789–1792. Dudley Tunnel and Canal, Parkhead–Dudley Junction, Engineer, £50,000

1791–1795. Herefordshire and Gloucestershire Canal, Engineer, Gloucester–Ledbury, £106,500, completed 1845 to Hereford

1791 1795. Worcester and Birmingham Canal, Surveyor and Consultant, completed 1815, William Cartwright Engineer, £610,000

1792. Survey of Gloucester/Berkley Canal, estimated cost £102,108, started by Robert Mylne to new survey

1792–1795. Shrewsbury Canal, Engineer £65,000–70,000, completed by Thomas Telford in 1796

1792–1795. Dudley Number 2 Canal (Blower's Green Junction–Selly Oak/Worcester–Birmingham), Engineer, £140,000, completed 1798

1792–1795. Stratford Canal (Kings Norton–Kingswood), Engineer, completed 1816, £153,771

COATES, Richard (fl. 1792–1822), civil engineer and timber merchant, is known through his work on river navigations in East Anglia. When John Rennie replaced William Jessop (q.v.) as Engineer on the Ipswich–Stowmarket Navigation in December 1791, Coates was appointed surveyor at a salary of £200 p.a., as somebody with knowledge of bricklaying and masonry work. He left in October 1793 with a £50 gratuity. In the eighteenth century several schemes were considered for improving the navigation on the Chelmer were generally opposed by Maldon which felt that its position as a port was under threat. Finally, in 1792, with John Rennie (q.v.) as engineer, surveys were carried out by Charles Wedge, and in the following year by Matthew Hall. The navigation bypassed Maldon and linked Chelmsford to Collier's Reach. An Act was obtained and work began in July 1793 near Chelmsford, with Coates acting as resident engineer, on £240 p.a.

Rennie had almost certainly met Coates as a result of being asked to report on the Ipswich and Stowmarket navigation in late 1791—another river improvement scheme—in this case to the Gipping. William Jessop (q.v.) had been the engineer but problems had occurred with the contractors and Rennie discovered that the original survey was defective and so in 1792 recommended various improvements; Coates was involved with these works.

Works on the Chelmer included a 2½ mile cut between Beeleigh and Heybridge basin, a cut to a terminal basin at Chelmsford and thirteen locks, including a sea lock at Heybridge. Work was finally completed in 1797 but at the end of the year problems were already evident due to shoaling and flooding. In 1799 Rennie was called in to advise and thereupon acknowledged faults in the original design.

Coates is not known of in connection with any further civil engineering works, and extant correspondence with Rennie only covers the years 1793–1799. He became a successful merchant in Chelmsford, dealing in coal and timber. On his death on 12 April 1822 he left £17,750. His business was continued by his brother George, a masonry contractor on the Stowmarket and Chelmer Navigations. It is not known if he was related to the 'Cotes' mentioned in connection with the Oxford Canal.

MIKE CHRIMES and P. S. M. CROSS-RUDKIN

[EM 400/2 and 400/11, Suffolk County RO; D 1236, C/2/9/1 and C/4/4/3, Essex County RO; Rennie mss, National Library of Scotland; Rennie reports, ICE Library; J. Boyes and R. Russell (1977) *The Canals of Eastern England*, 66]

Works

1792–1797. Ipswich and Stowmarket Navigation, Surveyor, 17 miles, £35,000

1793–1797. Chelmer and Blackwater Navigation, resident engineer, 14 miles, £50,000

COLBY, Thomas Frederick, FRS, FRSE, MRIA, Major-General (1784–1852), military surveyor, was born in Rochester, Kent, into an established Pembrokeshire family on his father's side and a military family on his mother's. He was educated at Northfleet School and the Royal Military Academy at Woolwich, in 1801 receiving his commission as a Second Lieutenant in the corps of Royal Engineers and then becoming assistant to Major William Mudge, Superintendent of the Trigonometrical Survey. A promising career was nearly ended in December 1803, aged nineteen, Colby lost his left hand and sustained a serious head injury when a pistol exploded; part of the pistol was permanently lodged in his skull, causing lifelong difficulties and ultimately his death. Recovering sufficiently to resume his duties, Colby participated in survey work in Wales, Cheshire and Yorkshire, spending intervals at the ordnance map office in the Tower of London supervising the publication of maps.

In 1809, Mudge was appointed lieutenant governor of the Royal Military Academy Woolwich and Colby became chief executive officer of the survey. In 1813 he was charged with the principal triangulation and the measurement of the arc of the meridian between the Shetland Islands and the Isle of Wight. On the death of Mudge in 1820 and on the recommendation of Sir Joseph Banks, Colby, by then a Major, became Superintendent of the Trigonometrical Survey. The House of Commons order for a Survey of Ireland was to provide the great work of Colby's life, commencing in 1824. He moved to Dublin in 1828, supervising personally for the next ten years the massive effort required to produce accurate detailed maps. Dissatisfied with existing methods, Colby invented a superior system—known as Colby compensation bars and subsequently used throughout the world—for correcting base measurements to allow for actual conditions at the time of measurement. By 1846 he had been promoted to Major-General, the last sheets of the Survey of Ireland being by this time nearly complete.

In 1828 Colby married Elizabeth Hester Boyd of Londonderry, and had four sons and three daughters. He was an original Fellow of the Astronomical Society from its foundation in 1820, a Member of the Board of Longitude, a Fellow of the Royal Societies of London and Edinburgh, of the Geographical, Geological and Statistical Societies of London, Member of the Royal Irish Academy, LLD of the University of Aberdeen and a Knight of Denmark. He was a member of the Smeatonian Society from 1822, in the second class of honorary members. He joined the Institution of Civil Engineers in 1820, proposed by Thomas Telford (q.v.), and was transferred to the class of Honorary Member; based in London until commencing the Irish Survey, Colby frequently attended meetings and presented Ordnance maps and documents to the library.

After retirement from the Survey in 1846, Colby devoted himself to the education of his sons and lived for some time in Bonn and in Belgium. He died on 2 October 1852 in New Brighton, Wirral, and is buried in St. James's Cemetery, Liverpool. A pencil portrait of him exists; it is by Sir William Brockedon and is dated 1837.

TESS CANFIELD

[Membership records, Institution of Civil Engineers; W. Yolland (1847) *An Account of the Measurement of the Lough Foyle Base, with its Verification and Extension by Triangulation*, London; obituary (1853) *Annual Report Royal Astronomical Society 1852–1853*; obituary (1853) *Min Proc ICE*, **XII**, 132–137; DNB; J. E. Portlock (1855) Memoir of the late Major-General Colby, *Papers … of the Royal Corps of Engineers*, **iii, iv, v** (reprinted 1869); C. Close (1969) *The Early Years of the Ordnance Survey*; a lost portrait formerly at Southampton is reproduced in Close; T. Owen and E. Pilbeam (1992) *Ordnance Survey: Map Makers to Britain Since 1791*]

Publications

1811. An account of the Trigonometrical Survey, extending over the years 1800–1809, with W. Mudge in *Trigonometrical Survey of England*

1854. *Letters in Ordnance Maps*, correspondence respecting the scale for the Ordnance Survey and upon contouring and hill delineation, London

COLLINGE, Charles (1792–1842), engineer, of London, was engaged in mechanical pursuits from an early age. His father was John Collinge (q.v.), also an engineer.

A founder of the Institution of Civil Engineers, Collinge became a Member on 2 January 1818 at the inaugural meeting at the Kendal Coffee House in Fleet Street and on 25 June 1820 he moved the motion that requested the Secretary of the Institution to prepare an invitation to Thomas Telford (q.v.) to become its President. He was an active member, frequently taking the chair at meetings, but he resigned in 1822, then informing the Council that he could no longer remain a member.

In March 1820, Collinge proposed a design for a conical wooden roof of 66 ft. diameter, with iron connectors, a drawing of which survives in the ICE Archives. He held patents on axletrees (1833) and on spherical hinges. Under the title of Collinge and Mainwaring, he occupied a site developed for letting by George Smart (q.v.) from before June 1824 and, later, the manufacture of his hinges was continued by Arthur Collinge and by John Imray (1820–1902), who took over from Chas Collinge and Co., at 65 Westminster Bridge Road in about 1860. In July 1835, Charles Collinge gave evidence for John and Samuel Seaward (qq.v.) in a case concerning a paddlewheel patent.

He was re-elected a Member of the ICE in 1833 when he took up residence at Westminster Bridge Road and joined the firm of Maudslay, Son and Field. He was one of the five founding members who signed the ICE's newly instituted Register of Members in 1840.

TESS CANFIELD

[Membership records, ICE; John G. James Collection, Archives ICE; *Description of a Circular Roof* (1820), ICE OC 8; obituary (1843) *Min Proc ICE*, **II**, 13; G. Watson (1988) *The Civils*]

COLLINGE, John (fl. 1787–1830), engineer and iron founder, of Lambeth Road, London. He was the father of Charles Collinge (q.v.) and held many patents. In his earliest, for wheelboxes and axles (November 1787), he describes himself as 'cabinet maker'. Subsequent patents include further wheel boxes and axles (1792 and 1811), bearings for sugar mills (1794), ball and socket hinges (1821), door springs (1825) and ships rudders (1830).

He left Bridge Road, Lambeth, for Buxton Place, but by 1811 had returned to (Westminster) Bridge Road, initially as 'axletree manufacturer', but by 1821 he referred to himself as 'engineer'. In 1804, he erected a second-hand pumping engine for the Rotherhithe basin construction work for the Grand Surrey Canal Company.

As an experienced maker of wrought iron boilers, Collinge gave evidence in the 1817 inquiry on steamboat explosions. By 1824, in which year he became a member of the Institution of Civil Engineers, the firm of Messrs. Collinge and Mainwaring had rented premises from George Smart (q.v.). At that time he had just supplied lock gates and machinery for the entrance basin at the Custom House Dock, Dublin, to designs of Thomas Telford. Collinge's tender proved much below cost, and he spent a considerable effort in attempting to recover his costs.

Collinge's date of death is not known.

TESS CANFIELD

[Membership records, ICE; John G. James Collection, Archives ICE; Telford mss, ICE archives; Committee on the Highways of the Kingdom (1808) *1st Report*, 64; *Report of the Select Committee on Means of Preventing the Mischief of Explosion ... on Board Steam Boats*; *Phil Mag*, 1817, **50**, 332, 95–100 (repr. from HC 422, 17–20); letter from W. H. Dewhurst (1831) *Mech Mag*, **14**, 392–394; G. Dodd (1876) *Dictionary of Manufacture Mining Machinery and the Industrial Arts*, 55–68]

COLLIS, William (fl. 1830), road surveyor, of Sturry, was surveyor for several turnpike trusts in the Canterbury area, receiving a salary of £10–50 p.a. for each post. His main function appears to have been to deal with the application to the roads of the usual surface dressing of flint or gravel; he was also responsible for receiving and recording payments from relevant parishes in lieu of statute labour. He is typical of the many road surveyors active in the late eighteenth and early nineteenth centuries.

MIKE CHRIMES

[F. H. Panton (1985) Turnpike roads in the Canterbury area, *Archaeologia Cantiana*, 102, 171–191]

Works

1833. Canterbury and Sandwich Turnpike, 11½ miles, surveyor

1833. Canterbury and Barnham Turnpike (Act 1791), 7½ miles, surveyor

1833. Whitstable Turnpike (Act 1736), 5.2 miles, surveyor

1833. Canterbury–Ramsgate Turnpike, first district (Act 1802), surveyor

1833. Herne Bay Turnpike (Act 1814), 5½ miles, surveyor

1840. Canterbury–Ramsgate Turnpike, first district (Act 1802), clerk and surveyor

COLTHURST, Edmund (*c*. 1545–1616), originator and Overseer of the New River water supply to London, was the eldest son of Matthew Colthurst, MP for Bath, a considerable landowner in Kent and Somerset and treasurer of ordnance at Boulogne in 1544. In the 1540s, after the Dissolution, he acquired the site of Bath Priory and the Barton, its home farm estate. On his death in July 1559 Edmund, then a boy of about fifteen, inherited the Bath property. In 1572 he gave the Abbey Church, still only a shell, to the Corporation of Bath, but retained Abbey House as his residence, and in 1591 sold Barton Grange. In the late 1590s he saw military service in Ireland where he defended a castle in County Wexford; thereafter he is often referred to as Captain Colthurst.

How he became interested in London's water supply is not known. It was certainly an important matter. The population quadrupled during the sixteenth century to about 200,000. The conduit system had been extended, Peter Morris (q.v.) in 1582 and Bevis Bulmer in 1594 provided pumped supplies from the Thames and in 1590 the Hampstead Ponds reservoirs were opened. But more was needed. Colthurst's scheme involved bringing water to the City by an open channel from springs in Hertfordshire along a winding course almost on a level and nearly 40 miles in length. Although the first accurate survey of the route was made by Edward Wright (q.v.) in 1609, Colthurst must have been able to show that there was a small but adequate fall between the springs and the proposed terminus in Islington, north London.

His proposal was referred to the City authorities and the sheriffs of Hertfordshire and Middlesex. After local enquiries they approved the scheme in principle and in 1604 James I granted Colthurst a licence by Letters Patent to cut a channel or 'river' for the purpose. Places mentioned on the line (springs 'toward Hertford', Broxbourne, Cheshunt, Theobalds and Edmonton) indicate that his route was essentially that followed by the New River. Moreover he actually began work. Early in 1605 he had brought the channel forward 3 miles at a total expenditure of about £700, before running into financial problems.

With backing from the Privy Council he tried to persuade the City to contribute to the expenses, without success. However they decided to submit a Bill to Parliament for powers to carry out either his scheme or two others for taking water from the rivers Lea or Colne. After the Bill had its second reading in 1606 it went to Committee. Colthurst was invited to attend. On the committee as MPs were Hugh Myddelton (q.v.) and several others who later became Adventurers in the New River undertaking. The scheme as it emerged from this committee was much on the lines of Colthurst's Patent. On its third reading, the Bill was passed, and the City representative promised that such recompense as the Lord Chancellor should set down would be paid to Captain Colthurst 'for his travail and pains in the work'. The Bill soon passed into law as an Act enabling the Lord Mayor and citizens of London to bring a stream of fresh water in a channel not more than 10 ft. wide from springs at Chadwell and Amwell in Hertfordshire to the north parts of the City, which 'upon view is found very feasible'. No mention is made of the rivers Lea and Colne.

Financial aspects still had to be settled and delays ensued when the City, advised by the engineer William Ingelbert, submitted an alternative proposal for a brick tunnel, theoretically a good idea but immensely more expensive than an open channel. Nothing came of this apart from a fruitless second Act in 1607. Then, in October 1608, Colthurst petitioned the City stating that he, his friends and partners were ready to construct the river at their own cost under his Patent of 1604. After further thought the partners considered it would be better to have the City's power transferred to themselves. To this the City agreed in March 1609 and constituted Hugh Myddleton, 'citizen and goldsmith of London', as their lawful deputy to exercise their statutory powers under the 1606 Act. There is an indication that the estimated cost of the undertaking at this stage was £3,600. If so this turned out to be optimistic; the expenditure up to completion in 1614 amounted to over £18,000.

It is probable that Myddleton was one of the 'friends and partners' mentioned in October 1608 and that the City preferred to convey the powers of their Act to a relatively wealthy man well known to them rather than to Colthurst, who they knew had not the means to undertake the work. Myddleton took on managerial control of the project but Colthurst was in no way forgotten. He became overseer or resident engineer when work began in May 1609, at a salary of 14s per week, and in 1612 as his 'recompense' Myddleton arranged that a life-interest in four of the thirty-six Avdenturers' shares in the New River should be assigned free of charge to 'Edmund Colthurst of the Cittie of Bath, gent. In consideracon of the great labour and endeavour by him bestowed about the worke'. Colthurst 'the overseer, this try'd Man, an Ancient Soldier and Artizan', is the first to be mentioned at the ceremony on opening of the New River in September 1613. He continued to receive his salary during the next year while the water mains were being laid in the streets, and on to August 1616 when he probably died. His will has not been found, but by the 1612 arrangement his heirs were to have two of the New River shares.

A. W. SKEMPTON

[H. W. Dickinson (1954) *Water Supply of Greater London*; G. C. Berry (1956) Sir Hugh Myddleton and the New River, *Trans Hon Soc Cymmrodorion*, 17–46; J. W. Gough (1964) *Sir Hugh Myddleton, Entrepreneur and Engineer*; Roger Virgoe (1982) M. Colthurst, *The House of Commons 1509–1558*, **1**, 679–680; A. J. Keevil (1996) The Barton of Bath, *Bath History*, **6**, 25–53]

Works

1609–1614. The New River, Overseer, £18,520

COLVIN, John, Colonel (1794–1871), 'Father of irrigation in Northern India', was born on 20 August 1794 in Glasgow, the son of Thomas Colvin, a merchant. He attended Addiscombe

College as a cadet in 1809, the inaugural year of the East India Company's College for training its own officers, and subsequently served in the Bengal Engineers from 1810 until 1839, retiring in 1839 with the rank of lieutenant colonel. He was awarded the CB in 1838 and made honorary colonel on 28 November 1854. In 1838 he married Josephine Baker, the sister of his friend and fellow Bengal Engineer (Sir) William Erskine Baker.

Colvin's early duties in India were surveying work, for a lighthouse at Palmyras Point in 1812 and at East Sagar Island in 1813 as an assistant to G. R. Blane (q.v.). From 1817 to 1827 he was Garrison Engineer at Hansi under Richard Tickell (d. 1855), but was called away from this in 1818 to the Mahrata (Pindari) War, where his conduct at the siege of Mundela was mentioned in despatches. He also surveyed the route for Brigadier Arnold's force on the Bhuttee frontier. He returned to military action again in the siege of Bhurtpur near Agra (1825–1826), but it was his irrigation work which mainly brought him recognition.

In 1820 he was appointed to survey and reconstruct the oldest of the Jumna (Yamuna) Canals, the fourteenth century Feroze's Canal on the Western Jumna, 240 miles in length. The first stage of work was completed in 1825 but work on both this and the Delhi Canal, which branched from it, continued into the 1840s. In 1827 he was appointed Superintendent of Canals, Delhi Territory, in succession to Tickell, a post he retained until the end of 1836 when he went on furlough to Europe, studying canals in Italy. At the end of his furlough he retired.

Colvin's career made him an experienced irrigation engineer, whose benign personality earned him the description ' the father of irrigation in Northern India'. The Canal works were subject to constant attack from seasonal torrents crossing their path and Colvin was forced to develop methods of overcoming these difficulties. One innovation was his adaptation of the native use of wells in foundations, whereby he used large blocks of masonry, typically 10–15 ft. long and 4 ft. wide, undersunk in the river bed to a depth of 10 ft. and pierced with wells. At Dadupur, the lower head of the canal, Colvin built a large dam 777 ft. wide with sixty openings of 10 ft., connected by a revetment to a bridge across the canal. Its difficult position in a knot of rivers involved foundations of masonry blocks sunk 8–11 ft. and connected front and rear by lines of curtain walls. Both it and the bridge on the canal could be closed with sleeper planks to control water entering the canal. This dam, built in 1830, was one of his main achievements on the canal.

Colvin was interested from the 1830s in providing a canal in the doab between the Jumna and the Ganges and this was finally to see fruition with his encouragement, with the building of the Ganges Canal by Sir Proby Cautley. It was opened in 1854 and was to be one of the first great irrigation works of British India.

Colvin died at Leintwardine, Herefordshire, on 27 April 1871.

JOYCE BROWN

[P. T. Cautley (1839) On the use of wells etc. in foundations, *Journal of the Asiatic Society of Bengal*, **8**, 331; W. Baker (1849) *Memoranda on the Western Jumna canals in the North Western Provinces of the Bengal Presidency*; *Times*, 29 April 1871; T. Thackeray (1900) Biographical notes of officers of the Royal (Bengal) Engineers, 33–35; V. C. P. Hodson (1922) *List of the Officers of the Bengal Army, 1758–1834*, **I**, 368–369; E. W. C. Sandes (1935) *The Military Engineer in India*, **I**, 2615, **II**, 4–6; E. W. C. Sandes (1948) *The Indian Sappers and Miners*, 100–103; R. H. Phillimore (1950) *Historical Records of the Survey of India*, **II**, 391; J. Brown (1978) Sir Proby Cautley (1802–1871), a pioneer of Indian irrigation, *History of Technology*, **3**, 43–49]

Works

1820–1827. Western Jumna Canal (Feroze's Canal), 240 miles
1830. Dadupur Dam

Publications

1833. On the restoration of the ancient canals in the Delhi territory, *Journal of the Asiatic Society of Bengal*, **2**, 105–127

COOPER, John (*c.* 1744–1792) and his brother **James Cooper** (d. 1801) were sons of John Cooper, millwright of Poplar. They became leading London master millwrights, particularly well known for their association with the newly invented Boulton and Watt rotative steam engines. Seventeen of these engines were installed between 1784 (the first in London at Goodwyn's brewery) and 1799 in breweries, foundries, lead mills, etc., for which one or other of the brothers designed and executed the millwork.

John Cooper served an apprenticeship with Blagrave Gregory 1758–1765. As 'John Cooper, Engineer', he joined the Society of Arts in 1768, on the proposal of Thomas Yeoman (q.v.). He took his first apprentice in 1770, to be followed by four others in the next twelve years. He was admitted a member of the Society of Civil Engineers in 1776, attending meetings with regularity up to a year before his death in June 1792.

James Cooper also joined the Society, in 1780, but seems to have left soon after 1786. A patent for a new form of watermill was granted in 1787 to 'James Cooper of Poplar, Engineer'. Asked by a client to recommend someone to act in addition or conjunction with Cooper in the engineering and mill part of a project, John Smeaton (q.v.) in 1792 suggested John Rennie (q.v.) 'Engineer and millwright on Surrey Street over Blackfriars'. The status of expert industrial millwrights as what would later be called mechanical engineers is clear.

The firm 'John and James Cooper, millwrights, Poplar' is recorded in London directories to 1792 and afterwards as 'James Cooper, millwright, Poplar' until his death in April 1801. James lived there; John's address from 1774 was Portpool Lane, Holborn. John became a Liveryman of the City of London. He owned land and houses at Danbury, Essex, and property in Holborn, the rents from which were to be paid to his widow, Lucy (she died in 1807). James, a widower, left considerable holdings in government stocks to members of his family. The business appears to have been continued by his son John, but the 'extensive millwrights manufactory, large yard and premises' and an adjoining 'spacious residence' were sold in 1805. Thomas Cooper, millwright of Old Street, Finsbury (fl. 1794–1816) was probably a relative.

A. W. SKEMPTON

[Records of the Worshipful Company of Turners, Guildhall Library; Minute Books of the Society of Arts; correspondence between the Cooper brothers and Boulton and Watt, Birmingham Reference Library; John Smeaton's letters, ICE Library; advertisement of auction sale of premises in Poplar (1805) *Morning Chronicle*, June; Peter Mathias (1959) *The Brewing Industry in England 1700–1830*; Jennifer Tann (1970) *The Development of the Factory*; A. W. Skempton and Esther Wright (1971) Early members of the Smeatonian Society of Civil Engineers, *Trans Newcomen Soc*, **44**, 23–42]

COOPER, Robert (fl. 1757–d. 1766), civil engineer, in 1757 surveyed Rye New Harbour and made an estimate of the works necessary to complete it; in the same year he was appointed to direct the works at a salary of £50 p.a. with a rent free house. Operations at Rye had started in 1725 to plans by John Perry (q.v.). He left in 1730, to be succeeded by Edward Rubie (q.v.). By 1743, after a long interruption due to lack of funds, a canal had been dug about a mile in length from Winchelsea Channel to the sea with piers at the entrance and a great sluice rather more than half way along the canal towards the sea. Wharfing had been built, and the canal deepened to its full depth of 20 ft. below high water spring tides, for a length of 830 ft. from the piers. After another interruption to allow funds to accumulate, during which Rubie left in 1748, Stephen West (q.v.) reported in October 1756. His report was approved by Justly Waston (q.v.) and Edward Collingwood, Master Attendant at Deptford Dockyard, and work could now proceed.

The canal required to be deepened to 20 ft. throughout, a jetty 200 ft. long was needed as an extension to the west pier at the entrance, and finally the shingle bank left as a cofferdam between the piers had to be removed. These works were carried out by Cooper at an expenditure of £12,000 and the harbour was opened in 1762. The total cost of works since 1725 amounted to £42,000.

In December 1762 the Commissioners petitioned Parliament for powers to raise extra duties on shipping to cover the cost of works to enlarge Winchelsea Channel. Cooper gave evidence and John Smeaton (q.v.) reported in February 1763. But nothing happened until, after a new petition, an Act was passed in 1764. Even then delays ensued. Cooper again gave evidence to the House of Commons in February 1765 saying that he had accompanied Smeaton in his survey, that in his opinion the method proposed for completing the project 'is the most eligible and that it cannot be done in a more proper Manner', and he explained the proposed works in detail.

Work did indeed begin soon afterwards, but now under William Green, Cooper having been discharged with a years salary. He was consulted on Shoreham harbour, but before he could report he died in June 1766 and was buried at Rye Parish Church, leaving his widow, Judith.

A. W. SKEMPTON

[Rye Harbour Commissioners Minute Book (1724–1766) East Sussex Record Office; Robert Cooper (1762, 1765) Evidence, *JHC*, **29**, 404, 783]

Works

1757–1765. Rye New Harbour, further works, £12,000

COTTAM, George (fl. 1818–1857), ironmaster, was in partnership with George Hallen (probably a member of the West Midlands family of ironmasters), the proprietor of an iron foundry, seemingly founded about 1818 with premises in Southwark; the firm was still trading as Cottam and Williams in 1890. It produced all kinds of ornamental ironwork, perhaps most famously the gates for the South Transept of the Great Exhibition building. Cottam was elected an Associate of the Institution of Civil Engineers in 1825 and remained in membership until 1852, when he presumably retired. He was a regular attender at meetings, contributing to discussions on topics relating to iron, steam engines and fireproof buildings.

Although many of the firm's activities were involved in the decorative side of ironwork, they did produce structural ironwork. William Turnbull, who in 1832 produced an early *Treatise on the Strength of Cast Iron Beams and Columns*, had carried out calculations for Cottam, and Cottam himself was concerned that beams in buildings were being overloaded and recommended the more extensive use of wrought iron. The firm was involved in the export of iron buildings, including lighthouses, to the colonies, possibly encouraged by Ambrose Hallen (q.v.), who went to Sydney on the recommendation of Thomas Telford (q.v.), and his brother Edward, who returned to the firm in England. This connection was no doubt responsible for Cottam's brief note on the dimensions of some of the indigenous trees of New South Wales and van Diemans Land, presented to the Institution of Civil Engineers in 1833.

Cottam's sons, George Hallen and Edward, also joined the firm, Edward later going into business with Francis Robinson of Bramah and Robinson fame. The family took out a number of patents, mostly relating to ironware.

TESS CANFIELD

[Membership records, ICE archives; Minutes of conversations, ICE archives; J. G. James collection, ICE archives; On hot water systems (1828) *Gardeners Magazine*, 19 February; W. Turnbull (1831) *A Treatise on the Strength, Flexure and Stiffness of Cast Iron Beams and Columns*; W. Turnbull (1832) *A Treatise on the Strength and Dimensions of Cast Iron Beams*; W. Turnbull (1833) *An Essay on the Construction of the Five Orders of Architectural sections of Cast Iron Beams*; On testing iron (1843) *Min Proc ICE*, **2**, 128; On steam vessels (1844) *Min Proc, ICE*, **3**, 86; On Stirling's engine (1845) *Min Proc ICE*, **4**, 356; On dimensions of beams (1845) *Min Proc ICE*, **4**, 348; On fireproof buildings (1849) *Min Proc ICE*, **8**, 146–163; Obituary of E. Cottam (1905) *Min Proc ICE*, **162**, 424; G. Herbert (1978) *Pioneers of Prefabrication*]

COUCHMAN, Henry, Sr. (1738–1803), architect, was the eldest of five children of Henry Couchman, a carpenter of Ightham, in Kent, and his wife Sarah Luck. His grandfather Luck paid for his education and he was sent to a dame school at Borough Green when five years old, remaining there for about four years before going to day school at Wrotham until the age of fourteen. His grandfather intended that he become a surgeon or a shopkeeper and although he had learnt carpentry from his father, he obtained a succession of jobs in shops in Kent and London before becoming so ill that he was unable to work for more than a year. Recovered sufficiently, he worked as a foreman for a Mr. Dixon on building works where the architect was Matthew Brettingham. After some initial disappointments, Brettingham in 1766 employed him as Clerk of Works at Packington Hall, in Warwickshire, where he stayed for ten years. The client there was the Earl of Aylesford and Couchman soon developed a good working relationship with him, as a result of which he obtained a lease of Temple House, Temple Balsall and an appointment as Bailiff to the Trustees of Balsall Hospital. Lord Aylesford was also prominent at Quarter Sessions and it was presumably through his influence that Couchman was appointed Bridgemaster of Warwickshire in 1790, following the death of Francis Hiorne. He retained the office until his own death at Temple Balsall on 21 January 1803.

Couchman was consulted occasionally by canal companies in north Warwickshire. In 1786 he gave advice to the Oxford Canal Company about the best mode of repairing Brinklow

Aqueduct and in the later 1790s, as a member of committee of the Warwick & Birmingham and Warwick & Napton Canal Companies, he undertook some supervision of construction while their Engineer, William Felkin's, conduct was under scrutiny. He also provided a design for the centres of the aqueduct over the river Avon at Warwick.

P. S. M. CROSS-RUDKIN

[Manuscript autobiography, to 1776, in Warwickshire RO; PRO RAIL 855 & 881; *Coventry Mercury*, 24 Jan. 1803; Colvin]

Works

1786–1787. Ryton Bridge, Coventry, designed and built for £700
1792–1795. Barford Bridge, designed and built for £2,341
1800–1801. Wixford Bridge, £1,521

COUCHMAN, Henry, Jr. (1771–1838), architect, was the son of Henry Couchman Sr., noted above, and was baptised at Great Packington, Warwickshire, on 15 July 1771. He married Elizabeth Short at Solihull on 16 January 1809 and they had nine children over the next twenty-two years.

He succeeded his father at Balsall Hospital, as Chief Constable of Hemlingford Hundred in Warwickshire and as County Bridgemaster. During his tenure of this last position, the average annual expenditure on County bridges was about £1,300 p.a., almost twice as much as it had been during his father's time, though remaining constant at about 7% of the County total budget. In 1813 he succeeded in obtaining an increase in his salary from £40 p.a., which it had been since Francis Hiorne assumed office in 1781, to £60 p.a., but when he resigned in the following year his successor, John Docker, received only the lower figure.

P. S. M. CROSS-RUDKIN

[Extract from will at Warwickshire RO; Colvin]

Works

1807–1809. Leamington Priors Bridge, £2,900
1809–1812. Salford Bridge, Birmingham, £3,800
1812–1813. Aston Bridge, Birmingham, £1,750

COWPER, William, Captain (1774–1825), military engineer, was the son of William and Mary Cowper and was born in London on 12 April 1774. He served as an ensign with the Bombay Engineers at Cannanore in 1795 under Major John Conrad Sartorius and like many other military engineers he was seconded to civil works, following the defeat of the French and the British victory in the fourth Mysore war in 1799.

The first dry docks facilities at Bombay had been built by Lavji Nasananji between 1748 and 1766 and it was not until the early nineteenth century that the Bombay Engineers took an interest. Captain R. B. Crozier began work on a new dry dock parallel to the three earlier docks in 1805 but Cowper soon took over and was

confronted with problems of hard rock, difficulties with tides, and a lack of skilled workmen. Nevertheless, by 1808 he had completed the first new dry dock, the first vessel to be built there being the *Minden*. Cowper had completed a second dock by 1810 and these docks, known as the Duncan Docks after Jonathan Duncan, governor of Bombay, remained as built until the 1840s.

Cowper died on 27 May 1825.

MIKE CHRIMES

[E. W. C. Sandes (1933–1935) *The Military Engineer in India*, **I**, 198, **II**, 172–173]

Works

1805–1810. Bombay dry docks

COX, Lemuel (1736–1806), carpenter, inventor and bridge-builder, was an American who had a brief career as bridge designer and contractor in Ireland in the years 1789–1796. He was born in Boston or Malden, Massachusetts, in 1736, the youngest son of William Cox and Thankful Maudsley. He served an apprenticeship as a carpenter but was later described variously as millwright, inventor, architect and bridge-builder. He married Susanna Hickling at Boston in 1763 and had three sons and six daughters.

Cox's inventions for textile technology included a machine for cutting card wires; he claimed to have devised and operated equipment for the drawing of steel wire 'from the bigness of half an inch down to the size of a hair', and in 1776 he erected the first powder mill in the state of Massachusetts. He had also, however, been imprisoned as a royalist in 1775, perhaps for only a short time. His patriotism was apparently unquestioned in later years.

When a piled wooden bridge from Boston to Charlestown was projected in 1785, Cox became the 'master workman'; by one report he is named as contractor for the bridge in partnership with one Captain John Stone. The design and construction method were provided by Major Samuel Sewall, brought from Maine to do so; but credit for building the bridge was accorded unequivocally to Cox—whether as contractor or simply as salaried supervisor is not clear—and he himself claimed the 'invention' of the draw span and of the draw spans of two later bridges, each of the three 'upon different constructions whereby the conveniency of vessels are greatly expedited in passing thro'. There were seventy-five 'piers' or 'bents' of piles in the Charles River Bridge with spans between them of about 20 ft. It was opened to traffic on 17 June 1786.

To build a bridge of such length over water up to 42 ft. deep at highest tides was unprecedented and plans for other estuaries soon began. The first, in 1787, was for a crossing of the Mystic River only a few miles to the north where Cox at least designed the draw span and the lighting, but may have built the whole bridge. The second was the Essex Bridge of ninety-three piers and a draw span of 30 ft. from Salem to Beverly, where Cox was the salaried superintendent of construction from March to July 1788; after some disagreement with the proprietors he left before the bridge was finished.

Between then and June 1789 he travelled to Ireland at the request of the mayor and council of Londonderry, who had heard of the bridge at Boston. For them Cox sounded the river Foyle estuary beside the city and contracted with them to bridge it for £10,000. In spite of objections from navigation interests an Act of the Irish Parliament was passed by April 1790. Cox shipped oak piles from Sheepscot, Maine, with twenty 'skilled workmen' and the Foyle Bridge, of about sixty spans of 16½ ft. each and a draw span, was opened in 1791 at a cost of £16,594. It was 40 ft. wide and built, like all Cox's bridges of which any description survives, by the method he had learned from Sewall: each pier or bent of piles, most commonly seven piles all parallel in one plane, was assembled on shore with a capping beam, two or three parallel walings or 'girths' across both sides and diagonal braces also on both sides, the whole bent then towed to its position and upended to stand on the river bed. In general the bents sank into the bed up to 6 ft. by their own weight, but in at least one instance the bents were also hammered in by dropping repeatedly on the capping beam the free end of a heavy timber balk hinged at its other end to a supporting frame and the frame presumably mounted on a barge. The piles at Derry were up to 40 ft. long. When the pier was set, a raking pile was driven adjacent to each end of it to form cutwaters, the rake helping with the break-up of floating ice, one of the great hazards to bridges in America and which also damaged several of Cox's Irish bridges.

For a new bridge at Waterford, Cox was engaged in 1793 by the commissioners, first appointed by an Act of the Irish Parliament in 1786. Cox built for them in eight months of site work a bridge across the Suir of thirty-eight piers and 832 ft. long; it was opened in January 1794. It continued in use until 1911, the longest life of any of his bridges. Details of it were well recorded by William Friel, the resident engineer on site when it was dismantled. The setting out had been poorly controlled, the spans between piers varying randomly from a maximum of 22 ft. to a minimum of 9 ft. The width was 40 ft., including a 7 ft. footway at each side. The oak piles were made from pieces averaging 20 ft. long, connected by scarf joints held by oak treenails as well as iron bolts, submerged bolts having largely rusted away but the treenails still holding. To protect the piers a bed of small stones 'averaging 56 cubic inches in size' (by Friel's meticulous measurements!) had been dumped round the feet of all piles. Of the 4 ft. bed of stones, about 2 ft. had sunk into the mud.

Within the next three years (1794–1796) Cox built five more bridges in Ireland. One was truly an estuary crossing at Wexford, and at 1,554 ft. his longest in Ireland; the bridge at New Ross was at the entry of the Barrow to its estuary; shorter bridges were built a few miles upstream from both of these but still in tidal water, at Ferrycarrig and Mountgarret respectively; and lastly a non-tidal bridge well up the Shannon at Portumna, but still extending to 766 ft. in two sections, one each side of an island.

Although primary documents have not survived for all Cox's bridges in Ireland, it is almost certain that the piles of all were of American oak and deck beams and boards of several are stated to be of the same material. There is at least one mention of a purchase of fir which may have been native Irish or Baltic timber and was probably used for deck timbers, as it was certainly used frequently for repairs in later years.

All the bridges were authorised as toll bridges, which was necessary to raise enough capital and later to finance regular maintenance and major repairs after damage to most of them by floodwaters and ice. Whether from the drain on their funds for repairs or from shareholders' greed for the excellent dividends which some of the bridges yielded them, all remained in private hands and charged tolls up to the day of their final removal.

Lemuel Cox also earned a lot of profit, but a longstanding petition which he had submitted to the state parliament of Massachusetts in 1790, claiming recompense for his inventions and other services which, he said, had left him 'yet a poor man', evoked a response when he returned there in 1796 in the form of a grant of 1,000 acres of land in Maine. He sold the land soon after and lived for the remainder of his life in or near Boston.

Two of his sons, William and Lemuel, had been with him in Ireland. William was married in Dublin but died later in Savannah, Georgia, while Lemuel became a sailor and was lost at sea before 1799. William wrote to a Boston newspaper in 1794 to explain why his father had been accused—and acquitted—in a Dublin court of enticing tradesmen to emigrate to America, which was punishable by a twelve-month jail sentence. It is also reported by W. K. Watkins that Cox was asked 'about 1795' by the Corporation of London to take down the Monument, which was causing alarm by its vibrations but his price was too high and the Monument therefore survived.

When after his death in 1806 his will was proved, his wealth was only $20, but some thirteen years later, in consideration of the value of his house near Charlestown Bridge and debts due from others, $2,555 was distributed to his ten grandchildren.

TED RUDDOCK

[Minutes of New Ross Bridge Commissioners 1795–1868, MS 91 at NLI; Wexford Bridge, MS 11,102 at NLI; W. Priest (1802) *Travels in the United States of America ... year 1793* [to] *1797;*

W. T. Colby (1837) *Ordnance Survey of the County of Londonderry*, 2nd edn., 117–118, 130–131; S. Lewis (1837) *Topographical Dictionary of Ireland*, 2 vols.; Walter K. Watkins (1907) A Medford tax payer, Lemuel Cox, *Medford Historical Register*, **10**, 33–48, 57–64; W. Friel (*c*. 1911) Measured drawing, MS, Waterford Bridge, with annotations, at Waterford Corporation Offices; Edmund Downey (n.d., 1920s?). *Waterford's Bridges*; Peter O'Keeffe and Tom Simington (1991) *Irish Stone Bridges*, 258–268]

Works
Of Cox's works in America, only the bridge works which led up to his work in Ireland are listed.

1785–1786. Charles River Bridge from Boston to Charlestown, Massachusetts, superintendent (and contractor?), 1,503 ft. long, 42 ft. wide, seventy-five piers, draw span, cost $50,000
1787. Mystic River Bridge near Malden from Charlestown Neck to Sweetser's Point, design of draw span (possibly design and/or contractor for whole bridge). 2,420 ft. long, 32 ft. wide, draw span, cost £5,300
1788. Essex Bridge over North Ferry from Salem to Beverly, Massachusetts, superintendent of construction March–July, design of draw span, 1,484 ft. long, ninety-four piers, draw span, cost $16,000
1790–1792. River Foyle Bridge, Londonderry, design and contractor, 1,068 ft. long, 40 ft. wide, sixty-two piers, draw span, cost £16,594, partially destroyed by ice floes in 1814 and rebuilt, replaced in 1863
1793–1794. River Suir Bridge at Waterford, design and contractor, 832 ft. long, 40 ft. wide, thirty-eight piers, cost of bridge structure £20,000, replaced by concrete bridge in 1913
1794. River Barrow Bridge, New Ross (design and contractor). 358 ft. long, 40 ft. wide, 24 piers, draw span. Replaced 1869.
c. 1794. River Barrow Bridge at Mountgarret (design and contractor?), 209 ft. long, 18 ft. wide, seven piers, replaced by a concrete and steel bridge in 1929
1794. River Slaney Bridge, Ferrycarrig, design and contractor, about thirteen piers, draw span (drawing on company crest), cost *c*. £7,000, replaced in 1912
1794–1795. River Slaney Bridge, Wexford, design and contractor, 1554 ft. long, 34 ft. wide, seventy-five piers, draw span, cost £17,000, shortened in 1827–1832, replaced in *c*. 1867
1795–1796. River Shannon Bridge, Portumna, design and contactor, cost £8,000 (?), Galway side destroyed by flood in 1814, rebuilt and the whole repaired in 1818, rebuilt in 1834 with a swing span 40 ft. wide

CRAMOND, Robert (fl. *c*. 1767–1790), harbour engineer, in 1767 undertook work at Eyemouth, Scotland, where the west pier was severely damaged, and John Smeaton (q.v.) was called in for one of his first harbour works in Scotland.

Smeaton designed an additional pier, opposite the west pier, and suggested that some parts of the old pier should be re-built and others protected. He also recommended that the basal rocks on which the new pier was to be founded should be cut in such a manner that the masonry of the pier walls would slope in towards the core of the pier. However, he only prepared the outlines of this plan, arguing that it was better to leave the precise details of the pier's position and length to the contracting surveyor. Robert Cramond 'of Dunbar' was the contractor, completing the new pier in 1773; it remained virtually unchanged until the mid-1960s.

Cramond was almost certainly the person of the same name, although now said to be 'of North Sunderland', Northumberland, employed *c*. 1786 to build a 'pier', almost certainly implying a complete harbour, at what later became known as Seahouses. The present inner south pier there shows well the form of Cramond's construction, utilising large, squared blocks of red and grey sandstones, some of which were won by quarrying within the harbour itself; they appear to have been lifted into place with lifting dogs rather than Lewis bolts. He used a sloping-course technique similar to that at Eyemouth, and while some thought this 'an abuse of Gravity' no problems appear to have arisen over the next half century.

The harbour thereby created at North Sunderland was quite small, being only a little larger than its near contemporary at Beadnell, Northumberland, where Cramond was again involved. He was asked by John Wood, the landowner, to estimate for a new pier (harbour) in 1788, and thinking it the 'most eligible' situation for a harbour of most places he had seen, being in effect land-locked on all sides, Cramond recommended that the harbour walls should be constructed to stand 3 ft. above high water mark, have a 6 ft. high parapet, and be 21 ft. broad to allow carts to turn when goods were being shipped or landed. Nothing further was done until early in 1790 when Cramond produced a reduced scheme, a breakwater constructed of limestone which could later be faced with ashlar. The final arrangement was rather similar to that which Cramond devised at the same time for North Sunderland.

Wood was clearly uncertain whether to proceed with this proposal and he contacted a William Forster at Maryport about the matter. Forster sent details and sketches of piers at Maryport and Whitehaven and also criticised the manner in which Cramond was laying the stones at North Sunderland. Wood was also worried about the cost, for Cramond was charging somewhat less per linear yard of pier at North Sunderland, but he was informed that the reasons for this were 'too tedious to mention'. Cramond did, however, explain that the stone needed at North Sunderland was all to be found within the harbour, which greatly reduced the cost of the work, and that Beadnell had no such advantage. It can only be presumed that work did begin along the lines suggested by Cramond, and although there are certain similarities with his work at North Sunderland, it was certainly not finished to his specification.

Cramond was said to be possessed of spectacular strength, stopping an enraged bull with nothing but a steely stare before beating it off with a pick.

STAFFORD M. LINSLEY

[Craster papers, Northumberland County RO, colln. ZCR; Crewe Trust papers, Northumberland County RO; notes taken at Eyemouth Museum, April 1991; *Smeaton's Reports* (1812); [Warner?] (1826) A tourist, *The Border Tour*; Anon. (n.d., but purchased 1991) *Eyemouth Town Trail Guide*]

Works
1767. Eyemouth harbour, contractor
1786. North Sunderland (Seahouses) harbour, contractor
1788. Beadnell harbour, contractor

CRAVEN, Hiram (1779–1842), stonemason, civil engineering contractor and mill owner, was the eighth of twelve children of John Craven (c. 1743–1791), stonemason of Colne, Lancashire, and Jennet Whitaker (c. 1747–1785). Craven was baptised in Colne on 9 January 1780 and lived there until the age of fifteen when he was apprenticed to a mason, possibly a member of his mother's family, as the Whitakers were masons in the Keighley area to which he moved in 1795. He married Alice Calvert (c. 1778–1844) of Bingley in 1798, and made his home at Dockroyd, Oakworth. They had twelve children, several of whom joined him in business or became otherwise involved in the construction industry.

On completion of his apprenticeship Craven set up in business and came into association with Samuel Whitaker, probably his cousin, and Joseph Nowell (q.v.); although originally concentrated in the Yorkshire area the firm was to acquire national importance. Craven's earliest work was a small mill, known as Forks House Mill near the Brönte waterfall, Haworth; it was followed by work on various roads in the area, their Pennine location necessitating embankments and retaining walls. According to Whitaker the partnership was responsible for twenty to thirty bridges, the earliest attributed to Craven being the widening of Spinksburn Bridge (1809–1810). This was followed over the next decade by work at Hazlewood Bridge near Harrogate, Goose Edge Bridge (near Keighley), Wreaks Bridge (Birtsworth), Ramsgill Bridge (Pateley Bridge), Collingham Bridge, Hansworth Bridge (Cleckheaton), Pickle Hill Bridge (Bradford) and Saltersbrook Bridge. These contracts totalled c. £12,000 in value, additional to his other contracts with Joseph Nowell and Samuel Whitaker. From Whitaker and Craven's tender for Wreaks Bridge it would appear they may have worked on

the Leeds–Liverpool Canal and Hull Docks before 1811.

Their most important work was in York, notably the Ouse Bridge (c. 1810–1820), associated with works at Foss Bridge and to the Castle and City Walls. Craven made use of bales of wool to control the flow of the river during construction. This contract brought them to national prominence and led to their being consulted over London bridge. They also took out masonry contracts on the Edinburgh and Glasgow Union Canal and it was on his return from Edinburgh, possibly en route to London, in June 1821 that Whitaker fell from the roof of the London–Edinburgh coach which was speeding over Sunderland Bridge and died from his injuries. His death perhaps restricted the partnership's activities and although it was continued until 1828 by Joseph Nowell and Craven, little civil engineering work is known of, and the partnership was dissolved, presumably because both parties now wished to expand as family businesses.

There is abundant evidence that by this time Craven had diversified his business activities. Members of the family had been active in the local textile industry from the late eighteenth century. A John Craven bought Walk Mill, Keighley, in 1776 and it remained in the family, operated by John and Joseph Craven, worsted spinners and manufactures, until 1841. John acquired East Greengate Mill, Keighley, for cotton spinning in 1795 although it appears to have been given up by the family in c. 1810—after his death—and also had a share in Goose Eye Mill, Keighley.

At that time, most of the industry was still water powered, and mill buildings were relatively modest structures. With a little capital, technical knowledge and access to water a mill could readily be built. Generally, partnerships of local farmers and businessmen proliferated and with Craven's extensive family and business connections it would have been easy to get started. It is unclear what first prompted Craven's interest in the textile trade, but from around 1819 he was very active. His brother, Edward (1783–1861), was trained as an engineer and in 1808 he became partner in a cotton mill, the Ellam Carr Mill at Cullingworth. His original business associate was John Haggas of Oakworth Hall but the firm failed in 1816 and John Greenwood became Edward Craven's new business partner. He was one of the most active mill owners in the area and Edward acted as engineer on the other Greenwood mills, as well as marrying into the family. Hiram Craven appears to have acquired some, if not all, of Edward's previous partner's property; soon after he began exploiting water power in the Oakworth area, as well as later installing his youngest son, Thomas (b. 1814), as a farmer at Oakworth Hall. Around 1819 Craven built Ebor Mill to manufacture worsted, with Ebor House alongside for the mill manager. The reservoir and a three arch bridge over the sluice survive alongside the original water powered mill building, a three storey structure of seven bays, originally fitted with a water-wheel of c. 23 ft. diameter. The Mill was originally leased to Town(s)end & Co. and to provide access to it, and others, Craven acquired and enlarged in the 1820s Mytholmes, Upper and (possibly) Lower Providence Mills, building an occupation road, Providence Lane. This activity probably explains why his interest in masonry contracts waned.

His eldest son, John 'one eye' Craven (1800–1872), was now well established having worked on the Union Canal and may have taken the lead in the contracting business. He married Joseph Nowell's sister Frances (d. 1870) in 1819. In 1832 he was a shareholder in the Mytholme Mill, originally used as a cotton mill, with John Greenwood and William Newsholme and may have been responsible for the installation of a 40 ft. diameter water wheel in 1843. He also became a partner in the Hepworth Iron Company of Penistone. The third son, Abraham (1804–1832), became an architect in York where he died of cholera. At the time of his death the family had a yard at Skeldergate, York and John was resident there c. 1828–1834. Another son, Edward (1811–1833), died on a contract at Whitby docks; he was married to Joseph Nowell's daughter, Mary Willans. Of Hiram Craven's other children Benjamin (1800–1869) and William (1803–1847) are known to have worked with Hiram as contractors, Benjamin also having shares in the Hepworth Iron Company. After 1828 the firm operated as Hiram Craven and Sons, and subsequently as John Craven and Sons.

The family was involved in railway contracting from the first, tendering unsuccessfully for Contract No. 3 on the Leeds–Selby Railway in December 1830 and making unsuccessful enquiries about the Tees bridge on the Stockton and Darlington Railway in 1832. Their first work was on the London and Birmingham Railway (1834–1839), followed by the Manchester and Leeds Railway (1838–1842), the Manchester and Birmingham Railway (1838) and the York and North Midland Railway (1839–1841). While working on the latter Hiram's son, Jonas Craven (1809–1839), died; he was possibly acting as subcontractor on the Shade bridge.

Some impression of the high estimation in which the firm was held can be obtained from their success on the London and Birmingham Railway. The contracts at either end of Wolverton viaduct—Contract Nos. 2C and 3C—were originally awarded to William Soars but it soon became apparent that he was unable to fulfil both and, at Robert Stephenson's suggestion, Craven took over Contract 3C. Progress was generally good and in 1836 the earthworks on Contract 2C were re-divided so that Craven took on more of the work, engineering problems being experienced with the peculiar qualities of the ground in the area. Soars gave up his contract in 1837.

Craven's epitaph in Keighley churchyard reads that 'though his public works may testify his professional skill, [his family's] hearts only can appreciate his tenderness as a husband, affection as a father, uprightness as a master, and firmness as a friend'. He died on 26 August 1842 and his son John continued his success, leaving a fortune of £140,000.

MIKE CHRIMES

[Frank Smith papers, ICE archives; PRO RAIL various minutes; RAIL 384, London–Birmingham Railway; RAIL 454, Manchester & Birmingham Railway; RAIL 1175, Manchester & Leeds Railway; RAIL 667/1038, Stockton & Darlington Railway; RAIL 770/1-15, York & North Midland Railway; Q/AB/32/131, Bucks County RO; QAR 22, Cheshire RO; quarter session records, QD3/10, QD3/367, West Yorkshire RO, Wakefield; National Monuments Record, Royal Commission Historical Monuments, Swindon; *Report of the Committee of the House of Commons ... the State of London Bridge* (1821); P. Lecount (1839) *History of the London and Birmingham Railway*; C. Giles and I. H. Goodall (1992) *Yorkshire Textile Mills*; G. Ingle (1997) *Yorkshire Cotton*]

Works

Pre-1809. Forks House/Brönte Falls Mill, Haworth

With Whitaker and Nowell:

1803–1809. Humber Dock, Hull?
1809–1810. Spinksburn Bridge, near Fewston, widening, £392 14s 10d
1810. Hazlewood Bridge, near Harrogate, £607 3s 1d
1810–1820. Ouse Bridge, York, £80,000
1811–1812. Foss Bridge, York
1811. Willshill Bridge, £302–8–11
1811. Goose Edge Bridge, Keighley, widening, £90
1812–1814. Holme Bridge, widening, £2,089 10s 8d
1812–1814. Wreaks Bridge, Birtsworth, £3,791 12s 8d
1812–1814. Rams Gill Bridge, Pateley Bridge, rebuilding, £1,983 1s 2d
1814. Hebden Bridge–Leeds (Keighley) Turnpike, c. 8.5 miles
1814–1815. Collingham Bridge, rebuilding, £1,031 2s 3d
1814–1815. Tamewater Bridge, parapets, £553 7s 7d*
1815. Hansworth Bridge, Cleckheaton, £504 10s 3d
1816. Overwike, Beck Bridge, Bradford, widening, £1,003 12s 3d
1816. Threshfield Bridge, Skipton, £643 8s 9½d*
1817–1818. Saltersbrook Bridge, rebuilding, £1,557 –s 4d
1818–1821. Contracts on Edinburgh Glasgow Union Canal: Lot 18 (including six bridges) £28,990 10s; Slateford Aqueduct £12,500; Avon Aqueduct, £14,000

1818. Eastburn–Kildwick Bridge Road, c. 2 miles (Keighley and Kendal Turnpike)

1818–1820. Hodder Lower Bridge, £1802 –s 4d

With Nowell:

1823–1825. St. Matthews Church, Wilsden, £4,026
1823–1825. St. Paul's, Shipley, £7,961
1823–1825. St. Mary's, Quarry Hill, Leeds
1823. Staly Bridge, reconstruction, £2,325 (since rebuilt)
1826. Knaresborough Bridge, widening

Messrs. [H] Craven and Sons (operate as Craven & Sugden 1834–):

1819. Ebor Mill, Haworth
1819. Mytholmes Mill, enlarged and owner
1825. Providence Mills (2), Oakworth, rebuilt and owner, Providence Lane 'Occupation road', c. 1 mile
1826–1829. Hull Junction Dock
1828–1829. Layerthorpe Bridge, Foss, York, £1,150 (John Craven in York 1828–1834)
1833. Swing Bridge, Whitby Harbour
1835–1839. London–Birmingham Railway, Contract No. 3C, Castlethorpe, tender £49,735, payments £35,818
1836–1839. London–Birmingham Railway: Contract No. 2C, Wolverton Embankment; Contract No. 7C, Bletchley bridges
1838–1842. Manchester–Leeds Railway: Hallroyd Junction to Walsden, £63,000; Todmorden Tunnel (Benjamin under John Stephenson)
1838–. Manchester–Birmingham Railway, Contract No. 4, Heaton Norris, tender £57,246 12s 5d
1839–1841. York and North Midland Railway: Fairburn contract, 1 mile, fifty-two chains, £39,472; Alltofts contract, tender £36,344

CRAWSHAY, Richard (1739–1810), ironmaster, was born in Normanton, Yorkshire, the son of a farmer. In 1755 he was apprenticed to a London iron stockholder and made a successful impression as on his master's retirement in 1763 his business was passed to him. Crawshay married the same year and his eldest son, William (1764–1834), was born the following year and was to join him in business.

In 1765 Anthony Bacon MP (d. 1786), a shipowner and merchant with iron interests near Colchester, with his partner William Brownrigg, acquired mineral rights to 4,000 acres near Merthyr Tydfil. His agent, Charles Wood, erected the first blast furnaces for the Cyfartha ironworks in 1766–1767. In 1774 Bacon obtained a government contract for cannon and John Wilkinson (q.v.) originally supplied these under the terms of his patent. Following the cancellation of Wilkinson's patent, which was held to be detrimental to the public interest on 1 June 1776, Bacon was free to manufacture cannon and in the following year Crawshay, who had been associated with him since 1774, became his partner and Brownrigg was bought out. Crawshay proved a very vigorous partner; in 1783 Francis Homfrey

was leased the cannon boring mill to supply the cannon using Bacon's iron and after Bacon's death in January 1786 Crawshay bought out Homfrey the following year.

A crucial event took place in 1787 when Crawshay visited Richard Cort's works at Gosport with his partner James Cockshutt (1742–1819) and agreed to pay him 10s a ton for iron made in the puddling furnace according to Cort's method. At that time the Cyfartha works were only producing 10 tons of bar iron a week, but Cort's process was perfected and production rapidly expanded. Cockshutt, originally of Wortley ironworks near Sheffield, designed the first bar rolling mill for Cyfartha in 1787.

Crawshay took on Watkin George (q.v.) as manager and he built a new blast furnace with two 'tungeres'. In 1796 there were three blast furnaces at Cyfartha, producing 48 tons of bar iron a week; by 1803 this had doubled. In 1813 the works were producing 220 tons of bar iron a week, nearly double the production at Penydarren, and Cyfartha was regarded as the largest ironworks in the world.

Crawshay's capital was important in forming the Glamorgan Canal (1791–1794) and his control over it was an important factor in the construction of the Penydarren tramroad. On his death in 1810 Crawshay left a reputed fortune of half a million pounds. Regrettably his grandson, William (1788–1867), and Samuel Homfrey refused to credit Cort with his contribution to developing the puddling of wrought iron and opposed Cort's son's application for government recognition, despite the crucial part he played in making Crawshay's fortune. Cockshutt, in contrast, supported Cort Jr., and he was in the best position to know. Cockshutt, who died in 1819, was elected a member of the Smeatonian Society in 1795 and became a Fellow of the Royal Society in 1804, in recognition of his scientific and technological abilities. It is probably he and Watkin George who, in an engineering sense, made Crawshay the ironmaster he was.

MIKE CHRIMES

[J. G. James collection, ICE archives; DNB; R. A. Mott (1977) Dry and wet puddling, *Newc Soc Trans*, **49**, 153–158]

CREASSY, James (c. 1740–1807), surveyor and civil engineer, was 'born and bred in the midst of the Fens'. He first comes to notice as 'James Creassy of Boston', employed as a freelance surveyor on the Witham Drainage and Navigation scheme under the direction of Langley Edwards (q.v.) from 1762 to 1768. His tasks included setting out most of the 9 mile new cut from Boston to Chapel Hill, surveying Holland Fen, and (with James Hogard) measuring the quantity of excavation in the new cut. On the Black Sluice Drainage, also under Edwards's direction, he made a survey (1766) and measured up the new outfall sluice (1768).

In March 1774 Creassy reported on a project for a navigation from Sleaford to the Witham at Chapel Hill, having made the survey in November 1773. His report is notable for minutely detailed specifications and estimates for the locks. Nothing came of the scheme for the time being. The navigation was eventually carried out on a somewhat different plan between 1792 and 1794 by William Jessop and John Hudson (qq.v.).

Creassy retained a connection with the Witham. He deputised for one of the Commissioners at a meeting in March 1775. Later that year he drew up plans with estimates for improving drainage of the fens adjacent to the river between Lincoln and Chapel Hill, by cutting drains on each side of the river. The high and low lands water would thus be kept separate. John Smith Jr. (q.v.) gave his opinion on the proposals in a report signed by both men and dated January 1776.

Next year, at the request of Lord Townshend, Master General of Ordnance and a Board Member of Bedford Level Corporation, Creassy 'took a view' of the Middle and South Levels to see how their drainage could be improved. On the same principle of separating high and low lands water, he recommended the provision of drains on both sides of the river Ouse downstream from Denver with outfall sluices into the river just above King's Lynn, the drain on the east side to continue upstream to the higher ground at the river Cam.

This report was one of several in response to a set of proposals submitted to Parliament by the Corporation in February 1777. The Bill failed on its second reading. The solution finally adopted was to effect a major improvement in the river by the Eau Brink Cut, sanctioned by an Act of 1795.

Creassy came to the attention of Colonel Henry Watson (q.v.) while the latter was in England before returning to India in 1777 as Chief Engineer, Bengal. Several years earlier Watson had drawn up plans for docks and a yard for fitting out ships at Calcutta. He was now prepared to start work, seemingly at his own expense, and materials were arriving on site. At Watson's request, Creassy came out to superintend the work, at what is said to have been a large salary. Details of the operations are not known, but the first ship, the Nonsuch, was launched from the dock in 1781. Unable to obtain any financial compensation, Watson sent Creassy back to England to represent the case, but without success.

Creassy, who made his return journey via Egypt, Italy and the Netherlands 'viewing all the great works of drainage, both ancient and modern', was back in Lincolnshire before November 1782. He spoke of undertaking a 'voyage of discovery', though it is unlikely this ever took place. The next reference to him is in 1785 when he had moved to London and reported, in March, on the river Nene outfall and, in April, on the drainage of low grounds in Wainfleet, north of Boston.

In India, Watson's health declined. He resigned the service in January 1786 and several months later embarked for England. Creassy visited

Dover to greet his return but Watson died immediately after arrival on 19 September, as Creassy informed Sir Joseph Banks in a letter written next day.

In 1791 Creassy reported on Moulton and Holbeach sea banks, writing from London. Next year he was proposing to publish a map of the Fens. In 1800 he carried out a small but very successful job for William Madocks (q.v.): a sea bank 2 miles long and varying in height from 11 to 20 ft. to reclaim about 1,100 acres of sand and salt marsh in Traeth Mawr on the coast of north Wales at the site of what was to become Portmadoc. Some two-hundred men were employed, with one-hundred and fifty barrows, and the bank was built in six months at a cost of about £3,000. In addition Creassy cut two catchwater drains and there were sluices in the bank 'for venting the inland waters'. Two years later a large tract of the reclaimed land was under wheat and rape, followed in 1803 by barley and grass.

By the end of 1800, if not earlier, he moved to Burwood Copse, near Crawley, Sussex. From that address he reported in 1801 on the drainage of Keyingham Level and in 1802 on the Leven Canal, both in the East Riding. In 1806 he surveyed the river Adur in Sussex regarding improvements in drainage and navigation. Also in 1806 he prepared plans for a new sea bank to enclose an area of land in Traeth Mawr, much more extensive than that reclaimed in 1800, but before anything could be done he died in March 1807 at the age of sixty-seven and was buried at Crawley.

A. W. SKEMPTON

[Minute Books of Witham Drainage and Navigation Commissioners (1762–75) and Black Sluice Drainage (1764–68), Lincolnshire Archives Office; John Smith (1776) *Report on the Drainage of Low Lands on both Sides of the River Witham*; James Creassy (1785) Reports on Wisbech Channel and Wainfleet Drainage, MSS in Banks Collection, Sutro Library; Anon. (*c.* 1800) *Address to the Proprietors of Fens Lands near Spalding* (with biographical note on Creassy), Lincolnshire Archives Office; W. H. Wheeler (1896) *History of the Fens of South Lincolnshire*; Article on Henry Watson, DN.; R. H. Phillimore (1945) *Historical Records of the Survey of India*; W. R. Dawson (1958) *The Banks Letters*; Charles Hadfield (1973) *Canals of Yorkshire and North East England*, J. H. Farrant (1974) Civil engineering in Sussex around 1800, *Sussex Industrial History*, **6**, 2–14; Elisabeth Beazley (1985) *Madocks and the Wonder of Wales*]

Publications
Reports marked by an asterisk include an engraved plan.

1771. *Report respecting the Drainage of the South and Middle Levels of the Fens**
1774. *Report respecting the Opening of a Navigable Communication from Sleaford to the River Witham**

1791. *Report concerning Embanking the Salt Marshes adjoining the Parishes of Moulton, Waplode, Holbech and Gedney*
1801. *Report on the Drainage of Keyingham Level*

CROSLEY, William, I (d. 1796), was a land surveyor in West Yorkshire. He was usually described as of Brighouse, although he may have been based earlier in Huddersfield. He had been involved in surveys for the Rochdale Canal already when, in 1791, he assisted John Rennie (q.v.) in taking borings, but the Bill was lost in 1792 due to opposition by millowners. Crosley now surveyed sites for additional reservoirs but a second Bill foundered in 1793. The third attempt, with the survey by Crosley and advice from William Jessop (q.v.), was successful in 1794 and Crosley was appointed Resident Engineer, for a short time jointly with Henry Taylor and then alone.

Crosley and Rennie seem to have had a warm regard for each other and Rennie employed Crosley elsewhere. In 1792 he was an assistant surveyor for the Lancaster Canal, which had recently gained its Act, and he also surveyed an extension of the Arun Navigation to Horsham in Sussex. Despite initial enthusiasm the project was abandoned in the face of opposition from the Navigation proprietors and he and Rennie were still waiting for their accounts to be settled in 1795. In 1793 he was working on part of the proposed Kennet & Avon Canal. Rennie then offered him the survey for the Crinan Canal but it seems that matters at Rochdale were more pressing and he did not go. The next year Rennie had undertaken to report within four months on a proposal for a canal from Isleworth to Reading to bypass the lower Thames and expected eight weeks of Crosley's time, to start in one week; Crosley could not, or would not, comply and Rennie then offered to employ him at two guineas per day plus expenses, the terms which he then charged for his own services.

In 1793 Crosley and another Yorkshire surveyor, Joseph Lindley, published a map of Surrey which was notable in being the first to draw upon data derived from measurements taken in the National Survey. They agreed to share the profits and expenses equally, though it seems that Lindley had occasionally to subsidise his partner. That year must have been a very busy one for Crosley, as he surveyed the Gloucester & Berkeley Canal for Robert Mylne (q.v.), whom he had met in 1791–1792 when the latter was consulted by the Rochdale subscribers; he also published a report on a line for a canal in the Esk valley in North Yorkshire although it was not built.

Crosley died late in 1796, still employed on the Rochdale Canal.

P. S. M. CROSS-RUDKIN

[Rennie collection mss 19936, 19782, NLS; Rennie reports, ICE; Rochdale Canal ms, ICE; *Geographical Journal*, 1966, **132**, 372; RAIL/384, RAIL/812, RAIL/850, RAIL/886, RAIL/1075, PRO]

Publications
1791. *A Plan shewing the Line of the proposed Rochdale Canal ...*, for John Rennie; revised version, 1793
1793. *Line of the proposed Canal, between Whitby and Pickering*
1793. *Map of the County of Surrey*, with Joseph Lindley

Works
1794–1796. Rochdale Canal, Resident Engineer, with Taylor, then alone

CROSLEY, William, II (fl. 1802–1838) was a canal engineer who became one of Robert Stephenson's assistants on the construction of the London & Birmingham Railway. He was the son of William Crosley I (q.v.) and in 1802 he succeeded Thomas Townshend as Resident Engineer on the Rochdale Canal, the post which his father had held.

The canal was completed in 1804 and Crosley stayed on, although it seems that by 1809 he had left and was engaged on surveys for John Rennie (q.v.) for the Manchester and Salford Waterworks. In August that year he was appointed Engineer to the Brecknock & Abergavenny Canal, his late employers, the Rochdale Company, providing satisfactory references about his ability and honesty. This canal had been opened in stages from Brecon to Llanfoist, which served Abergavenny, and Crosley undertook to devote his whole time to the remaining 11 miles to link to the Monmouthshire Canal and to complete it within three years; his salary was to be £500 p.a. The company had powers to build tramroads within 8 miles to serve the canal but in 1810 Crosley made the surveys for two longer railways, from the canal at Brecon through Hay-on-Wye to Eardisley, and from Llanfoist to Llanvihangel Crucorney respectively. Both lines were subsequently modified by John Hodgkinson (q.v.) and built, the latter line forming the first stage of a through route between Abergavenny and Hereford.

The canal was opened within the time stipulated but Crosley had moved on in November 1811, though apparently retaining responsibility for the canal and devoting about one-quarter of his time to it. He was now engaged upon the Worcester & Birmingham Canal, which had fallen out with its previous Engineer, John Woodhouse (q.v.), because of problems with the lift at Tardebigge. Rennie had produced an adverse report on it and it was he who recommended Crosley. Woodhouse obtained the contract for the next length of canal, including forty-one locks. Crosley was ill in 1812 and unable to work for a time; his place was filled by Samuel Hodgkinson (q.v.). Crosley returned to see the works almost to completion but was replaced by Hodgkinson in July 1815, five months before the line was fully open. The Company now proceeded to arbitration against Woodhouse and there were suggestions that Crosley had authorised additional expenditure without informing his employers. He seems to have been able to defend his actions.

For the next two years Crosley spent his time as a surveyor, working mostly for Rennie. In August 1815 he was at Falkirk, one of several eminent surveyors working on alternative lines for a canal from Edinburgh to Glasgow. He was paid for yet another survey of a Cam to Stort canal in Essex—the actual work may have been carried out earlier—and then in December he was in Scotland again, preparing an estimate for completion of the Glasgow, Paisley and Ardrossan Canal. This amounted to £143,604 and he offered to undertake the whole contract himself. Rennie was asked for his advice and was mildly critical, so no action was taken; instead a railway, based on a survey by James Jardine (q.v.), was built under an Act of 1827 but only over part of the route

During this period, Crosley appears to have been based in Rochdale. As he sent the results of his surveys to Rennie, he added gentle reminders that he would soon be 'out of employ'. On one such occasion he was offered the superintendence of the Hincaster to Kendal section of the Lancaster Canal, taking over part of Thomas Fletcher's (q.v.) work, at £550 p.a. Having checked with Rennie that there were no better prospects available, he accepted in May 1817. He completed the northern end of the main line in 1819 and from 1820 he was Superintendent of the whole canal and in that capacity he constructed the Glasson Dock branch. He also found time to lend his name to a cast iron arch bridge over the river Dee at Eaton Hall, south of Chester, but as this was an ornate variant of Thomas Telford's (q.v.) standard 150 ft. span, and was cast by William Hazledine (q.v.), the extent of Crosley's input is not clear.

In 1826 Crosley became more closely associated with Telford who, with Thomas Brown (q.v.), had successfully steered the Act for the Macclesfield Canal through Parliament. Telford remained as consulting engineer and appears to have provided some of the designs, for example the cast iron aqueduct at Congleton which was clearly the forerunner of the one at Nantwich which he illustrated in his autobiography. Crosley was appointed Engineer to the Committee of Management and bore the brunt of supervision of construction. By this late stage in the canal era competent contractors were available, able to commit adequate resources on lengths of three or 4 miles, and the canal was open by 1831. One of these contractors was William Seed, who had worked with Crosley on the Brecknock & Abergavenny and Lancaster canals; another was William Soars, who would work for him again with less success on the London & Birmingham Railway.

Crosley left the canal for the railway in June 1833 and was responsible, as one of four Assistant Engineers reporting directly to Robert Stephenson (q.v.), for the Tring to Wolverton section. (It may be noted that the contemporary book on the railway by Lieutenant Lecount is wrong in naming John Crossley in this post; subsequent authors have repeated the error. Conversely, the William Crossley mentioned in Hadfield's *Canals of the East Midlands* was probably John Crossley, though not the man who later became Chief Engineer of the Midland Railway). His salary and that of George Buck (q.v.) was £600 p.a., which was more than that of the other two Assistants, Frank Forster and Thomas Gooch, and reflected their long experience of canal construction. The principal work in Crosley's section was the Tring Cutting, where about 1½ million cu. yd. of chalk were removed; its maximum depth was 60 ft. Crosley's salary in 1837 had increased to £700 p.a. and he was still employed in April 1838, when his section opened, but left later in the year as the Company restructured its operations.

Crosley died at some time before 1850 but the exact date has not been ascertained.

When employed on the railway construction, Crosley had as one of his assistants his nephew, William Crosley III (1819–1874), the son of his brother, Richard. He later worked on the Great Western Railway and in Scotland for the contractors Brassey, Mackenzie and Stephenson, as well as undertaking work which would now be described as mechanical engineering.

P. S. M. CROSS-RUDKIN

[Rennie collection, mss 19936, 19782, NLS; Rennie reports, ICE; Rochdale Canal, ms, ICE; *Min Proc ICE*, **41**, 224–225; *Geographical Journal*, 1966, **132**, 372; RAIL/384, RAIL/812, RAIL/850, RAIL/886, RAIL/1075, PRO]

Publications
1810. Llanvihangel Railway, HLRO Q/RW/T1

Works
1802–1809. Rochdale Canal, Resident Engineer
1809–1811. Brecknock & Abergavenny Canal, built Llanfoist–Pontymoile section, Engineer, £500 p.a.
1811–1815. Worcester & Birmingham Canal, Engineer, built Tardebigge–Worcester section and Cofton Hacket (Bittell) reservoir
1813. Finished Kingsnorton reservoirs
1817–1820. Lancaster Canal, Hincaster–Kendal, Resident Engineer, £550 p.a.
1820–1826. Lancaster Canal, Superintendent of whole canal
1824. Eaton Hall Bridge, Aldford, Surveyor, 150 ft. cast iron arch over river Dee
1826–1833. Macclesfield Canal, Engineer, supervised construction
1833–1838. London & Birmingham Railway, Assistant Engineer in charge of Tring–Ashton section, £600 p.a.

CROWLEY, Sir Ambrose (1658–1713), iron and steel manufacturer, nail maker, millwright and engineer, was born at Stourbridge. He was the only child of Ambrose Crowley (1635–1721) and his first wife, Mary Hall, but had seven step-brothers and six step-sisters by his step-mother, Sarah Morris. He was the third generation of iron forgers and nail makers and was probably received into the Quaker religion as embraced by his father. In 1674 he was apprenticed to an iron-monger (then a dealer in iron products on a large scale), Clement Plumstead, at the Minories in London. He completed his apprenticeship in 1681 and set up as an ironmonger in Carey Lane, London. In 1682 he married Mary Owen of Shrop-shire and moved his home and business to Thames Street. She bore him eleven children: one son, John, five daughters and five who died in infancy.

Believing the best method of transporting iron products to be by water Crowley set up work-shops in Low Street, Sunderland, to manufacture nails and iron goods. He transported them to his London warehouse by sailing colliers, ultimately using three of his own ships, at the same time im-porting his raw material, Swedish iron, arriving as ballast in returning empty colliers. Not satisfied with the local workers' productivity, he brought in nailers from Liege but an influx of Catholics into a Protestant community was unsuccessful and after a failed petition to King Charles II he moved his works to the village of Winlaton, on the river Derwent in the north part of County Durham.

Here he set up a water powered iron and steel works and a community of nailers. Crowley himself was responsible for the design of the river dams, leats, waterwheels and pits together with the arrangements of the cementation furnaces (and their operation), rolling and slitting mills. Where he obtained his knowledge is not re-corded but his father had been involved in similar work. The works were extended to other areas in north Durham, particularly at Swalwell (by buying out a rival company), where the river Derwent was navigable from the Tyne; this re-sulted in the formation of the largest iron and steel works in Europe.

Crowley never lived anywhere but in the London area, now with his warehouses and mansion on the quayside at Greenwich. Apart from his visits to the north, which were not fre-quent, all instructions to his works were sent by post, leaving London on Thursday afternoon, ar-riving at Newcastle by mail coach on Sunday af-ternoon, picked up and taken to Winlaton and acted upon by the 'Works Council', meeting at 5.30 on Monday morning. Crowley was a prolific writer of rules and instructions to his workers and his *Law Book of the Crowley Ironworks* is extant in the British Museum. Whilst he demanded long hours of work and exemplary behaviour from his workers he was a caring employer and provided many social benefits, including medical and spiritual care, schooling, sickness and unemploy-ment pay, pensions and widows' pensions and funeral expenses. These payments were contrib-uted to by those in work and recipients of welfare payments were required to wear a badge an-nouncing them to be 'Crowley's Poor'.

Works production was principally nails for use by the Admiralty in sheathing the wooden hulls of its vessels and small iron items for domestic, agricultural, mining and similar activities. Steel manufacturing enabled the making of edge tools and many of these were exported to the Colonies. Crowley achieved a high reputation for service and quality, particularly with the Admiralty, and he was soon providing them with all iron goods including anchors and ordnance.

The extent of Admiralty work involved Crowleys in being owed substantial sums but the Government itself was suffering a simultaneous lack of funds, so much so that Robert Harley, first Earl of Oxford, formed the South Sea Company which took over the National Debt. Subsequently Crowleys received payment from the Admiralty in South Sea Company shares and so concerned was Crowley that he contrived to be elected to Deputy Governor of the Company and must be presumed to have successfully guarded his company's interests.

Crowley was a Freeman of the Drapers Company, and its Master in 1707; a Common Councillor for Downgate Ward; Alderman for Greenhithe Ward; and Sheriff in 1706–1707. He was Knighted in 1707 and was MP for Andover in 1713.

He was a hard, demanding man, too forceful to win friends and was almost certainly the subject of Addison's satire in the *Spectator* as Sir John Envil. He died on 7 October 1713 and was buried in Mitcham Parish Churchyard; he was commem-orated by a plaque in the church. His son, John, continued the business and he was followed by his widow, Theodesia (d. 1782), who managed affairs for many years; in all, the business lasted for more than one-hundred and fifty years.

J. B. ARMSTRONG

[*Spectator*, 12 February 1712; W. Bourne (1893) *Whickham Parish, its History, Antiquities and In-dustries*; *Victoria County History of Durham* (1907) vol. 2; M. W. Flinn (1962) *Men of Iron*; T. S. Ashton (1963) *Iron and Steel in the Industrial Revolution*, 3rd edn.; Winlaton and District Local History Society (1975) *A history of Blaydon*]

CUBITT, William, FRS, MRIA (1785–1861), civil engineer, was born in 1785 at Dilham, Norfolk, the elder son of Joseph Cubitt (born *c.* 1760), a miller at Dilham, and his wife Hannah, *née* Lubbock.

He gained a rudimentary education in the village school at Dilham and, when his father moved to another mill at nearby Southrepps, took advantage of an opportunity for self-improve-ment in the library of the parish curate, the

Reverend Erasmus Drury; he later extended his self-taught studies in the library of another tolerant clergyman, the Reverend J. Humphrey of Wroxham. William's brother, Benjamin, born at Dilham in 1795, was later to join him in railway projects as a locomotive engineer. Another brother, Joseph, practised as a stockbroker.

From an early age he was employed in his father's mill where he exhibited considerable aptitude in the repair of machinery. At the age of fifteen he was apprenticed to James Lyon, a cabinet maker and joiner at Stalham, and with some relief left this stern service after four years, having developed marked skills in carpentry including the manufacture of furniture. During this period his father moved to Bacton Wood mill, Norfolk, where William rejoined him in 1804.

His great interest in machinery led him to join Cook, an agricultural machine maker at Swanton, Norfolk, where he took part in building horse-powered threshing machines and other agricultural implements. Here, the rewards of his apprenticeship were fulfilled in the accuracy he achieved in making patterns for iron castings used in making the machines. He was frequently employed in repairing windmill mechanisms and, having identified a common problem in controlling the mill sails during stormy weather, he invented a device to act automatically in self-regulating the sails. The mechanism consisted of linking the shutters to a weighted central 'spider' which kept the shutters closed until the wind pressure overcame the weight and the shutters then opened to spill the wind. The device was patented in 1807 and was progressively adopted throughout the country; it is still used on most of the few remaining active windmills.

By this time he had established himself at Horning, Norfolk, as a millwright but continued to diversify his interests by making machines mounted on cast-iron tripod frames for draining the marshes in that locality; some of them were still in use fifty years later. His only son, Joseph, was born at Horning in 1811 and he was later to work with his father on railway projects and was the engineer in the construction of the Great Northern Railway.

William struck up a fortuitous friendship with James Ransome, an ironmaker of Yarmouth, who made castings for him whilst he was in business at Horning. Ransome moved to Ipswich in 1809 to develop an agricultural machinery works there and in 1812, at the age of twenty seven, William joined the firm as the Engineer. This was the start of Cubitt's long and distinguished career as a civil engineer.

At Ransome's, his experience in ironwork and his skills in the correlated craft of pattern-making quickly enabled him to widen the scope of work undertaken by the firm to include cast-iron bridge units as is illustrated in a contemporary list of works carried out under his direction between 1812 and 1816. This document is in the archives of the Rural History Centre at the University of Reading. Of particular note was the design and construction of three cast-iron road bridges, at that time not yet widely used in bridge construction.

The first bridge, a single span over the river Brett near Brent Eleigh, in Suffolk, was built in 1813; in 1953 it was bypassed and is now disused and scheduled as an ancient monument. The next bridge, still in use, is a three span structure over the River Stour at Clare on the Essex/Suffolk border, also built in 1813. In the following year he built an improved version of the Brent Eleigh bridge at Witham, Essex, spanning the river Brain; this was still in use in 1998 but subject to a weight restriction. His next bridge enterprise was in 1818, a much larger cast-iron span at Stoke Bridge over the river Orwell in the Ipswich docks area, still using the semi-elliptical profile of his earlier bridges.

Although Cubitt was a partner in Ransome's firm from 1821 to 1826, he also undertook outside consultancy work involving the gasworks and docks at Ipswich and a paper read much later to the Institution of Civil Engineers by George Hurwood on the Port of Ipswich stated: 'In 1821, under the advice and superintendence of Sir William Cubitt (Past-President, ICE) a new channel was cut from the upper end of Lime Kiln Reach to the lower end of Hog Island Reach'. Cubitt had first been approached in 1819.

During his period of employment at Ransome's, Cubitt developed the treadmill, evidently intending it as a means of grinding corn or pumping water by utilising the labour of convicts. The device took the form of a large horizontal cylinder, free to revolve, with steps fitted externally so that the cylinder could be rotated by treading the steps; it was soon adopted in British prisons, commencing at Brixton in 1817, as a form of punishment rather than a useful activity.

Cubitt had been engaged in the preparation of three reports on the Norwich Navigation between 1814 and 1822 and on the third occasion he came into contact with Thomas Telford (q.v.). In January of the following year Cubitt became a Member of the Institution of Civil Engineers. His work on the Norwich and Lowestoft Navigation resulted in Parliamentary action in 1825–1826 and an Act was finally passed in 1827. Improvements were carried out on the Yare, a cut dug from Reedham to Haddiscoe of 2½ miles, a new lock built at Oulton Broad, and a new harbour at Lowestoft. John Timperley (q.v.) acted as resident engineer. The sea lock at Lowestoft was on a scale sufficient to attract the favourable comments and attention of Telford.

The demands for his services had now widened to such an extent that he moved to London in 1826 to establish a consultancy and his career developed over the next three decades to place him in the forefront of his profession at a time when civil engineers could achieve great prominence and fame. His early contributions to the Institution of Civil Engineers' *Minutes of*

Conversations reveal his variety of interests: hot air blast, gas manufacture,, gas vacuum engines, steam navigation, locomotives, hydraulics of weirs, etc., and knowledge of works at home and abroad. His subsequent career will be described in detail in a second volume.

His work on the Norwich navigation had clearly demonstrated he was a safe pair of hands and as old age made it difficult for Telford to adequately deal with the problems of construction on the Birmingham and Liverpool Junction Canal, Cubitt was asked by the proprietors in 1833 to report on the problems there. Shortly afterwards he also effectively replaced Telford on the Ellesmere canal, although Telford continued to give advice on both works until his death. Cubitt continued to act for the two companies and supervised the final stages of construction on the Ellesmere canal, its works carried out by William Alexander Provis (q.v.).

His early experience of harbour works at Ipswich and elsewhere in East Anglia led to frequent consultancies and when he gave evidence to the Tidal Harbours Commission in 1845 he claimed to have replaced more than forty-eight tidal harbours. Of all the engineers whose practices had commenced before the railway age he perhaps was the most successful. He gave evidence in support of the Liverpool and Manchester Railway in 1825, albeit critical of the original survey. Although Henry Robinson Palmer (q.v.) had done the early work on the South Eastern Railway, Cubitt was appointed Engineer-in-Chief on its construction and this involvement led in time to work on the Boulogne–Amiens railway. He then became the Consulting Engineer for the construction of the Great Northern Railway in association with his son, Joseph, as the Engineer.

He became a Fellow of the Royal Society in 1830, a Fellow of the Royal Irish Academy, a Member of the Society of Arts and President of the Institution of Civil Engineers (1850–1852). He was appointed by Prince Albert to supervise the construction of the Crystal Palace in 1851, for which work he was knighted by Queen Victoria. He finally retired in 1858 and died at his home in Clapham Common on 13 October 1861; he was buried at Norwood Cemetery. A portrait of him by W. Boxall hangs in the Institution of Civil Engineers.

TED LABRUM

[DNB; SK; membership records, ICE archives; minutes of conversations (1827–1838), esp. 18, 82, 83, 96, 118,119, 138, 167 [On Lock gates and sluices], and T. Telford (1831) *Norwich Lowestoft Navigation* (MC 131), ICE archives; *Practical Rules for Ascertaining the Discharge of Water over Weirs* (1831), OC 165, ICE archives; *Observations upon the Tides in the River Thames* (1836), OC 203, 1836, ICE archives; *Account of Work which has arisen from the Employment of William Cubitt and which would most probably not have been undertaken without him* (1812–1816), Ransome's archives, Rural History Centre, University of Reading; House of Commons, Committee on the Liverpool and Manchester railroad (1825) *Evidence*, 295–302; House of Commons, Committee on ... Navigable Communication ... between ... Norwich and Lowestoft (1826), *Minutes of Evidence*, 5–25; House of Commons, Committee on the London and Dover (South Eastern) Railway ... (1836), *Minutes of Evidence*, 4 May; House of Commons Select Committee on Dover Harbour (1836), *Minutes of Evidence*, 139–144, Commissioners upon ... Harbours of Refuge (1845), *Report*, 165–176; Tidal Harbours Commission (1845) *Report 1*, 124–129, (1847) *Report 2*; G. Hurwood (1860) On the River Orwell and the Port of Ipswich, *Min Proc ICE*, **20**, 4; obituary (1861) *Min Proc ICE*, **21**, 554–558; J. Boyes and R. Russell (1977) *The Canals of Eastern England*, 84, 94, 113–115, 117; J. Marshall (1975) *The Guinness Book of Rail Facts and Feats*, 34; J. Marshall (1978) *Biographical Dictionary of Railway Engineers*, 61–62; G. Biddle and O. S. Nock (1983) *The Railway Heritage of Britain*, 14, 193; E. Labrum (ed.) (1994) *Civil Engineering Heritage: Eastern and Central England*, 160, 161–162, 166; L. G. Booth (1996) The Cubitt's on the Great Northern Railway: one family or two, *Railway and Canal Historical Society Journal*; H. Hobhouse (1999) in lit.]

Publications
Prior to *c*. 1835:

1814. *Report to the Committee for Improving the Navigation from Norwich to Yarmouth* (in: *The Speech of Alderman Crisp Brown ...*, 1818)
1820. *The Second Report of Mr William Cubitt ... the Best Means of making Norwich a Port ...*
1822. *Description of the Treadmill invented by William Cubitt*
1825. *Map of the proposed Navigation for Ships, from Norwich to the Sea at Lowestoft* [1820; resurveyed by Cubitt and R. Taylor, 1825]
1829. *A Report and Estimate on the River Waveney, between Beccles Bridge and Oulton Dyke, towards making Beccles a Port ...*
1832. *Description of a Plan for a Central Union Canal ... between London and Birmingham*
1833. *Description of a plan for a Central Union Canal ... between London and Birmingham ...*, 2nd edn.
1834. *A Report on the Financial state of the Birmingham and Liverpool Junction Canal*
1835. *Dublin Harbour: Report to the Commissioners*
1835. *Report on Ashworth Harbour*
1835. *Report on the Improvement of the Port and Harbour of Belfast*

Works
Prior to *c*. 1834:

1811. Brent Eleigh Bridge, cast iron arch, £648
1813. Hempstead Mill, £1,220

1813–1814. Clare Bridge, three spans, masonry, £700
1814. Witham Bridge, cast iron arch, £700
1818. Stoke Bridge, Ipswich, cast iron arch
1821. Ipswich harbour, new cut
1827–1833. Norwich–Lowestoft Navigation improvements, 32 miles, c. £150,000
1831. Ouse lower navigation and drainage
1833. Birmingham and Liverpool Junction Canal, consulting engineer
1833–. Ellesmere Canal, consulting engineer

CUNDY, Nicholas Wilcox

CUNDY, Nicholas Wilcox (b. 1778–c. 1837), architect and engineer, was the son of Peter Cundy of Restowick House, St. Dennis, Cornwall, and his wife Thomasine, *née* Wilcocks. His eldest brother was Thomas Cundy Sr. (1765–1825), the architect and builder. In 1793 Cundy was articled to a civil engineer and subsequently moved to London where he was articled to an architect for whom he worked for nine years. His brother Thomas was in London from the 1790s and this may have swayed his decision to move there. Cundy's most famous architectural work, not without some criticism of its execution, was the reconstruction as a theatre of the Pantheon, Oxford Street, in 1811–1812.

Around 1823 Cundy appears to have taken up civil engineering again. He became involved in expensive proposals to build a ship canal between London and Portsmouth, becoming involved in some controversy over an alternative scheme proposed by the Rennies (qq.v.); although this all came to nothing, it occupied him for the next five years. He claimed the support of leading figures like the Duke of Wellington for this proposal, albeit without their agreement.

In the early 1830s Cundy was an advocate of several ambitious railway schemes, mostly in southern England, which again came to nothing. The most important was his proposal of 1833 for the Grand Southern Railway, which was his route for the London–Brighton Railway, on which he gave parliamentary evidence in 1836. His proposed Great Northern Railway was resurveyed by James Walker (q.v.) as the Northern and Eastern Railway. Walker was also involved in 1837 with surveys for his Central Kent Railway.

Nothing is known of his subsequent life other than the fact that he was married to a Miss Stafford-Cooke, and stood unsuccessfully as MP for Sandwich, where his family had historical links.

MIKE CHRIMES

[Colvin (3); Skempton; House of Commons and House of Lords (1836) *Minutes of Evidence … on the London and Brighton Railway Bill*]

Publications

1824–1827. *Reports on the Grand Ship Canal from London to Arundel Bay and Portsmouth*
1828. *Imperial Ship Canal from London to Portsmouth*
1832. *Proposals for an Equitable Reform in Taxation and Finance, etc.*
1833. *Inland Transit. The Practicability, Utility and Benefit of Railroads*, 1st edn. (2nd edn. 1833)
1833. *Great Northern Railway*
1834. *Grand Southern Railway*,
1836. *Plan of the Central Kentish Railway from London to Maidstone, Canterbury and Sandwich with a Branch to Gravesend*
1836. *London, Shoreham and Brighton Railway without a Tunnel* (prospectus)

CURR, John

CURR, John (1756–1823), mining engineer, was born at Kyo, County Durham, and must have been acquainted with coal mining operations from an early age. Around 1774 he became agent for the (9th) Duke of Norfolk's collieries near Sheffield and was responsible for a number of innovations, the most important of which was his introduction of flanged iron plateways. In 1774 Curr was responsible for construction of a 2 mile tramroad with wooden rails from the colliery to the centre of Sheffield, a work which provoked riots as it was seen as a subterfuge to increase the cost of coal.

Curr wrote his *Coal Viewer and Engineer Builder's Practical Companion* in 1794 and it remains the major source on his activities and knowledge. Although best known for its account of plate rails it also details the atmospheric or Newcomen-style steam engine in its most advanced form and it provides an early detailed account of boiler making, discussing manufacture. Curr was a leading engine builder of the time and one innovation he supported was the removal of the boiler from the engine house, developing the idea of side boilers.

With regard to his development of rails, it appears that around 1774 he experimented with replacing the wooden sledge tram used underground, without guidance, with some kind of wheeled trams running on flanged timber barrow ways. This seems to have been an unsuccessful development but Curr must have persevered as around 1787 he introduced at Sheffield Park Colliery the use of wheel-curves and flanged iron plate rails. The system was inspected by John Buddle (q.v.), who reported favourably upon it, and in 1788 it was introduced on the surface at Joseph Butler's ironworks at Wingerworth, Derbyshire. The system spread rapidly, being particularly associated with Benjamin Outram's (q.v.) Butterley ironworks set up in 1790, although Curr seems to have produced them at his own foundry from c. 1792. They were adopted in all the coalfields outside the North-East; it remained loyal to the edge rail and flanged wheels. The inherent weakness of the flanged plateway with its thin section led to it being superseded from the mid-1820s by iron edge rails following their success on the Stockton and Darlington Railway.

Curr took out ten patents, mostly concerned with ropes and rope haulage in mines (No. 1660,

1788; No. 1924, 1792; No. 2270, 1798; No. 2891, 1805; No. 2914, 1806; No. 2947, 1806; No. 2960, 1806; No. 3157, 1808; No. 3502, 1811; No. 3711, 1813). He remained employed by the Norfolk family until 1801 when he was dismissed, perhaps due to lack of profitability in the pits. He also had an interest in Queen's Foundry, in Duke Street, Sheffield, and was consulted by the Coalbrookdale Company, among others.

He appears to have retired to Paris around 1820 but died in Sheffield on 27 January 1823; he was buried in St. Marie's Roman Catholic Church. His son, also named John Curr, emigrated to New South Wales, where he wrote a book which referred to his father's achievements.

MIKE CHRIMES

[J. Curr Jr. (1847) *Railway Locomotion and Steam Navigation*; F. Bland (1931) John Curr, originator of tramroads, *Trans Newc Soc*, **xi**, 121–134; C. F. D. Marshall (1938) *A History of English Railways down to the Year 1830*; C. E. Lee (1960–1961) Some railway facts and fallacies, *Trans Newc Soc*, **33**, 1–16; M. J. T. Lewis (1970) *Early Wooden Railways*; P. J. Riden (1973) The Butterley Company and railway construction 1790–1830, *Transport History*, **6**, 1, 30–52; R. L. Hills (1989) *Power from Steam*; L. T. C. Rolt and J. S. Allen (1997) *The Steam Engine of Thomas Newcomen*]

Publications
1797. *The Coal Viewer, and Engine Builder's Practical Companion*

CURTIS, Robert (fl. 1686–1736), stonemason, was one of many men from the Lyme Regis area who found regular employment on the Cobb, a structure with its origins in the thirteenth century. Originally a timber structure, perhaps packed with loose stones, successive repairs and improvements meant that by the late seventeenth century greater reliance was being placed on masonry rather than a heavy timber framework. Men working on the Cobb were regularly consulted or found employment about pier work elsewhere, as at St. Peter's Port (1590) and Hastings (1596). In the mid-seventeenth century extensive coastal works were carried out in the area by Samuel Hoyte (q.v.). Following this, in recognition of its importance as a port from 1684, Lyme received an annual grant of £100. Curtis appears to have been a leading beneficiary of this regular source of funding for the maintenance of the harbour, working on the Cobb from 1686 until 1736; this included major work in 1697 and 1736. Initially Curtis worked with Be(a)rnard Brown but the 1736 work was carried out with Nicholas Lackey.

MIKE CHRIMES

[R. A. Otter (1991) *Civil Engineering Heritage: Southern England*; Keystone (Historic Buildings Consultants) (1993) *The Cobb, Lyme Regis, Dorset*]

Works
1686–1736. The Cobb, Lyme Regis

D

DADFORD, family of canal engineers, the careers of four of whom are detailed below.

Thomas Dadford Sr. (1730–1809) was a canal engineer; his birthplace and parentage are not known. However, he had strong Wolverhampton connections and may have been born in the area. He was married at St. Peter's, Wolverhampton on 9 September 1759, aged twenty-nine, returned to Wolverhampton in retirement, *c.* 1800, and died there in 1809. His wife was Frances Brown, daughter of Samuel Brown, a Wolverhampton toymaker. Thomas and Frances were Roman Catholics and it was a marriage by licence, his occupation being then recorded as joiner and carpenter. Dadford had four sons, Thomas, James, John and William, the first three of whom followed in his footsteps as canal engineers. There was also a daughter, Mary.

His known canal career started on 17 March 1767 when he was appointed Carpenter and Joiner to the Staffordshire and Worcestershire Canal Co. 'to provide the persons necessary to work under his Direction but to have no profit whatsoever from them'. He constructed a substantial masonry aqueduct across the river Sow under a separate contract in 1771, suggesting that by that time he had an established reputation as a builder. The large canal basin at Stourport and the wide locks down into the Severn were reported to be among his works. The canal was open throughout by May 1772 and in September 1773 he was appointed Engineer to the company at £100 a year.

In 1774, probably at the instigation of the Staffs. & Worcs., which had an eye to the growth of its coal trade, he surveyed the river Stroudwater with John Priddey (q.v.), their scheme to make it navigable recommending new cuts to avoid the many mills. In April 1776, while working out the notice he had given to the Staffs. & Worcs., he enquired of the Stroudwater Navigation whether they wished to treat with him to be their 'Surveyor'. Simultaneously, he offered his services to the Stourbridge Canal Co., stressing his expertise in lock design: 'Experance as taught me every Imperfection of Locks; therefore I should make your Locks mutch improved and to continue with little Repares or Indarance to your Canal …'. The Stourbridge company appointed him their Engineer, a position he held until he resigned in 1781. From 1776 to 1783 he was also Engineer to the Dudley Canal and from 1785 he was consulting engineer for the extension of the Dudley to join the Birmingham Canal via a 2942 yd. tunnel but was paid off in 1787. In 1778 he surveyed and reported on the state of the Stroudwater Navigation but was too 'engaged in sundry Navigations' to take on the post of engineer to complete the navigation, recommending instead 'Mr. Clews' (Josiah Clowes (q.v.)).

Dadford was engineer for the Trent & Mersey Canal Company from c. 1780 to c. 1787 and, together with his son, Thomas (q.v.), he surveyed the river Trent for improvements, reporting in November 1781. In 1784, with Robert Whitworth (q.v.), he advised the Coventry Canal Company on their proposed aqueduct over the river Tame. From 1785 to 1787 he surveyed the Fazeley to Fradley section of the Coventry Canal and was initially in charge of the part of it from Fradley–Whittington Brook for the Trent & Mersey Co. It was possibly during this work that Dadford first met his future partner, Thomas Sheasby Sr. (q.v.), of Tamworth, who was the contractor for the Whittington Brook to Fazeley section of the Coventry Canal (1785–1790) and for the Atherstone to Fazeley section (1785–1789).

Dadford was by now well known in canal circles, and in September 1788, in a letter to John Sparrow, Clerk of the Trent and Mersey, he referred to an intended visit 'to Mr. Powel' at Abergavenny, the earliest indication that thoughts were being given to a canal from Abergavenny to Newport.

In November 1789 Dadford and Sheasby were appointed contractors for the construction of the Cromford Canal in Derbyshire, William Jessop (q.v.) being the principal engineer and Benjamin Outram (q.v.) the resident engineer. Dadford and Sheasby evidently underestimated the job, for in November 1790 they were reported to be in financial difficulties and in January 1791 they declared their inability to proceed and withdrew. Outram completed the work for the company, employing direct labour.

By November 1790 certainly—and probably as early as August—Dadford was already at work with his son, Thomas, and Sheasby as engineer-contractors for the Glamorganshire Canal. They were thus simultaneously engaged on the construction of two different canals over 100 miles apart for a period of some months. One wonders whether withdrawal from the Cromford had more to do with the prospect of greater earnings on the Glamorganshire than with real losses on the Cromford. Whatever the reason, Dadford seems to have escaped from the Cromford Canal works with little damage to his reputation and he was clearly held in high regard by Richard Crawshay (q.v.), the ironmaster and principal proprietor of the Glamorganshire Canal. In April 1791 Crawshay wrote of his 'Surveyor & Navigation Maker' to Count de Reden in Berlin: 'On my return from Wales I found your Favour of 26th Febry.—when your Engineer appears here I will put him under the Tuition of the ablest Man who has promised me he will teach him all he knows I mean Dadford our Contractor, he has made so great Progress that had I not lately been an Eye

Witness of any report would have fall'n far short of …'.

Crawshay's opinion may have been less good later as costs began to exceed estimates. In 1794 the canal company sued Dadford and Sheasby after they refused to repair a breach without extra payment and withdrew from completion of the works and management of the canal. However, independent arbitration found in the contractors' favour.

In 1796, following the resignation of his son, John, he took over as engineer of the Montgomeryshire Canal, Thomas Dadford Jr., recommending his father 'under whom they were bred, and who has had great experience in Canal works …'. He completed the section from the junction with the Ellesmere Canal at Carreghofa to Garthmyl south of Welshpool in 1797. This was his last major canal work, though he carried out a survey for a railway at Penrhyn slate quarries, in north Wales, c. 1799.

Dadford was a significant engineer of the first period of canal building in Britain. As well as his work on the canals of the Black Country and the 'Grand Cross', the canals and associated tramroads constructed by the Dadfords and the Sheasbys down the Welsh valleys to Swansea, Neath, Cardiff and Newport made a major contribution to industrial growth in South Wales. The Glamorganshire was conspicuously successful, embarrassing its proprietors for years with the problem of surplus revenues. Sound canal engineers of the second rank, Dadford and his sons were well able to carry out demanding work according to contemporary custom and practice but perhaps showed limited innovative engineering ingenuity. They purchased shares in the canals they helped to build. Both Thomas Dadfords were at various times allowed or instructed to build boats for canal company use. James and John Dadford were carriers on the Glamorganshire canal, 1792–1794.

Dadford also acquired skills as an architect. In the 1770s he advised on the rebuilding of the parish church tower at Chaddesley Corbett, Worcestershire. In 1780 he was closely involved with the rebuilding of Stafford Gaol, though in what capacity is not clear. In 1780–1781 he contributed to the cost of building St. Peter's Chapel (Roman Catholic) at Rushton Grange, Cobridge (demolished 1936), for which he may well have provided the designs. As Thomas Dadford of Cobridge he was then appointed Surveyor of Bridges to Staffordshire Quarter Sessions at Easter 1785. He was succeeded by James Trubshaw (q.v.) in 1792. In 1800–1801 he designed a chapel for the noted Roman Catholic school at Sedgley Park, near Wolverhampton, which his three younger sons had earlier attended.

In 1783 he was elected a member of the Smeatonian Society of Civil Engineers, a distinction he shared with other contemporary canal engineers. A self-made man, whose spelling remained idiosyncratic to the last, he seems to have been possessed of considerable charm and a persuasive personality.

Thomas Dadford died at Wolverhampton in October 1809 at the age of seventy-nine and was buried at St. Peter's, church on 28 October. He outlived at least three of his sons and survived his wife, Frances, by a few months.

Thomas Dadford Jr. (c. 1761–1801), canal engineer, was the eldest son of Thomas Dadford Sr. (q.v.) and Frances Brown. His birthplace is not known but it was probably in the Wolverhampton area. He married Ann, daughter of James and Ann Parker, of Bluntington Green, Chaddesley Corbett, Worcestershire on 15 August 1797, then aged thirty-seven; both Thomas and Ann were Roman Catholics. There are no known children of the marriage.

Trained by his father, he assisted him in the construction of the Stourbridge Canal from 1776, until the committee decided in 1777 that 'his Service be discontinued'. He also assisted his father in 1782 on a survey of the river Trent, for improvements.

In 1790 he worked with his father and Thomas Sheasby (q.v.) as engineer-contractors on the construction of the Glamorganshire Canal. In the same year he surveyed the route of the proposed Neath Canal, assisted by his father and brother, John. In 1791 he was appointed engineer ('general surveyor') to superintend the Neath over Jonathan Gee as engineer-contractor. Also in 1791 he was appointed engineer for the Leominster Canal. He left the Neath Canal in 1792 to become engineer of the Monmouthshire Canal, for which he was contracted to give three-quarters of his time, one-quarter being left for the Leominster, for which he remained engineer until 1795. Sheasby took over the completion of the Neath. Construction of the Monmouthshire canal commenced in 1792 and the main line was completed in 1796. Dadford's work included six 'edge-rail' tramroads: the independent Trevil Rail Road and five for the canal company, connecting the canal with quarries, collieries and ironworks, the latter including Beaufort, Blaenavon, Sirhowy and Blaendare. On behalf of the Monmouthshire he also surveyed an alternative 'high level' line, subsequently adopted, for the southern part of the proposed Brecknock and Abergavenny Canal, by which the canal would be lock-free for the greater part of its length. He remained engineer of the Monmouthshire canal until late 1798.

In 1794 Dadford undertook to assist his brother, John, in the latter's work as engineer for the Montgomeryshire Canal. He was already overstretched and was criticised by the company for non-attendance, though he did act as contractor for a short length in 1795–1796. In 1797 he inspected the whole line of the canal with his father and reported upon it.

From 1 January 1796 he was Engineer of the Brecknock & Abergavenny Canal, initially at a salary of £133 6s 8d, on condition he devoted a quarter of his time to the job. Perhaps because of the amount of work he already had in hand he appears not to have been the Company's first choice; they advertised the post in 1794 even though he was by then de facto engineer. By 1797 he was being paid £100 a quarter, and was still in post when he died in 1801. Amongst numerous other works, he assisted William Jessop (q.v.) in 1793 in reporting on alternative lines and branches for the Ellesmere Canal, surveyed the line of the proposed Brecon, Hay & Whitney Canal, and surveyed a line from the Montgomeryshire Canal to the Leominster Canal via Montgomery, Bishop's Castle and Ludlow. In 1798 he surveyed an extension to the Neath Canal to Giant's Cave and in 1800 he re-surveyed the route of the Aberdare Canal.

Amongst his major works may be counted the flight of fourteen locks at Rogerstone, on the Monmouthshire Canal, the massive embankment carrying the Brecon & Abergavenny Canal over the river Clydach at Gilwern, and the fine four-arched masonry aqueduct carrying the Brecon & Abergavenny over the river Usk at Brynich. Tunnels seem not to have been his strong point. He was criticised by John Rennie (q.v.) in 1795 following the collapse of the Southnet Tunnel on the Leominster Canal and some of the blame for the collapse of the 375 yd. long Ashford Tunnel on the Brecon and Abergavenny during construction must also be laid at his door.

Dadford died at Crickhowell on 2 April 1801 and was buried at Llanarth, Mons. The cause of his early death is not known but overwork may well have been a contributory factor. He may have had a weak constitution—his illness apparently delayed a start on the Glamorganshire Canal by nearly a month in 1790—and his brother, James, also died young. He died intestate, letters of administration for his 'goods, chattels and credits' amounting to £2,000 being granted to his widow.

Of his character traits little is known. There is some slight evidence that he spoke with a Black Country accent, which would not be surprising. He could be obsequious when occasion demanded: '... I perfectly Agree with Mr. Golborn in all his remarks which shew him to be a Man of Judgement & Experience'. He may have been somewhat headstrong and was certainly imprudent in 1791 when he used his men to cut ground on part of the Glamorganshire canal through Lord Cardiff's land, where it had been specifically agreed that Lord Cardiff's own workmen would be employed. There is the suspicion that at times he showed a readiness to distance himself from awkward problems for which he bore at least partial responsibility.

James Dadford (c. 1768–1804), canal engineer, was the second son of Thomas Dadford (q.v.) and Frances Brown. His birthplace is not known but is likely to have been in the Wolverhampton area. He attended Sedgley Park Roman Catholic School from April 1777 to June 1781. It is not certain that he was brought up as an engineer but he assisted with the construction of the Glamorganshire Canal and in June 1793 he was engaged by the Aberdare Canal Company as engineer for the survey and construction of an edge-rail tramroad from Aberdare via Hirwaun to Penderyn, linking colliery, ironworks and limestone quarry. He was a carrier on the Glamorganshire canal, 1792–1794.

In 1795 he was appointed resident engineer for the Gloucester and Berkeley Canal, under Robert Mylne (q.v.) as Chief Engineer, continuing as sole engineer after Mylne's dismissal in 1798. He was himself given notice in November 1800 when funds became exhausted. It was no doubt James that the displaced Mylne had in mind when he wrote irascibly to the Company Secretary in 1801: 'He [William Jessop (q.v.), who had been consulted on the canal] and Dadford are mere drudges … The Irish are advertising for a Resident Engineer, I think Dadford and they would fit one another to a T, for wrong heads and deficiency of knowledge' (Gotch (1993), 32). For all Dadford's relative inexperience, this judgement seems unduly harsh. His canal-related career continued. He surveyed and reported with Charles Roberts in 1799 on the need for strengthening the Dudley Canal and in 1800 he reported on a rail road from Windmill End to Cabbage Hall.

James Dadford died at Stourport on 20 February 1804 after 'a long and painful illness', his death 'lamented by a large circle of friends and acquaintance'. At the time of his death he was described as engineer to the Staffordshire & Worcestershire Canal.

John Dadford (c. 1769–c. 1800?), canal engineer, was the third son of Thomas Dadford (q.v.) and Frances Brown. His birthplace is not known but is likely to have been in the Wolverhampton area. He attended Sedgley Park Roman Catholic School from May 1778 to June 1781.

His first noted canal work was assisting his father and brother, Thomas (q.v.), to survey the course of the proposed Neath Canal in 1790. He probably also assisted with the construction of the Glamorganshire Canal. In 1792 he surveyed a route for a canal and tramroad link connecting the Glamorganshire and Neath canals, simultaneously surveying a road route up the Aberdare valley. He was a carrier on the Glamorganshire canal in 1792–1793.

Dadford was effectively the first engineer to the Brecknock and Abergavenny Canal and the scheme as presented to Parliament in January 1793 was his. It was a modified version of his original line with the canal now joining the Monmouthshire Canal instead of the river Usk as first intended. It had been read twice when a successful petition was entered for the alternative line favoured by the Monmouthshire Canal Company, surveyed by his brother Thomas. In 1793–1795 John engineered the first sections of the edge rail Clydach Rail Road, from Llangroiney to Gellifelen and from Gellifelen to Fossalog, near Nantyglo, totalling 5½ miles. His bridge carrying

the railroad across the river Usk collapsed in the floods of February 1795; his failure to provide for flood relief culverts in the approach embankment was judged a contributory factor.

In July 1794, still relatively inexperienced, he was appointed engineer for the Montgomeryshire Canal, with his elder brother, Thomas, to assist him 'with his opinion and advice as often as it shall be thought necessary by the Committee'. In the event he received little assistance, being often in demand himself to assist Thomas with the latter's projects. He resigned in July 1796, probably as a result of partial failures of aqueducts over the Vyrnwy and the Rhiw for which he felt responsible, and emigrated to America.

He is not mentioned in his father's will in 1809 and it seems likely that he died in America before that date. That he was highly regarded is indicated by a Monmouthshire Canal Committee minute to the effect that Thomas Dadford Jr. be requested 'to send down a proper person to take up the Mason's and other Work on the different Lines of Canal And that the Committee wd. prefer Mr. John Dadford's coming down in preference to any other person …'.

J. NORRIS

[Crawshay letter book; Maybery Papers, vols. 1 and 2, 1720–1885, National Library of Wales; Marriage Registers and Marriage Bonds, St. Peters Wolverhampton, c. 1750, Lichfield RO; Staffordshire & Worcestershire Canal, Committee Minutes, 1766–1785, PRO; Staffordshire & Worcestershire Canal, Proprietors Minutes, 1766–1845, Staffordshire RO; Staffs. & Worcs. Canal, Orders from Mr. Brindley 1767–1768, Receipt Book 1766–1767, Daybook 1769–, etc.; Clifford Papers, c. 1760–1790, Staffordshire RO (copies of originals held at East Riding County RO (ERCRO), Beverley); Letter, Thomas Dadford to the Committee of the Stourbridge Navigation Co., 15 April 1776, ERCRO; Letter, John Davenport to Thomas Clifford, 22 September 1773, ERCRO; Letter, Thomas Dadford to John Sparrow, 13 September 1788, ERCRO; Stroudwater Navigation, Committee Minutes, Gloucestershire RO, c. 1775; Stourbridge Canal, Committee Minutes, 1776–1781, PRO; Dowlais Iron Co., Letter Books 1792–1794, Glamorgan RO; Aberdare Canal Company, Proprietors Minutes 1793–, Glamorgan RO; Brecknock & Abergavenny Canal, Committee and Proprietors Minutes, 1793–1795, PRO; Parliamentary Committee Minutes, 1793, House of Lords RO; Ellesmere Canal, Committee Minutes, 1793–, PRO; Montgomeryshire Canal, Committee Minutes, 1794–1797, PRO; Monmouthshire Canal, Minutes, 1795, PRO; Diary of Elizabeth Prowse, 1774, Sharp Papers, Gloucestershire RO; School Registers, Sedgley Park School, c. 1781–1787 (in lit. Father Petroc Howell, Diocesan Archivist, RC Archdiocese of Birmingham); Letter Book of Richard Crawshay, 1790–1797, Gwent RO; Marriage Registers, Belbroughton, c. 1797, Worcestershire RO; Quarter Sessions records, D141,

Gloucestershire RO; Berrows' *Worcester Journal*, 1 March 1804; Rev. F. C. Husenbeth (1856) *History of Sedgley Park School, Staffordshire*; C. Hadfield, *Index of Canal Engineers, Surveyors and Contractors*, LSE; C. Hadfield (1967) *The Canals of South Wales and the Border*, 2nd edn.; A. W. Skempton and Esther Clark Wright (1971–1972) Early members of the Smeatonian Society of Civil Engineers, *Newc Soc. Tr.*, **xliv**; J. Roper (*c.* 1978) *A History of St. Cassian's Church, Chaddesley Corbett*, 2nd edn.; C. Hadfield & A. W. Skempton (1979) *William Jessop, Engineer*; C. Hadfield (1981) *The Canals of the East Midlands*, 3rd impression; C. Hadfield (1985) *The Canals of the West Midlands*, 3rd edn.; R. B. Schofield (1985–1986) The design and construction of the Cromford Canal, 1788–1794, *Newc Soc Tr*, **57**; J. Garth Watson (1989) *The Smeatonians*; S. Hughes (1990) *The Brecon Forest Tramroads*; C. Hadfield (1993) *Thomas Telford's Temptation*; E. Paget-Tomlinson (1993) *The Illustrated History of Canal and River Navigations*; C. Gotch (1993) *The Gloucester & Sharpness Canal and Robert Mylne*]

Evidence to Parliamentary Committees
Thomas Dadford:

Trent & Mersey Canal, House of Lords Committee 11 April 1783
Trent & Mersey Bill for Coleshill agreement, Lords Comm. 6 June 1785
Glamorganshire Canal (with James Cockshutt), Commons Comm. 14 May 1790

Thomas Dadford Jr.:

Glamorganshire Canal, House of Lords Committee 12 May 1790
Newport (i.e. Monmouthshire) Canal, Lords Comm. 11 May 1792
Brecknock & Abergavenny Canal, Commons Comm. 26 February 1793, Lords 14 March 1793
Monmouthshire Canal (extension), Lords Comm. 19 June 1797

John Dadford:

Aberdare Canal, Commons Comm. February 1793, Lords 5 March 1793; Brecknock & Abergavenny Canal, Commons Comm. 12 February 1793, Lords 14 March 1793; Montgomeryshire Canal, Commons Comm. 4 March 1794, Lords 25 March 1794]

Publications
Entries marked with an asterisk are originals, not engraved or printed, and therefore fall outside the strict definition of 'printed' reports.

Thomas Dadford Sr.:

c. 1781. *Plan of the Canals between the Ports of Liverpool, Bristol and Hull, the Towns of Manchester and Birmingham, and the Cities of Coventry and Oxford*
1790. *Plan of a Canal from the Parish of Mirthyr Tidvil to or near the Town of Cardiff in the County of Glamorgan*

1799.* *Plan of intended railroad from new engine on Essington Wood to join the Staffordshire and Worcestershire Canal*, Staffs. RO
1799.* *Plan of Intended Canal from the Staffordshire and Worcestershire Canal at Radford Bridge to Stafford Town*, Staffs RO

Thomas Dadford Jr.:

1789. *A Plan of an Intended Canal from Kington in the County of Hereford to the River Severn near Stour Port in the County of Worcester*, BM
1789. *A Plan of an intended Canal from Kington in the County of Hereford to the River Severn near Stourport in the County of Worcester*, Hereford City Library
1792. *Plan of a Canal from Newport to Pontnewynydd with a Branch to Crumlin Bridge and Rail Roads to Blaen-Afon, Beaufort, Nant-y-Glo, Sorwy [sic], Trosnant and Blaendir Iron Works in the Counties of Monmouth and Brecon*, Gwent RO
1793. *Plan of a Canal from the Town of Brecknock to join the Monmouthshire Canal near the Town of Ponty Pool in the Counties of Brecknock and Monmouth with a Plan of the Monmouth Shire Canal as taken in 1792, and also of the dif.t Rail roads that communicate with it*, PB, Brecon Museum
1796.* *A Plan of the Intended Extension of the Monmouthshire Canal below the Town of Newport*, T. Dadford Junr. Engineer, Gwent RO
1797.* *Report to the Commissioners of the Court of Sewers for the Levels of Caldicott & Wentlooge*, Gwent RO
1798. *Brecknock & Abergavenny Canal. The Engineer's Report*, 26 April, B, Powys RO

John Dadford:

1792. *Plan of a Canal and Rail Road, for forming a Junction between the Glamorganshire and Neath Canals in the County of Glamorgan*
1792.* *Plan Of A Canal from the Town of Brecon to near New Brige With a Branch of a Rail Road From Abber Clydach to Llwd coad In the Counties of Brecon and Monmouth*, Powys RO
1792.* *Plan of the Propos'd Branch of Canal From The Abergavenny and Brecon Canal Near Pontypool*, Powys RO
1792.* *Plan Of A Canal from the Town of Brecon to join the Monmouthshire Canal near Ponty Moil with Rail Roads to Llwydecaed, Wain Dew to Langrina Forge & Village to Abergavenny in the Counties of Brecon and Monmouth. By John Dadford*, Powys RO

Works
Thomas Dadford Sr.:

1768–1774. Staffs. & Worcs. Canal Carpenter, construction contractor and, from 1773, resident engineer
1776–1781. Stourbridge Canal, engineer
1776–1783. Dudley Canal, engineer
c. 1780–1787. Trent & Mersey Canal, resident engineer; surveyed the River Trent for improve-

ments, reporting in November 1781; in 1784 advised Coventry Canal Company on proposed aqueduct over the River Tame; from 1785 to 1787 surveyed the Fazeley to Fradley section of the Coventry Canal and constructed the part from Fradley–Whittington Brook for the Trent & Mersey Co.

1785. Maisemore Bridge, over River Severn, Glos.
1785–1787. Dudley Canal, consulting engineer for the extension of the Dudley to join the Birmingham Canal via a 2942 yd. tunnel
1789–1791. Cromford Canal, contractor, with Thomas Sheasby; withdrew in financial difficulties
1790–1794. Glamorganshire Canal, engineer & contractor, with Thomas Sheasby, £103,000
1796–1797. Montgomeryshire Canal, engineer, for the completion of the first section, from Carreghofa to Garthmyl

Thomas Dadford Jr.:

1776. Stourbridge Canal, assisted his father
c. 1781. River Trent, assisted his father with a survey of the River Trent for improvements on behalf of the Trent & Mersey Canal Co.
1789–1791. Cromford Canal, assisted his father and Thomas Sheasby, £103,000
1790–1794. Glamorganshire Canal, engineer & contractor, with his father and Thomas Sheasby
1790–1792. Neath Canal, engineer, £40,000
1791–1796. Leominster Canal, engineer, £93,000
1792–1798. Monmouthshire Canal, engineer, £220,000
1794–1796. Montgomeryshire Canal, assisted his brother John as engineer
1796–1801 Brecknock and Abergavenny Canal, engineer

James Dadford:

1795–1800. Gloucester and Berkeley Canal, (resident) engineer

John Dadford:

1793–1794. Brecknock and Abergavenny Canal, engineer, Clydach Rail Road, 5½ miles
1794–1796. Montgomeryshire Canal, engineer, for the first section, Carreghofa to Garthmyl

DAER, Lord Basil William Douglas (1763–1794), was the second, but eldest surviving son and heir of the fourth Earl of Selkirk. He was involved in radical politics. In 1786 his father delegated to him the management of his estates in Galloway and, in the course of this work, he realised the importance of good roads to agriculture and instigated improvements to and reconstruction of those on the estates. In this he was assisted by a land surveyor, John Gillone, whom he trained in his methods. Lord Daer took up the suggestions of Sir George Clerk of Penicuik for the improvement of roads through hilly country and developed them further, in particular by fixing the alignment strictly by chain and level so

as to maintain a uniform gradient suitable for horse-drawn vehicles. A road had been constructed north-east from Durisdeer on similar principles in 1770, but to a gradient of 1 in 13, whereas Lord Daer favoured 1 in 25.

As well as the roads on the Selkirk estates, he pioneered this method throughout Galloway, firstly on part of the road between Newton Stewart (then known as Newton Douglas) and New Galloway. He was also instrumental in obtaining a Turnpike Act for the Stewartry of Kirkcudbright, although this was not passed until 1796, after his death. Under this Act, John Gillone was appointed County Engineer and carried on Lord Daer's work on several roads, particularly that from Dumfries to Castle Douglas (now the basis of A75) and A712. Lord Daer predeceased his father. John Gillone died in 1810.

A. D. ANDERSON

[Minutes of the Road Trustees of the Stewartry of Kirkcudbright; Rev. Samuel Smith (1810) *General View of the Agriculture of Galloway*, 313–318; Balfour, ed. (1910) *The Scots Peerage VIII*, 521–522]

Works
Road from Kirkcudbright towards Gelston
Road from Kirkcudbright towards Gatehouse of Fleet
Road north-east of Newton Stewart

DAGLISH, Robert (c. 1779–1865), mining engineer and ironfounder, was allegedly born on 21 December 1779, although there is conflict regarding census information and details of his death. His brother John (?1781–1851) was born on 27 May 1781 in Gateshead, the son of Alexander Daglish, and Robert may be the son of Alexander and Jean Daglish, born in Glasgow on 12 December 1777.* Like his brother he was probably trained as a coal viewer in the north-eastern coalfield and then in Leeds. He became an agent to Alexander Lindsay (1752–1825), Earl of Crawford and Balcarres, and from c. 1804 became his manager at the Haigh Foundry, Wigan, and Brook Mill Forge (James Lindsay and Company), building pumping, winding and blast engines for the local collieries, such as that at Arley, and establishing a reputation for the quality of his engineering. Throughout his subsequent career he was heavily involved with coal mining. His work as manager at Orrell Colliery, the most important on the Lancashire coalfield, led to his involvement in early railways.

He built a railway from Orrell to the Leeds-Liverpool Canal in 1812 and the next year built and introduced into service a locomotive of the type evolved by John Blenkinsop (q.v.). This was so successful, saving £500 p.a. compared to the use

* It is possible that John (d. 1851) and Robert were sons of John Daglish (1756–1812), a millwright, and his wife, Ann (Nancy) John.

of horses, that he built two more; they remained in service for more than thirty-six years. Although the bulk of his activity was in mining and foundry work he maintained his interest in railways and was involved in the Bolton and Leigh Railway which obtained its Act in 1825. Although George Stephenson (q.v.) was nominally the engineer, Daglish prepared the original survey and was responsible for the line's construction. Because of the wishes of local landowners it undulated with the countryside, unlike most early lines which avoided any kind of gradient; as such it was heavily criticised by Stephenson.

Daglish's involvement with the Liverpool–Manchester scheme was small, but he rebuilt the Novelty engine in 1833 for the St. Helens and Runcorn Gap Railway on which his son Robert Jr. was working, erecting machinery for inclined planes. They subsequently operated this line from 1839 until 1848. He was consulted about the Newcastle–Carlisle Railway in 1832 and in 1834 he won a prize offered by the London–Birmingham Railway for the best form of pedestal (chair) for rails. This may have led to his being consulted by various early North American railway companies: the Baltimore and Susquehanna; the Boston and Providence; the New York and Harlem; and Norwich and Worcester; as well as the Great North of England Railway.

Daglish's son, **Robert Jr.** (1809–1883), also became an engineer, a source of confusion during their active lifetime. After training with Hick (q.v.) and Rothwell, he joined the St. Helens Iron Foundry of Lee Watson and Company where Robert Sr. had an interest. Under his stewardship this foundry expanded greatly and supplied a variety of machinery for mills, mines, waterworks and glassworks, as well as for railways where early iron lattice truss bridges were supplied to the Liverpool-Bury Railway in 1846. He also became a general railway contractor.

Robert Daglish Sr. became a Member of the Institution of Civil Engineers in 1830. He died at his residence at Orrell, Lancashire on 28 December 1865 and is buried at All Saints, Wigan. He had a number of children by his marriage to Margaret Twizel apart from Robert Jr. The eldest surviving, George (1805–1870), became a surgeon but it is believed that his other sons, John (b. 1811), John (b. 1812) and Charles (b. 1814), died young.

MIKE CHRIMES

[Membership records, ICE; Minutes of conversations, ICE archives; F. Smith Archive, ICE; Obituary (1867) *Min Procs ICE*, **26**, 561–562; C. F. D. Marshall (1938) *A History of British Railways down to the Year 1830*; I. Hedley and I. Scott (1999) The St. Helens iron foundry, *I A Rev*, **xxi**(1), 53–59]

Works

1812–1813. Orrell Railway, Engineer, 2 miles
1825–1827. Bolton & Leigh Railway, 5 miles

DARBY, Abraham, III (1750–1789), ironmaster, was born on 24 April 1750 in Dale House, Coalbrookdale, Shropshire, the eldest son of Abraham Darby II (1711–1763) and his second wife Abiah Sinclair (*née* Maude).

Confusion often arises due to the fact that there were four different members of the Darby family who bore the same name. Abraham Darby I (1678–1717) moved from Bristol to take over the operation of a blast furnace in Coalbrookdale, Shropshire, in 1708. Soon after, he pioneered a new method of smelting iron in the furnace, using coke as fuel in place of the traditional charcoal. He son, Abraham Darby II (1711–1763) was responsible for numerous technical improvements at the works, and their rapid expansion, including the building of new furnaces at nearby Horsehay and Ketley in the 1750s. Abraham Darby IV (1807–1878), a member of the fifth generation, having helped the Coalbrookdale Company through a difficult period of change in the 1830s and 1840s, moved away from Shropshire in 1851 to pursue interests in the Ebbw Vale area of South Wales.

Abraham Darby III is best remembered for his involvement with the Iron Bridge. Having served an apprenticeship under Richard Reynolds, he took over the management of the works on his eighteenth birthday in 1768. His diary reveals that a bridge across the Severn was suggested as early as 1773, though the first meeting of interested parties took place in Broseley in 1775, and he was elected treasurer. Thomas Farnolls Pritchard, a Shrewsbury architect, was appointed to design the bridge, and an Act of Parliament was obtained in 1776. Darby, Pritchard and the ironmaster John Wilkinson convinced the other subscribers that the bridge should be built of iron, since some were suggesting alternatives made of stone, brick or timber. Preparatory work was undertaken in 1777–1778, and the ironwork of the bridge was erected during a three month period in the summer of 1779. Following completion of the approach roads on either side, it was opened to traffic on 1 January 1781. Although details have survived concerning the cost of the bridge, opinions differ as to how it was put together and, despite the fact that the Coalbrookdale Furnace was enlarged in 1777, exactly where the bridge components were cast. It was a triumph in engineering terms, but financially it was a disaster, leading the Darby family, who were major shareholders, to near ruin and years of debt. Abraham Darby III died on 20 March 1789 at The Haye, a prestigious house on the northern slopes of the Severn Gorge from which he had a splendid view of the bridge he helped to create. He is buried in the Quaker Burial Ground in Coalbrookdale.

There are no known portraits of the first three Abraham Darbys, probably due to their strong religious beliefs as practising Quakers. Abraham Darby IV was an Anglican, and there is an oil painting and an engraving of him in the collection of the Ironbridge Gorge Museum Trust.

JOHN POWELL

[Minute Book of the Proprietors of the Iron Bridge, 1775–1799, Shropshire Records and Research; various Darby family papers in the collections of Shropshire Records and Research and the Ironbridge Gorge Museum Trust; A. Raistrick (1953) *Dynasty of ironfounders: the Darbys and Coalbrookdale*; B. Trinder (1974) *The Darbys of Coalbrookdale*; B. Trinder (1979) The first iron bridges, *Industrial Archaeology Review*, **III**(2), 112–121; N. Cossons and B. Trinder (1979) *The Iron Bridge: Symbol of the Industrial Revolution*; J. Smith (1979) *A Conjectural Account of the Erection of the Iron Bridge*; B. Trinder (1981) *The Industrial Revolution in Shropshire*, 2nd edn., esp. chap. x 'The phenomenon of the age'; H. Hodson (1992) The iron bridge: its manufacture and construction, *Industrial Archaeology Review*, **XV**(1), 36–44; D. de Haan (1999) *Abraham Darby's Iron Bridge of 1779: Construction and Restoration*; E. Thomas (1999) *Coalbrookdale and the Darby Family: the Story of the World's First Industrial Dynasty*]

Works

1777–1781. The Iron Bridge, major shareholder, Treasurer to subscribers, manufacturer of iron used in construction

DARGAN, William (1799–1867), civil engineering contractor, was born in Co. Laois, not far from Carlow town, on 28 February 1799, the son of a farmer on the estate of the Earl of Portarlington.

He received his early education in Carlow and commenced his training in a local surveyor's office. His potential was recognised by Major Alexander of Milford and by Sir Henry Parnell MP, supporter of the Holyhead Road and thus a close associate of Thomas Telford (q.v.). Parnell wrote a letter of introduction to Telford, which led, in 1820, to Dargan being appointed as overseer of a portion of the Holyhead road contract. Amongst his responsibilities was the embankment over the Stanley Sands joining Holy Island to Anglesey.

Returning to Ireland, Dargan took on a number of small contracts on his own account, the most important being sections of the road from Dublin to Howth, including the long sea wall protecting the road towards Sutton. He also acted as surveyor to the Dublin to Carlow and Dublin to Dunleer road trusts and carried out some work on the River Barrow navigation.

In 1831, he took on his first large and prestigious project when he became the contractor for the construction of Ireland's first public railway, the Dublin and Kingstown. This contract provided the foundation upon which Dargan built his subsequent contracting enterprises. These included the Kilbeggan branch of the Grand Canal, the Ulster Canal, and contracts for many of the Irish railway companies, notably the routes from Dublin to Cork, Dublin to Drogheda, Dublin to Wicklow and Mullingar to Galway.

By 1853 he had constructed over 600 miles of railway and his name became synonymous with the development of Ireland's transport infrastructure. In the same year, he sponsored the Dublin Exhibition of Industry and was offered a knighthood, which he declined.

In 1866 Dargan was seriously injured in a fall from his horse and died in Dublin on the 7 February 1867 and was buried in Glasnevin cemetery in the city. He was survived by his wife, Jane. An oil painting by Stephen Smith the Elder hangs in the National Gallery of Ireland, Dublin, an institution which he helped financially to establish, and a fine bronze statue to him stands outside the building.

RON COX

[DNB; Holyhead Road Commissioners (1819–1830) *Reports*; (1853) *Irish Tourists Illustrated Handbook*, 12, 41, 148; *Illus Lond News*, 14 May 1853, 390; J. Sproule (1854) *Irish Industrial Exhibition of 1853*, ix–xiv; *Times*, 8 February, 1867, 12; *Gent Mag*, March 1867, 388–389; K. A. Murray and William Dargan, *Jn Ir Rlwy Rec Soc*, **2**, 1950, 94–102]

Works

Works prior to 1830:

1820–1821. Western end of Holyhead road, including Stanley embankment between Holy Island and Anglesey, surveys and supervision of construction
1821–1830. Road construction and maintenance contracts in Dublin and surrounding districts, for grand juries

DAVIDSON, James (1798–1877), civil engineer, was born in 1798, the youngest of three sons of Matthew Davidson (q.v.), master mason of Langholm.

Matthew Davidson's three sons were all born in Wales and were welcomed as apprentices of Telford in London, but in the end Thomas and John gave up engineering and became surgeons. James, however, was already sufficiently experienced and well thought of by Telford to be appointed to succeed his father on his death in 1819. He was elected a member of the Institution of Civil Engineers in 1820. He described the works carried out at Clacknaharry under his father's superintendence, where the ground for the lock was preloaded, a notable case study in the early development of practical soil mechanics.

The Caledonian Canal was opened to shipping in October 1822, but for many years afterwards it was necessary to carry out major repairs and improvements. James succeeded Alexander Easton (q.v.) in 1823 as resident engineer for the whole canal, at a salary of £300 p.a., but in 1829 he was advised for the sake of his health to seek a warmer climate, and he remitted his responsibilities to George May, the toll collector at Clacknaharry; when it became clear that he would be unable to return for some time, May became resident engineer.

George May died in August 1867 and was succeeded by James Davidson once more. He had returned to live with his wife, Eliza Green, at Burnfoot, near Inverness. After his death at Inverness on 30 September 1877 his estate was recorded as £990.

RON BIRSE

[Sir Alexander Gibb (1935) *The Story of Telford*; Jean Lindsay (1968) *The Canals of Scotland*; Skempton; W. T. Johnston (1998) *Scottish Engineers and Shipbuilders*, published as computer database]

Publications
1826. A general report of the state and progress of the Caledonian Canal. Works, from May 1825 to November 1826, in: *Twenty-third Report of the Commissioners for the Caledonian Canal*. House of Commons, 1827, pp 18–21. Subsequent progress reports by Davidson are in the 24th (1828) and 25th (1829) Commissioners' Reports

Works
1819–1823. Caledonian Canal, eastern section Resident Engineer
1823–1829. Caledonian Canal, Resident Engineer
1867–1877. Caledonian Canal, Resident Engineer

DAVIDSON, Matthew (1755–1819), stonemason and civil engineer, was born on 14 January 1755 at Langholm, Dumfriesshire, where he grew up with Thomas Telford (q.v.). When Telford was appointed Surveyor of Public Works for Shropshire he asked his old friend to act as resident engineer for the construction of his first bridge in the county, Montford Bridge (1790–1792). Work on other Shropshire bridges followed.

Davidson worked for Telford for the remainder of his life, acting as resident engineer for the erection of the masonry on Pontcysyllte Aqueduct, before moving to Scotland in June 1804 to act as superintendent of work on the Clacknaharry Division of the Caledonian Canal. Telford regarded him as the senior superintendent on the canal, appointing him as his deputy when he was absent in Sweden in 1808.

Davidson had fallen in love with Wales, as well as the Welsh girl he married, and when he was sent to the Scottish Highlands he made no secret of his low opinion of the country and its inhabitants. He persuaded many who worked for him to follow him north from Wales. His greatest achievement, not as spectacular as the Welsh aqueducts but more difficult to build, was the sea lock on the Beauly Firth at Clacknaharry. Its entrance had to be 400 yd. from the shoreline, founded on more than 50 ft. of soft mud, and when it was successfully completed in 1812 it was the largest lock in the world at 170 ft. by 40 ft.

Davidson had an important role in selecting the masonry contractors for the Caledonian Canal, *viz*. John Simpson, John Wilson and James Cargill (qq.v.), and possibly the earthworks contractors, Thomas Davies and William Hughes (q.v.), based on experience on Shropshire bridges and the Ellesmere Canal. He also advised on suitable workmen for the Gotha Canal.

There is eloquent testimony to Davidson's character. He had no love for his fellow countrymen from the Highlands and was described by Southey as 'a strange cynical humorist'. He was well-read, and in 1809 sent a box of books of value £17 12s to his eldest son, Thomas (1794–1839), studying to be a doctor in Oswestry. He had married Janet Irvine on 29 December 1780 and, aside from Thomas and six children who died in infancy, had two more sons. John (1797–1843), also to become a surgeon, spent some time in London with Telford, as did James (q.v.) who succeeded his father as superintendent on the canal. Davidson died at Clacknaharry on 8 February 1819.

MIKE CHRIMES

[Telford mss, ICE; Gibb papers, ICE; J. Davidson (1820) Foundations of the sea lock, Clacknaharry, ICE OC110; T. Telford (1814) Navigation, inland, *Edinburgh Encyl*; A. Gibb (1935) *The story of Telford*; A. D. Cameron (1972) *The Caledonian Canal*; A. Penfold (ed.) (1980) *Thomas Telford: Engineer*; A. Blackwall (1985) *Historic Bridges of Shropshire*]

DAVIES, Robert (d. 1723), shipwright, leased a shipbuilding yard in Leith at the start of the eighteenth century. At that time there was already a small dock, 60 ft. by 24 ft., known as Lynn's Dock, built at the end of the previous century. Davies was well connected by his marriage to Janet Livingstone, the daughter of the Court of Sessions Clerk. In April 1709 he was made a Burgess of Edinburgh and in 1710 successfully renewed his petition to build a dock capable of the repair of large vessels. The dock was similar to previous docks in the area and used a right-angled kink in the bank of the Water of Leith to provide two sides of the dock, with a third wall built out parallel with the bank in the direction of the flow.

In 1718 Davies remarried, being styled as Honourable Shipbuilder in Leith. At his death in 1723 he left his family of three daughters and a son (who died in childhood) his yard, a small vessel worth Scots £300, and property in Leith.

MIKE CHRIMES

[S. Mowat (1994) *The Port of Leith*]

DAWSON, William Francis, Captain (d. 1829), military engineer, was born in Newmarket, the son of Francis Dawson. A Royal Engineer, he arrived in Colombo, Sri Lanka, in 1819 as 'Second Captain' RE, and acted as private secretary to Sir Edward Barnes (1776–1832), the Lieutenant General. Dawson was the engineer responsible for the construction of several of the military roads built in Ceylon in the 1820s and that over the Kadugannawa Pass presented such great difficulty that in 1832 a posthumous memorial was erected to his memory at the summit. He died at Kandy on 28 March 1829 having contracted dysentery while surveying the Paumben and Mannaw Channels and was buried at St. Peters, Colombo.

MIKE CHRIMES

[P. M. Bingham (1921–1923) *History of the Public Works Department, Ceylon, 1796–1913*, 3 vols. (includes portrait)]

Works

1822. Road over Kadugannawa Pass, Kandy District, Sri Lanka

1822. Road from Kospetta-oya over the Galagedera Pass, Sri Lanka

DEBBIEG, Hugh, General (1731–1810), military engineer and surveyor. No career covering sixty years and rising from Matross (Assistant Gunner) (1742) to General (1803) can be considered anything but successful, but Hugh Debbieg's might have been more so were it not for a series of acrimonious disputes with the Duke of Richmond, Master General of Ordnance, leading to two court martials, in 1784 and 1787, and a premature and enforced transfer to the Corps of Invalids (1799). During his lengthy career Debbieg worked on a variety of civil engineering projects, albeit generally from a military perspective.

Debbieg began a cadetship at the Royal Military Academy in 1744; his precocious talents being rewarded with a 'work experience' place on what proved a disastrous expedition to Lorient. In 1746 he returned thence with considerable personal credit to continue his studies in engineering and drawing and in 1748 was sent to work with William Roy (q.v.) on the survey of Scotland where he eventually organised and conducted one of six survey parties, covering parts of Peebles, Lothian and Lanark as far south as the Edinburgh–Glasgow line.

Most of the next decade was spent working on roads where he surveyed an almost continuous route from Newcastle through to Stranraer, although it is unlikely it was ever planned as a single line. The first section—the military road from Newcastle to Carlisle—on which he worked, with Dougal Campbell (q.v.) from 1748 to 1749, was a retrospective solution to General Wade's (q.v.) inability to move his army across country to counter the Jacobite threat in 1745. The second section from Sark Bridge near Gretna Green to Port Patrick (1757) was a forward looking attempt to speed the progress of troops to Ireland. The sections differed in construction—the Newcastle–Carlisle being an entirely new road whereas the Dumfries-shire route improved existing roads wherever possible, although an estimate exists of 'the bridges found necessary to be built on a survey by Captain William Rickson, Deputy Quartermaster General

for North Britain and Hugh Dibbieg [*sic*], Lieutenant of ingineers' (£629 18s 1d for twenty-one bridges). The road was actually constructed by Rickson in 1763–1764.

By then Debbieg had returned to active military service in north America under General Wolfe as his Assistant Quartermaster General, at Louisberg, where he was in charge of four hundred miners, and at Quebec where he was almost the only officer portrayed in West's painting of 'the death of Wolfe' whose presence can be historically proven.

In 1762 Debbieg produced a plan of the town and harbour of St. John's, Newfoundland, with plans for new defences built of masonry costing £72,000. This proved too expensive for the Ordnance Board who finally, ten years later, agreed to his revised scheme built of earth and sods at a saving of £7,000. In 1763 he produced surveys of Grace and Carboniere Harbours in Conception Bay.

During his time in Newfoundland he received 50s a day reduced to 40s 'secret mission' allowance when in 1767 he was sent to France and Spain to survey harbours. He returned with detailed plans of Barcelona, Cartagena, Cadiz and Corunna written up as 'remarks and observations on several seaports in France and during a journey in those countries from 1767–1768'. He not only observed military activities, but the state of roads, noting that the Spaniards were building three new roads in the Pyrenees to bring iron and timber down to the port of Corunna. The book was presented to George III who rewarded him with a pension of 1s a day for life.

From 1777 to 1784 Debbieg was Chief Engineer on the staff of the Commander-in-Chief Lord Amherst in charge of all military operations in England including field works, bridges and surveys. He was placed in charge of the defence of public buildings in London during the Gordon riots of 1780. This included the Bank of England who asked him to determine necessary measures for a permanent defence. He remarked that 'I cannot make it a fortress but (can procure) it a greater degree of security than has been done by the frippery designs of a civil jobbing architect'. His estimate was £30,000.

In 1778 he was also Chief Engineer at Chatham at an extra salary of 25s per day where he improved and extended the earth ditch and rampart defences of J. P. Desmaretz (q.v.) (with whom he had worked there in 1755–1756) with massive brick walls coped on the high ground with stone and an underground tunnel system. Work at Chatham included a road from Rochester to Chatham with a three arched span brickwork bridge crossing Rome Lane begun in 1781 (demolished in 1902). The finished road was handed over to the Commissioners of the Rochester Trust in 1783. This was his only fixed bridge, but in 1779 he had constructed the 'favourite work of mine', a bridge between Tilbury and Gravesend comprising barges supporting a platform running the entire length, arranged to allow passage of ships. Its demolition by Richmond was one of the causes of their acrimony although 'I have the comfort of knowing that in the judgement of many it was deemed a masterpiece'. He also designed a new, more portable, pontoon for bridging small rivers and tidal inlets. It weighed 3 cwt and was 3 ft. broad and 16 ft. long. This continued in service until 1815 when it was replaced by a design of Charles Pasley (q.v.). While at Chatham he also built a new hospital, and wrote to Lord Sandwich proposing the closure of Gillingham Creek both for the land reclamation of St. Mary's Isle and to improve the navigation of the Medway. As usual Debbieg was ahead of his time the Admiralty finally closing the creek in 1863. As a corollary to this Debbieg also presciently considered that Chatham should be developed solely as a dockyard the current state of the Medway rendering it useless as a harbour.

His professional troubles began in 1782 when the Duke of Richmond became Master General of Ordnance and introduced a series of changes which Debbieg saw as a personal affront, *viz.* restricting his services to the Commander-in-Chief, removing officers from his command and reducing his pay from 25s to 17s a day. This was even further reduced to 9s with Richmond's reorganisation of the Royal Engineers 'the death warrant of the Corps'.

Debbieg's area of activity was reduced from virtually the whole of the south east to Chatham itself which he considered was being starved of funds to finance Richmond's vastly expensive plans for fortifying Portsmouth and Plymouth. Consequently in 1784 he wrote the Duke a letter accusing him of partiality, offering him slights and oppressing and humiliating him by his reforms, his changes and his crudities. This lead to his arrest and court martial, but a court full of generals who had requested his services in the past merely reprimanded him and wrote a minimalist apology for him to utter in court.

Unsurprisingly matters did not improve. In 1785 Richmond refused him the position on the Board of Land and Sea Defences recommended by Parliament. In 1789 Debbieg once again put pen to paper this time describing the Duke's plans as 'a system only calculated to invite the enemy into the very bosom of Britain'. For good measure he published a copy in the Gazette. This time he was suspended for six months and despite the sympathy of both the King, who after his retirement granted him an extra 12s per day pension, and political satirists (see the *Rolliad*) it was effectively the end of his career. Despite subsequent promotion to General he was virtually unemployed for the last twenty years of his life. His wife Jannet died in 1801. They had three sons: Hugh, Henry and Clement, the first two christened at Swallow Street Scottish Church, Westminster, in 1767 and 1769 respectively. Henry subsequently became a Lieutenant-Colonel, and Fort Major at Dartmouth Castle.

Debbieg died at Margaret Street, Cavendish Square, London, on 27 May 1810.

SUSAN HOTS

[PRO, Minutes of the Surveyor General (Thomas Lascelles) WO 47/34; SRO 9010/546/1&2; ZAN.MS.M.5, Northumberland County RO; Conolly Papers, Royal Engineers Library; Porter, vols. 1 and 2; Bendall; DNB; (1784) *Correspondence between the Duke of Richmond and Colonel Debbieg and between Major General Brabham and Colonel Debbieg on which the Charges exhibited ... are founded*; (1789) *Copy of a General Court Martial held at Horseguards for the Trial of Colonel H. Debbieg on Three Charges exhibited against him by Charles, Duke of Richmond*; Captain T. W. J. Conolly (1887) General Hugh Debbieg, *Royal Engineers Journal*, **17**, 245–280, 266–269; T. Hodgson (1902) Military roads in Cumberland, *Transactions of the Cumberland and Westmoreland Antiquarian and Archaeological Society*, **11**, 274–281; G. H. Griffith (1903) Old records of Chatham garrison, *Royal Engineers Journal*, **23**, 168–169, 194–195; *Cambridge History of English and American Literature* (1907–1921) **XI**, chapter 2, part 2, The Rolliad; M. C. Arnott (1949) Military road to Portpatrick, 1763, *Transactions of the Dumfries-shire and Galloway Natural History and Antiquarian Society*, 1949–1950, 120–134; W. Lawson (1966) The origin of the military road from Newcastle to Carlisle, *Archeologia Aeliana*, **4**, 44; J. Hamilton-Baillie (1974) The fixed defences of the 16th to 18th centuries illustrated by the defences at Chatham, *Royal Engineers Journal*, **88**; W. Taylor (1976) *Military roads in Scotland*, 1st edn.; D. W. Marshall (1976) *British Military Engineers 1741–1783*, unpublished dissertation, University of Michigan; W. Lawson (1979) Construction of the military road in Cumberland, *Cumberland and Westmoreland Antiquarian and Archaeological Society, New Series*, **79**, 109–119; A. Saunders (1986) *Fortress Britain*, 120; A. A. Anderson (1997) Some notes on the old military road in Dumfries-shire and Galloway, *Transactions of the Dumfries-shire and Galloway Natural History and Antiquarian Society*, **72**, 79–88; IGI]

Reports

1757. *Report on the Road from the River Sark to Portpatrick in North Britain by W. Rickson and Lieutenant Debbieg, Engineer*

1767–1768. *Remarks and Observations made upon Several Seaports in France and Spain during a Journey in those Countries*, BL King MS.41

Maps and surveys

Most of the following are unpublished (in the BL).

c. 1748–1749. *Plan or Survey of all the Country between these Places (Newcastle to Carlisle) and some Miles further on each end extending in all 60 mile and about 6 miles in Breadth* (with Dougal Campbell)

1755. *Plan of the Lines designed to inclose his Majesty's Dock-yard and Ordnance Wharf at Chatham, including Gillingham Fort, and the Country for Half a Mile in front of the Lines, distinguishing the Part not yet executed, and shewing the extent of the King's Land*

1756. *Plan of the Entrenchment inclosing his Majesty's Dock-yard and Ordnance Wharf at Chatham* (drawn by Joseph Heath)

1757. *Plan of Loch Eil*

1765. *Sketch of the Town and Harbour of Toulon, with the adjacent Coast and Country to the Westward, as far as Cassis, in which an Attempt is made to shew the Posts that ought to be previously occupied by an Army and a Fleet, whose Object might be the Siege of the Place*

1767. *Plan of the City of Barcelona … together with its Port and two Citadels*

1767. *Plan of the Town and Harbour of Cartagena, in the Province of Murcia in Old Spain, with its Dock Yard and Bason, Forts and Batteries, as they stood in the month of October, 1767; together with a Sketch of the Works intended by the Court of Spain for its better Security, as they are at present traced upon the spot; the Soundings of the Harbour, and the Bason, and without the Harbours Mouth, …*

1768. *Plan of the City, Bay, and Harbour of Cadiz, the Caraccas or Dock Yard, the Island of Leon and Country adjacent, from the Town of La Rotta to the Castle of St. Pédré, taking in the Puenté de Suacé and all the Ground relative to the Strength of the Place*

1780. *Plan of the Harbour of Ferrol, with the adjacent Country, and the Bay and City of Corunna*

1781. *Plan of Quebec and Environs, with its Defences and the occasional Entrenched Camps of the French commanded by Marquis de Montcalm; shewing likewise the principal Works and Operations of the British Forces, under the command of Major General Wolfe, during the Siege of that Place in 1759. Surveyed … by Lieutenant Colonel Debbieg … by Major Samuel Holland … by Captain Des Barres*

Works

1762–1770. Defences for St. John's Harbour, Newfoundland, £65,000

1778–1789. Great Lines, Chatham, Brompton, Fort Amherst, c. £45,000; £27,000 spent, 1778–1780

1779. Temporary military bridge across the Thames between Tilbury and Gravesend

1781–1783. Road from Rochester to Chatham (including three-span brick arch bridge)

DE BUDE (DEBUDÉ), Henry, Major

(1800–1843), military engineer, was born on 3 November 1800, the son of Lt. General Jacob de Bude and Mary (*née*) Lambert. He joined the Bengal Engineers in 1818 and reached the rank of Major on 31 March 1840.

In his first years in India, Debudé was seconded to the Surveyor-General's Department to carry out route surveys such as that of the road from Kashipur to Almora in the Himalayan foothills in 1820. At that time British military engineers in India were beginning to look at irrigation in some detail. Work had already begun under Lieutenant Blane (q.v.) on restoring the Western Jumna Canals, and in July 1822 Debudé was sent to survey an old Mughal Canal, the Eastern Jumna (Yamuna) or Doab Canal, which the British administration was keen to restore. From 1823 to 1826 he worked on this under Richard Tickell (q.v.) and was succeeded in this task by Captain Robert Smith (q.v.). Along with other canal officers he saw service in the siege of Bharatpur near Agra in 1825–1826.

From 1827 to 1828 Debudé was Executive Officer for the fortress of Aligarh. The possibility of extending irrigation into the Ganges plain interested officers such as John Colvin (q.v.), who had succeeded Tickell as superintendent of Canals in the Delhi Territory in 1827. In 1828 Captain Debudé carried out a survey from Muzafarnagar to Aligarh in the lower part of the Doab between the Jumna and Ganges to explore the possibility of irrigating this tract of land by damming two natural watercourses on the site of Muhammad Abu Khan's eighteenth-century 11 mile long canal. His proposal, costing 300,000 rupees, was not carried out, however, as it was soon realised that the watercourses would not be able to keep up a steady supply. His survey did, on the other hand, establish that the Ganges would have to be the source of water for any irrigation scheme; this was eventually achieved with the building of the Ganges Canal scheme by (Sir) Proby Cautley between 1842 and 1854. In 1831 Debudé was appointed Garrison and Executive Engineer at Delhi; Superintending Engineer in the Public Works Department in the Central Provinces in 1835, and Secretary to the Military Board in 1841.

Debudé was married three times: to Mary Anne at Meerut on 12 July 1825; secondly to Miss J. A. Royle; and third to Margaret Davidson at Calcutta on 14 April 1842. He died on 8 November 1843 at Calcutta.

JOYCE BROWN

[Sir P. T. Cautley (1860) *Report on the Ganges Canal Works*, vol. I, 6–8, 15–17; E. W. C. Sandes (1933) *The Military Engineer in India*, vol. I, 253–260; E. W. C. Sandes (1935) *The Military Engineer in India*, vol. II, 5–6; R. H. Phillimore (1954) *Historical Records of the Survey of India*, vol. III, 24, 437; J. Brown (1978) Sir Proby Cautley (1802–1871). A pioneer of Indian irrigation, *History of Technology*, **3**, 44, 56]

Works

1823–1826. Delhi Canal, assistant engineer

DE GOMME, Sir Bernard

(1620–1685), military engineer, was born in 1620 at Terneuzen, Zeeland, son of Peter de Gomme. De Gomme was first brought to England by Prince Rupert at

Sir Bernard de Gomme

engineering employments are brief and disappointingly poorly documented but, nevertheless, there are aspects in his work as a military engineer which have relevance to the field of civil engineering.

His detailed drawings specify the plans of the timber framework and piling for Charles Fort and James Fort in Portsmouth Harbour and earlier for the fortifications at Dunkirk. The pattern of piles for the abandoned water bastion at Tilbury Fort can still be seen at low water. His Dutch experience led to the adoption of double moats and schemes for inundations at Portsmouth and Tilbury. Perforce he dealt with problems of coastal erosion at Sheerness. Architectural skills were employed in designing barracks, magazines and storehouses at Plymouth, Tilbury, Sheerness and Portsmouth. Ultimately, de Gomme became the leading influence in the building works of the Ordnance Office, particularly upon his promotion to Surveyor General in 1682, a post which he combined with Chief Engineer. De Gomme is best known for his surviving works: Plymouth Citadel and Tilbury Fort but as Chief Engineer at the Ordnance Office (1661–1685) he also designed fortifications at Portsmouth, Gosport, Sheerness and Harwich as well as working at Dunkirk and Tangier.

At his funeral in November 1685 his achievements were honoured by a sixty-gun salute and he was buried in the Chapel of St. Peter ad Vincula, Tower of London.

ANDREW SAUNDERS

[Portfolio of maps and plan of fortifications in The Netherlands by de Gomme (*c.* 1637–1651) BL King's Top. Col., **cii**, 21; plans and drawings, estimates, surveys, reports, etc., relating to de Gomme's fortifications, Add MSS 16370, 16371, etc., British Library; Board of Ordnance papers, PRO; Dartmouth Collection, Staffordshire RO; Dartmouth Collection, National Maritime Museum; Bodleian Library; Portsmouth City Library; *Cal State Papers Domestic, Charles II* (1661–1662) 155, (1665–1666) 57, (1670), 373; H. C. Tomlinson (1979) *Guns and Government: The Ordnance Office under the later Stuarts*; Andrew Saunders (1996) Sir Bernard de Gomme: a Dutch military engineer in English service, *Vestingbouw overzee: Militaire architectuur van Manhattan tot Korea*, Stitching Menno van Coehoorn, Utrecht, 9–17; DNB; Colvin (3)]

Works

1661–1662, 1670. Dover Pier, repairs
1665–1666. River Cam, navigation
1665–1683. Plymouth Citadel, etc.

the start of the Civil War, eventually becoming the Royalists' chief engineer. He had been knighted by 1645 but little is known of his early life. He was certainly in the army of Fredrick Henry, Prince of Orange, as early as 1637 and there are signed and dated siege maps and plans of Netherlands fortifications, including a portrait miniature, in a large portfolio of his drawings in the British Library. Returning to England and royal service in 1660, he lived for many years in Bury Street, Bevis Marks, and registered as a member of the Dutch Reformed Church; he obtained denization in 1667. He was married twice with a daughter, Katherine, and a stepson, Adrian Beverland, by his first wife.

A seventeenth-century military engineer had to be omnicompetent in all branches of mathematics, land surveying, civil and military architecture and quantity surveying. De Gomme had experience in drainage and land reclamation besides and in 1636, a 'Bernhardt de Gomme' was commissioned as a surveyor in his native Zeeland. De Gomme was ordered to Dover in 1661 to advise on repairs to the harbour pier and then, between fortification commissions in 1663, he was again appointed surveyor in his native Zeeland.

Two years later he was required to assist the commissioners responsible for making the river Cam navigable and establishing a communication with the Thames, but almost immediately he was sent to Plymouth to design the Royal Citadel. Subsequently, in 1670, lack of progress with the repairs at Dover and suspected misappropriation of funds led to a commission of inquiry being established, consisting of de Gomme, Christopher Wren, Jonas Moore (qq.v.) and others. These civil

DELAVAL family, members owned estates in Northumberland, including the harbour of Hartley, between North Shields and Blyth.

Sir Ralph Delaval (1622–1691), the son of Robert Delaval and his wife Barbara Selby, was born on 13 October 1622 and having been trained in the law he married Anne Leslie in 1646, the union producing a family of 12; he was created baronet in 1660. As the owner of the port of Hartley his first improvement was the protection of its entrance in 1661 by building piers of both timber and stone. He found that the harbour silted up and to remedy this defect he placed, *c.* 1675, 'a new strong Sluice with Flood-gates upon his Brook; and these being shut by the Coming-in of the Tide' so impounded water which at every low tide 'scoured the Bed of the Haven clean'. From this time the port became known as Seaton Sluice. Delaval died on 29 August 1691 and was buried at Seaton Delaval.

Sir John Hussey Delaval (1728–1808), the second son (of a family of twelve) of Francis Delaval (1692–1752) and his wife, Rhoda Apreece, was born on 17 March 1728. In 1750 he married Susan Potter, *née* Robinson, the marriage resulting in a family of seven. After her death he married again but this union produced no children. John Hussey Delaval became a Member of Parliament in 1754 and was created baronet in 1761; at about the same time, as the port's nominal owner, he became responsible for the second phase of its improvement. As a result of the development of the family's nearby coal mines, the trade of the harbour had increased and it was considered expedient, a century after the initial work, to improve it further.

John Smeaton (q.v.) visited Seaton Sluice in 1758, but there is no evidence that he then recommended improvement works. Consultations involving the Delavals and William Brown (q.v.) followed and the harbour was then improved further—to whose designs is not known—by the formation of a cut through rock to the south of the established entrance. The new channel was some 800 ft. long, 30 ft. wide and 52 ft. deep and, protected by timber stoplogs at each end, formed what was, in effect, the first wet dock in northeast England; shipping was further protected by a substantial pier at the seaward end. To permit easier loading, staiths were built so that vessels could lie in the protection of the new channel. Constructed over a period of some three years, the new cut was brought into service on 20 March 1764.

Delaval was made baron in 1783. He died on 17 May 1808 and was buried in Westminster Abbey.

Thomas Delaval (*c.* 1730–1787), the brother of John, was the fourth son of Francis Delaval and was, with—or perhaps in place of—his brother, responsible for the port's operations. Born *c* 1730, he lived for a time in Hamburg where, as a merchant and speculator in coal and iron, he was said to have amassed considerable wealth. By 1763 the manufacture of copperas and glassmaking—principally the manufacture of bottles—had been established at Seaton Sluice; development of the family's Hartley colliery had included the installation of Newcomen engines. Thomas Delaval married Cecelia Watson in 1768; within two years he was on the verge of bankruptcy and his ownership of the glassworks was offered for

sale to his brother, John. He died on 31 August 1787 when 'as he was taking an airing on horseback in Hyde-park, he dropped from his horse in a fit, was carried home, and expired immediately'.

R. W.RENNISON and S. M. LINSLEY

[*Gents Mag* (1787) 57, 839; R. Welford (1895) *Men of Mark Twixt Tyne and Tweed*, 49; H. H. E. Craster (1909), *A History of Northumberland*, 126–129; J. U. Nef (1932) *The Rise of the British Coal Industry*, **1**, 378, 379; Northumberland Record Office, ZDE/12 and ZDE/41]

Works
1661. Hartley, construction of sluices to scour harbour
1761–1764. Seaton Sluice, formation of new cut, cost £10,000

DENTON, William Smith (see: *NIXON family*)

DESAGULIERS, John Theophilus, FRS (1683–1744), lecturer and writer on experimental philosophy, was born on 12 March 1683 at La Rochelle and came as a child to England with his Huguenot parents, Jean and Marguerite Desaguliers. For a time his father was minister of the French Chapel in London and then opened a school in Islington; after his father's death in 1699 he taught at the school. In 1705 he entered Christ Church, Oxford, graduating in 1709 and the following year he transferred to Hart Hall (later Hertford College) where he took over the pioneer lectures on experimental philosophy previously given by John Keill, FRS (1671–1721). Still at Oxford, he married Joanna Tuckey in October 1712. Soon afterwards he began a course of lectures on 'mechanical and experimental philosophy' given at a booksellers near Temple Bar, London, for which he charged an attendance fee of 2 guineas, they proved to be successful. By 1715 he settled in Channel Row, Westminster, where he delivered his regular courses for more than twenty years.

Desaguliers had already been elected a Fellow of the Royal Society in 1714, and in 1716 he became curator of experiments there, a post he held until 1743, receiving £30–50 a year according to the number of experiments performed. He published more than fifty papers in *Philosoph ical Transactions* and was awarded the Copley Medal for distinguished research three times, in 1734, 1736 and 1741. The University of Oxford awarded him the degree of DCL in 1719.

In 1716, having been ordained some years earlier, he was appointed chaplain to the Earl of Carnarvon (later Duke of Chandos) who was at that time building his great house at Canons near Edgware. The project involved a waterworks supplying both fountains and the house and Desaguliers made observations on the flow of water in a long pipeline. This led to his interest in hydraulics, and in 1718 he translated Mariotte's

Traité du Movement des Eaux, first published in 1686.

He met Henry Beighton (q.v.) through a shared interest in the steam engine erected in 1714–1715 at Griff colliery in Warwickshire by Thomas Newcomen (q.v.). Beighton went on to build a similar engine, with improved valve gear, at Washington Fell, Co. Durham, in 1718; it incorporated a lever safety valve suggested to him by Desaguliers the year before. At the same time Desaguliers devised an improved form of the steam engine invented by Thomas Savery (q.v.). Seven of these were erected to his design for pumping water to fountains, the first in the gardens of Peter the Great at St. Petersburg.

Desaguliers was consulted by the Edinburgh authorities in 1721 on the Comiston aqueduct, a lead pipe 3 miles long from the springs to a receiving reservoir in the city. He realised that the problem of insufficient supply resulted principally from air trapped in the 'eminences' of the pipeline, and solved it by introducing hand-operated air valves. Further thought led to the invention of an automatic valve in 1726; in its development he collaborated with Newcomen and his 'operator', Joseph Hornblower.

Desaguliers had the honour of presenting a course of lectures before George I at Hampton Court. Soon after the accession of George II in 1727 he was appointed chaplain to Frederick, Prince of Wales, and it was under his patronage that Desaguliers published in 1734 and 1744 the two volumes of the *Course on Experimental Philosophy;* both were reissued in 1745 and again in 1763. Meanwhile he had translated from the Latin the *Mathematical Elements of Natural Philosophy* by W. J. Gravesande (1682–1742), Professor at Leiden University, in successive editions of 1720, 1721, 1726 and 1737.

As for Desaguliers's contributions, they are summed up in his own words, written towards the end of his life: 'I am still acting in my Province, which has been for many years to explain the Works of Art, as well as the Phaenomena of Nature'. In 'explaining the Works of Art' he became one of the founders of engineering science: the first English writer to analyse machines on the basis of statics and elementary dynamics; to give experimental results on the flow of water through an orifice and through a rectangular notch (useful in gauging streams); and to give a formula for the velocity of flow under a sluice gate under a head ($v = \sqrt{2gh}$, where g is the gravitational constant (32.15 ft./s^2)). He gives rules for the size of elm pipes to sustain various heads and, similarly, for lead pipes. Characteristically, he also describes an apparatus to test the bursting pressure of a lead pipe and advises a factor of safety of four in design. Well aware of the losses in a pipeline due to fluid friction, he gives rules for the up-sizing of pipes according to length. He also describes and illustrates many machines from actual examples: cranes, pile-drivers, a Newcomen engine, and so on. A valuable feature of the *Course* is the provision by 'my ingenious and very good friend Mr Henry Beighton' of descriptions of several of the best watermills then in existence. These accounts are well illustrated by Beighton's own drawings and include an analysis of their performance.

Desaguliers refers to 'Mr. Charles Labelye (q.v.), formerly my Disciple and Assistant, and since that time appointed Engineer of the Works of Westminster Bridge'. The two men met in 1725, soon after Labelye came to England. Following the latter's submission to Desaguliers of calculations regarding the 'fall' in water level between the piers of his proposed Westminster Bridge, both men gave evidence on the bridge to the House of Commons in February 1736, Desaguliers also to the Lords in April. In a similar manner John Grundy (q.v.) submitted in 1734 to Desaguliers for comment a paper on fen drainage.

Though not a great originator, Desaguliers played an important role in the dissemination of scientific knowledge and the application of science in engineering. In this capacity he was joined by his friend Petrus van Musschenbrock (1682–1761), Professor at Utrecht, who in 1729 published the first accurate tests to measure the strength of timber and iron in tension, compression and bending, and in France by Bernard de Forest Belidor (1698–1761), whose *Science des Ingenieurs* also appeared in 1729.

When construction of Westminster Bridge eventually started in 1738 with Labelye as engineer, Desaguliers had to move, owing to the demolition of Channel Row to make room for the approach to the bridge. From then on he lived in lodgings over the Piazza in Covent Garden where he carried on his lectures. He died on 29 February 1744 and was buried in the Savoy Chapel. His elder son held a living in Norfolk after graduating from Hart Hall, Oxford, and his younger son, Thomas, became lieutenant-general of the Royal Artillery.

A. W. SKEMPTON

[John Grundy (1738) *Some Propositions ... on the draining of Lands*; Charles Labelye (1736) The result of a calculation made to estimate the fall of the water at the intended bridge at Westminster, in: Nicolas Hawksmoor, *A Short Historical Account of London Bridge, with a Proposition for a new Stone-bridge at Westminster*; DNB; A. Wolf (1938) *A History of Science ... in the Eighteenth Century*; E. G. R. Taylor (1966) *Mathematical Practitioners of Hanoverian England*; A. E. Musson and E. Robinson (1969) *Science and Technology in the Industrial Revolution*; Margaret Rowbottom (1968) John Theophilus Desaguliers, *Huguenot Soc Proc*, **21**, 196–218; Rupert Hall (1971) Desaguliers, *Dictionary of Scientific Biography*, **4**, 43–45; L. T. C. Rolt and J. S. Allen (1977) *The Steam Engine of Thomas Newcomen*; R. J. B. Walker (1979) *Old Westminster Bridge*; E. H. Winant and E. L. Kemp (1997) Edinburgh's first water supply: the Comiston aqueduct, *Proc ICE*, **120**, 119–124]

Selected publications

1718. *The Motion of Water*, translated from the French of Edme Mariotte

1720. *Mathematical Elements of Natural Philosophy*, translated from the Latin of W. J. Gravesande (later editions 1721, 1726, 1737)

1734. *A Course of Experimental Philosophy*, vol. 1

1744. *A Course of Experimental Philosophy*, vol. 2 (both vols. reissued 1745 and 1763).

DESMARETZ, John Peter (*c.* 1686–1768), military engineer. Desmaretz's sixty-year career was evenly divided between mapping and fortifying most of the coast of south-east England, eventually leading to his proposals for the harbours at Ramsgate and Shoreham, which are his main civil engineering contributions. At one period (1758) he was at the same time Clerk to the Fortifications (£60 p.a.), Architect to the Ordnance Board (£120 p.a.) and Master Draughtsman (£100 p.a.), but his engineering training and priorities are impossible to determine as nothing is known of his life before he took service with the Duke of Marlborough in 1709. A Huguenot background can be postulated. His epitaph tantalisingly states 'though born a foreigner he early adopted every sentiment of civil and religious liberty and exerted his active abilities for the service of this nation in quality of an engineer'.

Desmaretz stayed on after the end of the war in 1713 working with John Armstrong (q.v.) on the demolitions at Dunkirk where he was employed to survey the works and chart the coast between Ostend and Gravelines. His first work in England was also as a surveyor, charting the river Medway in 1724.

He was first employed by the Ordnance Board as a draftsman in the Tower in 1725, where he helped to train the next generation of military draftsmen, such as the Durnfords, Andrew Frazer and William Twiss (qq.v.). Despite being officially based in the Tower he still provided charts, reports and estimates and was involved in the construction of every major Board of Ordnance project in the south of England. These included Landguard where he surveyed Harwich Harbour (1732), the Brompton Lines (1755) built to defend the dockyard at Chatham, Dover Castle (1756) where he made the first major alterations in the castle's defences in 500 years, employing 734 men and spending £3,658, and the Gosport Lines (1757) built to defend Priddy's Hard and the dockyard at Portsmouth where a large part of his career was spent, becoming overseer in 1748 and Commanding Engineer 20 years later. He drew up numerous plans of the harbour, town, dockyard and defences, one of the last (1762) with an estimate of £1,761 for repairs to the harbour for damage caused by extraordinary tides and winds.

His major work there was the design and construction of Fort Cumberland, an irregular star-shaped fort, built to defend Langstone harbour. Desmaretz's estimate of £3,787 was agreed on 18 November 1746 and work began the following new year's day. His contradictory instructions were 'to see that everything is done in a frugal but substantial manner with the best materials and all convenient dispatch'. The earthworks alone cost £1,874 involving the removal of 52,000 cu. yd. of earth.

In 1749 while still employed on Fort Cumberland Desmaretz seems to have answered an advertisement by the Board of Trustees for Ramsgate for engineers to deliver sealed plans for piers of stone or wood to enlarge the harbour. He attended a meeting to give his opinion on the manner and materials for the building of these piers, and was recalled for a second meeting on the 15 December for taking into consideration all the plans and proposals. In February 1750 work began, the overall plan of the harbour being only slightly modified from the original prepared by Edward Gatton, of the Ordnance Board, in March 1749.

By 1754 the piers had been built to about one-third of their final length when William Ockenden, one of the harbour Trustees, advocated reducing the width by 286 ft to 1,200 ft. Work on this revised plan actually started, but so annoyed local merchants and seafarers that in February 1755 the House of Commons was petitioned to put a stop to it. After a vast amount of evidence had been heard, Desmaretz and Admiral Sir Peircy Brett were commissioned to report. They proposed to retain the original width but to extend the piers into deeper water than originally intended. For the better protection of ships liable to be damaged by lying aground they also suggested a deepened basin on the east side of the harbour. The total estimate for the works was £195,906 7s 6d. Their report of December 1755, was accepted by Parliament in February 1756, when the works done for contracting the harbour width were ordered to be taken up. Work eventually resumed in 1761 but when the piers were completed by Thomas Preston (q.v.) in 1772 they were very much along the original 1750 lines.

Having acquired a taste for harbour engineering Desmaretz had also been commissioned to survey Shoreham in 1753 and provide a scheme for a new entrance to the harbour, since the mouth of the river had shifted 4 miles to the east of the old village of Shoreham and was blocked with shingle. At the first meeting of the Harbour Commissioners in 'The Star' at Shoreham on 25 June 1760 it was ordered 'that a cut be made through the beach at the place fixed by a plan made by Mr Desmaretz ... and piers erected and built in order to make a useful harbour at the said town'. This was done, and for the first few years the harbour was successful, witnesses stating 'that while the piers remained the harbour was safe and good having ... between 24 and 25 ft. depth' and that 'there was no bar or lodgements at the mouth of the harbour'.

Unfortunately, unscrupulous contractors had only driven half the piles into the marl and chalk foundations, the rest being cut off at 8 ft. and the timber sold. Unsurprisingly, the piers failed and the harbour mouth drifted back east again. When work began on restoring the harbour in 1817 the entrance was again made at the same location as Desmaretz had proposed in 1753. This time, with William Chapman (q.v.) as engineer and William Clegram (q.v.) as resident engineer, the works were properly carried out and proved entirely successful.

Desmaretz's prolific military career continued with a report and estimates for Senegal (1758), and in 1759 the building of sea defences on sites on the south coast most vulnerable to French attack from Littlehampton to Folkestone, completed in 1761 at a combined cost of £14,764. The scale of these works, involving the use of 120,000 'exceedingly well burnt bricks' at Brighton alone, caused a shortage of bricks and timber along the entire coast.

He also worked on most of the Board of Ordnance powder mills and stores, beginning in 1733 with a porch for the west front of the magazine at Greenwich. In 1743 he reported on repairs to the gunpowder store at Woolwich and obviously by now considered an expert, worked on a project for a powder magazine in Jamaica. In 1755 he produced a report and estimates for the grand magazine at Purfleet and in 1762–1763 designed both horse mills and powder mills at Faversham; the same year he also produced reports and estimates for an infirmary at Sheerness.

In 1763 Dunkirk once again came into British possession and Desmaretz returned as First Commissary and along with Andrew Frazer (q.v.) and William Roy (q.v.) drew up plans for the two new channels at Mardyke in 1766. He died there in 1768, but was buried alongside his wife in the garrison church in Portsmouth where the tablet erected by his daughter Mary luckily survived the firebombing in 1941. Mary was married to Stillingfleet Durnford who worked for the civil branch of the Ordnance as Clerk of Deliveries from 1758–1768, and had a son Desmaretz Durnford who served as an engineer in America, and a daughter Charlotte who married his former assistant at Dunkirk, Andrew Frazer. Their son Sir Augustus Simon commanded Wellington's Horse Artillery in the Peninsular War, also continuing a family tradition, since Desmaretz himself, although never holding military rank in the Corps of Engineers had been made a Lieutenant-Colonel of Artillery in 1761.

SUSAN HOTS

[Trustees of Ramsgate Harbour (1749–1772) General Meeting, Minutes and Report books, PRO; Plans (1724–1738) of the Medway, Harwich, Sheerness and Plymouth, PRO; Plans (1743–1766) of Eastney, Portsea, Gosport, Portsmouth, Purfleet and Mardyk, BL map room; A plan of the river Medway from Rochester Bridge to Sharpness Point, Royal Engineers Museum; Royal Engineers Library, Conolly papers; John Smeaton (1791) *An Historical Report on Ramsgate Harbour*; W. Chapman (1815) *Report on the Harbour of new Shoreham*; A naval officer (1837) *A Brief History of Dover and Ramsgate Harbours*; W. H. Tregallis (1886) Historic sketches of the coast defences of England, *Professional Papers, Corps of Engineers*, **12**, 67–86; H. Brookfield (1949) A critical period in the history of Shoreham harbour, *Sussex Archaeological Collections*, **88**, 42–50; G. R. Taylor (1966) *Mathematical Practitioners of Hanoverian England*; A. H. W. Robinson (1962) *Marine Cartography in Britain*; D. W. Marshall (1976) The British military engineer 1741–1783, unpublished Ph.D. dissertation, University of Michigan; R. B. Matkin (1977) Construction of Ramsgate Harbour, *Trans Newcomen Soc*, **48**, 53–71; G. H. Williams (1979) Western defences of Portsmouth Harbour 1400–1800, *Portsmouth Papers*, No. 30; J. G. Coad and P. N. Lewis (1982) Later fortifications of Dover, *Post Medieval Archaeol*, **16**, 141–200; R. Fox (1991) Portsmouth's ramparts revisited, *Fortress*, **11**, 29–38; F. Kitchin (1991) Aspects of the defence of the south coast of England 1756–1805, *Fortress*, **19**, 11–22; P. Magrath (1992) Fort Cumberland 1747–1850, *Portsmouth Papers*, No. 60; J. Douet (1998) *Barracks 1600–1914*]

Publications

1755. *A Plan for Making a Harbour at Ramsgate*, jointly with Sir Peircy Brett

1755. *Report and Estimate Subjoined, Relating to the Harbour of Ramsgate*, jointly with Sir Peircy Brett

Works

Works include:

1747–1750. Fort Cumberland, estimate £3,787
1753. Plan for new entrance to Shoreham Harbour
1755–1763. Powder and horse mills at Faversham
1759. Batteries on coast of Sussex and Kent, £15,000

DEVILLE, James (d. 1846), manufacturer, was elected a Member of the Institution of Civil Engineers on 25 February 1823 and until the early 1830s was a regular attendee of its meetings. Originally a manufacturer of plaster casts, he branched into lamp manufacture and then more general architectural metal ware, including lighthouse lanterns and gas fittings. In this capacity he supplied lights for the Menai Bridge. His contributions to discussions at the Institution covered heating, buildings and the properties of cements and grouts.

His general scientific interests included phrenology; he also collected ancient bronzes.

MIKE CHRIMES

[Membership records and Minutes of Conversations, ICE archives; Annual report (1847) *Min Procs ICE*, **6**, 5]

DIGGES, Thomas (*c.* 1546–1595), mathematical practitioner, MP and General Surveyor of Dover harbour was the eldest son of the mathematician Leonard Digges (d. 1559) of Wootton, Kent, and his wife, Bridget Wilford, and is said to have been educated at University College, Oxford. In 1571 he published *A Geometrical Practice named Pantometrica* based on a manuscript by his father, with additions by himself, on practical geometry and surveying; it includes descriptions of the circumferentor and theodolite. Elected MP for Wallingford in 1572, he sat on the parliamentary committee which in 1576 recommended setting up a commission to report on Winchelsea, Rye and Dover harbours. The Commission, under the leadership of William Borough (q.v.), reporting to the Privy Council, recommended Dover as the most promising for development and produced a scheme for doing so.

The chief problem at Dover had been the northward drift of shingle along the coast. In an effort to prevent the harbour entrance being choked by it a pier was built on its south side by John Thompson (q.v.) in 1535–1540 and an extension, the King's Pier, followed in 1541–1551. Although taken into water 21 ft. deep at low tide the King's Pier failed to stop shingle moving round its end and forming an off-shore bar to Dover Castle. Behind the bar was a salt-water lagoon into which the little river Dour flowed. Borough's plan involved building a long wall on the bar down to the harbour entrance and a cross wall, with a sluice, back to the shore about two-thirds of the way down the lagoon to form a large basin. The basin, later known as the Pent, would fill at high tide and the water thus stored could be released thorough the sluice at low tide to scour the harbour south of the cross wall.

Owing to the large estimated cost of £30,000 nothing was done for the time being. A violent storm in 1579 destroyed what was left of the King's Pier, leaving only its foundation of massive rocks known as the Mole. Consequently a new commission was set up under Lord Cobham (Warden of the Cinque Ports). The Commission called on Borough for advice and with him Matthew Rickworth, sluice-master of Dunkirk, and Peter Pett, master shipwright of Deptford. A plan emerged, revised in detail from the 1576 proposal, which Borough explained to the Privy Council. The estimate was now reduced to £21,000 though a group of Flemish engineers brought in to carry out some first aid works apparently offered to execute the scheme for £18,200.

This still seemed an excessive charge. As an alternative the Commission entered into agreement with John Trew (q.v.) to rebuild the King's Pier on its existing foundation, but now in masonry instead of the original timber framing with an infill of large stones. This more solid and enduring structure in a highly exposed situation was estimated to cost between £8,000 and £10,000. By 1580 Trew was well under way with

100 workmen quarrying stone at Folkestone. Borough had not been consulted on Trew's proposal and early in 1581 he submitted on his own initiative *Notes on Dover Haven* in which, after a dissertation on littoral drift, he demonstrated that, contrary to the view held in certain quarters, reconstruction of the King's Pier would not solve the problem.

It was at this stage that Digges came on the scene at Dover, in July 1581. He seems to have had no connection with the proceedings there since 1576 but is reputed to have made a survey of Winchelsea harbour in 1577. In 1579 he published *An Arithmeticale Militare Treatise named Stratioticos*. About 1580 he married Agnes, daughter of William St. Leger of Leeds Castle, Kent. In the early months of 1581 he was busy in Parliament.

Digges later claimed he was the first to 'discover that grosse error of [John] Trew', having been called to examine the question, but more probably he was consulted as a result of Borough's report. However that may be the Privy Council now had to admit that Trew's plan 'is not likely to take that good effect we looked for' and informed the Commissioners that he would have to be discharged.

Attention then returned to Borough's plan. Before the end of the year Digges made an accurate survey of the old harbour and the offshore bar. In December preparations were in hand to obtain 1000 ton of timber and a pile driver for the timber-framed walls of the Pent envisaged by Borough. Early next year Fernando Poyntz, who had recently completed repairs to the Thames bank of Plumstead Marshes, offered to execute the scheme for £14,600 using less costly walls composed of shingle mixed with partially dried mud. The Council then sent Borough and Digges to Dover for a conference, and at about that time (probably in February 1582) Digges presented to the Queen his *Brief discours … [on] making Dover Haven*.

This repeats the criticism of Trew's project, backed by short accounts (dated December 1581 and January 1582) of the history and present state of the harbour by ship masters and towns people of Dover, and includes a detailed description of his own proposals. These amount to a major revision of Borough's plan, still with the long wall but with the sluice and an adjacent lock gate placed close to jetties at the (new) harbour mouth, thus creating a much larger basin which in effect becomes the harbour (or, strictly, a dock) to be entered at slack water around high tide. The cost is kept down to £13,365 by building the wall as a bank of shingle and chalk with a central core of mud; the banks are to be about 10 ft. high, rising 3 ft. above high water level. He also proposes to remove the inner part of the Mole and use the rocks to raise the outer part to form a breakwater: cost £1,760.

In March 1582 a new commission was appointed, again under Lord Cobham but now with

Digges, Sir Thomas Scott (q.v.) and Richard Barrey (Lieutenant of Dover Castle) among its members. In April they wrote to the Council saying 'by the good advice of Sir William Wynter, Mr. Digges and Mr. Borrowe we have finally decided on such a plat as we judge most convenient'. Wynter at that time was Surveyor and Master of Ordnance of the Navy, brought in, like Borough, as a consultant to the commission.

In the plan then adopted the long wall would finally be built to its full length of about 3,300 ft. down to the north entrance jetty, but in the first instance it would be only some 2,000 ft. in length and from its southern end the cross wall would run about 600 ft. back to the shore with a lifting-gate sluice near its landward end at the position shown in Borough's 1579 plan. The Pent, thus created, would have an area of 17.5 acres; the harbour, south of the Pent, would cover 15 acres at high tide. Subject to a decision on the exact position of the new harbour mouth, the outer part of the Mole would become a breakwater as proposed by Digges. Later the long wall would be extended to its full length and entrance jetties built.

After deciding on this sensible plan, progress was rapid. Digges was appointed General Surveyor, Scott become Treasurer for the first year, and John Hill acted as the local administrator. The officers reported to Francis Walsingham, Secretary of State, who controlled the project on behalf of the Privy Council. In June Poyntz and his Plumstead men were employed to open a cut through the old harbour entrance in order to drain the lagoon as far as possible, to facilitate construction of the walls, and to build (temporary) protective groynes. Thomas Bedwall supervised these operations which were completed in January 1583 at a cost of £1,200.

Meanwhile, methods of building the walls were under discussion. The master shipwrights, Matthew Baker and Peter Pett, favoured timber construction, as proposed by Borough. Poyntz continued to press for his banks of shingle and mud. Digges preferred his shingle and chalk banks with a central core. Then, in a decisive intervention, Scott suggested the method long used in building and repairing the sea wall of Romney Marsh at Dymchurch. This involved a bank of earth fill, tipped from horse-drawn carts and levelled in layers by workmen, protected on the seaward face by faggots fastened down by short oak piles. The question was finally settled in March when Walsingham brought two of the Romney experts to London and arranged a conference, under Wynter, with them and Digges, Borough and Poyntz.

In April the Commissioners accepted the Romney principle and announced that work would begin in May. On the appointed day no fewer than two-hundred carts arrived on site. By the end of June the number had risen to above five hundred, implying a workforce of a least a thousand men and boys. Sir Thomas directed the work with enthusiasm and personally supervised the cross wall; Barrey supervised the long wall. Both walls were completed by August, with a improvised sluice in the cross wall. Pett and Baker then proceeded to built a permanent, timber-framed, sluice.

The next question related to the position of the new harbour entrance. Opinions were sought from Martin Frobisher and John Hawkins, as well as Digges, Borough and Barrey among others. The problem was resolved in June 1584. It involved the removal of the inner half of the Mole, a length of 500 ft. Digges, in London, sent Robert Stickells to supervise this difficult operation as 'work master'; he completed the task by the end of 1585. Another engineer, Paul Ivye (q.v.), was brought in to construct two groynes with the object of protecting the bar from erosion in front of the long wall. To obtain the benefit of expert advice from the Low Counties, Walsingham called on Humphrey Bradley (q.v.). He took part in discussions on the harbour entrance (in June 1584) and was consulted on the groynes. He also advised on the sluice, which turned out to be leaky and difficult to operate.

Among those employed in this period was John Symonds (Surveyor of Works at the Castle) who took soundings within and around the harbour in December 1583 and in September 1584 he was called in to confer on the design of jetties and the suitability of different types of local stone for the purpose. He also produced a plan of the works as they existed c. 1584. By 1586, with repairs or perhaps a rebuilding of the sluice, the first stage of the scheme envisaged in the 1582 plan had been completed at a cost of £12,000. Meanwhile Digges was busy in the new Parliament in 1685, having been elected MP for Southampton, and in The Netherlands under the Earl of Leicester, a post he held until March 1594.

At Dover little happened during the years 1586–1591. In the latter year a new administration was established with a Receiver and Paymaster in the person of Sir Thomas Fludd, who had a clerk, a sluice-keeper and a clerk of works under him, with an overseer brought in when there was work to be done. The long wall was extended to the harbour mouth in 1592–1593. John Stoneham, a carpenter from Rye, was brought in to construct the entrance jetties which were built in the traditional way: timber framing with an infill of hard stones. Finally the sluice was rebuilt in masonry by William Taylor in 1598, he having been dispatched beforehand to inspect sluices in Zeeland and Holland. This brought to a conclusion the most successful civil engineering work of Elizabeth's reign. Its total cost is not known, but probably did not exceed £20,000.

On his return to England Digges seems to have been involved again at Dover and a very fine plan of the harbour dated 1585 is always attributed to him. When he died on 21 September 1595 he was not more than about fifty years of age. Besides about a dozen published works he left uncompleted manuscripts on astronomy, navigation and fortification. He had two sons, Sir Dudley (1583–1639) and Leonard the younger (1588–1635), and also two daughters.

A. W. SKEMPTON

[T. W. Wrighte (1794) Transcript of Thomas Digges [1582]. A brief discourse … [on] the making of Dover Haven, *Archaeologia*, **11**, 212–254; T. Coope (1888) Thomas Digges, DNB, **15**, 71–73; Alec Macdonald (1937) Plans of Dover harbour in the sixteenth century, *Archaeologia Cantiana*, **49**, 108–126; E. G. R. Taylor (1957) Cartography, survey and navigation, in: C. Singer (ed.) *A History of Technology*, vol. 3, 536–557; A. H. W. Robinson (1962) *Marine Cartography in Britain*; John Summerson (1982) Dover harbour, *King's Works*, **4**, 755–764; Stephen Johnston (1994) *Mathematical Practice … in Elizabethan England*, unpublished Ph.D. thesis, Cambridge University; Dorothy Beck (1995) The drainage of Romney Marsh and maintenance of the Dymchurch Wall in the early seventeenth century, in: Jill Eddison (ed.) *Romney Marsh, the Debatable Ground*, Oxford University Committee for Archaeology Monograph No. 41, 164–168]

Works
1583–1585. Dover harbour, the Pent, £12,000

DODD, Barrodall Robert (*c.* 1780–1837), civil engineer, was the eldest son of Ralph Dodd (q.v.). Details of his early life are lacking, but, like his younger brother George (q.v.) he was employed as his father's assistant on the Grand Surrey Canal scheme from which all three were dismissed in April 1802. In 1805 he took out a patent (No. 2839) for a fireplace. About 1808 Dodd became involved in water supply schemes in Portsmouth, and was engineer for the Farlington scheme which was one of two schemes to obtain an Act of Parliament in 1809. At the end of the year he resigned and was succeeded by Ralph Walker (q.v.); G. W. Buck (q.v.) was the resident engineer. It can be assumed that he accompanied his father to the north-east around 1810, where the family had probably originated. In that year Dodd produced his earliest known report, on a proposed canal between Newcastle and Hexham, a proposal that had been looked at over the previous twenty years by his father and William Chapman (q.v.), among others. In early 1814 he became involved in a steamer service on the Tyne. The first vessel, the *Tyne Steam-Packet* was launched at Gateshead in February, and was in service in May. In November it was announced that he was going to start a similar service on the Thames but, in fact, it seems that it was George who took on this venture. Two years later, however, it appears that Dodd was 'constructing' steam vessels on the Tyne.

There then follows a hiatus in his known career until 1826 when he became involved in a proposal for a high-level bridge across the Tyne at

Newcastle, a proposal which was revived in 1836 with Samuel Brown (q.v.) as a chain suspension bridge; this was his last known scheme. He died of asthma at Old Fish Street in the City of London on 26 September 1837, his son Robert having predeceased him in 1816. At the time of his death he was campaigning for posthumous recognition of his father's achievements.

MIKE CHRIMES

[J. G. James collection, ICE Library; *New Monthly Magazine*, 1816, **5**, 469; M. Hallett (1971) Portsmouth's water supply, *Portsmouth Papers*, **12**, 3–25; J. G. James (1976) Ralph Dodd, the very ingenious schemer, abridged in *Newc Soc Trans*, **47**, 161–178]

Publications
1810. Report … on the proposed canal navigation between Newcastle and Hexham, in: W. Thomas (1825) *Observations on Canals and Railways*, 31–44
1826. *Important/improvements of the Great North Road, by a Proposed Bridge over the River Tyne*

Works
1809. Portsmouth (Farlington) Waterworks, Engineer

DODD, George (*c.* 1783–1827), civil engineer, was the second son of Ralph Dodd (q.v.). Details of his early education are lacking, but once his father's career in civil engineering had begun in earnest Dodd can be assumed to have worked under him. He is first mentioned as giving evidence in 1801 on his father's Grand Surrey Canal scheme which connected Rotherhithe (and its docks) to Vauxhall, with a branch from Kennington to Mitcham. Ralph was appointed engineer, with Dodd and his brother Barrodall Robert (q.v.) as assistants. All were dismissed in April 1802 with work under way, allegedly because Ralph was not devoting sufficient time to the project.

The same year Dodd invented a form of gun lock (Patent No. 2825, 1805), proof against accidental discharge, which was awarded a Society of Arts medal in 1803. It was approved by the Ordnance Board in 1804 and patented in 1805. With this success he moved on to his next scheme, the forerunner of Waterloo Bridge. The original proposal envisaged raising £100,000 to erect a timber bridge and, from tolls, eventually to finance a masonry structure. The City of London opposed the idea of a temporary structure, and a more ambitious scheme for a masonry bridge, involving the raising £500,000 capital, after failing to obtain Parliamentary approval in 1808, obtained its Act on 20 June 1809. Dodd issued his first report in 1806, more detailed plans in 1807, with a revised report appearing in 1809. His design for a masonry arch structure with *corne de vache* details at the lower edges, emulating Perronet's Neuilly bridge, was exhibited at the Royal Academy in 1810.

By then Dodd's involvement was under threat. No doubt concerned about his lack of experience with such a major structure, William Jessop and John Rennie (qq.v.) were consulted about its feasibility. Rennie pointed out vagueness in the scheme and requested further details, which Dodd was probably unable to provide, and in June 1810 his design was rejected and Rennie made chief engineer. Although Dodd acted as resident engineer at £1,000 per annum, and received £5,000 from the company, his active involvement ceased the following year, when he was also declared bankrupt; Rennie developed his own design.

It is unclear what George did next but he was involved with paper-making machinery and took out patents for artificially making wine (Patent No. 3522, 1812) and telescopic umbrellas (Patent No. 3668, 1813). In August and September 1813 he surveyed a proposed Portsmouth breakwater for Alexander Lamb and then became interested in steamboat services on the Thames. His brother and father also became involved in steamboats on the Tyne at the beginning of 1814 and in 1815 Dodd was appointed engineer to a company intending to run steam services from London to Richmond, Margate and Gravesend. In April he brought their vessel, *The Thames*, from Port Glasgow via Dublin to London, arriving in June. By 1817 he had five vessels, all powered by low-pressure wrought iron boilers, in his charge and gave evidence to the Select Committee investigating steamboat explosions. He clearly had a great deal of knowledge about the vessels then in service although his evidence in favour of low pressure boilers was essentially conservative. The vessels under him—*The Thames* (74 ton, 14 hp engine), *Majestic* (90 ton, *c.* 20 hp engine), *Richmond* (60 ton, 8 hp engine), *London* (70 ton, 14 hp engine) and *Sons of Commerce* (80 ton, 20 hp engine)—cost on average £2,000 to £4,000, and his experience was incorporated in his *Historical and Explanatory Dissertation on Steam Engines and Steam Packets* (1818), largely a reprint of the Parliamentary evidence. In the same year he supervised the construction of another vessel, the *Viking* (160 ton, 40 hp engine), for the Margate service, and this probably marked the height of his fame and prosperity. By the mid-1820s competition was becoming cut-throat, with larger, faster, high-powered vessels requiring more capital, to which Dodd lacked access.

Dodd turned increasingly to drink and, although he took out a patent in October 1824 for a method of extinguishing fires at sea using water jets (Patent No. 5023), this was a last, unsuccessful flourish. On 16 September 1827 he was turned away, drunk, from some lodgings, appeared in court the next day, and died at Giltspur Street, Compter, on 25 September 1827, having refused all medical aid; he left a son and daughter.

MIKE CHRIMES

[Rennie reports, ICE archives; Strand bridge (1808), ICE Works of art catalogue, No. 250; Memorandum (1808) Select Committee on Highways of the Kingdom, 2nd report, 163; *Journal des mines*, 1815, **38**, 175–199; Evidence (1817) *House of Commons Select Committee on steamboats*, 29–31, 37, 45; Obituary (1827) *Gentlemen's Magazine*, **97**, 468–469; W. Baddeley (1828) *Mechanics Magazine*, **10**, 217–218; *Mechanics Magazine*, 1834, **22**, 65–76; C. Hadfield (1969) *Canals of South and South East England*; J. G. James (1976) Ralph Dodd, the very ingenious schemer, abridged in *Trans Newc Soc*, **47**, 161–176]

DODD, Ralph (*c.* 1756–1822), was the second of three sons of Alexander Dodd, a burgess of Berwick, resident in the City of London. His family is believed to have come from the North East. Ralph's brothers, Robert and George, became artists, Robert achieving fame as a marine painter, and Ralph himself, after attending school and receiving some education in mechanics, spent five and a half years in the Royal Academy Schools and practised as a portrait painter.

In the early 1790s he became interested in engineering, no doubt inspired by the growth in activity at the time, and around 1794 developed some kind of mechanical aid to excavation which was put to trial use on the Grand Junction Canal at Dawley. Full implementation of the equipment never took place, allegedly because of the luddite attitude of the workforce, and this rather inconspicuous beginning was characteristic of Dodd's subsequent career. Ralph Dodd had numerous ideas but saw virtually none through to reality.

To deal with his canal engineering exploits first, following his Grand Junction experiences Dodd turned his attention to his native North East and, after reporting on improvements to the river Wear, in mid-1794 he surveyed and promoted a canal from Newcastle to Carlisle later that year. This was followed by publishing a plagiarised version of John Phillips's (q.v.) *General History of Inland Navigation* (1792). While this may point to a lack of originality in Dodd's thought, it is at least indicative that he was making up for lack of practical experience as an engineer by reading up on the subject.

Dodd continued surveying and reporting on canal and harbour improvements in the North East over the next few years, at Hartlepool (1795 and 1802), Tynemouth Harbour (1796)—where he may also have designed a dry dock—and canals from Stockton to Winston via Darlington, with a branch to Durham, and from Redheugh on the Tyne to Picktree, near Chester-le-Street. The Stockton-Winston canal had already been surveyed by Joseph Brindley and Robert Whitworth (qq.v.) and Dodd merely updated the estimates and developed the Durham branch.

None of these early canal schemes were realised and Dodd turned his attention to mining and tunnelling schemes. His 1796 report on Tynemouth had contained a suggestion for a tunnel between North and South Shields beneath the Tyne. Citing underwater mine working at Wylam and Whitehaven, he developed this 1797–1798 report using knowledge of the original estimates of the tunnels on the Grand Junction Canal to estimate costs for a 400 yd. tunnel. In 1798 this was followed by a proposal for a tunnel beneath the Thames between Gravesend and Tilbury and marked a move to the London area. An Act was obtained for this tunnel in July 1799 but, although a shaft was sunk on the Gravesend shore, work was abandoned in 1802 due to flooding of the workings and lack of finance.

As with his North Eastern proposals of the 1790s a flood of civil engineering schemes in the south-east of England now followed, few of which progressed beyond the pamphlet stage. Linked with his Thames tunnel was a proposed Thames–Medway Canal incorporating a 700 yd. tunnel. An Act was obtained in 1800 and, following some suggested amendment by John Rennie, work progressed until 1802 when Dodd was dismissed. He was succeeded by John Rowe, at a reduced salary. In 1799 Dodd suggested a canal between Croydon and Rotherhithe. This developed into the Grand Surrey Canal, for which Dodd, assisted by sons Barrodall and George (qq.v.), became engineer in 1801, and also the Croydon canal, where Rennie replaced him as engineer, a position which was to be repeated. Dodd was spending too much time elsewhere to warrant his £600 p.a. salary. Although Dodd was involved in some further canal proposals—the North London Canal from London to Cambridge (1802), a link between the Basingstoke and Andover canals (1810) and a continuation of the Surrey Canal to Vauxhall (1815)—none was executed, and his attention turned elsewhere.

His interest in port and harbour improvements also turned from north-east to south-east England and he became interested in improvements to the Port of London. He suggested in 1798 that a new, high-level, five-span cast-iron bridge be built to replace London Bridge, improving access up the Thames to Blackfriars. This was received by the House of Commons in 1799–1800 and became one of a number of schemes suggested for a new London Bridge, most famously that of Thomas Telford (q.v.). Bridges now became a major interest of Dodd, and he only produced two more harbour schemes, Brighton (1806) and Grimsby (1811). In 1806 he produced a preliminary proposal for what was to become Vauxhall Bridge, a thirteen span masonry structure, whilst his son George developed proposals for what was to be Waterloo Bridge. After three unsuccessful Bills (1806–1808), in 1809 Vauxhall bridge received its Act and, almost immediately, following reports by Robert Mylne (q.v.), Dodd was replaced by Rennie as engineer.

Typically Dodd was undismayed. In 1808 he took out a patent (3141) for *Improved Bridge Floorings or Platforms, Fireproof Floorings and Fireproof Roofings*, using wrought iron tubular beams, columns, piles and iron arched plates in conjunction with earth or concrete. It would appear that the patent was not exploited, with the exception of tubular iron piling which, made of cast iron, was used at Springfield Bridge over the Chelmer in 1819–1820. This structure, one of the few engineering works for which Dodd can actually take any credit, was on his system of tenacity, developed in 1814, which appears to have been a form of trussed girder combined with tubular beams; this system was first suggested for a crossing of the Thames between Ratcliff and Rotherhithe in 1816–1817. A trial span was erected, and an example possibly exported to the West Indies. Following a proposal for a bridge across the Fleet at Holborn, work began at Chelmsford. Completed in 1820, by February 1821 it had been reinforced with cast iron dovetails, Henry Maudslay and Bryan Donkin (qq.v.) reporting on its many defects of construction. Reports of this bridge's fate can have done nothing for Dodd's proposals for a pier on his system at Brighton in 1819 and a replacement for London bridge the following year.

While Dodd promoted his bridge schemes he was also involved in a number of proposals to improve water supply in London and elsewhere. In 1804 he proposed waterworks in south-west and north-east London, following them up with his *Observations on Water* to encourage investors; further proposals followed for west Middlesex and Kent. Subsequently Dodd was involved with waterworks promotion in Birmingham, Portsmouth, Manchester and Colchester (all in 1808). The South London Waterworks received its Act in 1805 and on its completion in 1807 became one of Dodd's few successful engineering ventures, subsequently known as the Vauxhall Waterworks. Dodd took out a patent (No. 2397, 1806) for applying a Trevithick style high pressure engine to raising water in 1806, possibly building on his experience there. Although the West Middlesex (1806), East London (1807) and Kent Waterworks (1809) Companies received their Acts, in all cases Dodd was replaced as Engineer almost immediately.

With the end of his waterworks promotions he became interested in steam boats and his sons, George and Barrodall, were both to become pioneers of steamer services. Ralph's own involvement, however, was ultimately fatal as he was injured in a boiler explosion on the *Sovereign* steamboat at Gloucester in 1822. He was advised by his doctor to visit Cheltenham Spa but by then he was so impoverished that he had to make the journey on foot; clearly in a weakened physical state, the next day he called for a doctor but it was too late and he died at 10 a.m. on 11 April 1822. With only £2 5s on his person he left his seventy-year-old wife of forty-six years standing in straightened circumstances, a daughter Fanny, and two surviving sons. A third son, Thomas Alexander, also alleged to have practised as a civil engineer, is believed to have predeceased him.

Dodd was a prolific promoter of engineering schemes but lacked the civil engineering experience, and possibly business contacts, to take them through to fruition. Often a plagiarist, it would be easy to dismiss him as no more than a pamphleteer were it not for the fact that several of his schemes in the London area in particular became a reality, albeit under other engineering guidance.

A namesake Ralph Dodd(s) (1792–1874) was head viewer at Killingworth Colliery. While there he trained his relative, Isaac Dodds (1801–1882), who subsequently had an impressive career as a railway engineer and ironmaster, and also Nicholas Wood (q.v.). He was a mentor to George Stephenson (q.v.) and with him took out a locomotive patent (No. 3887, 1815) which incorporated features of Trevithick's *Catch-me-who-Can* locomotive. Details appear in the *Repertory of Arts and Manufactures.*

MIKE CHRIMES

[J. G. James collection, ICE archives; Rennie reports, ICE archives; Royal Academy Archives; *Nicholson's Journal*, 1798, **2**, 239–240, 473–476; Evidence (1801) *Sheerness Pier Bill*, 11 June, HLRO; Evidence (1805) *Select Committee on Camberwell Waterworks*, 5–6; Evidence (1806) Brighton, *J House Commons*, **19**, 93–368; Evidence (1808) Chichester Waterworks Bill, 15 June, HLRO; Colchester waterworks, Drawing 2174, list of (Telford) drawings, ICE; Evidence (1808) Portsmouth Waterworks Bill, HLRO; Obituary (1822) *Gents Mag*, **92**, 474; R. Stuart (1829) *Historical and Descriptive Anecdotes of Steam Engines*, vol. 2, 534–536; S. A. Snell (1921) *A Story of Railway Pioneers*; H. W. Hart (1964) Ralph Dodd and the Gravesend and Tilbury tunnel scheme, *J Transp Hist*, **6**, 249–257; J. G. James (1976) Ralph Dodd, the very ingenious schemer (abridged in *Newc Soc Trans*, **47**, 161–178; A. W. Skempton (1980) Telford and the design for a new London bridge, in: A. Penfold (ed.) *Thomas Telford, Engineer*; A. W. Skempton (1981–1982) Engineering in the Port of London, 1789–1808, *Newc Soc Trans*, **53**, 100–101; M. Chrimes (1985) Bridges: a bibliography of articles published in scientific periodicals 1800–1829, *Hist Technol*, 10, 217–257; DNB]

Works

1800–1802. Grand Surrey Canal, engineer
1801–1802. Gravesend-Tilbury tunnel, shaft sunk, engineer
1805–1807. South London Waterworks, engineer
1819–1820. Springfield (iron) Bridge, Chelmsford, engineer

Publications

1794. *Reports to the Honourable Commissioners of the River Wear, on the state ... of the Port of Sunderland*

1795. *A Short Historical Account of the Great Part of the Principal Canals in the Known World*
1795. *Report on the First Part of the Line of Inland Navigation from the East to the West Sea by way of Newcastle and Carlisle*
1795. *Report on the various Improvements, Civil and Military that might be made in the Haven ... of Hartlepool*
1796. *Report on the present state of Tynemouth Harbour*
1796. *Report on the Line of Inland Navigation, from the City of Durham, to the Navigable part of the River Wear*
1796. *Report on the Line of Inland Navigation from Stockton to Winston*
1797. *A Report, and Estimate, on the Projected dry Tunnel ... under the River Tyne, to Communicate with the Two Towns of North and South Shields*
1798. *Reports ... of the proposed Dry Tunnel ... from Gravesend ... to Tilbury ... also on a Canal from near Gravesend to Strood*
1798. *Letters to a Merchant, on the Improvement of the Port of London; demonstrating its Practicability without Wet Docks*
1799. *Report on the proposed Canal Navigation, forming a Junction of the Rivers Thames and Medway*
1799. *Plan of the Intended Thames and Medway Junction Canal*
1799. *Introductory Report on the proposed Canal Navigation from Croydon to the River Thames at Rotherhithe*
1799. *Plan of ... the General Line of the proposed Croydon Canal*
1799–1800. Mr. Dodd's account of London bridge, with ideas for a new one, *Third Report, Select Committee upon the Improvement of the Port of London*, 47–50, 51–54
1800. *Report on the Intended Grand Surrey* [sic] *Canal Navigation*
1800. *Plan of Part of the proposed Grand Surrey Canal*
1800. *Introductory Report ... on the proposed Bridge across the Mersey at Runcorn*
1802. *Report on the Intended north London Canal Navigation ...*
1804–1805. *Observations on Water*, includes reports on South and East London waterworks
1807. *Introductory Report on the Intended Bridge across the River Thames ... near Vauxhall ...*
1809. Specification of the patent ... for improved bridge floorings ..., *Rep Arts Manufs, 2nd series*, **14**, 145–147
1815. *Practical Observations on Dry Rot*
1816. Tontine Waterloo bridge, *New Monthly Mag*, **5**, 168–169
1817. Iron bridges on the principle of tenacity, *Phil Mag*, **50**, 389–391
1817. Proposed iron bridge of tenacity, from Holborn Hill to Snow Hill, *Eur Mag*, **72**, 120–121
1819. *Engineer's Report on the Brighton Sea Jetty*
1820. *Report ... on London Bridge, with descriptive Plans for a New One*

DODSON, William (fl. *c.* 1630–1665), civil engineer, was born at Ely, probably about 1600. As a young man he visited The Netherlands where, in his own words, 'by my observations ... I gained some experience to serve my native country'. It can be assumed that this service included employment on the drainage of the Fens in the first campaign (1631–1636) under Francis (fourth) Earl of Bedford, since in his report on the Great Level of the Fens, written in 1664, Dodson refers to 'my industry and thirty years experience and more' as the basis for his detailed description of that region. Whether he worked in the second campaign (1640–1641), in which Sir Cornelius Vermuyden (q.v.) was Director of Works, is uncertain.

During the Civil War Dodson served in the Parliamentary army, attaining the rank of Lieutenant-Colonel, and was Governor of Crowland in 1646. In June of that year a committee set up by Parliament gave notice that they were ready to receive petitions for and against further work in draining the Fens. During the autumn Dodson, with two others, was involved in discussions on the numerous petitions submitted and in reporting to the committee. Vermuyden and Andrewes Burrell (q.v.) gave evidence early in 1647 and, after interruption by the second Civil War, the Act of 29 May 1649 was passed confirming William (fifth) Earl of Bedford as undertaker for completing the drainage of the Great Level begun by his father. Within days of the Act being passed the 'Company of Adventurers', headed by Bedford, asked Vermuyden to become Director but sought advice as to the best plan to be adopted. Vermuyden submitted his proposals on 11 June; the Dutch engineer Jan Barents Westerdyke was consulted; and, on a petition from the Isle of Ely, so was Dodson. He presented his 'Designe' on 14 June. To do so he must have been considering the matter for some time and was clearly drawing on previous experience.

The principal works proposed in Dodson's scheme were: (1) to deepen the old South Eau for 2 miles north of Guyhirn to Murrow and thence to make a new embanked cut 50 ft. wide running 5 miles to the river Nene below the sluice at the Horseshoe, a mile or so downstream of Wisbech. This would benefit the fens north of Wisbech and act as slaker, or flood relief channel, for Morton's Leam, the channel of the Nene from Guyhirn up to Stanground near Peterborough; (2) to make a new embanked cut 80 ft. wide and 6 ft. deep and about 8 miles long from the Horseshoe to the river Ouse at St. Germans, where (3) a great sluice would be built in a cut alongside the river with its floor 2 ft. below low water level, which there was several feet lower than low water at Wisbech; and (4) a pound lock (sasse), near the sluice for navigation from King's Lynn up to Ely and points beyond. A similar sluice and lock on the Ouse 2 miles upstream at Magdalen had been proposed by Burrell in 1642.

Dodson's scheme appears to have received little support; the company could not agree with Vermuyden as to his terms of employment and Westerdyke had not yet reported. In the hope of getting something started Sir Miles Sandys and two other Adventurers, with assistance from Dodson, submitted on 22 June 'a Designe and estimate of the works ... thought fit to be done this summer'. A few weeks later Dodson, as acting Director with Robert Burton (q.v.) as Overseer, began operations only to be frustrated by lack of funds and the necessity, in September, of having to dismiss the workmen. At about this time Dodson himself became an Adventurer by purchase of 256 acres from Sir Miles. There followed a confused period of further negotiations with at least one other scheme coming under consideration until, on 25 January 1650, Vermuyden was appointed Director of Works on conditions less favourable than he had originally been willing to accept.

Dodson was asked to submit his expenses in February. Vermuyden began setting out the first works in March, again with Burton as Overseer or resident engineer, and Jonas Moore (q.v.) became Surveyor General in August. Vermuyden's scheme, which was carried out in the third and final campaign in 1650–1656, differs materially from Dodson's and in almost every respect from that eventually submitted by Westerdyke in September 1650.

Contrary to what has been assumed there is no evidence of Dodson being employed on Vermuyden's staff or succeeding him as Director. At some time in the 1650s he returned to The Netherlands where between 1657 and 1660 he took out several patents relating to drainage mills. In 1662 he was back in England as one of the Adventurers petitioning against certain aspects of the proposed formation of the Bedford Level Corporation.

When the Corporation was established in July 1663 drainage of the Level had been formally recognised as 'complete'. But much remained to be done by way of improvement and in this connection Dodson was called for advice as an independent consultant. His report, delivered in June 1664, is essentially a repeat of his 1649 proposals, presented in detail with estimates of cost together with many general observations. Among these is his recognition (expressed here for the first time) that a fundamental problem in the Fens is the shrinkage of peat land after drainage. He also exhibited a model of the proposed stone-built sluice on the Ouse at St. Germans and in the text gives all the leading dimensions. Its cost is estimated at £10,000 (including £200 for tarras) plus £200 for the foundation pit, taken 4 ft. below floor level to good solid clay, about £300 for the dam across the river and £400 for the cut.

When published in 1665 the report included answers to objections to his scheme, raised since its submission, and his opinion, given in November 1664, on another scheme then under consideration. This resembled Westerdyke's plan of dispensing with sluices on the main rivers thus leaving them open to the tides and requiring much higher banks; an idea haunting Fens engineers up to the early nineteenth century. Dodson argued against it, for good reason. His own scheme was not adopted though elements of it were incorporated in some later proposals.

A. W. SKEMPTON

[Andrewes Burrell (1642) *A Briefe Relation discovering ... how* [the Fens] *may be drained*; Thomas Badeslade (1725) *The History of the ... Port of King's Lynn ...* [and] *... of Draining the Great Level of the Fens*; Samuel Wells (1830) *The History of the Drainage of ... Bedford Level*, a reprint of Dodson's report is given in vol. 2 of this work; H. C. Darby (1940) *The Draining of the Fens*; G. Doorman (1942) *Patents for Inventions in The Netherlands during the 16th, 17th and 18th Centuries*; L. E. Harris (1953) *Vermuyden and the Fens*; Keith Lindley (1982) *Fenland Riots and the English Revolution*; Frances Willmoth (*in lit* 1996) additional information]

Publications

1665. *The Designe for the perfect Draining of the Great Level of the Fens, called Bedford Level ... As also, several objections answered since the Delivery of the said Design*, with folding engraved plan.

DONKIN, Bryan, FRS (1768–1855), Vice-President of the Institution of Civil Engineers, was born at Fountain Hall, Sandoe, Northumberland on 22 March 1768, the third son, in a family of eight, of John Donkin, a land surveyor. His father was agent for the Errington family and the Duke of Northumberland.

From an early age Donkin displayed an interest in mechanical contrivances and built his own workshop. He was initially trained as a surveyor by his father, before working as a land agent for four years—from the age of twenty—to the Duke of Dorset at Knowle Park, Sevenoaks, Kent. His father knew John Smeaton (q.v.) from works on the river Tyne, and on Smeaton's advice in 1792, Donkin was apprenticed to John Hall (1755–1836), millwright of Dartford.

At that time Hall was involved in installing machinery in various paper mills, and it was in that area that Donkin first made his name. At that time paper was still largely made by hand but in 1799 Louis Robert took out a French patent for a paper (mould) making machine. John Gamble, then resident in Paris, introduced the idea to a leading London stationery firm, the Foudriniers (Walker, Bloxham and Foudrinier), who approached Hall about the idea. Donkin was entrusted with making a model and in 1802 he was established in a new works at Fort Place, Blue Anchor Road (Southwark Park Road), Bermondsey. When giving Parliamentary evidence in 1807 in support of the extension of the patent, he was still described as foreman rather than proprietor.

Hall had already advanced Donkin £250 in capital to set himself up as a paper mould maker in 1798, when he married Hall's wife's sister, Mary Brome. The Foudriniers advanced further capital to enable Donkin to work on a practical machine and the first was installed at a mill at Frogmore in 1804, with a more successful machine at Two Waters Mill, Hertfordshire in 1805. Around 1808 the Foudriniers leased the Bermondsey works to Donkin and by 1810 he had installed 18 machines and established his reputation. At the time of his death over two-hundred had been installed at home and abroad.

The early paper patents (extended by Act of Parliament in 1807) were taken out by the Foudriniers and Gamble as they had funded the developments but from 1803, for nearly forty years, Donkin took out a variety of patents covering various aspects of machinery and gearing, steel pen nibs, textile manufacture, printing and, latterly, paper making.

Donkin's reputation was established in what would now be considered the sphere of mechanical engineering. He was familiar with all aspects of steam engine and machine tool design and knew all the leading engineers in these fields. His work on dividing engines, a development of that of his friend Henry Maudslay (q.v.), was pursued for more than thirty years (1810–1846). Donkin helped manufacture Babbage's calculating machine.

With Hall and Gamble, Donkin took out a patent in 1810 for the preservation of food, based upon the process of a Frenchman, Nicholas Appert. This was the origin of the canned food industry and, after a period of experimentation by Donkin, from 1812 the Bermondsey works were supplying the Royal Navy.

As an eminent engineer Donkin was frequently consulted on civil engineering matters. He supported Telford's proposals for a massive suspension bridge at Runcorn using bar cables (1814) and assisted in the pioneering tests using hydraulic equipment at Bruton's (q.v.) manufactory in 1817. By the 1830s he was convinced of the need for vertical and horizontal trussing to stabilise suspension bridge decks. In 1821 he and Maudslay reported on an iron bridge erected to the design of Ralph Dodd (q.v.) at Springfield, Chelmsford; although critical of the design, they supported the payment of the contractors who had fulfilled their obligations. In the 1820s Donkin was involved with the Thames Tunnel scheme. He had got to know Marc Brunel (q.v.) when he supplied equipment for his machinery at Chatham Dockyard. He was one of the original committee and in 1829 his eldest son, John, married Caroline, the daughter of the Tunnel Chairman, Benjamin Hawes. In 1825–1827 Donkin supplied equipment for pumping water from the tunnel and also workmen for modifying the shield; at one time it was suggested that he should replace Brunel as engineer. In 1826 he built a model of a landing stage proposed by Brunel for use at Liverpool.

Donkin's Works regularly supplied equipment for use on civil engineering schemes, including dredging machines for the Prussian Government in 1817 and for the Calder and Hebble Navigation in 1824. Much was supplied for the Caledonian Canal, including stationary steam engines for use in construction of the locks. Thomas Telford (q.v.) subsequently commissioned Donkin to convert these for use in steam tugs on the canal but Donkin advised against their use. In 1833 Telford employed Donkin in his surveys of rivers in the London area for the Water Supply report completed shortly before his death. In 1834, with William Cubitt (q.v.), Donkin reported on the Montgomeryshire Canal's 'Western' branch.

Donkin exercised a great influence, directly or indirectly, on the development of the civil engineering profession. In 1805, with Hall and others he had formed the Society of Master Millwrights, acting as its Treasurer. He was a member of the (Royal) Society of Arts, becoming a Vice-President, Chairman of the Committee of Mechanics, and receiving two Gold Medals. Among his pupils was Henry Robinson Palmer (q.v.) the key figure in the formation of the Institution of Civil Engineers in 1818. The Donkin family soon became heavily involved in the Institution and Donkin himself joined on 4 December 1821. He regularly chaired meetings, initially as 'Chairman in ordinary' and subsequently as a Vice-President; on Telford's death in 1834 he stood unsuccessfully as his successor. It was his influence, together with Telford's, which enabled the Institution to obtain its Royal Charter in 1828 and he advanced 100 guineas towards the costs.

His paper, presented in 1824, *On the Construction of Carriageway Pavements*, recommending the use of a subgrade of broken stone, was the first Institution paper to be published. Another early paper, on hydraulic experiments conducted at Woolwich in 1823–1824 with Peter Barlow (q.v.), was subsequently published in the Institution's first volume of *Transactions*. Donkin retired in 1848 and resigned his membership.

His eminence as an engineer was reflected in his election as Fellow of the Royal Society in 1838, where he became a member of its Council. He was elected a member of the Smeatonian Society of Civil Engineers in 1835. A keen amateur astronomer he was a founder member of the Royal Astronomical Society. Of his six sons, John (1802–1854), Bryan (1809–1893), and Thomas (1819–1894) all followed him into the engineering profession. He died on 27 February 1855 at his home, 6 The Paragon.

MIKE CHRIMES

[Donkin Diaries, etc., Donkin Archive, Derbyshire RO; Membership records, ICE Archives; Council minutes, ICE Archives; Minutes of conversations and OC14 and OC60, ICE Archives; Smeatonian Society Archive, ICE Archives; Telford mss, ICE Archives; Thames Tunnel papers, ICE Archives; Springfield Bridge, D/TX 4/13, Essex RO; Parliamentary evidence (1807) (Halls mills), HLRO; T. Telford (1833) Diary, SRO; House of Commons Select Committee on Steam Boats (1817) Evidence, 5–10; 1824; House of Commons Select Committee on Artizans and Machinery (1824) 28–42; Obituary (1856) *Proc R Soc London*, **7**, 586–589; S. B. Donkin (1950) Bryan Donkin FRS, MICE, 1768–1855, *Trans Newn Soc*, **27**, 85–95; D. M. Henshaw (1953) *A Brief Account of Bryan Donkin FRS*; A. W. Skempton (1975) A history of the steam dredger, *Trans Newc Soc*, **47**, 111–114; H. Miller (1985) *Halls of Dartford 1785–1985*]

Publications

1803. Radius of spur wheels, *Nicholson's Journal of Arts and Sciences*, **6**, 86–88

1808. Table of the radii of [water] wheels, from 10 to 300, the pitch begin two inches, in: R. Buchanan, *An Essay on the Teeth of Wheels*

1810. An instrument to ascertain the velocity of machinery (tacheometer), *Trans [R] S A*, **28**, 185–191

1819. A counting machine, *Trans [R] S A*, **37**, 116–123

1834. *A Paper read before the ICE, on the Construction of Carriage-way Pavements* (1824) Montgomeryshire Canal (Western Branch), with W. Cubitt

1836. An account of some experiments made in 1823. and 1824, for determining the quantity of water flowing through different shaped orifices, *Trans ICE*, **1**, 215–218, pl. xxiii

1837. Description of Mr. Henry Guy's method of giving a true spherical figure to balls of metal, glass, agate or other hard substances, *Min Proc ICE*, **I**, 22; *Trans ICE*, **2**, 47–49

1837. Description of a spring level, *Trans R Astronom Soc*, 31

Works

1817. Steam dredger for Caledonian Canal, built 1814–1816, began work 1817

1822. Steam dredger for Gotha Canal, built 1821–1822, began work 1822

1824. Steam dredger for Calder and Hebble navigation

DOUGLAS, Sir Howard, General (1776–1861), third baronet of Carr, Perthshire, soldier, was born at Gosport in 1776, the son of Vice-Admiral Sir Charles Douglas (first baronet) and Sarah Wood, his second wife. The premature death of his mother when he was three, and his father's naval career meant he was brought up by his aunt Mrs Helena Baillie, in Musselburgh. He was educated at the local grammar school, but spent much of his childhood with local fishermen. His father's sudden death in 1789 led to the abandonment of the idea of a naval career and he was sent to the Royal Military Academy, Woolwich. Although he initially failed the entrance examinations, he soon passed, and distinguished himself in mathematics, becoming a favourite of Charles Hutton (q.v.). He passed out, as a second lieutenant, Royal Artillery, in 1794 and the following year was transferred to service in Canada, where he remained until 1799. On his return he married Anne Dundas (d. 1854) of Edinburgh. They were to have three daughters and six sons, of whom the eldest survivor, Sir Robert Percy Douglas, became lieutenant governor of the Cape of Good Hope. Over the next five years he served in the Royal Artillery in England, the most interesting part of his service being ten months in General Congreve's mortar brigade (1803–1804).

In 1804 Douglas was promoted to Captain, but soon afterwards he was appointed Commandant of the Senior Department of the recently founded Royal Military College, at High Wycombe, subsequently becoming Inspector General of Instructions, until 1820. His work at the College led to his being put on half pay and placed on the retired list of the York Rangers, of which he eventually became Major General in 1821.

Douglas made various contributions to the teaching and administration at the Royal Military College before, in 1808, returning to active service in Spain and Flushing. He was recalled from the Peninsular campaign in 1812 as he was sorely needed at the Military College. On 23 May 1809 Douglas succeeded to the baronetcy on the death of his half brother.

Following his work at the RMC he became Governor of New Brunswick from 1823–1831 where he improved the roads and navigation lights along the coast. He stood, unsuccessfully, for Parliament for the Tories in 1832 and 1835 and was then appointed Lord High Commissioner of the Ionian Islands where he again promoted public works, serving there until 1840. His military promotion continued and he became General of the 15th Foot Regiment in 1851. As a Conservative he was elected MP in 1842 but opposed Peel's line on the Corn Laws and stood down from Parliament at the 1847 election. He was awarded a host of honours, and continued to advise the government on military matters, and published a succession of important treatises.

It is by this published work that he made his best remembered contribution to civil engineering. The first, and in civil engineering terms most important, publication was his textbook on military bridges which he first drafted in 1808 although it was not published until 1816. Based on his teaching at the Royal Military College and to support his campaign for a separate Corps of Pontonniers, the first edition dealt with all aspects of military bridge design—bridges of boats, pontoons, carriage bridges, and trestle and truss bridges as well as a section on bridge hydraulics. It was the first textbook in English to be published on the subject, with the exception of Hutton's work on bridges. Of particular interest is his description of the use of rope suspension bridges, as the book appeared at the time Thomas Telford (q.v.) was investigating the idea of a suspension bridge crossing of the Mersey at Runcorn. It has been suggested that Douglas

influenced Telford, but there is no evidence for this, and in any event the book was published after Telford began his investigations. The second edition of the work, which appeared in 1832, is in some ways a more useful text, with the sections all considerably expanded, and an appendix detailing properties of materials. It is a testament both to Douglas's own knowledge of the subject, and to the quality of information available to the Royal Engineers active abroad in the 1820s and 1830s. It drew on foreign works and Douglas's personal knowledge of military engineering, as well as private communications to the author. A third edition appeared in 1853, again expanded.

Douglas died at Tunbridge Wells on 9 November 1861 and was buried beside his wife at Boldre, near Lymington, Hampshire. He left an estate of c. £16,000.

MIKE CHRIMES

[DNB; Commissioners upon ... Harbours of Refuge (1845–1847) *Reports*, etc.; S. W. Fulton (1862) *A Life of Sir Henry Douglas*]

Publications
1816. *An Essay on the Principles and Construction of Military Bridges and the Passage of Rivers in Military Operation*, new editions 1832, 1853
1817. *Observations on the Motives, Errors, and Tendency of M. Carnot's System of Defence ...*
1820. *Treatise on Naval Gunnery*

DREWRY, Charles Stewart (1805–1881), barrister, of 77 Chancery Lane, London, was born on 16 September 1805 in Harlesden, Middlesex, the son of Samuel Drewry, Paymaster of HM Forces. He was the author of the first textbook in the English language on suspension bridges, and indeed the only textbook devoted to the subject to be produced in the British Isles before the Second World War. His work, entitled *A Memoir on Suspension Bridges* was published in 1832, by which time the form was well established in Britain. The early section, comprising an historical account, was derived in part from travel literature and the works of the French engineer, Navier, and Robert Stevenson (q.v.) in particular. It does, however, reveal a familiarity with primary and secondary sources which suggests Drewry was as abreast with the subject as any of his contemporaries, and his work provides a useful distillation of developments down to 1830 when British engineers dominated the field. It describes in detail many of the early British suspension bridges and correctly ascribes to Thomas Telford (q.v.) the credit for making this form of bridge acceptable to the engineering community, with his proposals for bridges at Runcorn and the Menai. Having said this, the work contains much about the bridges designed by Samuel Brown, Marc Isambard Brunel and William Tierney Clark (qq.v.). It describes foreign bridges, information which would probably have been otherwise inaccessible to many English readers ignorant of French and German. Perhaps unfortunately,

Drewry is critical, for practical reasons, of the continental preference for wire cables, which were regarded as more prone to corrosion, more difficult to handle for large spans and prone to developing kinks which were difficult to remove. Drewry had himself conducted experiments with small scale catenaries of wire cable which had reinforced his opinion and had also conducted experiments into the use of wooden chains. The final section of Drewry's work deals with formulae for the design of suspension bridges and strength of materials, and can be described as a simple exposition of these subjects.

It is unclear what prompted Drewry's work on suspension bridges. When he joined the Institution of Civil Engineers as an Associate in 1827 he was described as an engineering draughtsman and as such he helped Brown with his plans for Clifton Suspension Bridge in 1829 and dedicated his book to him. He had been educated at the Brussels Lycee and had the education and intellectual capacity to come to terms with some of the mathematics involved. Soon after writing his work he must have begun legal training, being enrolled at Inner Temple on 10 June 1836. He resigned from the Institution in 1842, presumably because of his legal career. He married Laurentia Buschman on 11 November 1835 at St. Pancras Old Church. He died at 1 Westbury Terrace, Paddington on 3 September 1881.

MIKE CHRIMES

[Membership records, ICE archives; Pigot's directory (1839)]

Publications
Drewry also wrote on legal matters.

1832. *A Memoir on Suspension Bridges*

DUCART, Davis (c. 1735–c. 1785), civil engineer and architect, in his report on the extension of the Coalisland Canal, dated 17 November 1767 and published in the Irish parliamentary journal for November of that year, wrote that he was born and bred an engineer in 'the hilly parts adjacent to the Alps ... so often visited by the English nobility and gentry'. It appears that he was of Franco-Italian descent and that his real name, as given in his will, was Daviso de Arcort. In fact, the only biographical information of any substance is to be found in his will, dated 30 November 1780, but not proved until 29 March 1786.

Ducart (variously spelt Duchart or Duckart) came to Ireland from the Sardinian service. The earliest reference to him in Ireland is in 1761, when Cork Corporation ordered a payment of £25 'to Mr Davis Ducart, Engineer, for his trouble in taking the level of the river Lee, and drawing several plans of waterworks to supply this City with water'. In 1768, the Cork Pipewater Company, established in 1761, installed the first waterwheel and pumps at the Lee Road works in accordance with Ducart's plan and recommendations.

The first building in Ireland to be attributed to Ducart is the Custom House in Limerick, which now houses the Hunt Museum. The building, with a frontage of 175 ft., was commenced in 1765 and completed in 1769 at a cost of £8,000. It also seems likely that he was responsible for the general grid plan and layout of Newton Pery, Limerick's new development, which dates from around the same time and is due to Edward Sexton Pery, the Speaker of the House of Commons. Pery was instrumental in pulling down the old city walls and building new roads, a bridge and quays. In 1765, Ducart entered into an agreement with Cork Corporation to carry out his design for the new Mayoralty House in Cork (now the Mercy Hospital) for a sum not exceeding £2,000, being paid five per cent on the money expended.

Although Ducart received several substantial architectural commissions in the second half of the 1760s (e.g. country houses at Kilshannig and Lota in Co. Cork and Casteltown-Cox in Co. Kilkenny) no further work is recorded after 1770. In fact, nothing is known of Ducart between 1774 and his will in 1780. An attack on him in the *Freeman's Journal* for 3–4 February 1773 states that he had given up architecture by this time and that he (Ducart) was 'actually ignorant of the common rules and proportions of architecture'.

In his will, Ducart states that he was employed as an engineer on the Newry canal and on the Tyrone and Boyne navigations. He remained connected with canal work and the collieries in county Tyrone for most of his life, possessing some property at Drumtea in that county. In 1768, we find Ducart estimating the expenses of an underground canal at the county Tyrone collieries and in 1771 he is mentioned in a contemporary newspaper as closing a colliery and operating another.

Ducart is probably best known for his design of the stretch of canal linking the Coalisland Canal (opened 1733), part of the Tyrone Navigation, with the adjacent collieries. This was the first and only canal in Ireland to use a system of inclined planes rather than locks to raise or lower boats from one level to another. Coal was conveyed in small wheeled tubs, which were run on to lighters on the canal. The canal, constructed in the late 1770s, consisted of three level sections connected together and to the existing canal by stone arched ramps, known as 'dry hurries', down which the tubs were run to the next level. This innovation was not a success and the canal fell into disuse after only a few years. The most prominent structure on Ducart's canal still extant is the ashlar stone aqueduct at Newmills, built c. 1778, where the canal was carried over the river Torrent.

The date of Ducart's death is not known, but is likely to have been around 1785.

RON COX

[Anon., *Freeman's Journal*, 2–4 February 1773; F. Sloane (1786) Map of Ducart's canal, Royal Irish

Academy, Map; A. R. Caulfield (ed.) (1876) *The Council Book of the Corporation of the City of Cork,* 752, 815–816; will of Davis Ducart, dated 30 November 1780, proved 29 March 1786, abstract in Genealogical Office, Dublin, MS 424, 237–238, summarised in *Georgian Society Records,* 1913, **5**; Knight Of Glin (1967) A Baroque Palladian in Ireland: the architecture of Davis Duckart, *Country Life,* **CXLII**, 735–739, 798–801; W. A. McCutcheon (1980) *The Industrial Archaeology of Northern Ireland,* 58–65; R. C. Cox and M. H. Gould (1998) *Civil Engineering Heritage: Ireland,* 167–168; Irish Architectural Archive, *Index of Irish Architects: Biography,* in preparation]

Works
Works include:

1761–1768. Cork waterworks
c. 1778. Ducart's canal

DUDLEY, Sir Robert (1573–1649), naval engineer and mathematician, was the son of the Earl of Leicester and was first married to a granddaughter of Sir Richard Cavendish (q.v.). He used two ships which he inherited from her brother, Sir Thomas, to explore Guyana and the Orinocco River in 1594. This served as the basis for a chart of the coastline in volume 6 of his *Dell Arcano Del Mare.* This was the first sea atlas using the Mercator projection for all its charts, which included an early one of the coastline of Virginia showing soundings, banks and reefs, magnetic variation and notes on winds and currents. Volume 1 covered means of determining longitude including a method of his own using a water clock. Volume 2 has plans of ports and harbours. This work was published in Florence in 1646–1647 as he never returned to England after a precipitate departure in 1606.

In exile he was employed by the Grand Duke of Tuscany on the design and construction of a new port at Livorno, with a famous mole. He was later given the task of draining the marshes between Pisa and the sea. He was also in charge of the Livorno Arsenal. His invention of an azimuth dial is in the British Museum.

SUSAN HOTS

[DNB; *Dizionario Biografico degli Italiani*; J. T. Leader (1895) *Life of Robert Dudley*; R. A. Skelton (1952) *Decorated Printed Maps of the 15th to 18th Centuries,* 62–63; E. G. R. Taylor (1954) *The Mathematical Practitioners of Tudor and Stuart England,* 191–192; A. H. W. Robinson (1962) *Marine Cartography in Britain*; http://es.rice.edu/ES/humsoc/Galileo/Catalog/Files/dudley.html]

Works
1611. Cosimo Mole, Livorno
1621. Livorno marshes, drainage

DUDGALE, Sir William (1605–1686), historian, was born at Shustoke, Warwickshire, on 12 September 1605, the son of John Dugdale, lawyer and bursar of St. John's College Oxford. After his father's death he bought Blythe Hall near Coleshill and this remained his country home until the end of his life. In London he held various posts at the College of Heralds from 1638 to 1644, from 1660, becoming Garter King-of-Arms; in 1677 he was knighted. In 1665 was published volume 1 of *Monasticon Anglicanum,* with Dugdale as editor, and in 1656 appeared his archaeological and topographical masterpiece *Antiquities of Warwickshire,* on both of which he had been working for many years. Then followed *The History of St. Paul's Cathedral* but before its publication in 1658 he was commissioned to write a history of the draining of the Great Level of the Fens.

This project, involving the drainage of some 300,000 acres of fenland, was carried out in three stages: 1631–1636, undertaken by Francis, (fourth) Earl of Bedford and his associates; 1640–1641 under King Charles I; 1650–1656 by William, (fifth) Earl of Bedford and his Company of Adventurers. The Company saw the need for an objective account of the whole undertaking and of the legislation upon which the authority for action in each stage had been based.

Dugdale's diary of his visit to the Fens in May 1657, which is transcribed by Darby (1940), probably marks the beginning of his research, after completion of the St. Paul's Cathedral manuscript. The results are given in *The History of Imbanking and Drayning of divers Fenns and Marshes* published early in 1662 at the expense of the Company and at the request of Richard, Lord Gorges its principal officer, for which Dugdale received £150. The book doubtless played a part in the successful application by the Company for a royal charter, granted by the Act of 1663, as the Bedford Level Corporation. Dugdale's book is a 242-page folio with eleven engraved plates. As indicated by its size and by the title it is more than a history of the Great Level from the negotiations in 1629 leading to the start of work in 1631 and on to completion in 1656. In the first place he takes the history of attempts at general draining of the Fens back to the late sixteenth century, including an account of the important planning and (small) amount of work carried out 1604–1606. Secondly he outlines the history of other drainage schemes in the early seventeenth century such as Hatfield Chase, Deeping Fen and Lindsey Level. Thirdly, with characteristic devotion to scholarship, he traces the local works of fen drainage and marshland reclamation on a county basis from medieval times (being the largest section of the book) and fourthly he briefly treats work in other countries.

Dugdale has been criticised for a lack of impartiality; certainly he expresses opinions in favour of draining and against its opponents. But by far the greater part of his book is straightforward narrative history based throughout on primary sources, all referenced, of which many are now difficult to find and some no longer exist. He took care to provide maps and, similarly, several are the only known copies. They comprise: Romney Marsh; Sedgemoor; Hatfield Chase, by Josias

Aerlebout (1639); Ancholme Level, by Francis Wilkinson and John Fotherby (1640); Deeping Fen; South Holland; Norfolk Marshland, after William Hayward 1591; The Great Level 'as it lay drowned' after Hayward 1604; the Great Level 'drayned' after Jonas Moore 1654; Lindsey Level; and East & West Fen.

A stock of five-hundred copies of the book, possibly half of the original issue, perished in the Great Fire of 1666 along with many documents held in the Fen Office at the Inner Temple. By the mid-eighteenth century it had become a rarity and the Corporation decided to publish a new edition. Charles Nalson Cole (1723–1804), their Register and himself the author of works on Bedford Level, undertook the task of making any necessary typographical and other corrections, from Dugdale's own annotated copy, modernising the layout and spelling, except for proper names, and reprinting the maps from the original copper plates which were found to be in good condition. He also wrote a new Preface and provided an index. The copyright belonged to Richard Geast, of Blythe Hall, a descendant on his mother's side from Sir William. He generously offered not only to forego any royalties but to meet the expense of publication and in this form the second edition appeared in 1772. It remains an indispensable source for the early history of fen drainage in England.

Dugdale continued to publish learned works on various subjects up to 1685, still in that year 'having his sight and memory perfect'. He died at his home on 10 February 1686, leaving an only son, Sir John Dugdale (1628–1700), who also worked in the College of Heralds.

A. W. SKEMPTON

[C. N. Cole (1772) Preface to Dugdale's *History of Imbanking and Drayning*, 2nd edn.; Francis Espinasse (1888) Sir William Dugdale, DNB, vol. 16, 136–142; H. C. Darby (1940) *The Draining of the Fens*; articles on Richard Atkyns, John Hunt, John Liens and Sir Cornelius Vermuyden in the present volume]

Publications

1662. *The History of Imbanking and Drayning of divers Fenns and Marshes ... and of the Improvements thereby. Extracted from Records, Manuscripts and other Authentick Testimonies*; by Sir William Dugdale, 2nd edn., revised and corrected by Charles Nalson Cole, Esq., of the Inner-Temple, 1772

DUMMER, Edmund (fl. 1682–1712), Surveyor to the Navy, began his career in the Navy in the early 1680s. In the years following the Restoration of 1660 there had been relatively little investment in dockyard facilities and the Navy had been humiliated by the Dutch in 1667. In the 1680s the position began to change and following the Glorous Revolution of 1688 the French replaced the Dutch as the chief threat, causing reconsideration of the deployment of dockyard facilities.

As Surveyor, from 1692 to 1699, Dummer was responsible for the design and construction of warships and for their maintenance. He supervised the Master Shipwrights in the Royal Dockyards as well as being responsible for the buildings and engineering works of the dockyards themselves. Dummer's contribution to warship building has yet to be studied in detail but his chief claim to fame lies in his work in the royal dockyards, notably Portsmouth and Devonport, or Plymouth Dock as the latter was known.

Dummer's remarkable ability as a draughtsman can be seen in the meticulous studies of Plymouth Dock and in his earlier illustrated journal of '... remarkable Lands, Cities, Towns and Arsenals' in the Mediterranean, the fruits of a three year study begun in 1682 when he was sent to Tangier as a midshipman-extraordinary. It was probably this draughting ability that led Dummer, then assistant master shipwright at Chatham, to be selected by the Navy Board in 1689 'to goe to Dartmouth and Plymouth, and taking notice of all the parts adjacent thereunto, to represent what were found most suitable to the design and building one single dry-dock'. As a result of Dummer's recommendations, the Navy Board selected the site of the present Devonport Dockyard, making Dummer responsible for the layout of this new yard and all the buildings and engineering works.

The original intention had been simply to construct a dry-dock just to service the western squadron but this modest aim was revised and expanded to create a full-fledged dockyard with both wet and dry docks, ropery, smithery and storehouses. Dummer's dry dock was constructed in the solid rock and was the first in England to have stepped sides; it was drained by gravity and was protected from tidal surges and storms by a wet-dock, 452 ft. in length and 1.3 acres in area, in front. Although both were to be substantially altered and modernised in the nineteenth century, they still form the focal point of what is now South Yard, Devonport. Work began in the spring of 1691 and was substantially completed by the contractor, Robert Waters, by the end of 1692. The docks came into use in 1693.

Contemporary with his work at Plymouth, Dummer was busy designing and constructing another dry dock at Portsmouth. This too, was protected by an outer wet dock but the Portsmouth scheme was although more ambitious in that it included a second wet dock to the north, approached by a long channel. The wet docks, both 1.6 acres in extent, had entrance gates 51 ft. wide. The engineering works at Portsmouth were also far more complex than those associated with the Plymouth wet and dry docks. At Portsmouth, much of the work was on reclaimed land, involving very extensive foundation work. This was to give trouble later, but the *Royal William* became the first warship to use the new dry dock—now known as No. 5 Dock—in June 1698. An innovation here was the horse-driven pumps, apparently capable of lowering the water in the 190 ft. long dry dock 6 ft. in one night; extant is an ingenious drawing by Dummer proposing a water-wheel as an alternative to the horses. This was powered on the tide-mill principle, water from the two wet docks turning the wheel at low tide. However, his North Basin, used as a reservoir since the 1770s, lies beneath Brunel's Blockmills, while most of the stonework of the north wall of what is now No. 1 Basin probably dates from Dummer's time. The only partly surviving building attributed to Dummer, who may have been influenced in its design by Hooke's Bethlehem Hospital, is the monumental Officers Terrace at Devonport.

In 1698 Dummer completed a survey of all the royal dockyards as well as a survey of the south coast but the following year he was dismissed from his post after accusations of bribery involving the contractor at Portsmouth, John Fitch. For a while Dummer ran a packet-boat service to Spain and the West Indies but it was not a success and in February 1711–1712 he went bankrupt, dying shortly afterwards in the Fleet Prison. Dummer's engineering works at Plymouth Dock and Portsmouth were bold in concept and execution, and his step-sided dry docks formed the models for all subsequent ones.

J. COAD

[BL, King's MS, 40; BL, Lansdowne, 847; BL, King's MS, 43; J. G. Coad (1989) *The Royal Dockyards 1690–1850*; J. M. Collinge (1978) *Navy Board Officials 1660–1832*; Colvin (3); J. B. Hattendorf, R. J. B. Knight, A. W. H. Pearsall, N. A. M. Rodger, G. Till (1993) *British Naval Documents 1204–1960*; R. D. Merriman (1961) *Queen Anne's Navy*]

Works

1691–1692. Plymouth Dockyard, *c.* £11,000
1691–1698. Portsmouth Dockyard, £16,000
1697. Chatham Dockyard, masthouse and pond, *c.* £4,000

DUNCOMBE, John (d. 1810), civil engineer, came from Oswestry and is best known for his association with Thomas Telford (q.v.). In 1789 Duncombe was involved, as engineer, with the earliest proposals for what was to become the Ellesmere Canal. In 1791 he produced an estimate for the canal, with several branches, of £171,098, with Joseph Turner (q.v.), another local engineer, who was county surveyor for Denbighshire. In August 1791 he was asked by the promoters to look at an alternative line proposed by the Whitchurch engineer William Turner (q.v.) and in November William Jessop (q.v.) was called in to provide experienced advice to the promoters, in the following August recommending a modified version of the original line. It was the time of the canal mania and there was no problem in raising the necessary subscription capital and in 1793 the canal Act was obtained.

The original engineering staff comprised Jessop, with Duncombe, his assistant, Thomas Denson (d. 1811), and William Turner (q.v.) but in October 1793 Telford was appointed General Agent, and played an increasingly important role. Duncombe appears to have been responsible for the detailed surveys of the line and Turner drew up the original designs for the aqueducts, which were subsequently superseded by later proposals of Telford and Jessop. Duncombe himself designed a two-tier masonry aqueduct for Pontcysyllte in 1794. Despite Telford effectively supplanting Turner and Duncombe, it is evident that a reasonable working relationship prevailed. Duncombe was paid a guinea a day while working from home, and one and a half guineas (£1 11s 6d) when away from home as from April 1795, and was appointed 'engineer' on 29 July 1796; his clerk, Denson, was paid half this rate. At the end of June 1803 Denson replaced Duncombe as Engineer, with a salary of £150 p.a. At that time Duncombe had been paid £4402 2s 5d by the Company since 1791. Denson laid out the Pontcysyllte–Ruabon railway in 1804–1805 and by the end of 1805 had been paid £1782 3s 2d, commencing January 1798. These salaries reflected the amount of work both men devoted to the Canal, with Duncombe's exceeding that of Telford. Denson was succeeded as Engineer by Thomas Stanton (q.v.) in November 1811; by then Duncombe had died.

In 1806 Duncombe had gone to join Telford's team in Scotland to work on the Highland Road Surveys where he succeeded James Donaldson as Superintendent of Roads; this role was not a success as by 1809 Telford was very dissatisfied with his work.

His association with Jessop probably led to his election to the Smeatonian Society of Civil Engineers in 1793. He wrote an unpublished treatise on 'railways' while at work on the Ellesmere Canal, the highpoint of his career. He died in prison in Inverness in 1810.

MIKE CHRIMES

[Ellesmere Canal papers, RAIL 827, PRO; Gibb papers, ICE Library, esp. Rickman–Telford correspondence; M. Davidson, Treatise on canals and rivers, etc. (mss), including Duncombe's treatise on railways and inclined planes, ICE Library; Reynolds' collection of drawings, Science Museum; Reports (1803–1821) *Commissioners for Highland Roads and Bridges*; J. Mitchell (1883) *Reminiscences of my life—the Highlands*, vol. 1, (repr. 1971); C. Hadfield (1969) *The Canals of the West Midlands*, 167–170; A. W. Skempton and E. Wright (1971) Early members of the Smeatonian Society of Civil Engineers, *Trans Newc Soc*]

Works

1793–1803. Ellesmere Canal, resident engineer
1808. Glenmorriston Road, Inverness, survey, 14 miles 469 yd.
1809. Lochy-Side Road, Argyll & Inverness, survey, 12 miles 540 yd.

1809. Asheik Road, Isle of Skye, survey, 4 miles 360 yd
1809–1810. Laggan Road, Inverness, survey, 42 miles 568 yd., with John Mitchell

DUNN, Matthias (c. 1790–1870), colliery viewer and Mines Inspector, was born c. 1790, probably in the north-east of England. His father, also Matthias (c. 1750–1825), was a viewer for the Jarrow and Hebburn collieries in County Durham c. 1800, and a minor colliery lessee. Dunn served his time with the viewer, Thomas Smith, and later was an assistant to John Buddle Jr. (q.v.). Dunn became a viewer in his own right, notably for the Hetton Coal Company in County Durham, where there were particular difficulties in shaft-sinking through waterlogged strata and in dewatering the mines. He accompanied George Stephenson (q.v.) on his survey for a proposed railway between East Hetton and Hartlepool in c. 1831 and, in 1835, Dunn himself drew up a prospectus for a railway from Newcastle to Morpeth. In the following year he, with others, argued for a line between Newcastle and Dunbar, a route which eventually came to pass. He was called before the Select Committee on the Coal Trade as a leading expert in 1836, and in 1839 he joined Buddle and others as a partner in the Stella Coal Co. He introduced the use of wire rope into British mining in 1840.

A calamitous inundation of Workington colliery, where Dunn had previously been viewer, led to his becoming, in the 1840s, a leading advocate for government inspection of mines. In this he was sharply at odds with the likes of Nicholas Wood and Robert Stephenson (qq.v.), the latter suggesting that it was absurd to imagine that any practical good might result from government inspection. Perhaps not surprisingly therefore, Dunn was appointed Government Mines Inspector for Durham, Northumberland, Cumberland, and Scotland, under the Mines Act of 1850. Following his role as Inspector in the enquiry into the Seaham Colliery explosion of June 1852, where he severely criticised the ventilation arrangements and reliance on naked flame lights, he entered into discussions with George Elliott which led to the formation of the North of England Institute of Mining and Mechanical Engineers in the following month. Dunn was thereafter an honorary member of the Institute. He ceased to be a Mines Inspector in 1864–1865, and in 1866 he left his home of some years at 5 St. Thomas Place, Jesmond, Newcastle upon Tyne, for retirement at Highland Villa, Central Road, Upper Norwood, London. He died there in 1870, aged eighty years.

STAFFORD M. LINSLEY

[*The History of the Viewers* (1811) MS, NEIMME collection East/3B, Northumberland RO; Journal, 1831–1836, MS, Central Library, Newcastle upon Tyne; R. L. Galloway, (1898) *Annals of Coal Mining and the Coal Trade*, vol. 2; W. W. Tomlinson (1914) *The North Eastern Railway*; T. V. Simpson (1930–1931) Old mining records and plans, *Trans Inst Mining Engrs*, **LXXXI**, 75–108, 292–294; R. Church (1986) *The History of the British Coal Industry: 1830–1913*; M. Sill (1986) The journal of Matthias Dunn, 1831–1836: some observations of a colliery viewer, in: R. W. Sturgess (ed.) *Pitmen, Viewers and Coalmasters*, North East Labour History Society, 55–80; P. E. H. Hair (1988) *Coals on Rails ...*]

Publications
1831. Extract from a Minute Book on the Explosion at Harraton Colliery, in the year 1808, *Transactions of the Natural History Society of Northumberland, Durham, and Newcastle upon Tyne*, **1**, 241–243
1835. *Prospectus of a Railway from Newcastle to Morpeth to be called the Northumberland Railway, ultimately to be constructed by Branches with Blyth and Shields*
1838. Notice on the Gravel Bed of St. Lawrence Colliery, *Transactions of the Natural History Society of Northumberland, Durham, and Newcastle upon Tyne*, **2**, 285–287
1838. Narrative of the Sinking of Preston Grange Engine Pit ..., *Transactions of the Natural History Society of Northumberland, Durham, and Newcastle upon Tyne*, **2**, 226–235
1844. *An Historical, Geological and Descriptive View of the Coal Trade of the North of England ... appended a Concise Notice of the Peculiarities of certain Coal Fields in Great Britain and Ireland*
1852. *A Treatise on the Winning and Working of Collieries*, 2nd edn.

DUNNELL, Daniel (fl. 1694–1702), carpenter and engineer, is first noticed as Dan. Dannell on a piece of masonry surviving from the 1694 waterworks at Exeter. The works raised water using a waterwheel on a leat of the Exe. From a waterworks lease of 1695 it would appear Dennell [*sic*] was a carpenter from Gloucester, substantiated by the baptism of a son, Edward, there the same year. Work on the water supply continued until at least 1699; one of his partners in the scheme was Ambrose Crowley (q.v.).

Evidently well regarded locally he was appointed in 1699 to carry out works on the Exeter Canal. He had previous experience of similar work as in 1697 Dunnell was employed to survey the Lugg Navigation, work for which he was paid £1 2s. The Exeter Canal as originally built in the 1560s by John Trew (q.v.) was about 1¾ miles long from Exeter to a point on the tidal river Exe, with a timber weir across the river, three pound locks and a navigable sluice at the lower entrance. It had a width of 16 ft. and a depth of 3 ft. and, after some improvements in 1581, was navigable by lighters of 16 tons burden. Richard Hurd (q.v.) extended the canal by about a mile towards Topsham in 1676 and built a large sluice and transhipment basin at the new entrance. The sluice could admit ships of at least 60 tons burden, at high water spring tides. He also built a stronger weir.

In July 1698 Exeter Chamber resolved to enlarge the canal to carry ships throughout its length and in September they made an agreement with William Bayley, an engineer from Winchester, to widen and deepen the canal, to replace the three locks by a single very large one, to build a new weir and to enlarge the quay at Exeter; the works to cost £9,000. Specifications for the new lock show two pairs of gates, 26 ft. wide, with masonry abutments, enclosing a turf-sided chamer or 'pool' 350 ft. long and 75 ft. wide (at the water surface) over a length of 200 ft. and fall of 6 ft. It became known as the Double Lock. Work began, but in May 1699 Bayley embezzled some of the funds and fled whereupon the city immediately reimbursed the labourers and in June appointed Dunnell as engineer at a salary of £3 per week.

Estimates were then prepared for completing the works. Adding up to £5,985 they include £1,563 for finishing the upper gates of the new lock, making the lower gates and excavating the pool; £800 for a new sluice or flood gate (King's Arms Sluice) at the head of the canal; £307 for finishing the quay; £407 for deepening the canal from the new lock to the lower sluice; and £1,000 for land purchase and associated costs. Provision was made for building a dock at Exeter (£750) and making the lock (£605) but as these were not carried out they are not included in the above figure.

Whether Bayley had built a new weir is uncertain. It seems clear, however, that he had started work on the quay and on the upper gates of the new lock and had deepened the canal from its head to the lock, a length of 1¼ miles.

Work under Dunnell's direction began with help from voluntary labour given by the people of Exeter but by June 1700 more funds had been raised, enabling work 'to be carried out with Vigour', and the canal was re-opened in November 1701. It was 50 ft. wide a 10 ft. depth of water. Allowing, say, £2,000 for work by Bayley the total expenditure would have been approximately £8,000.

The entrance lock was rebuilt, and probably enlarged, in 1707. Exeter now had a ship canal 3 miles long capable of accommodating vessels of 120 tons burden. It remained essentially unchanged until completely rebuilt and extended by 2 miles in 1825–1827 by James Green (q.v.).

A. W. SKEMPTON

[P. G. De la Garde (1845) Memoir of the canal of Exeter from 1563 to 1724, *Min Proc ICE*, **4**, 90–102; T. S. Willan (1936) The Exeter lighter canal, *Journal of Transport History*, **3**, 1–11; E. A. G. Clark (1960) *The Ports of the Exe Estuary*; K. R. Clew (1984) *The Exeter Canal*; A. Brian (1994) As to the River Lugg, *Trans Woolhope Naturalists Field Club*, **48**(1), 37–96; V. R. Stockinger (1996) *The Rivers Wye and Lugg Navigation*, 121–167]

Works
1694–1699. Exeter Waterworks
1699–1701. Exeter Canal enlargement, £6,000

DUNTHORNE, Richard (1711–1775), astronomer and Superintendent General of Bedford Level, was born in 1711 at Ramsey, Huntingdonshire, the son of a gardener. He attended Ramsey Grammar School and came to the notice of Dr. (later Professor) Robert Long, astronomer and Master of Pembroke College, who brought him to Cambridge as a college servant with opportunities to study mathematics. Later he undertook for a few years the management of a preparatory school but was recalled to Cambridge as assistant to Dr. Long. Dunthorne published *The Practical Astronomy of the Moon* in 1739 and four papers on astronomical subjects in the *Philosophical Transactions* of the Royal Society between 1747 and 1761.

In June 1753 'Richard Dunthorne of Cambridge' was elected Surveyor of the South and Middle Levels of the Fens with a salary of £80 p.a. Two years later this was increased to £100 and in 1764 he became Superintendent General for the whole of Bedford Level, probably at a salary of £150 p.a.

His duties were chiefly concerned with maintenance of the rivers, banks, drainage channels and sluices in the Fens but he was also involved in any new works. Of these the principal was the improvement of the river Nene outfall beyond the River's End 4.5 miles below Wisbech. The first attempt at improvement had been made in 1721–1722 under the direction of Nathanian Kinderley (q.v.) by cutting a new channel 1.5 miles long from River's End past Gunthorpe Sluice towards Peter's Point but the Wisbech townspeople opposed the scheme and organised the destruction of the diversion dam before it had been completed. Little was achieved for several decades afterwards and conditions in the North Level continued to deteriorate. In 1753 a great part of the Level lay under a foot of water. In 1763 a violent flood breached the north bank of the river in several places between Peterborough and Wisbech with the loss of lives and many cattle. The bank was repaired but further breaches occurred in 1765 and 1767. The river banks were at or close to their maximum height on the soft ground foundations. The only solution was a radical improvement of the outfall to increase the gradient in the river and so to increase the rate of discharge of flood water.

Driven to this conclusion the Commisoners of the North Level in 1767 consulted John Smeaton (q.v.) and employed William Elstobb (q.v.) to prepare a map and longitudinal section of the river from Peterborough to the sea. Dunthorne was present during these investigations and in his journal of 1767 expressed the view that the most practical way of improving the outfall would be to re-open the cut made by Kinderley and preferably to extend it through reclaimed land to the sea near Sutton Leam (a total distance of 6 miles from River's End) though he also referred to the rather curious plan advocated by Kinderley himself of making a cut from the Nene through Norfolk Marshland to the Great Ouse near King's Lynn.

Smeaton reported in August 1768 with a very different proposal: to build a great scouring sluice a short distance above River's End. This would have been regarded as a controversial suggestion. In July 1769 John Wing (q.v.) one of the North Level Commissioners and Steward of the Duke of Bedford's estate at Thorney, wrote to Smeaton desiring him to take a further view. He also wrote to James Brindley, Thomas Yeoman and John Golborne (qq.v.). Smeaton appears to have declined the invitation but Golborne reported in October 1769. He rejected the sluice, and proposed a cut to Sutton (as Dunthorne had already suggested) with an extension through the sands to a permanent deep-sea channel at the Eye. He estimated the cost at £26,000. A month later Yeoman reported in the same vein, as did Brindley in August 1770 but with the much higher estimate of £42,000.

Queries published in September on Brindley's and Golborne's reports were answered by Dunthorne in a carefully reasoned response. A terrible breach in the river bank in the following winter cost £6,000 to repair and further advice was sought from Langley Edwards (q.v.). His report, in April 1771, recognised the merit of the long cut but recommended as an initial step making a cut only to Peters Point, though with very generous dimensions (top width 200 ft. and 12 ft. deep) which brought his estimate to £17,000.

Dunthorne having reported to the Commissioners in May, they set up a Committee in July to bring the matter to a conclusion and Edwards made a new survey. In September Golborne was recalled, when he, with Edwards and Dunthorne, set out what they considered to be the most convenient line for the cut from River's End to the Eye. The Committee, including John Wing, 'took a view' with the three engineers and decided that the cut as now marked out 'is the best method of improving the outfall from Wisbech to the sea'.

Before the end of the year an Act was obtained enabling the Commissioners to make the cut and raise a sum not exceeding £20,000 for doing so, this was the correct decision. The Nene Outfall Cut eventually completed in 1830 followed the line proposed in 1771. But Wisbech Corporation strongly objected to any tax levied for the purpose and it was not until June 1773, after negotiations in which John Grundy (q.v.) played a role, that a new Act was obtained by which the North Level proprietors, largely at their own expense, undertook the more limited task of restoring the cut from River's End past the outfall of Gunthorpe Sluice and on towards Peter's Point. It would be named Kinderley's Cut in honour of its first Director.

On 5 July 1773 Dunthorne was appointed to direct the works: Wing was to act as Receiver. Three weeks later Dunthorne specified the dimensions: 100 ft. top width, 8 ft. deep with 4 in 1 side slopes, and a foreland 60 ft. wide between the cut and a bank on the east (sea) side. The end of the cut was to be protected by stone flagging, as was the outer side of the bank. Two-hundred barrows and some other equipment were ordered, and advertisements placed for 'spade and barrow men'. On 9 August Elstobb was appointed to superintend the work, as resident engineer, with Thomas Pear (q.v.) as assistant.

Work began very soon afterwards. Excavation of the cut had been completed by June 1774 and the 'great dam' across the old channel, to divert flow down the cut, was finished by September. Dunthorne died in March 1775. Writing next month to Golborne, John Wing said 'we completed the works last autumn ... and it was opposed by some of the Wisbech people ... but to their shame the first fresh that came down in October last ground the bottom [of the cut] to a depth of 15 or 16 ft.'. The cut was also being widened by erosion and, as anticipated, much slope protection had to be done.

By the summer of 1776, with Elstobb still in charge, the whole scheme was effectively completed at a cost of about £9,000. The results exceeded expectation; a lowering of low water levels as recorded in the early 1780s (compared with those in 1767) of 6 ft. at Gunthorpe Sluice, 3 ft. at Wisbech, and nearly 2 ft. as far upstream as Peterborough, with great benefit to drainage of the North Level.

It was quite usual for the Bedford Level Surveyors or Superintendents to be consulted from time to time on other projects and early in 1773 Dunthorne was asked to advise on draining low grounds adjacent to the river Stour in Kent. At his request levels were taken from Fordwich to the sea by Thomas and Henry Hogben. He visited the site again in May 1774. In June he was back on Kinderley's Cut, and reported on the Stour in September. Dunthorne died at Cambridge on 10 March 1775 and Thomas Yeoman took over the Stour scheme while James Golborne (q.v.), nephew of John Golborne, succeeded him in 1776 as Superintendent General of Bedford Level.

Dunthorne never married. He was esteemed for his integrity and kindliness and is known to have made settlements for life on his younger, humble relatives.

A. W. SKEMPTON

[North Level Commissioners Order Books (1754–1782) and Committee Book for Kinderley's Cut (1773–1786) Cambridgeshire Record Office; BL, Add. MS 5489 (river Stour); William Elstobb (1767) Map of Wisbech River & Channel. MS in Smeaton's collection, Royal Society; printed reports by J. Smeaton (1767), J. Golborne (1769),

T. Yeoman (1769), J. Brindley (1770), L. Edwards (1771); W. Watson (1827) *An Historical Account of ... Wisbech and the Fens*; N. Walker and T. Craddock (1849) *The History of Wisbech and the Fens*; H. C. Darby (1983) *The Changing Fenland*; DNB; Skempton]

Publications

1770. *Remarks on certain Queries*
1771. *Report on Wisbeach Outfall*
1775. *Report on river Stour, in Kent*

Works

1773–1775. Kinderley's Cut, completed 1776, £8,900

DURNFORD, family of military engineers who provide an enduring name in Royal Engineering history from Quebec (Augustus, 1759) to the Second World War.

Elias Durnford Jr. (1739–1794), was the son of Elias Durnford Sr., Deputy Treasurer of the Ordnance in 1720. He trained under John Peter Desmaretz (q.v.) at the drawing office in the Tower of London during its brief period as *de facto* alternative to the Royal Military Academy, afterwards being commissioned in the Corps of Engineers. He was probably the Durnford who worked with Desmaretz on the south coast defences in 1759, building two batteries at Seaford for £2,000.

From 1762 to 1781 he worked in America, first as Surveyor General of West Florida and later as Lieutenant-Governor. In both capacities he was involved with Robert Hutchins in the Iberville canal project, a plan to bypass French territory in New Orleans by connecting the Mississippi to Mobile via the Iberville river and lakes Maurepas and Ponchartrain. Durnford's plan was to burn all the obstructions in the river Iberville during the dry season, so enabling flood waters to increase the depths of the passage by cutting into the loam bottom. Then Hutchins canal could be cut from a point on the east bank of the river Mississippi 300 yd. above the mouth of the Iberville causing the flood waters to raise the level in its relatively dry channel. Plans were also drawn up for a town near Fort Bute. The scheme was feasible in engineering terms and would have been the first canal in America, improving trade and settlement and possibly even affecting the course of the war; unfortunately it fell foul of military and political changes of personnel and the reports were never acted upon, neither was his recommendation that £2,500 needed to be spent on the defences of Mobile, which Durnford was subsequently forced to surrender to the French in 1781 after a three-month siege. Returning to England he worked on the 'Maker Heights' defences of Plymouth Dockyard before becoming Chief Engineer in the Caribbean where he died at Tobago in 1794.

Elias Walker Durnford (1774–1850) was the son of Elias Durnford Jr. and his wife, Rebecca Walker. He began his career with his father in the

Caribbean, eventually becoming a Lieutenant-General. His career is treated separately below.

Andrew Durnford (1744–1798), the younger son of Elias Durnford Sr., also trained under Desmaretz at the drawing office in the Tower of London and, like his brother, was afterwards commissioned in the Corps of Engineers. He began his career in 1770 working on the demolitions at Dunkirk before moving to America where he served throughout the war beginning with Burgoyne's Expedition of 1777, where he served with his cousin Desmaretz Durnford and William Twiss (q.v.), moving on to become Deputy Assistant Quartermaster-General to his brother in 1779. He was in New York in 1783 when he was sent to Bermuda, where, other than a few years as Chief Engineer at Chatham, the rest of his career was to be spent.

The loss of the American colonies had drastically altered the significance of Bermuda which became the only British held harbour between Newfoundland and Jamaica. Andrew Durnford's first commission was to examine the defences and recommend necessary improvements, but the entrance to Castle Harbour became so dangerous that it had to be closed to shipping. An alternative was urgently needed so Durnford returned in 1788 to carry out a special survey to include lands, shores and soundings. He also assisted with the hydrographic survey of the reefs undertaken by the navy. These surveys opened up an anchorage north of St. George's Island and laid the groundwork for the establishment of the dockyard.

At the same time he worked on most of the forts: Fort Devonshire; repairs to Castle Island; Fort Popple and Fort St. Catherine. In 1791 he built a new military road linking the latter with St. George's Town and Town Cut Battery which he also strengthened. Unfortunately in 1796 he was suspended for padding his payroll and held for court martial. Specific charges were that he had diverted funds to build Queen's Warehouse (still in use) and a house in St. George's Town, of which he was the first Mayor. He died on 10 September 1798 still awaiting trial and protesting his innocence.

Durnford had married Jemima Margaret Isaacson in 1772 and his grandson, Edward (1803–1889), was one of Charles William Pasley's (q.v.) pupils; his career included test boring in 1825 to check the strata beneath proposed new fortifications at Sheerness, working on a survey of Ireland in 1827 and being one of the first engineers to work in Hong Kong in 1841. His sons continued the tradition, Anthony being killed at Isandzhana in 1878 and Arthur (1838–1912) rising to Deputy-Inspector-General of Fortifications.

SUSAN HOTS

[PRO WO55/928; Durnford papers, Bermuda Archives, Hamilton; Royal Engineers Library, Conolly papers; Porter, **2**, 400–401; Bendall; T. W. J. Connolly (1857) *History of the Royal Sappers* and Miners ...; Mary Durnford (1863) *Family Recollections of Lieutenant General Elias Walker Durnford*; D. S. Brown (1946) The Iberville Canal project ... 1763–1775, *Mississippi Valley Historical Review*, **32**, 491–516; D. W. Marshall (1976) The British military engineer 1741–1783, dissertation, University of Michigan, MS; A. Mardis (1987) A battery in Bermuda, *Fortress*, **15**, 21–38; J. Weiler (1987) *Army Architects*, **3**, 373–374 (Ph.D. thesis, University of York); E. Harris (1986) Bermuda fortifications, *Fortress*, **1**, 37–48; E. Harris (1989) American spies at Bermuda's forts, *Post-medieval Archaeology*, **20**, 311–331; D. Evans (1990) The Duke of Richmond as designer, *Fort*, **18**, 92; E. Beerman (1990) El Diario de Bernardo de Galvez de la Batalla de Mobila (1780) *Cvadernos de Investigacion Historica*, **13**, 125–144; F. Kitchen (1991) Aspects of the defence of the south coast of England 1756–1805, *Fort*, **19**, 12–13; E. C. Harris (1997) *Bermuda forts, 1612–1987*; M. Andrews (1998) *For King and Country, Lieutenant Colonel John By*; G. Napier (1998) *The Sapper VCs*, 85–87]

Plans
Elias Durnford:

c. 1767. Plan of the River Iberville from Lake Maurepas to the forts with part of the Comit and Amit (MS copy, William L. Clements Library, 1772)

1770. The communication between the Iberville and the Mississippi, MS map, William L. Clements Library

Andrew Durnford:

1788–1792. Survey of Bermuda, including lands, shores and soundings

Works
Andrew Durnford:

1791. Military road between St. George's Town and Fort St. Catherine, Bermuda

DURNFORD, Elias Walker, RE, Lieutenant-General (1774–1850), military engineer, was born on 28 July 1774 in Lowestoft, the son of Elias Durnford (q.v.) and Rebecca Walker. After attending a preparatory school, he attended the Royal Military Academy at Woolwich and was commissioned a Second Lieutenant in the Royal Artillery in 1793 and later transferred to the Royal Engineers. His first posting was to the West Indies where he served alongside his father. Starting in 1794 he supervised the construction of fortifications in Guadeloupe where, for seventeen months, he was a prisoner of the French. After being released he continued his engineering duties first in England and later in Ireland. In 1808 he was appointed Commanding Royal Engineer in Newfoundland. Here he was primarily responsible for the construction and maintenance of the coastal batteries and various blockhouses. In 1813 he was made a Major in the army and

Lieutenant-Colonel in the Royal Engineers and added garrison duties to his responsibilities.

From 1816 to 1831 Durnford was stationed in Quebec City as the Commanding Royal Engineer for the Canadas (now Quebec and Ontario). His most significant work during this period was the design and construction of the Quebec citadel. Among other fortification works for which he was responsible during this period, the Quebec citadel was the most important first defence for British North America against both American attack and potential local uprisings. Durnford was also involved in the construction of canals on the Rideau and Ottawa rivers although the significant Rideau Canal was not under his direct command but rather under his friend, John By (q.v.). One contribution he made in 1829 concerned testing a local rock for its pozzolonic qualities. He found this local source, named after the contractor, Ruggles Wright, as Wright's Hull cement, was an economic substitute for imported Harwich cement for pointing the locks and advised on its manufacture.

Durnford returned to England in 1831 and retired in 1837. In 1846 he was made a Colonel Commandant in the Royal Engineers and Lieutenant-General in the Army. He has been characterised as a hard worker, rising early and inspecting construction sites before breakfast, then working in his office each day until late in the evening. Various contemporary authors have indicated that he regularly won the esteem of his colleagues and superiors through his honesty and strong professional principles. He had married Jane Sophia Mann in October 1798 and they had thirteen children. Their six sons each became officers in the British army and two, Elias and Viney, joined the Royal Engineers. Durnford died on 8 March 1850 at his home in Tunbridge Wells.

MARK ANDREWS

[M. Durnford (ed.) (1863) *Family Recollections, Lieut. General Elias Walker Durnford, A Colonel Commandant of the Corps of Royal Engineers* (privately published); R. F. Leggett (1955) *Rideau Waterway*; J. Weiler (1987) *Army Architects*, **3**, 373–374 (Ph.D. thesis, University of York); DCB (1988), **VII**, 265]

DU VERNET, Henry Abraham, Colonel (1787–1843)—in 1842 he changed his name to Henry Abraham DuVernet Grosset Muirhead—military engineer, was born on 4 April 1787, the eldest of ten children of Abraham DuVernet and Miriam Grosset Muirhead. His father was an officer in the Royal Artillery and an aide-de-camp to Prince William Henry. After his father's accidental death in 1806, the prince befriended the family and helped to obtain commissions for the sons. Without any formal military education, DuVernet was appointed ensign in the Royal Staff Corps in 1803. He served on the Mediterranean and, under Lieutenant-General Sir John Moore, participated in the retreat from Coruna, Spain. He was invalided home in 1809, promoted to Captain in that year, to Major in 1825 and to Lieutenant-colonel in 1828.

In 1818 a proposal was put forward by the British military to construct an inland canal system up the Ottawa and Rideau rivers in the Canadas in order to avoid the established, but vulnerable, transportation route along the St. Lawrence River. The Americans missed their opportunity to cut the vital St. Lawrence route during the War of 1812 and the British feared it would the first focus of attack should hostilities recur. In 1819, DuVernet was sent to Canada to oversee the construction of the required canals and locks on the Ottawa River portion of the inland route. Six years later, John By (q.v.) was sent to Canada to undertake the Rideau River part of the scheme. DuVernet began construction in 1820 on a 6-mile canal with five locks using Staff Corps forces with a limited annual budget. The canal was to bypass the Long Sault Rapids on the Ottawa River (near present day Hawksbury, Ontario). In 1826, in conjunction with the approval of the Rideau Canal, the Duke of Wellington ordered that the work on the Ottawa River canals be accelerated and additional works be undertaken downstream at Carillon and Chute-a-Blondeau. DuVernet continued with expanded responsibility on the works until they were completed in 1833. He returned to England in that year and in 1834 was placed on half-pay. Like many of his engineer colleagues, he was an accomplished and prolific watercolour artist.

DuVernet married Martha Maria Iqualin Van Kemper and with her had three children. Shortly after his mother's death he changed his name and thereby inherited the Muirhead family estate, Bredisholm, near Coatbridge, Scotland, where he died on 16 December 1843.

MARK ANDREWS

[DCB (1988), vol. VII, 268; R. F. Legget (1988) *Ottawa River Canals and the Defence of British North America*]

DYSON (fl. 1760–1850), a family of civil engineers and contractors, originating from Austerfield in Yorkshire, had careers which spanned the formative years of the civil engineering profession in the British Isles. With the Pinkertons (q.v.), with whom they were on occasion associated, they can claim to be the earliest civil engineering contractors.

John Dyson Sr. (d. *c.* 1804) is first known for his excavation work on the Adlingfleet Drainage, Driffield Navigation and Laneham Drainage schemes in the late 1760s, where he was the partner of James Pinkerton (q.v.). At that time he was described as yeoman of Austerfield, Yorkshire. While the Pinkertons seem to have concentrated very much on the contracting side of civil engineering, members of the Dyson family seem to have alternated far more between undertaking contracts and acting as assistant engineers and surveyors.

It is unclear what he was doing in the 1770s, but one can assume that, as at Laneham Drainage, he was assisting engineers like John Grundy Jr. surveying land drainage works and subsequently tendering for work. There was certainly no shortage of the latter with at least sixteen Fen enclosure schemes in south Lincolnshire alone at that time.

In 1785 Dyson was appointed a Commissioner for the embanking and drainage of the Timberland Fen. By this time he was clearly well-known and respected as an engineer and surveyor. He was called upon to give Parliamentary Evidence in support of the Bills for the Gainsborough Bridge and Road (1787), Bawtry Road (1793), Epworth Drainage (1795), Everton Drainage and Gringley Common Drainage (both 1796); in 1792 he surveyed the Sleaford–Tattersall Turnpike.

John Dyson Jr. (1761–1849) and **Thomas Dyson Sr.** (1771–1852), were involved in civil engineering as the 1790s progressed and a more complicated work pattern becomes evident. In the 1780s the Dysons were generally based at Newington, near Bawtry, but by the early nineteenth century they were in the London area.

On the Everton, Gringley Common and Misterton drainage schemes John Dyson Sr. had drawn up plans and estimates in 1787/8; following the Acts of Parliament he is believed to have acted as (resident) engineer from 1796 until February 1801. Despite these engineering appointments Dyson had continued to work as a contractor. In 1786 he had built the Tattersall Navigation with John Gibson and shortly afterwards he was involved with the construction of a lock on Valentine's Drain on the Witham at Nordyke Bridge and also the construction of the Misterton sea outfall sluice. In 1790 he was again working with the Pinkertons on the Stowmarket Navigation and this was followed by seven contracts, with Peter Tyler and John Langwith on the Sleaford navigation (1792–1793). Although there were some problems with defects in the construction of the locks at Haverholme in 1794, this did not affect Dyson's standing locally as he was appointed engineer for the design of the works for the Horncastle navigation on 25 February 1794 and the next month appointed surveyor, jointly with John Hudson. It is in connection with this work that, in early 1796, John Dyson Jr. is first identified, writing about settling his father's accounts.

There follows a period of uncertainty regarding the careers of the two John Dysons. From 1796–1798 a John Dyson was acting as (resident) engineer for the Isle of Axholme Drainage Commissioners. This would have been an undemanding appointment which did not preclude him from acting as resident engineer to the Everton, Gringley and Misterton Drainage, 1796–1801. This was almost certainly John Dyson Sr.'s last job, as on 14 February 1801 he was paid the balance of his account and by April 1805 he had died, confirmed by the fact that his executor was still trying to settle his account on this job.

If John Dyson Sr. had died by 1805, his namesake was developing into an important contractor. On 24 July 1800 John Dyson 'the younger' of Newington near Bawtry obtained a contract for excavations on the City Canal in London's docks and by December 1802, when he was working on the canal's lock walls, he had clearly moved to London and was living in Prospect Place, Mile End. He had by now a family comprising his wife, Elizabeth (b. c. 1780), and son, John William (1799–1843); his daughter, Elizabeth, and son, Thomas, were born there c. 1806. In addition to his contracts for the City Canal he also had excavation contracts on the West India Docks and the Grand Surrey Canal. Whether he was responsible for managing all of this work more or less simultaneously is unlikely and an explanation is probably to be found in a reference in the West India Dock Committee Minutes of January 1801 to a Mr. Dyson of Hull. This is almost certainly Thomas Dyson who was working in the Hull area at that time. John also had as partners William Bough and James Hollinsworth (qq.v.) on work on the Royal Military Canal in 1805; it was a project from which they were all dismissed.

As work on the docks wound down Dyson, with a young family to support, began to seek work elsewhere, possibly of a more permanent nature. He gave evidence on a tow path arbitration for the Kennet and Avon Canal Company in the summer of 1808, having been recommended by John Rennie (q.v.). The Company 'found Mr Dyson a very intelligent, good witness ...' (Rennie MS, NLS, MS 19781 f. 122). In September Dyson discussed possible employment on the canal with John Thomas (q.v.), but the salary of £150 p.a. was evidently insufficient as in June 1809 he gave evidence on the St. Thomas and Christchurch (Southwark), Surrey Sewers Bill, where he had taken levels; this is his last connection with the London area. At the Midsummer Sessions he was appointed County Surveyor for Dorset at a salary of £500 p.a. together with £50 for a clerk's services 'for such time as the Magistrates should think proper'. Unfortunately for Dyson this level of salary proved unsustainable to the Magistrates and he was replaced by the architect, George Allen Underwood (c. 1793–1829). In early 1817 he attempted to return to contracting by tendering for work on the New Custom House Wall; John Rennie advised against acceptance of his tender as being too low, particularly when he had no men or materiel from other contracts to carry forward.

On 8 July 1824 John Dyson Jr. was appointed on (Sir) John Rennie's recommendation as engineer to the Bedford Level Commissioners at an annual salary of £250 p.a. inclusive of expenses. He was provided with accommodation at Downham Market, Norfolk, and moved there with his family. There was a large body of work to deal with at that time, involving the

Thomas Dyson, 1771–1852

reconstruction of locks and sluices at Welmore Lake, Salter's Lode, Old Bedford River, Tongs Sluice, Hermitage Sluice, Nordelph Sluice, Stanground, Great Dyke, Welch's Dam and Reach Lode and 'a great many other works of great importance', to a value of £50,000 to £60,000. At Salter's Lode, Dyson carried out the work himself there being no suitable tenders. However, the bulk of the work was carried out by the 1830s and in 1832 the Commissioners considered discontinuing his employment with effect from April 1833. Although this decision was rescinded the end came in June 1838 when it was decided to do away with his services from April 1839 and a final gratuity of £50 was paid in view of his advanced age.

In 1839 his address at Bridge Street, Downham Market, was shared with Thomas Dyson (q.v.), his son, also practising as a civil engineer. John died of jaundice on 8 September 1849, at the age of eighty-seven. His son John William had predeceased him.

Thomas Dyson (1771–1852), assumed to be the brother of John Dyson Jr. is first mentioned as contractor, of Newington (near Bawtry), for works on the Everton, Gringley and Misterton drainage in the late 1790s where his father was engineer. This was followed by work on the Beverley and Barmston Drainage and Driffield Navigation, and elsewhere. He almost certainly assisted his brother with the London docks' contracts. After acting as contractor on the Portsmouth and Arundel Canal (1818–1821) there is some uncertainty about Dyson's subsequent career. He acted as resident engineer on the Leeds and Selby Railway, resigning in July 1832, although continuing to aid his successor. He left in the hope of making more money as a contractor on the Aire and Calder Navigation. He gave evidence on the London-Brighton Railway in 1836, before acting as a resident engineer on the northern section of the North Midland Railway. In 1840 he was resident at Sandal, near Wakefield.

Thomas Dyson (1806–1851) was the son of John Dyson Jr. who assisted his father as engineer to the Bedford Level Commissioners and from 1830 there is a possibility of confusion between the two men. This Thomas Dyson was a Member of the Institution of Civil Engineers from 1832 until 1844 and practised with John Dyson Jr. as a civil engineer at Bridge Street, Downham Market, until c. 1845, latterly at Lynn Road. It was almost certainly he who was acting as railway contractor in the 1840s and went bankrupt in 1850.

Thomas Dyson, railway contractor, of Hardinge Terrace, Albert Square, Newington, Surrey, died on 17 January 1851.

MIKE CHRIMES

[Membership records, ICE; ICE works of art catalogue; J. Grundy (1768–1772) Reports, **X** and **XI** (Laneham Drainage), Brotherton collection, Leeds University Library; Sir Joseph Banks Papers, vol. 841, Spalding Gentleman's Soc.; Estimates, correspondence, etc., Cherry Cobb embankment, East Yorks RO; Beverley and Barmston drainage commissioners minutes, East Yorks RO; Sleaford Navigation Co. (1792–1794) Proprietors' minutes; J Dyson Jr. (1809) Minutes of evidence committee of the House of Lords on the Surrey sewers bill, HLRO; Dorset Quarter Sessions Minutes, 1806–1815, 1822–1824, Dorset RO; J. Rennie (1817) Reports, 7, 70, ICE archives; Bedford Level Corporation, Proceedings 1824–1850, R 59.31.12(3–8), Cambridge RO; S. Wells (1830) *History of the Drainage of the … Bedford Level*; Leeds–Selby Railway committee minutes, PRO RAIL; T. Dyson (1836) *Minutes of Evidence … London-Brighton Railway*; Pigot (1839) *Norfolk*; F. Whishaw (1840) *Railways of Great Britain and Ireland*; White's directory (1845) *Norfolk*; *London Gazette* (1850–1851); F. McDermott (1887) *The Life and Works of Joseph Firbank*, 26–27; T. Mackay (1900) *The Life of Sir John Fowler*; C Hadfield (1972–1973) *The Canals of Yorkshire and North East England*; A. W. Skempton (1974) William Chapman (1749–1862), civil engineer, *Trans Newc Soc*, **46**, 45–82; J. Boyes and R. Russell (1977) *The Canals of Eastern England*; C. Hadfield and A. W. Skempton (1979) *William Jessop, Engineer*; A. W. Skempton (1985) The engineering works of John Grundy, *Lincolnshire History and Archaeology*, **19**, 65–82; W. M. Hurst (1995) The Sleaford navigation and Benjamin Hadley's private bank account, *RCHS Jnl*, **31**, 458–477]

Publications
John Dyson Jr.:

1824. *Report of Mr. John Dyson, Engineer to the Bedford Level Corporation*
1824. *Specification for a Sluice near Welney*
1825. *Map of the Courses of Rivers passing through the [Bedford] Level*

Works
John Dyson Sr.:

1767–1768. Adlingfleet Drainage, with James Pinkerton, 170,000 cu. yd., c. £8,000
1768–1770. Driffield Navigation, with James Pinkerton, c. 5 miles, c. £13,000
1769–1772. Laneham Drainage, with James Pinkerton, eight contracts
c. 1770. Cherry Cobb Sands embankment, with James Pinkerton
1785. Timberland Fen Drainage, Lincolnshire, Commissioner
1786. Gibson's Canal, contractor with Gibson
1787. Everton, Gringley and Misterton Drainage survey
c. 1788. Misterton Soss Sluice and Drain contract
1787–1788. Witham Drainage, lock on Valentine's Drain at Nordyke
1788–1789. Malltraeth embankment, Anglesey, including sluice £442, with John Pinkerton
1790. Ipswich–Stowmarket Navigation, with George Pinkerton, c. £1,500, completed by James Smith
1792–1794. Sleaford Navigation, bridges and locks, six contracts, c. £4,000, with Peter Tyler and John Langwith
1794–1798. Horncastle Navigation, surveyor

John Dyson Sr. or Jr.:

1795–(1798). Isle of Axholme Drainage, resident engineer
1796–1801. Everton, Gringley and Misterton Drainage, resident engineer

John Dyson Jr.:

1800. City Canal excavation, contractor
1801. West India Dock, Limehouse Basin excavation, contractor, with T. Dyson
1801–1805. City Canal, gravel excavation, contractor
1802–1803. Grand Surrey Canal excavation, contractor, 3¼d/cu. yd.
1803–1804. West India Dock, Export Dock, wall excavation, contractor
1803–1805. City Canal, West Lock excavation, contractor
1809–1817. Dorset County Surveyor
1813. Longham Bridge repairs, Engineer
1813. Canford Bridge, Wimbourne Minster, Dorset, Engineer
1824–1840. Bedford Level, engineer, c. £60,000 work

Thomas Dyson:

1796–1801. Everton, Gringley and Misterton Drainage, contractor for mother drain, triple sluice, etc.

1799–(1806). Beverley and Barmston Drainage: Hull Clough outfall sluice, Beverley Tunnel, Walton Beck–Driffield drain, contractor (other contractors included John Pinkerton)

1802–1805. Driffield Navigation, contractor, with T, Atkinson

(1804–1806). Muston and Yedingham Drainage: tunnels and culverts, contractor

1808 Bridge near Scorborough, Yorkshire, £170

c. 1812. Lavenham drainage survey

1818–1821. Portsmouth and Arundel Canal, contractor

1830–1832. Leeds-Selby Railway, resident engineer

1833–1836. Aire and Calder Navigation, contractor

1837–1840. North Midland Railway, northern section, resident engineer

1846–1848. Midland Railway, Syston and Peterborough, Melton–Oakham contract, £139,227 3s 6d

1848–1850. LNWR, Rugby–Harborough Branch, 13½ miles, completed by J. Firbank

E

EASTBURN, Henry (1753–1821), civil engineer, was baptised on 7 February 1753, the son of Michael Eastburn, apothecary of York, and his wife, Faith Jenkinson, sister of John Smeaton's wife, Ann. He became an apprentice or pupil of Smeaton in 1768 and worked as his assistant from 1775 until 1788, following William Jessop (q.v.) in both capacities. Eastburn's work during this period included a survey of Hatfield Chase (1776), drawings of the winding engine at Walker Colliery (1783) and of Aberdeen Harbour (1788). Later, in 1788, he was working with Jessop on some aspects of a scheme for bringing water from the river Colne to Marylebone; in March 1789 he and Jessop had a consultation with Smeaton on this subject.

Eastburn's first big job came in the next year when he was appointed resident engineer on the Basingstoke Canal (37 miles long from the river Wey to Basingstoke with twenty-nine locks, a cutting 1000 yd. long up to 70 ft. deep and a 1200 yd. tunnel). An initial survey had been made by Joseph Parker and an Act obtained in 1778 but nothing was done until 1788 when Jessop was appointed engineer with William Wright as 'assistant surveyor' or resident engineer. Jessop made some modifications to the original plans and in October 1788 agreed a contract with John Pinkerton (q.v.). Work began immediately at the east end of the canal at Woodham, where it meets the river Wey about 2 miles south of Weybridge, and on the tunnel near Odiham.

Jessop made some further revision, and Pinkerton's contract was accordingly revised. By the summer of 1789 doubts arose about the accuracy of levels taken by the principal surveyor, George Pinkerton, and the ability of Wright to exercise sufficient control on the contractors. At Jessop's suggestion John Rennie (q.v.) checked the levels (finding serious errors) and Eastburn replaced Wright. Eastburn arrived on site about April 1790, with William Cartwright (q.v.) as assistant. Thereafter Jessop seems to have left the direction of the works entirely to Eastburn, who settled at Odiham.

In June 1791 he reported that fifteen locks had been completed; five-hundred and fifty men were now employed on the line with forty-eight horses besides the teams bringing materials. By August 1792 all but five of the locks were built, together with fifty-two bridges and nearly 900 yd. of tunnel, as well as the Deep Cut and a long embankment. In 1793 progress slowed due to financial maladministration but by the end of that year practically all the engineering works had been finished and the canal was finally opened throughout in September 1794. Between February 1790 and April 1793 expenditure on works under Eastburn's direction amounted to some £65,000

In August 1793 Rennie, as Engineer of the Lancaster Canal, asked Eastburn if he would come as resident engineer on the southern section of the canal (between Garstang and Wigan) bringing Cartwright with him as assistant. Archibald Millar (q.v.) was already Resident Engineer on the northern section (27 miles from Borwick to Garstang) on which work started in late 1792. Eastburn accepted the post but after leaving the Basingstoke Canal about November 1793 he surveyed the Yorkshire Derwent at the request of Earl Fitzwilliam with a view to extending the navigation from Malton to Yedingham.

Eastburn arrived in Lancashire in March or April 1794, settling at Preston, and began work on the southern section with Thomas Fletcher (q.v.) as assistant, Cartwright having meanwhile gone in January to be assistant engineer on the foundations of the great Lune Aqueduct, under Millar. After completing his work on the aqueduct (for which he received a silver cup in October 1795) Cartwright joined Eastburn as assistant, Fletcher having moved on to a position on the Aberdeen Canal. In January 1797 the senior engineering staff comprised: Rennie, Principal Engineer (250 p.a.); Millar, Resident Engineer north section (£400) with four Superintendents; Eastburn, Resident Engineer south section (£400) with Cartwright, Assistant Engineer (£250) and two Superintendents (£90 and £100).

The south section of the canal, as built, consisted of three parts: a 15 mile length from Garstang to Preston; a 13 mile length from Clayton Green to Bark Hill near Wigan (the South End); and a 5 mile tramway linking the two parts. The Garstang–Preston part was opened in November 1797 and by February 1798 most of the South End was completed. Soon afterwards, when it was no longer necessary to have two resident engineers, Cartwright took over from Millar and Eastburn and it was he who, *c.* 1800, completed the South End and from 1801 to 1803 built the tramway.

After 1798 there is little to record of Eastburn's career. In June 1801 Rennie recommended him as resident engineer on the London Docks, saying 'he has acted for me on the Lancaster Canal for about four years, in which he acquitted himself much to may Satisfaction'. However, James Murray (q.v.) was appointed to this job in September 1801 and Eastburn, surprisingly, seems to fade from the engineering scene.

As for his personal life, in October 1779 he married Elizabeth Simon, daughter of the vicar of Whitkirk (Smeaton's parish church). They had one daughter, Elizabeth, who in 1808 married the Rev. William Jenkinson at York. Eastburn was elected to the (Smeatonian) Society of Civil Engineers in March 1789 and attended meetings up to April 1790, thereafter being listed as a 'country member' and last appearing in the 1797 list.

The will of Henry Eastburn, gentleman, of York is dated December 1812, leaving everything to his wife or, if she dies, to his (recently widowed) daughter. The date of his death has not been found but presumably it is not long before probate was granted in August 1821.

A. W. SKEMPTON

[John Smeaton (1781–1792) Letter Book, ICE library; Smeaton's 'Designs', Royal Society; Basingstoke Canal Minutes (1790–1793) Hampshire RO; John Rennie (1793–1801) letters in NLS; C. T. G. Boucher (1963) *John Rennie*; B. F. Duckham (1967) The Fitzwilliams and the navigation of the Yorkshire Derwent, *Northern History*, **2**, 45–61; Charles Hadfield and Gordon Biddle (1970) *Canals of North West England*; A. W. Skempton (1981) *John Smeaton*; P. A. L. Vine (1994) *London's Lost Route to Basingstoke*]

EASTON, Alexander (1787–1854), civil engineer, having been born near Carron, in Stirlingshire, was baptised on 5 July 1787 at Bathgate in West Lothian. His father, Alexander, of Falkirk (b. *c.* 1768?), and grandfather (James?) were contractors for the construction of parts of the Forth and Clyde Canal and after receiving 'a good plain education' he acquired upon their works and under their supervision, the first rudiments of professional knowledge. He received further practical training under Andrew Brocket (q.v.) a contractor of considerable eminence in the west of Scotland.

He was appointed by Thomas Telford (q.v.), at the early age of eighteen, as one of the surveyors of roads in Argyllshire and two years later, in 1807, was promoted by him to the position of resident engineer for the works of the western district of the Caledonian Canal. In May 1814, Easton having complained that his income was 'insufficient for maintaining a due respectability in his situation', his salary was raised to £250 p.a. He

remained in that post until the opening of the canal to traffic in 1822 and in the following year he was elected a member of the Institution of Civil Engineers. At about the same time he resigned and removed to Ireland where he was engaged for two or three years on works connected with the construction of a gaol in Queen's County. On his departure from the Caledonian Canal it was resolved, in view of 'the approaching completion of the work', that no successor should be appointed and James Davidson (q.v.) thus became resident engineer for the whole canal.

On his return in 1824 he was again employed by Telford, firstly on surveys for proposed improvements to the mail road between Chepstow and Milford Haven and later on supervision of the extension of the Bude Canal in Cornwall. In 1826 he was appointed resident engineer on the construction of the Birmingham and Liverpool Junction Canal at a salary of £700 p.a. and, later, on some of the works of the Ellesmere and Chester Canal. They involved very difficult ground conditions, particularly at the Shelmore embankment, which delayed completion of the canals. He subsequently became the resident engineer for the care and maintenance of the whole of these two canals, an appointment he held for more than twenty years until his retirement in 1851.

He was a man of great energy of purpose, active habits and sound practical talent; from long experience he had acquired considerable skill in designing, laying out and constructing the canal works in which for many years he specialised. He was enjoying an active retirement when unfortunately, in assisting in the felling of a tree near his home at Sutton in Staffordshire, he was injured so severely that he died soon after, on 19 March 1854. He was remembered as a man of strict integrity, yet of a kind and cheerful disposition, and an excellent temper; his decease was lamented by a circle of old and devoted friends. He was probably the brother of John Easton (1785–1826) (q.v.).

RON BIRSE

[Plans in the SRO; RAIL 808/1, PRO; Memoir (1855) *Minutes of Proceedings of the Institution of Civil Engineers*, **14**, 131–133; J. Lindsay (1968) *The Canals of Scotland*; Skempton; W. T. Johnston (1998) *Scottish Engineers and Shipbuilders*, published as a computer database]

Publications

1810. *Plan of the Intended Road commencing on Loch Lochy ... up the East Side of the River Spean to the Junction with the Loch Laggan Road*
1820. Description of the Outlet at Strone, in: *Seventeenth Report of the Commissioners for the Caledonian Canal*, House of Commons, 20

Works

1805–1807. Argyllshire, road surveyor
1807–1823. Caledonian Canal, western district, Resident Engineer

1826–1835. Birmingham and Liverpool Junction Canal, and Ellesmere and Chester Canal, Resident Engineer (construction)
1835–1851. Resident Engineer (maintenance)

EASTON, John (*c.* 1785–1826), road surveyor, was probably the son of Alexander Easton of Falkirk, and Mary Simpson, christened on 24 April 1785, and brother of Alexander, noticed above. He lived at Langholm, Dumfriesshire, before being employed by Thomas Telford (q.v.) on the Glasgow to Carlisle road from 1816.

By 1820 Easton was Telford's assistant-in-charge on the English part of the London to Holyhead road, superintending improvements and issuing instructions to the various Turnpike Trusts between London and Chirk regarding the selection of road-making materials and the breaking and placing of stone to the required shape and width. He prepared the drawing *Tools for Making and Repairing Roads* and probably much of the text in detail for Telford's important and widely circulated *Rules for Repairing Roads*, first published in a parliamentary report in June 1820. These *Rules* which were subsequently published separately in several octavo editions and also in Sir H. Parnell's *Treatise on Roads* (1833 and 1838), and later publications, were still being applied well into the twentieth century. Easton died in post in 1826. His salary was £200 p.a.

R. A. PAXTON

[T. Telford (1820) *Reports ... to the Commissioners for the Improvement of the Holyhead Road upon the State of the Road between London and Shrewsbury*, Holyhead Roads Commissioners (1820–1830) reports]

Publications
1822. Report ... Improvement and repairs of the road from Shrewsbury to London, in: *Select Committee on the Road from London to Holyhead*, 84–114
1823. Specification for Holyhead Road St. Albans and South Mimms, in: T. Telford (1838) *Life*

EASTON, John (1788–1860), surveyor, was born on 18 December 1788, the second of twelve children of Josiah and Hannah Easton of Hele, near Taunton, Somerset.

Josiah Easton (1761–1848) was well-known in Somerset as a surveyor and land agent who undertook civil engineering work such as canal, land drainage and road surveys in the area. Several of his sons followed him into engineering and surveying, perhaps the best known being **James Easton** (1796–1871) who, as part of Easton and Amos, practised as a mechanical engineer.

In 1794–1796 William Jessop (q.v.) employed Josiah to survey an early version of the Grand Western Canal scheme, between Taunton and Exeter via the Exe. This scheme, which marked the start of a long period of co-operation between the Jessop and Easton families, included a link

onward from Taunton to Uphill which Josiah first surveyed in 1793, and the same year Josiah suggested a further link to Axmouth, on the south coast.

From 1795 until 1797 he acted as engineer to the Ivelchester and Langport Navigation, which ran out of funds before it could be completed. Possibly John's first civil engineering survey was carried out for John Rennie (q.v.) when the latter was asked to look at the Bristol & Western Union Canal scheme (later called the Bristol & Taunton) in 1810. William White (q.v.) surveyed the section from Bristol to the river Parrett upstream of Bridgwater, and Easton from there to Taunton. Easton's survey included an aqueduct over the Parrett near Huntworth and a link down to riverside quays in Bridgwater. In 1811, he, or his father, drew up a scheme for a ship canal linking the Parrett at Combwich, downstream of Bridgwater, to a floating harbour in the middle of Bridgwater. The ship canal and floating harbour idea was revived by Henry Jessop, William's fifth son, in 1828–1829 but no action was taken. A further report by Henry Habberley Price (q.v.) in 1835 drew on this and Easton's work.

As agent to the Grand Western Canal in Devon *c.* 1814–1818, John was responsible for fifteen road bridges. Although he surveyed revived Taunton to Bridgwater scheme in 1822, it was in fact James Hollinsworth (q.v.) and Charles Hodgkinson who acted as engineers following the passage of an Act in 1823. The revised line terminated at a basin connecting with the Parrett at Huntworth. In 1841 the canal was extended 1 mile downstream from Huntworth to a dock basin against the Parrett below Bridgwater Bridge, again with advice from Easton.

While his work as canal and river navigation surveyor seemed doomed to frustration, Easton had more success with his work for Local Boards and Turnpike Trusts, where Josiah was already well established.

In 1813 Easton was appointed surveyor by the Trustees of the Hazelstone–Cullompton Turnpike which had just obtained its Act. The following year, at a salary of £50 p.a., he was put in charge of the works which he estimated at £2,000. The Trust got into financial difficulties and when he presented his final account of £8,000 in June 1817 he had to accept an arrangement whereby he would receive the final payment in two years. While this work proceeded Easton was engaged on other surveys such as a road from Rawridge to Chard and a railroad link to the Grand Western Canal. He was also acting as contractor for a number of bridges, the largest being that over the Culm on the Cullompton Turnpike. His various experiences were detailed in his unsuccessful application to the Somerset magistrates in 1818 to be appointed County Surveyor.

Among his brothers, **Josiah** (1790–1848) and **Edward** (1799–1898) were both involved in road surveying and county bridge maintenance and Abel (1807–1848) assisted him in surveys of

King's Sedgemoor in 1829. There John reported on a possible navigation between the Parrett at Dunball, below Bridgwater, and Somerton, Ilchester and Yeovil, and an associated drainage scheme costing £48,000. (Sir) John and George Rennie (qq.v.) also reported on the scheme. In 1830 the Othery, Middlezoy and Weston Zoyland drainage appointed Abel their surveyor.

John Easton remained active into the mid-1850s. He had a twenty-one years maintenance contract on the Taunton turnpike between Minehead and Milverton from 1828, and continued working for them until 1857. Easton died on 19 August 1860 and his family memorial is in St. Giles Church, Bradford-on-Tone.

MIKE CHRIMES, DAVID GREENFIELD
and BRIAN GEORGE

[Turnpike Trust Records, Quarter session minutes, and deposited plans (some redrawn by Haskoll, see below), Somerset and Devon County RO; C. Hadfield (1967) *The Canals of south west England*; T. Haskoll (1994) *By Waterway to Taunton*]

Works

c. 1814–1818. Grand Western Canal, including fifteen bridges, agent
1813–1817. Hazelstone–Cullompton New Turnpike Road, Devon, surveyor to the Trustees
c. 1815. Baulk Bridge, Cullompton, contractor
c. 1815–1817. Knapp Bridge, North Curry, Somerset, design and contractor, £380 (for the River Tone Coservators)
c. 1815. Crawley Bridge, Yarcombe, Devon, contractor
c. 1817. Honiton–Sidmouth New Turnpike Road, Devon, surveyor to the Trustees

EDGEWORTH, Richard Lovell, MRIA, FRS

(1744–1817), educationalist, inventor and engineer, was born at Bath in 1744, the seventh child of Richard and Jane Edgeworth (nee Lovell). The family returned to Ireland in 1747 and settled on the family estate at Edgeworthstown, in county Longford. A visit, at the tender age of six years, to a mechanics workshop in Dublin, appears to have made a deep impression on Edgeworth. Throughout his long life, he devoted much time, thought and study to the construction of carriage wheels and the making of roads; it was an absorbing interest of his life, and he ranks, in the opinion of many later observers, as one of the most far-seeing road engineers of all time. This honour, however, eluded him, and the credit for his achievements was given to others.

Following schooling at Warwick, Drogheda and Longford, he entered the University of Dublin but, in 1761, was removed by his father to Corpus Christi college in the University of Oxford. Before he was twenty years of age, he had married for the first time and had a son. A daughter by a later marriage was the celebrated writer Maria Edgeworth. After pursuing a somewhat tempestuous lifestyle in England, Edgeworth returned to Ireland in 1782 to manage the family estate and became intensely interested in agricultural improvements. His reputation in this area made him an obvious choice to assist in the work of the Bog Commissioners (1810–1814).

As a landowner he had the resources and the time to dabble in what came to be regarded as mechanical engineering, and his developments in carriage design are noteworthy. He is also credited with making the first practical telegraph and was deemed to have been more than twenty years ahead of the Frenchman, Chappé.

When in later life he turned his mind to road-making, he was able to draw on almost half a century of experimentation. His fundamental contribution to the development of road-making arose from his view, unique at the time, that road construction and carriage design were linked. In *An Essay on the Construction of Roads and Carriages*, published in 1813, Edgeworth described in detail what became accepted as the best method of road construction, in preference to that propounded by John Loudon Macadam (q.v.). Edgeworth's book on road construction was a thoroughly comprehensive work embodying all aspects of the subject and his ideas and suggestions were very much in advance of his contemporaries. He advocated, for example, that a national roads authority should be responsible for the maintenance of Britain's highways.

Edgeworth's road construction method was, in fact, what later became known as 'macadamising'. He and Macadam followed the same pattern of road construction up to a point, that is, a foundation of large flat stones upon which were laid stones of six or seven pounds weight, with a final covering of small stones an inch or two in size. Here, however, their views diverged. Macadam used no further covering of the stones on the basis that the subsequent traffic would break up the stones and the chippings so produced would fill the interstices between the larger stones. Edgeworth, on the other hand, insisted on a final covering of 'clean angular gravel that may insert itself between the interstices of the stone' and thus bind or metal the roadway. Again, Edgeworth differed on the question of the proper curvature of the road profile, insisting on a cambered surface, whereas Macadam, at least initially, preferred to maintain a flat profile. The principle of metalling a road, as introduced by Edgeworth, had not been known to Thomas Telford (q.v.), when in 1802 he issued his specification for a Scottish road. There is no evidence to suggest that Macadam used the 'macadamising' method prior to 1811, and Edgeworth was the first to describe the method, his book appearing some seven years prior to that by Macadam.

Lay and Clarke, perhaps the foremost authorities on the history of road-making, justly credit Edgeworth with priority, and point out that Edgeworth's use of binding material became adopted as the preferred method, whereas it was condemned by Macadam. According to John Joly, writing in 1913, 'Edgeworth formulated a thoroughly scientific procedure, based on correct first principles. He stands as one of the most far-reaching and far-seeing scientific road engineers of all time.'

Edgeworth died peacefully, surrounded by his immediate family, at Edgeworthstown on 13 June 1817.

His son William (q.v.) also became involved in civil engineering.

RON COX

[*Commissioners … into Bogs in Ireland* (1810–1814), 1st to 4th reports; M. Edgeworth (ed.) (1820) *The Memoirs of Richard Lovell Edgeworth*, 2 vols.; J. Joly (1913) *Reminiscences and Anticipation, Richard Lovell Edgeworth*, 164; D. Clarke (1965) *The Ingenious Mr. Edgeworth*; N. McMillan (1991) The first Macadam road, *Technology Ireland*, May, 30–31; M. G. Lay (1992) *Ways of the World*; DNB]

Publications

Selected writings of R. L. Edgeworth relevant to engineering matters:

1788. Essay on springs and wheel carriages, *Transactions of the Royal Irish Academy*, **2**
1795. Essay on the telegraph, *Transactions of the Royal Irish Academy*, **6**
1801. Essay on rail-road. *Nicholson's Journal*, **1**
1810. On telegraphic communication, *Nicholson's Journal*, **26**
1812. Report on District 7, *Commissioners into Bogs in Ireland, Reports*, **2**, 173–193 (dated 1810)
1813. *Essay on the Construction of Roads and Carriages* (2nd edn. 1817)
1814. Report on District No. 15, *Commissioners into Bogs in Ireland, Reports*, **4**, 102–107 (dated 1813)
1816. On wheel carriages, *Nicholson's Journal*, **48**
1817. *Experiments on Carriage Wheels*

EDGEWORTH, William (*c.* 1794–1829), of Edgeworthstown, County Longford, Ireland, was a son of Richard Lovell Edgeworth (q.v.) by his third marriage, to Elizabeth Sneyd (d. 1797). His father, after a somewhat dissipated youth, had settled down both to mechanical pursuits and the management of his Irish estates.

As a young man William Edgeworth is said to have spent much time with William Strutt (q.v.) the son of Jedidiah Strutt, the cotton manufacturer, and to have worked as an engineer in the Strutt factory. He also assisted his father in compiling bog drainage reports in 1810–1813, as a surveyor receiving a guinea a day. In 1814 his father was struck down with a dangerous illness and Edgeworth, now a young engineer, planned bridges and devoted a considerable amount of time to experimenting with carriage wheels and springs, in fact carrying on his father's work. He is noted as assisting with the construction of a bridge at Longford in 1814.

His subsequent career is not known in full but he carried out some work in Ireland for Alexander Nimmo (q.v.). This included the Oughterard–Clifden road (County Galway). In 1823 he was working on roads in County Waterford, and in 1825 he was 'laying out the road to Glengariff'. That year and in 1827 he carried out surveys of the River Tees in England for Henry Habberly Price (q.v.) who had been commissioned by the Tees Navigation Company to report on improvements to the river's navigation. Price suggested that a suitable means of effecting improvements was to construct a canal from Stockton to the sea, eliminating the need for the use of the river, and Edgeworth's surveys show the line of the canal in two versions with lengths of 4½ miles and 2¼ miles. Neither was built but one was used by the Stockton and Darlington Railway for its extension to Middlesbrough.

Edgeworth was elected a Member of the Institution of Civil Engineers in 1829, probably through the interests of Marc Isambard Brunel and Richard Beamish (qq.v.) but he died of consumption that May, his death commented upon in 1851 by William Bald (q.v.), later to be involved in affairs relating to the Tees and with whom Edgeworth may have worked. He then referred to him as 'a young Gentleman of great Talent, who died early in Life, to the great Regret of all his scientific Friends … (and) had it pleased Providence to spare him, his Career promised to be great and brilliant'.

RON COX, MIKE CHRIMES,
R. W. RENNISON and TED RUDDOCK

[Membership records, ICE archives; Catalogue of prints and drawings, ICE; *Commissioners … into … Bogs in Ireland* (1810), 1st report; R. L. Edgeworth (1812) Report on District No. 7, Appendix 8, *Commissioners … into Bogs in Ireland*, 2nd report, 173–193 (dated 1810); R. L. Edgeworth (1813) *An Essay on the Construction of Roads and Carriages* (2nd edn. 1817); R. L. Edgeworth (1814) Report on District No. 15 … east of the river Shannon; and observations on Mr. Roscoe's improvements of Chat Moss in Lancashire, Appendix 7, *Commissioners … into Bogs in Ireland*, 4th report, 102–107 (dated 1813); R. L. Edgeworth (1816) *A Letter to the Dublin Society, relative to Experiments on Wheel Carriages*; C. B. Noble (1938) *The Brunels*; DNB; W. Bald (1851) *Report of the Admiralty*; E. Inglis Jones (1959) *The Great Maria*; D. Clarke (1965) *The Ingenious Mr. Edgeworth*. London]

Surveys

1825. *The River Tees below Stockton Bridge …*, for H. H. Price
1827. *Survey of the River Tees below Stockton Bridge …*, for H. H. Price

EDWARDS, David (fl. 1780–1803), bridge-builder and farmer at Beaupré, near Cowbridge, Glamorganshire, was a son of William Edwards (q.v.) and his wife Elizabeth. He inherited some of his father's reputation as a bridge designer and contractor and is believed to have worked initially on William's contracts. He went on to both design and contract for masonry bridges, in which he maintained the simplicity of style and construction that characterised the bridges built by his father, but two of David's structures surpassed any of his father's bridges in scale and cost. These were the three-arch bridge over the Tywi at Llandilo-yr-Ynys and the bridge over the Usk at Newport (Mon.). The former is his finest monument, having been restored to exactly its former appearance after a partial collapse in 1931. In the Newport Bridge, by much his largest project, he was joined as partner in the contract in 1794 by his brother Thomas (q.v.), but David finished the contract himself after Thomas's death in 1800. Two of his sons, William and Thomas, worked with him on it.

Another son of David's, also called David, became a mason and was superintending the building of locks and bridges on the Kennet and Avon Canal when Malkin made his tour of Wales in 1803, and was considering taking a contract for the rebuilding of a bridge at Caerleon. A bridge of three segmental arches was built there a few years later.

TED RUDDOCK

[Minutes, letters and drawings concerning Newport Bridge 1791–1801, Gwent County RO; accounts re. same, Newport Public Library; B. H. Malkin (1804) *The Scenery, Antiquities and Biography of South Wales*, 82–93; E. Jervoise (1936) *The Ancient Bridges of Wales and Western England*, 72–96, pls. 36–37; T. Ruddock (1979) *Arch Bridges and their Builders 1735–1835*, 51–53; H. P. Richards (1983) *William Edwards, Architect, Builder, Minister*]

Works

1786. Llandilo-yr-Ynys Bridge (R. Tywi), three arches, 45 ft. span
?. Pont Loerig (Afon Taf), three arches, 27 ft. span, widened in concrete cast to the shapes of masonry features in 1931
1786. Llandilo-yr-Ynys Bridge (R. Tywi), three arches, 45 ft. span
?. Edwinsford Bridge (Afon Cothi), no details known, demolished for replacement before 1936
1786. Llandilo-yr-Ynys Bridge (R. Tywi), three arches, 45 ft. span
?. Bedwas Bridge (R. Rhymney), two arches (possibly undertaken in partnership with his brother Thomas)
1794–1801. Newport Bridge (R. Usk), five arches, 62–72 ft. span, cost £15,941, demolished 1927

EDWARDS, Langley (fl. 1748–d. 1774), civil engineer, was Steward of the Duke of Bedford's estate at Thorney from 1748 until replaced by John Wing (q.v.) in 1750. In June 1751 he and the surveyor, John Aram, surveyed and produced an estimate for making the river Nar navigable to Narborough. Nothing happened for six years, nor

is it known what Edwards was doing during this period, except that in 1754 he is listed among the newly appointed Commissioners of the North Level of the Fens. In 1757 he became involved with three river navigation schemes, giving evidence to a Parliamentary Committee in March on the river Ivel, making a re-survey of the Nar in May, and a survey of the river Blyth in June.

The Act for the Ivel Navigation having been obtained in May 1757, Edwards and Thomas Yeoman (q.v.) were appointed Surveyors of the Works (engineers). Work began soon afterwards and was completed (from Tempsford to Biggleswade) in October 1758, with five locks, at a cost of £6,000. Meanwhile on the Nar, Edwards agreed to undertake the work for his estimated cost of £2,500, starting in September 1757. The navigation was opened in August 1759; it was 12 miles in length with one lock and nine staunches. He received £100 for his survey, paid on completion.

On the Blyth scheme, following the Act of 1757, sufficient funds had been raised by June 1759 for proceedings to begin. Edwards made a re-survey and estimate and work started in November, his assistant Samuel Jones (q.v.) overseeing construction. With three locks and eight short new cuts the 9 mile navigation was opened in July 1761 at a cost of £3,820.

At this time Edwards was living at King's Lynn and in 1760 was busy surveying and preparing a report on the River Witham, so he managed only two site visits to the Blyth, in Suffolk. The Witham became the scene of his greatest achievement. John Grundy and his father (qq.v.) had proposed a scheme for improving navigation of the river and drainage of the adjacent fenland in 1744, the essence of which was a new cut 7 miles long from Chapel Hill to Anton's Gowt. This proved too expensive for the landowners to contemplate but in 1753 Grundy was asked to re-examine the proposal. He reported with an improved line of the cut, an extension 2 miles long towards Boston, and a large outfall sluice, with an adjacent navigation lock, near Anton's Gowt. This plan was favourably received, with one exception: the sluice should be placed nearer to Boston. Nevertheless, again no action was taken. Five years later Grundy suggested that the sluice might be made at Boston Bridge. Edwards was asked to comment on this idea and the scheme in general and reported in 1760 proposing that the sluice be placed on a short cut just upstream of Boston. In other respects he agreed with Grundy's 1753 plan, though recommending that the main cut for the Witham should be 10 ft. wider.

Encouraged by what seemed to be the emergence of an acceptable scheme, the landowners asked John Smeaton (q.v.) to join Grundy and Edwards in a re-examination of the whole question. They took 'a fresh view of the River and Fens' in October 1761 and reported on 23 November. An engraved plan by Grundy appeared early in 1762. The estimated cost was rather more than £54,000.

All three engineers gave evidence on the Bill in Parliament, and the Act passed in June 1762. Edwards received £90 for his contributions, Grundy charged £254 (probably dating back to 1753) and Smeaton £73. The scheme finally adopted involved Grundy's 1753 cut, the Grand Sluice (as it was called) exactly as proposed by Edwards, a new 14 ft. sluice for the outfall of the drain at Anton's Gowt and three locks in the river between Chapel Hill and Lincoln; the total length from Lincoln to Boston was 32 miles of which 9 miles would be in new cut, bottom width 50 ft., 9–10 ft. deep with side slopes of 1.5 : 1 and banks 10 ft. high set back 40 ft. from the top edge of the cut. The Grand Sluice would have three openings each 17 ft. wide, the adjacent lock to be 15 ft. wide and 58 ft. between gates, with 10 ft. of water on the sills at high water springs.

Commissioners for drainage and navigation held their first meetings in August. Edwards was ordered to survey the ground from the site of the Grand Sluice to Chapel Hill and 'mark out'. In November he delivered his plan and was appointed Surveyor of the Works (engineer) at an annual salary of £225. He was to make drawings and estimates for the Grand Sluice and submit them to Grundy and Smeaton for approval. This they gave in January 1763 and advertisements were placed for contractors. Work began in April, with Samuel Jones assistant to Edwards at 52 guineas p.a. A second assistant, John Quince, joined soon afterwards and James Creassy (q.v.) was employed from time to time as a freelance surveyor.

In April 1764 Edwards wrote to Smeaton saying that a cargo of stone quoins and copings had arrived, an order for them having been placed on Smeaton's recommendation. He was by this time resident in Boston and had hoped to visit Smeaton, but floods and a heavy fall of snow prevented him leaving the Witham. Even so, upwards of one-hundred men were kept at work all winter and a third of the 'spade work' between Boston and Chapel Hill had been completed. Smeaton replied eight days later from Austhorpe, giving a graphic account of the floods on his works on the Calder Navigation, and adding a postscript: 'the progress you mention in the Spade work surprises me'. As well it might, for a third of the total excavation amounted to over 300,000 cu. yd. and had been completed in a year.

Edwards had the honour of opening the Grand Sluice on 15 October 1766 'in the presence of a very large concourse of spectators'. By July 1767 all work up to Chapel Hill reached completion. Deepening the river upstream and building the three locks took until 1771; total expenditure on the whole scheme was £48,000. Edwards presented his last report in July 1772 but Samuel Jones had been effectively in charge from 1770 and was kept on as Surveyor until 1776. Improvement of the river led to improvements in the drainage of some 30,000 acres of adjacent fenland.

The floods in the winter of 1763–1764, referred to by Edwards in his letter to Smeaton, were the worst ever remembered. They submerged the entire fenland west and south of Boston, an area of 58,000 acres comprising Lindsey Level and Holland Fen, known collectively as the Black Sluice District.

After years of neglect it was now decided to take urgent action. The Witham Commissioners on request granted Edwards leave of absence to advise on the problem. He reported to the landowners in April and again in July and September. By this time he had produced a plan and a scheme with preliminary estimates. John Landen, FRS (1719–1790), Steward to Earl Fitzwilliam's estate near Peterborough and a distinguished amateur mathematician, was then asked for his opinion and help by taking levels. He agreed with Edwards's proposals. The scheme was approved and application made for a Bill. The Witham authorities gave Edwards permission to attend the Parliamentary proceedings, provided the Black Sluice proprietors paid his fees and expenses. The Act, duly passed in May 1765, lists the works to be done: principally a new sluice at the outfall on the South Forty Foot Drain just south of Boston; this drain (8 miles long) to be deepened; the old Main Drain, continuing the South Forty Foot 14 miles down to the bank of the river Glen, to be enlarged in a new cut, and all side drains to be improved. These works would reinstate the drainage of Lindsey Level completed by John Liens (q.v.) in 1639, destroyed by rioters in 1642 and never properly restored. As for Holland Fen, the bank of the new Witham cut, when completed, would give protection from river flood water; its main drain, the North Forty Foot (dating from the 1720s), was to be improved and led into the new sluice.

At the first meeting of the Commissioners in June 1765 Edwards and Landen were appointed joint Surveyors of the Works, later referred to as joint Engineers, Landen at an annual salary of 40 guineas, Edwards to give what time he could without detriment to the Witham. John Chapman and Richard Strattard were appointed (assistant) Surveyors of the Works at annual salaries of 50 guineas each. Edwards could employ Samuel Jones occasionally to assist him in the early stages.

Contracts for carpentry, brickwork and masonry of the new sluice, to Edwards's design, were let in September. Known as the Black Sluice, it had three openings to give a waterway of 40 ft. Early in 1766 work began on the South Forty Foot and the new Main Drain cut. In July Edwards received £243 for his fees to date. All the works were nearing completion by December 1768. Edwards and Landen retired in July 1769, total expenditure having reached about £24,000. Edward Hare (q.v.) then took over as Surveyor, Chapman and Stattard stayed in post. What little new work remained to be done seems to have been finished by 1771, perhaps another £4,000 having been spent.

W. H. Wheeler, the historian of the Lincolnshire fens, says that 'the works were efficiently carried out and, being well-designed, entirely answered expectation. The fen, which, before the drainage, was ... growing a coarse herbage and affording a scanty pasture during the summer months, became rich arable and grass land, and the annual value increased tenfold'.

In April 1771 Edwards presented a decisive report on improvement of the river Nene outfall. The history of this project begins in 1721 with the making of a new cut by Nathaniel Kinderley (q.v.) from River's End (4½ miles below Wisbech) past Gunthorpe Sluice toward Peter's Point. But rioters destroyed the diversion dam in 1722 before it was completed. Nothing was done until several engineers submitted reports between 1768 and 1770, the outcome of which was a decision to re-open Kinderley's Cut and extend it a total distance of 6 miles from River's End. Edwards recognised the merit of the long cut but recommended as a measure more likely to be sanctioned the restoration of Kinderley's Cut, with larger dimensions that those originally adopted. Nevertheless plans to carry out the long cut were put in hand and Edwards, Richard Dunthorne and John Golborne (qq.v.) were employed in setting out what they considered to be the best line. However, objections were raised to the cost and eventually, when an Act passed in June 1773, it was for the cut which Edwards had proposed in 1771, though with smaller dimensions. Kinderley's Cut, as it was called after its originator, was carried out under Dunthorne's direction with William Elstobb (q.v.) as resident engineer. When opened in 1776 it proved highly beneficial to drainage of the North Level.

Edwards joined the (Smeatonian) Society of Civil Engineers in March 1773. He attended another meeting on 22 April 1774 but died before 5 July; in August 1774 his daughter, Ann Edwards, was granted Administration of the estate of the late Langley Edwards of Boston. Next year Ann received £100 from the Black Sluice Commissioners for her father's outstanding fees.

Edwards is said to have a portly figure and as 'being a very good hand at a bottle'. He is mentioned in warm tones by Yeoman in a letter to Smeaton.

A. W. SKEMPTON

[North Level Commissioners Order Book (1754–1775) Cambridgeshire RO; Minute Books of Witham Drainage and Navigation Commissioners (1762–1775) and Black Sluice Drainage (1764–1775) Lincolnshire Archives Office; John Smeaton (1764) Letter Book, ICE Library; W. H. Wheeler (1846) *History of the Fens of South Lincolnshire*; John Boyes and Ronald Russell (1977) *Canals of Eastern England*; A. W. Skempton (1996) *Civil Engineers and Engineering in Britain, 1600–1630*]

Publications

1761. *Reports ... concerning ... the River Witham* (with Grundy and Smeaton)
1764. *Plan of the Low Fen Lands and Marshes lying between the Rivers Witham, Kyme Eau and Glen*
1771. *Reports for Amending the Outfall of the Nene*

Works

1757–1758. River Ivel Navigation (with Thomas Yeoman), £6,000
1757–1759. River Nar Navigation, £2,600
1759–1761. River Blyth Navigation, £3,800
1763–1771. Witham Drainage & Navigation, £48,000
1765–1769. Black Sluice Drainage, £24,000

EDWARDS, Thomas (d. 1800), bridge-builder and innkeeper of the Three Cocks, Aberllynfi near Glasbury, Radnorshire, was a son of William Edwards (q.v.) and his wife, Elizabeth. After learning the business of bridge-building in his father's employment, Thomas became contractor for a number of bridges built in mid-Wales and South Wales. For one of the largest, the seven-arch bridge at Glasbury, he contracted to build a bridge his father is said to have designed and the same attributions are likely to apply in the case of Dolauhirion Bridge which has a stone in the parapet carved with 'THOMAS EDWARD 1773'. However, for other bridges built by him, Thomas Edwards himself is likely to have made the designs.

Between 1781 and 1791 Edwards was consulted repeatedly by the Justices of the Peace of Breconshire on all their difficult bridge repairs, and also by other authorities. He joined his brother, David (q.v.), in the contract for the family's biggest bridge at Newport (Mon.) but died in 1800 before it was finished. One of his larger contracts also ran into difficulty at about the same time: he contracted to repair and widen the old bridge over the Usk at Brecon in 1793, with the usual clause requiring him to keep it in repair for seven years afterwards, only for the catastrophic flood in the river in February 1795 to cause a partial collapse of the bridge. When he died—between 15 July and 7 October 1800—his wife, Sarah, with his brother, David, paid £150 for the cancellation of his unexpired bond to maintain the bridge.

TED RUDDOCK

[Breconshire Quarter Session Orders 1773–1801, NLW, Aberystwyth; see also sources named for *Edwards, David*]

Works

1773. Dolauhirion Bridge (R. Tywi), contractor, one arch, 84 ft. span, with one cylindrical void through each spandrel
1777. Glasbury Bridge (R. Wye), contractor, seven segmental arches, contract price £3,000, destroyed by flood in 1795

1781–1782. Talybont Bridge (R. Usk), rebuilt, design and contractor, contract price £159, rebuilt in 1931

1783. Pont Ithel (Afon Llynfi), rebuilt, design and contractor, three arches, max. 23 ft. span, cost £160, widened in 1952

1793. Brecon Bridge (R. Usk), widened and repaired, design and contractor, seven arches, cost £1,000, further partial collapse in 1795

EDWARDS, William (1719–1789), bridge-builder and minister of the Independent Church in Wales, the third surviving son of Edward David, a farmer, was born at Ty Canol, Groeswen, Caerphilly, and baptised at Eglwysilan Church on 8 February 1719. His father died by drowning seven years later when fording the river Taff, and his mother then moved with her family to Bryn Tail, another farm near the church. Bryn Tail was to be William's home to the end of his life.

He had little or no school education but is presumed to have been taught by his mother 'all she knew', and more by local church ministers. He lived at the time in entirely Welsh-speaking communities and only in his twenties was he taught English by Walter Rosser, a blind baker with whom he lodged when building a forge in Cardiff. His son David (q.v.) explained to Benjamin Malkin many years later that while working on the farm in his youth William Edwards undertook mason work, learning his trade by observation of other masons' work and of the ruins of Caerphilly Castle. There was also the substantial Caerphilly Furnace, producing pig iron, to engage his interest. Concurrently, he became deeply involved in a religious revival in the district, which led eventually to his ordination as minister of the Independent Chapel at Groeswen, about 4 miles from Pontypridd, in 1745 at the age of twenty-six. He also had a third parallel career as farmer of Bryn Tail, which passed into his ownership from his mother.

When he began contracting for buildings on his own account, his early work included houses, forges and a water mill. In 1746 he was commissioned to build a bridge across the river Taff less than 2 miles from his home, at the place now called Pontypridd, and in Welsh Pont-tŷ-pridd, 'the bridge of the cob house', a name obviously referring to a previous bridge undoubtedly of timber. For £500 Edwards built a three-arch bridge and guaranteed to keep it in repair for seven years but in fact it was destroyed by a flood in the river after about two years. Edwards then proposed to replace it with a single arch, eliminating the danger from floods, but when it was approaching completion the centring was carried off by another flood and the arch collapsed. When he built it again, with better centring, it stood for a short time but then collapsed yet again by breaking upwards at the crown of the arch, due to the large weight of stone on the haunches and the lightness and slenderness of the crown. After consultations and the raising of extra funds, Edwards built it once more of the same overall

William Edwards

shape but with three open cylindrical voids through each of the haunches of the bridge to achieve a satisfactory distribution of weight on the arch. The result, when it was completed in 1756, was a stable arch which has stood to this day, albeit without ever carrying heavy traffic because it was too steep for the passage of large vehicles, especially when drawn by horses.

The span of the arch, 140 ft., was the longest yet built in Britain and remained so for forty years; its stability and structural 'equilibrium' has remained a subject for study by structural theorists up to the present day. Edwards undertook contracts for about seven more bridges, all built, so far as is known, to his own design and of similar materials to those of Pontypridd; in three of them he repeated the device of one or two cylindrical voids in each spandrel. Only in one instance, Dolauhirion Bridge near Llandovery, can the voids still be seen. The spans of all three were much shorter than Pontypridd and the roadways rose at gentle gradients, showing a maturity of design which made Edwards a respected authority on masonry bridges in Wales. He was consulted by several roads authorities for advice on special problems and about the work of other masons.

He also continued to undertake other building work in stone but few facts about his works have been discovered. There is clear evidence, however, of his assisting John Morris, the proprietor of the copper industry by the river Tawe, in many ways. Edwards drew the layout plan of the town built to house the workers as a right-angled grid and remained in touch with the development for many years. The town was christened Morriston.

Edwards had four sons by his wife, Elizabeth, two named David and Thomas (q.v.), both of whom became significant bridge-builders, and William and Edward. He died on 7 August 1789 after a long illness, having survived his wife by seven months. He was still minister of the Groeswen meeting house but was buried in the graveyard of his 'home' parish church of Eglwysilan. His very modest possessions were divided by his will amongst his children and grandchildren, but his six-volume biblical commentary was bequeathed to his church for the use of future ministers.

A miniature painting of William Edwards, shown half-length, was commissioned by John Morris from Thomas Hill Jr., and exhibited at the Royal Academy in 1779. It is now in the National Library of Wales with a copy printed in Richards' *William Edwards*. A copy of it is said to have been made for the Empress Catherine the Great of Russia and a relief in bronze was also copied from it to form part of a memorial plaque to Edwards placed in Groeswen Chapel in 1906. Hill's portrait is the basis for the engraving reproduced in Smiles.

TED RUDDOCK

[B. H. Malkin (1804) *Scenery, Antiquities and Biography of South Wales*; T. M. Smith (1846) Account of Pont-y-tu-Prydd, *Minutes of Proceedings of Institution of Civil Engineers*, **v**, 474–477; S. Smiles (1904) *Lives of the Engineers; Smeaton and Rennie*, 87; E. I. Williams (1945) Pont-y-typridd: a Critical Examination of its History, *Transactions of the Newcomen Society*, **xxiv**, 121–130; E. C. Ruddock (1974) William Edwards's Bridge at Pontypridd, *Industrial Archaeology*, **11**, 194–208; T. Ruddock (1979) *Arch Bridges and their Builders 1735–1835*, 48–53; H. P. Richards (1983) *William Edwards, a Builder for Both Worlds*; Colvin (3), 336–337; MS sources quoted by the above]

Works
Edwards was both designer and builder for most of the bridges listed, but in some of the later instances he may have acted only as designer.

c. 1740. Constructed a forge in Cardiff
1746. 'New Bridge' at Pontypridd, Glamorgan, three arches, price £500, destroyed by a flood after about two years
1748–1756. 'New Bridge', built again as a single-arch bridge, span 140 ft., collapsed due to destruction of centring by a flood, rebuilt on stronger centring, after standing for some time it failed by breaking upwards at the crown in 1755, rebuilt 1756, total cost (1746–1756) £1,154
?. (attrib.) New Bridge at Tredunmock, Glamorgan (R. Usk), three segmental arches
1746–1752. R. Usk bridge at Usk, five arches, 42–52 ft. spans, cost £980, widened 1836

?. Beaufort Bridge (R. Tawe), three arches, demolished 1968
?. Pontardawe Bridge (R. Tawe), one arch, span 80 ft., with (formerly) one void through each spandrel
?. Betws Bridge (R. Amman), one arch 45 ft. span
c. 1768. Aberavon Bridge (R. Avon), one arch, 70 ft. span, demolished 1842
1773. Dolauhirion Bridge (R. Tywi), one arch 84 ft. span with one cylindrical 8 ft. diameter hole through each spandrel; Thomas Edwards, named on a stone in the parapet, was probably contractor, independently or in partnership with his father
c. 1777. Glasbury Bridge (R. Wye), seven arches, contractor Thomas Edward, contract price £3,000, destroyed by flood in 1795
1778. Wychtree Bridge (R. Tawe), near Morriston, one arch, 95 ft. span, two cylinders in each spandrel, 8 ft. and 2 ft. 6 in. diameter, demolished 1959
?. (attrib.) Addition to Pontcymer Bridge (R. Pelena), one short arch, formerly a packhorse bridge
1760s–. Copper industry workshops at Morriston ('superintendent' of planning and construction)

Non-engineering works:

c. 1768. Plan for the new town of Morriston near Swansea
1782. Chapel of Libanus at Morriston (part patron, designer and builder)

ELMES, James (1782–1862), architect and writer, was born in London, the son of Samuel and grandson of John Elmes, both builders. Elmes was educated at Merchant Taylors' school before becoming a pupil of George Gibson, an architect with important connections in the City business community. In 1804 he entered the Royal Academy school and exhibited regularly at the Academy from 1801 until 1842.

Elmes's work as an architect has been overshadowed by his writings which include the first documented biography of Sir Christopher Wren (q.v.) and his work on the Port of London, produced for the publisher John Weale, and generally now known as part of *Public Works of Great Britain* (1838), a compilation generally attributed to F. W. Simms. This work was doubtless commissioned by Weale because of Elmes's position as Surveyor to the Port of London, an officer of the Corporation of London, and a post created following the 1799 *Port of London Act*. His work is an important source of published information on London's early docks.

For the publication, Weale borrowed Telford's drawings for St. Katharine's Docks from the Institution of Civil Engineers, to whom they had been presented on Telford's death. Elmes was himself briefly a member of the Institution, elected an Associate on 3 March 1829. He was a Vice-President of the London Architectural Society and when applying for election as Surveyor to the Port of

London in 1828 he described his qualifications as an architect and civil engineer. Among his pupils was his only son, Harvey Lonsdale Elmes (1814–1847), whose brilliant career as architect for St. George's Hall, Liverpool, was cut short by early death.

Elmes retired from practice due to failing sight in 1848 and died at Greenwich on 2 April 1862. The Royal Institute of British Architects has a portrait of Elmes by his friend James Lonsdale.

MIKE CHRIMES

[ICE membership records, ICE archives; Correspondence and Council Minutes, ICE archives; Obituary (1862) *Builder*, 275; Skempton; Colvin (3)]

Publications
A full list of publications is given in Colvin (3).

1808. Essay on foundations, *Essays of the London Architectural Society*, 169–189
1823. *Memoirs of the Life and Works of Sir Christopher Wren*
1827. *A Practical Treatise on Architectural Jurisprudence*
1827–1829. *Metropolitan Improvements on London in the Nineteenth Century* (for T. H. Shepherd)
1831. *London Bridge, from its Original Formation of Wood to the Present Time*
1831. *A Topographical Dictionary of London and its Environs*
1838. *A Scientific, Historical, and Commercial Survey of the Port of London*

ELSTOBB, William (fl. 1730–1781), was a land surveyor and civil engineer. From 1730 to 1765, but not in the following years, he styled himself 'William Elstobb Jr.'. In this manner he may be distinguished from 'William Elstobb', perhaps his father, who in 1725 was living at King's Lynn, subscribed to Badeslade's *History of the Port of King's Lyn*, advertised in 1739 as a teacher of mathematics at Wells, and published in 1742 *Some Thoughts on Mr. Rosewell's Proposal for amending Lynn Channel and Harbour*; presumably he died *c.* 1765.

'Wm. Elstobb Jr.' first comes to notice in July 1730, writing from Lynn, as the author of a note communicated to the Peterborough Society on experiments on hydrostatic pressure at a depth of 55 fathoms (100 m) in the sea off the Norway coast. He joined the Society in 1733, submitting observations on a lunar eclipse made with 6 ft. telescope in November 1732. He also contributed two mathematical problems, and another in 1734.

In a published survey of Sutton and Mepal Levels in the Fens (1750) he advertises that 'Any Gentleman wanting to have Estates survey'd in any part of England, Scotland or Ireland, may have them done with Accuracy and Integrity, and map'd in the most elegant and natural Manner, upon very reasonable Terms by William Elstobb Jr. of King's Lynn. By whom young Gentlemen

may be Taught the Principles of Geography, Astronomy, the Use of the Globes, Mensuration, Surveying, Algebra and other Parts of the Mathematics, in a familiar way'. Another, longer, advertisement in the same vein appeared in the *Norwich Mercury* in June 1756 by 'William Elstobb Jr., Land Surveyor and Teacher of the Mathematics, King's Lynn'.

Apart from his surveying and maths teaching he took an interest in engineering matters, to the extent of publishing in 1745 a tract on *The Pernicious Consequences of Replacing Denver Dam and Sluices*. But it was in 1767 that he became well known in engineering circles, when the Commissioners of the North Level of the Fens asked him to survey the river Nene from Peterborough to the estuary below Wishbech, a distance of 32 miles. His map of the river and adjacent topography (scale 1 in. to 1 mile) remains in manuscript copies. His longitudinal section showing water levels at low tide and the river bed was published as an engraving.

At that time measures to improve the river, and particularly its outfall, were becoming necessary as the only practical means of preventing frequent floods in the North Level. The Commissioners consulted John Smeaton, Thomas Yeoman and John Golborne (qq.v.) among others. Their reports were submitted between 1768 and 1771, those by Smeaton and Yeoman including Elstobb's section. Eventually in July 1773 it was decided to re-open the 1½ mile cut from River's End (4½ miles below Wisbech) past Gunthorpe Sluice towards Peter's Point started by Nathaniel Kinderley (q.v.) in 1721 but abandoned after the diversion dam was destroyed by rioters before completion in 1722. The new cut was to be named Kinderley's Cut in honour of its originator.

Richard Dunthorne (q.v.), Superintendent General of Bedford Level, who had taken part in all the planning, determined the dimensions of the cut and directed operations. Elstobb was appointed Superintendent or resident engineer in August 1773, to be paid 1.5 guineas (£1 11s 6d) per day 'for time, expenses and attendance on the works'. He set out the cut, and work began. Excavation of the cut had been finished by June 1774, and the great dam across the old channel, to divert flow down the cut, was completed in September. As expected, erosion deepened the cut, and much slope protection was needed to prevent it becoming too wide. Dunthorne died in March 1775, after which Elstobb took full charge, the whole scheme being effectively completed by the summer of 1776 at a cost of some £9,000. Elstobb spent about 370 days on site and presented the accounts. He also made a detailed survey of the estuary, for which he received £172. The results exceeded expectations, proving of great benefit in flood control and drainage of the fens.

In the meantime attention turned to the Middle and South Levels and the river Great Ouse. Bedford Level Corporation produced an *Address* to the *Public* and a draft of a Bill in 1775. Elstobb published *Observations* on this in 1776 and in the same year surveyed sections of the Ouse from Lynn up to Denver Sluice and from thence up the Hundred Foot River. The Corporation submitted a petition to the House of Commons in February 1777, but the Bill failed on its second reading. John Golborne was then consulted. He reported in December 1777, strongly advocating the cut (later known as Eau Brink Cut) first advocated by Kinderley to bypass a long shallow loop in the Ouse above King's Lynn. Elstobb again entered the discussion with *Remarks* on Golborne's report, published in 1778. He also produced another carefully surveyed section of the river from Denver up to Clayhithe, and reported on the navigation of this stretch in 1779.

Elstobb approved the cut, but argued against keeping Denver Sluice (he also thought the recently built Grand Sluice on the Witham at Boston was a mistake). Others argued against both the cut and the Sluice. Bedford Level Corporation wanted the cut and intended leaving the Sluice as it was, while Hyde Page (q.v.) insisted that the cut should be wider than proposed and the Sluice should be rebuilt with a much wider waterway. Finally it was not until 1795, after fierce debate in Parliament, that the Eau Brink Act became law. Denver Sluice was retained.

As early as 1772 in a manuscript report on levels taken between Guyhirn and Knights Gool, Elstobb described himself as 'Land Surveyor and Engineer' and did so in nearly all his subsequent surveys and reports. At about that time he seems to have moved to London, or at least to have had a base there, and it is as William Elstobb of London that he appears as a Commissioner of several small land drainage schemes in the south Lincolnshire fens in the early 1770s. An advertisement added to his 1776 *Remarks* states that 'he is already to be consulted and engaged in any Canal, Navigation, Improvement of Harbours, Drainage, Surveying, Valuation, Inclosing of Lands &c'. Letters addressed to him could be left at an address in Soho.

Elstobb's last commission was to survey and report on Wells harbour, on the Norfolk coast. His engraved map was published in 1779 with his assistant John Turpin. In May 1781 he attended a meeting at the Temple (in London) on the harbour with Robert Mylne (q.v.) who had been consulted a few months earlier. But Joseph Hodskinson (q.v.) in his report on Wells harbour in July 1782 refers to the late Mr. Elstobb. The exact date of his death has not been found.

According to Gough (1780), Elstobb in 1771 proposed printing by subscription *An Historical Account of the Great Level of the Fens*. This was eventually published in 1793 by Whittington of Lynn, clearly as background to the Eau Brink discussions then actively in progress. The numerous subscribers include most of the engineers involved: Mylne, James Golborne, John Watté, Rennie, John Hudson and Hodskinson. At the back of the book Whittingham advertises the printing of *A Philosophical Treatise on Rivers* by the same author, 'the late W. Elstobb, Engineer'. But this was not published.

A. W. SKEMPTON

[North Level Commissioners Order Book (1767–1778) and Committee Book for Kinderley's Cut (1773–1778), Cambridgeshire RO; William Elstobb (1767) *A Map of Wisbech river and channel ... with Part of the North Level*, MS, Smeaton Collection, Royal Society; William Elstobb (1772) *Report relating to Levels taken from Guyhirn to Knight's Gool*, MS, Northants RO; Thomas Hyde Page (1777) *Observation upon the Draining of the South and Middle Levels of the Fens*; Richard Gough (1780) *British Topography*; Minutes of meeting on Wells Harbour (1781), Sutro Library; Joseph Hodskinson (1782) *Report on Wells Harbour*; N. Walker and T. Craddock (1849) *History of Wisbech and the Fens*; W. H. Wheeler (1896) *History of the Fens of South Lincolnshire*; M. J. J. Winter (1939) Scientific notes from the early minutes of the Peterborough Society, *Isis*, **31**, 51–59; H. C. Darby (1940) *The Draining of the Fens*; R. V. Wallis and P. J. Wallis (1986) *Bibliography of British Mathematics*; Sarah Bendall]

Publications

1745. *The Pernicious Consequences of Replacing Denver Dam and Sluice*

1750. *A Book of References to the Map of Sutton and Mepall and A Map of Sutton and Mepall Levels ... and Byal Fen.*

1754. *Remarks on a Pamphlet intitled An exact Survey of the River Ouse*

1767. *A Chain and Scale of Levels along Wisbech Rivers Ouse from the Old Bar Beacon below Lynn up to Denver Sluice and from Denver Sluice up to Over Cote and Lateral Sections of the River Ouse*

1776. *Observations on an Address to the Public ... and on a Plan and Draught of a Bill... for preserving the Drainage of the Middle and South Levels*

1778. *Remarks on the Report of Mr John Golborne*

1779. *Report on the State of the Navigation between Clayhithe and Denver Sluice with Longitudinal Section of the River* (1778)

1779. *A Map of the Harbour & Haven of the Port of Wells ... with the Marshes*

1793. *An Historical Account of the Great Level of the Fens*

Works

1773–1776. Kinderley's Cut, £9,000

ELWELL (ALWALL), John (fl. 1780–1808), ironmaster, was resident in Rotherham in 1780 with his wife, Sarah, and son, William. He became an early shareholder in the Bowling Ironworks near Bradford where the value of the local coal and ironstone deposits had been recognised by John Sturges Sr. (d. 1814), a Wakefield ironmaster, in 1784. In December 1787 he took shares to set up the ironworks: Sturgess, £700; John Sturgess Jr (d. 1823), £1,050; Richard Paley, £700; and Elwell, £350. In September 1788 they were joined by William Sturgess (d. 1811), £700, of Datchett, Berkshire. In October the following year they all signed a partnership agreement for works at Fall Ings, Wakefield, where Elwell was then based, and at Bowling. Pig iron was produced at Bowling and then sent to Fall Ings for conversion to wrought iron until 1792, when the partnership was dissolved and Elwell took over the Fall Ings concern, with the help of a £1,000 mortgage from Paley. At that time John Green Paley (1774–1860) joined the firm and later became manager and partner at Bowling. In the Napoleonic period they concentrated on ordnance.

At this time Elwell became involved with John Crawshaw (formerly a partner of the Walkers (q.v.)) with whom he bought the iron rights at Shelf in 1793. Crawshaw, then of Burton Hills, Sheffield, bought a £2,000 half share in Fall Ings in January 1794, enabling Elwell to pay off his mortgage and later the same year Elwell sold half of his remaining share to Samuel Aydon (q.v.) together with a 25% share in the Shelf enterprise. Around this time they began building Newcastle style wagonways which included rails for the Lake Loch Railroad in 1798, as well as steam engines for local mills; from 1795 they were using iron beams for steam engines. Cast iron beams from the Shelf Works are among those for which the earliest strength test results survive, the tests carried out by Reynolds at Coalbrookdale in April 1795.

By 1808, Crawshaw had been joined at Fall Ings by his brother William. Elwell and Aydon increasingly concentrated their activities at Shelf where from 1810 they established a reputation as iron bridge manufacturers, specialising in moveable spans.

It would appear that by 1810 John must have died as in *Monthly Magazine* the Shelf firm of Samuel Aydon, William Elwell and his mother, Sarah Elwell, is listed as bankrupt. Apparently it was revived as there followed a period of major bridge building activity at the works down to 1821 when the firm was again in financial difficulties and in 1822 William Elwell and Samuel Aydon were again declared bankrupt. The firm was taken over by the Low Moor Company in 1824.

The Elwell family were well known in the Black Country iron trade. William Elwell (1709–1790), ironmaster, purchased a Newcomen style steam engine cylinder from the Coalbrookdale Company in 1746. He had three sons William (1736–1793), Edward (1737–1809) and John (1739–1784), and the family operated furnaces in both towns in West Bromwich and Walsall, and from 1752 William and Edward leased a mill at Hateley Heath, West Bromwich. In 1775–1778 they were involved in a property deal in Madeley, letting it to Abraham Darby III; William was involved as a Commissioner for the Iron Bridge. He also served as Mayor of Walsall in 1778 and 1787. His grandson Edward

Elwell of West Bromwich, Staffordshire, had a forge at Wednesbury when he took out a patent (No. 4836, 1823) for manufacturing spades and shovels. This reflected the firm's dominance in the edge tool industry of the area.

MIKE CHRIMES

[J. Banks (1803) On the power of machines, *Monthly Magazine*, 1810, **30**, 263; *European Magazine*, 1817, **70**, 558; *New Monthly Magazine*, 1817, **6**, 567; *European Magazine*, 1821, December; Reeves (1836) *History of West Bromwich*, 24, 28, 51–53; W. Cudworth (1891) *Histories of Bolton and Bowling*, 205–213; R. A. Mott (1963) The Newcomen engine in the eighteenth century, *Transactions of the Newcomen Society*, **35**, 69–86; W. L. Norman (1969) Fall Ings, Wakefield, *Industrial Archaeology*, **6**(1), 74–79; J. Goodchild (1968) The Ossett Mill Company, *Textile History*, 46–61; M. J. T. Lewis (1970) *Early Wooden Railways*; J. Goodchild (1977) *The Lake Loch railroad*; B. Trinder (1979) *The Industrial Revolution in Shropshire*, 37, J. G. James (1987–1988) Evolution of early iron arched bridge designs, *Transactions of the Newcomen Society*, **59**, 153–186; C. J. D. Elwell (1992) *The Iron Elwells*, 2nd edn., 152–153]

Publications
1780. Patent 1251, *Making Utensils and Articles of Grain Tin alone or with another Metal, for the purposes of Cooking*

Works
With Aydon:

1798. Lake Lock Railroad, rails
1803. Ossett Mill steam engine, £1,750
1804. Bristol, cast iron bridges tender
1804. London Docks, Wapping entrance, swing bridge
1807–1808. Humber Dock, Hull, swing bridge
1808. Bristol, cast iron bascule bridge
1810. Queen's Dock, Liverpool, swing bridge
1811. London Docks, communication lock bridge
1811. West India Docks, Limehouse lock, west end, one bridge
1812–1813. West India Docks, Blackwall lock, east end bridge
1812–1813. King's Dock, Liverpool, swing bridge
1813. West India Docks, Blackwall lock footbridge
1814–1815. Old Dock, Hull, bascule bridge
1814. London docks, swing bridge
1815–1816. Old Dock, Liverpool, swing bridge
1816. Gasworks Bridges, Sowerby Bridge, 92 ft. span
1816. Holyhead Harbour swing bridge, £1,750
1818–1820. Princes Dock, Liverpool, north entrance, swing bridge
1818–1820. George's Dock, Liverpool, north entrance, swing bridge
1819. Newlay Bridge, Horsforth, 75 ft. span

EMERSON, William (1701–1782), mathematician, was born at Hurworth, near Darlington, County Durham, on 14 May 1701, the son of Dudley Emerson, a schoolmaster. William was educated first at home and later at schools in Newcastle-upon-Tyne and York, after which he returned to Hurworth where he lived for the rest of his long life. While at the schools he had shown strong interest and ability in mathematics, which was developed in the isolation of his rural home, and stimulated particularly by an experience just after his marriage, when he was about thirty-two. Before the marriage his wife had lived with her uncle, a wealthy surgeon and ordained clergyman, who had promised her a 'marriage portion' of £500 but refused to pay it when requested by Emerson and treated him 'with contempt, as a person of little consequence, and beneath his notice'. In response, Emerson sent back to him all the possessions his wife had brought from her former home and swore to prove himself a better man than the uncle, by becoming a famous mathematician. He succeeded eventually in winning national renown, but throughout his life spurned the normal rewards of his success, such as membership of the Royal Society in London. His ordinary dress could be called plain, shabby or eccentric, his wigs were few and dirty and all his habits of life remained rustic, or even uncouth, to the date of his death.

His attempts at private teaching only confirmed that he was temperamentally unsuited to it, and he lived for a long time on an income of only £70 or £80 p.a., the profit of a small estate he inherited from his father. He scarcely ever travelled by horse or carriage, or more than a few miles from Hurworth.

He was also critical, if not contemptuous, of many other mathematicians of his day, including Euler, J. Bernoulli and Thomas Simpson, but idolised Sir Isaac Newton and his invention of fluxions, or, in modern terms, calculus. Not surprisingly, Emerson's first book, published in 1743, was *The Doctrine of Fluxions* and was a purely mathematical textbook. However, he viewed the calculus as the description of movements, which perhaps suggested a relationship with the subject of mechanics, and towards the end of the book his 'doctrine' proceeds to such matters as centres of gravity, projectiles, and the equilibrium of arches. Within ten years more, he produced a whole book entitled *The Principles of Mechanics*, in which he dealt with many matters, from Newton's basic laws of motion to principles of the motion of animals and of windmills, in the textbook manner; but then went on to describe and illustrate pumps, waterwheels, pile-driving, simple steam engines and many other machines and structures, all of which were subjects in the nascent theory of civil and mechanical engineering. The importance of the book was that it contained good engraved plates of all these devices in a form readily usable by engineers and tradesmen in practice. Most of them were described without theory or calculations, but with clear explanation of their parts and their operation.

Both the books mentioned were issued in second editions during the 1750s but it appears that his writing was not contributing much to

Emerson's income until he was given an introduction to a London publisher named John Nourse who was himself a mathematician. Emerson then made, in 1763, his only recorded journey to London, walking the whole distance, and agreeing with Nourse to write a series of monographs on mathematical subjects. These books flowed from the press in subsequent years, on the mathematical principles of geography, astronomy and optics and on purer mathematical subjects such as trigonometry, algebra, and the arithmetic of infinities. Only two, *The Art of Surveying, or Measuring Land* and *The Laws of Centrifugal and Centripetal Force*, would appear to be directly useful in civil engineering. That the *Mechanics* went through repeated editions up to 1836 attests its usefulness in practice, though its author was guilty, at least in the early editions, of one or two errors, for instance in his theory of beams.

No records have been found of Emerson being consulted by practising engineers or their clients, but he offered unsolicited advice publicly on at least two occasions. In 1758 he sent an article to *The Gentleman's Magazine* on the form of arches; and after the committee for Blackfriars Bridge in London had, in 1759, invited eight 'gentlemen of the most approved knowledge in building, geometry and mechanics' (not including Emerson) to advise on the designs submitted to them, he wrote his brief and uncompromising views on the issues at stake in a letter published in the *London Chronicle*. He included a statement that 'elliptical arches are good for nothing', but the bridge committee a few days later adopted a design which used elliptical arches and proved a successful structure (see: *MYLNE, Robert*).

Emerson died on 20 May 1782 and was buried in the churchyard of Hurworth. His wife survived him by two years. They had no children.

TED RUDDOCK

[W. Bowe (1793) Some account of the life and writings of Mr. William Emerson, prefixed to W. Emerson, *Tracts*; DNB; E. C. Ruddock (1974) Hollow spandrels in arch bridges, *The Structural Engineer*, **52**, 281–293; Ted Ruddock (1979) *Arch Bridges and their Builders, 1735–1835*]

Publications

In addition to those listed below, Emerson published about eighteen other books on mathematical subjects.

1743. *The Doctrine of Fluxions* (later editions 1757, etc.)
1754. *The Principles of Mechanics* (later editions 1758, 1773, 1794, …, 1836)
1758. The best way of constructing the arches of bridges, *The Gentleman's Magazine*, **28**, 307–309
1760. Letter, *The London Chronicle*, 21 February
1770. *The Art of Surveying, or Measuring Land*

ESSEX, James (1722–1784), was the son of James Essex (d. 1749), a carpenter and joiner of Cambridge. He was educated at the grammar school attached to King's College, and afterwards studied architecture under Sir James Burrough, an amateur architect and Master of Gonville and Caius College. They collaborated in designing several Cambridge buildings, while Essex continued to ply his father's trade. In 1749–1750, after the latter's death, he erected the wooden bridge over the Cam at Queens' College to the design of William Etheridge (q.v.). In 1769 he designed a similar structure to replace Garret Hostel Bridge, which he called a mathematical bridge, and the name has clung to the bridge at Queens' College.

He had made a design for Corpus Christi College in 1748, which the Master of that college tried to appropriate, but from about the mid 1750s Essex was able to practise independently and see his works executed. He was remarkable in his time for his understanding—obtained by antiquarian enquiry—of Gothic architecture as a structural form, rather than merely a decorative style.

From 1757 he restored the lantern at Ely Cathedral, a fine example of fourteenth century carpentry. It consisted of eight vertical oak posts, each 63 ft. high and weighing about 10 tonnes, 100 ft. above the floor, and braced together by a complex framework of timber struts and curved rakers. Essex replaced some of the rakers and added other members, which enabled him to remove the external flying buttresses (since replaced by Sir George Gilbert Scott). He then went on to replace the roof of Lincoln Cathedral Chapter House to a new, lower profile which involved an innovative and ingenious support structure. The sketches he made as his ideas developed are preserved in the British Library, and are discussed by Yeomans. Even where he used conventional king post trusses, the members are lighter and more economical than contemporary practice.

Although a carpenter by training, Essex used stone when called upon to rebuild the Great Bridge, which till then had been of timber. This structure was not entirely successful, as it required repairs in 1799 and was replaced in 1823. A more lasting stone bridge is the one at Trinity College, which made use of the newly fashionable elliptical arches and elegant use of two types of limestone, from Dorset and Northamptonshire, to differentiate the elements of the structure.

He married a daughter of a Cambridge bookseller called Thurlbourne, and they had one son, who predeceased him, and a daughter. Essex's mother died only four months before his own death, of a paralytic stroke, on 14 September 1784.

He was elected a Fellow of the Society of Antiquaries in 1772. A list of those of his architectural writings published in his lifetime is given below; a fuller list of his works is given in Colvin.

P. S. M. CROSS-RUDKIN

[*Gentleman's Magazine*, 1784, **ii**, 718; Willis and Clark (1886) *The Architectural History of the University of Cambridge*; DNB; Colvin; David Yeomans (1992) *The Trussed Roof*]

Publications
Publications include:

1776. Remarks on the antiquity of different modes of brick and stone buildings in England, *Archaeologia*, **iv**
1776. Some Observations on Lincoln Cathedral, *Archaeologia*, **iv**
1782. On the origin and antiquity of round churches and of the round church at Cambridge in particular, *Archaeologia*
1784. Observations on Croyland Abbey and Bridge, in Nichols, *Bibl Top Brit*, **xxii**
1785. A Description and plan of the ancient timber bridge at Rochester, *Archaeologia*, **vii**

Works
1749–1750. Mathematical Bridge, Cambridge, contractor, rebuilt twice, most recently in 1904
1754. Great (Magdalene) Bridge, Cambridge, design, £1,609, now demolished
1757–1762. Ely Cathedral, restored Lantern
1762–1765. Lincoln Cathedral, design for Chapter House roof
1763–1765. Trinity Bridge, Cambridge, design, £1,500
1764–1769. Clare College Chapel, Cambridge
1769. Garret Hostel Bridge, Cambridge, designed and built, now demolished
1772. Winchester College Chapel, repaired tower

ETHERIDGE, William (fl. 1744–d. 1776), was foreman in 1744 to James King, the master carpenter at Westminster Bridge, and when the latter died he succeeded to the post. There had been no bridge on this scale built in England for the past century or more and there were many problems to be solved anew. After a proposal to build a wooden superstructure of King's design (1738) on stone piers had been disposed of, the carpenters' role was to erect and remove the centres for the stone arches designed by Charles Labelye (q.v.). These were made at first to a design by King, but proved difficult to strike and in 1745 Etheridge developed a system of straight wedges and battering rams which proved much easier to use. There were also problems in withdrawing from the river bed the piles on which the centres had been supported, and Etheridge designed instead a pontoon-mounted machine for sawing them off underwater. For these inventions, which saved the bridge commissioners £1,500, he was awarded a premium of £200. However, when one pier settled substantially in 1747 and it became necessary to reinstate centres under the adjacent arches, Labelye rejected Etheridge's design in favour of his own. Etheridge appears to have taken umbrage and, being paid by day work, took far longer to construct them than Labelye had expected. Friction between the two men continued and led in 1749 to Etheridge's dismissal from the works.

At the time that the dispute arose, Etheridge had been engaged in designing a wooden bridge over the river at Walton-on-Thames for the promoter, Samuel Dicker. When it was opened in 1750 its central span of 130 ft. was the longest in Britain until William Edwards (q.v.) succeeded in building his single arch at Pontypridd. The arrangement of the timbers in this bridge, and in the footbridge at Queen's College, Cambridge, is remarkably like King's design for Westminster Bridge. In 1754, Dicker and Etheridge made estimates for stone or timber bridges at Blackfriars, London, to cost £150,000 or £60,000 respectively. Dicker admitted that they had made a mistake in the form of Walton Bridge, which should have had side spans of the order of 60 or 70 ft. rather than the 44 ft. actually built. He also claimed that the timber would last for two-hundred years, but in this he was hopelessly optimistic. Water had penetrated some of the joints near the centre of the main span, causing it to subside and requiring propping sometime before 1778. Although it might have remained in this condition for some years, a decision was taken to rebuild in stone and the work was done between 1783 and 1786.

In February 1750 work began on the piers at Ramsgate Harbour, with Thomas Preston (q.v.) as Master Mason and by the end of 1751 the piers were approaching low-water mark. Discussions took place on the best way of founding them underwater. Etheridge submitted proposals for using caissons, as at Westminster Bridge, and a tool for digging trenches into the chalk and levelling the ground for the caissons. His scheme was adopted and he was appointed Surveyor of the Works in January 1751 at a salary of £200 p.a.

Work proceeded satisfactorily. By 1754 about a third of the total length of some 1,600 ft. of each pier had been built. However, in December 1753 William Oxenden, one of the harbour Trustees, advocated reducing the width of the harbour to save costs. Work started on the realigned west pier. But the idea met with strong opposition and a petition was submitted to Parliament early in 1755. Preston and Etheridge gave evidence in March. Work came to a halt and Etheridge left. Sir Peircy Brett and Captain J. P. Desmaretz (q.v.) reported in December, recommending a plan differing little from the original. When work eventually resumed in 1761 Preston was in charge. He completed the piers in 1772.

Nothing has been found of Etheridge's career after 1755. He died on 3 October 1776.

P. S. M. CROSS-RUDKIN

[Gents Mag, 1750, 587; Gents Mag, 1754, 116; Evidence on Ramsgate Harbour (1755) JHC, **27**, 213–230; John Smeaton (1791) An Historical Report on Ramsgate Harbour; R. Willis and J. W. Clark (1886, reprinted 1988) Architectural History of the University of Cambridge; R. B. Matkin (1977) The construction of Ramsgate Harbour, Trans Newcomen Soc, **48**, 53–72; Ted Ruddock (1979) Arch Bridges and their Builders; R. J. B. Walker (1979) Old Westminster Bridge; Colvin]

Publications

1745. A Perspective View of the Engine made use of for Sawing off under Water the Piles which helped to support the Centres for the turning the Arches of Westminster Bridge
c. 1750. A Plan of the Bridge [at] Walton-upon-Thames

Works

1747. Walton Bridge, design, built 1748–1750
1749. Queen's College Bridge, Cambridge, design, built 1749–1750 by James Essex
1752–1755. Ramsgate Harbour piers, Surveyor of Works, £45,000

EVANS, Richard, MRIA (fl. 1766–d. 1802), was born in Wales, where he was working on Kymer's Canal in the mid-1760s. He went to Ireland in the mid-1770s to seek employment on the inland navigations. He seems to have received a sound early education and training as he rose rapidly to become one of Ireland's leading canal engineers and consultants.

In 1778, Evans prepared estimates for a canal running parallel with the river Erne, from Ballyshannon to Lough Erne and Ballyconnnell, but the scheme was not proceeded with. In the 1780s, he carried out improvements to the Shannon Navigation between Jamestown and Killaloe and later, in 1800, he was called upon by the Board of Inland Navigations to make revised estimates for the completion of the navigation on the Shannon. Around 1783, he took over from Captain Charles Tarrant as resident engineer on the Grand Canal and oversaw the completion of the Leinster aqueduct, which carries the canal over the river Liffey; he remained with the Grand Canal company until 1789.

Charles Vallencey's (q.v.) site for the Leinster aqueduct had been adopted by the Grand Canal company and work had commenced in the Spring of 1780. Evans took over active control of the canal works, with Tarrant submitting occasional reports; Evans received a gratuity of IR £70 'as a proof of our opinion of his integrity and attention to the works of the company.' He built two special boats to transport material on the river Liffey and, when this work was completed, he erected platforms on them to form a temporary bridge.

Evans surveyed a branch of the Grand Canal towards Naas in county Kildare in 1782 and produced plans for a line with three locks, but the line was built some years later by another company with William Chapman (q.v.) as engineer, his first canal project.

By May 1784, Evans was working on the Barrow Line of the Grand Canal through Ballyteague Bog and finding the going tough. Although he was in charge of the works, all matters of importance still had to be referred to Tarrant. Evans was also increasingly in demand by other navigation companies and eventually was dismissed by the Grand Canal company in December 1789, subsequently becoming engineer to the Royal Canal and other navigations. In 1790, he took over as chief engineer to the Boyne Navigation with Daniel Monks (q.v.) as his assistant and deputies John Brownrigg (q.v.) and Henry Walker. The section from Slane to Navan was completed, including Somerville Bridge (1792). During this period, Evans is recorded as having an office at 25 Queen Street, Dublin.

Evans worked on the Royal Canal between Dublin and Mullingar, including the construction of aqueducts over the Rye Water and the river Boyne but he became involved in a dispute over levels with John Brownrigg (q.v.), the surveyor responsible for the original surveys. Thomas Hyde Page and William Jessop (qq.v.) were called in and the dispute developed into a full-scale enquiry. Large cost over-runs, occasioned by the Rye Water aqueduct and quarries on the chosen line of the canal southwards towards Maynooth, resulted in Evans being criticised and accused of failing to give proper attention to the works. However, there is no evidence to suggest that he was an incompetent engineer. Nevertheless, although his dismissal was recommended, he remained with the company in some capacity or other until his death in January 1802. The line of the Royal Canal to the Shannon, as originally laid out by Evans, was the one eventually adopted and the main line of the canal was finally completed in 1817.

RON COX

[M. B. Mullins (1859) Presidential Address, *Trans ICEI*, **6**, 1859–1861; C. Ellison (1983) *The Waters of the Boyne and Blackwater*; R. Delany (1992) *Ireland's Royal Canal*, **31**, 33–38, 40, 44, 50, 54; R. Delany (1995) *The Grand Canal of Ireland*, 22, 25, 28, 57]

Publications
c. 1778. *Proposal and Estimate for Cutting and Completing a Navigable Canal from Lough Erne to the Tide Water at Ballyshannon* (undated report in IEI Archives, but approx. date 1778)

Works
1765–1766. Kymer's Canal, surveys
1783–1789. Grand Canal, resident engineer, Leinster Aqueduct and Barrow Line
1790–1792. Boyne Navigation, engineer
c. 1792–1802. Royal Canal, engineer, Rye and Boyne aqueducts

EVERTON, Thomas (fl. 1532), carpenter, was contractor for a new weir on the river Avon, installed for a watermill at Ringwood, Hampshire. The indenture, dated 4 February 1532, is between Thomas Everton, carpenter of Battramsley (near Brockenhurst), Nicolas Harding, Surveyor to Margaret Countess of Salisbury, and Oliver Franklin, her Receiver General. Everton is to build a weir with eighteen 'floodgates' each 2 ft. 6 in. wide

between the posts and two 'tumbling bays' 4 ft. deep, the length of apron to be 24 ft. between the upstream and downstream cill beams. He is also to rebuild the mill wheel. The whole job is to be done for £20, including cost of timber, and finished before the end of September.

Thus the Surveyor, on behalf of the client, lays down the general dimensions but leaves all details of construction to the experienced contractor-craftsman. This would have been usual practice in early days of civil engineering for weirs, and locks on rivers, sluices on fen drainage channels and for bridges, but very few contract documents for such works have been published.

A. W. SKEMPTON

[L. F. Salzman (1952) *Building in England down to 1540*]

EWART, Peter (1767–1842), civil engineer and mill owner, was born at Troquaire Manse, near Dumfries on 14 May 1767, the youngest of the ten children of a Church of Scotland minister, several of whom had distinguished careers; his eldest brother, Joseph, became British Ambassador in Berlin and William was an important Liverpool merchant. Peter, on the other hand, showed an early interest in mechanical devices and by the age of twelve had built his own clock. He was educated at Dumfries Parish School where Dr. Dinwiddie was a gifted teacher of mathematics and at the age of fifteen he went to Edinburgh where he attended a course of lectures, and probably met Robison (q.v.) and Playfair, both of whom got to know Ewart well. He was apparently related to Robison, who wrote to Matthew Boulton in 1781 looking for employment for him.

In the event Ewart became John Rennie's (q.v.) first apprentice, paying a premium of £21 in August 1784. Their first job was the erection of a small water mill, Knows Mill, on Phantassie Farm. Ewart accompanied Rennie to London, and worked with him on Albion Mills and at this time he must have come again to the attention of Boulton and Watt as in 1788 he was sent by Rennie to Soho to erect a waterwheel for a mill for rolling copper, etc. He then worked for Boulton on machinery for his new mint, clearly distinguishing himself as an engineer of some ability as he then became (1790–1792) Boulton and Watt's agent in the Manchester area, installing steam engines in the growing number of textile mills, including a 40 hp rotative engine at Simpson's Shudehill Mill in 1792, the largest Boulton and Watt engine in a cotton mill to that date.

His success as an engineer brought Ewart to the attention of those engaged in the textile industry and in 1792 he was persuaded to become the partner of Samuel Oldknow in a bleaching and calico works at Stockport, although by 1793 a trade depression meant that much of the labour force was laid off. In 1795–1796 Ewart was again working for Boulton and Watt helping to establish a foundry and engine shop there to enable them to fabricate their own engines. He designed their boring mill, which suffered from problems.

In 1798 Ewart became involved once more in the textile industry, becoming the partner of Samuel Greg of Styal Mill; after the dissolution of the partnership in 1814 he continued cotton spinning on his own until 1835. At Styal he made extensive improvements to the water power arrangements; a new dam was built costing £456 in 1799, two new waterwheels were connected in 1801 and in 1807 he introduced the first iron waterwheel there. Throughout this period he maintained his contacts with the engineering and scientific community; he devised an automatic expansion gear for the steam engine (1788) and presented seven papers to the Manchester Literary and Philosophical Society between 1808 and 1828, of which only one was published. He took out three patents (1813–1833), two related to textile machinery and the third, for iron cofferdams (1822) was employed by Jesse Hartley (q.v.) in Liverpool Docks. This was the earliest patent for iron sheet piles although David Mathews (q.v.) had used iron piles at Bridlington prior to this. His work also appeared in the *Annals of Philosophy*. He maintained his contact with Boulton and Watt and in 1809 assisted Watt with his machine for copying sculpture. In 1814–1815 he was offered a partnership by Watt Jr.

In the 1820s he was involved in discussions regarding methods of traction for the Liverpool and Manchester Railway and in 1835, on the recommendation of James Watt Jr., Ewart became Chief Engineer and Inspector of HM Dockyards, taking responsibility for the steam machinery of the Navy. He was fatally injured in an accident at Woolwich Dockyard when he was struck by a broken chain and died on 15 September 1842.

Ewart was greatly respected by his contemporaries. He recognised the importance of Eaton Hodgkinson's work on cast iron beams and introduced him to William Fairbairn (qq.v.). Hartley felt his support was essential in him securing the post of Dock Engineer at Liverpool and James Walker remarked: 'I never met with a man who had so general an acquaintance with engineers and mechanical men of his own time as Ewart had ... His knowledge of machines ... was very intimate'. His career spanned a period between the time when the early cotton manufacturers had a first-hand knowledge of mechanics and when they became largely businessmen employing experts to design, install and maintain machinery.

MIKE CHRIMES

[DNB; Rennie mss 19934, NLS; Specification of the patent ... for a new method of making cofferdams, *Rep AM*, 2nd ser., **43**, 193–202, ill.; *Select Committee on Exports of Tools and Machinery* (1824); J. Walker (1843) Presidential address, *Min Proc ICE*, **2**, 25–28; W. C. Henry (1846) A Biographical notice of the late Peter Ewart, *Memoirs of the Literature and Philosophical Society of Manchester*, 2nd series, **7**, 113–156; J. P. Muirhead (1859) *The Life of James Watt*, 283, 411, 456; H. W. Dickinson and R. Jenkins (1927) *James Watt and the Steam Engine*, 118, 267–268, 277, 288–289, 295; J. Tann (1981) *The Selected Papers of Boulton and Watt*, 1, *passim*; R. L. Hills (1989) *Power from Steam*; M. Williams and D. A. Farnie (1992) *Cotton Mills in Greater Manchester*; A. Calladine and J. Fricker (1993) *East Cheshire Textile Mills*]

EYES, John (fl. 1725–d. 1773), was a land surveyor practising in Liverpool from 1725. In 1740 he published a survey of the river Dee estuary which included the ship canal recently completed by Nathaniel Kinderley and John Reynolds (qq.v.) from Chester down the south side of the estuary, together with details of the diversion dam and sluices (drawn by J. Willson), the only known source of information on these structures. Eyes surveyed the river Calder from Wakefield to Salterhebble in 1740–1741 and, together with Thomas Steers (q.v.), planned the intended navigation. Nothing happened for the time being, but in 1757, when John Smeaton (q.v.) was asked to act as engineer, the Eye–Steers plans were sent to him and they formed the basis for his own scheme, presented later that year. Discussion ensued regarding the western end of the navigation and Eyes extended the former survey to Sowerby Bridge in 1758. Work proceeded from 1760 under Smeaton's direction and was largely completed by 1764.

The first survey to prove the practicability of a canal to join the Trent and Mersey was made by Eyes and William Taylor in 1755 at the expense of the Liverpool merchants. A survey by Eyes in 1759 shows the Sankey canal, engineered by Henry Berry (q.v.), completed from Sankey Bridge, half a mile from the Mersey, up to Gerrards Bridge near St. Helens. The canal was opened in 1761, but at neap tides there was insufficient depth for navigation in Sankey Brook, downstream of the bridge. Eyes surveyed for an extension down to the Mersey at Fiddler's Ferry. The Act for this extension passed in April 1762 with Eyes as principal witness for the Bill.

In 1769 the Liverpool committee of the proposed Leeds & Liverpool Canal challenged the route chosen through Lancashire and they asked Eyes and Richard Melling to undertake a survey. The work seems to have been done too hastily and when Robert Whitworth (q.v.), at Brindley's request, went over part of the line with Eyes he found a serious levelling error. However, as finally built the part of canal from Colne to Wigan followed the line quite closely, on a route different from that sanctioned in the Act of 1770.

In addition to his survey of the Dee estuary, Eyes produced four charts of the Lancashire coast, surveyed in 1736–1737 and published 1738. They were the first to show Greenwich as the prime meridian and were claimed as the first marine surveys made 'with instruments other than the magnetic needle', possibly with Hadley's octant. The nautical directions on the charts were compiled by Samuel Fearon, a shipbuilder and mariner. Eyes issued a re-survey of the Liverpool Bay chart in 1764.

He was concerned, though in what capacity is not clear, in the building of St. Thomas's Church, Liverpool, in 1750. His nephew Charles Eyes (c. 1754–1803) succeeded him as Town Surveyor in 1781.

The will of John Eyes is dated April 1773 and was proved in August that year.

A. W. SKEMPTON

[R. Stewart-Brown (1910) Maps and plans of Liverpool and district by the Eyes family of surveyors, *Trans Hist Soc Lanc Cheshire* [*THSLC*], **62**, 143–173; F. Webster (1930) The River Dee reclamations, *Trans Liverpool Eng Soc*, **51**, 63–92; S. A. Harris (1938) Henry Berry, Liverpool's second dock engineer, *THSLC*, **89**, 91–116; T. C. Barker (1949) The Sankey Canal, *THSLC*, **100**, 121–155; A. H. W. Robinson (1962) *Marine Cartography in Britain*; C. Hadfield and G. Biddle (1970) *Canals of North West England*; Jean Lindsay (1971) *The Trent & Mersey Canal*; Mike Clarke (1994) *The Leeds and Liverpool Canal*, Colvin (1995), Sarah Bendall (1997)]

Publications

1740. *Survey of the River Dee*
1741. *Plan of the River Calder from Wakefield to … Salter Hebble Bridge* [and] *the intended Navigation* (with Thomas Steers)
1742. *Plan of the Liverpool Docks*
1758. *Plan of that part of the River Calder that lies between Sower Bridge and Halifax Brooksmouth*

F

FADEN, William (1749–1836), map maker and geographer to the King, was born in London on 11 July 1749, the sixth of eight children of William (1711–1783) and Hannah Faden. His father was a well-known painter and in 1764 Faden was apprenticed to James Wigley, a Fleet Street engraver. He rose rapidly to prominence in the map-making world and in 1773 became a partner in the business of Thomas Jeffreys in St. Martin's Lane, taking it over completely in 1783 on the death of his father. In that year he became 'Geographer in Ordinary to his Majesty', George III. His map work in the American War of Independence brought him widespread fame and in 1776 he was elected to the (Smeatonian) Society of Civil Engineers, no doubt in recognition of the quality of his maps and charts, which were of the highest standard. Initially his business encompassed the production of land plans.

Faden worked for the government, producing the first Ordnance Survey map (of Kent) in 1801, having already commissioned his own surveys

of Hampshire (1791) and Norfolk (1797), both published at 1in to a mile. Some of his charts were officially adopted by the Admiralty. As publisher, he produced Mudge and Dalby's *Account of the Operations carried on … for … a Trigonometrical Survey of England and Wales, 1784–1799* (1799–1801). He was responsible for some of the finest plans made to accompany late eighteenth and early nineteenth century engineering reports and was particularly associated with the work of London-based engineers such as John Rennie (q.v.). For the Smeatonian Society he had already undertaken work of a minor nature such as the plates for their summons card when, in 1797, he was asked to publish the first volume of Smeaton's *Reports*, which became available for sale to the public in March 1798.

On his retirement in 1823 his business passed to James Wyld (q.v.). He died in Shepperton on 21 March 1836 and was buried in the parish church. His relationship to James Faden, gentleman, of Woolwich, elected an Associate Member of the Institution of Civil Engineers in 1826 is unknown.

MIKE CHRIMES

[ICE archives; ICE membership records; A. W. Skempton (1971) The Publication of Smeaton's reports, *Notes and Records of the Royal Society*, 135–155; J. G. Watson (1989) *The Smeatonians*; Smeatonian Society Archives; DNB]

Publications

Publications include:

1811. *Plan showing the Present and Projected Navigation Communication between the Cities of London, Bath, Bristol and Exeter*

FANSHAW(E), Edward, Lieutenant-General (fl. 1801–1850), military engineer. Edward Fanshaw's military prospects were adversely affected by his participating at two of the worst military disasters of his time: Montevideo (1807) and Walcheren (1809), but he still rose to Assistant Inspector General of Fortifications *c.* 1830–1850, was involved for nearly ten years on repairs to the Cobb at Lyme Regis (1817–1826), and became an acknowledged expert on construction in the West Indies, where he carried out missions for the Duke of Wellington.

Fanshaw began his career as a military engineer at Plymouth in 1801 and then spent a year working on Cork Harbour (1804–1805). In 1806 he went to the Cape of Good Hope where he was promoted to Captain and made the first survey of the Salt River. He also took possession of a small island off the coast which he fortified to provide a safe harbour for mercantile shipping.

In 1807 he was sent thence to South America where according to General Burgoyne he was the only engineering officer present during the siege and capture of Montevideo where he was mentioned in dispatches. Unfortunately he was also present during its subsequent incompetent loss.

In 1808 as a military secretary on the staff of Sir Hew Dalrymple he was involved in the negotiations which removed the French from Portugal. However, since all three commanders involved were recalled to face a court of inquiry this did little for his immediate prospects. However, it did gain him the 'useful interest' of Wellington one of the other commanders and a wife—he married Dalrymple's daughter Frances Mary on 15 June 1811.

His bad luck continued when he was sent on the expedition to the Schelde in 1809 where according to Charles Pasley (q.v.) 'the Corps of Engineers is disgraced and damned for ever'. Flushing (captured in August) had to be evaluated and it was decided to demolish the sea defences and basin by destroying the piers at the floodgates. Fanshaw later wrote an account from notes kept at the time. Since the Royal Military Artificers had been decimated by a fever worsened by excavations in marshy and unhealthy ground, a body of nearly four hundred civil artificers and navigators was sent from England to assist—a workforce organised by Hugh McIntosh (q.v.). The main demolition had to be carried out without affecting the surrounding town necessitating placing the charges so that the foot of the wall would be blown into the entrance and the upper part collapsed inward rather than thrown upward. This was achieved so successfully that not even a pane of glass was blown out in the neighbouring lock house.

One of the engineers excoriated by Pasley at Walcheren was Robert D'Arcy with whose earlier work on the Cobb at Lyme Regis Fanshaw was to become acquainted when he was sent to report on damage from the storm of 20 January 1817. Repairs to the Cobb were the reluctant responsibility of the Board of Ordnance since the mid-eighteenth century despite persistent and accurate complaints that 'the Cobb was a work in no way connected with any fortifications or defence ... [and] the Board were not aware of any reason why the expence should be borne ... anymore than in other ports'. D'Arcy had carried out repairs in 1792 and according to an unacted upon report by William Jessop (q.v.) in 1805 these had already been damaged to a length of over 120 ft.

Fanshaw's report to the House of Commons considered the recent repairs secured by bolts of iron or dovetails of oak to be adequate except for the coping, but a breach in the old part had completely opened the harbour to south western gales. His original recommendation was that 100 ft. of the pier should be taken down and rebuilt, enabling an estimate of the damage to the original foundations to be made—if secure they were to serve as footings for the new work—if not the lower part of the old wall was to be removed and a new foundation sunk, enabling the coping front wall and interior walls of the main breach to be rebuilt. The foundations of the outer pier were to be underpinned with large blocks of Portland cap stone dovetailed to each other and secured in front by a row of short piles sunk 5 ft. into the ground. His estimate for this was £13,224 12s 9½d. That these repairs took nearly ten years was down to the Treasury's refusal to pay anything over the original estimate and a contractor Edward Taylor whose slow progress and inflated prices showed that little had changed in relationships between the Board of Ordnance and its contractors since the Fitch brothers (q.v.) one-hundred and fifty years earlier. On 18 December 1818 the Inspector General of Fortifications Gother Mann wrote informing them that if they did not make greater progress the Board would be 'reduced to the necessity of ordering their solicitors to prosecute them for failing in the performance of the contract'. By June 1819 Fanshaw was taking statements from masons confirming Taylor had totally ignored his instructions and that £8,856 18s 4½d had so far been spent for no results. The case went to arbitration in September 1819 when despite the Ordnance Board's arbitrator insisting that the contractors were wholly culpable the vote went against them. However, the matter was rendered academic by the Treasury's refusal to pay and Taylor ended up in a debtors prison from which he was extricated by the Board's reluctant payment of £3,263 2s 8d; thereafter Fanshaw personally oversaw the work 'without benefit of contractor'.

The Treasury paid for temporary repairs, but another storm in August 1824 brought him back the following year. This time the old wall had been driven on to the new works. Fanshaw's detailed report stated that no damage had been done to the 1787 works on the outer pier, but the landing wharf had been breached in two places and the old works had a breach of 230 ft. × 50 ft. and needed to be taken down. His recent work had survived as well as could be expected since 300 tons of stone from an unsupported parapet built by the Corporation of Lyme Regis had descended on it. These repairs were finally completed in 1826 the occasion marked with a brass plate stating that 'the repair of this Cobb was commenced and finished under the direction of Lieutenant-Colonel Fanshaw RE by order of the Master-General and under the immediate superintendence of Captain Savage. Length of pier rebuilt 232 ft., length of parapet rebuilt 447 ft., amount of estimate £19,193 19s 10d, amount of expenditure £17,333 ¼d, date of completion of work 18 November 1826'.

In between trips to Lyme Regis Fanshaw had been specially chosen by the Duke of Wellington to report on the defences of the West Indies, particularly Bermuda. The dockyard in Bermuda had been established at Ireland Eye in 1809 and attempts had been proposed for defending its 25 acres for nearly twenty years including Thomas Cunningham (Martello towers) in 1811 and Thomas Blanchard in 1823. On Fanshaw's visit in 1826 he proposed two new works, Fort

George and Fort Albert, and altered the plan for local defences. Fanshaw's report was approved by Wellington on 26 November and resulted in the last bastion style fortifications to be built by British military engineers. These are still intact save for the Ravelin and Couvre Port and included a bastioned rampart along the seaward side and a new breakwater to shelter the harbour. He also designed a cast iron type building for use there, at the same time planning the iron work experiments for a new barrack design for the West Indies. He also later wrote about problems with the use of Yorkshire paving in the tropics. In his opinion deterioration was not from climate but from using inferior stone from the upper levels of the quarries.

In 1827 he reported on the defences of Pembroke and a proposed a new pier at Milford Haven. More unusual was a report on Captain Manby's apparatus for extinguishing incipient fires in buildings.

In 1830 he was made First Assistant Inspector General of Fortifications, which entailed regulating, revising and reporting on all activities in the fortification and barrack branches of the Ordnance Department. He also served on a variety of committees including the state of the Royal Military Academy (1835). Reviving his brief diplomatic career he also reported on the ownership of the Port of Antwerp which was disputed between Belgium and Holland in 1832. He resigned as First Assistant Inspector General of Fortifications in 1850 as 'he had passed a full period in the office and (thought) that a change might be beneficial to the service'.

A memoir by Sir John Burgoyne appeared in the Corps papers 1860, but gives no details of when he actually died.

SUSAN HOTS

[MT/19/73/304, 1818–1824, PRO; MPH/604, fols. 1–4, PRO; WO 44 134, PRO; WO SS 1551/3, PRO; F/ 1805–1871, Dorset Record Office; F8 1–6 Dorset Record Office; Conolly papers, Royal Engineers Library; Porter, vol. 1; Memoir (1860) *Professional Papers (new) Corps of Royal Engineers*, **xix**, 62; Some RE works 80 years ago, *Supplement RE Journal*, **5**, 1906, 46; E. Harris (1986) Bermuda forts, *Fortress*, **1**, 37–48; J. Weiler (1987) Army architects, Ph.D. thesis, University of York, 416–417; E. Harris (1989) American spies at Bermuda's forts 1842–1852, *Post Medieval Archaeology*, **20**, 311–331; J. Coad (1989) *Royal Dockyards 1690–1850*, 3/4–3/5; A. D. Harvey (1991) Captain Pasley at Walcheren, August 1809, *Journal of the Society of Army Historical Research*, **69**, 6–21; Keystone Historical Buildings Consultants (1993) *The Cobb Lyme Regis*, vol. 2; IGI www.familysearch.org]

Publications
1817. *Report on Re-committed report on Lyme Regis Harbour Petition*
1825. *Repairs to the Cobb at Lyme Regis and Sketch of the Cobb of Lyme Regis shewing the*

Breaches made in the Gale of the 23 November 1824
1838. A short account of the demolition of the piers of the entrance chamber of the large basin at Flushing, 1809, *Professional Papers, Corps of Royal Engineers*, **21**, 33
1839. Report on the effect of climate on Yorkshire paving, *Professional Papers, Corps of Royal Engineers*, **3**, 180

Works
1817–1826. Repairs to the Cobb at Lyme Regis, £17,000

FAREY, John, Jr. (1791–1851), mechanical engineer and polymath, was the eldest of the seven children who reached maturity of **John Farey Sr.** (1766–1826) and Sophia Hubert (1770–1830). He was a brother of Joseph Farey (q.v.).

Farey was born at Lambeth, London, on 20 March 1791 but the family moved in 1792 to Pottsgrove, Beds., following the appointment of his father to the post of Agent to the Woburn estates of the fifth Duke of Bedford; a return to London was made in 1802 on the death of his father's patron. In the course of his work as a surveyor and land agent, John Sr., had acquired a great deal of knowledge about the soils and agriculture of Britain and it is as a geologist that his posthumous fame has endured. He was a prolific author of some one-hundred and eighty papers and articles but is best known in civil engineering for his articles on *Canals* for Rees's *Cyclopedia* and his *General View of the Agriculture of Derbyshire* (1811–1817).

Nothing certain is known of the education of John Jr., but it is possible he had training at the school of William Nicholson (1753–1815), with whom he later worked on the production of patent specifications. From 1804 he spent two years investigating manufactories around London, taking notes about machines and processes.

Farey had a precocious talent for mechanical draughtsmanship, as did his brothers and sisters, and from 1804 provided illustrations and descriptive texts for scientific and technical books and encyclopaedias. Of particular interest are the articles on *Cotton Manufacture*, *Steam Engines* and *Water* for Rees's *Cyclopedia* of 1818. The latter two drew heavily on the Smeaton manuscripts to which the family had access, John Sr. having known both Smeaton and Sir Joseph Banks, who had acquired the Smeaton papers. They were used by John Jr. and Joseph to prepare the illustrations for Smeaton's *Reports*, published under the auspices of the Smeatonian Society of Civil Engineers. Subsequently the Smeaton papers seem to have passed into John Jr.'s care. In addition to these publishing activities the family business assisted inventors in developing new machines and obtaining patents for them and during his career Farey acted for many prominent inventors; not only did he advise on specifications, but he also produced models for use in patent trials.

In 1819 Farey spent a month in St. Petersburg investigating ironworks and whilst there learned about the operation of Watt's steam engine indicator which he introduced to engineers on his return to Britain. Farey gave up his interest in the family business in 1821 on taking an appointment with John Heathcoat in his lace manufactory at Tiverton, Devon, and in that year he began writing his *Treatise on the Steam Engine*. On 1 February 1822 he married Mary Taylor at St. Pancras Church, London. In 1824 he took an appointment at Marshall's flax mills, near Leeds, but gave this up following both the death in childbirth of his wife in February 1825 and the serious illness in the summer of that year of his brother, Joseph, who had been running the family business.

Farey was elected a Member of the Institution of Civil Engineers in 1826 and he subsequently played an active role in the reform of the bylaws, advocating a more rigorous approach to the training of engineers. He married Elizabeth Pugsley on 24 February 1827 and she helped him in running the business from then on, the first volume of his *Treatise* being published later that year.

Farey was an important witness at the Parliamentary Committee on the Law relating to Patents for Inventions of 1829 and in the same year his plans for the publication of the second volume of his *Treatise* were abandoned and his publisher entered into litigation. From this time the thrust of his work changed direction and he dropped much of his writing in favour of engineering consultancy. He was a witness to various Government enquiries, appeared at inquests as an expert witness and undertook a wide range of work for patentees and inventors.

A disastrous fire occurred in his house at 67 Guildford Street, London, in December 1844, in which four men were killed. He was able to salvage much of the contents of his office and library and quickly resumed his profession from premises nearby. His second wife, Elizabeth, died on 15 November 1850 and six months later Farey was appointed a juror to the machinery section of the Great Exhibition but on 17 July 1851 he died of heart failure at Sevenoaks, Kent. He was buried with Elizabeth in Highgate Cemetery. A Daguerreotype portrait of 1842 is in the National Portrait Gallery.

A. P. WOOLRICH

[Membership records and correspondence, including annotated copies of his publications, ICE archives; House of Commons, Select Committee on the law relating to patents for inventions (1829) *Minutes of Evidence*, 16–38, 131–155; A. P. Woolrich (1985) John Farey and the Smeaton MSS, *History of Technology*, **10**, 181–216; A. P. Woolrich (1997) John Farey Jr. (1791– 1851), engineer and polymath, *History of Technology*, **19**, 111–142; A. P. Woolrich (1998) John Farey Jr.: technical author and draughtsman;

his contribution to Rees's *Cyclopaedia*, *Industrial Archaeology Review*, **xx**, 49–67; A. P. Woolrich (forthcoming) John Farey and his Treatise on the Steam Engine (1827), *History of Technology*]

Publications

Publications include:

1808–1818. Contributions to Rees *Cyclopedia*, etc.
1827. *Treatise on the Steam Engine*

FAREY, Joseph (1796–1829), mechanical engineer and draughtsman, was the third son of the seven children who reached maturity of John Farey Sr. (1766–1826) and Sophia Hubert (1770–1830). He was a brother of John Farey Jr. (q.v.), the well-known engineer, and was born at Pottsgrove, Beds, the family home, when his father was agent to the Woburn estates of the fifth Duke of Bedford. The family moved to London in 1802 when John Sr. began a career as a consulting mineralogical surveyor and writer.

Nothing is known of Farey's education. He was a talented engineering draughtsman and, together with his brothers and sisters, joined in the family business of assisting inventors in developing mew machines and preparing patent specifications for them, and for publishers by providing descriptive texts and drawings. Examples of their work may be seen in the original patent specifications in the Public Record Office and in the many important scientific and technical publications of the time, such as Rees's *Cyclopaedia* (1802–1819), Smeaton's *Reports* (1812), *Pantologia* (1808) and *British Encyclopaedia* (1808). Farey was awarded a prize by the Society of Arts in 1808 for some drawings he submitted and in 1821 he took over as head of the business when his elder brother, John Jr., moved to Tiverton, Devon, where he worked with John Heathcoat on the development of new lace machines.

Farey became a member of the Institution of Civil Engineers in 1822 but does not appear to have been very active. In July 1825, whilst he was taking out a patent for a lamp he had designed (Woodcroft No. 5214), he became paralysed following a cut from a barber's infected razor and appears never to have worked again. He died in 1829 in Paris in the house of his brother-in-law, the engineer Charles Antoine Gengembre.

Joseph Farey is a very shadowy figure, whose work tends to be obscured by his more famous brother. Engraved drawings are variously signed Farey, J. Farey, J. Farey Jr., and very, very, occasionally, Jos. Farey; they are always modified by the craft of the engraver, such as Wilson Lowry (1762–1824).

No portrait has been found and he never married.

A. P. WOOLRICH

[Membership records, ICE archives; A. P. Woolrich (1997) John Farey Jr. (1791–1851), engineer and polymath, *History of Technology*, **19**, 111–142; A. P. Woolrich (1998) John Farey Jr.:

technical author and draughtsman; his contribution to Rees's *Cyclopaedia*, *Industrial Archaeology Review*, **xx**, 49–67]

FAVIELL, Mark, Sr. (1786–1861), of Kearby near Wetherby, contractor, was the second child and eldest son of William Favell, mason and Ann Linfoot. There was a family story about him changing his name in order to receive an inheritance willed to him as Mark Favill, but it was spelt Faviel at the time of his baptism on 4 June 1786 at Kirkby Overblow. His father had widened Hebble End Bridge in Calderdale in 1806/7 and at first he followed this tradition of local bridge building. In 1821, however, in consortium with Abraham Pratt for masonry and Simon Hamer for earthworks, he won a contract on the Knottingley & Goole Canal. By 1822 he had moved to Amcotts, near the mouth of the river Trent in Lincolnshire, where he died on 22 June 1861.

He and his wife, Maria, whom he married in 1808, had eleven children, of whom three followed their father as contractors. In the 1830s these sons, Mark Jr., Charles and William Frederick were operating from West Bromwich and Stratford-upon-Avon and Mark Sr. provided the necessary sureties for the contracts they obtained; on these documents he is described as a farmer. He was also surety on a contract for Abraham Pratt, who had been his partner on the Knottingley & Goole Canal, when the latter died in 1838. From 1835, Mark Sr. and his sons obtained many and large railway contracts throughout England, and later on the Indian subcontinent, occasionally tendering in various partnerships both within and outside the family.

P. S. M. CROSS-RUDKIN

[Warwickshire RO; West Yorkshire RO; H. Speight (1903) *Kirkby Overblow*; Frank Smith collection at ICE]

Works

1810–1812. Bawtry Bridge, contractor, £1,874
1818–1819. Potts Bridge, Killinghall, contract for widening, £852
1821–. Knottingley & Goole Canal, contractor with Abram Pratt (q.v.) and Simon Hamer; value £21,236

FAWCETT, James (fl. 1718–d. 1721), was engineer of Sunderland harbour. Commissioners of the River Wear and Port of Sunderland were appointed by an Act passed in June 1717 and soon afterwards they asked Fawcett to survey the harbour and draw up a scheme for its improvement. His report, accompanied by a map and description of the harbour, is dated 19 August 1718.

Except from some earlier and not very successful attempts to deepen the north channel in the estuary, the harbour in 1718 was still in its natural state. Fawcett proposed to build a masonry pier 700 ft. long projecting in a north-north-west direction from a point on the south shore and a 100 ft. long pier from the north shore,

leaving an entrance 300 ft. wide between the pier heads centred on the north channel. He also proposed a quay wall running some 300 yd. back from the south pier to join the Custom House quay and another, longer, wall on the north shore; the area so enclosed would be 20 acres in extent, protected from storms by the piers. The river's flow would be concentrated in the north channel (instead of being divided between two channels as hitherto) thus inducing a sluicing effect. He recognised the need for dredging the channel but wisely refrained from guessing the cost involved.

In preparing the estimates, Fawcett remarks 'I have carefully compared the prices allowed in other parts of the Kingdom where I have been concerned in the Public works (or that have otherwise come to my knowledge) with observations and information I have received from the best workmen in these parts'. And he goes on to say with regard to the 'executive part' of the works at Sunderland, that 'if the Commissioners shall think me in anyway qualified to serve them, I am willing to be retained on as moderate and reasonable Terms as they can expect, and shall act therein with the Utmost Diligence and Fidelity, to which I hope to will please God to add Success, and I shall count it a peculiar happiness to be an instrument of advantage to this Place'.

Though contemporary accounts have long been lost, it is known from Parliamentary evidence given by Mark Burleigh and Isaac Thompson (q.v.) in 1747 that works costing £1,024 were carried out in 1719–1720 and there can be little doubt that they were carried out under Fawcett's direction since he was living in Sunderland when he made his will in March 1720. Almost certainly the costs relate to the southern quay wall.

Nothing has so far been discovered concerning Fawcett's career, other than the fact that he and his wife, Elizabeth, had at some time lived in Barbados. Clearly he was a skilled surveyor and from the quotation given above he had previous experience of public works in Britain. 'James Fawcett, Ingeneer' was buried at Sunderland on 23 March 1721 and was succeeded as harbour engineer by William Lellam (q.v.).

A. W. SKEMPTON

[James Fawcett (1718) *A scheme and Proposal for the Improvement of Sunderland Harbour, with Estimates and A Brief Description of the Harbour*, Tyne & Wear Archives, Mark Burleigh and Isaac Thompson (1747) evidence, *JHC*, **25**, 296; John Murray (1847) An account of the progressive improvement of Sunderland harbour, *Min Proc ICE*, **6**, 256–277; J. W. Summers (1858) *The History and Antiquities of Sunderland*; A. W. Skempton (1977) The engineers of Sunderland harbour, *Indust Archaeol Rev*, **1**, 103–125]

FEATHERSTONE, Joseph (fl. 1746–1771), worked as Assistant Surveyor of works in Deeping Fen under John Grundy Sr., John Grundy Jr. and

Thomas Hogard (qq.v.) at a salary of £40 p.a. During the twenty-six years of his employment works of maintenance and improvement were carried out at a cost of around £40,000. He also made several surveys, including a fine map of the Fen, the river Welland and its estuary and adjoining land. The drainage mills are shown (about forty-five in total) and lengths of all the drains and banks are listed.

A. W. SKEMPTON

[Minute Books of the Adventurers of Deeping Fen (1746–1771), Lincolnshire Archives Office]

Publications
1763. *Map of Deeping Fen*

FIELD, Joshua, FRS (1786–1863), engineer, was born at Hackney in 1786, the son of John Field, a corn and seed merchant in the City of London who later became Master of the Worshipful Company of Merchant Taylors. From 1794 Field was educated at a school at Harlow, Essex, kept by the Rev. J. Brown. In 1803 he started an engineering pupilage at Portsmouth dockyard under Simon Goodrich (q.v.) as a draughtsman in the office of Samuel Bentham (q.v.), and later transferred to the Admiralty at Whitehall. In the following year Field was transferred by the Admiralty to Henry Maudslay's (q.v.) workshop in Margaret Street (Cavendish Square, off Oxford Street), London, where he worked on the drawings for the Portsmouth block-making machinery, which Maudslay was then building. In 1810 he moved with Maudslay to a new works at Lambeth and, two years later, became a partner in the firm of H. Maudslay & Co.: in 1822 the firm's name changed to Maudslay, Son & Field.

In 1817 Field was one of the founders of the Institution of Civil Engineers and the only one to become President (1848–1849). With Henry Robinson Palmer he was the mainstay of the Institution in its formative years, regularly chairing meetings, and he wrote various papers for the Institution, most of which remain unpublished. In 1821 Field spent several months making a tour of engineering works in the Midlands and his illustrated diary contains valuable information on them.

Marine engines became the firm's speciality. The first set of ship's machinery which the firm built was for the Thames vessel the PS *Richmond* (1813), and their first naval contract was for HMS *Lightning* (1825). They also made stationary steam engines of various patterns; machinery for sugar, rice and flour mills; flour and saw mills; minting equipment; machine tools; waterworks pumping machinery; railway locomotives and fixed haulage engines for railway inclines. They made the first shield for M. I. Brunel's Thames Tunnel.

Field was a well-known consulting engineer, working on such projects as helping to plan the machinery for laying the Atlantic Telegraph Cable, and advising the Metropolitan Board of Works on the machinery for the sewage disposal works at Deptford. He advised I. K. Brunel on the machinery for his steamships and was one of the committee appointed to deal with the Metropolitan local committees of the Great Exhibition. He was granted five patents under the Old Law (pre-1851), including one granted in 1824 concerning a method of reducing the concentration of salt in marine boilers, but this was superseded later by the widespread introduction of the surface condenser.

Field was elected a Fellow of the Royal Society in 1836, joined the Institution of Mechanical Engineers in 1862 and was also a Member of the Society of Arts. He was by religion nonconformist. Details of his family life are sparse. His wife was Matilda (1797–1874), who is thought to have been a niece of John Medenham, one of the partners in the firm, 1812–1820. Of their three sons, Joshua Jr. (1829–1904) and Telford Field, became partners in the company during the 1860s.

Field died at Balham Hill House, Surrey on 11 August 1863, and was buried at Norwood Cemetery. A portrait by James Andrew hangs in the Council Room of the Institution of Civil Engineers.

A. P. WOOLRICH

[Field papers (NRA 9524), Science Museum; Membership records, Council minutes, etc., ICE archives; Original communications 1, 4, 10, 35, 54, 111, 182, ICE archives; Obituary (1863) *The Engineer*, **16**, 112; Obituary (1864) *Min Proc ICE*, **23**, 488–492; Obituary (1864) *Proc Institution of Mechanical Engineers*, 15–16; John W. Hall (1926) Joshua Field's diary of a tour in 1821 through the Midlands, with Introduction and Notes, *Transactions of the Newcomen Society*, **vi**, 1–41; Captain E. C. Smith (1932–1933) Joshua Field's diary of a tour in 1821 through the Provinces. Part 2, *Transactions of the Newcomen Society*, **xiii**, 15–50; J. Foster Petree (1934–1935) Maudslay, Son & Field as general engineers, *Transactions of the Newcomen Society*, **xv**, 39–61; Cyril. C. Maudslay (1948) *Henry Maudslay (1771–1831)*; J. Foster Petree (ed.) (1949) *Henry Maudslay, 1771–1831 and Maudslay, Sons & Field, Ltd*; J. Foster Petree (1950) Henry Maudslay and the Maudslay Scholarship, *Journal & Record of the Transactions of the Junior Institution of Engineers*, 1–18; K. R. Gilbert (1971)*Henry Maudslay, Machine Builder*; K. R. Gilbert (1971–1972) Henry Maudslay, 1771–1831, *Transactions of the Newcomen Society*, **xliv**, 49–62; J. G. Watson (1988) *The Civils*]

FINLEY, James (1756–1828), was born in County Antrim, Northern Ireland, in 1756 and emigrated to America in the latter half of the eighteenth century along with numerous other Presbyterians from Ulster who settled in western Pennsylvania. The date of his arrival in Uniontown, Fayette County, Pennsylvania, and the extent of his education is unknown. He served as a judge of the

Court of Common Pleas in Uniontown and to-gether with Albert Gallatin (1761 1849), Swiss born financier and statesman, served as Secretary of Treasury in Washington. During his term in this office Gallatin published a report in 1808 on the federal role in internal improvements. He mentioned Finley's ingenious invention of a chain suspension bridge in laudatory terms, as a near neighbour to Finley in Fayette County being aware of his early bridges. In the same year as Gallatin's report, Finley published a seminal paper on his chain link suspension bridge incorporating details of all of the aspects of the modern suspension bridge.

He refers to an encyclopaedia which provided information on the mathematics of the catenary curve, strength of materials, and primitive suspension bridges in the Andes and Himalayas, but little else is known about his sources. He built the first modern suspension bridge over Jacob's Creek, near Uniontown, in 1801. This diminutive bridge of 70 ft. (21.3 m) span was the harbinger of more than forty such bridges built from 1801 until his death in Uniontown in 1828. He appointed agents such as John Templeman in various states and he served as a consultant for the designs; Templeman developed a two span design. Most bridges had spans of just over 100 ft., but that over the Merrimac was 244 ft., and those at Lehigh Gap and Newbury Port formed parts of longer viaducts. Several had a short life—that at Schuylkill Fall collapsed in 1811 and 1816, and those at Brownsville and Essex–Merrimack failed after eleven and seventeen years respectively, under snow and vehicle loads.

At the end of his description of an experimental design method, Finley said: 'I know the young mathematician, with mind half matured would smile at my mode of testing the relative force and effect of several ties and bracings on any piece of framing, but the well informed will not so lightly treat any information obtained or supposed to be obtained by actual experiment'. The essence of his method was to erect to scale a model of a chain loaded with weights at each suspender point and with the ends passing over pulleys and loaded with weights. By mounting the model on a wall he could determine not only the geometry of the loaded chain but also the force in the chain for various sag/span ratios. In the days of empirical structural engineering this was a remarkable achievement.

With regard to the long link chain, he discusses how to lap forge the wrought iron bars at their ends to achieve maximum resistance to direct loads. For anchorages he recommended masonry of sufficient weight to resist the force in the back stays. In order for the towers to receive vertical compression loads and not induce lateral bending, he pointed out that the slope of the back stays must match the slope of the chain at the tower. The vertical suspenders were hung from the joints of the chain to prevent bending in either component. The most notable feature of

the deck was Finley's insistence upon trusses on each side of it to provide stiffness for moving live loads and the forces of wind. Later in the century, engineers were to learn to their chagrin that unstiffened decks often produced unstable behaviour under certain wind loadings.

It was intended that the ironwork would be produced by local blacksmiths and hence, like the iron bridge in Shropshire of 1779, the suspenders were attached with a cotter system without the use of bolts. As an economy measure, it was expected that the towers would be framed in heavy timber and not with either masonry or iron. As a result, they were enclosed with clapboards to protect them from the weather.

Although the 'eye' bar chains featured in Britain and the parallel wire cables employed in France and America were to replace the less efficient link chains, Finley had succeeded in producing the first modern suspension bridge capable of carrying vehicular traffic and utilising iron in a most efficient manner, i.e. direct tension. Mention of his invention appeared in several publications and came to the attention of not only American bridge builders but also engineers in Britain and France. Thomas Telford (q.v.) owned a copy of Pope's work which described Finley's bridges and mentioned his work in connection with the design of Runcorn Bridge.

E. L. KEMP

[Dobson (1798) *Encyclopedia*, Philadelphia; Albert Gallatin (1802) *Report of the Secretary of the Treasury on Roads and Canals*, S. Doc. No. 250, 10th Congress; T. Pope (1811) A treatise on bridge architecture; J. Cordier (1820) *Histoire de la Navigation Interieure*, Paris; C. S. Drewry (1832) *A Memoir on Suspension Bridges*; Theodore Cooper (1889) American railroad bridges, *ASCE*, **21**(418); A. P. Mills (1911) Tests of wrought iron links after 100 years of service, *The Cornell Civil Engineer*, **19**(7), 256–262; W. H. Boyer and I. A. Jelly (1937) An early American suspension bridge, *Civil Engineering*, **7**(5), 338–340; E. L. Kemp (1979) James Finley and the modern suspension bridge, *ASCE*, pre-print 3590, 2–31; E. L. Kemp (1979) Links in a chain: the development of suspension bridges, 1801–1870, *The Structural Engineer*, **57A**(8); E. Kranakis (1998) *Constructing a Bridge*]

Publications

1810. A description of the patent chain bridge, *The Port Folio*, **3**(6), 441–453

Works

All the following are suspension bridges.

1801. Jacob's Creek Bridge, Uniontown, $600
1807–1808. Potomac Bridge, $5,000
1807–1810. Cumberland Bridge
1807–1810. Brownsville Bridge (Dunlap's Creek)
1809. Brownsville Bridge
1809. Neshamy Creek Bridge, two spans
1809. Schuylkill Bridge, two spans, $20,000
1810. Brandywine Bridge, $7,500

1810. Essex–Merrimack Bridge, $25,000
1810. Frankfurt Bridge, Kentucky river, two spans
1814. Lehigh Bridge, two spans, $20,000
1826. Lehigh Gap Bridge, Dalmerton
1827. Newbury Port Bridge, two spans, $66,000

FITCH, John (1642–1706), contractor, a son of William Fitch of Barkway, Herts, held a small contract in connection with the Fleet Canal (1671–1674) on which his elder brother, Sir Thomas Fitch (q.v.), was the main contractor. Between 1675 and 1678 John Fitch worked on three buildings in London designed by Robert Hooke (q.v.): the College of Physicians, Bethlehem Hospital and Montagu House. From 1679 he was in partnership with his brother on Portsmouth fortifications and from 1682 he held the post of Workmaster to the Ordnance Office.

In the Fens he carried out the brickwork and masonry of Denver Sluice, 1683–84, the carpentry being by John Hayward. Fitch's biggest contract, agreed in October 1691, was for two wet docks and the Great Stone (Dry) Dock at Portsmouth Dockyard to plans by Edmund Dummer (q.v.). Trouble occurred with the foundations of the gates of the Upper Wet Dock but the whole scheme was completed, after bitter arguments, in 1698 although some repairs were still going on in 1702.

While working at Portsmouth Fitch built the ropehouse and other buildings at Plymouth Dockyard, 1693–1697. Later works included the west front of Chatsworth House, Derbyshire, and in *c.* 1702 the contract for all the masons', bricklayers', carpenters' and other building work for Buckingham House, London, for £7,000.

In 1698 he married Melior, daughter of William Russell of Kingston Lacy in Dorset. He subsequently acquired High Hall near Wimborne, which still belongs to his descendants and where his portrait is preserved. He died early in 1706.

A. W. SKEMPTON

[Bedford Level Corporation Orders and Accounts (1682–1684) Cambridgeshire RO; J. F. Reddaway (1940) *The Rebuilding of London after the Great Fire*; John Ehrman (1953) *The Navy in the War of William III*; J. G. Coad (1989) *The Royal Dockyards 1690–1850*; Colvin (3)]

Civil Engineering Contracts
1683–1684. Denver Sluice, £3,000
1691–1698. Portmouth docks, £16,000

FITCH, Sir Thomas (1637–1685), contractor, was a son of William Fitch of Barkway, Herts, and the elder brother of John Fitch (q.v.). His earliest recorded work is the brickwork contract for Kingston Lacy, the mansion in Dorset designed by Sir Roger Pratt and built between 1663 and 1665. After the Great Fire of London (1666) he rebuilt the Thames waterfront stairs at Three Crane Wharf and Blackfriars and undertook the large contract by which he is chiefly remembered: cutting and wharfing the Fleet Canal between 1671 and 1674.

The idea of transforming the lower reach of Fleet River, or Ditch, as it was usually called, into a canal with roads or wharves on either side and bridges across it appeared in several plans for rebuilding the City after the Fire. As a matter of necessity Fleet Bridge was rebuilt in 1688–1689; it had a semi-circular masonry arch of about 20 ft. span with abutments running back on a level above the valley sides to Fleet Street and Ludgate Hill. In March 1670 a definitive plan for the canal was approved by the City authorities in conference with Sir Christopher Wren (q.v.). It would have a length of about 700 yd. from the Thames to Holborn Bridge—about to be rebuilt by the mason Thomas Cartwright—the channel to be 40 ft. in width with 30 ft. wide wharves and two footbridges.

The lines having been staked out and most of the land purchased, a contract for the work was awarded jointly to Fitch and John Ball in August 1671. Wren, 'his Majesties Surveyor General', directed the project and the works were supervised by Robert Hooke (q.v.) and John Oliver (c. 1616–1701) as Surveyors to the City; measurements of work done were made by William Leybourn and Richard Shortgrave (qq.v.) acting as quantity surveyors for new buildings and works in the City. Fitch had William Boulter as 'head man' or agent on site until February 1673 but his successor appears to be unknown. The first work was to remove silt and rubbish from the Fleet and dig the channel to a depth of 5 ft. below mean tide level (at 2s per cubic yard). Work then began on the walls. Designed by Wren and Hooke, they were to be in brickwork for 50 ft. below Holborn Bridge, and the same distance above and below Fleet Bridge, and in timber elsewhere. They soon proved to be inadequate, and Ball turned out to be unreliable. In consequence a new contract was made in May 1672 with Fitch alone and at the same time Wren revised the wall design. The brickwork was now to extend 200 ft. from the bridges, in best stock bricks with the outer courses set in tarras mortar, and where the walls exceeded 20 ft. in height they were to have piled foundations. Moreover, for 120 ft. from the bridges there would be storage vaults with cross walls under the streets descending from the bridge abutments. Soon afterwards it was decided to have brick throughout and the base width of the high walls was increased from 7 ft. to 9 ft. 4 in. Under a separate contract with John Fitch additional works were carried out above Holborn Bridge to prevent silt and rubbish washing into the channel but they proved only partially successful. For example, Hooke and Oliver reported after a storm in January 1674 that as much as 4,700 cu. yd. of silt would have to be dredged to restore the channel to its former state.

Despite various troubles and changes in design Fitch proceeded calmly and efficiently. Drawing £500 a week on account, he kept two-hundred men fully employed on site and completed the contract in October 1674 to everyone's satisfaction at a cost of £51,300. One of the footbridges, in Venetian style, was built by John Young and the others by Thomas Cartwright, masons. The total cost amounted to £56,000 exclusive of land purchase.

It may be noted that the canal was arched over in 1733 between Holborn and Fleet Bridges to provide a site for the old Stocks Market. The lower reach, depicted by Samuel Scott and other artists, was covered in 1765 for the new road leading to Blackfriars Bridge, then under construction.

To mark its appreciation of Fitch's work the City presented him with plate valued at £200. A knighthood, discussed at the time, was to follow in December 1679. Meanwhile his next big contract, in this case in partnership with John Fitch, was to build the fortifications at Portsmouth to designs by Sir Bernard de Gomme (q.v.). He then undertook the fortifications at Hull (1681–1683) designed by Sir Martin Beckman.

Fitch next appears as a consultant rather than contractor. By 1682 it had become necessary to rebuild the three-arch sluice on the Great Ouse at Denver. Timber-framed, this was built 1655–1656 with a clear waterway of 30 ft. to designs by Jonas Moore (q.v.) as the final work in draining the Great Level of the Fens, known as Bedford Level, under the general direction of Sir Cornelius Vermuyden (q.v.). Ralph Peirson (q.v.), Surveyor of the South Level since 1671, drew up plans for rebuilding the sluice in brick with a wider waterway. Work on the foundations began in June 1682 and was completed by May 1683 at a cost of £1,774. Fitch had been consulted on designs for the superstructure in November 1682 and in December a contract was let by Bedford Level Corporation to John Hayward of Southwark, Carpenter, to build the sluice 'according to the Platt or Draft hereunto annexed or otherwise as Sir Thomas Fitch shall direct'.

Work began as soon as the foundation had been finished and reached completion in March 1684 at a cost of £3,059 (making a total of £4,833), approved by Sir Thomas. It was probably he who suggested that the total waterway should be increased from 42 ft. (Peirson's original plan) to 54 ft., and he also proposed a small increase in dimensions of the piers. As built, the sluice had three arches each 18 ft. wide, and the 'sea doors' were 18 ft. high. John Fitch had been in partnership with Hayward as subcontractor on the brickwork.

Sir Thomas married Anne, daughter of Richard Comport of Eltham. He acquired an estate at North Cray, Kent, where he owned and perhaps built a mansion called Mount Mascall. He designed the Court House at Windsor c. 1687 and supervised its construction, but before completion he died on 16 September 1688, nine days after being created a baronet.

A. W. SKEMPTON

[Bedford Level Corporation Orders and Account Books (1682–1684) Cambridgeshire RO; H. W. Robinson and W. Adams (1935) *The Diary of Robert Hooke 1672–1680*; T. F. Reddaway (1940)

The Rebuilding of London after the Great Fire; L. E. Harris (1953) *Vermuyden and the Fens*; Colvin (2); Richard Trench and Ellis Hillman (1984) *London under London*]

Works

1671–1674. Fleet Canal, contractor. £51,300 (total cost £56,000)
1683–1684. Denver Sluice, consultant, £3,000

FLETCHER (fl. 1790–1850) is the family name of at least four families engaged in canal construction. Definite links between the several holders of the name have not been established.

George Fletcher (fl. 1790–1794), originally a bricklayer, worked for John Pinkerton as site agent on the Birmingham Canal works and also on the Dudley Canal Tunnel before working for Thomas Sheasby (q.v.) on tramroad work in Monmouthshire. He became the Inspector of the Buildings and Works for the General Committee of the Kennet and Avon Canal in October 1794.

Following the passage of the Act for the Leicestershire and Northamptonshire Union Canal in 1793, with John Varley (q.v.) as its engineer, **John Fletcher** (fl. 1793–1797) was appointed Clerk to the Company in July 1795. Problems arose with, in particular, the Saddington tunnel on the canal which was intended to link the Soar and Leicester Navigation to the Nene at Northampton. Christopher Staveley (q.v.) was originally acting with Varley as joint surveyor but by July 1796 Fletcher had replaced him and discovered that the tunnel was not correctly aligned. A succession of small contractors had been involved with its construction—James Gladwell, Thomas Packer Batchelor, Robert Biggs, Valentine Harrison, Henry Ludlam, Thomas Walker, Thomas Hill, Joseph Parkin and Peter Tipping—which cannot have helped supervision. Varley had reported in February that progress was satisfactory and now offered to pay to rectify the error. James Barnes (q.v.) was called in and he confirmed Fletcher's findings. Fletcher was given the contract for constructing the canal from the tunnel to Debdale, while retaining the post of general supervisor, and the section was opened in April 1797.

John Fletcher (fl. 1797–1820), of Chester, was engineer for the Chester Canal in 1797 in succession to Joseph Turner (q.v.). He carried out work for the Ellesmere Canal, becoming contractor on the Hordley–Weston Lullingfields section in 1795, following this with a contract for the Whitchurch Branch in 1796 and Church Bridge, Norbury–Wrenbury section in 1801. With John Simpson (q.v.) he was contractor on the Ellesmere Canal extension from Whitchurch to Hurleston.

Fletcher was employed as a surveyor by the Post Office 'for a great number of years' on the Chester to Holyhead road, suggesting improvements to it in 1810. He submitted a design for a bridge across the Mersey at Runcorn c. 1814 and was present with Thomas Telford and Bryan Donkin (qq.v.) when tests were carried out on wrought iron cables at Brunton's manufactory in May 1817. In April 1817 he wrote to Telford regarding improvements to the Holyhead Road at Rhyalt between St. Asaph and Holywell; Telford referred his recommendations to William A. Provis (q.v.).

Fletcher evidently enjoyed a considerable local reputation among his contemporaries and was described by Telford, when crediting his observations on mills, as an 'able mathematician and mechanic'.

Samuel Fletcher (d. 1804), civil engineer, apparently from Bradford, with others of his family were involved with the construction of the Leeds–Liverpool Canal from 1790. A **James Fletcher** was appointed overlooker of masonry on the Yorkshire end while Samuel worked between Burnley and Accrington. In 1795 Samuel Fletcher became Inspector of the canal works and, following the death of Robert Whitworth (q.v.) in 1799, he became engineer for the works on the canal, a position he held until his death in 1804, at which time the post of engineer was discontinued.

James Fletcher (d. 1844), his son, was then appointed as principal overlooker and surveyor jointly with Samuel's brother, **Joseph**, who had been employed by the Leeds and Liverpool Canal Company since 1801; their salaries were some £350 and they were provided with a house at Gannow. In 1805 James and Joseph Fletcher reported on links with the Manchester, Bolton and Bury Canal at Redmoss, estimating links from Enfield to Redmoss (25¾ miles) at £245,275; and Wigan to Redmoss (6¼ miles) at £101,725, but nothing was done immediately. In 1808 agreement was obtained for the Blackburn–Wigan section, and, in 1810 for a link to the Lancaster canal at Kirklees. Construction was completed to Blackburn in 1810 and a grand opening of the Leeds–Liverpool line, including the Lancaster link, took place in October 1816; a Leigh Branch was completed in 1820. Throughout this period, and indeed down to his death in 1844, James Fletcher was engineer to the canal; he was paid a gratuity of 100 guineas and his salary increased for his work on remedying the failure of the Foulridge Tunnel in 1824. For much of the period he had as his assistant at Liverpool Walmsley Stanley, who had succeeded Albinus Martin (q.v.) in 1827. They had reported on improvements to the Douglas navigation shortly after this and in 1839 they reported on a proposal for a link to the Aire at Leeds which was finally executed in 1845. The **Thomas Fletcher** (fl. 1844), who acted as an assistant engineer on the Yorkshire side of the Leeds–Liverpool Canal from 1844 and was later assistant to the engineer, Walmsley Stanley, was presumably a relative. The links of this family with Thomas Fletcher, noticed below, are unclear.

Thomas Fletcher (c. 1770–1851), civil engineer, may have come from the East Midlands, the son of Thomas Fletcher (d. 1796) a land and estate surveyor of Whitwell, Derbyshire. In October 1792 he was working for the Pinkerton (q.v.) family on the Leicester Navigation, when John Rennie approached him about work on the Lancaster Canal. He was appointed surveyor from January 1793 at a salary of £100 p.a. In 1794 he was working, under Henry Eastburn (q.v.) on the Lancaster Canal south of the Ribble, at a salary of £150 p.a. Subsequently he was appointed resident engineer under John Rennie (q.v.) on the Aberdeen Canal following its Act of Parliament in 1796. Progress on the canal was affected by shortage of funds and compensation for landowners and, in view of this, Fletcher became concerned about his employment prospects and in April 1800 asked Rennie if he had alternative work; Rennie hoped to recommend him for a post in the West India Docks. In the event Ralph Walker (q.v.) selected the assistant engineers there, but Fletcher successfully applied for the post of Superintendent of the construction of Union Bridge, Aberdeen, in 1801.

There were problems identified by Fletcher with the original bridge design by Hamilton and alternative designs were sought from both Rennie and Telford. Fletcher developed his own design for a main span of 130 ft. and this was eventually approved; construction was completed in 1805. Fletcher then worked as contractor on Conon Bridge near Dingwall. In 1812 he tendered unsuccessfully for Cree Bridge, Newton Stewart.

When Conon Bridge was completed schemes had been revived for the completion of the Lancaster Canal and its extension to Kendal. Fletcher became engineer, drawing up the specifications, examining the tenders, and supervising construction down to 1819, although in May 1817 William Crosley II (q.v.) was appointed to complete the work north of the Tewitfield Locks. The canal was opened to Kendal on 18 June 1819.

At this time Telford (q.v.) was looking for a reliable engineer to supervise the completion of the Gloucester and Berkeley Ship Canal and Sharpness Docks. Fletcher was appointed in 1820 at a salary of £500 p.a. but such was the parlous state of the canal's finances that he was owed money in 1821 and the Exchequer Bill Loan Commissioners took over the administration of the works. Fletcher remained in charge of the canal until its completion, his presence in the area leading to his involvement with other works of Telford, the Over Bridge at Gloucester and drainage works for Colonel Berkeley. At this time, in 1825 and then resident at Frampton, Stroud, Gloucester, he was elected a corresponding member of the Institution of Civil Engineers,

Details about the later stages of Fletcher's career are sparse: he was involved in surveying an early version of the Grand Junction Railway in the West Midlands in the early 1830s for John Urpeth Rastrick (q.v.); in 1832 he proposed a ship canal between Lytham and Preston; and in the late 1830s he was involved with the Manchester and Salford Junction Canal for which John Gilbert was the original engineer. He then retired to

Horningsea, near Cambridge, where he was in receipt of an annuity from William Mackenzie (q.v.), the contractor. He died there in December 1851, apparently a widower with a surviving daughter, Margaret.

MIKE CHRIMES

[**John Fletcher (fl. 1793–1797)**: C. Hadfield (1970) *The Canals of the East Midlands*, 2nd edn.; J. P. A. Stevens (1972) *The LeicesterLine*. **John Fletcher (fl. 1797–1820)**: Ellesmere Canal Minutes, PRO Rail 827; T. Telford (*c.* 1798) Report on mills, & Telford's Runcorn bridge notebook, & T/HO/15–16, Telford mss, ICE Library; *Committee on Holyhead Roads and Harbour* (1810) 2nd report, 7, 11. **James (d. 1844), Joseph Samuel (d. 1804) and Thomas (fl. 1844) Fletcher**: C. Hadfield and G. Biddle (1970) *The Canals of North West England*, 1; M. Clarke (1994) *The Leeds and Liverpool Canal: a History and Guide*. **Thomas Fletcher (*c.* 1770–1851)**: Gloucester & Berkeley Canal correspondence, PRO Rail 829/32; Lancaster Canal correspondence, etc., PRO Rail 844/262 and 844/230; Lancaster Canal specifications, ICE Library; Mackenzie collection, ICE Library (Ledgers 1835–1850, 1851–1852; Diaries 1840–1841); Telford mss, ICE Library; ICE membership records; Rennie papers, NLS (93784; 93797); Telford collection, Ironbridge Gorge Museum Trust; *Reports of the Highland Roads and Bridges Commissioners*; C. Hadfield (1969) *The Canals of South and South East England*, 346–347; C. Hadfield and G. Biddle (1970) *The Canals of North West England*, vol. 1, *passim*; T. Ruddock (1979) *Arch Bridges and their Builders, 1735–1835*, 175; G. N. Crawford and C. Gotch (1993) *The Gloucester and Sharpness Canal and Robert Mylne*; T. Day (1995) Telford's Aberdeenshire bridges, *Industrial Archaeology Review*, **xvii**(2), 193–207]

Publications

Thomas Fletcher (*c.* 1770–1851):

1813–1817. *Lancaster Canal Specifications*
1831. *Basford and Wolverhampton Railway, Section Ending at Basford Township*

Works

John Fletcher (fl. 1793–1797):

1795–1797. Leicestershire and Northamptonshire Union Canal: resident engineer; and Debdale, cutting, contractor, £902 14s 6d

John Fletcher (fl. 1797–1820):

1795–1805. Ellesmere Canal extension, Whitchurch to Hurleston, contractor with J. Simpson

Samuel Fletcher:

1790–1804. Leeds–Liverpool Canal, superintendent of work (–1799) and resident engineer (1799–1804)

James Fletcher:

1804–1844. Leeds and Liverpool Canal, principal overseer and then engineer

Thomas Fletcher (*c.* 1770–1851):

1794–1796. Lancaster Canal, Assistant Engineer
1797–1800. Aberdeenshire Canal, Resident Engineer
1801–1805. Union Bridge, Aberdeen, Engineer
1807–1809. Conon Bridge, five arches, Contractor, £6,854
1812–1819. Lancaster Canal, 14½ miles, Engineer, *c.* £100,000
1820–1827 Gloucester and Berkeley Ship Canal, Resident Engineer, *c.* £160,000
1826–1830. Over Bridge, Gloucester, Resident Engineer
1830–1831. Basford and Wolverhampton Railway, surveys
1836–1841. Manchester and Salford Junction Canal, Engineer (?)

FORSYTH, John (fl. 1794–1821), contractor, of Avoch near Fortrose, was responsible for the British Fisheries Society harbour at Loch Bay, Skye. There was concern about the softness of the local stone and this prompted Thomas Telford (q.v.), the Society's engineer, to carry out trials there with Parker's cement in 1796, leading to its adoption elsewhere. Work continued across the Highlands and elsewhere in Scotland. With George Forsyth he built Newhaven Pier near Edinburgh and its extension, and the Bridge of Earn in Perthshire to the designs of John Rennie.

MIKE CHRIMES

[Rennie mss, NLS; J. Dunlop (1978) *The British Fisheries Society 1786–1893*]

Works

1794–1796. Loch Bay Harbour, Skye, contractor
1814–*c.* 1820. Newhaven Pier, near Edinburgh, contractor (extension contract 1817–1820), original contract £10,000
1821. Bridge of Earn

FORTREY, Samuel (1622–1682), landowner and fen drainer, was born on 11 June 1622, the only son of Samuel Fortrey, a London merchant who came to England from France in about 1604, and his wife, Katherine de Latfeur. As a young man he worked with his father, living at the family home on Kew Green. He married Theodora Josceline, from Essex, in February 1648 and soon afterwards they moved to Oakington, a village near Cambridge, where four of their six children were baptised between 1650 and 1660. Before 1653 he purchased 400 acres in Byall Fen, adjacent to Bedford River 6 miles west of Ely. The land is shown on the map of the Fens (1658) by Jonas Moore (q.v.) as 'belonging to Mr Forte'. He also built a 'commodious habitation' nearby at Mepal.

In October 1653 Fortrey attended his first of many meetings of the Company of Adventurers for draining the Great Level of the Fens, the Company formed by William (fifth) Earl of Bedford in 1649 to complete the immense project of draining some 310,000 acres of fenland begun by his father in the

1630s. The whole area was partitioned into three divisions: the North Level between the Welland and Morton's Leam; the Middle Level between Morton's Leam and Bedford River; and the South Level between Bedford River and the border of the Fens. Main works in the North and Middle Levels were completed by 1651 and those in the South Level by the end of 1653. But much remained to be done in making roads, bridges and division dykes and in the perennial task of maintaining the banks, rivers, drains and sluices.

In a reorganisation in August 1654 Edmund Welch (q.v.), the former principal Overseer or resident engineer, took charge of the direct labour force at a salary of £60 p.a. Jonas Moore, previously the Surveyor General, became Overseer at £100 p.a. and as such let out the last major contract in April 1655, for Denver Sluice. He spent most of 1656 and 1657 producing his great map of the Fens. Sir Cornelius Vermuyden (q.v.), Director of Works since 1650, was now rarely seen on site. The principal change, however, was the appointment of three officers to the new posts of Conservators of the three Levels; chosen from Adventurers who hitherto had been actively supervising works in the field, they received an annual salary of £150 and each had assistants, known as Bailiffs.

The Conservators appointed in August 1654 were: Alexander Blake, of Peterborough, in the North Level; Richard (later Lord) Gorges in the Middle Level; and Anthony Hammond of Wilburton, previously Superintendent or project manager, in the South Level. After a brief interregnum Fortrey succeeded to this last post on 3 December 1655. George Barnes and Captain Lane, former assistant overseers, were the South Level Bailiffs.

Following the Restoration of Charles II in May 1660 the Company's chief concern was to obtain an Act of Incorporation to provide a firm legal foundation. Many disputes had to be settled, petitions were submitted to Parliament, and in May 1662 a special commission was established under which the Company could carry on, pending the Act. Lord Gorges and Fortrey were among its members. In July, in another reorganisation Gorges, who had been Governor of the Company since 1657, became Surveyor General; Edmund Welch, who had taken over from Blake as Conservator of the North Level in 1658, became Surveyor of that district, and Fortrey was appointed Surveyor of both the South and Middle Levels at £156 p.a. with Barnes as his assistant. Finally the Act of 27 July 1663 established the Bedford Level Corporation under which the affairs of the Great Level were conducted for the next two-hundred and fifty-seven years. The Board comprised the Governor (the Earl of Bedford), Bailiffs and Conservators, old titles but now with quite different connotations. The officers included three administrators together with the Surveyor General and Surveyors of the Levels. Fortrey continued as Surveyor of the South and Middle Levels until 1667

and he also served on the Board as Bailiff or Conservator to 1679.

In 1663 Fortney published a forty-two page tract, *England's Interest and Improvement*, reprinted in 1673 with several later editions. On the title page he is described as 'one of the Gentlemen of his Majesties most Honourable Privy Chamber'. How he obtained an appointment at Court is not known, though the family seems to have enjoyed a higher social status than might be supposed. His aunt married Sir Thomas Trevor in 1647 and his youngest son, James (1656–1719), was page of honour to the Duchess of York, afterwards groom of the bedchamber of her husband James II and a major in the Horseguards. Fortrey himself was appointed Clerk of the Delivery of Ordnance in November 1670. He held this post until he died and would have been rather fully occupied during the Third Dutch War (1672–1674); otherwise he devoted part of his time to the pursuit of his engineering interests.

Chief among these was the drainage of Deeping Fen. This area of some 30,000 acres in south Lincolnshire between the rivers Glen and Welland had been drained during the years 1632–1636 by Francis (fourth) Earl of Bedford and his associates but rioting in the Civil War period destroyed much of the works. The land remained 'drowned', with no one willing to undertake a rescue operation until 1665 when the Earl of Manchester and other Adventurers obtained an Act to restore the drainage in return for one-third of the land in the Fen.

For reasons which are now obscure nothing was done until June 1669 when the Adventurers of Deeping Fen held their first meeting. It was then agreed that 'Samuel Fortrey esq. shall from time to time appoint, direct and cause such works to be done, as in his Judgement shall be thought meete for the dreyning of Deeping Fennes, and shall also issue out such summe or summes of money for the said worke … as he shall think requisite'. Also that 'Mr. Andrew Browne shall be overseer of the works … who is to observe and follow such orders and directions as he shall receive from Samuel Fortrey esq. who is the Surveior and Director of the said works, and for his recompanse he shall receive the summe of £40 p.a. from 29 September next'.

Work on Deeping Fen probably started in the spring of 1670. At a meeting in June Fortrey received orders to employ a land surveyor 'to survey, plott out and discribe the fenns'. In March 1671 the Act was amended to allow a further three years to completion as progress had been hindered by 'the unseasonableness of the weather and other unavoidable accidents'. Details are not available but at the meeting in April 1675 Fortrey was to be 'director and supervisor of the works for the ensuinge year' and Browne was to continue as overseer. It may be assumed that the works were completed by 1676, since orders made at the 1677 and 1678 meetings appear to relate only to maintenance and improvements, and no further mention is made of Fortrey, or anyone else, as Director. He had been an Adventurer from the start, however, and continued to attend the annual meetings up to and including April 1681.

While the Deeping Fen works were in progress Fortrey and Sir Samuel Morland were consulted in 1671 on making the Little Ouse navigable from Brandon to Thetford. Jonas Moore commented on the estimated cost a few months later. The work was undertaken by the Earl of Arlington and was completed by 1675 but it is unlikely that Fortrey was involved.

Also in 1671 the Corporation of Lincoln obtained at Act for restoring navigation on Fossdyke, the 11 mile Roman canal from Lincoln to the Trent at Torksey, and for improving the river Witham. Nothing was done at the time regarding the Witham, but Fortney undertook to work on Fossdyke. After the Corporation had spent £500 he was to bear one-third of the cost and receive one-third of the profits and he was to 'afford his best advice, direction and assistance, according to the best of his skill and knowledge'. Work began immediately and reached completion by the end of 1672, the principal feature being a lock or navigable sluice at Torksey as originally proposed c. 1632 by Simon Hill (q.v.). The canal could accommodate boats of 18 tons burden.

Fortrey's last work was on the Hampshire Avon. In 1665 the Earl of Clarendon obtained an Act for making the river navigable from Christchurch to Salisbury. He did some work on Christchurch quay, but none on the river. In 1675 Andrew Yarranton (q.v.) surveyed the river at the request of Lord Salisbury. He thought it could be made navigable. As a result Salisbury Corporation decided to make a start, at an estimated cost of £2,000, and appointed Fortrey to direct the works at a salary of £50 a quarter for five quarters and a lump sum of £250 when work was finished. The first spit was cut by the Bishop of Salisbury in September 1675 but by 1677 funds were running out and certain private individuals agreed to take over. This may have been the end of Fortrey's connection with the enterprise, which was completed after a fashion several years later at a cost of £3,500.

The will of 'Samuel Fortrey esq. of Salisbury Court in the City of London' is dated 30 March 1680. He leaves his land in Byall Fen together with all the tenements, houses, barns and stables, and all the household goods and furniture to his wife Theodora. The exact date of his death has not been found but probate was granted on 16 February 1682.

His interests in fen drainage were continued by his eldest son, Samuel Fortrey Jr. (1651–1689), an Adventurer of Deeping Fen, 1675–1685, and a Bailiff or Conservator of Bedford Level, 1680–1686. In or before 1684 he was appointed Surveyor of the South Level. To him is traditionally attributed *The History or Narrative of the Great Level of the Fens called Bedford Level*, written in 1684 and published in 1685 by Moses Pitt who at the same time issued the second printing of Jonas Moore's map of the Level. *The History* is notable for a detailed list of all the banks, rivers and drains in each of the three Levels. In Mepal church is a memorial to 'Samuel Fortrey esq. of Byal Fen in this parish: ob. 10 Feb. 1688 [1689 NS] aet. 38'.

A. W. SKEMPTON and FRANCES WILLMOTH

[Proceedings and Accounts of the Company of Adventurers of the Great Level (1653–1662) and Bedford Level Corporation Orders and Accounts (1663–1668), Cambridgeshire RO; Minute Book of the Adventurers of Deeping Fen (1669–1685), Lincolnshire Archive Office; Sir William Dugdale (1662) *The History of the Great Level of the Fens, called Bedford Level*; Edwin Canan (1889) Samuel Fortrey, *Dist Nat Biography*, **7**, 50; L. Gaches (1890) Fortrey of Byal Fen, *Fenland Notes and Records*, **4**, 353–358; W. H. Wheeler (1896) *A History of the Fens of South Lincolnshire*; T. S. Willan (1936) *River Navigation in England 1600–1750*; J. W. F. Hill (1956) *Tudor and Stuart Lincoln*; Charles Hadfield (1969) *The Canals of South and South East England*; John Boyes and Ronald Russell (1977) *The Canals of Eastern England*; H. C. Tomlinson (1979) *Guns and Government*; Frances Wilmoth (1993) *Sir Jonas Moore*]

Works

1655–1667. Surveyor of the South Level, to 1662, and of the South and Middle Levels to 1667
1669–1676. Director of Works, Deeping Fen
1671–1672. Restoration of Fossdyke Navigation
1675–1677. Hampshire Avon Navigation, not completed

FOSTER, John (1759–1827), architect, contractor and civil engineer, was born in Liverpool in 1759, a son of John Foster, master joiner.

Foster is not so much a forgotten engineer as one who escaped notice in the first place. This is not surprising, for he served his time as a joiner and seems never to have worked outside Liverpool; he certainly could not be described as a qualified engineer and yet in 1799 he was appointed Surveyor to the Trustees of Liverpool Docks at a salary of £300, three times that of his predecessor, Thomas Morris (q.v.), who was a trained engineer. He was now in charge of the engineering of the fastest growing port in the country. In 1802 he was appointed Corporation Surveyor as well, in which role he was responsible for designing a number of important buildings on the Corporation Estate, including St. John's Market, the Public Dispensary and the New Gaol.

As a young man, he had successfully taken charge of major municipal reconstruction under the *Liverpool Improvement Act* of 1786 but was still no engineer within the accepted meaning when he first undertook work on the Corporation-managed parts of the docks—Chester Basin—possibly as early as 1795.

In 1800 he sought, or was required to seek, the advice of William Jessop (q.v.) and later, in 1809, of John Rennie Sr. (q.v.) on the subject of Princes Dock, the next major development in Liverpool. It is clear, however, that day-to-day management of the construction of Princes Dock was under Foster's control.

The reason we know this is that Foster was profoundly corrupt and those who investigated him left their records. Contracts for dock works often ended up with his own family firm or with those of relatives or associates. He had established a dynasty in Liverpool, which was governed not by the ballot box but by the realities of the way one did business at the time. One son, Thomas, was Town Clerk; another, William, was Secretary to the Dock Committee; John Jr., was Borough Architect; James a partner in a foundry. Using his insider knowledge, he invested in speculative building in a fairly large way; the houses he built on Seymour Street still stand. All this and more was investigated by the Dock Audit Commissioners, whose Minute Book has survived. Furthermore, by his marriage to Jane Gregson, a member of a prominent family of merchants, bankers and manufacturers, he acquired a brother-in-law, Mathew Gregson, who loathed him and kept a number of incriminating documents which also survive. Eventually his malpractice became public knowledge and he was forced to resign in March 1824.

None of this concerns the quality of his work, which was high. Princes Dock, his largest and most expensive job, was an immediate success and continued to be a good earner for the port into the 1870s. Modest modernisation schemes in 1895–1896, 1904–1905, 1929–1930, 1954 and 1964–1966 kept it usable by the Irish trade until 1981.

He also undertook large and complex modernisation of existing docks, including Canning Graving Docks, built in 1765 and modernised in 1812–1813. Kings Dock (1788) was deepened and Queens Dock (1796) was extended by 4 acres. He laid the groundwork for the unpopular infilling of the obsolete Old Dock, opened in 1715, and the sale of its valuable site, though one may note that John Jr. secured the contract for building the new Custom House thereon. These substantial achievements, involving every dock in the estate, could be equated to the construction of at least two major new docks.

At Princes Dock, he was willing and able to apply the latest technology: Rennie-type 'banana' dock walls, Walker-pattern iron swing bridges and, perhaps most notably, a high degree of mechanisation in construction. The French engineer, J. Dutens, has left us a description of the centrally-powered system in which one Boulton & Watt engine, the third which Foster had bought, drove pumps, cranes, mortar mills and a spoil-raising railway extending some 3 miles. The know-how almost certainly came from Rennie but he spent little time in Liverpool, so Foster deserves much of the credit for its successful application. He was also responsible in 1806 for the introduction of large cast iron pipes as a cheaper alternative to brick culverts.

Despite these money-saving methods, Princes Dock cost roughly twice as much per acre as it should have done and the reason was Foster's inadequate control of supplies and contract work which made it almost impossible to account accurately for work or materials. Serious defalcation was proven, with far more 'known about' but not proven. This sloppiness was probably deliberate, for two things about John Foster are certain: he was not stupid and his will was proved at 'Under £60,000', a sum which does not include his significant real estate and might be considered quite high for the son of a joiner.

Foster died in 1827. He had successfully carried through complicated and extensive works, which greatly benefited both the port and the town at large. During his tenure the tonnage using the port more than doubled and revenues multiplied almost tenfold. Regardless of any critical findings by the Dock Audit Commisson he played a considerable part in that remarkable success.

ADRIAN JARVIS

[*Minutes of the Dock Committee, 1793–1824*, Minutes of the Dock Audit Commissioners, Correspondence, Maritime Archives and Library, Merseyside Maritime Museum; Gregson papers and Liverpool City Council Minutes and Accounts, Liverpool RO; J. Dutens (1819) *Memoires sur les Travaux Publics d'Angleterre*; J. Longmore (1982) The Development of the Liverpool Corporation estate, 1760–1835, Ph.D. thesis, Reading University; N. Ritchie Noakes (1984) *Liverpool's Historic Waterfront*; A. Jarvis (1990) The interests and ethics of John Foster, Liverpool dock surveyor 1799–1824, *Trans Hist Soc Lancs Ches*, **140**, 141–160; A. Jarvis (1991) *Liverpool Central Docks 1799–1905*; Colvin (3)]

Works
For municipal and architectural works see Longmore (1982), and for dock reconstruction works see Ritchie Noakes (1984) and Jarvis (1991).

1810–1821. Princes Dock, Liverpool, £650,000

FOULDS, John Torr (1742–1815), came from North Wingfield, Derbyshire, where he was trained as a millwright and in this trade he began work at London Bridge Waterworks in 1763. At that time there were five waterwheels in use, in the first, second and fourth arches at the north end of London Bridge. The wheels had been installed 1726–1742 and they powered sixteen engines to raise water. Such were the constrictions imposed by the bridge that a considerable head of water resulted and this body of water powered the works. In 1761–1762 a larger arch was opened in the centre of the bridge to facilitate navigation but, although this reduced the damming effect, it also reduced the power available to the works.

In 1761, in anticipation, the Waterworks Company was awarded the lease of the third arch; it was inadequate and in 1766 it was allowed the lease of the fifth arch, and a wheel on the south side in the second arch. Foulds therefore arrived in a period of rapid change at the waterworks, not least the design and installation of John Smeaton's (q.v.) 'Great Engine' in the fifth arch. The millwright responsible was Joseph Nickalls (q.v.) and Foulds could not have been exposed to better masters.

The waterworks probably employed three full-time millwrights who were fully occupied in the maintenance of the complex machinery. Foulds, who had married Lydia Powell at St. Saviour's, Southwark in 1765, had by 1776 become the chief millwright at the works, his salary then £7 11s a quarter and the lease of a company house in Churchyard Alley, near the waterworks. Around 1780 his practical experience was resulting in inventiveness and in 1779 he had dealt with the loss of the works water tower in a fire by installing equipment to enable the waterwheels to pump water directly into the mains.

In 1784 London Bridge Waterworks ordered a Boulton and Watt steam engine to replace the atmospheric engine they had as standby for pumping and Foulds appears to have supplied the drawings for the engine, installed in 1786. This additional power enabled Foulds to embark on a rebuilding program at the works and, beginning with the second arch wheel in September 1786, six wheels were replaced, including Smeaton's. Foulds was suitably rewarded, his salary being increased in 1792 to £100 p.a. paid quarterly; this was increased to £120 in September 1796 and his salary was regularly supplemented by gratuities.

Foulds, as a master millwright, appears to have been a leading figure in the Master Millwrights' Association and become embroiled in labour disputes with the journeymen millwrights, a large number of whom must have been engaged in the renewal work at London Bridge. He was recognised as one of the leaders of the profession and in 1793 was elected to the Smeatonian Society of Civil Engineers. By this time his work was becoming more diversified. In 1791 he was appointed Assistant Engineer to the City Corporation and so was responsible for the repair of London Bridge. His 1795 device for cutting off piles below water was presumably a result of this experience.

His work for the Corporation led to involvement in the schemes for the development of the Port of London from the start. He became most heavily involved with the plans for West India Docks, taking soundings, drawing up designs, and making estimates. This was accompanied by work on the neighbouring City Canal, for which he acted as Assistant Engineer in the initial stages. In the late 1790s Foulds was also involved with the Shadwell Waterworks Company, in 1798

designing and directing the installation of a new Boulton and Watt engine.

Foulds's first wife predeceased him and in 1805 he was remarried, to Jane James. From this time it is clear that he was increasingly troubled by ill health. He had trained his two sons, John Powell (1774–1796) and William (1782–1814), as millwrights but both predeceased him. In 1808 Foulds retired, working for the waterworks only on a consultancy basis, and William became engineer. Foulds died on 10 February 1815, only a few weeks after being consulted about the Company's fifth arch wheel, which was renewed in 1816–1817. In addition to his two sons he had three daughters who survived into adulthood.

MIKE CHRIMES

[Smeatonian Society Records, ICE; Description of the plate of the model of Mr. John Fould's machine for cutting off piles under water, *Transactions of the Society of Arts*, 1795, **13**, 241–242, plate 3; Committee appointed to enquire into ... the Port of London (1796) *Minutes of Evidence*, 107–108, appendix Cc; Select Committee for better regulating the Port of London (1800) *Report*, 39; J. G. Moher (1987) John Torr Foulds (1742–1815): millwright and engineer, *Trans Newc Soc*, **58**, 59–73]

Works
1763–1815. London Bridge Waterworks
1798. Shadwell Waterworks engine
1800–1802. City Ship Canal, London docks, assistant engineer

FOULSTON, John (1772–1842), architect, was a pupil of Thomas Hardwick, father of Philip Hardwick (q.v.). Foulston was an admirer of Grecian architecture and was most proud of what he called the 'most acceptable part of his work, *viz.* its Constructive Architecture'. He practised first in London, beginning in 1796, and in 1811 he won the competition for a group of important civic buildings in Plymouth, including the Royal Hotel, the Assembly Rooms and Theatre. He was particularly interested in the theatre's stage machinery and published it in detail in his 1838 book. His design was distinguished by the extensive use of ironwork, in roofs and in the structure, with the intention of making the buildings fireproof; his roof design was a development of the ideas of Thomas Pearsall (q.v.). Foulston claimed to have suggested, some years before Thomas Tredgold (q.v.) published his book on cast iron, that his, Tredgold's, *Treatise on Carpentry* be used as a guide for king and queen post trusses of iron in order to avoid problems of shrinkage and warping.

His reputation made, he moved to Plymouth where he was for twenty five years the leading architect in the area. His most striking group of buildings is the picturesque group of 'Egyptian' library, 'Hindoo' nonconformist chapel and 'primitive Doric' town hall at Devonport, formerly Plymouth Dock, plus the massive commemorative Column of 1824. Foulston developed a novel method of raising and setting the stones for the column involving the use of a derrick, which he detailed in his book. In his later years he was briefly in partnership with George Wightwick, who succeeded to his practice in about 1836.

In retirement, Foulston amused himself with landscape gardening. A Grecian to the end, his gig was built in the form of an antique war chariot, mounted from behind in correct fashion. He died on 13 January 1842 and was buried in St. Andrew's New Cemetery, Plymouth.

His engraved portrait and signature are reproduced in *Public Buildings*.

TESS CANFIELD

[G. Wightwick, (1852–1892) Memoir, in: W. Papworth (ed.) *Dictionary of Architecture*; F. Jenkins (1968) John Foulston and his public buildings in Plymouth, *Journal of the Society of Architectural Historians*, **xxvii**; DNB; Colvin(3)]

Publications

1838. *The Public Buildings, erected in the West of England, as designed by John Foulston, FRIBA*

FOWLER, Charles (1792–1867), architect, was born in Cullompton, Devon, where his family had strong local connections. At the age of fifteen he was articled to a local builder and architect, John Powning of Exeter, and on completing his apprenticeship in 1814 he moved to London and worked for D. Laing (q.v.) before setting up in practice on his own account in 1818. From the start he was carrying out work for the Earl of Devon's family, who were his major patrons, and much of his work was based in the south-west.

Fowler established a reputation for market design beginning with Gravesend market (1818–1822), and his Covent Garden (1828–1830) and Hungerford markets (1831–1833) were widely admired as opening a new phase in market design, imposing order and openness on what were by their nature chaotic assemblages. Hungerford was particularly noteworthy for its early and successful use of iron for its roof, supplied by Joseph Bramah (q.v.). Fowler's use of iron here, and earlier for the Syon House conservatory, Isleworth (1820–1827), has led to comparison with the work of Thomas Telford (q.v.) as an rare example of an architect who successfully embraced iron to produce structural forms of outstanding quality and innovation. As with Telford, Fowler's works at Syon and Hungerford attracted favourable comments from British and foreign observers. At Syon he combined Palladian concepts with a 59 ft. high central glass and iron dome supported on arched cast iron ribs 29 ft. 6 in. long; cast iron ribs had hitherto been used only in more traditional buildings. The free-standing cantilever roof structure at Hungerford fish market (1834) has been described as a double butterfly roof, and preceded the railway station roofs of the 1840s.

The origins of Fowler's innovative approach may lie in his provincial origins which meant that he was not restricted by the more conventional approach of the London architectural establishment: perhaps also he was more open in his youth to the ideas of ironmasters. Aside from markets at Tavistock (corn market, 1835) and Exeter (lower and higher markets, 1835–1838), Fowler was responsible for a number of churches, hospitals and other buildings which he carried out in a more conventional manner. He won the prize for London bridge in 1822, but John Rennie's (q.v.) scheme was preferred, and he designed Totnes bridge in Devon (1826–1828).

Fowler was a founder member of the (Royal) Institute of British Architects. He retired due to ill-health in 1852 and died at Great Marlow on 29 September 1867.

MIKE CHRIMES

[Obituary (1867) *Builder*, 761; *RIBA Trans,, 1867–1868, 18; Hungerford market in London, Allgemeine Bauzeitung*, 1838, plates 246–253; *RIBA Trans*, 1862, **13**, 54–57; J. Taylor (1964) Charles Fowler: master of markets, *Architectural Review*, March, 174–182; R. Thorne (1980) *Covent Garden Market*; G. Kohlmaier and B. von Sartory (1986) *Houses of Glass, passim*; R. Thorne (ed.) (1990) *The Iron Revolution*; Colvin (3)]

Works

Selected works:

1818–1822. Gravesend market
1820–1827. Syon House, glasshouse
1822–1823. Silk factory, Great Yarmouth
1826–1828. Totnes Bridge
1828–1830. Covent Garden market
1831–1833. Hungerford market
1835. Corn market, Tavistock
1835–1837. Lower market, Exeter
1835–1838. Higher market, Exeter

FOWLS, Samuel (1778–1847), civil engineer, was born at Allostock, (now Lostock) and baptised at Lower Peover, Cheshire, on 9 August 1778, the son of Joseph and Martha Fowls. Shortly afterwards they removed to a farm at Wilton, Northwich. From the age of nine Samuel worked on the farm, having been educated by the local parish clerk. He was soon apprenticed to a local joiner and builder and began to study the sciences after work. As a consequence of his study of mechanics he began to obtain work on machinery in local saltworks.

In 1807 the Weaver Navigation obtained an Act of Parliament for an extension from Frodsham to Weston Point on the Mersey. Their engineer, John Johnson, who had been employed since the 1780s on various improvements including tramways and chutes at Anderton (1799–1800), had first surveyed the line to Weston Point in 1796. In 1806 Thomas Morris (q.v.) had reported on improvements at Frodsham and the new cut, offering to do the work for £38,000, but nevertheless the trustees had decided to put Johnson in charge. By 1809 he was clearly not up to the job, which ran over budget, and Fowls was appointed resident engineer with Thomas Telford (q.v.) as consultant; Fowls remained engineer to the Weaver Navigation until his death.

Under the guidance of Fowls the Weston Basin was opened in mid-1810, with piers added in 1817 and 1820, and revenue soon paid for the improvements. In the 1830s further cuts were carried out and on-going improvements to the Weston Basin followed. On William Cubitt's (q.v.) recommendation, the locks and weirs were upgraded in the 1840s. Fowls's reputation locally was high and he became Deputy Bridgemaster for Cheshire in 1825 at £300 p.a. and County Surveyor at £400 p.a. in 1829. As Bridgemaster he was involved in the replacement and repair of a number of bridges, including the design of a temporary bridge at Warrington while the Victoria Bridge was under construction (1835–1837) in masonry; Thomas Harrison's (q.v.) 1814 design in timber needing remedial work almost immediately, and it was apparently a Harrison design in masonry which replaced it. He left office in 1844.

Doubtless because of his connection with Telford, Fowls became an early member of the Institution of Civil Engineers, elected in 1820, and he attended meetings whenever in London. Telford would pay his subscription and Fowls refunded the money when they next met.

On his death, the Weaver Trustees wrote to his family that they 'desire to record their high sense of his integrity, zeal, and the extreme ability with which he conducted their works during such a lengthened period; and which, coupled with his uniform correctness of conduct, secured to him their respect and unlimited confidence'. His son, John (d. 1832), assisted him in his work and was also elected to the Institution but his career was tragically cut short in a bathing accident. Fowls left an annuity of £60 to his wife Mary and the bulk of his remaining property to his daughter, Ellen Johnson. He died on 1 September 1847.

MIKE CHRIMES

[QAR; Letter book of Samuel Fowls, 1831–1840, Minutes of Quarter Sessions, Cheshire RO; Cheshire wills and parish records, Cheshire RO; ICE membership records, ICE archives; Detailed account of the building of Saltersford Weir, OC 107, ICE archives; On a sea wall, OC113, ICE archives; Obituary (1849) *Min Proc ICE*, **8**, 11; C. Hadfield and G. Biddle (1970) *The Canals of North West England*, vol. 1, 131–142]

Works

Weaver navigation:

1809–1820. Weston cut and basin (total cost c. £82,000)
1832–1835. Barnton cut, £19,400; Crowton cut; Aston Grange cut
1833–1844. Weston basin improvements
1844–1845. Lock and weir improvements (under Cubitt)

Cheshire bridges:

1829. Saughall Massey Bridge
1832. Broom Stair Bridge
1834–1837. Victoria Bridge, Warrington, £6,000
1841. Baguley Bridge

FRANCHE, Thomas (fl. 1520–1540), master mason, was son of John Franche (d. 1489), a mason who lived at Linlithgow and had worked on the royal palace there. Thomas appears to have moved to Aberdeen and become master mason to Bishop William Elphinstone, presumably in the latter's late years. The bishop had founded the university in Aberdeen, erected part of the cathedral, and both dedicated funds and gathered materials for the building of a stone bridge over the Dee near the city. Benefactions in his will at his death in 1514 amounted to £10,000. Construction of the bridge had not begun and there was no progress until 1518 when Gavin Dunbar assumed the Bishop's chair. Building began in 1520 with Thomas Franche as master mason and practical supervisor, but subject to Alexander Galloway (q.v.) as director of the work. The bridge, an ashlar structure of seven ribbed semicircular arches and a horizontal road across all of them, was finished in 1527 and in 1529 was accepted into ownership by the town together with 'the lands of Ardlar to be held in feu for maintenance of the structure'. A year or two later Franche left Aberdeen. In 1535 he was made Master Mason to King James V and worked on the palaces of Linlithgow and Falkland. He died before 1551.

TED RUDDOCK

[R. S. Mylne (1893) *The Master Masons to the Crown of Scotland*; G. M. Fraser (1913) *The Bridge of Dee*]

FRASER, John, Lieutenant-Colonel (1791–1862), military engineer, was born on 25 January 1791. He served in the Peninsular War where he led the forlorn hope at the siege of Burgos before being gazetted to the 1st Ceylon Regiment as Captain in 1813. He participated in the Kandyan War of 1815, the suppression of the Uva rebellion and in 1820, with the rank of Assistant Quartermaster General, he became Director of the Island (Public) Works, a position he held until 1832 when the public works departments were reorganised and Fraser relieved. In the 1820s, under the governorship of Sir Edward Barnes (1776–1838), a great deal was done by Fraser, in association with Royal Engineers such as Captain W. F. Dawson and Major Thomas Skinner (qq.v.) to survey the Island of Ceylon and improve the road network. Much of the works was carried out using forced local labour. Fraser completed a series of maps commencing with one of the mountain area at a scale of 1 in. to a mile and culminating in a map of the island, published by Arrowsmith in 1826 at a scale of 4 miles to an inch.

Fraser's most famous civil engineering work was the design of a timber arch bridge of 205 ft. span over the Mahaweli Ganga at Peredenia. There were four arched ribs of treble planks at 5 ft. centres, the whole cross-braced, and made of satin wood; work began in July 1832 and was completed by the end of the year.

In 1826 Fraser was appointed Deputy Commissary General and in the following year Deputy Quartermaster General and became a Lieutenant Colonel. He remained Deputy Quartermaster General until 1854, was made Colonel of the 37th Regiment in 1858, and Lieutenant General in 1859. After residing with his family 'in patriarchal fashion at Rangboda' he died in Kandy on 29 May 1862.

MIKE CHRIMES

[Peradenia Bridge, OC105, ICE archives; P. M. Bingham (1921–1923) *History of the Public Works Department, Ceylon, 1796 to 1913*, 3 vols., incl. portrait; S. Xavier (1999) in lit.]

Works
1820–1821. Kurungale–Kandy Road
1820–1825. Ambepussa–Kurungale Road (with above), 48 miles
1820–1825. Colombo–Kandy Road, 72 miles
1820–1827. Kurungale–Dumbulla Road, 33 miles
1822. Bridge of Boats, Grand Pass
1825–1831. Kandy–Mantale Road
1831–1832. Mantale–Dumbulla Road
1832–1833. Peradenia Bridge, Sri Lanka, 205 ft. span timber arch

FRAZER, Andrew, FRS (1746–1792), was a military engineer whose life marks an interesting family progression from the purely civil career of his father, George (1701–1774), Deputy Auditor of Excise in Scotland, who produced an unbuilt design for the North Bridge over North Loch, Edinburgh, and the military career of his son, Sir Augustus Simon (1776–1835), who commanded Wellington's field artillery in the Peninsula and at Waterloo. Andrew interested himself in canals, harbours and land drainage as well as ecclesiastical and military architecture.

He was trained in the drawing office in the Tower of London before becoming a practitioner engineer in 1759. He worked in Portugal and Liverpool after which he became John Peter Desmaretz's (q.v.) assistant at Dunkirk in 1763, taking over as Chief Commissioner on his death in 1768. To continue the connection, he also married Desmaretz's granddaughter, Charlotte Durnford, in 1773. During this period he drew up plans and reports on the town and harbour of Dunkirk and made a joint report on three new canals at Mardyke.

Frazer left Dunkirk in 1778 when France allied itself with the American colonies and was sent to Scotland as Chief Engineer. Since John Paul Jones's raids had demonstrated the vulnerability of isolated areas his main task was to modernise the northernmost fort in Britain, at Lerwick, in 1781. This involved rebuilding the west wall, north gate and main gate and building a new transverse wall, rainwater cistern, magazine, ammunition store and barracks for two-hundred and seventy men. The new fort was (supposedly) renamed after Queen Charlotte, and survives relatively unaltered.

The same year he built the Governor's House in Edinburgh for £302 and won a competition for the design of St. Andrew's Church there, emulating his father who had designed the High Church in Inverness. The church was built between 1782 and 1785 and is in the background of a portrait of Andrew Frazer by Zoffany. In 1783 he also built a battery and stables in Leith. He was made a Burgess of Edinburgh 'in recognition of his services to King and Country in general, to this city in particular in the line of his profession'. Unfortunately his service in the West Indies, where he was sent in 1785, was to have a less happy outcome and ended in his being cashiered in 1792.

Ironically one of his first actions in Dominica was possibly his most important civil engineering commission, when he removed the principal objection to occupying Prince Rupert's (Scotts Head?) by draining the morasses which had previously made the site too unhealthy and recommended its development as a port. He also provided a plan of the main town at Roseau and built temporary works costing £1,000. In 1788 he became a Colonel and worked on the fortifications of Barbados, Grenada and Antigua. Unfortunately he had received no written orders for the work on the latter two islands and was summoned back to England to be tried in 1791.

It was the opinion of his fellow engineers that Frazer was an excellent engineer who had only followed established precedent but the Duke of Richmond, the Master General of Ordnance, was implacable in his prosecution and Frazer was cashiered in 1792, leaving the country almost immediately. He died and was buried at Chalons on his way to Geneva. The same year, unsurprisingly, his son chose to join the artillery rather than the engineers. After Waterloo he was appointed British Commissioner for taking over the French fortresses and was elected FRS in 1816.

SUSAN HOTS

[PRO W0/55; surveys 1783–1786, mss 1646, 1649, plan Z3.48, NLS; Conolly papers, Royal Engineers Library; D. W. Marshall (1976) The British military engineers 1741–1783, Ph.D. thesis, University of Michigan, 148; J. Gifford (1992) *Buildings of Scotland, Highlands and Islands*; C. Tabraham and D. Grove (1995) *Fortress Scotland and the Jacobites*; D. Pringle *et al.* (2000) An old pentagonal fort … excavations at Fort Charlotte, *Post Medieval Archaelogy*, **34**, 105–143; DNB; Colvin (3); Bendall]

Plans
Plans include:

1772. Report and plans of Dunkirk harbour [BL Add. mss, 17779-82]

Works

Works include:

1781. Fort Charlotte, Lerwick, Shetlands
1787. Drainage of marshes in Dominica

FROME, Sir Edward, General (1802–1890), military engineer and artist, was born on 7 January 1802 in Gibraltar, the son of the Reverend J. T. Frome, a military chaplain. Orphaned at the age of two, Frome received an early education at Bexley and Blackheath. He entered the Royal Military Academy at Woolwich in 1817 and was commissioned a second lieutenant in the Royal Engineers in 1825. Frome was posted to Upper Canada (now Ontario) in 1827 on the Rideau Canal where he was assigned as officer-in-charge of the Kingston Division. In this regard his most significant work involved the construction of a dam and three locks at Kingston Mills. He returned to England in 1833 upon completion of the canal and for the next seven years was an instructor of surveying and engineering at Chatham. In 1840 he was appointed Surveyor-General of South Australia and later Colonial Engineer. In these posts he was responsible for surveying most of the known parts of the country and for exploring large tracts of unknown territory. After returning to England he was appointed, in 1852, Surveyor-General of Mauritius. He again returned to England and in 1859 was promoted to Colonel and made Commanding Royal Engineer first in Scotland, then in Ireland and later in Gibraltar. In 1868 he was appointed Inspector-General of Fortifications for the Board of Ordnance and in 1869 made Lieutenant-Governor of Guernsey.

Edward Frome was an enthusiastic artist who was able to combine his analytical engineering mind with a sensibility which lifted his art above mere pedestrian work. He completed an impressive collection of pencil sketches, pen-and-ink drawings and watercolours, all of which are now housed at the Library of Cambridge University.

Frome died on 12 February 1890 in Ewell.

MARK ANDREWS

[T. W. J. Conolly (1898) *Roll of Officers of the Corps of Royal Engineers*, Chatham; *Frome Papers*, Commonwealth Collection, Cambridge University Library and Archives, 1957; E. F. Bush (1976) *The Builders of the Rideau Canal*, Parks Canada, MSS Report No. 185, Ottawa; M. E. Andrews (1998) *For King and Country, Lieutenant-Colonel John By, R.E. Indefatigable Civil-Military Engineer*]

FROST, James (*c.* 1774–1851), cement manufacturer, was a builder from St. Faith's Lane, Norwich. Following the experiments of John Smeaton (q.v.), interest in cements other than traditional lime mortars grew and in 1796 James Parker patented his 'Parker's' or Roman cement; almost immediately Thomas Telford (q.v.) reported favourably on it for the British Fisheries Society. Parker's cement was a consequence of his discovery on the Isle of Sheppey of 'nodules of clay' or septaria, or cement stone, which on firing produced a hydraulic cement. A short time afterwards, his patent was purchased by Samuel Wyatt (q.v.) and exploited by his cousin Charles (d. 1819). The patent, although renewed, ran out in 1810 opening the way for other manufacturers and once the 'secret' of the septaria had been identified the search was on for similar geological formations. Frost identified these on the Essex coast, possibly *c.* 1807, but more probably as a consequence of his contracts for the construction of Martello towers (1808–1812). His other early known work was the St. Miles (Coslany Bridge), Norwich, a cast iron structure.

It would appear that Frost began experimenting in the Harwich area *c.* 1810 with various combinations of clay and lime and other ingredients to develop his own patent cement. They were not immediately successful and his first cement patent of 1822 was of no practical value. Frost, however, visited Vicat, the world authority on cements, in France, and by 1825 had embarked on a new phase in his life, moving to Finchley, in London, where he built (1825–1827) two houses incorporating fire-resisting floors and iron ribs. At the same time, in October 1825, he leased land at Swanscombe, Kent, to erect a cement works. In December 1828 they were visited by Charles Pasley (q.v.), who described Frost's works in some detail in his books, although not to Frost's satisfaction. At the same time he acquired premises at 6 Bankside, Southwark, and it was from that address that he joined the Institution of Civil Engineers as an Associate in 1828. He was a regular contributor to discussions until his resignation at the end of 1830, speaking on subjects as varied as road foundations, hardness of stone, steam engine construction—no doubt based on their installation at his works—and iron casting techniques, as well as cement and his own fire-proof floor system. His statements reveal his knowledge of the work of others such as Vicat and James Aspdin (q.v.). The first major use of Frost's cement was for Hungerford Market in 1830 but in 1833 Frost resolved to sell up and emigrate to the United States of America. His works were purchased by Francis and White, another early cement manufacturer. In the United States he wrote on cements for the Franklin Institute.

He died on 31 July 1851 in New York, described as an architect.

MIKE CHRIMES

[Membership records, ICE archives; Minutes of conversations, 38, 53, 54, 58, 61, 62, 88 (1828–1830), ICE archives; C. W. Pasley (1838) *Observations on Limes, Calcareous Cements ...*, iii, appendix 13–16; S. Sutcliffe (1972) *Martello*

Towers; A. J. Francis (1978) *The Cement Industry, 1796–1914: a History*; E. A. Labrum (1994) *Civil Engineering Heritage: Eastern and Central England*, 118–119; R. M. Telling (1997) *English Martello Towers*; Colvin (3)]

Publications

1835–1836. Essays on calcareous cement, *Franklin Institute Journal*, 1835, **16**, 217–219; 1836, **17**, 234–239; 1836, **18**, 179
1842. On limes and cements. *Mechanics Magazine*, **36**, 310–311

Works

1804. St. Miles Bridge, Norwich, cast iron arch, 36 ft. 2 in. span, £1,100
1818. St. Hellesdon Bridge, contractor, £1,140

FROST, James (1791–*c*. 1861) contractor, was baptised at Sedgley in October 1791, the son of Samuel and Mary Frost. His father was a labourer and he and his brother, Matthew, were agricultural labourers and tenant farmers who became contractors in the West Midlands. James was responsible for the Sandhills branch of the Birmingham Canal, five furlongs in length, *c*. 1838–1840, and 2 miles of the Stourbridge extension canal in 1838. Matthew built several tramroads in the Black Country as well as the Hatherton branch of the Staffordshire and Worcestershire canal and, in the 1840s, several of the extensions to the Birmingham canal. He later turned to railway contracting while James seems to have become a mine agent and surveyor (1841–1861).

MIKE CHRIMES

[Frank Smith collection, ICE archives]

FRYER, John (1745/1746–1825), was a surveyor employed almost exclusively in north east England. His origins have not been ascertained but he was described as having been pupil and assistant of Charles Hutton (q.v.) 'when the latter occupied the school-house near the foot of Westgate Street' in Newcastle. He married Margaret Stephenson (1745–1777) in Newcastle in 1770, at that time being assistant to Robert Harrison at the Trinity House school in Newcastle. The following year he was appointed master at a salary of £20 p.a.

In 1772 Hutton published, in his name only, a map of Newcastle, the survey for which had been completed two years earlier, seemingly with the involvement of Fryer. Almost contemporaneously, Fryer began a survey of the river Tyne, at first only in its lower reaches; soundings were included. Initially the survey had been at the behest of Newcastle Trinity House but at the request of Newcastle Corporation, it was extended, first to Newcastle, where in 1772 it included the damage incurred by the bridge in the disastrous flood of the previous year, and then to the river's tidal limit. Due to other work, the survey—'justly admired for accuracy and beauty'—was not completed until 1782. Retained by a committee of

shipowners, he again inspected the river in 1787 and then found that since his earlier survey it had deteriorated appreciably; afterwards he had no further involvement in navigation matters.

On 17 November 1771 an unprecedented flood on the river Tyne destroyed all the bridges on it except that at Corbridge. Fryer came to be involved peripherally in the rebuilding of those at Chollerford and Hexham. At Chollerford he prepared two designs for a bridge to be built on the site of the former structure, both drawings showing five spans varying in their dimensions; one of them indicated a maximum span of 60 ft. while the other showed the maximum span to be 64 ft. Due to the fact that John Wooler (q.v.) also prepared a design it is uncertain as to whose design was adopted but the new bridge, as built, incorporated five varying spans with a maximum of 56 ft., the centre span being the only one differing from Fryer's design; its cutwaters are different from his drawing.

At Hexham the bridge carried away in the flood had been built only a year before. A drawing prepared by Fryer for the County magistrates in 1773–1774 shows the ground conditions on the site of the bridge then proposed, loose material underlain by clay, in turn underlain by hard sand; a second drawing gives an elevation of a bridge with nine spans varying from 70 ft. to 50 ft. Work did not proceed and Wooler was then employed to provide the replacement bridge, apparently initially adopting Fryer's design, then amending it and finally abandoning the project on account of unsuitable ground conditions.

Fryer's first wife died in 1777 and in 1779 he married Elizabeth Harrison (1750–1792) of Warden, near Hexham. From this time Fryer undertook many estate surveys in Northumberland and County Durham: Corbridge, 1779; Washington, 1783; Boldon, 1784; Blackhall, 1790; Jarrow, Barnes, Stranton, Newsham, Westholme, Allensford, Gibside and Coxlodge, from 1803 to 1817.

In 1794 and again in 1807 Fryer produced drawings for a turnpike from North Shields to Morpeth together with a road from Newcastle to North Shields and in 1820 published a map of Northumberland, which 'for the first time shows the county correctly placed as regards longitude'. He also undertook, with others, surveys for several enclosure and division awards: Beamish, 1803; Thorneyburn, Greystead and Stannersburn Commons, 1804; Alston Common, 1805; Elingham Rig and Shitlington Common, 1805; Beamish South Moor Division, 1806; and Framwellgate and Witton Gilbert Division, 1808. In 1817 he entered into a bond regarding the collection of taxes in the northern counties, becoming Receiver General. Fryer lived, and had his office, in Newcastle, first in Westgate Street and then in Northumberland Street.

Fryer, who possessed 'soundness of judgement, unremitting industry, and the most unbending integrity', died in Newcastle and was buried at St. Andrews church on 8 October 1825, in his will leaving the bulk of his estate, including 'the best horse which I shall die possessed of' to his son William (q.v.). Of his nominal £7,000 estate, legacies of £600 were left to each of his five surviving daughters and two other sons, Charles and Joseph Harrison Fryer (d. 1855), the latter a geologist.

R. W. RENNISON

[*Hexham Bridge, 1767–1789*, Northumberland RO, M17/57; *Northumberland Quarter Sessions: Orders, 1772–1775*; *Newcastle Courant*, 1803–1808; E. McKenzie (1827) *A Descriptive and Historical Account … of Newcastle*, 332, 444; H. Whitaker (1949) *A Descriptive List of the Maps of Northumberland, 1576–1900*, 112; Estate Plans held in Durham and Northumberland Record Offices]

Drawings

Plan and Elevation of a Bridge designed to be erected over the River Tyne at Chollerford in Northumberland, Newcastle Central Library, L624.G546

Plan and Elevation of a Bridge to be Erected over the River Tyne at Hexham, Newcastle Central Library, L624. H614

Plan of Several Roads between North Shields and Morpeth, 1794, Newcastle Central Library, L625.7

Major surveys

1772. *Plan of Such Parts of Newcastle and Gateshead as lie Contiguous to the River*, Tyne and Wear Archives, 285/9/1

1773. *A Plan of the Lower Part of the River Tyne … from an Accurate Survey finished in 1772*, Northumberland RO, 3410/WAT/24

1820. *Map of the County of Northumberland … John Fryer & Sons, Newcastle*, Newcastle Central Library, L912.2

FRYER, William (1788?–1864), was the son of John Fryer (q.v.), a prominent Newcastle land surveyor and his second wife, Elizabeth. The second son of a family of twelve, he was baptised a non-conformist and was registered at St. Andrews church, Newcastle on 6 January 1789.

Fryer worked with his father and in addition to his surveying work was, in 1827, described as 'receiver general of the land revenues of the Crown for the counties of Northumberland, Cumberland, Durham, Lancashire and Westmorland'. Fryer had his office in Scafe's Court, Pilgrim Street, Newcastle. His residence was at first in Newcastle and, later, in Gateshead and then Ryton. He was a member of the Literary and Philosophical Society of Newcastle and was described as 'a man of artistic tastes'.

In addition to estate surveys, Fryer was involved with his father in the production of a map of Northumberland, the first to show 'the county correctly placed as regards longitude'. He also undertook the first survey of the eastern end of the Newcastle and Carlisle Railway, carried out under the direction of Benjamin Thompson (q.v.) in 1825 but, perhaps as a result of errors in levels, he was not involved in the preparation of the Parliamentary drawings in 1828.

It is possible that Fryer did not marry and in 1852 he moved to Tynemouth and died when living there with his sister, Mary. He was buried in Preston Cemetery, Tynemouth, on 30 January 1864.

R. W. RENNISON

[*Proceedings of the Society of Antiquaries*, **1**(3S), 41; Edward Hughes (ed.) (1963) *The Diaries and Correspondence of James Losh, 1811–1833*, Surtees Society, vol. 174, 44; H. Whitaker (1949) *A Descriptive List of the Maps of Northumberland, 1576–1900*]

Publications

1808. *Survey of Hilton Ferry Estate*

1822. *Map of the County of Northumberland, John Fryer and Sons, 1822*

1825. *Plan and Section of an Intended Iron Rail Way or Tramroad from the Town and County of Newcastle upon Tyne to the Canal Basin near to the City of Carlisle in the County of Cumberland, with seven Branches therefrom, by Benjamin Thompson*

FULTON, Hamilton (*c*. 1780–1834), civil engineer, was born in Renfrewshire *c*. 1780 and worked under John Rennie and Thomas Telford (qq.v.) in the early nineteenth century. He assisted Telford in surveys of the Highlands Roads, receiving over £700 between 1803 and 1812, and accompanied him to Sweden in 1808. In 1809–1810 he surveyed the Stamford Junction Canal for Telford, marrying Sarah Collins Martin, on 25 April 1810 in Stamford. He then worked on a survey of mail roads in North Wales as part of the Holyhead Roads survey for Telford and the Post Office, the latter in 1813 supporting Telford's recommendations.

Subsequently, in July–August 1815, Fulton went out to Bermuda on behalf of the Admiralty, with Rennie's support. In 1809 the Navy had acquired Ireland Island as a site for a dockyard, and from 1810 to 1816 £145,000 had been spent on buildings, many of a temporary nature, as all labour and materials had to be imported. Larger structures were probably designed by Edward Holl (q.v.). It was probably on the basis of Fulton's reports on these facilities that Holl and Rennie developed a plan for larger facilities, including a breakwater and dry dock, which were executed after their deaths, in the 1820s. In 1817 Fulton made another dockyard report, again at Rennie's instruction, at Malta. The dockyard here was of ancient origin, but in 1811 it had been agreed to build a dry dock. Unfortunately the site was one of fissured rock, and in 1816 the Admiralty decided to abandon it; Fulton's findings confirmed the decision and it was not until the 1840s that a dry dock was built there

In 1819 Fulton was appointed state engineer, on Telford and Rennie's recommendation, to North

Carolina and Georgia, and worked in the U.S.A. for ten years. His appointment, at a salary of £1,200 p.a., and an assistant, Mr Brazier, at £300 p.a., came after a near year-long search by Peter Brown of the General Assembly of North Carolina, who came to Britain seeking a state engineer in the summer of 1818. He discovered that none of the 'first Engineers' nor most of the 'second rate' engineers would entertain his offers of work, 'so well situated' were they. Fulton was temporarily unemployed and prepared to travel.

On his arrival in July 1819 he set about surveying the state with a view to making recommendations about navigation, harbour facilities, canals and river transport, roads, and land drainage. Unfortunately he was soon affected by poor health brought on by some locally contracted fever. He returned to England in 1829 with his family of eight children and took his son Hamilton Henry Fulton on as a pupil.

In 1832, acting for the owner of Togston colliery, near Amble, Northumberland, Hamilton Fulton reported on the feasibility of constructing a harbour there. He recommended that a new outlet be formed for the River Coquet immediately to the south of that existing and he also proposed that breakwaters be built to protect the harbour's mouth. Sir John Rennie (q.v.) later became responsible for the forming of a new harbour but in a form different from Fulton's. Fulton also surveyed the London–Brighton Railway line for Rennie Jr., and on Fulton's death in 1834 his son worked for Rennie before embarking on a career in civil engineering.

MIKE CHRIMES

[Drawings register, 153, 158, 162, ICE Library; Rennie reports, 1815, ICE Library; *Reports of the Commissioners for Roads and Bridges in the Highlands of Scotland*, 1803–1821; Select Committee on Holyhead Roads (1815) *Minutes of Evidence*, 8–12; Evidence of Sir John Rennie (1836) *House of Commons Select Committee on London and Brighton Railway Bill*; Obituary of H. H. Fulton (1886–1887) *ICE Min Proc*, **87**, 418–422; *Engineer*, 1887, **62**, 158; N. Fitzsimons (1985) Wanted: one civil engineer, *Engineer as Historian*, **61** (from *Civil Engineering*, Jan. 1971); J. G. Coad (1989) *The Royal Dockyards 1690–1850*; SK]

Publications

n.d. *Report on the proposed Canal between the Rivers Heyl and Helford*
1810. *Plan of the proposed Stamford Junction Canal* (for Telford)
1813. *Report on the Mail Roads in North Wales*

Works

1809. Thurso Road survey, 20 miles 475 yd.
1810. Mail roads in North Wales (for Telford)
–1810. Lochieside (Loch Arkegg) Road survey, 12 miles 540 yd.
1811. Trotternish Road survey, Skye, 21 miles 638 yd.

1811–1812. Loch Awe Road survey, 30 miles 1,089 yd.
1811–1812. Morvern Road survey, 35 miles 541 yd.
1812. Huna Road survey, 17 miles 854 yd.
1812. Kerneachtrack Road survey, Jura, 22 miles 480 yd.
1812. Fleet Mound survey
1812–1813. Holyhead–Shrewsbury, and Llandegai–Chester Road surveys (for Post Office)
n.d. Loch Feachan Road survey

FULTON, Robert (1765–1815), artist, canal engineer, and naval inventor was born on 14 November 1765 at Little Britain, Pennsylvania. Of Ulster–Scottish descent, he was eldest son of Robert Fulton, a farmer, and Mary (*née* Smith).

Fulton was educated at a Quaker school in Lancaster. An indifferent scholar, he nevertheless had early interests in science and mechanics but a talent for painting led to an artist's career before he eventually departed for England in 1787. A tall, handsome and socially-gifted man, he enjoyed success as a portrait painter among the aristocracy and became acquainted with Earl Stanhope, with whom he corresponded on ship propulsion and canals.

He was to abandon art in favour of engineering and in 1793 he advocated the use of inclined planes on the Bude Canal. During 1794 Fulton lived in Manchester, attracted by the steamboat experiments in progress at Worsley on the Bridgwater Canal. He unsuccessfully applied for work on the Gloucester to Berkeley Canal for which he proposed to design a rudimentary earthmoving machine. Through friendship with Robert Owen and other industrialists, Fulton was recommended as a contractor on the Peak Forest Canal, where he interested the proprietors in his ideas for tub-boat canals, and inclined planes in place of locks. He also proposed an iron aqueduct in place of the Marple aqueduct then under construction, promising substantial savings. He published articles and a book on his designs, most of which were rejected by engineers of the day.

Fulton returned to America in 1807 but not before his fertile mind had invented submarines, mines and torpedoes, which were demonstrated to both British and French governments. After building the first successful steamboat, which sailed on the Hudson River, Fulton continued with his remarkable naval warfare inventions.

His works were ahead of their time and influenced designs of a later generation and his status in America is reflected in his statue in the Hall of Statues in the US Capitol building. He married Harriet Livingston on 7 January 1808; there were four children of the marriage. Fulton died of pneumonia on 24 February 1815, in New York. Two self-portraits, one bust, and a vignette of Fulton on a memorial stone, are known to exist in addition to the statue. All are in the United States.

R. SCHOFIELD

[Peak Forest Canal Company. Committee Minute Book, 1/94/1795, PRO RAIL; W. Chapman (1795) *Observations on the various Systems of Canal Navigation*; R. Owen (1857) *Autobiography: The life of Robert Owen*, vol. 1; H. W. Dickinson (1913) *Robert Fulton: Engineer and Artist*; Wallace S. Hutcheon Jr. (1981) *Robert Fulton: Pioneer of Undersea Warfare*]

Publications

1796 *Treatise on the Improvement of Canal Navigation*
1796. *Report on the proposed Canal between the Rivers Hayl and Helford* [Helston Canal]

Works

1794. Peak Forest Canal, earthworks contract, Hyde

G

GALLOWAY, Reverend Alexander (fl. 1514–1530), priest and architect-construction director, is believed to have designed, and certainly directed the construction of the Bridge of Dee at Aberdeen in 1520–1527. He is described as the 'parson of Kinkell' and also as 'the most faithful friend [to whom] Bishop William [Elphinstone, d. 1514] had entrusted all his fortunes, to … employ them in the way destined in his will'. Materials had been prepared before Bishop William's death and when the bridge project was again set in motion by Bishop Gavin Dunbar in 1518, he placed Galloway in charge of the work again. It is therefore assumed that he designed the bridge, which was built by Thomas Franche (q.v.). Galloway performed the ceremony of handing it over to the town in 1529. In that year he also provided 'the townsfolk' with a scheme of fortification by an encircling wall and fosse, which they formally approved but failed to carry out.

TED RUDDOCK

[G. M. Fraser (1913) *The Bridge of Dee*]

GANDON, James, MRIA (1742–1823), architect, was born in London on 20 February 1742 to Peter and Jane Gandon (*née* Wynne) of French Huguenot stock. At seven years of age he was sent for two years to a boarding school at Hatfield in Hertfordshire. For the next five years he was at school in Kensington, London. Early in his development he showed a great taste for drawing, mathematics, chemistry, landscape and ornamentation. Although his father had intended his son to enter the Corps of Engineers, he became, at the age of fifteen, a pupil in the office of Sir William Chambers, who had recognised his abilities. Gandon began to practise independently around 1764 and was married in 1770.

He undertook some commissions for public company buildings in London before being appointed about 1775 by the East India Company to be Architect and Civil Engineer to Fort Malborough on the island of Sumatra in Indonesia. However, owing to the nature of the climate, he declined the appointment and in 1776 he, together with an Irishman named Woolfe, Architect to the Board of Works, formed the intention of publishing a continuation of *Vitruvius Brittanicus*, begun by Colin Campbell and published in three volumes.

Gandon moved to Ireland in 1781 and was retained to supervise the construction of the now celebrated Custom House and its associated docks and warehouses, on which John Rennie (q.v.) was later employed. The foundations of the Custom House proved difficult owing to the swampy nature of the river-side site. Workers digging the trenches for the south front met a 'boiling spring' with sand at about 4 ft. below the surface, but by the judicious use of piling, pumping and sodding, and a constant relay of masons, a solid foundation was eventually obtained. The Custom House was completed in 1791 and the first dock opened in 1796 at a time when Gandon was working on the design and construction of the Four Courts (first stone laid 1 August 1795), one of the chief architectural glories of Dublin. Another notable achievement was the portico of the Parliament House, now the Bank of Ireland. Other commissions in Dublin included the Military Infirmary in the Phoenix Park, the King's Inns and Carlisle Bridge.

In 1782 the Wide Streets Commissioners in Dublin decided to erect a bridge to connect Sackville Street (now O'Connell Street) on the north bank of the river Liffey with D'Olier and Westmoreland Streets on the south bank and commissioned Gandon to undertake the work. His original design was for a triumphal bridge to commemorate the Army and Navy achievements during the reign of George III, but the design as executed was much more modest. The first stone of the bridge was laid in March 1791. It was named Carlisle after the Lord Lieutenant of Ireland at the time, and comprised three 50 ft. semicircular masonry arches with steep approaches and a width of 40 ft.; it was replaced by the present O'Connell Bridge in 1880.

Gandon was one of the first to be elected to membership of the Royal Irish Academy. He died of gout at Lucan, County Dublin, in 1823 and is buried in the churchyard at Drumcondra, in Dublin. He left one son and two daughters.

His portrait is reproduced in McParland.

RON COX

[J. Warburton (1818) *History of the City of Dublin*, 516–519; *Dublin Builder*, **1**, 1859, 79, 101–102; J. C. Kinoulty (1970) *A Biographical Dictionary of Ireland from 1500*, unpublished manuscript, Trinity College, Dublin, 253; E. McParland (1985) *James Gandon: Vitruvius Hibernicus*]

Works

1781–. Custom House dock and warehouses, Dublin c. £140,000 spent on dock by 1794
1791–1795. Carlisle (now O'Connell) Bridge, Dublin

GARSTIN, John, Major General (1756–1820), was born in 1756, the son of Margaret Garstin of Half Moon Street, Westminster. He had a brother, Edward (d. 1779). He joined the Bengal Engineers as an Ensign in 1778 and rose to become Surveyor-General in Bengal (1808–1813). His early civil engineering work included a survey of the navigability of the Cossimbazar River above Murshidabad in 1780 and the production of a large scale map of Calcutta (c. 1783), which he worked on with several colleagues.

He was then stationed at Patna (1784–1793) and was responsible for the construction of a brick dome granary at Bankipore. The walls were 12 ft. thick, the internal diameter 108 ft. and the height 94 ft. It is believed that the initial design was due to Henry Watson (q.v.). He moved to Chunar in 1793 where his second son, Edward (1794–1871), also to be an engineer, was born. He moved to Calcutta in 1797 and for health reasons returned to Britain with his family the following year, then being based at Fort William.

On his return to India he surveyed a trunk road route from Cuttack to Calcutta in 1804. In 1807, while in temporary charge of the Surveyor General's drawing office, he was made responsible for the construction of Calcutta Town Hall. Work was completed in 1813 but subsidence problems led to disturbance of several pillars in the portico of the building and the structure had to be rebuilt at Garstin's expense. In 1806–1807 he had been acting Chief Engineer and held this appointment, and that of Surveyor General, after the departure of Colonel Kyd (q.v.) until April 1813, when Charles Crawford was appointed Surveyor General.

Under Garstin's charge, James Tod and John Macartney (qq.v.) undertook surveys of the ancient canals alongside the Jumna. He took a great deal of interest in their surveys and was concerned about the accuracy of the levels. His interest in these, and other works such as his proposed cut (1812) between the Ganges and Cossimbazar rivers, to prevent the flooding of Murshidabad, led him to translate Colonel Henry Watson's copy of Frisi's *Treatise on Rivers and Torrents* during a voyage home in 1815. While away, he visited some of the works in Italy described by Frisi. He returned to Calcutta and resumed his command in the autumn of 1818, remaining there until his death on 16 February 1820.

He married Mary Lufftie in Singapore on 21 November 1789. She died in Calcutta in 1811 and was buried in South Park Street, Cemetery, Calcutta, where Garstin was interred; they had seven children.

MIKE CHRIMES

[E. W. C. Sandes (1935) *The Military Engineer in India*; R. H. Phillimore (1951) *Historical Records of the Survey of India*, **I** and **II**]

Publications
1818. *A treatise on Rivers and Torrents* [translation of Frisi]

Works
1784–1786. Bankipore dome, 175,640 rupees
1807–1813. Calcutta town hall

GASCOIGNE, Charles (c. 1745–1806), ironmaster, was the son of Woodgrave and Grizel Gascoigne, of Parlington, Yorkshire.

Employed by the East India Company he met and in 1759 married the daughter of Simon Garbett (1717–1803), a Birmingham merchant and industrialist. In that year Garbett became a shareholder in the Carron ironworks, near Falkirk, a company set up to produce cast iron-ware, particularly cannon and steam engine cylinders. The works was only the fourth in Scotland to be equipped with 'modern' blast furnaces and the first designed to be fuelled with coal; the other works, most notably that at Taynuilt (started 1753), were set up by English concerns to exploit local sources of timber for charcoal. At that time charcoal was still required for the production of high quality wrought iron, but wrought iron production was a secondary consideration at Carron. Garbett's chief partners were Dr. John Roebuck, a scientist with whom Garbett already had business links in metal refining and sulphuric acid production, and who is probably best known for his early support for the steam engine experiments of James Watt (q.v.), and William Cadell, a local Scottish businessman. It was Cadell who first proposed the erection of a cupola furnace but his leading partners were more ambitious; they imported skilled workers from the Birmingham area and began building a series of furnaces. The first blast furnace was completed in December 1760 and the second in September 1761. The first air furnace was completed in March 1760 and the first boring mill in September 1761. At this stage the management of the works was largely in the hands of Cadell's son, William Jr., although John Smeaton (q.v.) was consulted about the means of producing blast for furnaces in 1766, 1767 and again in 1769, two more furnaces having been built; by this time Gascoigne was in charge of the works.

Gascoigne first arrived at Carron in 1763 to act on behalf of his father-in-law who was concerned about the financial state of the company. Gascoigne gradually ousted the original leading partners. Roebuck was bought out in 1768 and after Gascoigne replaced Cadell Jr. as manager in 1769 Cadell's family severed all connection with the company in 1773. By this time a financial crisis in 1772 laid the foundations for financial problems for Gascoigne and the Garbetts. From 1776 until 1786 Gascoigne was in sole charge of the company and learnt all there was to know about managing a large ironworks of the period. He undertook four main projects: securing the works against theft, increasing the boring capacity in the works to manufacture ordnance, obtaining an adequate supply of water, and providing canal access into the works.

In dealing with the engineering problems the advice of Smeaton was constantly sought. In July 1769, when he was investigating blast to the furnaces, he was also asked to look at the boring mills, which were unable to work to capacity due to the shortage of water. Smeaton's solution was to design, in 1770, a new boring mill using the tail water from the furnaces. After further modifications in 1772 this mill worked well but water supply remained a problem as the river Carron only provided directly a quarter of the supply required by the works. The solutions adopted were: to raise an earlier dam at Larbert, the water storage in the furnace pools was increased, water rights were leased from Lord Elphinstone at Fannyside Loch, and, following reports by Smeaton in 1776 and 1777, it was decided to purchase a Boulton and Watt pumping engine to deal with summer shortages. The question of water transport into the works was resolved in 1780 by building a cut from the Carron, attempts to obtain a direct link to the Forth Clyde Canal having failed.

While, with Smeaton's help, Gascoigne sorted out the basic infrastructure of the Carron Works, he was also addressing the problems of its products; chief among them was the production of quality ordnance. From the start the works produced a whole variety of ironware but the production of cast iron pipes, cylinders and ordnance had been a principal aim from its first, and all required accuracy and quality that was initially missing. Although reasonable success was achieved in fulfilling small orders for cast iron pipes to Scottish waterworks in Edinburgh and Dundee, and even, in 1765, for 1,500 yd. of pipes for Blenheim Palace, defective workmanship prevented a breakthrough into the London market. Similarly, lack of accuracy in boring cylinders prevented the firm from supplying Boulton and Watt in the early years. For engineering expertise Smeaton was relied upon and it was he who provided the advice for the Kronstadt engine for Catherine the Great in 1773–1774, the start of an important Russian connection for Gascoigne.

With regard to ordnance, the capacity of the works made it an attractive supplier to the Board of Ordnance. When the works was opened cannon were cast hollow and bored to finish but from the start there was a high rejection rate. Cannon production was a specialist skill which had been developed to a fine art using charcoal filled furnaces in the Weald of Kent and Sussex. Although it was possible to hire some skilled labour this could not be accompanied by knowledge of the metallurgy of the iron castings being produced at Carron. Attempts were made through industrial espionage to discover more about production methods at John Wilkinson's (q.v.) works, and elsewhere, both as regards to the properties of the coal and iron used and boring

methods. The Carron works seemed unable to produce guns of the correct calibre or quality of iron and by 1773 the Ordnance Board were so dissatisfied that Carron guns were embargoed. Gascoigne responded to this crisis by erecting an assay furnace, banning the use of inferior iron for guns, and seeking Smeaton's advice on boring guns from the solid on Wilkinson's method. These measures were largely successful and a new lighter type of ordnance, known as the Carronade, was developed between 1776 and 1778. Despite the continuing ban by the Ordnance and Admiralty Boards these guns proved very successful in the private market and in 1779, following successful trials at Woolwich, Government orders resumed. Gascoigne's own role in the development of this type of gun, which was a most successful piece of ordnance in the Napoleonic wars, is unclear. It would appear that General Robert Melville may have first suggested such a gun to Gascoigne, and an Edinburgh banker and merchant, Patrick Miller, suggested that light guns would be suitable for use on one of his vessels, the *Spitfire*. Gascoigne must have supplied the technical knowledge to make their ideas a reality.

Although principally concerned with cast iron production, the works also produced wrought iron and, as with cast iron, experienced problems with quality. Gascoigne experimented with new methods of production through the 1770s and by the mid-1780s had developed a method, using charcoal in the process, of producing bar iron from Carron pigs. One unfortunate consequence was that when Henry Cort was allowed in 1786 to try out his method of wrought iron production in the Carron forge, Gascoigne found his process wasteful and Carron rejected it.

Despite the increasing technological success of the works, financial problems, which had been present from the start and no doubt aggravated by the continuing need for investment, came to the fore again in the 1780s. Although the concern was profitable in the American War of Independence, the affairs of both Garbett and Gascoigne were not so. Garbett was bankrupted in 1782 and Gascoigne, who had fallen out with him, followed. Garbett had his supporters in the firm and moves were on the way to oust Gascoigne when in April 1786 it was decided he should go to Russia, from where he never returned.

The success of the carronades had led Admiral Samuel Greig, already working for the Russians at Kronstadt, to obtain Catherine the Great's approval for an approach to be made to Carron for their help in improving cannon founding in Russia. Gascoigne was sent, initially—apparently—on a temporary basis with machinery and some workers, having received British government approval. He arrived at Kronstadt on 26 May 1786 and then proceeded to the Olonetz works at Petrozavodsk in September 1786, where he was put in charge. He formed the head of an important Scottish group of emigré workers including

Charles (q.v.) and James Baird, and George and William Clark, and William Wilson, many of whom came from Carron. New furnaces and boring mills were rapidly brought into production and cannon were produced from the end of 1786.

Following his success at Petrozavodsk, he was awarded the Order of St. Valdimir (4th class) in 1789 and became a 'Collegiate councillor' in July 1790. Further awards followed in 1795, 1797 and 1798 when he was 'Active councillor' and Order of St. Anna (1st class). In 1794 he rebuilt the Konchezersk works and in 1795–1797 set up the Lugansk works; he was then put in charge of the Alexandrovsky works in St. Petersburg, where he reorganised a large cotton mill. In addition to cannon these works produced a range of machinery. By 1799 he was earning an annual salary of 30,000 roubles (£2,500 p.a.). In 1801–1804 he organised the Kolpino works with Hastie (q.v.) and some early cast iron bridges were cast in his works.

Gascoigne fell ill in 1805 and died in St. Petersburg on 19 July 1806, although he was buried at Petrozavodsk. He was succeeded in the management of the Olonetz works by Adam Armstrong (1762–1818), a fellow Scot whose sons followed in Russian service. The works at Kolpino and Alexandrovsk were taken over by Alexander Wilson (q.v.).

MIKE CHRIMES

[J. G. James collection, ICE archives; W. Tooke (1799) *View of the Russian Empire*, vol. 3, 550; J. E. Norberg (1805) *Ueber die Produktion des Rohreisens in Russland ...*; K. B. (1843) Brief account of the life of ... K. K. Gascoigne, *Manufakturniya i Gornozavodskiya Izvestiya*, **7**, 52–55; A. V. Armstrong, *Manufakturniya i Gornozavodskiya Izvestiya*, **12**, 94–96; J. M. Morris (1958) The struggle for Carron, *Scottish Industrial Review*, **37**, 136–145; R. H. Campbell (1961) *Carron Company*; E. Robinson (1968) *The Transfer of British Technology to Russia, 1760–1820*; R. P. Bartlett (1977) Scottish cannon-founders and the Russian navy, 1768–1785, *Oxford Slavonic Studies*, **x**, 51–72; J. G. James (1983) *The Application of Iron to Bridges and other Structures in Russia to about 1850*; S. Fedorov (1996) Early iron domed roofs in Russian church architecture, 1800–1840, *Construction History*, **12**, 41–45]

Works

1806. Narodny (Politseisky) cast iron arch bridge, castings, total cost 92,325 roubles
1806. Kazan cathedral, St. Petersburg, ironwork for outer dome

GAYFERE, Thomas, the Elder (*c.* 1721–1812), mason, was involved in the construction of Westminster Bridge as clerk and foreman of Andrews Jelfe (q.v.), the joint contractor for masonry. Gayfere is described as one of Andrews Jelfe's foremen in about 1736, when he made a copy drawing of Labelye's (q.v.) first design for Westminster Bridge. He prepared detailed drawings of the timber superstructure proposed by James King (q.v.) for Westminster Bridge *c.* 1738 and was again described as one of Jelfe's foremen.

In 1747 he was clerk in the office of Jelfe and Tufnell, when he witnessed the contract for construction of the masonry pedestrian recesses over the cutwaters of the piers of the bridge. In November 1747, Labelye declared Westminster Bridge and its approaches complete, but by spring serious problems began to be apparent. A copy of Labelye's *Short Account* (1739) contains a sectional drawing by Gayfere of the proposed repair of the sinking pier presented to the Bridge Commissioners in December 1748. Gayfere was also involved in the improvements to the approaches to Westminster Bridge. In 1746 the Bridge Commissioners drew up a plan for a 70 ft. wide street, now Parliament Street. It was built by Jelfe and Tufnell, with paviour Thomas Phillips under Labelye's direction, and finished in 1756. Subsequently, Gayfere and Tufnell, with Thomas Phillips, contracted to pave the footway of a further improvement, the widening of Whitehall from the Admiralty to Spring Gardens. The work was completed in 1759 but aggrieved landowners forced the commission to defend their mason, Gayfere, in court, with legal action continuing into 1762.

In 1767, with William Jelfe and John Herobin, Gayfere failed to win the contract for resurfacing the roadway of the bridge. Worries about the bridge continued and in 1795–1796, at the recommendation of 'Mr. Gayfere the Mason', who had been present at the building of the bridge, one of the piers was opened for inspection, and found to be in good order. In June 1811, John Rennie (q.v.) reported on the condition of the bridge, using Gayfere's illustrated manuscript *Account Book*, which was presented to the Institution of Civil Engineers in 1821 by Rennie's widow. It was used again in 1823 by Thomas Telford (q.v.) in his investigations into the bridge's condition.

Gayfere seems the likely author of *Gephyralogia*, an historical account of bridges and a history and a detailed description of the rebuilding of Westminster Bridge. The author is generally in agreement with Labelye's work, although he is critical of the choice of caissons rather than piling, ascribing to it subsequent problems with settlement of a central pier. Gayfere rose to the top of his profession: he was appointed Master Mason to Westminster Abbey in 1766, and was Master of the London Masons' Company in 1773. His son, **Thomas the Younger** (1755–1827), followed his father into the mason's trade. He was an excellent draughtsman, exhibiting a view of Norton Church, Leicestershire, at the Society of Artists in 1774, and another drawing in 1777. He exhibited topographical views at the Royal Academy in 1778, 1779 and 1780.

He was appointed joint Master Mason to Westminster Abbey along with his father, who had held the post, until his retirement in 1823.

Gayfere the Younger executed the precise and highly accurate restoration of Henry VII's Chapel at Westminster Abbey for which Parliament voted funds for restoration in 1807–1808, with James Wyatt, Surveyor to the Abbey, nominally in charge. Gayfere, however, travelled to visit quarries throughout the country, selecting Combe Down Bath Stone for the repairs and he bore the main responsibility for the work, which resulted in a remarkably correct restoration. He took plaster casts of surviving fragments, made full size working drawings from them, and trained his masons to work in the historical style.

From 1819 until 1822, Gayfere was engaged in the restoration of the north front of Westminster Hall in Bath stone under the direction of J. W. Hiort of the Office of Works. While working on the Abbey, he had lived at 1 Abingdon Street, London but he died at Burton-upon-Trent on 20 October 1827 and is buried at Newton Solney, Derbyshire.

TESS CANFIELD

[*An Account of all the Stone used in the new Bridge at Westminster* (1739) containing thirty-seven watercolour drawings, ICE Archives; forty-four watercolour illustrations and mss. notes on stone in Labelye, *A Short Account ... of Laying the Piers of Westminster Bridge*, ICE Archives; Obituary [Thomas junior] (1828) *Gentleman's Magazine*, **XCVIII**(I), 275; T. Ruddock (1979) *Arch Bridges and their Builders*; R. J. B. Walker (1979) *Old Westminster Bridge*; Colvin (3)]

Publications

attrib. 1751. *Gephyralogia. An Historical Account of Bridges Antient and Modern ... including a more particular History and Description of the New Bridge at Westminster*, London

GEORGE, Watkin (fl. 1790–1811), foundry manager, of Merthyr Tydfil, was trained as a carpenter before being taken into partnership by Richard Crawshay (q.v.) at Cyfartha ironworks in 1792. With his technical advice the works expanded into the largest in the world by 1806, developing Richard Cort's puddling process. The 50 ft. diameter 50 hp cast iron water wheel he erected there allegedly cost £4,000 and became a local landmark. The two blast furnaces erected at Ynysfach ironworks in 1801 bore his initials. Around 1805 George left Cyfartha for the Pontypool ironworks owned by Hanbury Leigh.

George's chief interest in the history of civil engineering is due to his involvement in the design of several early iron bridges. These included the Pont y Cafnau and a Cyfartha works bridge, unusual for early designs in cast iron in being built to form an A frame heavily influenced by timber construction. Pont y Cafnau has been provisionally dated to 1792–1793 and was a combined aqueduct and tramroad bridge. A bridge there was sketched by William Reynolds (q.v.) in 1794 but it has been suggested, somewhat dubiously, the existing

bridge and its near neighbour may be somewhat later (c. 1810–1820).

George's Ynysgau bridge has a far more primitive appearance, each rib comprising three main frames 22–24 ft. long and 5 ft. high. There were a number of cast iron bridges erected over the Glamorgan Canal in the early 1790s of beam–frame design rather than arch form; their designer is unknown although it has been suggested George may have been involved.

George's last known involvement with bridges was for a design for a cast iron bridge on existing piers at Chepstow in 1811. He left Cyfartha with a considerable fortune, variously described as £40,000 and £100,000 by his contemporaries.

MIKE CHRIMES

[J. G. James collection, ICE archives; D25, 1961, Gwent RO; W. L. Davies (1992) *Bridges of Merthyr Tydfil*]

Works

1793. Pont y Cafnau, Merthyr Tydfil, design
1793–1796. Wooden aqueduct, 80 ft. above river, 606 ft. long
1799–1800. Ynysgau bridge, Merthyr Tydfil, design

GETHIN, John (1757–1831), stonemason and County Surveyor, was born at Kingsland, near Leominster, Herefordshire and was baptised there on 30 November 1757, the son of John and Jane Gethin. His father was also a stonemason who contracted for bridge construction and repairs; his younger brother Benjamin Gethin (c. 1770–1833), was a mason too and was associated with him in at least one of his works. His wife, Esther (d. 1824), bore him one son (another John) and four daughters.

In the 1790s there are frequent references in the Herefordshire Quarter Sessions records to contracts with Gethin for repairs and maintenance of bridges, for sums up to £100. Following a severe flood in February 1795 which did much damage, the Gethin brothers contracted to re-build Aymestrey Bridge for £440. Then on 8 October 1799 under the heading Surveyor of the County Bridges, the Quarter Sessions minuted 'it is ordered that John Gethin of Kingsland be directed to survey the same and make his report at the next Sessions', the first holder of that office in the county.

Over the next few years, when a bridge required to be rebuilt, Gethin would produce a plan and estimate, after which tenders would be sought in the local press. When, as not infrequently happened, there were no offers, Gethin undertook the work himself as contractor, the price being his earlier figure. By these means, more than sixty bridges of moderate size were built within the county to his plans, and many others repaired. His larger bridges derived their style from his father's work, seen for instance at Mortimer's Cross, and his single arch structures are recognisable, with arch and spandrels slightly recessed, the parapets often supported by dentilations. In 1815–1818, he was responsible for a series of three larger bridges, each of 54 ft. span, with circular voids in the spandrels, at Burrington, Bodenham and Kentchurch. At the time of his death he was supervising the construction, by another contractor, of Hampton Court Bridge (Herefordshire) but there is no evidence that he was its designer.

Gethin also produced designs on occasion for knee braced timber trestle bridges and in 1803 he was invited to inspect Caerleon Bridge, which was then a timber structure, for the Monmouthshire justices. On this occasion he prepared a remarkable design of three large cast iron spans, clearly derived from the Buildwas Bridge of Thomas Telford (q.v.); however, it was rebuilt in stone by John Hodgkinson (q.v.) in 1805. In 1820, having prepared plans for wooden bridges at Criftins Ford and Little Hereford church, he was instructed to consider the use of iron, but nothing came of this either.

In the first twenty years of the 19th century, the county rate levied in Herefordshire more than doubled, to pay for projects such as the new county gaol and shire hall, with which Gethin had little to do. The Quarter Sessions, who were intimately involved in the administration of bridge repairs, were increasingly reluctant to allow Gethin to commit monies and in 1820 he was given notice of dismissal, apparently to save costs. At the next Sessions the notice was withdrawn but his salary was reduced from £100 to £50 p.a. and his expenses curtailed. He continued in office until his death on 24 May 1831.

Gethin was responsible for no great technical innovations, unlike his contemporary, Telford who was County Surveyor in neighbouring Shropshire. However his structures were generally more economical and with a county bridge stock of very similar size, Gethin rebuilt roughly half as many again as Telford. More than half have survived to the present day, though widened to cater for the needs of modern traffic.

In 1809 he was one of the trustees named in the Rivers Wye and Lugg Navigation Act, which was very precise about the use of 'gentleman' and 'esquire'; Gethin was one of the former. In the Quarter Sessions minutes he was usually dignified by the title of Mister, though not always so.

P. S. M. CROSS-RUDKIN

[G. H. Jack (1931) John Gethin, Bridge Builder, of Kingsland, Herefordshire, 1757–1831 *Transactions of the Woolhope Naturalists' Field Club*, 86–97, Herefordshire RO; P. S. M. Cross-Rudkin and P. T. Shaw (2001) John Gethin, County Surveyor *Transactions of the Woolhope Naturalists' Field Club*]

Works

Gethin was responsible for the construction, either as designer or designer-builder, of over sixty bridges in Herefordshire and the repair of many more. A complete list of those designed by him is included in the last source above.

GIBB, John (1776–1850), civil engineer and contractor, was born at Gartcows, near Falkirk, the home of his father, William Gibb (q.v.), and was baptised on 13 October 1776. The youngest of seven children of whom three died in childhood, he was fifteen years old when his father died, insolvent, and soon afterwards the family moved to Larbert. He was first apprenticed to a mechanical trade and received some instruction in the rudiments of civil engineering from a brother-in-law employed under John Rennie Sr. (q.v.) on the Lancaster Canal, the first phase of which was largely completed in 1797. From about 1801 he worked at Leith on the construction of Rennie's Old East Dock with a contractor, Alexander Easton of Falkirk (see: *EASTON, Alexander*), whose daughter, Katherine, he married. The date and place cannot be determined as the parish records have been lost but it was possibly about a year before the birth of his son, Alexander, at Larbert on 21 September 1804.

He worked from 1805 to 1809 for the contractor Andrew Brocket (q.v.) under John Rennie Sr. on the new harbour for Greenock. While there he caught the eye of Thomas Telford (q.v.); when work began in 1809 on the reconstruction of the harbour at Aberdeen, Telford sent for Gibb to be his resident engineer. It was a task on which he laboured for the next six years, with the grand title of Permanent Superintendent of the Harbour Works at a salary of £250 p.a. He had first to restore the south pier head, a work of John Smeaton (q.v.) destroyed in the winter of 1807, then extend the north pier and, finally, build the breakwater on the south side. All these works, on which he introduced the first steam dredging machine into Scotland, still stand substantially unaltered.

From that time on Gibb was perhaps Telford's most trusted associate on canal, bridge, lighthouse and harbour works in Scotland. He removed to Aberdeen from Greenock in 1809 taking with him his wife and six-year-old son, Alexander. The family settled in a modest house at Footdee and John's only daughter, Janet, was born there in 1814. Alexander studied at the Grammar School and then at Marischal College in Aberdeen before enrolling as an apprentice with Telford in London.

Meanwhile his father carried out repairs to the Crinan Canal in 1817, completing a year's contract in nine months to the great satisfaction of Telford and the Parliamentary Commissioners for the improvement of the canal. In 1819 John Gibb 'of Aberdeen' became the first Corresponding Member of the Institution of Civil Engineers which had been founded in London in 1818. At that time he was supervising a number of harbour projects in north-east Scotland for Telford. For the next thirty years he worked closely with Telford, Robert Stevenson (q.v.) and other engineers, initially as a resident engineer and later mainly as a contractor, after about 1825 with his son, Alexander, as his partner. Although generally active in

Scotland he assisted Telford with surveys of the Dee at Chester, and of the Nene Outfall in the early 1820's.

At some time between 1825 and 1830 the Gibbs left Footdee and moved to 8 Canal Terrace, Aberdeen, then a charming little secluded row of three-storey Georgian houses. On 20 February 1845 Katherine Gibb died at Grahamston, Falkirk, at the age of sixty-one; she was buried in the Parish Churchyard there. She had not been in good health for some time, suffering from 'nervous depression'.

The poet, Southey, accompanied Telford and Gibb in 1819 on part of a tour of north-east Scotland and described how 'At Cullen we took leave of that obliging, good-natured, useful and skilful man, Mr. Gibb'. Apparently John Gibb, although in other circumstances he could display an irascible temper, was also well liked by his workmen. When building the Dean Bridge he had an elevated perch made from which he could survey their progress and bellow criticisms through a sort of megaphone. On occasions he descended amongst them 'to exchange blows, banter and snuff' to the delight of all.

One of the most impressive works on which he was engaged with his son from 1833 to 1836 was Telford's Broomielaw (or Jamaica Street) Bridge in Glasgow, a seven-span segmental arch bridge with an overall length of 560 ft. and a carriageway no less than 60 ft. wide. He completed the four-year contract in two years and eight months in January 1836 at a cost of £34,427. Considered by some to have been Telford's most beautiful bridge, it unfortunately had to be demolished in 1899 to allow the river Clyde to be deepened. Its replacement was considered to be a replica.

His last important contract was for the Almond Viaduct on the Edinburgh & Glasgow Railway. It was an impressive structure with thirty-six arches of 50 ft. span, averaging 70 ft. in height, completed with extraordinary expedition in 1842, but unfortunately it involved the firm in a loss of about £40,000. A mistake had been made in tendering which he noticed but refused to amend because the tender had already been submitted.

Soon after its completion Gibb suffered a decline in health and retired from active employment. In his later years he had suffered from extreme deafness, an affliction he passed on to many of his descendants. He died at Aberdeen on 3 December 1850 and was buried in St. Nicholas Churchyard there. He was at the time of his death one of the oldest members of the Institution of Civil Engineers, in whose affairs he had always taken a lively interest, and by whose members he was held in high regard.

RON BIRSE

[Tidal Harbours Commission (1845–1847) *Reports* and *Appendices*; Obituary, *Min Proc ICE*, **10**, 1851, 82–85.; DNB; A. Gibb (1935) *The Story of Telford*; G. Harrison (1950) *Alexander Gibb: the Story of an Engineer*; L. M. Rae (1961) *The Story of the Gibbs*; J. Lindsay (1968) *The Canals of Scotland*; Skempton; W. T. Johnston (1998) *Scottish Engineers and Shipbuilders* (computer database)]

Publications

1838. Aberdeen Harbour, in: *Life of Telford*
1839. Observations on the Crinan Canal, in: *Report for the Select Committee on the Caledonian and Crinan Canals*

Works

1801–1805. Leith, Old East Dock, contractor under Easton
1805–1809. Greenock New Harbour, contractor under Brocket
1809–1816. Aberdeen Harbour, reconstruction, resident engineer under Telford, £100,000
1811–1814. Potarch Bridge, Aberdeenshire, over the Dee, resident engineer, £20,000
1816–1822. Peterhead North Harbour, detailed designs, for Telford, supervised construction 1818–1822, £27,000, Gibb was paid a percentage fee of c. £1,800.
1817, 1835. Crinan Canal reconstruction, contractor under Telford
c. 1818. Sheerness Dockyard, supplier of granite from Aberdeenshire quarries, for Jolliffe and Banks, contractors (see: *BANKS, Edward*)
1818–1821. Harbours at Cullen, Banff and Nairn, resident engineer under Telford
1821–1822. Lanarkshire, Cartland Crags and other bridges, Gibb surety for Minto (q.v.) as contractor under Telford
1823–1826. Rhinns Of Islay Lighthouse, John and Alexander Gibb, contractors under Robert Stevenson, £9,000
1825–1827. Buchan Ness Lighthouse, John and Alexander Gibb, contractors under Robert Stevenson
1825–1828. Cape Wrath Lighthouse, John and Alexander Gibb, contractors under Robert Stevenson, £14,000
1827. Duror Church, with Minto (q.v.)
1827–1830. Don Bridge, Aberdeen, John and Alexander Gibb, contractors under Telford, much of the bridge had to be taken down and rebuilt in 1828–1829
1829–. Aberdeen Harbour Wet Dock and other works, John and Alexander Gibb, contractors under Telford, Gibb was also resident engineer to the harbour board
1829–1830. Suspension Bridges at Boat O' Brig and Findhorn, resident engineer, under Captain Samuel Brown, RN (q.v.)
1829–1832. Edinburgh, Dean Bridge, John and Alexander Gibb, contractors under Telford, £34,000
1830. Dundee Harbour improvements, design, Telford consultant, James Leslie (q.v.) resident engineer
1833. Girdle Ness Lighthouse, John and Alexander Gibb, contractors under Robert Stevenson
1833–1836. Glasgow, Broomielaw (or Jamaica Street) Bridge, John and Alexander Gibb,

contractors under Telford, a four-year contract, completed in two years and eight months
1836. Ayr Harbour breakwater, design
1836. Penshaw, Victoria Bridge over the Wear, John and Alexander Gibb, contractors under Thomas Harrison
1836. Lossiemouth and Elgin Harbours, consulting engineer
1839–1842. Kirkliston, Almond Valley Viaduct on Edinburgh & Glasgow Railway, John and Alexander Gibb, contractors under John Miller
1842. Wick Harbour, consulting engineer

GIBB, William (1736–1791), civil engineering contractor, was born on 2 June 1736 at Bo'ness in West Lothian, son of William Gib, the last of several generations of maltmen in that town, and his wife Janet Ker. He was the first 'civil engineer' in the Gibb family. William's father died at the age of thirty-one when he was six years old and soon afterwards his mother left Bo'ness and settled in Falkirk where in due course he was apprenticed to a stone-mason. He is recorded as having been admitted to the Brotherhood of Masons at Falkirk in September 1757, at which time he was living at 'Sunnyside', a farm on the outskirts of Falkirk.

On 27 November 1760 he married Alison Galbreath, daughter of William Galbreath of Larbert, and it was at about this time that the spelling of the family name was changed from Gib to Gibb. Over the next few years he is known to have been responsible, along with John Moir, a relation of his wife's, for the construction of many engineering works of some local importance, including the Kirkintilloch Aqueduct on the Forth & Clyde Canal, completed about 1773. Of their other works at this time little evidence remains apart from a stone carved with 'WG-JM' and the date '1775' from a bridge that spanned the river Carron near the famous iron-works.

In August 1772 he purchased a small mansion-house in its own grounds on the outskirts of Falkirk, known as 'Gartcows'. By that time he and his wife had three surviving children (out of six born between 1761 and 1771), and the last of their family, John (q.v.), was born there in 1776. Between 1775 and 1780 William, no doubt mindful of the family tradition, established a malthouse and distillery at Gartcows, but it was for him a financial disaster. In 1785 the house had to be sold along with the malthouse and distillery in order to meet his creditors and he was left with very little to show for all his work up to that time.

His partner John Moir stood by him, however, and when tenders were invited for the Kelvin Aqueduct on the Forth & Clyde Canal theirs was accepted. The foundation stone was laid on 15 June 1787 and they were contracted to complete it before the end of 1789. The aqueduct consisted of four masonry arches of 50 ft. span giving it a total length of 350 ft., a width of 57 ft. and a maximum height of 83 ft. above the bed of the river. Their work was, as always, of a very high standard, and the aqueduct survives in use today,

but the contract involved Gibb and Moir in a loss of £2,600, and it was not finally completed until 1791.

William Gibb died suddenly of a seizure, in Glasgow, on 10 January 1791.

RON BIRSE

[G. Harrison (1950) *Alexander Gibb: the Story of an Engineer*; L. M. Rae (1961) *The Story of the Gibbs*; J. Lindsay (1968) *The Canals of Scotland*]

Works
Known works, in association with John Moir:

c. 1773. Kirkintilloch Aqueduct on the Forth & Clyde Canal
1775–1777. Masonry bridge over the River Carron, Stirlingshire
1787–1791. Kelvin Aqueduct with associated locks and ship basin in Maryhill, Glasgow, on the Forth & Clyde Canal

GIFFORD, William (fl. 1827–1836), of 5 Great St. Helens, City of London, was elected an Associate of the Institution of Civil Engineers in 1827. He was responsible for the drawings of Telford's Dean Bridge, but no more is known about him.

MIKE CHRIMES

[ICE membership records, ICE archives; W. Gittin (1836) Inventory of Telford drawings, ICE archives]

GILBERT, Davies Giddy, FRS (1767–1839), mathematician and natural philosopher, was born at Tredrea, St. Erth, Cornwall on 6 March 1767, the son of the Rev. Edward Giddy and Catherine Davies. He was first educated at home, largely supervised by his father. He later attended Penzance Grammar School until, in 1782, the family moved to Bristol. In 1785 he entered Pembroke College, Oxford, where the Master described him as 'the Cornish Philosopher'. Giddy studied a range of subjects including anatomy, mineralogy, chemistry and botany. Whilst at Oxford he was elected FRS in 1791, and returned to Cornwall in 1793. In 1804 he was elected MP for Helston, and in 1806 for Bodmin, which he represented until 1832. In Parliament ha became an ardent representative of scientific and technical interests. In 1808 he married Mary Ann Gilbert of Eastbourne, the niece of Charles Gilbert from whom he inherited large estates in Sussex in 1816, and as a condition of inheritance he then changed his surname to Gilbert.

His involvement with early nineteenth century mechanical engineering was notably in conjunction with his fellow-Cornishman Richard Trevithick (q.v.), who always referred to him as Davies Giddy. Trevithick regularly consulted Gilbert over many years, saying that his 'theory on every occasion I have relied on, and found by practice to be correct'. Gilbert advised Trevithick on the mathematical and theoretical aspects of his design work on boilers, locomotives, steam navigation and many other topics, carrying out engine trials and giving evidence in connection with the Boulton & Watt lawsuits in 1798.

In civil engineering Gilbert is perhaps best known for his mathematical work on the complex properties of the catenary. His equations, or more probably, his simplifying tables, brought the analysis of the loaded catenary chain within the scope of most engineers. Gilbert's work influenced Thomas Telford (q.v.) in changes to the design of the Menai suspension bridge, and I. K. Brunel in the original design of the Clifton suspension bridge chains. Davies Gilbert was one of the two assessors of the Clifton bridge designs and he advised that 'all the parts' should be 'sufficiently divided so as to afford a power more than equal to supporting the bridge notwithstanding a failure in any one part, and also to afford ample means for replacing any part that has failed'. This advice was heeded by the young Brunel and subsequent generations of engineers.

Davies Gilbert was also involved with Committee for Rebuilding London Bridge, the Plymouth Breakwater and he was a Holyhead Road Commissioner. Gilbert was elected an Honorary Member of the Society of Civil Engineers (Smeatonian) in 1811, when he was living in Hollis Street close to Drury Lane. He was awarded an honorary DCL by the University of Oxford.

Gilbert retired to his house in Eastbourne in November 1839, where he died on 24 December of the same year.

DENIS SMITH

[W. Walker (1862) *Memoirs of the Distinguished Men of Science of Great Britain Living in the Years 1807–1808*, 79–82; F. Trevithick (1872) *Life of Richard Trevithick with and Account of his* Inventions; H. W. Dickinson and A. Titley (1934) *Life of Richard Trevithick: The Engineer and the Man*]

Publications

1805. Letter describing a singular fact of the invisible emission of Steam and Smoke together from the chimney of a furnace, though either of them, if separately emitted, is visible as usual, *Nicholson Journal*, **XII**, 1–2

1821. On some properties of the catenarian curve with reference to bridges by suspension, *Quarterly Journal of Science, Royal Institution*, 230–235

1822. On the ventilation of rooms, and on the ascent of headed gases through flues, *Quarterly Journal of Science*, **XIII**, 113 120

1823. An investigation of the methods used for approximating to the roots of adfected equations, *Quarterly Journal of Science*, **XIV**, 353–358

1823. On the vibrations of heavy bodies in cycloidal and in circular arches, as compared with their descents through free space, *Quarterly Journal of Science*, **XV**, 90–103; **XX**, 1826, 69–73

1825. On the general nature and advantages of wheels and springs for carriages, the draft of cattle, and the form of roads, *Quarterly Journal of Science*, **XVIII**, 95–98

1826. On the mathematical theory of suspension bridges with tables etc., *Phil. Trans.*, 202–218 [reproduced in W. A. Provis (1828) (q.v.)]

1827. On the expediency of assigning specific names to all such functions of simple elements as represent definite physical properties; ... also by some observations on the Steam Engine, *Phil. Trans.*, 25–38

1828. On the regular or Platonic solids, *Phil Mag*, **III**, 161–165

1830. On the progressive improvements made in the efficiency of steam engines in Cornwall, *Phil Trans*, 121–132

1830. On the nature of negative and of imaginary quantities, *Phil Trans*, 91–98

1831. A Table for facilitating the computations relative to suspension bridges, *Phil Trans*, 341–344

1836. Description of an improved mode of water tanks, *Polytechn Soc Trans*, 79–80

GILES, Francis John William Thomas (1787–1847), civil engineer, was born on 10 October 1787 and baptised on 21 October 1787 at Walton-on-Thames, the son of Samuel and Mary Giles. He came from a large family, was trained as a surveyor by his brother Netlam (q.v.) from c. 1803 and, after six years pupillage, became his partner. In his early career, like Netlam, he worked for John Rennie (q.v.) on a number of important civil engineering surveys.

Giles's first job was to assist in the survey of the London and Portsmouth Canal in 1803. This was followed by work on the Weald of Kent Canal survey (1803), Dymchurch Wall survey (1804), Shorncliffe Canal survey (1804), the Royal Military Canal (1804–1805), Dover Harbour (1805), the proposed Loch Earn–Perth Canal (1806), Birmingham Canal improvements (1807), and a proposed canal between Basingstoke and Winchester. Giles also carried out surveys of Ramsgate, Rye (1804–1805) and Margate Harbours for Rennie, together with Dublin Bay (1815–1816) and Wexford (1814–1815). He carried out surveys of the river Thames in connection with Vauxhall, Waterloo and Southwark Bridges and soundings under the bridges as far up river as far as Westminster and was responsible for setting out the foundations for Waterloo Bridge. Much of his early work was carried out jointly with his brother, such as their surveys of the Seaton-Beer Bay (1810), Weald Canal (1811), Bristol Harbour (1811), London and Cambridge Junction Canal (1811), Bridport Harbour (1809) and the Arundel and Portsmouth Canal (1815). After his brother's death he carried out survey work for Rennie on the Mersey (1820–1821) and at Holyhead (1821) and in the North East he produced detailed surveys of the river Wear (1819–1822). This appointment disappointed James Mills (q.v.) who had hoped to get the work.

Giles was entrusted with training Sir John Rennie (q.v.) on surveying, Rennie Jr. assisting him in the surveys of the coasts of south west

Scotland and Ulster (1814–1815) to identify suitable sites for artificial harbours, involving surveying the ports themselves as well as preparing charts of the intervening channels. The rugged terrain made it difficult to establish a base line and it is a measure of his ability that Giles overcame all difficulties. The younger Rennie also assisted in the surveys of the Tyne and Shields harbours (1813), Eastern Extension of the Kennet and Avon Canal, Stockton and Darlington Canal (1813) and the Thames, near Woolwich dockyard (1813–1814). Later, when in practice himself, Rennie employed Giles to survey the Witham in 1827 in connection with his report on the Witham outfall for Boston Corporation and he also assisted him with work on the Ancholme navigation where, together, they explored three lines, Giles taking the levels and drawing up the surveys, for a canal 100 yd. wide, 24 ft. deep and costing £7 million. Giles, around 1826, also surveyed Cundy's (q.v.) proposed line for a London to Portsmouth Canal on Rennie's behalf.

Following Rennie Sr.'s, death in 1821 Giles embarked on an independent career with offices in Salisbury Street, Adelphi, initially as a civil engineering surveyor, and carrying out surveys of the Thames in connection with London Bridge, on which he gave evidence in 1821. His first independent commission was the survey for the Ivel Navigation extension to Shefford (1821), having already worked for Rennie on estimates for an extension to Langford in 1819. In 1822 he was working on Rennie's plans for the improvements to the Aire and Calder Navigation.

Giles's early works were almost exclusively canal proposals. They included the Hertford (Lea) Union Canal (1823–1824, 1830), the Basingstoke Canal extension to Newbury (1825) and, later, the extension of the Sankey Brook or St. Helens Canal (1830–1833) to Widnes. Giles was also engineer for some schemes which were never built, including work for Sir George Duckett, a proposal for an aqueduct across the Mersey at Runcorn (1825), the 'Union' Canal (1830) and the Hampshire and Berkshire Junction Canal (1824). Perhaps his last canal work was to report on improvements to the Lee Navigation (1844), carried out following the Act of 1850.

In the 1820s Giles carried out a number of coastal and harbour surveys and works, building on the investigations he and his brother had carried out for Rennie. At Bridport, Netlam had carried out a thorough survey for Rennie and Francis carried out a number of improvements to the piers which had largely been unimproved since John Reynolds (q.v.) work of a nearly a century earlier; a workforce of over three-hundred men was involved. At Dublin, Giles was responsible, with George Halpin (q.v.) for the extension of the North Bull Wall. Throughout the 1820s he was employed by Liverpool Corporation, surveying the Mersey and investigating the sea approaches to Liverpool. He was responsible for the sea embankment at Leasowe, and similar

works in Lincolnshire and in 1831–1832 he was involved in the Sunderland Docks controversy, working with Isambard Kingdom Brunel. In the 1830s he completed Courtown Harbour, originally designed by Alexander Nimmo (q.v.). Here, problems were occurring with the south pier and it was decided to rebuild it parallel to the north pier, deepen the harbour and provide quays, sluice gates, and an overfall. Giles directed the Owenavorragh River into the harbour by an artificial cut and employed the Breanoge river to scour the harbour on the ebb tide. In 1825 he was appointed County Surveyor to Bedfordshire, a post he retained until his death, and in addition to his responsibilities for roads and bridges in the county, he also designed road bridges in Dorset and Cumberland.

When the Thames Tunnel was proposed, Giles was employed by the City Corporation to carry out soundings of the river along the proposed alignment. He was called in again by some of the directors of the Company who were losing confidence in M. I. Brunel (q.v.) following the first major irruption into the Tunnel workings in May 1827. He then carried out three borings, one of which he claimed to reveal quicksand, but it transpired that there were problems with the sampling methods, so revealing the limitations of many site investigations of the time. Giles was one of several engineers mooted to take over from Brunel when progress on the Tunnel became difficult but his own workload makes it unlikely that he would have been able to take on such a challenge.

With his substantial experience Giles inevitably obtained work on the early railways. He gave evidence against George Stephenson's (q.v.) Liverpool and Manchester Railway proposals in 1825, criticising the proposed route across Chat Moss. He carried out, possibly in c. 1824, an early survey of the London–Birmingham railway route which followed a line similar to that eventually adopted by Robert Stephenson (q.v.). He then became involved in work on the Newcastle–Carlisle Railway, originally proposed by Benjamin Thompson (q.v.) in 1828. Following the passage of the Act the Company requested an independent survey, which Giles supplied, and on 16 October 1829 he was appointed engineer.

This can probably be regarded as his engineering triumph, the line including the 102ft deep Cowran cutting and impressive bridges across the Eden, Gelt and Corby Beck. When Thomas Telford (q.v.) was asked to advise the Exchequer Loan Commissioners in 1832 he supported Giles's proposals and his report on the line in 1834 must have been among his last. Giles's printed specifications for works on the line may be the first such documentation for railway contracts (1830–1832). Giles became consultant to the Company from July 1833, before completion of the work, as his other commitments prevented him from devoting as much time to the work as the Company required. His principal assistant since the late

1820s, John Blackmore (1801–1844), remained as engineer.

Giles other railway work was less successful. In 1830 a proposal was advanced for a railway from London to Southampton and a preliminary survey was carried out by Doswell. One of the supporters, Lieutenant Colonel George Henderson (RE) was not very impressed and on the recommendation of the Royal Engineers it was decided to approach Giles who knew the area from his work on the Basingstoke Canal. He carried out his first survey in 1830 and was appointed Engineer the following year. He drew up plans and estimated the costs at £1,033,414 but in the event no progress was made until 1833 when a further survey was carried out; the Bill was passed in 1834. Henderson was appointed general superintendent of works at a salary of £1,000 p.a. and Giles was appointed Engineer on 1 September at an annual salary of £1,500 p.a., plus £500 expenses.

Progress on the Southampton line was mixed, and Giles's preferred method of contract procurement and supervision was investigated at length when he gave evidence in support of the Basing–Bath Railway, intended to connect with what became the London and South Western Railway, and against the Great Western Railway, in 1835. Giles preferred a system of advances to smaller contractors, who were advanced part of the contract sum to meet initial capital expenses, rather than dealing with large 'capitalists' who would not be so dependent upon the Railway Company if there was a contractual dispute. This said, two of the contractors used, David McIntosh and Francis Treadwell, were contractors of some reputation and when conflicts arose between Giles and the Company it is likely that there were other concerns at work. These included lack of progress partly caused by factors beyond Giles's control, such as delays in getting land, and rising costs caused by the railway boom. Another likely factor was that Giles was, again, so heavily committed that he was unable to devote a satisfactory amount of time to the project. A group of shareholders, most notably those from Lancashire, pushed for another engineer, specifically Joseph Locke, to report on progress and on 13 January 1837 Giles resigned on six months notice. It was intended he should help the Company to get a Supplementary Act enabling it to raise more money, but in giving his evidence Giles was critical of the intended method of raising the funds, which must have made the Company more determined in its resolve to dispense with his services.

Giles's commitments in the 1830s obliged him to take his son Alfred (b. 1816) out of education at the age of seventeen to help his already considerable practice, much of which was railway survey work. It included schemes associated with the London and South Western Railway, such as the Portsmouth Junction Railway (1836), and Basing–Bath Railway, an early survey of the London–Greenwich scheme (1832), and reports on the Great North of England Railway (1836), and the Chester–Holyhead line (1839); some of his lines were never built. He also carried out surveys on behalf of Sir John Rennie for a line from London to Brighton in 1833 and his proposed South Wales Railway (1845) led to him developing a design for a cable stayed railway bridge across the Bristol Channel at Aust Passage, with four main spans of 1,100 ft. and two side spans of 550 ft.

The London–Southampton line was linked with the development of Southampton as a port, and Giles drew up plans for the development of the docks. Work began in 1838 but financial problems meant that little was achieved until 1840 and the scheme was largely carried out by Alfred Giles. Giles's reputation as an engineer has suffered from his replacement as the engineer on the Southampton line, an event regarded as a triumph for Joseph Locke and Thomas Brassey (qq.v.). It should be remembered that the route was selected by Giles and on a cost per mile basis it was the cheapest of the early lines, carefully avoiding unnecessary gradients; Locke may have benefited from Giles's skills as a surveyor. His opposition to the Liverpool–Manchester railway proposal of George Stephenson may have played a part in the diminution of his reputation as it brought him into conflict with the rising star of the Stephensons and their circle. A circular, heavily critical of Giles's fees for his proposed Chester–Liverpool railway (£1,500) was in circulation in 1831 and Robert Stephenson resisted attempts to have Giles involved in the London–Birmingham railway at that time.

Giles had other interests outside civil engineering. He and Netlam published a map of Leeds in 1815 and they were contemplating a map of Lancashire at the time of Netlam's death. Giles was a shareholder in the Birmingham Canal and had mining and property interests at Dudley Port, near Tipton, Staffordshire, which he shared with Netlam's son Frederick. As civil engineering surveyors, Giles and his brother, Netlam, enjoyed a high reputation and the continuing confidence of John Rennie. It is apparent that at an early stage Netlam set up his own office, independently of Rennie, and they had their own staff and assistants. Of these the earliest known, and one of the most loyal, was Alexander Comrie (1786–1855), who first met Netlam in 1803 and was persuaded by Francis to work with them, which he did until c. 1824; his last work for them was possibly on the Liverpool–Manchester Railway survey. He appears to have been succeeded as principal assistant by Blackmore, who was working for Giles on the Lea in the late 1820s, before becoming his resident engineer on the Newcastle and Carlisle Railway. Other Assistants included Thomas Dodd, designing the Newcastle–Carlisle Railway bridges for Giles, and William Lindley (1808–1900) who became Giles's pupil in 1827 and worked on the Newcastle and Carlisle and the London–Southampton railways.

On Netlam's death Giles took over their practice and trained his brother George's sons, Francis (1806–1884) and George (1810–1877), as engineers. In due course Giles's own sons, Francis George (b. 1815) and Alfred (1816–1895), joined him. Alfred became President of the Institution of Civil Engineers and William Lindley became one of the most important British engineers working in Europe, a distinguished legacy for Francis to have trained. Giles died on 4 March 1847 at his home at (9?) Adelphi Terrace, leaving his estate to his wife Mary Ann, *née* Wyer, whom he had married at St. Martin's, Birmingham on 14 April 1814. They had several young children in addition to the two engineer sons, *viz.*: Mary Ann, Sarah Wyer, Maria Eliza, Edwin, Helen and Toby. Giles is buried in Kensal Green Cemetery.

MIKE CHRIMES

[Probate 11/2059/574 Middlesex, July 1847; Lindley notebooks and correspondence, West Yorks RO; Rennie papers, National Library of Scotland; Rennie Reports, ICE archives; PLA Archives, Museum of London; Hekekyan papers, vol. 1, BM mss. 37448; Q/RUm/54 Q/RUm/62, Staffs RO; House of Lords, Select Committee … London Bridge (1821) *Minutes of Evidence*; Select Committee … Liverpool and Manchester Railway (1825) *Minutes of Evidence* …; Select Committee London and Southampton Railway (1832) *Minutes of Evidence*; Select Committee Great Western Railway (1835) *Minutes of Evidence*; *Min Proc ICE*, **9**, 1848, 8–9; S. Smiles, *Lives of the Engineers*; DNB (1889); Hadfield, Canals, *passim*; P. S. Richards (1965) A geographical analysis of some of the surveys made for the London and Birmingham railway line, *Trans Birmingham Archaeological Soc*, **80**, 17–25; W. O. Skeat (1973) *George Stephenson and his Letters*, 140–141; P. L. Bell (1985) Surveying the country, *Bedfordshire Magazine*, **20**(153), 3–10; R. Zelichowski (1997) *Lindley owie: saga pionierow higeny*; R. W. Rennison (2001) The engineers of the Newcastle and Carlisle Railway, *Trans Newc Soc*, preprint]

Publications
Publications marked with an asterisk are for John Rennie.

1810. *Plan of Seaton and Beer Bay* (with Netlam)*
1811. *Plan of the Harbour and Part of the Town of Boston* (with Netlam)*
1811. *Plan of the Proposed London and Cambridge Canal* (with Netlam)*
1811. *Plan of the Proposed Weald of Kent Canal* (with Netlam)*
1815. *Plan of Leeds* (with Netlam)
1817. *Plan of the Intended Arundel and Portsmouth Canal … 1815* (with Netlam)*
1822. *Map and Chart of the River Witham and Boston Harbour*
1824. *Plan of the proposed Berkshire and Hampshire Junction Canal from the River Kennet at Midgham to the Basingstoke Canal at Basing*

1825. *Plan of the Stone–Baswich link with Staffordshire and Worcestershire Canal*
1826. *Map and Survey of the River Wear, 1819–1822*
1830. *Report on the Parliamentary line of Railway from Newcastle to Carlisle … 1829*
1830. *Second Report on the Line of Railway, from Newcastle to Carlisle … 1829*
1830. *Newcastle-upon-Tyne and Carlisle Railway. Report upon the Comparative Qualities of a Line between Scotswood and Gaw Cook Mill*
1830. *A Map of the Hundred of Wirral in Chester on which are delineated the two Lines of Railway*
1832. *Plan of the Entrance to the Port of Sunderland … proposed Dock on the South Side according to the Design of F. Giles (1831–1832)*
1832. *Estimate of the probable Expense of … a Double Railway … from … Southampton to Vauxhall*
1832. *Southampton and London Railway*
1833. *Plan of the Intended London and Southampton Railway*
1834. *Railway from Basing, in the County of Hampshire, to Bath* (with W. Brunton)
1834. *South Midland Counties Railway*
1834. *Plan der Intendirten Eisenbahn zwischen Altona, Hamburg, and Lubeck*
1836. *Plan of the Intended Portsmouth Junction Railway*
1836. *Plan and Sections of the intended Alterations and Denotations from the original Plan of the London and Southampton Railway*
1836. *Plan of the intended South Midland Railway*
1841. *Calcutta and Diamond Harbour Railway and Dock* (with J. Vetch)
n.d. *Proposed Dock at Thames Haven*

Works
The following exclude surveys for John Rennie.

1806. Lady Perth's pleasure pond, Resident Engineer, for Netlam
1821–1823. Ivel navigation, five locks, £7,700, Engineer
1821–1825. North Bull Wall, Dublin, Engineer, with G. Halpin, 7,500 ft.
1822–(1829). Aire and Calder navigation improvements, Engineer
1823. Bridport Harbour reconstruction, Engineer
1823–1824. Hertford Union Canal survey
1825. Bromham Bridge repairs, Bedfordshire
1825–(1829). Basingstoke Canal extension, survey
1828–1829. Wallasey (Leasowe) embankment, Engineer
1829. River Conway improvements
1829. Rye Harbour, Engineer
1829–1830. Hertford Union Canal, 1¼ miles, Engineer
1829–1833. Newcastle–Carlisle Railway, Engineer
1830. Birmingham–Warwick Junction canal survey
1830–1833. Sankey Brook extension, Engineer

1830–1837. London–Southampton Railway, Engineer
1832. London–Greenwich Railway survey
1833–1835. Warwick Bridge, Cumberland, Engineer
1834–1847. Courtown Harbour, Engineer
1838–1846. Southampton Docks, Engineer
1846. Reading–Reigate Railway, Engineer

Undated works:

n.d. Hayward Bridge, Dorset, Engineer
n.d. Grand Trunk Canal
n.d. Blyth Harbour
n.d. Pembroke
n.d. New Quay, Cardigan Bay
n.d. River Medway
n.d. Waterbeach Fen
n.d. Freiston embankment, Lincolnshire
n.d. Boston Harbour
n.d. Harrold Bridge, Bedfordshire
n.d. Luton Bridge, Bedfordshire

GILES, Netlam (c. 1775–1816), civil engineering surveyor, a son of John and Mary Giles, was baptised at Walton-on-Thames on 31 December 1775. He was one of a large family, having five brothers: George, Edward, Peter, Samuel and Francis (q.v.), and two sisters, Frances (Larner) and Mary (Willis).

Giles is best known for his work for John Rennie (q.v.). Their connection began early in the nineteenth century and Rennie used Giles to carry out civil engineering surveys on his behalf. It would appear that generally Giles was paid by the client rather than by Rennie himself and from 1804 Giles is known to have had offices at 27 Upper King Street, Bloomsbury. Here he took on his younger brother, Francis, as his pupil c. 1803 and shortly afterwards Alexander Comrie (1786–1855), whom he first employed as a field assistant on works in Scotland; Comrie remained with the brothers until c. 1825. After serving six years pupillage Francis became his partner c. 1809. At the time of Netlam's death they also employed another assistant, James Brazier.

Netlam Giles is known to have been in practice on his own account from c. 1799, when he purchased a share in the survey of Surrey by Joseph Lindley and William Crosley I (q.v.); Lindley, if not already a family friend, became one. Most of Giles's known work was, however, due to Rennie. His earliest known surveys relate to the Portsmouth and Weald of Kent canal (1802) and a report on Dymchurch Sea Wall (1804) and the Portsmouth Canal proposal of 1802–1803.

Around 1806 he was responsible for the supervision of the construction of an ornamental lake for Lady Perth, involving an embankment dam, sluice and associated buildings. At Bridport Harbour, although Rennie was the engineer, it was Giles who reported on the failure of recent works in 1809, and recommended to Rennie what should be done to improve the sluices and the east pier; he also surveyed a proposed canal route and measured the tidal range. In 1812

he was working on a survey of a sewer in Baker Street, London, which had been adversely affected by the Regent's Canal works. His overall responsibility for many of the surveys in which his brother was involved is likely, if uncertain.

In 1815 he produced, with Francis, a *Plan of the Town of Leeds and its Environs* at a scale of 1 in. to 550 ft. The following year they were involved in a proposal for a new map of Lancashire at a scale of 1 in. to a mile and were seeking subscribers when Netlam's premature death halted work.

His last civil engineering works were surveys of various harbours and channels around the coast of Ireland. Following Rennie's appointment as Engineer at Dun Laoghaire in October 1815, Giles carried out surveys the following winter and spring. This work followed surveys he and his brother carried out in the North Channel and along the coasts of Ulster and South West Scotland for suitable harbour accommodation to improve communications between Ireland and Scotland. This work was undertaken in the spring and early summer of 1815 with (Sir) John Rennie as their assistant, together with Alexander Comrie and James Brazier.

Netlam Giles died on 26 December 1816. He had three children, Frances (b. c. 1811), Frederick and, tragically, a posthumous son, Netlam John. His brother Francis, whose career overshadowed that of his elder brother, seems to have assumed responsibility for them, and was a partner, with Frederick, in mines at Dudley Port, Staffordshire.

MIKE CHRIMES

[Probate 11/1592/244 Effingham, 1817; Rennie Reports, ICE Archives; Rennie papers (National Library of Scotland); Hadfield, *passim*; Skempton; Keystone (in lit.); *Gentleman's Magazine*, 1816, December, 627; J. B. Harley (1964) A proposed survey of Lancashire by Francis and Netlam Giles, *Trans Hist Soc, Lancashire and Cheshire*, **116**, 197–206; J. B. Harley (1966) English counties map marking in the early years of the Ordnance Survey: the map of Surrey by Joseph Lindley and William Crosley, *Geographical Journal*, **132**, 372–378]

Publications
Publications marked with an asterisk are for John Rennie.

1010. *Survey ... of Seaton and Beer Bay* (with Francis)*
1811. *London to Cambridge Junction Canal* (with Francis)*
1815. *Weald of Kent Canal* (with Francis)*
1815. *Arundel and Portsmouth Canal* (with Francis)*
1815. *Plan and Sections of the Reservoirs and Feeders in the Parishes of King's Norton and Northfield ... for the Use of the Worcester and Birmingham Canal and Mills on the River Rea*
1815. *Plan of the Town of Leeds and its Environs*

Works

The following do not include printed surveys. Works marked with an asterisk are for John Rennie.

1802–1803. Weald of Kent, survey*
1803. London–Portsmouth Canal, survey*
1804–1805. Dymchurch Wall, plans and sections*
1804. Shorncliffe Canal, survey*
1804–1805. Royal Military Canal, survey*
1804–1805. Rye Harbour
1805. Dover Harbour, survey*
1805–1806. Loch-Earn–Perth Canal, survey*
1806. Lady Perth's ornamental lake, engineer
1807. Birmingham Canal improvements, surveys*
1809. Bridport Harbour surveys and east pier improvements, £164 12s*
1812. Baker Street sewer collapse*
1815. North Channel soundings and tidal current surveys*
1815–1816. Dun Laoghaire railroad, survey
1815–1816. Dublin Bay, survey, £300.00*

GLYNN, Joseph, FRS (1799–1863), civil and mechanical engineer, was born on 6 February 1799 at Hanover Square, Newcastle-upon-Tyne, one of the sons of James Glynn of the Ouseburn Iron Foundry. He was educated at the Percy Street Academy where Robert Stephenson, four years younger, was later a pupil. As a schoolboy he was severely burned by an explosion of gunpowder on board one of his father's ships. His convalescent trips at sea gave him a valuable knowledge—later to be used professionally—of ship construction.

At first he worked as his father's assistant at the foundry and in 1820 he erected a steam engine at Talkin colliery in Cumberland. The following year he designed and built the gas works in Berwick-upon-Tweed, which served that town for nearly forty years without alteration. He also designed gas works for Aberdeen but they were constructed by others as he had meanwhile joined the Butterley Company in Derbyshire, with whom much of his important engineering work was done.

At that time steam power in ships was in its infancy but during the next twenty-five years Glynn developed engine size to two hundred horsepower. However his name is most widely remembered for his work in fen drainage, providing engines of up to eighty horsepower to drain lands in all of the eastern counties of England from Yorkshire to Suffolk, as well as in The Netherlands and Germany. In British Guiana (now Guyana) he designed a scheme which could be used for drainage in the wet season and irrigation in the dry. During this period the Butterley Company supplied much machinery, mills and iron roofs from Glynn's design to three continents. Butterley had already cast some major iron bridges and Glynn built several more, including a bascule bridge for the railway at Selby; all of them have now been replaced.

As a practising engineer Glynn had little to do with railways but he was involved in the direction of some, becoming chairman of the Eastern Counties Railway, forerunner of the Great Eastern Railway, shortly after the downfall of George Hudson in 1849.

Glynn was elected a Member of the Institution of Civil Engineers in 1828 and contributed several papers to its meetings; for one of these he was awarded the Institution's Telford Medal. He became a Member of the Society of Arts in 1836 and subsequently a Vice-President and in 1838 he was elected a Fellow of the Royal Society. He died in London on 6 February 1863. A Full account of his career will follow in volume 2.

P. S. M. CROSS-RUDKIN

[Obituary, *Min Proc ICE*, 1863; Philip Riden (1990) *The Butterley Company 1790–1830*]

Publications

1836. Draining land by steam power, *Trans Soc Arts*, **51**
Others also published after 1830

GOLBORNE, James (1746–1819), civil engineer, was born in 1746, the nephew of John Golborne (q.v.) and probably the son of John's elder brother James (b. 1716). Writing to the Earl of Hardwicke in 1791 concerning the possibility of a job as Agent of his estate in Cambridgeshire, he says 'I am a Cheshire man, my Lord, who after seven years Apprenticeship to my Uncle Mr. John Golborne of Chester (deceased) travelled with him in various parts of these three Kingdoms, to assist him in executing works in the Engineering line, which he had undertaken in such places; until March 1776, when I was recommended to the Honourable Corporation of Bedford Level, as Superintendent General of all their Works, in which Employ I have continued ever since'.

His apprenticeship began in 1760 when John Golborne was engineer of the River Dee Company with responsibility for maintaining the 8 miles ship canal (completed 1740) from Chester to Connah's Quay and reclaiming land in the upper parts of the Dee estuary. Of the 'works in the Engineering line' the most important was the deepening of the river Clyde for 14 miles below Glasgow, a very successful scheme devised and directed by John Golborne and carried out 1770–1775 with James as resident engineer at a salary of £100 p.a. On its completion in December 1775 he received a gratuity of £100.

Three months later he arrived at Ely to take up the post of Superintendent General of Bedford Level, in succession to Richard Dunthorne (q.v.), though this seems at first to have been a provisional appointment confirmed in 1778 with a salary of £200 p.a. In 1782 he married Ann Marshall of Ely and they had two daughters and a son, James (b. 1785). From 1790 he held the additional post of Receiver General at £80 p.a.

As Superintendent he directed maintenance works on the rivers, banks, drains and sluices in this vast area of fenland—over 300,000 acres—and submitted reports from time to time on new works. One of these, in 1784, concerned improvements of navigation in the 19 mile length of the river Nene between Peterborough and Wisbech. Another referred to a proposed extension of Kinderley's Cut, which had been completed in 1774 by Dunthorne from River's End (4½ miles below Wisbech) towards Peter's Point. Golborne also gave evidence in 1777 to a Parliamentary Committee on a Petition presented by Bedford Level Corporation for a Bill allowing additional powers to raise taxes and a scheme to improve the drainage of the South and Middle Levels. The Bill was dropped after its second reading, but gave rise to the famous report of December 1777 by John Golborne, in the preparation of which James took part in the field investigations.

His most significant contributions, however, were a report and subsequent involvement in plans to implement the Eau Brink Cut. Intended to improve the outfall of the Great Ouse, it was to run 2¾ miles from Eau Brink (a short distance below St. Germans) nearly to King's Lynn in order to bypass a long, wide and shallow bend in the river; the cut would increase the gradient and achieve a lowering of low-water level of at least 4 ft. as far upstream as Denver Sluice, with great benefit to drainage of the South and Middle Levels. The cut was proposed by Nathaniel Kinderley (q.v.) in 1721 and again by him, with more detail, in 1730. The idea was publicised in a book by his son in 1751 and further advocated with great authority by John Golborne in his 1777 report. Yet it met with strong opposition from the merchants of King's Lynn who feared that the consequent loss of tidal indraught would be disastrous for their harbour.

As a result nothing was done until 1791 when a 'Committee of Land Owners and others interested in the Improvement of the Outfall of the River Ouse' decided to take action. They called on John Watté (q.v.) to take levels and soundings along the river from St. Germans to the deep sea channel 2 miles below Lynn. He did so, and also produced a map, and reported in April giving reasons in favour of the cut. The Committee then asked Golborne to prepare estimates for the cost of the cut, of a bridge over it, and of the necessary re-alignment of drains from Norfolk Marshland. His report is dated 30 August 1791. He, too, gives good reasons in favour of the cut. He made borings to determine the nature of the soil (chiefly a firm clay) through which it would be made and took tidal observations, and measured the width of the river at several places upstream. From the latter he deduced that the cut should have a width of 208 ft. at the upper end increasing to 240 ft. at the lower end, and be 5 ft. deep at low water (and about 16 ft. below ground level); the banks, spaced well apart, should have a top width of at least 30 ft. With these dimensions the cut would cost £38,000 including the

diversion dam across the old river channel and £1,300 for the bridge, to which he added £2,000 for a new drain from Marshland and its outfall sluice.

In what should have been an end to the investigations, the consulting engineer, Robert Mylne (q.v.), reported in October, strongly supporting the case for the cut and indicating what consequential works might be required upstream of Denver. Instead, the affair dragged on for years. Four more engineering reports appeared in 1792–1793, two in favour and two against the proposal; John Golborne's 1777 report was reprinted and several pamphlets, mostly against the scheme, were published. A petition went to Parliament in February 1793. Mylne had meetings with Watté and Golborne at London and on site; the Bill was withdrawn; a new petition was submitted in January 1795; Golborne gave evidence, from which it transpired that he had made further borings in 1794; an Act for the cut passed in May 1795; taxes for financing the work began to be collected; in 1801 Golborne and Mylne, the latter now acting as engineer jointly with Sir Thomas Hyde Page (q.v.), revised the estimate for the cut to £40,000 with an increased cost of land purchase. After a lengthy dispute between Mylne and Page the width of cut was settled in 1804, Golborne again meeting with Mylne, this time at his home in Ely. The centre line and spacing of the banks were to remain exactly as set out in 1801, the width of the cut at the upper end was to be nearly as proposed by Golborne in 1791, increasing to 270 ft. at the lower end.

Golborne's duties as Superintendent were severely tested in January 1809 by the worst floods in living memory. The bank of the Hundred Foot or New Bedford River suffered four breaches totalling 200 yd. in length, the Old Bedford bank slipped in two places, and many other banks in the South Level suffered damage. Large areas of fenland lay under water. All this is described in letters by Golborne in the Hardwicke Papers; at the end of one of them he adds a note that 'harassed and fatigued in both mind and Body, I am going to Rest, having never sat down except in a boat, or been under cover for seven days'.

This disaster might well have been prevented had the Eau Brink Cut been made. But arrears in payment of taxes, coupled with a further large increase in the estimated cost of land purchase and huge expenditure (£12,000) already incurred on the 1795 Act, meant that insufficient funds were available for work to begin. However, the floods persuaded many hitherto reluctant landowners to support the scheme and a renewal Act was passed in August 1809. In that month John Rennie (q.v.) began the investigations which culminated in his report of August 1810 on drainage and navigation in the South and Middle Levels. Golborne accompanied him in viewing the Ouse from Ely to Lynn and the Nene outfall. By 1818 work could start on the Eau Brink Cut. It was completed under Rennie's direction in 1821 in accordance with the 1804 plans, and proved highly beneficial to drainage and to Lynn harbour.

Although his later career was centred on the Fens, he was responsible for work elsewhere, including an embankment scheme at Malltraeth, at the mouth of the Afon Cefni in Anglesey. He was approached about this in 1781, and drew up plans, which were enacted in 1788. The embankment was completed the following year but suffered storm damage, and further acts followed, Benjamin Wyatt and William Provis (qq.v.) being involved on separate occasions. He may also have become involved in his uncle's proposals to bridge Menai.

In 1793 Golborne had been elected to the (Smeatonian) Society of Civil Engineers and he retired in 1813, then finding it difficult to ride on horseback and to execute his office to his satisfaction. He was buried at Holy Trinity, Ely, on 1 August 1819, aged seventy-three. He left an annuity to his wife and property in Ely, as well as 40 acres in Southery, to his daughters Ann and Susan, both of whom were married to clergymen. His personal effects were valued for probate at £450.

A. W. SKEMPTON

[Bedford Level Corporation Orders (1776–1813) Cambridgeshire RO; House of Commons (1777) *Minutes of Evidence, Bedford Level Petition*; Hardwicke Papers (1791–1809) British Library; House of Commons (1795) *Minutes of Evidence, Eau Brink Cut*; Robert Renwick (1912) *Records of the Burgh of Glasgow*; H. C. Darby (1940) *The Draining of the Fens*; A. E. Richardson (1955) *Robert Mylne*; Tidal Harbours Commission (1846) 2nd report, appendix B, 235–239; Tidal Harbours Commission (1847) 2nd report, appendix C, 391–395; D. Gwyn (2000) in lit.]

Publications

1791. *Report on the Outfall of the River Ouse*
1792. *Report on the Outfall of the River Welland* (with John Hudson and George Maxwell)
1799. *Report on the Drainage of Keyingham Level in Holderness*

Works

1770–1775. River Clyde improvement, resident engineer
1776–1813. Bedford Level, Superintendent General
1788–1789. Cob Malltraeth, Anglesey, consultant

GOLBORNE, John (1724–1783), civil engineer. The Golborne (or Golbourne) family lived in Chester from the mid-seventeenth century. John was baptised on 19 March 1724, a younger son of William Golborne, schoolmaster, and can be identified with John Golborne, gentleman, elected Freeman of the City of Chester in October 1759. He attended King's School, Chester, from 1735 until 1739. His elder brother, James (b. 1716), was one of the two Overseers of Works on the New Cut Canal of the river Dee from 1741 and John may

have worked with him and later, probably in 1754, become engineer of the River Dee Company. It is unfortunate that the Company records were lost or destroyed long ago, and details of his appointment are not known. However, an old workman speaking of the period around 1770 refers to 'Mr. Golborne of Chester who hath, with great reputation, effected works on the river Dee' while Thomas Yeoman (q.v.), in the following year, said that his 'experience has been very great upon the river Dee'.

The canal promoted by Nathaniel Kinderley (q.v.) was completed in 1743. It ran for nearly 8 miles along the south side of the Dee estuary from Chester to Connagh's Quay. The Company maintained the canal and progressively reclaimed land from the sea in the upper parts of the estuary. Maintenance involved the use of dredging 'ploughs' to ensure a depth of water of at least 15 ft. at moderate spring tides, and the construction of groynes or 'jetties' to prevent the channel spreading to a width of more than about 250 ft. at low tide.

The canal was by far the largest work of its kind in Britain in the eighteenth century. Its high cost (£56,000 had been spent by 1743) was to be repaid by tonnage dues on shipping and by rents from the reclaimed land. Reclamation works began in 1744 by building an embankment 2800 yd. in length from the canal bank straight across the estuary to the opposite (Cheshire) shore but in 1749 it was breached by wave action, portions crossing the old channels of the river being entirely washed away. A new bank, 4,780 yd. in length and following a more oblique line, was completed in 1754, so enclosing some 1,600 acres. A further bank 4,440 yd. long, completed in 1763, enclosed an additional 660 acres and this was followed in 1769 by another bank 500 yd. in length enclosing 350 acres. The banks were substantial: rising 12 ft. above ordinary spring tides with a top width of 6 ft. and side slopes of 4:1 (outer) and 2:1 (inner). They are shown, together with the canal, on a 1771 map by Thomas Boydell (c. 1729–1795).

Golborne remained engineer of the Dee Company to the end of his life—he is mentioned as such in his brief obituary notice in the *Gentleman's Magazine*—but like others in long-term jobs of this kind he was free to take on outside work from time to time on leave of absence. Thus in 1759 he was engaged on an extension of the Newry Canal in Ireland from the first lock towards the sea. Work began in April under his direction but was abandoned in the following year. In November 1759 he was consulted on the failure of a lock and weir on the river Weaver Navigation, and in March 1760 he gave evidence in the House of Commons on the cost of widening all the locks on this river. Also in 1760 he took as an apprentice his nephew James Golborne (q.v.).

Two small commissions followed in 1766. In February, Golborne and the surveyor, Hugh Oldham, reported on the proposed Macclesfield

Canal, with an estimate of cost, and later that year he attended Parliament in the Weaver Navigation interest during hearings on the Trent & Mersey Canal Bill. Then in 1768 he was recommended by James Brindley (q.v.) as 'the properest person for considering the navigation of the rivers Carron and Clyde, and deepening their beds'. This was in connection with a scheme for the Forth & Clyde Canal, alternative to that planned by John Smeaton (q.v.) on which work had already started. Golborne went to Scotland and reported on 30 September 1768. His report and those of Brindley and Yeoman on the alternative scheme are of minor interest since nothing came directly from them: work proceeded on Smeaton's plan. But Golborne's presence at Glasgow encouraged the city authorities to seek his advice on improving the Clyde, quite independently of the canal, and this led to the work by which he is chiefly remembered.

There were ten shoals in the 14 miles of river from Glasgow to Dumbarton. At the shoal nearest to Glasgow Quay the depth of water at high water springs was little more than 3 ft. while at the next three shoals in a 4 mile stretch the high water depth was 4 ft., with only 18 in. at low tide. These soundings had been taken by Smeaton in 1755 and 1758 in relation to a proposal to provide a lock and weir on the river at Marlin Ford, 4½ miles below Glasgow. Attempts to build these structures between 1760 and 1762 (not under Smeaton's direction) were a dismal failure. With his Dee experience in mind Golborne adopted a radically different approach: in his own words, 'by removing the stones and hard gravel from the bottom of the river where it is shallow, and by contracting the channel where it is worn too wide'. By these means he was confident of achieving a depth of 6 ft. at neap and 9 ft. at spring tides at Glasgow. Based on a map of the river by James Barrie and Alexander Wilson, Golborne estimated the cost of jetties and dredging at £8,640 with an upper limit of £10,000. He submitted his report on 30 November 1768 and received a fee of £85. The report was printed and circulated and on 5 January 1769 the Glasgow authorities decided to apply for an Act enabling the work to be carried out. Before leaving Scotland, Golborne recommended that James Watt (q.v.) should check the soundings. This he did, reporting in October 1769 and confirming Smeaton's observations.

Meanwhile in July 1769 Brindley, Yeoman and Golborne were requested to advise on improvements of the river Nene outfall. In 1721 work had begun under Kinderley's direction on a new cut 1½ miles long from River's End (4½ miles below Wisbech) past Gunthorpe Sluice towards Peter's Point. But the project met with fierce opposition and in June 1722 rioters destroyed the diversion dam then under construction across the old channel. Nothing more was done for several decades and the outfall continued to deteriorate. Golborne reported in October 1769,

recommending that the cut should be re-opened and extended for 6 miles through reclaimed marshes and farther to a deep sea channel at the Eye, at an estimated cost of £26,000. Yeoman and Brindley both agreed, through Brindley, reporting in August 1770, gave a much higher estimate. Golborne was briefly recalled to Wisbech in September 1771 and, with Langley Edwards (q.v.), actually set out the line of the proposed cut. For various reasons it was finally decided to restrict the scheme to re-opening the original cut, the work (known as Kinderley's Cut) being carried out 1773–1774 under the direction of Richard Dunthorne (q.v.). The very favourable results almost immediately achieved were communicated in letters to Golborne from John Wing (q.v.), Steward of the Duke of Bedford's estate at Thorney and as a Commissioner of the North Level responsible for bringing in Brindley, Yeoman and Golborne as consultants.

Back on the Clyde little time been lost in obtaining the Act, in January 1770. On 29 June a contract between Golborne and the city was approved. He undertook to create a depth of 7 ft. of water, at least, at neap tides in every part of the river within four years at a cost of £11,000. If he achieved the stipulated depth at a lower cost the surplus was his, but if the work cost more an additional £2,000 would be provided; after this the work was at his risk. He would be paid £120 p.a. and was expected to devote at least six months each year to the project. His full-time assistant would receive £100 p.a.; Golborne appointed his nephew James to this post.

Work began in July 1770. Golborne drew money weekly from the city to pay wages, purchase equipment and materials. By the end of the 1772 season more than a hundred jetties had been built, some up to 500 ft. in length, and with help from the dredging ploughs a depth of 5 ft. existed throughout the river at high tide. The ploughs were cast iron bucket-scrapers 4 ft. long and 3 ft. wide dragged across the river bed using capstans in punts moored at the banks.

The main obstacle to further improvement remained the Dumbuck Shoal 12 miles downstream where the wide river ran in two channels. Here jetties proved to be inadequate, and Golborne proposed to build a training wall 800 yd long at an estimated cost of £2,300. The management committee approved this additional expenditure, and the wall, known as Lang Dyke, was built between June and November 1773. Elsewhere work continued, dredging and building jetties where experience showed them to be necessary; and by October 1775 the results were so satisfactory that the city ordered that a two-handle silver cup be made for presentation to Golborne. A survey in December showed more than 7 ft. of water throughout at an ordinary neap high tide. Finally, with regard to what 'consideration' should be given to him 'for his labours, pains and attendance' in the works, the city decided to pay him a gratuity of £1,500 together with £100 to James Golborne for superintending the work.

In the period leading up to the Clyde works Golborne had doubtless been busy on the bank in the Dee estuary, completed 1769 at about the time of his Wisbech report. In 1770 he reported on a scheme for a road across north west Wales, including an embankment at Traeth Mawr, and via Porth Dinllaen, a link on to Dublin. This anticipated the dreams of William Madocks (q.v.) by half a century. Early in 1771 he acted for the City of Chester regarding questions on the route of the proposed Chester Canal. In March 1771 he became a member of the Society of Civil Engineers, later known as the Smeatonian Society, attending its second meeting with Robert Whitworth (q.v.) and four of the founder members Yeoman, Smeaton, Grundy and Mylne (qq.v.), a notable coming together of six leading members of the profession in its early days. In September he was back at Wisbech, as mentioned above, and in the summer of 1772 work began on a new quay wall at Ayr Harbour according to his plans, apparently drawn up in conjunction with James Watt. He also arranged for some workmen and dredging ploughs to be sent there on loan from the Clyde.

In 1777 Golborne was back in the Fens at the request of Bedford Level Corporation 'to take a View, and make a Report, of the Middle and South Levels, and to give an Opinion on the proposed Plan for a General Draining of those Lands; and what omissions, additions or variations of the Plan I shall think of most benefit'. He arrived on 24 June and spent five weeks on site accompanied by his nephew, recently appointed Superintendent General of Bedford Level, and for part of the time by John Wing.

The proposed Plan, embodied in a Petition submitted by the Corporation to Parliament in February 1777, involved extensive works estimated to cost over £150,000 and the granting of powers to raise additional taxes. Not surprisingly the Bill encountered strong opposition and was dropped after its second reading. The Corporation then called on Golborne. They would have been better advised to have do so before rather than later.

In his report dated 2 December 1777 Golborne points out that the essential requirement for better drainage of the Middle and South Levels is an improved outfall of the Ouse; that the proposed works are 'interior', within the fenland itself, and can contribute nothing to improving the outfall; and without such improvement the interior works will be of limited value quite disproportionate to their cost. He identifies the main cause of trouble as the wide and shallow bend, 6 miles long, in the Ouse above King's Lynn and says it is necessary to bypass this bend by a new channel cut from Eau Brink to ½ mile above Lynn 'nearly in a manner as formerly proposed by Mr. Kinderley'. This would be 2¾ miles in length and should result in a lowering of low water level by

at least 4 ft. as far upstream as Denver and Salters Lode sluices, to the great benefit of drainage in the South and Middle Levels. He is referring here to the suggestion first made by Kinderley in 1721, and again with more detail in 1730, and again by his son in 1751, but opposed by those who thought the long, wide bend provided a valuable indraught for the tides, essential in their view for preservation of Lynn harbour. Golborne's report includes recommendations for several other works, in particular on the river Nene above Wisbech, but is renowned for its reasoned advocacy of the Eau Brink Cut, as it was later called. The report was reprinted c. 1793 as a key document in the discussions leading to an Act of Parliament in 1795 under which the Cut was eventually made between 1818 and 1821, with results even more beneficial than had been anticipated.

In August 1781 Golborne was recalled to Glasgow to advise on possible improvements on the Clyde. 'It gave me great pleasure and satisfaction (he says) to find the general work in such good order'. The conditions at Dumbuck were especially pleasing. The new requirement was to deepen the river for ships of 200 or 300 tons burden up to Newshot, 7 miles below Glasgow. This could be achieved, he says, by dredging combined with eight new jetties at a relatively small cost. In concluding his report Golborne remarks: 'The generous treatment I met with in completing my work (in 1775) has made a lasting impression on my mind, and I hope that the present small tribute of my endeavours to promote the trade and navigation of the opulent and flourishing City of Glasgow will be accepted as the overflowing of a grateful heart'.

In April 1783 he studied the flow through the Menai Straits and drew up a combined embankment and timber bridge scheme for crossing of the Straits. Golborne carried out his last consulting job in May 1783, when he reported on the river Boyne and harbour of Drogheda. Work began but was not completed as he died five months later. He was buried at St. Oswald's, Chester, on 12 October 1783, age fifty-nine. He never married but had a natural son John Jones to whom he left the bulk of his estate, including his River Dee stock. Thomas Boydell of Trevalyn was one of the executors. Other bequests went to his sister, his niece and four nephews.

A. W. SKEMPTON and MIKE CHRIMES

[Telford Collection, ICE Archives; John Grundy (1750, 1754, 1759) West Chester Navigation, Reports, vol. 4, Brotherton Collection, Leeds University; Parish Records, Cheshire RO; Ayr Harbour Board Minutes (1772) Scottish RO; Tidal Harbours Commission (1846) On Drogheda Harbour; Admiralty Inquiry (1840) Chester, Dee Navigation, Minutes of Evidence; N. Walker and T. Craddock (1849) The History of Wisbech and the Fens; Clyde Navigation Trust (1854) Reports on the River Clyde and Harbour of Glasgow from

1755 to 1853; Robert Renwick (1912) *Extracts from the Records of the Burgh of Glasgow, 1760–1780*, **7**; F. Webster (1930) River Dee reclamations, *Trans Liverpool Eng Soc*, **50**, 63–92; T. S. Willan (1951) *The Navigation of the River Weaver in the Eighteenth Century*; W. H. McCutcheon (1965) *The Canals of the North of Ireland*; Ruth Heard (1975 in lit.) on Drogheda harbour; John F. Riddell (1979) *Clyde Navigation*]

Publications

1768. *Report relating to a Navigable Communication between the Friths of Forth and Clyde*
1768. *Report relative to the Deepening of the River Clyde*
1769. *Report concerning the Drainage of the North Level of the Fens, and the Outfall of Wisbeach River*
1770. *Proposals for making a Turnpike Road from Llangynog, in Meironyddshire, to Traeth Mawr …*
1777. *Report on a View … of the Middle & South Levels, and their Outfalls to Sea; with a Plan for the effectual Draining the said Levels* (reprinted c. 1793)

Works

1754–1783. River Dee Company, Engineer
1770–1775. River Clyde, deepening, £13,000

GOODRICH, Simon (1773–1847), engineering manager and civil servant, was born on 28 October 1773. He was the son of Isaac Goodrich of Suffolk. His education and training is not known, but in December 1796 he was appointed a draughtsman to the mechanist in the office of Sir Samuel Bentham (q.v.), Inspector General of Naval Works; he was appointed mechanist in October 1799 with an annual salary of £400. He acted as Bentham's deputy during the latter's absence in Russia (1805–1807). Bentham's office was abolished and the staff dismissed in December 1812, following investigations into Bentham's business ventures. Goodrich continued working on a freelance basis, as mechanist without warrant, until April 1814, when he was re-appointed engineer and mechanist to the Navy Board at an annual salary of £600. In the meantime he had been asked to report on Bridlington Harbour, but his renewed naval duties precluded him carrying out this work, which was entrusted to David Matthews (q.v.). Bentham also tried to get him to supervise work on Vauxhall Bridge, but Goodrich, perhaps wisely, refused to get fully involved. Thereafter Goodrich acted as a consultant to the Navy Board on engineering matters and managed the engineering works of the dockyards. Among his installations at the dockyard there was a testing machine powered by a hydraulic ram. He lived at Portsmouth, until his retirement in 1831, when his annual pension was £400.

The Navy had 800–900 ships of the line between 1805–1816 and to keep them afloat required repair facilities and stores backup of considerable sophistication. Bentham's goal was to streamline the production of marine stores, particularly ships' blocks, the annual consumption then being estimated at 100,000. Bentham's chief accomplishment was the construction at Portsmouth and other dockyards of mills for working metal and wood, the block-making machinery mills, rope and cordage mills and the millwrights' workshop. There seems little doubt that much of his achievement can be attributed to the ability of subordinates like Goodrich.

Goodrich's papers and drawings were presented to the Science Museum after his death and they include a daily journal which, though not complete, covers 1802–1845. His circle included Trevithick, Brunel Sr., Maudslay and Matthew Murray (qq.v.); many of their letters are preserved with his papers. Goodrich visited many of the major engineering manufactories as he travelled throughout the country on Naval business and his papers include notes, often illustrated, about machinery, and advertising leaflets, with details of prices, weights and dimensions.

In 1834 Goodrich moved to Lisbon and died there on 3 September 1847. He was elected a corresponding member of the Institution of Civil Engineers in December 1820 and transferred to Member in December 1837.

Goodrich did have a family, but rarely mentioned them. He received no public recognition in his life-time, and no contemporary biographies or obituaries appear to have been published. He was occasionally referred to as Goodrick.

A. P. WOOLRICH

[Goodrich papers, Science Museum; Admiralty papers PRO; membership records, ICE Archives; Tidal Harbours Commission (1846) *2nd Report, Appendix B*, 391–395; E. A. Forward (1922–1923) Simon Goodrich and his work as an Engineer, Part 1, 1796–1805, *Transactions of the Newcomen Society*, **III**, 1–15; E. A. Forward (1937–1938) Simon Goodrich and his work as an Engineer, Part II, 1805–1812, *Transactions of the Newcomen Society*, **XVIII**, 1–27; A. S. Crossley (1961) Simon Goodrich and his work as an Engineer, Part III, 1813–1823, *Transactions of the Newcomen Society*, **XXXII**, 79–91; R. C. Riley (1985) The evolution of the docks and industrial buildings in Portsmouth Royal Dockyard 1698–1814, *Portsmouth Papers*, 44; J. G. Coad (1989) *Architecture and Engineering Works of the Sailing Navy*]

GORDON, Alexander (1802–1868), civil engineer, was born in New York on 5 May 1802, the second son of David Gordon, an inventor who patented a portable system of compressing gas. In 1807 the family returned to Scotland, and Alexander was subsequently educated at Edinburgh University. He joined the Institution of Civil Engineers as an Associate in 1827 and was already well-known to several senior members, notably Telford, Donkin, Field, the Bramahs (qq.v.) and

James Simpson, through his father's work with gas engines. In the early 1820s he worked with his father on a number of early road locomotive designs. In 1819 they had experimented with the use of compressed gas as motive power and published in 1820 and 1824 designs for steam powered locomotives, one of which emulated the action of horses legs. He was familiar with Sir Marc Isambard Brunel's carbon gas engine and followed the work of Cayley and Sterling with interest.

Gordon was familiar with the work of Gurney and others on steam locomotion on roads and gave evidence, despite his relative youth, at Parliamentary inquiries on road traction in 1831. His evidence at this and further inquiries reveals a familiarity with key sources like Macneill (q.v.) and Parnell and many aspects of the subject, including early experiments, and wear on road surfaces by carriages. He was responsible for some experiments himself and his work in this field is best-known through his text on locomotion which first appeared in 1832 and subsequently went through three editions and several translations. Both he and his father were involved in setting up associations to promote road locomotives.

Gordon became an agent of Robert Napier and an expert in marine engineering, patenting a scientific propeller in 1845. This was only one of a growing number of interests and he is perhaps best-known for his work on lighthouses. In 1833 he introduced his systems of lighting apparatus, the polygonal arrangement of dioptic and cadioptric systems of Maritz, and Fresnel's cadioptric arrangement. The latter was used for the first time at the Town Pier, Gravesend, in 1833, three years before Trinity House adopted it; having displayed it to Trinity House he exhibited it at the British Association in 1834. Gordon took out his own patent for a holophotal apparatus. He specialised in the design of lighthouses for the British colonies, generally prefabricated structures of cast iron or timber; the first was the Morant Point lighthouse, Jamaica of 1842. This aspect of his career more properly belongs in Volume 2.

Gordon was clearly something of a polymath, reflected both in his contribution to meetings of the Institution of Civil Engineers and in his varied engineering and scientific activities. He presented the first paper on the application of photography to engineering as early as 1840, identifying its importance for progress records. He was a Fellow of the Royal Geographical Society and a founder of the Polytechnic Institution. He died at Sandown, Isle of Wight on 14 May 1868.

MIKE CHRIMES

[Membership records, ICE Archives; Publications index to ICE Minutes of Proceedings 1837–1879; account of a voyage made by the late Mr. David Gordon in his canal steam boat [1828], OC 71 (1831), ICE Archives; Prospectus (1833) *Institution*

of Locomotion for Steam Transport; Minutes of evidence (1831) *House of Commons, Select Committee on Steam Carriages* (HC 324), 74–78; Minutes of evidence (1836) *House of Commons, Select Committee on Turnpike Trusts and Tolls*, 51–54, 89–91, 110–112; Minutes of evidence (1839) *Select Committee on Turnpike Trusts* (HC 218), 26–31; Commission … into Condition and Management of Lights, Buoys and Beacons (1861) *Report*, 2 (appendix), 643–650 plus plates; D. B. Hague and R. Christie (1975) *Lighthouses*, 103–105; 218, 224; G. Herbert (1978) *Prefabrication*; M. [M.] Chrimes (1991) *Civil Engineering: a Photographic History*]

Publications
The following are a selection of publications prior to 1840:

1832. *Historical and Practical Treatise upon Elemental Locomotion*
1832 (and 1837). *Rapport … sur les Moyers d'Ameliorer la Navigation de la Dyle …*
1832–1833. *Journal of Elemental Locomotion*, editor
1835. *The fitness of Turnpike Roads and Highways for the most Expeditious, Safe, Convenient and Economical internal Communication*
1836. *Treatise on Elemental Locomotion*
1837. *Observations addressed to those Interested in either Roads or Turnpike Roads*
1840. *Photography, as applicable to Engineering*

GOTT, John (1720–1793), civil engineer, of Woodhall, Calverley, near Leeds, is the central figure in a family who through three generations made material contributions to the design, building and maintenance of county bridges in the West Riding of Yorkshire. His father, also John Gott (1696–1746), was a builder at Calverley, and John II was born his second son in 1720 and baptised on 2 October. The firstborn, one year older, was called William. This William had two sons, William (b. 1749) and John (b. 1755), while John II also had sons William (1745–1810) and John (1747–1751) by his first wife Mary Parker, who died in 1752. By his second wife, Susanna Jackson of Bradford (1723–1778), he had three daughters and one son, Benjamin (1762–1840), who became a prominent woollen manufacturer in Leeds, and also in later life a director of at least one railway company.

The members of the family who participated in bridge work were John I, his sons, John II and William I, and two Williams, sons of John II and William I. All are believed to have been trained initially as masons.

From 1740 to 1777 the repairing and some rebuilding, all on a modest scale, of county bridges of the West Riding, which numbered one-hundred and ten in all, were made by groups of 'undertakers' who entered into contracts for terms of seven years, each time for a fixed sum which varied from £300 down to £230 p.a. The undertaker group at the outset comprised John Gott I, John II, William I and John Westerman. It became John II and William I for the years 1756–1770 and from 1770 John II with his son and nephew, both called William. Throughout those years the justices had also been served by two 'surveyors', who were paid small salaries presumably for part-time duties and none of whom were Gotts. That changed in 1777 when John Gott II was appointed as sole surveyor for a salary of £250 p.a. and under a bond that he would not contract for any of the Riding's bridge work. (During his years as surveyor he was also called upon to deal with other buildings owned by the Riding, such as the House of Correction, and was referred to by the broader title of 'Surveyor of public works'). He was also allowed £50 p.a. for assistants, presumably to report on and supervise repairs in distant parts of the Riding. When John II died he was succeeded in the post by his son with salary of £150. William resigned four years later, to be replaced by Bernard Hartley (q.v.) at the same salary, but it was raised to £250 again in 1803, and the surveyorship remained in the Hartley family for more than eight decades.

The annual spending on bridge works, stated in the accounts of the Riding's treasurer, varied during the Gotts' period as undertakers from £420 in 1748 to £4,882 in 1776, the excess over the undertakers' annual fees being accounted for by major rebuildings, and possibly a few new bridges, which were the subjects of individual contracts. In the sixteen years of John II's surveyorship the overall spend varied between a maximum of £7,857 in 1779 and a minimum of £804 in 1781, but only in three years was it less than £2,000. During these years John II did all, or nearly all, of the designs.

John Gott II's work included design and/or construction of about ten large West Riding bridges, as well as many smaller ones, not all of which can be identified from the records. His biggest contract was for the four-arched Ferrybridge over the Aire, built to his own design for £3,950 in 1764–1765. For the repair and widening of Castleford Bridge, of seven arches and costing £1,284, he was joined in the contract by his son, William. He contracted in partnership with the young Williams, his son and nephew, and one Anthony Blanchard, to widen Ripon North Bridge to his brother William's design, for £1,797. For Bingley and Otley bridges, his responsibility was to design and supervise the contracts for widening, because he had by then become the Riding's salaried Surveyor.

He had other work as consultant and resident engineer. He was appointed by Peter Birt as engineer of the Aire and Calder Navigation, and is said by Hadfield to have spent in the 1760s '£4,000 on capital improvements and £9,000 on maintenance', chiefly rebuilding old weirs to restore their original heights. He appears to have retained a working connection with that navigation for most of his life. He went on to act as resident engineer for William Jessop (q.v.) on the construction of the Selby Canal connecting the

Aire to the Ouse, in 1775–1778, and designed and built coal staiths on the same canal in 1781–1782. He also served as engineer to the Barnsley Canal in conjunction with Jessop and Elias Wright.

About 1766 he was engaged by Sir Walter Blacket, a substantial landowner near Hexham, to design a bridge across the river Tyne after John Smeaton (q.v.) had refused the commission due to pressure of other work. Gott drew a seven-arched bridge very similar in style to his bridges in Yorkshire, and his brother contracted to build it for £3,459 in partnership with George Brown, surveyor of bridges for the southern part of Northumberland. It was opened for use in September 1770 but little more than a year later it was destroyed by the 'Great Inundation' which also broke down all the other bridges on the lower Tyne except Corbridge. The flood was so severe that Blacket absolved the contractors from the bond they had made to rebuild the bridge in the event of an early collapse, considering that such a flood was unforeseeable.

John Gott II was elected a member of the Society of Civil Engineers in London in 1773, probably due to his long acquaintance with Smeaton, the accepted leader of the profession and the Society. Gott died in 1793 and was buried on 7 June.

P. S. M. CROSS-RUDKIN and TED RUDDOCK

[JHC 22 Feb. 1774; QD 1/514, QD3, QS Orders of the West Riding at West Yorkshire RO; A. W. Skempton and E. C. Wright (1971–1972) Early Members of the Smeatonian Society of Civil Engineers, *Trans Newc Soc*, **44**, 23–47; Charles Hadfield (1972) *Canals of Yorkshire and the North East of England*, vol. 1; E. C. Ruddock (1977) The foundations of Hexham Bridge, *Géotechnique*, **27**(3), 385–404; S. M. Linsley (1994) Tyne crossings at Hexham up to 1795, *Archaeologia Aeliana, Fifth Series*, **22**, 235–253; Christopher Chalklin (1998) *English Counties and Public Building 1650–1830*, **70**, 113, 215–221]

Works
1740–1773. General repairs of county bridges of West Riding of Yorkshire, undertaker (with various members of Gott family and John Westerman), annual fee varied between £230 and £300
1760–1761. Bridge at Nostell Priory, near Wakefield, contractor. Client Sir George Savile
1760–1792. Aire and Calder Navigation, resident engineer
1764–1765. Bridge at Ferrybridge, River Aire, for West Riding, design and contractor, four arches, 47 and 60 ft. spans, cost £3,950, replaced 1804
1767–1770. Hexham Bridge, River Tyne, for Sir Walter Blacket, seven arches, 40–70 ft. span, contract price £3,459, destroyed by flood 1771
1768. Lindley Bridge, River Washbourn, resident engineer
1769. Wetherby Bridge, River Wharfe, widening, design, contractor William Johnson, cost £830

1770. Ripponden Bridge, design, contract price £350
1770. Glusburn Bridge, design, contract price £215
1770. Castleford Bridge, River Aire, repair and widening, contractor with his son William, contract price £1,284, replaced 1808
1770. Hebble End Bridge, River Calder, design, fee £49
1771. Keighley Low Bridge, design, contract price £412
1773. Brearley Bridge, design, contract price £340
1774. Ripon North Bridge, River Ure, widening, contractor with son and nephew (both William) and Anthony Blanchard, design by William Gott (brother?), price £1,797
1775. Stock Bridge, River Aire, design, contract price £345
1775–1776. Elland Bridge, River Calder, repairs and widening, contractor with son and nephew (both William), price £689
1775–1776. Bingley Bridge, River Aire, design and supervision, price £535
1775–1776. Otley Bridge, River Wharfe, widening, design and supervision, contractor Henry King, price £751
1775–1778. Selby Canal, resident engineer, engineer William Jessop
1776. (attrib.) Calverley Bridge
1776–1779. Methley Bridge, River Calder, design
1779. Hebden Bridge, design

Other bridges on which some work is attributed to William Gott:

1771. Higher Hodder Bridge, cost £312
1772. Burholme Bridge, two arches, 60 ft. spans, cost £1,100, one arch rebuilt 1845
1773. Cow Bridge, River Ribble, cost £899
1774. Slaidburn Bridge, River Hodder, widening, cost £352
1775. Long Preston Bridge, Priest Gill
1775. Clough Bridge, Clough Beck
1775. Swillington Bridge, River Aire, repair and widening, £700
1776–1777. Harewood Bridge, River Wharfe, widening, £797

GRAHAM, David (fl. 1550s), master mason of Edinburgh, was in charge of the masons who worked on the reconstruction of the quay at the 'Shore' at Leith between March 1553 and September 1555. The masonry was quarried at Granton under the supervision of Alexander Reid and dressed by a team of masons before being transported by boat to Leith where it was placed. Costs in the first year were £774 2s 9d, and less in the following two years. Further work, in timber, followed at New Haven, but it is not known whether Graham was involved.

MIKE CHRIMES

[S. Mowat (1994) *The Port of Leith*]

GRAHAM, James Gillespie (1776–1855), architect, was born at Dunblane on 11 June 1776, the son of Malcolm Gillespie, a solicitor in the Register of Births. He may have originally been trained as a mason and in the first decade of the nineteenth century was active in Skye as superintendent of the 2nd Lord Macdonald's extensive improvement works, including some small piers. By 1810 he had begun to obtain architectural commissions and in 1815 he married an heiress, Margaret, daughter of William Graham, Laird of Orchill in Perthshire, and assumed her name. Gillespie Graham's practice was largely gothic church and country house architecture, where he benefited from a friendship with Pugin, but he was also involved in important examples of urban planning on the Moray Estate in Edinburgh and in Birkenhead, where he laid out the original grid street plan and Hamilton Square. He corresponded with Telford (q.v.) about the Dean Bridge.

He died in Edinburgh on 21 March 1855.

MIKE CHRIMES

[J. Macaulay (1974–1975) James Gillespie Graham in Skye, *Bull Scot Georg Soc*, **iii**, 1–14; J. Gifford, *The Buildings of Scotland: Highland and Islands*; Colvin (3)]

Works
1800. Portree pier
1801. Kylerhea pier, Skye, £93
1801. Sconser pier, Skye, c. £30
1803. Broadford pier, Skye, c. £180

GRAINGER, Thomas (1794–1852), civil engineer, was born on 12 November 1794, at Gogar Green, Ratho, near Edinburgh, the son of a tenant farmer.

He became a pupil of John Leslie, an Edinburgh land surveyor, at the age of 16 and in 1817–1818 also attended classes in Natural Philosophy at Edinburgh University. A drawing from this early period survives of Leith Walk, dated May 1819. The following year he surveyed the Forth from Dysart to Kirkcaldy for John Rennie (q.v.), his fees being £44 18s 6d. In 1822, he set up in business on his own account as a civil engineer and surveyor and was responsible for a large number of road surveys, including the Hamilton–Lanark Road via Cartland Crags. He acted as co-surveyor on John Loudon McAdam's (q.v.) Edinburgh–Newcastle road improvement scheme in 1828 and reported on roads from Glasgow to Ayrshire.

Grainger was an early supporter of railways and this would have made him an obvious choice as engineer for the Monkland and Kirkintilloch Railway among its supporters. From 1823 he became involved in railways and over the next thirty years was engineer for a number of important lines in Scotland and northern England.

The Monkland and Kirkintilloch Railway obtained its Act in 1824. A mineral line intended as a canal feeder, it was envisaged that it would take locomotive traffic in due course. Its success was followed by neighbouring lines at Ballochney, and Wishaw and Coltness. From 1825, with a growing body of work, Grainger was in partnership with John Miller who had joined him in 1823 at the age of eighteen. In 1826 they produced plans for the Edinburgh–Dalkeith Railway. Their most famous early collaboration was survey work for the Edinburgh and Glasgow Railway. The failure of this line to get Parliamentary approval at that time led them to take separate responsibility for their own work. This arrangement continued until 1845 when the partnership was dissolved. It was generally Miller who undertook the most important lines, including the successful Edinburgh–Glasgow scheme (1838–1842).

Grainger was elected a Member of the Institution of Civil Engineers in 1829. He was President of the Royal Scottish Society of Arts, travelling to Holland so that he could give an informed address on Harlem River. He also served on Edinburgh Council. Still active as an engineer he suffered fatal injuries in a railway accident near Stockton-on-Tees on 25 July 1852. His obituary notes that 'he was regretted as a firm friend, a kind master, and a sincere, consistent Christian'. His portrait, owned by the Institution of Civil Engineers, hangs in the Department of Civil and Offshore Engineering at Heriot Watt University.

MIKE CHRIMES

[Membership records, ICE Archives; Drawings register, ICE Archives; J. T. Naylor (1847–1851) Notebook, ICE Archives; Memoir (1858) *ICE Min Proc*, **12**, 159–161; Skempton; Bendall]

Publications

1828. *Report relative to the proposed Railway, to connect the Clydesdale, or Upper Coal-field of Lanarkshire, with the City of Glasgow, and the East and West Country Markets* (with Miller)
1829. *Report of a Survey undertaken for the Purpose of ascertaining the Line by which the best Road from the City of Glasgow to Ayrshire is to be obtained* (with Miller)
1829. *Report to the Proprietors of, and Traders on the Canals and Railways terminating on the North Quarter of Glasgow, and other Gentlemen interested therein and in the Harbour and Streets of Glasgow* (with Miller)
1829. *Glasgow Railway and Tunnel*, reports by Grainger and Miller and Telford
1829. *Report and Estimate of the probable Expense of the proposed Extension of the Garnkirk and Glasgow Railway to Port Dundas* (with Miller)
1830. *Observations on the Formation of a Railway Communication between … Edinburgh & Glasgow* (plus plan)
1831. *Reduced Plan and Section of the proposed Edinburgh and Glasgow Railway*
1831. *Edinburgh, Glasgow and Leith Railway. Reports by Messrs Grainger and Miller of Edinburgh, and Mr. George Stephenson of Liverpool*
1834. *Report relative to the formation of a Harbour and Dock in Trinity Bay, on the southern Shore of the Firth of Forth* (with Miller)

1834. *Map showing the Line of the Glasgow, Paisley, Kilmarnock and Ayr Railway with its immediate Connextions* (with Miller)
1835. *Glasgow Water Supply. Report of 1834 …*
1835. *Report … relative to … Edinburgh … Eligibility for Manufacturing Establishments*
1844. *Edinburgh Northern Railway Prospectus*
1846. *Edinburgh and Bathgate Railway (Parliamentary) Section*
1849. *Address on the Desirableness of obtaining Communications relative to the Construction and details of Engineering and other Public Works, accompanied by the necessary Models and Drawings*

Selected Works
1824–1826. Monkland and Kirkintilloch Railway, 10½ miles, £25,000
1825–1826. Ballochney Railway, 5 miles, £18,500
1826–1831. Glasgow and Garnkirk Railway, £35,000
1829–1834. Wishaw and Coltness Railway, £60,000
1834. Paisley and Renfrew Railway, 3¼ miles
1835–1839. Arbroath and Forfar Railway, 15 miles
1836. Glasgow and Greenock Railway, survey only
1836. Edinburgh, Leith and Newhaven Railway
1836. Trinity Harbour scheme
1844–1848. Edinburgh and Northern (later Edinburgh, Perth and Dundee) Railways
1845–1847. Leeds, Dewsbury and Manchester Railway, 16 miles
1845–1852. Leeds (Thirsk) Northern Railway, including Morley and Bramhope Tunnels, 46 miles, Wharfe Viaduct, twenty-one arches, Yarm Viaduct, forty-three arches, Knaresborough Viaduct, four arches
1846. Edinburgh and Bathgate Railway, 11 miles
1846. East and West Yorkshire Junction Railway, 15 miles
1849. Broughty Ferry Harbour
1849. Ferry Port-on-Craig

GRANTHAM, John (1775–1833), civil engineer, was baptised on 28 November 1775 at Kirton in Lancashire, the son of James and Martha Grantham. Possibly trained as a surveyor, extant surveys relate to Kent, Berkshire, Oxfordshire and Warwickshire. By 1805 he had moved to Croydon where the Canal was under construction with Dudley Clark (q.v.) as Resident Engineer. In 1811 Grantham produced a plan of its route. He was employed c. 1819 by John Rennie (q.v.) on surveys of Lynn Harbour, for which he was paid £200, and between 1819 and 1823 as his assistant in making the first comprehensive surveys of the river Shannon in preparation for the improvement of the navigation and the prevention of flooding. In 1823 he was joined in Ireland by one of his sons, Richard Boxall Grantham (1805–1891), who assisted him on a number of projects, including the partial rebuilding of the bridges over the Shannon at Limerick, Killaloe, and Portumna. His son also assisted him in introducing steam navigation in 1825 between Limerick and Shannon Harbour, the entrance to the Grand Canal system. For much of that year Grantham, with his brother Edward, surveyed a route for a railway from London to Birmingham for George and Sir John Rennie (qq.v.). The Report was presented to the Railway Committee in 1826, but no action taken until 1830 when Sir John Rennie was again approached and Grantham and his sons reviewed the route. In 1831 Grantham took over responsibility for the construction of Wellesley (now Sarsfield) Bridge over the river Shannon in Limerick city. Construction work had commenced in 1824 under Alexander Nimmo (q.v.) but was not completed until 1835. He died at Killaloe, county Clare, in 1833 and a plaque in the cathedral at Killaloe commemorates his work on the Shannon Navigation. His brother Edward, based in Croydon, continued to work for Rennie on the London–Brighton Railway, etc.

RON COX

[Sir John Rennie (1826) Reports, 2, ICE archives; Plan of the Croydon Canal and Railway, 1811, ICE Archives; J. Walker and W. C. Mylne (1825) *Report on Eau Brink Cut*; S. Grantham (1835) London and Birmingham Railway, *Mech Mag*, **22**, 436–438; J. Rennie (1875) *Autobiography*; Bendall (1891) Obituary of R. B. Grantham, *Min Proc ICE*, **108**, 399–403]

Works

1819–1823. Shannon Navigation Survey, Limerick, Killaloe and Portumna Bridge reconstruction, etc.
1831–1835. Sarsfield Bridge, Limerick, resident engineer
1834. Portumna Bridge, reconstruction
1843. Killaloe Bridge, reconstruction

GRAVATT, William, FRS (1806–1866), civil engineer and scientific instrument maker, was born on 14 July 1806 at Gravesend, the son of Colonel William Gravatt, RE (1771–1851), Assistant Inspector of the Royal Military Academy, Woolwich, from 1814 until 1828. Gravatt Jr. thus had access to the Royal Arsenal where he studied the steam and other engines. He also showed a mathematical ability and about the age of fifteen was placed with Bryan Donkin (q.v.); during this period he became friends with scientific instrument maker Edward Troughton (q.v.). After his time with Donkin, Gravatt worked for Marc Isambard Brunel (q.v.) where he was employed on the Thames Tunnel. In 1832, after work on the tunnel ceased, Gravatt was appointed to the Calder and Hebble Navigation where he designed some bridges, before being replaced by William Bull (q.v.). He then worked for Henry Palmer (q.v.) on the London and Dover railway in the course of which, following on from some work of Troughton's, he devised a level which entered into widespread use and for which he was awarded a Telford Medal of the Institution of Civil Engineers, of which he was elected an Associate in 1826 and a Member in 1828.

In 1834 Gravatt was employed by Isambard Kingdom Brunel on surveying railways in South Wales, on the Bristol and Exeter Railway and on the design of bridges for the Great Western Railway. During the railway mania of 1845–1846, Gravatt unwisely became involved in contracting, which permanently altered his prospects. After this period he turned his attention to scientific instruments, building a large refracting telescope and a calculating machine. During the 1850s he took a prominent part in controversies at the Institution of Civil Engineers relating to the effectiveness of a propeller and on the Atlantic Telegraph cable. In all his work, Gravatt's mathematical approach was always in evidence. Following an accidental overdose of morphia, he died on 30 May 1866.

FRANK A. J. L. JAMES

[Membership Records, ICE Archives; F. Smith Collection, Ice Archives; Thames Tunnel Papers, ICE Archives; Obituary (1866–1867) *Min Proc ICE*, **26**, 565–575]

Works

1826–(1832). Thames Tunnel, assistant engineer
1832. Calder and Hebble Navigation, engineer

GREATOREX, Ralph (*c.* 1625–*c.* 1675), mathematical instrument maker, is first heard of when he was bound an apprentice clockmaker for nine years from 1639, aged about fourteen. His first master was Thomas Dawson; then he was turned over to Elias Allen. Although Greatorex's apprenticeship expired in 1648, and he took his own property in the Strand in 1650 and set up his own business there, he was not made free until 1653, after Allen's death.

Greatorex continued Allen's traditional making of fine mathematical instruments but he also worked as a mechanical engineer, making many pieces of experimental apparatus and machinery, and was probably the first professional maker of such equipment. He had a particular interest in hydraulics: he experimented with a diving bell in the lead mines of Derbyshire in 1655 and, interested in the variation of water pressure with depth, tried to adapt it for submarine use in the Thames. The bell was also tested in 1664 with the aim of using it to help construct defences at Tangier. In 1656 Greatorex developed a pump to provide a constant supply of water for firefighting and in 1660 his engine for lifting water, probably an Archimedean screw, was demonstrated in St. James Park. An air pump was commissioned by Robert Boyle in 1658 but Robert Hooke considered it unsatisfactory.

Greatorex's apprenticeship with Elias Allen led to future connections and employment. Thus Greatorex corresponded with William Oughtred, a client of Allen, about the current comet in 1652. Oughtred had invented the equinoctial ring dial and 'double horizontal dial' and versions of these instruments made by Greatorex survive today. Another contact was Jonas Moore (q.v.) who lodged with Allen during the 1640s. Moore directed the Tangier project, a concern of the Royal Society, one of the contractors being Sir Hugh Cholmley (q.v.), who asked Greatorex to devise a way of lifting stones; Cholmley thought him 'very ingenious' but not to be trusted with money. Greatorex went to Tangier himself, returning to England in early 1666 and, later in the year, he and Moore collaborated in surveying parts of London destroyed by the Great Fire; in 1674, Greatorex was involved with Moore's surveying of the fens in Cambridgeshire. Other surveying work was carried out for the Crown: in 1664 Greatorex went to Hampshire; in 1668–1669 he surveyed the Royal Arsenal at Woolwich. Probably as a result of this work, and Christopher Wren's (q.v.) knowledge of him, Greatorex surveyed Whitehall in 1670 and went on to survey the town and castle of Windsor two years later.

Described by Samuel Hartlib as having 'a most piercing and profound wit', Greatorex had many other contacts with the scientific communities in London and Oxford during the 1650s and 1660s and attended early meetings of the Royal Society, although he never became a member of it. Samuel Hartlib mentions many of the hydraulic inventions as well as other instruments made or designed by him, including dial clocks in 1651. In 1653, he invented a new form of coinage to prevent counterfeiting, he designed a new surveying instrument, he intended to copy a device for sowing corn and he manufactured Wren's perspectograph. In 1654, Hartlib talks about Greatorex's new way of cutting tobacco, his new art of dialling and his writing instruments. A new kind of brewing vessel and interests in beekeeping and the use of clover are reported in 1655 and in 1657 he maintained a garden for experimenting with the cultivation of exotic herbs. In 1655–1666, Robert Boyle exchanged letters with Greatorex about practical chemistry; the latter's reputation probably led to his advising the lexicographer Edward Phillips on the terminology of scientific instruments in 1658. Greatorex's friendship with Samuel Pepys probably started in 1660 when Pepys was shown Greatorex's 'sphere of wire' (probably an armillary sphere), in 1663, Greatorex made a thermometer and a ruler with special engravings for him and the two men discussed Greatorex's new type of varnish.

As a member of the Clockmaker's Company, Greatorex took at least five apprentices from 1654. The first, Henry Wynne, went on to have a distinguished career as an instrument maker. Attempts were made to involve Greatorex more closely in the affairs of the Company in 1666 but he was resistant until 1668, when he was elected an Assistant although he was never sworn into office. Neither, from 1671, did he pay quarterage to the Company as one residing within 10 miles

of the City in spite of the fact that he had stayed in the Strand until 1663 and had been living in St. James Street (south end, east side) since 1665 and was still there. In 1673 Greatorex was commissioned captain of miners and pioneers in the expedition of the Third Dutch War.

He made his will on 26 October 1674 and left one shilling to his son Ralph; the sole executrix was his wife, Anne, to whom was left the rest of the estate. The will was proved on 4 May 1675, so Greatorex was probably the man with this surname who was buried in St. Martin-in-the-Fields on 28 April 1675.

SARAH BENDALL

[MS 2710.1, 2715.1, 6939, Clockmakers' Company Records 1639–1676, London Guildhall Library; Hartlib Papers 1651–1658, Department of History, University of Sheffield; Parish records, St. Clement Danes and St. Martin-in-the-Fields 1651–1675, Westminster Libraries; Cholmley Papers 1663–1666, North Yorkshire Record Office; PCC Wills 1674–1675, Public Record Office; T. Birch (1756–1757) *A History of the Royal Society of London*; E. G. R. Taylor (1954) *The Mathematical Practitioners of Tudor and Stuart England*; H. W. Robinson and W. Adams (eds.) (1968) *The Diary of Robert Hooke*; R. Latham and W. Matthews (1970–1983) *The Diary of Samuel Pepys*; A. V. Simcock (1993) An equinoctial ring dial by Ralph Greatorex, in: R. G. W. Anderson, J. A. Bennett and W. F. Ryan (eds.), *Making Instruments Count*, 201–215; F. Willmoth (1993) *Sir Jonas Moore*; G. Clifton (1995) *Directory of British Scientific Instrument Makers 1550–1851*; S. Bendall (ed.) (1997) *Dictionary of Land Surveyors and Local Map-Makers of Great Britain and Ireland 1530–1850*; S. Thurley (1998) *The Whitehall Palace Plan of 1670*; DNB New]

GREEN, James (1781–1849), civil engineer, was born in Birmingham. His father, possibly named James or Thomas, was a civil engineer and contractor in Warwickshire and the adjoining counties and it was from him that James received his early experience until 1800, when he was employed by John Rennie (q.v.), for whom he worked as an assistant from 1801 to 1807, leaving his service after a year in Devon. Rennie had employed Green on extensive surveys, canal works, the drainage of bogs and fens and the design of engineering works generally, both in England and Ireland. At that time the repair and replacement of Dymchurch Wall came particularly under Green's care and the reconstruction of the sea lock of the Chelmer and Blackwater Navigation was entirely entrusted to him by the Earl of St. Vincent.

Probably it was from here that Green came to Devon for in July 1806 he took over from Rennie, instructing a local surveyor, Charles Tozer, at Totnes, Rennie then reporting to the Duke of Somerset on ways of improving the navigation below Totnes bridge. Rennie also employed

Green in 1807 on a survey of the rock at Cattewater intended for use for the construction of Plymouth breakwater, which scheme was commenced in 1812. Meanwhile, in a report to Lord Boringdon of Saltram in December 1805 Rennie had proposed an embankment from Pomphlett Point to Saltram Quay. This had a favourable reception for in the spring of 1806 Lord Boringdon contracted with Green for the construction of the embankment 970 yd. long to enclose 175 acres. Following the collapse of the newly rebuilt Fenny bridges near Honiton in 1808, Green contracted for the design and construction of a replacement bridge across the river Otter; it had three spans of 42, 48 and 42 ft. spans in brickwork and was 20 ft. between parapets.

At the Midsummer Sessions of 1808 a committee of magistrates had been reminded of a letter of July 1800 from the Clerk of the Peace of Shropshire giving information on their conditions of appointment of Thomas Telford (q.v.) as their county surveyor. The Devon magistrates decided to dispense with their six surveyors and appoint a civil engineer as their county bridge surveyor; Green was appointed in 1808 at a salary of £300 p.a. At the same time a sum of £500 was voted towards the cost to the Plymouth Eastern Turnpike for a bridge over the river Yealm, at Lee Mill, to be designed and supervised by Green.

Green therefore became Devon's first county bridge surveyor, a title which was quickly to become county surveyor when he took responsibility for the county buildings. As surveyor, he was contracted to inspect over two-hundred and thirty bridges every year, to report deficiencies to Quarter Sessions and to obtain the magistrates' sanction to carry out repairs for a particular sum of money. He was allowed to seek outside work and Green put a series of advertisements in the *Exeter Flying Post* informing the noblemen and gentlemen of Devon and the adjoining counties that he had taken up residence in Exeter and soliciting their patronage.

By 1820 some thirty-six bridges had been built or widened to take the rapidly expanded traffic of the day. Three span bridges were Fenny, New at Tawstock, Cadhay over the Otter, New at Kingsteignton, Emmets over the river Dart, Hele at Hatherleigh, Head over the Mole, Cowley near Exeter, Steps at Dunsford, Weston near Honiton, and Brightly north Okehampton. Standard widths between parapets were agreed with the justices: for the most important turnpike roads 18–20 ft., for the lesser turnpike roads and roads much used 15–18 ft., and for other roads 12 ft. clear.

In 1811 Green commenced work on a canal from Exeter to Crediton but this project was halted almost immediately. For Lord Rolle and others he carried out land reclamation of Braunton marsh on the estuary of the river Taw where, with John Pascoe as his surveyor, an embankment 4,000 yd. long enclosed 1,300 acres and was completed in 1814. At Budleigh Salterton in the estuary of the river Otter, Lord Rolle

commissioned Green to reclaim an area approximately 2,000 yd. long by 330 yd. wide, enclosing over 140 acres. However, in October 1813 he had joined Ralph Walker, Joseph Whidbey, John Rennie, William Chadwell Mylne, Josiah Jessop and William Chapman (qq.v.) in advising the Admiralty Solicitor that enclosing a creek at Alverstock, near Gosport, would interfere with the tidal flow near Portsmouth!

Green's most important architectural assignment had come in 1810 when he transformed Buckland House, damaged by fire in 1798. His work there led Pevsner to say that his work showed him to be an accomplished innovative practitioner in the neo-classical style which was at this time becoming popular in Devon. His construction of St. David's Church, only 100 yd. from his home Elmfield, was commenced in 1816. Although it was replaced in 1897, the appearance of this church was well known in Exeter from its distinctive octagonal tower with eight Doric pillars surmounted by a rounded dome.

In 1819 Green reported to the trustees of three turnpike roads, the Plymouth Eastern, the Ashburton and the Exeter, concerning the road from Exeter to Plymouth. As always in those days the problems were the need to reduce unnecessary ascents and descents, the width and rolling resistance. In all, some 14 miles of road were realigned.

Early ideas for a canal from Bude to Launceston had surfaced in the 1770s and Robert Fulton (q.v.) had already suggested that inclined planes would be more suitable than locks for the 362 ft. rise from the sea to the river Tamar. In 1817 the fourth Earl Stanhope commissioned Green to prepare a plan of a possible line for the canal and Thomas Shearm was appointed surveyor. Work began in 1819 and Green built 35 miles of canal with six inclined planes fed from a dam across the upper reaches of the river Tamar; a reservoir was included. Green invested £3,000 of his own money in the canal but the shares produced no return in his lifetime.

During the decade from 1821 to 1830, one important scheme followed another. Some forty-six bridges were built or widened, including the magnificent five-arched Beam aqueduct north of Torrington and three-span bridges at Clyst Honiton, Gosford over the Otter, Long at Cullompton, Otterton, Tinhay—over the Wolf—and Newnham, over the Taw. In 1823–1824 Green combined with Underwood, the Somerset County Surveyor, to produce plans for a new House of Correction to stand alongside the County Gaol and Green became responsible for the construction of the £12,700 building. At this time his salary was £550 p.a. but he insisted that the County also paid him the fee of £87 16s paid to Underwood. The magistrates eventually agreed but this matter caused resentment that was to surface in 1831 and result in a reduction in salary.

Green became heavily involved in canal work. The Bude Canal was completed in 1824, a fine example of the use of 6 ton narrow boats and inclined planes. In 1824 he commenced the Torrington canal for Lord Rolle, extending from downstream of Weare Gifford to a point alongside the river south of the town. Here Rolle also employed him to build new grist mills and erect the machinery. Meanwhile the City of Exeter in 1820 had asked him to advise them on improvements to their canal and work proceeded on rebuilding the entrance sluice, providing a uniform depth of 10 ft., lowering the cills of the Double Locks and constructing a culvert under the canal to drain land fed by the Alphin brook. In 1824 he proposed that the canal should be extended 2 miles from opposite Retreat House to Turf, where vessels drawing 12 ft. could navigate the estuary at all tides.

The canal was then further deepened but at Exeter there was solid sandstone below the quays on the river. Green therefore proposed the construction of a basin, independent of the river; Telford was consulted and work proceeded on this project, the canal being opened to Turf in 1827 and the new basin completed in 1830. Green was voted the Freedom of the City of Exeter on 20 October 1820, his recognition significant in view of the fact that he was a Quaker. The idea of linking the Bristol and English channels had been alive since 1768 and in 1821 Green was asked to make a survey. Naturally he proposed a tub-boat canal to run from the existing canal near Taunton to Beer and in 1824 Telford was engaged to make a survey. Although an Act was obtained in July 1825 no more was heard of the scheme.

In 1823–1824, in conjunction with Joseph Whidbey (q.v.) of the Admiralty, Green surveyed and reported on the harbours of St. Ives and Ilfracombe and in 1827 he surveyed the bay for a harbour at Coombe Martin. Also in 1823 he proposed improvements for Bridport harbour but instead a scheme proposed by Francis Giles (q.v.) was carried out in 1824. In 1829 a scheme prepared by Green for a dock at Cardiff was adopted by Lord Bute and submitted to Parliament in 1830 but on the advice of William Cubitt (q.v.) it was altered and West Bute dock was opened in 1839.

Now having firm links with the Exeter Turnpike Trust, Green was invited to make a survey of the road between Exeter and Crockernwell, the Trust's limit on the way to Okehampton. He produced a new route from Pocombe Bridge to Tedburn St. Mary using the valley of the Alphin Brook and it was opened in 1824. For the Countess Wear Committee of the Trust he rebuilt the swing bridge over the canal in 1823.

In conjunction with a proposal to build Laira bridge, the Plymouth Eastern Turnpike turned to Green to improve the roads from the eastern bank towards Totnes. Green produced a plan for a direct road to Yealmpton, some improvements to Ermington and then a new road up the Ludbrook valley, bypassing Ugborough to Ladydown; this required new bridges over the river Yeo, the Erme and the Lud. In 1825 a three-mile diversion was made just north of Sandwell to run

directly to Totnes and again he was asked to supervise the building of a new bridge over the river Harbourne. In 1827 he was responsible for a new road to Yarcombe for the Chard Turnpike Trust. In 1824 Green had built Eggesford Bridge over the river Taw, and the following year he built a new road down the valley of the Taw. This road saved over 1,000 ft. of unnecessary ascents and descents and provided Green with the opportunity to build four more bridges with a view to them being taken over by the county.

As result of a complaint that too much was being expended on the maintenance of the prisons it was proposed in 1830 that Green's salary should revert to £300 p.a. and a letter from Mr. Rendel was produced, offering to perform all Green's duties for £300. Green accepted this reduction in salary but was forced to look outside the county for as much consulting work as he could command; besides building another twenty-seven bridges during the next decade he turned his attention once more to canals and other proposals. For the Barnstaple Bridge Trust in 1834 he widened the existing 16 span bridge by cantilevering delicate and attractive footways 4 ft. wide on each side using ironwork from the Neath Abbey Iron Company. In 1832 he proposed water supply, sewerage and railway schemes for Torquay.

Rennie had built over 11 miles of canal from Tiverton to the Devon–Somerset border to convey limestone from the Canonsleigh quarries, and this had been opened in August 1814. The Grand Western canal proprietors wished to extend their canal to Taunton to join the Bridgwater and Taunton Canal. The distance was only 13 miles but the difference in level was 262 ft. Green had presented a report to the company in 1829, advocating boats 20 ft. by 6 ft. carrying 5 tons, six of these to be drawn by one horse. In a further report in 1830 he suggested one inclined plane and seven perpendicular lifts, with boats of 8 tons capacity, at an estimated cost of £61,324. Work commenced in June 1831 but both the lifts and the inclined plane had difficulties when working was commenced and by 1836 Green had ceased to be engineer; work was completed in 1838 at a cost of £80,000. In 1831 a canal for Chard was proposed and Green was consulted. He proposed the use of two lifts, two inclined planes and two tunnels, all at a cost of £57,000. Work got under way in June 1835 but soon Green ceased to be engineer, no doubt because of troubles with the Grand Western Canal; the canal was completed by May 1842.

The silting of the Gwendraeth estuary in South Wales in the early nineteenth century had caused Kidwelly to lose its facilities as a port for the coal of the valley. The Kidwelly and Llanelly Canal and Tramway Company had obtained powers by an Act of 1812 to extend a canal up the valley to beyond Cwm Mawr about 250 ft. above sea level and in 1832 the company called in Green to report on extending the canal beyond the point reached in 1824. In 1833 he recommended two locks and then three inclined planes at an estimated cost of £35,845. By 1834 work was well advanced but the following year Green informed the directors that he was unable to finish his inclined planes. He was dismissed in 1836, in the same year being dismissed as engineer to Burry Port as a result of the collapse of a dock wall.

In 1830–1831 Green's home was recorded at 38 Southernhay Place while in 1833, 1834 and 1836 it was in Magdalen Street. A notice of bankruptcy appeared in the *Exeter Flying Post* on 9 March 1837 following an entry in the *London Gazette*. By 1838, Green had moved out of the city to Alphington, no doubt to economise. The sums involved in his failed contracts were probably so large that he had no opportunity to recover them from his income during the closing twelve years of his life. This would have affected his status in the Religious Society of Friends, who might have disowned him because of his bankruptcy.

A contract for a dock at Newport had been let in 1835, but within two years the contractors were in trouble and some time around 1840 Green was appointed to take over from the previous resident engineer to complete the works. He took up residence there but in 1840 the Devon Justices were told that Green had said he could not continue his work in Devon satisfactorily without deputising the minor matters to his son. Some magistrates complained that they were having to do the work of the surveyor and Green was given twelve months notice from the Midsummer 1840 Sessions. So Green left the County's employment and in 1841 was listed as living in Heavitree with his son as 'Green, James and Son, Civil Engineers and Land Surveyors, Portview Cottage, Heavitree'. Green brought the work at Newport Dock to a successful conclusion and then it was in 1843, too late in his career, that he settled in London, 'for owing to the active competition of younger men, he was not so extensively employed as he might have been'.

In 1844, because of his knowledge of the estuary of the river Exe, Green was consulted on the building of the South Devon Railway. Exeter City opposed the bill to safeguard its navigation rights in the estuary and Green made a report in 1844. The essence of his evidence was that the embankment alongside the estuary would enclose 102 acres which would make a significant difference to the movement, and hence the scour, of the water of the estuary as it crossed the bar.

When the floating harbour of Bristol was formed by placing locks on the river downstream of the docks and diverting the river Avon along a new channel to the tideway below the locks, no thought had been given to intercepting and carrying off the sewage of the city away from the harbour. Further sewage was brought in by the tributary river Froome which passed through a populous part of the city. In 1846 Green was instructed by the Council to advise on the measures necessary for abating the nuisance and he reported in October. He recommended straightening the river Froome, making it of uniform width to give the greater scour of the bed and intercept the sewers that discharged into it. The Council considered that it would not proceed because it did not have the legal powers but further instructed Green to advise on action to be taken between Stone Bridge and Castle Moat. The report was made in March and during the summer works were carried out at a cost of £4,537 to clear this area of accumulated sludge. Green presented a paper on these reports and work carried out to the Institution of Civil Engineers on 8 February 1848; he had been proposed as a Corresponding Member by Telford in 1824.

In May 1805 Green had married Elizabeth Dand at St. Martin's, Birmingham. A son, Thomas, died aged three in 1815, but another son, Joseph, was born in 1817. James Green died from a heart attack on 13 February 1849 in his 68th year at 67 Manchester Buildings, Westminster, and was buried on 28 February at Bunhill Fields as a non-member of the Religious Society of Friends, though his connection was enough for a Quaker burial.

His death was noted in the *Bristol Mirror* which commented that his son Joseph D. Green was resident engineer at Bristol docks.

A. B. GEORGE

[Chelmer & Blackwater Navigation letterbook, DZ 36/40, Essex CRO; Quarter Sessions minutes, Devon County RO, Exeter; Exeter Turnpike Trust minutes, Devon County RO, Exeter; Deposited plans for canals and turnpike roads, Devon County RO, Exeter; Obituary (1849–1850) *Min Proc ICE*, **IX**, 98–100; C. Hadfield (1967) *The Canals of South West England*, 151; Helen Harris (1996) *The Grand Western Canal*; A. B. George (1999) *James Green, Civil Engineer*]

Publications
1838. Description of perpendicular lifts, *Transactions ICE*, **II**, 185–191
1845. (with P C. de la Garde) Memoir of the canal of Exeter, *Min Proc ICE*, **IV**, 91–113
1848. Account of the recent improvements in the drainage and sewerage of Bristol, *Min Proc ICE*, **VII**, 77–84

Works
1806–1807. Reclamation of Land for Lord Borington, 1807
1808–1841. Devon County (Bridge), Surveyor, over one-hundred bridges for the Devon Quarter Sessions and for various Turnpike Trusts, major works detailed below, for a full list see George
1808. Yealm Bridge, Lee Mill
1808. Fenny Bridge
1810. New Bridge, Kingsteignton, three spans, replaced in 1845
1810. Reconstruction of Buckland House, Buckland Filleigh
1810–1811. Emmett's Bridge, three spans

1810–1812. Hele Bridge, Hatherleigh, three spans
1812. Long Bridge, Membury, three spans
1813. Head Bridge, three spans
1813–1814. Cowley Bridge, three spans
1813–1814. Reclamation of Land for Lord Rolle and others at Braunton
1814. Reclamation of Land for Lord Rolle at Budleigh Salterton
1814. Improvements for the Exeter Turnpike Trust
1814. Improvements for the Ashburton Turnpike Trust
1816. Steps Bridge, Dunsford, three spans
1816. St. David's Church, Exeter
1817. Weston Bridge, three spans
1819. Exeter–Plymouth road improvements, 14 miles
1821. Clyst Honiton Bridge, three spans, £1,200, now widened
1823–1824. Exeter–Crockernwell road improvements
1823–1824. Exeter House of Correction, £12,700
1823–1825. Improvements for the Plymouth Eastern Turnpike Trust
1823–1825. The Torrington Canal, £40,000, including Beam Aqueduct, Torrington, five spans, now a road bridge
1824–1825. Gosford Bridge, three spans, estimate £1,200
1825. Long Bridge, Collompton, three spans
1826–1830. Exeter Canal, £110,000
1827. Otterton Bridge, 3 spans, estimate £1700
1827. Tinhay Bridge, three spans, estimate £1,200
1827–1828. Chard–Exeter road, improvements
1828. Dipper Mill Bridge, single span, £799
1829–1836. The Grand Western Canal, Lowdwells to Taunton, £80,000
1830. Newnham Bridge, three spans
1833. Polson Bridge, three spans, cast iron central span, since rebuilt
1833. Drainage of Westmoor, Kingsbury Episcopi, Somerset
1834. Barnstaple Bridge, sixteen spans, widening with 4 ft. footways in cast iron
1835–1836. The Chard Canal
1835–1836. The Kidwelly and Llanelly canal (part)
1835–1836. Bury Port
1836. Crocombe Bridge, three spans
1836. Long Bridge, Plympton, three spans
1838–1843. Newport Dock
1840–1842. Newton Poppleford Bridge, three spans, estimate £2,500

GREEN, John (1787–1852), civil engineer and architect, was the son of John Green and his wife, Ann Armstrong. He was born at Newton, north east of Corbridge, Northumberland, on 20 June 1787, the youngest of four children. His father was a carpenter and a maker of agricultural implements and Green Jr. joined him in the business, which was subsequently transferred to Corbridge. In 1810 he married Jane Ellis (1783–1846) in Newcastle and while living in the Tyne valley had two children.

In the early 1820s Green, described as 'a plain, practical, shrewd man of business', moved into Newcastle and established an architectural practice. His work as an architect has been detailed by Colvin who lists some twenty churches which were either designed or restored by him, together with fifteen secular commissions. In some instances the work was carried out in conjunction with his son, Benjamin (1813–1858), and from approximately 1835 onwards the work of father and son is virtually indistinguishable. Somewhat confusingly, his nephew John Green (1807–1868), was also an architect.

As a civil engineer, Green's first work was in connection with a suspension bridge over the river Tyne between North and South Shields, designing and specifying the masonry piers of the bridge proposed by Captain Samuel Brown, RN (q.v.). He also designed and this time built a similar bridge, again over the river Tyne, at Scotswood where the client was a company promoting a road westwards from Newcastle. With a main span of 353 ft. the bridge was completed in 1831. The same year he completed a suspension bridge over the river Tees at Whorlton and it remains as the oldest suspension bridge still supported by its original chains. Green was involved in the construction of Blackwell bridge, again crossing the river Tees, where designs for a suspension bridge, a three-span iron arch and a laminated timber structure were submitted, together with others. None of these proposals was adopted and it was a masonry bridge, incorporating cornes-de-vache and designed by Green, which came to be built in 1832.

The Durham Junction Railway was formed to enable coal to be brought from collieries south of the river Wear for shipment on the Tyne. Its line included a crossing of the Wear and in 1834 Green was invited to provide 'a Plan and Estimate for a Bridge of malleable Iron of 170 ft. span' and a height of some 100 ft. His proposal was not accepted but the same year two masonry bridges were constructed to Green's designs, at Lintzford, crossing the river Derwent, and at Bellingham, crossing the North Tyne. He also designed and built a masonry bridge for the Great North of England Railway, for which company Thomas Storey (q.v.) was the engineer; completed in 1840, it was located at Nether Poppleton and crossed the river Ouse north of York.

Green's most important contribution to civil engineering was his work on laminated timber bridges and viaducts. Although father and son worked together, John would seem to have been responsible for the timber bridge designs undertaken by the practice and Percy L. Addison, perhaps a grandson, writing from Whitehaven in 1887 maintained that they were John's work alone. Green had first investigated this type of structure when called upon to design the road bridge at Scotswood, although eventually recommending a suspension bridge, and in 1833 had entered a competition promoted by the

Newcastle and Carlisle Railway for a rail bridge to cross the Tyne near his earlier structure. Although a 1:12 scale model was exhibited and Green was awarded a premium of £10 work did not proceed.

The first laminated timber structures built by him—and the first such in Britain—were two viaducts for the Newcastle and North Shields Railway, opened in 1839. In 1835 the use of timber *vs* masonry for the viaducts had been investigated and, with cost the deciding factor, masonry was rejected and the design of laminated timber structures was put in hand, albeit with masonry piers. Green, the company's architect, estimated that the savings effected by the use of timber had reduced the cost of the two structures by some £20,000.

Green continued his interest in the application of laminated timber with proposals made in 1839 for a high level bridge to cross the river Tyne at Newcastle. Initially, no progress was made but in 1841 the High Level Bridge Company was formed to promote his plans for a four-arch crossing, its maximum span 280 ft. Again, no progress was made but two years later Green submitted designs for a bridge to take dual traffic; this time the design was for a seven-span masonry viaduct with an iron superstructure. His bridge did not materialise and the river was eventually crossed at the same point by the High Level Bridge, the work of Robert Stephenson (q.v.) and Thomas Elliot Harrison.

Green was elected a member of the Institution of Civil Engineers in 1840 and in the same year designed two further laminated timber structures. The first was a viaduct some 1000 ft. long over the river Esk at Dalkeith and the second a two-span bridge over the river Wear for the West Durham Railway. Like his earlier proposal for a bridge at Scotswood, but dissimilar from the viaducts already noted, the three arches of the Wear bridge were above deck level and the twin rail tracks were suspended from them by iron hangers.

Green's laminated timber spans were also used in roof construction. The roof of the trainshed of the North Shields station of the N&NSR comprised laminated arches of 12½, 25 and 12½ ft., forming a roof which extended to a length of 140 ft., supported on side walls and cast-iron columns. Another roof—still partially surviving—is that of the church at Cambo where 29 ft. span laminated arches are almost semicircular in form.

Green, like several other Newcastle engineers, was a member of the Literary and Philosophical Society and maintained an office in the Royal Arcade, Newcastle. He died on 30 September 1852 at his house in Oxford Street, Newcastle, and was buried in Jesmond Old Cemetery. His estate was modest and it passed principally to his daughter, Isabella, married to John Addison, a civil engineer. Benjamin Green outlived him for only six years, dying in Dinsdale Lunatic Asylum.

R. W. RENNISON

[*Durham Junction Railway Company, Minutes of Meetings* (1834) PRO, RAIL 165/1; *Newcastle and North Shields Railway, Minute Book* (1834–1839) PRO, RAIL 507B; B. Green (1846) On the arched timber viaducts on the Newcastle and North Shields Railway…, *Min Proc ICE*, **5**, 219–232; *Newcastle Weekly Chronicle*, 8 January 1887; R. Welford (1895) *Men of Mark Twixt Tyne and Tweed*, vol. 2, 326–330; L. G. Booth (1971–1972) Laminated timber arch railway bridges in England and Scotland, *Newc Soc Trans*, **44**, 1–21; H. Hagger (1976) The bridges of John Green, *Northern Architect*, **8**, 25–31; Colvin (1978) 361–363]

Reports and drawings

c. 1825. *Specification for Masonry, North and South Shields Suspension Bridge*, Newcastle Central Library, L942.8

1838. *Sketches illustrative of Richard Grainger*, Newcastle Central Library, L942.82

1843. *Coble Dene Wet Docks, North Shields* (with R. Nicholson)

Works

1829–1831. Scotswood suspension bridge, 353 ft. span

1830–1831. Whorlton suspension bridge, 183 ft. span

1831. Wynyard Park suspension bridge

1832. Blackwell bridge, masonry segmental arches, spans 68, 78 and 68 ft.

1834–1835. Lintzford bridge, masonry, two segmental arches, each 38 ft.

1834–1835. Bellingham bridge, masonry, four segmental arches of 50 ft.

1837–1839. Viaducts for Newcastle and North Shields Railway: Ouseburn, timber arches on masonry piers, five spans of 116 ft., 108 ft. high, £24,000; Willington, timber arches on masonry piers, seven spans of 120 ft., 82 ft. high, £23,500

1838–1840. Great North of England Railway bridge, Nether Poppleton, masonry, three semi-elliptical arches of 66 ft.

1840. West Durham Railway bridge, timber, two spans of 79 ft.

1840–1841. Dalkeith railway viaduct, timber arches, four spans of 120 ft. and two of 110 ft.

GREEN, Sir William, Bt. Major General

(1725–1811), military engineer, whose forty-year career, beginning as a cadet gunner in 1738 and including Fontenoy (1745), Quebec (1759) and twenty-two years on Gibraltar (1761–1783), is mainly of interest for his organisational reforms.

From 1786 until 1802 he was in fact the last bearer of the title Chief Engineer, his tomb in Plumstead churchyard describing him as Chief Royal Engineer, a distinction arising from the change in designation in 1787 from Corps of Engineers to Royal Engineers. This was a realistic acceptance of the change in perception of the term 'engineer' from the beginning of the eighteenth century, when it still had mostly military connotations, to its almost total 'civilian' usurpation by the end.

His other more practical reform was the creation of the 'Soldiers Artificers Company' in 1772. This removed ordnance construction projects from dependence on unreliable local labourers and unscrupulous contractors which had bedevilled them for over a century. They were first used in Gibraltar to implement Green's report of 1770 in which he estimated that £51,126 needed to be spent on remedial work on the King's and Princes lines and new works on the Prince of Orange Bastion, Montague Bastion and the King's Bastion, with casemates for 800 men defending the new mole. For the old mole he proposed a loose rubble dyke with two rows of piles driven into the sea bed in front.

It was Green's artificers, in particular Sergeant Major Henry Ince, who were responsible for the major engineering development during the siege of 1779–1783; the tunnelling of galleries into the rock originally to improve the angle of flanking fire on the Spanish. Green's earlier report on similar mines and galleries in Luxembourg obviously proved useful. Also in Gibraltar he designed a hospital (1771) and built a reservoir for his home, later converted into a bombshelter as described in his wife's diary of the siege.

He had married Miriam, daughter of Justly Watson (q.v.), in 1754 and they had a son, Justly Watson Green, and five daughters. Miriam unfortunately died of a chill caught in a 'bombproof' in 1782 but William survived her by thirty years, receiving the thanks of Parliament (the first engineer to be so honoured), promotion to Major General, a Baronetcy and the position of Chief Engineer. He died in 1811 and his portrait is hung in the Engineers Mess at Chatham.

SUSAN HOTS

[Royal Engineers Library, Conolly papers; M. Green (1912) A lady's experience in the great siege of Gibraltar, *Royal Engineers Journal*; DNB (1896) Porter, vols. 1 and 2; Bendall; T. H. McGuffie (1965) *The Siege of Gibraltar*; Q. Hughes and A. Migos (1995) *Strong as the Rock of Gibraltar*]

Reports
Reports, plans, etc., include the following:

1747. *Drawings of the Galleries and Mines of the Fortress of Luxembourg* (BL)
1770. *A Critical Review and Report of the State of Fortifications of Gibraltar* (National Army Museum)
1778. *Considerations upon the present State and Strength of Gibraltar*, plus SEVEN plans

GREENWOOD, John (1791–1867), surveyor of Gisburn, Yorkshire, is believed to be the 'surveying engineer' who was elected an Associate of the Institution of Civil Engineers in 1827. Greenwood was born at Gisburn on 25 March 1791 and, like his elder brother, Christopher (1786–1855), took up surveying. At one stage they were involved with Netlam and Francis Giles (qq.v.) in publishing maps. The Greenwoods were

responsible for an important series of early maps of Lancashire (1818), Cheshire (1819) and Shropshire (1826–1827) which formed part of the Telford collection left to the Institution of Civil Engineers on Telford's death in 1834. John Greenwood appears to have returned to Yorkshire, where he carried out a number of local surveys, and died in Gisburn in September 1867.

MIKE CHRIMES

[Membership records, ICE Archives; Catalogue of Telford drawings, ICE Library; Bendall]

GREGORY, Dr. Olinthus Gilbert (1774–1841), mathematician, was born at Yaxley, Huntingdonshire, on 29 January 1774. Despite humble origins he attended a local school and was then educated by Richard Weston, a Leicester botanist. He was clearly gifted in mathematics as his first published work, *Lessons, Astronomical and Philosophical* appeared in 1793 when he was only nineteen. The following year he began contributing to *Ladies diary*. His work on the sliding rule, although unpublished, brought him to the attention of Charles Hutton (q.v.), who subsequently became his patron.

From 1796 until 1802 he lived in Cambridge, teaching mathematics, and for a short time ran a bookshop. In 1802 he published *A Treatise on Astronomy* and the same year became a teacher of mathematics at the Royal Military Academy, Woolwich, with Hutton's support, and in 1807 succeeded Hutton in the Chair of Mathematics there; he held this post until his retirement in 1838.

From his arrival at Woolwich until his death there on 2 February 1841 Gregory combined his teaching with writing a succession of textbooks which seem to have been largely aimed at aspirant engineers and teachers of mathematics. They included a revision of Hutton's *Course of Mathematics*. These were generally very popular, most going through three or four editions, although his religious work *Evidences, Doctrines, and Duties of the Christian Religion,* originally published in 1811, was his most successful work, going through nine editions down to 1857.

Gregory's most important contributions to science were probably his *Treatise of Mechanics* (1806) and his paper on the *Velocity of Sound* (1833). His *Treatise on Mechanics,* of which the account of steam engines was reprinted in 1807 and 1809, took the form of a two volume handbook or encyclopaedia. It was much maligned by Todhunter and Pearson for reproducing Galileo's findings on strength of materials, then known to be erroneous: 'Nothing shows more clearly the depth to which English mechanical knowledge had sunk at the commencement of this century'. This harsh verdict totally overlooks the importance of Gregory's work in explaining to mechanics and engineers much of value in mechanics, including topics such as arch and catenary theory. His most valuable work for engineers was his *Mathematics for Practical Men* (1825), which was dedicated to Thomas Telford (q.v.), and generally well thumbed down to mid-century by engineers such as Edwin Clark. It was explicitly intended 'for the use of the younger members of the Institution of Civil Engineers'; it contains reference to Prony's earth pressure theory. Aside from more general writings, and contributions to Peter Nicholson's *Journal*, he edited *Gentleman's diary* (1802–1819) and *Ladies Diary* (1819–1840), and helped edit the *Pantalogia*, a twelve-volume technical encyclopaedia. Gregory's general involvement in education included support for University College; his name appeared on its original foundation stone.

Among Gregory's children was Charles Hutton Gregory (1817–1898) who had a distinguished career as a railway engineer and rose to the Presidency of the Institution of Civil Engineers; Gregory was himself elected an honorary member. He was a member of many learned societies overseas, and Secretary of the Astronomical Society.

FRANK JAMES

[Gregory papers, Royal Artillery Institution, Woolwich; Membership records, ICE; Obituary (1842), MPICE, **2**, 12–13; DNB; I. Todhunter and K. Pearson (1886) *The History of the Theory of Elasticity*, **1**, 88]

Publications

1806. *A Treatise of Mechanics*, two volumes plus plates; 4th edn., 1826
1807. *An Elemental Treatise on Natural Philosophy*, translation of work by R. Huet
[1813]. *Pantalogia*
1817. *Description and Use of a Ballistic Pendulum ... in the Royal Arsenal at Woolwich*
1825. *Mathematics for Practical Men*, 2nd edn., 1833; 3rd edn., 1848, enlarged by Henry Law
1840. *Hints ... for the Use of Teachers of Mathematics*

GRIFFITH, Sir Richard John, Bt., FRSE, MRIA (1784–1878), civil engineer and geologist, first baronet, son of Richard Griffith, of Millicent, Naas, County Kildare, by his first wife Charity, daughter of John Bramston of Oundle in Northamptonshire, was born on 20 September 1784 at 8 Hume Street, Dublin. His father had made a modest fortune with the East India Company before returning to Ireland to take up residence at Millicent. His marriage to Charity Bramston in September 1780 was short-lived as she died in 1789. Griffith senior was a Director of the Grand Canal and played an active role in its affairs.

The young Griffith attended school at Portarlington and Rathangan before completing his early education in Dublin. In 1800 he became an ensign in the Royal Irish Regiment of Artillery stationed at Chapelizod near Dublin, but, at the Act of Union in 1801, he retired from the army to pursue a career as a civil and mining engineer.

Around 1802 he went to London, where he studied chemistry, geology and mineralogy for two years, subsequently going to Cornwall to study practical geology. He visited Matthew Boulton at Birmingham and also the various mining districts in England and Scotland. Between 1806 and 1808 Griffith studied geology at Edinburgh University and in 1807, at the age of 22, was elected a Fellow of the Royal Society of Edinburgh. It was in Scotland that he met his future wife, Maria Jane Waldie (1786–1865), eldest daughter of George Waldie of Hendersyde Park, Kelso, whom he married in Kelso on 21 September 1812.

Griffith returned to Ireland in the Spring of 1808 and for the next fourteen years devoted himself to an investigation of Ireland's geology, including the search for the mineral wealth which it was hoped would trigger an Irish industrial revolution. He was commissioned by the Royal Dublin Society (RDS) in May 1809 to survey the Leinster coalfield and this he completed in 1813. In November 1812 the RDS had appointed Griffith as their Mining Engineer at a salary of £300 p.a. Having visited Newcastle upon Tyne, between 1814 and 1824 he completed surveys of coalfields in Connaught, Ulster and Munster. He lectured on geology and mining at the RDS during the period 1814 to 1829, when he resigned his position with the Society. In 1812 he was appointed Inspector-General of His Majesty's Mines in Ireland. During all this time, Griffith collected information for his *Geological Map of Ireland*, the first edition of which appeared in 1815. For this and other like achievements, he was awarded in 1854 the Wollaston Medal of the Geological Society of London.

On 28 September 1809 Griffith was one of seven engineers appointed by the Bog Commissioners to undertake surveys of the main bogland areas of Ireland. Between 1809 and 1813 he surveyed over 206,000 acres of bogland and made preliminary surveys of a further 267,000 acres of mountain bog. He also reported on the attempts being made to drain Chat Moss in Lancashire. These surveys represented 843 days of work for which he was paid £2,466.

Griffith commenced his road and bridge building activities in 1822 when he was sent by the government to Munster to carry out a programme of public works as a means of alleviating famine and opening up parts of the counties of Cork, Kerry and Limerick. Up until 1830, he was responsible for the construction of some 240 miles of roads, notable those connecting Abbeyfeale with Castleisland, Newmarket with Listowel and Newmarket with Charleville. Of the ten masonry road bridges designed by him prior to a report published in 1831, Listowel, with its five 50 ft. spans, and Wellesley (or Feale), with a single arch of 70 ft. span, are noteworthy.

In 1825, Griffith was appointed Director of the Boundary Survey, a land and property valuation survey which ran in parallel with the general mapping being carried out by the Ordnance Survey. He returned to reside permanently in

Dublin in 1828 at 2 Fitzwilliam Place, where he died on 22 September 1878 and was buried in Mount Jerome Cemetery in the city.

His later career embraced many public duties, including acting as a Commissioner enquiring into the development of Ireland's railway system, and as a Commissioner making recommendations for the improvement of the Shannon Navigation. He was Chairman of the Irish Board of Works 1850–1864. He was twice President of the Institution of Civil Engineers of Ireland, during the period 1850–1855 and again in 1861. He received an honorary degree of LLD from the University of Dublin in 1849 and an honorary master of engineering (MAI) degree from the same university in 1862. He was created a baronet on 8 March 1858. His portrait is reproduced in Herries-Davies and Mollan (1980).

RON COX

[Commissioners ... into Bogs in Ireland (1810–1814) *Reports*, 1–4; Select Committee on Public Works (Ireland) (1835) Minutes of Evidence, 220–240; *Report of the Select Committee on the General Valuation of Ireland* (1868–1869) Report IX, evidence 199–200 [HC 362]; Anon. (1874) *Dublin University Magazine*, **83**, 432–437; *Quarterly Journal of the Geological Society*, 1879; *The Irish Times*, 24 Sept.1878; *The Freeman's Journal*, 24 Sept.1878; *The Times*, 27 Sept.1878; Anon. (1878–1879) Memoir of Sir Richard Griffith, Bart, *Min Proc ICE*, **55**, 317; J. S. Crone (1928) *A Concise Dictionary of Irish biography*, 83; H. Boylan (1978) *A Dictionary of Irish biography*, 129; G. L. Herries-Davies and R. C. Mollan (eds.) (1980) *Richard Griffith 1784–1878*; P. O'Keeffe and T. Simington (1991) *Irish Stone Bridges*]

Publications

Twenty-six papers are listed in the *Royal Society Catalogue of Scientific Papers*, mostly on geological subjects in *Transactions of the Geological Society, Proceedings of the Geological Society, British Association Reports* and *Philosophical Magazine*.

1810. Report on the practicability of draining and improving a part of the Bog of Allen. *Commissioners on ... Bogs ... in Ireland, Reports*, **1**, 15–55

1812. Report on the practicability of draining and improving a part of the Bog of Allen or district of the River Barrow, *Commissioners on ... Bogs ... in Ireland, Reports*, **2**, 31–76

1814. Reports on Bogs in the counties of Galway and Roscommon which discharge into the river Suck, *Commissioners on ... Bogs ... in Ireland, Reports*, **4**, 109–162

1814. Report on the mountains of the counties of Dublin and Wicklow, *Commissioners on ... Bogs ... in Ireland, Reports*, **4**, 163–174

1814. Report Bogs in ... County ... Mayo and County Sligo, *Commissioners on ... Bogs ... in Ireland, Reports*, **4**, 175–186

1814. Report on bog improvements at Chat Moss, *Commissioners on ... Bogs ... in Ireland, Reports*, **4**, 211–216

1818. *Geological and Mining Surveys of the Western Coal Districts of Tyrone and Antrim*

1821. Report relative to the moving bog of Kilmaleady in the King's County, *Phil Mag*, **58**, 70–73

1829. *Report and Section of Tyrone collieries*, Royal Dublin Society

1836. On the Geological map of Ireland, *British Association Report*, 56–58

1838. On the leading features of the geology of Ireland, *Br Association Report*, 88–90

1859–1861. Presidential Address, *Transactions of the Institution of Civil Engineers of Ireland*, **6**, 193–221

Works

Works prior to 1830:

1822–1830. Roads and bridges in Cork, Kerry and Limerick, engineer, 240 miles, *c.* £120,000

GRIMSHAW, John (1763–1840), an active partner in a ropeworks in Sunderland, was born at Guiseley on 18 August 1763, the eldest of five children of William Grimshaw (b. 1731) and his wife Hannah Adamson (b. 1743), of Thirsk.

Initially a joiner and turner, Grimshaw entered into a partnership with Rowland Webster and Ralph Hill in 1793, resulting in the formation of Grimshaw, Webster & Co, ropemakers. Soon afterwards he 'opened what was probably the world's earliest factory for machine-made rope', housing in a four-storey building manufacturing plant powered by a 16 hp Boulton and Watt engine; his system of ropemaking eliminated the need for a ropewalk. Between 1796 and 1802 Grimshaw was granted three patents for improvements in ropemaking and by the end of the century was producing some 500 tons p.a. of cordage; in 1804 production was 800 tons.

Grimshaw was responsible in 1805 for undertaking repairs to the cast-iron bridge completed at Sunderland under the direction of Rowland Burdon (q.v.) in 1796, only the second built in Great Britain. As a result of a lack of bracing, the bridge had become unstable and the ribs had moved as much as 19 in. out of alignment. Grimshaw inserted iron diagonal braces between the arch ribs and Marc Brunel (q.v.) later wrote that, had the work not been done, 'it must ultimately—and I should add, very shortly too—have shared the fate of many cast iron bridges'. In the year during which the repairs were carried out Grimshaw became Bridge Treasurer.

Grimshaw took an active part in the controversy regarding the choice of lines for a canal to be formed to carry coal from south west Durham to the river Tees, writing in 1818 in opposition to the more northerly line under the pseudonym *Alexis*. By this time Grimshaw was a part-owner and manager of Fatfield colliery and had experimented on railway traction with Benjamin

Thompson (q.v.) and he became, in Tomlinson's words, 'the strongest intellectual force at the back of the railway movement'. He championed the cause of the proposed Stockton and Darlington Railway and the prospectus for its formation was drawn up largely in accordance with one of his letters to the press.

The ropemaking partnership was dissolved in 1817, from that time only Grimshaw's name being used in the works' title; his active management is confirmed by the fact that in 1822 he was granted a further patent for the manufacture of flat ropes. His rope-making activities perhaps account for the fact that he was a member of the provisional committee for the company established in 1825 to construct a suspension bridge between North and South Shields. It was soon after this time that he supplied the haulage ropes used on the S&DR's inclined planes; later, similar ropes were provided for the Canterbury and Whitstable Railway. His works continued after his death, a notice published in 1847 stating that John Grimshaw & Co. were able to supply ropes for marine and other purposes. In addition to his other interests, Grimshaw was a shipowner.

In 1795 Grimshaw had married Elizabeth Miller (1767–1841), his housekeeper; he died without children at his residence in South Street, Sunderland, on 22 May 1840 and was buried in the Quaker burial ground five days later. His estate amounted to some £7,000 and in his will he directed that it should pass to the daughter of his brother-in-law, 'my Wife and Myself having adopted the said Isabella Miller as our own daughter'; his wife Elizabeth was to have the house rent-free, together with an income of £200 p.a. Legacies of £150 each were paid to the two sons of his brother, William, and to the daughter of his sister, Hannah.

R. W. RENNISON and J. G. JAMES

[*Sunderland Herald*, 3 June 1840; W. W. Tomlinson (1914) *North Eastern Railway*, 46–56, 134; *Histories of Famous Firms: County Borough of Sunderland* (1954); J. G. James (1986) *The Cast Iron Bridge at Sunderland (1796)*, 33–37; Skempton; *The Corder Manuscripts, Vol. 1: Friends Pedigrees*, Sunderland City Library, L289.6C; J. F. Clarke (1997) *Building Ships on the North East Coast, 1640–1914*, vol. 1, 68]

Publications

1796. Patent 2089, *Vegetable Substances to bleach … Cloths, or other Materials*, 17 Feb.
1799. Patent 2335, *Manufacture of Ropes and Cordage*, 2 Aug.
1802. Patent 2651, *Machinery for laying Ropes*, 5 Oct.
1818. *A Report of the Repairs given to Wearmouth Bridge, in the Year 1805*, Sunderland
1822. Patent 4669, *Stitching, Lacing or Manufacturing Flat Ropes …*, 16 Apr.
1822. *Specification of a Patent granted to John Grimshaw … for an improved Method of making Flat Ropes …*, Bishopwearmouth, reprinted 1826

Works

1805. Wearmouth Bridge, repairs

GRUNDY, John, Sr. (*c.* 1696–1748), land surveyor and civil engineer was born *c.* 1696 at Bilstone, Leicestershire, the son of Benjamin and Mary Grundy, and for most of his first forty years or so lived at the neighbouring, larger village of Congerstone. He married Elizabeth Dalton and their first child, John, was baptised at Congerstone on 1 July 1719.

By the mid-1720s Grundy had established himself as a land surveyor and teacher of mathematics to private pupils, and could be 'inquired for' at addresses in Leicester, Derby, Market Bosworth and Congerstone. For surveying he used a theodolite, circumferentor, Beighton's improved plane-table and Gunter's chain. In 1731 he was employed by the Duke of Buccleugh to survey his estates in south Lincolnshire, a commission which led to a change in Grundy's career.

He arrived at Spalding in April, spent six months on the survey, and observed with interest the work in progress on the great sluice being built under the direction of John Perry (q.v.) on the river Welland at the south end of the town. In June he joined the Spalding Gentlemen's Society. Next year he presented the Society with a map of Spalding, and spent another six months completing the Buccleugh survey. This provided him with an opportunity for studying drains, banks, sluices and outfalls. He became convinced that the proper drainage of low grounds depended upon the applications of what he called mathematical and philosophical principles.

In May 1733 the Commissioners of Sewers (the land drainage authority) asked him to survey, map and take levels in the parish of Moulton bordering the sea coast 3 miles from Spalding, and to draw up a scheme for improving its drainage. He obtained from Jonathan Sisson (*c.* 1690–1747), the leading instrument maker of the period, a precision spirit level with telescopic sights, with which he could achieve an accuracy better than an inch per mile. He demonstrated the level and his method of using it to Henry Beighton (q.v.) in January 1734, to the latter's satisfaction.

Three months later Grundy published a paper in which, among other things, he makes an early attempt to tackle the problem of flow in open channels. By analogy with free fall on an inclined plane he calculates the time required for a particle of water to flow along a drain 4 miles in length with a gradient of 3 in. per mile. The result, 1 hour and 28 minutes, gives an average velocity of 4 ft. per second. But this ignores fluid friction, and from field observations he finds the actual velocity to be less than half the theoretical value. He also tackles the problem of flow through a sluice, using in this case an experiment to measure the flow through an opening in a tank under a small constant head. He suggests that the full scale flow can be found by simple proportion. Finally, he insists on the need for accurate levels and mapping, and observations on rivers and drains, in deciding on the best method of draining fenland. This general approach, and the experimental result, met with approval from J. T. Desaguliers (q.v.) who Grundy seems to have known; certainly he was a subscriber to his forthcoming book on *Experimental Philosophy*.

In April 1734 he set off again for Spalding, this time at the request of the Adventurers of Deeping Fen, to prepare a scheme for improving the drainage of this area of 30,000 acres of fenland west of Spalding. John Perry, the former engineer of the Fen, had died in 1733. Grundy took levels and soundings along 22 miles of the Welland, produced a map, and proposed as a principal new work a reservoir 10 acres in extent behind a sluice (to be rebuilt) at the outfall of the river Glen into the Welland estuary. By releasing water from the reservoir through the sluice at low tide the outfall would be scoured; a method, he says, used recently with success (at Gunthrope Sluice) in Bedford Level by 'the celebrated Drainer Humphry Smith Esq.' (q.v.).

In March and April 1735 Grundy was back again, making observations on Vernatt's Drain, the main drain of the Fen. He received a fee of £50 for 'his attendance and expenses in London, at Spalding and other places, and for Drawing up a Scheme by him proposed for Draining the Fenn'. All this was to bear fruit three years later.

It would have been widely known in engineering circles that Nathaniel Kinderley (q.v.) had started work in 1733 on a New Cut (ship) Canal about 7 miles long from Chester through salt marshes on the south side of the Dee estuary to improve navigation and, after its completion, to make possible the reclamation of large areas of land from the sea. Purely out of interest, Grundy visited the works in May 1735. He found some aspects of the project to criticise, but was more upset by a diatribe published by Thomas Badeslade (q.v.) in November 1735 which was highly critical of Kinderley's scheme and included adverse comments on Deeping Fen; Grundy replied early next year. Badeslade responded in March with another polemic which drew a spirited rejoinder from Grundy in May 1736. Apart from correcting factual errors concerning Deeping Fen and the river Welland, he defended the case for relatively narrow and deep channels, to give a high velocity of flow, as against those (like Badeslade) who promoted the case for maximum tidal indraught and therefore opposed new cuts, sluices on tidal rivers and the inclosing of salt marshes.

This argument reverberated throughout the eighteenth century. Grundy was the first to support his case by explicit reference to authorities such as Castelli's *Mensuration of Running Water* (in Salisbury's 1661 translation). In this respect he can be linked with Charles Labelye (q.v.) as a pioneer under the influence of Desaguliers in the application of scientific principles to civil engineering problems in England.

By 1737 the Adventurers of Deeping Fen decided to take action. They called on Humphry Smith and Grundy to 'take a View' of the Fen. This they did in July, producing a scheme for improving the drainage, partly based on Grundy's 1734–1735 proposals, and estimating the cost at £15,000. A Bill was submitted to Parliament for powers to raise taxes, and in April 1738 Smith and Grundy were appointed 'Surveyors and Agents … of Deeping Fen … for the putting into Execution of Scheme and Estimate by them delivered in for draining all the Fenns therein', Smith to have a salary of £150 p.a. and Grundy £150 p.a. for himself and an assistant. The Act was duly passed, and work began well before October 1738. At about that time the Grundy family settled in Spalding.

Initially Smith directed the operations with Grundy acting in effect as resident engineer but later they separately directed individual works. The principal works, completed by 1742, included repairs to the 12 mile long Deeping Bank, south of Spalding, deepening the Welland along this length, building the reservoir (8 acres) and sluice at the Glen outfall near Surfleet, erecting two drainage mills with 16 ft. diameter scoop wheels on Vernatts and Hills Drains, regrading the bed of the Glen and raising its banks. In addition, John Scribo (soon to become Steward of the Duke of Bedford's Thorney estate) was brought in to direct the strengthening of the 6 mile Country Bank.

Surfleet Sluice had a 24 ft. waterway in three openings with pointing sea doors and lifting land doors. It carried the inscription 'This Sluice was erected by order of the Honourable Adventurers of Deeping Fen, according to the model and direction of Messrs Smith and Grundy. W. Sands, Bricklayer. Samuel Rowel, Carpenter. 1739'.

Smith retired in April 1742 and Grundy then took full responsibility for further works, at a salary of £100 p.a. These were mostly of a routine nature costing on average about £1,200 a year, but they included a navigation lock on the Westlode (1743) and the difficult task of widening and deepening the Welland through Spalding (1744–1745). He also submitted a proposal for extending Vernatt's Drain to a new outfall 2½ miles further downstream. This was eventually carried out under a new Act of 1774.

While holding the Deeping job, Grundy was free to take on a certain amount of outside consultancy on leave of absence. The most important of such commissions related to the river Witham. He and his son John (q.v.) produced a very detailed survey of the river from Lincoln to the estuary below Boston in 1743 and a report, dated 1744, on a scheme for improving navigation and draining the adjacent fens. Both the map and the report were signed 'John Grundy Sen. and Jun. of Spalding, Engineers'. The main feature of the scheme was a new cut 7 miles in length bypassing a 'prodigious mendering course of the River' upstream of Boston. The estimated cost of £16,200 proved too high for the interested landowners at that time but the scheme served as a basis for future plans and work carried out in the 1760s.

Grundy's appointment on Deeping Fen was not renewed in 1747, perhaps due to ill health. He died at Spalding on 30 December 1748, age fifty-two, and was buried at Congerstone. His son erected a memorial in the church bearing the inscription: 'In memory of John Grundy, late of Spalding, in Lincolnshire, who without the advantage of a liberal education had gained by his industry a competent knowledge in several of the learned sciences and lived by all ingenious honest men deservedly beloved and died by all such truly regretted'.

Trained by his father, John Grundy Jr. succeeded him on Deeping Fen in May 1748 and in due course became one of the leading civil engineers in England.

A. W. SKEMPTON

[Derby Postman (1728) Grundy's advertisement of his services as 'Land Surveyor and teacher of Mathematicks'; Deeping Fen Minute book and Letter Book (1736–1748) Lincolnshire Archive Office; John Grundy Jr. (1770) Report on Deeping Fen, BL, MS 54784; W. H. Wheeler (1896) *A History of the Fens of South Lincolnshire*; J. H. Hopper (1980) The two John Grundys, *Lincolnshire Life*, **20**, 24–27; Neil R. Wright (1983) *John Grundy of Spalding*]

Publications

1734. *Some Proposition ... (on) the draining of Lands that lie near the Sea or low Fen-Lands*
1736. *Philosophical and Mathematical Reasons ... to prove that the Present Work ... to recover and preserve the Navigation of the River Dee most intirely Destroy the same. With some Remarks on Mr. Badeslade's Reasons thereon*
1736. *An Examination and Refutation of Mr. Badeslade's New Cut Canal... (and) Observations and Experiments made upon the River Welland*
1743. *A Map of the River Witham ... from Wiberton Roads to the Bradon above Lincoln* (with John Grundy Jr.)
1744. *A Scheme for the Restoring ... the Navigation of the River Witham from Boston to Lincoln, and also for draining the Low-Lands and Fenns contiguous thereto* (with John Grundy Jr.)
1745. *A Further Illustration of Messrs Grundy's Scheme for the Witham*

Works

1738–1742. Deeping Fen drainage (jointly with Humphry Smith), £15,000
1742–1747. Works on Deeping Fen, £6,000

GRUNDY, John, Jr. (1719–1783), civil engineer, was baptised on 1 July 1719 at Congerstone, Leicestershire, the elder son of John Grundy Sr. (q.v.) and his wife Elizabeth Dalton. He was trained by his father, a land surveyor and teacher of mathematics, who became engineer of

John Grundy Jr.

Deeping Fen in south Lincolnshire in 1738, at first jointly with Humphry Smith (q.v.) and from 1742 in sole charge. In 1739 the family moved to Spalding. John joined the Gentlemen's Society there in December 1739. His first recorded engineering work, doubtless under his father's direction, was Pinchbeck sluice built 1739–1740 at the outfall of Blue Gote Drain into the river Glen. In 1741 he published his first report, supporting objectors to a proposal to divert the Trent Navigation from its main course to a narrow channel close to Newark. Between August 1742 and January 1743 he executed repairs to the breached sea wall embankment on the Norfolk coast near Happisburgh in accordance with plans drawn up by his father when acting as a consultant on leave of absence from Deeping Fen.

Soon afterwards the two Grundys undertook a detailed survey of the river Witham from Lincoln to the estuary below Boston. This was published as an engraved map dated 1743, and a 48-page printed report appeared in 1744, both signed 'John Grundy Sen. and Jun. of Spalding, Engineers'. As a means of improving navigation on the Witham and drainage of the adjacent fenlands their main proposal was a 7 mile new cut to bypass the 'prodigious meandering Course of the River' between Chapel Hill and Boston. The scheme proved too expensive at the time but served as a basis for future planning in which John Grundy junior played a leading role.

In January 1743 he married Lydia (1721–1764), daughter of the Rev. and Mrs Knipe. They had two daughters: Mary (1742–1809) known as Polly before her marriage to William Thompson in 1769,

Lydia (1849–1878) who remained unmarried, and four children who died in infancy. Both girls had lessons in Spalding before going to boarding school in Grantham when they were 14. They left school at 18, Lydia having spent her last two years at a school in Stamford. Portraits of Grundy and his wife, probably painted at the time of his marriage, and of Grundy in about 1770, have been handed down through Mary Thompson's family. All three portraits are reproduced in Wright (1983).

His wife died in 1764. Two years later he married Ann Maud, widow of the Rev. John Maud who had been vicar of St. Neots. There were no children of this marriage.

Works at Grimsthorpe 1746–1752

Between 1745 and 1752 Grundy carried out several commissions for the Duke of Ancaster on his estate at Grimsthorpe in Lincolnshire. They mark effectively the start of his career as an independent consulting engineer. The outstanding work is an earth dam with a central clay core wall, built 1746–1748, to impound an ornamental lake of 36 acres extent known as the Great Water. The dam has a maximum height of 18 ft., a length of 420 ft. and a crest width of about 20 ft. No drawing of the dam has been found, but a year or so later Grundy produced a plan for extending the lake with another dam further downstream, for which he gives a section, specification and estimate. Except for a greater height it probably represents closely the Great Water dam: inner slope 3½:1, outer slope 2:1, a central clay core 6 ft. wide to be 'well Rammed and Watered' and taken down through a porous stratum to rock, the inner and outer banks to be of earth rammed in layers. The extension was not carried out but the Great Water and its dam still exist, little changed after two-hundred and fifty years.

Other works at Grimsthorpe include regrading a slope to allow a better view of the lake and an improved water supply. This involved pumping from a deep well into a cistern and taking the water in lead pipes from the cistern to the house, Grimsthorpe Castle, a distance of 1 mile with a difference in head of 10 ft. The water was raised from the well by three pumps in series operated by a triple crank driven by a 'horse engine' or gin. An elegant circular engine house was erected next to a square building for the cistern beside the well. Accounts record payments in 1751 to Joseph Smith, engineer, for pumps and engine and to labourers for digging trenches and laying pipes. Grundy supervised the work.

From time to time during the next twenty years he was called back to Grimsthorpe to advise on various improvements and to provide 'book maps' of the Duke's lands. In one of these, dated 1759, he includes a perspective view of Grimsthorpe Castle, displaying his skill as a topographical draughtsman; a skill also revealed in a view of his own house and premises in the High Street, Spalding, which he purchased about 1760.

Deeping Fen 1748–1764

Grundy had assisted his father at several times on Deeping Fen and in May 1748 succeeded him as 'Agent' or Surveyor of Works at a salary of £50 p.a. with Joseph Featherstone (q.v.) continuing as assistant at £40 p.a. Grundy resigned this post in April 1764, to be succeeded by Thomas Hogard (q.v.) but he was retained as consulting engineer to the Adventurers (proprietors) of the Fen.

His duties as set out in 1748 were 'to make all estimates of the several works necessary to be done, and to let out and bargain with workmen for the same, and to take up or measure them when finished or once a month, and to pay the workmen. To survey banks and dykes at least once a week, when at Spalding, during the time such works are going forward. To survey all sluice works and carpenter's work as often as necessary. To make his winter residence constantly at Spalding so never as to be absent from the attendance of the banks during floods. To keep accounts fairly. And to have a person of ability and integrity to follow his instructions in his absence and be constantly with the workmen or surveying and taking care of all the works which will require daily attendance all the year round'.

Maintenance of the river Welland banks for which the Adventurers were responsible (the 12 mile Deeping Bank and 6 mile County Bank) cost nearly £10,000 in the period 1748–1764 out of a total expenditure of £24,000. In the early 1750s Grundy almost completely rebuilt the Six Door Sluice on the river near Spalding. Originally constructed in 1730–1731 by John Perry (q.v.) as a scouring sluice, Grundy extended the floor, heightened the existing doors and provided another set of doors to stem the tides. A little later, in 1754, he rebuilt the adjacent navigation lock. Next year three drainage mills were erected on Hill's Drain and in 1758 a new sluice on the Forty Foot Drain. In 1759–1761 he deepened the Welland by more than 2½ ft. between the Six Doors and Spalding bridge, using a 'hedgehog', a spiked roller dragged along the river bed to loosen the silt so that it could more easily be sluiced away to sea.

Consulting work 1749–1764

Grundy was able to undertake consulting work on leave of absence from Deeping Fen. Thus in 1749 he reported on the drainage of the Earl of Lincoln's estate in Borough Fen near Crowland. In the same year he prepared estimates for a new outfall sluice in the south bank of the Humber at Stallingborough. By 1751 the decision was made to proceed. Grundy took levels and produced working drawings which show three rows of dovetailed sheet piling protecting the foundations against the effects of underseepage. The sluice was then built with William Hodgson as foreman.

In 1750 Grundy made the first of his visits to works in the river Dee estuary begun by Nathaniel Kinderley (q.v.) in 1733 and completed by John Reynolds (q.v.) in 1743. These involved cutting a ship canal 8 miles long from Chester through salt marshes on the south side of the estuary and then building a bank 1.6 miles long across to the north shore, behind which some 2000 acres of land could be reclaimed from the sea. The Cross Bank with a sluice at its northern end was breached in February 1750 by strong westerly gales combined with an exceptionally high tide. Grundy arrived on site in July and with Reynolds (who had left Chester in 1743) drew up plans for closing the breach and raising the bank 3 ft. Little seems to have been done, however, until a new Act was obtained in 1753 enabling more capital to be raised. John Golborne (q.v.) was then appointed engineer to the River Dee Company and Grundy was consulted in March 1754 to examine three proposals. He approved Golborne's preferred plan, with some modifications, and work began immediately on this new bank, following a line very different from the 1743 original and proving to be entirely successful. Grundy's third report on the Dee, in 1759, is particularly interesting as it includes some history of the works not otherwise available, the Minute Books having been lost many years ago.

By this time Grundy had established a reputation as a consulting engineer with well over a dozen reports to his name. He charged fees of 1 guinea (£1.05) a day plus all expenses and half a guinea a day plus expenses for an assistant. One of the most important schemes with which he was involved related to the river Witham. Landowners interested in the drainage of fenland adjacent to the river held meetings in 1752 and 1753. They asked Grundy to reconsider the proposals which he and his father had put forward in 1744 and also a scheme by Daniel Coppin in 1745 which included the provision of a 'Grand Sluice', on the river upstream of Boston. In a report of November 1753 Grundy came up with a revised (and better) line for the proposed new cut and a two-mile extension of the cut down to Boston on which the Grand Sluice would be built to stem the tides; the sluice was to have a 50 ft. clear waterway and a navigation lock alongside.

With a modified position of the sluice, nearer Boston than Grundy had proposed, his scheme was approved at a meeting held in Lincoln and it was decided that an application should be made to Parliament to impower Commissioners to levy a tax for carrying out the works. However, no action was taken for several years. Grundy reported again in 1757. Langley Edwards (q.v.) reported in 1760 chiefly on the location of the Grand Sluice, which seems to have been in dispute, but otherwise he agreed in essentials with Grundy's proposals. Finally the landowners asked John Smeaton (q.v.) to join Grundy and Edwards to take 'a fresh View of the River and

Fens'. Their report, dated 23 November 1761, is a classic. It sets out the principles involved and gives detailed recommendations with estimates totalling £38,000 for drainage and £7,400 for navigation.

The report was approved though the proposed line of the new cut, theoretically the best, was changed nearly to Grundy's 1753 alignment at a further meeting in January 1762. The Grand Sluice was to be exactly where Edwards had suggested and in the river from Chapel Hill up to Lincoln there would be three locks, not stanches as originally proposed. Grundy produced a fine engraved map soon afterwards. He and Smeaton gave evidence in Parliament and the Act passed in June 1762. Edwards was appointment engineer in charge with two assistants. He drew up working drawings for the Grand Sluice, which were approved by Grundy and Smeaton with some alternations. Their involvement then came to an end. A measure of Grundy's contribution is that when the engineers' fees were paid in March 1763 he received £254, Edwards £90 and Smeaton £73. Work began in April 1763 and the principal drainage works, including the Grand Sluice, were completed in 1768 at a cost of £42,000; navigation works continued to 1771, at a cost of £6,000.

On another scheme, for a canal from Louth to the north-east Lincoln coast, Grundy was consulted in 1756. He selected Tetney Haven as the best point of entry and submitted a report and plan in April 1757, for which he received £31. Little happened in the next three years until in January 1760 a subscription list was opened. In March the Town Clerk wrote to Smeaton asking him for a second opinion on Grundy's proposals. The two engineers met on site in August (their first meeting); some letters passed between them and their reports were published in August 1761 along with an engraved plan by Grundy. As would be expected, Smeaton made several useful and perceptive comments but in all essentials agreed with his colleague's scheme. After a further delay the Act was obtained in March 1763, both Grundy and Smeaton giving evidence on the Bill.

Also in 1763 Grundy was asked to prepare a scheme for the drainage of some 11,000 acres of low ground in Holderness, north-east of Hull and here again he was joined with Smeaton. They visited the area in November and Grundy completed his report on 30 December. It was sent to Smeaton who replied in February that he concurred with Grundy on every essential point, suggesting only minor modifications, and an engraved plan by the land surveyor, Charles Tate, was issued. Grundy attended the Parliamentary proceedings and the Act passed on 5 April 1764. Holderness Drainage was to be his first big work in the next phase of his career.

Major works 1764–1775

To supplement his salary from Deeping Fen and earnings as a consultant (£71 in 1757, for example) Grundy with two partners started trading in timber from the Baltic, a trade showing a profit by the late 1750s. In 1762 he also took on a part-time job, requiring only a few weeks work a year but paying more than his Deeping salary, as Collector of the Land Tax in the Spalding District. Thus by 1764 he had secured a degree of financial independence and, with the prospect of becoming engineer to the large new Holderness Project, he decided to resign from the Deeping Fen post, a decision made easier by the generous offer by the Adventurers to retain him as a consultant.

Holderness drainage. In a personal letter to Smeaton, dated 10 May 1764 and signed 'Your affectionate Friend', Grundy relates the sad news of his wife's recent death, after a marriage of 21 years, and refers to the 'teazing and peevishness' of the opponents of the Holderness Bill, which he presumed Smeaton knew had passed through Parliament. A fortnight later the Holderness Trustees ordered that 'the two Engineers Mr. Grundy and Mr. Smeaton be wrote to desiring they come over to view the low grounds and Carrs'. Soon afterwards Smeaton wrote a letter of sympathy to his 'Dear Friend', saying he was hard at work on the Calder Navigation. On 22 June Grundy wrote that he hoped they could meet on site on 4 July. This they did, and reported ten days later. In September Grundy sent working drawings of the outfall sluice to Smeaton who made some valuable comments, chiefly on the foundations, and prepared, in a report of December 1764, a detailed bill of quantities with estimates totalling £1,800,. The sluice had a waterway of 20 ft. in two 10 ft. arches.

This marks the end of Smeaton's involvement in the Holderness project. Grundy, it seems, had been in charge of operations since July 1764 when work began on the 17 miles of 'barrier bank' along the east side of the river Hull, with John Hoggard as Superintendent. Contractors (bricklayers, masons and carpenters) for the sluice were appointed in March 1765. Joseph Page (q.v.) of Hull became 'Surveyor' or resident engineer on the sluice and the main drains with a salary of £80 p.a. Charles Tate acted as land surveyor throughout the proceedings.

Grundy made four site visits in 1765 and several more before the sluice and main drains were completed by October 1767. He and Page then left the scene, but work continued on the banks and drains for five or six years under Hoggard. The scheme had practically been completed by 1772 at a cost around £24,000.

Louth Navigation. The Act for building a canal from Louth to the Lincolnshire coast at Tetney Haven, in accordance with plans by Grundy, was obtained in 1763 but it was not until February 1765 that he was appointed engineer at a salary of £300 p.a., out of which he had to pay the resident engineer, James Hogard. The latter, James Hogard, took up his duties in the following month and work began.

By the summer of 1767 the outfall sluice and adjacent navigation lock had been built and the 7 mile cut from the lock formed through flat fenland, the depth of the cut being such that water level was nowhere less than 2 ft. below ground surface. This ensured that the cut could act both as a canal and a main drain. Hogard now took over as engineer, at £140 p.a. and it was under his direction that seven locks were built on the length of 4 miles from the edge of the fenland up to Louth. The navigation, suitable for Yorkshire Keels, was opened in May 1770.

Adlingfleet Drainage. In a report of December 1764 Smeaton proposed a scheme for draining 5000 acres of low ground known as Adlingfleet Level, south of the junction of the rivers Ouse and Trent. The landowners accepted his proposals and obtained the necessary Act in March 1767. Commissioners were appointed and in June they asked Grundy to take charge of design and construction, doubtless on Smeaton's recommendation, with David Buffery (q.v.) as resident engineer (at £50 p.a.) and Charles Tate continuing as surveyor.

Grundy followed Smeaton's plan with only minor variations. He produced drawings and specifications for the outfall sluice in July; contracts for building it were let four weeks later. In October James Pinkerton and John Dyson (qq.v.) contracted to excavate the main drain 'agreeable to Mr. Grundy's dimensions'. By April 1769 'all the material work of the Drainage being completed', Grundy was paid £302 for his fees and expenses. Tate likewise received £275 and Buffery was retained on an *ad hoc* basis to let contracts for minor works required to finish the scheme. The total cost was approximately £7,000.

Driffield Navigation. This canal, about 6 miles in length, enabled ships for 40 tons burden to navigate between a basin in Great Driffield and the river Hull, itself navigable down to the Humber. Grundy described his scheme in a report dated December 1766, accompanied by a plan and section made by Isaac Milbourn under his direction. Grundy gave evidence in Parliament, the Act passed in May 1767 and in August he was appointed engineer. He produced working drawings of the four locks and two swing bridges. Samuel Allam (q.v.), of Spalding, arrived in October as Surveyor of Works. Work began on the first lock soon afterwards and in May 1768 Pinkerton and Dyson contracted to excavate the canal. The navigation was opened in May 1770 at a cost of about £13,000.

Laneham Drainage. In December 1768 Grundy was asked by a group of Nottinghamshire landowners to consider the draining of an area of low ground about 2 miles wide west of the river Trent and extending 5 miles north from Laneham to West Burton (2 miles south of Gainsborough). He visited the site with his 'clerk', Samuel Goodhand, and concluded that it would be essential: (1) to provide a Catchwater Drain along the edge of the higher ground to the west to intercept run-off from various streams, the drain to discharge above flood level in the Trent at West

Burton; (2) to build a bank along the Trent some 7 miles long from Laneham to West Burton to prevent flooding from the river; and (3) to carry off rain water falling on the low ground by a main or Mother Drain (with numerous side drains) leading to an outfall sluice in the Trent Bank at Sturton Cow Pasture, ½ mile east of West Burton.

His scheme meeting with approval, Grundy was then authorised to make detailed surveys, levels and plans, to prepare estimates and to attend the proposed Bill during its passage through Parliament. After six weeks field work, assisted by George Kelk (q.v.) and Buffery (brought from Adlingfleet to check the levels), he reported at the end of February. The area to be drained had now been determined as 5,900 acres and his estimates of the principal works were: Catchwater Drain, £2,700; Trent Bank, £6,800; Mother Drain, £2,400; Sturton Sluice, £900; and side drains, £1,200. Grundy spent from 8 March to 27 April in London on the Bill. He received £329 for his work up to the latter date. Kelk was paid £20 for thirty days on the survey. Their plan on the scale of 3 in. to a mile was published.

Commissioners appointed under the Act held their first meeting on 29 May, at which they appointed Grundy as engineer, David Buffery as Surveyor of works and Kelk as land surveyor. Grundy prepared specifications and working drawings, including a plan and elevation of Sturton sluice with its 12 ft. waterway. Orders were placed for bricks, lime, triass and timber and for a chain pump and a pile engine. Work began in August with local contractors for brickwork and masonry (quoins and coping) and a carpenter from Spalding, while Dyson and Pinkerton contracted for spade work at unit costs ranging from 3s per floor (400 cu. ft.) for the Catchments Drain, to 4s 3d for the 'well bedded and rammed' banks and 5s for the sluice foundation pit. Grundy later decided that it would be necessary to have a drainage mill with a 15 ft. diameter scoop-wheel at the Sturton outfall. Henry Bennett of Spalding contracted to build this 'Great Water Engine' in April 1770 for £458.

The works were completed in May 1772, as planned, at a cost of about £15,000. During the course of the work Grundy made at least seven site visits.

Weighton Drainage and Navigation. The feasibility of draining the low ground of Hotham Carrs and Walling Fen into the Humber, and making a navigation up to Market Weighton, was looked into by Smeaton in 1765. At his request levels were then taken by Tate. Grundy made a more detailed field study in September 1767 on the basis of which the landowners decided to proceed. Funds having been raised, John Smith (q.v.) was employed to draw up a scheme with estimates. This he presented in February 1772, accompanied by a plan surveyed by Smithson Dawson. A petition went to Parliament, Smith gave evidence, and the Act passed in May. At their first meeting, in June, the Trustees

appointed Smith as engineer and Allam as Surveyor of the Works at an annual salary of 52 guineas.

At this meeting consideration was given, 'for greater Security of the Execution of the Works', to employ Smeaton or Grundy 'occasionally to view and correct the proceedings of Mr. Smith'. But the expense was judged too great. However, by August some doubts arose concerning Smith's proposal and Grundy was called in for advice. He reported on 2 September 1772 with a radically different (and far better) method of constructing the outfall sluice and a revised scheme for the navigation together with a comprehensive set of levels taken by Allam and Goodhand.

Grundy's plans were immediately accepted, and he became the engineer. Contracts for excavation (James and John Pinkerton), masonry (Joseph Smith) and carpentry (John Everingham) were let 'according to the plans of Mr. John Grundy' and work began on 29 September.

The works were carried out in three well-defined stages. (1) 1772–1775: outfall sluice and the adjacent navigation lock and the main Mother Drain. Grundy was in charge, with Smith as assistant (to July 1773) and Allam as resident engineer. The cost was about £26,000 and Grundy's fees to December 1775 came to £160. (2) 1776–1780: side drains (total length 43 miles) and deepening the streams. Allam was in charge with Pinkertons as contractors. The cost was about £14,000. (3) 1781–1783: canal from the Mother Drain towards Market Weighton on a revised plan by Luke Holt (q.v.), now in charge, with new contractors. The cost was about £8,000.

The Mother Drain, 6 miles long, was cut with a bottom width of 30 ft., widened towards the sluice; it had a minimum depth of 7 ft. below lowest ground level in Walling Fen and an average depth of 14 ft. The sluice had a 30 ft. waterway in two arches with 18 ft. of water in the sills at spring tides; the lock had a width of 15 ft., a length of 70 ft. between the gates, and walls rising 23 ft. above sill level. The foundation pit was 18 ft. deep, excavated behind the Humber bank. After completing the structure new banks were built curving back to the side walls of the lock and sluice, and the old bank removed down to floor level. In this way the use of cofferdams was avoided. The area of drained land amounted to 16,400 acres. During repairs to the sluice in 1826 a tablet was placed on the parapet recording the names of Grundy and Allam and the mason and carpenter; the date was given as A.D. 1775.

Consulting work 1764–1775

During the period in which these six works were in progress Grundy continued to report on various projects, typically three a year. They include sea banks in Essex, Lincolnshire and Yorkshire; small drainage schemes, some as far away as Warwickshire and Worcester; the river Swale navigation, on which he gave evidence in Parliament in 1767 (though the work was not completed); a plan for the Chesterfield Canal in 1770, and a long report on Deeping Fen. Some idea of the scale of his activities is given by his income from engineering work in 1767 which amounted to £591.

It was, then, as one of the leading engineers in England, that Grundy became a founder member of the Society of Civil Engineers, attending the first meeting on 19 March 1771 at the King's Head Tavern, Holborn, along with Thomas Yeoman, John Smeaton, Robert Mylne and Joseph Nickalls (qq.v.). He attended the next meeting ten days later when John Golborne and Robert Whitworth (qq.v.) joined the Society. Later, he twice took the chair, and attended his last meeting in May 1780.

Deeping Fen. Grundy's report of February 1770 on Deeping Fen was requested by the Adventurers following a proposal by Thomas Hogard in 1767 to make a new cut for the Welland 7½ miles in length to the Witham at Wyberton, and its rejection in 1768 by Thomas Tofield (q.v.) who proposed instead a new cut from Vernatt's Drain (the main drain of the Fen) about 5 miles long to the north shore of the Welland estuary near Fosdyke. Grundy also rejected the Wyberton cut, claiming that a better and less costly scheme would be to extend the Welland in a cut 115 ft. wide and 1½ miles long to a new outfall near Fosdyke. He accepted the idea of extending Vernatt's Drains but disagreed with Tofield's suggestion that it should be cut to a depth of 2 ft. below the level of the outlet and at a dead level right up into the Fen.

The two engineers asked Smeaton for his opinion. In his reply dated March 1770 he outlined the theory of flow in open channels, showing that water below the level of the outlet was not 'dead' as Grundy supposed but, on the contrary, the extra depth is beneficial and inferring that where, as in this particular case, the fall has a very low gradient—less than 4 in. per mile—the additional cost is justified. In other words Tofield was right on this point.

Grundy had begun work on his report in September 1769. He includes a history of Deeping Fen from John Perry's time (1730), provides a detailed plan, and gives estimates of cost. He received £250, indicating that by this time he was probably charging 3 guineas per day on consulting work.

The subsequent history is briefly as follows. (1) an Act of 1774 sanctioned an extension of Vernatt's Drain (but only 2½ miles in length) and new outfall sluice and various improvements of the other drains in the Fen; (2) an Act of 1794 established the Commissioners of the River Welland who were empowered to carry out the Wyberton Cut; but (3) nothing happened until under an Act of 1801 Grundy's proposed Welland cut to Fosdyke was carried out, the Wyberton Cut being quietly forgotten.

Hull Dock. One of Grundy's last major consulting jobs was the planning and design of Hull's first dock. The need for extra space for shipping in Hull harbour and for a Legal Quay had long been recognised, but it was not until 1773 that a satisfactory solution to the problems emerged. Asked for his opinion on the effects of building the Quay as a solid structure projecting 40 ft. into the harbour Grundy reported in October 1772 that it would oppose too much obstruction to flood water in the river, that a flood relief channel could however be provided by enlarging the old moat around the town, but that with great advantage the north part of the moat would provide a suitable site for a large wet dock (i.e. a dock with lock gates to maintain water at high tide level) to accommodate ships, thereby relieving pressure on space in the harbour.

This idea was quickly taken up. The dock would have sufficient depth for fully laden ships so that the Legal Quay could be located on one side of the dock, leaving the need only for a 15 ft. quay on open piled foundations in the harbour. Acting for the Board of Customs in January 1773, John Wooler (q.v.) prepared outline plans and estimates. In April Smeaton reported favourably on the timber quay. Further to satisfy critics Smeaton and Grundy examined the problem again, reporting (separately) in December. The way now lay open for an application to Parliament and Grundy drew up detailed plans and estimates, totalling £68,000, in March 1774. The Act was obtained in May and in November he received £300 for his work.

Construction began in March 1775 with Luke Holt as resident engineer and Henry Berry (q.v.), the Liverpool dock engineer, as consultant. In January 1775 Berry had recommended that the dock should be made wider by 17 ft. and this was adopted. He visited the site in March and August, on leave of absence from Liverpool. Thereafter his contact seems to have been by occasional correspondence with Holt, in effect, carrying full responsibility for construction.

As completed in September 1778 at a cost of £73,000 the dock was 1703 ft. long and 254 ft. wide, giving an area of nearly 10 acres. It had brick walls rising 24½ ft. above dock bottom, with a massive stone coping. The entrance lock had a timber floor and stone-faced brick walls (the facing stones set in pozzolana cement on Smeaton's recommendations) with a length of 120 ft. between gates and width of 36 ft., retaining 20 ft. of water on the sills. Up to 100 ships could be accommodated in what was the largest commercial dock of the eighteenth century in England.

The last years, and the Report Books

From 1775 there is a distinct slowing down in Grundy's activities: no further site supervision, apart from finishing the Weighton job, no more than six reports between 1775 and 1777, only two between 1778 and 1779, and a note on Wells Harbour in 1782.

By about 1775 he completed, with the aid of one or two clerks, the compilation of the first twelve quarto volumes of what are known as his *Report Books*, totalling some 4,000 pages and covering the period 1741 to 1772 except for his work as Surveyor on Deeping Fen, a summary of which is, however, given in the 1770 report. They preserve fair copies of his reports, many with carefully drawn diagrams and plans, together with field observations, estimates, notes on progress of works, minutes of meetings, copies of letters and so on. They were clearly intended to ensure a detailed, formal record of his career. He also kept copies of all subsequent reports, up to 1779, but for some reason did not have them bound into volumes.

Grundy's will is dated 15 January 1783. The extent of his prosperity is indicated by bequests of £1000 to each of his four grandchildren and ample provision for his widow. He died on 15 June 1783 and was buried in Spalding parish church where a commemorative plaque was placed on the wall by the north door.

In his will Grundy left his books and manuscripts to his eldest grandson, the Rev. J. G. Thompson, who later became vicar of Belton near Grantham. In 1793 Sir Joseph Banks, President of the Royal Society, purchased from Mr. Thompson the twelve *Report Books*, the later reports (which he bound into five volumes), printed reports by Grundy and other engineers and thirteen Acts of Parliament. The seventeen *Report Books* remained in the Banks family until sold at Sotheby's in 1918. They then disappeared from view until 1955, when the Institution of Civil Engineers bought Volume 2 from a London book dealer, and 1988 when the other sixteen volumes were discovered in Leeds University library as part of the large collection of books and manuscripts bequeathed by Lord Brotherton in 1930. Of particular interest as the only or main sources of information, in the absence of Minute Books, are: Volume 2 on the Grimsthorpe works and other schemes around 1750, Volume 4 on the river Dee, and Volumes 10 and 11 on Laneham Drainage.

A. W. SKEMPTON

[John Grundy (1741–1779) Reports Books. Leeds University Library, except vol. 2, ICE; Minute Books of Deeping Fen (1746–1774), Louth Navigation (1760–1770) and Witham Drainage & Navigation (1762–1768) Lincolnshire Archive Office; Minute Books of Holderness Drainage (1764–1775), Driffield Navigation (1767–1771) and Weighton Drainage & Navigation (1772–1783), East Riding RO; Adlingfleet Drainage Minute Book (1767–1772) Adlingfleet & Whitgift Commission, Crowie; John Grundy (1770) Report on Deeping Fen, Report Book, vol. 7 and BL. MS54784; John Timperley (1836) An account of the harbour and docks at Kingston-upon-Hull, *Trans ICE*, **1**, 1–52; John Farey (1845) Letters of Grundy, Tofield and Smeaton 1770, *MP ICE*, **4**, 205–209; W. H. Wheeler (1896) *History of the Fens of South Lincolnshire*; Esther Wright (1938) The early Smeatonians, *Trans Newc Soc*, **18**, 101–110; A. W. Skempton and Esther Wright (1971) Early members of the Smeatonian Society of Civil Engineers, *Trans Newc Soc*, **44**, 23–42; Mark Baldwin (1974) The engineering history of Hull's earliest docks, MS, Science Museum Library; J. M. Hooper (1983) Lincolnshire Worthies: the two John Grundys, *Lincolnshire Life*, **20**, 24–27; N. R. Wright (1983) *John Grundy of Spalding, Engineer (1719–1783). His Life and Times*; A. W. Skempton (1985) The engineering works of John Grundy. *Lincolnshire History & Archaeology Soc*, **19**, 65–82; G. M. Binnie (1987) *Early Dam Builders in Britain*; O. S. Pickering (1989) The re-emergence of the engineering reports of John Grundy, *Trans Newc Soc*, **19**, 65–82; G. M. Binnie (1987) *Early Dam Builders in Britain*; O. S. Pickering (1989) The re-emergence of the engineering reports of John Grundy, *Trans Newc Soc*, **60**, 137–143]

Selected publications

1744. *A Scheme for the Restoring ... the Navigation of the River Witham from Boston to Lincoln and also for draining the Low-Lands and Fenns contiguous thereto* (with John Grundy Sr.), engraved plan published separately

1753. *Mr. Grundy's Plan* [for improving the River Witham], with folding engraved plan

1761. *A Scheme for executing a Navigation from Tetney-Haven to Louth*, engraved plan published separately

1761. *Report ... concerning ... the River Witham ... and the Navigation thereof ... and of the Fen Lands on both sides of the said River* (by Grundy, Langley Edwards and John Smeaton)

1762. *A Plan of the River Witham and adjoining Fens and Low Ground from Lincoln to Boston, with the New Works proposed*

1763. *Repor ... concerning the Drainage of Low Grounds and Carrs ... in Holderness*, engraved plan by Charles Tate published separately 1764

1766. *Observations ... and Report, Scheme and Estimate for making a Navigation from Great Driffield to Emmetland*, engraved plan and section published separately

1769. *A Plan of the Ings, Meadows, Marshes and other Low Ground in Laneham ... Leverton ... Stourton ... and West Burton* (with George Kelk)

1772. *Report ... respecting the Drainage and Navigation proposed for Walling Fens*, with folding engraved plan of Weighton Drainage and Canal

1772. *Report ... respecting the Proposition of making Quays and Wharfs ... in the Haven of the River Hull*

1774. *Observations ... on the East Fen ... and Schemes for the Draining thereof*, with folding engraved plan

Principal works

1746–1748. Grimsthorpe Great Water dam
1748–1764. Deeping Fen, improvements, £24,000
1753, 1761. River Witham, planning, works 1763–1768, £49,000

1764–1767. Holderness Drainage, completed 1772, £24,000

1765–1767. Louth Navigation, completed 1770, £27,000

1767–1769. Adlingfleet Drainage, £7,000

1767–1770. Driffield Navigation, £13,000

1769–1772. Laneham Drainage, £15,000

1772–1775. Weighton Drainage, completed 1780, £40,000

1774. Hull Dock design, works 1775–1778, £73,000

GWILT, George (1746–1807), surveyor, was born in Southwark on 9 June 1746, the son of Richard Gwilt, a wigmaker. At the age of fourteen he was apprenticed to a Southwark mason, Moses Waite, and subsequently worked for George Silverside, a leading London surveyor, and it is as a surveyor rather than architect or engineer that Gwilt is probably best categorised, given the nomenclature of the time.

Around 1770 he was appointed County Surveyor by Surrey, a post he retained until his retirement in 1806. In 1771 he became Surveyor to the Commissioners of Sewers for Surrey and in 1774 District Surveyor for St. George's Parish, Southwark. He was subsequently appointed Surveyor to the Clink [Southwark] Paving Commissioners, and the Kent Sewers Commissioners. Given the growth of London at that time much of his work would have been within the boundaries of modern London.

As County Surveyor, Gwilt was responsible for a number of bridges, as well as public buildings such as the County Sessions House in Southwark (1795–1799) and Kingston-upon-Thames House of Correction (1775). He also carried out surveying work for clients in Southwark, such as the Anchor—later Courage's—Brewery. Although architect for some private houses—including his own home, 18 Union Street, Southwark—it is his work as Surveyor to the West India Dock Company for which he is best known. Its massive five storey warehouses, which he designed in association with his son, George, suited his utilitarian style of architecture.

The West India Docks scheme drawn up by the engineers William Jessop and Ralph Walker (qq.v.) obtained its Act in 1799 and in February 1800, when work began, Jessop was appointed engineer and Gwilt and his son were appointed jointly as architects and surveyors at £1,000 guineas p.a. They remained until May 1804 when Thomas Morris Jr. (q.v.) took over their responsibilities once the major building was completed. Gwilt was to act as Surveyor, and his son as Clerk of Works. Some original design work had been done by George Dance Jr., and Ralph Walker had proposed ten warehouses on each side of the quays, but these plans were ignored and Gwilt's design brief was to provide ten warehouses on the North Quay, 224 ft. × 114 ft., with five 8 ft. tall storeys for sugar storage, basement vaults for rum, and attic floors for lighter goods, to be erected in four years. Gwilt altered this to nine warehouses 223 ft. long, with the three central blocks 125 ft. deep, and the others 116 ft. deep. The plans for the first six were approved in May 1800 and work began on 12 July 1800 with William Adam—of the famous Adam dynasty (q.v.)—and Alexander and Daniel Robertson as contractors; the work was to be completed in two years. This was a massive building contract and work was complicated by the proximity of the dock excavations, for which the contractors were supplying bricks. As a consequence of the rush, lime mortar was damaged by frost as work continued in the winter. Despite the consequent rebuilding the first three warehouses were ready for the official opening on 27 August 1802 and the remainder by January 1803. Because of a shortage of capital, the remaining three warehouses were redesigned as 'low' warehouses for coffee storage, with walls and foundations suitable for heightening, if necessary. They were completed in August 1803. In the meantime linking blocks were designed to make an overall frontage over a mile in length; the scale of these buildings was unprecedented. Originally in brick with timber internal framing, work was subsequently carried out from c. 1811 under John Rennie (q.v.) involving the use of cast iron columns, etc. By then Gwilt had died. The warehouses represent a tremendous achievement and despite their lack of architectural pretension made an imposing sight. It is of interest to note that Gwilt had previously worked with the Jupp family; Richard Jupp had been involved with the design of the East India Company's warehouses around Cutler Street and it may have been Gwilt's familiarity with these structures that helped him get the dock job. In addition to the warehouses, the Gwilts designed the permanent dock walls, office building and lodges.

Gwilt trained his sons, George (1775–1856) and Joseph (1784–1863). The elder son, aside from his work in the docks, carried out relatively little practical work with the exception of Southwark cathedral, including the erection of an iron roof above the choir. Joseph succeeded his father as Surveyor to the Commissioners of Sewers for Surrey—he resigned in 1846—and carried out a number of architectural commissions. He is best known for his writings which included a translation of Vitruvius (1826). His *Treatise on the Equilibrium of Arches* is essentially a mathematical treatment of the subject. It was dedicated, interestingly, to John Narrien, Professor of Mathematics and Fortification at the Royal Military College, Sandhurst. The second (and third) editions of this work included his prize winning, although unexecuted, design for London Bridge. Engineers may well have relied on the works of Hutton and Gregory (qq.v.) for a theoretical understanding of arch behaviour, and the third edition would have been effectively overtaken by the work of Henry Moseley and others.

Joseph's *Encyclopaedia of Architecture* (1st edition, 1842) became a standard reference work; it described the history of architecture and the properties of materials, gave a glossary of terms, and contained standard specifications. It provides a useful introduction to building practice of the Victorian period.

The Gwilt family have been to a degree neglected by architectural historians and their use of the new engineering materials iron and concrete might bear further examination. The use of iron in the roof of Southwark cathedral is little known, and Pasley describes Joseph's(?) use of concrete made from small broken stones cemented with 'hot lime grout' in foundations. James Walker, in an unpublished discussion of 1830 at the Institution of Civil Engineers refers to the use of a very weak concrete across the whole of the West India dock warehouse site; if this refers to Gwilt's work this would predate Smirke's use of the material by a decade.

George Gwilt died in Southwark on 9 December 1807. There is a family monument in Southwark Cathedral.

MIKE CHRIMES

[West India Dock drawings and minutes, Museum of London in Docklands; ICE Minutes of conversations, 1830, ICE Archives; J. Britton and A. Pugin (1825) *Public Buildings of London*; C. W. Pasley (1826, repr. 1862) *Practical Architecture*; House of Commons (1834) *Minutes of Evidence, Select Committee on Metropolitan Sewers*; J. Gwilt Jr. (1848) *Evidence, Metropolitan Sanitary Commission*, **1**, 259–273; LCC (1950) *Survey of London*, xxii: Bankside; D. G. C. Allan (1974) William Adam and Company, *RSA J*, **122**, 659–678; C. Hadfield and A. W. Skempton (1979) *William Jessop, Engineer*, 184–221; D. Cruikshank (1989) Gwilt complex, *Arch Rev*, **clxxxv**(1106), 55–61; R. A. Otter (1994) *Civil Engineering Heritage: Southern England*, 175–176; RCHME (1994) *Survey for London*, xliii: Poplar, Blackwall and the Isle of Dogs; Colvin (3)]

Works

See Colvin for complete list of architectural works.

George Sr.:

1782. Cobham Bridge, reconstruction, nine arches
1782. Leatherhead Bridge, reconstruction, fourteen arches
1782–1783. Godalming Bridge, reconstruction, five arches

George Sr. and Jr.:

1800–1803. West India Docks warehouses

George Jr.:

1822–1825. Southwark Cathedral, restoration

Joseph:

1819. Southwark Bridge, approaches

Publications
Joseph:

1811. *A Treatise on the Equilibrium of Arches*, 2nd edn., 1826; 3rd edn., 1839

GWYN, John (*c.* 1733–1789), civil engineer, in 1747 as the son of Henry Gwyn, deceased, was apprenticed to George Gwyn, carpenter and citizen of London. No doubt he served the usual seven years apprenticeship. By 1760, and probably from several years earlier, he was working for Joseph Nickalls (q.v.) millwright of Lambeth. John Smeaton (q.v.), recently appointed engineer of the Calder & Hebble Navigation, brought Nickalls up to Wakefield in December 1759 as assistant or resident engineer. Writing from Lambeth in January 1760, Gwyn applied for a foreman's job on the navigation works at 18s per week. He arrived on site in June. Meanwhile he appeared before a committee of the Society of Arts in April to explain a model of a windmill submitted by Nickalls.

Work proceeded on the Calder & Hebble. Nickalls' salary was increased to £120 p.a. in August 1760. But for reasons which are not clear, he was dismissed in November 1761. Advertisements were placed for two 'surveyors', and in March 1762 Gwyn was appointed in charge of carpentry at 50 guineas p.a. with Matthias Scott (q.v.) in charge of masonry and digging at 60 guineas p.a., they to work directly under Smeaton.

By July 1764 after an expenditure of £56,000 the navigation was open from Wakefield to Brighouse: 18 miles with seventeen locks. Not a great deal remained to be done, but disputes arose. A new management committee appointed James Brindley (q.v.) in January 1765 to complete the work on a revised plan and Smeaton was discharged, along with Gwyn and Scott.

A year later Gwyn went to Scotland to make borings in the river bed of the Tay at Perth for the bridge to be built there to Smeaton's design. Gwyn reported his findings in February 1766 together with an estimate of the amount of timber required for the cofferdams, piling, and 'stages and runs', for the barrows in excavations. He was to communicate his reports to Smeaton immediately on his return to England, and promised 'with the blessing of God, if I continue in good health' to be back at Perth in six weeks time to take up his appointment as Overseer of construction. John Adam (q.v.) had already been put in overall charge of operations, both at the quarries and the bridge.

Work began soon after Smeaton's site visit in May. By the end of 1766 the first pier was finished. Smeaton approved everything that had been done, but was alarmed at the cost, and in April 1767 recommended that the rest of the bridge be built by direct labour (not by contract) under Gwyn 'the very best man for executing orders I have ever met with in the whole of my practice'. In October 1767 Gwyn's industry was rewarded by a salary increase to 104 guineas p.a. and by being made responsible for all work in the absence of John Adam who in any case seems to have left soon afterwards. By August 1770 with the bridge nearing completion and £20,000 having been spent, Gwyn was allowed to leave, and 'sensible of the Merit he has Justly acquired in the executing of Mr. Smeaton's designs', he was given a gratuity of 25 guineas.

He then moved to Portpatrick as resident engineer on the harbour when work was starting, again to Smeaton's plans. By 1773 the main pier 270 ft. long had been finished. Next, work started on an extension to be 150 ft. long, founded by the *pierres perdues* method. This was well advanced by the end of the year, when Gwyn left for Aberdeen. James Kyle finished the pier and deepening the harbour three years later, total expenditure about £10,000.

Gwyn had already visited Aberdeen early in 1774, on Smeaton's recommendation, to make preliminary arrangements for quarrying stone, building a lighter and the purchase of machinery no longer required at Portpatrick. He explained that 'it will be the Latter End of this year before I can possibly be finally acquitted from my present Engagement at Portpatrick'. His appointment as 'Superintendent and Director of Works' at Aberdeen Harbour dates from January 1775, with an annual salary of £120, and he received £15 for the expense of removing his family from Portpatrick. Work proceeded by direct labour, contracts only being made for the supply of materials, and reached completion in October 1780. The pier had a total length of 1200 ft., containing some 45,000 tons of stone: total expenditure £16,000.

Smeaton reported on Peterhead harbour in 1772. Funds were raised. Gwyn prepared detailed estimates in 1774, probably as an additional task on his trip to Aberdeen that year, and the foundation stone of the new South Pier was laid in July 1775. While working at Aberdeen Gwyn visited Peterhead from time to time. The pier and deepening of the harbour at Peterhead reached completion in 1781 at a cost of about £6,000.

Gwyn then moved to another of Smeaton's projects in Scotland: Cromarty harbour. It is interesting that the extent to which Smeaton and his resident engineer in this remote spot kept in touch can be gauged from their letters between January 1782 and June 1783 (recorded in Smeaton's Journal for that period). During this period Gwyn wrote nine times and received four letters in reply. Gwyn's letters took eight to twelve days to arrive in London or Austhorpe.

With work drawing to a close at Cromarty in the summer of 1784 Gwyn became concerned about his future prospects. Fortunately, Smeaton was able to arrange for him to take a job at Hull as engineer for a new bridge, adding in his letter of recommendation that Gwyn 'can both Draw and Design very sufficiently, especially if he has a

Single Hint, from his old Master which course to steer'. Gwyn wrote from Cromarty in November that he had sent his designs to Smeaton in London for comments, and hoped to be in Hull at the end of December. Christmas Eve, however, found him stormbound at Peterhead and he resolved to travel overland, but three days later he was writing from Aberdeen that the roads were 'Impassable from this to Edinbro by either fly Chaize or horse, the Mail being for numbers of miles carried on mens backs'. He also wrote to Smeaton, worried about the effect his lateness might be having on his future employers. Smeaton immediately wrote a kind letter to the Hull authorities, assuring them again of Gwyn's ability and asking that they should not judge Gwyn on 'the style of his letters, or indeed of his discourse ... for having served his apprenticeship in London he both Speaks and Writes the London Vulgar; and withal has such a Knack of misapplying an heap of fine Words and Phrases as by no means do justice to his Mechanical abilities; and this I think it is but justice to him to say; as we are very apt to judge from first impressions'.

All went well at Hull. With more than 'a Single Hint', and indeed with a good deal of help, from Smeaton, Gwyn designed and supervised construction of the bridge. It had four stone arches, two of 20 ft. and two of 24 ft. span, and a central two-leaf drawbridge 38 ft. in span. The bridge was opened in September 1787.

Once again things turned out fortunately for Gwyn, as he was soon to secure a job at Ramsgate Harbour. Smeaton had acted as consulting engineer to the harbour authorities on several works since 1774 but when the question of building an extension to the east pier arose in 1787 it was decided that he should become Engineer with an experienced 'Resident Surveyor' working full time on site. Gwyn was duly appointed, his salary from January 1788 to be £100 p.a. The 'Advanced Pier', as it was called, was a massive structure founded 7–10 ft. below low water springs on timber caissons, 36 ft. wide at the base, 40 ft. high and 26 ft. wide at the top.

Replying on 27 February 1788 to a letter from Gwyn, Smeaton said he was very glad that he was getting settled in his new occupation and hoped he would soon 'be getting your wife and family along with you'. And the next day he wrote to the Secretary of the Harbour Trustees remarking that Gwyn 'having got a house to his mind, seems to think himself very happily situated, and I doubt not but as soon as the Season comes favourably every thing will go on Chearfully at Ramsgate'.

Smeaton with Gwyn and Henry Cull, the master mason, visited quarries at Purbeck and Portland for stone and at Lyme Regis for Lias hydraulic limestone. Work started in the pier in July, using a diving bell to level the chalk sea bed for the caissons, and by November 40 ft. of the pier had been built. Work resumed in April 1789 but, tragically, Gwyn died on 11 June after a short illness, aged fifty-six, and was buried at St.

Lawrence Church, Ramsgate, 'A real loss to the public', Smeaton said 'as well as lamented by his family and friends'. Gwyn's wife Mary died at Ramsgate in 1810, age sixty-seven.

A. W. SKEMPTON

[Apprenticeship Register Book (1747) Worshipfull Company of Turners, Guildhall Library; John Gwyn (1766) Report on Perth Bridge investigations, in Smeaton's *Reports*, **1**, 184–187; John Smeaton (1782–1783) Journal, Trinity House, London; Trevor Turner (1970) John Gwyn and the building of the North Bridge, Kingston upon Hull, *Transport History*, **3**, 154–163; Charles Hadfield (1972) *Canals of Yorkshire and North East England*; Robert Matkin (1997) The construction of Ramsgate Harbour, *Trans Newc Soc*, **48**, 53–71; Trevor Turner (1978) The works of John Smeaton, a chronological survey, *Trans Newc Soc*, **50**, 37–52; A. W. Skempton (1981) *John Smeaton*]

Works, as resident engineer
1766–1770. Perth Bridge, £20,000
1770–1774. Portpatrick Harbour, completed 1778, £10,000
1775–1780. Aberdeen Harbour, North Pier, £16,000
1775–1781. Peterhead Harbour, South Pier, £6,000
1781–1784. Cromarty Harbour pier.
1785–1787. Hull North Bridge.
1788–1789. Ramsgate Harbour, Advanced Pier, completed 1792, £52,000

GWYNN, John (1713–1786), writer, architect and bridge-builder, was a native of Shrewsbury. Nothing is known of his parentage or his formal education, but the diarist, Joseph Farington, wrote that Gwynn was 'originally a carpenter' who became self-educated in architecture. He seems to have used the knowledge for many years only as background for a series of writings about London and its monuments, chiefly St. Paul's Cathedral, and also about the needs and status of artists, including architects. His publications of the second kind contributed materially to the movement which resulted in the formation of the Royal Academy of Arts in 1768, Gwynn's first statement of the case having been published in 1749. He was rewarded by being one of the first 'architect' academicians, in spite of having little or no practical design experience at the time.

He was well known and respected in London, particularly amongst writers, and he figured prominently in the competition for the design of the intended bridge over the Thames at Blackfriars in 1759–1760. After a process of short-listing only three designers were under consideration, namely Robert Mylne (q.v.), John Smeaton (q.v.) and Gwynn. Gwynn's design, of semicircular arches and copious decoration, was supported in letters to a newspaper by Samuel Johnson, quite convincingly to lay readers (though the letters were unsigned) but on the single ground that semicircular arches were stronger and better than arches of any other shape. In the wider public controversy the much greater practical knowledge of Mylne and

Smeaton weighed against Gwynn and Mylne won the commission with his design of elliptical arches.

Six years later Gwynn's most important book, *London and Westminster Improved,* appeared with a dedication to the King composed by Johnson. It was a plea for control of building in London to be subject to a detailed 'general plan' of improvements, and therefore proof against frustration by strong private interests; and various important developments of later years, such as the formation of Trafalgar Square and the Embankment, and the building of Waterloo Bridge, were foreseen by it.

At the end of August 1767 Gwynn was invited to Shrewsbury by the committee planning an upgrading or replacement for the English Bridge over the Severn. They had already approached Smeaton, who refused the commission, and then employed Mylne for a year and a half. Much preparatory work had been done and when Gwynn attended the committee in mid-September he presented two alternative designs, as well as introducing a London mason, Richard Buddle, as prospective contractor for the work. The design which had semicircular arches was chosen and after delays, mainly due to the inadequacy of funds available, the foundation stone was laid in June 1769. Gwynn was named as 'surveyor' with William Hayward (q.v.), a local tradesman and aspiring architect, as assistant surveyor or clerk of works and Buddle as contractor. But Buddle's progress was very slow and he was too often ill or absent in London, and in April 1771 Gwynn made an estimate for the remaining work and offered to undertake it as contractor. After considerable delay, his offer was accepted and he completed the bridge in 1774.

Gwynn's work on a second large bridge at Atcham followed a similar pattern: initial appointment by the County Justices of Salop to design it, Buddle obtaining a contract for construction and failing to perform, and Gwynn then contracting to complete the bridge, which he did in 1776. He was also working concurrently as 'surveyor'—a term equivalent to 'engineer'—for two other large bridge projects, at Worcester and at Oxford, both involving considerable new planning of the adjacent streets. His appointment at Oxford was as 'Surveyor to the Commissioners of the Oxford Paving Act' and the salary was £150 p.a. To control projects in several cities at once meant that he must have able agents or assistants at each site. Hayward was 'clerk of works' at Shrewsbury and perhaps at Atcham, William Spiers was 'deputy surveyor' at Oxford, and there is limited evidence that Gwynn was also using a young artist who was a pupil of his London neighbour, Samuel Wale, presumably as a drawing assistant.

All of Gwynn's four large bridges were fine classical compositions and richly ornamented in a baroque idiom, the arches semicircular and of modest spans. The finest ornamentation was the carving of two dolphins on cutwaters of the English Bridge. All served the traffic for long periods though not without some adjustments of gradient and width. All except the Worcester bridge still display Gwynn's architectural decoration.

In his years of bridge design, Gwynn was generally paid, according to surviving documents, 'by poundage', clearly meaning a percentage of the costs of the projects he designed and 'surveyed', and so at that time he must have been moderately well off but never wealthy. He was described by acquaintances as a jolly companion and lively in conversation, and 'of quaint appearance and odd manners.' A portrait of him by Zoffany is in the Shrewsbury Museum and is reproduced in Ruddock, *Arch Bridges.*

Gwynn died at Worcester on 27 February 1786 and was buried there. He provided in his will for the education of his natural son, Charles, who survived him only by ten years, and he made bequests to the Royal Society and to the offspring of his early endeavours for the profession of artists, the Royal Academy.

TED RUDDOCK

[*The King's Maps*, xxxiv.33 and xliii.67, MS drawings of Magdalen Bridge and Worcester Bridge; W. Papworth (1863–1864) John Gwynn, RA, *The Builder,* **21**, 454; **22**, 27; DNB; A. W. Ward (1935) *The Bridges of Shrewsbury*; J. Greig (1923) *The Farington Diary,* **1**, 180; J. L. Hobbs (1962) The parentage and ancestry of John Gwyn [*sic*], *Notes and Queries,* Jan., 21–22; T. W. M. Jaine (1971) The building of Magdalen Bridge, 1772–1790, *Oxoniensa,* **36**, 59–71; T. Ruddock (1979) *Arch Bridges and their Builders 1735–1835*; Colvin (3)]

Publications

1749. *An Essay on Design, including Proposals for erecting a Public Academy ...*
1749. *A Plan for Rebuilding the City of London after the Great Fire in 1666, designed by ... Sir Christopher Wren.* Engraving
1755. *A transverse Section of St. Paul's Cathedral decorated according to the original intention of Sir Christopher Wren,* engraving
1758. *Plan of St. Paul's Cathedral* (with S. Wale), engraving, showing the dimensions
1766. *London and Westminster Improved, to which is prefixed a Discourse on Public Magnificence*
1781. *Mr. Gwynn's report to the Trustees for building a new Bridge at Worcester, 5 March 1781*

Works

1767–1774. English Bridge, Shrewsbury, River Severn, seven arches, cost £15,710 including property purchases, etc., designer-'surveyor' from Oct. 1768, contractor from May 1772 for contract value £5,789
1768–1776. Atcham Bridge, Salop, River Severn, seven arches, price £7,238, designer-architect from 1768, contractor from 1772 for contract value £5,998
1770–1781. Worcester Bridge, River Severn, five arches, also new streets adjacent

1772–1790. Magdalen Bridge, Oxford, River Cherwell, two ranges of three arches each, with long high road structure between, and adjacent new street works, mostly complete before Gwynn resigned in 1778, decorative works continued largely to his design

H

HADLEY, John (fl. 1689–1701), engineer. The earliest reference to him yet discovered dates from 1689, when Worcester Corporation granted a lease of the 'decayed waterworks' to 'John Hadley of West Bromwich', conferring on him powers to lay pipes in the streets and to erect a cistern, in the market place, holding 200 hogsheads (about 12,000 gallons) of water. He built a new waterwheel and pumps to raise water from the river Severn to the cistern, from which it would be distributed to houses in the city. Most, if not all, the pipes had been laid by 1691.

In this scheme, which initiated a series of about a dozen similar ones during the next fifteen years, Hadley was associated with John Hopkins. As partners, they were granted leave by the city of Chester in April 1692, at their own cost, 'to set up...and use new Waterworks engines, cisterns, pipes and other instruments'. The work was probably completed during the next two years. The 'waterhouse' and wheel are shown in the engraved *View of Chester* (1728) by S.& N. Buck. Hopkins and Hadley conveyed the works to other proprietors in 1698.

Meanwhile in March 1693 'John Hadley of Worcester, Engineer' took out a Patent for various devices including a mechanism for raising and lowering a waterwheel according to the level in a river, to maximise its efficiency. It is highly probable, and indeed Desaguliers (1744) positively states, that Hadley first used this device at Worcester. George Sorocold (q.v.) also used it in his waterworks at Derby between 1692 and 1693 and there can be little doubt the two engineers were acquainted by then. Certainly when Sorocold met Ralph Thoresby at Leeds in 1694 he referred to Hadley's work at Chester.

Hadley is next heard of in June 1696 when Sir Godfrey Copley MP, FRS (1653–1709), writing to his friend Thomas Kirke, says 'I have been this day and am to meet tomorrow Mr. Saracole and Mr. Hadley. I have seen his Engine ... which raises Tems water ... I do think it the best piece of work I have seen and I find Hadley to be a man of Mathemat and Bookish'. It is generally assumed that Copley is referring to Merchant's Waterworks in Westminster, built under Letters Patent of 1695 and 1696 to pump water through pipes from the Thames. The work appears to have been finished in 1697. Sorocold and Hadley were the leading 'engine-makers' of their day, but both of them worked in a wider context.

Hadley's last and by far his largest commission was as engineer of the Aire and Calder Navigation. In the seventeenth century boats of 30 tons burden could go up the Aire at high water springs as far as Knottingley. The first serious proposal for extending the navigation from Knottingley to Leeds came in 1679 but received inadequate financial support. So did another proposal of 1685 which included its tributary, the river Calder, from Castleford up to Wakefield. However, the scheme was successfully revived in 1697. Financed chiefly by the woollen cloth merchants, Leeds could count on subscriptions of £5,500 and Wakefield some £4,000.

Samuel Shelton was employed to survey the Calder. The same Thomas Kirke to whom Copley had written, and who had recently designed an improved surveying wheel, surveyed the Aire. Hadley was consulted as engineer. Ralph Thoresby FRS (1658–1725) of Leeds, also a friend of Kirke and Copley, records in his diary: 'Dec. 3 (1697) ... accompanied the Mayor and Mr. Hadley to view the river. Mr. Kirke and I followed the windings of the water and measured it with his surveying wheel. Viewed the several mills and shoals, and with much ado finished as far as Ferry-bridge against night, 10 miles by land, twenty by water. Dec. 4, down the river to Welland, observed the sands, &c ... and upon the whole the ingenious Mr. Hadley questions not its being done, and with less charge then expected, affirming it the noblest river he ever saw not already navigable'.

Hadley estimated £5,200 for works on the Aire: five locks between Leeds and Castleford (17 miles, 44 ft. fall) and two locks between Castleford and Knottingley, though improvements would be required in the river as far downstream as Welland House near Haddlesey (18 miles from Castleford, 34 ft. fall). On further thought he proposed a lock at Haddlesey, making eight in all.

In January 1698 a petition went to Parliament for leave to bring in a Bill. This was granted, but the Bill met with strong opposition from interests dependent on the Yorkshire Ouse. Hadley gave evidence in May in the House of Lords. Two officers of Trinity House were called in to report. They did so in July with a sensible assessment of the problem, concluding that the navigation works would not be detrimental to the Ouse. The petition was re-submitted in the next session when it again met with opposition. Hadley again gave evidence, in March, and finally the Act for the Aire and Calder passed on 4 May 1699.

Five days later Hadley was appointed to direct works on the Aire. He would receive 200 guineas (£210) in the first summer and a further 200 guineas on completion, expected in two years time. Work proceeded rapidly and by December 1699 locks at Knottingley and Castleford had been built, at the joint charge of the Leeds and Wakefield proprietors or 'Undertakers', and the first of the five locks upstream. The remaining four were completed and the navigation opened to Leeds by November 1700. Soon afterwards two additional locks were built on the Leeds line, one at Methley and the other on the 1½ mile Cryer Cut, newly excavated to bypass a very circuitous stretch of the river at Woodlesford.

In July 1701 work started on four locks on the Calder (13 miles from Castleford to Wakefield, 28 ft. fall). They were built by James Willans, mason of Dewsbury, and James Mitchell (q.v.), carpenter of Bradley, near Huddersfield, who had been employed on the Aire locks. In April 1702 Mitchell and George Atkinson (q.v.) contracted to build the lock at Haddlesey and another which had been decided on at Beal. These were completed in 1703 and the navigation was judged 'near perfected' in 1704, the total expenditure amounting to about £25,000.

The locks were 15 ft. wide and 60 ft. between gates, with masonry walls and timber floors, admitting boats carrying 25 tons on average and drawing 3 ft.: more in winter, less in summer. A view of the river included in Thoresby's *Ducatus Leodiensis* (1715) shows a lock at the tail of a cut some 250 ft. long with a weir across the river near the head of the cut. Other locks were located in cuts alongside mills. In places the river banks had to be raised and a good deal of dredging was needed.

In July 1699 Hadley had been consulted by York city authorities for a second opinion on a report by Thomas Surbey (q.v.) on improvements to the Ouse. He broadly confirmed Surbey's recommendations.

Hadley was active on the Aire in 1700 but almost certainly died before 23 April 1701. On that date Mary Hadley, presumably his widow, wrote to one of the Leeds proprietors of her concerns 'about the reference of our business betwixt the Undertakers of Leeds and myselfe'. This sounds very much as if she were seeking the promised second payment of 200 guineas due on completion of the Aire works.

A. W. SKEMPTON

[Ralph Thoresby (1715) *Ducatus Leodiensis*; J. T. Desaguliers (1744) *A Course of Experimental Philosophy*; Joseph Hunter (1830) *The Diary of Ralph Thoresby*; John Noake (1877) *Worcester Relics*; F. Williamson (1936) George Sorocold of Derby, *J Derbyshire Archaeol Nat Hist Soc*, **57**, 43–93; W. H. G. Armytage (1953) George Sorocold and Sir Godfrey Copley, *J Derbyshire Archaeol Nat Hist Soc*, **73**, 105–107; G. Ramsden (1954) Two notes on the history of the Aire and Calder Navigation, *Thoresby Soc Miscellany*, **12**, 388–395; T. S. Willan (1954) in lit.; G. Taylor of Chester Waterworks (1954) in lit.; R. W. Unwin (1964, 1967) The Aire and Calder Navigation, *Bradford Antiquary*, **42**, 53–85; **43**, 151–186; B. F. Duckham (1967) *The Yorkshire Ouse*; Charles Hadfield (1972) *Canals of Yorkshire and North East England*]

Works

1689–1691. Worcester waterworks, probably about £1,000
1692–1694. Chester waterworks

1699–1701. Aire and Calder Navigation. £25,000 (to 1704)

HAKE, Matthew (fl. 1500–1502), was a Flemish engineer from Gravelines commissioned to build a large sluice on the river Witham at Boston. Construction of relatively small weirs and locks on rivers and sluices on fen drainage channels could be left to experienced local craftsman, such as Thomas Everton (q.v.) and Joseph Hunt (q.v.). Large sluices on tidal rivers and for scouring harbours were uncommon in England before 1700 and they presented problems which in several cases led to the employment of experts from Flanders and the Netherlands. An early example is the 'Hollander', Gerard Matheson, brought in to build the Great Sluice for New Romney harbour (1409–1413).

The tidal range at Boston varies from 10 ft. at neaps to 21 ft. at springs and in its natural state the Witham was tidal for a distance of 9 miles above the town. The combined effect of the tide and fresh-water floods caused frequent drowning of the surrounding fenland. An attempt to build a timber-framed sluice at Boston appears to have made in medieval times, but with what success is not clear and there is no doubt that the natural state of affairs persisted throughout the fourteenth and fifteenth centuries. Eventually in 1495 Lady Margaret Beaufort, a powerful landowner in Lincolnshire (and elsewhere), persuaded her son King Henry VII to appoint a council to consider the matter. The council included John Morton who, when Bishop of Ely (1479–1486), planned and supervised construction of a new 12 mile channel for the river Nene in the Fens (still known as Morton's Leam) and in 1488 was consulted by the men of New Romney on a new cut to prevent silting up of their harbour. Another council member was Richard Fox, Keeper of the Privy seal, to whom in 1492 the King had delegated the task of improving the haven of Calais by the construction of a new sluice. From the proximity of Gravelines to Calais it is possible that Matthew Hake was employed on the Calais works and that Fox, having met him on this occasion, recommended him to the council. Two other members, Sir Richard Guildford and Sir John Fyneux, had both been involved in improvements to the drainage and sea defences of Romney Marsh

This high-level group 'communed together several times with long deliberation' and by February 1500 they had drawn up a plan of action and were ready to put it into effect: (1) to build a sluice on the river at Boston, (2) to employ Matthew Hake for its design and construction on terms already agreed with him, (3) to obtain a loan from the Crown to cover the estimated cost of £1,000, (4) to repay the loan by taxing the landowners of ground affected by the resulting improvement, on an acreage basis, and (5) to establish a commission to assess the acreage and collect the tax. The commission would operate on the system long-established in Romney Marsh, with a Bailiff and his servant paid 2s 4d a day and jurats at 1s 4d a day. Hake and 'his man' were to receive jointly 4s a day, masons 5s a week and labourers 4s a week.

An indenture with Hake was signed on 19 February by Sir John Hussey, a former Steward of Lady Margaret and Sheriff of Lincoln, and John Robinson, a wealthy merchant of Boston trading with Calais. On 21 February royal letters were dispatched to Boston concerning the loan. John Morton recommended John Clerk (q.v.), builder of the first harbour pier at Dover, as a specialist adviser on the project. He and Sir John Hussey went to Boston on 16 March to make plans for Hake's arrival, including the provision of stone and sending a ship to Calais for supplies assembled by Hake: nearly 2 tons of ironwork, two dozen water bailers, five mortar troughs, etc. Hake arrived on 10 May with fourteen masons and twenty-four labourers. Three days later a further indenture, confirming the earlier one, was made between Lake Margaret and Hake. She provided money for 'the great ramme of brass' for the pile driver and was also responsible for contracting with additional craftsmen including a master carpenter and his men from London. It is probable that one of her staff, James Morice, acted as Clerk of Works and an Expenditor was appointed to look after payments.

Work began almost immediately (in May 1500) and reached completion in the summer of 1502. A masonry pier 13 ft. wide and 44 ft. long was constructed near the middle of the river and piers built out from either bank, all on piled foundations, leaving a clear waterway of 65 ft. Massive timber gates were hung between the central and the land piers, and a timber deck provided a bridge. After numerous repairs and replacements the bridge lasted until the new single-arch iron bridge was built alongside 1802–1807.

The total cost of the sluice appears to have been more, but not much more, than the estimate: say about £1,100. The sluice proved to be successful. The gates were repaired in 1543 and several times later up to 1642. Thereafter it seems the sluice was allowed to fall into decay; no further repairs are recorded, and by the early seventeenth century the river was certainly tidal once again above Boston as it had originally been. After many reports and discussions, the Grand Sluice was built a little further upstream to the design and under the direction of Langley Edwards (q.v.) in 1764–1766, as part of a general scheme for improving drainage of the fenland and navigation on the river.

A. W. SKEMPTON

[Pishey Thompson (1856) *History and Antiquities of Boston*; W. H. Wheeler (1896) *A History of the Fens of South Lincolnshire*; H. M. Colvin (1975) Calais … 1485–1558, *King's Works*, **3**, 337–360; Michael K. Jones (1986) Lady Margaret Beaufort, the Royal Council and an early Fenland drainage scheme, *Lincolnshire History and Archaeology*, **21**, 11–15; Eleanor Vollans (1988) in Jill Eddison and Christopher Green (eds.), New Romney … in the later Middle Ages, *Romney Marsh, Evolution, Occupation, Reclamation*, Oxford Committee for Archaeology Monograph No. 24, pp. 128–141]

Works

1500–1502. Boston sluice, *c*. £1,100

HALL, Francis Benjamin (1792–1862), civil engineer, was born on 26 December 1792 in Clackmannan, the son of James Hall and Jean Russell. In 1808 he became a pupil of Robert Bald (q.v.), the mining engineer. He attended classes in chemistry, mathematics and natural philosophy at Edinburgh University, *c*. 1812.

Hall is known for a number of estate surveys on the Isle of Lismore, dated 1815, and an undated survey of the Glenfuir estate near Falkirk. as well as some surveys in the Glasgow area. He was with Hugh Baird (q.v.) 1813–1816 and surveyed the Edinburgh and Glasgow Union Canal for Baird in 1813. He moved from 636 Argyle Street, Glasgow, in 1814, to Edinburgh (there in 1818), consistent with work on the Union Canal.

His work on the Union Canal brought him into contact with Thomas Telford (q.v.) with whom he visited the Union Suspension Bridge of Captain Samuel Brown (q.v.). He later claimed he assisted in work on the Runcorn and Menai suspension bridge projects, as well as the Esk Bridge on the Glasgow–Carlisle Road near Rockcliffe. W. A. Provis (q.v.) is normally credited as chief assistant on these works but it is known there were others such as John Pollock (q.v.). He certainly surveyed a deviation on the road in 1820. Hall is also said to have worked on the Caledonian Canal, but no proof of this has been found.

In 1823, preceded by his parents and siblings, he emigrated to British North America where he remained for the rest of his life. Settling at Queenston in Upper Canada (now Ontario), he threw himself into a range of construction activity. Initially he was engaged to provide assistance on two canals. In 1824–1825 he prepared and published reports and surveys containing a preliminary design and initial cost estimate for the now famous Welland Canal between Lakes Erie and Ontario. He also designed the Burlington Bay Canal with an associated drawbridge and lighthouse, and supervised the canal's construction from the spring of 1824 until 13 April 1826. Then, having extensive commitments elsewhere, he resigned to the great displeasure of the Canal Commissioners. Meanwhile in 1824, he was architect for the Brock Monument at Queenston, erected 1825–1827, and proposed a chain suspension bridge across the Niagara at Queenston, anticipating better known proposals by twenty years. At York (now Toronto), the provincial capital, he was contractor for the Bank of Upper Canada (1825), begun but not completed to his designs, and proposed plans for a Parliament House there (1826).

Hall was soon involved in three canal projects in Nova Scotia. He carried out a survey and completed the detailed plans for the Shubenacadie Canal between Halifax and the Bay of Fundy in 1825. He moved to Halifax in the spring of 1826 as construction started. Telford, to whom Hall submitted his plans for review; strongly endorsed the project in August 1829. Unfortunately Hall's project for a 44-mile canal with fifteen locks designed for 8 ft. of draught was never completed, and when his contract as engineer was not renewed in the spring of 1832 he moved back to Upper Canada, probably to Lanark County where his parents lived. In 1825 he also had completed a detailed survey for a 6-mile canal with seven locks between the Gulf of St. Lawrence and the eastern end of the Bay of Fundy, and Telford was again asked to review the plans. Construction of this canal, later named the Chignecto Ship Canal, was not started until 1891 and by then it had been modified to a marine railway. Also in 1825 (a date of 1821 can be discounted) Hall prepared plans for the construction of a 2,700-ft. canal between the Atlantic Ocean and Bras d'Or Lake in Cape Breton Island on Nova Scotia's coast. Construction of this canal was also delayed, until 1854, but it was finally completed in 1869, after Hall's death.

In addition to his canal works, Hall was involved throughout the 1830s and early 1840s in various municipal works such as roads and land surveys in the Niagara region, harbour works on Lake Ontario at Cobourg, a revival of his suspension bridge proposal in 1836, and on the Peterborough and Port Hope Railway, for which he issued a substantial report in 1847. From around 1835 until 1850 he was living in St. Catharines, but then returned to Nova Scotia. His first wife Wilhemina Denham died in January 1826; although he was remarried, to Mary Albro of Dartmouth, Nova Scotia, he died childless, at Dartmouth, in 1862.

MARK ANDREWS

[Catalogue of plans and drawings, National Archives of Scotland; ICE (1834) Inventory of Telford Bequest, ICE archives; Telford collection, Ironbridge; T. Telford (1833) Diary, NLS; *Colonial advocate*, 27 May 1824, 22–23, 27; *Novascotian* (1825) 1 June; [Upper Canada] *Journal of the House of Assembly* (1826–1827) B: Report of the Burlington Bay Commissioners; *Bathurst Courier*, 29 Aug. (1834), 2; [Upper Canada] House of Assembly (1836) *Appendix to Journals*, 7–8; [Canada] Canal Commission (1871) *Improvements of the Inland Navigation of the Dominion of Canada*; Canada (1891) *Alphabetical Record: Engineers and Superintendents, etc. Principal Public Works*, Sessional Papers, **9**, Appendix 19; F. Staton and M. Tremaine (eds.) (1934) *A Bibliography of Canadiana* and *Second Supplement*, 1983; H. G. J. Aitken (1954) *The Welland Canal Company*, 46–47; J. P. Heisler (1973) *The Canals of Canada*, Occasional Paper, **8**, Parks Canada; R.

Legget (1976) *Canals of Canada*; J. Winearls (1991) *Mapping Upper Canada 1780–1867; An Annotated Bibliography of Manuscript and Printed Maps*; R. A. Styran, R. McAfee, R. R. Taylor (1992) *Mr. Merritt's Ditch*; S. Otto (2001) in lit.; Bendall]

Publications

1813. [Reduced] *Plan of the proposed Edinburgh and Glasgow Union Canal* (repr. 1816)
1820. *Plan and Section of intended Deviation from the Parliamentary Road between Stonehouse and the Glasgow Carlisle Road*
1825. Report relative to the Burlington and Dundas Canal, *Colonial advocate*, 22 Dec. 2
1825. *Welland Canal; Estimates for completing a Canal, between the River Welland and 12-Mile-Creek or Niagara, York* (now Toronto)
1826. *Map of proposed Canal* [in the Atlantic Provinces], Halifax, 1826?
1834. *Treatise upon hydraulic Architecture* (advertised but probably not published)

Works

1824. Brock Monument, Architect
1824–1826. Burlington Canal, Engineer, paid c. £400
1825. Bank of Upper Canada, Toronto, Contractor
1826–1832. Shubenacadie Canal, Engineer

HALL, John (d. 1828), civil engineer, was possibly trained as a mason. A John Hall was masonry contractor on the Leeds–Liverpool Canal in 1799. As 'Inspector', John Hall supervised the construction of Cree Bridge, Newton Stewart, 1812–1814. While there he submitted a three arch design, referred to Telford, for New Galloway Bridge. In the event John Simpson (q.v.) was contractor for the bridge, which collapsed under construction in 1815. Hall was still at Newton Stewart in 1816, but subsequently, in partnership with John Straphen (q.v.), Simpson's nephew, undertook some small contracts on the Holyhead Road.

He established a practice as the first professional architect in Bangor, acting as Clerk of Works at Bangor Cathedral 1824–1828, and repairing Beaumaris Church in 1825. Engineering work continued with surveys and design work for the Nantlle Railway. Prospects must have been limited in Wales because he succeeded Peter Logan (q.v.) as resident engineer on the construction of St Katharine's Dock in early 1828, when work was nearing completion. He died in November 1828, shortly after the opening ceremony and was succeeded by Thomas Rhodes (q.v.). Hall was elected as a Member of the Institution of Civil Engineers in 1828, along with his pupil Thomas Okes. Okes subsequently removed to Northampton.

Hall had two near contemporary namesakes. **John Hall (1764–1836) of Dartford** was the second of four sons of William Hall, a millwright. He originally set up in business as a smith in Dartford in 1785, establishing a foundry, and eventually building steam engines which were

widely exported in the early nineteenth century. His most famous early associate was Bryan Donkin (q.v.).

John Hall (fl. 1778–1803) of Flemington, New Jersey, worked for Boulton and Watt in the late 1770s, and installed steam engines for Walkers of Rotherham (q.v.), and at Wilkinson's Snedshill works in 1778. He emigrated to Philadelphia in 1785, meeting Thomas Paine (q.v.) with whom he became involved. He returned briefly to England 1791–1792, but returned to the United States where he established an ironworks in New Jersey, and was buried in Flemington.

MIKE CHRIMES

[**John Hall (d. 1828)**: ICE membership records, ICE Library; St. Katherine's Dock Board of Directors Minutes, Museum of London in Docklands; M. L. Clarke (1954) A Bangor architect of the early nineteenth century, *Trans Caernavon Hist Soc*, **15**, 49–50; A. W. Skempton (1981–1982) Engineering in the Port of London, 1808–1834, *Trans Newc Soc*, **53**, 82–87; D. Gwyn (2001) Transitional technology: the Nantlle Railway, in A. Guy and J. Rees (eds.) *Early Railways*, 50–51. **John Hall (1764–1836)**: Evidence (1817) *Select Committee on Steam Boats etc*, 27–28, 33–34 [HC 422]; H. Miller (1985) *Halls of Dartford*. **John Hall (fl. 1778–1803)**: J. G. James Collection, ICE archives]

HALLEN, Ambrose (fl. 1826–1834), architect of New South Wales, went to Sydney on the recommendation of Thomas Telford (q.v.) to the colonial government. He, and his brother Edward, who initially accompanied him, were responsible for the mansion, Telford Place, in Sydney. Hallen was the son of George Hallen, partner of George Cottam (q.v.). Cottam and Hallen were iron founders who became involved in the export of iron buildings to the colonies and it is possible that the Hallens were related to the Hallen ironmasters of Shropshire and Staffordshire, which may explain the Telford connection.

Hallen was elected a Corresponding Member of the Institution of Civil Engineers in 1826 but remained in membership only until 1831.

MIKE CHRIMES

[Membership records, ICE archives; P. Alsop (in lit.); J. G. James collection, ICE archives]

HALPIN, George, Sr. (c. 1775–1854), builder and civil engineer, was born about 1775. He was a builder by trade, with no academic qualifications, but contributed significantly to the development of the port of Dublin and the provision of lighthouses around the coast of Ireland.

The Act setting up the Corporation for Preserving and Improving the Port of Dublin (commonly called the Ballast Board) received Royal Assent in 1786. In September 1800 the Board appointed Halpin to the position of Inspector of Works in succession to the first supervisor of works, Francis Tunstall.

In 1810, responsibility for Irish lighthouses, beacons and seamarks was vested in the Ballast Board and Halpin took on the additional role of Superintendent of Lighthouses. At this time there were only fourteen Irish lighthouses but by 1867 there were as many as seventy-two. Halpin, in his reports to the Board, stated that many of the lighthouses were in a deplorable condition, badly maintained and badly managed. He commenced a programme of repair and replacement, whilst also establishing a considerable number of new lights. A large number of these early lighthouses were constructed to the designs of George Halpin, Sr., including the Bailey (1813) at the entrance to Dublin Bay and the Tuskar Rock (1815) guarding the approach to Rosslare Harbour; a full list is given below.

In 1818 Halpin was asked to carry out a new survey of the outer harbour at Dublin and of the site of a proposed northern breakwater, the North Bull Wall. He visited Plymouth, but found little to assist him in his calculations. He recommended that Francis Giles (q.v.) be commissioned to survey the outer harbour and bar. Giles commenced this survey in June 1818, with assistance from Halpin, and their joint report and proposed design were placed before the Board in May 1819. The design of the breakwater was based on earlier proposals of William Chapman (q.v., 1786) and two members of the Board, Maquay and Crosthwaite (1801). When built, the north and south breakwaters enclosed a large volume of water and this was employed on the ebb tide to scour the sand deposits from the bar, thus allowing Dublin to develop as a deepwater port.

In 1830, Halpin was joined by his son, George, and it is sometimes impossible to distinguish which of the Halpins actually designed the later lighthouses; most of the designs are basically similar and are usually attributed to the elder Halpin.

George Halpin, Sr., died in Dublin in his 80th year and is buried in Mount Jerome Cemetery in the city.

RON COX

[Records Of Dublin Port Company, Port Centre, Alexandra Road, Dublin; H. A. Gilligan (1988) *A History Of The Port Of Dublin*; B. Long (1993) *Bright Light, White Water* (revised edn. 1997)]

Works
Prior to 1830:

1814. Old Head of Kinsale tower (replaced 1846), Engineer
1815. Copeland Island Lighthouse, Engineer
1815–1817. Roche's Point Lighthouse (Tower, later dismantled to become Duncannon light), Engineer
1816–1818. Wicklow Head Lights (Towers), Engineer
1817. Fanad Head Lighthouse, Engineer
1817. Inishtrahull Lighthouse (original tower), Engineer
1818. Clear Island tower, Engineer

1818. Howth Harbour Lighthouse, Engineer
1818. Proposed tower for Coningbeg Rock (not built), design engineer
1819–1820. Redesign and rebuilding of Poolbeg Lighthouse, Dublin, Engineer
1824. Tramore Beacons, Engineer
1826. Skellig Michael towers, buildings and roadways, Engineer
1827. Inishgort Lighthouse, Engineer
1829. The Maidens Lighthouse, Engineer

HAMILTON, George Ernest (fl. 1825–1844), architect, civil engineer and surveyor, was living in Shrewsbury when elected a Corresponding Member of the Institution of Civil Engineers in 1827. A turnpike road surveyor for trusts such as the Sandon–Ireland Cross Road, the Drayton–Newcastle under Lyme Road and the Preston Brockhurst District, he was involved in road surveys in the area, and thus became acquainted with G. Buck and J. Provis (qq.v.) who proposed him for membership. In 1828 he was living in Stone, and subsequently at Wolverhampton (1836–1844). He claimed to have built, as contractor, ten bridges in Staffordshire and is known to have tendered for bridges in neighbouring counties. In addition to his road work, Hamilton undertook some railway surveys, most famously that of the Liverpool–Manchester railway for W. James (q.v.). In 1827 he reported with Thomas J. Woodhouse on the cast and wrought iron rails then in use on various railways for the Cromford and High Peak Railway; they discovered that many of the early wrought iron rails suffered from lamination. In 1828 he was asked to survey Knipersley Reservoir on the Trent and Mersey Canal, which had suffered problems due to settlement and the fracture of an outlet pipe, a job, he later confessed to Telford (q.v.), that he wished he had never taken on, so acrimonious had the subject become. He was employed by the Fairbanks on their railway surveys and surviving correspondence gives an insight into his character. Reporting on the 'Boston line' on 6 November 1836 he wrote: 'To level amid the rain, as you know, is impossible, for the Glasses become so dimmed, so as to be useless—all we can do is watch the weather and take advantage of any cessation of rain—for it has been little else since I have been here ... the Dykes all filling fast, every ditch becoming a dyke, and every dyke a river, would it be advisable to put some person at Peterboro' and commence levelling to meet me?—for if such weather as we now have continues, where shall we be even a fortnight hence'. On 13 November he followed up 'What a contract, what weather. This country is only fit for ducks and geese and human amphibians. In fact we are all amphibious here'.

His architectural work was mostly confined to church architecture, and his designs were criticised by his contemporaries. As a surveyor he developed an improved spirit level which he described to ICE members in 1831. He appears to have left ICE membership *c.* 1842.

MIKE CHRIMES

[Shropshire RO, DP13, DP87; Staffordshire RO, QS orders and treasurers accounts; Warwickshire RO, QS 24/28xv; ICE membership records; T/TR 80, 106, Telford mss, ICE archives; Minutes of conversations, 97 (1831), ICE archives; A. Blackwall (1985) *Historic Bridges of Shropshire*; G. Biddle (1990) *The Railway Surveyors*, 62–63, 75, 78; SK; Colvin (3)]

Publications

1825. *Report to the Commissioners of the Second District of the Turnpike Road from Coleshill, through the City of Lichfield and Town of Stone, to the end of the County of Stafford*
1827. *Report on the Turnpike Road from Sandon through Leek to Hugbridge, in the County of Stafford*

Works

1822. Liverpool–Manchester railway survey
1829–1830. Stableford Bridge, contractor, £172+
1830–1831. Bearstone Bridge, near Woore, contractor, £334
1831. Mill Dam Bridge, Eccleshall, contractor, £200+
1836. Gainsborough–Sheffield–Chesterfield railway survey (for Fairbanks)

HAMMOND, John (W.) (fl. 1828–1847), civil engineer, was probably a pupil or assistant of Richard Beamish (q.v.) when he was working in Cork following the halting of work on the Thames Tunnel in 1828. Although elected an Associate of the Institution of Civil Engineers while resident there in 1829, he was evidently very hard up, as Marc Isambard Brunel (q.v.) told the Institution of Civil Engineers he could not afford the subscription in 1831. He may have assisted Brunel with the Cork Jail Bridge in late 1833, but had moved to England to assist Isambard Kingdom Brunel with the site investigations for the Great Western Railway in time to give evidence for the Great Western Railway's second (successful) bill in the spring of 1835. Following its passage, Hammond became resident engineer for the western end of the Great Western Railway and, after completion of the line, he became Brunel's assistant in charge of civil engineering on the line until his death in 1847.

MIKE CHRIMES

[Membership records, ICE archives; House of Lords (Commons) (1835) *Minutes of Evidence on Great Western Railway Bill*, HL81, 564–568; I. Brunel (1870) *The Life of I. K. Brunel*, **65**, 92; L. T. C. Rolt (1957) *Isambard Kingdom Brunel*, **75**, 105–106; E. T. MacDermot and C. R. Clinker (1964) *History of the Great Western Railway*, **I**, 21, 30, 69, 363; P. B. Clements (1970) *Marc Isambard Brunel*, 204; G. H. Gibbs (1971) *The Birth of the Great Western Railway*, 34, 37, 48, 52; A. Vaughan (1991) *Isambard Kingdom Brunel*]

HANDLEY, Charles (*c.* 1750–1812), of Barford, Warwickshire, canal engineer. His birth is not recorded in the parish registers but when he married Sarah Adkins by licence at Kenilworth on 14 October 1782, he is described as 'of Barford'. They had four sons and seven daughters over the next seventeen years.

The Warwick & Birmingham Canal gained its Act in 1793 with assistance from the engineers of the Birmingham Canal, Samuel Bull (q.v.) and William Felkin in particular. Handley must have provided local input as he was paid £300 three months later 'as a gratuity for his loss of time in attending to the business of the canal'. Felkin was appointed Engineer with a land surveyor called James Sheriff to assist him, so Handley's appointment in December 1794 as surveyor to assess damages done by the contractors to property was a relatively minor one.

Handley's survey of the Warwick & Braunston Canal in 1793 was one of three, the others being done by Felkin and Sheriff. When its Act passed in 1794, Felkin was appointed Engineer, in addition to holding the same post with the Warwick & Birmingham, and Handley became land surveyor. This time he was paid £350 'as a gratuity for his loss of time in attending to the business of this canal' (the Clerk received 200 guineas). Construction was proceeding when, in August 1795, Handley proposed that the connection with the Grand Junction Canal at the eastern end be changed from Braunston to Napton. As a saving of £50,000 could be made, a revised Act was obtained and the Company renamed the Warwick & Napton Canal. The Company meanwhile was dissatisfied with Felkin as Engineer and after John Turpin of Wisbech resigned after only six weeks in the post, Handley was appointed Engineer from 1 November 1796. Construction proceeded uneventfully and the canal was opened to traffic in March 1800 at a cost well within the original share capital.

Handley was occasionally consulted by other bodies, such as the Nene Commissioners in 1802 on the subject of a proposed branch from the Grand Junction Canal to Northampton, and by the Grand Junction itself in 1804 when problems became apparent with the Ouse Aqueduct. It failed in 1805.

He died on 31 March 1812; the record in the parish register at Barford of his burial four days later describes him as a yeoman. His widow lived on for another forty years.

The Charles Handley who in 1819 built a private cut from the Oxford Canal at Long Itchington was presumably his son of that name.

P. S. M. CROSS-RUDKIN

[Warwickshire RO; RAIL 881 and 882, PRO]

Works

1796–1800. Warwick & Napton Canal, Engineer during construction at salary 350 guineas p.a. plus incentive shares; after opening in 1800, retained at £150 p.a.

HANDYSIDE, William (1793–1850), engineer, was born on 25 July 1793, the eldest child of Hugh Handyside, ironmonger, and Margaret Baird. His mother's family were well-known in the Scottish iron trade and although Handyside served two years pupillage to a Mr. White, a local architect, from the age of fifteen, it was as an engineer of iron that Handyside made his reputation.

In 1810 Charles Baird (q.v.), his uncle, offered him a position at his works in St. Petersburg. Baird had been there since 1786 and by the time of Handyside's arrival was one of the most important manufacturers in the Russian Empire. Handyside soon revealed he was a talented engineer and had the leading role in a number of Baird's most important engineering contracts.

He first worked on the installation of machinery at the Imperial Arsenal and Imperial Glassworks in St. Petersburg. He became particularly concerned with the manufacture and installation of steam engines; by 1830 about 130 steam engines had been produced by the works. In 1815 he helped build the first Russian steamship, the *Elizabeth*, and over the next ten years developed steamship design in a further ten vessels. Other works for Baird included developing a sugar refining process, gas lighting for the factory, and plant for armaments manufacture.

Handyside's activities led to his involvement in the construction of a number of important civil engineering structures, most notably his collaboration with the architect Montferrand. For years he worked on St. Isaac's Cathedral with its magnificent iron dome: he developed specialist machinery for the erection of the cathedral; he assisted in the design of the dome and its gilding; and he supervised the casting of the decorative bronze work. Montferrand and Handyside worked together again on the Alexander Column, Handyside casting the reliefs as well as organising the erection of the column. He described the methods of hauling large monoliths in a paper to the Institution of Civil Engineers.

Handyside was also involved with the construction of the chain suspension bridges erected in St. Petersburg from 1824 to Wilhelm von Traitteur's designs. He developed a testing machine for proving the iron links which is described in works by Traitteur.

Handyside was elected a Member of the Institution of Civil Engineers in 1822 but although he presented details of some notable Russian engineering achievements he did not attend meetings until his return to Britain after his retirement. When Handyside first arrived in St. Petersburg he lived with his uncle and in 1829 he married Sophia Gordon Busch, a fellow member of the British community in St. Petersburg; they had four children. Following Charles Baird's death, Handyside returned to Britain and retired from business. His brothers had joined him in St. Petersburg and although some remained there his younger brother, Andrew, returned with him to Britain and took over the Britannia ironworks at Derby.

William Handyside died in his mother's house, 48 India Street, Edinburgh on 26 May 1850, apparently of overwork.

MIKE CHRIMES

[Membership records, ICE archives; J. G. James collection, ICE archives; Methods used by the Russians to extract from the rock, pieces of granite of any size possible to be transported (OC12), ICE archives; Pantaleymon bridge (OC38), ICE archives; A. R. de Montferrand (1845) *Eglise Cathedrale de Saint Isaac*; Obituary (1850) *Min Proc ICE*, **10**, 85–87; DNB; T. Tower (1867) *Memoir of the late Charles Baird*; J. G. James (1983) *The Application of Iron to Buildings and other Structures in Russia to about 1850*; S. G. Federov, Early iron domed roofs in Russian church architecture, *Construction History*, **12**, 41–66]

HANKEY, William Alers (1771–1859), banker and (Honorary) Treasurer of the Institution of Civil Engineers, was born in London on 15 August 1771 and in early life was a neighbour in Hackney of the father of Henry Robinson Palmer (q.v.), founder of the Institution. Hankey was educated at Edinburgh University before embarking on a career in banking, becoming head of the firm of Hankey and Company, Fenchurch Street, London. When Palmer resolved on the formation of the Institution, Hankey offered every encouragement and in 1820 was elected an Associate, and subsequently was elected an Honorary Member. Later, in 1820, he was appointed Treasurer and Banker to the Institution, retaining that position until 1845 when it was decided to place the account with a Westminster bank, Cockburn's.

Aside from his banking, Hankey took a great interest in religious matters, and was involved in the Religious Tract Society and the British and Foreign Bible Society. He died at his home in Hyde Park Gardens on 23 March 1859.

MIKE CHRIMES

[ICE membership records, ICE library; *Min Proc ICE*, **20**, 1860, 444–445; J. G. Watson (1988) *The Civils, passim*]

HARBORD, Sir Charles (fl. 1633–d. 1679), Surveyor General, was employed by Charles I and Charles II as Surveyor General of Land Revenues. He is known to have carried out survey work at Audley End for both monarchs but in a civil engineering context he is best known for an early map of Chatham Dockyard (1633) and as Surveyor for the Commissioners of Sewers of Romney Marsh.

In the fourteenth century the Knelle Dam had been built alongside the Rother, directing the main river to the north of the Isle of Oxney. By the early seventeenth century the river had silted badly and 3,000 acres were 'drowned lands', perennially flooded, and 2,000 acres in Shirley Moor and Ebony were fit only for use as summer pasture.

In 1623 the great freshwater sluice was built across the channel below Appledore to regulate the flow of tides and river. These measures were not entirely successful in draining the 'upper levels' and in 1631 and 1633 agreements were signed between the Romney Marsh and Wittersham Level Sewer Commissioners to divert the river along a new course through Wittersham Level. The waters were directed through the newly embanked water course on 4 October 1635 but on Lady Day 1644 a very high tide broke through the walls and flooded much of the upper marsh. A new sluice was relocated far inland near Blackfriars Bridge in 1646 and much land lost to the flooding; it was not until the late 1670s that the reclamation of Wittersham Level was again seriously addressed.

In 1633 Harbord, as Surveyor General, produced a map, *The Topographical Description of ye Parte of ye River of Medway ... that lyeth betwene Rochester Bridge and the lower Mouth of St. Marie Creek, wherein are exactly noted the Ships of his Maties Navg ... The Olde and Newe Dockes with their particular Buildings, Storehouses and Offices.* As well as detailing the river defences of a barricade and chain, and vessels, the dock facilities are described, and the whole forms the most detailed survey of naval facilities at Chatham to that date. Between 1625 and 1636, £22,000 was spent on facilities there and this map was clearly made to record progress.

Harbord was knighted by Charles I on 29 May 1636. He married Mary van Alst and had two distinguished sons, Sir William Harbord MP (1635–1692), and Sir Charles, who died at the battle of Southwold in 1672. After the Civil War he remained Surveyor General to Charles II and served as MP for Launceston (1661–1679). One of his first acts on the Restoration was to commission Jonas Moore (q.v.) to survey the river Thames in 1662.

SUSAN HOTS

[DNB; J. Eddison and C. Green (eds.) (1988) *Romney Marsh: Evolution, Occupation and Reclamation*; F. Willmoth (1993) *Sir Jonas Moore*; R. Baldwin (1999) in lit.; Bendall]

HARDWICK, Philip, FRS (1792–1870), architect, was the son of Thomas Hardwick (1752–1829), a distinguished architect who had been a pupil of William Chambers. He was born on 15 June 1792, at 9 Rathbone Place, London, and educated at Rev. Dr. Barrow's School in Soho Square. In 1808 he entered the Royal Academy Schools, and became a pupil in his father's office. In 1815 he visited Paris, and spent a year in Italy (1818–1819).

He began independent practice in 1816 when he was appointed Surveyor to the Bethlehem and Bridewell Hospitals. His first major commission was as architect and surveyor to the St. Katharine's Dock Company at its foundation in 1825 and, as Surveyor, he was responsible for negotiations to purchase the more than 1,200 dwellings which covered the site; Thomas Telford (q.v.) was the engineer. The innovative design proposed the removal of goods directly from the ship to the warehouse, with no quayside space and lithographed plans dated 1824 and 1828 carry the names of both Telford and Hardwick. For the warehouses (1827–1829, demolished 1971–1981), Hardwick produced a severely functional design of brick walls, articulated only to allow the positioning of cranes, and reducing in thickness as the buildings gain height. Although Hardwick's interest in cast iron had been stimulated by his visit to France, he used brick vaults and timber beams to support the floors. Cast iron columns and beams were used only on the waterside frontages in order to provide open working spaces.

In 1828 Hardwick became Surveyor to the Goldsmith's Company and, from 1826 to 1856, to St. Bartholomew's Hospital in succession to his father. He was Surveyor to Westminster Bridge Estates from 1829.

Hardwick was one of the most influential architects of railway buildings, at a time when railway architecture was being invented stylistically. He was surveyor to the London and Birmingham Railway for whom he designed symbolic triumphal arches at each end of the line. At Birmingham an Ionic arch incorporated offices and booking hall and, in London, he built the famously elegant Greek Doric Euston Arch (1836–1840), demolished in 1962 to huge public outcry. At Euston he was also responsible for the Euston and Victoria hotels, although the Great Hall of the station was by his son, Philip Charles Hardwick (1822–1892).

Hardwick was often responsible for difficult negotiations with landowners whose property was required for railway purposes. In addition to his railway buildings, other notable works include Goldsmith's Hall, London (1829–1835), the City-Club House, London (1833–1834), and a house in Belgrave Square for Lord Sefton (1842–1845); he designed the Albert Dock Traffic Office, Liverpool (1846–1847), with a cast iron portico.

Hardwick was elected a member of the Institution of Civil Engineers in 1824 and a Member of Council in 1825, 1828 and 1829. He became a Fellow of the Royal Society in 1828 and was a Council Member. He was a founder member of the Institute of British Architects in 1834, Vice President in 1839 and 1841, and received its Royal Gold Medal in 1854. In 1839 he was elected an Associate of the Royal Academy, made RA in 1841, and was appointed Treasurer and Trustee in succession to Sir Robert Smirke in 1850.

In 1844 his health failed and he subsequently limited his activities to those he could carry out from his home, excepting only his activities on behalf of various societies. In 1861, his health became so poor that he withdrew entirely from architectural practice and from social life. He died on

28 December 1870 at his son's home in Wimbledon and is buried in Kensal Green Cemetery.

In 1819 he had married a daughter of the architect, John Shaw, by whom he had two sons, Thomas (1820–1835) and Philip Charles (1822–1892), who succeeded to his practice. His portrait is reproduced by Hobhouse (1976).

TESS CANFIELD

[DNB; Obituary (1871–1872) *Min Proc ICE*, **XXXIII**, 215–216; A. & P. Smithson (1968) *The Euston Arch*; H. Hobhouse (1976) Philip and Philip Charles Hardwick, in J. Fawcett (ed.), *Seven Victorian Architects*, 23–49; Skempton; C. Fox (ed.) (1992) *London World City 1800–1840*, 63, 64, 86 n76, 91 n91, items 12, 698; Colvin (3)]

Publications
1842. *Drawings of the New Hall and Library of Lincoln's Inn*

Selected works
1827–1829. St. Katharine's Docks, London, warehouses and offices, demolished 1971–1981
1836–1840. Euston Station, London, the Euston and Victoria Hotels, Doric Gateway and Lodges, demolished 1962
1838–1842. Birmingham, Curzon Street Station
1846–1847. Liverpool, The Dock Traffic Office, Albert Dock

HARE, Edward (*c.* 1741–1816), of Castor near Peterborough, was a surveyor and enclosure Commissioner. He was married twice, first to Mary Knowlton, who died aged twenty-six on 18 July 1771, giving birth to a son, Edward Knowlton Hare, and, later to Frances Layng in 1793.

He was one of typically two or three Commissioners named in an Act of Parliament for the enclosure of a defined area, their salaries fixed by the Act, usually at 2 or 3 guineas per day; for his work on Holland Fen at Frampton, Hare's remuneration was fixed at £63. He is known to have been involved in fifteen enclosures in Cambridgeshire, three of them with George Maxwell (q.v.), twenty-three in Northamptonshire, ten in Rutland, eleven in Lincolnshire and others in Hertfordshire and Nottinghamshire. In Cambridgeshire he earned £979 over fourteen years.

Some of these enclosures involved extensive drainage and sea defences. In 1793 the South Holland District comprised 19,400 acres between the rivers Welland and Nene. Hare, Maxwell and one other Commissioner were empowered to raise £38,400 to construct the New South Eau, 14 miles long, with ancillary drains, bridges and a tidal sluice; they were executed under the superintendence of Thomas Pear I (q.v.). By another Act of the same year, the South Holland Embankment was built north of the Old Sea Bank, 15 miles long, reclaiming about 4,500 acres from the sea. The enclosure award was made in 1811, at a total cost of £45,227. In 1800 Hare and Maxwell were joined by William Jessop and John Rennie (qq.v.) in presenting a report on draining

Deeping Fen, an area of 34,000 acres. An Act was obtained the following year under which Hare and Maxwell, with two other Commissioners, were obliged to carry out extensive works, but the steam engines at Pode Hole which had been recommended by Jessop and Rennie were not built and the scheme was not entirely successful.

Hare died on 4 April 1816 and is buried in the south transept of Castor church.

P. S. M. CROSS-RUDKIN

[W. H. Wheeler (1896) *A History of the Fens of South Lincolnshire*; M. W. Beresford (1946) Commissioners of Enclosure, *Economic History Review*; inf. ex. Dr. S. Bendall; inf. ex. F. Sheppardson; *Peterborough & District Family History Society Journal*, **17**(4), 1998]

Publications
1800. *Drainage of Deeping Fen* (with William Jessop, John Rennie and George Maxwell)

Works
1793–1796. New South Eau
1793–1811. South Holland Embankment
1794. Improvement of 2½ miles of river Welland outfall
1801. Deeping Fen

HARRAP, John (fl. 1769–d. 1812) was resident engineer on the Humber Dock at Hull. Although his signature is very clear, his surname is often misspelt Harrop or sometimes Harrup in contemporary records. He first appears in 1769 as a carpenter working under John Smeaton's resident engineer Luke Holt (qq.v.) on the Calder & Hebble Navigation and then in 1775 as a carpenter, also under Holt, on the construction of Hull (Old) Dock. After its completion he became in February 1779 a salaried employee of the Dock Company at £60 p.a. and effectively succeeded Holt as engineer from 1780, being variously described as 'overseer of the works', 'engineer' or 'surveyor'. This seems to have been a part-time appointment and, presumably on leave of absence, he could undertake other work. Thus in June 1781 he signed a contract drawing for a swing bridge on the Market Weighton Canal where Holt was now the engineer; he also carried out work on the locks. At Hull he rebuilt the walls of the entrance lock of the dock in 1785–1786. In 1790 a new swing bridge on the Driffield Canal was built under his direction.

The Dock Company raised his annual salary to £105 in 1792 and to £126 in 1800. During this period discussions took place on the possibility of building a second dock. Several engineers were consulted in 1793 and after a delay occasioned by the war with France the Company in 1800 resolved to send a Bill to Parliament as soon as peace returned. The enabling Act passed in June 1802, whereupon the Company appointed John Rennie (q.v.) as its engineer and Hull Corporation employed William Chapman (q.v.) to work with him. Harrap was the obvious choice for resident engineer but Chapman, realising that the work would be too much for him to handle on his own, recommended taking on an assistant and George Miller was accordingly appointed in this capacity.

Work on the Humber Dock, as it was called, began in 1803. Construction proceeded behind a large cofferdam segmental in plan and spanning 280 ft., built to withstand 30 ft. head of water. The dock had an area of 7.2 acres, providing accommodation for seventy ships of 180 tons burden, with an entrance lock 158 ft. long and 42 ft. wide with 26 ft. of water on the sills at spring tides. There was an entrance basin of 2.7 acres with two jetties projecting 100 ft. in front of a new river wall. The works were completed in 1809 at a cost of £233,000.

Harrap died at Hull in December 1812. A widower, he left his personal estate (valued for probate at £3,500) to his three married daughters.

A. W. SKEMPTON

[Market Weighton Drainage & Navigation Records (1781–1782) and Driffield Navigation Minute Book (1789–1790), both in East Riding RO; all other references from Mark Baldwin (1974) *The Engineering History of Hull's Earliest Docks*, TS, Science Museum Library]

HARRISON, Henry (fl. 1795–1813), land surveyor of Chester, was a superintendent under Thomas Wedge (q.v.) for the construction of the embankment at Rhuddlan Marsh from March 1795. Work continued over eighteen years, with Harrison's participation. In 1807 he was employed by local landowners to erect an embankment to enclose an area of salt marsh, known as Winemoor, at Cartmel, Cumberland. He designed an embankment 4,714 yd. long, between 6 and 14 ft. in height, with a slope between 1 in 5 and 1 in 7 on the sea face, and 1 in 2 on the landward face. In place the embankment was 130 ft. wide at the base, 4 ft. wide at the top, and covered in 'marsh soil', 4 in. thick on the seaward face, and 2 in. thick in the landward side. An area of 564 acres was enclosed and a total of 600 acres of pasture provided.

There are references to a Remmy [sic] Harrison providing advice and parliamentary support for William Madocks' (q.v.) embankment at Traeth Mawr in 1806–1807; it seems likely that this may be a misprint for Henry.

MIKE CHRIMES

[Rhuddlan embankment papers, Flint County RO; *Transactions of the Society ... of Arts*, **28**, 1810, 42–45; E. Beazley (1967) *Madocks and the wonder of Wales*, 138–138]

Works

1795–1813. Rhuddlan Embankment, resident engineer
1807–1808. Winemoor Embankment, 4,714 yd., £5,500

HARRISON, Thomas (1744–1829), architect and bridge designer, was the son of a joiner, born at Richmond in North Yorkshire and baptised on 7 August 1744. Having shown a talent for 'arithmetic, mechanics and drawing' he was sponsored by Sir Lawrence Dundas of Aske to go to Italy to study, his stay extending to seven years and including an unsuccessful submission for the annual architectural prize of the Academy of St. Luke in 1773, but he obtained access to membership of the Academy by the patronage of the Pope.

Soon after his return to England in 1776 Harrison made designs for a 'triumphal bridge' over the Thames at Somerset House, the site where Waterloo Bridge was built by John Rennie in 1811–1817. In the early 1780s he was living at Lancaster and in 1782 won a design competition for a bridge, Skerton Bridge, to span the River Lune, which when completed six years later exhibited important innovations in British bridge-building. His initial submission to the promoters included several alternative elevations with discussions of the arch forms and of the widths and cross-sections of waterway provided by each, quoting the best-known bridges in France and Italy, both old and recent, in support of his arguments. He then settled for elliptical arches as 'the lightest and perhaps the most beautiful; and when properly constructed, as strong as any other'. He designed an elevation of five arches with each of the spandrels between arches and over the abutments adorned with a pair of columns carrying a small pediment and framing the entry to a tunnel (which mimicked the flood passages in some ancient Roman bridges) through the spandrel. The classical details on the facades, together with his adoption of horizontal roadway and balustrades over the whole length of the structure, relate the design to the drawings of the Italian Renaissance architect Andrea Palladio and to a fashion for drawing 'triumphal bridges' which occurred in Rome in 1740–1760. That fashion had inspired several such drawings by other English architects as well as Harrison. The horizontality was an innovation in British bridge-building which was soon adopted by other designers of large bridges, including John Rennie. Rennie actually visited Lancaster while Harrison's bridge was under construction.

Harrison supervised the construction of Skerton Bridge closely and gained the reputation of an expert in bridge design and building. For a smaller public bridge at Derby he designed form and details similar to Skerton Bridge, and in the years 1792–1793 at least he appears in the Lancaster Quarter Sessions records to have fulfilled the functions of a bridge-master or county surveyor, attending the sessions to report on quite a number of (presumably small) bridges for which he directed repairs and/or rebuilding. Having transferred his residence to Chester, he began to undertake similar duties for the Sessions of Cheshire, being first named as 'county architect' in their records in 1800 although it was not until 1815 that he received a formal appointment with that title and a salary of £100 p.a. He was asked in 1800 to give a design for a new Stockport Bridge, for which he was paid in 1804. In 1805 he supervised construction of a new bridge at Craden Brook, and also in that year submitted a plan for the new Eden Bridge at Carlisle, but it was not adopted. In 1808 he offered advice on the proposed Cumberland Navigation Canal, intended to bring coasting vessels right into Carlisle.

He was subsequently asked for comment on a plan by George Dodd (q.v.) to bridge the Thames in London at the site adjacent to Somerset House for which he had made his 'triumphal bridge' design in 1776. Along with his critical comments on Dodd's design he sent two elevations, one of which is alleged to have corresponded in everything but the type of columns used on the spandrels to the bridge actually designed by John Rennie and built there in 1811–1817, when it was named Waterloo Bridge. However, Rennie had already built his bridge at Kelso in a similar style in 1800–1804 and can hardly be accused of copying Harrison's proposal. More interesting is the fact that Harrison extended his brief to suggest a scheme of continuous quays along the river from Westminster Bridge to Blackfriars, a proposal which can be seen now as a preview of the Victoria Embankment constructed several decades later.

After an Act of Parliament had sanctioned works to improve the old Ouse Bridge at York in 1809, Harrison was called on for a survey and advice; he recommended demolition and replacement with a new bridge, which was designed and built under the direction of the local architect Peter Atkinson junior (see Ruddock (1974) Hollow Spandrels in Arch Bridges).

His bridge design went seriously wrong when in 1813 the counties of Cheshire and Lancashire consulted him about a replacement bridge to cross the Mersey at Warrington. To obtain satisfactory gradients in a very built-up part of the town he designed a wooden bridge of a single arch of two ribs spanning 140 ft., each rib made up of five square balks of American red pine to a height of 5 ft. He created the curve of the arch by sawing halfway through each balk at every 6½ ft. and inserting wedges of iron or hardwood; the curved balks were then bolted and strapped together to form each rib. The ribs lacked stiffness and very soon after completion the bridge had to be propped from piles driven under the middle of the span. It was taken down and replaced in 1837. Another wooden bridge, and possibly of similar design but smaller span, was built at Cranage in Cheshire in 1816; how long it survived in use is not known.

Harrison made a second major contribution to the history of masonry bridges with his design for Grosvenor Bridge over the Dee at Chester. He is said to have made his first design some twenty-five years before work began in 1827. The design

then used was his but he declined, having reached the age of eighty-two, to take any part in the construction. The contract was supervised by his former pupil William Cole and Jesse Hartley (q.v.), the engineer of Liverpool Docks. The bridge was of a single arch of 200 ft. span, the longest masonry arch in the world. The architecture, with massive pilasters at the abutments topped with pediments and large niches at the level of the springings, is worthy of the huge arch, and Harrison deserves praise for such a grand conception, though credit for the interesting methods of construction must lie with Cole, Hartley and the contractor James Trubshaw.

Harrison's surviving correspondence (see especially Ockrim, chap. 9) reveals strong views about masonry, the influence of which is evident in his best bridges. He believed that architectural masonry should be made of the largest possible stones, such as he had seen in classical structures in Italy; and in expressing his belief he made no distinction between excellence of construction and of appearance. He commended and in some buildings succeeded in using columns of a single stone and monoliths up to 20 ft. long and weighing up to 20 tons such as he had seen in the bases of temples and in the pedestals of classical monuments. When using large stones he demanded of contractors the engineering skill to carry and place them, and high quality of mason work in dressing and laying them.

Harrison moved from Lancaster to Chester in 1793 and lived there until his death on 29 March 1829, aged eighty-four. He left a widow and two daughters. According to contemporaries quoted by Colvin, he was an architect of high professional skill and energy, but his career had suffered by his reserved and diffident behaviour. In the engineering of masonry bridges his few large works and the reports by which he explained them witness to wide knowledge and good judgement, together with the will and confidence to design bridges of unprecedented form and dimensions.

His extensive and important architectural works are listed and summarised by Colvin.

TED RUDDOCK

[D 533 A/TT 17, Derbyshire RO; *The Gentleman's Magazine* (1829) 468–470; S. Lewis (1835) *Topographical Dictionary of England*; J. B. Hartley (1836) Account of the New Bridge over the river Dee at Chester, *Transactions of the Institution of Civil Engineers*, **1**, 207–214; S. Blomfield (1863) Memoir on Harrison, *The Builder*, **21**, 203–205; J. W. Clarke (1958) The Building of Grosvenor Bridge, *Chester Archaeological Society Journal*, **45**, 43–55; A. C. Taylor (1969) Thomas Harrison and the Stramongate Bridge, Kendal, *Transactions of the Cumberland and Westmoreland Archaeological Society*, NS, **69**, 275–279; J. M. Crook (1971) A neoclassical visionary: the architecture of Thomas Harrison III, *Country Life*, **149**, 1088–1091; E. C. Ruddock (1974) Hollow spandrels in arch bridges, *The Structural Engineer*, **52**, 281–293; Ted Ruddock (19/9) *Arch Bridges and their Builders 1735–1835*, 123–124, 145–147, 186–188, and sources cited; M. A. R. Ockrim (1988) The life and work of Thomas Harrison of Chester 1744–1829, Ph.D. thesis, Courtauld Institute of Art, London, and sources cited; G. Woodward (1997–1998) The Brunels and the Grosvenor Bridge, Chester, *Transactions of the Newcomen Society*, **69**(1), 129–145; Colvin (3); DNB]

Works

1783–1788. Skerton Bridge, Lancaster, River Lune, five elliptical arches, each 68 ft. span with Palladian aedicules framing the entries to tunnels through all the spandrels; the road, cornices and balustrades all horizontal from end to end, contract price £10,400

1789–1790. Harrington Bridge, River Trent near Sawley, replaced by a steel girder bridge before 1932

1789–1794. St. Mary's Bridge, Derby, River Derwent, three elliptical arches, each *c.* 40 ft. span; road, cornices and balustrades horizontal; all ashlar, plain and rusticated, niches over piers

1793–1794. Stramongate Bridge, Kendal, River Kent, four segmental arches, shallow niches over piers

1812–1814. Warrington Bridge, River Mersey, single wooden arch, 140 ft. span, of two ribs 5–7 ft. deep of American or Baltic pine, contract price £3,000 (propped soon after construction due to structural deformation, taken down in 1837)

1814–1816. Shrewsbury, monumental column commemorating Lord Hill, adviser to designer, Edward Haycock, involving small modification to Haycock's design

1816. Cranage Bridge, River Dane, Cheshire, design, a wooden structure, obviously small but of unknown form, cost £250, apparently still in use 1835, date of subsequent removal unknown

1825–1826. Old Dee Bridge, Chester, widening, design

1827–1833. Grosvenor Bridge, Chester, River Dee, single arch, 200 ft. span, design

HARTLEY, Bernard (*c.* 1745–1834), architect and county surveyor, designed the Town Hall at Pontefract, built in 1785, and in 1796 he was of sufficient standing to be appointed arbitrator between the County and the contractor when it was decided to abandon the widening of Ferrybridge Bridge. The following year his was the successful tender for the rebuilding of this important structure to the design of John Carr (q.v.), which occupied seven years and cost £29,678. Having been appointed Surveyor of Bridges to West Riding of Yorkshire in January that year, on the resignation of William Gott Hartley employed Richard Allison as his agent at Ferrybridge.

As Surveyor of Bridges, Hartley's salary was £300 p.a., which was increased to £400 in 1803 and £600 in 1811, the latter amounts being significantly higher than those paid in other counties. From 1819, he held office jointly with his son, Bernard II (1779–1855), and most of the official reports were signed by both of them. In an extensive county with several large rivers, expenditure on bridges was quite high, being typically of the order of £10,000 p.a. Hartley lived in Ropergate at Pontefract, towards the centre of the county, but with large distances to travel he employed deputies such as William and Thomas Anderton of Gargrave to supervise the contractors in the more distant areas. That he himself undertook a lot of journeys may be gauged by the payment to him in 1801 of an extra £50 'for travelling expences on account of the high price of hay and corn'.

Hartley developed a distinctive architectural style for the elevations of his bridges: in the tooling of the stonework, radial courses extending through the spandrels and particularly in the sophisticated shape of the cutwaters. His contract drawings were much more detailed than those produced by other county surveyors of his day, detailing the sizes and shapes of the individual stones to be included. He also developed a standard bill of quantities which contractors were required to submit with their tenders, in order to make comparisons more easy.

In 1818 he produced a plan for a cast iron bridge with a single 85 ft. span over the river Calder at Mytholmroyd, quite forward looking for its day, but it was not used and the existing bridge was widened in 1823–1824.

He died on 5 February 1834 and as there is no record of his burial in the parish registers, it may be guessed that he belonged to a non-conformist church.

P. S. M. CROSS-RUDKIN

[West Yorkshire Archives Services; E. Baines (1822) *History, Directory and Gazetteer of the County of York*; *Gentleman's Magazine*, 1834, **1**, 342; Colvin (3)]

Works

1797–1804. Ferrybridge Bridge, contractor

The following bridges were widened or rebuilt by others under contract at a cost exceeding £1,000 during Hartley's Surveyorship:

1800–1801. Tamewater Bridge, £1,136

1801–1803. North Bridge, Ripon, widened, £1,255

1802–1804. Cock Bridge, Tadcaster, £1,409

1804–1805. Thurne (Turn) Bridge, Cowick, £1,168

1804–1806. Silsden Bridge, £3,529

1805–1808. Castleford Bridge, designer, three arches of 57 ft., 67 ft. and 57 ft. span, built by his son Jesse Hartley (q.v.), £14,160

1808. Collin Bridge, £1,350, supervised by Jesse Hartley (q.v.)

1809–1811. Low Gill Bridge, £1,017

1810–1812. Bawtry Bridge, £1,874

1810–1812. West Tanfield Bridge, widened, £3,485

1811–1813. Hubberholme Bridge, £1,681

1812–1814. Wreaks Bridge, Birstwith, £3,791

1813–1814. Skell Bridge, Ripon, £1,659
1812–1814. Ramsgill Bridge, £1,983
1812–1814. Holme Bridge, widened, £2,089
1812–1815. Pool Bridge, widened, £5,083
1813–1814. White Bridge, £1,039
1814–1815. Collingham Bridge, over Collingham Beck, £1,031
1816–1817. Gargrave Bridge, £2,951
1817–1818. Saltersbrook Bridge, £1,557
1818–1820. Lower Hodder Bridge, Hurst Green, £1,802
1824–1827. Wetherby Bridge, widened, £4,442
1826–1830. High Bridge, Knaresborough, widened, £1,395

HARTLEY, Jesse (1780–1860), dock engineer, was born 1780 in Pontefract, Yorkshire, a son of Bernard Hartley, noticed above, bridge builder. His early training was as a mason and bridge builder, at least in part under his father, and the bridge at Castleford (1805–1808) bears the names of Bernard Hartley as designer and Jesse as contractor. Sir Robert Rawlinson was later to describe him as 'the greatest stonemason England or any other country has [ever] produced'. After working in Ireland c. 1808–1818, during which time he both designed and built bridges and married Ellenor Penney of Dungarvan in 1809, he returned to England and worked as Bridgemaster in Salford.

In 1823 the Liverpool Dock Trustees were becoming seriously embarrassed by the improprieties of their Surveyor, John Foster (q.v.) and advertised for a Deputy Surveyor: writ large between the lines were the words 'Surveyor-designate'. The appointment of a bridge builder with no known dock or harbour experience seems a strange choice, and the *Liverpool Mercury's* suggestion that the Trustees were seeking honesty, force of character and managerial rather than engineering skills seems plausible. Be that as it may, within a fortnight of his appointment, the new Deputy was indeed Surveyor, albeit on a probationary contract.

Initially, his engineering work was largely a matter of continuing previous practice, for he rightly identified the management and control problems as more pressing. Existing safeguards, rendered nugatory by Foster, were enforced and new ones added. The drawing office expanded to produce the detailed progress drawings he required and work formerly done by contractors was increasingly taken in-house. By 1828 we see the recognisable form of the Dock Yard, the first civilian integrated civil and mechanical engineering works in the country, possibly the world. Princes Dock had been completed, and Clarence and Brunswick docks begun.

In 1826 Hartley leased a granite quarry at Craignair and in 1830 another one at Kirkmabreck. At one level he was continuing the elimination of corruption by means of vertical integration but at another he was initiating the most distinctive visual feature of his work, the so-called 'Cyclopean Granite'. These quarries had dislocated beds, which meant that for every decent ashlar block there was a great deal of rubble, and Hartley perfected a method of facing sandstone rubble cores with dressed granite rubble which produced retaining walls which were cheap, durable and effective. It was imitated only by his successors and a handful of his pupils but proved as successful in their hands as in his.

The first dock Hartley saw through from conception to completion was Clarence (enabled 1825, completed 1830), originally set apart for steamships and constructed at a distance from the sailing ship docks to prevent the former setting fire to the latter. Here, as in all his later schemes, he designed an ensemble rather than just a dock: there was the wet dock, two graving docks, a graving dock basin (with grid) and a half-tide entrance dock. But such was the pressure of trade that before Clarence was opened, Hartley was already designing Waterloo, Victoria and Trafalgar docks, which bridged the 'fire-gap' between it and Princes Basin.

At the south end of the estate, he completed Brunswick in 1832. Originally enabled in 1811, this had been a problem project and was, at over 12½ acres, the largest Liverpool dock to date; it was geared to the timber trade and had characteristically low quay walls. These works were not sufficient fully to occupy the energies of the man, dour, thickset, short-necked and noted for his powers of vituperation. He undertook (with permission) a number of outside works, including the Grosvenor Bridge, Chester, then (1826) the longest masonry span in the world. He also reported on other dock projects (see below) and surveyed the Manchester, Bolton and Bury Railway and the Carlisle Canal. As consultant to the Liverpool and Manchester Railway he took on the daunting task of telling George Stephenson that his proposed 'Gothic' Sankey Viaduct was impracticable and his alignments of the Wapping Tunnel wrong; he was successful in both interventions. A great turning point in Hartley's career occurred in 1840 when he was given a salary increase in return for waiving the right to undertake outside work; this made him Britain's first full-time civilian dock engineer.

In so far as one can or should appraise Hartley's early career, his most important achievements in dock engineering were probably managerial and organisational. His determination to grub out corruption brought about better supervision of work and clearer reporting lines which not only resulted in his employers getting more for their money, but brought desirable side effects, including a very low accident rate on his working sites. The most striking characteristic of his later career—the ability and willingness to innovate when, but only when, necessary—was already noticeable.

ADRIAN JARVIS

[Ladies Diary owned by Hartley, Minutes of the Docks Committee, their subcommittees and successors 1823–1860; autograph drawings and (larger) numbers of authorised drawings; MS reports; specifications; contracts; correspondence, etc.; copy material from his other works; his will, etc. (his personal papers were burned, on his orders, shortly after his death in 1860), Maritime Archives and Library, Merseyside Maritime Museum; his *Memoir* in the *Minutes of Proceedings of the Institution of Civil Engineers* was published as an afterthought to that of his son John Bernard, in **XXXIII**, 1871–1872, 216–222; N. Ritchie-Noakes (1984) *Liverpool's Historic Waterfront*; A. Jarvis (1991) *Liverpool Central Docks 1799–1905*; I. Weir (1993) Port of Liverpool Quay Walls, unpublished M.Sc. dissertation, University of Liverpool (History of Science and Technology); A. Jarvis (1996) *The Liverpool Dock Engineers*; A. Jarvis (1998) Theory and Practice in Dock Engineering, *Transactions of the Newcomen Society*, **69**, 57–68]

Publications

Publications prior to 1830:

1825. *Littlehampton Harbour*
1831. *Swansea Harbour*
1836. *The Surveyor's report to the Dock Committee on the General State and Progress of the Dock Works*

(N.B. In 1836, just before the Municipal Reform Act came into effect, Hartley published a massive report on his stewardship of the Dock Estate since 1824. Thereafter, his report was published annually, but for the period covered by this volume the, 1836 Report is all there is)

Works

Works, prior to 1830:

1805–1808. Castleford Bridge
1813–1816. Causeway Bridge, Dungarvan, Ireland
1824–1825. Completion of Princes Dock (sheds, etc.)
1825. Improvements to Georges Dock
1826. Infilling and reclamation of The Old Dock
1826–1828. First phase of Dock Yard
1826–1830. Grosvenor Bridge, Chester, completion
1829. Canning Dock

HASCHENPERG, Stephen, Von (fl. 1539–1544), military engineer, known as Stephen the Almaine and a Gentleman of Moravia, worked on most of Henry VIII's transitional castles from 1539–1543 including Camber, Deal, Sandgate, St. Mawes, and Pendennis where he is variously described as deviser, engineer and overseer. He was also sent to prepare a survey of the Calais Pale in 1540.

From 1541–1543 he worked on the fortifications at Carlisle only to be suspended for spending 'a great treasure to no purpose', a somewhat ironic comment given the siege of Boulogne cost £587,000. He wrote to Henry from Lubeck in 1545 proposing a hydraulic invention to convey water

economically for the Palace of Nonsuch, but ended his life as steward to the Bishop of Olmutz.

SUSAN HOTS

[B. H. St. J. O'Neil (1945) Stefan von Haschenperg ..., *Archaeologia*, **91**, 137–155; H. M. Colvin *et al.* (1975–1982) *King's Works*, vols. 3 and 4; J. Harvey (1987) *English Mediaeval Architects*]

Works

Total under Henry VIII:

1539–1543. Deal Castle, *c.* £27,000
1539–1543. Sandgate Castle, £5,500
1539–1543. Camber Castle, £15,800
1539–1543. Pendennis Castle, £5,500
1539–1543. St. Mawes, £5,000
1539–1543. Calais, £120,700

HASTIE, William (*c.* 1755–1832), architect and town planner, was trained as a mason in Scotland before responding to an advertisement in the *Edinburgh Evening News* of 21 January 1784 for Scottish artisans prepared to work in Russia. The advertisement was placed on the advice of Charles Cameron (1745–1812), the Scottish architect who had been invited to Russia in 1779 by Catherine the Great, probably on the strength of his book *The Pathos of the Romans Explained and Illustrated* (1772). The scale of Cameron's commissions in the early 1780s made urgent the recruitment of additional skilled workers and Hastie was one of sixty to seventy men engaged in 1784. Cameron was working at the Palace of Tsarskoe Selo, the adjoining town of Sofia, and Pavlovsk, as well as Bakchiserai in the Crimea. It is most probable that Hastie was chiefly engaged in design work at Tsarskoe Selo as this is the location for many of the designs for villas, summer houses, etc., which have survived in an album of his drawings at the Hermitage.

By 1794 Hastie had come to the attention of Catherine as an architect in his own right and from 1795–1799 he worked for her favourite Prince, Platon Zubov, on the restoration of the Kharn's Palace at Bakchiserai and other projects. For reasons which are now unclear, he then worked with Charles Gascoigne (q.v.) on the construction of the Kolpino ironworks (1801–1804).

Whatever the explanation, this was an important contact for Hastie as his next patron, Count Rumianzov, obtained work for him in the Office of Waterways. Apparently he discussed the practicability of constructing cast iron arch bridges with Gascoigne, developing a system which used cast iron voussoir blocks bolted together to build up very solid arches; the system, similar to that patented in England by John Nash (q.v.), was first used by Hastie to replace the existing timber Polizeiski bridge across the Moika in St. Petersburg. The bridge, 27 m long and 21 m wide was built up of voussoirs 3 m ×1½ m, nine such sections making an arch, and fourteen such arches making up the bridge. The castings were made at Gascoigne's works and the bridge was opened on

14 November 1806; it proved the successful prototype for a number of similar designs of 15–30 m
spans and the system was used by E. A. Adam and
P. P. Bazaine between 1824 and 1840.

Following on this success Hastie was appointed chief architect for the development of
Tsarskoe Selo in 1808. His plans for the town
were executed by V. P. Stasov and others,
Hastie's own brief being extended to planning
and developing towns throughout Russia. His
works included many provincial towns: Kiev,
Vilna, Smolensk, Ekaterinoslav, Tomsk, Saratov,
Viatka, Penza, Krasnojarsk, Omsk, Schluessel-
burg, and proposals for the reconstruction of
Moscow after the 1813 fire which were incorporated in the plans agreed in 1818. Hastie died at
Tsarskoe Selo on 4 June 1832

MIKE CHRIMES

[(N.B. Many of Hastie's drawings survive in
Russian archives, notably the Hermitage in St. Petersburg) M. Korshunova (1974) William Hastie in
Russia, *Society of Architectural Historians of
Great Britain, Journal*, **17**, 14–19, figs. 12–16; M.
Korshunova (1977) *Architekhtov V Geste* (1755–
1832) *Trud 'gosudarstvennogo Ermitazha 18*,
5, 132–143; A. L. Punin (1982) *Arkhitektura
otechestvennykh mostov*, 9–11; J. G. James (1983)
*The Application of Iron to Bridges and other
Structures in Russia to about 1850*; A. L. Punin
(1990) *Arkhitektura Peterburga seredin, xix veka*,
91–92; Colvin (3)]

Works
1806. Police Bridge, Moika, St. Petersburg, cast
iron arch, engineer, 92,235 roubles
1814. Alexandrovskii Bridge, Vvedensky canal,
St. Petersburg, cast iron arch, engineer
1814. Krasnyi Bridge, St. Petersburg
1816. Zhelti (Potzeluev) Bridge, St. Petersburg,
cast iron arch, engineer, 134,062 roubles
1818. Sini Bridge, Moika, St. Petersburg, cast iron
arch, engineer, 326,434 roubles
1818(?). Obvodni Canal Bridge, Moskovski
Prospekt, St. Petersburg, cast iron arch, engineer

**HAVILLAND, Thomas Fiott, de, Lieutenant-
Colonel** (1775–1866), military engineer, was born
on 10 April 1775 at Havilland Hall, Guernsey, the
eldest son of Sir Peter (d. 1821) and Cartarette de
Havilland. In 1792 he joined the Madras engineers
as a cadet; by 1824 he had been promoted to lieutenant-colonel. He 'amused himself in surveying'
to compile a map of Coimbature and Dindigul
(1800) and, two years earlier, had produced *A
Sketch from Tanjore, west to the Sources of the
Colleroon*. He served on a number of military campaigns including the siege of Pondicherry (1793),
the Columbo (1795–1796), Tippoo campaigns
(1799) and the Egyptian expedition of 1801, after
which he had been briefly captured by the French
in January 1804. While in Egypt he had surveyed
Lake Moeris (Birkat Qarun) and the Cairo–Suez
desert areas for water.

He then returned to India and completed a
survey of the Deccan begun by Colin Mackenzie
(1754–1821). In January 1807 he was put in charge
of repairs at Seringapatam and in 1809–1810 was
involved in a mutiny there, occasioned by the appalling treatment of the Army by the Government.
He was temporarily dismissed from the service in
1810 and not restored until 1812 when he was on
leave in Guernsey. On something of a busman's
holiday, he was commissioned to build Jeybourg
Barracks while there.

This Commission heralded something of a
change in direction in Havilland's career as on his
return to Madras in 1814 he was appointed civil
engineer and architect for the Presidency, his
building work including the completion of the
construction of St. George's Cathedral (1814–1815)
to the designs of Captain James Caldwell (q.v.) and
the design and construction of St. Andrews Presbyterian Church (1818–1821). In 1822 he compiled a
report on building limestones and urged that
samples be collected from the various Madras districts for analysis and testing.

As a civil engineer Havilland's achievements included the brick-built Seringapatam bridge over
the Cauvery River, the low rise (11 ft.) of its five
110-ft. spans leading him to construct a model of
a single arch of the structure to demonstrate its
practicability. At Madras he built the Mount road
and began to seriously address the problems of
the harbour. In 1770 Warren Hastings had sought
the views of Brindley, Smeaton and Mylne (qq.v.)
on a pier at Madras and in 1798 Captain W. C.
Lennon, a Madras engineer, drew up plans for a
harbour, estimating the cost of the works at
£450,000. Little practical progress had been
achieved before Havilland was appointed civil
engineer but, after six months observations, he
fixed the mean sea level and then designed a sea
wall, known as the Bulwark, to protect the 'Black
town' shore from damage. In 1836 he was consulted by (Sir) Arthur Cotton about a proposal for
a breakwater there; he supported Cotton's proposals although they were not acted upon.

Havilland was clearly a dynamic figure and
more or less his final act which followed his appointment as chief engineer in February 1821 was
to compile a report on the organisation of the
Madras Corps of Engineers, the existing problems
in its administration, its relationship with civil and
military authorities, the fact that the 'pioneers' did
not report to the engineers, and aspects of the
need for skilled officers to take responsibility for
public works. Havilland's original report was
dated 23 November 1821 and there followed a
fiery exchange with the Quartermaster-General,
Colonel R. B. Otto, who clearly saw some of his
own responsibilities threatened. Havilland was
particularly emphatic on the need for trained engineers for civil works: 'I cannot admit that road
making which comprehends bridge building and
the directing of waters, needs neither skill nor science'. Some at least of Havilland's suggestions
were acted upon over the following decade.

Havilland's belief in the importance of a scientific approach to the engineers' work was reflected in his translation of Dubuat's *Principes d'Hydraulique*. This text, one of the earliest in English on the subject, was all that was readily available to engineers like Captain John Colvin (q.v.) who, in the 1820s, began work on the Western Jumna canal.

Following the death of his father, Havilland left for England in January 1824 and retired in April 1825, then returning to Guernsey where he immersed himself in local affairs. In 1808 he had married Elizabeth de Saumarez; they had two sons, Thomas and Charles Ross, who both predeceased him, and two daughters. Following Elizabeth's death in 1818 he had married Harriet Gore in 1828. He died at Beauvoir, Guernsey on 23 February 1866.

MIKE CHRIMES

[Papers, 1805–1850, Add. 9548, Dept. of Mss, Cambridge University Library; DNB; H. M. Vibart (1883) *The Military History of the Madras Engineers and Pioneers*, **1**, 281, 288, 327, 442–448, 587–602; **2**, 1–35; E. W. C. Sandes (1933–1935) *The Military Engineer in India*, **1**, 171, 200, 228; **2**, 170–171; R. H. Phillimore (1945–1954) *Historical Records of the Survey of India*, vols. I and II, *passim*; E. W. C. Sandes (1948) *The India Sappers and Miners*, 81–83, 117; J. Brown (1978) Sir Proby Cautley (1802–1871), a pioneer of Indian irrigation, *History of Technology*, **3**]

Publications

1822. *Chevalier Dubuat's Principles of Hydraulics*, vol. 1
1836. Madras breakwater, *Reports of the Corps of Engineers Madras Presidency*, **1**, 1839, 4–6

Works

1799. Seringapatam Bridge, five spans
c. 1815–. Mount Road, Madras, and St. Andrews Bridge (13,000 pagodas)
c. 1815–. Bulwark sea wall, Madras, c. 3,000 ft.

HAWES, William (fl. 1817–1851), soap manufacturer, was the son of Benjamin Hawes (d. 1861), soap manufacturer of Old Barge House, Bankside. His father was a close personal friend of Sir Marc Isambard Brunel (q.v.), acting as a director and eventually Chairman of the Thames Tunnel Company. The friendship was reciprocated amongst the children, Sophia Macnamara Brunel marrying (Sir) Benjamin Hawes (1792–1862), who worked in the family business before becoming an MP and, eventually, Permanent Under-Secretary for War. William was, from childhood, a close personal friend of Isambard Kingdom Brunel and was an important source of information for the *Life of Isambard Kingdom Brunel* (1870). He was well aware of the Brunels' experiments with liquefied carbonic acid gas for engines, his father having given evidence on oil gas lighting to the House of Commons. In 1851, as Chairman of the Australian Royal Mail Company, he asked Brunel Jr. to advise

on the design of two steamships to provide a fast regular service between England and Australia, ships that were built by John Scott Russell. He worked in the family business and was described as a 'soap manufacturer' when he joined the Institution of Civil Engineers in 1829; he remained a Member only for a short period.

MIKE CHRIMES

[Membership records, ICE; Minutes of Evidence (1824) *Committee on the London and Westminster Oil Gas Bill*, 187–197; I. Brunel (1870) *The Life of Isambard Kingdom Brunel*; C. B. Noble (1938) *The Brunels, Father and Son*; G. S. Emmerson (1977) *John Scott Russell*, 56–64]

HAWKINS, John, FRS, Major (1783–1831), military engineer, was born on 19 April 1783, the son of Richard and Mary Hawkins of Kingsbridge, Devon. He went to Bombay as an Ensign in 1798, and while serving as a Lieutenant, in 1810, Hawkins helped carry out a Revenue survey of the Island. This work was interrupted by a head wound caused by being thrown from his carriage in 1812. He was elected a Corresponding Member of the Institution of Civil Engineers in 1822. Subsequently he was responsible for the design and supervision of the construction, between 1824 and 1829, of the Bombay Mint, a quadrangular building on two floors erected on land reclaimed from the sea.

Hawkins died at sea in April 1831, acting as both 'Mint Engineer' and Inspecting Engineer of the Bombay Presidency. He was married twice, to Frances Schutz Drury (d. 1818) who predeceased him, and Susan, who survived him.

MIKE CHRIMES

[ICE membership records and minutes conversations, ICE archives; E. W. C. Sandes (1935) *The Military Engineer in India*, **2**; R. H. Phillimore (1950) *Historical Records of the Survey of India*, **II**]

HAWKINS, John Isaac (1772–1854), engineer and inventor, was born at Taunton, Somerset on 14 March 1772, the son of Isaac Hawkins, a watch and clock maker. He went to the United States when young and for a while studied medicine at the University of Pennsylvania, but gave it up. In the early 1790s he interested himself in gas manufacture and, with a friend, demonstrated it in Philadelphia. For a time Hawkins lived at Bordentown, New Jersey, working with a local inventor, the Revd Burgess Allison. He was a good musician and while in America patented the upright piano as we now know it (English patent No. 2446, 1800).

He returned to England in August 1803 and set himself up as a consulting engineer and patent agent. He was involved from 1809 with The Thames Archway Company's scheme for a tunnel under the Thames near Rotherhithe incorporating the construction of cylinders of brick rings which were to be sunk in a trench excavated in the bed of the river. The scheme foundered in November

1811 due to the failure of the joint-stock company, but not before an experimental section of the tunnel was built and tested by Hawkins. Needless to say problems were experienced in correctly aligning adjoining cylinders on the river bed.

Hawkins had twelve English patents under the Old Law, the first two being taken out under his father's name since he was then still living in America. He was a versatile inventor; as well as the piano, he patented a claviole or finger-keyed viol which could imitate various musical instruments. Other patents included the ever-sharp pencil and a pentograph, a device for making multiple copies of correspondence. His numerous other inventions included gold pens with hard iridium nibs and a method of condensing coffee.

Hawkins's work in England is not well documented, working as he did as an adviser to inventors and patentees, and much of what is known has to be derived from his numerous contributions to discussions at the Institution of Civil Engineers, which he joined in 1824. From then until the summer of 1827 he was one of its most prolific contributors. His interests encompassed all branches of the profession of that time: puddle clay linings, steam power, drills, bearings, heating, coal gas etc. and they reflected his travels in North America and Europe. It would appear that he was an intimate of Jacob Perkins (q.v.). For some years he was involved in the erection and fitting up of sugar refineries at home and abroad, including one in Vienna. From 1827 until 1829 and again in 1830–1831 he was principally engaged in Austria–Hungary and was consulted about a proposed 300 mile canal from Vienna to Trieste. He was able to inspect von Mitis's steel suspension bridge in Vienna and gave an account of it to the Institution and to the British Association.

Hawkins published a translation of Camus's study of gearing teeth in 1806 (anonymously) and again in 1837. He was the first to appreciate the value of the involute curve in tooth outline. Camus, of course, dealt only with the epicycloid form and Hawkins only came to realise the error of this at the very end of the translation, and after much had been typeset; as a result his conclusions appear as an appendix to the work.

Hawkins's fortunes waned in the railway age. His changes of address were frequent and not for the better: Hampstead Road (1830), Quality Court (1839), Judd Place West (1843) and Charles Square, Hoxton (1847). In the late 1840s he was nearly destitute and with the support of friends returned to America in 1849. He started *The Journal of Human Nature and Human Progress* in 1852. This is something of a self-exposition of his accomplishments in London. His hopes of obtaining a receptive audience appears to have been unsuccessful and he died at Elizabeth Town, New Jersey, on 24 June 1854; it was nearly ten years before the Institution learned of his death. Hawkins once said of himself: 'The Creator has constituted me an inventor, and I consider every useful invention given me as a commission from Him in trust for the benefit of mankind'.

A. P. WOOLRICH

[Membership records and minutes of conversations, ICE archives; original communications 98, 125, 436, ICE archives; *Repertory of Arts and Manufactures* 5, 2nd series (1810–1813) **20**, 2, 141; **24**, 18; T. Gill (1827) On La Riviere's half round drills, *Technical Repository*, **XI**, 314–315; C. Knight (1857) Biography, *The English Cyclopaedia*; Obituary (1866) *Min Proc ICE*, **25**, 512–514; F. Boase (1892–1921) *Modern English Biography*; E. Pauer (1896) *A Dictionary of Pianists*; J. Wickham Roe (1916) *English and American Machine Tool Builders*; Rosamond Harding (1978) *The Piano-Forte*, 2nd edn.; Stanley Sadie (ed.) (1988) *The Piano*; M. Friesen (in press) Forthcoming thesis-friesen@mc.net]

Publications

1808. *Plan and Catalogue of Hawkins' Museum of useful and mechanical Inventions*

1812. Observations on building the piers and abutments of bridges ..., *Repertory of Arts, Manufactures*, 2nd series, **20**, 141–146

1826. Letter to Pennsylvania Society for promotion of internal improvement, *Franklin Institute Journal*, 1st series, 93–97

1827–1828. On sugar refining, *Repertory of Patent Inventions*, 3rd series, **5**, 219, 271, 340, 396

1828. Description of an hydraulic weighing machine, *Franklin Institute Journal*, 2nd series, **1**, 336

1828. On the valuable properties of the genuine emery stone, and on a superior process of washing, *Emery Franklin Institute Journal*, 2nd series, **1**, 88

1829. On the German polish for wood, *Gill's Technological Repository*, **V**, 103–105

1830. On a domestic steam bath, *Gill's Technological Repository*, **VI**, 294–304

1832. On the steel suspension bridge recently built over an arm of the Danube at Vienna, *British Association Report, 2, Trans of Sections*, 608–609

1833. An investigation of the principles of Mr. Saxton's locomotive differential pulley *British Association Report, 3, Trans of Sections*, 424–426 (and *RPI*, **16**, 223)

1837. Translation of Camus on gearing, *A Treatise on the teeth of Wheels*

1839. On wood paving in Vienna, *Mech Mag*, **31**, 307

1843. On the formation of concrete, etc., *British Association Report, 13, Trans of Sections*, 99–100 (printed separately, essentially a distillation of material presented to ICE ten or more years earlier)

1845. *The History and Resuscitation of the Claviole*

1847. On Trevithick, *Mech Mag*, **46**, 308

HAWKS family (fl. 1750–1863) members were involved in one of the major iron manufacturing concerns in the Gateshead area, the others being that of Ambrose Crowley (q.v.) and that of the

Abbot family. The firm existed under a variety of names from c. 1750 until 1889.

William Hawks Jr. (1730–1810) was the eldest child—one of twelve—of William Hawks (1708/ 1709–1755) and his wife, Jane England, and was born in Gateshead, being baptised there on 29 June 1730. He served an apprenticeship as a blacksmith at the works of Ambrose Crowley, for whom his father had worked before establishing his own works c. 1747, and then joined his father where he 'soon displayed remarkable ability in everything that concerned the welfare of the firm'. Following the death of Hawks Sr. in 1755 he became head of the firm and brought about its extraordinary expansion. Initially, the output of the firm was general ironmongery, products similar to those produced by Crowley, Millington & Co. at Swalwell.

In 1779 Hawks, with Thomas Longridge, began operating a forge at Lumley and two years later, again with Longridge, purchased the Bedlington ironworks; both works were subsequently extended. He continued the philosophy of his former employers in extending his works to places where water power could be harnessed, setting up in 1829 a new rolling mill and forge at Beamish where a water wheel was the source of power. The works at Bedlington were sold in 1809 but those at Beamish operated until 1877 when the mill dam failed. At Gateshead, where a lease from 1770 had included a dam and reservoir, Hawks installed a 12 hp Boulton & Watt engine in about 1790 and it was soon followed by a second engine of 22 hp.

What added greatly to the reputation and profitability of the firm was the securing of Government contracts for all types of ironwork. One of the firm's principal products was anchor chains; Hawks invented a machine for turning the links and in 1804 obtained a patent (No. 2776) for making chains. His reputation was such that he was at times called upon as consultant and it is perhaps through these activities that in 1804 he came to manufacture the castings of a bridge for the Marquis of Cornwallis at his estate, Culford Park, Suffolk. With a span of 60 ft. the bridge—designed to a patent of Samuel Wyatt (q.v.)—contained some 80 tons of cast iron and the castings cost approximately £1,457. There could well have been other bridges cast during the same period.

The firm later consolidated much of its work at Gateshead but in 1803 a proposal was made to build a canal from the Beamish ironworks to the river Tyne. No work ensued but it is difficult to imagine that this scheme, a modification of an earlier proposal by Ralph Dodd (q.v.), was not promoted by Hawks, especially as the firm appreciated the advantages of transport by water and actually owned four ships for the carriage of their products.

In 1759 Hawks had married Elizabeth Dixon (1736–1808) and it was written that 'much of his subsequent success in life was owing to her

energy and talent'. They had a family of eight boys and five girls, born in Gateshead between 1760 and 1780; four of their sons, George (1766–1820), William (1772–1807), Robert—to become senior partner—and John (1770–1830) became partners in the firm. After the death of his first wife, Hawks married Elizabeth (d. 1831), the widow of Joseph Atkinson.

William Hawks died on 4 December 1810 and was buried in Gateshead, the *Newcastle Courant* noting the 'extensive iron manufactory which had for its origin his personal exertions'.

Sir Robert Shafto Hawks (1768–1840) was the fifth child of William Hawks Jr. and his wife, Elizabeth, and was born in Gateshead on 17 May 1768. At the time of his marriage he was described as a 'clothier' and later as a 'woollen draper' but, nevertheless, he probably maintained an interest in the firm as on the death of his father he took over as head of the partnership. It was written that he 'was a most active and industrious man and carried on the works as successfully as his father had done'. His accession to management was accompanied by the erection of a testing machine—said to have been the first in England—for proving anchor chains to a loading of 90 tons. In 1814–1815 Hawks was responsible for the manufacture of parts—excluding the boiler and chassis—for a three-axled locomotive designed by William Chapman and John Buddle (qq.v.) for use at Whitehaven Colliery and later, c. 1830 and again through the same men, for a steam operated crane at Seaham Harbour.

Hawks married Hannah Pembroke Akenhead (1766–1863) at Gateshead in 1790 and they had two sons, neither of whom became involved in the partnership. Active in the Northumberland Volunteers, Hawks 'showed so much zeal in suppressing the riots in the North in the winter of 1816' that he was knighted by the Prince Regent the following year. Hawks left the partnership in 1838 and died on 23 February 1840 at his home in Clavering Place, Newcastle; he was buried at Gateshead, survived by his wife and sons.

George Hawks (1801–1863) was the eldest son of John Hawks and his wife, Jane (1777–1841), daughter of Thomas and Jane Longridge; he was one of a family of eleven. In 1827 he married Elizabeth Clark Wright at Wallsend and the union produced a family of seven.

In either 1838 or 1841 Hawks became senior partner in the firm and in 1839 obtained a patent (No. 8243) for the construction of carriage wheels for railway use. By 1843 the firm had assumed the title of Hawks, Crawshay & Sons as a result of the death of Frank Stanley, a partner, and the transfer of his interest in the company to his brothers-in-law, George and Edmond Crawshay, the great-grandsons of Richard Crawshay (q.v.).

The firm was by this time owned by the Crawshays with George Hawks as manager or, perhaps, as a figurehead; the division of responsibility, however, is unclear but it was at his time that its greatest expansion took place, in 1844 a

Nasmyth steam hammer being installed, an early example. George Hawks's interest in the firm virtually ceased with the building of the High Level Bridge at Newcastle, where the placing of its 5050 tons of ironwork culminated in 1849 with the last link being driven by him. In 1835 he had become the first Mayor of Gateshead; an ardent Liberal, he was again mayor in 1848 and 1849 and is commemorated by a statue in Gateshead.

Hawks died at his country house at Pigdon, near Morpeth, on 15 October 1863, survived by his widow, three sons and two daughters. By then the company employed some 1500 men at a works occupying some 47 acres in Gateshead. Employment was to reach 2000 by 1889, the year of the firm's sudden closure 'in somewhat mysterious circumstances' which resulted in George Crawshay (1821–1896), at his death seven years later, leaving estate valued at only £25.

R. W. RENNISON

[*Newcastle Courant*; *Newcastle Weekly Chronicle*; *Gateshead Observer*; J. Buddle, *Colliery Memoranda* (1804) Northumberland RO, BUD/24/34; *Proceedings, Newcastle Town Council* (1842); *Industrial Resources of ... Tyne, Wear and Tees* (1864); R. E. C. Waters (*c.* 1872) *Genealogical Notes ... of Longridge, Fletcher and Hawks*; W. Brockie (1887) The Hawks family, *Monthly Chronicle of North Country Lore and Legend*, 28–31; *The Industries of the Tyne; their Growth and Importance* (*c.* 1888) 170–171; D. J. Rowe (1984) George Crawshay, *Dictionary of Business Biography*; Information from Beamish Museum (1973) 268]

HAWTHORN, Robert (1796–1867), engine manufacturer, was born on 13 June 1796 at Newburn, near Newcastle, the son of Robert Hawthorn (1760–1842)—'the Duke's Engineer'—and his wife, Alice Burn. He was the eldest surviving son of a family of five boys and three girls, born between 1792 and 1809.

Hawthorn was educated in village schools at Dinnington, Newburn and Walbottle and then entered into an apprenticeship at Walbottle colliery, where his father was engineer. Whereas his father had been consulted on civil engineering matters, among them the Caledonian canal, Hawthorn's abilities were such that in 1817 he established a business for the manufacture of 'steam-engines, millwork and other machinery' in Forth Banks, Newcastle, and was joined there by his brother, William, the following year. In 1820, William became a partner and in the same year Hawthorn married Jane Taylor (1797–1857) of Lemington, the marriage taking place in St. John's church, Newcastle, where he was later a churchwarden.

At first, the Forth Bank works produced stationary engines and also devices to enable the removal of ballast from ships to be more easily effected. The first Cornish engine to be made in the North East was supplied by him to the Newcastle Subscription Water Company in 1834 and, later,

further pumping plant was manufactured for the successor company and for other water undertakings; by 1848 the firm had supplied engines to Derby, Wolverhampton, Nottingham, and Brighton. The manufacture of locomotives began in 1831, when two-hundred and twenty men were employed, and although it was later written that the 'style of engine was not so distinctive as that of some other makers', nevertheless by 1850, when the staff numbered almost one thousand, more than a thousand locomotives had been completed.

Hawthorn became a member of the Institution of Civil Engineers in 1839, in the same year being elected to the Newcastle Town Council, representing the Westgate ward; he remained a member until 1848. In 1845 he became a director of the Whittle Dean Water Company but this appointment was of short duration. As the result of entering into a contract to supply a pumping engine to the company he was forced to resign. Hawthorn also owned a half share in the Blaydon Chemical Company, later selling his share on the condition that all new manufacturing plant be made at the Forth Banks works.

On 26 June 1867 Hawthorn died at his home, Elswick Lodge, Newcastle, 'after a long and severe illness' and the funeral took place at Newburn. His interment was attended by more than six-hundred of his employees, to whom Hawthorn had 'ever acted with great kindness, always evincing a desire to promote their welfare', together with many men from the adjoining works of the Stephensons; both establishments had been closed for the occasion. He had no family and the works were continued by his brother, William (1799–1875).

R. W. RENNISON

[*Minutes, Newcastle Subscription Water Company*, Tyne and Wear Archives; *Minutes, Whittle Dean Water Company*, Tyne and Wear Archives; *Proceedings of the Town Council of Newcastle upon Tyne*; *Industrial Resources ... (of the) ... Rivers Tyne Wear and Tees*, British Association, 1863; *Newcastle Daily Journal*, 3 July 1867; *Engineering*, 5 July 1867; *Min Proc ICE*, **27**, 1868, 590–592; J. F. Clarke (1977) *Power on Land and Sea*]

Publications

1839. Patent 8277, 21 November, *Boilers for Locomotives and other Steam-engines.*.
1843. Patent 9691, 7 April, *Locomotive engines ...*
1851. Patent 13533, 24 February, *Locomotive engines ...*
All patents in conjunction with William Hawthorn.

HAYWARD, William (fl. 1591–1637), was a land surveyor, well-known for his surveys of the Great Level of the Fens and with an extensive practice. It is probable that he lived at Outwell, a village 5 miles south of Wisbech. His earliest map, existing as a copy, is of Norfolk Marshland and the fens as far south as Ely, drawn on a scale of 1 in. to 3 furlongs (2 2/3 in. to a mile) and ascribed

'Guiliemus Haiwarde descripsit 1591'. The marsh-land portion appears to be based on the map of *c.* 1582 showing the drainage mill of George Carleton (q.v.). Several surveys by Hayward in the next decade are known in Norfolk, Lincolnshire and Essex and of the Tower of London. Then, in 1604, he became involved in the first comprehensive scheme for draining the Great Level along with John Hunt (q.v.) who acted as the engineer and Richard Atkyns (q.v.). Hayward's contributions were a general map of the Great Level drawn on the scale of 1 in. to a mile and measuring 52 in. × 37 in. overall, and a written survey on 'The True Content or Number of acres in the Fenns, described in the general Plat ... as they were delivered by Will. Hayward, gent, Surveyor, upon his Oath at Wisbech the 13 July 1605'. This occupies two folio pages as printed in Dudgale (1662) and Badeslade (1725). It gives the area of each fen, the total being 307,242 acres.

In producing the map Hayward would have made use of his 1591 survey and the 1589 map of John Hexham (q.v.) of the northern fens. But much had to be added to complete the survey, which he accomplished in the early months of 1605 (1604 OS). The original no longer exists but in Cambridgeshire Record Office is a map, recently transferred from the Fen Office at Ely, entitled *An Exact Copy of a Plan of the Fenns as it was taken Anno 1604 by William Hayward, carefully coppy'd from the original by Mr. Payler Smith Anno Dom. 1727'.* In the left-hand top corner is a long descriptive title, presumably the original, reading *'A Generall Platte & Discription of the Fennes & other Grounds within the Isle of Ely & in the Counties of Lincolne, Northampton, Huntingdon & Marsh. Between the Sea Bankes on the North, the River Welland & the high grounds of Northampton & Huntingdon Sheires on the West, the Highland of the said Sheires & the River of Ouse on the South & the said River of Ouse on the East. The Hards or High Grounds lying within the said Fennes are Compassed about & shadow'd with Popingue Greene for the readie distinguishing of them from the Fennes. The Sea Banks, Fenne Banks & the Banks between which the Waters have Passage to the Sea are Coloured Light Greene. The Meares, Rivers & Principall Rivers ... are Coloured Blew; and the Highways & Cawsays Yeallow ... Compassed by Mr. Wm. Hayward Anno 1604.*

Smith's map, for which he received 15 guineas, is taken from a copy of 1618, a date added beneath the scale. A smaller variant made for Sir Robert Cotton appears to have been used by Hondius in the English edition of his Atlas published in 1633, with the addition of Deeping Fen north of the Welland and off-shore sand banks. The best known version is the map in Badeslade (1725), reproduced by Harris (1953) and Darby (1983), with the title *A Plan and Description of the Fens and other Grounds within the Isle of Ely and in the Counties of Lincoln [etc.] ... Surveyed by Wm. Hayward 1604. Copied by T. Badeslade 1724.* On the scale of 3 in. to 10 miles, this has a

colouring system almost exactly as in the original title (except that roads are omitted) and includes the Hondius additions.

The 1618 copy was probably made in connection with the investigations by Sir Clement Edmondes and Richard Atkyns in that year. Hayward himself briefly took part in subsequent proceedings in 1622 associated with proposals for draining by Cornelius Vermuyden (q.v.), from which nothing emerged at the time, But, more importantly, he served on the Commission granting powers under the so-called Lynn Law of 1631 to Francis (fourth) Earl of Bedford to undertake the draining of the Great Level. This first campaign was completed in 1636 and Hayward carried out the detailed survey of each area of drained land. His report, submitted to the Commissioners in 1636, is endorsed 'This booke was delivered in upon oath unto us ... by William Hayward, Esq. Surveyor, conteyning a Survey made by him of the Fennes and surrounded grounds called the Great Levell ... and done in pursuance of ... Lynne Law'. As printed by Wells (1828) it occupies ninety-five pages. The total area is now 312,000 acres. With some minor changes and corrections, following local disputes, the Commissioners made their final award to the Earl and his fellow Adventurers at St. Ives on 12 October 1637.

Surveyors employed under the Earl include Thomas Wright (q.v.) and Benjamin Hare. Owing to the paucity of contemporary records it is not certain whether, as seems possible, Hayward acted as Surveyor General, a role comparable to that of Jonas Moore (q.v.) in the third campaign of Fens drainage in the 1650s. What is clear, however, is that Hayward provided a link between the scheme put forward by Hunt in 1605 and works in the campaign of 1631–1636 which were largely based on Hunt's proposals. Hayward served on the 1637 St. Ives Commission; this is the last known reference to him.

A. W. SKEMPTON

[Sir William Dugdale (1662) *The History of Imbanking and Drayning of divers Fenns and Marshes*; Thomas Badeslade (1725) *The History of the ... Port of King's Lyn ... and of Draining the Great Levell of the Fens*; Samuel Wells (1830) *The History of the Draining of the Great Level of the Fens, called Bedford Level*, and vol. 2 (1828); H, G. Fordham (1908) *Cambridgeshire Maps, a descriptive Catalogue of the Maps of the County and of the Great Level of the Fens*; Edward Lynam (1936) Maps of the Fenland, *Victoria County History of Huntingdonshire*, **3**, 292–306; L. E. Harris (1953) *Vermuyden and the Fens*; H. C. Darby (1983) *The changing Fenland*; R. J. Silvester (1989) William Haiwarde and the Fens, *Fenland Research*, **6**, 38–42; Frances Willmoth (1993) *Sir Jonas Moore*]

HAYWARD, William (*c.* 1740–1782), architect and bridge-builder, was a son of John Hayward of Lincoln, a mason who died in 1778 and who was himself the son of a mason who had moved to Lincoln from Shropshire. By 1766 Hayward was resident in Shrewsbury and employed as Deputy to the appointed Surveyor of the Works for widening and repairing the town's Stone Bridge (later called the English Bridge) over the Severn. The Surveyor was Thomas Farnolls Pritchard, a local architect, but the work moved rather slowly until in August 1767 John Gwynn (q.v.) was invited to produce a design for full replacement of the old bridge; his proposal was quickly adopted. In October 1768 Pritchard was replaced as Surveyor by Gwynn and thereafter Hayward and Gwynn were often to be found working together.

Hayward became involved in work on Gwynn's other commissions, for the Severn Bridge at Atcham and for pavings and the building of Magdalen Bridge at Oxford. In 1777 he had full responsibility for minor finishing works to complete the construction contracts Gwynn had assumed for the bridges at Shrewsbury and Atcham. Also in 1777 Hayward designed and contracted to build by June 1780 an elegant bridge of a single large arch over the river Tern at Atcham, the design being chosen from nine alternative drawings, all by Hayward, and the decoration being paid for entirely by Noel Hill Esq. because it adorned the view from his house at Attingham Park. The structure is distinguished by the use of several cylindrical voids through each spandrel in the manner of William Edwards's (q.v.) bridge at Pontypridd but with the voids hidden behind the outer walls.

Hayward also designed a small bridge over the Tern at Walcot, 3 miles upstream from Atcham, which was funded by local subscriptions and completed in 1782. It bears an inscription that it was 'the last edifice erected by that ingenious architect William Hayward'. In fact, however, Hayward's biggest bridge and biggest contract, the Thames bridge at Henley, was not completed until 1786. It has been surmised by Francis Sheppard that his involvement in the Henley project resulted from acquaintance with General Henry Seymour Conway who lived at Park Place, near the site, and took a prominent part in the work of the committee which initiated and managed the project. When they advertised for plans in June 1781 several were considered and in mid-September the committee gave its approval to Hayward's design of five semi-elliptical arches. He was introduced as 'William Hayward of the Parish of St. Julians in the town of Shrewsbury'. With a recorded but unspecified alteration in 'the curve of the arches', the work began, the contract price being £9,727. But within two months Hayward died and the bridge was finished by John Townesend (d. 1784) of Oxford and his son, Stephen (1755–1800).

Hayward was buried on 16 January 1782 in St. Mary's Parish Church in Henley where he is commemorated by 'a beautiful small monument' at the base of the tower. His death was attributed to 'his kindness in giving up his place inside a coach to a woman on a stormy day, when he contracted a cold and fever'. He left no will and his estate did not exceed £300. Letters of administration were granted to his brother, John, a mason of Lincoln. His mother, Ann, also survived him.

P. S. M. CROSS RUDKIN and TED RUDDOCK

[Salop County Archive, MS drawings and contract for the Tern Bridge, and Quarter Session records Oct. 1777 and Jan. 1782; The British Library, *King's Maps*, **34**, 33, 3a–c, MS drawings of work in Oxford; S. M. Johnson and C. W. Scott-Giles (1933) *British Bridges: an Illustrated Technical and Historical Record*, 247; A. W. Ward (1935) *The Bridges of Shrewsbury*; J. L. Hobbs (1960) The Hayward family of Whitchurch, *Shropshire Magazine*, Jan.; N. Cossons and B. Trinder (1979) *The Iron Bridge*, 79–82; Francis Sheppard (1984) Henley Bridge and its architect, *Architectural History*, **27**, 320–330; Anthony Blackwall (1985) *Historic Bridges of Shropshire*; Colvin (1995), 483–485]

Works

1766–1777. English Bridge, Shrewsbury, deputy surveyor (see: *GWYNN, John* (1713–1786))
1768–1777. Atcham Bridge, assistant surveyor to J. Gwynn
1772–1777? Magdalen Bridge, Oxford, and paving contracts, assistant to Gwynn
1777–1780. Tern Bridge, Atcham, architect and contractor, one arch, 90 ft. span, approx. cost £2,100, widened 1932
1779. Cound Stank Bridge, Shropshire, designer (inscription on bridge)
1780. Preens Eddy Bridge at Coalport, designer, bridge over River Severn of two framed timber arches on stone abutments and pier
1781–1786. Thames Bridge, Henley, designer and contractor until death in January 1782, five elliptical arches of 40 ft. and 34 ft. span, contract price £9,727
1782. Mill Bridge, Walcot, designer, three small arches over the River Tern

HAZLEDINE, John (1760–1810), ironfounder, was the eldest of four sons of William Hazledine, a millwright. He was apprenticed to his father for seven years from 29 May 1774. By 17 October 1782, however, when he married Ann Davies of Eardington he was resident on the other side of the river Severn from Bridgnorth and by 1795 was the managing engineer of the Bridgnorth Foundry, which lay on a site by the river at the north-east end of Bridgnorth Bridge; his third and fourth brothers, Robert (b. 1768) and Thomas (b. 1771), were employed in the foundry.

The foundry began by making millwright's castings and later was a pioneer in the production of threshing machines. Hazledine developed improved methods of ploughing and of pig breeding and took out patents for them. By 1804 the foundry was associated with Richard Trevithick (q.v.), the pioneer of high pressure steam engines, and an engine built by Hazledine for him is now in the Science Museum. From about 1806 John Urpeth Rastrick (q.v.) was a partner. The foundry also built the second Stourport Bridge, a cast iron span of 155 ft., after the first had been washed away in the flood of 1795, but the contract by Hazledine, Rastrick & Company for the unusual cast iron bridge at Chepstow, which is sometimes attributed to John Hazledine, was executed after his death. Hazledine's other commercial interests included a lease of Leebotwood colliery, south of Shrewsbury.

He was made a freeman of the borough of Bridgnorth in 1803. He died on 28 October 1810 and was buried under a cast iron grave slab, from his own foundry, in St. Mary's churchyard, Bridgnorth. After sizeable legacies to his wife and sisters, the residue of his estate went to his brother, Robert, who continued to manage the foundry, and Thomas Davies, his wife's brother.

P. S. M. CROSS-RUDKIN

[Copy of will in Shropshire RO; Victoria County History of Shropshire; S. M. Tonkin (1945) Trevithick, Rastrick and the Hazledine Foundry, Bridgnorth, *Trans Newc Soc*, **xxvi**, 171–183; Barrie Trinder (1996) *The Industrial Archaeology of Shropshire*]

Works

1805–1806. Stourport Bridge, replaced in 1870

HAZLEDINE, William (1763–1840), ironfounder and engineer, was born at Shawbury, Shropshire in 1763, second son of William Hazledine.

William Hazledine was an engineer and ironfounder of considerable importance and influence during the last decade of the eighteenth century and the first four decades of the 19th century, but he has not received the recognition he deserves due to the fact that information about him is fragmentary and widely scattered, and a biography of him has yet to be written. He was born in Shawbury, Shropshire, though the family moved soon afterwards to a house at Sowbatch, near Moreton Corbet. He and his elder brother John (q.v.) were trained as millwrights under their uncle John. It is possible that their two other brothers Robert (b. 1768) and Thomas (b. 1771) also received similar training. John, Robert and Thomas went to Bridgnorth in or before 1795 at the foundry where Richard Trevithick carried out some of his early experiments with steam engines and subsequently they entered into partnership with John Urpeth Rastrick.

William, who had supervised the erection of machinery at Upton Forge, owned by the Sundorne family, when he was only sixteen or seventeen years of age, moved to Shrewsbury and thereafter had little to do with the Bridgnorth enterprise. He entered into partnership with a Mr. Webster in The Mardol, described as a clockmaker, though later he became an ironmonger. William Hazledine then set up on his own, and his first foundry was in Cole Hall, between Claremont Street and Barker Street, not far from the Welsh Bridge. That he was going up in the world is indicated by the fact that he joined

William Hazledine

the Salopian Lodge 262 of the Freemasons in 1789, where he made the acquaintance of a fellow member, Thomas Telford (q.v.). In 1794 he supplied new machinery for Broadstone Mill, Shropshire at a cost of £350. By 1796 he had outgrown the premises he was then occupying, and he purchased land in Longden Coleham, just south of the River Severn near the English Bridge, and it was here that he was to establish the large, well-equipped, steam-powered workshops which were to serve him for the rest of his career. In the same year he supplied cast iron beams and columns for Bage's Mill, now recognised as the world's first iron framed factory building, and it was also in 1796 that Telford referred to him by that most memorable phrase 'the Arch conjuror himself, Merlin Hazledine'.

In 1793 the Ellesmere Canal had obtained its Act of Parliament, and Thomas Telford was appointed engineer. Two years later, work began on the celebrated Pontcysyllte Aqueduct across the River Dee near Ruabon. Though William Reynolds and John Wilkinson (qq.v.) had held hopes of supplying the ironwork for this magnificent structure, the contract was awarded to William Hazledine on 17 March 1802, he having established a foundry at Plas Kynaston specially for the purpose c. 1800. The aqueduct was completed in 1805, and it seems likely that Hazledine was responsible for the erection of the ironwork, as well as merely supplying it, though his role in the venture, like that of William Jessop, was later played down. Hazledine also supplied the ironwork for the bed of the nearby Chirk Aqueduct on the same canal.

By this time, Telford and Hazledine, described by the latter's obituarist as 'kindred spirits' had struck up a firm friendship and business partnership, and Hazledine was to supply ironwork for numerous bridges, canal and dock projects for Telford over the next thirty years. The most notable examples would undoubtedly be the Menai Suspension Bridge, completed in 1826 (Hazledine was a passenger in the first carriage to cross the completed bridge), the Conway Suspension Bridge (also 1826) and the extremely attractive Waterloo Bridge at Betws-y-Coed (1815), all in Wales. In Scotland, the bridge at Craigellachie is one of the most attractive, whilst a similar structure at Eaton Hall, near Chester, and the Mythe Bridge across the Severn near Tewkesbury are probably the outstanding English examples. A trusty foreman by the name of William (S.) Tuttle assisted on a number of projects, and his name actually appears on one or two main works at Coleham and Plas Kynaston, Hazledine had increasing interests in various other enterprises towards the end of his life. He purchased the Calcutts Ironworks at Jackfield, in the Ironbridge Gorge in 1817, though according to one observer 'even his master talents could not ensure (their) success'. He operated a limeworks at Llanymynech and also had timber-yards, brick-yards and coal-wharves in various places as well as a substantial amount of property in Shrewsbury itself. He owned shares in canal, gas and insurance undertakings.

William Hazledine died on 26 October 1840, his wife having predeceased him in somewhat tragic circumstances: she died of grief and shock when he broke his arm in several places in a coaching accident in Wyle Cop, Shrewsbury. Many shops closed for his funeral on 6 November, and he was interred in St. Chad's Church, where there is a bust of him by Sir Francis Chantry.

A portrait of William Hazledine, oil on canvas 34 in. × 27½ in., is in the collection of Shrewsbury Museums, but is currently on loan to the Ironbridge Gorge Museum.

JOHN POWELL

[Watton's Newspaper Cuttings, vol. 3, SBL Col. 852 Shrewsbury Records and Research; J. Rickman (1838) *Life of Thomas Telford, Civil Engineer*; E. T. Burne (1936) On mills by Thomas Telford, *Newc Soc Trans*, **xvii**, 205–212; S. M. Tonkin (1945) Trevithick, Rastrick and the Hazledine Foundry, Bridgnorth, *Newc Soc Trans*, **xxvi**, 171–183; L. T. C. Rolt (1953) *Thomas Telford*; C. T. G. Boucher (1963–1964) Broadstone Mill, Corvedale, *Newc Soc Trans*, **xxxvi**, 159–163; B. Bracegirdle and P. Miles (1973) *Great Engineers and Their Works: Thomas Telford*; E. A. Wilson (1975) *The Ellesmere and Llangollen Canal: an Historical Background*; J. R. Hume (1977) *The Industrial Archaeology of Scotland II The Highlands and Islands*; N. Cossons and B. Trinder (1979) *The Iron Bridge: Symbol of the Industrial Revolution*; A. E. Penfold (ed.) (1980*) Thomas Telford: Engineer. Proceedings of Seminar*; A. E. Penfold (1981) *Thomas Telford: Colossus of Roads*; B. Trinder (1981) *The Industrial Revolution in Shropshire*, 2nd edn.; A. Blackwall (1985) *Historic Bridges of Shropshire*; W. J. Sivewright (ed.) (1986) *Civil Engineering Heritage: Wales and Western England*; B. Trinder (1988) *'The Most Extraordinary District in the World': Ironbridge and Coalbrookdale*, 2nd edn.; C. Hadfield (1993) *Thomas Telford's Temptation: Telford and William Jessop's Reputation*; B. Trinder (1996) *The Industrial Archaeology of Shropshire*]

Works
1794. Broadstone Mill, Shropshire
1796–1797. Bage's Mill, Shrewsbury, supplier of iron beams and columns
1796–1801. Chirk Aqueduct, supplier of ironwork
1802–1805. Pontcysyllte Aqueduct, supplier and erector of iron trough; he was also contractor for the original masonry plan with John Simpson (q.v.) and William Davies after John Varley (q.v.) withdrew
1804–1822. Caledonian Canal, supplier of lock gates, etc.
1811. Meole Brace Bridge, Shrewsbury, supplier of ironwork
1812. Long Mill Bridge, Shropshire, supplier of ironwork
1812. Bonar Bridge, supplier of ironwork
1812–1815. Craigellachie Bridge, supplier of ironwork
(c. 1815–1823). Pembroke, supplier of iron roofs for stores
1816. Waterloo Bridge, Betws-y-Coed, supplier of ironwork
1818. Cound Bridge, Shropshire, supplier of ironwork
1818. Sweden, supplier of swivel bridges for the Gotha Canal and one set of lock gates
1818–1821. Kington Railway, contractor for construction (with Morris Sayce), £10,000, and ten year maintenance contract
1820. Esk Bridge, near Carlisle, supplier of ironwork
1821–1822. Liverpool New Market, supplier of columns
1822–1824. Dublin Custom House Dock, supplier of iron roofs, etc., £20,000
1824. Plymouth, supplier of ironwork for Laira bridge for Earl Morley
1824. Eaton Hall Bridge, near Chester, supplier of ironwork
1825. Menai Bridge, supplier of chains and ironwork
1826. Holt Fleet Bridge, supplier of ironwork
1826. Mythe Bridge, Tewkesbury, supplier of ironwork
1826. Conway Bridge, supplier of chains and ironwork
c. 1827. London Docks, supplier of swivel bridges (St. Katharine's Dock)
1827. Cleveland Bridge, Bath, supplier of ironwork

c. 1828–1829. Montrose, supplier of ironwork for chain bridge

c. 1829. Marlow, supplier of ironwork for chain bridge, total contract value £22,000

1840. Newport Docks, Monmouthshire, supplier of lock gates

n.d. Liverpool Docks, supplier of iron swivel bridges

HEATHCOAT, John (1783–1861), machinist, inventor and lace manufacturer, was born in Duffield, near Derby, in August 1783, the son of Francis Heathcoat, a grazier, and Elizabeth Burton. Apprenticed as a frame smith or textile machine engineer in the East Midlands, he set up in business in Nottingham and in September 1802 married Ann, the widow of John Chamberlin of Hathern, and daughter of William Caldwell, also involved in the lace trade.

About 1803 he became interested in developing lace-making machinery, and attempted unsuccessfully to take out his first patent the following year. In partnership with his brother-in-law, Samuel Caldwell, he removed his business to Hathern in 1805. His first patent was taken out in 1808, and a more successful one in 1809, capable of making lace curtains. In partnership with Charles Lacy and a hosier named Boden, they established a lace mill, with fifty-five frames to his designs, at Loughborough which, on the night of 28 June 1816 was wrecked by a gang of Luddites. The year previously Heathcoat had begun operating in a six-storey textile mill at Tiverton, Devon, and he thereupon moved his plant and staff there. In this venture he was joined by John Farey (q.v.). They had three-hundred frames in operation

Heathcoat launched a lace enterprise in France at about the same time. A Brevet d'Importation for one of his machines was granted in 1817 and between 1820 and 1828 he took out six French patents for them. He was elected an Associate of the Institution of Civil Engineers in 1826, resigning in 1855.

In 1832 he patented a steam plough and in the same year he was elected MP for Tiverton, holding the seat until 1859. Heathcoat died at Tiverton on 18 January 1861. His wife had died in 1831, and they were survived by three daughters. His portrait is reproduced by Gore Allen (1958), but its current whereabouts is unknown.

A. P. WOOLRICH

[Membership records, ICE archives; DNB; W. Felkin (1868) *History of Machine-wrought Lace*; W. Gore Allen (1958) *John Heathcoat and his Heritage*]

HEDLEY, William (1779–1843), colliery viewer, was born at Newburn, Northumberland, on 13 July 1779, the son of William Hedley and his wife, Ann Atkinson. Hedley was educated at nearby Wylam and from 1801 worked as a viewer at Walbottle colliery before moving to Wylam

where, in 1805, he became viewer to Christopher Blackett, the owner.

The 4½ mile long waggonway transporting coal from the Blacketts' pits at Wylam had been first constructed in 1763 and comprised wooden rails with a gauge of 5 ft. Under Hedley, the waggonway was rebuilt using cast-iron plate rails in 1808. As a result of the rising cost of horse-haulage, Blackett wished to adopt steam haulage and in 1812 Hedley put in hand experiments using a man-powered test carriage to ascertain the practicability of using smooth-wheeled locomotives on the level waggonway, rather than stationary engines. His work led to a patent in 1813 and the subsequent modification of his test carriage to accommodate a boiler and cylinder. The engine worked—but not well—and was superseded by another locomotive of Hedley's design; it had two vertical cylinders believed to be 8 in. diameter and 3 ft. long. This engine was probably built in 1814–1815 and was closely followed by another; the locomotives were later converted to run on eight wheels. One of these engines, the *Wylam Dilly*, was put to work during a miners' strike in 1822 when it was mounted on a barge and modified by Robert Hawthorn (q.v.) to drive paddlewheels; as a tug-boat it was used on the river Tyne, first to tow loaded keels and, later, to tow ships to sea.

In 1824 Hedley became a coalowner and leased several collieries in County Durham. He left Wylam in 1828 and later concentrated his efforts on collieries at South Moor (Lanchester) and Coxhoe (West Hetton). After leasing further collieries, Holmeside and Craghead, he became involved in 1824 in investigating the quality and availability of coal in relation to the promotion of the Clarence Railway. For a period of three years after its construction, it carried only the produce of Hedley's mines.

When at Wylam, Hedley was a member of the Duke of Northumberland's Yeomanry Cavalry and later became an active member of the Literary and Philosophical Society in Newcastle. Having shipped coal on the rivers Tyne, Wear and Tees, he became a shipowner and at the time of his death owned four ships.

Hedley married Frances Dodds, of Berwick, in 1803 and between 1804 and 1809 they produced four sons, Oswald Dodd, Thomas, George and William. Hedley moved to Burnhopeside Hall in 1837 and died there on 9 January 1843, being buried at Newburn. His four sons continued his colliery interests.

Hedley's engines remained at work at Wylam until after his death and in 1862 *Puffing Billy* was acquired by the Patent Office Museum and remains on display in the Science Museum.

R. W. RENNISON

[O. D. Hedley (1858) *Who Invented the Locomotive Engine*; M. Archer (1882) *William Hedley, the Inventor of Railway Locomotion on the Present Principle*; R. Welford (1895) *Men of Mark 'twixt*

Tyne and Tweed; C. F. Dendy Marshall (1953) *A History of Railway Locomotives Down to the End of the Year 1831*; L. G. Charlton (1974) *The First Locomotive Engineers*]

Publications

1813. Patent 3666, 13 March, *Mechanical means of conveying Carriages laden with Coals, Minerals, and other things*

Works

1808. Rebuilt Wylam waggonway using iron rails
1812. Experiments on smooth-wheeled traction
1814–1815. Built first smooth-wheeled twin-cylinder locomotive

HENDERSON, David (*c*. 1735–*c*. 1788), mason, architect and bridge-builder, initially of Sauchie in Clackmannanshire, had a rather chequered career working mostly in south-eastern Scotland. In the 1760s his public pose on a number of occasions suggested that he was well educated in architecture, including technical and aesthetic aspects of bridge-building, and also in cost estimation, a style of self-promotion similar to that of some renowned eighteenth-century pamphleteers in London, such as Batty Langley and William Halfpenny. Unlike those men, however, he was seeking employment as a building contractor. His success in obtaining work in both design and building, though somewhat limited, and the time and energy it absorbed may explain why his self-promotion seemed to decline over the years.

In 1765 he described himself as 'architect and mason in Sauchie', a village a few miles from Alloa in an area of mining activity and some industrial development. Presumably it was his place of birth and first training as a mason, but where he acquired his announced skill at figures, writing and drawing and his apparent knowledge of architectural texts is not evident. The first contract he is known to have taken up on his own account was for a substantial stone bridge at Forteviot in 1761, and serious partial failure occurred soon after it was built. In 1763 he agreed to be contractor for the important bridge at Coldstream (see: *REID, Robert*) but after a year of dilatory attendance he was asked to leave. He was at the same time receiving some attention in Edinburgh and in December 1764 announced in a newspaper his intention to publish by subscription a book of house plans of all sizes and costs from £100 to £4,000, and illustrated with sixty copper plates in quarto. The book never appeared but the announcement implies that he understood something also of printing and publishing.

He was known to leading men in the Edinburgh Town Council, having been one of six tradesmen who in 1763 took part in and reported on a trial excavation for foundations of a bridge to link the Old Town with the proposed New Town to the North; Henderson was paid £30 odd for 'expenses debursed' relative to the construction of a trial pier in the excavation though he was not the builder of it. When designs for the bridge were invited in January 1765 Henderson's was chosen out of twenty-two submitted and he was awarded the premium of £30. He was then asked to tender for the construction both to his own design and to those of William Mylne (q.v.) and of Sir James Clerk. He could not find security enough to satisfy the Council and the job went to Mylne. When Mylne's bridge suffered a partial collapse in 1769, Henderson was one of those who offered schemes for repair, involving substantial rebuilding of the structure. It has been argued that his winning design in 1765 was probably a bridge with cylindrical voids through all the spandrels, as was his scheme for repair. It was a device intended to avoid instability and reduce weight in the spandrels of typical eighteenth-century arches, which were larger than those of earlier bridges.

His output of buildings and bridges in the twenty years following the competition success and the announcement of his book was disappointing in quantity, but it included some works of architectural quality and, in bridge-building, of sound construction and originality. As consultant on several occasions to the Edinburgh Town Council during their struggle to complete and sustain Mylne's damaged North Bridge between 1769 and 1784, and in the design and building of the tall viaduct across the Pease Dean on the Berwick–Edinburgh road in 1783–1784, he seems to have acted with considerable skill and some flair. The viaduct was designed and built with open cylindrical voids through all the spandrels. In 1780 he was invited by the Commissioners of Supply for Dumfriesshire to report on an unfinished bridge at Auldgirth and recommended that it be taken down. Four days later he provided a design accompanied by extensive construction details set out in a written specification. It was a design of original architecture, three equal segmental arches framed over the cutwaters with pairs of thick ashlar pilasters which at parapet level carry a cornice running round the base of a refuge which is covered by a semi-dome giving shelter from rain for travellers on foot. Another of his activities was the making of 'plans and estimates' for the Bo'ness Canal, probably when active work was under way in 1782–1783.

He was listed as 'architect' at a variety of addresses in Edinburgh trade directories from 1774 (when they were first published) until 1786 and not thereafter. In about 1774 he sent his son, John, to Rome to study and on his return in 1779 David Henderson worked sometimes as architect in partnership with John; he also undertook some building contracts with an Edinburgh mason named John Wilson.

Following the death of his son, John, in February 1786 Henderson initially took over his contracts, but his own death took place less than two years later, though the date is not known. Another son, David, who was a millwright at

Alloa, sought settlement of some accounts due to both of the deceased in the Commissary Court of Edinburgh on 18 June 1789.

TED RUDDOCK

[Minutes of the Trustees for Coldstream Bridge 1762–1774, MS, Northumberland County RO; Edinburgh Testaments, Court of Commissary 18 June 1789, National Archives of Scotland; T. Telford (1830) Bridge: history, *The Edinburgh Encyclopaedia*, **4**, 486; G. W. Shirley (1940–1944) The Building of Auldgirth Bridge, *Transactions of the Dumfries and Galloway Natural History and Antiquarian Society,* 3rd series, **23**, 71–75; A. Graham (1963–1964) Archaeology on a great Post Road, *Proceedings of the Society of Antiquaries of Scotland*, **96**, 318–347 (and sources quoted); J. Lindsay (1968) *The Canals of Scotland*, 190–192; E. C. Ruddock (1974–1976) The building of North Bridge, Edinburgh 1763–75, *Transactions of the Newcomen Society*, **47**, 9–33 (and sources quoted); James Robertson (1993) *The Public Roads and Bridges of Dumfriesshire 1650–1820*, 221–237; Colvin (1995) 487–488]

Works

1761. Forteviot Bridge, River Forth, unsymmetrical elevation of four arches, cost £700 after correction of one failed arch (Petition to the Commissioners for Forfeited Estates, read 4 December 1780, ref. E728/29/27, at SRO)

1763–1764. Work as 'undertaker' for Coldstream Bridge, discharged

1763–1765. Preparatory works and competition design for North Bridge, Edinburgh, winning design, not built

1769–1784. Repairs and alterations to North Bridge for Edinburgh Town Council on two occasions

1781–1782. Auldgirth Bridge, River Nith, design and specification, three arches all 56 ft. span, cost *c.* £1,486, contractor William Stewart (q.v.) (Photograph in S. M. Johnson and C. W. Scott-Giles (1933) *British Bridges*, 445)

1783–1784. Pease Bridge, four semicircular arches, spans 41–55 ft., cylindrical voids in three spandrels, height 124 ft., cost £1,500 (John Knox (1789) *A View of the British Empire* ..., 3rd edn., 568)

For architectural works, see Colvin.

HENDERSON, William (fl. 1810–1847), contractor, was related to Hugh McIntosh (q.v.) by marriage and acted as agent on a number of his contracts, most notably Pater (Pembroke) Dockyard where Henderson continued to work after McIntosh's death. In the 1830s he was McIntosh's agent on the Great Western Railway and in 1847 gave evidence in the Equity Suit between the Great Western Railway and McIntosh over unsettled claims. In 1838, described as of Bath, he became contractor on the Horley–Balcombe Tunnel section of the London–Brighton Railway (Contract 7), with David Harris and Thomas Gregory. He retired to Bangeston House, Pembroke.

Henderson was one of a number of McIntosh agents whose careers continued after the latter's death but whose careers never attained the first rank in contracting.

MIKE CHRIMES

[F. Smith archive, ICE Library; M. M. Chrimes (1995) Hugh McIntosh (1768–1840), national contractor, *Newc Soc Trans*, **66**, 175–192]

HENRY, David (fl. 1802–1843), civil engineer, worked as an assitant engineer on the Rochdale Canal from 26 August 1802 until 17 May 1805. He then was appointed resident engineer on the Glasgow Paisley and Ardrossan Canal where work began in 1806 to Thomas Telford's (q.v.) designs. The money ran out because of unforeseen costs arising from difficult rock excavation. Henry was also involved at linked improvements at Ardrossan Harbour. In 1810 and 1815 John Rennie (q.v.) reported critically on Telford's proposals here but such work as was carried out followed Telford's design. At the time of his election as a Corresponding Member of the Institution of Civil Engineers in 1820 Henry was resident at Ardrossan but, following the passage of the 1821 Act, became resident engineer for work at Portpatrick Harbour under Sir John Rennie (q.v.). Works here were continually hampered by lack of financial support from the government arising from uncertainty as to the value of Portpatrick as a harbour. Work had to be extensively revised in 1825 to accommodate steam vessels and in 1837 and 1843 there were further inquiries as to its value, James Walker (q.v.) urging completion of the work in 1843.

Henry may or may not have been related to the contractor David Henry (II).

MIKE CHRIMES

[Glasgow, Paisley & Ardrossan Canal papers, National Archives of Scotland; Rennie mss, NLS; ICE membership records; D. Henry (1820) A description of sundry parts and operations in the formation of Ardrossan Harbour, OC 108, ICE archives; Sir John Rennie (1854) *The Theory, Formation and Construction of Harbours*, **2**, 187]

Works

1802–1805. Rochdale Canal, assistant engineer

1806–1820. Glasgow Paisley & Ardrossan Canal, resident engineer

1821–1843. Portpatrick Harbour, resident engineer

HENRY, David (fl. 1802–1836), contractor and architect, was born in Dublin. In about 1802 Henry and Bernard Mullins (q.v.) took over the contract for the extension of the Grand Canal from Tullamore to the river Shannon after the death of Michael Hayes. In 1806 Henry was responsible for building the canal company's hotel at Shannon Harbour in county Offaly. Not long afterwards, about 1808, Henry, Mullins and John MacMahon (q.v.) came together to tender for the restoration and extension of the Kildare (Naas

Line) Canal for the Grand Canal Company; they were awarded the contract for the work, completed by 1810. The following year, the firm undertook the lock-less extension of the Naas Line to Corbally.

The partnership of Henry, Mullins & MacMahon became leading contractors for government projects in Ireland. Notably, they built from 1824 the extension of the Grand Canal to Ballinasloe in county Galway and, in 1827, the Barrow Aqueduct on the Barrow line at Monasterevan in county Kildare. Although successful in tendering for work on the Ulster Canal, in which they had shares, the firm became involved in a dispute with Telford and William Dargan (qq.v.) took over the works. The partnership was dissolved in 1827, the dissolution appearing to have been a business arrangement rather than a parting of the ways, and Mullins & MacMahon continued for many years to be a major force in Irish contracting.

David Henry was also an architect and in 1825 applied unsuccessfully to succeed Francis Johnston as architect to the Board of Works. In his letter of application, written from 92 St. Stephen's Green, Dublin, and dated 16 January 1826, he wrote:

The qualifications upon which ... I ... ground my pretensions, are, a long and intimate acquaintance with the various departments of Civil Architecture in which I have been employed on an extensive scale, planning, superintending, executing, or directing several of the most important Public works and private buildings in Ireland, during the last thirty years; namely, the Queen's County Gaol and Infirmary under the auspices of Lord Maryborough and Mr. Cosby; Palaces for the Bishops of Meath and Clogher; the Cathedral, Law Courts, Gaol and Town Hall of the City of Londonderry, for which I gave Designs and executed the Works; I have also had the execution of the Government Docks and Warehouses in Dublin and Sligo; and the Extension of the Grand and Royal Canals to the Shannon.

David Henry died on 24 November 1836, leaving a widow, Margaret. He was buried in the Pro-Cathedral, Marlborough Street, Dublin. His eldest son Thomas Henry (1807–1876) became a distinguished magistrate and barrister in London, and a younger son, David I. Henry became a civil engineer, elected an Associate of the Institution of Civil Engineers in 1837, resident in Dublin. By 1841 the latter was at West Regent Street, Glasgow, his last known address before being removed from the register of members in 1845. Their links to David Henry above are unclear.

RON COX

[Telford Correspondence, Dublin Docks, ICE archives; R. Delany (1992) *Ireland's Inland Waterways*; R. Delany (1995) *The Grand Canal Of Ireland*; Irish Architectural Archive, *Index Of Irish Architects:* Biography (in preparation)]

Works

1802–1803. Grand Canal, Shannon line, completion, contractor (with Mullins)
1806. Grand Canal Co., Shannon harbour hotel, contractor
1808–1811. Grand Canal, Naas line & Corbally extension, contractor (with Mullins and MacMahon)
1813–1817. Royal Canal, Shannon extension, including Whitworth Aqueduct, contractor (with Mullins and MacMahon)
1815–. Dublin docks
c. 1823–1830? Sligo harbour improvements
1824–1827. Grand Canal, Ballinasloe line, contractor (with Mullins and MacMahon)
1826. Clarendon Dry Dock 2, Belfast, contractor
1827–. Barrow aqueduct, contractor (with Mullins and MacMahon)

HENSHALL, Hugh (1734–1816), surveyor, engineer, canal proprietor, canal carrier and pottery manager was born in 1734 in North Staffordshire, probably at Bent House or Farm, Newchapel, in the parish of Wolstanton. His parents were John Henshall, a land surveyor, and Anne Cartwright who had married in Wolstanton on 6 January 1730 and five surviving children were born between 1731 and 1747. Henshall's sister, Jane, married William Clowes, a local landowner, with strong mining interests; he was also the elder brother of Josiah Clowes (q.v.), the early canal and tunnel engineer, who was later to form a close friendship and working partnership with Henshall and Robert Whitworth (q.v.). Another sister, Anne, married James Brindley (q.v.).

Nothing is known of Henshall's early life and education. The latter is likely to have been at Newchapel Grammar School, within sight of Bent farm. His vocational training was received at the hands of his father and, significantly, he became one of Brindley's pupils; it seems likely his father assisted Brindley on early surveys for the Trent and Mersey Canal, the families evidently became very close. As a result of his surveying work, Henshall met Thomas and John Gilbert and Josiah Clowes with whom he became great friends. In 1760, Brindley, John Gilbert and Henshall bought the Golden Hill Estate on the route of what would later become the Trent and Mersey canal.

As a young man, Henshall worked as a surveyor in North Staffordshire, Cheshire and Gloucestershire. In 1717 a proposal was made by Thomas Congreve to link the rivers of the Trent and Mersey and in 1758, Brindley, John Smeaton (q.v.) and Henshall made a survey of such a navigable waterway.

In 1765 Henshall was involved with Robert Pownall (q.v.) on a survey of the river Weaver from Winsford to Lowton via Middlewich and Nantwich. This work was to be of great value to Henshall a decade later when he came to engineer the northern section of the Trent and Mersey canal through the Saltersford and Barnton tunnels. In the summer of 1765, Henshall was requested to plan a level route across Cheshire between Harecastle and the Duke's Navigation and in the following year he was also involved in a survey of the river Severn. It is possible that this survey was conducted with help from Clowes as the latter was to assist with the building of locks from the Stroudwater Navigation to the Severn as well completing the Thames and Severn canal. It is also clear that the various navigation links that became part of the 'Great Cross' were already fermenting in the minds of Brindley and Josiah Wedgewood and must have been known to Henshall.

On 14 May 1766 the Trent and Mersey canal, or Grand Trunk Canal as it was then called, became a joint stock company, giving it the right to purchase land compulsorily and also conferring the right to levy tolls on goods carried along the canal. The canal, unlike the Bridgewater Canal which was financed by one wealthy aristocratic landowner, was to be a precedent for future canal and railway construction. Designed to link the Duke of Bridgewater's canal near the Mersey at Preston Brook to the river Trent at Wilden Ferry, north of Lichfield, the canal was financed by businessmen who operated along its route, particularly Josiah Wedgewood and the potters in Stoke.

Henshall drew the parliamentary map of the Trent and Mersey as built and in 1766 was appointed Clerk of Works to the premier engineering project in the country; his annual salary was £150 for himself and an assistant. In the same year Henshall was surveying roads in Newcastle under Lyme, almost certainly linked to the engineering and construction of the Trent and Mersey Canal, constructed with a width of 12 ft. wide at its base. It had a depth of 3 ft. and could accommodate boats 70 ft. long and 6 ft. wide drawing 2 ft. 6 in. when carrying a full load of 20 tons. On the 26 July, Josiah Wedgewood, in the presence of Brindley, Henshall and 'many respectable persons of the neighbourhood, who each cut a sod to felicitate the work', cut the first sod to inaugurate construction.

In 1767 Henshall was involved in the purchase of land for the new canal and, in the same year the canal was laid out in front of Eturia Hall. This part of work gave some insight into Brindley's irascibility as Wedgewood and Henshall were anxious to abide by Brindley's instructions or 'Mr. Brindley would go mad'. At this time it would seem that Henshall was still living at Newchapel, very possibly with his elderly parents and grandparents. Brindley had moved close by to Turnhurst Hall shortly after his marriage to Henshall's sister.

The watershed of the new canal lay on the high ground near Ranscliff which Henshall, Brindley, Gilbert and Clowes owned, north west of the potteries. The canal tunnel at Harecastle, when built, passed through the coal, ironstone and millstone grit under much of their land; the quartet of families clearly had interests in mining. At 2,880 yd. the tunnel, which had subsidiary tunnels branching from it to coal measures, reflected the practice at Worsley, and was the longest canal tunnel built in the world at that time. After Brindley's death in 1772, Henshall was left to complete the task of the engineering of the tunnel and the difficult northern section of the canal. The Trent and Mersey descended into the Cheshire plain after navigating the tunnel at Harecastle and formed a junction with the Duke of Bridgewater's canal at Preston Brook near Runcorn. Henshall was also responsible for the cutting of the Bridgewater Canal through to Runcorn thus finally completing the link between the rivers Trent and Mersey.

In 1768 Brindley, assisted by Henshall, surveyed the Staffordshire & Worcestershire canal, designed to link the Grand Trunk at Great Haywood to the river Severn at Stourport. The final part of James Brindley's 'Great Cross' came in 1768–1769 when the Coventry and Oxford canals were authorised to link the Trent and Mersey at Fradley to the Thames at Oxford. The Oxford was completed by Brindley's other brother-in-law Samuel Simcock (q.v.), Robert Whitworth (q.v.) having done the survey. Although Whitworth was to become a more prominent figure in the field of civil engineering than Henshall, it was Henshall to whom Brindley entrusted the engineering and completion of the Grand Trunk Canal.

In February 1768 there is abundant evidence of how busy Henshall was: on 12 February Brindley was asked to send him to survey the Birmingham Canal, Simcock (q.v.) went in his stead as on 19 February the Clerk of Works on the Coventry Canal was sent 'forthwith … to Mr. Henshall the Clerk of the Staffordshire Navigation works for instruction, Mr. Brindley having engaged that he shall be at liberty to remain with Mr. Henshall a month for that purpose upon a gratuity to Mr. Henshall of 10 guineas'. Clearly Henshall was training canal engineers for Brindley's works.

After Brindley's death in 1772, Henshall became his chosen heir. In Brindley's obituary it is said that 'Mr. Brindley had long been sensible of the precarious situation of his Health and wishing to be succeeded in his Profession by his brother-in-law, Mr. Henshall, Clerk of Works of the Grand Trunk Navigation, he spared no pains to qualify him for that important Trust'. Henshall, who had never married, also took over the support of Brindley's remaining family. In addition, Wedgewood commented that Henshall's 'knowledge of mechanics, unwearied industry, and strictest integrity' rendered him well qualified to complete the canal, which was now open from 'the Trent to the Pottery'.

The Chesterfield Canal had been surveyed by Brindley in 1769. The line ran 46 miles from Chesterfield to the river Trent at Stockwith via sixty-five locks and a long summit tunnel, 2,850 yd. long, at Norwood; the canal was built to the same narrow dimensions as the southern section of the Trent and Mersey. Trade focused on iron, coal, timber and lead. After Brindley's death, Henshall took over as supervising engineer to the Company in November 1772 and in 1774 he was appointed chief engineer as John Varley (q.v.), the incumbent engineer, was finding supervision of the project difficult; additionally, Varley's brothers had been found guilty of overcharging the company for their contract work in the tunnel. The brothers were dismissed but Varley stayed on to work for Henshall.

Henshall completed the difficult section of engineering between Acton and Middlewich on the Trent and Mersey Canal. In September, Wedgewood wrote that 'Mr. Henshall deeming it impracticable to make the canal along the high sloping banks on the side of the Weaver has cut those new tunnels' (Saltersford and Barnton). It is noteworthy that Henshall surveyed and engineered this difficult northern section of the Grand Trunk Canal completely independently of James Brindley.

In order to get the northern end of the canal completed Clowes, clearly working for Henshall, advertised for labourers on the Trent and Mersey canal at Middlewich, the transhipment point for barges allowing for broader and heavier vessels belonging to the Duke of Bridgewater to navigate the northern section of the canal. This section had been redesigned with greater dimensions to allow for the Duke's craft. On the 7 March 1775 Henshall gave evidence to a parliamentary committee requesting permission to raise more money and indicated that 75 miles of the canal were complete and that 'the great Tunnel through Harecastle Hill of the length of about 2,900 yd.' was made.

Early in February 1776 Henshall suffered a personal blow when his mother died. On the 5 September 1776 Henshall was criticised by the Chesterfield Canal Company for not attending the works as agreed but in this year the canal was virtually completed. It opened fully for trade on the 4 June 1777; in the same year the Trent and Mersey Canal was completed. Henshall had, despite many engineering and personal difficulties, eventually brought Brindley's major and enduring work to fruition.

In 1778 Henshall purchased the farm and adjoining acreage at Greenway Bank where horses could be raised and stabled, and his energies now turned primarily to his business interests although the canal carrying trade bearing his name had been developed during his busy years as an engineer. This canal carrying business, 'Hugh Henshall and Co.', became the Trent and Mersey Canal proprietors carrying company. The canal trade between South Lancashire and the Midlands became extremely profitable and later links were made with London on the completion of the Coventry and Oxford canals. Henshall's Company was eventually taken over by Pickfords almost certainly on a regional basis and in 1786 Pickfords boats were commuting between Manchester down the Bridgwater and Trent and Mersey canals to Shardlow. Goods were then taken by road to London.

In 1790 Henshall was asked to survey and prepare estimates for the building of the Manchester, Bolton and Bury Canal, linking the river Irwell in Manchester with growing cotton town of Bolton. The canal was 11 miles long with a 4¾ mile branch to Bury. In July the following year, Matthew Fletcher and Henshall were organising contracts for cutting and on 10 September 1791 Henshall was acting as chief engineer. The narrow canal, with seventeen locks, was opened completely in 1808, with John Nightingale as the Company's (resident) engineer.

Henshall's business developments were complemented by a further move into the manufacture of pottery with Robert Williamson at Longport. In 1793 the firm was entitled Henshall, Williamson and Clowes and specialised in the production of black basalt ware. By 1815 it was known as Henshall and Williamson. Despite these developing entrepreneurial interests, Henshall was available as a consultant to advise on canal schemes and surveys. Work in Staffordshire meant that such requests were not always followed up and in June 1785 Henshall declined to be involved in the arbitration of a dispute between Charles Jones (q.v.), the contractor of the Thames and Severn Canal, and the Company itself over his fees for tunnelling; as an excuse, he pleaded illness.

Henshall gave advice to the Mersey and Irwell Navigation on damage caused by flooding, his report received by the Company in December 1787. However, in the autumn of 1792, Henshall, now living in Longport, re-surveyed Josiah Clowes's plans for the Hereford and Gloucestershire canal reporting to the committee in 1792. He suggested a shorter route with the Newent branch incorporated into the main line of the canal. He also advocated that the canal be more deeply excavated to allow for boats with 4½ ft. draft as opposed to 3½ ft. The canal was built to these plans and the first section from Gloucester to Oxenhall was completed in 1795.

In 1793 Henshall was involved with canal engineering further afield. On this occasion, he worked with William Jessop (q.v.) on a survey of the Grand Western Canal, to have been part of an ambitious scheme to link the English and Bristol Channels thereby obviating the dangerous voyage around Lands End. The Grand Western was to link the Bristol and Taunton Canal to Topsham, south of Exeter. Initial surveys had been conducted the previous year by Whitworth and John Longbotham (q.v.) and Jessop was asked by the Company to comment on the two surveys. Interestingly, he appointed Henshall to the task. Jessop's report to the committee on 28 November 1793, based on Henshall's work, recommended Longbotham's line but was unenthusiastic about the Tiverton and Cullompton branches.

The significance of the choice of Henshall as surveyor should not be ignored. Jessop was clearly one of the outstanding engineers in the country at this time and by his choice he recognised Henshall as one of the country's most skilled surveyors of waterways. The canal, designed to foster trade with Hereford, Gloucester and the Forest of Dean, where Henshall's other work had taken place, was not built until the beginning of the next century. In 1793 Henshall also found time, with the assistance of Charles McNiven (q.v.), to survey the Mersey and Irwell Navigation as there were continued problems with flooding that had occurred six years earlier.

In 1794 between April and July, Henshall was assisting another of Brindley's old pupils, Thomas Dadford (q.v.), in a survey of a short tramroad to serve the Brecon and Abergavenny Canal. The tramroad, 1 3/8 mile long ran from Llam-march 'Coal and Mine Works' to the Clydach ironworks; it opened in June 1795. That year Henshall was also responsible for completing the survey of the Caldon Canal the last enterprise that Brindley had worked on prior to his death. The Canal linked the Trent and Mersey to Froghall and neighbouring limestone quarries, with a branch to Leek. The Canal was approximately 17 miles long with the same number of locks and a short tunnel on the Leek Branch.

Henshall's business interests left him a wealthy man. On his death he left a large number of properties, business premises and shares. These included extensive properties in Greenway Bank, Ranscliffe and Wolstanton and mines at Golden Hill. Businesses, listed in his will included potworks, shops and land in Longport, Dale Hall and Burslem; also included were farms at Biddulph and limeworks at Newbold Astbury. Further properties belonging to Henshall included warehouses and premises associated with the Trent and Mersey Canal and the Hendra Company which brought clay from Cornwall. His will stipulated that the proceeds of these assets were to be dispersed to fourteen members of the Brindley, Henshall and Williamson families with the bulk of Henshall's assets left to his sister, Anne Williamson, and her sons, Hugh Henshall Williamson and Robert Williamson. Certainly the Henshall/Williamson families did very well financially and socially from the canal, trading and pottery businesses. Anne Williamson's son, Robert, eventually purchased Little Ramsdell Hall in Cheshire, a striking Georgian building overlooking the Macclesfield canal and close to the limeworks at Astbury.

Hugh Henshall was not devoted to the civil engineering of canals in the same single minded manner as his two brother-in-laws, Brindley and Clowes. However, following Brindley's death, Henshall was to spend a further five years engineering the most challenging sections of the Trent and Mersey, namely the Harecastle Tunnel, and the northern section of the canal, including three further tunnels, thus completing the Grand Trunk Canal as set out by Brindley and himself. He was to survey eight other canals successfully and engineer two further canals. Throughout his long life he never abandoned the profession of engineer.

Henshall was an intelligent, hard working man of high probity. He was also blessed with his family's strong constitution and business sense. His character appears to have been generous and avuncular, given his devotion to supporting his family throughout his lifetime. Henshall died at Longport on 16 November 1816 having reached the remarkable age of eighty-two and was buried at St. James's, Newchapel. Although he chose the epitaph 'Esquire' for his tomb, and his interests were later in business and trade, the vital part that Henshall played in establishing the foundations of the early canal network and its carrying trade should be fully understood.

CHRISTOPHER LEWIS

[Norton-in-the-Moors, Parish Records (1924–1943) Staffs. Parish Record Soc (SPRS), vol. 1, 1574–1751; vol. 2, 1754–1837; Norton-in-the-Moors Parish Records; Burslem Parish Records 1578–1812 (1913) SPRS, 3 vols.; Stoke-upon-Trent Parish Records 1629–1812 (1914–1927) SPRS, 3 vols.; Wolstanton Parish Records 1624–1812 (1914–1920) SRPS, 2 vols.; Minutes of the Chesterfield Canal Co. January/April 1774, PRO, 817/1; The Henshall Map 1779, PRO, E97.52; Plan of the Manor and Parish of Norton-in-the-Moors belonging to Charles Bowyer Adderley with his estates lying thereon, surveyed 1771 by Hugh Henshall, City Central Library, Hanley, Stoke on Trent; Advertisement for labourers on the Trent and Mersey Canal (1775) *Aris Birmingham Gazette*, 12 June; Will of Hugh Henshall of Longport, Burslem, PRO, PROB 11/1593 CP 3438; documents relating to the Burslem Potter John Wedgewood, Account Book. 23 May 1772; information on hire of Boats from Mr. Henshaw, City Museum and Art Gallery Stoke on Trent; minute books of the Chester Canal Company, PRO RAIL (3 December 1779, 20 January 1780); boat hire; 1735/1779–1791 Accounts and advice, twenty-one items, Hugh Henshall & Co & Josiah Wedgewood, including a voucher for the use of a pleasure boat; Wedgewood papers, University of Keele; Mortgage of Greenway Bank (1809) Staffordshire RO. Stafford; memo of an agreement 27 January 1767 between N. Kent and John Sparrow and Hugh Henshall, British Waterways Archive BW No. 1058.95; memo of an agreement 1 February 1767 between George and John Whiston, Sarah Barker and Wm. Cross, purchase of lands at Kings Bromley, BW No. 1061.95; letter 2 June 1770, Wm. Wyatt to Hugh Henshall, BW No. 1068.95; letter E. Lingard to Hugh Henshall, BW No. 1080.95; extract from indenture between Dorothy Norton and Hugh Henshall, BW No. 1105.95; letter from Hugh Henshall to W. Harrison, 10 May 1773, BW. No. 1127.95; letter from Thos. Harrison to Hugh Henshall, 1 June 1773, BW No. 1128.95; indentures/Hugh Henshall Esq. and others to Messrs. Bancks and Barker/Lease for ironstone mining at Golden Hill, cancelled by Henshall, 1820, reinstated by Anne Brindley *et al.*, 1840, Spode Collection. University Of Keele, 229g; C. Hadfield, Notes on individual engineers, LSE; *Encyclopaedia Britannica* (1778–1793) 2nd edn.; J. Priestley (1830) *Historical Accounts of the Navigable Rivers, Canals* and Railways throughout Great Britain; S. Smiles (1861) *Lives of the Engineers*, vol. 1, 1861; C. Hadfield (1962) *Canals to Stratford*; C. Hadfield (1966) *Canals of the West Midlands*; C. Hadfield (1966) *Canals of the East Midlands*; C. Hadfield (1967) *Canals of South Wales and the Border*; Greenslade and Jenkins (eds.) (1967) *Victoria History of the County of Stafford*, vol. 2; H. Household (1969) *The Thames and Severn Canal*; L. T. C. Rolt (1969) *Navigable Waterways*; C. Hadfield (1970) *Canals of North West England*; J. Lindsay (1970) *The Trent and Mersey Canal*; D. Holgate (1971) *Newhall and it's Imitators*; R. Russell (1971) *The Lost Canals of England*; A. W. Skempton. and H. C. Wright (1972) Early Members of the Smeatonian Society, *Trans Newc Soc*, **XLIV**; H. Malet (1977) *Bridgewater, the Canal Duke 1736/1803*; C. G. Lewis (1978) Josiah Clowes, 2 parts, *Waterways News*, **79–80**; C. G. Lewis (1978–1979) Josiah Clowes, *Trans Newc Soc*, **50**; G. L. Turnbull (1979) *Traffic and Transport. An Economic History of Pickfords*; C. G. Lewis (1980) Summit Problem. [Sapperton Tunnel], *Waterways News*, **100**; P. Lead (1980) *The Trent and Mersey Canal*; P. J. G. Ransom (1984) *The Archaeology of Transport Deelopment, 1750–1850*; P. Lead (1990) *The Caldon Canal and Tramroads*; W. Klemperer and P. Sillitoe (1995) James Brindley at Tunstall Hall, an archaelogical and historical investigation, *Pubs Staffs Arch Studies*, **6**; C. Richardson (1997) Minute books of the Chesterfield Canal; K. M. Evans (1997) *James Brindley Canal Engineer. A New Perspective*; Pickfords, Birmingham City Council, internet 10 April 1996, http://birmingham.gov.uk/html/tourist-info/canals/canalcp.html]

Publications

1765. *Plan of the intended Trent and Mersey Canal*
1779. *A plan of the navigable Canals now made in England ...*
1794. *Report to the Committee of the Brecon and Abergavenny Canal*

Works

1766–1777. Trent and Mersey Canal, Clerk of Works and Chief Engineer/Surveyor in succession to Brindley, approx. value £300,000
1772–1777. Bridgewater Canal, link to Runcorn, Engineer, in succession to Brindley
1772–1777. Chesterfield Canal, Chief Engineer, approx. value £150,000
1776–1779. Caldon Canal, Engineer
1790–1791. Manchester Bolton and Bury Canal, Chief Engineer, Surveyor, value £115,500 when completed in 1808
1792. Herefordshire and Gloucestershire Canal, Gloucester–Ledbury, survey, value £106,500
1806. Glamorgan Canal, survey

HEWES (HERVES), Thomas Cheek (1768–1832), millwright and textile machinery manufacturer, was born in Beckenham, Kent, on 23 February 1768, the eldest known child of Thomas Herves and his first wife, Sarah; his father is known to have had eight other children. Hewes 'imbibed a love of mechanics at an early age, from the circumstances of a corn [water] mill being situated near his father's residence'. It has been assumed Hewes served an apprenticeship as a millwright and in 1790 he was setting up machinery in a textile mill in Belfast. At the age of twenty-four he moved to the Manchester area which was in the throes of a building boom associated with the growth of the cotton industry and by 1797 had set up in business. In these early Manchester years he met and married a local girl, Ann, and his only child, James Thomas, was born.

Hewes appears to have supplied all kinds of machine to textile manufacturers; although his name has been associated with certain innovations it is unclear what his precise role was beyond supplier/contractor. His practice was, however, extensive and covered the whole of the British Isles.

Hewes's first known work in the Manchester area was for McConnel and Kennedy (q.v.) at their 'Old Mill' in Union Street, Ancoats. In September and October 1797 Hewes and probably four workmen were involved in installing a Boulton and Watt engine in the mill and he was still working on 'rebuilding the engine' a year later, having in the meantime installed the associated millwork. This was followed by work on the Long Mill, where building work (by Richard Coppack, Bateman and Kennet branch) commenced in 1801, presumably on the machinery from 1803–1805. Hewes was also involved with the installation of millwork or machinery at three other Manchester mills where Boulton and Watt were involved at this time: Marsden and Neild's (1800), Jackson's Mill (1803) and Daniel Holt's. Hewes was supplying other mills with textile machinery at this time, such as breaker cards for J. & R. Simpson's Shudehill Mill, and a variety of machines for Darley Abbey Mills, Derby (1804).

Hewes most impressive early job may have been George Allman's cotton spinning mill at Bandon, county Cork, where he may have been involved in the structural work as well as the supply of a 40 ft. iron waterwheel and spinning machinery. Work began in 1802 and Hewes supplied additional machinery in 1810, although he turned down a similar contract in 1823. Early mill buildings on the site were of masonry and timber but those with which Hewes was probably involved contained cast iron columns of cruciform and circular cross-section, and sheet iron to protect the timber floors against fire.

This impressive contract is an indication of the extent to which Hewes had established a reputation by this time. His fame has generally been associated with water rather than steam powered mills, particularly with the introduction of waterwheels of the suspension type and governors for waterwheels, together with the use of iron in their construction. This work was usually accompanied by a dramatic increase in the power available to the mills. A major problem with wooden waterwheels was their durability. Hewes may have first used iron at Bandon (1802) and in 1805 he used a cast iron axle to replace a timber shaft at Darley. He became involved in improvements to Belper Mills, originally established by Arkwright and Strutt (1776–1778). The West Mill, erected by Strutt (q.v.) between 1793 and 1795, originally had two wooden waterwheels but by 1811 these had been replaced by two iron wheels; they were described by contemporaries as the first (wrought) iron wheels. Hewes was responsible for their construction and although Strutt was probably responsible for the design innovation, it was Hewes who went on to exploit the system elsewhere. The wheels were of 21 ft. 6 in. diameter, 15 ft. long with cast iron shafts and slender wrought iron spokes 1½ in diameter, with 40 ft. float boards. The wheels were rim driven, permitting a lightness of construction, and were perhaps half the weight of a conventional wheel with cast iron spokes. Around the same time a governor was employed for the first time on a waterwheel at Belper. At this time Peter Ewart (q.v.) introduced an iron wheel at Styal (1807), probably of the suspension type, which Hewes repaired in 1815; it is possible that Hewes supplied the original to Ewart's designs. In 1819 Hewes supplied a new wheel at Styal.

It is clear that by the 1820s Hewes was running a major enterprise. From perhaps four employees in the late 1790s, by 1824 he was employing between one-hundred and forty and one-hundred and fifty men, forty on 'heavy mill work', and one-hundred and ten 'on machinery, the heavy gearing, and the fireproof mills'. Much of his iron production was involved in the production of waterwheels, the heavy mill work. They were producing all kinds of machinery and castings for fireproof construction, as well as installing millwork. In 1800 he was described as a machine maker of Portland Street, Piccadilly; by 1804 he had moved to New Islington, Union Street, still as a machine maker; by 1808, in partnership with Henry Day, another machine maker or joiner, he was at 16 Dale Street; and he remained in that street, at number 12, until his death, then described as an engineer, in partnership with Henry Wren, who seems to have joined his staff in 1812 and became his partner in 1821. An example of their work is a large (restored) crane at the Huddersfield Canal basin beside the University in Huddersfield.

From 1816–1817 Hewes had briefly employed William Fairbairn, initially on a mill at Macclesfield and then as a draughtsman. The experience must have been useful as soon after he left, Fairbairn was able to set up in business on his own and soon rivalled and surpassed Hewes's own enterprise, although Fairbairn was building waterwheels to Hewes's design in 1827. Fairbairn allegedly split with Hewes over rival designs for a replacement for a timber bridge over the river Irwell, Blackfriars Bridge, Salford. Fairbairn proposed a single cast iron arch but details of Hewes's proposal are lacking. He is known, however, to have built a small suspension bridge at Armley, Leeds, in 1825 and may have been responsible for the Broughton suspension bridge (1827) which failed under a load of marching troops. According to local sources he was also involved, at least on the committee, with the crossing of the Mersey at Stockport for which the Nowells (q.v.) were contractors.

Hewes's mill work in the 1820s showed a remarkable geographical spread: Gloucestershire; Tiverton, Devon; Knottingley Mill, Leeds (1822); Thwaite Mill, Hunslet (1823–1824), rebuilt to (Sir) John Rennie's designs by Hewes and Wren; Bakewell Mill, Derbyshire (1827); Tutbury Mill (1829) and Tutbury mills, Rochester, Staffordshire (c. 1830–1831); Aberdeen, where he had supplied iron for fireproof construction; and work for Milne Cruden and Company in 1818, probably Woodside Mills and Gordon Mill.

Without detailed research into mill account books Hewes's significance can only be estimated. He had risen to eminence in Manchester life by the 1820s when he joined the Literary and Philosophical Society and the Board of the Mechanics Institute, as well as playing a role in the early gas supply of Manchester. In 1824 he was chosen by the Chamber of Commerce to represent Manchester engineers at the Parliamentary Select Committee where he gave some of the most interesting evidence on the organisation of textile mill building and machine making of the time. Hewes died in February 1832, resident at 52 Lever Street, Manchester.

MIKE CHRIMES

[*Select Committee on Export of Tools and Machinery* (1824); *Mechanics Magazine*, **8**, 1827, 204, 21 July, 12–13; S. B. Smith (1969) Thomas Cheek Hewes: an ingenious engineer and mechanic of Manchester, M.Sc. thesis, UMIST; S. B. Smith (1998) in lit.]

Works
Works excluding textile machinery:

1802, 1820. Bandon Mill, County Cork, mill and 40 ft. waterwheel
1805–1807. Belper Mill, waterwheels
1819. Styal Mill, £3179 9s 4d
1820. Milne Couden and Co.'s mill, Aberdeen
1821–1824. Knottingley Mill, Leeds, rebuilt, £1,439
1823–1824. Thwaite Mill, Leeds, rebuilt, £15,876
1826. Armley suspension bridge
1826. Stockport (masonry) Bridge
1827. Kirkthorp weir, Aire and Calder Navigation
1827. Bakewell Mill, 25 ft. waterworks, c.£2,000
1827. Turbury Mill, two 70-hp waterwheels
1830–1831. Rochester, two 30-hp waterwheels

HEXHAM, John (fl. 1579–1596) of Huntingdon, land surveyor, is best known for his work in the Fens though his surveys extended also to Essex and Suffolk. In 1586 he is referred to as an 'expert in Serveyenge [who] hath alreadie taken Paynes in drawing of a Platt of the [East and West] Fennes', while in August 1588 it is recorded that there are 'two persons of good faculty and skill, John Hexham and Ralph Agass, appointed ... to make a perfect platt of South Holland'. This refers to Ralph Agas (q.v.) a prominent land surveyor of his day, and the 'platt' was related to a scheme by George Carleton (q.v.) for draining land in that part of south Lincolnshire. Hexham is next heard of in company with Agas and the engineer Humphrey Bradley (q.v.) in a reply from the Privy Council to request from a Commission of Sewers (the land drainage authorities) that 'some special good service might be performed for the drowned and decaied grounds in the counties of Northampton, Lyncolne, Cambridge and Huntingdon'. The reply from the Council, dated 21 March 1589, states that they had 'received information by men of good credyt, skill and judgement that Humfrey Bradley of Bergen op Zome in Brabant, John Hexhame in Huntingdon and Ralfe Agasse of Suffolk, be known sufficient and of great experience in their faculties, were able to make viewe and platt for the severall fennes, marshes and decaied grounds in the said fower shires, and to observe the true dysentes of waters and quallyties of the soile through which the waters should be carryed'. And the Council commends these three persons to the Commissioners.

Thus began the first attempt to tackle the general drainage of the Fens. Part of the survey is shown in a map by Hexham kept in Lord Burghley's collection at Hatfield House. Covering the area from the Welland to the old Nene and from Peterborough to Wisbech, it shows, on a scale of 1 in. to a mile, the watercourses and banks with clarity and a good degree of accuracy. Whether the 1589 survey extended further is uncertain, but in any case a map of the southern fens from Norfolk Marshland to Ely was made by William Hayward (q.v.) in 1591.

Nothing came of the general scheme, on which Bradley reported to Lord Burghley in December 1589, but quite a lot of local activity took place in that year. In May, Bradley and Hexham were appointed to design the course and outfall of a new drain in South Holland along the lines of the proposal by George Carleton and in August they were again consulted, this time on the north bank of the Witham near Boston. Meanwhile Bradley was working on a great sluice on the river Nene at the Horseshoe, about a mile below Wisbech, and Hexham was appointed to be 'the Clarke attendant upon the works' of improving the Wisbech outfall.

Dugdale refers to Hexham as surveyor to Philip (first) Earl of Arundel. 'Mr. Hexham', surveyor to the second Earl, took part in preparing a scheme for general draining of the Fens submitted in 1619 by the Earl, Sir William Ayloffe and Anthony Thomas. This may be John Hexham or more probably a son.

A. W. SKEMPTON

[Sir William Dugdale (1662) *The History of Imbanking and Drayning of divers Fenns and Marshes*; Thomas Badeslade (1725) *The History of the ... Port of King's Lynn and Draining the Great Level of the Fens*; L. E. Harris (1933) *Vermuyden and the Fens*; H. C. Darby (1940) *The Draining of the Fens*; L. E. Harris (1961) *The Two Netherlanders, Humphrey Bradley and Cornelis Drebbel*; R. A. Skelton and John Summerson (1971) *A Description of Maps and Architectural Drawings in the Collection made by [Lord] Burghley now at Hatfield House*; Peter Eden (ed.) (1976) *A Dictionary of Land Surveyors*]

HICK, Benjamin (1790–1842), locomotive builder, was born in Leeds in 1790, and trained in the works of Fenton and (Matthew) Murray (q.v.) who were one of the leading steam engine manufacturers of the time. His brother, John, was also trained as an engineer and worked at the Bowling works.

Hick was soon given responsibility for the installation of engines but declined the offer of partnership when it was offered, moving instead to Bolton in 1810 to become manager at the Rothwells' Union Foundry. This firm became known as Rothwell Hick & Rothwell and, after the death of Peter Rothwell Sr., in 1824, Rothwell Hick & Co. In that year Hick was elected a member of the Institution of Civil Engineers.

Although Hick is generally associated with locomotive building, the works with which he was associated produced all kinds of machinery, including cast iron cranes for docks and when, in the 1820s, John George Bodmer (1786–1864) developed his patterns and textile machinery near Bolton, he made use of the Rothwell–Hick workshops. They began work on a specially designed waterwheel which was completed by William Fairbairn after Bodmer had temporarily returned to Europe.

Around 1814 Hick married Elizabeth Routledge. They had two sons, Benjamin and John (1815–1894), who their father trained as engineers, and with whom, in 1833, he set up a his own works in Bolton, the Soho foundry, under the name of Benjamin Hick and Sons. Hick had shown an early interest in locomotives, commenting on Blenkinsop's (q.v.) invention in 1822. Rothwell Hick and Co. built their first locomotive, the *Union*, in 1830, for the Bolton and Leigh Railway, in which Hick was a shareholder, and the advent of railways may have encouraged him to branch out on his own. Hicks built their first locomotive, the *Soho*, for goods traffic on the same railway. Hick took out three patents in 1833 and a final patent in 1839 and by the time of his death in Bolton on 9 September 1842, the firm was well established as locomotive manufacturers with British and foreign railway companies. Benjamin Jr., had by then left the firm and predeceased his father whilst John took it on to greater strengths and became a leading light in Bolton society, serving as MP in 1868.

MIKE CHRIMES

[Membership records, ICE archives; Bolton and Kenyon Railway (1830) *Mechanics Magazine*, 18 December; Obituary (1843) *Min Proc ICE*, **2**, 12–13; *Engineer*, **20**, 1865, 249; E. L Ahrons (1927) *British Steam Railway Locomotives 1825–1925*; E. L. Ahrons (1930) *Engineer*, **129**, 598; C. F. Dendy Marshall (1953) *A History of Railway Locomotives before 1831*]

HIGGINS, John Lane (fl. 1824–1850), of 370 Oxford Street, was elected an Associate of the Institution of Civil Engineers in 1824, nominated by John Farey Jr. Higgins took out a number of patents which is doubtless how he came to Farey's attention; they included a method of providing navigation lights on steamships. He remained in membership only for a few years.

MIKE CHRIMES

[ICE membership records, ICE archives; Evidence (1831) *House of Commons Select Committee on Steam Navigation*, 61–62; R. Smith (1998) in lit.]

HILL, Simon (fl. 1632–1636), was a fen drainage engineer working in Lincolnshire. It can be assumed that Hill's Sluice, built in 1632 at the outfall of Shire Drain into the river Nene north of Wisbech, was named after him and that he had been in charge of its construction under Francis, Earl of Bedford, the principal undertaker of draining the Great Level of the Fens, 1631–1636. Similarly, Hill's Drain and its outfall sluice into the river Welland near Spalding can be attributed to him. This formed part of the works on Deeping Fen (1632–1636), also under the Earl of Bedford. The Sluice dates from 1635 following the widening and deepening of the river which was still in progress during the summer of 1634. The name 'Mr. Hill' appears on several small parcels of land shown on the map (c. 1642) of the Fen in Dugdales's *History of Imbanking and Drayning*.

In April 1633 the King (Charles I) recommended the Earl of Lindsey, Sir Robert Killigrew, his son Sir William and Robert Long as undertakers for draining all other fens in Lincolnshire except those in the undertaking of Sir Anthony Thomas (the East and West Fens north of Boston) and the Eight Hundred Fen (west of Boston). This area of some 70,000 acres includes land northwest of Lincoln draining into Fossdyke, fens adjacent to the Witham between Lincoln and Boston, and fens (later known as Lindsey Level) north of Deeping Fen between Kyme Eau and the river Glen. A document in *State Papers Domestic* (SPD 339/28) endorsed 'Mr. Hills answer to Sir Rob. Killigrew his demands' is a preliminary report on this project. It is undated but must be earlier than the death of Sir Robert which occurred before 12 May 1633. The main points are: (1) to make Fossdyke a perfect channel to drain the adjoining land and to place a navigable sluice at its outfall into the Trent [at Torksey]; the work may cost £2,500; (2) the fens between the Glen and the Eight Hundred Fen can be drained reasonably well by several drains falling into the Welland estuary; and (3) for the Witham Fens a new cut should be made 30 ft. wide at the bottom from the Witham at Boston to meet the same river again at Dogdyke, with a navigable sluice at the outfall. The cut, about 8 miles long, may cost £4,000 and the sluice £2,500. Cleansing old drains and enlarging their sluices may cost £10,000.

In a slightly later report (SPD 339/29) Hill recognises that the 30 ft. cut is suitable only for drainage and considers the alternative of a 50 ft. cut suitable for drainage and navigation by vessels of 50–60 tons burden. In either case he now recommends that the cut should extend up to Washingborough, near Lincoln. The 30 ft. cut, 5 ft. deep would cost about £5,000 and a timber-framed sluice would suffice, costing £1,500. The 50 ft. cut, 7 ft. deep, might cost £12,000 and would require a stone-built navigable sluice costing £4,500.

Two years later the project had advanced to the point where the Commissioners of Sewers (the land drainage authority) of Lincolnshire decreed that Lindsey, with Sir William Killigrew as principal partner, should undertake the drainage in return for 24,000 acres, the work to be completed within six years. A document (SPD 315/44) dated 4 March 1636, records the appointment of Hill as Director of Works. Signed by Lindsey and William Killigrew, it states that 'the Earl of Lindsey ... hath agreed with Simon Hill gent. To be Director of all his works for the draining of the Fennes' and sets out his duties. Hill is to 'take a view' of all places that 'do charge those fennes with water' and 'to find the best and aptest outfall'; to decide on the placing and dimensions of all drains, sluices and bridges; to direct the Paymaster and Overseers, as well as the carpenters, masons, smiths and other workmen; and to set out all new works. He is expected to be in attendance at least three days a week and would receive a salary of £4 per week (say £200 p.a.) and 200 acres of fenland.

All seemed ready to begin operations but within a few weeks the scheme had been greatly reduced in scale to cover only Lindsey Level: an area of 36,000 acres of which the undertakers would receive 14,000 acres after drainage. Next, from two documents, undated but probably of late 1636 (SPD 339/31) and 339/44), it emerges that 'Mr. Wright' (? Thomas Wright (q.v.) who had worked on the Great Level) had been appointed General Overseer, with 'Mr. Walpole' as Surveyor, and that Hill had drawn up estimates for a scheme proposed by John Liens (q.v.) who was now to be Director of Works. The estimates, which are given in considerable detail, clearly relate to Lindsey Level and it is certain that Liens acted as Director up to 1638, if not later.

Liens in 1635 had completed the Dutch River, the 5 mile embanked channel for the river Don from Newbridge, near Thorne, to Goole on the Yorkshire Ouse, having previously worked under Sir Cornelieus Vermuyden (q.v.) on Hatfield Chase drainage. Apparently he was consulted after March 1636 for a second opinion on Lindsey Level and his proposals were preferred to those by Hill, though the latter was still employed to make the estimates, on which he seems to have been regarded as an expert. No later reference to Hill's career has come to light.

A. W. SKEMPTON and MARGARET KNITTL

[State Papers Domestic (1633–1636) PRO; Sir William Dugdale (1662) *The History of Imbanking and Drayning*; William Killigrew (1705) *The Property of all English-Men asserted, in the History of Lindsey Level*; W. H. Wheeler (1896) *A History of the Fens of South Lincolnshire*; H. C. Darby (1940) *The Draining of the Fens*; Margaret Albright (1955) The entrepreneurs of fen draining in England under James I and Charles I, *Explorations in Entrepreneurial History*, **5**, 50–65; Keith Lindley (1982) *Fenland Riots and the English Revolution*]

HODGKINSON, John (1773–1861), railroad engineer, was a member of the family of Overton Hall near Ashover, Derbyshire, which was connected by marriage and business interests to that of Benjamin Outram (q.v.). He probably worked for Outram for some time before January 1796, when he deputised for him in presenting a report to the proprietors of the Nutbrook Canal. From 1799 he supervised the construction of the tramroads of the Ashby Canal in Leicestershire, for which Outram was the Engineer, and also assisted him in surveys and reports for the Brecknock & Abergavenny Canal, the Monmouthshire Canal and railways in the Forest of Dean.

Hodgkinson's connection with the Monmouthshire Canal developed in 1801 when he took over estimating and engineering responsibility for the Sirhowy Railway, for which the canal company gained an Act in 1802. The line was 24 miles long, running north from Newport, Monmouthshire, and was designed on 'Outram's improved plan'. It contained substantial earthworks and a thirty-two arch viaduct, 1000 ft. long across the valley at Risca, easily the largest such structure on the early railway system (though demolished in 1901). He was then engaged at 12 guineas per week to supervise construction, which continued to 1805. During this period, he converted the company's tramroad to the Beaufort Works from edge rail to plate rail, the first recorded example of such a change. In the same year, 1803, he was consulted by the proprietors of the Leominster Canal in Herefordshire, who had built the central portion of their line but run out of funds. On the basis of his report, which recommended railways to fill the remaining lengths to Kington and Stourport, an Act was obtained but no work was done. Interestingly though, when he was again consulted by the company in 1812, he proposed a canal on a new route to join the Severn lower down, at Worcester.

In December, 1803 he bought the property of Alteryn (now Allt-yr-ynn), near Newport, which

was to remain his home for the rest of his life (though in 1811 he was described as 'of Cheltenham' and he seems to have lived for a time near Llanvihangel Crucorney, where the Llanvihangel Railway met the Grosmont Railway). In 1805 he surveyed a short extension of the Monmouthshire Canal at Newport southwards to a new terminus at Pillgwenlly (built probably in 1808 on a slightly different line) and contracted to build a substantial stone bridge over the river Usk at Caerleon, 5 miles to the north. In this he was associated with William Jessop (q.v.) who was active at Bristol then. In 1811 the Monmouthshire justices asked him to report on remedial work to Chepstow Bridge, but their colleagues in Gloucestershire would have no more patch and mend, and it was rebuilt by John Urpeth Rastrick (q.v.) in 1816.

In 1808 he was employed to sort out the conflicting schemes for the Gloucester and Cheltenham Railway or Tramroad, which gained its Act the following year; his estimate of £25,261 14s is said to have been significantly low. In 1809 the Act was passed for the Bullo Pill Railway in the Forest of Dean, which included the Haie Hill Tunnel, an early example for a railway. The contractor here was Robert Tipping, who in 1810–1812 was the contractor for the Severn Tunnel from Newnham to Arlingham. Hodgkinson appears to have possessed a copy of the Act, which however makes no mention of the name of the Engineer. The tunnel was flooded when about half of its length had been constructed, and the work abandoned. Tipping later worked for Hodgkinson again on the Hay Railway. On the other side of the Severn, Hodgkinson was one of several engineers who were consulted about the southern terminal of the Gloucester & Berkeley Canal, in his case in 1810.

Later that year or early the following, he was asked to resurvey the Llanfihangel Railway, for which the initial plans had been drawn up by William Crosley (q.v.). In the subsequent discussions Hodgkinson's ideas prevailed and he took over responsibility for the extension to Llangua (the Grosmont Railway) and subsequently to Hereford. For these surveys Hodgkinson employed David Davies of Llangattock, as he had done in 1801 for the Sirhowy Railway, and would do for an abortive railway from Monmouth to Skenfrith in 1813 and the Hereford Railway in 1825; they also worked on turnpike roads together. From 1811 Hodgkinson was Engineer to the Hay Railway also. He reported to the Brecknock & Abergavenny Canal on the problems with the aqueduct at Pontymoile and in 1814 to the Somersetshire Coal Canal on the conversion of the towpath of their Radstock branch to a railway. For the Worcester & Birmingham Canal he made an estimate of the cost of removing Worcester Bar, Birmingham, but seems to have overstretched himself as the Hay Railway dismissed him in September 1814, apparently for non-attendance to his duties. Subsequently he contracted for sections of the line but was unsuccessful with his tender in 1818 for its extension, the Kington Railway. He continued to obtain work as Engineer or contractor as opportunity arose, including a return to his home ground on the Cromford & High Peak Railway in Derbyshire, where Josias Jessop was the Engineer.

In his early railways Hodgkinson followed Outram's method of laying track with stone blocks and plate rails spiked directly to them. By the late 1820s, however, on the Hereford Railway he had been converted to the use of combined tie-bars and chairs (dovetailed sills), which had been popularised by George Overton (q.v.). The Duffryn Llynvi & Porthcawl Railway was almost the last line in Britain to use the cast iron fish belly rail. After using Outram's gauge of 4 ft. 2 in. between the flanges on the Sirhowy Railway, Hodgkinson specified 3 ft. 4 in. (3 ft. 6 in. over the flanges) for most of his lines, but 4 ft. 7 in. for the late ones in West Glamorgan.

He was still active in 1842, when he made a survey with Thomas F. Marsh for a railway from Newport to Blaina, which they estimated at £240,000; it was not built until 1855.

He had married Sophia Gunter in 1830 and one daughter, Helen, married Benjamin Outram Hodgkinson, her cousin. John Hodgkinson died on 12 February 1861 and was buried at St. Woolos Church, Newport, now the cathedral.

P. S. M. CROSS-RUDKIN

[PRO; Maybery, Harpton Court and Powis Collections, NLW; Gwent RO; Hereford RO; Rennie manuscripts, NLS; *Monmouthshire Merlin*, 16 February 1861; H. W. Paar (1963) *The Severn & Wye Railway*; H. W. Paar (1965) *The Great Western Railway in Dean*; Bertram Baxter (1966) *Stone Blocks and Iron Rails*; Kenneth R. Clew (1970) *The Somersetshire Coal Canal and Railways*; P. Stevenson (1970) *The Nutbrook Canal*; C. R. Clinker and C. Hadfield (1978) *The Ashby-de-la-Zouch Canal & Railway*; R. Cook and C. R. Clinker (1984) *Early Railways between Abergavenny and Hereford*; David Bick (1987) *The Gloucester & Cheltenham Tramroad*; Gordon Rattenbury and Ray Cook (1996) *The Hay & Kington Railways*; R. B. Schofield (2000) *Benjamin Outram*; inf. ex. J. van Laun; inf. ex. J. G. Calderbank]

Publications

1801. *A Plan of an Intended Rail Way, or Tram Road, from Sirhowy Furnaces ... by Tredegar Iron Works ... to communicate with the Monmouthshire Canal and the River Usk, at or near the Town of Newport ...*
1811. *The Hay Railway*

Works

1802–1805. Sirhowy Railway, Engineer, 16 miles
1803. Conversion of Monmouthshire Canal tramroad to Beaufort Works from edge rail to plate rail
1805–1807. Caerleon Bridge, contractor, for £9,500, three spans of 60, 70 and 60 ft. in tidal water

1808. Monmouthshire Canal, Newport Extension, Engineer
1809. Bullo Pill Railway, Engineer, including Haie Hill Tunnel, 1100 yd. long
1809–1811. Gloucester & Cheltenham Railway, Engineer, 3 ft. 6 in. gauge
1811–1814. Hay Railway, Engineer; estimate £52,743 18s, 24 miles, 3 ft. 6 in. gauge
1811–1814. Llanfihangel Railway, Engineer, 6½ miles; 3 ft. 6 in. gauge
1814. Llangattock Tramroad
1815–1818. Hay Railway, contractor for sections
1818–1819. Grosmont Railway, Engineer, 7 miles
1818–1820. Kington Railway, Engineer, 12 miles, £10,000
1819–1821. Glanllyn Reservoir, Glamorganshire Canal, Engineer
1825–1828. Duffryn Llynvi & Porthcawl Railway, Engineer, 16 miles; 4 ft. 7 in. gauge
1826. Cromford and Sheep Pasture inclines, Cromford & High Peak Railway, contractor
1826–1829. Hereford Railway, Engineer, 11 miles; 3 ft. 6 in. gauge, estimate £24,383 14s
–1829. Monmouthshire canal tramroad at Pont-newynydd, £2,140
–1830. Bridgend Railway, Engineer, 4 miles; 4 ft. / in. gauge
1820–1830. Hay Railway, contract to repair, £5 per mile per year

HODGKINSON, Samuel (fl. 1792–1825), canal engineer, offered his services to the Worcester & Birmingham Canal in April 1792 but was turned down because the salary he requested was too high. In October 1792 he won a contract for cutting part of the canal and in December was taken on trial as a Clerk at 1 guinea per week, to start as soon as he completed his contract. He was still with the company in July 1794, when he was given a gratuity of 20 guineas and his salary increased to 60 guineas a year. Thomas Cartwright (q.v.), the Sub- (i.e. resident) Engineer, had his salary raised to 200 guineas. Work on the canal wound down in 1795 as funds ran out, and Hodgkinson was transferred as an assistant on the Hungerford to Newbury section of the Kennet & Avon Canal in December 1795. In 1796 he was supervising two lots by direct labour after the contractor had been discharged, but more normally his duties were to inspect the contractors' works. In 1808 when Burbage Tunnel was starting, his health was not good enough to permit his appointment there and he was moved to Rowde, near Devizes, with an assistant to help him. He left the concern shortly afterwards when John Blackwell was appointed over him to supervise the whole of the Middle District. John Thomas, the Superintendent, commented that the work was now better managed, but the Company paid Hodgkinson a 100 guinea gratuity in consideration of his good conduct in its business.

He spent the next six and a half years working on the Birmingham Canal but it cannot have been a full time commitment as he was tendering unsuccessfully for a contract on the northern extension of the Lancaster Canal in November 1814. He was regularly consulted about work on the Worcester & Birmingham Canal, 1811–1815 but was still living at Summer Row, Birmingham, when on 17 July 1815 he was appointed to complete the Canal where his brother John was accountant clerk; it was opened on 3 January 1816 and later in the year he was given three months' notice to leave. He had moved to Bath Row, Birmingham, by August 1817 when both he and his son were out of work and seeking employment, and at St. Martins Street, Fiveways, Birmingham in 1820 when he sought unsuccessfully to be appointed to the Gloucester & Berkeley Canal. During these years he undertook occasional work for the Dudley Canal and made a survey for a Stourbridge Extension Canal; nothing was done about the latter until 1836. He was described as 'an eminent engineer' when in 1819 he was employed by the Stratford-upon-Avon Canal in a dispute with the Warwick & Birmingham Canal about water at Kingswood. He was consulted by John Urpeth Rastrick (q.v.)—in a letter dated 6 February 1823—about rails for the Stratford & Moreton Railway, and employed by (Sir) John Rennie (q.v.) in surveys of railways in Surrey and near Bristol in 1825.

P. S. M. CROSS-RUDKIN

[RAIL 829, 842, 875, 886, PRO; Rennie mss, NLS; Rennie reports, ICE; Gloucestershire RO]

Works

1792–1795. Worcester & Birmingham Canal, contractor and clerk
1795–1808. Kennet & Avon Canal, assistant engineer
1808–1815. Birmingham Canal, assistant engineer
1815–1816. Worcester & Birmingham Canal, Engineer, £350 p.a.

HODSKINSON, Joseph (fl. 1765–d. 1812), land surveyor and civil engineer, first comes to notice as the engraver of a map of Bedfordshire surveyed by John Ainslie (q.v.) and Thomas Donald and published by Thomas Jefferys, Geographer to the King, in 1765. Between 1767 and 1770, together with Ainslie and Donald, he was employed by Jefferys in surveying the whole county of Yorkshire; the map was published in 1772. Hodskinson married Ann Haynes on 22 July 1770 at St. George's, Bloomsbury; they were both in their thirties. Ann seems to have been a woman of some wealth. Within a year they moved to 29 Arundel Street, off the Strand, a house previously occupied by her or her family.

It was from this address in 1774 that Hodskinson engraved and published the map of Cumberland, surveyed by Donald. He had by now established a considerable surveying practice; at least a dozen of his estate surveys are known from 1773 to 1790 in Surrey, Kent, Bedfordshire, Cheshire and Worcestershire. Clearly he was a skilled draughtsman, exhibiting architectural drawings and designs at the Free Society of Artists, 1774–1779, and at the Royal Academy, 1784–1786. His greatest achievement in this period is in the map of Suffolk, surveyed c. 1777–1782 initially by Hodskinson and Andrew Dury and after the latter's death c. 1779 by Hodskinson alone. It was published in 1783 by William Faden (q.v.) on the scale of 1 in. to a mile and won for Hodskinson the Gold Medal of the Society of Arts; he was a member of the Society from 1775 until 1790.

He had been elected to the (Smeatonian) Society of Civil Engineers in 1777, undoubtedly as a surveyor (the Society was broadly based in its early days). He attended the meetings with great regularity and was Vice-President from 1781 until 1789. Probably on the recommendation of John Grundy (q.v.), a founder member, Hodskinson was asked by the Commissioners of Wells Harbour to survey and report on the state of the harbour. He spent March and April 1782 on site. His report, dated 5 July 1782, was published together with a report by Joseph Nickalls (q.v.) and it includes a plan of the harbour with the adjoining marshes. Grundy and Thomas Hogard (q.v.) accompanied Hodskinson in his survey and approved his report.

In 1788 the Commissioners of Shoreham Harbour decided to petition Parliament for increased powers to raise funds. Hodskinson and Nickalls reported jointly in January 1789 and Hodskinson gave evidence to the House of Commons in March. The Bill passed in May, but little was done until the work by William Chapman (q.v.) following a new Act of 1816.

Now established as an engineering consultant, Hodskinson advised on improvement of the River Cam Navigation and examined the Great Ouse down to Ely in October 1790. A year later he was again in the Fens, on behalf of the Mayor and Corporation of King's Lynn, to consider what effect the proposed Eau Brink Cut might have on their harbour. He reported in January 1792 that the cut, bypassing a long bend in the Ouse upstream of Lynn, would lead to silting up of the bend with a reduction of tidal reflux and disastrous consequences to the harbour. He came back in March 1793 to make detailed observation and calculations which he presented in evidence to the House of Commons in March 1794, estimating that the reflux would ultimately be reduced to one eighth of its present volume. A footnote in the printed proceedings records that 'this Witness having a Defect in his Voice, so that he could not be heard by the House, was permitted to communicate his Answers to his Nephew, Mr. Joseph Hodskinson, Junior, who stood near him at the Bar, and repeated his Answers to the House'. The nephew, a surveyor, had assisted in the field observations. Hodskinson gave further evidence a week later and subsequent proceedings continued into the next session. Those who supported the proposal, including Robert Mylne and John Rennie (qq.v.), eventually won the argument and the Bill for the Eau Brink Cut passed in May 1795.

Meanwhile, Hodskinson had been busily engaged elsewhere. In May 1792 he put forward a scheme for improving navigation on the river Stour from Sandwich to Canterbury. Later that year he advised Mr. William Constable on protecting the sea banks of Cherry Cob Sands, on the north shore of the Humber. The works recommended by him in May 1793 had been partially carried out when he reported again in February 1796, and were completed by 1799.

Hodskinson gave evidence to the Committee on the Rochdale Canal Bill in February 1793 and in February 1794, with particular reference to water supply to the summit level. In November 1793 he reported to the Participants (proprietors) of Hatfield Chase on problems arising from the proposed Stainforth & Keadby Canal.

An Act for improving the drainage of low grounds in the Isle of Axholme (adjacent to Hatfield Chase) having been obtained in June 1795, William Jessop (q.v.) and Hodskinson were appointed engineers. They reported jointly in October and their scheme was carried out with John Dyson (q.v.) as resident engineer, reaching completion by 1803.

During his second visit to Cherry Cob Sands, Hodskinson reported (again to Mr. Constable) on draining Keyingham Level, an area of low ground north of the Sands. Several features of his scheme were incorporated in the works carried out between 1803 and 1807 by Chapman.

Hodskinson gave evidence, in March 1796, to the Parliamentary Committee on improving accommodation for shipping in the Port of London, with a plan for extending the Legal Quays and deepening the river down to Deptford and in August 1797 he reported to Scarborough Harbour Commissioners with recommendations for carrying on the works and extending the South Pier. They thanked 'Mr. Hodskinson for his great Attention to a work of such public utility' but once again it was Chapman who in 1801 took over direction of the works. With one exception, noted below, there is no record of any further engineering work by Hodskinson. The surveying practice continued; his nephew died in 1800 and a little later George Bassett became a partner.

The Hodskinsons moved from 29 to 35 Arundel Street in 1783. He was still there in 1811 as 'Joseph Hodskinson, Surveyor'. In September of that year he produced his last report: a plan and estimate for a new harbour at Shoreham.

His wife had died about 1800; the widow of his nephew and her children came to live with him. His will is dated February 1811 and he died on 2 February 1812, leaving property and a small fortune to numerous relatives.

A. W. SKEMPTON

[Society of Art Minutes and Lists (1775–1790). Royal Society of Arts; Shoreham Harbour Commissioners Minute Book (1789) West Sussex RO; Joseph Hodskinson (1793) Report to the Participants of Hadfield Chase Level, MS, Nottingham University Library; *Minutes of Evidence on the Bill for improving the Drainage of the Middle and South Levels … and the Navigation of the River Ouse from … Eau Brink … to the Harbour of King's Lynn* (1794) House of Commons; William Jessop & Joseph Hodskinson (1795) Report on Isle of Axholme Drainage, MS, Nottingham University Library; Report from the Committee on the Port of London (1796) House of Commons; *Sundry Papers and Reports on the Defence of Cherry Cobb Sands, against the Humber* (1796); Scarborough Harbour Commissioners Minute Book (1797) Scarborough Library; Joseph Hodskinson (1811) Estimate and Plan of proposed Harbour at Shoreham, MS, West Sussex RO; British Museum (1967) *Catalogue of Printed Maps, Charts and Plans*; D. P. Dymond (1972) *The County of Suffolk, surveyed by Joseph Hodskinson*, Suffolk Records Soc., 15; Elizabeth Rodger (1972) *Large Scale County Maps of the British Isles*; Charles Hadfield and A. W. Skempton (1979) *William Jessop, Engineer*; Bendall]

Publications

1782. *Report on the State of Wells Harbour, Norfolk* (with engraved plan)

1792. *Report on the Probable Effect which a New Cut from Eau Brink to a little above Lynn will have on the Harbour and Navigation of Lynn*

1792. *Plan of the Outfall of the River Ouse*

1792. *Plan and Estimates for Improving the Navigation of the River Stour from Sandwich to Canterbury* (with engraved plan)

1796. *Report on Keyingham Drainage* (with engraved plan)

HOGARD, Thomas (fl. 1754–d. 1782) is recorded as attending meetings of the Adventurers of Deeping Fen from 1754. In April 1764 on the resignation of John Grundy (q.v.) as Surveyor (of Works) Hogard wrote to his fellow Adventurers saying 'as I am greatly concerned in that Undertaking as Proprietor and Occupier, and have the welfare of the Fen greatly at Heart, I do desire to succeed Mr. Grundy in that Office'. He was duly appointed at a salary of £50 p.a.; Joseph Featherstone (q.v.) continued as Assistant Surveyor.

In November 1767 Hogard submitted a scheme for improving drainage of the Fen and navigation from Spalding to the sea by widening the river Welland from Spalding to its junction with the river Glen, and thence to cut a new channel 100 ft. wide and 7½ miles long to Wyberton on the river Witham estuary (near Boston) where there would be a massive sluice and navigation lock. He estimated the works at £30,000 apart from large costs for land purchase. This scheme follows in a practical form an idea put forward by Nathaniel Kinderley (q.v.) in 1720 and again by his son in 1751. The Adventurers sought a second opinion from Thomas Tofield (q.v.). He reported unfavourably in November 1768, giving a full summary of Hogard's proposals and putting forward an alternative scheme for a new channel about 5 miles long from Spalding to the north shore of the Welland estuary at Fosdyke. Finally Grundy was asked to examine the whole question of improving the Fen drainage. Reporting in February 1770 he also rejected the Wyberton Cut and made two main proposals: to extend the outfall of Vernatts Drain further downstream, and to cut a new channel for the Welland 1½ miles long to Fosdyke. The subsequent history is briefly: (1) an Act of 1774 which sanctioned the extension of Vernatts Drain, (2) an Act of 1794 establishing the Commissioners of the River Welland who were empowered to carry out the Wyberton Cut, but (3) nothing happened until, under a new Act of 1801, the Welland Cut to Fosdyke was made.

Hogard retired from the Deeping Fen job in 1771 but continued to attend all the Adventurers' meetings up to 1781. Between 1764 and 1773 he served as Commissioner on several inclosures of fens in south Lincolnshire, from the records of which he appears to have moved from Deeping (St. James) to Spalding by 1767. He is almost certainly 'Mr. Hogard of Spalding' who in 1776 was appointed to superintend the rebuilding of Gunthorpe sluice at the outfall of Shire Drain into the Nene estuary north of Wisbech, and the 'Thomas Hogard, an engineer conversant with the management of low lands for near forty years' who was one of the witnesses to the Parliamentary Committee on a Petition from Bedford Level Corporation in March 1777.

He was elected a member of the (Smeatonian) Society of Civil Engineers in May 1772. Towards the end of this life he returned to Deeping. His will is dated 7 January 1782 and he was buried at Deeping St. James on 24 April 1782.

'Thomas Hogard junior' drew up plans for the new outfall sluice of Vernatts Drain in 1774 and attended meetings of the Adventurers from 1777 to 1781. He is presumably a relative but he continued to attend the meetings from 1783 to 1793, and is the Thomas Hogard of Deeping St. James who died in May 1794.

A. W. SKEMPTON

[Deeping Fen Adventurers Minute Book (1754–1793) Lincolnshire Archive Office; Thomas Tofield (1768) *Report on Deeping Fen*; Committee Book for Kinderley's Cut, North Level (1776); *Bedford Level* (1777) House of Commons; W. H. Wheeler (1896) *History of the Fens of South Lincolnshire*]

Works

1764–1771. Deeping Fen, improvements, £13,000

HOLDSWORTH, Arthur Howe (1780–1860), MP and Governor of Dartmouth Castle, was elected a Corresponding Member of the Institution of Civil Engineers in 1825, resigning in 1837. He was the seventh of seven Holdsworths who became virtually the rulers of Dartmouth between 1710 and 1832. Arthur Howe was the son of the fifth Arthur (1757–1787), who was also MP for Dartmouth and Governor of Dartmouth Castle. He lived at Mount Galpin on Clarence Hill, which was the town

house and office and also owned the country estate, Widecombe House, near Torcross. He was MP from 1802 until 1820 and from 1829 to 1833. He later left Dartmouth and lived at Brookhill, Kingswear, from 1840 and then at the Beacon.

Holdsworth wrote a number of pamphlets relating to naval affairs over a period of thirty years. Among his printed works, The Revolving Rudder described a rudder shaped to the inclination of the stern of the ship so that if the ship were hit by a very strong sea from the rear the rudder could turn forward without damaging either rudder or supporting brackets. He took out his first patent in 1817, on gasholders, and gave evidence on patent law.

In 1815 he had passed through Parliament an Improvement Act which gave the town's Corporation powers to build a market, widen, pave and light streets and over the next twelve years the old mill pool was filled in. Meanwhile Howe had supported a new turnpike scheme to link the town to Kingsbridge. In 1825, the line of the Butterwalk was extended west with a road, later called Victoria Road, alongside the filled-in pool and then up the hill to Townstal, and thence to Stoke Fleming. It was wide enough and at a suitable gradient to enable carriages for the first time to reach the town.

On the other side of the river he had already backed a similar turnpike leading from Kingswear to Torbay. His rival, John Henry Seale, promoted the Floating Bridge Act of 1830 where James Meadows Rendel (q.v.) provided his first floating bridge to solve the problem of crossing the river Dart, to be followed by others across the river Tamar and other estuaries.

The road to Kingsbridge was not completed until 1855 when Holdsworth, then aged seventy-five, stirred up his fellow trustees to raise the funds to carry out the bold plan which he laid before them for a new road out of Stoke Fleming, a long slant down to Strete Gate and then the laying of a macadam road along the shingle bank of Slapton Lea to Torcross; this constituted a noteworthy piece of civil engineering.

Holdsworth was married twice, to Elizabeth Were Clark (d. 1804) and to Catherine Henrietta Eastbrook. A portrait of Arthur Howe Holdsworth is reproduced by Freeman (1983).

MIKE CHRIMES and BRIAN GEORGE

[Membership records, ICE archives; ICE library catalogue; House of Commons (1829) *Select Committee on the Law relative to Patents for Inventions*, HC332, 119–126; P. Russell (1950) *Dartmouth, A History of the Port and the Town*; R. Freeman (1983) *Dartmouth: A New History of the Port and its People*; R. G. Thorne (1986) *The House of Commons 1790–1820*, **IV**, 214]

Publications

1832. *The Revolving Rudder* (reprinted 1850)
1841. *Dartmouth: the Advantage of its Harbour as a Station for foreign Mail Products*

1842. *Improvement in constructing certain parts of Ships … in order to arrest progress of Fire*
1852. *Water Bulkheads for reducing the Temperature and arresting Fire in Ships*

HOLL, Edward (d. 1824), Architect to the Navy Board, may have been the son of Edward Holl, a mason from Beccles in Suffolk. As architect to the Board from 1804 until his death he succeeded Samuel Bunce, who had been appointed as the first salaried architect on the staff of Brigadier General Sir Samuel Bentham (q.v.), the first and only Inspector General of Naval Works. Bunce was appointed in 1796, the year following Bentham's appointment, but only the Portsmouth Block Mills can be attributed to him and he probably spent most of his time vetting building designs sent in by the dockyard officers who, traditionally, were responsible for these. Holl was considerably more involved as an architect, probably because by the time of his appointment Bentham had succeeded in exerting more control from London.

In the course of twenty years, Holl's title was to change. On the abolition of the Inspector General's department in 1812, Holl came under the direct control of the Navy Board and his original title of Civil Architect was changed to Surveyor of Buildings. This marked the beginnings of a proper centralised Architect's Department that remained little altered until the abolition of the Navy Board in 1832. Holl's period in office was marked by considerable building works, although by then the main expansion programmes for the fleet bases of Portsmouth and Plymouth Dock (Devonport) were nearly over. Probably for that reason, his main work was to be in the eastern dockyards.

At Devonport, only one surviving building can be identified with him. This is the ropery spinning house, gutted by fire in 1812 and rebuilt by Holl to a fireproof design with stone slab floors carried on cast and wrought iron columns and joists, and with a metal-framed roof; this 1200 ft. long building was partly destroyed by enemy action in 1941. At Portsmouth, he was responsible for the School of Naval Architecture (1815–1817) and he also designed three dockyard chapels: at Plymouth Dock (1814–1817), Chatham (1808–1810) and Sheerness; the latter two still stand. At Chatham he was also responsible for the surviving offices, a dock pumping station—probably in collaboration with John Rennie (q.v.)—the fireproof lead and paint mills and, in collaboration with Marc Brunel (q.v.), the sawmills. Holl was also closely involved in the design for the naval hospital at Great Yarmouth (1809–1811), built to cope with casualties from the Eastern Squadron. For the new Bermuda dockyard he designed the commissioner's house; this too used cast and wrought iron in its construction, not for fireproofing but to aid prefabrication. Apart from the stone for the walls, which was quarried locally, virtually everything else was prefabricated in London. This building was to be completed by his successor, George Ledwell Taylor.

Holl's chief work after the end of the Napoleonic wars involved the early planning stages for the Royal William Victualling Yard at Stonehouse, Plymouth, although the main works here were to be the responsibility of Sir John Rennie (q.v.). Bentham by 1811 had a poor opinion of Holl's architectural abilities but this in an unduly harsh judgement. All his buildings are well-proportioned and still essentially Georgian in character. However, his use of cast and wrought iron, whether for fire-proofing or to simplify prefabrication, show him to be well abreast of contemporary building technology.

Little is known about the Holl family other than the fact that he died in 1824.

J. COAD

[J. G. Coad (1989) *The Royal Dockyards 1690–1850*; Colvin (3)]

Works
For a full list see Colvin.

1804–1824. Chatham Dockyard, Officers' offices, Dockyard chapel and sawmills (in conjunction with M. I. Brunel); lead and paint mills and pumping station (in conjunction with John Rennie)
1806–1807. Bermuda Dockyard, Commissioner's House
1812–. Devonport Dockyard, ropery spinning house

HOLLAND, Samuel (1728–1801), surveyor and military engineer, was Surveyor-General of Canada from 1764 until 1801. Born Samuel Jan Hollandt in Nijmegen in 1728, Holland trained as a draughtsman and military engineer in the Dutch army, in 1755 joining a British regiment raised to enable 'foreign Protestants' to serve as officers and engineers in America. As Captain-Lieutenant, his first tasks there were to provide plans of New York, Fort Ticonderoga and Louisberg, where he was entrusted with leading the attack on the north east harbour in the siege of 1758, a rare honour given James Wolfe's notorious dislike of engineers. He was with him when he died at Quebec the following year and it was a matter of lifelong complaint that he was omitted from the romanticised painting of the event. He also worked with James Cooke, whom he had earlier instructed in the use of the plane table, on surveys of the gulf and river of St. Lawrence.

In 1759 he constructed a pallisaded fort at the mouth of the St. John River on the Bay of Fundy and in 1760 he was made Acting Chief Engineer although in 1761 he returned to surveying when he was asked to make a survey of all the settled parts of Quebec. His plans were recommended, both by the military, General Murray, 'an intelligent engineer … [who] … merits to be advanced' and by local merchants who described him as an 'able engineer and skilful surveyor' and approved his estimate of £1,117 to carry out the survey. In March of that year he was made Surveyor-General of the Province of Quebec and the northern district of North America north of the Potomac at a salary of £365 p.a.

His first task was to survey the newly acquired territories of the Isle de St. Jean, the Magdalenes and Cape Breton which included Louisberg, of which he presciently noted that robbed of its military function the soil and climate were so bad that no one would want to remain. The survey of Prince Edward's Island, the renamed St. Jean, recorded for the first time the principal rivers and harbours and recommended suitable sites for new towns, particularly the main harbour of Charlottetown which he designed and sited on Point La Joue. He also noted the timber and soil conditions, in particular 'a soft red clay which when exposed to the air for some time becomes harder and not unfit for building'.

In 1770 he moved south, surveying 3,000 miles of the New England coastline, estuaries and islands, emphasising coastal waters, sandbars and rocks. This ended in 1775 when he was forced to flee after refusing a commission in the rebel army.

Holland spent the next six years as one of the few military engineers with the army in America before being recalled to his surveying duties in 1781, first to survey all the ordnance land in Montreal and then in 1783 to make the first detailed survey of Ontario. This was necessitated by the post-war influx of united empire loyalists, who had all been promised free land. He was first sent to survey the condition of Fort Frontenac, constructed at Cataraqui, where he reported that 'the harbour is in every respect good and most conveniently situated to command Lake Ontario' and 'there was fine timber fit for building vessels (and) the soil is rich and fit for all purposes of agriculture'. Temporary barracks, warehouses and wharves were built and the still extant town plan of Kingston was laid out. The Cataraqui survey is also of interest as it was the first survey of the Lake Ontario end of the Rideau Canal.

Eventually this pressure of work led to the administrative division into Upper and Lower Canada. He remained Surveyor General of the latter in an increasingly supervisory capacity until his death in 1801. During this period he originated the system of land surveying in use throughout Canada until the early twentieth century.

He had married as his second wife Marie Joseph Rolette and had 10 children, the eldest of whom, John, worked with him on the Cataraqui surveys. His widow returned to Prince Edward Island and died there in 1824. As a posthumous tribute, the public of the island voted to name a new college of technology after him.

SUSAN HOTS

[Joseph F. W. Des Barres (1774–1780) *Atlantic Neptune* (includes all of Samuel Holland's New England charts); H. Scadding (1895) A notice of Samuel Holland ... based on a hitherto unpublished manuscript letter addressed to him by Lieutenant-Governor Simcoe in the year 1792, *Canadian Magazine*, October; W. Chipman (1924) The life and times of Major Samuel Holland, Surveyor-General 1764–1801, *Ontario Historical Society Papers and Records*, **21**, 11–90; W. Foster (1965) *Post-occupational History of the old French Town of Louisburg 1760–1930*; N. Shipton (1968) Samuel Holland's plan of Cape Breton, *The Canadian Cartographer*, **5**]

Publications
Publications include:

1762. *Description of the Island of Cape Breton and its Dependencies*
1765. *A Map of the island of St. John in the Gulf of St. Lawrence divided into Counties and Parishes to which are added Soundings around the Coasts and Harbour* (revised version published 1775).
1768. Observations made on the islands of St. John and Cape Breton to ascertain the longitude and latitude of these places ..., *Philosophical Transactions, Royal Society*
1781. *Plan of Quebec and Environs* (with Lieutenant-Colonel Debbieg (q.v.) and Captain Des Barres)

HOLLINSWORTH, James (fl. 1787–d. 1828) was a mason who, despite some setbacks early in his career, became the trusted assistant of John Rennie (q.v.). He is first recorded as a subcontractor on the final length of the Oxford Canal, south from Banbury, where he built Dashwood Lock at Northbrook for £253 12s 8d. He also undertook stone bridges at £50 each in partnership with one Coates, who may have been Richard Coates (q.v.) or his brother, George.

While working on this job he lived at Adderbury and it was there on 8 March 1787 that he married a local girl, Ann Stanfield, probably a relative of the Stanfield who was his competitor for work on the canal and built Twyford Bridge there for £65. A daughter, Mary, was baptised at Banbury on 6 January 1792 and was followed by five brothers and a sister born as far apart as Wiltshire and Argyllshire as their father moved around.

The Oxford Canal was opened in 1790 but Hollinsworth must have found other work as in 1792, when being awarded the contract for Town Bridge, Lechlade, he was described as a mason of Banbury. The scope of his work was extended in 1793 to include the flood arches over the fields south of the river. He is next heard of in 1795, when a problem arose with the alignment of Braunston Tunnel at the north end of the Grand Junction Canal. John Biggs had been the contractor for this tunnel and on giving up the contract, had been appointed to supervise its construction by direct labour. When William Jessop (q.v.), the company's Chief Engineer, noticed in February that year that the tunnel was not following the line set out above ground by

James Barnes (q.v.), Hollinsworth was charged by Jessop to see that the fault was corrected. For some reason he failed to do so and by June, Barnes had to take remedial action; a small chicane remains in the tunnel to this day. Jessop placed the principal responsibility on Biggs but also wrote: 'I am sorry to have been under the necessity for blaming Hollingsworth for excepting in this instance I have had reason to think he has been a valuable servant to the Company; unaccountable as his neglect has appeared to me I have yet full confidence in his future attention, and as I hardly know anyone that can so well do his duty I hope the Committee will endeavour to forget what is past'. In fact the Committee warned Hollinsworth that only Jessop's recommendation had saved him from dismissal and they retained half of his salary as a guarantee for better performance. (It might be noted that Hollinsworth's name was often misspelt in contemporary documents; in the record of his daughter's baptism it became Hollandsworth)

Work on the Grand Junction continued for another ten years but Hollinsworth may have felt that he had little future there, or that he would be more successful as a contractor once more. By May 1796 he was living at Seend in Wiltshire and working on masonry for the swing bridges of the Kennet & Avon Canal. By August next year the Company was looking for an additional inspector of masonry and Rennie, the Company's Engineer, recommended Hollinsworth. He offered his services at £150 p.a., over 30% more than he had been receiving on the Grand Junction. The Company demurred a little and it was only in October that he was appointed, at the salary he had requested, but to inspect earthwork and carpentry as well as masonry. The Company's books continue to show payments to him as a contractor, but trouble was in store again, for on 19 April 1798 it was minuted that 'On examining Contractors' accts It appeared many of them had extra charges to make, wch must be deferd till the Principal Engineer attends—But James Hollingsworth of Seend a Contractor for mason work whose contracts amount to £1,668 has been paid £2,397 wch is £729 over—he cannot by his own acct make out any extras of consequence besides what he has been allowed in the statement—the Committee therefore came to a conclusion to arrest him for the amount of the surplus & though he says he cannot pay it—yet they think it their duty to enforce the repayment if possible—and if the money is not recovered they think some example should be made of him to deter others from shifting in any way from the punctual performance of contracts in future or from entering into others they are not likely to perform ...'. He was arrested briefly but no further action was taken and the Company continued to pay small sums to a J. Wilson for Hollinsworth. Rennie seems to have taken him under his wing almost immediately, as a payment

to Hollinsworth of 1 guinea is noted in his accounts on 1 June 1798.

By 1799 Rennie was employing Hollinsworth on several commissions. Northam and Bursledon Bridges, east of Southampton, had been designed by George Moneypenny but there had been problems with poor foundations. Robert Mylne (q.v.) had been consulted and a new Act obtained but when the abutments continued to move, Hollinsworth made the survey for Rennie to make a report. While in the area he was also involved in measuring the work done by the contractors on the Salisbury & Southampton Canal, to which Rennie had been consultant since the previous year and which would shortly slide into bankruptcy without completing its works. Then at the end of the year he was in Berwickshire, viewing the remains of Whiteadder Bridge, at Paxton, which had been swept away for the fourth time in about thirty years. He also appears to have done work for the Crinan Canal, in Argyllshire, as he was chasing payment in 1800.

In April 1800 Rennie asked Hollinsworth to state his terms for a survey of what became the Croydon Canal but sent him instead to survey a new road at Datchet. He was instructed to bring a supply of clothes as Rennie would then send him to Lincolnshire for two to three months. This prospect turned out to be the Grimsby Haven, where he replaced George Joyce as Resident Engineer. The job lasted for more than a year and saw the start of seventeen years of almost continuous salaried employment on major projects where Rennie was Engineer. At Grimsby, the site had not been well chosen and the lock walls were too heavy for the ground. Rennie designed and Hollinsworth supervised the construction of a novel system of cellular walls on piled foundations which proved entirely successful. Sir Samuel Bentham (q.v.) later queried the priority of invention of his method of placing foundations under water, but Hollinsworth assured him that at Grimsby the work had all been done in the dry inside an earthen cofferdam. From Grimsby Hollinsworth went in June 1801 to London Docks as Assistant Engineer, James Murray (q.v.) being the Resident Engineer. Here he evidently gave satisfaction as his salary was raised from £250 to £350 p.a. after eighteen months in post.

Proposals to deal with the possibility of an invasion by Napoleon's forces through the low lying lands around Romney, in Kent, had been considered and rejected by the politico/military establishment when, in 1804, the danger had become very pressing. John Rennie was consulted and a scheme which became the Royal Military Canal was agreed. With backing at the highest level, Rennie proposed Hollinsworth with William Bough and John Dyson (qq.v.), two of the contractors from the London Docks, as a consortium to undertake its construction. They contracted to complete this 28 mile canal, wider and deeper than most navigation canals, with flanking drains and a military road in less than eight months. This would require at least two-thousand men to be recruited but half way through the contract period only four-hundred and fifty had been employed. There were problems, too, with water in the excavation and two steam engines ordered from Boulton & Watt for pumping were not delivered because of non-payment by Bough & Co. At a meeting on 6 June, five days after the works ought to have been completed, the consortium agreed to give up the contract and the army took over. Some nine-hundred and sixty soldiers were engaged by the middle of July and one and a half thousand one month later. In October the battle of Trafalgar removed much of the threat but work continued and it was only in April 1809 that the canal was more or less complete. With the advantage of control over military wages, the canal cost only about 15% more than the original estimate and Hollinsworth and his partners escaped without financial penalty. Taken at face value there was an extraordinary lack of realism about this project and it is hard to escape the belief that the documents do not tell the whole story.

After this strange episode, Hollinsworth returned to the Crinan Canal, this time as Resident Engineer. Work on the canal had stopped for a while in 1804 for lack of money but a Government loan had enabled it to restart. The line was declared open in 1809 but repairs and improvements to previous work continued to be necessary and Hollinsworth remained until January 1811. During this period he found time to make a survey of Tarbert Harbour for the Commissioners of Highland Roads and Bridges and another for the Berwick and Kelso Railway, for which an Act was obtained in 1811 but which made no further progress. After leaving Crinan he surveyed the navigable river Nith in Dumfriesshire and proposed improvements to cost £16,000. The Act for this work was passed in 1811 too; part of the work was carried out shortly afterwards under Walter Newall as Engineer.

The next six years were spent supervising the construction of Strand Bridge, London, renamed Waterloo Bridge in 1816, the year before its opening. At that time its nine spans of 120 ft. each were exceeded in masonry in Britain only by that of Pontypridd. The bridge itself was built within Rennie's estimate by Jolliffe & Banks (q.v.), though two further Acts of Parliament were necessary to raise additional capital for the construction of the approaches. During this period Rennie continued to employ Hollinsworth occasionally to undertake the necessary investigations for his own reports, such as the state of the Old Dock lock at Hull in 1813. At other times Hollinsworth was employed directly, probably at Rennie's instigation, such works being: the failure of Margate Pier in 1813; the arbitration between John Woodhouse (q.v.) and the Worcester & Birmingham Canal in 1815–1816; the arbitration between Thomas Baylis (q.v.) and the Gloucester & Berkeley Canal in 1819; and as a witness in the court case about the collapse of the first New Galloway Bridge in Kirkcudbrightshire.

The six years from 1818 to 1824 saw Hollinsworth acting as Engineer in his own right. The Portsmouth & Arundel Canal as built was a development of an earlier scheme by Rennie, but suffered from problems of construction and of the natural tidal channels which connected the canalised lengths. He also surveyed improvements to the neighbouring Arun Navigation. In order to supervise these works effectively he moved to Rumbeldswyke, near Chichester, and so it was as a local engineer that he provided a design for a long timber viaduct to connect Hayling Island to Havant. In 1821 he found time to travel to Dumfriesshire to report to Rennie on problems with an embankment at Annan and in 1822 he was involved away from home on a truncated version of another Rennie canal. The Bristol & Taunton Canal had been envisaged as a link in a continuous inland navigation from London to Exeter, using the Kennet & Avon, the Bristol & Bath and the Grand Western canals but substantial portions had not been built. In 1822 Hollinsworth prepared an estimate of £34,145 for a scheme to bypass the navigable river Tone from Taunton to Bridgwater but in 1823 he raised this figure to £60,000 and on this basis a revised Act was obtained in 1824.

The canal was built between 1824 and 1827 but now Hollinsworth was back in London on his last great project, the new London Bridge. John Rennie had provided the design for this before his death in 1821 and it fell to his second son, who became Sir John Rennie (q.v.) on the bridge's opening in 1831, to fill the post of Engineer. Its central span of 152 ft. was the largest in Britain when it was completed, though soon to be overtaken by a substantial margin by Grosvenor Bridge, at Chester. Hollinsworth was Resident Engineer and personally involved in the setting out of the piers and arches. He was recognised now as an authority on large bridge foundations and he reported on the continuing problems with Westminster Bridge in 1825 and again in 1827. His health began to fail, however, and by February of the next year he was unable to continue. He died three months later, on 11 May 1828. Sir John Rennie, who had a flattering phrase for almost everyone with whom he was associated, later referred to him as a 'worthy and excellent man'.

Of his sons, James Jr. succeeded him as Resident Engineer on the Crinan Canal in 1811 but was dismissed in 1812. John was co-signatory with John Rennie of London Dock accounts in May 1821 but was unemployed in 1822. His father wrote to John Rennie Jr. seeking work for him but he does not appear to have held any post of responsibility. George was first assistant to his father at London Bridge in August 1824 and became joint Resident Engineer with William Knight when his father could no longer act. For John Rennie Jr., Thomas made a survey of Eau

Brink Cut south of King's Lynn, in 1825 to determine what changes had taken place since its opening. In the same year and for the same employer he surveyed the Kensington Canal in London. This led to a new Act in 1826 for a canal of larger dimensions and Thomas supervised its construction. The youngest son, William, does not seem to have taken up civil engineering.

P. S. M. CROSS-RUDKIN

[Rennie reports, ICE; Rennie mss, NLS; RAIL 830/39, 842/2, 842/3, 842/4, PRO; Cambridgeshire RO; Hampshire RO; *Weale's Quarterly Journals*, **6/IV**, 1849; Edwin Welch (1966) *The Bankrupt Canal*; I. Donnachie (1971) *Industrial Archaeology of Galloway*; G. Phillips (1981) *Thames Crossings*; Tony Haskell (1994) *By Waterway to Taunton*; inf. ex. P. A. L. Vine]

Publications

1801. *Plan of the Town and Harbour Of Great Grimsby ...*
1820. *Plan of the River Arun Navigation from Burpham to New Bridge*

Works

c. 1787. Oxford Canal, subcontractor
1792–1793. Town Bridge, Lechlade, Gloucestershire, one arch, 50 ft. span, contractor, £480
c. 1795. Grand Junction Canal, Assistant Engineer, salary 2 guineas per week
1796–1798. Kennet & Avon Canal, contractor for masonry; from August 1797, Inspector of Works from Whaddon to Foxhanger, salary £150 p.a., but still being paid for contract works
1800–1801. Grimsby Docks, resident engineer
1801–1804. London Docks, assistant engineer, £250 p.a. then £350 p.a.
1804–1805. Royal Military Canal, contractor (with William Bough and John Dyson)
1805–1811. Crinan Canal, resident engineer
1811. River Nith, consultant
1811–1817. Waterloo Bridge, London, resident engineer, cost £937,393, including approaches, rebuilt 1937–1942
1818–1820. Portsmouth & Arundel Canal, Engineer, salary £500 p.a. plus £50 expenses
1820–1823. Arun River, Sussex, survey and improvements
1822. Langstone Harbour Bridge, Hampshire, design, built 1823–1824, now rebuilt
1822. Bridgwater & Taunton Canal, Engineer
1824–1828. London Bridge, resident engineer, completed 1831, rebuilt 1967–1973, original removed to Arizona

HOLT, Luke (*c.* 1723–1804), of Middlestown, West Yorkshire, carpenter and Resident Engineer, at first confined himself to contracts for timber work only but by 1766 he was in partnership with Robert Carr Jr. (q.v.) to provide carpenter's and mason's work. They were jointly appointed Surveyors of the Calder & Hebble Navigation on 18 May 1769, when John Smeaton (q.v.) had resumed his oversight of that concern. Holt was

dismissed in 1772, presumably because he was by then working for the Ramsden interest in Huddersfield; his survey in 1773 for Ramsden's Canal led to the successful passage of its Act in 1774.

He did not see construction through to completion, though, as on 24 March 1775 he was appointed Resident Engineer for the first enclosed dock to be built in Hull, on the recommendation of his old superior, Smeaton. There were problems with variability of the ground beneath the dock walls and Holt advised extra piling. The Commissioners were against this but within nine months Holt had been proved right and remedial action had to be taken. The Act allowed seven years for construction but this, the largest commercial dock in the country at the time, was completed in almost exactly half that period.

He remained in the East Riding and when the timber North Bridge at Hull required repairs he supervised the work. It cannot have been more than a palliative as the bridge was replaced in stone by 1787. In 1781 he provided a design for the drainage of the area around Market Weighton between Hull and York, which was carried out successfully in the two years following. By 1786 he was back in the West Riding and providing a complete design for Ossett Mill, which was probably the first company mill in the county.

Luke Holt had three sons and three daughters, born between 1758 and 1770, of whom the eldest, also Luke, died in infancy. His youngest son, Elias (born 1762), was associated with him in the contract to build a bridge at Mirfield but does not seem to have continued in that line.

P. S. M. CROSS-RUDKIN

[West Yorkshire RO; Hull Dock Company Letter Book 1774–1788, DDMW/7/404, Hull RO; RAIL/815, PRO; *Trans ICE*, **1**; and *Proc ICE*, **xli**, 87; J. H. Goodchild (1968) The Ossett Mill Company, *Textile History*, **1**, 46–61; M. W. Baldwin (1973) The Engineering History of Hull's Earliest Docks, *Trans Newc Soc*, **46**; Skempton (1984) The Engineering Works of John Grundy (1719–1783) *Soc Lincolnshire History Archaeology*, **19**]

Publications

1773. *A Plan of the intended Navigable Canal from Cooper Bridge to Huddersfield, in the County of York taken November the 6th 1773*

Works

1762. Calder & Hebble Navigation, contract (with John Topham), all timber work for five locks and three weirs, £1,240
1766. Shipley Bridge, contract to repair (with Robert Carr Jr.)
1766–1768. Wakefield House Of Correction, contract (with Robert Carr Jr.) to build, to design of John Carr and John Watson
1769–1772. Calder and Hebble Navigation, Surveyor (Resident Engineer) (with Robert Carr Jr.)
1773. Sir John Ramsden's (Huddersfield Broad) Canal, survey

1775–1778. Hull Dock (New Dock, later called Old Dock), Resident Engineer, cost £73,000

1780. Hull, North Bridge, repairs

1781–1783. Weighton Drainage And Navigation, Resident Engineer, to his own plan, top 3½ miles; built by contract; cost *c*. £8,000

1786. Ossett Mill, design, including ancillary works in River Calder

1798–. Ledger Bridge, Mirfield (with Elias Holt, his son), contract to build, to design of Bernard Hartley I (q.v.)

HOOF, William (*c.* 1788–1855), contractor, began working on civil engineering contracts, probably as miner, at an early age. He worked for Daniel Pritchard (q.v.) and in due course became his son-in-law and partner, and in the 1820s, as Pritchard and Hoof, they became known as specialist tunnelling contractors on canals, successfully completing some of the more important canal tunnels of the time.

In the 1830s the partnership split up, and Hoof continued as a railway contractor, initially as Hoof and Sons, and following the death of his eldest son, James Hoof, and William Hoof the younger, as Hoof and Hill, with Thomas and James Hill. Their works on the London–Croydon railway were notoriously difficult due to the cuttings.

Hoof died of apoplexy at his home, Madeley House, Kensington, on 11 August 1855 at the age of sixty-seven. At that time he had four surviving children, Egmont, Alfred, Mary Ann and Fanny Elizabeth.

MIKE CHRIMES

[Telford mss, ICE Library; RAIL 635/20, RAIL 414/272, RAIL 45/1, RAIL 615/2, RAIL 620/1, RAIL 623/48, PRO; 623/49; Prob. II 2219 (1855); House of Lords/House of Commons (1836) Minutes and evidence; London and Brighton Railway; *Railway Times* (1844) 24 February; 1 March, 44; 16 March, 44; 26 March, 48; *Min Proc ICE*, **x**, 1850–1851, 97–98; *Civil Engineer and Architects Journal*, 1855, 326]

Works

1819–1824. Thames and Medway Canal, Strood Tunnel (as Pritchard and Hoof)

1824–1827. Trent and Mersey Canal, Harecastle Tunnel (as Pritchard and Hoof)

1827–1829. Hall Green branch, Macclesfield/Trent and Mersey canal, 1½ miles

(?1828). Oxford Canal, Coventry (as Pritchard and Hoof)

1829–1834. East London Waterworks, Lea Bridge, Reservoir, etc.

1830–1833. St. Helens and Runcorn Gap railway, 12 miles (as Pritchard and Hoof)

1830–1832. Wigan Branch Railway, 7 miles (as Pritchard and Hoof)

1832–1835. Wigan branch maintenance contract (as Pritchard and Hoof)

1835–1838. London and Croydon Railway, 8¾ miles

1838–1841. London–Brighton Railway, Merstham, Clayton Tunnels

1843–. South London and South Western Railway, Chandlers Ford contract

1844–. Bristol and Exeter Railway, Whiteball Tunnel (Hoof and Hill)

1845. South Eastern Railway, Tunbridge Wells branch, £50,000

1845–1847. London–Brighton and South Coast Railway, West Croydon and Epsom railway, £88,000

1845–1848. London and South Western Railway, Salisbury–Bishopstoke, value of tender £89,000

1846. London and South Western Railway: Weybridge–Chertsey, value of tender £19,359

1847–1849. Shrewsbury and Birmingham Railway, Oakengates–Wolverhampton

1847–1850. South Staffordshire Railway, Walsall–Dudley (Hoof and Hill), 6 miles

1847–1852. Birmingham, Wolverhampton and Stour Valley Railway, contract 2

1848–1850. London and South Western Railway, Salisbury–Bishopstoke maintenance of permanent way

1851. South Eastern Railway, Hastings–Priory–Bunker Hill, including four tunnels (Thomas Hoof and Hill)

HOOKE, Robert, FRS (1635–1703), scientist, was born on 18 July 1635 at Freshwater, Isle of Wight, the son of John Hooke, a curate on the Isle.

From a list of his achievements in many fields, Robert Hooke certainly deserves to be better known and acclaimed than he generally is. One misfortune was being talented in too many fields—mechanical engineering, strength of materials, materials science and structural mechanics (as we would now call them), as well as astronomy, microscopy, chemistry, biology, physics, surveying and architecture. Another was being born a contemporary of Huygens (1629–1695), Wren (1632–1723) and Newton (1642–1727) who also excelled in some of the same fields. Towards the end of his life he became increasingly cantankerous and bitter that his considerable contributions to these many fields were not credited by many of those who built on his work. He had long and famous disputes with Huygens, and with Newton who retaliated by excluding any mention of Hooke in his famous Principia mathematica.. Unlike Wren he left no children with a vested interest in championing his achievements. The one compensation is that he is known today, in the Law that bears his name, throughout the world to everyone who has studied science at school.

Hooke left home for London aged thirteen following the death of his father. As a boy he had shown talents both for making things and drawing, and he began as an apprentice to the painter Peter Lely but this did not stimulate him sufficiently. He used his small inheritance to attend Westminster School, and progressed rapidly, learning Latin, Greek and Hebrew and the works of Euclid. In 1653, at the age of eighteen, he gained a place at Christ Church College, Oxford as a servitor—a student who earned his keep as the servant to another, wealthier student—and later became a chorister. He gained his Master of Arts in 1662. While at Oxford he met Christopher Wren, Robert Boyle (1627–1691) and John Wilkins who later were among the founder members of the Royal Society. Hooke first found an outlet for his remarkable inventiveness while working with Boyle. In 1655, at the age of twenty, he had devised and made a vacuum pump for Boyle's experiments with gases and it was on his recommendation that Hooke was later appointed Curator of Experiments at the newly formed Royal Society in 1662.

By this time Hooke had undertaken experiments on pendulums, invented the anchor escapement for clocks and the balance wheel escapement (though Huygens is often mistakenly given the credit), all in his quest to devise a mechanism for a clock accurate enough to enable mariners to establish their longitude. It was this work on springs that led him towards his famous law of elasticity, though at this time (he later claimed) he was not willing to publish it for fear of others learning to match the accuracy of his clock mechanisms. In the following year, 1663, at the age of twenty-seven, Hooke was elected a Fellow of the Royal Society and in 1665 was appointed Professor of Geometry at Gresham College where the Royal Society held its weekly meetings. It is worth noting that the title of this post referred not to geometry in the modern sense; this was in a time when geometry was seen as the basis for explaining the natural world and embraced much of what we might nowadays variously call physics, mathematics, astronomy and science.

Various surviving extracts of his diary over the next thirty years provide an extraordinary and vivid picture both of his work and of science at this crucial time in its development. Although not always obvious, there was indeed method underlying the bewildering variety of the experiments that were conducted and discussed—from making microscopes, to weighing air, to testing the strength of wires and wooden beams (sometimes under different atmospheric pressures), to blood transfusions in dogs. The founder members of the Royal Society had espoused the Baconian scientific method based on induction. Experiments were undertaken in sufficient number to reach the point of being able to generalise to predict the results of experiments not yet undertaken.. Hooke was one of the first experimenters in science to use the method that came to replace induction—the hypothetico-deductive method. A possible explanation or hypothesis is suggested before experiments are undertaken, and the experiments are designed and conducted especially in order to test the hypothesis. Hooke's long list of scientific achievements were based largely on his use of this method and made him not only one of the very first experimental scientists, but also one of the most successful. Throughout his life, and even while engaged on major engineering and architectural work following the Great Fire in 1666, Hooke was one of the world's most creative and productive scientific minds.

In 1665 he published his Micrographia in which he presented the hitherto unknown world under the microscope—Hooke's own new type of microscope offering unprecedented magnification and clarity—and his wonderful drawings of insects took the scientific world by storm. But this was not all he studied. He also drew and discussed the crystal structure of ice, snowflakes and quartz and drew an analogy with the way lead shot could be closely packed in a regular way. He discussed the possibility of making artificial fibres by a process similar to the spinning of the silkworm and he was the first to use the word 'cell' to describe the honeycomb cavities in cork. He was later one of the first to propose the idea of evolution in the development of species.

The list of Hooke's other achievements in science are astonishing. He observed interference fringes in mica and soap bubbles and deduced that light must behave as if it were a type of wave. In 1672 he observed and likewise explained the phenomenon of diffraction. He also observed and explained the phenomenon now known as Newton's rings at a time when Newton still adhered to the corpuscular theory of light. Hooke was the first to state that all matter expands when heated. He was also the first to suggest that air is made up of small particles, separated from each other by relatively large distances.

In 1666 he was the first to suggest that the force of gravity could be measured using a pendulum; he was also the first to propose a universal law of gravitation, and the inverse square law of gravity to describe the motion of the planets—some thirteen years before Newton. He also attempted to show the earth and moon followed elliptical paths around the sun, his only shortfall being not quite as good a mathematician as Newton to conclude his proof.

In 1674 Hooke invented the universal joint which is so fundamental to the transmission system in modern cars. He also devised the spring barometer, still so popular today, and invented the octant (similar to a sextant) which allows accurate observations of stars to be made for navigation purposes. When Charles II appointed him as one of the Commissioners for setting up the new Greenwich Observatory, Hooke also designed many of the instruments.

Hooke's experiments in 1662 and 1664 to determine the strength of timber beams and metal wires were in response to proposals by members of the Royal Society with an interest in ship-building and had no link to his earlier interest in elasticity. The aim was to test experimentally the theoretical ideas that Galileo had put forward in his book Two New Sciences (1638) which was just then being translated into English. One experiment proved Galileo's theoretical conclusion that the strength of a rope or wire was not inversely proportional to its length, as many people believed (or rather that it is, but only because a long rope or wire is statistically more likely to contain a weak point or flaw). The work on wooden beams was less robust from a modern viewpoint. As a result of altering more than one variable at a time, the chance influences of changing material, length, breadth and depth led the members to conclude that the breaking strength of a beam varies with the cross-sectional area, i.e. with the depth, d, not as d^2 which both Galileo and later science tell us. Although these experiments were to establish the breaking strength of timber beams, Hooke was asked on one occasion to observe the deflection immediately prior to fracture, but, as chance would have it, he observed no difference among the different woods.

Hooke returned to the elasticity of materials in his work of 1678 *De Potentia Restitutiva* ('Of Spring') in which he first published his famous law, though as an anagram—ceiiinosssttuv—for fear of divulging his secret—*Ut tensio sic vis* (as the extension, so the force). Hooke had used the same device a year earlier in A description of helioscopes and some other instruments (1676) when publishing his other major contribution to structural mechanics—the solution to finding the best form of an arch to span between two walls a given distance apart, with a given rise. It was, he said, the form of a hanging chain, though initially he had thought the shape was a parabola rather than a catenary—abcccdeeefggiiiiiiiillmmmmmnn-nnoorrsssstttttuuuuuuuvx or *Ut continuum flexile, sic stabit contiguum rigidum inversum* (as hangs the flexible line, so but inverted will stand the rigid arch).

With this list of scientific achievements it is hardly surprising, perhaps, that Hooke's architectural and surveying work has been neglected. Sadly, much of Hooke's best building work has been lost and it is also true that his work has been overshadowed by that of his close friend Christopher Wren whose better-developed and more effective social skills undoubtedly contributed to his greater acclaim among contemporaries.

If Wren's move from academic life into architecture had been rapid, Hooke's was virtually instantaneous. At the age of thirty-one, with no previous architectural experience, and less than two weeks after the Great Fire had been extinguished on 6 September 1666, he submitted to the City his plan for its rebuilding. The City's aldermen approved of it and, indeed, preferred it to that of the city surveyor, Peter Mills. On the strength of this model (now lost) he was nominated as one of three surveyors put forward by the City, to work alongside the three put forward by the Crown, one of whom was Wren. At this time the term architect was not current and the term surveyor applied not only to Hooke's work carrying out surveys of the damaged streets and properties, it also incorporated a variety of other work acting as the client's representative. These included negotiating contracts with land owners and contractors for the rebuilding and helping supervise much of this work on the City's behalf and directing the work of the builders and contractors for houses, drainage and other construction work. Within this scope would have been what we now call designs for buildings, and so began his career as what we would now call an architect. Within just a few years he was acting independently for private clients on a number of major buildings, notably: the Royal College of Physicians, Warwick Lane, London, 1672–1678; Bethlehem (Bedlam) Hospital, London, 1675–1676; London, 1675–1679, which later became the first home of the British Museum; and Ragley Hall, Warwickshire 1679–1683. Although often credited to Wren, a drawing in the British Library indicates that it was Hooke's design for the Monument to the Great Fire that was built in Fish Street Hill, London during 1671–1676.

To what extent Hooke was innovative in his designs is difficult to determine as much has disappeared. At Montague House he made use of inverted arches in the foundations, perhaps the first such 'modern' use, anticipating Wren's practice at Christ Church College Library and St. Paul's. His ingenuity is again evident in his involvement in the development and early use of the sash window. It cannot yet be stated with confidence that he originated the idea of counterbalancing the weight of a vertical sliding window using lead weights and pulleys built into the casement, but, at the very least, he did play a part in its improvement. As with so many things, he worked closely with Wren in this matter. Wren had seen counterweights fitted to existing sliding windows in the Queen's apartment at Whitehall in 1669 and a couple of years later had specified the counterbalance to be built into the casement in some new windows in the same building. It would have been uncharacteristic of Wren to have devised such a mechanism, whereas Hooke possessed such talent in abundance. The early sash windows (c. 1672) were divided by a central mullion and it was in Hooke's building for the Royal College of Physicians in London in 1673, that full-width sash windows were first used giving the proportions that were later widely used by Wren at the Royal Palaces and set the pattern which lasted for the following two centuries in English architecture. At his own Montague House Hooke records in his diary of 1675 instructing the joiner installing sash windows how to prevent them sticking.

Hooke's early work after the fire also included some major civil engineering work, the largest of which was his supervising the 'planning' and construction of the 40-ft. wide canalised section of the river Fleet which flowed south through the City into the River Thames. He made a model of his and Wren's scheme for constructing the wharves on each embankment, though this was deemed too weak by the contractor Thomas Fitch (q.v.) and Hooke supervised the four 100-ft. lengths of wharf that were constructed to compare the merits of four alternative designs.

Though no proof has been found, there is strong circumstantial evidence that Hooke also designed and supervised construction of two masonry arch bridges over the Fleet Canal.

In the 1670s he was earning some £500 p.a., equivalent to at least £50,000 today. As well as his own work, Hooke was constantly in discussion with Wren about his work, especially the London churches and, most importantly, the new St. Paul's cathedral. In his diary for 5 June 1675, Hooke noted that Wren had made use of 'my principle about arches' and revised his design for the dome accordingly.

As work in the City petered out, Hooke continued as an architect on many private houses, a church in Buckinghamshire and possibly at the Dockyard in Plymouth, although he never pursued the profession as his sole activity and never matched Wren's architectural genius. He was appointed as Surveyor to the Dean and Chapter of Westminster from 1691–1697. During all this time since the Great Fire he continued to hold his professorship at Gresham College and he died there on 3 March 1703, aged sixty-seven. He was buried in the church of St. Helen Bishopsgate.

BILL ADDIS

[R. T. Gunther (1930) *Early Science in Oxford*, vols. VI–VII, *The life and work of Robert Hooke*; M. I. Batten (1936–1937) The Architecture of Dr. Robert Hooke, *Walpole Society*, **25**; T. F. Reddaway (1940) *The Rebuilding of London after the Great Fire*; E. N. Da C. Andrade, (1950) Robert Hooke, *Proceedings of the Royal Society*, **201**, 439–473; S. P. Timoshenko (1953) *History of Strength of Materials*; M. Epinasse (1956) *Robert Hooke*; H. I. Dorn, (1970) The Art of Building and the Science of Mechanics: a Study of the Union of Theory and Practice in the Early History of Structural Analysis in England, Ph.D. thesis, Princeton University; M. A. R. Cooper (1997–1998) Robert Hooke's work as surveyor for the City of London in the aftermath of the Great Fire, *Notes and Records of the Royal Society*, **51**, 161–174; **52**, 28–38; **52**, 205–220; J. Heyman (1998) Robert Hooke, *Notes and Records of the Royal Society*, **52**(1), 39–50; H. Louw and R. Crayford (1998–1999) A construction history of the sash-window c. 1670–c. 1725, *Architectural History*, Part 1, **41**, 82–130; Part 2, **42**, 173–239; M. A. R. Cooper (1999) Robert Hooke—City Surveyor: an assessment of the importance of his work as surveyor for the City of London in the aftermath (1667–1674) of the Great Fire, Ph.D. thesis, City University, London; Colvin]

Publications

1665. *Micrographia* ('Small drawings')
1675. *A description of Helioscopes and some other Instruments*
1678. *De Potentia Restitutiva* ('Of Spring')

Works

As City Surveyor:

1670–1674. Fleet Canal and designs for Thames Quay

As architect—in addition to the following major buildings, Hooke designed many private houses:

1671–1676. The Monument, Fish Street Hill, London
1672–1678. Royal College of Physicians, Warwick Lane, London
1675–1676. Bethlehem (Bedlam) Hospital, London
1675–1679. Montagu House, Bloomsbury, London
1679–1683. Ragley Hall, Warwickshire
1680. Willen Church, Bucks
1690–1693. Aske's Hospital, Hoxton, London

HOPKINS, Roger (1775–1847), civil engineer, was the son of Evan Hopkin(s) of Llangyfelach, who was engaged in South Wales during the late eighteenth century in the construction of canals, tramroads and other works. Regrettably little is known about Evan, and indeed Roger's brother David. One assumes Evan had built up substantial experience associated with the mining industry of South Wales and trained his family. Evan was responsible for the design and construction of the steam powered inclined plane at Glynneath (1802–1805), more or less contemporaneously with the other early powered inclined planes on the Lancaster Canal. Such a work could only have been carried out by an engineer with substantial knowledge and experience of engineering. Early inclined planes connecting with the canal network were generally on the balance principle, but this utilised a Trevithick high pressure engine. The line was surveyed by Thomas Cartwright (q.v.) who would have been aware of that built by William Cartwright (q.v.) on the Lancaster Canal, and formed part of the tramroad that linked the Aber-nant ironworks and the Neath Canal. This was followed by a contract to build the Aberdare Canal in 1809 and, with his son David, a further tramroad on to the Aberdare ironworks,; the engineer was Edward Martin and Thomas Sheasby was resident engineer (qq.v.). The canal was linked by tramroad to the Abernant route. Aside from his near contemporary survey of the South Wales roads little else is known of Evan, and his son Roger had by this time emerged as an engineer in his own right.

Roger Hopkins married Mary Harris, daughter of the Reverend R. Harris of Pwllheli, Carnarvonshire, at St. Mary's Church, Swansea in 1806. In that year he became trustee to the Baptist Meeting House of the Swansea General Baptist Church. He was elected a Corresponding Member of the Institution of Civil Engineers in 1827.

Hopkins had apparently in 1804 been involved with the tramroad between Pen-y-darren and Abercynon in South Wales upon which Richard Trevithick (q.v.) tried the first railway locomotive steam engine. In 1810 he was engaged as engineer on the Monmouth Railway which was built partly through the Forest of Dean and not finally opened until 1816. While in this position in 1811 he was permitted to superintend work on the Severn and Wye Railway, where progress was poor. In 1814 he came to Bideford to plan a tramroad or railway for John, Lord Rolle, to run alongside the river Torridge to Great Torrington, but the project came to nothing.

On 9 April 1821 the Plymouth and Dartmoor Railway appointed Hopkins as assistant engineer. At that time William Stuart (q.v.) had been borrowed from Rennie and Whidbey (qq.v.) as part-time engineer. The company requested Hopkins to inspect and report on the state of the railroad between Crabtree and Jump (Roborough), where it appeared that Stuart had deviated from the agreed route. The findings of Hopkins's report were apparently so serious that it became necessary to amend the earlier Act. Hopkins was sent off to parliament to guide through a new Bill at the Lords' Select Committee stage and, with the Earl of Shaftesbury in the Chair, Hopkins stated to the committee 'that the necessity for the present application to Parliament for the Bill was not manifested until the month of April last, and originated in the impracticability the Railway found with proceeding with the work on the original line …'. A new Act was passed, 1&2 Geo IV c125, 1821, and William Stuart was dismissed on 8 October 1821. Hopkins completed the supervision of construction and the railway was opened on 26 September 1823. His eldest son Rice (1807–1857), born at Swansea in the year 1807, began his career as an engineer on the tramroad, at the age of fifteen, as a pupil of his father. He was elected a Corresponding Member of the Institution of Civil Engineers in 1836.

During 1823 Roger Hopkins and James Rendel (q.v.) competed for the approval of the Earl of Morley to be allowed to construct a bridge at Laira, Hopkins wishing to construct a multiple span wooden bridge and Rendel planning first a suspension, and then a five span cast iron bridge. In the event, Rendel was successful but at the same time Hopkins would build a wooden bridge between Shaldon and Teignmouth. Late in December 1823 Hopkins set off for an extended spell in London where for the next five months he assisted in the preparation of an estimate and tender to supply Dartmoor granite for the whole construction of the new London Bridge; thus the P&DR Company would benefit from this trade by transporting granite from Dartmoor to quays in the river Plym estuary. Still in London, he finalised the design in February 1824 for the proposed bridge between Shaldon and Teignmouth.

The bill to erect the bridge at Teignmouth received assent as 5 Geo IV c114 on 14 June 1824 and on 8 June 1827 the Teignmouth and Shaldon bridge was opened to traffic by the Duchess of Clarence. It cost £20,000 and measured 1,672 ft. in length, comprising thirty-three timber arches and masonry approaches; it was the longest wooden bridge in England and only surpassed in Europe by the Pont de Lyons. There was a swing bridge over the main channel.

In 1827 the Hopkins family was established in Plymouth at 5 Brunswick Terrace where Roger lived with his wife, Mary, and three sons; Rice, Thomas(d. 1848) and Evan. It is interesting to note that Evan, the only son not to become a civil engineer, in due course married the daughter of William Stuart, who Roger had displaced from the Plymouth and Dartmoor Railway. In 1828 Roger Hopkins designed and constructed the Plymouth Royal Union Baths which were opened on 1 May 1830 to much praise. However, within twenty years they were demolished to make way for the Millbay railway.

In 1831 he returned to north Devon to make a survey for the proposed Bideford and Okehampton Railway but this 21 mile route did not come to fruition. Another 1831 proposal that he prepared was a proposal for the formation of a floating harbour at Swansea together with a bridge across the river and the proposed new channel.

In 1831, Sir William Molesworth, a landowner, engaged Hopkins to survey a railway route from Wadebridge to Wenfordbridge with branches to Bodmin and Ruthernbridge. The Bodmin and Wadebridge line was Cornwall's first standard gauge railway and also the first with steam traction. It was opened from Wadebridge to Bodmin and Wenfordbridge on 30 September 1834.

In 1836 the partnership of Roger, Rice and Thomas Hopkins, based at Bath, owned coalmines in South Wales and built, owned and directed the Victoria Ironworks in Ebbw Vale. In March of that year they also proposed a railway from Tremoutha Haven to Launceston. In 1837 they built a 7 mile tramroad from their pit at Gwauncaegurwen in the Swansea valley to the Swansea Canal. However, by 1840, the Victoria Ironworks had failed and the works were handed over to the Monmouthshire and Glamorganshire Bank Company in repayment of a debt of £12,500.

By late 1842 Roger Hopkins had turned his back on South Wales and settled in France, at Boulogne. In March 1845 he wrote to David Mushet at Colford, Gloucestershire, who had recommended him to the Plymouth and Dartmoor Railway, asking him to join him in a new company to erect furnaces, not only at Boulogne, but all over France; he does not appear to have received Mushet's support. He returned to England and died at the home of his elder son, Rice, at 109 Upper Stamford Street, Lambeth, on 27 June 1847 in his seventy-second year.

BRIAN GEORGE

[DP 43 Railroad, Leigham to Jump, two plans, 1821; DP 65 Proposed roads, Dawlish, Kenton and Mamhead, 1825; DP 83 Alternative tunnel and pipeline to public baths in parish of St. Andrews, Plymouth, 1828, all at Devon County RO; Teignmouth and Shaldon Bridge, original design, February 1824, House of Lords RO; Membership records, ICE; E. Hopkin (1805) *An Abstract of the Particulars contained in a perambulatory Survey of above 200 miles of Turn-pike Road through the Counties of Carmarthen, Brecknock, Monmouth, and Gloucester*; Thomas Hopkins, Obit., (1849) *Min Proc ICE*, **VIII**, 11–12; Rice Hopkins, Obit. (1859) *Min Proc ICE*, **XVIII**, 192; Evan Hopkins (1854) Memoir of William Stuart, *Devon and Cornwall Society of Architects and Engineers*, Plymouth, 8 December; H. R. Paar (1963) *The Severn and Wye Railway*; R. A. Williams (1973) *The London and South Western Railway*, **1**, 101–106, plates 7–14; S. Hughes (1990) *The Brecon Forest Tramroads*; K. S. Perkins (1992) Opening up South Devon: the Hopkins Connection, *The Devon Historian*, **45**, 9–17; K. S. Perkins (1995) Roger Hopkins, Civil Engineer 1775–1847, *The Devon Historian*, **51**, 22–28; R. A. Otter (1994) *Civil Engineering Heritage: Southern England*]

Publications

1831. *The Report ... on the Works proposed ... for the Formation of a Floating Harbour ... Swansea*
1831. Map of the proposed railway from the proposed harbour ... to Launceston, in: *The Duke of Cornwall's Harbour and Launceston and Victoria Railway Prospectus*
[1831]. *Map of the proposed Harbour to be called by Command of His Majesty the Duke of Cornwall's Harbour*, ICE DRWG 2173

Works

1802–1805. Glynneath inclined plane and tramroad, *c*. 8 miles, £21,500
1804. Tramroad from Pen-y-darren and Abercynon
1810–[1816]. Monmouth Railway
1811–[1812]. Severn and Wye Railway
1821–1823. Plymouth and Dartmoor railway
1824–1827. Teignmouth and Shaldon bridge
1828. Plymouth Royal Union baths
1831–1834. Wadebridge to Wenfordbridge and Bodmin railway
1836. Victoria Ironworks, Ebbw Vale
1837. Bath and Weymouth railway (engineer)

HORE, John (1680–1763), engineer of the Kennet and Bristol Avon navigations. It is probably he who was baptised at Thatcham (near Newbury) on 13 March 1680 as John, son of John Hore,. possibly the yeoman of that name and parish whose will is dated 1708 leaving property to his wife Frances and (only) son John. 'John Hoare Sr.', one of the original proprietors of the Kennet navigation, is perhaps the maltster of Newbury—his will is proved 1721—who may be a relative but is not the father of the engineer.

Hore is thought to have been initially a millwright but nothing is known for certain about his early life before 1718 when, in his own words, he was appointed 'Surveyor and Engineer to make the Kennet navigable' from Reading to Newbury. In this length of 23 miles the river has a gradient of 66 in. per mile and a correspondingly swift current with many shoals. Under an Act of 1715 the proprietors wasted a good deal of money by employing an 'unskilful Person' to build locks at

the ten mills along the river, a method quite unsuited to the circumstances. Hore realised that it would be necessary to provide many more locks, most of them on long cuts. As completed by him in 1723 the Kennet navigation had a total length of 18½ miles, of which 11 miles were in cut, with twenty locks; there was also a wharf at Aldermaston and a wharf and barge basin at Newbury. His appointment carried a salary of £60 p.a. plus expenses. After completion he was to be Surveyor of the Works for life, at the same salary, and to have 'an additional Consideration' as Book-Keeper or wharfinger at Newbury.

Surveying, planning and setting out the cuts, and negotiations for land purchase, would have taken some time. Work began in 1719. In 1720 a mob from Reading destroyed part of the works, and in the winter 'extraordinary floods' caused further damage. An Act was sought, and obtained in March 1721, for a two year extension of the time limit imposed by the original Act.

In July 1721, reporting progress, Hore wrote that 'Sheffield and Widmead Cutts are opened, and the two large boats employed raising Colthorp Bank, the dams are made, the screw fixed, and great part of the lock at Aldermaston. I hope next week to begin at Padworth. The locks at Fobney and Southcote are framed, as are the timber work for the two brick locks to be placed in Padworth field except the gates which must be those from the Old Locks when taken up and altered'.

This refers to the Archimedes screw used in pumping out the lock pit excavations. The cuts, embanked where necessary, had a width of 54 ft. at the water surface and a depth of 5 ft.; the navigable part of the river was typically 70 ft. wide. Apart from three brick walled locks the rest were 'turf-sided'. The walls and aprons of the head and tail gates were framed in timber, and the sides of lock chamber were timber sheeted to about 2 ft. above low water level from which height the sides were sloped at 45° and turfed.

In the late eighteenth century the Kennet locks were 19 ft. wide, with an effective length of about 114 ft. (between the head apron and the tail gates) and 4 ft. water on the sills, admitting barges 109 ft. × 17 ft. carrying 110 tons at a draught of 3½ ft. It has been suggested that these dimensions apply only to a rebuild after 1767, but the original locks cannot have been very different and could well have been of this size since in 1725 barges of 100 tons burden are said often to have passed fully laden from Newbury through Reading to London.

In 1720 an almost entirely new set of Proprietors took over, and Bryan Philpott was appointed Accomptant and Receiver of Duties. However, in the initial stages it was Hore who had to negotiate with the landowners. In doing so he incurred considerable expenses but seems to have been unable to present a proper statement. The management committee of 'governing Proprietors', who held their meetings in London, used this as an excuse to dismiss him from his office as Surveyor in 1724. A sum of £840 was later agreed but even by 1730 the Proprietors had not paid a shilling to Hore; nor had they yet settled for damages caused during construction. Hore prepared a 'Case', the proceedings went to arbitration in 1731 and a satisfactory compromise was reached; in 1734 he was reinstated as Surveyor at an increased salary of £100 p.a. He retained his position as wharfinger at Newbury throughout. All this tends to obscure the fact that Hore had done an excellent engineering job on the Kennet. In evidence to a House of Commons committee in 1727 Philpott stated that £40,403 had been expended in making the navigation.

Within a year of his dismissal as Surveyor of the Kennet, 'John Hoare of Newbury' was employed in the 'Direction and Chief management of the Works' on the Bristol Avon Navigation. Bath Corporation obtained an Act in 1712 for making the river navigable from Hanham Mills, the head of (tidal) navigation from Bristol, up to Bath. Nothing was done, however, until 1724 when the Corporation granted the powers conferred by the Act to a new body of proprietors among whom Ralph Allen (1694–1764) played a leading role. They met in Bath on 31 December 1724, called up the first round of subscriptions and in March 1725 decided to bring in Hore as their engineer. They wrote asking him to bring 'Proper Instruments … for taking of a Thorough Survey of the River. To mark out all the Grounds which will be anyway made use of in this undertaking and to prepare an Estimate of the whole Expence'. He was to receive 9 guineas per week for attendance and journeyings. His assistant Mr. Downs, paid £1 4s per week, would be full time on site and in charge in Hore's absence. Edward Marchant was to be chief mason at the high rate of 2 guineas per week during his attendance.

With a river of moderately good depth and a gradient of just over 30 in. per mile (30 ft. fall in 11½ miles) the engineering problems were less formidable than on the Kennet. The existing mill weirs could be retained. Five locks would be built on short cuts beside the weirs, with falls of 3–6 ft. But near Twerton there were several mills, on both sides of the river, and two weirs about a ¼ mile apart. Here a cut some 600 yd. in length (known as Weston Cut) had to be made with a single lock at the lower end accommodating the total fall (9 ft.) of the two weirs.

The six locks were masonry-walled, about 98 ft. long by 18 ft. wide with 6 ft. of water on the sills, taking barges up to 140 tons. Work got under way in the summer of 1725 and the navigation was opened in December 1727 although the quay at Bath did not reach completion until 1729. The total cost was about £12,000. Receipts from tolls in the period 1730–1737 averaged £978 p.a. The two navigations were linked by the Kennet & Avon Canal built 1794–1810 under the direction of John Rennie (q.v.).

In 1729 a group of Gloucestershire gentleman and clothiers considered the possibility of obtaining an Act for making the Stroudwater navigable from the river Severn up to Stroud and called on Hore to advise. It was immediately obvious to him that the river itself was inadequate for the purpose. He therefore drew up a scheme for what was in effect a canal 8½ miles long from Framilode on the Severn to Wallbridge near Stroud with twelve locks (average fall 7½ ft.) and a series of long cuts 33 ft. wide and 5 ft. deep, suitable for 60 tons boats, at an estimated cost of £20,000. On this basis a Bill was examined by Parliament. Hore gave evidence to the Commons on 11 February 1730 and to the Lords a month later. The Act duly passed in May, but opposition threatened by millers on the river led to the scheme being laid aside. After two false starts the Stroudwater Canal was eventually built between 1776 and 1779.

The river Chelmer Navigation, from Maldon 14 miles up to Chelmsford, follows a rather similar history. In July 1733 Hore surveyed the river and prepared two schemes, the first to make the river navigable at a cost of £9,300, the second to cut a canal on a new line. The latter would cost £12,800 but he strongly advocated its adoption. Here again nothing happened, despite a new attempt leading to an Act in 1766, until finally the navigation was built 1793–1797.

Back on the Kennet in 1734, Hore found the navigation in poor repair after ten years in neglect. At that time Lady Forbes, widow of one of the early proprietors, began taking an interest in the management. Francis Forbes, a relation, was appointed Superintendent at Newbury and her business-like attitude soon helped to improve the fortunes of the undertaking. Hore carried out the necessary repairs and in 1740 a central carpentry depot and yard was established at Aldermaston. Tolls were now bringing in about £2,000 p.a., representing an annual carriage of some 10,000 tons.

Hore's connection with the Kennet ceases in 1761. He is presumably the 'John Hore of Ham Mills in the parish of Thatcham… gentleman' whose will is dated April 1753, leaving land and property and shares in the River Kennet to his sister-in-law, Ann Cook. He died on 12 April 1763 and was buried at Thatcham four days later.

A. W. SKEMPTON

[Bryan Philpott (1727) evidence, *JHC*, **20**, 708; John Hore (1730) evidence, *JHC*, **21**, 437; W. E. Mavor (1808) *General View of the Agriculture of Berkshire*; F. S. Thacker (1932) *Kennet Country*; T. S. Willan (1936) *River Navigation in England 1600–1750*; K. R. Clew (1968) *The Kennet & Avon Canal*; C. Hadfield (1969) *Canals of South and South East England*; J. Boyes and R. Russell (1977) *Canals of Eastern England*; M. Handford (1979) *The Stroudwater Canal*; A. W. Skempton (1996) *Civil Engineers and Engineering in Britain, 1600–1830*; Brenda J. Buchanan (1996)

The Avon navigation and the inland port of Bath, *Bath History*, **6**, 63–81; P. A. Harding and R. Newman (1977) The excavation of a turf-sided lock at Monkey Marsh, Thatcham, *Industrial and Archaeological Review*, **19**, 31–48]

Works

1719–1723. Kennet Navigation, £40.000
1725–1727. Bristol Avon Navigation, £12,000

HORNE, James (fl. 1721 d. 1756), became deputy to Nicholas Dubois, Master Mason in the Office of Works, in 1721 and was appointed in 1726 to direct the civil engineering works in the transformation of the deer park of Kensington Palace into one of the earliest landscaped gardens in England. Henry Wise (c. 1653–1738) and Charles Bridgeman (fl. 1709–d. 1738), chief gardeners to George I, prepared the general plans, which involved the planting of several thousand trees chiefly under Bridgeman's supervision. The engineering works were carried out in two stages: (1) the Round Pond, a 7 acre basin near the Palace, and that part of the Serpentine lake within Kensington Gardens properly known as the Long Water, 1726–1728; and (2) the dam and lake in Hyde Park, 1730–1731. Strictly speaking this part only is the Serpentine but commonly the name now applies to the entire lake, 40 acres in area.

Charles Wither, Surveyor General of H.M. Woods and Forests, had control of the scheme as a whole and submitted the accounts to the Treasury. These are summarised by Rutton (1903, 1904). Payments to Horne, additional to those for survey and estimates, were 'for his care and diligence in directing the Works, looking after the Workmen, making up and settling all the accounts'. He received fees of 12s per day as 'Clerk of the Works'. Labourers were paid 1s 8d and the foreman 3s per day.

The Gardens and Park are situated on gravels of an ancient river terrace of the Thames. A tributary stream, the West Bourne, has eroded its valley through the gravels into the underlying London Clay, leaving gentle side slopes and alluvium in the valley bottom. In the early years of the 18th century a linked series of the ponds existed in the valley.

In the first campaign 59,492 cu. yd. were excavated for the Round Pond, and for the Long Water the valley (hitherto occupied by the four uppermost ponds) was deepened and widened by excavating 31,479 cu. yd. from the bottom and 6,515 cu. yd. from slope on the east side, most of the material being used in levelling the ground for a lawn around the Pond and building a wide embankment across the valley as a temporary dam (at the site of the present bridge): the cost of these works was about £5,000 and Horne's fees £203.

The second campaign chiefly involved building the main dam, or 'Head' as it was called, at the furthest possible place downstream: 300 yd. north of Knightsbridge and ½ mile east of the Long Water embankment. Horne drew up the plans in July 1730 and work began in September, employing about two-hundred men; the opening took place in May 1731. The accounts were closed in November, the cost having been about £4,800 with Horne's fees £259.

The dam was raised about 1 ft. in 1827 when the Long Water embankment was replaced by the present bridge and water level in the Serpentine increased. Otherwise it exists essentially in its original form with an impermeable core of rammed clay. Though there may be earlier examples of this type of earth dam it is the first for which records are available. As built in 1731 the dam had a maximum height of 17 ft. above ground level and 19 ft. above stream level, with 3 ft. freeboard. As shown in a map by John Rocque of 1736 it had a length of about 515 ft. and a crest about 60 ft. wide forming a terrace and walkway.

From the valley upstream of the dam 74,644 cu. yd. earth were dug, carted and placed in the shoulders of the dam and banks along the sides of what was to be the lake, at a cost of 9d per cubic yard, the material being compacted simply by the carts and horses crossing back and forth. In addition, 6,287 cu. yd. of clay were dug and carried to the dam, laid down and rammed, to make the core or 'clay wall', at 3s 9d per cubic yard. The great difference in unit costs indicates the special nature of the core: selected (London) clay would have been dug, barrowed, laid in fairly thin layers and compacted by rams or 'beetles' similar to those used by paviours. From the quantities it can be shown the core was about 18 ft. wide.

The next known dam of this type was built by John Grundy (q.v.) in 1748 with a maximum height of 18 ft. and a core of clay 'well rammed and watered'. It impounded the 'Great Water', a lake of 36 acres extent in the Grimsthorpe estate (Lincolnshire) of the Duke of Ancaster.

Horne's later works include three or four churches designed in a handsome but rather pedestrian Palladian style (Colvin, 1978). He held the post of surveyor of Westminster Abbey from 1746 until 1752 and of the Naval Hospital at Gosport between 1752 and 1756. He died on 19 April 1756.

A. W. SKEMPTON

[John Rocque (1736) *Plan of the Palace, Gardens and Town of Kensington*; W. L. Rutton (1903) The making of the Serpentine, *The Home Counties Magazine*, **5**, 81–91, 183 195; W. L. Rutton (1904) The making of Kensington Gardens, *The Home Counties Magazine*, **6**, 145–159; Colvin (2)]

Works

1726. Kensington Gardens and Hyde Park, the Round Pond, and Serpentine lake and dam, £10,000

HOUGHTON, James (fl. 1776 on), was the name of two canal contractors, father and son, which occurs in the minute books of several canal companies, but whose full careers await more research. By his own evidence, James Sr. started work on canals from 1760 or even earlier, but the first mention found is as one of four people who made agreements with the Stroudwater Canal Company in 1776 to supervise gangs of cutters to dig out the bed of the canal and he appears in the accounts from August 1776 to March 1777. Another ganger there was James Bough (q.v.), who recommended Houghton to John Pinkerton (q.v.) in 1785, who engaged him as his Agent in charge of digging on the Fradley Heath line and subsequently on the Minworth section of the Birmingham and Fazeley Canal. He was employed as an additional Superintendent of Works on the Shropshire Union Canal in February 1789, when the Surveyor, John Lawden, was under threat of dismissal, but both managed to survive the criticism of the Committee. Early in 1793 he was undertaking small contracts in the preparations for King's Norton Tunnel in the Worcester and Birmingham Canal, but in July his was the lead name of a team of contractors who agreed to cut 6 miles at the eastern end of the Shrewsbury canal. Rather naively, they offered to do the work at their own tendered rates, or at rates to be fixed by William Jessop and Thomas Dadford (qq.v.), who were expected to be the Engineers to the canal. The rates agreed must have been satisfactory as the team soon undertook the remainder of the line to Shrewsbury and the minutes record frequent and substantial payments for work done. With one of his partners, Thomas Ford, he next won a 4 mile stretch of the Ellesmere Canal in 1794; they were unsuccessful later in the year in another tender but were awarded another, larger contract in March 1795. James Houghton Jr. was working with them but for some reason the canal company ordered in August that he be dismissed from the works.

In June 1795, between the last two events, Houghton and Son won the contract for the final length of the Somersetshire Coal Canal, so it seems that James Sr. must have travelled south to manage it. In any event it appears to have gone well for him because the canal company's accounts later show £8,145 paid to Houghton, on a contract valued initially at £3,650. In 1796, though, when he was appointed Engineer of the Warwick & Birmingham Canal after William Felkin had been dismissed, James Sr. was said to be 'of Old Heath, near Shrewsbury'. His salary at Warwick was £250 p.a. but he did not manage to see the project through to completion; he was replaced by the company's Clerk, Philip Witton, who rolled the two posts into one for the salary which Houghton had been earning.

Houghton is noted subsequently on contracts in the south-west of England. On the Kennet & Avon Canal he worked with Gregor McGregor at least as early as 1804 and took a sizeable contract in his own right in 1806. In the Grand Western Canal contract in 1832, by which time it must have been James Jr., his address was given as Bishopshull (near Taunton). The contract was drafted after the start of work, however, and he may have moved there to manage the works.

P. S. M. CROSS-RUDKIN

[RAIL/827, 842, 868, 869, 882, 886, 1019, PRO; D1180, Gloucestershire RO; Rennie mss, NLS; Rennie reports, ICE; J. Pinkerton (1801) *Abstract of the cause* ...; K. R Clew (1970) *The Somersetshire Coal Canal and Railways*; T Haskell (1994) *By Waterway to Taunton*; inf. ex. Dennis Dodd]

Works

1785–1786. Trent & Mersey (for Coventry) Canal, Fradley Heath line, Contractor's Agent
1786–1787. Birmingham & Fazely Canal, Minworth section, Contractor's Agent
1789. Shropshire Canal, superintendent of works
1793. Worcester and Birmingham Canal, a contractor at King's Norton
1793–1796. Shrewsbury Canal, Contractor (with John Houghton and Thomas Ford), Wombridge–Longdon section at a Bill of rates and Longdon–Shrewsbury section for £20,170
1794–1795. Ellesmere Canal, Contractor (with Thomas Ford); had been paid £22,660 on Llanymynech and Weston Lines by 1805
1795–1796. Somersetshire Coal Canal, Contractor (Houghton & Son) for 2 miles near Camerton, £3,650
1796–1798. Warwick & Birmingham Canal, Engineer
1806–1809. Kennet & Avon Canal, Contractor for Milton lot, £10,326
1813. Grand Western Canal, Contractor (with William Pritchard for Holcombe lot, by self for two Lots between Halberton and Tiverton)
1824–1826. Bridgwater & Taunton Canal, Contractor (with William Wawman)
1831–1832. Grand Western Canal, Contractor (with William Wawman and Thomas Lear) for five lots (8 miles out of 12), contract price for first three lots £6,497

HOYTE, Samuel (fl. 1655–1669), Cobb warden, was responsible for the extensive work on the Lyme Regis's coastal works in the period *c.* 1655–1669, including improvements to the Cobb, an important civil engineering work with origins in the thirteenth century. Originally a timber framework with loose stones for fill, in the seventeenth century increasing reliance was placed on masonry rather than timber. Masons such as Hoyte were in regular demand for coastal works elsewhere because of their expertise in pier construction.

MIKE CHRIMES

[R. A. Otter (1991) *Civil Engineering Heritage: Southern England*; Keystone (Historic Buildings Consultants) (1993) *The Cobb, Lyme Regis, Dorset*]

Works
1655–1669. The Cobb, Lyme Regis

HUDDART, Joseph, FRS, Captain (1741–1816), was born at Allonby, Cumbria, on 11 January 1741, the only child of William (1704–1762) and Rachel (*c.* 1705–1792). Huddart's father was variously a farmer and shoemaker with interests in the local fisheries; his uncle, Joseph Huddart, was a pawnbroker living in London and had close family contacts with the East India trade. It is said that from his mother 'he inherited a vigorous constitution, and a strong and determined spirit; from her he, moreover, inherited frugal habits and those correct and moral principles which he maintained through life, and which rendered him an upright man and essentially a gentleman'.

His parents were able to provide him with a reasonable education at the local boarding school under the tuition of the Reverend Thomas Wilson and his son, from whom he learnt a great deal about astronomy, an interest which he maintained throughout his life. Like several of his contemporaries who became involved in civil engineering, Huddart had an enquiring mind. He learnt singing from a travelling music master and developed this by teaching himself musical theory and learning several instruments. He began building models, including a scale model of a local mill then under construction, and later, having acquired Mungo Murray's Treatise on shipbuilding and navigation, made a model of a seventy-four gun ship. While looking after his father's cattle, Huddart pursued his mathematical and scientific studies, using a portable desk he had built himself.

While there were some early hopes of Huddart entering the clergy, or becoming a shoemaker, his life was transformed in 1756 by the arrival of a large shoal of herrings in the Solway Firth which caused his father, and other local leading citizens, to invest in the fishing industry. Huddart was able to master the art of navigation and became involved in the white fish trade, probably taking command of small vessels while a teenager. In 1762 his father died and he inherited his financial interest in the fisheries. In that year he took command of the ship Allonby, conveying fish caught by the company to Cork. His sailing experience made him increasingly aware of the shortcomings of existing charts, reflected later in the production of an important series of his own. At this stage his burning ambition appears to have been to build his own ship and in 1767–1768 he built a brig, the Patience, intended for the coal trade at Maryport, and in the late 1760s became involved in transatlantic trade with North America, expanding his practical knowledge of the science of navigation.

In 1771 Huddart visited his uncle in London and met Sir Richard Hotham, his cousin's husband, who was also involved in the East India trade. Hotham was clearly impressed by Huddart's knowledge of naval architecture and the sea and by late 1773 he was able to persuade him to enter the service of the East India Company as a ship's officer, initially on a trial basis. In the event this was to become the focus of Huddart's career at sea. Aside from a short break in 1777 to complete for publication a chart of St. George's Channel, the remainder of his career as a naval officer was spent in the East India trade.

By 1788 Hotham had determined to retire from shipping and sell the Royal Admiral of which Huddart had been in command; Huddart agreed to retire, taking £1,000 in compensation. He had already made a name for himself for his scientific pursuits, mostly concerned with astronomy and navigation, and in the East Indies trade this was associated with his determining the longitude of Bombay more accurately than before, surveying the coastline from Bombay to Coringo, surveying the river Tigris from Canton to Sankeet, and improving navigation guidelines from the East Indies to China. In the course of his voyages he went out of the way to observe a solar eclipse in Chinese waters. On his return to England he continued his scientific interests which included clock and instrument making.

Huddart's skill in marine surveying being well-known, he was increasingly consulted about navigation issues. In 1790 he surveyed the Hasborough Gatt shores off Norfolk for Trinity House and the following year he was elected to the Elder Brethren. He became one of the most active of the Brethren, and remained so until his death, carrying out various surveys for floating lights, as well as charts and sailing directions; locations included Goodwin Sands, Carn Braes (Lands End), Flamborough Head, South Stack (Anglesey), and Hurst Point. Generally the lighthouses themselves were the work of Samuel Wyatt (q.v.) but Huddart superintended the construction of an additional light at Hurst Point in 1812.

Huddart's hydrographic surveys for Trinity House formed only a small proportion of his output and his surveys of the West Coast of Scotland for the British Fisheries Society (1789–1794) and False Bay, Cape of Good Hope (1796) were recognised for their authority. He was clearly an accurate observer and was widely recognised as such by his contemporaries.

In a civil engineering context Huddart is best known for his harbour surveys and reports which brought him into contact with leading civil engineers of the time. His earliest reports were in connection with harbours with which he must have been familiar from an early age: Maryport (1786) and Whitehaven (1789). His recommendation as regards improving these harbours by extending and modifying piers appear to have been followed, involvement at Whitehaven continuing until 1804. Elsewhere Huddart was consulted about Humber Dock at Hull, St. Agnes, and Swansea Harbour. John Rennie (q.v.) was engineer at Hull but at Swansea Huddart seems to have acted as engineer. Huddart also reported on Leith Harbour, again apparently in collaboration with Rennie.

In 1808 Huddart surveyed Holyhead Island and the Lleyn peninsula for alternative sites for a crossing to Ireland, following an earlier report by Rennie on Holyhead. He recommended Holyhead and identified a suitable site for a lighthouse, that at South Stack, subsequently erected to designs by Daniel Alexander (q.v.). Rennie was commissioned to draw up designs for Holyhead and Huddart was then asked to survey a suitable harbour site at Howth, near Dublin, again with Rennie, who was also investigating the state of the Royal Dockyards, reporting in 1807.

With Robert Mylne (q.v.), Huddart was then asked to survey Portsmouth, upon which he had already reported in 1796, when he recommended a new entrance. In 1808 he also surveyed Woolwich Dockyard, where access was affected by silting, and in 1815 Sheerness; in both instances, Rennie was engineer for the works as executed. Huddart, like Rennie, was also involved in the development of London's Docks and was a director of both the London and East India Dock Companies.

Huddart became involved in the Fens in 1793. He was asked to report on the state of Boston harbour, recommending a cut to improve the navigation up to the port. His recommendations were the basis for subsequent reports by John and Sir John Rennie (qq.v.). More controversially Huddart, acting as umpire, became involved in the dispute between Sir Thomas Hyde Page and Robert Mylne (qq.v.) over the Eau Brink Cut and King's Lynn harbour. There he made his own recommendations, which seem to have favoured Mylne more than Page.

Considerable as Huddart's hydrographic surveying achievements were, his most enduring legacy was probably his invention of rope making machinery. He had become interested in rope manufacture during his East India voyages where he observed that the exterior strands of anchor cables bore most of the strain due to the method of twisting the yarns. Prior to twisting, they were of the same length, but the outer yarns were shortened in twisting and took most of the strain. Huddart worked on a method of proportioning the original length of the yarn so that it would be of equal length and bear an equal strain when twisted. After some experiments in his own garden, Huddart offered his invention, without success, to the East India Company and Admiralty, before in February 1800 setting up a business, Huddart and Company, in Limehouse. In this enterprise he had three partners: Sir Robert Wigram, Charles Hampden Turner and a Mr. Woolmore. In 1793 Huddart had taken out his first patent, which envisaged some of the work being done by hand in the traditional method of a ropewalk, but in his second patent of 1799 he devised a means of laying a rope with a stationary machine. The whole method involved four processes and the machinery for manufacture was designed by Rennie. Huddart's patent was contested but his invention was vindicated and ultimately the original machine was purchased by the Admiralty and installed in Deptford Dockyard. Such was the general interest among civil engineers at the time that the Institution of Civil Engineers invited descriptions of it as being suitable for the Telford Prize in the 1830s.

If the rope-making machinery was the most practical fruit of Huddart's labours he was also responsible for a number of scientific investigations. These included an early investigation into colour blindness and Observations on horizontal refractions which affect the appearance of terrestrial objects, and dip, or depression of the horizon of the sea, was effectively an early description of the phenomenon of a mirage. These were both reported to the Royal Society, of which Huddart was elected a Fellow in 1791. He also displayed an interest in ship design and after his retirement looked into ship resistance, building two experimental vessels to assist in his research, which continued almost until the end of his life.

Huddart clearly enjoyed a place of high regard and affection among his contemporaries. His partner in the rope works and other ventures, William Cotton, remarked that the 'distinguishing feature of Captain Huddart's mind was an understanding integrity of purpose. In his investigations he had no preconceived views, and no interested motives to serve.

The pure love of scientific knowledge appeared to actuate all his mental powers and conducted him to the best practical result...In the highest sense of the term Captain Huddart was an honest man'. After a long illness, Huddart died on 19 August 1816 at his home in Highbury Terrace and was buried at St. Martin's, Westminster; his only surviving son, Joseph, erected a memorial to him in the Chapel at Allonby.

His wife, Elizabeth, had died at Allonby in February 1786; of their five sons only three had survived infancy. William (1763–1787) had accompanied his father as a midshipman on the Royal Admiral in 1778 before taking command of the York in 1785 at the age of twenty-two. Unfortunately, his health was poor and he died in China in 1787. The third son, Johnson (1771–1795), also sailed with Huddart on his East Indian voyages and continued in the trade until forced to retire on health grounds in 1794. He joined the second son, Joseph, in Leghorn but died in January 1795. Joseph Jr., joined his father in London and later managed the estate his father acquired in North Wales. A portrait of Huddart, by Hooper, hangs in the Institution of Civil Engineers.

MIKE CHRIMES

[Drawings catalogue, ICE archives; W. D. Dempsey (1838) Description of Huddart's improvements of the rope manufacture, OC 272; E. Birch (1838) Huddart's rope machinery, OC 273; correspondence on Joseph Huddart, ICE archives; J. Huddart (1821) Memoir of the late Captain Huddart, FRS, etc.; W. Cotton (1842) A memoir of Captain Huddart, MPICE, 2, 56–59; W. Cotton (1855) A brief Memoir of the late Captain Joseph Huddart, FRS; W. Huddart (1989) Unpathed Waters: Account of the Times of Joseph Huddart, FRS]

Publications

1778. Chart of St. George's Channel, with Remarks, Telford Drawings 1079, a new and original hydrographic survey of the North and St. George's Channel

1781. *River Shannon* (repr. 1794)
1781. *West Coast of Scotland from Mull of Galloway to Dunan Point*
1786. *Tigris Survey*
1786. *?Maryport Harbour Survey*
1789. *Whitehaven Harbour*
1789. *A New Hydrographic Survey of West coast of Scotland*
1790. *North Coast of Ireland and West Coast of Scotland* (repr. 1813)
1791. *St. Agnes*
1793. *Copy of the Report ... relative to the making of a new Dock at Kingston-upon-Hull*
1794. *Report on Swansea Harbour*
1794. *West Coast of Ireland from Shannon Mouth to Urris Head* (repr. 1832)
1794. *West Coast of Ireland from Sligo bay to Tory Island* (repr. 1830)
1794. *A New Hydrographic Survey of the North Coast of Ireland and the West Coast of Scotland*, TD 1078
1795. *Haisborough Gat* [1790]
1795. *West and South-west coast of Ireland* (repr. 1829)
1796. *Isle of Wight*
1796. *Cape of Good Hope*
[1800]. *Report upon the Improvement of the Port and Harbour of Boston* (another edition with John Rennie), 1793
1801. *East Coast from Lowestoft to Cromer*
1804. *The Coasting Pilot for Great Britain and Ireland*
1804. *On the Improvement in Manufacturing of Cordage*
1804. *Eau Brink New River ... Deed Poll*
1804. *The Report ... accompanied by a Plan for the Improvement of Whitehaven Harbour* (repr. 1836)
[1804]. *Swansea Harbour. Extracts from the Report ...*, 1794
1804. *Swansea Harbour, Second Report*
1809. *North Sea*
[1810]. *Howth Harbour. Report* (with John Rennie), 1808–1809
1811. *North Sea*
1813. *Portland to Aberyswyth*
[1817]. *Copy of a Letter ... to the Right Honourable Nicholas Vansittart* (on Howth Harbour), 1815
[n.d.]. *Chart of the West and South West Coast of Ireland ...*

Works

1786–. Maryport Harbour, improvements
1789–1804. Whitehaven Harbour, improvements
1804. Swansea Harbour, improvements
1812. Hurst Point Lighthouse

HUDSON, John (fl. 1759–d. 1801), land surveyor and civil engineer working in Lincolnshire, was one of three commissioners for the inclosure and drainage of Kirton Fens (1772) and Quadring and Surfleet Fens (both 1775) in the North Holland Fens, and in 1777 for Billinghay and Walcot Fens on the west side of the Witham north of Chapel Hill. The extent of his involvement in the design of these schemes is not known but in 1784 'John Hudson and John Dyson, Engineers' reported on the drainage of Timberland and Timberland Thorpe Fens lying north of Walcot and Billinghay. Hudson gave evidence on the Bill in the House of Lords in April 1785 and under the Act passed in that Session he and Dyson (q.v.) were appointed commissioners. Works carried out in accordance with their report included a bank 1½ miles long parallel to the river, linking on to the Walcot–Billinghay bank, leaving a ¾ mile wide washland between itself and the river, and a cross or side back to Carr Dyke at the foot of higher ground to the west; the bank, 50 ft. wide at its base, was built 10 ft. high to allow for 2 ft. settlement. The main drain ran alongside the bank and there were to be two 'engines' at the river bank: windmills operating scoop wheels. The area amounted to 2,500 acres and the cost of the works was c.£4,000.

Work on the next block of Witham fens north of Timberland (Blankney and Martin) proceeded under an Act of 1787. Then in 1788 Hudson and the surveyor John Parkinson (fl. 1771–d. 1818) reported on the fens north of Blankney, an area of almost 10,000 acres. These were dealt with in two blocks: Dunstan and Metheringham, and Norton, Potterhanworth and Branston. Hudson gave evidence in the Lords on both in June 1789 and the (separate) Acts passed in that year. He and Parkinson were appointed commissioners for the Norton block: 5,850 acres were completed 1793 at a cost of £10,000, with a 3 mile river bank set a ¼ mile from the river, banks 2½ miles and 2 miles in length at the south and north boundaries respectively running back to Carr Dyke, and drainage mills at the river bank. The Commissioners were responsible for carrying out the works but there may well have been a resident engineer, possibly William Bonner, Surveyor to the River Witham Commissioners, who could have undertaken such supervision as was required.

Hudson next reported jointly with William Jessop (q.v.) in November 1796 on the proposed river navigation from the Witham to Sleaford: 12 miles with seven locks on Kyme Eau and the river Slea. Hudson supplied specialised knowledge on the means to avoid interrupting drainage of the adjacent fen land. His engraved plan of the scheme appeared early in 1792. The Act passed in June and the navigation was opened in May 1794; its cost was £16,000. Here the works were at first safely left in the hands of Bonner, who contracted for the digging, and from February 1793 to John Jagger who had previously been agent to John Dyson, contractor for the locks.

Hudson was also involved with the Horncastle Canal, on the opposite side of the Witham. Jessop reported in June 1791, Hudson and Bonner in February 1792 and the Act was passed in June. Jessop had difficulty finding a resident engineer.

After delays, work began in April 1793 but William Crawley, the resident engineer, proved unsatisfactory. John Dyson and Thomas Hudson (John's son) took over for a year followed from April 1795 by Hudson himself. He worked on the canal as 'surveyor and engineer' to the beginning of 1797 when operations ceased due to lack of funds; the relatively small amount of work left to be done was finished in 1801–1802.

Meanwhile in October 1791 Hudson and George Maxwell (q.v.) were appointed by proprietors of estates in South Holland to prepare plans for an improved drainage system. They 'took a view' in November, ordered a survey, met again at the beginning of August 1792 and Maxwell took levels along the line of the proposed new main drain, finding a fall of 8 ft. in a length of 14 miles from the fens adjacent to the east bank of the Welland near Spalding to low water at the proposed outfall sluice at Peters Point on the Nene estuary. Borings were also made to determine the nature of the soils. Their report is dated 18 August 1792. The estimates were: cutting the drain £8,450; purchases of land £2,400; outfall sluice £3,100; bridge £1,600 and contingencies £2,350; a total of nearly £18,000. The area of land to be drained was 19,400 acres. The scheme met with approval, an Act was passed in 1793 and the works, carried out under the superintendence of Thomas Pear (q.v.), were completed in 1796. The sluice, dated 1795, had three arches with a waterway of 26 ft.

Hudson and Maxwell were also involved in a revival of interest in a scheme first proposed in 1702 by Nathaniel Kinderley (q.v.) and again, in 1767, by Thomas Hogard (q.v.) for a new outfall channel of the river Welland 7½ miles long to the Witham estuary at Wyberton Roads. Thomas Thorpe made a land survey, Maxwell took the levels (finding an 11 ft. fall to low water to Wyberton) and Hudson prepared the estimates: £49,000 total including £7,000 for land purchase and £10,500 for the large outfall sluice with adjacent navigation lock. Together with James Golborne (q.v.), Superintendent General of Bedford Level, they reported in October 1792. The Act enabling the works to be carried out by the newly appointed Welland Commissioners passed in 1794. No action followed, however, due partly to the high cost but perhaps chiefly to persisting doubts as to the validity of the schemes. Finally under a new Act of 1801 the better solution of improving the Welland by extending its channel within its own estuary was adopted, an idea put forward by John Grundy (q.v.) in 1770 as an alternative to the Wyberton Cut (of which he strongly disapproved).

Another controversial proposal, also originating with Kinderley, was to by-pass a long, wide and shallow bend of the river Great Ouse by a new cut from Eau Brink to King's Lynn. After years of indecision and conflicting advice a Committee of Landowners in the South Level initiated an investigation. They called on John Watté (q.v.) to survey and make borings, and on Golborne to report with estimates. Golborne supported the proposal. The Committee then sought independent advice from Robert Mylne (q.v.), Hudson and John Rennie (q.v.). In this distinguished company Hudson, like them, approved the Cut. Opposition came from engineers consulted by King's Lynn, who feared the effect the cut might have on Lynn harbour. Finally, after debate in Parliament in three consecutive sessions, the Eau Brink Act was passed in May 1795. The Cut proved highly beneficial to drainage of the South Level.

In 1796 Hudson returned to the Witham fens with a report on reclaiming the wide washland, known as the Dales, left between the river and the banks of Walcot, Timberland and Blankney Fens, an area of 2,800 acres. He was appointed sole commissioner under the Act of 1797, receiving the usual fees of 2 guineas per day paid to commissioners in this period. The works, presumably carried out under his direction, included a river bank 6 miles in length and the side banks of Martin, Timberland and Walcot Fens were extended across the Dales to join the new river bank.

The will of John Hudson of West Ashby (near Horncastle), gentleman, dated 9 May 1800, leaves property to his sons, Richard and Francis, and to a son and two daughters of his late son, Thomas. Probate was granted in April 1801, the personal estate valued at £8,000.

A. W. SKEMPTON

[Plan by T. Thorpe and Estimates by J. Hudson of proposed Wyberton Cut (1792). Welland papers, House of Lords RO; W. H. Wheeler (1896) *History of the Fens of South Lincolnshire*; H. C. Darby (1940) *The Draining of the Fens*; J. Boyes and R. Russell (1977) *Canals of Eastern England*; C. Hadfield and A. W. Skempton (1979) *William Jessop, Engineer*; W. M. Hunt (1979) The Horncastle Navigation engineers, *Journal of the Railway & Canal History Society*, **25**, 2–11; Bendall]

Publications

1784. *Report on the Practicability of Imbanking and Draining the Fens Lands in Timberland* (with John Dyson)

1788. *Report to the Proprietors of Low Lands in Metheringham, Dunstan, Norton … and Branston* [and] *an Estimate of the Probable Expenses* (with John Parkinson)

1791. *Report on the Means and Expense of making Navigable the Kyle Eau and River Slea, from the Witham to the Town of Sleaford* (with William Jessop) and an engraved plan by Hudson (1792)

1792. *Report relative to the intended Navigation from Lincoln to Horncastle* (with William Bonner)

1792. *Report on the Improvement of the Outfall of the River Welland* (with James Golborne and George Maxwell)

1792. *Report on the Probable Effect of the Proposed Cut from Eau Brink to Lynn will have on the Banks and Drainage of Bedford South Level*
1793. *Report on the Drainage of South Holland* (with George Maxwell)
1796. *Report on the Means and Expense of Embanking and Draining the Dales of Walcot, Timberland … and Blankney*

Works

1785–1787. Timberland Fens drainage (with J. Dyson), £4,000
1789–1793. Norton, Potterhanworth & Branston Fens drainage, £10,000
1793–1797. South Holland Drain (with G. Maxwell), £18,000
1797–1799. Walcot, Timberland & Blankney Dales drainage, £6,000

HUGHES family (fl. 1800–1850) of contractors and civil engineers came to prominence in the first decade of the nineteenth century, led by two brothers, John and William. They possibly originated at Northrop, Flintshire.

John Hughes (fl. 1800–1840) possibly began contracting on the Lancashire Canal. In June 1803 he tendered, with his brother, for the Tilstock–Grindley Brook length of the Ellesmere Canal. Moving to Poplar he was taken into partnership by William Bough (q.v.) for the second phase, 1803–1806, of the construction of the West India Docks, the first phase of which had been largely carried out by Bough, in partnership with John Holmes and Thomas Clark between 1800 and 1803. In 1804–1805 they undertook the excavation of the Export Dock and in a little over a year they excavated 490,000 cu. yd. of earth at an average of 8,000 cu. yd. a week at 11d per cubic yard. This work rate was achieved using a 6-hp steam engine to haul wagons on an inclined plane, formerly used by Bough and Holmes at London Dock. Following this, in February 1806, Hughes began dredging contracts for the City Corporation in partnership with John Mills, a lighterman of Limehouse.

In August 1806 Bough and Hughes decided, on Hughes's suggestion, to fit out a 60 ton barge, the Brunswick, with this steam engine and bucket-ladder dredging equipment. Subsequently they obtained a contract from the City Corporation to dredge the East India moorings to a depth of 20 ft. below low water spring tides. The work, personally supervised by Hughes, was carried out between February and November 1807, while, although he began as partner of Bough on the removal of mud from Greenland Dock for the Commercial Dock Company in late 1807, the excavation work for the docks was completed by Bough alone. It would appear that Hughes took responsibility for removing mud dredged by the Admiralty at Woolwich and to assist in this work purchased the remodelled dredger, the Plymouth, in 1808; it was active at Woolwich between 1809 and 1816. Hughes became the contractor for the

ill-fated foundations of the Millbank Penitentiary, designed by Thomas Hardwick; he blamed the architect for the subsequent fate of the building.

It is possible that Hughes worked for Richard Trevithick at Penydarren. John Hughes (1814–1889) of Uskside Iron Works, Newport—he was bankrupted in 1858 and also set up an important ironworks in Russia—in 1857 referred to a meeting with Francis Trevithick and a conversation to that effect. Trevithick was involved with the early Hughes dredgers and Samuel Hughes worked in South Wales. Another John Hughes was a famous quarry master in North Wales, and builder of bridges in the early nineteenth century.

William Hughes (fl. 1800–1836), the brother of John and working in Scotland, had accompanied William Jessop (q.v.) in his survey of the line of the Caledonian Canal in October 1803 and subsequently began to take excavation contracts on the Clachnaharry section of the canal. In 1808 Hughes, then described as one of Thomas Telford's (q.v.) assistants, accompanied him to Sweden to survey the Gotha Canal and after his return in 1811 became contractor for one of the lots in the middle district of the Caledonian Canal near Fort Augustus.

In 1805 a dredging machine was supplied by Outram (q.v.) for use in dredging to Loch Oich and Loch Dochfour, the barge being built at Inverness. Work began in the spring of 1806 but little was achieved as the barge sank that autumn. In 1811 the idea was revived by Jessop and Telford and two dredgers were delivered, one from Butterley's, delivered for work on Loch Dochfour in 1814, and the other by Donkin (q.v.); it began work in Loch Oich in 1817, both contracts being awarded to William Hughes. His expertise was later recognised when he was consulted by the Clyde Trustees regarding a dredger in 1822.

In the mid-1830s William Hughes, in partnership with his brother, took contracts on the North Union and London–Birmingham Railways. Both contracts hit problems due to lack of progress and experienced strikes as a result of the workforce not being paid; in due course they were released of the contracts. It is believed that Hughes suffered a stroke and in January 1836 it was considered that there was no hope of recovery; he died shortly afterwards. There are, however, later references to a William Hughes, contractor, tendering for works on the Pensnett Railway (1844–1845), while a John Hughes tendered, unsuccessfully for work on the Birmingham–Gloucester Railway in January 1838. So common is the surname that any reference to William Hughes after 1836 must be to another individual.

Further details of his later career are lacking although he had taken a contract on the Holyhead Road in the late 1820s and is known to have had two sons. One was born *c.* 1797 and fell in love with Benjamin Outram's eldest daughter c.1818. Both sons were educated in architecture by a Mr. [John?] Smith of Aberdeen. He was well regarded in the Highlands, Robert Young (1871) remarking:

'We remember Mr. Hughes well. He was a native of Wales—of small stature, sharp, and active, of very considerable talents; quiet, modest and unassuming in his manners; very much of a gentleman. He lived long at Inverness, where he was well known and much respected'.

Thomas Hughes (fl. 1790–c. 1846), the son of John Hughes, is perhaps the most distinguished of the family. After assisting his father with dredging work on the Thames he went north to join his uncle on the Caledonian Canal and it was he who sorted out practical problems concerning use of the dredgers. In total, 170,000 cu. yd. of material were excavated in eight months.

After seven years work on the Caledonian Canal, Hughes took work on other canals and harbours in Scotland. He tendered with Thomas Williams for lots 10–16 on the Edinburgh and Glasgow Union Canal and on one lot he lost heavily due to misleading information on ground conditions. In the 1820s he undertook three contracts on the London–Holyhead Road, work which included the use of a concrete foundation for the section at Highgate Archway; he also built a bridge over the Buckinghamshire Ouse. This was followed by work as an engineer and contractor in East Anglia, giving parliamentary evidence on the Norwich and Lowestoft navigation in 1827.

Early in 1833 Hughes undertook the surveys of the eastern half of the Great Western Railway for Isambard Kingdom Brunel; although Hughes occasionally frustrated Brunel by his elusive movements, he appears to have done a competent job. He may have been involved in a more practical capacity with railways in advising on the drainage of cuttings on the London and Croydon Line. It would appear that Thomas Hughes retired in the late 1830s, living at that time at 22 Palace New Road, Lambeth, and then at Acre Lane, Brixton, and wrote a series of works based on his experience. He joined the Institution of Civil Engineers in April 1844, but had died by 1850 when its next membership list was published.

Edward Hughes (fl. 1830s), the brother of Thomas, probably assisted on the Scottish contracts. In the 1830s he was living at Winchburgh, making a living as a small scale contractor, sinking pits and building reservoirs for mills. These latter included a reservoir for a paper mill at Curwick (1836–1837) and another for Murray's mill at Harmony (1836).

In May 1837 he gave some extraordinary Parliamentary evidence on the borings on which he had assisted on the line of the Edinburgh and Glasgow Railway. This revealed that the contractor, Crosby, had deliberately falsified the records of the site investigation. Hughes went with this information to the Union Canal Company, who were opposed to the line and for whom he had carried out work, subsequent to his brother's contracts, in the late 1820s. It is not known whether this Edward is the same Edward

who unsuccessfully tendered for lot 27 on the Eastern Counties Railway in February 1841.

Samuel Hughes (c. 1816–1870), the son of Thomas, assisted him on some of his later contracts on the Holyhead Road, measuring his employees' work and assisting in estimating. He acted as Joseph Gibbs' assistant on the London–Croydon Railway and assisted with estimating and measuring many of the railways built in the 1830s. He also reported on Chepstow Bridge. He then became involved in mineral surveys in South Wales and the Forest of Dean before subsequently returning to London in the early 1840s where he worked for the Tithe Commission, with an office at 8 Duke Street, Westminster. Samuel Hughes became a leading gas and water engineer, from c.1850 onwards writing a number of books on the subject. His most important written work, however, was a memoir of William Jessop. He participated in all the Metropolis Gas Inquiries from 1858 onwards, his offices then being at 14 Park Street, Westminster. He died on 31 October 1870 at the age of fifty-four, leaving an estate of £16,000.

The failure of the Hughes family to make an impression as railway contractors is an indication of the difference in scale between railway civil engineering and what had gone before.

MIKE CHRIMES

[Membership records, ICE archives; Telford mss, ICE archives; Ellesmere Canal records, PRO RAIL 827/2; West India Dock Company; Commercial Dock Company minutes; Bridgewater Estate papers, BW/C/2/3, Preston RO; Museum of London in Docklands, *Caledonian Canal Commissioners Reports 1812–1828*; House of Commons (1827) *Minutes of Evidence ... Navigable Communications ... Norwich ... Lowestoft*, HLRO; T. Hughes (1835) *Description of a proposed Method of laying down Lines of Railway without Pedestals or Chains*, ICE OC161; S. Hughes (1844) Application for assistant surveyorship of sewers for Westminster, ICE Library; Holyhead Road Commissioners (1827–1836) *Reports*; House of Commons (1837) *Minutes of Evidence ... Edinburgh and Glasgow Railway Bill*; J. Weale (ed.) (1843) The steam dredger, *Quarterly Papers on Engineering*, **1**; House of Commons (1867) *Special Report from the Select Committee on the Metropolis Gas Bill ... Minutes of Evidence*; Probate records (S. Hughes) (1870); R. Young (1871) *The Parish of Spynie*; J. Mitchell (1883) *Reminiscences of my Life in the Highlands*, vol. 1; A. Gibb (1936) *The story of Telford*; J. Lindsey (1968) *Canals of Scotland*, 74–75, 184–185; A. D. Cameron (1972) *The Caledonian Canal*, 35, 55, 60, 77–78, 80, 90, 104; A. W. Skempton (1975) The steam dredger, *Trans Newc Soc*, **47**, 97–116; J. Boyes and R. Russell (1977) *The Canals of Eastern England*, 128; E. G .Bowen (1978) *John Hughes*; (Yuzouka) 1814–1889; A. W. Skempton (1979) Engineering in the Port of London, *Trans Newc Soc*, **50**, 87–108; K. H.

Vignoles (1982) *Charles Blacker Vignoles*, 17; D. Brooke (1983) *The Railway Navvy*; G. Biddle (1990) *The Railway Surveyors*; J. Gifford (1992) *The Buildings of Scotland: Highlands and Islands*]

Publications

John Hughes:

1813. Patent 3672: *Raising, Screening and delivering Gravel*
1819. Method of ascertaining distances from one station to another (in levelling or surveying) on level and variously inclined surfaces, to their exact length of horizontal base, by the addition of an angular bubble attached to the telescope of a spirit level, *Phil Mag*, **53**, 271–272

Thomas Hughes:

1838. *The Practice of making and repairing Roads; of constructing Footpaths, Fences, and Drains*
1839. The origin and use of the steam dredging machine, *Civil Engineer & Architects Journal*, **2**, 9–11
1842. Pile driving machines; on the causes and prevention of slips, *Surveyor Engineer and Architect*, 13–24
1843. A series of papers on the foundations of bridges, *Weale's Bridges*, vol. 1
1845. Description of the method employed for draining some banks of cuttings on the London and Croydon and London and Birmingham Railways, *ICE Min Proc*, **5**, 78–86

Samuel Hughes:

1839. Report on Chepstow bridge, in J. Weale (ed.) (1843) *Bridges*, vol. 2, cxii–cxviii
1843. The inventor of the dredging machine, *Civil Engineer and Architects Journal*, **6**, 16
1843. Memoir of James Brindley, *Weale's Quarterly Papers*, **1**(8)
1844. Memoir of William Jessop, *Weale's QuarterlyPapers*, **1**(2)
1853. *Rudimentary Treatise on Gasworks*
1856. *Rudimentary Treatise on Waterworks*

Works

John Hughes:

1803. Ellesmere Canal, Tilstock–Grindley Brook, contract, £6,989 (with William Hughes)
1804–1805. West India Export Dock, excavation, 490,000 cu yd., *c*. £22,000 (with Bough)
1805 1806. City Canal, works
1806. Dartmoor Prison (with Bough; Bough responsible)
1806–1807. East India moorings, 50,300 tons dredged, £9,346 (with Bough)
1806–1807. City Canal entrances, dredging (with Bough)
1807–1817. Woolwich moorings
1808. Commercial Dock excavation (completed by Bough)
1809. Dredging for the City Corporation
1812–1813. Millbank Penitentiary, foundation

1838. Rochdale Canal, excavation, Castlefield, £5,000

Thomas Hughes:

1815–1817. Dingwall Canal, Peffery diversion, and harbour, £4,365
1817. Sutherland Harbour, for Lord Stafford
1818–1822. Edinburgh & Glasgow Union Canal, lots 10–16 (with T. Williams), 14 miles, *c*. £54,000
1820s. Hornbush harbour, Inverness
1820s. Loch Spynie, Morayshire, embankments and land reclamation
1825–1826. North Walsham and Dilham Canal, engineer and contractor, 8¾ miles, six locks, £29,000
1827. Norwich–Lowestoft navigation, engineer
1828–1830. London–Holyhead Road, Highgate Archway, Crown Hill, Forty Mile Hill, Fenny Stratford
1830. Olney Bridge, Bucks
(1834–1836?). Bridge over the Ouse, Buckinghamshire, 250 ft. long?, five spans?, Stoney Stratford

William Hughes

1808–1812. Loch Spynie, drainage, £12,740
1808?–1817. Caledonian Canal, Clachnaharry excavation, Fort Augustus excavation, dredging of Lochs Dochfour, Loch Oich and Loch Ness, dredgings 1 million cu. yd., £15,100 (in part with T. Davies), Contracts 13 and 14
1811–1813. Brora coalmine, shaft sunk, 76.2 m
1813. Brora harbour
1815–1817. Dingwall Canal (with Thomas Hughes)
1825. Caledonian Canal, repairs and dredging, Muirtown–Dochgarroch, £1,186 13s 6d
1828. Holyhead Road, Bangor–Shrewsbury, maintenance
(1835–1836). London–Birmingham Railway, contract 1F, Blisworth cutting
1835–1837. North Union Railway

HUNT, John (fl. 1590–1611), landowner, acted as the engineer for the first comprehensive scheme for draining the Fens. After the death of George Carleton (q.v.) in 1590 John Hunt 'of London' purchased his lands in Holbeach, South Holland and Coldham Fen, south of Wisbech. Within a few years Hunt was appointed to the Commission of Sewers (the local land drainage authority) in both Cambridgeshire and Lincolnshire. In that capacity he attended meetings at which contracts were made with Thomas Lovell for draining Deeping Fen in 1599 and with Sir William Cockayne in 1602 for improving the drainage of some 3000 acres between Upwell and Denver. A year or so later he began exploring the possibility of draining the whole of what became known as the Great Level of the Fens, an ambitious project previously considered only in broad terms by Humphrey Bradley (q.v.) in 1589, supplemented by an inadequate specific engineering proposal of 1593.

Hunt approached a number of landowners who offered part of their land 'to such as will undertake the recovering of the whole'. Encouraged, Hunt and Henry Totnall (of whom little is known) addressed the King, informing him there were more than 200,000 acres that could be greatly improved in value by drainage. James responded favourably and in July 1604 authorised Hunt and Totnall and 'such others as they shall think fit to employ' to 'take a view' of the Fens while requiring the Commissioners in the six counties involved (Suffolk, Norfolk, Cambridge and the Isle of Ely, Huntingdon, Northampton and Lincoln) to aid them therein.

An excellent map of the fens from the Welland to the Old Nene had been produced by John Hexham (q.v.) in association with Bradley in 1589, and William Hayward (q.v.) had mapped part of the southern fens in 1591. A survey of the entire area clearly had to be made and this task was tackled by Hayward, who completed his famous map early in 1605 showing the watercourses, existing land drains, banks and limits of the higher ground on a scale of 1 in. to a mile. Also, in July, he listed the area of each fen within the Great Level, the total being 307,242 acres. Meanwhile, Richard Atkyns (q.v.) reported on the nature of many individual fens, from observations made in the winter months of January and February 1605, and in April he explored the ground to a depth of 11 ft. in some fifty places using a hand auger. In the meantime, as events were soon to show, Hunt had been deciding what engineering works would be necessary for the general drainage.

Sir John Popham, the wealthy Lord Chief Justice, now took control if indeed he had not earlier been active behind the scene by desire of the King. In April 1605 he appointed a special Commission drawn from the six counties. Five weeks later, replying on 19 May to letters from the Privy Council, the Commissioners said that 'upon long debate, and all objections heard, they conclude this work of draining is feasible and not only so, but they did reckon it the most noble work, and most beneficial to the counties interested, that was ever taken in hand of that kind'.

Next month, to bring the scheme into sharp focus, several of the Commissioners accompanied by Hunt took 'a particular view' of the whole Level. Starting in the north they found the Welland outfall to be reasonably good but the river needed improvement from Spalding to Peakirk, on the edge of the Fens. Hunt proposed that it 'shall be amended … to 60 ft. wide and 8 ft. deep' and a new bank made on the south side from Crowland to Peakirk (4 miles); banks already existed from Spalding to Crowland. Also, for drainage of fens between the Welland and Nene, improvements of Shire Drain were essential. The Nene at Stanground, near Peterborough, was in good condition but in Morton's Leam there were 'wonderful defects in breadth and depth'. To redress which Hunt thought fit that, as well as

amending the Leam, there be a new river made 50 ft. wide and 8 ft. deep 'north of the Leam and some distance from it' and a bank made on the south side of the Leam; also a navigable sluice on the old Nene at Stanground. Similarly the river should be improved from Guyhirn past Wisbech to the estuary, making its channel 60 ft. wide, and Hunt further desired that a new cut be made over Tydd Marsh and Sutton Marsh 'into the deepe', with a dam to turn the water into the new channel. All this the Commissioners approved though leaving to others 'of longer experience' a final judgement on the 'sea workes'. Further south, Well Creek from Outwell to Salters Lode was almost choked and, as the upper part could not conveniently be amended, Hunt desired that a new river 80 ft. wide and 8 ft. deep be made from March river (the Old Nene) to join the lower part of Well Creek at Nordelph, and thence to Salters Lode 'where a great navigable sluice is to be placed'.

Travelling back to Ramsey and on to Earith, where the Ouse emerges on the Fens, Hunt desired a new river be made 'line right' from Earith bridge to fall into the Ouse at some apt place near Slaters Lode; the new river (21 miles long) was to be 120 ft. wide and 8 ft. deep with sluices at each end, and a navigable sluice was to be built on the Ouse at the Hermitage a short distance below Earith. This was on 25 June. Next day the party set off along the course of the Ouse for 40 miles past Ely and Littleport and arrived at Salters Lode on 27 June where they saw that low water in the river was 10 ft. below the soil of the fens. The necessity and feasibility of the new river became apparent. The excursion, thus briefly summarised, ended at Cambridge when Hunt who, as Dugdale says, 'was the Artist of the draining', represented what cuts, banks and sluices would, in his judgement, be further necessary to perfecting this work.

The Commission then met at Wisbech on 13 July 1605 and passed a 'law and ordinance' stating that 'Sir John Popham and others (named) shall undertake within seven years, at their own cost and charge, to drain all the Fens and surrounded (drowned) grounds between the old course of the river Ouse as it now runneth' and the Welland (an area of about 230,000 acres); the following works shall be made at an estimated cost of £120,000 'at the least'.

(1) A new river to convey the greatest part of the Ouse from some place at or near Earith bridge unto such place near Salters Lode as they shall think fit; the river to be between banks at least 30 poles (495 ft.) apart, the banks to have a height of at least 7 ft. (2) A like passage to be made from near Peterborough for the conveyance of the river Nene from thence to Wisbech. (3) And so likewise for the river Welland from its entrance into the Fens near Peakirk until it meets the Glen (below Spalding). (4) Land Eas (cut-off drains) from Earith to Peakirk to defend the grounds within them and between the new passages of

the Ouse and Welland from land floods falling from the upland countries (another of Hunt's proposals). (5) A navigable sluice to be made at Salters Lode on Well Creek which was to be enlarged and diked to Nordelph, and thence a new cut to be made 'as directly as may be' to the Old Nene above Upwell, with a sluice. (6) A new bank from Earith to Ely to defend the grounds between that and the new river from the overflowing of the Ouse. (7) To improve Shire Drain, hitherto the responsibility of the country.

Ten days later the King 'commended these endeavours'. On the 5th and 6th of August Hunt and Atkyns set out the cut 5½ miles in length, still known as Popham's Eau, from Nordelph to the Old Nene. On the 7th work began in the presence of Hunt (who cut the first sod), Atkyns, Totnall and some others. A large workforce must have been employed, since on 21 December the cut was completed and the river bank opened to allow water to run down to Well Creek and so to the Ouse, thus providing a better outfall for the Old Nene.

Proceedings had so far been conducted under the Wisbech 'law' of July 1065 but petitions against the general scheme were being received and even the legality of the undertaking was in question. Popham therefore sought an Act and in February 1606 submitted a Bill to Parliament. The principal works were to be very much as before, with the addition of banks in all needful places on the Cam, Lark, Little Ouse and Wissey (tributaries of the Ouse). Popham's Eau is not mentioned as it had already been made.

Popham and his fellow 'Adventurers' were now to drain the whole area of 307,000 acres and receive 112,000 acres, of which 12,000 acres would be reserved to provide income for maintenance, when drainage was completed, and a corporation would be established as 'Governors of the Fens'. Among those named as future Governors were the Bishop of Ely, several great landowners such as Sir Miles Sandys and some lesser ones including Henry Totnall, esquire, and John Hunt, gentleman.

It is highly significant that Hunt's proposals of 1605 (except the cut-off drains on the Fenland margins) formed the basis of the works eventually carried out under Francis (fourth) Earl of Bedford in the years 1631–1636 with which Hayward was closely involved both as Commissioner and Surveyor. The Hunt–Popham scheme probably had never been forgotten by 'the country' in general and certainly not by Hayward. Of special interest is the fact that Hunt's bold concept of a new 21 mile channel to carry the flood water of the Ouse was adopted as the 'Bedford River', made 1632–1635 though with the banks too close together, leading to a breach in the first major flood. The principle of providing a large 'washland' for the temporary storage of flood water, implicit in the 1605 proposals, was strongly advocated by Sir Cornelius Vermuyden (q.v.) in his Discourse of 1639 (published 1642)

and put into practice when he directed the works of the second (1640–1641) and third (1650–1656) campaigns of the Fens drainage.

However admirable the scheme may have been in engineering terms, the Bill failed to pass its third reading in May 1606. Meanwhile a great storm in March broke the newly made banks of Popham's Eau. Attempts to revive the Bill in the next session of Parliament proved unsuccessful and the death of Sir John in June 1607 effectively put a stop to the whole affair. The banks were in due course repaired and work on deepening and widening Well Creek from Nordelph to Salters Lode began in August 1610.

Thus an excellent plan ends with little achieved. John Hunt briefly appears once more. In June 1607 a Bill was drawn up for draining about 6,000 acres in Waldersey and Coldham and, on passing, became the first local Act in the Fens. After an unexplained delay, Commissioners in May 1611 sanctioned construction of the main drain to run from Coldham, passing under Well Creek and crossing Marshland for a length of some 6 miles, to an outfall sluice in the west bank of the Ouse near Stow Bridge. This is referred to as 'Hunt's Drain' and it can be assumed he was involved from the start in 1607. Masons began work on the sluice in August 1611. A cofferdam would have been required. It is mentioned as 'the dam built by Mr. Hunt' which failed in October and was repaired a few days later under the direction of Mr. Richard Hunt.

Early in 1611 John Hunt petitioned for protection against his creditors for a further year so that he could complete a drainage job which promised to yield enough profit to cover his debts, obviously the Waldersey–Coldham scheme. By October of that year he may have died or been seriously ill as in 1612 Sir John Peyton bought his interests in Coldham. His will has not been found but the will of Margaret, widow of John Hunt of London and Holbeach, proved in 1618, leaves only a modest estate.

Samuel Wells in his History of Bedford Level refers to 'Mr. Atkyns and Mr. Hunt, two of the most able and scientific fen men of that time'. Hunt, in particular, is truly the pioneer of draining the Great Level.

A. W. SKEMPTON

[Sir Cornelius Vermuyden (1642) *A Discourse touching the Drayning of the Great Fennes*; Sir William Dugdale (1662) *The History of Imbanking and Drayning of divers Fenns and Marshes*; Thomas Badeslade (1725) *The History ... of the Port of King's Lyn ... and the Draining of the Great Level of the Fens*; S. Wells (1830) *The History of the Drainage of ... Bedford Level*; Mark E. Kennedy (1985) So Glorious a Work as the Draining of the Fens ... in Elizabethan and Jacobean England, unpublished Ph.D. thesis, Cornell University; Professor Margaret Knittl (in lit. 1996 and 1997) provided valuable additional information from primary sources, including a transcript of the 1605 'view' by Hunt and the Commissioners, from Ely Diocesan Records]

HUNT, Joseph (*c.* 1652–1697 or later), of Godmanchester, was a carpenter working on the Great Ouse river navigation under Henry Ashley (q.v.). Speaking in 1689 Hunt refers to Ashley as being 'skilled in works of navigation'. However skilled in planning and directing such works, Ashley, like other river engineers and fen drainers, would depend on local craftsmen for details of construction and the building of sluices, locks and bridges. Hunt is one of few in the seventeenth century of whom a little more is known than the mere mention of their name in payment accounts. On two occasions in a lengthy dispute between Ashley and the owners of the navigation, Hunt, in evidence, gives his age (thirty-seven years in 1689), trade and place of residence and refers to two locks and three stanches (built 1680–1686) between St. Neots and Great Barford, saying they were built at the cost of Henry Ashley amounting to a thousand pounds (which is correct).

It is highly probable that Hunt was employed on some of these works. Certainly he was later employed on repairs, making and fitting new gates and so on in various locks and stanches on the navigation between St. Ives and Barford. In 1697 he is known to have charged 2s 6d per day or 15s a week. Labourers working under him were paid 1s a day.

A. W. SKEMPTON

[T. S. Willan (1946) The navigation of the Great Ouse between St. Ives and Barford in the seventeenth century, *Bedfordshire Historical Records Society*, **24**]

HUNTER, Walter (1772–1852), millwright, was born at Newbattle, near Edinburgh in 1772, the son of a wright. After training as a millwright he moved south and was employed in the Soho works of Boulton and Watt. He was then taken on by John Rennie (q.v.) as his principal foreman. Rennie's decision to send him to look at a Dutch-style bucket ladder horse dredger employed in Hull docks proved a crucial point in his career as he developed a forte in the application of steam power to dredging and was presumably involved in the construction of Rennie's Brunswick dock dredgers of 1802. Having tried unsuccessfully to set himself up in business he set up in partnership with William English (fl. 1807–1850) and in 1807–1808 set up their Bow works, which was in business throughout the nineteenth century.

One of Hunter and English's first jobs was the remodelling of Trevithick's (q.v.) dredger, Plymouth II, for John Hughes and with a (Matthew) Murray (qq.v.) engine; it was completed by July 1808 for use in dredging contracts on the Thames. Their later dredger contracts included the supply of the machines for Trinity House's first dredger on the Thames, in 1826, one for use on the South

Level (1827–1828), and exports to Hamburg in 1834 and Geneva in 1844.

Other dock related contracts followed. They included cast iron columns for a warehouse at Greenland dock (1811) and an iron swing bridge in Commercial dock to James Walker's design (1812). Walker was clearly impressed by their work as he persuaded William English to act as resident engineer on Vauxhall bridge (*c.* 1815–1816), remarking in his autobiography: 'I cannot speak too highly of the indefatigable industry, practical knowledge and sterling principles of Mr. English. I don't know that I ever knew a man possess these qualities to a higher degree'. English was awarded 150 guineas for this work.

Despite English's absence Hunter prospered. They carried out more than £2,000 worth of millwrights work and smith's work for the East India Dock Company between 1813 and 1816, including cranes and the millwork on a cast iron bearing bridge supplied by Horseley's. Their swing bridge work became as well-known as their dredging machinery and in 1827/8 they supplied two swing bridges for London docks. This was more or less contemporaneous with bascule bridges and lock gates for the Hull Junction Docks.

Their export of dredgers led to contact with the Hungarian Count, Szechenyi, who ordered a dredger, the Vidra (Otter), for use on the Danube. Their resident engineer on this work, Adam Clark, remained in Hungary and his presence, and the earlier contract with Szechenyi, no doubt helped them secure the contracts for the supply of the bed plates and roller carriages for the Budapest chain suspension bridge designed by William Tierney Clark (q.v.).

Around 1831–1833 Hunter supplied two canal boats to John Macneill for his experiments on canal boat traction on the Forth–Clyde canal. Aside from their civil engineering work Hunter and English supplied steam engines, pumping equipment and waterwheels, such as waterwheels installed at the Lea bridge mill of the East London Waterworks Company in 1837.

Hunter died on 8 February 1852; English had predeceased him. Hunter was assisted in his works by his son, James, who ran the works after his death and was in turn succeeded as works manager by his fifth son, James Bernard (1855–1899).

MIKE CHRIMES

[J. G. James collection, ICE archives; East India Dock Company, records, Museum of London in Docklands, J. Walker (*c.* 1860) Autobiography, ms, ICE archives; *Min Proc ICE*, **12**, 1853, 161–163; A. W. Skempton (1975) A history of the steam dredger, *Newc Soc Trans*, **47**, 97–116]

Works
1812. Commercial Docks, swing bridge
1813–1816. East India Docks, swing bridge, millwork, £416

1826–1829. Hull Junction dock, bascule bridges and lock gates
1827–1828. London Docks swing bridges (two), £5,180
1832–1833. Ironwork for Blackwall sewer outfall
1838. Dover harbour entrance, improvements, machinery
1847. Ironwork for Szechenyi chain suspension bridge, Budapest, £3,000

HURD, Richard (fl. 1676–77), of Cardiff, engineer, was engaged by the Exeter Chamber in 1676 to improve the Exeter Canal, for a fee of £100. In 1671 Exeter Chamber called a conference to discuss improvements of the Canal or 'haven' as it was called. A Dutch engineer was consulted but eventually Hurd was appointed.

The canal, originally built by John Trew (q.v.) in the 1560's, had a length of about 1¾ miles from Exeter Quay to a tidal creek of the river Exe, with three pound locks (the first in England), and a navigable sluice at the lower end to maintain water level in the canal at low tide. Just below the head of the canal Trew built a timber weir across the river. After some improvements in 1581 the canal could accommodate lighters of 16 tons burden. The city now required an extension of about a mile, nearly to Topsham, with a large sluice and transhipment basin at the entrance.

On 18 January 1676 the Chamber Act Book records that 'Whereas an Inginoor for the new haven is presented to this house as an experienced person in such works being at Cardiff in Wales who hath effected great work there as is affirmed and that neither time nor opportunity may be lost to accomplishing of that work here which is of so great concernment. It is this day agreed and ordered that Luke Falvey the surveyor of the city works shall make his speedy repair into Wales where the said inginoor liveth and to see what those works are there done by him'. Mr. Falvey duly made his visit and must have been satisfied with what he saw for a fortnight later the Chamber met and entered into an agreement 'with the Ingineere lately come from Wales'.

Work seems to have started almost immediately, and the canal was re-opened a year later. The 'very large sluice' at the new entrance, later named Trenchard's Sluice after its builder, had a single pair of gates capable of admitting vessels of at least 60 tons burden into the basin where goods could be transferred to or from the canal lighters. At the same time a stronger weir was built and the quay at Exeter thoroughly repaired and enlarged, there were extensive dredging operations in the existing canal channel and thorough repair of the banks. The works and land purchase cost upwards of £5,000. Hurd received a fee of £100.

Nothing has yet been discovered concerning his 'great work' in Wales before 1676, nor his later career.

A. B. GEORGE and A. W. SKEMPTON

[Exeter City Act Book 11, folio 163, Devon County RO; W. B. Stephens (1957) The Exeter Lighter Canal, 1566–1698, *Journal of Transport History*, **III**(1), 1–11; E. A. G. Clark (1960) *The Ports of the Exe Estuary*; K. R. Clew (1984) *The Exeter Canal*]

Works
1676. Exeter Canal, improvements, extension and new weir, c. £5,000

HUTTON, Charles, FRS (1737–1823), mathematician, was the younger son of Henry Hutton, pitman, and his wife Eleanor. He was born in Newcastle on Tyne on 14 August 1737 and after the death of his father five years later his mother remarried. Hutton at first sought work as a pitman but as a result of an injury sustained when young, he was unable to continue this work and became interested in scholastic pursuits. He attended several schools in Newcastle and later became a schoolmaster in Jesmond, then a village near Newcastle.

In 1760 he removed his school into Newcastle, an advertisement then placed stating that as well as instruction in writing and mathematics, surveying would be part of the syllabus. Such were Hutton's abilities as a teacher that pupils of the town's Royal Free Grammar School were taught by him in mathematical studies. In 1760 he married a distant relative, Isabella Hutton (1733–1785), and their marriage resulted in the birth of a son, George Henry, and three daughters.

As a schoolmaster, Hutton published tracts on arithmetic and mensuration and in 1770, at the invitation of the Corporation of Newcastle, he produced a plan of the town with 'an accuracy that has excited the surprise and approbation of subsequent surveyors', having been assisted in this work by John Fryer (q.v.); the plan was published in 1772. In the same year Hutton published *The Principles of Bridges...*, a book written as a result of the failure in a flood the previous year of the bridge connecting Newcastle to Gateshead; it covered all aspects of bridge design and construction.

In 1773 he applied for the post of Professor of Mathematics at the Royal Military Academy, Woolwich, and his successful appointment followed several days of rigorous examination of the eleven applicants. He took up the position in May, then leaving Newcastle, his wife and his family. At the Academy, Hutton was employed both in teaching and writing and it was said of him that 'his skill and patience as an instructor were generally acknowledged' and as an author he was 'the most popular of English mathematical writers ... at once concise and perspicuous'.

Hutton was elected a Fellow of the Royal Society in June 1774, among his sponsors being John Hunter, anatomist and surgeon; Nevil Maskelyne, Astronomer Royal; and Joseph Banks, to become President. He contributed significant papers to the Society and *The Forces of Fired Gunpowder and the Initial Velocities of Cannon Balls* gained him the Copley medal. He was

Charles Hutton FRS

developing 'a series of genteel houses'. In 1806 the Academy itself moved to Woolwich Common, there purchasing Hutton's property. Resigning the following year as a result of 'a pulmonary complaint' Hutton retired to a house in Bedford Row, London, with a pension of £500 p.a.

From 1796 until 1802 Hutton was a member of the Smeatonian Society and was twice called upon to report on London Bridge. In 1801 he was asked for his comments on the single 600 ft. span bridge designed by Telford and not long before his death he was again called upon, this time preparing a paper on 'the proper curve which should be adopted for the arches of the new design' for the bridge.

He never returned to Newcastle although it was said that at the time of his death he was planning such a visit, to be made by steamship from the Thames. Although he had left his wife and daughters in Newcastle, they were well-provided for by him, in spite of the fact that he fathered a daughter and married for a second time. His links with Newcastle were never completely broken; he maintained an interest in the Royal Jubilee School, the Schoolmasters Association and the Literary and Philosophical Society, the last-named still holding both a portrait and bust of him.

Hutton died on 27 January 1823 at his house in Bedford Row and was interred in a family grave at Charlton, Kent. As a young man he had shown leanings towards Methodism but this interest declined during his life although his first wife, who died in 1785, was buried in dissenters' ground in Jesmond cemetery, Newcastle. Hutton was survived by two of his daughters and by his son, George Henry (1761?–1827), who had been a cadet at the RMA and later continued his career by becoming a lieutenant general, dying in Ireland. One of his daughters, who died in Guadeloupe in 1794, was the mother of Charles Blacker Vignoles (1793–1875).

R. W. RENNISON

[*A General Index to the Philosophical Transactions* (1787); A Memoir of Charles Hutton (1823–1824) *Literary and Philosophical Society Reports, Newcastle*, **11**; E. McKenzie (1827) *A Descriptive and Historical Account ... of Newcastle*, vol. 5, 557–560; R. Welford (1895) *Men of Mark Twixt Tyne and Tweed*, vol. 2, 586–592; R. S. Watson (1897) *The History of the Literary and Philosophical Society, Newcastle*; Sk (1987); DNB; DSB, **6**, 576–7]

involved in experiments to ascertain the mean density of the globe and presented his report in 1778, the following year being appointed as the Society's foreign secretary and being awarded a LLD degree by Edinburgh University.

Hutton's literary output continued throughout his stay at the Academy. He wrote books or tracts on measurement, on the powers and products of numbers, on logarithms, on conic sections, on the teaching of mathematics and on cubic equations and infinite series. His most important work was, perhaps, his *Mathematical and Philosophical Dictionary* of 1795 but the biggest was his involvement, with others, on the eighteen volume quarto *Abridgement of the Philosophical Transactions of the Royal Society of London* in 1809, for which task he was said to have received £6,000.

In spite of Banks having been one of those who sponsored Hutton for Fellowship of the Society, differences between the two men became apparent after Banks had become President in 1778. As Foreign Secretary, Hutton found himself placed under pressure by Banks, who wished to secure for a protege the position held by Hutton. Because of the 'jealousies and dissentions that then prevailed' in the Society Hutton, with other mathematicians, resigned his membership in 1784.

While at the Academy Hutton, as a result of his health, had sought leave to remove himself to Shooters Hill where he had a house built *c.* 1784, at the same time obtaining water from a borehole. His purchase of land there enabled him to begin the manufacture of bricks, at the same time

Publications

1770. *Treatise on Mensuration ...*
1772. *A Plan of Newcastle ... and Gateshead taken from an Accurate Survey finished in the Year 1770 by Charles Hutton, Mathematician*
1772. *The Principles of Bridges*
1798. *Course of Mathematics ...*
1801. *Answers* [to the Questions respecting the Construction of a Cast Iron Bridge, of a Single Arch, 600 Feet in the Span, and 65 Feet Rise], 1801

1809. *Abridgement of the Philosophical Transactions, Royal Society*
1812. *Tracts on Mathematical and Philosophical Subjects, comprising ... the Theory of Bridges*
1815. *Mathematical and Philosophical Dictionary*

I

IRVINE, Archibald, Lieutenant-Colonel
(1797–1849), military engineer, was born in Westerkirk, Eskdale, in 1797; his father was a farmer of Hawcleugh, Roxburghshire. As a cadet in the East India Company he was trained at Addiscombe and subsequently at Chatham. He was one of the first of the East India Company's engineering cadets to have the benefit of a Chatham education. In 1816 he went to India and served in the Maratha War (1817–1818), surveying the route from Saugor to Bhopal. In the military sphere he distinguished himself at the siege of Bharatpur (1825–1826) as a Major of a Brigade of Engineers. He used for the first time in the field a method of ventilating extensive shafts and galleries, which he had been taught at Chatham. The use of this tactic was crucial in successfully undermining the fortifications and for his work Irvine was nominated a CB in 1831. Further promotion in India followed including, in September 1843, being appointed Superintendent of Marine at Calcutta. In 1846 he rejoined the army for the campaign in Sutlej for the battle of Sabraon, where he again distinguished himself.

Irvine was elected a Corresponding Member of the Institution of Civil Engineers in 1828. His civil engineering works in India included surveys in the early 1820s of the sources of the Nerbudda (Narmada), Slane (Sone) and Mahumuddie (Mahanadi) rivers. His work for the Committee of Embankments at Allahabad, involving surveys of the Jumna from Agra to Allahabad, was interrupted by military service and a return to England in 1826 prompted by severe wounds at Bharatpur. In the 1830s he supervised the construction of the Grand Trunk Road Benares–Allahabad Road. As an architect he was responsible for the design of churches in Cawnpore. After his return to England he was appointed Director of Works at the Admiralty in November 1846.

In 1835 Irvine had married Marianne Eliza Sparks in Allahabad. Following a riding accident, he died at Highgate, London, on 29 December 1849.

MIKE CHRIMES

[ICE membership records; *Min Proc ICE*, **x**, 1849, 87–90; E. W. C. Sandes (1933) *The Military Engineer in India*, **I**, 260–266; R. H. Phillimore (1954) *Historical Records of India*, **III**, 462]

Works
[1830s]. Grand Trunk Road, Benares–Allahabad

IVE (IVES or IVEY), Paul
(fl. 1584–1604), unlike other Tudor engineers, had no training other than as a military engineer, nor would he have considered it desirable as 'in this business the opinion of the soldier who hath experience of the defence and offence is to be preferred before the opinion of the geometrician or mason'. However, this opinion did not stop him encroaching on their territory by working on Dover Harbour nor from describing the use of pile foundations, retaining walls and earth reinforcements in his book on fortifications.

Ive was at Corpus Christie College, Cambridge, in 1560 but there is no evidence that he actually graduated. Certainly he knew Latin as he quotes Agricola's *De re metallica* in the section in his book on problems with water in ditches and foundations, and French as he translated *Instructions for the Warres* into English.

He spent the 1570s in the wars in the Low Countries, first in observation and then in exercise in fortifications. Ironically, the civil commission at Dover in 1584, where he was in charge of the new groynes, is his first known work. By this time the work on Dover Harbour was almost complete but it was felt that new groynes were needed on the seaward side of the shingle bank to stop erosion. Alternatively Minet considers they were intended to encourage the formation of the tidal harbour after the installation of the great sluice. In any event there was some debate over whether piers should be constructed of chalk stone or hard stone. Ive's groynes seem to have been of timber with chalk infilling and are described on a plan between 1584 and 1592 as 'Ives first groyne' and 'Ives second groyne'. Since it is described in the plan of 1595 as 'Paul Ives groyne decayed' this was possibly the wrong choice of material.

In 1589 Ive published *The Practise of Fortification*. From the walls up this is, as it claims, the first surviving original publication on fortifications in English. From the walls down it is a fascinating description of the knowledge of foundations and retaining walls current at the end of the Elizabethan period. He gave detailed specifications for the use of raft foundations for marshy ground, pile foundations for particularly unstable ground, and great chalk stones where quakes are liable to occur.

In what is probably the first description of a retaining wall in English he states that at 50 ft. high the walls should be 5 or 6 ft. broad at the ground; at 40 ft., 4 ft.; at 30 ft., 3 ft.; and at 20 ft., 2½ ft. or less. He also describes, as did Jacopo Aconcio (q.v.), the advantages of using reinforced earth instead of masonry. For problems with water he refers the reader to Agricola.

The book, or its dedication to Lord Cobham the Warden of the Cinque Ports and Sir Francis Walsingham with both of whom he would have come into contact at Dover, served its purpose as for over a decade he was responsible for building and updating the most important fortification works in the country. In 1593 he was made consultant for forts in the Channel Islands where he was responsible for Castle Cornet in Guernsey and Elizabeth Castle in Jersey, a totally new fortress where he was able to put his theories into practice. To build it, lime was brought from France, timber from the New Forest and stone quarried on the island. The project employed thirty masons and was financed by £750 in cash and the equivalent of £1,100 in labour, at four days work from every household.

In 1595–1596 he was in charge of modernising the defences of Portsmouth Harbour and the following year he was made the surveyor in charge of the repair of the Cinque Ports which cost £1,275. In 1597–1599 he undertook a similar task at Pendennis, encircling the Henrican castle with a bastioned enceinte to safeguard Falmouth Harbour.

In 1601 he was at the siege of Kinsale and stayed in the area to design James Fort and to provide charts of Kinsale Harbour which included information on the depths of the water over shoals and Youghal Harbour (1601). He has been suggested as the draughtsman of a small scale chart for the whole of Western Europe.

Ive died at Kinsale in 1604. The fort there was completed by Samuel Molyneux (fl. 1594–1625) at a cost of £2,050; Molyneux, the Clerk of the Royal Works in Ireland, also built Halbolin Fort, Cork (£1,777) after Ive's design.

In his will Ive described himself as 'servant to the King in fortifications'. He had a pension of £50 p.a. and his salary of 15s a day, interestingly, places him between Richard Lee (q.v.) and John Rogers (q.v.) on the Tudor salary scale. Sir Walter Raleigh, under whom Ive worked in Jersey, described him as having 'an excellent gift in these works and that which is rarely joined to such knowledge, as much truth and honesty as any man can have'.

SUSAN HOTS

[B. L. Cott ms, August, 1, 30; B. L. Cott ms, August 1, 46, plate iv; SP12/17; Cecil papers 209.5; DNB; William Minet (1921) Some unpublished plans of Dover Harbour, *Archaeologia*, **72**; A. Macdonald (1937) Plans of Dover Harbour in the sixteenth century, *Archaeologia Cantiana*, **49**, 108–120; A. H. W. Robinson (1962) *Marine Cartography in Britain*; Martin Biddle (1970) Introduction to reprint of *The practise of Fortification*; Rolf Loeber (1981) Architects in Ireland 1600–1720; Kings Works, vol. IV, 1982; A. Saunders (1989) *Fortress Britain*; P. M. Kerrigan (1990) Fortifications in Tudor Ireland, *Fortress*, **7**, 27–39]

Publications
1589. *The Practise of Fortifications*

Works
1584. Dover Harbour, groynes
1593–1601. Elizabeth Castle, Jersey
1601–1604. James Fort, Kinsale

IVORY, Thomas (*c.* 1732–1786), architect, was born in Cork around 1732. He was a self-educated man and, prior to practising in Dublin, he designed a 100 ft. span limestone masonry arch bridge over the river Blackwater at Lismore in County Waterford for the fifth Duke of Devonshire. It was completed in 1779 by the contractors Darley & Stokes. When built, the arch spanning the main river channel was the largest in Ireland; there was also an approach viaduct of eight smaller spans. In the early 1770s he designed a masonry arch bridge in classical style for the Duke of Leinster's estate at Carton in County Kildare.

As an architect, Ivory's most important commission in Dublin was the King's Hospital School in Blackhall Place, begun in 1773 and opened on 16 June 1777. He was Master of Architectural Drawing in the Schools of the Royal Dublin Society between 1759.

He had two daughters by his wife, Ellinor, and died in Dublin in 1786.

RON COX

[*Dublin Builder*, **1**, 1859, 115; E. McParland (1973) *Thomas Ivory, Architect*; J. C. Kinnoulty (1970) *A Biographical Dictionary of Ireland from 1500*, unpublished manuscript, Trinity College Dublin, 334–335; R. C. Cox and M. H. Gould (1998) *Civil Engineering Heritage: Ireland*]

Works
Early 1770s. Carton Demesne Bridge, architect
1776–1779. Lismore Bridge, architect
1773–1777. King's Hospital School, Dublin, architect

J

JAMES, Benjamin (*c.* 1746–1820), bridge builder, was the contractor in 1792 for a small aqueduct on the Monmouthshire Canal. He was described as 'of Blaenavon, Llanover' near Abergavenny in 1796 when he was selected to build the bridges between Llangattock and Gilwern on the Brecknock and Abergavenny Canal. By a further series of contracts over the next four years, in which he is now described as 'of Llangattock', he appears to have performed all of the subsequent mason's work on the canal up to Brecknock, including the lining of Ashford Tunnel.

In 1809 he contracted to repair Crickhowell Bridge, a thirteen-arch structure over river Usk which had been damaged by flood, and to widen it. In 1813 he was employed to certify payments for remedial works to the tower of Llywel Church and as the bridge on the present A40 road there was built in that year it is probable that he was its builder.

In the following year, Benjamin James the younger contracted with the County to rebuild Logyn Bridge on the same road at Trecastle, Pont ty Gwyn at Cray in 1816 and Aberclydach Bridge at Gilwern in 1817. These were smaller structures, of 10–21 ft. span, but in 1819 an Act of Parliament authorised a diversion of the A40 at Sennybridge, which necessitated a bridge over the Usk on a new site, and one over the Senni from which the village derives its name. Benjamin James the elder and the younger contracted together for them, the Usk bridge having a span of 50 ft. and the Senni 10 ft. less.

On the death of Thomas Waters in 1826 Benjamin James the younger was appointed County Surveyor of Monmouthshire. He produced regular quarterly reports on the County bridges but does not appear to have undertaken any major rebuilding.

P.S.M. CROSS-RUDKIN

[RAIL 500/5 and 812/3, PRO; QS P&R 0007, Gwent RO; B/D/BM A65 and Q/SO/6, Powys RO; Theophilus Jones (1809) *A History of the County of Brecknock*; Dewi Davies (1992) *Bridges of Breconshire*]

Works

1795–1796. Llangrwyney Bridge, contractor
1796–1800. Brecon & Abergavenny Canal, contractor for locks, bridges and tunnel linings
1799–1800. Brynich Aqueduct over river Usk, Brecon & Abergavenny Canal, contractor with Walter Walters, £2,200
1809–1810. Crickhowell Bridge repair and widening; £2,300
1819–1820. Usk and Senni bridges, Sennybridge, contractor; £1,000

JAMES, William (1771–1837), railway promoter, was born on 13 June 1771 at Henley in Arden, Warwickshire, the eldest son of William James, a local magistrate, and Mary Lucas.

James went to school at Warwick and Winston Green before going to study law at Lincoln Inn Fields in London. He returned to Henley in Arden after qualifying as a solicitor to share his father's practice in 1791. By now he had a brother and four sisters. He married Dinah Tarlton in 1793, an orphan of a local farming family, and moved into Yew Tree House.

William James bought his first share in the Stratford on Avon Canal Company in 1793. His father bought far more and started in canal company shares as 'canal mania' swept the nation. James began to act as a land agent; by 1798 he was agent for Mr. Dew in Wellesbourne and the following year he was appointed land agent for George Greville, Earl of Warwick. This soon involved him with turnpike roads between Warwick and London, and Warwick and Alcester. James expanded his transport interests with a projected rail road at Keynsham, Somerset.

By 1800 although James was one of Warwick's leading citizens, being Deputy Recorder and Commander of the local militia, his father had been virtually bankrupted by his speculation in canal shares, and with his health failing, asked his

son to look after his four sisters, and provided a dowry for them if they married with his approval; James agreed. The first three sisters married and each received a generous dowry. The last sister was to remain unmarried for a while longer.

James decided to enlarge his knowledge of the transport infrastructure and in 1802 he journeyed around the north of England inspecting railroads and canals. He met the Duke of Bridgewater and spent a day discussing transport with him. James began to form the opinion that railways would be the future for transport and in 1804 he was present at the Pen-y-Darren mine when Richard Trevithick's locomotive made its historic run.

James concentrated on his business and the Stratford Canal company for the next few years. He became its chairman and was responsible for the construction of Edstone Aqueduct, the largest in England. He had bought a small colliery at Wyken, near Coventry, and expanded this side of his business by opening the Balls Hill colliery at West Bromwich on the Earl of Dartmouth's land. He had identified the 10 yd. coal seam and moved to a new house on the site. He was an active man, acquiring shares in several other canal companies, leasing more collieries across the Black Country and running his land agency business from an office at New Boswell Court, London. He was involved with several turnpike roads and tramways and acquired the Upper Avon Navigation in 1813.

In 1818 James started to discuss plans for a trunk rail road with Lord Redesdale. He intended to create a steam powered railway from the terminus of the Stratford canal to Paddington, London, and produced a detailed survey of the route. Lord Redesdale persuaded him to start by building only the section to Moreton from Stratford.

The first public meeting for the Stratford and Moreton Tramway was on the 14 August 1820. James was dispatched to Newcastle to inspect a steam engine built by George Stephenson. James and Stephenson had a common interest although their social status was markedly different. Whilst they were discussing how they could work together, James was in correspondence with the promoters of other railways. The Stockton and Darlington was one such, and there were others being suggested. On 1 September 1821 James and Stephenson made a formal agreement that James and his son, William Henry, could have a quarter of any profit from engines sold below an imaginary line from Liverpool to Hull. Stephenson was to purchase the patent rights of a water tube boiler that William Henry had designed.

When the gold standard was re-imposed much of James's investment was devalued. His mines were run by managers whilst he pursued his railway interests and it is probable that they were incompetent. It was now that his last sister, Elizabeth, decided to marry a Mr. Muddie. James objected, and refused to pay the dowry and the matter was taken to court.

In 1821 James had been approached by Joseph Sanders to survey a rail line between Liverpool and Manchester, which he carried out at his own expense. The surveyors were Paul Padley, his brother in law, assisted by George Hamilton (qq.v.). Whilst surveying Chat Moss, William slipped into a bog and was saved from drowning in it by George reaching out to him with his surveying staff.

Unfortunately he was imprisoned for debt in August 1823. He owned thousands of pounds, part of which was for the disputed dowry. Joseph Sanders formed the Liverpool and Manchester Railway Company whilst James was in Fleet prison. Although James tried to keep the survey as a bargaining counter, Paul Padley took a copy to Stephenson. James managed to prove that the dowry allegedly owed to his sister was not legally due, but emerged from prison to find his affairs in the hands of receivers. Up to and including his time in prison he was involved in many different railway ventures.

From his early surveys of the Somerset railway in 1799 and the preliminary work on the Liverpool and Manchester in 1802 the importance of rail transport grew to overshadow his other surveying and legal work. After he became chairman of the West Bromwich Coal Masters Association in 1816 he proposed the Birmingham and Wolverampton, the Birmingham and Manchester, and the Birmingham and London railways. He was clearly thinking of a network rather than individual lines. His initial plans for the Stratford and Moreton Railroad in 1820 were entitled 'Plan of the lines of the Central Junction Railway or Tram-road ...' and showed the line extended to London. By 1823 his interests were on a national scale; the Bristol, Salisbury and Southampton Railway, and Bristol, Bath and Bradford Railway would have formed a network ready to couple with the London, Brighton, Portsmouth and Chatham. He took an active part in promoting all of them, even to the extent of publishing a report to illustrate the advantages of the London–Portsmouth route from his prison cell. He had also been involved with the Canterbury and Whitstable, the Truro and St Agnes, the Padstow Bodmin and Fowey, the Algavoar Moor, and the Bishops Stortford, Cambridge and Newmarket railway.

James was destitute when released from prison, his family living with his agent, Mr. Bill, at West Bromwich. He became the clerk to the Stratford and Moreton Tramway but by now he was embroiled in a feud with Stephenson as the board of the Stratford and Moreton railway was being filled with Stephenson's friends, such as John Urpeth Rastrick (q.v.), who engineered the line to completion. James was removed from his post as clerk in April 1826 and he retired to Plas Newydd Farm, Bodmin, Cornwall.

Here James carried on a legal practice and acted as land agent for the Cornish gentry. His wife, Dinah, and eldest daughter died just before the move. James remarried around 1830 and started to rebuild his life and in 1836 he was once again trying to promote new railways. In February 1837 he returned to West Bromwich to attempt to salvage his property from the receivers. He took a carriage journey back to Bodmin as his wife was giving birth to his last daughter Winfred, but he caught influenza during the rigorous trip. The illness was to prove fatal and he died on 10 March 1837. He was buried at Bodmin church.

N. BILLINGHAM

[Henley in Arden Parish Records, Warwickshire County RO; Estate papers of Earl of Darmouth, Staffordshire County RO; Minutes of the Stratford on Avon Canal Company, 1793–1816, PRO; Minutes of the Stratford and Moreton Tramway Company, 1821–1826, PRO; Kings Bench Record 1823, PRO; *The Two James's and the Two Stephensons* (1861) anonymous pamphlet by E.M.S.P., believed to be James' daughter Elizabeth Tranter Payne; C. Hadfield and J. Norris (1962) *Waterways to Stratford*; C. Hadfield (1969) *The Canals of the West Midlands*, 2nd edn.; G. Biddle (1990) *The Railway Surveyors*]

Publications

1798. *Two plans of London Dock, with some Observations respecting the River*

1820. *Plan of the Central Junction Railway*, copy in *The Two James's and the Two Stephensons*

1823. *Report to illustrate the Advantages of direct inland Communication through Kent, Surrey, Sussex and Hampshire to connect the Metropolis with the Ports of Shoreham, Rochester and Portsmouth*

Works

1806–1816. Southern section of Stratford upon Avon Canal, 13 miles, £143,000

1815–1823. Upper Avon Navigation, restoration, 17 miles, £18,000

1821–1826. Stratford and Moreton Tramway, 16 miles, £40,000

JARDINE, James, FRSE (1776–1858), son of a farmer of the same name, was born at Applegarth, Dumfriesshire on 13 November 1776. After having shown great ability in mathematics at Dumfries Academy, he went to Edinburgh with an introduction from his teacher to John Playfair, professor of mathematics at the university, where he attended classes. Playfair befriended him, obtaining employment for him as a mathematics teacher.

By 1809 Jardine, on Playfair's advice, had begun to practice as a civil engineer and by 1811 had opened an office in Edinburgh. In 1809 he observed water levels in the river Tay with reference to salmon stake nets. He was, according to Professor W. J. M. Rankine, the first to determine—by observations of the tides over a great extent of coast—the mean level of the sea and to show the symmetry of the undisturbed tidal wave above and below that level and the effect of a

James Jardine FRSE

river current in disturbing that symmetry; they were 'discoveries of high importance, both scientific and practical'.

In 1810 on the recommendation of Thomas Telford (q.v.), with whom he was to become closely associated during the next three decades, Jardine accurately determined the levels and produce of springs in the Pentland Hills, near Edinburgh. This work eventually led, in consultation with Telford, to his best known achievement, supplying Edinburgh with a plentiful supply of pure water. This water came from Crawley spring via an iron aqueduct and involved construction of a compensation water reservoir at Glencorse with one of the tallest earth dams of its time—nearly 120 ft. high—and a cut-off trench. In 1825 *The Scotsman* described these works, which had cost about £145,000, to be 'the most extensive, perfect and complete ever executed in modern times'.

From 1819 until 1846 Jardine was engineer to the Edinburgh Water Company, for whom his last major project was the harnessing via Clubbiedean reservoir of the Black Springs at the Bavelaw and Liston-Shiells estates on the North Pentland Hills and the construction of compensation water reservoirs at Threipmuir and Harlaw, near Balerno (1843–1848). Other water supply schemes upon which he is known to have been consulted included Perth (1828), Dumfries (1833), Glasgow (1834), Cobbinshaw reservoir for the Edinburgh & Glasgow Union Canal (c. 1818), and Leslie, Fife (1853). He was also employed on several important law cases involving hydraulics relating to mill dams. In 1830 he was consulted on the partial

drainage of Loch Leven by straightening the river flowing from it, building new bridges, and he successfully superintended the lowering of the loch by several feet. From 1831 until 1849 he acted as the Commissioner for this project which provided water power to many mills on the river Leven.

Jardine's other work in Edinburgh included road lines and levels, retaining walls and foundations, associated with the architectural projects of William H. Playfair and others, drainage of what remained of the North Loch (1813) and also, to the south, the 'Meadows' and heating the Signet Library. Jardine also had a considerable practice in the improvement of transport communications throughout Scotland. In 1813–1814 he surveyed and estimated the cost of the Annandale Canal project in Dumfriesshire, and the continuation of John Rennie's (q.v.) Union Canal line via the south and east of Edinburgh to Leith Docks. Neither proposal was executed. In 1818 he advised on the proposed deviation of the Union Canal line through Callandar Park, near Falkirk. From 1825 to 1830 he prepared extensive surveys and estimates for a railway between Edinburgh and Glasgow but his proposals, although seriously considered, were not adopted.

In 1826 Jardine was appointed engineer for the Edinburgh & Dalkeith, or 'Innocent' Railway as it was later known, which became operational in 1831 and cost, with the completion of its branches in 1838, about £133,000. At its Edinburgh end under Holyrood Park he built the first public railway tunnel in Scotland and a steam-engine operated inclined plane 1,160 yd. long. Near Dalkeith he built what is now his finest surviving bridge, Glenesk Bridge, to carry the railway over the river North Esk (c. 1830, conserved 1993). The 570-yd. tunnel and a now unique cast-iron beam bridge over the Braid Burn near Duddingston (1831) were conserved by Lothian Regional Council in the 1980s. Although operated by horse traction, the railway was commercially successful both as a mineral and passenger railway. For a number of years prior to its adaptation for steam locomotion in 1846–1847 it carried more passengers per mile than the Liverpool and Manchester Railway. In 1827 Jardine also became engineer for the Ardrossan and Johnstone Railway which, although underfunded, was opened from Ardrossan to Kilwinning in 1831. He also conducted road surveys, for example in Dumfriesshire in 1829 and on the 'Great North Road' to Perth from Milnathort to Glenfarg (1825–1832), probably at Telford's request.

Following the success of steam locomotion on the Liverpool & Manchester Railway, horse-traction became rapidly outdated and Jardine seems to have had little further involvement with railways, although he did survey the Nith valley line in Dumfriesshire in 1835. His river and maritime projects included Saltcoats Harbour (1811), river Tweed (1813, 1817), river Tay (1818–1833), river

Forth ferries, Earl Grey and King William IV docks at Dundee in consultation with Telford (1829–1832), Perth (1831), Leith (1833–1835) and Eyemouth (1837) harbours and an evaluation of metal lighthouse proposals for Skerryvore reef (1837).

Jardine was also a leading bridge engineer in Scotland. At Telford's request in 1821 he furnished calculations of chain strengths at different degrees of curvature for Menai suspension bridge. In addition to his railway bridges he is known to have built masonry bridges on routes in which Telford had an interest: at Threave, Kirkcudbrightshire (1825), Almond, near Perth (1827), and at and south of Dalkeith on the Edinburgh–Jedburgh–Newcastle Road (1827–1838). Jardine's un-executed masonry bridge projects included a five-span design for Dean Bridge, Edinburgh (1828) and a 160 ft. span at Coulternose (river Findhorn). He also erected a timber bridge over the river Tweed at Innerleithen in 1830 and a light iron truss bridge over the river Whitadder at Hutton Mill, Berwickshire, in 1837. Jardine's masonry bridge designs were influenced by Telford's practice but were more refined in some respects. He adopted notably small arch ring depths which reduced towards the crown, low-rise elliptical arches and longitudinal walls within hollow spandrels, all measures that economised on materials and reduced weight on the foundations. Rankine assessed Jardine's work as 'all models of skilful design and solid construction' and his masonry 'worthy of the study of every engineer'.

Jardine had a national reputation as a scientific engineer. He played an important part in determining the proportions of the old and diverse Scottish weights and measures to the imperial standards, conducting his enquiries 'with extreme precision'. In 1811 he determined the length of the ell as 37.0598 in. at 62°F. From 1812 to 1814 he measured water temperatures at different depths in Lochs Lomond, Katrine and the river Tay, work which was still considered of value when it was first published in 1871. He was elected a Fellow of the Royal Society of Edinburgh in 1812, the Geological Society in 1816, and to membership of the Institution of Civil Engineers and the Society of Civil Engineers in 1820 and 1827 respectively. He was also a director of the Edinburgh Astronomical Institution and its Astronomer from 1815 until 1825.

From 1826 Jardine operated his practice from his house at 18 Queen Street, Edinburgh, where he died on 20 June 1858. Although a bachelor, he had a daughter, Ann, by Margaret McGee. Rankine commented from personal knowledge, presumably having first met him when Rankine's father was manager of the Edinburgh & Dalkeith Railway, that although Jardine's manner 'was somewhat eccentric and cynical, he secured the warm regards of his intimate friends amongst whom were many of the highest eminence in science'.

R. A. PAXTON

[National Archives of Scotland (SRO), Register House Plans and E&DR minute books; W. J. M. Rankine, *Imperial Dictionary of Universal Biography*, **XII**; *The Scotsman*, Suppl. 252, 26 June 1858; F. Whishaw (1842) *The Railways of Great Britain …*; J. Colston (1890) *The Edinburgh and District Water Supply*; R. A. Paxton (1968) *Three Letters from Thos. Telford with Introduction and Notes*; Skempton; R. W. Jardine (1992–1993) James Jardine and the Edinburgh Water Company, *Trans Newcomen Soc*, **64**, 121–130; R. A. Paxton (1993) *Edinburgh & Dalkeith Railway—Glenesk Bridge, Dalkeith*; Private information]

Publications
1830. *Report … for a Survey and Plan of a Railway from Edinburgh and Leith to Glasgow*
1830. *Dundee Harbour, Report respecting the extension of the Docks*

Main works
1810–1846. Edinburgh water supply, £145,000 (to 1825)
1825–1838. Edinburgh & Dalkeith Railway, £133,000
1829–1832. Dundee Harbour
1831–1849. River Leven improvement

JEANS, Thomas (fl. 1829–1836), of 14 Manchester Buildings, Fleet Street, London, was elected an Associate of the Institution of Civil Engineers on 24 February 1829; at that time he was a pupil of Thomas Rhodes (q.v.). He may have been related to Thomas Jeans (c. 1775–1866) who, as architect to the Barrack Department of the War Office, had designed barracks during the Napoleonic Wars. An FSA, he was a founder member of the Architects' and Antiquaries' Club.

MIKE CHRIMES

[ICE membership records; Colvin (3)]

JEBB, David (fl. 1766–1788), canal engineer, is reported as being the engineer in charge of the Upper Boyne Navigation in the 1760s. He appears to have had a personal interest in the work as in 1766 he had completed a large mill at Slane. Under his direction a guard lock was built above Slane Bridge in 1785 at a cost of nearly £1,800.

In 1787, when the Commissioners of Inland Navigation were dissolved, local commissioners were appointed to take over the navigation and Jebb was retained as engineer, treasurer and secretary. The following year, he prepared a detailed report to the commissioners with estimates of the repairs to the navigation from Slane to Stackallan totalling £4764 8s and also made specific recommendations for work on a number of the locks on the navigation, most of which were eventually carried out. He was later entrusted with the duties of toll collector and paymaster of the lock-keepers.

Richard Evans (q.v.) took over as Chief Engineer on the navigation in about 1790.

RON COX

[C. Ellison (1983) *The Waters of the Boyne and Blackwater*, 16–17; R. Delany (1992) *Ireland's Inland Waterways*, 41–42]

Works
1760s. Upper Boyne Navigation, engineer
1780s. Boyne Navigation, engineer

JEBB, Sir Joshua, Major-General (1793–1863), military engineer and surveyor-general of prisons, was born on 8 May 1793, the eldest son of Joshua Jebb and Dorothy Gladwin. He graduated from the Royal Military Academy, Woolwich and was commissioned second lieutenant in the Royal Engineers on 1 July 1812. A year later he was posted to Lower Canada and served in the army during the War of 1812 against the Americans. He was commended for his service in the United States during the battle of Plattsburg on 11 September 1814.

After the war he continued active service in the Engineers Department in the Canadas, conducting land surveys as well as planning and designing various military works such as stores, wharfs, and other buildings. His most significant work, however, was the preparation of plans for a defensible, internal canal route partly along the Rideau River, through Upper Canada (now Ontario) from the Ottawa River to Kingston on Lake Ontario. As part of his 1816 report to Colonel Elias Durnford (q.v.), his Commanding Engineer, he recommended the use of a 5-mile long marine railway for transporting ships over the height of land between the rivers. This recommendation is one of the first documented marine railway proposals in British North America. In 1819 Jebb surveyed a preliminary canal route between Lakes Erie and Ontario and in March 1820 issued a report and maps showing the proposed route. This would prove to be the first survey and plan for the now-famous Welland Canal around the Falls of Niagara. Later in 1820 he returned to England and was stationed at Woolwich and then Hull. In 1827 he was posted to India and the following year was invalided home where he was appointed adjutant of the Royal Sappers and Miners at Chatham.

In 1839 Jebb was seconded to the Home Office as surveyor-general of prisons wherein he provided technical advice on the design and construction of prisons and on the management of penal institutions. He was strongly opposed to transporting convicts to remote colonies, favouring instead their employment on public works projects in England. Jebb developed a new design for the Pentonville prison which was long regarded as a model for prisons around the world. He was considered thereafter as the originator of modern prison architecture. He made a particularly significant contribution in the area of heating and ventilation, which unusually he varied to suit different climates, i.e. in the West Indies. In 1850 Jebb retired from military service and was appointed Chairman of the newly formed board which was to be responsible for the management of prisons throughout England. Jebb was made Colonel in 1854, Major-General in 1860 and KCB in 1859. He served the last two years of his life as a commissioner considering embanking the Thames in London and communications between Blackfriars Bridge and Mansion House and between Westminster Bridge and Millbank.

Jebb was married twice; first to Mary Legh Thomas, with whom he had four children, and second to Lady Amelia Pelham. He died suddenly on 26 June 1863 at Chesterfield.

A **William F. Jebb** (fl. 1820–1830), civil engineer, of Northants, was elected a Corresponding Member of ICE in 1820. Although he remained in membership until 1830, no family or other details are known.

MARK ANDREWS

[DNB; *Encyclopedia Brittanica*, 11th edn., **22**, 1911, 364; DCB, **IX**, 1976, 412, Toronto; W. Porter (1977 repr.) *The History of the Corps of Royal Engineers*, vol. II, reprinted 1977, Chatham; J. Weiler (1987) *Army architects*, vol. 3 (thesis, University of York, mss); J. Winearls (1991) *Mapping Upper Canada 1780–1867; An annotated Bibliography of Manuscript and printed Maps*; Colonel K. W. Dale (1994) *Sir Joshua Jebb (1793–1863) and the model prison—Pentonville*, *Royal Engineers Journal*, **108**, 1840–1842, 272–277]

Publications
1836. *A Practical Treatise on Strengthening and Defending Outposts, Villages, Houses, Bridges, etc.*
1844. *Modern Prisons: Their construction and ventilation*
1844. *Notes on the Theory and Practice of Sinking Artesian Wells*
1853. *Manual for the Militia, or Fighting made Easy: A practical Treatise on strengthening and defending Military Posts, etc.*
1855. Report on barrack accommodation
1860. *Observations on the Defence of London, with Suggestions Respecting the Necessary Works*
1863. *Reports and Observations on the Discipline and Management of Convict Prisons*

JELFE, Andrews (*c*. 1690–1759), master mason, son of William Jelfe of South Weald, Essex, was made free of the Masons' Company in 1711 having served a seven year apprenticeship with Edward Strong (1652–1724), one of the leading masons on St. Paul's Cathedral and Blenheim Palace. From 1719 to 1727 he was 'Architect and Clerk of Works' to the Office of Ordnance, working on military structures in Scotland and in 1720 supervising the construction of a new Gun Wharf at Plymouth. In partnership with Christopher Cass (*c*. 1678–1734) he carried out the masonry contracts for the entrance piers and Great Sluice at Rye New Harbour (1728–1733) to plans by John Perry (q.v.). Also at Rye, the Town Hall was built to Jelfe's designs in 1743. By far his largest work, in partnership with Samuel Tufnell (fl. 1719–d. 1765), was building the masonry of Westminster Bridge to designs by Charles Labelye (q.v.). Their first contract dates from June 1738 and further contracts were placed as work progressed, the last (of fourteen) being signed in February 1750. The last stone was laid in November 1750.

In 1742 he built thirteen houses on the north side of New Palace Yard and the south side of Bridge Street, later purchasing the freehold from the Westminster Bridge Commissioners; one of the houses he occupied himself. He bought a country house at Bletchingly, Surrey, in 1747. He died on 26 April 1759, leaving his house in New Palace Yard and £10,000 to his daughter, a similar sum to his younger son and the Bletchingley house to his elder son, a naval Captain.

A. W. SKEMPTON

[Rye Harbour Commissioners Minute Book (1728–1733) East Sussex RO; Patricia M. Carson (1954) *The Provision and Administration of Bridges over the Lower Thames, with Special Reference to Westminster and Blackfriars*, unpublished MA thesis, University of London; Colvin (1); R. J. B. Walker (1979) *Old Westminster Bridge*; R. Hewlings (1993) Hawksmoor's brave designs for the Palace, in J. Bold and E. Cheney (eds.), *English Architecture: Public and Private*]

Civil engineering masonry contracts
1728–1733. Cass & Jelfe, Rye New Harbour, piers (£3,920) and sluice (£4,310)
1738–1750. Jelfe & Tufnell, Westminster Bridge, £155,320

JESSOP, Josias (1781–1826) civil engineer, was baptised on 26 October 1781 at Fairburn in Yorkshire, the second son of William Jessop (q.v.) and his wife Sarah. Two or three years later the family moved to Newark, where they stayed until 1805. Trained by his father, he took his first job under him as assistant engineer on construction of the West India Docks in February 1801 at a salary of £100 p.a. In October 1802 William and Josias were appointed engineers of the proposed Croydon, Merstham and Godstone Railway, to prepare plans and sections. The petition for a Bill went to Parliament in December, William having prepared the estimates. Josias gave evidence and the Act passed in May 1803. By that time insufficient funds had been subscribed for the full length of line, so specifications were drawn up for building the railway from Croydon to Merstham (9 miles). The Butterley Company won the contract for the whole job with a bid of £36,350. Josias set out the line, George Leather Jr. (q.v.) checked the levels of the track as it was being laid, and Benjamin Outram (q.v.), manager of the Butterley Company, personally directed the contract. Subcontractors carried out the earthworks.

The line was opened in July 1805 but before that, in February 1804, Josias went to Bristol as resident engineer on the Floating Harbour works 'under the direction of his father' at an annual salary of £500, this to be a full-time appointment requiring 'his constant residence and attention to the works during their progress'. The entire scheme was certified as complete in May 1809 but Josias stayed on to the end of 1810 finishing various jobs, ordering a steam dredger, dealing with a slip in the river bank and so on. By March 1810 the official statement of expenditure showed a total of £611,000 of which the engineering works alone accounted for £457,000. The Floating Harbour provided Bristol with the largest area (70 acres) of impounded water for shipping in the world, the entrance lock was one of the largest yet built (45 ft. wide, 200 ft. between gates with 34 ft. of water on the outer sills at HWST) and the New Cut for the River Avon was 2 miles in length, 120 ft. top water width, crossed by two iron bridges. Prince's (Swing) Bridge over the Harbour seems to have been entirely the responsibility of Josias. He submitted estimates to Parliament, gave evidence in the Lords, and in May 1808 delivered plans and specifications. The bridge was opened in May 1809 at a cost of £14,300.

The Directors expressed their appreciation of 'Mr. Josias Jessop with whose accuracy and attention to the Interests of the Company they have every reason to be satisfied; and to whom it is no more the justice to state, that he acquitted himself in the important trust reposed in him with the most unsullied reputation and honor'.

Josias, rising thirty years of age, could now launch his career as an independent consulting engineer. In 1811 he was asked to select the best route for a canal linking the River Wey Navigation near Godalming to Newbridge on the River Arun. He spent four weeks with surveyors, submitting his report in August. A prospectus was issued in October. Further surveys followed and in May 1812 he produced a revised estimate of £86,000 for a partly re-aligned route. The Act passed in April 1813, and work began in July with May Upton as resident engineer. Jessop designed all the bridges, aqueducts, locks and lock-keepers' cottages. 18½ miles in length with twenty-three locks for 30 ton barges, the canal was opened in September 1816 at a total cost of £103,000.

Meanwhile, towards the end of 1813, Jessop was called in to estimate for a western extension of the Montgomeryshire Canal to Newtown. His report was accepted in July 1814 but the Bill, meeting strong opposition, was not passed until 1815. With John Williams as assistant engineer the canal (7½ miles long with 6 locks) was completed to Jessop's plans in 1819 at a total cost around £50,000.

After the death of Benjamin Outram in 1805 the Jessop family moved to Butterley Hall, and Josias' younger brother William (c. 1783–1852) took over as manager of the Butterley Company. From as early as 1794 one of the firm's specialities had been the manufacture of cast iron plate rails and in several cases, under Outram's direction, they acted as contractor to build the line. This practice continued, but in 1813 the company introduced I-section fish-bellied edge rails on the short (2 miles) railway from the Grantham Canal to Belvoir Castle, built for the Duke of Rutland. They were cast iron, 3 ft. long, weighing 40 lb. and laid on a 4 ft. 4½in gauge. Rails of identical type but heavier section were used by Josias on the 8 mile railway from Mansfield to Pinxton Basin on a branch of the Cromford Canal. The railway was authorised by an Act of 16 June 1817. A fortnight later Jessop wrote to John Rennie that he could not come to London for the time being as he had just begun setting out the line and would soon be letting contractors for the embankments. A newspaper account of the opening ceremony on 13 April 1819 refers to a large assembly of people 'at the beautiful five-arch bridge, constructed under the direction of Mr. Jessop, the engineer', proceeding thence to Mansfield market place accompanied by nearly three hundred workmen who had been employed on the railway. The bridge, known as King's Mill Viaduct, is now scheduled as an Ancient Monument and carries a plaque commemorating Josias Jessop.

In a report of October 1819 on Newhaven harbour Jessop recommended that the east pier should be extended by a jetty built to a little above low-water level, thus leading the ebb-tide over the shoals at the harbour mouth to increase the scour. The jetty was built in the following year and in a second report of August 1821 he stated that the channel had been improved. Further improvements were effected by extending the pier itself, following a report by William Cubitt (q.v.) in 1835. On other harbour works Jessop reported on Plymouth Breakwater in 1823 with Thomas Telford and the Rennie brothers George and John, and again in 1826 with William Chapman and John Rennie.

On 28 July 1824 Jessop received a request to ascertain the best route for a railway linking the Cromford and Peak Forest Canals, which required crossing the Derbyshire Pennines. He reported on 2 September, proposing a line 33 miles long rising 990 ft. from Cromford to the summit and falling 250 ft. to Whaley Bridge; the great majority of the rise and fall to be in steep inclines (typical gradient 1:8) operated by stationary steam engines. His bold scheme met with approval. He was then asked to make more detailed surveys and estimates. These were presented at a meeting on 1 December, the cost being estimated at £150,000 later modified by an increased allowance for contingencies to £155,000.

On this basis the enabling Act passed in May 1825. The line included two short tunnels and Burbage (or Buxton) Tunnel 580 yd. in length, several cuttings and embankments some of which had stone retaining walls on each side. Butterley supplied the rails (of the Belvoir type but 4 ft. long weighing 74 lb. and laid on a 4 ft. 8 in. gauge) and eight steam engines (two of 20 hp,

the others of 40 hp). Contracts in twenty-four lots were let for the earthworks, the last dating from June 1826. Thomas Woodhouse (1793–1855) was appointed resident engineer.

During this period there had been remarkable developments on the proposed Liverpool and Manchester Railway. Briefly, the Bill based on George Stephenson's survey of 1824 had been withdrawn in June 1825, partly due to serious errors in the survey and Stephenson's incompetent evidence. The Company than asked George and John Rennie (qq.v.) to plan a new line. This they did, with Charles Vignoles (q.v.) as assistant, and a petition went to Parliament in February 1826. Vignoles, Jessop and George Rennie gave evidence in support of the Bill in April and the enabling Act duly passed in May. The Rennies were asked to submit their terms to act as engineers. In doing so they stated they would be prepared to work with Jessop or Telford or indeed any member of the Society of Civil Engineers (to which Jessop had been elected in 1813) but not with Stephenson. Meanwhile Vignoles was appointed to set out the line and, the Rennies' offer having been declined, the Company appointed Jessop as consulting engineer on 21 June. A fortnight later Stephenson was appointed principal engineer to execute the works.

Jessop was consulted soon afterwards on a proposed modification of the line. How his relationship with Stephenson would have developed cannot be known, for Jessop died on 30 September 1826 following an illness supposedly from over exertion on the Cromford and High Peak Railway. Woodhouse then took over direction of the works and completed the line in 1831. In the north part of the summit tunnel is a stone inscribed 'Josias Jessop Engineer'.

A. W. SKEMPTON

[Rennie Papers (1817) NLS Box 23; Evidence (1817) Select Committee on Steam Boats, 37–38; Thomas Gray (1825) *Observations on a General Iron Rail-way* (for Jessop's reports on the Cromford and High Peak); Sir John Rennie (1848) *An Historical, Practical and Theoretical Account of the Breakwater in Plymouth Sound*; A. E. Carey (1886) Harbour improvement at Newhaven, Sussex, *MPICE*, **87**, 92–113; C. E. Lee (1938) The Belvoir Castle Railway, *Railway Magazine*, **72**, 391–394; J. A. Birks and P. Coxon (1949) The Mansfield and Pinxton Railway, *Railway Magazine*, **95**, 224–233; D. J. Hopkins (1963) The origins and independent years of the Cromford and High Peak Railway, *Journal of Transport History*, **6**, 39–55; P. A. L. Vine (1965) *London's Lost Route to the Sea* (for Wey and Arun Canal); Charles Hadfield (1966) *Canals of the West Midlands*; R. E. Carlson (1969) *The Liverpool and Manchester Railway Project*; R. J. Riden (1973) The Butterley Company and railway construction, *Transport History*, **6**, 30–52; Charles Hadfield and A. W. Skempton (1979) *William Jessop, Engineer* (for Croydon and Merstham Railway and Bristol

Floating Harbour); John Marshall (1982) *The Cromford and High Peak Railway*; J. K. Gardiner (1999) King's Mill Viaduct Commemoration, *PHEW Newsletter*, **82**, 1–2]

Publications

1813. *A Plan of a proposed Canal to connect the Rivers Arun and Wey*

1817. *Plan (and Section) of the Intended Railway from Mansfield ... to Pinxton Basin*

1824. *Report to the Committee of the Promoters of the intended Rail-way from Cromford to the Peak Forest Canal at Whaley Bridge*

1825. *Report ... on the most desirable mode of Improving the Communication between Newcastle and Carlisle*

1826. Evidence on Liverpool and Manchester Railway Bill, *House of Lords Sessional Papers*

Works

1804–1810. Bristol Floating Harbour, resident engineer, £457,000

1808–1809. Prince's Bridge, Bristol, £14,000

1813–1816. Wey and Arun Canal, £103,000

1815–1819. Montgomeryshire Canal, western extension, £50,000

1817–1819. Mansfield and Pinxton Railway, £32,000

1820. Newhaven Harbour, east jetty

1825–1826. Cromford and High Peak Railway, completed 1831, £160,000

William Jessop

JESSOP, William (1745–1814), civil engineer, was born in Devonport on 12 January 1745, the eldest son of Elizabeth (*née* Foot) and Josias Jessop. There were two younger brothers, Josias and Samuel, baptised in 1747 and 1749 respectively, and a sister, Mary. Josias Sr. was a shipwright who from 1756 until 1759 was assistant to John Smeaton (q.v.) during the construction of the Eddystone lighthouse.

William Jessop was educated locally. He proved a talented scholar, becoming proficient in French and the classics, and displaying a marked ability in mathematics and science; he also possessed good practical skills. Smeaton took him on as a pupil, at the age of fourteen to learn the theoretical and practical aspects of civil engineering. Jessop then left his home in Devon to begin a career with Smeaton at Austhorpe Lodge, near Leeds. He was to gain experience in a wide variety of work during his training, on harbour, drainage and canal engineering. After 1767 he continued as draughtsman and assistant to Smeaton for another five years and then became a consulting engineer in his own right.

Jessop was an able apprentice and excelled especially on surveys and designs for river and canal navigations which included the Ure, the Calder and Hebble Navigation, the rivers Wear, Witham and Yorkshire Ouse and the Ripon Canal, to mention but a few major works. Encouraged by Smeaton, Jessop first acted independently at the age of twenty-seven, in 1772, on the alignment of new river cuts for the Aire and Calder

Navigation. Then followed the design of a connecting waterway, the Haddlesey to Selby Canal, the Act for which was obtained in 1774, Jessop having given evidence on the Bill. Work began early in 1775 with Jessop as engineer—with a salary of £250—and the Pinkerton brothers, James and John (q.v.) as contractors. The Selby Canal was opened in 1778 and soon became a heavily used waterway.

In 1771 Smeaton was invited to visit the Grand Canal of Ireland (then under construction, since 1756) to advise on the unfinished line of the canal, as well as on proposed connections with the rivers Barrow and Boyne. Smeaton delayed his visit until June 1773 because of pressure of work on the Forth & Clyde Canal. Jessop then went with him and after two months spent viewing the route, Smeaton left him with instructions to make surveys. These were completed by the end of October. Smeaton then reported, making recommendations on channel construction, lock dimensions, and the route of the central section across the Bog of Allen.

During that first visit to Ireland Jessop went north to view the Tyrone Coal Canal, again on behalf of Smeaton. This tub-boat canal, designed by Davis Ducart (q.v.), was built with three inclined planes, known as 'hurries', in place of conventional pound locks. These planes, the first to be built in the British Isles, had rises of between 50 and 70 ft. One-ton boats were intended to run down the inclines over rollers but this system had not operated satisfactorily. Smeaton's suggestions to Jessop were to substitute slides for rollers and, if that method did not answer, then a horse-drawn tramway should replace the canal entirely. The planes were made to work, however, during the canal's short lifetime of ten years from 1777. Boats were first lifted over the ridges at the top levels with power provided by horse gins, before descending the planes on wheeled cradles, alternately down a double trackway. It is possible that Jessop devised this system, basing it on Vittoria Zoncas's published work of 1607 which described the Fusina inclined plane near Venice. This closely resembled the Tyrone hurries.

In 1773 Jessop became a member of the Smeatonian Society of Civil Engineers, founded two years previously; he was appointed Secretary the following year, a task which he retained until 1792. In 1777 he married Sarah Sawyer, of Haddlesey House, a lady ten years his junior, and after a period living in Pontefract they moved to Fairburn, near Ferrybridge. Their first child, John, was born there in 1779, their second, Josias (q.v.), in 1781. Seemingly, Jessop was unsure of a future direction for his career and for some time, in partnership with the Pinkertons, he had dabbled in local enterprises which included a lime-burning business, a coal-mine and a dry dock; he even tried canal contracting on the Chester Canal. But late in 1777 he was consulted on the canalisation of the Soar (the Loughborough Navigation); this was to lead him to the Midlands and to a new, significant phase in his career.

The Trent and Mersey Canal was opened in 1777, bringing increased traffic to the Trent which was then an unsatisfactory river navigation. The river was shallow and winding above the limits of the tidal reaches at Gainsborough and trading was restricted to vessels of 10–12 tons and 18 in. draft. Thomas Dadford (q.v.) had surveyed the river and recommended improvements estimated at £20,000. Although a Bill was introduced in 1782, it was withdrawn because the proposals were too costly. Jessop was invited to make a new survey and his recommended works, based on meticulous investigations, were more modest. His scheme was accepted and an Act was passed in 1783. Jessop was appointed engineer and the work was finished in 1784 at a cost of £3,100. In 1783 Jessop moved the family to Newark where, apart from his engineering, he continued with his business interests, investing in a cotton mill there which he retained until his death; it was bequeathed to his third son, William, born in 1783 also.

Shortly afterwards, Jessop was invited by the Staffordshire and Worcestershire Canal Company to survey the river Severn, down from the junction of their canal with the river at Stourport, with a view to improving the river navigation. His scheme included fourteen locks and a towpath, with reservoirs to collect flood waters for use in summer. Conflicting interests defeated his imaginative scheme and so the Severn languished, unimproved until the 1840s. Riparian industry, hampered by inadequate transport, had by then failed to develop its full potential.

Then followed in 1785 a succession of briefs for proposed (but unsuccessful) plans for extensions of the Soar and its tributaries, and reservoirs on the Don for Sheffield's domestic water supplies. The next year Jessop was designing harbour and river diversion works at Rye, a major drainage scheme in Holderness, near Hull, and improvements for the Thames Navigation to accommodate traffic from the Oxford and the Thames & Severn Canals. His practice was rapidly expanding, with work on other waterways too, including the Sussex Ouse and the Ipswich & Stowmarket Navigation. He also acted as engineer for the Basingstoke Canal between 1788 and 1790, leaving completion of the work to the resident engineer Henry Eastburn (q.v.).

Jessop's services were constantly in demand during the Canal Mania years of 1789 to 1796. He was regarded as the leading engineer of his generation and most promoting committees tried hard to get him to formulate their plans and to guide their Bills through Parliament. Hence he appeared before Lords committees for navigation Bills on twenty-seven occasions, far more than any other engineer of his day. Most of the schemes on which he was engaged were successful, probably because he could pick and choose his jobs. His works included the trunk routes of the Grand Junction Canal, the Grand Canal of Ireland, the Ellesmere Canal in North Wales, and the Rochdale Canal which crossed the northern Pennines. He developed an intricate network of waterways, centred on the Trent Navigation and extending into the Witham Navigation in Lincolnshire. Other canals on which he was consulted were as far apart as the Somersetshire Coal Canal, the Rother and Arun Navigation, the Lancaster Canal and the Newcastle to Solway Firth Canal.

The Cromford Canal linked Langley Mill on the Erewash Canal (which led down into the river Trent) with the industry and limestone quarries of Cromford and the coal mines of Pinxton. Jessop's imaginative scheme, designed in 1788, incorporated a 14-mile long summit pound which crossed some rugged, hilly country on one level to join Cromford and Pinxton. A line locked down from the summit into the Erewash Canal. The long pound necessitated driving the 2,978-yd. long Butterley tunnel, the construction of the Derwent masonry aqueduct of 80 ft. span and some 200 yd. in length, and the Amber embankment, also 200 yd. in length and 30 ft. high in places. Both structures partially failed shortly after completion, for which Jessop characteristically accepted the blame and even personally defrayed some of the costs of repairs. Jessop was assisted by Benjamin Outram (q.v.) in preparations for the Bill and also as resident engineer during the construction of the canal. Iron and coal seams were discovered when driving the tunnel and this was partly the reason why Outram and Francis Beresford, solicitor for the canal company, purchased the nearby Butterley Hall estate when it came onto the market in 1790. Two years later a partnership was established to include William Jessop and John Wright, a Nottingham banker, who was to provide most of the finance for the ironworks and coal-mining company established on the estate. This later became the Butterley Company.

In 1789, Jessop was instructed by the Trent Navigation Company to recommend further improvements to the river because of the increased traffic from the Cromford Canal. He wisely produced more sophisticated designs than previously and the company's very first side cut with locks was finished at Sawley by 1793. The Beeston cut which, with the Nottingham Canal (also engineered by Jessop) bypassed the notoriously difficult and dangerous Trent Bridge section of the river, was completed in 1796. This and other major improvements, which continued until 1801, were authorised by an Act of 1794, the result of a remarkably detailed report made jointly by Jessop and Robert Whitworth (q.v.) on the state of the 71 miles of river navigation down to Gainsborough.

Fresh interest for the Leicester Navigation was awakened in 1790 and Jessop was again invited to prepare a scheme and report. Detailed surveys were made by Christopher Staveley (q.v.) under Jessop's supervision and the Act was obtained in May 1791. The Navigation comprised two waterways, the River and Forest lines. The river Soar was made navigable by cuts and widenings, including a diversion up the Wreake, before it cut back to the Soar via the Sutton's Cut, and then to Leicester. The Forest line was intended to be a canal throughout in the 1785 scheme but it was later found that there was insufficient water for lockage. Hence it now began as a horse tramroad, rising 185 ft. over 2½ miles from Nanpantan, after which it ran for 7¾ miles as level canal, fed from a reservoir from the Blackbrook, continuing as a railway 2½ miles to Swannington and Coleorton. On this line he employed flanged wheels running on edge rails though all his later railways were built as plateways. The works were nearing completion by early 1795 but the reservoir was finished only two years later. Unfortunately, following a thaw in February 1799, the dam failed due to insufficient overflow capacity. Great damage was caused in the district, as well as to the structure of the waterway. The canal never reopened after that disaster and the line was abandoned.

Proposals for the Melton Mowbray Navigation were also revived during 1791 and the Oakham Canal the following year. Jessop was involved with both waterways although construction work was supervised by Staveley. The Wreake was canalised beyond the Sutton's Cut of the Leicester Navigation up-river to Melton Mowbray. There were twelve broad locks and construction was finished by 1797. The 15-miles long Oakham

Canal had nineteen broad locks and was completed by William Dunn, of Sheffield, in 1802.

During the first three years of the 1790s, Jessop was also involved in eight other East Midlands canals which had links with the Trent Navigation. He worked on the Derby and Nutbrook Canals with Outram, the Ashby with Robert Whitworth and on the Leicestershire and Northamptonshire Union Canal, the Grantham Canal, the Witham and Sleaford Navigations and the Horncastle Canal. Of particular interest was his review and modification of Outram's proposals for the Derby Canal. These resulted in 4 miles of railway, connecting the Little Eaton branch of the canal with Smithy Houses and collieries at Denby, an ingenious crossing of the Derwent in Derby and probably the cast-iron aqueduct which Outram built nearby in the Holmes.

The Grantham Canal was promoted as an adjunct of the Nottingham Canal to provide a market for coal transported through the latter and along the 33-mile route to Grantham. Originally, it was planned to begin at Radcliffe-on-Trent but the first Bill was withdrawn after strong opposition was encountered in Parliament on water supply proposals. Jessop re-designed the line to begin on the Trent, just opposite the entrance to the Nottingham Canal, and planned to provide water only from reservoirs, at Knipton and Denton. This was a new concept in canal building. The Bill was successful in 1793 and although James Green (q.v.) and William King built the canal, Jessop kept a watch on matters both during and after completion.

Improvements to the Fossdyke and Witham Navigations were already in hand when Jessop was asked to report on the Lincoln High Bridge, a medieval structure central to these interconnected waterways. This bridge impeded any improvements because only small vessels drawing one foot could pass through. Jessop first proposed a bypass canal but a public meeting preferred a route through the High Bridge. Hence he recommended lowering the wooden floor of the structure, together with an adjustment of weirs, to give a depth of 3 ft. 6 in. This was duly put in hand although Jessop's other proposals for the Witham were shelved through lack of funds. Further improvements were left until 1803 when John Rennie (q.v.) was called in to make recommendations which formed the basis for a new Act in 1808. Jessop was then invited to comment but the works proposed by both engineers, for a commercial centre in Lincoln and a floating harbour in Boston, were too advanced and expensive for the time.

Jessop and John Hudson (q.v.) had their proposals accepted for the 13-mile long Sleaford Navigation down to the Witham and an Act was obtained in 1792. The line followed the deepened Kyme Eau drain, partly along the river Slea and also a widened mill-stream. The waterway fell through 42 ft. via seven locks, each 60 ft. by 15 ft., the lower two with flood staunches. There was a series of small tunnels through the banks near the Witham used for watering cattle during dry seasons and these would be drowned out by the projected water levels. Objections largely centred on these feeders, as well as on the safety of the banks protecting Holland Fen from the waters of the Kyme Eau. The proprietors agreed to safeguard these interests and the waterway was operational by mid-1794.

An Act for the 11-miles long Horncastle Canal was granted on the same day as that for the Sleaford Navigation. Jessop had offered a choice of two schemes: a canalised river Bain from Dogdyke to Horncastle, or a new canal from Kirkstead on the Witham. Both would have twelve locks and costs would be much the same. The promoters chose the former. Progress of the works was impeded by incompetent contractors and their deviations from Jessop's plans, but a dearth of capital delayed completion until 1802 under John Rennie (q.v.).

Hence Jessop was concerned with the navigable Trent and all its associated waterways from Torksey to the Derwent Mouth, with the exception of the Erewash Canal and the Witham line. Most of his projections were designed during the ten years from 1783 when he was first briefed for work on the Trent and, for the most part, these waterways were finished within ten years of the authorising Acts. The total length of navigable cut was 316 miles, with one-hundred and sixty-five locks. All was broad waterway, with the exception of the Cromford Canal to the west of the Butterley tunnel, which was built to narrow boat standards. Thus Jessop established a remarkable transport base which permitted industry, mining, quarrying and agriculture to develop and flourish for decades to come.

In spite of the vast amount of work that Jessop undertook in the Midlands in 1793, he also accepted the appointment of chief engineer to the Grand Junction Canal, a broad waterway to link the industrial North and Midlands with London, and thus to bypass the established Oxford Canal route via the Thames. It was surveyed by James Barnes (q.v.); it ran from Braunston on the Oxford Canal through Blisworth, Tring, Watford and Southall to the Thames at Brentford, a distance of 93 miles. There were 31 miles of branch canals, more than one hundred locks, substantial earthworks and two major tunnels, at Braunston (2,049 yd.) and Blisworth (3,076 yd.).

With Barnes as full-time engineer, work began at both ends of the canal, as well as in the tunnels and at the Tring cuttings, and, by December 1793, three thousand men were employed. Braunston tunnel was unexpectedly driven through quicksand for 320 yd. at considerable extra costs but, in spite of this and some serious setting-out errors, it was finished ahead of schedule in June 1796. Blisworth tunnel fell behind schedule and, although treacherous strata and underground springs of water had been encountered, discovery of indifferent workmanship and above all,

a deliberate variation of the alignment without Jessop's knowledge, brought work to a halt in 1795. Jessop recommended locking over the hill rather than risk long delays to opening the canal but Barnes was opposed to this view, as were Whitworth and Rennie when called in to report. Jessop's recommendations were rejected and Barnes continued work in the tunnel, but further difficulties and burgeoning costs brought work to a halt in 1797. Jessop then proposed a railway across Blisworth Hill and this was accepted by the proprietors. A contract was undertaken by Outram and the railway was finished by late 1800; the canal was then opened from Braunston to London, although the tunnel was not completed until March 1805 under Barnes' direction.

For much the same period Jessop was concerned with the Rochdale Canal, another trunk line which linked the Bridgewater Canal in Manchester with the Calder & Hebble Navigation at Sowerby Bridge. Rennie had previously surveyed and reported for two Bills, both of which failed through strong opposition from mill-owners in the Pennines, who depended on the water of hill streams for power. Jessop was persuaded to pilot the 1794 Bill through Parliament and he succeeded, largely because he intended the canal to be fed from reservoirs; these would retain only surplus flood waters. All the mill feeders would pass under the canal and compensation reservoirs would be built for any streams cut off during construction.

William Crosley (q.v.) surveyed the line under Rennie's direction but Jessop planned the engineering construction, which incorporated ninety-two locks over the 33 miles of canal, with level branches to Rochdale and Heywood. To conserve water he made most locks with 10 ft. fall; gates were interchangeable. Rennie had intended to drive a tunnel 3,000 yd. long at the summit but, probably because of his experiences at Blisworth, Jessop substituted seven locks up and seven down, with a 30-ft. deep cutting between. There were major aqueducts at Hebden Bridge (four 25 ft. arches) and Todmorden (two 21 ft. arches), a massive embankment at Slack's Valley and reservoirs at Hollingsworth and Blackstonedge. Jessop built the first correctly-designed skew bridge in the country on this canal, at March Barn. The design was based on William Chapman's (q.v.) method, devised in 1787 for bridges on the Kildare Canal. The Rochdale Canal was fully operational in 1804.

The Barnsley Canal ran 10 miles, via fifteen locks, from the Aire & Calder Navigation in Wakefield to the Dearne & Dove Canal near Barnsley, with a branch of 5 miles rising by five locks to a terminal basin at Barnby. Railways connected this to five collieries. The principal engineering work was the aqueduct (five 30 ft. arches) across the river Dearne, just before the canal junction. Jessop appeared for the Bill, with Elias Wright, and the Act was obtained in June that year. The canal was opened to Barnsley by mid-1799 and the upper section followed by 1802.

Schemes had been proposed for an Ellesmere Canal, to join the Mersey with the Dee at Chester and the Severn at Shrewsbury, and in 1791 Jessop was invited to decide on the best route. Initially he considered a line across the Wirral to Wrexham and then on via a tunnel of 4,600 yd. to Ruabon, and across the Dee at Pontcysyllte. Another tunnel, 1,235 yd. long, followed at Chirk before the canal crossed the Ceirog at Pontfaen; it then passed by Frankton and a 476 yd. tunnel to the Severn at Shrewsbury. Eventually, Jessop proposed to avoid the tunnel at Ruabon by using a summit level and locks. The Act was obtained in 1793.

Thomas Telford (q.v.), then the Surveyor of Public Works for Shropshire, was appointed engineer with a brief to prepare design details and supervise construction. He also received instructions to 'submit such Drawings to the Consideration and Correction of Mr. William Jessop'.

The principal structures on the canal are the Pontcysyllte aqueduct across the valley of the Dee near Llangollen and another at Chirk. Masonry arches were originally considered for the former but Jessop finally decided to adopt an iron trough, 127 ft. above the river Dee, supported by masonry piers at 53 ft. centres, to give clear spans of 45 ft. The origin of this idea is uncertain; there was a growing interest at the time in iron structures and Telford was then building an iron road bridge at Buildwas, as well as an iron aqueduct at Longdon-on-Tern on the Shrewsbury Canal. Hence it is probable that he suggested the concept to Jessop. As built, the length of the aqueduct was 1,007 ft. with eighteen piers. The Chirk aqueduct was built in masonry although the channel base was formed in iron plates, tied into the side walls. It is 70 ft. high and 696 ft. long, crossing the river Ceiriog with ten spans of 40 ft. It was opened by 1801 but Pontcysyllte was not finished until late in 1805, entirely under Telford's direction from 1801.

In 1785, Jessop returned briefly to Ireland to advise on the Grand Canal and from 1790 onwards, until completion of the works in 1804, he was the consulting engineer for the company. His principal works included the Circular Line around Dublin to join the river Liffey at Ringsend, where extensive docks were built to his designs. These comprised three graving docks and a large wet dock with three entrance locks, one for barges and two for ships, the larger of which measured 149 ft. by 28 ft., with a depth of 18 ft. on the sills at high-water spring tides. Jessop then determined the best route for the unfinished canal to the Shannon and also surveyed the middle reaches of the river, from Lough Allen to Lough Derg, to formulate a comprehensive development plan. The most exacting of Jessop's problems was the crossing of several miles of undrained peat bog, notably the Bog of Allen, near Edenderry, upon embankment. His plan was to drain each side of the intended line of canal and then to build up the embankment, before excavating the central core and laying puddle clay for the channel. The method succeeded, although work was beset with difficult conditions and often poor workmanship. Jessop was highly regarded by the company; indeed the directors were to write 'that the Services of Mr. Jessop are of the utmost Value to the Company upon all Occasions'.

The enclosures of the eighteenth century, and the resulting land improvements, led to the drainage and reclamation of many thousands of acres of fen and low-lying ground. Jessop, in common with other engineers of the day, was consulted on such matters. He advised on at least eleven drainage schemes in East Anglia, Yorkshire and Nottinghamshire, Somerset and Sussex, at different periods between 1785 and 1812.

The Holderness Drainage scheme, north of the Humber, was typical of these. 11,200 acres had been improved under an Act of 1764 with John Grundy (q.v.) as engineer and Smeaton as consultant. Although the deepening and extending of drains and raising of river banks continued up to 1775, much remained to be done. Large areas near Leven in the north and by Weel, east of Beverley, were under water and winter floods were frequent. The drains were inadequate because they had to cope with rain from the carrs whilst full with water from the uplands. After a month of exploring some 40 square miles of ground, Jessop presented his report in July 1786. A plan was agreed the following year.

George Plummer did much of the survey work on which Jessop based his designs and specifications for drains, embankments, dams and tunnels. Essentially, the system adopted was based on the principle that drainage of the upland waters should be separated from internal drains. Jessop personally surveyed the river Hull the following year for improvements to the river outfall. Other drainage work at that period included the Norfolk Marshland Drainage, an area east of the river Ouse and south of Kings Lynn. The area comprised some 32,000 acres, of which 7,000 were imperfectly drained and 5,000 were permanently under water. Jessop's recommended solution was to install steam engines, capable of pumping a quantity of water equivalent to 2 in. of rainfall in a week over the area. Boulton and Watt designed suitable engines, recommending 8 hp for every 1,000 acres drained, requiring a lift of 8 ft. No action was taken at the time and, indeed, improvements did not take place until twenty-five years later. Subsequent experience has shown that Jessop's and Watt's design assumptions compare well with modern practice and, had their scheme been adopted in 1790, costs after thirty years would have been more than repaid and the land reclaimed would have greatly exceeded expectations.

When invited in 1799 to design a canal from Croydon to the Thames at Wandsworth, Jessop recommended a railway instead, because most water resources were already used by industry. Jessop noted in his report that 'it is but lately that they [railways] have been brought to the degree of perfection which now recommends them as substitutes for Canals; and in many cases they are much more eligible and useful'. And so the Act for the Surrey Iron Railway Company was obtained in 1801 and the world's first independent public railway was established. Jessop's estimate for the line was £33,000

The works comprised a substantial basin, with locks into the Thames at Wandsworth, from which the railway, using plate rails on a gauge of 4 ft. 2 in. and double-tracked all the way, ran to Croydon, with a branch to Carshalton. The total length was 9¼ miles and the costs were £2,950 per mile, only slightly higher than Jessop's first estimate. The line was opened in August 1803. Increased capital of £60,000 was authorised by a further Act of 1805, the increases being due to the enlargement of the basin over the original plans, as well as the inflationary advances in labour and materials. Initial enthusiasm for the railway led to consideration of a further extension to Portsmouth but eventually this was built only to Merstham, a further 8¾ miles of double-tracked line, total costs being about £50,000. Butterley provided all materials and labour for the works and Outram personally supervised the contract.

Another project devised by Jessop was the first public railway in Scotland, from Kilmarnock to Troon, which, with the associated harbour at Troon, was authorised by Acts of Parliament in 1808. The harbour was a wet-dock of 6 acres, from which a double-tracked railway of 4 ft. gauge ran 9¾ miles to Kilmarnock. For the first 2 miles the line ran on embankments up to 12 ft. high and, further along, crossed a mile of peat bog, followed by a masonry bridge of four arches, each 40 ft. span, across the river Irwine. The cost of engineering works was £42,000, compared with Jessop's estimate of £38,167. Unlike the Surrey Iron Railway, the line proved successful; coal was the principal commodity transported but other goods, as well as passengers, were also carried.

In 1801 and 1803 Telford was invited to report on the feasibility of a canal through the Great Glen of Scotland, the line of which had been surveyed by James Watt in 1773. He reported favourably for a canal of 100 ft. top width, 50 ft. at the bottom, and 20 ft. deep. Each lock, costing £5,000, would be turf-sided and rise 8 ft. There was an 'inexhaustible supply of water' on the summit and the entrances at Fort William and Inverness offered deep-water anchorages. He described the waterway as the Caledonian Canal and estimated construction costs as £350,000. Both Rennie and Jessop were examined for their views by government committee and a preliminary Act was obtained in 1803.

Telford was ordered to Scotland later that year to prepare surveys and to set out the line of the canal; he was also to begin excavations for the entrance basins, to open quarries, to order machinery and set up management procedures. The Commissioners appointed moved to obtain 'the opinion and assistance of Mr. William Jessop, another eminent and experienced Engineer' for the duration of the project. They made it plain that their engineers should take joint decisions on all policy matters. Jessop and Telford collaborated and their relationship was always amicable, although Telford invariably deferred to his senior colleague.

Jessop visited Scotland to examine the line in October 1803; later he took the responsibility of preparing the report and estimate (January 1804) for the enterprise, for which Telford contributed much information. Jessop's estimate was £474,531 for a canal of 50 ft. bottom width, 90 ft. at the top (later increased to 110 ft.) and 20 ft. deep. There would be twenty-three masonry locks, each 152 ft. by 38 ft. (later 170 ft. by 40 ft.). A sea lock was planned at Corpach, at the western end, the canal then rising by a flight of eight locks before running level by several aqueducts to Loch Lochy. The surface there was to be raised 12 ft. to bring it into line with the bed of Loch Oich, which needed dredging. The line would then follow and part-use the bed of the river Oich to Fort Augustus before locking down into Loch Ness. At the far end of the latter, Loch Dochfour would be dredged and its surface maintained at the same level as Loch Ness. From there, a canal would be cut to Muirtown, where four locks took it down to a large basin serving Inverness. The line would then continue to the sea lock, later built at Clachnaharry. These plans and estimates were prepared for the Bill of 1804, on which both Jessop and Telford were examined in Parliament.

Notable works built during Jessop's lifetime included the great summit cutting at Laggan, between Lochs Oich and Lochy, and the flight of locks at Banavie, known as Neptune's Staircase. The latter is probably considered the most spectacular work on the canal but in engineering terms, the sea lock at Clachnaharry was perhaps the greatest feat. A clay embankment was built on the site so as to consolidate, by its weight, deep mud which lay beneath. By this means, foundations for the sea lock were built without recourse to an expensive cofferdam.

By 1809 the two engineers estimated that some £300,000 would still be required to finish the canal, thus bringing costs to £550,000. This was not a great deal more than Jessop's original estimate and the Commissioners reported that 'we think that praise is justly due to Messrs Jessop and Telford for the accuracy of their Calculations'. Jessop's last visit took place in 1812 and much of the work remaining—almost all of it in the middle sections from Fort Augustus to Loch Lochy—was left to Telford to complete. The canal opened in 1822. In later years, Telford's writings made no mention of Jessop's participation and leadership during this great enterprise, nor did he acknowledge his senior colleague's contributions to the engineering of the Ellesmere Canal. Such surprising omissions did less than justice to Jessop's reputation and reflected no credit on Telford.

Jessop was experienced in dock and harbour work. His earliest commission was on proposals for improving Workington Harbour, the report for which was written in 1777 from his home in Pontefract. He was consulted on at least nineteen such projects during his lifetime. The degree of his involvement varied but it included works as far apart as Ardrossan and Portsmouth, Sunderland and Plymouth, Yarmouth and Liverpool, and notably the Ringsend docks in Dublin. None of these however, compared in scale with developments carried to fruition in the Ports of London and Bristol.

The Thames had provided excellent harbour, port and ship-building facilities since the earliest times. During the 1790s, the river was becoming overcrowded and the wharves and means for rapid handling of cargoes were proving inadequate. Major works, involving unprecedented capital expenditure, were necessary and miscellaneous schemes were reviewed by parliamentary committee. Jessop was engaged as consulting engineer by the City Corporation from 1796 to 1799 and participated in much of the early planning and designs for the Port of London.

Jessop was entirely responsible for the civil engineering works of the City Canal (the ship canal crossing the Isle of Dogs) from 1800 until its opening in 1805. The canal was 3,710 ft. long, had a surface width of 173 ft., a depth of 45 ft. and 76 ft. bottom width. Two locks, 193 ft. by 45 ft., were built with sills arranged to give depths between 20 and 24 ft. over high tidal ranges and vessels of 500 tons burden could be admitted. The surrounding land was below maximum high-tide levels, hence the canal was embanked 12 ft. high and 6 ft. above canal water level. The area to the dockside was filled to the same level, as was the land 50 ft. beyond the opposite side. The walls of the locks were of brickwork, curved in vertical profile and 6 ft. thick, faced with stone and backed with counterforts; the floors were built as inverted arches. Engineering costs were £187,600.

An unusual task on which Jessop was then engaged was the removal of the Blackwall rock, an outcrop in the bed of the Thames about 300 ft. long by 150 ft. wide, and only 2½ ft. below low-water spring tides. Explosives had been tried without success but Jessop's method was to use a heavy steel chisel, attached to a ram and operated from a barge, to break up the rock. A cylinder, sealed at the base with a skirt, was then lowered into the water and pumped dry to allow men inside to clean out the broken rock. The rock surface was eventually reduced by over 15 ft. Work began in 1804 and finished in 1808 at a cost of £42,000.

The outstanding project of Jessop's career was the West India Docks, built on reclaimed marshes of the Isle of Dogs between 1800 and 1806, at a total cost of £1,239,000. The system comprised separate, but parallel, import and export docks, both of which were just to the north, and also parallel to the City Canal. The import dock measured 2,600 ft. long by 510 ft. wide (30 acres) and water depth was 23 ft., the quay being 6 ft. above this. Massive walls, curved in vertical profile were 6 ft. thick at the top, backed with puddle clay and supported by counterforts. At the eastern end a lock, 162 ft. by 38 ft., with 24 ft. of water on the sills, led into the 6-acre Blackwall Basin and another lock, identical in dimensions to those on the adjacent canal, led into the river. At the western end of the dock, the 2-acre Limehouse basin had an entrance lock measuring 155 ft. by 36 ft. This was used by lighters and unloaded ships moving up river. Nine warehouses, 222 ft. long by 126 ft. wide and five storeys high designed by George Gwilt (q.v.), were built along the north quay, fronted by an open, timber-framed transit shed. Extensive sheds were also located on the south quay and cargoes of timber were handled at each end of the dock. This enormous undertaking was completed and operational in late 1802 and work on the export dock began one year later. That was also 2,600 ft. long but 400 ft. wide (24 acres). The locks into the basins were of the same dimensions as those into the import dock but there were no warehouses because cargoes went directly by lighter into outward-bound ships. The formal opening, on completion of the entire project, was in July 1806. Nothing on this scale had been built anywhere in the world before that date. The cost of the engineering works alone came to £515,000.

An Act of 1803, for improving and enlarging the Port of Bristol, was the culmination of investigations made at intervals over the previous forty years by Smeaton, Joseph Nickalls (q.v.), Jessop and others. Jessop's design of Bristol's Floating Harbour, completed between 1804 and 1810, was an outstanding achievement, comparable with the West India Docks.

The river Avon followed a sinuous, westerly course through Bristol and its tributary, the Frome, met the main stream in two confluences about 1 mile apart. In Jessop's scheme, the Avon was dammed at Temple Meads, ¾ mile west of the upper confluence and also 2 miles downstream at Redcliff, where the dam incorporated an overflow. A new river channel, 2 miles long with 120 ft. top width, was cut south of the city and this left a long, isolated loop of the Avon to form a floating harbour. The entire flow from the Frome passed through this loop which was also fed by a channel running from Netham weir, upstream on the Avon.

The Cumberland Basin at Rownham Meads connected the floating harbour with the Avon, below the dam at Redcliff. There were two entrance locks, 45 ft. by 200 ft. and 36 ft. by 185 ft. respectively, and the maximum tidal depth over the sills was 34 ft. The depth in the harbour was 16 ft., 22 ft. above low water at Rownham. Bathurst Basin, which was also a ship basin, linked the new cut with the harbour just east of the lower Frome confluence. Barge locks of 17 ft. width were located just below Temple Meads dam and also connected the harbour with the new cut. The floating harbour area was about 70 acres, the largest area in the world of water impounded for this purpose.

Jessop's estimate for the engineering works was £290,000 but unstable soils in the basins and the new cut, as well as progressive improvements to the design, had markedly increased costs. By 1810 total expenditure, including land purchase, legal and parliamentary expenses, had risen to £610,000; engineering works and contingencies amounted to £457,000. There can be no doubt, however, that the prosperity enjoyed by the Port of Bristol during the nineteenth century was largely due to Jessop's imaginative developments.

Following the death of Benjamin Outram in 1805, William Jessop and his family moved into Butterley Hall. By then the family had increased to eight children, including Charles, Sarah, Henry, George and Edward, all born between 1786 and 1797 at the former home in Appleton Gate, Newark. In spite of his busy and demanding professional life style, Jessop still found time to serve as Alderman in Newark for nineteen years, and even to hold the office of mayor in 1790–1791 and 1803–1804. He was also a Justice of the Peace for a short period before he left for Butterley. It is surprising that he was never elected to a Fellowship of the Royal Society, an honour which he surely deserved with several of his juniors, including John Rennie and Thomas Telford. The reasons for that remain a mystery although it has been suggested that he had once fallen foul of the then President of the Society, Sir Joseph Banks, who had seen fit to block his admission.

Jessop was not a wealthy man during much of his professional life and he had borrowed heavily to buy into the partnership at Butterley Ironworks, for which he was denied any profits until the debt was paid off in 1805. In his declining years, however, the Jessops seemed to live in comfortable circumstances. This was no doubt due to the increasing prosperity of the company, as well as the dividends accruing from his investments in several of the canals for which he was engineer.

William Jessop died on 18 November 1814, possibly following a stroke; he had been in poor health for much of the two preceding years. He was buried in Pentrich churchyard and his wife, Sarah, followed him on 21 December 1816. It was said of this worthy and most able man, who had towered over the civil engineering profession during the years from 1785 to 1805, that he was 'totally free of all envy and jealousy of professional rivalship, his proceedings were free from all pomp and mysticism and persons of merit never failed in obtaining his friendship and encouragement'.

Once established as a consulting engineer, Jessop charged professional fees of 3 guineas per day (plus expenses) up to 1793; thereafter to the end of his career 5 guineas per day. After planning a scheme and (more often than not) having given evidence on the Bill in Parliament, if appointed engineer to direct the works he usually charged the same fees or in some cases received an annual salary: e.g. on the Cromford Canal (1789) £300 p.a., for which he was expected to spend one-third of his time. There would always be a full-time resident engineer; names are given in the list of works appended in this article. In addition, in Ireland Jessop had Chapman as assistant engineer from 1789 until 1794 on the Grand Canal, and also in the early stages of Ringsend docks. Also on the Ellesmere Canal, Telford was engineer under Jessop as chief engineer from 1794 to 1800, while on the Grand Junction Canal James Barnes was full-time engineer throughout, first under Jessop as chief engineer from 1793 to 1797 and with Jessop as consultant from 1799 until 1803.

On relatively small jobs Jessop would visit the site once or twice a year during construction; thus seven visits on Holderness Drainage (1789–1792), six on Axholme Drainage (1797–1801) and on Yarmouth Harbour (1800–1806), and on average twenty-three days a year on the Trent Navigation (1794–1800). Quite often, having seen a job through the early stages he would leave its completion in the hands of the resident engineer. On the major works, however, his attendance on site was more extensive; about eighty days per year on the Grand Junction Canal while chief engineer and seventeen visits while consultant; no less than four-hundred and thirty-seven days on the West India Docks between February 1800 and October 1802 (average of one-hundred and sixty days per year), and a total of twenty site visits in the Bristol Floating Harbour job (1804–1808). As consultant on the Caledonian Canal he visited the works, with Telford, once each year from 1804 to 1812 inclusive.

Finally, such was the pressure on Jessop in the 1790s that several of the smaller canal schemes planned and taken through Parliament by him were carried out by the resident engineers without, or with little, nominal supervision by him; it was not a satisfactory arrangement but almost inevitable in the circumstances. They include the Melton Mowbray Navigation, 1792–1797 (Christopher Staveley); Nottingham Canal, 1792–1796 (James Green); and Nutbrook Canal, 1793–1796 (Benjamin Outram).

R. B. SCHOFIELD

[David Brewster (ed.) (1817) William Jessop, *Edinburgh Encyclopaedia*, **11**, 735–737; Samuel Hughes (1844) Memoirs of William Jessop, *Weale's Quarterly Papers on Engineering*, **1**(2), 32 pp.; Charles Hadfield (1966) *Canals of the East Midlands*; W. H. Slatcher (1968) The Barnsley Canal, its first twenty years, *Transport History*, **1**,

William Jessop: principal works

Date	Work	Cost (£)	Resident Engineer(s)
1775–1778	Selby Canal*	20,000	John Gott
1780–1785	River Calder, new cuts and locks		Samuel Hartley
1783–1787	River Trent Navigation*	13,000	Thomas Thompson
1787–1792	Holderness Drainage (completed 1805)	16,000	George Plummer (1787–1792) and Anthony Bower (1792–1795)
1788–1790	Basingstoke Canal (completed 1793)	150,000	William Wright (1788–1790) and Henry Eastburn (1790–1793)
1789–1794	Cromford Canal*	79,000	Benjamin Outram
1789–1800	Grand Canal, Ireland (Roberts-town–Tullamore)	174,000	James Oates and John Killaly
1790–1793	Circular Line, Dublin	52,000	Archibald Millar
1792–1796	Ringsend Docks, Dublin	113,000	Edward Chapman
1791–1794	Leicester Navigation* (completed 1797)	80,000	Christopher Staveley
1793–1797	Grantham Canal* (completed 1803)	118,000	William King and James Green
1793–1803	Grand Junction Canal* (with James Barnes, engineer, completed 1805)	1,500,000	James Hollinsworth and Henry Provis
1793–1802	Barnsley Canal*	95,000	Samuel Hartley
1794–1800	Ellesmere Canal* (with Thomas Telford, engineer, completed 1806)	459,000	John Duncombe (1794–1803), Thomas Denson (1803–1805) and Matthew Davidson (1794–1804)
1794–1802	Rochdale Canal* (completed 1804)	580,000	William Crosley (1794–1796), Thomas Townshend (1796–1802) and William Crosley Jr. (1802–1804)
1795–1800	River Trent (further works)*	20,000	Thomas Thompson
1797–1801	Isle of Axholme Drainage	20,000	John Dyson
1796–1803	Everton, Gringley and Misterton Drainage	c. 11,000	John Dyson (1796–1801) and William Gauntley (1801–1803)
1800–1805	City Canal*	126,000	John Foulds
1800–1805	West India Docks* (completed 1806)	515,000	Ralph Walker (1800–1802) and Thomas Morris (1803–1806)
1800–1806	Yarmouth Harbour, South Pier	18,000	John Druery
1801–1803	Surrey Iron Railway*	27,000	George Leather Sr.
1803–1805	Croydon and Merstham Railway	44,000	Josias Jessop (1803–1804) and George Leather Jr.
1801–1806	Axe and Brue Drainages* (completed 1810)	58,000	Robert Anstice
1804–1806	Blackwall Rock (completed 1808)	42,000	J. M. Warren (1804–1806) and James Mountague (1806–1808)
1804–1808	Bristol Floating Harbour* (completed 1810)	470,000	Josias Jessop
1804–1812	Caledonian Canal* (Jessop, consultant, Thomas Telford, engineer, completed 1822)	855,000	Matthew Davidson (1804–1819), James Davidson (1819–1822), John Telford (1804–1807) and Alexander Easton (1807–1822)
1808–1811	Kilmarnock and Troon Railway*	44,000	Thomas Hollis

48–66; Philip Riden (1990) *The Butterley Company, 1790–1830*; Charles Hadfield and A. W. Skempton (1979) *William Jessop, Engineer*; W. M. Hunt (1979) The Horncastle Navigation engineers, *Journal of the Railway and Canal Historical Society*, **25**, 2–11; Alastair Penfold (1980) Managerial organisation on the Caledonian Canal, *Thomas Telford, Engineer*, 129–150; R. B. Schofield (1986) The design and construction of the Cromford Canal, *Transactions of the Newcomen Society*, **57**, 101–122; A. W. Skempton (1990) Historical development of British embankment dams, *Proceedings of the Conference on Embankment Dams*, ICE, 15–52; R. B. Schofield (2000) *Benjamin Outram 1764–1805: An Engineering Biography*]

Publications

Jessop proposed numerous requests for proposed canal and river navigations, drainage schemes, dock and harbour improvements. Although many were printed, others can only be found as transcripts in various company minutes. The following are typical of the printed reports and plans:

1774. *Plan of the proposed Canal from the River Aire at Haddlesey to the River Ouse at Selby*
1777. *Report on the State and Means of improving and enlarging the Harbour of Workington*
1782. *Report on a Survey of the River Trent … relative to a Scheme for improving its Navigation*
1786. *Report on the State of the Drainage of the Low Grounds and Carrs … in Holderness … and on the Means of completing the Same*
1788. *Report on a Design for a Canal from Langley bridge to Cromford*
1792. *Plan of the intended Canal from the Town of Grantham … to the River Trent near Nottingham*
1792. *Report on the Grand Junction Canal*
1793. *Report and Estimate for improving the River Trent Navigation* (with Robert Whitworth)
1796. *Estimates for Floating Dock in the Isle of Dogs*
1800. *Report on the Drainage for Deeping inclosed Fens and the Commons adjoining thereto* (with Rennie, Maxwell and Hume)
1801. *Plans of the intended Iron Railway from … Croydon to the River Thames at Wandsworth*
1802. *Design for improving the Harbour of Bristol* (with William White)
1804–1812. [Annual] *Reports on the Caledonian Canal*
1807. *Report on the present State of the Pier of Sunderland Harbour and the Means recommended for their Improvement*

Works

Jessop's principal works are listed in the table with the dates of his involvement from the start of construction. Those on which he gave Parliamentary Evidence are marked with an asterisk. For canals the total cost is given; otherwise the cost applies to the engineering works alone.

JOHNSON, George (1784–1852), colliery viewer, was the son of George Johnson (d. 1801) and his wife Mary and was one of five children. He was born on 4 May 1784 at Byker, near Newcastle, while his father was viewer at the colliery there and was baptised at Wallsend three years later.

In 1826 Johnson became viewer at Willington colliery, in which concern he was a 4/64 owner, as he was of Benton. As a viewer, Johnson was involved with the operation of many collieries in the Tyneside area and, *inter alia*, produced reports on the following: a valuation of Hetton colliery made with Nicholas Wood (q.v.) in 1828; Hebburn colliery in 1840; the sinking of the Monkwearmouth pit in 1841 recommending that it be sunk to twice its depth, to 120 fathoms (720 ft.); the winning and lining of the shaft at Murton, again with Wood, in 1842; and the shaft at St. Hilda, South Shields, in 1843 where the

lining had failed. He also gave evidence to the Select Committee on Accidents in Mines in 1835.

Johnson was involved in the construction and operation of the region's railways and became a director of the Newcastle and Carlisle Railway in 1833, for which company Francis Giles (q.v.) was engineer. The major structures at the western end of the line were completed under Giles but differences between him and the company's directors brought about his resignation in 1833. The eastern section of the line was then completed under a committee of management comprising Johnson, Wood and Benjamin Thompson (q.v.), Tomlinson writing of the 'invaluable … engineering experience' of the three men, all colliery viewers; John Blackmore (q.v.) was appointed engineer in place of Giles. The full length of the line was brought into operation in 1839.

In 1834 a proposal was made for the formation of the Blaydon, Gateshead and Hebburn Railway, extending the N&CR eastwards. Johnson became a director of this company although, as a result of the formation of the Brandling Junction Railway, its length was principally constructed by the two last named companies. Johnson was involved, too, with the Newcastle and North Shields Railway, reporting on both its line and its bridges. The latter had been designed as laminated timber structures by John Green (q.v.) but Johnson's opinion, not agreed by the directors, was that they should be built in masonry.

When building Seaham Harbour, the third Marquess of Londonderry had arranged that a railway be built to transport coal from his collieries near Durham to the new port. Unwilling to finance it himself, it was built by others and leased by Londonderry but in 1840 he sought to purchase it. An arbitration ensued with Johnson one of the referees. Continuing his connection with the N&CR, Johnson, with Wood, published a report on the possible construction of a railway from the N&CR at Gilsland northwards into Scotland, a proposal viewed with concern by George Stephenson (q.v.), intent on providing a line which would reach Edinburgh via the east coast. In the same year an investigation was held concerning alleged malpractices in the affairs of the Brandling Junction Railway and a reply to the allegations was written by Johnson, then company chairman.

In 1845 Johnson became a member of the provisional committee of the Whittle Dean Water Company, later the Newcastle and Gateshead Water Company, which sought to build an impounding reservoir some 12 miles west of Newcastle and bring water from it to the towns of Newcastle and Gateshead, replacing the existing somewhat unsatisfactory supply. The company sought capital of £150,000 and Johnson subscribed for one hundred of the £25 shares; he became a director on the company's formation and remained on the board until his death.

In 1815 Johnson had married Margaret Short but he died childless. He and his wife brought up the two sons of his brother and one of them, John, became his assistant at Willington and later succeeded him. Johnson died at his home, Willington House, on 26 January 1852, leaving his not inconsiderable fortune in trust for his wife, to pass to his two nephews. In addition to the shares in the two collieries were two hundred shares in the N&CR, valued at £80 each, and one hundred shares in the WDWC.

R. W. RENNISON

[*Report from the Select Committee on Accidents in Mines*, 1835; W. W. Tomlinson (1914) *North Eastern Railway*; W. Richardson (1923) *History of the Parish of Wallsend*; R. W. Rennison (1979) *Water to Tyneside; Watson Collection*, 3410/JOHN/9, Northumberland RO]

Reports

1834–1841. *The Managing Committees Report to the Directors of the Newcastle and Carlisle Railway, Newcastle, 1834–41*, with Nicholas Wood and Benjamin Thompson
1843. *Report* [of the Directors of the Brandling Junction Railway]
1843. *Central Line of Railway into Scotland*, FOR 3/42/15, Works Northumberland RO

Works

1833–1852. Director, Newcastle and Carlisle Railway, Committee of Management, 1834–1841
1834–1836. Director, Blaydon, Gateshead and Hebburn Railway
1836–1845. Director and Chairman, Brandling Junction Railway

JOHNSON, Sir Henry, MP (*c.* 1623–1683), shipwright, the eldest son of Francis Johnson of Aldborough and his wife, Mary, *née* Pett, served his apprenticeship with his cousin, Phineas Pett (1570–1647), in the Royal Dockyards. He married Dorothy Lord (*c.* 1633–1664) in 1648 and was sufficiently established by 1649–1650 to be able to build two ships of *c.* 500 tons for the navy in his own yard at Deptford.

At that time shipbuilding was well-established on both banks of the Thames. In 1608 the East India Company's shipwright, William Burrell (fl. 1600–1620), had taken a lease on a yard at Deptford and in 1614, at his suggestion, land had been acquired to build a dry dock at Blackwall. This was progressively enlarged in 1615 and 1624, and a new dry dock built in 1618. The lease was assigned by Burell to the Company in January 1621. In 1630–1631, after Burrell's death, a third dry dock was built, which was widened in 1634. The yard became the major employer in the London area but the East India Company, in something of a financial crisis, resolved to get out of shipbuilding and in December 1652 initially let the yard to Johnson for twenty-one years. He benefited from the growth in trade and consequent demand for merchant shipping as well as the Anglo-Dutch wars which created a demand

for warships which outstripped the Royal Dockyards' capacity.

Johnson made a number of improvements at Blackwall, including building three storey warehouses, but the most significant was the decision to build a 1½ acre wet dock in 1659 for ship repairs. A carpenter, George Sammon of Wapping, was awarded the contract to build the dock, with sides *c.* 225 ft. by 310 ft. by 175 ft. The dock gates were modelled on latest practice at Deptford. By January 1661 Samuel Pepys was able to view the dock, which was completed that March, and was the largest on the Thames, before the construction of John Wells's (q.v.) Howland Dock in the 1690s.

Johnson, with a succession of partners, prospered as a shipbuilder; the wharves and docks were extensively rebuilt in 1677–1678 and in the latter year he was elected MP for Aldborough. In 1679 he was knighted by Charles II at his home. He left a successful business to his son, (Sir) Henry Johnson (d. 1719), on his death aged sixty in 1683. Henry Jr. failed to take the same active interest in its affairs, relying on occasion on his younger brother, William (*c.* 1660–1718). Both sons were also MPs for Aldborough. In the eighteenth century the yard passed into other hands.

MIKE CHRIMES

[J. Charnock (1801) *History of Marine Architecture*; S. Smiles (1884) Phineas Pett, *Men of Invention*, 1–49; H. Green and R. Wigram (1881) *Chronicles of Blackwall Yard*; India Office (1915) *A Calendar of the Court Minutes of the East India Company, 1650–1654*; S. Porter (ed.) (1994) Poplar, Blackwall and the Isle of Dogs, *Survey of London*, **XLIV**, 553–561]

Works

1659–1661. Blackwall wet dock, 1½ acres, excavation, etc., £1,057 10s; total expenditure £4,741

JOHNSON (JANSEN), Joas (fl. 1566–?1575), was a Dutch engineer brought over to direct work on Yarmouth harbour, known as the Haven. The early history of Great Yarmouth and its Haven was written by Henry Manship (a former Town Clerk) in 1612–1619. His manuscript, now lost, was transcribed in 1854 and edited, with useful notes, by Ecclestone (1971). From Manship's history it transpires that between 1350 and 1550 no less than six entrances to the Haven had been made at different places through the sand spit which in that period extended at least 6 miles south of the town. The entrance cut in 1408 lasted nearly a hundred years before silting up but the others had a shorter life. Between the sand spit and the shore is a tidal river typically some 300 ft. wide running from Breydon Water, a lake north of the town fed by the rivers Yare, Waveney and Bure. The Haven is that part of the tidal river between the entrance and the quays at Yarmouth.

Work began on the seventh (and final) entrance in March 1559. Practically the entire population of men, women and children laboured for three days to cut through the spit at a place opposite Gorlestone cliffs about 2 miles south of Yarmouth. Carpenters then built a 'stop' across the river to prevent water running south and direct its flow eastwards through the cut. The stop was strengthened by a backing of rubble stone. The combined energy of the river water and the ebb tide would enlarge the cut, no doubt assisted by further digging. By August funds were exhausted. A similar operation to open the fifth entrance (quite close to the seventh) in 1528 had cost £1,500.

Nothing more seems to have been done until December 1560 when advice was sought from 'a stranger from Emden, a man skilled in marine works'. The substance of his advice is not known, but in April 1561 the Duke of Norfolk successfully petitioned the Queen to give the town £1,000 and a loan of £3,000 repayable over fifteen years for 'the repair and maintenance of Yarmouth haven now sore decayed'. The cost was estimated to be about £7,000 by Adrian Harrison, who had been sent by the Privy Council to investigate the situation.

Clearly a plan was now in existence; from later events it involved the construction of a pier. Work proceeded, but very slowly. In 1564 two carts were ordered to carry sand 'to the new pier' and a small building was erected to store tools and shelter the workmen in bad weather. By 1566 it appeared wise to obtain more expert advice and it was then that 'a Dutchman, Joas Johnson, a man of exceptional knowledge of works of this nature', was brought over. Evidently he decided on building two long piers, one on the north side of the entrance, chiefly for heaving in of ships, the other on the south side; the latter (probably the one already started) to be 'solid' to direct the flow to the harbour mouth. It also served as a jetty.

In May 1566 Johnson started by constructing two groynes to protect the main works of the piers. They were probably advanced seawards ahead of the piers as work progressed. Next year the Queen granted the town a further £1,400 and in 1573–1574 Norwich and the counties of Norfolk and Suffolk contributed £500. By 1575 work reached completion. The north pier, 235 yd. in length, 40 ft. wide at the base and 20 ft. at the top, was built of huge tree trunks joined together and sloped up with beach, stone and shingle and with three rows of piles. The south pier, 340 yd. long and 30 ft. wide at top, was made of 12 in. square piles driven close together at an oblique angle and filled with stone. The pier rose 12 ft. above high tide level and when Manship wrote his account in 1613 it stood in 24 ft. of water. The harbour's mouth between the piers had a width of 114 yd. The north groyne, about 500 yd. from the north pier and 265 yd. in length (made of brushwood, planks and piles), served to hinder the transport of sand into the entrance and in doing so led to the building up of firm ground near the shore.

In the course of time a submerged sand bar began to form beyond the harbour mouth. Nevertheless Yarmouth Haven retained its position as the third or fourth largest port in England, in terms of tonnage of shipping, throughout the seventeenth century: behind London, Newcastle and, towards the end of the century, Bristol.

Assuming the preliminary works in 1559 cost £1,500 the total expenditure to 1575 would have been about £7,500. This includes the building of a timber jetty on the beach at the town for fishing vessels. Whether Johnson stayed at Yarmouth until the works were finished is uncertain. Presumably he returned to Holland, but nothing is known of his career there.

A. W. SKEMPTON

[A. W. Ecclestone (1971) *Henry Manship's Great Yarmouth*; Gordon Jackson (1953) *The History and Archaeology of Ports*]

Works

1566–1575. Yarmouth Haven, piers, total cost c. £7,500

JOHNSON, John (1732–1814), architect and county surveyor, was born on 22 April 1732 in Southgate Street, Leicester, the eldest known son of John Johnson (c. 1707–1780) and Frances Knight (c. 1708–1776); his father and grandfather were carpenters and joiners. By 1760 Johnson had moved to London with his wife Elizabeth, his eldest son John, who was to work closely with him, being born in Marylebone on 18 January 1761. At this time Johnson was working as a carpenter/builder around the West End, helping to develop the property of William Berners from c. 1766 onwards. By November 1767 he had moved to 32 Berners Street, where he lived until 1786. In this period he moved rapidly from a carpenter carrying out work for William Chambers at the German Lutheran church in the Savoy to an architect designing town houses and country residences. Johnson's architectural work is noted for its extensive use of the recently developed Coade stone, an artificial stone developed by Eleanor Coade who set up her works in Lambeth in 1769. Johnson used it in his works from 1774 to 1810. He was also involved in the development of an alternative form of stucco to that patented by John Liarder in 1773. In 1777 Johnson took out his own patent. He claimed to have improved upon Liarder's patent by the use of serum of ox-blood. Liarder took Johnson to court for patent infringement in two cases in February and July 1778 and Johnson lost on both occasions. The case became academic as Liarder's oil-based stucco began to fail.

In 1786 Johnson moved to 27 Charles Street, leaving his eldest son at Berners Street from where the business was run. He was by now well-established and through two of his clients who were magistrates in the county—T. B. Bramston and John Strutt—he had been appointed by Essex Quarter Sessions 'as Surveyor of the Gaol, Houses of Correction, Bridges and other Buildings' on 15 January 1782.

While Johnson's work as County Surveyor formed a relatively insignificant part of his total work, it is where his interest in civil engineering terms lies. He was paid 1s 3d (later 1s 6d) a mile expenses, and a 5% commission for his plans, etc., for works, together with a two guinea fee for attendance at the Quarter Sessions, subsequently increased to 5 guineas. His work involved the maintenance of the County's buildings such as the County Gaol, Shire Hall and Houses of Correction, and also necessitated designing a large number of County bridges which needed rebuilding.

Johnson is known to have been involved in the surveying, repair or rebuilding of nearly fifty bridges. During his period of office the County was taking on responsibility for an increasing number of bridges and the terms of Lord Elenborough's Act (43 GIII c5) these had to be in a good state of repair. Most of Johnson's bridges were modest affairs, generally of brick or timber, costing a few hundred pounds. The most important structure was Moulsham Bridge across the Can in Chelmsford. Johnson's only bridge in stone, it cost over £1,740 and replaced an existing fourteenth century bridge in poor condition. In addition to local tradesmen his son John Jr. worked as contractor on the bridge as he did regularly on his father's designs. The design involved the use of nearly 1,000 tons of Portland stone and 140 tons of arch stone from Swanage, and Dutch trass mortar; a notable feature was the decorative use of Coade stone. Work began in 1785 and was completed in 1787. There were complaints that Johnson's design was causing flooding by reducing the waterway from 54 to 36 ft. and Thomas Telford (q.v.) was consulted in 1824, after Johnson's death, then recommending the removal of the piers of the mediaeval bridge and replacing Johnson's bridge by a new cast iron structure. Only the first part of his advice was followed and flooding remained a problem until the 1960s.

Several of Johnson's brick bridges, including that at North Weald (1788–1786), had iron ties through the brickwork with cast iron tie plates, a very early use of the material in Essex. Johnson also made much use of iron ties in his timber bridges.

Around 1801 Johnson's health began to fail. Although he attended the Essex Quarter Sessions on 12 January 1802 he described himself as 'very unwell' four weeks later. In this situation his work for the County became increasingly reliant on his eldest son, who had prepared the accounts on his behalf in the autumn of 1799. He was not always able to attend the Quarter Sessions and surveys of bridges at Great Burstead (1812) and Dagenham Beam (1812) were carried out by his son. In this situation Johnson decided to resign at Michaelmass 1812, and retired to Leicester. His health problems had been compounded by his involvement with Dorset, Johnson, Wilkinson, Berners and Tilson, a London bank which failed in 1803; Johnson had been a partner since 1785. The failure was brought about by fraudulent behaviour by one of the partners, Wilkinson, of which Johnson was ignorant. Johnson was declared bankrupt on 5 May 1803 and suffered a stroke in the ensuing crisis. Although he was able to meet his liabilities and was discharged from bankruptcy in May 1804, he had to leave his home at 53 High Street, Marylebone, and move to Camden Town. Bankruptcy had cost him most of his wealth but fortunately he had already settled some on his family. He died on 27 August 1814 at his house in Southgate Street, Leicester, and was interred in the Chancel of St Martin, Leicester, near his parents. He was predeceased by his eldest son, John, who died on 21 February 1813. Of his three sons only Charles (1768–1841) survived him, the youngest, Joseph (1770–1802), dying at his Paddington home on 13 February 1802. A portrait is reproduced in Briggs

MIKE CHRIMES

[Court of Counter Sessions, Q/SO; Q/ABb3,4, Q/ABp2,3,7,47; Q/AB22/1, Essex RO; family papers; D/DOpE8, D/DP020, Q/FAc6/3/1,2, Q/FAc6/2/1–4, PRO; Stucco Patent infringement 1777 C12/1346/22; Bankruptcy of Dorset Johnson & Co., 1803–1805, B3/1268–1273; Will of John Johnson Sr., B11/1579, PRO; G. M. Benton, Fingringhoe bridge, *Essex Archaeological Society Transactions, New Series*, **xx**, 262–269; W. W. Hodson, Ballingdon Bridge, *Suffolk Institute of Archaeology Proceedings*, **viii**, 21–30; J. Newman (1973) The History of Battlesbridge, *Essex Journal*, **8**, 16–18; N. Briggs (1984) The Evolution of the office of County Surveyor in Essex 1700–1816, *Arch Hist*, **27**, 297–307; F. Kelsall (1984) Liardet versus Adam, *Arch Hist*, **27**, 118–126; F. Kelsall (1989) Stucco, *London Topographical Society*, **140**, 18–24; N. Briggs (1991) *John Johnson 1732–1814*; Colvin (3); A. Kelly (1990) Mrs. Coade's Stone]

Works

For a full listing of architectural work see Briggs (1991).

Bridges:

1784. Sutton Ford, rebuilt in brick

1784. Moreton, rebuilt in brick

1785. Passingford, rebuilt in brick

1785–1756. Weald, rebuilt in brick

1786. Dedham, rebuilt in brick

1787. High Ongar, rebuilt in brick

1787. Moulsham, Chelmsford, rebuilt in stone, £1,740

1788. Blackwater, rebuilt in timber

1788. Heybridge, Ingatestone, widened 1788, repaired 1801

1790. Cattawade, rebuilt in timber

1792. Chain Bridge, Mountnessing, rebuilt in brick

1793–1794. Pentlow, rebuilt in timber

1794. Two Mile Bridge, Wintle, widened
1795. Baythom, repaired
1795–1796. Battlesbridge, rebuilt in timber
1795–1796. Dedham, rebuilt in timber
1796. Half Mile Bridge, Ingatestone, rebuilt in brick
1797. Ongar, rebuilt in brick
1799. Blue Mills, culverts
1799. Wickham Mill Bridge, culverts
1800. Stifford, rebuilt in brick
1801. Ilford, repaired
1802. Roydon, rebuilt
1803–1805. Peet, rebuilt in brick
1804. Dagenham, repaired
1805. Widford, rebuilt in timber
1805. Ballingdon, rebuilt in timber
1806. Ackingford, rebuilt in timber
1808. Ramsey, rebuilt in timber
1808. Rod, rebuilt in timber
1808–1809. Langford, rebuilt in timber
1809–1810. Loughton, rebuilt in timber
1810. Leaden Wash, rebuilt in timber
1810. Shonks Mill, rebuilt in timber
1810–1811. Small Lea, Waltham Cross, rebuilt and widened in brick and stone
1812. White's Bridge, Great Burstead, rebuilt in brick
1812. Shellow, rebuilt in timber

JOHNSON, Rowland (fl. 1547–1584), master mason and military engineer spent most of his career in the north, latterly as underpaid and overworked assistant—'he earned the least with most travail of any officers'—to Sir Richard Lee (q.v.).

Johnson was first employed in 1543 and spent 1547–1550 as a Sapper in Scotland. In 1556 he was made master mason at Berwick earning 8d a day. This was increased to 2s 6d when he became Deputy Surveyor in 1560 and through a complicated deal to 8s 4d when he officially took over as Surveyor after Lee's death in 1575, but as he was wont to mention at length this was little compared to Lee's 20s per day and John Rogers (q.v.) 13s 4d.

He had a point, as he carried out the site work at Berwick from the beginning and was in *de facto* charge from 1567, as well as drawing up most of the surviving reports and plans. These include the scheme of comparing Lee's and Portinari's (q.v.) test walls and a plan showing a street plan of four-hundred and fifty houses within the walls as well as the planned fortifications, making it one of the earliest surviving town plans. When consulted on the progress of the work in 1567 Johnson declared to Cecil that the state of the walls left the town 'very bayr and ungardable' and the earthen ramparts had not even been begun. He proposed that they be reinforced with heather to speed the binding and also suggested building two traverse walls to retain water in the ditch and a timber bridge to cross it.

He was also concerned with the old timber bridge over the Tweed, repairs to which cost over £2,000 and occupied three years from 1570–1573 after which he took a sabbatical to besiege Edinburgh Castle. On his return to Berwick in 1576 he produced an estimate of £700 for a stone pier to stop the silting up of the harbour, using a wall of stone set in clay and strengthened (again) with heather and horse dung. The choice of a more elaborate and expensive scheme meant his last years at Berwick were rather acrimonious.

Johnson's duties as Deputy Surveyor also covered drawing up reports and plans for most of the Border forts, particularly Wark and Norham where his unbuilt design for a rigidly symmetrical enceinte was far more advanced than anything built at Berwick. In a rare excursion south he also drew up a plan of Portsmouth Harbour.

Johnson probably died in 1584, although his will was not actually proved until 1600.

SUSAN HOTS

[Cecil papers, maps 1/29, plan ii/59, SP 59/5, 9; *King's Works*, **iv**, 1982; I. MacIvor (1965) The Elizabethan fortifications of Berwick-on-Tweed, *Antiquaries Journal*, **65**, 64–96; A. H. W. Robinson (1962) *Marine Cartography in Britain*, 149]

Works
1570–1573. Berwick (timber) Bridge, repairs, £2,000

JOHNSON, William (1730–1807), stonemason and bridge surveyor, was baptised on 22 October 1730 at Hartburn, Northumberland. The eldest of five children, he was the only son of James Johnson (1699–1777), a stonemason, and his wife, Mary Caneck (1700–1776). At the termination of his apprenticeship in 1751 he married Mabel Moor (1731–1810) of Stamfordham.

Between 1752 and 1755 Johnson undertook repairs to some 9 miles of the Alemouth Turnpike, running from Hexham to Alnmouth, and it is possible that he also carried out other work of this nature. In 1765 the Northumberland Justices paid him £20 'on Account of Bridges by him building and repairing' and two years later he was awarded the contract for building the two-span Holywell Bridge, near Earsdon.

Following a catastrophic flood which in 1771 destroyed almost all of the Tyne bridges, Johnson—with Thomas Forster—was awarded the contract for building a new bridge at Chollerford, designed by John Wooler (q.v.) and/or John Fryer (q.v.). Of five spans, it was completed in 1776 at a cost of £2,707; the original tender price had been £2,500. In the same year Johnson, Forster and Thomas Allan were appointed jointly as undertakers for the County's bridges at a fee of £61 5s per quarter.

Johnson's next work for the County was the building of a cupola bridge at Whitfield. Designed by the County's bridge surveyor, Jonathan Pickernell (q.v.), the three-span bridge was built on a rock foundation at a cost of £733 and was completed in 1782. He was also involved in the building of a two-span crossing of the river South

Tyne at Ridley Hall where in 1779 Pickernell had designed a bridge. Some twelve months after its completion in 1781 it was destroyed by a flood and the County Justices sought the advice of Robert Mylne (q.v.) who was paid for work on the foundations although it is possible that the design of the superstructure was partly the work of Johnson. The bridge was built by contractors employed by the owner of Ridley Hall, probably under the supervision of Johnson and Robert Thompson (q.v.), surveyor for the southern bridges, and it was completed in 1792.

In 1788 Johnson was appointed bridge surveyor for the northern division of Northumberland at a salary of £25 p.a. and the following year was instructed by the County Justices—on the advice of Mylne—to proceed with the rebuilding of the bridge at Hexham, completed by John Smeaton (q.v.) in 1780 but destroyed in a flood two years later; supervision was to be the joint responsibility of Johnson and Thompson. Work on the new bridge began in 1789 and was completed in 1796, its superstructure built in accordance with Smeaton's design. So pleased were the Justices by the work of Johnson and Thompson that for their 'extraordinary trouble, care and attention' they were paid an honorarium of £50 each and saw their salaries increased to £40 p.a. The bridge would seem to have cost at least £7,500.

The last of Johnson's major works as surveyor of bridges was the rebuilding of a bridge at Wooler, apparently to his design and under his control. Work began in 1798 and was completed in 1801, the cost of the work having amounted to approximately £1,200.

Johnson retired from his position as county surveyor in 1802 and died in 1807 when the *Newcastle Chronicle* announced his death: 'On the 12th instant (March) at Stamfordham in his 77th year, Mr. William Johnson, who for upwards of 55 years, with much credit, kept an inn in that place; he was also for many years surveyor of bridges etc.'. He was buried at Stamfordham three days later, survived by his wife and several of his seven daughters.

R. W. RENNISON

[*Northumberland Quarter Sessions: Orders* (1765–1807); *Newcastle Chronicle*, 28 March 1807; S. M. Linsley (1994) Tyne Crossings at Hexham up to 1795, *Archaeologia Aeliana, 5th Series*, **22**, 235–253]

Works

1765. Minor bridges at Smallburn and Coldcoats
1767–1768. Holywell Bridge, construction, two spans of 20 ft., 10 ft. wide, cost £134
1772–1776. Chollerford Bridge, construction, five spans of up to 56 ft.; length 390 ft., width 16 ft., cost £2,707
1780–1782. Whitfield Bridge, construction, spans of 45, 49 and 45 ft., width 16 ft., cost £733
1788. Appointed bridge surveyor for Northumberland, northern bridges

1789. Ridley Hall Bridge, construction, two spans of 67 ft. 6 in.
1790–1796. Hexham Bridge, construction, with Thompson, nine spans of up to 50 ft., cost c. £7,000
1791. Akeld Bridge, construction, cost £331
1799–1801. Wooler Bridge, construction, cost c. £1,200

JONES family (fl. *c.* 1770–1835), contractors, probably originated as miners, possibly from the Shropshire or Gloucestershire coalfields.

Charles Jones Sr. was living at Preston on the Hill near Warrington in July 1773. His first known civil engineering work was on the Bridgwater Canal, followed by work on the Trent and Mersey and Chesterfield Canals. The original engineer, James Brindley (q.v.), died and was succeeded by John Varley acting with Hugh Henshall (qq.v.). Because of irregularities in the tunnel contracts Varley's brother, Thomas, who was a contractor for part of the work, was dismissed, but the work was redistributed between Jones and a third contractor, Samuel Fletcher (q.v.), in Jones's case at £3.00 a yard. With this background he came as well recommended by Josiah Clowes (q.v.) and others when he tendered for a contract for the Sapperton Tunnel on the Thames and Severn Canal in 1783, then settling at Coates, Gloucestershire. The story of his work on this canal is well documented by Household (1969) and is largely one of incompetence. In October 1783 he agreed to complete the tunnel, 3,817 yd. long, in four years from January 1784 at a price of 7 guineas a linear yard, having originally tendered at eight. After three months 'open jointy rock' was met and in these conditions Clowes insisted that the bore should be widened and the tunnel lined with brickwork, at no extra payment for the contractor. Jones had other problems as his sureties were not forthcoming: he was being pursued by creditors from Manchester and he was obliged to sign a further agreement whereby a company overseer would be responsible for payment of the workforce, including Jones. It is unlikely in these circumstances that Jones could have profitably executed the works. Progress was slow, punctuated on Jones's part by bouts of drunkenness and spells in prison for debt. The company attempted to dismiss him from the works in June 1785 but he was still on site as late as the summer of 1788 and in 1792 filed a bill against the company claiming that he had fulfilled the contract, completing the tunnel on time; in March 1795 his case was dismissed with costs. Jones would claim that he was harshly treated by the company and as he was able to continue to tender successfully for work one must wonder how typical an episode this was in his life. It would appear that the earliest recorded use of a contractors 'railway' was by Jones on this job, suggesting he had expertise beyond that displayed here. He then obtained a sub-contract for the Greywell Tunnel on the Basingstoke Canal, a source of constant complaint for John Pinkerton (q.v.) the main contractor.

Charles and George Jones, the sons of Charles Sr., assisted their father on the Sapperton tunnel and a son was agent on the Norwood tunnel on the Chesterfield Canal, where progress was not always satisfactory. The experience of the family in tunnelling enabled them to get work in the 1790s but they rarely completed these contracts, in what remains one of the most difficult of civil engineering arts. They took on the straightforward canal cutting job on the Rother navigation in 1791, but were unable to complete it. Their work on the Grand Junction Canal tunnels appears to have been taken on jointly with John Biggs and was eventually taken over by James Barnes (q.v.) for the Company. The Langleybury tunnel was in any case soon abandoned in favour of a diversion. Over the next thirty years the family was involved in a succession of civil engineering contracts, mostly for earthworks. By 1830, largely due to ill-health, Charles Jr. had effectively retired.

Samuel Jones, another son of Charles Sr., continued on the Rother navigation after his father's dismissal from a contract there in 1793, surveyed the Croydon–Reigate Railway in 1802 and built the larger part of the Grand Surrey Docks (1804–1806). In 1810 he surveyed the line for a canal from Canterbury to St. Nicholas Bay.

From Samuel Jones's parliamentary evidence on the London to Southampton Railway it is evident that the family never developed as large scale contractors, probably because they never made enough money to build up the necessary capital, and after the mid-1830s they fade away.

MIKE CHRIMES

[Thames and Severn Canal records, Gloucestershire County Records Office; Jones v. Corporation of Proprietors of the Thames and Severn Canal Navigation, 21 March 1795, PRO C33/488, p. 285d; Basingstoke Canal records, Hampshire County RO; House of Lords (1834) S Jones Jr., *Minutes of Evidence ... London to Southampton Railway*; H Household (1968, 1991) *The Thames and Severn Canal*; P. A. L. Vine (1965, 1994) *London's lost Route to Basingstoke*; P. A. L. Vine (1995) *London's lost Route to Midhurst*; C. Hadfield (1969) *Canals of South and South East England*, A. H. Faulkner (1972) *The Grand Junction Canal*; C. Richardson (ed.) (1996) *Minutes of the Chesterfield Canal Company 1771–1780*]

Works

Bridgwater Canal
Trent and Mersey Canal, Preston Brook
1773–1776. Chesterfield Canal, Norwood tunnel, 2,850 yd.
1784–1788. Sapperton Tunnel, Thames and Severn Canal, 1,487 yd., £14,335, dismissed
1788–1789. Greywell Tunnel, Basingstoke Canal, dismissed
1790. Andover section, Basingstoke Canal, survey

1791–1793. Rother Navigation, with Samuel, 12 miles, £7,148, dismissed
1793–1795. Braunston, £9.50 per linear yard (?), dismissed
1793–1796?. Blisworth Tunnel, £10.00 per linear yard, dismissed
1793. Langleybury Tunnels, Grand Junction Canal, with John Biggs
1795–1798. Aldenham Reservoir, Grand Junction Canal
1800. West India Docks
1801. Grand Surrey Canal
1801–1805. Surrey Railway, Croydon–Carshalton, 1½ miles
c. 1802. Samuel Whitbread's Railway
Contracts for Lord Camden:
1804–1806. Grand Surrey Docks, dismissed
1812–1819. Regent's Canal
1821–1822. New North Road
c. 1824. Battle Bridge, King's Cross, Holloway Road
1829. Devonport waterworks

JONES, Evan (fl. 1823), road surveyor, of Monmouth, acting for Thomas Telford (q.v.) carried out surveys of the South Wales Mail Roads from Moonall to Huntley, and from Huntley to Goodrich in 1823; he also produced a map of the road from Monmouth to Old Forge.

MIKE CHRIMES

[Gibb papers, ICE archives]

JONES, James (1790–1864), a founder member of the Institution of Civil Engineers, was born in London on 6 April 1790; his early career was as a copper and tinsmith. In the early 1820s he was involved in building models of steam engines etc for various engineers and patentees in the London area and built a locomotive for William James (q.v.) in 1826–1827. He assisted with equipment for the Cheshunt and Deptford railways built to a patent of Henry Robinson Palmer (q.v.) and when the latter was engaged in the Eastern Extension to the London Docks after 1826 he employed Jones to make models of dock machinery. Jones became Palmer's assistant at the docks.

On the recommendation of Thomas Telford (q.v.) and Palmer, Jones was appointed Resident Machinist and Engineer by the St. Katharine's Dock Company in 1831, where he remained until 1836. Jones went with Palmer in 1838 to build the Ipswich Docks, where he was employed until 1842. He become manager of Ransomes & Co., Ipswich, manufacturers of agricultural machinery, where he stayed until 1852, when he became engineer of the Oxford Water Works.

At the first meeting of the Institution of Civil Engineers on 2 January 1818, Jones was asked to be Secretary and Treasurer, and was duly elected on 10 January when the real activity of the Institution began. He read a Communication on an improvement to blocks, accompanied by models. With Palmer and William Maudslay (q.v.), Jones drafted the Institution's rules of order and, with William Provis (q.v.), he proposed that Telford patronise the Institution by becoming its President. In 1823 he proposed the first Commemorative Dinner, which became an annual event for the Institution. Jones continued as Secretary until January 1823 and became a Council Member in 1825. In February 1835, Palmer proposed that Jones be made a member exempt from annual contributions in recognition of his efforts on behalf of the Institution during its early years. He remained an active member and frequent visitor and was one of the five surviving founders who signed the Register of Members in 1840

He suffered a terrible death in the line of his professional duties: while inspecting Mr. Evans' brewery on 24 April 1864, he fell from a stage above the vats into boiling liquid and died the next day at the Radcliffe Infirmary. The Institution commemorated Jones with a tablet erected in the Oxford Church he attended, now part of the library of Lincoln College.

TESS CANFIELD

[J. Jones (1817–1827) Cash book, including ICE accounts, ICE archives; *Occasional Communication No. 3, Hudleston's Patent Lock* (1819) ICE archives; Obituary (1865) *Min Proc ICE,* **xxiv**, 523; R. Boase, Jones, James, in *Modern English Biography (1892–1921);* J. G. Watson (1988) *The Civils*]

Publications

1837. Account of the Youghal Bridge, *Min Proc ICE,* **I**, 11

JONES, Samuel (fl. 1759–1776), assistant to Langley Edwards (q.v.), acted as resident engineer on the River Blyth Navigation, 1759–1761, and on the River Witham Drainage and Navigation from the start of work in 1763 at an annual salary of 52 guineas. On the Witham, the Grand Sluice at Boston and a new 9 mile cut for the river from Boston to Chapel Hill were completed by 1767. Deepening the river and building three locks between Chapel Hill and Lincoln reached completion in 1771. The total expenditure was £48,000. Jones had been effectively in charge from 1770 and continued to be employed as Surveyor to 1776.

Meanwhile work began on the Black Sluice Drainage, south of Boston, in 1765 with Edwards and John Landen as joint engineers. Two Surveyors of Works were appointed but Edwards could only devote part of his efforts to the project and in 1766 he was allowed 1 guinea per week 'for the time he has employed Samuel Jones to assist him' in starting the works.

A. W. SKEMPTON

[Minute Books of the Witham Commissioners (1762–1776) and the Black Sluice Drainage (1764–1766) Lincolnshire Archives Office; John Boyes and Ronald Russell (1977) *Canals of Eastern England*]

JONES, Thomas, FRS (1775–1852), instrument maker, was a pupil of Jesse Ramsden a leading instrument maker of the late eighteenth century whose dividing engine revolutionised manufacture of scientific instruments at that time. Ramsden employed a workforce of fifty, and when Jones joined him in 1789 he could not have chosen a better master.

In due course Jones set up in business himself at Charing Cross and was one of the last generation of British instrument makers with the prestige to play an important role in the scientific community. He was involved in the establishment of the Astronomical Society in 1820, was elected a Member of the Institution of Civil Engineers in 1824 and elected a Fellow of the Royal Society in 1835. His reputation was particularly based on his skill in making large astronomical instruments such as the transit circle at Radcliffe Observatory, Oxford (1836). He also developed an improved hygrometer and published a guide to the use of mountain barometers. No doubt his work in such surveying instruments brought him into contact with civil engineers.

MIKE CHRIMES

[Membership records, ICE archives; DNB; J. A. Bennett (1987) *The Divided Circle*]

JOPLING, Joseph (1789–1867), architect and engineer, was the son of Joseph Jopling of Cothestone, near Barnard Castle. A. J. Joplin submitted a design for a 'Gothic' church at the Royal Academy in 1816 which indicates he had moved to London by this time. He got to know both Thomas Telford and Thomas Tredgold (qq.v.) who signed his membership proposal from the Institution of Civil Engineers, to which he was elected an Associate in 1824. At that time he was heavily involved in promoting his septenary system of drawing, particularly among naval architects.

Although best known for his works on the theories of proportion and architectural draughtsmanship, his contributions to the Institution of Civil Engineers suggest more diverse interests such as fireproof construction and blasting. The latter arose from his involvement with the Kirkby Ireleth slate quarries, which led to railway surveying work in the Furness area. He won a Society of Arts gold medal for his method of improving the construction of the ribs of groined arches. He also contributed to the *Builder* and *Mechanics Magazine.*

Jopling had at least three sons of whom Charles Michael (1820–1863), the second son, became a Member of the Institution of Civil Engineers and was predominately involved in railways. Jopling lived for much of his career at 24 Somerset Street, Portman Square, although from 1835–1843 he removed to Furness. He resigned his Institution membership in the 1850s.

MIKE CHRIMES

[Membership records, ICE; Original communications, ICE; Colvin (3); Obituary of C. M. Jopling, *Min Proc ICE*, **23**, 1863–1864, 508–511]

Publications

1823. *The Septenary System of generating Curves by continued Motion; including Sunday Observation*
1827. *Designs for Agricultural Buildings*, by C. Waistell
1827. *On the Construction of ordinary Dwelling Houses*, OC53
1830. *Smith's fireproof Building*, OC61
1833. *The Practice of isometrical Perspective*
1833. *Blackfriars Bridge and Thames Navigation*
1834. *On blasting Slate Rock*, OC124
1849. *Letter … on the Necessity of Numerous Examples of Mathematical Line*
1849. *An Impulse to Art*
1855. *A Key to Proportions of the Parthenon*

Articles:

1827. Septenary system, *Mechanics Magazine*, **8**, 36–39, 116–120, 281–263, (314), 357–362, 381–384, 389–390, 418–422, 435–438
1828. Mechanical geometry—septenary system, *Mechanics Magazine*, 9, 250, 394–395, 436–440
1828. Mechanical geometry, *Mechanics Magazine*, **10**, 8–10, 32, 74–76,156–151, 211–214, 226–229, 350–351, 391–392
1828. Designs for agricultural buildings, by C. Wastell [review], *Mechanics Magazine*, **10**, 60–63
1829. Radii of cumulative in a cycloid, *Mechanics Magazine*, **11**, 142–143, 392–393
1829. On Child's parabolic trammel, *Mechanics Magazine*,**11**, 297–298
1829. Septenary system, *Mechanics Magazine*, **12**, 329–331, 404–405, 469–472
1829. Naval architecture—the Septenary system, *Mechanics Magazine*, **12**, 239–240
1830. Improvements in ribbed groines, *Mechanics Magazine*, **13**, 241–244
1830. New Building Act, *Mechanics Magazine*, **14**, 317
1831. Smith's patent chimneys, *Mechanics Magazine*, **15**, 23–24
1831. Triangular figures and septenary system, *Mechanics Magazine*, **16**, 308–309
1833. Sliding catch for the swing frame for cheeses, *Mechanics Magazine*, **20**, 6–7
1833. Cracks in ceilings, *Mechanics Magazine*, **20**, 7 8
1833. Treatise on isometrical perspective [review], *Mechanics Magazine*, **20**, 289–294, 313–314, 350–351, 404–406, 442–444
1833. Smoky chimneys, *Mechanics Magazine*, **20**, 445
1833. Blackfriars Bridge, *Mechanics Magazine*, **20**, 25–29, 38–39, 280–283
1833. Isometrical perspective, *Mechanics Magazine*, **20**, [29–30], 45
1834. The practice of isometrical perspective [review], *Mechanics Magazine*, **20**, 235–238

1834. Metropolitan improvements—new road from Hyde Park to Clapham [etc.], *Mechanics Magazine*, **20**, 426–428
1834. On the most pleasing form of an ellipse, *Mechanics Magazine*, **21**, 51–52
1834. History on drawing isometrical perspective on wood, *Mechanics Magazine*, **21**, [186–187], 203, 280
1834. Isometrical perspective, *Mechanics Magazine*, **22**, 39–40
1834. The late conflagration of the Houses of Parliament, *Mechanics Magazine*, **22**, 168–169
1835 [1834]. The smoke nuisance, *Mechanics Magazine*, **23**, 298–299
1835. Railroad bars, *Mechanics Magazine*, **23**, 371–372
1835. Slate platforms for railways, *Mechanics Magazine*, **24**, 124–125
1835. Kirkby slate, *Mechanics Magazine*, **24**, 164–166
1836. Railway hints, *Mechanics Magazine*, **25**, 43
1836. Kirkby slate quarries, *Mechanics Magazine*, **25**, 56–59

JUBB, John, Sr. (d. 1808), and **JUBB, John, Jr.** (c. 1775–1816), millwrights and textile machinery engineers, were active in Yorkshire in the late eighteenth and early nineteenth centuries. Jubb Sr. may have originated in South Yorkshire where a namesake married Sarah Drabble of Sheffield at Doncaster in 1774. John Jr. was baptised at Wortley in April 1775.

John Sr. was apparently trained as a millwright, and was a partner in Churwell cotton mill in 1784, when he was also making textile machinery. By 1789 he was working in Leeds and had supplied John Marshall with various kinds of textile machinery. He made a name in this field but continued to practice as a millwright, although at the time of his second marriage in 1791 he described himself as a 'yeoman'; at his death in 1806 he was referred to as a millwright and machine maker. His business was continued by his son, who took over the Hunslet foundry in 1810; his premature death in April 1816 finished the firm.

The Jubbs are an example of the metamorphosis in skills taking place between those of the traditional millwright and the textile engineer at this time.

MIKE CHRIMES

[G. Cookson (1993) Early textile engineers in Leeds 1780–1850, *Publications of the Thoresby Society, 2nd series*, **4**, 40–61]

K

KANE, Richard (1666–1736), was Governor of Minorca and his memorial states that he 'had paved, fortified and adorned a truly royal road throughout the length of an island hitherto impassable'. This road, completed in 1720, was the first between the north and south of Minorca since the Roman period. The portion which remains between Mahon and Mercadel is still known as the Cami d'en Kane.

SUSAN HOTS

[Monumental memorial in Westminster Abbey; Archives of Mahon, Alayor Cuidadella and Minorca; J. Sloss (1995) *Richard Kane Governor of Minorca*; DNB]

KATER, Henry, Captain, FRS (1777–1835) scientist, was born on 16 April 1777 in Bristol, the son of Henry Kater, a sugar manufacturer of German descent. After two years working for a local lawyer, on his father's death in 1794 he resumed his education by studies of mathematics. In 1799 he purchased a Commission as Ensign in the 12th Regiment of Foot, then stationed in Madras where he became involved in surveying work, developed an improved hygrometer, and introduced improvements in the pendulum used in survey work before returning home in 1807. His surveys included that of the Pelar River, as well as on the triangulation of India. After attending the Royal Military College he served in Jersey, Uxbridge and Ipswich where he attained the rank of Brigade Major, finally being retired on half pay as Captain in 1814.

By this time Kater's scientific attainments were well-known and he was elected a Fellow of the Royal Society in 1815, acting as its Treasurer for several years. His work with pendulums led to him being awarded the Copley Medal in 1817 and becoming involved in the establishment of standard weights and measures for the United Kingdom, the standard foot being determined by the use of a pendulum. His work embraced astronomy, the work of the ordnance survey, variations in gravity, and the design of compasses.

Some indication of the interest in his work among civil engineers is given by the fact that offprints of nine of his twenty-five papers listed in the Royal Society's *Catalogue of Scientific Papers* are to be found in the Library of the Institution of Civil Engineers. He was elected a Member of the Institution of Civil Engineers in 1820, becoming an Honorary Member in 1824. Kater died at his home, York House, Regent's Park, on 26 April 1835.

MIKE CHRIMES

[Membership records, ICE archives; *Select Committee on Weights and Measures* (1821) Report HC571; R. H. Phillimore (1950) *Historical records of the survey of India*, **II**; DNB; *Royal Society Catalogue of Scientific Papers*]

Publications
Selected published works:

1807. Description of a very sensible hygrometer, *Asiatick Researches*, **IV**, 24–31; *Nicholson Journal*, **23**, 207–211

1807. Description of an improved hygrometer, *Asiatick Researches*, **IV**, 394–397; *Philosophical Magazine*, **27**, 322–325
1808. Description of a new compensation pendulum, *Nicholson Journal*, **20**, 214–220
1818. An account of experiments for determining the length of the pendulum vibrating seconds, in the latitude of London, *Phil Trans*, 32–109
1818. On the length of the French metre estimated in parts of the English standard, *Phil Trans*, 110–117
1819. An account of experiments for determining the variation in the length of the pendulum vibrating seconds, at the principal stations of the trigonometrical survey of Great Britain, *Phil Trans*, 336–508
1819. An account of experiments for determining the variation in the length of the pendulum vibrating seconds, at the principal stations of the trigonometrical survey of Great Britain, *Phil Trans*
1821. An account of the comparison of various British standards of linear measure, *Phil Trans*, 75–94
1825. An account of the construction and adjustment of the new standards of weights and measures of the United Kingdom, *Phil Trans*, 1826, 1–52.
1830. *A Treatise on Mechanics*, with D. Lardner
1830. *On the Error in Standards of linear Measure arising from the Thickness of the Bar on which they are Trained*
1831. An account of the construction and verification of a copy of the Imperial standard yard made for the Royal Society, *Phil Trans*, 345–348
1832. An account of the construction and verification of certain standards of linear measure for Russian government. *Phil Trans*, 359–382

KELK, George (fl. 1764–1800), was a land surveyor working in Yorkshire, Nottinghamshire and Lincolnshire. He assisted John Grundy (q.v.) in 1769 on a survey of some 6,000 acres of low ground west of the river Trent between Laneham and West Burton, a few miles south of Gainsborough. He received £20 for thirty days work, and was then employed from 1769 to 1772 as land surveyor on the drainage works. In May 1772 the proprietors of the River Don Navigation ordered their engineer John Thompson (q.v.) and Kelk to survey a line for a canal from Stainforth lock on the Don to Althorpe on the Trent. Thomas Tofield (q.v.) reported favourably, saying 'Mr. Thompson and Mr. Kelk have shown much Judgement in the choice of the Course', but the canal, on a slightly different line at its eastern end, was not built until the 1790s.

In 1773 Kelk reported on a scheme for draining Everton, Gringley and Misterton Carrs, an area of about 5,000 acres south of the river Idle. His proposals included a sluice 20 ft. wide in cut adjacent to Misterton Soss (navigation lock) on the Idle, an embankment up to 8 ft. high along the south side of the river, and a main drain running to a 12 ft. sluice through the bank into the river downstream of the Soss with side drains into the Carrs; the total cost was estimated at £9,000.

Apart from a small amount of work in the late 1780s, no action was taken until 1796 when an Act was obtained and work began in accordance with a plan by William Jessop (q.v.). In all essentials this was based on Kelk's scheme, the principal difference being a wider (30 ft.) sluice adjacent to the Soss.

A. W. SKEMPTON

[John Grundy (1769–1772) Report Books, vols. 10 and 11 on Laneham Drainage, Leeds University Library; George Kelk (1773) A scheme for embanking, draining, and preserving the low grounds in … Everton, Gringley and Misterton, ms, Nottinghamshire RO; Charles Hadfield (1973) *Canals of Yorkshire and North East England*; Charles Hatfield and A. W. Skempton (1979) *William Jessop, Engineer*; Bendall]

Publications
1769. *A Plan of the … Low Grounds in Laneham … Struton … and West Burton …which are frequently overflowed by the Trent and other Water*, with John Grundy
1772. *A Plan of the Intended Navigable Canal from Stainforth Cut to the River Trent at Althrop*
1773. *A map of the Low Grounds in … Misterton, Gringley and Everton, in the County of Nottingham*

KENNEDY, John (1769–1855), millwright and mill owner, was born at Knocknalling, Kirkcudbright, on 4 July 1769, the third of five sons of Robert Kennedy, a farmer. He received some elementary education in the winter months from a young cleric, and also from Alexander Robb, who instructed him in mechanics. His father died when he was still a child and he was brought up by his mother until the age of fourteen when he was apprenticed to William Cannan, machine maker at Chowbent, Lancashire, and in February 1784 he headed south. Cannan was clearly a mechanic of some eminence as his partner, Smith, was installing cotton machinery in Carlisle and brought Kennedy south. Adam and George Murray (q.v.), who came from the same part of Scotland, had already been apprenticed to Cannan and it is likely that this encouraged Kennedy.

Kennedy learnt all about the manufacture of the somewhat crude cotton machinery of the time and after a seven year apprenticeship moved to Manchester in 1791 and joined with Benjamin and William Sandiford, fustian warehousemen, and James McConnel, another millwright, in business as machine makers and mule spinners. Sandiford left the business in 1795. Crompton had built his first mules in 1780 and Kennedy's firm became leading builders of spinning mules, Kennedy himself improving the equipment, introducing 'double speed' machines which facilitated the spinning of fine yarns around 1793, initially via the Murrays at Drinkwater's Piccadilly Mill, then managed by Robert Owen and the first in

Manchester to be powered by a Boulton and Watt rotary beam engine. Their own works were small rented premises to which they added Savrin's Mill in 1793. A Wrigley steam engine was added in 1796 and steam power may have been applied to a mule at that time.

Kennedy proved an astute businessman as well as machine designer and McConnel and Kennedy built an important complex of mills alongside Murrays' in Ancoats, Manchester. Their 'Old Mill' was erected in 1796–1797 and was possibly the first to be erected to accommodate steam driven mules, using an internal layout that was to become standard throughout the Lancashire cotton industry. The steam engine was supplied by Boulton and Watt, and millwork by T. C. Hewes (q.v.). In 1801–1802 Long Mill was added alongside, followed by Sedgwick Mill (1818–1820), an eight storey structure on a U-plan of fireproof construction, like Chorlton New mill (1814), although with a timber roof, and powered by a Boulton and Watt engine in an internal engine house; millwork was by William Fairbairn. The Long Mill was built, presumably to Kennedy's designs, by Richard Copparts, with Bateman and Sherratt supplying the iron columns, they also supplied the columns for the Sedgwick Mill.

Kennedy's eminence in engineering circles was reflected in his appointment as an umpire at the Rainhill Trials in 1830. As a member of the Manchester Literary and Philosophical Society he contributed a number of important papers on the history of the cotton trade. He married Mary, daughter of John Stuart, and they had one son, John Lawson Kennedy, and several daughters.

Kennedy died at his home, Ardwick Hall, Manchester, on 30 October 1855 and was buried in the Rusholme Road cemetery, Ardwick. In his obituary notice Fairbairn notes that he was interested in all aspects of mechanical science and that he was well known in the scientific and engineering community of the day. John Kennedy's brother, James, was also trained as a millwright and by 1800 had a mill in Radium (German) Street, Manchester. After it was burnt down he built a new mill in Great Ancoats Street (1805), which used cast iron beams and jack arches, the beams following the model used by Bage in Leeds (1802–1803).

MIKE CHRIMES

[W. Fairbairn (1862) Memoir of the late John Kennedy Esq., *Memoirs of the Manchester Literary and Philosophical Society*, 3rd series, **1**, 147–155; DNB; S. B. Smith (1969) Thomas Cheek Hughes (1768–1832): an ingenious engineer and mechanic of Manchester, MSc thesis, UMIST; M. Williams (1992) *Cotton mills in Greater Manchester*]

Publications
1815. *On the Rise and Progress of the Cotton Trade*
1830. *Memoir of Crompton*

Works

1798. McConnel and Kennedy's Old Mill, Ancoats
1801–1805. New Mill, Ancoats
1818–1820. Sedgwick Mill, Ancoats

KILLALY, John A. (1766–1832), surveyor and canal engineer, was born in England and may, for a while, have worked on canals there. Killaly came to Ireland as a surveyor and in 1794 joined the Grand Canal Company as an assistant engineer at an annual salary of £150, having previously carried out some surveys for the company. William Jessop (q.v.) was consulting engineer at that time. Killaly also received a small payment for work on Daniel Augustus Beaufort's map of Ireland. Two years later, the directors praised his work and said that, from being a 'mere measurer and surveyor', he had become the 'complete superintendent of all kinds of work'. In 1798 he became the company's chief engineer with an increased annual salary of £400, this rising in due course to around £800. At about this time, Killaly married Alicia Hamilton (d. 1837) and three of their sons later graduated from the University of Dublin, two of them following their father in his profession. Along with a younger brother, John Sackville Killaly, Hamilton Hartley Killaly (1800–1874) emigrated to Canada in 1834 and later became the first president of the Canadian Society of Engineers.

The Canal was completed to Daingean (Philipstown) in 1797 and it was hoped the route would be completed to Tullamore shortly afterwards. However the early part of 1800 saw Killaly supervising the construction of the section near Edenderry. This was probably the most difficult part of the route of the canal, crossing as it does extensive deep bogs to the west of the twentieth lock at Ticknevin. Many engineers had been asked for their opinion as to the best way of taking the canal across the bog, including William Jessop, William Chapman, and Richard Griffith Sr. (qq.v.). Bernard Mullins and his son Michael Bernard (q.v.), in an important paper presented to the Institution of Civil Engineers of Ireland in 1846, described the engineering problems involved and offered a solution.

While Killaly overcame these difficulties, consideration was being given to the Canal's continuation on to the Shannon. In 1801 the Directors General of Inland Navigation, a body created in 1800, refused to approve the additional costs of a route via Frankford, and the Company therefore decided upon the Brosna Valley Route, both alternatives suggested by Jessop in 1797. Killaly soon realised it was easier to build a canal rather than improve the Brosna, despite the problems of building across a bog. Richard Griffith (q.v.) was appointed to inspect the work five days a month in recognition of its difficulty. As construction dragged on following the death of the original masonry contractor Michael Hayes, the contractors Henry and Mullins (qq.v.) agreed to complete the

work. The navigation to Shannon Harbour, in county Offaly, was officially declared open on 25 October 1803. Killaly also surveyed a line of canal from Monasterevan to Maryborough (now Portlaoise). Around 1803, with John Brownrigg (q.v.), he became an engineer under the Directors-General of Inland Navigation. Possibly at their request he visited a number of English Canals.

In May 1804 Killaly received a silver cup and 200 gold guineas from the Grand Canal company for his work on the canal. He continued to work for the company at a reduced salary (£630), Jessop's work as a consultant also ending. Various leaks were still outstanding and in 1804 he directed a general repair of the canal, which was not completed until the end of 1805. He also carried out surveys of various other projects such as the Ballinasloe line or Suck Navigation, considered in the 1790s, and put by the Company to Government 1802–1807 without success. In 1810 he was sent to Doonane, near Athy, to assess the situation in the Company collieries there and to report on their commercial viability. He reported on 6 April 1810 that prospects (for the collieries) 'were by no means flattering'. He threatened to resign if he would be asked to take over the position of manager but, during his short stay, he compiled a number of reports detailing the position at the collieries. He was shortly thereafter relieved by another of the company's engineers, Thomas Colbourne, who resigned in November 1810.

At that time Killaly then on a salary of £685 p.a., also left the Grand Canal company, as part of a programme of retrenchment under new directors. He was thus able to devote more time to his professional duties as an engineer to the Directors-General of Inland Navigation. He was paid a basic salary and was later allowed to carry out other government assignments. He did, however, continue to take an active interest in the affairs of the Grand Canal company.

Together with John Brownrigg (q.v.), Killaly inspected the state of the Shannon Navigation in July 1810 and in the following year he compiled a comprehensive report for the Commissioners on the state of the inland navigations of Ireland and laid down a number of proposals for action. He advised on the Corrib, Lagan, Newry and Suir Navigations, and on the Royal Canal. More specifically, he reported on the Upper Shannon Navigation, the Lough Allen Canal, and the Erne Navigation. Two years later, he turned his attention to Belfast where he made proposals for a dock and short ship canal. He also appears to have had a private surveying business in Dublin in partnership with James Oates. The services of Killaly ensured more accurate estimates and the completion of the Royal Canal was carried out within his estimate by a single contractor; additionally, the Directors-General were kept fully informed of progress.

As part of the proposed nationalisation of the Irish waterways, Killaly surveyed nine schemes and estimated the cost at a little over £1.5 million, but the idea was not proceeded with at the time.

A detailed map was prepared by Killaly to accompany a publication by James Dawson, *Canal Extensions in Ireland*, Dublin, 1819. It shows the Grand and Royal Canals with proposed extensions. Two years later, in a report on the Suir Navigation, he advocated the use of spur weirs and made proposals for the clearing of the navigable channels. He also put forward a plan for the canalisation of the river from Carrick-on-Suir to Clonmel and onwards to the Shannon.

Killaly acted as consulting engineer for the extension of the Grand Canal to Ballinasloe in County Galway, a project which was supervised by his son, Hamilton Hartley. An Exchequer Loan had finally become available in 1822 and work commenced in 1824, the contractors being Henry, Mullins and MacMahon. The canal was inspected in 1827 by George Halpin Sr. (q.v.), acting for the Loan Commissioners, and was opened on 29 September 1828. Work on the Mountmellick line, also financed by an Exchequer Loan, commenced early in 1827 under Hamilton Hartley Killaly; the route his father surveyed on to Portlaoise was not completed. In 1825 John surveyed a route for the Ulster Canal, on which his son began work in 1830. This was completed by Thomas Casebourne (q.v.) with William Dargan (q.v.) as contractor.

During the 1820s Killaly is reported to have employed a labour force of 9434 men on famine relief schemes. He was one of the government engineers appointed to carry out road works in the western counties at a period of distress as a means of affording relief by the employment of the peasantry in improving the road communications of the country. His main appointed district was County Clare, with smaller areas of South County Galway and North County Tipperary. The total mileage of roads built under his direction was 107, all but ten being in Clare. The main projects were the road from Ennis to Kilrush and a coastal road of 36 miles, including Bealaclugga Bridge at Spanish Point. All the roads were completed by 1828 using contractors. Reporting from County Clare in 1822, Killaly said 'the great destruction of morals and waste of public property which have taken place in the county of Clare from this cause (jobbing) are beyond my power to calculate'. In 1830 he prepared answers to a series of queries from the Chief Secretary, Leveson-Gower, and expressed the hope that the latter's labours towards eradicating the evils of the presentment system would be successful.

Killaly also urged the appointment of qualified county surveyors, something which indeed came to pass shortly after his death. He had a sound background knowledge of surveying and engineering and was simultaneously involved in a range of projects in different parts of the country. Due to this work load, he did not give the government roadworks the same personal supervision as Richard Griffith (q.v.), preferring to use contractors. He wrote fewer reports of his work and

furnished less detailed accounts, but he did succeed in completing a coherent plan of new roads in his area at a moderate cost, in fact taking over some of Alexander Nimmo's (q.v.) unfinished roads. The Loughrea–Derrybrien Road, Galway, laid out by him was not built before his death.

Towards the end of his life he enlisted the help of his son, Hamilton, to design a wet dock at Galway and discussed with John Rennie and John Brownrigg (qq.v.) plans for a ship canal from Dun Laoghaire Harbour to the Grand Canal Docks at Ringsend in Dublin.

He died at his residence, Williamstown, in county Dublin on 6 April 1832. A large memorial to him was erected in St Patrick's Cathedral, Dublin, by his widow, Alicia, who died in 1837.

RON COX

[Plan and section of Derrycooly supply (1802), Plan of Richmond harbour (1817), Section of the line of Lough Allen Canal (1818), Killaly Drawings (M. B. Mullins Bequest), Sketchbook, IEI archives; Plan for Richmond Bridge and harbour house (1817), Pakenham collection, Tullynally; Design for a ship canal from the Grand Canal Docks to the Royal Harbour at Kingstown (1825) Nat. Lib., Trinity College; Pembroke Estate papers, Irish National Archives; House of Commons (1812–1813); *Papers relating to Inland Navigations in Ireland*; House of Commons, Committee on Inland Navigation in Ireland (1813) *Reports*, **I–III**; Commissioners on Public Works, Ireland (1834) *Reports*, **1**, 1832; A. Nimmo *et al.* (1845) *Reports ... on the Harbour of Cork*; B. Mullins and M. B. Mullins (1848) On the origin and reclamation of peat bog, with some observations on the construction of roads, railways, and canals in bog, *Trans Inst Civ Eng Ireland*, **2**, 1–48; J. Hill (1851) On the maintenance of macadamised roads ..., *Trans Inst Civ Eng Ireland*, **4**, 12–23; R. Delany (1973) *The Grand Canal of Ireland*, 39–46, 53–55, 91–98, 149; R. Heard (1977) *Public Works in Ireland 1800–1831*, MLitt thesis, University of Dublin]

Publications

1800. *Report, Survey and Estimate of a proposed Canal from Monasterevan to Jenkinstown*
1808. *A Plan for making the Philipstown Level the Summit Level of the Grand Canal*
1811. *A Copy of a Survey and Estimate ... Canal from the Grand Canal at Monasterevan to Carrick on Suir*
1813. *Survey for Lines of Navigation ... between the Grand Canal, the River Suir, the Barrow and the Shannon* [181–1812]
1813. *Report ... on a Line of Canal from Gore's bridge to Kilkenny* [1812]
1813. *A Map of proposed Extensions of the Royal Canal surveyed ... in 1809 and 1813*
1818. *Survey and Section for a Line of Canal from Lough Allen to the River Shannon*
1819. *Map of the Grand and Royal Canals with proposed Extensions. Accompanying DAWSON J. Canal Extensions in Ireland. Dublin, 1819*
1845. *Report on Cork harbour* [repr.]

Works

1794–1810. Grand Canal, Ireland, Lowtown–Tullamore completion, Tullamore to Shannon Harbour construction, surveyor/assistant engineer (1794–1798), engineer (1798–1810), £135,000, total cost c. £340,000

1822–1828. Roads and bridges in county Clare, etc., engineer, 107½ miles

1824–1828. Grand Canal, Ballinasloe Branch, consultant, 14.5 miles, c. £40,000

1825. The Ulster Canal, surveyor

1827–1831. Grand Canal, Monasterevan to Mountmellick Branch, consultant, 11.5 miles, £33,000

KINDERLEY, Nathaniel (1673–1742), civil engineer, was baptised at Spalding on 23 May 1673, the younger son of Geoffrey Kinderley, a well-to-do north-countryman who had moved to Lincolnshire. The elder son having died young, Nathaniel inherited considerable property on his father's death in 1714. Apart from the recorded fact that he superintended repairs to a drainage mill near Moulton in 1705 nothing has been found about his engineering activities before 1720, though it is evident from subsequent events that he had by that time devoted much thought to fen drainage and was already known to the Bedford Level authorities.

In August 1720 Jacob Le Pla, Surveyor of the North Level of the Fens, reported to the Bedford Level Board that the channel of the river Nene in its estuary below Wisbech had swung across towards the Norfolk shore, leaving Gunthorpe Sluice stranded a quarter of a mile away. As the sluice was at the mouth of Shire Drain, the main drain of the North Level, something had to be done to restore its outfall. John Chicheley, Surveyor General of Bedford Level, was requested by the Board to consult a 'skilful engineer' and he brought in John Rowley (c. 1668–1728), a former leading London instrument maker and from 1715 Master of Mechanics to George I. Exactly what happened next is not entirely clear but it seems that by November the Board had decided on making a new cut to bring the river close to the sluice while Le Pla, contrary to orders, had begun work on some other plan which Rowley considered impracticable. A committee was then set up and three of its members, Chicheley and the Surveyors of the South and Middle Levels (Henry Saffery and James Fortrey), 'took a view' attended on site by Kinderley and Rowley taking levels. Their report, approved by the Board on 1 June 1721, advised that 'a New Cut be made through the Salt Marsh in Wisbech Channel above and below Gunthorpe Sluice with proper Jetties and Breakwaters in order to turn the Wisbech Channel over to the sluice for the more Effectual gaining a better outfall for the North Level'. This proposal was, however, strongly opposed by Wisbech Corporation on the fear that it would be harmful to navigation.

Nevertheless work began very soon afterwards under Kinderley's direction. Le Pla kept the accounts and presumably acted as resident engineer. The cut, some 120 ft. in width, was intended to extend from River's End (4½ miles below Wisbech) past Gunthorp Sluice towards Peter's Point, a length of 1½ miles. By June 1722 the cut was well advanced and work was proceeding on the diversion dam across the old channel, involving about one-hundred and forty men 'under Mr Kinderley's care', when it was attacked and destroyed by rioters from Wisbech.

Lengthy legal disputes ensued, but without result, and it was not until 1773 that work began on restoring the scheme, then named as Kinderley's Cut, under the direction of Richard Dunthorne (q.v.). Meanwhile conditions at Gunthorpe Sluice were improved in 1727–1728 by building a reservoir behind the sluice to provide water for scouring its outfall at low tide. This work and a new cut for the river upstream of Wisbech, between Stanground and Guyhirn, were carried out under the direction of Humphry Smith (q.v.).

It would have been early in 1721 that Kinderley wrote the paper published anonymously under the title *The Present State of the Navigation of the Towns of Lyn, Wisbech, Spalding and Boston*. His fundamental conclusion is that 'the main thing wanting for sure Draining is good Outfalls'. In this light he saw the proposed cut past Gunthorpe Sluice principally as a means of improving the Nene outfall, in contrast to a winding and changing channel through the estuary sands, and he put forward the practical suggestion of improving the Great Ouse outfall by making a new cut 3 miles long from St. Germans to King's Lynn, bypassing a 7 mile loop in which the river is far too wide and too shallow. This would be of obvious advantage to navigation, and to drainage, by increasing the fall in the river.

Kinderley also proposed a 'grand Design' of uniting the lesser rivers Nene and Welland with the Ouse and Witham, respectively, by cuts from Gunthorpe to Lynn (8 miles) and from Spalding to Boston (10 miles) and then to extend the cuts from Lynn and Boston to deep sea channels. He recognised that such a scheme might not be acceptable and pleads 'if these noble works can neither of them be done' at least the cut above Lynn should be made.

There followed almost a decade of arguments about the best way of improving the Ouse and the port of King's Lynn. Colonel John Armstrong (q.v.) in a report of May 1724 advised removing Denver Sluice (then in partial ruin), thereby giving free reign for the tides to run 25 miles or more above Lynn; and to widen and deepen the river where necessary and raise its banks. A few months later Charles Bridgeman went to the other extreme by adopting part of Kinderley's grand Design but with a sluice at the lower end of the cut near Lynn to prevent the tides running further upstream, while John Perry (q.v.) in 1725 suggested placing a scouring sluice at the upper end of the cut. Also in 1725 Thomas Badeslade (q.v.) published the results of three years research on the history of the Fens and their rivers, the practical conclusion so far as it concerned the Ouse being to endorse Armstrong's scheme. Finally in 1729 Humphry Smith proposed restoring Denver Sluice, with the addition of a navigation lock, and deepening the 4-mile slaker channel from Denver to Downham to provide a reservoir for scouring the river downstream. He also expressed a commonly held view that the new cut above Lynn could be disastrous to the harbour, by diminishing tidal flow in the large bend in the river upstream.

Smith's report prompted a stern reply from Kinderley, in which he gave more details of the proposed new cut. 'Having surveyed the Ouse (he says) from Denver to Lynn I found the river to be seven, eight and nine feet deep from Denver to St. Germans and Eau Brink, and not above two hundred feet broad; and that it had six inches Fall in a mile, so far, which is five feet; but below Eau Brink to Lynn, for 6 miles in a crooked course, I found the river expanded itself to above ½ a mile in breadth, and at Low-Water was but ten, twelve and fourteen inches deep in many Places in the Channel thence; and had eight feet Fall to Lynn'. By making a cut 200 ft. broad and 2 miles long from Eau Brink to ½ mile above Lynn a fall of 7 ft. would be gained, assuming 6 in. per mile in the cut. He estimated the cost at £15,000.

Here are the essentials of what became known as the Eau Brink Cut, sanctioned by Act of Parliament in 1795 and carried out (with an average width of 240 ft.) in 1818–1821 by John Rennie (q.v.). It proved to be one of the most beneficial works in the history of draining the Fens. In the meantime Kinderley was appointed Surveyor of the South Level in March 1722 at a salary of £85 p.a. in place of Henry Saffery, deceased. A month later he was elected to the Board of Bedford Level Corporation as Conservator. He held both positions for three years.

Kinderley's survey of the Ouse was probably made during or soon after this period. By the late 1720s he turned his attention to the river Dee where for many years past sea-going vessels anchored in the vicinity of Parkgate, on the north (Cheshire) side of the estuary, and goods were sent some 10 miles to and from Chester by land or by small boasts in a shallow channel through the sands. Under an Act of 1700 the city had attempted to improve and stabilise the channel, but with little effect, and early in 1677 Andrew Yarranton (q.v.) had put forward the rather impractical suggestion of making a canal across the sands which, as a by-product, would allow 3,000 acres of land to be 'gained out of the sea'. Kinderley's proposal, submitted in October 1730, was to make a canal about 6½ miles long from Chester through salt marshes on the south side of the estuary in return for tonnage dues and the right to reclaim the sands; the cut was to have a minimum depth of 15 ft. of water at common spring tides to accommodate ships up to 200 tons burden.

Chester Council liked the idea. They encountered opposition but eventually acquired an Act in June 1733. In evidence to the House of Commons in February Kinderley said he had surveyed the river for three years successively and estimated the cost between £40,000 and £50,000. After obtaining the Act he appointed a body of forty to finance the work and in April 1734 it was agreed to raise a joint stock of £40,000 divided into four-hundred shares.

The original Minute Books and Accounts were lost or destroyed long ago, and appear not to have been examined by historians; only copies of some Accounts for 1740–1742 survive. Minutes of the Dee Commissioners exist but they give no details of construction. From a few other contemporary sources, however, it is possible to outline the main sequence of events. Excavation of the canal probably got under way, with a large workface, in the late summer of 1733; it was cut to a depth of 8 ft. below marsh level, the excavated material being used to build embankments about 450 ft. apart with their crest 6 ft. above high water. A dam 480 ft. in length with a sluice 60 ft. long was built across the old channel close to Chester. Part of the dam and its sluice having been demolished by the tides, John Reynolds (q.v.) was consulted in June 1735 and advised that the dam should be widened and the sluice rebuilt 100 ft. in length with a deep cut-off of dovetail piles. This was done, almost certainly under the direction of Reynolds. The sluice gates were closed in April 1737, diverting flow down the cut. Details of the sluice and dam are given in drawings by J. Willson (dated 1739) inserted on a map of the canal and estuary by John Eyes (q.v.) in 1740. The combined force of the river and the ebb tide began scouring the cut, as anticipated, and by March 1740 the required minimum depth of 15 ft. had been attained at 'moderate' spring tides, defined as about 1 ft. below average high water springs. Where necessary the process of deepening was aided by dredging 'ploughs' hauled across the bed by hand-operated capstans in moored punts.

At some time after 1736 and before 1740 Kinderley proposed to extend the canal by at least a mile in the sands along the coast to place called New Quay (later Connah's Quay). The north bank was accordingly extended, with stone pitching on both sides and on the crest. Here, and in other places along the canal upstream, it became necessary to build groynes or 'jetties' to prevent the cut becoming too wide; a width of about 250 ft. at low water seems to have been the desired limit. The extension, the dredging and the jetties ranked as 'New Works' and were directed by Reynolds with two Overseers, William Brackley and James Golborne, elder brother of John Golborne (q.v.).

In December 1740 it was agreed that the joint stock should be increased to £52,000 and that proprietors be incorporated as the River Dee Company; this was achieved by an Act in March 1741. The New Works were considered to be complete by 1743 at a cost around £16,000, bringing the total expenditure to about £56,000. Reclamation works began in 1744.

Kinderley had retired by 1740. He spent his last years at Saltholme, near Stockton, where he died in 1742. He had two sons: John, the vicar of South Walsham, Norfolk, and Nathaniel who in 1751 published a greatly enlarged and up-dated version of his father's 1721 paper. It proved influential in the later decision to make Eau Brink Cut.

A. W. SKEMPTON

[Bedford Level Corporation Orders (1720–1722) Cambridgeshire RO; John Reynolds (1735) *Remarks on the Dam and Sluice making cross the River Dee*, in Thomas Badeslade (1736) *The New Cut Canal*; John Eyes (1740) *Survey of the River Dee*; River Dee Company Accounts (1740–1742) Chester City RO; Nathaniel Kinderley Jr. (1751) *The Ancient and Present State of the Navigation of the Towns of Lyn, Wisbech, Spalding and Boston*; Neil Walker and Thomas Craddock (1849) *The History of Wisbech and the Fens*; Admiralty Inquiry, Chester, Dee Navigation Improvement, Minutes of Evidence (1849) and Report by J. M. Bloxham and J. Albernethy (1850); W. H. Wheeler (1897) Nathaniel Kinderley, *Fenland Notes and Queries*, **3**, 112–118; F. Webster (1930) The River Dee reclamation, *Trans Liverpool Eng Soc*, **51**, 63–92; T. S. Willan (1936) *River Navigation in England 1600–1750*; G. W. Place (1992) The Rise and Fall of Parkgate 1686–1815, *Chetham Soc*, 39]

Publications

1720. *The Present State of the Navigation of the Towns of Lyn, Wisbeech, Spalding and Boston*
1730. *Mr Humphrey Smith's Scheme for Draining the Fens examined and compared*

Works

1721–1722. Wisbech outfall cut, not completed
1733–1740. New Cut Canal, River Dee, completed 1743, £56,000

KIRKHOUSE, William (c. 1776–1866), was the son of Bedlington Kirkhouse, a mining engineer, living near Neath. His grandfather, George, had been brought down c. 1750 to the lower Swansea Valley from Tyneside in order to develop the Llansamlet Coalfield and its waggonways.

Kirkhouse probably sank the first deep mine (620 ft.) in south Wales at Great or Main Pit, Bryncoch, Neath. In 1817–1818 and 1821–1824 he built the 9 mile Tennant Canal the second part of which, north of Aberdulais, included several large engineering works. The canal was conveyed through a deep cutting sunk into quicksand and reinforced with an inverted-arch bed and low retaining walls. In turn this was crossed by a high masonry tramroad causeway. Also on the canal were ten-arched masonry and single-span cast-iron aqueducts over the River Neath,

and an early canal, at Aberdulais. There was also a two-arched crossing of the Afon Clydach at Neath Abbey.

Kirkhouse built short branch-lines to Onllwyn colliery from the Brecon Forest Tramroad in 1835–1837 and the 8 mile long 'Glyncorrwg Mineral Railway' in 1839–1840 from the Neath Canal at Aberdulais to coal-levels at Blaen Cregan near Glyncorrwg. The major engineering works were deep rock cuttings on the Cefn Morfudd summit level and the head of the inclined-planes rising from the Vale of Neath were carried on a large masonry causeway built in typical early-nineteenth-century style; there was also a three-arched viaduct at Blaenpelenna. There were no fewer than one powered-incline and six gravity-worked inclines on the line and the level sections in-between were originally intended to be worked by horses. It is little wonder that the railway inspector, Joshua Richardson from Croydon, considered the line old-fashioned and wondered why a tunnel had not been built instead of using the more primitive expedient of inclined-planes rising to a summit level. In 1852 Kirkhouse was still acting as an agent for the Tennant estate and lived to be 90 years old, dying at his residence, Dulais House, at Aberdylais in the Vale of Neath.

STEPHEN HUGHES

[Journal and accounts of William Kirkhouse, Swansea Reference Library, MSS 464, 466–467; E. Phillips (1925) *Pioneers of the Welsh Coalfield*, **xxiv**; C. Hadfield (1967) *The Canals of South Wales and the Border*, 2nd edn., 77–81; H. Green(1977) Parsons' Folly—The Glyncorrwg Mineral Railway, *Transactions of Neath Antiquarian Society*; S. R. Hughes (1990) *The Archaeology of an Early Railway System: The Brecon Forest Tramroads*, 116–127]

Publications

1831. Mr Kirkhouse's Report, in *Reports on the Formation of a Floating Harbour at Swansea with reference to Plans submitted to the Trustees*

KYD, Alexander, Lieutenant-General (1754–1826), military surveyor, was born in Scotland on 14 March 1754, the son of Captain James Kyd RN and Hannah Bevis, his first wife. He entered the Bengal Engineers in 1775, reaching the rank of lieutenant general on 12 August 1819.

Kyd gained good experience as a military engineer, and from 1787 was involved in surveys of the island of Penang, or Prince of Wales Island, which had been ceded to the East India Company in 1786. He held the appointment of Surveyor General of the Fort of Budge in Calcutta from 1788 to 1794, but spent little time there. During 1789 and 1790 he was engaged in looking for new harbours in the Andaman and Nicobar Islands, but was called away from this during 1790–1792 when he was appointed ADC to the Governor General, Lord Cornwallis in the Third

Mysore War against Tipu, acting as Surveyor General at the siege of Seringapatam.

In 1792–1793 he returned to the Andamans and prepared estimates to fortify the new harbou Port Cornwallis on the North Island. He continued to be occupied with defences of the Andamans until 1796, when it was decided to rely solely on the harbour at Penang. Kyd was recalled to Calcutta and in 1798 was transferred to Allahabad to superintend the remodelling of the fortifications. He remained there until his furlough in 1804. His improvements there are commemorated in the name of the suburb of Kydgunj. He was promoted Colonel in 1805 and became Chief Engineer at Calcutta in 1807. He carried out some tidal observations on the Hooghly (Hughli).

Kyd had married Elizabeth Hay (*née* Wagstaff) on 13 November 1804. His sons James and Robert were sent to England to be trained in shipbuilding and on return took over the dockyard founded by Colonel Henry Watson (q.v.) at Kidderpore Dockyard. He died at Albermarle Street, London, on 25 November 1826.

JOYCE BROWN

[DNB; E. T. Thackeray (1900) *Biographical Notices of Officers of the Royal (Bengal) Engineers*, **6**; E. W. C. Sandes (1933) *The Military Engineer in India*, **I**, 166–170, 179; E. W. C. Sandes (1935) *The Military Engineer in India*, **II**, 175; R. H. Phillimore (1945) *Historical Records of the Survey of India*, **I**, 261, 345–347; Harfield (1988) Lieutenant Colonel John Macdonald, *Army Historical Research*, **66**, 205]

L

LABELYE, Charles (fl. 1725–1753), civil engineer, was born at Vevey in Switzerland where, by one account, he was baptised on 12 August 1705. His father was François Dangeau Labelye, a French refugee. From notes in some of his writings and recorded statements to committees it can be deduced that he came to England in the early 1720s, when he was not much less than twenty. Up to that time he had never heard English spoken. It was said by George Vertue that he worked first as a barber, but from his later accomplishments it must be presumed that he had had a good education with emphasis, perhaps, on mathematics and mechanics. In 1725 he joined a French Masonic Lodge in London, sponsored by John Theophilus Desaguliers (q.v.), another refugee Frenchman and an able 'natural philosopher', or physical scientist. Labelye was in Spain in 1727–1728 and founded a Masonic lodge in Madrid before returning to England.

In the years between his first arrival and 1734, Labelye gained experience which fitted him for the work of his subsequent career, but we have limited knowledge of his activity. He knew probably as much as any contemporary English engineer about construction works on the coast and in rivers and had observed such works in other countries, including Holland and Flanders and presumably France. Moreover, he had learned and was ready to use mathematical theories of hydraulics and of structures such as arches and presumably timber frameworks, theories of which few British engineers, if any at all, had any knowledge. The two articles which he contributed to Desaguliers's seminal two-volume *Course of Experimental Philosophy*, one of which is dated 1735, show him engaging in scientific discussion at an expert level; and in publishing his *Mapp of the Downes* in 1737 he was able to sign it as 'Engineer, late Teacher of the Mathematicks in the Royal Navy.' Doubtless his friendship with Desaguliers was helpful but there is also little doubt that he derived much of his engineering theory from Continental texts such as the writings of Belidor in France. In 1736, he made a critical contribution to the growing controversy about the proposed Westminster Bridge, using an original theoretical solution of the problem of 'the fall under bridges.' About two years later (1738–1739) he applied the equilibration theory of French writers in a mathematical check of the arches of his design for the fifteen-arch bridge. Each of these calculations was a first application in Great Britain—and one of the first in Europe—of quantitative mathematical analysis to the particular element of bridge design involved.

Labelye's use of English, in spite of apologies made in his writings of 1739 and 1745, had become by the latter date unusually accurate and trenchant, employing an admirably broad vocabulary. His engineering reports after 1735 on the engineering of rivers, ports and drainage show that by that date he had gained much knowledge of those subjects by observation if not also by practical civil engineering work. Clearly, if his attachment to the Royal Navy had lasted for some time and involved more than classroom teaching, it could have given him the knowledge and experience on which he seemed to draw. It is noticeable, for instance, that the confident analyses of the scouring actions of the ebb and flow of tides and of the land- or back-waters in the Fen drainage channels on which he reported in 1745 and in Yarmouth Haven, reported on in 1747, included principles and judgements which he had cited as early as 1736 in his proposal for a haven of refuge in the Downs. The late reports suggest that he was a hydraulic engineer of outstanding potential, yet one who never carried out a river, harbour or land drainage scheme. That was because virtually all his energies for some thirteen years—starting in 1738—were absorbed by the design and construction of Westminster Bridge.

The bridge was a national project for which Parliament accepted the responsibility of finding the vast capital sum required, thereby ensuring (though perhaps unintentionally) that the bridge would be free of tolls and free of debt as soon as it was completed. The sources of funds were lotteries set up by Acts of Parliament and grants from the public purse, and the management of the project lay with a body of commissioners appointed by Parliament. Their regular chairman and the conceived patron of the bridge was the ninth Earl of Pembroke.

Labelye was conveniently living in Westminster when planning for the bridge began, gave evidence in support of the Bill in Parliament in 1736, and was one of five men invited in June 1737 to present their surveys of the river as support for their propositions about the choice of site and type of foundations to be used. All were heard but the commissioners were undecided. Eventually in 1738 they accepted an offer by Labelye to direct the founding and building of the piers by the method he had proposed. That order and the successful completion of several piers gave them confidence and, after the severe winter of 1739–1740 with the Thames frozen over convinced them that proposals which they had entertained to build a wooden bridge for speed and economy were too risky, they appointed Labelye 'engineer' for design and supervision of the whole project, it being an early use of that title in civil engineering works in Great Britain. He was paid a salary of £100 p.a. plus 10s a day for subsistence when he was actually on the site but he claimed successfully when the bridge was finished that he had been promised a greater settlement at the start, perhaps by the Earl of Pembroke. The arrears were paid as a gratuity of £2,000 in 1751. From the start of work on the pier foundations the commissioners had also been served by other salaried officers of whom the highest paid was Richard Graham who, as 'surveyor and comptroller of the works', drafted and negotiated contracts, made or supervised all measurement and payments, and so left Labelye free to concentrate on engineering.

The chief design challenge was that of making adequate foundations for the piers in tidal water which was up to 24 ft. deep at spring tides, and on a river bed of very permeable gravel. Labelye decided on 'caissons', meaning open-topped wooden boxes which were floated out to their places and sunk by the weight of masonry for the pier; when a pier was built high enough the sides of the caisson were released and floated away and the pier completed by work which was all above the level of low tide. For the superstructure he designed semicircular arches with a maximum span of 76 ft., built externally of Portland limestone and with Purbeck Stone and Kentish Ragstone for internal work.

Although the commissioners required to see and hear of Labelye's proposals for each new phase of their unprecedented bridge, once he had been named as the engineer his authority for technical decisions was largely unquestioned; but he regularly sought the views of the master tradesmen whom he had met during the initial investigations, Andrews Jelfe (q.v.) and Samuel Tuffnell, masons, James King, carpenter, and Robert Smith, a 'ballastman' who dredged the river-bed to horizontal surfaces on which to found the piers. He himself undertook the boring of the bed at the site of each pier and decided that it was firm enough without driving piles. The caisson foundations were therefore economical. But piles for temporary and guard works were required—as the river was a busy water highway—and a sophisticated horse-powered piling machine was made from an idea provided by a watchmaker friend of Labelye's, a machine for cutting off pile heads under water was devised by James King's foreman, William Etheridge (q.v.), and a means of striking the centres from under the arches chosen by Labelye after a trial of several proposals. At all times he took the final decisions.

In one controversy the commissioners became deeply involved. One pier suffered serious settlement and progressive tilting when the bridge was almost complete in 1747 and widespread public discussion ensued, with criticisms and offers of remedies sent in by various sorts of men. The chosen method of repair involved taking down the two adjacent arches, stabilising the foundation and rebuilding with the spandrels lightened by large voids within them. It delayed the opening of the bridge for nearly three years. In fact the foundations continued to move less drastically—for they should have been piled at the start—and eventually after a century of use the bridge was demolished and replaced by a bridge of iron arches. One other misjudgement in the design was the use of Portland Stone, a famous material and simple to deliver to London by sea round the coast but which suffered progressive frost damage where it was regularly saturated in the range of tidal rise and fall.

When the bridge was finished Labelye still felt a duty to stay and maintain it, but after a year he begged leave on the grounds that he had 'contracted an asthma (daily growing upon me) by my constant attendance on the works, especially on the water in winterly and rainy weather' and was permitted to go to the south of France in order to recover his health. He never returned to Britain and only a few sightings of him were recorded in subsequent years. In 1753 he was seen by the British architect, Stephen Riou, in Naples and P.-C. Lesage reported in 1810 that Labelye had lived at Passy, near Paris, and bequeathed his papers and models of the bridge works to Jean-Rodolphe Perronet for the Département des Ponts et Chaussées. Belidor had published in 1752 a detailed drawing of the caisson (*Architecture Hydraulique*, **4**, plate xxviii) which was probably abstracted from the collection of drawings of all aspects of the construction made by Thomas Gayfere (q.v.) and now in the archives of the Institution of Civil Engineers in London; the collection significantly lacks any drawing of a caisson. Belidor's drawing is accompanied by a verbal description stated to be derived from a printed text by Labelye, presumably one of the publications listed below.

There is no evidence that Labelye ever married, nor that he had relatives in Great Britain. The voluminous manuscript records of the Westminster Bridge Commissioners suggest that he was engrossed in and satisfied with his work, and constantly observing and analysing the structures and natural forces it was his profession to control. One or two letters reveal comfortable relationships with Lord Pembroke and the commissioners and until the event of the sinking pier he seems to have worked in good harmony with the other paid officers of the commission and the contractors. Thereafter those relationships became cooler and sometimes difficult. In the preface to his report on the Fens, however, he briefly opens a window on his private life when 'in the summer of the year 1743 [he] had occasion to travel on horseback, and in company with some friends, from Cambridge to Lynn' and, without any professional involvement, he viewed the country and its drainage with interest, comparing it with similar parts of Holland and Flanders. The Fens were 'then in a most beautiful condition', unlike their state when he went there again two years later and found large areas inundated, about which he was engaged to report to the Corporation of the Fens. That commission was one of a very few undertaken during periods of special leave granted him by the Westminster Bridge Commission. His involvement in 1745 and 1748 in plans for the rebuilding of Denver Sluice in the Fens is detailed under *LEAFORD, John*.

With respect to bridges, he had two commissions other than Westminster. At Brentford he designed a new bridge of brick and stone which was built 1741–1742 and in 1746 he produced a design, illustrated in Maitland's 1756 *History of London*, for improving the old London Bridge by reducing the thickness of the starlings.

Labelye was accorded British nationality by a private Act of Parliament in 1746. His place of residence varied within Westminster, being given as Derby Court in 1739 and Crown Court, King Street, in January 1747. In the early 1820s, according to Joseph Mitchell, Thomas Telford (q.v.) was wont to tell guests at his house, 24 Abingdon Street, that Labelye had occupied the same house, which still had a painting of Westminster Bridge by Canaletto in a panel over the dining-room chimney-piece. As Telford had been acquainted with the ninety-year-old Gayfere who still lived in the same street in 1812, the information is probably sound but Labelye is likely only to have been tenant of or lodged in the house.

Labelye's death was reported by *The Gentleman's Magazine* as having taken place in Paris on 18 March 1762, by Lesage as taking place in 'about 1770' and by *Le Conservatoire Suisse* (1817) as having occurred on 17 December 1781.

TED RUDDOCK

[Commissioners for Westminster Bridge, MS, minutes of commissioners, of committees, accounts, contracts, reports, etc. (1736–1753), WORK 6/28–62, PRO; publications of Labelye listed below; Westminster Bridge papers, MS, Ninth Earl of Pembroke at Wilton House, Wilts.; B. F. de Belidor (1752) *Architecture Hydraulique*; P.-C. Lesage (1810) *Receuil de Divers Mémoires extrait de la Bibliothèque des Ponts et Chaussées*, part 2, 275; J. Mitchell (1883) *Reminiscences of my Life in the Highlands*, **1**, 87–88; DNB; *Ars Quatuor Coronatorum*, **40**, 1927, 37, 244; P. M. Carson (1954) Provision and administration of bridges over the lower Thames 1710–1801, with special reference to Westminster and Blackfriars, MA thesis, University of London (unpublished); Patricia Carson (1957–1958) The building of the first bridge at Westminster 1736–1750, *Journal of Transport History*, **3**, 111–122; Ted Ruddock (1979) *Arch Bridges and their Builders 1735–1835*; R. J. B. Walker (1979) *Old Westminster Bridge, the Bridge of Fools*; Colvin (1995); C. Chalklin (1998) *English Counties and Public Building 1650–1830*]

Publications

1/34. Description of the carriages made use of … to carry stone from [Ralph Allen's] quarries … to the River Avon … near the City of Bath, in J. T. Desaguliers, *A Course of Experimental Philosophy*, **1**, 274–279

1735. Dissertation concerning the paralogisms of F. W. Stubner relating to the forces of foreign bodies in motion, signed 15 April 1735, in J. T. Desaguliers, *Course of Experimental Philosophy*, **2**, 77–90

1736. Result of a calculation made to estimate the fall of the water at the intended bridge at Westminster, and … the effects … on the navigation', in N. Hawksmoor, *Short Historical Account of London-Bridge, with a Proposition for a New Stone-Bridge at Westminster*

1737. *A Mapp of the Downes … shewing the true Shape and Situation of the Coast between the North and South Forelands* [the east coast of Kent], with navigation lines, soundings and the new haven proposed in the plan which follows, engraved plan, ref. 3081 (1), BL

1737. *A Plan of the Intended Harbour between Sandwich Town and Sandown Castle* [to] *admitt 150 of the largest Merchant Ships*, engraved plan, ref. K.Top.17.11.1, BL.

1739. *A Short Account of the Methods made use of in Laying the Foundations of the Piers of Westminster-Bridge*

1739. *A Design of a Stone Bridge, adapted to the Stone Piers which are to support Westminster Bridge*, engraved plan

1743. (Attributed) *The Present State of Westminster Bridge … in a letter to a Friend*, the letter is dated 8 December 1742

1745. *The Result of a View of the Great Level of the Fens* [made for] *the Corporation of the Fens in July 1745*

1747. *The Result of a View and Survey of Yarmouth Haven* (re-published in *2nd Report of the Tidal Harbours Commission, Appendix*, 482–492, presented to both Houses of Parliament 1848)

1748. *An Abstract of Mr. Charles Labelye's Report relating to the Improvement of the River Wear and Port of Sunderland made in July 1748*

1748. *The Result of a Particular View of the North Level of the Fens, taken in August 1745*

1751. *A Description of Westminster Bridge*

1751. *The Geometrical Elevation of the North Front of Westminster Bridge*, with plan of the foundations and plan of the superstructure, engraved plan

1756. Proposals for improving London Bridge presented in 1746, in W. Maitland, *History of London*, 3rd edn., vol. 2, 829–830

Works

1735. Surveys of the Thames at Westminster for the promoting committee of Westminster residents, etc.

1738–1751. Westminster Bridge, fifteen arches, design and supervision, cost £198,323, demolished 1862

1741–1742. Brentford Bridge, replacement in brick and stone for an old bridge, £3,350, widened 1811, replaced 1824

LAING, David (1774–1856), architect and surveyor to the Board of Customs, became a member of the Institution of Civil Engineers in 1825. He was the son of a cork cutter of Tower Street, London, and was articled to Sir John Soane in 1790; he began independent practice in 1796.

In 1810, Laing succeeded Pilkington as Surveyor to the Customs. His design for the Plymouth Custom House (1810) showed his interest in contemporary French architecture. Beginning in 1812, he redesigned the London Custom House in an elegant neo-classical style, completing it in 1817. In 1825, part of the facade collapsed, a result of inadequate beech piling used in the foundations. After investigations, Laing's failure to supervise his contractor was ascribed either to incompetence or to collusion, ruining his career and forcing his retirement from practice. In 1821, he had been President of the Surveyors' Club, but subsequently he found himself dependent upon it and other organisations for financial support

One of his pupils was William Tite, the noted railway architect and he and Laing are jointly credited with the design of the church of St. Dunstan in the East (1817–1819, destroyed in 1941). Another pupil was Charles Fowler (q.v.), who won first prize in the 1823 competition for the rebuilding of London Bridge and designed the iron roofed Hungerford Market.

Laing died at 5 Elm Place, West Brompton, on 27 March 1856.

TESS CANFIELD

[Drawings in British Architectural Library Drawings Collection, RIBA; Obituary (1856) *The Builder*, **14**, 189; A. Stratton (1917) *The Custom House, Arch Rev*, **42**, July 1–4; J. M. Crook, (1963) *The Custom House Scandal, Architectural History*, **vi**; RIBA (1993) *Directory of British Architects 1834–1900*; Royal Commission on the Historic Monuments of England (1993) *The London Custom House*; Colvin (3); DNB]

Publications

1800. *Hints for Dwellings, consisting of original Designs for Cottages, Farm-Houses, Villas, etc.*

1818. *Plans of Buildings Public and Private Executed in Various Parts of England, Including the Custom House*

1830. *Appeal to the Hon. the Commissioners of H.M. Customs*

LAMPEN, Robert (fl. 1591), was a member of a family long resident at St. Budeaux. Robert Lampen is mentioned in 1566 in the St. Bude register as the father of another Robert who was baptised on 25 July. One of these men was unquestionably the surveyor by whom the Plymouth Leat was planned. He was accustomed to surveying for in 1592 he helped Robert Adams, sent down by the Privy Council to advise concerning the fortifications. Lampen was aided in work on the leat by his brother, James.

Drake's Leat was conceived as a water supply scheme for the town some years earlier but as no Parliament sat between 1572 and the end of 1584 it was in December that the Bill for the preservation of Plymouth Haven had its first reading. Royal Assent was obtained in March 1585 for a 'ditch or trench between 6 or 7 ft. wide between the town of Plymouth and any part of the said river Meavy'.

In 1590 the town agreed with Sir Francis Drake to bring in the water of the River Meve and gave him £200 in hand and £600 for which he had to compound with the owners of the land over which the river ran. In December 1590 work began and it was completed by April 1591.

Robert Lampen planned the course of the leat in detail and his brother, James, seems to have taken charge of the work with William Stockman and John Stevane as his foremen. The leat was 17 miles long and 2 ft. deep; a century later a hard floor and cut granite sides were provided. The head weir is now covered by Burrator reservoir. Around Yennadon, the ground was so rocky that a wooden trough carried the water across the ground.

Presumably Lampen, his brother, the foremen and the men were paid by Sir Francis Drake, but the town minutes made the following disbursements on completion: 'Robert Lampen in reward at the bringing in of the leat, 2s 6d; Given to two Lampens in reward touching their pains taken about the leat, 2s 8d; Given to Robert Lampen and his brother in reward about the pains about the water, 2s 8d; To William Stockman surveyor of the workmen of the leat in reward, 12s; To John Stevane one other surveyor of the said workmen, 14s'. Lampen's later career is unknown.

A. B. GEORGE

[L. Jewitt (1873) *Plymouth*, 123R.N. Worth (1890) *History of Plymouth*, 439–450; C. Gill (1966) *Plymouth, A New History*, 204–205; H. Harris (1966, 1972) *Industrial Archaeology of Dartmoor*, 136–137]

Works

1591. The Plymouth Leat (Drake's Leat), £200 (construction)

LANDALE, Charles (1764–1834), civil engineer of Dundee and Charlestown, Fife, was christened at St. Andrews on 12 February 1764. His father was David and his mother Christian Campbell. Landale, said to have been formerly an apothecary, had by 1814 obtained sufficient experience to be taken on and serve for two decades as engineer for Lord Elgin's extensive limestone quarrying and coal mining interests in and around Charlestown and Dunfermline. His work for Lord Elgin included a plate-way to Charlestown harbour on the Firth of Forth, via a long inclined plane with a three-span masonry viaduct, together with improvements to the harbour and, *c.* 1820, the upgrading of the 6-mile 'Elgin Railway' with edge-rails and the construction of the 'much admired' Pittencrieff and Colton inclined planes near Dunfermline following a survey and approval of plans by Robert Stevenson (q.v.). A passenger service operated between Dunfermline and Charlestown from 1833. Landale's work is characterised by ingenious conveyances, plant and mechanisms. From 1825, with his nephew, David Landale, he also superintended the Wemyss and Methil collieries.

An unusual assignment *c.* 1818 involved the relocation of the monument to Charles, Earl of Elgin (d. 1771) at the Abbey Church, Dunfermline. During operations the side of a stone vault collapsed to reveal the near 500-year-old remains of Elizabeth, Queen of Robert the Bruce. Landale obtained some of the Queen's long red hair which in 1841 was of great interest to Dunfermline antiquarians. Landale was engaged by Lord Elgin *c.* 1813 at an annual salary of £250 plus the use of a horse and expenses. From January 1825 to July 1834, when he was much employed elsewhere, his annual salary reduced to £150, but the accumulated sum of £1,425 was not paid until 1836, two years after his death.

Landale also improved Scotland's oldest wagonway, from Tranent colliery to Port Seton, Cockenzie, in East Lothian, the track of which had been re-laid with cast iron rails in 1815. In 1821 the line was extensively remodelled with two inclined planes under the direction of 'the ingenious Mr. Landale of Dundee'. By 1825, when he was employed to make proposals for a railway from Dundee across the Sidlaw Hills to the Vale of Strathmore, he probably had had more experience of horse-traction railways and steam-operated and self-acting inclined planes than any other engineer in Scotland.

With the approval and support of Mathias Dunn (q.v.), colliery viewer, implementation of Landale's Dundee & Newtyle railway proposal, with its three steam-operated inclined planes, commenced early in 1827. Unfortunately Landale was unable to manage the project within budget for reasons some of which were beyond his control and in July 1829 the directors replaced him as principal engineer by George Lish, who completed the line in 1832. The railway was extended to the harbour in 1837 and cost in all about £70,000.

In September 1829, following publication of a letter explaining his conduct, Landale seems to have been retained by the railway company in an advisory capacity, but his reputation and health declined. His railway layout, whilst satisfactory for horse-traction, proved inconvenient and expensive to operate when steam locomotion was introduced in 1833. Although the railway never made a profit it was much used. By *c.* 1840 it was carrying some 70,000 passengers and 50,000 tons of goods per annum.

In 1824 Landale, based on his experience of harbour work at Charlestown, put forward a £60,000 proposal for the improvement of Leith harbour involving a 1,450 yd. extension of the existing pier and a new west pier from Trinity over a mile in length. The piers would have consisted of deposited rubble breakwaters about 50 yd. wide at the base and terminated in a harbour entrance 167 yd. wide, flanked by lighthouses, much on the layout of the eventual development of the harbour by 1942. No work was executed.

Landale, who is believed to have been unmarried, had a brother, John, who was the father of David Landale (1806–1895), mining engineer of Burntisland. In summer 1834, although in poor health, Landale went to the continent to conduct railway surveys for a German nobleman. He died at London on 8 November 1834.

ROLAND PAXTON

[Landale's engineering notebook, drawings and other records held by Lord Elgin at Broomhall, Dunfermline; Dundee City Archives; D. McNaughton (1986) *The Elgin and Charlestown Railway*; N. Ferguson (1995) *The Dundee and Newtyle Railway*]

Publications

?. [*Report on improvement of Leith Harbour, Edinburgh*]
1825. *Report … to ascertain whether it is practicable and expedient to construct a Railway between the Valley of Strathmore and Dundee …*, 16 September 1825; also, *Observations upon the Line of Railroad by Matthias Dunn*, 1 October, 1825
1829. *Letter to the Committee of the Dundee and Newtyle Railway Company … and … Excerpts from Letters of the celebrated Smeaton,* relative to the Duties of Committees and Engineers on Public Works* [*to Redmond Morres, 1773]

Works

c. 1814–1834. Elgin Railway, Charlestown Harbour and other work for Lord Elgin
1825–1832. Dundee & Newtyle Railway

LAPIDGE, Edward (1779–1860), architect, was the eldest son of Samuel Lapidge. His father worked under Lancelot 'Capability' Brown (q.v.), eventually becoming head gardener at Hampton Court. Lapidge's architectural practice was concentrated in the Home Counties, including a number of gothic churches. His civil engineering work arose from his appointment as County Surveyor for Surrey in 1824. Shortly afterwards, in 1825, Kingston Corporation obtained an Act for a new bridge to replace the existing timber bridge which was in a very dilapidated condition and Lapidge became responsible for its design, a masonry structure, with Portland stone facings, of five elliptical arches and small flood arches at each end. The contractor was William Herbert. Work began in November 1825 and the bridge was opened by the Duchess of Clarence on 12 July 1828. It has been suggested the design was influenced by the style of Capability Brown. The bridge is Lapidge's chief claim to fame. Lapidge offered a similar design for a bridge at Staines but that of George Rennie (q.v.) was preferred. Lapidge's only other bridge was that at Betchworth (1842), although as County Surveyor he would have been consulted about the maintenance of others. In 1850 he also exhibited at the Royal Academy his own system for a suspension bridge. As Surrey County Surveyor he was responsible for a number of buildings and road surveys. He also entered a number of architectural competitions but with little success.

Elected a Fellow of the (Royal) Institute of British Architects in 1838 his pupils included George Wightwick and H. H. Russell. He died in February 1860 and is buried in Hampton Wick.

MIKE CHRIMES

[Quarter Session Records, Surrey County RO; G. Phillips (1981) *Thames Crossings*; Bendall; Colvin (3)]

Works

Works include:

1825–1828. Kingston Bridge, £31,000 (tender)
1842. Betchworth Bridge

LARKIN, Nathaniel John (fl. 1825–1827), teacher of mathematics and crystallographer, was born in London on 5 December 1781. He was appointed (second paid) Secretary of the Institution of Civil Engineers in January 1825. He was required to post £100 security on appointment and was entitled to 30 guineas per session plus 5% of all subscriptions collected. At that time his duties involved attending evening meetings in the session and Council meetings during the recess, but at the end of the year enlarged premises were taken over at 15 Buckingham Street with a view to having a resident secretary. Larkin, then living at Gee Street, Somers Town, and with other responsibilities elsewhere, resigned, being succeeded by William Rutt (q.v.). Larkin was elected an Associate in 1826 but remained in membership only for a short time. From 1820 he was a member of the Geological Society.

TESS CANFIELD

[Membership records, ICE archives; *Larkin, Nathaniel John*, in F. Boase (1892–1924) *Modern English Biography*; J. G. Watson (1988) *The Civils*, 199–200]

Publications

1810. *An Essay on a Species of Mosaic Pavement formed of Right angled Triangles of different Colours*
1820 *An Introduction to Solid Geometry, and to the Study of Crystallography, containing an Investigation of some of the properties belonging to Platonic Bodies independent of the Sphere*
?. *Rudiments of Linear, Plane, and Solid Geometry*

LATROBE, Benjamin Henry (1764–1820), architect and engineer, was born at Fulneck, near Leeds, the second son of the Reverend Benjamin Latrobe and Anna Margaretta, *née* Antes. His father was the head of the Moravian congregation in England and moved to London in 1768 but his son was educated at the Moravian school in Fulneck until, in 1776, he was sent to the Moravian seminary at Niesky (Silesia) and Barby (Saxony). His education included elements of mathematics and science which were to underpin his subsequent career in engineering. Around 1781 Latrobe briefly studied with the Saxon hydraulic engineer, Riedel, before touring Europe in 1783. In August that year Latrobe returned to London where he obtained a sinecure at the Stamp Office.

What is known about Latrobe's early training as an engineer and architect is largely derived from subsequent references in his correspondence. It has been suggested that Latrobe was a pupil of John Smeaton (q.v.) but this is unlikely as his return to England coincided with Smeaton's effective retirement. It would seem that Latrobe probably worked for Smeaton's best known former pupil William Jessop (q.v.). Latrobe's reference to securing work on the Fens could perhaps refer to work at Knights Gool on the Ouse, or on the Trent, or Hatfield Chase, where work was ongoing at the time. Clearer are references to work at Rye Harbour (1786–1787) and on the Basingstoke Canal (1787–1789). For reasons that are unclear, as there was an abundance of work for engineers at the time, Latrobe then began work as a draughtsman for the architect, Samuel Pepys Cockerell. By 1792 he was married to Lydie Sellon and in private practice as an architect working on private houses, and was also surveyor to the London Police Offices. It is as an architect, in England, and, more importantly in the United

States that he is best known, although his engineering career continued.

Latrobe's first wife died in 1793, an event which apparently had a major impact on him. In that year he gave Parliamentary evidence, for the first time, in support of his own scheme for improvement of the Blackwater and Chelmer Navigation. The first proposal was unsuccessful but Latrobe drew up a revised scheme which was presented to Parliament in April 1795. This scheme, also unsuccessful, marked Latrobe's last involvement in engineering in Britain as that autumn he emigrated to the United States where the young republic was desperately short of engineers and architects.

On his arrival in Norfolk, Virginia, Latrobe soon obtained architectural commissions and at the end of 1798 moved to Philadelphia in connection with one of the most important, the design of the Bank of Pennsylvania in the Greek revival style. His architectural reputation prospered and in 1803 he was appointed surveyor of public buildings to the Federal Government and designed the South Wing of the Capitol. His major contribution to the development of the neo-classical style in north America has made his reputation as an architect but from his arrival in the United States he was involved in engineering schemes.

In 1796 Latrobe reported on improvements to the Upper Appomattox Navigation. As with many of his schemes it was not executed. Frequently the financial resources were not available and, even when they were, they were inadequate for the task. This was true of his work on the Lower Susquenna River (1801), although his survey of the river is a geographical document of major significance. His proposals for a naval dry dock and Lower Falls Canal (1802) were frustrated by their cost. This work was followed by designs for a proposed Chesapeake and Delaware Canal; work began on a feeder for this canal in 1804–1805 but funds ran out before its completion. Further reports followed on canal schemes and navigation improvements: the Delaware River shoals (1807), Lower Appomattox Navigation (1809–1810), Washington Canal (1804), Potomac Navigation (1811), the New York Western Navigation linking the Hudson River to Lake Erne (1811), and the Jones Falls improvement at Baltimore (1817–1818). Of these only the Washington Canal was executed.

Latrobe took an interest in other schemes such as the National Road, but his best known engineering works were associated with water supply, Philadelphia Waterworks (1799–1801), and New Orleans. Both of these scheme made use of steam power to pump water from river sources. The Philadelphia scheme was based on his knowledge of London schemes, particularly Chelsea Waterworks, and was generally successful. In contrast his scheme for New Orleans was not completed at the time of Latrobe's death in 1820. Perhaps boosted by his success in Philadelphia, in 1809 Latrobe first put forward a scheme for a waterworks in New Orleans, and in 1811 the City Council gave Latrobe an exclusive contract to supply the city with water. He was supposed to have begun supplying water in 1813 but technical and financial problems bedevilled him and in 1817 Henry (b. 1792) his only son by his first marriage, died of Yellow Fever, depriving him of a reliable deputy on site. Latrobe lacked the money to build the steam engine and associated works, and in 1820 moved from Baltimore to New Orleans in a determined effort to complete them. His work there was cut short by his death.

Latrobe's problems at New Orleans reflected similar problems he experienced with other business ventures such as the Schuylkill rolling mill, Steubenville woolen mill, and the Ohio Steamboat Company (involving Robert Fulton (q.v.)). These relied on Latrobe's engineering designs but, generally, lacked capital for success.

Latrobe's career is of interest as an early example of technology transfer from Britain to the United States. Latrobe's knowledge of British engineering practice was reflected in his designs and the project organisation he adopted on those works he executed. A number of the contractors and mechanics also had British experience, such as Samuel Ellis (Philadelphia Waterworks and U.S. Navy Yard), James Smallman and John Herritt (ex-Boulton and Watt), John Davis (Philadelphia Waterworks), Charles Randall and John Cocksey (ex-Grand Junction Canal).

Latrobe exercised great influence through his pupils of whom perhaps the best known are his son, Benjamin Henry Jr., and William Strickland, who visited Britain in the 1820s, and in his own account of *Reports of Public Works of North America* (1841) gave details of Latrobe's work at Philadelphia and introduced the achievements of American engineers to a British audience, whilst reporting at home on British achievements in his earlier *Reports on Canals, Railways, Roads and other Subjects* (1826).

When attempting to expedite work at New Orleans Latrobe contracted Yellow Fever and died on 3 September 1820. His son, Benjamin Henry Jr., was a child of his second marriage to Elizabeth Hazlehurst, other children being John Hazlehurst Bonval and two daughters, one by each marriage. One of Latrobe's sisters, Mary Agnes, was the mother of John Frederic Bateman (1810–1889), Britain's most prolific designer of dams.

MIKE CHRIMES

[Note: Latrobe's papers are available in microfiche as *The Papers of Benjamin Henry Latrobe*, Maryland Historical Society, Baltimore, U.S.A. Blackwater navigation drawings, Essex County RO; T. Hamlin (1955) *Benjamin Henry Latrobe*; ASCE (1972) *A Biographical Dictionary of American Civil Engineers*, 77–78; D. H. Stapleton and T. C. Guider (1976) The transfer and diffusion of British technology; Latrobe and the Cheasapeake and Delaware canal, *Delaware History*, **17**,

127–138; D. H. Stapleton (ed.) (1980) *The Engineering Drawings of Benjamin Henry Latrobe*; R. E. Shaw (1990) *Canals for a Nation*; P. Way (1993) *Common Labour: Workers and the digging of North American Canals 1780–1860*; Colvin (3)]

Publications
1799. *View of the Practicability and Means of supplying the City of Philadelphia with wholesome Water*
1799. *Answer to the Joint Committee of the Select and Common Council of Philadelphia, on the Subject of a Plan for supplying the City with Water*
1799. *Remarks on the ... Delaware and Schuylkill Canal Company*
1809. First report ... what improvements have been made in the construction of steam engines in America, *Am Philosophical Soc Trans*, **6**, 89–98
1812. *Opinion on a Project for removing the Obstructions to a Ship Navigation to Georgetown*
1812. *Proposals for the Establishment of a Company for the Supply of New Orleans with Water*
1814. To the Editor of the Emporium [on the construction of the National Road], *Emporium of Arts and Sciences*, n.s., **3**, 284–297

Works
1799–c. 1801. Philadelphia Waterworks, engineer
1801. Susquenna River improvements, $10,000
1804–1805. Chesapeake and Delaware Canal, feeder, engineer, salary $3,500 p.a.
1810–1816. Washington Canal, 1½ miles, engineer, $85,000
c. 1811–1820. New Orleans Waterworks, engineer

LAW (LAWS), John (fl. 1770–1781), mason, of Lancaster, was responsible for the reconstruction of Walton and Ribchester bridges across the river Ribble in the 1770s and early 1780s. In 1775 Law may have been living in Hundersfield, Lancs., but in 1779 he was at Heywood. His namesake and partner, Robert, came from Todmorden, and Samuel from Heywood. The Laws worked with Robert Crabtree of Burnley on the Walton bridge, and with Robert Gudgeon, John Bradley, Edward Blackledge and Matthew Todall on the Ribchester bridge. Crabtree also worked on the repair of Stealey Bridge (Staley Bridge) in 1781.

Elsewhere they also worked with Samuel Lister (q.v.), Abraham Fielden and the Grindrod family.

MIKE CHRIMES and P. S. M. CROSS-RUDKIN

[Quarter session records, Cheshire, Lancashire and West Yorkshire County ROs; E. Jervoise (1931) *The Ancient Bridges of the North of England*, 137–138]

Works
1770. Hebble End Bridge. Yorks., reconstruction, £49 2s (contract price)
1771–1772. Higher Hodder Bridge, Yorks, £312 10s. (contract price), with Samuel Lister (q.v.) and Abraham Fielden
1774–1775. Ribchester Bridge, three segmental arches

1779–1781. Walton Bridge, near Preston, three segmental arches, c. £8,400
1780. Otterspool Bridge, Bredbury, three spans, with the Grindrod family

LE PLA, Mark (c. 1622–1697), fen drainage engineer, was born at Bouire in the county of Calais, France, and as a Huguenot was naturalised in 1662 at the age of forty having lived 'for divers years past' in the Cambridgeshire fens near Thorney. Presumably with some previous experience, of which no details are available, he was appointed Surveyor of the North Level of the Fens (Bedford Level) in 1665 on the death of Edmund Welch (q.v.). In this post, which he held for thirty years at the annual salary of £50, he would have been responsible for maintenance of the rivers, banks, drainage channels and sluices and for overseeing any new works.

On leave of absence from the Fens he reported to Lord Thanet in 1689 on improvements in the drainage of Wittersham Level in the western part of Romney Marsh. His report is accompanied by 'A Draft of Wittersham Levell and also of the Upper Levells, taken only on a view [not surveyed] March 18th 1688–1689 by Mark le Pla'. His recommendations include the installation of two 'engine mills', wind-driven scoop-wheel drainage mills which were becoming increasingly common in the Fens at the time.

In the Bedford Level account books Mark le Pla is Surveyor of the North Level up to and including the financial year 1693–1694. In 1694–1695 and 1695–1696 both he and his son, David (b. 1667), are listed jointly but from April 1696 only the name of David le Pla is given. Mark died on 19 April 1696 aged seventy-five and both he and his wife were buried in Thorney Abbey Church.

A. W. SKEMPTON and MARGARET FARRAR

[Bedford Level Corporation Orders and Accounts (1665–1696) Cambridgeshire RO; A. Shaw (1911) *Acts of Naturalization, 1603–1700*, Huguenot Society, **18**; Jill Eddison (1998) Changes in the course of the Rother and its estuary and associated drainage problems, 1635–1737, *Romney Marsh, Evolution, Occupation and Reclamation*, Oxford Committee for Archaeology Monograph No. 24,142–161; Bendall]

LEACH, Stephen (c. 1777–1842), was Engineer to the Corporation of London Navigation Committee. The City had jurisdiction over the Thames downstream of Staines (36 miles above London Bridge). Under an Act of 1774, with Robert Whitworth (q.v.) as engineer, the river was improved principally by dredging and building groynes to concentrate flow within the navigation channel. After Whitworth retired in 1784 the works were continued by his former assistant Charles Truss. In January 1810 Stephen Leach, who had been assistant to Truss since 1801, succeeded him as Clerk of Works or engineer with a salary of £300 p.a.

By 1790 the Thames Navigation Commission, the authority operating upstream of Staines, had built sixteen locks between Maidenhead and Oxford, mostly if not all in short cuts adjacent to mills, and in 1798 they added a lock at Romney (between Staines and Maidenhead) with a 160 ft. long weir, 120 ft. of which could be fully opened in times of flood.

The first intimation that the City might introduce locks below Staines came in a report by Truss in 1802. He proposed locks at Teddington and Molesey with solid weirs having long overfalls or 'tumbling bays' but further upstream he adopted an idea of John Rennie (q.v.) of making long cuts (without locks) to bypass shoals and bends in the river. Other schemes followed, for locks and weirs of the Romney type, or cuts well over a mile in length (opposed by landowners) or combinations of these. Eventually in 1810 the Committee put forward a practical scheme, drafted by Leach, for locks and weirs at Teddington, Sunbury, Shepperton and Chertsey, with no long cuts; the locks were to be timber-framed 150 ft. and 20 ft. wide to take the great 170 ton 'western' barges, and the weirs were to be in masonry up to 450 ft. in length with tumbling bays on each side of a central set of sluices, providing ample capacity for passing floods.

On this basis an Act was obtained in June 1810 and shortly afterwards Rennie and Leach visited the sites to decide on the exact position of the structures. Thereafter Leach had complete control, producing plans and estimates, letting contracts and supervising construction. Teddington and Sunbury were built 1810–1811 and 1811–1812, the contractors for both being Joseph Kimber and John Dows. Shepperton and Chertsey (at a revised site) followed in 1812–1813, both by the well known contractors, Mills and Bough (q.v.).

Encouraged by the success of these works, the City obtained Acts for building a lock and weir at Molesey, and a lock on a fairly short cut (without a weir) at Penton Hook. Both were built 1814–1815 by John Mills. These six very large locks and five weirs, costing something like £90,000 to build, constitute an outstanding achievement in river engineering.

Leach lived at Turnham Green, Chiswick. He died in November 1842, aged sixty-five, and was succeeded by his son, Stephen William Leach (1818–1881), who had been his assistant for several years.

A. W. SKEMPTON

[Minutes Books of the City Navigation Committee (1770–1843) Port of London Authority Archives; Brief obituary of Stephen Leach (1843) *Gentleman's Magazine*; Memoir of S. W. Leach, *MPICE*, **70**, 420–421; A. W. Skempton (1984) Engineering on the Thames Navigation, *Trans Newcomen Soc*, **55**, 153–176]

Publications

1815. *Map of the River Thames Navigation … from London Bridge to Lechlade*

Works

1810–1815. Locks and weirs on the Thames, £90,000

LEAFORD, John (fl. 1713–1753), was appointed a divisional officer of the South Level of the Fens in 1727. By his own account he had worked in this area since 1714, apparently as a contractor, and he remembered the failure of the three-arch western part of Denver Sluice in 1713 (originally built 1684); the separate two-arch eastern part, built by Richard Russell (q.v.) in 1700, remained.

After years of debate whether the sluice should be removed or rebuilt, and if so in what manner, the three-arch structure was rebuilt in its original form but with improved foundations incorporating dovetailed sheet piles and the eastern part was refurbished with a navigation lock in one of the arches. In this way four arches, each about 17 ft. wide, with traditional type pointing doors (rising 18 ft. above sill level) were provided to 'keep out the tides'. The work was carried out 1746–1750 under William Cole (Director of the South Level 1714–1750) to plans by Leaford, who supervised construction.

The discussions had centred on two opposing schools of thought: those following John Armstrong (q.v.) in favour of removing the sluice, allowing the tides to run many miles upstream, and raising the river bank, and those who wished to rebuild. Leaford in a memorandum to Bedford Level Corporation (1747) ends by declaring that the South Level 'must be intirely lost without a sluice'. A crucial point was that whereas quite high banks had been built on the clay soil downstream of Denver, it was a different matter to raise the river banks on peat foundations upstream.

A decisive move came in 1745 when the Corporation invited Charles Labelye (q.v.) to report. In principle he followed Armstrong but recommended that draw doors should be provided (four in each of the eastern arches and twenty in a timber framework at the site of the western part), the doors to be kept open in all but exceptional high tides; this seemed a good compromise. Although Labelye is always credited with the rebuilding of Denver Sluice the work was, in fact, essentially due to Leaford.

The opening of the Eau Brink Cut in the 1820s led to a considerable deepening of the river bed from King's Lynn right up to Denver. To gain full benefit for drainage in the South Level upstream a new three-arch sluice with a sill level 6 ft. lower was built 1832–1836 under direction of Sir John Rennie (q.v.). This left Leaford's three-arch western sluice intact; it remained so until replaced in 1932 by a very large lifting gate.

Leaford resigned his post in 1753. It is probable that he, rather than a son or relation, is the John

Leaford appearing as a Conservator of Bedford Level in 1756 and 1757.

A. W. SKEMPTON

[John Armstrong (1724) *Proposals for Draining the Fenns*, Cambridgeshire RO; Bedford Level Corporation Order Books and Conservators Proceedings (1714–1756) Cambridgeshire RO; Charles Labelye (1745) Plan of the Lock and Draw Doors proposed for Denver Sluice, Cambridgeshire RO; William Cole (1746 and 1747) South Level Journals, Cambridgeshire RO; Anon. (*c.* 1805) Drawing of Denver Sluice, Cambridgeshire RO; Charles Labelye (1745) *The Result of a View of the Great Level of the Fens*; Samuel Wells (1830) *History of the Drainage of the Great Level of the Fens, called Bedford Level*; H. C. Darby (1940) *The Draining of the Fens*; A. W. Skempton (1996) *Civil Engineers and Engineering in Britain 1600–1830*]

Publications

1740. *Some Observations made of the Frequent Drowned Condition of the South Level of the Fenns … with a Scheme for Relieving that Level*

Works

1746–1750. Rebuilding Denver Sluice, £5,000

LEATHER, George, Sr. (1748–1818), colliery and civil engineer, was the youngest son of Samuel Leather (1702–1776) and his wife Mary. He was born at Wood End Farm, Farnworth, near Liverpool and was one of a family of eight. He was baptised at Great Sankey, near Warrington, on 3 July 1748.

Little is known of his early career but it is probable that both he and his brother, Samuel (b. 1745), were employed as apprentices by James Brindley (q.v.) who often stayed at their parents' house at Farnworth when working for the Duke of Bridgewater and it is possible that both young men were employed on navigation works in the area. He is known to have contracted, as the junior partner of John Tickle, for excavating the Leeds and Liverpool Canal from Skipton to Bingley. The contract was abandoned following disagreements with the canal company, although the affair ended in the company paying over more money.

From about 1774 Leather worked for Thomas Fenton, the owner of collieries between Leeds and Wakefield. When in his employment he married Hannah Beaumont *c.* 1776 and they had a family of ten of whom George Jr. (q.v.) was the second surviving son. By 1795 Leather was chief engineer for William Fenton and was responsible for opening up his New Park colliery, near Ossett. He probably opened collieries at Rothwell Haigh, Wakefield, Lofthouse and Methley. When engineer to the Fentons, Leather devised a 'water machine' which lowered water down one shaft and enabled coal to be drawn up another adjacent; he also installed a waterwheel powered by the outflow of two reservoirs. For the Fentons he

built and operated colliery railways, work which by 1792 included the probable introduction of underground tramplates.

Leather was appointed as resident engineer under William Jessop (q.v.) for the construction of the Surrey Iron Railway, 8 miles long and running from Croydon to Wandsworth. It was completed in 1805 and although he was nominally in charge of construction it is probable that a portion of the work was undertaken by George Jr. He carried out work on the Derwent Navigation and, when employed there, was asked in 1812 to prepare plans for a canal to Pocklington. He declined the commission and the work was undertaken by George Jr.

In 1813, with two of his sons, George and James (1779–1849), Leather acquired the lease of Beeston Park colliery, on the outskirts of Leeds, and it was there that he died on 14 February 1818.

A. D. LEATHER and R. W. RENNISON

[PRO/RAIL/846/2/1; J. Batty (1877) *The History of Rothwell*; C. Hadfield (1972–1973) *The Canals of Yorkshire and North East England*; J. Goodchild (1978) *The Coal Kings of Yorkshire*; C. Hadfield and A. W. Skempton (1979) *William Jessop, Engineer*]

Works

1771–1773. Leeds and Liverpool Canal, excavation, Skipton–Bingley
1774–*c.* 1800. Engineer to Thomas and William Fenton, sinking and operation of collieries, construction and operation of colliery railways, construction of dams
1801–1805. Construction of Surrey Iron Railway, 8 miles long, £27,000
1808. Survey for improvements on River Don (for William Jessop)
1812–1816. Derwent Navigation, construction
1813–1818. Lessee of Beeston Park Colliery

LEATHER, George, Jr. (1786–1870), civil engineer, one of a family of ten, was the sixth child of George Leather (q.v.), also a civil engineer, and his wife, Hannah Beaumont. He was born at Stanley, near Wakefield, on 5 October 1786 and was baptised in Wakefield a month later.

As a youth, Leather was responsible, with his father, for the construction of colliery railways for the Fenton family at Rothwell, near Leeds, and also for the Surrey Iron Railway, 8 miles long, running from Croydon to Wandsworth, where Leather Sr. was nominally resident engineer under the general direction of William Jessop (q.v.); it was completed in 1805. In 1809 Leather married Sarah Wignall in Wakefield and the next year he both undertook a survey for the Derwent Navigation and became responsible for a canal to Pocklington, a venture for which his father had been first approached but had declined the work; the canal was completed in 1818, the year of his father's death. Leather assisted Thomas Wood in work on a canal from Haddesley, on the river Aire, to Dutch River but a different line was established by John Rennie (q.v.) in 1818. The next year he surveyed a possible branch to Womersley.

In 1820 discussion took place regarding the construction of a canal from the south west part of county Durham to the River Tees. The previous year Leather had drawn up a plan for a railway running from the same area to Stockton and it was natural that he should be called upon to survey a line for the canal's 29½ mile length. It was to have a bottom width of 24 ft. and, with fifty locks, was estimated at £225,283. At the time, two rival schemes were under consideration and Leather's scheme was dropped; in its place the Stockton and Darlington Railway was constructed.

Also in 1820, Leather was appointed consultant to the Aire and Calder Navigation and, in addition to navigational works, became responsible for the design of several cast-iron structures. Between 1827 and 1832 the Monk, Astley and Hunslet bridges over the river Aire were built. The Stanley Ferry aqueduct, a significant arched structure which carried the canal over the river Calder has been attributed to Leather but it is probable that much of its design was the responsibility of his son, John Wignall Leather (1810–1887); it was completed in 1839. The aqueduct formed part of a Leeds, Wakefield and Ferrybridge canal system and before work began the views of Thomas Telford (q.v.) had been sought. He had suggested that a cast-iron trough with overhead lattice trusses be used but it was the Leathers' single span arch of 155 ft. which came to be adopted.

Leather was also responsible for waterway improvements and for work on the Goole Canal, including the planning and construction of basins and wharves. The town of Goole was set out by him but his plans were not fully implemented. Following the completion of the canal, he had foreseen the need for docks at Goole and he was later responsible for the design and construction of two three-acre docks, one for ships and the other for barges; they were opened in 1828. A dock at Leeds was built under his control in 1840, as was the Crown Point Bridge there, the latter again with the significant involvement of his son, John.

In addition to his work on canals, Leather—one of the engineers whose career embraced the inception of railways—was consulted concerning his views on both docks and railways. When the Marquess of Londonderry was considering the establishment of a new harbour at Seaham in order to ship coals from his wife's collieries near Durham, Leather was asked for his views on the harbour proposed by William Chapman (q.v.). He was generally in agreement with Chapman's suggestions but expressed concern as to ships lying aground at low tide and asked if borings could be made to ascertain strata details, in the hope that softer materials would be revealed. His comments perhaps led to a somewhat

less ambitious harbour being built. Leather also became involved with work at Bridlington. In 1831 the Trustees of the port there approached both Chapman and Jonathan Pickernell (q.v.) of Whitby concerning the supervision of the construction of a new pier; both would seem to have declined to act for the Trustees and Leather was contacted, his brief being to inspect work twice a year.

In 1828 Leather became a member of the Institution of Civil Engineers and the following year he reported upon the line which had been proposed for the Clarence Railway but, although his views led to modifications, he was not responsible for its construction. Rivalry between the Stockton and Darlington and the Clarence railways led to Leather again becoming involved when the line was extended in 1833 to deeper water on the north bank of the Tees and he designed there a coal drop, differing from the more usual pattern in that the loaded chaldron wagons were lowered vertically to the waiting ship rather than falling in an arc. He was appointed engineer to the Stockton and Hartlepool Railway—in effect the extension of the Clarence line—to a new port, West Hartlepool, north of the Tees; the railway was opened in 1840. As a result of this involvement, he was consulted by the Tees Navigation Company in 1840 and, with Henry Habberley Price (q.v.), was made responsible for delineating an improvement line for that river. Two years later he became one of the three arbitrators appointed to resolve a dispute concerning the effect of the Monkwearmouth dock on the navigation of the River Wear.

Leather was responsible in 1826 for the enlargement of the Worsbrough reservoir of the Dearne and Dove Canal and in 1831–1832 he reported on the draining of Hatfield Chase. He also gave evidence when the Liverpool and Manchester Railway Bill was in Parliament and he spoke against the Newcastle and Carlisle Railway Bill in 1829, principally regarding the effect of floods on the line should the indicated levels be adhered to. As a continuation of his Yorkshire bridge work, between 1830 and 1832 he designed and built the four-span cast-iron Dunham bridge, crossing the River Trent south of Gainsborough.

The interests of Leather and, later, others of his family—namely his son, John, and his nephew, John Towlerton Leather (1804–1885)—extended to water supply. In 1837 the Leeds Waterworks Company engaged George Leather and Son as consulting engineers and this appointment led to modifications to the Eccup scheme, then under consideration. Leather eventually supervised the construction of the reservoir, a 1½ mile tunnel and a storage reservoir. Work began in 1840 and was completed in 1842 although such was the demand for water that by 1845 the scheme proved inadequate.

Almost contemporaneously, Leather was retained by the Holme Reservoir Commissioners to build reservoirs for supplying water to power a series of mills south of Huddersfield. One of the reservoirs was Bilberry and although Leather was engineer for its construction he seldom visited it during the period of its building. The contract for its construction ran from 1839 until 1843 at which time the formation of the 98 ft. high embankment was proving unsatisfactory and a repair contract was arranged, extending from 1843 until 1845. On 5 February 1852 Bilberry dam failed, causing extensive damage and 81 deaths. The subsequent inquest found that both the commissioners and Leather 'were culpable in not seeing to the proper regulation of the works' and this conclusion led, three years later, to Leather's retirement. He died, almost in obscurity, on 2 April 1870; the *Leeds Mercury* reported his death but did not mention his outstanding early career, embracing canals, land drainage, docks, railways, bridges and water supply.

From 1808 until 1825 Leather lived in Bradford, all his children being baptised in the cathedral there; between 1825 and 1837 he lived in Leeds and from 1837 up to his death he lived at Knostrop New Hall. From 1838 until 1856 the offices of 'George Leather and Son' were established in Wellington Street, Leeds. The fact that this partnership tendered for the construction of a dock on the River Tyne in 1845 would indicate that they were also involved in contracting and local directories show an interest in coal-mining, including the ownership, with his brother, James, of Beeston Park colliery.

Leather's estate amounted to some £10,000 and was left to his family. Of his nine children, four sons and two daughters survived him and the estate passed in equal shares to them and to the children of his daughter, Maria, the late wife of his nephew, John Towlerton Leather. Both he and Leather's eldest son, John, pursued careers in civil engineering.

R. W. RENNISON and A. D. LEATHER

[*Bridlington Harbour Commissioners Minutes*, 6 October 1831; J. S. Jeans (1875) *History of the Stockton and Darlington Railway*; W. W. Tomlinson (1914) *North Eastern Railway*; C. Hadfield (1972) *Canals of Yorkshire and North East England*; C. Hadfield and A. W. Skempton (1979) *William Jessop, Engineer*; G. M. Binnie (1981) *Early Victorian Water Engineers*; R. W. Rennison (1987) *The Development of the North East Coal Ports, 1815–1914*, unpublished Ph.D. thesis, University of Newcastle upon Tyne; Skempton (1987); J. D. Porteous (1988) *The Company Town of Goole*]

Reports and drawings

1822. [Report] *to Lord Stewart* (3rd Marquess of Londonderry) [on Seaham Harbour], 16 January
1829. *Report of Mr. Leather, Civil Engineer, on the Projected Line of Railroad between Newcastle and Carlisle*
1830. *Level of Hatfield Chase. The First and Second Reports of Mr. Leather on the Better Drainage of the Level ...*

1831. *Level of Hatfield Chase. The Third Report of Mr. Leather ...*
1831. [Report] *to the Proprietors of Fen Lands in the Parishes of Nocton ... in the County of Lincoln*

Plans

1811. *Plan of an Estate the Property of the Low Moor Company ... Geo. Leather, jnr, Bradford*
1814. *A Plan and Section of the line of an intended Canal from the River Derwent ... (to) Pocklington ... in the County of York*
1819. *Railway from Stockton to Evenwood Bridge ... by George Leather of Bradford*
1819. *Plan of the Line of the Proposed Canal from Knottingley to Goole*
1824. *Railway from Haverton Hill to Willington ...*
1824. *Luddenden Top to Skircoat Railway*
1836. *Plans and Sections, Hartlepool Harbour Improvements by George Leather of Leeds*
1836. *Sheffield Union Railway*
1837. *Plan of the Aire and Calder Navigation ...*
1838. *Leeds Water Works ...*
1839. *Plan and Section of Stockton and Hartlepool Railway*
1840. *Stockton and Hartlepool Railway from the Clarence Railway ... to Hartlepool Harbour*
1842. *Hartlepool West Dock and Railway*
1846. *Leeds Waterworks. General Plan ...*

Works

1801–1805. Surrey Iron Railway, with Leather Sr., 8 miles, £27,000
1810–1818. Pocklington Canal, construction
1820. Aire and Calder Navigation, Appointed Engineer
1822–1838. Goole docks and town
1820–1824. Knottingley and Goole Canal
1823. Long Sandall to Stainforth, waterway improvements
1826. Enlargement of Worsbrough reservoir for Dearne and Dove Canal
1827. Monk Bridge, Leeds, two spans of 112 ft. over river Aire
1829–1830. Hunslet Bridge, Leeds, 152 ft. span over River Aire
1831. Bridlington pier; supervision of construction
1831–1832. Astley Bridge, near Leeds, over River Aire
1837–1839. Stanley Ferry aqueduct, 155 ft. span over River Calder
1840–1843. Docks at Leeds
1830–1832. Dunham Bridge, four spans of 118 ft. over River Trent
1833. Clarence Railway coal drops
1837–1842. Leeds Waterworks
1838–1840. Stockton and Hartlepool Railway
1838–1845. Holme reservoirs
1839–1842. Birmingham and Derby Junction Railway
1840. Crown Point Bridge, Leeds, 120 ft. span

LEE, George Augustus (fl. 1799–1820), cotton manufacturer, is best known for his experiments on steam power, which began with a Boulton and Watt rotative engine at Philips and Lee's mill in 1792. He was owner, with Philips, of the Salford Twist Mill, erected 1799–1801 and only the second iron framed multi-storey building in Britain. Based upon designs by Charles Bage (q.v.), it is unclear, due to the demolition of this structure, how much the beams differed from the earlier structure in Shrewsbury. There was clear innovation in the columns which were hollow cylinders rather than being of cruciform cross-section and Lee may have been responsible for the variations from Bage's model. He was a close friend of Benjamin Gott, a pioneer of the textile factory in Yorkshire at his Beam Ing Mill, Leeds. For a new mill at Armley, Leeds (1805), Gott used inverted T-beams and cylindrical columns and the general arrangements show similarities to what is known about Salford. Lee's role may have been no more than one of general advice to his friend but that he was eminently capable of designing mills himself cannot be in doubt. In the early 1820s he was called in by J. & N. Philips of Tean Hall Mills, Staffordshire, to advise on erecting a factory to replace the current system of outsourcing in North Staffordshire, where were something like four-hundred looms were scattered across the county. A family link is assumed to be the reason why Lee was consulted. He recommended a mill costing £13,500 to £14,000 for the building, £3,600 for the steam engine, £2,000 for millwork, and £1,500 for gas lighting, for which Lee was a leading advocate. He anticipated that the outlay would be rapidly recovered by savings in labour works. While accepting Lee's overall conclusion, his designs were not followed, and two mills were built, the second at Cheadle.

MIKE CHRIMES

[A. W. Skempton and H. R. Johnson (1962) The First iron frames, *Architectural Review*, March; J. Tann (1970) The development of the cast iron frame in textile mills to 1850, *Industrial Archaeology Review*, **10**(2), 127–145; R. L. Hills (1989) Power from steam: a history of the stationery steam engine; C. Giles (1993) Housing the looms 1790–1850, *Industrial Archaeology Review*, **16**(1), 33–37]

LEE, Sir Richard (1513–1575), was the most famous military engineer and architect of his time, spending forty years on virtually every military structure built in England, Scotland and France from the Reign of Henry VIII to Elizabeth I. As a taste of his lifestyle, in 1544–1545 alone he provided plans for the attack on Leith and for the fortifications of Portsmouth, worked on new forts at Queenborough, Tynemouth, Berwick, Yarmouth, and block houses on the Medway and still managed to cross the Channel to advise in the Pale! The negative side of this eminence is that he never saw any work through to completion meaning that it can be difficult to assess his exact contribution to

any given project (even Berwick), making it hard to estimate his importance as an innovator. Also, despite his contemporary fame as a map maker few surviving examples from the period can definitely be attributed to him.

Lee was the son of Richard Lee and grandson of John Lee, arguably the masons of Ely and St. Albans respectively. Despite thus being a third generation master mason he chose the then more lucrative career of military engineer, consolidating his rise by marrying Margaret, daughter of Sir Richard Grenville, and eventually marrying his elder daughter, Anne, to Edward Sadler whose father, Ralph, was described as the richest commoner in England. He was knighted in 1544 (certainly the first architect to be so honoured, but as an engineer he was preceded by his colleague in Scotland, the designer of Haddington, Sir Thomas Palmer). He was also so well rewarded with the spoils of the dissolution of the Abbey at St. Albans that he was able to sit out the virtual moratorium on Crown building works during the reign of Mary without having to find private civil engineering commissions, as did less affluent colleagues, meaning that in strictly civil engineering terms his career is actually of less interest than others such as John Rogers (q.v.)

Lee's first known position was as page of the King's Cups in 1528. By 1533 he was serving as a soldier in France but, given his family connections to Thomas Cromwell, it is romantic rather than realistic to accept the description of him as a simple spearman; in any event from July to September of 1535 he was working for Cromwell on houses at Friars Austin, Hackney and Ewhurst. Perhaps in consequence of this, in 1536 he was made Surveyor and Paymaster of the Works of Calais at a salary of £20 p.a. Not yet being exclusively occupied as a military engineer he had time to advise on the conversion of Titchfield Abbey early in 1538, but in December 1539 Henry VIII was excommunicated and Lee was sent back to Calais with a commission to review the defences. The results of this necessitated the importation of timber and stone and the realisation that the harbour was virtually useless with no crane, a ruinous wharf, eroding jetties and out-of-order sluices which Lee reported that all the carpenters in Calais could not mend in less than three months.

The solution was an improvised harbour 1,650 yd. long and 44 ft. deep on the River Hamme, employing five hundred men. During the period 1539–1542 the monthly expenditure on all the works at Calais and Guynes was £1,000, including the military works, the most interesting of which was the Rysbank Fort which had involved digging underwater foundations on clay subsoil. Turning his attention inland in 1541 Lee worked with Rogers on building a great dyke to drain the marshlands around Calais for new settlements.

After returning to England in 1542 he was sent to Harwich the next year and received the sum of

£1,300 for works there of which £400 was allocated for the waterworks. According to local sources these are probably the new quay shown on the Elizabethan map and could have included walls and buildings used for shipbuilding, later taken over by the naval yard. His other civil contributions were few, mostly with Rogers: conferring on drainage in Boulogne in 1545, surveying the River Liane in 1546, being appointed to the Commission for New Harbour Works at Sandwich in August 1548 and inspecting the pier at Dover in October 1553. Also during his time as General Surveyor of the King's Majesties Works and Fortifications in the Northern partes in 1547, a period when he was more involved in administration than engineering, pioneers under his orders repaired the roads during and after the invasion of Scotland.

Of his considerable output as a military engineer three projects are of particular note: Eyemouth which he began in 1547, costing £1,908, bears a plaque describing it as the first *Trace Italienne* style fortification in Britain; Berwick 1558–1575, the best surviving example of this type of design in Northern Europe, eventually costing £130,000; and Upnor, which from 1559–1564 cost £3,621 13s 1d. Unfortunately, in all three cases changes had to be made to Lee's plans almost immediately: by the French at Eyemouth; by the Italians Aconcio (q.v.) and Portinari (q.v.) during the design phase at Berwick; and by William Spicer thirty years later at Upnor, casting doubt on his total understanding of the new fortification methods introduced from Europe.

There are mitigating factors in all three cases. The use of the *Trace Italienne* in Scotland was probably a political decision by Somerset (there is an account of Lee coming weeping from his presence) as these forts were quick and cheap to build with low thick walls which in the short term could be built of reinforced earth, timber and turf and needed no specialist builders and masons. The choice of Eyemouth was strategic as it commanded the main eastern route into Scotland; crucially however, it was overlooked by a smaller headland and Lee's protective earth curtain was too short, needing to be extended inland with the addition of two bastions at either end. The position of the Flankers also needed to be changed.

Both of these defects reoccur in the original plan for Berwick but once again Lee, made Surveyor of the Fortifications in Berwick in 1558 at a salary of £1 per day, did not have complete freedom in his designs as it was necessary to superimpose a modern artillery system onto the existing mediaeval walls. It is also relevant that after he was made Surveyor General at Berwick in 1559, he spent only a fortnight on site between 1560 and 1564. In fact Lee's contract both gave discretion to visit as he pleased and absolved him from blame for anything happening in his absence: 'no man can in all cases be so circumspect, but in some part he may be deceyved or at the least by alteration of time be misjudged'.

From 1560 Lee was also working at Upnor. The only entirely new fort built during this period, it shows no trace of a *Trace Italienne*, once again having too short a curtain wall and with the main building too high making it vulnerable to cannon fire, but even in Europe it was still not always considered necessary to build advanced bastions on a coastal defence work.

It would be useful to look to Lee's plattes (plans) for clarification of his design skills but, to the contrary, attribution of surviving plans depends on his skills as a military engineer. Merriman has on this basis identified plans of the Scottish fortifications in the Belvoir collection as Lee's. More controversial is the attribution of a plan of proposed defences of Portsmouth in 1545, annotated with an advanced bastion design in pencil. Among those which can be identified, a topographical chart of Orwell Haven in water colour, described as 'this plott made by Mr. Lee' in 1533–1534, is of interest as an early English description of an important harbour. In 1540 he provided a map of Calais and Boulogne. The only surviving plan of Berwick in his hand is an early one of 1558 among the Cecil papers.

But if his exact place in the introduction and development of modern artillery fortifications into Britain is debatable there can be no doubt of his talents as a project manager. According to a report by Portinari (not normally an admirer, who once accused Lee of purloining one of his alternative designs for Berwick) 'master Ley makes the groat of your Majesty do the work of 5d, so good a manager is he'. Interestingly, there were suspicions of his having enriched himself at the expense of the Berwick works! He had certainly by the time of his death in April 1575 acquired most of the estates of St. Albans Priory—except for the church—including the manor of Langleybury, which he sold back to the Crown, and buildings at Sopwell Nunnery where he built Lee Hall, inherited by his grandson, also Richard. The ruins still exist, with further remains at Salisbury Hall, Shenley.

He was buried in the church of St. Peter at St. Albans where a Latin slab in the chancel commemorated him and his two daughters. The ceremonial helmet originally on his tomb is in the St. Albans Museum. A portrait painted on board is documented up until 1719 but whether it survived long enough to be the source of Sir Walter Scott's description of a portrait of Sir (Victor) Lee—'a man of about 50 years of age in the ... manner of Holbein ... the face express(ing) forcibly pride and exultation' is unproveable.

SUSAN HOTS

[DNB; Whitworth Porter (1889) *History of the Corps of the Royal Engineers*, vol. 1, 22–32; A. H. W. Robinson (1962) *Marine Cartography in Britain*, 64–96; L. Weaver, *The Harwich Story*; I. MacIvor (1965) The Elizabethan fortifications of Berwick-on-Tweed, *Antiquaries Journal*, **65**; L. R. Shelby (1967) *John Rogers Tudor Military*

Engineer; L. White Jr. (1967) Jacopo Aconcio as an engineer, *American Historical Review*, 425–444; *King's Works*, vol. 3 (1975), vol. 4 (1982), *passim*; M. H. Merriman (1983) *Italian Military Engineers in Britain in the 1540s*; S. Tyake (ed.), *English Mapmaking 1500–1650*, 57–67; J. Harvey (1987) *English Medieval Architects*, 175–176; M. H. Merriman (1988) The forts of Eyemouth, *Scottish Historical Review*, **67**(2), 142–155]

Plans

1533–1534. *Plan of Orwell Harbour*, BL 135 Cott, ms
1538. *The Haven of Dover*
1540. *Calais*
1544. *Plan of Edinburgh*
1558. *Berwick*, PRO, MP5/137

Works

c. 1536–1542. Calais, various fortifications, total expenditure in 1552 £150,000
1539. New Harbour, Calais
1543. Waterworks, Harwich (£400),
1543. Fortifications of Berwick, £130,000, final cost £41,000 in 1540s
c. 1544–1545. Queensborough, fortifications, final cost £2,700
c. 1544–1545. Tynemouth, fortifications, final cost £2,650
c. 1544–1545. Yarnmouth, fortifications, final cost £2,400
c. 1544–1545. Eyemouth, fortifications, £1,908

LEGG, George (1799–1882), architect, was elected an Associate of the Institution of Civil Engineers in 1826 but seems to have ceased in membership before 1830. Based in London, where he was District Surveyor for Belgravia and Pimlico from 1858, he had two sons who also practised as architects.

MIKE CHRIMES

[ICE membership records; *Builder*, **42**, 1882, 309]

LEISHMAN, James (1800–1884), contractor, was an agent of Hugh McIntosh (q.v.) and was trained by his staff on the Edinburgh and Glasgow Union Canal where he began work in 1817; he worked for the McIntoshes until Hugh's death. In the 1820s he is believed to have been working at Dymchurch on the sluice and other works for which Hugh was responsible c. 1820–1832. In 1829 he married Sarah Waddell (1811–1894) at Dymchurch Parish Church and acquired property in the area which he leased to his wife's family. In 1827 he had been McIntosh's agent on the works at St. James Park lake. In 1832 he was in Yorkshire where his first son was born and he was probably active on McIntosh's Aire and Calder works. He was Hugh McIntosh's agent on the Great Western Railway contract 9L, and was apparently David's partner on the Derby contract of the North Midland Railway, moving to Duffield to supervise the firm's work in the area.

Leishman signed Hugh McIntosh's death certificate in 1840 when the latter was in Wakefield to meet all the agents in the area. Following the retirement of David McIntosh from contracting Leishman purchased an estate at Brownrigg, Dollar, where he was active locally in the Dollar Gas Company; he was a shareholder in the North British Railway. In the 1840s he was a partner of Edward Price, taking contracts on the North Staffordshire Railway and Londonderry and Enniskillen Railway. The partnership ended in 1849 and Leishman's last known active involvement with civil engineering was a visit to Turin on behalf of Mackenzie and Brassey to investigate various railway proposals.

Leishman was clearly one of the McIntosh's most dependable lieutenants and made a fortune out of contracting, leaving an estate of £120,154.

MIKE CHRIMES

[Frank Smith archive, ICE Library; Mackenzie collection, ICE Library; *Alloa Advertiser* (1884); M. M. Chrimes (1995) Hugh McIntosh (1768–1840), national contractor, *Trans Newc Soc*, **66**, 175–192]

LELLAM, William (fl. 1716–d. 1733), was engineer of Bridlington, Sunderland and Scarborough harbours.

At Bridlington harbour a north pier projected out from the foot of boulder clay cliffs and a south pier or breakwater ran parallel to the coast. The piers were of the traditional form: timber piles and planks with iron cross ties enclosing an infill of stones. They required frequent maintenance and on at least six occasions between 1537 and 1674 had to be more or less completely rebuilt as a result of storm damage, the work being funded chiefly by grants from the Crown. Following another violent storm in 1697 an Act was passed enabling duties to be levied on shipping towards the cost of reconstruction. As rebuilt the north pier was about 320 ft. long and the south pier approximately 700 ft. However, by 1716 the piers were 'by the Impetuosity of the Weather, and Violence of the Seas, shaken, disjointed and, if not timely prevented, in imminent Danger of being destroyed'.

A new Act was passed in June 1716, granting additional powers. Before the end of the year Lellam was put in charge of the works. In October 1717 he presented accounts for work on the north pier, totalling £744 on items such as oak timber, planks, iron work and workmen's wages, and his own 'wages and charges' (£38 14s 9d). A further £137, he said, would be needed to finish work for the winter and he estimated £1,133 to complete 'the lengthening and repairing' of the pier.

As for the south pier it was 'so far decayed that it cannot be repaired but that all or the greatest part thereof must of Necessity be rebuilt'. He estimated the cost at £8,272. The main items included: straight oak timber £3,696, 3-in. planks £341, ironwork £528, staging, ropes, ginns, etc., £320; sawyers' and carpenters' works £2,287; and filling with clay, chalk and stones £950.

By January 1719 a total of rather more than £1,800 had been spent and application was made to Parliament for a grant of £7,500 (additional to the levies). An Act was obtained accordingly in March 1719 and work began on the south pier.

Lellam spent another three years at Bridlington. Whether by the end of that time he had rebuilt the whole 700 ft. of the south pier is not known, but it would surely have been well on the way to completion. At the least we may assume that some £6,000 had been spent on works under his direction since 1716.

At Sunderland the history is different. Little was done to improve the harbour from its natural state until an Act of June 1717 gave powers to newly appointed 'Commissioners of the River Wear and Port of Sunderland' to raise the necessary funds for engineering works by a levy on coals shipped from the river. The Commissioners asked James Fawcett (q.v.) to survey the harbour and put forward a scheme for its improvement. He reported in August 1718 with a detailed map, his main proposals being a masonry pier 700 ft. long built in a NNW direction from the south shore and a 100 ft. pier from the north shore, leaving an entrance 300 ft. wide between the piers heads centred on the north channel in the estuary. He also proposed a quay wall from the south pier back to the Custom House and another on the opposite shore. He estimated the cost of the piers at £10,630 without contingencies. Work began and £1,024 was spent 1719–1720, probably on the south quay wall. But Fawcett was already ill in 1720 and he died in March 1721.

About a year later Lellam came from Bridlington as the new engineer of the Wear Commissioners. Work began on the south pier in 1723, on a very different plan. It started at the same point on the south shore but ran north-east with the object of concentrating flow of the river and tide in the southern channel. The pier formed the basis of all subsequent developments of the harbour, though dredging was required to deepen the channel.

In evidence to a House of Commons committee in March 1727 Lellam stated that £15,220 had so far been spent and a further £6,000 would be required to finish the work. Powers to borrow this sum on credit were granted by an new Act. Completed in 1730 or 1731 at a cost of £19,800 the pier was 1,000 ft. long and 30 ft. wide at the top, with a parapet, in masonry founded on rock or, where the overlaying gravel was too deep, on piles to driven to rock.

Lellam left Sunderland for Scarborough in 1731. He was not replaced. The Wear Commissioners, having exhausted their funds, could expect an annual net income of little more than £1,000. In 1737 Mark Burleigh and Issac Thompson (q.v.) produced a map of the river and harbour and the engineer John Thomas (q.v.) gave evidence in Parliament, seeking extra funds, but without success. The next Act was passed in 1747 and initiated a new phase of development with a report by Charles Labelye (q.v.) and the appointment of William Vincent (q.v.) as engineer.

Scarborough harbour provided the best haven for shipping between Tyne and Humber. A pier, timber-framed with stone infill, 800 ft. long and 20 ft. wide, was in existence and needing repairs in 1546. Repairs were again required in 1564 but this time it was decided to rebuild completely in stone. With 'great stones at the outer side thereof and well filled with stones in the middle, it is thought that it can never be moved nor much impaired by any Rage of water'. The great stones would be lifted from the shore and carried to the site at high tide by 'tuns', twenty or thirty of them with their iron chains, and then lifted into position by iron windlasses. This method of transporting large stones had been used before, for example at Dover for the foundations of the pier built in the 1530s by John Thompson (q.v.).

A petition submitted to the Queen resulted a year later in a grant of £500 plus 100 tons of timber and 6 tons of iron. In 1584 Dover asked for information on the pier. Scarborough replied that 'it was 800 ft. long, 60 ft. wide at the base, 22 ft. wide at the top and 30 ft. high, founded on (hard) clay under sand', and had cost £2,500 to build (to which sum most be added the value of the gift of timber and iron).

One of the Dover engineers went to see the pier, finding its construction to be 'crude'. Nevertheless the pier was still standing when Lellam arrived at Scarborough, though some damage had occurred in 1711, and a late seventeenth century extension was in bad repair.

Lellam's presence came about as a result of Scarborough Corporation asking for his advice on plans to enlarge the harbour. They had petitioned Parliament for additional power in March 1731, and again in November, but with no specific plan. Lellam started surveying in December and drew up a scheme, for which he received 20 guineas. On 10 February 1732 he gave evidence to a House of Commons Committee. He planned to remove the extension and build on from the Elizabethan pier 440 ft. in a more southerly direction into water at least 6 ft. deep at low tide, keeping the same 60 ft. base width, a wider top (30 ft.) and a height of 35 ft. at its extremity. He estimated the cost at £14,500, and gave details of the advantages to shipping.

On this basis the Act was passed in June 1732. Lellam received £36 5s for attending the Commons. He was then appointed Engineer (salary not stated but probably £70 p.a.) and Timothy Olby was appointed to collect the duties levied on ships sailing from Newcastle, Blyth, Seaton Sluice and Sunderland, all of which would benefit from an improved haven at Scarborough.

Lellam started assembling materials and equipment. His 'Book of Accounts for Building the new Pier' begins in December 1732. Early next year (on leave of absence) he and Mark Burleigh were asked by the Wear Commission to attend Parliament on their behalf. They agreed to do so for fees of 1 guinea (and expenses of 5s) per day each, plus travelling charges for themselves and a servant.

Back at Scarborough, the first payments to labourers date from April 1733. But two months later Lellam died and the burial of 'Mr. William Lellam, Engineer is recorded at Scarborough on 3 June 1733. He was replaced by Robert Wilkins who in turn was followed by William Vincent (q.v.) in December 1734. Vincent carried out Lellam's plan for the pier, acknowledging him as 'the first Engineer'.

A. W. SKEMPTON

[Petition (1716) and Report (1719) on Bridlington harbour, *JHC*, **18**, 390, and **19**, 78; James Fawcett (1718) A scheme and proposal for the improvement of Sunderland Harbour, MS, Tyne & Wear Archives; William Lellam (1727) evidence, *JHC*, **20**, 785; William Lellam and Mark Burleigh (1732) Weir Commissioners Old Papers, Tyne & Wear Archives; Scarborough Corporation Records (1731–1743) Scarborough Library; William Lellam (1732) evidence, *JHC*, **25**, 296; John Murray (1847) An account of the progressive improvements of Sunderland harbour expenditure, *Proc Min ICE*, **6**, 256–277; J. B. Baker (1872) *The History of Scarborough*; J. S. Baker (1926) *Bridlington Charters, Court Rolls and Papers*; Arthur Rowntree (1931) *History of Scarborough*; A. W. Skempton (1977) The engineer of Sunderland harbour, *Indust Archaeol Rev*, **1**, 103–125]

Works

1716–1722. Bridlington harbour, £6,000
1723–1730. Sunderland harbour pier, £19,800
1732–1733. Scarborough harbour, commencement of work

LESLIE, James, FRSE (1801–1889), civil engineer, was born on 25 September 1801 in Largo, Fife, the son of Alexander Leslie, architect and builder. Educated at local schools and Mackay's Academy, Edinburgh, from 1815–1817 he attended Edinburgh University where his uncle, Sir John Leslie taught.

In 1818 he began a pupillage with the architect W. H. Playfair, at that time working on Edinburgh University. The knowledge he acquired of architecture was put to good use in his subsequent career in civil engineering, which began in 1824 when he started with the civil engineers George and John Rennie Jr (qq.v.). In the next four years he assisted in works at London and Serpentine bridges, Plymouth Breakwater, Sheerness Dock yard, Royal William Victualling Yard, and West India Docks.

By then well qualified, in 1828 he was appointed resident engineer on William Chapman's (q.v.) scheme to extend the East Pier at Leith, and then superintended work on the west breakwater there. While working on these schemes he produced his own designs for a wet dock at Dysart, and a coal shipping pier at St. David's, on the Forth.

In 1832 he was appointed resident engineer on the harbour works designed by John Gibb (q.v.), with Thomas Telford (q.v.) as consultant, at Dundee. He remained there until 1846, designing harbour works elsewhere, and becoming involved in water supply schemes, initially with James Jardine (q.v.), whom he succeeded as Engineer to Edinburgh Waterworks in 1846.

His subsequent career, largely as a water engineer, involved him with the improvement of most towns in Scotland, as well as some overseas commissions.

He died on 29 December 1889; at the time he was the 'father' of the Institution of Civil Engineers, having joined in 1833.

MIKE CHRIMES and P. S. M. CROSS-RUDKIN

[Membership records, ICE; Tidal Harbour Commissioners (1845–1847) *Reports and appendices*; *Min Proc ICE*, **100**, 1890, 389–395; G. M. Binnie (1981) *Early Victorian Water Engineers*]

Works

1828–1832. Leith harbour, resident engineer
c. 1830. Dysart dock, engineer
1832–1846. Dundee harbour, (resident) engineer

LETHBRIDGE, John (Thomas) (fl. 1818–1827), founder member of the Institution of Civil Engineers, appears to have practised as an architect in the Exeter area, where he designed Baths at Southernhay (1821), housing in Exeter and a church at Torquay.

His practical involvement in civil engineering is unknown but he may have met some of the other early members through work in London's docks. Although he resigned membership of the Institution in November 1820 his name appears in its 1824 list of members.

MIKE CHRIMES

[Membership records, ICE; Colvin (3)]

LEYBOURN, William (*c.* 1626–1716), was a leading teacher of practical mathematics in the latter part of the seventeenth century. Among his many publications *The Compleat Surveyor*, first issued in 1653, became a standard text book; revised editions appeared in 1657, 1674 and 1679. He and Richard Shortgrave (q.v.) acted as quantity surveyors to the City of London after the Great Fire of 1666. As such they measured the works before the contractors accounts were passed for the payment on buildings and on the Fleet Canal works (1671–1674) by Sir Thomas Fitch (q.v.). Leybourn also surveyed the new street plan for Ogilby & Morgan's large-scale map of London (1672).

His portrait, age forty-eight, appears as the frontispiece to the 1674 edition of *The Compleat Surveyor*, reproduced by Taylor (1954).

A. W. SKEMPTON

[Charles Platts (1883) William Leybourn, *DNB*, **33**, 208–210; T. F. Reddaway (1940) *The Rebuilding of London after the Great Fire*; E. G. R. Taylor (1954) *The Mathematical Practitioners of Tudor & Stuart England*]

LIENS, Johan or John (fl. 1627–1641), civil engineer, was born at St. Maartensdijk in the isle of Tholen, Zeeland, about 1600, the son of Sir Cornelius Vermuyden's (much older) sister Cornelia, the wife of Joachim Liens who came to England in 1618 as a special ambassador of the Netherlands government and was knighted by James I in 1619. Vermuyden (q.v.) came to England in 1621, probably with Joachim's support, and in 1622 joined with his brother-in-law Cornelius Liens in submitting proposals for draining the Great Level of the Fens. Nothing came of this at the time but in 1626 Vermuyden signed an agreement with the King to undertake the drainage of Hatfield Chase and neighbouring areas totalling 73,000 acres. He quickly gained financial support from Netherlanders, who became 'Participants' in the undertaking and workforce and staff came from the Low Counties; among whom was John Liens. He is first heard of in November 1627 in relation to the sale of some land in the Chase.

Works in the adjacent Isle of Axholme formed part of the overall scheme and it seems they were carried out by Liens. The Axholme works began in August 1628. Exactly a year later, in connection with these, construction was due to start on Misterton Soss (Sasse), a navigable sluice on the river Idle near its outfall into the Trent at Stockwith, and the local Court of Sewers (the land drainage authority) record that 'John Liens and other Dutchmen of his party have lately undertaken to drain the Carrs in Axholme … and some necessary sluices have to be placed [on that river] in the parish of Misterton … It is agreed that John Liens may make the sluices and cut timber therefor on the commons'. In order to operate both as a navigation lock and a sluice the Soss had one or probably both gates of the lifting or 'guillotine' type. The lock chamber was 60 ft. long by 18 ft. wide. Commissioners appointed for the purpose judged Axholme drainage to be complete in 1631.

The next known reference to Liens is that in 1635 he was Director of Works for Hatfield Chase and had been for some years past. The work then in hand was a new channel for the river Don, cut and embanked for a length of 5 miles east from Newbridge, near Thorne, to Goole on the Yorkshire Ouse, in order to prevent flooding in the parishes of Fishlake and Sykehouse adjacent to the original channel running north from Thorne to an outfall on the river Aire. The new channel, later known as the Dutch River, was finished in 1635 and the work reached completion that year with the construction of an outfall sluice at Goole.

In July 1630 Vermuyden had transferred control of the Hatfield undertaking to a group of six leading Participants, of whom he was one, in expectation of becoming involved in a new proposal for draining the Great Level. In this he was again frustrated. Next year he took on two other projects: the reclamation of Sutton Marsh, north of Wisbech, and draining Dovegang lead mine in Derbyshire. Nevertheless he remained on the Hatfield management group and when the need for the Dutch River came into consideration it was Vermuyden who drew up the plan.

Work began in the summer of 1632. It can be assumed that Liens was is charge of construction with David Parolle (q.v.) as Overseer or resident engineer. Already by April 1633 upwards of £10,000 had been spent and the total cost is said to be about £33,000. It formed the final part of Hatfield Level drainage for the time being. Thereafter the Level was administered by a Court of Sewers appointed in February 1636. The name of John Liens, esquire, appears among its members.

An affidavit of June 1638 stating that 'John Liens, master-workman, and director of the Earl of Lindsey's undertaking of draining the Great Level, co. Lincoln, and the Eight Hundred Fen there, about January last obtained leave of the Earl to go into the Low Countries, upon promise to return in March or April. The Earl's undertaking hitherto depended upon Lyen's sole direction, and by reason of his absence the works have been much retarded'. The affidavit asks for, and got, more time to finish them. This document refers to fens in south Lincolnshire known as Lindsey Level, an area of some 39,000 acres between the river Glen and Kyme Eau, and to the contiguous Eight Hundred (originally Haute Huntre, later Holland) Fen covering 19,000 acres west of Boston.

Robert, Earl of Lindsey undertook to drain the Level in March 1636 but some time passed before he and his principal partner Sir William Killigrew reached agreement regarding their respective shares, and work probably did not begin until the spring of 1637. King Charles himself undertook to drain the Eight Hundred Fen in June 1637, only to sub-contract with Lindsey shortly afterwards. This was a sensible arrangement; both fens could readily be treated as a unit from an engineering point of view.

A paper, undated but presumably of late 1636 or early 1637, sets out Liens's duties and indicates the nature of the management structure. It reads as follows:

'To Mr. John Lyens Director of the works in the right Honourable the Earl of Lindsey his Undertaking.

1. He hath full power to direct all the work for the draining of the Levell.

2. To give dimensions to all drains and rivers and direct where and in what manner they shall be made.

3. To give dimension to all sluices, gotes and bridges, as also for the placing of sluices, gotes, bridges and tunnels.

4. To size and sett out the proportions of all highways, banks and Land Eaes.

5. To contract with all manner of workmen whatsoever shall be employed in or about the aforesaid draining work, with the assistance of the general overseer.

6. To direct the paymaster, the overseers and all the workmen

7. To set a value and rate of all works whatsoever.

8. To direct the Bookkeeper, Surveyors and all other officers whatsoever.

9. To entertain overseers as need shall require, and to discharge them as he shall think fit with the assistance of the general overseer.

10. To call to him Mr. Wright the general Overseer and Mr. Walpole, Surveior, to assist and advise him when and as often as need shall require.

11. To appoint certain days and places when and where all the aforesaid officers shall meet the said Mr. Liens to consult upon this business.

12. To take care of all things that concern the work, to see that all things be duly performed and at the best rate, and with the assistance of Mr. Wright to contract for all materials, and likewise to make reasonable provisions for the same, that they may be ready when the work shall need them'.

This document seems to have been drawn up shortly before construction began. The General Overseer Mr. Wright (? Thomas Wright (q.v.) who had worked on the Fens under Francis, Earl of Bedford) and the Surveyor Mr. Walpole were already in post and Liens had previously been consulted. An earlier paper entitled 'Mr. Hill his estimate of Lindsey Level' notes the appointment of Simon Hill (q.v.) on 4 March 1636 by Lindsey and Killigrew (at a salary of £200 p.a.) to take levels, 'view the places that doe charge these fennes with water', find the best outfall 'for venting all the water annoying the fennes' and act as Director.

No doubt Hill carried out these preliminary investigations and, though later replaced by Liens, he was still employed to make the estimates. They show the principal feature to be a main drain 10 miles long running north through the centre of the Level and joining Hammond Beck which flows north-east to the Witham at Boston; the Beck was to be widened and deepened with a new sluice at its outfall built in brick to replace the old one (built 1601). The completed scheme, however, shows a major modification. The drain was extended by 3 miles and continued in an easterly direction for 8 miles (this portion subsequently being called the South Forty Foot) along the southern edge of Eight Hundred Fen to an outfall sluice, close to the old one, with four pairs of pointing doors and an overall width of 55 ft. The revised system had the advantage of separating the 'living waters' of the Beck from the land drains and was better in all respects.

Other works, made in accordance with the 1636 propositions, include improved channels for three tributary streams of the Beck flowing off higher ground west of the Level; a new cut leading another stream further north to an outfall on the Witham; and a new cut 2½ miles long

linking Risedale Eau in the south to an outfall into Bicker Haven on the Welland. Finally a main drain nearly 6 miles long was taken from the South Forty Foot running north through the centre of Eight Hundred Fen. Commissioners declared Lindsey Level drainage to be complete in June 1639. No specific date for completion of the Eight Hundred Fen drainage is known but it is unlikely to have been much later. The total cost is reported as £45,000. Dugdale (1662) gives a map of the Level.

To Liens can be attributed the 16 page *Discourse on the Great Benefit of Drayning and Imbanking*, signed J.L. and published 1641. The author, a Netherlander, refers to 'my long practice and experience' in draining, apologises for his 'unexpertnesse' in the English language, and expresses his gratitude, duty, zeal and affection 'to his King and State' [of England]. After general remarks on the high cost of importing corn, cheese, flax and other commodities, and estimates of the value of crops which could be grown on land fit for drainage in the counties of Cambridge, Norfolk, Lincoln and York, he goes on to answer various objections against draining; objections which in recent year had given rise to disturbances or even rioting. The principal argument in favour is of course the improved quality of the land. In Hatfield Chase 'good sweet grass and clover' were growing two years after draining and the land was 'very fruitful for corn, flax and coleseed'. Similarly in the Earl of Lindsey's draining there were more horses and cattle than previously and barley, wheat, beans and oats were sown and proved extraordinarily good before the banks were cut [by rioters]. He adds that Lindsey and Killigrew and their partners made in their Level two highways along its length and ten in the breadth, where there was but one before, and many bridges, which they were not bound to make by their contract.

Nevertheless despite the undoubted improvements, represented in glowing terms by the author, local landowners and commoners felt resentment at having their way of life upset by outsiders with power at Court and by foreigners. Opposition was encountered in most of the early seventeenth century drainage schemes during construction; by 1640–1641 disturbances became widespread and in 1642, as the Civil War was breaking out, rioters severely damaged or destroyed the drainage works in all the fens except Hatfield Chase and Bedford Level which remained relatively unscathed, though work in the latter ceased after 1641. Orders were issued for restoration of Axholme drainage in 1645, with partial success; work in Bedford Level resumed in 1650; while redrainage of Ancholme Level and Deeping Fen took place in the 1660s. But despite repeated attempts to reclaim Lindsey Level and the East, West and Wildmore Fens (north of Boston) nothing was achieved until the second half of the eighteenth century. It is therefore improbable that John Liens found any employment in England after 1641; perhaps he returned to the Netherlands. The date of his death is not known.

A. W. SKEMPTON and MARGARET KNITTL

[No minute books exist for any of the pre-1640 drainage schemes. The 1635 reference, like many for this period, comes from State Papers Domestic (SPD), in the PRO; another principal source is Dugdale's *History of Imbarking and Drayning* (1662). In SPD is State Papers Domestic (1635–1638) 315/44, 339/44, 375/48, 398/25; Sir William Dugdale (1662) *The History of Imbanking and Drayning*; W. H. Wheeler (1896) *A History of the Fens of South Lincolnshire*; J. Korthals-Altes (1925) *Sir Cornelius Vermuyden*; Margaret Albright (1955) The entrepreneurs of fenland drainage in England under James I and Charles I, *Explorations of Entrepreneurial History*, **8**, 51–65; Keith Lindley (1982) *Fenland Riots and the English Revolution*; Pat Jones (1994) Vermuyden's navigation works on the river Don, *Railway & Canal Hist Soc*, **31**, 248–258]

Publications

1641. *A Discourse concerning the Great Benefit of Drayning and Imbanking, and of Transportation by Water within the Country …*

Works

1628–1631. Isle of Axholme drainage, in charge of works under Vermuyden, 13,000 acres, works include Misterton navigable sluice
1632–1635. Dutch River, Director of Works of Hatfield Chase, cost *c.* £33,000
1637–1639. Drainage of Lindsey Level and Eight Hundred Fen, Director of Works, 58,000 acres, cost *c.* £45,000

LINGARD, Edmund (*c.* 1725–1780), canal engineer, was probably born in Curdworth, Warwickshire, *c.* 1725, the son of Edmund and Mary Lingard. The family name was common in the West Midlands at that time. The Coventry Canal received its Act of Parliament in 1768, with James Brindley (q.v.) as its Engineer and Joseph Parker Clerk of Works. In September 1769 concern over labour costs and quality of supervision led to Brindley's dismissal and a reprimand for Parker. Thomas Yeoman (q.v.) acted briefly as Engineer, then Parker left and Lingard, apparently with no previous canal experience was appointed Engineer with Samuel Bull (q.v.) as his assistant. By the end of 1771 the original capital of £50,000 was exhausted and with the canal completed from Coventry to Atherstone, the onward link to the Birmingham and Fazeley was delayed, and not opened until 1790. In these circumstances it is unsurprising that Lingard sought employment elsewhere. In the spring of 1776 the Stroudwater navigation was looking to replace its Engineer, John Priddey (q.v.), and Lingard was offered the post in June 1776 on a temporary basis at 2 guineas a week. He evidently made a good impression as on 8 July 1776 he was offered the post of resident engineer at £300. Lingard

seems to have been responsible for locating, hiring and supervision of labour, most 'contracts' being for small amounts. The company made use of other trusted workmen such as James Bough (q.v.). Lingard was dismissed in November 1777, being suspected of connivance with one of the lock carpenters, John Pashley; at this time the navigation was still incomplete. Details of Lingard's subsequent career are lacking.

MIKE CHRIMES

[C. Hadfield (1969) *Canals of the West Midlands*, 2nd edn.; C. Hadfield (1970) *Canals of the East Midlands*, 2nd edn.; M. Handford (1979) *The Stroudwater Canal*]

Works

1770–1771. Coventry Canal, engineer
1776–1777. Stroudwater Navigation, resident engineer

LINGARD, Thomas (d. 1836) was agent or manager of the Old Quay Company, owners of the Mersey and Irwell Navigation, from 1814. He may have been a descendant of Edmund Lingard, (q.v.). By 1825, Thomas Lingard had general superintendence of the whole of the Old Quay Company, and lived on their premises at Manchester. In May 1825, when Lingard gave testimony to the Committee of the House of Commons on the Liverpool and Manchester Railway, the Old Quay Company had seven large warehouses at Manchester, the newest built in 1819. They were used chiefly for the storage of cotton to supply the textile mills of Manchester. During his superintendence, it was possible for a vessel to leave Manchester at 5 p.m. and arrive in Liverpool at 10 a.m.

Under Lingard's supervision, with Richard Phillips as the Company engineer, the Mersey and Irwell Navigation was improved by the Woolston Canal (1819–1821), which shortened the route by three quarters of a mile. Two further improvements which would eliminate 2 or 3 miles from the journey were planned, and property was purchased. He was also involved in making a large new dock at Runcorn, completed in 1825. Lingard testified against the railway which was proposed to pass through part of the Old Quay Company's 12 acre site in Manchester and would block a planned link to the Rochdale Canal, avoiding the town centre. In the 1830s Lingard designed an iron lighter decked over for timber cargoes.

Lingard's eldest son, Thomas Jr., gave testimony at the hearings along with his father. He lived with his father and had for the previous three years assisted him on the management of the Manchester facilities of the Company. He succeeded his father as Agent, and was involved in fierce competition with the railway and the Bridgewater Canal, which led to transfer of control of the Mersey and Irwell to the Bridgewater company in 1844. Thomas's fifth and youngest son, Frederick (1811–1847), born in Birmingham, was a very well known musician.

TESS CANFIELD

[Frederick Lingard, *DNB*; *Proceedings of the Committee of the House of Commons on the Liverpool and Manchester Railroad Bill*, Sessions 1825; E. Padget-Tomlinson (1978) *Complete Book of Canal and River Navigations*, 344–345]

Works

1819–1836. Mersey and Irwell Navigation, improvements, including Woolston Canal, £20,000, and Runcorn new lock and basin, 1822–1825

LISTER, Samuel (1746–1805), of Bramley near Leeds, mason, was the son of John Lister (1719–1775) and Hannah Farrer. He was baptised on 14 July 1776, the eldest of a family which included two sisters. His four brothers, John (b. 1753), James (1755–1798), Joseph (b. 1758) and William (1763–1811) were all practising as masons at Bramley in 1795, though William later had a foundry at Bramley and subsequently moved to Leeds where he joined Fenton, Murray (q.v.) and Wood.

An earlier Samuel Lister was a contractor with William and John Gott (q.v.) for Hebble End Bridge over the River Calder in 1748 but this subject's first contract appears to have been with his father and another man at Lindley, in Wharfedale in 1768. Three years later he was building Higher Hodder Bridge, west of Clitheroe, about 50 miles from home. In 1795 he was responsible for the design-and-build contract to replace Swarkestone Bridge over the River Trent, swept away in the floods of that year, but as the design bears some resemblance to that of other contemporary Derbyshire bridges, it may be that it was provided by Thomas Sykes (q.v.), the County Surveyor, who supervised the construction and determined the foundation details. Samuel's sureties for this job were his brothers, John and Joseph.

In 1797 Samuel Lister contracted, with his youngest brother, William, to build Bursledon Bridge but there were problems with the foundations and, when money ran short in 1798, work was suspended. A revised design in timber was prepared and work resumed with funds raised under a second Act of Parliament. Lister was buried at Bramley on 23 May 1805.

William Lister was the contractor for the widening of High Bourn Bridge in the North Riding, to John Carr's design, which the latter described for some reason as 'well executed, but the contractor is a bad workman'. Joseph Lister tendered unsuccessfully at £5,224 for Wolseley Bridge in Staffordshire in 1797 but he and his other brothers seem to have confined their activities otherwise to building bridges in the West Riding.

P. S. M. CROSS-RUDKIN

[West Yorkshire Archive Service; Derbyshire RO; Hampshire RO; Cheshire RO; Rennie notebooks at NLS; *Thoresby Society*, **23**, **29**]

Works

1768. Lindley Bridge, contract (with John Lister of Bramley and John Robinson of Horsforth) to rebuild, for £340 plus old materials

1771. Higher Hodder Bridge, contract (with Abraham Fielden and John Law (q.v.)) to rebuild to plan of William Gott, for £312 10s

1773. Sowerby Bridge, contract (with John Lister) to repair and widen to plan of John Gott, for £520

1789–1794. St. Mary's Bridge, Derby, contract (with J. Hanley) to build to design of Thomas Harrison (q.v.), for £2,900

1795–1797. Swarkestone Bridge, Derbyshire, contract to rebuild, for £3550

1796. High Bourn Bridge, near Masham, widened by William Lister to plan by John Carr, for £575

1796–1797. Ferrybridge Bridge, John and Joseph Lister contractors, for widening the structure of 1764, to plan by William Gott, for £3,415, contract was abandoned by the West Riding when a decision was taken to build on a new site further upstream

1797–1799. Bursledon Bridge, Hampshire, contractor (with William Lister), for £5,000, to plan of George Moneypenny; also, 1794 Monk Bridge, York, Joseph Lister contractor (with Christopher Dalton and King) to design of Peter Atkinson Jr., now widened

1800–1801. Marple Bridge, contractor to design and rebuild, £500

1801–1802. Colne Bridge, near Huddersfield, Joseph Lister contractor, for widening, for £726 4s 11d

LODGE, Ralph (*c.* 1730–1801), attorney and entrepreneur, was the only son of George Lodge (1695–1764) and his wife, Margaret Wilkinson (d. 1777), married in 1729 at Hubberholme, Yorkshire. He was admitted to student membership of Gray's Inn in May 1753 and subsequently would appear to have pursued a career in law.

In 1767 Lodge, described as residing in London, became a shareholder in the Carron Company, its works near Falkirk brought into production in 1760. Both John Smeaton (q.v.) and James Watt (q.v.) were at times associated with the company which by 1772 was experiencing financial problems and Lodge was then asked by management to 'obtain a copy of a draft charter' so that the company could be reconstituted.

In 1769 a petition was presented to the mayor of Newcastle by Lodge, with others, 'for supplying the Town … with good water'. He stated that he had undertaken surveys to find a suitable source and asked for 'a lease for a long term of years of the fountains, springs, reservoirs, pipes and other premises' so that his proposals could be implemented. Perhaps associated with this venture he subscribed to Armstrong's *Map of Northumberland* in the same year. The scheme was not given consent in full and as a result Lodge entered into an agreement with the Common Council to build a reservoir at the south end of the Town Moor, bringing underground water to it from Coxlodge, some 3 miles north of the town.

The proposals followed closely a scheme which had been put forward two years earlier by William Brown (q.v.), one of the region's leading colliery viewers. To eliminate competition, Lodge purchased the earlier works of William Yarnold (q.v.) and, as a result, the accrued profits (£1,612) of the older company passed into his hands. Some dissatisfaction as to the supply was expressed by the Council in 1785 and the town clerk was instructed to write to Lodge informing him of the Council's unease. Nothing further transpired and the works were eventually purchased in 1797 by the Newcastle Fire Office.

Lodge's next venture was in copper mining at Middleton Tyas, 4 miles from Richmond, Yorkshire. Lodge, then residing at nearby St. Trinian's Hall—bought by his father in 1731—took a lease of land at the Glebe for 21 years in 1775. Work began immediately and in 1776 a pumping engine was purchased to facilitate the mine's drainage. 'Between 13 April and 19 June Ralph Lodge paid £943 4s 11d in three instalments for the new engine, about half the cost of [the previous lessee's] engine twenty years earlier'. This earlier engine had been inspected by Brown in 1754, his report expressing some doubts as to its efficiency. So far as the later engine is concerned, only its cylinder came from the Carron Company. In spite of the expeditious start, consideration was given to disposing of the mines in 1779 but the outcome is not clear.

Lodge suffered financial problems in 1780 when some property of his was assigned to the rector of Melsonby, and in 1782 he was declared bankrupt, then described as 'iron manufacturer, dealer and Chapman'.

Lodge died, aged sixty-seven, at his home in Castle Street, Carlisle, and was buried on 17 October 1801 at St. Cuthbert's Church.

R. W. RENNISON

[*Newcastle Courant*; R. H. Campbell (1961) *Carron Company*; T. R. Hornshaw (1975) *Copper Mining in Middleton Tyas*; R. W. Rennison (1977) The supply of water to Newcastle upon Tyne and Gateshead, 1680–1837, *Archaeologia Aeliana*, 5th Series, **5**, 179–196]

Publications

1769. *The Humble Petition of Ralph Lodge ... for supplying the Town of Newcastle with good Water* (to John Baker, Mayor)

Works

1767–1772?. Carron Company
1769–1785. Water supply for Newcastle upon Tyne
1775–1780?. Copper mining at Middleton Tyas

LOGAN, David (1786/1787–1839), civil engineer, came of an engineering family living in the Angus area of Scotland. His father, Peter Logan, was a master builder who became foreman builder of the Bell Rock lighthouse. Another David Logan, probably the subject's uncle was an architect active in the Angus area 1791–1815.

David Logan's career was mainly in the field of dock, harbour and bridge works, first under Robert Stevenson, Thomas Telford and John Rennie Sr. (qq.v.) and later in his own right. His first known post was Clerk of Works for the construction of the Bell Rock lighthouse, carried out between 1807 and 1811 under Stevenson. Logan was based almost entirely ashore at the work yard at Arbroath and only worked for the last few months on the Rock itself. From 1811 to 1813 he was Inspector of Works for the Marykirk Bridge over the North Esk, designed and supervised by Stevenson whose employment he subsequently left to become, in August 1816, Superintendent of Works for Telford's improvements at Dundee Harbour, work which included the construction of a graving dock. In April 1820, on the proposal of Telford, he was elected a corresponding member of the Institution of Civil Engineers.

In January 1821 Logan resigned from Dundee to take up the position of Resident Engineer on Rennie's new Donaghadee Harbour project, the first stone of which was laid in August that year. Progress there was slow, due in large measure to the Government's reluctance to make adequate annual funds available, and was not finally completed until 1837. This afforded Logan the opportunity to undertake a number of other commissions both for John Rennie Jr., who had taken over the supervision of the work on the death of his father in October 1821, and on his own behalf.

In 1823 he submitted plans to the County Antrim Grand Jury for a five arch bridge across the river Lagan, and between that year and 1828 he supervised several works for the Belfast Ballast Board. His work included proposals for a steam dredger and plans for the disposal of dredged material ashore. Between 1829 and 1832, acting for Rennie, he supervised the extension of the pier at Ardglass Harbour. He also designed a single arch masonry bridge at Banbridge over the river Bann, completed in 1832.

In 1834, with the work at Donaghadee substantially complete, Logan was asked by Rennie to visit Whitehaven and the same year, having been appointed Engineer to the River Clyde Trustees, took up residence in Glasgow. In January 1839 he tendered successfully for the formation of a straight channel from the town of Belfast seaward, but before the work could commence his death was reported to the Belfast Ballast Board and the contract placed elsewhere.

David Logan died of palsy on 20 January 1839 at his home in St. Vincent Street, Glasgow, leaving an estate valued at £5,390. He was buried in the Kappa section of the Glasgow Necropolis and the memorial there records the death of his wife, Jane, and two of his sons: John, who died aged twenty-five, and the Rev. Robert Logan.

Logan was married three times: first, in 1808, to Jean Taylor of Arbroath; secondly, in 1822, to Mary Smith of Donaghadee, County Down; and lastly, in 1831, to Jane Hannay of Portpatrick. In all, he fathered four sons and four daughters. Only one son, David, seems to have followed in his father's footsteps; he was elected a Member of the Institution of Civil Engineers in 1867 and his career was mainly in railway engineering in India, where he became Chief Engineer of the Great Southern of India Railway. **Peter Logan**, David's cousin, succeeded him as Superintendent of Works at Dundee Harbour and later became Resident Engineer at St. Katharine's Dock, London.

R. SWEETNAM

[Dundee Harbour Commissioner's minute books (1816–1821) City of Dundee RO; D. Logan (1820) ICE Original Communication Paper No.109, Report on Graving Dock at Dundee; Papers of Robert Stevenson & Sons, deposit 216/92 Marykirk Bridge, NLS; Belfast Ballast Board Minute books (1823–1839) Belfast Harbour Commissioners]

Reports
1835. *Report on Improvements of the River Clyde and the Harbour of Glasgow*

Works
1807–1811. Bell Rock Lighthouse, Clerk of Works
1811–1813. Marykirk Bridge, Inspector of Works
1816–1821. Dundee Harbour, Superintendent of Works
1821–1834. Donaghadee Harbour, Resident Engineer, £145,000
1823–1828. Belfast Harbour, No 2 Graving Dock, etc., Superintendent of Works
1829–1832. Ardglass Harbour, Superintendent of Works
1831–1832. Banbridge Bridge, Engineer
1834–1839. River Clyde improvements, Engineer

LONGBOTHOM, John (fl. 1760–d. 1801), civil engineer, came from Halifax and in 1768 drew up a plan for a link from Halifax to the Calder Hebble navigation for which John Smeaton (q.v.) had been engineer; he subsequently designed two other proposed branches. He is best-known as Engineer of the Leeds–Liverpool Canal, his greatest civil engineering achievement.

The idea of a trans-Pennine canal, from the Ribble to the Aire, was raised in 1764 and John Stanhope, a Bradford attorney and landowner with links to the wool trade, employed Longbothom to carry out a survey, with his servants helping. Longbothom's 1765 proposals linked Preston to the Aire, with branches to Liverpool and Lancaster. By August 1767 he had completed a survey of a line from Preston to Leeds at an estimated cost of £101,831 and was developing a proposal for a canal—capable of taking 60 ton vessels—from Liverpool to Hull, by-passing the Aire and Calder Navigation.

In January 1768 the results of his Leeds–Liverpool survey were presented in Yorkshire and a committee was set up. Progress was slower in Lancashire where nobody had as yet taken the lead and it was not until August that the first meeting of supporters was held. It was decided that a second engineering opinion was required and James Brindley (q.v.) was called in.

Longbothom's line was then surveyed by Whitworth (q.v.) and Longbothom was generally supported, as was his view of the need for the canal to be large enough to accommodate Douglas/Aire and Calder navigation vessels. In December supporters decided to apply for an Act, based on estimates for a canal and branches of £259,777. It proved too late to get a Bill ready for the Parliamentary session and the delay offered Liverpool subscribers an opportunity to challenge Longbothom's route, which provided unsatisfactory links to the Wigan coalfield, a priority for Liverpool. An alternative route, surveyed on their behalf by John Eyes (q.v.) and Richard Malling, was rejected by the more powerful Yorkshire shareholders when Whitworth found faults in the levels. In July 1769 another Liverpool surveyor, P. P. Burdett, was commissioned to do a further survey; he recommended a route via Wigan. Subscribers thus had five routes to choose from as Longbothom modified his route and provided a link to the Wigan coalfield by a branch to Dean. In the event the lower cost of the main line of Longbothom's modified route—£174,324, cf. Burdett's £240,881—proved decisive and this provided the basis for the Parliamentary route of the 1770 Act.

Although Brindley was initially asked to be engineer, with Longbothom as clerk of works, he turned the offer down and Longbothom was appointed engineer at a salary of £400 p.a., later increased to £500. Longbothom supervised construction down to June 1775 when he was forced to resign. He had a packet boat business based in Liverpool and acquired coal mining rights at Upholland to which he built a branch canal, but he was in debt to the Company. His dismissal resulted from his debts and poor bookkeeping rather than lack of engineering ability. By this time 55 miles of canal at the Leeds and Liverpool ends were complete. Indeed after 1777 funds were almost exhausted and the central section of the canal was not completed for many years, work resuming only in 1790. One of Longbothom's last acts as engineer was, in 1775, to arrange to take water from the Ince and Hindley brooks to deal with water supply problems at Wigan.

Although well-known for his work on the Leeds and Liverpool canal, the remainder of Longbothom's career is not so well defined. He was an early member of the Smeatonian Society of Civil Engineers, being elected in 1773, but he attended irregularly and in 1784 was deleted from membership. He was clearly suffering financial problems after losing his job and in 1776 had to sell his coal mining rights. These problems seem to have continued until the end of his life as in 1800 he appealed to the Leeds–Liverpool Company for financial assistance and indeed they met his funeral expenses.

While working on the Leeds and Liverpool canal in 1767 he had drawn up a proposal for a canal between Leeds and the Ouse at Selby; it provided an alternative route to the sea avoiding the Aire and Calder. In 1772 the Leeds and Liverpool Canal Company gave him permission to work on the scheme, which went to Parliament but was defeated in 1774. There were doubts about the practicability of its water supply.

Following the Leeds–Liverpool debacle Longbothom is next heard of as engineer for a land drainage scheme near the River Alt in west Lancashire. In 1787 he was called on to survey a coastal canal near Morecambe, a forerunner of the Lancaster canal. Its proposed line was modified at the suggestion of John Wilkinson (q.v.) who pointed out the potential for land reclamation. Longbothom's scheme involved large areas of land reclamation and river navigation improvements as well as canal construction. In the end John Rennie (q.v.) was brought in and his recommendations formed the basis of the Act.

Longbothom was also consulted about the Bradford canal, which linked Bradford to the Leeds–Liverpool (1771–1774) and, in the 1790s, the Manchester, Bolton and Bury Canal's proposed Bury–Sladen extension (1792–1793). He was also consulted regarding proposals for the completion of the Leeds and Liverpool canal onward from Wigan (1792–1794). At this time he was resident in Chorley. Elsewhere he was consulted about the Grand Western Canal (1792–1793) and the Bristol and Western Canal (1793), the latter with William White (q.v.). Shortly after, in 1794, he was advertising in the *Exeter Flying Post* offering to undertake canal surveying work at three guineas a day plus expenses, with his own team of surveyors. His subsequent career is unknown but he died in April 1801.

MIKE CHRIMES

[Leeds and Liverpool canal minutes (PRO RAIL 846); F. M. L. Thompson (1968) *Chartered Surveyors: the Growth of a Profession*, **55**, 62; Hadfield, *passim*; Paget-Tomlinson; Skempton; M. Clarke (1994) *The Leeds and Liverpool Canal*]

Works
1770–1775. Leeds and Liverpool Canal, 55 miles, *c.* £160,000, engineer
1771–1774. Bradford Canal, engineer

Publications
1770. *A Plan of the Intended Navigable Canal, between Liverpool and Leeds*
1772. *A Plan of the Intended Navigable Canal from Leeds to Selby*
1796. Specification of the patent granted ... for ... a plan or method of supplying canals, or any other cuts, ponds or sluices, wanting the same, with water, *Repertory of Arts and Manufacturers*, **4**, 145–154

LOUDON, John Claudius (1783–1843), writer and landscape gardener, was one of the most important early innovators in the use of iron as a structural material. Born on 8 April 1783, in Cambuslang,

Lanarkshire, he was the eldest child of William Loudon of Kerse Hall, near Gogar, and Agnes Sommers. He attended school in Edinburgh. Interested in horticulture from childhood, Loudon obtained his father's consent to be trained as a landscape gardener and became draughtsman and assistant to John Mawer at Easter Dalry, learning particularly the management of hothouses. Mawer died before Loudon was sixteen, leaving him to spend several years with Mr. Dickson, nurseryman and landscape gardener of Leith Walk, Edinburgh. During this time he continued his education by taking classes at Edinburgh University.

In 1803 he arrived in London. He became a close friend of Sir Joseph Banks, and was elected a Fellow of the Linnaean Society in 1804. That year he was employed by the third Earl of Mansfield to lay out the grounds of Scone Palace, Perthshire, and returned to Scotland for several months, draining and improving estates and writing *Observations on the Formation and Management of Useful and Ornamental Plantations* ... . He continued writing, including the two volume *Treatise on Forming, Improving and Managing Country Houses* and practising as a landscape gardener until late 1806, when he contracted rheumatic fever while working in Wales. The disease resulted in a permanent disability curtailing his career as a travelling landscape gardener. Ill health dogged his life, often brought on by over work.

Loudon became interested in improving English farming, and, joined by his father in 1807, he set about the introduction of advanced 'Scotch' farming into England. His pamphlet on the topic resulted in an invitation to take up a part of Tew Park, Oxfordshire. There he trained young men in improved agricultural methods, and demonstrated the principles he promoted in *Observations on laying out Farms in the Scotch Style* ... . He also published an account on the use of tarred paper as a robust lightweight material suitable for agricultural and residential buildings. After the death of his father, Loudon left Tew in 1811 to return to London, having by 1812 accumulated £15,000 through his own labours he determined to travel, visiting the south of England in 1812–1813 and making the first of many continental visits 1813–1814.

On his return he designed a ducted hot air heating system for Coleshill, Oxfordshire (1814). In 1816, he moved to Bayswater, London, joined by his mother and sisters, and began experiments on the construction of hothouses, publishing his findings which included the earliest proposals for cast iron rectangular braced girders. In *Sketches of Hothouses* Loudon describes his wrought iron glazing bar, invented in 1816. The narrow bar could be bent along its length, allowing the resulting curved surface of the glasshouse to meet the sun's rays at nearly right angles for much of the day, in conformance with current theories of light penetration. Loudon's glazing bar was a pioneering event in the development of the rolled iron section. In Loudon's design, the wrought iron glazing bars and glass acted together structurally, producing the earliest useful shell structure of iron and glass. In 1818 he also published *A Comparative View of the Usual and the Curvilinear Process of the Roofing of Hothouses*, in which he invited commissions for hot houses, or for iron roofs, fireproof buildings, etc. Loudon worked jointly with a contractor, W. & D. Bailey, building elegant arching glasshouses and Bicton Park Palm House, Devon, (*c.* 1843) survives as a late rare example. Loudon made over his rights to Bailey in 1818, foregoing a likely fortune, but freeing himself for literary pursuits. Richard Turner's (q.v.) Palm House at Kew used a wrought iron glazing bar based on Loudon's designs. Loudon also proposed the ridge and furrow design to achieve maximum light penetration twice a day. There is no evidence that an example was built but the influence of his ideas on Joseph Paxton (q.v.) is undeniable.

In 1818, Loudon published his *Design for a Bridge across the Mersey at Runcorn*, a cast and wrought iron suspension bridge of 1,000 ft. clear opening with a minimum height above the water of 70 ft. It consisted of two giant cast iron piers with each part of the bridge deck suspended individually and directly from the piers. Loudon was friendly with Thomas Telford (q.v.), who also designed a bridge for the site.

Loudon made a long journey through Europe in 1819, gathering material for his *Encyclopedia of Gardening*; first published in 1822, it established his international reputation. His difficult health continued; but his enormous literary activity continued unabated. In 1826, he began the *Gardeners Magazine*, the first of five monthly publications he was to found, including the *Architectural Magazine* (1834–1839), the first periodical devoted entirely to architecture.

In 1828, Loudon reviewed an anonymous novel, *The Mummy*, set in the twenty-second century which postulated, among other things, a steam plough. It prompted him to arrange to meet its author who turned out to be a woman, Jane Webb (1807–1858), who Loudon married in September 1830, six months after they met. Jane joined her husband in his work, becoming a well known horticultural writer in her later years.

Loudon continued to tour Britain and the continent for the rest of his life, providing material for his extensive publications, including the influential *Encyclopedia of ... Architecture* the 1842 edition of which illustrated early intermediate railway station buildings as models of 'villa' style architecture. In 1833 he began his major work, *Arboretum et Fruiticetum Britannicum*; which left Loudon again in debt and he spent his final years in concentrated work attempting to clear this. He resumed activity as a landscape gardener, advising on country estates and laying out Derby Arboretum in 1839 for Joseph Strutt.

Loudon was almost alone in his concern for the improvement of taste and comfort amongst the middle classes, tirelessly promoting modern ideas and technology. In May 1818, he had proposed a scheme for decent housing for the poor, which eventually was published in 1832 as *Colleges for Working Men*. It is discussed in the 1844 parliamentary report on the *State of Large Towns*, to which he gave evidence.

Loudon died standing on his feet on 14 December 1843 while dictating to his devoted wife and is buried in Kensal Green Cemetery, London. Loudon's papers were destroyed in the Second World War but his house in Bayswater survives largely intact, but without its gardens and pioneering glasshouse.

TESS CANFIELD

[Biography file and MSS Collection of the British Architectural Library, RIBA; J. W. Loudon (1845) An account of the life and writings of John Claudius Loudon, in J. C. Loudon, *Self-instruction for Young Gardeners*; B. Howe (1961) *Lady with Green Fingers* (Jane Loudon); J. Hicks (1974) *The Glass House*; E. Macdougall (ed.) (1980) *John Claudius Loudon and the Early Nineteenth Century in Great Britain*; J. Sutherland (1983) *1780–1850, Structural Engineering*, 42–43; G. Kohlmaier and B. von Sartory, J. C. Harvey (trans.) (1986) *Houses of Glass* (first published as *Das Glashaus* (1981)); M. Simo (1988) *Loudon and the Landscape*; Colvin (3); DNB]

Publications

Publications include:

1804. *Observations on the Formation and Management of Useful and Ornamental Plantations; on the Theory and Practice of Landscape-Gardening and on gaining and embanking Land from Rivers or the Sea*
1806. *Treatise on Forming, Improving and Managing Country Houses*
1811. *Observations on laying out Farms in the Scotch Style adapted to England*
1811. *An Account of the Paper Roofs used at Tew Lodge*
1817. *Remarks on the Construction of Hot-houses*
1818. *A Comparative View of the Usual and the Curvilinear Process of the Roofing of Hot-houses*
1819. Design for a bridge across the Mersey at Runcorn, *Annals of Philosophy*, **11**, 14–27
1822. *Encyclopedia of Gardening*
1833. *Encyclopedia of Cottage, Farm and Villa Architecture*, and late editions
1833–1838. *Arboretum et Fruiticetum Britannicum*
1832. Colleges for working men, *Mechanics' Magazine*, **443**

Extensive bibliographies are included in Appendix B of Simo, and in MacDougall.

LOWE, George, FRS (*c.* 1788–1868), gas engineer, was born in Derby around 1788, the son of a brewer. Although trained as a brewer he took an early interest in lighting by coal gas, and wrote on gas purification in the *Philosophical Magazine*. This led to his appointment in 1821 to the Chartered [Gas, Light and Coke] Company as one of their engineers and he worked for them until his retirement in 1863. In addition, he acted as consultant to other companies such as the

Imperial Continental Gas Association, the European Gas Company and the Dublin Alliance Gas Company.

Lowe had developed a method for evaporating spent lime water under the retorts at a time when the Company was suffering from the problem of satisfactory effluent disposal. He was put in charge of the Curtain Road works in 1821 and demonstrated such ability that in 1832 he was appointed chief engineer. In 1831 the Company, on his advice, replaced traditional wrought iron retorts with cast iron, the Company marketing them to other firms. In 1835 he replaced the Brick Lane holders to telescopic principles and gradually other old holders were similarly modified. In the early 1840s naphtha was introduced into gas supplies for lighting purposes, first at the Reform Club (1841) and Buckingham Palace (1842).

Lowe was elected an Associate of the Institution of Civil Engineers in 1823 became a Member in 1829; he served on Council from 1837 until 1845. He regularly contributed to Institution discussions and more than twenty contributions are recorded in the *Minutes of Proceedings*, mostly related to gas engineering and chemical processes. His various scientific interests and attainments were reflected in his election as Fellow of the Royal Society in 1835 and membership of other societies such as the Geological Society, Microscopical Society, Chemical Society, and Franklin Institute.

From 1831 onwards he took out a number of patents relating to the gas industry: for long or 'reciprocating' retorts; methods of improving the illuminating power of coal gas; improvements to retorts; motive power meters; production of Prussian blue from liquid ammonia and refuse lime ligure; and the purification of gas.

Lowe married in 1818 but his wife, to whom he was devoted, died in 1843. He died aged 80 on Christmas Day 1868 at his residence at St. John's Wood Park, London.

MIKE CHRIMES

[Membership records, ICE archives; *Journal of Gas Lighting*, 5 January 1869; *Min Proc ICE*, **30**, 1869, 442–445; S. Everard (1949) *The History of the Gas Light and Coke Company, 1812–1949*, *passim*]

Publications
1819. On the purification of coal gas, etc., *Philosophical Magazine*, **53**, 262–266
1834. On some new chemical products obtained in the gasworks of the metropolis, *Brit Assn Rep*, 582–583

LOWE, Thomas (fl. 1760–1790), millwright, of Nottingham, was believed by William Fairbairn to have been responsible for the millwork in many of the early cotton mills, although describing the work as 'heavy and clumsy'. A dealer in timber he erected engine frames for Boulton and Watt in the 1770s, as well as acting as millwright to a number of cotton mills. Known examples include those

for John and William Elliott, cotton spinners, and John James (1787), all at Nottingham, and possibly the Robinson's mill at Papplewick.

Elsewhere he installed the millwork for Peter Drinkwater (1789), in Manchester, which was the first to use a Boulton and Watt rotary engine for cotton spinning. He worked for Charles Lees of Stockport (1795), Robert Owen in Chorlton (1795), the Salvins, in Durham (1796), Samuel Marsland (1798) and the Catrine Mills in Scotland. He was not exclusively concerned with cotton mills, working for Gott (Leeds, 1792) and Wedgwood (1801). His work for Gott followed a disagreement between Gott and John Sutcliffe (q.v.). Little is known of his life beyond his work.

MIKE CHRIMES

[W. Pole, *Life of Sir William Fairbairn*, 181; J. Tann (1970) *The Development of the Factory*, 97–99]

M

McADAM, John Loudon (1756–1836), road maker and administrator, was born on 21 September 1756 at Ayr, the son of James McAdam of Waterhead, near Carsphairn, Kirkcudbrightshire, and Susannah Cochrane, niece of the Earl of Dundonald. He was the youngest of ten children, having one brother and eight sisters. His father sold Waterhead about 1763 and the family moved to Straiton, from where McAdam went to school in Maybole, Ayrshire. He was still a schoolboy when his father died in August 1770, leaving an estate worth £6,000 or £7,000. Each of the surviving daughters received £500 and, his older brother being already dead, McAdam inherited the remainder.

In the autumn of that year he went to New York to join his uncle, William McAdam (d. 1779), a successful merchant, who adopted him. By 1776 John was a Commissioner of Naval Prizes and probably the owner of John McAdam & Co, auctioneers, and wealthy enough to undertake a tour of the West Indies, partly for business but also for pleasure. On his return to the growing turbulence in America he joined a militia unit, Delancey's Brigade, but it is unlikely that he took any part in the fighting. In 1778 he married Gloriana Nicoll, the daughter of a loyalist lawyer who presided over the Provincial Assembly and who was a relative of the De Lanceys.

McAdam stayed on in New York for a time after Independence, forming a new partnership called McAdam Watson & Co. in 1781, but many of his connections had left. He followed them on 4 May 1783, arriving in Scotland the following month, and in 1784 was one of the 'Sufferers' who petitioned for compensation under the Act of 1783. He had salvaged enough of his fortune to purchase the Sauchrie estate in Ayrshire in 1785 and to add to it in 1787.

In 1786 McAdam became the manager of the British Tar Company at Muirkirk, a venture of his cousin Archibald Cochrane, 9th Earl of Dundonald. In 1790, with £5,000 of his own money and a loan of just under £10,000, he bought the company, but a poorly drafted contract with the Muirkirk Iron Company for the supply of coke, and a downturn in demand for the product, left McAdam unable to repay the loan. The dispute with the Iron Company dragged on until 1803 and McAdam was still involved as a trustee of the lender's estate in 1805.

Apart from this, McAdam spent his time much like any other landed gentleman, undertaking public duties and attending county meetings. His first attendance at a road committee in Ayrshire was for Carrick district on 30 December 1785, and in October 1787 he is recorded as having been present at the general meeting of the Ayr Road Trustees in the county town. A year later he attended a Lanarkshire meeting to discuss a road between the two counties and in 1789 performed the same office in respect of a road into Dumfriesshire. He continued to take an active part in road affairs in Carrick until 1793 and in the wider county until March 1798. His growing public status was recognised by his being elected a burgess of Ayr in 1789, a member of the Council there in 1793, a major in the militia in 1794 and his appointment as a Deputy Lieutenant of the county on 6 February 1797. However his financial problems appear to have decided him to repair his fortunes elsewhere.

He was by now the father of three children. Ann and William had been born in New York, and James Nicoll in 1786, another James (b. 1784) having died in infancy. He left Scotland on Whitsunday 1798 and by 4 June was at Bristol, where another son, John Loudon Jr., was born on 29 September. Later, when he was publicly interrogated about the origins of his system of road making, he was deliberately vague about the next period in his life, claiming that he had no occupation during the eighteen years from 1798. However, on 21 November he had travelled on to Flushing, opposite Falmouth in Cornwall. An account published in Ayrshire soon after his death, where he still had relations, stated that he had been the superintendent of the Victualling Department for the west of England, but no record has been found of such an agent nearer than Plymouth. It is supposed, rather, that he was a prizemaster at Falmouth, organising the sale of ships captured during the Napoleonic war. He returned to Bristol in June 1802, during the temporary Peace of Amiens, and was based there for the following twenty-three years. Here he started again in public affairs, petitioning to be made a freeman in 1805 (though apparently not taking it up), becoming a Commissioner for Paving in 1806, chairman of the Committee of Citizens which agitated in 1816 for an improved gaol, and a commissioner of the Bristol turnpike roads.

John Loudon McAdam

In addition to this and his private occupations, in the fifteen years from 1798 he claimed to have travelled 30,000 miles at his own expense, gathering information about road making and administration. He found that there was a great deal of prodigality, which he castigated as worse than carelessness, and estimated that one-eighth of the money spent on roads in the country at large was misapplied, and more in London. By 1811 he was not always in good health and had been inclined to give up his quest, but was reinvigorated by a request from Sir John Sinclair, the President of the Board of Agriculture, to give information to a select committee of the House of Commons. His evidence was printed as an appendix to the committee's report and McAdam later published an edited version as his *Remarks on the Present System of Roadmaking*. Later editions were published in the United States and Germany.

He proposed 'to do what never yet has been done, to consider the making of the form and surface of roads scientifically'. This claim was somewhat inflated, as Robert Phillips had read a paper on the subject to the Royal Society in 1737 and Pierre Tresaguet in France had presented a memorandum on the subject in 1775. Already Thomas Telford and James Paterson in Scotland and Benjamin Wingrove (qq.v.) in England were putting Tresaguet's ideas into practice. McAdam, though, would have argued that the form was 'unscientific' insofar as it relied on a foundation of large stones, carefully laid and infilled, and a convex surface. He specified two layers of broken stone, which he expected to 'unite by its own angles, so as to form a solid hard surface'. He laid down in some detail the equipment and method to be used in breaking the stone, which had to be sufficiently small to pass through a screen of 1 in. opening. The surface should be level and not bound with any mixture of earth which might pass water through to the formation level or retain it and so damage the road. With the need for good side drainage and for care by trained roadmen he would have agreed.

His opportunity came on 15 January 1816 when, their roads having been indicted and McAdam having surveyed and reported on them, his colleagues on the Bristol Turnpike Trust elected him as their General Surveyor. He was now responsible for 149 miles of road and he managed to reduce expenditure by 9% in the first year and a further 6% the next, while improving their state. Thereafter, expenditure rose to its former levels, but considerable improvements were put in hand. Much of the reduction may be accounted for by the drop in labouring wages at the end of the Napoleonic Wars but his methods also played a part, as he was able to repeat the effect when he took over other trusts in later years. It may be noted too that the expenditure per mile in Bristol was significantly higher than the national average, as the roads were comparatively heavily trafficked.

McAdam believed that the general surveyors had to be 'persons of education, talents and be in the status of a gentleman. Without these qualifications, he can neither have the confidence of the Commissioners, nor have due weight in repressing abuses and controlling the working surveyors'; they should also be full time, 'so that trustees could demand an account of the manner in which their orders were carried into execution'. This concern with administration was a major theme of McAdam's programme, which he spread with evangelistic zeal. He also advocated amalgamation of trusts in order to have an adequate income from which to pay properly trained officers. At Bristol, already one of the larger trusts, he replaced fifteen sub-surveyors, whose annual salaries had amounted to £672, by ten only but he claimed to have paid them £100 p.a. each in order to attract competent people. Wherever possible he paid the roadmen by piecework rather than day rates, which latter were often little more than a kind of poor relief and correspondingly unproductive.

From Bristol his fame spread and, with the active support of Lord Chichester, the Postmaster General, nine trusts in Sussex were merged and a general surveyor who had worked at Bristol was appointed with a salary of £150 p.a. He was so successful that this was doubled after one winter's work. In Scotland, where the road trustees were generally organised on a county basis, the appointment of McAdam's nephew to the 253 miles of road in Midlothian in 1819 at the unprecedented salary of £700 p.a. was less well received. It was not renewed at the end of its term

of two years, more as a reaction to an incomer than because of any poor performance.

This appointment was part of a deliberate policy by McAdam to install members of his family, whom he had trained, as general surveyors, in order to promote his methods. His son, James, was the first, at Epsom in December 1817. Then in 1818 James added seven more trusts and his elder brother, William, took on his first seven. Altogether three sons, a nephew, four grandsons and a cousin by marriage managed more than 3,700 miles in one-hundred and forty-four trusts, before the last of them, William Jr., died in 1861.

An early setback occurred at Bath, where McAdam was asked in late 1816 to report. He spent a month with his son, William, whom he had brought from Scotland for the purpose, and a small team worked almost continuously on a survey. His report was accepted but when it came to a vote for the post of General Surveyor, one of the trustees, Benjamin Wingrove, was narrowly preferred to McAdam. Wingrove was an exponent of Tresaguet's methods and in 1821 published *Remarks on a Bill ... for regulating turnpike roads, in which are introduced strictures on the opinions of Mr. McAdam*. Like McAdam he took on multiple turnpike trusts, by 1823 including those at Taunton, Chippenham, Trowbridge and Bradford-on-Avon, and employed his son to assist him at Bath. However his administrative ability was less good and in 1826, after three years of internal opposition at Bath, he resigned and was succeeded by McAdam.

The aggregate salaries which the McAdam family drew from these activities seem very large by the standards of the day but they were intended to cover expenses such as travelling and, as the number of engagements grew, the costs of assistants to give the close attention which the general surveyor himself could not. In 1823, James drew £3,479 from thirty-nine trusts, covering 858 miles of road. In time this gave rise to occasional criticism, as some of the assistants were less well trained or grew more lax under infrequent supervision, and standards fell. In order to reduce travelling the family seems to have been geographically selective in accepting appointments. In 1824 McAdam had a house in Keswick, from which he visited a group of trusts in that area, and he persuaded the owners of the lead mines around Alston to combine 130 miles of road into one trust.

McAdam claimed that he was not an engineer and, in the sense that he designed no bridges or other great works for his roads, that may be so. But by any modern definition of the term he may rank as one. He derived his road specification from empirical yet systematic observations, and improved it in the light of experience. His initial insistence on a flat surface became a camber of 1 in 30, and the maximum size of the broken stone he used increased from 1 in. to 2½ in. By definition, his methods existed in places beforehand, but like so many men whose names have become inextricably linked with some development, it was his persistence, amounting at times to obstinacy, that led to the general adoption of his ideas. His control of finances was much superior to that of Wingrove who most closely rivalled him as a turnpike surveyor. McAdam's roads were less long-lasting than those of Telford but at a time when about £1½ million was spent annually on road maintenance, they were more economic yet adequate when maintained in accordance with his system. His promotion of the ideal of amalgamation in order to improve purchasing power and reduce administrative costs sounds decidedly modern, but in the face of vested interests was less successful than he had hoped. Probably the most successful scheme of this kind was the combination of fourteen trusts in Middlesex under the Metropolis Roads Act of 1826. James McAdam was appointed Surveyor to this, the richest of all of the trusts and an exemplar to all of the other trusts in London.

In 1824, McAdam surveyed a line for a railway from Bristol to London which would have passed through Wantage and Wallingford, much along the line which was chosen ten years later though going directly across the Cotswolds from Bristol to Wootton Bassett. It was only eight years since he had first been appointed at Bristol, but already a new system was developing which would deprive the turnpike roads of much of their long distance traffic and therefore of their income. The family does not seem to have had any further involvement with railways, staying with the turnpike roads which outlasted them. James McAdam in particular was an active protagonist and able administrator in his father's mould, and his work was recognised by a knighthood in 1834, an honour which his father had earlier declined.

From 1825 McAdam's principal home was Montague House, Hoddesdon, Hertfordshire, but he was accustomed to spend the summer in Scotland, where he was also employed. It was at Dumcrieff House, near Moffat, Dumfriesshire, that he died on 26 November 1836. Despite three grants of £2,000 each from Parliament in return for 'his great exertions and very valuable services', he was not a wealthy man when he died, and his widow later said that he left her only the amount of his salaries due.

P. S. M. CROSS-RUDKIN

[Accounts and Papers, 1805, HLRO; *Select Committee on the Highways of the Kingdom* (1819) Report and Minutes of Evidence, HC509, 17–37; *Select Committee on the Turnpike Roads and Trusts in England and Wales* (1820) Report and Evidence, HC 301; *Commissioners for Metropolis Turnpike Roads*, 1827–; *Select Committee on Turnpike Returns* (1833) Report and Evidence, HL 422; *Gentleman's Magazine*, Jan. 1837; R. H. Spiro (1950) *John Loudon McAdam, Colossus of Roads*, Ph.D. thesis, Edinburgh; *VCH Wiltshire iv*; L. A. Williams (1975) *Road Transport in Cumbria in the Nineteenth Century*; W. J. Reader (1980) *Macadam*; D. McClure (1994) *Tolls and Tacksmen*; DNB]

Publications

1811. *Observations on the Highways of the Kingdom*, reprinted with an addition 1816
1816. *Remarks on the Present System of Roadmaking … with a View to … the Introduction of Improvement …*, and eight further editions, with revisions and additions, to 1827
1817. *Memorial on the Subject of Turnpike Roads*, reprinted with revisions 1818
1819. *A Practical Essay on the Scientific Repair and Preservation of Public Roads*
1825. *Observations on the Management of Trusts for the Care of Turnpike Roads*
1828. *Plan of the Present and Proposed Road between Edinburgh and Newcastle by Jedburgh*
1829. *Report … to the Expense of the Survey of the Carter Fell*

Works

Surveyor to the following Trusts:

1816–1836. Bristol
1818–1823. Reading
1823–. Cockermouth–Penrith
1824–. Ambleside
1824–. Carlisle–Eamont Bridge
1824–1836. Alston
1824–1836. Heronsyke & Eamont Bridge
1824–1836. Milnthorpe & Levens
1825–1836. Carse of Gowrie
1826–1836. Bath
1830–1832?. North Queensferry & Perth
1832–1836. (Northern District)
1829–1836. Perth & Dunkeld
1831–1836. Perth & Crieff
1832–1836. Kinclaven
1833–1836. Perth & Blairgowrie
1833–1836. Whitehaven
1833–1836. Appleby & Kendal
pre-1834–1836. Haslingden & Todmorden
pre-1834–1836. Liverpool, Prescot & Warrington
pre-1834–1836. Rochdale & Burnley
pre-1834–1836. Sedbergh
pre-1834–1836. Abernant & Rhyd-y-blew

For surveyorships held by other members of the family, see Reader (1980).

McCRAKEN (McCRACKEN), John (fl. 1789–1825), was a mason of Dumfries who, like other tradesmen of that area and period, developed further skills. The most famous was, of course, Thomas Telford (q.v.), but the joiner, Thomas Boyd (c. 1753–1822), became an architect and the cabinet maker, Walter Newall (1780–1863), an architect and civil engineer. (For the latter two, see Colvin.)

McCraken designed and built the parish church of Kirkpatrick–Juxta in 1798–1800 (remodelled 1875–1877) and the Roman Catholic Pro-cathedral of St. Andrew, in Dumfries, in 1811–1813 (enlarged 1842–1843 and 1871–1872; destroyed by fire in 1961), as well as designing a church at

Kirkmichael (built 1813–1815). He also built the new courthouse and gaol at Dumfries, to plans by Boyd.

As a bridge builder he was more active in Kirkcudbrightshire than Dumfriesshire. When in 1798 it was alleged that Samuel McKean was not building Glenlochar Bridge to specification, the Commissioners of Supply asked McCraken to report, and part of the work was redone as a result. At Tongland Bridge, McKean was later the contractor who had to relinquish the work when his centres were swept away by flood in 1804.

McCraken was able to build bridges of moderately large span, his Ramhill Bridge (now bypassed) having a single arch of 72 ft. In 1805 he was the successful tenderer for rebuilding Eden Bridge, Carlisle, to Thomas Boyd's design, at prices up to £11,960 for various widths of road. Unfortunately for McCraken, the work was suspended late that year while the magistrates sought a Government grant and when work was resumed in 1812, it was Paul Nixon (q.v.) who obtained the work. McCraken continued to tender for bridge work, being unsuccessful at Newton Stewart in 1812 and was one of those consulted by the County in 1823 about the bridge over the river Dee at Airds, on the road north from Kirkcudbright.

P. S. M. CROSS-RUDKIN

[Commissioners of Supply minutes, Dumfries & Galloway RO; CQ AB/8/1, Cumbria RO; J. Gifford (1996) *Buildings of Scotland: Dumfries and Galloway*]

Works

1789–1790. Dee Bridge, Clatteringshaws, contractor
1798–1800. Ramhill Bridge, Urr, designer/contractor, for £550
1823–1825. Threave Bridge, Kelton, contractor, for £2,880

McILQUHAM, James (fl. 1788–1802), mason and bridge builder, was living at Old Kilpatrick, Dunbartonshire in 1788 when he was awarded the contract nearby for a canal lock and culvert at Dalnottar. The lock operated well for a year but later leaked badly, though there is no record that McIlquham was actually required to carry out repairs. He was one of several contractors who followed Robert Whitworth (q.v.) south to the Leeds & Liverpool Canal, working at first with his earlier partner Robert Watson and then with James Porteous. They were delayed in their contract for the masonry lining of Foulridge Tunnel and claimed £50 for their losses. The company must have been satisfied with their work as payment of £100 was made.

Porteous was asked by John Rennie (q.v.) in April 1793 to bid for construction of the great Lune Aqueduct on the Lancaster Canal, but lost the work to Alexander Stevens (q.v.). In 1795, McIlquham's was the lead name with Porteous in the successful tender for the Avoncliff and Semington Aqueducts on Rennie's Kennet & Avon Canal. They later also won the contract for the elegant Dundas Aqueduct.

P. S. M. CROSS-RUDKIN

[BR/FCN/1/12, Scottish RO; RAIL 842/2 and 846/4, PRO; L. J. Dalby (1971) *The Wilts & Berks Canal*]

Works

1788. Forth & Clyde Canal, contractor (with Robert Watson) for lock 37, £8 10s per rood
1791–1794. Leeds & Liverpool Canal, contractor (with Robert Watson) for three locks and a bridge at the west end of the summit, 8s 6d per cubic yard; then with James Porteous for bridges nearest Holme Bridge, and masonry in Foulridge Tunnel
1795–1799. Kennet & Avon Canal, contractor for Avoncliff, Semington and Dundas Aqueducts and bridges in lots 11–13
1799–1802. Wilts & Berks Canal, contractor for locks and bridges

McINTOSH, Hugh (1768–1840), civil engineering contractor, was born on 4 December 1768 in one of the hamlets named Kildrummie, Nairn, Scotland, the second son of David McIntosh (c. 1730–1790) and Margaret Tolmie. After a brief education at Inverness Academy, he began his working life as a navvy on the Forth & Clyde Canal and, under the engineer Robert Whitworth (q.v.), the Leeds & Liverpool. McIntosh's first contracts were awarded for work on the Lancaster Canal for John Rennie (q.v.), an engineer with whom he was associated to their mutual benefit for many years. Commencing in 1795, he and two partners in the next four years carried out six contracts on this canal for which the total payment was £14,294. In 1797, he is said to have taken the first of approximately a dozen contracts on the Grand Trunk Canal which eventually entailed work on the Leek, Burslem, Caldon and Lawton branches and Rudyard Reservoir. Probably as a consequence of new-found financial security, he married Mary Cross (1773–1833), the daughter of a farmer and agent of the canal company. At Cheddleton, Staffordshire, on the Caldon Branch of the Grand Trunk, their only surviving son, David (1799–1856), was born. Probably the single most important advance in his career occurred in 1803 when he received the first contract for the construction of the East India Company's Import and Export docks which by 1807 had earned him £65,634; the efficiency with which he organised traditional methods for the excavation of these docks especially impressed his contemporaries. Success with this assignment was the forerunner of twenty-one years of work in London's docks including the London Dock Company's Eastern Dock and a timber dock (Lavender Pond) for the Commercial Company. An appreciation of the requirements of the burgeoning population of East London was no doubt a factor in prompting McIntosh to engage in speculative building in the form of housing for the working classes in Commercial Road, Stepney, Bow and Bromley which at the time of his death was valued at £57,000.

It was in the early years of nineteenth century that this thriving business, notwithstanding its expansion into new fields of activity and parts of the country, came to be based in London where McIntosh took up residence in Charlotte Street, Bloomsbury, and later, 38 and 39 Bloomsbury Square. It was at these addresses that his clerks, in a frequently overlooked and underestimated aspect of contracting, compiled accounts which, on the evidence of the schedules presented to Chancery in the suit against the Great Western Railway of the 1840s, recorded building expenses down to the level of lamp oil and brooms.

London and its suburbs provided an abundance of opportunities for work by reputable contractors and some of that carried out by McIntosh included facilities of considerable benefit to the general public. He built sewers in Rotherhithe, Deptford and Southall (1835–1837), a bridge over the river Brent at Brentford (1824), and improvements to Vauxhall Bridge Road (1816) and Archway Road, Highgate (1810–1813). The most prestigious project for the public good which he undertook in London was, however, a programme of repairs and alterations to Blackfriars Bridge. In carrying out the terms of contracts of March 1834 and December 1835, he provided new parapets, renewed stonework in piers and protected these with wooden piles at a cost of £97,632. His services were secured by East London and Chelsea water companies for pipe laying, and more exalted patronage followed with commissions to carry out the same operation for Windsor Castle, Hampton Court, Kensington Palace and Brighton Pavilion. Developments in the commercial applications of new technology brought business from the Gas Light & Coke Company for the construction of retorts and installation of mains. Outside London, the expansion of the gas industry resulted in contracts to lay mains in Bristol (1817) and Shrewsbury (1823) and to construct and equip an entire gas works at Carlisle (1819).

The demands of the capital and new forms of industrial activity did not, however, lead to the abandonment of the type of operation which laid the foundations of his success. Canal construction and improvements continued to occupy a central place in McIntosh's business commitments. Thus he made valuable contributions to the building or improvement of the Regent's, Thames & Severn (1831–1822), Gloucester & Berkeley (with Sharpness Dock) (1822–1827), Great Western (1831–1835) and Grand Junction (1836–1838) canals and, in the final years of his life, the Aire & Calder Navigation; this last-mentioned programme alone earned him £174,806. In a letter of 1817 to Thomas Telford (q.v.) in which he sought work on the Edinburgh & Glasgow Union Canal, he also claimed to have played a significant part in the construction of the Thames & Medway,

Kennet & Avon, Stainforth & Keadby and Croydon canals. McIntosh's growing experience, resources and reputation in the early years of the century made it almost inevitable that the government would call upon his services for operations of national importance. Dockyard works of considerable magnitude became his responsibility at Pembroke (1813–1816), Plymouth (1823–1832) and Portsmouth (1824–1832). The Plymouth assignments, in particular, were onerous since they involved thirty-one contracts with a combined value in excess of £300,000. In an unusual episode in his life, he is believed to have accompanied the disgracefully organised Walcheren Expedition of 1809 in order to supervise the dismantling of the fortifications of Flushing in December of that year. If this is so, it probably makes McIntosh the first British contractor to work abroad, albeit in a negative role.

The volume and diversity of work carried out by this outstanding contractor preclude exhaustive discussion of his achievements in the space available here, but mention should be made of the improvements to the harbours at Shoreham (1817–1821), Whitstable (1831–1832) and Dover (1826–1829), and the sea defences at Dymchurch (1820–1832) and Gillingham. The construction of Junction Dock (1826–1829) provided Hull with its third enclosed dock. Although McIntosh was never a specialist bridge builder, he widened Clopton Bridge, Stratford-on-Avon, repaired the bridge at Rochester and built Folly Bridge over the Isis at Oxford and South Bridge, Northampton. The most elegant of his surviving bridges is that designed by Telford which he erected under the supervision of William Mackenzie to span the Severn at Tewkesbury (1823–1826). The contracts awarded to McIntosh early in 1834 on the London & Greenwich and Grand Junction (Birmingham to Warrington) railways marked the start of the most substantial transfer of resources and experience into the new form of transport by any contractor whose reputation and prosperity had been built on work in other branches of civil engineering. He was no doubt motivated by an assessment of the railway's potential to generate new business and thus to ensure the survival, under the leadership of his son, of the establishment he had founded. Advancing years and his deteriorating eyesight meant that his transfer of authority could not be indefinitely delayed. David McIntosh had assisted his father on several projects, including the construction of the Gloucester & Berkeley Canal, but had never been responsible for duties as onerous as that which came in February 1834 when he was put in charge of building Dutton Viaduct, the great viaduct in the Weaver Valley, for the Grand Junction Railway. In the event, David's precarious state of health led to retirement from business affairs shortly after his father's death.

It is not surprising that Hugh McIntosh did not make his first foray into railway construction at an earlier date since the Stockton & Darlington and

Liverpool & Manchester railways were built in small 'lots' or contracts which can have had no attraction for those who were accustomed to handling major assignments. Having entered the field, he competed with enthusiasm and in working for nine railway companies before his death in 1840 carried out contracts for which the tender price or actual payments amounted to approximately £1,265,000. The most valuable of these commissions were: for the Great Western Railway, the lines from Acton to Hanwell, and Brislington (Bristol) to the outskirts of Bath for which the company originally paid £221,148; for the North Midland Railway, the Altofts, Methley, Milford, Belper (with Thomas Jackson), Duffield and Derby contracts at a total tender price of £341,071; for the Midland Counties Railway, the line between Leicester and Rugby with a tender price of £258,629; and on behalf of the London & Southampton Railway, for bridges and culverts between London and the River Wey and for 'works below Winchester', he was paid £177,834. His other railway assignments included the Kirkthorpe Contract of the Manchester & Leeds Railway, the Northern & Eastern Railway's line from London to Broxbourne, and, for the London & Greenwich, 'the first railway in London', the 878 arches in the viaduct and bridges which comprised its original line. One of the few major English railway schemes of the 1830s on which he did not seek work was the London & Birmingham Railway.

Hugh McIntosh's final years were marred by an acrimonious dispute with Isambard Kingdom Brunel over payments for extra work on the Great Western Railway; McIntosh demanded £265,557 in addition to a similar sum which the company had already paid him. In the face of the rising costs of his line, Brunel opposed any settlement that would have satisfied McIntosh. After Hugh McIntosh's death, the matter went to Chancery and was pursued in a suit by David McIntosh and Timothy Tyrrell, the solicitor of, and occasionally surety for, the contract agreements of both McIntoshes. The matter was not settled until 1866 when their heirs (all the original litigants having died) received an award (£120,639) which substantially upheld Hugh McIntosh's original claim. Although Sir John Rennie once described him as having the 'character of being a very litigious person', it has to be noted that he worked on perfectly satisfactory business terms with many engineers, including Ralph and James Walker, Thomas Telford, George and Robert Stephenson, Joseph Locke and Charles Blacker Vignoles, and, apart from a bill before the Court of Exchequer against the Midland Counties Railway, this was the only occasion that his differences with a railway company could not be settled without recourse to law. The opposition of the young and inexperienced engineer must have come as an exasperating surprise to McIntosh in view of his leading position amongst the contractors of his time and widely-acknowledged reputation for probity.

On 30 August 1840 Hugh McIntosh died of apoplexy in a hotel in Wakefield, West Yorkshire, when inspecting the state of his works on the Aire & Calder Navigation and Manchester & Leeds and North Midland railways. It was his achievement to be amongst the first civil engineering contractors to have created a national organisation that could operate efficiently and, if necessary, simultaneously in several branches or skills of the profession. By the early years of the century his business had outgrown the management abilities of any single person. Thus many contracts were taken in the name of David McIntosh and above all, he was compelled to rely on the services of a number of agents. Amongst those who at one time or another occupied this position were James Leishman, the signatory on his behalf of the Derby Contract of the North Midland Railway, Edward Ladd Betts, his resident supervisor at Dutton Viaduct, William Betts, the negotiator of the terms of his contract with the Midland Counties Railway, James McIntosh, his brother and agent on the London & Southampton works, and Hugh Ross, William Mackenzie, William Henderson and William Radford. The last-named worked for him for 35 years and when asked by counsel in the suit against the Great Western to describe his duties replied that a proficient agent had to be able to estimate costs, produce plans and sections, supervise accounts on a day-to-day basis 'in fact to do whatever the Contractor would have himself to do with reference to the contract'. These men formed the essential foundations of any large business enterprise and their quality was a guarantee of the reliability required by companies and their shareholders. After 1840, some of them took the experience which they had gained under Hugh McIntosh forward into the next phase of British civil engineering and contributed to the success of other contractors, notably, to that of William Mackenzie.

In the second half of the nineteenth century whilst many contractors exceeded Hugh McIntosh in the value of the work which they had in hand at any particular moment, few equalled him in the diversity and geographical range of their construction operations within Britain. Amongst his obituaries was that which appeared in the press of the town in which he had died. 'Mr. McIntosh', wrote the *Wakefield Journal*, 'was one of that class of men whose energetic perseverance and natural talents raised him from the situation of an ordinary labourer to become one of the noted men of the time in which he lived, having undertaken to complete by contract some of the most stupendous works of the civil engineer ever contemplated in this kingdom'. The value of his personal estate was assessed at £300,000, to which one might add the amount of the final award against the Great Western. He was buried in the family vault at St. Matthias Church, Poplar.

DAVID BROOKE

[**PRO**: Aire & Calder Navigation—Undertakers' Minutes 1820–1841, Reports of the Auditor and Chief Clerk 1825–1853, Reports and Accounts 1834–1839, RAIL 800/10,/34, RAIL 1113/19–24; Gloucester & Berkeley Canal—Proceedings of the Company 1818–1827, Contracts for Construction, 1794–1826, Indenture for Completion by H. McIntosh 1823, RAIL 829/5,17,23; Grand Junction Canal–Directors' Minutes 1831–1840, RAIL 830/5,6,7; Lancaster Canal—Committee Minutes 1792 1813, Reports to the Canal Committee by Millar, Engineer, 1793–1797, and by Gregson and Eastburn 1794–1798, Copies of Articles of Agreements 1793–1795, Contracts and Estimates 1801–1803, RAIL 844/230, 231, 240, 241, 247, 248; Regent's Canal—Meetings of the General Committee 1812–1814, RAIL 860/5,6; Grand Junction Railway—Directors' Minutes 1833–1837, Contract with D. McIntosh for Dutton Viaduct 1834, RAIL 220/1, 22; Great Western Railway—Brunel's Reports to the Board, 1835–1842, London Committee Letter Books 1833–43, Bristol Committee Letter Books 1833–1841, Letters of Brunel 1839–1842, Brunel Special Collection: Letters 1835–1843, RAIL 250/82, RAIL 253/85–99, 100–103, 107, RAIL 1149/2-7; London & Greenwich Railway—Directors' Minutes 1831–1841, RAIL 389/1,2; London & Southampton Railway—Directors' Minutes, 1834–1841, Ledgers Nos. 1 and 2, 1833–1843, RAIL 412/1,7,8; Manchester & Leeds Railway—Directors' Minutes, 1836–1841, RAIL 343/2,4,6,8; Midland Counties Railway—Directors' Minutes 1836–1841, Finance Committee 1836–1841, Engineer's Reports 1835–1838, Committee of Works South of the Trent, 1836–1839, RAIL 490/3,8,9,17,15,16; Northern & Eastern Railway—Directors' Minutes 1836–1840, RAIL 541/3; North Midland Railway—Directors' Minutes 1836–1842, the Leeds Committee 1835–1842, the London Committee 1835–1841, Rail 530/1,2,4,6,7,8. **The Museum in Docklands Project**: Ledger and Minute Books of the East India Dock Company, 1803–1808. **Greater London RO**: Various Agreements and Contracts for Bridges, Sewers, etc., in the London Area, MA/D/Br/12 and 46, and SKCS/121,124 and 133. **City of London Record Office**: Bridge House Committee, Plan No. 4 and Associated Papers for Repairs to Blackfriars Bridge. **ICE Archives**: Reports of John Rennie, Telford mss; Edinburgh & Glasgow Union Canal, Letter H. McIntosh to T. Telford, 3 November 1817; Frank D. Smith Files. **Secondary**: J. Lindsay (1968) *The Canals of Scotland*; C. Hadfield (ed.) *Canals of the British Isles Series*; A. W. Skempton (1978–1979) Engineering in the Port of London, 1789–1808, *Trans Newc Soc*, **50**, 87–108; A. W. Skempton (1981–1982) Engineering in the Port of London, 1808–1834, *Newc Soc Trans*, **53**, 73–96; M. M. Chrimes (1994–1995) Hugh McIntosh (1768–1840), national contractor, *Trans Newc Soc*, **66**, 175–192; D. Brooke (1996) The Equity Suit of McIntosh v. Great Western Railway, *JTH*, Third Series, **17**(2), 133–149]

Contract works

Docks and harbours in London:

1803–1807. East India excavation contracts, contractor, payment £65,634
1806. Grand Surrey, entrance, contractor
1815–1822. Commercial, timber pond, contractor
1816. East India, extension dredging, contractor, payment £1,000
1819–1820. Regent's canal dock excavations, contractor
1825–1828. Eastern dock, excavations, contractor, payment £9,300
1832–1833. West India timber pond, excavations, contractor, payment £4,000
1832–1834. Brunswick Wharf, construction, contractor

Docks and harbours outside London:

1812–1815, 1833–1836. Dover Harbour, improvement and construction, contractor, contract west pier c. £19,300
1813–1816, 1830–1832. Pembroke Dockyard, improvement, contractor, payment c. £133,000
1817–1821. Shoreham Harbour, improvement, contractor, £58,000
1820–1832. Dymchurch sea defences, improvement, contractor
1820–1832. Gillingham sea defences, improvement, contractor
1824–1832. Plymouth Dockyard, improvement, contractor, payment c. £300,000
1824–1832. Portsmouth Dockyard, improvement, contractor, payment c. £200,000
1827–1829. Junction Dock, Hull, construction, contractor, contract £95,000
1830–1832. Pembroke Dockyard, improvement, contractor
1831–1832. Whitstable Harbour, construction, contractor, contract £10,000
1832–1833, 1837–1838. Southampton Harbour, pier construction, contractor, payment c. £12,375
1835–1838. Black Point Lighthouse, Beaumaris, construction, contractor

Canals:

1795–1799. Lancaster, Grand Trunk, Myerscough contract, contractor, payment £14,294
1797–1800. Grand Trunk Canal
1800–1801. Thames & Medway
1801–1809. Croydon
1812–1820. Regent's, improvement, contractor
1817–1822. Edinburgh & Glasgow Union
1821–1822. Thames & Severn, improvement, contractor, payment £1,169
1823–1827. Gloucester & Berkeley, construction, contractor, contracts c. £126,807 Kennet & Avon improvement?
1831–1834. Great Western, northern extension, contractor
1833–?. Stainforth & Keadby cut to River Don, contractor
1834–1841. Aire & Calder Navigation, improvement, contractor, payment £174,806

1836–1840. Grand Junction, improvement, contractor, payment c. £12,253

Bridges:

1814–1816. Clopton, Stratford-on-Avon improvement, contractor
1815–1816. South, Northampton construction, contractor, contract £9,600
1821. Rochester improvement, contractor, contract £8,578
1823–1826. Mythe, Tewkesbury, construction, contractor, contract c. £10,000
1824. Brentford, construction, contractor
1824–1825. Harmondsworth, construction, contractor
1827. Folly, Oxford, construction, contractor
1834–1836. River Itchen, Southampton, construction floating bridge, contractor
1834–1841. Blackfriars, London, improvement, contractor, payment £97,632
1838–1840. Portsmouth, landing for floating bridge, contractor, payment c. £6,800

Gas, water, roads, housing, etc.:

1801–1817. Chetney Hill Lazarette, Sheerness, excavation contract, dock and canal, contractor, payment £38,354
1807–1808. East London Waterworks, reservoirs
1809. Flushing Harbour, demolition
1810–1813. Archway Road, Highgate, London, improvement, contractor
1812, 1827–1832. Regent's Park, improvements, contractor, payment £11,706
1815. Gas Light & Coke Company, London, gas mains and equipment, contractor
1816. Vauxhall Bridge Road, London, improvement, contractor
1817. Bristol, gas mains, contractor
1819. Carlisle, gas mains and equipment, contractor
182?. East London water mains, contractor
182?. Chelsea water mains, contractor
182?. Windsor Castle, Hampton Court, Kensington Palace and Brighton Pavilion, water mains, con-tractor
1823. Shrewsbury, gas mains, contractor
1823–1828. British Museum, excavation contract, contractor
1825–1826. Buckingham Palace, excavation contract, contractor
1825–1835. Commercial Road, Stepney, Bow, Bromley house construction, contractor
1827. St. James's Park, improvement, contractor, contract £8.300
1827. Melville Hospital, Chatham, construction, contractor
1829. Hampton Court, water gallery, excavation contract, contractor
1829–1831. State Paper Office, excavation contract, contractor
1832–?. Windsor Castle, improvements, contractor
1835–1837. Rotherhithe, Deptford, Southall, Nine Elms sewer construction, contractor

Railways:

1819–1822. Plymouth–Dartmoor, light railway, construction, contractor, £28,000
1831–1832. Canterbury & Whitstable, harbour at terminus, contractor, contract £10,000
1834–1837. Grand Junction, viaduct, construction, contractor, contract £54,440
1834–1838. London & Greenwich, construction, contractor, payment c. £139,400
1836–1841. Great Western, construction contracts, contractor, contracts £160,953
1836–1842. London & South western, construction contracts, contractor, payment £177,834
1837–?. Birmingham & Derby Junction, construction contract, contractor
1837–1840. North Midland, construction contracts, contractor, payment c. £26,000
1837–1840. Manchester & Leeds, construction contract, contractor, contract £46,000
1837–1840. Midland Counties, construction contracts, contractor, contract £258,629

MACKELL, Robert (fl. 1762–d. 1779), is remembered as the resident engineer on the Forth & Clyde Canal from 1768 until 1779 but, according to a note made by John Smeaton (q.v.) in 1767, he was already well known for his engineering abilities, as a millwright in contemporary terminology. His works included a steam engine erected by himself and James Watt (q.v.) for Lord Kennet about 1766. They also proposed in December 1766 to erect a steam engine for the Carron Company. In 1767 Mackell built a horse engine and pumps (these designed by Smeaton) for dewatering the dry dock at Port Glasgow and in March 1768 he and the millwright, Andrew Meikle (q.v.), took out a patent for a machine for dressing and cleansing corn.

Mackell's connection with the canal began in 1762 when he and James Murray were employed by Lord Napier of Murchiston to survey a route linking the Forth, near Grangemouth, to the Clyde at Yoker, 6 miles below Glasgow. Smeaton was then asked to report on the proposal, at that time for a canal 40 ft. wide with a 5 ft. depth of water. At the request of the Glasgow merchants, Mackell and Watt surveyed a revised route early in 1767, bringing the west end of the canal to the Broomielaw; it was intended to be 4 ft. deep and 24 ft. wide. A Bill went before Parliament in March but proposals for a canal of much larger dimensions having received strong support, the Bill was dropped and Smeaton was again asked to report. He and Mackell carefully reviewed a modified course from Grangemouth to the Clyde at Dalmuir (2 miles downstream of Yoker) with a branch to Glasgow. To determine the nature of the ground Mackell made at least fifty-eight borings along the line to a depth of 12 ft., and Smeaton reported in October 1767. His plans met with approval. He and Mackell gave evidence on the new Bill in February 1768 and the Act passed on 8 March. The canal was now to be 48 ft. wide and 7 ft. deep with a large reservoir to supply water to the summit level.

A week later Smeaton was appointed Head Engineer at a salary of £500 and Mackell became Sub-Engineer at £315 p.a. These titles were soon changed, to Engineer-in-Chief and Resident Engineer. Alexander Stephen was appointed Clerk & Overseer under Mackell at £52 p.a. (raised in 1770 to £70) and John Laurie was employed as a free-lance land surveyor. Soon afterwards Smeaton and Mackell were paid £372 and £306 respectively for work done before their appointment.

Mackell and Laurie marked out the ground to be purchased. In June members of the management committee went with Smeaton and Mackell to view the east end of the canal as now staked out. The first two contracts were let in July. By November four contractors for digging were at work and two for masonry. Superintendents for masonry and carpentry were appointed. Smeaton sent in general plans for locks, aqueducts and cross sections, explaining that 'as the resident Engineer has proved himself to me and I trust in like manner to the Committee as a man of Judgement, Diligence and Ingenuity in the application of material put into his hands' it would not be necessary to send particular plans for each site unless specially requested.

Work proceeded westward from Grangemouth. By May 1770 over a thousand men were employed. Before the end of 1771 twenty locks (74 ft. long between gates with a clear width of 20 ft.) had been built up to the summit level, 156 ft. above sea level, and work started on the earth dam for the reservoir.

Meanwhile, in May 1770, Mackell submitted plans for a revised route for the canal west of Kirkintilloch, 19 miles from Grangemouth. It deviated south to Stockingfield, much nearer to Glasgow, and would then cross the river Kelvin on a large aqueduct, proceeding after 2 miles to join the original line to Dalmuir. He had found this route during a reconnaissance in 1768. It was pointed out to James Brindley and Thomas Yeoman (qq.v.) on the occasion of their rather curious intervention in the Forth & Clyde scheme in September 1768. At that time Smeaton rejected it, partly on grounds of cost. But on a site visit in July and August 1770, after careful consideration, he accepted that the advantage outweighed the extra cost and supported the plan both in committee and in Parliament, where the necessary Bill for the re-alignment was passed in March 1771. Concluding his report on the matter in August 1770 he wrote: 'In the whole of this Business Mr. Mackell has shown a great deal of Judgement and attention to the Interest of the Company. And I have the further pleasure to acquaint the Company that, having inspected the Works now going on, that they are in general and in things the most material, exceedingly well executed and much to my satisfaction'.

By 1772 the dam for Townhead reservoir had been completed and in September 1773 the canal was opened as far as Kirkintilloch. At this stage Smeaton retired as Engineer-in-Chief and Mackell took full charge, still with his £315 salary. He extended the summit level to Stockingfield, 26 miles from Grangemouth, by 1775 and the 2 mile branch from Stockingfield to Hamilton Basin in Glasgow in September 1777. Total expenditure up to March 1778 amounted to £164,000.

At times during the last five years Mackell displayed a rather quarrelsome nature and the fact that in most cases his complaints were shown to be justified did not improve relations with some of his colleagues and committee members. Nevertheless, despite the practical cessation of work due to the Company's funds being exhausted, he was retained until his death in Glasgow on 8 September 1779. Nicol Baird, the father of Hugh Baird (q.v.), was appointed Surveyor at a salary of £100 p.a. with responsibility for maintenance and it was not until 1785 that under Robert Whitworth (q.v.) work resumed on the canal westward from Stockingfield. A survey by Baird and Laurie showed a better terminal at Bowling, 3 miles downstream of Dalmuir, but otherwise the route followed Mackell's 1770 plan. The canal, including the great Kelvin Aqueduct, was finally opened throughout in 1790.

It remains to add that besides his work on the canal Mackell carried out two small commissions for the city of Glasgow: a plan for deepening the River Clyde between the old Glasgow and the Broomielaw bridges, and directing operations of taking down and rebuilding one of the arches of the old bridge to improve its ascent. For the former he received a fee of £33 in 1778; for the latter £25 in 1779.

Finally, two points of interest: Mackell was the first person to have the title Resident Engineer, a term introduced by Smeaton precisely for this post on the Forth & Clyde Canal, and his salary of 300 guineas (£315) was exceptionally high for the period. It reflected Smeaton's insistence on the necessity of having an experienced, able engineer in day-to-day charge of a work of such magnitude.

A. W. SKEMPTON and RON BIRSE

[Minute Books of the Forth & Clyde Canal (1767–1/80) Scottish RO; John Smeaton (1767) *Second Report touching the Practicability and Expence of making a Navigable Canal for the river Forth to the river Clyde*; John Smeaton (1768) *A Review of Several Matters relating to the Forth and Clyde Navigation*; article on Andrew Meikle, DNB; R. Renwick (1912) *Extracts from the Records of the Burgh of Glasgow, 1760–1780*, vol. 7; H. W. Dickinson and Rhys Jenkins (1927) *James Watt and the Steam Engine*; Jean Lindsay (1968) *The Canals of Scotland*; Jean Lindsay (1968) Robert Mackell and the Forth & Clyde Canal, *Transport History*, **1**, 285–292; A. W. Skempton (1981) *John Smeaton*]

Publications

1767. *An Account of the Navigable Canal proposed to be cut for the River Clyde to the River Carron* (with James Watt)

1770. *Plan of the Great Canal from the Forth to Clyde with the new intended line of Alterations*

Works

1768–1779. Forth & Clyde Canal, resident engineer 1768–1773, engineer 1773–1779, £164,000

MACKENZIE, William (1794–1851), civil engineer and contractor, was born at Marsden, now Nelson, Lancashire, to Alexander Mackenzie, a small contractor on the Leeds & Liverpool Canal. Having served an apprenticeship under a lock carpenter on this canal, he assisted John Clapham, Resident Engineer, with the building of a dry dock at Troon Harbour. He remained in Scotland for work under John Cargill (q.v.), a masonry contractor building the abutments of Craigellachie Bridge, Banffshire, as part of improvements to communications in the Highlands. Between 1816 and 1826, Mackenzie acted as an agent for the leading contractor, Hugh McIntosh (q.v.), on both the Edinburgh & Glasgow Union and Gloucester & Berkeley canals before becoming a Resident Engineer under Thomas Telford (q.v.) for the erection of Mythe Bridge over the river Severn at Tewkesbury. Mackenzie apparently gained Telford's confidence since in his next assignment he held the same appointment from 1826 to 1832 for the supervision of extensive improvements to the line of the Birmingham Canal between Tipton and Birmingham.

A change of far-reaching importance in his career occurred in November 1832 when he began to construct Lime Street Tunnel for the Liverpool & Manchester Railway and thus be- came both a contractor and railway builder. Success in this task was followed between 1835 and 1841 by work on the Grand Junction (Warrington Contract), North Union (Yarrow Contract), Glasgow, Paisley, Kilmarnock & Ayr (Kilwinning Contract), Glasgow, Paisley, & Greenock (Bishopton and Finlayston Contracts) and Midland Counties (Leicester–Nottingham and Derby Contract) lines.

The final ten years of Mackenzie's life were dedicated to spearheading the transfer of British expertise in railway construction, and the finance which accompanied it, into Continental Europe and especially into France. In 1840–1841, in partnership with Thomas Brassey and with the support of the engineer Joseph Locke, he obtained contracts for the building of the Paris & Rouen Railway between the St. Germain Railway (Le Pecq to the Gare St. Lazare) and the south bank of the Seine opposite Rouen. This line, as completed in May 1843, was then one of only three steam-powered trunk lines for passengers and goods in the country. Success with this assignment ensured favourable consideration in 1843 when contracts were awarded for the difficult rail route between Rouen and Havre, which

entailed building both tunnels and several large viaducts. In the strategy of the developing rail network, the Rouen and Havre lines were components in the formation of a new and swift highway between Paris and London which, in England, would include the London & South Western Railway. The efficiency of Mackenzie & Brassey's agents ensured that even the loss of time arising from the rebuilding of Barentin Viaduct following its total collapse in 1846 did not delay the opening of the line beyond March 1847. The Havre railway later included a branch to Dieppe (opened August 1848) for which this partnership was also responsible. The above lines were largely financed by private capital, approximately half of which came from England, and entirely built by the private contracting partnership of Mackenzie & Brassey. The overwhelming majority of lines in France in the 1840s came into existence under the terms of the Railway Law of June 1842 which guaranteed State financial and engineering assistance for railway construction. In December 1844, Mackenzie & Brassey obtained the contract to ballast and lay the rails of the track of the Orléans & Bordeaux Railway between Orléans and Tours; this task was eventually extended to include the entire 462 kilometres of line to Bordeaux. In July 1851, as his final appearance at a major public event in France, Mackenzie attended the opening of the section from Tours to Poitiers. A second route from Paris to the Channel coast came into existence in April 1848 with the completion of the Amiens & Boulogne Railway, via Abbeville. Mackenzie & Brassey were responsible for the excavation work and track laying on this line.

Mackenzie's participation in the railways of France was not confined to their construction since he also became enthusiastically involved in railway promotion in the boom years of 1845–6, invested heavily in the shares and bonds of at least nine companies and was a Director of the Tours & Nantes and Paris & Strasbourg lines. He continued to hold these assets and appointments even after the Revolution of February 1848 had led to a general reduction in British influence in the French railway scene. Also on the Continent, Mackenzie & Brassey invested in and served on the Boards of the Charleroi & Erquelinnes, Tournai & Jurbise and Landen & Hasselt companies of southern Belgium. In Spain, they provided the engineering and contracting expertise and some finance for the first railway in that country, the Barcelona & Mataro. In addition, their agents carried out surveys for several other schemes, including a railway between the Franco-Spanish frontier and Madrid, via Bilbao.

From the mid-1840s the day-to-day management of the Continental affairs of the partnership were left very much in Mackenzie's hands whilst Brassey concentrated on British affairs. Thus, although the latter's name alone is often associated with the building of the Eastern Union Railway from Colchester to Ipswich, the Buckingham-shire, the Great Northern between Kings Cross and Peterborough, the Trent Valley from Rugby to Stafford and much of the North Staffordshire line, in fact, these were formally part of the part-nership's affairs. With the assistance of a third partner, John Stephenson, Mackenzie & Brassey also built the Lancaster & Carlisle Railway and Caledonian (Carlisle to Glasgow and Edinburgh) with its northern extensions, the Scottish Central and Scottish Midland Junction lines. The final two British lines in which Mackenzie took a close in-terest were the Chester & Holyhead, on which the partnership had two contracts on the eastern half, and the Liverpool, Ormskirk & Preston. A survey of Mackenzie's work would be incomplete with-out reference to the contracts which he carried out on behalf of the Shannon Improvement Com-missioners in order to encourage the growth of trade on the river; these included the erection of a new bridge at Banagher.

William Mackenzie died in his house in Grove Street, Liverpool, after a prolonged illness, on 29 October 1851 and was buried in the ceme-tery of St. Andrews Presbyterian Church, Rodney Street, Liverpool. In partnership with Brassey and Stephenson, he had been responsible for approx-imately 620 miles of line in Britain and had either built or made a major contribution to 520 miles in France. France, in particular, benefited from the drive and experience which he brought to the task of major line construction, and from the confidence amongst English railway investors created by the names of Mackenzie, Brassey and Locke. His assets at the time of his death were worth approximately £350,000; most of this went to Edward, his brother and close associate in railway work. Mackenzie's second wife, Sarah, outlived him but by neither this marriage nor his first was there any offspring.

DAVID BROOKE

[**PRO—canals**: Birmingham Canal, RAIL 810; Committee Minutes, 1822–1838, RAIL 810/12,13; Reports to the Committee, 1826–1837, RAIL 810/99; Ledger, 1820–1831, RAIL 810/231; Glou-cester & Berkeley Canal, RAIL 829; Proceedings of the Company, 1818–1827, RAIL 829/5. **PRO—railways**: Birkenhead, Lancashire & Cheshire Junction, RAIL 35; Directors' Minutes, 1846–1862, RAIL 35/4; Buckinghamshire Railway, RAIL 35; Directors' Minutes, 1847–1874, RAIL 35/3; Con-struction Contract with T Brassey, 1847, RAIL 35/11; Chester & Holyhead, RAIL 113; Directors' Minutes, 1843–51, RAIL 113/3,4; Contract with Mackenzie, Brassey & Stephenson for the line Ryddlan to Llandrillo yn Rhos, 1845, RAIL 113/44; East Lancashire Railway (Liverpool, Ormskirk & Preston Railway), RAIL 176; Directors' Minutes, 1844–1847, RAIL 176/5; Works Committee, 1844–1847, RAIL 176/19; Eastern Union Railway, RAIL 187; Directors' Minutes, 1844–1846, RAIL 187/1; Grand Junction Railway, RAIL 220; Directors' Minutes, 1833–1837, RAIL 220/1; Warrington

Contract with W. Mackenzie, 1835, RAIL 220/29; Great Northern Railway, RAIL 236; Directors' Minutes, 1846–1847, RAIL 236/13; Committee of Works, 1846–1848, RAIL 236/271; Lancaster & Carlisle Railway, RAIL 346; Directors' Minutes, 1844–1848, RAIL 346/1; Liverpool & Manchester Railway, RAIL 371; Directors' Minutes, 1831–36, RAIL 371/2,3; Committee of Management and Sub-Committee, 1831–1833, RAIL 371/8; Midland Counties Railway, RAIL 490; Directors' Minutes, 1836–1841, RAIL 490/3; Finance Committee, 1836–1841, RAIL 490/8,9; Committee of Works North of the Trent, 1837–1840, RAIL 490/13,14; Contracts with W. Mackenzie, 1837 and 1838, RAIL 490/19,21; North Staffordshire Railway, RAIL 532; Directors' Minutes, 1846–1851, RAIL 532/4; Reports to the Board, 1847–1868, RAIL 532/26; North Union Railway, RAIL 534; Directors' Minutes, 1834–1839, RAIL 534/2,4; Shrewsbury & Chester and North Wales Mineral Railway, RAIL 616; Directors' Minutes, 1843–1848, RAIL 616/1; Accounts Ledger, 1844–1853, RAIL 616/14; Cash Book, 1844–1848, RAIL 616/21; Trent Valley Railway, RAIL 699; Directors' Minutes and Min-utes of the Trent Valley Committee of the London & Birmingham Railway, 1845–1848, RAIL 699/2; Contract for the line with Mackenzie, Brassey & Stephenson, RAIL 699/3; *Railway Times*, 1839–1850, ZPER2/2–14; *Herapath's Railway Journal*, 1841–1847, ZPER3/3–10; *Railway Record*, 1844–1850, ZPER6/1-7; *Bradshaw's Railway Gazette*, 1845–1846, ZPER41/1-2. **Scottish RO**: Caledo-nian Railway, Directors' Minutes, 1844–1848, CAL/1/7,8; Correspondence from the General Manager and Secretary, 1846–1847, CAL/4/74; Glasgow, Paisley & Greenock Railway, Directors' Minutes, 1837–1841, GPG/1/1; Works Com-mittee, 1838–1842, GPG/1/6; Glasgow, Paisley & Kilmarnock & Ayr, Minutes of the Direc-tors, Management and Shareholders Meetings, GPK/1/1–2; Minutes of Sub-Committees, 1838–1840, GPK/1/6; **Mackenzie Archive, ICE** (a se-lection of sources): Volumes of the Diary of William Mackenzie, 1840–1850; Letter Books of W. Mackenzie; Ledger Bill Books, Nos. 1, 2 and 3; Paris & Rouen Railway Waste Book; Estimates for Work on the Paris & Rouen and Rouen & Havre railways; Plans for the Reconstruction of Baretin Viaduct, 1846, Specifications for the Track, Tours to Bordeaux, May 1847; Proceedings at the Opening of work on the Tournai & Jurbise and Landen & Hasselt railways, 1846; Inventory of the Securities belonging to the Trustees of the late W. Mackenize, 31 October 1854; Agreement between W. Mackenzie and T. Brassey for a Divi-sion of British Shares, 12 February 1850; Agree-ment between W. Mackenzie and T. Brassey as to French Work, 15 October 1850; M. M. Chrimes *et al.* (1994) *Mackenzie—Giant of the Railways*; D. Brooke (1997) William Mackenzie and railways in France, *Construction History*, **13**, 17–28]

Works

c. 1804–1808. Leeds & Liverpool Canal, locks, construction, apprentice carpenter

c. 1809–1811. Troon Harbour, dock, construction, assistant engineer

1812–1815. Highlands of Scotland, roads and bridges, construction, assistant engineer

1816–1823. Edinburgh & Glasgow Union and Gloucester & Berkeley Canals, contractor's agent

1823–1825. Mythe Bridge, Tewkesbury, resident engineer

1826–1832. Birmingham Canal, construction and improvements, resident engineer

1840. Manchester & Salford Junction Canal, tunnel, construction, contractor, contract £47,000

1840. Peak Forest Canal, improvement, consulting engineer

1841–1848. Shannon River Navigation, improvement, contractor, contracts c. £100,176

1843–1844. Marne–Rhine Canal, Einville contract, contracting partner, contract £44,000

1845–c. 1851. Greenock Harbour, dock, construction, contracting partner, contract c. £100,000

Railway building in Britain:

1832–1835. Liverpool & Manchester, Lime Street tunnel, construction, contractor, payment £51,975

1835–1837. North Union, Yarrow contract, construction, contractor, contract £76,000

1837–1840. Midland Counties, Leicester to Nottingham and Derby, construction contracts, contractor, payment £206,469

1838–1840. Glasgow, Paisley, Kilmarnock & Ayr, Kilwinning contract, construction, contractor, contract £12,000

1838–1841. Glasgow, Paisley & Greenock, Bishopton and Finlayston contracts, construction, contractor, payment £164,837

1844–1846. Eastern Union, Colchester to Ipswich, construction contract, contracting partner, payment £171,010

1844–1848. Lancaster & Carlisle, line and stations, construction, contracting partner, payment £928,857

1844–1848. North Wales Mineral and Shrewsbury & Chester, contracting partner, contract c. £500,000

1845–1847. Trent Valley, contracting partner, contract £611,619

1845–1847. Kendal & Windermere contracting partner, contract £98,000

1845–1848. Chester & Holyhead, contract Nos. 2 and 3, contracting partner, contracts £361,753

1845–1848. Caledonian, construction partner, contract £1,275,000

1845–1848. Scottish Central and Scottish Midland Junction, contracting partner, contracts

1845–1849. Clydesdale Junction and Garnkirk, contracting partner, contract £349,000

1846–1847. Brikenhead, Lancashire & Cheshire Junction, planned construction, contracting partner, contract £538,508

1846–1849. Liverpool, Ormskirk & Preston, contracting partner, contract £200,698

1846–1849. North Staffordshire, Macclesfield to Colwich and branches, contracting partner, contract £960,000

1846–1850. Great Northern, Kings Cross to Peterborough, contracting partner, contract c. £1,500,000

1847–1850. Buckinghamshire, contracting partner, contract £515,566

Contracts outside Britain:

1840–1843. Paris & Rouen, construction, contracting partner, payment FF 21,741,620 (£869,665)

1843–1847. Rouen & Havre, construction partner, contract FF 26,520,000 (£1,060,800)

1844–1856. Orléans & Bordeaux, track laying, contracting partner, payment FF 65,561,257 (£2,622,450)

1846–1848. Dieppe & Fécamp, Dieppe branch, contracting partner, contract FF 7,750,204 (£310,008)

1846–1848. Amiens & Boulogne, excavations and track laying, contracting partner, payment FF 3,775,249 (£151,010)

1847–1848. Barcelona & Mataro, construction, contracting partner, estimate £112,812

MACMAHON, John (fl. 1800–d. 1851), contractor and engineer, was born in Ireland around 1780. In 1808 he went into partnership with David Henry (q.v.) and Bernard Mullins (q.v.) to bid for the contract for the restoration and extension of the Kildare (Naas Line) Canal for the Grand Canal Company. The Kildare Canal itself had been built earlier by William Chapman (q.v.).

The contracting firm of Henry, Mullins & MacMahon went on to undertake many canal, navigation and drainage projects during the early years of the nineteenth century, including large sections of the Grand and Royal canals. From later papers in the Transactions of the Institution of Civil Engineers of Ireland, it is evident that MacMahon developed a sound knowledge of the design and construction of temporary works, in particular centring for bridges and caissons. A major contract was the construction in 1827 of the substantial masonry aqueduct carrying the Barrow Line of the Grand Canal over the river Barrow near Monasterevan in county Kildare. This was followed in 1831 by the completion of the Anglesea Bridge over the river Dodder in Dublin. Following the withdrawal of David Henry from the partnership on completion of the aqueduct, the firm of Mullins & MacMahon continued to play a major role in Irish contracting and, on occasions, working outside the country.

MacMahon took personal charge of the railway contracts undertaken in Lancashire on the North Union, and Lancaster and Preston Junction railways. On the North Union, apparently as 'Henry Mullins and MacMahon', they tendered successfully for contract No. 1, which included the Ribble Bridge, in January 1835. This was followed by the Douglas contract, originally awarded to William Hughes (see: *HUGHES family*) in February 1837, and smaller contracts, ending with the maintenance contract on the line in 1840–1841. Work on the Ribble contract proved contentious, and in 1842 MacMahon won an arbitration award of nearly £14,000 from the North Union.

In 1838 Mullins and MacMahon successfully tendered for work on the Lancaster and Preston Junction line which was opened in June 1840. At the time MacMahon was living in Fishergate, Preston, and in July 1838 his twenty-five year old daughter Kate married a local gentleman, Charles Saxton Ottley, at St. John's Church, Preston.

In October 1842, following the completion of the main contracts associated with the improvements to the Shannon Navigation, MacMahon terminated his partnership in the firm and became Engineer to the Drainage Department of the Board of Public Works (Ireland) in the following year.

When he resigned from the Council of the Institution of Civil Engineers of Ireland in 1851, the Vice-President, William T. Mulvaney, said of him that 'he had been so long and so favourably known to the profession, and the public, and in conjunction with the firm of which he was a member, [and] had filled so large a space in connection with some of the most extensive and important works in the country'.

RON COX

[Frank Smith collection, ICE archives; *London Gazette*, 1842, 3071; W. T. Mulvaney (1851) Chairman's remarks, *Trans ICEI*, **4**(III), 2; M. Mullins (1859) Presidential address, *Trans ICEI*, **6**, 66]

Works

1808–1811. Grand Canal (Naas Line), Corbally extension, contractor (with Henry and Mullins)
1813–1817. Royal Canal, Shannon extension, including Whitworth Aqueduct (with Henry and Mullins), £150,000
1815–1823. Custom House Dock, Dublin, cofferdam (£4,000), entrance lock and basin (with Henry and Mullins)
1824. Royal Canal, Ballinamore extension, contractor (with Henry and Mullins)
1826. Clarendon Dry Docks, Belfast (with Mullins)
1827–1831. Barrow Aqueduct, contractor (with Henry and Mullins)
c. 1830. Ulster Canal (with Mullins)
1831. Anglesea Bridge, Dodder, contractor (with Mullins)
1835–1841. North Union Railway: contract No. 1, Ribble viaduct, c. £48,000; contract No. 3, Douglas contract, completion, c. £50,000; engine shop, £2,600; Preston branch; maintenance contract North Union Railway, contractor (with Mullins; with Henry? until his death in November 1836)
1838–1840. Lancaster & Preston Junction Railway, one of four contractors
1839–1843. Shannon navigation, Athlone Bridge, contractor (with Mullins)

1846. Ballinamore and Ballyconnell navigation, engineer
1848. Eglington Canal, engineer

MACNEILL, John Benjamin, FRS (1793?–1880), civil engineer, son of Captain Torquil Parkes Macneill of Mount Pleasant, near Dundalk, in County Louth, was born at Dundalk about 1793. He first worked as a civil engineer on roads and bridges around the town, mainly under the direction of John Foster, the last Speaker of the Irish House of Commons.

Macneill then worked in the west of Ireland for the Scottish civil engineer, Alexander Nimmo (q.v.), before obtaining employment in late 1826 with Thomas Telford (q.v.). Macneill became one of Telford's principal assistants or 'deputies' and was appointed superintendent of the southern division of the Holyhead road from London to Shrewsbury, having his headquarters at Daventry in Northamptonshire.

From his arrival with Telford he seems to have acted as something of a think tank. He carried out important experiments for Sir Henry Parnell relating to traction on roads, using a dynamometer, and was initially opposed to the introduction of railways. His dynamometer tests were used to relate gradient, surface roughness and pulling resistance, and in 1833 he used these as the basis for assessment of the economics of the Holyhead Road. As regards road maintenance, Macneill arrived at the conclusion that the iron-shod feet of horses were more destructive to roads than any other form of contemporary transport. Claiming experience of more traditional methods of road foundation from 1816, he gave evidence to the Select Committee on the Holyhead and Liverpool Roads in 1830 in support of Telford's system of a paved bottom, i.e. unbroken stone laid on edge as a foundation. This, he felt, was cheaper as no breaking of stone was involved; also, it was more durable than a system using smaller stones on earth, such as that apparently advocated by John Loudon McAdam (q.v.). Macneill devised an instrument to be drawn along roads to indicate their state of repair by monitoring the deflections produced by irregularities in the road surface, tracing them on paper as a continuous curved line. On the Highgate Archway section of the Holyhead Road he laid a trial section of road using a concrete foundation, having earlier placed a small section of concrete paving in Pall Mall.

His research was not confined to roadmaking and he became involved in Telford's canal boat experiments which had begun under Henry Robinson Palmer (q.v.) and were intended initially to demonstrate the relative superiority of canals over railways, particularly for moving heavy loads. Tests were carried out into tractive power and hydraulics using various vessels and the dynamometer at the Adelaide Gallery and on the Grand Junction Canal in 1832–1833, and on the Monkland, Forth and Clyde, and Paisley canals after Telford's death. Macneill also devised tables for computing earthwork quantities in canal cuttings and these were published in 1833. Under Telford, Macneill acquired great technical and parliamentary experience in engineering matters and was the logical choice to succeed him as engineer to the Holyhead Road on his death in 1834.

In his later career, Macneill built up a considerable practice as a consulting engineer and was engineer to many of the railways in Ireland. He was responsible in the early 1840s for introducing into the British Isles iron latticed girders for bridge construction. He was appointed in 1842 to be the first holder of the Chair of the Practice of Civil Engineering at Trinity College, Dublin, and was knighted in 1844. He became an Associate Member of the Institution of Civil Engineers in 1827 and was transferred to Membership in 1831; he was elected a Fellow of the Royal Society in 1838.

Macneill died at 186 Cromwell Road, South Kensington, London, on 2 March 1880 and is buried in Brompton Churchyard. Two daughters survived him; his sons Torquil and Telford, both became engineers, Telford retired to his father's Mount Pleasant home at Dundalk, where he died aged one hundred in 1934. Macneill's portrait at the Masonic Lodge, Dundalk, County Louth, Ireland is reproduced in R. C. Cox, *Engineering at Trinity* (1991) with his signature.

RON COX

[Macneill papers, IEI archives, Dublin; Telford MSS, ICE archives; Gibb Collection, ICE archives; J. Macneill (1829) On the method of improving turnpike roads and what materials, cheapness and durability considered, are the best for the carriageways and roads, in clay countries, particularly about London, OC 44, ICE archives; On the means of ascertaining the comparative merit of roads from the power required to draw a given weight over them; by the application of a parameter or pull measure, OC 45, ICE archives; Holyhead Road Commissioners (1826–1843) *Reports*, esp. **7**, 1830, 21–45, and **12–20**, 1835–1843; House of Commons Select Committee on Whetstone and St. Alban's Turnpike Trusts (1828) *Minutes of Evidence*, 37–44; *Select Committee on Holyhead and Liverpool Roads* (1829); House of Commons Select Committee on Steam Carriages (1831) *Minutes of Evidence*, 93–118; House of Lords, *Select Committee … Turnpike Trusts* (1833), vol. 2, 127–138; Commissioners into the Post Office (1835) Further report; H. Parnell (1833) A Treatise on roads; J. Macneill (1836) Recent canal experiments, *Trans ICE*, **1**, 237–284; Obit. of J. B. Macneill, *The Times*, 5 March 1880, 7, and 8 March 1880, 8; *The Dundalk Democrat*, 6 March 1880; *The Irish Times*, 6 March 1880; *Engineering*, 12 March 1880, 203; *The Engineer*, **29**, 19 March 1880, 215; Obit. of J. B. Macneill (1882–1883) *Min Proc ICE*, **73**, 361–367; R. P. Dod (1853) *Dod's Knightage*; F. Boase (1897) *Modern English Biography*, **2**, 672; DNB; M. G. Lay (1992) Ways of the world]

Publications

Only publications relating to his work under Telford are included.

1833. *Road Indicators*
1833. *Canal Navigation. On the Resistance of Water to the Passage of Boats upon Canals*
1833. *Tables for calculating the Cubic Quantity of Earthwork in the Cuttings and Embankments of Canals, Railways, and Turnpike Roads*

Works

Works prior to 1830:

c. 1822–1824. Dundalk, roads and bridges, assistant engineer
c. 1824–1826. W. Ireland, roads and bridges, assistant engineer to Alexander Nimmo
c. 1826–1837. London–Holyhead Road, Southern Division, London to Shrewsbury, resident engineer

McNIVEN, Charles (*c.* 1746–1815), surveyor, nurseryman, and iron founder, of Scottish origin moved south with his brother, Peter (fl. 1776–d. 1818), and set up business in Alport Lane, Manchester around 1776.

Around 1790, possibly because of prior contact with Hugh Henshall (q.v.), McNiven became involved in canal engineering and was associated with the Mersey and Irwell navigation for many years. He undertook the Parliamentary survey for a canal to link Bolton and Bury to the river Irwell. This scheme, proposed by Matthew Fletcher, a mine agent, was resurveyed by Henshall. Following the passage of the Manchester, Bolton and Bury Canal Act in 1790 work was carried out under Charles Roberts (q.v.), with Henshall as consultant. Shortly afterwards, in 1793, McNiven surveyed the Mersey and Irwell navigation with Henshall, recommending various improvements. At that time, with Fletcher, he surveyed a preliminary line for a canal between Bury and the Leeds and Liverpool Canal at Church. He then briefly flirted with canal contracting, in 1794 with Robert Fulton (q.v.) taking a contract on the Peak Forest Canal, a venture from which he soon extricated himself.

The one work which McNiven is believed to have executed was a cut on the Mersey and Irwell between Warrington and Runcorn Gap which shortened the distance and inconvenience of that section of the Mersey. This was completed in 1804.

In addition to his surveying work in Lancashire and Cheshire, McNiven carried out some building development in Manchester and on occasion acted as an architect. He was involved in an iron foundry in the latter part of his life, and by the time of his death in 1815, was a wealthy man.

MIKE CHRIMES

[Obituary (1815) *Gentleman's Magazine*, **i**, 569; C. Hadfield and G. Biddle (1970) *The Canals of North West England*, 2 vols., *passim*; Bendall; Colvin]

Works
–1804 Warrington–Runcorn cut, Mersey & Irwell Navigation

MACQUISTEN, Peter (fl. 1828–1841), surveyor, of Glasgow, was responsible for a number of road surveys in Scotland. He was elected an Associate of the Institution of Civil Engineers in 1830, remaining in membership until 1836. He was succeeded in his practice by his assistant, William Low (1814–1886). His relationship with Bryce Macquisten [sic] another road surveyor (c. 1799–1818) is unclear.

MIKE CHRIMES

[Membership records, ICE archives; Plans of Gogo Water, Largs, and Ibert near Balfron, Stirling, Nos. 280/1-2, 12807-8 and 15926, SRO]

Publications
1828. Report ... relative to a New Line of Road, from the Confines of the Countie of Ayr and Renfew, at Drumby and Floak, to the City of Glasgow
1829. To the Trustees and Others interested in the Proposed New Approach from the Two Great North Roads to the City of Glasgow

MacTAGGART, John (1791–1830), civil engineer and author, was born on 26 June 1791 in the parish of Borgue, Kirkcudbrightshire, one of 11 children of a Galloway farmer. His early education included private tutoring and a few years at an academy in Kirkcudbright where he showed a strong aptitude for mathematics. He later attended classes in mathematics and physics, for one year, at the University of Edinburgh.

Early in the 1820s he moved to London where he taught mathematics and established a short-lived newspaper, the London Scotchman. His well-known Scottish Gallovidian Encyclopedia was published in 1824. This substantial work, which included descriptions of Scottish culture, antiquities and nature was quickly withdrawn from circulation due to local sensitivities but republished in 1876, long after his death. A short autobiographical sketch which is included as one of the entries in the encyclopedia makes no reference to any of MacTaggart's engineering training or experience. Nonetheless, two years after the book was published, reportedly on the recommendation of John Rennie Jr., he was appointed by the Board of Ordnance as clerk of works for the construction of the Rideau Canal in Upper Canada. Although there is no information regarding his engineering background, several authors have suggested that he served an indentureship under Rennie on the Plymouth breakwater.

His initial assignment on the Rideau Canal involved the design and supervision of construction of a series of seven bridges across the Ottawa River near the entrance to the planned canal. The largest of these was a 212 ft. wooden triple-arch truss across an excessively deep and fast flowing

channel. Subsequently he was made responsible for the preliminary surveys of the entire route and had overall direction of the civilian staff in the Engineers Department at the base camp, Bytown (now Ottawa). During construction, MacTaggart was reprimanded on several occasions for insubordination and drunkenness. The final occasion occurred in June 1828 when he offended Sir James Kempt, Lieutenant-Governor of Nova Scotia, while on an inspection tour of the canal. MacTaggart was dismissed from duty and ordered to return to England. Late in 1829, he published a two-volume account of his experiences in Canada entitled Three Years in Canada.

MacTaggart died on 8 January 1830, aged thirty-one, at his father's home, Torrs, Kirkudbright. One author described him as 'a tall handsome man of 6 ft. 2 in., and of stout figure, his proportions being majestic'.

MARK ANDREWS

[John MacTaggart (1981) Scottish Gallovidian Encyclopaedia, new edn., Strath Tay; H. V. Morton (1933) In Scotland Again; E. F. Bush (1976) The Builders of the Rideau Canal, Parks Canada MSS Series No. 185, Ottawa; E. Welch (1979) Sights and Surveys. Two Diarists on the Rideau, DCB, **VI**, 1987, 480]

Publications
1824. The Scottish Gallovidian Encyclopaedia; or, the Original, Antiquated, and Natural Curiosities of the South of Scotland (2nd edn., 1876)
1829. Three Years in Canada: an Account of the Actual State of the Country in 1826–7-8, 2 vols.

MADOCKS, William Alexander, MP (1774–1828), entrepreneur, was born in London on 17 June 1773, the third surviving son of John Madocks, MP and barrister, and Francis née Whitchurch. The Madocks family had long connections with North Wales and after being educated at Oxford University, William purchased property there. He became responsible for the reclamation from the sea of a large area of land at Traeth Mawr, north of Cardigan Bay, at a cost of £100,000 or more. Such a scheme had been proposed to Sir Hugh Myddleton (q.v.) in 1625, revised in 1718, and considered by John Golborne (q.v.) as part of a new road route to Ireland via Porthdynlleyn in 1770, but remained unfulfilled until Madocks arrived.

Madocks purchased the estate of Tan-yr-Allt, which comprised eight small farms, in 1798 and the first area to be enclosed, along the west side of Traeth Mawr, was the adjacent Penmorfa Marsh, amounting to 1,000 acres. A sand embankment, covered in 11 acres of turf, was built to the designs of James Creassy (q.v.). The embankment was 2 miles long and from 11 to 20 ft. in height. A large, three-arch sluice was provided for drainage from two main dykes. With the completion of the embankment Madocks set about improving his house and estate and, in 1802, successfully stood for Parliament in a by-election at Boston. His activities so far had been costly and financed in part by his mother. He inherited much of her property on her death in 1804 and sold this on to his eldest brother to enable him to further develop his Welsh property.

One obvious development concerned the crossing of Traeth Mawr which formed part of the route to Dublin via Porthdynlleyn on the other side of the Lleyn peninsular. Improvement of this route had attracted renewed interest as a result of the political changes in Ireland and a desire to improve communications with Dublin. For Madocks this was an opportunity not to be wasted as adoption of this route as the main link to Ireland would place his property on a principal artery of communication. As a first stage the Porthdinllaen Turnpike Trust Act was passed in 1803 and in 1807, with Madocks as its sponsor, the Porthdinllaen Harbour Bill was passed, a scheme for its improvement having been drawn up by the engineer Thomas Rogers (q.v.).

Porthdinllaen was of little immediate value to Madocks's estate and he considered developing a harbour at Ynys Cyngar, building a canal, effectively an enlargement of the Great Sluice drainage channel recently completed, which would provide a link to the new town Madocks was now planning. For all this activity Madocks needed an agent and he selected John Williams (c. 1778–1850), the son of an Anglesey farmer, who had previously worked on the gardens of the Marquis of Anglesey at Plas Newydd. Williams had worked on the embankment and then the Tan-yr-Allt gardens and proved his worth to Madocks. He was to be a crucial figure in the success of Madocks's subsequent engineering schemes. His first task was supervising the construction of Madocks new town, Tremadoc; it would appear that Madocks planned it all. One of the first major projects was the construction of a water powered wool mill, the 'manufactory', in 1805–1806, one of the first in North Wales. An engineer, Fanshaw, was employed to manage this but Madocks found him unsatisfactory and Williams soon took over.

While work at Tremadoc progressed Madocks was active again in Parliament in March 1806 with a bill for a dam across Traeth Mawr itself. The scheme was again designed by Creassy. At this stage Madocks suffered three blows. His eldest brother John died in March 1806, leaving his affairs in the hands of Trustees for his son, a minor, depriving William an obvious source of financial support. Perhaps unsurprisingly his Bill failed, soon after Creassy died, leaving Madocks without an engineer. Despite this in early 1807 Madocks was again petitioning Parliament for an embankment Bill. The first attempt floundered, but the second was successful that summer providing that Madocks should undertake the embankment at his own expense, in return for the reclaimed sands and some income (one fifth of the annual rent) from reclaimed marshlands. It would appear that, aside from the surveyor Renny (Henry?) Harrison (q.v.) who prepared the scheme for Parliament, and a report by John Claudius Loudon (q.v.), the responsibility for the Traeth Mawr embankment devolved upon Madocks and Williams. Williams was responsible for determining the final alignment of the cob, which was intended to be 1,600 yd. long, 21 ft. high, with five gate sluices 15 ft. high. Stone for the Cob was transported from local quarries by rails supplied by Brymbo ironworks in which Madocks had a share, and placed using a timber platform supported on 'stilts' (piles). Clearly as work progressed the channel for the passage of the Afon Glaslyn and tides would be restricted, making progress more hazardous, but Madocks also faced financial hazards. Problems were faced in recruiting and retaining the workforce which was two to three hundred in number by the summer of 1808; no doubt Madocks's precarious finances encouraged labourers to move to the neighbouring embankment scheme of R. Williams at Harlech. Horses were also scarce, although 104 were being employed by 1810. The problem of building a stone embankment on a sand foundation hampered progress and this was resolved by using reed matting as a foundation. Problems were also encountered as a result of not training the river in a new course from the start. In mid-1811 it appeared the scheme was going to be realised and in September a 'Jubilee' was held to celebrate the completion of the scheme. By then Madocks's financial affairs were in extremis with both Welsh and English creditors pursuing him through the courts; almost inevitably calamity ensued. In February 1812 the embankment was breached but with substantial support from local landowners, four-hundred men, two-hundred and twenty-two horses and sixty-seven carts in one week, and 892 men and 733 horses in another, Williams was able to bring the breach under control. Madocks's own resources were all but exhausted, his creditors seized most of his personal property, but he was able to retain control of his real estate. This could only fund piecemeal repairs, although the income from that summer's harvests helped, and Williams was eventually triumphant.

In straightened financial circumstances Madocks achieved little, but nature now lent a hand. A 1,400 yd. long embankment, now known as the Cob, was then erected across the estuary, enabling a further 6,000 acres to be reclaimed. It was a rockfill structure 90 ft. wide at its base, founded on rush matting. The restriction of the Glaslyn had scoured out a deep channel at its entrance to the sea facilitating the creation of a new port, Portmadoc. The 1807 Act for the reclamation of Traeth Mawr had permitted the construction of a tramway and the formation of Portmadoc in 1821–1824 encouraged Madocks to look at a rail link with the slate mines in the Festiniog area. The first proposal was surveyed by George Overton (q.v.) and an alternative route by William Provis (q.v.) in 1824. These failed to get Parliamentary approval, as did a second Provis scheme of

November 1825. By this time Madocks was in financial difficulties and concerned that a rival scheme sponsored by Nathan Rothschild, the Moelwyn and Portmadoc Railway, would result in his loss of control over the Portmadoc harbour.

Financial problems and ill health led to Madocks abandoning his Festiniog and Portmadoc railway scheme and in 1826 moving to France, where he died in Paris in September 1828. He had married a Mrs. Eliza Gwynn of Tregunter, a wealthy widow, in 1818 and they had only one daughter.

The promotion of the Festiniog railway became very much associated with James Spooner (1790–1856) who had come to North Wales on Ordnance Survey work and carried out the survey of Rothschild's line. When the Festiniog railway obtained its Act in 1832 he was engineer, construction work being carried out by James Smith between 1832 and 1836.

MIKE CHRIMES

[DNB; J. K. Boyd (1962) *Festiniog Railway*; R. Cragg (1997) *Civil Engineering Heritage: Wales and Western England*, 33–36; G. Biddle and J. Simmons (1997) *Oxford Companion to Railway History*, 466; E. Beazley (1985) *Madocks and the Wonder of Wales*]

Works

1800–1801. Penmorfa marsh, reclamation, 1,082 acres
1808–1812. Traeth Mawr, reclamation, 6,000 acres, £60,000
1821–1824. Portmadoc Harbour

MALLET, John (*c.* 1573–164?), promoter of the Tone Navigation, was born *c.* 1573, the son of Thomas Mallet (1546–1580) who died when he was seven. A wealthy landowner whose family had been proprietors of estates at Enmore in western Somerset for centuries, he was variously MP for Bath and Sheriff of Somerset. He was able to use his wealth and influence to obtain, in 1638, a Commission, under the Great Seal of Charles I, to improve the River Tone to Ham Mills, at his own expense, in return for which he was granted sole navigation rights on the river between Bridgwater and Taunton, and the right to levy bills on cargo.

There is documentary evidence of attempts to improve the Tone from the early fourteenth century. By the end of the sixteenth century cargo could be moved 7 miles up river from Burrowbridge, the confluence of the Tone and the Parrett, to Ham Mills, the limit of the tidal influence. The difficulties and limits of the navigation meant that the costs of some goods, such as coal were beyond the reach of the poorer sections of the community in Taunton. It would appear that Mallet's motivation in seeking to improve the river was to reduce the costs of goods, as well as encourage the local economy. The full extent of improvement is unclear.

The Tone above Ham Mills to Taunton was navigable only with great difficulty, being narrow, winding and obstructed by mills. As a result expensive land haulage was then necessary for the 4 miles from the 'Coal Harbour' at Ham to Taunton

An early navigation scheme, and typical of the time in its reliance on local gentry for support, the Tone was not complete by the time of Mallet's death, and responsibility passed to his son, John (d. 1656), and descendants. Little more appears to have been done until the end of the century. Plans were drawn up by the engineer John Nation, and by an Act of 1699 the family were bought out and the Conservators for the River Tone Navigation established. The work was carried out by John Wilson 'ingineer' that summer.

MIKE CHRIMES and DAVID GREENFIELD

[P. W. Hasler, *The House of Commons, 1558–1603*; T. Haskell (1994) *By Waterway to Taunton*]

Works

1638–. Tone Navigation, 7 miles of river improvement

MANBY, Aaron (1776–1850), engineer, was the father of Charles (1804–1884), who was Secretary and Honorary Secretary of the Institution of Civil Engineers for forty-five years from 1839, and of three younger sons, all of whom were civil engineers. The elder Manby was born at Albrighton, Shropshire, on 15 November 1776, the second son of Aaron Manby of Kingston, Jamaica, and his wife Jane Lane of the Lanes of Bentley. Manby's early years are obscure. The birth of his son, Charles, at West Cowes, Isle of Wight, has led to speculation that he may have been involved in banking there. The source of his engineering training is unknown.

In 1812 he became managing partner of the Horseley Coal and Iron Company, Tipton, Staffordshire, a firm that had existed since the late eighteenth century. He took out a patent (No. 3705) for using blast furnace slag as material for bricks and building blocks in 1813, describing himself as in business at Wolverhampton as an ironmaster. The Horseley Company included coal mines, blast furnaces and engineering workshops. In 1814, Horseley tendered for the castings of a large water tank for the East India Docks and in 1815 supplied a cast iron swing bridge; by 1816 they were advertising their services as providers and erectors of gas-light appliances. In 1821 Joshua Field noted three blast furnaces there, and also noted that Manby had been working several iron barges on the canal for the previous seven years.

In 1822, or possibly as early as 1819, Manby established an engineering works at Charenton near Paris, for some time under the management of Daniel Wilson (q.v.) of Dublin, a chemist. Captain (later Admiral Sir) Charles Napier went to live in Paris in 1819 and supervised the construction of engines for ships for the Société des Transports Accélérés par Eau, in which Manby apparently had an interest.

In 1821 Manby patented (No. 4558) a steam engine for marine use of a type he called oscillating. It was the first practical application of this type of engine, proposed by William Murdock, and patented by R. Witty in 1811. Manby patented it in France as well, and included a claim for making ships of iron and an improved paddlewheel. His first iron steamship was the *Aaron Manby*, 120 ft. long and 18 ft. beam, which was built in pieces at the Horseley works and assembled at the Surrey Canal Dock on the Thames. Built with the assistance of his son Charles, it was tested on the Thames on 9 May 1822. With Captain Napier in command, and probably with Charles Manby as engineer, the *Aaron Manby* set off for Paris, arriving on 11 June 1822, to the amazement of the French. It was the first iron ship to go to sea and the first vessel of any type to make the journey from London to Paris. By June 1822, Manby had apparently parted with the Société and established an English partnership, Manby & Co., owners of the iron steamship *Aaron Manby*. In the winter of 1822–1823, Manby established a French company to compete with the Société, the Compagnie des Bateaux à Vapeur en fer Entre Paris et Havre, which until 1830 operated four iron steam packets, including the *Aaron Manby*.

His second vessel for Paris, the *Commerce de Paris*, was built at Horseley in the winter of 1822–1823 and sent to France for assembly at Charenton. Manby himself left the Horseley Works in 1823, taking many workmen with him to France for his works there. In the same year he was operating a rolling mill of an English type at Charenton. He was back in 1824, recruiting more men with offers of high wages. Manby earned the hostility of British ironmasters both for 'seducing' workmen and for capturing the formerly lucrative French market for engines. By 1826, seven-hundred men and boys, half of them English, were employed at the works. Some said that the Select Committee on Artizans and Machinery of 1824 was in danger of becoming the Select Committee on the activities of Aaron Manby.

The Charenton Works was of great importance in the development of French industry. Its founders were awarded a Gold Medal by the Société d'Encouragement in 1825 and were alternately credited and blamed for France's independence from English engine manufacture. By 1825, Manby and his associates had fitted out at least six complete engineering or iron works in France, at Châtillon-sur-Seine, Abainville (Meuse), Raismes, Imphy (Nievre), Audincourt (Doubs), and La Chamdeau (Haute Saône). A description of the works at Charenton was published by A. Ferry in 1826, a result of a visit by the French Cabinet.

Manby was also interested in gas manufacture and the lighting of towns. In the same year as his marine engine patent he, with Wilson and another, took out a patent for manufacture and purification of gas, and for gas compressed in containers, called 'portable gas'. In 1819 Manby sent Wilson to Paris to seek commissions for building gas works and in May 1822 the Manby-Wilson Company (Compagnie Anglaise) obtained a concession for lighting Paris with gas, against strong French opposition. He was the first to provide Paris with gas light and was influential in the establishment of the Imperial Continental Gas Association which erected gas works in Germany and elsewhere. The Manby-Wilson company existed until 1847, when all the Paris gas companies were amalgamated.

In 1826, Manby and his associates took over the Creusot Ironworks and modernised it with new machinery made at the Charenton works. They manufactured the rails for Marc Seguin's St. Etienne–Lyons railway there in the same year. By 1828, the two had been amalgamated as the Société Anonyme des Mines, Forges et Fonderies du Creusot et de Charenton. Difficult economic conditions forced the liquidation of the company in 1833. Although ultimately a failure for Manby, the subsequent success of the Creusot Works under the Schneider brothers relied on Manby's updated machinery.

Manby returned to England in about 1840, first to Fulham, London, and later to Ryde, Isle of Wight. He married twice, firstly to Julia Fewster, mother of Charles, and secondly to Sarah Haskins with whom he had one daughter and three sons, John Richard (1813–1869), Joseph Lane (1814–1862) and Edward Oliver (1816–1864). All were civil engineers, mostly practising outside Britain. Manby died in his final home in Shanklin, Isle of Wight, on 1 December 1850. He was a Corresponding Member of the Institution of Civil Engineers from December 1820 and remained in membership until 1836. A portrait of him was exhibited in 1868.

TESS CANFIELD

[John G. James Collection, ICE archives; Membership records, ICE; *Specifications and Letters Patent concerning an Oscillating Engine* (1821) mss, ICE archives; *Report from the Select Committee of the House of Commons on Artizans and Machinery* (1824) (*Parliamentary Papers, Reports from Committees*, Session 3 Feb. to 25 June 1824, vol. 5); *Bulletin Société d'Encouragement ...*, Paris (1825) 123, (1826) 295, (1828) 204; A. Ferry (1836) Sur les fonderies, forges et ateliers de M.M. *Manby*, Wilson et Compagnie à Charenton, *Annales de l'Industrie Nationale et Etrangère*, **23**, 5–16; J. Grantham (1842) *Iron as a Material for Shipbuilding*, 6–9; J. Wilkinson (1842–1843) On iron sheathing, broad headed nails and inner sheathing for ships, *Min Proc ICE*, **II**, 168; Address of the President (John Rennie) (1846) *Min Proc ICE*, **V**, 89; A. De Barante (1849–1850) Rapport sur les etablissements du Creusot, *Min Proc ICE*, **IX**, 102–103; Obituary John Richard Manby (1869–1870) *Min Proc ICE*, **XXX**(II), 446; M. du Camp (1873) L'éclairage à Paris, *Revue des deux Mondes*, 780; DNB; S. Everard (1949) *The History of the Gas Light and Coke Company*; W. H. Chaloner and W. O. Henderson (1958) Aaron Manby, builder of the first iron steamship, *Trans Newc Soc*, **XXIX**, 77–91; W. H. Chaloner and W. O. Henderson (1956–1957) The Manby's and the Industrial Revolution in France, *Trans Newc Soc*, **30**,

63–75; W. O. Henderson (1967) *The Industrial Revolution on the Continent*, 98–100; W. O. Henderson (1972) *Britain and Industrial Europe (1750–1870,)* 3rd edn., 49–58]

Publications

1822. Au Roi, en Conseil d'Etat. Les Sieurs Manby, Henry et Wilson, formant la Compagnie d'Eclairage par le Gaz Hydrogene … contre les Sieurs Bourienne, Caret st Chaptal …

MARTIN, Albinus (1791–1871), architect, was born in Beckington, Somerset, on 21 March 1791, the son of a surgeon who was so strong an advocate of the French Revolution that he was forced to flee to America, leaving his son in the care of relatives. The death of his father in America ended the son's intended medical career and he was instead articled to Joseph Woods (1776–1864), a Quaker architect of adventurous spirit and literary abilities. Woods proposed a 600 ft. span single arch stone bridge over the Thames in 1802 and a continuous building some 2,500 ft. long from Charing Cross to Westminster Bridge in 1814. He gave up architecture in about 1814. Martin spent some time in the office of James Savage, where he was involved in preparing drawings of a scheme to salvage Trevithick's (q.v.) failed tunnel under the Thames.

With Samuel Beazley (1786–1851), an architect and playwright, Martin built the first English opera house, on the site of the Lyceum Theatre, Strand, London. Renamed the Theatre Royal English Opera House in 1815, the new theatre opened on 15 June 1816 but was destroyed by fire in 1830, its site claimed by the construction of Wellington Street; it was rebuilt by Beazley between 1831 and 1834 on a site farther west.

Martin gradually abandoned architecture and took up engineering. He was Surveyor of the Cromford Canal, Resident Engineer of the Leeds and Liverpool Canal (1824–1826), manager of the collieries of the Earl of Balcarres, and Bridge Master of a division of the County of Lancaster. As Engineer of the Leeds and Liverpool Canal, Martin defended the interests of the canal against the proposed Bill for the Liverpool and Manchester Railway. By the time of the reintroduction of the Bill in 1826 and no longer connected with the canal, Martin promoted the proposed railway. He joined the London & Southampton Railway Company in 1836, becoming eventually Manager and Resident Engineer, resigning after thirteen years in 1849. He then took up practice as a consulting engineer specializing as a referee and on colliery questions and retired from active practice in 1864, suffering from debilitating attacks of gout.

Martin became a Member of the Institution of Civil Engineers in 1849 and in 1851 he presented five volumes of pamphlets, chiefly on canals, to its Library; they had formerly belonged to William Vaughan (1751–1831). In 1864 he tendered his resignation on grounds of ill health and absences from London but his resignation was not accepted; he agreed to allow his name to remain as a Life Member. He died on 7 October 1871, aged eighty, a highly esteemed, popular individual known for his generosity.

TESS CANFIELD

[Membership records, ICE; Photographic archive of the ICE; Membership records, ICE; *Letter to C. Manby re: gift of 5 vols. of Pamphlets to the Institution of Civil Engineers* (1851), ICE archives; Obituary (1871–1872) *Min Proc ICE*, **xxxiii**, 223–226]

MARTIN, Edward (fl. 1780–d. 1818), mining engineer, was born in Martyndale, Cumberland, and became one of the most important late eighteenth century engineers on the South Wales coalfield. He acted as agent for John Smith who had mining rights to the Llansamlet coalfield on the eastern side of the lower Swansea valley and may have been engineer for Smith's canal (1783–1785) which served the coalfield. He also acted as the Duke of Beaufort's colliery agent in Wales and clearly developed a deep knowledge of the South Wales coalfield. He advocated the use of tunnels and adits, using both canals and railways to exploit the coalfield rather than more expensive shafts and pumping and winding gear. This was practicable in the deep Swansea valley and under his management the Gwannclawdd colliery was very productive. Martin was very much an entrepreneur and in partnership with his father-in-law, Thomas Lott, a Swansea timber merchant, operated three of the principal collieries along the Swansea canal.

Martin's civil engineering work can be presumed to derive from his interest in exploiting the coalfield. He made the first survey for the Swansea Canal (1780), and remained involved in its affairs, building the Cwm Clydach and Twrch tramway lines which linked to the canal (1796–1798) together with the link to Gwannclawdd, all using plate rails. He appears to have been the first engineer to realise the potential of the canal as a source of water power. In 1810 he designed Felin Ynys Dawe corn mill, exploiting a waste water weir constructed to supply forges and mills on the Tawe as part of the Swansea Canal Act; he owned the mill until 1814. He remained consultant to the Swansea canal until his death.

Martin's other canal work included surveys for the Oystermouth Canal (1803), the Aberdare Canal (1809), the Penclawdd Canal (1810) and the Kidwelly and Llanelly Canal (1811–1812). These works were frequently associated with plate ways or tramroads and Martin acted as engineer for the Oystermouth tramroad, which he used himself to transport limestone, and surveyed the Bridgend railway. In several of these works Martin was associated with the surveyor David Davies (c. 1786–1819), who became his partner in 1807 at the age of twenty-one. In 1815 they applied to survey the Swansea–Brecon road.

Martin was involved in various harbour schemes along the coast. In 1802 he was involved in setting out and organising the construction of the east pier of Swansea Harbour and jointly proposed its first floating dock. He carried out embanking work for Swansea Corporation and reported on the condition of the Mumbles lighthouse.

Martin was a successful businessman, investing in property near his homes at Ynys Tawe and Morristown, and was described in one obituary as 'a man of sterling integrity and worth'.

MIKE CHRIMES

[G. F. Gabb (1979) Edward Martin, *South West Wales Industrial Archaeology Society Newsletter*, **22**, 3; S. R. Hughes (1979–1980) The Swansea canal: navigation and power supplies, *Industrial Archaeology Review*, **4**(1), 51–69; S. R. Hughes (1990) *The Brecon Forest Tramroads*, 112–113]

Works
1783–1785. Smith's Canal
1796–1798. Cwm Clydach and Twrch tramroads
1802. East pier, Swansea Harbour
1802. Oystermouth tramroad
1811–1812. Kidwelly and Llanelly Canal

MARTINEAU, John (fl. 1817–1832), engineer, was a leading manufacturer of machinery at the start of the nineteenth century. The Martineaus were a leading Norwich family of Huguenot extraction; at the end of the eighteenth century Thomas Martineau was a textile manufacturer and the family had mining interests in Devon. These concerns gave the family an interest in the manufacture of steam engines, other kinds of machinery and chemical production. Around 1799 John Taylor (1779–1863), of Norwich, who became a distinguished mining engineer, was asked to manage their Wheal Friendship mine and between 1801 and 1805 was joined by his brother, Philip (1786–1870). Philip returned to Norwich in 1805 and set up a drug manufactory. They both got to know Arthur Woolf (q.v.) and his improvements in high pressure steam engines. In 1812 John joined Philip in chemical manufacturing, taking out various patents, with new works in Stratford near London. By the time of the 1817 *Report on Steam Boats*—whether on Taylor's advice or not—the Martineaus had introduced into their works high pressure steam boilers, manufactured by John Braithwaite (q.v.). By 1815 John Martineau Jr., then resident at Stamford Hill, had joined Taylor in business and over the next ten years in various guises Philip Taylor and Martineau took out a number of patents relating to chemical engineering, gas manufacture and steam engines. John Taylor had left the firm by 1820 but, operating as engineers and chemists, (Philip) Taylor and Martineau had established premises in City Road by 1825.

According to Martineau's evidence to the *Select Committee on the Export of Tools and Machinery* (1825) and *Artisans and Machinery* (1824) the firm's principal product, for export at least, was apparatus for making gas, including oil gas for lighting, although they built steam engines of which, according to J. C. Fischer, eighteen were under construction in July 1825.

Among the firm's customers were Marc Brunel (q.v.), for steam engines at his Battersea saw mill and also the Thames Tunnel, and the French engineer and manufacturer, Marc Seguin. Taylor and Martineau were leading exponents of oil gas lighting in 1824–1825.

In 1827 Martineau became involved with R. C. Smith and J. C. and C. Fischer in a steel making process for which Martineau had earlier taken out an English version of their patent (No. 5259, 1825). This may well have led Martineau into financial problems leading to Martineau and Taylor's failure in 1828. Taylor moved to France, becoming an important manufacturer in the Marseilles area.

Martineau joined the Institution of Civil Engineers in 1821, then resident at Bury Court, St. Mary Avenue, and he remained a Member until his death on 6 January 1832. He played a leading role in the establishment of the London Mechanics Institute, later Birkbeck College of the University of London.

MIKE CHRIMES

[Membership records, ICE archives; Report from the Select Committee on Steam Boats with Minutes of Evidence (1817) HC422; Select Committee on Artizans and Machinery (1824) *First Report*, 5–14; Select Committee on Export of Tools and Machinery (1825) *Report (and Minutes of Evidence)*, HC504, 19–26; House of Commons Committee on the London and Westminster Oil Gas Bill (1824–1825) *Minutes of Evidence*; *Mechanics Magazine*, **1**, 1823, 186–187, 189, 229, 256, 264, and **2**, 1824, 10–12, 24; O. Raveaux (2000) Un technicien Britannique en Europe Médionale: Philip Taylor (1786–1870), *Histoire, Économie, et Société*, **19**(2), 253–266]

MATHIESON, Kenneth (fl. 1801–1832), of Glasgow, contractor, was noted as a mason in Gallowgate in 1801, but moved several times over the next few years. He was described as 'of Inchinnan', Renfrewshire in 1812, and it may be presumed that he was employed in some capacity on the bridge being completed there, for which Robertson Buchanan and Andrew Brocket (qq.v.) were respectively the Engineer and Contractor. He was then the successful tenderer for the bridge over river Cree at Newton Stewart, Wigtownshire. He built it within the time stipulated and must have been pleased enough with it, for in 1813 he offered to build a similar bridge at New Galloway, where Telford's design was considered too expensive, and prepared a design based on it for a new bridge over river Doon at Alloway, Ayrshire in 1815. Fortunately for Mathieson, the Kirkcudbrightshire Commissioners persevered with Telford, for his bridge collapsed during construction, and a lengthy lawsuit against the contractor, John Simpson's (q.v.) sureties ensued.

In 1816 he took a feu of quarries at Hopehill near Port Dundas in Glasgow and obtained a contract for a substantial bridge over the river Clyde on the road from Edinburgh to Ayr via Kilmarnock. He was back in Glasgow in 1818, where he was having difficulty in making a watertight foundation for a gasholder pit, but after 1820 he disappears from the city directories and appears again in south-west Scotland where, with Rennie dead and Telford not much in favour, he was commissioned to prepare plans and specifications for a bridge to bypass the awkward Granyford Bridge over river Dee at Kelton. The Commissioners asked James Jardine (q.v.), who was passing through the area, to review the design and he modified it after tenders had been obtained, and again during construction. Mathieson submitted an offer but was unsuccessful.

The harbour at Buckhaven, Fife was built by Messrs Mathieson of Glasgow in 1838 at a cost of about £4,200. Some time later, c. 1859, a Kenneth Matthieson built the Bridport Railway. He may have been a son.

P. S. M. CROSS-RUDKIN

[Glasgow directories, Mitchell Library, Glasgow; Kirkcudbrightshire Commissioners of Supply minutes, Ewart Library, Dumfries; Rennie reports, ICE; Rennie mss, NLS; *New Statistical Account*, **v**]

Works

1812–1814. Cree Bridge, Newton Stewart, contractor, five arches, cost £8,234
1816. Newton Stewart, weir across river Cree below Cree Bridge, contractor
1816–1817. Garrion Bridge, Dalserf, contractor, three arches, 65 ft. span
–1820. Stranraer Harbour, west pier, contractor (remodelled 1978–1979)
1820–1824. Ken (or Kenmore) Bridge, New Galloway, contractor, five arches up to 90 ft. span, cost £10,960
1827–1830. Leith Harbour, eastern pier, contractor
1831–1832. New Bridge, Stirling, contractor, cost c. £17,000

MATTHEWS, David (fl. 1800–1834), was Engineer to the East Country Dock Company who were responsible for the construction of their dock adjacent to the Greenland Dock in Rotherhithe c. 1807–1811. Following its completion, Matthews was appointed engineer to Bridlington Harbour in 1815 following a report on the piers by Simon Goodrich (q.v.).

There he may have been responsible for the first use of (cast) iron sheet piles in the construction of the foundations of the North Pier around 1820. He appears to have died in Bridlington by 1835.

MIKE CHRIMES

[Tidal Harbours Commission (1846) 2nd Report, Appendix B, 389–396, M. A. Borthwick (1836) Memoir on the use of cast iron in piling, *Trans ICE*, **1**, 195–205; J. G. James Collection, ICE library; PLA archives; A. W. Skempton (1981–1982) Engineering in the Port of London, *Trans Newc Soc*, **53**, 73–94]

Works

1807–1811. East Country Dock, London, 4½ acres, £40,000
1815–1834. Bridlington Harbour, improvements

MAUDSLAY, Henry (1771–1831), mechanical engineer, was perhaps the greatest machine tool builder of the early nineteenth century. Born at Woolwich on 22 September 1771, the son of William Maudslay, he began working in Woolwich Arsenal as a powder monkey, filling cartridges at the age of twelve. His father had also worked in the Arsenal as a storekeeper after being severely wounded serving in the Royal Artillery in the West Indies and had died when Henry was only nine. Maudslay soon demonstrated that he was capable of more skilled work and after a period in the carpenters' shop was transferred to the blacksmiths' shop when he was about fifteen.

Maudslay's experience at Woolwich was essentially one of hand-working materials and gave him an intuitive understanding for the shaping and forming of wood and metals which underlay his talent for machine tool design. His skill in the workshop became well-known among smiths in the London area and he was brought to the attention of Joseph Bramah (q.v.), whom he joined in 1789 and helped make the Bramah lock a reality. Despite his youth he soon became Bramah's foreman and in 1791 married his employer's housekeeper, Sarah Tindale (1762–1828); their first child, Thomas Henry, was born in 1792. The pressures of a growing family encouraged Maudslay to seek a higher salary than 30s a week, but Bramah could not oblige. As a consequence, in 1797 with his reputation well-established, Maudslay set up in business on his own at 64 Wells Street, off Oxford Street. His knowledge of the machines developed at Bramah's was to be applied again in his own works and machinery.

The Maudslay lathe and slide rest was developed soon after setting up on his own; no doubt a refinement of lathes used at Bramah's, it perhaps displayed most clearly his three dimensional imagination. It was all metal, ensuring a more consistent accuracy of workmanship, and heralding the modern age of mechanical engineering. He realised the importance of accurate screw cutting and was continually refining his machinery until the end of his life. He developed a micrometer, nicknamed the Lord Chancellor, capable of measurements to an accuracy of one ten thousandth of an inch. Thousands of screws were cut and the most accurate used as a guide screws. Such screws were not only used by Maudslay but by his pupils who took them with them when they moved on.

Maudslay's first commission at his new business had been the ironwork for an artist's easel. From this modest beginning his business rapidly expanded and in 1802 he moved to 78 Margaret Street, where eventually eighty men were employed. This move may in part have been occasioned by the large amount of work arising from his involvement in making Marc Brunel's (q.v.) block making machinery which revolutionised the production of pulley blocks for rigging for the navy. On the recommendation of Agustin Betancourt, a fellow refugee, Brunel initially approached Maudslay in connection with parts for a model of block making machinery around 1800. It soon became apparent that Maudslay was an ideal collaborator and for seven years, from 1802 until 1809, Maudslay was largely occupied with making the forty-five machines for Portsmouth Dockyard's block making machinery. This was described by Gilbert (1965) as the first large scale plant to employ machine tools for mass production.

In addition to machine tools Maudslay and his firm became important builders of steam engines. The accuracy of his machine tools meant that he could build more compact and neat engines than many of his contemporaries. His table engine, patented in 1807 (No. 3050), proved ideal for driving machinery for factories and workshops.

Maudslay's work at Portsmouth brought him into contact with Joshua Field (q.v.) who worked as a draughtsman in the Dockyard. In 1804 Field joined his staff and by 1810 the firm had so prospered that it relocated to Lambeth Marsh, where the works remained until their closure in 1900. Around September 1812 the firm became known as Henry Maudslay and Company, and Maudslay took his son, Thomas Henry, Josuha Field, and Field's wife's uncle, John Meedham, into partnership. In 1820 Meedham retired and the firm adopted an approximation to its familiar name Maudslay, Son and Field; in 1831 with Maudslay near to death his third and fourth sons, John and Joseph, were admitted into the partnership and the title 'Maudslay, Sons and Field' adopted.

The firm became particularly well-known as marine engineers. In 1815 Maudslay supplied his first engine, of 17 hp, for the *Richmond*, the first passenger steamer on the Thames; there followed machinery for forty-five ships in Maudslay's lifetime, including in 1823, HMS *Lightning*, the Royal Navy's first steamship, and the 1825 *Enterprise*, the first steamer to voyage to India. At the end of his life in 1830 he built a pair of 200 hp marine engines, the largest hitherto built, before a firm order had been received, such was his belief in the marketability of his workmanship. They were fitted in a mail packet paddle steamer, *Dee*, completed in 1832, after his death.

In addition to steam engines Maudslay supplied a whole range of general engineering work including saw mill machinery, gun boring machines, a draw bench and melting house equipment for the Mint, portable cranes and hydraulic presses. Such machines were not just for the United Kingdom market and engines and machinery were exported to Belgium, France, Prussia, and even Brazil. Some machinery was manufactured in association with Maudslay's patents such as those for Calico printing (No. 2872, 1805; No. 3117, 1808) a differential gear hoist (with Bryan Donkin (q.v.), No. 2748, 1806), aeration of water (with Robert Dickenson, No. 3338, 1812) and for changing water in marine boilers (with Field, No. 5021, 1824).

Maudslay's works was employing over two-hundred men in the 1820s—they later employed well over a thousand and had subsidiary works—and the scale of the enterprise meant that he could not possibly have been solely responsible for the design work and management of the works. His most important collaborators, apart from Field, were his family. Thomas Henry (1792–1864) was particularly known for his work on engines supplied to the Royal Navy. The second and third sons, William Nicholson (1795–1818) and John (1799–c. 1835), died young and appear to have played a relatively small role in the firm, William never even becoming a partner. Joseph (1801–1861), on the other hand, seems to have displayed much of his father's ingenuity. Trained as a shipwright with William Pitcher of Northfleet, and intended to be a shipbuilder, he joined the family firm and was able to combine his knowledge of naval architecture with expertise in marine engineering. He adopted Murdoch's (q.v.) oscillating (rotary) engine in a patent of 1827 which utilised ancillary slide valves, worked by an eccentric, to engines with oscillating cylinders and a pair of these engines was fitted to the Thames steamer, *Endeavour* in 1828. There was much prejudice against the use of oscillating engines and Joseph spent much of his career in developing alternative forms of direct acting marine engines and screw propulsion.

Maudslay's works were a fertile training ground for the next generation of mechanical engineers including Samuel Seaward (q.v.), Joseph Clement, Richard Roberts, James Nasmsyth and Joseph Whitworth. The Maudslays also made an important contribution to early nineteenth century civil engineering. Maudslay's collaboration with Brunel continued in the 1820s when the firm supplied the first shield for the Thames Tunnel and a 24 hp steam engine. They built the only example of Congreve's patent lock for the Regent's Canal, fitted at Hampstead Road. Pumping engines were supplied for Haddenham level drainage, Lambeth waterworks, and, after Maudslay's death, for the Grand Junction and Chelsea waterworks. At the Lambeth works an important early iron roof was erected and although it collapsed on 24 May 1826 it was subsequently rebuilt and was widely commented on.

Perhaps most importantly the firm played an important role in the formation of the Institution of Civil Engineers. Thomas and William together, with Josuha Field, were founder members of the Institution, with William taking the chair at the first meeting. William died later that year and

Thomas resigned in 1824 due to pressure of work; Henry was elected in 1829 and Joseph in 1833.

Henry Maudslay, like Field, lived at the Westminster Bridge works and there he died on 15 February 1831. He was buried under a cast iron tomb at St. Mary's, Woolwich. His epitaph, penned by Field, read: 'A zealous promoter of the arts and sciences, eminently distinguished, as an engineer for mathematical accuracy and beauty of construction, as a man for industry and perseverance, and as a friend for a kind and benevolent heart'.

MIKE CHRIMES

[Brunel diaries, ICE archives; Maudslay collection, Science Museum; Membership records, ICE archives; Evidence (1817) *Select Committee on Steam boats etc.*, HC 422, 23–24; Accident (1826) *Mechanics Magazine*, **6**; *Verein zur Beförderung des Gewerbleissess in Preussen* **12**, 1833, 248–249, plate xx; C. Holzapffel (1846) *Turning and Mechanical Manipulation*, **2**, 641–642; S. Smiles (1863) *Industrial Biography*, 183–197; R. Beamish (1862) *Memoir of the Life of Sir Marc Isambard Brunel*; The late Joseph Maudslay (1861) *Mechanics Magazine*, 29 November, 351–352, *Min Proc ICE*, **21**, 1862, 526–529; Thomas Henry Maudslay DNB, p. 82; *Practical Mechanics Journal*, 1864, 81; J. Nasmyth (1883) *An Autobiography*; W. A. S. Benson (1901) The early machine tools of Henry Maudslay, *Engineering*, **71**, 64–66, 134–136; J. F. Petrie (1934) Maudslay, Sons & Field as general engineers, *Trans Newc Soc*, **16**, 39–61; J. F. Petrie (1949, repr. 1956) *Henry Maudslay, 1771–1831 & Maudslay Sons & Field*; K. R. Gilbert (1965) *The Portsmouth Block-making Machinery*; L. T. C. Rolt (1965) *Tools for the Job*; K. R. Gilbert (1971–1972) Henry Maudslay, *Trans Newc Soc*, **44**, 49–62; K. R. Gilbert (1971) *Henry Maudslay, Machine Builder*; J. G. Watson (1988) *The Civils*; R. C. Hills (1989) *Power from Steam*, 70–94; F. J. Evans (1994) The Maudslay touch, *Trans Newc Soc*, **66**, 153–174; DNB]

Works
For a more complete list see Petrie (1934).

1800–1801. Models of block-making machinery for M. I. Brunel
1802–1808. Portsmouth Dockyard, block-making machinery
1814–1815. Regent's Canal, Congreve patent lock
1824–1825. Thames Tunnel, shield and steam engine
1826. Lambeth works (1) and (2), cast iron roof, 55 ft. span
1830. Haddenham Level drainage engine and scoop wheel, £3,500
1830. Belvedere Road, pumping engine reconstruction for Lambeth Waterworks
1831. Lambeth Waterworks, 200 hp pumping engine
1835. Engine and boiler for Dance's steam carriage
1836–1838. Kew Bridge, Grand Junction Waterworks, engine

1837. Chelsea Waterworks, beam engine
1838. Euston incline, London–Birmingham Railway, two condensing engines for rope haulage
1840. Minories, London–Blackwall Railway, two pairs of rope haulage engines

MAXWELL, George (*c.* 1744–1816), land agent, was born near Fletton, Huntingdonshire, into a farming family. He lived in the neighbourhood all of his life except for six and a half years in Hertfordshire and, for a time during 1792, at Houghton while his house, Fletton Lodge, was being built.

He first became interested in the Bedford Levels of the fens in the early 1760s but it was not until 1772 that he undertook his first drainage scheme, at Titchmarsh in Northamptonshire. In 1783 he was appointed land agent to Lord Eardley, who owned substantial estates at Fletton and in the Middle Level. During the next twelve years Maxwell was responsible for erecting ten large windmills and for lining the drains with clay to improve his employer's lands. In either 1782 or 1783 he examined the drainage at Lakenheath in Suffolk, and about 1787 he was responsible for the drainage of Castlemartin Corse in Pembrokeshire, which he considered to be his first commission as an engineer.

He was an active enclosure Commissioner, claiming in 1800 that he had worked in that capacity more than one hundred times during the previous twenty-seven years. In enclosures where significant drainage or embankment works were involved he seems to have worked with Edward Hare (q.v.), under whose entry some of them are described. He was an enthusiastic supporter of the idea of improved outfalls of the major rivers as an aid to better drainage of the Fens, citing Kinderley's Cut on the river Nene as an example. Under an Act of 1794 he and Hare were Commissioners for the improvement of the Welland outfall but much of the work authorised was not carried out and had to wait until the passage of another Act in 1837. Much of Maxwell's *General View of the Agriculture of the County of Huntingdon* (1793) is an essay on the need for the Eau Brink Cut, which was carried out by Rennie and Telford (qq.v.) in 1820.

Maxwell was buried on 1 January 1817 at Fletton, aged seventy-two.

P. S. M. CROSS-RUDKIN

[*Minutes of Evidence on Second Reading of Eau Brink Drainage Bill*, 1794; *House of Commons Committee Reports*, 1800]

Publications
1788. *Reasons offered to the Proprietors of the North Level against a new Tax*
1789. *An Address to the Proprietors of the North Level*
1791. *Observations as to the Present State of the Lands in the Neighbourhood of Spalding with a view to an Improvement in the Outfall of the River*

Welland, and to the Drainage of the Lands on both sides thereof

1792. *An Essay on Drainage and Navigation, occasioned by the Scheme now in Agitation for Improving the Outfall of the River Ouse*

1793. *Observations on the Advantages to be derived from an Improved Outfall at the Port of Lynn; with Answers to the Objections which it is supposed will be used against that Measure*

1793. *Reasons attempting to shew the Necessity of the Proposed Cut from Eau Brink to Lynn; with Extracts from the Reports of Engineers and other Writers on the Subject … Addressed to all Persons interested in the Drainage or Navigation of the River Ouse*

1793. *The General View of the Agriculture of the County of Huntingdon*, Board of Agriculture

1800. *The Report of Messrs Maxwell and Hare, on the Drainage of Deeping, Langtoft, Baston, Crowland, Cowbit, Spalding & Pinchbeck Commons, in the County of Lincoln*, with Edward Hare

1800. *Drainage of Deeping Fen*, with William Jessop, John Rennie and Edward Hare

Works

1787. Castlemartin Corse, drainage

MAY, George (1805–1867), engineer, was the son of Andrew May, clerk to John Mitchell (q.v.), Thomas Telford's (q.v.) Chief Inspector of Highland Roads, based in Inverness, Scotland. Andrew was the first Collector of Dues at Clachnaharry on the Caledonian Canal, serving from June 1824 to October 1827 on a salary of £150 per year, half that of the resident engineer's salary. His son, George, born in Inverness on 14 September 1805, was educated at Inverness Academy and at King's College, Aberdeen for three years.

Both May and James Davidson, then Superintendent to the Canal Company, had been pupils of Telford in London. He lived in Telford's London house in Abingdon Street at the same time as Joseph Mitchell, author of *Reminiscences of my Life in the Highlands*. May remained in London from 1823 to 1827, when he returned to Inverness on account of his father's death.

May succeeded his father as Collector at Clachnaharry, a position he held from October 1827 until March 1830. In 1828 he accompanied Telford on his steamboat tour of inspection of the canal works. May became Engineer for the Caledonian Canal in 1829, succeeding James Davidson, whose health had failed. He remained as Engineer until his death in 1867, when he was succeeded by Davidson, who had recovered his health and been employed as Collector at Clachnaharry.

The quality of construction of parts of the Canal left much to be desired. In November 1834, substantial damage was caused to the canal by severe flooding, causing water to reach levels three ft. higher than the lock gates at Gairlochy, and threatening local residents. Just before the floods, May had completed a report criticising the masonry work at the west end of the canal, and the Banavie

and Corpach locks as 'execrable'; James Walker (q.v.) supported May's opinions but was more moderate in his language. May raised the canal banks and the stonework of the Gairlochy Lock. In 1837 a collapse at Fort Augustus caused many large vessels to be blocked. It took May three weeks to repair the damage.

A Select Committee of the House of Commons examined many witnesses on the subject of the canal and its problems, including George May in 1839. Major repairs to the Canal were undertaken in the 1840s, under May's supervision. According to Walker and Burges' Report of 1846 on the state of the Caledonian Canal works, 'the character and valuable services of Mr. May are too well known to you to require our confirmation'. By 1847 the defects of the Canal had been fully corrected, creating the Canal as it should have been at its opening in 1822. It was re-opened on 1 May 1847.

In 1850 George May toured the Baltic ports, seeking to publicise the canal in Riga particularly. He returned in 1851, with notices, plans and charts, visiting Oslo, Copenhagen, Stockholm, Hamburg and Stettin. These efforts were curtailed by the Crimean War, which stopped trade from the Continent. May continued as Engineer and Superintendent to the Caledonian (and later Crinan) Canal until within a few weeks of his death on 28 August 1867, in Inverness.

He was an Associate of Institution of Civil Engineers from 1824 until 1837 when he was transferred to the class of Member. His younger brother, James (1818–1876), was also a civil engineer, elected to membership of the Institution in 1864. His career included harbour works at Alderney, Alexanrdia, Egypt and Madras, India.

A portrait of May in the possession of British Waterways Board is reproduced on page 109 of Cameron.

TESS CANFIELD

[Obituary (1867–1868) *Min Proc ICE*, 595–596; Obituary (James May) (1876) *Min Proc ICE*, 244–246; Skempton; A. D. Cameron (1972) *The Caledonian Canal*; C. Hadfield and A. W. Skempton (1979) *William Jessop, Engineer*; C. Hadfield (1993) *Thomas Telford's Temptation*]

Publications

1830. General Report of the State of the Caledonian Canal Works to May 1830, signed and dated Inverness 1 May 1830, in *Twenty-sixth Report of the Commissioners for the Caledonian Canal*, House of Commons, 10–13 (May's annual reports to the Commissioners continue until 1867, excepting the years 1844–1847 by Walker and Burgess)

1839. Mr. May's Report in reply to Mr. Borron, signed George May, Inverness, 1 Nov. 1837, in *Report from the Select Committee on the Caledonian and Crinan Canals*, House of Commons, 120–144]

MEAD, John Clement (1798–1839), architect, was the son of C. Mead and his second wife. His father was a London surveyor and Mead was trained in his father's office being admitted to the

Royal Academy School in 1815. He entered various architectural competitions in the early 1820s and was responsible for Cambridge Observatory (1822–1824). He visited and worked in France and was interested in abattoir design and casting bronze sculpture.

He was elected an Associate of the Institution of Civil Engineers in 1828 and remained a member until his death on 15 January 1839. He had suffered a long illness and had retired to Piddletrenthide, Dorset. He left a widow and three sons.

MIKE CHRIMES

[ICE membership records, ICE archives; On grouting & brickwork (1834) Minutes of Conversations, ICE; A section of borings at Wood House, Shepherds Bush, ICE archives; Colvin (3)]

MEIKLE, Andrew (1719–1811), came of a long line of versatile millwrights whose careers stretched from the latter years of the seventeenth century through to the early years of the nineteenth. The Meikle family are chiefly associated with the construction of mill buildings and machinery, and for the development of machinery for threshing and winnowing grain. Their work, much of it of an ingenious nature, received favourable comment from contemporary engineers.

The first of the lineage, John Meikle, a founder, has been credited with the introduction of iron-founding into Scotland having in 1686 received encouragement in this enterprise by an Act of the Scottish Parliament.

James Meikle, a wright of Keith in Haddingtonshire, was the next notable member of the family. In 1710 he was sent to Holland by Andrew Fletcher of Saltoun to learn the construction of a mill to produce pot-barley and fanners to separate chaff from barley. On his return to Scotland he erected a barley mill at Saltoun using imported ironwork. The first set of fanners he constructed did not come into use at Saltoun until 1720, and religious suspicion delayed the completion of the next winnowing machine, built in Roxburghshire, for a further seventeen years.

Andrew Meikle, son of James Meikle, was a farmer, miller and millwright who, as a skilful mechanic, is best known for the development of the drum threshing machine. His ingenuity was widely regarded. For much of his life, Meikle carried on his various businesses from Houston Mill on the Phantassie estate that was leased by George Rennie, father of John Rennie (q.v.). Much of Meikle's work as a millwright was based locally in and around the town of Haddington, and his wright's business appears to have flourished to the extent that he was unable to fulfil all his obligations. This did not prevent him on several occasions offering advice about their mills, gratis, to Haddington Town Council: acts that were subsequently acknowledged in 1763 by his admission as a burgess of the town. In 1750 he surveyed the Tyne and Colstoun Waters for the Haddington Tarred Wool Company, and in that

year visited England for the same company to inspect waulk mill machinery. Meikle reported and worked on a number of mills: at Bonnington, Edinburgh (1749 and 1759), Kilmarnock (1745), Thornton (1776), Dumfries (1781), and, in conjunction with his son George, the burgh mills at Linlithgow (1785). His work was not entirely that of a millwright constructing machinery within mills but also included the assessment of the hydraulics required to power mills. In 1771 John Smeaton (q.v.) recorded the work done by Meikle at Dalry Mills, Roseburn, where a third mill was to be built adjacent to two existing mills with all three mills powered from the same water course. Meikle proposed that the existing mills should be powered by a single wheel and the third by a new wheel.

In 1747 Andrew and Robert Meikle (Mackell?) began their association with the Board of Trustees for Manufactures, being engaged as millwrights to the Board in 1751, and in the following year they received an annual payment of £20 from the Board to train apprentices. Andrew Meikle's reputation as a competent craftsman acted as inspiration to two future engineers, John Rennie and Peter Nicholson (q.v.). Rennie, having been educated locally until the age of twelve, spent two years, 1773–1775, working at Meikle's millwrights shop before leaving to continue his education at a school in Dunbar. In 1777 Rennie returned to work part-time at Houston Mill before leaving again in 1779 to study at the University of Edinburgh. During the summer while still at university he took 'overflow' work from Meikle. It has been suggested that Rennie whilst working with Meikle 'acquired a mastery of the various crafts of the carpenter, the mason and the blacksmith' and that his subsequent engineering career evolved from this training. Nicholson was apprenticed to Meikle, and later, between 1800–1808, practised as an architect in Glasgow, laid out the town of Ardrossan and subsequently became the Surveyor to the County of Cumberland. He is best-known as the author of some thirty publications, written in the years 1792–1844, on mathematics, architectural draughtsmanship and building science.

Meikle's ingenuity is best illustrated by the developments he made to agricultural machinery for winnowing and dressing grain and to milling. In 1768 he, in conjunction with Robert Mackell (q.v.), registered a patent for a machine to dress grain. It was based on the use of riddles and fanners, and although practical was thought to be 'of no great novelty'. In 1772 Meikle designed a mechanism to control the speed of windmill sails. The hinged wooden shutters of the sails could be trimmed automatically by a tensioned control spring when they were subject to excessive wind pressure. The originality of Meikle's work was questioned as a similar patent had been registered earlier in 1745 by Edmund Lee. Smeaton, although aware of Meikle's proposal, did not use it in his own work.

Meikle's patent for the drum threshing machine, registered in 1788, was the culmination of many years work. Earlier Francis Kinloch of Gilmerton, East Lothian, had seen threshing machines built by Ilderton and Smart in Northumberland. Kinlock modified these designs and had an improved model built, but this failed when powered by Meikle's water-wheel at Houston Mill. The failure caused Meikle to re-examine the design but his re-designed machine of 1776 did not meet his expectations, and it was not until 1786 that he and his son George erected their first drum threshing machine for a Mr. Stein, a distiller and farmer in Clackmannanshire. The second machine was built on the Phantassie estate for George Rennie in 1787, and by 1789 Meikle and his son were advertising as manufacturers of threshing machines. It is estimated that, in the next twenty years, about three-hundred and fifty threshing mills were built in East Lothian at a cost of some £40,000. As Meikle had built mills to his design before the patent was registered he was unable to gain financially from his work.

Towards the end of Meikle's life Sir John Sinclair raised £1,500 to be invested on his behalf. Amongst those who subscribed were the wright's friends James Watt and John Rennie. In his obituary Meikle was described as a man of great genius and 'of original cast', and Smeaton, who was acquainted with his work, was quoted as saying 'that if Andrew Meikle had a better address, and greater confidence in his own powers, he would out rival the whole of his contemporaries, and be considered the first engineer in the Island'.

Until 1788 George Meikle (d. 1811) worked in partnership with his father as a millwright and on the development of the drum threshing machine. Later he worked on mills at Alloa (1789) and Lanark (1795) and on his own account built threshing mills. His most notable work was the drainage in 1787 of Kincardine Moss, Blair–Drummond, by the use of a 'water-raising wheel'. Meikle used a wheel, 28 ft. diameter and 10 ft. broad, to raise water 17 ft. to provide a supply of 40–60 hogsheads per minute to wash away the layer of peat that, overlying a clay substratum, formed the moss. Robert Whitworth (q.v.) had previously submitted a plan to drain the moss, but on inspection of Meikle's model of the wheel 'declared that, for the purpose required, it was greatly superior to the machine recommended by himself; and advised it to be immediately erected'. George Meikle died on 29 November 1811, two days after his father.

TOM DAY

[*Scots Magazine*, **51**, 1789, 211, and **74**, 1812, 79; *Farmers Magazine*, **12**, 1811, 566; S. Smiles (1874) *Lives of the Engineers*, vol. 2, 105–117; DNB; C. T. G. Boucher (1963) *John Rennie, 1761–1821*, 2–3; J. Shaw (1984) *Water Power in Scotland, 1550–1870*, 102–106, 156–157; S. Jamieson (ed.) (1984) *The Water of Leith*, 58–59]

Patents

1768. No. 896: *Machine for dressing Wheat, Malt, and other Grain, and cleaning them from Sand, Dust and Smut*, 14 March 1768 (with Robert Mackell)
1788. No. 1645: *Machine, which may be worked by Cattle, Wind, Water or other Power, for the purpose of Separating Corn from the Straw*, 9 April 1788

MELVILLE, Robert (fl. 1780–d. 1812), fishery agent, was the nephew and erstwhile partner of Robert Fall, who owned a fishery business. Melville was acting on behalf of the firm in Caithness and Sutherland when it went bankrupt in 1788, at a time when the British Fisheries Society were first becoming active. Melville offered to take on the role of agent and contractor at Ullapool where a pier was required. Work was delayed as another contractor, Roderick Morrison, farmer, fish merchant and businessman of Stornoway and Tarbert, thought, correctly, that the intended pier was badly positioned. The question was referred to John Smeaton (q.v.), but pressure of work prevented him acting, and a director of the Society, John Call, an East India Company engineer, drew up plans for a pier and breakwater. Morrison, originally contracted to build the pier, refused the work because of its location and Melville went ahead. He lacked the necessary experience. The pier was built on soft ground with inadequate foundations which caused it to spread and so required repositioning the breakwater. James Maclaren, a surveyor, reported on the work in July 1789 but specified only minor improvements and the pier and breakwater were finished in early 1790, having required far more material than originally envisaged. Smeaton was consulted and blamed the poor workmanship on Melville who, he felt, should bear the extra costs. In something of an impasse, on William Pulteney's advice, Thomas Telford (q.v.) was asked to report. He felt that Melville should bear the additional cost of the breakwater but not the pier. Melville refused to accept any responsibility for bearing the extra costs. The works themselves were suffering problems to do with shingle encroachment, as predicted by Morrison. On the advice of John Bain, with Telford's approval, defenders were added to the pier.

In the early years of the nineteenth century Melville suffered financial problems was declared bankrupt in 1808 and died in 1812. Telford remarked of him: 'His activity was entitled to great profits if he had been content with the reward of his real merit without aiming at extra advantages'.

MIKE CHRIMES

[J. Dunlop (1978) *The British Fisheries Society 1786–1893*]

Works

1788–1796. Ullapool Harbour, including pier (c. 136 ft.) and breakwater (137 ft.), £6,000

MEREDITH, Michael (d. 1865), architect, was elected as an Associate of the Institution of Civil Engineers in 1824 but does not appear to have renewed his membership. At that time he was resident at White Hart Court, Bishopsgate. As an architect he appears to have practised largely in the London area. He died at 99 Guildford Street on 16 June 1865.

Benjamin Meredith was active as a contractor in the 1820s, carrying out work at Highgate Archway in north London.

MIKE CHRIMES

[ICE membership records; Colvin (2)]

METCALF, John (1717–1810), entrepreneur and roadmaker, known as Blind Jack of Knaresborough, was sufficiently famous to be the subject of a ghosted autobiography which ran to more than one edition in his lifetime. Subsequent accounts, including one by Samuel Smiles in his *Lives of the Engineers*, have drawn heavily on this book, which was written and published by Edward Beck at York in 1795. The picture which is painted there of Metcalf is of a 'rum lad'. The narrative is designed for popular consumption and does not pay strict attention to completeness or chronology. Nevertheless it is clear from this and other sources that, after a varied career, he made about 180 miles of road over a period of forty years and subsequently branched out again in other spheres.

Metcalf was born in Knaresborough on 15 August 1717. At the age of six, when he had been attending school for two years, he suffered an attack of small pox which left him blind. Possessed of a remarkable degree of self confidence, he taught himself the pursuits of normally sighted people. He was tall by the standards of his time, being about 6 ft. 1½ in. in height, and well built and he claimed to have engaged in several wild adventures, mostly for a challenge or a bet.

From 1732 he earned a living by playing the fiddle, in the fashionable rooms of Harrogate and Ripon, and spending two seasons in London. His income was sufficient by 1739 to enable him to marry Dorothy Benson, the daughter of the landlord of the Royal Oak in Harrogate. Characteristically, he eloped with her on the eve of her wedding to another man. There were four children of the marriage. He had a house built in Knaresborough and started a chaise hire business in Harrogate, the first of its kind there, but when competition grew, he gave it up and developed a business selling fish from Whitby in Leeds and Manchester. This did not prosper and he was again working as a musician when the rebellion of 1745 broke out. He enlisted as a volunteer, his company forming part of General Wade's army, which proved so ineffectual. Metcalf was present at the Battle of Falkirk on 17 January 1746, playing his company to the field; the battle ended in defeat for the Hanoverian army. After the final Jacobite defeat at Culloden, he started to trade between Scotland and England, taking horses north and returning with cloths which he could sell in Yorkshire. By 1751 he was running a stage wagon regularly between York and Knaresborough but in 1754 he sold that business to take the first of his contracts for turnpike roads. Smiles, who is normally fairly careful with the facts, unless they interfere with the story he wishes to tell, says 'about 1765', but that is too late.

In 1752 an Act had been passed to repair roads leading from Leeds to Harrogate and thence to Ripon and Boroughbridge. Metcalfe contracted to build 3 miles of this extensive scheme, from three to 6 miles north-east of Knaresborough, through easy country. It was soon followed by a contract for a bridge of 18 ft. span at Boroughbridge, 7 miles away. Another contract for the same road on the other side of Knaresborough was followed by stretches nearer Leeds, then by lengths for a different trust in the west of the county and almost the whole length of the Huddersfield to Standedge road. On two of these jobs, Metcalfe supported a road over bogs by laying the road stone on bundles of heather carefully placed. On the second occasion he obtained the contract—in the face of opposition from the turnpike surveyor who wished to excavate to a solid foundation—by offering to try his own method and, if it failed, to redo the work at his own expense to the surveyor's specification. He employed sixty men on this job in order to complete on time; on an earlier job he had taken on four hundred men for the same reason.

Metcalfe's wife, in need of medical treatment, went to stay with a married daughter in Stockport and she died there in 1778, aged sixty-one. He subsequently lived for some years in his daughter's house, transferring his roadmaking to the adjacent parts of Lancashire, Cheshire and Derbyshire, but also joining his son-in-law in a cotton manufactory. On a trip back to Knaresborough to sell some of his goods, he contracted for a mile and a half of road at Marsden, near Huddersfield. He lost money by it and undertook no more road contracts in Yorkshire.

Metcalfe finished roadmaking in 1792 and returned to Yorkshire, buying a house at Spofforth, near Knaresborough. He continued to trade occasionally in an active old age and died aged ninety-two on 26 April 1810.

P. S. M. CROSS-RUDKIN

[DNB; Edward Beck (1795) *The Life of John Metcalf*; W. B. Crump (1949) *Huddersfield Highways Down the Ages*]

Works
1754–1755. Harrogate–Boroughbridge, 3 miles from Minskip to Ferrensby, with a small bridge at Boroughbridge
1754–1755. Knaresborough–Harrogate, 1½ miles, including bridge over Starbeck, £400
1754–1755. Harrogate–Harewood Bridge, 6 miles, £1,200

1754–1755. Chapeltown–Leeds, 1½ miles, widened Sheepsar Bridge, almost £400
1754–1755. Skipton–Colne–Burnley: 2 miles Burnley; 2 miles from Broughton to Marton; 2 miles through Addingham, £1,350
1756. Wakefield–Halifax, 4 miles from Mill Bridge to Belly Bridge, 3 miles from Belly Bridge to Halifax
1756. Wakefield–Dewsbury, 5 miles from Wakefield to Checkingley Beck
1756. Hag Bridge–Pontefract, 3½ miles
1756. Wakefield–Doncaster, 1½ miles from Crofton to Foulby
1759–. Wakefield–Austerlands: 9 miles from Blackmoorfoot through Marsden to Stanedge Foot; 2½ miles from Lupset Gate to Horbury; 3 miles from Stanedge to Thurston Clough; 6½ miles to Huddersfield; 1½ miles from Huddersfield to Longroyd; including three tollhouses, around 21 miles, for £4,500
1759–. Guide Bridge: 18 miles, Docklanehead to Ashton-under-Lyne, Guide Bridge–Stockport and Stockport toMottram-in-Longdendale; 8 miles, including embankment 9 yd. high, £2,000; 4 miles between Whaley Bridge and Buxton; £1,100
1759–. Huddersfield–Sheffield, 1½ miles at Highflats, £300
1777–1778. Huddersfield–Halifax, 8 miles, £2,711
1777–1778. Congleton–Red Bull, 6 miles, £3,000
1781. Knaresborough–Wetherby (Ribston through Kirk Deighton to Wetherby), £380
1781?. Marsden, New Mount Road, 1½ miles, including Ottiwell's Bridge £1,000
1782. Wilmslow–Congleton
1789–1791. Bury–Haslingden, Haslingden–Accrington, branch to Blackburn, £3,500

Various other roads, including parts of the A5002 and B5470 near Buxton and the B6265 in the Yorkshire Dales, not mentioned in the Life, are attributed to him.

MILLAR, Archibald (fl. 1785–1830), canal engineer, began his engineering career on the Forth and Clyde Canal, and in 1785, with the surveyor John Laurie, proposed the change of terminus from Dalmuir to Bowling. He then worked as an overseer with the Grand Canal Company in Ireland and progressed to being in charge of the construction of the canal from Monasterevan to Athy in the early 1790s. He left the company in 1793 and went to England to become resident engineer on the Lancaster Canal. Problems were experienced with the contractors Pinkerton & Murray (qq.v.) and finally in September 1795, following arbitration by Robert Whitworth (q.v.), their lots were re-let under Millar's superintendence. By that time the supervisory staff had been increased to include Henry Eastburn, Thomas Fletcher and William Cartwright (qq.v.). Millar was paid £400 p.a. for his role as resident engineer on the Tewitfield–Calder section. In 1799, by which time the major work, the Lune aqueduct was completed, Millar's position was not renewed, leaving Cartwright to complete the outstanding works.

RON COX

[BR/FCN/11, Scottish RO; C. Hadfield and G. Biddle (1970) *The Canals of North west England*, **1**, 186–188; R. Delany (1973) *The Grand Canal of Ireland*, 35]

Works
c. 1790–1792. Grand Canal, Barrow line, resident engineer
1793–1799. Lancaster Canal, resident engineer

MILLINGTON, John (1779–1868), engineer, lecturer and writer was born on 11 May 1779 at Hammersmith, Middlesex, the son of Thomas Charles Millington, attorney, and his wife Ruth, née Hill. He entered the University of Oxford to study law but was forced to withdraw due to the poverty of his father. From then until about 1819 he lived at Hammersmith, when he moved to Bloomsbury.

In the period after 1803 Millington had a practice as a patent agent and indeed in 1816 took out a patent for propelling boats. In 1806 he was elected a Fellow of the Society of Arts and in 1815 was engaged by the Royal Institution to deliver a course of lectures on natural philosophy which he repeated the following year. From then until 1829 he delivered a course of lectures annually at the Royal Institution on a wide range of subjects including agriculture, the uses of steam power, mechanics, magnetism and electricity. He was appointed Professor of Mechanics there and was the first person to deliver the Christmas lectures for children at the Royal Institution in 1825–1826.

Millington was appointed County Surveyor of Bedfordshire in 1816 and this possibly brought him into contact with McAdam (q.v.), with whose work he was familiar. As County Surveyor he was responsible for work on the cast iron bridge at Broom. He owned a foundry at Webb's Lane Hammersmith. He also briefly worked for the West Middlesex Waterworks, based in Hammersmith, in the latter part of 1810.

His work for Bedfordshire brought in fee income and expenses-£153 in 1822, and £115 in 1823. Although he received a testimonial in 1821 he was dismissed in 1825 as he was so often absent due to other commitments. Aside from this little is known of any practical engineering commissions although, following the decision in 1828 to replace the existing bridge at Marlow, he drew up plans for a suspension bridge there. It featured eye bar chains suspended over cast iron towers on stone pedestals. The drawings were prepared by his pupil William Carpmael (1804–1867) but Millington was replaced as engineer by William Tieney Clark (q.v.) before the bridge was built; he had found a more promising situation.

The 1820s were years of intense activity for Millington. He married Emily, the daughter of William Hamilton, the painter. Out of some of his lectures he published in 1823 his *Elementary*

Principles of Natural Philosophy. He was elected a member of the Linnean Society in 1823 and was one of the founder members, in 1820, of the Royal Astronomical Society and served as one of its Secretaries between 1823 and 1826. He also lectured at Guy's Hospital, at the London Mechanics' Institution and, in 1827, he was appointed Professor of Engineering and the Application of Mechanical Philosophy to the Arts at the newly founded University of London (now University College London). However, he resigned this position, before delivering any lectures, the following year due to the inability of the university authorities to guarantee him a minimum annual income. He also provided expert evidence in a number of court cases and occasionally to Parliamentary committees.

In late 1829 Millington accepted, on a three year contract, the position of Chief Commissioner of the Anglo Mexican Mining Association at Guanaxuato where he oversaw the mining and minting of silver there. While working there he prepared the second edition of his *Elementary Principles of Natural Philosophy*. At the expiry of his contract he moved to Philadelphia in 1832. His wife died on her way to join him there leaving him with several children. In Philadelphia he set up in business as a philosophical instrument maker, took the degree of MD at Jefferson Medical College and married Sarah Ann Letts as his second wife. In 1836 he was appointed Professor of Chemistry and Natural Philosophy at the William and Mary College, Williamsburg, Virginia, a position which entailed a heavy lecturing load. There he wrote his 'Elements of Civil Engineering' published in 1839, an encyclopaedic text reflecting the polymathic knowledge he had acquired over the previous thirty years. It was only the second such general civil engineering textbook to appear in English, and anticipates that of W. J. M. Rankine by twenty years.

In 1848 he was appointed Professor of Natural Sciences at the newly founded University of Mississippi at Oxford and was appointed state geologist there. In 1853 he was appointed Professor of Chemistry and Toxicology at Memphis Medical College, Tennessee. He retired at the age of eighty in 1859 and lived in La Grange, Tennessee until the Civil War forced him to flee to Philadelphia. He then lived with his daughter in Richmond, Virginia. He died on 10 July 1868 and was buried in Bruton Parish churchyard, Williamsburg.

FRANK A. J. L. JAMES

[Millington's papers and books, Swem Library, William and Mary College; West Middlesex waterworks papers, acc 2558/wm/1/2, GLRO; House of Commons(1829) *Select Committee on the Law relating to Patents for Inventions, Report*, HC 332, 98–103; C. S Drewry (1832) *A Memoir on Suspension Bridges*; F. A. J. L. James (1991–1993) *The Correspondence of Michael Faraday*, **I–II**; DNB; DAB]

Publications

1816. Account of an hydraulic machine for raising water, called the 'water ram', *Quart J Sci*, **I**, 211–215

1818. On street illumination, *Quart J Sci*, **V**, 177–181

1823. *An Epitome of the Elementary Principles of Natural and Experimental Philosophy* (2nd edn. 1830)

1839. *Elements of Civil Engineering* (2nd edn. 1843)

Works

Broom Bridge, Beds, 20 ft. span

MILLS, James (fl. 1770–1842), civil engineer, was the son of Thomas Mills (c. 1735–1820), a Quaker bookseller in Bristol. His sister Seline married the radical Zachary Macaulay (1768–1838) in 1799; their eldest child was Thomas Babington (Lord) Macaulay. Despite these distinguished relations much about Mills is obscure although his acrimonious personality is not.

He was trained at a young age in civil engineering and from 1795–1802 he worked on the Western District of the Kennet and Avon Canal, possibly acting as clerk to the superintendent of works (resident engineer), Dudley Clark (q.v.). His first report of 4 June 1795 was in connection with the line around Bath, and the tunnel there. He is known to have acted as clerk of works/assistant engineer certifying work done by the contractors on the canal, such as Edward Bushell, contractor for the Bath tunnel and also claimed to have been involved in the design of the inclined plane from the Canal Company's Quay at Corkwell (1800–1801). In July 1800 Dudley Clark was dismissed and although Mills hoped to get his job John Thomas (q.v.), a fellow Quaker and wholesale grocer from Bristol, who was on the Committee, was appointed. Mills claimed he had trained Thomas, and evidently left the Canal Company's employment with some resentment. It was originally intended to give Mills a golden handshake of 100 guineas but Mills dragged his heels over handing over documents and financial records and in 1803 the Company called in a solicitor, Simon Ward, from Bristol, to sort out matters, the affair dragging on for many months, and it transpired Mills had collected canal tolls and not passed them on to the Company. Mills was succeeded as clerk by Benjamin Davis. His relations with the Canal Company were to come back to haunt him.

Mills later claimed involvement with the Somersetshire Coal Canal although whether this was carrying out surveying and levelling work on its original construction or in connection with the Radstock Tramway (1814–1815) is uncertain from his Parliamentary Evidence on the London–Brighton Railway (1836). He is known to have worked several years for John Rennie (q.v.) and although based in Bristol, he surveyed part of the proposed Portsmouth canal in 1803 for him.

Around 1806 he was once more involved with the Kennet and Avon working on the designs for bridges and aqueducts. He also carried out work for William Jessop (q.v.), in December 1802 reporting on the sluice at Highbridge on the Brue Drainage. He surveyed for the Trustees the turnpike road between Redruth and Penzance (1806). Work continued with Rennie who in 1817 provided him with a testimonial for the post of engineer to the River Wear Commissioners. Mills prepared a plan of the river and was so incensed when he was unsuccessful in his application that he published a broadsheet attacking the Commissioners and their refusal to meet his expenses. He scathingly described the experience of the successful applicant, Thomas Milton (q.v.): 'this young man never did anything but keep a public house aboard the Alexander frigate to supply the workmen...(at Sheerness dockyard)'. At that time Mills was still living near Bristol. However, in 1819 he may have fallen out with Rennie when Francis Giles (q.v.) was asked to survey the Wear, in preference to himself. Mills claimed to have defended Rennie's character against others, and expected more. At that time his address was 4 Church Passage, Guildhall, London.

Around 1822 Mills began working for Thomas Telford (q.v.) and claimed in 1834: 'I have made during the last ten years I think, all the principal surveys for the Government myself; Mr. Telford from his age and infirmities has merely signed his name to the surveys when they were completed'.

Mills was in fact only one of a number of engineers to whom Telford delegated work in the last decade or so of his life, other examples being William Mackenzie, John and William Provis and Henry Welch (qq.v.). His assertion that he was effectively doing Telford's work was clearly an exaggeration given the number of other assistants Telford used at that time.

Mills' earliest known work for Telford was a survey of the Thames between Teddington Lock and Putney Bridge (1823) for the New London Bridge Enquiry. This was followed by surveys of the Girvan–Stranraer Road (1824), Edinburgh–Morpeth and London–Retford Roads (1825–1826). He claimed involvement in the design of a bridge at Lanark, presumed to be Cartland Crags (1821–1822). His best-known work was his investigation of the state of the Liverpool and Manchester Railway, which formed the basis for Telford's report to the Commissioner for the Loan of Exchequer Bills (1829), strongly critical of George Stephenson's conduct of the works; this report was one of many that Telford provided as Engineer to the Exchequer Board from 1817 onwards. Mills claimed involvement with all those after c. 1824, as well as 1,400 miles of Post Office road surveys.

Mills and Telford fell out in October 1833; Telford was shown correspondence between Mills and the Kennet and Avon Canal Company and immediately broke off relations. It happened in the midst of the compilation of a report on London's water supply and the evidence presented to the House of Commons Select Committee the following year illustrates the breakdown in their working relationship. While Telford claimed to have employed Mills 'in perambulating the Countryside, taking trial levels', along with other engineers and surveyors, Mills claimed that 'from the very first commencement of this inquiry [1831] I believe I have conducted the whole under the name of Mr. Telford'. The radical MP, Francis Burdett, the chief mover in the establishment of the Inquiry saw Mills as the key figure and indeed Mills had been calling for action on the state of London's supply since around the time of his survey of the Thames. Mills felt Telford had disregarded his advice in compiling the report but his own proposals were generally very heavily criticised by the Water Company Engineers who gave evidence. Following Telford's death Mills put in a claim for £1,200 against his estate. His bitterness was still evident when he gave evidence on the London–Brighton Railway in 1836, in support of Cundy's scheme, evidence which reveals weaknesses in his apparently rushed survey. In the following year the preamble of his own scheme failed at the standing orders stage.

In the early 1840s Mills moved from Battersea Fields, where he had lived for many years, to Paris, becoming involved in assessing the alternative routes from Calais and Boulogne to Paris. He joined the Institution of Civil Engineers in February 1825 and was deleted for non-payment in 1844. He was described by Joseph Mitchell (q.v.) as a radical, 'very clever and a great talker'.

MIKE CHRIMES

[Minutes of the Kennet and Avon Canal Company, PRO RAIL 842/1–7; Rennie papers, National Library of Scotland; Telford mss, ICE Library; Thomas Telford (1833) Diary, National Library of Scotland; *Royal Cornwall Gazette*, 2 August 1806; *Report from Select Committee on Metropolis Water, with Minutes of Evidence* [571], 1834; *Minutes of Evidence taken before the Committee on the London and Brighton Railway Bill*, House of Lords and House of Commons, 1836; *Railway Times*, 7 March 1840; J. Mitchell (1883) *Reminiscences of my Life in the Highlands*, repr. 1971; A. Gibb (1935) *The Story of Telford*; G. G. Hopkinson, Road development in South Yorkshire and North Derbyshire 1700–1850, *Trans Hunter Arch Soc*, **10**, 1971, 14–30; Skempton]

Publications

1828. *Report of the Commissioners ... into the State of the Supply of Water in the Metropolis*, HC 267
1829. *Report on the Sheffield to Barnsley Turnpike*
1834. *Report from Select Committee on Metropolis Water*

Works

1795–1806. Kennet and Avon Canal, assistant engineer

1802. Brue Drainage, survey
1803. Portsmouth Canal, survey
1806. Redruth and Penzance Turnpike, survey
1820–1823. Great North Road, Edinburgh–Morpeth survey
1823. River Thames, survey between Teddington Lock and Putney Bridge
1824. Girvan–Stranraer Road, survey
1825. Birmingham–Liverpool (Rock Ferry) Railroad, estimates, fees £52 (Glasgow Portpatrick Road)
1825–1826. Great North Road, London–Retford Road, survey
1827–1828. Dee and Mersey Ship Canal, survey
1828. East and West India Docks railroad, survey
1828–1829. Liverpool–Manchester Railway, inspection and report
1828–1829. Clarence Railway, survey and report
1828–1834. London water supply, surveys and reports
1829. Sheffield–Barnsley Road
1836–1837. London–Brighton Railway, surveys

MILNE, James (fl. 1800–1840), civil engineer, worked as a resident engineer on a number of projects designs by John Rennie and his son (Sir) John (qq.v.). The earliest of these were the ferry piers at Queensferry on the Forth.. This was followed by work on the pier at Newhaven near Edinburgh. He almost certainly produced a survey of the Dee–Mersey Ship Canal in 1818. In the early 1820s he was resident engineer on Cramond Bridge, designed by Rennie (Sr.), but completed posthumously by (Sir) John.

In the late 1820s it is not known where he was working, although probably practising in the Edinburgh area. In 1832 he was appointed resident engineer on the harbour improvements at Hartlepool. Unfortunately, he came into conflict with the engineer, Sir John Rennie, over the design and execution of the work and was dismissed the following year and replaced by James Brown (q.v.). It seems possible the he is the James Milne identified by Colvin as a mason and architect of Edinburgh (fl. 1809–1834) who was living in Newcastle in 1834, and who wrote *Elements of Architecture* (1812), a work owned by Telford, and a tract on bridges (1816). James Milne reported on Newcastle Quay wall c. 1838.

A paper by a James Milne was presented on a piling engine at Montrose Harbour works at the Institution of Civil Engineers in 1844; as he was not a member it is impossible to verify his background. Unfortunately it is a common name. Colvin identifies two other examples, a Leith builder (d. 1826), and **James Milne** (c. 1792–1863), Northamptonshire County Surveyor, 1826–1863. Moreover a **James Milne** (fl. 1829–1832), stone and marble merchant, of 117 Millbank, Westminster was proposed as an Associate of the Institution of Civil Engineers in 1829 remaining in membership until 1832

A relationship with **John Milne** (fl. 1822–1842), the Leith civil engineer and architect, who

reported on Leith Harbour (1835–1839), and specialised on the ventilation of buildings is similarly elusive.

MIKE CHRIMES and P. S. M. CROSS-RUDKIN

[Rennie reports, ICE archives; membership records, ICE archives; Tidal Harbours Commission (1847) *2nd Report*, appendix C; RIBA Library catalogue; Bendall; Colvin; Skempton]

Publications
1816. *An Enquiry into the Theory and Principles of Bridges and Piers ...*
1844. Description of the piling machine used at Montrose harbour works, *Min Proc ICE*, **iii**, 197–200

Works
1809–1818. Queensferry Piers, Firth of Forth, resident engineer, £34,000
1814–1820. Newhaven Pier, Firth of Forth, near Edinburgh, resident engineer, £18,000
1821–1824. Cramond Bridge, contractor
1832–1833. Hartlepool Harbour, improvements, resident engineer
[1837–1842]. Montrose Harbour, improvements, including 3¼ acre wet dock, James Leslie (q.v.) engineer

MILTON, Thomas (1780/1781–1857), civil engineer, was born in Rochester, Kent, his father perhaps a contractor. Nothing is known of his early years other than the fact that he married Sarah Smith in 1808, the marriage resulting in at least three children, born between 1809 and 1820.

He was appointed as engineer to the River Wear Commissioners in May 1817, in succession to Matthew Shout (q.v.). His appointment caused some acrimony on the part of one of the unsuccessful applicants, James Mills (q.v.), who later wrote to the Commissioners to the effect that Milton was 'well known at Newcastle as an actor on the stage under the name of Kendal' and had done little engineering work other than in a very minor role at Sheerness. Milton's first task at Sunderland was to report on the cost of providing a dredger with a double set of buckets to further deepen the river. In 1821 the Commissioners called upon John Rennie Sr. (q.v.) to suggest harbour improvements and he subsequently recommended the extension of both north and south piers in masonry. Milton supervised the building of a 230 yd. long extension of the southern pier, 40 ft. wide and 10 ft. above high water; comprising stones of up to 7 tons in weight it was paved in freestone with a promenade and parapet and its construction involved the use of a diving bell. While employed at Sunderland, Milton's salary was £300 p.a. and in 1824 he was granted permission by the commissioners to purchase a share in a ship.

In 1826, while at Sunderland, Milton undertook dredging work for the Hartlepool Port and Harbour Commissioners and when, under the influence of Christopher Tennant, a prospectus for the

Hartlepool Dock and Railway Company was published in 1831 he became responsible, with Sir John Rennie (q.v.), for the design of its docks; George Stephenson (q.v.) was engineer for the railway to the port. Construction followed authorisation but not to either Milton's or Stephenson's designs although Rennie remained as consultant.

Following the death of his first wife, in 1823 Milton married Ann Whinnern, the union producing a daughter, born in 1828. While at Sunderland, the family first lived in the Pier House, later moved to Vine Place and, from 1830, lived in a new house adjacent to the south pier, also used for meetings of the commissioners. In May 1830 he was elected a Member of the Institution of Civil Engineers, having been nominated by James Walker (q.v.). Milton resigned his appointment with the RWC in January 1832 and was succeeded, very briefly, by James Leslie (q.v.) who, in turn, was followed by John Murray (q.v.).

In April 1834 Milton was appointed as Bridgemaster for the county of Cumberland where for thirty years the supervision of the county's bridges had been undertaken by two district surveyors, with responsibilities delineated East and West by Ward boundaries. His arrival there occasioned some ill-feeling concerning the two men whose work he took over in that they were reluctant to hand back to the Justices papers considered to be county property.

His first major task was to implement the reconstruction of the bridge over the River South Tyne at Alston where there had been a history of failures and its completion was achieved in 1836; the rebuilding of several other county bridges followed, as did general maintenance and repair work. In 1850 Milton agreed, in addition to his bridge responsibilities, 'to undertake the superintendence of the whole of the County Property', so embracing court-houses, gaols and police 'lock-ups'.

Milton married a third time in 1847, his marriage then to Elizabeth Pattinson (b. c. 1822) resulting in another daughter, born in the same year. In 1851 he was living in the village of Botcherby but he later moved to Spencer Street, Carlisle. He resigned his appointment at Easter 1856—by which time his salary was £400 p.a.—but he continued in post for some six months, being then succeeded by John A. Cory, of Durham. He died on 16 July 1857 and was buried in Carlisle Cemetery five days later, aged seventy-seven.

R. W. RENNISON

[*River Wear Commissioners: Minute Book* (1817–1832) T&WAS 202/1003 and 1004; *Hartlepool Port and Harbour Commissioners: Minutes* (1817–1832); Rennie ms, 19804 f.155 NLS; *Minutes of the Hartlepool Dock and Railway Company* (1831, 1832) RAIL/294/14, PRO; *Public Orders* (1834–1856) Q/7/6 and 7, Cumbria RO; Membership records, ICE; J. Murray (1847) An account of the

progressive improvement of Sunderland Harbour and the River Wear, *Min Proc ICE*, **6**, 272]

Works

1822–1832. South pier, Sunderland; supervision of building
1834. Crofton Bridge, 28 ft. single span, 18 ft. wide
1834. Ennerdale Bridge, 36 ft. span, 16 ft. wide
1834. Threlkeld Bridge, two spans of 46 ft., 20 ft. wide
1834–1836. Alston Bridge, two spans of 35 ft., 20 ft. wide
1838. Irt and Bull Gill-brow bridges, rebuilding
1840–1841. Rack Bridge, three spans of 33 ft., 16 ft. wide; Stakes Bridge, two spans of 24 ft.
1841. Workington Bridge, design, three arches
1844–1846. Low Wiza Bridge, two spans, 25 and 12 ft.
n.d.. High Gelt Bridge, single span, 36 ft., 21 ft. wide
n.d.. Southwaite Bridge, two spans, 24 and 21 ft.

MILTON, William (1749–fl. 1793), canal contractor, was born near Blackstone Edge, Lancashire, c. 1749. His first job was on the construction of the Blackstone Edge Turnpike under Alexander Midgley, a work which required special construction techniques to deal with the ground, involving the use of small reservoirs to wash away unwanted earth. He then began a peripatetic career in canal construction over a period of twenty-eight years, returning to his home c. 1790. In that period he was involved in the construction of five reservoirs, beginning on the Staffordshire and Worcestershire Canal, and latterly on the Thames and Severn Canal.

Milton carried out borings for the Blackstone Edge reservoir on the Rochdale Canal and reported in favour of the scheme to Parliament.

MIKE CHRIMES

[Minutes of Evidence on the Rochdale Canal, 23 February 1793; Rennie Reports, ICE Library]

MINTO, William (1761–1847), civil engineer and contractor, son of John Minto and Barbara Murray, was born at Inverkeithny, Banffshire. In 1809 he married Jane Morison by whom he had eight children, the last being Farquharson Minto (d. 1915) who was for forty years a road surveyor in the counties of Aberdeen and Forfar. Circa 1810 Minto became domiciled at Alford, Aberdeenshire, where he subsequently conducted his activities as architect, engineer and contractor. His architectural practice does not appear to have been significant and his main professional involvement with civil engineering was as a masonry contractor employed on the construction of bridges and harbours, and to a lesser extent as a road contractor and as a builder.

His association with bridge building commenced in 1805 when he was linked with the construction of a bridge at Marnoch, Banffshire, but it is not known whether he was employed as the contractor or as a working mason. In 1810 having removed to Alford, his career as a contractor became established when he built bridges at Alford and Potarch for the Commission for Roads and Bridges in the Highlands of Scotland. Thomas Telford (q.v.), the Commission's engineer, became aware of Minto's prowess as a contractor by the diligence and resolution he displayed when completing these contracts. This occurred despite the unforeseen fall, caused by logs becoming jammed against the piers and temporary works, of the almost completed Potarch bridge. On three occasions whilst these two bridges were under construction arches were turned within twenty-four hours; the largest of these arches was a 65 ft. masonry span at Potarch where the total weight of the masonry voussoir blocks in the arch ring was 300 tons.

In 1816 Minto completed his contract for the construction of Keig Bridge, Aberdeenshire. This bridge had a single span of 100 ft. and it was constructed to the design specification used by Telford for Highland bridges. Prior to this, Minto had widened his engineering activities by taking the contract to construct the Commission's road between the bridges of Alford and Potarch. After the completion of Keig bridge he constructed piers at the Moray coast ports of Cullen and Fraserburgh, the latter being designed by Robert Stevenson (q.v.).

Before work at Fraserburgh was complete, Minto, as the Commission for Roads and Bridges had been unable to attract sufficiently competitive offers, was invited to tender for the construction of a bridge at Cartland Crags, Lanarkshire. This bridge over the Mouse water had three spans of 52 ft. with a clearance to the crown of the central arch in excess of 120 ft. At the time of completion it was recorded that 'Its altitude is... superior to any other bridge in Great Britain'. Whilst working in Lanarkshire he also constructed bridges at South Calder and Fidlers Burn and constructed the Carluke section of the Lanarkshire roads.

Returning to Aberdeenshire he designed the East Side Bridge at Turriff (built by William Smith of Montrose) and then, after the disastrous floods of August 1829, was employed reinstating bridges in north-east Scotland. The major contract associated with this work was the rebuilding of the Spey Bridge at Fochabers. Two arches of this four-span bridge (designed by George Burn (q.v.)) had been destroyed and Minto, in partnership with William Leslie of Aberdeen, replaced the fallen spans by a single span comprising three timber arch ribs of 184 ft. span. The replacement span was designed by Archibald Simpson, the Aberdeen architect. Each rib had an average depth of 3 ft. and was fabricated from six timbers, laid three deep and two wide in the manner proposed by Thomas Tredgold (q.v.).

Between 1827 and 1829 Minto, acting alone and in partnership with John Gibb, Aberdeen, built a number of manses and churches in the Highlands of Scotland and on the island of Islay. These were designed by Telford, and the Commissioners for Roads acted as the agents for governmental expenditure. After 1832 Minto became less active professionally and he died at Alford in 1847.

TOM DAY

[*Reports of the Commissioners for Roads and Bridges in the Highlands of Scotland*, 1811, 1813, 1815, 1817 and 1821; T. Day (1987) The Old Spey Bridge, Fochabers, *Industrial Archaeology Review*, **10**(1), 75–81; A. MacLean (1989) Telford's Highland Churches; T. Day (1995) Telford's Aberdeenshire Bridges, *Industrial Archaeology Review*, **17**(2), 199–202; T. Day (1997) Did Telford rely, in Northern Scotland, on vigilant inspectors or competent contractors?, *Construction History*, **13**, 101–112]

Works

1805–1806. Marnoch Bridge
1810–1811. Bridge of Alford
1812–1814. Potarch Bridge, 65 ft. span
1816–1817. Keig Bridge, 101 ft. span arch
1816–1819. Alford–Potarch Road
1817–1819. Cullen Harbour, improvements
1818–1821. Fraserburgh Harbour, improvements, c. £6,000
1821. Fidlers Burn Bridge, three arches, 105, 150 and 105 ft.
1821–1822. Cartland Crags Bridge, three 50 ft. spans
1821–1822. Carluke Road
1826. Turriff, Eastside Bridge
1827–1829. Turriff, Churches and Manses
1830. Carr Bridge, repairs
1831–1832. Spey Bridge, Fochabers, repairs

MITCHELL, James (fl. 1699–1710), carpenter of Bradley near Huddersfield, worked on the locks on River Aire navigation from Knottingley to Leeds 1699–1701 under John Hadley (q.v.). In July 1701 he and his son Joshua, jointly with James Willans and his two sons, masons of Dewsbury, contracted to build four locks on the River Calder to extend the navigation from the Aire at Castleford up to Wakefield. They completed the work in April 1702 at a cost of £2,300, and thereafter held the maintenance contract for seven years. Also in April 1702 Mitchell and George Atkinson (q.v.) agreed to build locks and weirs at Beal and Haddlesey on the Aire below Knottingley. This work was completed about a year later at a cost or rather more than £1,600. Mitchell then engaged in the coal and lime trades on the rivers. He seems to have retired after 1710.

A. W. SKEMPTON

MITCHELL, Joshua (fl. 1701–1723), worked with his father, as mentioned above. The maintenance contract on the Calder was held by Joshua Mitchell and the sons of James Willans. Mitchell also continued his father's trade as a carrier on the rivers, at least to 1716.

In 1720 the Hon. Thomas Wentworth (later Earl of Malton) agreed with Mitchell and Mark Andrews of Knottingley that they would undertake to make the Yorkshire Derwent navigable. This they succeeded in doing from its junction with the Ouse up to Malton, a distance of 38 miles with a towing path and five locks large enough to take Yorkshire keels 55 ft. by 14 ft. The work was completed in 1723 at a cost to Wentworth of about £4,000. He then leased the tolls to Mitchell and Andrews.

Meanwhile in 1722 William Palmer (q.v.), employed as a surveyor on the Derwent, and Joseph Atkinson (q.v.) were commissioned to survey the River Don with a view to improving the navigation to Doncaster and extending it to Sheffield. They brought in Mitchell to help. The job was finished in November, after four weeks. Mitchell and Palmer took the levels or 'falls' and Atkinson the distances. The map, by 'Will. Palmer & Partners', with tables of the falls and distances, was published in 1723. Nothing has yet been found of Mitchell's later activities.

A. W. SKEMPTON

[Will, Palmer & Partners (1723) *A Survey of the River Dunn*; T. S. Willan (1965) *The Early History of the Don Navigation*; R. W. Unwin (1967) The Aires and Calder Navigation, *Bradford Antiquary*, **43**, 151–186; Charles Hadfield (1972) *Canals of Yorkshire and North East England*; W. N. Satcher (1972) in lit.]

Works

1701–1702. James and Joshua Mitchell with James Willans, River Calder Navigation, £2,300

1702–1703. James Mitchell and George Atkinson. Beal and Haddlesey locks, £1,600

1720–1723. Joshua Mitchell and Mark Andrew, Yorkshire Derwent Navigation, £4,000

MITCHELL, Joseph, FRSE (1803–1883), civil engineer, was born in Forres, Morayshire, on 3 November 1803, the eldest of eight children of **John Mitchell** (1779–1824), a stonemason, and his wife Margaret Philip. John Mitchell was employed on the building of the Caledonian canal, where his diligence and skill were noticed by Thomas Telford (q.v.), who afterwards described him as 'a man of inflexible integrity, a fearless temper and indefatigable frame'. In 1804 he was recruited by Telford to superintend the building of some of his more important bridges and for the last fourteen years of his life he was principal inspector of the roads and bridges in the Highlands of Scotland. He was said to have covered upwards of 10,000 miles on his tours of inspection every year, which proved too much in the end even for his 'indefatigable frame'.

John Mitchell and his family moved to Inverness in about 1810 and Joseph attended the Academy there for three and a half years. His education was completed by a year of studies in Aberdeen and, being determined to follow his father into engineering, he too began as an apprentice stonemason. In 1820 he was sent to work on the locks at Fort Augustus on the Caledonian canal, where he also had the opportunity to study various other aspects of practical civil and mechanical engineering.

Within a year he had so impressed Telford that he was invited to become his assistant in London and he spent the next three years there, becoming an Associate of the Institution of Civil Engineers in 1824. He had the distinction of being the first to take notes of the proceedings of the Institution at their meetings and, Telford having discovered them, he read out the digest at the next meeting, this practice leading in due course to the published *Minutes of Proceedings*. He was transferred to Member of the Institution in 1837 and in 1843 he was elected a Fellow of the Royal Society of Edinburgh.

The illness of his father in the summer of 1824 prompted Telford to send Joseph back to Inverness, where John Mitchell died two months later. Telford thought that his son should succeed him, but as Joseph was then only twenty-one years old the commissioners insisted that he was appointed initially on six months' trial, at the end of which he was confirmed as general inspector and superintendent of the Highland roads and bridges, a post he held for almost forty years. In addition to these responsibilities, the commissioners employed him to build forty new churches throughout the highlands and islands of Scotland.

As well as his official appointments, he had in his early professional life a large private business in Perthshire, constructing, within the space of a few years, bridges, embankments and roads in that County to the value of £180,000. For many years after his father's early death Joseph supported his mother and her seven younger children. On 27 January 1841 in Inverness he married Christian, daughter of James Dunsmure, formerly secretary to the Fisheries Board in Edinburgh; they had three children, two daughters and a son who became a minister in the Church of Scotland.

From 1828 to 1850 he acted as engineer to the Scottish Fisheries Board, building and improving numerous harbours. Mitchell also played a major part in the planning and construction of the railway system in the Highlands, latterly in partnership with William and Murdoch Paterson, from about 1845 until his retirement in 1867, following a stroke he suffered in 1862 and from which he made only a partial recovery. The second volume of his *Reminiscences* was in the printer's hands when he died on 26 November 1883 at 66 Wimpole Street, his residence in London. His estate amounted to £99,524 2s 6d. There is a marble bust of Joseph Mitchell by Alexander Munro (1825–1871) in the Town Hall, Inverness.

R. BIRSE

[Correspondence, etc., ICE archives; SRO; Obituary notice (1884) *Proc ICE*, **76**, 362–368; J.

Mitchell (1883–1884) Reminiscences of my life in the Highlands; Sir Alexander Gibb (1935) *The Story of Telford*; J. Marshall (1978) *A Biographical Dictionary of Railway Engineers*; A. W. Skempton (1987) *British Civil Engineering 1640–1840: a Bibliography*; DNB; W. T. Johnston (1998) *Scottish Engineers and Shipbuilders*, computer database]

Publications

Publications (up to 1830):

1828. *Answer … to Lord Colchester,* [and] *Notices of the improved State of the Highlands of Scotland in 14th Report of the Commissioners for repair of Roads and Bridges in Scotland*
1830. *Report and Estimate of the Several Bridges and other Works injured by the Floods of August last in 16th Report of the Commissioners for repair of Roads and Bridges in Scotland*
1830. *Report and Description of Churches and Manses in 6th Report of the Commissioners for Building Churches in the Highlands and Islands of Scotland*

Mitchell's reports continue to feature in these Commissioners' reports after these dates.

Works

Works (up to 1830):

1824–1863. For the Commissioners of Highland Roads and Bridges, general inspector and superintendent of road and bridge works
1824–1835. For the Commissioners of Highland Roads and Bridges, surveyor of the forty new Churches built in the highlands and islands of Scotland
1828–1850. For the Scottish Fisheries Board, building, maintaining and improving harbours throughout Scotland
pre-1830. Bridges, embankments and roads in Perthshire to the value of £180,000

MONKS, Daniel (fl. 1759–1810), canal engineer, was the son of William Monks, overseer on the Grand and Royal canals. Daniel was engineer-in-charge of the Boyne Navigation and later the Coalisland Canal on the Tyrone Navigation. In 1801 he reported on a proposed still-water canal from Dublin to the northern counties of Ireland via Trim and Navan. Daniel died in 1810 and the administration of the Coalisland Canal was taken over by father William.

RON COX

[*Trans ICEI*, **6**, 1859 1861, R. Delany (1992) *Ireland's Inland Waterways*]

Publications

1801. *Report to the Committee appointed to Report on the Practicability and Advantages of a Still Water Navigation, to extend from the City of Dublin, through the Northern Counties of Ireland. Navan, January 26th*

Works

c. 1790s. Boyne Navigation, assistant engineer
1800–1810. Coalisland Canal, engineer-in-charge

MOORE, Sir Jonas, FRS (1617–1679), Surveyor General of drainage works in the Fens and later of the Royal Ordnance, was born on 8 February 1617 at White Lee near the village of Higham in Pendle Forest, Lancashire, as the second son of Hugh Moore, a prosperous yeoman, and his wife Mary (Aspinall). It is probable that Jonas went to a local grammar school, perhaps at Burnley. He appears not to have attended a university but continued his education by reading mathematical books; he was assisted and encouraged in this by friends, especially members of the Shuttleworth family.

In November 1637 he was appointed clerk to the Chancellor of the diocese of Durham. The post proved short-lived as the courts ceased to function when the region was invaded by the Scots in August 1640, and formally ceased to exist in October 1642. He married Eleanor Wren at St. Giles, Durham, on 8 April 1638. Their elder daughter, Mary, was baptised in March 1639 and the younger, Helen, in January 1641. A son, Jonas, was the youngest of the three children; the date and place of his birth are unknown.

In 1640 Moore furthered his mathematical studies by visiting William Milburne at Brancepeth. He probably left Durham before the end of 1642 and his whereabouts during the Civil War are not known. He emerges from obscurity in 1647, acknowledged in the preface to *The Key of the Mathematicks* (the first English edition of William Oughtred's famous *Clavis Mathematicae*) for the care he had taken in correcting and proof-reading the volume; he was also tutor in mathematics and geography to the young Duke of York (later James II). During the next three years he had other well-connected pupils. His *Arithmetick* was published in 1650. Its preface is dated 30 October 1649 from 'my chamber' in the house of Elias Allen, a leading instrument maker and close friend of Oughtred.

Moore later attributed 'his rise' to his work as Surveyor General to the fifth Earl of Bedford's Company of Adventurers for draining the Fens. His appointment to the post, on 26 August 1650, undoubtedly resulted from his presence in this circle of mathematical practitioners and their patrons in London.

Much had been accomplished in the first campaign (1631–1636) of fen drainage under the fourth Earl of Bedford and in the second (1640–1641) under Charles I with Sir Cornelius Vermuyden (q.v.) as Director of Works. Completion of the project required further works on an extraordinary scale in the third and final campaign. They were carried out between 1650 and 1656 under an Act passed on 29 May 1649.

Moore is briefly mentioned in the Company's minutes towards the end of June, offering to draw a map of the Level, but this was merely an incident in a long and confused period of conflicting schemes, a false start and negotiations with Vermuyden, finally resolved by the adoption of the latter's plan and his appointment as Director of Works on 25 January 1650. During the

Sir Jonas Moore FRS

negotiations (in the previous December) Vermuyden told the Company that a Surveyor and four Overseers would be required; this is possibly an indication of the staff he had employed in the second campaign. Within days of his own appointment Vermuyden reminded the Company that a Surveyor should be found 'to sett out the works and see that they bee done'. Why it took until August to seek one is not clear.

Vermuyden lost little time in giving 'directions' to the principal Overseer, Robert Burton (q.v.), who began organising the first works, in the vicinity of Wisbech. Burton was joined in March by Sir Cornelius himself 'to stake out the worke' and during the first season he may have been personally involved in tasks that were later carried out by the Surveyor. Moore did not arrive on site until the beginning of September. At first he probably stayed in lodgings close to wherever he happened to be engaged but in February 1652 he brought his family to live in the Fens, at March; in 1653 they moved to Southery on the Fenland edge of Norfolk and remained there until 1657.

Vermuyden's duties as Director were to plan the scheme; to decide what works should be done, where and when, and their dimensions; and to issue directions accordingly to the Surveyor and principal Overseer. The Surveyor, constantly on site, set out the works, measured land for compensation and, with his assistants or 'undersurveyors', divided the land after drainage; he dealt with boundary divisions, they with the internal subdivisions. Moore was also responsible from January 1653 for planning and setting out 'high wayes and Lanes throughout the whole

Levell'. The Overseer 'let out' contracts. He and his assistants had to oversee construction and 'take up' or measure quantities for payment of work done. Overall control resided in a group of Adventurers which included the Treasurer, Expenditor, Comptroller and the Superintendent, Anthony Hammond, who acted in effect as project director. In addition, several of them were deputed to 'husband' or supervise works in the field on behalf of the Company. Moore, who had a salary of £200 a year, was closely associated with Hammond in seeing 'that the works bee done'. Hammond received an allowance of £20 per month for what was virtually a full-time job and the Overseers and undersurveyors were paid £1 a week.

At its maximum, in the summer of 1652, the workforce numbered ten thousand men and payments to them peaked at £8,000 a week, as later reported by Hammond. Total expenditure was about £250,000, the greatest amount spent on any civil engineering work in England during the seventeenth century; it should be compared with £97,000 on the first campaign and £23,000 on the second.

Drainage of the North and Middle Levels (170,000 acres between the Welland and 'old' Bedford River) was declared complete by Commissioners meeting in March 1651 and hearing evidence from, among others, Moore and Burton. Work then proceeded in the South Level (140,000 acres) and was adjudged complete in March 1653, the main evidence being given by Vermuyden. Works in the South Level included the Hundred Foot or New Bedford River 21 miles long from Earith to Salters Lode with barrier banks on either side half a mile apart; a dam and navigable sluice on the Ouse at Denver; a slaker channel 114 ft. wide running 4 miles from Denver to outfall sluices (with 72 ft. waterway) at Stow, and banks along the Ouse tributaries, Cam, Lark, Little Ouse and Wissey.

Marshland Cut or Tongs Drain was made in 1653, after the adjudication, when Moore was granted extra powers in Vermuyden's not infrequent absence. From January 1653 he had a deputy, Gabriel Elliott, one of the five land surveyors then employed, their number increased from four in 1652. Edmund Welch (q.v.) was now the principal overseer, having in 1652 replaced Robert Burton who died early in the next year. Under Welch there were three overseers including George Barnes and Jeffrey Hawkins.

On 17 August 1654, in a reorganisation of the Company, the post of Surveyor was terminated and Moore became the Overseer at £100 a year; Hammond was appointed Conservator of the South Level at £150 a year with Barnes as his assistant (Bailiff); other Adventurers were Conservators of the North and Middle Levels. Welch became 'principal contractor' at £60 a year, presumably in charge of the direct labour force engaged on the relatively minor works still in progress such as highways, bridges, division dykes and so on. However, experience soon showed the need for an additional sluice at Denver and Moore drew up the contract in April 1655 for a timber-framed structure providing a 30 ft. waterway. It was confirmed by Edward Barber (supervisor), certified by Barnes and awarded to John Savery; the work was to be completed by June 1656 for £600. The cut and foundation pit were probably made by direct labour and may have cost as much as the sluice itself.

Vermuyden issued his last set of 'directions' in December 1655 and completion of Denver Sluice effectively marked the end of the undertaking. The company was again reorganised in September 1656, this along the lines of the long-established Romney Marsh administration with governors, bailiffs and jurats. Among the Jurats were Jonas Moore and the former overseers, Welch and Hawkins, as well as Elliott and John Savery; all ranked as 'gentleman'. Moore qualified in this respect by his purchase in 1653 of a 100-acre piece of Methwold Severals near Southery. He almost immediately used the property as security for a loan but before long lost it to a mortgagor. Later, however, one of his daughters bought back the same land.

So far as mapping the Fens is concerned, dozens of plans of parcels of lands were produced, but the first 'general' map since the pre-drainage survey by William Hayward (q.v.) of 1605 is one attributed to Moore and now existing only as an eighteenth century copy of an original probably dating from 1651. It is on a scale of about ¾ in. to a mile and shows the drains made up to that year. An engraved *Mapp of the great Levell of the Fenns at it now lyeth drayned ... Described by Jonas Moore* can be dated to 1654. It shows all the drains but on the small scale of ½ in. to a mile and is to some extent diagrammatic. Clearly an accurate large-scale map was desirable and this Moore proceeded to make, with permission and support from the Company but to be published at his own expense. On the scale of 2 in. to a mile, printed on 16 sheets and measuring about 6 ft. by 4½ ft. overall, the map is a landmark in cartography; it is unrivalled by anything previously done in precision, clarity and wealth of detail. Moore brought the engraver, John Goddard, to Southery to begin the work under his supervision; Moore and Elliott were allowed travelling expenses to London 'about the Mapp' several times in 1657. It must have been published early in 1658.

The great map was reissued in 1684 by Moses Pitt, 'pasted and coloured' for 30s, using the original plates with one or two very minor alterations and the coats of arms of mostly Commonwealth notables appropriately removed. The first and second editions now exist only as single copies, located in the PRO and Bodleian respectively. The third edition, ordered to be printed in July 1705 and issued without change by Christopher Browne, is well known.

Moore marked his return to London in 1658 by publishing a mathematical paper, followed by a pamphlet on algebra (1660) and a revised edition of his *Arithmetick* (1660). He was now living in Stanhope Street and remained, at two further addresses, in the fashionable west side of London for the next five years. During this period he derived his income from teaching and a variety of free-lance commissions, such as surveying the manor of Woburn (1661) and mapping the Thames from Westminster to the sea for the Navy Board (1662). Between June and September 1663 he visited Tangier as part of an expedition to deliberate upon the design of the Mole, the large breakwater wall planned to create a fortified harbour there, before work began under contract by Hugh Cholmley (q.v.). Moore was involved 'in setting out the Bounds of the Mole' and on his return published 'a brave draught of the Mole to be built', in the form of a view of the town with the proposed harbour. On the engraving, which appeared in 1664, he is termed Surveyor to the Duke of York.

In June 1665, during the Second Dutch War (1665–1667), Moore was appointed Assistant Surveyor of the Royal Ordnance. He succeeded to the post of Surveyor General in July 1669, when he took up residence in an official house in the grounds of the Tower of London. In November 1669 he reported on Dover harbour and in 1670, with Bernard De Gomme, Christopher Wren and others, he served on a commission requested by Parliament to ensure that the 1662 Act for repairing the harbour was implemented. On 28 January 1673 he was knighted in recognition principally of his services in the Third Dutch War (1672–1674); also in that year he published *Modern Fortifications*.

After the War, Moore played a very prominent role in founding, constructing and equipping the Observatory at Greenwich and in the appointment of Flamstead as Astronomer Royal. Indeed the initiative for the whole project was largely his, and more than anyone else he was responsible for seeing it through to completion; even some of the key pieces of equipment were paid for by Moore himself, at a cost of several hundred pounds. He was elected a Fellow of the Royal Society on 3 December 1674.

Moore continued in office as Surveyor General until his death on 27 August 1679 at Godalming on a return journey from Portsmouth. He was buried on 2 September in the Tower chapel of St. Peter ad Vincula, where a monument was later set up by his elder daughter. His outstanding mathematical library was sold at auction in 1684. There are two portraits, engraved as frontispieces to the 1650 and 1660 editions of *Arithmetick*.

FRANCES WILLMOTH and A. W. SKEMPTON

[Proceedings of the Company of Adventurers (1649–1656) Cambridgeshire RO; Samuel Wells (1830) *The History of the Drainage of the Great Level of the Fens, called Bedford Level*; E. G. R.

Taylor (1954) *Mathematical Practitioners of Tudor and Stuart England*; Frances Wilmoth (1993) *Sir Jonas Moore, Practical Mathematics and Restoration Science*]

Selected publications

1650. *Moores Arithmetick. In two Bookes*, 2nd edn. 1660, 3rd edn. 1688
1654. *A true Mapp of the Great Level of the Fenns as it now lyeth drayned ... Described by Jonas Moore*, scale ½ in. to a mile
1658. *A Mapp of the Great Levell of the Fenns ... as it now drayned, described by Jonas Moore Surveyor Generall*, 2nd printing 1684, 3rd printing 1705, scale 2 in. to a mile
1664. *A Mapp of the City of Tangier*
1673. *Modern Fortification of, the Elements of Military Architecture*, 2nd edn. 1689

MORGAN, James (*c.* 1773–1856), architect and engineer, was born in Wales *c.* 1773. Most of his career was spent in the office of the architect John Nash (q.v.), and he remained his principal assistant for nearly forty years. From 1806 to1815 he was Nash's partner as Architect to the government's Office of Woods and Forests. In September1811 Morgan produced a Plan and Sections for the promoters of the proposed Regent's Canal. Nash's influence helped Morgan to secure a senior post with the Regent's Canal Company which obtained its Act in July 1812. On 10 August 1812, it was resolved unanimously that 'Mr. James Morgan be appointed Engineer, Architect and Land Surveyor to this Company'. On the same occasion Morgan was asked to prepare a plan of the intended canal for the next meeting of the Directors, and to stake out the line of the canal. Perhaps because of Morgan's lack of civil engineering experience, the Company decided to advertise for designs of the locks and tunnels, with a premium of 50 guineas, to be adjudicated by William Jessop, William? Nicholson and James Walker (qq.v.). In the event the submitted designs were rejected and on 17 December 1812 Morgan was empowered to 'take the entire conduct of making and perfecting the canal ... he shall furnish the drawings, plans, and specifications, and give all necessary Direction relative thereto'. He was awarded a salary of £1,000 a year and from this date he earned the right to be recognised as an engineer. Morgan undertook the survey of the route of the canal in 1812. Daniel Pritchard (q.v.) was the contractor for the Maida Hill and Islington tunnels, two major works on the canal, whilst Hugh McIntosh (q.v.) undertook much of the earthwork. In 1817 the company were short of funds and, following a report by Thomas Telford (q.v.) to the Exchequer Loan Commissioners, successive loans up to £250,000 were granted to complete the work. The canal was opened to traffic in 1820. Morgan also designed the revolving machinery for the *Diorama* erected in Regent's Park in 1823. The *Diorama* was still in use in 1851 when a description said 'The pictures are suspended in separate

rooms, and a circular room containing the spectators is turned round, much like an eye in a socket', so the machinery must have been substantial. Morgan was subsequently engineer to the 'United Committee for the Stour Navigation and Sandwich Harbour', producing a report dated 11 March 1824, together with 'Estimates; and a Statement of the Revenue'. Following this he appears to have returned to architectural work. He died in Hammersmith in February 1856.

DENIS SMITH

[Regent's Canal Minute Books, RAIL 860/5, PRO; Colvin; Skempton]

Publications

1811. *Regent's Canal. Plan and Sections of an intended Navigable Canal from the Grand Junction Canal at Paddington to the Thames at Limehouse ... September 1811*
1818. *Plan of the Regent's Canal in the County of Middlesex. Shewing such parts of the Works as are Finished, In hand* [or] *Not Begun*, repr. 1819
1824. *Report to the United Committee for the Stour Navigation and Sandwich Harbour* [with] *a Plan*

Works

1812–1820. Regent's Canal, survey of canal route, Engineer, £720,000, including, Maida Hill Tunnel (370 yd.), Islington Tunnel (960 yd.), Macclesfield Bridge, Regent's Park, brick arches supported on cast iron columns supplied by the Coalbrookdale Company. gauging surveys for summit level feeder reservoirs (75 acres) on the River Brent, and at Ruislip
1823. Regent's Park, Diorama machinery

MORRIS, Evan William (fl. 1819–1875), civil engineer, was elected a Member of the Institution of Civil Engineers in 1819, along with John Provis (q.v.), whose address at 16 Union Street (Middlesex Hospital) he then shared. He attended several early meetings. He was probably under training, and as an engineering draughtsman he was responsible for the drawings of John Rennie's (q.v.) London Bridge and drew some of the plates for William Alexander Provis's (q.v.) account of the Menai suspension bridge. With the Provis connection he may have assisted W. A. Provis on his canal contracts. Certainly in the 1830s he turned to contracting.

Morris' earliest known tender was for the Stoke Hammond contract on the London-Birmingham Railway, Provis acting as one of his sureties. He was one of the few original contractors to successfully complete his contract. Unsuccessful tenders on other early railways such as the Great Western, Manchester and Birmingham, and London and South Western railways followed. He was successful on the Preston Wyre Railway, and on the Godstone and Mersham contracts of the South Eastern Railway (SER), where he was reprimanded for his use of the notorious 'Tommy shop' system. This had been in use by contractors on the

Holyhead Road, and Morris may have regarded it as normal practice.

Around 1838 Morris had been involved with surveying the abortive Central Kent Railway with Thomas R. Crampton (1816–1888) for Sir John Rennie (q.v.). Around 1850, by which time he had purchased Folkestone Harbour for £10,000 and sold it on to the SER for £8,000 profit, they felt the time was ripe to revive it. They brought in George Burge (q.v.), whom Morris must have known many years, and organised the creation of the East Kent railway, which obtained its Act in 1853. This, which metamorphosed into the London Chatham and Dover Railway, was perhaps the archetypal contractors' speculation of the era, with Samuel Morton Peto and Edward Ladd Betts (see: *BETTS, William*) both becoming involved. Morris continued to be involved in contracting for the work until the 1860s, apparently hitting financial difficulties in 1863; Burge had already pulled out.

Interestingly, aside from his Provis connections, he also worked with another contractor of the Telford era, William Hoof (q.v.) on the West Croydon and Epsom Railway (1845). In 1847 he was living in Pembury, and was subsequently described as of Smeeth, both in Kent. Regrettably no more is known of him except he was allegedly in poverty in 1875.

MIKE CHRIMES

[Frank Smith archive; ICE membership records, ICE archives; South Eastern Railway Minutes, RAIL 635/15, PRO]

Works (all after 1830)

1835. London Birmingham Railway, Stoke Hammond contract
1840. Preston & Wyre Railway (there were other contractors)
1842–1843. South Eastern Railway, Godstone contracts 1, 4, 6 and 7, Merstham contract
1845–1847. West Croydon & Epsom railway, with Hoof
1853–1860. East Kent Railway, Strood–Faversham–Canterbury railway, with Crampton and Burge
1855–1861. East Kent Railway, Dover extension, with Crampton and Burge
1856–1860. London, Chatham & Dover Railway (East Kent Railway), Sittingbourne to Sheerness, with T. R. Crampton(1816–1888)
1857–1863. Forest of Dean Railway?
1860–1861. Faversham–Whitstable–Herne Bay contract, with Crampton and Burge

MORRIS (MORICE), Peter (fl. 1575–1584), hydraulic engineer, described as a Dutchman (or German) by birth but a free denizen of England, was in the service of Sir Christopher Hatton in 1575 when he submitted a petition for the granting of a monopoly on the use of 'divers engins and instruments by motion whereof running streames may be drawen farr higher than their naturall levells ... and also dead waters [may] be drayned ... whereby the ground under them will prove firm ... and more fertil'. A Patent was duly

granted by royal consent on 24 January 1578 to 'our wellbeloved subject Peter Moris (who) hath by his great labor and charge found out and learned the skill to make some newe kynde and manner of engynes to drawe and raise up waters ... out of any manner of fenne groundes or other places not nowe or heretofore practized or used by any other ... and none but he ... or such as he shall lycence shall make or putt in practice any such worke within the space of twenty one yeres'.

The nature of the engines is not disclosed and the work for which Morris is remembered had nothing to do with fen drainage. This was to supply water from the Thames to part of the City of London, the plan being to acquire a long lease on the first arch of London Bridge, at its northern end, and there to build an undershot water wheel operating twin reciprocating pumps which would raise the water into a cistern on an adjacent tower from whence it would be distributed in wooden pipes. Work began on the piled foundations before July 1580 and the supply of water started on 24 December 1582.

Meanwhile George Carleton (q.v.) adopted the idea of improving the drainage of his land in Coldham Fen, near Wisbech, by means of an engine. To achieve this he and a partner agreed with Morris that his Patent be conveyed to them. In June 1580 delays were experienced 'for want of workmen and labourers' but the engine was completed within a year or so since it is depicted in a little sketch of a windmill, titled in translation from the Latin 'watermill with sails' on a map of *c.* 1582. The site can be identified at the north end of Coldham Fen near Friday Bridge 3 miles south of Wisbech. It is also shown in a sketch on the 1589 map of John Hexham (q.v.) as 'The Engyne', a windmill discharging water into Elm Leam.

Neither sketch shows the mechanism for raising the water. Possibly it was a paddle wheel, which would be more suitable for lifting large quantities of water through a small height than pumps, which lifted small quantities through a large height. Carleton presumably had to take over the Patent as it could be interpreted as covering the drainage of fenland by any (unspecified) type of engine.

The Coldham 'engyne' is the first drainage mill in England for which positive evidence exists, just as the London Bridge waterworks mark the beginning of London's pumped supply. The waterworks proved to be successful. Morris took a lease on the second arch in 1584 to build another engine, but no details are available. Nor is the date of his death known. His descendants, however, continued to enjoy a considerable income from the waterworks up to the time of the Great Fire of 1666, when the works were destroyed. They were rebuilt by Thomas Morris in 1668 and finally sold in 1701 to a new company with George Sorocold (q.v.) as their engineer.

A. W. SKEMPTON

[Rhys Jenkins and E. Wynham Hulme (1895) Notes on London Bridge Waterworks, *The Antiquary*, 243–246, 261–264, reprinted in *The Collected Papers of Rhys Jenkins* (1936) 131–140; H. W. Dickinson (1954) *Water Supply of Greater London*; Richard L. Hills (1967) *Machines, Mills and Uncountable Costly Necessities*; J. S. P. Buckland (in lit. 1997) further information]

Works

1580–1582. London Bridge waterworks

MORRIS, Thomas, Sr. (b. ?1727; fl. 1763–1782), canal engineer, is first heard of in November 1763 working under James Brindley (q.v.) on the Bridgewater Canal. Early references suggest he may have been a carpenter. After 1765 Brindley was chiefly concerned with the Trent and Mersey, and Staffordshire and Worcestershire canals and it is possible that Morris thereafter had charge of construction on the Bridgewater onwards from Altrincham. Certainly in September 1767 the 'ingenious man Thomas Morris' caught the attention of a visitor to the works then proceeding in the Bollin Valley. The canal, including the great flight of locks at Runcorn, was completed (apart from a very short section) in December 1772 and little over a year later, in May 1774, Morris took over from Samuel Weston (q.v.) as engineer of the Chester Canal. He had already, in 1770, worked with Weston on gauging the water supply for the canal. In April 1775 he was asked to survey an alternative route to Middlewich but, like others before and after him, he ran into trouble with the proprietors and left in August 1777 to be followed by Josiah Clowes (q.v.) and then by Morris's former assistant.

In March 1777 'Mr. Thomas Morris of the City of Chester, Engineer' was elected to the (Smeatonian) Society of Civil Engineers while staying in London at St. Martin's Lane. At that time he was giving Parliamentary evidence on behalf of the Chester Canal Company to seek extra funds and approval for his alternative route to Middlewich. In the event construction was halted at Nantwich in 1779 and Middlewich was not looked at further until Thomas Telford (q.v.) revived the idea in 1813.

His name is entered on a 1782 list of Smeatonians but was later crossed out and does not appear in the next list of 1784. This suggests that he died c. 1783 but the date of his death has not been found. He may have been the Thomas Morris who married Mary Matthews at Chester St. Peters on 9 February 1753. It is probably correct to assume that Thomas Morris junior (c. 1754–1832), the Liverpool dock engineer (q.v.), was his son.

A. W. SKEMPTON and I. WEIR

[J. Brindley, Day book (diary) 2 November 1763, ICE Library; Minute Book of the Society of Civil Engineers (1771–1792); facsimile copy printed by the Society (1893); Charles Hadfield (1966) *The Canals of the West Midlands*; C. T. G. Boucher (1968) *James Brindley, Engineer, 1718–1772*; Charles Hadfield and Gordon Biddle (1970) *The Canals of North West England*; E. A. Shearing Chester canal projects, *RCHS J*, part II, **28**(4), 1985, 141–154; I. Weir (1998) Thomas Morris— the unknown hero, *CHS Newsletter*, **50**, 1–4]

MORRIS, Thomas, Jr. (c. 1754–1832), civil engineer, is presumed to be the son of the canal engineer, Thomas Morris Sr. (q.v.), and may have married Ann Neale at St. Nicholas Church, Liverpool on 6 November 1780. There is a reference to a Thomas Morris travelling over from Ireland in September 1771 in the Leeds and Liverpool Canal Minutes, but nothing definite. Nothing is known of his early life until October 1783 when he was consulted by the Commissioners of the Port of Lancaster concerning the pier and a proposed dock at Glasson near the mouth of the Lune estuary. The west wall of the recently completed pier had bulged out and Henry Berry (q.v.), the dock engineer of Liverpool, was asked in August 1782 to arbitrate with the contractor and to estimate costs of making a dock alongside the pier. He was too busy at Liverpool to do so and the Commissioners called on Morris who produced his plan in November 1783. It involved rebuilding the wall and building a short pier out from the opposite shore to form the abutment for a dock gate. A new contractor, Mr. Fisher, estimated the cost at about £2,700. Morris was appointed to superintend the work at a salary of £100 p.a. and the dock was opened to shipping in March 1787. His attendance at Glasson ceased in December.

Fifteen months later, on 19 February 1789, 'Mr. Thomas Morris of Lancaster, Engineer' was appointed 'to succeed Mr. Berry (who had retired) as Engineer of the Docks and other buildings' at Liverpool, with a salary of 100 guineas (£105) p.a. In the ten years during which he held this post his principal work was the construction of Queen's Dock. This had been planned by Berry, along with the neighbouring King's Dock (completed 1788), but Morris would have made the working drawings, let the contracts and supervised construction. The dock had an area of 7 acres, the entrance was 42 ft. wide with a single pair of gates 25 ft. high, and the walls were built of sandstone. Work began about 1792 and was completed by May 1795 at a cost of £35,000 (excluding the cost of land). From February 1793 his salary was doubled, but payment of the extra amount was deferred until he resigned soon after October 1799 following a dispute over his claim for a further increase. By that time he had become involved with the Parliamentary Bill for what was to be Princes Dock, built by his successor, John Foster (q.v.), the former surveyor. Foster's salary at appointment was £262.

While holding the post at Liverpool Morris undertook several small consulting jobs while on leave of absence. He was responsible jointly with Joseph Turner (q.v.) for the survey of the 'Eastern Canal', the forerunner of the Ellesmere Canal, producing the deposited plans in 1792. In July 1793 he reported on a proposed new dock at Hull and in 1796 he was consulted by the Mersey and Irwell Navigation on proposals for a long cut from Latchford (near Warrington) towards Runcorn. His broad estimate was approved in June 1797 and work began in 1799 on what became the Runcorn and Latchford Canal. In November 1797 he produced a plan for a new lock and cut on the River Weaver Navigation from Northwich weir to Anderton, for which he received 40 guineas and a further plan followed next year. In February 1799 he was asked by his old employers, the Lancaster Commissioners, to survey the estuary and report on the best place for a new graving dock; for this he received £30. Finally in November 1799 'Mr. Thomas Morris of Liverpool' was called in to advise the West India Dock Company on the entrances to their proposed docks. This he did, with a report and plan for which he was paid 100 guineas. He then returned to Liverpool from where he was called in April 1801 to advise on a sluice at Clawydd Llwyd for the Rhuddlan Embankment Commissioners on the advice of their surveyor, Thomas Wedge.

At that time there was no question of Morris joining the permanent staff of the West India Docks. Ralph Walker (q.v.) was already Resident Engineer; William Jessop (q.v.), principal consultant since 1797, became Engineer in February 1800 and at the same time George Gwilt (q.v.) was appointed Surveyor to design and take charge of building the warehouses; each of the principal officers had an assistant. The great Import Dock (30 acres in area) and its warehouses were completed by July 1803. Walker had resigned a little earlier; he was about to become joint engineer with John Rennie (q.v.) of the East India Docks. For a year there was no successor, the work being carried on by an assistant, but on October 1803 Morris was appointed Resident Engineer at a salary of 800 guineas with a rent-free house at Blackwall, an assistant (at £50 guineas p.a.) and a clerk. Moreover in May 1804 he took over from Gwilt as Surveyor (without an increase in salary). The work in hand from 1803 was the Export Dock (24 acres), its locks communicating with the Blackwall and Limehouse entrance basins and three stacks of warehouses. The dock was formally opened in July 1806, the engineering works alone having cost £205,000.

Jessop left the West India Dock Company in June 1805 and thereafter Morris acted as Engineer (and Surveyor) still at the same salary of £800 guineas p a until his retirement in March 1811. He then returned to Liverpool where it seems he had kept possession of his house in St. James' Place. He remarried in 1819 and died on 10 April 1832, aged seventy-eight, leaving his widow, Bathsheba, and a step-daughter. His personal property was valued for probate at under £3,000.

A. W. SKEMPTON and I. WEIR

[Leeds and Liverpool Canal Minutes, PRO RAIL 846/2/1; Minutes of the Commissioners of the Port of Lancaster (1776–1787), Lancaster Ref. Library;

Liverpool Docks Minute Books (1793–1800) Mersey County Museum; Minutes of the Commission of the Rhuddlan Embankment, Flintshire County RO; West India Docks Company Minute Book (1799–1811) Port of London Authority archives; J. A. Picton (1873) *Memorials of Liverpool*; S. A. Harris (1937) Henry Berry (1720–1812) Liverpool's second dock engineer, *Trans History Soc Lancashire and Cheshire*, **89**, 91–116; S. A. Harris and T. C. Barker (1960) Henry Berry, an inventory of his professional papers, *Trans History Soc Lancashire and Cheshire*, **112**, 57–63; Charles Hadfield and Gordon Biddle (1976) *The Canals of North West England*; A. W. Skempton (1979 and 1982) Engineering in the Port of London, *Trans Newcomen Soc*, **50**, 87–106, and **53**, 73–96]

Publications

1793. *Report relative to the making of a New Dock at Kingston-upon-Hull*
1797. *A Plan to improve the River Mersey Navigation by an inland Cut from Howley to Hempstones*

Works

1784–1787. Dock at Glasson, about £3,000
1792–1795. Queen's Dock, Liverpool, £35,000
1803–1806. Export Dock, West India Docks, resident engineer (1803–1805) and engineer (1805–1806), £205,000

MORTON, Thomas (1781–1832), engineer and inventor, was born at Leith, the port of Edinburgh, on 8 October 1781, younger son of Hugh Morton, wright and builder. Other details of his early life are not known, but he evidently acquired some knowledge of engineering and shipbuilding, either in his father's business or elsewhere. His elder brother, Samuel (1776/1777–1842), was an agricultural implement manufacturer in Leith who with his son, Hugh (1812–1878), later carried on Thomas Morton's shipbuilding business as S. & H. Morton.

He married Dorothy Gutzmer on 20 September 1810 and they had at least two children, sons Samuel and Thomas. At about the same time Thomas established a ship-building yard on the north side of the Water of Leith and here he developed Morton's Patent Slip (Patent No. 4352; 1819) which enabled ships in need of repair to be drawn up out of the water along an inclined plane, a much cheaper and, in many places, a more practical expedient than having to use a dry dock. The ships were supported on a strong carriage running on iron rails and were kept upright by a system of wedges on transverse slides drawn in under the bilges as the vessel was being raised. The incline of the slip was generally between 1 in 15 and 1 in 20, and a capstan, or similar device, enabled ships of several hundred tons to be raised by six men for every 100 tons.

The first slip was built at Bo'ness in about 1822, partly at Morton's own expense in order to demonstrate the advantages of his invention. Within five or six years slips were in use at most of the principal dockyards in the UK, as well as in

France and Russia. Morton's patent was infringed by the construction of a slip on the same principle at Stobcross, in Glasgow, but he was successful in the High Court at Edinburgh in a restraining action in 1824.

Steam engines came into use to draw ships up some of the larger slips but as the size and weight of ships increased the limitations of the patent slip became more serious and in the longer term its use was restricted to smaller shipyards.

In about 1830 Morton removed from his original shipyard and on 24 December 1832 he died at his home in Pilrig Place, Leith, and was buried in South Leith Burial Ground. His wife died on 16 January 1842. For several decades the firm was carried on by his family and it is recorded that the 1300 ton SS *Leith* was built by Hugh Morton & Co. in 1862 at their shipyard on the western edge of Leith Docks, where launches could take place directly into the waters of the Firth of Forth.

RON BIRSE

[*Encyclopaedia Britannica* (9th edn., 1880), **11**, 470a; DNB; S. Mowat (1994) *The Port of Leith: its History and its People*; W. T. Johnston (1998) *Scottish Engineers and Shipbuilders*, computer database]

Publications

1819. Patent No. 4352: *Dragging Ships out of Water on to dry Land*
1824. *Infringement of a Patent. Notes of a trial before the Jury Court at Edinburgh, 15th March 1824. T. Morton, Pursuer versus I. Barclay and others ... for an Infringement of a Patent ... for an Invention called a Slip by which Ships are hauled out of the Water, etc.*

MOSER, Richard (fl. 1810–1830), ironmaster, possibly the son of Robert Moser (1739–1785), also an ironmaster, was elected Associate of the Institution of Civil Engineers in 1828 but only remained a member for a year. The Moser family was involved in the London iron trade for much of the nineteenth century, with premises in Southwark.

In 1814 Moser took a £3,000 share in the business of William Holner, a Southwark ironmonger, the stock being valued at £9,000. Each of three partners was supposed to reside at the premises of 165 Borough High Street for seven years. Moser was related to Richard Crawshay (q.v.) who provided the firm with loans in 1816 following the slump after Waterloo. By 1823 the firm was turning over £44,000 in stock a year.

The family acted as London agents of the Crawshays ironworks. In July 1825 they successfully tendered for the ironwork for Buckingham Palace, supplying it from Toll End Ironworks, Staffordshire, at a great profit; Robert [*sic*] Moser gave evidence on this to the Parliamentary Select Committee in 1831.

In the 1830s Moser became sole partner until in 1845 his son, Richard, became managing partner.

The family remained in charge of the business until 1924.

MIKE CHRIMES

[ICE membership records, ICE archives; *Select Committee on Windsor Castle and Buckingham Palace* (1831) 2nd report, 129; W. A. Young (1938–1939) Ironmongers and engineers, *Trans Newc Soc*, **19**, 231–232; Portrait of Robert Moser (1988) Sotheby Old Masters sale, lot 256]

MOSES, Moses (fl. 1825–1838) of Abercarne Colliery, Newport, Monmouthshire, was elected a Corresponding Member of the Institution of Civil Engineers in 1825 and remained in membership until 1838. Little is known of his activities aside from a letter he wrote to William Bull (q.v.) regarding the use of blue lias lime.

MIKE CHRIMES

[ICE membership records, ICE archives; On lime, OC 20, February 1825]

MOYLE, Samuel (1784–1857), mining engineer, was born at Chacewater on 17 January 1784. He was elected a Corresponding Member of the Institution of Civil Engineers in 1825 and remained a member until 1830. He drew up several proposals for early railways and harbour improvements in Cornwall, while practising chiefly as a mining engineer.

He died on 5 January 1857 in a house in Ferris Town, Truro only a few days after removing from his home of many years at Bosvigo. His death was noted by the *Royal Cornwall Gazette:* 'He possessed considerable talent as an engineer and was the projector of many scientific inventions of which others were too apt to take advantage, leaving him without the reward to which he was entitled... for several years he was engaged as chief manager in the prosecution of mining adventures in South America, chiefly in the Province of Guatemala, and was accompanied throughout his perilous sojourn abroad by Mrs Moyle, his devoted wife'.

M. M. CHRIMES and R. P. TRUSCOTT

[Membership records, ICE archives; Report (1834) Committee of House of Commons, on Sir G. Gurney's claim to the invention of the locomotive; Obituary, *Royal Cornwall Gazette*, 9 January 1857, Royal Institution of Cornwall; Boase and Courtney (1878) *Bibliotheca Cornubiensis*, 378, Royal Institution of Cornwall]

Publications

1813. *Report for a breakwater in the Plymouth Sound*, submitted to the Board of Admiralty, 7 September 1812
1836. *Truro, Redruth and Penzance Railway ... with a breakwater at Penlee Point, Mount's Bay*
1838. *On the Ventilation of the Cornish Mines and the Deterioration of the Air in the lower Levels*, Report RIC, 26–29
1839. *On the Ventilation of Mines*, Report RIC, 57–60

1840. *An Account of some Specimens of Aurif-erous Pyrites*, Report RIC, 76–77

MULLER, John (1699–1784), mathematician and educator, was born in Germany in 1699. By 1736 he was established in London where he published his first book, on conic sections. In 1741 he became head master of the Royal Military Academy, Woolwich and, with Thomas Simpson, Muller transformed an undistinguished school filled with ill disciplined youth into an enlarged academy with a cadet corps. Muller was appointed Professor of Fortifications, a post he kept until retirement in 1766. He was described as the scholastic father of all the great engineers this country employed for forty years. He died in April 1784. His portrait was painted by J. Hay, and engraved by T. Major

TESS CANFIELD

[DNB; *Gentleman's Magazine*, **i**, 1784, 475; Ted Ruddock (1979) *Arch Bridges and their Builders*, 38, 46, 70–71]

Publications
Publications include:

1736. *A Mathematical Treatise, containing a System of Conic Sections and the Doctrine of Fluxions and Fluents applied to Various Subjects*
1744. *A Treatise containing the Practical Part of Fortification, for the Use of the Royal Military Academy, Woolwich*, new edn. 1755
1746. *A Treatise concerning the Elementary Part of Fortification*, 2nd edn. 1756, 2nd edn. 1774, 4th edn. 1782, 5th edn. 1799, 6th edn. 1807
1747. *The Attack and Defense of Fortified Places*, 2nd edn. 1757, 3rd. edn. 1770
1756. *A Treatise on Fortification, Regular and Irregular. With Remarks on the Constructions of Vauban and Coehorn*
1757. *A Treatise of Artillery … with a Theory of Powder applied to Firearms*, 2nd edn. 1768, 2nd edn. 1780
1759. *The Field Engineer*, translated from the French of La Mamye Clairac (1760); 2nd edn., with additions (1773)

MULLINS, Bernard (1772–1851), civil engineering contractor, was born in 1772. He apparently gained his first employment in the late 1780s as a young overseer on Grand Canal contracts. He met William Chapman (q.v.) and had frequent communications with him during his residence in Ireland. In about 1802 he joined with David Henry (q.v.) to take over the contract for the extension of the Grand Canal from Tullamore to the River Shannon. Not long afterwards, about 1808, Mullins joined Henry and John MacMahon (q.v.) to tender jointly for the restoration and extension of the Kildare (Naas Line) Canal for the Grand Canal Company. They were awarded the contract for the work, which was completed by 1810. The following year, the firm undertook the lock-less extension of the Naas Line of the Grand Canal to Corbally.

The contracting firm of Henry, Mullins & Mac Mahon went on to undertake many of the canal, navigation and drainage projects during the early years of the nineteenth century. These included large sections of the Grand and Royal canals. Their familiarity with the work led to a joint report on expenditure on both canals in 1823.

Henry removed from the partnership following the completion of the aqueduct carrying the Barrow Line of the Grand Canal over the river Barrow at Monasterevan, but Mullins & MacMahon continued to play a major role in Irish contracting and, occasionally, undertook work outside the country, e.g. the 120 ft. span bridge on the North Union Railway over the river Ribble near Preston, built for Charles Vignoles (q.v.) in 1838. This appears to have been the responsibility of MacMahon.

Mullins set out his views on inland navigation in a brief treatise on the subject, published in 1832. He claimed thirty years professional experience of designing and constructing navigations. He considered that agriculture and its dependencies, being the stamina of society in Ireland, easy communication to its different districts would extend the market and facilitate the transfer of goods. He suggested a number of extensions to the system, for example the extension from Ballinasloe to Galway and the opening up of the East Galway and Roscommon area. None of these came to fruition as, within a few years, the railways had begun to compete for water-borne passengers and freight.

Drawing on his experience as a contractor, when he frequently met some of the most demanding working conditions, Mullins, together with his son Michael Bernard, presented an important paper to the Institution of Civil Engineers of Ireland in 1846 'On the origin and reclamation of peat bogs, with some observations on the construction of roads, railways, and canals in bogs' in which he described the difficulties encountered when driving the Ballinsloe Canal through the bogs of county Offaly (12 of the 15 miles of the canal were constructed though deep bog, on average from 26 ft. to 46 ft. in depth, and bounded on two sides by the River Shannon and the River Suck). He gained further experience when extending the Royal Canal through the bog at Clonbrainey in county Longford.

Michael Bernard Mullins, in his presidential address to the ICEI in 1859, referred to his father Bernard as 'one of our late Vice-Presidents' and as 'assistant engineer to Mr. Evans in the execution of the Royal Canal' He recorded that Bernard Mullins 'having found the field preoccupied, almost to the exclusion of Irishmen, looked out for a combination to enable him to embark extensively in the execution of public works, and in the early part of the present (nineteenth) century commenced with his partners, Messrs Henry and MacMahon, a career which, associated with engineering, will lose nothing by comparison with the most successful period of the profession in this (Ireland) or possibly in any other country'.

In 1832, he is recorded as living at 1 Fitzwilliam Square South in Dublin, an address still used by his son in 1854. In October 1842, when the partnership with MacMahon was dissolved Mullins had an office at 9 Mabbott Street, Dublin.

RON COX

[Telford mss, ICE archives; Irish Railway Commission (1838) *2nd Report*, 73–74; M. B. Mullins (1863) Presidential address, *Trans ICEI*, **6**, 66]

Publications

1823. *Observations on the Grand and Royal Canals*, repr. 1838
1832. *Thoughts on Inland Navigation, with a Map, and Observations upon Propositions for Lowering the Waters of the Shannon and of Lough Neagh*, accompanying map exhibiting the lines of still water and river navigation now in operation in Ireland
1846. On the origin and reclamation of peat bogs, *Trans ICEI*, 2, 1–48 (with M. B. Mullins)

Works

1790s. Royal Canal, assistant engineer
1802–1803. Grand Canal, Shannon Line, construction, contractor (with David Henry)
1808–1811. Grand Canal, Naas Line Corbally, extension, contractor (with Henry and MacMahon)
1813–1817. Royal Canal, Shannon Extension, including Whitworth Aqueduct (with Henry and MacMahon), £150,000
1815–1823. Custom House Dock, Dublin, cofferdam (£4,000), entrance lock and basin
1824. Grand Canal, Ballinasloe extension, contractor (with Henry and MacMahon)
1826. Clarendon Dry Dock, Belfast (with Henry and MacMahon)
1827–1831. Barrow aqueduct, contractor (with MacMahon)
1830. Ulster Canal (with MacMahon)
1831. Anglesea Bridge, Dodder, contractor (with MacMahon)
1835–1842. North Union Railway, Lancashire (with MacMahon)
1838–1840. Lancaster and Preston Junction Railway (with MacMahon; one of four contractors)
1839–1843. Shannon Navigation, contractor (with MacMahon)

MURDOCK (MURDOCH), William (1754–1839), engineer and inventor, was born on 21 August 1754 at Bellow Mill near Cumnock in Ayrshire, the oldest of the four surviving sons of John Murdoch, miller and millwright, and his wife, Anna Bruce. William attended the parish school in Old Cumnock until he was ten years old and then the school in Auchinleck for another two or three years before assisting his father full-time at the mill. He displayed a natural ability and inventiveness in practical mechanics which for the next ten years he was able to exercise while working for his father, although by the end of that time he knew that he would have to seek further advancement elsewhere. He seems to have been aware of the work then being done by James Watt (q.v.) on his steam engines in Matthew Boulton's Soho manufactory and in August 1777 he set off to walk the 250 miles from Cumnock to Birmingham.

Boulton was favourably impressed and gave Murdock a job at fifteen shillings a week in the pattern shop, where a year later he was already esteemed by Watt as the best of the pattern makers. He was also by that time gaining experience in the field, supervising the erection of Watt's cumbersome steam engines in different parts of the country. Recommending him to the mine-owners at Wanlockhead in 1779, Watt described him as '… a very sober, ingenious young man who has a good deal of experience under us in putting engines together…'. and there is no doubt that he was a great asset to the firm of Boulton & Watt in the erection and maintenance of their engines, especially in Cornwall where they were extensively used for drainage of the copper and tin mines. For twenty years from late 1779, Murdock lived in Cornwall and devoted his considerable energies and talents to furthering the engine business of his employers; in that time about fifty Watt engines were supplied and erected, commissioned and maintained. In addition, more than two-thirds of them were subsequently dismantled and re-erected on another site, many of them more than once. The work was hard and included as much building and civil engineering as mechanical improvisation, training of erectors and operators, and trying to meet the conflicting demands of Watt the perfectionist, Boulton the businessman and the often irascible Cornish mine captains. Murdock was by nature somewhat stubborn and uncommunicative and, in the eyes of Boulton and Watt, too keen to develop his own inventive ideas, but resourceful and extremely hard-working. They knew that they would be unlikely to find a better man for the job.

In the summer of 1783 Murdock was struck down by a serious illness from which he was to suffer intermittently for the rest of his life. Never clearly diagnosed at the time, it is now surmised that it might have been a form of malaria then found in some parts of Britain. It was at this time that he met Anne Paynter, the young daughter of the mine captain at Polgooth. They were married by special licence on 27 December 1785 and at the end of July the following year Anne was delivered of twins, both of whom died before they were two years old. William and his wife made their home in Redruth and two surviving sons were baptised in Redruth Church, William on 22 July 1788 and John on 1 June 1790. Sadly, Anne herself died shortly after giving birth to John, and the boys and their father were, until his return to Birmingham, looked after by Anne's recently widowed mother. The boys were then sent to their Scottish grandparents where they had the advantage of a good secondary education at Ayr Academy.

Throughout his working life Murdock found the greatest satisfaction in solving engineering problems and bringing new ideas to practical fruition. Griffiths, in *The Third Man*, lists more than forty of these, ranging from major inventions such as the sun-and-planet gear, the D-slide valve, the oscillating cylinder engine, model steam carriages, small steam engines and coal-gas lighting, to relatively minor devices such as an open-cast excavator powered by a water wheel, pneumatically-operated house bells and improved methods of boring out wooden water pipes, as well as schemes for dry docking machinery, compressed-air engines, a steam gun and other ideas which were never, or only partially, realised. Murdock himself took out only three patents. The first in 1791 for the production of various chemicals, required the use of a retort, and it was the extension of these experiments to different types of coal that led Murdock to the use of coal-gas for lighting. The second in 1799 for 'manufacturing and constructing steam-engines' was in four sections, including a boring machine operated by an endless screw, and a method of cutting out and boring a cylinder and steam case from a single casting. The last in 1810 for boring pipes, cylinders etc out of solid blocks of stone. The sun-and-planet gear was incorporated, almost as an afterthought, in Watt's patent of 1781, but the surviving correspondence and most of the contemporary references point to its having been Murdock's idea.

His experiments with model steam carriages showed a far-sighted appreciation of the possibilities of steam locomotion that was not shared by either Boulton or Watt, who did all they could to dampen his enthusiasm. Nor did Watt see much future in the small self-contained rotating steam engines he developed originally as whims or winding engines for the Cornish mines, but here Murdock was supported by Boulton and James Watt Jr., who were well aware of the huge potential market for rotative steam engines. The D-slide valve, also known as the long slide valve, was included in his 1799 patent, but again he encountered opposition from Watt because it was a modification of the four-valve arrangement Watt had devised for his original steam engines. The long slide valve in its various forms was widely used in reciprocating steam engines for most of the nineteenth century.

The success of Murdock's experiments in the use of coal-gas for lighting is another example of his determination to make things work. The existence of ignitable gases in coal, wood and other substances was well known and had been demonstrated before the end of the eighteenth century by Lord Dundonald and others, but none of them put the gas to any practical use. In the words of the paper presented to the Royal Society in 1808, Murdock concludes with the words 'I believe I may, without presuming too much,

claim both the first idea of applying and the first actual application of this gas to economical purposes'. He was awarded the Rumford gold medal of the Royal Society, inscribed *Ex Fumo Dare Lucem*, in 1808.

The first large-scale application of gas lighting was carried out by Boulton & Watt, under the supervision of Murdock, in 1805–1806 in the Manchester calico mill of Philips and Lee. The initial installation of fifty lights enabled trials to be made to determine the best form of gas burner and the final total of 904 lights (equivalent to 2,500 candles) was approximately four times more economical than candles. John Southern (q.v.), another Boulton & Watt employee, suggested that fewer retorts would be required if gas could be produced and stored when the lights were not burning and a number of small gas-holders were duly incorporated in the scheme. The fundamental problems in the use of coal-gas for lighting having thus been solved, Murdock preferred to leave it to others such as his assistant Samuel Clegg (q.v.) to exploit its commercial possibilities and turned his attention elsewhere.

The last phase of his career with Boulton & Watt centred round their new Soho Foundry which was opened with due ceremony on 30 January 1796, although Murdock himself was still living in Cornwall until the end of 1798. He took a leading part in the planning and equipping of the foundry and machine shops under the general direction of Boulton and Watt's sons, and he continued to be responsible for the installation and commissioning of their larger engines in mines, mills and steamships. Once in full production the Foundry had a regular output of about fifty engines a year, varying from four to fifty horsepower.

In 1815 he bought some land not far from the Foundry and two years later moved into the newly-completed Sycamore House, which was lit by gas piped across the fields from the Foundry and heated by a ducted hot air central heating system. In the meantime his own two sons had also joined the firm, John in the foundry and William Jr. in the counting house at Soho. John lived with his father and was greatly concerned by the increasing frequency of recurrences of the illness he had first suffered from in Cornwall.

In the last two decades of his working life Murdock was particularly concerned with the design of engines for steam-ships, following on from the pioneering work of William Symington (q.v.) and others. An experimental installation in 1795 was followed in 1804 by an order for a twenty horse-power Boulton & Watt engine to be fitted in Robert Fulton's (q.v.) *Clermont* which operated the world's first passenger steam-ship service on the Hudson river from 1807. In 1814 two engines were fitted in a Clyde steamer, then in 1817 came a series of competitive trials in which the *Caledonia,* engined by Boulton & Watt, emerged triumphant.

This was followed over the next few years by a large, though short-lived, increase in the number of orders for marine engines, in sizes ranging from twenty to a hundred horse-power, the designs being progressively improved by Murdock and some of his colleagues. At the same time it became clear that the firm of Boulton & Watt was no longer so competitive, compared to such rivals as Matthew Murray (q.v.), Maudslay Son & Field (qq.v.), and the Napiers (qq.v.) on the Clyde. By the late 1820s the engine business had declined to such an extent that it was no longer profitable and at the end of September 1830 Murdock's contract with Boulton & Watt was terminated. Shortly before he died in 1819 Watt had written to Murdock: 'I shall always retain a due sense of the zealous friendship with which you have furthered my views, and the invaluable assistance I have derived from you'. The relationship between the sons of Watt and Boulton, and Murdock and his sons, was never so amicable, and sometimes distinctly strained.

William Murdock's health had been slowly deteriorating for many years, not helped by his tendency to overtax his strength in the physical exertions of engine building. In his retirement he became something of a recluse, and eventually died in Sycamore House on 15 November 1839. He was buried in Handsworth Church, where there is a bust of him by Chantrey, executed in 1828. He left no will, but is thought to have been worth some £30,000 to 40,000.

A portrait of Murdock in oils by John Graham-Gilbert is in the possession of the Royal Society of Edinburgh, and there is another by the same artist in the Art Gallery, Birmingham.

RON BIRSE

[J. P. Muirhead (1854) *The Origins and Progress of the Mechanical Inventions of James Watt*; S. Smiles (1884) *Men of Invention and Industry: William Murdock*; A. Murdoch (1892) *Light without a Wick: a Century of Gas-lighting*; H. W. Dickinson and R. Jenkins (1927, repr. 1981) *James Watt and the Steam Engine*; D. Chandler and A. D. Lacey (1949) *The Rise of the Gas Industry in Britain*; J. Griffiths (1992) *The Third Man: the Life and Times of William Murdoch, Inventor of Gaslight* (with comprehensive bibliography); J. Tann (1996) Riches from copper: the adoption of the Boulton & Watt engine by Cornish mine adventurers, *Trans Newc Soc*, **67**, 1995–1996; W. T. Johnston (1998) *Scottish Engineers and Shipbuilders*, computer database; DNB]

Publications
1791. Patent No. 1802: *Making from the same Material (pyrites), Copperas, Vitriol, and different sorts of Dye-stuffs, Paints and Colours; also a Composition to preserve Ships' Bottoms*
1799. Patent No. 2340: *Manufacturing and Constructing Steam Engines*
1808. On the use of coal gas for lighting, *Phil Trans*, **xcviii**
1810. Patent No. 3292: *Boring and forming Pipes, Cylinders, Columns, and Circular Discs, out of Solid Blocks and Slabs of Stone*

Principal works
1779–. Supervising the erection, commissioning and maintenance of more than fifty Boulton & Watt's steam engines
1805–1806. Supervising the first large-scale application of gas lighting in Philips and Lee's Manchester cotton mill

MURRAY, Adam and **George** (fl. 1790–1820), millwrights and mill owners, came from Kirkcudbright. After training by William Cannan, machine maker of Chowbent, Lancashire, they set up in the emerging cotton trade of Manchester. After renting a succession of premises they began to specialise in the spinning of fine cotton.

In 1798 their 'Old Mill' beside the recently built Rochdale Canal in Ancoats reached an unprecedented seven storeys in height and was followed in quick succession by the adjacent Decker Mill (1802), New Mill (1804) and the Murray Street and Bengal Street Blocks of 1804–1806, the latter two being four storey warehouses. These mills were not fireproof but made use of cruciform cast iron columns and large timber beams and joists, possibly reflecting the conservative nature of the Murrays' own designs. Power was provided by Boulton and Watt engines in external blocks and renewal of the millwork there in 1817–1818 was William Fairbairn's first job as an independent engineer. This complex and the adjacent mills of McConel and John Kennedy (q.v.) are among the most important monuments of the industrial revolution to have survived in the Manchester area.

MIKE CHRIMES

[M. Williams and D. A. Farnie (1992) *Cotton Industry in Greater Manchester*]

MURRAY, John (fl. 1780–1810) and **James** (d. 1807), were civil engineers and contractors. A John Murray is recorded as contractor, with Alexander Sutherland, for the Glasgow branch of the Forth & Clyde Canal. He later successfully tendered for a lot on the western extension with James Gray, in August 1788. In November 1790 John Murray of 'near Glasgow' contracted for Lot 3, near Colne, on the Leeds and Liverpool Canal, work continuing on subsequent lots 1791–1792. He was paid at 3d a cubic yard, or slightly more. John Murray of Colne and his son, James, are first mentioned in John Rennie's (q.v.) account book in the 1790s where they are regularly paid for survey work for Rennie. John acted as contractor, with John Pinkerton (q.v.), for earthworks on the Lancaster and Ulverston canals in the early 1790s, being removed from these contracts due to poor workmanship (Lancaster) and lack of progress (Ulverston). He then worked as an assistant on the Kennet and Avon Canal between Hembridge and Devizes, appointed at a salary of £200 pa in December 1795.

His son, James, was also carrying out surveying work for Rennie, for example on the Kennet and Avon Canal (1794), the projected Polbrock Canal, near Bodmin (1796), the Tamar Manure Canal (1797), in the Boston area of the Fens (Wildmore Fen scheme) (1799) and for an early version of the Regent's Canal (1802). At that time John Murray, in partnership with Robert Lees, was building Kelso Bridge, one of Rennie's most famous designs, and James assisted him briefly. There were some financial problems and James disappeared with the firm's books at the time of his marriage in 1801.

On Rennie's recommendation, James Murray was then appointed resident engineer at London Docks in September 1801, at a salary of £500 p.a. By 1804 this work was drawing to a close and in August Murray was offered a post on the Royal Canal in Ireland, Rennie agreeing that he could be spared. As one of the resident engineers on these works he died at Mullingar on 19 March 1807.

John Murray is known to have tendered successfully for an army building contract in Dublin c. 1804 but absconded when his estimates proved wrong and payments were withheld due to a dispute over prices for hammer dressed and ashlar masonry. In 1806 Rennie was attempting to find Murray to settle his account but it proved no barrier to obtaining work as in February 1807 John Aird (q.v.) reported to Rennie that he had successfully tendered for the West Dock at Greenock.

Rennie remarked of the father 'he possesses but little judgement in these matters (bridge building), a very accommodating conscience and a D___d hard mouth'. He was kinder about the son: 'a young man of good disposition & possesses integrity'

John Murray Sr. had a grandson, John Murray (1804–1882), born at Kelso on 12 December 1804; he was trained by William Chadwell Mylne before becoming dock engineer at Sunderland in 1831.

A James Murray was employed in 1762 to survey the Forth Clyde Canal, with Robert Mackell (q.v.), while another James Murray, no known relation, chronometer maker of 30 Cornhill, in the City of London, was elected an Associate of the Institution of Civil Engineers in 1825 but resigned in 1826.

P. S. M. CROSS-RUDKIN and MIKE CHRIMES

[BR/FCN/1/12, SRO; RAIL 846/4, PRO; Rennie account book, 1794–1800, ICE Library; Rennie reports, ICE Library; Rennie papers, NLS; Rennie reports, ICE Library; London Dock Co., Court of Directors Minute Book 1, PLA Library; J. Rennie (1800) Report concerning the Drainage of Wildmore Fen and of the East and West Fens; European Mag., 51 (1807) 317; Obituary notice of John Murray (1883) Min Proc ICE, 71, 400–402; S. T. Miller (1982) John Murray: a bold and skilful engineer, IAR, 7(2), 102–111; R. Philpotts (1993) Building the Lancaster Canal]

Publications

James Murray:

1796. Plan of the proposed Polbrock Canal … near Wadebridge

1802. Plan of the proposed London Canal from the Grand Junction Canal at Paddington to the London Docks …

Works

John Murray:

1792–1795. Lancaster Canal, two contracts: Tawilfield–Ellel, £52,000; Ellel to Ray Lane
1793–1795. Ulverston Canal, excavation work
1800–1804. Kelso Bridge, five span masonry, c. £12,876
1804. Government Building,, Dublin, contract
1807–1809. West dock, Greenock harbour, contractor, £44,000 total for East India Docks scheme

James Murray:

1802–1804. London Docks, resident engineer
1804–1807. Royal Canal, Ireland, resident engineer

MURRAY, Matthew (1765–1826), steam engine manufacturer, was born in Newcastle-upon-Tyne, one of three known children. After starting an apprenticeship with a smith or millwright there he moved to Stockton where he completed his apprenticeship as a whitesmith and worked as a journeyman and flaxmill mechanic in Darlington. Already married to Mary Thompson (1764–1836) he moved to Leeds to work for John Marshall, a leading flax manufacturer, in his water-powered mill at Adel in 1789. Murray rapidly established himself, becoming Marshall's chief mechanic, and set up home at Black Moor, near Adel.

In 1790 Marshall, in partnership with Benyon, established a new mill at Holbeck, initially powered by a waterwheel, with water supplied via a Savery-type steam engine; in 1793 this equipment was supplanted by a Boulton and Watt steam engine. From 1790 Murray had been taking out patents for improvements to textile machinery and he now began to apply his inventiveness to steam engines. In partnership with James Fenton, who supplied most of the capital, David Wood (1761–1820), his original partner and responsible for textile machine design, and William Lister, he established a steam engine works in Holbeck. The firm rapidly rose to prominence and Murray took out a series of patents which attracted the attention of Boulton and Watt. They successfully challenged two of his patents and purchased adjacent land to stop the works expanding. Despite these efforts, and the use of 'spies' inside the works, Murray's 'Round Foundry' became Boulton and Watt's most serious rival. As well as beam engines, other types of steam engines, millwork and machine tools were supplied, and in 1804 the firm's first overseas order (to Sweden) was received, and Samuel Owen (q.v.) went there initially as Murray's agent.

In 1811 the firm built a high pressure engine on Trevithick's model, fitted the following year to the paddle steamer L'Actif, based at Yarmouth. At that time John Blenkinsop, manager of Brandling's colliery at Middleton, near Leeds, had patented (No.

3431, 1811) a method of handling coal wagons on the colliery 'railway' there. The patent was vague about the design of the steam locomotive but Murray successfully built one for Blenkinsop; it operated on the rack system. Three more locomotives were supplied for the colliery and one for the Coxlodge railway near Newcastle.

In 1816 Francis B. Ogden, the U.S. Consul in Liverpool, ordered several engines for steam vessels, one of which worked in a steam tug on the Mississippi. In 1815 Tsar Nicholas, then a Grand Duke, visited the works and subsequently much material was exported to Russia, where several of Murray's descendants were to work. His son, Matthew, died in Moscow 22 July 1835, aged forty-two. Murray is known to have visited Moscow in 1818 when he brought back a letter for Thomas Telford (q.v.) on behalf of General Alexander Wilson (q.v.), then in charge of the Admiralty Izhora (Kolpino) ironworks near St. Petersburg. Among Murray's exports was a large hydraulic testing machine for chain cables, supplied to Wilson in 1824.

Murray was involved in supplying machinery for all types of engineering applications including land drainage pumps, gas and waterworks. Following Bramah's death and the exposing of his patent for the hydraulic press, Murray improved it with a rack and pinion device which effectively doubled the stroke of the press cylinder. He was instrumental in the establishment of the first gasworks at Leeds and his works there was the first in the town to be lit by gas.

It had been thought that Murray was structurally illiterate but research by the Royal Commission on Historical Monuments (England) has revealed interesting insights into the mill buildings with which his firm and John Marshall were variously involved, suggesting that someone must have been thinking about the mill structures. The supply of the ironwork for Benyon's mill at Meadow Lane, Leeds, is well-known. The Barracks Flax Mill (1809) at Whitehaven contained a central rank of paired cruciform columns and similar arrangements were found at Broadford, Aberdeen (1808), Bell Mill, Dundee (1806) and Whitby, North Yorkshire. The quatrefoil cross sections and decoration detail are similar to each other but differ from the earlier work by Charles Bage (q.v.) at Ditherington, although found in Marshall's 'C' Mill at Leeds (1815–1826).

Murray died on 20 February 1826 and was buried in a vault at St. Matthew's Churchyard, Holbeck, Leeds. He had three daughters: Ann (1790–1874) who married Charles Gascoigne Maclea; Mary (1797–1864) who married Joseph Ogdin Marsh, of the eponymous firm of Leeds machine toolmakers; and the eldest, Margaret (1784–1840), who married Murray's partner, Richard Jackson. Murray's firm continued in operation until 1843 and was responsible for the training of several notable engineers including Benjamin Hick, Richard Peacock, Luke Longbottom, Murray Jackson and David Joy.

MIKE CHRIMES

[Goodrich mss, Science Museum; Telford mss, ICE; S. Smiles (1861) *Industrial Biography*; E. K. Scott (ed.) (1928) *Matthew Murray Pioneer Engineer*; A. W. Skempton and H. R. Johnson (1962) The First iron frames, *Architectural Review*, March; K. A. Falconer (1993) Fireproof mills—the widening perspectives, *IA Review*, **16**(1), 11–26; G. Cookson (1994) Early textile engineers in Leeds, *Publications of the Thoresby Society*, 2nd series, **4**, 40–61]

Publications

1809. A machine for hackling hemp or flax, *Trans* [R] *S A*, **27**, 148–153

Works

1802–1803. Meadow Lane Mill, Leeds, ironwork
1820. Steam dredger, Aire and Calder Navigation

MUSCHAMP family (fl. 1770–1785) of masons were active in Lancashire and Yorkshire in the second half of the eighteenth century.

In 1770 Joseph Muschamp of Goole was one of the contractors for Ripponden Bridge in the West Riding of Yorkshire. The following year John Muschamp was a contractor (with Thomas Muschamp) for masonry work on the Leeds and Liverpool Canal; he became an Overlooker in 1774. A Mr. Muschamp was appointed 'Overlooker of Masons' on the Chester Canal in March 1775, on the recommendation of Joseph Longbothom (q.v.), engineer of the Leeds and Liverpool Canal. John Muschamp 'of Harwood' superintended the widening of Harewood Bridge in the West Riding for John Gott (q.v.) in 1776–1777.

In 1783 John Muschamp, Thomas Muschamp of Carlton, Joseph Muschamp of Otley (Yorkshire) and Edward Exley (carpenter) of Otley, Robert Thompson (carpenter), William Muschamp of Liverpool, and Thomas Hawkesworth, contracted to build Skerton bridge over the Lune in Lancaster. The bridge designed by Thomas Harrison (q.v.) had five elliptical arches, and is said to have been the first large bridge to be built with a completely level roadway. Benjamin Muschamp was also involved in this contract as he found it impossible to drive the sheet piles for the coffer dams into the cobbles of the river bed and had to expend £1,400 on pumping; he put in a claim of £600 for this work. Benjamin was later consulted about the Red Bush Bridge in Lancashire, 1797. In the West Riding he was contractor for Silsden Bridge in 1804–1806, the designer being Benjamin Hartley (q.v.).

MIKE CHRIMES and P. S. M. CROSS-RUDKIN

[Articles and bond, etc., Skerton bridge, QAR/2/5, QSP 2203/6 and 2211/32, Lancs. County RO; Correspondence on Red Bush Bridge, QSP 2392/5, Lancashire RO; Chester Canal papers, RAIL 816/2, PRO; E. Jervoise (1931) *The Ancient Bridges of the*

North of England; E. C. Ruddock (1979) *Arch Bridges and their Builders*; Colvin (1)]

Works

1770. Ripponden Bridge, repair, contractor
1771–1774. Leeds and Liverpool Canal, masonry work, mason and overseer
1775–. Chester Canal, overseer of masonry
1776–1777. Harewood Bridge widening, superintendent of works
1784–1788. Skerton Bridge, Lancaster, five spans masonry, £10,400
1804–1806. Silsden Bridge, Yorks, contractor

MYDDLETON, Sir Hugh (*c*. 1560–1631), entrepreneur, was born *c*. 1560 at Galch Hill in the parish of Henllan, near Denbigh, the sixth son of Richard Myddleton, governor of Denbigh Castle, and his wife Jane Dryhurst. Sir Thomas Myddleton (1550–1631), Lord Mayor of London, was a brother. Hugh went up to London in 1576 to be apprenticed to a goldsmith. He was granted the freedom of the Goldsmiths' Company in 1585, a year in which he spent three months in Antwerp and set up shop in Cheapside. His name appears as a liveryman of the Company in 1592. His first wife having died childless, he married Elizabeth Olmstead in 1598 and a few years later they moved to Wood Street and at some time after 1614 to Basinghall Street. In 1603 Myddleton was elected MP for Denbigh, and served in every Parliament up to 1628. He became a Warden of the Goldsmiths' Company in 1604; in 1610 he was Prime Warden, and held this office again in 1624. As was usual at this time, Myddleton combined his trade with the business of banking and also took part with his brother in several overseas financial 'adventures'.

A member of the London merchant aristocracy in the early 1600s he would have been aware of the efforts by Edmund Colthurst (q.v.) to bring water to the City by an open channel or 'river' from springs at Chadwell and Amwell near Ware in Hertfordshire, a work which was started in 1604 under Letters Patent from the Crown but early in 1605 had been temporarily abandoned due to lack of adequate funds. Conscious of the need for an additional water supply for a rapidly expanding population (200,000 in 1600, having quadrupled in the sixteenth century and set to double by 1650) the City then prepared a Bill for statutory powers to carry out the scheme. After the second reading in January 1606 the matter was referred to a Committee which included Myddleton among its members. Colthurst was invited to attend. The Bill duly passed into law in June but the City authorities were not willing themselves to finance the undertaking. Negotiations continued and finally in March 1609 they agreed with Myddleton that he would act as their deputy to exercise their statutory powers and finance the work, then apparently estimated at £3,600, in return for the profits.

Myddleton had some partners in the enterprise, but from start to finish he had managerial control

Sir Hugh Myddleton

of the entire operation. While superintending the works he lived in a house at Bush Hill, near Edmonton, which he afterwards made his country residence. Work began in May 1609, with Colthurst as Overseer or resident engineer at a salary of 14s per week. Myddleton employed the mathematical practitioner Edward Wright (q.v.) as 'my Arts man' to carry out the levelling, a delicate task since the fall over the 40 miles of the route from Chadwell spring to Clerkenwell was only 18 ft. or 5 in. per mile. For three surveys back and forth he received £20 and from May was paid a weekly allowance of £2, doubtless for setting out the work. Myddleton kept the accounts up to August; thereafter they were entered up by his clerk Edward Hughes—paid 12s per week—each page being signed by Myddleton.

The original account books, which run from 1609 to 1631, show that labourers were paid 10d per day (5s per week). Most of them worked for the excavators who usually operated in gangs at piece work rates. Bricklayers earned 18d per day (9s per week) and carpenters were paid 1s 4d per day (8s per week). The 'master of timber-work', William Parnell, supplied timber and worked on contract. There is no evidence of Dutch or other foreign experts being employed. Commissioners were appointed to settle landowners, compensation.

By September 1609 work was well under way with some one-hundred and thirty labourers engaged but from February 1610 operations came almost to a standstill. The river had reached nearly to Cheshunt, about a quarter of its total length, yet rather more than £1,100 had been spent and the costly business of making and laying pipes in the City had not been started. Moreover the project encountered strong opposition from several local landowners and against diversion of the springs which might adversely affect mills and navigation on the river Lea.

Eventually in September 1611 Myddleton reached agreement with the Lord Treasurer that King James would pay half the expenditure, in return for half the profits, and command the cessation of all unlawful opposition. Myddleton also split the other half into thirty-six shares, which he assigned to the Adventurers in proportion to their contribution. A life interest in four shares was assigned to Colthurst free of charge 'in consideration of great labour and endeavour by him bestowed'. With the future of the undertaking assured, work resumed in November 1611. Edward Pond took the place of Wright, at the slightly higher salary of £2 6s 8d per week. William Lewyn replaced Hughes as Clerk, and from January 1612 Myddleton received a weekly allowance of £2 6s 8d for himself, 'his man' and two horses. Colthurst continued as before.

Rapid progress was now made with up to two-hundred men at work, and sometimes well above this number. In twenty-two months the river was brought from the northern boundary of Cheshunt to New River Head at Clerkenwell, a winding route of some 30 miles. The formal opening took place in September 1613, witnessed by the Lord Mayor accompanied by Sir Thomas Myddleton (Lord Mayor elect) and Sir Henry Montague (the Recorder). A column of sixty labourers carrying their 'spades, shovels, pickaxes and such like instruments of laborious employment' marched round the reservoir and a speech alluded to the leading personalities involved, beginning with the Overseer (Colthurst), the Clerk (Lewyn), the Mathematician (Pond) and Master of the Timber work (Parnell).

The work of making and laying the elm pipes had begun some months before the opening ceremony and was completed in November 1614 (though additional work amounted to £18,525). The King paid his half share of this in instalments, the last being made in 1617. Myddleton had borrowed £3,000 from the City in 1614 to keep going before any substantial receipts were forthcoming. By March 1615 there were only three-hundred and eighty-four households paying water rates (typically 5s or 6s a quarter); by 1618 the number had passed the one thousand mark and by 1622 profits are recorded.

The length of the New River, as previously mentioned, was about 40 miles. Four sluices or weirs along its course reduced the water gradient to 2 or 3 in. per mile. The channel was 10 ft. wide and about 4 ft. deep, with boarded sides, crossed by numerous timber bridges and a few in brickwork. There were shallow cuttings in places, and low embankments; only at one situation, near Edmonton, was a wooden trough aqueduct built over a stream, although in 1618 another was constructed on a fairly large embankment at Highbury to shorten the course by eliminating a loop of about a mile in length.

In 1619 the New River Company was incorporated with Myddleton as the first Governor, a Deputy Governor (Robert Bateman) and Treasurer, together with twenty-six Adventurers. In 1622 Myddleton undertook to manage the whole undertaking for £800 p.a., raised to £1,000 the next year. He continued in this post until his death.

On 19 October 1622 he was created a baronet in recognition of his work on the New River and for two other achievements: the reclamation of Brading Haven and the developments of silver mines in north Wales. In the early seventeenth century Brading Haven was a tidal mud flat on the east coast of the Isle of Wight covering about 700 acres, which Myddleton decided to reclaim from the sea by building an embankment with a sluice across the relatively narrow entrance. Dutch workmen were brought over, presumably for their skill in sluice making. Work began in December 1620. Myddleton seems to have incorporated some ideas of his own, since in July 1621 he took out a patent for 'a new invention or way for the wyninge and drayninge of anie grounds overflowen with water' which 'he hath by his great paynes and charges devised and found out'. But no details are given. The job was completed in 1622 at a cost of about £4,000 plus £2,000 for purchase of the land. While supervising the work he stayed at Nunwell Park near Brading. The quality of reclaimed land proved rather poor. Myddleton and his partners Robert Bateman and nephew Richard Myddleton sold the land to Sir Bevis Thelwall in 1624; fortunately for them as events were to prove, since in March 1630 the embankment failed, probably by seepage erosion in the underlying sand. Within four days the sea had so widened and deepened the original breach that repairs were considered not to be economically feasible.

By contrast the Welsh mining enterprise was highly successful. In May 1617 Myddleton secured a 21 year lease from the Society of the Mines Royal to work silver-bearing lead ore in Cardiganshire. The richest mine, at Cwmsymlog, yielded upwards of 70 oz. of silver per ton of ore. It had been worked by open cast and shallow shafts. Myddleton overcame the problem of draining it by means of pumps and an adit, probably employing foreign mining engineers. The gross takings in 1623 are estimated to have been £1,500 a month. Much of the silver was sent to London for coinage. In 1625 he secured from the King an extension of his lease and the mines were put under the control of a commission of eminent persons including the Lord Treasurer, Sir Hugh himself, his partner Sir Bevis Thelwall, the Cornish mining expert, Sir Francis Godophin, and the German engineer, Daniel Hochstetter.

In September 1625 Sir John Wynn wrote to Myddleton, on one of his periodic visits to Wales, inviting him to consider undertaking the reclamation of Traeth Mawr, an expense of sand lying between Aberglaslyn and the sea in North Wales. Sir Hugh replied from Cardigan saying that 'I am grown in years, and full of business here at the mynes, the river at London, and other places ... which maketh me verie unwillinge to undertake anie other worke ... My desire is great to see you ... yet such are my occasions at this tyme here for the settling of this great worke, that I can hardlie be spared one howre in a daie. My wife being also here, I cannot leave her in a strange place. Yet my love to publique works, and desire to see you, may another tyme draw me into those parts'. Nothing came of this proposal until the land was eventually reclaimed in the early nineteenth century by William Alexander Madocks (q.v.).

The City of London in 1623 awarded Myddleton a gold chain and jewelled pendant for 'his worthy and famous worke', the New River. He is shown wearing the chain in his portrait painted by Cornelius Johnson (or Jannsen) in 1628; the original is in the Museum of Art at Baltimore, Maryland. There are several nearly contemporary copies, one being in the Court Room of the Goldsmiths' Hall, together with a portrait of Lady Myddleton; another is in the National Portrait Gallery.

Sir Hugh died on 7 December 1631 and was buried three days later at St. Matthew's, Friday Street, where he had served as a churchwarden and where his children were baptised. A monument to his memory designed by Robert Mylne (q.v.) was set up on an island in the New River at Amwell in 1800 and a statue of Myddleton was erected in 1862 on Islington Green.

In his will Sir Hugh left £2,300 to his four sons and three daughters, in rather unequal amounts as some of them had already received bequests, and £20 or £30 to several named employees at the mines and the New River Company. To his wife, the sole executorix, he left a life interest in twelve shares in the New River, the dividend from which at the time yielded about £160 p.a. (and more than double that amount in 1640). She was also authorised to dispose of her interest in the mines. Lady Myddleton spent her widowhood at the Bush Hill house. She died on 19 July 1643 and was buried in Edmonton church.

A. W. SKEMPTON

[Samuel Smiles (1861) *Lives of the Engineers*, vol. 1; G. C. Berry (1956) Sir Hugh Myddleton and the New River, *Trans Hon Soc Cymmrodorion*, 17–46; J. W. Gough (1964) *Sir Hugh Myddleton, Entrepreneur and Engineer*]

Works
1609–1614. The New River, £18,520
1620–1622. Brading Haven, reclamation, £4,000

MYLNE family (sixteenth to eighteenth centuries), a Scottish family from which came the architect/engineers Robert Mylne (1733–1811) and William Mylne (1734–1790), noted individually. Earlier generations of the family were known as master masons but were responsible for some major and minor engineering works as well as for many buildings.

Alexander Mylne (1474–1548), the first of the family to be noted, was Canon of Dunkeld and Master of the Work of the Bridge thereof' for the years 1513 and 1514, therefore presumably designer of the bridge, and subsequently auditor of accounts for 1515 and 1516. He then left to become abbot of the important monastery of Cambuskenneth which was near Stirling, and was the burial place of King James III; the bridge was finished by others. In later years he was very active in parliament and the law, becoming the first President of the Court of Session when it was instituted in 1532.

John Mylne (d. 1621) was Alexander's great-nephew and was made a burgess of Dundee in 1587 for services to the town, among which special mention was made of his 'renewing of the whole of the harbour works'. From 1605 to 1617 he was 'Master Mason to the Brig of Tay' at Perth, an eleven-arch replacement of the old bridge, much of which had fallen in 1582. He was made a freeman of Perth in 1607. Some months after his death in 1621 the bridge was destroyed by a flood. In 1774 a memorial stone to John and his bridge was erected near his grave in Perth by Robert Mylne (q.v.).

John Mylne II (c. 1590–1657) worked in early life with his father, John, on construction of the Brig of Perth. By his appointment as Master Mason to King Charles I in 1631 he became responsible for all royal castles in Scotland and he later repaired the fortifications of the town of Dundee in 1645–1651. For various works he was made burgess of Edinburgh, Dundee and Aberdeen. In 1610 he married Isobel Wilson.

John Mylne III (1611–1667), the eldest son of John II, was Master Mason successively to Charles I, from 1636, and Charles II. He took charge of the defence of Edinburgh Castle in 1639–1640 and undertook repairs to the pier of Leith in 1643–1645. In 1660 he reconstructed a dam on the Water of Leith. He was three times married but failed to produce a male heir. He died on 12 December 1667 and is the first of several members of the family to be commemorated on a fine monument which stands in Greyfriars Cemetery, Edinburgh.

Robert Mylne (1633–1710), of Balfargie, was the nephew of John III and was much involved in the building of Holyrood House, the royal residence in Edinburgh, from 1672 to 1679. Robert was made a burgess of Edinburgh in 1660. In 1665 he built a fort and barracks at Lerwick, Shetland, but it had a very short life. A design and estimate for a bridge at Drumlanrig Castle, made in Robert's hand before 1673, survive at the Castle but it was not built. He made five 'cisterns and fountains' for water supply in the High Street of Edinburgh in 1674–1676 and several still survive. In 1679 he was contractor for the rebuilding—at a cost of £1,150—of fortifications at Edinburgh Castle and in 1682 he contracted to build for £300 a single-arch bridge of 45 ft. span over the Clyde at Rommellweill Crags, north of Abington; it was replaced in 1769. Among building developments at the Shore of Leith he was responsible for maintaining the quay and is likely to have built the windmill of which the tower still stands. He was Master Mason to King Charles II and to Queen Anne, from 1668 to 1710 in all. He married Elizabeth Meikle in 1661 and had eight sons and six daughters; his eldest daughter, Janet, became the second wife of James Smith (q.v.), an important architect and owner of coal mines.

Portraits of John Mylne II, John III and Robert are reproduced in R. S. Mylne, *Master Masons to the Crown of Scotland*.

TED RUDDOCK

[Robert Mylne's book of family history (MS) *c.* 1775, British Architectural Library (microfilm, Mylne Papers, SRO); R. S. Mylne (1893) *Master Masons to the Crown of Scotland*; Colvin (3); J. Gifford (1996) *The Buildings of Scotland: Dumfries and Galloway*, 223; DNB]

Works

For architectural works of all the above, see Colvin.

MYLNE, Robert, FRS (1733–1811), architect and civil engineer, was born in Edinburgh on 4 January 1733, the eldest surviving son of Thomas Mylne (d. 1763), Surveyor to the City of Edinburgh, burgess and magistrate, and his wife Elizabeth Duncan (d. 1778).

In 1747 he began a six-year apprenticeship to a wright (or carpenter) called Daniel Wright and was admitted to Mary's Chapel, the lodge of building craftsmen in Edinburgh, in 1754. He was employed in making wooden carvings for the Duke of Atholl at Blair Castle until he left, with his father's support, to pursue his education for architecture by travel on the Continent, first joining his brother William (q.v.), who was already in France, in Paris in October 1754. By late December they were in Marseilles and soon afterwards reached Rome. They travelled and lived cheaply on a very modest joint allowance of £60 p.a. from their father. Robert evidently worked hard and made lasting friends of Italians, including Piranesi, and English and Scottish gentlemen for whom he made drawings and taught some of them architectural drawing. In 1757 he went with a Welshman, Richard Phelps, to survey the antiquities of Sicily, about which he proposed to publish a book of prints on his return to Britain, an intention never fulfilled because important employment fell into his hands immediately on his arrival.

Robert Mylne FRS

By entering in 1758 in the *Concorso Clementino* of the Academy of St. Luke and winning the Silver Medal, never before won by a Briton, Mylne gained publicity both in Scotland and London, but he reached London in July 1759 apparently unaware that the City authorities were about to advertise for designs for Blackfriars Bridge, the second bridge over the Thames within the City. Mylne very quickly established contact with the leader of the City's committee, and continued developing his design and lobbying aldermen and others until late November when his design was publicly announced to be the leading contender. A pause of three months ensued while eight 'experts' were consulted about his proposal to use elliptical arches, which were new in Britain. His design was finally adopted on 22 February 1760 and he was appointed 'surveyor' to direct the whole project at a salary of £350 p.a. At the end of construction the annual payments were subsumed into a fee of 5% of the whole construction cost, plus 1% of the cost of property purchases which he managed; he was forced to take his case to court to obtain this settlement.

He had won the commission over fifty submitted designs, including those of famous architects such as William Chambers and the engineer, John Smeaton (q.v.) who had previously been the City's favoured candidate. Whether before or after his appointment, Mylne had to learn a lot about bridge-building, and did so, though it cannot be doubted that he had previously applied his mind to some extent to the study of heavy masonry and construction in water. His demonstration of skill in these matters during the ten years it took to build the bridge brought him many other commissions not only for bridges but for hydraulic engineering, ranging in succeeding years from public water supplies and arterial drainage to canal and river navigations.

It is perhaps more surprising, and contributes greatly to his professional stature, that his architectural practice developed in parallel, and just as quickly; before the bridge was completed he had become surveyor to both St. Paul's and Canterbury Cathedrals and had designed major additions to at least five country houses and many properties in London. This architectural practice, which cannot be further noticed here, continued throughout his lifetime, spreading throughout England and Scotland as far north as Inveraray and the Isle of Mull. His engineering reports and correspondence reached to equally distant places, including harbours in the Western Isles of Scotland and design of a major bridge for Londonderry in the northwest of Ireland, though the latter was not built.

After winning the Blackfriars commission, he moved quickly into a new manner of life, renting a 'charming apartment' in Arundel Street near the Thames in September 1760 at 50 guineas p.a., and employing a clerk, a valet de chambre and a waterman with gondola which would be vital for his supervision of the bridge-building. Clerks became his regular aides and some, perhaps all of them, were aspiring professional architects. Robert Baldwin was one of the first, from 1763 to 1766, and was paid £50 p.a. Thomas Cooley, engaged in 1764, received £40 with breakfast and lodgings. House servants came and went fairly frequently.

There is evidence of methodical ways of managing his work. From 1762 he kept diaries in which he made notes on most days of where he had been, of letters and drawings received and dispatched, the fees to be charged for each piece of work, and also occasional notes of his technical observations at work sites and references to important family events. Substantial reports have either been preserved in the manuscript archives of his clients or have survived in print. Drawings, of which he made many, have not survived in great numbers and are widely scattered. All his directions to the several contractors working on Blackfriars Bridge, including some construction drawings, were copied into a folio volume which provides a thorough record of what was his largest single project.

There is no doubt that his behaviour and speech were also well ordered and forthright, but within ten years of his establishment in London the family in Edinburgh were finding him overbearing on his visits there. His brother, William, spoke of 'his Honour of London' and sister, Anne, called Robert 'the Bashaw'; on more than one occasion she was deliberately absent when he came to visit. There is a scatter of notes on Robert Mylne in the letters and diaries of other acquaintances which convey similar impressions of autocratic, sometimes angry

and always self-confident behaviour. However, his relationships with many important people who were his clients seem to have been cordial, and his status amongst fellow-professionals, both engineers and architects, was high. In business, it is clear that he consciously set himself a high standard of integrity once he had become established. At the start, his engaging in the controversy about the designs for Blackfriars Bridge by an unsigned pamphlet and similar letters to newspapers was, at the least, lacking in candour.

From 1760 until 1764 the work at Blackfriars occupied most of Mylne's time during the construction seasons; from then on he was able to travel frequently throughout the year. Two early concerns were bridges in the parks of the Earl of Warwick, a very fine single arch, and the Duke of Portland, a large three-arch bridge at Welbeck Abbey in Nottinghamshire, both of them engineering structures but designed to beautify the estates. The Welbeck bridge suffered movements before the centres were struck, culminating in February 1768 in the collapse of the middle arch of 90 ft. span, and the termination of Mylne's employment by the Duke. Other commissions for new bridges came later, for the Bishop of Durham, the Duke of Argyll, and for local authorities at Tonbridge, Kent, and Romsey, Hampshire. A very interesting design which never came to fruition was made in 1774 as replacement for an old stone bridge which connected Inveraray Castle with its old estate village. With the village being rebuilt further away from the Castle, leaving the bridge in estate parkland, Mylne drew a bridge of two iron (presumed wrought iron) arches with its parapets and the gate at midspan all in the current Chinese style. It is his only known design of an iron bridge and was made five years before the first substantial iron bridge in Britain was built.

A broader spectrum of civil engineering came under his care in works for the supply of water to London. One supplier was the London Bridge Waterworks, operating water wheels in the flow through several arches of the bridge to drive pumps which raised water to pipes supplying some of the streets; Mylne reported to the Common Council in 1767 on these works. Earlier in his diaries he had recorded visits to, and reports on, several parts of the works of the New River Company which supplied clean water to another part of the city by their long channel from the springs near Ware in Hertfordshire. Their system involved dams, embankments, bridges and reservoirs, and all the necessary control valves, sluices, etc., and the distribution mains and branches in the streets. At the end of 1767 Mylne was appointed 'joint surveyor' to the Company, to work with their long-standing surveyor, Henry Mill, for an annual salary of £200. When Mill died in 1771 Mylne became chief engineer to the New River and held the post for over thirty-five years. Near the end of his life, in 1804, he accepted his son William Chadwell Mylne (1781–1863) as joint engineer and in 1810 took a pension, so leaving his son as sole chief engineer for the next forty-nine years.

From working at Blackfriars amongst the Thames river traffic, and also on the whole New River system, Mylne gained a practical knowledge of hydraulic engineering that was unchallengeable. He was therefore consulted by large numbers of owners, managers and new promoters of navigations, water supplies and sewerage authorities, and owners of mills who competed for use of watercourses. At New River Head, the London end of the River and headquarters and centre of the Company's distribution network, he was also involved in 1767 in the installation of a steam engine designed by Smeaton, to pump water from the River to the service reservoir called the Upper Pond. He was given use of a house there in 1771 which he used as his office and also to some extent as a residence. In 1782 it was refronted and extended to his design. He also had the use of a house at Greenwich during his tenure of the office of Clerk of Works to Greenwich Hospital from 1775–1782, and his family spent summer months there. In the 1790s he built himself a house at Amwell near the source of the New River in Hertfordshire, and it became the family's normal home. In 1800, after the deaths of his wife, two daughters and his eldest son, a mausoleum was built in the local churchyard for them, while at the same time he erected on an island in the New River at Amwell a memorial urn to the founder of the Company and original constructor of the New River, Hugh Myddleton (q.v.).

Throughout the 1770s the diaries record frequent consultations, surveys, letters and reports on rivers, harbours and bridges which were other engineers' responsibilities, Mylne giving advice and 'second opinions' about particular problems or attending the sittings of parliamentary committees, and occasionally court hearings in various parts of the country, always representing a client either in support of or in opposition to proposed developments. These actions were additional to his work on the projects for which he was himself the engineer or architect. In the 1780s the number of his engagements of this sort increased and in almost every public argument concerning a bridge in Britain he played a role. For instance, in the litigation which dragged on for six years after the destruction by flood of Smeaton's bridge over the Tyne at Hexham, Mylne advised the county justices, making several visits of inspection and attending assizes at Carlisle; when the Irish Society of London considered many plans for a bridge across the estuary of the Foyle at Londonderry, he attended their meetings and the parliamentary committee, spoke to individuals, wrote letters and finally in July 1787 delivered to them 'a design of many drawings, reports, papers and estimate', and charged them the considerable sum of 100 guineas, including fifteen for his brother, William, who, being resident in Dublin, must have visited the site and given other assistance; in

1785–1786 he several times met the projectors of a bridge over the Menai Straits; and in 1791–1792 he consulted with Rowland Burdon (q.v.) more than once about the Sunderland Bridge proposal and gave 'long evidence' to the parliamentary committee which approved it.

The 'canal mania' years of the early 1790s also brought him a lot of business. He reported in 1790 on surveys for navigational improvements on the Severn and gave evidence in favour of a Bill in Parliament. In 1791–1793 he made personal surveys of almost the whole of the Thames from Lechlade to Staines and recommended many improvements including new pound locks, bypass cuts, weirs and dredging of shoals. The estimated cost of works was £28,000. Much of what he recommended was carried out, but not by Mylne. (He made further proposals for new cuts above Oxford in 1802.) He attended committees in Parliament on the Worcester and Birmingham Canal, the Staffordshire and Worcestershire Canal, the Grantham Canal, the Nottingham Canal, and several other navigations during 1791–1792. Also in 1791, he made a report (published 1792) on the perennial problems of adequately draining the Fens and proposed the making of a long straight channel which, when eventually adopted more than twenty years later and then called the Eau Brink Cut, resulted in major improvements.

A great opportunity came when he was invited to take charge of planning, including obtaining an Act of Parliament, for the Gloucester and Berkeley Canal. In September 1793, six months after the Act was passed, he was appointed chief engineer. It was to be a ship canal about 18 miles long and 18 ft. deep to take sea-going vessels of 300 tons from the Bristol Channel up to Gloucester Docks. Mylne's estimate of the cost was £121,000. Excavation began late in 1794 but proceeded slowly, and the problems of an unsatisfactory resident engineer, escalation of the estimate and frictions in the committee, some members of which exhibited a settled prejudice against Mylne, led eventually to his dismissal in 1798 with only 5 miles of the canal made. Further progress was slight and only long after Mylne's death was it restarted and the whole line opened in 1827 as the Gloucester and Sharpness Canal.

In his late years his diaries show his employment as 'surveyor', or property advisor and valuer, which had always been a feature of his practice, taking even more prominence, with frequent service given to clients such as the Stationers Company of London, the banker Thomas Coutts, and the Marquess of Bute.

Robert Mylne was proud both of his ancestry and of his own professional status. Items in the history of the family as far back as the sixteenth century which he had collected were placed in a manuscript notebook, apparently about 1775, and added to in subsequent years, so starting an interest and pride in family history which continued through three more generations culminating in the publication of the grand book

Master Masons to the Crown of Scotland, by the Rev. R. S. Mylne in 1893. Robert Mylne himself was married in 1770 to Mary Home, daughter of a regimental surgeon, taking a girl without money which, he said, he 'always despised, because [he] could procure it by industry'. At that time he had 'an income of £1,000 per annum (fixed and flying)', meaning, respectively, his fixed annual fees or salaries, and the one-off sums earned for consultancies and design of works, architectural as well as engineering. He had six daughters and three sons, only one son surviving to maturity. His eldest daughter, Maria, who died in 1794 aged twenty-two, made a miniature painting of him—still held by the family—from which was engraved the best-known portrait of him for inclusion in *Master Masons to the Crown* .... There is also a head in profile of him at age twenty-four, engraved by Vangeliste in 1783 from a drawing made by Brompton in Rome, and printed in Richardson's book *Robert Mylne, Architect and Engineer 1733–1811*; and a pencil portrait at age sixty-nine made by George Dance is in the National Portrait Gallery, London.

He was a very healthy man. He reported in letters to the family the long walks by which he and William covered much of the distance across Europe to Rome in 1754–1755. After his return to Britain he often travelled to Scotland by coach and chaise in about four days (and nights), and journeys to other cities and estates throughout the country were made every few weeks. His river surveys on the Severn and the Thames, carried out when he was near to or over sixty years old, involved very long walking or possibly riding. The quantity of engagements recorded in his diaries seems not to diminish as the years passed, right up to the year before his death. He had a great liking for claret.

Robert Mylne was elected Fellow of the Royal Society of London in 1767 but made no contributions to its *Transactions*. In 1771 he was one of seven men who gathered in a London tavern on 15 March to form the Society of Civil Engineers and at their second meeting he was chosen as Vice-President, which post he held until 1778. He wrote the first printed description of the early years of the Society in his preface to the *Reports of the late John Smeaton FRS*, published in 1797, and is credited with taking a key role in the survival of the Society in 1793 after a crisis of personalities had threatened its existence; it was reorganised with Mylne as treasurer, the only office in the Society at that time and which he filled until 1810, a year before his death. His view of the value of the Society is given in his description of the meetings: 'Conversation, argument, and a social communication of ideas and knowledge, in the particular walks of each member, were, at the same time, the amusement and the business of the meetings'. When a somewhat parallel club for professional architects was founded in 1791 he was elected a member but was never an officer of the club. He had, however, sustained a standard

of practice, in combining the roles of civil engineer and architect, that was not seen again as the separate professions became formalised in the years after his death.

Mylne died at New River Head on 5 May 1811 and was, at his own request, buried in the crypt of St. Paul's Cathedral near the tomb of Sir Christopher Wren, to whom he had erected a monument there in 1807–1810.

TED RUDDOCK

[Mylne Papers, GD1/51, SRO, containing MS letters etc. of William (brother) and Ann (sister), various other family papers and microfilms of Robert's book of family history (RH4/82) and of his diaries (1762–1810); Mylne Papers, MS, British Architectural Library, London, containing letters of Robert and William from France and Italy 1753–1760, later letters of Robert, and originals of his book of family history and his diaries 1762–1810, also his book of orders made during construction of Blackfriars Bridge, etc.; J. Smeaton (1760) *Mr. Smeaton's Answer to the Misrepresentations of his Plan for Black-Friars Bridge*; Rev. R. S. Mylne (1893) *The Master Masons to the Crown of Scotland*; C. Gotch (1951) *The missing years of Robert Mylne*, The Architectural Review, **110**, 179–182; H. W. Dickinson (1954) *Water Supply of Greater London*, 38–42; A. E. Richardson (1955) *Robert Mylne, Architect and Engineer*, containing an incomplete and inaccurate transcript of the diaries; D. Stillman (1973) British architects and Italian architectural competitions 1758–1780, *Journal of the Society of Architectural Historians* (American), **32**, 45–50; A. W. Skempton (1971) *The Smeatonian Society of Civil Engineers*; Ian Lindsay and Mary Cosh (1973) *Inveraray and the Dukes of Argyll*; Ted Ruddock (1979) *Arch Bridges and their Builders 1735–1835*, and references quoted therein; Garth Watson (1989) *The Smeatonians*; Ted Ruddock (1993) *Travels in the Colonies in 1773–1775*; Colvin (1995) 679–85; Roger Woodley (1998) Robert Mylne, 1733–1811: the Bridge between Architecture and Engineering, Ph.D. thesis, University of London; DNB]

Publications
1759. (attrib.) Letter 'in defence of elliptical arches and iron-rails', *London Chronicle*, 4 December 1759
1760. (attrib.) *Observations on Bridge Building and the several Plans offered for a New Bridge* [Blackfriars] and *Postscript*, published separately but with continuous pagination and both signed 'Publicus'
1760. *View of Mr. Mylne's elegant Design of a New Bridge, to be built from Black Fryers to the opposite Shore, approved by the Committee in Common Council*, elevation and pictorial foreground; the elevation was also published in *The London Magazine*, March
1767. *Mr. Mylne's Report and Opinion of the Design for keeping the Ships afloat at all Times in the Harbour of Bristol. And also, Mr. Smeaton's*

Opinion upon Plans laid before him for that Purpose, signed at Bristol, 12 January 1767

1767. Report relating to the Petition of the Proprietors of the London-Bridge Water-Works, *To the right Honourable the Lord Mayor, Aldermen, and Commons, in Common Council Assembled*, signed at Arundel-Street, 27 February 1767

1772. *Mr. Mylne's Report respecting Tyne Bridge. with his Plan for a Temporary Bridge: also Mess. Rawlings and Wake's Abstract of the Borings into the Bed of the River Tyne*, signed at Gateshead, 12 March, 1772

1781. *Mr. Mylne's Report, on his Survey of the Harbour, etc. of Wells, in Norfolk*, signed at New River Head, London, 28 April 1781

1783. *Mr. Mylne's Opinion and Report ... respecting the Practicability of building a Permanent Bridge at Hexham*, signed at Hexham, 24 April, 1783. In: John Smeaton (1812) *Reports on Various Occasions*, **3**, 296–298. *Mr. Mylne's Second Report ...*, signed at Edinburgh 30 September, 1783. In: Smeaton (1812) *Reports on Various Occasions*, **3**, 322–324

1789. *A Letter from Robert Mylne, Esq., to the Right Worshipful John Patteson, Esq., Mayor of Norwich, on the State of the Mills, Water-Works, etc., of this city, commonly called the New Mills*, signed 1 January 1789

1791. *Mr. Mylne's Report to the Commissioners, for improving the Navigation of the Rivers Thames and Isis*, signed at New River Head, 8 May 1791. *Report the Second, by Mr. Mylne, Surveyor and Engineer on the Navigation of the River Thames, between Lechlade and Whitchurch*, signed 10 August 1791, [and] *An Estimate of the Works proposed to be done agreeable to Mr. Mylne's Opinion and Report*

1792. *The Report of Robert Mylne, Engineer, on the Proposed Improvement of the Drainage and Navigation of the River Ouze, by executing a Straight Cut, from Eau-Brink to King's Lynn*, signed at New River Head, 26 October 1792

1793. *Report on a Survey of the River Thames from Boulter's Lock to theCity Stone near Staines and on the best Method of improving the Navigation of the said River, and making it into as compleat a state of perfection as it is capable of*, signed at London, 20 August 1793

1793. [Report] *To the Gentlemen of the Committee of Subscribers to the Proposed Canal from Bristol to Cirencester*, signed 27 September 1793

1794. With Robert Whitworth (q.v.) [Report] *To the Gentlemen Subscribers for the London Canal, from Boulter's Lock to Isleworth on the River Thames*, signed 17 July 1794, In: *Extract from Minutes of Proceedings of Subscribers ... presented to the Commissioners of the Thames Navigation*, 1 August 1794

1794. With Robert Whitworth (q.v.) *Plan of the proposed London Canal from a place called Hog Hole in the River Thames in the Parish of Datchet to a place called the Rails Head in Parish of Isleworth*, surveyed 1794

1797. Preface, *Reports of the late John Smeaton made on Various Occasions in the course of his Employment as an Engineer*, **1**, iii–xi; reissued with vols. 2 and 3 added, 1812

1800. *Report of Mr. Mylne, on the state of the River Thames and its bed; on the structure of London Bridge, and as to the navigation of the River above and below it etc.*, with an appendix and drawings, signed at New River Head, 30 May; printed as Appendix (A.1), *Third Report from the Select Committee upon the Improvement of the Port of London*, House of Commons, 28 July, 25–38; also in *Reports from Committees of the House of Commons (1793–1802)*, **14**, 1803, 550–554, pls. 29–30

1800. *A Scheme or Outline of a Plan for a new Bridge, adapted to the Situation and Circumstances of London Bridge*, signed 23 June; printed as Appendix (B.2), *Third Report from the Select Committee ...*, 28 July, 51–56; also in *Reports from Committees ... (1793–1902)*, **14**, 562–564

1800. *Inscriptions for a Votive Urn at Amwell in Hertfordshire*, by Robert Mylne FRS; reprinted in John Nichols (1815) *Library of Anecdotes of the 18th Century*

1801. *Mr. Mylne's Answers. To the Select Committee of the Bridge-House Lands, appointed to consider a Report of the Select Committee of the House of Commons, for the Improvement of the Port of London, by rebuilding London-Bridge, etc.*, signed Robert Mylne, 15 May [and] 30 October; printed in *At a Sub-Committee ... held at Guildhall, on Wed. the 19th day of December 1801*

1802. *Report by Mr. Mylne, Engineer, on Three New Cuts, for the Improvement of the Navigation of the River Thames, above Oxford*, dated 15 May

1802. *Report, by Mr. Mylne on the Present State of the Navigation of the River Thames, between Maple-Derham and Lechlade*, signed 1 June 1802

1802. Second Part of the *Report on Three New Cuts ...*, signed 16 June 1802

Works

1760–1769. Blackfriars Bridge, London, designer and 'surveyor' for construction, nine elliptical arches, max. span 100 ft., cost £160,000, demolished 1864

1764–1768. Welbeck Abbey Bridge, Notts, three arches, spans 75, 90 and 75 ft., middle arch collapsed before completion

1765. Survey for Leafield Bridge in the estate of the Earl of Warwick, constructed 1772–1776, one high arch, 80 ft. span, very decorative, greatly damaged by vandalism 1975–1995

1767–1810. New River Company for water supply of London, directed extensive works of maintenance and development as Joint Surveyor (1767–1771) and Surveyor (1771–1810) to the Company pre-1768 (uncertain). Early involvement in his brother's design of the new bridge over the Clyde at Glasgow; drawing inscribed *Original design by which the New Bridge ... at Glasgow was constructed after the scale was enlarged for a wider part of the River* ..., signed R. Mylne, British Architectural Library, RAN1/J/1

1768. Dumbarton Bridge, River Leven, report on correction of large settlement which occurred during construction, repair probably executed by his brother William (q.v.)

1772–1776. Aray Bridge, Inveraray, design of bridge and directions to Col. Skene, military Inspector of Roads, and design of ballusters for the Duke of Argyll

1773–1776. Tonbridge Great Bridge, River Medway, design and supervision for Kent County Justices, three arches, each 22 ft. span, built on the piers of the former bridge (?), cost £1,036, demolished 1887 (bills, etc., at Kent County Archive)

1774–1780. Southern end of the Tyne Bridge, Newcastle-upon-Tyne, design and supervision for the Bishop of Durham, three segmental arches, the longest 50 ft. span, joined the northern part, built for the Common Council of Newcastle, in midstream, widened *c.* 1800, demolished *c.* 1876

1782–1784. Middle Bridge, Romsey, River Test, design and supervision for Hampshire County Justices, single masonry arch, 48 ft. span, cost £3,039, demolished 1930 to make way for a 'lookalike' bridge of concrete arch with stone facings (contract, bills, etc., at Hants. County Archive)

1783–1787. Dubh Loch Bridge, River Garron, Inveraray Estate, design and directions for the Duke of Argyll, single masonry arch, in rustic style, 60 ft. span, cost £312 (tender) with some materials free from the owner (documents and drawings at Inveraray Castle; copies at NMRS)

1785–1787. Ridley Hall Bridge (rebuilt after collapse), design and directions for Northumberland County Justices, two arches, asymmetrical, cost £1,129 (Quarter Session Records at Northumberland County Archive)

c. 1790. Clachan Bridge, Seil, Argyll, a high arch over the tidal channel between island and mainland, revised the design of John Stevenson for the Earl of Breadalbane and others, built 1791 (Royal Commission on the Ancient and Historical Monuments of Scotland, *Argyll*, **2**, 294–295)

1790?. Pinford Bridge, Sherborne Castle Estate, Dorset (Colvin (1995) 575)

MYLNE, William (1734–1790), master mason, architect and civil engineer, was the second son of Thomas Mylne (d. 1763), Surveyor to the City of Edinburgh, burgess and magistrate, and his wife, Elizabeth Duncan. Both Thomas and William were therefore members of the family six of whose members had held the office of Master Mason to the Crown of Scotland between 1481 and 1707.

William was entered as apprentice mason in the lodge of Edinburgh (Mary's Chapel) on 27 December 1750 and admitted as freeman mason and 'operative master' on 20 December 1758 (but his burgess ticket is dated 12 April 1758). He had in the meantime travelled to the Continent at his father's expense, living in Paris and Rouen from March to December 1754, in Rome until mid 1757, and visiting Florence, Vicenza and Rotterdam before returning to Edinburgh about March 1758.

He began a business which continued for fifteen years as architect and builder in Edinburgh, often producing reports and designs as well as undertaking construction contracts, for buildings, bridges, roads, the pier at Leith, and water supply and drainage, many of his commissions coming from the town council and from road trustees in the district and further afield. He designed in 1759 and may have built (about 1764) the three-arch bridge at Yair (or Fairnilee) over the Tweed; in the outskirts of Edinburgh he repaired the road to the west at Bell's Brae, designed a bridge at Canonmills in 1767 and advised on repairs to Cramond Bridge in 1769; removed and re-erected a water conduit head at Niddry's Wynd when a new supply was routed into the city in 1760; he built the masonry of St. Cecilia's music hall, designed by his brother Robert (q.v.); and reported in 1766 that the roof of the chapel of Holyroodhouse was unsafe and would collapse, as it duly did in December 1768.

More importantly, perhaps, he was commissioned early in 1759 by the Lord Provost, George Drummond, the prime mover of the expansion of Edinburgh, to 'make out a plan of the new intended north passage immediately'. This involved a crossing of the marshy low-lying ground on the north side of the city to reach a ridge on which an extension of the burgh, soon to be known as 'the new town', would be developed. Six years later, when the city advertised for designs and tenders for the bridge, Mylne obtained the contract with a high bridge of three large arches and a smaller one at each end. On the evening of 3 August 1769, when the bridge was virtually complete and already in use by members of the public on foot, three vaults in the high approach from the south end collapsed together with their supporting walls and part(s) of one or both of the side retaining walls of the approach; as the earth filling subsided five people were engulfed and killed and one or two others injured. Mylne's contractual responsibility for repair was confirmed immediately, but John Smeaton (q.v.) was called to report to the Town Council, which he did jointly with the local men, John Adam and John Baxter (qq.v.). With feuing (sale for building) of sites in the 'new town' already going on, meetings of the feuars began demanding partial demolitions of other walls of the bridge for inspection and repairs, the Council stopped payments to Mylne and he fell heavily into debt. When facing a further demand for inspection in February 1773, when he believed his work was again completed, he protested formally and left Edinburgh. The repairs had involved the formation of large cylindrical voids through each of the spandrels between the arches.

While working on the bridge he had obtained other commissions, most notably as designer and overseer of a contract for a seven-arch bridge over the Clyde at Glasgow, built in 1768–1772, in which there were, from the initial design, open cylindrical voids through all the spandrels. Some collusion with his brother, Robert, about the design may be indicated by Robert having signed an earlier drawing (RIBA, BAL, RANI/J/I), similar to the contract drawing. When Robert gave advice about serious settlement of the unfinished bridge of Dumbarton in 1768 it appears that William was employed to undertake or supervise the work and he was made a burgess of the town in the following year. He was also admitted burgess of the far-away town of Banff in October 1769 'on account of the singular respect which the … magistrates [bore] to him'.

William Mylne also played a significant part in the planning of the Edinburgh 'new town', plans being invited in open competition in 1766. The winning premium was awarded to the young James Craig in August, though without committing the Council to adopting his plan. Although the names of the other five competitors were not made public, it would be surprising if Mylne was not one of them. In October, the Council received a 'rectified plan of the improvements' from him and passed it to the group which was assessing Craig's plan; amendments to Craig's scheme then continued until July 1767 when it was adopted. It is fairly certain that Mylne's suggestions had had considerable influence.

Another indication of his good standing is that from March 1770 to July 1772 he directed and supervised minor but important works at Inveraray Castle, seat of the Dukes of Argyll and one of the great houses of Scotland.

Before he left Edinburgh in May 1773 Mylne had already suffered illness which he associated with the strains of conflict and debt. Long before, on his way home from Rome, he had confessed his lack of ambition: 'I have little hopes … about making a fortune as … I can be contented with very little and frequently reflect over the mending a hole in my old shirts or coat how bainful ambition is to the happiness of man'. From London in September 1773 he took ship for Charleston in South Carolina and in January 1774 retired to the backlands near Augusta in Georgia where for some months he occupied a log cabin, grew corn and melons, kept chickens, shot some game and fished, and hoped that with the payment of what was owed to him for the bridge he might set himself up as a planter. But his presence became known, immigrant Scots took him away in the heat of summer to see their new land grants and advise them on buildings, and his bouts of illness returned. In the winter of 1774–1775 he left on horseback to seek architectural work in the cities to the north, but with the unsettled state of life and business as the colonies drifted towards revolution nothing was offered and he returned incognito to London in September 1775 and later to Edinburgh.

In 1776, while planning for a large increase in their public water supply, the 'pipe water committee' of Dublin's town council asked the advice of Robert Mylne, who had been surveyor to the New River Company, suppliers of much of London's water, since 1767, in seeking a person to direct the new works and superintend the whole of the city's supply. Robert recommended William, who was appointed engineer at £140 p.a. with a house provided, the salary increased a year later to 225 guineas.

The large increase of available water arose from the completion of the navigable Grand Canal connecting Dublin with the river Shannon; the eastward flow of the canal was purchased cheaply by the corporation and diverted into their storage 'basin'. In the prosperous centre of the city there were many house owners ready to pay for connections and Mylne's task was to design the network, purchase material—mostly elm trees for which he scoured both Ireland and Britain, but also some other species, and, for a few large mains, cast iron—lay the new mains, branches and house and industrial connections, and organise the back-up services of maintenance, collectors and turncocks, while monitoring the activities of the canal company and its users to avoid contamination of the supply, and similarly controlling other existing sources in the river Liffey and smaller streams. In spite of feeling frustration with his masters on the committee amounting at times to despair, and which led him several times during his tenure of thirteen years to tender his resignation, and also of continuing indifferent health, he was each time persuaded to take some leave and then return to his post. By the time of his death in March 1790 he had presided over a huge change in the water supply, in both scope and quality, and a capital investment of over £30,000. He took particular pride in the provision of 'conduits, or fountains for supplying the poor with water'. The whole network he had projected soon after his arrival was still unfinished, by a few years' work, at the time of his death. A silver salver presented to him in 1786 testifies to the city's 'entire approbation of the laudable exertion of his great abilities in rescuing those important works from the very bad condition they were in, and bringing them to a state of perfection'.

He was buried in the churchyard of St. Catherine's Church, Dublin, and is commemorated in the south aisle of the church by a marble plaque erected by his brother, although Robert had a sharp disagreement with the corporation about the ownership of a plan of the system drawn by William and for which Robert asked, and the corporation refused to pay, a fee of £100!

He was unmarried, and had one natural son named Willy, whose education was supervised for some years by his sister, Anne (1745–1822), and who later served on ships trading in the East. William was greatly attached to Anne, at least up to the time of her marriage just before his return from America, and his letters to her are testimony to his habits (and hers) of reading and graphic letter-writing, but also to his professional ambitions which seem always to have been modest in scale but demanding as to the quality of his work. In temperament he was probably best suited to the public service role to which he devoted his last years.

All that survives of his will is a tiny abstract in which the two names mentioned are those of his brother, Robert, and his nephew, Robert's son and heir, William Chadwell Mylne. They were presumably residual legatees after provision had been made for others, in particular his elderly house-servant.

TED RUDDOCK

[MS letters of Robert, William and Anne Mylne, and Robert's book of family history, Mylne Papers, British Architectural Library, London.; letters of the same and other family documents, Mylne Papers, GD1/51, SRO; Edinburgh Town Council Minutes 1760–1776 and Bridge Committee Minutes 1764–1770, Edinburgh City Archives; Minutes of Edinburghshire Road Trustees, Cramond and Dalkeith Districts, SRO; MS contract with drawing for Broomielaw Bridge Glasgow, 1768, Glasgow City Archives; R. S. Mylne (1893) *The Master Masons to the Crown of Scotland*; Sir John and Lady Gilbert (1905–1909) *Calendar of Ancient Records of Dublin*, **12–14**; Ian Lindsay and Mary Cosh (1973) *Inveraray and the Dukes of Argyll*; E. C. Ruddock (1974–1976) The building of North Bridge, Edinburgh 1763–1775, *Transactions of the Newcomen Society*, **47**, 9–33; Ted Ruddock (1979) *Arch Bridges and their Builders 1735–1835*; S. Harris (1992) New Light on the first New Town, *Book of the Old Edinburgh Club*, new series, **2**, 1–13; Ted Ruddock (1993) *Travels in the Colonies in 1773–1775 described in the letters of William Mylne*, and sources quoted; Kitty Cruft and Andrew Fraser (eds.) (1995) *James Craig 1744–1795*]

Works

1759, 1764?. Yair Bridge (river Tweed), Selkirkshire, design and contractor (attrib.), three arches (letter William to Robert Mylne 27 Feb. 1759; RCAHMS (1957) *Inventory of Selkirkshire*, 26–27, 71)

c. 1760. Removal and re-erection of a water conduit head at Niddry's Wynd, Edinburgh

1764. Reconstruction of the road at Bells Brae, Edinburgh, advice and supervision, for Cramond District Road Trustees

1765. Lugton Bridge, design, for Dalkeith District Road Trustees, one arch, 55 ft. span, built by Alexander Stevens (q.v.)

1765–1773. North Bridge, Edinburgh, design and contractor, for Edinburgh Town Council, five arches, cost £17,000

1766–1767. Canonmills Bridge (Water of Leith) for Cramond District Road Trustees, one arch, built by Alexander Stevens, cost £180, demolished 1840

1768–1772. Broomielaw Bridge, Glasgow (river Clyde), design and arbiter of contract, replaced in 1833–1835

1768. Repairs to unfinished bridge of Dumbarton (river Leven) (Robert Mylne Diary 31 May 1768 and 11 Sept. 1772)

1769. Proposed repairs to Cramond Brig, consulted by Cramond District Road Trustees

1777–1790. Engineer to the waterworks of Dublin, expenditure £30,000

N

NAPIER, Sir Charles James, General (1782–1853), Governor of Scinde, had a military career which lasted from 1794 until 1850, covering the Peninsular War—Corunna (1809), Busaco (1810) and Fuentes del Onoro—the infamous conquest of Scinde (1843) and the first Sikh War of 1845.

As a Captain in the Staff Corps in 1805, assisting the Royal Engineers, he worked on the construction of the Royal Military Canal. In 1815 he attended the Military College at Farnham where his studies included building construction, useful in his career as an administrator in the Ionian Islands from 1819–1830 and as Governor of Scinde.

He first visited the Islands as inspecting field officer in 1819 and three years later was appointed Resident of Cephalonia. While there he was in charge of public works and built a network of roads covering the island. In 1825 he wrote *A Memoir on the Roads of Cephalonia*.

In 1840 Napier was asked to report on the situation of barracks in the manufacturing towns of the North of England. His report was the first to factor railways into the strategic situation and recommended the building of large defensible barracks in Preston, Bury and Ashton.

While Governor of Scinde (1843–1847), he investigated the mouths of the Indus and had plans to make Karachi the second port of the Indian Empire. These included construction of a breakwater and the supply of fresh water from Komaru, channelled through the 'Napier mole'. The plans were described by James Walker (q.v.) as 'large and enlightened'.

He returned to England and was working on *Defects Civil and Military of the Indian Government* when he died on 29 August 1853. His statue, by G. G. Adams, stands in Trafalgar Square.

SUSAN HOTS

[Report of Sir Charles Napier on the barracks in the Northern District, HO55/451, PRO; J. Walker(1858) *2nd Kurrachee Harbour*; E. W. C. Sandes (1933) *The Military Engineer in India*, **1**, passim; E. W. C. Sandes (1948) *Indian Sappers and Miners*, 111–114; J. Douet (1998) *Barracks*]

Publications

1825. *A Memoir on the Roads of Cephalonia accompanied by Statistical Tables*

NASH, John (1752–1835), architect, was probably born in London in 1752, the son of a Lambeth millwright of Welsh origin, who died when Nash was a child. He worked for the architect Sir Robert Taylor before setting up on his own as an architect and builder in 1775.

In 1783 he was declared bankrupt and for the remainder of his career he was to be dogged by financial and professional controversy. He retreated to Carmarthen, where he continued to rebuild his practice whilst maintaining a metropolitan contact via the architect, Samuel Saxon. As an architect in Wales he developed the picturesque style which has generally been associated with him and was able to return to London in 1796 with his reputation re-established.

The main aspects of his subsequent career are well-known and require little elaboration. For some years he had a partnership with Humphrey Repton, the landscape architect, and executed work all over the country. The partnership dissolved in 1802 by which time it had brought him into contact with the Prince Regent through work at Brighton. He became his favourite, and the architect of the Regency style. In 1806 Nash became attached to the Department of Woods and Forests which led to his planning the development of the Crown estate, centred on what is now Regent's Park, including Regent Street. From 1813 he was, with Sir John Soane and Robert Smirke (qq.v.), one of the architects attached to the Office of Works. His specific responsibility was for Carlton House, Kensington, and St. James Palace, and the lodges at Windsor, but from 1815 until 1822 he was personally selected by the Prince Regent to work on Brighton Pavilion. Following George IV's accession in 1820 he was put in charge of the reconstruction of Buckingham House, now Palace. Nash's association with George IV, who became increasingly unpopular, provided much work but ultimately tarnished his career. Because of expenditure on Buckingham Palace a succession of Parliamentary Inquiries investigated his work and, although exonerated, much about his modus operandi was criticised by his contemporaries, the related Parliamentary reports providing a fascinating insight into procurement at that time.

Nash, unsurprisingly given the scale of his practice and speculative interests, became involved in civil engineering. Whilst in Wales he designed two masonry bridges in Cardiganshire, both subsequently demolished. Of some interest is his patent (No. 2165, 1797) for iron bridges, which made use of hollow box cast iron voussoirs, possibly based on ideas of Robert Fulton (q.v.), which were most fully developed by William Hastie (q.v.) in Russia. His first essay in iron bridge construction, over the Teme at Stanford, Worcestershire, was a light affair and soon collapsed. It was rebuilt in 1797, really a unique contrivance with cast iron arch ribs and road plates, but with hollow brick cylinders through the spandrels. It was replaced in 1905.

Nash's work in London included involvement in the planning of the Regent's Canal, which crosses Regent's Park which he was laying out at the time, and for which his assistant, James Morgan (q.v.), acted as engineer once it had obtained its Act. Nash was also brought in when Robert Vazie's (q.v.) Archway Tunnel collapsed in 1811 and was responsible, with James Morgan, for the bridge and road which replaced it.

Perhaps of greatest interest is the way in which Nash made use of cast iron in his buildings, which included private commissions such as Corsham Court, as well as public works such as a conservatory at Windsor and the framing for the dome at Brighton Pavilion. The views of his contemporaries on his practice are revealed in the relevant Parliamentary reports. The engineers Sir John Rennie, John Urpeth Rastrick and Francis Bramah (qq.v.) were severe in their criticism, although their own insight at times appears inadequate. Nash himself stated that he was 'the principal user, and perhaps I may add, the introducer of cast iron in the construction of the floors of buildings'. Surviving drawings show how extensive and unorthodox his approach was. At Buckingham Palace, although the founders, Crawshay, would have carried out all necessary tests, he felt confident that he could detail how iron should be used.

With George IV's death in 1830 Nash's own career was almost over. He died on 13 May 1835 at his country residence, East Cowes Castle, and is buried in the local churchyard. He was married twice; to Jane Kerr and Mary Anne Bradley, but his main legacy, apart from his buildings were his many pupils and assistants.

MIKE CHRIMES

[*Select Committee on the Office of Works* (1828); *Select Committee on the Works at Windsor Castle* (1830); *Select Committee on Windsor Castle and Buckingham Palace* (1831); H. M. Colvin (ed.) (1973) *The History of the King's Works*, **VI**, 1782–1851; T. Ruddock (1979) *Arch Bridges and their Builders*, 140–141; J. Summerson (1980) *The Life and Works of John Nash*; R. Thorne (ed.) (1990) *Iron Revolution*; Colvin]

Publications

1797. Patent No. 2165: Constructing bridges of plate iron, etc., *Repertory of Arts and Manufacturers*

Works

See Colvin for full details.

1792–1793. Tre-Cefel Bridge, Cardiganshire, masonry arch
1795–1797. Stanford Bridge, cast iron arch, 96 ft. 8 in. span
1797–1800. Aberystwyth Bridge, over the Rheidol, masonry arch
1812–1813. Archway, Highgate
1815–1822. Brighton Pavilion
1825–1830. Buckingham Palace

NASMYTH, Alexander (1758–1840), painter and occasional architect, landscape consultant and bridge designer, was the second son of Michael Naesmyth (1719–1803), wright and builder in Edinburgh, and Lilias Anderson. Alexander, who dropped the 'e' from the surname when he first set up in business as a painter, was born on 9 September 1758 in the family home at Anderson's Land, Grassmarket. He was educated at home and later at the High School of Edinburgh, with emphasis on mathematics until his natural interest and skill in drawing and painting led him, at the age of fifteen, into an apprenticeship to a coach-builder for whom he decorated the panels of the coaches. He also took lessons in drawing and when the famous painter, Allan Ramsay, noticed his work he invited him to join his studio in London; Nasmyth worked there for four years, then returned to Edinburgh and set up as a portrait painter. After another four years, in 1782, Patrick Miller, a prosperous banker in Edinburgh, paid for him to go to Italy for two years of study and painting. On his return he married, on 3 January 1786, a distant cousin called Barbara Foulis. At this time he was expanding his work from portraiture to the landscape paintings on which his fame now chiefly rests.

Nasmyth made drawings for a study by Miller of the propulsion of ships which culminated in the successful trial in October 1788 of a steam-powered, paddle-driven and twin-hulled pleasure boat on the loch at Dalswinton, Miller's estate near Dumfries.

Nasmyth mixed with men of science in Edinburgh, including (Sir) David Brewster of the University and Sir John Hall of Dunglass, a competent amateur geologist with wide interests in science and arts who was president of the Royal Society of Edinburgh. Nasmyth equipped a workshop in his house at York Place with a lathe and other tools and made models and experiments as relaxation from his painting. He published proposals for developments in steam engines and ship propulsion of his own invention and made some small drawings in or about 1794 of bow-string girders for bridges and roofs, types of structure proposed by others, e.g. James Jordan and Robert Fulton (q.v.), at about the same date but not yet used. His surviving sketches also included studies of construction techniques such as building masonry piers in water with men in diving suits.

Between 1801 and 1808 Nasmyth was employed in the early stages of design of works on at least eight major estates, giving advice on the siting, form and style of houses and farm buildings, and on the landscaping of their surroundings, by means of sketches, paintings—both in water-colours and oils—and a small number of ruled drawings and occasional models. In general, an architect was appointed later to carry each project through, and some adhered closely to his sketches.

Engineering significance attaches to his designs for a lighthouse by the harbour of Inveraray (1801) and for the Nelson Monument on Calton Hill, Edinburgh, neither of which was executed. A small bridge with a pointed arch, and an adjacent cottage named 'Hoolity Ha', on the Culzean estate in Ayrshire, have long been attributed to Nasmyth, but without documentary confirmation.

However, his involvement in the design of two fine stone bridges is well attested. For Tongland Bridge over the Dee near Kirkcudbright (1803–1808), Thomas Telford (q.v.) was the engineer but Telford's design 'required a smoother style of masonry than could be easily executed in the country, or than seemed to accord with the bold and rugged scenery [and] it was, with his approbation, altered, in so far as related to the external architecture … by Mr. Nasmyth' (Smith, 1813). Nasmyth's sketch survives, corresponding to the bridge as built, one large arch over the river with flanking turrets and castellated parapets.

The second bridge was built for Nasmyth's friend, Henry Erskine, sometime Lord Advocate of Scotland, over the river Almond on the approach to his house at Almondell, West Lothian, and is now called Nasmyth Bridge. It is of a clever asymmetrical design, with one small and one large arch of roughly squared sandstone and castellated parapets crowned with short balustrades over the middle of the large arch. He made a sketch and an oil painting of it after completion, the painting exhibited in 1811 with a catalogue note to the effect that it was built to his design. Whether his design included any responsibility for the bridge's construction must be doubted; it suffered a partial collapse in 1973 and was faithfully restored in 1996–1997.

The styles of masonry, parapets and other details in these two bridges are very similar to those of Dunglass New Bridge by James Burn (q.v.), 1798, and Bilsdean Burn Bridge by George Burn (q.v.), 1799, on Sir John Hall's estate which suggests that Nasmyth might have helped in the styling of those bridges also.

Nasmyth's enthusiasm for invention and engineering had a strong influence on his youngest son, James (1808–1890), who became a highly successful mechanical engineer, credited with the invention of the industrial steam hammer; a second son, George, also became an engineer. His eldest son, Patrick (1787–1831), and at least two of his daughters became painters. Alexander Nasmyth made useful paintings of many bridges, including the Tweed bridge at Kelso after the collapse of two arches in 1797 (painting in Ednam House Hotel, Kelso) and the Union suspension bridge near Paxton, by Samuel Brown (q.v.); the latter was exhibited in 1820.

Eleven portraits of Nasmyth are listed by Cooksey and several reproduced, one by Andrew Geddes c. 1825 in colour as frontispiece. A sketch at age sixteen by Philip Reinagle is in Smiles, p. 25.

Nasmyth died in Edinburgh on 10 April 1840. For his architectural works see Colvin, and Cooksey.

TED RUDDOCK

[National Gallery of Scotland. Collection of drawings by A. Nasmyth; drawings and paintings in other public and private collections (some listed and reproduced by Cooksey); The Rev. Samuel Smith (1813) *General View of the Agriculture of Galloway*, 324–327; S. Smiles (ed.) (1883) *James Nasmyth, Engineer. An Autobiography*; J. C. B. Cooksey (1991) *Alexander Nasmyth H.R.S.A. 1758–1840*; Ted Ruddock (2000) Telford, Nasmyth and picturesque bridges, in D. Mays *et al.* (eds.), *Visions of Scotland's Past: looking to the Future*; Colvin.]

NEWCOMEN, Thomas (1664–1729), inventor of the atmospheric steam engine, was born at Dartmouth, Devon, and was baptised at St. Saviours Church on 28 February 1663 (OS). He was one of five children—two boys and three girls—of Elias Newcomen (d. 1702), a Freeholder, ship owner and merchant, and his wife, Sarah (d. 1666–1667). His grandfather was Thomas Newcomen (d. 1652), a merchant venturer and ship owner, who may have led the Dartmouth Newcomens away from the established church to join the Baptist faith, a matter which was to be most important in the life of Thomas the inventor. Shortly after the death of Thomas's mother, his father, Elias, married Alice Trenhale of Kingwear on 6 January 1668 and it was she who brought up young Thomas.

At the age of forty-one, Thomas married Hannah, the daughter of Peter Waymouth of Malborough, Devon, on 13 July 1705. Thomas and Hannah had three children: Thomas (d. 1767), who became a Sergemaker at Taunton; Elias (d. 1765), an Ironmonger; and Hannah. Thomas Newcomen is said to have served an engineering apprenticeship in Exeter, but no trace of this has been found. It appears that non-conformist apprentices were excluded from the lists of Freemen of Dartmouth due to the prevailing political situation. Nevertheless he commenced to trade as an Ironmonger in Dartmouth about 1685 and described himself thus throughout his life.

All Dartmouth references to his trade are for small sums for nails, locks and latches but there is also reference in 1704 to his having mended the Town Clock. The references also confirm that Newcomen was acting as Overseer of the Poor. These trivial dealings should not obscure the much larger general trade with which as ironmonger he was involved. For example, he traded with the Foleys in the Midlands at least between 1694 and 1700, purchasing iron in quantities of up to 10 tons from mills at Cookley, Wolverley, Wilden and Stourton.

In 1707 Newcomen either renewed or took out new leases for a number of properties in Dartmouth. The chief of these was where he lived in North Town, abutting Higher Street in the west

and Lower Street in the east. There were also some 'cellars or ground rooms' that may have been the site of his workshops, where he was to experiment. Newcomen's partner was John Calley (c. 1663–1717), a glazier and member of an ancient Dartmouth family who served a lawful apprenticeship.

Thomas Savery (q.v.) was a member of a family of prosperous merchants in Totnes. On 25 July 1698 he was granted a fourteen-year patent (No. 356) for 'Raising water and imparting motion to all sorts of mill-work by the impellant force of fire, useful for draining mines, serving towns with water and working all kinds of mills in cases where there is neither water nor constant wind'. It was in 1699 extended by twenty-one years to 1733. Savery's pump relied on a vacuum to raise water to a vessel and the use of high pressure steam to force this water to an appropriate height. The device was limited to atmospheric pressure in order to raise water to the vessel and by the available technology to build vessels and pipework to withstand the high pressures necessary to force water to any great height. Nevertheless Savery developed the principal using two receivers and demonstrated this to the Royal Society in June 1669. In 1702 he published *The Miner's Friend* in which he described and illustrated his pump in its final form. He appears to have abandoned his efforts at mine drainage about 1705 although pumps for other purposes were installed after this date.

Newcomen's engine was essentially different from that of Savery and comprised a large rocking beam with arch heads to which were connected chains. At one end, these chains were connected to a piston able to move in an open topped cylinder fixed to the pump's rods descending into the mine where they were fixed to the pumps. The boiler produced steam at just above atmospheric pressure which entered the cylinder when the piston was raised. Cold water was then injected directly into the cylinder from an overhead tank, condensing the steam and creating at least a partial vacuum beneath the piston. Atmospheric pressure then pushed down the piston, raising the pump rods, which depending upon the type of pump, raised the water or, when the vacuum was broken by the admission of more steam, the pump rods descended and acted upon a force pump.

The injection water was supplied from an overhead tank, itself fed by an in-house pump operated by a plug-rod suspended from a small arch head on the beam. This plug rod also served, by a series of plugs and catches, to operate the levers which moved the valves and caused the engine to operate in irregular but continuous motion. The engine was thus capable of construction by available technology and was enlarged in capacity by the increasing ability to cast and machine larger cylinders and to build larger boilers. The cylinders and valves were at first of brass but, from the early 1720s, many were of iron.

Savery had in June 1699 demonstrated his pump with two receivers to the Royal Society. It would seem that a demonstration of Newcomen's engine took place in London in October/November 1712 for in correspondence between John Spedding (q.v.) of Whitehaven and Sir James Lowther in London there is reference to 'Capt. Savery's invention' and 'ye engines for raising water'. Spedding, on 26 November 1712, wrote 'I am glad you have found so great a satisfaction in the experiments you have seen for raising water'. At this date mine engineers were well aware of Savery's pump and the reference must be to the Newcomen engine.

The date at which Newcomen commenced his experiments is not certain. John Theophilus Desaguliers (q.v.), writing in 1744, referred to what is recognised as Newcomen's first successful engine: 'About the year 1710 Thomas Newcomen, Ironmonger and John Calley, glazier, of Dartmouth, ... Anabaptists, made then several experiments in private, and having brought [their engine] to work with a piston, &c, in the latter end of the year 1711 made proposals to draw the water at Griff, in Warwickshire; but their invention meeting not with reception, in March following, throu' the acquaintance of Mr. Potter of Bromsgrove, in Worcestershire, they bargain'd to draw water for Mr. Back of Wolverhampton where, after a great many laborious attempts, they did make the engine work; but not being either philosophers to understand the reason, or mathematics enough to calculate the powers and to proportion the parts, very luckily by accident found what they sought for'. He then proceeds to state that the condensation by injection of water inside the cylinder (instead of outside, according to Savery's practice) was discovered accidentally and that the engine was rendered self-acting by the ingenuity of Humphrey Potter, a boy employed to mind the engine, who contrived a series of catches and strings worked from the beam, by which the several valves were opened and closed in due order. The accuracy of some aspects of Desaguliers accounts has been questioned and extensive research has failed to establish the identity of 'Mr. Back'. The engine was certainly fully self-acting by 1717 as depicted in a print by Henry Beighton (q.v.) that year.

Marten Triewald who came to England from Sweden in 1716 and met both Newcomen and Calley, was impressed with their engine and assisted in the erection of an engine at Byker Colliery, near Newcastle-upon-Tyne in 1717. He later wrote: 'Now it happened that a man from Dartmouth named Thomas Newcomen, without any knowledge whatsoever of the speculations of Captain Savery, had at the same time also made up his mind in conjunction with his assistant, a plumber by the name of Calley, to invent a fire machine for drawing water from the mines. He was induced to undertake this by considering the heavy cost of lifting water by means of horses, which he found existing in the English tin mines.

These mines Mr. Newcomen often visited in the capacity of a dealer in iron tools with which he used to furnish many of the tin mines ... For ten consecutive years Mr. Newcomen worked at this fire-machine...'.

It is known that an engine was built in 1712 on a site in or near Lady Meadow, Coneygree Coalworks, Tipton, Staffordshire, within sight of Dudley Castle, becoming known as the 'Dudley Castle Engine'. A copper plate print of the engine, 'Delin and sculp. by T. Barney, 1719', a file maker of Wolverhampton, is headed 'The Steam Engine near Dudley Castle. Invented by Capt. Savery and Mr. Newcomen. Erected by ye later 1712'. The print has explanatory matter printed in letterpress on the side. From this, the brass cylinder was 21 in. in diameter and 7 ft. 10 in. high. The engine made twelve strokes a minute, each stroke lifting 10 gallons and raising water 51 yd. perpendicularly.

The Netherton Cinder Bank Baptist Church, Dudley has records which foreshadow the building of the engine. In February 1711 they include John Dunford of Tiverton (a Baptist family known to Newcomen) and Elias Newcomen, the nephew of Thomas Newcomen by his brother John Newcomen, Apothecary of Chard. They are both recorded as 'Received by a letter' and later as 'Gon [sic] from us'. Following the undoubted success of this engine a number of others were built in which Newcomen himself was involved. These include Griff Colliery, April 1714; Woods Mine, Hawarden, Flint, 1714–1715; Moor Hall, Austhorpe, Leeds 1715–1715 (where John Calley was to become ill and die, being buried on 29 December 1717); and Stone Pitts, Whitehaven, November 1715.

Following the death of Savery on 15 May 1715 his friend, John Meres, Clerk to the Workshipful Society of Apothecaries of London, launched a joint stock company known as 'The Proprietors of the Invention for raising water by fire'. The nature of Newcomen's association with Savery within his broad patent becomes clear from an opinion dated 1 July 1720 by Sir Thomas Pengelley, a leading lawyer who stated: 'Thomas Savery ... divided the profit to arise by his invention into 60 shares and sold divers of them ... and after his death ... his executors sold all the rest. One Mr. Newcomen having made considerable improvements to the said invention, the Proprietors in 1716 came to an agreement amongst themselves and with the said Newcomen and by indenture in persuance of articles made between Savery and Newcomen in full of his improvement and of the said agreement'.

Shortly afterwards the London Gazette for 11–14 August 1716 included an advertisement as follows: 'Whereas the invention for raising water by the impellant force of fire, authorised by Parliament, is lately brought to the greatest perfection, and all sorts of mines, &c., may be thereby drained and water raised to any height with more ease and less charge than by the other methods hitherto used, as is sufficiently demonstrated by diverse engines of this invention now at work in the several counties of Stafford, Warwick, Cornwall, and Flint. These are therefore, to give notice that if any person shall be desirous to treat with the proprietors for such engines, attendance will be given for that purpose every Wednesday at the Sword Blade Coffee House in Birchin Lane, London ...'. All the counties mentioned are known to have had Newcomen engines, although the site of that in Cornwall remains uncertain. The Proprietors also licensed others for royalties to build engines including in the United Kingdom, Stonier Parrott, George Sparrow, Henry Beighton, members of the Potter and Hornblower families and Henry Lambton, together with others abroad. The spread of the use of the engine was both remarkably rapid and widespread, some 104 being known before the expiry of the patent in 1733.

Newcomen was in Bristol in 1722 where he went 'designing to turns his engines or part of them into cash'. In 1725 he wrote from Dartmouth to Lord Chief Justice King concerning 'a new invented wind engine'. He spent much time in London during 1727–1728 and is also recorded as being at the foundry of Harrison and Waylett, London, on 29 January 1727–1728. Perhaps his last official documentary involvement was on 5 July 1729 when William Baddiford was bound to Arthur Holdsworth and Thomas Newcomen for £61.

Newcomen died in London on 5 August 1729 at the house of Edward Wallin. He was buried on 8 August in the Nonconformist Burial Ground at Bunhill Fields, Finsbury, London. Letters of administration were granted to his widow, Hannah, by the prerogative court of Canterbury on 29 November 1729 and she moved to Malborough and died twenty-seven years later in 1756. Newcomen was strong in his Baptist faith throughout his life as indicated by his acting as Pastor at Dartmouth. That he was well respected can be seen by references in correspondence between James Lowther and John Spedding at Whitehaven: 'There is nothing that will do our business so well and be less liable to accidents than the engine, and, 'tis the cheapest, safest and best way of keeping the colliery dry'. Lowther's further comments, made in February 1727–1728, include: 'We have been very successful from first to last in the timing of things about the Fire Engine, which I should hardly have ventured upon if I had not mett with such a very honest good man as Mr. Newcomen who I believe would not wrong anybody to gain ever so much'.

Memorials have been erected at Dartmouth to Thomas Newcomen, including a granite obelisk in 1921, and later a wall plaque. In 1964 a small late eighteenth century atmospheric engine was moved there from Coventry to commemorate the tercentenary of his birth and is known as the 'Newcomen Memorial Engine'. In 1994 a further plaque was placed by the Town Council in Newcomen Road at a position shown by research to be the location of Newcomen's Workshop. A full sized working replica of the 1712 'Dudley Castle' engine has been built at the Black Country Museum, Dudley, West Midlands, about a mile from the original engine site.

Although improvements were made to the details and workmanship, Newcomen's engine remained little changed for almost three quarters of a century until it was gradually superseded by that of Watt, with a separate condenser, patented in 1769. Even then, many Newcomen type engines continued to be built.

J. S. ALLEN

[Dartmouth Borough Receiver's Accounts and the Mayor's Accounts; Death notice (1729) Monthly Chronicle II, 169; Register records, Mr. Newcomen from St. Mary Magdalens, buried in a vault. 00-14-00; J. T. Desaguliers (1744) *Experimental Philosophy*, **ii**, 532.; Martin Triewald (1734) *A Short Description of the Fire and Air Engine at the Dannemora Mines*; John Calley, cf. Whitkirk Parish Register; J. Farey (1827) Treatise on the Steam Engine, 155 n.; J. R. Harris (1967) *The Employment of Steam Power in the Eighteenth Century, History*, **LII**; G. J. Hollister-Short (1976–1777) The introduction of the Newcomen engine into Europe, *Trans Newc Soc*, **48**; A. Smith (1977–1778) Steam and the City. The Committee of Proprietors of the Invention for raising water by Fire, 1715–1735, *Trans Newc Soc*, **49**; A. Smith (1994–1995) Engines moved by fire and water, *Trans Newc Soc*, **66**; Svante Lindqvist (1984) *Technology on Trial, The Introduction of Steam Power Technology into Sweden, 1715–1736*; I. H. Smart (1988–1989) The Dartmouth Residences of Thomas Newcomen and his Family, *Trans Newc Soc*, **60**; J. S. Allen, Thomas Newcomen, New DNB (to be published); L. T. C. Rolt and J. S. Allen, *Thomas Newcomen and the Steam Engine*, 2nd edn.]

NICHOLLS, Nathaniel of Bethlem (Bethlehem) Hospital, London, appears in membership lists of the Institution of Civil Engineers from 1824 to 1841, and again in 1866. He was elected auditor with James Howell in 1837. Nicholls was steward of Bethlem Hospital from 1819 until at least 1851. Until 1948, Bethlem was jointly administered with Bridewell House of Occupation, where Nicholls was steward from 1830 to 1853. Bethlem Hospital, Southwark, was completed in 1815, to the design of James Lewis (c. 1751–1820), who preceded Nicholls as steward. It appears that Nicholls' work was mainly administrative, involving ordering and paying for provisions and tradesmen's services, and paying staff.

TESS CANFIELD

[Membership records, ICE; Archive of the Bridewell Royal Hospital; Archive of Bethlem Royal Hospital]

NICHOLSON, George (1736–1793), stonemason, architect and bridge surveyor, was the son of Richard Nicholson and his wife Elizabeth Clark and was baptised in Durham on 25 October 1736, the second of at least four children. In 1763, then described as a mason, he married Elizabeth Watson of Ackworth, Yorkshire.

He became Surveyor, possibly on an *ad hoc* basis, to the Dean and Chapter of Durham cathedral, a position he held until his death, and during his tenure of office carried out several restoration works on the cathedral, some of them in conjunction with John Wooler (q.v.).

Nicholson was responsible for the construction of Prebends' Bridge, crossing the river Wear in Durham and replacing an earlier structure destroyed in the calamitous flood of 1771. Robert Mylne (q.v.) advised on the location and form of the bridge but the design is considered to be Nicholson's. Comprising three semicircular spans each of 67 ft., the bridge was completed in 1778.

In 1776 Nicholson, again described as a mason, was appointed as surveyor of bridges in County Durham; his initial salary was £20 p.a. but within a year it was doubled. His responsibility was the upkeep of bridges in the southern part of the county, those in the north coming under the supervision of John Bell (d. 1784), a Durham architect. In 1784 Nicholson's responsibilities were extended to cover the whole of the county, among other works completing the new bridge at Witton-le-Wear in 1789.

Nicholson died in Durham and was buried at St. Giles church on 31 January 1793, then noted as an architect; he was survived by his wife and two married daughters, two children having predeceased him.

R. W. RENNISON

[*Chapter Acts, 1729–1777*, Dean and Chapter, DCD, University of Durham; *Durham Quarter Sessions: Orders* (1777–1793) Q/S/OB, Durham RO; Colvin (2)]

Works

1772–1778. Prebends Bridge, Durham, design and construction, cost c. £3,250

1776–1784. Stockton and Darlington Wards, County Durham, Bridge Surveyor; widening or rebuilding of Wolsingham and West Auckland bridges

1784–1793. Bridge Surveyor for County Durham; work at Witton-le-Wear, Cockerton, Yarm, Barnard Castle, Eggleston and Rushyford bridges

NICHOLSON, Peter (1765–1845), author and architect, was born on 20 July 1765 at Prestonkirk, East Lothian, the son of a stonemason. He attended the local school where, although the instruction was very limited, he showed considerable mathematical ability, particularly in geometry. From the age of twelve he assisted his father but after a year he turned to cabinet making and was apprenticed for four years in Linton. Throughout his apprenticeship he studied algebra in his leisure hours before and after work and, having served his apprenticeship, he moved to Edinburgh to work as a journeyman and continue his mathematical self-education in calculus.

According to Smiles he also had some training with Andrew Meikle (q.v.).

Nicholson moved to London c. 1785 where he worked for an uncle as a builder. During his leisure hours he returned to his study of geometry, which now became the subject of lectures for his colleagues and a source of extra income. The lectures attracted a wide audience and the fees gave him more freedom to study the application of geometry to building technology. They also provided finance for his first book in 1792, entitled *The Carpenter's New Guide*. The book was devoted to the application of old and new solutions to problems in solid geometry, including curved twisted handrails and the intersection of domes by prisms; a second edition was published in 1793. *The Carpenter and Joiner's Assistant* (1797) was considered by Nicholson to be a supplement to the *Guide;* it included some elementary mechanics of beams and the strength properties of various species of timber that had been established mainly in France by Belidor, Buffon and Duhamel. Nicholson found the test data to be variable and did not explain how the information could be used for design purposes, but he did assist engineers by giving drawings of existing roof trusses with sizes of the members and details of the joints.

He then became a prolific writer on a wide range of topics: for example architecture (*The Architectural Dictionary*, 1812–1819), mathematics (*The Rudiments of Algebra*, 1819), carpentry and joinery (*A Treatise on the Construction of Staircases and Handrails*, 1820), building (*New Practical Builder and Workman's Companion*, 1823), masonry (*Treatise on Masonry and Stone Cutting*, 1827), and civil engineering (*The Guide to Railway Masonry, containing a Complete Treatise on the Oblique Arch*, 1839). The DNB gives the titles and dates of all editions of twenty-four books spanning the years 1792–1839. Nicholson also wrote numerous articles on architecture, carpentry, masonry, perspective, etc., for Rees's *Cyclopaedia,* and on carpentry, etc., for Brewster's *Edinburgh Encyclopaedia* and for the *Encyclopaedia Metropolitana.*

He practised as an architect in Glasgow from c. 1800 to 1809, details of his work in this field are listed in Colvin. His work in 1808 included a timber footbridge consisting of nine 42 ft. spans over the river Clyde at Hutchesontown. Nicholson laid out the town of Ardrossan where the harbour had been designed by Thomas Telford (q.v.) and it was on Telford's recommendation that in 1808 he became Surveyor to the county of Cumberland, working on Telford's designs for the new Courts.

In 1810 he returned to London where he gave private lessons in mathematics, land surveying, mechanical drawing etc. and produced the two volume *Architect's Dictionary* (1812–1819). The usefulness of his geometrical discoveries was recognised by the award of a gold medal by the Society of Arts for an improvement in hand railing in 1814: there were further medals and awards in 1815.

In 1827 he began a work entitled *The School of Architecture and Engineering* which was designed to be completed in twelve parts. The publishers went bankrupt after five numbers and Nicholson lost heavily, a burden that he appeared to carry for the rest of his life. He was, according to Thomas Sopwith writing at this time, a 'veteran Mathematician and a kind and amiable man. He possesses great talents but is of too easy a temperament to cope with the roughness of the world'. It was, perhaps, a result of his amiable nature that a fund was established to assist him financially.

In 1829 he went to live in Morpeth and in 1832 moved to Newcastle on Tyne where he opened a school. It was when he lived there that in 1834 it was proposed that a 'general subscription' should be raised for 'this venerable and talented individual'. Almost every engineer in the town spoke of his merits but unfortunately subscriptions were not as forthcoming as had been the eulogies. He remained in high esteem, as is shown by his election in 1835 as president of the Newcastle Society for the Promotion of the Fine Arts. He continued to write until 1839, including *The Guide to Railway Masonry* (3rd edn., 1846) which was notable for a history of oblique arches. The memoir in *Carpenter and Builder's Complete Measurer* emphasises that the whole of his active and scientific labours had been directed towards applying science to useful purpose. However, the memoir also points out that Nicholson was not 'in the highest sense of the word *an architect*'. He did not design 'grand building(s) ... nor did he understand the different styles of architecture ... but as a mechanician and mathematician, and as a guide in the construction of the different parts of a building, he was, if not unequalled, at least unsurpassed'.

Nicholson will be remembered for his long career as an author of books on building technology, where his great ability as a mathematician, especially in the field of solid geometry, enabled him to simplify and generalise many old methods, particularly in carpentry and joinery, and to invent new ones.

Nicholson was twice married and had two sons and a daughter. His elder son, Michael Angelo, collaborated with him during the production of some of his books. Nicholson died in Carlisle on 18 June 1844 and was buried in Christ Church graveyard. Portraits are to be found in *Mechanic's Magazine* (1825) by an unknown artist, and in several of his works from 1824 onwards, based on paintings by Heaphy, Train, and Derby.

L. G. BOOTH

[ASPD; Colvin (3); DNB; T. Sopwith, Diary of Thomas Sopwith, 1828–1879, Newcastle University Library; Memoir (1844) *Civil Engineer and Architect's Journal*, **7**, 425–428; S. Smiles (1861) *Lives of the Engineers*]

Selected publications

1792. *The Carpenter's New Guide* (13th edn., 1848)
1797. *The Carpenter and Joiner's Assistant* (6th edn., 1886)
1804. *The Student's Instructor*
1811. *Mechanical Exercises*
1812–1819. *Architectural Dictionary*
1818. *Essays on the Combinatorial Analysis*
1820. *Essay on Involution and Evolution*
1820. *Treatise on the Construction of Staircases and Handrails*
1823. *New and Improved practical Builder and Workman's Companion* (repr. 1837)
1825. *The Mechanic's Companion*
1827. *Carpenter and Builder's Complete Measurer*
1827. *Popular and Practical Treatise on Masonry and Stone Cutting*
1828. *The School of Architecture and Engineering*
1830. *Practical Masonry, Bricklaying and Plastering*
1839. *Guide to Railway Masonry* (3rd edn., 1846)
1846. *The Carpenter, Joiner and Builder's Companion*

Works

Architectural works not included (see Colvin (3)).

1808. Glasgow, footbridge over River Clyde at Hutchesontown (illustrated in *Edinburgh Encyclopaedia*, **4**, 1830, 537 and plate 102)

NICHOLSON, William (1753–1815), technical author, was born in December 1753, the son of a solicitor. At the age of sixteen he left school and made two voyages in the service of the East India Company, but his father's death led to him leaving the Company at the age of twenty. There followed a succession of jobs including work for Josiah Wedgwood on the continent. In the 1770s he was evidently developing his literacy skills and in 1781 began his career as a writer of technical and scientific works with his two volume *Introduction to Natural Philosophy*. A succession of popular works followed, including a translation, in five volumes of Fourcroy's *Chemistry* (1787 and new edition 1799). Topics ranged from navigation to bleaching wool. In the 1780s Nicholson began to contribute to the *Philosophical Transactions* of the Royal Society on a variety of scientific topics and was to carry out various chemical investigations, including the chemical properties of galvanised iron. With Wedgwood's support he was made Secretary to the Committee of the General Chamber of Manufacturers of Great Britain in 1784, and in 1791 became an active member of the Society for the Encouragement of Naval Architecture.

In 1792 he began to publish one of the earliest scientific and technical magazines the *Journal of Natural Philosophy, Chemistry and the Arts*, generally known as 'Nicholson's Journal'. One of its earliest issues contained Nicholson's translations of Venturi's paper on 'the lateral communication of motion in fluids'. A later issue contained a description of Thomas Telford's (q.v.) earliest proposal for bridging the Menai Straits using suspended centering.

In 1799 Nicholson established a school at 10 Soho Square, London where he lectured on natural philosophy and chemistry, with the aid of a grant of £1,500 from Thomas Pitt (1775–1804), Lord Camelford. His name was a sufficient selling point for it to be applied to the *British Encyclopaedia* a six volume work published in several editions 1808–1814, but completed by Jeremiah Joyce (*c.* 1764–1816).

In the early nineteenth century he took out a number of patents—for file cutting machinery, and steam operated blast furnace machinery, and became more involved in engineering.

Possibly consulted about London Bridge Waterworks, he became involved in the water supply of west London and Portsmouth, both schemes involving the Dodd family (q.v.). In 1808 Nicholson proposed a water supply scheme for Portsmouth, Portsea, and neighbouring parishes at Wymering, Widley and Farlington; Barrodall Robert Dodd was engineer to a rival scheme which also encompassed Gosport. Despite local support for Nicholson's scheme the group backing Dodd were able to buy out the proprietors of an earlier scheme authorised by an Act of 1740. In 1809 Nicholson's scheme obtained Parliamentary approval and the Portsea Island Company, as it became known, began work on a well at White Swan Field. After some technical problems with pumping, and rivalry over laying mains, water supply was commenced to Portsmouth in April 1811. Unfortunately the supply from the well was inadequate in the summer months and the company was in fairly continuous problems, although Nicholson was no longer involved, until its amalgamation with the other scheme in 1840.

At West Middlesex Nicholson was hardly more successful. The original scheme was devised by Ralph Dodd, but he resigned and in early 1807 Nicholson was appointed engineer. He installed two steam engines to pump water from the Thames at Hammersmith, and proposed a more ambitious area of supply, necessitating a further range of capital by an Act of 1810. The scheme was not complete by the time of Nicholson's death and he was succeeded by John Millington (q.v.); in the event it was William Tierney Clark (q.v.) who brought the scheme to maturity. Nicholson's work as a civil engineer led to a number of consultancies, including the Regent's Canal. It would appear after a period of prosperity he fell on hard time and died in poverty in 1815.

MIKE CHRIMES

[Memoir (1812) *European Magazine*, **62**, 83–87 and portrait; *New Monthly Magazine*, **4**, 1815, 76–77]

Selected publications
1784. *Navigator's Assistant*
1797–1814. *Journal of Natural Philosophy*, 2nd series

1807–1808. *British Encyclopaedia*
1810. *A letter to the proprietors ...*, Portsea Island waterworks

Works
1807–1815. West Middlesex Waterworks, engineer
1808–1810(?). Portsmouth (Portsea Island), Waterworks

NICKALLS, Joseph (fl. 1758–d 1793), was a millwright and civil engineer. As a leading London millwright, based at Lambeth, Nickalls was asked in November 1758 by the Society of Arts to serve jointly with Thomas Yeoman (q.v.) as an expert on a committee on mill design. He would therefore have been well known to John Smeaton (q.v.) who had recently been appointed engineer of the proposed Calder & Hebble Navigation. The Commissioners soon appointed Nickalls as 'deputy Surveyor', or resident engineer, with a salary of £100 p.a. and £21 for removal expenses. Nickalls arrived in Wakefield in December 1759 and in June 1760 his former employee, John Gwyn (q.v.), joined the staff as foreman. Two months later Nickalls's salary was increased to £120 p.a. but he was dismissed in November 1761. Gwyn was then put in charge of carpentry and Matthias Scott (q.v.) of masonry and digging; both men were to work directly under Smeaton.

Smeaton had a high regard of Nickalls. He employed him to build the waterwheel and pumping machinery of the Stratford mill, east London, 1762–1763, and it was Nickalls who built the great 32 ft. diameter waterwheel and pumps, to Smeaton's design, in the fifth arch (the second from the south end) of London Bridge in 1767–1768.

In 1762 a dispute arose between the Trustees of the River Lea Navigation and Thomas Walton, owner of the Waltham Abbey Gunpowder Mills. To settle the matter a joint survey was ordered in November to inspect the dams near the mill tail, with Nickalls, 'Millwright and Surveyor of Navigations', acting for Walton and Thomas Yeoman for the Trustees. Nickalls was involved with Yeoman in surveys of Tottenham and Enfield Marshes in 1766 and he acted again for Walton in April 1771.

In 1770 the Thames Navigation Commissioners asked Nickalls to make a survey with a view to improving the navigation in that part of the river under their jurisdiction upstream of Staines. He concluded that it would be necessary to build pound locks in cuts adjacent to the watermills as preferable alternatives to the existing flash locks in the mill weirs. This was in response to a proposal by the town of Reading for a canal bypassing the river from Reading to Maidenhead and a further proposal, supported by the City of London, for an extension of the canal down to Isleworth; both schemes had been put forward by James Brindley (q.v.).

Petitions were submitted to Parliament by Reading in November 1770, by the Commissioners in December and by the City in January

1771. Nickalls and others gave evidence, the upshot being that Bills for the canals failed and the Commissioners obtained an Act in April 1776 granting power to build pound locks between Lechlade and Maidenhead, to dredge the navigation channel and to extend the towpath over the full length.

Nickalls was appointed General Surveyor, or engineer, in May 1771 at an annual salary of £200 with an inspector of works between Reading and Maidenhead, where eight pound locks were to be built as a first stage, and an assignment on channel improvements downstream to Staines. By the end of June agreement had been reached on the siting and design of five locks. Plans for the next two locks followed about a month later and, after some hesitation, for the last lock in October.

Contracts for each lock were let within a few weeks of the plans being agreed; construction was to be in accordance with a 'model' by Nickalls. The first lock, Boulters, was opened in December 1772; the other seven followed from January to April 1773. Dredging started in July 1771 and continued for several years, gravel raised from the riverbed being used to form the towing path. The locks were timber framed with turf slopes, 133 ft. between gates, a clear width of 18 ft. 3 in. and a depth of about 3½ ft. on the sills at low water.

In June 1774 the Thames Commissioners 'returned their thanks to Mr. Nickalls for his great care and fidelity in the Execution of the Works' but in 1776 he was discharged, it being thought he had recently been spending too much time on other work. By that year £26,000 had been spent. Thomas Yeoman took his place for two years, on a freelance consulting basis, to be followed by John Clarke (q.v.) as Surveyor.

Nickalls was a founder member of the (Smeatonian) Society of Civil Engineers, attending the first meeting on 15 March 1771 along with Yeoman (President), Smeaton, John Grundy, and Robert Mylne (qq.v.). He attended future meetings with great regularity, and became President in 1783. He also joined the Society of Arts, elected as 'Mr. Joseph Nickalls, Engineer, Southwark' in January 1771. In 1782 he became joint chairman of the Mechanics Committee, and remained a member of the Society to the end of his life.

In September 1775 Nickalls received a request to advise on improvements of Dover Harbour. He devoted considerable time to the subject, probably the reason for his absence from the Thames, and his printed report, together with a plan of the harbour, is dated 1777. Nothing came of this for the time being, but his report was reprinted in 1782 and he was subsequently employed as engineer on various works, notably building two new sluices in the lower crosswall (between the tidal harbour and the Basin), refacing the wall in stone, enlarging and lowering the sill of the gates in the upper crosswall (between the Basin and the Pent) and deepening the Basin. Those works

appear to have been well executed, but Nickalls is said never to have been accurate in estimating the costs or the time required for completion, and he was eventually dismissed in 1791. Henry (later Sir Henry) Oxenden then took charge of the harbour works and from 1792 James Moon was employed as resident engineer.

In a rather similar way Nickalls was employed as engineer on the River Wey Navigation, being called on from time to time to advise on specific problems and giving directions for work to be done. There are records of payments to him of about £50 or £60 in 1780 and 1781, and in September 1793 he was paid £150 on account, the balance (£100) being paid to his family after his death. In 1782 he reported on Wells Harbour and then in 1783 on Yarmouth Harbour, receiving £200 for his survey and fees. He went on to direct some works there, besides producing a model for an improved dredging 'engine', before being discharged in 1786. Meanwhile in April 1785 he reported on the River Severn, presenting reasons against proposals by William Jessop (q.v.) for building locks and weirs on the river, a scheme defeated in the 1786 session in Parliament.

Nickalls then became involved in the scheme first put forward in 1767 by William Champion, a Bristol shipbuilder, for converting the Avon into a 'floating harbour' by building a dam and navigation lock across the river at Redcliff, about a mile below the quays of the city. Little happened for the next twenty years but in 1786 the Society of Merchant Venturers asked Smeaton to recommend an engineer to examine the matter and if possible to come to Bristol himself; in January 1787 he named Nickalls and Jessop. Nickalls reported in November when he came up with a still more ambitious proposal of his own, for damming the Avon at Black Rock about 3 miles below the quays. This had the great advantage that ships could enter the locks at any state of the tide but the cost and difficulties of construction would be considerable.

Jessop reported in May 1788, praising Nickalls's concept but coming out in favour of Champion's plan, modified by placing the lock a little further downstream. Smeaton was now prevailed upon to consider the whole matter and in July 1789 he produced a masterly report. Not without regret he decided against the Black Rock scheme and emphasised the practical advantages of Jessop's scheme, which he investigated thoroughly and modified to some extent.

Smeaton, Jessop and Nickalls visited the site together in September. Borings and flood observations were made. Jessop reported again in February 1790, and Nickalls in August reported on the mills on the river. He had no further connection with this great project which was eventually carried out by Jessop between 1804 and 1809, after further revisions.

In the Minutes of the Society of Civil Engineers for December 1791 there is a curious entry: 'Mr. Nickalls, on a representation of the Offence given

Mr. Smeaton expressed himself in the following terms; that he was mistaken, that he was sorry for the offence, and that he begged his pardon as a Member of this Society'. The nature of the offence will probably never be known. In any case his pardon was accepted and meetings proceeded with Nickalls still President of the Society until May 1792.

Nickalls's last report, dated February 1793, to the Mayor of King's Lynn, refers to the proposed Eau Brink Cut. He opposed the scheme on the grounds that by eliminating a long, wide bend of the River Ouse upstream of Lynn the Cut would be a disadvantage to the harbour by reducing tidal reflux. Hodskinson held the same view but those in favour of the idea, including Robert Mylne and John Rennie (q.v.), won the day. After long Parliamentary debates the Act for making Eau Brink Cut was passed in 1795.

Nickalls died in November 1793 and was buried at Christ Church, Southwark. In June 1794 Administration of the estate (valued at £5,000) of Joseph Nickalls, late of Gravel Lane—where he had been based from 1786—in the Parish of Christ Church, widower, was granted to his daughters and only children, Elizabeth and Mary.

A. W. SKEMPTON

[Society of Arts, Minutes of Committees (1758–1782) and Subscription Books (1771–1793) Royal Society of Arts; Lea Navigation Minute Books (1762–1771) PRO; Thames Navigation Commission Minute Book (1771–1776) Berkshire RO; Joseph Nickalls (1776) Plan of Dover Harbour, in Smeaton's *Designs*, f. 139, Royal Society Library; Joseph Nickalls (1786) Plan of a timber bridge with three drawbridges, mss 19771, NLS; Wey Navigation Records (1778–1794) Surrey History Service; Yarmouth Harbour Minute Book (1783–1786) Norfolk RO; Joseph Nickalls (1785) Report on the River Severn, in T. Nash (1799) *History and Antiquities of Worcestershire*; Joseph Nickalls (1787 and 1790) Reports on Bristol Floating Harbour, Bristol RO; John Lyon (1813) *History of the Town and Port of Dover*; Charles Hadfield (1972) *Canals of Yorkshire and North East England*; Charles Hadfield and A. W. Skempton (1979) *William Jessop, Engineer*; Alec Hasenson (1980) *History of Dover Harbour*; A. W. Skempton (ed.) (1981) *John Smeaton*; A. W. Skempton (1984) *Engineering on the Thames Navigation, Trans Newcomen Soc*, **55**, 153–176]

Publications

1771. *Plans of the Thames Navigation, with the Improvements as intended by the New Bill*
1777. *Report on Dover Harbour, with engraved plans*, reprinted 1782
1782. *Report on the State of Wells Harbour*
1793. *Report on the Consequences which the New Cut from Eau Brink would be attended with to Drainage, Navigation, and the Harbour of Lynn*

Works

1762–1763. Stratford pumping engine

1767–1768. London Bridge Waterworks engine
1771–1776. Thames Navigation, locks and dredging, Maidenhead to Reading, £26,000
1783–1786. Yarmouth Harbour, improvements
1783–1790. Dover Harbour, works

NIMMO, Alexander, FRSE, MRIA (1783–1832), civil engineer, was born in Kirkcaldy in 1783 (within the first twenty days of January, if his reported age at death is accurate). His father, also Alexander, was first a watchmaker, one of several Nimmos to practise that trade, with his business in the High Street of Kirkcaldy, but was later described as a merchant and as proprietor of a hardware shop. Young Alexander had a brilliant career at the town's grammar school and then went to St. Andrews University where he attended classes from 1796 to 1799, including Latin, Greek, mathematics, logic, ethics and natural philosophy. For one further year, 1799–1800, he attended the University of Edinburgh, taking courses in physics, ethics and mathematics, and doubtless attending lectures by such eminent professors as John Robison (q.v.) and John Playfair.

His first known employment was as second master (of three) and teacher of mathematics in Fortrose Academy. The records show him settled in this post in February 1802 and he is likely to have taken it up in mid-1800 or 1801. His teaching embraced 'arithmetic, book-keeping, and drawing; the elements of Euclid, navigation, land surveying and other mensurations; architecture, fortification and gunnery; also the elements of chemistry and of natural philosophy'. He had a salary of £35 p.a. and class fees from fifteen students, plus boarding fees from boys whom he housed in an 'excellent and commodious house' provided for him. When the harsh regime of a new rector of the Academy caused the number of students to decline he left discontented in June 1805 to become Rector and first master of the Academy of Inverness, a much bigger school in the largest town of northern Scotland. He obtained the post after interview conducted by three professors of Edinburgh University, one of whom was John Playfair, over three days. It was a well-endowed school with a large public hall, six spacious classrooms, a library and some 'philosophical [i.e. scientific] apparatus'. Nimmo taught mathematics and natural philosophy and also, to small numbers, Spanish and Italian. By 1811, when he left, the Academy had eight masters and almost three hundred boys.

His practical abilities soon became known in the Inverness area and when Thomas Telford (q.v.) made known, apparently in the early months of 1806, that the Commissioners for Highland Roads and Bridges had need of a man to undertake a difficult job of surveying, he was directed to Nimmo. Consequently, Nimmo was engaged to spend his first summer vacation (1806) trying to establish and map the boundaries between counties in which road-building schemes were projected or already in hand. It involved him in almost two

Alexander Nimmo FRSE, MRIA

months of travel, carrying instruments supplied from London, viewing the ground, measuring, interviewing residents and resolving disputed boundaries, all for a fee of £150.

His contribution was towards a map of Scotland prepared for the Commissioners by Aaron Arrowsmith (q.v.), a mapmaker of nationwide reputation. Three or four years later, Nimmo made for Telford a survey to establish the route of a proposed new road southward from the Great Glen along Loch Treig and across Rannoch Moor to Glen Lyon—a road which was never made, though a report was published. It is likely that he had other tasks related to the current engineering works in the Highlands during his six years in Inverness. He was concurrently gaining status as a man of scientific knowledge and skill; for instance in 1804, while he was still at Fortrose, he had sent to John Playfair an 'account of the removal of a large mass of stone to a considerable distance along the Murray Firth'—presumably an observation of large-scale coastal erosion and drift—and which Playfair read to a meeting of the Royal Society of Edinburgh. Nimmo was elected to Fellowship of the Society on 3 December 1810, proposed by Sir George McKenzie Bart. and seconded jointly by Alexander Christison, Professor of Humanity in the University of Edinburgh, and Thomas Allan, a banker. He was also, in 1809, elected an honorary Member of the Geological Society of London.

As he was known to John Robison (q.v.) who wrote many articles on 'mechanical philosophy' for the *Encyclopaedia Britannica* (3rd edition) and to Playfair who succeeded to the Chair of Natural Philosophy on Robison's death in 1805, it is not surprising that Nimmo turned his scientific mind to some of the subjects which had occupied Robison. Before he left Inverness for a much busier life in engineering practice he had already written most of the learned articles which he contributed to volumes of *The Edinburgh Encyclopaedia*. Compilation of the *Encyclopaedia* had been initiated in 1807 with David Brewster (q.v.), who was a probable classmate of Nimmo's at Edinburgh University, as editor. Other articles published later dealt with subjects which had probably also engaged his attention before 1811, although their content shows that they were not completed at that early date. It is apparent that being in Inverness was not an obstacle to Nimmo's studies, as he referred in his early articles to all relevant sources whether they were written in English, Latin or modern European languages; his remarkable linguistic skills being used to good effect. Moreover, in the same articles he showed good understanding of the nature and limitations of practical civil engineering and their impact on application of the theories he developed.

He resigned his post at Inverness in June 1811 to begin his career in engineering. It was claimed many decades later by Joseph Mitchell that he had incurred censure by the governors of the Academy for his failure to accompany the boys regularly to church and that his resignation was a response to the criticism. However, his departure to work which would be much better paid and would engage his growing interests is quite natural. It seems that the offer of a new job came many months before he left Inverness; for though his resignation was not made public there until June 1811, his appointment as engineer to the Commissioners on the Bogs in Ireland was recorded in the Commissioners' Fourth Report as dating from 5 January. The appointment ran to 31 December 1813 and he was paid for 720 days work at 2 guineas per day of actual employment. The time could have been worked without his having actually started before July 1811 or even later.

He was one of nine engineers engaged by the Commissioners, each assisted by several surveyors, to assess and report on large tracts of the Irish landscape which were covered with peat bogs. The method was to take levels of the surface and borings down to 'hard' ground, along a set of parallel traverses spaced at intervals of a quarter-mile; and the purpose was to identify areas which could be drained and brought into agricultural production thus easing the endemic problems of over-population and poverty in the Irish countryside. The areas Nimmo surveyed were in west-coast counties, especially Galway, and parts of the counties of Kerry and Cork in the south-west of the country. In all, his surveys covered nearly 2,000 sq. miles and were reported to Parliament in London in 1814.

Thereafter Nimmo was never without substantial civil engineering work in Ireland, but his work was increasingly varied and involved him in frequent visits to England and occasionally elsewhere. Early in 1814 his first report on a project for a major harbour in the south-east of Ireland was published. It was designed as a terminal port for packet-boats of the General Post Office, and sited in a cove near Dunmore (now Dunmore East) near the seaward end of the wide inlet called Waterford Harbour. He became engineer for the project, which involved heavy pier and breakwater construction.

In October 1814 Nimmo visited his home town of Kirkcaldy in Scotland and was witness to the baptism of a daughter of his brother John. In 1815 and 1816 he wrote letters at different times from Dublin, Cork and Galway, and he also went 'immediately after the peace' with France in 1815 to France, the Low Countries and Germany to see famous civil engineering works of such eighteenth-century engineers as Belidor and De Cessart. It was probably on the same trip that he collected documents which gave thorough descriptions of the French government's establishment of engineers for the construction and maintenance of roads and bridges, inland navigation, and water supply for industry at two different dates, 1760 and 1811. He also learned the details of the educational provision for engineers in the Ecole des Ponts et Chaussees, and in due course he passed copies of the documents to ministers responsible for organising the service of government engineers in Ireland.

These actions show that Nimmo, in his early thirties, was already thinking further than the practical works he was engaged in to the desirable staffing, education and organisation of engineers for local and national public works. Although the national structure of engineering service he envisaged was not set up in his lifetime, he was engaged by government to lead works which spread across large tracts of Ireland and required him to create networks of officers with graded responsibilities (a more permanent structure was initiated in 1831 when a government Board of Works was established). In December 1817, however, when he was appointed by the Lord Lieutenant of Ireland to be one of a board of engineers led by Telford to examine candidates for office as county surveyors in Ireland under a new Grand Jury Act, Telford objected quietly to the Chief Secretary because Nimmo was 'yet but a young engineer, and … a candidate for a surveyorship'.

The construction of Dunmore Harbour was not completed until 1821, when it was assessed by Telford who praised the 'improved diving machine', or diving bell, which had been used in deeper water than had been done before. In 1820 Nimmo was appointed by the Commissioners of Irish Fisheries to make surveys of all the coasts of Ireland, and within the next three years he directed surveys of most of the coastline of the country from which maps and charts were plotted at large scale. Most of them survive and some, chiefly those depicting the approaches to the larger ports, were engraved for publication. A particular outcome of the work was Nimmo's publication in 1824 of a paper on the relevance of geology to practical navigation, and later on his completion of a chart and extensive piloting directions for navigation in the Irish Sea, which were published in 1832 after his death. All this work was of great importance in improving the safety of navigation in the Irish Sea and round the coast of Ireland. A list of piers round the coast which were built, repaired or improved up to June 1830 named sixty-one piers, of which fifty-four were finished or under construction, those unfinished numbering eighteen. A large majority of the fifty-four were projects of Nimmo, either for the Fisheries Board or for the Western District (see below).

With the danger of famine never far away, sporadic partial failures of the potato crop during the first twenty years of the century had brought various actions from the government in London aimed at relieving poverty amongst the Irish peasantry. Some of the government actions concentrated on improving employment opportunities, and another crop failure in 1821 resulted in the designation of several areas to which subsidies were to be directed to provide work and create public amenities, under the direction of engineers. Nimmo was given charge of 'the Western District' in 1822 and public works in that area occupied him until the end of the decade and his life. Not a few projects were unfinished when the programme was stopped at the end of June 1831. Nimmo himself died suddenly early in 1832.

The Western District comprised five counties, Galway, Mayo, Sligo, Roscommon and Leitrim, and within the years 1822 to 1830 roads and bridges were made in all of them, harbours and piers on the coasts of the first three named, and piers and other navigation works on inland loughs and rivers. The total length of roads made or repaired was several hundred miles and the overall expenditure up to mid-1830 was reported to Parliament as £167,000, while in 1831 Nimmo estimated that £60,471 would be needed to complete the intended works, the great majority of it to be spent on roads and bridges.

Management of the programme required an office in Dublin for dealings with government bodies—and perhaps to expedite visits to London—and the Western District's office occupied a three-storey terrace house at 78 Marlborough Street built for the purpose by George Nimmo, a brother of Alexander who was by that time resident in Ireland. Alexander also lived there when in Dublin. For a year before the building was ready, the office had been at 56 Marlborough Street. In the West Alexander built, as a personal speculation and to his own design, a substantial inn at Maam Bridge, by an important road junction at the head of Lough Corrib; but while the Western District works were going on it was used as office and residence by some of his supervisors. Stores and workshops were also built there. The house has been known as Corrib Lodge and is said to have been occupied by John Nimmo, another brother, after Alexander's death. It is now a public-house.

A third personal initiative of Nimmo concerns the village and harbour of Roundstone on the coast of County Galway. To overcome the objections of a local landholder Nimmo himself took on the lease of the land required to build the harbour. It made Roundstone almost his property as well as his project, and the name of Nimmo survives on a house which may have been the first fish store. His brother John went to live there, during or after the construction work in the district, and both John and his wife Marion died there in 1850 and 1849 respectively.

While he was directing the work in the Western District, Alexander undertook many other commissions. They included his most famous and most costly structure, the Wellesley Bridge and Docks at Limerick, which was started in 1824. It was a five-arch bridge of sophisticated structural form and architecture with an opening span of two iron swivel girders, quays and a wet dock, all included in the gross cost which had risen to £89,000 by its completion in 1835. The engineer for completion of the work was C. B. Vignoles (q.v.). 'Mr. L. Grantham' was credited with 'immediate superintendence' (House of Commons (1833) *Accounts and Papers*, **17**, 373).

In 1823 Nimmo offered a design for another bridge over the Shannon a few miles upstream, to be built by the unprecedented principle of three underslung suspension spans, but the proposal was not accepted. In 1826 he reported to the harbour authority of Drogheda with proposals for improvement by dredging and river training, based on his geological knowledge as well as hydrographic survey. At an unidentified date he designed a bridge over a gorge at Pollaphuca on the turnpike route from Dublin to Baltinglass, a bridge which shows his desire for fine architectural effects, being a large, ornamental Gothic arch spanning over the very deep gorge. It cost £4,074 and a second pointed arch bridge less than half a mile to the south is assumed to be also of Nimmo's design.

By the middle of the 1820s he was being consulted about most of the major engineering developments planned in Ireland. He planned a railway from Limerick to Waterford in the south, with his proposals published in 1825 and again with some revision in 1826, the line to be used first by horse-drawn trains but suitable in the future for locomotives. He planned a rail link from Dublin to Kingstown in 1831 and construction followed,under Vignoles's direction after Nimmo's death. Another commission, which reached the construction stage in 1830, was for a crossing of the estuary of the Blackwater near the town of Youghal, on the south coast between the counties of Waterford and Cork. Here Nimmo designed a very long wooden bridge approached by a causeway of stone and earth. The bridge was of Memel pine and comprised fifty-seven piers, each of five timber piles, and deck spans of 30 ft., plus a draw span of 40 ft. for ships to pass. His brother George contracted to build it in partnership with John Bennett of the City Foundry in Dublin, for £17,000. The final cost was £22,000.

During these years he must often have passed through Liverpool in transit from Dublin to London and back, and in the late 1820s he was frequently employed by men of Liverpool to assist with their projects. In a case which was heard at Lancashire Summer Assizes in September 1827 the mayor and council of Liverpool prosecuted the Mersey and Irwell Navigation Company for causing a nuisance in the Mersey by diverting some of the river water into the Navigation's channel. Nimmo's evidence that the extractions of water were insufficient to alter significantly the river flow and its scouring action downstream was based on personal observations over four years, measurements made at critical stages of the tides, and calculations. The surviving verbatim record of the case shows him cool under the hectoring of lawyers in cross-examination, studious in requesting pauses to consult his notes and make extra calculations, and both judge and jury and even the Lord Chancellor Henry Brougham as counsel for the plaintiff surrendered to his logic.

Other commissions in the area followed quickly. In partnership with Robert Stevenson (q.v.) he reported in 1828 on a scheme to cut a ship canal through the Wirral peninsula from the estuary of the Dee, which was easily accessible to sea-going ships, to avoid problems of constant silting at the mouth of the Mersey. In two further reports they were joined by Telford, representing the Exchequer Loan Commission from which the promoters of the scheme might obtain loan finance. The scheme was deemed viable, but made no further progress.

In the same year Nimmo gave his only recorded advice to an overseas project. He and Telford jointly answered a request for views about aspects of the planned Welland Canal, a ship canal joining the Lakes Erie and Ontario in Canada.

Having made good friends in Liverpool, he was engaged by some of them to assist in the establishment of companies and route planning for several nascent railways in the North-West of England. They were striving for quick results, in the aftermath of the Rainhill Trials and the opening of the Liverpool and Manchester Railway in 1829. The meetings of most or all of the boards and committees took place in Liverpool. A 'provisional committee' of the *Liverpool and Lancashire Northern Railway*, soon to be renamed the *Liverpool and*

Leeds Railway, appointed Nimmo their engineer in September, 1830, to survey a line for their intended railway and help them with the formation of the company, the design of the railway and the parliamentary hearings in London. For these services he drew from the committee about £1,350 in the first four months of his service—clear evidence of intensive work by himself and his assistants. Within the first twelve months of their existence, the company purchased a controlling interest in the Manchester, Bolton and Bury Canal. Nimmo had attended a special general meeting of the Canal Company in September 1830 to assure them of the wisdom of converting at least some parts of the canal to a railway, and records infer that he then became their engineer, though a year later it was Henry Eaton who was 'continued as Engineer' and after some delay the company's substantive proposals for railway and canal were drafted by Jesse Hartley (q.v.) at the end of 1832.

Nimmo had also served the Preston and Wigan Railway and they ordered fees of £200 to be paid to him in June 1831, but Vignoles was their engineer and was paid £3,000 in fees at the same date. Nimmo's obituaries recorded that he had been 'consulting engineer' at about the same time to the Duchy of Lancaster, the Mersey and Irwell Navigation, the St. Helens and Runcorn Gap Railway ... and Birkenhead and Chester Railway'. His responsibilities to at least some of these must have been small. For instance, it was Vignoles who undertook the main surveys and planning for the St. Helens and Runcorn Gap line although Nimmo visited the site twice in 1830.

Within the last two years of his short life Nimmo became subject to criticism on several counts concerning the works he had directed in the Western District. Firstly, too many of the roads which had been made were unfinished or were already in need of repairs. In the existing circumstances of work ceasing for months in the winter, and a wet and windy climate, delays and difficulty of making roads with adequate surfaces and drainage could be expected, but it is also apparent that the transfer of finished roads to the county authorities did not occur smoothly. The need to work largely with unskilled labour—for relief of the poor as well as lack of experienced men available—would be a valid excuse for some of the problems. Once criticism began and reached leading officials of the Irish government, allegations were made that Nimmo's engineers and supervisors had been allowed to take materials such as good timber and iron from the stores for their private use, and that control of contracts and payments was lax. Payments to some contractors and salaries to his staff were long overdue and funds from government to pay them were not forthcoming. In the early months of 1831 he made a report on the work of the previous year and another in summary of the whole eight years of his appointment. He answered letters from government officers, but they felt they had to stop the work. His staff were sacked at the end of June.

Nimmo then had to plead for payment of the arrears due to his staff and contractors, amounting to £2,382. The salaries due to his staff totalled £933, the largest sum being due to his brother John for six months salary. John had lived for some time in the west as 'resident inspector and paymaster-general' and earned 7 guineas per week.

The government had to act and sent John Killaly (q.v.) to Mayo to report on the state of the works. Nimmo's accounts were also referred to the 'Board of Accounts' for examination. Before they reported he had died. They found that the anomalies in the accounts, as presented to them by his brothers George and John, had been sufficiently explained and that the amounts unaccounted for should be allowed. They also noted that there had been great difficulty in meeting the responsibilities placed on Nimmo in the first years (1822–1825) of the work, and were convinced that his dealings had all been entirely honest. Remarkably, the final accounts reveal that even in the year 1831, while he was acting for at least one railway company in Liverpool, he had devoted three hundred and twenty-six days to the affairs of the Western District.

He had been writing letters from addresses in England at least until the end of October 1831, and the cause of his death in Dublin on 20 January 1832, aged forty-nine, is not known. His brothers both stayed in Ireland. George had first become a contractor in Dublin, but later moved to the West and died at Ballinahinch in 1851. John remained at Roundstone till his death in 1850. John's son Alexander was trained as an engineer, and worked in Canada, then again in Ireland until his death at an early age in Dublin in 1839.

It was not until 1828 that Nimmo became a member of the Institution of Civil Engineers. He had never married but clearly made many personal friends. The novelist Maria Edgeworth, whose father had been one of the engineers employed with Nimmo on the bog surveys of 1811–1813, and whose brother William (q.v.) was assistant to Nimmo on some road-building in County Waterford in 1823, wrote of cheerful calls made by Nimmo to her house at Edgeworthstown in County Longford; also of his having given 'a course of engineering' to the prodigiously clever daughter of Thomas Martin of Ballinahinch Castle in Connemara. The young William Rowan Hamilton, already blossoming as a mathematician and astronomer, enjoyed a rapid trip through Ireland, the North of England and Scotland with Nimmo in 1827, including a descent in the diving bell which was in use for the foundations of Wellesley Bridge in Limerick. A picture of Nimmo as a highly sociable man emerges and he must have been welcomed in many well-to-do households at the same time as he was working so hard for improvements in the lives of the Irish poor.

One other probable member of his social circle was John Edward Jones (1806–1862), who had been an art student in Dublin but also trained as an engineer under Nimmo and later worked as an engineer in London. When about forty he gave up engineering to become a sculptor and he carved in marble the only extant likeness of Nimmo, a bust which is now in the library of the Royal Dublin Society.

TED RUDDOCK

[St. Helens & Runcorn Gap Railway minutes, PRO RAIL 593/1; Reports of the Commissioners on Bogs in Ireland (2 vols.) (1814) in *House of Commons Papers*; MS letters of Alexander Nimmo (various dates) BL; MS plans of bog surveys, map room, NLI; MS survey plans of Irish coast and harbours (1820–1830), maps classes 15 and 16, NLI; MS papers re. public works in the Western District (1822–1832) National Archives of Ireland, Dublin; MS records of NE England railway companies (1828–1832) PRO; *The Edinburgh Encyclopaedia* (18 vols.) (1830); Obituary (1832) *The Freeman's Journal*, 24 January; Will books and Testamentary Index, PRO, Dublin; S. Lewis (1837) *Topographical Dictionary of Ireland*; R. P. Groves (1882) *Life of Sir William Rowan Hamilton* (3 vols.) vol. 1, 232–281; Joseph Mitchell (1883, repr. 1971) *Reminiscences of my Life in the Highlands* (2 vols.), vol. 1, 42–47; E. Inglis-Jones (1959) *The Great Maria*; E. C. Ruddock (1974) Hollow spandrels in arch bridges, *The Structural Engineer*, **52**, 281–293; Ted Ruddock (1979) *Arch Bridges and their Builders 1735–1835*, 196–200, and sources quoted; P. O'Keeffe and T. Simington (1991) *Irish Stone Bridges*, 274–277; R. C. Cox and M. H. Gould (1998) *Civil Engineering Heritage: Ireland*; DNB]

Publications

1807. Boundaries of the Northern Counties … , Appendix U to *Third Report of the Commissioners for Highland Roads and Bridges*, 68–84
1810. Report on survey for a drove road from Kyle-Rhea to Killin, in T. Telford, *Report and Estimates, Relative to a Proposed Road from Kyle-Rhea to Killin by Rannoch Moor*

1830. Contributions to *The Edinburgh Encyclopaedia* (D. Brewster, ed.) dates of issue of volumes as follows:

1811. *Boscovich's Theory of Natural Philosophy*, vol. 3, 749–768
1811. *Bridge: Theory*, vol. 4, 489–519
1812. *Theory of Carpentry*, vol. 5, 495–512
1814. *Draining*, vol. 8, 63–81
1821. (attrib.) Part of *Navigation Inland*, vol. 15, 209–315

1814. Reports on the Bogs in Iveragh in Roscommon, parts of Kerry and Cork, and in Galway West of Loch Corrib, signed 1811 and 1812, all in Fourth Report of the Commissioners on Draining and Reclaiming the Bogs in Ireland, *House of Commons Papers*, 27–57, 59–102, 175–186 and pls. 1–8
1814. *Report of Sir Charles Coote Bart and Alexander Nimmo, Civil Engineer, respecting the Harbour of Dunmore*, dated March

1814. Estimate of a packet harbour at Portcullin Cove near Dunmore, dated 25 May, in *House of Commons Journal*, 25 May, 2 pp.
1815. *Report on Improvements to Cork Harbour*, dated 11 September, 30 pp.
1818. *Plan of the Works of Dunmore Harbour*, dated May, engraving
1819. Evidence on roads, harbours, drainage of the bogs, in *First Report of the Select Committee on the State of Disease and the Labouring Poor in Ireland*, 101–109, App. 223–253

Charts and plans made for the Commissioners for Irish Fisheries:

1821. *The Bay and Harbour of Sligo*, engraving
1821. *Carlingford Lough*, engraving
1821?. *The Coast of Down from the Lee Stone to St. John's Point*, engraving
1821. *Killough and Ardglass*, engraving
1821. *Strangford River, or the Entrance into Lough Cone*, engraving
n.d. *Belfast and Larne*, engraving
1822. *Design for a Pier at Crossfarnogue*, engraving
1823. *The Harbours of Roundstone and Birterbuy, with part of the Coast of Galway*, engraving
1823. *Dublin Bay*, engraving

1824. *Map of the New Road and Bridge between the City of Limerick and Parteen* (the bridge is Athlunkard Bridge, here designed as a three-span underslung suspension bridge, spans 136 ft., not built)
1824. On the application of the science of geology to the purposes of practical navigation, *Transactions of the Royal Irish Academy*, **14**, 39–50; also published as a pamphlet, 1825
c. 1824. *Elevation of the New Bridge of Limerick over the Shannon*, engraving, n.d. (the drawing is of Wellesley Bridge, and the elevation is accompanied by a sectional plan)
1825. On railways, *The Dublin Philosophical Journal and Scientific Review*
1825. *Report on Proposed Railway from Limerick to Waterford*, reprinted (1826) in *The Dublin Philosophical Journal*
1826. *Report on the Improvement of the Harbour of Drogheda*, dated 10 May, 12 pp.
1826?. *Plan of the River Boyne from the Bridge of Drogheda to the Sea with a Design for Improving the Navigation*, engraving
1827. Evidence in *The King … v. Samuel Grimshaw, John Rowles et al. … for a Nuisance, in diverting the Water from the River Mersey, at Woolston, in the County of Lancaster*, 162–178, at Lancashire Summer Assizes, September
1828. Papers in: David Stevenson, *Life of Robert Stevenson* (1878); (with Robert Stevenson) Preliminary Report … on the Proposed Improvements at Wallasey Pool, 23 February, 132–135; (with Thomas Telford and Robert Stevenson) Report … recommending two extensive new sea ports … with a floating harbour or ship canal to connect them, 16 May, and Further Report respecting the Proposed Two New Ports, etc., 14

July, 132–150; Wasting Effects of the Sea on the Shore of Cheshire between the Rivers Mersey and Dee (with Robert Stevenson), 8 March, 225–229
1830. *Accounts and Papers of Parliament*, vol. 27: Report on public works in the Western District of Ireland in the year 1829, 10 March; return of all the public money expended under direction of Mr. Nimmo on public works in Ireland, 7 July; account of works in the Western district commenced under the direction of Alexander Nimmo, civil engineer, in the summer of 1822, 18 June
1832. *The Harbour of Valentia, by Alexander Nimmo, 1831*, Admiralty chart
1832. *New Piloting Directions for St. George's Channel and the Coast of Ireland*, 209 pp. and chart

Works

1806. Survey to define the boundaries of northern counties of Scotland, for Commissioners for Highland Roads and Bridges, fee £150
1809?. Survey of route for a drove road from Kyle-Rhea to Killin, for same
1811–1813. Surveys of bogs in the South-West and West of Ireland, for the Commissioners on Draining and Reclaiming the Bogs in Ireland, area surveyed *c.* 2,000 sq. miles
1814–1821. Dunmore (East) Harbour, design and construction, for the General Post Office, cost £42,500
1815. Survey, report and proposals for the river and harbour of Cork (not executed)
1818. Portumna Bridge (River Shannon), western half rebuilt (timber trestle bridge)
1820–1823. Surveys of Irish harbours with proposals for improvements, etc., for the Commissioners for Irish Fisheries. The following harbours and sections of the coast were included: Dublin Bay, Sligo Bay and Harbour, Bray, Carlingford Lough, the coast of Down from the Lee Stone to St. John's Point, Killough and Ardglass, Strangford River, Belfast Lough and Larne, Portballantrae, Clare Island, Roundstone and Birterbuy, Carrigaholt, Brenogue Head, the bay of Balcheline, and Barra, Cleggan, Clifden and Galway
1822–1831. As Engineer for the Western District of Ireland, extensive works for new development and for employment and relief of the poor: chiefly construction of roads and bridges, piers and harbours, quarries and an iron mine and smelting furnace. Works included, but not all completed:
In Galway town: four lines of road; Slate Pier; Claddagh Pier; the market house. In County Galway, the Connemara coast road; Galway to Killary Harbour road, including bridges at Toombeola and Clonisle; branch road to Kilkerran Bay, including bridge at head of Lough Corrib (i.e. Maam Cross); road Spiddal to Moycullen; road Killeries to Cong; road Headfort to Galway; road Oughterard to Clifden; the Central Connemara road; road Creggs to Galway; road Tuam to Galway; road from the coast road to Costello Pier; branch roads (various); quay and cut at Corofin; bridge at Ballinderry; sewer at Tuam; piers (3 or 4) on Lough Corrib; quay at Leenane; quay at Derryinver; quay at

Clonisle; quay at Greatman's Bay; quay and fishing pier at Clifden; fishing pier at Cleggan; fishing pier at Roundstone; fishing pier at Costello; fishing pier at Barna; fishing pier at Burrin.
In County Mayo: road Westport to Killary Harbour; road Westport to Claremorris; roads Newport to Achill Head; road Hollymount to Claremorris and Ballyhaunis; the Erris Road, comprising Castlebar to Blacksop Point, Binghamstown and Tarmon piers, including nine bridges; road Killala to Ballina, Foxford and Swineford; road Ballina to Moyne; road Ballyscanlon to Swineford; road Killala to Broadhaven; bridges at Bandurra, Annies and Kilnacarra.
In County Sligo: roads in Sligo town; roads to Catron, Ballyfarnon, Drumcliffe, Gartlownane Circuit and Geevah colliery; road Bonnyconlon to Pullagheeny pier; pier at Raughly.
In County Roscommon: road Castlerea to Ballymoe; road Boyle to Tarmonbarry and Slievebawn; road Carrick-on-Shannon to French Park; road Arigna ironworks to Mount Allen.
In County Leitrim: road from Cloone to Mohil; road Ballinamore to Belturbet, and to Dromahair colliery. Total expenditure up to mid-1830, £167,000

1824–1835. Wellesley Bridge and Docks, Limerick (River Shannon), five arches, masonry, iron opening span, with wet dock and ship lock, for Directors of Inland Navigation, contractor Clements and Son, cost £89,000 (docks never finished)
1825. Bridge at Cahir, Co. Tipperary, for road trustees
1820s. Two bridges at Pollaphuca on the Dublin–Baltinglass turnpike route, each of one pointed arch, one 14 ft. span, one 65 ft.
1830. Youghal Bridge, wooden trestle bridge of fifty-seven spans of 30 ft., 1,787 ft. overall length, contractor John Bennett and George Nimmo, cost £22,000
c. 1830. (attrib.) The Spring Rice Monument, Limerick, a tall fluted Doric column topped by a statue of Thomas Spring Rice, MP for Limerick 1820–1832
1830–1831. Liverpool and Leeds (previously Liverpool and Lancashire Northern) Railway, Engineer during planning, obtaining an Act, etc.
1830–1831. Preston and Wigan Railway, brief appointment (for survey?)
1830–1831. Manchester, Bury and Bolton Railway, brief appointment
1831–1834. Dublin–Kingstown Railway, planned (construction after Nimmo's death, engineer C. B. Vignoles (q.v.))

NIXON (NIXSON) family (fl. 1812–1838) members were in evidence in civil engineering work in the Carlisle area, both on bridges and railways.
  William Nixon (1760–1824) was the son of William Nixon and was baptised on 22 June 1760 at Dalston, near Carlisle. Perhaps initially a stonemason, he became one of the two Bridge-Masters

of the county, responsible for Allerdale Wards above and below Derwent, i.e. West Cumberland; he held the appointment from Midsummer 1816 until his death. His salary was £30 p.a., but it was not a full-time appointment. When he was required to attend every day at the construction of Caldew Bridge, Carlisle, in 1820, he was paid an additional 2 guineas per week. The following year at Cockermouth, further from his home, he was paid 4 guineas per week. In 1823 he was also appointed—jointly with Christopher Hodgson—as Inspector and Clerk of the Works for the new gaol in Carlisle. In 1789 he married Elizabeth Carr and the union produced at least four children; a daughter, Anne (b. 1813), married John Blackmore (q.v.), engineer to the Newcastle and Carlisle Railway. Nixon died at Chalk Lodge, Dalston, on 2 March 1824.

Paul Nixon (b. 1768) was a brother of William Nixon and was baptised at Dalston on 7 March 1768. In 1795 he married Ann Slack of Carlisle and the marriage resulted in at least three daughters. He was the contractor for Eden Bridge, Carlisle and was uncle to William Nixon Jr., also a contractor. As a matter of general policy, in 1823 both of these men were barred from working for the County in the Allerdale wards. Nixon gave evidence in 1829 to the Parliamentary Committee on the Newcastle and Carlisle Railway, at which time he was forming the foundations for a bridge carrying the London Road into Carlisle. Later, with his son-in-law, he contracted for Warwick and Alston bridges in the east of the county. Their partnership would appear to have ended c. 1835.

William Smith Denton (b. 1799) was the son of Samuel and Eve Denton and was baptised on 20 January 1799 at Withernwick, in the East Riding of Yorkshire. In 1824 he married Sarah (b. 1798), the daughter of Paul Nixon, his partner, with whom he worked on Warwick, Alston and Irt bridges. Denton gave evidence in 1834 to the Parliamentary Committee on the London and Southampton Railway. By this time—perhaps with Nixon—he was building 6 miles of the Newcastle and Carlisle Railway; its several major works included the 95 ft. high viaduct crossing the river Eden. The fact that only his name appears on the commemorative plaques on the individual bridges would seem to indicate that Nixon was not involved, at least after 1835.

His eight children—at least—were baptised either in or near Carlisle between 1826 and 1835 but the place and date of his death have not been ascertained.

P. S. M. CROSS-RUDKIN and R. W. RENNISON

[CQ/7/5 & WQ/A/B 41, Cumbria RO; *Carlisle Journal*, March 1824; House of Commons (1834) *London and Southampton Railway: Minutes of Evidence*; House of Commons (1839) *Evidence ... on the Newcastle and Carlisle Railway*]

Works
William Nixon:

1818–1819. Isel Bridge, design and supervision, cost £2,000
1821–1822. Derwent Bridge, Cockermouth, design and supervision, cost £2,800
1823. Westlinton Bridge, design and supervision, cost £1,650

Paul Nixon:

1812–1816. Eden Bridge, Carlisle
1829–1830. Harraby Bridge, Carlisle
1833. Alston Bridge
1833–1835. Warwick Bridge
1836–1837. Irt Bridge

William Smith Denton:

1833. Alston Bridge
1833–1835. Warwick Bridge
1835. Bampton Grange Bridge, over river Lowther, for £1,030
1836–1837. Irt Bridge
1830–1836. Newcastle and Carlisle Railway, 6 mile length of track including Wetheral and Corby viaducts

NOWELL families, stonemasons and contractors, originated in Dewsbury, Yorkshire, in the eighteenth century and carried out works all over Britain throughout the nineteenth century. They were related by marriage and through business interests with the Cravens, Hemingways and other Yorkshire contractors. The similarities of names, careers and family links make it difficult at times to establish which branch of the family was responsible for work undertaken.

Jonathan Nowell (1757–1810) was the second of three sons in the family of six children of Richard Newill (1718–1768), a mason, and Jane Johnson (d. 1799); he was baptised in Dewsbury Parish Church on 6 February 1757. Presumably trained by his father he is known to have built a bridge over the Calder, carried out work on Dewsbury Market Cross and built property in Dewsbury. On 10 October 1781 he married Hannah Chadwick (1762–1834). They had ten children, and the four sons who survived into manhood became involved in construction. Of their five daughters Hannah (1782–1854) married Benjamin Cooper (1778–1854), and Frances (1800–1870) married John Craven (q.v.), both husbands being involved in building work.

'Joseph' Nowell (c. 1784–1836), Jonathan's eldest son, was baptised as Joshua in Dewsbury on 24 February 1784. He later had a mason's yard at Greaves Road, Dewsbury, and owned property in the town. One assumes he worked with his father until the latter's death in 1810. About that time he became involved in a partnership with Samuel Whitaker (d. 1821) and Hiram Craven (q.v.). Together they were responsible for between twenty and thirty bridges in the Yorkshire area, of which the most important was Ouse Bridge, York, erected 1810–1820. This work led to Nowell giving evidence on the state of London Bridge in 1821. The first bridge contract directly attributed to Joseph Nowell was the widening of Holme Bridge, Gargrave (1812–1814), followed by parapets for Tamewater Bridge, Saddleworth, (1814–1815) and the new Threshfield Bridge near Skipton.

In 1818 the partners obtained some important masonry contracts on the Edinburgh and Glasgow Union Canal, including an aqueduct over the Avon, and Slateford Aqueduct. While this work was in progress in June 1821 Whitaker died in a coaching accident at Sunderland Bridge, near Durham. Nowell and Craven continued to work together, mostly on church contracts—then in abundance under the recent Parliamentary grants—until 1828 when the partnership was dissolved. Joseph was involved in other building contracts at Liversedge Church (1812–1816) and at Manchester Infirmary, as well as at his home, Quarry Hill House. His last known building contract was Holy Trinity Church, Ripon (1826–1827), on which he was joined by his eldest son **John Willans Nowell** (1806–1851), who married Elvia Simpson (d. 1837) there in 1828.

Joseph had married his first wife Alice Willans in 1802. They had three children, the second son being **Jonathan Willans Nowell** (1809–1846); their daughter was Mary Willans, who married Edward Craven (q.v.). From 1828 Joseph and his sons were concerned exclusively in civil engineering work.

The first important contracts of Joseph Nowell and Sons were on the Macclesfield Canal: the Dane Aqueduct and the locks at Bosley. At this time they began to establish a loyal team of foremen including Joseph Longbottom, in charge of carpenters at Ripon, and James Briggs, Inspector of Works, whose family worked for the Nowells for many years. For the Macclesfield Canal works stone was transported over rails from local quarries. It was on these works that Jonathan met his wife Martha Percival (1806–1879). Following the completion of the canal in 1831 John Willans moved to Farnworth to supervise the construction of Widnes Dock for the St. Helens and Runcorn Gap Railway, and Jonathan Willans moved to Elleby Lane, Leeds, to supervise their earliest railway contract, the first section of the Leeds and Selby Railway (1831–1833). This included Marsh Lane Tunnel and henceforth Jonathan established a reputation as a specialist tunnelling contractor, describing himself as 'the great borer'. In the meantime John supervised work on the Leicester–Swannington Railway. By the standards of the time Joseph Nowell and Sons were now well-established contractors.

The family successfully tendered for two contracts on the London and Birmingham Railway, contract 2B Harrow and 7F Kilsby Tunnel. Joseph moved to Hatch End near Pinner, John Willans to Bushey, while Jonathan Willans took charge at Kilsby. Problems soon became apparent at Kilsby where quicksand made conditions all but impossible. In these circumstances Robert Stephenson (q.v.) advised the railway company to take over the contract. On his return from Birmingham, where he was given the news, Joseph fell ill at Jonathan's house and died shortly afterwards on 12 January 1836; he was buried in Dewsbury Parish Churchyard. His sons completed the other contract, but not without concern. On several occasions Joseph's brother, **James Nowell** (1793–1859) and Hiram Craven (q.v.)—having acted as surety for Joseph Nowell—were involved in discussions with the railway company over progress on the works and it was through their advice and support that Joseph's sons were allowed to successfully complete the contract. Problems were also experienced with progress on the Lodge Hill contract on the North Midland Railway in 1838.

The skew bridge they erected over the Bushey–Watford Road was the first to be built on the lines of George Watson Buck (q.v.). They also introduced tip wagons on their works and this, and the general efficiency of their works, were commented on by Conder who noted a 'stout, hale, well-humoured Yorkshireman (presumably John Willans Nowell) stood or sat in from early morning till sunset and inspected the progress of the wagon with a never flagging interest'. Jonathan broke his leg in a coach accident on Harrow Hill and walked with a limp for the rest of his life.

With the completion of these contracts the brothers took a succession of railway contracts before forming a partnership in the mid-1840s in association with Richard Hattersley, who had married Frances Percival, Jonathan's sister-in-law, and William Shaw who had worked for them at Kilsby. Jonathan died of a chill in 1846 and John retired shortly afterwards to the 'Elms', Macclesfield, following losses on their Stalybridge–Huddersfield contract. By this time seven of Joseph's children by his second marriage to Jane Cowling (née Barnes) were involved in contracting and they continued in that capacity until the end of the century.

Of Joseph's brothers **Benjamin** (1785–1823), **Jonathan** (1789–1825) and **James** (1793–1859) on 6 February 1819 formed a partnership which was effectively continued as James Nowell and Sons after his brothers' deaths. They built churches across Central England, including several designed by Thomas Rickman. James married Tabitha Hemingway (1794–1864) and resided at Stonefield, Dewsbury, which house James built. Tabitha's cousins, John (1795–1854), James (b. 1802) and Joseph Benjamin (b. 1812) Hemingway were trained as masons by the Nowells. James and Joseph Benjamin worked for James Nowell on jobs in the Midlands in the 1830s, such as Stoneyhurst College (1830–1832) and Birmingham Grammar School (1832–1837). James Nowell's own children began to become involved in his projects at this time and four of his five sons worked as engineers or contractors at various times, **Michael** (1818–1840), **Benjamin**

Jonathan (1820–1891), **James** (1830–1902) and **Jesse** (b. 1833).

James Nowell and Sons' first major civil engineering job was Rea (Lawley Street) viaduct of 27 arches on the London–Birmingham Railway (1834–1837), followed by Wolverton Viaduct and the Leighton Buzzard (Mentmore–Linslade contract) including Linslade tunnel. On 11 June 1838 they were awarded the contract for Gauxholme Viaduct on the Manchester–Leeds Railway. Michael Nowell worked for his father on the Tamworth Viaduct (1837) on the Derby–Birmingham Railway. In the 1840s, with James Nowell's support, Benjamin Jonathan tendered for contracts with John Hemingway and Charles Pearson, most famously for the masonry on the Britannia Bridge.

Of the same generation as Jonathan Nowell (1757–1810) were the four children of **Benjamin Nowell**, also of Dewsbury. Both **George** (c. 1772–1826) and **John** (c. 1773–1844) were masons. George married Frances Rhodes (1773–1858) and had nine children, most of whom became masons and contractors. Their oldest son **George** (1795–1852) was injured in the construction of Cob Wall Viaduct, Blackburn, and died on 13 April 1852. **Benjamin** (1799–1837) was a mason who married Mary Hemingway. **Abraham** (1801–1845) and one of his sons (1825–1845) were killed in an accident at Ashton Aqueduct in April 1845. **John** (b. 1805), **Thomas** (b. 1811) and **Jonathan** (1814–1873) were also masons, operating as Nowell and Company on the Liverpool–Crosby and Southport Railway (1850). **Joseph** (1817–1867) and **Jacob** (1820–1866) were railway contractors operating as Nowell Brothers of Watford. Two of Joseph's sons, **Joseph Dore** (1847–1897) and **Herbert Mason** (1859–1935), became major railway contractors in the late nineteenth and early twentieth centuries.

MIKE CHRIMES

[Frank Smith papers, ICE Archives; Briggs papers, National Railway Museum, York; Birmingham & Derby Junction Railway Committee Minutes, 1836–1844, RAIL/36/1–6, PRO; Birmingham & Gloucester Railway, RAIL 37/1–, PRO; Blackburn Railway, RAIL 52/1, 52/9, 52/14, 52/15, 52/21, PRO; Bristol & Gloucester Railway 1841–1847, RAIL 4/76/1–37, PRO; Cheltenham and Great Western Union Railway Minutes etc., RAIL 109/1, 109/19, PRO; Great Western Railway General Committee and Board Minutes 1833–, RAIL 250, PRO; London–Birmingham Railway Minutes, RAIL 384/1–146, PRO; London–Birmingham Railway contracts etc., RAIL 384/163, 169, 176, PRO; Macclesfield Canal, RAIL 850/1–2, PRO; Manchester–Leeds Railway, RAIL 1175/76–79, PRO; North Midland Railway, London Committee Minutes 1837–1840, RAIL 530, PRO; South Eastern Railway Minutes, etc., 1845–1846, RAIL 635/18, 635/20, PRO; St. Helens Canal and Railway, RAIL 593, PRO; The Nowells: a remarkable Dewsbury family, *Dewsbury Reporter*, 30 July 1932, West Yorkshire RO, Dewsbury; Wakefield: QD 3110,

West Yorkshire RO; *Mining Journal*, 11 January 1840; P. Le Count (1839) *History of the London and Birmingham Railway*; [F. R. Conder] *Personal Recollections of English Engineers* (1868), 17–19]

Works

Benjamin Nowell and sons:

1819–1822. Birmingham, St. George's Church
1821–1825. Sheffield, St. George's Church
1822–1823. Erdington, St. Barnabas Church
1823. Dewsbury Parish Church

James Nowell and sons:

1826–1829. Birmingham, St. Trinity Church
c. 1831–1832. Lissadell Mansion, County Sligo
1832–1835. Stonyhurst Church
1833–1837. Birmingham Grammar School
1834–1837. London–Birmingham Railway, contract IG (RAIL 384/241), Lawley Street (Rea) Viaduct, £15,500
1834–. London–Birmingham Railway, contract 4C (RAIL 384/176), Wolverton viaduct, £25,964 tender
1835–1839. London–Birmingham Railway, contract 5C (RAIL 384/219, 310–312), Leighton Buzzard, 7½ miles, £43,162
1838–1840. Manchester–Leeds Railway, Gauxholme viaduct ,eighteen arches, £22,000
1838–1840. Derby–Birmingham Railway, Tamworth viaduct
1838–1840. Ashby de la Zouch, Holy Trinity Church
1839–1840. Ancoats, All Saints Church
1839–1840. Hawarden Church

Joseph Nowell, Craven and Whitaker:

For full list see *CRAVEN, Hiram* (q.v.)

Joseph Nowell and sons:

1811. Willshill Bridge, £302 8s 11d
1812–1814. Holme Bridge, widening, £2,089 10s 8d
1814–1815. Tamewater Bridge, walls, £553 7s 7d
1818–1820. Hodder Bridge, £1,802 4d
1824–1826. Wellington Road Viaduct, Stockport, eleven arches
1824–1826. Brighouse Bridge, £3,952 10s (estimate)
1825–1828. St. Martins Church, Liverpool
1826. Dewsbury Market Cross, £650
1826–1827. Holy Trinity Church, Ripon
1828–1831. Macclesfield Canal, locks, etc.
1830–1833. St. Helens and Runcorn Gap Railway, Widnes Dock, £37,000, six contracts for rails, waterside, etc., £20,000
1830–1834. Leeds and Selby Railway, first contract including tunnel 170,000 cu. yd., £25,244
1833. Coleorton Railway, £2,373
1834–1831. London–Birmingham Railway, contract 7F (RAIL 384/231), Kilsby Tunnel (completed by Company)
1834–1838. London–Birmingham Railway, contract 2B (RAIL 384/163, 213, 297, 298), Harrow contract, £144,574 tender

1837. North Midland Railway, Altofts contract, £42,000 tender

O

OATES, James (fl. 1778–1821), canal engineer, built—with a partner named Delahunty—some of the early locks on the Barrow Navigation, as well acting as an assistant engineer. Following the appointment of Archibald Millar (q.v.) as Resident engineer on the Grand Canal he was appointed an assistant engineer in late 1789 or early 1790. He became involved with the link between the Grand Canal and the Barrow at Athy and William Jessop (q.v.) approved his suggestion for the location of the junction between the two in 1791. When this branch was completed Millar was put in charge of the Circular line of the Canal, leaving Oates on the Shannon line where he worked with John Killaly (q.v.). They surveyed a line to join the Shannon near Athlone in the early 1790s, which was not adopted. He worked as an assistant engineer for the Grand Canal Company until his death in 1821. He was also employed as a draughtsman and worked elsewhere with Killaly.

RON COX

[C. Hadfield and A. W. Skempton (1979) *William Jessop, Engineer*; R. Delany (1992) *Ireland's Inland Waterways*]

Works
1780s. Barrow Navigation, contractor (with Delahunty)
1780s–1790s. Barrow Navigation, assistant engineer and surveyor
c. 1790–1804. Grand Canal, construction, assistant engineer, £174,000

OGBOURNE, Sir William (c. 1662–1734), a master carpenter, was—after serving an apprenticeship—admitted a freeman of the Carpenters' Company in 1693. He built Trinity Almshouses, Stepney, for the Corporation of Trinity House 1694–1697 and in 1697 'William Ogbourne of Stepney, house carpenter', contracted for the timber work of Howland Great Wet Dock in Rotherhithe, north of Deptford Dockyard, and wharfing along the river front.

The dock, designed by the shipwright John Wells (q.v.), had an area of 10 acres and a 17 ft. depth of water, with an entrance lock 44 ft. wide and 150 ft. long between two pairs of gates. The walls of the dock and lock, at least 20 ft. high and some 3000 ft. in length, consisted of timber piles and planking which would have been supported by anchored land ties. Other contractors carried out the excavations. The dock was opened in 1700 at a cost of about £12,000, the greater part of which can be attributed to the timber work.

The dock was built on land leased from Mrs Elizabeth Howland, a widow whose daughter and heir became after marriage the Duchess of Bedford. Records relating to the dock are among the Bedford estate papers. Ogbourne went on to build Streatham House for Mrs Howland in 1706. For many years he held the post of Master Carpenter to the Board of Ordnance. He was knighted in 1727 after serving as Sheriff of London. He died on 13 October 1734 and was buried at St. Olave's Church, Hart Lane, where there is a monument to his memory.

A. W. SKEMPTON

[Bedford Estate Papers (1695–1706) Greater London Record Office; Joseph Broodbank (1926) *History of the Port of London*; W. Crawford Snowden (1948) *London 200 Years Ago*; Mike Clarke (1993) Thomas Steers, in Adrian Jarvis (ed. 1998) *Ports and Harbours*; Colvin (3)]

Works
1697–1700. Howland Dock, Rotherhithe, £12,000

OKES, T. (see: *HALL, J.*)

OMER, Thomas (fl. 1750–1770), canal engineer of Dutch origins, was sent to Ireland by the government to work on the navigations. A Thomas Omer is recorded as engaged as a master carpenter on the Kennet Navigation in 1740. Omer, directed by the Commissioners of Inland Navigation, was to be involved with almost all the mid-eighteenth century Irish inland navigation schemes, many of which became notorious for their lack of progress and expense.

Omer worked for a short while on the Newry Canal in 1754 and the following year for the Grand Canal Company. His surveys and recommendation for a southerly line between Dublin and the Shannon were adopted, and with Omer in charge work began on the Grand Canal in 1756. By 1763, with 12 miles completed from Clondalkin to the west, including three locks, six bridges, and seven aqueducts, Ir. £57,000 had been spent. Omer was now busily engaged on other schemes and responsibility was taken over by Dublin Corporation. Rather than building a continuous navigation he began work wherever he had easy access to land, leaving much incomplete and much of Omer's work had been rebuilt or altered by the end of the century by Trail (q.v.) and others.

In 1755 Omer and William Ockenden were appointed to take over the supervision of the construction work on the Tyrone Navigation but, writing in 1977, Heard notes that 'the unfortunate choice of Omer and Ockenden as the principal engineers compounded the weaknesses, through their technical limitations and the multiplicity of navigation works superintended by them, which created opportunities for deceits by the local overseers.' These deficiencies were compounded by technical difficulties associated with the ground conditions which saw construction drag on from 1733 until 1787.

In 1760 Omer reported on an alternative line from the Newry navigation to the Tyrone collieries, in connection with the Newry Ship Canal which was itself completed in 1769.

Omer began the work on the Lagan Navigation in 1756 and with adequate funding work proceeded rapidly between Belfast and Lisburn, but after about 1765 work more or less ground to a halt at Sprucefield as no money was available.

In 1759 Omer is recorded as working on both the Barrow and Boyne navigations. The initial grant on the Barrow was for £2,000; with the lock at Behann and section from there to Ballyverra open in 1761, £5,263 was authorised for the St. Mullins–Graiguenamonagh section. In 1765, 3 miles were opened and a further grant of £13,240 was made, followed by £6,184 in 1767. By 1769 only 4 miles were left to complete.

During the period from 1755 to 1767, with Ockenden, Omer surveyed and superintended the construction of parts of the Nore Navigation from Kilkenny to Inistioge in Co. Kilkenny and continued with the work following Ockenden's death in 1761. However, 'ten years of effort did not produce one complete navigation'. Omer was examined before a House of Commons Parliamentary Committee in 1761.

Around 1760 Omer was involved with proposals to improve Drogheda Harbour. In 1767 he reported to the House of Commons on the Lagan Navigation. He died in 1770.

RON COX

[M. B. Mullins (1859–1861) Presidential Address, *Trans ICEI*, **6**, 1–186; R. Heard (1977) Public works in Ireland, M.Litt. thesis, Dublin; E. W. Paget-Tomlinson (1978) *The Illustrated History of Canal and River Navigations*; R. Delany (1992) *Ireland's Inland Waterways*]

Publications
1767. Proposals by Omer and Ducart for the Tyrone Navigation, *IRISH H.C.*, **8**(II), 179–180, 17 November

Works
1754. Newry Canal, superintending engineer
1755–1763. Grand Canal, engineer, 12 miles, c. £53,000
1755–. Limerick (Shannon) Navigation, to Killaloe, engineer, c. 12 miles
1755–1769. Upper Shannon Navigation, c. 80 miles of improvements
1755–1767. Nore Navigation, surveys, £10,000
1756–1767. Lagan Navigation, engineer, c. 10 miles, £40,000
1758–1769. Newry Ship Canal, Newry–Fathom, with C. Myers, £23,000
1759. Boyne Navigation, engineer, 9 miles, £75,000 by 1789
1759–1761. Drogheda navigation and harbour improvements, engineer, over £2,000
1759–1769. Barrow Navigation, c. £30,000

ORROCK, Robert (fl. 1540–1542) was paymaster and master of works for the 'new havin' of Burntisland, built on the Forth for James V, who established a Royal Burgh there in 1541. At that stage the works included a harbour wall and pier, the quays being largely of timber.

Orrock may have been a relative of James Orrok, mason, who was one of the masons responsible for Fordell House at Inverkeithing (1567).

MIKE CHRIMES

[J. Gifford (1988, 1992) *The Buildings of Scotland: Fife*]

OUTRAM, Benjamin (1764–1805), engineer and managing partner of the ironworks of B. Outram & Company—now the Butterley Company—was baptised on 1 April 1764 at Alfreton, Derbyshire, the eldest son of Joseph Outram (1732–1810), Estate Agent and Commissioner for Enclosures, and Elizabeth (nee Hodgkinson). Benjamin married Margaret, daughter of Dr. James Anderson FRS (1739–1808), on 4 June 1800. The third of their five children became Lt. General Sir James Outram Bt.(1803–1863), the 'Bayard of India'.

Although nothing is known of his early education, Benjamin Outram was already an experienced land surveyor and estate agent in his father's practice before he became, in 1788, assistant to William Jessop (q.v.), engineer for the Cromford Canal Bill before a Committee of the House of Lords, particularly on water supplies and the operation of mills on the river Derwent. Subsequently he was appointed superintendent of works under Jessop. The canal had a summit level of 14 miles and linked Cromford with coalfields to the east and, through the Erewash Canal, to the Trent. Major works included a masonry aqueduct of 80 ft. span across the Derwent, massive earthworks and the Butterley Tunnel, 2978 yd. long. Iron and coal were discovered during construction of the tunnel which led to Outram, in partnership with Jessop and others, establishing the ironworks during 1792.

During this period the Nottingham Canal was promoted, with Jessop as engineer and Outram associated with the construction of reservoirs, although he later took over Jessop's responsibilities. Then followed the Nutbrock Canal, designed by Jessop but constructed entirely under Outram. By 1792 Outram was commissioned to survey and estimate for the Derby Canal. Jessop advised on the plans but Outram designed this waterway which joined the Trent and Mersey Canal, at Swarkestone, to the Erewash Canal at Sandiacre via a crossing of the Derwent in Derby. A short branch off the Sandiacre line ran to Little Eaton, from which Outram's first railway led to the Denby collieries.

This railway, built as a more economic alternative to a canal, consisted of angle-iron rails supported on stone sleepers; it was constructed on Outram's 'improved plan'. The line was used by horse-drawn carriages, with detachable bodies, which ran on flat-rimmed wheels. In later years these railways became known as tramroads, or plateways. A recurring myth claims that 'tram' derives from Outram's surname but the word had been in use before his time to describe the wooden sub-frames of small carriages used in coal-mines and can be traced to the old Scandinavian word for a timber baulk. 'Tramroads' was never used by Outram to describe his railways although he was familiar with its use in mining practice.

A unique feature of the Derby Canal was the Long Weir, built across the Derwent in Derby below the location of the crossing. It contained a conduit connecting both sides of the canal system such that water could be fed from one side to the other, both pounds being at the same level, but higher than the river. On the west bank, a 40 ft. long iron aqueduct carried the canal across a stream at the Holmes. This trough was prefabricated at the Ripley ironworks and completed by late 1795, although its inadequate design resulted in partial failure during 1802. It was the second to be completed in this country after Ponty Cafnan. In 1794 Outram was appointed engineer to the Huddersfield Narrow Canal. This 20 mile long waterway rose 339 ft. from the Ashton-under-Lyne Canal to a summit level at 650 ft. AOD, where the Standedge Tunnel was driven 3 miles through the Pennines, before descending 439 ft. to Huddersfield. Outram's second prefabricated iron aqueduct of 55 ft. clear span crosses the river Tame at Stalybridge. Financial problems delayed the canal's opening until 1811 but Outram withdrew in 1801 during a lengthy stoppage of work. Telford advised on construction and planning during 1807; his programme was diligently followed until completion.

Much of the Ashton-under-Lyne Canal was finished by 1797 when Outram was engaged to construct the final half-mile in Manchester, as well as to lay out the terminus and nearby streets for housing and industry. An important feature is Store Street Aqueduct, a skew structure built during 1798 on principles devised by William Chapman (q.v.); it remains the oldest bridge of its type on the canal system.

Interlinked with these narrow canals is the Peak Forest Canal which runs from Dukinfield to Whaley Bridge, where a short branch to Buxworth connected with a 6-mile long railway—it included a spectacular self-acting inclined plane near Chapel-en-le-Frith—to quarries at Doveholes. The 15 mile canal was designed on two levels with sixteen locks at Marple, rising 212 ft., immediately below which is the Grand Aqueduct, a three-arch masonry structure 368 ft. long and 95 ft. high.

After 1801 Outram concentrated on the management of the ironworks, besides the promotion of his railway systems. The 12-miles long railways of the Ashby-de-la-Zouche Canal Company were outstanding examples of his method. Outram subsequently advised on railways from collieries in the Forest of Dean to the rivers Wye and Severn and also made recommendations to the Monmouthshire, and the Brecknock and Abergavenny Canal Companies on their feeder systems. He undoubtedly influenced the introduction of plateways in South Wales.

Outram was a man of much ability but in most aspects of his works he seemed not so much an original thinker as an engineer who recognised a good idea when he saw one and developed it to the full when time and circumstances were right, so much is evident in his railways and his iron aqueducts, to mention but two of his engineering successes.

He was said to be a fine-looking, active and spirited person of great engineering ability. Accustomed to command, he had little toleration for stupidity or opposition and even errant clients did not always escape his hasty temper. Serious illness during 1796–1797 probably led to his early demise from 'brain fever' in May 1805 during a visit to London, where he was buried in the churchyard of St. Paul's Cathedral on 18 May. Outram died intestate and the affairs of the ironworks were left in some disarray. The company was renamed the Butterley Company and the management of it was taken over by William Jessop Jr.

R. B. SCHOFIELD

[Cromford Canal, BTHR, PRO; Nottingham Canal, RAIL, PRO; Huddersfield Narrow Canal, RAIL, PRO; Manchester to Ashton-under-Lyne Canal, RAIL, PRO; Peak Forest Canal, RAIL, PRO; Ashby-de-la-Zouche Canal, RAIL, PRO; Monmouthshire Canal, RAIL, PRO; Brecknock and Abergavenny Canal, BTHR, PRO; Derby Canal, Derby Borough Library; Evidence on the Cromford Canal Bill, HLRO; Chandos-Pole-Gell Mss, Derbyshire RO; J. Anderson (1803). *Recreations in Agriculture, Natural History, Arts and Miscellaneous Literature*; M. F. Outram (1932) *Margaret Outram 1778–1863*; P. Stevenson (1970) *The Nutbrook Canal: Derbyshire*; P. Riden (1972) *The Butterley Company 1790–1830*; R. B. Schofield (2000) *Benjamin Outram 1764–1805: An Engineering Biography*]

Publications

1792. *A Plan of the intended Derby Canals and Railways with a Sketch of the adjacent Canals, Rivers and Roads*

1793. *Report and Estimate on the Proposed Sheffield Canal*

1793. *Report on the Proposed Canal from Sir John Ramsden's Canal, at Huddersfield, in the County of York, to join the Canal at Ashton-under-Lyme, in the County of …*

1794. *Estimate of expence [sic] to make a Canal and Railway from Dukinfield Aqueduct to Whaley Bridge and Loads Knowl*

1795. *Report on the Proposed Macclesfield Canal*

1796. *Particulars of some Very Singular Balls of Stone found in the Works of the Huddersfield Canal, Philosophical Transactions XIV*

1798. *Observations on Mr. Bagshawe's Estate on the Banks of the Leeds and Liverpool Canal in Craven*

1799. *Report on the Worcester Canal and Railways*

1799. *Observations on the Brecknock and Abergavenny Canal and Railways*

1800. *Report on the Somersetshire Canal and Railways*

1800. *Report on the Railways of the Monmouthshire Canal Company*

1801. *Report and Estimate of the Proposed Railways from the Collieries in the Forest of Dean, to the Rivers Severn and Wye,*

1803. *Minutes to be Observed in the Construction of Railways,* in J. Anderson (1803)

Major works

1789–1795. Cromford Canal, superintendent, £83,055

1792–1796. Nottingham Canal, resident engineer, later principal engineer, £75,000*

1793–1796. Nutbrook Canal, principal engineer, £20,000*

1793–1796. Derby Canal and Railways, principal engineer, £100,000*

1794–1801. Huddersfield Narrow Canal, principal engineer, £402,653; Outram resigned from the company in 1801 and the works were not completed until 1811

1794–1801. Peak Forest Canal, principal engineer, £177,000; Outram resigned in 1801 and the works were finished, according to his designs, in 1805

1798–1799. Ashton-Under-Lyne Canal, construction of the Piccadilly section, canal terminus and warehousing, engineer and contractor, £7,000*

1799–1803. Ashby-De-La-Zouche Railways, engineer and contractor, £30,160

1799–1800. Blisworth Hill Railway (Grand Junction Canal), contractor, £9,750

N.B. After 1800 Outram devoted most of his time to the development of the Ripley ironworks, although he occasionally reported on various canals and railway systems under construction by others, but the extent and costings of these ventures is unknown.

OVERTON, George (fl. 1795–d. 1827), mining engineer, of Llandetty Hall, near Brecon, was an early engineer of railways in the era before steam locomotives were widely introduced.

In his *Description … of South Wales* (1825) he claimed experience of tramroads extending over the greater part of 30 years, favouring flanged rails or 'tramways' over flanged wheels and rails or railways, as he described them, and cast iron rails and sleepers. He also claimed to have developed a tram or wagon subsequently generally adopted, initially of wood, and from *c.* 1815 largely of iron.

Overton's early training and experience is not known, but was evidently in mining. His earliest

known tramroad connected the Dowlais (Iron) Company's limestone quarries to their blast furnaces at Merthyr Tydfil. On this horses allegedly hauled wagons of 9–10 tons on a 1 in 53 incline. This was followed (1799–1802) by the Penydarren Tramroad, a project which arose from the dispute between Richard Crawshay (q.v.), of the Cyfartha ironworks, and his neighbours, over the way the Glamorganshire Canal was being run, essentially to Crawshay's benefit and their detriment. The tramroad is the best known of the early Welsh railways as the scene of the pioneering locomotive experiments of Richard Trevithick (q.v.). He then became involved in a partnership—Bowzer, Overton and Oliver—as proprietor of Hirwaun Ironworks, and built a connecting tramroad (1806–1808) with a high causeway and impressive span over the Afon Cynon (N.B. not to be confused with the earlier tramroad crossing by E. Hopkin, 1803–1805). Work followed on the completion of the Aberdare Canal and the Llwydcoed tramroad. In 1813 the Bowzer ironworks ceased trading but Overton was as busy as ever, working on a cast iron arched beam replacement bridge over Morlais Brook in Merthyr. His next tramroad was the 12 mile long Brinore (Bryn oer) which connected the Bryn-oer colliery, in which Overton had an interest, to the Brecknock and Abergavenny Canal with a connection to the Rhymney Iron Company's tramroad.

Overton's reputation spread beyond South Wales as a consequence, almost certainly, of his relationship, by marriage, with Jeremiah Cairns, the steward of Thomas Meynell, squire of Yarm, and a leading figure in moves to connect Stockton and Darlington by a canal or railway to the Durham coalfield around Winston. John Rennie (q.v.) had recommended a canal route in 1813, and George Leather Jr. (q.v.) also surveyed a canal line but Meynell and others favoured a tramroad. Rennie looked at a rail route in 1814 and in September 1818 Overton was invited to report on such a line. His estimate for the 51 miles of main line and branches was £124,000. Despite some misgivings, particularly over the route which crossed the land of opponents, it was decided to proceed to Parliament with a Bill, Overton forcibly arguing that in South Wales tramroads were more profitable than canals. On 5 February 1819 he proceeded to London to offer Parliamentary evidence while attempts were being made to overcome opposition. George Stephenson (q.v.), meanwhile, reported somewhat critically on Overton's line. Parliamentary progress was delayed through 1819 and 1820 and Overton resurveyed his route, which in its 1820 form was less than 37 miles. In this form it received Parliamentary approval in 1821. By then concern was being expressed about what were seen as high charges by Overton and Edward Pease, in particular, favoured selecting a different engineer for construction. Pease's name is generally connected with the Stephensons but other engineers were considered, notably Robert Stevenson (q.v.). While George Stephenson was asked to report, Overton remained in waiting until December 1821 when he realised the game was up. One area in which he had failed to keep abreast of developments was in the application of locomotive power, of which Stephenson was a convincing advocate, although this did not prevent him being busy in South Wales. Overton's last project was to survey the Rumney railway, from the Sirhowy railway to Rhymney ironworks, which featured a 4 span ashlar arch structure at Bassaleg. Overton died in 1827 and is buried at Llandetty, Breconshire.

MIKE CHRIMES

[W. W. Tomlinson (1914) *The North Eastern Railway*; C. F. Dendy Marshall (1938) *A History of British Railways down to the Year 1830*; Bertram Baxter (1966) *Stone blocks and Iron Rails*; C. Hadfield (1967) *The Canals of South Wales and the Border*; S. Hughes (1990) *The Brecon Forest Tramroads*; W. L. Davies (1992) *Bridges of Merthyr Tydfil*; M. W. Kirby (1993) *The Origins of Railway Enterprise*]

Publications
1825. *A Description of the Faults or Dykes of the Mineral Basin of South Wales*, part I

Works
–1793. Dowlais tramroad, engineer, 3 miles, 4 ft. 2½ in. gauge
1799–1802. Penydarren tramroad, engineer, 9½ miles, 4 ft. 2 in. gauge; bridges over Taff
1802–1808. Hirwaun ironworks and tramroad
–1811. Llwydcoed tramroad, engineer, 2 miles, 4 ft. 2 ft. gauge; cast iron bridge at Robertstown
1813. Pontmorlais, Merthyr Tydfil, design; cast iron; 27 ft. span, demolished 1980
1814–1815. Brynoer tramroad, engineer, 12 miles over Black Mountain
1826. Rumney railway, Bassaleg viaduct; the tramroad itself was completed in 1836, after Overton's death

OWEN, Jeremiah (fl. 1800–1850) was one of the seventeen children of Jacob Owen (1778–1870) an engineer trained by William Underhill (q.v.), whose daughter he married; Underhill was resident engineer on the Dudley Canal in charge of the Lappal Tunnel 1793–1798. From 1804 Jacob was attached as Clerk of Works to the Royal Engineers department at Portsmouth before being transferred from 1832 until 1856 as principal engineer and architect to the Irish Board of Works, a post in which he was succeeded by his son James Higgins Owen (d. 1891).

No doubt his father's work provided Jeremiah Owen with the opportunity for dockyard work and he became metallurgist to the Admiralty and storekeeper at Woolwich Dockyard, having almost certainly gained experience at Gospel Oak ironworks, Tipton, where his father had contacts. He served on the Admiralty Committee on Metals in the late 1840s and organised tests associated with these. Owen was elected a Member of the

Institution of Civil Engineers on 17 February 1829 and resigned in January 1831, when he was working for the Admiralty at Somerset House.

MIKE CHRIMES

[ICE membership records, ICE; DNB article on Jacob Owen]

Publications
1845. *Admiralty Committee on Dockyard manufacturing ... of Articles of Iron, Copper or other Metals*, 1st report (copper sheathing), 2nd report (iron)
1847. *Report on the Results of Experiments made ... on the Relative Strength of Common Cast Iron and toughened Cast Iron*

OWEN, Samuel (1774–1854), manufacturer, was born on 12 May 1774 at Norton-in-Hales, Shropshire, the son of a farmer, George Owen, and Catherine Owen. At the age of nine and a half he attended school for a short time but left at the age of ten and a quarter to work on a local estate as a shepherd or swineherd. He did this for two years, followed by other similar work until at the age of seventeen he got a job leading horses pulling canal barges. A year later he was apprenticed to a carpenter and began to attend evening classes. At the age of twenty or twenty-one he began working for Boulton and Watt in their Soho works, where he remained for four years, working as a pattern maker. While there he got married. With the help of the foundry master, Abraham Storey, he studied steam engines and developed his own ideas. One suspects he was then headhunted as after four years he moved to Leeds to work for Boulton and Watt's major rivals Fenton, Murray and Wood. Owen went to Sweden in 1804 to install four (steam) engines, which Baron Abraham Niclas Edelcrantz had ordered from Murray, at factories in Lidingo, Kungsholm, Ladeurgardsland, and Dannemora iron mines. On his return to England Owen worked for Arthur Woolf (q.v.). Meanwhile Edelcrantz ordered another steam engine from Murray for use at Eldkvarnen in Stockholm. Owen was approached by Edelcrantz to install this but when he arrived in Sweden he discovered Murray had brought three workmen to do the work. Owen, however, decided to remain in Sweden and in 1806 found work at the foundry of Wilcke and Erdman in Bergsund. This works had been established by the Scottish engineer, Thomas Lewis, c. 1769–1770. Lewis had been trained at the Carron Ironworks, but had advanced production at Bergsund to such an extent that in 1775 he was advising Charles Gascoigne on cannon boring technology for Carron. At Bergsund Owen designed the first Swedish built steam engines for Dannemora mines, and a sheet rolling mill for Kloster works in 1808 which established Owen's reputation as a machine designer.

In August 1809 Owen set up an iron foundry and machine works on Kungsholm. He obtained some English workmen and rapidly developed the works which became the leading manufactory in

Sweden; by the 1820s Owen was exporting goods as far afield as Rio de Janeiro. The products of the works comprised various foundry goods, steam engines, dredgers, and castings and iron work for bridges. The first iron bridge in Sweden had been imported in 1813 from England for the Gotha Canal. In 1815 Owen built the first Swedish iron bridge between Djurgården and Maniela in Stockholm; it was 5½ cu. wide (c. 20 ft.) and the top of the arch measured 13 cu. (c. 59 ft.). He built an arch bridge over the canal at Stromvatterren in 1832 and designed a suspension bridge of 186 m span over Norrstrom, 1830–1833. Although the iron towers were erected the proposal was abandoned and the towers sold for scrap in 1844.

Owen introduced iron boat construction to Sweden and in 1816 designed the first Swedish steamer, the *Witch*, with screw propeller propulsion, followed in 1817 by a paddle steamer. By the time he left his works in 1843 they had built thirty steamships, including several ironclads. In January 1843, virtually bankrupt, he gave up his works to his creditors and retired to property near Kalmar where he built a threshing machine and installed a steam engine to drain a bog. In 1847 he was appointed works manager by his friend, Martin von Wahrendroft, at Åker gun foundry where he remained until 1851 when ill-health forced him to retire.

Owen married three times. His first (English) wife Ann Spen Toft (1796–1817) died in 1817; they had six children of whom two sons and two daughters survived. His second marriage in 1817 was to Britta Carolina Svedell (d. 1821) from Hedemora; they had two children of whom a daughter survived. His third wife, Johanna Magdalena Elisabeth Strindberg, survived him. They were married in 1822 and of their nine children, six sons and two daughters survived. Owen died in Stockholm on 15 February 1854.

MIKE CHRIMES

[Manuscript autobiography in Gothenburg University Library; *Kongl Vetenskapsakademiens handlingar for ar 1853* (1855) 469–475; F Schütz (1975) Samuel Owen (1774–1854), *Daedalus*, 93–140]

Works

n.d. Dredger for the Waldo canal
1815. Manillaholm bridge, cast iron arch, c. 59 ft span
1832. Stromvatteren arch bridge
1830–1833. Suspension bridge over Norrstrom, Stockholm, 186 m (iron towers only erected)

OXLEY, Joseph (c. 1724–1813) was a miller, agent and engineer. Details of his early life and work have not yet emerged, but by 1753, he was based at Russel's Factory, seemingly a woollen mill by the Tyne at Benwell, Newcastle. He was also advising the Northumberland landowner and coalowner, John Hussey Delaval (q.v.), on a wide variety of topics ranging from mining and milling, to the planting of willows and rape seed. It is not known how Oxley's association with Delaval

came about, but it was to last a further forty-five years, with only minor differences ever seeming to appear between them.

While based at Russel's factory Oxley developed a system whereby a Newcomen pumping engine could be modified to give rotary output for shaft-winding purposes. Oxley, describing himself as a miller, patented this rather complicated device on 29 July 1763, (1763, No. 795), it being popularly described as a 'machine for drawing coals out of a coal pit by the help of fire'. William Brown (q.v.), then Delaval's viewer, had built a Newcomen engine at Hartley in 1760 and the first public demonstration of Oxley's device, presumably attached to Brown's engine, took place in October 1763.

The device was far from ideal, and Oxley had recognised some its defects, so that by early 1764, while still based at Russel's factory, he was already 'a good way on' with a second machine. Such modifications were clearly sufficient to catch the attention of James Brindley (q.v.), who expressed an interest in using and promoting it, and of many others. By March 1765, his second coal-winding machine was at work 'with incredible success', drawing coals out of the mine at the rate of a corf a minute. The mining companies continued to show great interest for, as Oxley informed Delaval, their horses were 'much Distress'd and Corn Dear'.

One of Oxley's machines, probably the second, was seen by the Frenchman Gabriel Jars in 1765 and James Watt, who later developed his own rotative systems, also visited the engine. Some forty years later he informed his son that the first reciprocating engine adapted to rotary power that he knew of, was one he 'saw at Hartley Colliery about 1768 ... It was employed to draw coals out of a pit, had no flywheel, and went sluggishly and irregularly'. Jars, after visiting Oxley's machine, suggested that it would be much better if the pumping engine simply delivered its water to a water wheel, which could in turn drive a winch for bringing up the coal, and whether by independent decision or influenced by Jars or some other experience, this was precisely the method of shaft winding upon which Oxley now embarked, seemingly the first person to do so.

A memorandum of October 1766 assigned Oxley's invention of a new engine to Delaval for the small sum of 10 guineas and he enrolled a patent on 13 March 1767, unfortunately without specification, for a 'Machine or Water Wheel for Drawing Coals or other Purposes by Water; which Machine also Counterballances Ropes made use of in Drawing Coals or Other Things' (1767, No. 871). By February 1767, a new winning was underway at Hartley, and a new fire engine had been built by Brown, making it the first to deliver its pumped water direct to a water-wheel winder. This system, sometimes known as a 'water whim' or 'water gin', was soon extensively adopted in the north of England.

Delaval persuaded Oxley to become chief agent at Ford, but in spite of his professed avowal of things mechanical, he continued his experiments

with winding engines and other devices and shortly after his appointment he began to construct a
horse-driven threshing machine, one of the first
people to do so; it was not so successful as the
1780 design by Andrew Meikle. Oxley was also
very much concerned with other estate matters:
the relocation of a fulling mill; the construction of
an iron forge; the development of the colliery; the
reconstruction of a lime works; the building of a
new brewery; the redevelopment of three water
mills; and the design of a new corn mill for the
Corporation of Berwick.

From 1774, his main field of activity moved to
the Hartley estates where, among other things, he
had certain responsibilities for the harbour and its
coal trade. By September 1800, however, Oxley
was in financial trouble and was declared bankrupt. He continued to live at Hartley until his
death in 1813, at the age of eighty-nine years.
His wife, Ann, had died in 1802 at the age of
eighty-two, and two of his five sons had also predeceased him; the survivors were spectacularly
unsuccessful.

STAFFORD M. LINSLEY

[*Delaval Papers*, Northumberland County RO; M.
G. Jars (1765) *Voyages Metallurgiques*, J. Bailey
and G. Culley (1794) *General View of the Agriculture of Northumberland* (repr. by Frank Graham,
1972); Rev. John Baillie (1801) *An Impartial History of Newcastle upon Tyne and its Vicinity*, Newcastle upon Tyne; John Farey (1827) *A Treatise on
the Steam Engine* (repr. in 2 vols., David &
Charles, 1971); J. P. Muirhead (1859) *Life of James
Watt with Selections from his Correspondence*,
2nd edn.]

P

PADLEY, Paul (1778–184?), surveyor, was born in
1778, the son of William and Elizabeth Padley. The
family were Quakers, and his father was a merchant in Swansea. Padley was the brother-in-law
of William James (q.v.), well known as an early
railway promoter and he is best known for his involvement in the preliminary surveys of the Liverpool and Manchester Railway; Padley had shown
early ability as a draughtsman and worked in
James's office. In 1821 James first became involved
in proposals for a railway between Liverpool and
Manchester and in 1822 Padley led a team of surveyors, including George Ernest Hamilton, Robert
Stephenson (qq.v.) and Hugh Greenshields, who
surveyed a possible route to demonstrate its
feasibility. Padley then began work on detailed
surveys. Unfortunately, James's financial affairs
were in a critical state and in early 1823 he was declared bankrupt. In these circumstances the provisional committee turned to George Stephenson
(q.v.) as engineer and in June 1824 he engaged
Padley to carry out the surveys. This led James
to claim later that he had in some way been

defrauded of his surveys. While Padley may have
given Stephenson access to work carried out for
James, he was only one of a number of surveyors employed by Stephenson and the earliest
Stephenson plan to be lithographed, of 20 November 1824, names Thomas Oswald Blackett
(q.v.) as surveyor. From the summer of 1824
Stephenson was involved in a number of railway
schemes and Padley was employed on the London
and South Wales scheme, as the Liverpool and
Manchester scheme was so well advanced. He is
last mentioned in connection with a proposal of
John Urpeth Rastrick's (q.v.) in the Birmingham
area of 1830.

Padley was a well-known surveyor and examples of his work survive covering parts of Cambridgeshire, Cheshire, Glamorgan, Kent, Lancashire, Staffordshire, Surrey and Warwickshire. He
was one of the most experienced surveyors to be
employed on the early railway schemes. Perhaps
his most famous pupil was John Errington
(1806–1862).

MIKE CHRIMES

[EMSP (1861) *The Two James's and the Two
Stephensons*; R. E. Carlson (1969) *The Liverpool
and Manchester Railway Project (1821–1831)*,
40–98; P. R. Reynolds (1977) Paul Padley, *Journal
RCHS*, **23**, 21–23; R. H. G. Thomas (1980) *The Liverpool and Manchester Railway*, 11–32]

PADMORE, John (fl. 1710–1739), mason and
engineer, of Bristol, is well-known through his
works which attracted the attention of his contemporaries in print. He was variously described
as 'ingenious'—by Ferguson in 1764—and having
a 'natural genius for mechanics'.

Padmore was possibly trained as a mason, as
he is so described with reference to work in the
Swansea area *c*. 1728, and this would fit his earliest known work, unlikely to have been his first
because of its scale: the construction of a dock
4 miles below Bristol at Sea Mills in 1712. Here
the River Trym enters the right bank of the Avon.
This dock, with a 35 ft. entrance lock, may have
been the first British enclosed dock to be completed for general cargo, although by 1720 its
poor transport links had led to its use being restricted to fitting out privateers. Extant remains
include two sluice gate openings and the entrance piers.

Details are lacking of Padmore's career over
the next few years. In 1717, Dr. John Lane, a
Bristol physician with business interests in non-
ferrous mines in Cornwall and Devon, and copper and lead works at Kidwelly and Neath Abbey,
established a new copper smelting works at
Llangyfelach, Landore, near Swansea. The buildings were erected by the mason, Daniel Rauleigh,
who had previously worked at Neath Abbey, and
Richard Kelick carried out the carpentry work.
Padmore spent at least fifty-six days on site; the
existence of a masonry aqueduct supplying the
water mill and sophisticated millwork suggests

John Padmore

the handiwork of an engineer of ability. He was recalled to South Wales by the new owners of the Landore mills, Morris Lockwood and Company, to erect a copper mill at Lower Forest or Fforest Isaf in 1728.

Before then he was involved on the Avon Navigation. In 1720 he was involved in a proposal for a floating harbour at Bristol with a dam and lock on the Avon below Hotwell House. Proposals for improving the Navigation itself had been revived in 1724, after receiving its Act in 1712. John Hore (q.v.) was engineer but Padmore supplied the design for the lock gates at Keynsham and was consulted about the location of the neighbouring wharf there; he remained involved through to the completion of the navigation in 1727.

The leading proprietor on the navigation was Ralph Allen (1694–1764), and it is through him that Padmore became involved in the scheme largely responsible for his posthumous fame: the Combe Down Tramway, the first English 'railway' to be described in any detail in print. Allen, while based at the Bath Post Office, developed the cross post system which avoided the need for mail to be sent via London. By farming the 'cross-post' contracts himself he made a vast fortune, allegedly £500,000, and invested the proceeds in a variety of concerns aside from the Avon, including Bath Stone quarries at Bathampton and Combe Down.

Perhaps as early as 1724 a wooden tramway was built to Bathampton and the opening of the Avon navigation in 1727 led Allen to consider a tramway from Combe Down to a basin on the Avon. Padmore improved on original designs by the recently deceased 'Mr Hedworth' and although Allen is known to have consulted north eastern mine owners about the design of wagons and tramroads, the design as executed, with narrow gauges, cast iron wheels and axles, and a long wheel base, owed much to Shropshire wagonways. Whether there were a direct borrowing of ideas or whether the scheme relied on Padmore's knowledge of Sir Humphrey Mackworth's Shropshire style wagonways is unclear. The wagons had cast iron spoked wheels with flanges and an ingenious braking system, which attracted the attention of Charles Labelye (q.v.), who described the system in Desaguliers' *Course of Experimental Philosophy* (1734).

Padmore displayed his ingenuity elsewhere on the tramway system with a number of cranes and it is as a crane builder that his contemporary reputation was probably most firmly based; a skill he had possibly developed at Sea Mills. What is known of this aspect of his work is concentrated in the 1730s. At Combe Down there was a crane, powered by a capstan, with a jib swinging through a semi-circle, to move heavy blocks. Padmore developed a reducing gear to control its slew. Another crane at Combe Down was used to place the stone for squaring while a 'rat-tailed' crane was used for lowering blocks at Dolemead wharf on the Avon. This contained safety devices to control lowering. Perhaps his most famous crane was the 'Great Crane' at Mud Dock in Bristol, a treadmill crane with three independent jibs supported on fourteen cast iron columns, with a ratchet to stop it moving backwards when loaded. It is of interest to note that a crane is featured in the portrait of him at the Victoria Art Gallery, Bath. It was erected in 1735.

It can safely be assumed Padmore carried out other, now forgotten works. He replaced the original waterwheel pump near Cobham erected *c.* 1698 by George Sorocold for Bristol waterworks, allegedly by an atmospheric steam engine, but more probably a rebuilt waterwheel. John Smeaton (q.v.) was consulted about the machinery in July 1786, and commended Padmore as 'a very able engineer', and felt it improper to alter it. Another example of his ingenuity was his attempt in 1731 to deal with the subsidence of one of the piers at St. Nicholas Church, Bristol, by using a cast iron 'machine' to screw up neighbouring pillars. Although unsuccessful, and more conventional remedies were applied, it is indicative of the man's ingenuity, although it can have done little to enhance his reputation.

Unfortunately, little is known of his personal life save that he was married, and had died by 1740, when his wife sold off another of his cranes.

MIKE CHRIMES and BRENDA BUCHANAN

[J. T. Desaguliers (1734) *A Course of Experimental Philosophy*, **1**, 120–122, 179–181, 274–279; (1744) **2**, 417; J. Wood (1749) *An Essay towards a*

Description of Bath; J. Ferguson (1764) *Lectures on Select Subjects*, 55; W. H. Tregellas (1885) *Ralph Allen*, DNB; D. E. Gibbs and R. O. Roberts (eds.) (1955) The copper industry of Neath and Swansea, *South Wales and Monmouth Records Society*, **IV**, 125–162; A. Elton (1963) Prehistory of railways, *Proceedings of the Somerset Archaeological and Nautical History Society*, **107**, 31–59; P. McGrath (1975) *The Merchant Venturers of Bristol*, 192–193; M. Watts (1975) John Padmore's cranes at Bath and Bristol, *BIAS Journal*, **8**, 17–19; W. J. Sivewright (1989) The ingenious Mr. Padmore, *PHEW Newsletter*, **42**, 2–4; B. J. Buchanan (1996) The Avon navigation, *Bath History*, **VI**, 62–87; S. R. Hughes (1999) *Coppropolis: the Early Industrial Landscape of Swansea*]

Works

1712. Sea Mills Dock, Bristol, £9,600
1718. Landore copper mills
1725–1727. Avon Navigation, lock design
1728. Lower Forest Mills, Swansea
1731. Combe Down Tramway, wagons, cranes, £10,000
1735. Mud Dock crane, Bristol

PAGE, Joseph (*c.* 1718–1776) trained as a bricklayer and became one of the leading master builders in Hull from the mid-1740s. Soon after work began on the Holderness Drainage scheme, designed and directed by John Grundy (q.v.), he was appointed in May 1765 to superintend the building of the outfall sluice and making the main drain, with a salary of £80 p.a. In other words he acted as resident engineer on these works and continued to do so until their completion in August 1767. When work was due to start on Hull Dock (also designed by Grundy, with Henry Berry (q.v.) of Liverpool as consultant on construction) it was hoped that Page would act as resident engineer, but the Commissioners had to inform Berry in February 1775 that 'from Mr Page's ill state of health and the many other Employments he has to attend to, he will not be able to give you that constant and daily attendance which the importance of this great work will require'. Instead, on John Smeaton's recommendation, Luke Holt (q.v.) was appointed. Page was admitted a member of the (Smeatonian) Society of Civil Engineers in 1774 but attended no further meetings. He died on 23 April 1776, aged fifty-eight, and was buried at Barton-on-Humber, his birthplace.

A. W. SKEMPTON

[*Minute Book of Holderness Drainage Trustees (1765–1768)*, East Riding RO; Mark Baldwin (1974) *The Engineering History of Hull's Earliest Docks*, TS, Science Museum Library; Colvin (3)]

PAGE, Sir Thomas Hyde, FRS (1746–1821), military engineer, was born in Harley Street, London in 1746, the son of Robert Hyde Page. Famous in his day—he was knighted by George III, and granted the freedom of Dover, Dublin and King's

Sir Thomas Hyde Page FRS

Lynn—he has been neglected by historians ever since. Trained at the Royal Military Academy, Woolwich, Page won the King's gold medal in 1765, and in 1769 was appointed Practitioner Engineer and Sub-Lieutenant, becoming a Lieutenant in 1774.

His first civil engineering commission came as early as 1770 when he produced a scheme for embankments at Langstone Harbour, near Portsmouth, using fascine work and involving the repair of a breach 150 ft. long and 10 ft. deep at a cost of £2,000. Attempts had already been made to repair the breach by the then conventional method of piles or planks, but this had blown up at the foundations, in Page's opinion because the softest part of the mud and water had been too much obstructed and confined within the bank. His technique was awarded a (Royal) Society of Arts Medal in 1782. He later applied his method to building breakwaters at Dover, Sheerness and Margate roads and to stopping a breach in the canal docks at Dublin. Less successfully he advocated its use for Dublin Harbour and the Fens.

He also had theories about types of sluices, particularly those with outfalls across moveable shingle or sand, blocked at certain times of the year by changes in prevailing winds or currents. His solution for such a sluice in the Isle of Wight was to construct a covered tunnel instead of an open channel with an outfall beyond the reach of the beach. Versions of this were later built throughout Hampshire.

In 1775 he was asked by Lord Townshend, the Master General of Ordnance, to 'take a view of the Bedford Level', which view turned out to be decidedly similar to that of his fellow military engineer John Armstrong (q.v.) fifty years earlier, namely that the basic solution lay in the restoration of the River Ouse to its function as the 'great and natural sewer of the South Level of the Fens'. He recommended confining the river to narrower bounds so that the increased power of the water would scour the river to the proper depth. Building new banks within the existing ones to the height of half ebb tide, with fascines and stakes was his chosen method.

Page volunteered for the war in America and fought at the battle of Bunker Hill as ADC to General Howe, where he was so badly wounded in the leg that it led to an amputation. His career in America was summarised by John Montressor, Chief Engineer in America, as 'Page served 11 days and was then wounded and returned home and had 10/– per diem settled for life'. However, within 3 years he was able 'to saunter without a stick' and it was not until twelve years later that he applied for a transfer to the Invalid Corps. On his return to England in 1777 he published an engraving of the battle of Bunker Hill from a ground plan by Montressor.

In 1778 Page was requested by Lord Townshend to consider the problems of water supply in forts and garrisons where a plentiful supply of fresh water was not available 'within the command of the cannons'. In Page's opinion the three cases which most signally failed this test were Sheerness, which had to have water brought from Chatham at a cost of £2,000 p.a., Landguard which had water piped in from a spring 2 miles away, and Harwich where there was a total lack of 'wholesome water' as the existing supplies were brackish from the infiltration of sea water and rendered bad by the presence of copper and other minerals. Page suggested in 1781 that the solution was to sink wells to the proper depths in each place and was ordered by the Board of Ordnance to carry out the task.

The work at Sheerness began on 17 April 1781 and was completed 4 July 1782. His first attempt, using established methods, failed and was described in the House of Commons as 'not a well for fresh water but a sink for the money of the public'. He changed sites, and used a new method wherein he marked out a circle of ground 22 ft. diameter, excavated to a depth of 5 ft. and placed a timber rib with boards pressing upon it. The well was slowly excavated through the treacherous ground using successive ribs to support the sides. Once firm marl was reached at 36 ft. a smaller diameter timber framed shaft was constructed, lined with brick, and the space between the two shafts backfilled with clay to make the whole watertight. The well was then taken down to a depth of 330 ft. when water was hit, rising 189 ft. in six hours and settling within days to a level of 8 ft. from the surface. The machinery to draw the water was powered by wind or, in the event of calm weather, by horses.

Page then moved on to Landguard Fort despite doubting whether it would be possible, given the sandier soil, to find a hard bottom as at Sheerness. Luckily while excavating a ditch for one of the batteries concurrently being erected to cover the approaches to Harwich, a small quantity of fresh water was found just a few feet from the surface; this water continued fresh to a depth of 18 ft. when it became entirely salt. Page had sand thrown in to raise the depth to 12 ft. and filtered out the salt water. He then had another well sunk 40 ft. away with the two connected by a horizontal brick drain with holes in the side for filtration. The bottoms of both wells were secured with hard materials so that the whole supply of water had to pass through the drain, filtering out any remaining sand.

The problem at Harwich was the reverse: the upper surface water was of bad quality and Page's solution was to cut through rock to avoid the contaminated land drains, hitting moist sand which, when bored through, produced a spring powerful enough to supply 'every public and private purpose in even the driest season'. It was this provision of a good water supply at Harwich which instigated the development of the garrison, dockyard and town.

Page was elected a Fellow of the Royal Society in 1783 and his account of the wells was published in the society's *Philosophical Transactions* in 1784, arguably the first account of such methods of construction. He had been knighted in 1783 and despite his opinion that it was for his active service in the war, it seems more likely that it was for his present engineering feats rather than past heroics, especially as George III had taken a personal interest in the work at Sheerness, visiting the site in 1781 and ordering him to continue as long as there was any possibility of success.

He was also Commanding Engineer at Dover from 1778 until 1782, when he was promoted to Engineer-in-Charge of the coastal districts. This involved drawing up plans and reporting upon the construction of batteries on the coasts of Suffolk, Kent and Sussex, in particular at Tilbury, Gravesend, Chatham and Sheerness, where he also applied his fascine theories to the breakwater. At the same time he was specifically In 1778, his first year in charge, he was allocated £4,100 for works at Dover, £2,000 for Dover Castle itself, and the rest for construction of new defences around the harbour and work on the Western Heights. There he designed a large obtuse angled bastion on the site of the Citadel. The allocation was originally to cover the construction of field works to protect port and town, which under successors such as William Twiss (q.v.) grew into the huge complex of fixed defences and earthworks estimated by the 1820s to have cost £30,000 and described as the result of 'either madness the most humiliating or profligacy the most scandalous'. The work proceeded so slowly, however, that in 1794, seven years after leaving Dover on health grounds, Page was instructed to work with Twiss to complete the works planned on the west hill; the work was considered to be of such national importance that in 1795 he was refused three weeks leave to attend to the Eau Brink disputes.

Page's interest in the harbour was a natural extension of his work on the harbour batteries, which were in a very exposed position and needed protection. He designed a fascine jetty or breakwater and considered this far easier than his earlier work on Langstone embankment because of the lack of backwater. Extra protection was given by sea beach, shingle or small stones thrown in among the fascines using the power of the waves to bond with the brushwood. The slopes of these jetties were graded to offer as little resistance as possible to the 'surf'. This work, completed in only six weeks, saved not only the batteries but a large part of the shore used for shipbuilding.

In 1784 he extended his interest to the rest of the harbour, publishing his *Considerations upon the State of Dover Harbour*. His solution was to reopen the Dour to navigation by dredging which would stop it being forced through the Pent and to make a new haven near the Maison Dieu, with new piers being constructed to guard the Dour to the east at Lord North's battery, using timber and fascines. During the height of the Napoleonic War he produced a revised proposal to drain the marshy ground by The Ropewalk to make the whole length of the Pent navigable. This was taken sufficiently seriously for measurements to be taken by engineers on site but, once again, it was rejected in favour of more work on the bar.

In 1792 'The King graciously approved his going to Ireland to assist the Engineer on the Royal Canal'. As well as stopping the dock breach, Page's work on the canal included drawing up plans for floating docks in the North Lotts (land parcels behind the quays) near the Custom House and for converting the bar of Dublin into an island. His initial involvement with the Royal Canal Company seems to have ended with his criticism of William Jessop's (q.v.) report for the Irish House of Commons in 1793 and he moved on to working on the Newry Canal and planning a lighthouse on the South Rock off Co. Down. However, his most prolific period in Ireland was 1800–1801 when he was commissioned to provide plans and reports on the harbours of Dalkey, Howth, Dublin and Wicklow, a harbour of refuge in Dublin Bay and a ship canal. He advised on the sinking of a well at the Pigeon House, inspected the Pipe Waterworks and 'took a view' of the Royal Canal to Mullingar.

For Dalkey he proposed a curved breakwater pier carried 300 yd. to the south-west into Killiney Bay to provide shelter against dangerous south south-east winds with a pier from the sunken rock in Dalkey Sound to Lamb Island to act as a quay for small ships and fishing boats. However, the area which he considered most suitable for the development of a harbour and ship canal was Sandycove Point which gave a square mile of anchorage in the best ground in the bay of Dublin, enabling ships trading with Dublin to enter the connecting ship canal from half flood to half ebb tide. His estimate for the works at Dalkey and Sandycove Bay was £1,014,600; it is interesting that he considered the harbour at Howth less practical because of 'the great expenditure of money necessary to execute works'.

Page was also consulted over the planned construction of a north wall or embankment from the east end of the north wall quay in Dublin as far as Poolbeg and his estimate for his scheme was £246,000 in comparison to John Rennie's (q.v.) £655,872, both well above the eventual cost of £95,000 when it was completed to the design of Francis Giles (q.v.) twenty years later. Page's plans were approved by Lord Cornwallis, the Lord Lieutenant, and the townsfolk of Dublin voted him the freedom of the city on 10 August 1801, and petitioned Cornwallis and his successors to implement his plans; no action ensued because of their cost. In these circumstances Page resigned, stating the Navigation Board 'having got his designs felt no desire for further services'.

His involvement in English affairs continued. In 1794, his views on the river Ouse having changed little in the intervening decades, he was called upon to give evidence against the Eau Brink Cut. As a sop to the Corporation of King's Lynn, it was agreed at a meeting with the Committee of Landowners that each side would appoint an engineer to 'be authorised and empowered to determine, mark and stake out the dimensions ... and direction of the Cut from the River Ouse at Eau Brink to the Harbour of King's Lynn'. Unsurprisingly Page was hired by the Corporation; Robert Mylne (q.v.), with whom he had previously worked on Margate Harbour, by the Committee. Specifically they were to draw up plans and schemes for drainage and navigation, with particular attention to the present Ouse between St. German's Bridge and Eau Brink and the outfall below the harbour and river at Lynn; £80,000 was allocated for the task. The unlikelihood of both engineers ever actually agreeing to anything was prophetically covered by the arbitration clause which allowed the employment of an 'able and experienced engineer by way of umpire'. In the event they could not even agree over the umpire but, surprisingly, they did manage to work in harness for almost a decade and agreed that the Ouse from Earith to Denver Sluice should be cleansed and scoured to the greatest possible depth consistent with the safety of the banks and bridges. They also agreed that the width of the forelands and the base of the embankment at the lower end of the Cut should be 278 yd. from bank to bank at the top. The line of the middle cut was staked out by December 1801 and this, unfortunately, was the last thing on which they could agree. Sadly more ink and money were devoted to the subsequent personal and legal aspects of the case than to any engineering progress made. In 1802 Page resigned from the Smeatonian Society, which he had joined in 1793, and of which Mylne was Treasurer, citing his being too busy to attend meetings. Mylne combined this resignation with aspersions as to Page's engineering skills, declared him to be incompetent for such a complicated question and for good measure accused him of duplicity. The matter came before the Court of King's Bench in January 1804 and a judgement was made in February 1805 but there was no agreement even over this, Mylne stating in his diary that all four judgements favoured his views whereas a more neutral view records the judge speaking strongly of Mylne's obstinacy. It was the view of Charles Hutton (q.v.) that Sir Thomas had acted with perfect candour in his proposals.

Page eventually agreed to Joseph Huddart (q.v.) as umpire after Mylne had turned down his counter proposals of Rennie, Jessop and Daniel Alexander (qq.v.). Huddart agreed to most of the line of the middle Cut, but then imposed his own solution on the points of controversy, but that it was not favourable to Page can be inferred from the fact that while Mylne and Huddart visited Limehouse together only four days later, Page resigned his appointment soon after; it was a sad end to thirty years involvement in the Fens.

In the meantime Page had kept his hand in as a military engineer by constructing a hard, a sloping roadway, across the foreshore of the Medway at Chatham dockyard and Upnor in 1803 and had been acting as a consultant for the Harbour Board of Jamaica since 1802. With more leisure he was therefore able to investigate plans for the Harbour at Montego Bay, for which the Board voted him a cask of rum. By May 1805 £26,000 had been spent on the Harbour but no more details are known of his involvement.

He visited the Thames 'Archway' tunnel works of Richard Trevithick (q.v.) in September 1808 with Daniel Alexander (q.v.) who recommended him that November as the person best able to advise on Trevithick's proposals to complete the tunnel following its failure. In February 1809 the Archway Company thanked him for his distinguished and zealous services.

Page died on 30 June 1821, at Boulogne, survived by his third wife, Mary Everett. He had married his first wife, Susannah Bastard—a widow—in 1777 and his second, Mary Woodward (d. 1794), in Canterbury Cathedral in 1783; he had five children by his second marriage. A portrait of Page, of c. 1780, exists in Boston. Formerly attributed to Joshua Reynolds it is now thought to have been the work of James Northcote.

SUSAN HOTS

[Roll of Officers of the Royal Engineers 1660–1898, PRO; Connolly papers, Royal Engineers Library; A plan of the present state of Sheerness, Royal Engineers Museum; DNB; *Burke's Landed Gentry*; *Minutes of Evidence … on the second*

reading of the *Eau Brink Drainage Bill* (1794); J. Huddart (1804) *Eau Brink New River, or Cut. Deed Poll, stating the Opinion of J. Huddart, in Pursuance of the Reference to him by Sir Thomas Hyde Page, and Robert Mylne, under the Powers of an Act of Parliament, 35th George III C.77*; Civis (1805) *A Letter addressed to the … Directors-General of Inland Navigation … respecting the Harbour of Dublin*; Royal Canal Company (1812) *Papers relating to the Royal Canal Company*; Anon. (1814) *A Letter to … Lord Viscount Whitworth … concerning the Harbour of Dublin, by 'A Dublin Merchant'; with additional Remarks by 'An Observer'*; H. Wells (1843) An account of the reclaiming and draining of land in the Bedford Level commonly called the Great Level, *Architect, Engineer and Surveyor*, **4**; J. P. Griffith (1878–1879) The Improvement of the Bar of Dublin Harbour by Artificial Scour, *ICE Min Proc*, **58**; W. Porter (1889) *History of the Corps of Royal Engineers*, **1**, 200–205; Major-General C. S. Akers (1886) Historical sketch of the fortifications of Dover; *Professional Papers of the Corps of Royal Engineers, Occasional Paper*, **12**; A. E. Richardson (1955) *Robert Mylne Architect and Engineer, 1733–1811*; W. Y. Carmen (1955) Sir Thomas Hyde Page engineer, *J. Army Hist. Res.*, **33**; V. T. M. and D. R. Delany (1966) *The Canals of the South of Ireland*, 94–95; D. W. Marshall (1976) *The British Military Engineers 1741–1783*, unpublished Ph.D. dissertation, University of Michigan; Trinity College, Dublin (1978) *Engineering Ireland 1778–1878: Exhibition Catalogue*; A. Hasenson (1980) *The History of Dover Harbour*; J. Coad and P. Lewis (1982) The later fortifications of Dover, *Post-Medieval Archaeology*, **16**, 141–200; H. A. Gilligan (1988) *A History of the Port of Dublin*; A. Saunders (1989) *Fortress Britain*, 123; R. Delaney (1992) *Ireland's Royal Canal 1789–1992*]

Publications
Publications include:

1775. *Observations on the Present State of the South Level of the Fens, with a Proposed Method, for the better Drainage of that Country, made by the desire of the Master General of the Ordnance, and the Honourable the Corporation of Bedford Level*
1775. *Observations on the Present State of the South Level of the Fens*
1777. *Observations upon the Draining of the South and Middle Levels of the Fens*
1782. Account of a cheap and effectual method of making or repairing Banks to prevent the overflowing of lands by the sea, or great rivers, *Royal Society of Arts, Memoirs of Agriculture*, **3**
1784. *Considerations upon the State of Dover Harbour, with its Relative Consequences to the Navy of Great Britain*
1794 *Estimate of the Expence of carrying into Execution the Plan of Embankment, recommended … for the Improvement of the Drainage of the South and Middle Levels, and other Lands*

having their Drainage through the River Ouse, and for the Improvement of the Navigation of that River ... Together with the Survey and Measurement of the Several Rivers upon which the Estimate is founded

1797. An account of the commencement and progress in sinking wells, at Sheerness, Harwich and Landguard Fort. London: John Stockdale. Originally published in *Philosophical Transactions*, Royal Society of London, **74**, 1784

1801. *Reports relative to Dublin Harbour and adjacent Coast*

1801. *Observations upon the Embankment of Rivers; and Land inclosed upon the Sea Coast*

Maps

1775. *Boston, its Environs and Harbour, with the Rebels Works raised against that Town in 1775, from the Observations of Lieutenant Page of His Majesty's Corps of Engineer, and from those of other Gentlemen.* Pen and ink and watercolour: scale 1:24,000; relief shown by Hachures and depths by soundings; in Library of Congress

1777. *A Plan of the Action at Bunkers Hill on the 17th of June 1775*

1783. *Chart of the Thames Estuary*

c. 1800. *A Survey of Dalkey Island and Coast from Killiney Bay to Dunleary upon a large Scale showing the proposed Works at Dalkey and Sandcove Point*

c. 1800. *Survey on a Large Scale of the Island called Ireland's Eye, with the Town of Howth and adjacent Coast showing proposed Piers, etc.* These and others relating to the Dublin coastal area were deposited in the office of the Director General of Inland Navigation; Their present whereabouts is unknown

Works

Works include:

1770. Langstone Harbour, repair of breach, 150 ft. × 10 ft , £2,000

c. 1780. Tunnel outfall, Isle of Wight

1781–1782. Wells at Sheerness, Landguard, Harwich

c. 1783. Breakwater, Dover Harbour

1792. Dublin canal wall; repair of breach

1795–1804. Eau Brink Cut, consultant

PAINE, James (1717–1789), architect, practised first in south Yorkshire and later in London. He underwent a seven-year grounding in construction work as supervisor of the building of Nostell Priory, a large house in Yorkshire, and in his later practice was required repeatedly to add estate bridges for the country houses which he designed or extended, experience which led to his building four considerable public bridges over the Thames.

As his biography and his architecture have been the subject of a long entry in Colvin's *Dictionary* (1995) and a detailed monograph (1988) by Peter Leach, this note deals strictly with the bridge works which are his contribution to civil engineering. Paine's own lavishly illustrated

account of the country houses and the bridges he had built, published in two volumes in 1767 and 1783, provides much of our knowledge of the bridges.

All the bridges we know of were built of ashlar. They were typically built with segmental arches, but in two instances he used semi-ellipses. All the bridges are symmetrical in elevation with their highest points over the crown of the middle arch; classical balustrades predominate, but there is some solid parapet in the larger, public bridges; there are niches over piers, some stone balls on parapets, and sculptures were placed on the cutwaters of the park bridge at Chatsworth. In two instances, the three-arch Cavendish Bridge over the Trent and the single-arch Beeley Bridge at Chatsworth in Derbyshire, the arches are built ribbed in the manner of medieval bridges, a surprising and probably unique way of building for the eighteenth century.

Paine's important bridges were four public bridges over the Thames, all linking Middlesex with Surrey. By the time they were built, from 1774 onwards, he was a justice of the Peace in both counties. He became effectively the county bridge surveyor for Middlesex and the designs of all four bridges appear to have been made by him. Surrey had previously adopted many more bridges as 'county bridges' than had Middlesex and so already had a county surveyor, an architect named Kenton Couse (1721–1790) who within his lifetime filled many public offices. Paine and Couse shared responsibility for supervision of Richmond and Chertsey bridges, which were county bridges. Those at Kew and Walton were privately owned and Paine seems to have been solely responsible for them, though he was assisted by his son, also James.

Surviving specifications and other documents of the four large bridges show a thorough understanding of the best current practice in bridge foundations. Paine clearly favoured the use of caissons, which had been introduced by Charles Labelye (q.v.) for Westminster Bridge and also used by Robert Mylne (q.v.) for Blackfriars Bridge, and also the placing of clay seals over arches to prevent slow saturation and decay of the masonry. His choice and construction of decorative elements was also of high quality.

TED RUDDOCK

[MS minutes of Richmond Bridge Commissioners 1773–1786, Richmond-on-Thames Central Library; Drawings of Richmond Bridge, BL King's Maps; Chertsey Bridge Papers, Middlesex County RO and Surrey County RO; J. A. Stonebanks (1969) *The Thames Bridges at Walton*, Walton and Weybridge Local History Society Papers, No. 4; J. Paine (1767, 1783) *Plans, Elevations and Sections of Noblemen and Gentlemen's Houses*, 2 vols.; T. Ruddock (1979) *Arch Bridges and their Builders 1735–1835*, 107–109; P. Leach (1988) *James Paine*, and refs. quoted; Colvin (1995); DNB]

Publications

1767. *Plans, Elevations and Sections of Noblemen and Gentlemen's Houses*, vol. 1
1783. *Plans, Elevations and Sections of Noblemen and Gentlemen's Houses*, vol. 2

Works

1755. Bridge at Wallington Hall. River Wansbeck, design and supervision, for Sir Walter Blacket, main arch semi-elliptical, others semicircular cost £300
1756–1757. Beeley Bridge, River Derwent at Chatsworth, Derbyshire, design and supervision for the Duke of Devonshire, one ribbed arch
1758–1761. Cavendish Bridge, River Trent at Shardlow, Derbyshire, for bridge trustees, three main arches, segmental and ribbed, contractor John Chambers, John Spratley and Joseph Bannister, cost £3,300, collapsed in flood in 1947
1760. A 'Gothic bridge' in the estate of Syon House, Isleworth, demolished
1760–1764. Bridge in Chatsworth Park, River Derwent, for the Duke of Devonshire, three arches, fine architectural detail including sculptures by Cibber, some since removed
c. 1770. Bridge in Chillington Hall Park, Staffordshire, one arch, small
c. 1770. Bridge in Weston Park, Staffordshire, one arch, small
1772–1774. Brocket Hall Park Bridge, Hertfordshire, River Lee, for Lord Melbourne, three arches over a weir
1774–1777. Richmond Bridge, River Thames, for the bridge commissioners established by Act of Parliament, joint surveyors Paine and Kenton Couse, five arches, the largest 60 ft. span, contractor Thomas Kerr and Edward Chapman, cost £15,000, widened 1937
1780–1785. Chertsey Bridge, River Thames, for counties of Middlesex and Surrey, joint surveyors Paine and Kenton Couse, five arches, the largest 42 ft. span, contractor Charles Brown, contract price including approaches *c.* £10,600, altered in the nineteenth century
1783–1786. Walton-on-Thames Bridge, for M. D. Sanders, design and supervision by Paine, four stone arches on one new and two existing piers, long approach of brick arches, collapsed in a flood in 1859
1784–1789. Kew Bridge, River Thames, for Robert Tunstall, design by Paine, supervision by Paine and his son James Jr., seven arches, the middle one 66 ft. span, contractor Charles Brown and John Haverfield, cost *c.* £16,500, replaced in 1903

PAINE, Thomas (1737–1809), political pamphleteer and promoter of iron bridges, was born at Thetford, Norfolk, in 1737. Raised as a Quaker and apprenticed as a stay-maker, his father's trade, he moved to London at the age of twenty. Eschewing his stay-making, he is reputed to have been a privateer, but later settled down as an excise man and shopkeeper. It was through his interest in science that he met Benjamin Franklin

and later emigrated to America in 1774. Two years later he published *Common Sense*, the first in a trilogy of political pamphlets upon which rests his controversial reputation. His politics and interest in science led to his association with Jefferson, Franklin, Whitney, Fulton, Rittenhouse and other natural philosophers.

By 1786 Paine conceived the idea of an iron bridge across the Schuylkill River in Philadelphia, having first considered a wooden bridge across the Harlem in 1785. He discussed technical details with Jefferson and others regarding bridge design, including the mathematics of the catenary. With the aid of John Hall, also of English birth and who had worked for Boulton and Watt, three models were built in 1786 to promote his bridge design. Failing to secure support in America, Franklin urged him to seek endorsement for his bridge model and plans by the French Academie des Sciences. Accordingly he sailed to France in 1787 and won the approval, in general terms, of the Academie. The next year he received a British patent (No. 1667, 1788), which showed some influence of his French experience. With an agreement with the Walker Brothers of Rotherham, a 90 ft. test span was constructed at their works in 1788–1789 and, following its success, they erected a 110 ft. span model iron bridge in London (1789–1791). Details were probably worked out by the Walkers and their foreman William Yates (qq.v.). It was intended to be sent to America, but the impecunious Paine could not raise the necessary funds. Walker Brothers repossessed the model bridge and wrought iron bars from the model were utilised in their great Sunderland Bridge, completed in 1796.

In 1791 Paine produced the first part of his *Rights of Man* and the consequent storm led to him returning to France where he remained for most of the 1790s. In 1797–1798 he developed two further bridge models, but details are lacking. In 1802 Paine returned to the United States to face a storm of public opposition for his *The Age of Reason*; dubbed an anti-Christ, he received a cool reception from Jefferson and others. The following year he made one last effort to have an iron bridge built and this marked the end of his career as a bridge builder. He died in poverty and without honour on 8 June 1809, in New York City.

EMORY KEMP

[W. H. G. Armytage (1951) Thomas Paine and the Walkers: an early episode in Anglo-American co-operation, *Pennsylvania History*, **18**, No. 14, 16–30; A. Williamson (1973). *Thomas Paine*; David Freeman Hawke (1974) *Paine*; E. L. Kemp (1978) Thomas Paine and his pontifical matters, *Newc Soc Trans*, **49**, 21–40; J. G. James (1987–1988) Thomas Paine's iron bridge work, 1785–1803, *Newc Soc Trans*, **59**, 189–221]

PAKENHAM, William (d. 1823), of the Ordnance Department, was elected a Member of the

Institution of Civil Engineers in 1820. He attended a few early Institution of Civil Engineers meetings.. No more is known of his involvement in civil engineering before his death on 11 February 1823. A Pakenham was involved in consultations over Dalkey Sound, near Dublin, with Sir Thomas Hyde Page (q.v.) in 1800.

MIKE CHRIMES

[ICE membership records, ICE archives]

PALMER, Henry Robinson, FRS (1795–1844), founder of the Institution of Civil Engineers, was born in Hackney in 1795, the son of Reverend Samuel Palmer, a non-conformist minister. His father ran an academy which Palmer attended before being apprenticed about 1811 to Bryan Donkin (q.v.) where he learnt draughtsmanship and mechanical engineering skills. In 1817 he began working for Thomas Telford (q.v.) and became his chief assistant in his London office. He remained with Telford for around seven years and helped with surveys for many of his works. An incisive portrait of him at this time is provided by Joseph Mitchell, one of Telford's pupils, who described him, c. 1821, as 'a good mechanician and surveyor, self complacent and indolent. He imagined himself a genius, but had no experience practicably of engineering works. Palmer was very kind to me ...'; Mitchell's remarks provide a good guide to Palmer's subsequent career.

Palmer's surveys for Telford included the Knaresbrough Canal and Railway (1818), Burnham marshes (1819), Archway Road, London (1820), Loose Hill and Valley Road improvements (1820), York–Bawtry Road, Barnsdale–Pontefract Road, Croft Bridge–Northallerton Road, Cockett–Lougher Road, Ely–St. Fagans Road, Derby–Stockport Road, Portishead Harbour (1824), Uphill Bay (1824) and Scilly Islands (1824). At Loose Valley, Palmer may have acted as Resident Engineer with Hugh McIntosh (q.v.) the contractor.

Palmer was heavily involved in the formative stage of the British railway network between 1818 and 1840 although his practical contribution to railway construction was minimal. In 1821 he took out a patent (No. 4618) for a monorail system with iron rails fixed to timber beams supported on columns of cast iron or timber. A model, built by James Jones (q.v.) was initially displayed at George Smart's (q.v.) manufactory. The system was employed in Deptford dockyard and to transport stone from a quarry near Cheshunt to the river Lea c. 1825. Such a system could only be successful in short lengths over a level (and stable) surface and was viewed at the time as rather impractical.

Of rather more importance was the associated work Palmer undertook on the resistance to tractive effort. Initially his study was restricted to data on early tramroads; this work was cited in Tredgold's work on railroads. However, in 1825 Telford sent Palmer to the north east to visit the Stockton and Darlington and Hetton Colliery railways. Palmer studied the performance of the locomotives and subsequently carried out a series of experiments to compare the resistance to traction of locomotives and horses on the Hetton Railway and also on boats, initially on the Ellesmere Canal. Subsequently a whole series of experiments was carried out on the Mersey and Irwell Navigation and the Grand Junction Canal, with Bryan Donkin, Peter Barlow, Benjamin Bevan and William Chapman (qq.v.) in attendance. These experiments were widely quoted and Palmer used them in Parliamentary evidence in support of navigation interests opposed to the Liverpool and Manchester Railway in 1825. They claimed to demonstrate that for speeds under 4mph resistance was less on canals, but above that the advantage lay with railways. In the primitive stage of railway traction at that time this could be cited as a concern, but by 1830 these findings had been overtaken by events, although Telford commissioned further experiments on a model canal in the Adelaide Gallery.

There seems little doubt that these studies were in response to Telford being approached about a number of early railway proposals, of which the Liverpool and Manchester was the best known; proposals also included lines from London to Birmingham and from Birmingham to Birkenhead, for Liverpool. Palmer was involved in 1825 in reporting on these for Telford but by then seems to have been striving to set up his own career.

Despite his evidence against to the Liverpool and Manchester Railway, repeated in the successful 1826 Parliamentary passage, Palmer does not appear to have held any animus against railways *per se*, and more or less simultaneously was acting for two important early proposals. The supporters of the Kentish Railway scheme initially persuaded Telford to act as their engineer when they launched their prospectus in 1824 but when he stepped down a short time later Palmer took over and surveyed the route, which began north of the Thames, crossed over near Woolwich, and continued close to the Thames Estuary in the Erith area, before continuing via Strood and across the Medway to Dover.

There were clear problems with the break in journey across the Thames and considerable misgivings about the bridging of the Medway, and money was never forthcoming for a complete survey of the route, although Palmer did survey it thoroughly to Strood, and also branches from Tunbridge Wells to Maidstone and Snodland. His knowledge of the route made him an obvious engineer to approach when the idea was revived in the 1830s. More or less contemporaneously with the Kentish scheme, Palmer was approached by the committee of the Norfolk and Suffolk Railway, subsequently the Norfolk, Suffolk and Essex Railway, which was an early proposal for a railway from London to Ipswich and onward to East Anglia. He was asked to survey the line in 1825 and by August 1826 had billed the committee for £1,573 in engineering fees. This scale of expense was unsustainable when few subscribers had come forward and the scheme was abandoned. Palmer contented himself with £750 out-of-pocket expenses. In these early surveys he employed the young George Parker Bidder (q.v.) as assistant.

Bidder acted as his assistant again when he was appointed Resident Engineer for works at London Docks in June 1826. The company had a scheme prepared by William Chapman for their Eastern Dock extension in 1824 but the original resident engineer, J. W. Hemingway, died in 1825, leaving an opening for Palmer whose position was further enhanced when Daniel Alexander (q.v.), who had overall responsibility, retired in April 1828; Palmer's salary then rose to £800 p.a. As engineer, Palmer was responsible for the supervision of the construction of the Eastern Dock and warehouses, Shadwell entrance—with communicating lock and basins—and a number of swivel bridges. In 1833, with the works substantially complete, Palmer's salary was reduced to £300 p.a. By the time he left the company in 1835 his work was under something of a cloud as there were problems with the stability of the entrance lock walls, the new entrance itself being difficult to navigate. These problems were resolved by George Rennie, with John Smeaton (qq.v.) acting as resident engineer; one explanation for the problem may have been that Palmer was too busy elsewhere to exercise proper supervision. Certainly by 1835 he was working as a consulting engineer with an extensive practice in Westminster from 18 Fludyer Street, and then Great George Street. Of the large number of canal, railway and harbour schemes he investigated there were relatively few which he saw through to execution.

While working at London Docks Palmer had taken out a patent (No. 5786, 1829) for the application of corrugated iron 'to the roofs and other parts of buildings'. This was apparently the first use of the term 'corrugated' in relation to iron. Although Palmer did not claim any originality in the idea of corrugating iron, practical problems in manufacture had delayed its mass production until the late 1820s. One can imagine that Palmer considered the idea while contemplating warehouses at London Docks. Palmer sold the patent on to Richard Walker of Grange Road, Bermondsey, who over the next two years perfected the manufacture of corrugated iron, subsequently employed in the docks' warehouses. As a development of the idea Palmer applied it to bridge decks. This may have been first developed for a bridge at Yarmouth where an Act was being considered in 1829; a suspension bridge was subsequently adopted. Examples of decks of this type are known to have been built at Swansea and, after Palmer's death, Peter William Barlow (q.v.) designed one for a road bridge over the South Eastern Railway at Tunbridge Wells. Aside from these examples, little is know of Palmer's bridge designs although he reported on cracks in Sydenham Bridge over the Croydon Canal.

Palmer had been approached frequently about other work when employed at London Docks. He was called upon to give evidence on most of the important railway proposals of the 1830s, generally in support of the successful parties. On the London and Birmingham Railway (1832), which he had surveyed with Bidder for the Company, he was asked about the prices and estimates, as well as his knowledge of relevant ground conditions on the neighbouring Holyhead Road. Much of his evidence on the London–Southampton Railway (1834) was concerned with his experience of contract procurement; his estimates were higher than those of Francis Giles (q.v.), the Engineer. This implied criticism of Giles became explicit when, in 1835, Palmer gave evidence in support of Isambard Kingdom Brunel's Great Western Railway and criticised Giles's survey of the Basing–Bath Railway. Palmer had recently left London docks and some attempt was made to discredit his evidence by questioning him about the failure of the entrance lock.

The railway with which Palmer is most commonly associated is the South Eastern Railway. Perhaps unsurprisingly, given his involvement with the Kentish Railway in the 1820s, Palmer was approached by supporters of a railway to Dover in 1833, and provided a rough route. With no capital forthcoming no detailed surveys were undertaken until 1834–1835 when a route was surveyed following the London–Croydon Railway, and continuing on to Oxted where the line divided, part going on to Brighton while the main line continued on to Dover. Palmer thus became involved in the scramble for Parliamentary approval for a route to Brighton.

The supporters of the Dover route soon realised that seeking Parliamentary approval for their Brighton branch was likely to delay their scheme and the South Eastern Railway, as it became known, concentrated on the route from London to Dover. Although Palmer gave evidence as the Engineer in 1836, and organised the surveys by P. W. Barlow, Froude, William Gravatt (q.v.), Montague Harrison, and F. W. Simms, it is clear from his evidence that his health had broken down. For his estimates he relied on calculations by other former Telford assistants, W. A. Provis and J. Macneill (qq.v.), possibly an indication of how busy he was, or incapable.

Soon after the Act was obtained William Cubitt (q.v.) was appointed Engineer, presumably because of Palmer's state of health. At the same time Palmer was involved in the ambitiously named England and Ireland Union, and the Ireland, South Wales, Gloucester, Cheltenham and London Junction Railways, in excess of 120 miles of railway surveys, taking a railway from the main Great Western line, with branches to Cheltenham, through to Cirencester, and from Gloucester to Llandovery via Hereford, and a branch from Brecon to Merthyr. These schemes floundered, presumably because of lack of capital for what was generally a rather unpromising route, although Palmer's state of health may have played a part.

Despite these setbacks in the railway arena, Palmer continued to work until *c*. 1842. From *c*. 1836–1839 he was involved in improvements to Penzance Harbour and from 1837–1842 at Ipswich Harbour. At Penzance he recommended a northern pier of granite 30 ft. in width with a cross wall enclosing a 30 acre floating dock. He designed the pier at an angle of 5° to the prevailing current, in accordance with his view on the effect of tidal currents. The approaches to Ipswich along the Orwell had been improved by Chapman and subsequently by Cubitt earlier in the century but by the mid-1830s there was a growing demand for dock accommodation. In 1837 Palmer designed a wet dock scheme and acted as Engineer until 1842 when his resident engineer, G. Horwood, replaced him and completed the work. At Port Talbot there was a shortage of capital and, rather than employ a large workforce to excavate the basin, at the suggestion of one of the proprietors Palmer excavated a channel through which land (flood) water could be directed to scour out the necessary waterway.

He also carried out surveys of harbours at Neath and Swansea, Briton Ferry, Dover, Folkestone (1826), Ramsgate, Fishguard, and Lowestoft. His harbour work provided further opportunities for Palmer to indulge his interest in scientific investigation. As engineer of London Docks he was concerned about the effects of the removal of old London Bridge on the behaviour of the river at the entrance to the Docks and he developed an instrument to help monitor the rise and fall of the tides, reporting on the results to the Royal Society. His near contemporary investigations at Folkestone led him to report again to the Royal Society on the movement of shingle along the south coast, based on observations at Sandgate, Folkestone, Dover and Sandwich. His canal schemes included a proposed improvement to the Mersey–Irwell navigation to enable it to accommodate sea going steamers (1840).

For such a prolific surveyor his engineering output was small and his most enduring achievement was the foundation of the Institution of Civil Engineers. As a pupil of Donkin he displayed an interest in self-improvement by setting up an early mechanics institute among his fellow workmen in Bermondsey in 1813–1814. By 1817 he was contemplating the more ambitious project of an institute for younger civil engineers where they could meet regularly in the evenings for discussions and professional self-improvement. With the support of his friend and late assistant, James Jones (q.v.), and several other young mechanical engineers he established the Institution of Civil Engineers in 1818 and at the first meeting on 2 January laid down objectives which remained fundamentally unaltered, with one notable exception, for the next two hundred years. The Institution—a model for similar professional engineering bodies world-wide—was essentially set up to encourage the study of civil engineering, with full membership restricted to practising civil engineers. Initially, to prevent reserve among younger members, membership was restricted to engineers under thirty but, while this answered a very real need for access to knowledge at a time when even a pupillage system was in its infancy, the age restriction was too burdensome, and was relaxed to enable Telford to be invited to take up the Presidency. Under his leadership, with Palmer as Vice-President for much of the time, the Institution began to prosper.

Palmer died on 12 September 1844 and his widow presented his papers, including more than four hundred drawings, to the Institution of Civil Engineers. Their subsequent loss has perhaps precluded a fair assessment of his work as an engineer but the foundation of the Institution is, in itself, a fitting memorial to him.

MIKE CHRIMES

[DNB (missing persons); Skempton; ICE membership records, ICE Council minutes, Minutes of Conversations, ICE archives; ICE Drawings register, ICE archives; Telford mss, ICE archives; Experiments on the resistance of barges moving on canals (OC 93, 1833), Minutes of Conversations, **44** (1833), ICE archives; London dock minutes and drawings, Museum of London in Docklands; P. W. Barlow, Palmer's system used at Tunbridge Wells (OC 764, 1844) ICE archives; *Proceedings of the Committee of the House of Commons on the Liverpool and Manchester Railway Bill* (1825); T. Tredgold (1825) *A Practical Treatise on Rail-roads*; *Minutes of Evidence taken before the Lords Committee … Railway from London to Birmingham* (1832); *Minutes of Evidence taken before the Lords Committee … Railway from London to Southampton* (1834); *Minutes of Evidence … House of Commons … Great Western Railway Bill* (1836); *Great Western Railway Bill; Minutes of Evidence … House of Lords* (1836); House of Commons (1836) *Minutes of Evidence … South Eastern Railway Bill*; Port of Ipswich (1837) *Civil Engineer and Architects Journal*, **1**, 75; F. Whishaw (1837) *Analysis of Railways*; F. Whishaw (1838) *Analysis of Railways*; F. Whishaw (1840) *The Railways of Great Britain and Ireland*; S. C. Brees (1840) *Railway Practice*, 2nd series, plates 12–13; E. Evill (1844) Description of the iron shed at the London Terminus of the Eastern Counties Railway, *Min.Proc ICE*, **3**, 288–290; Obituary (1845); *Min Proc ICE*, **4**, 6–8; H. W. Dickinson (1943) A study of galvanised and corrugated sheet metal, *Trans Newc Soc*, **24**, 27–36; C. Von Oeynhausen and H. Von Dechen (1971) *Railways: England 1826 and 1827*, 75–77; G. Herbert (1978) *Pioneers of Prefabrication*, 34–36; A. W. Skempton (1981–1982) Engineering in the Port of London, 1808–1834, *Trans Newc.Soc*, **53**, 87–88; R. Malster and R. Jones (1992) *A Victorian Vision: the Building of Ipswich Wet Dock*; Lewisham Public Libraries, 1998 (in lit.)]

Publications

1818. *Map of a Proposed Line of Navigable Canal from near the Town of Knaresboro … with a Line of Rail-road from Knaresboro … to Pateley Bridge*
1823. *Description of a Railway on a New Principle*, 2nd edn., 1824
1824. *Report on the Comparative Advantages of Canals and Railways*
1830. *Plan of the Harbour of Swansea showing the Proposed Improvements*
1831. Description of a graphical register [*sic*] of tides and winds, *Philosophical Transactions*, 209–213
1831. *Plan of London Docks on Completion in 1831*
1834. Observations on the motion of shingle beaches, *Philosophical Transactions*, 567–576
1835. *Map of the Line of the South Eastern Railway*
1836. *Reports on the Proposed Improvements of the Port of Ipswich*
[1840]. Report on the improvement of the Harbour of Penzance, *CEAJ*, **3**, 21–22
1840. *Report on the Improvement of the Rivers Mersey and Irwell, between Liverpool and Manchester, describing the Means of Adopting them for the Navigation of Sea-going Vessels.*

Works

1826–1835. London Docks, extension works
1836–1839. Penzance Harbour, improvements, *c.* £35,000
1837–1842. Ipswich Dock, *c.* £55,000
1840. Port Talbot Harbour
1840. Swansea Bridge

PALMER, William (fl. 1700–d. 1737), is probably the William Palmer who from 1700 worked as a watchmaker in partnership with Robert Warter of Crayke, near York. He appears to have invented the hydraulic ram and advertised as a builder of mills and 'engines'. His introduction to civil engineering and the first certain reference to 'William Palmer of the City of York, Engineer', as he describes himself in his will, occurs in 1720 to 1722 when he was employed as a surveyor on the Yorkshire Derwent navigation, work on which was being carried out by Joshua Mitchell (q.v.) for the Hon. Thomas Wentworth (later Earl of Malton). In October 1722 Palmer was engaged by Doncaster Corporation to survey the River Don below the town and Joshua Mitchell was brought in to help. Palmer and Mitchell took the levels or 'falls' and Joseph Atkinson (q.v.) measured the distances; the job took about a month. The original sketch map on a scale of about 2½ in. to a mile still exists. An engraving of a map on a reduced scale by 'Will. Palmer & Partners', with tables of the falls and distances, was published in 1723 together with a printed account of the methods proposed for making the navigation. The proposals had been presented by Palmer and Atkinson at a meeting in November 1722. During the next two years Palmer came to further

meetings in Sheffield, twice from York and once from London, receiving a total of £43 for fees and expenses. After lengthy negotiations the Cutlers petitioned Parliament for a Bill in March 1726, it having been decided to terminate the navigation at Tinsley, as the last 3 miles up to Sheffield were obviously going to be very difficult. Palmer gave evidence to the Committee, though he was prevented from doing so on a second occasion as he was 'in a feavor … and dangerously ill'. The Act was passed on 24 May and within a month work began on the first two locks above Doncaster. Payments between June and November 1726 were made to 'workmen about the locks, to labourers digging and to masons, and to William Palmer at several times' of £106 16s 9d. In fact he had made five site visits and was clearly acting as engineer directing the works.

Detailed records thereafter appear to be missing, but it may be assumed that Palmer continued on the Don navigation for the next three years. By 1731 at an expenditure of about £8,000 five locks, three of them on a 2½ mile cut, had been built on the 12 miles above Doncaster to Kilnhurst. They had masonry walls and timbers floors, an average fall of rather more than 6 ft., and could accommodate boats 55 ft. × 14 ft. carrying 20 tons.

Meanwhile Doncaster obtained an Act in April 1727. Atkinson gave evidence, supported by Richard Ellison. Presumably he went on to act as engineer for the lower part of the navigation. By 1731 Doncaster had spent about £3,500. The two sets of proprietors were amalgamated as a single company in August 1731 with John Smith (q.v.) as their engineer. Under his direction the navigation was completed in 1751: 33 miles from Fishlake Ferry to Tinsley with seventeen locks. An early nineteenth century canal extended the navigation from Tinsley to Sheffield: 3 miles with twelve locks.

Before work began on the Don, Palmer received instructions from York Corporation to survey the river Ouse with a view to improving its navigation. They had consulted Thomas Surbey (q.v.) in 1699 and sought a second opinion from John Hadley (q.v.). Both engineers recommended a lock and weir at Naburn as the principal requirement. York sent a petition to Parliament but the Bill faltered on its second reading and nothing more was done until 1725. Palmer produced an elaborate map of the river in December, accompanied by tidal data. The City succeeded in obtaining an Act in May 1727. Palmer received £50 for his survey and 10 guineas for attending the Parliamentary Committee in 1726.

No decision had yet been made as to exactly what works were required. The City again sought a second opinion, this time from John Perry (q.v.). His main proposal centred on several long cuts which would undoubtedly have been beneficial but their cost was beyond the available resources. The lock at Naburn was also a costly affair and by itself would not be sufficient without improvements in the river. Palmer also proposed a cut, about a mile long but even this was too expensive and he had to be content with measures to improve the river by making the tides scour a deeper channel. This he intended to achieve by groynes or training walls to make the flow more rapid, narrowing the channel at low water to 90 ft. from the original 200 ft. width.

Work on these lines started under Palmer's direction, he receiving a salary of £40 p.a. from May 1728 in addition to other special fees. Considerable improvements had been effected by 1732, when the navigation was producing £600 p.a. in tolls. But in places dredging had to be carried out and in 1734 Richard Wilkins, then in charge of Scarborough harbour, brought an 'engine' from Bridlington to 'plough' the river bed.

Francis Drake in his *Eboracum* of 1736 refers to these works by 'Mr Palmer, an engineer of our own growth, as I may call him' and says they cost £4,000 to £5,000. Eventually it became necessary to build the lock at Naburn. This was done by John Smith between 1753 and 1757. It provided a depth of 7 ft. up to York, but shoals downstream still limited navigation to Humber keels of 60 tons burden with a draught of 5 ft. The final improvement was brought about by a steam dredger in the 1830s.

Palmer's other consulting work included advice on Beverley Beck in 1726. With Richard Ellison he surveyed the rivers Swale and Ouse from Richmond to York in 1735 and finally, in 1736, he was consulted on improvements on the Aire and Calder Navigation.

His will is dated 12 January 1737, signed in a very shaky hand. He was buried on 24 January at St. Olave's Church, York.

A. W. SKEMPTON

[William Palmer (1722) MS map of river Don, Doncaster Council Archives; Anon. (1723) *The Methods proposed for making the River Dunn Navigable*; Francis Drake (1736) *Eboracum, or the History and Antiquities of the City of York*; J. D. Leader (1897) *Records of Burgery of Sheffield*; T. S. Willan (1936) *River Navigation in England 1600–1750*; T. S. Willan (1965) *The Early History of the Don Navigation*; R. W. Unwin (1967) The Aire and Calder Navigation, *Bradford Antiquary*, **43**, 151–186; B. F. Duckham (1967) *The Yorkshire Ouse*; Charles Hadfield (1972) *Canals of Yorkshire and North East England*; R. V. Wallis and P. J. Wallis (1986) *Biobibliography of British Mathematics and its Applications*]

Publications

1723. *A Survey of the River Dunn in order to Improve the Navigation ... to Doncaster and ... to Sheffield*
1725. *A Map of the River Ouze, from its Rise ... to its Falling into the Trent and Humber*
1735. (with Richard Ellison) *A Survey of the Rivers Swale & Ouze from Richmond to York, in order to Improve the Navigation thereof*

Works

1726–1731. River Don Navigation, £8,000
1728–1734. Yorkshire Ouse Navigation, £5,000

PARKES, Josiah (1793–1871), civil engineer, born in Warwick on 27 February 1793, was the third son of John Parkes, owner of a wool mill. Both he and his younger brother, Joseph (1796–1865) a radical politician, attended Dr. Burney's school at Greenwich, but while Joseph trained as a solicitor, Josiah, at the age of seventeen began work in his father's mill.

Parkes familiarised himself with mill machinery but in 1820 the mill was closed and he moved to Manchester. There he moved in the local scientific circles and investigated means of smoke prevention. He had already met leading members of the Institution of Civil Engineers and was elected an Associate of the Institution in 1823 while living in Manchester. He now abandoned his smoke research in favour of developing a salt refining process at Woolwich. In 1825 he moved to France, to Puteaux, on the Seine, where he set up a works and became involved in opposition politics. In the 1830 revolution, although his side was triumphant, his business was ruined and he returned to England.

The next phase of his career, in land drainage, brought him fame, although not necessarily personal success. He was interested in developing a system of 'steam cultivation' and was employed by John Heathcoat (q.v.) to drain part of Chat Moss, Lancashire. To do this he employed—unsuccessfully—steam power but his experiments in deep cutting of the bog were more successful. From these he concluded that the optimum minimum depth of drains beneath the surface was 4 ft., by which arrangement the surface layer would be drained, water rising by capillary action from below, leaving the surface layer dry, warm and friable. There ensued a prolonged discussion of the merits of Parkes's system and an earlier shallow drainage system developed by Smith of Deanston and it was not until 1843–1844 that suitable drain cutting tools and cheap, cylindrical clay pipes—invented by John Meade—became available and made Parkes's system practical. With the aid of government loans the drainage of heavy clay soils became widespread. Much discussion of his ideas is to be found in the publications of the Royal Agricultural Society.

Parkes himself was intolerant of advice and a poor manager of men; his own early schemes were poorly executed and other engineers like W. B. Denton made his ideas more affordable and practicable. His shortcomings, however, did not preclude him from acting as consultant to the Enclosure Commission. He was still actively engaged on government drainage work at Yaverland and Warden Point, Isle of Wight, from 1862 until 1869.

Parkes's other great contribution related to steam engine theory. His early (unpublished) papers to the Institution of Civil Engineers had dealt with bogs (1824) and salt manufacture but from 1837 he presented a number on steam power, including one on fuel consumption which was awarded the Telford Gold Medal in 1841. He was also a frequent contributor to the meetings of the Institution.

Parkes died at Freshwater, Isle of Wight, on 16 August 1871.

A. P. WOOLRICH

[Membership records and original communications, ICE archives; Obit. (1871–1872) *Min Proc. ICE*, **33**, 231–237; DNB]

Publications

Publications include:

1822. On the economy of fuel, *Q. J. Sci.*, **XIII**, 58–61

1822. Observations on the economical production of steam, and the consumption of smoke

1838. On the evaporation of water from steam boilers, *Min Proc ICE*, **I**, 17–20; *ICE Trans*, **II**, 160–150

1839. On steam boilers and steam engines, *Min. Proc. ICE*, **I**, 54–58; *ICE Trans*, **III**, 1–48

1840. *Report on A. M. Perkins Patent Steam Boiler*

1840. On steam engines, principally with reference to their consumption of fuel, *Min. Proc. ICE*, **I**, 6–14; *ICE Trans*, **II**, 49–160

1840. On the action of steam in Cornish pumping engines, *Min Proc ICE*, **I**, 75–78; *ICE Trans*, **III**, 257–294

1841. On the percussive or instantaneous action of steam and other aeriform fluids, *Min Proc ICE*, **I**, 149–150; *ICE Trans*, **III**, 409–439

PAROLLE (PERIOLE), David (fl. 1633–1649), civil engineer, was consulted in 1633, as one of the 'skilfullest Dutchmen in Marshland', by the City of York on improvements to the Ouse navigation; he reported in July. 'Marshland' in this context is the area of low ground through which the new 5 mile channel of the river Don (later known as the Dutch River) was made from Newbridge to Goole on the Ouse between 1632 and 1635. It formed the final part of the drainage of Hatfield Level by Sir Cornelius Vermuyden (q.v.). Periole was probably the Overseer, or resident engineer. In June 1649, during negotiations preceding the third campaign for draining the Great Level of the Fens, 'Mr Parolle' is mentioned as 'an Artist in Hatfield Chase' who might be of use in the project and from a letter written in September it appears that 'Mr Prowle' actually worked there for a short spell under William Dodson (q.v.). All these various spellings refer to the same man, since Abraham De La Pryme, the historian of Hatfield Chase, wrote in 1703 to a descendant of David Parolle, who had been a Surveyor in the Chase, for information on Vermuyden's son and two daughters so that he could get in touch with them. It seems that Periole (or Parolle in the adopted anglicised version of his name) had remained as Surveyor of the Level and may have been in charge of reparations of the sluices and banks, destroyed by rioters during the Civil War, which were ordered to be carried out in 1645.

A. W. SKEMPTON

[Samuel Wells (1830) *The History of the Drainage of the Great Level of the Fens, called Bedford Level*; J. Korthals-Altes (1925) *Sir Cornelius Vermuyden*; L. E. Harris (1953) *Vermuyden and the Fens*; R. F. Duckman (1967) *The Yorkshire Ouse*; K. Lindley (1982) *Fenland Riots and the English Revolution*; F. Willmoth, in lit., 1996]

PASLEY, Sir Charles William, FRS, General (1780–1861), Royal Engineers, was born on 8 September 1780 at Eskdalemuir, Dumfriesshire, Scotland, the son of a London merchant. He was educated in the school of Andrew Little of Langholm and later at Selkirk. Pasley entered the Royal Military Academy at Woolwich in 1796 and graduated the next year, receiving a commission in the Royal Artillery. In 1798 he was transferred to the corps of Royal Engineers and posted to Portsmouth. During the period 1799 to 1809, Pasley served on the east coast of England as well as in Minorca, Malta, Naples, Sicily and various European theatres of the Napoleonic Wars. He was severely wounded at the siege of Flushing during the Walcheren expedition and this incapacitated him for further combat duty. Thereafter, Pasley's career in the corps was to focus on the education of military engineers and building technology research.

He had already expressed his concerns about the state of Engineers' education when in 1810 Pasley joined a group of six Royal Engineers in forming a society for procuring useful military information. The next year, while in command of a company of royal military artificers at Plymouth, he instituted at his own expense a course of instruction for the non-commissioned officers and men in order to improve their knowledge of fortifications and fieldworks. In 1812 a Royal Warrant established Pasley's school permanently at Chatham, making him director, and opened instruction to junior officers of the Royal Engineers as well as the military artificers, now renamed the Royal Sappers and Miners. This new military educational institution was called the Royal Engineer Establishment. Pasley began with a teach-yourself system of training in which non-commissioned officers led the lessons. This method was influenced by Pasley's visits to the schools of Joseph Lancaster and the Reverend Andrew Bell, the two dominant figures in the primary education field in Britain. Pasley produced a three volume textbook, *Course of Instruction Originally Composed for the Use of the Royal Engineer Department* and began a professional library at the Establishment.

Pasley married Harriet Cooper in 1814 but she died after a only a few months and five years later he married Martha Matilda Roberts, by whom he had six children. His son, Charles Pasley (1824–1890), also served in the Royal Engineers, including an appointment as Commissioner of Public Works for Victoria, Australia and as Director of the Admiralty Works Department. As early as 1816, Pasley was elected a Fellow of the Royal Society and three years later he was appointed to a committee of experts to consider the Thames tunnel scheme of Sir Marc Isambard Brunel (q.v.). Well recognised as a man of practical science and possessed of high social standing as an army officer, Pasley was asked by Thomas Telford (q.v.), a life-long friend, to become a member of the recently founded Institution of Civil Engineers in 1820. For the rest of his career, Pasley participated regularly in discussions and presented papers at meetings of the Institution.

In 1825, following upon orders from the Master General of the Ordnance, the Duke of Wellington, Pasley began an architectural course at the Establishment. He developed another textbook, *Outline of a Course of Practical Architecture*, and appointed a civilian clerk of the works as instructor. Pasley was to serve for twenty-nine years as director of the Royal Engineer Establishment and from 1839 to 1855 acted as a public examiner at Addiscome, the military college which trained engineer officers for service in India. He thereby became a major formative influence on the education of British military engineers in the nineteenth century.

Pasley's position as director of the Establishment induced him to pursue research into building technology. Initially, he undertook experiments with wooden models to determine the stability and most efficient form of retaining walls. Pasley published the results in the textbook for the architectural course and in his *Course of Elementary Fortification*, the century's first fully comprehensive English work on the subject. In 1826 he began experiments in an attempt to make an artificial Roman cement and four years later succeeded; he tried his cement in structures at the Brompton Barracks, Chatham. He also printed an essay on his research and distributed it to all Royal Engineer stations at home and in the colonies, and to engineer officers of the East India Company; by 1836 his cement was equal in quality to the best of its kind. Pasley's achievement was founded on a prodigious programme of testing in which he used a variety of standard methods of the time for determining the tensile strength of mortar. In 1838 Pasley published his master work, *Observations on Limes, Calcareous Cements, Mortars, Stuccos and Concrete etc.*, wherein he described his experimental findings and gave an historical account of cementitious materials unequalled in the nineteenth century. The book was highly commended by the technical press and reprinted in 1847. Pasley's principal influence on the manufacture and use of cementitious materials was in the promotion of Roman cement, both natural and artificial, establishing its superiority over ordinary lime and supporting an English preference for its use over hydraulic limes in civil engineering work.

General Sir Charles William Pasley FRS

As an adjunct to his cement experiments, Pasley was involved in the testing and promotion of hoop iron reinforced cement bond brickwork as a substitute for bond and chain timbers in brick walls and in door and window lintels. His experiments, which began in 1837, were modelled on the testing to destruction of brick beams undertaken respectively by Marc Isambard Brunel and Messrs. Francis and Sons, cement manufacturers. Pasley was credited with resolving the debate between the two over the effectiveness of hoop iron in reinforcing the test beams; he concurred with Brunel that the material greatly increased strength. Pasley's 1838 book became the recognised authority on the subject.

Another achievement of Pasley's experimental work at the Establishment was with the use of military mines in blasting under water. Between 1838 and 1844 he carried out the removal of shipwrecks which obstructed navigation at Gravesend, Spithead and St. Helen's. In 1841 Pasley was appointed Inspector-General of railways. This government post had been created only the year before and he became an easy target for criticism by civil engineers who objected to such state regulation. Nevertheless, Pasley kept a diary during his five-year term and it demonstrates that he overcame his lack of practical knowledge of railways by great industry and conscientiousness. His reports on slips in clay cuttings are particularly valuable.

Pasley was made a KCB in 1846, became Colonel-Commandant of the Royal Engineers in 1853 and a general in the army in 1860. A portrait of him, by Eddis, hangs in the Royal Engineers Headquarters Officers Mess at Brompton Barracks, Chatham.

Pasley died on 19 April 1861 at his home, 12 Norfolk Crescent, Hyde Park, London, from an ailment of the lungs.

JOHN WEILER

[Pasley Papers, Add MS 41766, British Library; Pasley Papers, Add MSS 41989–41992, British Library, Sir Charles Pasley's Diary as Inspector General of Railways 1841–1846; Obit. (1861–1862) *Min Proc ICE*, **21**, 545–550; Obit. (1863) *Proc R Soc*, **12**, xx–xxv; W. Porter (1889) *History of the Corps of Royal Engineers*, **2**; G. R. Redgrave and C. Spackman (1905) *Calcareous Cements*; J. C. Tyler (1929) *General Sir Charles William Pasley*; P. H. Kealy (1930) *Sir Charles William Pasley*; H. Parris (1960) A civil servant's diary 1841–1846, *Public Administration*, **38**, 369–380; H. Parris (1965) *Government and the Railways in Nineteenth Century Britain*; R. J. M. Sutherland (1976) *Pioneer British Contributions to Structural Iron and Concrete: 1770–1855*, in C. E. Peterson (ed.) *Building early America*, 96–118; A. J. Francis (1977) *The Cement Industry 1796–1914: A History*; J. R. E. Hamilton-Ballie (1980) Nineteenth century concrete and the royal engineers, *Concrete*, **14**, 12–16, 18–22; J. M. Weiler (1987) *Army Architects: Royal Engineers and the Development of Building Technology in the Nineteenth Century*, unpublished D.Phil. thesis, University of York; A. D. Harvey (1991) Captain Pasley at Walcheren, August 1809, *Journal of the Society of Army Historical Research*, **69**, 6–21; A. W. Skempton (1995) Embankments and cuttings on the early railways, *Construction History*, **11**, 33–49; DNB]

Publications

1814. *Course of Instruction Originally Composed for ... the Royal Engineer Department*, vol. 1
1817. *Course of Military Instruction Originally Composed for ... the Royal Engineer Department*, vols. 2 and 3
1818. *Standing Orders of an Establishment for Instruction ... the Royal Engineer Department in their Duties in the Field*
1822. *A Course of Elementary Fortification*, 2nd edn., 2 vols.
1826. *Outline of a Course of Practical Architecture Compiled for ... the Royal Engineers*
1830. *Observations Deduced from Experiment Upon the Natural Water Cements of England and on the Artificial Cements ... Used as Substitutes for Them*
1836. *Improved Systems of Retaining the Earth of Military Mines and Blasting Under Water*
1838. *Observations on Limes, Calcareous Cements ... the Same Subjects*
1839. Description of the state of the suspension bridge at Montrose ... gales of wind, *Transactions of the Institution of Civil Engineers*, **3**, Pt. 3, 219–227

PATERSON, James (fl. 1807–1826), of Montrose was one of the proponents in the often vitriolic debates which took place in the 1820s about the best methods of making roads. In 1807 he surveyed the Edinburgh to Musselburgh road, where there was a dispute between the trustees and the contractor. He built thirty-eight bridges and 21 miles of road for Commission for Highland Roads and Bridges, most probably the Glendaruel and Riddan roads in Argyllshire, and then worked as a surveyor at Tynebank, near Edinburgh, where he obtained a reference as honest, sober and active before being appointed Surveyor of the Dundee–Montrose road in 1815.

Within eighteen months of his arrival at Montrose, Paterson had spent £7,000 in improving the road. He was unusual in his time in that he applied different specifications, distinguishing between roads on level ground and inclines and on dry or wet formation. His construction depth varied from 7 to 14 in. or more, with a bottom layer of larger stones in the deeper cases only. He agreed with J. L. McAdam (q.v.) that gravel was unsuitable and with Benjamin Wingrove (q.v.) that hard stone broke down too much. With R. L. Edgeworth (q.v.) he was an advocate of wider wheels, as doing less damage to roads, an idea which found official favour for a time but proved difficult to enforce in the long term.

He subsequently was appointed to four other roads so that by 1826 he had charge of 109 miles of road. In that year the Metropolis Roads Act was passed, and Paterson applied for a position as one of the four General Surveyors proposed. In the event a single appointment was made, being filled by (Sir) James McAdam. Having already published a withering attack on the McAdam system, denying him with some justification any priority of invention but using also words like 'quackery', this must have been most galling.

P. S. M. CROSS-RUDKIN

[Tracts 8vo 4, 8vo316, ICE]

Publications

1819. *A Practical Treatise on the Making and Upholding of Public Roads ... and a Dissertation on the Utility of Broad Wheels and other Improvements*
1822. *A Series of Letters and Communications, addressed to the Select Committee of the House of Commons, on the Highways of the Kingdom*
1824. *M'Adam and Roads*
1826. *A Circular Letter for distribution amongst the Honourable Commissioners of the Metropolis Roads*

Works

?1811–1812. Highland Roads and Bridges, Surveyor and Contractor
1815–. Montrose–Dundee Turnpike, Surveyor

PATERSON, John (fl. 1794–d. 1823), civil engineer, is first mentioned in the 1790s as resident engineer under John Rennie (q.v.) on the Crinan Canal. The idea of a canal for small sea-going vessels from Loch Crinan to Loch Gilp was first advocated in the 1780s by John Knox, and then Robert Fraser for the British Fisheries Society. In

1792 Rennie was asked to report on a suitable route and the following year an Act was obtained. From the first the scheme was bedevilled by lack of funds and Paterson's reports also draw attention to problems in obtaining satisfactory labour and materials. Intended to be completed in 1796 construction continued throughout the 1790s. Although opened in July 1801, the canal was incomplete and as the company looked for alternative funds Rennie obtained for Paterson another post at Leith. Following the 1804 Act, James Hollinsworth (q.v.) was appointed in 1805 as Paterson's successor.

Paterson was resident engineer in charge of the works at Leith docks from *c.* 1803. Although for many centuries Scotland's leading port, little significant engineering work had been carried out there aside from improvement of quays and piers established in the sixteenth century, together with the construction of small shipbuilding docks in the eighteenth century. Proposals were made by John Adam and Robert Whitworth (q.v.) in 1786 and 1787 and an Act of Parliament was passed in the latter year to sanction Whitworth's scheme. Whitworth's proposals languished, however, and it was not until Rennie reported on a scheme involving docks in 1799 that anything was achieved. Rennie's proposals involved an extended north pier, docks of 31½ acres in the sands off North Leith, and an outer harbour of 18 acres at Newhaven. This ambitious scheme was restricted to the construction of two docks of 10½ acres, two dry docks, and extensive warehouse facilities at North Leith. Paterson was placed in charge of this work (and has been erroneously credited with the design of the dockside warehouses). The works were completed in 1817 and. Although his eyesight was then failing, he retained the post of dock engineer until his death.

From May to December 1809, Paterson was Superintendent of Public Works in Edinburgh and was Surveyor to the Heriot Trustees in 1809–1810. He should not be confused with the contemporary architect, John Paterson (d. 1832), an associate of Robert Adam, with an extensive practice in Scotland and northern England.

Paterson seems to have acted as Rennie's senior 'resident' engineer in Scotland; their frequent correspondence was almost social in its nature, and displayed Paterson's fierce criticism of Thomas Telford and Robert Stevenson (qq.v.).

Paterson had married Elizabeth Jameson and he died on 4 January 1823; he left an estate of £4,096 13s 6d.

MIKE CHRIMES

[Crinan canal records, SRO; Rennie letters, NLS; Rennie reports, ICE Library; Rennie sketchbook, ICE Library; *Scots Magazine* (1823) 256; Sir J. Rennie (1854) *British and Foreign Harbours*, **2**; Skempton; Colvin (3); J. Lindsay (1968) *The Canals of Scotland*; W. Johnstone (1997) *Scottish Engineer and Architects*, 2nd edn.]

Publications

1810. *Report to the Lord Provost of Edinburgh and the Joint Committee for Superintending the Construction of the Wet Docks at Leith*
1814. *Observations occasioned by the Mineral Survey and Report of Robert Bald, on the Proposed Level Line of Canal between Leith, Edinburgh, and Glasgow, surveyed by Mr. Rennie in 1798, in which a Comparison is attempted to be drawn between the Utility of that Canal and of the proposed Union Canal*

Works

1794–1802. Crinan Canal, 9 miles, £141,000, Resident Engineer
1803–1817. Leith docks, 10½ acres, £260,000, Resident Engineer

PEACAN, William (fl. 1826–1830), civil engineer, of Southampton Row, London, was elected an Associate of the Institution of Civil Engineers in 1826 and remained in membership only for a few years.

It is not known if he was related to J. Peacan who worked for the Ordnance Survey in 1832.

MIKE CHRIMES

[ICE membership records, ICE archives; Bendall]

PEAR, Thomas, was the name of three successive generations of fenland surveyors and civil engineers, who originated in Cambridgeshire but whose principal business was conducted from Spalding, Lincolnshire.

Thomas Pear I (fl. 1773–1795) was Assistant to William Elstobb (q.v.) during the construction of Kinderley's Cut on the River Nene in 1773. He also superintended the construction of the New South Holland Drain, a channel 14 miles long to drain much of the land between Spalding and Sutton Bridge. It was built in 1793–1796 and completed by his son, which appears to fix the date of his death, though the record of it has not yet been found.

Thomas Pear II (*c.* 1770–1858) was born in Cambridgeshire, possibly at Elm near Wisbech, though his name does not appear in the parish registers. He married Katherine Howard on 25 April 1793 at Wisbech, where a son, Thomas III, and two daughters were born. By 1826 he was living at London Road, Spalding, while in 1830 he was described as a builder and architect of London Road and High Street there.

As Superintendent of the North Level of the fens, he was much involved in drainage schemes though ultimate responsibility was often assumed by national engineers. He reported on the drainage and navigation of the River Welland and drainage of Deeping Fen in 1815, proposing a new cut from below Fosdyke to cost £50,000, but this was abandoned because of opposition by William Chapman (q.v.). A further report on the same area in 1820 recommended using a steam engine for drainage and this was executed by Benjamin Bevan (q.v.) in 1823. His scheme of 1817 for a navigable canal along the line of Ouse

Mere Lode from Billingborough to the South Forty Foot Drain in Lincolnshire was never made.

Pear's principal work was the North Level Main Drain, a line 8¼ miles long from Clough's Cross past Tydd to the recently completed Nene outfall, to replace the 12 miles of the Old Shire Drain. On the basis of his report of 1828, an Act was obtained the following year with Thomas Telford (q.v.) as consultant. Working in collaboration with William Swansborough (q.v.) Pear supervised the works which, when completed, provided for the first time within living memory a natural drainage from much of the North Level.

Pear 1842 took his son into partnership by 1842 and had retired some time before his death at London Road on 14 October 1858.

Thomas Pear III (1796–1873) was baptised at Wisbech on 14 November 1796 and began working for his father at the age of 16. He also worked for Sir John Rennie, Thomas Telford, James Walker and William Cubitt (qq.v.) at various times. He died on 14 August 1873 and his will was proved at £450. He was unmarried and the succession to the business fell to his nephew, John Kingston, who lived with the Pear family at Spalding while under training and later became Superintendent to the River Welland Trustees (by 1866) and County Surveyor of Parts of Holland on the creation of County Councils in 1888.

P. S. M. CROSS-RUDKIN

[Pigot's Directory (1830); HLRO 1841; W. H. Wheeler (1896) *A History of the Fens of South Lincolnshire*; Cambridgeshire RO; Lincolnshire RO]

Publications

1820. *The Report ... on the Improvement of the Drainage of Deeping Fen and adjoining Commons*
1825. *Survey of the Shire Drain*
1826. *Plan of Bare Sands and inclosed Marshland between Kinderley's Cut and the Sea*
1828. *Preliminary Report and Estimate of the Means of Improving the Discharge of the North Level Waters from Clow's Cross to the Sea*
1828. *Plan of the Marshes, Bare Sands, Channel and Lands lying between Kinderley's Cut and the Sea made by Thomas Pear and referred to in the Nene Outfall Act of 7th and 8th Geo 4th c85 1828* [with W. Swansborough[*Report on Conducting Waters of North Level and Portsand to Clows Cross*
1828. *Plan and Section of Proposed New Drain from Clow's Cross to the New Outfall near Buckworth Sluice*
1829. *Plan and Section of New South Eau Drain from Black Horse Sluice to Clows Cross Sluice and also of the Proposed New North Level Drain from Clows Cross Sluice aforesaid to the Nene Outfall Cut*
1834. *Plan of Marshes ... allotted by Thomas Pear under Nene Outfall Act*
1835. *Plan for Tunnels and Slackers at White Hart Bridge*

Works
1806. Lutton Leam Inner Sluice
1829. North Level Sluice, now demolished
1830–1834. North Level Main Drain, including Clough's Cross Bridge, 1833

PEARSALL, Thomas (*c.* 1759–1825), was the son of John Pearsall (1716–1777) and grandson of John Pearsall (1683–1762), Quakers and owners of an iron rolling mill and foundry at Willsbridge on the (Bristol) Avon. John Pearsall Sr. had come from Rowley Regis to Bristol in 1712; apparently after losing money gambling he underwent something of a religious conversion and married a Quaker in 1714, purchasing land at Willsbridge around 1716. He developed his mill producing iron sheets and iron bars *c.* ½ in. in width for supplying forges and farriers in the Bristol area, using imported Russian iron. The firm was closely associated with the Winwood family, ironmasters of Bristol.

In 1811 Thomas Pearsall took out a patent (No. 3503) entitled *A Method of Constructing Ironwork for certain Parts of Buildings*; it described a wrought iron roof truss comprising 'rafters' of wrought iron *c.* 3¾ in. wide (deep) and 1/8 in. thick, with diagonally intersecting stays *c.* 1¾ in. wide and ¾₁₆ in. thick for a 25 ft. span roof; for spans above 40 ft. an additional arched stay was used. The patent was most famously and successfully used for the 53 ft. span roof of the Theatre Royal, Plymouth (1811–1814), designed by John Foulston (q.v.). It had trusses at 7 ft. centres and 3 in. × ³⁄₁₆ in. rafters at 1 ft. centres.

In February 1812 Pearsall issued a pamphlet promoting the system and listing a number of structures where it had been used. His system clearly came to the attention of leading engineers and architects as Pearsall and Winwood, in late 1813, were supplying ironwork for John Rennie (q.v.) for the rum sheds at West India Docks. The eastern roof failed in October 1813 and the western roof was propped up; in January 1814 Pearsall and Winwood were instructed by Rennie to repair the roof or remove it. Although it was strengthened, Rennie was still concerned a year later as some parts showed signs of buckling, as a result of which he could not certify payment. In 1816 the roofs were replaced by structures designed by Rennie and supplied by Butterley (and Horseley). These were large roofs and unsurprisingly put Pearsall under a large financial strain with the result that on 12 April 1815 he was declared bankrupt. John Winwood was a major creditor and tried to sell Pearson's works in May 1816, together with the patent rights. The mill sale was unsuccessful and Winwood later converted the works to flour mills.

Pearsall himself slid into obscurity.

MIKE CHRIMES

[J. G. James Collection, ICE Library; Rennie Reports 1813–1816, **7–9**, ICE Library; West India Dock Company minutes and drawings, Museum of London in Docklands; *New Monthly Magazine* (1815) **3**, 389; A. P. Woolrich (1986) *Ferrier's Journal, 1759–1760*; M. Tucker (1989) Thomas Pearsall's wrought iron construction, *CHS Newsletter*, **19**, 3–6]

Publications

1812. *Wrought Iron Roofs, and other Inventions of Wrought Iron Work for divers Purposes in Buildings*

Works

c. 1811–1814. Theatre Royal, Plymouth
c. 1811. H. Mandrell's premises, Heddington, Wilts., including roof joists
c. 1811. J. Merewether's Blackland Mill, Calne, Wilts., roof
c. 1811. Reynolds and Company, iron warehouse, Bath Quay
1811. Mangeon's riding house, Clifton, Bristol
1811. Winwood and Company, iron foundry roof
1811. Pearsall's iron works, Willsbridge
c. 1813–1815. Rum warehouse roof, West India Docks

PEIRSON, Ralph (fl. 1671–1709), Surveyor and Expenditor of the South Level, part of the Great Level of the Fens, known as Bedford Level. He was appointed to this post in 1671 at a salary of £100 p.a. with responsibility for maintenance of the rivers, drainage channels, banks and sluices, and for overseeing any new work. In April 1682 he reported that the sluice on the Great Ouse at Denver should be rebuilt in brickwork with three arched openings, the middle one to be 18 ft. wide and the other two 12 ft. wide. The old sluice, timber-framed with three arches giving a total waterway of 30 ft., had been built 1655–1656 to designs by Jonas Moore (q.v.) as the last major work in the drainage of the Great Level under the general direction of Sir Cornelius Vermuyden (q.v.) Peirson's proposal was approved, and work began in June 1682. The cofferdams and piled foundations were built by direct labour or 'day work' with Nathaniel Browne in constant attendance on site as supervisor at 15s per week.

The superstructure was to be built by contract, and Sir Thomas Fitch (q.v.) was consulted on its design. He delivered his 'proposals' in December 1682. The main recommendation appears to have been that all three arches should be 18 ft. wide. A week later a contract was made with John Hayward of Southwark, carpenter, to build the sluice 'according to the Platt or Draft hereunto annexed, or otherwise as Sir Thomas Fitch shall direct'. Work continued on the foundations, now presumably being enlarged. Fitch intervened to say that the piers should be 1 ft. wider and 4 ft. longer than in the contract articles. The foundations were completed by the end of May 1683, at a cost of £1,774. Work then began on the brick piers and abutments and making the six pairs of mitre-gates. John Fitch, younger brother of Sir Thomas, was in partnership with Hayward as sub-contractor on the brickwork. Browne continued as

supervisor. Work reached completion in March 1684 at a cost of £3,059, giving a total expenditure of £4,833. Peirson had been in charge from start to finish.

He resigned in 1685 to take up the post of Steward of the Duke of Bedford's estate at Thorney, where he stayed until 1709.

Denver sluice was a massive structure. The foundations rose above the river bed to a level about 7 ft. below low water. The 'sea doors', pointing downstream to keep out the tides, rose to a height of 18 ft. above the cills and 1½ ft. above high water springs; the 'ebb doors' were 10 ft. high, pointing upstream; the piers had a width of 13 ft. and were about 60 ft. long. However well built by the standards of the day, the sluice failed in 1713. Badeslade (1725), who gives a plan and elevation, says it was first undermined and afterwards 'blown up' and destroyed by the tides. An additional sluice 60 ft. to the east was built 1699–1700 under the direction of Richard Russell (q.v.),

A. W. SKEMPTON

[Bedford Level Corporation Orders and Accounts (1671–1685) Cambridgeshire RO; Russell Estate Collection, Bedfordshire RO; Thomas Badeslade (1725) *The History of the … Port of King's Lyn … and the Draining of the Great Level*; L. E. Harris (1953) *Vermuyden and the Fens*]

Works

1682–1684. Denver Sluice, western part, £4,833

PENN, John, FRS (1805–1878), was born at Greenwich, the son of the engineer, **John Penn** (1770–1843). He was apprenticed to his father in the family business (founded at Greenwich in 1800) which specialised in millwrighting and general engineering; the first prison treadmill, designed by William Cubitt (q.v.), was made by the firm in 1817.

He spent some time in France in 1826 superintending experiments by the French Government on the Perkins Steam Gun, which the company had made for the inventor. Later the firm was much involved with the French Government on marine matters, including making a number of steam-engine indicators for engine testing.

John Penn and Son branched out into marine engine manufacture and in 1838 they made the oscillating engines for HMS *Black Eagle* and later introduced the trunk engine. By the time of his death, 230 vessels were fitted by the firm with that pattern engine. In the Crimean War, Penn was responsible for organising the fitting out of ninety-seven gunboats, arranging manufacture of components by other manufacturers.

Penn wrote about experiments on bearing materials for screw-propeller shafts, in particular the suitability of lignum vitae; superheated steam in marine engines; and the construction of steam boilers.

He joined the Institution of Mechanical Engineers in 1848, becoming President in 1858–1859 and 1867–1868. He became as Associate of the Institution of Civil Engineers in 1826, a Member in 1845 and member of Council from 1853–1856. He was elected Fellow of the Royal Society in 1859.

Penn married Ellen English in 1847 and of his children, two sons became active in the management of the firm after his retirement in 1875. He died at his residence, Lee, Kent, on 23 September 1878.

A. P. WOOLRICH

[Memoir (1878) *Proc Inst Mech Engrs*, **13**; S. Smiles (1863) *Industrial Biography*, 182–228; DNB]

PENSON, Thomas, Sr. (*c.* 1760–1824), practised from Wrexham as a mason, surveyor and architect; he was appointed 'Surveyor of Bridges and other Public Works' in Flintshire in 1805 and was *de facto*—there appears to be no record of a formal appointment—County Surveyor of Denbighshire *c.* 1808–1820. In 1787 he married Charlotte Brown (d. 1824), of Wrexham, by whom he had three children including Thomas Penson Jr. (q.v.) and John William Todd Penson the artist (1796–1826). Thomas Penson, and his elder son and namesake, practised at a time when there was no clear distinction between the professions or callings of civil and mechanical engineer, surveyor and architect and practitioners moved with relative ease between these disciplines and others.

Thomas Penson Sr.'s public works included bridges, roads, gaols and building in the counties of Caernarfon, Denbigh and Flint. His first recorded public work was the design and construction of a new county gaol in Caernarfon (1784), the magistracy anticipating John Howard's third Report on the Condition of Prisons (1784), which gave a damning account of the facilities at Caernarfon. Penson was also responsible for improvements to the Wrexham house of correction (1795) and the county gaol, Flint (1806–1807). From 1785–1790 he worked on the construction of the county hall, Ruthin (architect Joseph Turner (q.v.)), but is best known for his work on improving the roads and bridges in the counties of Denbigh and Flint.

From the 1780s, Penson's relationship with the magistracy appears to have been uncomfortable owing to his inability to keep costs within estimate and by 1810 the Flintshire magistrates had resolved to put all work out to competitive tender and to pay Penson a fixed salary. The difficulties came to a head in August 1813 with the collapse, during construction, of the single-arch Overton Bridge spanning the River Dee. Penson subsequently produced plans for a two-arch bridge, also of stone, but his failure to provide adequate security for the work resulted in his dismissal from the contract by the Flintshire magistrates and the appointment in his place of his son, Thomas.

The elder Penson's other work included a fine single-arch bridge, Pont Llanfihangel Glyn Myfyr,

Denbighshire (1798–1799), a theatre adjoining his house at the Beast Market, Wrexham (1818), and the construction (to a bizarre design by Thomas Harrison (q.v.)) of the Jubilee Tower on Moel Fammau, Denbighshire, to commemorate George III's jubilee, 1810. In 1808 he was consulted about the drains and sluices on the Rhuddlan embankment.

Penson died on the 30 March 1824 and was buried in the old cemetery, Wrexham.

ROBERT ANTHONY

[*Quarter Sessions Minutes, Orders and Reports for the Counties of Flint, Denbigh and Montgomery (1785–1830)*, Flintshire CRO, Denbighshire CRO and Montgomeryshire CRO, respectively; Minutes of the Commission of Rhuddlan Marsh, DC1140, Flintshire CRO; W. H. Jones (1889) *Old Karnarvon*, **115**; A. N. Palmer (1893) *History of Wrexham*, 123, 128, 215; A. H. Dodd (*c.* 1968) *The History of Caernarvonshire 1284–1900*, 407–408, 412; E. Beazley and P. Howell (1975) *The Companion Guide to North Wales*, 362, *passim*; Colvin (2)]

Works

T. Penson Sr.—masonry arch bridges:

1788–1799. Pont Llanfihangel
c. 1805. Pont David, Flintshire, £658
Penson was also responsible for a number of bridge repair projects in both Flints. and Denbighshire.

T. Penson Sr. and Jr.—masonry arch bridges:

1813 Overton Bridge

PENSON, Thomas, Jr. (1790–1859), was an architect, civil engineer and surveyor; he was County Surveyor of Denbighshire from about 1820 and of Montgomeryshire from 1817 until 1859. Born on the 5 May 1790 at Wrexham, he was the elder son of Thomas Penson Sr. (q.v.) and Charlotte Brown and became a pupil of the distinguished neo-classical architect, Thomas Harrison (q.v.) of Chester. In 1814 he married Frances Kirk (b. 1793), the daughter of Richard Kirk (later 'Kyrke'), the Wrexham ironmaster. They had ten children, including Richard Kyrke Penson (1815–1886), surveyor and architect (County Surveyor of Montgomeryshire, Carmarthenshire and Cardiganshire) and Thomas Mainwaring Penson (1817–1864), surveyor and architect (County Surveyor of Cheshire and Flintshire, and designer of Shrewsbury Railway Station).

Following marriage, Penson moved to Overton Cottage close to the end of Overton bridge over the River Dee and, after his father's dismissal from the Overton bridge contract in 1815, constructed (under the supervision of the architect John Carline II (q.v.)) the stylish two-arch stone structure which still stands. By 1823, Penson had moved to Oswestry, Shropshire, where he set up in practice and where he lived until his move, in 1839, to Richard Kyrke's house, Gwersyllt Hill, Wrexham. Penson, a versatile man interested in new materials such as iron and terracotta, designed and constructed houses, churches, schools, gaols, workhouses and other buildings. He also erected many bridges, especially in Montgomeryshire, and was responsible for the improvement of many lines of road. Although Penson's stone and iron bridges were not innovative either in design or construction, his contribution, as County Surveyor, to the development and improvement of communications, particularly in Montgomeryshire over a period of forty years, was immense.

Penson's earliest Montgomeryshire bridge at Pont Llogel (1818) was similar in design to his father's bridge at Llanfihangel Glyn Myfyr, Denbighshire; he also built Caersws Bridge (c. 1820) and Llanidloes Long Bridge (1826), both over the river Severn, and Llanerfyl Bridge (c. 1820) over the river Banwy. A contrast in design—and revealing the influence of Harrison—is provided by the superb neo-classical bridges over, respectively, the rivers Vyrnwy at Llanymynech (1826–1828) and Severn at Newtown (Long Bridge, c. 1827, with later ironwork). His Flannel Exchange building (1830–1832) still stands adjoining the Newtown end of the Long Bridge. Other works at this time include the construction of The National School, Welshpool (1821), the extension to the County Gaol, Ruthin (1824–1825) and the remodelling of the town hall, Montgomery (1828). Most of Penson's other major works, particularly on iron bridges and churches, were post-1830.

Penson became an Associate of the Institution of Civil Engineers in 1839 and a Fellow of the Royal Institute of British Architects in 1848. He died at Wrexham on the 20 May 1859 and was buried at Gwersyllt Church (which he had designed in 1850–1851), near Wrexham.

ROBERT ANTHONY

[Quarter Sessions Minutes, Orders and Reports for the Counties of Flint, Denbigh and Montgomery (1785–1830), Flintshire CRO, Denbighshire CRO and Montgomeryshire CRO, respectively; W. H. Jones (1889) Old Karnarvon, 115; A. N. Palmer (1893) History of Wrexham, 123, 128, 215; A. H. Dodd (c. 1968) The History of Caernarvonshire 1284–1900, 407–408, 412; E. Beazley and P. Howell (1975) The Companion Guide to North Wales, 362, passim; Colvin (2); C. R. Anthony (1995) Penson's progress: the work of a nineteenth-century county surveyor, Mont. Colls., 83, 115–175 (this includes a Gazetteer of Thomas Penson II's bridges and other works in Montgomeryshire)]

Publications

T. Penson Jr.—Deposited plans, etc.:

1824. *Denbigh to Llanwrwst Road*
1825. *Four Crosses, Montgomery to Rolly House, Salop Road*
1835. *Map of Turnpike Roads in Montgomeryshire*
1836. *Tal y Cafn–Llanwrwst Road*

Works

T. Penson Sr. and Jr.—masonry arches (all single span masonry arches unless otherwise indicated):

1813. Overton Bridge

T. Penson Jr.—masonry arches (all single span masonry arches unless otherwise indicated):

1817–1818. Pont Llogel, 44 ft. span
1819. Pont Llanerchemrys
1821. Caersws Bridge, reconstruction, three arches
1823. Newtown Short Bridge
1826. Llanidloes Long Bridge, three arches, £3,000
1826–1828. Llanymynech Bridge, three arches, max. span 72 ft., £14,000
1827. Newtown Long Bridge, three arches, 60 ft. span, £4,000
1827. Pont Farrog, reconstruction, £254
c. 1820s. Berriew Bridge, reconstruction
1829–1831. Abergwydol Bridge, reconstruction
1830. Abermule Bridge, reconstruction
1830–1831. Pentyrch Bridge, reconstruction
1831. Abergwydol Bridge, reconstruction
1831. Wigdwr Bridge, reconstruction
1832. Pontgwenwynfarch, reconstruction
1832. New Mills Bridge, reconstruction
1832–1833. Millbrook Bridge, £374
1833. Llyfnant Bridge, reconstruction, £400
1833. Pont Cedig, reconstruction, £222
1835. Aberhiriaeth Bridge
1835. Cwmllinau Bridge, reconstruction, 24 ft. span
1835. Heniarth Bridge
1835. Pentre Bridge
1835. Pontafongarno, reconstruction
1835. Pont Sycoed, reconstruction, two arches

PERKINS, Jacob (1766–1849), was born at Newburyport, Massachusetts, on 9 July 1766. His family originated in the Forest of Dean, but migrated to America.

He was apprenticed to a goldsmith but quickly turned his attention to mechanical inventions and became the holder of twenty-one American and nineteen English patents.

By 1815 he had moved to Philadelphia where he became interested in a scheme to improve the engraving of banknotes to prevent forgery. He moved to England in 1819 in the hope of interesting the Bank of England. His printing machine attracted the attention of engineers and he joined the Institution of Civil Engineers in 1820. Perkins established a printing company which supplied many country banks with banknotes and foreign countries with postage stamps. Stamp production started with the 1840 1d black and 2d blue postage stamps for the Crown.

Perkins investigated high pressure steam at pressures up to 2,000 lb/sq. in., developing boilers and engines which were not hitherto a practical proposition. This excited hostility within the engineering world as his critics were still wedded to the low-pressure philosophy of the single-cylinder Watt engine. His concepts were revived a century later as the uniflow engine, often named the 'heat extraction engine'.

About 1829–1830 various business problems led to lawsuits and Perkins was forced to close his newly-opened engine manufactory. With his son, Angier March Perkins (1799–1881), he then set up a prosperous business manufacturing their patent central heating system utilising the hermetic tube method of steam raising. In 1836 a number of locomotives were built for the London and South Weston Railway utilising the hermetic system.

In about 1832 Perkins opened the National Gallery of Practical Science in Adelaide Street, West Strand. An exhibition centre with lectures devoted to modern inventions it was here that Thomas Telford (q.v.) carried out his canal boat experiments.

Perkins retired in 1843 and died in London on 30 July 1849.

A. P. WOOLRICH

[*Trans Soc Arts*, **38** (1820) 80–82, 102–110; **39** (1821) 133–135; Obituary notice (1866) *Min Proc ICE*, **25**, 516–519; Letter to George Watkins about Perkins from Charles R. King, 3 April 1926, Watkins collection, RCHME; C. R. King (1924) Uniflow engines were built in the year 1827, *Power*, October 2 (includes portrait); Anon. (1925) Supersteam Practice a Century Ago, *Cheap Steam*, May, 34–36. Anon. (1925) Jacob Perkins and his steam engine, *The Engineer*, **140**, 6–9; Anon. (1926) Fairlie–Perkins super-pressure locomotive, *The Engineer*, **142**, 580; Anon. (1931) Origin and evolution of the indirect system of heating steam boilers, *The Engineer*, **151**; Anon. (1931) Perkins hermetic tube boilers, *The Engineer*, **152**, 405–406; D. Bathe and G. Bathe (1943) The contributions of Jacob Perkins to science and engineering, *Trans Newc Soc*, **XXIV**, 49–53; E. Ferguson (1965) *Early Engineering Reminiscences (1815–40) of George Escol Sellers*, 12–25 (has an early portrait of Perkins)]

PERRY, John (*c.* 1669–1733), naval officer and civil engineer, was born *c.* 1669 at Rodborough in Gloucestershire, the second son of Samuel Perry, gentleman, and his wife Sarah, daughter of Sir Thomas Nott. He entered the Royal Navy in his teens and in January 1691 he was serving with the rank of Lieutenant in HMS *Montagne* when, at the beginning of an action, his Captain was killed. Perry stayed at his post for over an hour with an undressed wound in his right arm. The delay endangered his life and though he recovered, he lost his arm. In 1693 he superintended repairs to the *Montagne* in Portsmouth Dockyard and built an 'engine' for rapidly draining the dry dock. Later that year he sailed with the rank of Captain in command of HMS *Cygnet* in consort with HMS *Diamond* under the command of Capt. Henry Wickham, Perry's superior officer. In September both ships surrendered to French privateers. In April 1694, after release from St. Malo, the Captains were court-martialed, fined and sentenced to ten years imprisonment. While in Marshalsea prison Perry published in 1695 a pamphlet criticising the press gang system. In 1797 Wickham was pardoned and Perry's pardon followed soon afterwards. He retained his title and in his subsequent engineering career is nearly always referred to as Captain Perry.

It is remarkable that within a short time of his release he was commissioned to design the first dry dock in the Netherlands, at Flushing (Vlissingen), which was eventually built 1704–1705. In April 1698 he was introduced by Lord Carmarthen and Edmund Dummer (q.v.), Surveyor (of Works) to the Navy, to Tsar Peter the Great, then in England, as a person suitable to superintend the naval and engineering works in progress or planned in Russia. The Tsar engaged him immediately at a salary of £300 p.a. with a subsistence allowance of 25 roubles per month. It may be supposed that Perry, while at Portsmouth in 1693, met Dummer, who was impressed by his practical ability and intelligence: qualities amply demonstrated in Perry's later writings and works.

Perry stayed in Russia from 1698 to 1712. He gives an account of the works carried out under his direction in his book on *The State of Russia under the Present Czar* published in 1716. Major works were started but not completed before he was ordered to begin the next project; officials resented the presence of a foreigner; the workforce was inadequate in number and skill; and apart from the occasional gratuity and his subsistence allowance he rarely (if ever) received his salary. Finally he fled the county under the protection of the British Ambassador. Nevertheless he achieved a good deal including surveys and about half of a canal with several locks to link the Black and Caspian seas via the Don and Volga; a dam with sluices and a navigation lock for large vessels on the Veronezh at its junction with the Don, and surveys for a canal link from St. Petersburgh to the Volga.

These achievements were not unknown in England. Within a fortnight of his return he was taken to see the work then in progress to close a breach in the Thames embankment at Dagenham. The breach occurred in 1707. Originally only 14 ft. wide, it had been neglected and soon became much wider, flooding some 1,000 acres of reclaimed marshland in Dagenham and Hornchurch Levels. Attempts to close the breach, first by local labourers and later by contractors, were nearing completion in 1713. Perry thought they were unlikely to be successful but refrained from becoming involved at this late stage in the proceedings.

In July he went to Dublin on the recommendation of Sir Alexander Cairns to advise on improvements of the harbour. He completed his surveys in September and shortly afterwards presented his proposals. They included a basin behind the harbour with a sluice which could be closed at high tide and opened at low tide to discharge a rush of water to scour the channel, together with a breakwater to narrow the harbour entrance. Political problems in local government unfortunately meant that the work could not be authorised. Perry received a consulting fee of £67, and was retained in Dublin for several months in the hope that something could be accomplished.

Meanwhile at Dagenham the works, allegedly completed in October 1713, had been swept away by a storm in February 1714, and the problem was compounded by an extensive build-up of silt from the breach which threatened navigation in the river. The situation was now beyond the resources of the local landowners. The Havering Commissioners (with jurisdiction over the marshes and banks form West Ham to Hornchurch), the City of London and Trinity House joined forces to submit a Bill to Parliament. An Act was passed authorising a ten year levy on shipping using the port of London and creating a Trust to deal with the breach. The Trust held its first meeting in July 1714.

Hearing of these developments, Perry left Dublin and drew up proposals for closing the breach. They were considered by the Trust in August, but a contract was awarded to William Boswell who submitted a tender for carrying out the job at £16,300, a much lower figure than Perry's estimate. Boswell proved to be unequal to the task and his contract was cancelled in November 1715. Perry re-submitted his proposals in December and they were accepted by the Trustees in January. As his scheme had been criticised by Boswell and others Perry was forced to justify it, successfully, to the House of Commons before the contract was signed in June 1716.

By now the breach had widened to 400 ft. and at its deepest was 17 ft. below low water level. Perry undertook to close it permanently, to repair the bank along the whole 2 mile river front of the two Levels, to make a 40 ft. wide sluice for discharging water from the Levels, and to clear away the silt in the river channel. Payments of £15,000 would be made for work in the breach up to low water, a further £6,000 on successfully turning the tides out of the Levels, and £4,000 on completion of the whole contract: £25,000 in total.

Work began immediately under Perry's direction. His foremen were John Reynolds (q.v.) who had previously been employed by Boswell and Edward Rubie (q.v.). There were three other assistants and what must have been a moderately large workforce. Apart from the sluice, itself a considerable structure, the main work was the earth-fill dam across the breach. This had a deep cut-off of dovetailed timber piles 8 in. thick, a wide clay core with chalk shoulders up to low water level, and clay placed in 3 ft. layers above, forming an embankment with its crest 25 ft. above low water, 5 ft. above HWST or 2 ft. above the highest known tide.

During construction the top of the bank had to be protected twice every twenty-four hours when it became submerged. In September 1717 the bank, nearly at its full height, was breached and

severely eroded. It was rebuilt but again breached in September 1718. This caused a crisis which would have been disastrous, as Perry says, 'had not my good friends the Russia Merchants stood firmly by me'. Thanks to their support he was able to proceed and the tides were finally turned out of the Levels on 18 June 1720. The dam was raised to allow for settlement. It withstood a great tide in November 1720 which caused damage to several banks along the Thames.

In May 1721 Perry could report all works completed at the breach and on the 2 mile length of bank, and that much of the silt had been dredged from the river. By then the expenditure amounted to £40,473. The Trustees quickly obtained an Act empowering them to pay out an additional £15,000. Dredging continued, rather slowly, for the next two years.

Repayments to Perry's backers probably included interest at a high rate. They could afford to write off part of their loan and still make a profit. They were also prepared to write off Perry's personal debt, and probably did so. Fourteen in number, the backers included a Russian merchant, the deputy master of Trinity House and several members of Perry's family.

The closing of Dagenham Breach, a prominent and difficult undertaking, established his reputation which he enhanced by publishing a full account of the operation in 1721. Already in 1718, at the request of Lord Aylmer, he visited Dover harbour. His report, dated November 1718, is printed as an appendix in the Dagenham book. In October 1721 he reported again on Dublin harbour.

In February 1723 petitions were referred to Parliament for the granting of revenue from Passing Tolls on shipping for completing repairs to Dover Harbour and for 'restoring' the harbour at Rye, which was becoming blocked by shingle driven eastward along the coast. Perry and Samuel Walton gave evidence on Dover and representatives of the Navy and Trinity House reported on Rye. An Act having been obtained, the Commissioners then consulted Perry on proposals for restoring Rye harbour. He came up with a plan for a new harbour about 2 miles to the west, to be formed by a wide canal from Winchelsea Channel down to and through the beach ridge. This had the advantage that ships coming out would enter deep water, thereby avoiding the great flat and bar off the entrance of the old harbour.

His idea met with approval. A petition was submitted to Parliament in February 1734 for a Bill 'making more effectual' the previous Act. Perry gave evidence, briefly describing his scheme and laying a plan of the harbour before the Committee with an estimate of £13,732 for carrying out the works. He said that he 'had tried the Ground where the intended Canal is proposed to be cut, and finds it very good and proper for the same, and that such a Canal may be made without any Danger of flooding the neighbouring Lands, and will securely carry off the Waters from the said Lands'.

The new Act was duly passed. At their first meeting thereafter, in May 1724, the Commissioners approved various payments including £105 to 'Captain John Perry for his Attendance at Rye three times and in several Committees of the House of Commons, and for his Judgement in the Affair of the Harbour'. Next day he delivered his proposals: the canal to run about a mile from Winchelsea Channel to the coast, with a scouring sluice and navigation gates rather more than half way down the canal, and piers 120 ft. apart at the entrance; the canal to have a depth of 20 ft. at high water springs so that ships of 300 tons burden drawing 15 or 16 ft. may enter safely.

The Bank of England advanced a loan of £4,000 so that work could begin. In June Perry was appointed to direct the works and attend for at least three-quarters of the time at a salary of £170 p.a.for himself and a 'clerk'. By August he had marked out the land to be purchased. He ordered a drawbridge to be built over Winchelsea Channel, essential for access from Rye to the new harbour. This was constructed in 1725 by John Reynolds.

While land purchase was still in progress Perry applied for leave of absence to visit Ireland to report (again) on Dublin harbour. He brought down Edward Rubie 'whom he proposes to leave as his Agent in his absence, to whom he hath given full and particular directions'. Rubie received £20 from Perry and in July was appointed by the Commissioners as his foreman at 25s per week.

Perry had now returned to Rye and work began on the canal, using four contractors. It was to be 8 ft. deep initially, with a top width of 140 ft. above the sluice and 180 ft. below, widths later increased to 150 ft. to 200 ft. respectively. Having designed the piers Perry applied in September 1726 for six weeks leave of absence to report on the North Level of the Fens. This led to the construction, with some modifications, of the reservoir and scouring sluice at Gunthorpe under the direction of Humphry Smith (q.v.) in 1727–1728. Perry was also consulted in early 1727 by York Corporation regarding the Ouse navigation. He recommended a flash lock above Cawood and five cuts on the river to secure a navigation depth of 5 ft. up to York. He was paid £40 guineas for his report but the scheme was beyond their resources.

Back at Rye, work began on the west pier in 1728. It was founded on a timber grillage 2 ft. above the level of low water neaps, supported on piles and surrounded by dovetailed sheet piling 20 ft. long. Built of Portland stone with rubble masonry infill set in hydraulic lime mortar, the pier was 62 ft. long, 27 ft. broad and 18 ft. high, rising to the level of high water springs. Christopher Cass and Andrews Jelfe (q.v.) contracted for the masonry. The foundations were built by direct labour under Rubie's supervision. He also

measured up the masonry on completion of the pier in September 1730.

By that time Perry had gone to Deeping Fen, Rubie having been given a permanent appointment in July 1730 at a salary of £65 p.a. with a rent-free house. Rubie, speaking in 1738, said he 'had the management of the Works under Captain Perry and the care of them since his death' (in 1733). This suggests that Perry continued to take an interest in the proceedings at Rye. Although the east pier (1730–1731) and the Great Sluice (1731–1733) were built under Rubie's direction and to his working drawings, they undoubtedly followed plans made earlier by Perry. The largest structure of its kind in England at that time, the Sluice had massive abutments, five lifting gates each 6 ft. wide, separated by a central pier 24 ft. and 58 ft. long from the 40 ft. navigation opening with a pair of gates pointing upstream. The whole structure was founded on a timber grillage supported on piles with three rows of dovetailed sheet piling and was designed to have a depth of 20 ft. of water on the sills at high water springs. The expenditure on works up to 1733 was about £20,000. After a pause due to lack of funds work resumed in 1734 under Robert Copper (q.v.), the total cost of the works then amounting to about £42,000.

Deeping Fen is an area of some 30,000 acres of fenland west of Spalding between the rivers Welland and Glen. Limited success in draining the land had been achieved in the 1630s and 1680s but the return on capital barely sufficed for adequate maintenance of the banks, drains and sluices. Perry would have gained an idea of the state of the Fen during his 1726 visit to the neighbouring North Level. In April 1729 he wrote to the Deeping Fen Adventurers (proprietors) saying that the best means of improving drainage was to improve the outfalls of the Welland, of Vernatts Drain (the main drain of the area) and of the Glen near Surfleet. This he proposed to do by scouring sluices built on the Welland in Spalding and at the outfall of Vernatts Drain and at Surfleet.

The Adventurers approved the scheme. They offered to convey nearly 6,000 acres of land to Perry as payment; this he accepted. On 16 April 1730 he attended their annual meeting and on the same day joined Spalding Gentleman's Society (which had been founded in 1710, one of the earliest provincial learned societies). In June he sold about 2,000 acres to raise funds for the work. He probably moved to Spalding in July, and began work before the end of 1730. By April 1731 the foundations and floor of the Great or Six-Door Sluice at Spalding had been completed.

The sluice had six lifting gates each 6 ft. wide adjoining a navigation lock. As at Rye, the structure was founded on a timber grillage supported on piles, with rows of dovetailed sheet piling beneath the gates and at each end of the floor or apron. All six gates could be raised or lowered simultaneously by chains connected to a horizontal axle operated by a treadwheel. The sluice retained water in Cowbit Wash, an area of about

1,000 acres between widely spaced banks of the Welland upstream. With the Wash full at high tide level the gates were first raised, at low tide, in September or October 1731. The river had previously been diked out to the level of the sluice apron for a length of 180 yd. and loosened by 'porcupines', wooden rollers fitted with iron spikes dragged along the bed. The rush of water on opening the gates produced a scour, the effect of which extended up to 3 miles downstream as reported by John Grundy (q.v.) in 1736.

Perry started other works but before they were completed he died at Spalding on 11 February 1733. He was buried two days later at Spalding Parish Church where a mural tablet bearing a coat of arms and a long inscription was placed by 'William Perry of Penshurst, Kent, Esq., his Kindsman and Heir'.

He is the best known civil engineer of the early eighteenth century in England. John Smeaton (q.v.) referred in 1763 to 'the famous Captain Perry' and he is the only one of the period to receive more than the briefest mention by Samuel Smiles in *Lives of the Engineers*. Apart from his works, Perry has the distinction of being the first engineer to have an appreciable number of publications. His report on the North Level (1727), a fourteen-page folio with folding engraved plan, set a standard for all subsequent printed reports.

A. W. SKEMPTON

[Rye Harbour Commissioners Minute Book (1724–1733), East Sussex RO; Deeping Fen Adventurers Minute Book (1730–1733), Lincolnshire Archives Office; John Grundy (1736) *An Examination … of Mr. Badeslade's New Civil Canal … and Captain Perry's Proceedings*; E. Rubie (1738) Evidence on Rye harbour, *JHC*, **23**, 38; S. Smiles (1861) *Lives of the Engineers*, **1**, 72–82; G. P. Moriarty (1896) John Perry, *DNB*, **44**, 35–36; Anon. (1900) Captain John Perry, *Fenland Notes and Queries*, **4**, 293–296; J. J. Bootgezel (1930) The first dry-dock in the Netherlands, *Trans Newc Soc*, **27**, 241–253; J. G. O'Leary (1966) Dagenham Breach, *VCH Essex*, **5**, 285–289; B. F. Duckham (1967) *The Yorkshire Ouse*; A. Hasenson (1980) *Early Maps from the Dublin Port Collection*; M. A. Gulligan (1988) *A History of the Port of Dublin*]

Publications

1716. *The State of Russia under the Present Czar*
1721. *An Account of the Stopping of Dagenham Breach*
1721. *The Description of a Method … for the making of better Depth coming over the Barr of Dublin. As also for the for the making of a Bason within the Harbour*
1725. *Remarks on the Reasoning and different Opinions … in some late Printed Proposals for Restoring the Navigation of the Ports of King's Lynn and Wisbich, and Draining the great Levels*
1725. *A Method proposed for making a safe and convenient Entrance to the Port of Dublin*
1727. (Report) *To the Gentlemen Land Owners in the Parts of South Holland, in the County of*

Lincoln. With A Map of the North Level and of the Marshes

Works
1698–1712. Works in Russia
1716–1723. Stopping Dagenham Breach, £40,000
1724–1730. Rye New Harbour, to 1733, £20,000
1730–1733. Deeping Fen, £4,000

PHILLIPS, John (*c.* 1709–1775), master carpenter, was the son of Matthew Phillips (d. 1777) of East Hagbourne, Berkshire, and nephew of Thomas Phillips (q.v.), who probably trained him and to whose extensive London business he succeeded following his death in 1736. He was the erstwhile partner of George Shakespeare (d. 1797), Clerk of Works at the Radcliffe Library where Phillips was responsible for the wooden dome and other fine joinery. Their work there was arguably surpassed by their joinery at Christchurch College Library.

Like his uncle, John acted as a speculative builder in the London area and also had pretensions as an architect. In the civil engineering sphere he successfully contracted for a temporary bridge and the centering for the new arch designed for London Bridge by George Dance Sr and Sir Robert Taylor. Although he submitted eleven designs for Blackfriars Bridge he was unsuccessful, one reason being alleged extravagance of timber in the centering at London Bridge. He was consulted about work at Bristol Bridge in 1760 and was the successful contractor at Battersea, where he built the timber bridge (1771–1772) to the designs of Henry Holland; it had 19 spans, varying in width between 15 and 32 ft.

He lived in his uncle's house in Brook Street, dying there on 28 December 1775, aged sixty-six. He was childless and his wealth was ultimately inherited by his brother, William (d. 1782), and his descendants, who lived at Culham House, Oxon, which John had rebuilt.

MIKE CHRIMES

[Publicus (1760) *Observations on Bridge Building*; J. Dredge (1897) Thames bridges; T. Ruddock (1979) *Arch Bridges and their Builders, passim*; Colvin (3)]

Works
1759. London Bridge, centering for replacement arch
1771–1772. Battersea Bridge, timber, £16,000

PHILLIPS, John (fl. 1760–1803), author and surveyor, is believed to have been born in Essex. He trained as a builder and surveyor but little is know about his life beyond his own writings. In the 1760s he became acquainted with James Brindley (q.v.) and helped survey a canal route from Liverpool to Hull, probably part of Brindley's Grand Cross scheme delineated in his 1765 *Plan of the Navigable Canals … for Opening a Communication to the Ports of London, Bristol, Liverpool and Hull*, rather than the Leeds–Liverpool proposal of

1769. From Phillips's writings it would appear he knew Brindley personally and his general history devotes much space to the man and his works. He became a firm advocate of inland navigation and perhaps its most important publicist, as well as surveying his own proposals in East Anglia.

Probably subsequent to his work with Brindley, Phillips was taken as a 'prisoner on remand' to America; perhaps he was captured in the American Wars of Independence. He then went to Russia, possibly in response to the advertisements being placed in Britain for experienced surveyors, and worked briefly, 1783–1784, on an attempt to complete the Don-Volga Canal, a proposal on which John Perry (q.v.) had worked at the start of the eighteenth century. The intention was to link the great rivers by their tributaries, the Ilovna and Camnshinkshka, but problems encountered included unreliable water supplies and lack of depth. Phillips's efforts were frustrated by lack of men and all he was able to do in nineteen months was to fell trees. Following his return to St. Petersburg in 1784, the following year he returned to England via Sweden where he observed the various inland navigation proposals there. His outward journey had been via Germany and Poland.

The publication of his *General History* in 1792 coincided with a growth in canal construction and the work was substantially enlarged and reprinted. It became the main source of information on British canals but the rapid growth of the canal network precluded any detailed description of the engineering works involved. Its status is reflected in the way in which Cordier, the French engineer, translated large sections of it in his *Historie de la Navigation Interieure* (1819, vol. I). Phillips was also the author of the *New Builder's Price Book*, published by Crosby, which appeared in its 13th edition *c.* 1809. It is possible he had a hand in I. and J. Taylors's *Builder's Price Book* (1st edition, 1776).

Phillips may have died *c.* 1803 as the 5th edition of his *General History* has no updates later than that date, and the 1817 *Price Book* was 'corrected by C. Surman'.

There were at least four other men by the name of John Phillips active in the late eighteenth and early nineteenth centuries. Aside from those above, a John Phillips (1771–1843) was a London architect and another was surveyor of Westminster sewers in the 1840s. The most likely source of confusion could be with John Phillips (1800–1874), the geologist and nephew of William Smith (q.v.), who wrote more than a hundred papers and books from 1829 onwards.

MIKE CHRIMES

[DNB; Colvin (3); Bendall; RIBA Library catalogue; M. Baldwin and A. Burton (1984) *Canals: A New Look*]

Publications

1785. *A Treatise on Inland Navigation: Illustrated with a Whole Sheet Plan, delineating the Course of an Intended Navigable Canal from London to Norwich and Lynn*
1792. *A General History of Inland Navigation, Foreign and Domestic: Containing a Complete Account of the Canals already executed in England*
1793. *Addenda*, 2nd edn.
1795. *2nd Addenda*, 3rd edn.
1803. 4th edn.
1809. 5th edn. (reprint of 4th edn.)
[1819. Translation, J. Cordier, *Historie de la Navigation Interieure*]

PHILLIPS, Thomas (*c.* 1689–1736), master carpenter, a son of Matthew Phillips, a mason of East Hagbourne, Berkshire, was apprenticed to Jeremiah Franklin, an Oxford carpenter, in 1705. In the 1720s he was active in London both as a speculative builder and also as a carpentry contractor on churches such as St. Martin-in-the-Fields. This church contains interesting stiffening details in the truss above the portico, possibly introduced by Phillips. As a master carpenter he obtained the contract to build a timber bridge across the Thames, the first Fulham Bridge, 1729–1730. A plethora of designs from architects such as John Price (d. 1736) were considered by the Bridge Commissioners between 1726 and 1729 before the designs of an amateur, Sir Jacob Acworth, were approved. Phillips erected the timber beam structure, with spans varying between 15 ft. and 30 ft., in nine months. Supervision was probably by another proprietor, Dr. William Cheselden, surgeon at St. Thomas's Hospital

Phillips's last work was possibly a carpentry contract in the Treasury buildings, Whitehall, erected 1733–1737. He died in 1736, and is buried in East Hagbourne churchyard.

MIKE CHRIMES

[A. Chasemore (1825) *History of the Old Bridge at Fulham and Putney*; J. F. Wadmore (1890) Old Fulham bridge, *Trans London Middlesex Arch Soc.*, **VI**, 401–448; J. Dredge (1897) *Thames Bridges*; [E. C.] T. Ruddock (1979) *Arch Bridges and their Builders*, 28–29; D. Yeomans (1992) *The Trussed Roof*, 148, 156; Colvin (3)]

Works
1729–1730. Fulham Bridge, timber, £12,000

PICKERNELL, Jonathan, Sr. (1738–1812), civil engineer, was the son of James Pickernell (b. 1698) and his wife, Martha Overin, and was baptised at East Woodhay, Hampshire, on 27 December 1738. He was one of at least four children. Nothing is known of his early life other than the fact that in 1796 he stated that he had been employed as an engineer since 1766.

In 1757 Pickernell married Catherine Batter (b. 1733) at East Wellow, Hampshire. At the instigation of John Smeaton (q.v.) he moved with his family to the North East in 1775 to act as resident engineer under John Wooler (q.v.) for the construction of a bridge over the River Tyne at Hexham. Plans for a bridge were initially drawn up by John Fryer (q.v.) and approved by Wooler but, later, this project was abandoned and a new design was prepared by Wooler, the bridge to be built 'by days work ... under the direction and inspection of Mr. John Wooler and Mr. John Pickernell'. It was when undertaking site investigations that Pickernell pushed down by hand an iron bar into the river's bed a distance of 46 ft., finding 'no more resistance than chaff'.

The result of this finding was the siting of a replacement bridge approximately a mile down river and the departure of Wooler from the project. A new bridge was built to Smeaton's designs with nine river arches up to 51 ft. span and a total length between abutments of 518 ft. With Pickernell acting as resident engineer, construction began in 1777 and the bridge was certified as complete in 1781; it collapsed in a flood the following year and a new bridge was built between 1790 and 1795. Payments of approximately £750 were made to Pickernell on account of his work at Hexham but details are scant.

Pickernell was appointed as bridge surveyor to Northumberland County on 17 July 1776 and, in addition to his work at Hexham, was subsequently responsible for the design of bridges at Whitfield in 1779 and Featherstone in 1780, both built under contract. He prepared a design for a bridge at Ridley Hall in 1780 and it was completed by Isaac Dodds and Thomas Allen the following year, only to be carried away in a flood in March 1782. For his work as County bridge surveyor Pickernell was paid £40 p.a. but in addition he received several honorariums, one such being £32 14s 8d for 'his extraordinary trouble and expence in drawing Plans and making Estimates for Work to be done to several Bridges in this County'. He resigned his appointment in 1781 and in that year became Harbour Engineer at Whitby with a salary of £100 p.a.

Pickernell produced a plan of Whitby and his principal work there was the rebuilding and strengthening of the West Pier, 338 yd. long. He also investigated the condition of the East Pier, in 1812 providing estimates for its rebuilding. As Harbour Engineer, his salary rose to £121 by 1811 and, in addition, he received annual payments of up to £40 for his 'extraordinary Diligence'. Between 1798 and 1812 the expenditure on the harbour totalled £28,317.

Pickernell was also able to undertake work other than that for the Harbour Trustees. In 1787 he produced plans for a dock at Grimsby—where in 1796 he was consultant for a period—and in the following year designed the Town Hall at Whitby. Four years later he was called upon to investigate the condition of the River Tees and produced a plan for a dramatic shortening of the course of the river at Mandale by providing a cut some 220 yd. long in order to eliminate approximately 2 miles of the channel; his plan was later put into effect by William Chapman (q.v.). In 1794 he devised centring for the iron bridge under construction at Sunderland under the supervision of Rowland

Burdon (q.v.) and Thomas Wilson (q.v.) and in 1808 carried out an inspection of the foundations of the bridge crossing the River Tees at Stockton.

Pickernell died in Whitby and was buried there on 5 August 1812, his estate valued at less than £5000. He was survived by his wife and two sons, Peter Giles (1772–1859) and Jonathan (q.v.). The former attained the rank of Commander in the Royal Navy and the latter became a civil engineer.

R. W. RENNISON

[*Northumberland Quarter Sessions: Orders* (1775–1781) Northumberland RO, QSO/11 & 12; *Whitby Harbour Trustees: Minute Book* (1781–1812); J. Smeaton (1812) *Reports of the Late John Smeaton, FRS*, vol. 2; G. Young (1817) *A History of Whitby*, vol. 2, 535–536; S. M. Linsley (1994) Tyne crossings at Hexham up to 1795, *Archaeologia Aeliana*, 5 Series, **22**, 235–253]

Drawings

1791. *A Plan for altering the Course of the River Tees between Stockton and Portrack*
1791. *Plan of the Town of Whitby*, inset in *Draught of part of the Coast of Yorkshire*

Works

1775–1781. Hexham Bridge, resident engineer, cost £5,700
1776–1781. Northumberland county, bridge surveyor
1779. Whitfield Bridge, design, cost £733
1780. Featherstone and Ridley Hall, bridge designs, cost £538 and £1,300
1781–1812. Whitby Harbour, Engineer, principal work the rebuilding of the West Pier
1788. Town Hall, Whitby, design and construction

PICKERNELL, Jonathan, Jr. (1764–1814), civil engineer, was the elder son of Jonathan Pickernell (q.v.) and his wife, Catherine Batter. He was baptised on 27 May 1764 at Mottisfont, Hampshire, but nothing is known of his early life although it is possible that he worked with his father at Whitby. In 1794 he married Catherine Watson of Whitby and following her death married Peggy Ridley of Lythe, near Whitby, in 1795, this marriage resulting in three children.

Pickernell, 'a gentleman of cheerful and urbane manners', was appointed as Engineer to the River Wear Commissioners in 1795 at a salary of £120 p.a. At Sunderland he was responsible for the purchase of dredging machinery from Boulton and Watt, the 4 hp engine being mounted on a barge and set to work in 1798, the first steam dredger engaged in river and harbour works.

In the same year Pickernell reported on measures to be adopted for improving the harbour. He considered the cost of forming a dock too great and recommended, instead, that dredging would be more appropriate for accommodating bigger ships in the river. Surplus revenue would be better expended on improvements to the piers and he suggested that the north pier should be lengthened 250 ft., at the same time forming a training wall up-river from its base. The south pier should also be extended so as to provide a distance between the pier heads of 400 ft..

These works were begun in 1799 and within two years the north pier had been extended 100 ft. with work on the south being put in hand in 1802. A further work was the erection of a lighthouse, an elegant masonry structure 76 ft. high; it was completed in 1803 at a cost of £1,400. In 1803 Pickernell received a gratuity of £50 for his 'extra attention as Engineer' but in spite of this he was dismissed the following year for appropriating for his own use timber and other materials. Seemingly, he then obtained work on the construction of the Crinan Canal, from there writing to John Rennie in 1804 seeking alternative employment as work on the canal had been suspended.

In 1808 Pickernell unsuccessfully applied for the post of Engineer at Falmouth Harbour, following which he was offered a position on a naval brig commanded by his brother, Peter Giles Pickernell (1772–1859) who had served on *Revenge* at the battle of Trafalgar. The ship in which he was travelling was captured by the French and Pickernell was taken as a prisoner to Longwy where, after initial privations, 'his situation became comparatively comfortable' and he was able to inspect and make drawings of the principal buildings of the town. He also 'drew plans and gave instructions for erecting a bridge in the vicinity' and was said to have been offered an engineer's commission in the French army, which he declined. In 1814 he was taken, with some 1700 other prisoners, to Beauvais, where he died on about 20 January 1814. His only son, Francis (1796–1871), became Engineer to Whitby Harbour, holding the position from 1822 until his retirement in 1861.

R. W. RENNISON

[*River Wear Commissioners: Minute Book (1795–1804)*, Tyne and Wear Archives, 202/1102,3; *Sunderland Navigation: Orders of the Commissioners* (1795–1804), Tyne and Wear Archives, 202/1003; Letter from Pickernell to John Rennie (20 July 1804) NLS; Brief Account of Mr. Pickernell (n.d.), bound with *A Plan for altering the Course of the River Tees* ... Lit & Phil Soc., Newcastle, Acc. No. 1904; A. W. Skempton (1977) The Engineers of Sunderland Harbour, 1718–1817, *Industrial Archaeology Review*, **1**, 2]

Reports

1798. *A Report on the Present State of the Harbour of Sunderland, in the County of Durham, with Remarks on the Measures to be adopted for its Improvement*

Works

1795. Appointed Engineer, Sunderland Harbour
1798. Provided steam dredger, Sunderland
1799–1801. North pier extension, Sunderland
1802–1803. Built lighthouse, Sunderland
1802–1804. Extended south pier, Sunderland

PINCH, John, Jr. (d. 1849), architect, was elected a Corresponding Member of the Institution of Civil Engineers in 1821 but was a member only until 1826. Pinch was the son of John Pinch (*c*. 1770–1827), a builder and architect of Bath, to whose practice he succeeded. His father was surveyor to the Pulteney Estate from 1793 and consequently father and son would have been known to Thomas Telford (q.v.).

MIKE CHRIMES

[ICE membership records, ICE archives; Colvin (3)]

PINKERTON family (fl. 1760–1840) was a family of at least two generations of civil engineering contractors believed to have originated in Lincolnshire or the East Riding of Yorkshire. With a continuous succession of contracts from the 1760s onwards they can perhaps be regarded as the earliest contracting business in civil engineering.

James Pinkerton (*c*. 1736–1784) was the first of the family to take civil engineering contracts, on the Adlingfleet Drainage in 1767, on the Driffield Navigation the following year and on the Laneham drainage in 1769, all with John Dyson (q.v.) as his partner. There must have been some division of responsibility as Dyson appears to have been in charge of the Laneham job. The engineer for several of their jobs was John Grundy Jr. (q.v.). At this time Pinkerton was living in Cawthorne, near Barnsley, Yorkshire, and was described variously as Yeoman or Gardener, descriptions which his brothers (?) also used. His children, Hannah (b. *c*. 1764), **James** (*c*. 1766–*c*. 1840), Anne (b. *c*. 1767), Margaret (b. *c*. 1770), and Jane (*c*. 1772–1838) were all born there. In 1779 he removed to North Cave in the East Riding, buying the White Hart public house, although shortly afterwards he was imprisoned in York for debt and it had to be taken over on his behalf by his brother. His son, James, also became a contractor and one cannot be sure who was active after *c*. 1780. James Sr. died on 25 March 1784 at the age of forty-seven; his wife, Janet, had perhaps died between 1777 and 1781.

James Sr. appears to have had three younger brothers **John** (d. 1813), **Robert** (fl. 1775–1796), and William, a nurseryman based in Wigan. Of these, John was by far the most important and best known. He worked on the Driffield Navigation as agent for James Sr., and contracted with his brother on the Bishop Soil Sluices in 1770. They continued to work together for the next decade, although in 1777 John had sole management of the Billinghay embankment. From 1780 John, sometimes with other partners, appears to have become the chief contractor in the family and his standing is reflected in his election to the Smeatonian Society in 1777. In 1791 he was consulted by Loammi Baldwin, then embarking on the Middlesex Canal at the dawn of the canal era in the

United States. He shared business interests with William Jessop (q.v.) and appears to have been well regarded, at least initially, by John Rennie (q.v.) who advised him of possible work on the Lancaster Canal. John Pinkerton married Ann Lakin in 1786, at which time he was living in the Birmingham area, at Sparkhill. Ann's father, Henry, was involved with the Birmingham Canal, and was a contractor for bridges in Warwickshire. John and Anne had four children, **Henry Lukin** (b. 1792), **John Lukin** (b. 1793)—both born in Odiham, Hampshire—Caroline, and Harriett. They had moved to Tottenham by the time of John's death in Bath on 21 March 1813. By this time two other nephews of John, **Francis** (fl. 1785–1810) and **George** (fl. 1785–1805)—perhaps the younger sons of James Sr.—had joined the profession. It would appear that Francis continued the firm after his uncle's death and the last mentions of it are some unsuccessful tenders by **William Pinkerton** for the Great Western and Bristol and Gloucester railways between 1839 and 1841.

The bulk of the Pinkertons' work was as muck-shifters, either digging drainage ditches or as canal cutters. It would appear that they operated as a primitive form of management contractors, supplying labour and organising subcontractors for canal companies. Some indication of their aspirations can be obtained from a circular issued in connection with Grimsby Dock in January 1800 where they stated 'As you know we are enjineers [sic] in our turn', and offered to complete the work, which had been bedevilled by construction and other problems to alternative methods than those proposed by John Rennie (q.v.) then acting as engineer; they were unsuccessful in their bid. In the 1820s Marc Brunel (q.v.) got in touch with the Pinkertons for labour on the Thames Tunnel, as well as for a possible supply of bricks. In general terms their drainage work appears to have been more successful than their canal work, which on several occasions ended in court action.

The Pinkertons' early contracts were carried out successfully and they made a favourable impression on Jessop when he encountered them on the Market Weighton Canal and the Mirfield Cut on the Calder and Hebble Navigation in the 1770s. One result was for Jessop and John Pinkerton to become involved in various enterprises in the area: Whitley Wood Colliery, Mirfield Dry Dock, and lime burning. In 1778 Jessop, along with James and Robert Pinkerton, obtained the contract for the Beeston Brook–Nantwich section of the Chester Canal. Jessop also acted as surety on some of John Pinkerton's contracts, such as Rye Harbour and the Dudley Canal.

From the mid-1780s the scale of the enterprise grew, and so did complaints about scamping, general lack of progress and lack of adequate supervision by John Pinkerton, by then the leading member of his family. To some extent this must have been the result of stretching slender financial resources and, indeed, as the pace of canal construction quickened the financial viability of

some of their clients was also dubious. An indication of the economic circumstances of the time are the tokens issued by Pinkerton on the Basingstoke Canal in 1789–1790. John's need to manage several contracts simultaneously meant that he had to delegate, and to do this meant relying largely on his nephews.

We first hear of James Jr. on the Rye Harbour contract in 1786–1787; he then moved to Odiham to supervise the Basingstoke Canal work with Francis who had earlier worked at Rye and on the Dudley Canal Tunnel (1785–1787). Francis, James Jr. and George were also involved briefly as contractors on the Leicester Navigation, including 8 miles of railway (1791–1792), George signing extant drawings. Francis and James, Jr. were contractors on the Beverley and Barmston Drainage (1800), one of William Chapman's (q.v.) finest works, and the Muston and Yedingham Drainage, another Chapman job. Francis was evidently dissatisfied as in 1802, when working under John on the Fens near Boston, he applied to Rennie for a job as a resident engineer on the Kennet and Avon Canal; earlier, in 1794, George had made a similar offer. Francis, in 1805, and James Jr. between 1809 and 1813, are known to have worked on John's contracts in South Lincolnshire.

George was trained by William Jessop, his father James paying a premium of 100 guineas. He was paid £557 6s 7d for his work on the Basingstoke Canal, down to February 1790, almost as much as all the other surveyors and professionals combined, and must be assumed to have carried out the canal survey. In contrast, in the same period Jessop, the engineer, was paid only £197 18s and the resident surveyor, William Wright, £194 19s. In 1791, on Jessop's recommendation, Henry Eastburn (q.v.) took over as resident engineer until the completion of the canal. The canal is of great interest as John Pinkerton undertook the whole work for a sum of £76,690 6s 8d by a contract of 3 October 1788. For a canal of over 35 miles this must have been the largest civil engineering contract to that date to be awarded to a single contractor. Pinkerton appears to have been responsible for the hire of the sub-contractors as in August 1789 the company ordered him to sack Charles Jones (q.v.), contractor on the Greywell tunnel.

The Pinkertons first had major problems on the Dudley Canal tunnel (1785–1787). It was their first tunnel contract and they soon found themselves in difficulty with the 'extreme hazards of the strata' and the quantity of groundwater. In 1787 the complaints about lack of progress, poor quality bricks in the lining and rising costs led to John Pinkerton being forced to pay £2,000 to be released from his contract. By this time he was having problems with his other contracts in the Birmingham area. The first, the Broadwater extension of the Birmingham Canal (1783), was troubled by disputes between Pinkerton and James Bough (q.v.), the Company superintendent, over the cement Pinkerton was using. The next dispute, over the

Minworth–Fazeley section of the Birmingham Canal proved even more contentious. Pinkerton and James Watt (q.v.) were asked to supply estimates before tenders were invited and Pinkerton began work in January 1787, before the contract had been signed (May 1787). Some work was delayed over problems of access to land. At the end of 1786 George Pinkerton had discovered that Bough's levels were wrong, and Samuel Bull (q.v.), the Company Engineer, confirmed that Pinkerton was correct. This can hardly have helped relationships between Bough and the Pinkertons, and the quality of their workmanship and materials was heavily criticised. By late 1788 costs were exceeding the estimates and payments were stopped. Although John Pinkerton had largely completed the canal when relieved of his contract in February 1789, he was called back in December to repair two locks at Curdworth. In 1792, the company attempted to settle accounts, but Pinkerton delayed submitting his own. At dispute were the 'extras' for which he had been paid but which the company claimed were overpayments. In part these were due to rising labour costs, which Pinkerton claimed had cost him £2,750 15s 9d. Eventually, as a result of arbitration in 1801, Pinkerton was paid £436 of his claim, but he was not satisfied and issued a vitriolic justification which included a libellous attack on the Company Clerk, John Houghton, resulting in Pinkerton being fined and sent to prison. Pinkerton's work on the Coventry Canal was also criticised. He began work on this in 1784 before the Act had been passed and although due for completion by the end of 1786 it was not finished until December 1788.

Pinkerton's work at this time was under criticism elsewhere. The banks on the Basingstoke Canal collapsed in two places within six months of opening and the Greywell tunnel was found to be poorly built. The family's contracts on the Leicester Navigation, where Jessop was the engineer, were cancelled in 1792 within a year of being awarded. Progress had been rapid at first but within 9 months the Company had to make a cash advance of £200 to the Pinkertons and they were subsequently dismissed due to due to lack of progress, and the works completed by direct labour under Christopher Staveley (q.v.). Similarly, the Pinkertons were sacked from the Upper Ouse (Sussex) Navigation in 1792 for lack of progress. On the Warwick and Birmingham Canal the firm lost the contract on the Rowington–Shrewley embankment in July 1794, almost as soon as it was awarded, when Robert made an improper financial suggestion to the Engineer. In 1793 Pinkerton, with Murray, was given the contract for the Ulverston Canal, Edward Banks (q.v.) cutting the first sod. They were obliged to abandon the contract in 1795 due to lack of money to pay the workers and at the same time were discharged from the Lancaster Canal for lack of progress. Rennie remarked of Pinkerton's claims that 'nothing seems extraordinary from

that quarter'. With perhaps eight projects simultaneously in hand all over the country in the early 1790s problems of cash flow and adequate supervision are unsurprising. Contract conditions were harsh-on the Leicester Navigation not only were they subject to 5% retention on payments for work done, but also a £1,000 retained guarantee for two years after completion of work. If their estimates gave only marginal profits, with a large number of contracts going awry they could soon find themselves in financial difficulty.

In difficult ground such problems were inevitable. In 1793 John Pinkerton was given the contract for the Barnsley Canal after agreeing to Jessop's prices. As work progressed the Resident Engineer, Samuel Hartley, became dissatisfied with Pinkerton's progress and concerned about overpayment. There was, however, a major problem with excavating the Cold Hiendley cutting, where Pinkerton spent £540 on gunpowder for blasting, and costs were 2s a yard rather than the expected 6d. There were also problems with the puddle, in part caused by delays in filling. Jessop's report was sympathetic to Pinkerton, who was not forced to make good the deficiencies, carried out by James Rhodes, but the company wanted his bond. The case went through Chancery and although the judge criticised Jessop, he ws legally obliged to rule against the contractor and in 1812 Pinkerton was obliged to pay £3,137 to the company.

Despite these several problems the Pinkertons continued to win contracts and their work for Rennie at East, West, and Wildmore Fen, and Chapman on drainage works in the East Riding were major successful civil engineering projects. Even here, Pinkerton's litigious nature was in evidence, Sir Joseph Banks remarking to Rennie on 18 October 1804: 'We have not redressed Pinkerton, but we all incline to do something for him. The worst of all is that his claims which originally were for £4,300 I think, they, now he has observed some inclination in us to assist him seems swelled into a demand of £7,200'. Nonetheless, Rennie was inclined in 1806 to make allowance for the problems of indurated clay and allow Pinkerton some profit.

As an example of their later work, in April 1807 James Pinkerton, in partnership with a Mr. Ormond, successfully tendered for Lot 4 on the Glasgow, Paisley and Ardrossan Canal. The contract, which involved the excavation of 84,370 cu. yd. of material and the construction of eleven bridges, was probably the easiest on the canal, where opening was delayed because of the hard rock encountered in some areas. Nevertheless the contract was forfeited in 1808 with only 12,417 cu. yd. of excavation completed. Pinkerton successfully tendered to complete the work with a new partner, Orr, and by May 1809 the work was virtually finished. In November 1810 it was still the only contract to have been completed and at £4,938, only marginally over the original tender of £4,641, by far the most satisfactory progress on the canal.

John died in 1813 and left c. £7,500 in his will; he also left a fine reputation 'distinguished by sincerity of manners and gentleness of temper. As a husband, uniformly affectionate: as a parent, laudably indulgent, as a friend invariably constant ... He was attached to liberty [and] continually embraced the Unitarian doctrine'.

Although the Pinkertons have been overshadowed by the activities of Sir Edward Banks and Hugh McIntosh (qq.v.) in the two decades following John's death, they continued working. Perhaps their most substantial works of this period were those in South Wales associated with the Kidwelly and Llanelly Canal, again, regrettably, distinguished by controversy over poor workmanship.

MIKE CHRIMES

[Basingstoke canal papers, Hampshire RO; PRO/RAIL minute books, 800, Aire & Calder Canal; 806, Barnsley Canal, 810, Birmingham Canal; 815, Calder Hebble Navigation; 816/3, Chester Canal; 818, Coventry Canal; 828/1, Erewash Canal; 829, Gloucester & Berkeley Canal; 844, Lancaster Canal; 880, Ulverston Canal; 881, Warwick and Birmingham Canal; Rennie reports, ICE Library; Rennie papers, NLS; Glasgow, Paisley and Ardrossan Canal minute book, C96/2002, Scottish RO; Grundy reports, vols. x–xi, Brotherton Collection, Leeds University; Cherry Cobb embankment papers, East Yorks. CRO; Tender for Lechlade summit, Thames & Severn canals, 1783, TS 193/1, Gloucs CRO; Market Weighton Drainage papers, Hull University Library; *Gentleman's Magazine*, **1** (1813) 390, 672; House of Lords (1836) Select Committee ... Railway ... between London and Brighton, *Evidence of James Pinkerton, jnr*, G. Jackson (1966) *Grimsby and The Haven Company*, 17–21; W. N. Slatcher (1968) Barnsley Canal, *Transport History*, **1**, 48–66; D. F. Gibbs and J. H. Farrant (1970) The Upper Ouse navigation 1790–1868, *Sussex Industrial Archaeology*, **1**, 22–40; S. R. Broadbridge (1971) John Pinkerton and the Birmingham canals, *Transport History*, **4**, 33–49; C. Hadfield (1972–1973) *The Canals of Yorkshire and North East England*; A. W. Skempton (1974) William Chapman (1749–1862), civil engineer, *Trans Newc Soc*, **46**, 45–82; J. Boyes and R. Russell (1977) *The Canals of Eastern England*; C. Hadfield and A. W. Skempton (1979) *William Jessop, Engineer*; A. W. Skempton (1985) The Engineering works of John Grundy, *Lincolnshire History and Archaeology*, **19**, 65–82; P. A. Stevens (1992) *The Leicester and Melton Mowbray Navigations*; P. A. L. Vine (1994) *London's lost Route to Basingstoke*; P. A. L. Vine (1995) *London's lost Route to Midhurst*]

Publications

John Sr.:

1801. *An Account of the Cause lately arbitrated between the Birmingham Canal Navigation and J. Pinkerton*

Works

James Sr.:

1767–1768. Adlingfleet Drainage (with John Dyson), 170,000 cu. yd., c. £5,000
1768–1770. Driffield Navigation (with John Dyson), c. 5 miles, c. £13,000
1770. Cherry Cobb embankment (with John Dyson)
1769–1772. Laneham Drainage, excavations and earthworks, c. 5,000 acres, c. £11,000
1778–1779. Chester Canal (Beeston Brook–Nantwich) (with Robert P. Jessop), c. 6 miles, £3764 13s

James and John Sr.:

1770. Bishop Soil Sluice
1772–1775. Market Weighton Drainage, sluice lock and mother drain, c. 6 miles, 870,000 cu. yd., c. £8,000
1774–1776. Hedon Haven and Navigation
1775–1778. Selby Canal, £20,000
1775–1778. Aire & Calder, £30,000
1776–1780. Market Weighton Drainage, drains and deepening Fowley River, etc., 61 miles, c. £14,000
1776–1780. Calder & Hebble Navigation
1777. Billinghay Embankment
1777–1779. Erewash Canal, excavation, 11¾ miles, c. £21,000

John Sr.:

1783–1785. Birmingham Canal, Broadwater extension
1783–1785. Thames and Severn Canal, Lechlade summit and Cirencester branch, tender
1784–1788. Coventry Canal, Fradley–Dennisbrook–Fazeley, c. 8 miles, c. £9,000
1785. Calder–Hebble Navigation, Battye Ford Cut
1785. Marshland Fen, Norfolk, Knights Gool Sluice
1785–1787. Dudley Canal Tunnel, 3,172 yd.
1786–1787. Pinkerton's (Union) Sluice Rye Harbour, c. £3,400
1787–1789. Birmingham Canal, Minworth–Fazeley, c. 10 miles, £12,300
1788–1789. Malltraeth Embankment, Anglesey, including sluice, £442, (with J. Dyson)
1788–1794. Basingstoke Canal, 37½ miles, £153,462
1790–1792. Ouse Navigation, Lewes–Laughton, upper Ouse, Thomas, James & Francis, £15,000
1791. Rother Navigation, tender, £8,740 (by George Pinkerton)
1792–1795. Lancaster Canal (with John Murray), Keer Aqueduct, £844 1s 6d
1793–1795. Ulverston Canal (with John Murray)
1793–. Gloucester and Berkeley Canal
1793–1799. Barnsley Canal
1801–1814. East, West and Wildmore Fen Drainage, c. 9,000 acres, £51,448, expenditure by J. P. (Boston)
1811. Black Sluice and Sea Banks
[1812. Wade embankments, Essex?]

Francis:

1792–1793. Lower Ouse navigation

[1794–1798. Witham–Horncastle navigations? (with John Dyson), resident at Wragby, Lincs.
1799–(1810). Beverley and Barmston Drainage (with James Jr.), 25 miles cuts, 20 miles embankment, total cost £115,000 (there were other contractors, including Dickinson, Dyson, Nicholson and Smith (qq.v.))
1800–1804. Muston and Yedingham Drainage, Sea Cut, (with James Jr. and Thomas and Simon Harmer), c. £2,000, 3 miles

George:

1790–1793. Ipswich–Stowmarket Navigation (with John Dyson), c. 10 miles, £1,500, discharged

James Jr.:

1791–1792. Leicester Navigation, river line (with George & Francis)
1792. Leicester Navigation, Forest line and railways (with George and Francis)
1807–1809. Glasgow, Paisley and Ardrossan canal, Lot 4, £4,938
?1808–1809. Butterwick Marsh Enclosure?

Robert:

1794. Warwick and Birmingham Canal, Rowington and Shrewley embankments

Pinkerton and Allen:

1812–1818. Kidwelly and Llanelly Canal and Tramroad (and works in 1820s)
1817–. Pembrey Old Harbour
1818–. Kidwelly Sea Defences
1825–(1838?). Pembrey New Harbour and Tramroads

Pinkerton and Hemming:

1827–1831. Leith Harbour, western breakwater

John (Lukin?) Jr.:

[1819–1825. Thames–Medway Canal, 6½ miles, £300,000 total]
1823. Yantlet Creek

PLEWS, John (1795–1861), contractor, was born on 1 May 1795 at Thornton Steward, Yorkshire, the son of a farmer. Having received some education and mathematical training, in 1812 he moved to London and obtained work with the contracting firm of Jolliffe and Banks, Edward Banks (q.v.) being his uncle. His first post was at Waterloo Bridge where he remained until its completion in 1817.

Plews remained with the firm until 1835, when both partners died. He became a trusted employee, having for the last twelve years the general superintendence of their accounts and acting as one of the executors of Banks's will. As such, it was his responsibility to wind up the contracting business. Plews was associated with several specific projects including Southwark Bridge, Customs House quay wall, Hermitage Entrance to London Docks, the river wall of the Victualling Office, Deptford, extensive works at Woolwich

and Sheerness Dockyard, the Serpentine Bridge, Staines Bridge, London Bridge, and the Nene Improvement at Sutton Wash for the North Level Drainage. He gave evidence regarding the foundations of London Bridge to the London Bridge enquiry of 1831. He was heavily involved for five years (1817–1822) with the quarry owned by Banks at Merstham.

Following the settlement of Banks's affairs, Plews worked as a docks contractor and engineer, acting as engineer for Cardiff Docks for about sixteen years (1837–1852). He was joined by his son, John Plews Jr., who acted as resident engineer on work there and at the Northumberland Dock on the north bank of the River Tyne. Following the completion of Northumberland Dock in 1857, Plews retired to his residence in New Road, Kennington, and became involved in the affairs of the parish of St. Mary's, Lambeth. He joined the Institution of Civil Engineers in 1856 and died on 23 June 1861.

His son worked in New Zealand and Melbourne, Australia before returning to Brighton in 1877. He died around 1879.

MIKE CHRIMES

[Telford mss, ICE Library; Minutes of Evidence (1831) *London Bridge*; Minutes of Evidence (1836) *London–Brighton Railway (Stephenson's Line)*; *Min Proc ICE*, **21** (1861) 564–567; I. Gall (1984) *A Budapesti Duna-hidak*; R. W. Rennison (Ph.D., 1987) *The Development of the North East Coal Ports*]

Works
1835–1836. London Docks, Shadwell entrance reconstruction
1837(?). Danube Bridge, Buda-Pesth, feasibility report for Hungarian Government
1837–1852. Cardiff Docks, Engineer; Bute Docks 1837–1839, 19½ acres, £222,757
1852–1857. Northumberland Dock, Tyne, engineer, 15 acres, £200,000

POLLOCK, John (fl. 1815–1832), road surveyor, superintended the Glasgow and Carlisle and Lanarkshire roads, whose improvement had been reported on by Thomas Telford (q.v.). In 1832 he surveyed a section of the Great North Road, Telford remarking that he was the most judicious and experienced road surveyor he knew.

MIKE CHRIMES

[Gibb papers, ICE archives; House of Commons, *Select Committee on the Great North Road* (1832), Report, appendix III]

POOLE, Moses (fl. 1827–1849), Patent Agent, of the Patent Office, was elected a Member of the Institution of Civil Engineers in 1827 and resigned in 1849. He gave evidence to the Parliamentary Select Committee of 1829 on the costs and delays of the existing Patent system. According to John Farey (q.v.), Poole was the most important patent agent active in the early nineteenth century.

MIKE CHRIMES

[Membership records, ICE archives; Minutes of Evidence (1829) *Select Committee on the Law relative to Patent Inventions* (HC 332), 82–88; *Min Proc ICE*, **9** (1849), 97]

POOLEY, Henry, Captain (*c.* 1799–1843), military engineer, probably attended the Royal Military Academy at Woolwich and was commissioned a Second Lieutenant in the Royal Engineers in 1816. He was promoted to Lieutenant in 1825 and posted to the Rideau Canal project in Upper Canada (now Ontario). Early in his posting, Pooley was tasked with constructing a temporary timber king-post bridge over a stream in what is now downtown Ottawa. To this day the current bridge and area are referred to as Pooley's Bridge. As a junior officer, he worked closely with the Commanding Engineer, John By, RE, and in fact was dispatched to England as the latter's emissary to explain the planned changes and cost increases for the canal project. On his return to Canada, Pooley was responsible for the supervision of construction of the locks and weirs at Smiths Falls. He was promoted to Second Captain in 1837, retired with full pay in 1840 and died at Bath on 6 November 1843.

MARK ANDREWS

[T. W. J. Conolly (1898) *Roll of Officers of the Corps of Royal Engineers*; E. F. Bush (1976) *The Builders of the Rideau Canal*, Parks Canada MSS Series No. 185, Ottawa]

POPINJAY, Richard (fl. 1560–1587), surveyor, spent most of his career at Portsmouth, rising from mason's clerk to surveyor. His area of responsibility also included surveying fortifications and harbours in the Channel Islands and the Isle of Wight.

Included in his work at Portsmouth was the new quay, built in stone rather than wood, possibly as a result of his own estimate in 1563. His plan shows a waterfront of 157 ft. with side quays of 42 ft. and 80 ft. with positions indicated for three cranes; the total cost was £1,000. In 1568 he was commissioned to rebuild 'the great platforme where the great ordinance Lyeth,' as well as bridges and a barbican. Popinjay's plan for the platform shows a rectangular masonry structure with the seaward side joisted and planked and the landward side filled with gravel and earth. In 1584 a new program of works was proposed, lasting nine years and costing £9,000. However, he retired as surveyor in 1585 and most of the work was carried out by William Spicer as task work.

Popinjay's main claim to interest is as a map maker where he has been described as the most notable English exponent before Saxton. Of particular note is a survey of St. Helier (1563) which shows anchorages, submarine rocks, quays and roads as well as a scale, a feature still unusual in England at the time. He also produced surveys of Castle Cornet (1566), Hurst (1567), Southsea (1567)- where he also repaired the breach in the

sea wall—and Carisbrooke (1583). Even after his official retirement he produced plans of Portsmouth Harbour and the Isle of Wight in 1587.

Popinjay's will was proved on 9 March 1595, most of his property in Portsmouth being left to his wife, Bridget.

SUSAN HOTS

[*King's Works*, 4; E. Lynam (1950) English maps and map makers of the 16th century *Geographical Journal*, **116**, 9–28]

PORTINARI, Giovanni (fl. 1525–1566), was an Italian military engineer who worked both for Henry VIII, possibly designing the fort at Sandown and on mining operations in Boulogne, and Elizabeth I, who gave him a pension of 500 crowns and employed him at Portsmouth (1560), Le Havre (1562) and as a consultant on the design of Berwick fortifications from 1560–1566. This involved a scheme by William Cecil to set a standard for the design of masonry walls for fortifications by building test walls to the specifications both of Portinari and Richard Lee (q.v.). The work was begun by Richard Popinjay (q.v.) to ensure neutrality but there is no record of a completion or judgement in either's favour. Nothing further is heard of Portinari after this date.

SUSAN HOTS

[*Cal SP, 1553–1558*, 196; *King's Works*, **IV**, 1982]

POTTER, James (1801–1857), was the youngest son of Joseph Potter (q.v.) and was articled to William Brunton (q.v.) of the Eagle Foundry, Birmingham. After a spell in his father's office, he was elected a Member of the Institution of Civil Engineers in 1824 and then was Resident Engineer for Thomas Telford (q.v.) during the construction of the second Harecastle Tunnel, which was completed in a remarkably short time—February 1825 to March 1827—for the Trent & Mersey Canal. His salary, in his first post of responsibility, was £500 p.a., which was twice his father's salary as County Surveyor; his ability in conducting these works earned him a glowing testimonial from Telford to the Canal Committee:

I have, in a former letter … stated, that the proper management of the tunnel works, joined with those of the Knypersley reservoir, including the expenses to which any person, having this charge, is unavoidably subjected, is a service deserving a salary of £500 a year, and this is what would have been required by any other duly qualified person I could have provided, and this more especially when there is such an unusually great demand for persons of this description.

Although it has so happened that Mr. James Potter, a very young man, has been appointed to this charge, and that this salary may appear large for his first essay, yet, as by the due exertion of talents and unwearied assiduity aided by the experience and judgement of his father, the whole of the complicated, and in many instances dangerous operations, have been arranged and in every aspect carried on in a well regulated and successful manner under the management of persons in whom the company have cause to place confidence, I do not see any reason why less remuneration is due, than what a stranger would have required. (6 August 1825, TR/48, ICE)

As is clear from this, Potter was also Telford's Resident Engineer, in 1825–1829, on the construction of Knypersley Reservoir for the Trent and Mersey Canal Company. This project was not entirely successful and led to the appointment of James Trubshaw Jr. (q.v.) to undertake remedial works. Potter was employed on the improvement works on the northern section of the Oxford Canal (1830–1834), where Frederick Wood was Engineer and John Ferguson Resident Engineer. He appears to have been responsible for the design of the cast iron aqueduct over the old Lutterworth road out of Rugby. Later he was involved in the supervision of the construction of several railways, particularly those which required tunnelling, until in 1852 he obtained his final appointment, as Engineer-in-Chief of the Manchester, Sheffield and Lincolnshire Railway. He died in 1857.

P. S. M. CROSS-RUDKIN

POTTER, Joseph (*c.* 1755–1842), was an architect and builder of Lichfield with a considerable practice in Staffordshire and the neighbouring counties. Early in his career he was employed by James Wyatt to supervise the latter's alterations to the cathedrals at Lichfield (1788–1793) and Hereford (1790–1793). He was also associated with Wyatt in the repair of St. Michael's Church, Coventry (1794)—now the old cathedral—and at Lichfield he became the established cathedral architect after Wyatt's employment had ceased.

Potter was employed in 1798 by the magistrates of Staffordshire to prepare plans for a House of Correction at Wolverhampton and in 1800 John Rennie (q.v.) was writing to him to ask him to use his influence with the magistrates about the problems with the contractor for Wolseley Bridge. In that year too it was decided to dispense with the services of James Trubshaw Sr. (q.v.) as Surveyor of the County Bridges and in 1801, after advertisements had been placed in the Stafford and Birmingham newspapers for a replacement, Potter was chosen from a shortlist of four, the others being relatively unknown to posterity. During the first years of his Surveyorship, he was paid large sums of money for work on many bridges, rising in 1808 to £4,045 (out of a bridge account for that year of £7,495 and a total County expenditure of £13,953), presumably executing the works by direct labour. At Easter 1809, his remuneration was changed to £80 p.a. plus 5% of costs of all new bridges over £300. By 1830 it would seem that these additional fees amounted to more than

£300 and at Michaelmas 1831 his salary was again changed, to a flat rate of £250 p.a.

Potter assumed responsibility in 1800 for completing Wolseley Bridge to John Rennie's design after the successive failures of James Trubshaw Sr. and John Varley (q.v.) to do so. For the next few years, the design of new County bridges was done by Rennie (Radford at Stafford and Darlaston at Stone) with local input by Potter. In 1806 the magistrates ordered Potter to make designs for Stretton but it seems that Rennie provided four alternatives for this, too. From about 1810, however, Potter was providing designs of his own, though his architectural treatment of major bridges such as Hopwas and Fazeley was a rather thin version of Rennie, probably on grounds of cost. For smaller bridges he provided simple yet elegant designs, such as the series in upper Dovedale, which Jervoise calls the Staffordshire type, and describes as exceptionally well built.

The rate of expenditure on County bridges fell off very sharply in the years 1833–1835 and remained low thereafter, before the arrival of railways caused a change in traffic patterns. It is to be presumed that Potter had brought the bridge stock to a state of tolerable repair. In 1836 the Surveyorship was split and James Smith of Repton was appointed for the northern division of the county. Potter continued to be responsible for the southern division, together with the County buildings in Stafford, and his salary was divided 100:150 until his resignation in January 1842, when Smith assumed his duties and his total remuneration. During Potter's tenure of office, more than 60 masonry bridges were rebuilt or widened under his supervision.

About 1822, he had been joined by his third and youngest son, James Potter (q.v.) and the following year, with his son's help, Potter prepared the plans for Chetwynd Bridge, Alrewas, a cast iron structure whose central span is 75 ft. In its general design and detailing, it also derives quite heavily from Rennie's Southwark Bridge, completed in 1819. When in 1828 a replacement was required for High Bridge, Handsacre, Potter provided the plans for a single, cast iron span of 140 ft. and 14 ft. rise, comparable in size to the series of bridges which Thomas Telford (q.v.) had built. The ironwork for both of the Staffordshire bridges was provided by the Coalbrookdale Company.

In his son's obituary notice, Potter is described as Engineer to the Grand Trunk (Trent & Mersey) Canal Company. Unfortunately the Trent and Mersey minutes are no longer available and the dates of his appointment cannot be found. In a surviving letter of 1820, when he was involved with John Rennie in selecting a site for a new reservoir at Knypersley, he is called Surveyor. It would appear that he felt himself unable to take sole responsibility for the major work then required, the duplication of the Harecastle tunnel. The deposited plans bear Potter's signature but first Rennie, and after his death, Telford was employed as consulting engineer, and his son acted as resident engineer. Potter continued to act as engineer to the canal company until 1837, when he was replaced by James Trubshaw Jr. (q.v.).

He was recognised sufficiently as a canal engineer to act for the Worcester and Birmingham Canal Company in their dispute with John Woodhouse (q.v.) in 1815, though as James Hollinsworth (q.v.) acted for Woodhouse and William Crosley (q.v.) had been the Resident Engineer, the hand of John Rennie may be detected. Hollinsworth described Potter as obstinate and surly, though in this opinion he was betraying his partisanship.

Potter lived in St. Johns Street, Lichfield, and was sufficiently respected to be elected a Town Councillor in the first election after the Municipal Reform Act of 1835. He died at Lichfield on 18 August 1842, aged eighty-six. His elder sons, Robert (c. 1795–1854) and Joseph (c1797–1875), practised as architects, the former in Sheffield and the latter taking over his father's practice in Lichfield.

P. S. M. CROSS-RUDKIN

[Staffordshire RO; QS Order Books and Treasurer's Accounts; Warwickshire RO; Telford mss, ICE; *ICE Proceedings*, **xvii** (1858); Jervoise (1932) *Ancient Bridges of Mid and Eastern England*; G. M. Binnie (1987) *Early Dam Builders in Britain*; Colvin]

Works

1800–1802. Wolseley Bridge, Colwich contractor or supervisor of direct labour for completion, after dismissal of John Varley (q.v.)

The following masonry bridges in Staffordshire designed by Potter cost over £1,000 and were built by contractors under his supervision:

1805–1814. Cheddleton
1808–1811. Bridgeford
1811–1814. Weston
1811–1815. Hulme End
1812–1817. Hopwas
1815–1817. Tutbury
1816–1817. Froghall
1823–1825. St. Thomas, Stafford
1826–1828. Alton
1826–1829. Fazeley
1830–1832. Beamhurst

Cast iron bridges:

1824–1827. Chetwynd, Alrewas
1829–1831. High, Handsacre

His architectural works are listed in Colvin.

POWNALL, Robert (fl. 1735–d. 1780), was engineer of the Weaver Navigation. The River Weaver had been made navigable from Frodsham via Northwich up to Winsford (18 miles) by building eleven locks between 1730 and 1735 under the direction of Thomas Robinson (q.v.). The timber-framed locks and adjacent timber weirs were constructed within locally widened sections of the river, not in cuts. Repairs were made from time to time but by the late 1750s improvements were becoming necessary in a navigation then carrying some 40,000 tons annually.

In December 1757 Pownall, who had been employed as Clerk of the Tonnage at Winsford since 1735, with four others, surveyed and reported on the state of the works. In March 1758 he was promoted to the post of Inspector and Superintendent of the navigation at a salary of £50 p.a. and Henry Berry (q.v.), the Liverpool dock engineer who had recently completed the Sankey Canal, was brought in as consultant.

Under new plans four of the locks downstream of Northwich were to be rebuilt, in brickwork, on cuts up to a mile in length and to rather larger dimensions (about 70 ft. long and 17 ft. wide with 4½ ft. of water on the sills). However, in March 1759 Northwich lock collapsed after a rock-salt pit had given way. In the same month a flood breached the banks of the newly made Pickerings cut; a few months later the weir across the river blew up immediately after construction; and in March 1760 one of the walls of Pickerings lock suffered a foundation failure.

John Golbourne (q.v.) was consulted in November 1759, Berry's services were no longer retained, and in July 1760 Pownall became Director and Surveyor of the Works at £70 p.a. This was raised to £130 p.a. in July 1761 though he had the additional duty of collecting tonnage dues at Northwich.

Pownall successfully built a new lock at Northwich. He rebuilt the weir and lock at Pickerings and by 1762 completed a new cut and lock at Saltersford. He also planned and built (with approval by James Brindley (q.v.)) a short tributary extension of the navigation to Wilton Mill in 1764–1765.

In 1766 with Hugh Henshall (q.v.) he surveyed a line from Harecastle to Winsford for a possible connection with the proposed Trent & Mersey Canal and in the same year he and Golborne represented the Weaver Navigation interest in parliamentary proceedings on the Trent & Mersey Bill.

Back on the Weaver, Pownall built a new cut and lock at Winnington (1768–1771) and at Acton Bridge (1773–1778), in each case receiving an extra 30 guineas 'for his extraordinary trouble and attendance'.

Pownall resigned his post as collector of tonnage in 1778 due to ill health but remained inspector and manager of the works at £100 p.a. 'in consideration of his long services and faithful discharge of his duty'. These services came to an end with Pownall's death in 1780.

By that time tolls were bringing in almost £5,000 p.a. from an annual carriage of some 100,000 tons. Pownall was succeeded by John Johnson who remained in post until 1807. He built a lock and weir at Frodsham and replaced the original locks above Northwich by larger ones in new cuts following the pattern established by Pownall.

A. W. SKEMPTON

[T. S. Willan (1951) *The Navigation of the River Weaver in the Eighteenth Century*, Chetham Soc., 3rd series, vol. 3; C. Hadfield and G. Biddle (1970) *Canals of North West England*]

Works

1760–1778. Weaver Navigation, improvements, £24,000

POWSEY, John (*c*. 1734–1799 or later), lived at Blackwall and was appointed District Surveyor for Wapping and Limehouse in 1774. During a period of nearly thirty years from 1771 he worked on occasion as an engineer at the famous Blackwall shipyard for its owner, John Perry (1743–1810). In this period works included repairs to the Old Wet Dock, 1½ acres in area built between 1659 and 1661 by the then owner, Sir Henry Johnson (q.v.), a new dry dock (the fourth in the yard) in the 1780s, and the large Brunswick Dock.

In preparation for the new dock, Perry in 1788 obtained permission to make an embankment along the river in front of 17 acres of marshland, which he already owned, adjacent to and east of the yard. Powsey designed the dock and directed construction, assisted by Thomas Williams, an 'officer' of the shipyard, who probably acted as clerk of works.

Work on the dock began in March 1789 and the opening ceremony took place on 20 November 1790. The dock comprised two linked basins, each with its own entrance: the 8 acre western or Great Basin, with a single pair of gates 44 ft. wide and sills 16 ft. below high water spring tides, capable of admitting 1,250 ton East Indiamen unladen, and the eastern or Little Basin, 3 acres in area with an entrance lock 160 ft. between gates, a width of 32 ft. and 13½ ft. of water on the sills at HWST, used by smaller ships. The Great Basin was equipped with a 120 ft. high mast house, the first ship being masted there in October 1791, while the east quay of the Little Basin was leased for the use of whalers in the Greenland trade.

The docks' walls were built with timber piles and sheeting supported by land ties, Powsey later commented that the project was undertaken 'at a very great expense' and 'we had great difficulties to struggle with, and some unforeseen'. One problem was the discovery of fossilised trees 12 ft. below the surface in the Thames alluvium, which must have hampered construction of the dock walls. The cost is not known, but by comparison with the Howland Dock at Rotherhithe, built 1697–1700 by John Wells (q.v.) in a similar manner, and allowing for increases in wages during the eighteenth century, the expenditure on Brunswick Dock is likely to have been not less than £25,000. William Daniell published an excellent view of the dock in 1803.

In 1804 the dock was incorporated in the new East India Docks, engineered by John Rennie and Ralph Walker (qq.v.). The dock company paid £35,600 for the eastern portion of the yard, including Brunswick Dock, and a large piece of ground to the north. The mast house remained in use until 1862.

By the early 1790s congestion of shipping in the Thames had become intolerable. William Vaughan in 1793 proposed as a solution the creation of enclosed wet docks equipped with warehouses. He envisaged docks at Rotherhithe, the Isle of Dogs, and St. Katharine's, but thought a site of Wapping to be the most suitable. His ideas received strong support. In February 1794 Powsey was commissioned to survey an area of some 34 acres in Wapping, and shortly afterwards produced plans which formed the basis of what eventually became the London Docks, built between 1801 and 1805 under the direction of John Rennie. Powsey gave evidence to the Committee on the Bill before Parliament for the London Docks in March 1799. Nothing later is known of his career.

A. W. SKEMPTON

[Minutes of evidence (1799) on the Bill for Rendering more Commodius and for better Regulating the Port of London (City Plan) House of Commons; H. Green and R. Wigram (1881) *Chronicles of Blackwall Yard*; P. Banbury (1971) *Shipbuilders of the Thames and Medway*; A. W. Skempton (1979) Engineering in the Port of London 1789–1808, *Trans Newc Soc.*, **50**, 87–108; S. Porter (1994) Poplar, Blackwall and the Isle of Dogs, *Survey of London*, **44**, 556–563]

Publications

1794. Plan of the river with the proposed docks, in William Vaughan (1794) *Plan of the London Dock, with some Observations*

Works

1789–1790. Brunswick Dock, *c*. £25,000

PRESTON, Thomas (fl. 1750–d. 1777), was appointed master mason in February 1750 when work began on the piers of Ramsgate Harbour. As completed, the piers were 1,486 ft. apart, running south from the shore and then turning inward to enclose an area of 46 acres with an entrance 300 ft. wide in water 6½ ft. deep at LWST (tidal range 16 ft.); both piers were about 1,600 ft. long. The first plan had been drawn up in 1749 by Edward Gatton, of the Ordnance Board, who proved by borings that the sea bed consisted of chalk with a thin covering of sand. After much discussion the harbour Trustees decided, in January 1750, that the east pier should be entirely of stone, the west pier would be timber framed with chalk infill for a length of 615 ft. (to LWNT) followed by stone. The plan then adopted was essentially the same as Gatton's.

By December 1751 the piers were approaching low water mark and a meeting was held on how best to found the stone piers below water. Of several suggestions the most promising was submitted by William Etheridge (q.v.). He had worked on Westminster Bridge, the foundations of which were laid in caissons devised by Charles Labelye (q.v.), engineer of the bridge. Etheridge suggested a similar method, coupled with a tool of his own design for making a trench in the chalk and levelling the ground on which the caisson would rest. His scheme was adopted and he was appointed Surveyor of Works in January 1752.

Work proceeded steadily. By September the timber portion of the west pier had been completed and Preston had built some 300 ft. of the east pier. That summer George Semple (q.v.), on a visit from Dublin seeking information on building foundations under water, was most impressed by the Ramsgate works and refers to Preston and Etheridge as two very courteous gentlemen. By 1754 the stone extension of the west pier had been advanced 68 ft., plus 138 ft. of foundations, and the east pier was now 757 ft. long. Thus about one-third of the work had been done and there was every hope of completing the pier in the next five or six years.

However, in December 1753 William Ockenden, one of the harbour Trustees, proposed to divert the west pier inwards, thereby decreasing the harbour width to 1,200 ft. on the misguided view that this would save cost. The proposal was accepted and work began on the new alignment of the west pier in the summer of 1754. But the idea met with strong and widespread opposition, leading to a petition being submitted to Parliament in February 1755. Etheridge, who supported the revised plan, and Preston both gave evidence, along with many others. Work came to a halt. Etheridge left. Asked to look into the matter, Sir Peircy Brett and Captain J. P. Desmaretz (q.v.) reported in December 1755. They recommended a plan retaining the original width, but with piers extended into water 8 ft. deep at low tide, and a deepened part of the harbour adjacent to the east pier. A Parliamentary committee approved their plan in February 1756, ordering that the work done for contracting the harbour width be taken up.

It was not until 1760, however, that the Trustees resumed activities, and then on a plan differing little from the 1750 original, with Preston now acting in effect, if not in name, as engineer. Work began in January 1761 and by June 1762 the abortive works had been removed and good progress was being made on the piers. They were completed in 1772.

Nevertheless, not all was well. As the piers advanced seawards so an ever-increasing quantity of sand accumulated within. By 1773 the harbour had been choked up to levels in places above low water, and this despite the removal by spade and dredger of more than 50,000 tons of sand during the past three years.

In the summer of 1773 the Trustees sought advice from John Smeaton (q.v.). He could not come immediately as he was about to visit Ireland. Another call for help came early next year. He made his site visit in April and reported in October 1774, proposing a basin in the north-west corner of the harbour from which, having been filled at high tide, water could be discharged through sluices to scour out the sand.

Soon after Smeaton's report, Preston proposed an important modification: a wall going right across the northern part of the harbour and a pair of gates (as well as sluices) in the wall to permit access for ships into the basin. The Trustees approved his plan and work began in 1776 under Smeaton's direction with Preston as resident Engineer.

Preston died in 1777. Edmund Hurst, who had been his foreman since the beginning, took over and Henry Cull was appointed foreman. The basin wall, gates and six sluices were successfully completed in 1781 and by 1783 the sand was almost entirely cleared away. Up to 200 ships of 200 to 400 tons burden could now be accommodated in what was the largest enclosed harbour of the eighteenth century in England.

A. W. SKEMPTON

[E. Gatton (1749) Evidence on Ramsgate Harbour Plan, *JHC*, **25**, 764–765; Evidence to the Committee on Ramsgate Harbour (1755) *JHC*, **27**, 213–230; G. Semple (1776) *A Treatise on Building in Water*; J. Smeaton (1791) *An Historical Report on Ramsgate Harbour*; R. B. Matkin (1977) The construction of Ramsgate Harbour, *Trans Newc Soc*, **48**, 53–71; A. W. Skempton (1981) *John Smeaton*]

Works
As Master Mason:

1750–1755. Ramsgate Harbour, piers, £45,000
1761–1772. Ramsgate Harbour, piers, *c.* £100,000
1776–1777. Ramsgate Harbour, basin, completed 1781

PRICE, Francis (*c.* 1704–1753), surveyor and author, held the post of Surveyor to Salisbury Cathedral from 1737 until his death. Nothing is known of his family and training and his name first occurs when his *Treatise on Carpentry* was published in 1733; the book was recommended by Hawksmoor as 'a very Usefull and Instructive Piece'. This view was confirmed by the publication of a further five editions in 1735, 1753, 1759, 1765 and 1768. The second edition carried a new title, *The British Carpenter: or, a Treatise on Carpentry*, and was enlarged from twenty-eight to forty-four plates and from thirty-three to fifty-two pages. The topics included floors, partitions, girders and domes; trussed roofs were extensively illustrated; and scarf joints for beams and ties in trusses were described. Small bridges were illustrated, one of which used vertically laminated arched ribs. The book concluded with a table of required sizes of oak and fir for components such as posts, joists, rafters, etc. in 'small' and 'large' buildings. Seasoned timber was required; strength properties were not given and no attempt was made to calculate the required sizes of the members. It was a pattern book. Although subsequent issues were described on their title pages as editions, the only change made was that the type was reset in 1753. This meant that the second edition effectively had an impressive thirty-five year life with four reprints.

Prior to Price's treatise, timber had been treated as one topic in multi-material builders' manuals. Price and James Smith (1733) were the first to devote complete books to carpentry and joinery. Price's treatment pointed the way for Peter Nicholson's (q.v.) books and for Thomas Tredgold (q.v.) to treat timber as an engineering material.

Price made considerable repairs to Salisbury Cathedral; his book represents the first serious attempt to analyse the structure of a major Gothic building. He also acted as architect for some buildings around Salisbury. A portrait by George Bear (1747) is in the National Portrait Gallery.

L. G. BOOTH

[J. Smith (1733) *The Carpenter's Companion*; D. T Yeomans (1992) *The Trussed Roof: its History and Development*; Colvin (3)]

Publications

1733. *Treatise on Carpentry*
1753. *A Series of particular and useful Observations ... upon ... the Cathedral Church of Salisbury*

PRICE, Henry Habberley, MRIA (1794–1839), civil engineer, was born in 1794, the son of Peter (1739–1821) and Anna Price, Quakers. Peter Price had been born at Madeley, Shropshire, and was apprenticed at Coalbrookdale before working at the Carron Ironworks. He then set up furnaces in Pennsylvania, Maryland, North Carolina and Virginia before returning to England in the War of Independence. On his return he worked as Agent for the Coalbrookdale Company in London and in 1781 he married Anna Tregelles. He and his father-in-law, Samuel Tregelles, were original shareholders with the Fox family in an iron foundry at Perran Wharf, near Falmouth (1791), and Neath Abbey Ironworks (founded 1792); Price Sr. became manager at Neath Abbey from 1800, the works enjoying a very high reputation and becoming important builders of Cornish engines. From 1810 Henry Price's brother, Joseph Tregelles Price, was involved in running associated collieries and, increasingly, the ironworks. Price Sr. died in 1821 by which time the works was dominated by the Price family interests, Henry Habberley Price joining his brother in its management.

Price had begun his apprenticeship as an engineer in 1808 and subsequently worked on inland navigation improvements in Ireland, on the Suir and the Lee at Cork. It seems likely he was working for Alexander Nimmo (q.v.), who later gave him some of his books. He is claimed to have been particularly involved in the building of steamships, but there may have been some confusion with his namesake son, who with his brother, Edwin, carried on the company with

their uncle after Henry I's death. As well as steam engines the firm built gasworks throughout South Wales and the West Country; from 1829 it built locomotives and also had interests in collieries and other works across South Wales.

While South Wales was one focus of his activity, Price seems to have practised widely as a civil engineer, although many of his schemes remained unfulfilled. For the Tees Navigation Company in 1824, however, he produced two reports on the river, recommending its shortening downstream of the Mandale Cut, in 1810 completed under William Chapman (q.v.). To facilitate work, two surveys were carried out under his direction by William Edgeworth (q.v.), following which the Portrack Cut was completed in 1831, so making the channel from the sea to Stockton much more easily navigated. Attempts were also made to increase the river's velocity downstream of the new cut and so reduce silting but the construction of groynes to do so, although having a beneficial effect on part of the river, actually caused the silting up of the coal staithes of the Stockton and Darlington Railway—located in a wider section of the river at Middlesbrough—and so led in 1835 to a fierce argument between the two parties. Price eventually reached agreement with the railway company the following year, by which time his association with the river Tees would seem to have ended. Again in the North-East, Price had initially been appointed consulting engineer to the Clarence Railway—it obtained financial support from the Exchequer Bill Loan Commissioners on the recommendation of Thomas Telford (q.v.)—but this appointment lasted only until 1833. He became involved, too, with its branch, the Durham South West Railway.

Price had previously met Telford in connection with the South Wales Mail Road improvements, probably through Alexander Nimmo In early 1823 he had been representing local interests in South Wales, arguing for improvements to the Mail Road. Telford was appointed engineer and although few improvements were made Price became engineer in 1827 for new ferry piers at Old Ferry Passage across the Severn Estuary.

In 1831 Price reported on Swansea Harbour and his South Wales connections led to his becoming involved in railway schemes between London and South Wales. With William Brunton (q.v.) he proposed a rail line between Bristol and London, terminating at Paddington. He opposed in 1834 the original Bill for the Great Western Railway on behalf of Lord Jersey, a major landowner in South Wales but subsequently gave support in 1835 for the successful Bill, attending several public meetings in preparation for the Act. As engineer to Bridgwater Corporation he opposed the line of the Bristol and Exeter Railway where it affected the interests of the Parrett Navigation and in 1835 drew up plans for a ship canal and dock for Bridgwater. He also drew up a plan for a railway between Gloucester and Hereford.

Almost contemporaneously, he became involved in the controversy regarding the rail route between London and Brighton, giving evidence on behalf of the Croydon Railway. He also worked on a scheme to link the Croydon Canal by canal to Vauxhall and the Grand Collier Dock Scheme in Deptford and Rotherhithe, with rail connections to the South Western and London and Greenwich lines. After the canal Bill had failed, the proposals involving rail connections were revived as the grandiosely entitled Westminster Bridge, Deptford and Greenwich Railway,

Like many engineers involved in railway schemes in the 1830s, few of his projects came to fruition. Price joined ICE in 1832 as a corresponding member. He died in 1839.

MIKE CHRIMES

[Membership records, ICE; ICE Annual report (1840) *Min Proc ICE*, **1**; House of Lords (1829) *Minutes of Evidence … Clarence Railway Company* [HL42], 74–83; Tees Navigation Company, Committee Orders, 1824 1839, J. Whishaw (1837) *Analysis of Railways*; C. Hadfield (1967) *The Canals of South West England*, 55; L. Ince (1977) Neath Abbey Ironworks, *Industrial Archaeology*, **11/12**, 21 37]

Publications

1824. *First Report of a Survey of the River Tees* [Stockton]
1824. *Second Report…*
1831. *Report on the Improvement of the Harbour of Swansea*
1835. *Report on the Harbour of Falmouth*
1835. *Report on the Establishment of a Ship Canal and Docks at the Port of Bridgwater*
1835. *Report … respecting the Completion of the Grand Surrey Canal … and a Railway … to … Deptford*
1836. *Map of the Gloucester and Hereford Railway*
1837. *Plan of the Grand Collier Docks*

Works

1827. Old Passage Ferry Piers, River Severn
1828–. Tees Navigation
1829–1833. Clarence Railway

PRIDDEY, John (*c.* 1731–177?), canal engineer, was probably christened at St. Andrews, Droitwich, on 16 May 1731, the son of Joseph and Elizabeth Priddey. The Priddey family were well established in the Droitwich area and in December 1767 Priddey gave Parliamentary evidence regarding the poor state of roads in this area.

Following the passage of the Droitwich Canal Act in January 1768 the company appointed James Brindley (q.v.) Inspector of the Works, and John Priddey was appointed Clerk of Works, or resident engineer, at £90 p.a. He arranged for the lock contractor, John Bushel, to receive some training on the Trent and Mersey Canal and set about obtaining labour and organising construction. Although the initial capital was exhausted by March 1771, the works were well executed,

much of the credit due to Priddey's supervision, and he was able to move on to further works, initially as assistant Clerk of Works on the Oxford Canal in April 1772. As Brindley had died, Samuel Simcock (q.v.), the Clerk of Works, seems to have been acting as engineer on the canal, work progressing very slowly. Priddey was dismissed in April 1773.

In June 1774, together with Thomas Dadford Jr. (q.v.), Priddey was asked to survey a line for a possible Stroudwater Canal, for a fee of 15 guineas (Dadford was to be paid 11 guineas). Their plans were accompanied by an estimate of *c.* £16,500. Following further consultation with Thomas Yeoman (q.v.) a prospective committee appointed Samuel Jones (q.v.) of Boston as resident engineer, planning to use powers granted in a 1730 Act. Jones's working relationship with the proprietors broke down and in February 1775 Priddey was appointed resident engineer and surveyor at £100 p.a. For this he was expected to spend at least a quarter of the year on site. Priddey found the committee as difficult as his predecessor and after some argument he was finally paid off in August 1776. His subsequent career is not known.

MIKE CHRIMES

[Minutes of Oxford Canal Company; RAIL PRO; Evidence regarding repairing the roads from the City of Worcester… (1967) *Journal House Commons*, **31**, 476–477; C. Hadfield (1969) *The Canals of the West Midlands*; C. Hadfield (1970) *The Canals of the East Midlands*, 2nd edn.; M. Handford (1979) *The Stroudwater Canal*]

Works

1768–1771. Droitwich Canal, resident engineer
1772–1774. Oxford Canal, resident engineer
1775–1776. Stroudwater Canal, resident engineer

PRITCHARD, Daniel (*c.* 1777–1843), contractor, emerged as a specialist tunnelling contractor by the 1820s. It is likely he began as a miner and began working on civil engineering in the first decade of the nineteenth century. He was probably a relative of John Pritchard, who contracted for work on the Lancaster Canal. Daniel's contracts included the construction of Bosworth and Crick tunnels on the Grand Union Canal, Maida Hill and Islington tunnels on the Regents Canal, Kendal tunnel on the Lancaster Canal, and the Thames and Medway Canal Strood tunnel near Rochester. The latter was notable for the first use of transit instruments in its alignment as well as its great size. His most famous tunnel contract was perhaps the second Harecastle tunnel (1825–1827) on the Trent and Mersey Canal, described by Thomas Telford (q.v.) as a 'Tunnel more perfect than any other that was hitherto constructed'. The Grand Union Canal tunnels were also notable achievements, unveiling difficult geological conditions. Pritchard had as his partner on several of these works his son-in-law, William Hoof (q.v.).

Surprisingly, given this experience, Pritchard never really became involved in railway contracts. His most important was the Berkswell contract on the London–Birmingham Railway, on which he used a *Planet* type locomotive, the *Sun*, for moving spoil.

The 1841 Census records that at his home were Sarah Pritchard, aged thirty-five, of independent means—presumably his daughter—and Mary W. aged six, Sarah aged four, and Thomas P. Pritchard aged four months, presumably grandchildren. Pritchard died on 7 July 1843 at the age of sixty-six at his home, Broad Meadow House, Kings Norton, Worcestershire. His sons Daniel Baddeley Pritchard (1827–1872), Samuel (b. *c.* 1817) and John, were also contractors. Daniel worked in Australia and John tendered for several railway contracts including (successfully) contract 2C on the Bristol and Gloucester Railway in May 1843.

Pritchard was arguably the first specialist civil engineering contractor with a high reputation in the 1820s for his tunnelling work.

MIKE CHRIMES

[Telford mss, ICE archives; J. S. Tucker (*c.* 1830) Specifications, ICE Archives; PRO; F. W. Simms (1838) *Public Works of Great Britain, ICE Min. Proc*, **38** (1874), 293–295; P. A. Stevens (1972) *The Leicester Line*; P. K. Roberts (1977) *British Canal Tunnels*, D.Phil., Salford University]

Works

1811–1813. Grand Union Canal, Husbands Bosworth Tunnel, 1,070 m
1812–1814. Grand Union Canal, Crick Tunnel, 1,397 m
(*c.* 1812–1820). Regents Canal, tunnels, Islington 878 m, Maida Hill, 249 m
1815–1817. Lancaster Canal, Hincaster Tunnel, 378 yd. (346 m), *c.* £11,500 (as John Pritchard)
1815–1818. Lancaster Canal, Farleton embankment
(?1820). Regents Canal, Chelsea engine house
1819–1824. Thames & Medway Canal, Strood Tunnel, 3,931 yd. (3,608 m) (with W. Hoof)
1825–1827. Trent & Mersey Canal, (second) Harecastle Tunnel, 2,926 yd. (2,676 m), *c.* £115,000 (with W. Hoof)
1827–1829. Macclesfield–Trent & Mersey Canal, Hall Green branch, 1½ miles (with W. Hoof)
1830. King's Scholars Pond, London, sewer, £5,200
1830–1832. Wigan branch railway, 7 miles
1834–1836. London–Birmingham Railway, Berkswell contract (with W. Hoof?)

PROVIS, Henry (1760–1830), civil engineer, the son of Alexander and Catherine Provis, was baptised on 26 July 1761 at St. Gluvias, Cornwall. He probably received his earliest training as a surveyor before joining Sir John Soane as an 'outdoor clerk' or Clerk of Works in April 1791; he continued in this employment until the spring of 1802. Initially he superintended the building work at

Wimpole Hall, Cambs., moving to Tyringham Hall, Bucks, in February 1795. In February 1802 he was corresponding with Soane regarding his salary— 2 guineas a week if in the office or 4s 6d a day if in the country. Tyringham is close to Newport Pagnell and Provis must have been aware of the work in progress on the Grand Junction Canal. Evidently dissatisfied with his salary he found employment on the Canal under James Barnes (q.v.) in June 1802, supervising all of the works south of Uxbridge to Brentford and Paddington, at a salary of £200 p.a. In August 1803 he was appointed to superintend the construction of Wolverton bank and aqueduct, locks at Stoke Bruerne, and the superintendence of Blisworth tunnel.

Following the opening of the canal and the retirement of Barnes in 1805, Provis was made responsible for the Southern District of the canal. When Wolverton Aqueduct collapsed in February 1808 he supervised the contract for a temporary wooden trunk. His work included responsibility for the design and construction of its Aylesbury branch between 1811 and 1815. For this branch he was paid £500 p.a. initially, rising to £1,000 p.a., including—from December 1812—a Clerk. Thomas Telford (q.v.) was impressed with this work, which he reported in May 1815 as being 'in a very perfect state'. Shortly after its completion, in 1816 he left the Canal's service.

While working on the canal Provis was involved in other activities such as his prize-winning entry for the Regent's Canal tunnel of September 1812, which he copied from a William Jessop (q.v.) design. He acquired some experience in water supply, surveying for a feeder to the Paddington Branch in 1809, and constructing a reservoir for the Hampstead Waterworks. In 1809 he also surveyed the proposed Western Canal from the Grand Junction at Marsworth to Abingdon, this with William Whitworth and John Barker. Of greater importance was his work as engineer for Tickford cast iron bridge (1810), a 25 ft. wide, 58 ft. span structure over the river Ouse, based on the patent of Thomas Wilson (q.v.) and cast by Walker of Rotherham (q.v.). The bridge was strengthened in 1900 and 1976 but can be regarded as a fine example of cast iron architecture of the period. Provis would have had experience of bridge construction under Soane at Tyringham and he designed a masonry bridge at Newport Pagnell, the North Bridge, at the same time. It appears that his own design for a masonry structure at Tickford was overruled by the local committee who negotiated with Wilson, and Provis's role became largely supervisory.

Shortly before his resignation from the Grand Junction Canal Provis was appointed by the County of Buckingham to design a new bridge over the Ouse at Sherington, completed in 1818. Although urged by the local committee of Justices to bid for the work he clearly had reservations about the propriety of being both Engineer and Contractor. A compromise was reached wherein his son John was formally awarded the contract although Henry actually undertook the work. During the maintenance period of this bridge Provis was appointed County Surveyor for the northern district of Buckinghamshire, at Epiphany Sessions, 1819, at a salary of 2 guineas for each day engaged on county work. When the County Surveyor for the southern district died in 1822 Provis was appointed for the whole county. In that capacity he carried out a number of bridge repairs and also designed a chapel and new rooms for the gaol in Aylesbury, and other building works in Buckingham. In 1827 he submitted plans for a new bridge at Padbury for the Wendover to Buckingham Turnpike Trust and acted as Surveyor to the Trustees during construction. He was working on a new bridge at Olney at the time of his death.

Provis had married Mary Marshall at St. Leonard's, Shoreditch, in 1788 and his sons, William Alexander and John (qq.v.) both practised as civil engineers under Telford. A further son, Henry Thomas (fl. 1809–1836), was elected a Member of the Institution in 1825, then resident in Paddington. He presumably had assisted his father in Grand Junction Canal works but details of his career are uncertain. Drawings, no longer extant but listed in the Institution of Civil Engineers inventory of drawings, were attributed to him c. 1850. They relate to the Abingdon branch of the Grand Junction Canal (1809–1810), and the Holyhead–Chester Road between Tal-y-pont and Penmaenmawr, and Pwll Crochon [sic] and Llandulas, suggesting that he, like John, assisted William Provis on the Holyhead Road. Henry Provis Jr. married Caroline Barrett on 25 May 1813 at St. Martin's-in-the-Fields and is known to have had three children: Ellen Sargeant, William Alexander, and Mary Marshall, all baptised at Somerset Street Baptist Chapel, Bath, between 1818 and 1823. This is consistent with a survey of the Shaftesbury–Honiton Road carried out in 1825. Either may have carried out a survey of the Macclesfield canal for Telford in 1825. Henry Thomas was presented with a silver snuff box by the Shropshire Union Railways and Canal Company in 1851, suggesting a continuing involvement in civil engineering.

Henry Provis joined the Institution of Civil Engineers in 1820. Resident at Bridge House Sherrington, Buckinghamshire, from 1815, he died there on 23 August 1830, aged seventy.

J. B. POWELL

[Membership records, ICE; Shaftesbury–Honiton Road papers, Telford mss, ICE archives; Tide in the Thames, OC 120, (1827); Notebook 15, Sir John Soan's Museum; Q/AB/4, 16, 32, 35, 69, Q/AG/10/11, A/AM/7, Q/SO/34, T/3/9 and 31/9, Bucks RO; RAIL: 830/ 41, PRO; *Gents. Mag.*, **100**(2) (1830) 380; A. T. Bolton (1923) *The Works of Sir John Soane, Architect*; A. H. Faulkner (1968) The Aylesbury branch of the Grand Junction Canal, *J. RCHS*, **14**(1), 12–15, **14**(2), 26–29; A. H. Faulkner (1972) *Grand Junction Canal*; J. G. James (1979) The cast iron bridges of Thomas Wilson 1800–1810, *Newc Soc Trans*; S. Evelergh

(1994) The building of Padbury Bridge, *Records of Buckinghamshire*, **36**, 165–172; E. A. Labrum (1994) *Civil Engineering Heritage: Eastern and Central England*, 183; A. Provis (2000) in lit.]

Publications

1829. On a supply of water for the metropolis, *Mechanics Magazine*, **11**, 43–44

Works

1802–1816. Grand Junction Canal, resident engineer, southern district
1810. North Bridge, Newport Pagnell, engineer
1810–1811. Tickford bridge, cast iron arch
1811-1813. Aylesbury branch canal, 6 miles
1816–1818. Sherington Bridge, engineer
1819–1822. Buckinghamshire, Northern district, County Surveyor
1822–1830. Buckinghamshire, County Surveyor
1823–1824. Ickford Bridge, repairs
1824–1825. Aylesbury gaol improvements, architect
1826–1827. Castlethorpe Bridge, repairs, engineer
1827–1828. Padbury Bridge, Engineer, £1,800
1830. Olney Bridge, design

PROVIS, John (1801–185?), civil engineer, was born in 1801 in Middlesex, the son of Henry Provis (q.v.) and younger brother of William Alexander Provis (q.v.), whose assistant he became. Initially trained by his father, whom he assisted on Sherrington Bridge (1816–1818), in 1819 John Provis's address was in Paddington, as was his brother's, doubtless arising from their father's past involvement on the Grand Junction Canal, but in 1822 he removed to Shrewsbury so he could take charge of supervising the testing of the ironwork for the Menai Suspension Bridge, then being made by William Hazledine (q.v.). It is likely that he had already assisted his brother in some of the Holyhead Road surveys, not least because his brother was in charge of roads over a very large area in north Wales and the borders, where communications were poor, a view supported by a statement given in Parliamentary Evidence that he had worked under Thomas Telford (q.v.) for fourteen years.

Provis remained in a junior capacity until 1826 when, following the completion of the Menai Bridge, his brother resigned and John was appointed Assistant Engineer to the Commissioners, with similar responsibilities for road improvements in north Wales, at a salary of £150 p.a. This salary was raised to £250 in 1829 when he assumed responsibility for the Holyhead Harbour works. Following Telford's death in 1834 John Provis was appointed Engineer for the Roads from Chester and Shrewsbury, to Holyhead, John Macneill (q.v.) being made Engineer for the English section of the Holyhead Road. This situation continued until 1844 when Macneill resigned and Provis took over the whole route. While acting as Engineer the Menai bridge was severely damaged by gales in 1836 and 1839 and it is of note that it was William, in association with Thomas Rhodes (q.v.), who reported on the damage and organised the repairs. In the mid-1840s the question of improvements to Holyhead Harbour was raised and outside consultants were again called in, suggesting that John Provis was a competent surveyor and administrator but lacked the engineering ability of his brother and other contemporaries.

When living at 16 Union Street, Marylebone, Provis was elected a Member of the Institution of Civil Engineers in 1819 and he remained a member until the late 1830s. He married his sister-in-law, Emily Stanton (b. 1809), the daughter of Thomas Stanton (q.v.), in 1833. They had 5 children: John (b. 1834), Thomas John (1836–1913), William Alexander (b. 1838), Mary (b. 1839) and George Stanton (1840–1894). The family lived at Salt Island, Holyhead until the early 1850s but after that time the career of John Provis is unknown; he probably retired to Ellesmere.

MIKE CHRIMES

[Census records, Anglesey RO; Q/AB/32 and Quarter session records, Bucks CRO; Membership records, Notebooks of John Provis, ICE archives; Membership lists, (Royal) Society of Arts (1819–1825); T. Telford (1833) Diary, NLS MS9157; W. A. Provis (1828) *A Historical and Descriptive Account of the Suspension Bridge constructed over the Menai Straits ...*; Commissioners for ... Holyhead Road (1826–1850) *Reports*; Minutes of Evidence (1836) *House of Commons Select Committee on Turnpike Trusts and Tolls*, 91–102]

Publications

1829. *Report* [on] *Swilly Rocks, Appendix to Sixth Report, Commissioners Holyhead Road*
1836–1850. *Report(s) to Commissioners ... London to Holyhead Road*

Works

1826–185?. Shrewsbury–Holyhead Road, (assistant) engineer, *c*. £60,000
1827–1833. Menai Straits, navigation improvements, resident engineer, £9,400
1829–1850. Holyhead Harbour, resident engineer, *c*. £30,000
1844–1850?. London–Holyhead Road, engineer, *c*. £24,000

PROVIS, William Alexander (1792–1870), civil engineer, was born at Wimpole, Cambridgeshire, on 5 May 1792 where his father, Henry, was employed by Sir John Soane as a Clerk of Works. There is a strong possibility that Provis was the first pupil of Thomas Telford (q.v.), for whom he worked briefly in 1805, probably when Telford reported on his father's work on the Grand Junction canal at Wolverton, but then spent three years as his father's pupil before rejoining Telford in 1808, accompanying him on his Highland tours. Provis was responsible for the civil engineering drawings of a number of important Telford projects, including those for Bonar Bridge, and the Menai Suspension Bridge.

The earliest dated survey for which Provis was responsible was the Stamford navigation (1810) to Oakham, probably working with Hamilton Fulton (q.v.). In 1810 Parliament was considering the question of road communications with Ireland via Holyhead and Telford was asked to report on the roads from London to Holyhead, with Hamilton Fulton surveying the existing roads in England while Telford looked more closely at the Welsh roads. This included a crossing of the Menai Straits, for which John Rennie (q.v.) had already proposed cast iron bridges. A major problem was the issue of navigation under any bridge and Telford proposed (1811) either a single 500 ft. span cast iron arch at Ynys Moch or a three 260 ft. span cast iron arch bridge by Swilly Rocks. The design of these bridges was largely identical to the Bonar Bridge, designed on the model of one of Telford's 1801 proposals for a replacement London Bridge, with the cast iron lozenge spandrels which became Telford's characteristic design. The published drawings for the bridges, and a similar design for Conway, were drawn by Provis. Progress on the Menai Bridge stalled and Telford turned his attention to a 1,000 ft. suspension bridge across the Mersey at Runcorn (1814). Provis assisted in the experiments associated with this work and was responsible for the published drawings for the proposal of 1817.

At the same time he assisted in the original survey for the Glasgow–Carlisle Road of 1814, which was presented to Parliament the following year. In that year Telford was appointed Engineer to the Holyhead Roads' Commission and a short time later Provis was appointed as Assistant Engineer. Early progress, with John Stanton and John Straphen (q.v.) as contractors, was hampered by lack of funds. His first reports were on the London to Shrewsbury Road, via Coventry, and that from Stone Bridge to Shrewsbury (1817), but soon Provis was based in Wales, carrying out surveys of Rhyall Hill (1817), the Menai Straits (1818) and the Chirk–Gobowen Road. Although he carried out a survey of the Prior's Lee section in Shropshire in 1821, his attention was largely focused in Wales for the next eight years. He was made resident engineer for the Chester–Bangor Road and in addition to these responsibilities was involved in the evolution of the Menai suspension bridge design, drawing up the 1819 proposal with stiffened catenary chains. To adequately supervise all these tasks was clearly impossible given the poor communications the road was intended to remedy. His brother, John Provis (q.v.), took over the main supervising role, leaving Provis with the opportunity to improve the design.

On completion of the Menai Bridge in 1826 Provis took over responsibility for the Birmingham and Liverpool Junction Canal proposals, which included the associated extensive dock works at Ellesmere Port, and for which Telford was the engineer. Having drawn up the designs, Provis later acted as contractor on part of the canal. The drawings for the original line for the canal had been deposited in November 1825 but a revised plan, for which Provis was largely responsible, was deposited in June 1826, necessitated by landowners' objections. Further modifying legislation followed in 1827, including the Newport branch laid out by Provis. The idea of the canal, as suggested by Telford to the Birmingham Canal and other canal interests, was to meet the challenge of the railways by shortening the links between Birmingham, Liverpool and Manchester by twelve miles. The main line went from Wolverhampton to Nantwich, and thus on to Ellesmere Port via the Ellesmere and Chester Canals' new branch, and to Middlewich where it joined the Trent and Mersey; the Newport branch linked with the Shrewsbury Canal. The resident engineer for the canal was Alexander Easton with John Wilson (qq.v.) acting as contractor for the first three lots, and subsequently that between Church Eaton and Autherley.

Telford was at the same time looking at a southward link from Birmingham to London, but attempts to get a Bill through Parliament in 1825 and 1828 failed. Possibly the potential of this scheme led Provis to turn down Telford's offer to act as engineer on the Birmingham and Liverpool Junction and it was not until June 1829 that he tendered for lot 4 on the Birmingham and Liverpool line. He subsequently took on much of the Newport branch. Unfortunately lot 4 involved a very high embankment at Shelmore and deep cuttings at Grub Street, which proved very difficult to stabilise. The original tender price of £114,678 was soon exceeded and, rather than two years, the construction period extended to five and a half years; in 1843 problems were still being experienced. Because of Telford's poor health William Cubitt (q.v.) was consulted by the canal company to advise on the work; he was generally supportive of Provis's methods and recommended continuing to place material until the slopes stabilised. Progress was not improved until work had been completed on Provis's other contracts on the canal, freeing up his resources. The canal eventually cost c. £800,000, more than double Telford's original estimate of £388,474 for the main line.

It is unclear why Provis took up contracting. From 1824–1829 he was busy with a number of surveys: the Nantlle Railway an early proposal for the Festiniog Railway, the link of the Wyrley and Essington Canal; the Staffordshire and Worcestershire Canal and reservoir (1826); a canal from Worcester to Gloucester along the Severn Valley (1826); the Wiltshire and Buckinghamshire from Abingdon to Aylesbury (1828); and, with John Urpeth Rastrick (q.v.), the Stourbridge extension canal (1829). The fact that most of these schemes, and the London–Birmingham Canal, came to nothing may have encouraged Provis to change direction. Notwithstanding this, Provis was still consulted about other schemes such as the improvement of the Forest line and railway of the Leicester Navigation (1832–1833)

Following Telford's death in 1834 Provis, to whom Telford had bequeathed £400, took over

his London house at 24 Abingdon Street, Westminster. He had already acquired a property, The Grange, at Ellesmere, where he spent much of his leisure time in searching for fossils and other geological pursuits. This was retained in the family until c. 1903.

After the completion of the Birmingham and Liverpool Junction Canal Provis designed a timber bridge at Poole Harbour and, as contractor, carried out further improvements to Ellesmere Port. He was asked at short notice to advise on the South Eastern Railway and a 'South Eastern' route to Brighton (1836–1837), in which he was associated with Henry Robinson Palmer (q.v.), and gave Parliamentary Evidence. Cubitt became the engineer for the line. Associated with this was the 11 mile South Eastern and Maidstone Railway for which Provis was engineer. His proposal for an extension of the Glasgow & Garnkirk Railway along the line (to be converted) of the Union Canal, was not realised. These railways were followed by proposed improvements to canal communication between Birmingham and Manchester (1837–1838) and a partial railway conversion of the Shropshire Union Canal system (1845), also not realised. The latter scheme was transmuted into an arrangement with the London and North Western Railway by which the Shropshire Union shared income with the railway. As a result of his efforts with this work Provis's health broke down.

The lack of success in the railway age was not untypical of the career of the engineers of the 1820s, but Provis's reputation remains high, principally for his work in strengthening the Menai suspension bridge following damage in 1836 and 1839, work which secured the bridge for another sixty years service. He was called on again by the Commissioners of Woods and Forests, then responsible for the Bridge, to comment on the implications of the proposed Chester–Holyhead Railway. He was critical of the proposal to use the bridge for railway traffic, and supported the idea of a new bridge as suggested by James Walker (q.v.)

Provis's last project concerned the Dee navigation, where in the 1840s he had succeeded Telford as consulting engineer to the Company. He retired following the passing of an Act of Parliament in 1851.

Provis married Harriott [sic], the daughter of Thomas Stanton (q.v.), on 18 July 1825 at Ellesmere, Shropshire. They had no children. He had joined the Institution of Civil Engineers in 1819 and bequeathed £500 to its Benevolent Fund. Provis died on 29 September 1870 and was buried in Kensal Green Cemetery.

MIKE CHRIMES

[Membership records, ICE archives; On canal locks (OC36) and On Lime (OC37), ICE archives; Caledonian Canal sketchbook, Telford mss, ICE Library; Holyhead Road papers, Telford mss, ICE Library; Gibb papers, ICE archives; Plan of Nantlle Railway, XM/Plans/R/1, Caernarvon RO; House of

Commons, *Committee on Holyhead Road* (1811) Report (HC197); House of Commons, *Select Committee on Holyhead Road* (1815) 1st report (HC313), 14–15; House of Commons, *Select Committee on Holyhead Road* (1817) 4th report; House of Commons, *Select Committee on Holyhead Road* (1817) 2nd report (HC217), 16; House of Commons, *Select Committee on Holyhead Road* (1819) 3rd report; *Menai Bridge* (HC256); 6th report, *Turnpike Trusts*; Commissioners of Holyhead Road (1816–1826, 1836–1840) reports; House of Commons, *Committee on Brighton Railway* (1836) *Minutes of Evidence* (HC195.2) 687–707; House of Commons, *Committee on the London and Dover (South Eastern) Railway Bill* (1836) *Minutes of Evidence*, 4–21; J. Rickman (ed.) (1838) *Life of Thomas Telford*; F. Whishaw (1837) *Analysis of Railways*, 219–220; House of Commons, *Select Committee on ... Navigation of the River Severn* (1841) *Minutes of Evidence*, 1–16; H. Tuck (1846) *Railway Shareholders Manual*, 266; Tidal Harbours Commission (1846) 2nd report, Appendix B, Dee, 292–319; Admiralty (1849) *Chester–Dee Navigation Improvements*; Memoir (1870) *Min Proc ICE*, **31**, 225–230; L. T. C. Rolt (1958) *Thomas Telford*; M. Hughes (1966) A truck shop at Dinas, near Betws-y-Coed, on Telford's London to Holyhead Road, *Trans Caern Hist Soc, 139–148;* E. *Beazley (1985) Madocks and the Wonder of Wales*, 222; E. A. Shearing (1990) The Birmingham and Liverpool Junction Canal: planning and construction in the Norbury district, *RCHS J.*, **30**(2), 73–85; **30**(3), 138–146; P. A. Stevens (1992) *The Leicester and Melton Mowbray Navigations*]

Publications

1810. *Plan of the proposed Stamford Junction Canal* (by H. Fulton, drawn by W.A.P.)
1811. *Design(s) for a Bridge proposed to be erected over the Menai* (drawn by W.A.P.)
1811. *Plan of Improvements in the road between Carlisle and Portpatrick as proposed by Mr. Telford and Mr. Rennie* (drawn by W.A.P.)
1813. *Bonar Bridge. General elevation* (drawn by W.A.P.)
1815. *Map of the Mail Road between Glasgow and Carlisle with the proposed Improvements* (surveyed by W.A.P. for Telford)
1817. *Design for a Bridge over the River Mersey at Runcorn* (drawn by W.A.P. for Telford)
1817. *Report on the present State of the Road from London by Coventry to Shrewsbury*
1817. [second] *Report on the Road from Stone Bridge to Shrewsbury*
1818. *Design for a Bridge over the Menai Straits* (by Telford, drawn by W.A.P.)
1820–1821. *Report(s) to Commissioners of ... Shrewsbury and Bangor Ferry Road* (1st and 2nd)
1828 *An Historical and Descriptive Account of the Suspension Bridge constructed over the Menai Straits in North Wales*
1836. *Report ... relative to the Menai and Conway Bridges* (with T. Rhodes, in 13th report of Holyhead Road Commissioners)

1837. *Suggestions for improving the Canal Communications between Birmingham, Wolverhampton, Shropshire, Cheshire, North Wales, and Manchester, by means of a New Canal*
1839. Observations on the effects of wind on the suspension bridge over the Menai Straits ..., *ICE Trans.*, **3**, 357–370
1844. *Report on the Menai Bridge*, 1843 (HC 212)

Works

1815–1826. Holyhead Road, Shrewsbury–Holyhead, resident engineer, 106 miles, £197,000
1818–1826. Holyhead Road, Menai Suspension Bridge, resident engineer, £185,346
1822–1826. Chester–Bangor Road, resident engineer, 54 miles, completed 1829, £16,500
1822–1826. Chester–Bangor Road, Conway Suspension Bridge, resident engineer, £51,000
1826–1835. Birmingham and Liverpool Junction Canal and branches, engineer and contractor, c. 25 miles, £300,000
1835–1836. Poole harbour bridge
1836. Menai Bridge repairs, engineer, £6,100
1839–1840. Menai Bridge, strengthening, engineer
1839–1843. Ellesmere Port, works, contractor, £70,000

R

RABBARDS, Ralph (fl. 1550–1600), military engineer, appears to have been born overseas and brought to England as a child. As a young man he was caught up in the persecution of the Protestants in the reign of Queen Mary and was imprisoned and tortured, only escaping death by the accession of Queen Elizabeth. From 1550 he was engaged in developing 'warlike engines', not only in Britain but also abroad. His experiments are listed, ranging over medicinal and toilet preparations to the preparation of fire-works and 'A Firy Chariott without horses to runne upon the battaile and disorder it, that no man shall be able to abide or come nighe the same, and will be directed even as men will to tourne, to staye, or elles to followe and chase the enemye in their flighte.' There are no illustrations and the document is simply a catalogue, similar to the *Century* of the Marquess of Worcester of seventy years or so later. In 1592 Rabbards caused to be printed George Ripley's *The Compound of Alchemy*.

In 1588, at the time of the Spanish Armada, he compiled an album of drawings nearly all relating to military affairs with written descriptions on the facing pages. There are seventy drawings of various engines and implements followed by twenty-three diagrams of fort-ifications. In this album are some notes about his foreign military service in the Netherlands, Germany and Italy. The topics covered include eight sketches of lifting and forcing jacks, a crane weighing machine, an air mattress, a diver, eight sketches of bridges, two sketches of pumps, two sketches of a pile driver, two sketches of a powder mill, three sketches of printing presses and two sketches of a hand cornmill.

A. P. WOOLRICH

[Rhys Jenkins Collection, Science Museum Library; Lansdowne MSS,121 No. 14, BL]

RADFORD, William (c. 1790–1856), contractor's agent, was one of the many agents employed by Hugh McIntosh (q.v.). He is believed to have originated in Buglawton, Cheshire. His earliest known association with McIntosh was as agent for work at Pembroke dockyard (1813–1816). In the 1820s he was agent at the King William Victualling Yard, Plymouth dockyard, and was possibly involved with work from 1834 on Blackfriars bridge, where his second son, also called William (1816–1854), is known to have been employed.

Radford joined the Institution of Civil Engineers as an Associate in 1841 'because his engagements as a contractor for public works qualify him to concur with the civil engineer in the advancement of professional knowledge.' He was then resident at Upper Ground Street, Blackfriars. His son worked on the Hanwell Viaduct for McIntosh, and various works for the Grand Junction Waterworks. This may have provided him with the contacts to succeed William Anderson (q.v.) as engineer to the Regent's Canal. The latter part of his career was largely involved in work for James Meadows Rendel (q.v.). Radford had three more sons. George Kent (c. 1825–1908), worked with his older brother on the Regents Canal, Birkenhead Docks, and other Rendel projects before emigrating to Canada where he worked on the Grand Trunk Railway. He worked eventually in the U.S.A. as did Edward (1811–1920) the fourth son.

MIKE CHRIMES

[Rennie reports, ICE Library, ICE membership records; Obituary notice of W. Radford (Jr.); *ICE Min Proc*, **14** (1855) 136–137; M. M. Chrimes (1995) Hugh McIntosh (1768–1840), national contractor, *Trans Newc Soc*, **66** (1994–1995) 175–192]

RANGER, William (1800–1863), civil engineer, was the son of William Ranger, a builder and surveyor of Brighton. From 1823 until 1833 he was active in Sussex. His early work included drainage and coastal defence contracts at Pevensey and Worthing. His contracts in the Brighton area included the approaches to the suspension chain pier designed by Samuel Brown (q.v.) and Shoreham suspension bridge designed by William Tierney Clark (q.v.). He removed to Dean's Yard, Westminster in 1833, shortly after he had taken out a patent for architectural stone (4 December 1832, amended 1834).

In the 1820s a concrete sea wall at East Cliff, Brighton, was erected using formwork to support the sea face of the wall, with grey chalk lime mixed with shingle and sand from the local beaches. The wall was up to 60 ft. high, at which

point it was 22 ft. 6 in. thick at the base, with a slope of 1:3. It was designed by Thomas Cooper, with John Wright as resident engineer and William Lambert as contractor. Ranger must have been aware of this and developed a system of precast concrete units using timber moulds. The concrete was mixed alongside with boiling water to produce a vigorous slaking action and then packed into the moulds, which generally took the form of blocks, although more decorative moulds were used. Experience suggested that the blocks needed to be rested to gather strength for two to three months after the forms had been struck.

The most famous applications of Ranger's system were for the Royal College of Surgeons, Lincoln's Inn Fields, and Wellington Barracks; it was also used for wharves at Woolwich and Chatham Dockyard. Ranger's concrete generally comprised eight parts sandy gravel to one part Malling lime. Problems were experienced with Ranger's work. For the foundations of a storehouse in Chatham dockyard the concrete was only as wide as the brick superstructure and underpinning was only necessary. The wharf walls at Chatham and Woolwich were vulnerable to water and frost action, and Charles William Pasley (q.v.) urged the use of facing masonry on such structures. Ranger's agent on the works at Chatham and Woolwich was Charles Nixon (1814–1873), whose father, William, had worked for John Nash (q.v.).

When railway construction began in earnest, Ranger took on contracts 1B, 2B, 3B and 3B extension (March–August 1836) on the Great Western Railway in 1836 but by October 1837 it was apparent that he would be unable to complete these contracts. In June 1838 the company determined to relet them and there followed a notorious series of court cases between Ranger and the GWR, not resolved until 1855. Nixon remained as agent on the contract for the bridge over the Avon near Bristol, acting for McIntosh (q.v.) who took over the work.

In this situation Ranger sought more regular employment and worked for the Board of Health on more than sixty inquiries into the condition of towns. This was combined with lecturing on civil engineering to the Royal Engineers, at Chatham, and the College of Civil Engineering at Putney.

He died on 12 September 1863 at his home in St. George's Square.

MIKE CHRIMES

[F. Smith archive, ICE archives; C. W. Pasley (1838) *Observations on Limes, etc.*, 18–25; C. W. Pasley (1847) *Observations on Limes, etc.*; Evidence (1848) Metropolitan Sanitary Commission, **1**, 360–369; *Builder*, **21** (1863) 672; E. T. Macdermot and C. R. Clinker (1964) *History of the Great Western Railway*, 54–55, 94; M. M. Chrimes (1996) Concrete foundations and substructures: a historical review, *Structures and Buildings*, ICE *Proc*, **116**, 344–372; Corps RE, QS, **1**]

Works

c. 1823. Brighton, chain pier approaches

1830–1833. Shoreham, suspension bridge
c. 1833–1836. Chatham and Woolwich Dockyards, improvements
1836–1838. Great Western Railway, contracts completed by McIntosh

RASTRICK, John (1738–1826), millwright and civil engineer, was the fourth son of a family of seven, born between 1726 and 1743 to William Rastrick (1695–1772)—a millwright—and his wife, Sarah Donnison. He was baptised on 15 October 1738 at Mitford, Northumberland, and lived at nearby Morpeth for much of his life; little is known of his early years. In 1766 he married Catherine Reed by whom he had three daughters and, following her death, he married Mary Urpeth in 1774; by her he had three sons and three daughters. In 1781 he was described as 'engineer, millwright, patent chain maker' and was at that time noted as being 'High Constable for the West Division of Morpeth Ward'.

Rastrick became a member of the Worshipful Company of House-carpenters and Millwrights of Newcastle in 1764 and in 1777 he patented the 'Imperial barrel-churn', fitted internally with ribs and having a handle for rotation. Later he entered into a long and bitter argument with Andrew Meikle (q.v.) as to the invention of a threshing machine, patented by Meikle in 1788 but claimed by Rastrick as having been invented by him in 1784–1785; the dispute was seemingly resolved to Rastrick's advantage. In 1788 a lighthouse notable for its 'very large and distinguishable' lights was brought into operation at Blyth, built 'under the direction of Mr. Raistrick'; its funding was provided by Sir Matthew White Ridley who was in effect the harbour's owner. Rastrick also claimed an involvement in the design of the iron bridge at Sunderland, completed in 1796 to the design of Rowland Burdon (q.v.). He maintained that he had discussed the use of iron voussoirs with Joseph Walker in 1791 and later had had talks with Burdon. Rastrick also claimed an involvement in the design of the iron bridge at Sunderland, completed in 1796 to the design of Rowland Burdon (q.v.). He maintained that he had discussed the use of iron voussoirs with Joseph Walker (q.v.) in 1791 and later had had talks with Burdon.

In 1798 he married again, this time to Anna Tewling, 'the widow of a jeweller at Charing Cross', but this marriage did not result in further children. Soon afterwards he became associated with the bridge at Weldon Mill, 9 miles north west of Morpeth. In April 1802 he made an inspection of the bridge there and found its condition such that the County authorities decided that a new structure was required. Contractors were appointed for its building and at the same time Rastrick was ordered to 'immediately erect a temporary bridge' for £100. He was later paid £150 for the provision of the permanent bridge's foundations and although records do not reveal if the bridge itself was to Rastrick's design the fact that drawings of it were available for inspection by contractors at his office in Morpeth would make his involvement likely. In 1807 he was paid £30 'for an Engine for driving Piles which the County wanted to Purchase' rather than keep it on hire for an extended period.

Rastrick died in Morpeth on 8 June 1826, leaving his estate to four daughters and three sons, one of them John Urpeth Rastrick (q.v.), also a civil engineer. His will provided for an 'annual supper' for his daughters and in the will he commented that he had devoted much of his time 'to Study the Principles of Mechanics and [had] Invented many useful Machines'. It was later written of him that he 'was a man of very strong and original mind but the want of education was readily perceived and was probably the cause of many eccentricities in his conduct'; otherwise his talents would have 'placed him in a very high rank as a civil engineer'.

R. W. RENNISON

[Freemen of Newcastle upon Tyne: Admissions (1764) Tyne and Wear Archives, GU/NCF/1/7; Northumberland Quarter Sessions: Orders (1802–1809) Northumberland RO, QSO/15 and 16; Newcastle Courant; Will (1824) Northumberland RO, ZMD 168; Northumberland RO (1799–1831) ZAN M.16/B.2; C. E. Baldwin (1929) The History and Development of the Port of Blyth, 28–30; J. G. James (1986) The Cast Iron Bridge at Sunderland (1796), 9, 10]

Publications
1777. Patent No. 1166: Barrel-churn. 'Imperial barrel-churn', 8 August

Works
1784–1785. Invention of a threshing machine
1787–1788. Construction of Blyth Lighthouse
1802–c. 1804. Weldon Bridge, provision of foundations, and probably superstructure; two segmental river spans of 51 ft. with 8 ft. circular pierced spandrel at the river pier together with segmental arch of 23 ft. over mill race
1817. Blyth, construction of ballast crane

RASTRICK, John Urpeth, FRS (1780–1856), civil engineer, was one of the most important engineers of his generation; his work has often been overshadowed by that of the generation of Thomas Telford, which preceded it, and that of Robert Stephenson which followed. He was born on 26 January 1780 at Bullers Green, Morpeth, the fourth child and eldest son of John Rastrick (q.v.), millwright, and Mary whose maiden name he took. At the age of fifteen he became apprenticed to his father and after six years pupillage moved to Shropshire where he began working at Ketley Ironworks. It is unclear how long he stayed there but in 1811, and probably earlier, he was the 'Engineer' at the Bridgnorth Foundry established by John Hazledine (q.v.) and his brothers, Robert (1768–1837) and Thomas (1771–1842). John died in 1810 which may have given Rastrick more responsibility

From around 1802 the Bridgnorth works had been doing work for Richard Trevithick (q.v.), mostly for his high pressure steam engines, and most of the references we have to Rastrick's work over the following decade involve Trevithick. Surprisingly, this included some involvement in the Thames Archway scheme, Trevithick's attempt to tunnel under the Thames from Rotherhithe to Wapping which was effectively abandoned in 1809 when the workings were flooded. Perhaps Rastrick was on hand to supervise engines but, at all events, he was there in temporary charge in the late spring and early summer of 1808 and he attended a meeting of proprietors on 16 November 1809 in support of Trevithick's plans to overcome the problems of quicksand, etc. Generally, however, their collaboration concerned the supply of steam engines to Trevithick's designs, mostly for Cornish mines, some for export and use in the sugar works of the West Indies, some prototype marine engines, but also, according to Rastrick's later evidence, the Catch-Me-Who-Can locomotive which ran trips in London in 1808.

In June 1817 Rastrick's partnership at Bridgnorth was dissolved. Through his work with Trevithick he now had experience at the cutting edge of steam engine technology, he was well experienced in working with iron, and had knowledge of millwork. He had also begun to practice as a civil engineer. In 1811 he unsuccessfully tendered to supply the cast iron work for Meole Brace Bridge. In 1814 the Bridgnorth Foundry obtained the contract to build an iron bridge at Chepstow where, in the opening decade of the nineteenth century, concern was being expressed about the state of the existing stone and timber bridge. In 1811 Watkin George (q.v.) was asked to estimate for iron arches to replace some of the masonry arches and in 1812–1813 John Rennie (q.v.) also reported. Rastrick's design was for a five span cast iron structure with a central span of 110 ft. The bridge was completed in 1816, with Rastrick being awarded over £1,100 in extras.

In 1817–1818 Rastrick produced reports for a number of mine and mill owners before becoming involved in his most famous collaboration, dating from mid-summer 1819 until mid-summer 1830, as managing partner of the Foster and Rastrick works at Stourbridge. This works had been set up in 1800 by John Bradley (c. 1770–1816), and by 1814 was employing four-hundred and fifty men. Expansion continued and shortly after Rastrick arrived in 1820 a 'new foundry' was opened; it was described by Joshua Field (q.v.) in 1821 as the best organised works he had visited.

The Stourbridge works produced all kinds of engineering products—cable-testing machinery, gas retorts, steam engines, roofs, boilers, gas holders—for Stourbridge and Manchester—furnaces, waterwheels, millwork. Of particular interest is the design of the large span 36 ft. cast iron beams that Rastrick supplied for the King's Library in the British Museum. He supplied these in preference to some kind of combination of cast iron beams with wrought iron ties which Robert Smirke (q.v.) had originally intended to use. They were among the largest beams hitherto cast and it is regrettable that little is known of Rastrick's other structural work. He did, however, give an indication of his experience at government inquiries on Buckingham Palace (1831) and elsewhere. He was for example consulted about the dimensions of the ironwork of the tied arch of Stanley Ferry Aqueduct designed by George Leather Jr. (q.v.).

The most famous products of the Stourbridge works were undoubtedly steam locomotives, particularly the three ordered by the American engineer, Horatio Allen, for the Delaware and Hudson Railway in 1828. These, of which the best known is the Stourbridge Lion, arrived in America in 1829 although only the Lion seems to have seen service. Their construction was preceded by that of the Agenoria, and indeed Rastrick's involvement in railways was fairly continuous from the time of his connection with Trevithick.

In 1818 he was involved with the proposed Hay Railway, designing an unexecuted suspension railway bridge with diagonal stiffening between the hangers and deck. Although for horse-drawn traffic, this is the earliest known design for a railway suspension bridge. In January 1825 the promoters of the Liverpool and Manchester Railway consulted Rastrick about the railways of the North East coalfield; his report on this was followed by giving authoritative Parliamentary evidence in support of the Company's Bill later that year. As one of the few engineers with locomotive engineering experience he reported with James Walker on the use of locomotive and stationary engines in 1829; surprisingly, the report favoured the latter. He was called upon again to judge in the Rainhill trials and later, with Jesse Hartley (q.v.), reported on alternative types of carriage wheels for the company.

This consultancy work for the Liverpool and Manchester company took place against a background of railway projects executed by Rastrick elsewhere. In 1826–1827 he built the Moreton-in-Marsh railway, first proposed by William James (q.v.). While it was allegedly the first to use wrought iron edge rails to John Birkinshaw's (q.v.) patent, this was probably on an experimental section as Rastrick's correspondence refers to more traditional fish-bellied track.

Of rather more historical significance is the Shutt End or Kingswinford Railway which was built to link the new works which Rastrick built for his firm at Shutt End, on land acquired by his partner in 1823, to the Astwood Basin on the Staffordshire and Worcestershire Canal. Built between 1827 and 1829 it was the first locomotive powered railway in the Midlands. His locomotive, the Agenoria, featured the sandwich frame rapidly adopted by Robert Stephenson & Co., the spring safety valves, wheel-mounted balance weight and the first mechanical lubricator, and was probably the most advanced locomotive to have been built hitherto. In the wake of this triumph, further work followed. Already known to Lancashire investors he was one of the engineers involved in the early proposals for a rail link between Birmingham and Lancashire, that became the Grand Junction Railway. In 1830 he was put in charge of the surveys for the southern section, George Stephenson (q.v.) having overall responsibility. The 1830 proposal failed in Parliament but a second proposal, for which Rastrick prepared detailed estimates between Birmingham and Basford, and a Wolverhampton branch, was successful in 1833. Thereafter Rastrick's interest faded, perhaps because he saw Stephenson in overall charge, and the impending disagreement with Locke looming. At all events he found an abundance of work elsewhere.

Of this, that for which he is best remembered is the London and Brighton Railway. Controversy over the best route for a line between London and Brighton involved most of the leading civil engineers in the mid-1830s. Rastrick supported the most direct route, generally determined by Sir John Rennie (q.v.). However, following the passage of its Act in 1837, it was Rastrick who was appointed engineer. Undoubtedly his civil engineering masterpiece, the widely admired Ouse viaduct, tunnel portals and station buildings were the result of collaboration with the architect, David Mocatta, who was responsible for the architectural treatment, although Rastrick was responsible for engineering issues such as the ironwork of the train shed at Brighton. The line was completed in 1840 and Rastrick went on to engineer the associated lines to Lewes, and on to Hastings, to Shoreham and Portsmouth, and to Horsham, East Grinstead, and Epsom. In total, the value of these works exceeded £2½ million. Elsewhere in the south east from around 1842 Rastrick was involved in a scheme to convert the Thames and Medway Canal to a railway, and it was he who devised the scheme, carried out in 1845, to partially convert the Strood tunnel to rail traffic. His last work was the Parliamentary preparation for the Lynn and Ely and Lynn and East Dereham Railways, the construction being carried out under his assistant J. S. Valentine.

Simultaneously with the Brighton job Rastrick was engineer for the Bolton and Preston Railway, depositing plans in November 1836. The Act was passed in 1837 and construction, which included a very difficult tunnelling section near Chorley was not completed until 1843.

Unsurprisingly for an engineer operating in the years of railway mania, Rastrick was engineer for, or consulted about, a large number of railway schemes which never came to fruition under him: a Stourbridge-Birmingham line (1835), Manchester and Cheshire Junction Railway (1835–1837), South Metropolitan Railway (1838), the proposed West Coast Main Line route to Scotland across Morecambe Bay (1838), a Manchester-Derby line (1839), the Potteries and Leek Railway

(1840), a Stockport-Macclesfield Railway (1840), an East Anglian Railway from Bishops Stortford to Yarmouth (1841), a link line from Croydon to Vauxhall (1842), the Stourbridge Railway (1843), the London and Manchester Direct Railway (1845–1846), the Ambergate and Nottingham, and the Boston and Eastern Junction Railway. In 1845 he had seventeen railways under his direction. Similarly Rastrick was regularly consulted about various canal schemes, and was engineer for a time to the Staffs. and Worcs. Canal. These proposals included the Stourbridge extension canal (c. 1829) from the Stourbridge canal to Shutt End, a tramroad connection to the Leicester Navigation (1833), a tramroad link between the Leominster Canal and Stourbridge, the Parliamentary survey for the Stourbridge, Wolverhampton and Birmingham Canal (Act 1837), a canal scheme be- tween Newhaven and Lewes (1838), a revived Stourbridge extension canal (1837), the Birmingham and Stourbridge Junction Canal (1840) and consultancy work for the Birmingham and Liverpool Canal (1838).

Although mostly concerned with railways after his departure from the Stourbridge works, he continued to be employed by ironworks, designing a rolling mill at Chillington (1833) and one at Montalaire in France (1835–1836). His work at Dowlais (1833–1836) included a rolling forge engine and a throttle valve for a winding engine. He also worked on an engine for the forge at Stourbridge (1835). He was an impressive Parliamentary witness, bringing to bear his years of practical experience in ironworks at inquiries such as that into the use of iron in railway structures (1849).

He effectively retired in 1847 and spent his latter years in occasional consultancies and sorting out his papers, most of which have survived. His membership of learned societies followed his increasing Parliamentary work in London. He was elected to the Royal Society of Arts in 1833, the Institution of Civil Engineers in 1834, and became a Fellow of the Royal Society in 1837.

Rastrick married Sarah Jervis on 24 December 1810 and of four known children he was predeceased by his son, John, also an engineer, who died on 5 July 1853. Rastrick himself died on 1 November 1856.

MIKE CHRIMES

[Rastrick collection, University of London Library; Membership records, ICE archives; Mackenzie collection, ICE archives; PRO RAIL; House of Lords (1825) *Select Committee on Liverpool and Manchester Railway Bill*; *Select Committee on Windsor Castle and Buckingham Palace* (1831) 2nd Report; House of Lords (House of Commons) (1836–37) *Minutes of Evidence on London— Brighton Railway Bill*; J. Weale (comp.) (1839– 1843) *Bridges in Theory, Practice and Architecture*, cxii–cviii, cxxiv–vii, clxxx–iv, plates 36–38, 44–45, 51–52, 113–116; *Commissioners appointed* to *Inquire into the Application of Iron to Railway Structures* (1849) Report and minutes, 286 287, Memoir (1857) *Min Proc ICE*, 128; H. W. Dickinson and A. Lee (1923) The Rastricks, *Trans Newc Soc*, **4**, 48–63; C. F. D. Marshall (1930) *Centenary History of the Liverpool and Manchester Railway to which is Appended a Transcript of the Relevant Portions of Rastrick's 'Rainhill Notebook'*; H. J. Gough (1934) Tests on cast iron girders removed from the British Museum, *ICE SEP*, **161**; S. M. Tonkin (1949) Trevithick, Rastrick and the Hazle-dine Foundry, Bridgnorth, *Trans Newc Soc*, **26**, 171–183; N. W. Webster (1972) *Britain's First Trunk Line*; R. J. M. Sutherland (1990) Who sized the beams?, in R Thorne (ed.), *The Iron Revolution*, 25–32; J. Crompton (1998) Rewriting the record—John Rastrick's locomotives, in *Perceptions of Great Engineers*, **II**, 51–58; (1995) The British Museum: upgrading the floor over the King's Library, *Structural Engineer*, 20 June]

Publications

1829. *Liverpool and Manchester Railway. Report to the Directors on the Comparative Merits of Loco-motive & Fixed Engines, as a Moving Power* (with James Walker; 2nd edn., separately published)
1836. *Report* (to the Manchester and Cheshire Railway)
1836. *The Direct London and Brighton Railway* (another edn. 1837)
1836. *Bolton and Preston Railway*, deposited plan
1837. *Report of the Directors with Copies of the Reports of J. U. Rastrick etc.* (London & Brighton Railway)
1837. *Reports on the London and Brighton Railway*
1838. *West Cumberland, Furness and Morecambe Bay Railway* (further reports, 1839)
1842. *Gravesend and Rochester Railway Company*

Works

1815–1816. Chepstow Bridge, five cast iron arches
1822. Kidderminster gasworks, ironwork
1824–1825. King's Library, cast iron beams, British Museum
1824–1825. Ironwork for British Museum
1825–1827. Stratford and Moreton-in-Marsh railway, construction, 16 miles
1827–1829. Shutt End Railway
1830. Leeds and Selby Railway, ironwork contractor
1830–1833. Grand Junction Railway
1831–1840. London and Brighton Railway
1836–1843. Bolton and Preston Railway, c. 15 miles, £164,000
1840–1847. London and Brighton Railway, extensions
1845–1846. Thames and Medway Canal, railway conversion

RAWLING family (fl. 1730–1820) members were associated with boring for coal over an extended period and were described in 1795 as 'the only respectable and professional Borers in the North'.

Records show that they had been in the trade from at least as early as 1723 when the men then involved were George, John and Thomas, thought to be brothers although relationships have not been established. The name continued with at least two successors, **Thomas** (1733–1809) and **George** (b. *c.* 1730), the sons of **George Rawling** (b. *c.* 1700) and his wife, Margaret, daughter of Edward Liddell of Ravensworth castle; the Liddell family was in 1726 one of the signatories to the 'Grand Alliance'.

The term 'Messrs. Rawlings' would seem to have been used between 1750 and 1780 while 'Thos & Geo Rawling' was used from the 1770s until the end of the century. It was these two men who were responsible for the borings taken in 1772 for site investigation purposes as a precursor to the rebuilding of bridges on the river Tyne, destroyed in the flood of the previous year. Work was undertaken at Newcastle, Chollerford and Haydon Bridge and, in 1786, for Ridley Hall bridge. The partnership was responsible for at least a further thirty recorded boring contracts in Northumberland and Durham.

In 1775 Thomas Rawling (1733–1809) married Elizabeth Maddison, their residence being at Marshall Lands farm, Whickham, near Gateshead; the marriage produced two sons, **Thomas** (b. 1777) and **George** (b. 1780), and two daughters. George Rawling (b. *c.* 1730) married Elizabeth Smith and lived at Ravensworth Hill Head, also near Gateshead; their union resulted in at least three sons born between 1765 and 1775, one of them also named George.

Thomas Rawling—a member of the Newcastle Literary and Philosophical Society—died on 6 February 1809 and was buried at Whickham three days later, described as 'farmer and borer'. His will was made in 1807 when he described himself as 'indisposed in body but quite perfect in mind' and it provided for his farm at Whickham to pass to his son, Thomas, and his property at Murton to pass to his other son, George, who also received his boring rods. George, also a member of the Literary and Philosophical Society, later inherited money from his uncle, Thomas Maddison, a builder, and following Maddison's death in 1811, his name was adopted by Rawling.

The date of death of George Rawling (b. *c.* 1730 and noted both as 'borer' and 'farmer') has not been established but his son, also George, died in Bristol in early 1796, noted as being from Ravensworth. As a further son was apprenticed as a skinner and another as a plumber, the Thomas and George Rawling who continued boring operations, together and individually, until *c.* 1820 would seem to have been the sons of Thomas (1733–1809).

R. W. RENNISON

[A Collection of Boring Notes, Northumberland RO: 3410/BUD/26/1; *Newcastle (Weekly) Courant*; *Newcastle Chronicle*; *Guild Apprenticeship Indentures*, T&WAS; Will, Durham University Special Collections, 1809/R.3; *List of Members, Literary and Philosophical Society*; (Northumberland) *Quarter*

Sessions: Orders, Nothumberland RO: QSO/11 & 12; R. Mylne, *Mr. Mylne's Report respecting Tyne Bridge ...*, Gateshead, 1772; M. W. Flinn (1984) *The History of the British Coal Industry*, **2**, 1700–1830]

REED, John (1749–1817), stonemason, was a son of Matthew Reed (d. 1790) and his wife, Mary Thompson (1726–1789). He was born at Elrington and was baptised at nearby Haydon Bridge on 15 January 1749, one of at least seven children. On 12 December 1771 Reed married Ann(e) Wear (1749–1831) at Warden and their marriage seemingly produced only a single daughter, Ann, baptised at Newburn on 10 May 1772 although Reed and his wife were then resident in Newcastle.

Reed acted as foreman to Edward Hutchinson for the construction of the northern section of the Tyne Bridge, Newcastle, taking over the contract on the death of Hutchinson in 1780 and so becoming responsible for the bridge's completion in 1781. The design of the northern six spans of the bridge had been undertaken by John Wooler (q.v.) while Robert Mylne (q.v.) had designed the three southern arches. The *Newcastle Courant* noted that Reed 'finished that great work for the benefit of (Hutchinson's) orphan children, in a way to give universal satisfaction to his employers'.

In 1800, Reed built the pedimented Tuscan doorcase of the entrance to the Trinity House chapel to the design of John (?) Stokoe and was said to have been the 'only regularly-bred mastermason' in Newcastle at this time; Newcastle directories show him as a stonemason only from 1790, his premises then in the Close. He also built quay walls at several unidentified locations on the river Tyne and in March 1804 he completed a graving dock for William Row at St. Peters Quay, Newcastle. The opening of the dock, able to receive vessels drawing 12 ft. of water, was celebrated by the entry of two ships of 300 tons.

Reed's last major work was 'the stupendous bridge over Pandon Dean ... carried on and finished by him under difficulties which would have appalled any other man'. Although the bridge, forming part of the North Shields Turnpike, was a major structure, the difficulties were mainly financial; Reed was never fully paid for his work. With a central span of 50 ft. and a flanking span of 45 ft. at each side, the bridge was 78 ft. high and 30 ft. wide. It had been designed by John Stokoe (c. 1756–1836) and was completed in 1812 at a cost of £7,448. The bridge was commended in 1831 by John Dobson, Newcastle's leading architect, who described Reed as 'one of the most judicious and substantial builders [he] ever knew'.

In 1817 Reed became a member of the Literary and Philosophical Society in Newcastle but died at his house in the Close on 9 June of that year and was buried at St. Nicholas's church, Newcastle, survived by his wife and daughter.

R. W. RENNISON

[*Newcastle Courant*, 20 December 1817; E. McKenzie (1827) *A Descriptive and Historical Account ... of Newcastle upon Tyne*; T. Oliver (1831) *A New Picture of Newcastle upon Tyne*, Newcastle; *Calendar of Common Council Book, Newcastle, 1824–1835*, Tyne and Wear Archives Service; J. Sykes (1833) *Local Records*, Newcastle; N. Pevsner (1992) *The Buildings of England: Northumberland*]

Works
1772–1781. Tyne Bridge, Newcastle, rebuilding
1800. Trinity House, Newcastle, extension
1803–1804. Dry dock, Newcastle, construction
1810–1812. Pandon Bridge, Newcastle, construction

REID, Robert (fl. 1735–1790), mason, builder and resident engineer was employed in his youth on the construction of the Tay Bridge, Aberfeldy, designed by William Adam (q.v.) for General Wade (q.v.). It was built in 1733–1735, the crowning achievement of Wade's programme of road-building in the Highlands. At some later time Reid became resident in Haddington and in 1742 he contracted as mason, in partnership with a wright named George Peirie, to build the new Town House there, which was also designed by Adam.

He next appears in 1762 in the records of the Coldstream Bridge project. It was to be a major bridge across the Tweed at a site adjacent to Lennel House, a property of the Earl of Haddington, and Haddington, being a prominent member of the road trustees who promoted the bridge, may be assumed to have introduced Reid to the project. He was engaged as 'overseer' to make preparations for a contract and, with a modest workforce, he investigated two alternative sites and several quarries, extracted some stone and began hewing it. During this period he directed the making, chiefly by his carpenter James Blaikie, of a bottomless caisson for use when making borings in the river-bed, a machine for raising stones in the quarry, sheds and carts, and a float for carrying stones and from which to drive piles in the river; Blaikie later made centering on which to build the arches.

Reid had also made a design for the bridge, which he had taken, in attendance on some of the trustees, to London to explain the project to a committee in Parliament. John Smeaton (q.v.) also accompanied the trustees, bearing his own design for the bridge. Parliament awarded a grant of £4,000. Later, David Henderson (q.v.) was accepted as contractor but having failed to start serious work he was discharged a year later, Smeaton having meantime been appointed engineer for the bridge. Reid was continuing his work at the site, with an expanding workforce, and thereafter directed the whole project at Coldstream, subject to Smeaton's instructions issued on visits to the site and by letters. As the job was nearing completion in 1767, Reid agreed with the trustees to build a tollhouse to his own design for £27, but incurred their displeasure for the first time by building the substructure which was necessary to support the tollhouse at road level as a two-storey basement and moving into the basement as a house for himself; up to that time he had kept his family at Haddington. The trustees were soon persuaded by Smeaton that the basement residence formed a useful buttress to the side wall of the bridge, and Reid remained resident.

When rental of the tolls was first advertised in 1767 Reid made a bid and became the toll-collector, but gave it up after one year. For more than twenty years, however, he continued to supervise the maintenance of the bridge, reporting to Smeaton on a number of occasions about dangerous scour round the foundations, and suggesting that a small weir across the river just downstream (which he had seen used effectively in similar circumstances at Aberfeldy) would be beneficial. Smeaton eventually agreed and the weir was built in 1785–1786; it has survived and, in a much-enlarged form, protected the foundations throughout the twentieth century.

Reid must also have found other employment, of which one recorded example was a stone bridge built in 1770 at Easter Greenlaw. By that date he had also become a bailie (i.e. magistrate) of Coldstream and remained so for at least fifteen years. The last evidence of his activity is a 1790 report to Smeaton about continuing maintenance of the bridge's foundations.

The trustees' minutes at Coldstream and Smeaton's correspondence with Reid and with his employers reveal their implicit and constant trust in his knowledge of complex bridge construction and his management of the workforce to complete a bridge of very high quality.

TED RUDDOCK

[*Minutes of Trustees for Coldstream Bridge, 1762–1772*, MS, CS Misc. 3, Northumberland RO; *Engineering Designs 1741–1792 of John Smeaton*, MS, **4**,150–157, 160v, Royal Society of London; John Smeaton (1812) *Reports on Various Occasions*, **3**, 235–251; R. Gibson (1905) *History of Greenlaw*, 230 ff.; D. Walker (1972) Haddington, *Country Life*, 10 August; T. Ruddock (1979) *Arch Bridges and their Builders 1735–1835*]

Publications

1784. Report of Robert Reid on the State of Coldstream Bridge to the … Trustees … for the Roads and Bridges in Berwickshire, signed 19 October 1784, in Smeaton (1812) *Reports*, **3**, 245–6.

Works

1762–1767. Coldstream Bridge, overseer of preparatory works and of construction of the bridge by direct labour, design by John Smeaton, cost c. £6,000
1767–1790. Coldstream Bridge, direction of works of maintenance
1770 Easter Bridge, Greenlaw, probably designed and built, cost £275 6s 8d

RENDEL, James Meadows, FRS (1799–1856), civil engineer, was the only son of James Rendel,

country surveyor and farmer of Okehampton, and the grandson of an architect, John Meadows FRS. He passed his youth in the neighbourhood of Teignmouth, receiving his education at a country school, and was initiated into the practical operations of a millwright by his uncle who resided there. From his father, who had charge of a district of roads, he obtained a certain degree of familiarity with the rudiments of civil engineering. He was elected a corresponding member of the Institution of Civil Engineers in 1824.

At an early age, in about 1817, he went to London and obtained an appointment with Thomas Telford (q.v.), who employed him on surveys and experiments for the proposed suspension bridge across the river Mersey at Runcorn. About the year 1822 he settled at Plymouth and commenced practice on his own account, then being chiefly employed on the construction of roads in north Devon. In September of that year, having commented on a proposal for a suspension bridge for crossing the Tamar at Saltash, he came under the notice of Lord Morley, who as Lord Boringdon had employed another civil engineer, James Green (q.v.), some fifteen years earlier.

In 1823 Lord Morley entrusted to Rendel the design and construction of a suspension bridge for the crossing of the river Plym at Laira. When an Act was obtained Samuel Brown complained that Rendel had 'made an exact transcription of his plan for the Tamar' and the idea of a suspension bridge was dropped. Roger Hopkins (q.v.) competed to provide a wooden bridge at Laira but at the last moment Rendel won the day by proposing an elegant cast iron structure with five spans and with the ironwork provided by William Hazledine (q.v.). He completed his bridge in 1827 and it lasted until 1962. This fine bridge was described in the *Transactions of the Institution of Civil Engineers*, volume I, 1836, and for this paper he gained a Telford medal.

Rendel's experience of suspension bridge design with Telford was not wasted. He appreciated the importance of longitudinal stiffening girders to provide aerodynamic stability, advising on this for Menai and Montrose bridges. He rebuilt the latter in the 1830s and later designed suspension bridges in St. James' Park, London, and Inverness. It is unclear when he first developed the idea of a deep longitudinal truss as his drawings for the Laira proposal do not exist; the illustrations of his design for Clifton Gorge suggest it may have been in place by 1830, although no mention is made in the accompanying text.

Soon after the completion of Laira bridge, Rendel constructed some roads for Lord Morley, the Cann Quarry tramway and a sluice of peculiar construction at the northern end of James Green's Chelson Meadow embankment along which Lord Morley had built a roadway to join Saltram House to Laira bridge. He also improved several turnpike roads, including a southern route between Sequers bridge and Totnes, the road from Plymouth to Cornwall via Saltash and the road from

Devonport to Liskeard via Torpoint. In 1826 he constructed Bowcombe bridge over a creek of the Kingsbridge estuary with four masonry arches and an opening span which originally was a drawbridge and where hydraulic power was first applied to machinery for bridges.

The Cann Quarry tramway was a short branch of 4 ft. 6 in. gauge off the Plymouth & Dartmoor Railway leading to the quarry and built for Lord Morley. A two span cast iron tramway bridge crossed the river Plym on the Cann quarry route. The bowstring girders of 25 ft. span are 9 ft. 6 in. apart, have cast iron cross girders carrying a longitudinal sleepered deck with a 4 ft. 6 in. gauge railway. In 1828 Rendel commenced a survey for a suspension bridge across the river Dart at Dittisham but this project was blocked by the landowner, James Elton. Rendel then turned his attention to a proposal for pulling a boat along a fixed chain using steam power and in 1831 a floating bridge was constructed for crossing the Dart at Dartmouth. This, now known as the Higher Ferry, also required 1½ miles of new road to Hillhead, where the road from Brixham meets the Churston to Kingswear road.

After building a similar ferry at Saltash in 1832–1833, which lasted until the suspension bridge was built in 1961, he established another floating bridge across the Tamar at Torpoint in April 1834. A paper published in *Transactions*, volume II, 1838, earned Rendel a second Telford Medal from the Institution of Civil Engineers. Two more ferries were built to his designs, one at Woolston, Southampton, and the other at Gosport. While these two are no longer working, such ferries can be found today at Cowes, Poole harbour and Trellisick, near Truro.

In January 1830 he applied for the post of County Surveyor of Somerset, without success, and in January 1831 he tried in Devon by offering to do the work for £300 against James Green's salary of £550. Green retained his post but at the reduced salary of £300. During his time in Plymouth, Rendel reported on nearly every harbour in the south west of England which founded his mastery of this branch of civil engineering on which his fame largely rests. In 1829 he designed the harbour at Par, in Cornwall, and in 1835 he enlarged the sea lock and basin of the Bude canal.

In 1836 he designed the harbour and breakwater at Brixham, in Devon, using the rock obtained from Berry Head; the breakwater has since been lengthened twice. In 1839 he was engaged in preparing various schemes for a railway from Exeter to Plymouth over Dartmoor, via Dunsford, Chagford, Princetown, Sheepstor and Roborough Down, and in 1841 he constructed the Millbay pier, Plymouth, a work of considerable difficulty, owing to the great depth of water. Here he first introduced the method of construction, since employed with so much success at the great harbours of Holyhead and Portland.

In about 1838 he moved to London leaving Mr. Beardmore as his partner in Plymouth. Rendel then concentrated on harbour works, although he acted as consultant on Indian railways. He was President of the Institution of Civil Engineers in 1852 and 1853 but died in November 1856.

A. B. GEORGE

[Membership records, ICE archives; India Office Library and Records, BL; Duchy of Cornwall Archives, London; Deposited plans, Devon RO: DP50, Roads Teignmouth to Teignbridge and Thornes, 1822; DP53, Roads from Start Place, Plymouth to Laira, 1823; DP56, Roads in Okehampton, 1823; DP59, Roads Dartmouth to Modbury and Kingsbridge to Frogmore, 1823; DP84, Higher Ferry road to Brixham Cross, 1832; DP97, Sutton Pool, Plymouth; DP103, Turnpike roads of the Saltash Trust, 1832; DP108, Kingsbridge Turnpike Trust improvements, 1833; DP139, Milbay Pier, Plymouth, 1839; DP148, Plymouth and Exeter Railway proposal over Dartmoor, 1840; *Exeter Flying Post*, 6 January 1831; QS 1/27, Epiphany Sessions 4 January 1831, Minute re. salaries, Devon RO; drawings of Montrose bridge, Angus RO; discussion on the causes of injury to suspension bridges (1841) *Min Proc ICE*, **1**, 77–80; G. C. Dobson, (1842) Description of a drawbridge at Bowcombe creek near Kingsbridge, *Min Proc ICE*, **2**, 68; T. Judd (1842) Chelsea meadow sluice, *Min Proc ICE*, **II**, 62–63; Obit. (1857) *Min Proc ICE*, **16**, 133–142]

Selected publications

Note: Rendel made more than fifty contributions to the *Minutes of Proceedings of the ICE*, and more than fifty printed reports are held by ICE, most after 1830.

1823. *View of the Proposed Suspension Bridge at Saltash*

1829. *Clifton Suspension Bridge*

1830. Particulars of the construction of a cast iron bridge over the Lary ..., *Transactions, Plymouth Inst*, **I**

1835. *A Report to Subscribers to a Survey of proposed Turnpike Road from Plymouth & Devonport to St. Austell ...*

1836. *Report and Plans for a Breakwater at Brixham in Torbay*

1836. Particulars on the construction of Lary Bridge, *Trans ICE*, **I**, 99–108

1837. *Report on the Practicability of forming a Harbour at the Mouth of the Loe Pool, in Mount's Bay near Helston*

1838. Particulars of the construction of the floating bridge lately established across the Hamoaze ..., *Trans ICE*, **II**, 213–217

1838. On the floating bridge lately established across the Hamoaze ..., *Min Proc ICE*, **I**, 21–24

1841. Memoir of the Montrose suspension bridge, *Min Proc ICE*, **1**, 122–129

Early works

1822–. Various improvements to turnpike roads in Okehampton, the South Hams, for the

Kingsbridge Turnpike Trust, and from the ferries over the Tamar to Liskeard in Cornwall
1823–1827. Laira bridge, Plymouth
1825. Chelsom Meadow Sluice, Devon
1826. Bowcombe bridge, near Kingsbridge, Hydraulically operated opening span
1829. Plymouth & Dartmoor Railways, Cann Quarry Branch, opened 20 November
1831. Higher Ferry, Dartmouth, and access road from Brixham Cross
1832–1833. Saltash Ferry, Plymouth
1834. Torpoint Ferry, Plymouth
1836. Brixham Harbour
1841. Millbay Pier, Plymouth

RENNELL, James, Major, FRS (1742–1830), geographer, is famed as 'the father of [our] Indian geography [Warren Hastings], godfather of the Royal Geographical Society [founded three months after his death], and creator of the entire science of oceanography. However, his engineering interest is mainly confined to his period in India, 1764–1777, as both surveyor and the puzzled recipient of a commission in the Bengal Engineers, 'I must confess I was never more surprised in my life'.

Rennell was born in 1742, the son of Captain John Rennell, Royal Artillery (d. 1747). His precocious interest in both surveying and engineering is evidenced by a plan of 1757 showing a profile of 'Clitter Road' with planned improvements to ease gradients and incorporate a new bridge. Of more relevance to his career in the navy which he joined as a fourteen-year-old midshipman in 1756 was a plan of Milford Haven. Throughout most of the Seven Years' War he served under Captain Hyde Parker, a veteran of Anson's circumnavigation who encouraged him in making surveys which included the Philippines, Madagascar and the harbours of Port Mathurin and Trincominalle. While in Ceylon he also suggested that the sandbars separating it from the Coromandel Coast could be made navigable by dredging the Straits of Ramisseran—as was in fact done sixty years later.

The end of the war limited his prospects in the navy, but he had already been loaned to the East India Company to survey Manila, Nicobar, Malacca and the Pamben Channel and was in the right place when a land surveyor was needed for the Company's newly acquired territories. This was accompanied by the 'surprising' commission as Practitioner Engineer on 9 April 1764.

At this time the most important means of transport for communication and commerce in Lower Bengal were the waterways and Rennell's main task was to survey the Ganges to find a channel navigable from Calcutta all year round. In 1765 Robert Clive extended his remit to cover the making of 'a vast map of Bengal' and in 1767 made him Surveyor General of Bengal and Captain of Engineers at a combined salary of £1,000 p.a. Over the next ten years he provided workable maps of 150,000 square miles of previously unsurveyed

territory which later enabled the Indian army to build the infrastructure on which much of India was to depend. Half a century later many had still not been superseded. The area covered was from the Assam Frontier to the Himalayas and included a report on the roads of Bengal and Bihar and a survey of the Brahmaputra River. These were mostly route surveys using a perambulator with compass bearings checked by astronomic observations. Unfortunately his continued adherence to this system meant that the later surveys of India owed more to the methodology of his contemporaries Michael Topping and William Roy (qq.v.) whose methods of triangulation were used by Rennell's successor Frederick Lampton (even down to a Jesse Ramsden theodolite) in the great trigonometrical survey of India in 1818.

Rennell also earned his engineering commission with a plan of the new citadel at Fort William and a report on damage caused by river floods at Chandernagore in 1772 (four years after he had worked on the destruction of fortifications which the French had claimed were only embankments to prevent just such a misfortune).

By 1777 his work on the *Bengal Atlas* was complete and as he claimed 'my health is so bad I am advised ... to quit India [and] am disabled from pursuing active employment', the East India Company begrudgingly agreed to a pension of £600 p.a. which he lived to draw for another 54 years. But they continued to make use of his expertise, as in 1796 when he was consulted over proposals for cutting a canal between the Ganges and the Hooghly. His arguments against were never refuted.

In 1779 Rennell published the *Bengal Atlas* and in 1783 the *Memoir of Hindustan*, a copy of which was owned by Thomas Telford (q.v.). These established his reputation as Britain's greatest geographer, consolidated by his work on Mungo Park's explorations for the African Association (one of the ancestors of the Royal Geographical Society). He became FRS in 1781 (recipient of their Copley Medal in 1791) and Associate Institut Francaise in 1801.

His position also permitted him to use ships' charts and journals provided by the Admiralty for the scientific study of winds and currents which enabled him to form the system which was his main legacy. His view that 'the winds with a very few exceptions are to be regarded as the prime movers of the currents of the ocean' is with the single insertion of 'surface winds' still acceptable today. His views were remarkably present on the Gulf Stream, the North West Passage, the currents of Greenland and the 'Rennell' current in the Isles of Scilly which stopped notorious shipwrecks such as that of Cloudsley Shovel. An investigation of the currents of the Atlantic Ocean was published posthumously as (in another parallel with Roy) was his work on the geographical system of Herodotus—the latter edited by his daughter Jane. He was also survived by two sons.

He died on 29 March 1830 and was buried in Westminster Abbey, surmounted by a bronze bust by Hagbolt. Other portraits exist by E. Scott and J. Opie.

SUSAN HOTS

[India Office, Records, BL; DNB (1896); Porter, **2**, 401–404; C. R. Markham (1895) *Major James Rennell and the Rise of Modern English Geography*; Sir H. Yule (1887) *Royal Engineers Journal*; R. Rodd (1930) Major James Rennell, *Geographical Journal*, **75**, 4, 289–299; R. H. Phillimore (1950) *Historical Records of the Survey of India*, **I**; E. G. R. Taylor (1966) *Mathematical Practitioners of Hanoverian England*, 271; G. Griffiths (1993) *James Rennell and British Arctic Expeditions, 1818–1829*; Soldiers of the Raj; M. H. Edney (1994) British military education, mapmaking and military map-mindedness: the later enlightenment, *Cartographic Journal*, **31**, 1, 14–20; M. E. Edney (1997) *Mapping an Empire: the Geographic Construction of British India*, 1765–1843]

Publications

1778. *A Description of the Roads in Bengal and Bahar*
1778. *Chart of the Banks and Currents at the Lagullas*
1779. *Bengal Atlas*
1781. *Map of Hindustan or the Moguls Empire*
1781. *Chart of Abai Harbour and the North-west Coast of Borneo*
1793. *Memoirs of a Current that often Prevails Westward of the Isles of Scilly*
1809. *On the Effects of Westerly Winds in raising the Level of the Bristol Channel*
1809. *Atlas of the Comparative Geography of western Asia*

Papers for the Royal Society *Philosophical Transactions* include: An account of the Ganges and Brahmaputra Rivers, **71** (1781) 87–114; and Observations on a current that often prevails to the westward of the Isles of Scilly endangering the safety of ships that approach the English Channel now known as Rennell's current.

Charts in the Royal Geographical Society include:

1757. Profile Clitter Road from Beggars Bush towards Newton, planned improvements to ease the gradients and incorporate new bridge, plan Milford Haven
c. 1764. Plan of Calcutta Citadel

RENNIE, George, FRS, MRIA (1791–1866), civil and mechanical engineer, was the eldest son of John Rennie (q.v.) and his wife Martha Ann (*née* Mackintosh). He was born on 3 December 1791 while his parents were resident at Stamford Street, Blackfriars, London. His early education was first at a school run by a Dr. Greenlaw at Isleworth, West London, then for two or three years at the better known St. Paul's School. His father groomed him from the start to succeed him in his business as an engineer and in 1807, after

taking him on a tour of his current projects round England, Ireland and Scotland, settled him at the University of Edinburgh where he studied for four years, attending courses in mathematics, mechanics, natural philosophy, chemistry and classics. He made personal friendship with professors known to his father, most importantly John Playfair, Professor of Natural Philosophy and successor to John Robison (q.v.) whose influence had been important in setting the elder Rennie on his road to great success in engineering.

When George returned to London in 1811 he joined his father's firm and gained experience both of large civil engineering works and of the manufacture of machinery and structures which was carried on in a large workshop in Holland Street, Southwark, premises which had been acquired in 1810 after the firm had outgrown its former workshop in Upper Ground Street. Both were close to the family home in Stamford Street. He was free to observe the many large projects in his father's practice, including three naval dockyards, other harbour works and breakwaters, and bridges. In particular, he was a regular visitor to the construction site of Waterloo Bridge from its start in 1811, and recorded in a manuscript notebook, from 1813 until the work was finished in 1817, both the sequence of construction and all the methods used. He was absent only for two or three periods, the longest one of four months, when he was travelling in Europe, an activity by which his father encouraged both his sons to increase their knowledge and their business potentialities.

In the same years Rennie was involved in the work of engineering manufacture in Holland Street and his experience in that business was doubtless what qualified him for appointment in 1818 as Inspector of Machinery and Clerk of the Irons (i.e. the dies) at the Royal Mint, then located in the Tower of London. He held the post for almost eight years, and within that time his tenure of a public office was found to debar him from appointment as Engineer for the building of the new London Bridge. The appointment passed therefore to his younger brother, John, and earned him a knighthood when the bridge was opened, but George claimed that he himself had made the design for his father (see Cooke, *Views of the Old and New Bridges*, 9); he probably also worked with John during its construction.

On the death of Rennie Sr. in 1821 George Rennie and his brother inherited both the works in Holland Street and the task of completing all their father's extensive projects. They formed a partnership called 'George and John Rennie' in which George must be considered the senior partner but, in what was common practice at the time, only one of them was named as Engineer in their commissions by individual clients, even if they were both involved in the work, as was often the case. The list of works given below represents an attempt to identify works in which George was either the appointed Engineer or made an important contribution. In the largest works, such as the government dockyards, both brothers are likely to have been effectively the 'engineer' for different parts of the whole complex.

It is clear that George Rennie took the larger responsibility for the workshop, his father having expressed the wish that his 'millwright business' should be continued by George, who should take into partnership the 'trustworthy foreman John Walker.' Arising from this, it has been suggested that he was mainly a mechanical engineer. He also had more interest in—and presumably more knowledge of—theoretical aspects of engineering than had John, and wrote a number of useful papers on such matters, using his workshop equipment and staff to make experiments, for instance on the strengths of materials and on mechanical friction (see list of publications below). The workshops, moving with the times, developed special expertise in the manufacture of steam engines for ships and pioneered the use of screw propulsion. One or two whole iron ships were fabricated and dispatched to be assembled elsewhere, one even in the Caspian Sea, and about twelve railway locomotives were built in the years 1836–1843.

The 'equilibration of arches' was a process developed by eighteenth-century mathematicians to check the form and dimensions of arches, and George Rennie was one of the few practising engineers who could make the checks himself while designing his bridges, a fact which gives him special importance in the history of arch bridge design As mentioned, he designed the new London Bridge in 1820, including 'equilibrating the arch', and went on to engineer—including design and supervision of construction—the firm's best masonry bridges, namely the Serpentine Bridge in Hyde Park, London, and Staines Bridge over the Thames, although John claims some credit for each in his autobiography. Both designs employed segmental arches of low profile and restrained classical details like those commonly used on the bridges of Rennie Sr. John Jr. allowed full credit to George for important changes made to the design of both the arch and the foundations of Grosvenor Bridge, Chester (see: *HARRISON, Thomas*), in the mid-1820s when the unprecedented scale of its arch and cost of the bridge inspired the promoters to invite the Rennies to visit Chester and assess the design.

The Rennie brothers were jointly appointed 'engineers-in-chief' of the Liverpool and Manchester Railway in 1825 after the company, with George Stephenson as its engineer, had failed to obtain an Act of Parliament for the line. Employing Charles Blacker Vignoles (q.v.) as assistant in the field, George Rennie re-surveyed the line—his brother being unfit due to an accident at the London Bridge site—and both brothers and Vignoles helped the company to a successful outcome in Parliament in the following year. The Company then offered to make George 'consulting engineer' for the construction phase with Stephenson as an assistant, but the Rennies wanted full control with freedom to choose their collaborators and failed to get the appointment on their terms. Although George directed initial surveys for several other lines of railway, including a line from Birmingham to Liverpool in 1825, the Rennies obtained relatively little work on British railways. George Rennie's most significant completed railway work was as engineer-in-chief of the Namur and Liège railway in Belgium, to which he was appointed in 1846 and which included four substantial stone bridges.

He had frequent contact with scientists throughout his life and was respected for his publications. He was elected to Fellowship of the Royal Society in 1822 and held office there for a time both as a vice-president and as treasurer. In 1841 he became a member of the Institution of Civil Engineers, of which his brother was president in 1845–1847. George Rennie was also a Fellow of the Royal Irish Academy, and of the Academies of Rotterdam and Turin. He was elected a member of the Institution of Mechanical Engineers in 1857 but is not known to have made any contributions to its proceedings, as he did to those of the Civil Engineers.

He married Mary Anne, daughter of Sir John Jackson, Bart., in 1828, and had two sons, George Banks and John Keith, and one daughter; all survived him. In his later years both sons were in partnership with him, each owning one-sixth of the manufacturing works then operating in premises at Greenwich as well as Blackfriars. The firm's total turnover up to 1861 was £200,812. The partnership was dissolved at the end of that year, when George yielded control, although he still went sometimes to the works, as noted below. Presumably he had earlier terminated the partnership with his brother.

Although obviously firm, efficient and reliable in work situations, George Rennie is believed to have been gentle and sympathetic to members of his family. A letter he wrote when he was twenty-four, to his aunt Henrietta who had kept house for the family in London subsequent to his mother's death, but later returned ill to Scotland, shows this aspect of his character and is printed in full by Boucher.

Though he is said to have been 'partially crippled' and to have suffered to some extent from epilepsy (Boucher, *John Rennie*, 29), he lived until his seventy-fifth year. He died at his home, 39 Wilton Crescent, London, on 30 March 1866 as the result of an accident which had occurred on his way home from the works several months earlier. He was buried in the churchyard of Holmwood Common, near Dorking, Surrey.

A portrait of George Rennie in 1825, painted by J. Linnell, is in the National Portrait Gallery, London, and reproduced in Ruddock, *Arch Bridges and their Builders*.

TED RUDDOCK

[Rennie Papers, MSS 19777-19968, NLS, including correspondence, journals and family papers;

Rennie drawings, ICE archive; MS report and estimate on proposed Grosvenor Bridge, in *Reports of Sir John Rennie*, **2**, 292–293, ICE archive; MS plan for railway from Birmingham to Liverpool, 29 November 1825, by George Rennie and Josias Jessop, Shropshire CRO; Sir John Rennie (1846) Presidential address, *Min Proc ICE*, **5**, 19–122; Stone bridge over the R. Meuse for Namur and Liège Railway (1847) *The Civil Engineer and Architect's Journal*, **10**, 348; memoir of George Rennie (1868–1869) *Min Proc ICE*, **28**, 610–615; Sir John Rennie (1875) *Autobiography*; DNB; J. W. Clarke (1958) The building of Grosvenor Bridge, *Chester Archaeological Society Journal*, **45**, 43–55; C. T. G. Boucher (1963) *John Rennie 1761–1821*; T. Ruddock (1979) *Arch Bridges and their Builders 1735–1835*; K. H. Vignoles (1982) *Charles Blacker Vignoles: Romantic Engineer*]

Publications

1818. Account of experiments made on the strength of materials, *Philosophical Transactions of the Royal Society of London, Part 1*, 118–136

1829. Experiments on the friction and abrasion of the surfaces of solids, *Philosophical Transactions of the Royal Society of London, Part 1*, 143–170

1833. E. W. Cooke, *Views of the Old and New London Bridges*, with 'A concise essay on bridges from the earliest period etc. etc., derived from information contributed ... by George Rennie'

1835. On an instrument for taking up water at great depths, *Report of the British Association for Advancement of Science*, 595–596

1839. On the dimensions and performance of the Archimedean steamer, *Minutes of Proceedings of the ICE*, **1**(3), 70–72

1840. Experiments to determine the resistance of a screw when revolving in water at different depths and velocities, *Minutes of Proceedings of the ICE*, **1**, 69–171

1840. On the expansion of arches, *Minutes of Proceedings of the ICE*, **1**(4), 4–6

1842. Practical examples of modern tools and machines ..., supplement to R. Buchanan, *Practical Essays ...* , George Rennie (ed.), two vols. (one text, one plates)

1848. *The Architect's Pocket-book for the Year 1848 ... to which have been added Numerous Experiments by George Rennie*

1850. *Report on the Supply of Water to be obtained from the District of Bagshot*

1855. The bridge aqueduct of Roque-Favour, *Minutes of Proceedings of the ICE*, **14**, 190–235

1855. Description of the Pont du Gard, *Minutes of Proceedings of the ICE*, **14**, 236–238

1856. *Suggestions for the Improvement of the River Danube between Isatcha and the Sulina Entrance* (may be by George Banks Rennie)

1857. Rubble béton, or concrete, in engineering and architecture, *Minutes of Proceedings of the ICE*, **16**, 423–448

1857. On the quantity of heat developed by water when violently agitated ..., *Report of the British Association for Advancement of Science*, 169–171

Works

Mechanical engineering works are not listed.

1820. New London Bridge, design, five elliptical arches, largest span 152 ft., constructed 1824–1831, engineer Sir John Rennie, cost £495,000

1824–1827. Hyde Park Bridge, River Serpentine, London, design and construction (with Sir John Rennie?), five segmental arches

1825. Birmingham to Liverpool Railway, initial survey and plan (with Josias Jessop)

1825. Liverpool and Manchester Railway, survey and route planning (with (Sir) John Rennie), constructed 1827–1829, engineer George Stephenson

1829–1834. Staines Bridge, River Thames, design and construction, three segmental arches, cost *c*. £30,000

1846–1848. Namur to Liège Railway, engineer-in-chief

RENNIE, John, FRS, FRSE (1761–1821), was one of the leaders of the civil engineering profession in the first two decades of the nineteenth century, and was one of the subjects chosen by Samuel Smiles for his *Lives of the Engineers*.

He was born on 7 June, 1761, the youngest of nine children of James and Jean Rennie. His father was a moderately prosperous tenant farmer at Phantassie, which Rennie himself spelt Fantasie, in the parish of Prestonkirk in East Lothian. When James Rennie died in 1766, his eldest son George was only seventeen, but he was already the manager of a brewery which his father had built. Later to achieve a degree of fame as an agricultural improver, George now proved an effective head of the family and John Rennie was able to attend the parish school. While there he spent much of his spare time in the nearby workshop of Andrew Meikle (q.v.), the celebrated millwright who was then still working on the development of the successful threshing machine which made his name, though not his fortune. At the age of twelve, having learnt all that the school at Prestonkirk could teach him, Rennie went at his own request to work full time for Meikle. This lasted for about two years, until it was decided that he should continue his education at the Public School in Dunbar, 5 miles away. There he stayed for two more years until he had completed the curriculum, and famously earning a mention in David Loch's *Essays on the Trade and Commerce, Manufactories and Fisheries of Scotland* '... his master could not propose a question either in natural or experimental philosophy, to which he gave not a clear and ready solution ...'. He then returned to work for Meikle, building a flour mill in Angus in 1778 and others in Fife and Midlothian in 1779, until in September 1779, with the assistance of his brother George, he commenced in business on his own account. With the money thus earned, and by working during vacations, he was able to support himself during three years (1780–1783) of higher education at Edinburgh University.

John Rennie FRS, FRSE

One of the professors under whom he studied was John Robison (q.v.), who provided him with an introduction to James Watt (q.v.) at Birmingham. Accordingly, Rennie made a leisurely study tour during May and June 1784 to Boulton & Watt's Soho Works, visiting and making notes on bridges, canals, docks, iron works and several different types of mill on the way. Watt was then involved in supplying a steam engine to power the Albion Mills designed by Samuel Wyatt (q.v.) in London. This was the first such use, and Watt needed to employ an agent to supervise the local fabrication, erection and maintenance of the machinery. Impressed by his visitor, and confirmed later in this opinion by a favourable reference from Robison, he offered a seven year contract to Rennie, who had by now returned home. The offer was accepted, probably during July as he took Peter Ewart (q.v.) into apprenticeship on 4 August. They returned to Soho in September 1784, travelling on to London in November. It was largely due to Rennie's skill and persistence that the mill was made operational, in February 1786, and he was responsible for the installation of a second engine with improvements in 1788. The management of the mill was poor however, so that Rennie was forced to complain of possible damage caused by overworking of the machinery and neglect of maintenance. In the event, a bearing overheated in March 1791, causing a fire which raged for three days and gutted the building. Commercially a failure, it was a notable technical success, Matthew Boulton noting afterwards 'The Albion Mill was an example of mechanism

that hath been copied and introduced into different manufactories, and daily experience proves that it will be the principal cornerstone of all the great fabrics and manufactories of the Kingdom.'

Rennie's employment at Albion Mills was not full time. He was involved in the installation of early Boulton and Watt rotative engines such as that at the Whitbread brewery in Chiswell Street, London and at the rolling mill at the King and Queen foundry in Rotherhithe in late 1784 and early 1785. In 1785 he prepared designs for an engine house for a hosiery factory in Nottingham. By February 1786 he was tendering for the supply of machinery to Goodwyn's works and negotiating for the lease of the premises in Southwark where he would set up his works. The Albion Mills became something of a social spectacle and soon Rennie was escorting titled visitors around the works. Watt deprecated this in a letter to Boulton, pointing out that dukes, lords and noble peers would not be his best customers. Had Rennie remained solely a manufacturer that might have been true, but as his career as a civil engineer engaged in large public works developed, he enjoyed the acquaintance of several powerful patrons.

Rennie had come to London armed with letters of introduction from influential contacts in Scotland to compatriots in the capital, one describing him as 'of great integrity & of equal genius in the science of mechanics—he has not much fortune, but great knowledge as an Engineer ...' In June 1786, conscious of his ability but unable to have his way with the managing shareholders at Albion Mills, Rennie had wished to leave the concern. Obliged by his contract to remain, he won contracts for machinery for most of the large breweries in London and a flour mill in Cadiz. He remained on good professional terms with Boulton & Watt, always recommending the use of their engines when steam power was required, and they often employing him to design, supply and erect machinery. He was consulted about the supply to watermills, although he recorded in a letter to his old professor that he had gained little business from them. In 1798 he was consulted about the machinery in the old Royal Mint, which would bear fruit later when it was rebuilt on a new site. An unsolicited offer of a quarter share in a patent for a cotton packing machine in return for help to develop it was declined. Later specialities included sugar mills for the West Indies, saw frames for various parts of the world, and dredgers for harbours. His works flourished and he moved to larger premises in 1800 and again in 1806.

In 1790 he had been consulted about the drainage of the lead mines at Wanlockhead in Dumfriesshire, where William Symington (q.v.) was a mechanic, and noted the experiments made by the latter in steam boat haulage, more than ten years before the successful launch of the *Charlotte Dundas*. From about 1818 Rennie was actively involved in developments, collaborating with Boulton & Watt in ship design and experiments to determine the most suitable positioning of the engines. A full scale trial on the Thames was successful and his son credits him with having encouraged the Navy to adopt steam tugboats.

Bridges

There was a belief in Rennie's family that he superintended the building of the bridge over the Water of Leith at Stenhouse Mills in Edinburgh in 1784, but research by Ted Ruddock has shown that the designer was Alexander Stevens (q.v.), the contractor William Stodart and the Resident Engineer Alexander Laing when it was built in 1783–1785. It is quite possible that Rennie was employed in some capacity, as his old professor, Robison was interested in the theory of masonry arches, and the notes of Rennie's tour in 1784 are evidence of his own interest in the subject.

In fact his first bridges to be constructed were those which he designed for the Lancaster Canal, including the splendid Lune Aqueduct in 1794. Here he introduced as a modification to his original design the inverted arch between adjacent arches, which would feature in all of his major multi-span masonry bridges. The contractor for this structure was Alexander Stevens and when he and Rennie disagreed on the best mode of construction, it was Rennie who gracefully gave way. He nevertheless held decided opinions on the quality of work required, preparing long and detailed specifications and insisting on adherence to them. On his first road bridge, Wolseley Bridge over the river Trent, not one but two experienced contractors successively suffered significant financial losses and were forced to give up their contracts. The problem lay mainly in the foundations and Rennie expressed a preference for building these by direct labour, as those of the major Kennet & Avon Canal and the Lune aqueducts had been. Although Wolseley Bridge took five years to build, the Staffordshire county authorities continued to commission designs from Rennie to 1805 when their own Surveyor, Joseph Potter (q.v.) was sufficiently experienced to undertake the work himself.

In 1799 Rennie designed replacements for two bridges which had been destroyed by floods. That over the Whiteadder between Paxton and Berwick upon Tweed, was not built but its history demonstrates something of Rennie's approach. He visited the site himself and sent an assistant, James Hollinsworth (q.v.) to survey the area in more detail. He proposed bridges at new locations, which would have incurred significant expenditure on new roads but which would have improved communications more generally. He made a design in cast iron but finding it considerably more expensive than stone, prepared three alternatives in masonry. One would have an arch of 150 ft. span, larger by 10 ft. than Pontypridd which was then the longest stone span in Britain.

The cost, which Rennie recognised was high, was beyond the resources of the road trustees and a bridge was built on the old site. The other, over the Tweed at Kelso has elliptical arches, a shape which he favoured to give maximum free waterway while preserving the strength of the arch.

Rennie's design in 1803 for a bridge at Musselburgh, where the old bridge still stands but is inconveniently steep and narrow, reverted to segmental arches and a gently curving roadway, but the end spans had the large span/rise ratio of 10:1.

Rennie's finest bridge is generally considered to be the Strand Bridge over the Thames in London, designed in 1808 and renamed Waterloo Bridge shortly before its opening in 1817. This bridge was easily the most expensive to have been built then in Britain and entailed significant temporary works. The foundations were laid within cofferdams and the piles driven by steam engines. The cutwaters in plan were Gothic arches, the shape which Rennie had come to prefer, though some of his earlier bridges had other shapes. The voussoirs increased in size as they descended from the crown to the springing, a point which Rennie insisted on though he was aware of large spans where that was not the case, and the centres were preloaded to minimise deflection when they were struck. Many of the details were repeated in his design for London Bridge, which was built after his death by his son, Sir John.

Occasionally Rennie would produce a design in timber, where funds were inadequate for a more permanent material. At Fosdyke in Lincolnshire an Act had been obtained for a bridge in 1794 but nothing had been done. He therefore suggested in 1810 the use of oak or pitch pine, which might be expected to last forty years with only moderate repairs, and the saving of £1,800 for pine compared with oak could form a sinking fund to rebuild in stone. In fact the road trustees opted for oak and the structure lasted for almost a century. For a similar structure at Elmley Ferry over the Swale in 1816 he suggested cast iron casings to give protection against decay, but the bridge was not built.

Rennie had no decided preference for bascule or swing bridges over his canals, which he usually designed in timber for economy. For docks however he usually provided swing bridges and at London Docks in 1803 introduced the first of that type in cast iron, a development of a form which had been proposed by Ralph Walker (q.v.) in 1800 but not executed.

Rennie had produced designs for a cast iron arch bridge in 1791, along very similar lines to Thomas Wilson's celebrated Wearmouth Bridge, for which an Act had been obtained the previous year. It would have had a span of 110 ft. and rise 13¾ ft., with six triple-membered ribs and cross frames to connect them. He did not submit any proposals in the London Bridge competition in 1800 and so was one of the panel which reported on the spectacular span of 600 ft. put forward by Telford and Douglass. Although Rennie claimed later that he preferred radial or vertical spandrel supports instead of circular, his first designs for Town Bridge, Boston in 1801 and his early free-hand sketch for Menai in 1802 both show the latter. His cheapest estimate for a span of 450 ft. across the Menai Straits was £259,140 so nothing was done then, and Boston was his first cast iron bridge to be built. Doubts were expressed soon after its completion about its stability and it did indeed need strapping across fractures in the ironwork. Rennie blamed these on the founders and the bridge survived until 1912. Originally it was to have had a span of 72 ft., but when this was changed to 86 ft. the rise of 5½ ft. was not altered. The resulting span/rise ratio was greater than any previous cast iron bridge in Britain and was only surpassed by that of the footbridges which Rennie, with the assistance of William Tierney Clark (q.v.) provided, also at Boston, in 1811.

The estimate for Southwark Bridge across the Thames was made in 1811, and its central 240 ft. span became the longest of cast iron in Britain. The design for a span of 250 ft. across the Wye at Chepstow in 1812 was almost identical in form with his earlier essay at Menai but by the time that the working drawings for Southwark were produced in 1814, he had adopted solid arch ribs and lozenge panels in the spandrels. These latter gave significant problems in the casting to the Walkers (q.v.), due to differential rates of cooling where the cross-section changed, but the calculations of rise and fall in the finished structure due to changes in the ambient temperature were found to be quite accurate.

In 1802 Rennie had considered use of chains to erect his cast iron arch at Menai, as did Telford in 1811. Interest in suspension bridges was aroused by the Runcorn Bridge proposals of 1813–1814, so when he was asked to consider a bridge at Elmley Ferry in 1816 he suggested one as an alternative to the wooden structure mentioned above. He reported positively on proposals for suspension bridges over the Tweed at Upsettington and Norham in 1817 and in 1818 he proposed improvements to Captain Brown's (q.v.) intended Union Bridge, which were incorporated into the structure as built.

In 1820, to replace Telford's bridge over the river Ken at New Galloway, he designed a self anchored suspension bridge, a concept far ahead of its time. The deck would have been a strut to take the horizontal forces at the anchorages, which would have been masses of stone. As he distrusted the quality of forged eyes in the chains then in use, he preferred steam engine chains with alternate single and double links. They would have passed over the towers on rollers to allow thermal movement. In the event a conventional masonry structure was built.

In 1820 also, Rennie proposed a similar but smaller self anchored bridge across the Aire in Leeds or a slightly cheaper alternative of the arch-suspension type. This latter principle was not new, some having been built in France and a British patent using laminated timber having been granted in 1796. Here the arch would have been of cast iron. The project did not proceed, but George Leather Jr. (q.v.) who was working as Rennie's assistant on the Knottingley-Goole Canal at this time, later constructed several large examples.

Canals

On his journey south to meet Watt in 1784, Rennie had made time to view the Bridgewater Canal and its famous Barton Aqueduct. As early as February 1789, while still involved with the Albion Mills, he went over ground which had been covered previously by Robert Whitworth (q.v.) and reported on possible routes for a canal north from Bishops Stortford to join the River Great Ouse or one of its tributaries, and so into the fens. In August he was employing Thomas Fletcher (q.v.) and Daniel Warner to make further surveys. For the rest of his life he had an extensive practice in canal engineering, though by no means all of those projects came to fruition.

In September 1789 he was asked by William Jessop (q.v.) to resurvey the Basingstoke Canal, then under construction, and reported that George Pinkerton's levels were wrong, some by several feet. This led to a commission to survey possible extensions of the canal to Andover, from which there was a navigation to Southampton, and he was careful to ask Jessop's permission to proceed. Canal mania was starting and in October he was investigating a line from Bury St. Edmunds to Ipswich and others to Hadleigh and Mistley, as well as a navigation to Thetford. The next month he made his first survey of the Western Canal, but was prevented by atrocious weather from completing it. By March 1790 he was preparing four schemes for Parliament, of which the Bury-Mistley scheme had most priority. In 1791 he was heavily involved with the Rochdale Canal and complaining that they expected too much of his time, to the detriment of his other commitments. There were problems with competition for water supply from the numerous mills along the route and from alternative routes via Huddersfield and Bury, so that the Bills were lost in 1792 and again in 1793. Jessop was called in and praised Rennie's work, but it was the weight of the older engineer which procured the Act in 1794. Rennie had by now dropped out of Rochdale affairs as, by a strange reversal, he had taken over the much smaller but badly managed Stowmarket Navigation in Suffolk from Jessop late in 1791. Earlier that year he had also been intensively involved in surveys for the Lancaster Canal, 75 miles long, and was appointed Engineer shortly after it gained its Act in 1792. Success with the Stowmarket Navigation led to appointment to the neighbouring Chelmer & Blackwater Canal, for which Rennie's survey book still exists, and he also found time in 1792 to report on the Crinan Canal. In 1794 the Western Canal, now renamed the Kennet & Avon, was authorised by Parliament and Rennie was responsible for constructing canals in Essex, Somerset, Lancashire and Argyllshire.

Two of these canals, the Lancaster and the Kennet & Avon, were on the whole well built, with sizeable resident staffs to supervise the works. Financial stringency caused by the canal mania and by war inflation prolonged the construction and the Lancaster was completed by Thomas Fletcher and William Crosley Jr. (qq.v.). At Crinan, Rennie's first estimate had been £63,628 for a canal 12 ft. deep, but this had been increased to £107,512 for 15 ft. Many shareholders defaulted and obtaining competent labour and even supplies of food in a sparsely populated location proved troublesome. The line traversed deep mosses and cut through whinstone to a depth of 60 ft. Even so, the canal had cost only £105,000 when it was opened, albeit with an effective depth of just 8 ft. £40,000 more was spent on improvements under Rennie, and Thomas Telford subsequently put in hand a programme of repairs.

The next years brought mixed fortune, with Acts for the Grand Western Canal and Tamar Manure Canal in 1796, on the first of which only 11 miles were completed in Rennie's lifetime and only 2¾ miles ever of the latter. The Polbrock Canal of 1796 in Cornwall and the Glenkens Canal of 1802 in Galloway were never started, despite obtaining Acts. Schemes to extend the Kennet & Avon west to Bristol and east to London, to connect the Bristol and English Channels, or London to the south coast at Portsmouth or Southampton and the Medway to the Sussex Rother, in extensive areas of Ireland and Perthshire and between Edinburgh and Glasgow were surveyed and resurveyed, with no tangible outcome. In 1810 a scheme to cross the Peak District would have cost £644,316, another to connect the Stratford Canal to the Thames via the Cotswolds £499,646. By now Rennie had an extensive connection with competent surveyors on whom he could call, particularly Netlam and Francis Giles (qq.v.), but lest it might be thought that he was content to provide reports on hopeless schemes and pocket his fees, it may be noted that he invested in shares of at least a few of them.

One extraordinary scheme which did proceed was the Royal Military Canal around Romney Marsh. After considerable delay caused by conflicting views amongst the different branches of the military, this line of defence against Napoleonic invasion was authorised by the Privy Council at a meeting in October 1804, at which Rennie was present. He was offered the contract to construct the works but declined, citing conflict with his professional status. Instead he introduced Bough, Hollinsworth and Dyson (qq.v.), who started work at the end of the month and undertook to complete within eight months, with a one month grace period. It is not clear why Rennie, by now well experienced in works of this kind, ever thought it could be completed within that time, or

anything like it. In any event, after an argument during a site visit with Lieutenant-Colonel Brown in February 1805, he was effectively excluded from the job; the contractors followed in June and four more years were needed for the army to complete the work.

When planning the Croydon Canal in 1801, and concerned about a potential lack of water, Rennie provided two estimates, one of which included inclined planes. He had seen these at work in Shropshire, though for tub boats only, and had been impressed,. The canal was built with conventional locks. He was less happy about vertical lifts. His report with Jessop on the experimental lift on the Ellesmere Canal has not survived, but it may be assumed to have been critical, as the Company did not proceed with it. In 1810 he was equally unconvinced by the lift at Tardebigge on the Worcester & Birmingham, noting 'the intricacy of [lifts] construction & their liability to get out of order'. Again the experiment was discontinued, though it is surprising that, with his mechanical expertise, he appears not to have tried to develop a workable example.

Rennie was consulted about problems with other engineers' canals. In 1796, when problems with running water had led Jessop to consider giving up Blisworth Tunnel on the Grand Junction Canal, Rennie and Robert Whitworth made a joint report. Jessop, who had already eliminated the tunnel which Rennie had proposed for the Rochdale Canal and shortened that on the Kennet & Avon, at the expense of increased lockage, now wished to take the canal over the hill here too. Rennie must have felt some satisfaction when the Company agreed to tunnel and although lack of funds delayed the work for some years, it was duly completed. Rennie also criticised the shape of the original tunnel at Blisworth, preferring an upright ellipse truncated at the invert, with hard stone springers at the junction, as the present tunnel is built. On another occasion, asked to comment on Ralph Dodd's (q.v.) plans for a deep cutting at Higham on the Thames & Medway Canal, he showed that an economic crossover point from open cut to tunnel occurred between 40 and 45 ft. He repeated his views when consulted about the Highgate Archway, a road tunnel, proposed by Robert Vazie (q.v.) in 1809 and was consulted about repairs when poor materials and bad workmanship caused a collapse in 1811.

Drainage

Although Rennie's bridges and canals have captured the imagination, his talents were displayed to at least as great effect in less visible works of land drainage and sea defence. He was consulted at least as early as 1790 by the Clerk of the Sewers for the County of Norfolk about draining Marshland, but it was the year 1800 which saw two of his seminal reports. Deeping Fen in Lincolnshire had been inundated in 1798, and he with William Jessop had been called in to consider the problem with the local engineers George Maxwell and

Edward Hare (qq.v.). As the falls of the existing drains were inadequate, and a natural drainage could only be obtained at great expense by a completely new outfall, he and Jessop recommended installing steam engines at Pode Hole. These would keep the internal main drains clear and allow the existing windmills to work the local drains. Steam power was as yet untried in the fens and although some of the recommendations of the main report were carried out under an Act of 1801, Deeping had to wait for Benjamin Bevan (q.v.) in 1823 to install steam.

Rennie's other report, on the East, West and Wildmore Fens north of Boston led directly to an Act in 1800 and the works were carried out. The principle of catchwater drains to intercept and carry off the water from higher land around was not new, but had not been applied on such a scale in Britain in modern times. New drains were constructed, many of them navigable, with special attention to the levels to keep the upland and fen waters separate. Anthony Bower (q.v.), the Resident Engineer, later calculated the increased annual value of the 60,481 acres which had been improved was £110,561; that £433,905 had been spent on the drainage works and a further £146,800 on roads and dividing the lands; the interest on this money at the prevailing rate of 5% p.a. was £29,035, so that the increased annual income was £81,256. He might have added that therefore the payback period was seven years. Rennie was one of those to benefit, as he bought 379 acres.

In 1809 Rennie produced an even more extensive scheme, for the South and Middle Levels in Cambridgeshire. The catchwater drain would have run round the south and east boundaries of the fens, but the estimate was the enormous sum of £1,188,189 and it was only in the 1960s that a very similar scheme was constructed.

There had for many years been arguments about the means of reconciling the conflicting demands of navigation and drainage in the rivers flowing into the Wash. Following the success of the first improvement of the River Nene outfall, pressure grew for a similar scheme on the Great Ouse above King's Lynn. Rennie was appointed by the drainage interests and Telford by the navigation, and after arbitration by Joseph Huddart (q.v.) on the size of the channel, the great bend in the river at Eau Brink was eliminated. An indirect consequence of this was the installation of steam power to drain Bottisham and Swaffham Fen, further up the river. After discussion with William Murdock (q.v.) and William Creighton of Boulton & Watt on the relative merits of pumps and scoopwheels, Rennie elected to supply the latter, which was less effective but familiar to the fenmen who would have to maintain it.

Rennie was not committed exclusively to steam power. In a report on Borough Fen, part of the North Level furthest from the outfall, he discussed how lands drained thus might cause others still drained by windmills to be flooded, and advocated a balanced approach. He also recognised

the perils of too effective a drainage. Reporting on the Somerset Levels in 1818, he noted 'There are two objects to be attended to in making out a scheme for the drainage of low lands. The first is that all water which may prove injurious to these lands should be carried off and the second is that such water as is advantageous should be retained. I have in many cases found that as much injury has been done by draining the lands too dry as by their not being sufficiently drained.' His scheme made provision for deliberate flooding of the lands as and when required. In 1803 he had designed a system of low dams at intervals down the river Lea, north-east of London, to be used to flood the valley as a line of defence in the event of Napoleonic invasion. He proposed two plans, one with twelve dams, which would be cheaper to build but require more men to defend; on this point he deferred to military advice. Despite including a premium in his estimate for speedy construction, only some were built, and those not completed until 1807, some time after the immediate scare had receded.

In 1804 Rennie was consulted about repairs to the sea wall on the south coast at Dymchurch. There had been a long history of reclamation, but the existing wall was irregular, formed of soil overlaid by brushwood faggots about 6 in. thick. The bank was too steep, being about 2:1 on the seaward face, and irregular, presenting obstructions to the free flow of the tide along it. In addition there were timber groynes up to 6 ft. high. Rennie proposed to face the slopes with stone, and he sent Netlam Giles as far away as Dover in a fruitless search for a suitable source. The following year he was asked to prepare a design for a bank to reclaim some new land from the river Plym. He proposed an embankment 13 ft. high, with central impermeable core tapering from 4 ft. to 2 ft., its outside slope 57 ft. long, to be pitched with stone and inside 19 ft. 6 in. long, to be sown with grass and a ditch 2 ft. deep at the inside toe. Both of these works were executed by James Green (q.v.).

A larger project than either of these was the lazarette, or quarantine establishment for immigrants, at Chetney Hill on a creek beside the river Medway. James Wyatt was responsible for the buildings and Rennie was to provide a canal round the site to form an island, together with the necessary protective works, docks and landing places. Building work commenced soon after the passing of the Act in 1800, but the landowners were clearly worried by the possibility of diseases spreading and there were protracted delays in purchasing some of the land. Civil engineering work did not start until 1806; the site proved most unhealthy and several workmen died after prolonged illness in 1807, which not unnaturally led to difficulties in recruitment; and work continued until 1816. The buildings were never completed and it is not clear whether any use was made of them at all.

Harbours

Rennie's introduction to harbour engineering appears to have been surveys of Wick, Loch Bay on Skye and Ullapool carried out for the British Fisheries Society in 1792. Although 400 acres of land were purchased at Wick, nothing more was done then and it was left to Thomas Telford fourteen years later to make physical progress. A pier had already been constructed by Smeaton at Aberdeen when Rennie submitted a report in 1797 on further developments. An Act was passed to put them into effect but by a strange coincidence, fourteen years elapsed before any work was done, again by Telford.

Harbours of refuge on the east coast of Scotland were then few and far between, and of small capacity, and Rennie investigated several possibilities in 1800. He submitted a report on Fraserburgh eighteen months later but a town, then, of 2,200 inhabitants could not find the £15,000 to execute the work. The harbour entrance would be too shallow for naval vessels to enter, so approaches to Government for funding were unsuccessful. Further, the surveys which the town had provided in 1801 proved to be inaccurate and it was only after a further survey by Giles that work started in 1807 on a reduced scheme. At Peterhead, for some reason Rennie did not submit his report until 1806. Again the scheme was beyond the funds available locally, but it is probable that the delay had allowed some political spadework to be done, as a grant of a little over half of the cost of the South Harbour was quickly forthcoming.

In these early schemes it is possible to find some of the themes which appear throughout Rennie's harbour works. There is concern that the entrance should be placed so as to be readily negotiable in time of storm by the sailing vessels then in use. He would take considerable pains during his preliminary reconnaissance to consult the harbour master and mariners on this point. Within a harbour he aimed to have a sloping beach where the waves might be stilled. Where funds were limited he would spell out the ideal solution but give advice on an affordable sequence of works, while recognising that the interim solution would not be ideal. At Fraserburgh and Peterhead the full scope of Rennie's ideas were realised later as revenue from increased traffic made funds available. Sometimes a client would modify Rennie's plans, as at Torquay, Margate and Ardglass, which led to structural failure in the first two cases and extensive silting in the last.

In order to deepen the harbour basin of the north-eastern ports, he proposed to employ men working within cofferdams, or if these would cause obstruction to the shipping, in diving bells. Smeaton had used these at Hexham Bridge and Ramsgate Harbour, but when Rennie first used one at the latter place, his model was the French practice at Toulon. As confidence grew he used bells often in places where caissons or cofferdams would be inadequate, in one case in a depth of 45 ft. of water, and by 1820 had ten of them at work, for blasting and quarrying under water, cutting off piles as well as laying masonry.

Rennie's first report on Holyhead as a packet port to Ireland was submitted in 1802, shortly after the Act of Union. Opposition from commercial interests in Liverpool delayed the passing of an Act until 1809. In 1807 he reported critically on plans by Major Taylor for a corresponding port on the Irish side at Howth, but the work was put in hand until in 1809 a section of a pier collapsed and Rennie took over control. For these jobs, as for most of his marine works, Rennie invested significant effort in identifying suitable yet economic stone to provide long term protection from the might of the sea. At Howth the inner foundations were of stones laid at 40° to the vertical, as had been his practice before, but to seaward he used rubble stone of large dimensions imported from Runcorn, and tipped into the sea to find its own slope. This technique, called *pierre perdu*, had been used by the French for some time and described in their books, and also used by Smeaton, so Sir John Rennie was mistaken in suggesting that this was its first use in Britain. The harbour was sufficiently advanced by 1819 for the packet boat service to commence but, as Rennie had forecast, the harbour was prone to silting and the boats were transferred to his original choice, Dun Laoghaire only seven years later.

Meanwhile he had reported on the possibility of protecting the naval base at Plymouth by a breakwater in The Sound. His estimate was the vast sum of £1,102,440 and despite showing how benefits would accrue as the project proceeded, it was six years before work was put in hand. The original sea slope was 3:1 and landward 3:2, but a storm in January 1817 caused the top to subside, so Rennie and Whidbey (q.v.) altered the design to follow the slope which the waves had given it. Sir John Rennie (q.v.), who took over after his father's death, introduced dovetailed granite blocks at low water to strengthen the work and further flattened the seaward slope above this level to 5:1. These changes pushed the cost to £1½ million and financial stringency caused the six years of the estimate to increase to thirty-eight. It spawned several similar schemes for harbours of refuge, such as Mount Bay in the Scillies and Port Nessock in Galloway, which foundered on grounds of cost, but the work itself has proved long lasting and effective.

Grimsby Haven was set up to rival Hull as an east coast port, but was bedevilled by poor management and inadequate funds for such an ambitious scheme. The original Engineer, Jonathan Pickernell (q.v.), had been dismissed and work was proceeding under the direction of the shareholders when there were geotechnical problems. Rennie was consulted twice before he was formally appointed Engineer. He considered that the position of the lock was wrong but that the expense of moving it would not be repaid by the benefits. He therefore contented himself with altering the side slopes of the canal, but made extensive changes to the form of the lock. His resident engineer, George Joyce, varied these plans without authority, which led to more failure in April 1800 and Joyce's dismissal, to be replaced by James Hollinsworth. Over the next three months, in correspondence with Hollinsworth, Rennie devised a new form of quay wall, which proved adequate for the job and would be used frequently elsewhere where ground conditions were poor. By building cellular walls on inverted arches, the width of the base was increased from 7 ft. to 20 ft. and the ground pressure correspondingly reduced.

The new century saw a proliferation of schemes for wet docks. The largest of these was the West India Dock at London, built in 1800–1806 under the direction of Jessop, who was also responsible for the Floating Harbour at Bristol in 1804–1810. Rennie was responsible, alone or with a colleague, for all of the other major docks during these years—Leith for the City of Edinburgh, London Docks, Humber Dock at Hull (initially with William Chapman (q.v.)), East India Docks, London (with Ralph Walker) and the harbour at Greenock. These were works on a massive scale which required innovation in construction as well as design. Rennie was particularly well served by his Resident Engineers, only the works at Hull giving rise to problems in use.

In 1803 Rennie criticised some of Jessop's proposals in the Bill for the Bristol Harbour; Jessop responded by claiming that he was the more interested in preparing an economical design, citing some of the details of Rennie's specification for London Docks. Clearly touched at a sensitive point, Rennie issued a strong rebuttal, stating that permanence had been a prime requirement of the client, and proving to his own satisfaction that the amount of extras such as pozzolana in the mortar at London had been far outweighed by the consequential costs of failures at West India Docks. Rennie took over responsibility at West India after the opening of the second phase and by an ironic twist, had to report the failure of a cofferdam in 1810.

A similar accident occurred at Sheerness, the most expensive of all Rennie's works. In 1807 he had produced a magisterial report on the state of the naval dockyards, showing how their haphazard growth and changes in requirements had rendered them inefficient. His solution was an entirely new yard at Northfleet. Unfortunately, it would cost £8,927,874. Little more was heard of this and instead a major improvement of Sheerness was put in hand in 1813. The major difficulty was the soft mud and running sand which underlay the site to a considerable depth, a problem solved by extensive piling and the use of cellular quay walls of the type built previously at Grimsby. In order to explain the works to the many illiterate workers, including convicts, employed, Rennie had a large model made, to one-sixtieth scale. Nevertheless the works demanded

much of his attention, with many more frequent visits and reports than his other jobs required.

Other works

Rennie's practice over the years included a number of interesting projects outside the normal scope of his work. One of the more challenging, and subsequently controversial, was the Bell Rock Lighthouse, off the coast of Angus. There had been agitation from naval quarters for a long time for a light there. Robert Stevenson, the Northern Lights Commissioners' Engineer, had been exploring the possibility since at least 1794, but the formidable problems and the estimated expense had been enough to put off any action. In August 1805 Rennie and Stevenson visited the rock together and Rennie's subsequent report started to convince the Commissioners of the practicability of a permanent stone tower. Rennie had, in his usual fashion when confronted by a new problem, consulted other works of a similar nature, consulted published works such as Smeaton's *Narrative ... of the Eddystone Lighthouse* and Belidor's *Architecture Hydraulique* and consulted other practitioners in the field. Inevitably the design was strongly influenced by Eddystone, but Rennie altered the shape, 'being much more extended in the base, to prevent the waves from cutting the rock at the foundation ...'. In December 1806 Rennie was appointed Engineer and he obtained the appointment of Stevenson as Assistant Engineer. After an initial flurry of correspondence between the two on methods and materials, Stevenson spent the first year in setting up work bases on shore and on site, where the rock was fully covered at each high tide, and in levelling the rock when conditions allowed. Progress was slow but Rennie was fully satisfied with Stevenson's efforts, as his reports to the Commissioners showed, and the work was completed within the four years mentioned in Rennie's initial report.

Afterwards Stevenson was pressed to write an account of the construction, which was already recognised as a great achievement, but it was not published until 1824, by which time Rennie was safely dead. In it Stevenson claimed sole responsibility for the design as well as construction, a claim which was disputed by Sir John Rennie in his *Autobiography* in 1875, when Stevenson also was dead. The dispute rumbled on between the families until the 1950s. Paxton has shown how Rennie's design, and the structure as built, differed from Smeaton's Eddystone and Stevenson's proposal. Equally, Stevenson was given and took considerable latitude in the detailed design, as the following extract from a Rennie report in 1808 makes clear: 'Mr. Stevenson's directions were to make the moulds or templates as much as possible to suit their quarries, so that the kind of stone supplied might be done with as little expence as possible to the Contractors, & for this purpose the plan of each course of stones conforming to a principle laid down was left to him,

& it appears from the extract of his report sent to me that he has fully complied with my directions, and so far amply vindicated the confidence placed in him.' Each may take a fair share of the credit for this remarkable building.

As Bell Rock was proceeding, Rennie was involved with a company which would bring him much less credit. As a supplier of pumps driven by steam or water, he was consulted about water supply to various towns and cities in England and overseas. For two of these projects, where one Samuel Hill was a shareholder, he agreed to use distribution pipes supplied by the Stone Pipe Company, of which Hill was also a shareholder. Rennie visited the quarry in the Cotswolds early in the proceedings and expressed misgivings about the mode of preparing the pipes, but failed to pursue the problem. When installed, the joints in the pipes proved woefully inadequate and the whole had to be replaced by cast iron.

Rennie did direct a survey in 1811 for the improvement of the road from Dumfries to Newton Stewart, on the way to Portpatrick, proposing many miles on a new alignment which would both ease the gradient and avoid the worst of the winter snowfalls. It was not built but later the railway followed the line very closely. The report gives no indication of the road construction he might have used.

Rennie used railways in several of his dock and harbour works for transporting the large quantities of earth and stone involved. He also built a line from Caldon Quarries to the Leek branch of the Grand Trunk Canal. These lines were worked by horses, with inclined planes where the direction of traffic provided most of the motive power. He was aware of Chapman's and Brunton's patents for steam engines, but advised against the use of steam haulage at Howth as being less economic than a traditional system. He proposed railways over sizeable distances, such as Kelso to Berwick and Stockton to Winston. This latter was an alternative to a suggested canal and he recognised that the nature of a railway would dictate a different route. The proposal matured ten years later as the Stockton & Darlington Railway, the Act for which was based on the unexecuted Kelso & Berwick Act of 1811.

Less usual requests received equal consideration. He was consulted about the possibility of raising the ship Royal George, which had sunk in the approaches to the Isle of Wight and for which he proposed a method based on that used by his colleague Captain Whidbey at the Nore some years earlier. Another report dealt with cast iron dams, of which he admitted the practicability in principle but pointed out the practical problems. Three steel dams were in fact built in the United States later in the century, but the idea never found general favour. An inventor of a new kind of rotative steam engine, an object which Rennie called the *ne plus ultra* in the steam engine line, was advised to contact Boulton & Watt to take out a patent, but there is no evidence that this ever bore fruit. Nor could an idea from the brother of a naval officer to use a water wheel to pump up water, which could then be used to augment the power of the wheel.

Professional practice

Rennie was consulted on many schemes, and almost invariably would present a written report to his client. Early in his career he would visit the site himself, but he soon came to rely on the advice of surveyors, whom he would employ as and when the need arose. Occasionally he would base his report closely upon the advice of a resident engineer, but generally the engineering input was his alone. He was prepared to give advice on projects overseas, relying on plans and replies to a list of queries which he would submit. His reports were often redrafted, sometimes as much as half of the first draft being deleted and rewritten. At least once he submitted a report to James Watt junior to correct for grammar and style. The results were models of clarity, and were copied fair into bound volumes. Rennie could thus refer to lapsed proposals, sometimes observing tartly that he could think of no cheaper ideas than his earlier ones, or giving detailed history to show how costs had escalated because of changes made by a client.

Rennie did employ clerks and draughtsmen, and on occasion required help from outside in order to fulfil his obligations. The drawings for major works such as the Southwark Bridge took several months to complete, and would if necessary be produced as work progressed.

For new schemes the report was usually accompanied by one or more detailed estimates of cost. For the Aberdeenshire Canal he explained 'It cannot therefore be imagined that the sums mentioned in any article of the estimate will agree with the costs. It is sufficient if they agree on the average which if managed properly I have no doubt they will do.' The estimates usually included an addition of 10% as a contingency item, though for the Bell Rock, where he admitted that his site costs were based on guesswork, he added 25%. Much of his work was done during the Napoleonic War, when there was significant price inflation, and he would load his estimate if he felt that the work would not proceed quickly. Nevertheless, many of his contracts were delayed by financial stringency and cost more than expected, the overrun often being due also to variations to the original scheme agreed by the client. The contracts which ran on after 1815 were much closer to the mark.

As has been mentioned, Rennie was closely involved with several influential patrons. In 1793, Sir John Jervis, the future Earl St. Vincent and First Lord of the Admiralty, was a shareholder in the Chelmer & Blackwater Navigation, and remained a staunch friend in Staffordshire, where he also had an estate, and at the Admiralty, when Jessop was called in to give a second opinion on the Plymouth Breakwater. In 1794, although not directly involved, Rennie assisted Sir Joseph Banks to find a Resident Engineer for the Horncastle Canal, and they were thenceforth closely allied in the promotion of fen drainage schemes. It was probably due also to Banks, who was a leading promoter of the Act for the Bell Rock Lighthouse, that Rennie obtained his appointment there.

Rennie worked in an age when it was still quite usual for the committee members of public works to be active in the day to day management. At Grimsby this led directly to the problems which led to Rennie's appointment, and at Leith to a strained relationship with the City Council. On the Royal Canal in Ireland, he spelt out his position: 'I do not mean to say (the Directors) should not settle the works of any dimensions most agreeable to themselves, but when that is done, they ought to leave the execution to me, or appoint some person more capable ...' and insisted on acceptance of his terms before any further work was done.

On another occasion, when challenged to explain why his design for the Custom House Wharf in London was more expensive than a cast iron alternative which had been proposed anonymously, he complained '... as I am not informed who the person is that has suggested the project, I have no means of estimating the value of his professional reputation which he offers as a pledge for the success of his scheme. I must therefore request the honourable Board will be pleased to obtain the requisite information and transmit the same to me and in doing this I need scarcely caution the honourable Board against any charges that may be made for the same as projects of this sort are too frequently made to the public by needy adventurers who are incapable of carrying their pledges into effect while the temporary employment suits their purpose.'

Rennie sometimes worked jointly with other engineers, as with Ralph Walker at the East India Docks in 1803–1806 and William Chapman at Humber Dock in 1802–1809. Several times he reported jointly with Captain Whidbey on maritime problems and co-operated with Matthew Boulton on machinery. His disagreement with William Jessop about the Bristol Docks, which clearly hit a raw nerve, has been noted, and some of the strain still showed when the latter was asked to give a second opinion on the proposal for the Plymouth Breakwater. After Jessop retired, the other leading civil engineer of the day was Thomas Telford, and it is well known that he felt that Rennie spurned his acquaintance. This has been ascribed to their differing social backgrounds, but may also be due to events in late 1803–early 1804, when Telford made a critical report on the Crinan Canal works and submitted an unsolicited estimate for the Bell Rock Lighthouse, both of which offended Rennie's sense of professional propriety. He was scornful too of Telford's use of other engineers, but his own preference for providing the engineering input himself certainly restricted the amount of work he

was able to accept, and ultimately limited his influence on the development of the profession.

Often though not always, part of Rennie's obligation as Principal or Consulting Engineer when a project started was to find a suitable person to act as Resident Engineer. The skills required were quite varied, and obtaining suitable people with ability, integrity and experience could be difficult. The Resident Engineer was employed by the project client and Rennie made it clear that he would not be responsible for his nominee's conduct. In time Rennie was able to rely on a corps of competent people, though even some of the most senior slipped from grace. Early on at Sheerness he had recommended that the supervisors should have nothing to do with purchasing, but John Thomas (q.v.) was found to have at least condoned some of the peculation which was rife there.

On widespread works such as the Kennet & Avon or the Lancaster Canals, Rennie preferred to keep the number of senior staff to a minimum and to use the money saved to pay good salaries to attract people of good calibre. He prepared lists of their duties, specifying where they ought to live and the accommodation which ought to be provided. Resident Engineers on canals were given authority to make improvements in the line, as they would be more intimately aware of the nature of the ground, and could determine the positions of bridges in consultation with landowners and the shareholders. On occasion the Resident Engineer departed from Rennie's plans without his authority, leading to failure of part of the works, as at Grimsby in 1800 and at Margate in 1813. At the former, the Resident Engineer was George Joyce, whom Rennie had recommended from being a supervisor of earthworks on the Lancaster Canal, and he was dismissed with a reference which bore testimony to his integrity but made no mention of his technical skill.

In the early days of the Kennet & Avon Canal, when the canal mania made it difficult to find enough or adequate staff, the contractors' work was not valued properly as they progressed and payment was made on the basis of costs expended rather than contractual entitlement (a practice not uncommon on other canals). In his instructions Rennie laid some stress on the need for regular measurement and valuation of the works. It was also an important part of the Resident Engineer's duties to keep a record of and evaluate any extra works ordered. Rennie was consulted on several contracts where inadequate records had been kept, and found considerable difficulty in coming to a fair decision on the monies due to the contractors. On one contract worth £70,000, the extras had amounted to £40,000.

It was Rennie's practice to make at least one extended tour each year, to visit the sites under construction and to obtain data for use in reports on new projects. In 1798 he spent seventy-five days on one trip, going as far as Argyllshire and in 1801 he made three tours aggregating one-hundred and five days, reaching Ireland on each occasion. Towards the end of his life, when work on the naval dockyards occupied much of his attention and he was turning away almost as much work as he accepted, because of his commitments, he made more frequent but shorter visits.

Rennie's career coincided with the emergence of contractors with the financial muscle and managerial skills to manage several contracts in different locations at one time. John Pinkerton (q.v.) was contractor for the Ulverston and Lancaster Canals, but was required to give them up because of poor performance. For some time afterwards Rennie generally let works in small packages to single trade contractors, though in the fens he employed Pinkerton again much more successfully. Later, for the large dockyard projects, he preferred to employ men such as Edward Banks and Hugh McIntosh (qq.v.), though he complained of the former that his work became much less reliable as his firm expanded and he was no longer in personal charge of a particular project.

Rennie believed strongly that the lowest tender was not necessarily the best, and on several occasions counselled clients strongly against accepting the cheapest offer. In 1798 he noted 'I have generally found contractors' workmen apt to slight their work, where the price is low and no superintendence; however, vigilance can make them do it well—the first matter therefore is to allow them a fair price for the work to be performed & then, whatever orders are given may be enforced with rigour but without this they never can.' and on another contract at the same period 'In all the Specifications of the work I have retained in the hands of the Company's Engineer a power of explaining his own meaning of discharging the Contractor and of valuing the work himself. These may appear arbitrary and improper powers to be lodged in the hands of any man; granted, but necessity requires it. Without such a power being lodged somewhere there is no possibility of doing a work of this kind properly. Suppose for instance a mason was to contract for a single lock, culvert or bridge and was not proceeding properly either in the way of executing his work or in the time of doing it. He might by this means delay the opening of the canal for many months after all the other works were finished, and by going into the Court of Session when the business would be settled. The settling the work by arbitration is little better; the time lost is often great and arbitrators universally lean to the side of the Contractor ... . I hold it as a principle that every contract should be mutually beneficial to the parties. Unless it is so, no Contractor can be bound to perform it, and the only way to have work to proceed with expedition and be well executed is to let the Contractors have a reasonable profit, and the directors of the work a full compulsory power over them ... and I find no difficulty in getting good contractors to enter into these terms.' He did on occasion recommend clients to make ex gratia payments to contractors who had performed well but who had sustained a loss by the strict terms of the contract, or to allow them a further contract at an advantageous price to recoup the loss.

He seems to have been much less aware of contractors' need for an acceptable cash flow. It was common practice at the time to accept bonds from third parties for the proper performance of contracts, but he came to prefer security in money instead. The sums involved could be significant: for a contract in Dublin worth a little over £4,000, he proposed to take £1,000. As it was also intended to pay the contractors interest at 5%, they would presumably have borrowed the money and the effect might not have been very different from the old system.

Rennie was quite an imposing man, standing almost 6 ft. 4 in. tall and weighing 15 stone. However his health in middle and later life was not always good, almost certainly as a result of the nature and extent of his professional life. He suffered badly from rheumatism, as early as 1800 being scarcely able to walk because of it. In August 1811 he had an accident which prevented him from working for a month and in 1813 was obliged to spend two months at Cheltenham to take the waters.

The steady flow of reports, some dated 25 December, with only occasional breaks due to illness, show a life dedicated to his work. His only known holiday was a two month tour in the autumn of 1816 through parts of the Netherlands and France 'for the sole purpose of examining the public works in those countries, ... in case of a future war.' His principal relaxation appears to have been his excellent library, which included first editions of Shakespeare and mediaeval histories of England as well as engineering volumes in English and French. He maintained contact with French engineers during the war, receiving a gift from Prony in 1811 and he returned the compliment by sponsoring his friend for Fellowship of the Royal Society in 1816. It has been claimed that he also learnt German while at university, but he cited ignorance of the language as a reason for declining an offer of membership of the Academy of Sciences at Munich.

In 1788 he was engaged to be married to a Miss Durnford of Betchworth in Surrey but the wedding did not take place and it was on 6th November, 1790 that he married Martha Mackintosh of Inverness at Christchurch, Blackfriars Road, Southwark. She died in March 1806, shortly after the birth of the youngest of their nine children, of whom seven were still alive in 1808 and five survived their father's death in 1821. The two eldest, George and John junior (later Sir John) are noticed separately.

He lived in the parish of Christchurch, Southwark, from the time of his arrival in London, and moved to Stamford Street, now called Rennie Street, in 1793. The house, now demolished, was a three and a half storey, end of terrace building.

He was clearly well established in business by then, as he had been able to withstand the loss of £3,000 in the previous year owed to him by bankrupt clients. His manufactory was close by and continued to grow, moving to larger premises in 1800 and again in 1806, and being expanded by the purchase of the Albion Stone Wharf in 1820. In 1790 he charged his time as a consulting engineer at 1½ guineas per day, which increased to 2 and 2½ guineas in 1792, 3 guineas in 1797, 5 guineas in 1800 and finally 7 guineas. In 1793 William Jessop had increased his daily fee from 3 guineas to 5 guineas, which remained his standard charge In 1808 Rennie drew up a draft will, which put his worth at about £40,000. In 1809 his net income from millwork was £806, from engineering £975 plus salaries and he received interest on canal and dock shares as well as Exchequer Bills. In 1810 he drew up what proved to be his final will. It was not attested but when finally proved in 1821, was valued at £110,171 12s 6d, a large sum for the time.

During his life Rennie received several notable honours. He was elected a Fellow of the Royal Society of Edinburgh in 1788 and of the Royal Society of London ten years later. He became a member of the Smeatonian Society of Civil Engineers in 1785, being then the youngest person to have been elected and remaining its youngest practitioner until 1811. With Robert Mylne, Robert Whitworth and William Jessop, he reformed the Society in 1793 and when Mylne died in 1811 he took over its leadership. In 1806 he was elected to the Linnaean Society and in 1811 a Governor of the Royal Hospitals. In 1817 he was offered a knighthood at the opening of Waterloo Bridge, but declined it.

He died on 4 October 1821 after a short illness and was buried in St. Paul's Cathedral. His memorial there asserts that the many splendid and useful works he superintended are the true monuments of his public merit

P. S. M. CROSS-RUDKIN

[Rennie reports, ICE; Rennie manuscripts, NLS; Diaries of Lt.-Col. Brown, mss 2868 and 3269, NLS; GD/171/177/32, Falkirk Museum; C. Hutton (1812) History of iron bridges, *Tracts on Mathematical and Philosophical Subjects etc.*, **1**, 144–166; (1846) *Second Report of the Commissioners appointed to inquire into Tidal Harbours*; Sir John Rennie (1854) *The Theory, Formation and Construction of British and Foreign Harbours*; J. F. LaTrobe Bateman (1884) *History and Description of the Manchester Waterworks*; W. H. Wheeler (1896) *A History of the Fens of South Lincolnshire*; H. W. Dickinson (1954) *Water Supply of Greater London*; C. T. G. Boucher (1963) *John Rennie*; R. L. Hills (1967) *Machines, Mills and Uncountable Costly Necessities*; J. Mosse (1968) The Albion Mills, *Trans Newc Soc*, **40**, 47–60; A. W. Skempton (1971) Samuel Wyatt and the Albion Mill, *Architectural History*, **xiv**; P. A. L. Vine (1972) *The Royal Military Canal*; M. W. Baldwin (1973) The engineering history of Hull's earliest docks, *Trans Newc Soc*, **46**, J. Taylor (1977) John Rennie's reconstruction of Sheerness Dockyard, *Industrial Archaeology Review*, **I**, 3; W. M. Hunt (1979) The Horncastle Navigation engineers, 1792–1794, *Journal of the Railway and Canal Historical Society*, **XXV**, 1; A. R. Buchan (1979) The engineers of a minor port, *Industrial Archaeology Review*, **III**, 3; A. W. Skempton (1979) Engineering in the Port of London, *Trans Newc Soc*; P. Lead (1980) *The Caldon Canal and Tramroads*; G. M. Saul (1987) John Rennie (1761–1821): one of his contributions to waterworks technology, *Trans Newc Soc*, **59**, 3–14; G. Watson (1989) *The Smeatonians*; Margaret Bradley (1994) *Prony the Bridge-builder*; J. Leslie and R. A. Paxton (1999) *Bright Lights*; Skempton]

Publications

1790. *A Plan showing the Line of the Proposed Navigation from Bishops Stortford … to the Brandon River, on the border of Norfolk*

1791. *A Plan showing the Line of the proposed Rochdale Canal between the Calder Navigation near Sowerby Bridge in the County of York and Manchester … with its several Branches*, and revised plan, 1794

1792. *Plan of the Proposed Lancaster Canal from Kirby Kendal in the County of Westmorland to West Houghton in the County Palatine of Lancaster*

1792. *Plan of the Proposed Crinan Canal between the Lochs of Crinan and Gilp, in the County of Argyll*

1792. [Report] *To the Committee of the Proposed Chelmer Navigation*

1793. *Plan of the Proposed Navigation from Chelmsford to Colliers Reach with a Branch to Maldon in the County of Essex*

1793. [Report] *To the Committee of Landowners, and others, interested in the Improvement of the Outfall of the River Ouze*

1794. *Plan of the proposed Navigable Canal, between the River Kennet at Newbury … and the River Avon at Bath … Likewise of a Branch from the said Canal near a place called Marsh Barn to the Towns of Calne & Chippenham, to which is added a Plan of the proposed Somersetshire Coal Canal*

1794. *Report of a Survey of the River Thames, between Reading and Isleworth; and of Several Lines of Canals projected to be made between those Places: with Observations on their Comparative Eligibility*

1794. *Plan of the proposed Grand Western Canal, from Topsham in the County of Devon, to Taunton in the County of Somerset; with the Branches to Tiverton and Cullompton*

1796. *Plan of the Proposed Polbrook Canal, from the Tide-Way of the River Camel, near Wadebridge, to Dunmeer. With a Collateral Branch towards Ruthern Bridge, in the County of Cornwall*

1796 *Estimate of the proposed London Docks in Wapping and of the proposed Canal from … Wapping to Blackwall*

John Rennie FRS, FRSE: works

Date	Work	Cost (£)	Resident Engineer
1792–1793	Stowmarket Navigation	35,000	Richard Coates
1792–1803	Lancaster Canal (Lune Aqueduct 1793–1797)	490,000 (48,000)	Archibald Millar (1793–1799), Henry Eastburn (1794–1799) and William Cartwright (1799–1804)
1793–1797	Ulverstone Canal	9,000	Hugh Baird?
1793–1797	Chelmer and Blackwater Canal		Richard Coates
1794–1809	Crinan Canal	165,000	John Paterson (1794–1802) and James Hollinsworth (1805–1811)
1794–1810	Kennet and Avon Canal	860,000	Dudley Clark (1795–1800) and John Thomas (1802–d. 1827)
1797–1800	Grimsby Haven	60,000	George Joyce (1798–1800) and James Hollinsworth (1800)
1797–1802	Grand Trunk Canal, Leek Branch (Rudyard Dam 1798–1800)	(8,000)	(Joseph Potter)
1797–1805	Aberdeenshire Canal	48,000	Thomas Fletcher
1798–1802	Wolesley Bridge, Colwich	12,000	Joseph Potter
1800–1804	Kelso Bridge	13,000	John Duncan
1801–1805	London Docks	320,000	James Murray
1801–1806	East Dock, Leith	105,000	John Paterson
1801–1809	Croydon Canal	127,000	Dudley Clark
1802–1804	Caldon Low Railroad		
1802–1810	East, West and Wildmore Fens	430,000	Anthony Bower
1802–1817	Royal Canal, Ireland	200,000	George Knowles, Thomas Townshend and James Brown(e)
1803–1806	East India Docks, London	250,000	Ralph Walker
1803–1809	Humber Dock, Hull	230,000	John Harrap
1804–1805	Royal Military Canal (completed 1809)		George Jones
1804–1806	Radford Bridge, Stafford	6,200	
1804–1807	Lea Valley, inundation dams	c. 20,000	
1804–1807	Town Bridge, Boston	22,000	John Watson
1805–1809	East India Dock, Greenock	44,000	John Aird
1806–1808	Musselburgh Bridge	11,000	
1806–1816	Chetney Hill, Lazarette		John Timperley
1807–1809	Darlaston Bridge, Stone, Staffordshire	2250	
1807–1811	Fraserburgh Harbour, south-east pier	11,000	William Stuart
1807–1811	Bell Rock Lighthouse	61,000	Robert Stevenson
1807–1811	Grand Trunk Canal, Uttoxeter Branch		
1807–1821	Ramsgate Harbour, improvements		George Louch (1807–1841) and Nathaniel Gott (1814–)
1808	Margate Harbour		
1808–1809	Stretton Bridge, Staffordshire	2,250	
1808–1811	South Harbour, Peterhead	8,000	William Wallace
1809–1818	Queensferry Piers	34,000	James Milne
1809–1818	Howth Harbour	347,000	John Aird
1809–1821	West India Docks, improvements		Thomas Morris (1809–1811), William Pilgrim (1811–1816) and Thomas Shadrake (1816–)
1809–1821	Manchester Waterworks		David Chrisholme and Freemantle
1809–1821	Berwick Harbour (completed 1824)	63,000	John Fox
1810–1814	Grand Western Canal	145,000	John Thomas (d. 1827) and William Wallace
1810–1817	West Dock, Leith	70,000	John Paterson
1810–1821	Admiralty Pier, Holyhead (completed 1824)	80,000	James Brown

John Rennie FRS, FRSE: works (contd.)

Date	Work	Cost (£)	Resident Engineer
1811–1817	Waterloo Bridge	480,000	James Hollinsworth
1811–1821	Grand Junction Waterworks, London		William Anderson
1812–1814	Cree Bridge, Newton Stewart	8,000	Adam Blane
1812–1814	Lucknow Bridge, India, designed and cast; erected *c.* 1846		
1812–1815	Fosdyke Bridge	6,000	
1812–1817	Witham navigation, including replacement of Tattershall Bridge	110,000	Thomas Townshend
1812–1821	Plymouth Breakwater (completed 1850)	1,500,000	Joseph Whidbey (1812–1830) and William Stuart (1830–1850)
1813	Abbey Bridge, Stoneleigh		
1813–1821	Sheerness Dockyard	1,600,000	John Thomas (d. 1844)
1814	Ardglass harbour, south pier		
1814–1820	Newhaven (Edinburgh) Pier	18,000	James Milne
1815–1819	Southwark Bridge	660,000	Charles Meston
1815–1821	Pembroke Dockyard	130,000	William Wallace
1815–1821	Custom House Dock, Dublin, completed by Telford, 1824		John Aird, William Wilkie and Mr Archer
1816–1819	Deptford Dockyard	67,000	William Lewin
1817–1819	Custom House, River Wall, London	73,000	
1817–1819	Wellington Bridge, Leeds		John Timperley
1817–1821	Kingstown Harbour (completed 1836)	700,000	John Aird
1818–1821	Eau Brink Cut (completed 1823)	250,000	Thomas Townshend
1819	Helmsdale Harbour		
1819–1820	Stranraer Harbour, west pier		
1820–1821	Yarmouth haven, improvements		
1820–1821	Chatham Dockyard, dry dock	165,000	Phillip Richards
1820–1821	Bottisham and Swaffham Fen	6,400	Thomas Townshend
1821	Bridge of Earn		
1821	Ken Bridge, New Galloway (completed 1824)		James Gould

1796. *First* [and Second] *Report by John Rennie on the Harbour of Aberdeen*
1796. *Plan of the Intended Navigable Canal from the Harbour of Aberdeen to the Bridge over the River Don at Inverury*
1797. *Report concerning the Different Lines surveyed by Messrs. John Ainslie & Robert Whitworth, Jun, for a Canal proposed to be made between the Cities of Edinburgh and Glasgow ... with an Account of a Running Level, taken for a New Line by Linlithgow and Falkirk*
1798. *Report concerning the Practicability and Expence of the Lines suggested by Messrs.. John Ainslie & Robert Whitworth, Jun, for a Canal proposed to be made between the Cities of Edinburgh and Glasgow ... with the Improvements that have been made on these Lines; and also concerning ... the New Line by Linlithgow and Falkirk*
1800. *Report* [and Second Report] *concerning the Drainage of Wildmore Fen, and of the East and West Fens*
1800. *The Report of Messrs.. Jessop, Rennie, Maxwell and Hare, on the Drainage of Deeping Inclosed Fens, and the Commons adjoining thereto*
1800. *Report concerning the Improvement of Boston Haven, by John Rennie, Esq.*

1801. *The Report of William Jessop and John Rennie* (at a General Meeting of the Company of Proprietors of the Lancaster Canal Navigation ...)
1801. *Answers* (to the questions respecting the construction of a cast iron bridge, of a single arch, 600 ft. in the span, and 65 ft. rise)
1801. *Mr. Rennie's Reports on the state of the Harbour of Port Patrick,* and 1802
1801. *The Report of John Rennie, Esq. on the Ancholme Drainage and Navigation* (and Further Report, 1802)
1802. *The Report of John Rennie, Esq. on building Bridges over the Straits of Menai, and over the Conway River*
1802. *Plan of the Proposed London Canal from the Grand Junction Canal at Paddington to the London Docks in the County of Middlesex*
1802. [Reports] *To the Directors General of Inland Navigation in Ireland*
1802. *The Report for Improving the Harbour, and making Wet Docks, at Greenock*
1802. [Report] *To the Commissioners for Navigation of the River Witham*
1803. *A Plan of the River Witham between Boston and Lincoln, with the Intended Improvement of the Navigation*

1803. *Report and Estimate of a Canal proposed to be made between Croydon and Portsmouth. By means of which, and the Croydon Canal, an Inland Navigation will be opened between London and Portsmouth*

1803. *Plan of a Canal proposed to be made betwen Croydon and Portsmouth; and which, by means of the Croydon Canal, will form a Communication with the River Thames at Rotherhithe*

1803. *Report of John Rennie, Esq. Civil Engineer ... upon a Survey & Plan of a Canal from the River Clyde at the City of Glasgow, to the West Coast ... at or near the Harbour of Saltcoats, by John Ainslie, Surveyor*

1803. *Plan of a Canal, Proposed to be made between the River Clyde at the City of Glasgow; and the Harbour of Saltcoats, with a Branch to Paisley*

1803. *A Plan of the East India Docks at Blackwall* (with Ralph Walker) (and revised scheme, 1804)

1803. *Mr. Rennie's Report to the Commissioners for the Navigation on the River Witham, and to the Proprietors of the Foss Navigation, and the Landowners draining through the same*

1803. *Specification of the Locks proposed to be built on the Royal Canal of Ireland, on the Lands of Porterstown, Riverstown (etc.)*

1804. *Plan of Rennie's proposed improvements to Dublin Harbour*

1804. *River Lee. Mr. Rennie's Report*

1805. *Report of John Rennie, Esq. on the State of the Works of Drainage of the East, West and Wildmore Fens*

1805. *Report and Estimate of the Expence of Enlarging and Improving the Harbour of Peterhead, agreeably to a Plan and Sections theirof*

1806. *The Report of Mr. John Rennie and Mr. Jos. Whidbey, to the Right Honourable the Lords Commissioners of the Admiralty* (on Plymouth Sound), (and further report, 1806)

1807. [Report] *to the Trustees of the Holderness Drainage*

1807. *Letter from John Rennie, Esq. to be added to certain Resolutions of the General Commissioners for Drainage, and the Commissioners for Navigation, by the River Witham*

1807. *Plan of the River Clyde at the City of Glasgow; with the proposed Dock at the Broomielaw*

1808. *Reports by Mr. Rennie, Civil Engineer, respecting the proposed Improvements at Pettycur, and intended Ferry-Boat Harbour at Newhaven* [and] *relative to the Improvements proposed to be made upon the Harbour of Burntisland*

1808. *Messrs. Rennie and Huddart's first Report on the best means to be adopted for improving the Harbour of Howth* [and] *second Report on their plan for the completion of the Harbour, and 1809*

1809. *Mr. Rennie's Reports, presented to the Lords of the Admiralty* (on the Bay and Harbour of Ardglass, with the Proposed Pier)

1809. *The Report of John Rennie, Civil Engineer, FRS &c concerning the Practicability and Expence of making a Navigable Canal through the Weald of Kent, the join the Rivers Medway, Stour and Rother*

1809. *Report, by Mr. John Rennie, Engineer, respecting the proposed Rail-Way from Kelso to Berwick*

1810. *Estimate of the probable Expense of building a Pier at Holyhead, extending 650 Feet beyond the Perch, and forming a Road, and excavating Part of the Interior Harbour*

1810. *Report and Estimate of the Grand Southern Canal, proposed to be made between Tunbridge and Porsmouth: by means of which and the River Medway, an Inland Navigation will be opened between the River Thames and Portsmouth*

1810. *New Shoreham Harbour. The Report of John Rennie, Esq.*

1810. *Report and Estimate on the Improvement of the Drainage and Navigation of the South & Middle Levels of the Great Level of the Fens*

1810. *A Plan of the Proposed Canal from Bristol to Taunton. Together with Collateral Branches to Cheddar and Nailsea in the County of Somerset*

1810. *Line of the proposed Ship Canal from Bridgewater to Seaton*

1811. *A Plan of the Proposed Extension of the Kennet and Avon Canal from the City of Bath to the City of Bristol*

1811. *Plan of the Docks at Liverpool, with the Proposed Alterations & Additions*

1811. *Mr. Rennie's Survey of the proposed Line of Road from Dumfries to Port Patrick*

1811. *Report of John Rennie, Esq. Civil Engineer, FRS &c on the Proposed Ship Canal between the English and Bristol Channels*

1811. *Plan of the Frith of Forth with the Improvements at the Queensferry Piers and Landing Places forming the Great Line of Communication between the North & South of Scotland*

1811. *Plan of the Proposed London & Cambridge Junction Canal from the Head of the Stort Navigation at Bishops Stortford to the River Cam near Clay Hithe Sluice ... and of the proposed Branch from Sawston to the North Road near Whaddon in the County of Cambridge*

1811. *Plan of the Proposed Weald of Kent Canal, to Unite the Rivers Medway and Rother, and of the Proposed Branches to Lamberhurst, Cranbrook and the Chalk Hills near Wye*

1812. *A Plan of the River Witham between Lincoln and Boston, with the Intended Improvements of the Navigation*

1812. Report on the Hundred Foot Wash

1812. Report on St. John's Eau

1812. *Copy of Report made by John Rennie, Esq. respecting Rye harbour, and the Upper Levels, on the Banks of the River Rother*

1813. *The Report of John Rennie, Esq. on the Drainage of Hatfield Chace*

1813. *A Plan and Elevation of the New Bridge now building over the River Thames near Somerset Place in the Strand*

1814. *Mr. Rennie's Report and Plan for the Improvement of the Outfall to Sea through Moreton's Leam Wash and Wisbeach; and fr the Drainage of the North Level, South Holland, and the low Lands adjacent*

1814. *Report by Mr. John Rennie, Engineer* [on the intended Edinburgh and Glasgow Union Canal]

1814. *The report of John Rennie, Esq., Engineer, for the Enlargement and Improvement of the Harbour of Whitehaven* (printed 1836)

1815. *Specification* [and Tender] *of the Masonry, Brick Work, Piling &c of a New Entrance to the Basin at the Royal Dock Yard at Deptford, and also for the Building of a River Wall from the said entrance to the Second Slip on the South*

1815. *Report to the Commissioners for the Construction of a Pier and other Works at Holyhead*

1815. Report on the Black Sluice Drainage

c. 1815. *Plan of the Intended Arundel and Portsmouth Canal, Aqueducts &c*

1816. *Mr. Rennie's report* [relating to the River Tyne] (printed 1836)

1816. *Copy of a Letter from John Rennie, Esq. to Henry Yeo, Esq. Secretary to the Commissioners for the Harbour of Howth*

1817. *Design for the New proposed Asylum Harbour at Dunleary* (with John Aird)

1817. *Specification of the manner of quarrying Rock ... for the Works of Dunleary Harbour* [and] *Specification of Horse Work proposed to be done by Contract in drawing the Waggons on the Railroad for Dunleary Harbour*

c. 1817. *Dimensions of the Southwark Cast Iron Bridge, erected over the Thames under the direction of John Rennie, Esq.*

1818. *Communication from Mr. Rennie, to the Commissioners of the Navy, on the best Method of preventing the general Shoaling of the River Medway*

1818. *Report ... to the Commissioners of the Haven and Piers at Great Yarmouth, Norfolk, on the State of the Bar and Haven, and the Measures advisable to be adopted for improving the same*

1819. *Letter* [and Paper] *from Mr. Rennie on the subject of a new Bridge at Rochester*

1819. *The Reports of John Rennie, Esq. Civil Engineer, on the Proposed Bridge over the River Nene, to effect a direct communication from Norfolk and Suffolk, with Lincolnshire ... and a Letter thereon from the Right Hon. Lord William Cavendish Bentinck, M.P. to His Grace the Duke of Bedford*

1821. *Copy of a Letter from John Rennie, Esq. to W. G. Adam, Esq.* [on the Improvement of the Outfall of the River Nene]

1821. *A Map and Survey of the River Wear, as far up as Biddick Ford, and of the Port and Haven of Sunderland* (printed 1826)

Works

See table (pp. 566–567).

RENNIE, Sir John, FRS (1794–1874), civil engineer, was the second son of John Rennie Sr., George Rennie (qq.v.) being the eldest. He was born in London on 30 August 1794 and was educated at home and at private schools, mainly in the Classics, until he was fifteen. Then, at his father's insistence, he underwent four years of training in the various departments of his father's business, while extending his studies in mathematical sciences and modern languages in the evenings. In 1813 he was given a post of some responsibility on the site staff at Strand (later Waterloo) Bridge, London, for which his father was the Engineer. A similar appointment to the first Vauxhall Bridge scheme in London was cut short when inadequate funds forced the abandonment of the project in the early stages of construction. In 1814, construction of another of his father's major bridges over the Thames in London, Southwark Bridge, was started. This bridge had the largest span of any cast iron bridge constructed in Britain and Rennie was a member of the site staff under the Resident Engineer. In this case, however, he reported directly to his father and was responsible for the working drawings. In his autobiography he claims to have had the major responsibility for supervision of the construction, but elsewhere he relates that under Francis Giles (q.v.) he was involved at the same time in several lengthy surveys for canals and dockyards outside London.

On completion of the Southwark Bridge, Rennie, at his father's desire again, travelled abroad to study architecture and engineering. He spent more than two years touring through Switzerland, Italy, Greece and Turkey to Egypt, returning home in September 1821. Eleven days later, his father died and Rennie, who was now twenty seven but had not yet had an independent appointment, soon set about obtaining the succession to his father's posts. His first success was at the Admiralty, where he became responsible for the completion of the Plymouth Breakwater and the dockyards at Woolwich, Deptford and Sheerness. He continued his father's improvement of the Eau Brink Cut in the river Ouse above King's Lynn, in conjunction with Thomas Telford (q.v.) and took over as Engineer to the Trustees of Ramsgate Harbour. The works at Portpatrick and Donaghadee harbours for the Irish packet ships went on spending about £15,000 to £20,000 p.a. At Whitehaven, where only preliminary work had been done before financial constraints stopped his father's scheme, he altered the design and brought John Fox from Berwick-on-Tweed to supervise the works. A commission which was Rennie's own, though carried out for the Admiralty, was the Royal William Victualling Yard at Plymouth. It replaced four, much smaller establishments and is an early and most impressive example of planned industrial buildings. In 1823 a competition was held for a design for a new London Bridge, to replace the mediaeval structure. Rennie, with his brother, submitted a design that their father had made shortly before his death. It was not one of the three prize winners (nor was a design by Telford) but after a trial of political strength between the Corporation of London and Parliament, the design was chosen and Rennie appointed Engineer. Construction of the bridge, with its approaches, occupied six years and on its completion in 1831, Rennie was knighted, the first consulting engineer to be so, though his father had declined the honour fourteen years earlier.

In 1825, the Liverpool and Manchester Railway Company failed to obtain its Act of Parliament, in no small part because of the shortcomings in the evidence of their Engineer, George Stephenson (q.v.). Within a month, he had been superseded by the Rennie brothers, who employed Charles Vignoles (q.v.), to make a new survey. A new route, to the south of the old one and going through Chat Moss, was selected and it was this line which was authorised by Parliament in 1826, George Rennie and Vignoles being the principal witnesses for the Bill. The Rennies had planned the railway to be operated by horses but a substantial party of the proprietors wished to have locomotives. The Rennies declined to have Stephenson associated with them in the construction of the railway and so, after two weeks of discussion, the Company turned down the Rennies' terms and appointed Stephenson as their Engineer. He spent more than one third above the sum for which the Rennies undertook to build the line but, when it opened, it was worked by steam locomotives.

At this period Rennie surveyed a line for a railway from London to Birmingham, via Banbury, but nothing was done then and the route chosen in 1833 was Robert Stephenson's through Rugby. Rennie was also organising a survey of railways in Surrey, Sussex, Hampshire, Wiltshire and Somerset, which included a direct line from London to Brighton, employing Vignoles at the same time to investigate the longer route through the Dorking Gap. The proposals were revived in 1829 and 1833 and led to an epic battle in Parliament in 1835, when Robert Stephenson favoured the Dorking route and Rennie the direct line. The resulting compromise was based largely on Rennie's route but he no longer had the entire confidence of the Proprietors and the railway was built by John Urpeth Rastrick (q.v.). Although he was much engaged in surveys during the railway mania of 1843–1844, Rennie had to wait until 1852 to construct his first line, in Sweden.

Rennie was elected a Fellow of the Royal Society in 1823 and a Member of the Smeatonian Society in 1822. He was not elected to the Institution of Civil Engineers until 1844 but became its President in 1845–1847. He was also a Fellow of the Geological, Geographical and Antiquarian Societies. He continued to practice as a harbour engineer, but spent much of the latter part of his life in compiling his book on harbour engineering and his autobiography. He died in 1874 and is buried in Kensal Green cemetery.

P. S. M. CROSS-RUDKIN

[Rennie reports, ICE archives; Sir John Rennie (1854) *The Theory, Formation and Construction of British and Foreign Harbours*; Sir John Rennie (1875) *Autobiography of Sir John Rennie*]

Publications

1822. *Report concerning the Improvement of Boston Haven*

1822. *Map and Chart of the River Witham & Boston Harbour from the Grand Sluice at Boston to Hob Hole Sluice & Clay Hole*

1825. *Report … on the Ancholme Drainage and Navigation*

1826. *Map of the Country between Liverpool and Manchester, exhibiting the proposed New Line of Railway* (with George Rennie)

1826. *Specification for sea wall and West Pier, Leith Harbour*

1826–1830. *Reports by Mr. Rennie on the Works in Progress at Whitehaven Harbour*

1826. *Report … relative to the Approaches to the new London Bridge*

Early works

Those marked with an asterisk were started by John Rennie Sr.

*1821–1823. Sheerness Dockyard

*1821–1823. Pater Dockyard, Pembroke

*1821–1824. Cramond Bridge

*1821–. Chatham Dockyard

*1821–1831. Woolwich Dockyard west basin; cost £340,000

*1821–1850. Plymouth Breakwater

*1822–. Ramsgate Harbour improvements

*1822–. Sunderland Harbour

*1822–1836. Donaghadee Harbour, cost within £145,000

*1822–1836. Kingstown (Dunlaoghaire) Harbour; cost nearly £700,000

*1822–1837. Portpatrick Harbour, cost £115,000 × 1.3

*1822–. West India Dock, London improvements

1824–1827. Hyde Park Bridge, London (with George Rennie), five arches of 40 ft. span, cost £45,464

1824–1831. London Bridge rebuilding, five arches to 150 ft. span, cost £1,458,311 including approaches

1824–1834. Royal William Victualling Yard, Plymouth

1825–1844. Ancholme Drainage and Navigation

1826–1829. Eau Brink widening (with Thomas Telford)

1827–1830. Nene River Outfall, (with Thomas Telford), cost £200,716

1827–1833. Witham River Outfall, cost £27,262

1828–1836. Portrush Harbour

RENTON, Amhurst Hawker (*c.* 1804–1889), engineer, was born in Pimlico in 1804, his mother being Maria Knight. He was baptised at St. Luke's, Chelsea, in 1811. He joined the Institution of Civil Engineers as an Associate in 1828. Among his proposers was Francis Bramah, and it is likely he worked with the Bramahs from an early age. He assisted Francis Bramah with a series of experiments on the strength of cast iron beams which followed Bramah's investigation into iron work in

Buckingham Palace and corresponded with Thomas Tredgold (q v.) on various forms of wrought and cast iron roof trusses in 1827, developing an early form of trussed iron girder. He evidently had a position of some responsibility at Bramah and Robinson's works and between 1829 and 1831 was present at various tests on stone using a Bramah Press. He gave an early (1838) paper to the Royal Institute of British Architects on ironwork and was involved in the fabrication of the cast iron lighthouses designed by Bramah's works for export. His later writings and patents are involved with marine boilers. In 1860 he gave evidence on behalf of the Universal Lime Light Co. to the Commission on Navigation Lights. The system had been used at South Liverpool Landing Stage and at Westminster Bridge.

Renton was resident at Brewer Street, Golden Square, London, when he joined the Institution, but subsequently gave his address as Belgrave Cottage and 49 Cambridge Street, Pimlico, where he remained until he left membership c. 1860. His last known address was 1 Hanover Chambers, Buckingham Street, Adelphi. He was married twice, to Mary (b. c. 1813) and to Eliza Emma Walton whom he married at St. Andrews, Cambridge, on 5 January 1836. Eliza bore him at least one son and two daughters.

Renton died in Hastings in 1889.

MIKE CHRIMES

[Census returns, 1851; J. G. James Collection, ICE archives; ICE membership records; ICE original communications 191, 238, ICE archives; Tredgold correspondence, ICE archives; *Mechanics Magazine*, **10** (1828) 97–98; *CEAJ*, **1** (1838) 171, 174, 200, 230, **2** (1839) 7, **4** (1841) 333–334, **53** (1850) 466–472, **64** (1856) 110, **67** (1857) 214; *Min Proc, ICE*, **1** (1839) 8; *Artizan*, **13** (1855) 220–221; Minutes of evidence (1861) *Commissioners … into the Condition and Management of Lights, Buoys, and Beacons*, **2**, appendix, 621–622; IGI has different wife, etc.]

RESTALL, John (fl. 1824–1827), was resident in Wanstead and was surveyor to the Great Essex Road when elected to the Institution of Civil Engineers in 1824. He remained in membership only for a short time.

MIKE CHRIMES

[ICE membership records, ICE archives]

REYNOLDS, John (fl 1713–1753), civil engineering contractor, began his career in civil engineering working for John Perry (q.v.) and by the 1730s had acquired a reputation for his expertise in timber work associated with river and harbour improvements.

In the early eighteenth century members of the Reynolds family were active as carpenters and joiners in the Stepney area; John Reynolds was a carpenter and is known to have been living in Poplar 1713–1717 before moving, perhaps to Dagenham. In August 1714 William Boswell was awarded the contract to repair Dagenham Breach and he employed a number of shipwrights and carpenters from the Stepney area 'having had great experience in water-works, dams, sluices, etc.', among whom was John Reynolds, who was appointed one of two foremen. Boswell ran into financial difficulties which were compounded by allegations of sabotage and his repairs failed. Reynolds was implicated in these allegations. Although he signed a letter supporting Boswell's efforts it appeared that he sensed that Boswell was going to be replaced and started to neglect his work. It was even claimed that he got the workers drunk when they should have been on watch. Whatever the truth of the matter, when Perry obtained the Dagenham contract Reynolds was kept on as his foreman and must have benefited greatly from this practical experience. As work at Dagenham was completed Reynolds followed Perry, possibly to Dover, and began to work on his own.

His first known contract, as 'John Reynolds of Poplar', was to rebuild a timber sluice, the Scots Float Sluice, for the Commissioners of the Kent and Sussex Rother Levels in 1723. This work incorporated a dovetailed sheet pile wall beneath the foundations, a feature of Perry's work at Dagenham which became a characteristic of Reynolds's own work. The work was evidently carried out satisfactorily as in July 1725 he was offered a permanent position at an annual salary of £65 to look after the maintenance of the Levels. He moved to Iden, Sussex, and retained the post until 1739, when the pressure of work elsewhere obliged him to resign. His wife, Mary, was buried at Iden parish church on 20 August 1732. In the 1720s siltation of the Rother Estuary was causing increasing problems for all the sluicing arrangements on the Rother Levels, rendering much work obsolete. Scots Float Sluice required more attention by 1729 and in 1732 Reynolds, due to changes further up the estuary, rebuilt it to provide an additional outlet. In the early 1730s new channels were dug in the Levels, with varying success, to direct all the waters of the upper levels through Scots Float.

In addition to work on the Rother Levels Reynolds was active elsewhere in Sussex. An early work was a timber bridge at Winchelsea in 1725. In March 1730 he reported to the House of Commons on the state of the timber pier at Newhaven, estimating the cost of repairs at £3,000, and carried out the works, subsequent to the 1731 Act, for the Harbour Commissioners. These involved the reopening of an existing outfall beneath Castle Hill and the repair and extension of piers to control the channel. These works were completed in 1735. In association with this he built a sluice at Newhaven-Piddinghoe (1731–1733) which was intended to hold back the waters and help scour out the channel. This work, carried out for the Drainage Commissioners, was removed in 1736, due to damage.

While work was proceeding at Newhaven, Reynolds gave evidence in March 1733 to the House of Commons on the state of the harbour at Littlehampton, recommending the erection of piers and locks to increase the depth of water in the harbour, at an estimated cost of £5,000. This work, of a similar nature to that at Newhaven, was along lines advocated by Perry in his report on Kings Lynn (1725), and here carried out by Reynolds.

From the mid-1730s the scale of Reynolds's operations increases, possibly as a consequence of his appearances as a Parliamentary witness, which would have brought him to the attention of a wider group of potential clients. One consequence was for him to be asked in June 1735 to visit the Dee navigation works, then under construction to the designs of Nathaniel Kinderley (q.v.). Kinderley's scheme, incorporated in an Act of Parliament of 1733, involved diverting the Dee from its natural channel along the north (Cheshire) shore into an artificial cut along the Welsh shore in order to increase the depth of the navigation from Chester to the sea and enable vessels to reach Crane Wharf, Chester. The new cut was across several miles of consolidated salt marshes, in preference to an earlier eighteenth century scheme— which had failed—which intended to improve the natural channel. The Act also transferred the City of Chester's rights to the 'white sands' of the Cheshire side of the river estuary to Kinderley if he restored the navigation. Whatever the engineering merits of the location of Kinderley's cut, its commercial advantage was obvious as it considerably increased the potential for land reclamation.

The scale of the scheme attracted the immediate interest of contemporaries like John Grundy and Thomas Badeslade (qq.v.), who became embroiled in an argument over the soundness of the hydraulic principles underlying the scheme; Reynolds's input was of a more practical nature. Kinderley had closed off the old channel with a sluice whose foundations were cut off using rows of square piles with rocks between. Reynolds correctly predicted this would be unable to withstand the dramatic changes in hydrostatic pressure beneath the sluice arising from the high tidal range and recommended the use of dovetailed sheet pile walls.

Kinderley's scheme was nominally completed in April 1737 as regards the navigation but many problems were already evident and for the next century engineers' attention was being directed to securing the navigation and enclosures against the action of the tides.

The most serious problem was the failure of Kinderley's sluice; indeed from Badeslade's remarks it probably never operated successfully. Reynolds told John Grundy Jr., in 1750 'that in attempting to make a stoppage of ye said fresh waters to force them down the new cutt ye said sluice blew up, as he apprehends from its being too weak in the foundation having only square piles to support it'. It is uncertain when this

occurred but problems were apparent in early 1735. As Reynolds had predicted the failure that June it is unsurprising that he was brought in, but not until some unsuccessful attempts had been made to deal with the problem. By the time he arrived there was a 30 ft. hollow beneath the sluice. Perhaps the delay was caused by Kinderley's effective retirement.

Reynolds's first act, one assumes in 1736, was to build a new, ten-gate, sluice, 40 ft. in waterway, with five rows of dovetailed piles, filling the foundations between with stones, alongside the breach. Once this had been done a seventeen gate sluice with a 68 ft. waterway was constructed on the site of Kinderley's sluice, driving three lines of dovetail piles—40 ft. long in the middle and 20 ft. at the wing walls—one beneath the gates and the others at front and back of the wing walls, and rock beneath the foundation. These sluices were marked on a survey made in 1739 by John Eyes (q.v.) and published in 1740. Their construction by no means marked the end of Reynolds's involvement at Chester. The rise of tides and river water had to be closely monitored to preserve the sluices as the new channel was not sufficient to accommodate the flow. 'Many expedients were used to deepen the New Cut', including 'hedgehog' and 'porcupine ploughs'.

Further down river at the seaward extremity of the New Cut the North Bank was under threat. Within a year of its completion by Kinderley the final half mile to Pentre Rock had been destroyed by a high tide (c. 1736), and the 'old' channel began to move eastwards up river, threatening to erode the North Bank. For four or five years nothing was done until Reynolds, presumably having secured the works near Chester, was able to turn his attentions down river. A stone causeway about 6 furlongs long (1,320 yd.) was erected (1741–1742) at the Flint end of the North Bank with groynes to protect it and direct the channel.

Having secured Kinderley's work Reynolds was then asked to organise the first land reclamation scheme and a cross bank, 17 ft. high, was erected 1742–1744 across the Dee from near Sandy Croft to near Blacon where a sluice was erected to Reynolds's designs; his commitments elsewhere meant that the practical supervision of these works was often in the hands of two overseers, James Golborne and William Brackley. Reynolds's contractual relationship is unclear but possibly he was paid a management or consultancy fee with work carried out under the two foremen by direct labour.

Reynolds's local reputation led to his being consulted by Edward Williams in April 1742 about a land reclamation scheme in Anglesey. It is unclear when Reynolds retired from acting as Engineer for the Dee Company as few papers have survived but he was there in 1750 after a high tide that February had breached the Cross Wall near the sluice. Both he and John Grundy Jr., ascribed the extent of the breach to the lack of stone pitching to the landward side of the Cross Wall, but neither his nor Grundy's recommendations can have been acted on as no repairs were effected by 1754 when John Golborne (q.v.) was Engineer to the Company. One can believe that financial problems were a factor in this delay as the proprietors did not receive a dividend until 1770, and in 1743 Reynolds reported to Parliament that over £56,000 had so far been expended, well in excess of Kinderley's initial capital of £40,000. One could also speculate that Reynolds's death may have removed an obvious director of works prior to Golbourne's appointment, but we do not know when either of these events took place from existing documentary evidence.

While based in Chester in 1741 Reynolds was awarded the contract for improving Bridport Harbour by the construction of timber piers out to low water mark, and directing the river via a sluice between the piers and creating a harbour capable of taking 150 ton vessels, a figure later modified to 100 tons. Work continued at Chester while this was carried out. The construction of the seven-gate sluice and navigation gates permitting access up river was by a second contract of 1743. Further work was necessary following storm damage in 1744–1745 and by the early nineteenth century the piers had to be extensively repaired and extended. Records of the 1744–1745 work mention a John Reynolds Jr., presumably a son. In view of Reynolds's commitments on the Dee it may be him who is mentioned as 'of Bridport' in the 1743 contract.

In the late 1740s Reynolds is again found to be resident in Poplar but he also carried out work around the country. He surveyed Southwold Harbour in 1747 following its improvement Act of 1746 and built the north pier he recommended and in 1751–1752, following this by construction of the south pier. In 1753 there was a proposal to improve the Blyth navigation to Halesworth and in May Reynolds submitted an estimate for £4,614. The Act was not obtained until 1757 but the work was carried out between 1759 and 1761 by Andrew Chandler and others, with Langley Edwards as consultant and Samuel Jones (qq.v.) as resident engineer. In the meantime Reynolds had tendered unsuccessfully for work on the piers at Yarmouth (1750), work which was in the end carried out under Edward Rubie (q.v.). The 1753 proposals at Blyth may represent Reynolds's last work. According to Grundy he had died by 1759 and his absence from Chester in 1754 and disappearance from Poplar Rate Books by that date suggest he may have died around then.

Given this uncertainty one may have to turn to his son for some of the other work ascribed to 'John Reynolds' at this time. A John Reynolds is reported to have been acting as 'overseer and director of works' at Dover in 1747 and the same phrase is used to describe the Reynolds acting as Clerk of the Cheques there in June 1758. A John

Reynolds also worked at Shoreham Harbour where in 1760 he drew up plans for a cut through the banks; in 1761, while working on the piers to train the river, he was dismissed. Later in the 1760s a John Reynolds is again found working at Bridport. Clearly, much remains to be learnt about this early engineer.

Reynolds's career is of interest in a number of respects. He clearly came from a craft background, as a carpenter, and offered craft-based solutions to the problems of the minor harbours of the early eighteenth century. The characteristics of his work—dovetailed sheet piled walls and the use of sluices and piers to improve channels—had been advocated by Perry, who can be seen as his mentor. His approach was essentially that of the design and build contractor; the independent approach of Smeaton and many of the late eighteenth century engineers, who provided engineering advice to clients, but left the client to procure the contractor, was quite alien. Reynolds offered his advice to obtain work, not as an end in itself. Of the quality of his work at its peak there can be little doubt; in 1750 John Grundy Jr. remarked that his sluices 'equal if not exceed any I have ever met with'. As fine an epitaph as any eighteenth century engineer could seek.

MIKE CHRIMES

[Hamlet of Poplar and Blackwall, Land Tax Books 1712–1760, London Borough of Tower Hamlets, Local History Collection; J. Boswell (1717) *An Impartial Account of the Frauds and Abuses at Dagenham Breach*; *Journal House of Commons*, **21** (1730) 461, 492–493, 615, 625, **22** (1733) 55, 73; **24** (1743) 599; Rye Harbour Commissioners Minutes, 1725; J. Perry (1725) Remarks ... on the Ports of Kings Lynn and Wisbech; T. Badeslade (1736) The New cut; DAP, *passim*; SRA 6/11/1, East Sussex RO; Littlehampton Harbour Commissioners Minutes 1733–1736; Dee navigation accounts 1740–1742, C/CR/134, Chester RO; Wardens of Dover Harbour Minutes 1747; Yarmouth Ports and Haven Commissioners Minutes 1750; Southwold Harbour Commissioners Minutes 1752–1753; J. Grundy (1750–1762) West Chester navigation, *Reports*, **IV**, Brotherton Library, Leeds University; Dover harbour papers, DHB AP 11, Kent RO; Shoreham Harbour Commissioners Minutes 1760–1761; H. Symonds (1912) Bridport Harbour through seven centuries, *Proceedings of the Dorset Natural History and Antiquarian Field Club*, **33**, 161–199; D. Swann (1968) Engineers of British port improvements 1600–1830, *Transport History*, **1**, 153–168; J. H. Farrant (1972) Newhaven Harbour and Lower Ouse, *Sussex Archaeological Collection*, **110**, 44–60; J. H. Boyes and R. Russell (1977) *Canals of Eastern England*; J. Eddison (1988) Drowned lands: changes in the course of the Rother and its estuary and associated drainage problems, 1635–1737, *Oxford University Committee for Archaeology Monograph*, **24**, 142–161; Keystone (1997) *West Bay Dorset Historical Report*]

Publications

1735. Remarks on the dam and sluice making across the River Dee, in Thomas Badeslade (1736) *The New Cut Canal*

Works

1723–1724. (Scot's New Float) sluice, Rother Levels, £750
1725. Winchelsea Bridge, £50
1725–1739. Rother Levels, maintenance work
1730–1735. Newhaven Harbour, piers, £2,000
1731–1733. Newhaven-Piddinghoe sluice, £650
1732. Scot's Float sluice, Reconstruction
1733–1736. Littlehampton Harbour improvements, £5,000
1736–1744. Dee navigation; total expenditure by 1743, £56,460; 1740–1742, new contract, c. £16,000
1741–1744. Bridport Harbour, piers, sluice, etc., £4,000
1747. Southwold Harbour, north pier
1751–1752. Southwold Harbour, south pier

REYNOLDS, William (1758–1803), ironmaster and industrial entrepreneur, was born at Bank House, Ketley, Shropshire on 14 April 1758, eldest son of Richard Reynolds (1735–1816) and his first wife Hannah Darby.

William Reynolds is generally considered to have been the most able and innovative of the Shropshire ironmasters and, had he not died so young, it is likely that his achievements would have far outweighed those of the Darbys, John Wilkinson and most of the other of his contemporaries. His father, the Quaker ironmaster and philanthropist Richard Reynolds, had moved to Coalbrookdale from Bristol in 1756 to look after the affairs of Thomas Goldney, a merchant who had a financial interest in the works. In 1757 Richard Reynolds married Hannah Darby, daughter of Abraham Darby II, and his son William was born the following year: thus, William Reynolds was the grandson of Abraham Darby II. He studied chemistry and took an interest in many other fields of science, maintaining a laboratory and a library at the family home, Bank House in Ketley. He also established a sizeable geological collection, offering rewards to local miners who brought interesting specimens to his attention. His father Richard was manager of Coalbrookdale between the death of Abraham Darby II in 1763 and the coming of age of Abraham Darby III in 1768, and it was during this period, when many new developments—such as the introduction of the first iron rails—were occurring that the young William was serving his apprenticeship at the works. He became familiar with all aspects of the company's operation, and by the 1780s he held a position of some importance at Ketley, and was a close friend and advisor to his cousin Abraham Darby III. In 1782, he and his father were considering the possibility of building a steam-powered corn mill, whilst in 1783 he was involved with his brother-in-law in the establishment of a new ironworks at Donnington Wood. Amongst his friends and contacts, many of whom

were frequent visitors to Coalbrookdale, were Matthew Boulton, James Watt, John Wilkinson, Lord Dundonald, Thomas Telford, the Darwin family of Lichfield and the circle of intellectuals encouraged by Joseph Plymley.

In 1786, William Reynolds was responsible for driving a canal tunnel into the north bank of the River Severn with the intention of bringing coal out on the level, instead of having to haul it up shafts. This tunnel later became the Tar Tunnel, which survives as one of the sites in the care of the Ironbridge Gorge Museum Trust. In the following year, 1787, he built the short Ketley Canal to take goods and raw materials to and from Ketley ironworks. Although it incorporated a short tunnel, its most significant feature was an inclined plane which carried boats 73 ft. down to the works. This was a self-acting incline, with two sets of parallel rails, whereby a loaded boat going down on its cradle counterbalanced a boat going up, the ropes and chains connecting them using a single winding drum built across the top of the slope. There were lock chambers at the top of the incline, with lock gates at the incline end: once the boat was on its cradle, water was pumped out of the chamber into a reservoir before the gates were opened and the descent began. This inclined plane was the first to be used on any canal in mainland Britain. An illustration of its appears on one side of the Coalbrookdale tokens issued in 1789, and it can also be seen in the background of the Wilson portrait. In 1788, an Act of Parliament was obtained for the construction of the Shropshire Canal, from Donnington Wood on the northern side of the Coalbrookdale Coalfield to Southall Bank, from where there were to be two branches, one via Coalbrookdale and the other to reach the river Severn near Haye Farm. Reynolds was responsible for surveying the route and, when completed in late 1792, the canal included several inclined planes, though unlike the one at Ketley which had not been a great success, they had reverse slopes at the top of the incline in place of a lock chamber and lock gates.

Reynolds was also closely involved with the Shrewsbury Canal, which connected the East Shropshire tub boat canal network to the county town. The engineer Josiah Clowes (q.v.) worked under his supervision and, when a masonry aqueduct across the river Tern at Longdon was destroyed by a flood in 1795, it was Reynolds who encouraged the new engineer Thomas Telford (q.v.) (Clowes having died) to build the replacement structure from cast iron, supplied from his own Ketley furnaces.

Following the completion of the Shropshire Canal, William Reynolds devoted much of his time and energy to the development of the area where its eastern branch met the River Severn. In 1792, he constructed a large warehouse which spanned the canal and the land between it and the river: he developed wharves and built housing, and also became a partner in one of the china works which were established alongside the canal. By 1794, the area had acquired the name Coalport, and attracted other trades and industries such as chainmaking and boatbuilding. In 1800, Reynolds moved from Bank House, Ketley to The Tuckies, a large house in Jackfield on the south side of the river opposite Coalport. From 1796 he was also occupied with the running of the Madeley Wood furnaces at Bedlam which, together with Ketley, were sold by the Darbys to the Reynoldses in that year. Shortly before his death, he collaborated with Richard Trevithick in the design and construction of the world's first steam railway locomotive, built at Coalbrookdale, and he also experimented with steam propulsion for one of his own boats. William Reynolds, who had married his cousin Hannah Ball in 1789, died at The Tuckies on 3 June 1803, and is buried in the Quaker Burial Ground, Coalbrookdale.

A portrait of William Reynolds, oil on canvas, 37 in. by 55 in., is in the collection of the Ironbridge Gorge Museum Trust. At least one other miniature is known to be in a private collection.

JOHN POWELL

[J. Plymley (1803) *A General View of the Agriculture of Shropshire*; J. Randall (1880) *History of Madeley, including Ironbridge, Coalbrookdale and Coalport*; H. W. Dickinson (1921) An 18th century engineer's sketch book, *Trans Newc Soc*, **2**, 132–140; A. Raistrick (1953) *Dynasty of Ironfounders: the Darbys and Coalbrookdale*; C. Hadfield (1966) *The Canals of the West Midlands*; B. Trinder (1974) *The Darbys of Coalbrookdale*; Shropshire County Library (197?) *Two Shropshire Ironmasters: Biographical Extracts, Letters and Memoranda relating to Richard and William Reynolds*; Ironbridge Gorge Museum Trust (1978) *Coalport New Town of the 1790s: a Guide to its Buildings and Monuments*; S. Smith (1979) *A View from the Iron Bridge*; Shropshire Libraries (1980) *Shropshire Canals: Articles from the Shropshire Magazine 1950–1965*; B. Trinder (1981) *The Industrial Revolution in Shropshire*, 2nd edn.; H. S. Torrens (1982) The Reynolds–Anstice Shropshire geological collection, 1776–1981, *Archives of Natural History*, **10**(3), 429–441; E. Thomas (1999) *Coalbrookdale and the Darby Family: the Story of the World's First Industrial Dynasty*]

Works

1786. Tar Tunnel, engineer
1787. Ketley Canal and inclined plane, engineer
1788–1792. Shropshire Canal, promoter, surveyor and engineer
1793–1797. Shrewsbury Canal, esp. Longdon-on-Tern Aqueduct

RHODES, Thomas, MRIA (1789–1868), civil engineer, was born at Apperley Bridge, near Bradford, Yorkshire, the son of James Rhodes, a carpenter and millwright who worked on the Leeds–Liverpool Canal. His eldest brother, William (b. 1779), was trained by his father while working on the Rochdale Canal and, following education at Calverley School, Thomas was apprenticed to William before he was fourteen. His father had been seriously injured in an accident on the Leeds–Liverpool Canal which disabled him for life. A younger brother, James (1791–1865), also worked in civil engineering. After working for William for about eight years, where he learned about all aspects of canal construction and developed great skill in engineering drawings, Thomas began work on the Glasgow and Paisley Canal in 1810. This was a short-lived employment and in 1811 he moved further north, to the Caledonian Canal.

Work on the locks of the Caledonian Canal, for which the joint engineers at the time were William Jessop and Thomas Telford (qq.v.), had been hampered by construction difficulties, particularly poor ground conditions and a shortage of timber of suitable dimensions for the locks. To deal with the latter problem cast iron was introduced into the framing and when Rhodes arrived at Corpach in September 1811 the ironwork had just been delivered. After successfully installing the lock gates and machinery for the sea lock at Corpach, Rhodes was transferred to Clachnaharry and repeated the process there. The continuing shortage of timber led to Rhodes further experimenting with the use of iron for the other lock gates on the canal, and developing a standard design. Rhodes was fully occupied on the canal until December 1822, working on the locks, swing bridges, dredgers and barges, although for eight months he acted as a foreman on the temporary works for the Bonar Bridge (1812–1813), for which John Simpson and William Hazledine (qq.v.), were the contractors. His spare time was taken up by study at the Inverness Academy.

By the end of 1822 Telford's scheme for the improvement of the Holyhead Road was well advanced and William Alexander Provis (q.v.), the assistant in charge in north Wales, was being sorely stretched as work had begun on the Menai Suspension Bridge, a bridge of unprecedented scale. Telford asked Rhodes to take on responsibility for supervising the erection of the ironwork and timber both there and at Conway. These works were completed in 1826 and Rhodes was then transferred to St. Katharine's Dock where he took charge of inspecting the civil engineering work. Telford's view of his work there deserves to be quoted in full:

... with regard to the rapidity with which the works were performed and the risk and difficulty of laying foundations, so much under the level of low water, the chief merit is due to Mr. Thomas Rhodes, who had been during many years employed by me as a complete master in carpentry, and who at my recommendation acted ... as Resident Engineer in the formation of St. Katharine's Docks where his dexterity as a mechanic and fertility of resource overcame every obstacle in constructing cofferdams under a lateral pressure of 40 ft. of water; and the water-tight security of the west wall foundations, as well as the perfect accuracy of the locks, their sills, gates and bridges, afford especial evidence of his superior skill and unremitting attention.

Initially Rhodes reported to Peter Logan and John Hall (qq.v.) but in November 1828 he became resident engineer and so was responsible for supervising the construction of the Eastern Dock.

St. Katharine's Dock was completed in 1830 and Rhodes then acted as resident engineer on the timber pier at Herne Bay, where the same contractor, George Burge (q.v.), was employed. Rhodes had now begun to be recognised as an engineer in his own right, such was his experience of major civil engineering work. He reported on a graving dock near the Commercial Dock in London, Llanelli Railway and Docks, and the London and Dover Road. He also probably prepared Telford's designs for Clifton suspension bridge. His standing was also reflected in the fact he was taking on pupils, such as Thomas Jeans (fl. 1829–1836) who joined the ICE while his pupil in 1829.

In 1833 he designed improvements in the Ouse, with locks at Naburn and Linton, jetties, ship repair slips, and dredgers. He also acted as Engineer at York Waterworks and in January 1834 he was asked by a Belfast committee to report on a proposed ship canal and floating dock. His scheme, estimated at £20,000, included a swing bridge across the Lagan but it was subsequently rejected in Parliament. In 1834 he also reported on the Hartlepool Dock and Railway for the Exchequer Loan Commissioners and subsequently the Clarence Railway, for which he became engineer. He reported on the Ulster Canal (perhaps with William Cubitt (q.v.)) where another Telford associate, Thomas Casebourne (q.v.), became engineer. In 1835 he reported on Berwick Harbour, recommending various improvements to the entrance, and his involvement continued until 1838 although the costs of schemes for a floating dock and other improvements precluded action. Rhodes also surveyed the Derwent and drew up plans for the Stockton and Darlington Railway's Byer's Green branch, the Stockton and Hartlepool Union Railway and Docks, and in 1836 improvements in the Severn between Stourport and Gloucester. The latter would have linked the Staffordshire and Worcester Canal to the Gloucester and Berkeley Canal by a lockage navigation.

Many of Rhodes's proposals were, as typical of the time, never realised but his involvement in the Shannon, which began in 1831, was a major triumph. In 1831 he was asked to report on the Shannon and its tributaries for the Irish Commissioners of Public Works. Subsequently, in 1835, he became a Commissioner for the improvement of the river and, finally, in 1839, engineer for the improvement works. For the next decade he was largely employed in major works in Ireland, the contractor on many of these being William Mackenzie (q.v.). On the Shannon itself were major

bridges, weirs and locks, but Rhodes also acted as engineer for Limerick Bridge and Docks and surveyed Loughs Neagh, Erne and Corrib, the rivers Bann, Blackwater and Banner, and reported on the sites suitable for piers and harbours on the Kerry coast. These were major achievements and, possibly with a view to a more leisurely schedule, in 1849 he agreed to act as Resident Engineer for James Walker (q.v.) on the harbour works at St. Catherine's, Jersey, and at Alderney, where, involved with its major breakwater, he remained until his retirement in 1860.

William Rhodes had a son, William, who worked with his uncle, became an agent for Thomas Brassey and eventually became engineer on the Caledonian Canal; he acted as executor for his uncle's will.

Thomas and William Rhodes, sons of Samuel Rhodes (d. 1802), supplied brickwork for Islington tunnel on the Regents Canal in September 1814. This William Rhodes also developed De Beauvoir Town from 1818. The family were regarded as the largest brickmakers in London. It is not believed they are related to Thomas. Similarly the William Rhodes (fl. 1790) and Israel Rhodes (fl. 1802–1804), involved with Irish Canals, appear unrelated.

Rhodes was elected to the Institution of Civil Engineers in 1827 and for the next four years contributed to meetings while based in London, giving details of his experiences at St. Katharine's Dock and elsewhere. His Institution obituary states that 'had it not been for his unobtrusive and retiring disposition, he might have taken a more prominent position in the profession, for which he possessed the highest qualifications'. While one could see that his career was overshadowed in the 1840s and 1850s by that of engineers like Brunel and Stephenson, his involvement in the major works of the preceding era—the Caledonian Canal, Menai Bridge, London's docks and his widespread activity in the 1830s—indicate he was at the top of his profession. His work in Ireland was of major importance for that country. Rhodes's career is an indication of how much could be achieved in the early nineteenth century by largely self-taught men; such careers had almost disappeared by the time of his death at Paignton on 6 June 1868.

MIKE CHRIMES

[Membership records, ICE archives; Minutes of conversations, 51, 71, 91, 104, ICE archives; Holyhead road papers, Telford mss, ICE archives; Mackenzie collection, ICE archives; Drawings of Caledonian Canal, BWB (see Cameron, 1972); Drawings of St. Katharine's Docks (ex-ICE), Museum of London in Docklands; W. A. Provis (1828) *An Historical and Descriptive Account of the Suspension Bridge ... over the Menai Strait*; House of Commons (1832) *Papers relating to the River Shannon Nagivation*; J. Rickman (1838) *The Life of Thomas Telford*; F. Whishaw (1840) *The Railways of Great Britain and Ireland*; Tidal Harbours Commission (1847), second report, *Appendix B, Limerick,* 130–145, and *Appendix C,*

Berwick, 115–138; Admiralty (1849) Report on the Severn Navigation Bill; Admiralty (1852) *History of the Harbours of the UK: Belfast*, 15–16; Obituary (1869) *Min Proc ICE*, **28**, 615–618; A. D. Cameron (1972) *The Caledonian Canal*, 91–92; A. W. Skempton (1982) Engineering in the Port of London, 1808–1834, *Trans Newc Soc*, **53**, 82–87; B. Cherry and N. Pevsner (1998) *Buildings of England: London, North*, 507; A. H. Faulkner (1998) A Tale of a tunnel, *RCHS Journal*, **37**, 614–620; R. Bowdler, English Heritage (1998)]

Publications

1832. *Report upon Canal and River Navigation from Limerick to Killaloe*
1832. *Second Report upon the Means of improving the Shannon Navigation …*
1832. *Report upon the Lower Brosna River*
1833. *Third Report upon the present State of the River Shannon*
1834. *Report upon the Ouse Navigation* (and longitudinal section)
1836. *Report on Menai and Conway Bridges* (with W. A. Provis)
1837–1839. *Reports* [1–4] *of the Commissioners for the Improvement of the river Shannon*

Works

1811–1822. Caledonian Canal, superintendent of installation of lock and bridge equipment
1811. Bonar bridge, resident engineer
1823–1826. Menai Suspension Bridge, assistant engineer for erection of iron and timber work
1826–1830. St. Katharine's Dock, assistant and resident engineer
1830. Herne Bay pier, resident engineer
1831–1848. Shannon Navigation, improvements, consultant and engineer, £650,000
1833. York Waterworks, engineer
1834. Clarence Railway, engineer
1835–1838. Berwick Harbour, improvements, including Carr Rock jetty, consultant
1849–1860. Alderney breakwater, resident engineer
1849–1860. St. Catherine's Harbour, resident engineer

RICHARDS, family of military engineers, **Jacob** (1660–1701), **John** (1669–1709) and **Michael** (1673–1721), Chief Engineer of England. Collectively described as the nearest English equivalent of Vauban and of seminal importance in the development of all branches of military engineering, the Richards brothers' diaries also throw light on aspects of civil engineering of the time.

The diaries originally arose through Jacob being the recipient of arguably the first travelling engineering scholarship, when in 1685, he was given a Royal Warrant instructing him to improve himself in foreign parts so that he might by improvement render himself fit to be employed as one of his Majesties engineers for his service in England; to which end he was to keep an exact journal of every days proceeding. Passing through Holland he commented on the comparative state of geo-technical engineering in the two countries 'I observed that throughout all Holland … they take great care to plant their works with trees and in bringing up their earth or turf work, be the soil never so good, they interlace every floor of earth with willow boughs and grass seed, which extremely binds and secures their works, which practice we have wanted in England and has been the greatest reason for many of our earth works falling and giving way.' That Holland was also ahead in tendering for contracts is shown by another entry', 'He that undertakes at the lowest rate has the work and their money paid them according to contract'.

On his return to England in 1686 he was made 3rd Engineer of England at a salary of £150 p.a., but despite being made Chief Engineer in Ireland and Flanders in the 1690s and placed in charge of the first regular train of ordnance set up in peacetime, he was passed over for promotion in England which may have been the reason that he was serving in Poland when he died in 1701. Such was his reputation that Frederick Augustus III appropriated his books instruments and designs, to the annoyance of John Richards who had also been serving there.

John's career is of least interest as being a Catholic he was barred from English service after William III invaded in 1688. However, he was employed as an engineer by Britain's ally Portugal from 1704 until his death during the siege of Alicante where he was Governor in 1709.

The youngest brother **Michael** had the most successful career rising from 3rd Engineer in Flanders in 1692 to Chief Engineer from 1711–1714 with a salary of £300 p.a. and a travelling allowance of 20s a day, Surveyor General of Ordnance on 19 November 1714 and Superintendent General and Inspector of Foundries in 1716 with a salary of £500 p.a. Milestones on route were commanding the pontoon train in Flanders, from 1692–1696 training which he presumably put to good use when he later superintended the construction of bridges during Marlborough's campaigns in 1704. In the intervening period from 1696–1703 he was Engineer in Charge of Building Coastal Defences, Batteries and Barracks at St. Johns in Newfoundland.

He made a brief foray into civil engineering after the war when he was asked to comment on schemes of John Perry (q.v.) and W Boswell for filling the Dagenham Breach. The latter considered that his 'peremptory' choice of Perry's scheme, apparently just on the basis of a sketch or draft rather than a full model, did not give his plan proper consideration.

As Surveyor General he was responsible for all the Board's buildings and those built during his period in office have certain unique stylistic characteristics, although functionally as diverse as the arsenal at Woolwich, dockyards at Chatham, Portsmouth and Plymouth, barracks at Berwick-on-Tweed and Upnor and magazines at Tilbury and Hyde Park. However, despite his presenting the draughts to the Board the designs have variously been attributed to John Vanburgh and Nicholas

Hawksmoor, both friends and neighbours indebted for Richards's peacemaking efforts with the Duchess of Marlborough.

He was instrumental in setting up a regular Corps of Engineer on the 26 May 1716. A memorandum explained as part of his reasons that 'whenever anything has been done in our fortifications it has been from pure necessity and the performance in a hurry whereas ... they require the most mature deliberation'. It originally consisted of twenty-eight members: a Chief Engineer, three directors and a mix of engineers, sub-engineers and practional engineers. These last were to be trained in providing surveys and plans. Similar establishments were set up in Minorca and Gibraltar in 1717.

This interest in the training of engineers was also applied to the Royal Military Academy, which he planned with his successor John Armstrong (q.v.) and the setting up of the Royal Regiment of Artillery.

He died in 1721 and was buried in the church of St. Luke, Old Charlton near Woolwich. An engraved print of a portrait by Kneller is in the National Army Museum.

SUSAN HOTS

[British Library, Stowe mss; Whitworth Porter (1889) *History of the Corps of Royal Engineers*, **1**; H. T. Dickinson (1968) The Richards brothers: exponents of the military arts of Vauban, *Journal, Society Army Historical Research*, **66**, 186, 76–86; F. J. Herbert (1976) The Richards brothers, *The Irish Sword*, **48**, 12, 200–211; R. Hewlings (1993) Hawksmoor's brave designs for the police, in J. Bold and E. Cheney (eds.), *English Architecture: Public and Private*; C. Duffy (1985) *The Fortress in the Age of Vauban and Frederick the Great*; DNB, Bendall]

RICHARDSON, Thomas (fl. 1756–1777), a 'master mason of Plymouth', was chosen by John Smeaton (q.v.) as foreman of one of the two companies of masons working on the Eddystone Lighthouse under the direction of himself and Josias Jessop Sr., his 'general assistant' or resident engineer. Richardson acted in that capacity during the whole period of construction, 1756–1759. Jessop was then appointed Surveyor in charge of maintenance; after his death in 1760 Richardson took over.

He next appears in February 1765 as contractor for building a new pier at Penzance Harbour, the pier to be in masonry, 170 ft. long, 27 ft. high, 40 ft. wide at the base and 30 ft. at the top. Richardson left in June 1767, the work being completed a year later by direct labour under the Corporation's Trustees.

In 1767 an Act was obtained for a new pier at St. Ives Harbour, in accordance with designs and a report by Smeaton of October 1766. The pier was to be founded by the method of *pierres perdues* 9 or 10 ft. below low water springs on sand proved by probing to be quite dense at this depth, about 6 ft. below the sea bed. It was 360 ft. long, 40 ft. wide at the base, 24 ft. at the top and 36 ft. high, surmounted by a 9 ft. parapet wall rising 15 ft. above high water springs. On the large random stones of the foundation the core of the pier would be built of rubble masonry encased between external walls of coursed masonry. Richardson was awarded the contract, and seems to have directed the work. Constructed in this manner, and containing some 35,000 tons of stone, the pier was completed in 1770.

On his visit to the Eddystone in 1787 Smeaton notes that 'since my last visit in 1777 the very careful and diligent Mr. Richardson departed this life'.

A. W. SKEMPTON

[John Smeaton (1791) *Narrative of the Building ... of the Edystone Lighthouse*; P. A. S. Pool (1974) *History of the Town and Borough of Penzance*; A. W. Skempton (1981) *John Smeaton*]

Works
1765–1767. Penzance Harbour, pier, £2,900
1767–1770. St. Ives Harbour, pier, £9,500

ROBERTS, Charles (fl. 1771–1799), is mentioned in connection with several canals, though it is not certain that all of the references are to the same person.

The first mention occurs in 1771 in the levelling account of the Birmingham Canal. In Parliamentary evidence given in 1797, it was said that Roberts had superintended and set out the Caldon branch (1776–1778) of the Trent & Mersey Canal. He left the company some time before 1783 but was consulted by them on extensions to the Caldon reservoirs in 1796. From 1783 there is a gap until evidence is given on the Andover Canal in 1788 but in 1791 Roberts was appointed by Hugh Henshall (q.v.) to superintend construction of the Manchester, Bolton & Bury Canal. Progress was not as rapid as the Company wished and in 1793 he was dismissed, to be replaced by the nephew of one of the more influential Committee members. The next year he was appointed Engineer of the Swansea Canal to build the canal by direct labour; Thomas Sheasby senior initially assisted him and later took over from him. In 1797 he made a survey of the Newcastle-under-Lyme Junction Canal, a short line in Staffordshire. Finally a Charles Roberts, mining agent to Lord Dudley, was involved with the Dudley Canal after the opening of its Lappal extension in 1798 and the subsequent resignation of his employer from the Committee.

P. S. M CROSS-RUDKIN

[RAIL 810/344 and 878/113, PRO; House of Lords Committee Book 41 and Witness Books, House of Lords RO; C. Hadfield (1960) *The Canals of South Wales and the Border*; C. Hadfield (1966) *The Canals of the West Midlands*; V. I. Tomlinson (1969) *The Manchester, Bolton and Bury Navigation*]

ROBINSON, Thomas (fl. 1730–1753) was engineer of the Weaver Navigation. The Act for making the River Weaver navigable from Frodsham to Winsford (18 miles) dates from 1721 but nothing was done until, with new undertakers, work began in 1730. Robinson was appointed 'Surveyor General' of works at a salary of £40 p.a. He planned and supervised construction of 11 locks and adjacent weirs built by local contractors such as Peter Walley and John Wood. The navigation was opened in April 1732. Robinson's employment ceased in 1735 when all the works had been completed and he then received a gratuity of 60 guineas for 'his faithful services'.

There appear to have been no mills on the river, and the locks were built not in cuts but in locally widened sections to give an adequate weir length. The weirs extended from the outer wall of the lock across to the opposite river bank, the whole structure of lock and weir being in timber. The locks, about 16 ft. wide and 60 ft. between gates with 4 ft. of water on the sills, could accommodate boats of 48 tons burden. By the early 1740s tolls averaged about £2,000 p.a. from the carriage of some 30,000 tons.

The office of Surveyor General lapsed until 1758 when the next phase of construction began under Robert Pownall (q.v.). However, from 1735 Robinson was granted 10 guineas p.a. for 'his attendance in the stating and settling the Undertakers' accounts'. What other employment he had is not known. His name last appears in the Weaver records in 1753.

A. W. SKEMPTON

[T. S. Willan (1951) *The Navigation of the River Weaver in the Eighteenth Century*, Chetham Society, 3rd series, vol. 3; C. Hadfield and G. Biddle (1970) *Canals of North West England*]

Works
1730–1735. Weaver Navigation, £18,000

ROBISON, John, FRSE (1739–1805), scientist, was born at Boghall, Stirlingshire in 1739, the son of John Robison, a Glasgow merchant. He was educated at Glasgow Grammar School and University from where he graduated in 1756. In 1758 he went to London and as the result of an appointment as a tutor to the midshipman son of Admiral Knowles, went on General Wolfe's expedition to Quebec. Robison assisted in surveys of the St. Lawrance River and surrounding countryside. He evidently impressed his contemporaries as an individual of scientific aptitude as, on his return to England in 1762, he was appointed by the Board of Longitude to take charge of John Harrison's chronometer on its trial voyage.

Back in Britain Robison returned to Glasgow and in 1766 was appointed Lecturer in Chemistry at the University in succession to Dr. Black, perhaps the leading teacher in the subject in the country at that time. In addition to his own scientific investigations he resumed his friendship with James Watt (q.v.), then instrument maker to the University, whom he had met when a student. Robison encouraged Watt's interest in the steam engine, and even suggested the idea of a steam locomotive. In 1796 Robison was to give supporting evidence in Watt's case against infringement of his patent. The two were regular correspondents throughout their lives.

In 1770 Robison accompanied Admiral Knowles to St. Petersburg as his private secretary following Knowles's appointment as President of the Russian Board of Admiralty, and in 1772 had his own appointment teaching mathematics to the Corps of Sea Cadets. The following year he returned to Scotland to take up the Chair of Natural Philosophy at Edinburgh University; the breadth of his teaching there, encompassing hydrodynamics, astronomy, optics, electricity and magnetism provides an indication of his learning. This was also reflected in his appointment as General Secretary to the Royal Society of Edinburgh on its foundation in 1783. The Royal Society of Edinburgh provided a vehicle for his literary output, which also included editing Dr. Black's 'Lectures', in 1799, and, most importantly, a number of articles for the Third Edition of the *Encyclopaedia Britannica*. These articles were considered sufficiently valuable to be republished in a posthumous edition by David Brewster (q.v.) as his *System of Natural Philosophy*. This work, which also contained many of his scientific papers, consolidated his reputation among his contemporaries and established it for posterity.

In a practical sense Robison is best known for his contribution to the development of the steam engine. He was, however, consulted about a number of engineering subjects. These included advising the Northern Lighthouse Board, in 1787, on the reflectors to replace the open fires currently in use for lighthouse illumination. He was one of the first authors in English to suggest a method for analysing timber trusses; he recognised the contribution of continuity to the strength of beams, and similarly, the strength of encastré vis-à-vis simply supported beams. His method of calculating arch thrust was displayed when he was consulted about the proposed 600 ft. span cast iron arch bridge proposed by Thomas Telford (q.v.) to cross the Thames at London Bridge. On the basis of the evidence of the Select Committee Report he was one of three experts who offered a practical method for calculating the stresses, the others being George Atwood and Charles Hutton (qq.v.).

Robison's articles on subjects such as 'strength of materials' and 'carpentry' reveal he was well aware of a range of relevant sources by British and continental authors. Although his work on the 'steam engine' had, unsurprisingly given technological advances, to be revised by the younger Watt shortly after his death, later authors such as Peter Barlow and Thomas Tredgold (qq.v.) quoted widely from his work. Todhunter and Pearson were very critical of his dismissal of Euler's work on columns, although it was based on observation of the behaviour of materials under load, and of his failure to correctly identify the position of a neutral axis in a beam. This overlooks his understanding of other aspects of beam behaviour. It is of some interest that Claude Navier, the French engineering scientist who perhaps more than any other individual was responsible for developing 'modern' methods of engineering analysis quoted his work. Robison was arguably the most important British engineering scientist of his generation.

Robison died on 30 January 1805. He had married Rachel Wright (*c.* 1759–1852) in 1772, and had four children: John (1778–1843), Euphemia (d. 1819), Charles (d. 1846) and Hugh (d. 1849). Of his children (Sir) John followed something of an engineering career, working for a Mr. Houston a Scottish millwright and machinist, and becoming a lifelong friend of Watt the younger before moving to London in 1802. There he established an artillery corps for the Nizam of Hyderabad, and made a fortune. Following his return to Scotland in 1815 he became involved in the Scottish scientific community. This included acting as Secretary of the Royal Society of Edinburgh from 1828 until 1840, being a founder member of the Scottish Society of Arts. His papers included a description of the failure of the Invalides suspension bridge in Paris designed by Navier (*Edinburgh Journal of Science*, **6** (1827) 240–242), and an account of the experiments on the resistance of vessels moving with different velocities conducted on the Forth and Clyde Canal, a continuation of the canal boat experiments fostered by Telford.

MIKE CHRIMES

[Select Committee upon the improvement of the Port of London (1800–1801) *3rd–4th Reports*; E. Todhunter and K. Pearson (1884) *A History of Elasticity and of the Strength of Materials*, **1**, 86–87; A. W. Skempton (1980) Telford and the design for a new London Bridge, in A Penfold (ed.), *Thomas Telford, Engineer*, 62–83; T. M. Charlton (1982) *A History of Theory of Structures in the Nineteenth Century*]

Publications
Publications include:

[1797]. Articles for the *Encyclopaedia Britannica*
1797. *Outline of a Course of Lectures on Mechanical Philosophy*
1804. *Elements of Mechanical Philosophy*
1815. Articles 'Steam' and 'Steam engines', written for the *Encyclopaedia Britannica*, edited by David Brewster with notes and additions by James Watt
1822. *A System of Mechanical Philosophy*, 4 vols

ROBSON, Joseph (*c.* 1704–1780), was engineer of Sunderland harbour. After the death of their engineer, William Vincent (q.v.), in April 1754, the Commissioners of the River Wear and Port of Sunderland instructed his foreman, Joseph Taylor, to carry on with the work in hand (his wage being increased from 15s to 1 guinea per

week) and began searching for a replacement. In December they made enquiries in London concerning Joseph Robson. A month later they asked him to come to Sunderland and survey the harbour. He arrived in March 1755 and in June was appointed harbour engineer at £120 p.a.

He is almost certainly the Joseph Robson 'late surveyor and supervisor of buildings of the Hudson's Bay Company' and author of *An Account of Six Years Residence in Hudson's Bay from 1733 to 1736 and 1744 to 1747*, published in 1752.

At Sunderland, Taylor was chiefly involved in dredging the south channel in the estuary, begun under William Vincent's direction in the autumn of 1752. Robson continued this operation and also cut through an outcrop of rock, giving an extra 3 ft. depth of water. Completed before 1759, deepening of the channel cost about £4,500 and enabled ships of 300 tons burden to leave the harbour fully laden.

Soon after his arrival Robson advised taking down the outer part of the south pier and building an extension in a more easterly direction. Proposing to build the extension in stone-filled timber framing, he went to see the piers in this form of construction at Bridlington, built in the 1720s by William Lellam (q.v.), who subsequently became harbour engineer at Sunderland and built the south pier there. Robson later decided to use masonry. The Commissioners approved the extension in April 1757 and it was completed, 320 ft. in length, probably in 1762. By the end of that year total expenditure on the harbour since the original Act of 1717 amounted to £48,000. Of this some £9,000 had been spent on works under Robson's direction.

In 1759 Robson gave evidence to a House of Commons Committee on the state of the harbour and on the question of making the river navigable from Biddick Ford (the current limit) up to New Bridge, Chester-le-Street, a distance of rather more than 2½ miles. This could be done at great cost by deepening the river bed, but the more practicable way would be to build a lock and weir about half way along the stretch. Evidence was also given by the land surveyor, Isaac Thompson (q.v.) and John Smith (q.v.), an expert in river navigation. The scheme was abandoned on the grounds that the cost, £7,000, would exceed the benefits.

Like other engineers in long-term appointments, Robson was able to take on outside work subject to approval from the Commissioners. The most important job of this kind was Stockton Bridge. In August 1760 he obtained leave of absence 'to go to Stockton to give his opinion on building a bridge there over the River Tees'. John Smeaton (q.v.) also visited the site in that year, producing a plan and estimate, and it was he who attended Parliament a year later; the Act was passed in March 1762. However, the Bridge Trustees then turned to Robson. His design was accepted in May 1763 but before construction began, Robert Mylne (q.v.) was consulted. He remarked: 'I have in investigating these circumstances been greatly obliged to the good Nature and Candour of Mr. Robson' and went on to say that he and Robson found themselves in agreement. Their joint report, *A Method of Laying the Foundations of the Pier of Stockton Bridge*, dated 16 October 1763, is copied into the Trustees Minute Book.

The contract for building the bridge was signed by Robert Corney, Robert Shout of Helmsley (q.v.) and Robert Nellson. Work began in 1764, the last arch was completed in September 1768 and the opening ceremony took place in April 1769. The bridge had five quasi-elliptical arches with a central span of 72 ft. and side spans of 60 ft. and 44 ft. Robson's fees and expenses in connection with the design amounted to £39 (Smeaton had charged £26 plus £42 for attending Parliament); after completion of the bridge, Robson received a further payment of £105, which indicates he had made several site visits during construction.

Meanwhile he gave advice on the building of a dry dock at Tynemouth in 1758 and in October 1765, it is recorded that 'a new machine ... built under the direction of Mr. Robson, a very able and experienced engineer, for cleaning the River Tyne, was launched off the Quay at Newcastle'.

Robson and Mylne again worked together when they were consulted on the best position for the proposed legal Quays at Hull. Their report is dated April 1767.

Back at Sunderland, Robson recommended building a pier or mole on rocks south of the harbour, with the object of reducing littoral drift of sand into the harbour entrance. The Commissioners sought advice from John Wooler (q.v.). He broadly agreed with the proposal. Considerable progress was made in the work but it was finally abandoned, presumable on grounds of cost. John Murray, writing in 1847, criticised the scheme as over ambitious and concludes 'there can be no doubt that the enthusiasm of the engineer carried him far beyond the bounds of prudence'. This may be so, but it is only fair to add that Robson's plan for a mole on the south rocks, and for a similar structure in the north, which he contemplated but never began, foreshadow the great outer piers (completed in the early years of the twentieth century) which form such important features of the modern harbour.

At a meeting of the Commissioners in September 1779, Robson signified that 'from his Age and present State of Health it is necessary for him to have some Assistance in the Execution of his Office, and recommended Mr. James Shout of Gateshead as a Proper Person to be concerned with him for that purpose'. The Commissioners agreed to appoint James Shout as Assistant Engineer but it is clear that he immediately took over all responsibility and that Robson retired.

Robson moved to Bishopswearmouth, where he died in November 1780 aged 76; he was buried at Sunderland. Administration of his estate was granted to his eldest son, William Robson, Mariner.

A. W. SKEMPTON

[Wear Commissioners Minute Book and Accounts (1754–1782) Tyne & Wear Archives; Joseph Robson (1759) evidence, *JHC*, **28**, 477–478; Stockton Bridge Trustees Order Book and Accounts (1762–1770) Durham RO; J. Skyes (1833) *Local Records* (of Newcastle); John Murray (1847) An account of the progressive improvement of Sunderland harbour, *Min Proc ICE*, **6**, 256–277; T. Richmond (1868) *Local Records of Stockton*; M. W. Baldwin (1973) The engineering history of Hull's earliest docks, *Trans Newc Soc*, **46**, 1–12; A. W. Skempton (1977) The engineers of Sunderland harbour, *Indust Archaeol Rev*, **1**, 103–125]

Publications
1766. *To the Commissioners of the River Wear ... on reasons against ... lessening the flux of water in navigable rivers*

Works
1755–1762. Sunderland harbour improvements, £9,000
1764–1769. Stockton Bridge, £6,300

ROGERS, John (fl. 1533–d. 1558), military engineer, was an important figure in the transitional period between the mediaeval master mason and the diverging professions of architect and military and civil engineer. His career, well documented from 1541 until his death in 1558, involved works in all these areas. As a mason he worked at Hampton Court and as an architect on Henry VIII's manor at Hull. As a military engineer it can be argued that he was instrumental in replacing the rounded bastions and quatrefoil shapes of the Early Henrican Forts on the south coast with the straight-sided angle bastions later used at Hull and Berwick. His career as a harbour engineer seems to have originally occurred almost as a footnote to his military career, when improvements needed to be made to the harbours which he was fortifying in France. He was later sent to report on progress with the harbour at Dover and hired purely as a civil engineer by the town of Sandwich in an attempt to solve the centuries old problem of the silting up of the harbour. Unlike his colleague and contemporary, Sir Richard Lee (q.v.), whose fortifications at Berwick remain virtually intact, the military works of Rogers were soon rendered obsolescent by new developments in artillery (not to mention invasion by the French) and few of his civil works reached fruition. Luckily his plattes/plans or surveys of many of the above have survived in the Cottonian collection in the British Museum and give a good idea of his engineering versatility. According to Shelby these plans also saw the same transition in style between mediaeval draftsmen and modern topographers as is demonstrated in his career as a whole, notably in being among the first Englishmen to draw maps to scale, as evidenced by his plan of the defences of Guines Castle and of Hull.

Little is known of either his early career or his family life. There is mention of a freemason John

Rogers as a member of the Duke of Buckingham's household in 1514 but as he was active as an engineer up until his death in 1558, this seems a little early to refer to the same John Rogers. To complete a triumvirate of John Rogers, there is also mention of a military engineer working in Ireland as late as the 1570s which has confused Porter and Robinson as to the exact date of his death.

Probably the first reliable reference to Rogers's work is as a mason at Hampton Court from 1533–1535 for 3s 4d per week. There is then an unfortunate gap in the details of his career until he was appointed master mason of Guines in 1541. The plans there called for the fortification of the town and the building of four new bulwarks. The plan chosen shows definite similarities to later works at Hull and Boulogne (Rogers) and Berwick (Lee). Rogers's job and position as master mason would have involved design and technical supervision of the building and directing the construction itself, as well as reporting directly to Henry VIII on progress (three times in the summer of 1541). His only rival to the title of foremost English engineer of the mid-16th century, Richard Lee, had been Surveyor of the Works at Calais since 1536 and their paths were to cross many times over the next decade, raising questions as to who was ultimately responsible for projects under their joint control. Among the first of these was the completion of the new mill tower and the raising of the town ramparts. They were also both concerned with digging an extensive new dike, 3 miles long, 28 ft. wide and 28 ft. deep to prevent the French tampering with the water supply.

In 1542 he was promoted to Surveyor of the Fortifications of Hull where he worked as an architect on the King's manor house as well as an engineer on the fortification of the town. A programme was drawn up to modernise the town defences of Hull by Henry VIII after his visit in 1541. This scheme was modified by Rogers to include block houses on the north and south sides of Hull Haven, linked to Hull Castle in the centre by a curtain wall to protect from attack by sea and land. These were later incorporated in the seventeenth century fortifications by Bernard De Gomme and Martin Beckman (qq.v.) and were demolished in 1864. The north bastion was demolished in 1802. All of these were later used as prisons, magazines and hospitals. Excavations show the walls to have had a thickness of 7 ft. 7 in. at a height of 11 ft. and to have been constructed of a wall core of chalk and limestone rubble. The brick curtain wall from the castle to the south block house was 10 ft. 5 in. wide with a limestone rubble core. A plan exists, probably by Rogers, and the whole scheme resembled work previously undertaken at Guines. Rogers was the chief technical supervisor of the project, the total cost of which was £21,056 and which involved twenty carpenters, sixty bricklayers, ten plumbers, thirty brickmakers and three-hundred labourers.

At the same time changes were being made to Hull manor which Henry acquired in 1538. Four plans were drawn up, also probably by Rogers, which show plans to rebuild the manor along lines similar to Hampton Court, Windsor and Nonsuch. Sadly, as this would have been Rogers only non-military building commission, there is no evidence as to what was actually built.

In 1544 Rogers was sent to view the works at Berwick and Wark. In 1543 he was granted an annuity of £36, more than Lee from whom he was his superior at Guines. In 1544 he was granted the office of Clerk of the Ordnance in England, which he held for the next thirteen years.

More relevant to his engineering career is the fact that in the same year Rogers was promoted to Surveyor of the Works and Fortifications of Boulogne at a salary of 6s 8d per day with an allowance for two clerks and four servants. This involved works on fortifying the 'Haute Ville' and a citadel at the Basseville (together costing over £82,000) and protecting shipping in the Liane Estuary by building detached works. Originally responsibility was shared by engineers such as Lee and Richard Cavendish (q.v.) and military commanders, but Rogers was eventually made entirely responsible for all the works in the Boulonais with directions to both civil and military officer not to interfere. 'One may say that considering the uncertainties of the opinions of your Lordship (Surrey) and Sir Thomas (Palmer) his majesty debated with Rogers and conceived certain plattes the execution of which he committed to Rogers … the man is plain and blunt which must be borne withal as long as he is well meaning and mindeth the service of the King's Majesty'.

His first order with regard to Haute Ville was to confer with Lee to 'draw the water and moisture out of the streets to render (it) more salubrious'. His work on the harbour involved the building of a stone pier or jetty 26 ft. high, 300 ft. long and 16 ft. wide; this was begun in 1545. Work on the town fortifications was halted by the Treaty of Campo, made with the French in 1546. (Rogers was sent to survey the head of the Liane River as part of the boundary negotiations.) This forbade the building of new fortifications but in 1547 work was nonetheless begun on a fortified mole opposite the harbour entrance, ostensibly to keep back drifting sand and to prevent the blocking of the channel. It could also, as the French were quick to point out, be used to mount cannons. This mole consisted of two parallel walls of masonry divided by cross walls into compartments filled with sand. On top were barracks, artillery platforms and other buildings; by 1549–1550 400 ft. had been completed. It was intended to extend it to 1,000 ft. and link it to the harbour pier with a five arch bridge. When Henry II of France attacked it in 1550 he described it as 'a great long jetty of large coarse stones brought together and constructed with great skill. Since the time of the Romans there has not been built in this place a sturdier or more superb edifice than this which they call the Dunette'.

As the harbour in Boulogne was causing problems, Rogers had been sent with Sir Thomas Wyatt to survey the newly acquired harbour of Ambleteuse in March 1546. His plattes show the problem: the depth of water in the river near the town was only 5 ft. The proposed solution was to divide the river into two channels each controlled by a sluice gate. Since there was no short term solution to be found, work was also begun at Blackness (Cap Gris-Nez). There were complaints about Rogers's slow progress on the earthwork fort there, but that he was giving considerable attention to the harbour is shown by his plan for a cut 'through the land ... from sea to sea, compassing the Blackness in and making him an island. For so shall both the flood and ebb run through the Haven and make it always deep'. The total cost of the works undertaken during his time at Boulogne was £122,696 3s 6d.

Following the death of Henry VIII Rogers lost much of his influence and for the next 10 years spent most of his time as a peripatetic engineering consultant without an official position. In 1548 he was concerned with harbours at Folkestone, Dover and Sandwich; from 1549–1551 with the fortifications on Guernsey; at Alderney (where £5,000 was spent on works at Longes Bay, traces of which still survive); at the Isles of Scilly (costing £3,787); in 1551 with fortifications in Ireland and Sheerness; 1552–1553 saw him back in Dover.

There is then a three year gap in his career before he was appointed Surveyor of the Queens Works in Berwick in 1556. He held this position for barely a month before being given the even more prestigious post of Surveyor of the Works of Calais. The major interest in this period is obviously the harbour works, but to round up his career as a military engineer it is worth mentioning that plans of both Harrys Fort in the Scilly Isles and Corkbeg at the entrance to Cork Harbour continue the development of the Italian Bastion style already noted at Boulogne and Hull.

Rogers's work on harbours in England began in 1548 when the Privy Council instructed him to go to Dover to report on the state of the pier then under construction. In February 1552, in company with Sir Richard Cotton, he was instructed to produce a plan showing the damage to the pier. In October 1553, this time with Sir Richard Lee and Benjamin Gonson, he was once again sent to see whether the pier could be repaired for use and at what cost, and also to consider any other method for improving the harbour. This report does not survive and nothing seems to have been done as a result of it. In 1548 he was also sent to Folkestone to consult with other officials as to the best means of constructing a new mole there.

The most substantial results of his work at this period were at Sandwich where, in 1548, he assisted the Mayor with a survey of the Haven. This study concluded that both the North mouth and the mouth of the river Stour were too silted up for dredging to be the solution. Instead it was proposed that a 'new cut' should be made from Sandwich Strait south east to the Sea, a solution reminiscent of his plans for Blackness. Rogers was not officially on the commission appointed to investigate proposals for new harbour works in Sandwich but a later commission of 1559 makes his responsibility clear by describing 'the newe cutte at Sandwich lately begun by devise of John Rogers'.

This cut was to run from the south east corner of the loop in the river Stour to Sandwich Bay, bypassing the silted channels of the Lower Stour and Pegwell Bay and reaching the coast between Sandown and Sandhills. Work began in April 1549, using large gangs of shovelmen with a foreman to every ten, but progress was slow due to lack of financial support from the Crown. It was kept going by the citizens of Sandwich throughout 1550 but work ceased permanently in November 1551. The completed portion, still visible, was the basis of later schemes of 1559, 1573 and 1737 although Andrian Andrison, a Dutch engineer working on the 1573 survey, felt it should have been set further upstream.

During the brief period when Rogers was the last English Surveyor of Calais, there was time to do very little, but of interest is the making of 'two sluices by Lanterne Gate into the new molle'. It is indicative of the high regard in which Rogers was held by his contemporaries that he was one of the fifty of the military officers and officials of Calais held for ransom by the French after the fall of Calais on 7 January 1558. It seems likely that he died while a prisoner of the French. His successor at the Board of Ordnance received back pay from (March) 1558 and it was the custom of the time to commence payment from the death of the previous office holder.

His will, however, was not probated until 10 May 1559. It left his house in London—St. Giles Hall—the lease of the Lordship of Streat in Somerset, the lease of the Priory of Folkestone and lands and goods within the town of Calais to his wife, Agnes, and her heirs.

There are no portraits or verbal descriptions of his physical characteristics, but his 'plain and blunt' nature is attested to in numerous acerbic letters from colleagues. However, this was balanced by, in his own words, 'having no trade or occupying in the world, but the King's business where in I travail night and day', a view endorsed by the granting of an annuity of 20 marks to his widow in recognition of his long service to Henry VIII, Edward VI and Queen Mary.

SUSAN HOTS

[SP68/13–15, PRO; SP1/215, PRO; Maidstone, Kent Archives Office; Sandwich Old Red Book 1527–1551; W. Porter (1889) *History of the Corps of Royal Engineers*, **1**, 24–29; A. H. W. Robinson (1962) *Marine Cartography in Britain*, 18–19, 147–148; L. R. Shelby (1967) *John Rogers Tudor Military Engineer*, *King's Works* (1975), **III** and

IV; M. H. Merriman (1983) Italian military engineers in Britain in the 1540s, in S. Tyake (ed.), *English Mapmaking 1500–1650*, 57–67; M. Foreman (1989) The defences of Hull, *Fortress*, **2**, 36–45.; A. Saunders (1989) *Fortress Britain*; P. M. Kerrigan (1990) Fortifications in Tudor Ireland, *Fortress*, **7**, 27–39; Bendall (1997); J. Tomkinson (1998) The Henrican Bastions at Guines Castle Fort, **26**, 121–142.]

Plans

All the following are in the British Library, Cotton mss:

Guines (four)
Hull fortifications
Survey of Hull area
Hull Manor (four)
County of Guines and Boulonais
Ambleteuse (four)
Survey of Boulogne (four)

Principal works

1542–1544. Hull fortifications, £21,000
1544–1548. Works at Boulogne, £122,700
1549–1551. New Cut at Sandwich

ROGERS, Thomas (fl. 1786–1812), engineer, is first noticed as a glass cutter near the Strand, Westminster, in 1786 when he was granted, with George Robinson—who supplied lighting apparatus to Trinity House—a patent for attaching glass ornamentation to furniture and mirrors.

Around 1787 an inspection of lighting apparatus in French lighthouses was made by Trinity House. Following experiments on Blackheath with reflectors and lamps Trinity House invited persons with proposals for improved lighting to send in their apparatus. Rogers submitted a cadioptric lens light, apparently comprising a plain convex lens with a reflector. The lens was, after experiments, installed at Portland High lighthouse. Although this was the only example of his lights employed by Trinity House, Rogers installed others more successfully elsewhere, examples being fitted at Howth (1790), Hook (1791) and North Foreland (1792), the last-named privately operated.

It was in Ireland that Rogers emerged as an engineer, severing his connection with Robinson in 1792 and moving to Dublin. From 1793–1797 he erected an important lighthouse at South Rock; The Rock, with one side only above high water was 3 miles off Newcastle, County Down. Charles Vallancey (q.v.) had drawn up a scheme for a light there in 1767 and although the proposal was considered again in 1783 nothing was done before Rogers began work. Conical in form the tower was 37 ft. in diameter at its base and 17 ft. at a height of 70 ft.; the first 22 ft. are of solid masonry. It was crowned with a 8 ft. lantern with 6 ft. tall glazing and was built of granite blocks with the end joints of tiers joggled into each other. Nine vertical iron bars 3 in. square run within the wall, keyed into circular cast iron plates. Work began in September 1793 and by the summer of 1794 a masonry platform had been prepared for the dressed stone. Rogers built a small harbour nearby. The chief problems encountered were with delivery and placement of the stones and also with disintegration of the foundation rock. Robert Stevenson (q.v.) visited the lighthouse several times to help him with the design of Bell Rock and although John Rennie (q.v.) was very critical of the work, it has survived for two centuries.

In the 1790s Rogers installed fixed versions of his lights at Copeland and Arranmore, in addition to those mentioned above. In 1796 responsibility for all lighthouses around the coast of Ireland passed to the Revenue Commissioners. The work was entrusted to Rogers, as 'Inspector' for the Revenue Commissioners, also referred to as 'the contractor'. By 1800 his work at South Rock had indicated that he had genuine talents as a civil engineer. In 1802 a further granite tower 130 ft. high was completed at Cranfield, Carlingford Lough, with eleven of his lights, and another at Loop Head with twelve lights. He also built a new tower at the Aran Islands. In 1803 the Revenue Board took over the lights at Waterford and Kinsale, which were also then equipped with Rogers's lights.

There were, however, concerns being raised about the whole management of Irish Lights and in 1810 responsibility for ten lighthouses and three harbour lights was transferred to the Ballast Board. Three inspectors were appointed to visit the lighthouses without the knowledge of Rogers and they came out with a report highly critical of his lights. At that time Rogers was contracted to manage ten lights but the Ballast Board restricted his contracts to six months, with a tight specification, and from July 1812 all his contracts were terminated. It appears that Rogers had assumed he had a contract for life and had abused his position, appointing unsuitable lighthouse keepers and making excess profits; his contract for nine lighthouses in 1810 was £5,899 p.a. whilst the same lights were being maintained in 1832 for £3,363 p.a. Rogers was superseded as engineer by George Halpin (q.v.).

The work of Rogers as a lighthouse engineer led to his being consulted about other schemes, notably improvements to Dublin Harbour and a design for a harbour at Porthdynllaen (1807) as part of improvements to London-Dublin communications. The latter report was carried out at the instigation of William Alexander Madocks (q.v.). In 1800 he suggested a new harbour at Howth, with a canal to Dublin. His 1807 report on Dublin contains a large number of tidal observations. He was one of several engineers, including Hyde Page (q.v.), considering the possible use of Howth as an alternative harbour but the scheme adopted was that of Captain George Taylor, the work being eventually completed to John Rennie's (q.v.) designs.

Although Rogers is known to have had twelve children by 1812, some still dependent on him, details of his career after that time are lacking.

MIKE CHRIMES and RON COX

[Sir John Rennie (1844) *Report on the Harbours of Holyhead and Porthdynllaen*; D. A. Stevenson (ed.) (1946) *English Lighthouse Tours, 1801–1818*; D. A. Stevenson (1959) *The World's Lighthouses before 1820*; E. Beazley (1985) *Madocks and the Wonder of Wales*, 2nd edn.; M. A. Gilligan (1988) *A History of the Port of Dublin*; R. C. Cox and M. H. Gould (1998) *Civil Engineering Heritage: Ireland*]

Publications

1807. *Observations on the Reports … for the Improvement of Dublin Harbour*
1807. *Documents … relating to Porthdynllaen Harbour*

Works

1788. Portland High Lighthouse, lights
1790. Howth Lighthouse, reconstruction
1791. Hook Lighthouse, light
1792. North Foreland Lighthouse, lights
1793–1797. South Rock Lighthouse, £16,372
1790s. Copeland Lighthouse, lights
1790s. Arranmore, Aran Islands, Lighthouse, lights
c. 1800. Aran Island Lighthouse, contractor
1802. Cranfield Lighthouse
1802. Loop Head Lighthouse, Co. Clare, reconstruction
1803–1804. Waterford Lighthouse, lights
1803–1804. Kinsale Harbour Lighthouse, lights
1803–1804. Old Head of Kinsale Lighthouse, lights

ROMER, John Lambertus (1680–1754?), military engineer, was the son of Wolfgang William Romer (q.v.) and was born in 1680. Father and son worked together only in Portsmouth in 1708–1710, but their careers followed similar patterns: mapping, surveying and building forts in recently conquered territory, the father in America and the son in Scotland.

Romer progressed through the ranks from Engineer on the establishment in 1714 to Director in 1742. In 1714 he was Engineer-in-Charge at Sheerness, his duties also covering the Medway and Thames area; a plan of 1716 describes a cistern he constructed there. He was also one of very few engineers actually present in Scotland during the 1715 rebellion and perhaps because of this he was made Engineer-in-Charge and Surveyor of the Northern Districts and Scotland, succeeding Andrew Jelfe (q.v.) in 1719. A few years later he was placed in charge of the administration in the same area, theoretically based in London, but still spending most of his time on site, providing surveys and estimates, and building forts and barracks including Hull, Clifford's Fort, Berwick and Carlisle. His surveys from the period 1723–1742 survive and show that, despite in some cases possessing the outward attributes of four square angle bastioned fortresses, most of his buildings were designed more as grand civic mansions than

defensible forts; consequently few of them survived intact or unaltered after the 1745 Rising.

His first commission was to complete barracks at Ruthven, Bernera Kiliwhimen and Inversnaid; the first was finally built by Sir Patrick Strachan at a cost of £1,555 in 1724. In 1722 Romer designed a pier to protect shipping at Inverkeithing. In 1724 he accompanied General Wade (q.v.) on his mission to report on the condition of the Highlands of Scotland. As a result of this it was decided to replace Kiliwhimen barracks with Fort Augustus (1729–1742) and to enlarge Ruthven barracks, where 'three of the roads lately made through the Highlands' met, to include stables for thirty horses (1734). He also modernised the walls surrounding the old castles at Edinburgh and Dumbarton (1735). Most of these survive, including his signature pepper pot sentry posts.

His new forts were less lucky. Fort Augustus held out for only 2 days before the Jacobites hit the exposed powder magazine. However, traces of it can still be discerned in the monastery now occupying the site. Fort George, at Inverness, built by Romer at a cost of £50,000, surrendered at once and was then blown up by the Jacobites. Perhaps annoyance at having twenty years work go up in smoke accounts for the unlikely presence of a sixty year old engineer at the Battle of Culloden on 16 April 1646. He was badly wounded, but did not retire until 1751.

Romer is believed to have died in 1754 and been buried in St. Margaret's, Westminster, but there is no documentary evidence of either fact. He was married to Mary Hammond. His son, John (1713–1775), also had a military career.

SUSAN HOTS

[National Library of Scotland, John Romers surveys 1723–1742; Royal Engineers, Conolly papers; Military engineers of the Stuart Dynasty compiled by J. C. Tyler (n.d.), ms; DNB; Vetch; Colvin, 833; Bendall, **2**, R247/247–1; W. Porter (1889) *History of the Corps of the Royal Engineers*; D. W. Marshall (1976) *The British Military Engineers 1741–1783*, unpublished Ph.D. dissertation, University of Michigan; C. Duffy (1985) *The Fortress in the Age of Vauban and Frederick the Great*; A. Saunders (1989) *Fortress Britain*; J. Gifford (1992) *Buildings of Scotland: Highlands and Islands*; C. Tabraham and D. Grove (1995) *Fortress Scotland and the Jacobites*, J. Douet (1998) *British Barracks 1600–1914*]

Publications and maps

1712. *Report on Portsmouth Harbour*
1716. *Plan and Section of a Cistern built at Sheerness in the Year 1716*

ROMER, Wolfgang William (1640–1713), was born on 23 April 1640 while his father, Matthias, was an Ambassador with the Exiled Court of the Elector of the Palatine in The Hague; the Elector, Charles Louis, was godfather at his baptism on 17 May 1640. Romer took service as a military engineer and was already a Colonel when he accompanied William III on his invasion of 1688. Until 1697 his commissions were all by Royal Warrant, placing him in an ambiguous position with the Board of Ordnance. These included fortifying Cork, Longford and Thurles, drawing up drafts, designs and estimates for the defences of Guernsey and reporting on the defences of Plymouth.

In 1697 he was ordered to go to New York and New England. He refused as the appointment was still at his existing salary of 20s per day and was suspended. However, his illustrious connections proved useful as the King agreed to a salary of 30s per day, to the fury of the Master General of Ordnance, Lord Romney, who complained that it was a higher salary than the Chief Engineer of England. He was recalled after only six months—the order was only rescinded on the personal intervention of the Governor of New York—but then kept there for nearly nine years despite his pleas to return as he had 'a distemper not curable in these parts for want of experienced physicians'. On his return to England (via a sojourn in St. Malo as a prisoner of war of the French) he was suspended again for staying too long in America! His work in America validated the salary. In 1697 he drew up a plan of the Hudson River and adjoining country and in 1698 he recommended the building of two stone forts, at Albany and Schenechtady, at a cost of £2,000. In 1700 he produced a map of his journey to the five Indian nations. Although charmingly adorned with vignettes of elks, bears and Indian villages this is still recognisably the earliest map of the Hudson and Mohawk rivers leading through Lake Oneida to Lake Ontario, later the site of the first major canal system linking New York to the great lakes. Less accurate, but still identifiable, are lakes Ontario and Erie with 'the great fall' connecting the two, an early mention of Niagara, first visited only thirty years previously by the French.

From 1701 until 1703 he built defences for Boston Harbour, in particular Fort William on Castle Island. When this was demolished after the British evacuation of 1776 a modest plaque was found stating that it was constructed by Romer, a military architect of the first rank. Romer was captured by the French on his way back from America in 1706 but allowed home on parole to arrange his ransom or exchange. It was therefore two years before he was able to work again as an engineer at Portsmouth, originally designing Block House Fort, but extending his interest to the harbour where he proposed the construction of breakwaters to protect it.

He remained in charge at Portsmouth until his death in 1713, with occasional excursions to report on the defences for Harwich and towns in Flanders. He died on 15 March 1713 was buried in Dusseldorf, complaining to the end that the Board of Ordnance had never recompensed him for moneys spent in America. His son, John Lambertus (q.v.), was also a military engineer.

SUSAN HOTS

[Royal Engineers, Conolly papers; Military engineers of the Stuart Dynasty, compiled by J. C. Tyler (n.d.), ms; DNB; Vetch; Colvin, 833; Bendall, 2, R247/247–1; W. Porter (1889) *History of the Corps of the Royal Engineers*; D. W. Marshall (1976) *The British Military Engineers 1741–1783*, unpublished Ph.D. dissertation, University of Michigan]

Publications
1694. *Plan of Castle Cornet*
1700. *Mappe of Colonel Romers Voyage to ye Five Nations Confederated with His Majesty on the Continent of America*
1712. *Report on Portsmouth Harbour*

ROSS family (fl. 1800–1840) members were contractors, various of whom were associated with civil engineering in the early nineteenth century, some with the Lancaster Canal, the earthworks of the Caledonian Canal and the improvements to Highland roads and bridges.

Hugh Ross (fl. 1820–1840) and his nephews, **Hugh** (c. 1809–186?) and **Alexander Mackenzie Ross** (1805–1862), were perhaps the best known and all worked for Hugh McIntosh (q.v.). Hugh Rose (Ross) is first known to have acted as McIntosh's agent on the Gloucester–Berkeley Ship Canal in 1823, based at Slimbridge, at which time his nephew, Alexander Mackenzie Ross, became his pupil. Later in the 1820s they worked on the London Dock extension works. Alexander was involved in the early 1830s with work carried out by McIntosh at West India Docks (1832–1833) and c. 1836 he joined Robert Stephenson's staff to work on surveys for the London–Brighton Railway. He was associated with Stephenson for much of the rest of his career, both on the Chester–Holyhead Railway and the Grand Trunk Railway in Canada, his role on which was the subject of some controversy. He was involved with William Mackenzie's (q.v.) schemes in North Wales and Spain, carrying out detailed surveys for a railway between Madrid and Bilbao.

In the 1830s Hugh (U.) Ross was McIntosh's agent on several Great Western Railway contracts (2B, 3B, 3B extension, 6B) and after McIntosh's death, Hugh Ross pursued claims on the North Midland Railway contracts. Hugh Ross Jr. remained involved with David McIntosh, being at his house at the time of the 1851 census. It is almost impossible to establish which Hugh Ross did what work for McIntosh in the 1830s, although it seems likely that it was the elder whose name features most often. Hugh, Alexander's brother, was still alive in the 1860s.

MIKE CHRIMES

[Frank Smith collection, ICE Library; Telford mss, ICE Library; Joseph Mitchell, letterbooks, ICE Library; Mackenzie collection, ICE Library; Telford collection, Ironbridge Gorge Museum Trust; *Reports of the Highland Roads and Bridges Commissioners* (1816); A Penfold (1980) Managerial organisation on the Caledonian Canal 1803–1820,

Thomas Telford, Engineer, 129–150; M. M. Chrimes (1995) Hugh McIntosh (1768–1840), national contractor, *Trans Newc Soc*, **66**, 175–192]

Works

1802–1805. Union Bridge, Aberdeen (W. Ross)
1803–1821. Caledonian Canal, cutting (Ross family?)
c. 1810. Fearn Road, Highland Roads, Mackenzie and Ross, £5,966 17s 9d
c. 1831. Sutherland Coast Road (Hugh Ross), contractor

For contracts with Hugh McIntosh, see McIntosh.

ROY, William, FRS, Major General (1726–1790), military engineer and surveyor. Despite a thirty year career in the Corps of Engineers William Roy's current reputation rests more on the posthumous results of his twin passions for archaeology and surveying, notably his *Military Antiquities of the Romans in North Britain*, published in 1793 and in many cases the sole surviving record of Roman engineering in Scotland, and in the Ordnance Survey of Great Britain, finally established the year after his death, after 40 years of advocacy. In light of this it is interesting that he himself told Hugh Debbieg (q.v.) that he did not consider himself an engineer. Possibly the remark refers to a lack of formal training, since while his birth at Carluke in Lanarkshire on the 4 May 1726, son of John Roy of Milton Head is documented, as is his education at the local grammar school, there is then no trace of him until he appeared as Assistant Quartermaster to David Watson (q.v.) in 1747, seemingly fully qualified in the use of surveying instruments, sketching, map mapping, topographical drawing and military engineering. Theories have abounded virtually from the moment of his death, some fanciful in the extreme such as the obituary in the *Gentleman's Magazine* of July 1790 which claimed he was already (aged twenty-one) a Colonel of Artillery. Equally fanciful is his descriptions as Watson's nephew (a confusion with David Dundas). Possible, but unproveable are suggestions that he worked for the civilian branch of ordnance, was a pupil of Colin Maclaurin, a Professor of Mathematics at Edinburgh University and that he worked for the Post Office in Edinburgh. Since the Post Office Act of 1711 required regular road surveys to be made it would have been a logical career move once the '45 rebellion had 'convinced government that a country so very inaccessible by nature should be thoroughly explored and laid open by establishing military posts ... and carrying roads of communications to its remotest parts'.

At first Watson's sole assistant, Roy was soon in charge of six parties for the work in the north and two when it was later extended to the south of Scotland. The courses of all rivers and streams were followed to the source and measured and all roads and lakes surveyed, using a plain theodolite with no sites, according to Roy 'Instruments of common or inferior kind, the sum

annually allowed for it being inadequate for so great a design'. It was drawn on a scale of 0.568 miles per inch (1:36,000), in comparison to Charles Vallence's (q.v.) similar works in Ireland on a scale of 0.635 miles per inch, Samuel Holland's (q.v.) of north America at 0.757 and James Rennell's (q.v.) of Bengal at 5 miles to the inch. The work continued until 1755 (coincidentally the area surveyed in that year during which Roy was in sole charge was 'the course of the Roman wall (Antonius) called Grimedyke'), but war with the French left the survey unfinished when Roy, Watson and Dundas were sent to survey invasion routes in the south of England. Plans from this period include the Lewes Road, the coast of Kent, a sketch of Dorchester, (with a diversion to survey Maiden Castle), Salisbury, Gloucester and Pembroke and proposals for a harbour at Milford Haven.

Despite having been in charge of engineering officers it was not until December 1755 that Roy was officially enrolled as a Practitioner Engineer with the Corps, 'fast tracking' through to Deputy Quartermaster-General in Germany by 1762. The peace of 1763 saw him returning to former preoccupations proposing 'a general survey of the whole island at public cost to which the map of Scotland (all 94 rolls) was to have been subservient'. But a combination of cost and distrust of the military meant nothing was done and he was instead sent to 'inspect, survey and make reports from time to time on the coasts and districts ... adjacent to the coasts of this kingdom', for which he was paid 20s per day. The same year (1765) saw him in Ireland reporting on the state of roads, harbours and natural features and in Dunkirk with J. P. Desmaretz (q.v.) reporting on the state of the demolitions and work on the Mardyke Channel.

In 1770 in his capacity of Surveyor-General of the coasts he was the first to point out the importance of the Marker Heights for the defence of Plymouth Dockyard. By 1783 he was both a Director and Lieutenant-Colonel with the Royal Engineers and Major-General in the army and a member of the Board on Fortifications (to the disgruntlement of his former colleague Debbieg). For his 'own private amusement' and as 'a hint to the public for the revival of the now almost forgotten scheme of 1763' he was surveying a base line of 7,744 ft. between Marylebone and St. Pancras, when in 1783 Cassini de Thury Director of the French Royal Observatory suggested to the government—who passed it on to the Royal Society—the need for an accurate measurement of the distance between Greenwich and Paris (a discrepancy existed of 11° longitude and 15 seconds latitude) Roy, a member of theirs since 1767, was therefore the obvious candidate.

A first step towards the triangulations was the establishment of an accurate base line measured on Hounslow Heath. For this Roy provided a 'Table of the expansion of metals', comparing different types of brass, steel, cast iron and glass to find measuring rods least affected by humidity. He eventually settled on glass tubes, with a total expansion of only 0.279 in., the original measurement of 27,404.7219 ft. was only 1.672 in. out.

The continuation of the triangulation was delayed by Roy's insistence (this time) on a specially designed theodolite from Jesse Ramsden, but it was eventually completed exposing errors on existing maps of both countries and providing the basis for typographical surveys of Surrey, Middlesex, Kent and Sussex.

Roy died suddenly on 1 July 1790 while preparing a paper for the Royal Society which had given him their Copley Medal for his work on the base line in 1785. The Duke of Richmond, the Master-General of Ordnance and a long term patron of Roy finally persuaded George III to authorise the extension of the national survey in 1791, utilising the technical standards laid down by Roy during his lengthy career.

SUSAN HOTS

[Considerations on the propriety of making a general military map of England with the method proposed for carrying it into execution and an estimate of the expense (1776), in J. W. Fortescue (ed.) (1927–1928) *Correspondence King George III 1760–1783 from the Original Papers in Royal Archives at Windsor Castle*, 328–334; WO 30/55 and WO 55/365, PRO; RE Library, Conolly papers; DNB; Porter, **1**; R. H. Vetch (1888) Colonel Hugh Debbieg, *Royal Engineers Journal*, 2 Feb, 33; **1**; G. Macdonald (1917) General Roy his military antiquities of the Romans in north Britain, *Archaeologia*, **68**; E. G. R. Taylor (1966) *Mathematical Practitioners of Hanoverian England*, 24; A. H. W. Robinson (1962) *Marine Cartography in Britain*, 95–96; R. A. Skelton (1967) *The Military Survey of Scotland 1747–1755*, Royal Scottish Geographical Society, SP No. 1; D. W. Marshall (1976) *British Military Engineers 1741–1783*; M. H. Edney (1994) British military education, map making and military map-mindedness in the later enlightenment, *Cartographic Journal*, **31**, 14–20]

Publications

1765. *Military Description of the South East part of England, July 1765*, BL maps C7d12 (plus a large map), BL maps, K Top v197

1770. *Memoranda concerning Plymouth made in a Tour thro' that Part of the Country in August 1770*

1774. *Mappa Britanniae Septentrionalis Faciei Romanae*, drawn from Thomas Chamberlain

1785. An account of the measurement of a base on Hounslow Heath, *Royal Society Philosophical Transactions*, **75**

1787. An account of the mode proposed to be followed in determining the relative situation of the Royal Observatories of Greenwich and Paris, *Royal Society Philosophical Transactions*, **77**

1790. An account of the trigonometrical operations whereby the distance between the meridians of the Royal Observatories of Greenwich and

Paris has been determined, *Royal Society Philosophical Transactions*, **80**(1)

1793. *Military Antiquities of the Romans in North Britain, and particularly their Ancient System of Castramentation illustrated from Vestiges of the Camps of Agricola existing there*, Society of Antiquities (posthumous publication)

RUBIE, Edward (fl. 1716–d. 1753), master carpenter and civil engineer, was employed as a master workman or foreman under John Perry (q.v.) on the Dagenham Breach works, probably from the start in 1716 to completion in 1720. In 1724 Perry was appointed to direct works at Rye New Harbour on his plan to make a wide canal about a mile in length from Winchelsea Channel down to and through the beach ridge, with a great sluice rather more than half way down and piers at the entrance 120 ft. apart. Having set out the canal, and while land purchase was still in hand, Perry brought Rubie down from London in May 1725 to act as his 'agent' during a short absence in Ireland. He received £20 from Perry and from July he was engaged by the Harbour Commissioners as foreman at 25s per week.

Work had by then started on the canal, dug initially 8 ft. deep with widths of 140 ft. above the sluice and 180 ft. below. Work on the west pier began in 1728. It was 62 ft. long, 27 ft. wide and rose 18 ft. above the foundations which were 2 ft. above low water level neaps. They comprised a timber grillage supported on piles surrounded by dovetailed sheet piling 20 ft. deep, and were built by direct labour under Rubie's supervision. Christopher Cass and Andrews Jelfe (q.v.) contracted for the masonry: Portland stone with a mortared rubble masonry infill. When completed in September 1730, Rubie measured up the masonry for payment to the contractor.

By that time Perry had gone to Spalding as engineer of Deeping Fen. From July 1730 Rubie was in charge at a salary of £65 p.a. with a rent free house, though later he modestly says he had 'care of the works' since Perry's death in 1733. The east pier was built from October 1730 to July 1731, exactly as the west pier. He then prepared working drawings for the sluice, built September 1731 to October 1733, undoubtedly in accordance with plans made earlier by Perry. The largest structure of its kind yet seen in England, it had massive abutments, with five lifting gates each 6 ft. wide, separated by a pier 58 ft. long by 24 ft. wide from the 40 ft. wide navigation opening with a pair of gates pointing upstream. The whole structure was founded on a timber grillage, on piles, with three rows of dovetailed sheet piling, and was designed for a head of 20 ft. of water on the sills.

By 1733 about £20,000 had been spent. There followed a pause while revenue from Passing Tolls on shipping could accumulate. In February 1738 Rubie gave evidence to Parliament, describing what had been done and estimating that a further £27,000 would be required to complete the project. The Commissioners received authorisation to go ahead. Work was resumed in 1739, the main tasks being to deepen the canal to a depth of 20 ft. below the level of high water springs and to build wharfing along its sides. Stephen West (q.v.) contracted for the wharfing: timber piles and planks supported by anchored land ties, the top being 3 ft. above HWST, forming a retaining wall 23 ft. high. By 1743 the wharfing had been brought 830 ft. from the piers along both sides, the canal having been widened to 200 ft. The expenditure on works since 1725 amounted to £29,700.

This figure comes from a report of December 1743 to the Lords of the Admiralty which shows that the total expenditure since 1723 was £40,000 including land purchase and interest on the capital borrowed to make up the shortfall of £10,000 on the £30,000 of revenue received. With such a large debt work again came almost to a halt while revenue accumulated. Rubie was kept on the pay roll, as the works had to be maintained but from December 1748 his salary was reduced to £30 p.a.

He had to look for other employment and this came in April 1749 when he was recommended to the Commissioners of Westminster Bridge by Charles Labelye (q.v.) as 'a water carpenter who had been several years employed in the Works of Dagenham Breach and Rye Harbour'. In his book on the bridge (1751) Labelye says he had known Mr. Rubie 'many years for a Man of strict Probity, and of Great Skill and Experience in all Work of this Kind'. The job was to construct the foundations for a new pier to replace the pier which had subsided. The old pier and the two adjacent arches were taken down by May. Work on the foundations started in July and by December some two hundred piles (12 in. square and 30 ft. long) had been driven and sawn off under water and the surrounding dovetailed sheet piling completed. The Commissioners noted in the Minutes that 'their Officers and Mr. Rubie ... had performed their Duty in this work with great Application and Judgement'. By February 1750 Rubie had completed the centring for the two new arches.

He then returned to Rye, but in July asked for leave of absence to go to Yarmouth till works at Rye 'shall again be taken in hand'. This was agreed, and the Rye Commissioners continued to pay him £30 p.a. salary as a retainer,

The job at Yarmouth arose from a report by Labelye in January 1748 and a successful application to Parliament for funding. The newly appointed Commissioners at their first meeting in July 1750 wrote to Labelye desiring him to inform Rubie (the Engineer recommended by him) that they wished him to attend them on 23 August, and to be in Yarmouth some days before. Rubie duly attended and was appointed as Director and Supervisor of the work on the south pier in the manner proposed by him, if approved by Labelye, and that he be allowed £200 p.a. His foreman would receive 1 guinea and his clerk £1 per week.

In October Rubie produced a model and plan for the pier, which had been approved by Labelye, and the Commissioners ordered that the work be put into execution. This involved a rebuild of the outer part of the pier on the south side of the harbour 30 ft. wide now 'much decayed' which was to be made solid with new piles, planking and cross ties and a mortared chalk infill. It reached into water at least 6 ft. deep at low tide (12 ft. deep at high water springs).

Contracts were placed in February 1751 for Riga timber, chalk, and oak piles. The carpenters received 13s and the labourers 10s per week. Good progress had been made by the winter of 1753, when Rubie died. The exact date of his death and his age have not been found, and his burial is not recorded at Yarmouth; presumably he was taken, or had returned, back home to London.

His son, John, took over his post at Rye for a few years. Work there resumed in 1757 under Robert Cooper (q.v.) and the harbour was opened to shipping in 1762. At Yarmouth the pier was completed in 1755 by the same Stephen West who had worked on the wharfing at Rye and who was appointed engineer at Yarmouth from December 1753. Another Edward Rubie became assistant to Thomas Yeoman (q.v.) on the River Lea Navigation works in 1769. He moved to this job from Portsmouth.

A. W. SKEMPTON

[Rye Harbour Commissioners Minute Book (1724–1762).East Sussex RO; E. Rubie (1738) Evidence on Rye harbour, *JHC*, **23**, 38; C. Labelye (1751) *Description of Westminster Bridge*; Yarmouth Port and Haven Commissioners Minute Book (1750–1756) Norfolk RO; J. Smeaton (1763) *Report on the Harbour of Rye*; R. T. B. Walker (1979) *Old Westminster Bridge*]

Works

1725–1748. Rye New Harbour, 1/25–1/30 foreman, 1730–1748 engineer, £30,000
1749–1750. Westminster Bridge, foundations
1750–1753. Yarmouth Harbour, south pier extension

RUDYERD, John (fl. 1703–1709), silk merchant and engineer of Eddystone Lighthouse, was born in Cornwall. His parents were described as labouring class but Rudyard apparently found a benefactor in a Plymouth gentleman who acted as his patron and provided for his education. He was thus able to practice as a silk mercer in London, although never part of the Mercers Company.

After Winstanley's lighthouse on the Eddystone rock collapsed in 1703, Henry Whitfield's interest of 1696 was assigned to John Lovett. He employed Rudyerd as his designer and work began in 1706, following a new Act of Parliament. The Admiralty provided a ship to give protection against French privateers and also to save the workmen from the naval press gangs. There is no indication of how Rudyerd developed his design, although he was helped in its execution by Messrs. Smith and Norcutt, shipwrights of Woolwich Dockyard.

The new lighthouse was circular in shape, with a basement of oak and granite ingeniously keyed down to the rock, this time with thirty-six iron stakes instead of the twelve used by Winstanley. The light was shown on 28 July 1708 but the building was not completed until 1709 at a cost of £10,000. It was a tall, slender structure, with a uniform taper from base to the light. Its construction involved 500 tons of stone, 1,200 tons of timber, 80 tons of iron and 35 tons of lead, etc.

When occasional repairs were required, the Admiralty placed the resources of the Plymouth dockyard at the disposal of the Trinity House, or of the lessees. This second lighthouse came to a sudden end by fire on 2 December 1755, the lantern catching fire from the candles at 9am. The flames were seen from Plymouth about noon and assistance sent at once, but by then it was too late. The three lightkeepers were taken off by a fishing boat at 1.30 p.m., one dying from his injuries.

A. B. GEORGE

[DNB; J. Smeaton (1791) *A Narrative of the Building of Edystone lighthouse*; M. M. Oppenheim (1968) *The Maritime History of Devon*, 96–98; Smeaton's Tower, *Maritime History*, **V**(2) (1977) 136–137]

Works

1706–1709. The second Eddystone lighthouse, £10,000

RUSSELL, Sir Henry (1783–1852), was a civil servant and resident at Hyderabad. While living there Russell is alleged to have designed a remarkable multiple arch masonry buttress dam, *c*. 1806, the earliest of the type. The dam is up to 40 ft. high and 2,500 ft. long, laid out in a long curve, with the wall divided into twenty-one semicircular arches varying in span from 70 ft. at the edges to 147 ft. near the centre. The arches are supported on buttresses 42 ft. long and 24 ft. thick. The reservoir capacity was *c*. 8,000 acre-ft., and took its water from the river Issa to supply Hyderabad.

Russell was resident at a time when French influence was ended and the actual design may have been by either a French or English military engineer, most obviously Thomas de Havilland (q.v.).

MIKE CHRIMES

[E. Wegman (1922) *The Design and Construction of Dams*, 7th edn.; H. M. Wilson (1904) *Irrigation Engineering*, 452–453; N. A. F. Smith (1971) *A History of Dams*, 189–191]

Works

c. 1806. Meer Allum dam, Hyderabad

RUSSELL, Richard (fl. 1685–d. 1700), was Surveyor of the South Level, part of the Great Level of the Fens known as Bedford Level. When Ralph Peirson (q.v.) retired in April 1685 Russell and William Gallaway were appointed jointly in his place, Russell with £70 p.a. and Gallaway with £30 p.a. Peirson had been responsible for rebuilding in brickwork the old timber-framed sluices at Denver, 1682–1684, but the adjacent navigation sluice (Sasse), also timber framed and dating from 1653, remained. In November 1685 Russell reported it to be in bad condition. Temporary works were carried out to ensure its safety and finally a completely new structure was built 60 ft. east of Peirson's sluice. It was to have two openings each 15 ft. wide with two pairs of mitre-gates, the abutments and central pier to be in brickwork with stone quoins and splays. The 'sea doors' pointing downstream were to be 18 ft. high, rising 1½ ft. above high water spring tides, and the 'ebb doors' 10 ft. high.

Bedford Level Corporation account books give details of payments for 'the New Sluices at Denver under the care of Richard Russell Esq.' who directed work from the start in May 1699. Materials, purchased by the Corporation, include £195 for bricks and mortar, £78 for Ketton stone and £152 for timber. Sums paid for 'day work', not by contract, include £587 for bricklayers and labourers under supervision of Russell's site assistant, Walter Frost; £55 to the master mason, William Hanning; £340 for timber work by the carpenter, William Newling; and £53 iron work by Emanuel Hall.

Most of the brickwork and masonry had been finished by November, as well as three of the sea doors. Russell was ordered to get the fourth one finished and all four hung as soon as possible to prevent damage should the cofferdam fail in winter floods. The whole work was completed shortly before Russell died in July 1700. Payments for June and July were handled by Gallaway. The total cost, finally settled in December, came to just over £3,000. The sluices, which came to be known as 'Russell's two Eyes', were converted to a single sluice and a navigation lock in the rebuild of Denver Sluice by John Leaford (q.v.) in 1746–1750.

Russell was buried on 8 July 1700 at Ely St. Mary. He left no will and administration was granted to his widow, Susanna. An inventory of the 'Goods and Chattles of Richard Russell Esq. of Ely' exists. Dated October 1700 it values his estate (exclusive of land and buildings) at £275. The house comprised the Hall (main living room), two Chambers (bedrooms), Kitchen, Bakehouse and two Cellars. In the Yard were stables and a barn, three horses, a cart and a small 'chairiott'. Among the silver were seven forks and a dozen silver-handled knives. The Russells were clearly gentry. He seems to have married in middle age; three children of 'Captain Richard and Susan Russell' were baptised between 1691 and 1696, two at St. Mary and one at Ely Cathedral. He may have been a retired army officer or perhaps a Captain in the militia.

A. W. SKEMPTON and MARGARET FARRAR

[Bedford Level Corporation Orders and Accounts (1685–1700) and Inventory of Richard Russell, Cambridgeshire RO; Thomas Badeslade (1725) *The History of the ... Port of King's Lynn ... and the Draining of the Great Level*; John Leaford (1740) *Some Observations ... on the South Level of the Fenns*]

Works

1699–1700. Denver Sluice, eastern part, £3,050

RUTT, William (1784–c. 1836), engineer and surveyor, was interested in surveying, drainage, and navigation. He was resident in Watlington (1792–1798), Sotwell, Berkshire (1813–1815), and in Hackney, London, in 1826. In that year he was elected Secretary of the Institution of Civil Engineers but resigned after two years because his residence was distant from the Institution. He was elected an Associate of the Institution in 1828 and remained a member until 1836. He worked with Edward Kelsey (1784–1839?) from 1784 to 1804, Rutt and Kelsey providing surveys for Robert Mylne's (q.v.) report of 1802 on three new cuts for the improvement of the navigation of the Thames above Oxford. His plan for the Mersey and Irwell Navigation may indicate an association with Thomas Lingard (q.v.).

TESS CANFIELD

[Bendall; J. G. Watson (1988) *The Civils*; Skempton]

Publications

1802. R. Mylne, *Report by Mr. Mylne, Engineer, on Three New Cuts, for the Improvement of the Navigation of the River Thames, above Oxford*, plans, surveyed by Rutt and Kelsey

1825. *Plan of the connection of the Maesey and Irwell Navigation with the River Mersey near Runcorn showing the proposed improvements* in *Proceedings of the Committee of the House of Commons on the Liverpool and Manchester Railroad Bill*, Session 1825, plan drawn by Wm. Rutt of Hackney

S

SANDBY, Thomas (1721–1798), Architect of the King's Works, was born in Nottingham, the son of Thomas Sandby, gentleman. He and his younger brother, Paul, were trained by Thomas Peat, a local land surveyor, before obtaining posts in the drawing office at the Tower of London about 1741. From 1743 Sandby was with the Duke of Cumberland, the Commander-in-Chief of the British Army, and accompanied him in the campaign against the Jacobites in Scotland, and subsequently on the continent. With the Duke's patronage, possibly in 1746, he became Deputy Ranger of Windsor Great

Park, which provided him with an entry into landscape gardening, and subsequently architecture.

At Windsor *c.* 1748–1752 he was involved with work at Virginia Water, then the largest artificial lake in the country, over 1½ miles in length. Much of the work was carried out by troops, disbanded after the Culloden campaign and transferred by Cumberland. Sandby's personal role is unclear; Henry Flitcroft was carrying out a lot of work in the park, and is generally credited with the design of the timber 'High' bridge there. With Sandby's lack of experience it is unsurprising that the main dam, of sand and clay, failed in a storm in 1768, earning him the nickname 'Thomas Sandbank'. Although attempts were made to repair it, these were not immediately successful, and it was not until 1782 that a full reinstatement, or rather reconstruction began. It was a much more ambitious project, and several hundred men were involved, including some soldiers. The main dam was 30 ft. high and 200 yd. long. Work was carried out with advice from Charles Cole. Associated with the lake were an enormous rock cascade, various grottoes, completed in 1785, and a stone bridge, replacing the earlier work by Flitcroft. Sandby's designs came under much scrutiny, and the design of the penstock was entrusted to a Colonel Dansey due to concerns about Sandby's engineering competence. Completion of the clay cut off wall was delayed until 1789, due to financial shortages.

Although Sandby drew up designs for a bridge at Somerset House his only other executed design was Staines Bridge for which he was appointed architect in 1792. The bridge was opened in 1797, the contractor being Stephen Townesend, the son of John Townesend, the contractor for Maidenhead (1772–1777) and Henley (1782–1786) bridges. Despite their experience settlement soon threatened the bridge, which was initially rebuilt in cast iron by Thomas Wilson (q.v.); it was not until George Rennie (q.v.) was called in that a satisfactory solution was found to the problems of the site.

Despite his lack of success in the civil engineering field Sandby prospered as an architect with the support of Cumberland, his successors and George III. In 1777 he became (joint) Architect of the King's Works, subordinate to James Adam (q.v.) and in November 1780, Master Carpenter, on the Board of Works. He was one of the original Members of the Royal Academy and its first Professor of Architecture where despite his lack of practical success his inaugural lecture included designs for a 'bridge of magnificence'. He died at his home, the Deputy Ranger's Lodge on 25 June 1798 and is buried in Old Windsor Churchyard.

His brother, Paul's, career paralleled his in certain respects. He maintained the Scottish connection acting as Chief Draughtsman on the military survey of Scotland under David Watson (q.v.) from 1747 to 1751, returning in 1753 to reduce the immense map to a scale of 12,000 ft. to an inch. He worked with Thomas at Virginia

Water where he etched eight plates for the prospectus, but his engineering interests remained theoretical, culminating in his appointment as Chief Drawing Master at the Royal Military Academy from 1768 to 1797. He was succeeded in the position by his son Thomas Paul, who had married Thomas's daughter, Harriet.

MIKE CHRIMES

[DNB; Colvin (3); T. Ruddock (1979) *Arch Bridges and their Builders*; R. A. Skelton (1967) The military survey of Scotland, 1747–1755, *Royal Scottish Geographical Society*, SP, 1; R. South (1983) *Royal Lake: the Story of Virginia Water*; J. Roberts (1997) *Royal Landscape: the Gardens and Parks of Windsor*; IGI www.familysearch.org]

Works
1748–1752. Virginia Water
1782–1790. Virgina Water, reconstruction and extension
1783–1788. Great (High) Bridge, Virginia Water, five stone arches, replaced in 1820s
1792–1797. Staines Bridge, demolished 1798–1799

SANDYS, William (*c*. 1607–1669), Member of Parliament and river engineer-undertaker, was the second son of Sir William Sandys of Fladbury, Worcestershire, and his wife Margaret Colepeper. Many of the family were distinguished in public life and his cousin Sir Miles Sandys (1563–1644) joined the Earl of Bedford as one of the original Adventurers in draining the Great Level of the Fens. William followed this tradition, as MP for Evesham and JP for Worcestershire, but is best remembered for his work on the Warwick Avon navigation from Tewksbury to Stratford; he was called 'Water-work Sandys, from his taste in improvements for that kind'. He was educated at Oxford, entering Gloucester Hall (later Worcester College) in June 1623, aged sixteen, followed after three years by a studentship in the Middle Temple. In 1633 he married Cecily, daughter of Sir John Stede of Kent. They had two sons and a daughter.

In January 1636 he received from the Corporation of Stratford 'our approbation, commendation and allowance' for making the Avon navigable to that town, and on 9 March 1636 the King granted Letters Patent to 'William Sandys of Fladbury, gentleman [who] had undertaken at his own costs to make the River Avon ... passable for boats of reasonable burthen from Severn near Tewksbury unto or near the City of Coventry, and hath already been at great charge in his endeavours to bring that work to perfection ... and [he] shall have the benefit of the water-carriage as in such cases is usual'. On the same day, the Privy Council appointed a Commission to take evidence and fix reasonable rates for land purchase. The Commission met at Evesham in August. They agreed on the price to be paid for land for the towing path, reported that there were to be thirteen cuts to 'sluices' between Tewksbury and Stratford, and recommended the price to be paid

for this land (which was not to exceed an acre in any one place) and for trees which must be cut down.

A contemporary account of the undertaking by the Worcestershire antiquary, Thomas Habington (d. 1647), reads in part as follows: 'The Avon never bore a boat of any burden before industrious Mr. Sandys beginning his unexpected design in March 1635 [1636 NS] did in three years make it passable for barges of 30 tons from the mouth thereof where it enters Severn, near Tewkbury, to Stratford-on-Avon ... purchasing with excessive charge mills, meadows, and other grounds to cut in some places a course for this watery work, to have a way through the firm land beside the main channel; and towards the accomplishing hereof he made sluices at Tewksbury [and eleven other named places], and so wrought by his sluices keeping up the water in summer ... He erected also locks and placed many weirs in the quickest streams. Neither meaneth Mr. Sandys to set at Stratford the bounds of his labours, but intendeth (if it may be) to extend the same up to Warwick. And for expense which he hath hereupon bestowed, it cannot be valued so little as twenty thousand pounds'.

All thirteen 'sluices' (Habington forgot to mention one at Stratford) were situated in new cuts at the existing watermills. Most of them would have been flash locks, similar to those on the upper Thames, but at least four were pound locks (one at Cleeve and three below Evesham at Chadbury, Wrye and Pershore) with unusual diamond-shaped lock chambers designed to minimise erosion of the turf sides by water discharged from the gate paddles. There were also two stanches at shallows downstream of Fladbury and Pershore mills. It is doubtful if Habington was correct in saying that 30 ton barges could travel to Stratford; navigation above Evesham may have been possible only for smaller vessels at that time. And he might have exaggerated the cost.

By 1639 Sandys had done as much as he could on the river. In 1640 he became MP for Evesham but was expelled in 1641 as a monopolist for charging a duty of a shilling per chaldron of Newcastle coal, although power to do so had recently been granted by Letters Patent. As a result he lost the navigation and William Say, one of his sureties, took it over. He probably did little to improve the river during the Civil War, but later replaced the remaining 'sluices' between Evesham and Tewksbury by pound locks, thus completing a seven-lock navigation in this 26 mile length. As a regicide MP, however, Say lost his lands and the river at the Restoration in 1660; they were forfeited to the Crown and granted to James, Duke of York.

Since 1636 Sandys had been living at Pershore. Following the death of his father in 1640 (the eldest son having predeceased him) he conveyed the manor of Fladbury to his sister, Jane, and her husband. During the Civil War he engaged in purchasing arms at Dunkirk and shipping them to the Royalist army in England. After the execution of

Charles in 1649 Sandys travelled widely in Europe raising funds to finance the new (exiled) King's expedition to Scotland. On the Restoration he returned home, was appointed a JP for Worcestershire in July 1660 and re-elected MP for Evesham in 1661; he remained active in both capacities for the rest of his life.

Sandys hoped to regain the Avon navigation and fulfil his original intention of extending it up to Coventry (or Warwick), but nothing came of this. By an Act of 1662 he and two relations as partners became undertakers of the river Wye navigation from Monmouth to Leominster. The chief difficulties were the swiftness of the stream and the large weirs (one at least was 11 ft. high) supplying power for mills and iron forges along the river. Sandys made new cuts with pound locks to provide passage for boats past the weirs. However, the scheme was not a success and by the end of the century the locks had fallen into decay. Later efforts were made, at considerable cost, but the navigation was never perfected.

As for the Avon, the trustees in 1664 assigned the river to Thomas, Lord Windsor. He in turn granted shares in the upper part (above Evesham) to a syndicate including Andrew Yarranton (q.v.) and it was the latter who replaced five 'sluices' between Evesham and Stratford by pound locks, and added another. The river now had seven locks in the upper, as well as seven in the lower part, and the whole length of 43 miles could be navigated by 30 ton barges.

Sandys became one of the most active committeemen in the Cavalier Parliament. He was concerned with seven river navigation projects and twelve Bills for land reclamation and drainage, among many others. He was appointed to his last committee on 24 November 1669, a month before he died. It is known that he left some shares in the Avon navigation, but his will has not been found.

A. W. SKEMPTON

[Thomas Habington (c. 1640) An account of Mr. William Sandys making the River of Avon navigable, printed by John Amphlett (1899) A Survey of Worcestershire, 2, 468–469; T. R. Nash (1781) Collections for the History and Antiquities of Worcestershire; O. Moger (1913, 1924) Victoria County History of Worcestershire, 3, 354, and 4, 178; W H Willan (1936) River Navigation in England 1600–1750; C. Hadfield and J. Norris (1962) Waterways to Stratford; E. Rowlands (1983) William Sandys, The House of Commons 1660–1690, 3, 390–391]

Works

1636–1639. River Avon navigation, Tewkesbury to Stratford, 43 miles, thirteen 'sluices', of which at least four were pound locks, and two were stanches; probably less than £20,000

SAVAGE, James (1779–1852), architect, was born in Hackney on 10 April 1779. He was a pupil of Daniel Alexander (q.v.) and worked under him as Clerk of Works at London Docks. In 1798 he became a student of the Royal Academy.

While Savage's career would now largely be considered that of an architect, for much of his early career he was involved with civil engineering. He won second prize in 1800 for improvements to the City of Aberdeen and in 1805 won the prize for a design for rebuilding Ormonde Bridge, Dublin. Although this was not carried out, his design for Richmond Bridge there in 1808 was successfully implemented and he assisted George Knowles in the construction of Lucan Bridge, the largest single span masonry arch in Ireland. This was followed by another successful design for a three span masonry arch bridge over the Ouse at Tempsford, Bedfordshire. Another unsuccessful design, for London Bridge (1823), resulted in Savage issuing a pamphlet highly critical of the Rennies' (q.v.) design. Savage had also written an essay on *Bridge-building* for the London Architectural Society all indications that at this stage in his career he was taking bridge design very seriously. He used his method of proportioning the arches and determining the lines for London Bridge for his design for St. Luke's, Chelsea (1820–1824), a scholarly work of gothic construction. Henceforth it was as an architect, particularly of churches, that he was chiefly engaged. He did, however, in 1825, prepare plans for a 'Surrey Quay' along the Thames from London Bridge to Lambeth, and in 1835 entered the competition for the new Houses of Parliament.

One of his pupils was Albinus Martin (q.v.) who assisted him with plans for the completion of Richard Trevithick's (q.v.) Thames Tunnel, c. 1814. Savage was elected a Member of the Institution of Civil Engineers in 1828, while resident at 31 Essex Street, Strand and resigned in the late 1830s. He was President of the Surveyors' Club in 1825, Vice-President of the London Architectural Society, a founder member of the Graphic Society, a Freeman of the City of London and member of the Skinners Company. He was, briefly, a Fellow of the Institute of British Architects but resigned over their regulations. He was for many years chairman of the Society of Arts Committee on Fine Arts. He was much employed in legal cases and provided much support to Henry Peto over the Custom House affair (1827–1830). He died at North Place, Hampstead Road, London on 7 May 1852 and is buried in St. Luke's, Chelsea.

MIKE CHRIMES

[ICE membership records; *Builder*, **x** (1852) 377; *Civil Engineer and Architects Journal*, **xv** (1852), 226; T. Ruddock (1979) *Arch Bridges and their Builders*, 196; R. C. Cox and M. Gould (1998) *Civil Engineering Heritage: Ireland*; Colvin (3); Bendall]

Publications

1810. Bridge building, *London Architectural Society, Essays*, **2**, 119–167
1823. *Observations on the Proposed new London Bridge*

Works

1813–1816. Richmond Bridge, Dublin
1814. Lucan Bridge, County Dublin
1815–1820. Tempsford Bridge, Bedfordshire
c. 1825. Crown Estate, near Reading, two road bridges

SAVERY, Thomas, FRS, Captain (*c.* 1650–1715), military engineer and inventor, was probably born in Shilston, Devon, around 1650, the son of Richard Savery. The family was well established in the area of Totnes. In 1691 Savery was commissioned as an ensign in the Duke of Bolton's Regiment of Foot. Despite his age he served in the wars against the French in the Low Countries, and learnt sufficient Dutch to be able to translate Koehoorn's work on *Fortification* in 1705. This work he dedicated to Prince George of Denmark who sponsored him in the Office of Treasurer to the Sick and Wounded the same year. Savery had been promoted to Trench Master in 1696 and Captain in 1702.

Savery is best known for his role in the development of the steam engine. His first invention was for millwork to polish glass and stone, for which he first petitioned for a patent in June 1694. At the end of 1695 he was granted a patent for this and 'for rowing ships with great ease and expedition'. In November 1697 he petitioned for a patent for 'a new invention for raising water ... by the impellent force of fire'.

Interest in the potential of using steam to raise water is evident in England from the start of the seventeenth century; Saloman de Caus, David Ramsay, Edward Somerset (Marquis of Worcester), and Sir Samuel Holland were all active in this field. Although these early experiments can be characterised as curiosities, there was a growing need for a reliable alternative to existing pumping equipment, particularly as the number and depth of mines grew, as well as the desire of landed gentry to ornament their estates with fountains and cascades, and existing systems of town water supply were enhanced. Water and wind power were the usual motive forces and engineers at the close of the seventeenth century, like George Sorocold and John Hadley (qq.v.), developed better engines and pumps.

In the second half of the seventeenth century there was increasing awareness of the potential of atmospheric pressure in pumping, using cylinders and pistons. Robert Hooke (q.v.) was demonstrating the possible use of steam to Fellows of the Royal Society in the 1660s. The Huguenot, Denis Papin, came to England in 1675; his investigations of steam included, in 1695, a demonstration of an engine with a cylinder partially filled with water which was then boiled, forcing a piston and rod attached to it to the top of the cylinder, air being driven out via a non-return valve. The rod was held in position by a catch, the fire removed, as the cylinder cooled the steam condensed, a vacuum was created, and when the catch was released the piston was driven down by atmospheric pressure. Despite his involvement with the Royal Society he never fully developed his invention.

In 1698 Savery was granted his patent; originally for fourteen years the patent was extended by Act of Parliament for a further twenty-one years, down to 1733. In Savery's engine, steam was passed from a boiler into a closed vessel (receiver) filled with water; steam pressure forced the water through a non-return value and up an ascending pipe. When the water had to be expelled, the steam supply was shut off and a cock on a pipe from the main delivery pipe was opened, releasing cold water which poured over the exterior of the receiver, cooling it and condensing the steam. The vacuum thus created forced water up a suction pipe and through a second non-return valve into the receiver by atmospheric pressure. Once the receiver had been refilled the cycle was recommenced. When Savery demonstrated his invention to the Royal Society in June 1699 he used a pump with 2 receivers with the aim that one could always be under steam pressure, although complicated to operate. The engine was further developed by using a sector valve before his work, *The Miner's Friend,* was published in 1702, describing his pump. The sector valve was a feature of most eighteenth century steam engines.

Savery had a workshop in Salisbury Court, off Fleet Street, where he advertised the manufacture of his pumps. Unfortunately, although they were used for private water supply, as at Campden House, Kensington, they were impractical for mines, the most pressing need. His attempts to introduce a higher pressure steam engine for York Building Waterworks failed.

Despite the limitations of Savery's device, John Theophilus Desaguliers (q.v.) persevered with improvements including internal water injection for cooling the receiver. Later in the century Joshua Wrigley (q.v.) and others used Savery-type pumps to raise water into a cistern to drive waterwheels.

Following his 1705 appointment, Savery was obliged to travel around the country, an itinerary which included Dartmouth; this would have provided an opportunity for him to meet Thomas Newcomen (q.v.). Evidence suggests that Newcomen had already made much progress in developing a practical steam engine but there were strong arguments for sheltering under Savery's extended patent protection, giving the 'Proprietors' as they became known, plenty of opportunity to recoup their investment. Savery, with the help of contacts in the City of London, set up an unincorporated joint stock company to exploit steam engines, initially with sixty shares, many of which he sold. In 1716, after his death, Newcomen was granted a further twenty shares.

Savery was dismissed from his post as Treasurer to the Sick and Wounded in 1713 but, with the patronage once more of Prince George, he became Surveyor to the Waterworks at Hampton

Court. He proposed a £1,000 scheme, using a waterwheel to improve the 'several fountains' there.

He died on 30 May 1715, leaving a wife Martha, but no children.

MIKE CHRIMES

[Royal Engineers archives, Chatham; An Account of Mr. Tho Savery's Engine for raising water by the help of fire (1699) *Phil Trans*, **21**, 225; R. Bradley (1/18) *New Improvements of Planting and Gardening*, 75; R. S. Hills (1989) *The Stationary Steam Engine*; A. Smith (1994–1995) Engines moved by fire and water, *Trans Newc Soc*, **66**, 1–25; L. T. C. Rolt and J. S. Allen (1996) *The Steam Engine of Thomas Newcomen*; DNB]

SCHALCH, John Augustus, Major (1793–1825), military engineer, was the son of Andrew Schalch, a Captain in the Royal Artillery. His family, of German Swiss origins, had a long tradition of service with the British monarchy and his great uncle, Andrew Schalch (1692–1776), had been master founder at Woolwich Arsenal. Schalch, like his brother, Philip, joined the Bengal Infantry as an Ensign in 1809, after training at the Royal Military College, Marlow, in 1807/8; his poor health had curtailed his studies there. Soon after his arrival in Bengal, with (Sir) George Everest's support, Schalch was attached to surveying duties, initially at Etawah and then at Murshidabad (1813). On the latter survey he so pleased his commanding officer, Colonel Fleming, with his competence as a surveyor that he wrote 'I really think you might pick and choose out of any thousand men (old and young) in the Service, and could not have lighted on one that from all appearance would, or could, have been more agreeable, not only to me in a public point of view, but to us as an intimate of our Family ...'. Schalch came fully equipped with his own theodolite, which he clearly mastered, and his competence was no doubt enhanced by his interest in astronomy, with which he occupied his leisure time.

Schalch continued with his surveying work in the Sundarbans in 1814 before joining General Wood in the field at the end of the year, mapping his campaign on the Nepal Frontier in 1815. Further survey work and promotion followed until in 1820 he was appointed by the Lottery Committee to improve Calcutta's waterways. The committee, funded by lottery income, was an early improvement commission for the port, and Schalch's title was Superintendent of Canals and Bridges in Bengal. Schalch, now a Lieutenant, was given an allowance of 1,000 rupees a month to cover the cost of boats and other equipment, and Lieutenants Taylor and Prinsep were appointed his assistants. In addition to detailed surveying and levelling work, Schalch prepared a scheme for navigable canals between the Hooghly and waterways to the Sunderbunds, Tolly's Nullah being inadequate for the traffic. He proposed a canal from Chitpore, on the Hooghly,

to the old Eastern Canal, parallel with the Circular Road. Work was begun under Schalch and completed in 1831. Other canals, to Hoseinabad on the Jaboona River, were authorised in March 1823 but Schalch's proposal to use the 'Circular Canal' to take Calcutta's sewage was not acted upon, neither was his proposal for docks off Tolly's Nullah.

Schalch's canal work was associated with a number of bridge designs of which the most important was a bar chain suspension bridge over Tolly's Nullah at Kalighat (Khallee Ghat). It was 141 ft. long but only 8 ft. wide and was intended for pedestrian and bullock traffic. Although military engineers had made use of rope suspension bridges in India and elsewhere before this time, this appears to have been the first iron suspension bridge to have been designed by the British in India. The design showed that Schalch was familiar with the work of Samuel Brown (q.v.) on the Union Bridge over the river Tweed.

Schalch's civil engineering work was interrupted by the outbreak of war with Burma in 1824. Major Schalch, as he now was, was put in charge of a Corps of Pioneers and Surveyors. He made use of some of the boats he had acquired for service in Calcutta, including the *Pluto*, one of the earliest steamboats to be used in India, which he had used as a dredger around Calcutta and converted for use as a floating battery.

Schalch was wounded on 23 February 1825 in action at Kiungpala and died on 25 February 1825. His work on the canals of Calcutta was largely completed by Captain Thomas Prinsep (1800–1830). Schalch's surveys were widely admired and remained standard sources into the second half of the nineteenth century.

MIKE CHRIMES

[Iron bridges of suspension in India, *Asiatic Journal*, **15** (1822) 60–61; Suspension bridge to be erected over Tolly's Nullah, *Asiatic Journal*, **16** (1823), 347–349; H. Douglas (1832) *An Essay on the Principles and Construction of Military Bridges*, 340–345; E. W. C. Sandes (1935) *The Military Engineer in India*, **2**; *Selections from Bengal Government Papers* (1905) *Canals 1865–1904*, BL IOLR; R. H. Phillimore (1950) *Historical Records of the Survey of India*, **2**, 440–441; **3**, 12–14, 495–496, 500–501]

Publications

1823(?). *Plan for Opening Water Communications from Calcutta to the Upper and Eastern Provinces of India*
1825. *Plan of the City of Calcutta and its Environs*

Works

1822. Kalighat suspension bridge, Tolly's Nullah
1822–(1825). Circular Canal, Calcutta,
1823–(1825). Hoseinabad Canal, Calcutta

SCHNEIDER, Gaulterius, Captain (1772–1841), military engineer, was born at Jaffna in 1772, the son of Lieutenant Johan Hendrick Schneider of Kirchheim, and was serving in the Dutch military as an engineer when the British took over Sri Lanka. In 1810 he was placed in charge of the civil engineer and surveyor general's office, a position he held until the arrival of F. B. Norris in 1833. Among his reports were those on the Wanny Tanks (reservoirs) in 1807, on the Galle, Matara and Hambuntota Districts in 1808, and the first map of the island produced under British rule. Schneider's predecessor had been George Atkinson (fl. 1802–1811) who had assisted Joseph Joinville, civil engineer, 1802–1805, before taking over from him. Atkinson did one of its earliest reports by a British engineer on irrigation, on Multuragewalla, in 1802. The first civil engineer under the British was Gerrit Joan Nagel (1799–1805) also a Dutch engineer.

MIKE CHRIMES

[P. M. Bingham (1921–1923) *History of Public Works Department, Ceylon*]

SCOTT, Matthias (fl. 1762–1783), was a surveyor of works. In November 1761 Joseph Nickalls (q.v.) was dismissed from his post as assistant or resident engineer on the Calder & Hebble Navigation under John Smeaton (q.v.). To take his place John Gwyn (q.v.), already on site as a foreman, was appointed superintendent of carpentry and Matthias Scott of masonry and digging; their appointments date from March 1762 at annual salaries of 50 guineas and 60 guineas respectively, they to work under Smeaton's direction. Work had started early in 1760. By July 1764, after an expenditure of £56,000, the navigation was open from Wakefield to Brighouse, a distance of 18 miles with seventeen locks. Not a great deal remained to be done, but disputes arose. A new management committee appointed James Brindley (q.v.) on 31 January 1765 to complete the work on a revised plan and Smeaton was discharged, along with Gwyn and Scott.

In 1765 and 1766 work began in accordance with plans drawn up by Smeaton for the drainage of Potteric Carr, an area of 4,250 acres of wetland south of Doncaster. Records of the first stage of operations no longer exist. Nevertheless it is almost certain that Thomas Tofield (q.v.) directed the works and, possibly on Smeaton's recommendation, Scott was employed as Surveyor. Certainly he succeeded George Forster as Surveyor of Works on the nearby Hatfield Chase in September 1771 at an annual salary of £40, and when work resumed on Potteric Carr in April 1772 he took on a part-time appointment there at 10 guineas p.a. His first task was to design the sluice at the outfall of the main drain into the river Torne.

He could combine the jobs as works on Hatfield Chase were mostly routine maintenance. However, repeated flooding of the Torne where it crossed the Chase (downstream of Potteric Carr) caused much trouble, and after years of

delay the Participants (proprietors) of the Chase decided to take action. They called on Tofield for advice. He reported in September 1773. Next month the Participants sought to employ Scott full-time at £70 p.a. from May 1774 'provided he gives up his farm at Wellingley (6 miles south of Doncaster near Tickhill) and the place he holds under the Trustees of Potteric Carr'.

Scott duly resigned his post in April 1774, to be succeeded by Henry Cooper, and by 1775, if not earlier, he moved to Thorne. Tofield proposed diverting the Torne outside the Chase to the river Don near Thorne. As an alternative Thomas Yeoman (q.v.) in December 1774 proposed a new cut for the drainage water to an outfall into the Trent 4 miles north of Althorpe where the drains and the Torne had their combined outfall. This idea was further investigated by Scott in a report of October 1775. Both schemes were logical but suffered from the disadvantage of requiring the purchase of a considerable amount of land outside the Chase. Then in October 1776 Smeaton reported in great detail on the Torne problem. His proposals represented a major improvement. Work began on parts of his scheme but two modifications were made before the matter was satisfactorily resolved. In the first of these, proposed by Scott, the northern drain was taken to a new outfall at Keadby, a mile north of Althorpe. The necessary Act was obtained in March 1783, with Scott giving evidence both in the Commons and the Lords, before he retired in June that year.

The work was carried out by his successor, Samuel Foster, appointed in August 1783 at the same salary of £70 p.a. It was Foster who put into effect the second modification, namely the provision of separate outfalls at Althorpe for the Torne and the southern drain, which became possible after making the Keadby drain. All these works were completed by 1789.

A. W. SKEMPTON

[Potteric Carr Trustees Proceedings (1772–1778) Potteric Carr Drainage Board, Bawtry; Court of Sewers for Hatfield Chase Record Book (1771–1789) Nottingham University Library; C. Hatfield (1973) *Canals of Yorkshire and North East England*; A. W. Skempton (1981) *John Smeaton*]

SCOTT, Sir Thomas (c. 1535–1594), a leading Kent landowner, MP and commissioner of Dover harbour, was the eldest son of Sir Reginald Scott of Scot's Hall and his wife, Emmeline, daughter of Sir William Kempe. Educated at the Inner Temple, Scott inherited large estates from his father in 1556. He became a JP in early 1560s, at about the time of his marriage to Elizabeth, daughter of Sir John Baker of Sissinghurst, and was knighted in 1570. He was elected MP for Kent in 1571, became Sheriff in 1576 and Deputy Lieutenant of the County before 1582. In March of that year he was appointed a Commissioner of Dover harbour. A month later the commission decided to go ahead with a plan for reconstructing the harbour, largely in accordance with a scheme proposed by William Borough (q.v.). Thomas Digges was appointed General Surveyor and Scott became Treasurer for the first year; they reported to Francis Walsingham, Secretary of State, who controlled the project on behalf of the Privy Council.

An off-shore bar extended more than half a mile north from the harbour mouth to the shore beneath Dover Castle, enclosing a salt-water lagoon. The scheme involved transforming the northern part of the lagoon into a large basin, known as the Pent, by building a wall some 2,000 ft. long on the bar, joined at its southern end by a wall about 600 ft. in length running back to the shore, with a sluice near its landward end. The basin would fill at high tide and the water thus stored could be released through the sluice at low tide to scour the harbour (south of the cross wall) and its entrance, which had become choked with shingle.

The first step was to open a cut at the entrance to drain the lagoon as far as possible in order to facilitate construction of the walls. This was done by January 1583. But discussions were still taking place on methods of building the walls; some were too expensive, others not satisfactory. Scott intervened to suggest that the method long used for repairing the sea wall of Romney Marsh at Dymchurch provided an economical and proven solution to the problem. This comprised a bank of earth fill, tipped from horse-drawn carts, levelled in layers by workmen, and protected on its seaward face by faggots fastened down by short oak piles.

As a Commissioner of Romney Marsh, Scott was well acquainted with Dymchurch wall, and his cousin Reginald Scot (c. 1537–1599) was Surveyor of the Marsh. The question was settled at a meeting held in March 1583 by Francis Walsingham. The harbour commissioners, accepting the decision, announced on 10 April that work would begin on 13 May. This left little more than four weeks to allocate sites for collecting the earth and chalk for the bank fill (as recommended by Reginald Scot) and to send out proclamations inviting men to bring in horses and carts.

On the appointed day no fewer than two-hundred carts arrived. By the end of June the number had risen to more than 500, implying a workforce of at least 1,000 men and boys. From the start Scott directed operations and personally supervised work on the cross wall while Richard Barrey, Lieutenant of the Castle and another of the harbour commissioners, supervised the long wall. Scott captured the allegiance of the men in the task of creating the Pent from the sea. They worked with enthusiasm and almost military precision against time and marine erosion. Carts came from so far afield as Maidstone and Sevenoaks; in the early stages work could proceed only a low tide; the logistics must have been formidable. Yet the whole job, including a temporary sluice in the cross wall, was completed in three months by August 1583. The walls rose to a height of 3 ft.

above high tide level, the tidal range at Dover being 19 ft. at springs and 11 ft. at neaps.

Much remained to be done: a permanent sluice; foundations of an old pier removed for a new harbour entrance; and groynes to protect the seaward face of the shingle bar. All was complete by 1586, at a total cost of £12,000.

Scott remained active at Dover until 1585, having served a second term as Treasurer. In 1586 he was again elected MP for Kent. At the time of the Armada (1588) he was appointed chief of the Kentish forces which assembled at Northbourne near Dover; he equipped four thousand men himself within days of receiving orders from the Privy Council. Renowned for his hospitality and public spirit he died on 30 December 1594, aged about fifty-nine. The eldest son Thomas (c. 1563–1610) inherited Scot's Hall and seven manors, leaving the property to his younger brother Sir John Scott (1570–1616).

A. W. SKEMPTON

[J. Summerson (1982) Dover harbour, *King's Works*, **4**, 755–764; M. R. Pickering and P. W. Hasler (1982) Sir Thomas Scott, *The House of Commons 1558–1603*, **3**, 355–358]

SEAWARD, John (1786–1858), civil engineer, of the Canal Iron Works, Limehouse, London, was born in Lambeth, London, son of a successful builder. He was educated at the Mansion House, Kennington, in classics and mathematics. He worked first for his father as architect/surveyor, followed by several years with Messrs. Doulle, Government contractors, as surveyor and works manager. Seaward directed the building of Vauxhall Bridge for Messrs. Grillier and Co., who had established an early cement works at Millwall. He then went to Wales for several years to manage lead mines, where he gained considerable knowledge of chemistry. He became highly knowledgeable about the machinery and processes of Welsh lead mining, and knew Arthur Woolf and Richard Trevithick (qq.v.), with whom he maintained a correspondence. He returned to London and superintended the construction of several private docks on the river Thames, including Gordon's and Dowson's. He became agent for the Gospel Oak Iron works in Staffordshire and, simultaneously, for the Imperial and Continental Gas Company. During this period he provided lighting for several towns and cities, including some in France, Belgium and Holland. In 1823 he submitted a design for a new London Bridge of three arches of 230 ft. span, with a treatise on the feasibility of constructing it. In his 1828 paper on iron roofs, Seaward claimed to have constructed many iron roofs. An advocate of mixed cast and wrought iron, he describes a 43 ft. span iron roof he built about 1819, and a 50 ft. diameter gasholder of about 1823. In 1830, he was one of the judges of the second Clifton Suspension Bridge Competition.

In 1824 he established the Canal Iron Works, Millwall, Poplar, London, for the construction of every sort of machinery. The following year, he was joined by his younger brother, Samuel (q.v.). They designed and built large swing bridges, hoisting shears, dredging machines, cranes and equipment for docks in London, Liverpool, Southampton, Lowestoft, and for foreign ports, and produced equipment for mills of various sorts. Seaward and Co. provided the swing bridge for Thomas Telford's (q.v.) St. Katharine's Dock (1827). Their outstanding expertise was in the building of marine engines. In 1829 they helped form the Diamond Steam Packet Company, building many of the boats which ran between London and Gravesend. In 1836, Messrs. Seaward and Co. introduced the direct-acting engines known as Seaward's Engines or Gorgon engines, first used in the *Gorgon* and the *Cyclops*. Large savings of fuel resulted from the use of double slide valves for both steam and exhaust. They went on to build engines for 24 of the British Navy's largest ships, and for many smaller ones. They introduced tubular boilers to the Royal Navy, and many other improvements. Their business extended to East India Company, steam navigation companies and foreign companies and governments. In the late 1830s the works was employing 1000 men.

After the death of Samuel Seaward in 1842, the elder brother continued to run the Canal Iron Works until his own death on 26 March 1858, at 20 Brecknock Crescent, Camden Town, London. Between 1819 and 1846, Seaward obtained eight patents alone and three jointly with his brother for improvements to steam engines and propelling machinery, The 800 hp engines for the *Amazon*, built in 1851, were a notable achievement which he intended as his memorial, but the ship was lost on her first passage.

In 1826, John Seaward joined the Institution of Civil Engineers as a Member, in 1829 taking the place of John Isaac Hawkins (q.v.) on the Council. Until his last years he was an active participant at meetings and wrote papers on professional subjects.

TESS CANFIELD

[*On the Construction of Iron Roofs* (1828) ICE OC133, ICE Archives; *Explanation of the Mechanical Power employed in Propelling the Model Coach exhibited at the Scientific Repository in Adelaide Street* (1833) ICE OC 95, ICE Archives; ICE membership records, ICE Archives; J. G. James Collection, ICE Archives; Obituary (1858–1859) *Min Proc ICE*, **18**, 199–202; Obituary (1858) *Civ Eng Arch J*, 28 March, 153; Obituary of William Morris, *Min Proc ICE*]

Publications

1823. *Observations on Suspension Chain Bridges; with an Improved Method of forming the Supporting Chains or Rods* ... (reprint from *Phil Mag*, **62**, 1823)

1824. *Observations on the Rebuilding of London Bridge*

1829. *Observations on ... Employing Steam Power in Navigating Ships between this Country and the*

East Indies (1829, 1834) London (with S. Seaward) (extract in *Mech Mag*, **21**, 1843, 298–303)

1836. On procuring supplies of water for cities and towns by boring, *Trans ICE*, **I**, 1836, 145–150

1837. *On Steam Communication to India; with a Comparison of Two proposed Routes, by the Red Sea and by the Cape of Good Hope*

1837. *Description of the Engines on board the Gorgon and Cyclops Steam Frigates, with remarks on Comparative Advantages of Long and Short Stroke Connecting Rods, and Long and Short Stroke Engines*

1846. (Seaward and Capel) *Copy of Letter to the Hon. H. L. Corry MP on the use of High Pressure Steam in the Steam Vessels of the British Navy*

1858. (Fairbairn, Forrester, J. Seaward & Co.) *Steam Navigation, Vessels of Iron and Wood ...*

Works

1827–1828. St. Katherine's Dock, cast iron swing bridge, £4,540

SEAWARD, Samuel, FRS (1800–1842), engineer, was born in Lambeth London, younger brother of John Seaward (q.v.). In 1814 he entered the service of the East India Company as a midshipman. After two voyages to India and China, he returned to London and was placed by his brother as an apprentice with Henry Maudslay (q.v.). After five years with Maudslay, Seaward joined Messrs. Taylor and Martineau and worked in Cornwall under Arthur Woolf (q.v.) erecting several large pumping engines. He then became superintendent of part of the Harvey's Hayle Foundry, where he worked with Richard Trevithick (q.v.). In 1825 he returned to London to join his brother at the Canal Iron Works, Limehouse, manufacturers of steam engines and machinery. Seaward was particularly interested in the adaptation of direct action engines to marine purposes and, jointly with his brother, made great advances in steam powered navigation. He held four patents relating to steam engines and navigation, and another three jointly with his brother.

He became an Associate of the Institution of Civil Engineers in 1828 and was a Member of Council in 1841 and 1842. He was awarded a Telford Medal in Silver in 1842 for his paper on auxiliary steam power for sailing vessels. He was also a Fellow of the Royal Society. He died on 11 May 1842 at Endsleigh Street, London, aged only forty two.

TESS CANFIELD

[*Calculations, Tenders and Memorandums* (c. 1835) mss, ICE archives; J. G. James Collection, ICE archives; Obituary (1843) *Min Proc ICE*, **2**, 11–12; DNB]

Publications

1824. Description of an improved gauge for ascertaining with precision the pressure of highly compressed steam gasses and fluid bodies ... on a hydro-pneumatic pump, *Phil Mag*, **63**, 36–60

1824. Description of a hydro-pneumatic pump for compressing gasses and other elastic fluids, *Phil Mag*, **64**, 12–17

1824. On a hydro pneumatic pump, *Phil Mag*, **64**, 441–442

1829. Observations on...employing steam power in navigating ships between this country and the East Indies (1829, 1834) London, (extract in *Mech Mag*, **21**, 1834, 298–303) (with J. Seaward)

1841. Remarks on the comparative advantages of long and short connecting rods and long and short stroke engines, *Min Proc ICE*, **6**, 53–55

1841. Upon the application and use of auxiliary steam power for the purpose of shortening the time occupied by sailing ships upon distant voyages, *Min Proc ICE*, **1**, 63–67

1841. Supplementary account of the use of auxiliary steam power, on board the 'Earl of Stonwick' and the 'Vernon' Indiamen, *Min Proc ICE*, **1**, 129–130

1842. Memoir on the practicability of shortening the duration of voyages by the adaptation of auxiliary steam power to sailing vessels, *Trans ICE*, **3**, 385–408

SEED, William (fl. 1775–1837) was a canal contractor who was also successful in the early days of railway construction. A man of this name is mentioned in the Leeds & Liverpool Canal minutes in 1775, probably a minor employee rather than a contractor. No mention of the name has been found thereafter until 1809 when he, or possibly a son, was described as a canal cutter while being awarded a contract on the Brecknock & Abergavenny Canal, where William Crosley (q.v.) was Engineer. Seed subsequently won substantial contracts on two more of Crosley's projects, the Lancaster and Macclesfield Canals. On the second of these he also found time to contract to rebuild a bridge over the river Goyt for the County of Cheshire.

The contracts on the Macclesfield Canal were taken in the name of William Seed and Sons. Presumably Benjamin Seed was one of the latter, as he appears alongside William in two early railway contracts. While working on the St. Helens and Runcorn Gap they were living at Prescot, from which they moved directly to a section of the Grand Junction Railway in north Cheshire, including the Vale Royal Viaduct. Other sections of this line were taken by the new breed of larger contractors, such as David McIntosh, James Trubshaw, Thomas Townshend, William Mackenzie and Thomas Brassey (qq.v.), and nothing more is known about William Seed after this contract was finished. However John Errington, who had been the Resident Engineer there, subsequently employed Benjamin on the Glasgow, Paisley & Greenock Railway to supervise part of the work by direct labour.

P. S. M. CROSS-RUDKIN

[RAIL 220/23, 593/1, 812/4, 844/232 and 846/2, PRO; Lancaster Canal records, ICE; QJB 4/5, Cheshire RO; in lit. Graham Cousins; inf. ex David Brooke]

Works

1809–1812. Brecknock & Abergavenny Canal, Mamhilad to Pontymoile, 2½ miles, contractor, £17,500

1814–. Lancaster Canal, cutting at Farleton, contractor, £24,854

1826–1829. Macclesfield Canal, Marple to Lyme Hanley (3 miles), Tytherington to Sutton (3 miles), contractor

1829. Haigh Bridge, Cheshire, £650 plus?

1830–1833. St. Helens & Runcorn Gap Railway, lot 2, contractor (with Benjamin Seed), £24,700

1834–1837. Grand Junction Railway, Dutton–Wharton section, contractor (with Benjamin Seed), £92,066

SEMPLE, George (c. 1700–1782), builder and engineer, was born in Dublin around 1700. His earliest known work is the steeple of St. Patrick's Cathedral, completed in 1749. In May 1751 he was building by contract St. Patrick's Hospital in Bow Lane and several houses in Dublin, when he was asked to construct a timber deck across part of Essex Bridge over the river Liffey, the bridge having failed four months previously. This he did in July of that year and was then induced to consider rebuilding the entire bridge in stone. He searched for information on bridges in his own 'fine and valuable collection' of books and travelled to London to purchase more. There he saw the newly completed Westminster Bridge by Charles Labelye (q.v.).

On his return to Dublin, Semple made twenty-eight borings at the site of Essex Bridge. He found rock at depths of 22 ft. to 26 ft. below high water level, covered by 10 ft. to 15 ft. of sand, much of which was loose and, in places, a little gravel. He modelled his design closely on Westminster Bridge but nevertheless departed from this precedent in one important respect, namely, to build the piers on piles driven to rock within cofferdams instead of founding the piers by the caisson method, as Labelye had done.

In view of the ground conditions, this was the correct decision, but Semple could find no details of cofferdam construction. He visited London again in May 1752, met Labelye, ordered various pieces of equipment and a quantity of *tarras* and then went to Ramsgate, where William Etheridge and Thomas Preston (qq.v.) were building the massive harbour piers. Neither they, nor Labelye, could help him.

Returning to Dublin, Semple heard that a bookseller's son was in Paris and earnestly requested that he procure at any expense all the books and plans that could help in the matter of laying foundations in deep water. One of the books brought back was the very recently published fourth volume of Belidor's *Architecture Hydraulique*, which, to Semple's delight, contained illustrations of the cofferdams used in building the bridge at Orleans. 'My drooping spirits were instantly revived (he writes) and I immediately went on with my Work with Vigour. and entertained the most sanguine Hopes of Success.'

Progress thereafter was rapid. Semple had designed his bridge with five semi-circular arches of 36, 41, 46, 41 and 36 ft. spans, with piers 8 ft. to 9 ft. thick, and an unusually large overall width of 51 ft. He now undertook to complete the bridge within two years at a cost of £20 500. He built a pile-driver and a pumping engine. Demolition of the old bridge began on 19 January 1753. The first piles of the north cofferdam were driven a month later and the bridge was open to the public on 10 April 1755, the final cost being £20 660. To prevent scour, Semple provided a masonry 'pavement', not less than 5 ft. thick, right across the river bed. Essex Bridge remained in use until replaced in 1863–1874 by Grattan Bridge, although the Engineer to Dublin Port, Bindon Blood Stoney, incorporated the original foundations into the new structure.

Soon after completing the bridge, Semple's health broke down and he was obliged to retire to the country. He recovered to finish St. Patrick's Hospital in 1757. He received a gratuity of £500 from the Irish Parliament in 1761. The following year, he presented to the Ballast Office (the forerunner of the Dublin Port & Docks Board) a set of eight charts of Dublin Bay. In these he 'collected and carefully laid down some of the most authentic surveys of this harbour for some hundred years past' and added his own ambitious proposals for developing the mouth of the rivers flowing into the bay, and a plan for a harbour at Dun Laoghaire.

Later, when in his seventies, he wrote his *Treatise on Building in Water*, one of the classics of eighteenth century civil engineering literature, giving an account of his work at Essex Bridge and, amongst other things, a plan for a harbour in the South Downs in Sussex.

Semple died about 1782 and was at that time in receipt of a pension of £100 p.a. from Parliament.

RON COX

[J. P. Griffith (1880) The lowering and widening of Essex Bridge, *Trans ICEI*, **13**, 34; J. C. Kinoulty (1970) *A Biographical Dictionary of Ireland from 1500*, unpublished manuscript, Trinity College, Dublin, 759; T. Ruddock (1979) *Arch Bridges and their Builders*, 40–45; G. Daly (1993) George Semple's charts of Dublin Bay, 1762, *Proc R Ir Acad*, **93c**, 81–105; J. W. De Courcy (1996) *The Liffey in Dublin*, 354–356]

Publications

1776. *A Treatise on Building in Water*
1780. *Hibernia's Free Trade or Plan for the general improvement of Ireland, peculiarly adapted to a Free Trade …*

Works

1755–1757. Essex Bridge, Dublin, rebuilding, engineer/architect

SEPPINGS, Sir Robert, FRS (1767–1840), naval architect, was born in 1767, the fourth child in a family of six. His father, also Robert, was a cattle dealer in Fakenham, Norfolk. In 1782 his uncle, Captain Thomas Milligen, placed him as an apprentice shipwright with John Henshaw, then Master Shipwright at Plymouth Dockyard. Henshaw, later knighted, was Surveyor of the Navy, 1784–1806, and a powerful supporter of Seppings.

On completion of his apprenticeship in 1789 Seppings was steadily promoted, becoming Second Assistant to the Master Shipwright in the dockyard 1797–1803. From his subsequent career one can deduce that these promotions were based on merit, and an inventive approach to dockyard work.

In 1803 he was awarded the Society of Arts Gold Medal for his invention of 'Seppings's blocks', which he had developed in 1800. They comprised triple sets of adjustable wedges, which made it easier to lift ships in dry dock to facilitate repair, reducing the labour required from 500 to 20 men and the time involved by 30%. He was awarded £1,000, by the Admiralty, for this. The Navy Board calculated they had saved £11,000 in three years, but the savings in time were worth far more than that.

In 1804 Seppings was appointed Master Shipwright at Chatham and began to concern himself with developing a new method of ship construction to address a number of fundamental problems associated with traditional ship design in timber. Wooden sailing vessels were known to hog, viz. their ends began to droop relative to the midship section, after a period of service. This was, in part, caused by the relative buoyancy of the midship and end sections, exacerbated by the even distribution of armament in warships, and by the motion of the ships in water. The ships also faced the more obvious problem caused by the dynamic loads imposed by waves which had to be resisted by the hull and its internal structure. The result was for ships to lose their shape, for joints to open up, and to become less seaworthy, necessitating their being taken out of service for reconstruction. The problems were obvious, but the solution had not been found when the French Revolutionary and Napoleonic wars increased pressure on the navy to maximise their active service strength. Shortage of shipbuilding timber was an added complication.

The problem had been considered by shipwrights before, and aspersions had been cast more generally on British vis-à-vis French ship design. In 1791 and 1796 Gabriel Snodgrass, employed by the East India Company, suggested renewing old vessels by the addition of iron knees and cross bracing to increase the transverse strength, and fastening an outer skin planking on top of the existing planking for longitudinal strength. These ideas were adopted but when referred to Seppings in February 1805 he decided to try out his own ideas on the vessels, Orion and Kent, instead. Over the next six years he developed a system, first fully employed on the Albion in 1811, and, for a new vessel, the Howe, in 1815.

Seppings's system involved filling the spaces between the transverse frames up to the orlop deck with old timber, inserting an internal trussed frame the length of the vessel, replacing transverse beam knees with shelf piece and waterways to form an inserted near horizontal ring, and replacing the fore and aft deck planking with planking running at forty-five degrees to the centre line. Of these innovations the most famous was the longitudinal trussing was first fully developed on the Tremendous when it was repaired in 1810. The Admiralty had intervened, through Charles Yorke, to overrule the Navy Board and have Seppings's trussing used for the 74s, Tremendous, Ramilles and Albion.

There followed something of a dispute between Seppings and Thomas Young (q.v.) recorded in successive papers in the *Philosophical Transactions* of the Royal Society where Young attempted a theoretical analytical approach to the problems of ships' strength, hogging and transverse strength. Although Young ultimately came down in support of Seppings's system, Seppings and his supporters felt he had to justify himself, and also disprove allegations of plagiarism, which tended to detract attention from both the merits of his innovations, and Young's attempts to provide a theoretical basis for ship design. Seppings (1813) recognised his debt to Yorke, then First Lord. It is of some interest that Seppings claimed he had drawn his inspiration for the trussing from the Schaffhausen Bridge rather than from any British practice.

Seppings's work was accompanied with attempts to measure deflection, and in a trial on the *Justica*, about to be broken up, in 1817, temporary bracing placed in the opposite, compressive, direction to that usually adopted, still produced a reduced deflection, but confirmed Seppings in his view that his system was better suited as its members were in tension. These observations, although in the field of naval architecture, were among the earliest on such large structures and reflect Seppings's practical approach to engineering design. In modern parlance he developed a reinforced membrane capable of resisting shear forces, although much of the theoretical discussion was focused on beam behaviour.

Seppings's next involvement in structural design also stemmed from concerns about the longevity of ships, in this case the problem of dry rot which affected ships laying uncovered on building slips for long periods. At the end of the eighteenth century several continental navies were building ships under cover, although not specifically to deal with the problem of dry rot which was the most immediate British inducement. Samuel Bentham (q.v.) a Commissioner of the Admiralty saw such covered dockyards abroad and advocated them to the Admiralty in 1812, but the adoption of the idea of covered slips was generally the work of Seppings after Bentham's dismissal and Seppings's appointment as Surveyor of the Navy in 1813, although it is possible a small span roof had been built in 1812. By 1814, probably initially at Plymouth, Seppings had developed a roof design suitable for wide spans, which had been approved by John Rennie (q.v.). In form they were something of a hybrid—braced arch, portal frame or balanced cantilever—clearly rooted in Seppings experience as a shipwright, and probably evolved in a short time frame as a result of experience. By the early 1820s almost all of the dockyard slips had been covered, most in Seppings style roofs. The form was discussed by Thomas Tredgold (q.v.) in his work on carpentry. The roofs are of particular interest in view of their large spans—29 m at Plymouth and 30 m at Chatham—rarely exceeded before that time. Several of these have survived at Chatham Dockyard, an indication of the soundness of the system if properly waterproofed and maintained.

In 1811 Seppings introduced the standard round bows for large vessels. The eighty gun *Worcester* had a distinctive circular stern which Seppings had a part in designing in 1819, although others claimed priority for the idea.

Seppings's contributions to the navy were widely acknowledged at the time, and in 1819 the Finance Committee of the House of Commons remarked '... Your committee deem it their particular duty to notice Mr. Seppings, one of the surveyors of the navy, to whose abilities and exertions this country is mainly indebted for many of its most valuable improvements in naval architecture'. In 1827 he approved the refitting of older warships with elliptical sterns, after ideas developed in Bombay in 1819 for the Ganges, which used teak to resist dry rot; Richard Blake laid claim to the idea.

Following reforms of the administration of the navy and the transfer of direct control of the dockyards to the Admiralty in 1832 Seppings, then aged sixty-five, retired. He had been elected to the Royal Society in 1814 and knighted in 1819. In 1836 he was awarded an Honorary DLL by Oxford University for his contribution to naval architecture and education. In 1811 he had been influential in the founding of a School of Naval Architecture at Portsmouth, giving evidence to the Commission of Naval Revision (1806–1809). A model of his method of ship construction survives, which originated in the Admiralty collection of ship models which Seppings established, another example of his approach to design. He married his cousin Charlotte (d. 1834), and they had two sons, John Milligen, and Edward. He died at Taunton, where he had retired, on 25 September 1840.

MIKE CHRIMES and ROBERT BALDWIN

[YOR/16a, Yorke mss, NMM; ADM/BP/24A, 5 May 1840, and ADM/B217, 16 February 1805, NMM; ADM/2236, 5 January 1805, PRO; Seppings models, Science Museum and NMM; *Soc Arts Trans*, **22** (1804) 275–292; T. Tredgold (1820) *Elementary Principles of Carpentry*; J. Knowles

(1821) *An Inquiry into the Means which have
been taken to Preserve the Royal Navy from ... Dry
Rot*, annotated copy in ICE Library; J. J. Packard
(1978) Sir Robert Seppings and the timber
problem, *Mariner's Mirror*, **64**(2), 145–157; D. K.
Brown (1979) The structural improvements in
wooden ships investigated by Robert Seppings,
Naval Architects, 103–104; T. Wright (1981)
Thomas Young and Robert Seppings: science and
ship construction in the early nineteenth century,
Trans Newc Soc, **53**, 55–72; R. Morriss (1983)
*The Royal Dockyards during the Revolutionary
and Napoleonic Wars*, Leicester University Press,
24–25, 36–37, 144, 155, 159, 163, 222; J. Coad
(1988) *The Architecture of the Sailing Navy*; R. J.
M. Sutherland (1989) Shipbuilding and the long
span roof, *Trans Newc Soc*, **60**, 107–126]

Publications
1814. On a new principle of constructing His
Majesty's ship of war, *Phil Trans*, **54**, 285–302
1817. On the great strength given to ships of war
by the application of diagonal braces, *Phil Trans*,
57, 1–8
1820. On a new principle of constructing ships in
the mercantile navy, *Phil Trans*, **60**, 133–143

Works
c. 1814–1821. British naval dockyard slip roofs,
in timber, *c*. 100 ft. spans, at least thirty-six roofs
at Woolwich, Chatham, Portsmouth, Devonport,
Sheerness, Pembroke and Deptford

Note: The Science Museum has a fine collection
of Seppings's models—perhaps rather more revealing than the National Maritime Museum collection. Both have published catalogues

SHEASBY, Thomas, Sr. (*c*. 1740–1799), civil engineer and contractor, was probably christened
on 28 October 1740 in Tamworth, Staffordshire.
Described as a builder of Tamworth, he carried
out bridge repair work for the Warwickshire
Quarter Sessions from 1775–1787; he contracted
to design and build Polesworth Bridge in 1776
and Duke's Bridge, Coleshill, in 1780. His subsequent career, in its early phase, was closely intertwined with that of the Dadford family (q.v.), as it
was with that of his son, also Thomas (q.v.).
Sheasby Sr. was a contractor in the late 1780s
on the Birmingham–Fazeley Canal. There was
some dispute over the Coventry Canal's completion to Fazeley and in June 1785 Sheasby was
awarded this contract; Thomas Dadford Sr. reported on the aqueduct he was building over the
Tame. Work was more or less completed by 1789
and Sheasby and Thomas Dadford Sr. decided to
tender for work on the Cromford Canal. This
work was short-lived as in 1790 they received a
more tempting offer of work on the Glamorganshire Canal. On 30 June 1790 Sheasby and
Thomas Dadford and his son tendered to build
the canal for £48,258, giving a bond for £10,000.
There was no engineer and the contract was run
by a management committee, an arrangement

which may have contributed to subsequent difficulties between the contractors and the company.
The canal was opened in February 1794, but
not before the contractors had accumulated over
£17,000 in payments for extra work. Unfortunately, there was a breach in the bank and the
contractors were called back to repair it. They
refused to act without a cash advance and the
company had Sheasby and Dadford Sr. imprisoned so as to recover the £10,000 surety, arguing
that they had already been overpaid £17,000.
Although subsequently Robert Whitworth (q.v.)
awarded them all but £1,500 of the extras, the imprisonment meant that both Dadford and Sheasby
had their next projects interrupted and the next
phase of the Glamorganshire Canal was built
under Patrick Copland.
Thomas Dadford Jr. had avoided arrest as,
shortly after they had begun work on the Glamorganshire Canal, he had been asked to survey the
Neath Canal and had been appointed as its engineer and contractor once it got its Act in 1791.
By mid-1792 he had completed work as far as
Ynysbwllog Aqueduct when an offer to build the
Monmouthshire Canal enticed him away. Sheasby
was taken on as engineer and contractor to complete the canal to Glynneath, including the aqueduct, by 1 November 1793 for £14,886; £2,500
was to be retained for three years. Sheasby was
unable to complete on this schedule but as discussions proceeded to resolve this problem in
1794 he was arrested on the Glamorganshire
writ and the company had to complete the outstanding work.
Despite these problems, Sheasby was soon at
work again, assisting Charles Roberts (q.v.) as engineer on the Swansea Canal, to be built by direct
labour. Sheasby had made the original survey in
1793 and might have been appointed engineer
from the start had it not been for his other problems. From 1796 he became engineer, with his
son assisting him, and the canal was partially
opened in 1796 and completed in 1798. After his
death in 1799 he was succeeded by his son as engineer to the company. Aside from his contribution to the construction of canals in South Wales
he was also responsible for a number of surveys,
including those for the Shropshire Canal (1788),
for a Brecon Forest tramroad and a Llandeilo–
Llandovery canal project in 1793.

MIKE CHRIMES

[Quarter Sessions records, Warwickshire RO;
RAIL 869/1, PRO; H. W. Paar (1963*) The Severn
and Wye Railway*; C. Hadfield (1967) *The Canals
of South Wales and the Borders*, 2nd edn.; C. Hadfield (1969) *The Canals of the East Midlands*;
Paget-Tomlinson]

Works
1776. Polesworth Bridge, £364
1780. Duke's Bridge, Colehill, £306
1786. Birmingham–Fazeley Canal, Aston Junction, Minworth, contractor

1785–1789. Coventry Canal, Athestone–Fazeley, including Tame Aqueduct
1786. Birmingham Canal, Narn, Farmer's Bridge, Digbeth
1789–179(1). Cromford Canal, contractor, with Thomas Dadford
1790–1794. Glamorganshire Canal, contractor, with Thomas Dadford, c. £65,000
1792–1794. Neath Canal, completion, contractor, £14,886
1795–1799. Swansea Canal, (assistant) engineer

SHEASBY, Thomas, Jr. (fl. 1794–1847), civil engineer and contractor, was christened on 1 April 1766 in Tamworth, Staffordshire. Like his father, Thomas (q.v.), his career was intertwined with that of the Dadford family (q.v.).

Sheasby's career began by helping his father on the Swansea Canal and he acted as its engineer from 1794 until c. 1802, when it was open and trade was expanding rapidly. He then acted as engineer to the Warwick and Birmingham Canal until June 1804. He is next heard of in 1810 when he was appointed resident engineer for the construction of the Aberdare Canal. The canal had been authorised in 1793 and surveyed at various times by John and Thomas Dadford and Edward Martin (qq.v.) but construction had been postponed while work progressed on tramroad connections. The canal was completed by August 1812 but in February 1811 Sheasby had been put in charge of work on the Severn and Wye Railway, variously as manager, clerk and engineer, succeeding Roger Hopkins (q.v.).

The Severn and Wye Railway had its origins c. 1799 when the idea of a tramroad in the Forest of Dean coalfield was first raised. Benjamin Outram and John Rennie (qq.v.) were consulted before the Act for the Lydney and Lidbrook Railway was obtained in 1809, its name changed to the Severn and Wye in 1810. In addition to its rail route it included the Lydney Canal and harbour.

The company found it difficult at first to obtain an engineer with all the various attributes required and problems were experienced with the work of the contractors who included James Barnes (q.v.). Financial problems were evident by the time Sheasby arrived, but he proved a vigorous manager and engineer. Eventually twelve branches were built in addition to the main line, most under Sheasby's charge, linking the many small collieries and foundries in the Dean coalfield. As well as supervising construction, Sheasby managed the traffic and investigated rail quality. By the time of his retirement due to ill-health in 1847, the tramroad was suffering from competition from the main line railway system. Despite this, in association with the Dadfords, the Sheasbys made a major contribution to opening up the industry of South Wales and the Borders by building its infrastructure. This involved approximately £500,000 of work, more than 70 miles of canals, and one-hundred and fifty

locks, with the canals rising up to five hundred above sea level.

MIKE CHRIMES

[H. W. Paar (1963) *The Severn and Wye Railway*; C. Hadfield (1967) *The canals of South Wales and the Borders*, 2nd edn.; C. Hadfield (1969) *The Canals of the East Midlands*; A. Faulkner (1985) *The Warwick Canals*]

Works
1795–1799. Swansea Canal, assistant engineer
1799–1802. Swansea Canal, engineer
1801–1804. Warwick and Birmingham Canal, engineer
1811–1847. Severn and Wye Railway works, engineer, 3 ft. 6 in. gauge, 28 miles, c. £75,000
–1813. Lydney Canal, wide lock and sea gates, £20,000 total cost
–1811. Ivy Moor Head branch
1811. Brookhall ditches branch, 800 yd.
1811–1814. Bishopswood extension, 1 mile
1812. Churchway branch completion, 1½ miles
1812. Birches branch
1812. Milkwall branch
1814. Monmouth Railway, branch, 1 mile
1819. Darkhill furnace extension, 300 yd.
1821. Churchway branch, relaid
1824. Brookhall ditches, relaid
1824–1826. Parkend branch and cut and cover tunnel
1837–1841. Foxes Bridge, extension, £930
1841. Wimberg Slade branch, 1 mile
1841–1844. Kidnall's Mill branch, 2¾ miles, £3,570
1845–1846. Foxes Bridge extension

SHERES, Sir Henry, FRS (1645?–1710), military engineer, was the son of Henry Sheres of Deptford, a naval captain. Following the Restoration Sheres obtained office at Court and in a diplomatic capacity accompanied Edward Montague, the first Earl of Sandwich, the ambassador, to Spain in March 1666. Prior to his return to England, in 1668, Sandwich was asked to report on the state of affairs at Tangiers. Sheres drew up a 'plat' of the fortifications for which he was subsequently paid £100. On his return to England in 1667 Sheres became friendly with Samuel Pepys, then Treasurer to the Tangiers garrison, and his wife.

At Tangiers an enormous breakwater was under construction by Sir Hugh Cholmley (q.v.) and others. Work had begun in 1663 using *pierres perdues* (loose boulders) as a foundation for an upright masonry sea wall. An alternative method of construction, employed in the mid-seventeenth century by Italian engineers at Genoa when building another large breakwater, involved sinking large open timber chests which were subsequently filled with masonry to make up the breakwater. Cholmley felt this approach was unsuitable for the rough conditions and high tidal range off Tangier, however practical it had proved at Genoa. From the start there had been supporters of the Genoan method. Signor Jacomo, a Genoan, had offered to carry out the work in 1664 but his offer was not accepted.

By 1668 the project was experiencing a variety of problems. Cholmley's two original partners had died and family circumstances had meant he had been unable to remain in Tangier constantly to supervise the project. Progress had been hampered by the slow release of funds, by the military situation there and at home, and by the engineering difficulties. In these circumstance Sheres's arrival was timely. He became, for reasons not entirely clear, the advocate of the 'chest' party. In April 1669 he was sent out to Tangier as a Clerk of Works and in August he visited Genoa. Shortly after this, on 27 August 1669, an order was finally made to replace the original contractors of the mole, and appointing Cholmley as Surveyor-General, Major Samuel Taylor, who had been in charge in Cholmley's absence, Deputy Surveyor, and Sheres as Clerk Examiner, at a salary of £250 p.a. (Taylor returned to England in 1670 due to ill health.) The work was now effectively a direct labour project. In Cholmley's absence (he did not return to Tangiers until April 1670) Sheres had been developing the idea of using Genoan-style chests in the mole and this had become a definite proposal by 10 March 1670, and had the support of the new Governor, Lord Middleton.

After months of preparation the first chest was ready to be placed in September 1670. For a period of two weeks attempts were made to position it but weather conditions made this very difficult. When it was originally positioned it slipped and was not got into position for several days. The actual work was supervised by Chomley, Sheres having taken to his bed, probably due to the stress. Despite the costs and the problems encountered, the chest method was persevered with and slowly the mole advanced.

In April 1670, 380 yd. of the mole had been completed; by the end of February 1673 it was 440 yd. long. At this stage the weather intervened and the winter of 1674–1675 caused several breaches. Cholmey offered to carry out the mole a further 100 yd. and make good the repairs in six years. This was poorly received by a Government hard pressed for cash and in June 1675 he offered to carry out the mole to 500 yd., and complete a south east return of 100 yd. in four years, for £30,000 p.a. Cholmley's proposal was turned down, as was another by another consortium including Thomas Fitch (q.v.), and Sheres offered to do the same work as Cholmey for £10,000 less.

Sheres took over as Surveyor-General in June 1676 and in the following sixteen months placed eleven chests. By November 1677 the mole had still only reached 457 yd. in length. Although critical of Cholmley's early work, Sheres was experiencing similar problems with payment and supplies, and in 1680 work was effectively halted by the 'Great Siege' by the Moors. Despite the conclusion of peace Charles II, embattled at home, was determined to cut costs as much as possible, and have troops loyal to him at home.

Paradoxically Sheres had a hand in the report of 1683, which sealed the mole's fate, exaggerating its costs and problems of siltation and navigation within the harbour. (All charges Sheres confided to Pepys he could refute.) The overall cost of maintaining the colony was estimated at nearly £4.8 million. At the time of its demolition in 1683 the mole was 479 yd. long, on average 110 ft. wide, and 18 ft. above low water. Despite Sheres's criticism of the early work this proved hardest to demolish. Demolition of the mole took around three months and was not complete until January 1684. Sheres himself took a leading role and displayed greater skills in mining than either the Master Gunner or Master of Ordnance.

The structure contained 2,843,280 cu. ft. of material weighing 167,250 tons. It is difficult to assess the relative importance of the roles of Cholmley and Sheres in what was one of the greatest English civil engineering achievements of the Stuart period. Something like £220,000 was spent under Cholmley to 1676. Sheres's accounts do not survive, but could only account at most for £78,000 of expenditure. Cholmley was, however, often in London, and the actual superintendence owed much to his assistants Taylor, Bolland, and Sheres.

In 1682 Sheres, at home to help resolve the issue of the Mole, first reported on the state of Dover Harbour, which he described as 'entirely useless, being filled and choked up with sand and mud, and... a bank of beach at the mouth of the harbour'. The condition of the port reflected the general lack of investment since Elizabethan times.

The Act which had provided maintenance revenue from 'passing tolls' was not renewed after 1610, although between 1605 and 1613 Samuel Elfreth, the master carpenter, 'brought the harbour to perfection', it having been found necessary to relay the sluice at the bottom of the Great Pent, and rebuild the parapet on Clark's pier. By the time of Charles II's accession, the Paradise Pent had silted up and its backwater was no longer effectively sluicing the harbour; storm damage had also affected the various harbour walls, which had been repaired only in a piecemeal fashion.

Following Charles II's Restoration, Bernard De Gomme (q.v.) had carried out some improvements between 1660 and 1662, with a new cross wall, built across the main harbour, of timber filled with chalk, and a new sluice. Following the completion of these works passing tolls were granted to Dover for eight years, but by 1670 when they ran out, no more than £9,000 had been raised which could only have covered the cost of routine works. In 1676 a Commission of Enquiry had been appointed to survey Dover harbour, with which Jonas Moore (q.v.) was involved, but no action was taken and so when Sheres reported in 1682 it had suffered from twenty years neglect.

Following his 1682 report, Sheres was ordered to return to Dover and make a plan for improvements, but on its receipt Charles remarked 'it was a noble project indeed, but that it was too big for his purse, and would keep cold'. Sheres was sent back to Tangiers and it was not until William III was on the throne that more was done. In 1698 the Admiralty was asked to report on the state of the harbour and Sir Cloudesley Shovel's report was submitted on 28 April 1698, recommending improvements to the piers and a more substantial cross wall and sluices; the cost was estimated at £7,850. These were relatively modest proposals but in the following year Sheres petitioned Parliament with his own proposals costing £30,000: £5,000 for dredging, £3,000 for tools, and £2,000 wages, and the remainder on works. In 1700 the passing tolls were renewed for nine years, and then for another nine years. Approximately £20,000 was raised and works began to Sheres's plans. They were, however, hardly successful as in 1718 a report by John Perry (q.v.) revealed many shortcomings and in 1723 two-thirds of the passing tolls revenue, still being renewed, was directed to Rye.

Sheres, as an officer of artillery, participated in the campaign against the Duke of Monmouth in 1685 and was knighted for his services. He succeeded De Gomme as Surveyor-General of Ordnance in 1685. Although professing loyalty to James II at the time of the Glorious Revolution, he was conveniently sick. Once William had established his position he made his peace with him, although replaced as Surveyor General in July 1689; he was arrested in June 1690, and again in 1696, on suspicion of conspiracy for the Jacobite cause. In 1700–1701 he acted as a Trustee for the regulation of Irish land grants, reporting to Parliament on this matter.

Sheres took a general interest in science, and was elected a Fellow of the Royal Society in February 1676. He wrote a number of works on scientific and naval issues including a transcript of the 1582 report on Dover harbour by Thomas Digges (q.v.), as well as poetry. He died on 21 April 1710. When he made his will in 1709 he referred to his 'poor fortune'.

MIKE CHRIMES

[Archives, Royal Engineers, Chatham; Sir C. Shovel (1698) *Report to Admiralty regarding Dover Harbour*; DNB; E. M. G. Routh (1912) *Tangier; England's lost Atlantic outpost 1661–1684*; J. B. Jones (1916) *Annals of Dover*; Q. Hughes and A. Migos (1995) *Strong as the Rock of Gibraltar*, 7–11; E. Chappell (1935) The Tangier papers of Samuel Pepys, *Publications of the Navy Records Society*, LXXIII; E. G. R. Taylor (1954) *The Mathematical Practitioners of Tudor and Stuart England*; R. Latham and W. Matthews (1976) *The Diary of Samuel Pepys*, **9**, 1668–1669; H. C. Tomlinson (1979) *Guns and Government, the Ordnance Office under the later Stuarts*; A. Hasenson (1980) *The History of Dover Harbour*]

Works
1676–1683. The Mole, Tangier, £78,000

c. 1700. Dover Harbour, works, £20,000

SHONE, Richard (fl. 1790–1820), surveyor of St. Asaph, was possibly a relative of Charles Shone, appointed one of the two supervisors of works on the Rhuddlan embankment designed by Thomas Wedge (q.v.) in June 1795. In October 1800 Shone himself was appointed to superintend the completion of the embankment to the west of Rhuddlan bridge at 1½ guineas a week. He clearly showed some ability as in February 1801 it was his design for an embankment along the Deserth River which was approved.

Shone appears to have subsequently taken up contracting on the scheme, in March 1805 taking on work on the Clawydd Llwyd sluice and embankment on the eastern side of the Clwyd and eventually completing most of it in 1807–1808 for £226. He seems to have been responsible for the report which formed the basis for the supplementary 1813 Act for completion of the Rhuddlan works, and set out the Parliamentary line for the Rhyl Marsh-Clwyd drain. By this time Shone had undertaken surveys for turnpike roads from Colwyn to Llandulas, from St. Asaph to Holywell (1811), from Ruthin to Holwell, and from Llandyrnoy Village to Brookhouse bridge (all with Huw Jones, 1819).

MIKE CHRIMES

[Quarter session records, QS/D/DJ/4, 8 and 10, Denbigh RO; Minutes of the Commissioners of Rhuddlan Marsh, DC/140, Flint RO]

SHORTGRAVE, Richard (fl. 1659–d. 1676), instrument maker and surveyor, was appointed 'Operator' or technician to the embryonic Royal Society at Gresham College in 1659 and assisted Robert Hooke (q.v.), the curator of experiments from 1662. He and Willaim Leybourn (q.v.) were quantity surveyors to the City of London after the Great Fire of 1666; in this capacity he was involved in the works of Sir Thomas Fitch (q.v.) on the Fleet Canal (1671–1674). In 1674 Shortgrave devised a new water-level in a closed tube fitted with telescopic sights and also undertook road surveys for John Ogilby's *Britannia,* published 1675. He died in October 1676.

A. W. SKEMPTON

[H. W. Robinson and W. Adams (1935) *The Diary of Robert Hooke 1672–1680*; T. F. Reddaway (1940) *The Rebuilding of London after the Great Fire*; E. G. R. Taylor (1954) *The Mathematical Practitioners of Tudor & Stuart England*]

SHOUT, Matthew (1774–1817), civil engineer, was the youngest child of Robert Shout Jr. (q.v.) and his wife, Mary Moorhouse; he was baptised on 6 October 1774 at Helmsley, Yorkshire. In 1781 the family moved to Sunderland when Robert Shout was appointed harbour engineer there; they lived in the Engineer's house on the Quay. Whether Matthew stayed on at Sunderland after his father's retirement in 1795 is uncertain, but in 1801 he was living at Blyth, Northumberland, and might have been working on Blyth harbour for several years. However that may be, in May 1801 he was appointed resident engineer on Scarborough harbour works under William Chapman (q.v.). He arrived in October to take up the post at a salary of £85 p.a., increased from January 1803 to £105 on Chapman's recommendation.

Work on the East Pier at Scarborough, a massive masonry wall founded at low water level on a mound of loose rocks, had been proceeding since 1746 according to plans by William Vincent (q.v.). From time to time several distinguished engineers were consulted regarding the best direction of the pier and on details of construction, but with little effect until Chapman reported in September 1800. By that time the pier extended 920 ft. out from the shore. Thereafter Chapman acted as chief engineer, but as resident engineer Shout carried responsibility for construction and supervision of the direct-labour workforce.

Meanwhile at Sunderland, Robert Shout's successor, Jonathan Pickernell Jr. (q.v.) was dismissed in May 1804 for alleged misconduct. The Commissioners of the River Wear and Sunderland Harbour advertised for a new engineer and ten days later appointed Matthew Shout. On leaving Scarborough, the harbour Commissioners recorded their 'thanks to Mr. Shout for his strenuous, efficient and indefatigable services'.

Shout took up his duties at Sunderland in June at a salary or £120 p.a. with free rent of the Engineer's house. A year later he received £80 in recognition of his diligence and from January 1811 his salary was increased to £300 p.a. It would have been soon after arriving in Sunderland that he married Jane, daughter of Thomas Dougal, a London merchant, and his wife, Louisa. Matthew and Jane Shout had two children: Louisa (b. 1805) and Robert (b. 1807); the latter died in 1824 as a midshipman in the East India service.

From the start, Shout realised that the existing harbour entrance was too wide, exposed as it was to all easterly gales. He therefore began extending the South Pier, from the point reached by Joseph Robson (q.v.) in 1762, in a more northerly direction. By the end of 1807 after little more than three year's work the pier had been extended in masonry construction for a length of about 350 ft. Moreover, in the same period he extended the North Pier by at least 100 ft. in timber framing filled with stonework up to high tide level.

Costs were obviously mounting. In May 1807 the Commissioners made preparations for a new Act enabling them to increase their revenue. Shout reported in June on proposed improvements, which the Commissioners accepted in part. But they decided to consult William Jessop (q.v.) on the whole question of future plans for the harbour. He reported in December, approving what Shout had so far achieved and making specific proposals for the directions and lengths of the piers. From 1808 work proceeded accordingly. By 1813 Shout had built a further

300 ft. of the South Pier in stone-filled timber framework and a masonry end 120 ft. in length. Also, an 8 hp bucket-ladder steam dredger was obtained from the Butterley Company, built in 1810–1811 to Shout's specifications at a cost of about £1,500. The total cost of the works in the period 1804–1816 is not known, but expenditure on the piers alone probably could not have been less than £15,000. Like all previous work at Sunderland, as at Scarborough, the piers were built by direct labour, the only contracts being for the supply of stone and other materials. In 1810 masons were paid £1 a week and labourers received 16s.

On leave of absence, Shout reported in 1809 and again in 1810 on rebuilding Hartlepool pier, the outer 200 ft. of which was in a ruinous condition. His masonry specifications probably reflect the practice at Sunderland: facing stones not less than 4 ft. on bed and 2 ft. between joints, laid in courses backed with rubble stonework, the upper 6 ft. of which was to be mortared. Work began in 1811, supported by local subscriptions. He visited the site again in 1812. Further funds were raised under an Act of 1813 and the work appears to have been completed soon after 1815 at a cost within the estimate of £3,500.

This was his only commission as an independent consultant, though he visited Aberdeen in 1814 to inspect the piers there. Had he lived longer other consultancy would no doubt have come his way, but he died on 14 March 1817, aged forty-two, and was buried in the churchyard of Holy Trinity, Sunderland, next to his father. He left no will. Probate on his estate, valued at under £3,000, was granted in 1818 to Jane Shout who, presumably with her children, was then living in Lambeth with her brother, Commander George Dougal, RN.

Matthew Shout's exceptional ability and devotion to duty were recognised at an early age by the Scarborough authorities. The chairman of the Hartlepool Commissioners in 1815 refers to him as 'a gentleman of great practical experience and acknowledged judgement' while the Wear Commissioners, in advertising for a successor, go out of their way to refer to 'Mr. Shout, their much respected Engineer, lately deceased'.

A. W. SKEMPTON and R. W. RENNISON

[*Scarborough Harbour Commissioners Minute Book* (1752–1833) Scarborough Ref. Library; *Commissioners of the River Wear and Port of Sunderland Minute Book* (1795–1820) Tyne & Wear Archives; C. Sharp (1816) *A History of Hartlepool*; J. Murray (1889) An account of the progressive improvements of Sunderland harbour and the River Wear, *Min Proc ICE*, **6**, 256–277; A. W. Skempton (1977) The engineers of Sunderland harbour, *Ind Archaeol Rev*, **1**, 103–125]

Works

1801–1804. Scarborough Harbour, resident engineer

1804–1817. Sunderland Harbour, engineer, extended south pier (780 ft.) and north pier (270 ft.), c. £15,000

1811–1815. Hartlepool pier, c. £3,500

SHOUT, Robert, Sr. (1703–1774),

was a stonemason and Surveyor of Bridges to the North Riding, Yorkshire. It is reasonably certain that he was born in 1703 the elder son of William Shout, mason, of Welburn near Malton in the North Riding who was born and brought up at Weston near Otley in the West Riding but on or about the time of his marriage, at York in 1702, to Mary Thompson of Welburn had settled there. He worked at nearby Castle Howard and at Hovingham Hall. Robert, presumably trained by his father, moved to Helmsley, in the same district, and married Margaret Ellis (1708–1780). Of their six sons, Robert (q.v.) became a mason and later a civil engineer, John (1738–1781) a mason, and James (1751–1780) a carpenter.

Nothing has yet been found on Robert Shout's early career before 1752 when he and Bryan Fenwick, also of Helmsley, contracted to build Newsham Bridge over the river Rye about 4 miles from Malton. It has three semi-circular arches of 22, 32 and 22 ft. spans high above normal river level to accommodate floods—the river has massive flood banks—and was completed in October 1753. Shout's next job, this on his own, was to build Scawton Bridge, also on the Rye and close to Rievaulx Abbey. The first payment was made in October 1755 and work reached completion in April 1757, costing £679. John Carr (q.v.) in his *Book of* [North Riding] *Bridges* says 'this bridge was planned and executed by Robert Shout'. It is almost identical to Newsham Bridge, with spans of 21, 31 and 21 ft. Both were therefore probably planned by him in the 'design and build' tradition of the master mason. They remain in good condition with the addition of a masonry pavement 'framed and sett' on the river bed at Scawton to protect the foundations from scour in floods.

In April 1757 Shout was appointed Surveyor of the Bridges in that part of the North Riding east of the Hambleton Hills, at a salary of £20 p.a.; John Wrather held the comparable post in the western part. As Surveyor, Shout would be responsible for advising on repairs and the need for rebuilding but was free to undertake work on contract. Thus in 1760 he contracted to rebuild two arches and one pier of the three-arch Kirkham Bridge on the Derwent, 4 miles south of Malton, for £450. In July of the same year he and John Bennison took on the contract for building Egton Bridge in Eskdale for £572. Completed in July 1761 this had three segmental arches of 24, 36 and 24 ft. spans. Carr notes that it was 'designed and executed by Robert Shout of Helmsley'.

In April 1762 an advertisement in the York newspaper announced that a meeting would be held in July to receive plans and estimates for a bridge over the Esk at Sleights. The contract, for £580, was awarded to Robert Shout the younger and William Ellis but the design is very similar to Egton Bridge and the first £50 payment was made to 'Robert Shout one of the Surveyors of the Bridges'. Sleights Bridge, with three segmental arches of 35, 45 and 35 ft. spans, was completed in October 1763. Both bridges survived for nearly two hundred years until severely damaged by a flood in July 1930.

In July 1766 both 'Bridge Masters', Shout and Wrather, were discharged, for no very obvious reason, and no new appointment was made until 1772 when John Carr became Surveyor of the Bridges in the whole Riding, at £100 p.a. Carr had now become the leading architect in Yorkshire and, like Shout, designed the new bridges in the Riding, either built by others or, in a few cases, by direct labour under the Surveyor's supervision.

This separation of function between architect or engineer and contractor is exemplified in Shout's last, and largest, job: the building of Stockton Bridge on the river Tees, designed by the engineer Joseph Robson with Robert Mylne (qq.v.) as consultant. It had five arches of 44, 60, 72, 60 and 44 ft. spans. The contract for £5750 was signed on 26 October 1763 by Robert Shout and Robert Nellson (masons) and Robert Corney (carpenter). Payments began in April 1764 and work reached completion in April 1769. In all, 547 men were employed, among them 79 masons (including Robert Shout senior and junior, John Shout and four members of the Nellson family), 32 carpenters, 3 blacksmiths and 416 labourers. The costs, settled in December 1770, came to £6,320 inclusive of fees and expenses, with a few later payments totalling about £150.

Finally, on 15 March 1774, Robert Shout the elder and Robert Shout the younger of Helmsley, masons, John Shout of Stockton, mason, and James Shout of Helmsley, carpenter, contracted to build the southern part of the Tyne Bridge, Newcastle, to plans by Robert Mylne. The northern part, with John Wooler (q.v.) as engineer, was built by other contractors. The work was carried out, but not by Shout himself. He died three months later and was buried at Helmsley on 11 June 1774. He had been churchwarden there in 1753 and as a juryman frequently signed the church accounts between 1735 and 1762. His will, dated 28 March 1774, leaves land and property in Easingwold and the house with 2 acres, in Helmsley, to his wife, Margaret, bequests of about £50 each to his sons, smaller amounts to two granddaughters, and the residue to his wife.

A. W. SKEMPTON and R. W. RENNISON

[*North Riding Quarter Sessions Order Books* (1731–1763) North Yorkshire RO; John Carr (c. 1800) *Book of Bridges*, North Yorkshire RO; *Stockton Bridge Order Book* (1762–1776) Durham County RO; *Tyne Bridge Contract* (1774) Special Collections, University of Durham; E. Jervoise (1931) *The Ancient Bridges of the North of*

England; J. H. Harvey (in lit. 1975) information on William Shout; Colvin]

Works

1752–1753. Newsham Bridge, designed and built (with B. Fenwick), £650
1755–1757. Scawton (Rievaulx) Bridge, designed and built, £679
1760. Kirkham Bridge, reconstruction, £450
1760–1761. Egton Bridge, designed and built, £572
1762–1763. Sleights Bridge, designed (built by R. Shout Jr. and W. Ellis), £580
1764–1768. Stockton Bridge, built (with R. Nellson & Corney), £5,750

SHOUT, Robert, Jr. (1734–1797), stone-mason and civil engineer, was baptised on 30 December 1734 at Helmsley, Yorkshire, a son of Robert Shout (q.v.) and his wife, Margaret Ellis. In view of his later career there can be little doubt that he worked on several bridges designed and built by his father in the North Riding, such as Newsham (1752–1753), Scawton (1755–1757) and Egton (1760–1761). In April 1762 he married Mary Moorhouse (1739–1806) of Helmsley. They had five (or perhaps six) children of whom Matthew (q.v.), born 1774, was the youngest. Robert served as churchwarden at Helmsley in 1769 and as juryman signed the church accounts on five occasions between 1773 and 1779.

His first responsible job was the contract, undertaken jointly with William Ellis, for building Sleights Bridge (1762–1763), a handsome three-arch structure with a central span 45 ft. over the river Esk, to designs by his father. For about a year from April 1764 he worked with his father and brother, John (a mason), on Stockton Bridge over the river Tees, designed by Joseph Robson (q.v.). With Ellis and Robert Corney (carpenter) Shout then took on the contract in July 1765 to build a new bridge at Whitby. Completed in 1768, probably to his plans, it had five stone piers and a timber deck. Next he designed and built two single arch bridges on the road from Pickering to Helmsley: Tilehouse Bridge, a rather ornate structure, and the simpler Sinnington Bridge with a 47 ft. span. They were completed in July and October 1769 respectively.

In June 1771 advertisements appeared for the work of repairing and widening the bridge at Ayton over the river Derwent. Shout and Ellis got the job, for £608, and work began in October. However, in June John Carr (q.v.) was asked 'to consider Mr. Shout's plan'. A month later Carr became Surveyor of the Bridges in the North Riding and shortly afterwards a new contract was drawn up with Shout and Ellis for building the bridge 'agreeable to the Plan and direction of Mr. Carr ... for £1,200'. This amounted to a new project. The first payment under the revised contract dates from October. A year later £1,150 had been spent with about 20 per cent of the work still to be done. Unable to finish the job at the contract price, Shout and Ellis nevertheless

carried on for another year but from October 1774 work was completed by direct labour under the supervision of the Surveyor. Opened in 1775, at a cost of £1,420, the bridge still exists. It has arches of 27, 36 and 27 ft. span and displays excellent workmanship throughout.

In October 1772, only three months after signing the Ayton contract, Shout agreed 'to rebuild Rutherford Bridge for £450 before Christmas 1773 according to Mr. Carr's plans and direction'. The old bridge, on the river Greta, a tributary of the Tees near Barnard Castle, was one of several destroyed by the great flood in November 1771. The new bridge comprised a single arch of 82 ft. span founded on rock; it was completed on time and within the remarkably low contract price.

'Robert Shout the younger of Helmsley' signed his last and by far his largest contract in March 1774 for building the southern part of the Tyne Bridge, Newcastle, in partnership with his brothers, John of Stockton (mason) and James of Helmsley (carpenter). Their father is also named but he died three months later. Robert Mylne (q.v.) acted as engineer on behalf of the Bishop of Durham with John Mylne as overseer. John Wooler (q.v.) was engineer for the northern part—the property of Newcastle Corporation—with different contractors. The old bridge had been destroyed in the 1771 flood and a temporary timber bridge was erected in 1772. Piling for the foundations of the southern part of the new bridge began soon after the contract date and the first stone was laid in October 1774. The first arch was closed in July 1775. Meanwhile work had started on the northern part. The last arch—on the northern section—was closed in September 1779 and the bridge became open to the public in April 1781. In round figures, the southern part (three arches) cost £9,000 and the northern part (six arches) about £26,000.

In June 1781 Shout was appointed harbour engineer at Sunderland, a post under the Commissioners of the River Wear and Port of Sunderland, at a salary of £100 p.a. with a rent free house. Joseph Robson, harbour engineer from 1775 to 1779, would have known the Shout family from their work on Stockon Bridge. His retirement, from a temporary appointment to December 1780 was to have been followed, as if by pre-arrangement, by John Shout but John died in June 1781 and Robert's appointment was made a fortnight later.

Work had already started on an extension of the South Pier planned by James Shout and approved, with some modifications, by John Smeaton (q.v.). Funds were running out, however, and construction ceased in 1782 with little accomplished. The Commissioners submitted petitions for an Act granting powers to increase the revenue for harbour works, without success. Shout reported in 1784 and gave evidence to Parliament in 1785, and finally in April of that year an Act was obtained. Work resumed, but in the winter of 1785 a strong northerly storm caused the harbour entrance to be blocked by a large sand bank. Shout immediately began building a temporary pier on the north side to contract the channel and enable the ebb tide to scour a deeper bed. By the winter of 1787 he had completed a pier 1,000 ft. long formed of stone-filled timber caissons; it had the desired effect and work then started on replacing the outer part by a permanent masonry structure. By 1793 a length of 700 ft. had been built. A massive structure founded on piles, typically driven through 12 ft. of sand and gravel to rock, it had a top width of 34 ft. (10 ft. above high water springs) a height of 24 ft. and a base (at low water springs) 48 ft. wide. It was built by direct labour under Shout's supervision, probably at a cost of about £12,000.

Though not his only work at Sunderland, the North Pier was Shout's major contribution to improvement of the harbour. Recognised as an engineer of ability he was consulted on other projects, in each case on leave of absence from his duties at Sunderland. In 1784 he was asked for his opinion on the line to be followed in building a pier at Bridlington; in the following year he visited Stockton to advise on improving the navigation of the Tees; and in 1792 he reported on means to be employed for removing a sand bank from the river Tyne. Meanwhile at home in 1784 he prepared a design for a masonry bridge with a span of 170 ft. to cross the river Wear, a precursor of the iron bridge erected by Rowland Burdon and Thomas Wilson (qq.v.) in 1796; for this design he received 50 guineas.

As a result of ill health Shout resigned from his post at Sunderland in August 1795. He died on 25 May 1797 and was buried at Sunderland parish church three days later. By his will, dated March 1797, he left a third share in the ship *Affinity* and the major part of his estate to his wife and son, Matthew, with small legacies to his four married daughters.

A. W. SKEMPTON and R. W. RENNISON

[*North Riding Quarter Sessions Order Book* (1762–1773) North Yorkshire RO; John Carr (*c*. 1800) *Book of Bridges*, North Yorkshire RO; *Stockton Bridge Order Book* (1762–1776) Durham RO; *Tyne Bridge Contract* (1774) Special Collections University of Durham; *Sunderland Harbour Minute Books* (1779–1797) Tyne & Wear Archives; A. W. Skempton (1977) The engineers of Sunderland harbour, *Industrial Archaeology Review*, **1**, 103–125; J. G. James (1986) *The Cast Iron Bridge at Sunderland*, Newcastle upon Tyne Polytechnic, Papers in the History of Science and Technology No. 5]

Works

1762–1763. Sleights Bridge, built (with W. Ellis) to design by Robert Shout Sr., £580
1765–1768. Whitby Bridge, built (with W. Ellis and R. Corney) to design, probably by himself, £2,860
1768–1769. Tilehouse Bridge, designed and built, £160

1768–1769. Sinnington Bridge, designed and built, £120
1772–1773. Rutherford Bridge, built to design by John Carr, £450
1772–1774. Ayton Bridge, built (with W. Ellis) to design by John Carr, £1,420
1774–1780. Tyne Bridge, southern section, built (with brothers John and James) to design by Robert Mylne, £9,345
1781–1795. Sunderland harbour, Engineer
1787–1795. Sunderland harbour, North Pier, 700 ft. long, Engineer, £12,000

SIBLEY, Robert (1789–1849), architect, was the son of a builder and surveyor. He was pupil to S. P. Cockerell in London, and later practised from Great Ormond Street. In 1818 he became County Surveyor of Middlesex, a post he held until 1828 when he became District Surveyor of Clerkenwell. As County Surveyor, Sibley provided large amounts of technical information for the Magistrates' report on the bridges of Middlesex (1826). The report by William Dickinson (d. 1822), George Saunders, and Samuel Perkins, was commissioned to recommend the correct division of maintenance responsibilities between the County, parishes, and private individuals. It includes an inventory of the County's bridges and culverts along with an analysis of relevant legislation. Sibley's works include numerous roads and bridges in Middlesex, Brentford Bridge (1828), modifications to Chertsey Bridge (1821), the Sun Brewery and Wharf at Wapping, a warehouse in the London Docks, and several houses. Two of his drawings, now destroyed, were included in Telford's bequest to the Institution of Civil Engineers, the *Great North Road* (London, 1822), a continuation of Farringdon Street, and the *Uxbridge–Brentford Turnpike* (1825). He supervised the erection of the Hanwell Lunatic Asylum (1829–1831) to the advanced designs of William Alderson. In 1832–1833 he adopted a system of constructing wharf walls using grooved cast iron piles and backing plates, backed with concrete. The earliest examples were at the Island Leadworks and at the Thames wharves at London Bridge. From 1839 he was surveyor to the Ironmongers' Company.

A friend and supporter of Thomas Telford (q.v.), he became a Member of the Institution of Civil Engineers in 1824, and served on its Council, elected in 1825, 1828 and 1829. He served on the committee which drafted the new By Laws and Regulations considered at a Special General Meeting in June 1834. Noted for his strict integrity, he was engaged in arbitration and valuations toward the end of his career. He died on 31 March 1849, aged fifty-nine. His son, Robert Lacon Sibley (1818/1819–1882), continued his father's architectural practice, and took up the post of District Surveyor of Clerkenwell on his father's unexpected death. His younger son, George (1824–1891), was a civil engineer of considerable distinction; another son, Septimus (1831–1893), was a physician.

TESS CANFIELD

[Membership records, ICE archives; *Description of a Bridge which was erected over the United Streams of the Cran Brook and the Isleworth Mill River near the Hanworth Powder Mills on Houslow and Hanworth in the year 1821 by the Magistrates of the County of Middlesex* (1834) ICE OC 114, ICE archives; *Plan of Building Walls on inverted Arches* (1834) ICE OC 116, ICE archives; *Motives which induced the Adoption of Cast Iron Piles and Panels to face the Wharf of the Leadworks at Limehouse* (1834) ICE OC 140, ICE archives; *Heating by Warm Water to the Middle-sex Pauper Asylum* (1834) ICE OC 147, ICE archives; *On Lighting and Ventilating Tunnels* (1836) ICE OC 188, ICE archives; List of Buildings by Sibley, Greater London RO; *Report of the Committee of Magistrates … respecting the Public Bridges of the County of Middlesex* (1826); Obituary (1849) *Builder*, **7**, 160; Obituary (1849–1850) *Min Proc ICE*, **ix**, 101–102; J. G. Watson (1988) *The Civils*; Colvin]

Publications

1828. *Observations on the Choice of Ground for the Middlesex Lunatic Asylum and Remarks on the General Principles required to be considered in the Selection of a Design for People labouring under Mental Derangement*

1828. *Observations on the Means of improving Counties by the Labour of Criminals*

Principal works

1821. Hanworth Bridge, £1,500
1821. Chertsey Bridge modifications
1828. Brentford Bridge

SIMCOCK, Samuel (fl. 1749–179?), civil engineer, is believed to have come from Staffordshire. Possibly trained as a millwright he worked as a carpenter and surveyor on the Bridgwater Canal and is first mentioned in the diary of James Brindley (q.v.)—he married one of his sisters, Esther, at Prestbury in June 1749—on 31 December 1759 when he was working on the Congleton Silk Mills. While work progressed on the Bridgwater he took levels for Brindley on the Sankey Navigation.

From the mid-1760s Brindley's workload exploded as his 'Grand Cross' scheme obtained Parliamentary sanction. The only way he could see the schemes through to completion was by relying on his assistants, of whom Simcock was but one. He surveyed the line of the Staffordshire and Worcestershire Canal with Hugh Henshall (q.v.) in 1766 and following the passage of its Act that year supervised its construction with Thomas Dadford (q.v.); John Baker acted as Clerk of Works. From 1768 he was also involved with Robert Whitworth (q.v.) on the Birmingham Canal, reporting, for example, on deviations and water supply; incredibly from 1769 he was appointed assistant on the Oxford Canal.

Simcock's involvement with the Oxford is best known, and most enduring, partly because its construction dragged on whilst the Staffordshire and Worcestershire Canal was completed in 1770 and the Birmingham in 1771. Following Brindley's

death in 1772 he was appointed engineer for the canal at £200 p.a. Construction continued until 1779, at an overall cost of £205,148. In 1786, after a further Act, a new period of construction, from Banbury to Oxford, commenced together with connections via the Coventry Canal to Birmingham; this was largely complete by 1790. Simcock, with Samuel Weston (q.v.) and four others, was the contractor, initially for the 7 miles to Aynho and subsequently the remainder. As work progressed he, at times with Weston, reported on the use of windmills to pump water for the locks, a diversion via Kidlington Green, and the Oxford terminus at Hythe Bridge. His final involvement, again with Weston, was the survey work for the Hampton Gay Canal, a 60 miles direct link between the Oxford canal and the Thames near London, first suggested in 1791. The scheme was intended to protect Oxford Canal traffic in the face of the proposed Grand Junction Canal. Simcock gave Parliamentary Evidence in support of the scheme in 1793 but by 1795 the proposal had been abandoned and Simcock disappears from canal history.

As an engineer he is difficult to assess. His lines on both the Birmingham and Oxford canals were criticised as being too winding; in 1771 he and Brindley had defended the Birmingham line on the grounds that only thus could it service its potential market. A generation later it was shortened by Thomas Telford (q.v.)

MIKE CHRIMES

[Minutes of Staffs & Worcs Canal, Rail 871/1, PRO; Section of Oxford Canal, 1786, CR 1590, P2, Warwks RO; S. Smiles (1861) *Lives of the Engineers*; C. T. G. Boucher (1968) *James Brindley, Engineer*; C. Hadfield (1969) *Canals of the West Midlands*; C. Hadfield (1970) *Canals of the East Midlands*; H. J. Compton (1976) *Oxford Canal*]

Works

1766–1770. Staffordshire and Worcestershire Canal, 46⅛ miles, forty-three locks, assistant engineer, with Dadford
1768–1771. Birmingham Canal, 22 miles, twenty-nine locks, assistant engineer, with Whitworth
1769–1771. Oxford Canal, assistant engineer
1772–1779. Oxford Canal, engineer, 63¾ miles, £137,000
1786–1789. Oxford Canal, contractor, with Samuel Weston, 27 miles, total cost c. £100,000

SIME, John, Sr. (d. c. 1778) and **SIME, John, Jr.** (d. 1796), shipbuilders, were among the leading shipwrights working in Leith in the eighteenth century. John Sime Sr. served his apprenticeship c. 1720, possibly with James Beattie (d. 1733), who was the owner of Lyon's Dock, a late seventeenth century dock at Leith. Sime apparently married Beattie's widow, Margaret (d. 1797), a few months after his death, but there is some doubt about the legality of the relationship as he married Margaret Hogg in 1748, without evidence of a divorce. He took John Jr. his son by this marriage, into partnership around 1766 and in July 1770 they petitioned the Council for permission to build a dry dock in their yard next to their property in Glasshouse Quay.

The scheme, approved in 1771, was Leith's first 'modern' dry dock, capable of handling vessels of 150 tons. Shortly after, in 1774, another shipbuilder, Robert Drybrough (d. 1793), built a rival dry dock. Through illness he was unable to continue and John Sime Jr. took the dock over, becoming by far the largest shipbuilder in Leith, although his property and business were already affected by improvements to the harbour carried out following an Act of Parliament of 1787. Plans had been drawn up by Robert Whitworth (q.v.), although only a small part was carried out, including the construction of a drawbridge to replace the existing Abbot's bridge. Sime supplied the timber platforms at the south end of the bridge.

Following his death his foremen, Strachan and Gavin, took over the yard. He left shares worth £2,000, as well as property in the town; he had also been involved in the establishment of the North Leith Iron Manufactory.

MIKE CHRIMES

[S. Mowat (1994) *The Port of Leith*]

SIMMS, William, FRS (1793–1860), was born on 7 December 1763, in Birmingham, the second of nine children of William (1763–1828) and Sarah Simms. His father was an instrument maker in Birmingham where the family was involved in jewellery and instrument making. Soon after William's birth they moved to London to enable William Sr. to help his ailing father, James, who had a jewellery business in Whitecross Street. This business was rapidly turned over to the manufacture of optical instruments. Generally William Sr. prospered and in 1804 he was elected a Freeman of the City. The family also moved in 1796 to Doby Court, Monkwell Street, and in 1802 to Bowman's Buildings, Aldersgate.

William Jr. was sent in January 1806 to be educated in mathematics by a Mr. Hayward who had taught in the navy. After two years education, in January 1808 he was apprenticed to Thomas Penstone, a member of the Goldsmith's Company. His interests lay elsewhere and in 1809 he was apprenticed to one of Jesse Ramsden's former workmen, a Mr. Bennett of Charles Street, off Hatton Garden. Following this he was elected a Freeman of the Goldsmith's Company in 1815 and set up in business. Initially he worked in his father's new premises at 4 Broadway, Blackfriars, and then from 1818 at Bowman's Buildings. His elder brother, James (1792–1857), was already establishing a reputation for navigation instruments.

Simms was interested in the division of the circle, the accuracy of which was vital to the manufacture of accurate measuring instruments. He entered into correspondence with Thomas Jones (q.v.) who brought him to the notice of Edward Troughton (q.v.) and persuaded him to join the Society of Arts. Here he met Bryan Donkin (q.v.), and also Colonel Colley of the Ordnance Survey. These contacts were to be the foundation of his career, culminating in Simms becoming Troughton's partner in 1826; on his retirement, Simms took over the whole business, which enjoyed the highest of reputations for the manufacture of scientific instruments.

Simms specialised in surveying instruments and from around 1817 began supplying theodolites for the Ordnance Survey and then the East India Company, including those supplied to Colonel Everest, c. 1830. On a larger scale he supplied telescopes, training instruments, mural circles and other astronomical instruments for observatories at Krakow (1828), Madras (1830), Cambridge (1832), Lucknow (1832), Calcutta (1829–1832), Edinburgh (1834), Brussels (1834) Greenwich and elsewhere. By the end of his career he had supplied most of the leading observatories in the world with equipment. His work formed the basis of the treatise on mathematical instruments written by his younger brother, Frederick Walter's (1803–1865), and published in 1836; F. W. Simms became an important writer on civil engineering.

The reputation of Simms was enhanced by the improvements he made to graduating instruments, and his self acting circular dividing engine reduced the labour involved in manufacture from weeks to hours. He was elected an Associate of the Institution of Civil Engineers in 1828, regularly attended meetings, and in 1834 was persuaded by Robert Stevenson (q.v.) to present a paper on the Edinburgh Mural Circle, and Graduating Astronomical Instruments. He was a Fellow of the Royal Astronomical Society, joining in 1831 and describing his dividing engine to them; in 1852 was elected a Fellow of the Royal Society. He helped develop both the standard yard and standard chains for the Admiralty.

Simms married Ann Nutting (1798–1839) in 1819 and had nine children by his first marriage; he also had three children by his second marriage, to Emma Hennell (1811–1888). From his partnership with Troughton in 1826 he had premises in Fleet Street, initially at 136, and from 1841 at 138 Fleet Street, addresses which achieved worldwide fame as the home of Troughton and Simms, instrument makers.

Simms died at his home in Carshalton, Surrey, on 21 June 1860. His family, notably his son, James, carried on the instrument making tradition.

MIKE CHRIMES

[Membership records, ICE archives; Minutes of Conversation (1834) ICE archives; E. Troughton (1809) An account of a method of dividing astronomical and other instruments, *Phil Trans*, 105–145; F. W. Simms (1836) *A Treatise on ... Mathematical Instruments*; Obituary (1860–1861) *Min*

Proc ICE, **20**, 167–168; E. Mennim (1994) *Transit Circle: the Story of William Simms (1793 to 1860)*]

Publications

1843. On a self-acting circular dividing engine, *Astronomical Society Memoirs*, **xv** (1846) 83–90

SIMPSON, James (1799–1869), was the fourth son of Thomas Simpson (q.v.) and became involved with Chelsea Waterworks at an early age. Due to his father's infirmity he was acting engineer for Chelsea Waterworks from 1820 and succeeded his father as engineer in May 1823. Similarly he acted as consultant to Lambeth Waterworks.

Around 1825–1826 he began important experiments on slow sand filtration. He visited filter beds in operation in the Glasgow area, and elsewhere, and designed in 1828 a filter bed 40,000 sq. ft. in area supplying two service reservoirs. In many ways this represents the start of modern water supply engineering. At much the same time he was surveying water wells in the London area and contributed to the contemporary Parliamentary investigations into London's water supply.

Under Thomas Simpson, the Lambeth company had invested its early profits in capital works and from 1802 all wooden mains were replaced by iron; a new steam pumping engine was added in 1815. By this time the company was the most profitable of the London concerns. A new Act in 1832 enabled Simpson to build new reservoirs at Brixton and Streatham.

By the 1830s Simpson was effectively a consulting engineer and had carried out work on the water supply for Windsor Castle and Hampton Court. His brother, William (1809–1864), Thomas's sixth son, whom James had trained in civil engineering, ran the family's engine manufactory in Pimlico.

Simpson's involvement with the Lambeth Company led to his recommending it in 1849 to take its supply from up river and removing its works to Seething Wells, near Kingston, which the company did in 1850, in advance of the 1852 Act which obliged all the London companies to take their water from above the tidal flow.

Simpson was elected to the Institution of Civil Engineers in 1825 and was on its Council from 1826 to 1844 when he was elected Vice-President. He regularly contributed to the meetings from the 1820s contributing to discussions on all manner of subjects, but particularly water supply, heating and ventilating, and steam engine design. In 1854 he became the first water engineer to be elected President of the Institution. His brother, William, was also a Member of the Institution.

After a career as a consultant, advising extensively on water supply, and other matters, he died in 1869.

MIKE CHRIMES

[Obituary of William Simpson (1865) *Min Proc ICE*, **24**, 539–540; Westminster Archives; ICE Membership Records; ICE Minutes of conversations, esp. *The Best Method of Filtering Water in Large Quantities for the Supply of a City*, MC57, 17 March 1829; MC59, 24 March 1829; *On Shaft Sinking*, MC60–61; *On Wells*, MC66, 26 May 1829, 68, 9 June 1825; *Select Committee on the Supply of Water to the Metropolis* (1821); J. Farey, *Treatise on the Steam Engine* (1827); *Min Proc ICE*, **30**, 1869–1870, 457–461. H. W. Dickinson (1954) *Water Supply of London*; G. M. Binnie (1981) *Early Victorian Water Engineers*; J. Tann (ed.) (1981) *The Selected Papers of Boulton and Watt*, vol. 1]

Works

1828–1829. Chelsea Waterworks, filter beds, £6,796 (F. Richman, Contractor)

SIMPSON, John (1755–1815), stonemason and contractor, is best known for his work for Thomas Telford (q.v.) who described him in a letter to his childhood friend, Andrew Little, as 'a treasure of talents and integrity'.

Apparently a stonemason from Stenhouse, Midlothian, he came to Shrewsbury in 1790 to supervise restoration work at St. Chad's. He tendered successfully for one of Telford's masonry arch bridges at Chirk in 1793 and over the next twenty years built a large number of bridges in the county. The original contractor for Pontcysyllte aqueduct was James Varley of Colne but by 1796 he was behind with his contract although by then an iron aqueduct was under consideration. Simpson joined him on the work in September 1795 and in 1800 tendered to complete the piers and abutments for the iron aqueduct design, then finalised. In the meantime he had, on 20 January 1796, successfully tendered for Chirk aqueduct.

Clearly Telford was impressed by his work, as was William Jessop (q.v.). In 1804 they agreed to let the masonry work on the Caledonian Canal to Simpson, with John Cargill and John Wilson (qq.v.) as his supervisors at the east and west ends respectively. They were clearly well organised and the housing they built as accommodation for their masons exists as Telford Street, Inverness. They also undertook excavation contracts on the canal, such as that at Corpach Moss, under Wilson, and completed works there after Simpson's death in 1815.

There was much work on the Highland Roads going on at this time and Simpson undertook the contracts to build the road from Banavie along the north side of the canal to Culross Burn, fifty-six bridges on the road being built by Colonel Cameron of Lochiel.

Earlier, in November 1806, Simpson had accompanied Telford to a meeting of the committee of the newly formed Glasgow, Paisley & Ardrossan Canal Company, subsequently successfully tendering £4,700 for Lot 3 with a Mr. Ashworth. By November 1808 it was apparent that work was not progressing well on the canal as great problems were being experienced with the costs of rock excavation. Fortunately for Simpson his lot involved the least cutting but a number of bridges for which

John Simpson

he was well equipped. In 1808 it was agreed he should do the works connected with the basin at Paisley and also the Cart Aqueduct. Because of spiralling costs the canal was not completed to Ardrossan and the first section was opened in October 1811. Lot 3 cost £5,780, but the increase here was modest compared with that at Paisley—£8,637 compared with the estimate of £4,088; even the Cart Aqueduct was higher than budgeted for: £5,440 compared with £3,459. Simpson did well.

Examples of the ingenuity of Simpson's work as a contractor are described by Telford in his article on Bridge practice for the *Edinburgh Encyclopaedia*. For the Corpach Lock on the Caledonian Canal a massive cofferdam was constructed by Simpson and Wilson, secured to the bedrock by iron dowels installed using a wooden cylinder and piling engine. Another cofferdam by Simpson, on this occasion with Cargill, was employed to underpin a pier and the eastern abutment of Pulteney Bridge, Bath, where all previous attempts had failed.

Simpson was persuaded, against his better judgement, to erect the masonry for Telford's famous iron bridges at Bonar and Craigellachie; with their remote location he lost money, but these bridges and the multi-span arch bridge at Dunkeld which he built for Telford were acknowledged with pride on his memorial at St. Chad's, Shrewsbury. He died on 16 June 1815, at his home in Belmont, Shrewsbury; his business in Shropshire was continued by his nephew, John Straphen (q.v.). One of his last contracts, the New Galloway Bridge, collapsed after his death.

Simpson's son, James, accompanied Telford on his second visit to Sweden in 1813 and remained there until 1815. Unlike John Wilson, who accompanied him, his work was not successful and to his father's shame he was sent back somewhat under a cloud.

MIKE CHRIMES

[Glasgow Paisley & Ardrossan Canal Minutes, SRO; Gibb papers, ICE Library; Telford mss; *Reports of the Commissioners for the Caledonian Canal*, 1804–1815; *Reports of the Commissioners of Roads and Bridges in the Highlands of Scotland*, 1–9; (T. Telford) (1814) Article on Bridge practice, *Edinburgh Encyclopaedia*; *Gentlemen's Magazine*, **85**(11), 1815, 572; A. Gibb (1935) *The Story of Telford*; A. Blackwall (1985) *Historic Bridges of Shropshire*]

Works

1790–1792. St. Chad's, Shrewsbury, restoration
1793. Chirk bridge, 50 ft. masonry span, £1,093
1794. Grindleforge bridge causeway, *c.* 100 yd. long
1795. Two timber bridges near Longdon
1795–1805. Pontcysyllte aqueduct, masonry piers and abutments
1795–1796. Vyrnwy aqueduct, Montgomeryshire Canal
1796–1801. Chirk aqueduct, masonry
1798–1799. Bewdley bridge, £8,512
1799. St. Mary's Church, Shrewsbury, restoration
1799–1802. Pulteney Bridge, Bath, underpinning (with Cargill)
1803. Standford bridge, masonry
1803–1804. Ironbridge abutment repairs
1804–1815. Caledonian canal, masonry work and some excavation (with Wilson)
1804–1805. Ellesmere canal, extension, Whitchurch–Hurleston (with Fletcher)
1805–1809. Dunkeld bridge, seven spans, £29,361
1807–. Glasgow, Paisley & Ardrossan Canal, Lot 3, Paisley Basin, & Cart Aqueduct, *c.* £20,000
1808. Wem Mill bridge, 24 ft. masonry span, *c.* £700
1809–1810. Banavie Road
1809–1810. Lochieside Road, 12 miles 540 yd., £5,178 14s (with Wilson)
1809–1811. Laigh Milton viaduct
1810. Lochie Ferry Piers, £154 9s 11d (with Wilson)
1809–1810. Pont Rhyd Meredyth, 80 ft. masonry span, £4,303
1811. Tern bridge, masonry, 41 ft. span
1811(?). Ardrossan harbour
1811–1812. Puleston bridge, three masonry spans
1811–1812. Corran ferry piers, £415 14s 2d (with Wilson)
1811–1812. Bonar bridge, masonry, £6,985 12s 3d (with Cargill)
1811–1812. Meole Brace bridge, abutments
1812. Long Mill bridge, masonry abutments
1812–1813. Cantlop bridge, abutments
1812. Wayford bridge, masonry, 24 ft. span, *c.* £550
1812–1815. Craigellachie bridge, masonry, £4,100 (with Cargill)
1812–1814. Tobermory harbour (with Wilson)
1812–1813. Creech Road, £1,616 (with Cargill)

1814–1815. Tenbury Bridge, widening with cast iron ribs, £1,200

1815. New Galloway Bridge, collapsed after Simpson's death

SIMPSON, Thomas (c. 1754–1823), civil engineer, was trained as a millwright in Cumberland before moving to London in 1778.

He was appointed Engineer to the Chelsea waterworks in 1783 and remained in office until his death. His arrival in London coincided with the installation of what was only the second of James Watt's (q.v.) steam engines for a water supply in London, Chelsea waterworks, in 1776. This engine was experimental in its nature, being one of Watt's earliest, and probably needed more skilled attention than the two atmospheric engines installed by John Wise of Hawkesbury Colliery, Bedworth, in 1741–1742. These were used to pump water from the company's works, on the site of what is now Victoria Station, to reservoirs in Hyde and St. James's Parks, work that had originally been carried out by a tide mill erected in the early 1720s. It may have been to look after the Watt engine that Simpson was first appointed.

Simpson's experience of the use of iron and wooden pipes was detailed to the *Select Committee on the Supply of Water to the Metropolis* of 1821, he coming down decidedly in favour of iron. The Chelsea company had been using iron pipes with success since 1734. Around 1785 Simpson invented the spigot and faucet joint, conducting some experiments before using it on a long main of c. 1,500 yd. in length for Chelsea Waterworks c. 1794.

While at Chelsea he became involved in the promotion of, and was appointed Engineer to, the Lambeth Waterworks Company, which commenced operations in 1785. This was the first to supply water in that area of London, taking its supply from the Thames near the present Hungerford Bridge, with its works in Belvedere Road, Lambeth.

Simpson was responsible for the installation of Watt steam engines for Chelsea Waterworks c. 1803–1804, 1809–1812 and 1818, and the installation of many water mains, including an 18 in. main from the New Engine House to Pimlico in 1811. The Watt engine installed at Pimlico Wharf pumping station in 1803 was a single acting engine and was regarded by Dickerman as 'an example of the high water mark of Watt's inventiveness and mechanical construction'. It replaced the 1742 engine and supplied water to the Hyde Park reservoir at the rate of 175 cu. ft. a minute. Outside London Simpson was Engineer to the Liverpool waterworks c. 1799.

Thomas Simpson died in April 1823 and was buried at St. George's, Hanover Square, on 2 May 1823 at the age of sixty-nine. His sons James (q.v.) and William were also engineers.

MIKE CHRIMES

[Westminster Archives; ICE Membership Records; ICE Minutes of conversations, esp. *The Best Method of Filtering Water in Large Quantities for the Supply of a City*, MC57, 17 March 1829; MC59, 24 March 1829; *On Shaft Sinking*, MC60–61; *On Wells*, MC66, 26 May 1829, 68, 9 June 1825; *Select Committee on the Supply of Water to the Metropolis* (1821); J. Farey, *Treatise on the Steam Engine* (1827); Obituary of William Simpson (1865), *Min Proc ICE*, **24**, 539–540; *Min Proc ICE*, **30**, 1869–1870, 457–461.; H. W. Dickinson (1954) *Water Supply of London*; G. M. Binnie (1981) *Early Victorian Water Engineers*; J. Tann (ed.) (1981) *The Selected Papers of Boulton and Watt*, vol. 1]

SIMS, William (1762–1834), mining engineer, was born near Chacewater on 29 December 1762 and became one of the leading designers of steam engines in Cornwall in the early nineteenth century. He was agent for Michael Williams of the United Mines and engineer to the Eastern Mines and as such responded to Arthur Woolf's advertisement in 1811 and went to London to see his engine 'now brought into such a high state of perfection'. He took a half share in Trevithick's pole engine patent (No. 3922, 1815) shortly before Trevithick's departure for Peru in 1816 and together they compounded engines at Wheal Chance and Treskerby by adding pole cases to the existing Boulton and Watts engines. Sims alone later compounded three more at United Mines Williams, Wheal Damsel and United Mines Polding, and by the time of the 1817 Parliamentary Inquiry into high pressure steam engines had emerged as one of the leading designers.

He died at White-Hall, in Kenwyn, Truro on 16 October 1834, aged seventy-two. In his address in the *Journal of the Royal Institution of Cornwall* in 1872, W. J. Henwood included the comment 'To the courtesy of this intelligent and excellent gentleman, I owe much information, which it might now be impossible to obtain, regarding the early application of steam-power in Cornwall'.

William Sims's second son, James (1795–1862), continued his father's work on steam engines with specifications for 'Certain improvements ...', published in 1841 and 1847.

R. P. TRUSCOTT

[House of Commons (1817) *Select Committee on Steam Boat Explosions*; *West Briton & Journal of RIC*, April. 1872, Courtney Library of the Royal Institution of Cornwall; Boase & Courtney, *The Bibliotheca Cornubiensis*, vol. II; H. W. Dickinson and A. Titley (1934) *Richard Trevithick, The Engineer and The Man*; T. R. Harris (n.d.) *Arthur Woolf, 1766–1837, Cornish Engineer*]

SINCLAIR (SINCLAR), George (c. 1625–1696), philosopher, scientist, civil and mining engineer, and inventor, is thought to have been born in East Lothian about the year 1625. At one time he owned property in the town of Haddington, and on the title-page of his *Ars nova ...* he refers to himself as 'Scoto-Lothiani'. His brother, John, was from 1647 to 1682 the minister at Ormiston, in East Lothian.

In 1654 George Sinclair is known to have been acting as a 'pedagogue' in the University of St. Andrews but he was admitted a Master at Glasgow University in October of that year and was then appointed regent of philosophy at Glasgow. He was one of the first in Scotland to make a study of the science of physics, then held, as he put it, 'of little account'. In 1655 he experimented with a primitive diving-bell in exploring a ship of the Spanish Armada wrecked off the island of Mull. He later described in his *Hydrostatical Experiments ... of* 1680 a kind of diving-bell of his own invention, which he called an *Ark*.

He remained in Glasgow for about ten years until he was forced to resign because of his refusal to declare his adherence to the episcopal form of church government being imposed on Scotland at that time. In 1665 he is noted as one of the regents in the less authoritarian University of Edinburgh, and he seems to have stayed in that city for more than twenty years, engaging in a variety of occupations where his scientific knowledge was applied to practical problems. For example, he made use of the barometer to measure relative heights both above and (in coal mines) below ground, though he based his calculations on the erroneous assumption that the atmosphere is a homogeneous fluid.

On many occasions he was consulted by coal-owners in the Lothians for advice on overcoming difficult geological conditions and on improving drainage of the lower seams at a time when mechanical pumps or other water-raising devices were not in general use in the coal-fields. In his writings he displays a knowledge of English as well as Scottish collieries. In 1672 he published his observations as an addendum to his *Hydrostaticks*, a work which was reprinted in 1683 as *Natural Philosophy improven by New Experiments ... and a True Relation of an Evil Spirit; also a large Discourse anent Coal ....* The 'true relation of an Evil Spirit' was an account of the witches of Glenluce and it was an early manifestation of Sinclair's preoccupation with such metaphysical phenomena, culminating in 1685 in his best-known work *Satan's Invisible World Discovered ...*, which was reprinted many times over the next two-hundred years.

In 1672 he was asked by the Magistrates of the City of Edinburgh to design and superintend the construction of the works for a new water supply from springs some 3 miles south of the Old Town. This work was successfully accomplished in 1673–1676 using lead piping of 3-in. bore, and for his 'attendance and advyce in the matter of the waterworks' Sinclair received £66 13s 4d, and in 1681 a pension of £100 Scots p.a. from the Town Council.

After the Glorious Revolution of 1688 he was able to return to his post of regent in Glasgow where two years later, in March 1691, he was appointed Professor of 'Mathematicks and Experimentall Philosophy'. The last notice of him in the records of the University of Glasgow was in April 1696 and he appears to have died later in that year. The college treasurer observed that he died poor, but 'was ane honest man'. The biographical notice of 1871 said of him 'It is curious to find science and superstition so intimately mingled in the life of this extraordinary person... and in company with such a zeal for the propagation of real knowledge.'

RON BIRSE

[G. Sinclair (1685, reprinted 1871 with biographical notice) *Satan's Invisible World Discovered*, ix–liii; OUP (1897) DNB; R. Birse (1994) *Science at the University of Edinburgh, 1583–1993*, 13–14; W. T. Johnston (1998) *Scottish Engineers and Shipbuilders* (computer database) DNB]

Selected publications
1661. *Tyrocinia Mathematica, sive Juniorum ad Matheses addiscendas Introductio in Quattuor Tractatus ...*
1669. *Ars Nova et Magna Gravitatis et Levitatis, sive Dialogorum Philosophicorum Libri Sex de Aeris Vera et Reale Gravitate*
1672. *The Hydrostaticks, or the Weight, Force and Pressure of Fluid Bodies made evident by Physical and Sensible Experiments: together with A Short History of Coal* (reissued with additions, 1683)
1683. *Natural Philosophy improven by New Experiments: touching the Mercurial Weather Glass, the Hygroscope, ... the Pressure of Fluids, the Diving-Bell, and all the Curiosities thereof. Together with a True Relation of an Evil Spirit ... and a large Discourse anent Coal, Coal-Sinks, ... Levels, Running of Mines, Gaes, Dykes, Damps, and Wild-fire*
1685. *Satan's Invisible World Discovered, ... proving evidently ... that there are Devils, Spirits, Witches and Apparitions, from Authentick Records ...* (frequently reproduced up to 1871)
1688. *The Principles of Astronomy and Navigation; ... etc. ... also A New Proposal for Buoying up a Ship of any burden from the Bottom of the Sea*

Works
1672–1676. Edinburgh, design and supervision of construction of new water supply

SINCLAIR, John (c. 1789–1863), civil engineer, apparently originated from Banff. It seems probable he met Thomas Telford (q.v.) in Scotland and was encouraged to move south to work on the Holyhead Road, probably as one of four inspectors under William Alexander Provis (q.v.). He is known to have surveyed the Shrewsbury–Chirk and Shrewsbury–Wellington sections of the Holyhead Road in 1818 for Telford. He was resident engineer on the cast iron bridge at Bettwys y Coed and suggested it be called Waterloo Bridge.

Soon after he became for 'upwards of 42 years resident engineer on the Coventry Canal' at a salary of £200 p.a. Construction of the canal had begun in 1768 and was completed by 1790, and by the end of the first decade of the nineteenth century had become a very prosperous concern, which position it maintained in the face of increasing railway completion. Sinclair's work, therefore, was essentially one of maintenance. Of his other works, in 1837 he designed an iron towpath bridge at Hawkesbury, Warwickshire, an early Handyside of Derby structure (see: *HANDYSIDE, William*), as part of the improved Coventry–Oxford canals connection.

Sinclair joined the Institution of Civil Engineers in 1829 but was a member only until 1832. At that time he was resident at Atherstone, on the Coventry canal. In 1841 he considered applying for the post of Warwickshire County Surveyor. He died on 17 March 1863 and is buried at Mancetter. John Sinclair Jr. of Coventry and Robert Cooper Sinclair of Hartshill were almost certainly his sons.

MIKE CHRIMES

[ICE membership records, ICE archives; Register of (Telford) drawings, ICE archives; Colvin 3; C. Hadfield (1970) *Canals of the East Midlands*, 2nd edn.]

Works

1815. Waterloo Bridge, Bettws y Coed, resident engineer
c. 1821–1863. Coventry Canal, engineer
1837 Hawkesbury bridge, Warwickshire

SKINNER, Thomas, Major (1804–1877), military engineer, was the son of an army officer and originally intended to join the navy. He received his first commission, in the army, at the age of fourteen, and was posted to Ceylon and placed in charge of a detachment of riflemen. In 1820 he was appointed to command two-hundred Sri Lankans intended for labour on the military roads to Kandy via the Kadugannawa Pass and Seven Korales. For the remainder of his career he was largely employed on Sri Lanka's roads, becoming Commissioner in 1841 and carrying out a great deal of survey work on the island.

He retired on 1 July 1867 and died on 24 July 1877.

MIKE CHRIMES

[P. M. Bingham (1921–1923) *History of the Public Works Department, Ceylon, 1796–1896*, 3 vols.]

Works

1820–. Sri Lanka Roads

SKINNER, William, Lt. General (1700–1780), military engineer, was the Chief Engineer of Great Britain from 1757 until 1780. The best summary of Skinner's career was his epitaph in St. Alphage churchyard, Greenwich: 'To the memory of Lieutenant General William Skinner who died on the 25 day of December 1780, having served sixty-one years as an engineer 23 of which Chief Engineer of Great Britain'. William Skinner was born on St. Kitts and was brought up by his uncle,

the military engineer Talbot Edwards, after the death of his father, Thomas, a merchant. After an education at military colleges in Paris and Vienna, Skinner became a practitioner engineer in 1719, subsequently rising to Chief Engineer in Scotland (1747) and Great Britain (1755) and Major General (1761).

In 1720 he worked on the gun wharf at Devonport and in 1722 at Minorca, before beginning a lifelong association with Gibraltar in 1724. He produced the first general survey of the defences of the Rock followed by a plan of the north front and several water-colour views now in the British Library. In 1727 he was rewarded with additional pay for his zeal and intrepid exertions during the siege. His time there gave him unique experience with the climate and ground conditions of what he described as 'the most peculiarly situated place in Europe'. This was utilised when as Chief Engineer he was asked to arbitrate in a dispute over works carried out by the Governor of Gibraltar, Lord Tyrawley. Skinner's report was so unfavourable to him that he was summoned to the bar of the House of Commons to substantiate it. His main criticism was the old argument between fixed masonry defences and temporary earthworks. Skinner was not opposed to the latter in principle but considered them wrong for Gibraltar as the earth would dry and crumble in the summer sun and then be washed away by winter floods, as demonstrably happened to the works on the south bastion. For good measure, he maintained that the landguard defences were also wrongly sited to defend the harbour and mole. Skinner was thereafter recognised as an authority on Gibraltar, producing a report in 1759 and giving his observations in 1769 on rival projects.

In 1746 the Duke of Cumberland requested 'an engineering officer of high standard to be sent to Scotland for the purpose of constructing forts, etc., to control disaffected Highlanders' and Skinner was sent, complaining about the vile state of the roads, the unhealthy climate and the intolerable nastiness of Edinburgh: 'Our hogs there (Gibraltar) being more clean', views hopefully unknown to the citizens of Edinburgh when they voted him the Freedom of the City ten years later, as also did Stirling, Perth, Aberdeen and Inverness. His first few months were prolific. He drew up plans for the restoration of Fort Augustus (dry moat, glacis and pier) put his experience to good use building barracks at Corgarff and Braemar and built a magazine at Dumbarton.

Most importantly he decided to resite Fort George—variously described as both a massively expensive white elephant and 'the most considerable fortress and best situated in Great Britain' [General James Wolfe]—on an isolated windswept promontory at Ardersier, 9 miles from Inverness. It was planned as an impregnable base for the army after the failure, during the 1745 Rebellion, of the small forts built by General Wade and John Romer (qq.v.). It was the largest construction project in Scotland prior to the

Caledonian Canal. One of the best preserved artillery fortresses in Europe, Skinner's designs survive virtually intact.

The buildings, including barracks for 1,600 men, were mostly two and three storey masonry blocks with suspended timber floors and roofs of single or dual pitched timber and slate with king or queen post trusses, as necessary. The casemates had brick arch masonry vaulted roofs with rubble infill; the magazine for 2,500 barrels of powder had a brick vaulted roof. The sandstone fronted walls comprised broad ramparts packed with earth excavated on site and stone cross walls. The ditches could be flooded by opening sluice gates in the batardaux (small dams) holding back water. The three span bridge included a drawbridge operated by upper counterpoise beams. All this was surrounded by the most complete range of outworks (counterscarp covered way lunettes, traverses and glacis) ever built in Britain.

Construction was undertaken by members of the Adam family, John, William, Robert and James (qq.v.) and Skinner personally supervised the work whenever possible. As the road to Blairgowrie was not completed until 1754 all men— 1,000 worked on the earthworks alone—and materials had to be brought in by sea, originally under constant threat of attack. This meant that a pier and harbour had to be built. The original estimate was £92,673 but the fort eventually cost over £200,000, more than the Scottish GNP in 1750.

In 1755–1756 Skinner reported on fortifications and garrisons in Ireland, but no action was taken until Charles Vallancey (q.v.) resurrected the plan twenty-eight years later. He also surveyed Jersey, Milford Haven and Plymouth providing new defence works and repairing old ones. In 1761 he was sent to Belle Isle after General Hodgson had complained about 'the set of wretches I have for engineers' to survey the defences, restore the works and provide accommodation and shelter for 4,000 men from the prodigious rains.

As Chief Engineer all plans and projects were submitted for his opinion and he came to be considered the ablest critic and authority in matters of fortification. Among these projects were plans for the harbour at St. Johns in Newfoundland, underground works in Minorca (1773) and revisions of John Archer's plans for Priddy's Hard, Gosport, to include a sluice to scour the basin. He also commented favourably on a cut in the river at Iberville, proposed by Robert Hutchins and Elias Durnford (q.v.) which would be of the greatest utility for British settlers being short, safe, and answering every communication problem. He considered the estimate of £2,500 to be far to low, but it was in any event academic as the war in America began before work on it could commence. Also in 1779, returning almost full circle, he approved fortifications for the Grenadines, Grenada and Tobago.

One of his first actions as Chief Engineer in 1757 had been to gain substantive army rank for the Corps and he also gained for them comparable salaries when his own pay, miserly for a Lieutenant General, was raised from £1.70 per day to £2.40. His rare financial rectitude was recognised by the granting of extra annuities of £100 to his widow, Margaret, and £50 to his granddaughter 'in consideration of his long and faithful service of sixty-one years (when) he had not earned above £60 a year for his family when he had frequent opportunities of assigning large sums'. His will only mentioned £500 in annuities and a house in East Greenwich.

His only son, also William, drowned in 1761 but his grandson, William Campbell Skinner (d. 1787), served as an engineer in America. His descendants retained connections with the Corps of Engineers, donating to it both a copy of his portrait, now in the Officer's Mess, from the Governors House in Gibraltar and his model (now missing) of Fort George.

SUSAN HOTS

[British Library, mss, 1749–1750; NLS surveys, 1747–1769, mss 1645–1649; Reports, c. 1746–1770, including full reports and estimates for Fort George, Royal Engineers Museum; Royal Engineers Library, Connolly papers; DNB (Colvin 3), 873; Bendall, 467; Porter, **2**, 382–384; Notes and queries, 2nd series, **1**, 168; Report from the committee appointed to examine and consider of a book presented to this house (1758) *Reports, Plans and Estimates for Fortifying Milford Haven, together with a Survey of the said Harbour*; R. Southey (1929) *Journal of a Tour in Scotland in 1819*, 108; D. S. Brown (1946) The Iberville Canal project … (1763–1775) *Mississippi Valley Historical Review*, **32**, 491–516; R. Whitworth (1958) *Field Marshall Lord Ligonier*; NCE, 22/29 December 1983, 12/13; D. W. Marshall (1976) *The British Military Engineers 1741–1783*, unpublished Ph.D. dissertation, University of Michigan; J. Coad (1989) *The Royal Dockyards 1690–1850*, 260–261; A. Cruden (1989) Fort George rehabilitation, *Structural Survey*, **8**, 31–43; A. Saunders (1989) *Fortress Britain*; F. J. Herbert (1990) Belle Isle and the siege of 1761, *Fort*, **18**, 69–81; J. Gifford (1992) *The Buildings of Scotland, Highlands and Islands*; Q. Hughes and A. Migos (1995) *Strong as the Rock of Gibraltar*; C. Tabraham and D. Grove (1995) *Fortress Scotland and the Jacobites*; I. MacIvor (1996, rev.) *Fort George*; J. Douet (1998) *British Barracks 1600–1914*]

Publications

Publications include:

1757. *Observations on the New Works at Gibraltar performed in 1757* (BL add ms 10034, includes four drawings)
1770. *Report and Opinion on Lieutenant Colonel Green's Report on Gibraltar*

Charts and plans

Charts and plans include:

1724. Survey of the defences of the Rock of Gibraltar

1758. Survey with particular soundings of the mole at Malaga

1771. Book of thirty-three original plans of Fort George

Works

1747–1769. Fort George, estimated £92,673 19s 1d, cost over £200,000

SMART, George (d. 1834), engineer and timber merchant, of College Street, Pedlar's Acre, London, became a Resident Member of the Institution of Civil Engineers in 1822. His membership nomination was written on the back of an envelope which survives in the Institution archives, its original address to Henry Robinson Palmer (q.v.) crossed out. He paid his subscription until April 1828 and was carried in membership until his death in 1834.

He appears as a member of the Society of Arts in 1796, residing in Camden Town until 1802. In 1800, he patented a design for a hollow mast for sailing ships and incidentally included a 'bow and spring' rafter, proposed as a temporary yard arm; that bowstring truss won a Silver Medal from the Society of Arts in 1819. From 1803 until 1833 he was at Pedlar's Acre, Westminster Bridge, and also, by 1806, at Ordnance Wharf, Westminster, until at least 1819. Smart received a Gold Medal from the Society of Arts in 1806 for his extendable chimney cleaner (the Scandiscope of 1803), instrumental in the campaign to eliminate the use of children as sweeps. In 1812 he patented (No. 3562) a method to prevent timber from shrinking and invented a method of coopering in 1813. In 1814 he described himself as a timber merchant and patented (No. 3796) machinery for grinding corn. Joseph Gibbs (1798–1864), on his return from the West Indies, worked with Smart in Lambeth, erecting corn and saw mills in the London area. A description of Smart's own sawmill of about 1812 states that it was driven by a horse wheel but, by 1825, steam power was provided by one of Charles Woolf's (q.v.) double cylinder expansive engines, fuelled by sawdust from the mill. Smart is credited with using circular saws as early as 1790 and with using 4 ft. diameter saws in about 1827. He may have contributed to Marc Isambard Brunel's (q.v.) invention of the large segmented veneer saw patented in 1805.

Smart was active in property development and leased large areas from the Cooke family, in 1819 being allowed to build on 'John Cooke's land', roughly from the bank of the Thames extending across the present York Road, Lambeth. By 1823, he had an engine house and workings on the site, along with another area allocated to Messrs. Collinge (q.v.) and Mainwaring. In that year, a specimen of Palmer's monorail could be seen at Mr. Smart's, Pedlar's Acre near Westminster Bridge. He patented a design for an iron truss in 1822 (No. 4688, for 'mathematical chains'), which was notably light, strong, and portable. This may have been the earliest example of an iron lattice truss. He

exhibited a timber lattice bridge at his works in 1824, a demonstration of the principles of his patent iron truss and its applicability to timber construction. He is believed to have introduced the idea of Tile Creasing, an early method of fire proofing construction used by Charles Fowler (q.v.) at Hungerford Market.

At his death in 1834 his wife, Eliza, survived him, as did his son, Nevil, farmer of Finchley, and grandchildren.

TESS CANFIELD

[ICE membership records; John G. James Collection, Archives of the ICE; *Monthly Magazine*, **22**, 1806, 10–11; Gold medal for chimney cleanser, *Trans Soc Arts*, **25**, 1807, 97; G. Gregory (*c.* 1812) *Shipbuilding, Dictionary of the Arts and Sciences*, 2nd edn., 661–662; T. Martin (1813, 1818) *Coopering, Circle of the Mechanical Arts*, 236–238; Account of brush work (1816) *European Magazine*, **70**; Chimney sweeping (1816) *New Monthly Magazine*, **6**; Bowstring rafter (1819) *Soc Arts*, **37**, 100–102, pl. 13; Description of Smart's timber bow string rafter (1823–1824) *Herbert's Register of Arts and Sciences*, **1**, 357–358; Mr. George Smart's patent iron bridge (1824) *Herbert's Register of the Arts and Sciences*, **II**, 49–51; Chimney sweeping (1834) *Mech Mag*, **21**, 420; G. D. Dempsey (1850) *Rudimentary Treatise on Tubular Bridges*; Discussion of G. L. Molesworth paper (1857) *Proc ICE*, **17**, 17–51; J. G. James (1982) The Evolution of the Iron Truss Bridge to 1850, *Trans Newc Soc*, **52**, 67–101]

Publications

1804. *Description of the Chimney Cleanser, and Method of using it*

1808. Account of experiments on sweeping chimnies, *Nicholson's Journal*, **xxi**, 170

SMEATON, John, FRS (1724–1792), civil engineer, was born in the parish of Whitkirk, Leeds, Yorkshire, on 28 May 1724, and was baptised in Whitkirk church on 22 June. He was the eldest of three children of William Smeaton (1684–1749) and Mary Stones. William Smeaton had a successful practice as an Attorney at Law in Leeds. John Smeaton's great grandfather, John, lived in York where he was well known as a watchmaker and a Freeman of that city. It was Smeaton's grandfather, also John, who left York to set up in business in Leeds. He prospered and built Austhorpe Lodge, the family home, in Whitkirk in 1698. It was here that John Smeaton was born and which was to be the base for his subsequent lifelong consulting engineering practice. At the age of ten he entered Leeds Grammar School where he remained until he was sixteen years old. By this time he already had a workshop at home and two years after leaving school he had built himself a lathe on which he could cut a screw thread, he could also melt metal, forge iron and had a range of tools at his disposal. He had made the acquaintance of Henry Hindley of Leeds, a talented clock and scientific instrument maker, and they conversed at length on scientific

John Smeaton FRS

matters. The fact that Hindley treated the young Smeaton as a friend, and even as a colleague, from their first meeting is evidence of Smeaton's ability to attract confidence and respect in those who knew him. When Smeaton was eighteen years old his father became somewhat anxious about his son's career prospects and in the autumn of 1742 he sent John to London to receive a legal education at Gray's Inn. Smeaton returned home in the summer of 1744 and Richard Beamish (q.v.) later said of this period of Smeaton's life that 'mechanical design and the construction of models had more interest than the drawing of deeds, or the engrossment of parchment, to which he had been destined by his father, but against which his nature rebelled'. Smeaton spent the next three or four years at Austhorpe developing his practical and experimental skills in constructing a set of mathematical instruments with a view to becoming a scientific instrument-maker. At the same time he was in correspondence with friends about many scientific and technical topics and particularly on astronomy which was to become a life-long interest.

In 1748, he again left home and returned to London, with his parents' blessing, to set up in business as a scientific instrument-maker. He first gained notice with his improved vacuum pump completed in 1749. In July 1750 he presented a paper to the Royal Society on a new mariner's compass that he had developed in conjunction with Gowin Knight, whom he accompanied on a sea trial in HMS *Fortune* in 1751. By this time Smeaton was employing three craftsmen in his London workshop and in December had moved to larger premises in Furnival Inn Court. It was in this building in 1752 that he conducted his important model experiments on the power of water and windmills. His intelligence and clarity of mind were clearly demonstrated in the design of the test rigs and the manner of dealing with the results. The following year he began to apply his findings to practical mill design. In 1754 his engineering experience was broadened by a consultation about the drainage of Lochar Moss in Dumfrieshire. In 1756 he was first approached about the design of a new Eddystone Lighthouse. His success there made his reputation as a civil engineer, and whose talents were beyond the ordinary.

Just before he began his civil engineering career proper, Smeaton undertook a five-week long technical *Grand Tour* of the coastal lowlands of Belgium and Holland. He left London on 15 June 1755 Here again his acute observation, powers of technical description, and intelligent critical assessment of all he saw is revealed in his diary of the journey. On 1 July he was in Leyden and met Professor Mussenbroek whom he described as 'truly an experimental philosopher'. In the diary he describes pumping machinery, wind and water-powered mills in a variety of industrial applications, sluices, dredgers, details of mechanisms of all types, and the construction of ships and military fortifications. It is obvious that he had read the technical literature in advance as on visiting mill sites he typically states that a construction 'conforms minutely to the description in the Dutch Mill Books' and on visiting a hemp mill he critically described the machinery as 'sufficiently ill-applyed, as well as clumsily executed'. But he also saw much to admire and the tour must have greatly helped to equip him for his subsequent civil engineering career. He returned to Britain on 19 July, landing at Blyth on the Northumberland coast.

On 7 June 1756 Smeaton married Ann Jenkinson, of York, in St. George's Hanover Square in London. John and Ann had three daughters, Hannah born in 1757, Ann born 1759, and Mary born 1761.

In 1768 Smeaton, with his usual clarity, defined his view of the consulting civil engineer saying, 'I consider myself in no other light than as a private artist who works for hire for those who are pleased to employ me. ... They who send for me to take my advice upon any scheme, I consider as my paymasters; from them I receive my propositions of what they are desirous of effecting; work with rule and compass, pen, ink, and paper, and figures, and give them my best advice thereupon. ... I endeavour to deliver myself with all the plainness and perspicuity I am able.'

Smeaton ran his consulting engineering practice largely from his home at Austhorpe Lodge in a purpose-built square tower, four stories high. The basement contained his forge; the first floor his lathe; the second his models; the third was his drawing office and study, and the fourth he used for astronomical observations. Smeaton considered engineering drawing as an important aspect of his work saying 'the rudest draft will explain Visible things better than many words' and that 'I do not think it within the compass of human knowledge, to form the best possible Design *at once*. Things are better finished, by touching and retouching as is usual, and necessary to the greatest painter'. Smeaton employed only two draughtsmen during his career. The first was William Jessop (1745–1814) (q.v.) who was the son of Smeaton's Resident Engineer at the Eddystone Lighthouse. Smeaton employed the fourteen-year old Jessop after the death of his father, and he stayed with Smeaton for fourteen years. Smeaton's second draughtsman was Henry Eastburn (q.v.) who entered the Austhorpe office in 1769. In discussing Smeaton's prodigious engineering work it will be best to treat the subjects thematically rather than chronologically.

Scientific work

Although best known as a civil engineer Smeaton was untypical in that he made a number of interesting experimental scientific studies out of pure intellectual curiosity. His scientific work is closely connected with the Royal Society with which he was associated for forty or more years, being elected a Fellow in 1753. The Society published eighteen papers by Smeaton in the *Philosophical Transactions* between 1750 and 1788. The subjects fall into three main categories, namely: scientific instruments, experimental mechanics and astronomy. He experimented and wrote on a range of topics including the mariner's compass, the air (or vacuum) pump, the ship's log, a new pyrometer, and the thermal expansion of metals. In addition he experimented on subjects that were more directly concerned with civil engineering, namely wind and water mills, mechanic power and impact, friction loss in pipes, the practical measure of horsepower and the strength of hydraulic limes Undoubtedly his most important work was the series of elegant experiments he made on model water wheels and windmills in 1752. He designed and made the apparatus himself and revealed his awareness of the fundamental problem in model experimentation by stating at the outset, 'in this case it is very necessary to distinguish the circumstances in which a model differs from a machine in large; otherwise a model is more apt to lead us from the truth than towards it'. He compared the performance of undershot and overshot waterwheels finding that the latter were about twice as efficient as the former. The results of these experiments, made a year before he built his first waterwheel, were not published in the *Philosophical Transactions* until 1759 as he was concerned not to publish the results until he had 'an opportunity of putting the deductions made therefrom in real practice, in a variety of cases, and for various purposes'. His results achieved a wide circulation and were being cited in engineering literature well into the twentieth century. His experiments on a whirling-arm windmill model were of somewhat less practical value and dealt with such issues as the most efficient aerodynamic warping, or 'weathering' of the sail surfaces, and the optimal area of canvas. Nevertheless he greatly influenced millwright practice and his results were still being quoted in the technical literature a century and a half later. To study friction effects in pipes Smeaton experimented with a pipe 100 ft. long and with a bore of 1¼ in. and showed that the relationship between the head loss (h) in inches and the velocity of flow in inches/second (v) was of the form: $h = 0.1v + 0.035v^2$. Smeaton's scientific works distinguishes him from his contemporary civil engineers.

Mills and millwork

Millwork practice was a regular feature of Smeaton's long engineering career and is therefore an important aspect of his work. Smeaton was not a traditional millwright but considered himself a designer of works that others would build. He designed nearly sixty new millworks during his career, the earliest in 1753 and the last in 1791. He also was much involved in supervising modifications to existing mills and was occasionally retained to arbitrate in disputes between rival mill-owners and others. The majority of Smeaton's mills designs were for watermills, he built very few windmills, and even fewer horse-driven mills. The first stage in a new watermill design is to investigate the site to determine whether there is an adequate supply of water and Smeaton would sometimes delegate this work to a local surveyor, but on other occasions would undertake the survey himself. The site survey would also determine whether an undershot. overshot, or breastshot waterwheel would be suitable. Unlike the traditional millwright, Smeaton had a professional attitude to his design work saying that his main objective was 'to get the most Power possible out of a given quantity of water in dry seasons' and that 'The construction of Mills, as to their Power, is not with me a Matter of *Opinion* it is a Matter of *Calculation* and I should draw the same result from the *Data* this year, that I did Twenty years ago'. In a Report of 1780 Smeaton stated his philosophy of the design of mill machinery saying, 'I have endeavoured to assemble the following properties, that is, strength and durability of all the matter, materials, and fundamental parts, ease of repair of such as are liable to decay by weather or wear; to have all possible simplicity, … and the whole to be of such form and construction as to be capable of being adequately repaired and kept in proper order by the artificers in the country'.

Smeaton designed mill machinery for a wide range of applications including corn, oat and snuff mills, oil mills, water-pumping, water-powered machinery for ironworks driving rolling and slitting mills, tilt hammers, and blowing engines for blast furnaces. He also designed edge-runner mills for incorporating the ingredients of gunpowder. His designs followed common practice in having a vertical wheel carried on a horizontal axle. On this axle a vertical 'pit-wheel' drove a horizontal 'wallower' on a vertical shaft Smeaton had a scale of fees covering travelling and subsistence, but for the preparation of drawings at Austhorpe he said in August 1782, 'As making the Designs for the Erection of Mills for the direction of Workmen is a work of Time he always does it at home. His constant Price for water Mills for Corn, Oil, &c. for thirty years past has been, &Is 25 Guineas each sett, … Wind Mills 30 Guineas'.

Smeaton was a pioneer in the application of cast iron to water and windmill mechanisms, and frequently suggested the use of iron to his clients. Writing to the Carron Company in February 1782 he said:

> … in the year 1755, that is 27 years ago, for the first time, I applied them as totally new subjects, and the cry then was, that if the strongest timbers are not able for any great length of time to resist the action of powers, what must happen from the brittleness of cast iron? It is sufficient to say … that the good effect has … drawn them into common use, and I never heard of any one failing.

In order to improve the supply of water to a waterwheel Smeaton also designed six river dams associated with ironworks during the 1770s. His largest dam, built in 1776, was for an ironworks on the river Coquet in Northumberland. It was an arched masonry structure, 162 ft. wide with a vertical air face 8 ft. high and built to a radius of 170 ft. Referrring to this dam Smeaton said, 'There is not a more difficult or hazardous piece of work within the compass of civil engineery than the establishment of a high dam upon a rapid river that is liable to great and sudden floods'. Nevertheless, his dam on the river Coquet survives today.

He built only four windmills, although there are surviving design drawings of others. The windmills built were at Wakefield (1755), Leeds (1774), Sykefield (1781) and Newcastle-on-Tyne (1782). The typical Smeaton windmill had five sail arms and a windshaft which projected in front of the plane of the sails. Adjustable rope stays, or chains, attached to the end of the windshaft ran diagonally to the whips to restrain deflection.

Smeaton designed a few horse-driven mills during his career. In 1761 he designed a horse-driven archimedean screw at Kew for the Dowager Princess of Wales. The circular horse track, 25 ft. diameter was contained in an octagonal building. The screw was 24 ft. long, 2 ft. 6in. diameter and lifted 13½ gallons per turn, through 11 ft. After a trial Smeaton wrote 'The horses at easy work went 3 turns per minute, but if they were at all pushed, they caused the screw to make 20 turns' which he described as 'light work for 2 light horses'. Having been commissioned in

1779 by HM Victualling Office to examine the pumping of water at a brewery near Gosport he designed a horse-driven well pump and advised his clients:

> I expect that two horses at the New well engine will perform the ordinary service … but, if the full quantity of 200 tons be required to be daily delivered, then, to keep the service going during twenty-four hours, will require three sets and one spare horse in case of accidents and to rest, that is, in the whole seven horses.

But, apparently keen to suggest an alternative technology, he said 'If thought necessary, by way of easing the horses, a wind engine might be raised upon the same building …'. Smeaton designed four waterwheel-driven pumping plants for water supply. His designs comprised an ensemble of waterwheel, crank, rocking beam and reciprocating pumps. The largest of such schemes was installed in the fifth arch of old London Bridge in 1767–1768.

The Eddystone Lighthouse

In December 1755, just after Smeaton had returned from his tour of the Lowlands, the Eddystone lighthouse was destroyed by fire. The lighthouse was on a wave-swept rock some 14 miles south of Plymouth. Robert Weston was the principal shareholder of the Proprietors of the Lighthouse, leased from Trinity House, and he approached the President of the Royal Society for advice who recommended John Smeaton. Smeaton had only recently embarked on civil engineering work and this project was to prove a major turning point in his career, becoming one the busiest consulting civil engineers during the second half of the eighteenth century. Smeaton was in Scotland when Robert Weston invited him to rebuild the lighthouse and they first met in February 1756. Smeaton was first in Plymouth from 27 March to 15 May where he met Josias Jessop who was to be his resident Engineer on the project. During this period ten trips were made out to the rock and landing was possible on only four occasions. Smeaton made major design decisions during this summer. Portland stone was to be used for the inner work with a cladding of the more-durable Cornish granite. A survey of the sloping rock was made and Smeaton designed an ingenious instrument for this purpose, He later wrote 'I think my reader will expect some general idea of what had employed me upon the rock for 3 successive tides, amounting in the whole to full 19 hours; and this was the taking such dimensions as would enable me to make an accurate model of such part of the surface of the rock as we were likely to have any concern with the rebuilding'. Smeaton paid particular attention to the anchoring of the structure to the rock and cut a series of steps to take the various courses of interlocking stones. At the same time he was making experiments to improve the lime/Pozzolana mortar he proposed to use. This work, which became generally known in the 1790s after the publication of Smeaton's *Narrative* on the lighthouse, aroused the interest of engineers in the properties of lime and cements, and anticipated the explosion in the use of concrete in the nineteenth century. Stone was being quarried, delivered to Plymouth where it was dressed ready for delivery to the rock. During August 1756, work on setting out the steps to be cut in the rock began as well as the cutting and levelling. Work during the first season on the rock ceased for the winter on 26 November. The winter was spent in making drawings and models. On 12 August 1757 the first stone, weighing 2½ tons was laid in position. The whole project required the cutting, transporting and fixing of 1,493 pieces of stone, weighing about 1,000 tons in total. Work continued during the summer of 1758. The summer working season of 1759 began on 12 May and by 17 August the main structure was up to cornice level. On 17 September the cupola of the iron lantern was lifted in one piece and fixed in position. The light was first exhibited on 16 October and Smeaton's work was complete. Smeaton played a particularly individual part in the work, personally setting out the work on the rock, making the moulds for cutting the stones and supervising every new or critical operation.

Inland navigation

From 1757 Smeaton was regularly consulted about improvements to navigable rivers and the design of new canal navigations. A common problem with all such works was reconciling the needs of those who regarded the rivers as a source of drinking water, the millers requiring waterpower and those who saw it as a transport route. He was involved with five major inland navigation schemes, the first of which was the Calder and Hebble Navigation close to Smeaton's home. The river was to be made navigable by making cuts and locks. Smeaton was first consulted in 1756, but pleaded that he too busy with the Eddystone works. But asked again in 1757 he agreed and made his survey in October and November. Smeaton was appointed Engineer from November 1759 and after most of the work had been done Smeaton left the work in January 1765 when he was succeeded by James Brindley. In 1766 Smeaton designed a scheme for upgrading the ancient River Lee and the work was carried out in 1767–1771. At the same he was working on the Ure Navigation and the Ripon Canal carried out in 1767–1772. But, undoubtedly, Smeaton's greatest canal work was the Forth and Clyde Canal designed to provide an east-west navigation joining two seas. Smeaton had reported on this proposed scheme in 1755 and 1758. After many routes were discussed the canal as built was 35 miles long with a 16 mile long summit level 155 ft. above low water in the river Forth. A flight of twenty locks was required to reach the summit level from the Forth and a further nineteen locks to descend to the Clyde. When most of

the canal was built Smeaton left his 'great canal' in September 1773. When, in 1785, he was requested to return to the job he said 'it will be better to make a fresh beginning with a fresh engineer' and Robert Whitworth (q.v.) was appointed. Smeaton's last canal work was on the Aire and Calder Navigation, constructed in 1775–1779, although his advice was sought on many other proposed works up to June 1791. Smeaton's pupil and assistant, William Jessop became the leader of the next generation of canal engineers.

Fen drainage

Land surveyors had undertaken the drainage of land for agricultural purposes for many centuries before Smeaton's time. Promoters of land improvements, particularly in the low-lying east coast land of East Anglia, Lincolnshire and south Yorkshire were keen, from the mid-eighteenth century, to consult the newly emerging civil engineers. Land drainage and navigation work were often related and in reporting to a client in 1789 Smeaton said 'For my own part, I am professionally as great a friend to drainage as to navigation ... If I can shew ... how the drainage may be very materially improved ... by means that will improve the navigation also, and that without any new charge, I shall be doing an evident service to both'. Smeaton was first consulted on drainage in 1754 about problems at Lochar Moss near Dumfries—his first civil engineering project. He arrived in Dumfries in August and submitted his Report on 21 September, although his proposed scheme was not carried out. Smeaton was involved with fen drainage schemes, which were put into practice, certainly from 1762 and his last such work was completed in 1789. Typically, Smeaton's design work was informed by accurate surveys and methods of gauging flow in rivers and channels and of calculating the discharge through sluices. He designed fenland drainage schemes for Potteric Carr, south-west of Doncaster, implemented in two stages; 1765–1768 and 1772–1777; for Adlingfleet, the work directed by John Grundy (q.v.); and in Hatfield Chase.

Bridges

Bridge repair was the subject of Smeaton's first engineering study, written in 1748, when he devised a method of repairing a sunken pier of Westminster Bridge. Smeaton built ten road bridges during his career, (excluding his two canal aqueducts). He built in stone or brick. Cast iron bridge structures emerging only in the 1770s, rather late in his career, he never adopted metal for bridge building. His four major bridge structures are Coldstream (1763–1767) over the Tweed, Perth (1766–1771), over the Tay, Banff (1772–1779) over the Deveron, and Hexham (1777–1780) on the Tyne. Three of these structures survive. His first bridge design was for Coldstream on the river Tweed. This handsome stone structure has five segmental arches, the largest being 60.7 ft. The Bridge at Perth, of similar design, has seven arches, the largest of 75 ft. Banff Bridge also has seven arches, the largest being just 50 ft.

His smaller bridges in stone are at Plymouth (1773), Altgran, Ross & Cromarty (1777)—both single arch spans—and Amesbury, Wiltshire (1775). In addition his two brick-built structures are the 74-arched bridge over the river Trent flood plain at Newark, and the handsome red-brick bridge (1778) on the Whitbread Estate at Cardington near Bedford. Earlier in his career Smeaton had been consulted about Blackfriars Bridge in London, and in July 1776 wrote a lengthy Report on the reduction of water to the waterwheels of London Bridge.

Smeaton designed his bridge at Hexham, spanning the river Tyne, in 1777 and it was completed in 1780 under Jonathan Pickernell (q.v.) as resident engineer. It was his largest bridge structure having nine arches. Soon after it was completed one of the piers subsided and repairs were made. However, in the spring of 1782 an unprecedented and violent flood swept away some of the pier foundations and arches. After hearing of the accident Smeaton wrote to Pickernell saying 'All our honours are now in the Dust! It cannot now be said, that in the course of 30 years' practice ... not one of Smeaton's works has failed. Hexham Bridge is a Melancholy witness to the contrary'.

When requested to rebuild the bridge he said, in October 1783, 'I would beg you to consider ... employing some other able Engineer who has not got the Horrors of the River Tyne painted upon his imagination'.

Steam engines

Thomas Newcomen (q.v.) built his first 'atmospheric' beam engine in Staffordshire in 1712. This type of engine dominated pumping technology during the first two-thirds of the eighteenth century, and Smeaton designed several such Newcomen-type engines making improvements that greatly increased their performance. His first steam engine to be built was for the New River Company in London when he was consulted about the pumping arrangements at New River Head in Islington. Horse-driven pumps were being used to pump water from the round pond to the upper reservoir at Claremont Square. After investigations Smeaton recommended a Newcomen-type beam engine. This he designed in 1767 and it was at work in Islington early in 1769. An unusual feature of the engine was asymmetrically-pivoted laminated timber beam. However its performance was disappointing and Smeaton said 'By good luck the Engine performed the work it was expected to do, as to the raising of Water, but the Coals by no means answered calculation'. Typically, Smeaton built an experimental engine. It had a 10-in. cylinder and developed about 1 hp. Beginning in 1770 he made about one-hundred and thirty tests and as a result was able to optimise the various features of the engine resulting in considerable improvements in performance. He went on to make a number of large engines for coalmine pumping

and winding, dry-dock pumping, and returning engines for overshot waterwheels.

In 1772 Smeaton designed a pumping engine for Longbenton Colliery in the Tyne coalfield. The engine had a laminated timber beam 22 ft.-long and 50 in. deep at the centre. The cylinder was 52 in. in diameter with a stroke of 7 ft. During a test in May 1774 when working at 12 stokes per minute it developed about 40 hp. This was a landmark in power engineering and was 25% better than the most efficient engine yet built. Five years later, for the same colliery, he designed a winding engine and to achieve rotary motion he used the beam engine to return the tail water from a 30 ft. diameter overshot waterwheel to the top of the wheel. In 1773 Smeaton received his only overseas commission for a pumping engine. This was for a dry dock at the Russian naval dockyard at Kronstadt on the River Neva near St. Petersburg. The engine, which was built by the Carron Company near Falkirk, had a 66-in. diameter cylinder with a stroke of 8 ft. 6 in. The engine replaced two wind-pumps and was set to work in 1777.

Smeaton was always aware of what his contemporaries were doing and when James Watt invented his improved steam engine with a separate condenser Smeaton entered into a long correspondence with Boulton & Watt. Writing to them in January 1778 he said:

> Ever since I have been acquainted with your method of condensing out of the cylinder I have been much struck with the ingenuity of the thought; & the promising Probability of Improvement; and still more with your last method by an Air Pump.. But it always appeared to me equally applicable to Newcomen's Engine as to your own ... I have now an opportunity of making trial of your Condenser as applicable to a common Engine; and as this may probably be adopted upon the Coal Pit Hill more readily than your double Cylinder, I shall be glad to know if you are mindful to encourage an Application of this kind.

Writing again in March 1778 he said 'I have only now to add, Gentlemen, that I wish you all the Success that your ingenious discoverys and indefatigable Labours have deserved, and shall endeavour to promote your interest'. This Smeaton did by recommending his clients, on several occasions, to approach Boulton & Watt and only built his own improved Newcomen engines at collieries where the cost of fuel was 'of little consequence upon the coalpitt hill' and ultimate thermal efficiency was not an issue. His last engine was designed for the Walker colliery in 1783.

Harbours

Smeaton's drawings and reports on the subject of harbours are the most numerous of the range of engineering subjects he tackled, and cover thirty sites in England and Scotland. between 1776 and his death in 1792. Based in Leeds, he had to travel great distances, by carriage or on horseback to visit harbour sites from Banff in the north of Scotland to St. Ives in Cornwall. Although Smeaton was consulted frequently about harbour works, this only led to construction work at eight sites. These harbours were St. Ives in Cornwall, Eyemouth, Rye in Sussex, Portpatrick, Aberdeen, Peterhead, Cromarty, and Ramsgate.

Smeaton published his Report on St. Ives Harbour in 1766 for the design of a pier, 360 ft. in length, containing about 35,000 tons of stone. It was built to replace an existing pier, 1767–1770, and is still known as 'Smeaton's Pier'—the only work commemorating his name.

At Eyemouth a pier had been built in 1747 but the harbour was still vulnerable to northerly gales. Smeaton produced a design in 1767 for a technically-innovative pier, 340 ft. long, to give the desired protection. This was built in 1768– 1770.

In 1768 The Lord Provost of Aberdeen wrote to Smeaton asking for his advice on the improvement of the harbour. Smeaton reported in 1770 suggesting the removal of the bar at the river mouth and the building of a pier on the north side of the mouth of the river Dee. The work was carried out in 1775–1780 and cost £16,000. He was consulted again in 1787. After visiting Aberdeen he reported, in 1788 recommending building a 'catch pier' projecting at an angle from the north pier. This was built 1788–1790 at a cost of £1,500.

But Smeaton's major harbour work was undoubtedly that at Ramsgate on the Kent coast. The Trustees of the Harbour consulted Smeaton in the summer of 1773 but he was unable to visit Kent immediately. However he made his investigations on site in April 1774 and reported in October. The problem was the siltation of the harbour. Sand was being deposited at the rate of about 400 tons per week and Smeaton estimated that about 300,000 tons of sand was already in the harbour. In his report Smeaton says 'This is the natural tendency of all harbours; for wherever there is Mud or Matter to deposit, an Addition to the Soil is the natural Consequence of a Place of Repose'. His proposed solution involved building a basin in the inner harbour, which filled at high tide, and the water discharged at low tide through sluices in the basin wall, thus scouring the sand out to sea. Smeaton reported again in 1777 and, with modifications, the scheme was completed in 1781.

The Trustees of the Harbour approached Smeaton again and he produce two further Re-ports in 1783 and1788. These Reports include the construction of a dry dock and the Advanced Pier which projects from the East Pier. Smeaton was involved with Ramsgate Harbour right up to 1792, the year of his death.

Professional practice

John Smeaton was one of the first to describe himself as a *Civil Engineer* and, more importantly, he regarded himself as being a *professional*, person and frequently used the word in correspondence with his clients. He didn't enter the engineering world from a craft apprenticeship and almost certainly gained his professional concept from his lawyer father. In a Report of 1764, for example, he said 'I think it necessary to give my opinion with that freedom that becomes me in a matter wherein I am consulted, and … where my opinion is desired in the way of my profession' and again in 1782, 'I take the opportunity of reporting my opinion, as a professional man'. His understanding of what constituted a professional attitude included such matters as a scale of fees, training the next generation of engineers, sound estimates with contingency allowances and an ethical basis to relationships with clients and other professionals. It is clear that Smeaton considered himself in a kind of partnership with the craftsmen who would execute his designs saying 'It is unnecessary to enter into a minute description of the operation of those parts which are made as common, and are obvious to all practitioners in this kind of work' and 'that where I have given no particular directions, as I suppose myself addressing my designs to ingenious engineers, acquainted with the usual practice, I suppose the thing to be done in the usual way, or subjected to convenience and their discretion'.

In January 1783 Smeaton wrote 'In the civil engineering business many are self taught; many come into it gradually, from being led to assistant business: I am one of the self taught:' and early in 1788 he corresponded with Professor Copland of Aberdeen University concerning the education of engineers and said:

> I very much like your plan of a set of lectures upon the practical part of Mechanical Philosophy. I have very long thought it a great pity, that Nothing of that kind was ever set on foot in any of our British Universitys. I have often thought the practical mechanical sciences might be taught in Colleges, with the same success that the chemical and anatomical sciences have been, provided there were but professors who were themselves equally men of practice: Here however was the rub.

Smeaton clearly stated his scale of fees in a letter of August 1782 saying, 'For 14 years past Mr. Smeaton's Terms of Business abroad have been and are 2½ Guineas p. day during the whole time of being out from home, with the addition of 2½ Guineas upon each Day actually employed upon the Business in hand; The above includes his personal Expenses but to be allowed what he pays out of pocket in Coach & Chaise Hire'.

In 1768,with reference to the Forth and Clyde Canal, he produced a model framework of the management structure for carrying out civil engineering works in which he describes meticulously his views on the duties of the *Engineer in Chief*, the *Engineer Resident*, and of the *Surveyors of Particular Districts.* He remarked on the importance of communication on management and, in particular, the need to 'preserve a good understanding between the Resident Engineer and the Committee that directs him' and obviously drew on his own experience when he added

… but the great difficulty is to keep Committees from doing either too little, or too much; too little when any case of difficulty starts, and too much where there is none.

An interesting feature of Smeaton's correspondence with clients is that it contains a litany of reasons why he should decline taking on further work. Smeaton explained that 'By a forced march in the year 1754 I caught a violent cold which brought on an Asthmatic Complaint, that still remains with me; and has ever since left me in an habit far too tender a delicate for my profession'. Which explains why, as early as 1764, he was minded to reduce his workload and wrote to Robert Weston saying:

To tell the truth I am rather looking out for small jobbs than large ones, as being attended with less care and anxiety of mind, and as least as much profit. Know then, that the renowned builder of the Eddystone Lighthouse does not think it below him to undertake the making of a pump: Indeed I begin to want a respite from high and mighty achievements ….

In 1783 he wrote that he 'had totally declined the entering upon any new subject, that I may dedicate the Remains of my life to … publishing my Works that others of the Profession may be aided by my Experience'. He began his writing in the winter of 1783–1784 with his account of the building of Eddystone Lighthouse (published in 1791) In February 1785 he considered this project as:

A work that I expect will do more towards the forming of Civil Engineers in the future generation, than any thing that has yet appeared in favour of the Profession; an indeed of much more real consequence to the public (if it is supposed that I am myself of any consequence) than any fresh examples I can now give in the executive part; the fundamental principles whereon I proceed, not appearing, but in the effect.

In March 1771 a group of the leading engineers of the period met in Holborn and 'agreed that the civil engineers of this Kingdom do form themselves into a Society'. Smeaton was a founder member and remained a member of the Society until his death.

In 1789 Henry Eastburn, Smeaton's assistant at Austhorpe was elected a member. After Smeaton's death the Society added the prefix 'Smeatonian' to its title.

Conclusion

John Smeaton died of a stroke on 28 October 1792 in Austhorpe Lodge. He was buried on 1 November in the parish church of St. Mary's, Whitkirk where his memorial on the north wall of the chancel is crowned by a relief model of the Eddystone Lighthouse. The inscription predicts that his lighthouse 'seems likely to convey to distant ages, as it does to every Nation of the Globe, the Name of its constructor'. His Will, dated 16 April 1792, makes dispositions to his kinsman John Holmes and others including 'John Reynolds whom I have acknowledged as my natural son' who was born in August 1775.

Smeaton was a first-rate engineer with a professional view of his chosen occupation, whose design principles were informed by a scientific outlook and to which was added good practice. His obituary in the *Gentleman's Magazine* said:

To the Publick, in whose service this gentleman spent the most valuable Part of his life, his death may be, eventually, a serious inconvenience; mechanical knowledge equal to his being very rare … As a civil engineer, Mr. Smeaton was not equalled by any of the age he lived in; it may, perhaps, be added, by none of any previous age.

And in all his professional work he was said to have 'distinguished himself by his modesty, punctuality, and undeviating integrity'.

DENIS SMITH

[Machine Letters and other correspondence, ICE Archives; Smeaton drawings, Royal Society archives; *Reports of the Late John Smeaton* (1812); S. Smiles (1861) *Lives of the Engineers*; A. W. Skempton (ed.) (1981) *John Smeaton FRS*]

Publications

1754. *Report on the Drainage of Lochar Moss, near Dumfries; Drawn up for Charles Duke of Queensberry and others* (published 1812)
1757. *From a Survey of the River Calder, from Wakefield to Brooksmouth, and from thence to Salter–Hebble Bridge, near Halifax, taken in the Months of October and November 1757, by John Smeaton, it appears …*
1757. *A Plan of the River Calder from Wakefield to Brooksmouth and thence to Salter Hebble Bridge, laid down from a Survey taken in October and November 1757; with a Projection for continuing the Navigation from Wakefield to Salter Hebble Bridge near Halifax in the County of York*
1758. *A Plan of the River Clyde in the County of Clydesdale from Dumbarton to Rose-Bank, laid down from a Survey 1758, by James Barry, with a Projection of the Navigation*
1759. *A Plan of the River Wear, from Biddick Ford to the City of Durham, with a Projection of a Navigation thereon*
1760. *An Experimental Enquiry concerning the Natural Powers of Water and Wind to turn Mills, and other Machines, depending on a circular Motion*
1760. *Mr. Smeaton's Answer to the Misrepresentations of his Plan for Black-Friars Bridge, contained in a late anonymous Pamphlet, addressed*

to the Gentlemen of the Committee for building a Bridge at Black-Friars

1761. *The Report of John Smeaton, Engineer, concerning the Practicability of a Scheme of Navigation, from Tetney Haven to Louth, in the County of Lincoln, from a View taken thereof, in August 1760; As projected by Mr. John Grundy of Spalding, Engineer*

1761. *Report on the River Witham* (with Grundy and Edwards)

1763. *The Report of John Smeaton, upon the Harbour Rye in the County of Sussex*

1763. *The Report upon the Question proposed to him by the Committee for improving, widening, and enlarging London Bridge*

1764. *Plan of part of Holderness with a Scheme for Draining the same* (with Grundy)

1765. *Proposal for laying the Ships at the Key at Bristol Constantly Afloat, and for Enlarging Part of the Harbour by a New Canal through Cannon's Marsh*

1766. *The Report of John Smeaton, Engineer, upon the New-making and completing the Navigation of the River Lee, from the River Thames, through Stanstead and Ware, to the Town of Hertford*

1767. *A Plan of the River Lee from Hertford to the River Thames, with a Profile of the Fall*

1767. [Report relative to the Petition of the Proprietors of the London-Bridge Water-Works] *To the Right Honourable the Lord Mayor, Aldermen, and Commons, in Common-Council Assembled*

1767. *A Report upon the Harbour of King's Lynn, in Norfolk* (2nd edn., 1767)

1767. *A View of the Bridge over Tweed at Coldstream. Finished in December 1766*

1767. *The Report of John Smeaton Engineer, and FRS concerning the Practicability and Expence of joining the Rivers Forth and Clyde by a Navigable Canal, and thereby to join the East Sea and the West* (2nd edn., 1768)

1767. *The Second Report of John Smeaton, Engineer, and FRS touching the Practicability and Expence of making a Navigable Canal from the river Forth to the river Clyde ... for vessels of greater burden, and draught of water, than those which were the subject of his First Report* (repr.)

1768. *A Review of Several Matters relative to the Forth and Clyde Navigation, as now settled by Act of Parliament: with Some Observations on the Reports of Mess. Brindley, Yeoman, and Golburne*

1768. *Mr. Smeaton's Report on Lewes Laughton Level*

1768. *The Report of John Smeaton, Engineer, concerning the Drainage of the North Level of the Fens and the Outfall of Wisbeach River* (and 2nd edn.)

1769. *The Report of John Smeaton, Engineer, upon the Harbour of Dover*

1769. *Report on the new bridge at Edinburgh* (with Adam and Baxter)

1770. *The Report of John Smeaton, Engineer, upon the Harbour of the City of Aberdeen* (printed 1772)

1770–1774. *The Report of John Smeaton, Engineer, upon the Harbour of Port Patrick* [and] *Explanation of the Plan for completing the Interior Harbour* (printed 1809)

1772. *The Report of John Smeaton, Engineer, upon the means of improving the Navigation of the rivers Aire and Calder, from the free and open Tides-way to the Towns of Leeds and Wakefield respectively*

1772. *Plan of the Rivers Air and Calder, from the Towns of Leeds and Wakefield to Snaith*

1772. *Aire and Calder Navigation—7 plans*

1772. *Report on Tyne Bridge* (with J. Wooler)

1772. *The Report of John Smeaton, Engineer, upon the Plan and Projection of a Canal, upon the North-Side of the River Air from Haddlesey to ... Brier Lane End in the Township of Newlands; as laid down by Mr. Jessop, Engineer*

1773. *Plan of the Rivers Air and Calder* (with W. Jessop)

1773. *Letters between Redmond Morres, Esq.; one of the Subscribers to the Grand Canal, and John Smeaton, Esq.; Engineer, and FRS in 1771, and 1772, relative to the Manner of Carrying on that Navigation* [etc.]

1773. *The Report of John Smeaton, Engineer, upon his view of the Country through which the Grand Canal is proposed to pass, and in answer to the several Matters contained in the Queries ... agreed to on the 3d of August 1773*

1776. *The Report of John Smeaton, Engineer, upon the Means of improving the Drainage of the Level of Hatfield Chace*

1776. *Estimate for the Excution of the New Works proposed for the Improvement of the Drainage of the Level of Hatfield Chace ...*

1778. *A Report of John Smeaton, Engineer, upon the state of the Bridlington Piers*

1780. *The Report of John Smeaton, Engineer, upon Mr. James Shout's Plan for rebuilding and extending the old pier of the harbour of Sunderland, in the County of Durham*

1782. *The Report of John Smeaton, Engineer, upon the state and Condition of Wells Harbour in the County of Norfolk*

1783. *Mr. Smeaton's Memorial Concerning Hexham Bridge*

1788. *Second Report by John Smeaton, Engineer, upon the Inrun of the Seas into the Harbour of Aberdeen, in Easterly Winds* [with] *Letter from Mr. Smeaton respecting the Harbour of Aberdeen* (printed 1834)

1791. *A Narrative of the Building and a Description of the Construction of the Edystone Lighthouse with Stone: to which is subjoined, an Appendix giving some Account of the Lighthouse on the Spurn Point, built upon a Sand* (2nd edn., 1793; 3rd edn., 1813)

1791. *An Historical Report on Ramsgate Harbour: written by Order of, and Addressed to the Trustees* (and 2nd edn.)

1794. *Experimental Enquiry concerning the Natural Powers of Wind and Water to turn Mills and other Machines* (and two other papers; 2nd edn., 1796)

John Smeaton FRS: civil engineering works

Date	Work	Cost (£)	Resident Engineer
1756–1759	Eddystone Lighthouse	16,000	Josias Jessop
1760–1770	Calder and Hebble Navigation	75,000	Joseph Nickalls, 1760–1761
			John Gwyn, 1762–1765
			Matthias Scott, 1762–1765
			Luke Holt, 1769–1770
			Robert Carr, 1769–1770
1763–1767	Coldstream Bridge	6,000	Robert Reid
1765–1777	Potteric Carr Drainage	9,000	Thomas Tofield, in charge
			Matthias Scott, 1765–1774
			Henry Cooper, 1774–1777
1766–1771	Perth Bridge	23,000	John Gwyn
1767–1770	St. Ives Harbour	9,500	Thomas Richardson (engineer-contractor)
1767–1771	River Lea Navigation	25,000	Thomas Yeoman, in charge
			Edward Rubie RE
1767–1772	Adlingfleet Drainage	7,000	John Grundy, in charge
			David Buffery RE
1767–1772	Ure Navigation and Ripon Canal	16,500	John Smith
1768–1773	Eyemouth Harbour		Robert Cramond (engineer-contractor)
1768–1770	Newark Flood Arches	12,000	
1768–1777	Forth and Clyde Canal I	164,000	Robert Mackell
1769–1773	Rye Harbour, new channel	19,000	William Green
1771–1776	Spurn Lighthouse	8,000	William Taylor (engineer-contractor)
1771–1778	Portpatrick Harbour	10,000	John Gwyn
1772–1779	Banff Bridge	9,000	James Kyle
1775–1779	Aire and Calder Navigation	30,000	William Jessop, in charge
			John Gott RE
1775–1780	Aberdeen Harbour, north pier	16,000	John Gwyn
1775–1781	Peterhead Harbour	6,000	John Gwyn
1775	Dunipace Dam		
1775	Queensberry Bridge, Amesbury	2,000	
1776–	Nent Force Level		
1776–1781	Ramsgate Harbour, basin and sluices		Thomas Preston, 1776–1777
			Edmund Hurst, 1777–1781
1776–1789	Hatfield Chase Drainage	20,000	Matthias Scott, 1776–1783
			Samuel Foster, 1783–1789
1777	Altgran Bridge, Evanton		
1777	Amesbury Bridge		
1777	Coquet Dam		
1777–1780	Hexham Bridge (failed 1782)	9,500	Jonathan Pickernell
1778	Cardington Bridge		
1781–1784	Cromarty Harbour		John Gwyn
1785–1787	Hull North Bridge		John Gwyn
1788–1790	Aberdeen Harbour, catch pier	1,500	Alexander Gildavie (engineer-contractor)
1788–1791	Ramsgate Harbour, dry dock		John Gwyn, 1788–1789
			Henry Cull, 1789–1792
1788–1792	Ramsgate Harbour, advanced pier	52,000	John Gwyn, 1788–1789
			Henry Cull, 1789–1792

1797. *Reports of the late Mr. John Smeaton, FRS made on Various Occasions in the Course of his Employment of an Engineer* (vol. 1)
1812. *Reports of the late John Smeaton, FRS made on Various Occasions, in the Course of his Employment as a Civil Engineer* (3 vols.)

1814. *The Miscellaneous Papers of John Smeaton, Civil Engineer, FRS Comprising his Communications to the Royal Society, printed in the Philosophical Transactions*

Works
(For civil engineering works, see table above.)

Watermills:

1753. Halton, Lancashire
1754. Wakefield
1761. Colchester
1761. Hounslow Heath
1763. Stratford, London
1763. Thornton, Fife
1765. Kilnhurst Forge
1765. Carron Ironworks
1768. London Bridge
1768. Wandsworth
1769. Carron Ironworks
1771. Carron Ironworks
1771. Dalry, Edinburgh
1771. Waltham Abbey
1771. Worcester Park
1774. Griff Colliery
1776. Woodhall, Northumberland
1776. Coquet Ironworks
1777. Carron Ironworks
1778. Hull
1778. Carshalton
1779. Deptford
1780. Cardington
1780. Wanlockhead Mine
1782. Newcastle
1783. Carshalton
1784. Deptford Dockyard
1785. Waren, Northumberland
1787. Loose, Kent
1790. Carshalton
1790. Wandsworth
1791. Waddon, Surrey

Windmills:

1755. Wakefield
1774. Leeds
1781. Austhorpe
1782. Newcastle

Steam engines:

1769. New River Head
1774. Long Benton Colliery
1775. Chacewater Mine
1777. Kronstadt Docks
1777. Long Benton Colliery
1778. Lumley Colliery
1779. Gateshead Park Colliery
1780. Carron Ironworks
1781. Middleton Colliery
1781. Seacroft Foundry
1782. Beaufort Ironworks
1783. Bourn Moor Colliery
1783. Walker Colliery

SMEATON, John (1806–1842), civil engineer, worked for the Rennies in the 1820s, on London Bridge and elsewhere. Following the resignation of Henry Robinson Palmer (q.v.) as their Engineer in 1835 George Rennie (q.v.) was asked by the London Docks Company to report on the state of the Shadwell entrance and he prepared specifications for the repair of the wing wall, the contract

for which was awarded to John Plews (q.v.) and Slater in August 1835. Smeaton was appointed Resident Engineer and, on completion of this work, he was made Engineer to the company at a salary of £300 p.a. in 1837.

Smeaton was elected a Member of the Institution of Civil Engineers in 1842 but died, suddenly, shortly afterwards.

MIKE CHRIMES

[Membership records, ICE; London dock warehouse, *Surveyor Engineer and Architect*, **1**, 1840, 137; ICE Annual Report, 1842, *Min Proc ICE*, **2**, 1843, 13; A. W. Skempton (1982) Engineering in the Port of London 1808–1834, *Trans Newc Soc*, 53, 87–88; E. F. Clark (1983) *George Parker Bidder*]

Works
1835–1842. London Dock Company
1835–1837. Shadwell entrance works, warehousing, £21,600

SMIRKE, Sir Robert, FRS (1780–1867), architect, was born in London on 1 October 1760, the second son of Robert Smirke, RA, an artist. He was educated at Apsley Guise, Bedfordshire, a private school, and was trained as an architect under Sir John Soane, George Dance Jr. and Thomas Bush. After studying at the Royal Academy Schools from 1796 until 1799 he went on an architectural tour of Europe, returning in 1805. On his return to London he embarked on an immediately successful career as an architect, his practice being one of the largest in the nineteenth century. In part, this was due to his father's contacts opening him to a wide circle of Tory patrons. In addition to his private work he was responsible for a large number of public buildings, being appointed one of the three architects to the Office of Works. In recognition of these services he was knighted in 1833. Smirke's architectural style, firmly rooted in the classical Greek revival, was criticised even in his lifetime for its lack of adventure. His designs were, however, reliable, and as a safe pair of hands he dealt with the failures of others at Millbank Penitentiary and the London Custom House. Both of these featured the use of concrete in the foundations and as a pioneer of this material, and also in the use of iron in his buildings, he is held in the highest regard.

Structural ironwork was employed by architects in churches in the eighteenth century and by engineers in multi-storeyed mill buildings before 1800. Smirke appears to have been a pioneer of its use in domestic architecture and public buildings, the latter almost certainly as a means of fireproofing. He first used iron in the floors at Cirencester Park (1810) and then at Worthy Park (Hampshire) in 1816. There he used a composite trussed girder of wrought and cast iron, apparently developed by Thomas Tredgold (q.v.). When Smirke suggested their use at the King's Library (British Museum, 1824) the iron founder, John Urpeth Rastrick (q.v.), was dismissive of the system and persuaded Smirke to use single castings, 39 ft. (12½ m) in span. This was more or less contemporary with his work at the General Post Office where Smirke used cast iron beams, iron floor plates and brick arches to provide a fireproof construction. Rastrick also supplied 30 ft. (9 m) beams, 18 ft. 6 in. columns and wrought iron fireplates for Smirke's reconstruction of the Custom House (1825–1828). The fireproof King's warehouse and Long Room represent a more advanced design than the dock warehouses of the time.

Smirke's work at the Custom House was occasioned by the subsidence induced failure of the original design by David Laing (q.v.). Smirke was brought in to investigate and his meticulous report revealed the incompetence of the contractors and the inadequacy of Laing's supervision. The chief cause of the failure was the inadequacy of the foundations; the piles had not been installed as specified. Smirke rebuilt the foundations as a bed of lime concrete 12 to 15 ft. thick covered with York stone landings and surmounted with brick footings. His success was widely publicised by George Godwin, Charles Pasley (q.v.) and others, and marked the acceptance of lime concrete as a foundation material, although Smirke had already been using it for ten years. His pioneering use was for remedial measures to the foundations of the Millbank Penitentiary. The prison had begun cracking up under construction and Smirke was brought in during September 1816; the contractors, including Jolliffe and Banks (q.v.) were dismissed and replaced by Samuel Baker (q.v.) and Sons.

Smirke reported with John Rennie (q.v.); they had discovered the foundations to be inadequate. Smirke developed the idea of a foundation raft of what he described as a stratum of grouted gravel. The idea of concrete had been suggested to him by a chance reference of Rennie to a stratum of calcareous concrete beneath Waterloo Bridge formed by a load of hydraulic lime mixing with the gravel of the riverbed. Other engineers had been experimenting with 'puddled gravel' and other proto-concrete mixes, but Smirke was apparently the first to specify a properly mixed concrete of (Dorking) lime and Thames gravel, in proportions of 1:7 or 1:8. At Millbank this was laid on a bed of gravel.

His success there was followed by other applications by him at Lancaster Place, Sir Robert Peel's house in Whitehall Gardens, in the 1820s, and the British Museum and Oxford and Cambridge Club in the 1830s. His favoured contractor, Samuel Baker, married to his sister, Sarah, made use of concrete elsewhere from the time of the Penitentiary job, beginning with Lancaster Place (1820–1823). Smirke's success was soon widely emulated. His experience of foundations led to his being consulted about Dover Harbour in 1845.

His most important civil engineering work was the Eden Bridge, Carlisle. The original design for a new bridge over the Eden had been developed by Thomas Boyd (1753–1822), a local Dumfries architect who had designed a bridge there (1791–1794).

Boyd worked on the Eden designs 1801–1805, but work was halted after preliminary borings had been sunk. Thomas Harrison (q.v.) developed an alternative design in 1805, and Thomas Telford (q.v.), who had already reported on road communications in the area, agreed to look over all the plans. In 1806 Telford put forward his own scheme. Little more was done until The Report of the *Committee upon the Roads between Carlisle and Port Patrick* was published in 1811. This supported Telford's plans for a bridge and associated embankment works to close off a branch of the Eden. Telford estimated the costs at £20,000, and it was agreed half should come from central government. Smirke had been brought to Carlisle in 1810 to deal with problems in the County Courts. The original designs were by Thomas Telford and Peter Nicholson (q.v.) but the work was badly executed and the foundations were giving way. His competence in dealing with these problems may have encouraged some supporters of the bridge to turn to him. At all events by the end of 1811 he had been appointed designer by the county magistrates. He was encumbered by the limitations of Telford's estimates and was criticised as these were rapidly exceeded once work had begun in late 1812. The fact that he chose a curved approach rather than straight causeway suggested by Telford and Harrison did not help. With the aid of the government grant the five span elliptical masonry arch structure, built by Paul Nixson (q.v.), was completed in 1817, although payments were still being made in 1822. It was by far the most expensive bridge to be erected outside London at that time.

Smirke married Laura Freston in 1819, and had one child, a daughter. His pupils included many leading architects of the Victorian period. He died in 1867 leaving an enormous corpus of work. His library, sold soon after his death, included many engineering treatises, reflecting his interest in all aspects of construction.

MIKE CHRIMES

[C. W. Pasley (1826) *Outline of a Course of Practical Architecture*; G. Godwin (1836) Essay upon the nature and properties of concrete, *(R)IBA Trans*, **1**, 1–39; C. W. Pasley (1838) *Observations on Limes, Calcareous Cements* ...; Harbours of Refuge Commission (1847) *Report*, 42; J. U. Rastrick (1849) Evidence, *Royal Commission on Application of Iron to Railway Structures*, 286–287; J. M. Crook (1965–1966) Sir Robert Smirke: a pioneer of concrete construction, *Trans Newc Soc*, **38**, 5–22; M. I. M. MacDonald (1971) The Building of the New Eden Bridge at Carlisle, *Cumberland & Westmorland Arch Soc Trans*, N.s., **lxxi**, 248–259; H. Colvin *et al.* (1982) *History of the King's Works*, **6**; R. Thorne (ed.) (1990) *The Iron Revolution*; RCHME (1993) *The London Custom House*; M. M. Chrimes (1996) Concrete foundations and substructures: a historical review, *ICE Proc, Structures and Buildings*, **116**, 344–372; Colvin; DNB]

Works

Works notable for their use of concrete or iron:

1810. Cirencester Park
1812–1817. Eden Bridge, Carlisle, £70,000
1813–1818. Eastnor Castle, 30 ft. cast iron beams
1814–1817. Westgate Bridge, Gloucester
1815–1816. Bolton (chain suspension) Bridge, Westmorland
1816–1819. Millbank Penitentiary, reconstruction, £320,000
1816. Worthy Park, Hampshire
1817–1823. Savoy Precinct, including 1820–1823 Lancaster Place
1823–1846. British Museum (NB: dome by Sydney Smirke)
1825–1827. Custom House, London, reconstruction
1830–1833. General Post Office
1833–1839. London Bridge, approaches (buildings)
1834–1837. Shrewsbury Shire Hall (in consultation with Telford (q.v.), d. 1834)
1835–1838. Oxford and Cambridge Club, Pall Mall

SMITH, Humphry (*c.* 1671–1743), was a fen drainage engineer and his first known employment was as Steward to the estate of the future Earl of Leicester at Holkham, Norfolk, from about 1708 to 1718. He held a similar post on the estate of the Duke of Bedford at Tavistock, Devon, from 1724. However, in January 1727 he presented a report to the Board of Bedford Level Corporation on his scheme for restoring the outfall of Gunthorpe Sluice, near Wisbech, and it seems that by that time he had moved to the Duke's estate at Thorney in the North Level of the Fens.

Gunthorpe Sluice in the west (Lincolnshire) bank of the river Nene estuary was at the mouth of Shire Drain, the main drain of the North Level. Shortly before 1720 the river channel swung towards the opposite (Norfolk) shore, leaving the sluice stranded a quarter of a mile away. It was then decided to make a new cut 120 ft. wide from River's End, 4½ miles below Wisbech, past the sluice and on towards Peter's Point, for a length of 1½ miles. Work began in 1721 under the direction of Nathaniel Kinderley (q.v.) but met with fierce opposition, and in June 1722 rioters attacked and destroyed the diversion dam then under construction across the old channel. The next plan was to transfer the Shire Drain outfall 3 or 4 miles further downstream, to deeper water, by a cut 70 ft. wide with a new sluice at its end. Thomas Badeslade (q.v.) took the levels and Colonel John Armstrong (q.v.) approved the scheme. Nevertheless it was not carried out, perhaps on grounds of cost, and conditions in the North Level continued to deteriorate. Finally in January 1726, with water lapping over the banks in Thorney and elsewhere in the North Level, the silt (up to 10 ft. thick) in front of the sluice was removed and a cut made through the sands to the river channel. This improved the situation but provided only a temporary remedy.

Humphry Smith

In September 1726 landowners in the district called on John Perry (q.v.) for advice. He reported in February 1727, suggesting that a new cut 2 miles in length and 100 ft. wide should be made from the sluice through the salt marshes, the sluice to be rebuilt with draw-doors on the land side as well as the usual pointing sea doors, and that Hills Sluice (on Shire Drain 1.2 miles above Gunthorpe) should similarly be rebuilt. In this way a reservoir would be created, between the two sluices, from which water could be discharged at low tide to scour and keep open the cut. The estimated cost came to about £11,000. It may be noted that, as a preferred alternative, Perry also proposed re-opening, widening (to 150 ft.) and extending the cut started by Kinderley with a scouring sluice above River's End, at a cost estimated around £18,000.

Though Smith's report was submitted a month earlier than Perry's he probably knew of the latter's proposals. At any rate there can be little doubt that he was brought in by the Duke of Bedford to resolve the problem. He did so by recommending the creation of a reservoir between Hills and Gunthorpe Sluices and rebuilding them, Hills with doors pointing in both directions and Gunthorpe with draw-doors to scour the outfall, scouring to be aided by 'porcupines': wooden rollers fitted with iron spikes dragged along the channel to harrow the sands. He makes no mention of a cut through the salt marshes.

Smith's scheme was approved by the Board and work began under his direction. In May 1728 the Board reported that the Duke of Bedford and the Earl of Lincoln (Surveyor General of Bedford Level) had advanced £3,400 as a loan to cover the cost of improving the outfall, which 'has proved to be of very Great Service to the North Level'. Gunthorpe Sluice now had a 20 ft. waterway, in two openings.

At the same meeting in May 1728 the Board considered 'a proposal by Mr. Humphry Smith for the more Effectual Draining and preservation of the North Level by making a new Channel in the Wash near the North Bank of Mortons Leam for strengthening and raising that Bank, the Charge whereof will amount to £6,600'. Here again Bedford and Lincoln were willing to advance the capital and to undertake the 'care and management' of the works; in other words with Smith as director. The channel, later called Smith's Leam, ran 12 miles from Stanground, near Peterborough, to Guyhirn, a few miles above Wisbech. The date of completion has not been found but it can be taken as 1730.

While the Gunthorpe works were in progress Smith was elected a Conservator of Bedford Level in May 1727. He became a Bailiff in May 1729. In that year he took up the post of Receiver to the Cambridgeshire, Norfolk and Suffolk estates of the Duchy of Lancaster and in 1738 he became Receiver and Expenditor General of Bedford Level.

During the 1720s various proposals and counter proposals emerged regarding drainage of the South and Middle Levels with particular reference to the river Great Ouse, and also to improvements of King's Lynn harbour. Smith promised the Earl of Lincoln to look into the matter. His report, of August 1729, was considered by the Bedford Level Board a few months later, but like the other reports of this period it had no immediate effect.

In April 1738 Smith was appointed 'Surveyor and Agent' of Deeping Fen at a salary of £150 p.a., jointly with John Grundy Sr. (q.v.), for 'putting into Execution the Scheme by them delivered in for draining all the Fen Land therein'. They had drawn up the scheme in 1737, on the basis of which an Act was obtained enabling taxes to be raised to defray the estimated cost of £15,000. Operations began in the summer of 1738, initially with Smith directing the works and Grundy acting in effect as resident engineer, though later they separately directed individual works and for one John Scribo (soon to become Steward of the Duke of Bedford's Thorney estate) was brought in as director. The principal works carried out during the next four years included repairs to the 12 mile Deeping Bank and the 6 mile Country Bank, deepening the rivers Welland and Glen, providing an 8 acre reservoir and sluice at the Glen outfall in Surfleet parish 5 miles below Spalding, a new sluice at the outfall of Lords Drain, and two drainage mills, one on Vernatt's and one on Hill's drains, both with 16 ft. diameter scoop wheels.

Surfleet Sluice had three openings giving a waterway of 24 ft., with pointing sea doors and lifting land doors. It carried the inscription: 'This Sluice was erected by order of the Honourable Adventurers of Deeping Fen, according to the model and direction of Messrs. Smith and Grundy, W. Sands, Bricklayer, Samuel Rowel, Carpenter, 1739'.

Smith retired in April 1742, when Grundy took over full responsibility for future works. In Ely Cathedral is a portrait bust and memorial tablet to 'Humphry Smith Esq., who departed this Life the 27 day of March 1743 in the 72 Year of his Age. A Man ... of a Competent Knowledge in the most Useful Arts and Sciences, but most Eminent for his Superior Abilities in Draining Fenny and Marsh Lands. Witness his Performances in Thorney Level in the Isle of Ely & Deeping fens in Linconshire, in which Places his Memory will be dear to all Generous Minds'.

A. W. SKEMPTON

[Bedford Level Corporation Order Book (1727–1729) Cambridgeshire RO; Deeping Fen, Adventurers Minute Book (1738–1742) Lincolnshire Archives Office; Thomas Badeslade (1725) *The History of the ... Port of King's Lyn ... and of Draining the* [Great Level] *of the Fens*; John Perry (1727) *Report to the Landowners in the Parts of South Holland*; Nathaniel Kinderley (1730) *Mr. Humphrey Smith's Scheme for Draining the Fens examined and compared*; W. H. Wheeler (1896) *A History of the Fens of South Lincolnshire*; H. C. Darby (1940) *The Draining of the Fens*; Sarah Bendall (1997) *Dictionary of Land Surveyors*]

Publications
1729. *Mr. Humphry Smith's Scheme for the Draining of the South and Middle Levels of the Fens*

Works
1727. Gunthorpe Sluice and Reservoir, £3,400
1728–1730. Smith's Leam, £6,600
1738–1742. Deeping Fen, drainage (with John Grundy), £15,000

SMITH, James (*c.* 1645–1731), architect, builder and industrialist, was the son of James Smith, master mason, first of Tarbat in Ross-shire and from 1659 a burgess of Forres in Morayshire, where he died *c.* 1684. It was presumably a Catholic family and the young James went to Rome and enrolled at the Scots College in 1671, where it was expected that his studies would continue to ordination as a priest. In fact he left in 1675 and returned via Paris to Scotland. At some time during his time in Italy he had visited northern cities and made drawings of buildings by Palladio, the famous sixteenth-century architect. Those drawings are now recognised as seminal documents of the eighteenth-century revival of Palladian architecture in Great Britain.

Within four years of his return to Scotland Smith was in touch, in spite of his Catholic faith, with leaders in society and in architecture, and had married Janet, the eldest daughter of Robert

Mylne (1633–1710) (q.v.), Master Mason to the Crown of Scotland, and been made a burgess of Edinburgh in consequence. Smith himself was invested in 1683 with an official title of Surveyor of the Royal Works, the appointment lasting until the Union of Parliaments in 1707. He later served as surveyor for the construction of forts in the Scottish Highlands for the Board of Ordnance until 1719 when Andrews Jelfe (q.v.) replaced him.

Although an architect—as such his works, many of which he built as well as designed, are described by Colvin—his involvement in engineering had begun very early when he contracted for building (and probably designing) the four-arch bridge across the Ness at Inverness in 1680–1682. It continued with his engaging in quarrying and coal mining and presumably constructing associated harbours for shipping out the coal. By an agreement made in 1699 he was granted a patent by Thomas Savery to use his invention of an early steam engine for the drainage of water from mines. In 1696 he had been involved in 'a scheme to supply Scottish towns with water'. He worked frequently in partnership with another architect/builder, Alexander McGill, and both men were joined with John Adair (q.v.) in a survey and report on the Bridge of Dee at Aberdeen in 1713. Probably at a similar date the same trio, with George Sorocold (q.v.), made surveys to assess the best line for and the cost of a canal from the Forth to the Clyde.

Smith had made his home in the Edinburgh area from early in his career and built a mansion house on an estate called Whitehill, near Musselburgh, which he had purchased in 1686. A personal financial crisis ('he ruined himself by a drowned colliery', wrote Robert Mylne) forced him to sell it some years later, but he needed substantial accommodation and income to house and educate his family of thirty-two children, eighteen of them born to his first wife, Janet, by the time she was thirty-seven (but she had also helped him in the making of drawings!) and fourteen more to a second wife. He was a justice of the peace, had sat in the Scottish parliament prior to 1707 and offered himself unsuccessfully in 1715 for election to the London parliament as member for the City of Edinburgh.

He died at Edinburgh on 6 November 1731, aged eighty-six.

TED RUDDOCK

[*Acts of the Parliament of Scotland*, **10**, 80, 257; Robert Mylne's book of family history (MS) *c.* 1775, British Architectural Library (microfilm in Mylne Papers, SRO); John Crawford (1874) *Memoirs of the Town and Parish of Alloa*, 90–91; H. M. Colvin (1974) A Scottish origin for English Palladianism?, *Architectural History*, **17**, 5–13; A. Roberts (1991) James Smith and James Gibbs: seminarians and architects, *Architectural Heritage*, **2**, 41–55; P. G. Vasey (1992) The Forth–Clyde Canal: John Adair, progenitor of early schemes, *Journal of the Railway and Canal History Society*, **30**(Pt. 7, No. 150), 373–377, and sources quoted; Colvin (3), 893–896, and sources quoted]

SMITH, John (fl. 1731–1767), of Attercliffe, near Sheffield, was carpenter and engineer of the River Don Navigation. Acts were passed in 1726 and 1727 granting powers to the Company of Cutlers of Hallamshire (Sheffield) and Doncaster Corporation, respectively, to make the river navigable from Tinsley (3 miles from Sheffield) 21 miles down to Doncaster and a further 8 miles downstream to Wilsick House. William Palmer and Joseph Atkinson (qq.v.) drew up the plans in 1723 and they probably directed the works until 1731. By that time six locks, three of them on a 2½ mile cut, had been built at a total expenditure of £12,000. The two sets of proprietors were amalgamated in August 1731 and the new management committee appointed John Smith as engineer of the whole undertaking, a post he was to hold for more than thirty years.

By 1740 the navigation reached Rotherham, 26 miles from Wilsick. Under powers of a new Act the navigation was then extended 4 miles below Wilsick. This involved two locks on a 2 mile cut and deepening the river channel to Fishlake. Finally, by 1751, three locks and a mile long cut above Rotherham completed the navigation to Tinsley, a length of 33 miles with 17 locks, achieved at a total capital expenditure of about £40,000.

Writing in 1767 John Smeaton (q.v.) said he had seen several of Mr. Smith's locks 'which have stood and answered very well'. They had masonry walls and timber floors, 60 ft. between gates and 16 ft. wide to accommodate boats of 30 tons burden. Trade on the navigation increased rapidly; by the 1760s tolls brought in some £6,000 p.a.

Smith gained a considerable reputation for his work on the Don. In 1748 he was consulted by York Corporation on improvements of the river Ouse. He recommended 'vigorous dredging'. Doubtless this proved beneficial and followed previous efforts in the 1730s. But in 1752 the decision was made to implement a long-standing proposal originated by Thomas Surbey (q.v.) to build a lock at Naburn, 6 miles below York. Though costly, this provided the best means of securing an adequate depth of water at all times up to the city.

The lock would be on a cut, with a weir—or 'dam' in local terminology—across the river. Detailed specifications were drawn up and a contract was signed on 28 September 1752 by 'John Smith of Attercliffe, carpenter', and his son, John Smith Jr. Work started in 1753. John Smith Jr. acted as site engineer and payments were made to him but it can be taken that his father was responsible for design and expert supervision. On a river subject to large floods work proceeded rather slowly, reaching completion in 1757 at a cost not exceeding the contract price of £5,500.

Naburn lock had a clear width of 20 ft. and a length of 90 ft. between gates, counterforted

masonry walls 23 ft. deep and a timber floor. The
masonry weir was, and still is, V-shaped in plan,
180 ft. in length overall with three 4 ft. sluices at
each end, sustaining an upstream water level
10 ft. above ordinary low tide. A depth of 6 ft. on
the lower lock sills at high water neaps allowed
the passage of Humber Keels of 80 tons burden.

In February 1758 commissioners of the river
Nene navigation invited 'Mr. Smith of Attercliffe'
to submit proposals for making the river navigable from Thrapston up to Northampton, a
length of 26 miles. A month later he and his son
attended a meeting with plans and estimates.
Other proposals were considered but on 22 June
1758 the contract was awarded to John Smith |Jr
for the sum of £14,070. Here again it can be
assumed that Smith Sr. was chiefly responsible for
the planning. Reference is made to 'his Integrity
and superior Experience in such affairs'. However, the works were certainly carried out by his
son and successfully completed in 1761.

John Smith of Sheffield was also consulted by
the Commissioners of the river Wear and Port of
Sunderland when they contemplated extending
the navigation for 2½ miles from Biddick Ford,
8 miles above Sunderland, up to New Bridge at
Chester-le-Street, the limit of their jurisdiction. He
arrived on site in October 1758 and reported on 14
November, having viewed the river in company
with Joseph Robson (q.v.), the Sunderland harbour engineer. Smith concluded that a lock and
weir should be built about half way down and the
river downstream deepened by 2 ft., the total cost
estimated at nearly £7,000. He gave evidence to
this effect before the House of Commons committee in March 1759 on the basis of which, the
cost being considered disproportionate to the benefits, the limit of the Commissioners' jurisdiction
was removed to Biddick Ford. In the same year
Durham Corporation looked into the possibility of
making the river navigable from Biddick up to the
city. Smeaton drew up a scheme but it was never
carried out.

All this time improvements were being made
on the Don. In 1766 John Thompson (q.v.) was
appointed engineer under Smith's supervision
and the following year he took over the job.
Smith, 'now elderly', would be consulted when
necessary. This is the last reference to him so far
discovered.

A. W. SKEMPTON

[Naburn Lock and Dam contract (1752) and Ouse
Navigation Trustees Account Book (1752–1758)
York City Archives; Nene Navigation Commis-
sioners Minute Book (1758) Tyne & Wear Ar-
chives; John Smith (1759) Evidence, *JHC*, **28**, 478;
T. S. Willan (1965) *The Early History of the Don
Navigation*; B. F. Duckham (1967) *The Yorkshire
Ouse*; C. Hadfield (1972) *Canals of Yorkshire and
North East England*; A. W. Skempton (1996) *Civil
Engineers and Engineering in Britain, 1600–
1830*]

Works
1731–1751. River Don Navigation, £28,000
1753–1757. Naburn Lock and Dam (with John
Smith Jr.), £5,500

SMITH, John, Jr. (fl. 1752–1782), was a civil engineer and a son of John Smith (q.v.) of Attercliffe
near Sheffield, engineer of the River Don Navigation from 1731 until 1767. In 1752 father and son
signed a design-and-build contract for Naburn
lock and weir on the river Ouse, 6 miles below
York. The design would have been by the elder
man; Smith Jr. acted as site engineer. The works
were completed in 1757 at or near the estimated
cost of £5,500.

In April 1756 an Act was passed for making
26 miles of the river Nene navigable from Thrapston up to Northampton, Thomas Yeoman (q.v.)
having drawn up a sketch plan in 1754 and given
evidence on the Bill. Within a year the necessary capital had been raised. In November 1757
Ferdinando Stratford (q.v.) made a more detailed
survey. In March 1758 the Smiths submitted proposals. Yeoman and Stratford examined these
and other tenders and in June the contract was
awarded to John Smith Jr. His estimate came to
£14,070 for building 20 locks and various ancillary works, including nearly 4 miles of cuts. The
locks, with stone-faced brickwork and timber
floors, were 100 ft. long between the gates so as
to accommodate two barges in tandem.

Smith acted as engineer and contractor. He
made some variations and was asked, and paid, to
carry out a few additional items. Yeoman inspected the works on two occasions on behalf of
the Commissioners, signifying his approval. The
navigation was opened in August 1761 at a total
cost of £15,600. However in September 1763 it was
reported that Smith had overspent by £2,700 on
items not covered in his contract. An appeal was
launched to meet his costs. By this time he was operating as a coal and timber merchant at South
Bridge Wharf, Northampton. Lady Day the following year he took up an appointment on the
Wey Navigation, leaving the Nene in 'good order'.

When Smith next comes to notice he was back
in Yorkshire. In April 1767 Acts were passed for
making navigable: (a) the upper Ouse from
Linton, 10 miles above York, to Swale Nab, its
junction with the rivers Swale and Ure, a length of
9 miles, with a lock at Linton; (b) the Ure with
two locks on the river and three on a 2¼ mile cut
up to Ripon, a total length 16 miles; and (c) the
Swale to Bedale, a length of 30 miles with five
locks. John Smeaton (q.v.) had prepared plans for
the first two sections and John Grundy (q.v.) for
the third following a survey by Smithson Dawson. Smith was appointed engineer for all three
schemes, with John Jackson as resident engineer
on the Swale.

Work began on Linton lock early in 1768. Smith
based his design on Naburn but sought advice
from Smeaton on some points. The lock, with a
10 ft. fall and large enough for boats 60 ft. long

and 15 ft. 4 in. beam, was completed by August 1769. Work then proceeded on the weir in the river. The navigation was opened in August 1771 and Smith received a salary from the Ouse Commissioners of £2 per week.

Meanwhile, on the Swale, Smith and Jackson encountered serious trouble from floods, difficulties in finding skilled workmen and lack of adequate funding. After building two locks and starting another, the works were abandoned in December 1769.

On the Ure, work started soon after the Act; a contract with a Halifax mason for building the locks, witnessed by John Smith, Engineer, is dated 18 June 1767. The first river lock at Milby, near Boroughbridge, on a 1,100 yd. cut, together with a bridge for the Great North Road over the cut, had been finished by October 1769. The bridge was probably in timber, later replaced by an iron beam structure. Smeaton visited the site in April 1770 and seems then to have taken over direction of the works. He found the first lock on the long cut, later called the Ripon Canal, to be almost completed. He personally set out the upper part of the canal and his assistant, William Jessop (q.v.), took levels and soundings. On a second visit in January 1771, Smeaton advised on repairs to the weir on the river adjacent to the second river lock at Westwich Wath. In May 1771 he found all five locks built and was chiefly concerned with the removal of shoals. Jessop took soundings in the summer. Smeaton made his last visit in September 1771.

Although it appears that Smith left the Ure works in 1770, he certainly stayed on the Linton Lock Navigation until its completion: his salary was paid to July 1771. Not long afterwards he was employed to draw up a scheme for improving the drainage of low grounds south of Market Weighton and for a canal from the Humber up to the town. Smeaton had shown the feasibility of the project in 1765 and Grundy reported in more detail in 1767. Smith gave evidence on the Bill in March 1772, his proposals illustrated in a plan by the surveyor, Smithson Dawson. The Act passed in May and at the first meeting of the newly appointed Trustees in June John Smith was appointed Engineer, with Samuel Allam (q.v.) Surveyor of the Works (resident engineer). However, by August some doubts had arisen concerning Smith's proceedings and Grundy was called in for advice. He reported on 2 September with a different, and better, method of building the outfall sluice and with a revised scheme for the navigation. His plans were immediately accepted. He became chief engineer with Smith as assistant. Work began on 29 September and continued under Grundy's direction until 1775.

Smith was admitted as a member of the (Smeatonian) Society of Civil Engineers in March 1772 while in London on the Weighton Bill. He attended meetings in 1776 and 1777.

Presumably on leave of absence from the Weighton job, he reported in October 1772 on a proposed canal from Wakefield to the river Don. Nothing came of this and he continued at Weighton until September 1773, or perhaps a little later. He then appears in June 1774 as engineer when work started on the River Bure Navigation in Norfolk, 9 miles long from Coltisham to Aylsham, with five locks. In October 1777 he reported that £3,600 had been spent and another £3,000 was required. More funds were raised. Work resumed in February 1778 and the navigation was opened in October 1779.

The leisurely pace of these operations left time for other activities though Smith took on more than he could properly handle. In January 1776 he reported jointly with James Creassy (q.v.) on improvements to the drainage of fenland adjacent to the river Witham downstream of Lincoln. In April 1776 he undertook to make 9 miles of the river Soar navigable from the Trent at Redhill to Loughborough. He built a stanch (not a lock) at Redhill and another further upstream. Later that year, or early 1777, he drew up plans for the proposed Erewash Canal from the Trent to Langley Mills, 12 miles with a rise of 108 ft. The Act was passed in April 1777 and John Varley (q.v.) was appointed engineer. The Soar shareholders were now considering a more ambitious scheme and they called on Jessop for advice. He reported in December 1777, recommending six locks. Soon afterwards John May was put in charge, employing new contractors to carry out Jessop's plan.

Apart from finishing the Bure Navigation in 1778–1779 nothing more is known of Smith's career. He was in correspondence with Smeaton in July 1782 and died before June 1784 when his name is listed among those members who had died since the first meeting of the Smeatonian Society. He is not to be confused with the land surveyor, John Smith (c. 1742–1820), who worked with Jessop on the Trent in 1786 and with Jessop and Robert Whitworth (q.v.) on the Ashby-de-la-Zouch Canal from 1792 until 1794.

A. W. SKEMPTON

[Minute books of the River Nene Commissioners (1756–1762) Northamptonshire RO; *Northampton Mercury* (1761–1764); Contract (1767) for building River Ure locks, Halifax Public Library; Market Dreighton Drainage and Navigation Trustees Orders (1772–1775), East Riding RO; John Smeaton (1782) Journal, Trinity House, London; C. Hadfield (1966) *Canals of the East Midlands*; C. Hadfield (1972) *Canals of Yorkshire and North East England*; J. Boyes and R. Russell (1977) *Canals of Eastern England*; P. Jones (1996) John Smith and England's first iron bridge, *Journal of Railway & Canal History Soc*iety, **32**, 178–185; D. Bates (2001) in lit.]

Publications

1772. *Plans of the Intended Drainage and Navigation from the Humber to Market Weighton*
1772. *Report on making a Navigable Canal from Wakefield … to the River Dun*

1776. *Report on the Drainage of Low Lands on both sides of the River Witham from Lincoln to Boston* (with James Creassy)
1777. *Plan of the Intended Canal from the River Trent to Langley Bridge*

Works
1758–1761. River Nene Navigation, £15,600
1767–1769. Milton lock, River Ure
1768–1771. Linton lock and weir, River Ouse, £8,400
1774–1779. River Bure Navigation, £6,000

SMITH, John (1782–1864), architect, builder and occasional sculptor, of Darnick near Melrose in Roxburghshire, was the second son of John Smith (c. 1748–1815), mason and builder at Darnick, and Mary Williamson. The family had been resident there since before the middle of the seventeenth century.

John had three brothers, one of whom died in infancy, and five sisters. His elder brother, James (1779–1862), became established as an architect and builder in Edinburgh and bore the title 'HM Master Mason for Scotland,' while his younger brother, Thomas (1785–1857), worked with John in the family business at Darnick. An obituary on John's death in 1864 noted the extent of their work: 'they executed nearly all the principal buildings of the district for many years, [John] from his meditative and ingenious nature, devoting his attention principally to the devising of plans and the preparatory portions of the work, and [Thomas], from his active and pushing habits, being admirably qualified for carrying these ideas and schemes to a practical issue.' Transcripts from John's diaries show, however, that he was much involved in the supervision of site work done to his designs and in finding and purchasing the necessary materials. The Smiths were contractors for most of the works which they designed, as well as some works designed by others, and John is quoted as saying that the firm employed eighty men in 1822.

John was on intimate terms with the owners of great houses in their area and undertook substantial work at Abbotsford for Sir Walter Scott, at Bowhill for the Duke of Buccleuch and at Dryburgh Abbey for the Earl of Buchan, who described the brothers early in their careers as 'my pupils the Smiths of Darnick'.

The Smiths were deliberate in adopting what they thought to be best practice in their construction of bridges of two kinds, and in publishing their methods for the benefit of others. One of these types was arch bridges of rubble whinstone masonry. Their views on these bridges comprised most of a paper which they sent to the Institute of British Architects in response to a book of questions distributed by the Institute in the early 1830s. In the paper, which was printed in the Institute's *Transactions* for 1835–1836, they describe five bridges they had built between 1831 and 1834 with arches of span 51 to 76 ft., all built with rough quarried whinstone in blocks of small size and irregular shapes, and to which they applied various amounts of facing and architectural features in hewn stone (generally sandstone) of larger blocks. They confessed to annoyance with the use of common lime mortar because of its slowness to set and the consequent need for centring to stay in place for a long time, but claimed that 'with proper precautions, [whinstone] arches might be constructed of a much longer span than any which have yet been attempted'.

Their opportunity to prove that claim came in 1846 when Sir James Russell of Peel proposed a new bridge for access to his house across the Tweed. John Smith designed a whinstone arch of 133½ ft. span, which has since been known both as Low Peel Bridge and Ashiestiel Bridge. Three weeks after the arch had been keyed and before the centring had been eased, the arch collapsed. Weeks were spent salvaging the timber and stones from the river and in the following season a re-designed centre was erected and the bridge rebuilt in about three months. It was claimed, perhaps correctly, to be the longest arch of rubble masonry in the world. It is of roughly elliptical form with a rise of 26 ft.

The Smiths' paper of 1835–1836 dealt more briefly with their historic contribution to the development of suspension bridges. Less than two months after the first iron suspension bridge built for regular use in Great Britain—a *wire* footbridge of 111 ft. span—had been erected at Galashiels, John Smith, on 17 January 1817, offered to the Earl of Buchan a design for a *chain* bridge over the Tweed at his estate of Dryburgh Abbey. Digging of foundations began within a month, erection of the ironwork began on 16 June and the bridge was completed by 2 August. With a span of 261 ft., a timber deck was suspended from wooden portals on the banks by eight pairs of chain stays, four from each portal, at different inclinations and each chain formed of ten-feet long wrought iron eyebars of half-inch to one-inch diameter. All the chains were made by the Smiths' own blacksmiths. On 15 January of the following year a powerful gale caused rampant oscillations and the bridge collapsed. John Smith revised the design and, with almost constant attendance by himself, the bridge was rebuilt in June–August 1818. This time it had catenary chains as the main suspension elements and it lasted for some thirty years.

During 1819 Smith visited Captain Napier at Thirlstane Castle in Ettrickdale, presumably to offer him some advice about a stayed iron wire suspension bridge which was built there in or before 1821. In 1823 he offered a design for a chain bridge over the Dee near Aberdeen and designed another for a site at Caddonfoot, in the Tweed Valley, in 1830. Neither was built, but in July 1825 he received an order to design a chain footbridge to connect Melrose with Gattonside across the Tweed and construction began in September. This time Redpath and Brown, a blacksmith firm in Edinburgh, contracted to build 'the chain bridge for

£600', but the masonry towers were almost certainly built by the Smiths. John had a leg injured in December when 'joining the chains of [the] bridge' and was confined to bed for a few days. The construction with catenary chains of eye-bars was very similar to that of the second Dryburgh Bridge, almost all hand-work by blacksmiths, unlike the more industrial methods used by Samuel Brown (q.v.) in his Union Bridge (1819–1820) over the river Tweed at Paxton, and elsewhere. The Smiths' bridge was opened to traffic in October 1826 and used with few repairs until 1991 when a misguided programme of strengthening destroyed most of the original ironwork, leaving only the main chains as non-functional elements.

As well as his many trips to Edinburgh and other towns in South-east Scotland, Smith occasionally travelled further, using a journey to Ireland in 1826 to view Telford's recently-completed Menai Bridge and the Minster at York, as well as to travel on the Stockton–Darlington Railway. He tried out the Liverpool–Manchester Railway on the way to London and Paris in 1835 and travelled to London again in 1851 to see the Great Exhibition.

His best-known sculpture is a huge statue of William Wallace, the thirteenth-century warrior patriot, commissioned by the Earl of Buchan and still standing on the hillside above Dryburgh. His extensive works of architecture, including many churches, are summarised by Colvin.

He was a man of great physical endurance, recording amongst his constant journeys to scattered work sites some very long rides and walks. On one occasion when he missed the coach at Edinburgh and could not get a horse, he simply walked the 37 miles home. On bridge sites, repeatedly, he recorded spending 'thrang days' with gangs of up to fifty men pumping and digging when founding bridge piers in the water. The diaries also record his courtship of Alison Purves, who lived at Housebyres, and marriage to her on 4 June 1818. He had seven sons, four of whom emigrated to Australia in the 1850s, and four daughters, the eldest of whom, Jeanie (1819–1911), followed her brothers and kept house for them for many years before returning to Scotland.

James Smith's estate was valued at £8,377 on his death in 1862. On John's death on 13 September 1864 his estate was valued at £5,717, including an estimated £900 in property in Australia and £2,806 already advanced to various members of his family. By a late codicil to his will he had determined that of his living children, the females should have equal shares of his heritable wealth with the males, the heirs of those already dead having similar equal portions.

TED RUDDOCK

[John Smith diary (1813–1864) selective transcripts, Acc. 8499, NLS; design drawing, Dryburgh first bridge (1817) MS, J8/7, British Architectural Library Drawings collection; sketch of Dryburgh first bridge (1817) by Lady Polwarth, MS, GD157/ 2009, SRO; sketch of Dryburgh second bridge (1818) by Sir David Erskine, MS, NMRS; letter, John Smith to Robert Stevenson, 25 May 1820, MS, Acc. 10706, No. 77/47, NLS; R. Stevenson (1821) Description of Bridges of Suspension, *The Edinburgh Philosophical Journal*, **5**(10), 237–256; J. Bower (1827) *Description of the Abbeys of Melrose and Old Melrose*, 108; Messrs. Smith of Darnick (1835–1836) On Whinstone Constructions etc., *Transactions of the Institute of British Architects*, **1**, 52–60; Sir D. Erskine (1836) *Annals and Antiquities of Dryburgh*, 170–174; John and Thomas Smith (1836) Description of Dryburgh Bridge, *The Penny Magazine*, 22 Oct., 415–416; Smith MSS, including letter book 1850–1858, diaries of 1822 and 1864, other journals and family trees, H2 SM1, NMRS; family testaments and other papers, MS, GD241/ 250, SRO; will of John Smith, MS, *Consistorial Record of Roxburghshire*, **40**, 480–545,SRO; D. M. Hood (ed.) (1978) *Melrose 1826*; Colvin, **3**; T. Ruddock (1999) Blacksmith Bridges in Scotland and Ireland 1816–1834, *Proceedings of Conference on Historic Bridges*, Wheeling, West Virginia, October 1999]

Publications

1835–1836. On whinstone constructions etc., *Transactions of the Institute of British Architects*, **1**(1), 52–60
1836. Description of Dryburgh bridge, *The Penny Magazine*, 22 Oct., 415–416

Works

Except where noted otherwise, all bridges were of whinstone rubble.

1815. Bridge at Crosslee, designed and built
1816. Bridge at Yarrow, designed
1817. Dryburgh Chain Bridge 1 (footbridge 261 ft. span), designed and built for Earl of Buchan for £560, collapsed in a gale in January 1818
1818. Dryburgh Chain Bridge 2, of same dimensions, for Earl of Buchan, extra cost £240
1820. Branxholm Bridge, repairs
1825–1826. Melrose–Gattonside chain bridge (footbridge), designed and supervised, fabricated and erected by Redpath and Brown, smithwork contractors for £600, total cost £726(?)
1827–1832. Newlands Bridge, designed and built
1831–1832. Two bridges over Tweed and Ettrick at Sunderland Hall, each of three elliptical arches approx. 55 ft. span, designed and built, cost £2,220 for structures and £430 for 'the ornamental part' of both
1831–1832. Hawick Bridge, repairs
1832. Wilton Bridge near Hawick, designed and built
1832–1833. Bridge at Yarrow, one arch 63 ft. span, designed and built
1833–1834. Falshope Bridge, one arch 76 ft. span, designed and built, for the Duke of Buccleuch
1833–1834. Bridge at Yetholm, designed and built
1834. Peebles Bridge, fifteenth-century bridge, five arches of 35–40 ft. spans, substantially widened on both sides (width doubled from 9 to 18 ft.), probably report or design only

1834. Ettrickbridge bridge, probably widening of an older bridge

1837. Merton Bridge, design and evidence on the Bill in Parliament, four spans, timber on stone piers

1841. Ashkirk Bridge, designed and built?

1846–1848. Ashiestiel Bridge over river Tweed, designed and built, whinstone arch of 133½ ft. span, collapsed 15 July 1847, rebuilt 1848

SMITH, Robert, Colonel (1787–1873), military engineer, was the son of James and Mary Smith of Bideford, Devon, baptised at Nancy, France, on 13 September 1787. He served in the Bengal Engineers from 1805 to 1830, when he retired with the rank of major. He was later made an honorary colonel and created CB.

In his first years in India, he helped to construct the lighthouse at Kirji (Kedgree) in Diamond Harbour, Calcutta, between 1808 and 1810. He saw service in the capture of Mauritius from the French in 1810, in the invasion of Java in 1811 and in the Nepal War of 1814–1815. He also took part in the siege of Bharatpur near Agra in 1825–1826.

Smith was a highly skilled surveyor whose detailed maps proved invaluable to the military as they settled newly acquired territories. He made a map of the wild jungle area of the Palamau, Shahabad and Mirzapur districts of the Ganges Plain in 1813–1814 and then spent some time in Penang (Prince of Wales Island). On return from his three-year furlough in 1822 he was appointed Garrison Engineer and Executive Officer, Delhi, and had among his duties the repair of great Mughal monuments in and around Delhi. From this post he was detailed in March 1823 to complete the survey of an ancient Mughal canal, the Doab or Eastern Jumna (Yamuna) Canal, in succession to Henry Debudé (q.v.), and in 1827 was appointed Superintendent of the Doab Canal. This canal had been constructed by Ali Mardan Khan in the seventeenth century, largely by joining watercourses together, but is thought not to have functioned for long because of the steepness of its slope and the interference it suffered from mountain torrents crossing its path in the north.

Under Smith the channel was excavated to a depth of 4 ft. below the surface of the country, following as far as possible the natural courses of rivers, but straightening out the most tortuous parts of the channel. A new head was established at Fyzabad using temporary bunds, and a new cut made from Gokulpur to Selimpur to deal with the tail-water. Various strategies were employed to deal with the seasonal torrents in the north. The worst of these, the Muskurra torrent, had a dam built across it, originally with eighteen openings of 7 ft., the centre six with gates; it later had to be increased in length. It operated in conjunction with a regulating-bridge at Kulsea, which had three spans, a centre one of 20 ft. and two side ones each of 15 ft., all controlled by gates and sleeper planks. Similar arrangements were made at Nogong and Nyashur; there was a slightly

longer dam at Nogong, and at Nyashur a dam and regulating-bridge arrangement. Smith usually used curtain walls front and rear in the foundations of his structures, but for works in difficult locations, such as at Muskurra, curtain walls were formed by parallel rows of wells sunk between 6 ft. and 12 ft., with the spaces between them filled by piling. This was an adaptation of a traditional Indian method of using wells in foundations. The lack of hydraulic knowledge in this period meant that irrigation work was largely empirical and much work had to be undertaken later to introduce falls to reduce the slope and to provide larger regulating-works.

From 1825, Smith had as his assistant Lieutenant Proby Cautley (1802–1871), whose delightful surviving sketch-book gives details of the structures and their foundations. Water was admitted in 1830 but Smith had already taken sick leave and retired at that time.

Smith combined his ability as an engineer and surveyor with the gifts of architect and artist. A number of his paintings survive in the India Office Records. Among other buildings in Delhi he repaired the Red Fort, the Qutb Minar, the Jami Masjid mosque, as well as producing designs for St. James's Church, Delhi (built in 1836), and Mughal-style palaces for himself in Nice and Paignton. He died on 16 September 1873.

JOYCE BROWN

[J. Colvin (1833) On the restoration of the ancient canals in the Delhi Territory, *Journal of the Asiatic Society of Bengal*, **15**, 116–117; P. T. Cautley (1833) Doab Canal sketches, Department of Civil Engineering Library, Imperial College, London; P. T. Cautley (1839) A description of the use of wells for foundations, *Journal of the Asiatic Society of Bengal*, **8**, 47–64; P. T. Cautley (1853) *Notes and Memoranda on the Eastern Jumna or Doab Canal*; E. W. C. Sandes (1933) *The Military Engineer in India*, **I**, 232, 240, 261; F. W. C. Sandes (1935) *The Military Engineer in India*, **II**, 4, 260; R. H. Phillmore (1950) *Historical Records of the Survey of India*, **II**, 442; M. Archer (1972) An artist engineer—Colonel Robert Smith in India (1805–1830), *The Connoisseur*, 79–88; J. Brown (1978) Sir Proby Cauley (1802–1871), a pioneer of Indian irrigation, *History of Technology*, **3**, 44–49; C. A. Bayley (ed.) (1990) *The Raj—India and the British, 1600–1947*, National Portrait Gallery Publications, 208–209]

Works

1823–1830. Eastern Jumna (Doab) Canal, 134 miles

1823–1830. Muskurra dam, Kulsea

1823–1830. Nogong dam, Budhi Jumna

SMITH, Thomas (1752–1815), tinsmith and lighthouse engineer, was born in Ferryport-on-Craig, Fife. In 1764 he was apprenticed to a metal worker in Dundee after which he came to Edinburgh, probably in 1770, at the time the building of the 'New Town' was gathering momentum. He was employed by an established metal worker and by 1781 was trading as a tinsmith in the city, manufacturing oil lamps, brass fittings, fenders and other household metal articles. His business prospered, particularly from 1786 when he took an enterprising interest in improving the illumination of lighthouses.

Smith's earliest known lighthouse illumination proposal was made in 1786 but not implemented until 1815–1816. He proposed that a lamp with metallic reflectors be substituted for the coal-fire light at the private lighthouse built on the Isle of May in the Firth of Forth in 1635. In support of this idea he wrote on 16 June 1786 'A comparative view of the superior advantages of lamps above coal light when applyd to light houses', in which he confirmed that he had 'constructed 2 small reflectors & lamp with a view to demonstrate by experiment what has been only laid down in theory'. Soon afterwards, at the expense of the Board of Manufactures in Edinburgh, he successfully tried out his reflector lamp experimentally at Inchkeith in the Firth of Forth.

Smith's pioneering work in illumination brought him to the notice of the Northern Lighthouse Board of Commissioners which had been formed later in 1786 to improve the almost non-existent lighting of Scotland's dangerous coast and on 22 January 1787 they appointed him as their first Engineer. Several weeks later he was sent to Norfolk to receive a short course of instruction in lighthouse construction and illumination from Ezekiel Walker of Kings Lynn.

On his return Smith enthusiastically set to work on the provision of new lighthouses for the Northern Lighthouse Board. At first this was in an honorary capacity and it was not until 1793 that the Board awarded him a salary of £60 p.a. over and above his expenses. The Board had appointed Smith primarily for his lighting expertise and did not regard his lack of building and architectural experience as an impediment as such skills could be and were brought in under his direction from Alexander Kay, architect and others. During the next two decades, commencing in 1787 with the conversion of Kinnaird Castle, into a lighthouse, Smith was responsible for providing or improving thirteen lighthouses. Independently of the Board, he was responsible for harbour lights at Leith and Portpatrick and lights on the rivers Clyde and Tay.

In 1789 Smith was elected to the Edinburgh Guild of Hammermen becoming its Master and a city magistrate in 1802. From 1797 he delegated and allowed almost complete autonomy in lighthouse matters to his apprentice and stepson Robert Stevenson (q.v.) who married his daughter Jane and formally succeeded him as Engineer to the Northern Lighthouse Board on 12 July 1808. This delegation of work enabled Smith to concentrate on lamp manufacture and the expansion of his shipping and other interests particularly his general and street lighting business. By 1800 his lamps were lighting much of eastern Scotland and the central belt as far west as Glasgow. In 1804 he was the public lighting contractor for both the Old

and New towns of Edinburgh and, by 1807, for lighting the streets of Perth, Stirling, Ayr, Haddington, Aberdeen. He retired from business in 1808 and died on 21 June 1815 at his home, 1 Baxter's Place, Edinburgh.

In terms of technical innovation Smith developed and made arrays of parabolic reflector oil lamps. Each lamp had a light source at its focus and a curved reflector formed of small pieces of mirror glass set in plaster that produced a beam of light. The adoption of glass rather than metal had the advantages of being more resistant to distortion and also to wear from frequent cleaning before the use of glass chimneys. Smith's expedients undoubtedly represented the most practicable means of achieving an immediate improvement in Scottish lighthouse lighting before the Argand lamp, patented in 1784, was eventually applied to this purpose.

Smith's first operational lighthouse light, at Kinnaird Head, had an intensity of about 1,000 candlepower, which although feeble compared with its modern counterpart of 690,000 candlepower, nevertheless represented a worthwhile improvement on coal lights. He continued to adopt glass-faceted reflectors for new lights until 1801, after which, because of Robert Stevenson's influence, he started to manufacture Argand lamps with silvered-copper reflectors. This improvement which produced a significantly brighter light was probably first installed in Scotland at Inchkeith lighthouse in 1804.

Details of Smith's reflectors became more generally known from an article 'Reflector for a lighthouse' in the supplement to the third edition of *Encyclopaedia Britannica* (1801). In it Smith is described as 'an ingenious and modest man [who] has carried [his inventions] to a high degree of perfection without knowing that something of the same kind has been long used in France'. This tribute was omitted from later editions, including the last carrying the article (1823), after the editor had learned of Ezekiel Walker's prior development of the glass facet reflector lamp concept. Nevertheless, Smith was the first to introduce brighter lights into Scottish lighthouses. More generally, he deserves to be remembered chiefly for his important contribution in improving both public and private lighting in Scotland and for laying the foundation of the Stevenson dynasty of engineers through his encouragement of Robert Stevenson.

Smith has a good claim to be regarded as Scotland's first lighting engineer.

R. A. PAXTON

[Private information; National Library of Scotland, Business records of Robert Stevenson & Sons, MS, Acc. 10706; Obituary notice, *Edinburgh Advertiser*, 30 June 1815; Obituary notice, *Edinburgh Evening Courant*, 1 July 1815; Obituary notice, *Caledonian Mercury*, 1 July 1815; D. A. Stevenson (1959) *The World's Lighthouses before 1820*; C. Mair (1978) *A Star for Seamen—The Stevenson family of engineers*; J. Leslie and R. Paxton (1999) *Bright Lights—The Stevenson Engineers 1751–1971*]

Works (all lighthouses)
1787. Kinnaird Head
1787–1788. Mull of Kintyre
1789. North Ronaldsay
1789. Eilean Glas
1789. Tay Lights (improved 1789)
1790. Pladda
1790. Leith (improved 1790)
1790. Portpatrick
1793. Little Cumbrae (improved 1793)
1794. Pentland Skerries
1797. Cloch
1802–1806. Start Point
1804. Inchkeith

Street lighting in various cities and towns in Scotland including Edinburgh, Glasgow, Perth, Stirling, Aberdeen, and Haddington.

The extent of Smith's private work is not known but it was probably considerable.

SMITH, William (1769–1839), the 'father of British geology', was the son of John Smith (d. 1777) and his wife Ann (*née* Smith) and was born on 23 March 1769 at Churchill, Oxfordshire; he was the eldest son of a family of three boys and a girl. After the death of his father, of farming stock and 'a very ingenious mechanic', at the age of eight, he was supported by his uncle, William, and educated at the village school. At the age of eighteen he became assistant to Edward Webb, surveyor of Stow-in-the-Wold, in whose house he lived. He worked for him on landscaping the grounds of Warren Hastings' house.

Webb arranged for him to survey the estate of Lady Elizabeth Jones at Stoney in Somerset in 1791. He settled in the area. After undertaking work on his own account for the High Littleton Coal Company and other clients (in Somerset), he became responsible in 1793—under John Rennie (q.v.)—for survey work on the Somersetshire Coal Canal, as a precursor to which he made a tour to inspect completed canal schemes, travelling as far north as Tyneside in the late summer of 1794. His salary was then a guinea a day plus expenses. After work began, Smith was appointed as sub-engineer in 1795 with a salary of £450 p.a., holding the post until 1799 when he left the company following a disagreement. By this time he had developed a keen interest in geology fostered by his work in the coalfield and dedveloped a stratigraphical theory. He had been elected to the Royal Bath and West of England Society in 1796, and by 1799 had produced a manuscript geological map of the Bath area and section of the strata. He became an independent consulting engineer and between 1800 and 1810 reporting on land drainage, sea defences, water supplies, canal routes and the siting of coal mines. His living was precarious because of debts he had accumulated and a large mortgage on Tucking Mill House. In the course of his work at this time he met some wealthy patrons, including Joseph Banks in 1801, and took in as an assistant John Farey (q.v.) while working on the Bedford Estate at Woburn. He moved to London in 1803, occupying 15 Buckingham Street, Adelphi, later the premises of the Institution of Civil Engineers in 1804. He married his wife Mary Ann (b. 1791/1792) in 1808. Aside from Farey, his chief assistant was his nephew John Phillips (b. 1800), who joined him in 1815.

In 1801 was published his 'tabular view' of the strata around Bath and in 1815, after financial difficulties, he published a geological map of England and Wales to a scale of 5 miles to the inch, dedicating it to Sir Joseph Banks. The cost of this work placed him under a heavy burden, as a result of which he sold his mineral specimens to the British Museum in 1818, and in the summer of 1819 suffered ten weeks in a debtors prison. On his release he left London for Yorkshire. Between 1819 and 1824, a period during which his wife's health failed, Smith produced a series of county maps and in 1828 he became land agent to Sir John V. B. Johnstone of Hackness, Yorkshire, remaining there for six years.

One of his most significant achievements was the identification of the presence of coal in the eastern parts of County Durham, beneath the overlying Magnesian Limestone. The advice he gave to the landowners there led to the establishment of new collieries, in turn resulting in the construction of port and dock facilities at both Hartlepool and Seaham Harbour during the 1830s; both ports were connected by rail to the new mines. The Geological Society of London, which had earlier ignored his work, awarded its first Wollaston Medal to Smith in 1831 and four years later he received an honorary doctorate from Trinity College, Dublin. In 1838 he became a member of the commission which advised the government on the type of stone to be used for the new Houses of Parliament but very soon after signing the commission's report, when travelling to the meeting of the British Association in Birmingham and staying with friends in Northampton, he died on 28 August 1839. He was buried at St. Peter's church and is commemorated by a tablet and bust. Little is known of his family life save that his wife survived him by five years, spent in an asylum in York. There were no known children

R. W. RENNISON

[J. Phillips (1844) *Memoirs of William Smith* ...; DNB; K. R. Clew (1969) *The Somersetshire Coal Canal and Railways*; D. A. Robson (1986) *Pioneers of Geology*; S. Winchester (2001) *The Map that Changed the World*]

Publications and drawings
1801. *General Map of Strata found in England and Wales*
1806. *Observations on the Utility, Form and Management of Water Meadows, and the Draining and Irrigating of Peat Bogs, with an Account of Prisley Bog*
1808. *Observations on the Utility, Form and Management of Water Meadows*

1815. *A Memoir to the Map and Delineation of the Strata of England and Wales, with part of Scotland ...*
1816. *Strata identified by organised Fossils*
1818. *Report on the Plan for draining the Low Ground north and south of the River Went, between the Rivers Aire and Don, in conjunction with the proposed Aire and Dun Canal, and Went Branch*
1819. *The Proposed Aire and Dun Junction Canal to drain the contiguous Lands and to shorten and connect the present Navigations*
1821. *Geological Map of Yorkshire*
1824. *Geological Map of Cumberland ...*
1824. *A Geological Map of Buckinghamshire*

Works

1791–1793. Surveying work, Somerset
1793–1799. Somersetshire Coal Canal, surveyor and sub-engineer, 10 miles in length
1799–1828. Consulting engineer
1799. Drainage work for James Stephens, Somerset
1799. Drainage work for Thomas Crook, Tytherington, Wiltshire
1800. Drainage work for Thomas Coke, Lord Leicester, at Holkham, Norfolk
1801. Drainage work for the 5th Duke of Bedford, Woburn
1801. Drainage of Hickling Marshes, Norfolk
1801. Drainage work at Dolymelynllwyn near Dolgellan
1801. Coal mine at Torbock, Lancashire
1801. Laugharne Coast works, Carmarthenshire
1801. Ouse Navigation, Sussex
1801. Kennet and Avon Canal, repairs
1801. Kidwelly Harbour
1806. Prisley Bog, Woburn, drainage
1810. Great Spring, Bath, repairs
1812. Minsmere Level drainage
1820. Scarborough, water supply
1828–1834. Land Agent, Hackness
1829. Scarborough Museum, design

SNAPE, John (1737–1816), land surveyor, was baptised at St. Chad's Church, Wishaw, Warwickshire, on 4 October 1737, the eldest of six children of Solomon and Mary. Solomon became schoolmaster of Lady Hacket's free school at Wishaw in 1745; John surveyed and mapped the estates of Andrew Hacket of Moxhull Hall in 1763 (two years after John's earliest known map), and was actively concerned in surveying and mapping land from this time. In 1767 he married Jane Lander, a widow, and they had two daughters, both named Mary. The elder one was buried in Sutton Coldfield in 1768; the younger was baptised in 1769 and outlived her father. From the year of his marriage until 1783, Snape gave Moxhull (near Wishaw) as his address. His estate surveying took him to Lincolnshire, Rutland, Staffordshire and Worcestershire for a number of landowners including Sir John Sheffield.

From 1777 Snape was heavily involved in surveying canals. In this year he drew a plan of the canal from Birmingham to Autherley, then he mapped several towns (Birmingham, Lichfield and Walsall) but returned to canals in 1782 and 1783 when he drew maps for extensions to the Birmingham Canal. He became a member of the Society of Civil Engineers in the latter year. Two years later, in 1785, he and Samuel Bull (q.v.) surveyed a proposed extension from the Birmingham Canal to the Dudley Canal. With Josiah Clowes (q.v.), Snape surveyed the line of the Worcester and Birmingham Canal in 1789; he then returned to the Dudley Canal. In 1785, he had admitted to the Parliamentary Committee enquiring into the proposed canal that he had no experience of tunnelling but in 1792 the line that he surveyed included the fourth longest tunnel in England, at Lappal. In the same year he surveyed for the Stratford Canal and from this time until 1807 he had an address in Birmingham, in Hospital Street.

In the first decade of the nineteenth century, Snape was involved in inclosure work (including Birmingham Heath and West Bromwich), but also continued his estate and canal surveying. He worked on the Grand Junction Canal in 1802, mapped the Birmingham Canal through Birmingham in 1805 and part of the Walsall Canal in 1808. This is the latest known date of map-making by him and he moved to Aston Road, Birmingham, in this year.

Snape died in Aston on 1 January 1816. In his will of 21 November 1815 he left his small estate to his second wife, Mary, with bequests to his daughter, Mary Woolfe, her husband John (a draper of Birmingham) and her six children.

SARAH BENDALL

[Will, 21 November 1815, Lichfield Joint RO; A. W. Skempton and E. C. Wright (1967) Early members of the Smeatonian Society of Civil Engineers, *Newc Soc Trans*, **44**, 1967–1968, 23–47; S. R. Broadbridge (1974) *The Birmingham Canal Navigations*, vol. 1, *1768–1846*; C. Hadfield (1985) *The Canals of the West Midlands*, 3rd edn.; A. F. Fentiman (1986) *John Snape Land Surveyor 1737–1816: A Comprehensive List of his Works*; A. W. Skempton (1987) *British Civil Engineering 1640–1840: a Bibliography of Contemporary Printed Reports, Plans and Books*; G. Watson (1989) *The Smeatonians: the Society of Civil Engineers*; R. V. Wallis and P. J. Wallis (1993) *Index of British Mathematicians, Part 3: 1701–1800*; S. Bendall (ed.) (1997) *Dictionary of Land Surveyors and Local Map-Makers of Great Britain and Ireland 1530–1850*]

Publications

1785. *A Plan of the Intended Extension of the Dudley Canal into the Birmingham Canal near Tipton Green in the County of Stafford: and also of the Dudley and Stourbridge Canals*
1789. *A Plan of the intended Navigable Canal from the Town of Birmingham into the River Severn near the City of Worcester*
1792. *A Plan of the Intended Navigable Canal, from the Worcester & Birmingham Canal at Kings*

Norton, in the County of Worcester; to Stratford upon Avon, in the County of Warwick; also the collateral Branches to Grafton &c
1792. *A Plan of the intended Navigable Canal from the present Dudley Canal Navigation at Netherton in the Parish of Dudley ... to the Worcester and Birmingham Canal at Selly Oak ... with the collateral Branches to communicate therewith*

SOROCOLD, George (1668–fl. 1716), engineer, was recognised by two of his contemporaries, Hatton (1708) and Thoresby (1715), as the 'Great English Engineer', almost certainly the first to receive such an appellation. Despite our knowledge of many of his works and their importance, much about Sorocold remains obscure, not least the circumstances of his death.

George Sorocold, son of James Sorocold (1627–1675) of Ashton in Makerfield and Elizabeth Barrow, was born in 1668. The name Sorocold was relatively common in Lancashire in the seventeenth century. In addition to the family's property in Lancashire, some was acquired by George's grandfather in Derbyshire. It would seem likely that this is where George was brought up. This raises the tantalising possibility that Sorocold may have been trained as an engineer by Cornelius Vermuyden II (see: *VERMUYDEN, Sir Cornelius*) or his assistants at Middleton near Wirksworth where George is known to have been active.

This remains speculation as our earliest knowledge of Sorocold's life is his marriage to Mary Francis (d. 1728) on 7 December 1684 at All Saint's, Derby. Their first child, Elizabeth, was baptised in 1687. Mary was the daughter of Henry Francis (Franceys), apothecary, and George described himself as a gentleman, reflecting a relatively distinguished social background. According to Thoresby Sorocold had had thirteen children by the age of twenty-eight; this has proved impossible to confirm. Whatever Sorocold's origins, Derby became the centre of his operations for the remainder of his life.

No details are known of Sorocold's reported involvement in a water supply scheme at Macclesfield in 1685–1687. Much better detailed is his work at All Saint's Church, Derby, where he rebuilt the frames and machinery for the bells in 1687–1688, for which he was paid £26. The work was carried out by John Baxter, of Northants, presumably to Sorocold's designs, but aside from this relatively modest start we are in ignorance of Sorocold's activities at this time. His next known project could only have been undertaken with much technical knowledge. On 5 March 1692 Sorocold took over Gunpowder Mill—by a deed with Derby Corporation—Byfleet Island and two adjoining sluices on the Derwent, with a view to supplying Derby with water on a three year contract, at the end of which he could give the works up to the Corporation. The waterwheel, which was employed to pump water from the river to a cistern near St. Michael's Church, for onward distribution by pipes, could be raised or lowered

according to the river level. The works were still in use in 1829. This type of device is generally believed to have been first used by John Hadley (q.v.) who patented it in March 1693. Hadley is known to have worked with Sorocold later, but this suggests an earlier connection. From a letter to the Earl (Duke) of Marlborough it would appear likely that this was Sorocold's first work of this type; it could well have proved his last. According to Thomas Surbey (q.v.), writing in 1699, while demonstrating 'his contrivances to a friend [Sorocold] chanced to drop in above a corn millwheel which going sucked [him] through but one of ye ladles breaking hee was taken out of ye water below without any harms'.

Instead of a fatal disaster Sorocold's success at Derby was followed by a large number of early water supply schemes, the majority involving waterwheels for pumping. Chronologically, the next scheme was that at Leeds where Sorocold in 1694, in partnership with Henry Gilbert of Netherseal, made an agreement with the Leeds Corporation for erecting an engine to take water from the Aire and supply it to Leeds householders; the partners were exempted from paying local rates. This was followed by a licence to lay 2,000 linear yards of lead pipes.

In September of the same year Sorocold for a payment of £200 and an annual rent of £25—with another partner, Richard Barry of Westminster, signed a 99 year agreement with Norwich Corporation to supply water and by a separate agreement acquired a lease of New Mills. These arrangements had to be confirmed by an Act of Parliament in 1700 but work had begun on the works by 1698. Barry was also Sorocold's partner for schemes at Portsmouth and Great Yarmouth in 1694. The work at Portsmouth included supplying the hospital and barracks and was completed *c.* 1697 but information is lacking on the Yarmouth venture. Of greater interest is Sorocold's next documented scheme, Marchant's Waterworks of 1696.

In 1695 Hugh Marchant and other partners had been granted letters patent to use 'sewer' water to drive 'mills' to raise water from the river Thames and elsewhere for public water supply. In 1696 these powers were extended and an overshot wheel installed which drove three mill wheels with small cranks at each end of the axle tree, so raising Thames water. The works were at Tom's Coffee House in Hartshorn Lane (Northumberland Street) and other pumps were powered by a windmill near Windmill Street. These works were carried out by Sorocold in association with John Hadley. The works were probably abandoned in the 1730s, once Chelsea waterworks was established, and by 1775 the mill wheel was being used to drive a corn mill.

Shortly after the Marchant scheme Sorocold was consulted about the waterworks at Sheffield. A Nottingham engineer, Peter Whalley, had begun work but he died before much was achieved; according to Defoe 'there is a fine Engine or mill also for raising water to supply the town, which was done by Mr. Seracal'.

By 1700 Sorocold was acquiring a national reputation and in the following year he commenced his most famous work, London Bridge waterworks. The original works had been built by Peter Morris (q.v.) and completed in 1582. His descendants had continued to operate the works, which were badly damaged by the Great Fire, until Michaelmas 1701 when they sold their rights to Richard Soame, a goldsmith. Soame formed a company with a share capital of £150,000 and obtained the lease on the fourth arch for new machinery and Sorocold was put in charge of reconstruction. It is uncertain where he began, but almost certainly he would first have installed the new wheel under the fourth arch. This was 20 ft. diameter with floats 14 ft. long and 1 ft. 6 in. deep. The wheel was connected by gearing to two 'four throw' cast iron beams attached to the pump rods, enabling the wheel to work sixteen pumps. The waterwheel was capable of pumping 880 gallons per minute against a head of 120 ft. of water (c. 32 hp). Once this was completed the two wheels in the third arch, powering twelve and eight pumps respectively, and the wheel in the first arch, powering sixteen pumps, were rebuilt. The waterwheels were all undershot and capable of being raised or lowered according to the flow. The works as a whole were capable of pumping at least 1,820 gallons per minute (66 hp) and were described by Henry Beighton (q.v.) in his paper to the Royal Society as superior to the Marly works which supplied Versailles. There seems little doubt that the fame of these works led to Sorocold's appellation, 'great'. Sorocold moved to London at this time and, building on his reputation, he petitioned the Earl of Marlborough for work in the Ordnance Department.

London Bridge was not Sorocold's last venture in the London water supply as he was also consulted in 1708 by the New River Company who needed a new, higher, reservoir to supply residents in Islington, an area which was being developed at that time. To supply the reservoir in Claremont Square, Sorocold built a windmill to pump the water but the intermittent nature of wind power meant that by the 1720s horses were being employed to power the pumps.

Sorocold's name has been linked with a number of other waterworks but confirmatory documentary evidence is lacking in some cases. At Bridgnorth a waterwheel and pumps were installed in 1705–1706 to distribute water from the river Severn to the top of Castle Hill through a 4 in. main to a cistern from where it could be distributed around the upper town; the scheme was financed by Sir William Whitmore, and Sorocold referred to an 'Enjoyable project at Bridgnorth, where I put my floating mill'. The works remained in use until 1857 when a steam engine was installed in a new works. According to Cox (1727), 'This wondrous piece of work was performed by those that erected the new waterworks at London Bridge'.

According to Thoresby, Sorocold was involved in waterworks installations at Wirksworth, Bristol (1696), Deal (1700–1701) and King's Lynn. At Deal two schemes proceeded, alternatively supplying water from a river and a well. Sorocold almost certainly provided advice on the former. In King's Lynn, a windmill powered scheme of 1682 was almost certainly replaced with a waterwheel by 1698. In 1694 Sorocold was consulted about Newcastle waterworks and possibly also the application of water power to the draining of mines but the eventual scheme was the work of William Yarnold (q.v.). The 1696 Nottingham waterworks used a system similar to London Bridge but may have been the work of Whalley rather than Sorocold.

In addition to his schemes for pumping water for public supply Sorocold worked on a number of private estates. His work for the Marchant's scheme brought him to the attention of Sir Godrey Copley, MP, FRS, whose family seat was at Sprotborough, on the Don near Doncaster. Copley had been impressed by the ornamental waterworks he had seen at Chatsworth and wanted to install similar features on his estate. To install fountains at Sprotborough Copley needed to pump the water up from the Don to a height of around 150 ft. He met Sorocold and Hadley in 1696 and inspected the Marchant's installation which he described as 'ye best piece of work I have seen'. Copley commissioned Sorocold to devise a scheme to provide water for a drinking water supply, and fountains on the estate, and an engine house was erected containing a 15 ft. diameter undershot wheel with 3 ft. wide float boards. These works, completed in 1703 and augmented in 1707 with the installation of a 35 ft. long heated swimming pool, mark a new phase in Sorocold's career. Definite evidence also exists for his involvement at Melbourne Hall, Derbyshire, where in 1705–1706 he raised the level of a mill pond by 2 ft. to give sufficient head of water to supply the pools and fountains in the garden of Thomas Coke. These works involved 1,000 yd. of limestone paving. It is possible that Sorocold carried out similar work at Calke Abbey in 1701 and at Alloa c. 1707, where work is ascribed to 'George S[illegible]'.

Some uncertainty attaches to Sorocold's work in Scotland although the fact that he worked there at all provides evidence enough of his nationwide reputation as an engineer. Sorocold was brought to Scotland to provide technical advice to Sir John Erskine, 6th Earl of Mar who, following his inheriting the title in 1689, had ambitious plans to improve his estates, including his coal mines. As early as 1694 it appears that Mar had rebuilt the Gartmorn Dam on his Alloa estates to provide water power for mines and mills in the area. It is unclear whether an engineer was consulted about this work but it is more likely that Sorocold was first consulted around 1709–1710 when it perhaps became clear that the technical problems associated with Mar's ambitions were beyond any local expertise. Although there is a possibility that Sorocold was consulted about 'water works' at Mar's Alloa property in 1707, it is generally agreed that in 1709 Mar sent an agent to Newcastle to try and find an engineer to advise on the drainage of the mines at Alloa and, in 1710, for a fee of £50 Sorocold arrived and gave his advice. He surveyed the line of a feeder for the dam from the Black Devon river using a 'large wooden quadrant set upon a tripod with brass lights along the upper radius, the index being a plummet suspended by a fine thread.' This feeder was presumably built to ensure a supply of water all through the year; with the Broughty Burn impounded already, the reservoir extended over 160 acres. Associated works included a 10 ft. high weir at Forest Mill to direct the water through a sluice to the feeder, and a lade, or channel, from the dam to drive a waterwheel intended to power a crank above the mine shaft to drive beams and pumps. It appears that Sorocold's advice was disregarded and an earlier chain and bucket arrangement continued. The water course continued to Alloa itself, powering mills and ultimately sluicing out the harbour. The pumping arrangement may not have been successfully completed as politics obliged Mar to go to London and, following the 1715 Rebellion, into exile. By 1723 the main coal seam appeared nearly exhausted, suggesting that Sorocold's proposals for pumps capable of draining deeper shafts had not been fully implemented.

Sorocold's work for Mar led to his involvement with James Adair (q.v.) and others in the earliest survey of a Forth–Clyde Canal, c. 1714. In early 1715 he was again in Scotland, in the Fifeshire coalfield at Torryburn, on unknown work. By April 1716 he was in France trying to contact Mar at Fontainbleau. To have done this so soon after the 1715 rising may indicate that he was a Jacobite; he certainly left behind debtors pursuing him in the Scottish courts.

The notion that Sorocold may have fled England as a Jacobite is also supported by the Mayor of Derby's statement of March 1717 in connection with the Derwent Navigation Bill, referring to the 'ingenious, unfortunate mathematician, Mr. Sorocold'. This has normally been taken to refer to the miraculous accident, described by Defoe, which befell Sorocold when showing visitors around the Derby silk mill. It is now known that this event must have taken place some twenty years earlier and the exact nature of Sorocold's involvement in the early Derby silk mills is uncertain, although involvement with the latter could suggest an alternative reason for his presence in France.

The silk industry was well established in the Macclesfield area by the late seventeenth century and Sorocold is certain to have been aware of this as he worked on the waterworks there. The development of the English silk industry was hampered by the monopoly the Italian silk industry held over the mechanical preparation, or throwing, of high quality silk threads such as organzine.

By the sixteenth century, in centres such as Bologna, waterwheels were being used to power winding and throwing machines, in the same building, on a primitive factory system. By the late seventeenth century British merchants were anxious to break the Italian technological monopoly and themselves produce organzine.

Around 1704 Thomas Cotchett, a barrister, obtained a lease on water rights on the river Derwent in Derby to enable him to set up a silk mill. It is generally assumed that Sorocold designed the waterwheel and associated machinery. The mill itself was not a great success as only part of the necessary machinery was installed, a fact referred to by Defoe. John Lombe (c. 1693–1722), the son of a Norwich worsted weaver and conjectured to have worked for Cotchett, travelled to Italy in 1716 with the express intention of accurately pirating Italian machinery. It would have been logical for him to have taken with him an engineer such as Sorocold, especially as Sorocold had Norwich (and London) business connections.

In 1717, with a reward for his capture from Italian authorities hanging over him, Lombe returned with Italian workmen and details of the most advanced—Piedmontese—throwing machinery. In September 1718 his half brother, Thomas Lombe, a London merchant and member of the Mercers Company, took out a patent (No. 422) for the manufacture of silk. In 1721 the brothers opened Lombe's mill alongside Cotchett's mill on the banks of the Derwent. The mill, 360 ft. long and five storeys high, was powered by an undershot waterwheel 23 ft. in diameter and built of wood. It powered twelve circular throwing machines occupying the two lower storeys and twenty-six winding machines on each of the top floors. Sorocold has been credited with the design of the waterwheel but it is unclear what his involvement was beyond that, although the importance of the mill cannot be doubted. It was followed by other similar mills in East Cheshire associated directly or indirectly with the Lombe family and was in many ways the forerunner of the late eighteenth century textile mills associated with the British industrial revolution. It was arguably the first successful use of the factory system in Britain.

Sorocold's own role in this episode must remain doubtful as there are no verified mentions of him after 1716. His association with another pioneering work, London's first important wet dock, Howland Dock, Rotherhithe (1697), although frequently quoted, seems even more unlikely. It appears that John Wells (q.v.), shipwright, and William Ogborne (q.v.), a carpenter of Stepney, were responsible for all the timber work and wharves. One reason for Sorocold's conjectured involvement is his well documented work on the design of the first (old) dock at Liverpool. Sorocold was first consulted in London by Thomas Johnson, MP for Liverpool, in January 1708 and in November the Town Council agreed that their MPs, Johnson and Richard Norris, should 'treat with and agree for a proper person to come to this town and view the ground and draw a plan for the intended dock'. On 7 March 1709 Sorocold and his Derby-based surveyor, Henry Huss, had the Freedom of the City conferred upon them. Their survey plans and estimates were completed by September 1709 and in the following January a Bill for the dock went before Parliament, obtaining Royal Assent in March. In May the Corporation agreed to follow the plans of Thomas Steers (q.v.) who was appointed engineer for the scheme.

The (Derbyshire) Derwent navigation was another scheme in which Sorocold was involved with the initial planning but there remains doubt about his final involvement. Attempts had been made to improve the navigation to Derby from 1664 onwards, and bills were prepared in 1696 and 1698. He was called upon to support Derby Corporation when they petitioned Parliament about their scheme. In January 1703 Sorocold attended to give evidence, following which he submitted a plan of his designs, with four new cuts and weir locks over a length of 10 miles. This scheme was unsuccessful but in 1717 Derby successfully petitioned for a Bill, the Mayor stating that the accompanying map had been prepared by Sorocold. These improvements were carried out and completed c. 1724.

Sorocold was involved with other navigation schemes in the North. In 1694, 1699 and 1704 he carried out surveys of the Yorkshire Derwent and the scheme completed in 1722 followed his plans. In 1706 he was consulted about improvements to the river Dee in connection with proposals by the City of Chester to revive the port and was then paid £20. Further South he was consulted about the Lea and Cam.

With so many unknowns in Sorocold's career a final assessment is difficult. It is likely that more works remain unattributed, although the lack of documentary evidence after nearly three centuries is unsurprising. Two possible areas of work which require further examination are the application of water power to landscape architecture and mine drainage, where Sorocold may have been an important figure. This is certainly suggested by surviving correspondence at Melbourne Hall, which refers to Sorocold's activities at Mill Close and Wirksworth, both lead mining centres. He designed pumps for the mine at Sheldon.

He was clearly a man of some substance and would not have relied on his engineering work for all his income. On some schemes he was simply consulted while elsewhere he acted as undertaker. He had an national reputation and if his involvement in schemes such as Lombe's mill and Howland's Dock could be confirmed his reputation would be of international significance.

Of his alleged thirteen children, the birth of six in Derby has been confirmed. The date and place of his death are unknown.

MIKE CHRIMES

[Papers of Dr. William Farrer, LI, Manchester Central Library; Committee for Dee Navigation, A/B/3/149, Chester City RO; Parish registers, Derby local history library; F. Williamson papers, Derby local history library; Melbourne Hall Muniments, Derby; T. Surbey (1699) A Journal from London to York upon a survey on the River of Ouze, May 1699, Acc. 65, York City RO; Ms, Journal of the House of Lords (Derwent Bill), 1 February 1703, HLRO; Mar family papers, SRO; E. Hatton (1708) A New View of London, **2**, 791; D. Defoe (1727) A Tour through England and Wales; H. Beighton (1731) Phil Trans Roy Soc, **37**, 5–12; J. T. Desaguliers (1744) A Course of Experimental Philosophy, **2**, 436–441, 528; R. Thoresby, Diary, vol. 1; R. Thoresby (1715) Ducatus Leodiensis; W. Hutton (1791) History of Derby, 14; R. Bald (1812) General View of the Coal Trade of Scotland, 10–11; New Statistical Account ... Scotland; R. Jenkins (1918) George Sorocold: a chapter in the history of public water supply, Engineer, **126**, 333–334; H. Peet (1930) Thomas Steers, Trans Hist Soc, Lancashire and Cheshire, **82**, 163–242; F. Williamson (1936) George Sorocold of Derby, J Derb Arch Nat Hist Soc, n.s., **10**, 43–96; R. Jenkins (1934) Historical notes on some Derbyshire industries, Trans Newc Soc, **14**, 167–171; Survey of London (1937) The Strand, **xviii**, 24–26; G. House, Old London Bridge; F. Williamson and W. B. Crump (1941) Sorocold's waterworks at Leeds, 1694, Pub Thoresby Soc, **37**, 166–182; C. Morris (ed.) (1947) The Journeys of Celia Fiennes; H. W. Dickinson (1954) Water Supply of Greater London, 24–28, 38, 51–52; W. H. Chaloner (1953) Sir Thomas Lombe (1685–1739) and the British silk industry, History Today, **3**, 778–785; A. W. Skempton (1953–1955) Engineers of the English river navigations, Trans Newc Soc, **29**, 43–44; D. H. Thomson (1957) History of the Portsmouth and Gosport Water Supply; B. F. Duckham (1967) The Fitzgibbons and the navigation of the Yorkshire Derwent, Northern History, **2**, 45–61; D. Swann (1968) Engineers of English port improvements, Transport History, **1**, 157; R. D. Mitchell (South Yorkshire County Council) (1978), in lit.; A. W. Skempton (1984) Engineering on the English river navigations, in M. Baldwin and A. Burton (eds.), Canals: a New Look, 23–44; P. Heath (1989) Melbourne Hall reconsidered (1988) Georgian Group Report, 48–60; A. Calladine (1993) Lombe's Mill: an exercise in reconstruction, Industrial Archaeology Review, **xvi**(1), 82–99; P. G. Vasey (1992) The Forth Clyde Canal: John Adair progenitor of early schemes, RCHS J, **30**(7), 373–377; A. Calladine and J. Fricker (1993) East Cheshire Textile Mills, 24–25; M. Craven(1998) Derbeians of Distinction, 187–188]

Publications

[1703]. A Survey of the River Darwent ... together with the intended Cutts for making the same navigable

[1716]. A Map with relation to the Navigation of the River Darwent *c.* 1716. *Reports on the Improvement of the Derwent near Derby*

Works
1685–1686. Macclesfield, water supply
1687. All Saint's Church, Derby, bell hanging
1692. Derby, waterworks
1694. Leeds, waterworks
c. 1694–1697. Portsmouth, water supply
1694–1700. Norwich, waterworks
1696–. Marchant waterworks, London
1696–1703. Spotborough Hall, engine house and fountains
c. 1699. Sheffield, waterworks
1702. London Bridge, waterworks
1704. Cotchett Silk Mill, waterwheel
1705–1706. Bridgnorth waterworks, £1,000
1705–1706. Melbourne House, fountains, ponds and weirs
1708. New River, waterworks, windmill
1709–1710. Alloa Mines, waterworks
[1720–1721. Lombe's silk mill, waterwheel, attrib.

SOUTHERN, John (*c.* 1758–1815), was a son of Thomas Southern of Wensley near Wirksworth, Derbyshire, a mining engineer. In June 1782, at the age of twenty-four he began a three year agreement with Boulton and Watt, nominally as a draughtsman but essentially to act as Watt's personal assistant, and remained with Boulton and Watt for the rest of his life. It would appear that Boulton knew his father as he recommended him to Watt, in terms which suggest intimate knowledge: 'If you have a notion that young Southern would be sufficiently sedate, would come to us for a reasonable sum annually, and would engage for a sufficient time, I should be very glad to engage him for a drawer, provided he gives bond to give up music, otherwise I am sure he will do no good, it being the source of idleness.'

Southern must have abandoned serious study of music and, although he interested himself in radical politics, he generally seems to have devoted his life to his professional work. In 1800 this was recognised in his being awarded a 2¼% commission upon all goods manufactured at Soho, with a minimum underpinning of £600, and by 1810 he was clearly involved in the improvement and refinement of Watt's engines and associated machine work. He is most famously associated with the introduction of the steam indicator used by George Augustus Lee (q.v.) in his experiments; he was responsible for much of the detailed engine design. He gave evidence in support of Watt's patent litigation, 1793–1796. Watt relied on him for revising the algebraic analysis in the *Encyclopaedia Britannica* article on the steam engine.

His ability as a mechanical engineer led to him being consulted about all kinds of engineering questions, most notably about Thomas Telford's (q.v.) 600 ft. span cast iron bridge proposal for London bridge in 1801. Southern himself proposed a variant design which employed a triangulated truss between two arch ribs. This, one assumes, was developed from the triangular Warren-style truss used for steam engine beams, developed at Boulton and Watt's works from 1784 to 1798, initially with wooden beams and then with cast iron. It is probable that Southern was involved with the development tests for these beams but there is no evidence that any bridges were built on such a basis at this time.

In 1793 Southern married the daughter of Thomas Dobbs, one of Boulton and Watt's customers and owner of a rolling mill at Kings Langley. Following his death at his home in Handsworth in July 1815 Southern was buried in the Dobbs family vault at Kings Langley.

MIKE CHRIMES

[J. P. Muirhead (1854) *The Mechanical Inventions of James Watt*, 2 vols., *passim*; H. W. Dickinson and R. Jenkins (1927) *James Watt and the Steam Engine*; A. W. Skempton (1980) Telford and the design for a new London Bridge, in A. Penfold (ed.), *Thomas Telford: Engineer*, 62–83; J. Tann (ed.) (1981) *The Selected Papers of Boulton and Watt*, vol. 1, *passim*; J. G. James (1987) The origins and worldwide spread of Warren-truss bridges in the mid-nineteenth century: part 1 (1986) *History of Technology*, **11**, 66–68; R. L. Hills (1989) *Power from Steam, passim*.]

Publications
1801. *Appendix to the Report of the Select Committee on the Improvement of the Port of London*, 64–70, plate 2
1801. On the equilibrium of arches, *Philosophical Magazine*, **11**, 97–107
1818. Letter on properties of steam [etc.], in J. Robison, *The Articles on 'Steam and steam engines'*, with notes and additions by James Watt

SPEDDING, Carlisle (1695–1755), coal viewer, 'the greatest of all the Whitehaven mining engineers', was the younger son of Edward Spedding (d. 1706), a farmer and carrier, and his wife Sarah Carlisle. He was baptised at Moresby, Cumberland, on 22 September 1695; his brother, John (1685–1758), also a viewer, became principal steward to Sir James Lowther. Intended first for a maritime career, Spedding was employed in the Lowther concerns from 1710 and in 1716 he married Sarah Towerson.

In 1730 Spedding took over the direction of the Lowther family's collieries and so became responsible for the sinking of some twenty-five pits both to the east and west of Whitehaven, those to the east drained by a level 1½ miles long. His major work was the sinking of the Saltom pit, one of the first of the undersea mines. Brought into production in 1732 and 'wainscotted all over', it reached a depth of 80 fathoms and was drained by a steam-driven pump. In 1732 staiths capable of storing 3,000 wagonloads of coal at a height of 37 ft. above quay level were built at Saltom, together with a quay, and in 1735 a wagonway to Whitehaven was constructed, its timber track carrying iron-wheeled wagons. Another major work was the King pit, sunk by Spedding in 1750 to a depth of 123 fathoms and connected to the port by the Whingill wagonway in the 1750s; at the time of his death Spedding had begun the Bransty arch carrying the line over the road into the town.

Spedding was a trustee of Whitehaven harbour from 1718 and, with his brother, established ship-building at the town in 1732. He was, however, concerned by the inadequacies of the harbour and in 1751 advocated the extension of the mole by 30 yd. Between 1738 and 1753 Spedding sub-scribed to several books on scientific matters, and by the latter year was described as a 'gentleman'. Such was the regard in which he and his brother were held that Lowther, at his death in 1755, left £1,000 to John and £500 to Carlisle.

Throughout his career, Spedding was involved in matters regarding safety in mines and in the 1730s adopted the 'coursing of air' through the workings. At this time, too, he invented the steel mill, used for the safer lighting of mines. He cor-responded with the Royal Society concerning mine gases and wrote in 1751 that he had been 'much engaged of late in preventing accidents by firedamp in which (he had) been successful'. Nevertheless, he was killed in a mine explosion on 8 August 1755 and was buried at Holy Trinity church.

He was survived by his wife and at least two of his five children. His son James (1720–1788) became Lowther's principal steward on the death of his uncle while Thomas (b. 1722) became the first incumbent of St. James's church, designed and built by his father in 1752 and where Sped-ding himself had been a churchwarden. At his death the *Newcastle Journal* paid tribute to his 'uncommon Abilities, assiduous Application and intrepid Conduct'.

R. W. RENNISON

[*Brown Collection*, Acc. 3410, Northumberland RO; *Newcastle Journal*, 16 August 1755; T. V. Simpson (1930–1931) Old mining records and plans, *Trans Inst Mining Engrs*, **81**; O. Wood (1988) *West Cumberland Coal*; B. Scott-Hindson (1994) *Whitehaven Harbour*]

Works

1718–. Whitehaven Harbour, Trustee
1730. Appointed Colliery Viewer to Sir James Lowther
1732–1735. Sinking of Saltorn Pit, 480 ft. deep, with wagonway
1750–1755. Sinking of King Pit, 738 ft. deep, and construction of wagonway

SPEDDING, James (fl. 1795–1810), was presum-ably a member of the well-known family of mining engineers in west Cumberland, of whom Carlisle Spedding (q.v.) was an earlier representa-tive. He is first noticed as a tunneller, living at Braunston, Northamptonshire, in 1795 where the Grand Junction Canal was under construction. In

April 1796, shortly before the Braunston Tunnel was completed, Spedding was employed by John Rennie (q.v.) to take borings on the summit level of the Kennet & Avon Canal and two years later he was one of those brought in to complete the Salisbury & Southampton Canal, on which Rennie had recently reported. He was still employed as a tunneller in 1802, when he was at work on the Thames & Medway Canal, but by 1804 he was engaged in constructing a cofferdam of novel design for Rennie at the East India Docks in London. A partial failure after four weeks' service was repaired successfully, probably by doubling the number of inner piles which supported the dam, suggesting a fault in design rather than workmanship. In 1805 he installed the coffer-dams for constructing the wing walls at the ends of the City Canal across the Isle of Dogs, where William Jessop (q.v.) was Engineer, this time without problems, and for the same Engineer he removed the rock outcrop in the river Thames at Blackwall. Experiments with blasting having failed, Jessop devised a system of heavy iron chisels operated like pile drivers from barges, and a diving bell in which miners could work to remove the fractured rock. Spedding completed the work to specification and on schedule.

The last two contracts which Spedding is known to have executed were again for Rennie. On com-pletion of the Blackwall job he returned to the summit level of the Kennet & Avon Canal to con-struct the Burbage Tunnel. Initially his agent was a 'Mr. Jones', who was possibly the Charles Jones who had been less than successful on the Grand Junction Canal. In any event, Jones left the works and Spedding carried on. At Chetney Hill on the Medway estuary, the unsuccessful attempt to con-struct a lazaretto in Britain, he undertook the piling, the earthworks forming a separate contract for Hugh McIntosh (q.v.).

P. S. M. CROSS-RUDKIN

[Rennie reports, ICE; Rennie collection, NLS; RAIL 842, PRO; A. W. Skempton (1979) *Engineering in the Port of London*, TNS 50, 87–108]

Works

1798. Salisbury & Southampton Canal, contractor for remedial works to Southampton Tunnel
1804. East India Docks, cofferdam, contractor
1805. City Canal, London, cofferdams, contractor
1806–1808. Blackwall Rock, River Thames, removal, contractor, £26,000
1808. Kennet & Avon Canal, Burbage (Bruce) Tunnel, contractor
1810. Chetney Hill, part of works, contractor, paid £2379 8s 9d

SPENCER, Arnold (1587–1658), gentleman and pioneer of river navigation works was baptised at Cople on 7 November 1587, a younger son of Robert Spencer (d. 1620). The family held Rowlands manor in Cople, a village 3 miles east of Bedford, from before 1530 to the early eigh-teenth century. How Spencer became interested in river navigation is not known but by 1618, at the latest, he had decided to undertake a scheme for making the Great Ouse navigable from St. Ives (the then head of navigation from King's Lynn and Ely) up to Bedford. To share the financial risk he went into partnership with Thomas Girton, vintner of Westminster.

John Gason of Finchley, a typical 'projector' of the period, secured a Patent in 1617 for making river navigations anywhere in England during the space of twenty-one years, paying an annual rent of £2 to the Crown. Spencer and Girton had therefore to negotiate with him, and in February 1618 he assigned to them his rights over rivers in the counties of Bedford, Huntingdon and Cambridge.

There can be no doubt that Spencer was the active partner, with local knowledge, and that he had already planned what works were required. In the first stage of operations these were 'sluices' in new cuts adjacent to each of the six watermills on the 15 mile length of river between St. Ives and St. Neots. Arrangements for purchase or rent of land for the cuts and towing path seem soon to have been completed and the work was finished three years later in 1621.

In a document of 1628 there is reference to the six sluices or locks upon the Ouse 'for the better raising the water in the river for the passage of boats through the said locks' and a diagrammatic plan of c. 1689 shows that the 'sluices' were indeed pound locks with two pairs of mitre gates enclosing a rectangular lock chamber. A 'sluice' is shown on a cut bypassing a mill and its dam, and alongside is a note stating that 'in this manner are the other ancient sluices between St. Ives and St. Neots'.

Details of one of these locks, at Godmanchester 'a little above Huntingdon', were recorded by Thomas Surbey (q.v.) in an appendix to his report on the Yorkshire Ouse 1699. He notes that two boats in tandem are drawn by a horse, the boats being 28 ft. long and 9 ft. 9 in. broad, each carrying 14 or 16 tons with a draught of 2 ft. He gives a plan of the site and drawings of the lock. It is timber framed, the sides being supported by overhead lintels (of the type familiar from John Constable's paintings of locks on the Essex Stour), the lock being long enough and just sufficiently wide to accommodate two boats in tandem, and with 3 ft. of water on the sills.

This lock is known to have been repaired about 1650. In 1674 the navigation was leased by Henry Ashley (q.v.) who in 1688 stated that he had not altered any of 'sluices' between St. Ives and St. Neots except Houghton sluice, which he rebuilt in 1678; instead his works were concentrated upstream of St. Neots to Bedford. It is therefore probable that Surbey's drawings depict Godmanchester lock in its original form apart from any (minor) improvements made during the c. 1650 repairs.

If that is the case, Spencer can be credited not only with creating the first river navigation in England with an integrated system of pound locks, but the locks themselves represent an advance on any built earlier in this country. The next 'modern' pound locks (with masonry walls) were three built 1632–1638 on the Thames.

As for the later history of the navigation, in 1625 Girton transferred his interest to Spencer who in turn assigned his rights to John Jackson of St. Neots in 1625. This was presumably a move to raise capital for the next stage of operations. Two months earlier Spencer had petitioned for a Patent to make rivers navigable, on the grounds that he had been 'at great charge' on works on the Ouse and wanted 'the sole use of his own engine'. The Patent was granted in January 1628 for a period of eight years, paying a rent of £5 p.a. Soon afterwards he built a lock at Eaton Mills a mile above St. Neots. He also began scouring shallows at eleven places in the river upstream to Great Barford, a 7 mile stretch on which there were no mills, using his 'engine'. This appears to have been a horse-drawn scarifier or some kind of plough to loosen the soil and so allow particles to be carried downstream to deeper parts of the river bed. In addition, he made two lateral cuts up to 200 yd. in length and 33 ft. wide, probably bypassing places where the river was exceptionally shallow.

All this appears to have taken ten years to accomplish, since the first boats to reach Great Barford are said to have done so about 1638. Before that time, and possibly as early as 1633, Spencer reacquired the original locks from Jackson. So he now owned the whole navigation from St. Ives to Barford. A considerable trade flourished on the river but the undertaking had been costly, and trading 'very much decayed' after outbreak of the Civil War in 1642. Spencer, far from being able to extend the navigation to Bedford, had by 1644 become involved in financial difficulties. Three years later he was compelled to mortgage the navigation to his son, Robert, and to Luke Spencer (whose relationship to them is uncertain). Later it passed to Robert's two daughters, Anne and Elizabeth, and their husbands. It was from them that Ashley leased the navigation.

For some time Spencer had been living at Eaton Socon, where he died in March 1658.

A. W. SKEMPTON

[Thomas Surbey (1699) *Report on the Yorkshire Ouse Navigation*, MS, York City Archives, transcribed by Paul Hughes 1992; T. S. Willan (1946) The navigation of the Great Ouse between St. Ives and Bedford in the seventeenth century, *Bedfordshire Historical Record Society*, **24**; D. Summers (1973) *The Great Ouse: the History of a River Navigation*; A. W. Skempton (1984) Engineering on the English river navigations before 1760, in M. Baldwin and A. Burton (eds.), *Canals: A New Look*, 23–44]

Works on the Great Ouse

1618–1621. Six locks between St. Ives and St. Neots, 15 miles

1628–1638. Lock at Eaton Mills, two cuts and scouring the river upstream to Great Barford, 8 miles

SPIERING, Hugo (fl. 1631–1633), civil engineer. In 1631, having been recommended to the Commissioners for improving the Thames navigation from Burcot to Oxford as 'a gentleman of understanding in such affairs', Spiering was invited to Oxford by Lord Dorchester, Secretary to the Commissioners, to take a view of the river and give his advice. He replied in September from Thorne apologising for having missed the appointment, explaining that he had charge of 'certain sluices and a Sasse' and added that as Oxford is 200 miles distant he hoped to be excused and promised to keep any further appointment at two months notice. Though he wrote again a month later nothing more is known of Spiering in this connection but the story shows him to be an engineer of some reputation in charge of Turnbridge Sasse (navigable sluice) and the adjacent Great Sluice, the largest structure of its kind in England at the time with seventeen openings each 6 ft. wide by 8 ft. high. These were built between 1629 or 1630 and 1631 on the River Don near its outfall into the river Aire, 3 miles north of Thorne, under the general direction of Sir Cornelius Vermuyden (q.v.), presumably with Spiering as engineer. Hugh Speyring, who asked Vermuyden in July 1633 how proceedings in the Chancery Court between him and his fellow Participants in Hatfield Level drainage were going on, is clearly the same man.

A. W. SKEMPTON

[F. S. Thacker (1914) *The Thames Highway*; F. N. Fisher (1952) Sir Cornelius Vermuyden, *J Derbyshire Archaeol Nat Hist Soc*, **25**, 74–118; P. Jones (1994) Vermuyden's navigation works on the river Don, *J Railway Canal Hist Soc*, **31**, 246–258]

SPITTAL, Robert (fl. 1515–1535), master tailor and bridge-builder (or merely patron), was named as having 'fundit' (i.e. founded) or 'built' several bridges in the reign of King James IV of Scotland and within the environs of his capital, Stirling. Another memorial stone within the burgh credits him as 'donor of the hospitall in this burgh for relief of decayed tradesmen' and he left land in trust for support of the institution.

Two bridges over the river Devon near Tullibody are attributed to Spittal, the remaining one having two pointed arches and long built-up approaches over the flood-plain with further flood arches, known to have been cut in war in 1560 when it was a vital link between Stirling and the eastern lowlands. Inscriptions naming Spittal exist on the bridge of Bannockburn, dated 1516, and Bridge of Teith (1535) near Doune, the latter an early and probably original stone on which he is described as 'tailor to the maist noble preces Margaret spous to King Iamis ye feird'. It is a large structure for its date, of two arches spanning

about 50 ft. each, 12½ ft. wide when first built, and carries the road high over the river flowing in a steep-sided valley, but the foundations are placed on a shelf of rock running almost flat right across the river.

While it is most unlikely that Spittal was a workman on the sites of the bridges, the concentration of his philanthropy on bridges justifies an assumption that he took an active part in the planning and design of the projects.

TED RUDDOCK

[Royal Commission on the Monuments of Scotland (1933) *Inventory of Fife, Kinross & Clackmannan*; Royal Commission on the Monuments of Scotland (1963) *Inventory of Stirlingshire*, vol. 2; A. Swan (1987) *Clackmannan and the Ochils. An illustrated Architectural Guide*; C. McKean (1994) *Stirling and the Trossachs. An illustrated Architectural Guide*]

Works
1516. Bannockburn Old Bridge, one arch, 34 ft. span, widened and/or repaired 1631, 1710, 1781, etc., bypassed.
n.d. Tullibody Bridge, now replaced by a concrete structure
c. 1535. Bridgend Bridge at Tullibody, two river arches and three flood arches, repaired 1560, 1697, etc., bypassed
1535. Bridge of Teith, two arches, c. 50 ft. span, widened 1866 (Stevenson drawings collection, ref. 5863/93, NLS Map Room)

STANTON, Thomas (1782–1846), canal engineer and deputy county surveyor, was baptised at St. Julian's, Shrewsbury on 3 March 1782, the son of a joiner. He was employed by Thomas Telford (q.v.) in 1798 to manage the accounts of the Ellesmere Canal Company and in the general re-organisation of that company at the end of 1805 he was promoted to General Accountant, with a salary of £150 p.a. In July 1806 he was described as Agent and it may be presumed that he then moved to the Canal Office at Ellesmere; certainly he was resident there by 1809. From 27 November 1811, following the death of Thomas Denson, he added the post of Engineer to that of Agent. Telford had not been based in Shropshire for a number of years but had retained the County Surveyorship, delegating to Denson and Stanton many of the duties. Stanton's salary on promotion was to be £400 p.a., plus expenses of a horse, for which he agreed to devote the whole of his time to the Canal Company and to give up his County work after his appointment had been ratified by a General Meeting. Despite this, he did continue to deputise for Telford until the latter's death in 1834, and to work for the Ellesmere (which amalgamated to become the Ellesmere & Chester) Canal. When he retired because of ill health at the end of 1845, it was as General Agent, the title which had earlier been enjoyed by Telford.

For the County, Stanton designed and supervised the rebuilding of perhaps sixty masonry arch bridges, but none with a span greater than 50 ft. In 1811, however, he produced the working drawings from sketches provided by Telford for a cast iron bridge at Meole Brace, south of Shrewsbury. With a light and elegant span of 55 ft., the design was used, with variations, several times in Shropshire and on the Birmingham Canal, an interesting example of standardisation. Stanton was also able to undertake a certain amount of work for other associates of Telford, acting for John Simpson (q.v.) in settling his accounts for work done to Bridgnorth Bridge in 1814.

Stanton married Harriet Hunt at Shrewsbury on 17 December 1804 and they had five children. Harriet, the eldest, and Emily married two Provis brothers, William Alexander and John (qq.v.). Their mother died in 1820, aged thirty-eight, and Stanton himself died on 17 January 1846 at Leamington Spa, of dropsy. He was buried one week later at Ellesmere; on the morning of his funeral, all of the shops in Ellesmere closed as a mark of respect. A window in the church is a memorial.

Thomas Stanton's brother, John (c. 1790–1870), was a contractor with John Straphen (q.v.) for sections of the Holyhead Road in 1815–1818 but by 1828 he was described as a maltster, of Scotland Street, Ellesmere. Thomas's younger son, George, was Engineer and Agent for the Middlewich branch of the Ellesmere & Chester Canal after its opening in 1833, and John and George were both committee members of the canal in 1835 at the time of its amalgamation with the Birmingham & Liverpool Junction Canal to form the Shropshire Union Company.

P. S. M. CROSS-RUDKIN

[RAIL 826 and 827, PRO; A. Blackwall (1985) *Historic Bridges of Shropshire*; Deposited plans, Shropshire RO]

Works
Although Stanton did most of the work, Telford retained nominal responsibility for the Shropshire bridges until his death in 1834. These bridges are described in Blackwall, though it should be noted that Bearstone Bridge was designed by Joseph Potter (q.v.).

1836. Hurleston Reservoir for Ellesmere & Chester Canal

STAVELEY, Christopher, Jr. (1759–1827), was the second child and elder son of Christopher Staveley (1726–1801), an architect and mason of Melton Mowbray, and his wife, Sarah Hill. Christopher Jr. followed his father's vocation and by 1799 he was practising as an architect and surveyor at Loughborough. In 1819 he moved to Leicester, where in addition he advertised himself as an engineer and land agent. He married a Miss Ella of Loughborough in 1793 and, like his father, he had two sons called Edward and Christopher who were associated with him in some of his business.

Staveley's involvement in canal engineering flowed from his surveys of projected canals for William Jessop (q.v.) from 1785. His first appointment was to the Leicester Navigation, at a salary of £200 p.a., Jessop as principal engineer being paid £350 p.a. For one year, from September 1791, construction was carried out under contract by the Pinkertons (q.v.) but their work was unsatisfactory. Their contracts were terminated and Staveley completed the works himself using direct labour.

While work on the Leicester Navigation was going forward, Staveley was also in charge of the construction of two navigations which, together, formed a branch from Leicester to Oakham by an end-on connection at Melton Mowbray. In 1795 however, he was removed from his post with the Melton Mowbray Navigation and, after a critical report on it and the Oakham Canal, he resigned also from the latter. He did much of the preliminary survey work for the Leicestershire and Northamptonshire Union Canal, which was intended to extend the Leicester Navigation southwards, but he left it soon after it gained its Act in 1793, leaving John Varley (q.v.) in sole charge.

The branch from the Leicester Navigation's main line to the coalfields was in fact a mixture of railways and canal, because of difficulties with procuring an adequate water supply for a canal throughout. Jessop's last act before leaving the company in 1795 was to recommend the building of a reservoir to help solve this problem and Staveley was put in charge of construction of the dam at Blackbrook. It was completed in mid-1797 but burst in 1799, causing considerable damage. Jessop attributed the failure to a fault in the design of the ancillary works, though poor workmanship may also have been to blame. The dam was repaired by James Barnes (q.v.) and Staveley continued to work for the company until 1825. The family's connection ended when his son, Edward, who had succeeded him, absconded in 1833 owing a large sum of money to the company.

Staveley's reputation was such that he was asked in 1803 to report on the tramroads of the Ashby Canal. He was guardian of John Sydney Crossley (b. 1812), later Engineer-in-Chief of the Midland Railway, and articled him to his son, Edward Staveley.

P. S. M. CROSS-RUDKIN

[RAIL 847/1&2, PRO; *Monthly Magazine*, 1801, 189; J. D. Bennett (1968) *Leicestershire Architects 1700–1850*; M. G. Miller and S. Fletcher (1984) *The Melton Mowbray Navigation*; D. Tew (1984) *The Melton to Oakham Canal*; G. M. Binnie (1987) *Early Dam Builders in Britain*; P. A. Stevens (1992) *The Leicester and Melton Mowbray Navigations*]

Publications
1790. *A Plan of the intended Navigation from Loughborough to Leicester and the proposed Water Level and Railways from Loughborough to*

the Coal Mines ... and Lime Works ... Also of the Intended Navigation from the Leicester Navigation to Melton Mowbray (with William Jessop)

1792. *A Plan of the intended Union Canal from Leicester ... to Hardingstone ... with a Collateral Branch to Market Harborough* (with John Varley)

1792. *A Plan of the Proposed Union Canal to join the Leicester Navigation ... to near Northampton: also of a Branch to Market Harborough* (with John Varley)

1793. *A Plan (and Profile) of the intended Navigation from Melton Mowbray ... to Oakham ...* (with William Jessop)

Works

1791–1825. Leicester Navigation, Surveyor, £200 p.a., from 1794, Engineer

1791–1795. Melton Mowbray Navigation, Resident Engineer

1793–1794. Leicestershire & Northamptonshire Union Canal, Engineer, jointly with John Varley (q.v.)

1793–1795. Oakham Canal, Engineer

STEEDMAN, John (fl. 1814–1834), civil engineer, of Edinburgh, became a Corresponding Member of the Institution of Civil Engineers in 1829. Steedman worked as an assistant (1814–1826) with Robert Stevenson (q.v.). This included his credited role as delineator for the *Report on the Improvement of the Harbour of Dundee, 1814* and as Surveyor for the *Reduced Survey & Section of a Line of Canal between Edinburgh & Glasgow upon One Level ...* (1817). The 1816 *Nautical Survey of the Frith of Tay by Robt. Stevenson, John Steedman Assistant*, formerly in the collection of the Institution of Civil Engineers, is now lost. A reference to his drawing for the Edinburgh–Dalkeith Railway for Stevenson (1818) has proved impossible to trace.

Steedman's most important work was as the contractor for Stevenson's Hutcheson Bridge, Glasgow, one of Stevenson's most impressive designs.

TESS CANFIELD

[ICE membership records; Skempton]

Publications

1817. *Reduced Survey & Section of a Line of Canal between Edinburgh & Glasgow upon One Level ... made in the Years 1814 & 1815, by Robert Stevenson, John Steedman Surveyor* (engr. plan)

1817. *Reduced Survey & Section of the proposed Canal into the Bale of Strathmore between ... Forfar and Aberbrothwick, Surveyed by ... Robert Stevenson, 1817, John Steedman, Surveyor* (engr. plan)

1818. *Reduced Survey of part of the Frith of Tay as seen at low water of Spring tides ... by Robert Stevenson, John Steedman, Surveyor* (engr. plan)

1818. *Reduced Plan of part of the Shires of Edinburgh & Haddington shewing the Lines of the Proposed Railways, by Robert Stevenson, 1818, John Steedman, Surveyor* (engr. plan)

1819. *Reduced Survey & Sections of several Lines of Railway from the Port of Montrose to the Town of Brechin ... by Robert Stevenson ... J. Steedman, Surveyor* (engr. plan)

1822. *Reduced Survey & Section of Part of the River Severn shewing the Proposed Improvement upon New Passage Ferry, by Robert Stevenson. Surveyed by W. H. Townsend of Bristol and J. Steedman of Edinburgh* (litho. plan)

1826. *Reduced Plan & Sections of the Valley of Strathmore shewing a Line of Railway from the Ports of Montrose & Arbroath to Perth, Surveyed under the direction of Robert Stevenson, Wm. Blackadder, John Steedman, Surveyors* (engr. plan)

1827. *To the Subscribers to the Survey of the proposed Railway from the Forth and Clyde Canal to Stirling ... and to Alloa Ferry, near Kersie Nook, the Report of John Steedman, Engineer*

Works

1831–1834. Hutcheson Bridge, Glasgow, contractor, £24,000

STEEL, Edward (1771–1833), colliery viewer and civil engineer, was the son of Edward Steel and his wife, Mary Stephenson. The eldest of at least four children, he was baptised at Whickham, near Gateshead, on 31 March 1771. Nothing is known of his early life but in 1796 he married by licence Isabella Rutter (1775–1816) of Lamesley, near Gateshead.

As a coal viewer Steel, in 1812, had become responsible for the construction of the Newbottle railway, or wagonway, running to the river Wear at Sunderland. The line was 6½ miles long and was designed to be operated by horses although in 1815 William Brunton's 'steam horse' was the subject of a trial. He resigned from his position as 'Superintendant Viewer' of the Grand Allies collieries in 1815 and from a similar post at Newbottle two years later.

In 1818 he surveyed coal deposits in south Durham, a work tied to the proposal for a canal to Stockton, and the next year worked on the projected development of Hetton colliery, including a survey for a possible line of railway. In 1823 Steel was involved in the presentation of a petition to Parliament in an attempt—it was successful—to force the Stockton and Darlington Railway to construct its Haggerleases branch so as to serve collieries owned by, amongst others, the Countess of Srathmore, for whom he worked extensively.

From 1827 Steel undertook work for the Clarence Railway from Simpasture—on the Stockton and Darlington Railway—to Stockton, shortening the route between collieries in south west Durham and the river Tees. On the formation of the company, intending to use both locomotives and horse traction, he became its engineer in 1828. He reported several times on the railway's progress but in May 1831, presumably because of unspecified shortcomings, the line was inspected by two other viewers 'versed in colliery railways'. Subsequently they reported that they were 'quite satisfied that the ability of Mr. Steel the Engineer is adequate to the work and that under his superintendence every difficulty will be eventually overcome'; in October 1833 the company decided to dispense with his services and Thomas King succeeded him.

Steel was also engineer to the Hartlepool Dock and Railway Company, at least so far as its railway was concerned; James Milne was engineer for the docks. Immediately after his appointment in July 1832 he was instructed to design the bridges on the line but he resigned in November, without recorded explanation, and was succeeded by Stephen Robinson.

On 19 December 1833, Steel died when his gig overturned in an accident and he was 'unfortunately thrown under the forewheel (of a wagon) ... which went over his head and caused immediate death'. Three days later he was buried at Lamesley, near Gateshead, survived by eight of his eleven children.

A. P. GUY and R. W. RENNISON

[Watson, Buddle and Forster Collections (1799–1832) Acc. 3410, NRO; Londonderry and Strathmore collections, DCRO; Matthias Dunn (c. 1820) *View Diary*, Beamish Museum; RAIL 117/1 (1831–1833) PRO; RAIL 294/3 (1832) PRO; *Newcastle Chronicle*, 28 Dec. 1833; W. W. Tomlinson (1914) *North Eastern Railway*; A. P. Guy and J. Rees (eds) (2001) *Early Railways ... First International Early Railways Conference*]

Drawings
1824. *Plan of part of the Ancient Inclosed Lands ... [at] Evenwood. Surveyed by Edward Steel, 1823–1824*
1827. *A Plan of the proposed Clarence Railroad from Simpasture to Haverton Hill ... November 30th, 1827*
1828. *Clarence Railway. From Haverton Hill to Durham ... with Bridge Elevation*
1831. *Plan of Marley Hill Colliery ... surveyed June 1831 by Edward Steel*
1831. *Clarence Railway additional Branches from Haverton Hill to Simpasture...*
1832. *Clarence Railway. Branches to Etherley ...*

Works
1812–1817. Newbottle Railway, viewer
1819. Hetton Colliery Railway, survey
1827–1833. Clarence Railway, Engineer
1832. Hartlepool Dock and Railway, Engineer

STEELE, Thomas Ennis (d. 1848), gentleman engineer and politician, was a descendant of an established family of County Clare, Ireland, where he inherited an estate. He took degrees at Trinity College, Dublin about 1817, and at Trinity College, Cambridge in 1820, having studied classics, mathematics, mechanics and chemistry. He became an Associate member of the Institution of Civil Engineers in 1827 where he repeatedly advocated the purchase and preservation of Sir Isaac Newton's birthplace. Perhaps as a result of his interest in geology, as well as his concern for the economic

effects of the difficult navigation of the river Shannon, Steele undertook a detailed survey of the river bed. He used original methods, including walking along the line of submerged reefs. His suggestions for improvements to the river were successfully carried out. He became interested in the diving bell, and made improvements. He also devised a method of lighting for divers while under water. In about 1825 he became a committed follower of Daniel O'Connell and became deeply involved in Irish politics. At O'Connell's death in May, 1847, Steele abandoned politics, a disappointed man, and himself died soon after. A copy of an engraved portrait is in the National Portrait Gallery, London.

TESS CANFIELD

[ICE Membership records; *Subaqueous Operations—Subaqueous Illumination* (1838) ICE Ms. OC 282; Obituary (1848) *Min Proc ICE*, **VIII** 16–19]

Publications
1828. *Practical Suggestions on the General Improvement of the Navigation of the Shannon between Limerick and the Atlantic, and more particularly of ... the Narrows*

STEERS, Thomas (c. 1672–1750), civil engineer, surveyor and merchant, is best known as the first engineer for the docks in Liverpool and as engineer for the Newry Canal, in Ireland.

Thomas Steers was probably born in Rotherhithe or Deptford around 1672. His handwriting was clear, and his calculations relating to surveying and land measurement show a familiarity with mathematics, which is suggestive of good education. Little is known of his early life, but he entered the army in his teens. A family story suggests that he was present at the Battle of the Boyne in 1690. He is mentioned, in the Army List of 7 July 1702, as a Quarter-Master in the 4th Regiment of Foot (The King's Own), though not on active service. This regiment was present at the Battle of the Boyne and also served in William of Orange's subsequent campaign against the French in the Low Countries. It was involved with the sieges of Huy and Namur, both situated on the navigable Meuse, before returning to England after the Peace of Namur in 1697. As William's army was poorly supplied with engineers and transport, Steers would have been in a good position to develop those skills, which he used in his later engineering work. He would also have been able to study recent developments in hydraulic engineering in the Low Countries.

Soon after returning to Kent, Steers married Henrietta Maria Barber in 1698 or 1699. Her father, Abraham Barber of Rotherhithe, settled property on the couple in Queen Street, Rotherhithe. At this time Howland Great Dock in Rotherhithe was already under construction. The dock was to be built on part of the Howland estate, Elizabeth Howland leasing the ground in 1695–1696 to John (q.v.) and Richard Wells who were originally to build a dry-dock and shipbuilding yard. Subsequently they were granted a

Thomas Steers

second lease and £12,000 to build a wet dock. There is no evidence that Steers was directly involved in the construction of this dock. However, a survey made by him in 1707 for Elizabeth Howland of the whole of the Wet Dock Field suggests that he may have been employed as a surveyor by the Howland Estate. Steers also leased part of the field from the estate and he is described as a house-carpenter in the lease.

Early in 1708, the merchants and politicians in Liverpool decided that a wet dock, similar to that at Rotherhithe, would help increase the town's trade. George Sorocold (q.v.) was approached, and by mid-1709 he and his associate, Henry Huss, had drawn up a plan and estimate for the Dock's construction. Probably because of other work, neither Sorocold nor Huss would agree to become the dock engineer. Then, on 17 May 1710, the Town Council were informed that Thomas Steers had arrived in Liverpool and had set out the works to his own design in which the dock was to be built on land reclaimed from the Pool. This location needed less excavation work, and was to set the standard for dock construction in Liverpool, only Stanley Dock being built on unreclaimed land.

A possible explanation for Steers's appearance in Liverpool is that his engineering work whilst in the army had come to the notice of the Hon. J. Stanley who had commanded the 16th Regiment of Foot in Flanders. Stanley became the 10th Earl of Derby in 1702, was Mayor of Liverpool in 1707 and Lord Lieutenant of Lancashire till 1710. Having such influential backing could certainly account for Steers subsequent rapid advancement in the town's hierarchy.

Steers, assisted by William Braddock, was responsible not only for the design of the Dock but was a contractor for the excavation work. Other contractors included Edward Litherland, responsible for the masonry, William Bibby for the lime and brickwork, and Thomas Hurst and Thomas Pattison provided timber. The Dock (3½ acres in area) had opened for shipping by mid-1715, but the works were still incomplete. A second Act, in 1717, authorised the construction of a tidal basin and three graving docks next to the entrance to the dock. Work was still in progress in 1720 when the north side was raised to stop flooding. Further expansion was envisaged and Steers had drawn up plans for a southward extension in 1718. He reported completion in 1721, the same year that the new Custom House was opened.

Steers's salary as Dock Engineer seems to have been about £20 p.a. The Council Treasurer's Accounts also show payments to him of £290 in 1721 and £429 in 1723 on account of the dock, presumably for construction work. From 1717, he was appointed Dock Master at an annual salary of £50, his assistant William Braddock being made Water Bailiff. This arrangement changed in 1724 when Steers took over the Water Bailiff's post as well, although from this time he ceased to be paid and had to rely upon the perks and fees associated with the post.

Besides being responsible for, and actively involved in building the Dock, Steers had time for other works. In 1712 he proposed and produced surveys for the Mersey and Irwell Navigation and the Douglas Navigation. He was named as an undertaker for both in their Acts of 1720. His actual involvement with the Mersey & Irwell is uncertain, but he certainly began work on the Douglas, building a lock and bridge at Rufford, straightening part of the river and commencing work on the tidal lock at Croston Finney. He estimated that he had spent some £700 out of an estimated £8,000 needed for the navigation. Unfortunately his associate, William Squire, who was raising money for the navigation in London, became involved with the South Sea Bubble and seems to have lost most of the money he raised.

These two navigations promoted by Steers, together with others opened in northern England at this period, were built to encourage new traffic and not simply to make existing trade easier. Steers's participation in both their promotion and construction suggests that he appreciated how vital a good transport infrastructure would be to the development of trade. Later in the century, it was one of the most important factors in establishing the Industrial Revolution in Britain.

Steers's other interests included land, and he leased an area along the foreshore to the south of the Dock, in 1715. In 1720 he became a promoter of the Liverpool Waterworks together with Sir Thomas Johnson and Sir Cleave Moore. His interests also included an anchor smithy near the Dock, a business which one of his sons, Spencer Steers, continued after his death. He was a joint owner of the vessel 'Dove' trading to the West Indies in partnership with Richard Gildart and Peter Hall, and in 1725 he was appointed as a commissioner for the turnpike road from Liverpool to Prescot. The same year, Steers, together with James Shaw, drew up plans and estimates for St. George's church. The construction took from 1725 to 1734 and cost £2,984. Steers was responsible for the foundations and building the steeple. By the early 1820s, the steeple had become unsafe and had to be rebuilt. In 1736 he surveyed the buoyed channel on Hoylake Bank, and the following year he presented a plan to the Corporation for a new dock and pier. He appeared for the Council before the Parliamentary committee considering the application for a new Act of Parliament for the Weaver Navigation in 1737. He leased, in 1727, the playhouse in Chorley Street, and later, in 1740, opened a new theatre on the site of the Old Ropery. A number of houses for poor and destitute seamen were built by him in 1739 on waste land belonging to the Corporation.

Steers was involved in political activity in Liverpool, becoming a Freeman in 1713 and a member of the Town Council in 1717. He was a Town Bailiff in 1719 and 1722, becoming Mayor for 1739. He was also an Out-burgess for Wigan by 1746, possibly through his work on the Douglas Navigation, though the politics of Liverpool and Wigan were interlinked. Liverpool was a Whig stronghold at this period.

Steers had seven children by his first wife, Henrietta Maria, who died in 1717. Only three, Philip Mell, Thomas and Gustavus Adolphus grew to adulthood. They probably became seamen and all three had died by 1732. Steers remarried in 1719, his new wife also coming from Rotherhithe. She was Ann Tibbington, nee Walton, the widow of Thomas Tibington, mariner of Rotherhithe. She already had one son, John Tibington, and a further four children were born during her marriage to Steers. Only two, Spencer and Ellen, survived into adulthood.

Apart from Liverpool's docks, Steers's most important work was the Newry Canal. 18 miles in length with 14 locks, it was the first summit level canal in the British Isles. He was asked to be the engineer in 1729, but his request for 100 guineas per month was not accepted. Instead an Irish architect, Edward Lovett Pearce, was appointed, his assistant, Richard Castle (q.v.) taking over on Pearce's death in 1734. In the spring of 1736, Steers was paid 50 guineas for a survey of work already undertaken on the canal, and in the following year he took over responsibility for its construction from Castle who had been dismissed in December 1736. Steers agreed to be in Ireland for four months in 1737 and for two months in each of the following two years. However, the work took considerably longer than at first envisaged and was not completed till 1741, by which time he had received £1,320, including payment for fifteen months extra work. Two of the locks were constructed on the 'French' pattern, with the sluices at the upper end built into the walls of the lock rather than in the gates themselves. It was an early use of this method, which makes it possible to build locks with much greater fall than with simple gate paddles, water being fed into the lock chamber close to the lock floor. The water supply for the canal came from streams and from Lough Shark, which was used as a reservoir, feeding the canal via a wooden aqueduct. In a contemporary published account of the canal, Steers was maligned for his work, possible because he was English, whereas his Irish deputy, Mr. Gilbert, was praised. What is certain is that the canal was not well constructed. The 'French' pattern locks with ground paddles, one of which was at Poyntz Pass at the southern end of the summit level, were reported as in poor condition as early as 1750 and they were rebuilt shortly afterwards. From descriptions, they were probably similar to those on the Canal de Briare, with both chamber and sluice walls being built on wood-covered pile work. A contemporary report suggests that the sluices had failed, leading to damage to the lock walls.

In Ireland, Steers was also employed, in 1738, on the Ballycastle harbour by Hugh Boyd, the local land and colliery owner. He constructed a pier using a framework, comprising three rows of wooden piles held together by iron straps, which was then filled with rubble. The pier was at least 200 ft. in length and 40 ft. in width, producing a harbour capable of accommodating around forty boats in 1743. It failed fairly quickly due to worms eating the framing and it was possibly his experience here which led him, in 1743, to rebuild the wooden pier at Liverpool in stone. Ballycastle pier was rebuilt in stone and extended in 1747, probably removing much of Steers's work. At Ballycastle he was both designer and undertaker, being assisted by William Needham on site. In 1746, he undertook a preliminary survey of the river Boyne. Two years later he made a complete survey, which was used subsequently when the river was made navigable. He was paid 250 guineas for this and was probably assisted by Henry Berry (q.v.). Steers suggested that the Boyne was the most difficult river he had surveyed for a navigation. Work started on the navigation in 1748 under John Lowe, though Steers may still have been involved. By the time of his death in 1750, the tidal lock at Oldbridge had been completed. The navigation was only completed to Navan in 1800.

During the period when he was working in Ireland, Steers also continued the improvement of dock facilities in Liverpool. An Act for a new dock and enlarged tidal basin (4 acres) was obtained in 1738, and Steers was once again called upon to oversee the work. He was to be paid at the same rate as when he built the first dock. The New Dock (4½ acres) eventually opened in 1753, Henry Berry overseeing the work after Steers's death in 1750. The old wooden pier at the dock entrance was extended in 1740 and rebuilt in stone in 1744, while, in 1746, a graving dock was ordered to be built at the north end of the new pier.

One characteristic of all Liverpool's eighteenth century docks was the use of a single pair of gates for dock entrances. In Steers's time, there was a seasonal nature to many of the trades in which large vessels were engaged, and the dock was used more for laying-up and re-fitting, resulting in the large number of ship builders around the dock edge. The tidal basin was provided so that large vessels could be loaded and unloaded in comparative safety without the time consuming process of entering and leaving the dock, and the construction of the stone pier could well have further enlarged cargo-handling areas. The New Dock, soon called Salthouse Dock, was used by coastal vessels in the salt trade. They would have had much less difficulty in entering and leaving the dock than the larger vessels in the colonial trades. The entrance gates of both docks were about 34 ft. wide with 22 ft. of water on the sills at HWST.

Steers continued to be involved in a wide variety of work. In 1740–1741 he surveyed the Calder & Hebble Navigation together with John Eyes (q.v.). Eyes then produced, in 1742, a 'Plan of the Docks and Piers in Liverpool', suggesting further collaboration. He also worked on the Douglas Navigation with Steers who was called in by Alexander Leigh, the new undertaker, for advice in the early 1740s. In Liverpool, Steers surveyed the Exchange building in 1740 and was involved in the design of the New Exchange in 1748. In the same year he advised on the supply of stone for Saint Thomas' Church and during the Jacobite Rebellion, in 1745, he was responsible for the fortification of the town. He continued to have an interest in shipping, being part owner, in 1746, of the sloop 'Hoadley' which traded with Ireland, where he also had a part share in a saltworks close to the Newry Canal.

Steers died late in 1750 and was buried in St. Peter's Churchyard on the second of November in that year. The sole acknowledgement found in the Town Records is the statement 'Whereas Mr. Alderman Steers is lately dead, it is ordered that Henry Berry, late clerk to him, be continued to oversee the works till further order'. The only other contemporary appreciation of his work is contained in a letter from John Smeaton to the Calder and Hebble committee in 1757 referring to Steers and Eyes survey saying '... as those gentlemen were generally esteemed men of character and ability in their profession, particularly Mr. Steers ...'.

Thomas Steers certainly made a lasting impression on British engineering practice with the design and construction of the world's first successful commercial dock. The Newry Canal, his other major work, is also of great importance, not just because it was Britain's first summit level canal, but also because of the two locks built after the French pattern; an example of the transfer of hydraulic engineering technology from the continent.

Examples of Steers's work have not lasted. The Old Dock was filled in during 1827, and the Newry Canal had to be rebuilt around 1800. The only remaining construction, which may possibly have been built by him, is a weir on the River Douglas Navigation close to the canal locks at Dean. Steers's forte was perhaps in surveying. He seems to have been a good teacher as his assistants often went on to fame in their own right. Amongst these, Henry Berry built the Sankey Navigation as well as being Liverpool's second Dock Engineer; William Needham became one of Ireland's early hydraulic engineers; and John Eyes (and others from his family) produced well-known county surveys in north west England. Liverpool's eighteenth and early nineteenth century importance as a centre for surveyors, such as the Eyes family and P. Burdett, is almost certainly the results of Steers's work.

Not only was Steers involved with engineering problems, but he also produced architectural designs. Unlike Liverpool's later Dock Engineers, he was actively involved in both the trade and politics of the town. The merchants of Liverpool, who needed someone with engineering skills, practical experience and ideas to develop their trade, could hardly have chosen a more suitable candidate for the position of Dock Engineer. The Dock and the navigations linking it with the port's hinterland, both the result of Steers's skills, enabled a small port to become the world's largest within one hundred years. The increase in north-west trade brought about by his, and Liverpool's, investment in the transport and industrial infrastructure is certainly one of the significant events which was to result in the industrial revolution of the late eighteenth century.

MIKE CLARKE

[Bedford Estate Papers re. Howland Dock survey and leases (1695–1709) GLRO; Survey of the River Douglas (1713) Lancashire RO; Account Book Ledgers and Treasurers accounts, 1720–1734, Liverpool RO; Accounts concerning the River Douglas (Leigh Papers), & Plans and Accounts of lands used for the Douglas Navigation, 1720–1721, Wigan RO; Letter from Mr. Knightley in Dublin, 24 May 1729, Add MS 21134, BL; Faulkner's Dublin Journal, extracts describing the improvements of the harbour at Ballycastle, 1737–1738, PRONI; River Boyne accounts, 1746–1822, PRONI; John McDowett (1754) *Mr. Thomas Steers, Map of the River Boyne coppyed in the year 1754. By order of the Corporation for promoting and carrying on an inland navigation in Ireland* [colour], Balfour Estate Maps, National Library of Ireland; Hugh Boyd (c. 1740) *An Account of the Collieries and Harbour at Ballycastle*, PRONI; W. Harris (1744) *The Antient and present State of the County of Down*; M. B. Mullins (1850–1863) Address on engineering in Ireland, *Trans Inst Civil Engnrs Ireland*, **6–7**, 89–90; H. Peet (1930) Thomas Steers, *Trans Hist Soc Lancs Ches*, **82**, 163–242; T. S. Willan (1936) *River Navigation in*

England, 1600–1750; E. R. R. Green (1963) *The Industrial Archaeology of County Down*; W. A. McCutcheon (1965) *The Canals of Northern Ireland*; N. Ritchie-Noakes (1984) *Liverpool's Historic Waterfront*; M. Clarke(1990) *The Leeds & Liverpool Canal, a History and Guide*, M. Clarke (1993) Thomas Steers, in *Dock Engineers*, papers presented at a Research Day, Merseyside Maritime Museum, 13 February 1993; R. Delaney (1998) *Ireland's Inland Waterways*]

Works

1710–1715. Liverpool, Old Dock, £11,000
1717–1721. Liverpool, tidal basin, £5,000
1737–1741. Newry Canal, total cost 1731–1741 £52,000
1738–1742. Ballycastle Harbour, pier, £10,000
1739–1750. Liverpool, Salthouse Dock and enlarged tidal basin, completed 1753, £21,000
1740–1741. River Douglas Navigation

STEPHENSON, David (1757–1819), architect, was the son of John Stephenson (d. 1796), carpenter, and his wife Ann Crawforth. The eldest of five children, he was baptised on 6 November 1757 at the Castle Garth Presbyterian chapel in Newcastle.

After completing an apprenticeship as a carpenter in 1778, Stephenson turned to architecture, working in London for two or three years before returning to Newcastle and marrying Margaret Gibbon (1755–1839) in 1783; the marriage brought four daughters and four sons, born between 1784 and 1800, all baptised in the (nonconformist) Hanover Square church, Newcastle.

Stephenson's architectural works are listed by Colvin and were mainly restricted to Tyneside and Alnwick; his major work was the elliptical All Saints church, Newcastle, completed in 1796. He also designed works of a civil engineering nature: in Newcastle, the formation in 1787 of the new thoroughfares of Mosley Street and Dean Street, the latter including the culverting of a stream running through the old town; in Gateshead he laid out Church Street in 1791, so improving the southern approach to the Tyne Bridge; and in North Shields he formed a new quay in 1806.

Stephenson, whose father had built the temporary timber bridge over the Tyne in Newcastle after the earlier structure had been destroyed in the flood of 1771, became responsible in some degree for three bridges in the North East: in 1787 he worked with Robert Mylne (q.v.) on the design and construction of Ridley Hall bridge over the river Tyne; in 1801 he became responsible, both as architect and contractor, for widening the masonry bridge at Newcastle from 21½ ft. to 33½ ft. by forming additional arches to both faces of the bridge, basing them on the existing piers and so obviating the need for extensive river works; and, lastly, for the Duke of Northumberland he designed a small three-span iron estate bridge which, fabricated by I. & T. Cookson of Newcastle and bearing the date 1812, stands in Hulne Park, Alnwick.

One of Stephenson's pupils was John Dobson, later responsible for undertaking many Newcastle improvements. Stephenson was a founder member of the Literary and Philosophical Society in Newcastle and was an early member of the Society of Antiquaries. In 1803 he commanded a volunteer corps in the town. He died suddenly in Alnwick on 29 August 1819, survived by his widow, two daughters and a son, two of his sons having died earlier in the service of the East India Company. He was buried in All Saints church, Newcastle, on 3 September 1819 and is commemorated by a plaque there.

R. W. RENNISON

[*Freemen of Newcastle upon Tyne: Admissions* (1778) GU/NCF/1/7, Tyne and Wear Archives; *Contract as to widening Tyne Bridge* (17 March 1802) 612/345, Tyne and Wear Archives; E. MacKenzie (1827) *A Descriptive and Historical Account … of Newcastle*; R. Welford (1895) *Men of Mark Twixt Tyne and Tweed*; Colvin]

Works

1787. Ridley Hall bridge, partial design
1787–. Mosley Street and Dean Street, Newcastle
1791–. Church Street, Gateshead
1801–1802. Tyne Bridge, Newcastle, widening
1806. North Shields Quay
1812. Iron estate bridge, Alnwick, design

STEPHENSON, George (1781–1848), enginewright, railway surveyor and builder, was born on 9 June 1781 at Wylam, Northumberland, the second son of Robert Stephenson (1748–1817), colliery artisan, and Mabel Carr. His public acclaim for bringing about the steam locomotive operated public railway is reflected in the rare distinction of his vignette featuring on the current Bank of England £5 note.

Stephenson began his colliery career without formal education; he subsequently received basic tuition but was otherwise self-educated. He had an aptitude for colliery engine maintenance and improvement and his innovatory ideas at Killingworth colliery, Northumberland, included, in 1812, the first underground haulage engine. He witnessed the 'travelling engines' of William Hedley (1779–1843) and Matthew Murray (qq.v.), and built his first locomotive at Killingworth in 1814, probably with advice from John Buddle (q.v.). Jointly with Ralph Dodds (1792–1874), his mentor and Colliery Viewer, Stephenson took out his first patent in 1815, for locomotive coupling rods and crank pins. In 1815 he gained acclaim in the Newcastle-on-Tyne Literary and Philosophical Society when, with the assistance of Nicholas Wood (q.v.), he presented the findings of experiments to develop a reliable colliery safety lamp. His development work was coincident with experimental work by Sir Humphrey Davy and so became controversial, with rival claims for precedence between Stephenson's Geordy lamp and the Davy lamp.

George Stephenson

Stephenson sought to reduce displacement and breakage of cast-iron rails on the Killingworth Colliery railway. Jointly with William Losh (1770–1861) of the Walker Ironworks, he took out a patent in 1816 for a new form or rail-joint and chair. In 1818, jointly with Wood, he carried out trials into rolling stock resistance using a simple dynamometer device.

In 1819, as engineer of the Hetton Railway, County Durham, Stephenson surveyed and supervised the construction of the 8-mile long colliery line. His brother, Robert (1788–1837), was resident engineer, the only one of his three brothers to distinguish himself as an engineer. Opened in 1822, the railway was a combination of self-acting inclines, stationary engine inclines and locomotive-worked stretches.

In 1821 Stephenson carried out a survey of the 27 mile long Stockton & Darlington Railway, between Witton Park and other collieries in County Durham; shipping staithes were to be provided at Stockton-on-Tees. He persuaded the railway's directors, particularly Edward Pease (1767–1858), to adopt locomotive haulage for the majority of the route. Stephenson was appointed as the railway's engineer in 1823.

Stephenson realised the limitations of cast iron track for locomotives and, disregarding his patent benefits, he recommended wrought iron rails, rolled at Bedlington Ironworks, Northumberland, under the patent of its Principal Agent, John Birkinshaw (q.v.). Anticipating the market for iron products, especially steam engines, for the Stockton & Darlington and other railways, Robert

Stephenson & Co. of which his son, Robert (q.v.), was managing partner, was established as a manufacturing business in Newcastle.

In 1824, the building of the Stockton & Darlington Railway stimulated interest in railways by financiers, particularly from London and Liverpool. Stephenson's services were much sought after and, to formalise this consultancy work, he and his son, Robert, set up the firm, George Stephenson & Son. Whilst overseeing the building of the Stockton & Darlington line he supervised teams of surveyors for several trunk routes, including the Liverpool & Manchester line.

Stephenson received a set-back in 1825 when the Liverpool & Manchester Railroad Bill was rejected by Parliament, during which survey discrepancies were highlighted. He then concentrated on the successful completion of the Stockton & Darlington line, the first public railway to use steam locomotive haulage, opened in September 1825.

The Liverpool & Manchester line was resurveyed, to meet landowners' objections, by Charles Vignoles (q.v.) under the direction of John and George Rennie (qq.v.). Following the railway's enactment in May 1826, Stephenson was appointed Chief Engineer to build the 31-mile line. It took over four years to build and was one of the largest civil engineering projects thus far undertaken and requiring substantial—and innovatory—works and structures. They included: two substantial masonry skew arches, at Rainhill and across the river Irwell (consultant Jesse Hartley (q.v.)); the Sankey viaduct formed of nine 50-ft. span arches (maximum height 70 ft.), of brick with stone facings (consultant Jesse Hartley); Olive Mount rock cutting (a mile long, maximum depth 70 ft.); Wapping tunnel, Liverpool (2,250 yd. long, through rock, 22 ft. wide and 16 ft. high on an incline of 1 in 48); and the 4-mile crossing of Chat Moss on a bed of wooden frames, brushwood, shingle and cinders, and with parallel drainage ditches.

Throughout the project, Stephenson campaigned tenaciously for the railway to adopt locomotive power against opposition from some railway directors. His motives were questioned and the Rainhill Trials of 1829 allowed the Stephensons to demonstrate that their locomotives were best, as well as demonstrate superiority over other forms of traction. Stephenson's son, Robert, undertook the research and development of the locomotives that led to *Rocket*, which won the Trials, and the *Planet* class, the first to operate services on the Liverpool—Manchester line.

Stephenson's fame stemmed from the opening of the Liverpool & Manchester Railway in September 1830, recognised as the beginning of a new transport era. Although he was not an innovator of mechanical or civil engineering principles, his strong beliefs, persuasive arguments and tenacious application to railway and locomotive technology were the attributes acknowledged by many who engaged his services. His strong presence and belief that his views should prevail upset others, however, and the remainder of his engineering career was often chequered with controversy.

Stephenson was a founder member and first President of the Institution of Mechanical Engineers. For his railway work he was appointed a Knight of the Belgian Order of Leopold in 1835. He was also offered a British knighthood several times, but declined it, preferring to remain identified with his working class origins. Stephenson became wealthy, having developed coal and limestone interests in the Midlands and, as a major investor in railways, was appointed a director of some of them. He was married three times: to Frances Henderson in 1802, Elizabeth Hindmarsh in 1820, and Ellen Gregory in 1848. He died of pleurisy on 12 August 1848, at his home, Tapton House, Chesterfield, Derbyshire, and is buried in the town's Trinity Church.

MICHAEL R. BAILEY

[Stephenson's letters and reports, National Museum of Science & Industry, Institution of Mechanical Engineers, Institution of Civil Engineers, Darlington Public Library, Liverpool RO; C. Von Oeynhausen and H. Von Dechen (1827, translated from the German text by E. A. Forward, edited by C. E. Lee in collaboration with K. R. Gilbert for the Newcomen Society, 1971), *Railways in England 1826 and 1827*; S. Smiles (1857, etc.), *The Life of George Stephenson, Railway Engineer*; J. G. H. Warren (1923) *A Century of Locomotive Building by Robert Stephenson & Co. 1823–1923*; L. T. C. Rolt (1960) *George and Robert Stephenson, The Railway Revolution*; R. E. Carlson (1969) *The Liverpool & Manchester Railway Project 1821–1831*; W. O. Skeat (1973) *George Stephenson The Engineer & His Letters*; M. R Bailey (1979) Robert Stephenson & Company, 1823–1829, *Trans Newc Soc*, **50**, 109–138; R. S. Fitzgerald (1980) *Liverpool Road Station, Manchester, An Historical and Architectural Survey*; R. H. G. Thomas (1980) *The Liverpool & Manchester Railway*; M. R. Bailey (1981) George Stephenson—locomotive advocate: the background to the Rainhill trials, in 'George Stephenson: A Commemorative Symposium for the 200th Anniversary of his Birth', *Trans Newc Soc*, **52**, 171–179; D. Beckett (1984) *Stephenson's Britain*; A. Guy and J. Rees (eds.) (2001) *Early Railways*]

Publications

1815. Dodds and Stephenson, *Specification, 28th February 1815 (No. 3887) for Locomotive Coupling Rods and Crank Pins*
1816. Losh and Stephenson, *Specification, 26th November 1816 (No. 4067) for a New Form of Rail-joint and Chair*
1817. *A Description of the Safety Lamp Invented by George Stephenson*
1822. *Plan and Section of the intended Railway or Tramroad from Stockton by Darlington to the Collieries near West Auckland*

1824. *A Plan and Section of an intended Railway or Tramroad from Liverpool to Manchester*
1825. *Report and Estimate of an intended Railway from Bolton le Moors to Eccles*
1830. *Report on the Line of the Edinburgh Glasgow and Leith railway*

Works (to 1830)

1821–1822. Hetton Colliery Railway Construction, Chief Engineer, 8 miles
1822–1825. Stockton & Darlington Railway, Construction, Chief Engineer, 27 miles, £167,000
1825–1830. Canterbury & Whitstable Railway, Construction, Chief Engineer, 6 miles, £83,000
1826–1830. Liverpool & Manchester Railway, Construction, Chief Engineer, 31 miles, £820,000
1827–1828. Bolton & Leigh Railway, Construction, Chief Engineer, 7 miles, £68,000
1827–1828. Nantlle Railway, Construction, Chief Engineer, 9 miles

STEPHENSON, Robert, FRS (1803–1859), railway, bridge, civil and mechanical engineer, was born on 16 October 1803 at Willington Quay, Northumberland, the only child of George Stephenson (q.v.), engine-wright and railway builder, and Frances Henderson (1769–1806). He was educated to the age of sixteen before undertaking a three-year apprenticeship in mining engineering with his father's friend, Nicholas Wood (q.v.), Head Viewer of Killingworth colliery. In 1821, Stephenson gained experience in railway surveying when assisting his father in the survey of the Stockton & Darlington Railway route. Stephenson's formal education was concluded when he attended the University in Edinburgh for seven months in 1822–1823.

In June 1823 he became Managing Partner of the manufacturing firm, Robert Stephenson & Co., established near Forth Street, Newcastle-on-Tyne, to meet the anticipated demand for iron products, particularly steam engines, for the Stockton & Darlington and other railways. His partners were his father, two influential Quaker investors, Edward Pease and Thomas Richardson, and the manager of the Bedlington Ironworks, Michael Longridge. The factory undertook general manufacturing business to supplement its early railway products and, in 1824, Stephenson showed his mechanical engineering capabilities by initiating the design of a paper-drying machine for high-quality bank-note paper.

In early 1824, he joined his father in the railway consulting firm, George Stephenson & Son, formed by financiers in London and Liverpool to meet the considerable interest in railway surveying and building. Later that year, Stephenson left England to work in Colombia as the manager of mineral mines on behalf of the London-based consortium, the Colombian Mining Association. He sought to gain experience, on his own account, of managing a substantial project and a large work-force. Although the mines did not produce a substantial profit, his three years in Colombia taught him much about project management which benefited him in his later career.

His return to England in November 1827 coincided with the urgent need to develop the steam locomotive into a robust form of motive power, suitable for goods and passenger services on the Liverpool & Manchester Railway route. George Stephenson had argued strongly for the line to be operated with locomotives rather than stationary engines and haulage ropes, but his commitments in supervising the line's construction took up his full attention.

Robert Stephenson returned to the Newcastle factory in January 1828, and with William Hutchinson (1792–1853), works manager, and George Phipps (1807–1888), draughtsman, he undertook a programme of research and development to advance locomotive design beyond the colliery-type used on the north-east coal-carrying lines. For inter-city operations, Stephenson sought to increase the quantity and rate of steam generation, improve thermal efficiency, provide a reliable form of suspension and a more direct means of transmission between piston and wheel, increase operating speed and adopt more reliable materials.

New boiler, suspension and transmission designs were each developed within general design arrangements, which were restrained by axle loading. Experimental locomotives incorporating innovations were built at the Newcastle factory and sold to customers in France and South Wales, and for the Stockton & Darlington and Liverpool & Manchester railways. The *Lancashire Witch* acquired by the Bolton & Leigh Railway in 1828, was a notable advance with steel leaf-springs and direct drive using cross-heads and slide-bars. The *Rocket*, the first to be equipped with a multi-tubular boiler, gained considerable fame as the winner of the Rainhill locomotive trials.

The completion of stretches of the Liverpool & Manchester Railway route during 1830 provided the Stephenson team with a test-track on which to try out and evaluate further innovations. In September 1830, just thirty-three months after the programme commenced, the *Planet* was completed as the prototype main-line locomotive incorporating all components subsequently employed in locomotive technology. These included, especially, a multi-tubular boiler incorporating a water-jacketed firebox and internal steam-pipe, and a smokebox with blast-pipe. *Planet*-class locomotives operated the first inter-city services in Britain, other European countries and in North America.

In 1828 and 1829 Stephenson carried out the first railway surveys in his own name for the Kenyon & Leigh Junction Railway and the Warring- ton & Newton Railway. He then supervised the construction of those railways. After 1830, Stephenson became a leading railway and bridge builder; the highlights of his career included the building of the London & Birmingham, Chester & Holyhead and Newcastle & Berwick railways, and the bridges at Conway, Menai, Newcastle-on-Tyne High Level, and the Royal Border Bridge, Berwick-on-Tweed. His overseas work included railways in Egypt, Norway and the Victoria Bridge near Montreal, Canada.

Stephenson married Frances (Fanny) Sanderson in June 1829. She died in October 1842, without issue, and Stephenson did not re-marry. He was first elected Member of Parliament for the Whitby constituency in 1847, and was made a Fellow of the Royal Society in 1849, President of the Institution of Mechanical Engineers 1849–1853, and President of the Institution of Civil Engineers 1856–1857. He received an Honorary DCL at Oxford University in 1857.

Stephenson was created a knight of the Belgian Order of Leopold in 1841 but, like his father, declined to accept a British Knighthood. He became a wealthy man and a much respected member of London society. He died of liver failure on 12 October 1859 at his home in Gloucester Square, London, and has the distinction of being one of only two engineers, with Thomas Telford (q.v.), to be buried in Westminster Abbey.

MICHAEL BAILEY

[Stephenson's letters and reports, National Museum of Science & Industry, Institution of Civil Engineers, Institution of Mechanical Engineers; S. Smiles (1862) *Lives of the Engineers*, vol. III, *George and Robert Stephenson*; J. C. Jeaffreson (1864) *The Life of Robert Stephenson, FRS*, 2 vols.; J. G. H. Warren (1923) *A Century of Locomotive Building by Robert Stephenson & Co. 1823–1923*; L. T. C. Rolt (1960) *George and Robert Stephenson, The Railway Revolution*; M. R. Bailey (1979) Robert Stephenson & Co., 1823–1829, *Trans Newc Soc*, **50**, 109–138; R. H. G. Thomas (1980) *The Liverpool & Manchester Railway*; M. R. Bailey (1984) *Robert Stephenson & Company, 1823–1836*, unpublished M.A. thesis, University of Newcastle on Tyne; Derrick Beckett (1984) *Stephenson's Britain*; M. R. Bailey (1989) Robert Stephenson & Co. and the Paper Drying Machine in the 1820s, *IPH Information, Bulletin of the International Association of Paper Historians*, **23**, Nos. 1 and 2; M. R. Bailey (1997) Learning through replication: the *Planet* locomotive project, *Trans Newc* Soc, **68**]

Publications (to 1830)

1830. *Observations on the Comparative Merits of Locomotive and Fixed Engines as Applied to Railways &c.* (with J. Locke)

Works (to 1830)

1828–1830. Canterbury & Whitstable Railway, Construction, Engineer, 6 miles, £83,000
1829–1831. Kenyon & Leigh Junction Railway, Construction, Chief Engineer, 2½ miles
1829–1831. Warrington & Newton Railway, Construction, Chief Engineer, 4¼ miles

STEVENS, Alexander (*c.* 1730–1796), builder, architect and engineer, was described in 1767 as 'mason at New Preston', on the estate of

Prestonhall in Midlothian, owned by Lord Adam Gordon who was both Stevens's landlord and his referee and guarantor in various architectural and engineering works over a long period. In 1775 Stevens held 38 acres on a long lease, the largest holding in terms of annual rent on the estate, and he was still there in 1788, a year before the estate was sold. Details of his family are scanty, but he was held responsible to the estate in 1787 for a small rent due on 'his brother's house', and a son named Alexander (fl. 1782–1812) became a recognised architect and joined him in several construction contracts in the 1790s. A daughter named Jean, noted in a legal document after his death as his 'only executrix dative qua nearest of kin', married James Fyfe who was agent for Lord Mountjoy at Dublin but resident in Edinburgh at the date of the document, June 1796. From her father Jean had also received on her marriage in 1794 the house in Ayr which he had built in 1786–1789 while he was building the New Bridge there; it was a tenement with a double bow front of four storeys facing the river.

While it seems likely that Stevens Sr. was born and bred in the district where he lived for most of his life, his education, his training as a mason, and how he acquired his knowledge of architecture and skill in design of buildings and bridges, are all an unsolved mystery. He answered a call to Moffat from the Earl of Hopetoun in 1784, within two years of the date when Raehills, a mansion built nearby for the Earl and attributed to Alexander Stevens Jr. was completed, suggesting that the father may have also been involved. In 1794 he was engaged in the building of The Burn, a large new house for Lord Adam Gordon near Edzell, while in the 1780s he designed a new manse and other buildings in his home parish of Cranston; but none of those works would have merited the lasting fame and honour which he achieved by his designs and construction of fine masonry bridges.

By 1766 he was being asked repeatedly for designs and estimates for bridge repairs and construction by the several districts of the Edinburghshire Road Trust. In 1773 and 1779/80 he designed and built large bridges over the Clyde at Hyndford and the Tweed at Drygrange respectively. In the former he adopted shapes of cutwater in plan—two curves meeting in a point—which were used in French bridges but in very few British bridges up to that time, and in the latter the creation of voids in the spandrels by longitudinal internal walls which had been first used by John Smeaton (q.v.) in his Perth Bridge in 1769; together they show that Stevens was familiar with the most up-to-date structural techniques of the time. In architectural design his largest bridges boast more ornament than those of any other contemporary designer of public bridges in Scotland, and as much as any in Britain; and his choice of material was good enough to ensure that their ornament would last.

Several manuscript documents indicate that he himself travelled to the distant sites of his bridges and sometimes spent considerable time there, the bridge at Ayr being a case in point. If, as seems likely, he discontinued his tenancy at New Preston when the Prestonhall estate was sold in 1789, he may have been resident in Ireland while Sarah's Bridge over the Liffey in Dublin was built in 1791–1793; it has been said that he also designed or built some locks on the Grand Canal but the evidence for it is not clear. At this time he was also called to travel far to the north in Scotland—a clear witness to his high status as a bridge designer—to report to residents of the area round Fochabers and give them a design for a bridge over the Spey, which he did by proposing a bridge of three long-span arches estimated to cost £14,000 (*Statistical Account*, 1792). The intentions were good but it took several years to raise enough money and the bridge was then built to a design by George Burn (q.v.).

The South Esk Bridge project next required Stevens's presence at Montrose—which he combined with building at The Burn not very far away. The bridge, built largely with timber imported directly to Montrose from Scandinavia, probably required his presence for substantial periods in 1792–1794, and overlapped at the end with his next contract at Lancaster. He was at Lancaster for much of the 1794 construction season while the foundations of the Lune Aqueduct were made by direct labour, and then spent two weeks or more of January 1795 in London to confer with John Rennie (q.v.), the engineer of the Lancaster Canal, in preparation for the start of building of the Aqueduct's superstructure by contract in the next season. It was a high structure of five semicircular arches in the building of which 'state-of-the-art' technology was employed. In particular, a steam engine had been used for the de-watering of the foundation pits, one of the first instances of the use of steam power in bridge-building.

When Stevens died at Lancaster on 29 January 1796 his son, Alexander, took over the work and finished it in 1798, possibly using the name Alexander Stevens and Son, which had been used for the South Esk Bridge contract. There is some difficulty in identification of the separate works of father and son, prior to the father's death, in cases where one name is given in documents; but several important works have always been attributed to Alexander Jr. One is the drawbridge at the harbour of Leith, with a central timber opening span, probably of two half-span bascules, and a stone arch each end. He also designed Wamphraygate Bridge, Dumfries-shire, a whinstone arch of 40 ft. span. Having served John Rennie (q.v.) as contractor of the Lune aqueduct, he made unsuccessful bids for two other large bridges by Rennie at Wolseley in Staffordshire and Kelso on the Tweed. In structural masonry he designed and built three church steeples, two of them for prominent churches in Edinburgh.

Raehills, mentioned above, was his largest known architectural commission.

Alexander Stevens Sr. was buried in the church-yard of Lancaster Parish Church, where a plaque on the south wall of the church reads: 'The many public works executed by him, especially the Aqueduct over the River Lune, near this town, are the best encomium of his professional merit.'

TED RUDDOCK

[Minutes etc. of the Edinburghshire (or Midlothian) Road Trustees, Calder, Cramond, Corstorphine and Dalkeith Districts 1757–1792, CO.2/1–8, SRO; Minutes of Ayr Town Council and committees (1780s), B6/29/7, SRO; MS drawings etc. re. Ayr New Bridge, Ayr Public Library (copies of some at NMRS); MS contracts and correspondence re. Drygrange Bridge and Bridge of Dun, and Prestonhall Estate papers, Lord Adam Gordon's Papers, GD244, boxes 18 and 19, SRO; MS Town Council Minutes and correspondence etc. 1782–1788, Montrose Library; Teviot Bridge papers, Duke of Roxburgh's Muniments, NRA (Scot.) 0179, section 3, box 3/18, bundle 46; correspondence re. Lune Aqueduct, Lancaster Canal Papers, PRO (Transport); W. Radford's commonplace book, QAR/5/39, Lancs. CRO; MS memorandum book of Alexander Stevens Sr. 1794–1795, with further notes after his death by his son Alexander, process paper No. 3842, CS 229/G.8/6, Edinburgh Commissary Court; facts and comment from the ministers of three different parishes, *Statistical Account of Scotland*, **14**, 1792, 271, 396–400, **15**, 103; Obituary notice (1796) *Gents Mag*; J. A. Morris (1912) *The Brig of Ayr*, pl. at p. 24; T. Ruddock (1979) *Arch Bridges and their Builders 1735–1835*, 120–123, 129–131 and refs. quoted; J. Robertson (1993) *Public Roads and Bridges of Dumfriesshire 1650–1820*; Colvin (1995) 923–924; W. T. Johnson (1996) *Scottish Engineers and Shipbuilders* (computer database)]

Works

1765. Lugton Bridge, by Dalkeith, Midlothian (contractor), for Dalkeith District of Edinburghshire Road Trustees, one arch, 55 ft. span

1766. Reconstruction of Colt Bridge (Water of Leith), Edinburgh (design and contractor) for Corstorphine District, Edinburghshire Road Trustees, one arch, 50 ft. span, 20 ft. wide, cost £232, bypassed

1766. Rebuilding of Gogar Bridge for Corstorphine District, 17 ft. span, cost £60

1767. Canonmills Bridge (Water of Leith), Edinburgh (contractor, to design of William Mylne (q.v.)) for Cramond District, Edinburghshire Road Trustees, cost c. £180, demolished 1840 (shown in the painting *A View of Edinburgh from Canonmills* (1820?) by John Knox at Scottish National Portrait Gallery]

1773. Hyndford Bridge (River Clyde) near Lanark (design and contractor), five arches, public road bridge (*Statistical Account of Scotland*, **15**, 1793, 26; T. Reid (1912–1913) Fords, ferries, floats and

bridges near Lanark, *Proceedings of Society of Antiquaries of Scotland*, **11**, 209–256)

1774. Nether Mill Bridge at Keith Marischal House, Humbie (Hopetoun House Archive, NRA (Scot.), 888/Box 51/16.2084)

1779–1780. Drygrange Bridge (River Tweed) for Roxburghshire Road Trustees, three arches, middle arch 105 ft. span, decorative urns in deeply-recessed circular panels in spandrels and other ornaments, cost £2,100, bypassed 1970s

1782. Bridge of Ae (Water of Ae), Dumfriesshire (design), two dressed sandstone arches, each 40 ft. span, built by William Stewart (q.v.), bypassed

1783. (attrib.) Glen Bridge, Oxenfoord Castle, Midlothian, three tall arches of short span over a burn, castellated parapets, estimate £180 (A. Rowan, *Country Life*, 15 Aug. 1974, 430–431)

1783. Stenhouse Bridge (Water of Leith) on road to Glasgow leaving Edinburgh (design, built by Wm. Stodart of Hamilton), for Calder District, Edinburghshire Road Trustees, three segmental arches, contract price £730, demolished 1927

1784. Ancrum Bridge (River Teviot), for Roxburghshire Road Trustees (design and contractor), three segmental arches, middle arch 57 ft. span, profiled parapet, etc., bypassed

1785. Stockbridge, Edinburgh (Water of Leith), (design and contractor) for Cramond District and subscribers, one segmental arch, 55 ft. span, with Gothic ornaments on spandrels and parapets, cost £ 500, widened 1900–1901

1785–1787. Bridge of Dun (River South Esk) near Montrose (design and contractor), for Burgh of Montrose and others, three segmental arches, middle arch 60 ft. span, Gothic and classical ornament, and profiled parapet, cost £ 3,128

1786. Brechin Bridge, rebuilt north arch, cost £350

1786–1789. New Bridge, Ayr, River Ayr (design and contractor) for Burgh of Ayr, five arches with profiled parapet, balustrade and decorated spandrels bearing cast lead sculptures, cost £4,063, demolished 1878

1780s?. (attrib.) Bridge on entrance drive to Carolside, by Earlston, Berwicks, one ashlar arch, 60 ft. span, with classical balustrades (info. from Miss C. Cruft)

1791–1793. Sarah's Bridge (River Liffey), Dublin (design and contractor?), for Dublin Pavings Commissioners, single elliptical arch, 104 ft. span, 40 ft. wide. (T. Telford (1812) Bridge, *The Edinburgh Encyclopaedia*, **4**, 487; *Gentleman's Magazine*, **63** (1793), 311)

1794–1795. (attrib.) Teviot Bridge, Kelso (designs 1784 and 1788, built by William Elliot), three segmental arches, profiled parapet, etc., cost c. £2,200

1794–1796. South Esk Bridge at Montrose, Timber trestle bridge of thirteen spans plus drawbridge, with stone arch at each abutment, length 700 ft. (design and contractor with his son Alexander), for Burgh of Montrose, cost £6,884 plus

government grants, demolished 1829 (*Scots. Mag.*, **73** (Feb. 1811) 83 and plate)
1793–1798. Lune Aqueduct, Lancaster (Alexander Stevens and Son, contractors, to design of John Rennie (q.v.)) for Lancaster Canal Co., five semicircular arches of 75 ft. span, cost £48,000

STEVENSON, Robert, FRSE (1772–1850), civil engineer, the only child of Alan Stevenson, a West India merchant, was born in Glasgow on 8 June 1762. Two years later his father died suddenly leaving his family in straightened circumstances and Stevenson was educated at a charity school in Edinburgh. In 1786 he was apprenticed to a gunsmith and was himself described as such about the time he began to work for Thomas Smith (q.v.), an Edinburgh tinsmith who in 1787 had been appointed engineer to the newly formed Northern Lighthouse Board. Smith became Stevenson's stepfather in 1792 and father-in-law in 1799 when Stevenson married his daughter, Jane Smith (1779–1846).

Stevenson was largely self-taught and of the practical school. He gained an engineering related education by part time attendance at Professor John Anderson's classes in natural philosophy at Glasgow University from 1792–1794 and was directed by him towards an engineering career. From 1800 to 1804 he attended classes at Edinburgh University: in natural philosophy by Professor Robison, mathematics by Professor Playfair, chemistry by Dr. Hope and natural history by Professor Jameson, but could not graduate because of his 'slender knowledge of Latin and total want of Greek'. Under Smith, before being formally apprenticed to him from 1796–1802, Stevenson had gained experience of lamp installation at Portpatrick harbour and on the erection and illumination of Pentland Skerries lighthouse. During his apprenticeship, but with much more responsibility than was usually the case for an apprentice, he specialised in the firm's lighthouse work, making reflectors, installing lamps and assisting with arrangements for the erection and maintenance of lighthouses. In 1800 Smith took him into partnership in the business.

The Bell Rock Lighthouse

Towards the close of the eighteenth century, with increased maritime trade and an intensification of shipwrecks in and around the entrances to the firths of Tay and Forth, Stevenson proposed building a lighthouse on the Bell Rock, 11 miles out to sea from Arbroath. This was an outstandingly difficult engineering challenge made even greater by the fact that the rock was submerged to a depth of about 12 ft. during every tide. In 1799 he envisaged a beacon-style lighthouse on cast iron pillars but in 1800, after seeing the rock and considering the possible damage to a beacon by ships, he abandoned this idea and prepared a design for a stone tower following the general concept of John Smeaton's (q.v.) Eddystone lighthouse (1759). As part of the design and promotional process for the

Robert Stevenson FRSE

project he had both proposals accurately modelled, a practice that he often employed subsequently on important maritime and bridge works.

Following the failure of a parliamentary Bill for the lighthouse in 1803, and because of the project's hazardous and unprecedented nature, the Northern Lighthouse Board on Stevenson's advice secured the services of an engineer with a national reputation, John Rennie (q.v.). With his support, the necessary Act of Parliament was obtained in 1806, based on Stevenson's design of 1800, and in December 1806 the Board resolved that the lighthouse be erected under Rennie's direction as 'Chief Engineer'; Stevenson was to execute the work under his superintendence as 'Assistant Engineer'.

The relative roles of Rennie and Stevenson in creating the Bell Rock lighthouse subsequently became a controversial issue between their families; it became public in 1848–1849 when each claimed that it was their father who had 'designed and built' the lighthouse. In fact both are entitled to this credit in differing degrees. Rennie approved of Stevenson's design and model of 1800 as a general concept but advocated, and later implemented, a closer adherence to Eddystone Lighthouse in the design of the as-built tower except for the greater width and height required. Their combination proved excellent for the success of the project. With Rennie's greater experience and direction and Stevenson's energy, ability and assiduous superintendence throughout the progress of the work, a remarkable achievement evolved which reflected great credit on both engineers, particularly on Stevenson for overcoming the exceptional difficulties of its execution in fulfilling the 'resident engineer' element of his remit.

It is now clear from family records and newly found evidence of the Clerk of Works, David Logan (q.v.), who later became Engineer for the Clyde Navigation and of John Paterson (q.v.), dock engineer at Leith c. 1803–1823, that Rennie's role was much more significant than it has always been portrayed in Stevenson publications. Stevenson's sons maintained that after the authorising act was obtained, Rennie's role was limited to that of 'an advising engineer to whom Mr. Stevenson could refer in case of emergency and who had suggested some alterations on Mr. Stevenson's design in which he did not see his way to acquiesce'. It is now evident that the as-built shape and internal dovetailed construction of the lighthouse tower were determined by Rennie, who for the duration of the work acted as a conventional Chief Engineer by means of meetings, occasional visits, reports and a considerable correspondence. The significance of Rennie's contribution was acknowledged by Logan who was of the opinion that if Stevenson's proposal for undovetailed horizontal courses at Bell Rock Lighthouse had been implemented 'not one stone of it would have been left standing upon another'. Paterson's opinion was similar, that the lighthouse 'would have shared the fate of the beacon on the Carr Rock'. In 1817 the sea destroyed this stone tower beacon designed by Stevenson in 1810.

In determining the external shape of the Bell Rock lighthouse tower Rennie adopted a much greater curvature at the base, its sides rising at about 40° from the horizontal compared with about 70° in Stevenson's design. Rennie's curvature was more effective in directing the waves upwards and dissipating their energy. Both Stevenson's and the as-built designs had base diameters of 42 ft. and Rennie's curvature resulted in a narrower tower at and near water level which had the advantage of reducing the surface area upon which the waves acted within, and well above, the tidal range.

Within Rennie's broad parameters Stevenson carried out almost all of the detailed design for the project, including conceiving and building the apartment floors cantilevered inwards from the outer wall to support a central core into which, at Rennie's insistence, they were dovetailed. This arrangement represented an improvement on the arched floors of Eddystone Lighthouse. These operations involved Stevenson taking many decisions on his own initiative. He undoubtedly had complete autonomy in the matter of fixtures and fittings and was also responsible for planning and executing the impressive shore base and signal tower at Arbroath. The lighthouse cost £61,331 to build and became operational on 1 February 1811.

Under Stevenson's direction several remarkable engineering innovations were introduced at the works that greatly expedited and facilitated construction. These included the temporary beacon barracks in which he and twenty-eight men were accommodated, elevated cast iron railways from the boat landings to the tower, ingenious and highly efficient moveable jib cranes and, for building the tower, the world's first iron balance crane. He may have conceived the idea of some of these innovations and certainly superintended their provision, modification and use, but there is strong evidence from Logan that the cranes were invented by foreman millwright, Francis Watt, and that he also designed the beacon and probably the railway. This attribution is consistent with a Stevenson letter to Watt in January 1808: 'So soon as you have got a proper draught of the crane, of the rock and railway, and of the wooden house for the beacon—come this way (to Edinburgh'. Stevenson did not specifically claim 'invention' of the cranes or acknowledge this to Watt, but he did state that the designs were 'his', at least in a proprietary sense and in time was credited with their invention.

Stevenson's definitive *Account of the Bell Rock Light-house*, a civil engineering classic, in which the various operations and intricate machinery and equipment used on the work were described and profusely illustrated by the best artists and engravers, was published in 1824 having taken 13 years to complete. It was perhaps intentional and certainly fortunate for Stevenson that the book was published after Rennie's death as he was most unlikely to have approved of it, not least for the omission of his report of 2 October 1809 made after a site visit demonstrating him exercising a key role in the tower's construction up to the floor of the lowest apartment at a height of about 45 ft. Although only three-hundred copies of the book were printed it was widely distributed to influential recipients and added considerably to Stevenson's reputation. He was perceived in some quarters as having not sufficiently acknowledged the contributions of Rennie and Watt and in 1841 this led to unfavourable comment in *The Surveyor, Engineer and Architect* and later in Smiles's *Lives of the Engineers*. Smiles stated that the credit for the lighthouse was 'almost exclusively' given to Stevenson because Rennie was 'in a great measure ignored' in Stevenson's book and that he 'should not be deprived of whatever merit belonged to him as chief engineer'. David Stevenson disputed Smiles's claim but to no effect.

Other lighthouse work

As Engineer and, in those days, chief executive of the Northern Lighthouse Board from 1808 until his resignation in 1843, Stevenson can be considered to have inaugurated the modern lighthouse service in Scotland. From 1806 he was responsible for the design and construction or improvement of at least twenty-five lighthouses, of which that at Cape Wrath, built by John Gibb (q.v.) of Aberdeen, is a typically fine example in the north. In lighthouse illumination he improved on Smith's work and brought the catoptric system, that is using silvered-copper parabolic reflectors and Argand lamps, to a high degree of perfection, so much so, that he was reluctant to adopt Fresnel's improved dioptric or lens system adopted in France in 1822 on grounds of operational economy. It was not until 1835 that the system was eventually introduced into a Scottish lighthouse at Inchkeith, the delay for which Stevenson and the Northern Lighthouse Board were criticised in the press by Sir David Brewster and others. This matter led to an acrimonious exchange of tracts in 1833 and 1859–1860. In order to distinguish between the ever-increasing number of lights Stevenson invented 'intermittent' and 'flashing' lights. For the latter distinction he received in 1829 a gold medal from the King of the Netherlands as a mark of his approval.

The firm of Robert Stevenson 1811–1843

The successful completion of the Bell Rock Lighthouse enabled Stevenson, from 1811, to establish in Scotland within a decade, an indigenous civil engineering business of sufficient importance to make modest inroads even into the work of the London-based practices of Thomas Telford (q.v.) and Rennie. To create an engineering dynasty, which flourished for nearly a century and a half through four generations is indicative of a man with exceptional qualities, particularly in the rigorous and exemplary training of his successors. Stevenson's success was based on his ardent acquisition, application and promotion of largely self-taught practical knowledge, combined with shrewdness, ambition, determination, hard work, and outstanding entrepreneurial flair and management ability, all combined with a good financial start. In engineering terms his strengths, which were experimental and practical, related to maritime work, river navigation improvement and inland communication, particularly bridges. His theoretical and mechanical engineering attributes, except with regard to lighthouse equipment, were less remarkable. In 1825 when invited by David Brewster to write the article 'Steam Engine' for the *Edinburgh Encyclopaedia* he replied 'I should be afraid of disappointing you every way' and offered him instead an article on suspension bridges which was not taken up.

Maritime and river navigation engineering

Stevenson's practice in these branches of engineering throughout much of Scotland and northern England on numerous harbours and rivers, of which the improvement of the Tay navigation was of particular note, represented a fundamental element of the firm's practice. He made recommendations and carried out detailed hydrographic surveys which the relevant authorities were often unable to act upon for financial reasons. These included, besides those noted as published below, Stonehaven (1812, 1830), North Berwick (1812), Cellar Dyke (1814), Kingbarns, Fife (1814), Rothesay (1815), Grangemouth (1815), Elie (1815, 1836), St. Andrews (1815), Methil (1815), Leith (1815, 1824), Montrose (1816), Newport (1818), Broughty Ferry (1818), Fraserburgh (1818–1830), Ferry-port-on-Craig (1818), Alloa (1826), Peterhead (1826), Lossiemouth, Ayr (1830–1841), Ballyshannon (1836), Chester (1839–1845), Fisherrow (1839), Musselburgh, Aberdeen and Portpatrick.

From 1838, when his son, David, joined his brother, Alan, and himself in the partnership, Stevenson's own contribution diminished and with it the firm's railway and bridge work. The reduction in railway and bridge work was more than outweighed by an expansion in the firm's maritime and river improvement business under David Stevenson's able direction. Robert Stevenson retired in 1846.

Stevenson's state-of-the-art maritime work included the design and construction in 1821 of a sea wall at Trinity, near Edinburgh, with a cycloidal curve vertical profile, experiments on the destruction of timber by the *Limnoria terebrans* that influenced the universal adoption of greenheart for marine timberwork and invention of the 'hydrophore'. This was an instrument he invented in 1812 for water sampling from different depths to support his hypothesis that in a tidal river salt water from the sea flows up its bed in a layer separate from the outgoing fresh water on top. A developed hydrophore was used in the Challenger Expedition oceanic investigations in 1872–1876. He had a consuming interest in coastal erosion and, from a study of the bed of the North Sea, maintained that its sandbanks were the result of this action. He published papers on this subject and, in 1820, a chart with cross-sections shown *in situ* and said to be the first use of 'sectio planography' in this context. Stevenson also measured Scottish coastal water depths and published important charts of the Firth of Tay and the coasts of Scotland and parts of Ireland and England. In 1819 he published his proposal for the *Dalswinton* steamboat, with paddles located on its centre-line fore and aft, intended for use at Leith Harbour and on the Forth & Clyde Canal. In 1838 'Robert Stevenson & Sons' specified and had built at Preston the steam operated bucket-dredger *Robert Stevenson,* which began operation on the river Ribble in 1839.

Canals, roads and railways

In the early years of the firm Stevenson was engaged extensively on canal, road and railway projects, often adopting a promotional role. In 1814 he made a £1½ million detailed proposal for a level canal between Glasgow and Edinburgh, locking down at its ends to Broomielaw Quay on the river Clyde and to Leith on the Firth of Forth. In 1817 he proposed a canal in preference to a railway from Forfar in the Vale of Strathmore to Arbroath. His ruling practice in canal design was to reduce lockage to the practicable minimum. By 1828 Stevenson's standing was such that he was working with Telford and Alexander Nimmo (q.v.) on a proposed harbour at Wallasey and a ship canal across the Wirral to the river Mersey, estimated to cost £1.4 million. His canal schemes were not executed because of lack of the necessary finance. Several substantial lengths of road were executed under his direction ranging from lighthouse accesses to main routes. His road making, probably influenced by the work of Charles Abercrombie (q.v.) was at the forefront of national practice. He adopted a mode of construction similar to that increasingly advocated by John Loudon McAdam (q.v.) from 1811, but more substantial. He also advocated pairs of stone tracks in town roads to facilitate traction and these were installed at gradients on several main roads in Edinburgh.

By 1818 Stevenson had become convinced of the superiority of horse-drawn railways over small canals for inland communication and proposed the 'Edinburgh Railway' to connect with the Midlothian coalfield. In 1819 he proposed a railway between Montrose and Brechin. His reputation as an advocate of railways was now such that he was called in as a consultant to advise on the proposed Stockton & Darlington Railway and also on the Elgin Railway, Fife, extension. About this time he edited for publication, with *Notes*, the numerous *Essays on Rail-Roads* submitted to the Highland Society. In 1823 when consulted by Sir John Sinclair about the best mode of inland communication between Edinburgh and London, he advised that a 'railway was not only much more practicable but more commodious and useful for general intercourse than a canal'.

By 1836 the railways proposed in Stevenson's various reports traversed Scotland from the Tweed valley north to Perth and Aberdeen and from Edinburgh across to Glasgow, more or less on the lines of the eventual railway network, but their estimated costs were considerable and the necessary finance not forthcoming. As steam locomotion developed, he lost his pre-eminence in Scottish railways to Thomas Grainger (q.v.) and Miller. His only railway proposal known to have been executed was the short Newton Colliery line from Little France on the Dalkeith road near Edinburgh, although his 'Edinburgh Railway' proposal of 1818 to some extent facilitated the successful Edinburgh & Dalkeith Railway that superseded it in 1831.

Stevenson's design practice for railways, similar to that of William Jessop (q.v.) and Telford, was to make them as near level as practicable, avoid the use of heavy rolling stock to reduce track damage and to adopt inclined planes with stationary steam engines for overcoming differences in level. As early as 1818, for the 'Edinburgh Railway', Stevenson advocated the adoption of 12 ft. malleable iron edge rails in preference to the short cast iron rails then in common use and influenced the development of John Cass Birkinshaw's (q.v.) epoch-making malleable iron forerunner of the modern edge rail. In 1821 this role was acknowledged by George Stephenson (q.v.), who paid him the tribute, 'you have been at more trouble than any man I know of in searching into the utility of railways'. A year later the young Robert Stephenson (q.v.) who

had met Stevenson inconnection with the Stockton & Darlington Railway project was less complimentary. After learning 'that Mr. Stevenson had surveyed an immense quantity [of railways] but had not had the good fortune to get them into action', he wrote to William James, projector of the Liverpool & Manchester Railway: 'If he has executed any railway it must be of very trivial consequence. I hope we shall be able to keep him out of the Liverpool concern'.

Stevenson's many transportation proposals involved more than one hundred bridges, most of which were never built. Of more than twenty bridges which he did erect or improve, the majority were of the masonry arch type, the most important and difficult to construct being the five-arch Hutcheson Bridge (1831–1834) over the river Clyde at Glasgow. In 1861 Hutcheson Bridge was considered by Fenwick, a structural analyst at the Royal Military Academy, as one of the 'best specimens of the segmental masonry arch' type in Britain, instancing it with Rennie's London and Waterloo bridges as the 'finest structures of the elliptical arch'. In Weale's *Bridges,* Hutcheson Bridge is described 'as one of the best pieces of bridge masonry in the kingdom'. This elegant bridge had a life of only thirty-four years because it was believed that the removal of a weir immediately upstream as part of a navigational improvement would render its foundations insecure.

Segmental arches characterise the style of Stevenson's masonry bridges. Another fine example, which still carries main road traffic, is the William IV, or New Bridge (1829–1832) over the river Forth at Stirling, for which he also said to have planned its town approach. This is not as imposing as his earlier London and Regent Road approaches into central Edinburgh skirting Calton Hill, which included proposals for housing terraces and the open 'triumphal' parapets which still exist at Regent's Bridge to enable its users to enjoy the fine views. From 1815–1819 he engineered this approach from east of Calton Hill to Waterloo Place and Princes Street. It was a work of particular difficulty and included the engineering input to Regent's Bridge, severing Old Calton cemetery, extensive rock blasting and a massive retaining wall in front of the new High School.

At Marykirk Bridge (1814), which is still in service and was Stevenson's first major bridge, some evidence of his design practice as early February 1811 is found in his letter to Logan, then the bridge inspector. He wrote, 'Your drawings are finely executed indeed, and the specifications are much more in the business way than any of the others. Your plan with the elliptical form, rather appears to be a little overdone with a view to keep the roadway level, and the same very laudable intention it strikes me has induced you to adopt too great a radius for the plan [proposal] with the segment of the circle—necessary to give the bridge a greater rise or increase the number of arches to at least four'. Segmental rather than elliptical arch profiles were adopted for the four arches. Similar profiles were used in his proposals for the Bridge of Don at Old Aberdeen (1823) and Canonmills Bridge, Edinburgh (1812–1834) with a span of 110 ft., but with variable depth voussoirs.

Stevenson did not neglect cast iron as a bridge building material. Influenced by the general concept of Telford's cast iron footway additions to Glasgow Old Bridge over the river Clyde in 1821 he proposed iron additions to existing major masonry bridges at Perth (1827), Newcastle-upon-Tyne (1828) and North Bridge, Edinburgh (1832).

Of at least five timber bridges erected under Stevenson's direction, basically to traditional design, the largest was the notably wide temporary fourteen-span structure (1832–1846), with innovative iron fittings, erected over the river Clyde at Portland Street, Glasgow. Its purpose was to accommodate traffic whilst Telford's Broomielaw Bridge was being built, but it proved so popular with the public that it was retained for pedestrian use until 1846. Although the work was to be constructed to Stevenson's 'entire satisfaction', reference was to be made to Telford in any dispute.

State-of-the-art unexecuted designs by Stevenson for other timber bridges included a slender laminated arch for Dornoch Firth in 1830 and, influenced by David Stevenson's visit to the U.S.A. in 1837, a Long's frame bridge for India and an eight-span 'Town' truss bridge crossing of the river Tweed at Norham in 1838. The Dornoch Firth bridge proposal, the firm's model for which is now in the National Museums of Scotland, consisted of a four-leaf arch stiffened at 5 ft. intervals by king-post roadway supports. Its design was probably influenced by the work of John Green (q.v.) from c. 1828.

Stevenson also contributed to the development of iron suspension bridges. In 1820 he proposed crossing the river Almond at Cramond near Edinburgh by means of a new type of underspanned wrought iron suspension bridge. The design was novel in that its roadway superstructure, a cast iron framework, rested on the chains rather than being suspended from them. This unexecuted proposal, together with his authoritative accounts of other Scottish suspension bridges based on correspondence with their designers, was publicised widely throughout Europe in his 'Description of Bridges of Suspension', published in the *Edinburgh Philosophical Journal* in 1821. He proposed designs on this or the suspension truss principle for other locations, possibly over the river North Esk at Melville Castle in 1822 and certainly for North Bridge, Edinburgh. His friend James Smith of Deanston (1789–1850), agricultural engineer, erected a bridge on this principle at Micklewood in 1831. Two years later Stevensons designed and built Abbey St. Bathans Bridge, Berwickshire, after which numerous short spans on this basic principle were erected in Britain. By, and almost certainly before, 1861 these bridges were being manufactured as a standard item by Charles D. Young and Co., Edinburgh.

By the 1830s Stevenson had become one of Britain's leading bridge builders and knowledge of his practice was disseminated in one of the most comprehensive and influential works of its day, Weale's *The Theory, Practice and Architecture of Bridges* published in London from 1839–1843. This work, which carried Stevenson's portrait as a frontispiece and five examples of his timber, masonry and iron bridges illustrated on eighteen plates, was of particular value to the newly emerging generation of railway engineers,

Tall structures other than lighthouses

Stevenson's successful erection of the Bell Rock Lighthouse brought him commissions for other tall structures: for example, to advise on the severely cracked tower of Montrose church (1811), safeguarding Arbroath Abbey ruins (1814–1815) and, one which furnishes a good example of the architect–engineer relationship at that time, the design and erection of the Melville Column, St. Andrew Square, Edinburgh. In March 1821, with the architect, William Burn, Stevenson examined the 31 ft. square foundation pit for the column, which was then 8 ft. deep. He reported that as the monument was upwards of 140 ft. high and weighed about 1,500 tons, it 'becomes necessary to obtain the best foundation … the pit should now be dug to the depth of 12 ft. before any final decision is entered into'. After this had been done and he had pronounced the foundation sufficient without planking or piling, he specified the form, dimensions and method of constructing the masonry base, staircase dimensions and the column wall thickness.

In order to fund these alterations Stevenson urged the committee 'to extend the funds to £500 or £600 a sum still too inconsiderable to be put in competition with the more certain stability of a building intended to perpetuate the memory of so illustrious a statesman as the late Viscount Melville'. The project went ahead and the scaffolding and tackling for the incredibly delicate task of raising and positioning the large statue on top of the column were carried out under Stevenson's direction using the Bell Rock iron balance crane.

Publications, learned society activities and death

Stevenson had a life-long interest in gaining and promoting knowledge and had formed an outstanding office reference library strong in the key English and French engineering works from 1737 to 1850. His writings, which were of a descriptive and practical character, appeared in more than sixty publications. Many were engineering reports, but about one-third achieved wider circulation through leading periodicals, text-books and ten articles in the *Edinburgh Encyclopaedia* and the *Encyclopaedia Britannica*. Many of his publications, because of their depth and authority, now represent a valuable historical resource.

Stevenson's professional and scientific interests are reflected in his membership of numerous learned societies, the earliest known being that of the Highland Society in 1807. By 1812 he was a member of the council of the Wernerian Natural History Society and, in the following year, a founder director of the Astronomical Institution of Edinburgh. In 1815 he was elected to Fellowship of the Royal Society of Edinburgh, the Geological Society and the Society of Antiquaries of Scotland. In 1821 he became a founder subscriber to, and soon afterwards a director of, the School of Arts in Edinburgh, Britain's first Mechanics Institute, from which Heriot-Watt University traces its origin. In 1827 and 1828 he was elected respectively to membership of the Smeatonian Society of Civil Engineers and the Institution of Civil Engineers. These elections, particularly the latter, for which he was sponsored by Telford, were a fitting recognition of Stevenson's acceptance into the first rank of British civil engineers.

Stevenson is now chiefly remembered for erecting the Bell Rock Lighthouse, for establishing the Scottish lighthouse service and for the useful contribution his surviving bridge, lighthouse and harbour work continues to make to mankind. He died in Edinburgh on 12 July 1850.

ROLAND A. PAXTON

[Stevenson papers, NLS; J. Weale (comp.) (1843) *Bridges*; Tidal Harbours Commission (1845–1847) *Reports and Appendices*, contains reports on Stonehaven (1812), Leith (1824), Alloa (1826), Elie (1826), Stirling (1828) Granton (1834), Fisherrow (1839), Lossiemouth, Fraserburgh, Peterhead, the Tay, Perth, etc.; A. Stevenson (1861) *Biographical Sketch of the late Robert Stevenson*; S. Smiles (1862) *Lives of the Engineers*; D. Stevenson (1878) *Life of Robert Stevenson*; C. Mair (1978) *A Star for Seamen*; J. Leslie and R. Paxton (1999) *Bright Lights—The Stevenson Engineers 1752–1971*; R. A. Paxton (1999) An Assessment of Aspects of the work of the Stevenson Engineers, Ph.D. thesis Heriot-Watt University]

Selected publications

1803. *Memorial and State relative to the Lighthouses erected on the Northern Part of Great Britain*

1810–1830. Articles in *Edinburgh Encyclopaedia*

1812. *Report relative to the Harbour of Stonehaven*

1814. *Report on the improvement of the Harbour of Dundee*

1814. *Reduced plan of the lands of Calton Hill* (Regent's Bridge, Edinburgh)

1816. *Nautical Survey of the Frith of Tay*

1816. Description of the bridge at Marykirk, *Scots Mag*, **78**, 883–885

1816. Observations upon the alveus or general bed of the German Ocean and British Channel, *Annals of Philosophy*, **8**, 173–182, etc.

1816–1824. Articles in *Encyclopaedia Britannica*, supplement

1817. *Report and Reduced Survey & Section of a Line of Canal between Edinburgh & Glasgow upon One Level … made in the Years 1814 & 1815*

1817. *Report and Reduced Survey & Section of the proposed Canal into the Bale of Strathmore between … Forfar and Aberbrothwick*

1818. *Reduced Survey of Part of the Frith of Tay as seen at Low Water of Spring tides …*

1818. *Reduced Plan of Part of the Shires of Edinburgh & Haddington shewing the Lines of the Proposed Railways*

1818. *The Report relative to the Improvement of Communication by the Ferries betwixt Fife and Forfar; and Observations* (1820)

1818. *Sketch of the Coast from Lincolnshire to Hampshire*

1819. *Report relative to the various Lines of Railway from the Coalfield of Midlothian to the city of Edinburgh and port of Leith*

1819. *Report and Reduced Survey & Sections of several Lines of Railway from the Port of Montrose to the Town of Brechin …*

1819. *Report relative to the Compensation Reservoirs for the Mills on the Water of Leith and Bevelaw Burn*

1820. *Memorial relative to Opening the Great Valleys of Strathmore and Strathearn, by means of a railway or canal*

1821. Description of bridges of suspension, *Edin Philos J*, **V**, 237–256

1821. [Dalkeith, Galashiels & St. Boswells Railway Report]

1821. [Report on the proposed railway in the Tweed valley between Roxburgh and Selkirk]

1821. [Report relative to proposed lines of railway from the coal-field of Midlothian to the rivers Tweed and Leader]

1822. *Report and Reduced Survey & Section of Part of the River Severn shewing the Proposed Improvement upon New Passage Ferry*

1824. *An Account of the Bell Rock Light-house*

1824. Notes … in reference to essays on railways, *Prize Essays Trans Highland Soc Scotland*, **6**, 130–146

1824. *Sketch Plan of a Design for obtaining new Access to the Cross of Edinburgh*

1825. *Plan for a Smooth and durable City Road*

1825. *To the Subscribers for the Survey of the East Lothian Railway*

1827. *Report and Reduced Plan & Sections of the Valley of Strathmore shewing a Line of Railway from the Ports of Montrose & Arbroath to Perth*

1827. *Excerpt from the Minutes of the Meeting of the Trustees upon the Turnpike Road from Crieff to Longcauswayhead …*

1827. *To the Commisssioners … Harbour of Sunderland*

1832. *Report relative to the Improvement of Ballyshannon Harbour*

1832. *A Chart of the Coast of Scotland with Part of England and Ireland* (with Alan Stevenson)

1833. *Survey of the River Tay—Harbour of Perth* (with Alan Stevenson)

1834. *Report relative to Granton Harbour* (with Alan Stevenson)
1834. *Report on the Navigation of the Tay, and the Extension of Perth Harbour* (with Alan Stevenson)
1835. *Plan of the Edinburgh and Glasgow Railway* (with Alan Stevenson)
1836. *Chart of Skerryvore Rocks* (with Alan Stevenson)
1838. [Forth navigation report] (proposals based on a survey of 1826–1827)
1839. [Dee navigation report] (with Alan and David Stevenson)
1845. *Tay Navigation, Effects of Perth … Railway*

Selected works
Lighthouses:

1797–1825. Cloch, Renfrewshire
1804. Inchkeith, modernised 1815
1806. Start Point, Orkney
1809. North Ronaldsay, Orkney (modernised 1809)
1811. Bell Rock, £61,000
1812. Toward Point, Argyllshire
1816. Isle of May
1817. Corsewall, Wigtownshire
1818. Point of Ayre, Isle of Man, two towers
1818. Calf of Man, three towers
1821. Sumburgh Head, Shetland
1824. Kinnaird Head, Fraserburgh, modernised
1820s. Eilean Glass, Harris, modernised
1820s. Pentland Skerries, Orkney, rebuilt
1820s. Mull of Kintyre, rebuilt
1820s. Pladda, Arran, modernised
1825. Rhinns of Islay, Argyll
1827. Buchan Ness, Aberdeenshire
1828. Cape Wrath
1830. Mull of Galloway
1830. Tarbat Ness, Ross & Cromarty
1831. Dunnet Head, Caithness
1832. Douglas Head, Isle of Man
1833. Barra Head
1833. Girdle Ness, Aberdeen
1833. Lismore

Harbours:

1812–1814. North Berwick, engineer
1818–1820. Fraserburgh, improvements, including south quay, engineer, £6,600
1825–1830. Stonehaven, engineer, £9,000
1826. Alloa, piers
1826–1828. Crail Harbour, west pier, designed by Stevenson in 1821, built by John Gosman, £1,100
1830. Fraserburgh, improvements, Middle pier, £5,700
c. 1832. Lossiemouth, c. £5,000
1834–1837. Granton, recommendations not fully implemented, but construction of 800 ft. steam boat pier supervised by David, 1836–1837
1835. Cockenzie Harbour, improvements

River navigation improvements:

1812. Parts of the Dee, Aberdeenshire
1834–. Tay, £62,000
1839–. Ribble improvements, work carried out by his sons

1843–. Stirling, improvements to the Forth to Alloa, etc., initial report 1828, Act 1843

Bridges:

1811–1814. Marykirk, four segmental arches, £10,000
1816. Lugton, Dalkeith, 50 ft. span arch
1819. Regent, Edinburgh, £17,000
1819. Clockmill tunnel, Edinburgh, £600
1824. Annan, timber temporary crossing, £500
1824–1827. Annan, three segmental arches in red sandstone, £6,000
1824. Stannochy Bridge, Brechin (1824, unattributed—if not by Stevenson, influenced by Marykirk Bridge design)
1825. Boddam, timber, £200
1828. At least nine on lighthouse access roads, e.g. Chearbaig (Kearvaig) (1828), near Cape Wrath, Glenmanuilt on the Mull of Kintyre road (1832)
1829–1832. Stirling, five segmental masonry arches, £17,000
1831–1834. Hutcheson, Glasgow, five segmental arches, £24,000
1832. Glasgow, temporary timber bridge, retained until 1846
1833. Abbey St. Bathans, Berwickshire, two 60-ft. iron truss spans
1836. Wardie Burn tunnel, Edinburgh, 100 ft. long
1842. Allanton, two masonry spans, £6,058

Other:

1814. Road at Marykirk
1815. Arbroath Abbey ruins
1818. Newton colliery railway, near Edinburgh
1819. Regent Road approach to central Edinburgh
1821. Melville Column, St. Andrew Square, Edinburgh

STEWART, William (fl. 1770–1806), was a bridge builder of Moniaive, Dumfriesshire. By 1793 he had built 20 bridges and was working on the New Bridge at Dumfries. In this contract he was described as a mason and his colleague, Thomas Boyd, as an architect, but in a subscription list of 1795, Stewart too is shown as an architect; by 1801 he had added the role of timber merchant. In the same year he subscribed 5 guineas to the new County gaol, for which Boyd had provided the plans.

P. S. M. CROSS RUDKIN

[Rennie collection, NLS; *Dumfries Weekly Journal, passim*; J. Robertson (1993) *The Public Roads and Bridges in Dumfriesshire 1650–1820*]

Works
1770. Nithsdale, three small bridges, cost c. £265
1777–1778. Nith Bridge, Thornhill
1781–1782. Auldgirth Bridge, contractor; cost £1,145
1782. Lochar Bridge, Collin, contractor for £166
1783–1784. Ae Bridge, contractor, £635
1785–1786. Shinnel Bridge, Tynron, contractor, £230
1789. Kirtle Bridge, Rigg, contractor, £200(?)

1791–1794. New Bridge, Dumfries, contractor (with Thomas Boyd), cost over £4,000
1796. Dalwhat Bridge, Moniaive, contractor, £200(?)
1798–1800. Brydekirk Bridge, contractor (with Thomas Boyd), £750
1806 (attrib.). Nethertack Bridge, Moniaive

STICKNEY, Robert (1772–1815), surveyor, of Riseholm, Rise, in the former East Riding of Yorkshire, was responsible for a number of surveys connected with navigation and drainage schemes in Lincolnshire and East Yorkshire in the late eighteenth century. These included the Witham, Horncastle navigation, Keyingham navigable drains, and Humber. Although the Horncastle and Keyingham schemes were carried out, there is no indication that Stickney was actively involved beyond the preparation of plans.

Stickney was a Quaker.

MIKE CHRIMES

[Bendall; SK; W. H. Wheeler (1896) *A History of the Fens of South Lincolnshire*; C. Hadfield (1973) *The Canals of Yorkshire and North-east England*, 311; J. Boyes and R. Russell (1977) *The Canals of Eastern England*, 268–280]

Publications
1792. *Plan of Part of the River Witham …*
1792. *A Plan showing the Course of the Rivers Nairn and Waring, and the Works proposed to be Executed thereon, to open a Navigation from Horncastle to the River Witham in the County of Lincoln* (with Samuel Dickinson)
1795. *A plan of the navigation, from the Humber to Great Grimsby … with the proposed alterations for the improvement thereof* (for M. Pilley and Jonathan Pickernell)
1802. *A map of part of … Holderness … comprising the levels of Keyingham to Burstwick and the drainages thereof and projected Improvements …*

STOREY, Thomas (1789–1859), civil engineer, was born at Ponteland, near Newcastle, on 7 December 1789, the son of Edward Storey and his wife, Alice Hindmarsh; he had two brothers, John (b. 1792)—a civil engineer—and Edward (b. 1805). His mother was the sister of George Stephenson's second wife, Elizabeth.

Storey was educated at Stamfordham and was apprenticed to John Watson, viewer at Willington colliery. He then moved to Lancashire, later working as a mining engineer with Clark, Roscoe & Co. in Wales and Shropshire. He married Elizabeth Scott (1786–1858), from Wragby, Yorkshire, c. 1810 and they had four sons and three daughters. He was appointed as assistant engineer to the Stockton and Darlington Railway in 1822 and took up residence at St. Helens Auckland where, under George Stephenson (q.v.), he was responsible for the construction of the western section of the line.

On completion of the S&DR in 1825 Storey was appointed as chief engineer to the company and four years later became a member of the Institution of Civil Engineers. He was responsible both

for the Haggerleases branch westwards from West Auckland and for the extension of the railway eastwards towards Middlesbrough, where staiths were built so as to provide better shipping facilities than had hitherto existed. He intended that an iron bridge be adopted to cross the river Tees but a suspension bridge, built by Captain Samuel Brown (q.v.), came to be used; it was not a success and was replaced in 1842. In 1829 he was called upon to survey the proposed road bridge at Whorlton, also crossing the Tees, where a suspension bridge was completed by John Green (q.v.) in 1831. In 1836 Storey relinquished his position with the S&DR although remaining as its consultant.

He then became involved with the proposed formation of the Richmond and Cleveland Railway, which did not proceed, and the following year was appointed as engineer to the Great North of England Railway, its line initially extending from Gateshead to York. It was written that Storey 'belonged to that class of old-fashioned engineers who considered bridge building the duty rather than of an architect than an engineer' and, as a result, the line's two major bridges were designed by architects, at Croft by Henry Welch (q.v.) and at Poppleton by John Green (q.v.). To undertake design work an office was established at Rushyford for the duration of the line's construction but staff were later transferred to Storey's office at St. Helens Auckland, where designs for timber bridges at Shincliffe, Chester-le-Street, and Durham were undertaken. The GNER was opened between York and Darlington in 1841 but its extension to Newcastle—with its timber bridges—was abandoned. Storey resigned immediately prior to the opening.

To obviate the difficult working of the S&DR's rope-hauled inclines, a tunnel was planned with an extension north-westwards towards Bishop Auckland. As consultant to the company, Storey became engineer-in-chief to the Shildon Tunnel Company, the joint venture established for construction. The engineer was Luke Wandless and the line opened in 1842. Storey also became engineer for the Bishop Auckland and Weardale Railway, a line extending the S&DR further westwards; it was completed in 1843.

Storey prepared plans for two further railways, the Northumberland and Lancashire Junction, from Newcastle, via Bishop Auckland to Kirkby Stephen, and the Newcastle and Leeds Direct, from Wath to St. Andrews Auckland. Both schemes were part of extensive arrangements, some of which were, in the words of Tomlinson, 'wildly impracticable [and] covered the surface of the northern counties in every direction with a complicated network of imaginary lines'. Neither railway was built at that time.

Storey's interests were not confined to civil engineering. In 1828 he was described as 'agent to the rail road Co' and as an 'iron and brass founder' in 1853, at which time he was employing seven men at St. Helens Auckland. He was also Secretary to the coal trade on the river Tees and, on its behalf, gave evidence in London to the Select Committee of 1836.

He was described as 'tall and athletic, and capable of undergoing great fatigue. He possessed great decision of character and was deservedly respected for his strict integrity and honesty of purpose'. As his health deteriorated in later life, he retired and 'died calmly, after a short illness' on 15 October 1859; three days later he was buried at St. Helens Auckland. He was survived by his son, Thomas, a civil engineer, another son, Robert, and his three daughters; to them he left effects of some £4,000.

R. W. RENNISON

[*Report of the Select Committee on the state of the Coal Trade*, House of Commons, 2 August 1836; *Great North of England Railway*, Northumberland RO; FOR/B/44, NRO; Memoirs, *Min Proc ICE*, **19**, 1859–1860, 182; *The Builder*, 7 April 1860; W. W. Tomlinson (1914) *North Eastern Railway*, 468, 9; Sk (1987)]

Publications

1827. *Branches of the Stockton and Darlington Railway showing Elevation of Tees Bridge and Docks at Marske*
1829. *Plan and Section of a Branch Railway from the Stockton and Darlington Railway to the River Tees*
1833. *Stockton and Darlington Railway from Egglescliffe to Stokesley*
1835. *Tramroad from Croft to Redheugh*
1836. *Railway from the Stockton and Darlington Railway ... to Frosterley, via Bishop Auckland*
1836. *Report on the Great North of England Railway connecting Leeds and York with Newcastle upon Tyne with plan and sections*
1836. *Plan and Section of the Great North of England Railway ... [from] Gateshead ... [to] the River Tees*
1837. *Plan ... of the Extension of the Great North of England Railway from the River Tees to the City of York*
1837. *Railway from Gateshead to Newcastle with Plan and Elevation of a Bridge*
1842. *Weardale Extension Railway*
1845. *Northumberland and Lancashire Junction Railway*
1845. *Newcastle and Leeds Direct Railway*
1853. *Bishop Auckland and Weardale Railway*

Works

1822–1825. Stockton and Darlington Railway, construction, total length 27 miles
1825–1836. S&DR, Engineer: Haggerleases branch, 4 miles; Middlesbrough branch, 4 miles
1836–1841. Great North of England Railway, Engineer: construction of railway York–Darlington, 43 miles
1836–1842?. S&DR, Consulting Engineer
1840?–1842. Shildon Tunnel, Engineer-in-chief, railway *c.* 3 miles, tunnel ¾ mile
1841–1843. Bishop Auckland and Weardale Railway, Engineer, 8½ miles

STRAPHEN, John (*c.* 1774–1826), mason and contractor, was the nephew of John Simpson (q.v.) and succeeded to his business on his death in 1815. It is likely that he was responsible for Simpson's work in Shropshire once Simpson had begun work in Scotland around 1804. Aside from some bridge work in Shropshire, he was contractor with his partners, John Stanton (of Ellesmere) and, latterly, John Hall (q.v.) for many of the early contracts on the Holyhead Road. He was originally awarded the contract for the masonry on the Menai Bridge but was replaced by John Wilson (q.v.). In 1820 he was requested by the Herefordshire Quarter sessions to submit a tender for an iron bridge and estimate for another bridge.

Straphen died at Shrewsbury on 27 July 1826, aged fifty-two.

MIKE CHRIMES

[Colvin, **1**; QS/SM/22, Herefordshire CRO; DP 26, Salop CRO; *Reports of the Commissioners ... Holyhead Road, 1822–1830*]

Works

1811–1812. Bromfield Bridge, £2,015
1815–1816. Lord Hill's Column, Shrewsbury
1815–. Holyhead Road: Ty Grewyn–Lake Ogwen, £3,281 6s; near River Lhigwy, £1,134; near Bettwys-y-Coed, £5,035 14s 6d; Glyn Conwy, £729 12s
1816–. Holyhead Road: at Llynin Bridge, £17,27 14s 5d; at Cerig-y-druidion, £536 5s; between Bangor Ferry and the City, £2,689 4s 3d
1817. Holyhead Road: at Glyn Dyffws, £1,662 2s 6d; Glynn Dyffws–Maes Mawrfachen, £4,355
1818. Cound Bridge, demolition
1818. Holyhead Road: near Rhyallt, £2,050; At Llan-issa, £1,720
1821. St. Alkmund's Church, Shrewsbury, repairs
1822. Stokesay Bridge, abutments for cast iron arch

STRATFORD, Ferdinando (1719–1766), civil engineer and surveyor, was christened Ferdinandus, the son of Walter and Francisca Stratford on 4 July 1719 at Guiting Power, Gloucestershire. He is known to have had an elder brother, Walter (b. 1716), and possibly a younger brother, William. Stratford was based for most of his career in Gloucestershire and is known for his estate and enclosure surveys in that country, and for survey work in Essex, Monmouthshire, Somerset and Wiltshire. He was resident in London in 1748, possibly working with Charles Labelye (q.v.), and moved to Bristol *c.* 1760.

His civil engineering work was mostly concerned with inland navigations, the earliest being surveys for the proposal to improve the western division of the Nene Navigation between Thrapston and Northampton, the subject of a revised Act in 1756. Other proposals were received, including those of Thomas Yeoman, and John Smith Jr. (qq.v.). In the event, Smith's scheme was adopted and Stratford's bill was settled at £46 2s in July

1758. This is unlikely to have been his first scheme as it was some way from his home base.

In the 1760s Stratford was involved in two further schemes, one, of 1765, was the idea of a Chippenham extension to the Avon navigation, which influenced the first proposal for a Kennet and Avon Canal. The other, the Chelmer navigation, literally had fatal consequences for Stratford. He, and his brother 'William' [*sic*] were brought in by opponents of a scheme of Thomas Yeoman. Stratford urged the use of stone walls for the locks, rather than the timber suggested by Yeoman. While Yeoman's scheme was incorporated in an Act of 1766, both Stratford brothers contracted malaria (ague) as a result of their surveys of the Chelmer and Stratford died on 28 April 1766.

Stratford's other civil engineering schemes included involvement in proposals of 1760 for a new bridge at Bristol. He submitted his design and, with James Bridges, inspected the old bridge as it was demolished. Bridges felt the existing foundations of the pier could be reused whilst Stratford proposed a more ambitious single arch design with new foundations. He discussed the form of the arch with John Smeaton (q.v.).

The trustees vacillated over a number of years until November 1763 when they decided on reusing the foundations. Stratford, along with Bridges and another aspirant designer, John Wood, helped draw up the contract and specification but the actual work was carried out by the contractor, 'Mr. Britton', under the supervision of the Trustees' supervisor, Thomas Patty.

As noted by Colvin, Stratford was responsible for a number of architectural works in Gloucestershire and Ulster.

MIKE CHRIMES

[Nene Navigation Commission minutes (1756–1758) Northants CRO; K. R. Clew (1968) *The Kennet & Avon Canal*; C. Hadfield (1969) *The Canals of South and South East England*, 226; J. Boyes and R. Russell (1977) *The Canals of Eastern England*, 63–64, 199; M. Handford (1979) *The Stroudwater Canal*, 104–105; T. Ruddock (1979) *Arch Bridges and their Builders*; Colvin, **3**; Skempton]

Publications

1760. *A Short Account of the Manner proposed for Rebuilding Bristol Bridge*

[1788]. *Observations on a Scheme for extending the Navigation of the Rivers Kennet and Avon ... by a Canal from Newbury to Bath* (includes extract from Mr. *Stratford's Plan for extending the Navigation from Bath to Chippenham*, 1765)

STROZZI, Piero de (fl. 1540–1550), military engineer, was employed by the French in 1548 to design fortifications at Leith which had been sacked by English forces in 1544 and 1547, Mary of Guise, the Queen Mother, and the French party then being in the ascendant in Scotland. Strozzi used the latest Italian designs, first employed at

Verona in 1530, with low earthen rhomboidal and angular bastion ramparts replacing the familiar medieval style wall. Further work was carried out to post-1550 Italian models in 1559. This work pre-dates defences on the Italian model south of the border at Berwick.

MIKE CHRIMES

STRUTT, William, FRS (1756–1830), a versatile innovator in the design of buildings and in several branches of engineering, was born at Blackwell, near Alfreton in Derbyshire, the eldest son of Jedediah and Elizabeth Strutt.

Jedediah Strutt patented machinery to knit ribbed stockings in the late 1750s and established a successful business in Derby. The Strutts were dissenters and initially William went to school at the academy in Findern, where some years earlier a four-year course had included such subjects as logic, mathematics and natural philosophy. His formal education was curtailed at the age of fourteen when he entered the family business, his father becoming one of Richard Arkwright's early partners at about the same time. Four years later, in a letter from his father on the importance of continued self-improvement and how to acquire the manners of a gentleman, Strutt was gently upbraided for a modesty and shyness that was to remain with him throughout his life. In replying, he accepted the usefulness of learning French, though not Latin, and wrote: 'Algebra & other Branches of the Mathematics, I only endeavour'd to learn enough of, to qualify me to read, & understand those Books which (in my opinion) treat on some subjects worth knowing'.

A shift in emphasis from making stockings towards spinning cotton followed the success of Arkwright's patent machinery. When the partnership with Arkwright ended in 1781, William Strutt was playing a key role in the family firm. In the late 1770s his father bought iron forges and a slitting and rolling mill and established machine-making shops downstream from his first mill at Belper. Before 1790 he was one of the first to try the self-acting mule for cotton spinning, although the Arkwright frames proved better suited to the water-powered mills at Belper. He invented the 'Devil', a machine that opened the tightly packed cotton bales. He also played a part in the development of the all-iron suspension waterwheel, introduced at Belper early in the nineteenth century. When Strutt was elected a Fellow of the Royal Society in 1817, without seeking the honour, one proposer was Marc Brunel (q.v.), himself the inventor of a knitting machine the year before.

His influence extended beyond the family firm to Derby itself and, from 1792 onwards, he chaired a series of Commissions set up by Acts of Parliament with powers to raise funds for improvements to the buildings, streets and infrastructure of the Borough. Strutt became abreast of the latest developments in bridge design. He was involved with the architect Thomas Harrison (q.v.) in the design

William Strutt FRS

of St. Mary's Bridge over the river Derwent, built from 1789–1794, and designed a number of smaller bridges over Markeaton brook himself. Writing in 1789 to his friend and fellow cotton spinner, Samuel Oldknow, who was then planning to build a bridge over the river Goyt, he advised him to 'consult some learned mathematician' and made reference to Thomas Paine's iron bridge patent of 1788.

Strutt inherited his father's inventiveness but his scientific approach to the solution of practical problems owed much to the influence of Dr. Erasmus Darwin. After Darwin moved from Lichfield to Derby he wrote in March 1783 to his friend and fellow member of the 'Lunar Society', Matthew Boulton, that 'we have established an infant philosophical society at Derby'. Two years later, Strutt's range of 'subjects worth knowing' had expanded and he was assisting Darwin in experiments concerned with adiabatic expansion that explained the formation of clouds. Derby benefited from this association when, with Darwin's help, Strutt applied his mechanical skills to the invention of a clock for use by the nightwatchmen of the Borough, which greatly improved their efficiency.

As the family business expanded, Strutt took the lead in technical matters. Large mills with timber internal structures were vulnerable to fire, which had claimed the firm's Nottingham mill in 1781. When Strutt planned a new mill at Derby and a warehouse at Milford a little over ten years later, the then recent destruction of Samuel Wyatt's famous Albion corn mills in London was perhaps foremost in his mind. In May 1793, Boulton wrote to Strutt: 'I understand that you

have some thought of adopting the invention of forming Arches by means of the hollow potts & thereby saving the use of Timber in making Floors, & guarding against Fire'. In Strutt's design the arches were paved with brick tiles and sprang from timber skewbacks that were covered with sheet iron and fixed to the sides of heavy timber beams, the soffits of which were plastered for fire protection. In the surviving Milford warehouse, two rows of cruciform section cast iron columns 9 ft. apart provide intermediate support to the timber beams, which are in one piece between the external supporting walls. The beams pass through a cast iron crush-box over the columns, through which are wedged wrought iron tie bars to complete the framing in the transverse direction. Both buildings were completed in 1793 and a third, on the same principle, was completed at Belper in 1795. They initiated a new era in structural engineering.

In early 1793 Strutt married Barbara, daughter of Thomas Evans, owner with his brothers of extensive cotton mills at Darley-Abbey near Derby. Two years later, with his father-in-law, he bought a 1,200 acre estate at Kingston-on-Soar. However, he continued to live in Derby, where he was a significant figure behind many developments, including the Derby Canal, completed in 1796. In 1797 his father died and the firm became W. G. and J. Strutt, with William at its head, his brother, George, managing the mills at Belper and Milford and another brother, Joseph, primarily concerned with commercial matters. The scale of the Strutt cotton enterprise rivalled Robert Owen's New Lanark and there were frequent visitors to the mills. One of these was Charles Bage (q.v.), who adopted fireproof floor arches in a Shrewsbury flax mill but substituted cast iron beams for Strutt's heavy timbers; the Shrewsbury mill was completed in 1797.

Another fire destroyed the early timber floored North Mill at Belper in January 1803. In August of that year, probably in response to a request, Bage sent Strutt his theorems 'On the strength of Cast Iron Beams' and the mill was quickly rebuilt on fireproof principles, but with improved cast iron beam and frame connection details. Strutt was also in correspondence, in 1803, with the very scientifically-minded George Lee, whose own mill, built in Salford, had followed the Shrewsbury mill in its use of cast iron beams. Strutt built further fireproof buildings, including the South Mill at Belper, rebuilt as fireproof in 1811–1812; it was his last and most refined building of this type.

The early cotton mills were heated to prevent frequent breakage of the fibres and later incorporated heated rooms for drying yarn. Strutt greatly improved the existing heating systems and developed the 'Belper stove'. Advanced warm air heating and ventilation systems were incorporated in his most important public building, the Derbyshire General Infirmary, opened in 1810. His friend Charles Sylvester (q.v.) oversaw the construction of the Infirmary, though Strutt was 'principal director in its arrangement and construction'. Strutt also designed houses for himself and his family and probably drew up the specifications for workers housing at Belper and elsewhere. One portrait shows him, perhaps as he wished to be remembered, at a table with building plans and drawing instruments.

The Strutts, for the times, were caring employers and William and his brother, Joseph, were interested in education at all levels. William Strutt succeeded Darwin as President of the Derby Philosophical Society in 1802. He supported his brother, Joseph, in the formation of a Lancasterian school in Derby in 1812 and in the establishment of the Derby Mechanics' Institute in 1825; among the latter's early lecturers was a surgeon, Douglas Fox, assisted by his younger brother, Charles, later Sir Charles Fox (1810–1874), the civil engineer, the Strutt and Fox families being linked by marriage.

William Strutt died on 30 December 1830, leaving a son, Edward (later Lord Belper), and many daughters. The daughters of his friend Richard Lovell Edgeworth (q.v.) described him as 'the ingenious, indefatigable and benevolent Mr. William Strutt'.

TOM SWAILES

[Charles Bage (1803) Letters to William Strutt, Shrewsbury RO; George Lee (1803) Letter to William Strutt, Derby Local Studies Library; H. R. Johnson and A. W. Skempton (1956) William Strutt's Cotton mills 1793–1812, *Trans Newc Soc*, **30**, 179–205: R. S. Filton and A. P. Wadsworth (1958) *The Strutts and the Arkwrights*; C. L. Hacker (1960) William Strutt of Derby, *J Derbyshire Arch Soc*, **80**, 49–70; G. Unwin (1968) *Samuel Oldknow and the Artwrights*; D. King-Hele (1968) *The Essential Writings of Erasmus Darwin*; R. L. Hills (1970) Power *in the Industrial Revolution*; A. F. Chadwick (1971) Derby Mechanics Institute, M.Ed. thesis, Manchester; A. Menure (1993) The cotton mills of the Derbyshire Derwent, *Ind Arch Rev*, **16**, 38–61; M. Craven (1998) *Derbeians of Distinction*]

Works

1792–1793. Derby Cotton Mill
1792–1793. Milford Warehouse
1793–1795. Belper West Mill, £7,000
1803–1804. Belper North Mill
1811–1812. Belper South Mill

STUART, William (1773–1854), mason and civil engineer, was born in North Leith, Edinburgh on 5 January 1773. His father was involved in the West India Trade but his death left Stuart an orphan at the age of six. With the assistance of his relatives he continued at school until the age of fourteen, when for seven years he was apprenticed as a mason to a builder active in southern Scotland. He then spent four years executing masonry works on his own account and for employers.

From 1798 he was employed for about eight years by Andrew Brocket (q.v.) the Glasgow contractor for whom he superintended all the locks built on the Crinan Canal, as well as extensive works in Glasgow and elsewhere. Among these were the opening and working of stone quarries and the erection of public buildings, including the erection of Nelson's monument at Glasgow. Through his work on the Clyde navigation and Greenock harbour, he became acquainted with John Paterson (q.v.).

Early in 1807 John Rennie (q.v.) wished to find a person to superintend proposed work at Fraserburgh harbour, of which he was chief engineer, and Paterson introduced Stuart to Rennie. On 16 March 1807 Rennie wrote to Stuart, telling him he was appointed to superintend the works, quarries were obtained and opened and on 1 September 1807 the foundation stone of the pier was laid. Here Stuart was occupied until August 1811, when the pier and harbour were complete. The Town Council presented him with a commemorative silver cup.

Thomas Telford (q.v.) visited the works as Engineer to the Commissioners of Highland Roads and Bridges, who had contributed half the cost of the works. During the construction of the harbour, Stuart made various surveys for Rennie, Telford and others, including a report on improving the harbour of Rosehearty (1809), and on the coast of Buckie where it was proposed to form a harbour.

In 1811 Stuart left Scotland for London to assist Rennie in some works there. The Admirality had now approved a report made by Rennie and Joseph Whidbey (q.v.) in 1806 for the construction of a breakwater at Plymouth and on 3 December 1811 Stuart received an appointment from the Lord Commissioners. Whidbey had been appointed superintendent of the works by the Admiralty the month before. Following Stuart's arrival in Plymouth suitable quarries were agreed in Oreston where also were set the offices. Construction of the wharves from whence the stone would be shipped, railways and buildings began. In March 1812 the House of Commons voted the sum of £80,000 for the commencement of the works and on 12 August the first stone was formally laid.

Moving the stone quarried at Oreston to the waiting ships had required Rennie and Stuart to set up an elaborate railway system at the Oreston quay.

Sir Thomas Tyrwhitt devoted much of his life and financial resources to a plan for the reclamation and agricultural development of the Forest of Dartmoor and in November 1818 made a proposition to the Plymouth Chamber of Commerce that Princetown should be linked to Plymouth by a railway. William Shillabear, a schoolmaster, had carried out the survey but for engineering expertise Sir Thomas had approached Stuart in September. After it had been made quite clear to Sir Thomas that Stuart could give his services only on a part-time basis, the Admiralty agreed to Stuart working for the Plymouth and Dartmoor railway in his leisure time. Stuart made an estimate of £27,783 for the works in May 1820 and a tender was accepted from Hugh McIntosh (q.v.). In August Stuart was engaged to carry out the entire superintendence of the works to completion and for twelve months afterwards for the overall sum of £1,000. However in March 1821 difficulties concerning the gradients of the lower part of the route became evident and in April Roger Hopkins (q.v.) was appointed as assistant engineer to resolve them. A further Act of Parliament was required and in October the committee resolved to dismiss Stuart and Hopkins was appointed engineer in his place.

On 4 October 1821 Rennie died and in 1822 some alterations in the original plan of the breakwater being proposed, Telford, Josias Jessop, George and (Sir) John Rennie (qq.v.) were consulted by the Admiralty and in 1823 visited and reported on the works. The most anxious period of Stuart's work on the breakwater was in November 1824 when the great storm around the coast disturbed the upper part of the whole work and remedial measures were carried out with the concurrence of William Chapman (q.v.), Jessop, and George and Sir John Rennie.

In 1828 the idea of constructing a breakwater in Torbay was revived by the Duke of Clarence, the Lord High Admiral. Whidbey had surveyed Torbay in 1798 and again in 1822. Now Whidbey and Stuart were required to visit Torbay and prepare alternative estimates for the breakwater using hired-men, hired-men and convicts, or convicts only. The length of the breakwater would be 1½ miles, the average width 255 ft. and the height 60 ft., requiring twenty-eight years to deposit 280,215 tons per year. No more of this matter was heard but the Duke of Clarence visited the Plymouth breakwater in July to witness the construction of that work.

In 1830 Whidbey retired from the Superintendence of the breakwater and the Admiralty directed Stuart to take over the entire management of the work, which he did and continued so to do until his death. In 1838 the work at the breakwater suffered greatly from storms and James Walker was called in by the Admiralty to confer with Stuart. After reporting, Walker continued to act as consulting engineer to the Admiralty and it was from the designs of Messrs Walker and Burges that Stuart commenced in 1841 the breakwater lighthouse, which he completed in 1844. In June 1845 the beacon at the eastern end was commenced and completed in October. When Stuart died, £1,516,144 had been expended and 3,834,957 tons of stone deposited.

Between 1811 and 1854, Stuart was called on to assist and report on many other public works. Among these were: the Government reservoir and pier at Staddon Heights, commenced in 1815 and completed in 1825, at a cost of £40,000; wharf walls in the Devonport dockyard in which a diving bell was used; harbour works at Brixham, Torquay and Bude; embankments and sea walls at Kitley and sea works at St. Mawes.

Stuart was elected a member of the Institution of Civil Engineers in 1828. In addition to his membership of the Institution, he was a founder member of the Devon and Cornwall Society of Engineers and Architects. Inevitably most of his publications concerned the breakwater, his model of which was awarded a medal at the Great Exhibition. Stuart was for many years a director of the Devon and Cornwall Banking Company. Some idea of the esteem in which he was held can be gained by the decision of the artist, Thomas Roods, to offer a mezzotint of his portrait to subscribers in 1848.

After having given forty-three years of service to the breakwater work Stuart died at his residence, Woodside, Plymouth, on 11 January 1854. In recognition of his part in what was one of the great public works of the early nineteenth century the Admiralty sent a lengthy letter of condolence to his family.

A. B. GEORGE

[Membership records, ICE archives; On limestone, OC 268, ICE archives; Drawings register, ICE archives; Rennie reports, ICE archives; Rennie papers, NLS; Anon. (1821) *Interesting Particulars, relative to that Great National Undertaking, the Breakwater now constructing in Plymouth Sound*; Minutes of Evidence (1844) *Commissioners … Harbours of Refuge*; E. Hopkins, (1854, repr. 1855) *William Stuart, a Memoir to the Devon and Cornwall Society of Architects and Engineers*; Obit. (1855) *Min Proc ICE*, **XIV**, 137–142; H. G. Kendall, (1968) *The Plymouth and Dartmoor Railway*]

Publications

1838. On the limestone … in the neighbourhood of Plymouth, *Min Proc ICE*, **1**, 35–36
1841. Account of the original construction and present state of the Plymouth breakwater, *Min Proc ICE*, **1**, 160–162
1841. On Plymouth Breakwater, *British Association, Reports (Mech Sect)*, **11**, 99–100

Works

c. 1800. Crinan Canal, locks, mason
1807–1811. Fraserburgh Harbour, resident engineer, £9,000
1811–1854. Plymouth breakwater, £1½ million

STUDHOLME, John (1787–1847), land surveyor, was the son of Joseph Studholme and his wife, Mary Moor, and was baptised on 31 October 1787 at St. Cuthbert's church, Carlisle. Nothing is known of his early life but he would seem to have married c. 1825, the union resulting in a family of eight.

In 1825 he was responsible, with William Fryer of Newcastle, in surveying a line for a railway between Newcastle and Carlisle, the work carried out under the direction of Benjamin Thompson (q.v.). Nothing then came of the proposal but in 1828 the line was resurveyed, this time with the eastern end surveyed by Thomas Oswald Blackett (q.v.). Their work, again with Thompson, resulted in the formation of the Newcastle and Carlisle Railway.

Studholme subsequently undertook many surveys for Enclosure and Tithe Awards and, in addition, had a financial interest in the Blenkinsopp Coal and Lime Company, its fortunes dependent upon the N&CR. He was presented by the Agricultural Society of Cumberland with a service of silver plate for his work on agricultural improvement, namely the introduction into the county of the use of tile land drains.

Variously listed in local directories as land agent, surveyor and valuer, Studholme occupied several premises in Carlisle between 1829 and 1847 and lived from 1837 at Morton Head, on the outskirts of the town. After his death, the notice for its sale recorded that, in total, Studholme had owned almost 400 acres of land in several lots.

Studholme died at Morton Head on 14 September 1847 and was buried at Grinsdale the following week. He disposed of his property to his widow, Elizabeth, and his family. His interest in the Blenkinsopp concern presumably passed to the family of John Blackmore—engineer to the Newcastle and Carlisle Railway—who was to have been his executor but who predeceased him.

R. W. RENNISON

[Copy of the Evidence taken before a Committee of the House of Commons on the Newcastle and Carlisle Railway Bill (1829); Carlisle Journal; Carlisle Patriot; Copy of will (1843) Cumbria RO]

Publications

1825. Plan and Section of an Intended Iron Rail Way or Tramroad from the Town and County of Newcastle upon Tyne to the Canal Basin near to the City of Carlisle in the County of Cumberland, with seven branches therefrom, by Benjamin Thompson

1828. Plan and Section of an intended Railway or Tramroad from the Town ... of Newcastle to the City of Carlisle ... Surveyed under the direction of Benjamin Thompson by T. O. Blackett and I. Studholme

SURBEY, Thomas (fl. 1699–d. 1703), engineer, was chosen by two MPs of the city of York following their being asked in February 1699 'to send an Ingineer as soon as possibly they can ... to view the River Ouze in order to make it more navigable'. Surbey's long-forgotten report, included in a journal of his journeys from and back to London and his survey of the river, was acquired by York Corporation in 1964; a transcript of the report, which he would have left at York, has not been found.

Surbey arrived in York in the evening of 4 May after a four-day journey by coach from Holborn. Next day, having received instructions from the Lord Mayor, he set off in a boat with two gentlemen, John Atty and Benedict Horsley, the

captain, two watermen and a boy. Soundings and other observations were made along the Ouse from York and on the Humber to Hull. The party returned to York in the evening of 13 May.

The main conclusions of the study were: (1) that an adequate depth of water existed downstream of Naburn (6 miles below York) for boats drawing 7 ft.; (2) the best way to provide this depth up to York would be to build a lock (and weir) at Naburn; (3) the alternative of dredging would be more costly and less certain in the long-term; (4) several proposed long cuts would do little to increase the depth of water and would also be costly; and (5) the destruction in 1688 of the sluice on the Dutch River at Goole had, contrary to general supposition, little if any effect on the tidal regime of the Ouse.

Surbey then spent six days drawing plans and sections of the lock and weir, preparing detailed specifications and estimates and writing the report. He had a meeting with the Lord Mayor, dined with Sir William Robinson, one of the MPs, and presented the report to the Corporation on 23 May. He received £24 in fees plus travelling expenses (£4 10s). His estimated costs were, in summary, land purchase £250, lock and lock cut £3,800, weir £960, and £450 for contingencies, a total of £5,460.

He next spent two days with John Aislabie (MP for Ripon) on a problem of river erosion on the Ure near Ripon, for which he received 2 guineas. After bringing his journal up to date he set off for London on horseback on 22 May via Bawtry, Nottingham, Derby, Warwick and Oxford, arriving back at his lodgings in Temple Bar on 2 June.

The report covers nineteen folio pages, including a map or chart of the river. One of the earliest practical civil engineering reports in England, it was known to Duckham (1967) and has been fully transcribed, along with relevant parts of the journal, by Hughes (1994). The report was well received by York Corporation. Soon afterwards they sought a second opinion from John Hadley (q.v.) who had just begun work on the Aire & Calder Navigation. He broadly agreed with Surbey's conclusions. York sent a petition to Parliament in January 1700 for an Act enabling funds to be raised for carrying out the works, but for no obvious reason the Bill never reached the statute books. Some improvements in the river were made by William Palmer (q.v.) in the period around 1730 but not until 1752 was action taken to build Naburn lock and weir. John Smith (q.v.) designed and supervised the work, completed in 1757 on lines similar to those proposed by Surbey.

Surbey's journal of his return journey to London includes comments and sketches of repairs to the piers of St. Mary's Church, Warwick, and the roof truss of the Sheldonian Theatre, Oxford. He also relates the remarkable escape from injury of George Sorocold (q.v.) when he fell under the waterwheel at Derby. This disproves the interpretation by most historians of the account given by Daniel Defoe in his *Tour through the Whole Island of Great Britain*, vol. 3 (1726) that the accident occurred at Lombe's silk mill in Derby, built 1718–1721.

Of particular interest is the 'Memoir' added by Surbey to his journal in which he gives drawings and notes on pound locks on the Great Ouse near Huntingdon, on the Wey at Weybridge and on the Welland near Stamford. They are the only known contemporary measured drawings of seventeenth century locks in England. He also gives an illustrated description, again the earliest known, of 'bag and spoon' dredgers used on the Thames. In this volume references to the Huntingdon and Weybridge locks are made in the articles on Arnold Spencer and Sir Richard Weston.

Surbey died within the London jurisdiction. Administration of the estate of 'Thomas Surbey, bachelor, late of Plymouth' was granted to William Thistleton, his half-brother on his mother's side; the inventory was to be produced by the last day of June 1703.

The report and the fact that he was chosen to go to York show Surbey to have been an experienced and able engineer. Nothing is otherwise known of his career. There is an important clue, however, in that he wrote to Edmund Dummer (q.v.), Surveyor to the Navy, no less than three times during his short time in Yorkshire. Dummer designed and directed construction of the dockyards at Plymouth and Portsmouth between 1691 and 1698. Moreover, as Hughes has pointed out, the writing and style of Dummer's chart of the South Coast (1698) is identical with Surbey's chart of the Ouse. It is therefore virtually certain that Surbey, 'late of Plymouth', worked under Dummer in some capacity on the dockyard and coastal survey. Finally it may be noted that Dummer served with John Aislabie on the fabric committee of Greenwich Hospital from 1695, and Aislabie was brother-in-law of Sir William Robinson, the York MP.

B. M. J. BARTON and A. W. SKEMPTON

[York Corporation House Book (1685–1700) York City Archives; Thomas Surbey (1699) *A Journal from London to York upon a Survey of the River Ouse ... showing how far it is Navigable and where it wants to be Improved*, also a *Report of the Survey with a Draught of Naburn Ings showing a place for a Cut to the River with a Lock thereon*, MS, York City Archives; Baron F. Duckham (1967) *The Yorkshire Ouse*; P. Hughes (1994) Thomas Surbey's 1699 survey of the river Ouse and Humber, *Yorkshire Archaeological Journal*, **66**, 149–190; P. Hughes (1996) Some civil engineering notes from 1699, *Local Historian*, May]

SUTCLIFFE, John (fl. 1780–1816), civil engineer, appears to have been trained and practised as a millwright in the Halifax area before becoming involved in the canal development of Lancashire and Yorkshire. He began work as a millwright *c.* 1781 and by 1790 had acquired a detailed knowledge of

the mills in the area and the rivers which powered them. He was responsible for Blackburrow and Marshall's Mills at Leeds. In the 1790s two of the largest Yorkshire manufacturers, Wormald, Fountaine and Golt, and Markland, Cooksen and Fawcett, made use of his services. His surviving mill drawings are of a very high standard, but his advice in recommending to both Marklands and Golt the use of 'common' steam pumping engines in combination with water wheels may be regarded as out of date; certainly both employers se lected Watt engines despite his advice. He was brought in as an expert witness on the impact of the Rochdale Canal proposals on the local mills and hydrology in 1793. From this it is clear he had been gathering data on the water requirements of canals for some time. About 1795 he was asked to report on improvements to the River Tyne.

His work, *A Treatise on Canals and Reservoirs* (1816), reflects his experience and strongly held views on canal navigations, cotton machinery, waterwheels, corn mills, etc. He was very critical of what he saw as waste of water on canals, the use of flights of locks, poor route selection and general poor design and construction evidenced on existing canals. He suggested his own method of lock design and location aimed at the more ef ficient use of water resources. These were themes which reflected his own experience of canal consultancy, as he had been asked to report on proposals for inclined planes and caisson locks on the Somerset Coal Canal in 1800.

Although Sutcliffe's practical experience of canal works was limited to the Somersetshire Coal Canal, he carried out several surveys and, from his experience of the water mills in the Pennines, would have been knowledgeable about reservoir and river design. He was consulted about the Axe drainage scheme in 1801, although the works were carried out by Robert Anstice (q.v.).

MIKE CHIRMES

[*Minutes of Evidence on Rochdale Canal, 1793–1794*, Rennie Reports, ICE Library; W. Chapman (1797) *Observations on Mr. John Sutcliffe's Report on a proposed Line of Canal from Stella to Hexham*; K. R. Clew (1970) *The Somersetshire Coal Canal and Railways*; C. Hadfield (1973) *The Canals of Yorkshire and North East England*, **2**, 354–358; J. Tann (1970) *The Development of the Factory*; Skempton]

Publications

1796. *Report on the proposed Line of Navigation from Stella to Hexham*
1797. *Report on the Line of Navigation from Hexham to Haydon Bridge, Proposal as a Continuation of the Stella and Hexham Canal*
1800. *A Letter to the Proprietors of the Somersetshire Coal Canal*
1816. *A Treatise on Canals and Reservoirs*

Works

1794–1795. Somersetshire Coal Canal

SWANSBOROUGH, William (fl. 1821–1833), engineer and builder—of Wisbech—was elected a Member of the Institution of Civil Engineers in 1821. In the 1820s he carried out some architectural and building work on local churches.

The problems of the Nene outfall had been investigated by a number of engineers and in the 1820s both Thomas Telford and Sir John Rennie (qq.v.) reported, before an Act was passed in 1827; it made provision for a new cut and the construction of the Cross Keys Bridge at Sutton Wash. The original contract price tendered by Jolliffe and Banks (q.v.) was £100,000, an indication of the scale of the works, and their eventual bill was £149,259. Swansborough was appointed resident engineer on Telford's recommendation as 'a man experienced in drainage operations.' Telford had previously employed him on his original survey of the river of 1822. In the spring of 1831 Swansborough described to the Institution of Civil Engineers experiments he had conducted on silt deposits on the river Nene. Whilst these works were in progress Swansborough was also co-operating with Thomas Pear II (q.v.) on the North Level drainage outfall sluice, doing the original survey in 1828.

Swansborough died in June 1833.

MIKE CHRIMES

[ICE membership records, ICE archives; Minutes of conversations, 103, ICE archives; T. Telford (1833) Diary, NLS; Maps (relating to Nene Outfall), 1822, 1838, 1864, 1826–1828, Northants CRO; J. Priestley (1831) *Historical Account of the Navigable Rivers, Canals, and Railways, throughout Great Britain*; J. Rickman (ed.) (1838) *Life of Thomas Telford*, 112–118; J Boyes and R Russell (1977) *The Canals of Eastern England*, 211–215; Colvin, **3**; Bendall]

Works

1827–1830. Nene outfall, resident engineer, £200,000
1830–1833. North Level drainage, assistant engineer, £150,000

SWITZER, Stephen (1682–1745), landscape designer and writer on gardening and water works, was baptised on 25 February 1682, the second of two sons of Thomas Switzer, a native of Hampshire. He was brought up in Stratton, near Winchester. His mother died soon after his baptism and his father died in 1697, leaving his estate, excepting £20, to the elder son. With his £20, the younger son was expected to purchase an apprenticeship. By 1698, he was at the famous Brompton Park Nursery, London, under George London (d. 1713) and Henry Wise (1653–1738), Royal gardeners.

Switzer's early projects while with London and Wise included the transformation of gravel pits at Kensington Palace into a garden amphitheatre (1704–1705), and Wray Wood at Castle Howard (1704–1705). As assistant to Wise, Switzer was park quarries supervisor at Blenheim (1705), planted 10,000 hedge yews, dug the foundations for Vanbrugh's bridge, and formed the river Glynde

into an irregular canal. By 1710, Switzer had left Blenheim and his association with London and Wise, and moved to Lincolnshire, where he wrote *The Nobleman, Gentleman and Clergyman's Recreation* (1715), a book on gardening, and designed and built gardens at Grimsthorpe, and probably Bresby and Belleau. After 1715, Switzer worked with Allen Bathurst (1683–1775) at Cirencester, Gloucestershire, and Riskins, Buckinghamshire. By 1718 Switzer was living in Newbury, Buckinghamshire, where he designed Caversham, near Reading, for William Cadogan (1675–1729). At Hampton, near Leominster (c. 1723), Switzer dug a new river in the park at a high level providing a head of water for the garden fountains and a reservoir for irrigating unwatered park lands to the north. Unlike Kent and many other garden designers, Switzer worked in plan and to levels.

Some time between 1718 and 1724, Switzer visited Holland and again, in 1729, he visited Italy, Holland and Denmark. By 1724 he was living and working at Spy Park, Wiltshire, by which time he had married Elizabeth. In the same year he designed extensive waterworks for The Drum, outside Edinburgh, and published his *Practical Fruit Gardener*, dedicated to Boyle. He undertook the design of a water garden for William Lord Brooke at Braemore, Hampshire, in the late 1720s. In 1726, Switzer, his wife and son Thomas were living in Pewsey, Wiltshire, but by August 1727 had removed to Kennington Lane, London, and opened a seed shop in Westminster Hall. Seed selling was a lucrative business and by 1731 Switzer had opened a nursery on Millbank. He published pamphlets to instruct the purchasers of his seeds and continued his design work with annual tours in England. Switzer was associated with agricultural improvers, and in 1731 was a corresponding member of the Society for Improvement of Knowledge in Agriculture, a mostly Scottish group founded by James Murray, Duke of Atholl. Switzer began a short lived agricultural magazine (1733–1734), the *Practical Husbandman and Planter*, which boasted a distinguished list of subscribers.

Switzer's *Hydrostatics and Hydraulics* (1729), reissued in 1734 as *An Universal System of Water and Water-Works*, is a work of over four-hundred pages with sixty plates including illustrations of experiments and technical information, as well as engines, machines, and garden water features. It includes views of the wheels of both London Bridge and Chelsea water works, along with Captain Savery's (q.v.) 'engine for raising water by fire'. Switzer intended to present 'the properest Methods of raising and distributing Water for the Use of Country Seats, Towns Corporate, &c.' as well as the theory and practice of hydrostatics. The book relies on the works of Boyle and Wallis, De Caus, Marriotte, Gravesande, Desaguliers (q.v.), Watts and Hawksbee as well as earlier sources and Switzer's own experience. Switzer was guided by Samuel Lindsey, chaplain to William, Lord Brooke, at Braemore, Hampshire, and had access to the library of Charles Boyle (1676–1731), Earl of Orrery, 1st Lord Marston, at Marston, Somerset, his 'Friend and Master'. It was the earliest work in English to attempt to address all the practical engineering aspects of the use and control of water, encompassing historical aspects, theoretical principles, and practical advice on the construction of fountains, distribution systems, and reservoirs. Only three earlier books on the subject were published in English (Thomas Salusbury's translation of the Italian Beneictus Castellus's works (1661), Desaguliers's translation of Marriotte (1718), and John Leak's 1659 translation of Issac De Caus's *New and Rare Inventions of Water-works*, largely based on Salomon De Caus's 1615 *Les Raison des Forces Mouvantes*.) None is so comprehensive as Switzer's and none is so firmly based on British experience.

TESS CANFIELD

[Plans at Nostell Priory, Yorkshire and at Leicester CRO; G. W. Johnson (1829) *A History of English Gardening*, 158–182; D. Green (1953) *Father of English Gardening* [Stephen Switzer] in *The Listener*, 3 Sept. 1953; D. Green, (1956) *Gardener to Queen Anne* [Henry Wise]; W. A. Brogden (1973) *Stephen Switzer and Garden Design in the Early 18th Century*, Ph.D. thesis, University of Edinburgh, 2 vols.; W. A. Brogden (1974) Stephen Switzer, in P. Willis (ed.), *Furor Hortensis*, 21–30; W. A. Brogden and P. Goode (1986) Stephen Switzer, in *The Oxford Companion to Gardens*, 545; DNB]

Publications
1715. *The Nobleman, Gentleman and Gardener's Recreation*
1718. *Ichnographia Rustica*, 2nd edn., 1742
1724. *The Practical Fruit Gardener*, 2nd edn., 1731; reissued 1752 and 1763
1727. *The Practical Kitchen Gardener*
1728. *A Compendious … Method … for the Raising Italian Brocoli …*, and subsequent editions to 1755
1729. *An Introduction to a general System of Hydrostaticks and Hydraulicks*
1731. *Dissertation on the True Cythisus of the Ancients … *, 2nd edn., 1735
1732. *The Country Gentlman's Companion*
1733–1734. *The Practical Husbandman and Planter*
1734. *An Universal System of Water and Water-Works … in four books*

SYKES, Thomas (1739–1816), of Chesterfield, County Surveyor of Derbyshire, was probably the person baptised at Holmesfield, near Dronfield on 26 December, 1739. There is no record in the Quarter Sessions order books of his appointment as County Surveyor, though Colvin states that it was in 1786. The first payment to him is mentioned in 1791. Throughout his tenure of office, committees of two to four magistrates were appointed for each bridge which needed repair or rebuilding and the records do not always mention Sykes's involvement. He does not appear to have been given any salary and his bills for superintendence were few and modest. In 1809 he received

£91 5s 2d for the four years since 1805. From 1808 he became 'one of the Surveyors of County Bridges', confining himself to the Hundreds of High Peak, Scarsdale and Wirksworth in the north of the county, John Welch taking over the remainder.

For his larger bridges he developed a distinctive style, which included semicircular piers with a spherical cap. The attributions of Swarkestone and Yorkshire Bridges below are based partly on stylistic grounds, though it is known that he was involved with both. He resigned his post at the Translation (July) Sessions, 1816 and was buried at Chesterfield, where he was a capital burgess, on the 21st of the following month. His reputation locally in 1794 was such that, when Robert Whitworth (q.v.) was not available to report on the completion of the Cromford Canal, he was engaged instead. His notebook for 1791–1813, which was noted in 1890 as being amongst the Quarter Sessions records, cannot be found now.

P. S. M. CROSS-RUDKIN

[Q/AH/1/1 and QS Order Books, Derbyshire RO; R. B. Schofield (2000) *Benjamin Outram*; Colvin]

Works
1795–1796. Whatstandwell Bridge, £1,161
1795–1797. Swarkestone Bridge, £3,550
1796–1798. Belper Bridge, £2,180
1800–1801. Calver Bridge, £1,323
1805. Duffield Bridge, widening, £1,100
1810. Yorkshire Bridge, Bamford

SYLVESTER, Charles (fl. 1804–1827), civil engineer and contractor, was born in Sheffield. He was the author of *The Philosophy of Domestic Economy* (1819), dedicated to William Strutt (q.v.), a work on the design of heating and ventilating, washing, drying and cooking arrangements for institutional buildings. It includes an explanation of Sylvester's work at the Derbyshire General Infirmary (c. 1810), notably its advanced system of hot air heating. Sylvester was a close associate of Strutt and worked with him on Derby Mill of 1792–1793, probably the first example of heating by warm air. Sylvester's execution of numerous works of heating and ventilation, as well as his knowledge of mathematics and chemistry (*An Elementary Treatise on Chemistry*, 1809, and many articles on galvanism, metals, gasses and mathematics) earned him a high reputation. Active in the community of engineers and scientists, he was a supporter of George Stephenson's (q.v.) views on the Liverpool and Manchester Railway, and author of a *Report on rail roads and locomotive engines* (1825). He never joined the Institution of Civil Engineers but his name appears in a list of contributors to 'a fund of the Institution' in June 1827. John Claudius Loudon (q.v.) had his Bayswater villa fitted out with the most advanced Sylvester system of hot air heating but was forced to replace it with a hot water system as he concluded that the rapid temperature changes in a small building were unhealthy. However, he thought the system very suitable for large buildings. He refers to 'the late Mr. Sylvester' in 1835. There is a pencil portrait of him by Sir Francis Chantrey at the Victoria and Albert Museum, London.

TESS CANFIELD

[John G. James Collection, ICE Archives; Obituary, John Sylvester (1852) *Min Proc ICE*, **12**, 165–167; J. Hix (1974) *The Glass House*]

Publications
Publications include:

1809. *An Elementary Treatise on Chemistry*
1819. *The Philosophy of Domestic Economy*
1825. *Report on Rail-Roads and Locomotive Engines addressed to the Chairman of the Committee of the Liverpool and Manchester projected Rail-Road*

SYLVESTER, John, FRS (1798–1852), engineer and contractor, was connected in business with his father Charles Sylvester (q.v.), executing large heating and ventilating projects. He built upon his father's work and extended it into ordinary household heating, greenhouses, conservatories and vineries, as well as Arctic Discovery Ships. He built forcing houses and vineries for Edward Strutt, MP, at Kingston Hall, Derby, of a ridge and furrow form. Joseph Paxton's (q.v.) New Victoria Regia House (1850) contained a Sylvester furnace for warming air. Sylvester was a Fellow of the Royal Society, and must have had a manufacturing connection, as he presented a Sylvester radiating stove grate, now lost, to the library of the Institution of Civil Engineers, of which he became an Associate in 1828 and was an active member. He was a talented amateur artist, and possessed considerable horticultural knowledge.

TESS CANFIELD

[John G. James Collection, Archives of the ICE; Obituary, John Sylvester (1852) *Min Proc ICE*, **12**, 165–167; J. Hix (1974) *The Glass House*]

SYMINGTON, William (1763/1764–1831), engineer and inventor, was born at Leadhills in the southern uplands of Scotland in October 1763 (according to a tablet placed near his grave in London in 1903) or in 1764 (according to the obelisk at Leadhills erected in 1891). Information on his parents is equally uncertain, but there is a reference in the biography written in 1862 by his son-in-law, J. Rankine, to William's father having been 'a practical mechanic and the superintendent of the mining company at Leadhills'; on his mother we have no information. It is known that his schooling was intended to take him to a university to study for the ministry, but he himself wrote that he was more attracted to engineering and in his teens he was working at the nearby Wanlockhead mine, where his brother, George, was the engineer under the manager, Gilbert Meason.

Drainage of the Wanlockhead workings had become a serious problem and Meason believed that steam-engined pumps would be more effective than the water-wheel pumps then in use. He met James Watt (q.v.) in Edinburgh in August 1776 and ordered from him an engine to pump from 600 ft. below the drainage adit. As no one at either of the mines had any experience of steam engines, George Symington was sent to Torryburn on the north side of the Firth of Forth where Boulton & Watt's erector, Logan Henderson, was building an engine for Andrew Colville.

By the time the parts for the Wanlockhead engine had been delivered in 1778 Meason decided that George had acquired sufficient experience to undertake its erection on his own. Watt, who at that time was short of skilled erectors, reluctantly agreed, and though there is no direct evidence it is very likely that, at about fifteen years of age, the mechanically gifted William Symington would have assisted his older brother. In March 1779 the engine was nearly complete and Watt sent William Murdock (q.v.) to supervise the largely trial-and-error process of commissioning; two months later the engine was working well, and more valuable lessons had been learned by the Symingtons.

Over the next few years Meason found that the lead veins in the mine were going deeper than expected and he had to order a larger engine which was working by 1787, again with the assistance of one of Watt's erectors. At the same time, Meason was apparently so impressed by William Symington's abilities that he arranged for him to attend Edinburgh University for one session of studies under such eminent Professors as Joseph Black and John Robison. This attendance at the university in 1786–1787 was afterwards regarded by Symington as 'a course of education and study' that entitled him to describe himself as a Civil Engineer.

His varied career in fact included not only engine-building but also, between 1794 and 1807, the management of coal mines at Kinnaird on the Bruce estate, at Grange near Bo'ness and at Callendar near Falkirk, as well as some small commissions as a consulting engineer. From 1795 until 1802 he lived in a house on the Kinnaird estate. Not all of his business ventures were successful and, especially after the collapse in 1803 of his plans for steam tugs on the Forth and Clyde and other canals, he seems to have become rather embittered and unreasonably litigious.

In the last quarter of the eighteenth century, when Watt's extended patent on the separate condenser was still in force, there were many inventive engineers trying to improve the steam engine without too obviously infringing his patent. From the time of his initial introduction to Watt's engine in 1779 Symington determined to do just that, but his first essay in utilising the power of steam was the construction of a model steam carriage, perhaps inspired by hearing of Murdock's experiments in Cornwall. No accurate drawing or description of his model, completed in 1786, has survived but it is known to have had a non-condensing steam engine with the cylinder placed horizontally, connected to the wheels by a rack and ratchet device. The model was taken to Edinburgh by Meason and successfully demonstrated on a number of occasions in his house in St. Andrew's Square. Symington's attempts to raise funds to build a full-scale steam carriage failed but in the meantime he had devised a scheme for the improvement of Watt's engine.

More of a simplification than an improvement, perhaps, it replaced Watt's entirely separate condenser with a condensing chamber in an extension of the working cylinder itself, thereby avoiding some of Watt's quite complicated pipework, pumps and valves. It was still an atmospheric steam engine without the possibility of further development but within its limits it was a practical success and Symington took out a patent in June 1787, the cost of which was met by Meason. More than twenty-five such engines were built by Symington between 1789 and 1808, almost all of them in Scotland. The cylinders and other finished parts were supplied by the Carron Ironworks, at Falkirk, and the majority of the engines were rotative, using the system of racks and ratchets illustrated in his patent.

Symington enjoyed a close and long-lasting relationship with Carron and there he met his wife, Elizabeth Benson, whose father was employed in the ironworks. They were married in about 1790 and a daughter, Elizabeth, was born the following year, and sons William and Andrew in about 1802 and 1803. A second daughter, Isabella Margaret, whose date of birth is not known, was married in London in 1816 to Dr. Robert Bowie; Symington and his wife lived with them in London from 1829 until his death in 1831. The Bowies emigrated to Australia in 1853 and were joined by William Jr. in 1855 and another William in 1862.

He continued to build atmospheric beam engines until 1808 but in 1801 he took out another patent which included a drawing of a double-acting horizontal cylinder steam engine connected directly to a flywheel. In principle it was a great advance on the beam engine but it would be a quarter of a century before materials and manufacturing processes allowed such engines to be made in large numbers. Symington himself, however, designed an engine of this type for the *Charlotte Dundas II* in 1803.

One man who had seen Symington's model in Meason's house was Patrick Miller, an Edinburgh banker who had become interested in the use of paddle-wheels as auxiliary power for sailing vessels. In 1786–1787 he built what would now be called a trimaran, and later a catamaran, both with paddle-wheels between the hulls that could be raised out of the water when not in use. Power was provided by teams of men turning cranks or capstans and, at the very least, it could be said that it was primitive, but it worked. He was assisted in this work by James Taylor, who had been at Edinburgh University at the same time as Symington, and was then appointed tutor to Miller's younger sons. Both Taylor and Miller

seem to have considered the possibility of applying the power of a steam-engine to the paddles, and after some hesitation Miller commissioned Symington at the end of 1787 to build a small engine that could be fitted to a boat he already had on Dalswinton Loch, on his estate in Dumfriesshire.

The engine built by Symington at Wanlockhead was indeed small, two vertical cylinders of 4 in. diameter and 18-in. stroke, estimated to have developed about one-half horse-power. Installed in the 25 ft. twin-hulled boat at Dalswinton, it was demonstrated in October 1788 and was said to have driven it at a speed of five knots, though, if true, the engine must have been assisted by a very favourable following wind; the partially reconstructed Dalswinton engine is now in the Science Museum, London. Although the power of the engine was inadequate, the most important outcome of the trial was that it showed the real possibility of using steam engines to turn paddlewheels and propel ships.

A few months after these trials Miller decided that a more powerful engine should be fitted to a larger boat so that it would be able to tow vessels along the Forth and Clyde canal. This time Symington used two cylinders of 18 in. diameter with a stroke of 36 in; the engine developed about ten horse-power. The boat was Miller's own 60 ft. catamaran with two paddle-wheels in tandem, previously tested using man-power. It was of very light construction, hardly suitable for an engine of that size. The first trials at the end of 1789 caused a sensation among the spectators but were certainly less successful in the eyes of Miller and some of his friends who were on board: the speed was disappointing and when efforts were made to increase it some of the paddle blades broke off. A few days later after the paddles had been strengthened and the gearing modified the boat performed well, reaching a speed calculated as 6¼ miles an hour, but by that time Miller and his friends had long since departed. Neither he nor Symington gave much thought to marine steam engines until more than ten years later.

Then in 1800, at the instigation of Lord Dundas, Governor of the Forth and Clyde Navigation Company and an old friend of Gilbert Meason's, Symington was asked to supply a vertical cylinder rotative engine for a small boat intended for use as a tug to assist sailing ships and tow barges through the Forth and Clyde canal. The *Charlotte Dundas* was launched in March 1801 and steamed with some success on the rivers Carron and Forth but it did not have sufficient power to tow other vessels on the canal and there was in any case a good deal of prejudice at that time against the use of steamships on canals for fear of the damage they would cause to the banks.

The enthusiasm of Lord Dundas was not shared by the other members of the Scottish canal committee and they voted in January 1802 to discontinue steam-boat experiments; by that time Symington had already begun to build another boat on which he could try out the horizontal single-cylinder engine outlined in his 1801 patent. Construction of this boat and its engine went ahead at Grangemouth and Carron in 1801 and 1802 and at the same time, at Lord Dundas's request, a model of the boat and its engine was made and taken by Symington to London, where it impressed the Duke of Bridgewater so much that he ordered eight tugboats for his English canals. Symington must have been quietly confident of success and an end to his financial worries, at last.

The *Charlotte Dundas II* made its maiden voyage through the canal to Glasgow in January 1803 and there, before a large crowd of spectators at Port Dundas, it successfully towed a sloop, but at a disappointingly slow speed. Neither drawings nor the original model have survived but it is reasonably certain that the engine consisted of a double-acting horizontal cylinder of 22 in. diameter and 4 ft. stroke, with a separate condenser, developing about ten horse-power. The connecting-rod acted through gearing on a single paddle-wheel inset at the stern. Again the ratio of the gears had to be altered and when that had been done the boat took two sloops of 70 tons burden in tow and travelled almost 20 miles along the canal in 6 hours in the face of a stiff breeze. The members of the canal committee were not impressed, however, and to make matters worse he learned that the Duke of Bridgewater had died suddenly, and his tugboat order had been cancelled.

Symington had designed and built what was without doubt the most compact and efficient marine steam engine up to that time and the *Charlotte Dundas II* had proved to be the first really practical steamship. Over the next few years he had to watch his pioneering experiments lead directly to the acclaimed success of Robert Fulton's (q.v.) *Clermont* (1807) and Henry Bell's (q.v.) *Comet* (1812); both men are said to have examined *Charlotte Dundas I* and *II* after they had been laid up in the Forth and Clyde canal. Symington could get no support for any continuation of his work and in fact over the next sixty years, both before and after his death in 1831, a succession of articles and pamphlets, some written by Taylor and members of Miller's family, attempted to belittle his part in the development of the steamship. A Parliamentary Report on steamships in 1822 mentioned the work of Miller and Symington but gave most of the credit to Bell.

With the support of Sir Ronald Ferguson, the MP for Kirkcaldy, and Sir John Clerk he presented a petition to the Lords of the Treasury in 1825 for recognition and reward as 'the first individual who ever effectively applied the power of the steam engine to the propelling of vessels'. Impoverished as he was by that time, he no doubt hoped for a modest pension, but all he received were two *ex gratia* payments of £100 and £50.

Symington seems to have had little or no remunerative work after 1808. He is known to have designed a lifting engine for Gilbert Meason's son at Wanlockhead in 1819 but there is no record of its

having been built. There is no record of any further work as a colliery manager, only a few drawings of devices such as a water clock and a diving bell, an improved paddle wheel and a new type of blast furnace, none of which so far as is known were ever built. Symington and his wife moved to London in 1829, where they lived in Wapping with their daughter, Isabella Margaret, and her husband, Dr. Robert Bowie. By that time he was suffering from a serious illness and died in March 1831 and was buried on the 22nd of that month at St. Botolph's, Aldgate.

R. M. BIRSE

[J. S. Russell (1841) *On Steam and Steam Navigation*; J. Rankine and W. H. Rankine (1862) *Biography of William Symington, Civil Engineer, Inventor of Steam Locomotion by Sea and Land ...*; DNB (1898) *Symington, William (1763–1831)*; H. P. Spratt (1958) *The Birth of the Steamboat*; W. S. Harvey and G. Downs-Rose (1980) *William Symington, Inventor and Engine Builder*; W. T. Johnston (1998) *Scottish Engineers and Shipbuilders* (computer database)]

Publications

1787. Patent No. 1610, 5 June: *A Steam Engine, on Principles entirely New* (Specified his methods of (a) condensing steam in the lower part of the working cylinder, and (b) producing rotatory motion by a system of racks and ratchets)

1801. Patent No. 2544, 14 October: *A New Mode of Constructing Steam Engines, and Applying their Power to ... Rotatory and other Motions, without the Interposition of a Lever or Beam* (Specified his method of transmitting power directly from the working cylinder to a rotating flywheel etc using a piston-rod and connecting-rod)

1829. *A Brief History of Steam Navigation*, repr. 1863 (The published version of his Petition to the Treasury in 1825)

T

TALBOT, Richard (fl. 1611), is noted in the Titchfield parish register for 23 June 1611 wherein it is recorded that 'Titchfield Haven was shut out by one Richard Talbottes' industry under God's permission at the cost of the Right Honourable the Earle of Southampton' (i.e. Henry Wriotheseley, 3rd Earl). This work refers to the River Meon Canal which runs adjacent to the river for a distance of some 3 miles south of Titchfield to the mouth of the Meon estuary. Its construction was associated with major estuary closure works completed in the early seventeenth century.

Closure was achieved by a shingle bank with two exits. A sluice controlled flow into the small Hill Head harbour to the East, the other permitted access to the canal via a sea lock. Remains of this lock are still visible but it is crossed by the Meon Road supported on twin masonry arches. The lock was 16 ft. wide and the surrounding walls indicate it was at least 36 ft. long. Soundings through the accumulated silt suggest that the lock floor was 7 ft. 6 in. below the coping level which is 3 ft. above the current water level.

The main purpose of the canal was to assist the drainage and irrigation of the estuary and to enhance the trade of Titchfield for it ended close to the corn mill and tannery and there was a bridle way leading to the market. However, it apparently did not bring great long term prosperity to the town and eighteenth century maps show several bridges across the canal which would have almost certainly prevented navigation. Certainly the sea lock was 'converted' to a bridge structure carrying the Meon Road.

Other problems had beset the works for a map dated 1753 shows that a second shingle bank had been constructed some 200 yd. in front of the original. Talbot's work is however an important early example of this type of navigation scheme.

MIKE CHRIMES AND BOB OTTER

[PHEW records, ICE archives]

Works
1611. Titchfield Haven

TALFOURD, Field (1809–1874), was christened on 22 October 1809, the younger son of Edward and Ann Talfourd. His father was a brewer in Reading. When living at 2 Derby Street, Westminster, Talfourd was elected an Associate Member of the Institution of Civil Engineers in 1827, transferring to full membership in 1837. He was proposed as a member by Thomas Telford (q.v.). In the late 1830s he was working on the Birmingham and Gloucester Railway, but perhaps availing himself of his skills as a draughtsman, became a portrait and landscape painter. He exhibited thirty-four works at the Royal Academy, 1845–1873. He died on 5 March 1874.

MIKE CHRIMES

[ICE membership records; ICE archives; Archives, Royal Academy of Arts; RAIL 37/15, PRO; Boase]

TAYLOR, Richard Cowling (1789–1851), mineral surveyor, was born on 18 January 1789, the third son of Samuel Taylor, a farmer in East Anglia. After attending school in Halesworth he became a pupil of Edward Webb, a land surveyor at Stow-on-the-Wold in July 1805. Following further training with William Smith (q.v.) he set up as a land surveyor in Norwich in 1813, operating as Browne and Taylor (1814–1824). In October 1826 he moved to 7 Wilmington Square, in London, and while there joined the Institution of Civil Engineers as an Associate in 1828.

He assisted in the Ordnance Survey of England, but increasingly became involved professionally with mining projects. His plaster model of the South Wales coalfield north of Pontypool, chiefly the property of the British Iron Company, was awarded the Isis medal of the Society of Arts in 1829. In July 1830 he emigrated to the United States of America where he carried out a number of mineral surveys in Pennsylvania and the ironfields. These included an early proposal for a railroad between Blossburg and Laurenceville (1833). In addition to his surveys in the United States he worked in Cuba (1836) and Canada.

Taylor's interests were not confined to geology and his writings and investigations encompassed natural history and antiquarian research, notably contributions on the monasteries of East Anglia (1821). He was a member of a number of learned societies in addition to the Institution of Civil Engineers, notably the Geological Society and American Philosophical Society, and is credited with thirty-eight scientific papers, published in British and American learned journals.

Taylor married Emily Errington, from Great Yarmouth in 1820, and had four daughters. He died in Philadelphia on 26 October 1851.

MIKE CHRIMES

[Membership records, ICE; DNB; Skempton; Bendall; *Royal Soc Cat Sci Papers*; Geological model (1829) *Trans Soc. Arts*, **xlviii**, No. 1, 45–46]

Publications
1823. Geological section of Hunstanton Cliff, *Phil Mag*, **61**, 81–83
1824. Remarks on the position of the upper marine formation exhibited in the cliffs on the north east coast of Norfolk, *Phil Mag*, **63**, 81–85
1825. Plan of the proposed navigation for ships, from Norwich to … Lowestoft (for W. Cubitt)
1827. On the geology of East Norfolk, *Phil Mag*, **1**, 277–290, 346–353, 426–432
1827. On the natural embankments formed against the German ocean on the Norfolk and Suffolk coast, and the silting up of some of its estuaries, *Phil Mag*, **2**, 295–304
1827. On the geological features of the eastern coast of England, *Phil Mag*, **2**, 327–331
1829. Progress of geology, *Mag Nat Hist*, **1**, 442–452
1830. Introduction to geology, *Mag Nat Hist*, **3**, 62–78
[1830]. Notice of two models and sections … of South Wales in the vicinity of Pontypool, *Geol Soc Trans*, **3**, 1835, 433–436
1833. Report on the surveys … of a railroad, from the coal and iron mines near Blossburg … to … Lawrenceville, and mineralogical report on the coal regions in the environs of Blossburg
1835. On the relative position of the transition and secondary coal formations in Pennsylvania …, *Trans Pennsylvania Geol Soc*, **1**, 179–193
1835. Memoir of a section passing through the bituminous coalfield near Richmond, in Virginia, *Trans Pennsylvania Geol. Soc*, **1**, 314–325
[1836]. Notice of a vein of bituminous coal, recently explored in the vicinity of Havana, *Phil Mag*, 1837, 161–167
1837. On the geology of Cuba, *Phil Mag*, **11**, 17–33

1841. Notice of a model of the Schuylkill … coal-field …, *Silliman J*, **xli**, 80–91

[1843]. On the geology of the north-east part of the island of Cuba, *Am Phil Soc Trans*, **ix**, 1846, 204–218

1845. On the authoritative and bituminous coal-fields in China, *Franklin Inst J*, **x**, 51–57

1848. *Statistics of Coal*

[1849]. *Table … in Regard to the Daily Temperature … Panama, Port Royal in Jamaica, and on the Return Voyage to New York … October 1849*

[1849]. Substance of notes made during a geological reconnaissance in the auriferous porphyry region next to the Caribbean Sea … Panama, *Philadelphia Acad Nat Sci J*, **2**, 1850–1854

[1854]. On a view of asphaltum at Hillsborough … New Brunswick…, *Am Phil Soc Proc*, **5**, 241–243

TELFORD, John (*c*. 1771–d. 1807), civil engineer, probably a kinsman of Thomas Telford (q.v.), worked for him in Shropshire in the 1790s. He acted as his draughtsman there, preparing plans for St. Mary Magdalene Church, Brignorth (*c*. 1792). In June 1797 he was appointed Toll Collector on the Wirral line of the Ellesmere Canal. He displayed sufficient competence and honesty to be recommended by Telford to William Jessop (q.v.) in June 1804, to superintend work at the Corpach (western) end of the Caledonian Canal 'as [has] to my own and your knowledge, for ten years past, been employed upon works of a similar nature whose abilities may be relied on and who [is] likely to enter with zeal into the spirit of the undertaking'. In August Jessop wrote 'Instructions … to Mr. John Telford for marking out the Caledonian Canal', and provided a number of drawings, some of which John Telford then copied into his notebook. This suggests John was relatively inexperienced in canal construction. He was perhaps fortunate that the massive sea lock for which he was responsible was founded on rock, rather than the more treacherous ground of the eastern end where Matthew Davidson (q.v.) was in charge. That said, much pumping was necessary to keep the excavation dry.

Telford remained in charge of the work at Corpach until his death in April 1807, aged 36. Much of his correspondence survives, revealing that management, particularly the availability of pay for the workforce, was as much an issue as the engineering of the canal. He was succeeded as Resident Engineer at Corpach by Alexander Easton (q.v.).

MIKE CHRIMES

[Telford collection, ICE archives; John Telford letterbook, transcript, Gibb collection, ICE archives; Commissioners for the Caledonian Canal (1804–1808) *Reports*; A. Gibb (1935) *The Story of Telford*; L. T. C. Rolt (1958) *Thomas Telford*; A. Penfold (ed.) (1980) *Thomas Telford: Engineer*; C. Hadfield (1993) *Thomas Telford's Temptation*]

Works

1804–1807. Caledonian Canal, western end, superintendent of works

TELFORD, Thomas, FRS, FRSE (1757–1834), first President of the Institution of Civil Engineers, second son, the first of the same name having died in infancy, of John Telford, an Eskdale shepherd, was born at Glendinning sheep-farm, Westerkirk, Dumfriesshire on 9 August 1757. Four months later his father died and Telford was brought up by his mother Janet (*née* Jackson, d. 1794). The closely knit Eskdale community, in particular his mother's brother Thomas Jackson, believed to have been factor to Sir James Johnstone of Westerhall, helped to support the family. He gained a good basic education at Westerkirk parish school and obtained occasional farm work whenever he could to help support his mother and himself. At school he met the younger generation of leading local families, and formed a close friendship with Andrew Little, schoolmaster in Langholm and later the postmaster there, James Little, his correspondence with whom, now in the ICE archives, represents the main source of information on his early life and his life-long connection with Eskdale.

On leaving school *c*. 1772 Telford was at first apprenticed to a stone-mason at Lochmaben, from whom he is understood to have run away after being ill-treated, and then to Andrew Thomson at Langholm working on the simple buildings of the area. Langholm Bridge built *c*. 1778 bears mason marks that are said to be his and he is reputed to have carved both the Pasley family memorial and the headstone to his father's grave in Westerkirk churchyard. From a recent study of mason marks in the locality the question has been posed that Telford's mark as illustrated in the earliest known published reference (Smiles, 302), might be incorrect and this subject requires further investigation. Whenever he could Telford diligently gleaned knowledge from any books he could find, for example, on literature and poetry from the elderly Miss Pasley of Craig who befriended him. In 1780 having mastered such mason-work as Eskdale could furnish he went to Edinburgh to improve his prospects and is reputed to have worked on its 'New Town'. Whilst at Edinburgh he learned to draw and studied the architecture of the locality. He particularly admired the Gothic splendour of Melrose Abbey and Rosslyn Chapel, a style which later influenced the elevations of much of his own work when it could be afforded.

In February 1782 Telford's restless ambition drove him to seek more challenging and better paid work in London where, through John Pasley an eminent merchant and relative of Miss Pasley, he met the architects Robert Adam (q.v.) and Sir William Chambers and obtained employment as a stonemason on the building of Somerset House. The following year he seriously considered, but eventually decided against, entering into business with a fellow stonemason, Mr. Hatton, to contract for work at Somerset House. Whilst in London

Thomas Telford FRS, FRSE

Telford was consulted by Sir James Johnstone of Westerhall about alterations to his house in Eskdale, possibly the present main entrance, and received his instructions from Sir James's brother William Pulteney (1729–1805). Pulteney, who had changed his name from Johnstone on marrying the heiress of the Earl of Bath, was impressed by Telford and his work and employed him on the restoration of Sudborough Rectory, Northamptonshire in 1783–1784. Other commissions followed and within a decade a close friendship had developed between them to the extent that Telford was known as 'young Pulteney'. His subsequent early career owed much to Pulteney's powerful patronage.

In 1784 the funding for building Somerset House became exhausted and Telford obtained employment in Portsmouth working on the Dockyard Commissioner's house and chapel designed by Samuel Wyatt (q.v.). Before long he was superintending the contract for this work, his first important position of independence and responsibility. Whilst at Portsmouth he widened his knowledge and experience by observing harbour and dock work under construction and by studying limes and mortars from copies of the lectures of Black and Fourcroy. He wrote down the information gleaned in his pocket-book compilation of useful data which became his vade-mecum, some of which was later published in text books and in his *Life*. Telford was by now a Freemason and on 1st February 1786 was about to direct the fitting up of a masonic lodge room to his plans at the George Inn, now demolished. On completion of the

dockyard buildings, later in 1786, Telford went to Shropshire at Pulteney's invitation, to undertake the restoration of Shrewsbury Castle for his use as an occasional residence in connection with his duties as an MP for Shrewsbury. In July 1787, with a reference from Robert Adam, Telford became clerk of works for the new county gaol at a salary of £60 p.a. Soon afterwards he became County Surveyor of Public Works, a post that he held for life, later through his able deputy Thomas Stanton (q.v.) at Ellesmere, with responsibility for public buildings and the provision of at least forty-two bridges.

During and after restoring Shrewsbury Castle in the Gothic style Telford lived in and practised as an architect from the castle. His church work included restoration of St. Mary's, Shrewsbury and All Saints', Baschurch, building the new churches of St. Mary Magdalen, Bridgnorth, which Pevsner (1958) calls 'a remarkable design, of great gravity inside and out, and apparently done in full awareness of recent developments in France', St. Michael's, Madeley and, almost certainly, the planning of St. Leonards', Malinslee. From 1787 to 1793 his other work included the county infirmary, private houses, and street improvement and drainage. In 1788 at Pulteney's request Telford advised on St. Chad's, Shrewsbury, accurately predicting its fall just before the actual event. He also superintended the excavation of the ruins of the Roman city of Uriconium, on Pulteney's estate near Wroxeter, the plan and sections for which, in *Archaeologia* 1789, are his earliest-known published drawings. In 1793 he added greatly to his knowledge of architecture and antiquities from a study tour of Bath, Oxford, London and other cities. The considerable extent of his architectural knowledge by the early nineteenth century is indicated by the content of his substantial 'civil architecture' article in the *Edinburgh Encyclopaedia*.

In 1790, at Pulteney's instigation as a director of the British Fisheries Society, Telford's life-long connection with the Society began. He advised on the improvement of numerous harbours and settlements in northern Scotland including Lochbay, Tobermory, Ullapool, Keiss, Staxigo, Broad Haven, Wick, Sarclet, Clyth, Lybster, Forss, Dunbeath, Helmsdale, Brora and Portmahomack. The largest, Pulteneytown, at Wick, executed to his designs over several decades, with its impressive Argyll Square, still survives as a fine testimonial to his architectural and planning skills. In 1796 Telford tested and soon afterwards used at Lochbay pier on Skye, an aluminous hydraulic cement patented by James Parker, later known as 'Roman cement', which set to a very considerable extent in about twenty minutes. His support for and extensive use of Roman cement to inhibit water penetration influenced its nationwide adoption for many years in facing, pointing and brick-jointing mortars.

Telford's honorary work for the British Fisheries Society led to his involvement in governmental surveys of the Highlands in 1801–1802 and to his wide-ranging recommendations for improvement of inland and maritime communications including harbours, partly with a view to stemming emigration, and to facilitate naval protection for the fisheries. His reports paved the way for the setting up in 1803 of Commissions for making the Caledonian Canal and many Highland Roads and Bridges, particularly north and west of the Great Glen. Recognition of Telford's outstanding contribution to the development of the Highlands was reflected in his election as a Fellow of the Royal Society of Edinburgh in 1803 when, in his own words, 'I had the honour of being proposed by three professors' (Little Mss, ICE). In 1834 the Society made Telford a present of inscribed silverware 'in grateful acknowledgement of the numerous and valuable professional services gratuitously rendered during a long course of years' (Dunlop, 59).

Canals

Telford's civil engineering career developed from 1793 on his appointment as 'General Agent, Surveyor, Engineer, Architect and Overlooker' to the important 68 mile Ellesmere Canal, intended to join the rivers Mersey, Dee and Severn. This canal, now a thriving leisure facility, still utilises many buildings and structures designed and built under Telford's direction. The most remarkable structure is Pontcysyllte cast iron aqueduct over the Dee, a development of his embryo sketch design dated March 1794, except for the piers, but not developed until after the novel iron trough concept had been proved operationally at Longdon-on-Tern aqueduct on the Shrewsbury Canal, in 1795–1796. At Pontcysyllte, with the support and approval of William Jessop (q.v.) the principal engineer for the canal, Telford deviated from traditional masonry aqueduct construction by building abutments and eighteen slender masonry piers joined by nineteen arches with cast iron ribs supporting a iron trough with 1 in. thick sides 1007 ft. long and 126 ft. above the river. The ironwork was made and erected by William Hazledine (q.v.), the masonry was built by John Simpson (q.v.), and the whole supervised by Matthew Davidson (q.v.). The result was the supreme structural achievement of the canal age. Sir Walter Scott thought it 'the most impressive work of art he had ever seen'. A misleading attempt by Hadfield in 1993 to question the traditional attribution of the concept and design of the aqueduct to Telford is incompatible with authoritative early evidence that it was designed and executed under his direction.

The 60 mile Caledonian Canal constructed from 1804 to 1822 was engineered by Telford and Jessop jointly until 1812, afterwards solely by Telford, basically with the same team that built the Ellesmere Canal aqueducts. Matthew Davidson superintended work at the eastern end and John Telford (q.v.), succeeded by Alexander Easton (q.v.), at the western end. John Simpson was the main masonry contractor with John Wilson and John Cargill (qq.v.) working as his partners. In

engineering terms the 100 ft. wide ship canal with its twenty-eight immense locks and deep summit cutting at Laggan, was then the most advanced of its kind in the world. Interesting features included the use of railways, comprising plateways formed with 3 ft. long Jessop pattern cast-iron rails, to facilitate excavation and banking, purpose designed and built machinery and equipment, steam engines for pumping and dredging, and ingenious and unprecedented lock construction. For example, at the Beauly Firth (Clachnaharry) entrance lock a 55 ft. depth of mud was pre-consolidated by the weight of imported clay fill loaded with stone from which, after removal of the stone, the lock-pit was excavated.

Despite the hard-won achievement of the Caledonian Canal and the much-needed work that it provided in the Highlands, (1385 men were employed in 1811), the project was less successful in other respects. Costs escalated with high inflation and unforeseen difficulties, additional funding was difficult to obtain, and further problems arose through defective workmanship, for example at Banavie and Fort Augustus locks, where ground water penetrated through the walls. Penfold attributed the defective workmanship to a lack of close site supervision arising from a management structure that favoured the contractors. The canal eventually opened in 1822 with an operational depth of 12 ft. instead of 20 ft. as planned at the outset. It had cost about twice the original estimate and taken eighteen instead of seven years to construct. By then, through no fault of Telford's, the reasons for creating the canal had been overtaken by events with the end of the Napoleonic wars and the development of the steamboat. Although important locally, relative to its capacity the canal has never been much used, except in 1918 when there were 6,254 passages associated with mine-laying in the North Sea. It is now a major tourist attraction and more traffic is expected on completion of the Forth & Clyde Canal regeneration in 2001.

In Sweden the Trollhätte Canal, comprising the western end of the Gotha Canal, had been completed in 1800 under the direction of its promoter Count von Platen and the Swedish engineer Samuel Bagge. From 1808 Telford, at the invitation of the King of Sweden, acted as consulting engineer for the canal's 114 mile eastwards extension from Lake Vänern to the Baltic at Söderköping, with at first Bagge and later Lagerheim, superintending operations on site. In 1808 Telford, with his assistants William Hughes and Hamilton Fulton (qq.v.), met and surveyed the line with von Platen. It was the start of a close friendship between Telford and von Platen which lasted until the latter's death in 1829. Construction commenced in 1809 and four years later seven thousand men were employed including John Wilson (q.v.) and James Simpson who had accompanied Telford on a visit in 1813 but did not return with him. There were delays in constructing the canal, which was not completed until 1832. Telford's advice was transmitted to Von Platen in a voluminous correspondence. In 1809, Telford was made a Knight of the Swedish Royal Order of Vasa in recognition of his valuable services, his letters from Sweden afterwards being addressed to 'Sir Thomas Telford'. His international reputation on canal construction was now such that he was also consulted by the Russian government on canal navigation schemes, as well as schemes such as the Darien and Welland Canals in central and British North America.

Telford can be considered the last of the great canal engineers of the Industrial Revolution. Of his later projects, Harecastle Tunnel more than 2,700 m long and constructed, in exact accordance with his plans under the supervision of the resident engineer, James Potter (q.v.), in less than three years from fifteen shafts, was one of the most remarkable feats in tunnelling history. The Birmingham Canal improvement engineered to Telford's characteristically direct line, and with a particularly deep cutting at Smethwick, saved 8 miles in length and thus maximised user-benefits. This canal was one of the finest achievements of the canal age. Similarly, the Birmingham & Liverpool Junction Canal, with a 12 mile saving in length, but requiring long cuttings up to 90 ft. deep, in what proved to be slip susceptible marl, and a diversion at Shelmore to avoid Lord Anson's game preserves involving a mile-long embankment up to 60 ft. high. Both features presented great problems as Telford's health declined but they were eventually overcome by the contractor, William Provis (q.v.), under William Cubitt's (q.v.) direction from 1833 to 1835. Telford's canal–seaport warehouse interchange at Ellesmere Port on the River Mersey that was greatly used for over a century represented a peak of efficiency of the canal age.

Telford acted as engineer on a number of other canals, beginning with the completion of the Shrewsbury Canal, designed by Josiah Clowes (q.v.) before his death. Others included the Glasgow, Paisley and Ardrossan Canal, not completed because of financial problems, the Weston Canal for the Weaver Navigation, the repair and upgrading of the Crinan Canal and work on the Trent and Mersey Canal and the completion of the Gloucester–Berkeley Ship Canal. Elsewhere he was regularly consulted about the Grand Junction (1805–1818), Carlisle (1808–1819), Bude (1818–1823), Edinburgh and Glasgow Union (1816–1822), and Ulster Canals (1826–1834). Telford's work for the Exchequer Bill Loan Commissioners led to further consultations (see below). He also reported on the Leicester and Northamptonshire (1803), Grand Union (1803), Aberdeenshire (1806), Forth and Clyde (1807), and Oxford (1830–1834) canals. His involvement with the Macclesfield Canal was largely confined to the Parliamentary survey; he was also responsible for surveying schemes which came to nothing such as the Stamford Junction to Oakham (1810), the English and Bristol Channel Ship Canal (1824–1826), the London–Birmingham (1828) and Dee and Mersey Ship Canal across the Wirral (1828).

Telford also worked on river navigation improvements on the Dee (Chester), Severn, Weaver, Thames, Mersey, Avon, Aire and Calder, Clyde, Dee (Aberdeen), Tay and Forth. He also investigated the development of fast canal boats and in 1832–1833 hundreds of experiments were made at the Adelaide Gallery, London by his chief assistant John Macneill (q.v.) in an unsuccessful attempt on behalf of canal interests to compete with railway travel. Further, full-scale, experiments were carried out on canals, and continued by Macneill after Telford's death.

Road making

Telford's main achievements in road-making were the London to Holyhead and Bangor to Chester roads as Engineer to the Holyhead Road Commissioners from 1815 to c. 1830, and the Glasgow to Carlisle, Lanarkshire, and Highlands of Scotland roads as Engineer to the Highland Roads Commissioners from 1803 to c. 1830. Although many of these roads declined in use from the 1840s as the railway network advanced, they entered into a new lease of life in the early twentieth century with the advent of the motor vehicle. Abroad Telford advised the Russian government on the 100 mile Warsaw–Brzesc road completed in 1825 on the route to Moscow. He also advised on the provision of a causeway at Bombay (Salsotte). Unexecuted or partially completed improvements on roads that he surveyed for the government or others included, the Carlisle to Portpatrick, Birmingham to Liverpool, Carlisle to Edinburgh, London to Milford Haven and South Wales, and the Great North Road from London via York and Edinburgh to Inverness.

In the Highlands, with the valuable assistance of John Rickman, James Hope and John Mitchell, respectively, Secretary, Agent and Chief Inspector to the Highland Road Commissioners, Telford was responsible for the provision of about 1200 miles of new or improved roads. These roads, authoritatively described in parliamentary publications and by Haldane, opened up Scotland west and north of the Great Glen, in Telford's own words, 'advancing the country at least a century' (Smiles, 389). His connection with the Holyhead and Scottish roads continued through inspections for most of the rest of his life. In terms of construction his major roads were commodious, well drained and incorporated a hand-pitched stone foundation beneath a layer of conventional road metal. Unlike McAdam's roads, they were properly engineered to improved lines and gentle gradients, and although more expensive initially, facilitated traction and reduced maintenance costs. On its completion the Holyhead Road was described as 'a model of the most perfect road making that has ever been attempted in any country' (Parnell, 35). Much of the road is still in use and in 1998 nearly a mile of the original retaining wall up to 40 ft. high and Pont Pen y

Benglog Bridge in the Nant Ffrancon Pass were economically and tastefully conserved by the Welsh Office. The road in North Wales is now designated an 'Historic Route' with information signing to encourage environmental and leisure use, which is entirely fitting for this 'long-lasting memorial to Telford's skill and vision' (Penfold, 58).

Masonry bridges

Throughout his lifetime Telford planned, built or advised on several thousand masonry bridges, including 1,100 to a standard specification on Highland roads alone. His bridges ranged from simple culverts to the sophisticated 150 ft. elliptical span of Over Bridge, Gloucester 1826–1830, influenced by Perronet's work. Telford's first major bridge was erected over the Severn at Montford from 1790 to 1792 using convict labour. Six years later it was followed by Bewdley Bridge, which, with its segmental arches and classical balustrades in a gentle arc, has been described as 'one of the most elegant bridges in England' (Ruddock, 154). Telford's most notable Scottish bridges include Dunkeld (1805–1809), also with its extrados on a large radius arc and, his finest development of architectural experimentation and excellence of construction, Dean Bridge, Edinburgh (1829–1832) with its intricately achieved slenderness. In terms of its elevation Telford's 'elegant' five-arch bridge at Pathhead, Midlothian, with its visually less pronounced 'ascititious' (Telford's *Life)* arches to some extent 'served as the prototype' for Dean Bridge (Paxton, 1998).

Telford's bridges at Broomielaw, Glasgow (1833–1835) and Dean have been described as 'a fitting crown to his creative life' (Gibb, 26). Both bridges were constructed by John and Alexander Gibb (q.v.) under the competent supervision of resident engineer Charles Atherton (q.v.). In construction terms, from the 1790s Telford developed, and widely influenced, the beneficial adoption of hollow piers and spandrels in large-span bridges by Robert Stevenson and others. This practice resulted in a stronger structure, facilitated its internal inspection and reduced the weight bearing on foundations. Telford's architectural experience enabled him to impart grace and beauty to the appearance of many of his bridges.

Cast iron bridges

Telford's innovative design practice was also effectively applied to cast-iron bridges carrying roads. Buildwas Bridge 1796, said to have been the second major cast iron bridge to be completed in Britain, just before Sunderland Bridge, differed considerably in concept from Coalbrookdale Iron Bridge, achieving a bridge of half the weight of that at Coalbrookdale with a considerably increased span of 130 ft. Four years later Telford, in association with James Douglass, made a very bold proposal for a 600 ft. cast iron arch over the Thames to replace London Bridge. Expert opinion on its practicability was widely canvassed by a parliamentary committee and varied greatly. For example, in respect of horizontal thrust, Dr. Maskelyne thought that there would not be any, Professor Playfair calculated it at 11,100 tons and Professor Robison at 20,550 tons (Paxton (1975), 80). Although the project was seriously considered for many years it was not implemented because of 'the unprecedented scale of the project, coupled with lack of knowledge of and agreement on the technical factors involved' (Skempton, in Penfold (1980), 79).

In 1810 Telford designed an economical pre-fabricated, lozenge lattice spandrel arch based on his practical precepts for use at locations where it would be more expensive or impracticable to construct a masonry bridge. At least nine arches with standardised spans of 105 or 150 ft. were cast and erected by William Hazledine (q.v.) from 1812 to 1830 of which those designed by Telford at Craigellachie and another at Tewkesbury of larger span, are still in use although strengthened. The prototype at Bonar Bridge over Dornoch Firth erected in 1812 lasted until 1891. The similar bridge, which still spans over the Birmingham Canal at Galton was cast by Horseley Ironworks, who also supplied a number of smaller bridges over the canal. Some of these were to a similar design to that developed by Telford with his assistant Thomas Stanton (q.v.) for small spans in Shropshire. Of Telford's cast iron bridges it has been aptly written, 'No other man has ever handled cast iron with such complete assurance and understanding, his exact knowledge ... enabling him to achieve that perfection of proportion which gives strength the deceptive semblance of fragility' (Rolt, xiii).

Suspension bridges

Telford's creation on the Holyhead Road of the elegant Menai wrought iron suspension bridge over the Menai Straits, with an unprecedented span of nearly 580 ft., was the most outstanding bridge development of the early nineteenth century, and sealed his international reputation among contemporary engineers. Its final form evolved from his experimentally based proposal of 1814–1818 for Runcorn Bridge, further experimental work and an almost continuous design process from 1819 to its opening in 1826. In 1814 Telford had correctly anticipated modern practice in envisaging parallel wire main cables and had erected and load tested a 50 ft. span model of the earliest known wire bridge, but he eventually opted for flat chain-bar links as being more practicable to achieve and maintain at that time. The masonry, which is of exceptional quality, was executed by John Wilson. Hazledine manufactured the ironwork, the testing and fixing of which with newly invented equipment under the supervision of resident engineer William Provis, his brother John, and Thomas Rhodes (q.v.), was at the forefront of technology of its day. Nothing that could be quantified was neglected and, although it could not quantified, the probability of deck trussing being required to counter undulation. Nearly 36,000 bars and plates, including all those used in the bridge, were tested to about twice their design load. The ironwork, which is believed to have been painted white originally to minimise temperature movement, continued in use until 1940 when the bridge was tastefully reconditioned by Sir Alexander Gibb to accommodate modern traffic, which it still does.

Telford's experimental results on the strength of iron wire and bars were widely propagated in leading text books from 1817 and in 1828 William Provis published the definitive account of the design and construction of Menai Bridge, then the world's largest bridge span. The project led to a surge of interest in suspension bridge building and exercised a fundamental influence on practice and development by Isambard Kingdom Brunel, William Tierney Clark, James Meadows Rendel (qq.v.) and others from 1819 to 1840. It established this type of bridge 'in its true role as the most economic means of achieving the largest spans' (Paxton, in Penfold (1980), 113).

A minor shortcoming of the design of Menai Bridge was that the then unquantifiable phenomenon of wind-induced undulation proved to be a greater problem than had been foreseen. This became apparent in 1825, as the bridge was nearing completion and caused Telford great anxiety. Measures were taken which proved partially effective and it was not until 1839, five years after Telford's death, that timber shrinkage contributed to the destruction of part of the deck in a gale. For a while the bridge was widely regarded as a 'complete failure' but this was 'an over-reaction' (Paxton (*ibid.*, 1980), 112). The bridge was fully reopened a fortnight later and then underwent strengthening under Provis's direction for about 5% of its initial cost. Nevertheless, from *c*. 1840, 'Telford's great achievement was mistakenly left unappreciated and greatly undervalued' (Roebling).

Telford's other suspension bridge projects included Conwy Bridge, also opened in 1826 which still has its original ironwork and his controversial Clifton suspension bridge proposal of 1830 for a three-span bridge with two ornate 'Gothic Revival' towers rising dramatically from the floor of the gorge. In order to understand Telford's reasons for allowing his innate admiration for the Gothic style full reign, apart from the fact that this style was then much more popular that it is now, it is necessary to know something of the project's background. This crossing of the Clifton Gorge had been contemplated for many years and was considered to offer an environmental and an artistic opportunity rather than one for a little-needed national road. Apart from its architectural appeal to Telford, this proposal also reflected his understandable concern about extending the spans of suspension bridges much beyond that of Menai Bridge, at an exposed location 200 ft. above the river, until more was known about undulation. His design is said to have received general approval at the time, but it failed to attract sufficient funding. To the modern eye Telford's most visually attractive bridge elevations are those in which the Gothic Revival influence was either muted or absent.

Railways and steam-carriages

Telford believed that steam power could best be applied to land transport in the form of self-propelled vehicles operating on roads rather than on railways. He was concerned that the railway companies would establish a monopoly to the detriment of road use. He supported the setting up of and gave evidence to a parliamentary Select Committee in 1831, which reported that steam carriages were practicable, safe and should be protected from high tolls to encourage their adoption and use. By 1833 Telford was a leading promoter in a steam carriage company intended to operate on the London to Holyhead Road and took part in an experimental journey on the London to Birmingham section in Dance's steam carriage. The size of the engine proved to be insufficient and the carriage only reached Stoney Stratford 57 miles from London, at an average speed of 7 m.p.h. High tolls, opposition from vested interests, mechanical shortcomings, and Telford's death in the following year, all contributed to the demise of this initiative.

Despite this perceived anti-railway bias Telford was involved, generally as consultant with the majority of the railway schemes of the 1820s, most of which were not realised. Railway projects upon which Telford advised generally in his capacity as consultant to the Exchequer Loan Commission included the Stratford & Moreton (1821–1826), operated with horse traction; Clarence (1828–1829); Newcastle & Carlisle (1829–1834); and the Liverpool & Manchester (1827–1829), whose directors had offered him the post of engineer in 1825, an offer that he declined. He was however instrumental in persuading the company to abandon the idea of fixed engines and inclined planes in favour of a level line suitable for locomotive haulage. Important railway proposals which he planned, but which were not executed, included the Glasgow to Berwick Railway (1810), intended to be operated with horse traction and steam-powered inclined planes; the locomotive operated London to Dover (1824); the East & West India Docks (1828) and the Glasgow, Forth & Clyde Canal to Broomielaw (1829), which was mainly in tunnel.

Fen drainage

From 1818 to 1821, Telford advised on behalf of the Navigation Commission on the execution of the Eau Brink Cut designed by John Rennie (Sr.) (q.v.) which bypassed the meandering Ouse above Kings Lynn. The Cut, the width of which had been specified by Joseph Huddart (q.v.), proved to be insufficient and soon afterwards it was widened by Telford and John Rennie (Jr.) (q.v.) with most beneficial effects. Telford also worked with Rennie (Jr.) on the Nene Outfall Cut from Wisbech to Crab Hole in The Wash, executed from 1827 to 1830 for about £200,000. It was on this work, whilst visiting Crab Hole with Rennie, that he was soaked to the skin in a storm and caught a severe chill. While returning to London, Telford was taken ill at Cambridge where he was confined for a fortnight and nearly died. His health never fully recovered.

On this work Rennie (Jr.) found Telford 'a most agreeable facetious companion' (Rennie, 201). This relationship contrasts with that of his father who wrote in a letter to Robert Stevenson (q.v.) in 1806 that Telford 'has no originality of thought and has all his life built the little fame he has acquired upon the knowledge of others which he has generally assumed as his own' (NLS. Acc. 10706, 73, 55). Rennie's comment provides substance for references in a letter Telford wrote to Watt in 1805. 'I am sorry to inform you', he wrote, 'that Mr. Rennie's conduct prevents me from benefiting by his acquaintance. Altho' I never had any connection with him in business or ever intentionally did anything to injure or interfere with him I, in every quarter, hear of his treating my character with a degree of illiberality not very becoming. This is so marked a part of his conduct, that I really believe it does him a serious injury ...' (Boulton & Watt Collection).

Telford's most important achievement in the Fens, made possible by the Nene Outfall Cut, was the drainage of about 48,000 acres of the North Level. For this major work, carried out from 1830 to 1834, he was the sole engineer.

Docks, harbours and piers
Telford advised on or acted as engineer for the improvement of more than 100 harbours, docks or piers, including many in Scotland for the Highland Road Commissioners. In addition to the British Fisheries Society harbours already mentioned, these included: Aberdeen, Peterhead, Ardrossan, Glasgow, Fraserburgh, Dundee, Leith, Belfast, Holyhead, Dublin, Howth and Dunmore, St. Katharine's Dock, Leith, Greenock and Dover. In some cases, such as Aberdeen this represented the design of several works over a period of more than twenty years. Elsewhere, such as Belfast and Glasgow, there were several consultations, but little work. Work at Fraserburgh, Leith, Holyhead, Howth and Greenock generally represented the overseeing of works designed by others, notably John Rennie. At St. Katharine's Dock, on a restricted and awkwardly shaped site which had required the demolition of over 1,250 houses and the excavation of 27 acres, Telford designed an entrance lock giving access to a basin interconnecting two irregular-shaped docks. This arrangement enabled each dock to be cleaned out separately without interrupting shipping operations. The loss of water through lockage was compensated for by an ingenious arrangement of pumps which delivered water from the river either into the lock or basin as required. This work, which was then the most advanced of its kind, was executed under the diligent supervision of Peter

Logan, John Hall (q.v.) and Thomas Rhodes as resident engineers.

Water supply
Telford advised on several water supply schemes, one of the earliest, from 1799 to 1802, being a piped supply to Liverpool pumped by steam engines from springs at Bootle. In 1806 he was appointed engineer of the Glasgow waterworks and, in association with James Watt (q.v.), completely reorganised its defective and impure supply. An innovative feature was a cast iron main with flexible joints specially invented by Watt to enable the water to cross the Clyde. From 1810 to 1822 Telford was consulted on Edinburgh water supply and in association with local engineer James Jardine (q.v.) was involved in the construction of Glencorse reservoir, with what was then one of the tallest earth dams in Britain. With characteristic attention to detail, every iron pipe in the main leading into the city was proved at a pressure equal to that of a column of water from 300 to 800 ft. high and they are still in service. *The Scotsman* in 1825 considered these works 'the most extensive, perfect and complete ever executed in modern times' (IX No. 603). From 1827 to 1834 Telford, with assistants John Macneill and James Mills (q.v.), was engaged on his largest water supply project, to supply London much-needed pure water. In 1834, Telford proposed to bring in water from the Verulam near Watford and the Wandle at Beddington at an estimated cost of £1.177 million. Although not implemented, according to Smiles, these proposals strongly stimulated the water companies and eventually led to great improvements.

Highland churches
From 1823 to 1834 Telford superintended the design and provision of Highland churches and manses at forty-three sites from Islay northwards to the Shetland Islands. As superintending surveyor to the Highland Churches Commission, he prepared plans, specifications and estimates for standardised structures based on the proposals of his three surveyors, of which those of William Thomson (q.v.) for the churches were closest to the form finally adopted. The basically austere structures, most of which still exist, are often enlivened by a touch of Telford's architectural artistry. He estimated that the churches were capable of containing about 22,000 persons without inconvenience.

Exchequer Bill Loan Commissioners
Telford's appointment as Engineer to the Holyhead Roads Commission confirmed his status as a leading consultant on civil engineering matters to the British Government. His only challenger in this regard was John Rennie, who dominated dockyard work. Telford's position was consolidated in 1817 when he became Engineer to the Exchequer Bill Loan Commissioners, established by the Poor Employment Act (57 Geo. III c.34). He was appointed their adviser 'on all works

requiring the information of a civil engineer'. In the period of high unemployment which followed the end of the Napoleonic Wars the Government was increasingly concerned at associated political and serious unrest; it was also being besieged by requests from various canal companies which lacked the funds to complete their navigations. To address these issues the Act, recognising the 'great Advantage ... in affording Employment for the labouring classes of the Community', authorised loans for public works, and other initiatives such as fisheries and mines. From 1817 to 1834 more than £4½ million in loans was authorised, the bulk involving public works, and on many of these schemes Telford's recommendations were sought. Aided by a growing team of assistants, notably James Mills, Telford produced a stream of reports. Nearly four-hundred applications were successful, and many more fell, either because Telford's report was critical, or the scheme was otherwise found wanting. Telford's role varied from recommending against such as that for the Wiltshire and Berkshire, and Lechlade canals, to putting forward schemes such as that for the ferries across the Tay where he was engineer. With some schemes he had had earlier involvement as a consultant, such as the Union Canal in Scotland, whilst on the Gloucester–Berkeley Canal he became Engineer when the Commissioners took over responsibility, and on the Ulster Canal he also he became the leading consultant. The overall result was for Telford to be consulted about almost every civil engineering scheme of any size undertaken outside the dockyards in the period 1817–1834, giving him an unprecedented, and probably unparalleled, view of civil engineering work. Coincidentally, the Commission changed its policies in 1834, the year of Telford's death, concentrating its resources on the construction of work houses, which meant that Telford's successors as engineering consultants to the Commission could never attain his overall insight into the profession in that way.

Publications

Telford's technical publications consisted mainly of engineering reports. His earliest known publications were poems. In addition, he wrote authoritative articles for the *Edinburgh Encyclopaedia*, of which he was a leading proprietor. These were 'Bridge' with co-author Alexander Nimmo (first published 1812), 'Civil Architecture' (1813), 'Navigation Inland' also with a contribution by Nimmo (1821, partly written in 1814), and almost certainly 'Jessop' (1817). Altogether they amounted to in excess of three-hundred pages with eighty-two plates and were particularly influential before 1830 when the whole encyclopaedia was issued. The drawings for at least forty-six of the plates were made by Telford's assistants and at least twenty-seven by outside draughtsmen. It is unlikely that he drew any himself.

Telford's periodical publications included articles in the *Philosophical Magazine* on the proposed iron arch replacement for London Bridge (1801), canals (1803), his Highlands report (1803), Stamford Junction Navigation (1810), and later, on the effect on the River Thames of rebuilding London Bridge (1825). Telford's proposal to suspend arch bridge centering by means of radiating iron stays in *Nicholson's Journal* (1813) encouraged the design of suspension bridges generally. His compilation *General Rules for Repairing Roads* widely circulated from 1819, and from 1833, Parnell's *Treatise on Roads*, both propagated his improved road-making practice as applied on the Holyhead Road and exercised a fundamental influence on road authorities for more than a century. It is evident from Telford's surviving diary that he spent considerable time on editing Parnell's book.

Several publications were dedicated to Telford of which the more important were W. A. Provis's *Account of the Suspension Bridge over Menai Strait* (1828), G. Bradshaw's 'Map of Canals ...' (1828–1833) and, of more general interest, that of the writer Robert Mudie in *A Second Judgement of Babylon the Great* (1829). Mudie wrote, 'To Thomas Telford, The Engineer, who in the intervals of an important, arduous and successful professional life, has found time to do more real kindness than any advertising philanthropist, who has rescued more genius from oblivion and turned more talent to the best interest of his country than any maecenas of any age'. He concluded, that Telford, by 'instituting the Royal Society of Civil Engineers' (the ICE) has given 'stability' to the nation's 'best art'.

Telford's most important publication, despite its deficiencies as a personal narrative, was his autobiographical *Life ...*, edited by John Rickman, 1838, written from 1831 as his health, hearing, and new commitments declined. Although its magnificent atlas of engraved plates, text and appendices constituted an invaluable record of his practice and achievement, the book was not very successful commercially at a price of 8 guineas. By the time it was eventually published, four years after his death, the nation was in the grip of railway expansion and it was to some extent outdated as a working manual. The unsold stock was bought by James Walker (q.v.), who succeeded Telford as president of the Institution of Civil Engineers, and was used for Telford premium prizes for many years, for example, to James Leslie in 1846.

Poetry

Telford's enjoyment of the books loaned to him by Miss Pasley marked the start of his life-long love of poetry and an almost excessive admiration for literary ability. This interest led to his friendship with the Rev. Archibald Alison, author of an essay on *Taste* in 1790, whom he had met at Sudborough rectory c. 1783, and who had introduced him to Thomas Campbell. In seeking support for a quarto publication of Campbell's

Thomas Telford: grants approved for the Exchequer Bill Loan Commissioners

Date of grant	Work description	Value of loan (£)	Surveyor (for Telford)	Engineer
1817	Southwark Bridge	78,000	–	John Rennie
1817	Folkestone Harbour	10,000	J. Upton	–
1817	Hay Railway	8,000	–	None (in 1817)
1817, 1820	Llanfihangel Railway	2,000, 3,000	–	None (in 1817)
1817	Llansamlet Tramroad	–	–	–
1817	North Stafford Railway	8,000	–	–
1817	Grosmont Railway	5,000	–	–
1817	Middle Level Drainage	350,000	–	John and Sir John Rennie
1818, 1821	Regent's Canal (completion of canal; Islington Tunnel)	250,000	–	James Morgan
1817–1833	Eau Brink Cut	15,000	Townshend, Casebourne, Swansborough	John and Sir John Rennie
1818	North Wiltshire Canal	–	–	William Whitworth
1818	Montgomeryshire Canal, Western Branch	6,000	–	George Buck
1818–1826?	Gloucester and Berkeley Canal (completion) *Note*: The Commissioners took over the canal in 1821	160,000	–	Thomas Fletcher, William Clegram
1820, 1823	Edinburgh and Glasgow Union Canal	100,000	–	Hugh Baird
1820, 1823	Plymouth and Dartmouth Railway	28,000	–	James Rendel
1820, 1823	Portsmouth and Arundel Navigation	40,000	–	James Hollinsworth
1820–1823	Holyhead Road, eight grants	44,000	John Easton	Thomas Telford
1821 (1826?)	Thames and Isis Navigation	13,000	–	–
1821	Black Sluice Drainage	5,000	–	–
1822	Tay Ferry	25,000	–	Thomas Telford
(1823), 1824, 1826	Bude Harbour and Canal	20,000	Alexander Easton	James Green
1823	Porthleven Harbour	23,000	–	–
1823	Isis Bridge, Oxford	6,000	–	–
1825	Kingston Bridge	40,000	–	Edward Lapidge
1826	Teignmouth Bridge	8,000	James Green	Roger Hopkins
1826	Bridgwater and Taunton Canal Navigation	15,000	–	James Hollinsworth
1826	River Witham Drainage	–	–	–
1826	New Shoreham Harbour	15,000	John Upton	William Clegram
(1826), 1827	Exeter Canal	10,000	–	James Green
1827	Liverpool and Manchester Railway	100,000	James Mills	George Stephenson
1827	Hackbridge and Wentbridge Railway	7,600	–	–
1827	Courtown Harbour	6,000	–	Nimmo and Giles
1828	Ulster Canal	120,000	Thomas Casebourne	Casebourne
1829, (1824)	Hertford Union Canal	20,000	–	Francis Giles
1831	Glastonbury Navigation and Land Drainage	5,000	–	Sir John Rennie
1832?	Norwich and Lowestoft Navigation	54,000	–	William Cubitt
1832	Clarence Railway	111,000 (second grant 1835)	T. Rhodes	–

Thomas Telford: grants approved for the Exchequer Bill Loan Commissioners (contd.)

Date of grant	Work description	Value of loan (£)	Surveyor (for Telford)	Engineer
1832	Newcastle and Carlisle Railway	160,000 (second grant 1835)	–	Francis Giles
1834	Ouse Navigation (Yorkshire)	10,000	T. Rhodes	–
1834	Bodmin and Wadebridge Railway	8,000	–	R. Hopkins
[1835]	Hartlepool Dock and Railway	30,000	(T. Rhodes)	–

Note: this list excludes works which were not executed following reports by Telford.

poems in 1802, Telford wrote to Watt, 'Have you any people guilty of a taste for fine poetry in your quarter. If so, I stand ready to whet them in their transgression by adding their names to a list I am trying to procure for this young Scot'. Rickman introduced Telford to Southey. All became his close friends. Telford's earliest-known printed work was an eight-verse poem in Ruddiman's *Weekly Magazine, or Edinburgh Amusement* on 5 May 1779, in traditional early eighteenth century style, which ended:

Lang may ye sing, weel may ye phrase,
Hae routh and plenty a' your days;
And I shall gar a' our green braes
Ken weel your name,
I'm sure ye still sall hae the praise
O' ESKDALE TAM.

This poem was closely followed by *Eskdale*, first published separately at London in 1781, probably for the Pasleys, prefaced by his introduction as a 'stonemason ... a young man of no education but common reading, assisted by some books lent him by neighbouring gentlemen'. It was reprinted several times and was well thought of by Southey. At least twelve poems are known to have been written by Telford during his lifetime, many of which are referred to by Smiles. The last known, in 1831, was a tribute *To Sir John Malcolm on receiving his Miscellaneous Poems* (ed. Paxton, 1971). The most widely circulated was a manuscript poem to Burns, which Dr. Currie considered of 'superior merit' to other poems found among the poet's papers, and he printed 26 verses of it in many editions of Burns's works from 1801. The original version begins:

To Robert Burns

How sweetly flow thy simply strains,
Dear Bard of Scotia's happy plains
Thou pride of a' the cottage swains,
None sings like thee.

Some of Telford's poems were signed with the initials 'TT'. His usual signature in correspondence and reports was reports was 'Thos. Telford' with a flamboyant underlining of his surname,

but a small but significant change occurred from *c.* 1806 after which the letters 'o' and 's' were not joined.

Professional societies

In 1820 Telford was invited to become the first president of the Institution of Civil Engineers, an office he held until his death. The Institution, the world's first professional engineering body, had been founded by a group of young engineers, led by Henry Robinson Palmer (q.v.) in 1818, but its youthful leadership lacked the gravitas to attract members; Telford provided the leadership, contacts and prestige to attract members from all over the country and even overseas. Soon after taking office his staff were able to provide continuity in secretarial support to establish firmly the valuable tradition of recording the proceedings and discussions of meetings and the substance of papers read. His diligent and invaluable fostering of the Institution as a forum for engineering knowledge included, encouraging membership, provision of a library of books and drawings, urging members to present papers, and even chasing up and making good the subscriptions of Gotha Canal engineers Lagerheim and Edström. He had considerable influence with the governmental establishment and was the driving force behind the obtaining of the Institution's royal charter in 1828. In his will he left the Institution his largest bequest of over £3,000, and his books, drawings and papers. The money provided financial security, enabling the Institution to find permanent premises and to plan its publications. Telford's contribution to the Institution was fundamental to its early development. His next two largest bequests were to the parish libraries at Westerkirk and Langholm. He was elected to fellowship of the Royal Societies of Edinburgh and London in 1803 and 1827 respectively.

Character and personal

Telford never married and according to John Rickman, 'lived life as a soldier always in active service' (Telford's *Life*, 283). From *c.* 1800 he needed a permanent base in London and lived in rooms at the Salopian coffee house, Charing

Cross, until taking possession of 24 Abingdon Street in 1821, where he lived until his death. This enabled him to change his office organisation. In his earlier career Telford relied largely on personal recommendations and acquaintance for his assistants such as Matthew Davidson. His responsibilities meant he had an office in Shrewsbury from much the same time, administered by Stanton and John Telford. Around 1808 he took on his first pupil, William Provis, but, lacking a permanent home, was unable to develop a formal London base. From 1821, however, he took as pupils the children of trusted deputies such as Mitchell, May, Davidson, Gibb, and others who applied such as Turnbull. To these were added office assistants such as Henry Robinson Palmer and Thomas Casebourne. In addition he continued to rely heavily on largely autonomous deputies in the field like Provis, Stanton and Macneill. There can be little doubt this was much closer to the model of consulting engineering practice that was to follow than that of John Rennie who had no pupils, and kept a strict control of decision making.

Telford's contact with Alison and his family from 1783 to 1834 was probably the nearest approach to home life that he ever enjoyed. Little of Telford's character is to gleaned from his publications, but the opinions of some of his contemporaries are informative. Rickman stated that Telford's most distinguishing character trait was a benevolence which made him accessible to all who came to him for information. His pupil, Joseph Mitchell, recounted his master's 'great delight in his work. The perfect good faith and honour of all his transactions, his clear conscience, and his cheerful temper cast a halo of happiness throughout our establishment' (Mitchell, 102). In 1793 Catherine Plymley wrote after meeting Telford that he was 'an excellent architect and a most intelligent and enlightened man. His knowledge is general, his conversation very animated, his look full of intelligence and vivacity' (Penfold 'Colossus', 2). She also praised the liberality and cheerfulness of his charitable donations.

Southey wrote of Telford, 'there is so much intelligence in his countenance, so much frankness, kindness and hilarity about him flowing from the never-failing well spring of a happy nature, that I was upon cordial terms with him in five minutes' (Southey, 7). Telford was not a lover of concert music. He wrote, 'the melody of sounds is thrown away upon me, one look, one sentence from Mrs. Jordan has more effect upon me than all the fiddlers in England' (Gibb, 294). Sir David Brewster wrote of Telford's apparent sternness of manner which created difficulties at times in his relationships, and belied the genuine benevolence and kindness of his nature. He also referred to Telford's 'quick perception of character, his honesty of purpose, and his contempt for all other acquirements, save that of practical knowledge and experience which was best fitted ... have

enabled him to leave behind him works of inestimable value ... which have not been surpassed either in Britain or in Europe' (Sir D. Brewster, 46). Telford died on 2 September 1834 at 24 Abingdon Street, London of a fatal bilious derangement. His remains rest in Westminster Abbey.

Of several portraits of himself Telford preferred that painted by S. Lane *c.* 1822, an engraving of which by W. Raddon was published by E. Turrell in January 1831 and used as the frontispiece to his *Life*. Telford also preferred the engraving to the painting as can been seen from the following letter written by him in February 1831 which was recently discovered on the back of Rickman's copy of the engraving: 'My Dear [Rickman]' he wrote, 'Accept a copy of my portrait made by knowing artists, a masterly performance. It was upwards of two years ago undertaken as a speculation by Mr. Turrell and another engraver, they have only just finished it. I prefer it to the painting. Place it among your friends' (private).

Conclusion

With hindsight, a more theoretical approach would have benefited Telford's structural design practice, but his reliance on experimental and practical procedures was the best available means of achieving the desired result at the time. An essential factor in his immense achievement was his sound judgement in selecting capable and reliable assistants and contractors, many of whom have been mentioned in this article, to whom he was able to devolve responsibility without losing control. Under his direction, they translated his designs into effect often at and even extending the frontiers of engineering technology and knowledge.

Telford's beneficial influence is most alive today through his many surviving works, contract procedures, and the Institution of Civil Engineers as a forum of engineering excellence. Of Telford's surviving works, although leisure interest in his canals is increasing, his roads and bridges, for which he was aptly dubbed by Southey as the 'Colossus of Roads' and 'Pontifex Maximus' (Smiles, 476), now make the greatest contribution to society. A welcome development in 1998 was the designation by the Welsh Office of the Holyhead Road across North Wales as an 'Historic Route' with appropriate information signing. Rolt, who was also the biographer of George Stephenson and Isambard Brunel, believed that Telford's 'achievement was as great as theirs and of equal historical significance' (Rolt, xi). Smiles, Rolt, Penfold and others have all helped to restore his reputation from its low ebb during the 'railway mania' era. Telford was undoubtedly one of the great civil engineers of all time.

ROLAND A. PAXTON

[NB: Original Telford material is to be found in most record offices in the British Isles. The main collections are to be found in the PRO; National Archives Scotland; Telford Collection, Ironbridge;

Telford mss, ICE archives. His large drawings collection, left to ICE on his death, was broken up in 1906, and effectively destroyed. J. Bennett (1830) *The History of Tewkesbury*; Sir H. Parnell (1833) *A Treatise on Roads*; T. Telford (1838) in J. Rickman (ed.), *Life of Thomas Telford*; R. Southey (1839) Review of Life of Thomas Telford, *Quarterly Review*, Jan.–Mar.; Sir D. Brewster (1839) Review of Life of Thomas Telford, *Edinburgh Review*, **LXX**; J. P. Muirhead (1854) *The Origin and Progress of the Mechanical Inventions of James Watt*; S. Smiles (1861) *Lives of the Engineers*, **II**; J. A. Roebling (1867) Report, 1867, *Annual Report of Covington & Cincinatti Bridge Company*; Sir J. Rennie (1875) *Autobiography of …*; J. Mitchell (1883) *Reminiscences of my Life in the Highlands*; R. Southey (1929) *Journal of a Tour in Scotland in 1819*; Sir A. Gibb (1935) *The Story of Telford*; ICE (1950) *A Collection of Works of Art and Objects of Historical Interest*; ICE (1957) *Thomas Telford Bicentenary Exhibition* (catalogue); L. T. C. Rolt (1958) *Thomas Telford*; N. Pevsner (1958) *Shropshire*; A. R. B. Haldane (1962) *New Ways through the Glens*; T. Telford (1968) in R. A. Paxton (ed.), *Three Letters from Thos. Telford*; T. Telford (1971) in R. A. Paxton (ed.), *To Sir John Malcolm upon Receiving his Miscellaneous Poems—A Poem by Thomas Telford*; J. Dunlop (1978) *The British Fisheries Society 1786–1893*; R. A. Paxton (1975) The influence of Thomas Telford (1757–1834) on the use of improved constructional materials in civil engineering practice, M.Sc. thesis, Heriot–Watt University; T. Ruddock (1979) *Arch Bridges and their Builders 1735–1835*; A. E. Penfold (ed.) (1980) *Thomas Telford: Engineer*; A. Maclean (1989) *Telford's Highland Churches*; A. E. Penfold (1989) *Thomas Telford 'Colossus of Roads'*, exhibition catalogue; C. Hadfield (1993) *Thomas Telford's Temptation*; R. A. Paxton (1993) Review of 'Thomas Telford's Temptation', *The Institution of Civil Engineers Panel for Historical Engineering Works Newsletter*, December No. 60, 6–7; R. Paxton (1998) *Road and Bridge Making on Main Routes in and around Dalkeith*; R. A. Paxton (1999) The early development of the long span suspension bridge in Britain, *Proc International Conference*, Wheeling, West Virginia, USA; Skempton; R. A. Paxton; DNB]

Selected publications
1796. *A Copy of a Letter to the Secretary of the British Society; containing a Course of Experiments on Mr. James Parker's Cement*
1796. Some account of the inland navigation of the county of Salop, in *General View of the Agriculture of Shropshire by Joseph Plymley* (1803)
1800. Plans and estimates for iron bridges of three and five spans over the Thames (with James Douglass), in *Third Report from the Select Committee upon the Improvement of the Port of London*
1800–1802. Plans and an account of a proposed single arch iron bridge of 600 feet span over the Thames (with James Douglass), in *Select Committee on Port of London*, reports and in a tract. Separately published large aquatint view by T. Malton with the bridge engraved by W. Lowry
1801. Mr. Telford's Reports of Cromarty, Aberdeen and Wick, made to the Lords of the Treasury in 1801, in *Third Report from the Committee on the Survey of the Coasts &c. of Scotland: Naval Stations and Fisheries*
1803. Report by Thomas Telford on the Harbour of Aberdeen, in *Reports upon the Harbour of Aberdeen* (1834)
1803. *A Survey and Report of the Coasts and Central Highlands of Scotland, Made by Command of the Treasury, in the Autumn of 1802* (roads and bridges, naval stations, Caledonian Canal, fisheries, emigration, Carlisle & Portpatrick Road)
1803. *A Survey and Report of the proposed Extension of the Union Canal; from Gumley-Wharf, in Leicestershire, to the Grand Junction Canal, near Buckby-Wharf in Northamptonshire* (2nd edn., 1804)
1804–1821. Reports &c., in *Reports of the Commissioners for Highland Roads and Bridges*; reports from 1815–1834 in *Reports … for maintaining … roads and bridges in Scotland*
1804–1830. Reports etc. (some with W. Jessop), in *Reports of the Commissioners for the Caledonian Canal*
1804. *Suggestions … relative to the Canal from Glasgow to the West Coast of the County of Air* [Ardrossan Canal]
1805. *A Report relative to the Proposed Canal from the City of Glasgow to the Harbour of Ardrossan on the West Coast of the County of Ayr, in Scotland*
1805. Report of the general state of the Grand Junction Canal, in *Report of the General Committee of the Grand Junction Canal Company*
1805. Report (with W. Jessop), in *Reports upon the Harbour of Aberdeen* 1834
1805. *A Report respecting supplying the City of Glasgow and its Suburbs, with Water*
1805. Report respecting the River Clyde, in *Reports on the Improvement and Management of the River Clyde* (1854)
1805. Report on improvements to Grangemouth harbour, in *State of Facts and Observations relative to the Affairs of the Forth and Clyde Navigation* (1816)
1806. Report on proposed harbour at Port Gower, in *Report from Committees … Forfeited Estates*
1806. *Report respecting supplying the City … of Glasgow with water …*
1806–1826. Reports etc. in *Reports on the improvement of the River Clyde and Harbour of Glasgow* (1854)
1808. *Mr. Telford's Report on the intended Cumberland Canal*
1808. Report respecting the line of communication between the north of England and Ireland (by the south-west of Scotland, with John McKerlie), in *Report from the Committee Appointed*

to *Examine into Mr. Telford's Report and Survey* (1809), with sixteen large plans separately bound

1808. Report of the Gotha Canal; being the result of a Survey made in 1808, in *Life of Thomas Telford* (1838)

1809. Report, in *Reports upon the Harbour of Aberdeen* (1834)

1810. *Report relative to the proposed* [cast-iron] *railway from Glasgow to Berwick-upon-Tweed; with Mr. Jessop's opinion thereon*

1810. *Plan and Sections of the Track of a proposed Cast Iron Railway, from the City of Glasgow, to Berwick-upon-Tweed, passing through the Counties of Lanark, Peebles, Selkirk, Roxburgh, Berwick, and Northumberland*

1810. *Report and Estimates relative to a proposed Road in Scotland from Kyle-Rhea in Inverness-shire to Killin in Perthshire by Rannoch-Moor*

1810. *Report of Mr. Telford, respecting the Stamford Junction Navigation*

1810. Report by Thomas Telford, Esq., in *Reports on the Means of Improving the Supply of Water for the City of Edinburgh and on the quality of the different Springs* (1813)

1811. Mr. Telford's surveys of New Galloway Bridge: Glenluce Road; and Carlisle and Garistown Road, in *Report from the Committee, upon the Roads between Carlisle and Port Patrick* (1811)

1811. Report to the Treasury, respecting the Great Roads, from Holyhead through North Wales, in *Report from Committee on Holyhead Roads* (1811); includes 'Method of constructing an iron bridge with one arch of extensive span proposed to be carried over the Menai' using iron stays'. Also in W. Nicholson (ed.) *A Journal of Natural Philosophy*, 1813

1812–1821. Contributions to *Edinburgh Encyclopaedia* [see p. 689]

1813. Report on the Crinan Canal (1813), in *Papers relating to the Crinan Canal 1813–1815* (1816); another in *Fourteenth Report of the Commissioners for the Caledonian Canal* (1817)

1814. *Report respecting the Harbour of Dundee in the County of Forfar*

1814. Design for a bridge over the River Mersey at Runcorn and reports, in *Report of the Select Committee appointed to Consider the most Practicable and Expedient Mode of effecting the Proposed Bridge at Runcorn*, Warrington (1817)

1815. *Report on the intended Edinburgh and Glasgow Union Canal*

1815. Report [Grangemouth Harbour], in *State Facts and Observations Relative to the Affairs f the Forth & Clyde Navigation* (1816)

1815–1830. Report respecting the Road from Carlisle to Glasgow, in *Report from Select Committee on Carlisle and Glasgow Road* (1815); later reports made to the Commissioners, … *for Maintaining and Keeping in Repair certain Roads and Bridges in Scotland*

1815. Reports, estimates, etc., in *Report of the Commissioners for Repairing the Roads between London and Holyhead by Chester and between*

London and Bangor by Shrewsbury (1816); numerous later reports in the publications of the Select Committee on the Roads from Holyhead to London, last known 9 August 1834

1815–1817. First (and second) report of Mr. Telford on the intended Edinburgh and Glasgow Union Canal, in *Observations by the Union Canal Committee, on the Objections made by the Inhabitants of Leith to this Undertaking*

1817–1819–1828. Dee Navigation reports, in *Local Acts. Copies of Admiralty Reports under Preliminary Inquiries Act 1850—Dee Conservancy Bill*

1818. *Map of a proposed Line of Navigable Canal from near the Town of Knaresbro' to the River Ouse at Ancaster Sailby with a Line of Rail-Road from Knaresbro' … to Pately Bridge in the County of York*

1818–1819. [Menai Bridge] Report, plan, and estimate for building a Bridge over the Menai Strait, near Bangor Ferry, in *Papers relating to the Building a Bridge over the Menai Straits near Bangor Ferry*, House of Commons 1819; [and] Design for a bridge over the Menai Straits at Ynys-y-Moch, in *Third Report from the Select Committee on the Road from London to Holyhead &c Menai Bridge*

1819. Report for Glasgow Dock, in *Reports on the Improvement and Management of the River Clyde* (1854)

1819. Some account of the drainages of Holland and short account of the drainage of the fens, in *Second Report for the Select Committees in the State of Disease and Condition of the Labouring Poor in Ireland*, 111–115

1819–1820. *Northern Roads. Copy of a Report and Estimate, of Two proposed Lines of Roads: The one leading from Catterick Bridge, in the County of York, to the Carter Fell, on the Borders of Scotland: The other leading from Catterick Bridge to New Castleton, in the County of Roxburgh; made under the Instructions of the Lords of His Majesty's Treasury*

1820. Report 5 June 1820, in *Holyhead Road. Reports of Mr. Telford*, including 'General Rules for Repairing Roads' (published separately in 8vo 1820; 4th edn., 1823; J. Weale 1837)

1820–1826. Reports on the extension of Broomielaw Quay, in *Reports on the Improvement and Management of the River Clyde and Harbour of Glasgow* (1854)

1821. Report and plan, in *Report by the Committee of Management, relative to the Plans of the New Ferry Harbours at Dundee and the opposite coast of Fife, proposed by Mr. Telford*

1821. *Mr. Telford's Report on the Points referred to him by the Corporation of Wisbech* (Fen drainage)

1821. *Dunmore Harbour. A Copy of the Report of Mr. Thomas Telford*

1822. Edinburgh and Morpeth Road. Report and estimate of a proposed line of road from Morpeth by Wooler, in *Report from Select Committee on Morpeth and Edinburgh Road* (1822)

1822. *Report on the Eau Brink Cut* (with (Sir) John Rennie)

1823. Mr. Telford's Reports [on the foundations of Westminster Bridge], in *Reports by Messrs. Telford, Cubitt, and Swinburne, Civil Engineers, as to the State of the Foundations, &c. of Westminster Bridge* (1836)

1823–1831. *Report of Thomas Telford, Esq. on the Effects which will be Produced on the River Thames by the Rebuilding of London Bridge.* Other reports from 1823 to 1831 in *Report from the Select Committee on Westminster Bridge* (1844)

1823–1824. Report(s) on roads from Glasgow to Port-Patrick, in *Report(s) from the Select Committee on Glasgow and Port-Patrick Roads* (1823–1824)

1824. Report respecting the street pavements &c. of the Parish of St. George, Hanover Square (London), in Parnell (1833) *A Treatise on Roads*

1824. *Report on the Practicability and Advantages of Opening and Improving the Navigation of the River Stour*

1824. Prospectus and Mr. Telford's preliminary report, in *English and Bristol Channels Ship Canal*

1824. *Ship Canal, for the Junction of the English and Bristol Channels. Reports of Mr. Telford and Captain Nicholls* (repr. 1825)

1824–1825. *Berwick and Morpeth Road. Report made by Mr Telford, to the Lords of the Treasury, respecting the Mail Road between the City of Edinburgh and the Town of Morpeth by the Towns of Berwick and Alnwick*

1825. *Map or Plan & Sections describing the line of an intended Turnpike Road to be made from the Parish of West Wycombe in the County of Buckingham, through Thame to Chilworth in the County of Oxford*

1825–1826. Mr. Telford's report on his Survey of Mail Roads and Steam-boat Stations in South Wales, in *South Wales Roads. A Copy of the Postmaster General's Letter to the Lords Commissioners of His Majesty's Treasury upon the subject of the Roads through South Wales*; other reports on the improvement of communication with Milford Haven across South Wales followed in 1827

1825–1834. Plans and elevations of a church and a manse, in *First Report of the Commissioners for Building Churches in the Highlands and Islands of Scotland. House of Commons*, 1825 (see also 2nd–7th reports)

1826–1829. *Liverpool and London Road. Report(s) on the state of the road from London to Liverpool*

1827. *Report respecting the Mail Road between London and the Town of Morpeth, made under the direction of His Majesty's Postmaster General* (and maps 1826–1827)

1827. *Mr. Telford's Reports, Estimates and Plans for improving the Road from London to Liverpool*

1828. *Report respecting the Lower Ferry between the Counties of Mid-Lothian and Fife*

1828. *Plan of the St. Katharine Docks*

1828. Report on the Queensferry Passage (across R. Forth), in *Report of the Committee*, Edinburgh

1828. Reports with R. Stevenson and A. Nimmo on the intended ship canal between the Rivers

Dee and Mersey, in D. Stevenson (1878) *Life of Robert Stevenson*

1828. Report with P. Roget and W. T. Brand, in *Report of the Commissioners appointed by His Majesty to inquire into the State of the Supply of Water in the Metropolis*

1829. *Liverpool and Manchester Rail-way. Mr. Telford's Report to the Commissioners for the Loan of Exchequer Bills*

1829. Report on the improvement of Swansea Harbour, in *Reports on the Harbour of Swansea*

1829. Report of the proposed new mail road, from Carlisle to Edinburgh (by Langholm) (and map), in *Report from the Select Committee on the state of the Northern Roads* (1830)

1830. [Report on] *Glasgow Railway* [to connect the Forth & Clyde Canal with Broomielaw Harbour]

1830. Report respecting the port and harbour of Belfast, in *Admiralty–Harbour Department—History of the Harbours of the United Kingdom* (1852)

1830. Report, in *Report from the Committee on Birmingham and London Junction Canal Petition*

1830. Report on state on the Holyhead and Liverpool Roads with report from Mr. Macneill on the labour of horses in drawing carriages on roads of different construction, in *Seventh Report of the Commissioners for the further Improvement of the Road from London to Holyhead and from London to Liverpool*

1830. Collected edition of articles on architecture, bridges, inland navigation and Jessop. First published *c.* 1810–1821, in *The Edinburgh Encyclopaedia* (see p. 689)

1830. Plan for improving the harbour of Aberdeen, in *Reports upon the Harbour of Aberdeen*

1833. Report, in *Report of the Committee of Management on the Glasgow Bridges; relative to the Building of Jamaica St. Bridge*

1833. Report, with others, of the result of an experimental journey … in Sir Charles Dance's steam carriage, in A. Gordon (1836) *A Treatise on Elemental Locomotion*

1834. *Metropolis Water Supply. Report of Thomas Telford, Civil Engineer, on the Means of Supplying the Metropolis with Pure Water*

1834. Report, in *Report from the Select Committee on Dover Harbour*

1834. *Report on the Present State of the New London Bridge* (with James Walker)

1834. Report on Firth of Forth Ferries (Newhaven–Burntisland), in *Forth Ferries, House of Commons* (1838)

1838. *Life of Thomas Telford, Civil Engineer, written by himself,* John Rickman (ed.)

Works

See table on pp. 696–697

THATCHER, Thomas (fl. 1799–1834), contractor, presumably had experience of canal construction when he applied for the post of Agent on the Kennet and Avon Canal in 1795. In the event

Thomas Telford: works

Date	Work	Cost (£)	Resident Engineer
1782–1793	Shrewsbury Gaol	–	
1790–1792	Montford Bridge	6,000	Matthew Davidson
1792–1794	St. Mary's Church, Bridgnorth	7,000	–
1794–1796	St. Michael's Church, Madeley	–	–
1794–1805	Ellesmere Canal*	460,000	John Duncombe, 1794–1803
	1795–1805: Pontcysyllte	47,000	Thomas Denson, 1803–1805
	1790–1701, Chirk Aqueduct	21,000	
1795–1796	Buildwas Bridge	6,500	–
1795–1796	Longdon Aqueduct	–	–
1795–1799	Bewdley Bridge	9,000	Matthew Davidson
1803–1821	Highland roads and bridges	430,000	James Donaldson,
	1806–1808: Dunkeld Bridge	30,000	1803–1806
	1811–1812: Bonar Bridge	14,000	John Duncombe, 1806–1809
	1812–1814: Craigellachie Bridge	8,000	John Mitchell, 1809–1823
1803–1822	Caledonian Canal†	900,000	Matthew Davidson,
			1804–1819
			James Davidson, 1820–1822
			John Telford, 1804–1807
			Alexander Easton,
			1807–1822
1805–1806	Tongland Bridge	7,710	Adam Blane
1806–1811	Glasgow–Paisley Canal and Ardrossan Harbour	130,000	David Henry
1806–1824	Highland harbours	100,000	
	1808–1811: Fraserburgh‡	11,000	William Stuart
	1814–1817: Fortrose	4,000	John Mitchell
	1816–1821: Banff	16,000	John Gibb
	1818–1823: Peterhead	23,000	John Gibb
1807–1812	Loch Spynie Canal	12,740	William Hughes
1807–1810	Weston Canal, Weaver Navigation	50,000	Samuel Fowls
1809–1833	Gotha Canal, Sweden	–	John Wilson and others
1810–1815	Aberdeen Harbour II	100,000	John Gibb
1815–1825	Dundee Harbour	120,000	David Logan, 1816–1821
			Peter Logan, 1821–1825
1815–1830	Holyhead Road	495,000	William Provis, 1815–1826
	1815–1830: Shrewsbury-Bangor	117,000	John Provis, 1826–1830
	1818–1826: Menai Bridge	185,000	William Provis
	1819–1829: Anglesey Road	61,000	William Dargan, 1820–1822
	1820–1830: London-Shrewsbury	132,000	John Provis, 1822–1829
			John Easton, 1820–1826
			John Macneill, 1826–1830
1816–1817	Crinan Canal, repairs	–	William Thomson
1815–1823	Glasgow–Carlisle Road	50,000	William Provis John Pollock
1818–1827	Gloucester–Berkeley Canal	330,000	Thomas Fletcher
1819–1823	Edinburgh Water Supply	145,000	James Jardine
1819–1826	Dee (Chester) Navigation improvements	26,000	Thomas Wedge
1820	Glasgow Old Bridge, widening with cast iron cantilevered walkways	–	–
1820–1821	Loose and Linton Hill Road, Kent, improvements	–	Henry Robinson Palmer
1820–1823	Lanarkshire Roads	19,000	John Pollock
	1821–1822: Cartland Crags Bridge	5,000	Henry Welch
1822–1826	Conway Bridge	51,000	William Provis
1822–1825	Dublin Docks completion	40,000	John Aird
1823–1828	Holyhead Harbour II	43,000	James Brown(e)
1824–1826	Mythe Bridge, Tewkesbury	35,000	William Mackenzie
1824–1827	Harecastle Tunnel	112,000	James Potter
1825–1827	Knypersley Dam	16,000	James Potter
1825–1830	Birmingham Canal III	24,000	William Mackenzie
	1825–1828: Rotton Park Dam		
	1829: Galton Bridge		

Thomas Telford: works (contd.)

Date	Work	Cost (£)	Resident Engineer
1826–1828	Over Bridge, Gloucester	43,000	Thomas Fletcher
1826–1828	Holt Fleet Bridge	–	William Mackenzie
1826–1829	Eau Brink Cut, widening‡	33,000	Charles Burcham
1826–1830	St. Katharine's Docks, London	250,000	Peter Logan, 1825–1828
			John Hall, 1828
			Thomas Rhodes, 1828–1830
1826–1835	Birmingham and Liverpool Junction Canal‖	800,000	Alexander Easton
1827–1830	Chester–Bangor Road	14,000	John Provis
1827–1830	Nene Outfall Channel‡	200,000	William Swansborough
1827–1834	Highland churches	54,000	Joseph Mitchell and others
1827–1830	Bridge of Don, Aberdeenshire	25,000	John Smith
1827–1830	Pathhead Road Lothian Bridge	6,500	Henry Welch
1829–1831	Morpeth Bridge	–	John Cargill
1829–1831	Dean Bridge, Edinburgh	19,000	Charles Atherton
1829–1832	Aberdeen Harbour III	60,000	John and Alexander Gibb
1830–1834	Dundee Docks	150,000	James Leslie
1830–1834	North Level Drainage of Fens	150,000	William Swansborough and Thomas Pear
1833–1836	Broomielaw Bridge, Glasgow	34,000	Charles Atherton

*William Jessop, principal engineer
†William Jessop, consultant 1803–1812
‡Sir John Rennie, joint engineer
§John Rennie, principal designer
‖Completed by William Cubitt

he was not appointed, but with John Clark was awarded three lots west of Devizes. These contracts were for earthworks only, the bridges in two of them being awarded to James Hollinsworth (q.v.). Thatcher, again with Clark was next awarded the first excavation contract on the City Canal in London's Isle of Dogs on 24 October 1799. This related to the top 6 ft. of earth, and was carried out January to July 1800. John Dyson (q.v.) completed the excavation. The engineer for the work was William Jessop (q.v.) and he was again engineer for the next contract for which we have information, that for the soil excavation at Bristol Dock in April 1804. Thatcher's initial (business) partner was James Sharp, who left in June 1805. Thatcher obtained further contracts in the docks for excavating the eastern half of the New Cut, and the Rownham and Bathurst contracts in 1804. Later in 1805, with a new partner, Richards, he obtained the excavation contracts for the western end of the New Cut, and the Cumberland entrance locks In June 1806 they followed this with a contract for walling the basin. Work generally progressed smoothly, although in June 1808 a high tide broke over the cofferdam and through the bank at Rownham, necessitating £15,000 of additional expenditure. By dint of extraordinary exertions, Thatcher was able to complete the basin and locks on schedule. More than one thousand men were employed on the construction of the docks, although other contractors were involved, such as Hodge and Langman. Richards was a mason who brought

additional expertise to Thatcher's enterprise, and built the Prince Bridge. The whole scheme was completed in May 1809.

Thatcher's subsequent career is unknown, until he reappears in the fens in 1825, undertaking work again involving cofferdams for Samuel Bower (q.v.). In the 1830s Richard Parr cited Thomas Thatcher as his surety when tendering for a contract on the London–Birmingham Railway.

P. S. M. CROSS-RUDKIN AND MIKE CHRIMES

[Frank Smith Files and Jessop papers, ICE archives; Kennet and Avon Canal papers, RAIL 842, PRO; R59/31/6a/1, Cambs CRO]

Works
1796. Kennet and Avon Canal, contractor for lots 11, 13 and 14, worth £641, £1148, £1500, respectively
1799–1800. City Canal, excavation, contractor
1804–1809. Bristol docks, excavation and masonry work, contractor
1825. Welmore Lake Sluice, Denver, contractor, over £9,000

THOM, Robert (1774–1847), civil engineer, was born in Tarbolton, Ayrshire, and spent his early years on his father's farm. In the winter months, and whenever he could be spared from agricultural employment, he attended school or pursued his studies on his own. He then, under the instruction of an elder brother, became an expert joiner or 'wright', at which trade he worked when he moved to Glasgow in the hope of advancing in life. There he was able to attend evening classes at the University and the Andersonian Institution, one of the first colleges of adult education in Britain, offering to artisans and tradesmen courses of lectures in mathematics, natural philosophy and other subjects.

About 1794 he obtained employment in the rapidly-expanding industry of cotton-spinning and after a few years was offered the management of the mills at Pollockshaws, in Glasgow. He next became the manager of, and later a partner in, Blackburn cotton mill in West Lothian and by 1813 he had purchased Rothesay cotton mills with William Kelly, the manager of the mills at New Lanark. As early as 1800 there were two steam engines at the Rothesay mills, producing about thirty horse-power between them, but when Thom required more power for an expansion of the mills he determined to achieve this by making more effective use of the available water power, coal being very expensive on the Island of Bute. By 1821 he had doubled the capacity of the mills' storage reservoir by raising and strengthening the dam, and greatly increased the catchment by a system of collecting aqueducts. He incorporated self-acting sluices of his own design to regulate the flow of water and by these means increased the available water power to seventy horse-power. An incidental benefit of the collecting aqueducts was the improvement they brought to the drainage of the whole upland catchment area.

He was then asked to submit proposals for increasing the power available from the River Almond, on which his Blackburn mill stood, and in 1824 he drew up plans for a storage reservoir and automatic control sluices, but despite the support of most of the mill owners the scheme came to nothing. Similar proposals made in 1827 for the Molendinar Burn in Glasgow were also abandoned.

His most ambitious and successful work, however, was the Shaws Water Scheme at Greenock on the south side of the Clyde estuary, which was built in several stages over about fifty years from 1825. When Thom first surveyed the area there was no water power in the whole of Greenock and water for drinking and industrial processes was in very short supply. A small water supply scheme had been surveyed and constructed by James Watt (q.v.) in 1773 but when the idea of utilising the Shaws Water was mooted Watt dismissed it as impracticable, a view supported by a number of other engineers, including John Rennie (q.v.).

Robert Thom, however, by then established as laird of Ascog on the Isle of Bute, replied that such a scheme was perfectly feasible and his offer to carry out the work was accepted. His proposals involved building a reservoir in the hills behind the town and leading the water through a 5-mile aqueduct with a fall of only 26 ft. into two partly artificial water-courses each falling more than 500 ft. through the town into the sea. He again made use of his automatic sluices to control both high and low flows and calculated that more than thirty mills could be supplied with a reliable source of power from such a scheme.

Another problem for Thom was to provide a supply of pure, filtered water to domestic premises and the sugar refineries which were a major source of employment in the area. The slow sand filters then in use were frequently choked by algal growths, so he devised and built self-cleansing sand filters in which algal scum could be removed by reversing the direction of flow of the water, a principle still in use today. For the supply of water to the filters a separate smaller stone-lined aqueduct was used and the filtered water was then conducted in a covered channel to the Town's Dam.

The Great Reservoir (later renamed Loch Thom) was an artificial lake of nearly 300 acres created by the construction of a 60 ft. high dam on the upper reaches of the Shaws Water. Together with six supplementary reservoirs, the compensation reservoir and one of the two water-courses—the western one was never built—the whole project had cost £90,000 when the first phase was completed in April 1827. The east falls had provision for nineteen mills, a typical water-wheel developing some fifty horse-power from a fall of about 28 ft. An outstanding iron wheel, one of the largest in the world at the time, was built by James Smith of Deanston about 1830 for the Shaw's Cotton Spinning Company; 70 ft. in diameter with a width of 13 ft., it produced over 200 hp from a head of 64 ft. By 1845 eleven falls were being utilised and it was not until 1881 that five of the highest remaining falls were combined to drive a turbine which replaced Smith's fifty-year-old water-wheel.

By that time demand for domestic and industrial process water had increased to such an extent that the embankment on Loch Thom had to be raised and two new reservoirs built on the Gryfe Water at a cost of £207,000. In the latter half of the nineteenth century many of the great water-wheels were abandoned and replaced by turbines and the eventual decline and closure of most of the mills was unconnected with their source of power provided by the foresight of Thom a hundred years before.

He retired from active engineering in 1840 but was afterwards induced to advise on many schemes for the supply of water to towns and cities in Britain and overseas. He died on 14 December 1847 at Ascog, on the Isle of Bute, and was interred in the High Kirkyard of Rothesay, overlooking the works of his first water power project in which his ideas and his skills in hydraulic engineering were fully realised. His writings reveal that he was concerned also in the social and environmental problems of his day. There is a marble bust by Patric Park (1811–1855) in the McLean Museum, Greenock.

His memoir in the *Proceedings of the Institution of Civil Engineers* notes that 'as he has left a mass of ms. papers, partially arranged, with a view to the publication of a work *On the Supply of Water to Towns*, it is hoped that the whole will be given to the world by his son, Mr. William Thom, who has been long engaged in carrying out the plans of his father'. Unfortunately, the work was never completed.

R. BIRSE

[Obituary, *Proc ICE*, **7** (1848) 7–9; R. M. Smith (1921) *History of Greenock*, 41–46; J. Ferrier, Robert Thom's water-cuts, *Trans Glasgow Archaeological Soc*, **15** (1960) 129–138; T. S. Reynolds (1983) *Stronger than a Hundred Men: A History of the Vertical Water Wheel*; J. Shaw (1984) *Water Power in Scotland 1550–1870*, 482–486; Skempton; W. T. Johnston (1998) *Scottish Engineers and Shipbuilders* (computer database)]

Publications
1825. Description of a new self-acting lever sluice, and of a waster sluice, *Mech Mag*, **III**, 34
1827. *Reduced Plan of Lands drained into the Reservoirs, and into the Aqueduct near Greenock …*
1829. *A brief Account of the Shaws Water Scheme* (includes *A Report on the Supplying of Greenock with Water*, 1824)
1829. *An Extract from A Report … upon the Plentiful and Economical Supply of the City of Edinburgh with Pure Water …*
1837. *A Report on Supplying Glasgow with Water …*
1843. On the formation of embankments and reservoirs to retain water, *Min Proc ICE* **2**, 191–192
Works
1821. Water power scheme, Rothesay, Isle of Bute
1825–1827. Shaws water scheme, Greenock, £90,000

THOMAS, John (fl. 1721–1737), engineer, is known at present only through two pieces of evidence given by him to committees of the House of Commons, in 1721 on Bridport harbour and in 1737 on Sunderland harbour. On Bridport in December 1721 'John Thomas, Engineer' reported that the haven had been 'neglected and is choked up with sands, and the piers ruined, but is capable of being improved and made commodious for shipping'. An Act was passed in February 1722 for restoring and rebuilding the piers and making a sluice, but nothing was done until 1741 when John Reynolds (q.v.) undertook to carry out the works.

At Sunderland the Commissioners of the River Wear and Port of Sunderland, appointed under an Act of 1717, were anxious to obtain a new Act before the term of the first expired in 1738. They called on Isaac Thompson (q.v.) to produce a survey of the harbour in 1736 and in March 1737 Thomas gave evidence during the committee stage of the Bill. He laid a map (Thompson's) before the committee, showing a large pier on the south side of the harbour, and said that another pier on the north side was absolutely necessary, ships having been stranded on the sand there for want of the pier, one instance of which he himself

had seen. He estimated the cost of the pier at £12,000 or thereabouts.

For no obvious reason the Bill failed and no further action was taken until the inhabitants of Sunderland petitioned successfully for an Act in 1747. Newly appointed Commissioners then sought advice from William Vincent and Charles Labelye (qq.v.) and work restarted in 1750.

Between December 1733 and June 1738 some £5,000 had been spent on the harbour but whether Thomas had been employed by the Commissioners during this period or part of it is uncertain. More probably he was brought in as a consultant towards the end of 1736.

A. W. SKEMPTON

[J. Thomas (1721) Evidence, *JHC*, **19**, 688; J. Thomas (1737) Evidence, *JHC*, **22**, 795; H. Symonds (1912) Bridport harbour through seven centuries, *Dorset Nat Hist Antiquarian Field Club*, **33**, 160–199; A. W. Skempton (1977) The engineers of Sunderland harbour, *Indust Archaeol Rev*, **1**, 103–125]

THOMAS, John (c. 1753–1827), wholesale grocer, of Bristol, was involved with the Kennet and Avon Canal from its inception until his death. A Quaker, he was a shareholder and member of the Kennet and Avon Western District Committee when construction began in 1794. He suggested to John Rennie (q.v.) that the canal contracts should be let in ½ mile lengths in September 1794. Following the dismissal of Dudley Clark (q.v.) in July 1800 Thomas spent an increasing amount of time in supervising the works at the western end. In August 1802, to the chagrin of James Mills (q.v.), the Clerk of Works, he was appointed Superintendent of Works at a salary of £600 p.a., with £150 p.a. allowance for expenses. Benjamin Davis was appointed his Clerk. Henceforth he played a leading role in directing operations, establishing a lime kiln in Devizes in January 1807, determining with Rennie how to build the locks in 1804, and keeping an eye on the activities of the contractors such as Charles (?) Jones (q.v.) on Savernake (Burbage) Tunnel. Following the completion of the canal in December 1810 Thomas retained an interest, and was consulted on occasion. It is possible Thomas then worked on the Grand Western Canal.

Thomas died at his home, Prior Park, near Bath in 1827 at the age of seventy-four. He had purchased this property in 1812. He was succeeded as engineer by John Blackwell (q.v.).

MIKE CHRIMES

[Rennie, mss 19781 and 19782, NLS; Rennie reports, ICE Library; K. R. Clew (1968) *The Kennet and Avon Canal*]

THOMAS, John (d. 1844), of Sheerness, became a Corresponding Member of the Institution of Civil Engineers in 1823 and was transferred to the class of Member in 1835, at which time he was living in Highgate. He was active in the affairs of the Institution and in June 1835, after the death of Thomas Telford (q.v.), President, Thomas nominated James Walker (q.v.) to succeed him.

Thomas may have worked as a surveyor under Thomas Townshend (q.v.) for the Irish Bog Commissioners (c. 1810–1812). His major work was as Superintendent of the Works to execute construction of the new Royal Dockyard at Sheerness, designed by John Rennie in 1813. Thomas's appointment took place sometime before 19 January 1814, when a letter from Rennie to the Commissioners of the Navy stresses the importance of a suitably experienced person, and anticipates the need for as many as six or seven men as assistants. For proper management of the works, Rennie proposed that Thomas should not be responsible for payment of money and that receipt of materials and progress of work be jointly certified by him and by an officer of the Dockyard.

On completion of the foundation of the river wall of the new works, Thomas received a promised increase in pay. Rennie applied on his behalf in a letter of 28 March 1818, describing him as 'indefatigable in his attention to every part of the Work, the best proof of which is the success that has attended a very difficult and hazardous work and the superior manner in which it is executed'. This high opinion stood Thomas well in August 1818, when misuse of materials and fraudulent payment of wages brought his position into question, although there was no evidence of his direct involvement. Rennie was asked by the Commissioners if Thomas was indispensable to the work. In reply he points out 'that it has been a most hazardous as well as an arduous task, notwithstanding which there has not been an accident of any amount since the work began. I am not aware that his place could be easily filled ... without considerable loss'. Thomas survived, and accompanied George and John Rennie (qq.v.) on the final inspection of the new works at Sheerness in July 1830, nine years after the death of their designer and sixteen years after his own involvement began.

Thomas died in 1844.

MIKE CHRIMES

[Membership records, Archives of the ICE; Rennie letter books, Archives of the ICE; Commissioners ... Bogs ... in Ireland (1810–1814) *Reports*; ICE (1845) Annual report, *Min Proc ICE*; G. Watson (1988) *The Civils*]

Works

1813–1830. Sheerness dockyard, resident engineer

THOMAS, Richard (1779–1858), civil engineer and surveyor, the son of Richard Thomas, was born at Falmouth on 27 December 1779. Usually referred to as Richard Thomas of Falmouth, it is recorded that he resided at Mellingye, Perran-ar-Worthal and his work was largely centred on this area of west Cornwall.

By contrast in 1815 he published *Hints for the Improvement of the Severn* ... and also a *Chart of the Severn ... to His Grace the Duke of Beaufort most respectfully inscribed by his Obedient Servant Richard Thomas Jnr ...* . This was followed by a geological map and report on the mining district of Cornwall from Chasewater to Camborne in 1819 and a second map of the northern mining district including the parishes of St. Agnes, Perranzabuloe and others in 1823. The same year saw the survey for and design of 'an intended Tram-Road or Rail-Way from Restrongett Creek to Redruth ...' which became known as the Redruth and Chasewater Railway. It was opened on 30 January 1826 with horse drawn wagons, converted to steam traction in 1854 and finally closed in 1915.

In 1826–1827, Lord Falmouth was pressing the Truro Turnpike Trustees to improve the road from Truro to Bodmin and presented them with a complete survey and estimates for a new line of road between Indian Queens and Truro that had been prepared for him by Richard Thomas. This motivated the Trustees to engage John Loudon McAdam (q.v.) to survey not only the line in question but 'any other that might present itself'. The result was a very decided preference for the one proposed. The work was started in 1828 to be finished in 1829 and with a few more recent modifications is still in use. Thomas continued to work with McAdam, this time for the Bodmin Trust who, in 1831, were engaged on improvements to the London road and in November 1832 his plan for proposed new works on the turnpike roads out of Helston towards both Penryn and Marazion was deposited with the Clerk for Quarter Sessions.

In the following November his plan, *Proposed line of Railway from Hayle to Helston, Redruth and Tresvean Mine, Richd. Thomas, Engineer* was also received by the Clerk and the same plan with amendments, deposited in November 1835, formed the basis for the Hayle Railway which opened on 23 December 1837. He continued to publish reports on local subjects, including a series of fifty-five articles for the *West Briton* on the 'Statistics of Cornwall', from December 1850 to June 1852.

Thomas died on 21 February 1858, aged 78, at the house of his son-in-law, T. J. Buxton, at St. Mary's, Isles of Scilly, and was buried at Budock, near Falmouth. His *History ... of Falmouth*, published in 1828, was specifically referred to in the notice of his death in the *West Briton*, where he was described as 'a thoroughly independent man who has left an honourable name for undeviating uprightness and integrity'.

R. P. TRUSCOTT

[Bibliotheca Cornubiensis, Boase and Courtney (1878) Royal Institution of Cornwall, Cornwall RO; *West Briton Royal Cornwall Gazette*]

Publications

1815. *Hints for the Improvement of the Navigation of the Severn*

1816. *Observations and Directions for Navigating the Severn*
1819. *Report on a Survey of the Mining District of Cornwall from Chasewater to Cambourne*
1828. *History and Description of the Town and Harbour of Falmouth*
1831. *Report on the proposed Line of Railway between Perranporth and Truro*

Works

1823–1826. Redruth and Chasewater Railway
1828–1829. Turnpike, Truro to Indian Queens
1835–1837. The Hayle Railway

THOMPSON, Benjamin (1779–1867), ironmaster, coal-owner, railway engineer, and inventor, was born at Whitely Wood Hall, Ecclesall, near Sheffield, on 11 April 1779, the seventh son of Anthony Thompson and Sabra Clark. He was educated at Sheffield Grammar School, but he left Sheffield when in his early twenties for Aberdare, in South Wales, where with his brother, John, he established blast furnaces and rolling mills and operated coal and ironstone mines in the vicinity. He married Ann, daughter of Samuel Glover of Abercarne, in 1806 and in 1811 he headed north to become a managing partner of the Bewicke Main (later called 'Ouston') colliery in County Durham, and Fawdon colliery near Newcastle. Thereafter he was ever identified with North East England. He introduced many innovations in his capacity as managing partner in the collieries, including a new design of edge rail for underground haulage, the case hardening of wagon wheels and, in about 1830, a lighter form of railway for collieries, where the wagons were half the usual size.

In May 1807, William Chapman (q.v.), the Tyneside engineer, had taken out a patent (No. 3030, 1807) for a new type of coal staith and he installed one at Benwell on the Tyne in that year. Known as 'Mr. Chapman's lowering down machine', it enabled a chaldron wagon to be lowered vertically from the end of a staith into the hold of a waiting vessel. In 1812, Benjamin Thompson introduced what could be seen as a variant on Chapman's device, probably at Wallsend. It differed from Chapman's in that the wagon was not lowered vertically, but had a long pivoted lever arm from which the wagon was suspended on a platform, such that it could be lowered in an arc to reach the hold of a vessel. This had certain advantages, particularly that the drop could reach deeper water, even if located on the river bank. Thompson claimed to have had the basic idea in mind when Chapman was building his drop at Benwell. The Thompson-type coal drop, with a few variations, came to be almost universally adopted in the north-east.

Bewicke Main colliery was served by a railway to the Tyne, which had horse-drawn wagons on a level section of the way and a stationary haulage engine to draw the wagons up a valley side, but Thompson replaced the horses with a stationary steam engine in 1821, using the longest rope then made—1½ miles long—and thereafter fixed engines came to be commonly used on level sections of mineral railways in the North East as well as on adverse gradients. Thompson then became a great advocate for the use of fixed engines, publishing a pamphlet in 1821 which extolled their virtues: this resulted in a prolonged argument between Thompson and George Stephenson's ally, Nicholas Wood (q.v.), in the pages of a local newspaper. Thompson used his 'reciprocating system' on the Fawdon wagonway c. 1822 and on the Brunton & Shields line ('Kenton and Coxlodge') of 1826, where five fixed engines worked the entire line at an average overall speed of about 6 m.p.h. So impressed by the latter line were the engineers deputed to advise on the best form of traction for the Liverpool & Manchester Railway by its provisional committee, that they recommended Thompson's reciprocating system for their line, on account of its 'economy, despatch, safety, and convenience'. In the event, of course, steam locomotives were used.

William Chapman and Josias Jessop (q.v.) commenced a survey for the Newcastle & Carlisle Railway in c. 1826, which was completed by Thompson. Initially this line was to be worked entirely by horses, for the intention then was to build a line for the use of the public, in the same way that a public road was used; such a line would rule out the use of stationary or locomotive engines. Moreover, Chapman and Thompson believed that horses would be less likely to antagonise the landowners along the route than locomotives. As time passed, however, and locomotives proved their superiority over other forms of traction on level track, Thompson became warmly supportive of their merits and, with Wood, was amongst the first directors of the Newcastle & Carlisle Railway; Thompson was also a director of the abortive Blaydon, Gateshead & Hebburn Railway. But where achieving a level track for a strictly mineral line would prove difficult or expensive, he continued to advocate the use of his reciprocating system, as on the successful Rainton & Seaham Railway in County Durham for Lord Londonderry in 1831, and for a projected but never implemented line at Ford, Northumberland, in 1852.

Thompson and his fellow partners in the colliery concern commenced the Birtley Iron Works in County Durham between 1828 and 1830, with two blast furnaces and a large foundry designed by Thompson and, in 1836, he and his brother, George, established the Wylam Iron Works in Northumberland, with access to the Newcastle & Carlisle Railway via a combined road and railway bridge over the Tyne, possibly designed by Thompson. The Thompsons built several steam locomotives at Wylam, in particular for the Newcastle & Carlisle Railway, amply confirming Benjamin's conversion to the new mode of transport. After 1844, the Thompsons leased the

Wylam works out to Bell Brothers, and Benjamin retired.

Thompson died at Gateshead on 19 April 1867, at the age of eighty-eight years, one of his sons, James, having become a prominent merchant and shipowner in Newcastle.

STAFFORD M. LINSLEY

[Delaval Papers, Northumberland County RO; M. Dunn (1852) *A Treatise on the Winning and Working of Collieries*, 2nd edn.; R. Welford, (1895) *Men of Mark 'twixt Tyne and Tweed*; W. W. Tomlinson (1914) *The North Eastern Railway*, repr. 1967; P. R. B. Brooks (1975) *Wylam and its Railway Pioneers*]

Publications

1821. *Copy of the Specification of a Patent Granted to Benjamin Thompson ... for his Invention of 'A method of Facilitating the Conveyance of carriages along Iron and Wood railways and Other Roads ... with Remarks thereon ...*

1847. *Inventions, Improvements, and Practice of Benjamin Thompson ...*

THOMPSON, Isaac (c. 1703–1776), land surveyor, son of Isaac and Hannah Thompson, was born at Kendal. By 1730 he had established a practice as a land surveyor based at Newcastle-upon-Tyne. He also gave courses of lectures on 'natural and experimental philosophy' and was closely associated with the River Wear and Sunderland harbour at various times between 1736 and 1759.

Commissioners of the River Wear and Port of Sunderland were appointed under an Act of June 1717, for a period of twenty-one years, with powers to raise funds for improving the harbour and navigation in the river. The principal achievement of their engineer, William Lellam (q.v.) was a masonry pier, 1,000 ft. long on the south side of the harbour, completed in 1731. But this left insufficient capital for building a north pier, considered necessary for further improvement of the harbour, before the term of the Act expired in 1738. Preparations were therefore made for submitting a petition to Parliament for a new Act granting increased powers.

As a first step Thompson was employed in 1736 to survey the harbour and the river 11 miles up to New Bridge, Chester-le-Street, the limit of the Commissioners' jurisdiction. He produced the finest engineering survey of the early eighteenth century. In association with Mark Burleigh (d. 1774), a Sunderland merchant and an excellent draughtsman, the survey was published in 1737 in the form of a large engraved map on the scale of 600 ft. to an inch. It included soundings in the river and for some distance out to sea. The Commissioners also consulted the engineer, John Thomas (q.v.). In March 1737 Thomas gave evidence to the House of Commons, laying a map (Thompson's) before the Committee, emphasising the need for a north pier and estimating its cost at £12,000.

The Bill failed, as did a renewed petition next year, and so the Commission ceased to exist after June 1738; nothing was done until finally in January 1747 the inhabitants of Sunderland and local landowners petitioned for a new Act. Evidence at the committee stage was given only by Burleigh and Thompson. They quoted figures of past expenditure (historically valuable as the original accounts were lost), described briefly what had been achieved and the poor condition of the harbour after ten years of neglect, and stated that the only proper method to improve it was by lengthening the pier already built on the south side and erecting a new pier on the north side.

This time the application to Parliament proved successful and led to the passing of an Act in June 1747. Burleigh and Thompson each received about £145 for 'attendance in London', a figure indicating they had done a good deal more than simply giving evidence on one day. The newly appointed Commissioners immediately sought advice from William Vincent (q.v.), the Scarborough harbour engineer and shortly afterwards they consulted Charles Labelye (q.v.). Too busy on Westminster Bridge to come at once, he arrived on site in June 1748 and presented a lengthy report, with a plan, in September. The Commissioners asked Thompson to prepare an abstract (a hundred copies of which were printed in December) and to make copies of the plan.

A period of almost uninterrupted activity began in 1750 and continued for over fifty years, by which time Sunderland ranked as one of the major ports in Britain. A problem remained, however, concerning the navigation of the river Wear from Biddick Ford (the current limit) from 2½ miles up to New Bridge, Chester-le-Street. Thompson gave detailed evidence to a Parliamentary committee in March 1759. Other evidence and estimates of work to be done were given by Joseph Robson (q.v.), the Sunderland harbour engineer, and by John Smith (q.v.), the experienced engineer of the Don Navigation. They all agreed that the only practical solution was build a weir and lock, but the expense proved too high in relation to the benefits. The limit of jurisdiction was therefore moved to Biddick. Durham Corporation then considered the possibility of making the river navigable from Biddick up to the city but nothing came of this scheme.

Thompson, a Quaker, became land agent to the Duke of Northumberland in 1766. He died on 6 January 1776, aged seventy-two, leaving two sons, Jonathan and John, the latter succeeding to his father's surveying practice.

A. W. SKEMPTON

[M. Burleigh and I. Thompson (1747) Evidence, *JHC*, **25**, 296; Wear Commissioners Minute, Book and Accounts (1747–1759) Tyne & Wear Archives; I. Thompson (1759) Evidence, *JHC*, **28**, 477–481; M. A. Richardson (1843) *The Local Historian's Table Book*; J. Murray (1847) An account of the progressive improvement of Sunderland harbour, *Min Proc ICE*, **6**, 256–277; A. W. Skempton (1977) The engineers of Sunderland harbour, *Indust Archaeol Rev*, **1**, 103–125; R. V. Wallis and P. J. Wallis (1986) *Biobibliography of British Mathematics and its Applications*; Bendall]

Publications

1737. *A Plan of the Mouth of the River Wear* [and] *Harbour of … Sunderland. And of the River Wear from Newbridge to Sunderland Barr* (drawn by Mark Burleigh)
1757. *A Course of Natural and Experimental Philosophy* (with Robert Harrison)

THOMPSON, James (1794–1851), Colliery Agent, was the third child of Thomas and Isabella Thompson, tenant farmers at Fairlam Hall, near Brampton, and was born on 21 August 1794.

In 1808 he began his service in Lord Carlisle's colliery offices at Kirkhouse under the Agent, Thomas Lawson. Concerned with reconstructing and extending wagonways in the Tindall Fell region of the coalfield he was early involved in the replacement of cast iron or wooden rails by wrought iron. Their effectiveness at Tindall Fell was noted by Robert Stevenson (q.v.) in his report on the proposed Edinburgh Colliery Railway of 1818, a copy of which was sent to George Stephenson (q.v.), then at Killingworth, who in turn sent it to Michael Longridge at his Bedlington ironworks. The development and patenting of the wedge shaped wrought iron rail by John Birkinshaw (q.v.) appears to have evolved from this report.

In 1819 Thompson replaced Lawson as Colliery Agent and set out to redevelop the Naworth coalfield, extending and improving the wagonways, opening up quarries and constructing associated industries including lime kilns, brick works, coke ovens and joinery shops. At that time the coalfield enjoyed a near monopoly in the supply of coal to the east Cumberland area, including Carlisle, Penrith and Brampton. Thompson was keen to maintain this, particularly in view of the revived proposals for a canal from Newcastle to Carlisle, soon followed in 1824 by the proposal for, and then the construction of, a railway to join the two towns. To counter this threat, Thompson prepared designs and estimates of cost—some £55,000—for a private railway connecting the Naworth coalfield to Carlisle in the west and possibly Alston in the south, so to tap the lead trade. His aim was to connect with the Newcastle and Carlisle Railway east of the coalfield in the region of Greenhead, where N&CR would terminate; Lord Carlisle would thus maintain control of the through route to Carlisle.

Carlisle's London Agent, James Loch, considered the proposal so radical that a second opinion was required and after some dissent from London based engineers, George Stephenson was appointed. After visiting the site and holding discussions with Thompson he sent his report dated 23 August 1824, from Liverpool, giving his approval to the scheme and concluding: 'I think it is only fair to state, in justice to Mr. Thompson, that I have never seen any coal works apparently better conducted than those under his care'.

Despite this approval the proposals were considered to be beyond the means of Lord Carlisle and were not proceeded with but Thompson continued to battle with the N&CR over the routeing and gradients of the western section of the railway. He believed that locomotives should provide motive power for railways, rather than the stationary engines and horses which were the current favourites of the N&CR directors and their engineer, Benjamin Thompson (q.v.), unrelated. With James Thompson's persistence the final alignment of the railway proved acceptable to the colliery company and it retained the branch line to Brampton which was relaid to a new alignment between 1834 and 1836, at which time the western section of the N&CR was opened.

Under Thompson's direction the colliery continued to improve and expand and locomotives were introduced; the first to be obtained, in October 1836, was *Rocket*, somewhat modified from its Rainhill Trials days and purchased in Liverpool by Thompson and his brother, Mark, for £300 then shipped to Carlisle, completing its journey on rails.

In 1837 a proposal that had long been under consideration came to a head; with the backing of Nicholas Wood (q.v.) and Matthew Liddel, now Lord Carlisle's London Agent, suggested that the whole of His Lordship's industrial interests be leased out so as to provide him with a regular, assured income from his estates. After negotiations, the lease was taken by Thompson and his two sons. The family continued to develop the enterprise, including the construction of its own locomotives (with advice from Robert Stephenson), all under Thompson's personal direction until his death on 14 July 1851, aged fifty-six.

His widow, Maria, nee Bell of Brampton, whom Thompson had married in 1822, then took over the company and traded as M. Thompson and Sons, continuing until her death in 1891; the family continued to trade until 1909.

J. B. ARMSTRONG

[W. W. Tomlinson (1914) *North Eastern Railway*; B. Webb and D. A. Gordon (1978) *Lord Carlisle's Railway*]

Works

1834–1836. Brampton Railway, diversion

THOMPSON (TOMSON), John (fl. *c.* 1530–1549), was a cleric and Surveyor of the works at Dover harbour. Certainly by 1533 and probably for several years earlier he was rector of St. James, Dover, and following the death of John Clerk (q.v.) in January 1535 he also became Master of the Maison Dieu, a religious house or 'hospital' providing maintenance for the poor and infirm of Dover.

There can be very little doubt that John Clerk, Master of the Maison Dieu from 1484, directed construction of the first harbour pier at Dover, completed c. 1495. The pier, later known as the Stone Jetty, extended some 500 ft. out to low-water mark on the south side of the harbour. It gave shelter against south-westerly storms and kept the harbour free from shingle driven north-wards along the coast. But in the 1520s shingle which had been banking up against the pier began encroaching around its end and by 1532 the harbour was completely choked.

The town, hitherto responsible for maintaining the harbour, could not afford the substantial works necessary to remedy this state of affairs. An attempt to obtain a grant from Parliament came to little effect. The mayor and jurats therefore decided to petition the King (Henry VIII), pleading the national importance of a good harbour at Dover. In November 1533 Thompson took the petition to London along with his own plans for greatly extending the pier and building a new pier or jetty on the north side. He received a favourable response; commissioners were sent to Dover; they approved the plans, and work began in July 1535.

Thompson was appointed Surveyor of the works with four Overseers, chosen from the best mariners of Dover, and a Paymaster (John Whalley, formerly Paymaster of the works at the Tower of London). A year later they were joined by Thomas Wingfield as Comptroller. Thomas Cromwell (Lord Privy Seal), to whom the officers reported, kept general control over the project. The King took a keen interest, visiting the works five times between 1536 and 1542. The workforce, at its maximum in the second half of 1536, numbered about five hundred, including twelve carpenters, six sawyers, fourteen tun men, one-hundred and twenty tide men, three-hundred and twenty labourers and fifteen clerks. The wages bill peaked at £360 a month. Blacksmiths are not included in these figures; they were paid by piecework.

The tidal range varies from 19 ft. at springs to 11 ft. at neaps. Piers situated on ground dry at low tide were framed by two rows of timber piles, the rows 24 ft. apart and the piles at about 15 ft. centres with diagonal bracing and some form of sheeting. The piles, iron-shod, were driven by 'a great engine called a ram' and held together by iron ties. The frame thus made was filled with large stones and pebbles. In deeper water a mound or 'mole' of rocks randomly deposited in the sea was brought above low-water level to form a foundation for the pier. This was framed and filled in the same manner as before, with timber posts set in holes hewed in the rocks taking the place of the piles. Rocks weighing up to 20 tons each were picked up from the shore beneath Dover Castle and at Folkestone, floated at high tide between casks or 'tuns', towed to the site and deposited in the mole. This ingenious and economical method of transport was devised by John Young, master of the tun men, but a similar technique may have been used earlier in building and repairing the pier (the Cobb) at Lyme Regis.

By September 1535 the south pier had been framed over a length of 350 ft. and partly filled, and work had started on the north jetty. In June 1536 Thompson reported that the latter was 380 ft. in length, two-thirds of which had been filled; the south pier, now heading in a southerly direction, was 500 ft. long and work was starting on the mole for the pier's further extension. The King made his first visit in July. He approved what he saw, ordered that an extra boat be acquired for towing the rocks, and proposed that a 'jetty' should be made running E.N.E. from the present end of the pier. Work began soon afterwards on the mole foundation for this jetty, which Thompson called 'the Kynges devyse'.

Following some dispute in December between the officers and doubts concerning the need to extend the pier much further, Cromwell insisted that work should continue and nothing was to be done without Thompson's advice and consent. By June 1537 the jetty had been framed over 120 ft. on its foundation of rocks and the greater part filled.

Henry visited Dover again in September 1538. The harbour seemed to be in good shape. He ordered Thompson to produce a 'platte of the workes' within two weeks. This he did; it formed the basis for a large drawing, perhaps by a Flemish artist at Court. It depicts several ships in the harbour; the north pier is complete; the south pier is shown with its southern end on a foundation of massive rocks, as also is the E.N.E. jetty. The piles and posts and diagonal bracing are shown, but the nature of the sheeting in not clear. Contemporary notes on the drawing state 'the south est jette' [the south pier] is 770 ft. long, with a groyne extending 60 ft. beyond its end, and 'the est northest jette' is 224 ft. long.

There had been ominous signs of shingle advancing around the pier and jetty in the winter of 1537–1538 and worse was to follow in 1538–1539. Henry's next visit in March 1539, so soon after his previous one, reflects this crisis. In April 1539 Wingfield reported that new work devised by the King and 'set forth by Mr. Cavendish' was going ahead satisfactorily. This refers to Richard Cavendish (q.v.) who arrived at Dover in the summer of 1537—in what capacity is uncertain— and was about to take over as Comptroller, Wingfield having been posted to the castles in the Downs (Deal, Sandwich and Walmer). The new work to which he refers can be identified as a northerly extension of the jetty terminating in the Crane Head. Meanwhile the jetty itself was being extended in the hope of reaching water deep enough to inhibit migration of shingle.

Expenditure from July 1535 to March 1539 amounted to about £8200. The new works begun in April 1539 accounted for a dramatic increase, bringing the total by the end of 1540 to approximately £15,000. In August 1540 there is a reference to 'the weakeness and age of the Surveior' and Cavendish seems to have taken over some of his duties. A kindly note from the Privy Council in October reassured Thompson of the King's favour and instructed him to come to Court, if he could without damage to his health, bringing platts of the works or to send others, having informed them of his plans.

This is the first indication that works on a grander scale were being considered. In January 1541 a commission was issued to Thompson, Cavendish, Auchar, Bartlet and A'Borough to oversee the works and procure materials and workmen. Antony Auchar had succeeded Whalley as Paymaster; John Bartlet and John A'Borough were experienced mariners and navigators. The King came to Dover in March. As a result a majestic plan emerged. This is shown in a drawing signed by Cavendish, Bartlet, A'Borough and Auchar. It depicts an extension of the south pier by 1017 ft. into water 21 ft. deep at low-tide and an entirely new north pier 1935 ft. in length.

Few details are available on progress of the work from 1541 but in summary the south pier extension, known as the King's Pier, was built almost to its planned length by 1551 at the enormous cost of about £36,000 (bringing total expenditure since 1535 to £51,000) and work never began on the proposed north pier. For the sad fact is that, against reasonable expectation, shingle continued to pass northwards around the end of the King's Pier, forming an off-shore bar which greatly restricted the harbour's entrance. Much later the bar formed the basis for a brilliant plan by William Borough (q.v.), son of John A'Borough, carried into successful effect by Thomas Digges (q.v.) and others in the 1580s.

Thompson seems to have retired as Surveyor early in 1541 and it that year he also resigned from St. James. He remained at the Maision Dieu until its dissolution in December 1544 (to become a victualling station). He was then granted a pension of £53 p.a. and is known to have been still in receipt of the pension in 1545–1546.

Finally 'Mr. Tompson of Dover' was consulted on Yarmouth harbour in 1549. Emden (1974) maintains that he died in July 1551 but may be a case of mistaken identity.

A. W. SKEMPTON

[R. C. Fowler (1926) The hospital of St. Mary or the Maision Dieu, Dover, *VCH Kent*, **2**, 217–219; A. Macdonald (1937) Plans of Dover Harbour in the sixteenth century, *Archaeol Cantiana*, **49**, 108–126; A. H. W. Robinson (1962) *Marine Cartography in Britain*; A. W. Ecclestone (1971) *Henry Manship's Great Yarmouth*; A. B. Emden (1974) *A Biographical Register of the University of Oxford 1501–1540*; M. Biddle (1982) Dover Harbour, *King's Works*, **4**, 729–755; Cotton, MS, Aug. I. i. 22, 23, British Library, and reproduced by Macdonald (1937)]

Works

1535–1540. Dover Harbour, north pier *c.* 400 ft., south pier *c.* 1,500 ft., about £15,000.

THOMPSON, John (fl. 1766–d. 1795) of Sheffield, civil engineer, was appointed engineer of the River Don Navigation in 1766, first under the supervision of John Smith (q.v.), who had been engineer since 1731, and from 1767 with full responsibility. He held this post until his death in 1795.

As early as 1772, in association with Thomas Tofield (q.v.), he surveyed the line of a proposed canal from Stainforth on the Don to Althorpe on the Trent, the idea being to bypass a long stretch of the Don from Thorne to Goole and of the Ouse to Trent Falls. Nothing came of this at the time but the scheme was revised in 1792, with an outlet to the Trent at Keadby a mile downstream of Althorpe. Also in 1792 the Don Company looked into the possibility of a canal from the Don at Swinton to the proposed Barnsley Canal, to be known as the Dearne & Dove Canal. Thompson and the surveyor, William Fairbank (1730–1801) of Sheffield, made the surveys and estimates of both lines and Robert Mylne (q.v.) was brought in as consulting engineer. He gave evidence on the Bill for the Dearne & Dove, Thompson and Fairbank giving evidence for the Stainforth & Keadby, the Acts were passed in June 1793.

Engraved plans were published by Fairbank and Thompson supervised construction on both canals, 1793–1795. Robert Whitworth (q.v.) took charge of the Dearne & Dove canal while Daniel Servant, who seems to have succeeded Thompson on the Don, completed the Stainforth & Keadby.

A. W. SKEMPTON

[C. Hadfield (1972) *Canals of Yorkshire and North East England*]

THOMPSON, Robert (1742–1806), stonemason and bridge surveyor, was probably the son of Robert and Ann Thompson, baptised on 14 December 1742 at St. Nicholas church, Newcastle. In 1767 he married Sarah Snowball (b. 1740) of Ovingham, their marriage producing a family of three daughters and three sons between 1768 and 1780. In 1781 he was appointed bridge surveyor for the southern division of Northumberland at the same time removing his family from Hedley Mill to Hexham.

In 1782 he carried out major repairs to Haydon Bridge, then impassable even by pedestrians, and two years later carried out work on the House of Correction at Hexham. He was responsible for repairs to many of the bridges in his area and was involved in the crossing of the river South Tyne at Ridley Hall where in 1779 Jonathan Pickernell (q.v.) had designed a bridge destroyed by a flood only twelve months after its completion. Foundations for the replacement bridge were designed by Robert Mylne (q.v.) and carried out under the direction of David Stephenson (q.v.) but the rebuilding of the superstructure was probably under the direction of Thompson and William

Johnson (q.v.), surveyor for the northern bridges; work was completed in 1792.

On the advice of Mylne, the County Justices decided that the bridge at Hexham, completed by John Smeaton (q.v.) in 1780 but destroyed in a flood two years later, could be rebuilt. Supervision was the joint responsibility of Thompson and Johnson and work began in 1789 and was completed in 1796, the superstructure built in accordance with Smeaton's original design. As a result of his 'extraordinary trouble, care and attention' he was paid an honorarium of £50 and his salary was increased from £25 to £40 p.a.

Thompson married for the second time in 1797 and retired from his position as bridge surveyor in 1803. He died at Hexham on 6 June 1806, the *Newcastle Chronicle* then noting him as 'master of the Black Bull inn, formerly surveyor of the county bridges'. He was buried at Hexham three days later, his widow, previously Margaret Reid, continuing to run the inn.

R. W. RENNISON

[*Northumberland Quarter Sessions: Orders* (1765–1807); *Newcastle Chronicle*, 14 June 1806; S. M. Linsley (1994) Tyne Crossings at Hexham up to 1795, *Archaeol Aeliana*, 5 Series, **22**, 235–253]

Works
1781–1803. Bridge Surveyor for Northumberland, southern bridges
1783. Designed and built Newbiggin Bridge
1789. With Johnson, built Ridley Hall Bridge; two spans of 67 ft. 6 in.
1790–1796. Hexham Bridge, construction, with Johnson, nine spans of up to 50 ft., cost *c.* £7,000

THOMSON, James (1787–1871), ironmaster, worked for Nicol and Hugh Baird (q.v.) for twelve years and gained experience in foundry work, with steam engines and the like, before working in Liverpool and London. In February 1817 he wrote, with Baird's support, to Thomas Telford (q.v.) regarding a possible post establishing a foundry for the Gotha Canal. At that time he was working for the Scotch Patent Cooperage Company. That May he met Telford in London and was entrusted with £50 and various books and drawing instruments which he took to Count von Platen, Telford's chief contact on the Gotha Canal, in Sweden. After looking over the line of the canal he and von Platen fixed on a site at Motala for a foundry for casting lock gates and iron bridges, beginning work on the foundry on 12 August. Thomson was negotiating for a salary of £450 p.a. and the scale of his demand, and political problems in Sweden, meant that work on the foundry fell through in the autumn, with Thomson returning to Glasgow.

Thomson was then employed in preparations for the Slateford aqueduct on the Edinburgh and Glasgow Union Canal, and in May 1818 he visited William Hazledine's (q.v.) works to discover details of how the Pontcystllte aqueduct ironwork was prepared for possible application there. Thomson got on well with Hazledine and by June 1818 was contemplating working for him, tendering jointly for the ironwork on the Union Canal. He also argued that Hazledine could supply ironwork for Sweden more cheaply than it could be manufactured there, although von Platen was still trying to persuade him to return. In 1819 Hazledine duly quoted for the Gotha Canal lock gates to Thomson's designs. Although von Platen was concerned at the cost, these gates were installed in 1820. The Swedes themselves found they were able to successfully imitate these 'models', ending Thomson's involvement.

When Thomson joined the Institution of Civil Engineers in 1820 he was employed at Hazledine's Calcutts Ironworks and he remained there until 1824, when he moved to Lightmoor Ironworks near Shifnal, Shropshire. He moved to Fenton Park Ironworks in 1836, and remained there until 1841; he then moved to Blair ironworks in Scotland. He was a Member of the Institution until 1836 although not struck off for non-payment of subscriptions until 1845. He served as a councillor and JP in Shropshire and retired to the Welsh Borders where he died, at Penybryn Hall, near Ruabon, in 1871.

MIKE CHRIMES

[Membership records, ICE Library; Gotha Canal, and Edinburgh and Glasgow Union Canal papers, Telford mss, ICE Library; Gibb papers, ICE archives; L. T. C. Rolt (1958) *Thomas Telford*; S. Purves (2000) Aqueduct construction in the canal age, *Millenium Link: Proceedings of the ICE Conference*, 23–31]

THOMSON, William (1783–1860), civil engineer, worked on the Crinan Canal from about 1804, when Thomas Fletcher (q.v.). At that time work on the canal had been virtually halted due to lack of funds and although some progress was made over the next 10 years, with James Hollinsworth (q.v.) and his son acting as resident engineers from 1805 to 1810, it was Thomson, who in 1816, estimated that more than £19,000 was still required to complete the canal. He was confirmed in post as resident engineer when the canal was taken over by the Caledonian Canal Commissioners in 1817. In addition to his work as engineer he acted as clerk and accountant. In the 1820s further attempts were made to raise funds for improvements but by the late 1830s there was still dissatisfaction amongst the potential users of the Canal.

Associated with Thomas Telford (q.v.) in a number of his Highland works he carried out surveys in Argyll for the Highland Church Commissioners and prepared the plans for churches at Duror, Appin, and Strontian. He was possibly related to James Thomson (q.v.). In 1805 he was noted as a lieutenant in the Canal Basin Volunteers at Ardrishaig. In 1839 he was being paid, in addition to a house and garden, £220 p.a. but by 1841 he had retired and was living at 46 Claremont Place, Glasgow. He died in 1860.

MIKE CHRIMES

[Crinan canal papers, BR/CRI/1/5, SRO; Gibb papers, ICE archives; *Commissioners for Building Churches in the Highlands and Islands of Scotland* (1825–1831) 1st–6th reports; *Select Committee on the Caledonian and Crinan Canals* (1839) Report and minutes of evidence, HC551; J. Lindsay (1968) *The Canals of Scotland*]

Works

1804–1839. Crinan Canal, resident engineer

THOROLD, William (1798–1878), farmer, millwright, civil engineer, surveyor and architect, was born on 9 October 1798 at Northwold, Norfolk, on the Bedford Level. He joined his parents in farming in 1812 and later took a farm at Melton, which he retained until 1831. In 1821 he attended the opening of the Eau Brink Cut and resolved to become a civil engineer.

In 1824 he exhibited a design for a county hall and corn exchange at Norwich. He was awarded two prizes by the Society of Arts in 1827, one for a machine for slicing turnips, and one for an improvement to Captain Manby's life saving apparatus. In 1828 he established himself as millwright and engineer at Norwich, building mills of all kinds, together with steam engines for draining, grinding, and for boats. He discovered that the circulation of water in a steam engine was essential and sent a paper to the Institution of Civil Engineers on the subject in 1831.

Thorold built roads and drainage works, designed and superintended the construction of buildings and built six Union Workhouses in Norfolk (1836–1837), and one in Chelmsford (1838). In 1835 he submitted plans for the Houses of Parliament. After the publication of his pamphlet on making Norwich a port (1826) he was hired by William Cubitt (q.v.) on the works for the locks at Mutford Bridge. He was Surveyor for many turnpike roads, including the Norwich, Swaffham, and Mattishall, the Norwich and Scole, the Norwich and New Backenham, the Norwich and North Wolshaw, and the Norwich and East Dereham roads.

His works in Norwich were destroyed by fire in 1840. Tragically he lost many of his papers, which included those of Thomas Tredgold (q.v.). In the new works he proceeded to build large quantities of contractors' plant, railway switches, drainage engines and a stationary engine for the Norfolk Railway. He continued to operate as a manufacturer until early in 1851, when he closed his works, having built a great number of turntables for railways. He then restricted his activities to those of a consulting engineer and to his many appointments as Surveyor to turnpikes. He submitted a list of his works to the Commissioners of the River Witham Drainage when he was a candidate for the post of Surveyor to the Board in 1863.

He was elected Associate of the Institution Civil Engineers in 1827 and transferred to the class of Member three years later. He was married and had at least one son who had little interest in his father's occupations. Although practically blind in his last years, Thorold maintained his interest in engineering until his death at Thorpe Hamlet, Norwich, on 17 December 1878.

TESS CANFIELD

[Thorold manuscripts, ICE: *Description of a Swing Bridge* (1830) ICE OC 62; *On Supplying High Pressure Steam Boilers with Water, and on their Duration* (1831) ICE OC 72; *On the Turnpike Road from Norwich to Walton* (1832) ICE OC 81; *Improvements. Norwich and Walton Roads* (1834) ICE OC 142; *A Trip to Ireland in June 1833* (1836) ICE OC 204; *Steam Engines at the Brewery of Sir E. Lacon and Sons* (1839) ICE OC 332; all ICE archives. *Design XIII. Farm House and Farmery for 100 acres of Lands to be Cultivated on the Norfolk System, with a Flour Mill driven by Wind, Design,* **XXXVI**; A public house and farmery, the publican being, at the same time, a small farmer and a butcher (1835, 1846) in J. C. Loudon (ed.), *Encyclopaedia of Architecture*, 471–473, 544–545; Obituary (1878–1879) *Min Proc ICE*, **55**, 321–322; Colvin]

Publications

1826. *An Inquiry into the Means of making Norwich a Port*
1829. *Remarks on the Present State of the Law relative to Patents for Inventors in a Letter to Davies Gilbert, esq. M.P.*
1845. An account of the failure of the suspension Bridge at Great Yarmouth, *Min Proc ICE*, **IV**, 291–293
1849. *Essay on the Present and Future Prospects of Farming in Great Britain*, 3rd edn.
1861. *Letters to the Proprietors of Land in that Part of the Bedford Level called the South Level, on the Expediency of the 'Middle Level Bill'*

TICKELL, Richard, Lieutenant-General (1785–1855), military engineer, was born 10 September 1785, the son of Thomas Tickell of Co. Kildare, Captain, 5th Royal Irish Dragoons, and Sarah Sparks, his wife. He was admitted to the service of the Bengal Artillery in 1803, but transferred to the Bengal Engineers on 21 September 1804. He reached the rank of lieutenant-general on 11 November 1851. He was made a CB in 1831.

Almost immediately on arriving in India, Tickell was called to military service in the 2nd Maratha War of 1803–1805. He then went as a surveyor with the Hon. Mountstuart Elphinstone's mission to Kabul from 1808 to 1809, and spent the next years from 1811 to 1816 on other surveys.

He served in the Nepal War of 1816 and the Maratha War of 1818, where he was mentioned for distinguished conduct. In April 1822 he was appointed to survey and prepare estimates for the restoration of an eighteenth century canal taking off from the eastern bank of the Jumna, but he was shortly afterwards made Superintendent of Canals, Delhi Territory, in succession to George Blane (q.v.) who had died prematurely. He held this post until 1827, during which time work went ahead on both the Eastern and Western Jumna (Yamuna) Canals under Robert Smith (q.v.) and John Colvin (q.v.) respectively. In 1827, Tickell was appointed Superintending Engineer for the South-West Provinces and in 1829 Superintending Engineer for the Lower Provinces. He took his furlough in 1837 and did not return to India. He was married twice, first on 1 February 1808 to Mary Anne Proctor. who died in 1833, and secondly on 18 June 1840 to Margaret Scott Walker. He died at Cheltenham on 3 August 1855.

JOYCE BROWN

[E. T. Thackeray (1900) *Biographical Notices of Officers of the Royal (Bengal) Engineers*, 26; V. C. P. Hodson (1947) *List of the Officers of the Bengal Army, 1758–1834*, **IV**, 274–275; R. H. Phillimore (1950) *Historical Records of the Survey of India*, **II**, 65–66, 446; R. H. Phillimore (1954) *Historical Records of the Survey of India*, **III**, 24, 433, 503]

Works
1822–1827. East and West Jumna Canals

TIMPERLEY, John (fl. 1796–1856), was trained as a land surveyor and first worked for William Bennet (q.v.) on the Dorset and Somerset Canal. He then worked as an assistant engineer on the Grand Surrey Canal before being employed by John Rennie (q.v.) from 12 March 1805 as a 'superintendent upon the works' at London Docks, which were nearing completion. This was the beginning of fifteen years employment on Rennie projects. In September 1806 he was transferred to Chetney Hill where canal and lock works were being carried out in association with the Lazarette, Hugh McIntosh (q.v.) being the contractor. Timperley undoubtedly made a great impression for his competence in dealing with this major public work which was abandoned without ever coming into use.

In 1817 he was appointed resident engineer under Rennie on the construction of the Wellington Bridge over the Aire in Leeds. In 1819 Rennie reported on the Aire and Calder navigation, George Leather Jr. (q.v.) also providing advice. No doubt Timperley would already have met the Leathers and, following the passage of the Act in 1820, Timperley was appointed resident engineer under Leather, with Jolliffe and Banks (q.v.) as contractors. Before his departure in 1825 several locks had been built and improved and new cuts formed at Goole. Unfortunately Timperley used the wrong lock sill depth (6 ft. rather than 7 ft.), which may have hastened his departure.

In 1826 James Walker (q.v.) employed Timperley in surveys for the Norwich and Lowestoft Navigation, Timperley drawing up estimates for the work, and in the following year, on Walker's recommendation, Timperley began his connection with Hull Docks, his best known work. Initially he was responsible for supervising the construction of the Junction Dock which Walker designed. Following this he was appointed engineer

for Hull Docks and wrote an important paper on the docks, awarded the first Telford Gold medal by the Institution of Civil Engineers.

In 1836 work began on the Hull and Selby Railway, again designed by Walker, and Timperley was appointed engineer at an annual salary of £250 p.a. Following its completion, no doubt on Walker's recommendation, Timperley was appointed harbour engineer in the Isle of Man where his career apparently ended. In 1846 Timperley was awarded a testimonial by the First Lord of the Admiralty, Lord Auckland, for his work in the island. At the time of his retirement, in November 1848, his salary was £210 p.a. and he continued to be consulted until 1854. There are mentions of a John Timperley advising the Isle of Man Harbour Board as early as 1772; perhaps this was his father.

Timperley's date of death has not been traced.

MIKE CHRIMES

[London Docks minutes (1805) Museum of London in Docklands; Rennie reports 4, 121; 6, 419; 7, 259; and 10, 33, ICE Archives; House of Commons (1805) *Minutes of Evidence on Camberwell Waterworks Bill*, 23–24; *Minutes of Evidence on Norwich and Lowestoft Navigation Bill* (1826) HC 369, 221–223; 225–230; *Minutes of Evidence Tidal Harbour Commissioners* (1846) 2nd report, Appendix B, 243–270; G. G. MacTurk (1879) *A History of the Hull Railways*; S. B. Smith (1970) Thwaite Mill, Hunslet, *Thoresby Society*, **53**, 82–90, M. Baldwin (1973) Hull docks, *Newc Soc Trans*; C. Hadfield (1972) *Canals of Yorkshire and North East England*, **1**, 130–143; Manx Museum (1998) in lit.]

Publications
1836. An account of the harbour and docks at Kingston-upon-Hull, *ICE Trans*, **1**, 1–52
1842. Description of the tanks for the kyanizing the timber for the permanent way of the Hull and Selby Railway, *Min Proc ICE*, **2**, 80–81
1844. Account of the building of the Wellington Bridge over the river Aire, at Leeds, *Min Proc* ICE, **3**, 104–106

Works
c. 1796–1803. Dorset and Somerset Canal, assistant
1803–1805. Grand Surrey Canal, assistant engineer
1805–1806. London Docks, resident engineer
1806–1814. Chetney Hill, Lazarette Canal, resident engineer
1817–1819. Wellington Bridge, Leeds, c. 91 ft. masonry arch, resident engineer
1820–1825. Aire and Calder navigation, improvements, Thwaite cut, etc., resident engineer, c. £17,000
1826. Norwich and Lowestoft navigation, surveys
1827–1829. Junction Dock, Hull, resident engineer, £15,000
1829–1836. Hull Docks, engineer
1836–1840. Hull and Selby Railway, resident engineer
1841–1848. Isle of Man harbour, engineer; Derby Haven breakwater; Ramsay north pier, extension;

Castleton pier; Douglas harbour, improvements; Peel harbour, south pier, widened, improvements 1842–1843. Ramsay improvements, north pier lengthening, 133 yd., £3,500, lighthouse erected 1845. Castleton pier, 268 yd., £2,700
1845. Douglas harbour, improvements, deepening and north pier erection
1845. Peel harbour, deepening harbour and widening quay, £1,054
1842–1843. Derby Haven breakwater, 260 yd., £2,520

TOBIN, Maurice (c. 1703–1773), whitesmith and ironmonger, of Briggate, Leeds, built the first documented iron arch bridge in Britain in 1769. From 1737 onwards Tobin features in local newspapers and church and estate accounts as a supplier of decorative metalwork and ironmongery: screws, locks, etc. His work brought him into contact with John Carr (q.v.) in the early 1750s and he provided high quality ironwork, such as balustrading, for the houses on which Carr was working.

In 1769 Tobin erected a bridge in the grounds of Kirklees Hall near Brighouse. According to a contemporary newspaper report it was 'a most curious bridge of one arch, 6 ft. wide, and 72 ft. in span; made entirely of iron ... It has also iron balustrades, which are ornamented with roses of the same metal ...'. Estate plans by William Crosley [the elder (q.v.)] of 1788 show this bridge crossing an ornamental pond with the (timber) deck resting upon the ribs of the arch. It has been generally assumed that it was a wrought iron structure.

Tobin married Elizabeth Todd in Leeds in October 1735 and they had several children, most of whom died in their infancy. Elizabeth died in 1756 and Tobin died after a stroke in September 1773. His son carried on the business for 12 months before moving to London.

MIKE CHRIMES

[Leeds Intelligence, 2 January 1770, 14 September 1773; D. Northcliffe (1980) The first iron bridge?, *Industrial Past*, **1**, 14–17]

Works
1769. Kirklees iron bridge, 72 ft. span, £157 10s

TOFIELD, Thomas (1730–1779), botanist and civil engineer, was born at Wilsic Hall near Wadworth (4 miles south of Doncaster) on 18 December 1730, the only surviving son of Thomas Tofield (1695–1747) and his wife Elizabeth Atkinson. He attended William Burrow's School at Chesterfield and in 1747 went up to Trinity College, Cambridge, taking his B.A. in 1751. Returning to Wilsic he began an intensive study of the local flora. This activity brought him into contact with William Hudson who in the late 1750s was working on a new British Flora based on the Linnean system. Tofield sent him many specimens and records. In Hudson's classic *Flora Anglica* (1762) Tofield, along with John Blackstone, was singled out for special acknowledgement. He

continued to send botanical data, much of which was included in the enlarged second edition of the *Flora* (1778).

He married Ann Smith in March 1761 and they settled in Balby, a village close to Doncaster and adjacent to an area of 4,250 acres of wetland known as Potteric Carr. Moves were afoot to drain the Carr, Tofield's father-in-law, Thomas Smith, being one of those involved. Tofield took a keen interest in the proceedings. He would have met John Smeaton (q.v.) who reported in September 1762, and again in March 1764. Tofield signed the Petition to Parliament and was appointed a Commissioner under the Act in May 1765. James Brindley (q.v.) then made a brief visit, apparently to check on a possible error in some of the levels which may have been pointed out by Tofield. Smeaton reported yet again with details of the scheme in the form finally adopted.

Work began in 1765 or 1766. The Commissioners' Minute Book no longer exists, however, there can be little doubt that Tofield played an active role in directing operations. By 1768 a new cut had been made for the river Torne, a new bridge built over the cut, and the first part of the main or 'Mother Drain' completed with two branch drains. Then came a pause while the northern parts of the Carr, belonging to Doncaster Corporation, were divided and inclosed. The Commissioners' task now finished, they made their Award in 1771 and were succeeded by the Trustees, whose minutes survive.

Work resumed in April 1772 with Tofield (one of the Trustees) effectively in charge and Matthias Scott (q.v.) as Surveyor or resident engineer; it was a part-time appointment as he also held a similar post on the nearby Hatfield Chase drainage works. Scott left two years later to become employed full-time on Hatfield Chase, his place being taken by Henry Cooper. With Tofield still directing, the whole scheme reached completion in 1777. By that time the total lengths of the cuts were: new River Torne 4.6 miles (embanked for about half its length), Mother Drain 4½ miles (with a 12 ft. outfall sluice into the Torne), and catchwater drains 3 miles. Joseph Colbeck, the land surveyor, produced a map of the drainage in 1782.

That Tofield had already gained a reputation from work on Potteric Carr by 1768 is indicated by an invitation early that year to examine a scheme recently put forward by Thomas Hogard (q.v.), Surveyor of Deeping Fen, for improving navigation from Spalding to the sea and drainage of the Fen. Tofield visited the site in April viewing the Fen, the River Welland (taking levels from Deeping to its outfall into Fosdyke Wash) and the land to the north through which Hogard proposed a new cut 100 ft. wide and about 7 miles long to Wyberton on the Witham estuary below Boston. Delayed by ill health for four months, Tofield finished his report in November 1768. He gives a full summary of Hogard's scheme—fortunately, as the original cannot be found—but rejected it on several well-reasoned grounds. He then outlines an alternative set of proposals, the principal one being a 5 mile extension of Vernatts Drain, the main drain of the Fen, to a new outfall sluice near Fosdyke; the bottom of the drain was to be carried on a dead level 2 ft. below low water in the Wash. He also proposed a barrier bank and catchwater drain along the western boundary of the Fen.

Faced with such different recommendations the Adventurers (proprietors) of Deeping Fen decided in March 1769 that Tofield, Smeaton, John Grundy and Langley Edwards (qq.v.) be 'desired to give their separate opinions on the best method of draining the Fen' in the hope of reaching a conclusion. Smeaton declined and Edwards was too busy on the fens near Boston, but Grundy reported at great length in February 1770, taking the opportunity of summarising the history of work in the Fen from 1730 successively by John Perry, by his father John Grundy Sr. (qq.v.) and by himself to 1764. His principal proposal is to improve the outfall of the Welland by making a new cut 115 ft. wide and 1½ miles long to a sluice and navigation lock near Fosdyke. Among other suggestions he accepts Tofield's western catchwater drain and the extension of Vernatts Drain, but considers as not justified the extra cost of digging below low water mark.

Tofield and Grundy met at Retford to discuss this difference of opinion and decided to seek Smeaton's advice. Each of them wrote to him and he replied in March 1770, saying 'I think myself much honoured by the reference of my brothers Grundy and Tofield, in a matter of art' and proceeding to explain that water in an open channel below the level of the outlet is not 'dead water', as Grundy seemed to think. On the contrary, the extra depth is beneficial since for a given gradient the flow increases with the ratio of cross-sectional area to wetted perimeter. Consequently, where, as in this case, the fall has a very low gradient (less than 4 in. per mile) the additional cost is justified. He himself had adopted this principle of the Mother Drain on Potteric Carr, with which of course Tofield was perfectly familiar. In May 1772 the Adventurers agreed to make the catchwater drain and to extend Vernatts Drain though eventually, under an Act of 1774, the extension was limited to a length of 2½ miles. Tofield's consulting fee and expenses amounted to £62.

His mother having died in 1770, Tofield moved back to Wilsic Hall with his wife and family. As well as resuming work on Potteric Carr in 1772 he was asked, in May that year, to consider the problem of the river Torne in Hatfield Chase. The river had been taken by Sir Cornelius Vermuyden (q.v.) in a new cut to an outfall in the Trent at Althorpe as part of his work for draining the Chase in the seventeenth century, but gave trouble from flooding. Lack of funds and the absence of any clear idea of the best remedial measures prevented any major improvement. Serious complaints, however, eventually led to a call for investigation. Tofield viewed the whole course of the river and surrounding ground and took levels, from which he concluded that the Torne could diverted out of the Chase to the river Don, near Thorne, with a fall of about 6 ft. in 7 miles. The engineering cost of this logical scheme was modest but negotiations for purchase of land would be necessary and an Act obtained, so nothing was done. The Torne problem was finally resolved by works carried out 1776–1789 by Matthias Scott and his successor, Samuel Foster, following, with important variations, a scheme devised by Smeaton in 1776. Tofield had reached his conclusions in the autumn of 1772 but due to pressure of work did not complete his report until September 1773. He received £43.

It was also in May 1772 that the proprietors of the River Don Navigation ordered their engineer, John Thompson, and the surveyor, George Kelk (qq.v.), to draw up a line for a canal from Stainforth lock on the Don to Althorpe on the Trent. When they had done so, Tofield was called in. He reported in October, approving the line, estimating the cost at £14,600 and showing how the canal could accommodate his proposed new course of the Torne. The canal was eventually made in the 1790s on a slightly different line at its eastern end to Keadby a mile downstream of Althorpe.

Tofield's third consulting job in 1772 concerned proposals for draining some 10,000 acres of low ground in the Vale of Pickering from Muston in the east to Yedingham and improving the river Derwent down to Malton. Landowners instructed the land surveyor, Isaac Milbourn, to prepare and deliver a plan to Tofield early in September 1772. This he did, and Tofield visited the site that month. His scheme having been accepted in principle, he submitted estimates and reported in September 1773. Here again nothing was done at the time and the drainage was finally carried out 1800–1808 by William Chapman (q.v.).

After 1773 and up to 1779 Tofield served as land commissioner on four inclosures in Yorkshire and Nottinghamshire, receiving the usual fee of about a guinea per day. He continued to work in Potteric Carr (he let the last contract in June 1777), pursued his botanical studies and added to his book collection. When sold at Leigh & Sotheby's in May 1780 'the elegant and valuable' collection fetched £430, augmented by £21 for mathematical and scientific instruments (including a theodolite and two levels). Among the books were: Belidor's *Architecture Hydraulique*, Desaguliers's *Experimental Philosophy*, Dugdale's *History of Inbanking and Draining*, several on surveying, many on mathematics and natural history.

Between March and mid-July 1779 Tofield was active on botanical field trips. His health then deteriorated rapidly and he died on 18 August 1779 leaving no will.

A. W. SKEMPTON

[Doncaster Corporation Minutes (1761–1771) Doncaster Archives; Potteric Carr Trustees Proceedings (1772–1778); Potteric Carr Drainage Board, Bawtry; Deeping Fen Adventurers Minute Book (1768–1772) Lincolnshire Archives Office; Court of Sewers for Hatfield Chase Record Book (1771–1789) Nottingham University Library; Meeting on Muston and Yedingham Drainage (1772–1773) East Riding RO; Leigh & Sotheby (1780) *Catalogue of the Library of Thomas Tofield*; J. Farey (1845) Letters from Grundy, Tofield and Smeaton 1770, *Min Proc ICE*, **4**, 205–209; W. H. Wheeler (1896) *History of the Fen of South Lincolnshire*; C. Hadfield (1973) *Canals of Yorkshire and North East England*; P. Skidmore, M. J. Dolby and M. D. Hooper (1981) *Thomas Tofield of Wilsic*; A. W. Skempton (1981) *John Smeaton*]

Publications

1768. *Reports on a Survey of Deeping Fen, for the more effectual Drainage thereof*

1772. *Report on the Practicability of making a Navigable Canal from the River Dun to the River Trent*

1773. *A Scheme for effectively Securing the Land of Hatfield Chase*

1773. *Report on the present State of a large Tract of Low Ground, extending from Muston to Malton* [with] a *Scheme for the Drainage thereof*

Works

1766–1777. Potteric Carr, drainage, £9,000

TOLL(E)Y, William, Lt. Col (d. 1784), military engineer, was trained at the Royal Military Academy and served as an Ensign in the Corps of Engineers from 1759 and participated in the siege of Belleisle of 1761 before resigning his Commission and returning to England in 1763. In that year he was appointed Principal Engineer at Fort Marlborough, where from 1765 until 1766 he commanded a train of artillery. In 1767 he was transferred as Captain to Bengal and there he remained for the remainder of his career, being ultimately promoted to the rank of Lieutenant-Colonel in the Artillery.

Tolly's chief contribution to civil engineering was the construction of what became known as Tolly's Nullah, a navigable canal linking the Hugli (Hooghly) near Kidderpore to the Bidyádhari at Samookpota. Previous to its construction it was difficult for boats to reach Calcutta and in certain seasons the Nullah became the only navigable link between Calcutta and the Eastern Bengal waterways. Around 1775 Tolly identified the potential of such a link, making use in part of an old channel, and obtained approval from the East India Company's Calcutta authority for what was nominally a private venture. In return for its construction Tolly was to receive a twelve-year lease with a toll of 1% on shipping using the cut. He obtained a loan of 100,000 [*sicca*] rupees for the work, which began in 1776 and was completed the following year.

Although it subsequently required deepening and widening the canal proved a tremendous success. Unfortunately, the line made use of land earmarked by Major Henry Watson (q.v.) for a similar scheme and dock, which he had been promoting since the 1760s and for which he obtained approval from the Court of the East India Company in London. Watson's scheme became bogged down in controversy not only over Tolly's scheme but also other landowners' rights in the area. Watson began negotiations with Tolly to take over his lease but unfortunately Tolly died on a voyage back to England in 1784 and his widow was able to get the lease extended a further fifteen years until 1804. At that time the land passed into Company control.

Tolly's first wife had died prior to his arrival in Bengal and he married Anna Maria Theresa Hintz (d. 1797) on 11 April 1768 in Calcutta. Four of their sons, Charles Edward (b. 1776), Henry Dunbar (1780–1837), John George (b. 1779) and William (b. 1775) pursued part of their careers in the army.

MIKE CHRIMES

[Biographical index, India Office collections, BL; India (1885) *Reports on Docks at Calcutta*; India (1904) *Selections from the Records of the Bengal Government … 1865–1904*]

Works

1776–1777. Tolly's Nullah, *c.* 18 miles

TOPPING, Michael (fl. 1785–d. 1796), was originally sent to Madras by the East India Company as an astronomer but his career is of equal interest for its surveying and engineering content. As a surveyor he was the equal of James Rennell (q.v.) but, owing to the Madras government's refusal to appoint an official surveyor general until 1810, he never held a position commensurate with his talents, having to settle for that of 'surveyor and astronomer on the coast'. He considered himself equally unappreciated as an engineer complaining that no account was taken of the 'office I virtually fill of civil engineer'. In response to this, in 1795 he was made 'superintendent of tanks and water courses' (in effect the first public works department in India) at a salary of 4,000 pagodas per month.

Nothing is known of his life before he arrived in India in 1785, having measured the longitudes of Ceylon and the Maldives en route. Immediately on arrival he did the same from Masulipatam to Calcutta, following it up in 1787 with a chart of the currents in the Bay of Bengal. Described as a person 'of very considerable mathematical and geographical knowledge', he was then hired to make an accurate survey of the 'sea coast from Madras to the southernmost extremity of the peninsular'. As no ship could be made available he, in his own words, 'conducted a series of triangles near 300 miles in length … travelling some thousands of miles', the first systematic triangulation used in India which he recommended should be extended to the whole of the country. However since the actual occasion of this survey was 'the late dreadful calamity', the sinking of an East India ship in Coringa Bay his main purpose was to examine the coast for a safe harbour and port of shelter for large ships. To this end he surveyed the mouth of the Godavary River but reported (18 September 1790) that this site was impossible, the previous survey in favour having been totally inaccurate. He preferred a safe and convenient harbour in the roads of Coringa after taking a survey of soundings and making systematic observations of the tides in Coringa Bay in 1789. The survey and report had obviously not been surpassed by 1855 when they were belatedly published by the government of Madras.

In 1793 Topping was sent to survey the Kistna and Godavary rivers to provide measurements for Alexander Beatson's (q.v.) irrigation scheme. Levels were taken of the Kistna 'with an excellent instrument of Mr. Ramsden's construction' from Masulipatam to Bezwada, a similar survey of the Godavery being prepared by his assistant James Caldwell (q.v.). A preliminary report of 14 February 1794 stated that it would be more economic to irrigate with a series of small cuts rather than a large dam. Topping also produced a scheme for the irrigation of Masulipatam itself. He considered (prophetically since he was to die of fever there in 1796) that it was the most unhealthy spot on the coast, being built on a swamp created by sea and land floods from which it needed to be secured by means of an embankment. Unlike the Kistna Godavary this plan was carried out and a map exists of the port and environs of Masulipatam showing the embankments, canal of navigation and other improvements executed in 1795–1796.

Also in 1795 he was sent to report on the repair of reservoirs and watercourses in the company's districts. He felt that the main problem was a lack of surveyors which he remedied by establishing a school of assistant revenue surveyors on 17 May 1794, the earliest establishment for engineering education in India.

That he had not neglected his original astronomic tasks is noted on his epitaph: 'here lies buried an honest gentleman Michael Topping of great intellectual gifts ... [and] of outstanding mechanical ability ... having been sent out to this country ... for the advancement of astronomy he designed and erected near Fort St. George an astronomical observatory ... he succumbed to a fever January 7 1796 aged 48'. This observatory remained in use until 1899. His versatility is demonstrated by the fact that his position of company's astronomer, geographical and marine surveyor-in-chief was divided among three successors.

SUSAN HOTS

[T. Jervis (1837) Memoir on the origin progress and present state of the surveys in India, *J R Geographical Soc*, **7**, 136; E. Buckle (1839) Wulloorpullum sluice, *Prof Papers, Madras Engineers*, **1**;

D. Sim (1839) Coleroon anicuts, *Prof Papers, Madras Engineers*, **1**; E. Buckle (1846) Report on the resources of the districts bordering the Kistna, *Prof Papers, Madras Engineers*, **2**, 1–13; R. B. Smith (1856) The Cauvery, Kistna and Godavary, 1–47; E. W. C. Sandes (1935) *The Military Engineer in India*, **24**, 190; R. H. Phillimore (1945) *Historical Records of the Survey of India*, **1** (with sketch by John Smart)]

Publications

1792. An account of the measurement of a base line upon the sea beach near Porto Novo on the coast of Coromandel, *Phil Trans R Soc*, **82**, 99

1798. *A Map of the Countries between the Kistna and Gaudavari Rivers with a Survey of these Rivers by the Late Mr. Topping and Captain Caldwell*

TOWNSHEND, Thomas (*c.* 1771–1846), was a civil engineer and contractor. The circumstances of his birth and early years are obscure but a payment from John Rennie (q.v.) in 1794 for duties on the Lancaster Canal indicates that by then he was in the service of that engineer. Between 1796 and 1802, whilst acting as Resident Engineer on the Rochdale Canal, he produced a plan for deviations in its line between Rochdale and Manchester. This was followed by a decade of association with Ireland in which he reported to Rennie on the state of the Royal Canal (1802) and on two occasions to the Commissioners appointed by the Government to enquire into the drainage of bogs and their subsequent cultivation. A leading feature of his contribution to this subject, in the report published in 1810–1811, was his belief that projected major new drainage channels might be used as canals or navigable drains. He was assisted in this work by his relative John Thomas (q.v.).

In 1812 Townshend was appointed Resident Engineer and Surveyor of the Witham Navigation Company with the specific task of implementing the scheme of improvement for this waterway drawn up by Rennie in the previous year. His responsibilities, which were extended in 1814 to include the supervision of a contract originally awarded to Francis Pinkerton, came to an abrupt end in 1817 when financial problems compelled the company to suspend its programme of new works. Townshend had been associated with his brother in his Irish work, but the latter died in 1818, while working on the Eau Brink cut, leaving his son, Richard (1807–1888), to be trained by Thomas.

Rennie's recommendation had assisted in securing Townshend's appointment with the Witham, and subsequently, no doubt, the Eau Brink Drainage, where he is noted as superintendent in July 1819. His salary was £100 a quarter. The works, for which John Rennie was engineer, although Thomas Telford (q.v.) was involved as adviser to the Navigation, had been delayed due to controversy over the design, with Joseph

Huddart (q.v.) acting as arbiter. Jolliffe and Banks (q.v.) were awarded the contract in May 1819, but previous to that Townshend had been directing a labour force, and evidently carried out other works on this basis outside the main contracts. Following Rennie's death in 1821 Townshend briefly acted as Engineer before Sir John Rennie (q.v.) was appointed at the end of October. He was awarded a cup of 20 guineas for these services. The main contracts were completed in July, although lesser work continued.

In 1825 Telford supported Townshend's tender for work on the Birmingham Canal which, with subsequent contracts over the following six years, finally resulted in payments of £243,248. His contribution to the very extensive schedule of improvements made by the Birmingham company at this time included work, usually in the form of excavations, at Sandy Turn, Rotton Park, Birmingham Heath, Winson Green, the Anson Branch, the Island line and, most notably, in lowering the summit of the canal by a huge cutting at Smethwick which accounted for approximately half of the above award. He also built the Galton Bridge, with ironwork supplied by Horseley Ironworks, the only one of Telford's standard large spans not supplied by William Hazledine. The completion of this work ahead of the time set in his contracts earned Townshend the praise of both Telford and the Resident Engineer, William Mackenzie (q.v.), and ensured favourable consideration for his tender to excavate a branch between Bloomfield and Deepfield. This extension included driving Coseley Tunnel, which gave rise to more difficulties than had been anticipated and delayed the opening of the branch. Although an inspection by James Walker (q.v.) in 1837 exonerated Townshend from any blame for the collapse of a tunnel shaft, this episode came as an additional problem in a period which culminated in the bankruptcy of his business.

The extent of Townshend's commitment to railway construction in the mid-1830s reflected both his desire for new work and, in a situation of huge railway expansion, the comparative scarcity of experienced contractors. Between September 1834 and August 1835, he took on railway assignments with a combined tender price of £362,063. This consisted of the Tring Contract on the London & Birmingham Railway, for which the sureties were Sir Edward Banks and John Thomas, and the Madeley, Wolverhampton, Hamstead and Birmingham contracts of the Grand Junction Railway (Birmingham to Warrington). During 1836, the works at Tring cutting, which required a labour force of about five-hundred men, began to suffer from a shortage of workers, partly as a result of the relative absence of existing accommodation between there and Mentmore. The London & Birmingham had assisted other contractors with loans for temporary housing but it refused to help Townshend and, finally, took over the contract when he abandoned it in October 1837. Robert Stephenson's (q.v.) revised estimate now

increased the anticipated cost of this contract by 38% in excess of Townshend's tender price of £104,496. This contractor's failure was unexpected, wrote Peter Lecount in *The History of the Railway connecting London & Birmingham* because he was 'a man of capital and talent, and had established a reputation for years as an able contractor'; the root of his problems, Lecount claimed, lay with events on the Grand Junction. At the same time as the Tring contract was running down, work on the Grand Junction continued only because the company made ad hoc payments beyond the original tender price. Things staggered on in this fashion until, in July 1837, on the eve of the opening of the line, he surrendered his contracts. The causes of is failure on this line may have been more complex than those on the London & Birmingham and included the disruption arising from changes made in the original specification by the Chief Engineer, Joseph Locke, and delays in obtaining possession of the land. It is also reasonable to speculate that Townshend suffered from the substantial rise in the costs of labour which occurred in the mid-1830s as a result of the demands of railway expansion; moreover, at his comparatively advanced age for a contractor, he may have been overwhelmed by the burden of a heavy and novel involvement in railway work.

Townshend was declared a bankrupt in October 1837 owing £24,212. Within seven months his creditors had been paid off and he returned to contracting with assignments in the early 1840s for the Tame Valley and Walsall Junction canal companies; the latter included the not inconsiderable task of building eight locks.

Apart for the year in which it occurred (1846), nothing is known about the death and burial of Thomas Townshend. He had a large family and his wife, Elizabeth had severe health problems, resulting in her confinement in a lunatic asylum in 1826. His children included daughters Susan, Catherine, Jane, Eliza and Hannah, and three sons, John, Benjamine and Thomas all who became involved in engineering. Thomas junior. was a contractor on the Hull and Selby Railway, the Bridgwater–Taunton section of the Bristol and Exeter Railway, and the Whitehaven and Furness railway *c.* 1837–1850. Townshend's nephew, Richard Townshend, who had joined the Institution of Civil Engineers when chief draughtsman of Sheerness Dockyard in the late 1820s, worked with Townshend on the Tring Cutting, a London–Birmingham Canal Survey and Rennie's London & Brighton Railway survey. He became a civil engineer with the Admiralty.

DAVID BROOKE

[In PRO: *Birmingham Canal, Committee Minutes* (1822–1838); *Reports to the Committee* (1826–1837); *Ledgers* (1820–1846); RAIL 810/12, 810/13, 810/99, 810/231, 810/232; *Witham Navigation—General Committee Minutes* (1762–1867), RAIL 885/1, 885/2; *London & Birmingham Railway, London Committee* (1833–1837); *Engineer's*

Reports (1834–1837); contract with T. Townshend for the line Tring–Mentmore (1834); *Contracts, Prices and Supervisors* (1838), RAIL 384/40–384/44, 384/101–384/103, 384/168, 384/203; *Grand Junction Railway, Directors' Minutes* (1833–1837); contracts with T. Townshend (1834–1835), RAIL 220/1, 220/24, 220/26, 220/27, 220/28; *Report to John Rennie by Thomas Townshend on the Condition of the Royal Canal* (5 October 1802). Reports of John Rennie, ICE archives; Reports on the State of the Wolverhampton and Madeley Contracts, Grand Junction Railway (undated), ICE archives; Frank D. Smith Files, ICE archives; Mackenzie collection and diaries, ICE archives; Eau Brink Cut Commissioners & Navigation Minutes, Cambs RO; Townshend correspondence, Gen. 835, Edinburgh University Library Special Collections; Second Report of the Commissioners appointed to enquire into the Nature and Extent of the Several Bogs in Ireland; and the Practicality of Draining and Cultivating them, *Parliamentary Papers*, **VI**, 1810–1811, 579; Third Report of the Commissioners—and Cultivating them, *Parliamentary Papers*, **VI**, Part 1, 1813–1814, 1; Obituary of Richard Townshend (1888) *Min Proc ICE*, **93**, 492–495; C. Hadfield and G. Biddle (1970) *The Canals of North West England*, **2**; Skempton]

Works
1794. Lancaster Canal, assistant engineer
1796–1802. Rochdale Canal, resident engineer
1802–1810. Royal Canal, Ireland, assistant engineer
1810–1812. Bog Surveys, Ireland, assistant engineer
1812–1817. Witham Canal, resident engineer and surveyor, £100,000
1818–1824. Eau Brink Cut, resident engineer, c. £300,000 (payments to contractor)
1820–1821. Bottisham and Swaffham Fen, resident engineer, £6,400
1825–1830. Birmingham Canal, improvement, contractor, payment:£243,248
1829. Hampton Lucy Bridge, Warwickshire, principal contractor, 60 ft. cast iron arch supplied by Horseley ironworks
c. 1832–1833. Yore Navigation, improvement, Carrow Bridge, Norwich, double leaf bascule, and Haddiscoe cut, contractor
1832–1837. Birmingham Canal, improvement, contractor, £66,738
1834–1837. Grand Junction Railway, construction contracts, contractor, contract £257,567
1834–1837. London–Birmingham Railway, construction, contractor, £104,496
1840–1842. Walsall Junction Canal, improvement, contractor, payment £24,942
1840–1843. Tame Valley Canal, improvement, contractor, payment £38,026

TRAIL, Sir John (c. 1725–1787), was a canal engineer working in Ireland. Work commenced on the Grand Canal in 1756 but little progress was made during the early years of construction. Dublin Corporation were interested in the canal as a source of water and in 1768 the pipe-water

committee of Dublin Corporation advertised for a new contractor to try to achieve some progress. John Trail laid his proposals for carrying on the works before them and was duly engaged. He was described as 'a judicious and intelligent person and has received the approbation of Mr. Edgar, the officer of the Navigation Board'. In the following year, Trail gave evidence before a committee of the House of Commons to the effect that the line of the Grand Canal was practicable, including the section through the Bog of Allen.

By 1772 a considerable amount of money had been spent but there was little progress and a controversy was continuing between Trail and Charles Vallencey (q.v.) as to what line the canal should take to cross the river Liffey. In that year the Company of the Undertakers of the Grand Canal was incorporated and Trail was appointed Engineer to the company. John Smeaton came over with his assistant, William Jessop (qq.v.), to report on the canal and Jessop remained to carry out the necessary surveys with Trail.

Difficulties arose with work on the canal, in particular the standard of workmanship and exceeding the estimates; as a result Trail was dismissed in 1777 and replaced by a Captain Tarrant. It was later recognised that Trail had been somewhat of a pioneer in planning and supervising the construction of such a major project and he was subsequently knighted for his services to Dublin Corporation. He was a member of the Royal Dublin Society from 1772 to 1787.

RON COX

[R. Delany (1970) John Trail—Grand Canal Engineer, *J Kildare Archaeol Hist Soc*, **14**, 626–630]

Publications
1771. *The Report of John Trail, concerning the Practicability and Expence of compleating the Grand Canal from Dublin to Tullamore*

Works
1756–1777. Grand Canal, surveys and construction, contractor/engineer

TREACHER, John, Sr. (c. 1735–1802), of Sonning, carpenter, founder of a firm of builders and Surveyor (of works) on the Thames Navigation, is first heard of in connection with the navigation in November 1771 when recommended by Joesph Nickalls (q.v.) to join John Smith of Sonning on the contract to build Sonning pound lock. Like the seven other locks built in this period between Maidenhead and Sonning it was framed in fir or 'Memel' timber, 133 ft. long between gates, 18 ft. wide, with 3½ ft. of water on the sills: finished in 1773 it cost about £1,000.

Treacher next began work on the towpath at Marlow in March 1773, and six months later at Shiplake, using gravel dredged from the river (about 120 tons per week). Then in 1786 Smith and Treacher contracted to rebuild Shiplake lock in oak under the direction of John Clarke (q.v.),

Surveyor of the Lower Districts (Staines to Mapledurham).

In August 1786 work began on the first of a new set of locks in the Upper Districts, at Whitchurch. It was built by direct labour to Clarke's plan and under his supervision, and opened in June 1787. Meanwhile in January and February 1787 Clarke had prepared plans and estimates for the next two locks, at Goring and Cleeve, and he was still working on the rebuilding programme of the 1773 locks lower down the river. Clearly he could not continue to superintend the Lower Districts while undertaking new works many miles upstream, so in March 1787 Treacher was appointed to take charge at Goring and Cleeve and become Surveyor of the Upper Districts, at a salary of 2 guineas per week 'for his constant attendance'.

Carpenters began making pile engines and piles at both sites and making lock gates in Treacher's yard in Sonning. Construction started in May with about thirty labourers and eight carpenters, and both locks were completed by February 1788. In the same manner, but now with Robert Treacher as foreman (at 1 guinea per week) Benson and Days locks were built in 1788–1789. Each of the locks, timber-framed in oak, length about 125 ft. and 18 ft. wide, cost around £1,100 but ancillary works such as lock cuts, modifications to (existing) weirs, dredging and making towpaths added a good deal more. Further works upstream at Sutton and Abingdon, including a new stone-built lock—contractors Edward Edge and John Owen—completed in 1791 brought the total expenditure to about £8,000.

In 1789, following a report by William Jessop (q.v.), work began on locks on the river from Oxford to Lechlade. On this length of 32 miles six masonry locks (110 ft. × 15 ft.) were built by 1792, nominally under Treacher's direction but with Daniel Harris and Josiah Clowes (q.v.) as engineers in charge of three locks each. The contractors included John Nock, Samuel Weston (q.v.) and Edward Edge. By 1792, then, the Thames was equipped with pound locks (twenty-four in number) from Maidenhead to Lechlade, a length of 94 miles.

In March 1791 Treacher was promoted to General Surveyor of the whole navigation under the Thames Commission (upstream of Staines) with an increase in salary of 20 guineas p.a., though Clarke remained as Surveyor of the Lower Districts. Both of them gave evidence in 1793 to a committee of the House of Commons. Clarke reported that navigation above Maidenhead was good, but from Maidenhead down to Staines it was bad and three locks and weirs should be built in this stretch. After long discussions, Treacher drew up plans for the first of these, at Romney, and Clarke supervised construction between 1796 and 1798 with Joseph Kimber as contractor.

Treacher's last work was Blake's Lock on the Kennet, used by Thames barges in getting to the Reading wharf. He submitted plans and estimates in February 1801. Contracts were let to Smith and John Treacher Jr. for the lock and Mordecai Brookes for the cut. Opened in the summer of 1802 at a total cost of £2,800 the lock had oak timber walls.

Long before starting work on the Thames, Treacher married Elizabeth Simmons in 1756. She died in 1791. Treacher was buried at Sonning on 13 January 1802 and his eldest son, John (b. 1760), succeeded him as Surveyor of the Upper Districts.

A. W. SKEMPTON

[*Minutes Books of the Thames Navigation Commission* (1771–1802) Berkshire RO; *Treacher Papers* (1773–1798) Reading Central Library; *Report from the Committee on the Navigation of the Thames* (1793) House of Commons; A. W. Skempton (1984) Engineering on the Thames Navigation, *Trans Newcomen Soc*, **55**, 153–176]

Works
1787–1791. Locks on the Thames, Goring to Abingdon. £8,000
1796–1798. Romney lock and weir, £2,200
1801–1802. Blake's lock, £2,800

TREACHER, John, Jr. (1760–1836), was Surveyor (of works) to the Thames Navigation Commission. He was baptised in February 1760 the eldest son of John Treacher of Sonning (q.v.) and his wife, Elizabeth Simmons. There can be little doubt that he worked in the building firm founded by his father and may well have headed the firm after 1787 when his father took on the Thames Navigation. In that year he married Elizabeth Collier. With John Smith of Sonning as partner he contracted in April 1801 to build Blake's lock on the Kennet, used by Thames barges in getting to the Reading wharf.

His father, who went on to become General Surveyor in 1791, died in January 1802. At that time John Clarke (q.v.) was Surveyor of the Lower Districts (Staine to Mapledurham) with James Millard as Surveyor of the Upper Districts (up to Lechlade). For two years Treacher worked with Millard, but the latter was dismissed in September 1803 and a few months later Clarke died. In February 1804 Treacher was appointed Surveyor of the Upper Districts at a salary of £140 p.a. with an allowance for an assistant in the navigation above Oxford, and Zachary Allnutt (q.v.) took over from Clarke. They shared responsibility for the navigation as a whole until Allnutt retired (to become General Receiver) in 1821. Treacher was then appointed General Surveyor with a salary of £320 p.a. His son George Treacher (1791–1863) was appointed Assistant Surveyor in 1824.

Treacher's first big job on the Thames was the 1400 yd. long cut and a masonry-walled pound lock at Culham, bypassing a very cumbersome arrangement at Sutton Courtenay mill. The survey had been made by Allnutt, but Treacher directed the works which were carried out from 1808 to 1810 at a total of nearly £9,000. He received a gratuity of £100 for his services.

Meanwhile he was consulted by Wallingford Corporation on their bridge over the Thames, the five river arches of which were destroyed by a flood in 1809. In place of these he designed three arches (the centre one to have a much increased span of 49 ft.) in a simple classical style. Construction was carried out between 1810 and 1812 by Richard Clarke as principal contractor under Treacher's supervision.

Treacher also held the post of Surveyor of Bridges for Berkshire from October 1812, receiving a retainer of £5 p.a. plus fees for any 'extraordinary trouble' on specific works. From 1814 his duties applied only in the eastern part of the County; Thomas Adey of Newbury thereafter had responsibility for the western part.

Following the successful Culham cut, the Commission considered applying a similar solution to problems presented by rocky shoals and an awkward bend in the river further downstream near Clifton Hampden. Treacher drew up plans but owing to legal complications over land purchase it was not until 1820 that work began on the 900 yd. cut and stone-built pound lock, completed in 1822 under Treacher's direction at a cost of £6,700.

Attention had now reverted to that part of the Lower Districts between Staines and Maidenhead where no locks had been built before the 1790s. John Clarke in evidence to Parliament in 1793 recommended the construction of three locks, with weirs. One of these, at Romney, was built between 1796 and 1798. It proved very successful but for no obvious reason nothing more was done until Bell and Old Windsor locks were built in 1817–1818 and 1821–1822. Treacher and Allnutt prepared plans and estimates, though it seems that Treacher supervised construction at Bell, and certainly at Old Windsor. George Treacher contracted for the lock and Ambrose Oliver for the 1,000 yd. cut at Old Windsor: the total cost of works there was £9,300.

Finally, Treacher planned three new locks on the river upstream of Maidenhead. (1) At Marlow a lock with masonry walls and planked floor, 135 ft. between gates and 20 ft. wide, replacing in a new position the old timber-framed lock, and a new weir; built 1825–1826 at a cost of £4,600 under supervision by George Treacher who was awarded £150 as testimony for his services. (2) Similarly a new Boulters lock (replacing the old one) on a new 600 yd. cut, built 1827–1828 at a cost of £11,800 with George Treacher in charge of construction; he again received a gratuity 'as testimony of the sense entertained by the Commissioners of the value of the services rendered by him upon various occasions and particularly in the construction of the new Boulters Pound Lock'. (3) A completely new lock and 900 yd. cut at Cookham, built 1829–1830 at a cost of £8,400.

The 108 mile long navigation from Staines to Lechlade, administered by the Thames Commissioners, was now approaching its present pattern; only two more locks, at Boveney and Bray, were added in 1838 and 1845. Treacher was buried at Reading in March 1836, aged seventy-six. George Treacher succeeded his father as General Surveyor in May 1862 and held this post until he retired in 1862.

A. W. SKEMPTON

[Minutes Books of the Thames Navigation Commissioners (1771–1836) Berkshire RO; Treacher Papers (1773–1836) Reading Central Library; Zachary Allnutt & John Treacher (1815) An Account of the several Pound Locks on the Thames Navigation, MS, Berkshire at Marlow and Ray Mill (Boulters), Berkshire RO; Colvin (2); G. Phillips (1981) *Thames Crossings*; A. W. Skempton (1984) Engineering on the Thames Navigation, *Trans Newcomen Soc*, **55**, 153–176]

Works
1809–1830. Wallingford Bridge, rebuilding, £7,000

TREDGOLD, Thomas (1788–1829), technical auth-or and consultant, was born on 22 August 1788 at Brandon, near Durham, the son of Thomas Tredgold and Rosamond Atkinson. He is now best remembered as the author of the definition of Civil Engineering in the first Charter of the Institution of Civil Engineers (1828). When he wrote this his reputation was at its peak; within thirteen months, his health broken by his industry, he died leaving his family to appeal to his fellow civil engineers for charity. Despite his financial failure he was the most influential technical author of his generation and possibly of the nineteenth century.

Little is known about Tredgold's family background in Durham. His mother was the sister of William Atkinson, who as an architect was a future employer of Thomas. He had two younger brothers; one, Ralph, later became the engraver of many of the plates in Thomas's books. At the age of fourteen Tredgold was apprenticed to a local carpenter for six years and then went to Scotland as a journeyman carpenter from 1808 until 1813. There is some uncertainty about the date of his subsequent move to London. The *Encyclopaedia Britannica* (1842) implies 1813: if, however, Tredgold was the 'contractor' for the New Barracks on Hampton Court Green he may have arrived in 1812, or even 1811.

He married Sally Burton on 30 December 1819. They had five children; of the two boys, one died early, while the other, Thomas, became an engineer. Two of the girls survived their mother (d. 1831) but one, the eldest, soon followed her. The other girl and Thomas Jr. were brought up by their aunt, Jemima Urquart (*née* Burton), in Southwark. Thomas Jr. was apprenticed to Bryan Donkin (q.v.) and later made some of the tests for Eaton Hodgkinson for the report into the *Application of Iron to Railway Structures*. He went to India as an engineer in the East India Company Office of Stamps and died there in 1853 aged twenty-eight.

From the time of his arrival in London up to 1820, Tredgold lived with Atkinson at 20

Thomas Tredgold

Bentinck Street, Marylebone, London; after his marriage his addresses were 2 Grove Terrace, Lisson Grove, and from September 1823 until his death, 16 Grove Place, Lisson Grove. In London, Tredgold joined the office of William Atkinson who, born in Bishop Auckland, began life as a carpenter. He then became a pupil of James Wyatt and moved his practice to London in 1813 when he succeeded Wyatt as Architect to the Board of Ordnance, a post he held until the abolition of the Office in 1829. During Tredgold's stay in Scotland, Atkinson was involved with the re construction of Scone Palace, Rossie Priory, Rosebery (Midlothian) and Bowhill. Although there is no evidence that Tredgold worked on any of these houses it is likely that it was through Atkinson that he found employment in Scotland. Tredgold described himself as Atkinson's principal assistant and as such it is probable that he acted as the practice's engineer with responsibility for structural designs in timber and cast iron, and, in later years, for heating schemes. Tredgold remained with Atkinson until 1823 when he began his own practice as a civil engineer and author.

Tredgold began his career as a technical author in a modest fashion. In July 1815 he wrote a two-page note for the *Philosophical Magazine* on the comparative strength of beams. A stream of six equally short notes in the same magazine followed until January 1816. Four of these were on the strength of materials, one on the elasticity of fluids and one on the effect of increasing temperature on the volume of water. Two years later, in

December 1817, he wrote his next contribution, *On the Resistance of Solids*, fifteen pages of standard theory with tables of strength properties compiled from various sources. The year 1818 was very productive with six notes on a variety of topics; in 1819 there were only three papers and in 1820, the year in which his first book, *Elementary Principles of Carpentry*, was published, just one.

Tredgold remained a regular contributor to the *Philosophical Magazine* with nine notes during the years 1822 to 1828. He also contributed for short periods to several other magazines. Between April 1818 and October 1819 he wrote six articles for the *New Monthly Magazine*. Henceforth his books were his main publications, but after 1825 he found time to make four contributions to the *Edinburgh Philosophical Journal* and two to each of the *Gardener's Magazine*, the *Repertory of Patent Inventions* and the *Mechanics' Magazine*. He also wrote articles on Joinery, Stone Masonry, and Engraving on Steel for the *Encyclopaedia Britannica*. A comprehensive assessment of the quality of his papers has yet to be made. Although the early notes were probably written partly for publicity purposes, they were also an outlet for the excitement he felt in expanding his knowledge and were an opportunity for him to help society.

Having worked as a carpenter, it was appropriate that Tredgold's first major publication, *Elementary Principles of Carpentry*, published in 1820, should deal with this subject. By 1820 cast iron was firmly established as a structural material and engineers were forced to reassess timber's role. Undaunted, Tredgold drew on experience and marshalled his extensive reading of previous publications on both carpentry and the strength of materials; the result was the first book on carpentry to be based on engineering principles and was the first timber engineering manual; its author could be described as Britain's first timber engineer. It became the standard reference work on the subject for the next century; two editions were published during his lifetime and subsequently some twenty-five editions from various publishers appeared, culminating in a final version in 1946, one hundred and seventeen years after his death.

The title page of the first edition indicates that Tredgold intended to cover all aspects of the mechanical principles of the art of carpentry. The book presents a design philosophy in addition to illustrating the design of building components. In size and scope it surpassed all its predecessors. The only omission, and that a deliberate one, was any mention of descriptive geometry, which he considered had been ably handled by Peter Nicholson (q.v.). Tredgold's book was the first British one devoted entirely, by title and contents, to carpentry, as distinct from carpentry and joinery.

Many treatises on carpentry had been written in the previous century but Tredgold criticised them on the grounds that 'in none of them is to be found anything on the mechanical principles of the art, except it be a few rules for calculating the strength of timber; and these are founded on erroneous views of the subject; and therefore not to be relied upon'. (Tredgold, 1820, vii) His philosophy of design is clearly defined in the Preface and in it he states that a 'just economy of materials should be one of the first objects of the builder's attention, and this desirable object is to be obtained only by judicious combinations of the materials to be used'.

Tredgold meticulously quotes his sources. His predecessors frequently emphasised architecture and joinery at the expense of carpentry; none concentrated on carpentry as he did. Nevertheless he owed, and acknowledged, a debt to five major authors: Francis Price (q.v.), James Smith, Peter Nicholson, John Robison and Thomas Young (qq.v.). Price (1733) and Smith (1733) had taken the first steps to show that carpentry was an art in its own right and not merely a facet of architecture.

When Tredgold discusses the strength of timber he quotes the well-known test data obtained by Emerson and Barlow; the theoretical behaviour in bending and compression is obtained from Robison and Young. These are a few of the British authors quoted, but they give a good indication of the breadth of his reading. When he quotes French sources it is difficult to know if he was making use of his own reading of them or whether he relied on summaries of the work in the various *Encyclopaedia* and in Barlow (1817).

Tredgold's arguments in his treatment of theoretical topics are difficult to follow even though they generally lead to the correct results. An example is his derivation of the deflection of a simply supported beam and he introduces—perhaps for the first time—the concept that the permissible deflection should be proportional to the length. Using load–deflection data provided by various experimenters (including himself) he tabulates the properties of oak and fir (from a number of different sources) together with values for another twenty species. Although the formula for the deflection is given, Tredgold works all his examples through a set of rules. He also shows how the constant in his formula must be modified for fixed-ended and cantilever beams of both rectangular and circular cross-section under central point and uniformly diffused (i.e. distributed) loads.

To determine the strength of beams Tredgold quotes the correct formula and refers the reader to Emerson, Gregory (1815) or Young for its derivation. He considers that the formula will give the greatest weight a beam will bear without fracture. The expression appears to have been meant for checking purposes only as he does not discuss the application of a factor of safety.

The main use of timber in Tredgold's time was for floors and consequently he devotes a complete section to this topic. The form of the rule given for establishing the sizes of single joists shows that the floor was designed to limit the

deflection (to a proportion of the span) and not the bending stress. The constants used depend on the deflection limit, the modulus of elasticity and the applied load. Tredgold, however, makes no attempt to calculate them theoretically: the constant numbers were derived from practical observations on bearings of different lengths.

The required depths of yellow fir joists at 12 in. centres are then tabulated for spans varying from 5 to 25 ft. and breadths from 1½ to 5 in. Similar tables are given for girders (at 10 ft. centres), binding joists (at 6 ft. centres) and ceiling joists (at 12 in. centres). Although various authors had previously given similar tables, it is thought that Tredgold's were the first to be based on the combination of practical experience and theory. Certainly no previous author had given a formula for the derivation of his tabulated sizes.

Tredgold's comments on floors are also based on the combination of practice and theory. He emphasises the advantage of using bridging joists in considerable lengths so that they extend over several binding joists and have increased strength and stiffness; to substantiate this increase he quotes theoretical work on fixed-ended beams by Belidor and Robison. He points out that to make a 'strong floor with a small quantity of timber, the joist should be thin and deep; but a certain degree of thickness is necessary, for the purpose of nailing the boards, and two inches is perhaps quite as thin as the joists ought to be made'. (Tredgold, 1820, 62).

Tredgold devotes several pages to the design of columns with rectangular, square and circular cross-sections (Tredgold, 1820, 48–59) and distinguishes between short columns that fail due to crushing and longer columns that suffer from bending failure and then gives design rules. He does not apply his two rules to develop design tables for timber posts (columns) to the same extent that he did for floor joists as he considers they are of limited value because the required size depends on the nature of the building. He therefore recommends that the load on each post should be estimated and the size calculated.

Tredgold also gives rules for the determination of the cross-sections of members in King and Queen post trusses. As with floors, the constants used in the rules are determined from a combination of practical experience and theory (Tredgold, 1820, 72–85) and an indication of his awareness of later practices is his analysis of Robert Seppings's (q.v.) recently introduced dockyard roofs.

The most difficult problem in carpentry is forming the joints, particularly in trusses and built-up beams. Tredgold recognises the importance of this topic and devotes a complete section to Joints and Straps. When it is remembered that the theory for shear was even less well understood than bending and compression, it is not surprising that the section is devoid of theory and concentrates on giving practical advice on the forms of joints which had proved successful in the past. (Tredgold, 1820, 136–148)

Although Tredgold generally writes clearly, his explanations of the derivation of various formulae can be very difficult to follow and it must be remembered that he inherited mistakes from Robison and Barlow. In writing his treatise, he cannot have imagined that it would be used for the next hundred years. He was attempting to fill an existing gap. He succeeded and set a high standard for those that followed

Tredgold published other books, his interest in the strength of materials leading to that on cast iron in 1822. By now he was established and in 1823 he left Atkinson's office to practise on his own. With more time, if not more money, to spare he continued his interest in building with the publication of *Warming and Ventilating* in 1824. A knowledge of heat naturally led to *Rail-roads and Carriages* in 1825 and finally, *The Steam Engine* in 1827. Even if the books failed to make his fortune, they certainly made his reputation with, during his life time, French and German editions of both *Cast Iron* and *Warming and Ventilating*, and French editions of both *Rail-roads* and *The Steam Engine*. After his death, his books were published in Spain (*Rail-roads* and *The Steam Engine*), Belgium (*The Steam Engine*) and America (*Carpentry*). In addition to his own books he edited an enlarged second edition of Robertson Buchanan's (q.v.) *Practical Essays on Mill Work and other Machinery* (1823), and *Tracts on Hydraulics* (1826).

Long after his death the name of Tredgold was still considered a valuable aid in selling a book. Although *The Steam Engine* was extensively revised by Woolhouse in 1838, with several appendices added by different authors in the 1840s, it was still known and advertised as *Tredgold's Steam Engine*.

Tredgold's book, *A Practical Essay on the Strength of Cast Iron* (1822), was intended for engineers, iron masters, architects, millwrights, founders, smiths and others engaged in the construction of buildings and machines. It contains practical rules for designers, accounts of recent tests and an extensive table of the properties of materials. This would suggest an identical work to *Carpentry*, but this was not the case. To compare the two it is necessary to consider the way in which the materials had been used in the past and the way in which they were used in 1822. Prior to 1779, timber and masonry were the major structural materials. Timber was a pattern book material with solutions that had slowly evolved with time and with fashion playing a major part in the changes; it also had the advantage of being a non-structural material capable of being used internally for joinery (particularly doors, windows and wall panelling) and furniture. The year 1779, with the building of Ironbridge, was a watershed; from that moment timber was, with the exception of housing, fighting to keep its structural market.

Nicholson (1797) had responded with the inclusion of strength properties of some species but had given no indication as to how they could be used to design beams and combinations of beams in floors. From the turn of the century the engineer could find calculations for simple beams, columns and trusses in articles in encyclopaedia. Tredgold had used this knowledge to produce the first book for carpenters based on engineering principles. The carpenter now had the choice of using the pattern book solution, which would be perfectly satisfactory for most buildings, or, if he had the knowledge, looking for improvements by recourse to the theory of mechanics. The timber content of *Carpentry* was virtually unchanged for the next fifty years; the main additions were expanded tables of strength properties and the use of metal tension members in roof trusses. Timber under Tredgold had made the change in 1820 from pattern book to design manual and had then stood still.

By 1820, cast iron had been used in churches for 50 years and for inner structures of mills for twenty years or so: scores of bridges had been built. However it was still an exciting new structural material and was attracting the innovative designer. *Cast Iron*, was about a new material looking for new markets. When the second edition appeared in 1824 it had doubled in size and the third edition of 1831 (two years after Tredgold's death) was improved and enlarged. The fourth edition, now consisting of two volumes and under the editorship of Eaton Hodgkinson, was published in 1842 but by then cast iron was no longer a new structural material: it was competing with its replacement, wrought iron. To have remained in competition in the 1820s and 1830s, timber would have needed more efficient fastening devices, improved preservative treatments, waterproof adhesives and structural plywood, techniques not available for another hundred years.

It was against this background that Tredgold wrote *Cast Iron*. In Sutherland's (1979) view he produced a book that was an assemblage of many things: load tables, test results, mathematical expressions and rules for the design of beams, columns and tie rods, and tables of strength properties. It was not a book for designers of complete buildings but a manual for iron component designers. Although Barlow revised *Carpentry*, Tredgold would have recognised immediately the 1840 edition, but he would have been in a different world with the 1842 edition of *Cast Iron*. His influence on timber was long-term; on cast iron it was short-term.

Cast Iron contains tables to aid the design of beams but it is not made clear what governs the design. Sometimes it appears to be deflection, sometimes the elastic limit (he warns against permanent set but appears to be happy to use the elastic limit as a working stress). Shear stresses in beams were not understood until the 1860s and are not mentioned. Tredgold spends a considerable amount of space dealing with columns but his arguments are difficult to follow. They did, however, lead to what became known as the

Tredgold column formula which was widely used during the rest of the nineteenth century. An overall view of the book on cast iron is that it is now open to criticism. The theoretical mechanics contains mistakes although the correct solutions were well known on the continent. Tredgold develops a fairly consistent set of principles for the design of beams and columns, although today they would not be considered safe enough. Where *Carpentry* dealt with an old material in a new way and was an outstanding book, *Cast Iron* was a first step towards the use of a new material.

Having considered Tredgold's book on timber, it is astonishing to find that no timber structure can definitely be attributed to him although doubtless he was responsible for the carpentry in the houses designed by Atkinson's practice and presumably he undertook work in connection with Atkinson's appointment as Architect to the Board of Ordnance. The only structures that can definitely be attributed to him are the floor and roof of the Riding School of the Horse Bazaar, King Street, Portman Square where he covered the 46 ft. span with cast iron trussed girders and timber secondary joists and purlins. (Nicholson, 1826, 24)

The same mystery applies to his work on warming. His book contains no references to any of his work but he did, however, design the system for the Duke of Northumberland's hot houses at Syon House, Brentford. At the time it was 'the most complete work of the kind and on the most extensive scale, that has hitherto been executed in this or any country' (*Gardener's Magazine*, 1829).

Apart from Tredgold's many publications, few details of his work are available after he left Atkinson's office. An exception is his work for John Stephen Langton (d. 1833). Tredgold's contact with Langton dates from the mid 1820s when Langton sought his opinion on a seasoning process he had invented. Tredgold investigated the theoretical and practical aspects of the process and produced *A Report on a Process for Seasoning Timber* which described the process briefly and concluded in its favour.

Soon afterwards, Langton took shares in the proposed Alford Canal and when he became doubtful about the practicability of the proposal he turned to Tredgold for advice. The canal was to join Anderby, on the Lincolnshire coast, to the town of Alford, about 7 miles inland. Three engineers were involved in the disagreement about the proposal: William Tierney Clark, Thomas Telford (qq.v.) and Tredgold.

Tredgold's practical experience on hydraulic problems had begun in his childhood with his father's involvement in scouring the river Tyne. Although knowledgeable on the theory, he does not appear to have had much practical experience other than a proposal to improve the River Lee Navigation, which had been referred to him by the Company's engineer, Bryan Donkin (q.v.) in consequence of his own ill health. Clark had

been appointed by the Canal Committee to prepare proposals for the haven and the canal and submitted a report which also obtained a favourable opinion from Telford via Clark. How ever, a major disagreement arose on the likely levels of the water in the canal due to the incoming tide. Langton thought Clark's scheme would not be able to fill the canal and persuaded the Committee that Tredgold's opinion should be sought. Tredgold's opinion was that the tide would not fill the canal as Clark predicted; indeed to achieve the levels Clark expected would require the tide to enter the canal with a speed that would damage the banks. As a means of resolving the conflict it was proposed that evidence from the two parties should be put to referees nominated by each engineer. Tredgold agreed and he suggested Thomas Young for any theoretical aspects and Benjamin Bevan (q.v.) for general engineering matters.

This was not quite the end of the story. Langton visited Telford and extracted from him a grudging acknowledgement that Tredgold had been right. In the event, no work was done on the canal but this episode illustrates Tredgold's confidence in his own ability and his annoyance that he was rarely given the opportunity to prove himself. Perhaps clients were happier with a rough practical approach and his theoretical text books did not count in his favour.

Tredgold was elected a Member of the Institution of Civil Engineers in March 1821 and during the next eight years played an important part in its development. He presented six papers between 1824 and 1828 covering a range of topics and his status within the Institution and the profession was recognised when in June 1824 he was elected an Honorary Member, 'having by his writings on various subjects connected with the Profession of a Civil Engineer conferred benefit on that Science'. In 1828 he was approached regarding the position of Secretary to the Institution; he accepted the offer, but never took up the post. The previous year, when the Council was preparing a petition to the Attorney General for the granting of a Charter, it was advised that this should contain a description of a Civil Engineer and Tredgold was asked to supply a description of the profession. His description began:

'Civil Engineering is the art of directing the great sources of power in Nature for the use and convenience of man; being that practical application of the most important principles of natural philosophy which has, in a considerable degree, realised the anticipations of Bacon, and changed the aspect and state of affairs in the whole world' (ICE, 1828).

He continued by defining the scope of civil engineering works (roads, bridges, railroads, aqueducts, canals, river navigation, docks and store houses, ports, harbours etc.) in order to distinguish between Civil Engineering and Architecture, and finally to eulogise upon the power to be gained from the control of water founded on the science of hydraulics. An abridged version was subsequently embodied in the Charter of 1828.

Although the Council failed to persuade Tredgold to become Secretary, it continued to use his literary ability and just before his death resolved that he should revise the rules and regulations of the Institution; he never completed the task. Tredgold was a dedicated professional and as his health deteriorated he worked from his bed up to his death on 28 January 1829. Although he had been ill for some time he made his will only the day before he died, giving to his 'dear wife Sally' the whole of his property. She was appointed the sole executrix and the will was witnessed by Bryan Donkin and Timothy Bramah; there is no indication of the value of the legacy. A sad little postscript completes his association with the Institution. On 2 June 1829 the Council resolved that Mrs. Tredgold's copy of the *Encyclopedie Methodique* should be purchased from her for forty pounds. Poor Mrs. Tredgold, left with four children and only a library to support them.

Tredgold's estate cannot have been substantial; by October 1829 an appeal on behalf of his family had been launched by the Donkin and Bramah families. The leading contributor was Langton, joined by engineers such as Telford and other of Tredgold's engineering friends. His widow, Sally, died in September 1831 leaving everything to Timothy Bramah, John Donkin and Matthew Habershon (architect of Mortimore Street, previously employed by Atkinson) to hold in trust for the children as their guardians. The appeal was unsuccessful and five years later the *Philosophical Magazine* (1834) wrote that Tredgold's early death had left the family in a state bordering upon destitution and, supported by Telford, exhorted its readers to contribute to the appeal.

It is difficult to assess the true nature of his contemporaries' opinions of his talents. The views that have survived have mostly come from obituaries and reviews, which tend just to list the contents of the books with very little criticism. For example, the obituary in the *Philosophical Magazine* read: 'Science and its practical application have thus lost in the prime of life a most zealous and able promoter' and the *Literary Gazette* wrote in a similar vein. The *Encyclopaedia Britannica*, writing about his books, considered that they 'have entitled him to an honourable place among those who have made important exertions for the advancement of the useful arts' and the *Philosophical Magazine*, in commending the 1834 appeal, noted that his books 'are sufficient proofs of his extraordinary talents, and of the benefits he has conferred upon his country'. The expressions that pervade contemporary views are ones of technical admiration and pity, tinged with a feeling of guilt. When George Rennie (q.v.) returned the volume of letters to Francis Bramah he was moved to write: 'I return to you your very valuable letters of the very talented and estimable Tredgold. That poor man seems to have been

little estimated during his life time except by such as yourself and a few others'.

Surprisingly, no memoir was written immediately after Tredgold's death but from the article in *Philosophical Magazine* (1834) it is clear that the characteristic that ruled his life developed during his apprenticeship and pervaded his years in Scotland: he was thirsty for knowledge and pursued it to such an extent that he damaged his inherently poor constitution. In return 'he acquired an extensive knowledge of Geometry, Chemistry, Geology, etc, and became a very excellent mathematician'. While admiring his technical skills the writer also praises his 'strong benevolent feelings for his fellow-creatures; an anxiety for the improvement of all around him, for their happiness and their good; with a readiness to communicate information, equalled only by his eagerness to acquire it'.

That he was very confident in his own ability was illustrated during the Alford episode where he was not concerned that Telford, initially, disagreed with him. From a few personal paragraphs in the Bramah and Langton letters one senses an ambitious man driven by his desire for knowledge and disappointed, if not bitter, that his efforts had not been financially rewarded. A further measure of the man is his portrait, which appeared as the frontispiece of the second edition of *The Steam Engine* and elsewhere, and shows him with a haggard face. His poor health is confirmed many times in his letters to Bramah and Langton but there is no mention of the cause. Even bearing in mind the perpetual burden of poor health, it is still curious that, in an era of great rewards for his contemporary engineering giants, Tredgold was unable to translate his undoubted wealth of theoretical and practical knowledge into financial security. Perhaps he loved learning a bit too much.

L. G. BOOTH

[N.B.: For full references see Booth (1979a,b,c; 1998). ICE (1828) Minutes of Council, 4 January, ICE documents, and *Letters to F. Bramah Esq. on Engineering Subjects*, including T. Tredgold (1823), letters to F. Bramah on engineering subjects and G. Rennie (1839) letter to Francis Bramah 13 July, ICE archives; S. P. Langton (1825) Langton Papers; T. Tredgold (1825) letters to J. S. Langton, Langton Collection, T. Tredgold (1827) Letter to J. S. Langton, 26 September, Langton 5/10; T. Tredgold (1828) *Report on a Process for Seasoning Timber Invented by John Stephen Langton Esq.*, Langton 3/6/9; T. Tredgold (1827a) *A Report on the Questions proposed by JOHN STEPHEN LANGTON, Esqr., respecting the intended Haven at Anderby, and the Canal from thence to Alford*, Langton 5/6; T. Telford (1828) letter to Alford Canal Committee, 6 March, Langton Collection 5/14; Bramah, Joseph and Sons (1829) letter to J. S. Langton, 11 February, Langton 3/7; B. Donkin and T. Bramah (1829) letter to J. S. Langton, 31 October, Langton 3/6/14, all in Langton Collection of documents,

Lincolnshire Archives, Lincoln; F. Price (1733) *A Treatise on Carpentry*; J. Smith (1733) *The Carpenter's Companion*; P. Nicholson (1792) *The Carpenter's New Guide*; W. Emerson (1794) *The Principles of Mechanics*, 4th edn.; P. Nicholson (1797) *The Carpenter's and Joiner's Assistant*; J. Robison (1797) *Encyclopaedia Brittanica*, 3rd edn.; J. Robison (1801) *Encyclopaedia Brittanica*, Supplement, 3rd edn.; T. Young (1807) *A Course of Lectures on Natural Philosophy and the Mechanical Arts*; O. Gregory (1815) *Treatise on Mechanics*, 3rd edn.; P. Barlow (1817) *An Essay on the Strength and Stress of Timber*; W. T. Clark (1825) *Report on the Practicability and Advantages of a Navigable Canal from the sea at Anderby Haven to the Town of Alford*, Langton 5/4; P. Nicholson (1826) *Practical Carpentry, Joinery and Cabinet Making*; T. Tredgold (1827b) *Steam Engine*; *Gardener's Magazine*, **5**, 1829, 240; Memoir on William Thorold (1878–1879) *Min Proc ICE*, **51**, 321–322; J. C. Loudon (1833) *Encyclopaedia of Cottage, Farm, and Villa Architecture*; Thomas Tredgold, *Mechanics Magazine*, **31**, 1839, 352; *Encyclopaedia Brittanica*, 7th edn., **21**, 1842, 348; Anon. (1834) The late Thomas Tredgold, Civil Engineer, *Philosophical Magazine*, **4**, 394–396; H. B. Hodson (1855) Letter, *The Builder*, **23**, 658, 437; Thomas Tredgold, *English Cyclopaedia*, **6** [1856–1862] 153–154; I. Todhunter and K. Pearson (1886) *A History of the Theory of Elasticity and of the Strength of Materials*; N. S. Billington (1979) Thomas Tredgold (1788–1829): some aspects of his work. Part 4. Warming and ventilating, *Trans Newcomen Soc*, **51**, 83–86; L. G. Booth (1979a) Thomas Tredgold (1788–1829): some aspects of his work. Part 1. His life, *Trans Newcomen Soc*, **51**, 57–64; L. G. Booth (1979b) Thomas Tredgold (1788–1829): some aspects of his work. Part 2. Carpentry, *Trans Newcomen Soc*, **51**, 65–70; L. G. Booth (1979c) Thomas Tredgold (1788–1829): some aspects of his work. Part 5. Tredgold's publications, *Trans Newcomen Soc*, **51**, 86–92; R. J. M. Sutherland (1979) Thomas Tredgold (1788–1829): some aspects of his work. Part 3. Cast iron, *Trans Newcomen Soc*, **51**, 71–82; L. G. Booth (1998) Thomas Tredgold (1788–1829): some further aspects of his life and work, *Trans Newcomen Soc*; Colvin (3); Skempton]

Publications

For a complete list of Tredgold's publications other than books, reference should be made to Booth (1979c, 1998). There are articles from magazines and journals (*Philosophical Magazine, New Monthly Magazine, Mechanics' Magazine*, etc.), original communications and others for the ICE, articles from *Encyclopaedia Britannica*, letters to clients and edited works.

Tredgold's most important publications were the five books he wrote between 1820 and 1827: the following are the English editions (for foreign editions see Booth (1979c)). When *Carpentry* was published in Weale's Rudimentary Series there was confusion between editions and impressions;

it would appear that the nomenclature changed at some time between 1902 and 1921.

1820. *Elementary Principles of Carpentry* (2nd edn., 1828) [Appendix to second edition, 1840; 3rd edn. by Peter Barlow, 1840; 4th edn., 1853; 5th edn., 1870; 6th edn., by E. Wyndham Tarn, 1885; 7th edn., 1886; *Elementary Principles of Carpentry, chiefly composed from the Standard Work of Thomas Tredgold*, Weale's Rudimentary Series, edited by E. W. Tarn, 1873; 2nd edn., 1875: 3rd edn., 1880: 4th rev. edn., 1885; 5th rev. edn. and extended edn., 1890; 6th edn., 1899; 7th? and 8th edns., 1902; 9th? and 10th edns., 1919; 11th edn., 1921; 12th?, 13th? and 14th edns., 1942; 15th edn., 1946. *Carpentry and Joinery*, an atlas of engravings to accompany and illustrate *Elementary Principles of Carpentry*, 1873; 2nd edn., 1878; 3rd edn., 1886. *Elementary Principles of Carpentry. Revised … by John Thomas Hurst*, 1871; 2nd edn., 1875; 3rd edn., 1880; 4th edn., 1883; 5th edn., 1886; 6th edn., 1888; 7th edn., 1890; 8th edn., 1892; 9th edn., 1895; 10th edn., 1899; 11th edn., 1904; 11th edn., new imp., 1914; 12th edn., 1919]

1822. *A Practical Essay on the Strength of Cast Iron … Containing practical Rules, Tables, and Examples; also an Account of some new Experiments, with an extensive Table of the Properties of Materials*, 2nd edn., 1824; 3rd edn., 1831; 4th edn. with notes by Eaton Hodgkinson, 1842–1846; 5th edn., 1860–1861

1824. *Principles of Warming and Ventilating Public Buildings, Dwelling Houses, Manufactories, Hospitals, Hot-houses, Conservatories, etc. … to which are added, Observations on the Nature of Heat; and various Tables useful in the application of Heat*, 2nd edn., 1824; 3rd edn. with an appendix by T. Bramah, 1836

1825. *A Practical Treatise on Rail-roads and Carriages*, 2nd edn., 1835

1827. *The Steam Engine, comprising an Account of its Invention and Progressive Improvement; with an Investigation of its Principles, and the Proportions of its Parts for Efficiency and Strength: detailing also its Application to Navigation, Mining, Impelling Machines, etc. And the Results collected in Numerous Tables for Practical Use*, 2nd rev. edn. edited by W. S. B. Woolhouse, 1838

TREVITHICK, Richard (1771–1833), mining engineer, was born near Carn Brea in the parish of Illogan, Cornwall, on 13 April 1771. He was the son of Richard Trevithick (b. 1735) and Anne Teague, who had married in 1760. He was employed as an engineer at Eastern Stray Park at 30s per month, and other mines, at the exceptionally early age of nineteen, having been a difficult scholar but intensely interested in mining.

In 1777, Boulton and Watt had introduced James Watt's (q.v.) patent engine into Cornwall, saving the Cornish miners great expense in the use of coal but soon dissatisfaction at the royalties that had to be paid to the Birmingham company caused the Cornish to try to avoid the charges incurred by the use of the patent. In 1795 Trevithick erected an engine at Ding Dong Mine, infringing the Watt's patent. With Edward Bull, Trevithick visited the Soho Foundry, Birmingham, when an injunction was served on both of them and so in 1796 Trevithick altered the Ding Dong engine to work as an atmospheric one.

In 1797 Richard Trevithick Sr. died and Trevithick married Jane Harvey on 7 November at St. Erth Parish Church. The year of his marriage marked 'the beginning of the harvest of his inventions and from then on until he left for South America was a period when he displayed his brilliance of inventions and improvements to existing ideas'. In that year he made models of high-pressure stationary and locomotive engines and the following year he made small high-pressure engines for winding and work on the mines. The first engine was sent to London in the charge of Arthur Woolf (q.v.). Two years later, in June 1800, Boulton and Watt left Cornwall on the expiration of Watt's patent and the following year Trevithick built his first steam road carriage at Camborne. His genius had provided an engine suitable for locomotion.

On 26 March 1802, with Andrew Vivian (q.v.), he patented the high pressure steam engine for stationary and locomotive use and William West became a partner in the patent. Next year he built a second steam road carriage and sent it to London and also commenced erection of his high-pressure engines in London and elsewhere. He met Samuel Homfray, ironmaster of Merthyr Tydfil, and built engines for ironworks in South Wales and an experimental locomotive engine at Coalbrookdale. There was a setback when one of his cast iron boilers exploded at Greenwich.

By 1804 many foundries were engaged in making his engines and Vivian sold his share in the patent to Robert Bill. Trevithick built a tramway locomotive at Penydaren (Penderyn, Rhondda) as a result of a bet with Homfray for 500 guineas that Trevithick would transport 10 tons of iron ore by steam locomotive from Penydaren ironworks to the Glamorganshire canal. He tried it out on a tramroad 9 miles long, built in 1804 by Roger Hopkins (q.v.) between Penydaren and Abercynon. In the event it also towed five wagons containing seventy men.

Also during this year he was sent for by the Admiralty to explain his engine and met Simon Goodrich (q.v.). The invasion of England was threatened by Napoleon and the suggestion was made that steam-driven fireships could be used against the French flotillas. In 1805 he built a railway locomotive engine at Newcastle-on-Tyne and drove a canal barge by means of a steam engine and paddle-wheels. This was followed in 1806 by the construction of the first of three steam dredgers which was put to work on the Thames and on 22 July a second dredger on the gun-brig *Blazer* also began work.

Meanwhile an Act of Parliament, 43 Geo.III cap.117, had been obtained on 12 July 1805 for making an archway (tunnel) under the river Thames from the Parish of St. Mary, Rotherhithe, Surrey, to the opposite side of the river in Middlesex. Robert Vazie (q.v.), who had been considering these works for over four years, commenced the works on behalf of the Thames Archway Company, 315 ft. away from the river by sinking an 11 ft. diameter shaft. The directors approached Trevithick for assistance and the latter was engaged on 10 August for a fee of £500 when the drift would be halfway and a further £500 on completion. The miners were accustomed to working in such tunnels, 5 ft. high and 3 ft. wide, but Trevithick and Vazie did not work easily together and Vazie left in October. By 21 December the drift had reached 947 ft. in length but it was noticed that the rock was dipping at 1 in 50 and on December the roof broke in; the situation was retrieved by dumping material in the river above the hole. By 26 January the drift had reached 1028 ft. Water broke into the drift again and this time the miners and Trevithick only just survived. Trevithick thereupon made proposals for restarting the drift 14 ft. lower but the directors were undecided and published in March 1809 a public request for plans to proceed. Finance prevented anything but experimental work being carried out. Although the geological findings of Trevithick were available to Marc Isambard Brunel (q.v.) in planning his ultimately successful Thames Tunnel, Trevithick himself moved on.

Early in 1808, Jane had brought the young family of four to London only to find conditions in Limehouse very rough compared with the home in Cornwall. In the same year Trevithick had made the acquaintance of Robert Dickinson and they formed a partnership, together patenting in July machinery for towing ships and discharging cargo, in October iron tanks for stowage of cargo on board vessels and in April 1809 for floating docks, iron ships, iron masts, bending timber, diagonal framing for ships, iron buoys and steam cooking. In that year Trevithick raised a sunken ship off Margate but, because he was not paid immediately, allowed it to sink again.

In May 1810 Trevithick suffered a serious attack of typhoid and gastric fever causing his brother-in-law, Henry Harvey, to suggest the family's return to Cornwall; in September they travelled to their home in Penponds via the Falmouth packet. As a result of his illness and absence from London, the partnership was declared bankrupt on 5 February 1811. Trevithick declared that he owed no one a penny but he appears to have been let down by Dickinson. Discharge was obtained on 1 January 1814. Trevithick had sold the tank patent to Henry Maudslay (q.v.), who put it to profitable use.

In 1811 Trevithick applied himself to the expansive use of steam and at Wheal Prosper he built the first Cornish engine and also his first plunger-pole steam engine. Next year he applied his high-pressure Cornish boiler to the pumping engine at Dolcoath mine and invented the single-acting high-pressure expansive engine and applied it to agricultural use. He supplied and engine and apparatus for rock boring for Plymouth breakwater and constructed a screw propeller for marine propulsion. In 1813 he built an engine for ploughing and next added plunger-pole high pressure cylinders to existing mine pumping engines, thus compounding them. In 1815 he built a recoil engine to be used with a screw propeller for ship propulsion and also built a high-pressure plunger-pole pumping engine at Herland mine; later in the year he patented the plunger-pole engine, the recoil engine and the screw propeller. In 1816 he proposed a tubular boiler and William Sims bought a half-share in the plunger-pole patent. Also in that year he sold shares in Wheal Francis, insisting that the buyer should obtain independent advice before accepting payment.

The ancient silver mines in the Cerro de Pasco district of Peru, situated at an elevation of over 14,000 ft. and about 160 miles from the capital city of Lima, had fallen into a bad state due to lack of dewatering equipment. In 1811 Francisco Uville was sent to England to find a suitable steam engine. When consulted, Boulton and Watt said that the low vacuum engine would be impracticable to construct in small enough pieces to be carried over the only available road, a mule track rising to over 17,000 ft. However, Uville found a model steam engine of Trevithick's in London and took it back to Peru where he and his partners found that it worked satisfactorily. Uville returned to England with the backing of his company in May 1813 and found Trevithick who, within a fortnight, had received an order for six engines and was preparing drawings for Hazledine, Raistrick and Co.

On 1 September 1814, 15 months after Uville's visit to England, nine engines were shipped to Peru; they included four pumping, four winding and one rolling mill engine. They arrived in Peru in January 1815 with William Bull, Thomas Trevarthen and Henry Vivian to erect the machinery. Trevithick followed on 20 October 1816 in the South Seas whaler *Asp*, arriving at Callao on 18 February 1817. Two engines were drawing water and two drawing ore but the installation was in a parlous state and Vivian died in May. In August 1818 Uville died; Bull had died some months previously and Trevithick found himself in full control of the Carro de Pasco mines. However the mining enterprise was ruined by the civil war, machinery being thrown down the shafts by the rebels.

In 1817 Trevithick visited Chile and opened up copper mines which were still working in 1872 when his son, Francis, published his *Life ...* in 1872. In 1824, at Guayaquil, the seaport of Ecuador, he met James M Gerard who was interested in mines on the mountain plateau of Costa Rica. Together, they set off across country to reach the Caribbean Sea and, after having become separated, Trevithick reached Cornwall

penniless on 9 October after being away for eleven years without ever communicating directly with his wife, Anne. With Gerard, who had arrived later, he set about raising money for Costa Rica mines but was unsuccessful; Gerard died in Paris.

Trevithick visited Holland in 1828 concerning the draining operations that were necessary after extensive flooding, and designed a suitable pumping engine for which Harvey and Company of Hayle received the order to construct. However, when finished, there was disagreement and the engine was not sent from Hayle. He drew up a petition to Parliament for a compassionate grant but, unlike other claims made by John Palmer and John Loudon McAdam (q.v.), it was unsuccessful; he did receive a small sum on behalf of his plunger pump patent. In 1831 Trevithick patented apparatus for heating apartments by warm air and in 1831 and 1832 patented a boiler and superheater for locomotives.

On 22 April 1833 Trevithick died at Dartford and was buried there four days later. After half a century, at a meeting of the Institution of Civil Engineers on 10 April 1883 it was resolved that funds should be raised to commemorate his life. As a result, a window was placed in Westminster Abbey; a Trevithick Scholarship at Owen College, Manchester, was inaugurated; and the Institution of Civil Engineers triennially awards a Trevithick Premium of £50.

BRIAN GEORGE

[F. Trevithick (1872) *Life of Richard Trevithick*; H. W. Dickinson and A. Tittle (1934) *Richard Trevithick, The Engineer and the Man*; L. T. C. Rolt (1960) *The Cornish Giant*; A. Burton (2000) *Richard Trevithick*]

Works
1805–1808. Thames Driftway

TREW, John (fl. 1563–1588), of Glamorganshire in Wales, gentleman, practised as a mining engineer and mineral surveyor in the middle of the sixteenth century. He developed expertise in various types of pumping and lifting devices associated with mining, which he described in a petition to Lord Burghley. This suggests that he worked in copper mines in Devonshire and built some kind of engine in Ireland, as well as working on early town water supply schemes. Precise details are lacking on his career before he was engaged, for £200, by the Exeter Chamber in 1563 to enable vessels of 10 tons laden weight to reach the city walls. His problem was that since the reign of Henry III (1216–1272) the Countess of Devon and other members of the Courtenay family had maintained stake weirs across the River Exe preventing vessels from reaching the city.

At first, when Trew arrived in Exeter, he was considering placing a canal on the eastern side of the river and a memorandum of 28 September 1563 made this clear saying that an agreement had been made with John Cove and Nicholas Crown, 'for and concerninge certeyan weares placed and fyxed yn the watercourse from St. James's Weare to the Weare Mylls'; that is to say they covenanted 'that one John Trewe, who hathe takn yn hande the conductinge of the ryver or haven, shall have all staks or other stuffe as he or his men shall pull up, for the conductinge of the said water, without any denyall'. And further they consented that neither they or any assignee should in any way intermeddle with St James's Weir

The plan might have been executed at a small cost in a short time, but Trew changed his mind and this was creditable both to his boldness and sagacity. The current of the Exe must have been difficult to stem, its rapidity was shown by the Tucking Mills, and the channel near Exeter was apparently shallow and rocky. Moreover, the purchase of the Wear mills, or at least compensation for damage, was avoided.

On 15 December 1563 it is recorded that Trew, after viewing the ground, had now changed his mind and thought it 'better to take the ground and way by the west side'. Instead of removing any weirs, Trew added the one which is known by his name below St. Leonard's church and was originally a timber structure

As built the canal contained three pounds locks or 'pools' and its length to the Pyll, or Matford brook, was 9,360 ft. The breadth of the canal was 16 ft. and its depth 3 ft. The length of the lower pound was 189 ft., its breadth 23 ft. and its depth 5½ ft. The middle pound had similar dimensions, but the breadth of the upper pound appears to have been 51 ft. wider, possibly to serve as a floating dock for vessels which, during floods, could not proceed at once to the quay. The gates were almost certainly of the lifting or guillotine type and the locks the earliest pound locks known in Britain. A sluice at the harbour entrance maintained water level in the canal at low tide. Mitre gates, shown in a drawing of c. 1630, may have been one of several improvements made in 1581, after which the canal could accommodate lighters up to 16 tons burden.

Having been commenced in February 1564, the canal appears to have been opened to navigation in the autumn of 1566 and work was finished, including a quay with a crane built by Mr. Smyth, receiver, in the spring of 1567. At the lower end of the canal it joined the pyll which entered the river at a position just above the Countess Wear Bridge which now carries the Exeter Ring Road. The canal was later extended in 1676 by Richard Hurd (q.v.), who rebuilt Trew's weir, and greatly enlarged in 1701 by Daniel Dunnell (q.v.). Despite his innovation Trew did not benefit by this work, which did not satisfy the City's expectations, and he had to petition Lord Burghley to help secure payment. His claims were not settled until 1573, when he received from the Corporation an annuity of £30 and the sum of £224. Aside from pursuing his claims little is known of his activities for a period of ten years.

Commissioners were appointed in 1575 to improve the Lee navigation in Essex. In 1576 a new pound lock with mitre gates (the first in England) was built at Waltham mill with timber walls and floor. Serious settlement required reconstruction of the lock on masonry foundations only two years later. The operation, completed in June 1579, was directed and supervised by Trew, who was paid £6 13s 4d.

Trew's name appears again in accounts of the improvement of Dover harbour. In February 1579 a violent storm destroyed what was left of the King's Pier. This was a long mole begun by Henry VIII consisting of larger stones deposited under water surmounted by timber framing with rubble masonry infill. Unfortunately the town of Dover had lacked the resources for proper maintenance and once it had begun to decay the process had been hastened by the poorer townsmen removing valuable timber and ironwork.

Earlier, in 1576, William Borough (q.v.) had been commissioned by the Privy Council to report on Winchelsea, Rye and Dover harbours and three years later he was brought in to advise a commission appointed to consider what could be done to improve the harbour at Dover. He produced a scheme that formed the basis of a most successful civil engineering work, but it was costly and was put to one side in favour of a proposal by Trew in the summer of 1580. Trew concentrated on rebuilding the mole on the existing foundations with two masonry walls enclosing chalk rubble infill.

In August 1580 the Council recommended Trew's proposals to the commissioners, but with the proviso that a 33 ft. trial section should first be made to assess the unit cost more accurately, as it would eventually be 1,000 ft. long. He was engaged at the high rate of pay of £3 10d per week. By December 1580 there were forty hewers of stone (paid 6s per week) and sixty labourers (at 4s per week) working in quarries at Folkestone. At the same time Trew requested two boats of 40 tons burden to transport the stone and ordered a hundred labourers to begin digging chalk at Dover. He proceeded with the trial sections, but by July 1581 costs were mounting to an ominously high level and the Council had to admit that Trew's plan 'is not likely to take that good effect we looked for!'

Attention then returned to Borough's scheme which, with some modifications, was adapted in April 1582. By that time Trew had been discharged having spent £1,300. The last reference to his career occurs in the Armada year. In December 1588 he petitioned:

'though an old man [I] desire to be employed in the wars. He has an invention which would do as much as 5000 men in time of extremity, and also an engine to be driven before men to defend them from the shot of the enemy'.

A. B. GEORGE and M. M. CHRIMES

[Chamber Act Books (1563–1567) Devon RO, Exeter; John Trew (*c.* 1570) Petition, Landsowne mss, 107 art. 73, BL Department of Manuscripts; P. C. De la Garde (1838) On the antiquity and invention of the lock canal of Exeter, *Archaeologia*, **28**, 7–25 (also Memoir of the canal of Exeter, *Min Proc ICE*, **4**, 1845, 90–102); S. Robertson (1876) Medieval Folkestone, *Archaeologia Cantiona*, **10**, 113–119; W. B. Stephens (1957) The Exeter Lighter Canal, 1566–1698, *Journal of Transport History*, **3**, 1–11; E. A. G. Clark (1960) *The Ports of the Dee Estuary*; K. R. Fairclough (1979) The Waltham Pound Lock, *History of Technology*, **4**, 31–44; *The History of the King's Works*, *IV*, 1485–1660 (Part II), 1982, 756, 757; K. R. Clew, *The Exeter Canal*; A. W. Skempton (1984) Engineering on the English river navigations to 1760 (from *Canals—A New Look*); S. Johnson (1994) *Dover Harbour*, unpublished thesis, Oxford University; *Calendar of State Papers (domestic), 1547–1580, 1581–1590*]

Works

1564–1567. Exeter Canal, £3,000
1576. Pontymoile Ironworks
1579. Waltham Mill, Essex, new lock on the Lee Navigation
1580–1582. Dover Harbour, trial pier, £1,300

TROUGHTON, Edward, FRS, FRSE (*c.* 1753–1835), philosophical instrument maker, was born in Corney, Cumberland. In 1773 he was apprenticed to his eldest brother, John (*c.* 1739–1807), an instrument maker established in Dean Street, Fetter Lane, London. In 1782, John purchased the family business of Benjamin Cole (1695–1766) at 136 Fleet Street, London, an address long known as a source of high quality scientific instruments. Edward finished his apprenticeship in 1789 and joined his brother in partnership until John's retirement in 1804. Between 1800 and his own retirement in 1831, Edward Troughton was the finest scientific instrument maker in England. He was invited by Thomas Telford (q.v.) to become an Honorary Member of the Institution of Civil Engineers soon after Telford became its President, in recognition of the 'vast services he had rendered to the profession of a Civil Engineer and science in general by the high state of perfection to which he has brought various mathematical and scientific instruments'. Troughton joined in 1820 and regularly attended meetings until 1823. His name appears in membership lists of 1824, 1827 and 1830.

Troughton's achievements include the 'pillar' sextant, which he patented in 1788, the result of an Admiralty order for a double sextant to the design of Graeme Spence. He perfected a method of dividing by roller which remained his trade secret until he made it public in 1809. He built an improved dividing engine in 1793, the 5 ft. standard scale for Sir George Shuckburgh, and developed a highly successful new type of surveyor's level before 1812. His theodolites were used for the American coastal survey (1815) and the Irish

and Indian arc measurements (1822 and 1829). Troughton and Simms built Thomas Colby's (q.v.) compensation bars. Examples of Troughton's work are in the Science Museum, London, and the Greenwich Observatory.

Troughton was a member of the Society of Civil Engineers (Smeatonian Society) from 1810 as an Honorary Member in succession to his brother, John, who had been invited to join in 1793. He was a member of the American Philosophical Society, a Fellow of the Royal Society of both London (1810) and Edinburgh (1822), and a founder of the Royal Astronomical Society (1820). In 1825 he visited Paris and in 1830 was awarded a gold medal by the King of Denmark.

Troughton never married. He took William Simms (q.v.) into partnership in 1826 and continued to reside at 136 Fleet Street with the Simms family until his death there on 12 June 1835. (Simms was the brother of Frederick Walker Simms (1803–1865), civil engineer and author of a standard work on tunnelling). He is buried in Kensal Green Cemetery.

MIKE CHRIMES

[ICE Membership records, ICE archives; Obituary (1835) *Gentlemen's Magazine* (New Series), **4**, 15–16; Edward Troughton, *DNB*, **57**, 259–260; D. Baxandall (1924) The circular dividing engine of Edward Troughton, 1793, *Transactions of the Optical Society*, **25**, 135–140; E. W. Taylor and J. S. Wilson (*c.* 1949, 1960) *At the Sign of the Orrery: the Origins of the Firm of Cooke, Troughton and Simms*, York; A. W. Skempton and J. Brown (1973) John and Edward Troughton, mathematical instrument makers, *Notes and Records of the Royal Society of London*, **27**(2), 233–262; A. McConnell (1992) *Instrument Makers to the World; A History of Cooke, Troughton and Simms*, York; E. Mennim (1995) *Transit Circle: the Story of William Simms*, London]

Publications

1788. Patent Specification AD 1788, No. 1644: *A New Method of Framing … to be used in the Construction of Octants, Sextants and Quadrants*
1800. Nachricht von neuen astronomischen Instrumenten und Beobachtungs-Methoden, Zach, *Monatliche Corresp*, **II**, 207–222
1804. Description of a tubular pendulum, having all the properties of the gridiron …, Nicholson, *J Natural Phil*, **IX**, 225–230; Gilbert, Annal., **XXV**, 1807, 255–266
1809. An account of a method of dividing astronomical and other instruments by ocular inspection …, *Phil Trans*, **99**, 105–145
1819. Comparison between the length of the seconds pendulum …, *Edinb Phil J*, **I**, 105–108
1819. Description of the nautical top, *Edinb Phil J*, **I**, 105–108
1830. Letter written *c.* 1814 to David Brewster describing his dividing engine of 1793, *Edinburgh Encyclopaedia*, **10**, 353–357
1822. An account of the repeating circle and of the altitude and azimuth instrument …, *Memoirs*

Astronom Soc, **1**, 33–54; *Phil Mag*, **LX**, 1822, 8–18; Hertha, *Z Frd- Völker- staaten- Kunde*, **XII**, 1828, 26–36

TRUBSHAW, James, Sr. (1746–1808), County Surveyor, was a son of Charles Cope Trubshaw, a mason and himself the son of a mason. By his first wife, Margaret Landor, whom he married in April 1768, he had two sons, Thomas (1771–1820) and Richard (b. 1771) who died shortly after. Margaret also died in 1771 and James took as his second wife Elizabeth Webb of Levedale, on 27 August 1775. They had seven sons, of whom the second, James Jr., is noticed below, and two daughters.

Trubshaw inherited his father's business in 1772 but little is known about his work. It cannot have been particularly successful as he had to dispose of the family house and a large part of his wife's inheritance in 1791 to repay debts. At Michaelmas 1792, however, he was appointed Surveyor of the County Bridges in Staffordshire, after advertisements had been placed in Birmingham and Wolverhampton newspapers. His salary was fixed at £52 10s p.a., in return for two days attention to his duties each week and he was required to provide a 'bond in the penalty of £200 with two different sureties ... for the due execution of his office'. In 1795 many of the County bridges were swept away or damaged by flood, including Wolseley, Halfhead, Blackbrook, Hopwas, Fazeley and Wichnor and Trubshaw was involved in the repair of many of them.

Wolseley Bridge over the river Trent on the great post road to the north west however was entrusted to John Rennie (q.v.) to design and James Trubshaw & Sons were the successful tenderers in 1797 for the contract to build it. By 1799 they were unable to complete the work and resigned the contract, the County agreeing to pay the subcontractors direct. Presumably as a consequence, Trubshaw gave up the County Surveyorship in January 1801, continuing on an informal basis until Joseph Potter (q.v.) had been appointed to succeed him. A committee of Justices appointed at Translation Sessions 1803 to examine Trubshaw's accounts took two years to report, during which time an attempt was made to call on his sureties to meet his alleged debts. Trubshaw made counterclaims and after a report by Potter and the County Treasurer, mutual release was given at Michaelmas 1805.

Trubshaw died on 13 April 1808.

P. S. M. CROSS-RUDKIN

[Staffordshire RO; William Salt Library, Stafford]

Works
1796. Lady Bridge, Tamworth design, built by contract for £1,260

TRUBSHAW, James, Jr. (1777–1853), was the second son of the second marriage of James Trubshaw Sr. (q.v.). He was born on 13 February 1777 at Colwich in Staffordshire. He attended school at Rugeley until the age of eleven, when his father removed him to enter the family building business. His lack of an extended education clearly rankled and is mentioned disapprovingly by his obituarist. Five years later, becoming dissatisfied with life at home, Trubshaw moved to London and obtained work there and elsewhere through introductions to Staffordshire connections. By 1797 he had returned to Little Haywood and joined his father and brother, John, in the contract for Wolseley Bridge nearby, which they gave up in 1799. On the 21 January 1801 he married Mary, daughter of Thomas and Mary Bott of Stone. They stayed for a short time only at Little Haywood before moving to Stone where Trubshaw set up in business on his own account. Lack of capital was initially an impediment but he made gradual progress as a building contractor. In about 1807 he was engaged to build Ashcombe Park, near Cheddleton, work which continued for about four years and which seems to have established him on a firmer financial basis. At the same time he undertook his first major bridge contract, building Darlaston Bridge to a design by John Rennie (q.v.).

Returning to Haywood on his father's death in 1808, he took over the latter's business as a builder and civil engineering contractor. He also claimed the title of architect but generally avoided that kind of work if possible. He confined his activities to Staffordshire and its neighbourhood—in 1836 he admitted that he had been to Sheffield often but never to Leeds—but there was sufficient work for him to prosper.

Although the bulk of his building work comprised country houses and churches, some of the buildings are of structural engineering interest. These include the Charles Barry-designed Royal Manchester Institution (1827–1829), now the City Art Gallery, with cast iron beams and columns, and jack arch floors, and St. Peter ad Vincula Church, Stoke-on-Trent (1826–1829), which used cast iron trusses in the nave. His name has also been linked to Philip's linen mill, of four storeys with iron framing, at Upper Tean, Chudley, Staffordshire, erected 1822–1824. Early in 1827 he ventured into Cheshire to contract for the largest masonry arch bridge built in Britain, Grosvenor Bridge over the river Dee. The architect, Thomas Harrison (q.v.) was by then aged eighty-two and his design was modified by George Rennie (q.v.) and the details by the clerk of works, Jesse Hartley (q.v.). Trubshaw's contribution to this great undertaking was to design a novel centering, of three stone piers in the river from the top of each of which a fan of timbers radiated to the soffit. By this means, and by the disposition of the wedges, much greater control of the shape of the arch during construction and striking was possible. Construction proceeded fairly slowly but was completed sufficiently for a formal opening by Princess Victoria in October 1832 and to the public on 1 January 1834. Trubshaw's status was reflected in his election to the Institution of Civil Engineers in 1827. When, in 1832, the magistrates at Shrewsbury wanted a third opinion on the problems of the Welsh Bridge there, it was to the person employed on the Dee Bridge at Chester that they turned.

Early in 1829, Trubshaw was asked by the Trent and Mersey Canal Company, whose head office was at Stone, to effect repairs to Knypersley Reservoir, which had not been brought properly into service because of leaks where the discharge pipes passed through the core of the dam. He made improvements but, when it was refilled in May, the results were still not entirely satisfactory and a complete remedy was not made until 1868. The records of the canal company being no longer available, it is not possible to determine when he became their Engineer, but he held the post in 1836 and it is probable that his appointment dates from the period of the Knypersley episode.

A more successful operation was the righting of the tower of Wybunbury Church; 96 ft. high and built partly on soft sandstone and partly on clay, it had developed a lean of more than 5 ft., due to brine extraction in the strata below. By digging a trench within the tower and boring a series of holes from it under the footings of the tower—they were filled with water to soften the ground—the material was removed in small stages until the tower regained the perpendicular. Trubshaw then demolished and rebuilt the body of the church to his own design; a subsequent rebuilding has now also been removed, leaving the tower free-standing.

One of his visits to Wybunbury was made on the way home from a visit of inspection to the Stratford & Moreton Railway, whose line had been leased out to Benjamin Baylis (see: *BAYLIS, Thomas*) and was in a poor state of repair. He surveyed a branch to Shipston, and built in 1833–1836. In 1834, he won a sizeable contract on the Grand Junction Railway in his home county. His railway work brought him into contact with George Stephenson (q.v.) and the two men of elementary education but energy and achievement formed a firm friendship. Trubshaw gave evidence to Parliament in support of Stephenson's proposals for railways in the north Midlands. Had his own life been spared, Trubshaw would have erected as a memorial to Stephenson—at his death—a monolith taller than Cleopatra's Needle.

Trubshaw, like several of his family before him, was a man of impressive physique and on one occasion won a race carrying the sculptor, Chantrey, on his back against an athlete of some repute. He remained rooted in the part of Staffordshire where his family had been stonemasons for generations. He was a churchwarden at Colwich and when he died on 28 October 1853 he was buried there. He had been married for nearly fifty-three years and his widow lived for another three. They had had three sons and three daughters. The eldest, Thomas, was an architect and the second, Mary,

married an architect. The youngest, Susanna, was the author of the family history.

P. S. M. CROSS-RUDKIN

[Watkins collection, National History of Wales; HLRO 1836, vol. C32, 33, day 9; 1837, vol. L3, day 4; 1846, vol. C47, day 5; will dated 18 March 1852, proved 23 January 1854, PRO; Chester City RO; Staffordshire RO; DNB; *Transactions of the Institution of Civil Engineers* (1836); *Gentleman's Magazine*, **1**, 1854, 97–101; S. Trubshaw (1876) *Family Records*; A. Bayliss (1978) *The Life and Works of James Trubshaw (1777–1853)*; G. M. Binnie (1987) *Early Dam Builders in Britain*; G. Woodward (2001) Trubshaw, Hartley and Harrison: early nineteenth century engineers and architects, *Trans Newc Soc*, **72**(1), 77–90]

Works

1807–1809. Darlaston Bridge, Stone, contract to build (with Charles Webster) at a schedule of rates

1811–1814. Cheddleton Bridge, contract to build

1813–1817. Hopwas Bridge, contract to build for £6,394 10s

1820–1822. Mill Bridge, Shenston, designed and built

1827–1833. Grosvenor Bridge, Chester, built under contract for £29,300

1829. Knypersley Reservoir, repairs

1832. Wybunbury Church, restored tower to plumb

1832–1834. Walton-On-Trent Bridge, designed and built of timber, replaced

1833–1836. Stratford & Moreton Railway, Shipston extension

1834–1837. Grand Junction Railway, contract for £83,100 to build Whitmore–Stafford section (14 miles)

1836–1839. Willington Bridge, designed and built

1843–1847. North Staffordshire Railway, contract for £14,289 to build Froghall–Caldon Low section (to replace Caldon Low tramway)

1848–1850. Exeter Bridge, Derby, designed and built, replaced in 1929–1931

1853. Burton-On-Trent Bridge, design and estimate £12,500, built to design of J. S. Crossley in 1863–1864

TURNBULL, George (1809–1889), civil engineer, was born on 2 February 1809 in Luncarty, the youngest of eleven children of William Turnbull who owned the Hunting Tower bleach fields near Perth. Turnbull was educated at Perth Academy and Edinburgh University. A friend, David Hogarth, became a pupil of Telford (q.v.) and Turnbull decided to follow him. Unlike Hogarth, who made a career in the Church, Turnbull took to civil engineering and in Telford's final years became his chief personal assistant.

He joined Telford in 1828 and worked on St. Katharine's Dock, then drawing to a close. He worked on a number of Telford's final projects including the Metropolitan Water Supply surveys, the Adelaide Gallery canal boat experiments and drawings and made preparations for Bute Docks, Cardiff, the Ulster Canal, and Belfast Harbour.

On Telford's death he became one of the beneficiaries of his will. There followed a period of working with William Cubitt, who took on much of Telford's consultancy work, at Folkestone and Dover Harbour, the Great Northern Railway and, under J. M. Rendel, Birkenhead Docks. In the 1840s he also worked on Middlesbrough port facilities and associated observations on the Tees. With Cubitt's and Rendel's support he became Engineer to the East India Railway Company in 1850, a post he held until 1863; on his retirement he continued to act as arbitrator for the Company. He purchased an estate near Abbot's Langley, where he became involved in local affairs.

He joined the Institution of Civil Engineers in 1829 and acted as secretary at some of its early meetings. By the time of his death on 26 February 1889 he was its longest-standing member.

MIKE CHRIMES

[Membership records, ICE archives; ICE Register of drawings, c. 1840–1852, ICE archives; T. Telford (1833) Diary, mss, India Office records, BL; Obituary (1890) Min Proc ICE, **97**, 417–420]

TURNER, Joseph (c. 1729–1807), architect, practised in Cheshire and north-east Wales in the last quarter of the eighteenth century. Initially resident at The Elms, Hawarden, he and his wife Elizabeth (d. 1794) had five children between 1755 and 1765 of whom only John (b. 1755) and Thomas seem to have survived childhood. At Hawarden he may have supervised work by his uncle/relative, Samuel Turner the elder of Whitchurch. Joseph was carrying out surveys for Chester Corporation from 1767 and by 1781 had moved his practice to Chester where he was made a Freeman in 1784 and elected Alderman in 1794, living at Paradise Row, Chester.

His work varied from church monuments, found in several parish churches, to country houses and public works such as Denbigh and Flint gaols and the Northgate Bridge arches, Chester. He also drew up a scheme for developing the Forelands, and a graving dock, in 1776. His local eminence led to his being consulted about a number of civil engineering schemes, the most famous of which was the Ellesmere Canal. After he had drawn up a preliminary plan for the canal in September 1791, he was responsible for the deposited plans for the *Proposed Eastern Canal from the Severn to the Chester Canal and Branches* of 10 November 1792, with Thomas Morris Jr., John Chamberlain and William Cawley.

Subsequently his cousin, William Turner of Whitchurch, became involved in early work on the canal and submitted early designs for the aqueduct at Pontcysyllte.

Joseph designed a proposed bridge for Menai Straits in 1785 and surveyed a turnpike from Chester to Parkgate in 1793. In 1794 he was described in the Ellesmere Canal Minutes as 'Agent and Engineer' of the Chester Canal. Another relative, possibly his son, John Turner, also of Whitchurch—Society of Arts prize winner for his novel church design in 1814—may have become his assistant. He was County Surveyor of Flintshire from 1815 until 1827.

Joseph Turner died in 1807.

MIKE CHRIMES

[Chester Street directories 1781–1789, Chester City RO; Eastern Canal plans (QS/DC/1), Ellesmere canal plan (D/BJ/432), Flint CRO; *Transactions of the Society of Arts*, **32**, 1814, 23; Pevsner, *Buildings of England*; Cheshire, *passim*; Colvin (3)]

Publications
c. 1791. *A Sketch of the Country through which it is Intended to make a Navigable Canal from Shrewsbury to Chester and Liverpool*
1792. *Proposed Eastern Canal from the Severn to the Chester Canal and Branches*

Works
1786. Pentre Bridge, Flint

TURRELL, Edmund (d. 1835), engraver, of 46 Clarendon Street, Somers Town, London, became an Associate of the Institution of Civil Engineers in 1823 on the recommendation of Thomas Telford (q.v.), with whom he had an association dating from before 1813. In 1826, Turrell proposed that the Institution award medals to recognize outstanding contributions to its Original Communications but the Institution then lacked the money to implement the idea and the medals came into existence only in 1835, funded by a bequest of Telford providing £2,000 in trust, the interest to be spent on annual premiums. Turrell himself was bequeathed £200 by Telford.

Turrell became involved in Telford's *Life of Thomas Telford*. Telford had limited his activities in order to concentrate on this work in 1830 and had engaged his friend, Turrell, to prepare the plates for the accompanying Atlas, had decided on printer and publisher, and on type and sizes; Turrell had been paid a fixed sum on the production of proof copies of each plate, usually 15 guineas for single plates and 30 guineas for double plates. Telford received proofs of fifty-eight plates prior to his death in 1834 and Turrell continued working on the plates but was himself incapacitated by illness within six months, and in September 1835, died, leaving the work incomplete. His widow attempted to persevere, but this was found unacceptable, and the plates were obtained by John Rickman in June 1837 when he described them as 'unfit for the press'. His account for the cost of the plates reveals, however, that only five of the sixty-six plates credited to Turrell required the services of other engravers.

TESS CANFIELD

[Gibb Collection, ICE Archives; ICE Membership records; ICE archives; *On Granite Paving* (1824) ICE OC Ms. 9, ICE archives; J. Rickman (ed.)

(1838) *Life of Thomas Telford*; G. Watson (1988) *The Civils*; Skempton; C. Hadfield (1993) *Thomas Telford's Temptation*]

Publications
1808. An improved method of making muffles for chemical purposes, *Phil Mag*, **31**, 187–192
1827. On an improved mode of forming and sharpening the points of etching-needles and dry-points, *Gill's Tech Rep*, **2**, 254–259
1832. Improvements in the microscope, *Trans Soc Arts*, **48** (with C. Varley and W. Valentine)

Engravings
1813. Bonar Bridge general elevation, drawn by W. A. Provis, E. Turrell sc. 1813, in *6th Report of the Commissioners for Highland Roads and Bridges*
1814. *Design for a Cable Suspension Bridge over the Mersey 1814, with a Central Span of 1000 ft.*, by Edmund Turrell from a drawing by W. A. Provis, 1814
1819. *Design for a Bridge over the Menai Straits at Ynns-y-moch*, by Thomas Telford, Drawn by W. A. Provis, Edm. Turrell sc. 1819, in *Papers relating to the building a Bridge over the Menai Straits near Bangor Ferry*, House of Commons, 1819
1820. *Holyhead Road. Tools for Making and Repairing Roads*, I. Easton del. E. Turrell sc. Plate X, in *Reports of Mr. Telford to the Commissioners for the Improvement of the Holyhead Road*
1838. *Atlas to the Life of Thomas Telford, Civil Engineer, containing eighty-three Copper Plates, illustrative of his Professional Labours*, Payne & Foss, London. Most plates were drawn by George Turnbull, with five by Thomas Casebourne. All but 16 of the 82 (No. 28 was not published) copper plate engravings are by Edmund Turrell

TWISS, William, General (1745–1827). William Twiss's career divides into three sections: a rather pedestrian beginning (1760–1776); a flurry of engineering activity in Canada (1779–1783), and twenty-five years in charge of the defences of the south-east coast (1785–1809), where he was instrumental in the introduction of the once ubiquitous Martello towers into the landscape.

Twiss was born in 1745 and trained for two years in the Tower of London Drawing Office (1760–1762) before going to Gibraltar, where the only noteworthy event was his rescue from drowning during a storm in 1766. He also worked on the dockyards at Portsmouth and Plymouth, but his chance came when he was sent to America in 1776. There was a dearth of engineering officers throughout the War of Independence and Twiss was made 'Comptroller of the Works to superintend (the moving) of the fleet to Lake Champlain'. This involved transporting about five-hundred boats up the Richelieu River and building a dockyard with launching slips for larger vessels. He was praised for his 'indefatigable attention', but unable to progress his

General William Twiss

modern-seeming design for a square bowed landing craft for infantry assault.

Unfortunately, his reward for his efforts in the defeat of the American navy at Valcour Island was to be named Chief Engineer on General Burgoyne's disastrous expedition the following year (1777). He contributed to its main success, the taking of Fort Ticonderoga, by deciding that it was practicable to build a road for artillery 16 ft. wide and 9 miles long up to the summit of Sugarloaf Hill, which overlooked it, in twenty-four hours. In consequence the enemy fled destroying roads and bridges as they went. Twiss was sent to survey the demolitions, measure streams, count the culverts and bridges to be rebuilt, and stake out sections and build causeways. His report lead to Burgoyne's decision to go overland to the Hudson River, arguably contributing to the defeat at Saratoga where Twiss was taken prisoner. After his exchange he was sent back to Canada where he found a patron in the Governor General Lord Haldimand who twice wrote requesting that he 'be appointed Chief Engineer in the province as I have found his zeal, activity and abilities equal to this important trust' as he had 'performed with great judgement and economy and I have such confidence in his abilities and integrity'. But the Ordnance Board were unwilling to give the position to someone of such low rank. (Twiss wrote complaining of having been eighteen years a Lieutenant) and it was not until 1781 that he was to be de jure as well as de facto Chief Engineer in Canada.

Twiss first task in 1779 was to build a naval establishment on Lake Ontario on Carleton Island, which he preferred to Cataraqui (later Kingston) because of its superior harbours. Twiss built a general hospital, prison, barracks, lime kiln and saw pit there, as well as the fort. Because of its ambiguous position, Carleton Island was handed over to the USA in 1817 and all that remains of one of the most important posts in Canada between 1778–1783 are a ditch, chimneys and rubble heap. Twiss also built new roads and hospitals at Sorel and Three Rivers, a prison at Montreal, windmills at Sorel and Lachine, bridges at Berthier and over the Riviere du Loup and Maskinonge, ironworks at Sorel and Three Rivers and a saw mill dam at the Chambly Rapids. In 1783 he assisted with the first saw mill and grist mill in Canada at Newark. He also build outworks and a temporary citadel on Diamond Hill at Quebec, but work ceased after the end of the war in 1783 and was only completed with the Twiss design incorporated with those of his successors Gother Mann and Elias Walker Durnford (q.v.) in 1825.

However Twiss's greatest historic contribution to engineering in Canada was the construction of the Coteau du Lac Canal, the first locked canal in north America. The war had seen a considerable increase in river traffic between Montreal and the Great Lakes, which was seriously impeded by the junction of the St. Lawrence and Ottawa rivers between Lakes Louis and Francis, where there was a drop of 100 ft. followed by a series of boiling rapids. To circumvent this Twiss proposed a fortified canal across a narrow peninsula of land at Coteau Du Lac on the north shore, situated so as to avoid damage by ice, and paid for by tolls.

Work began in 1779 when £1,520 was spent. The canal was 900 ft. long, 7 ft. wide and 2½ ft. deep with three locks. There is perhaps a touch of partiality in Twiss letter to Haldimand in 1780 'I wish your excellency could see this post … it will be formed into locks as useful to navigation as anywhere in the world', but others described it as 'a very complete canal' and considered that Twiss had 'solved questions which seemed impossible'. It was more than sufficient for the boats of the time which were 6 ft. wide with a 2 ft. drought, and was for twenty years the only artificial stretch of water on the St. Lawrence. Enlarged and deepened by Mann and Durnford it continued in use until 1845 when it was decided to replace 'the engineers canals' with one on the south shore (the Beauharnais Canal). One hundred metres of the northern canal section has been preserved. After the Coteau Du Lac Canal, Twiss was instructed to 'take all measures to facilitate navigation on the St. Lawrence' and built three more canals—at Trou Du Moulin (120 ft. long without locks), at Cascades (410 ft., one lock) and at Split Rock (one lock).

Twiss returned to England in 1783 and only worked on two other purely civil engineering projects. In 1793 a report with Samuel Wyatt (q.v.) proposed the replacement of the double dock at Portsmouth with a large single one. This was approved but never built. In 1798 he was sent to report on Ralph Dodd's (q.v.) scheme for a tunnel under the Thames between Gravesend and Tilbury. Twiss was so favourably impressed he joined the Company, but, although a shaft was sunk at Gravesend, flooding and insufficient funds lead to the project being abandoned in 1802.

Most of Twiss's later career was spent on the defences of the south coast first, as Secretary to the Board of Land and Sea Officers, reporting on the defences of Portsmouth and Plymouth 1785–1792, then as commanding Royal Engineer for the southern military district 1792–1809. He had only brief periods away; in 1799 as Commanding Engineer on the expedition to Holland where he was consulted over the destruction of the sluice gates and basin on the Bruges Canal at Ostend, in 1800 in Guernsey and Jersey, and in 1802 reporting on the defences on Ireland. During this time Twiss was involved with three of the largest, most extravagant and controversial construction projects of the time, Fort Cumberland at Portsmouth, Dover Castle in particular the Western Heights, and the Martello towers.

Work at Fort Cumberland was originally intended to be restricted to repairs to the fort built by J. P. Desmaretz (q.v.) but 'instead of improving old Fort Cumberland with money voted for that purpose [it was] destroyed and pulled down entirely'. It was the last angle bastion fort built in England, a compromise between the systems of Vauban and Montalembert and was begun in 1785. In 1792 Twiss complained that the convicts employed were insufficient, it would take four hundred to complete the work in seven years. In fact he underestimated considerably it was finally completed after twenty-five years at a cost of £106,966. Twiss involvement was mainly in a managerial capacity, suggesting in 1790 that work might be improved by paying the convicts from ½d to 2d per day 'productivity bonus' and being ordered to personally check the beer measures after violent complaints that they were being watered down—possibly contributing to Edward Gibbon's description of 'bulwarks erected at enormous price by heads of folly and by hands of vice'.

At Dover Castle Twiss rendered himself anathema to medievalists and archaeologists by reducing the height of the medieval curtain wall, which he backed with a massive earth rampart, to strengthen the eastern defences, and built four new bastions connected by tunnelling underground passages through the existing foundations. He also demolished the old entrance, building a new road (still in use) to the Canon's Gate. Repairs and new works at the castle cost around £150,000.

However, his main work at Dover was on the western heights, beginning in 1794 when he worked with Sir Thomas Hyde Page (q.v.) on field works which by 1807 had been turned into an irregular shaped permanent citadel with walls revetted to stabilise the slippery clay subsoil. It continued in use, in the 1990s as a young offenders prison, a better fate than that suffered by the rest of Twiss works which were mostly demolished by Dover Corporation after it bought the land for £20,000 in the 1960s (a bargain considering the £238,889 originally spent on it). Among the casualties were the grand shaft barracks for seven hundred men and the hospital (later the headquarters of the garrison engineer), all the records of which were also destroyed. The grand shaft itself, Twiss's most elegant and innovative engineering feature on the heights only narrowly escaped the same fate, despite a contemporary guide book claiming 'it would have reflected credit even upon the genius of Sir Christopher Wren [it] combines gracefulness with the utmost simplicity and accommodating usefulness'. Twiss insisted on a stairway rather than a road to connect the heights to the town because while the distance on horseback was 1½ miles over unstable ground it was only 300 yd. horizontally from the sea (180 ft. above high watermark). He bored two brick lined shafts, through solid rock, one inside the other with the inner acting as light well and the outer housing three concentric staircases with intertwined spirals on brick arches. It was 140 ft. deep with a 26 ft. diameter, a 180 ft. tunnel at the bottom connected it to the town. The treads were originally of purbeck marble and the landings and light apertures secured by iron rails and gratings. Sadly the description of one stairway for officers and ladies, one for sergeants and wives and one for soldiers and their women seems to be apocryphal.

Part of Twiss's responsibilities for coastal defence was to report on the possibility of flooding Romney Marsh in the event of an invasion. In his view 'great impediment might be offered to the enemy by opening the sluices under proper management', but that opening Dymchurch Wall and damming the Rother would not work if speed were essential as two tides would be necessary for complete inundation. This along with his report on the vast expense necessary to update the existing defences of Dungeness lead to an overall plan for the area consisting of a line of towers to defend the coast, partial flooding of Romney Marsh and the Royal Military Canal as a final defence. The latter was built by the Royal Staff College not the engineers, but Twiss was literally peripherally involved through defending either end—reprieving the battery at Rye (rendered useless for coastal defence by the silting up of the Rother) and proposing for the eastern outface by Shorncliffe Battery an (uncompleted) miniature version of the Western Heights budgeted at £30,000.

The exact ancestry of the traditional Martello Tower—(Corsica, Minorca, Ireland) is still a matter of debate, but Twiss is generally considered to be at least their godfather. While in Ireland he had recommended a round tower for a redoubt on an island in Bantry Bay and was consequently receptive to a memo from Captain (later Major General) William H Forde (1803) proposing a line of square towers along the coast of south east England, although he originally considered they would be more use restricted to areas such as Dymchurch or attached to existing batteries. He recommended the report to his superiors, Robert Morse the Inspector General of Fortifications and David Dundas (both of whom had been present at the seminal engagement with the French tower at Mortadella, Corsica, in 1794).

Two years later in 1804 after political and military arguments, conferences and committees Twiss was ordered (despite complaining 'old age creeps fast upon me') to make arrangement for the siting and construction of seventy-four (round) towers between Folkestone and Seaford, completed by 1808. The standard design for these was elliptical with the walls varying in thickness from 13 ft. at the base of the seaward side to 6 ft. at the top facing inland, with a taper of 4–5°. The entrance was 20 ft. above ground level, ventilation was into the thickness of the walls and most had water cisterns sunk into the foundations which were strengthened with inverted arches. The roof was a brick vault supported by a massive brick pier from the base of the tower. The bricks (200,000 to 250,000 per tower) were set in a special mortar of lime, ash and hot tallow which set phenomenally hard, with a cement and stucco finish. The contractor William Hobson, who also built the Old Bailey, apparently secured the contract by cornering the market in bricks. The towers were so strong that high explosives were needed to demolish one in 1986. Only twenty-five remain, most lost to beach erosion. Twiss was sent to report on how far the same system could be implemented on the east coast, but had retired by the time the twenty-seven trilobular towers, some built by James Frost (q.v.), had been completed there by 1812.

He had been promoted Brigadier General (1804), Major General (1805) and Lieutenant General in 1812. From 1794 to 1810 he was Lieutenant Governor of the Royal Military Academy and one of his last tasks in 1809 was to report on its state. He retired in 1809 and died on Harden Grange, Yorkshire on 14 March 1827. A copy of a portrait by Lawrence is held by the Royal Engineers.

No details of Twiss family life have been found beyond a Chancery devise summoning 'any person or persons claiming to be the heir or heirs of General Twiss to prove his or her descent [or] be excluded from benefit of said devise'.

SUSAN HOTS

[Add mss 21/84, 21/815, BL; WO 1/783, 28/6, 30/56, 30/62, 55/831, 55/778 (Engineers papers of the Dover district) PRO; Haldimand papers B/154/188/131, MG21, Public Archives of Canada; Conolly papers, Royal Engineers Library; DNB; *Biographical Dictionary of the History of*

Technology (1996) 719; R. Dodd (1798) *Report with Plans, Sections, etc., on the dry Tunnel from Gravesend to Tilbury in Essex*; L. Fussell (1818) *A Journey around the Coast of Kent*; W. H. Tresallis (1886) Historic sketches of the coast defences of England, *Professional Papers Corps of Engineers, OC*, **12**, 67–86; S. G. P. Ward (1949) Defence works in Britain, *Journal of Army Historical Research*, **27**, 18–37; C. Bond (1960) The British base at Carleton Island, *Ontario History*, **52**(1), 1–16; J. C. Kendall (1970) William Twiss, Royal Engineer, *Duquesne Review*, **15**, 175–191; J. C. Kendall (1971) The construction and maintenance of Coteau Du Lac, the first lock canal in North America, *Journal of Transport History*, **1**(1)(ns), 37–49; S. Sutcliffe (1972) *Martello Towers*; P. A. L. Vine (1972) *The Royal Military Canal*; R. Legget (1976) *Canals of Canada*; D. W. Marshall (1976) The British military engineer 1741–1783, unpublished Ph.D. dissertation, University of Michigan; B. Pegden (1980) The purchase of bricks for the Martello Towers in the year 1804, *Fort*, 8; J. C. Coad and P. N. Lewis (1982) Later fortifications of Dover, *Post Medieval Archaeology*, **16**, 141–200; J. E. Goodwin (1988) Fortifications against a French invasion of the east Kent coast 1750–1815, *Fort*, **16**, 83–96; R. Legget (1988) *Ottawa River Canals and the Defence of British North America*; A. Saunders (1989) *Fortress Britain*; J. Coad (1989) *Royal Dockyards 1690–1850*; P. M. Grundy (1991) The Martello towers of Minorca, *Fort*, **19**, 47–57; F. Kitchin (1991) Aspects of the defence of the south coast of England 1756–1805, *Fort*, **19**, 11–22; P. McGrath (1992) Fort Cumberland 1747–1850, *Portsmouth Papers*, No. 60; H. Kalman (1994) *History of Canadian Architecture*, **1**, 205, 224, 233; P. Wells (1995) *Brick Bulletin*, Autumn; R. M. Telling (1997) *English Martello Towers*; G. Demetri (1999) *Architects' Journal*, November, 18–19; R. M. Ketchin (1999) *Saratoga*, http://members.tripod.com/jeff_howe/grandshaft/; http://parcscanada; http://www2.prestel.co.uk/history/martello/bib.html]

Works
Canada, 1779–1783:
Canals: Coteau du Lac, La Faucilles (Cascades Point), Trou du Moulin, Split Rock
Bridges: Riviere du Loup and Maskinonge
Dams: Saw Mill Dam, Chambly River
Hospitals: Carleton Island and Montreal
Prisons: Coteau Island (military), Montreal (civil)
Ironworks: Three Rivers, Sorel
Mills: Combined saw mill and grist mill, Newark; windmills, Sorel, Lachine

Dover, 1783–1809:
Fortifications: castle (£68,225); Western Heights (£238,889); Grandshaft (£3,331)

Martello Towers, south-east coast, 1805–1808:
Seventy-four towers, *c.* £3,000–8,000 per tower

Reports, etc.
1794. *Report on the Dry Dock at Portsmouth Harbour*

1805. *Report on System of detached Redoubts and Towers which the Government has adopted for the Coasts of Kent and Sussex*

TYRRELL, Charles (1795–1832), architect of Aldermanbury, London, was the fifth son of Timothy Tyrrell (d. 1865), a solicitor and banker who practised law at Guildhall Yard, City of London, from 1823 to 1863. During the 1830s Timothy Tyrrell was the main business partner of the McIntosh family (see: *McINTOSH, Hugh*) and stood surety for many of their contracts.

Charles Tyrrell was articled in Sir John Soane's office from 1811 until 1817, and was a student at the Royal Academy. Having begun his own practice by 1820, he travelled to Italy and Sicily in 1821–1822 in the company of Henry Parks, a fellow pupil of Soane. In 1823, with William Anderson (q.v.), he published a design for the rebuilding of London Bridge on the old foundations. In 1828 he became Surveyor for the Eastern District of London and died in 1832.

TESS CANFIELD

[Italian Drawings, Drawings Collection of the British Architectural Library; Colvin, (3); D. Brooke (1996) The equity suit of McIntosh v. the Great Western Railway: the 'Jarndyce' of railway litigation, in *Journal of Transport History*, September]

Publications
1823. *London Bridge* (with W. Anderson)

U

UNDERHILL, William (fl. 1793–1804), canal engineer and contractor, was appointed Resident Engineer, with a salary of £150 p.a., on the Dudley Canal in 1793 after it had obtained an Act to extend its original line to the Worcester & Birmingham Canal at Selly Oak. This included Lappal Tunnel, the fourth longest in the English canal system, which proved extremely troublesome to construct. Initially not really up to the job, he gained by experience and when the Engineer, Josiah Clowes (q.v.) died in 1796, the Company promoted Underhill to his place, under the direction of their Committee of Works. Estimated originally to cost £90,000, a second Act was required in 1797 to raise further capital, and the final cost was about £140,000.

Early in 1801 Underhill was commissioned to report on the Dorset & Somerset Canal, whose shareholders had advanced less than half of the promised capital, and whose works therefore were largely uncompleted. His report was read on 23 March and led to a resolution to complete the branch to Frome. Despite this, money ran out in 1803 and an Act passed in that year was unsuccessful in raising more, so work ceased with the line unfinished.

In June 1802 he was in partnership with William Whitmore (q.v.) and Charles Norton, a Birmingham builder (see Colvin) to supply timber for cranes supplied by Whitmore to the Kennet & Avon Canal. He also undertook some erection work and had presumably been working for the Company for some time, as he was paid £100 for making a railroad at Foxhanger, west of Devizes where the Caen Hill flight of locks had not yet been started. In October Underhill and his son, also called William, contracted with the Company to dig almost three miles from Horton to All Cannings, east of the town. The railroad was put to work in June 1803 but by December Underhill's affairs back in Staffordshire were in a bad state and by March 1804 he had given up his contracts and settled with the Kennet & Avon.

He trained Jacob Owen (1778–1870) (see: OWEN, Jeremiah), who joined the Royal Engineers Department of the Ordnance in 1804, married his daughter and subsequently became an architect.

P. S. M. CROSS-RUDKIN

[RAIL 842/3 and 842/42, PRO; Rennie ms 19781, NLS; C. Hadfield (1966) *The Canals of the West Midlands*; K. R Clew (1971) *The Dorset & Somerset Canal*; Keystone (1998) *Kennet and Avon Canal Linings*; Colvin]

Works

1793–1798. Dudley No. 2 Canal (Selly Oak Extension), Resident Engineer, then Engineer, £140,000
1802–1804. Kennet & Avon Canal, contractor

UPTON, John (c. 1774–1853), civil engineer and contractor, was possibly the son of John Upton (d. 1812), surveyor to Lord Egremont. Upton is first noted in Gloucester, where he lived at 2 Spa Villas and is mentioned as Clerk to the Gloucester and Berkeley Canal in 1813. This canal, envisaged to permit ships to reach Gloucester avoiding the hazards of the River Severn, obtained its Act of Parliament in 1793, with Robert Mylne (q.v.) as its original engineer. As a ship canal it was a major undertaking and the costs of construction rapidly exhausted the financial resources of the Company, leaving the canal in an uncompleted state in 1800. Various options were tried to raise funds but little had been achieved when Upton was appointed. In the absence of another, Upton assumed the role of engineer to the canal, tidying up the basin as it existed and in 1815 issuing a report on the canal; it criticised earlier ideas of a terminus on the estuary at Hock Crib or Berkeley Pill, and instead recommended Sharpness Point. In his report he described himself as a marine engineer and there is some evidence that he also had some knowledge of the iron trade.

His proposal was looked at by the Canal Committee and Benjamin Bevan (q.v.) was asked to report on the suggestion; he also recommended Sharpness. The proprietors, following the passage of the Poor Employment Act in 1817, were able to make a successful application to the Exchequer Bill Loan Commissioners for aid in completing the canal. Despite his success in selecting the final line for the canal, Upton was forced to resign in March 1819 as a result of Thomas Telford (q.v.) and the Commissioners writing to the Company criticising his conduct, namely poor supervision of the work combined with allegations of selling materials to himself as Engineer. Upton gave pressure of work as the reason for his resignation. This is substantiated by the fact he was working for Telford reporting on a number of Exchequer Bill Loan applications namely the Thames Medway Canal (1817, 1822) and Folkestone (1817) and Shoreham harbours (1826). He clearly still had Telford's confidence.

At Midsummer 1818 Upton was appointed Surveyor to the Old Stratford and Dunchurch Turnpike where later, according to an auditor's report of May 1826, 'a system of the grossest fraud and deception has been practised (by Upton) from the beginning'. Upton had charged the Trust for labour and stone never supplied and it was estimated that he owed the Trust £1,163 9s 5d. The fraud was uncovered when it was noticed that payment had been made for breaking up more stone than had been supplied. The illiteracy of the workforce had made it easy for him to claim for extra labour. When his case came up at Northamptonshire Assizes he absconded. The Reports of the Holyhead Road Commissioners reveal the road to have been in a poor state and badly managed in the early 1820s. Upton, although the chief surveyor, was in charge of only part of the road and Telford in his 1824 report suggested he be put in sole charge, and move to Weedon. In his report of 1825 Telford remarks he had never seen 'the surveyor upon the road'; the explanation may lie in his extensive activities elsewhere.

In the 1820s Upton seems to have taken up contracting, building roads and bridges in South Wales, including a suspension bridge proposal for Llanvihangel Crucorney, near Abergavenny. Upton & Co. took six contracts on the Cromford and High Peak Railway in the late 1820s and there are references to their work there in the Butterley ledgers. This work was carried on despite the fact that he had skipped bail and it may be that it was further pressure from creditors arising from building work at Goodrich Court, Hereford, in early 1828 that persuaded Upton to flee the country.

The architect at Goodrich Court was Edward Blore and he had a commission around that time for a Palace at Alupka, in the Crimea, for the Governor of the province, Mikhail Vorontzov. Upton himself went to Russia where he found his various talents very much in demand. He worked on the construction of Sebastapol Dockyard and was made a Lieutenant Colonel in the Russian Army. His son, William, assisted in the work, married a Russian and was captured by the British in the Crimean War.

John Upton is believed to have died in October 1853, having retired to England, where he was living in the London area.

MIKE CHRIMES

[Quarter session records, Q/CB P&S and P&R 0007, Gwent CRO; Records of Old Stratford and Dunchurch Trust, Northants CRO; *Select Committee on the Holyhead Road*, 4th report, 1822; *Commissioners for the ... further improvement of the Road from London to Holyhead*, Reports 1824–1827; W. Shears (1847) Description of the iron gates for the docks at Sebastapol, *ICE Min Proc*, **6**, 47–56; *Willis's Current Notes* (1854) October, 80–81; C. Hadfield (1955) *The Canals of Southern England*; P. J. Riden (1973) The Butterley Company and railway construction, *Transport History*, **6**, 39; J Marshall (1982) *Cromford & High Peak Railway*; H. Conway Jones (1984) *Gloucester Docks: An Illustrated History*; G. N. Crawford (1989) Thomas Telford and the Gloucester and Berkeley Canal, *Industrial Archaeology Review*, **xi**, 2; C. Gotch (1993) *The Gloucester and Sharpness Canal and Robert Mylne*; R. Stiles (1998) in lit. Boase; Bendall]

Publications

1815. *Observations on the Gloucester and Berkeley Canal*

1816. *Plan of the Gloucester and Berkeley Canal, showing the different Lines*

Works

1813–1818. Gloucester and Berkeley Canal, engineer
1818. Ross and Abergavenny Road, 10 miles, contractor
1818–1826. Old Stratford and Dunchurch Road, surveyor
1824–1826. Llanelen bridge
1824–1826. Pant-y-Goitre bridge, Llanfihangel
1826–1829. Cromford & High Peak Railway, six contracts, *c*. 6 miles including Buxton tunnel
c. 1830–1846. Sevastapol dockyard (with W. Upton)

V

VALLANCEY, Charles, FRS, General (1725–1812), military engineer and antiquary, was born in Flanders, France, in 1725 and was brought to England as a child. It seems likely that he was educated at Eton before, at an early age, entering the army. He was attached to the Royal Engineers, rising to the rank of Lieutenant-general in 1798 and General in 1803. Vallancey must have been posted to Gibraltar shortly after his first commission; he served there seven years and a plan of Gibraltar, dated 1750, executed by him, is in the British Library.

In the late 1750s Vallancey was posted to Ireland, joining the Corps of Engineers there, and his *Essay on Fortification* was published in Dublin in 1757. His next work, *Field Engineer* was written at Kinsale. In 1760 he reported on the defences of Carrickfergus. He assisted with the military survey of the island and from 1774–1790

Charles Vallancey FRS

worked on his General Survey of the Southern part of Ireland, a highly commended work. In this he deliberately moved away from a map intended solely for military manoeuvring to a complete topographical map.

Vallancey made Ireland his adopted home, became Director-General of Engineering, and was consulted by government agencies on a number of public works. For Dublin Corporation, he designed the Queen's (now Liam Mellowes) Bridge in Dublin. This masonry bridge of three semicircular arches is the oldest of all the bridges spanning the river Liffey still extant, having been built between 1764 and 1768. The bridge has some architectural merit with alternate voussoir stones projecting from the arch ring. There are pilasters over the piers and a balustraded parapet.

In 1770 he surveyed the Grand Canal and the following year, acting for the Commissioners of Inland Navigation, Vallancey selected the site of the Leinster Aqueduct to carry the canal across the river Liffey near Sallins, in County Kildare. The location of the aqueduct was a matter of great debate and argument amongst the engineers of the day and both John Smeaton and William Jessop (qq.v.) were consulted before the final decision was made to follow Vallancey's recommendation.

Vallancey became something of an antiquary and published a number of works concerning the history, philology and antiquaries of Ireland and was an active member of many learned societies, including the Royal Dublin Society (1763–1800). He received an Honorary LL.D. from the

University of Dublin in 1781 and was elected a Fellow of the Royal Society in 1786. These distinctions were earned, not for his achievements as an engineer, but for his contribution to the understanding of the antiquaries of Britain and Ireland. However the biographer Webb wrote in 1878 that 'in the light of modern research his theories and conclusions (were) a fanciful compound of crude deductions from imperfect knowledge (and have been) shown to be without value, and such as would now not receive a moment's attention'.

Vallancey died in 1812. Several portraits exist including a portrait in oils by Chinnery in the Royal Irish Academy, Dublin, and a portrait in oils by Solomon Williams in the Board Room of the Royal Dublin Society, Dublin

RON COX

[Vallancey mss, *Scientific Notebooks*, 23 manuscript notebooks on various topics: chemistry (16), algebra (3), physics (2), climatology (2). Royal Irish Academy, MS.12K.34.35; *Plan of Gibraltar* (1750) MS Add. 21576.3F3h, BL; extract from a report made to the Navigation Board of the survey of the levels of the River Boyne and the then state of the Boyne Navigation in October 1771, MS.7352, National Library of Ireland; *Survey of 1776*, MS.2737, Royal Irish Academy; Report to Sir Richard Heron on navigation in, and defense of the bay and harbour of Dublin (1778?) MS.13.058 (Heron Papers), National Library of Ireland; *Proceedings of the Corporation for the Preservation of the Port of Dublin*, Vol. 1 (1793, 1799), National Archive of Ireland, Dublin; *Defensive State of Cork Harbour* (1797) MS, Add. 33.119, Add. 33.105, BL; *Letters to Colonel Brownrigg*, MS, 33.102, BL; FF 174; Connolly, Royal Engineers Library, Chatham; T. Jones (1813) *Catalogue (dated 1813) of a Valuable Collection of Books, Manuscripts and Irish History, the Library of the Late Celebrated Irish Historian* [sic] *General Charles Vallancey*. Thomas Jones, Auctioneer, 6 Dame Street, Dublin; A. Webb (1878) *Compendium of Irish Biography*, Dublin, 540–541; M. Nevin, General Charles Vallancey 1725–1812, *J R Soc Antiq Ir*, **123**, 19–58; J. H. Andrews, Charles Vallancey and the Map of Ireland, *Geog J*, **132**, 48–61; R. C. Cox and M. H. Gould (1998) *Civil Engineering Heritage: Ireland*, London, 41, 109; DNB]

Publications

1757. *Essay on Fortfication*
1758. *The Field Engineer*
1763. *A Treatise on Inland Navigation or the Art of making Rivers navigable, of making Canals in all sorts of Soils, and of constructing Locks and Sluice*
1766. *A Practical Treatise on Stone-cutting*
1771. *A Report on the Grand Canal or Southern Line*
1779. *An Essay on Military Surveys*

Works

1764–1768. Queen's (now Liam Mellowes) Bridge, Dublin, engineer
1771. Leinster Aqueduct (Grand Canal), surveyor

VARLEY, John (1740–1809), surveyor and engineer, was perhaps born at Heanor, Derbyshire, the son of Francis Varley. His wife, Hannah, was the daughter of Benjamin Patters of Dutton, near Runcorn. They had six children.

Varley was one of James Brindley's (q.v.) assistants. His Chesterfield Canal survey of 1769 was thought 'a very correct one' by John Grundy (q.v.) and his Greasbrough Canal survey of the same year was judged correct by John Smeaton (q.v.) but with insufficient water supply.

After Brindley's death Hugh Henshall (q.v.) was made Inspector of the Works (May 1773) for the Chesterfield Canal but Varley retained construction responsibilities. In late 1773 Varley's family was involved in contractual anomalies: some work in Norwood Tunnel had been allocated to his father, Francis, and his brothers, Francis and Thomas. As a result Henshall was appointed Chief Engineer in March 1774 and in May Varley's family was dismissed from the works, Varley being severely reprimanded.

Nonetheless, Varley completed the Chesterfield Canal with minimal involvement by Henshall. Along its 45 miles he organised the construction of six wide locks, fifty-nine narrow locks, one reservoir (Pebley) and two tunnels, one of which (Norwood) was the longest of the period (2,884 yd.*). This was achieved in a creditable six years with slender resources of experienced manpower, much of which was employed on the contemporary Trent & Mersey Canal. Of particular note was the completion of Norwood tunnel in only four years (1771–1775), the multi-staircase Thorpe flight of fifteen locks (1772), and the Norwood flight of thirteen locks in one-third of a mile (1775).

Following the completion of the canal he worked on the Erewash Canal and made other canal surveys in the East Midlands. An Erewash Canal lock was rebuilt in March 1780 'owing to his mistake in making the level'. His surveying on the Union Canal was mapped by John Varley Jr. (b. 1776) who acted as Resident Engineer on its construction.

Varley took over the contract for Wolseley Bridge from James Trubhaw, but after two seasons in which little progress was made he gave up the contract in May 1800. The work was continued by direct labour, Joseph Potter (q.v.) being responsible for supervision.

In May 1808 Varley was 'in a very infirm and distressed state and soliciting an Annuity from the [Erewash Canal] Company'. His home was at Pennyholme (Kiveton Park) from 1773 until his burial at Harthill, South Yorkshire, on 16 February 1809.

---

* The length of Norwood tunnel differs in many publications. There is no evidence that it was either shortened or lengthened at a later date. Recent calculations using the portal OS map references and Pythagoras result in a tunnel length of 2,884 yd.

James Varley of Colne was the original contractor for the piers at Pontcysyllte in 1794, but in 1795 he had to accept John Simpson (q.v.) as his partner; he subsequently withdrew completely.

CHRISTINE RICHARDSON

[Leeds & Liverpool Canal Company Minute Book, 1768–1769, RAIL 846/1, PRO; John Grundy (1770) *Report on the Chesterfield Canal Route to Bawtry*; Strutt Library, A66, Derbyshire RO; Harthill Church records (1772–1784); RL 942.741, Rotherham Central Library; John Smeaton (1775) Survey for the Greasbrough Canal, WWM. MP47. D, Sheffield Archive Office; Erewash Canal, Company Minute Book, 1777–1780, RAIL 828/1, PRO; Huddersfield Canal, Company Minute Books, 1794–1807, RAIL 838, PRO; C. Richardson (1992) *The Waterways Revolution—From the Peaks to the Trent, 1768–1778*; C. Richardson (ed.) (1996) *Minutes of the Chesterfield Canal Company, 1771–1780*; indexed transcription of RAIL 817/1, PRO; R. Schofield (2000) *Benjamin Outram*]

Publications

1769. *Map of a Superseded Route of the Chesterfield Canal, to Bawtry*
1769. *Map of a Superseded Route of the Chesterfield Canal, to Gainsborough*, D258. S/3C, Derbyshire RO
1769. *Map and Estimate for the Greasbrough Canal*, WWM. MP47. B, Sheffield Archives Office
c. 1770. *Map of the Chesterfield Canal, as built*, RAIL 817/2, PRO
c. 1782. Survey and estimate, Sleaford Navigation, Box 14/8, Spalding Gentlemen's Society, Banks & Stanhope Collection

Works

1769. Chesterfield Canal, survey, Brindley's assistant, 45 miles
1769. Greasbrough Canal, survey, 1½ miles, five locks
1769. Leeds & Liverpool Canal, survey, Brindley's assistant
1771–1777. Chesterfield Canal, construction, Clerk of the Works, 45 miles, £150,000, sixty-five locks, one reservoir, two tunnels
1777–1779. Erewash Canal, survey, Resident Engineer, 11¾ miles, £23,000
1782. Sleaford Navigation, survey, 13 miles, seven locks
1788. Erewash Canal, survey, Consultant
1791. Nutbrook Canal, survey, Consultant, 4½ miles, thirteen locks
1792. Nottingham & Beeston Canal, survey, Consultant, 6 miles, three locks
1792. Leicestershire and Northamptonshire Union Canal, survey, Jessop's assistant
1792. Erewash Canal, survey, Consultant, reservoir
1793. Leicestershire and Northamptonshire Union Canal, construction, Resident Engineer, 40 miles, £200,000, four tunnels, thirty-eight locks
1798–1800. Wolseley Bridge, Colwich, Staffs, Contractor

1800–1801. Huddersfield Narrow Canal, repairs, Subcontractor

VAZIE, Robert and William (fl. 1790–1830), civil engineers, were involved *c.* 1800 in some of the earliest proposals for underwater tunnels for roads rather than mining purposes.

Robert Vazie (b. 1758) was the second son of 'Mr. Robert Vazie' (b. 1727) and was baptised at Hexham, Northumberland, on 20 January 1758; he was one of a family of nine. Nothing is known of his early years. Vazie took out his first patent (No. 1384) in 1783 while resident in Hexham. This was for a 'Method of using steam in aid of fire, in obtaining salts of all kinds ...'. A succession of patents followed indicative of his varied interests: saddles (1931, 1793), guns (2466, 1801), brickmaking (2986, 1806), erection of arches (3516, 1812), metal compounds (4700, 1822) and food processing (5523, 1827). They also provide a clue to his movements: South Shields (1793), London, (1801–1812, 1827) and Chacewater mine, Cornwall (1812).

In 1807 Vazie spoke of his 'professional information in the art of mining' although from 1806 he was describing himself as a civil engineer. By then he had been involved for about three years in a proposal for a tunnel under the Thames from Rotherhithe to Limehouse. In 1798 Ralph Dodd (q.v.) had proposed a tunnel beneath the Thames at Gravesend, a rather uninspiring location. Vazie's scheme was more sensibly located in terms of the growth of London's docks at the time. He may have been contemplating the idea as early as 1802—it certainly was being discussed in the press in September 1803—and in July 1805 the Thames Archway Company obtained its Act of Parliament with the Vazie family amongst the shareholders.

Vazie began operations by attempting to sink an 11 ft. diameter shaft approximately 315 ft. from the river on the Surrey shore. Progress was slow. Although the shaft was only sunk 42 ft. the capital was exhausted at the end of the first year. The shaft was now reduced to 8 ft. diameter and caisson type linings employed, but when a depth of 34 ft. was reached quicksand and water ingress led to the Company consulting John Rennie and William Chapman (qq.v.). Vazie himself blamed his problems on the fact that the directors would provide only a 16 hp engine rather than the 50 hp steam engine he requested. After some delay Vazie introduced Richard Trevithick (q.v.) to the Board and in August 1807 they began to act as joint engineers on a driftway. Progress was rapid, in comparison with what had taken place before, using timber supports the driftway was excavated 5 ft. high, 3 ft. wide at the bottom, and 2 ft. 6 in. wide at the top. By 12 September they had reached 180 ft. from the shaft and by 19 October 394 ft. had been reached.

The conclusion of the directors, perhaps unsurprisingly, was that Trevithick was largely responsible for the about turn and Vazie was dismissed when further problems occurred. The driftway eventually reached c. 1,040 ft. at the end of February 1808. The workings were then again flooded, and advice taken from other engineers, including John Buddle and John Rastrick (qq.v.). It would appear that J. U. Rastrick had temporary charge in the late spring and early summer while Trevithick considered alternative means of completing the tunnel. The Company invited proposals for completion on 30 March 1809 but, although experiments were conducted by John Isaac Hawkins (q.v.), lack of finance precluded any progress until Marc Isambard Brunel (q.v.) revived the idea of a tunnel in the early 1820s.

Vazie himself rapidly moved on, acting as engineer to three road schemes in North London: Junction Road, from Kentish Town to Highgate, the New North Road from Shoreditch to Highbury, and that for Highgate Archway. The first scheme proposed, in 1808, a tunnel more than a mile long at Highgate to the north of London, with two forks at the southern end at the foot of Highgate and Westhills; this was modified, after the first Parliamentary Bill failed in 1809, into a 360 yd. long tunnel with approaches in cuttings which obtained Parliamentary approval in 1810. In April 1811, with work well advanced on the approach cuttings, Rennie was consulted about the tunnel and he recommended reducing its length to 211 yd. He was consulted again in early December 1811 and provided a specification for repairs using doweled brickwork. The following April, when it had been driven about 279 ft. the tunnel failed. This was partly due to poor construction and partly wet weather combining with difficult ground conditions. Rennie was exonerated from any part in the failure as his recommendations had not been fully followed, whilst Vazie was in charge of the actual construction. A second Act was obtained and with the architect, John Nash, acting as engineer, the road was built in a cutting with a bridge above. Vazie was still described as engineer to the Archway and Kentish Town Junction Road at the time of his daughter, Harriet's, marriage to Captain Henry Taylor, RN, in October 1814.

Despite the failure of these two tunnel schemes Vazie was involved with further road schemes in Essex and carried out tests on the vertical bonded brickwork he developed with P. Moore (1812–1813). His last civil engineering reports were on Shoreham harbour (1816) and Ralph Walker's (q.v.) proposal for the Romford Canal (1818–1824).

William Vazie (b. 1756), the brother of Robert Vazie, was baptised at Hexham on 25 August 1756. He was the eldest of a family of nine, born between 1756 and 1771. Like his brother, nothing is known of his early career but in 1805–1807 he became involved in a proposal to drive a tunnel under the Forth between Rosyth and Queensferry, with James Millar and John Grieve (fl. 1777–1824). This used geological information of the area gained from the mining activity and knowledge of the

underwater mines at Whitehaven, provided by John B. Longmine. In support of the proposal, Vazie obtained the views of Grieve and Robert Bald (q.v.), among others, who felt that it was practicable. Perhaps unsurprisingly, funding could not be raised. It is probable that he was involved with his brother in the Thames Archway Company as a letter survives, dated 1808, indicating that he sought the participation of John Buddle (q.v.) in the project, an invitation which was declined.

Robert and William Vazie are known only for their involvement in these early tunnelling schemes but, nevertheless, they must be regarded as pioneers of one of the most difficult of the civil engineering arts.

MIKE CHRIMES

[J. Rennie (1811) General reports, 237–242, Rennie mss, ICE archives; Rennis mss 19771, NLS; A tunnel under the Thames (1803) *European Magazine*, **44**, 244; Thames archway (1805) *Monthly Magazine*, **20**, 179; Some account of the archway or tunnel intended to be made under the River Thames (1805) *Repertory of Arts, Manufactures*, 2nd series, 7, 371–373; Tunnel under the Thames (1807) *Repertory of Arts, Manufactures*, 2nd series, **10**, 399–400; Letter from John Buddle to W. Vazie (1808) BUD/23/48, NRO; Appendix (1809) *Third Report Select Committee ... Broad Wheels and Turnpike Roads*, 119–121; Tunnel under the Thames (1809) *Repertory of Arts, Manufactures,* 2nd series, 14; (Highgate archway) (1809) *Monthly Magazine*, **27**, 328–329; **28**, 32; **31** (1811) 533–535; **33** (1812) 377; **34**, 459; **36** (1813), 357–358; **42** (1816) 210–211; History of the tunnel and archway (1813) *European Magazine*, **64**, 204–208; *Monthly Magazine*, **36**, 224–225; (Experiments on vertical brick bond) (1813) *Monthly Magazine*, **35**, 56, 99–100; **36**, 224–225; (Marriage of Harriet Vazie) (1814) *Monthly Magazine*, **38**, 377; The Thames Tunnel (1832) *Penny Magazine*, **I**, 257–258; (Highgate archway) (1833) *Mechanics Magazine*, **18**, 351, 384, 392; (Thames Tunnel) (1835) *Mechanics Magazine*, **24**, 68–72; F. W. Simms (1844) Practical tunnelling, 184–186; H. W. Dickinson and A. Tilley (1934) *Richard Trevithick*, 90–105; J. Boyes and R. Russell (1977) *The Canals of Eastern England*, 55–57; A. Lynch (1982) How the Archway Road was built, *Hornsey Historical Bulletin*, **23**, 1–5; S. Morris and T. Mason (2000) *Gateway to the City; the Archway Story*]

Publications

W. Vazie (with J. Grieve, J. Taylor, J. Millar, *et al.*):

1806. *Report of Surveys ... making a Land Communication by a Tunnel under the River Forth*
1807. *Observations on ... making Tunnels under Navigable Rivers ... proposed Tunnel under the Forth*

R. Vazie:

1812. Patent No. 3516: *Construction or Formation of Arches and other Erections and Buildings* (with J. A. Kelly)

1816. *Comparative Statement of the Specific Differences between forming a Detached Harbour near to New Shoreham*

Works
1805–1807. Thames Archway, shaft and driftway, c. 400 ft., Robert Vazie engineer
1810–1812. Highgate Archway Road and Tunnel, c. 150 yd., Robert Vazie engineer

VERMUYDEN, Sir Cornelius (c. 1590–1677), civil engineer and entrepreneur, was born c. 1590 at St. Maartensdijk in the Isle of Tholen, Zeeland, a son of Gillis Vermuyden and Sarah Werckendet, his wife; Gillis was High Sheriff of St. Maertensdijk. Sarah's father, Cornelius Werckendet, was a land drainage engineer and Vermuyden followed the same profession. In translation from the Dutch, a *Map and Project of the Drowned Lands lying in Brabant across from the Town of Tholen* by him is dated October 1615 and he reputedly worked in Beveland and Flanders before coming to England in 1621.

In July 1621, following various unsuccessful attempts made in the preceding two years to promote the general draining of the Fens, King James declared himself willing to undertake the project. This involved the drainage of rather more than 300,000 acres of fenland, later known as the Great Level, in the counties of Suffolk, Norfolk, Cambridge, Huntingdon, Northampton and Lincoln. The main problem was raising sufficient funds for such a vast undertaking and it may have been for this reason that the King brought in an engineer from the Netherlands who might be able to command financial support from his fellow countrymen, already well known as entrepreneurs of land drainage and reclamation at home and abroad. That the choice fell on Vermuyden is probably due to the presence of a 'friend at court' in the person of Sir Joachim Liens who had married Vermuyden's (much older) sister Cornelia and came to England in 1618 as a special ambassador of the Netherlands government to discuss the affairs of the Dutch East India Company and the Greenland fisheries. He was knighted in 1619.

In his *Discourse touching the Drayning of the Great Fennes*, written in 1639 (published 1642), Vermuyden says 'King James of blessed memory did undertake that great Worke, who for the Honour of this Kingdom (as his Majesty told me at the time) would not suffer any longer the Land to be abandoned to the will of the Waters, nor let it lye waste and unprofitable' and he goes on to say 'When … at that time I was come over into England, invited to this Worke, I took several Viewes thereof, went away, returned, and reviewed the same, took advice of Experienced men of the Low Countries, and from time to time did study how to continue that Worke for the best advantage, being at that time in proposition to have undertaken the doing of the said Worke together with my Friends at our own charge for a proportion of Land'.

As a result of this activity, by February 1622 the Privy Council accepted that a feasible proposition was now available and named as undertakers Cornelius Liens (brother of Sir Joachim) and Vermuyden, they no doubt being confident of raising capital from their 'Friends' in the Netherlands. The King would have been pleased at such an outcome from his initiative. Nevertheless for reasons which are now obscure the scheme was abandoned for the time being.

Vermuyden cannot have been unaware of the considerable amount of previous investigation and planning in the Fens. These notably involved the proposals by John Hunt (q.v.) in 1605 which included a new channel for the Ouse 21 miles long from Earith to Salters Lode, foreshadowing 'Bedford River' made in the 1630s and the maps of the Fens drawn up in association with Hunt by William Hayward (q.v.), who was also briefly involved in the 1622 proceedings, and the reports by Sir Clement Edmondes and Richard Atkyns (q.v.) in 1618. It may be supported that Vermuyden used much of this information and, coupled with his own observations, produced the plans considered by the Privy Council to be 'feasible'. However, no details of his plans are available.

Vermuyden remained in favour with the King and was given the job of trenching and draining Windsor Great Park, which he completed in 1623. In November of that year he married Katherine Laps at St. Mary's, the parish church of Rotherhithe, where presumably she was living with her mother and step-father, Joos Croppenburgh (whom her mother, as a widow, had married in 1614 at the Dutch Reformed Church, Austin Friars). Vermuyden would have known Croppenburgh in the Dutch community in London and when in June 1622 Croppenburgh contracted to reclaim 490 acres of drowned land in Erith Marshes beside the Thames, Vermuyden became his assistant or engineer. Incidentally, Croppenburgh was a London merchant and entrepreneur who between April 1622 and December 1623 created Canvey Island, a group of islets surrounded by salt marshes, in return for one-third of the 3,600 acres of reclaimed land. He employed Dutch workmen, some of whom settled on the island, but it is not certain that Vermuyden took part in the undertaking.

Dagenham Breach
In 1622, probably in August, Vermuyden himself took on a contract to repair a breach in the Thames bank at the outfall of Dagenham Creek, which had led to the drowning of large parts of Dagenham Marsh (550 acres) and Hornchurch Marsh (590 acres). A breach at or near this spot had occurred in 1591. After some delay it was repaired by John Legatt of Hornchurch under a contract of 1594, but a 'violent tide' in September 1621 again destroyed the bank. This time, after a shorter delay, the Commissioners of Havering Level (with jurisdiction over the marshes and

banks from West Ham to Hornchurch) entered into the contract with Vermuyden. Early in 1623 he was called before the Privy Council to answer complaints of arrears in payment to the workmen. He replied that though he had expended £3,600 he had received nothing from the Commissioners. In response they claimed that little had so far been achieved. The truth appears to be that the landowners involved had neglected to pay the tax levied for this purpose, the only source of funding available. The work was successfully completed, with a sluice in the bank, probably in 1624 since in the following year Vermuyden received a parcel of the reclaimed land by Letters Patent 'in recompense of his charges' at Dagenham. Some authors have assumed this to be the site of the more famous breach of 1707 eventually stemmed by John Perry (q.v.) in 1716–1720, but the scene of Perry's heroic efforts is half a mile west of Dagenham Creek.

Hatfield Level

The Crown owned large tracts of low-lying ground in Hatfield Chase and neighbouring districts and King James saw the possibility of great benefits if the land could be drained. He asked several leading landowners in the surrounding region to look into the question but they reported in July 1622 that 'considering how great the Levels were, and how continually deep with water, how many rivers run thereunto, and such like, they did humbly conceive it is impossible to drain and improve them'. Though the affair may have been put aside for the time being the King was not content to accept this verdict and two years later asked Vermuyden to examine the matter. This he did, and declared himself prepared to undertake the project, but before negotiations had been concluded James died in March 1625. However, his successor Charles I lost little time and on 24 May 1626 signed a contract with Vermuyden for the draining of some 60,000 acres of 'drowned' grounds in such manner as to be fit for tillage and pasture. Work was to start within three months of the King reaching agreement with persons claiming 'any estate, interest or common in the said grounds'. An amount of land not exceeding 3,000 acres could be left on each side of the rivers 'as a receptacle [washland] for the sudden downfall of waters'. Vermuyden was to make such 'channels, watercourses, banks, highways, sasses [locks], sluices … as he thought most necessary'. He and his associates would receive one-third of the drained land, the Crown also one-third, and income from the remaining third would be dedicated to maintenance of the works.

Clearly, all the essential planning had been done. The very day the contract was signed Vermuyden received permission from the Privy Council to proceed to the Low Countries with his wife, two children and four servants. There he quickly raised financial support form backers, who

thereby become 'Participants' in the underaking, and recruited a large workforce. Returning to England, he set up headquarters at Sandtoft, an 'island' in the Chase close to the junction of the rivers Idle and Don, and work began about August 1626. Vermuyden's success in raising capital abroad may have resulted in part from a promise of support from the influential Sir Philibert Vernatti, a Lombard by birth, a Hollander by adoption and a favourite of the English Court. He became a leading Participant along with prominent Dutch merchants and entrepreneurs.

Operations proceeded with remarkable expedition. The principal works, completed before the end of 1628, were as follows. (1) A dam across the Idle, at a place subsequently named Idle Stop, to divert flow out of its northern branch in the Chase along the eastern branch (originally a navigation channel cut by the Romans) known as Bycarrsdike to its outfall into the Trent at Stockwith. (2) A barrier bank about 5 miles long from Idle Stop to Stockwith along the north side of Bycarrsdike. (3) Ashfield Bank, 2 miles long on the south side of the Don from Fishlake to Thorne, to divert flow out of the Chase entirely along its northern branch (another Roman channel) known as Turnbridge-dike. (4) A navigable sluice in Ashfield Bank to permit navigation along the eastern branch to Sandtoft and places in the northern district. The sluice had lifting gates enclosing a lock chamber 50 ft. long by 15 ft. wide. (5) Dikesmarsh Bank along the east side of the north branch of the Don running 5 miles from Thorne to its outfall into the river Aire at Turnbridge. The bank protected land to be drained in Dikesmarsh north of Thorne and the Chase. It was set some distance from the river to allow space for washland between itself and the existing bank on the west side protecting land in Fishlake and Sykehouse. (6) A completely new embanked channel for the River Torne from Wroot, near its junction with the old north Idle, running for 6 miles in a north-easterly direction across the Isle of Axholme and thence due east for 3 miles to an outfall sluice on the Trent at Althorpe. (7) A main drain 8 miles long running north from Idle Stop, and passing under the Torne at Tunnel Pit, to Dirtness. There it was joined from the west by a 3 mile long drain and altogether they proceeded eastward for 5 miles to another sluice at Althorpe. The sluices, adjoining each other, had clear waterways of 11 and 14 ft. respectively. Feeding into the main drains were dozens of secondary ones.

It will be seen that the 'living waters' of the Idle and Don and also of the Thorne (by its banks) were excluded from the Level and kept separate from the land drains which had to cope only with rain or snow, a principle recognised by Vermuyden as the 'general Rule of drayning'.

Nothing on this scale had previously been seen in England, or possibly anywhere else and, in recognition, Vermuyden was knighted by King Charles at Whitehall on 6 January 1629. But more remained to be done. The King had experienced little difficulty in reaching agreement with the inhabitants of the Chase and the manors of Wroot, Crowle and Dikesmarsh, the areas so far drained. The manor of Epworth, composing most of the Isle of Axholme, was a different matter with freeholders and commoners holding land under ancient rights. Much of the Isle in the parishes of Haxey, Epworth and Belton was good land leaving only 13,400 acres 'drowned' and this they had no desire to see taken over in large part by the Crown and foreign undertakers even though receiving full compensation. Nevertheless, without agreement having been reached, work began in Haxey Carrs in August 1628, to be opposed immediately by rioters. The story thereafter is complicated, but at least a temporary peace was reached in September 1629 when Commissioners awarded the people of Epworth manor 6,000 of the 13,400 acres, and drainage was declared complete in 1631. Much of this land could be drained into the system recently established for the Chase. The only major work was a main drain for Haxey Carrs running 6 miles to an outfall sluice on the Trent at Owston.

In 1628 John Molanus is recorded as being 'agent' to the Participants, so it may be that Vermuyden, hitherto in total command of operations, now left most of the day-to-day management in his hands (as he later did in the Dovegang lead mine). Vermuyden's nephew, John Liens (q.v.), son of Sir Joachim, seems to have been in charge of the Axholme works and it was he who built the navigable sluice on Bycarrsdike known as Misterston Soss (or Sasse). This prevented flooding of the carrs south of Bycarrsdike in exceptional tides while allowing navigation in normal conditions up the Idle to Bawtry. Constructed in 1629–1630, about a mile upstream of Stockwith, it was timber-framed with lifting gates enclosing a lock chamber 60 ft. long by 18 ft. wide, and high banks ran from the Soss to the Trent bank.

In all, some 73,000 acres had been drained by 1631. The Participants received their share amounting to 24,000 acres, of which Vermuyden held 4,500 in return for planning and directing the works. However, the draining of the Chase and parts of Axholme was achieved at the expense of unforeseen flooding of land outside the Level on the west side of Turnbridgedike. Its prevention required heavy additional works which placed a severe burden on the Participants and damaged Vermuyden's reputation.

The problems here arose from the difficulty of discharging flood flows of the Don into the River Aire (which itself could rise 8 ft. above normal high tide level) after the eastern Dikesmarsh bank had been built. Vermuyden had provided the washland but it proved inadequate and floodings occurred in the western parishes of Fishlake and Sykehouse in the winter of 1628. Rioters tried to cut through Ashfield Bank. The west bank was restored and a navigable sluice built at Turnbridge (of the same dimensions as Misterton Soss) in 1629. And as flood flows were far greater in the Don than in the Idle, a large outfall sluice was built alongside with seventeen openings each 6 ft. wide and 8 ft. high.

The 'Great Sluice', as it was called, was probably built by Vermuyden's assistant, Hugo Spiering (q.v.), and completed before the end of 1630. Meanwhile in May 1630 the inhabitants of Fishlake, Sykehouse and nearby places petitioned the King, representing the losses they had sustained from repeated flooding, and in July of that year Vermuyden signed an agreement with the Hatfield Participants to give up his position as director of the drainage in return for a payment of £1000 for work he was responsible for but had not yet finished. Management thenceforth lay in the hands of a group of six principal Participants, of whom Vermuyden was one.

The Fens, 1629–1631

Vermuyden's withdrawal from direction of Hatfield drainage reflects growing unease in the relationship with his fellow Participants and, probably, also his anticipation of again becoming involved in the Great Level. Interest in this project, dormant since 1622, revived in June 1629 when the King recommended to a Commission of Sewers drawn from the six counties that a proposal by Sir Anthony Thomas and associates should be considered. Thomas had previous experience. Together with the Earl of Arundel and Sir William Ayloffe he was involved, with 'the assistance of some rare engineers', unnamed, in drawing up plans for draining the Fens in 1619. This time he headed a group including Henry Briggs, the famous mathematician, and Cornelius Drebbel, a military engineer and inventor. They may have provided little more than an appearance of scientific respectability, but Sir Anthony was soon to prove his ability in draining the East and West Fens, near Boston, completed between 1631 and 1634 under his supervision. However, the Commissioners politely turned down his scheme in October 1629, chiefly on the grounds that he was unwilling to reveal adequate details of the intended works. At about this time another proposal by Sir Thomas Crooke and William Burrell also proved unsuccessful. Sir Anthony returned to the attack in January 1630, but to no avail. Dissatisfied with these proceedings, Charles took the matter into his own hands in February 1630, requiring the Commissioners to play a more active role. This they did under the leadership of Francis, fourth Earl of Bedford, the largest landowner in the Fens. By July Bedford was in touch with Vermuyden and on 1 September 1630 the Commissioners, with Bedford presiding, declared their intention in principle of making a contract with Vermuyden to drain all the fens south of the river Glen, an area estimated at 360,000 acres. Vermuyden showed them a 'map [which] described the said fenny marsh, waste and surrounded [drowned] grounds, and the outfalls thereof, by lines and other descriptions, and also by writing expressed and set down what drains, sasses, sluices, banks, cuts and other works he intended to make'.

This leaves no doubt that Vermuyden had prepared a comprehensive design. Produced at short notice, it was presumably a version of his 1621–1622 scheme, but again no details are available. He was to receive 90,000 acres of the drained land of which the King, for his favour, would have 30,000; it was an excellent deal for him and for Vermuyden, provided the latter could raise the necessary (large) financial support. But storm clouds were gathering over the affairs of Hatfield Level and Vermuyden's Dutch friends appear to have lost confidence in him. Moreover 'the country' expressed opposition to the proposed contract and approached the Earl in the hope that he would undertake 'so great and so noble a work'. Faced with this situation and the possibility—or the certainty, perhaps—that Vermuyden would not be able to gain sufficient support, Bedford agreed to undertake the work at a meeting of the Commission at King' Lynn on 13 January 1631, when the so-called Lynn Law was passed in the presence of Sir Robert Heath, Attorney General. The land to be drained now comprised the fens south of the Welland, an area of about 310,000 acres defined as the Great Level of the Fens (later known as Bedford Level). Of this, the Earl and his fellow Adventurers were to receive 95,000 acres of which 40,000 would be reserved to provide an income for maintenance and the Crown would receive 12,000. Clearly much high level negotiation had been taking place.

Vermuyden's name is absent from those of the fourteen original Adventurers. Along with Bedford himself, they include Edward Lord Gorges, Sir Robert Heath, Sir Miles Sandys, Sir Robert Bevill, and Sir Philibert Vernatti. However, Vermuyden's plan was approved at the Lynn meeting, at least in broad terms, and it might reasonably be supposed that he would have been appointed to direct the works.

This assumption was first made by Samuel Wells in his *History of Bedford Level* of 1830 and has been repeated by subsequent authors up to and including Darby (1983). Nevertheless, as Professor Margaret Knittl has recently shown, it is not correct. Vermuyden certainly directed works in the second (1640–1641) and the third (1650–1656) campaigns in the Great Level but, after January 1631, played no part in the first. Nor is it certain that his plan was adopted.

An outline of what was done, and of the organisation, in the first campaign is given in the article on Andrewes Burrell (q.v.) who was employed throughout the years 1631 to 1636 under the Earl of Bedford. The Earl, with a mostly different set of Adventurers, also undertook the drainage of Deeping Fen (1632–1636), an area of 30,000 acres north of the Welland near Spalding.

Sutton Marsh
In May 1631 Vermuyden began embanking and reclaiming an area of 1,120 acres of saltmarsh outside the old (so called Roman) sea bank of Sutton township north of Wisbech. This seems to have been completed by September 1632.

Dovegang Lead Mine
In 1631 Sir Robert Heath acquired a thirty-one year lease on Dovegang mine in Derbyshire and in October of that year took Vermuyden into partnership; he was to receive two-thirds of the profits in return for his skill in draining the mine. The exceptionally rich Dovegang vein ran for about half a mile east of Middleton. By the early seventeenth century workings had reached or were approaching the water table. Attempts to lower the water in 1615 by an 'engine shaft' 240 ft. deep (probably a horse-operated chain pump) and again in 1626 (employing a 'skilled engineer', John Bartholemew) proved of little value and the mine had not been worked since 1630. Vermuyden's bold expedient was to drive a sough' or adit from the Derwent valley at Cromford for three-quarters of a mile in a southerly direction, at a gentle upward gradient, to reach the vein at a depth of 300 ft. or more beneath ground level.

Vermuyden brought John Molanus from Hatfield to Middleton as his agent in 1632. Driving the sough through hard rock by hand, without the use of explosives, would have been a slow process. Work began in 1632 and since the mine was showing good profits by 1637 the sough probably reached the vein in 1636 or perhaps a year earlier. During the Civil War, in which Molanus served with distinction in the Parliamentary army, the mine was not worked, or only to a small extent. Full-scale operations were later resumed and by 1651 Vermuyden's son, Cornelius II, had taken on the direction. A pumping engine was installed in the sough by the 'skillfull Ingineere', Edmund Wheatcroft, to effect a further lowering of the water table and almost unheard of quantities of ore were soon being produced. Molanus was still the 'agent'. Cornelius II lived for some time at Middleton Hall. He was joined in 1652 by his younger brother, John, and continued to hold an interest in the mine at least until 1688.

The Dutch River
North of Hatfield Chase, with the Great Sluice at Turnbridge discharging water into the River Aire at low tide and the (admittedly rather small) washland providing temporary storage while the sluice was closed at high tide, it might be expected that the flooding problem was solved. Such proved not to be the case, and the six-man management group of Hatfield Level was compelled to adopt a radically different solution in the form of a new embanked channel for the Don running 5 miles from Newbridge, near Thorne, to the Yorkshire Ouse at Goole. There, at times of high flow, water levels could be as much as 5 ft. lower at high tide, and 10 ft. lower at low tide,

than in the Aire at Turnbridge. As originally planned and built the 'New River', or 'Dutch River' as it later became known, was entirely for flood protection. Navigation continued to Turnbridge for the time being.

The management group called on Vermuyden to 'make a model' or plan of the new river. Work began in the summer of 1632 and John Liens, Director of Works of Hatfield Level from about this time, was in charge of construction with David Parolle (q.v.) as Overseer or resident engineer. Upwards of £10,000 had been spent by April 1633 and construction was finished, with a large outfall sluice at Goole, before the end of 1635 at a cost of about £33,000.

The drainage of Hatfield Level had now been completed, but at the expense of complex legal and financial disputes between Vermuyden, the Participants and local landowners and commoners. One dispute in particular followed a complaint to the King against Vermuyden by Sir Phillibert Vernatti and three other Participants in December 1632. After allegations had been heard, Vermuyden was committed to the Fleet prison in May 1633 for refusing further to cooperate with the court and he remained there for several months. In September Sir Robert Heath protested that Vermuyden's 'restraint hath fallen out unreasonably for our mines in Derbyshire where he should have been long since'. But it was not until February 1634 that, following a petition to the King, the case was re-examined and found to be one of wrongful imprisonment. Presumably he was released soon thereafter.

The Fens, 1639–1642

Little has come to light concerning Vermuyden's activities during the next five years. He and his family were living in the parish of St. Dionis Backchurch near St. Paul's in the City. His lands in Hatfield Chase had been sold by 1635. His main income probably came from the Dovegang mine but he was still in debt to a Zeeland financier from whom he had borrowed substantial sums in 1630. It would therefore have been with relief that he learnt in September 1639 of his appointment as 'agent' to the King for the next phase of Fens drainage at a salary reputed to be £1,000 p.a. The events leading to this appointment are briefly as follows.

Works under the fourth Earl of Bedford resulted in the Fens being dry from April or early May every year; in other words 'summer grounds' had been achieved. The next step was to prevent winter flooding and thereby secure 'winter ground, fit for arable'. This required funding on a scale beyond the powers of the Earl and his partners who had already expended £97,000 on the Great Level and £23,000 on Deeping Fen. Thereupon the King again declared himself willing to be undertaker and again in the larger area of 360,000 acres envisaged in the 1630 scheme though it is not clear how he proposed to raise the necessary capital.

Nevertheless the King's proposals were accepted in a decree by a new Commission meeting at Huntingdon in July 1638. Soon afterwards he requested 'divers gentlemen, expert in such adventure, to give their advice, how these lands might be made winter grounds' excepting 'forelands and receptacles for winter floods'. Andrewes Burrell was one of those responding, in October 1638, but little is known of his suggestions except for a sluice on the Welland near Spalding although he may well have put forward some of the proposals subsequently incorporated in an engineering report of 1642. Another respondent was Vermuyden who presented his 'Discourse' to the King in January 1639. Different in several respects from what had been done already, his principal measures were: (1) to divert the Welland from the point of its entry onto the Fens into Morton's Leam, the late fifteenth century channel of the River Nene running from Stanground, near Peterborough, to Guyhirn, a few miles upstream of Wisbech; (2) to build a navigable sluice on the Old Nene downstream of Stanground; (3) to provide banks along the 12 mile length of the Leam, placed wide apart to form a receptacle for floods 'and so keep the waters at a less height by far against the banks'; (4) to improve the Nene from Guyhirn to Wisbech and either make a new channel north of the town or bypass the Horseshoe Sluice (which had been built about a mile below Wisbech in 1634–1635); (5) to provide new, and higher, banks for Bedford River, the recently completed 21 mile channel of the Ouse from Earith to Salters Lode, the banks to be a great distance the one from the other, so that the water 'in time of Extremity may goe in a large room to keep it from rising too high … And those lands which are to be left for the water to Bed on will be good meadow ground'; (6) to make a new cut-off drain for the rivers Lark, Little Ouse and Wissey running near the eastern edge of the Fens and into a channel to the Ouse at Denver.

In a paper on 'Further Considerations touching the work of the fennes', written in August 1639, Vermuyden suggests that the work should be done in three stages, beginning with the Deeping area, and estimates the total cost to be £128,000. His proposals were provisionally adopted after debate, and referring to his appointment (19 September 1639) Vermuyden says 'His Highness … was pleased to put the Direction of that Worke upon me'

Nothing could be done in the field, however, without approval from a General Session of Sewers. For whatever reason that Session did not sit until August 1640, and then expressed serious misgivings. What they amounted to is not known, though diversion of the Welland is likely to have been a quite unacceptable proposition. Certainly it was not carried out, then or at any later time. And when operations began in the autumn of 1640, they were concentrated on improvements of the Nene.

Evidently Vermuyden had made all the necessary preparations for a rapid start and much was accomplished in the next twelve months. But work was slowing up by October 1641 for lack of funds and came practically to halt by the end of the year. In this short second campaign works in the Great Level include: (1) a navigable sluice, timber-framed, on the Old Nene at Stanground; (2) a bank on the south side of Morton's Leam running 2 miles from Stanground to the west side of Whittlesey and for 8 miles from the east side of Whittlesey to Guyhirn; (3) a bank on the north side of the Leam, begun but nowhere near completed, with the intention (realised in 1651) of creating a large washland; (4) deepening the Nene from Guyhirn to Wisbech and building along part of this length a new south bank which together with existing north bank formed a washland known as the Waldersey Receptacle covering nearly 500 acres; (5) a new cut 60 ft. wide from the Horseshoe extending 2½ miles towards the sea.

Records of the proceedings in the second campaign, like those of the first campaign, no longer exist; they were probably lost in the Great Fire of 1666 when the Fen Office in London was destroyed. Details of the works are taken chiefly from Burrell (1641 and 1642) and Dugdale (1662). From a remark by Vermuyden in 1650 it appears that his staff in the second campaign included a Surveyor and four Overseers. Expenditure is said to have been £23,500.

The Civil War Period

In January 1642 Vermuyden received an order to attend the House of Commons 'to give an account by what Authority he goeth on with his Workes in the Fens', not that much was going on by that time. Next month, the unsatisfactory nature of the Fens project having been referred to a Parliamentary committee, it was ordered that his 'Designe together with the Mappe be printed ... that all men whom it may concern to take notice of, may thereby inform themselves, and may offer any other designe, and they shall have notice to be heard'. Shortly afterwards Vermuyden's *Discourse* appeared in print, with its map, and Burrell presented a comprehensive scheme. This, too, was published, along with his second critical but informative pamphlet on Vermuyden's works in the second campaign.

Before any further steps were taken the Civil War broke out in August. During the next four years minor disturbances occurred in the Fens, but nothing comparable to the rioting in Deeping and in Hatfield Level where even the church at Sandoft, which had been built in 1638 for the foreign settlers, was seriously damaged.

Meanwhile the Earl of Bedford, who died in 1641, was succeeded by his son, William, the fifth Earl, later to play a dominant role in the Fens. Similarly, Sir Miles Sandys Jr. succeeded his father in 1664. Vernatti was impeached by Parliament in 1645. Vermuyden seems to have kept a low profile during the War, but a Colonel Vermuyden served in the Parliamentary army. Some doubt

exists regarding his relationship to Sir Cornelius; most probably he was Bartel Vermuyden, a nephew, who afterwards returned to Zeeland and died at Tholen in 1650.

Following the decisive battle of Naseby such of the Fenland Adventurers as were left, or their representatives, united with the Earl in a petition to Parliament in August 1645. A committee was then set up and as a first step called Bedford in order to examine the interest and title he had in the draining. Among other things he submitted a summary of the expenditure on the Great Level between July 1631 and August 1638. The committee decided that the Earl still be considered the undertaker for the drainage, despite the intervention by the King in 1638. Having arrived at this somewhat revolutionary decision, the next step was to consider whether the work of draining 'be favourable or not'. Several petitions were received against the drainage but in January 1647 Vermuyden delivered a paper 'purporting reasons for the feasibility of that Worke of draining the Great Level', while Burrell gave evidence in February of the success of the fourth Earl's work in creating summer grounds. Finally in November the committee reported that the Level could be drained and 'to the great benefit of the Commonwealth'. Maps were examined, boundaries of the Level confirmed, and in December the matter went to the House for legislation.

Before anything more could be done the second Civil War erupted, to be concluded by Cromwell's victory in August 1648 and the execution of the King in January 1649. Thereafter the 'Act for the draining the Great Level of the Fens' passed the Commons on 29 May 1649. Essentially this repeats Lynn Law but with the condition that the Level should be made winter ground, the works to be completed before October 1656, and the Huntingdon decree of 1638 declared null and void.

The Fens, 1649–1656

The fifth Earl of Bedford and his associates immediately formed themselves into the 'Company of Adventurers' for draining the Fens. The Company's *Proceedings* still exist as contemporary copies made to be kept at Ely. They therefore escaped the Great Fire and have now been transferred to Cambridgeshire Record Office. One of the first decisions was to appoint a Director of Works whose duties were to decide 'how and where and in what manner and when from tyme to tyme the workes or Banks, sluices, Sasses, River and Draynes are to be made and of what dimensions'. The company considered Vermuyden for the post but reserved the right to exercise general control, and at this stage were far from certain on the plan to be adopted. Vermuyden agreed to submit his design; Bedford himself consulted a Dutch engineer, Jan Barents Westerdyke, and, on a petition from the Isle of Ely, William Dodson (q.v.) was asked to present his scheme, which he did a few days later. Vermuyden then submitted his terms for 'recompense' and after discussion and modification they were agreed, but he could not accept the proposed control by what, in effect, would be a management committee. At this point, on 19 June, negotiations broke down. There followed a confused period, including a false start on work 'fit to be done this summer', which after a month had to be abandoned in September for lack of funds. Dodson's plan received little support and Westerdyke had yet reported (he eventually did so in September 1650). Vermuyden was again approached but again refused to compromise.

On 12 December the Company voted to adopt a scheme of limited extent submitted by Sir Edward Partridge. At the same meeting it was ordered that six Adventurers or any two of them 'be desired ... to be constantly upon the works to husband the same for their best advantage' and to be allowed £20 per month for doing so. Further, on 10 January, it was decided that nine Adventurers or any five of them 'shall have full powers for the management of the whole work for drayning', and that the Company would henceforth meet on quarters days to receive an account of all their proceedings.

There could now be no escape for Vermuyden. A management structure was firmly in place; he would have to accept it or leave. But the Company knew that a Director of skill and experience was required. So, at last, on 25 January 1650, Vermuyden 'declared his designe for the Company's approbation, and upon debate and consideration thereof his designe was approved in the generall ... as a possible way to effect the drayning according to the Act'. On the same day he was appointed Director of Works subject to observing 'such orders ... as he shall receive from the persons appointed on the tenth of January'. He would receive 4,000 acres of drained land, of which 2000 were to be held in trust for seven years, plus £1,000 (paid £300 down and the balance at £20 per month) subject to a stiff penalty clause should expenditure on works exceed £100,000, an exceedingly optimistic estimate.

Robert Burton (q.v.), already appointed Overseer, received 'directions' from Vermuyden and proceeded to Wisbech to organise work in that vicinity. In March Sir Cornelius joined him to 'stake out the work'. During the first season he may have been personally involved in tasks properly those of the Surveyor, for it was not until September that Jonas Moore (q.v.) arrived on site as Surveyor General (at £200 a year) despite the fact that Vermuyden, within days of his own appointment, had reminded the Company to employ a Surveyor 'to sett out the works and see that they bee done'.

The first jobs were probably the bypassing of Horseshoe Sluice (noted by Westerdyke in September as now deserted and useless) and resuming construction of the north bank of Morton's Leam. When completed, before February 1651, this bank together with the existing south bank, built in 1641, created a washland or receptacle for flood water covering some 3,000 acres. Other works in the first twelve months included the Forty Foot (Vermuyden's) Drain running 10 miles from Ramsey to Bedford River at Welches Dam and the Sixteen Foot Drain, 9 miles in length, to drain the fens near Chatteris.

For purposes of administration the Great Level was divided into three parts: the North Level from the Welland to Morton's Leam, the Middle Level from Morton's Leam to Bedford River, and the South Level from Bedford River to the border of the Fens. The first season's works, together with those already made, secured winter grounds in the North and Middle Levels, an area totalling some 170,000 acres. Commissioners in March 1651 examined witnesses, among them Moore and Burton, and on the 26th of that month declared the draining of this area to be complete.

Under Burton there had been two or three overseers and an 'overseer of materials'. Several Adventurers were frequently on site, in particular Anthony Hammond who throughout most of the campaign acted as Superintendent or project director. A land surveyor, known as an 'undersurveyor', had been on the staff since March 1650. He was joined by another after the adjudication to assist the Surveyor in dividing out the drained land. Nothing had yet been done since 1636 towards improving the South Level, an area of about 140,000 acres with problems greater than those in the fens further north. Immediately after the March adjudication the Earl called a meeting to consider the matter, at which 'after serious debate' it was decided that Sir Cornelius should present a specific plan. This he accordingly did, and the works were entrusted to him as Director.

Here is proof that the scheme for the South Level was Vermuyden's. From what was done it is clear that he had now abandoned the idea of an eastern cut-off drain. Not that it was a bad idea (it formed the basis of the successful flood protection scheme of 1954–1964) but the overriding necessity was remodelling Bedford River. Excellent in conception but inadequate in execution, its banks were too close together and the channel was too narrow (70 ft. instead of 120 ft. as proposed by John Hunt) with the result that its banks were breached in the first great flood. Once this had been made capable of safely carrying upland flood water of the Ouse from Earith, the downstream tributaries could be handled (or so it seemed) without much difficulty.

Vermuyden's treatment of the problem was impressive. He cut a new channel, known as the Hundred Foot or New Bedford River, parallel over most of its 21 mile length to the old one and about half a mile from it with banks outside each channel enclosing a washland of 5,800 acres. The banks were typically 10 ft. high with a top width of 10 ft. and a 50 ft. base, made of earth brought from Over (4 miles south of Earith) and not of the soft fenland soils. On the Ouse at Denver, just upstream of the New Bedford outfall, Vermuyden built a 24 ft. wide 'double sasse', that is, a navigable sluice with doors to keep out the tides and another set of doors to maintain river level for navigation from King's Lynn up to Ely and places beyond.

By these means the course of the Ouse from Earith past Ely to Denver received water only from its four tributaries. To cope with floods the lower parts of these rivers were embanked, with washlands covering a few hundred acres each, and a slaker channel 114 ft. wide with high banks (known as St. John's or Downham Eau) was cut from a point a short distance upstream of Denver sasse to an outfall sluice on the Ouse 4 miles downstream near Stow Bridge. The sluice, stone-built, had three openings each of 24 ft. clear waterway.

On 17 February 1653 Vermuyden submitted a memorandum declaring that the main works in the South Level were finished and fit for adjudication. Two days later the Company requested Commissioners to meet at Ely on 24 March and ordered the printing of notices to be read in all churches. Orders were also issued to Moore to have maps ready for the meeting, and to the Comptroller to make sure that all outstanding payments to workmen were settled. The Commissioners, having heard evidence and a lengthy report from Vermuyden, made their award on 24 March. The next day a service of thanksgiving was held at Ely.

In the preamble to his report Vermuyden says that the North and Middle Levels, adjudged drained in March 1651, are so far improved that there are now about 40,000 acres sown with cole-seed, wheat and other winter grain, beside innumerable sheep and cattle where never had been any before. He then outlines very clearly the problems in the South Level and their causes, and gives details of the works (from which the foregoing description is taken) together with their purpose. The total area of washlands, including that of the Nene, is 12,100 acres (individual areas are given in an appendix by Moore), well within the upper limit of 15,000 acres specified in the 1649 Act. In conclusion, Sir Cornelius is able to say 'The workes have approved themselves sufficient, as well by the great tyde about a month since, which overflowed Marshland banckes and drowned much land in Lincolnshire and other places, and a flood, by reason of a great snow and rayne upon it following soon after, and yet never hurt any part of the whole Levell ... [proving] a claere dreyning according to the Acte'. And he adds 'If any thing shall bee with reason objected against the dreyning, I am ready to answer the same, and as I hope to your lordshipps' satisfaction; and to make it appear to your lordshipps that the designe is ... every way sufficient for the intended dreyning, without prejudice to navigation in the rivers'.

In comparison with any other seventeenth century undertaking in England the works completed in the years 1650–1653 are immense. Cutting the Hundred Foot alone involved the excavation of about 2 million cu. yd. and the south bank required nearly 1.5 million cu. yd. of earth to be placed, while Denver Sluice and the sluice at Stow rank among the biggest structures of their kind yet built. The workforce at its maximum numbered 10,000 men, probably in the summer of 1652, and the wages bill peaked at £8,000 a week. Edmund Welch (q.v.) became Principal Overseer in 1652 in place of Robert Burton, who died early the next year. Under him were three overseers, and Moore had four assistant surveyors. Adventurers acting as managers in the field continued to play an important role.

After the adjudication of March 1653 Vermuyden took a less active part in the operations. Moore was given extra powers and now had a deputy, Gabriel Elliott. Dozens of miles of highways were built, as well as six new bridges, and a slaker channel known as Marshland Cut or Tongs Drain was made later in 1653, running 4 miles from Nordelph to a sluice in the west bank of the Ouse above Stow bridge. In August 1654, in a re-organisation, the post of Surveyor General was terminated and Moore became the Overseer at £100 a year. The three Levels were now put in charge of Adventurers acting as 'Conservator' at £150 p.a., each with an assistant. Hammond took control of the South Level.

Experience showed that a sluice was required, alongside the navigable sluice already built at Denver, to provide an additional 30 ft. waterway. Moore drew up the contract in April 1655, which was awarded to John Savery; it was to be completed by June 1656 for £600. In December 1655 Vermuyden issued his last set of 'directions'. The completion of Denver Sluice effectively marked the end of the third campaign, at a total cost of about £250,000. Drainage of the Great Level to give winter grounds had been achieved and remained essentially unaltered for the next hundred years apart from the introduction of drainage mills to combat the effects of peat shrinkage.

Vermuyden last attended a meeting of the Company on 4 February 1656. His departure occasioned no comment in the Proceedings, beyond a decision that the 2,000 acres of his 'land held by trustees should remain in their hands for a further seven years', presumably because the cost of the works had exceeded the £100,000 limit specified in the conditions of his appointment. Moreover it seems he had already disposed of the other 2,000 acres.

In September 1656 a governing body of the Great Level was established. Among its officers were Moore, Welch, Elliott, Fortrey and Hammond, but not Vermuyden. If, as appears to be the case, he had now severed all links with the Company, his family nevertheless maintained an interest. Cornelius II purchased land in the Fens during the early 1650s and retained the property until his death in 1693, when it was transferred to his brother, Bartholomew, at that time the only surviving son of Sir Cornelius.

Later years

In August 1655 Vermuyden petitioned the Protector concerning drainage of land in King's Sedgemoor (Somerset) which he had purchased in 1632. The petition received a favourable hearing and a Bill was introduced in December 1656 only to be rejected by local opposition. He held on to the land (it passed eventually to Cornelius II) but if he did make further attempts at gaining permission to drain it they met with no success.

By 1657, when his son Charles was admitted a student at Grays Inn, Vermuyden had moved to Westminster (next year Charles entered Christ Church, Oxford, where he graduated in 1661; he became a licentiate of the College of Physicians in 1662 and died 1673). In 1663, on the occasion of the marriage of Vermuyden's daughter, Susannah, his address is given as Channel Row, Westminster, a large house later inhabited by Cornelius II in somewhat lavish style, and it was in St. Margaret's, Westminster, that Sir Cornelius was buried on 15 October 1677.

A. W. SKEMPTON

[A. Burrell (1641) *An Explanation of the Drayning Workes which have lately been made for the Kings Majesty*; A. Burrell (1642) *Exceptions against Sir Cornelius Virmudens Discourse for the Draining of the Great Fennes*; J. Moore (1658) *A Mapp of the Great Level of the Fenns*; Sir W. Dugdale (1662) *The History of Imbanking and Drayning*; T. Badeslade (1725) *The History of the ... Port of King's Lyn ... and the Draining of the Great Level*; C. H. Cole (1784) *Extracts from the Report of a View of the South Level*; S. Wells (1830) *The History of the Drainage of the Great Level of the Fens, called Bedford Level*, and vol. 2 (1828); W. H. Wheeler (1896) *A History of the Fens of South Lincolnshire*; J. Korthals-Altes (1925) *Sir Cornelius Vermuyden*; H. C. Darby (1940) *The Draining of the Fens*; F. N. Fisher (1952) Sir Cornelius Vermuyden and the Dovegang lead mine, *Journal of Derbyshire Archaeology & Natural History Society*, **25**, 74–118; L. E. Harris (1953) *Vermuyden and the Fens*; M. Albright (1955) The entrepreneurs of fen drainage in England under James I and Charles I, *Explorations in Entrepreneurial History*, **8**, 50–65; L. E. Harris (1957) Land drainage and reclamation, *A History of Technology*, **3**, 300–323; H. Grieve (1959) *The Great Tide*; T. G. O'Leary (1966) Dagenham Breach, *Victoria County History of Essex*, **5**, 285–289; M. Williams (1970) *The Draining of the Somerset Levels*; K. Lindley (1982) *Fenland Riots and the English Revolution*; H. C. Darby (1983) *The Changing Fenland*; T. D. Ford and J. H. Rienwerts (1983) *Lead Mining in the Peak District*; F. Willmoth (1993) *Sir Jonas Moore*; P. Jones (1994) Vermuyden's navigation works on the river Don, *Journal of the Railway & Canal History Society*, **31**, 246–258; M. Knittl (1997) The design of the initial drainage of the Great Level (in press); Dr. F. Willmoth and Prof. M. Knittl (in lit., 1996 and 1997) additional information from primary sources]

Publications

1642. *A Discourse touching the Drayning the Great Fennels ... as it was Presented to his Majestie. By Sir Cornelius Vermuyden.*

Works

1622–1624. Dagenham Breach, rebuilding the Thames bank and sluice at Dagenham Creek, at least £3,600
1626–1630. Hatfield Level, drainage of 73,000 acres, and navigable sluices on the rivers Idle and Don, about £56,000
1631–1632. Sutton marshes, reclamation of 1,120 acres of salt marsh at Tydd St. Mary
1640–1641. Great Level of the Fens, third campaign of draining, under Charles I, £23,500
1650–1655. Great Level of the Fens, third campaign of draining, under William, Earl of Bedford, about £250,000, completing drainage of some 310,000 acres

VIGNOLES, Charles Blacker, FRS (1793–1875), civil engineer, was born on 31 May 1793 at Woodbrook, near Enniscorthy, in County Wexford. Descended from a notable French Huguenot family, his father, Captain Charles Henry Vignoles, was stationed in Ireland in 1793 and afterwards in the West Indies.

Following the death from yellow fever of both his parents in Guadaloupe, Vignoles became a prisoner-of-war until freed by his uncle, Captain Henry Hutton. Although still an infant, he was given a commission in the army as ensign and immediately placed on half pay. He was returned to England and placed under the care of his maternal grandfather, Charles Hutton (q.v.), a mathematical professor at the Royal Military Academy at Woolwich. Hutton 'adopted' Vignoles as his own and laid the foundations of a sound and liberal education.

Around 1807, Vignoles was articled to a law firm but having broken with his grandfather at the age of twenty was sent to the Military Academy at Sandhurst under the private tutorial care of Professor Leybourne. The following year he saw active service at the Siege of Bergen op Zoom and received a commission in the Royal Scots. After a spell of duty in Canada, Vignoles was sent to Fort William in Scotland and from there, in May 1816, to Valenciennes.

He married Mary Griffith in 1817, went on half pay from the army, and sailed alone for South Carolina where he carried out surveys and produced an account of the Dominion of Florida, accompanied by a highly acclaimed map.

Vignoles returned to England in May 1823 and after briefly working for James Walker and William Tierney Clark (qq.v.) was engaged in the summer of 1825 by the Rennies (q.v.) on surveys in Surrey and Sussex of a line of projected railway from London to Brighton. Shortly afterwards he undertook surveys for the proposed Liverpool and Manchester Railway, including the route over Chat Moss. He was appointed resident engineer

Charles Blacker Vignoles FRS

by the Rennies on 12 August 1825 and, having given Parliamentary evidence in its favour in 1826, drew up the first contracts and began work on draining Chat Moss. When George Stephenson (q.v.) resumed control of the line he regarded Vignoles as a Rennie man and made life difficult for him, blaming him for faults in the alignment of the Wapping Tunnel. As a result, Vignoles resigned his position with the railway on 2 February 1827; such rebuffs were typical of his career. He maintained an interest in the line and attended the Rainhill Trials, reporting on them to *Mechanics Magazine*. He was closely involved with the Braithwaite–Ericsson *Novelty* design and, indeed, later used their locomotives on the construction of his St. Helens line.

In the early summer of 1827 he was asked by the government to undertake the re-survey of the Isle of Man and he placed J. E. Terry in charge of the work, which was completed by 1828. In that year, following a preliminary survey by Marc Isambard Brunel (q.v.), he was approached to improve the Oxford Canal, which felt threatened by a scheme for a London and Birmingham Junction Canal which had Thomas Telford (q.v.) as its engineer. The latter scheme came to nothing but the Oxford Canal route was shortened by 13⅜ miles to the designs of Vignoles. His work was admired by both George Rennie (q.v.) and Telford, who visited the works on a number of occasions. Among the engineering works were Newbold Tunnel and a number of cast-iron bridges.

Vignoles sought to take over the Thames Tunnel project from Marc Isambard Brunel (q.v.) and put forward an alternative method of completing the work, which involved tunnelling deeper through the London clay. He also made proposals for a tunnel under the River Mersey at Liverpool.

In 1830, in conjunction with Captain John Ericsson, Vignoles devised a method of working railways up steep inclines by introducing, in the centre of the road, a third rail, which was nipped by two horizontal rollers actuated by a lever from the locomotive. This centre-rail system was essentially the same as that afterwards employed for the line over the Mont Cenis pass in Switzerland, but advances in mechanical engineering proceeded rapidly to the point where main-line locomotives were able to cope with any reasonable incline.

Despite his activities elsewhere, Vignoles maintained an office and contacts in Liverpool. This was rewarded with his appointment as engineer for the St. Helens and Runcorn Gap Railway in 1829 and a line from Parkside, on the Liverpool and Manchester Railway, to Wigan, the origin of the North Union Railway, with which Vignoles was to be involved for many years. At the same time, he was involved with proposals for a line between Barnsley and Goole, and with the Liverpool and Birmingham Railway, eventually realised as the Grand Junction Railway. In 1829 he resumed his links with Ireland when, with John Collister's assistance, he surveyed improvements on the Slaney Navigation.

The subsequent professional career of Charles Blacker Vignoles included many achievements, including being given credit for the invention of the eponymous 'Vignoles Rail', although it seems more likely that he introduced it to Europe from the U.S.A. He was responsible for the design and supervision of the construction of the first passenger railway in Ireland—the Dublin & Kingstown—and numerous other railways in Europe and South America. He also constructed a large suspension bridge over the River Vistula at Kiev, in Russia.

Vignoles was Professor of Civil Engineering at University College, London, between 1841 and 1843. He was elected a Fellow of the Royal Society in 1855 and was President of the Institution of Civil Engineers in 1870–1871, having been elected a Member of that body in 1827.

He had five sons and two daughters by his first wife but she died in 1834 and he remarried in the Spring of 1849. He died at Hythe, in Hampshire, on 17 November 1875 and was buried in Brompton cemetery. Reproductions of portraits of Vignoles are to be found in the biographies by his descendants; a bust is on display at the Institution of Civil Engineers

RON COX

[Membership records, ICE archives; North Union Railway correspondence and contracts, Mackenzie collection, ICE archives; Vignoles diaries, mss, British Library; Anon. (1829) The Thames Tunnel, *Mech Mag*, **11**, 203–207: Anon. (1830) Statement of experiments made with Winans railway wagons, *Mech Mag*, **12**, 461–463; Lord Cochrane and Mr. Alexander Galloway v. Messrs. Braithwaite and Ericsson, *Mech Mag*, 13, 1830, 308–312; Presidential address (1869–1870) *Min Proc ICE*, **29**, 272–318; *Min Proc ICE*, **43**, 1876, 306–311; O. J. Vignoles (1889) *Charles Blacker Vignoles*; H. J. Compton (1976) *The Oxford Canal*; R. H. G. Thomas (1980) *The Liverpool & Manchester Railway*; K. H. Vignoles (1982) *Charles Blacker Vignoles: Romantic Engineer*]

Publications

1823. *Observations on the Floridas*
1828. *Map of the intended Improvements along part of the Line of the existing Oxford Canal*
1829–1830. Competition for locomotive carriages on the Liverpool and Manchester Railway, *Mech Mag*, **12**, 114–116,129–142,145–152, 161–167, 180–181
1830. A table exhibiting a gross load which can be drawn by a twenty-horse locomotive engine, *Mech Mag*, **13**, 156–157
1830. Memorandum relative to the experiments made at Mr. Laird's works … with the new low pressure boiler on the exhausting principle of Messrs. Braithwaite and Ericsson, *Mech Mag*, **13**, 235–237 (with A. Nimmo)

Works (to 1830)

1822. London to Brighton railway, surveys of projected line
1825–1827. Liverpool and Manchester Railway, surveys and resident engineer
1828–1834. Oxford Canal, improvements, engineer, *c.* £168,000
1829–1833. St. Helens & Runcorn Gap Railway, 13 miles; Widnes Dock, engineer
1830–1832. Parkside–Wigan Railway, 7 miles, engineer
1830–. Wigan–Preston Railway (North Union)

VINCENT, William (fl. 1734–d. 1754), was the engineer of Scarborough harbour, which provided the best haven for shipping between Tyne and Humber. Its main feature was a stone pier built in Elizabethan times running south-west from the foot of Castle Hill for 800 ft., to which an extension in a more westerly direction had been added in the late seventeenth century. In 1732 Scarborough Corporation decided to enlarge the harbour. William Lellam (q.v.), formerly engineer of Sunderland harbour, drew up plans involving removal of the of the extension (which was now in poor repair) and building on from the Elizabethan pier 440 ft. in a southerly direction at an estimated cost of £14,500. The new pier was to have a width of 60 ft. at the base, 20 ft. at the top, and to be 35 ft. high at the extremity in water at least 6 ft. deep at low tide.

On this basis an Act was passed in June 1732. Lellam started assembling materials and equipment, and work began in April 1733, but two months later Lellam died. Robert Wilkins was appointed in his place in July. However on 3 December 1734 Scarborough Corporation Orders record 'that Mr. Vincent is thought a proper person for Carrying on the Works at the pier and it is … agreed that Mr. Vincent shall succeed Mr. Wilkins and shall be appointed Ingineer … and shall be allowed the sum of Seventy pounds a Year'.

Work proceeded until 1744, between thirty and forty men being regularly employed. Vincent was then ordered not to advance the work further than the foundations already laid. This meant closing the head of the pier in 1746 by which time the extension, known to the present day as 'Vincent's Pier' had reached a length of 320 ft. As Vincent later said, 'it was intended by the Plan of the First Engineer (Lellam) to be carried forty yards further' though he admitted that a depth of considerably more than 6 ft. of water had been reached. The cost is not recorded but on the basis of Lellam's estimate it may be taken as about £12,000.

In 1745 Vincent produced a map of the harbour and adjoining parts of the town. It was engraved and published in 1747. An indication of his intellectual attainments or aspirations is provided by the fact that he was one of the few engineers to subscribe to the *Course of Experimental Philosophy* published by John Theophilus Desaguliers (q.v.) in 1734, the very year in which Vincent took up his appointment at Scarborough. This raises interesting but as yet unanswered questions on his earlier career.

In 1746 the workforce, now reduced to about twenty men, was transferred to the great East Pier planned by Vincent to form an outer harbour and to act as a breakwater. Work on this massive structure continued under his direction until 1752, when he left for Sunderland. By that time it had a length of 170 ft. Work continued slowly under a succession of 'Surveyors' and by 1800 the pier reached almost exactly the position shown on Vincent's plan: a total length of 920 ft. During the period 1801–1826 it was extended and completed by William Chapman (q.v.).

At Sunderland harbour work came practically to a standstill after Lellam finished the South Pier and left in 1731. Isaac Thompson (q.v.) surveyed the River Wear and the harbour (1736), the engineer John Thomas (q.v.) reported on the need for a north pier (1737) and the Wear Commissioners petitioned Parliament but without success. Ten years later, a new petition was submitted. Mark Burleigh and Thompson gave evidence, and this time an Act was passed in June 1747.

The newly appointed Wear Commissioners asked Vincent for advice. He visited Sunderland and was paid £29 8s. for his journey from Scarborough in July. Possibly on his suggestion the Commission then consulted Charles Labelye (q.v.). Owing to commitments at Westminster Bridge it was not until June 1745 that he could set off for the north. He reported in September. Still with no engineer of their own, the Commissioners wrote to Vincent in July 1749 asking if 'he will be pleased to come here two or three days before the next meeting to give his further Opinion in regard the harbour'. This he did, advising the use of a dredger, for which he gave plans, and received 5 guineas 'for his trouble'. Then on 30 August 1749 the Sunderland Minute Book records that 'the Commissioners from the Good Opinion they have of Mr. Vincent and his judgement, agree to give him £120 p.a. ... so long as there will be any work to do for the preservation and improvement of the port and haven'.

Probably as he was committed to building the East Pier at Scarborough, and settled there with his wife and family, Vincent did not accepted this tempting offer. However, the Wear Commissioners ordered a dredging vessel to be built (it cost £213 with its 'machine' and came into operation in June 1750) and continue to consult him. In January 1751 they asked for his opinion on a plan for a 'jetty', and in December 1751 wrote to the Scarborough authorities requesting 'they will be pleased to give Mr. Vincent leave to come to give advice as to the situation of the proposed pier'. This refers to the north pier proposed both by Thomas and Labelye.

A little earlier Vincent and John Turner of Gosport were consulted on the construction of a masonry pier at Ramsgate, and when in January 1752 the Ramsgate Trustees advertised the post of harbour engineer, at £200 p.a., Vincent was one of the applicants. He produced ample testimonials but the job went to William Etheridge (q.v.).

Meanwhile a rapidly deteriorating state of affairs at Scarborough due to mismanagement, and even corruption, by the Corporation had reached such a pitch that the ship-owners and inhabitants petitioned Parliament in January 1752 for a Bill to remove authority for new works to an independent body of Commissioners. Vincent, who had in November 1751 suffered the indignity of an attempt to reduce his salary, gave evidence on progress of the works. The Bill duly passed, Commissioners were appointed under the Act of March 1732.

The Wear Commissioners, doubtless aware of these events, wrote again to Vincent in April repeating their offer of £120 p.a., now to be guaranteed for seven years. In May he attended at Sunderland and promised 'at midsummer next to enter upon the place of Engineer and Superintendent of Works'.

By August, Vincent was getting the dredger in good repair, procuring materials, opening a quarry and building a workshop. Hitherto, since 1750, dredging had been in the north channel. From the autumn of 1752 Vincent switched operations to the south channel, and continued dredging there throughout the following year. The opening of this channel for navigation was his important contribution to improvement of Sunderland harbour. But before he could complete it, or start work on the north pier, he died in April 1754. In his will, dated 12 March 1754, he leaves his wife, Mary, all personal estate and a house in West Sand Street, Scarborough. Presumably they and their two young sons lived there before moving into rented accommodation in Sunderland. The burial of 'William Vincent, Engineer' took place at Sunderland on 16 April 1754. His foreman Joseph Taylor carried on until Joseph Robson (q.v.) was appointed harbour engineer in June 1755.

A. W. SKEMPTON

[Scarborough Corporation Records (1731–1754) Scarborough Library; John Thomas (1737) Evidence, *JHC*, **22**, 795; Wear Commissioners Minute Book and Accounts (1747–1754) Tyne & Wear Archives; M. Burleigh and I. Thompson (1747) Evidence, *JHC*, **25**, 296; C. Labelye (1748) *Report relating to the Improvement of the Port of Sunderland*; W. Vincent (1752) Evidence, *JHC*, **26**, 459–460; J. Murray (1847) An account of the progressive improvements of Sunderland harbour, *Min Proc ICE*, **6**, 256–277; A. W. Skempton (1977) The engineers of Sunderland harbour, *Indust Archaeol Rev*, **1**, 103–125]

Publications

1747. *A Plan of Scarborough*

Works

1734. Scarborough harbour, Vincent's Pier, £12,000
1746–1752. Scarborough harbour, East Pier
1752–1754. Sunderland harbour, improvements

VIVIAN (fl. 1750 onwards) is a well established Cornish name and encompassed the whole range of society from working miners to Sir Richard Rawlinson Vyvyan, Baronet, of Trelawarren who graced the bench at Quarter Sessions in the 1820s. Their involvement in the copper trade in Cornwall led to involvement in developing the copper industry near Swansea.

Henry Vivian (d. 1817) was the third son of John Vivian of Vellansaundry near Camborne and has been considered worthy of note because he married Richard Trevithick's youngest sister, Thomasina. Described as a mine engineer working at Dolcoath, he was keen to go to Peru as the capable engineer required by M. Uvillé who sailed in September 1814 with the components of the engines to be erected there. Trevithick thought him to be suitable and it is recorded that he worked Trevithick's first high-pressure 'puffer' whim-engine in Stray Park mine. But Trevithick feared that he might be blamed by the family if any accident should befall him and, sadly, soon after arriving in Peru himself in 1817 Trevithick had to report home that Henry died on 19 May, adding, 'I believe that too much drink was the cause or it'! It does not seem to have been recorded whether Richard did suffer the family's displeasure as he feared or whether they were so pleased to see him after an absence of eleven years that he was forgiven.

Of **Andrew Vivian** (1759–1842), more has been recorded. He was born at Vellansaundry on 30 November 1759, the second son of John Vivian and his wife Anne, and he himself was married twice, having a total of eleven children. Although in the patent specification he registered with Trevithick in 1802 he described himself as 'Engineer and Miner', he was in fact also, banker, mine manager, merchant, farmer and from early in his career, the 'Confidential Agent' to the Vyvyans of Trelawarren and the Pendarves of Pendarves. In 1790, at the age of thirty-one he was in business supplying candles, leather and other mining requirements and by 1795 had extended his mining interests by becoming manager of Stray Park mine near Camborne. In the course of his career he was manager or purser of a number of mines and had dealings with the leading engineers of his day. At Dolcoath, in 1798, with Andrew Vivian as manager and Richard Trevithick, the younger, as engineer, the shareholders were induced to reopen the mine which, with its engines, had remained unused since 1788. So grateful were they that when he retired as manager in 1806, the Adventurers, through the Lord de Dunstanville, presented him with a handsome piece of plate with a suitable inscription. Likewise in 1811, when he was manager of Wheal Abraham, he offered the post of Engineer to Arthur Woolf (q.v.) who had but recently returned to Cornwall from London.

But it is his association with Trevithick's horseless carriage for which he is best known. He saw the potential of the compact high-pressure steam engine and was to buy one of the first of the high-pressure 'puffer' whim-engines for Stray Park and so he and William West, who then was with Henry Harvey at Hayle, were involved in the construction of the Camborne locomotive. Indeed, Andrew is recorded as having been steering when, on its third outing, it turned over on the atrocious road and was famously lost as its originators enjoyed their roast goose and proper drinks. This was Christmas 1801! Eventually a patent was secured, two-fifths to Trevithick and Vivian and one-fifth for West. But developing the engine proved more expensive than remunerative and eventually Vivian was forced to sell his share as he needed the money badly and wished to pay everyone their demands. His troubles were not over; his eldest son, also Andrew Vivian and also a mine Captain, died in 1805 at the age of eighteen and his second son in 1809. In 1811, the United Mine Company, which he had formed with others and Henry Harvey, was dissolved with Andrew unable to make the immediate payments required. But in August of the same year John Lean commenced the 'Lean's' monthly reports on the various pumping engines after first Andrew Vivian had been approached and, after declining, had suggested Trevithick. But he remained the manager of many mines. His obituary states that at one period of his life, in addition to his many other activities, he 'superintended the management of eighteen mines', including Wheal

Abraham. In 1817, with Trevithick now in South America, he gave evidence with Woolf and Thomas Lean before the Select Committee of the House of Commons. By 1827, Trevithick, now returned, and his friend Davies Gilbert M.P. were stating in correspondence that Captain Andrew Vivian would be able to supply dates and particulars that would assist Trevithick's petition to Parliament.

He was described as possessed of a great fund of wit and humour which, united to a remarkably cheerful and frank disposition, rendered him a most desirable companion. A measure of the respect with which he was viewed is the fact that although for several years his age prevented him from carrying out his duties, both Sir R. R. Vyvyan and Mr. Pendarves M.P. continued paying him his salary until he died.

In one of those little coincidences of life it happened that in September 1842, the obituary to Andrew Vivian and the report of the funeral of The Right Honourable Richard Hussey, Baron Vivian of Glynn and Truro, appeared on the same page of a Cornwall paper. Andrew Vivian was given 8 column inches. Lord Vivian was afforded 18½.

R P TRUSCOTT

[Cornwall RO; *West Briton, Royal Cornwall Gazette*; S. Vivian, *The Story of the Vivians*, The Courtney Library of the Royal Institution of Cornwall; House of Commons Select Committee on Steam Boats etc. (1817) Evidence, 40–41; F. Trevithick (1872) *Life of Francis Trevithick*; G. C. Boase and Courtney (1874) *Bibliotheca Cornubiensis*; G. C. Boase (1904) *Collectanea Cornubiensia*; H. W. Dickinson and A. Titley (1934) *Richard Trevithick, The Engineer and the Man*; T. R. Harris (1966) *Arthur Woolf, 1766–1837, The Cornish Engineer*; E. Vale (1966) *The Harveys of Hayle*; D. B. Barton (1966) *The Cornish Beam Engine*; T. R. Harris (1974) *Dolcoath, Queen of Cornish Mines*; A. Pugsley (1976) *The Works of Isambard Kingdom Brunel*; R. A. Griffiths (1988) *Singleton Abbey and the Vivians of Swansea*; L. A. Cook (1997) *An Examination of the Social Impact of the Vivians on Swansea, 1809–1894*]

W

WADE, George, Field Marshal (1673–1748), soldier and politician, was the third son of Jerome Wade of Kilavally, county Westmeath. His older brothers being destined for the church and the family estate respectively, he entered the army, being gazetted ensign in 1690. He was promoted quite rapidly, becoming lieutenant in 1693, captain in June 1695, major in March 1703 and lieutenant-colonel in October the same year. He had seen active service in Flanders from 1691 to 1697 and again from 1702. In Portugal from 1704 to 1708 and Spain from then to 1710 he achieved some distinction, but was placed on the retired list when peace was declared. In 1714, by now a major-general, he was given a command in Ireland but may not have taken it up as he was elected a Member of Parliament for the pocket borough of Hindon in Wiltshire. During the rebellion of 1715 he was sent with two regiments of dragoons to Bath, where there was active Jacobite sympathy, and enhanced his reputation by exposing plots to aid the insurgents.

The Disarming Act which followed the failure of the 'Fifteen had not succeeded in pacifying the Highlands and, following a memorial on the subject from Lord Lovat to the King, Wade was sent to Scotland to investigate the state of affairs and make recommendations for effective government. He set off on 3 July 1724 and his report was dated 10 December. Although he noted the want of roads and bridges in the Highlands, there was nothing in his proposals to remedy that lack. Appointed Commander-in-Chief, North Britain in December, Wade was still in London in April 1725, requesting £10,000 a year for two years for various purposes, including repairing the roads between the garrisons and barracks. Delayed in Edinburgh by the malt tax riots, he did not arrive in Inverness until 10 August. This first season was spent in disarming the disaffected clans, but it was immediately clear that the roads between the Government barracks were inadequate to their purpose. A start was made on the road to Fort William from Killihuimian (later Fort Augustus) and a boat was built to provide the link in the other direction to Inverness. In his report to Government at the end of 1725 Wade asked for a salary for the person employed as Inspector of Roads and such an official is mentioned in August 1726. It is probable that Wade enlarged his ideas about the scope of a roads programme about this time, for he gave orders for a road to be made along the south-eastern side of Loch Ness, to supplement the boat built the previous year. Back in London for the winter, he put forward the desirability of a road to Inverness from Perth, to be paid for out of the army contingency funds. The year 1727 was spent in 'completing' the road through Glen More and Wade proudly proclaimed its success in his report at the end of the year. Funds for the road from the south, which would start at Dunkeld, 15 miles north of Perth, were now made available as they were also for stone bridges over the rivers and streams on the Fort William to Inverness road.

The expenditure on roads and bridges in the first four years totalled about £1,000; in 1729 over £2,000 was spent, then about £3,500 for each of the years to 1733, a peak of £4,731 in 1734, with a rapid falling away thereafter. By 1731 there were six subalterns employed in charge of the working parties and a little over five-hundred non-commissioned officers and men. Two-hundred and fifty miles of road were built in eight years, a soldier being expected to complete 1½ yd. of road per day in reasonable ground. The men worked for six to seven months per year, being paid an additional allowance while so employed, and then returned to winter quarters. The construction of these roads was not dissimilar from Thomas Telford's (q.v.) practice eighty years later, in that a foundation of larger stones was laid to a depth of perhaps 18 in., covered by a layer of gravel ideally as much as 3 ft. deep, but sometimes only half that. Fascines were used in lieu of the base course where a road crossed boggy ground, anticipating John Metcalfe's (q.v.) famous use of them after he returned home from the 'Fortyfive. Banks, side drains and cross drains served to remove the copious rainfall from the road. The standard width was 16 ft. but 10 ft. was not unknown.

About 40 bridges were built to 1736, a few each year, under contract by masons, some of whom came from the Lowlands for the work. The smaller ones cost about £60 each, two arches £90 to £100 and the three largest considerably more. The largest single work on Wade's roads, the Bridge of Tay at Aberfeldy, was designed by William Adam (q.v.), Mason to the Board of Ordnance in North Britain. After preparatory work in 1732, a roadway was passable in 1733 and finishing works took until 1735.

General Wade's role in the construction of the roads which popularly bear his name was that of a promoter. There is some suggestion that he went over the route of some of the roads before they were built, but neither his expertise nor the time available would have allowed much engineering input. It would seem that the strategy developed only gradually, as the civilising influence of the programme showed itself, and that the very success which Wade claimed for it encouraged the paymasters in London to turn off the flow of funds before it was complete. No evidence remains of plans to improve the access to barracks such as Bernera, in Glenelg, or Inversnaid, beside Loch Lomond, and the survey in 1735 of the 'road' between Ruthven and Braemar was not followed by action. Only from 1741 was the network extended by William Caulfeild (q.v.) under Wade's successor, General Clayton.

Wade spent each winter and spring in London, where he attended Parliament first as Member for Hindon (1715–1722) and then, until his death, for Bath. He generally arrived in Edinburgh in July and travelled into the Highlands late in that month or early August, staying there for a couple of months before returning south. The roads were by no means his sole preoccupation. He had Fort Augustus rebuilt on a new site (see: ROMER, John Lambertus) and improved Edinburgh Castle, Fort William, Inverness Castle (which he called Fort George, before the later structure at Ardersier was planned) and Ruthven Barracks. The cost of Fort Augustus alone was slightly larger than the total spent on roads and bridges. He also conducted private business, being an adventurer in the lead mines at Strontian in Sunart.

His experience in North Britain may have led to his appointment as a Commissioner for Westminster Bridge in 1736. In Parliament he was involved with an Act to control gaming in 1739 and was an advocate of building barracks for soldiers, rather than quartering them, as was then the more usual practice. He voted consistently for the Whig government and seems to have been generally popular in Bath, polling all thirty votes in the election of 1734.

In 1727 he was promoted Lieutenant-General by the new king, George II, and General in 1739. He does not appear to have travelled north in 1738 or 1739 and resigned his appointment the following year, leaving the Highlands apparently at peace. He was appointed a Privy Councillor in 1742 and Field Marshal in 1743. By now seventy years old and asthmatic, his subsequent career as Commander-in-Chief in Flanders (1743–1745) and in England during the second Jacobite rebellion, until replaced by the Duke of Cumberland in December 1745, was not glorious. He lived on until 14 March 1748 and was buried in Westminster Abbey. His fortune amounted to more than £100,000. Although he never married, the bulk of his wealth was inherited by his two sons and two daughters, one of the latter of whom was married to Ralph Allen (see: *PADMORE, John*) the leading political figure in Bath and a close friend of William Pitt in his heyday.

P. S. M. CROSS-RUDKIN

[DNB; ms 295 and 7187, Wade Collection, NLS; J. B. Salmond (1934) *Wade in Scotland*; R. Sedgwick (1970) *The House of Commons 1715–1754*; W. Taylor (1976) *The Military Roads in Scotland*; T. A. Heathcote (1999) *The British Field Marshals: A Biographical Dictionary 1736–1997*]

Works

1725–1727. Fort William–Inverness road, 61 miles
1728–1730. Dunkeld–Inverness road, 102 miles
1730. Crieff–Dalnacardoch road, 43 miles
1731. Dalwhinnie–Fort Augustus road, 28 miles
1732–1733. Realignment of Fort William–Inverness road from Dores to Foyers
1732–1735. Bridge of Tay, Aberfeldy, £4,095
1735. Crubenbeg–Catlodge road, 4 miles
1735–1736. High Bridge, Spean Bridge, £1,087

WAKE, Henry (1740–1794), sinker and borer, was baptised at St. Mary's church, Gateshead, on 18 May 1740, the son of Henry Wake and his wife, Margaret Newton; he was the youngest of four children born between 1732 and 1740. The Wake family had been coal-borers from 1685, if not before and, like the Rawlings (qq.v.), their involvement in this aspect of the coal industry extended for more than a century.

Henry Wake Sr. had been associated with John Rawling at Longbenton c. 1750 and in 1772 Henry Wake Jr. became involved in the taking of borings prior to the Tyne Bridge rebuilding. Seemingly retained by Robert Mylne (q.v.), Wake's work was again undertaken in conjunction with the Rawlings. Wake is known to have carried out borings at Coxhoe in 1782 but many more contracts must have been undertaken.

Wake died in Gateshead and was buried on 12 October 1794, then being noted as 'Mr. Henry Wake, Sinker'. He remained unmarried and died intestate; his nephew, Ralph Wake (b. 1763), inherited his estate. Two further Wakes continued the same trade: in 1788–1789 Andrew Wake (1749–1811)—probably another nephew—undertook boring for the Ayrshire Coal Company, his work there extending over a year and reaching 378 ft.; another was Thomas Wake (1759–1798).

R. W. RENNISON

[*Probate Records*, Durham University Special Collections; Acc. No. 3410, NRO; *Newcastle Courant*; M. W. Flinn (1984) *The History of the British Coal Industry*, vol. 2, *1700–1830*, 73]

WALKER family of Rotherham were ironfounders, based near Sheffield. Joseph Walker (1673–1729) was a nail maker of Grenoside, near Sheffield, and he and his wife, Anne, of Hill Top, Ecclesfield, had three sons—**Jonathan I** (1711–1778), **Samuel I** (1715–1782) and **Aaron** (1718–1777)—and four daughters. His early death meant that to a degree the sons had to make their own way in the world, Samuel becoming a local schoolmaster. In 1741 the brothers attempted pot founding at their smithy but then a new foundry was built in 1744 and in 1746 when John Crawshaw, their brother-in-law, joined them a new works was established at Masborough, near Rotherham, with a casting house and two air furnaces. Production expanded rapidly from 96 tons of cast iron in 1747 to 1899 tons in 1782. In 1773 gun casting began and by 1782 this was more than half of their production. A steam engine was introduced in 1777 and in 1782 a Boulton and Watt engine. Their expansion was undoubtedly helped by the American War of Independence. At the time of his death Samuel I was the leading partner and was succeeded by his son **Samuel II** (1742–1792). He had three brothers **Joshua** (1750–1815), **Joseph** (1752–1801), and **Thomas** (1755–1828), and two sisters, Mary (1745–1803) and Sarah (1746–1813), all involved directly or through marriage with the family's ironworks or leadworks. The same was true of Aaron's son, **John** (1760–1804), and Jonathan's son, **Jonathan II** (1757–1807). On Samuel II's death the firm became known as Joshua Walker and Company, Crawshaw having been bought out in 1789. It was one of the most important firms in the country at the time with assets valued at £202,657 in 1794.

The firm's involvement with iron bridges began under Samuel II, almost certainly through the agency of John Hall, a former Boulton and Watt steam engine erector who had erected their second steam engine in 1782, before emigrating to the U.S.A., where he met Thomas Paine (q.v.). When Paine was seeking an ironworks to support his iron bridge proposals Hall suggested the Walkers and in 1788 they cast a 90 ft. test rib on Paine's system. This was test-loaded with 6 tons of pig iron, and temperature effects monitored, and the results sent to the Royal Society and Society of Arts. A surviving drawing of Paine's system depicts a spider's web arrangement of successive rings and spacers of cast iron and one imagines that details were worked out by Walker's foreman, William Yates (q.v.), and Thomas Walker, with whom Paine generally corresponded. Walkers must have hoped that interest in iron bridges would grow with the success of Ironbridge and they could demonstrate their own capacity in this field; at all events they agreed to fabricate a model arch bridge of 110 ft. span of cast and wrought iron which was erected in Marylebone in 1789–1790. Yates was unavailable to supervise erection on site and an American mechanic named Bull assumed the responsibility. Paine's other interests soon took over and in October 1791 the Walkers repossessed the ironwork and returned it to Rotherham.

By this time consideration was under way for a long span bridge across the river Wear at Sunderland and Walkers drew the attention of Rowland Burdon (q.v.) to their model, following which a Paine-style design was drawn up. Originally intended to be in masonry, the size of the span (240 ft.) and the attendant problems of erection and costs led Burdon to look seriously at the iron bridge proposal. By now Joshua Walker (1750–1815) had taken over the management of the firm. Paine's proposal was no basis for a satisfactory bridge; in 1791 John Rastrick (q.v.) had suggested to Walkers an alternative arrangement of iron voussoir blocks and this suggestion was probably fed into the final design developed by Burdon with Michael Scarth (d. 1805) and Thomas Wilson (q.v.), with Yates probably detailing the castings. After erecting a trial rib at the works in 1793 the final structure, with a nominal span approaching 240 ft., was erected in 1795–1796; it comprised ribs of cast iron voussoir blocks with spandrels of decreasing diameter cast iron rings.

Although the patent associated with the design of this bridge (No. 2066, 1795) was taken out by Burdon, it was Wilson whose name became more generally associated with the Sunderland type of design, and with whom Walkers became closely associated. Essentially the arch 'voussoirs' making up the ribs were open cast iron frames strapped together with wrought iron bands and laterally connected with cast iron tubes, with spandrels of diminishing circles supporting the decks.

Wilson's next executed design was Spanish Town Bridge, Jamaica (1800–1801), which showed certain modifications to the Sunderland design, as did the Stratfield Saye bridge (1802). In 1802 Burdon and Wilson took out a new patent (No. 2635) which reflected developments in the design and connection of the voussoir blocks which were incorporated in Wilson's next design

at Staines (1802–1803). Whether these refinements reflected input from the Walkers as fabricators is not known but they cannot have taken much pleasure in the failure of this bridge in 1804 and that at Yarm in 1806. Both failures were ascribed to inadequate abutments although at Yarm, possibly at the suggestion of John Rennie (q.v.), vertical struts replaced the diminishing circles in the spandrels.

The Walkers' next bridge, Boston, was the result of collaboration between Rennie and Wilson, with Rennie disapproving of some of the characteristic Wilson features, notably the form of cross-bracing and circular spandrel circles. Walkers were invited to tender and Wilson effectively out-manoeuvred Rennie by ordering the ironwork and supplying patterns before Rennie had fully appreciated what had been done. Although Rennie had proposed vertical struts in the spandrels, the cross-bracing did not take the form of his recommendation. Almost as soon as the bridge was erected cracks began to appear in the ironwork and Rennie and Wilson became involved in an argument about responsibility, Rennie arguing that as his advice had been ignored it was Wilson's responsibility to ascertain the cause. One consequence was that Walkers were not properly paid; as late as 1810 they had only received £2,126 of the £3,706 owed to them. Both Samuel Aydon (q.v.) and the Walkers advised on repairs and the affair dragged on until 1819.

These well publicised failures of Wilson's system did nothing to diminish the Walkers' reputation and in 1810 they supplied the ironwork for a modified version of a Wilson style bridge designed by Henry Provis (q.v.) at Newport Pagnell. This had diminishing circles in the spandrels and used circles in the casting of the voussoir ribs. More importantly, when considering his enormous Southwark Bridge design in 1811, John Rennie turned to the Walkers. Whilst he had been unable in the case of Boston to get Walkers to graduate the depth of the voussoir blocks, increasing them towards the springings, at Southwark he was largely successful. He also refused to give in to the Walkers' request to keep the castings below 5 tons and the thirteen rib segments varied between 6 and 8 tons, and were 6 to 8 ft. in depth. The bridge, with two side arches of 210 ft. and a main arch of 240 ft. span, erected 1815–1818 remained the record-breaking span in cast iron. While it served as a monument to their capabilities it also brought on the firm's bankruptcy following the Napoleonic Wars. Although Rennie generally supported their claims, delays in delivery exasperated the directors of the Bridge Company, who consequently disputed payments, and the structure became something of a liability rather than a showpiece for their works.

Joshua died at his home at Blyth, Nottinghamshire, at the age of sixty-five in 1815. (He had another house, designed by John Carr (q.v.) at Clifton, near Rotherham.) The iron tradition was carried on by **Samuel Walker III** (1779–1851), who completed Southwark bridge and carried out tests for George Rennie (q.v.) on cast iron beams in 1818. The financial burden of Southwark can hardly have helped him; in 1820 the firm was still owed £31,925 for its work on the bridge, an arbitrator had been appointed, and it was decided to wind up the iron concern. In 1821 they were offered 10s in the pound on the Southwark debt and two years later the Masborough furnace was closed down. Samuel joined with his cousin William Yates Jr. (c. 1781–1865), at Gospel Oak ironworks, Tipton, in the Midlands. They continued to produce structural ironwork of high quality and the business there was continued by John and Edward Walker in the middle of the century.

MIKE CHRIMES

[J. G. James collection, ICE archives; Rennie mss, NLS; Rennie reports and drawings, ICE archives; G. Rennie (1818) Account of experiences on the strength of materials, *Phil Trans*, **108**, 118–136; G. Rennie (1842) On expansion of arches, *Trans ICE*, **3**, 201–218; J. G. James (1979) The cast iron bridges of Thomas Wilson 1800–1810, *Trans Newc Soc*; J. G. James (1986) The cast iron bridge at Sunderland, *Newcastle-upon-Tyne Polytechnic, Occasional Papers in the History of Science and Technology 5*; J. G. James (1988) Some steps in the evolution of early iron arched bridge designs, *Trans Newc Soc*, **59**, 153–187]

Works

1789–1790. Paine model bridge, cast and wrought iron, 110 ft. span
1793–1796. Sunderland Bridge, cast iron, 240 ft. span, £6,130 ironwork
1800–1801. Spanish Town, Jamaica Bridge, cast iron arch, 82 ft. span, £4,000 (£1,000 ironwork)
1802. Stratfield Saye Bridge
1802–1803. Staines Bridge, cast iron arch, 180 ft. span, c. £3,000(?) ironwork
1803–1806. Yarm Bridge, cast iron arch, 180 ft. span, £5,560 ironwork
1805–1807. Boston Bridge, cast iron arch, 86 ft. span, £4,155 ironwork
1809–1810. Newport Pagnell Bridge, cast iron arch, 60 ft. span, £1,640 ironwork
1811–1820. Southwark Bridge, cast iron arches, 210–240 ft., total cost £384,000
1825–1827. Hammersmith suspension bridge, ironwork, 400 ft. central span, total cost £43,000
1828–1829. Montrose suspension bridge
1829–1831. Scotswood suspension bridge
1839–1849. Budapest (Szechenyi) suspension bridge

WALKER, James, FRS, FRSE (1781–1862), civil engineer, was born in the house built by his father in the Law Wynd in Falkirk, Scotland, the first of five children of John Walker and his wife Margaret. After attending the local school Walker went to Glasgow University in October 1794 where he studied for five years. He spent the first two years studying Latin and Greek, the third on Logic and the last two studying Natural Philosophy and Mathematics. He was an able student and took the first prize in Natural Philosophy and Mathematics. He left college in May 1799, aged eighteen and a half, and returned home to Falkirk. His future career was discussed and business and the law were considered. However, in the summer of 1800 James was asked to accompany his ailing brother-in-law on a sea journey to London where, on arrival, he went to his uncle Ralph Walker's house at Blackwall. It was intended that James would return to Scotland after a week or so, but Ralph (q.v.) quickly realised James's abilities when discussing his engineering work in the West India Docks with his young nephew with the result that James remained in London for the rest of his life to follow his sixty-year consulting engineering practice. He began his career by becoming a pupil of Ralph Walker in 1800. In April 1811 James took as a pupil Alfred Burges (1797–1886) who became a partner in1829 when the firm became known as 'Walker & Burges'. Walker was based in Limehouse in east London for the early part of his career, by 1832 he had an office in Westminster, firstly at 44 Parliament Street (now Whitehall) and subsequently at 23 Great George Street. In the 1830s the firm was described as 'the great nursery of civil engineers in England' and many of those who worked for the firm subsequently rose to eminence in the profession.

At the age of twenty-one he undertook his first engineering work in his own right. This was the construction of the Commercial Road in London, built under an Act of 1802 to convey goods from the West India Dock to warehouses in the City of London. In 1810 Walker was appointed Engineer to the Commercial Dock Company, and he remained as consulting engineer to the Commercial Docks until his death. Vauxhall Bridge was Walker's first major bridge design and was also the first cast iron bridge over the Thames. The Vauxhall Bridge Company obtained their Act in 1809 for a masonry bridge designed by John Rennie. However, when the Middlesex abutment was nearly complete the proprietors found that they could not afford the masonry structure. After some delay James Walker's design for an iron arched bridge was accepted and it was opened in July 1816.

The major part of Walker's extensive consulting engineering practice happened after 1830 and the firm was best known for such marine works as ports, harbours of refuge, lighthouses, canal and river navigations. Although Walker worked well on into the railway age his firm was not largely engaged in this field of work. In the first railway 'mania' of 1825 he was appointed Engineer, jointly with William Brunton (q.v.) for the 'Grand Junction Railroad'. Additional engineering expertise was bought in from (Sir) John Rennie and Josias Jessop (qq.v.). The scheme linked the Midlands to Yorkshire, and Walker was entrusted with the survey of a route from Goole to

Sheffield, Manchester, Wakefield and Leeds. When Walker was consulted by the Directors of the Liverpool & Manchester Railway about the best form of traction on the line he not only suggested that a comparative trial of stationary and locomotive engines should be made, he suggested a prize be offered, and Rainhill as the best site for the competition. He subsequently acted as one of the judges. Walker also acted as engineer to the Leeds & Selby Railway and its subsequent extension to Hull and the Northern & Eastern Railway in the 1830s.

Walker was an early member of the Institution of Civil Engineers and was actively involved with its development. After Thomas Telford's death in 1834 Walker became the second president and occupied the presidential chair from 1835 to 1844. By this time a younger generation of engineers was keen to play a major part in the Institution's affairs. Walker's presidency has been overshadowed by the means of his leaving office. Nevertheless during his presidency the Institution acquired a library, a permanent headquarters, began publishing its Proceedings, and introduced rules requiring candidates for membership to have had experience in civil engineering. He also endowed Walker prizes at the Institution and the University of Glasgow.

Walker was elected a Fellow of the Royal Societies of London and Edinburgh and was awarded an honorary degree of the University of Glasgow.

James Walker died at his home above the office in Great George Street on 8 October 1862. He was working right up to the end of his life and his weekly Report to Trinity House was laid before the Court the day before he died. He was buried in the churchyard of St. John's Episcopal Church in Princes Street Edinburgh alongside his wife Janet and his daughter Margaret. His epitaph describes him as 'Upright, Sincere, and Unostentatious' and *The Engineer* of 17 October said that Walker was 'Among the few remaining contemporaries of the elder Stephenson and Brunel. His long life was almost wholly devoted to the practice of his profession, of which he was certainly among the most eminent members.

DENIS SMITH

[J. Walker (Autobiography) and membership records, ICE Archives; Records of Trinity House; Minute Books of Commercial Road Turnpike Road Trustees, Metropolitan Archive; Minutes, etc., of London dock companies, Museum of London in Docklands; *Aris' Birmingham Gazette;* (1825); Houses of Parliament (1826–1827) *Minutes of Evidence on Norwich and Lowestoft Navigation Bill;* House of Commons (1834) *Minutes of Evidence, Select Committee on Metropolitan Sewers;* D. Smith (1997) James Walker (1781–1862): civil engineer, *Trans Newc Soc,* **69**, 1, 23–55; Skempton]

Publications
For a full listing see Smith (1997).

1804. *Plan of the late Improvements in the Port of London; with the Roads made, or projected, for connecting the same with the City* (repr. 1805)
1810. *Plan of the Commercial Docks at Rotherhithe with the Intended Improvements*
1816. *This Plate exhibiting a Picturesque Elevation of the Iron Bridge erected over the Thames at Vauxhall, under the Direction of James Walker, Engineer, & completed AD 1816*
1820. *Plan of the Commercial Docks at Rotherhithe* (repr. 1824)
1825. *Report on the Eau Brink Cut* (with W. C. Mylne)
1826. *Report … to the Commissioners of the Haven and Piers at Great Yarmouth, on the State of the Bar and Haven; and the Measures advisable to be adopted for their Improvement* (with Reduced Plan)
1826. *Second Report … to the Commissioners of the Haven & Piers, of Great Yarmouth, on the State of the Bar and Haven*
1826. *Report … to the Corporation of Great Yarmouth, on the Plan for a Ship Navigation from the Sea at Lowestoft to Norwich*
1826. *Report … to the Commissioners of the Haven & Piers, of Great Yarmouth, on the Practicability of making Braydon Navigable for Vessels drawing 10 ft. water, &.*
1829. *Liverpool and Manchester Railway. Report on comparative merits of locomotive and fixed engines* (with J. U. Rastrick; 2nd edn., 1829)
1829. *Report to the Committee of the Proposed Railway from Leeds to Selby* (with Plan and Section)
1831. *Report on the present State of the New London Bridge* (with T. Telford (q.v.))
1832. *Report to the Commissioners of the River Wear on the Formation of Wet Docks at Sunderland*
1835. *Northern and Eastern Railway. Report to the Committee for Promoting a Railway from London to York, with a Branch to Norwich, &c.*

Works
Works to c. 1830 include:

1802. Commercial Road, London
1810–1822. Commercial Docks, Rotherhithe, improvements, construction of Norway (1810–1811), Lady (1810–1811), Acorn (1811–1812) and Lavender (1815–1822) Docks/Ponds (1810–1813, £130,000 spent)
c. 1814. Surveyor to Poplar Sewers Commissioners
1816. Vauxhall Bridge, London, £175,432
1819. Lea Bridge, Lea Bridge Road, London
1825. Appointed Consulting Engineer to Trinity House (remained so until 1862)
1827. Canal boat resistance experiments
1828. Bow Bridge, London.
1829. Judge at Rainhill trials, Liverpool & Manchester Railway
1830. Belfast Harbour, work carried out 1837–
1830–. King's Scholars Pond Sewer, Westminster, Engineer
1832–1833. Sewer outlet, Blackwall, Engineer, fees £136 3s

WALKER, John (fl. 1807–1832), millwright, was principal foreman for John Rennie and his sons at their engineering works in Holland Street, Southwark, from *c.* 1807 until his retirement in 1832. It is possible that he was already working for Rennie when Walter Hunter (q.v.) left Rennie to set up his works in Bow *c.* 1807. In an 1827 report to the Admiralty on dredging he referred to upwards of 'twenty years' experience of building excavating machines and can therefore be presumed to have been involved with the dredgers built by the Rennies for use on the Witham (1810), at Hull (1811–1812), at Ramsgate (1815–1816) and (1820–1821), and Liverpool (1817–1818).

Walker was elected a Member of the Institution of Civil Engineers in 1825 and remained in membership until 1828. His origins and fate are unknown.

MIKE CHRIMES

[Membership records, ICE archives; Rennie reports, **4**, 12 September 1827, ICE archives; A. W. Skempton (1975) A history of the steam dredger 1797–1830, *Trans Newc Soc*, **47**, 97–116]

WALKER, Ralph (1749–1824), civil engineer, was born at Tullibody, in Clackmananshire, in 1749, the second son of James Walker and Helen May. He was brought up on his father's farm on the banks of the river Devon, and was first educated at the parish school in Dollar. Later his elder brother, James, sent him to an academy where he learnt the elements of marine navigation. He became Captain of a West India ship and subsequently settled as a planter on an estate in Jamaica belonging to his mother's family. Encouraged by the British Admiralty's support for his improved mariners compass, he returned to England in 1793 and settled in London just at the time when the proposal for a system of wet docks on the Thames was under discussion.

Walker became interested, and in 1796 submitted his proposals incorporating the forerunner of the City Canal across the Isle of Dogs. With the interest of West India merchants aroused he became their engineer and drew up plans for the Canal and West India Docks in the Isle of Dogs with William Jessop, John Foulds (qq.v.) and George Dance, surveyor to the City. Walker prepared the detailed plans enabling the promoters to obtain their Act of 1799 to construct both the docks and the City Canal. William Jessop (q.v.) was appointed Engineer and Ralph Walker as Resident Engineer under him in August 1799 at a salary of £600 p.a. Following a professional disagreement with Jessop about a structural failure at the works, Walker resigned his post in October 1802. However, his work there undoubtedly established Ralph Walker as a notable civil engineer. In 1804 Walker submitted his scheme for removing Blackwall Rock, which was an obstruction to navigation in the Thames off Blackwall

Ralph Walker

point. This work was carried out to a scheme by William Jessop.

The East India Dock Company was formed in 1803 and Ralph Walker was appointed Engineer with John Rennie responsible for the entrance lock. Walker decided that the bricks should be made on site and let the contract to a brickmaker who built accommodation for his workmen there. The Import and Export Docks were opened in 1806. Walker remained Consultant to the Dock Company until his death in 1824, designing improvements to the Blackwall Basin in 1815, which were carried out by William Bough (q.v.).

In 1807 an Act was obtained establishing the Commercial Dock Company to develop docks on the Rotherhithe site based on the Greenland Dock. Ralph Walker worked on the project from the outset and was assisted by his nephew James Walker (q.v.), from midsummer 1808 until Ralph was replaced by his nephew as Engineer to the Company in September 1810. His fees for the work amounted to £525.

In August 1807 Ralph was appointed Engineer to the newly-formed East London Water Works Company at a salary of £500 a year. He designed and supervised the construction of their original works at Old Ford on the River Lea, together with the distribution mains. Walker designed two low-level reservoirs to contain 4,266,000 gallons and an upper circular distribution basin with a pumping engine house. The contractor for excavating and puddling the reservoirs was Hugh McIntosh (q.v.). When the Company considered buying the steam engines in 1807 Walker expressed his philosophy by advising: 'I am of opinion that they ought not to be contracted for because those who would supply for the least money must necessarily supply you with the worst machinery and though they may work tolerably well at the first offset, they will want repairs much sooner than good ones would and consequently become in a short time the dearest. I am also of Opinion that neither Public nor private Bodies ought to run any risk by trying new experiments or inventions in the first instance'.

John Rennie, in conjunction with Ralph Walker, produced a report on improvements to Dover Harbour in 1802, and Walker also produced a solo report in 1812.

In April 1822 Walker was asked by Trinity House to survey and produce an estimate of repairs necessary to the Buoy Wharf at Blackwall. Walker's estimate for rebuilding the 130 ft. long timber wharf wall in brick with a stone coping was £1104 11s 6d. This wall survives and is the oldest structure at the Trinity Buoy Wharf. In December 1822 the Navy Board asked Trinity House to recommend the most skilful civil engineer regarding tides and currents and they replied 'were we desirous to consult a civil engineer, conversant with Tides and Currents ... Mr. Ralph Walker of Blackwall, would ... be the person, on whose advice we think we might rely'.

When James Walker, Ralph's young nephew, came to London from Falkirk in the summer of 1800, he went to his uncle's house in Blackwall and became a regular assistant of Ralph, who launched James on his own career in civil engineering.

Ralph Walker died in his house in the East India Dock Road in 1824 following a fall down some steps at home.

DENIS SMITH

[Minute Books of East London Water Works Company, London Metropolitan Archives; Trinity House By Minutes 1822, Guildhall Library; Account of Ralph Walker, Esq. (1801), *European Magazine*, 331–332; Mr. Ralph Walker, Civil Engineer (1862) *The Builder*, 13 September, 653; A.W. Skempton (1978–1982) Engineering in the Port of London, *Trans Newc Soc*, **50**, 87–108; **53**, 73–94]

Publications

1794. *A Treatise on Magnetism, with a Description and Explanation of a Meridional and Azimuth Compass* (repr. 1798)

?. *Memorial to the Board of Longitude* (instrument for showing magnetic variation)

1796. *A Proposed Plan of Wet Docks in Wapping*

1797. *A Comparative View of the Wet Docks in Wapping, and in the Isle of Dogs, for the West India Trade: also Hints respecting the bonding of all West-India Produce, with a Plan of the Docks*

1800. *An Accurate Plan of the Docks for the West India Trade, and Canal, in the Isle of Dogs; Begun 1800* (another edn. 1801)

1800. *Plan of the West India Wet Docks*

1800. *Plan of a Double Turning Arch Bridge*

1802. *Plan of the West India Docks, etc.*

1810. *Plan of the Intended London and Cambridge Junction Canal*

1810. *Plan and Section of the South Part of the Intended London & Cambridge Canal from the River Lea Navigation to Mile End*

1811. *Plan and Estimate upon ... a Canal from Canterbury to St. Nicholas' Bay* (with J. Rennie Sr.)

Works

1800–1802. West India Docks, Resident Engineer

1803–1806. East India Docks, Engineer (with Rennie), £250,000

1804. Blackwall Rock removal scheme, consultant

1807–1810. Commercial Docks, Engineer, c. £25,000

1807–. East London Waterworks, Engineer

1809–c. 1813. Poortsmouth (Farlington) Waterworks

1815–1816. East India Docks, enlargement of Blackwall basin, c. £10,000

1822. Trinity Buoy Wharf, £1,100

WALLACE, William (fl. 1810–1830), civil engineer, was appointed resident engineer for works at Peterhead Harbour in early 1808, on the recommendation of John Rennie (q.v.). Rennie had suggested various improvements at Peterhead in a report of February 1806, including alterations to the existing south harbour and the creation of a new dock and north harbour. This was endorsed by Thomas Telford (q.v.). Funding was released for the improvement of the south harbour, and a return to the west pier had been completed before Wallace arrived. Wallace's salary on appointment was £230 p.a., a sum increased by £50 in 1810, although he had been criticised for not building the west quay wall as instructed the previous year. The pay rise was probably an inducement to press on with the works vigorously.

Wallace's chief task was to deepen the harbour and create a quay wall on the west pier, and use any waste material to create a new embankment. In the event more than 39,000 cubic yd. of materials were excavated from the harbour, much from solid rock, the harbour deepened by 6 ft., and a 500 ft. embankment formed. Under Wallace's superintendence the main contract was completed by December 1810, a year ahead of schedule, and extra work on the west pier and the east quay wall completed by March 1812. By then £13,800 had been spent, £6,000 more than the original estimate, and completion of Rennie's plans stalled.

In the meantime work had begun on the Grand Western Canal, another Rennie project, in April 1810. John Thomas (d. 1844 (q.v.)) was the first resident engineer, but work had almost halted in 1811 before a new Act and a resumption of work the following year. Progress was slow although, with Wallace in charge, the canal was navigable in 1814, at a vastly inflated cost.

In October 1815 Pembroke was officially established as a royal dockyard, and improvements were soon underway to Rennie's designs, with Wallace acting again as Resident Engineer, and Hugh McIntosh (q.v.) principal contractor. Works included covered slipways to Robert Seppings's (q.v.) designs. Wallace was still there in 1825.

Wallace's subsequent career is obscure. He reported on the Ancholme drainage in 1828 for Sir John Rennie (q.v.). A Mr. Wallace supervised work on the Middle Jetty at Fraserburgh in the 1830s.

MIKE CHRIMES and P. S. M. CROSS-RUDKIN

[Rennie reports, ICE archives; Rennie correspondence, NLS; Rennie mss, NLS; Rennie reports, ICE archives; H. Harris (1973) *The Grand Western Canal*, 54; A. R. Buchan (1979) The Engineers of a minor port—Peterhead, Scotland, 1773–1872, *Ind Arch Rev*, **3**, 243–257; A. R. Buchan (c. 1980) *The Port of Peterhead*]

Works
1800–1812. Peterhead Harbour, improvements, resident engineer, £13,500
–1814. Grand Western Canal, resident engineer, £145,500
1815–1825. Pembroke Dockyard, resident engineer, £130,000
1830. Fraserburgh Harbour improvements, £5,654

WALLACE, William, FRSE (1768–1843), mathematician was born at Dysart, Fife on 23 September 1768, the son of a leather manufacturer. Following his father's removal to Edinburgh Wallace was apprenticed to a bookbinder and subsequently worked as a warehouseman in a printing office, and for a bookseller. In his spare time he educated himself in Latin, French and mathematics, attending university classes. He began giving private lessons and in 1794 was appointed assistant teacher of mathematics at Perth Academy. In 1796 he presented his first paper, relating to geometry, to the Royal Society of Edinburgh. Further contributions followed: to the *Transactions* of the Royal Society of Edinburgh; the *Encyclopaedia Britannica*; and, later, to the *Edinburgh Encyclopaedia*, *Edinburgh Philosophical Journal*, the Astronomical Society, the Cambridge Philosophical Society and the *Cambridge Mathematical Journal*, an indication of his eminence as a mathematician. In 1803 he became a professor at the Royal Military Academy, then at Marlow, and in 1819 he was elected Professor of Mathematics at Edinburgh University, a post he held until his retirement in November 1838. Several distinguished engineers attended his lectures, including Robert Stephenson (q.v.).

Wallace's eminence led to his election as Honorary Member of the Institution of Civil Engineers in 1824 and he made two important contributions to the engineering profession. Around 1821 he invented the eidograph, an instrument for making reduced copies of drawings, which was preferred by some to the pentograph; by a modification

it was possible to reverse copies for use by engravers. This device was exhibited at the Institution of Civil Engineers in 1839 and at about the same time he wrote a paper on curves of equilibration, from which he deduced a series for computing co-ordinates of the catenary and gave tables of the values of the computed co-ordinates, so providing engineers with a ready means of constructing catenaries with the forms of equilibrated curves.

Wallace died in Edinburgh on 28 April 1843.

MIKE CHRIMES

[ICE membership records, ICE archives; *Min Proc ICE*, **1**, 1839, 65; *Min Proc ICE*, **3**, 1843, 10–12; *Royal Society Catalogue of Scientific Papers*; DNB]

Selected works
1831. Account of the invention of the pantograph, and a description of the eidograph, *R Soc Ed, Transactions*, **13**, 1836, 418–431
1839. Solution of a functional equation, with its application to the parallelogram of forces and to curves of equilibrium, *R Soc Ed, Transactions*, **14**, 1840, 625–676

WALLER, Richard (fl. 1751–1758), was appointed surveyor to the Carlisle–Newcastle military road on 24 June 1751 at a salary of £40 p.a. Work was carried out on the road by the Boyers family. Waller died in 1758.

MIKE CHRIMES

[Bendall; Minutes of Carlisle–Newcastle Military Road, Ca/C10/8/5, Cumbria CRO]

WARCUP, W[illiam] (fl. 1817–1826), of Dartford and 7 Paradise Buildings, Stangate Street, Lambeth, developed a curvagraph as an aid to engineering draughtsmanship, which was commended by engineers such as Henry Maudslay and Marc Isambard Brunel (qq.v.). He was elected a Member of the Institution of Civil Engineers in 1819. He contributed some of the earliest original communications to the Institution in response to a discussion proposed by Joshua Field on 'the comparative advantages of the various canal locks' in April 1819. No more is known of him.

MIKE CHRIMES

[ICE membership records, ICE archives; Description of Woodhouse's lock (OC2), ICE archives (from RAM, 2, 8,405); Description of [R.] Fulton's canal lifts (OC5), ICE archives; Description of [Joseph] Stevens canal lifts (OC6), ICE archives; *Trans Soc Arts*, **25**, 1817, 109–112]

WATSON, David, Major General (1713–1761), military engineer, was born in 1713, the son of Thomas Watson of Muirhouse. Despite being involved in most of the major battles of the period: Dettingen (1743), Fontenoy (1745) and Culloden (1746), he is most noted for his 8 years, 6 months involvement with the military survey of Scotland from 1747–1755. William Roy (q.v.), his assistant throughout the project, is generally described as the father of the ordnance survey, but Roy himself acknowledged the primacy of his 'much respected friend the late Lieutenant-General Watson then Deputy Quartermaster-General in North Britain ... being himself an engineer active and indefatigable a zealous promoter of every undertaking [who] first conceived the idea of making a map of Scotland'. Watson submitted his scheme for a survey of North Britain in 1747 as a means of completing the subjugation of the Highlands. He had previously worked with Major Caufeild (q.v.) and combined his survey with the extension of General Wade's scheme for opening up the country with roads, bridges, culverts and channels, and fortified outposts such as Braemar and Corgarff.

The scheme had low priority and he was given only three assistant engineers, but luckily, as well as Roy, these were of the high calibre of David Dundas (1735–1820), Hugh Debbieg (q.v.) and Paul Sandby (see: SANDBY, Thomas). The former was the son of his sister, Margaret, but Watson can be exonerated from charges of nepotism since Dundas eventually rose to be the first engineer to become Commander-in-Chief of the British Army.

By 1755 the survey covered eighty-four rolls for the north and ten for the south of Scotland, one of which he appears to be holding in Soldi's portrait of 1756. The survey was never completed, leaving unpaid expenses of £4,690 16s 11d owing to Watson, as he and his team were diverted to survey the south of England for possible French invasion routes. This work included a survey and report on Milford Haven which described it as one of the finest natural harbours of the kingdom and recommended that 'nothing would more effectively contribute to the improvement of ye country than by establishing a port in a commodious part of the haven', but that 'unless some forts and batteries are erected there can be no security'. This report led to his giving evidence to a Parliamentary Committee on the defences of the haven where both he and William Skinner (q.v.) maintained that the harbour mouth was too wide to be worth fortifying and recommended building higher up river at Neyland.

Watson died in 1761. Roy's good opinion of his colleague was by no means universal and James Wolfe complained of 'his artifice and double dealing'.

SUSAN HOTS

[Board of Ordnance minutes, WO, vol. 77, PRO; Royal Engineers Library, Conolly papers; DNB; Porter, vol. 2; *Report from the Committee to whom the book entitled 'Report, plans and estimates for fortifying Milford Haven by Lieutenant-Colonel Bastide' ... November 1757, was referred together with a Survey of the Harbour of Milford*; W. Roy (1785) An account of the measurement of a base on Hounslow Heath, *R Soc Phil Trans*, **75**, 386; G. MacDonald (1917) General Roy his military antiquities of the Romans in north Britain, *Archaeologia*, **68**, 164–166; R. Whitworth (1958) *Field*

Marshall Lord Ligonier; A. H. W. Robinson (1962) *Marine Cartography in Britain*, 95–96; R. A. Skelton (1967) *The Military Survey of Scotland, 1747–1755*, RSGS, SP No. 1; D. W. Marshall (1976) The British military engineer, 1741–178, dissertation, University of Michigan; C. Tabraham and D. Grove (1995) *Fortress Scotland and the Jacobites*, 100–101]

Works
1747–1755. Survey of Scotland
1758. Survey, chart and report on Milford Haven

WATSON, Henry, Colonel (1737–1786), military engineer, was born at Holbeach, Lincolnshire in 1737, the son of a grazier. He attended Mr. Birk's school at Gosberton near Spalding and displayed a precocious ability in mathematics, contributing to the Ladies Diary in 1753 and coming to the attention of the local MP for Lincolnshire, Thomas Whichcot. Following a favourable reference from the master of Birk's school, Whichcot secured a place for Watson at the Royal Military Academy and an Ensign's Commission in the 52nd Foot. Watson developed a strong personal friendship with Thomas Simpson, Professor at the Academy and Editor of the Ladies Diary, to which he made further contributions down to 1761. So impressed was Simpson with his ability that on his death in 1760 he left Watson all his papers, including an unpublished treatise on bridges.

Watson graduated from the Academy in 1759 and embarked on a military career with a Commission in the Corps of Engineers. In 1761 he participated in the siege of Belleisle and the following year played a crucial role in the capture of Havana. Reports of his ability as an engineer brought him to the attention of Lord Clive who in 1763 arranged for him to be posted to India. Appointed a Captain in the East India Company's service on his arrival in 1764, he ultimately rose to the rank of Colonel and Chief Engineer of Bengal.

About 1765 the then Chief Engineer, Martin, suggested the idea of a dock at Calcutta. This was followed up by his successor, Lieutenant-Colonel Archibald Campbell (1739–1791), and at the end of 1768 the Court of Directors of the East India Company approved the idea. Watson assisted Campbell in drawing up the dock's plans, which were submitted in March 1769. The idea was discussed further in the following year. Watson acquired Campbell's rights in 1771, and in 1772 returned home on leave, presumably to lobby for political and financial support. By November 1776 Watson had secured the necessary support for his scheme, including docks at what became Kidderpore and a navigation cut from the Hooghly to the Ballee Ghat, and the appointment as Chief Engineer of Bengal. It was intended that he would concern himself not merely with the Docks but also with work at Fort William.

For the dock works Watson appointed as his agent the engineer James Creassy (q.v.), and ordered a considerable quantity of supplies, including

Colonel Henry Watson

one-hundred and fifty casks of pozzolana, most of which were on route to India in 1777. Unfortunately, as Watson himself discovered when he arrived in Calcutta in 1777, William Tolly (q.v.) had been granted a conflicting lease by the Calcutta authorities for a canal across part of the land earmarked for the docks. Watson also found that he had did not even have clear title to some of the land which he required for the docks. He began work on his scheme but spent considerable effort to secure his title to land and agreement with Tolly. While construction work at Fort William was progressing, frustration over the dockyard led Watson to contemplate returning to England. Eventually he sent Creassy back in his stead to present his case to the East India Company, assuming in January 1780 that he was going to give up the scheme entirely.

It is unclear what state the dockyard was in at the time, although it was alleged he sank over £100,000 in it, and there were sufficient stores and facilities to enable him to build a ship, the *Nonsuch*. From late 1781 this was employed in the opium trade, later joined by the *Surprise*, while Watson was still pursuing claims for losses on the dockyard and attempting to rescue it.

Watson's interests were not confined to commercial ventures; he maintained his interest in scientific subjects. He was responsible for the establishment of a mathematical school at Calcutta in 1784, with Reuben Barrow as the master. He had earlier translated Euler's work on ship design, probably while on furlough in England.

By 1786 Watson's health was breaking down and in January he resigned and returned to England where almost on landing, he died at Dover on 17 September 1786. His funeral was attended by three friends, Creassy, his shipwright George Louch, and John Barchard. He left a widow, Maria Theresa Kenman, whom he had married in 1780 and a daughter by a previous affair, to whom he left £6,000 p.a. His portrait is reproduced in *European Magazine*.

MIKE CHRIMES

[Conolly papers, Royal Engineers, Chatham; East India Company (1779–1780) Court minutes, 337, B/95, IOLR, BL; Colonel Watson's docks, H/68/HOME/MISC/f599, IOLR, BL; Deaths (1786) *Gentleman's Magazine*, 996–997; Anon. (1787) Sketch of the life and character of the late Colonel Henry Watson, *European Magazine*, December, 497–498; India (1885) *Reports on Docks at Calcutta*, V23/47/209, IOLR, BL; Fort William—India House correspondence, 1777–1781 (1981) *India Records Series*, **viii**]

Publications
1776. *Theory on the Construction and Properties of Vessels* (translation of Euler)
c. 1777. *Proposed Advantages arising to the East India Company and the Publick from the Establishment of Docks at Bengal* [and] *Plan of the Docks intended to be Built at Calcutta in Bengal* (repr. 1796) (The report is known only from a list of John Grundy Jr.'s library)

Works
176?. Burge Bridge Fort
176?. Melancholy Point Fort
1777–1781. Fort William, Calcutta, completion of works
1777–1780. Kidderpore Dockyard, £100,000?

WATSON, Justly (fl. 1726–1757), military engineer, did not enjoy a career as successful or eminent as either his father, Jonas, 'Chief Bombardier of England', or his son-in-law, Sir William Green (1725–1811), 'Chief Engineer of England', but his services as an engineer specialising in forts, harbours and coasts were in demand for over 30 years in the Caribbean, Newfoundland, West Africa and Rye Harbour, his main civil engineering contribution.

Watson began as a Cadet Gunner in 1726 and joined his father in Gibraltar during the siege of 1727. They were also together at Cartagena in 1741 where Jonas died at the age of seventy-eight. This expedition marked the nadir in the reputation of the British army and the engineers were not exempt from blame although Watson distinguished himself and was promoted to Lieutenant and Chief Engineer of Jamaica in 1742.

Over the next two years he made plans of Fort Charles and the Port Royal Peninsular in Jamaica and the towns' harbours and fortifications of Portobello, Darien and the Florida coast with the town of St. Augustine. However, this prolific period came to an end when he asked to be relieved of his engineering duties as he was afraid his absence from England was affecting his chances of promotion in the more prestigious and lucrative Harringtons Foot.

In 1744 he drew up the estimates for Gravesend blockhouse and Tilbury which included piling for the breach in the covered way. From 1748–1754 he was Chief Engineer of the Medway Division. This gave him responsibility for Gravesend, Tilbury, Sheerness, Harwich and Landguard, where an altercation between himself and his brother, Lovegood, and the Lieutenant Governor arising from 'liquor' led to his being court-martialled. This led to his suspension from Harringtons Foot but does not seem to have affected his position with the Board of Ordnance as he remained Chief Engineer and continued to provide plans such as that of Sheerness and the vicinity (1753) and Harwich, including the cliffs, town and encroachment of the sea since 1709 (1752) and a proposal for a breakwater (1754).

The same year he was sent to Annapolis in Nova Scotia as Chief Engineer but was almost immediately dispatched to West Africa to report on the defences, the scandalous state of which had been raised in Parliament. His reports and surveys of the state of the forts and harbours from Dixcove to Prampram were approved by the House of Commons and acted upon. His next report on the survey of Rye new Harbour, written jointly with Edward Collingwood, the Master Attendant of the Dockyard at Deptford, was produced in 1756.

Their opinion was that a safe and commodious harbour for ships drawing 16 ft. water could be made between the stone piers and a sluice up to the entrance to the Winchelsea Channel and that ships drawings 12/14 ft. could be navigated up to Rye itself. Their recommendations were that the sluice should be repaired, the West Pier excavated deeper in the water to prevent shingle choking the entrance mouth, and two timber breakwaters erected. The Winchelsea Channel needed 1,460 ft. of wharfing on either side. Their estimate of £29,622 probably explains the lack of action. He was not paid (£137 12s) until 26 July by which time he had returned to Newfoundland where he died later that year.

According to Conolly, citing a letter whose provenance and whereabouts are no longer known, this was the result of poison placed in his coffee by a black woman servant whom he had refused to bring to England. His widow, Susan (*née* Curtis), whom he had married on 15 November 1733 at Pagham was granted a pension of £40 p.a. in 1758.

His daughter, Miriam, maintained the family connection with Gibraltar, dying on 21 June 1782 from the after-effects of a chill caught in a bomb shelter during the siege of 1779–1783 where her husband, William Green, was the Chief Engineer. Her son, Justly Watson Green, became a Colonel

and 'military preceptor' to Queen Victoria's father, the Duke of Kent. He died around 1828.

SUSAN HOTS

[Rye Harbour minutes; Report from the committee to whom the petitions concerning the harbours of Rye and Dover were referred to which is subjoined a report with an appendix thereto, 25 February 1757; *Parliamentary Papers, Reports, 1744–1772*, vol. 31; Royal Engineers Library Conolly Papers; DNB (1899); Porter, vol. 1; Bendall, 540, 213; A. H. W. Robinson (1962) *Marine Cartography in Britain*; D. W. Marshall (1976) British military engineers, 1741–1783, Ph.D. thesis, University of Michigan, 56/58; IGI www.familysearch.org]

Selected charts, reports, etc.
1742. Portobello, plan of the town harbour and fortifications
1743. Florida, survey of the coast from Fort William near the St. Juan River to Mosquito River with plan of the town of St. Augustine
c. 1743. Darien, plan of the harbour and parts of the country on the Isthmus settled by the Scottish company
1752. Cliffe and town of Harwich with the encroachments of the sea since 1709
1754. Proposed breakwater at Harwich cliff
1755. Report and plans relating to James Fort on the River Gambia
1756. A report on the survey of Rye new Harbour (with Edward Collingwood)

WATT, James, FRS (1736–1819), engineer and scientist, was born in Greenock, one of eight children of James Watt and Agnes Muirhead, only one brother, John, and a sister, Jean, surviving to adulthood.

Watt's grandfather, Thomas Watt, was born sometime between 1639 and 1642, settling in Cartsdyke, a small seaport near Greenock. Thomas Watt was a mathematics teacher and married Margaret Shearer in 1679. Six children were born, of whom only two sons, James and John, reached adulthood. John taught arithmetic, book-keeping and navigation in Glasgow. He was a skilled cartographer, producing two plans of Port Glasgow and a survey of the river Clyde, thus anticipating his nephew, James's, work as a surveyor in Scotland. Thomas and Margaret Watt's other son, James, was born in 1698, settling in Greenock as a shipwright, chandler, builder, ship-owner and merchant. His wife, Agnes Muirhead, was said to be a woman 'of great sense, judgement and intellect'.

Childhood, education and training
James Watt was a delicate child, continuing to suffer from headaches when adult. He was taught at home by his mother at first and was then sent to M'Adam's school. He later went to the Grammar School where he was considered to be slow. However, on being introduced to mathematics, he showed both interest and ability. While still at school Watt assisted his father and on leaving school worked for him. It is likely that his

James Watt FRS

father intended his son to follow him in the shipwrighting and chandler's business but, on the failure of some commercial speculations, Watt went to Glasgow to learn the trade of mathematical instrument maker, taking a quadrant and some carpentry tools with him.

Watt lived in Glasgow with members of his late mother's family, through whom he came into contact with some of the leaders of the literary and scientific circles of Glasgow. His relative, George Muirhead, had recently moved from the Chair of Oriental Languages to the Chair of Humanity in the university. With John Anderson and Robert Dick, as well as Joseph Black and Gilbert Hamilton, he was a member of the Literary Society of Glasgow. Dr. Robert Dick, one of Watt's closest friends, become Professor of Natural Philosophy and it was he who advised Watt to go to London, giving him an introduction to James Short, a mathematician and optician. Watt was in his twentieth year and, never having served an apprenticeship, could not rank as journeyman and, moreover, in order to be the least possible financial burden on his father, he wanted to learn the greatest amount in the shortest possible time. Short introduced Watt to John Morgan, a mathematical instrument maker of Cornhill, and Morgan undertook to instruct Watt for one year in return for 20 guineas and Watt's labour.

Watt lived in considerable poverty in London, the draughty workshop, poor food, long hours and painstaking work taking their toll. By the year's end he noted 'I am now able to work as well as most journeymen' and, purchasing the bare necessities of materials and tools, he returned to Scotland. After a few months recuperation Watt went to Glasgow where he was given quarters in the college in order to assist Dr. Dick in renovating a collection of astronomical instruments which had arrived from Jamaica. Watt was permitted to open a workshop in the university and to call himself 'Mathematical Instrument Maker to the University' in 1757. Two years later he also opened a shop in the city. Watt came to the attention of Dr. Joseph Black, Professor of the Practice of Medicine, and made various instruments for Black's experimental work, including a perspective machine. In 1758 Watt met John Robison (q.v.) who later became Professor of Natural Philosophy at Edinburgh. Robison noted how Watt flourished on problem solving; 'everything became science in his hands'. He moved premises within Glasgow in 1763, by which time he was employing several journeymen and also taking on apprentices. The basis of his instrument business was the sale of quadrants, compasses, burning glasses and microscopes. In addition he also sold musical instruments and steel ornaments.

Early work on the steam engine

Watt's interest in the steam engine dates from around 1758 when Robison suggested that he consider the application of steam power to road carriages and mining. The idea was not developed but in the early 1760s Watt began some experiments on the force of steam 'and formed (a model of) a species of steam engine'. He abandoned the idea under pressure of other work, also recognising that an engine on the principle employed would be extremely inefficient.

In the winter of 1763–1764 Watt was asked to repair the model of a Newcomen engine belonging to the natural philosophy class at Glasgow University. This work led him to conduct experiments on the model as well as to understand the principles of steam power. He identified several problems, of which the wastage of steam during the ascent of the piston of the engine and the attempt to form a vacuum by cooling the system were two of the most fundamental. This led Watt to his well-known tea-kettle experiment in which he demonstrated the latent heat of steam, subsequently learning of the scientific principle from Joseph Black.

In 1765 and 1766 Watt erected several atmospheric engines in Scotland, probably incorporating parts made to his own designs. Work on steam power was abandoned for over a year while he undertook the first of his civil engineering engagements but in 1768 he recommended trials on model engines and began designing a colliery engine which was erected in 1769 for John Roebuck at Kinneil. Roebuck's financial situation was, however, precarious and he had begun to negotiate for the sale of a part or the whole of his share in Watt's engine to

Matthew Boulton of Birmingham in 1768. In the autumn of that year Watt initiated negotiations for a patent which was granted in 1769, two-thirds of the invention being assigned to Roebuck.

In June 1770 Roebuck was declared bankrupt and further engine development was delayed until Matthew Boulton took over Roebuck's share of the engine patent, as well as Watt's debts. In March 1774 Watt settled in Birmingham, thereafter devoting a large part of his professional life to his engine partnership with Boulton, an association which began in 1775.

Watt's family
Watt married his cousin, Margaret Miller, in 1764 and a daughter, Margaret, was born in 1767; a son, James Jr. was born in 1769 but he died unmarried in 1848,. It was in 1773 when Watt was surveying for a projected canal in the Highlands that his wife died in childbirth. In her, he wrote, 'I lost the comfort of my life, a dear friend and a faithful wife'. In 1776 he married again, his second wife being Anne, daughter of James McGrigor, a dyer of Glasgow. Two children were born of this marriage: Gregory in 1777 and Janet (known as Jessy) in 1779. Both died of consumption, unmarried, Janet in 1794 and Gregory in 1804.

Watt as civil engineer
Needing to provide for his wife and young family Watt began to undertake surveying work, while retaining his instrument business. His partner, John Craig, had died in 1765 and this put an additional financial burden on Watt, whose knowledge of scientific instruments, combined with his capacity and ingenuity as an instrument designer, clearly contributed to his success as a surveyor. Both were underpinned by a lifelong interest in measurement, something that he was later to share with several member of the Lunar Society of Birmingham. In 1765 Watt invented a machine for drawing in perspective. The idea was based on a machine acquired in India by an acquaintance. He made a portable version, employing a slide rule, the whole being capable of being put into an overcoat pocket, while the legs formed a walking stick. Altogether he made over fifty of these instruments, apparently designed by George Adams, a London instrument maker.

As civil engineering demands on Watt grew, he recognised the deficiencies of the surveying instruments at his disposal. In 1769 the surveyor's level came under scrutiny, one of the defects of which were the sightlines, besides the difficulty of ensuring that the instrument did not move in difficult terrain or adverse weather conditions. 'If I was to follow the business of levelling', he reported, 'I should certainly make some more alterations on levels'. In 1771 he developed a micrometer for measuring distance, showing it to the engineer, John Smeaton (q.v.), in 1772. He had had the idea of measuring distances with a telescope in 1769 and resolved the problem two years later. Working with a magnification of ×8 and using a 12 ft. rod he found that he could

measure 30 chains to an accuracy of 1 in 100. The practical limit to the instrument was the distance at which the rod could be read by the surveyor, a feature Watt addressed in 1772–1773, although he did not put the improvement into practice. As with the perspective machine, Watt did not patent the micrometer and in 1778 the Society of Arts made an award to W. Green for a similar instrument.

Watt's civil engineering career was wholly contained within the years of his residence in Scotland, before his move to Birmingham in 1774, and the projects were Scottish. The work consisted of surveying for river navigations and canals, roads, and bridges, as well as harbours and water supply. The waterways surveys composed the major part of Watt's work as a civil engineer. His first civil engineering projects were undertaken in partnership with the surveyor, Robert Mackell (q.v.). In 1767 they surveyed the route for a projected canal to join the Firth of Forth and the Clyde by the Loch Lomond passage. On the publication of their report, Watt travelled to London to promote the passage of the Bill in Parliament but it failed at the Committee stage, a setback that convinced Watt that he would 'not long to have anything to do with the House of Commons again'.

On his return from London he visited the Bridgewater Canal, James Brindley's (q.v.) early masterpiece, and the Aire and Calder Navigation, presumably to explore best practice in canal engineering. Others surveys undertaken in 1767 included the river Devon and the Forth. In 1769 he undertook a survey of the Clyde estuary at the initiative of the Glasgow magistrates who wished to see the development of Glasgow as a seaport. Watt's survey was published but the project was not implemented. In the same year he undertook a survey of the route for a 9 mile long canal between the Monkland collieries and Glasgow. He was also appointed resident engineer to this project, receiving a salary of £200 p.a. for two and a half years. In 1770 he undertook a survey for the Strathmore Canal. As Watt explained, 'I had to examine and survey a country of thirty 6 miles in length and to hunt about for a course for a canal through country where nature had almost done her utmost to prevent it; indifferent health and weather viciously cold and stormy were the attendants on my survey'.

Other canal surveys include the Bo'ness Canal to join the Forth and Clyde Canal and Bo'ness harbour (1771) as well as canals at Crinan and Tarbert (1771–1772), the latter being undertaken for the Commissioners for managing the Annexed Estates. In 1773 he made a survey for a coal-carrying canal from Macrihanish Bay to Campbeltown and a canal from Hurlet to Paisley.

Watt's charges for surveys were less than those of, for example, Smeaton. He charged £80 for forty-three days in the field (including travelling) and £30 for the preparation of the report and map for the Strathmore Canal. Besides parliamentary lobbying he disliked the management aspects of canal construction, commenting that he was 'terrifyed to make bargains and I hate to settle accounts'. Wisely recognising that he could 'by surveys and consultations, make nearly as much money with half the labor and realy think with double the credit' Watt did not undertake the supervision of construction work after the Monkland project. His skills as a civil engineer were, he recognised, 'to make an accurate survey and a faithful report of anything in the engineer way, to direct the course of canals, to lay out the ground and to measure the cube yards cut, or to be cut, to assist in bargaining for the price of work, to direct how it ought to be executed and to give an opinion of the execution to the Managers from time to time'.

The zenith of Watt's canal surveying was the commission, in 1773, to survey the practicality of constructing a canal between Fort William and Inverness through remote and wild country of lakes and rivers. Neither Watt's nor the subsequent proposal of Rennie were executed although the issue was revived in 1784, Watt being approached to be engineer, a task he declined. The project was successfully undertaken later by Thomas Telford (q.v.), who acknowledged Watt's contribution and credited him with it being 'just and masterly. I have introduced in my Report your general description, plainly saying that it could not be so well told in any other words'. Amongst Watt's port and harbour schemes were those for Port Glasgow between 1769 and 1772 (complementing his survey of the Clyde), recommendations for the improvement of Ayr harbour in 1771 and Greenock harbour and water supply in 1772. Contributions to road and bridge engineering include a survey and regulations for the contractor for a bridge over the Clyde at Hamilton (1770), Dumbarton road (1770) and streets in Glasgow (1773).

Watt's civil engineering career came to an abrupt end. Whilst engaged in the survey of the projected Inverness–Fort William canal he heard that his wife, who was expecting their fifth child, was dangerously ill. He hurried back to Glasgow to discover that she had already died in childbirth.

Later work on the steam engine
Since he had been in regular communication with Matthew Boulton and William Small at Soho, Birmingham, in connection with his steam engine and Boulton had already offered strong encouragement to settle in Birmingham to create 'a manufactory (from which) we would serve all the world with engines of all sizes', Watt took his family to Birmingham in May 1774 to direct his attention fully to the development of the steam engine business. During his time there Watt achieved international recognition as a scientist and engineer. He developed a rotative engine, patented in 1782, to generate power in factories and invented the parallel motion device in 1784;

he invented a mechanical letter copying device, patented in 1781. As a leading member of the Lunar Society, an informal philosophical society whose members included Matthew Boulton, Erasmus Darwin, James Keir and William Small, Watt enjoyed the stimulus of scientifically minded friends. It was, however, to his Scottish friends that he turned for assistance in the defence of his patents at law.

Watt gradually withdrew from day to day participation in the engine business after the opening of Soho Foundry in 1796, retiring in 1800. Watt died on 25 August 1819 and was buried beside Matthew Boulton in Handsworth church.

Watt's correspondence shows him to have been a man assailed by self-doubt. His strengths lay in achieving elegant solutions to scientific and engineering problems. He was generous in sharing ideas and did not patent his surveying inventions. While a number of his surveys for canal projects did not come to fruition, the testimony of Telford demonstrates that, in all probability, had Watt not gone into partnership with an able entrepreneur, Matthew Boulton, his civil engineering works would now be better known and he would have continued to develop in stature in the profession of civil engineer.

JENNIFER TANN

[Boulton & Watt Collection, Birmingham Reference Library, esp.: M1 Box 3, reports, surveys & some letters re. canals; MIV Boxes 18, 20, 21, canals; M1/15–21, notebooks concerning canals and other civil engineering projects inc. Bridges and Docks; M1/2/41, Report ... concerning the rivers of Forth and Leven; M1/2/42, notebook of expenses re. surveying; M1/2/52, report concerning the possible building and expense of a navigation to Strathmore. Muirhead (1854) *Origins and Progress of the Mechanical Inventions of James Watt*; Muirhead (1859) *The Life of James Watt*; S. Smiles (1861 etc.) *Lives of the Engineers*; H. W. Dickinson and R. Jenkins (1927) *James Watt and the Steam Engine*; H. W. Dickinson (1935) *James Watt*; R. L. Hills (2001) The Railway of James Watt, in *Early Railways, 1st Int Conf*, 63–81; Skempton; DNB]

Publications
1767. *Report on Proposed Canal from the Forth to the Carron* (with R. Mackell)
1770. *A Scheme for making a Navigable Canal from the City of Glasgow to the Monkland Coalierys*
1771. *Report concerning the Harbour of Port Glasgow*
1773. *Report concerning the Upper Part of the River Forth*

Works
1769–1771. Monkland Canal, Resident Engineer

WATTÉ, John (fl. 1769–1798), land surveyor, practised in the fens of Cambridgeshire, Norfolk and Lincolnshire, and in eastern Northamptonshire, from 1769. He was employed as an estate surveyor by several local landowners including the Duke of Bedford and was also involved in surveying for inclosures. From 1770 until 1783, he lived at Leverington in Cambridgeshire; he was also schoolmaster there in 1770.

Watté was involved in fenland drainage and navigation and in 1777 he surveyed and mapped the Nene estuary from the south end of Kinderley's Cut. He laid out and built the sea banks on the east side of the estuary. He advertised his levelling and draining work in 1778 and also published a sheet describing the forthcoming solar eclipse. In 1782, he was employed by the commissioners of the Bedford North Level to take levels and soundings of the river Nene from Peterborough Bridge to Gunthorpe sluice and in the following year he moved his land surveying practice to Wisbech.

While continuing to map estates and inclosures he became heavily involved in surveying for the proposed Eau Brink Cut. He was asked by landowners and other interested persons to give his opinion about Nathaniel Kinderley's (q.v.) proposals for the cut, and its likely effects, and started work in March 1791. His report, with a map of the cut, and line and scale of levels both along the present channel of part of the river Ouse and along the proposed cut, was published in April. In June, he, John Smeaton and James Golborne (qq.v.) independently estimated the expense of the works. In 1792 Watté extended the survey up to Denver Sluice and in 1794 he made ten borings up to 24 ft. deep along the line of the proposed cut. In 1795, together with Robert Mylne, John Rennie (qq.v.) and Golborne, he gave evidence in Parliament in support of the Bill, which passed in May. Meanwhile Watté and Golborne also worked together in 1794 on a survey and estimates for a canal between Wisbech and Outwell and both gave evidence that year on the Bill. The canal, 5 miles in length with two locks, was completed in 1796, seemingly under Watté's supervision, at a cost of £16,000. In 1795 he was elected an Honorary Member of the Society of Civil Engineers. He attended his last meeting in May 1798 but is not recorded in the 1799 list of members.

SARAH BENDALL

[H. C. Darby (1940) *The Draining of the Fens*; A. W. Skempton and E. C. Wright (1967) Early members of the Smeatonian Society of Civil Engineers, *Trans Newc Soc*, **44**, 23–47; J. Boyes and R. Russell (1977) *Canals of Eastern England*; [J.] G. Watson (1989) *The Smeatonians: the Society of Civil Engineers*; S. Bendall (1992) *Maps, Land and Society: AS History, with a Carto-Bibliography of Cambridgeshire Estate Maps c. 1600–1836*; Skempton; Bendall]

Publications

1777. *A Map of the Sands as are Vested in the Bedford Level Corporation by Act of Parliament 13 Geo.III.*

1791. *Report ... for the better Drainage of the South and Middle Levels of the Fens, and other Lands bordering upon each Side of the River Ouse, and amending the Outfall of the said River, by a New Cut or Channel from Eau-Brink to Lynn*
1792. *A Longitudinal Section with a line a Scale of Levels of the Rivers Ouse shewing the present State of its Channel ... from Denver Sluice to the Crutch about Two Miles below ... Lynn*

WEDGE, Thomas (*c*. 1761–1850), surveyor of Sealand, near Chester, was born *c*. 1761 and for much of his life, from 1788 onwards, acted as Agent for the River Dee Company. It is evident from evidence that he and others gave at the 1849 Admiralty Enquiry into the Dee that the affairs of the Company had, by the time of his appointment, become increasingly concerned with the management of reclaimed land rather than the improvement of the navigation as developed by John Golborne (q.v.). Indeed, from complaints by pilots and others, it appears that the groynes intended to deepen and control the channel were allowed to decay under Wedge's stewardship. As Wedge was not trained as an engineer himself questions of engineering import were referred to William Jessop and, later, to Thomas Telford and then on his death to William Provis (qq.v.). Wedge was responsible for carrying out the improvements recommended by Telford in the 1820s.

Wedge's local renown led to him, *c*. 1800, becoming responsible for drawing up an embanking and drainage scheme for Rhuddlan Marsh in North Wales. The canalisation of the Clwyd and associated embankments were an important medieval civil engineering work, giving access to Edward I's castle as Rhaddlan. The coastal and river defences had clearly, over the centuries, deteriorated and in February-March 1793 there were serious breaches by spring tides at Towyn. Wedge was invited by landowners to report and he provided an 'ocular survey' in June 1793, and more detailed reports in July and August 1793 recommending the construction of embankments, modifications to the beaches and cliffs, and improvements to the land drains. The local proprietors realised that the works, amounting approximately to £3,000 in value, could be financed only by the enclosure and sale of around 500 acres of wastes and common land. Following an Act setting up a Trust for this purpose in 1794 (34 Geo.III Cap(x)) Wedge was appointed by the Surveyor General of Land Revenue to establish the local parish and beach boundaries and also undertook to supervise the works for £120 p.a., including providing his own foremen. His plans were referred to Jessop for approval in September 1794.

Under Wedge's direction, the first phase of work, on the Abergele embankment, began in 1795, with two superintendents, Henry Harrison and Charles Shone. Work continued over the next eighteen years but it would appear that, although Wedge was consulted, much of the direction was by Harrison and Richard Shone. In 1800 work on the sea embankment was largely complete and Wedge became involved in the designs for the works along the Clwyd, calling in Thomas Morris Jr. (q.v.) in April 1801 to advise on a sluice across Clawydd Llwyd. Another engineer, Thomas Penson Sr. (q.v.) was consulted about sluices at Cwybyr and Morfa Llong Trunks in July 1808, although Wedge continued to receive fees.

By 1813 it was clear that additional funds would be required to complete the work, further breaches having taken place, and following the passage of a supplementary Act Wedge was again asked to supervise the setting out of the new works. The last record of payments for work for £186 14s 6d was made on 11 May 1819 and it may be that the works were substantially completed by then. By that time, November 1817, Thomas Harrison (q.v.) had already been consulted about work on the embankment and Edward Jones was appointed to carry out this work, which he continued to do until January 1822. Harrison and Telford were invited to report in January 1821, which suggests that Wedge had retired from the works by this time. A new superintendent, John Evans, was appointed in 1822 and he remained in post until 1843. When Francis Giles (q.v.) was asked to report on the state of the embankment in December 1826, following a breach on 2–3 December, he was very critical of the general state of the works, which may suggest that supervision was poor and that Wedge either did not supervise the work well or his original designs were defective. In view of Wedge's age his nephew, Henry, assisted him in the latter part of his life with river Dee business.

It is possible that Wedge was related to John Wedge (fl. 1760–1813) and a John Wedge carried out work for James Brindley. As an agent of the Earl of Aylesford, his description of land drainage works carried out on Aylesford's estates at Bickenhill, near Coventry, won the silver medal of the Society of Arts in 1792. Thomas Wedge also carried out work for the Aylesford family. It is unclear whether they were also related to Charles Wedge (1745–1842), another distinguished land surveyor, who worked for John Rennie (q.v.) and reported with Dudley Clark on the state of the Dee Navigation in 1805.

MIKE CHRIMES

[Minutes of the Commissioners of Rhuddlan Marsh, DC/140, Flintshire RO; J. Wedge (1792) Describing his manner of draining land, at Birkenhill, *Trans Soc Arts*, **10**, 130–150; Tidal Harbours Commission (1846) 2nd report; Admiralty (1849) *Chester–Dee Navigation Improvement*; Minutes of evidence, appendices, and plans, Bendall]

Works
1788–1849. Dee Navigation works, resident agent, *c.* 1,522 acres reclaimed (includes 1819–1826 stone causeway, 1,000 yd., *c.* £25,000

1793–1819. Rhuddlan Towyn, embankments and enclosure, *c.* 500 acres, 10 miles of embankments, consulting surveyor

WELCH, Edmund (fl. 1650–d. 1665), fen drainage engineer, was appointed in July 1651 to the staff of the (fifth) Earl of Bedford's Company of Adventurers in the third campaign of draining the Great Level of the Fens; perhaps he had previously been employed as a contractor on works in the Middle and North Levels (between the Welland and Bedford River) which began in March 1650. It seems that Welches Dam, built on Bedford River before Feburary 1651, was named after him. In March 1652 he replaced Robert Burton (q.v.) as principal Overseer or resident engineer under the Director of Works, Sir Cornelius Vermuyden (q.v.). It was in the summer of that year that the workforce reached its maximum number of 10,000 men. In 1652 and 1653 Welch had three (assistant) overseers working under him, mainly on the South Level.

By the end of 1653 most of the large works had been completed but much remained to be done in making roads, bridges and diversion dykes and the endless task of maintaining the banks, rivers, drains and sluices. In a reorganisation in August 1654 Welch became 'principal contractor', presumably in charge of the direct labour force engaged on such work, receiving a salary of £60 a year. At the same time Conservators were appointed to take control of each of the three Levels with assistants: one on the North and on the Middle Levels and two on the South Level. The Conservators were Adventurers previously active in field supervision; they received an annual salary of £150. Most of the assistants were former overseers. Jonas Moore (q.v.), formerly Surveyor General, became (principal) Overseer on £100 a year and, in this capacity, let out the last major contract, in 1655, for building Denver Sluice. Moore spent most of the next two years producing his great map of the Fens.

By 1656 Welch was second in command to the Conservator of the North Level, with an annual salary of £52; from April 1658 he was in sole charge, at the same salary. In July 1662 his post was retitled 'Surveyor of the North Level' and this he continued to hold until his death soon after February 1665.

A. W. SKEMPTON and FRANCES WILLMOTH

[Proceedings and Accounts of the Company of Adventurers of the Great Level (1650–1662) and Bedford Level Corporation Orders and Accounts (1663–1666), Cambridgeshire RO; Samuel Wells (1830) *The History of the Great Level, called Bedford Level*; H. C. Darby (1983) *The Changing Fenland*]

WELCH, Henry (1795–1858), civil engineer and architect, was baptised on 22 November 1795 at Strood, near Rochester, the eldest child and only son of Henry Welch and his wife, Abigail Weeks.

In 1824 he married Margaret Wilkin, born in Cumberland, but there were no children of the marriage. He was later described as 'a little, stout, fresh-coloured man, rather dignified in style, but a cheerful and amusing companion'.

Between 1821 and 1831 Welch was extensively employed on road surveys under the direction of Thomas Telford (q.v.). They included the Holyhead road, roads in South Wales, new roads linking London with Edinburgh via the eastern side of the country and roads connecting Carlisle with Edinburgh. As part of the improvement scheme for the Great North Road, Welch undertook the initial surveys for the bridge over the river Wansbeck at Morpeth—completed by Telford in 1831—and when, in 1834 and 1839, doubts arose as to its being subjected to damage as a result of a weir upstream, Welch reported on the matter.

In 1829, when resident at Carlisle, he became a member of the Institution of Civil Engineers—he resigned in 1854—and in April 1831 he was appointed Bridge Surveyor for Northumberland at a salary of £400 p.a. In 1839 he also became County Architect, his salary raised to £600 p.a. Welch also held minor appointments concurrent with his principal one and between 1834 and 1849, for example, he was surveyor to the Hexham to Alnmouth Turnpike Trust. In 1835 he carried out the survey for the Cow Causey and Buckton Burn Turnpike, near Belford.

In addition to his professional work, Welch operated a contracting business, a matter which led to his being taken severely to task as it was considered that he was in a position to award contracts to his own firm. In 1843 he resorted to a libel action against his accuser, Samuel Donkin, but was awarded only a token sum with no order made regarding costs.

He was responsible for the design and construction of the Great North of England Railway's bridge over the river Tees at Croft, a major four-span structure with a marked skew; it was completed in 1841, built by contractors Deas and Hogg. As Bridge Surveyor Welch supervised the construction and repair of several bridges in Northumberland, his major County work being that crossing the North Tyne at Falstone.

Many of Welch's architectural works were of a minor nature but the last was the building of the substantial lunatic asylum at Morpeth, completed in the year prior to his death and it was written that 'he died with the builder's account for extras in his hand'. His death took place on 21 December 1858 at his home in the fashionable Westgate area of Newcastle, where he had lived at two addresses since coming to the town. His will was proved at some £3,000, his estate passing to his wife with small legacies to a sister and his mother.

R. W. RENNISON

[List of road surveys, 1823–1827 (22 drawings no longer extant), Institution of Civil Engineers; *Northumberland Quarter Sessions: Orders*, QSO 23, Northumberland RO; Letters and reports, ZBS/2/42, Northumberland RO; *Local Biography*, **2**, 313; L920, Newcastle Central Library; S. Donkin (1886) *Reminiscences of Samuel Donkin …*, L920, NCL; Information from Highways Manager (1997) Northumberland County Council]

Drawings
c. 1826. *London–Morpeth–Edinburgh Road*, n.d.
1828. *Carlisle and Edinburgh Road … surveyed under the direction of Thomas Telford Esq., Civil Engineer*
1835. *Cow Causey and Buckton Burn Turnpike*

Works
1821–1822. Cartland Craggs Bridge, resident engineer, for Thomas Telford
1823–1824. South Wales, road surveys
1824. Road surveys in Scotland, Wales and for Great North Road
1825–1826. Surveys for Great North Road
1827. Surveys for roads in Scotland
1828. Surveys for Carlisle to Edinburgh road
1831. Sketch for Morpeth Bridge
1831. Survey for Gretna to Eskdalemuir road
1831. Breamish Bridge, repairs, four spans of 15 ft. (now demolished)
1832. Thornhope Bridge, rebuilding, 40 ft. span (now demolished)
1833. Kirkwhelpington Bridge, 25 ft. span
1833–1839. Felton Bridge, widening and repairs, spans 33, 41 and 31 ft.
1838–1841. Croft viaduct, Great North of England Railway, four spans of 59 ft.
1838. Stannington Bridge, rebuilt, two spans of 30 ft. (now demolished)
1841. Doddington Bridge, widened, 40 ft. span (now demolished)
1842. Knarburn Bridge, built, 32 ft. span
1843. Falstone Bridge, built, three spans of 45 ft.
1845. Warren Bridge, rebuilt, span 30 ft.
1850. Nest Bridge, rebuilt, 15 ft. span
1851. Gofton Bridge, widened, 27 ft. span
1852. Hartburn Bridge, widened, 34 ft. span
1856–1857. Morpeth lunatic asylum

WELLINGTON, James (fl. 1827–1828), of Bristol, ironmaster, supplied iron roofs in the 1820s, possibly developing the patents of Thomas Pearsall (q.v.) and Richard Jones Tomlinson (Patents No. 3750, 1813; No. 3883, 1815; No. 4556, 1821). He is best-known for supplying in 1827 the ironwork, using best Welsh iron, for the Brunswick Theatre, designed by T. Stedman Whitwell (q.v.); the roof failed on 28 February 1828, killing ten people. The roof form was a triangulated truss but, as was the habit of theatres of the time, equipment had been suspended from the tie-beam and overloaded the structure so causing its failure. Before this Wellington had supplied a roof at Deptford gas works and an earlier roof, designed to Tomlinson's patents, at Glasgow Cavalry Barracks; it failed in 1821.

MIKE CHRIMES

[J. G. James collection, ICE Library; ICE Minutes of Conversations, 11 March 1828, 13 May 1828; *Mechanics Magazine*, 1828, 8–10, esp. 9, 134; *Mirror*, **xi**, 1828, 161–163, 176, 187]

Works
?. Brunswick Theatre, roof

WELLS, John (1662–1702), was a member of a prominent family of Thames shipbuilders and the designer of Howland Dock.

The dock was built on land in Rotherhithe leased by John and Richard Wells from Elizabeth Howland, widow of a wealthy landowner. In 1695 Mrs. Howland's daughter and heir, while still in her teens, married Wriothesley Russell, Marquis of Travistock and grandson of the second Duke of Bedford. In February 1696 the Duke and Mrs. Howland petitioned Parliament stating that a dry dock had been built on the site for £2,500 and requesting powers to raise a further sum for making a wet dock, a dock capable of retaining a full depth of water at all states of the tide. A private Act was obtained in April 1696 and a Trust established for raising £12,000 for that purpose and afterwards for the Marchioness of Tavistock, or Duchess of Bedford as she became following the death of the second Duke in 1700. From the Bedford estate papers it appears that John Wells designed the dock and supervised construction and that William Ogbourne (q.v.), house carpenter of Stepney, held the contract for the timber work of the dock walls and wharfing on the river front. Other contractors nominated by two of the Trustees carried out the excavation.

Work began in 1697 and the dock was opened in 1700. It covered an area of 10 acres, having a depth of 17 ft. of water with an entrance lock 44 ft. wide and 150 ft. in length between two pairs of gates. The walls of the dock and lock, at least 20 ft. in height and some 3,000 ft. in length, consisted of timber piles and planking which would have been supported by anchored land ties. The dock was used by ships while laying up, and for fitting out; there were facilities for erecting masts. It could accommodate the largest commercial vessels such as 600 ton (unladen) East Indiamen.

A perspective view of Howland Great Wet Dock, as it was called, is depicted in a drawing by Thomas Badeslade (q.v.) engraved by John Kip c. 1718. By that time a second dry dock had been added nearby. A plan of the dock in John Rocque's large-scale map of London (1746) clearly shows the entrance lock with its two pairs of gates. It was then being used as a commercial dock by ships in the whaling trade. Later it became known as Greenland Dock and as such was incorporated, with an enlarged entrance, in the Surrey Commercial Docks in 1809.

Howland Dock had a far greater area than its predecessors: (1) the Old Wet Dock of 1½ acres built c. 1660 for the famous Blackwall shipyard,

(2) the Wet Dock at Plymouth Dockyard of 2 acres completed in 1693, (3) and (4) the two wet docks, both of about 1½ acres, completed in 1698 at Portsmouth Dockyard. Each of these had a single pair of gates, but the three naval docks, designed by Edmund Dummer (q.v.), had a greater width and depth (51 ft. and 23 ft. at Plymouth, slightly larger at Portsmouth) capable of admitting the largest first-rate warships.

There is no evidence to support the suggestion that George Sorocold (q.v.) acted as engineer of Howland Dock. However, it is possible that Thomas Steers (q.v.) was involved in some relatively junior capacity and he would certainly have been well acquainted with the dock as he lived nearby from 1698, if not a year earlier, after leaving the Army, and made a survey of the area for Mrs. Howland in 1709 (when he is described as a house carpenter) before moving early in 1710 to become Liverpool's first dock engineer.

A. W. SKEMPTON

[Bedford Estate Papers (1695–1709) Greater London RO; T. Badeslade (1718) *Howland Great Wet Dock*; J. Rocque (1746) *Plan of the Cities of London and Westminister and the Borough of Southwark*; J. Broodbank (1921) *History of the Port of London*; H. Peet (1930) Thomas Steers, the engineer of Liverpool's first dock, *Trans Hist Soc Lancs Cheshire*, **82**, 163–242; W. Crawford Snowden (1948) *London 200 Years Ago*; D. Swann (1968) Engineers of English port improvements, *Transport History*, **1**, 153–168; P. Banbury (1971) *Shipbuilders of the Thames and Medway;* J. G. Coad (1983) *The Royal Dockyards 1690–1850*; M. Clarke (1993) Thomas Steers, in A. Jarvis (ed.) (1998) *Ports and Harbours*]

Works
1697–1700. Howland Great Wet Dock, £12,000

WEST, Stephen (fl. 1741–1778), master carpenter and civil engineer, is first heard of at Rye New Harbour, building the wharfing 1741–1743. This comprised timber piles and planking supported by anchored land ties, forming a retaining wall 23 ft. high running for a length of 830 ft. on each side of the canal or dock. The work was carried out under the direction of Edward Rubie (q.v.).

In 1750 Rubie, now Director of Works at Yarmouth, began rebuilding the pier on the south side of the harbour entrance. This was timber framed with chalk infill. He died in the winter of 1753 and West was appointed his successor from December 1753 at a salary of £160 p.a. with William Dixon as his foreman. He completed the pier in 1755.

In June 1756 West applied for, and was granted, leave of absence to survey the works at Rye at the request of the Rye Harbour Commissioners. He gave evidence in Parliament, and his detailed report dated 19 October 1756 was printed by the House of Commons interleaved with comments by Justly Watson (q.v.) and Edward Collingwood.

West received £103 for his attendance. As a result of the reports and recommendations the Rye Commissioners gained extra funding; work proceeded under Robert Cooper (q.v.) and the harbour was opened in 1762.

Returning to Yarmouth in 1756, West continued as harbour engineer for many years, certainly to 1778 and possibly later. Stephen West mentioned in the Commissioners Minute Book up to 1800 may be his son. The date of his death has not been found.

A. W. SKEMPTON

[Rye Harbour Commissioners Minute Book (1741–1756) East Sussex RO; Yarmouth Port and Haven Commissioners Minute Book (1750–1800), Norfolk RO]

Publications

1756. Report on Rye Harbour, in *Report from the Committee ... concerning the Harbours of Rye and Dover*, House of Commons

WESTON, Sir Richard (*c.* 1591–1652), landowner, agriculturist and river engineer-undertaker, was the eldest son of Sir Richard Weston and his wife, Jane Dister. On the death of his father in 1613 he inherited the family estates at West Clandon and Sutton, in Surrey. He married Grace Harper, from Chestnut, and was knighted in 1622. Part of the Sutton estate had rich alluvial soil adjacent to the river Wey but much of the remainder was sandy heathland. Weston conceived the idea of increasing the area of water meadow by cutting an artificial channel 3 miles long from Stoke Mills, on the river north of Guildford, through his estate to an outfall on the river near Sutton Green. This would enable some 150 acres to be flooded in winter under controlled conditions, regulated by sluices. After negotiations with neighbouring landowners the cut was made in 1618–1619 at a cost exceeding £1500, with several bridges and a cartway alongside to facilitate maintenance. As a result 'his meadows watered with his new River' yielded two-hundred loads of hay per year, of which he sold about 120 loads 'at near £3 a load'.

This was one achievement in a lifetime spent on agricultural innovations, soil improvement and new crops. As a royalist and catholic his possessions were sequestrated during the Civil War and he seems to have been compelled to flee from the country. In 1644 he was at Ghent, Bruges and Antwerp. It was in the course of his exile that he made those observations embodied in his *Discours of Husbandrie used in Brabant and Flanders* which led to the adoption in England of a new system of crop rotation based on clover, flax and turnips. He sent the *Discours* to his sons *c.* 1645. It circulated in manuscript before Samuel Hartlib published an imperfect copy, without acknowledgement, in 1650 and a more correct copy, properly attributed, in 1652.

While in the Low Countries Weston also gained information about the working of pound locks and inland navigation. On his return to England in the late 1640s he sought to put this knowledge to use and by 1649 had drawn up a scheme for making the river Wey navigable from the Thames up to Guildford. Petitions for doing so had been submitted to Parliament by Guildford Corporation as early as 1621 and again in 1624, their case resting on the need for cheaper carriage of goods to enlarge trade and counteract the declining woollen cloth industry. Reference was made to the necessity of making 'new cutts' and 'weares', the latter presumably being flash locks as used from medieval times on the Thames navigation. Nothing came of these proposals nor is there evidence that a scheme had been worked out in detail with estimates and provision for an adequate source of capital. Nevertheless the idea was sound. It needed someone of Weston's skill and determination to carry it through.

He had first to find a person in favour with the Parliamentarians who would solicit the discharge of his sequestration and, even better, take an interest in the Wey scheme. Such a man was James Pitson, Commissioner for Surrey, who had risen to the rank of Major in Cromwell's army. He secured Weston's acquittal and applied to Parliament for an Act to authorise the making of the navigation. The Bill, submitted in December 1650, passed on 26 June 1651.

The original capital of £6,000 was raised in 24 shares; Pitson, Richard Scotcher and another put up £1,000 each and Weston £3,000. Work began in August 1651 and proceeded with great expedition under Weston's direction, evidently following a prepared plan. As many as two-hundred men were employed at a time. Sir Richard died on 7 May 1652 age sixty-one. In the nine months since the start of operations he had finished 10 miles out of the 15 miles length and spent £4,000 of his own money as well as supplying timber to the value of £2,000, in addition to the £3,000 raised by his partners.

Work continued under George Weston, one of the younger sons, who had bee instructed by his father; Richard Scotcher 'looked after the workmen' and kept the accounts. Pitson found other investors willing to support the enterprise. The navigation was opened in November 1653 at a cost of about £15,000. Of the total length, 9 miles were in cut, with 10 pound locks, four new weirs, twelve bridges and a wharf was built at Guildford. The locks were turf-sided and timber framed. Together they accounted for 60 ft. of the total fall of 72 ft. from Guildford to the Thames at Weybridge.

The navigation proved a success. The annual return is said to have soon been £1,500, chiefly derived from the carriage of timber and corn down to London and coal upwards. But its early history was clouded by disputes and litigation between Pitson, who seems to have been a rogue, Scotcher, John Weston (the heir of Sir Richard) and several others. In 1671, to clear up the muddle, an Act vested the river in six trustees and a board was set up to adjudge claims. The affair was finally settled in 1677, by which time several thousands of pounds had to be spent following neglect of proper maintenance and even wilful damage to some of the locks.

There is no reason, however, to suppose that any major changes were made. Thomas Surbey (q.v.), in an appendix to his 1699 report on the Yorkshire Ouse, gives a description of the Wey Navigation. He noted that barges carried 30 to 40 tons with a draught of 2½ ft. while his measured drawings of a lock show an effective length of 62 ft. (68 ft. between gates) and a width of 13½ ft. These dimensions may well apply to the original construction. In any case, the Wey navigation is certainly the finest piece of river engineering in seventeenth century England. Nor can there be any doubt that Sir Richard was its designer; if further proof is required it is provided by the report of a committee of the House of Commons in April 1662 which includes a resolution to this effect.

A. W. SKEMPTON

[R. Scotcher (1657) A short narrative of the proceedings concerning the making of the River Wey navigable; with an introduction by T. J. Lacy (1895) *The Origin of the River Wey Navigation*; T Surbey (1699) Report on the Yorkshire Ouse Navigation, MS, York City Archives, transcribed by P. Hughes 1992; O. Manning and W. Bray (1814) *History and Antiquities of Surrey*, vol. 3, App. 3, 54–58; E. Clarke (1899) Sir Richard Weston, *Dict Nat Biogr*, **60**, 367–369; T. S. Willan (1936) *River Navigation in 1600–1750*; C. Hadfield (1969) *Canals of South and South East England*; M. Nash (1969) Early seventeenth century schemes to make the Wey navigable, 1618–1651, *Surrey Archaeol Coll*, **63**, 33–40; A. W. Skempton (1984) Engineering on the English river navigations to 1760, *Canals: a New Look* (eds. M. Baldwin and A. Burton), 23–44]

Works

1618–1619. 3 mile cut from river Wey to create 150 acres of water meadow, at least £1,500
1651–1653. River Wey navigation (completed by George Weston), 15 miles long with 9 miles of new cuts and ten pound locks, about £15,000

WESTON, Samuel (fl. 1768–1804), civil engineer, became involved in civil engineering in the 1760s. His family probably came from Cheshire, possibly near Runcorn, and he married Mary Ankers at Aston-by-Sutton church on 2 October 1757. Like many of the canal engineers he appears to have operated variously as surveyor, engineer and contractor.

He worked as a staff holder and leveller for James Brindley (q.v.) who had surveyed a line for a canal from Chester to the Trent and Mersey at Middlewich; it failed to obtain Parliamentary sanction in 1769. Chester merchants and businessmen were keen to pursue the idea, in part because of concern for the impact of the Trent–Mersey canal on trade to the port of Chester. In March 1770 a revised scheme was launched and Weston, who may have worked for Brindley on the original proposal, was employed to survey a line, charging 10s 6d (52½ pence) a day. He was assisted in gauging the water at the summit level and Eaton Brook—a potential source of water—by Thomas Morris (q.v.). His estimates for a canal commencing at Boughton came to £32,395, with thirty-one locks and a reservoir, supplied from Eaton.

At the end of August 1770 meetings of the subscribers' committee became more formal and on 14 September Weston was asked to report on an alternative route to the north of the city joining the Dee near the sluices (see: *REYNOLDS, John*), via Bachers Pool or Northgate, the latter being adopted. Shortly after this, five hundred copies of 'observations' on the canal were printed and distributed to subscribers; they described the canal scheme, now nearly 21¾ miles long, with an aqueduct to supply the canal intercepting water from brooks near Tarporley. Preparations were made for a Bill, and John Longbothom (q.v.) was asked to look over the line. It seems unlikely that either he or Hugh Oldham, who was asked in December to survey the route, did so as in the end it was Thomas Yeoman (q.v.) who gave supporting engineering evidence to the House of Commons Committee.

Weston himself gave evidence, which revealed his lack of experience, but in the event the Bill failed in the House of Lords, largely because of the failure to obtain the agreement of the Trent and Mersey Company to a junction. Almost immediately, on 13 May 1771, Weston was asked to survey a branch from Tilstone Heath to Nantwich and in December a field plan for Parliament. A plan was published, nominally surveyed by Yeoman, drawn by Weston, and was accompanied by a pamphlet. The deposited plan and other surveys were signed by T. Weston [*sic*]. The new line followed a more southerly course from Tilstone Mill to Wardle Green with a summit level 40 ft. lower than the previous.

The new scheme, which did not specifically mention a junction with the Trent and Mersey, went before Parliament in January 1772, Weston again giving evidence. The Act was passed on 1 April 1772 but with a clause prohibiting the canal from crossing the Middlewich–Nantwich or Middlewich–London roads or approaching within 100 yd. of the Trent and Mersey without the permission of the proprietors. Weston was appointed engineer for construction. A son acted as his clerk, at a salary of £20 p.a., and was also assisted by John Lawton. Weston was paid off in January 1774, the month Lawton died. By then he had surveyed additional branches from the Dee to Gresford and 'Caegidog' in the North Wales coalfield, and to Shropshire. He was replaced by Thomas Morris (q.v.). This may have been due to problems with construction as in February 1774 the Committee ordered an aqueduct bridge to be taken down and rebuilt.

While engaged on the Chester Canal Weston was involved elsewhere. He was a contractor, with Lawton, on the Leeds–Liverpool Canal (1770) from Newburgh to Liverpool—a massive contract for the time—and was responsible for 4 miles of it.

Weston then moved to Oxford to work on the completion of the first phase of the Oxford Canal, probably as a result of contacts with other Brindley associates The canal formed part of Brindley's grand cross scheme to unite the rivers Mersey, Thames, Trent and Severn. Construction began in 1769 but had to be abandoned in 1778 when funds ran out. In the meantime he became involved in several schemes around Oxford. In 1778 he was appointed surveyor to the Stokenchurch, Wheatley Begbrook and New Woodstock Turnpike Trust, a post he retained until 1802.

Interest resumed in the completion of the Oxford Canal in 1785, with construction being carried out on the second phase 1786–1790. In September 1785 Weston was asked to survey the river Cherwell from Oxford to Banbury and it was decided by the Canal proprietors to seek Parliamentary approval for a scheme to make the Cherwell navigable to its junction with the Thames. In February 1786 Weston was asked to seek landowners approval for a towpath along the Cherwell. In the event this work was not adopted and Brindley's 1769 scheme was pursued through Parliament in 1786. Work began on the canal, supervised by James Barnes (q.v.) but Weston's involvement continued and in early 1788 he was asked, with Samuel Simcock (q.v.), to report on a new line through Oxford and in January 1789 he and Simcock drew up plans for a new wharf near Hythe Bridge, the final stage of the project. A year later Weston reported on the availability of land for the canal terminals. In addition to this engineering advice, Weston was a contractor on the canal, with Simcock and other partners. They were responsible for its maintenance from Longford to Banbury, and its construction south of Banbury.

No sooner was the canal opened than the limitations of the Thames as a navigable route to London became apparent. At more or less the same time a proposal to build the Grand Junction from the Coventry Canal to London was being launched, which could threaten the Oxford Canal's trade. Weston was asked to survey a route from the Oxford Canal near Thrupp to London in January 1792 with Simcock, and his plan was printed in September. This scheme became known as the Hampton Gay or London and Western Canal and was for a canal 61 miles in length leaving the Cherwell near Thrupp, proceeding on a level for 16 miles then rising 264 ft. with a branch to Aylesbury, followed by a fall of 474 ft. across Hounslow Heath, joining the Thames above Isleworth. To some extent this scheme was an attempt by the Oxford Canal to ensure that it could deal with the Grand Junction on reasonable terms. In 1793 the Bill came before

Parliament, Weston giving evidence on 20 February supported by Hugh Henshall (q.v.). The scheme was opposed by Thames navigation interests and the supporters of the Grand Junction scheme and the Bill was delayed in Parliament while the Grand Junction was approved; it effectively killed the Hampton Gay scheme.

Simultaneously with his work on the Oxford Canal, Weston was consulted in late 1788 about the 'Western Canal', a scheme to extend the Kennet navigation, which was the forerunner of the Kennet and Avon Canal. He was asked, along with Barnes and Simcock, to survey routes to join the Kennet and Avon. By the summer of 1789 surveys were completed. Weston's recommended route ran from Newbury via Hungerford, Ramsbury, Marlborough, Calne, Chippenham, Lacock, Melksham, Bradford-on-Avon to Bath. The report was referred to Robert Whitworth (q.v.) who queried the adequacy of the water supply at the summit level and at this stage John Rennie (q.v.) was brought in as engineer.

His interest in contracting evidently continued as Samuel Weston and Son, of Oxford, undertook the first two contracts on the Ellesmere Canal, which obtained its Act in 1793. The Wirral section from Netherpool (Ellesmere Port) and Chester was the first to be let on the canal and the second, the Llanymynech branch, was intended to connect with the Montgomeryshire Canal. The Wirral line presented no engineering difficulties and was completed in 1795. This was probably Weston's last job as both the Wirral line and the Welsh branch were apparently completed by Weston's partner John Fletcher (q.v.).

It is difficult to assess Weston's gifts as an engineer but that he was held in reasonably high regard by his contemporaries can be gleaned from the fact John Smeaton (q.v.) recommended him to the Ellesmere Canal Committee in September 1791, along with William Jessop and Eastburn (qq.v.), as being suitably qualified to survey their canal. Weston's reputation has tended to be overshadowed by that of his son William (q.v.) who went to the U.S.A. in 1793 to become a pioneer of inland navigation in North America.

His relationship with John Weston, surveyor (fl. 1774–1790), and Robert Weston, surveyor (fl. 1761–1808), of Brackley and Aynho, Northamptonshire, is indeterminate at present. It is even possible there were two Samuel Westons, as a Samuel Weston is listed in a Chester directory of 1781 as a civil engineer of Paradise Row and in 1793 a Samuel Weston of Halewood was involved in an enclosure award at Hope in North Wales, not necessarily compatible with his work in Oxfordshire.

Weston died between October 1804 and March 1805, when his will was written and probate granted.

MIKE CHRIMES

[S. Weston, Accounts, April–July 1770, TAV 2/55, Chester City RO; Minutes of the Chester Canal Company, 1770–, RAIL 816/2, PRO; Quarter session records, etc., (Robert Weston and son), Northants CRO; *Journal of the House of Commons*, **33**, 1771; *Journal of the House of Commons*, 31 January 1772; *Jackson's Oxford Journal*, 1 September 1781; Anon. (1806) *Report to the General Assembly of the Ellesmere Canal Propertors …* 1805; C. Hadfield (1969) *The Canals of the West Midlands*, 2nd edn., 43–44; E. A. Shearing (1985) Chester canal projects: part II, *Journal of the Railway and Canal Historical Society*, **28**(4), 146–154]

Publications

1770. *Observations on the Chester Canal*
1771. *Plan of the proposed Chester Canal* (with T. Yeoman)

Works

1770–1774. Chester Canal, engineer/surveyor
1770. Leeds Liverpool Canal, contractor
1785–1790. Oxford Canal, contractor/surveyor
1788–1789. Western Canal, survey, surveyor (not executed)
1792 1793. London and Western Canal Scheme, engineer/surveyor (not executed)
c. 1794–1795. Ellesmere Canal, contractor

WESTON, William (c. 1763–1833), civil engineer, is believed to have been born in or near Oxford in 1763, the son of Samuel Weston (q.v.). Little is known of William Weston's early career. It has been suggested that the may have served under James Brindley (q.v.), as Samuel is known to have done, but given the disparity in their ages this is clearly implausible. From details in his notebook he was probably involved in a junior capacity on the Oxford Canal, the first section of which was under construction from 1769 to 1778, and the second phase completed between 1786 and 1790.

Weston first came to prominence when he was responsible as both Engineer and managing contractor for the design and construction of the Trent Bridge at Gainsborough between 1787 and 1791. This is a handsome and substantial three span bridge in ashlar masonry across the tidal river Trent. The arches span 62, 70 and 62 ft. respectively and the overall width was originally 29 ft. although cantilevered walkways have been added. Weston's bridge still carries a busy main road without any sign of deterioration or distress, apart from superficial impact damage by barges, and is clearly the work of a confident and competent engineer.

This bridge is Weston's only known work in Britain, for in November 1792 he sailed from Falmouth for the United States of America with his bride, Charlotte, daughter of Richard Whitehouse, a merchant and brewer of Gainsborough, to take up a five year engagement as Engineer to the Schuylkill & Susquehanna Navigation Company of Pennsylvania. The Westons' address in America was initially 'c/o Honble Morris,

Philadelphia, North America'. This was Robert Morris who, besides being a promoter of the Company, was also a Member of Congress and Secretary of State.

At this time, at the start of the American canal era, there were few, if any, experienced engineers in the United States and Weston's services were obtained through the agency of Patrick Colquhoun, a London magistrate of Scottish and American origins. Weston was not Colquhoun's first choice although it seems that he was recommended to Colquhoun by William Jessop (q.v.). In this context it is noteworthy that Weston took with him what was probably the first engineer's level to be used in America, an instrument made for him by Edward Troughton (q.v.) of London.

Two years after Weston's arrival in America the Schuylkill Company became insolvent but by that time Weston had already become actively involved with three other canal projects: the Middlesex Canal in Massachusetts, the Potomac River Locks at Great Falls (at the personal request of President Washington) and the Western Inland Lock Navigation Company, linking the Hudson River with the Great Lakes in upstate New York and the precursor of the Erie Canal. Here Benjamin Wright, later to become America's leading canal engineer, was one of Weston's assistants.

In 1798–1799 Weston designed the two intermediate piers for the Schuylkill River Bridge at Philadelphia, a three span timber truss structure. Rockhead was encountered over 40 ft. below river bed level and massive cofferdams were necessary, among the largest ever constructed at that date. This bridge linked up with the 62 mile Philadelphia & Lancaster Turnpike with which Weston was also associated, the first major road to be built in the United States. Weston's final project in America was a proposal for a new water supply for New York city. By 1799 the city's population had reached 50,000 but was still largely dependent on shallow wells. Weston proposed to utilise the resources of the Bronx River, bringing its water to New York in a 14 mile open conduit, terminating in a filter bed and service reservoir in City Hall Park. The total fall along the conduit was a mere 23 ft., although 24 in. diameter cast iron pipes were to be used for the crossing of the Harlem Valley. In many respects Weston's proposals were well in advance of their time but they were not accepted and another forty years were to pass before any imported water was to reach the city.

Although Weston returned to England in late 1801 he retained his links with the United States and in 1811–1813 was acting as a consultant to the Erie Canal Commissioners, though he declined an offer of the post of Chief Engineer. Details of Weston's career after his return to England are just as shadowy as they had been before his departure from Falmouth seven years previously. He had been exceptionally well remunerated whilst working in America and, although still under forty, he could well have afforded to go into an early retirement, his engineering reputation already firmly established. He seems to have settled in Gainsborough, but died 'very ... suddenly ... from an ossification of the heart' on 29 August 1833, in London.

Weston was highly thought of as an engineer by his American contemporaries and their opinions are borne out both by the contents of his reports and by the skill and success with which he executed his works. The only criticism came from De Witt Clinton who remarked that Weston was, perhaps not surprisingly, totally ignorant of the country and its people and his designs were unnecessarily costly. Weston's reference notebook, a neatly hand-written compendium of technical information, survives and is preserved in the Institution of Civil Engineers archives. This is a fascinating document, revealing not only the extent of his technical expertise and understanding in such civil engineering topics as hydraulics (flow in pipes and open channels, including backwater curves and head losses through bridge piers, weir and aperture discharge characteristics), hydrostatics, wind and gravitational forces, the geometry of masonry arches and the design of sea banks, but also in subjects such as mechanical engineering (mechanics, frictional forces, pumps and steam engines), pneumatics, acoustics and optics—a remarkable breadth and depth of knowledge if the generally accepted idea that late eighteenth century engineers were essentially empirical men is to be believed.

BARRY BARTON

[William Weston's manuscript notebook (c. 1800) ICE Archives; R. Peters (1808) *Statistical Account of the Schuylkill Permanent Bridge*; E. Waston (1820) *History of the Rise, Progress and Existing Conditions of the Western Canals of the State of New York*; Obituary (1833) *Times*, 3 September; A. Stark (1843) *History and Antiquities of Gainsborough*; C. Eddy (1843) *Historical Sketch of the Middlesex Canal*; N. E. Whitford (1905) *History of the Canal System of the State of New York*; H. W. Hill (1908) *Historical Review of the Waterways and Canals Constructed in New York State*; American Council of Learned Societies (1936) *Dictionary of American Biography*, **X**, 21–22; R. S. Kirby (1936) William Weston and his contribution to early American engineering, *Trans Newcomen Soc*, **XVI**, 111–127; C. Roberts (1938) *The Middlesex Canal, 1793–1860*; J. S. English (1976) *Gainsborough: Some Links with America*, Pub Friends of the Old Hall Assn, Gainsborough 7–9.; D. H. Stapleton (1987) *Transfer of Early Industrial Technologies to America*, American Philosophical Society, Philadelphia; A. W. Skempton (1990) Letter, *ICE Panel for Historical Engineering Works Newsletter*, **48**; R. E. Shaw (1990) *Canal for a Nation, The Canal Era in the United States, 1790–1860*; R. Bourne (1992) *Floating West, The Erie and Other American Canals*; A. Jones (1997) in lit.]

Publications

[Two Reports by Wm. Weston are contained within in *An Historical Account of the Rise, Progress and Present State of the Canal Navigations in Pennsylvania*, Philadelphia, 1795]

1795. *Western and Northern Inland Lock Navigation Company—Report of Engineer ...*
1799. *Report on the Practicability of Introducing the Water of the Bronx into the City of New York*

Works

1787–1791. Trent Bridge, Gainsborough, Engineer and Contractor
1793–1795. Schuylkill and Susquehanna Navigation, Engineer
1793–1798. Delaware and Schuylkill Canal, Consultant
1793. Conewago Canal, Pennsylvania, Consultant
1793–1794(?). Philadelphia and Lancaster Turnpike, Consultant
1794–1796. Middlesex Canal, Massachusetts, Consultant
1795–1796. Potomac River Locks, Consultant
1795–1798. Western Inland Lock Navigation, New York, Engineer and Consultant
1796. James River Canal, Consultant
1800–1801. Schuylkill River Bridge, Philadelphia, piers and cofferdams, Engineer

WHIDBEY, Joseph, FRS (1755–1833), was a naval officer who gained his warrant as Master in 1779. In 1787 he was ordered to survey the harbours of Port Royal and Kingston, Jamaica, with George Vancouver. This marked the start of a fruitful collaboration in which he achieved a leading place as a coastal surveyor. Their survey of the ports was published in 1792 by which time they had set sail for the Pacific. He participated in Vancouver's famous surveying voyages along the North West Coast of America in 1792–1794, returning in 1795. In 1799 Lord St. Vincent commissioned him to make a survey of Torbay with a view to making it a completely safe fleet anchorage. His proposal that an artificial island across the eastern exposed extremities of the bay would make a safe anchorage for one hundred Sail-of-the-Line may have resulted from his experience on the voyage of 1792–1794 where he may have had memory of some island screening the entrance to an otherwise open bay.

In September 1799 Whidbey was appointed Master Attendant at Sheerness and his salvage work of a frigate near the Great Nore led to his election as a Fellow of the Royal Society in 1805, when among his proposers was John Rennie (q.v.). Meanwhile, in February 1804, he was appointed to the more prestigious post of Master Attendant at Woolwich. Problems of silting at Woolwich first brought him into contact with Rennie and when Lord St. Vincent, the newly appointed C-in-C of the Channel Fleet, became worried about the possible destruction of the fleet in a gale while at anchor, both Rennie and Whidbey, together with James Hermans from

Joseph Whidbey FRS

Chatham Yard, were sent to meet at Plymouth on 18–21 March 1806, the time of the greatest spring tides. Their report of 21 April to the Admiralty, proposing a breakwater sited on the Shovel shoals, ultimately led to the construction of Plymouth breakwater.

After a period in abeyance, a change of First Lord at the Admiralty in 1811 led to a flurry of reporting activity on Plymouth, with alternative schemes proposed by Sir Samuel Bentham, and Samuel Moyle (qq.v.) until Rennie and Whidbey's proposal for a detached rubble mound breakwater 1700 yd. long was adopted.

On 28 October 1811, Whidbey was appointed as the Admiralty Superintendent of the breakwater and construction began in 1812 with William Stuart (q.v.) at Oreston about a mile from Plymouth as the resident engineer. The first contracts, for quarrying the rock, were awarded to Smith and Austin, and Fox and Williams; Billings were awarded the contract to transport the stone to the breakwater, a contract they retained until 1830, when the Admiralty took over this work. The Admiralty also took over the main contract for building the breakwater from I. & W. Johnson in 1834; Jolliffe and Banks (q.v.) supplied the cranes. A succession of storms threw into question the whole design and Whidbey advocated a vertical sea wall. In this he was over-ruled and following a report by (Sir) John Rennie, Josias Jessop and William Chapman (qq.v.) in 1826 the design was modified to include a sea-slope of 5:1, the slopes and top paved with the largest blocks and overall form modified to take account

of experience. Greatest progress took place between 1812 and 1821, then pace of construction slackened until 1830 when Stuart took charge.

In 1822 Whidbey again surveyed Torbay and made an estimate for a breakwater, the cost of which he calculated as £1,630,000. In May 1823, with Sir John Rennie, he reported on means of improving Whitehaven harbour and, with Rennie, provided a design for its extension, while in September 1823 he reported on the harbour and pier of Newhaven. In 1823–1824, in conjunction with James Green (q.v.), whom he had met as Rennie's assistant at Plymouth in 1806, he surveyed and reported on the harbours of St. Ives and Ilfracombe. In September 1824 he reported on the improvement and management of the river Clyde and the harbour of Glasgow.

The idea of a breakwater in Torbay was revived in 1828 by the Duke of Clarence, Lord High Admiral, who required Whidbey to visit Torbay with William Stuart and go through the estimate for the breakwater in three different ways: with hired labour, with convict labour and with a mixture of both. Nothing further was done.

In 1809 Whidbey was elected to the Society of Civil Engineers as an honorary member. At that time Rennie was one of eight ordinary members that is, civil engineers—and, additionally, there were twenty-one honorary members. When John Rennie died in 1821 there was a considerable reorganisation of the Society and Whidbey was transferred to ordinary member in 1822; this no doubt reflected his involvement in the breakwater. Besides Whidbey, the full members were William Chapman, William Chadwell Mylne, Josias Jessop, James Watt, George Rennie and (Sir) John Rennie (qq.v.).

Whidbey was made a Freeman of the Borough of Plymouth in March 1815 and he retired from his position concerning Plymouth breakwater in March 1830 on a pension of £1,000 p.a., buying St. James House in St. James Street, Taunton, with Henry and Catherine Ogland to look after him. He died on 9 October 1833, aged seventy-eight years, and was buried in St. James churchyard, opposite the house where he lived, leaving over £2,000 to various friends and the residue estate to a great niece.

In Plymouth his contribution to the safety and growth of a famous harbour has been given a fitting memorial by the establishment in 1980 of the Whidbey Automatic Light (Occulting Green) on Andurn Point (SX 491 495), which commands the eastern entrance to Plymouth Sound. A portrait of him by J Ponsford (1815) was left to the Institution of Civil Engineers by George Rennie and a further portrait has been located in Portsmouth Dockyard.

A. B. GEORGE

[Rennie mss, Reports 7, December 1811 to February 1814, ICE archives; J. Rennie (1821) *Interesting Particulars Relative to … the Breakwater now constructing in Plymouth Sound* (two reports

to the Lord Commissioners of the Admiralty);
M. A. Borthwick (1836) Introduction, *ICE Trans*, **I**;
Whitehaven Harbour (1836) *Plans suggested as
different Periods for the Improvement of
Whitehaven Harbour*; Newhaven Harbour (1846)
*A Collection of Reports relative to the Harbour and
Piers of Newhaven*; Sir J. Rennie (1848) *An Histor-
ical Practical and Theoretical Account of the
Breakwater in Plymouth Sound*; James Green,
Obit. (1850) *Min Proc ICE*, **99**; Clyde Navigation
(1854) *Reports on the Improvement and Manage-
ment of the River Clyde and Harbour of Glasgow
from 1755 to 1853*; E. Hopkins (1854) William
Stuart, Obit., *Min Proc*; J. G. Watson (1989) *The
Smeatonians: The Society of Civil Engineers*; J.
Naish (1992) Joseph Whidbey and the Building of
the Plymouth Breakwater, *The Mariner's Mirror*,
37–56; W. K. Lamb (ed.), The Voyage of George
Vancouver 1791–1795, *Hakluyt Society Pubs*, 163–
166; D. Greenfield (1998) in lit.]

Publications

1806. *Report to the ... Admiralty* [on proposed
Plymouth Breakwater] (repr. 1812, 1821)
1823. *Report on the Means of improving the Harbour of Whitehaven* (with J. Rennie) (repr. 1836)
1823. On some fossil bones discovered in the
limestone quarries of Oreston, *Phil Trans R Soc*,
78–81

Works

1812–1830. Plymouth Breakwater, Superintendent
of Works, 2.8 million tons of stone (*c*. 70% of total,
completed 1850)

WHITE, William (*c*. 1749–1816), was a land surveyor based at Sand in the parish of Wedmore,
near Wells, with a practice extending through
Somerset, Gloucestershire, Devon and Wiltshire.
In 1791 an Act was passed for improving drainage
of peat land in King's Sedgemoor in the southern
part of the Somerset Levels. Five Commissioners
were appointed. Rather than simply improving
the existing drains and outfalls they decided to
make a new main drain (about 8 miles long)
leading through a 2½ mile cut in the coastal clay
belt to the site of an old sluice, called Dunball
Clyze, on the river Parrett estuary. White carried
out the survey and levelling, finding a very adequate fall from the peat moor to low tide at
Dunball.

The only objection to the scheme arose from
the cost of cutting to a depth of 15 ft. through the
high ground of the clay belt. John Billingsley,
one of the Commissioners, a farmer from Sutton
Mallet and author of *A General View of the Agriculture of Somerset*, was the leading proponent of
the scheme. As he explains: 'No alternative, however, presented itself. It appeared that this plan
must be adopted, or the works would be incomplete. Justified therefore by the concurrent
opinion of Mr. White and Mr. Jessop (whose
advice was taken) they proceeded boldy'. It is
probable that Robert Anstice (q.v.) acted as resident engineer. After much trouble with slips in

the cut and difficult foundation conditions at the
new sluice, work reached completion in 1795; the
area drained was 11,000 acres, the total cost was
£32,000 of which £15,000 was for engineering
works.

By 1794 a scheme had been drawn up for improving the drainage of low lands further north in
the Somerset Levels adjacent to the river Brue
(22,000 acres) and the river Axe (16,000 acres).
The scheme was proposed by Billingsley and surveyed and planned in detail by White. For the
Brue, an outfall sluice at Highbridge would
replace the old one, the sill to be 10 ft. lower, and
there would be a long cut (the South Drain) as
well as several short ones on the river. For the
Axe, principal works proposed were a sluice at
Hobb's Boat near Bleadon (no sluice on this river
having previously been built), a 1½ mile cut at
Loxton bypassing a long bend in the river, and
some other cuts.

Nothing much happened until 1800 when it
was decided to petition Parliament for a Bill for
Brue Drainage. William Jessop (q.v.) was appointed engineer. White gave evidence to the
Commons in March 1801, Jessop and White appeared before the Lords in May, White again in
June, and later that month the Act passed. Jessop
received £155 for his work up to that time and
White £147 'for measuring, levelling, preparing
maps and plans and for journeys and expenses'.
Jessop then sent in drawings for the sluice and a
new bridge. He and White marked out the sluice
and its cut in August. Anstice was appointed
Superintendent of Works (resident engineer) in
October at 4 guineas per week and work began,
first on the sluice and in 1804 on the South
Drain, reaching completion in 1806; the total cost
was £53,000, of which engineering works cost
£34,000. Jessop visited the site three times up to
1805, receiving £122 in fees. White's fees during
construction and for a map of the drainage as
completed came to £2,248.

Proceedings in respect of the Axe Drainage
started at about the same time in the Brue, with
reports by Josiah Easton (q.v.) in October 1800
and John Sutcliffe (q.v.) in May 1801 on the best
situation for the outfall sluice. Both preferred
sites further downstream than Bleadon. But when
the Act passed in May 1802, White having given
evidence in the Commons and Lords, the sluice
was to be at Hobb's Boat as originally proposed.
Jessop, appointed engineer, confirmed this location but discussion ensued on the need for a navigation lock alongside. Meanwhile work began on
Loxton Cut in July 1803, set out and probably supervised by White, and it was not until 1804 that
Jessop sent plans for the sluice (without a lock).
White and Jessop examined estimates, and excavation of the sluice pit began later that year with
Anstice now in charge of construction.

A deep bed of soft clay and artesian pressure in
the underlying sand caused problems. The whole
job was completed in 1810 at a total cost of
£41,000 of which engineering works totalled

£25,000. Jessop visited the site three times (one combined with a visit to the Brue) and wrote eight letters. He seems to have left direction of the works to Anstice from 1805, his fees amounting to £125. White must have been frequently on site, making his last visit to the works, with Anstice, in December 1809; his fees were £2,247 (plus £100 paid on account for attendance in Parliament in 1802) though this sum may include payment for a map of the completed drainage. Anstice received fees of £1,080.

While these three works on the Somerset Levels were in progress White was involved with Jessop in the planning of Bristol Floating Harbour. Between December 1792 and September 1793 he spent sixty-four days on surveying, levelling and river observations, twelve days attending meetings and forty-three days drawing plans, charging 2 guineas per day (including expenses) and £17 for assistants in the field: a total of £267. From White's drawings William Faden produced an engraved plan with longitudinal and cross sections of the Avon and details of the proposed dam on the river. In much the same period Jessop spent twenty-one days, charging £95 at 3 guineas per day in 1791 and 1792 and 5 guineas per day in 1793.

The scheme was later revised, leading to the Act of 1803. For work in 1801–1803 White charged £344 and Jessop £406. In September 1803 Jessop was appointed engineer and White surveyor, with Josias Jessop (q.v.) as resident engineer from February 1804. White was busily engaged in 1804 and on several occasions up to 1809, but only on surveying; all engineering work was supervised by Josias Jessop under the active direction of his father.

Finally, White was called in to survey the Weston valley between Clevedon and Partishead. He reported in June 1811 and was appointed engineer. But the Commissioners were dilatory and in a second report a year later White complained of a lack of progress. The work was eventually completed in 1815, with but little permanent improvement.

William White died in June 1816, aged sixty-seven, and was buried at Wedmore Parish Church.

A. W. SKEMPTON

[Documents on King's Sedgemoor Drainage (1791–1796) Proceedings of Brue Drainage Commissioners (1801–1807) and documents on Axe Drainage (1801–1812), Somerset RO; Bristol Dock Company Minute Books (1791–1810) Bristol RO; J Billingsley (1794) *General View of the Agriculture of Somerset*, 3rd edn., 1798; M. Williams (1970) *The Drainage of the Somerset Levels*; C. Hadfield and A. W. Skempton (1979) *William Jessop, Engineer*; Bendall]

Publications

1793. *A Plan of Part of the Rivers Avon and Frome from Rownham Ferry to Temple Bask ... Also Sections of the same together with the proposed Dam, Lock and Canals for the Improvement of the Harbour of Bristol* (with Jessop)

1794. *A Plan for more effectually Draining the Turf Bogs and Flooded Lands near the Rivers Brue and Axe*
1795. *A Plan of King's-sedgemoor Drains and Parochial Allotments*
1803. *Design for Improving the Harbour of Bristol* (with Jessop)

WHITMORE, William (*c.* 1748–1816), engineer and ironfounder, of Dudmaston Hall, near Bridgnorth, Shropshire was the son of Charles Whitmore, a wine merchant in Southampton, who was a cadet of an old landed family of Apley, near Bridgnorth. William Whitmore was twice married, to Frances Barbara Lister of Yorkshire who died in 1792 and then to Mary Thomas of Glamorgan, who died in 1813.

He is first noted in Birmingham in 1777 as a toymaker in Snowhill. By 1780 he had moved to Little Charles Street and in 1783 was a file maker there; in 1785 he advertised himself as a manufacturer of all kinds of scales, scale beams and steelyards, clock, watchmakers, jewellers and silversmiths tools and files, presses and stamps, rollers or flatting mills, machines for weighing wagons etc. and by 1812 he had added iron founder and manufacturer of steam engines to the list.

In 1799, Whitmore & Norton (for Norton see Colvin) offered to supply a 'forty foot lock of a new construction' to the Birmingham Canal to avoid the need for a proposed tunnel. It was not built and it was probably the same as the balance lift which they offered in 1800 to connect the lower and upper levels of the Somersetshire Coal Canal instead of locks or the inclined plane which had been proposed by Benjamin Outram (q.v.). They offered to do this at their own cost, recouping the expenditure by levying tolls on the traffic. Despite advice from William Underhill (q.v.) of the Dudley Canal and favourable reports by the SCC's engineers, this too was not accepted; the inclined plane which was built required to be replaced by conventional locks within a few years.

Despite apparently having no direct experience of this type of work, Whitmore was appointed Engineer to the Stratford upon Avon Canal on 23 January 1811. That company had effected a communication in 1802 between the Worcester and Warwick canals round the south side of Birmingham but, its funds being effectively exhausted, had made little subsequent progress towards Stratford. Completion of the canal was managed by a Board of Works, headed by William James (q.v.), to whom Whitmore was presumably known, but work was still hampered by lack of money. The surveys for the several alterations in the line were provided by Ebenezer Robins, a Birmingham surveyor. Whitmore's work was idiosyncratic and harked back to the 1790s and earlier, though the cast iron aqueduct at Edstone is the second longest in Britain. Whitmore was paid several thousand pounds in 1813, presumably for works executed by direct labour but possibly including ironwork for the aqueducts. He already had a substantial share-holding in the company and was paid by debentures on twelve months' notice. Whitmore died two months after the opening of the canal to Stratford in 1816 but his son continued to do work for them as a payment was made for balance of account to 31 May 1818.

The firm had been known as William Whitmore and Son since about 1812. This son was William Wolryche Whitmore (1787–1858); he had nine sisters, three half-sisters and one half-brother. He was educated at Shrewsbury School from 1799 and on 29 January 1810 married Lady Lucy Bridgman, daughter of the 4th Earl of Bradford; they had no children. He was Member of Parliament for Bridgnorth, as several members of his family had been before him, from 1820 to 1832 and after the Reform Act was Member for Wolverhampton, 1832–1834. He supported the Whig interest and was a strong advocate of free trade before that idea was accepted by the political parties. He was elected a member of the Smeatonian Society of Civil Engineers in 1825 and was Sheriff of Shropshire in 1838.

P. S. M. CROSS-RUDKIN

[RAIL/810 & 875, PRO; various Birmingham directories; H. T. Weyman (n.d.) *The Members of Parliament for Shropshire*; George Morris' ms pedigrees, Shropshire RO; inscription on Wootton Wawen Aqueduct; K. R .Clew (1970) *The Somersetshire Coal Canal and Railways*]

Works
1811–1816. Stratford upon Avon Canal, southern section from Kingswood to Stratford upon Avon, Engineer, salary 3 guineas per day plus expenses

WHITWELL, Thomas Stedman (d. 1840), architect, originated in Coventry and by 1806 he was in London and exhibiting at the Royal Academy. He is believed to have worked as an assistant to Daniel Alexander (q.v.) in the Architects Office at London Docks. Much of Whitwell's practice was in the Coventry and Birmingham area, although he was involved in Robert Owen's utopian project at New Harmony, Indiana.

Whitwell was based at Great Russell Street, Bloomsbury, in 1828 when he was elected Associate of the Institution of Civil Engineers, nominated by Thomas Telford, Robert Sibley and William Cubitt (qq.v.). At that time he was embroiled in the failure of the iron roof at the Brunswick Theatre, attributed by him to the hanging of theatrical equipment from the iron ties. The subject aroused considerable interest in the engineering profession and was discussed at meetings of the Institution on 11 and 18 March, and 6 May 1828, Whitwell describing the structure at the latter meeting. His association with this failure appears to have damaged his reputation as an architect but he remained an Institution member until his death in Gray's Inn in May 1840.

MIKE CHRIMES

[ICE membership records; *ICE Minutes of Conversation*, **2**, 1828–1834; *Mechanics Magazine*, **9**, 1828, 110–124; Colvin (3)]

Publications
1834. *On Warming and Ventilating Houses ...*

WHITWORTH, Robert (c. 1734–1799), land surveyor and engineer, was baptised at Sowerby, near Halifax, Yorkshire on 15 November 1734. He was the seventh and penultimate child of Henry Whitworth, a prosperous combsmith and Mary his wife. The family home, which is still standing, was interchangeably known as Waterside or Wheatleyroyd. It is probable that Robert continued to live there until his marriage to Sarah Irvin on 26 December 1765. For a short time while Whitworth was working for James Brindley (q.v.) they lived at Norton in the Moors close to Brindley's base at New Chapel. Here their son, William, was baptised in May 1772. Subsequently the family, including the elder son, Robert, whose baptism has not been traced, moved back to Sowerby and at some point took up residence in the family home at Waterside. Here they remained until the early 1790s when they removed to Hood House, near Burnley, so that Robert Whitworth could be closer to his work on the Leeds and Liverpool Canal.

Nothing is known of Whitworth's upbringing or education or, indeed, of any other aspect of his life until he emerges in 1761 as a qualified and practising land surveyor. The earliest known example of his work, a plan of an estate at Erringden, is from that year. During the next few years we find him engaged in a variety of professional tasks: measuring the stonework on the newly built Sowerby Church to ascertain what payments were due to the contractor, measuring the distances from various points in Sowerby chapelry to Halifax parish church, making a survey of the religious affiliations of the households in Sowerby, and, most significantly for the future, surveying and making a plan of part of the river Calder. We do not know why this survey was undertaken but it is quite possible that it was in connection with the Calder and Hebble Navigation and thus represents Whitworth's earliest documented involvement with navigable waterways. We can, however, reasonably surmise that his actual involvement goes back rather earlier. The construction of the Calder and Hebble Navigation had been progressing steadily upstream from Wakefield under its engineer, John Smeaton (q.v.), since 1759. There is no evidence that Whitworth was involved with this work but it is improbable that a young surveyor would not take an interest in such a large and unusual project taking place on his own doorstep and it is inconceivable that a man who was to became one of the leading canal engineers of his generation can have ignored it. Indeed it is more than possible that observation of the Calder and Hebble under construction aroused his first interest in becoming an engineer and indeed that through such observations of his methods Smeaton contributed more to his professional training than did Brindley.

The turning point in Whitworth's career came in 1765 when Smeaton was abruptly dismissed as engineer to the Calder and Hebble and replaced by Brindley. Again there is no direct evidence linking the two men but his eighteen months or so in this post was Brindley's only opportunity to assess Whitworth's capabilities and reach the conclusion that here was the man to provide the surveying and drawing expertise which his organisation lacked. The only hint at a connection between the two men at this time concerns the first (1766) survey of the line for the canal that was eventually built as Sir John Ramdens's Canal. Some secondary sources attribute this survey to Whitworth, others to Brindley. Do we have here the earliest example of what became their usual pattern of work whereby Brindley used his reputation to secure surveying commissions which he delegated to Whitworth to execute? Whatever may have been the extent of their collaboration on the Calder and Hebble, and associated schemes, the fact is that some time during 1767 Whitworth became established as Brindley's chief surveyor and draughtsman. During that year he produced survey drawings for the proposed canals from Birmingham to Aldersley on the Staffordshire and Worcestershire Canal which was then under construction, from Coventry to the Trent and Mersey Canal at Fradley Heath and from Droitwich to the river Severn. He also accompanied his employer during the 1767–1768 session of parliament, being described as his clerk. In 1768 Whitworth made the survey for the last of the group of narrow midland canals associated with Brindley, the Oxford Canal.

He then moved on to more distant projects, reporting in August on proposals for the Lagan Navigation in Ireland before moving on to County Durham where he drew a plan and wrote a report on a proposed canal from Stockton to Winston. The completion of this business then had to wait nearly a year until Brindley was free to collaborate on the production of the estimate. By Octoer, Whitworth was reviewing John Long-bothom's survey for the proposed Leeds and Liverpool Canal. He was called upon to do more work on this project in the following year. In particular, he was asked to assess an alternative, more southerly, route which had been proposed and was able to demonstrate that its advocates had made serious mistakes in their levels. He was also called upon to advise on the first version of a scheme for a canal from Leeds to Selby. In July 1769 he was obliged to move on to a project in Devon leaving the remainder of the Leeds and Liverpool work to John Varley (q.v.). The new project kept him occupied for over three months and entailed surveying various routes by which the English channel near Exeter might be linked with the Bristol Channel.

He remained in the south during 1770 making a survey for what became the Andover Canal before joining Brindley in London where he had been commissioned to advise on ways of improving the navigation of the river Thames. Brindley produced two reports predictably recommending that the river be bypassed with a canal, running, in this case, from Isleworth to Monkey Island. The proposals were illustrated on a plan and profile, described as surveyed by Brindley and drawn by Whitworth, the original of which was 7 ft. long and which was also doubtless drawn by Whitworth. Allied to this scheme was a plan for a second canal from Monkey Island to Reading. Whitworth surveyed this line and gave evidence in support of the unsuccessful bill to implement it early in the next year.

In 1771 Whitworth's status as a leading canal engineer was recognised in two distinct ways. Peer group recognition came in March when, at only its second meeting, he was admitted as a member of the Smeatonian Society. An indication of his standing among a wider public came when, in response to requests from his readers to provide information about canals under construction or proposed, the editor of the *Gentleman's Magazine* turned to 'Mr. Charles [sic] Whitworth, draughtsman to the celebrated Mr. Brindley' for the information which underlay the publication, between then and 1774, of a series of plans—mostly redrawn—of such waterways. Whitworth, meanwhile, was preparing a report on the river Trent at Newark in collaboration with John Grundy (q.v.) and surveying lines of canals from the Leeds and Liverpool to Kendal (what became the Lancaster Canal) and from Maldon to Chelmsford. The latter commission may well have owed something to the contacts made while he was working on the Thames a couple of years earlier. This was certainly the case with his main project for 1773. This was a survey of a canal from Waltham Abbey to Moorfields intended to bypass the lower reaches of the Lee Navigation and bring water transport right up to the outskirts of the City of London with a branch to Marylebone intended to link up with a future waterway to the Thames. Because it would approach so close to the City, that scheme gave Whitworth a rare opportunity to set his proposed canal in a wider context of urban design. The same awareness of the need to integrate canals into their environment appears in his discussion, some fifteen years later, of the plans for Port Dundas in Glasgow.

Nothing came of the Moorfields project but the London contacts bore fruit again in 1774 when Whitworth and Thomas Yeoman were commissioned to report on the implementation of the City Corporation's new statutory powers to improve the navigation of the river Thames up to Staines. They reported in June and Whitworth was immediately appointed Surveyor to the Navigation Committee at a salary of £250 p.a. with instructions to implement the proposals. He remained in this post until 1784 when the finance was exhausted although his salary had been reduced to £100 in 1780 in line with the falling workload. He was perhaps fortunate to secure this position when he did because canal work was at a low ebb during the mid 1770s and it was not until the very end of the decade that we again find him securing other commissions.

He reported on the lines of proposed canals in Herefordshire and Gloucestershire (1777), from Bishop's Stortford to Cambridge (1779–1780), and from Ashby de la Zouch to Griff, near Coventry (1781), as well as on the completion of the Oxford Canal from Banbury to Oxford (1779). In 1782 he was asked to review and comment upon rival proposals to build a canal to link the rivers Severn and Thames. He reported in favour of the more fully worked out proposal to extend the existing Stroudwater Navigation across the Cotswolds to the Thames near Lechlade in preference to the more sketchy plan for a canal direct from Tewkesbury. He gave evidence on the successful bill to implement this scheme but, contrary to a common assertion, he did not become engineer to the new company. This position fell to Josiah Clowes (q.v.), quite possibly on the recommendation of Whitworth who certainly helped him with some of the setting out and by drawing the profile for Sapperton Tunnel which was the most difficult engineering problem on the route. The following year the Thames and Severn Company commissioned Whitworth to survey the route for an extension of their canal to Abingdon which would have avoided some of the most difficult navigational conditions on the Thames. In the same year he also surveyed the still unbuilt section of the Coventry Canal from Atherstone to Fazeley and, while working for the City of London at Gravesend, was arrested as a French spy! In 1785 he was surveying the line of the proposed Oakham Canal and (again) the unbuilt portion of the Oxford Canal. This survey was preliminary to work being started.

In June 1785 he accepted the post of Chief Engineer to the Forth and Clyde Navigation. His brief was to complete the canal which, by 1777, had been brought from the Firth of Forth to the outskirts of Glasgow by Smeaton and Robert Mackell (q.v.). The plan now was to extend the main line to a basin nearer to the centre of Glasgow and to construct a branch down to the river Clyde so that the canal could be used, as its planners had intended, to take coastal shipping from one side of the country to the other. As a ship canal 56 ft. wide with locks 74 ft. × 20 ft. this waterway was on a much larger scale than any canal in England at the time. Much of the route was over rough terrain which presented the engineer with many problems of which the greatest was perhaps the crossing of the valley of the river Kelvin. Even Whitworth acknowledged in his typically understated way that he was breaking new ground: 'it (i.e. the Kelvin aqueduct) is in some respects new and out of the common road of bridge building. Indeed it is a new thing to carry a canal of 8 ft. depth of water over a bridge of such stupendous height'. However the most difficult problem was to keep what would probably be the most heavily used section of the canal fully supplied with water. Whitworth considered several options but eventually opted for a combination of making the canal a foot deeper throughout and tapping water from natural lochs and artificial reservoirs. Much of this water was to be delivered along the Monkland canal which was conveniently on roughly the same level as the Forth and Clyde and thereby became a navigable feeder. A basin and transhipment centre, which became known as Port Dundas, was built where the two canals met.

Whitworth was obliged to spend much of 1786 and 1787 in Scotland supervising these complex works but he was able to get away from time to time. He visited the Oxford Canal on several occasions, advising on water supply, especially the Wormleighton reservoir, and other matters. He also prepared proposals for various improvements to the port of Leith, including options for building new docks either adjacent to the existing harbour or on land to be reclaimed from the foreshore. Surveys of the lines for proposed canals from Arbroath to Forfar (1788) from the Forth & Clyde Canal to Borrowstouness (1789), and from Kelso to Ancrum Bridge (1790) were made.

In 1788 or 1789 he resurveyed the line of the Leeds and Liverpool Canal which, like the Oxford and the Forth and Clyde, had remained uncompleted since the 1770s. His first report was largely concerned with securing a better water supply than had been provided for in Longbothom's original plans of twenty-five years earlier. This was to be achieved by deepening the summit level so that it could serve as a reservoir and by lowering and lengthening it so that it could collect more water from higher land. This would entail making a tunnel under the summit at Foulridge with a reservoir nearby to collect the water. He was appointed engineer to the company with the task of implementing these proposals in May 1790. The first year of construction saw considerable progress on the summit level but in September the question arose of alternative routes on the Lancashire side. As was usually the case, Whitworth became very much involved advising on the merits of various options although the issues were in many ways more about canal politics and economics than engineering. The upshot was that Longbothom's original line on the north side of the Calder valley was abandoned in favour of a more southerly route which took the canal closer to Burnley, Blackburn and other expanding industrial towns. This change of plan gave rise to several substantial engineering problems not least among them being the need to carry the canal over the valley of the river Calder. This was achieved by the erection of an embankment 1,200 ft. long and up to 50 ft. high. It was punctured by an aqueduct over the Halifax turnpike road, by a culvert 5 ft. in diameter to carry the water for an engine and by another aqueduct over the river Calder.

While these and other works were in progress Whitworth's main priority became the establishment of his two elder sons in their careers as engineers. These were Robert, who was born around 1770, and had already helped his father during the final stages of his work on the Forth and Clyde Canal and William, who had been baptised in 1772 and who had assisted with setting out the Leeds and Liverpool and whose name appeared with his father's on the plan which they made in 1792 to show the line of the proposed deviation of his canal. Towards the end of 1792 Whitworth Sr. was asked to resurvey the line of the Ashby Canal and when this proposal received parliamentary sanction in 1794 he accepted the position of engineer to the company with responsibility for design work, setting out, and three months attendance each year while his son, Robert, was appointed full time resident engineer to provide day to day supervision of the works. These proceeded steadily with Whitworth Sr. submitting regular progress reports as well as advising on broader issues such as the substitution of railways for canals on some branches until, apparently without warning, both Whitworths were dismissed. There is a hint that the real problem may have poor attendance by the younger man: as a full time employee his absence from the meeting at which he was dismissed may not be without significance! It may well be that be was developing other business interests against the day when the canal was complete. Certainly other references to his engineering activities are few. In 1793, in collaboration with John Ainslie (q.v.), he surveyed a line for a proposed canal from Edinburgh to Glasgow, he is said to have succeeded his father as engineer to the Dearne and Dove Canal and at about the same time to have made a survey of a proposed ship canal to serve Canterbury.

Meanwhile, early in 1793, Whitworth Sr. and William had surveyed the lines of what became the Wilts and Berks Canal and in 1795 became its engineers on a basis similar to that adopted on the Ashby. Again, the father submitted regular progress reports until his death in 1799; William then became sole engineer. He saw the construction of this canal through to its completion in 1811 and served as engineer for the construction of the closely allied North Wilts Canal from 1814 to 1819. His lack of interest in the wider world of engineering was clearly signalled in 1810 when he indicated to the proprietors of the Lydney Canal that he was unable to recommend an engineer to them and that he could not advise them personally unless they were to visit him. By 1812 he was married to Rebecca Court and living with her and their two children close to the canal at Watchfield House near Faringdon. He was building up a range of business activities around his interest in the two canals. These included carrying on the canals and the ownership of collieries in the Forest of Dean.

Towards the end of 1795 Whitworth Sr. agreed to a reduction in his commitment to, and salary from, the Leeds and Liverpool Canal Company. This gave him more time to attend to his sons' projects and to the Dove and Dearne and Herefordshire and Gloucestershire Canal whose engineer he had become during 1795. However, since the day to day management of these projects was for the most part in the hands of capable and reliable subordinates, and he was only required to make periodic visits and reports, his final years saw what was perhaps the most varied workload of his career. We find him giving advice to the Gloucester and Berkeley Canal, the Grand Junction Canal, the Ilchester and Langport Navigation, and the Dudley Canal (all 1796), the Trent Navigation (1797) and the Don Navigation (1798–1799); arbitrating between the Lancaster Canal Company (1795) and the Glamorganshire Canal Company (1796) and their respective contractors; working on proposals for new canals such as the Commercial Canal (1795–1796), the competing routes on the north and south banks of the river Tyne (1797), and on a canal from the Tyne to the Wear with branches to Kio West Houses and via the river Wear to Durham (1797). As 1798 progressed the level of activity tailed off. Around this time he described himself as 'an old superannuated engineer' and, with hindsight, his last report to the General Assembly of the Proprietors of the Leeds and Liverpool Canal, written in September 1798 and in which he reviewed progress on the complex works being undertaken between Burnley and Henfield, reads like an apologia. In February 1799 he made his will, concluding with a spidery signature which was a shadow of its former bold and elegant self. On 30 March 1799, at the White Lion Inn in Halifax—still on his interminable travels—Robert Whitworth died. The cause of death was reported to be 'a mortification of his foot, brought on by wet and cold during the late severe storm'. He was buried at Sowerby where he and most of his children had been baptised. No words can sum up his life, work and personality better than those of his anonymous, but generally well informed, obituarist: 'He succeeded Mr. Longbottom in the direction of that grand undertaking, the canal from Leeds to Liverpool; which will be to his memory a monument more durable than marble or brass. When his person, his private worth, his social virtues, and his tender affections shall no longer be remembered, this together with his important works upon the Clyde ... and many other places will lastingly convey to posterity, with deserved celebrity, the extensive genius and scientific powers of this great and much regretted artist. He was extremely modest, unassuming and communicative [*sic*]; for what nature had denied him eloquence and colloquial rhetoric, she more than amply remunerated him in the lucid powers of his pen'.

G. W. OXLEY

[Parish registers of Halifax and Sowerby, West Yorkshire Archive Service, all branches; Plan of lands at Erringden and Sowerby (1761) YAS (DD99H1), West Yorkshire Archive Service; Papers relating to the rebuilding of Sowerby Church (1763–1764) Calderdale (STA215), West Yorkshire Archive Service; Records of the Calder and Hebble Navigation (1759–1767) RAIL 815, PRO; Collection of Canal Plans (1758–1774) ff 386 C9, Manchester Central Library; Records of the Coventry Canal (1768–1786) RAIL 803, PRO; Records of the Leeds and Liverpool Canal (1768–1769 and 1788–1799) RAIL 846, PRO; Records of the Oxford Canal (1769–1790) RAIL 855, PRO; Spencer Stanhope papers (1769) SpSt 13, West Yorkshire Archive Service, Bradford; Minute book of the Smeatonian Society (1771–1793) Institution of Civil Engineers; *Gentleman's Magazine*, 1771–1774; Records of the Lancaster Canal (1772, 1795) RAIL844, PRO; John Rennie's notebook re the Lancaster Canal (1772) Institution of Civil Engineers; Records of the Navigation Committee of the Corporation of London (1774–1787) MCFP303–9, Corporation of London RO; Records of the Thames and Severn Canal (1782–1819) TS, Gloucestershire RO; Records of the Forth and Clyde Canal (1785–1792) FCN, Scottish RO; Records of the Kennet and Avon Canal (1789) RAIL 842, PRO; Rennie papers (1789–1795) MSS 19834–19857, Manuscript Department, NLS; Published and MS plans by Whitworth, Maproom, NLS; Records of the Trent Navigation (1793) RAIL 829, PRO; Records of the Ashby Canal (1794–1797) RAIL 803, PRO; Canal reports and newspaper cuttings (1794–1799) L656.69, Newcastle Public Library; Records of the Herefordshire and Gloucestershire Canal (1795) RAIL 836, PRO; Canal Papers (1795–1799) M195, Carlisle Public Library; Records of the Grand Junction Canal (1796) RAIL 825, PRO; Will of Robert Whitworth (Prog. Apr. 1799) Borthwick Institute of Historical Research, York; F. S. Thacker (1914, 1920) *The Thames Highway*, 2 vols.; C. Hadfield *et al.* (1960–1977) *Canals of the British Isles*, 13 vols.; A. W. Skempton and E. C. Wright (1971–1972) Early Members of the Smeatonian Society of Civil Engineers, *Trans Newc Soc*, **XLIV**, 29–30; S. R. Broadbridge (1974) *The Birmingham Canal Navigation*, vol. I; L. J. Dalby (1986) *The Wilts and Berks Canal*; Humphrey Household (1987) *The Thames and Severn Canal*; R. Delany. (1992) *Ireland's Inland Waterways*; M. Clarke (1994) *The Leeds and Liverpool Canal*; G. Box, *The Construction of the Ashby Canal*, M.A. thesis, University of Leicester, 1997]

Publications
Robert Whitworth, Sr.:

1767. *A Plan of the Intended Navigable Canal from Birmingham to Aldersley near Wolverhampton*
1767. *A Plan of the Intended Navigable Canal from Coventry to Fradley Heath*

1767. *A Plan of the River Salwarp and of the Intended Navigable Canal from Droitwich to the River Severn*
1768. *A Plan of the Intended Navigable Canal from the Coventry Canal near Coventry to Oxford*
1768. *Reports, Estimates and a Plan of the Intended Navigable Canal from Lough-Neagh to Belfast*
1769. *A Plan of the Intended Navigable Canal from Taunton to Tiverton, Exeter and Topsham*
1769. *A Plan of the Navigable Canals now making … by James Brindley* (drawn by Robert Whitworth)
1770. *A Plan of the River Thames from Boulters Lock to Mortlake* (later editions extended to London Bridge, 1774, 1777)
1772. *A Plan of the Intended Navigable Canal from the Leeds and Liverpool Canal near Eccleston to Kendal*
1773. *Report, Plan and Profile of the Intended Navigable Canals from Waltham Abbey, to Moorfields and from Marylebone to Moorfields*
1780. *A Report, Plan and Profile of the Intended Navigable Canal from Bishops Stortford to Cambridge*
1781. *A Plan of the Intended Navigable Canal from the Coventry Canal near Griff to Measham, Oakthrop, Donisthorp, and Ashby Woulds,*
1783. *A Plan of the Intended Navigable Canal from Stroud to Lechlade*
1784. *A Plan of the Intended Navigable Canal from the Thames and Severn Canal at Kempsford to the River Thames at Abingdon*
1785. *Report to the Forth and Clyde Navigation relative to the Tract of the Intended Canal from Stokingfield westward, and Different Places of Entry into the River Clyde*
1785. *Report as the proposed Deepening of the Forth and Clyde Navigation to 8' and as to making Dolator Bog into a Reservoir*
1785. *Report respecting the South and North Lines for the Forth and Clyde Navigation westward of the River Kelvin*
1785. *Plan of the Great Canal from the Forth to the Clyde*
1785. *1st Report respecting bringing Supplies of Water through the Monkland Canal to the Forth and Clyde Canal*
1785. *Plan of the Intended Canal and River Navigation from Thrinkstone Bridge to Leicester* (with William Jessop)
1786. *A Plan of the Intended Navigable Canal from Melton Mowbray to Oakham and Stamford*
1786. *Plan for the Improvement of Leith Harbour* (n.d., *c*. 1786)
1786. *Report, and Estimates relative to the Enlarging of the Harbour of Leith, with Plans of a Bason above Leith Mills and between the North Pier and the Citadel*
1786. *Reports relative to Supplies of Water necessary to be taken into the Forth and Clyde Canal when 8' deep and the Junction of the Canal with the Monkland Canal*
1787. *Observations on Plans for Enlarging and Improving the Harbour of Leith*

1789. *Report to the Company of Proprietors of the Leeds and Liverpool Canal*

1789. *Report and Plan of the Proposed Navigable Canal from the Forth and Clyde Canal to Borrowstounness*

1789. *Plan of the Intended Navigable Canal from Andover to Redbridge surveyed in 1770 and revised in 1789*

1790. *Plan of the Great Canal from Forth to Clyde*

1791. *Report on the Practicability and Expence of making a Navigable Canal from Berwick to Kelso and Ancrum Bridge*

1792. *Plan of the Intended Navigable Canal from Ashby de la Zouche to the Coventry Canal near Griff*

1792. *Plan of the proposed Deviation Line of the Leeds and Liverpool Canal from near Colne to Wigan* (with William Whitworth)

1793. *Report and Estimate for improving the River Trent Navigation* (with William Jessop, who was almost certainly responsible for the printed text)

1793. *Plan of the Wilts and Berks Canal from Trowbridge to Abingdon* (with William Whitworth)

1794. *Plan of the Ashby de la Zouche Canal with the cuts or branches therefrom* (with William Jessop)

1794. *Report on the London Canal from Boulter's Lock to Isleworth on the River Thames* (with Robert Mylne)

1794. *Plan of the Proposed London Canal from Dachet to Isleworth* (with Robert Mylne)

1797. *Report on the Proposed Line of Navigation from Stella to Hexham, and from Hexham to Haydon Bridge upon the South Side of the River Tyne*

1797. *Report on the Proposed Canal on the North Side of the River Tyne*

1797. *Report on Mr. Dodd's projected line of navigation (i.e. from the city of Durham to the navigable part of the river Wear and from thence down the valley of team to the river Tyne with branches to Kyo West Houses)*

Robert Whitworth Jr.:

1794. *Report on a Canal between Edinburgh and Glasgow*

William Whitworth (see also Robert Whitworth, Sr.):

1810. *Plan of the Intended Severn Junction Canal*

1811. *Report on the Intended Severn Junction Canal from the Wilts and Berks Canal near Swindon to the Thames and Severn Canal at Latton*

1811. *Plan of the Intended Briston Junction Canal from the Wilts and Berks Canal near Wotton Basset to the Bristol Dock Company's Canal near Bristol Dock Company's Canal near Bristol*

Works

Robert Whitworth Sr.:

1774–1784. Thames Navigation, improvements, from Mortlake to Staines, Engineer

1785–1791. Forth and Clyde Canal, extension from Stockingfield to Bowling Bay on the river Clyde and from Hamiltonhill to Port Dundas, Chief Engineer, £140,000

1790–1799. Leeds and Liverpool Canal between Holmebridge near Gargrave and Henfield near Accrington, Engineer, £336,753

1794–1797. Ashby Canal, Engineer

1795–1799. Wilts and Berks Canal, Engineer

1795–1799. Dearne and Dove Canal, Engineer

1795–1799. Herefordshire and Gloucestershire Canal, Engineer

Robert Whitworth Jr.:

1790–1791. Forth and Clyde Canal, extension from Stockingfield to Bowling Bay on the river Clyde and from Hamiltonhill to Port Dundas, Sub-Engineer

1794–1797. Ashby Canal, Resident Engineer

William Whitworth:

1795–1810. Wilts and Berks Canal, Resident Engineer (1795–1799), Engineer (1799–1811), £255,262

1814–1819. North Wilts Canal, Engineer, £59,750

WILKINSON, John (1728–1808), ironmaster, the eldest of five children of Isaac Wilkinson (q.v.), ironmaster, was born at Little Clifton near Workington, Cumbria. Formerly a farmer working in a foundry for day wages, his father moved to Backbarrow, Furness, where he became 'potfounder' at the local blast furnace. There he prospered and was able to send his son to be educated at the Academy of Dr. Caleb Rotheran in Kendal until the age of seventeen. Wilkinson was then apprenticed to a Liverpool iron merchant for about five years before rejoining his family in Cumbria and acting as an ironmonger in Kendal. His father was now smelting iron ore at Lindale, and had a share in a blast furnace at Lowood. In 1753 he took a least on a charcoal furnace at Bersham near Wrexham. As an iron master Isaac took out a series of patents (No. 568, 1738; No. 675, 1751; No. 713, 1757; No. 723, 1758) relating to casting, crushing and bellows for blast furnaces.

As a result of the interests and abilities of his father Wilkinson would clearly have been well aware of developments in iron technology of the time. He joined his father in 1756 at Bersham but the following year sought work in the ironworks of the West Midlands, at Broseley, Coalbrookdale, where he eventually (1763) had sufficient capital to buy the works. By 1760 his father was in financial difficulties and in 1761–1762 Wilkinson and his younger brother, William (q.v.), took over the Bersham works and remodelled it. Their father moved to Bristol, there he died c. 1786.

The Bersham works gained a reputation for ordnance, which was on occasion exported to countries with which Britain was at war. At some stage capacity was added for the production of wrought iron and wire. In 1766 John built a new works at Bradley where in 1772 he was able for the first time to use his local pit coal for smelting ore without first coking it.

Wilkinson's first wife, Ann Maudesley of Kirkby Lonsdale, died in 1756 at the early age of twenty-three after a year's marriage, leaving him with a child, Mary (1756–1786), who was brought up by Mr. & Mrs. J. Clint of Shrewsbury. His second marriage, to Mary Lee (d. 1807) of Wroxeter, in 1763, brought him the additional capital with which he was able to take over the blast furnace and colliery at Broseley. It was here in 1774 that he took out his most important patent (No. 1063) for a 'New method of casting and boring iron guns or cannon', introducing the guide principle into machine tools, and boring cannon from solid castings.

He soon modified the design and developed a mill to enable it to bore other hollow articles, notably engine cylinders, the accurate manufacture of which was essential for the development of James Watt's (q.v.) improved steam engine; the first cylinder was made for Boulton and Watt in 1775. Details of the 'Bersham boreing mill' survive. Wilkinson was aware of the importance of Watt's invention and ordered for Broseley what was perhaps the first engine made commercially; it was installed in February 1776. This blowing engine was followed by an order for a forging hammer, which was not successfully developed until 1782–1783 using Watt's double-action rotative engine, at Bradley.

On 6 July 1787 Wilkinson launched another innovation, an iron barge, from Willey Wharf. Named the *Trial*, it was 70 ft. long and 6 ft. 8½ in. wide, made of riveted plates 5/16in thick, and was used for transferring castings on the Birmingham Canal. Earlier experiments on iron vessel construction at Bradley in 1786 apparently were unsuccessful. One assumes that Wilkinson was aware of the launch of an iron boat on the Foss at York in 1777 and may have had the idea in mind for some time. It has been claimed that he or his father built a vessel—abandoned in Helton Tarn—at Wilson House, Lindale, in the 1740s but confirmatory evidence is lacking. The *Trial* was a success, not least because of the shallow draught found possible with iron. Wilkinson built further canal and, less successfully, river barges.

Wilkinson's inventiveness continued with a patent for rifled ordnance (No. 1694, 1789) and 'A New method of making lead pipes' (No. 1733, 1790). This was, next to his boring mill, his most important invention and proved very profitable to him. His work attracted a European audience, and his brother, William, took the lead in transferring their technology overseas.

Wilkinson continued to expand his business activities, in 1780 starting a new works at Snedshill—it was sold c. 1793—and purchasing the 500 acre Brymbo Hall Estate in 1793, rich in coal and iron stone. The purchase brought to a head friction with his brother, William, who was a shareholder in Bersham. The result was for Bersham works to be sold, with John becoming sole owner. He subsequently built a new works at Hadley between 1800 and 1805.

Wilkinson's business success enabled him to purchase the estate of Castle Head on Morecambe Bay around 1779, adding further property to it in the 1790s. He died aged eighty on 14 July 1808 at his house at Bradley.

Wilkinson, whose brother-in-law was Dr. Joseph Priestley, was well respected by his contemporaries. James Watt Jr. was trained at Bersham although they subsequently fell out. He was High Sheriff of Denbighshire in 1799 and a member of the Society of Arts from 1787–1803. His second marriage was childless but he had three illegitimate children by Ann Lewis: Ann (b. 1802), Johanna (b. 1805) and John (b. 1806). Despite leaving precise instructions in his will regarding their inheritance the will was challenged by his nephew, Thomas Jones, and the legal process did much to dissipate his fortune. Although not a civil engineer his inventions played a significant part in the development of iron technology in the late eighteenth century, paving the way for its use as a structural material.

MIKE CHRIMES

[J. Wilkinson (1870) *A Biographical Notice of the late John Wilkinson*; J. Randall (1879) *The Wilkinsons*; W. H. Bailey (1886) *The First Iron Boat and its Inventor, John Wilkinson*; H. W. Dickinson (1914) *John Wilkinson, Ironmaster*; E. A. Forward (1924) The early history of the cylinder boring machine, *Trans Newc Soc*, **5**, 24–38; W. H. Chaloner (1951) John Wilkinson, ironmaster, *History Today*, 1 May 1964; R. Barker (1987) John Wilkinson and the early iron barges, *Journal of the Wilkinson Society*, **15**; *Wilkinson Studies* (1991–1992) **I–II**]

Works

1761–1762. Bersham Ironworks, remodelling
1766. Broseley, ironworks
1777–1780. Snedshill, ironworks, Oaken gates
1793. Brymbo Ironworks
1800–1805. Hadley Ironworks

WILKINSON, William (1738–1808), iron master, was the third son of five children of Isaac Wilkinson (q.v.). Although active in Britain, where he was in partnership with his eldest brother John (q.v.) at Bersham near Wrexham from 1763, he is best known for his role in introducing British technology to the continent. In late 1776 he emigrated to France to set up a cannon foundry for the French King. Work began in 1777 at Indret, near Nantes, involving air reverberating furnaces remolten iron and a green sand foundry, and was completed by 1780. In 1777 he also became involved in the proposal of the French engineer, Perier, to supply Paris with water from the Seine; this proposal received Royal approval in 1778. The Wilkinsons and some English friends provided capital for the scheme and obtained an order for 40 miles of cast iron pipe as well as supplying the first Boulton and Watt type steam engine in France in 1780. Both brothers attended a banquet to commemorate the opening on 14 July 1786.

While this work proceeded Wilkinson was involved elsewhere in France. He had visited Le Creusot, in 1780, a minor iron producing centre, but a valuable site because of local availability of coal and the following year he was engaged to remodel the works on British lines using coke for smelting. For around three years he personally directed operations and subsequently acted as consultant, probably until revolutionary turmoil made his stay impossible. At his peak he was earning 72,000 livres (£2,500) p.a. In that period 8 million livres was spent and four blast furnaces, five kilns, a foundry and a cannon boring mill were erected, as well as a Watt steam engine. The coke furnace he set up and had operational by 1785 was the first in France. By 1789 five steam engines were operation. Some of the earliest iron rails on the continent were laid in the works. Most of the ironwork and plant was supplied from Bersham. The works were not, however, a success at this time.

In 1782 Count von Reden, a Prussian aristocrat, visited English ironworks and gathered much information. As a consequence of this contact, John Wilkinson supplied a steam engine to the Prussians. In 1787–1790 von Reden attempted to smelt iron in a blast furnace of a Wilkinson type using charcoal rather than the coke for which it was designed. He contacted the Wilkinsons and asked for their help and in 1789 William visited Germany to introduce the use of smelting iron with coke there. During his stay at Friedrichshuette his efforts were successful. Reden returned to England with him and the Wilkinsons then provided plans for the erection of two English style blast furnaces for Silesia. William Wilkinson is generally believed to have invented the foundry cupola which was alluded to in his brother's patent of 1794.

There was increasing friction between John and William from around 1787 when William returned from France. On his return William lived at Plas Gronow, Wrexham, as well as purchasing property near his brother in Cartmel. In the early nineteenth century he became active in road improvements in the area. It has been suggested that he lacked John's originality and management ability although he appeared to get on better with Boulton and Watt than did John. Matters came to a head when John purchased Brymbo ironworks which could be seen as a threat to their joint concern at Bersham. In the consequent arbitration, Bersham was auctioned off in late 1795 and bought by John. At that time William disclosed to Boulton and Watt John's piratic dealings in the supply of engine parts and in November 1795 supplied John Watt Jr. with drawings of the Bersham equipment.

In 1791 William married Elizabeth Kirkes, a daughter of James Stockdale of Cartmel. They had two daughters, Mary Anne (b. 1795) and Elizabeth Stockdale (b. 1799). John Watt Jr. was appointed one of the guardians of the daughters in Wilkinson's will and Mary Anne married Matthew Boulton's son, Matthew Robinson Boulton. William Wilkinson died in Wrexham and was buried on 5 March 1808 in the Dissenters' Burial Ground there.

MIKE CHRIMES

[J. Wilkinson (1870). A biographical notice of the late John Wilkinson; J. Randall (1879). *The Wilkinsons*; W. H. Bailey (1886) *The First Iron Boat and its Inventor, John Wilkinson*; H. W. Dickinson (1914) *John Wilkinson, Ironmaster*; W. H. Chaloner (1951) John Wilkinson, ironmaster, *History Today*, 1 May 1964; R. Barker (1987) John Wilkinson and the early iron barges, *Journal of the Wilkinson Society*, **15**; *Wilkinson Studies* (1991–1992) **I–II**]

Works
1777–1780. Indret Ironworks
1777–1786. Paris Waterworks
c. 1780–1788. Le Creusot Ironworks

WILSON, Alexander, General (1776–1866), engineer, was born in Edinburgh on 27 February 1776, the eldest son of James Wilson, a smith. He was educated at Canongate High School before his education was interrupted in 1784 when his father was recruited by Charles Cameron to work in Russia. Cameron was the architect to Catherine the Great and had an enormous quantity of work in hand; sixty to seventy workers were recruited at that time, including William Hastie (q.v.).

While his father superintended the ironwork of the Cameron Gallery at Tsarskoe Selo, Alexander continued his education with Sir Robert Graham acting as his tutor for some lessons and at the same time attending a Russian school. Perhaps more importantly Hastie provided instruction in architectural and technical drawing. At the age of fourteen his formal education ceased and he joined his father as a draughtsman at the government small arms factory at Sisterbeck, in Finland. There he spent some of his leisure time in studying mathematics; this love of learning continued throughout his life and he built up a large private library.

In 1795 both Alexander and his father were given officer rank in the Russian Mining Corps. Subsequently, in 1800, Alexander became Secretary to Charles Gascoigne (q.v.) and in 1803 his father and the remainder of his family moved to Kolpino, where Gascoigne was building the works; James became assistant to the director of works there. In 1806 Gascoigne died and, in recognition of his talents, Alexander became director of works at Kolpino and at Alexandrosk in succession to Gascoigne. He probably was responsible for supplying the ironwork for the outside dome of Kazan cathedral, which was completed after Gascoigne's death and was one of the earliest iron domes (1806–1811), with a 17.7 m diameter.

Having served a long apprenticeship, Alexander was now able to display his own engineering ability. At Kolpino he rebuilt the existing sluices and waterwheels and designed new ones, as well as introducing steam power. At the works he built several steam engines, as well as supplying marine engines for other vessels. At both Kolpino and Alexandrosk Wilson was responsible for the planning of the small towns associated with the works, designing not just the workers housing, but schools and, at Alexandrosk, a foundling hospital with library, wash houses, etc. The Alexandrosk complex comprised all nature of textile mills and associated machines and even a factory for the production of playing cards. The management of these works was carried out by his younger brother, Lewis (d. 1847). Both these industrial complexes displayed features of structural engineering interest. The flax mill at Alexandrosk (1812) was a six floor rectangular building 56 ft. wide with cast iron columns at 14 ft. centres spanned by cast iron beams and brick jack arches, with a cast iron framed roof; it was re-roofed with a combination of 56 ft. span arched ribs and wrought iron trusses in 1845. This system had been used earlier for a 42 ft. span workshop roof at Kolpino in the late 1830s.

Wilson's work in Russia was not confined to his government factories. He reported on various Russian ports and other engineering works at Archangel, Nicolaeft and elsewhere, and managed two large cotton mills on behalf of a private consortium, its partners including Baron Stieghetz and Count Nasselrode. His knowledge of engineering was broadened not just by reading but by successive foreign trips. He visited Britain six times, meeting Telford on his first trip of 1814, and communicating regularly with him and other British engineers.

He was elected a Member of the Institution of Civil Engineers in 1824 and sent various communications on Russian engineering to Telford and to the Institution, including in 1818 details of Kulibin's timber proposal for bridging the river Neva and the construction of Peter the Great Monument. In 1830 he spoke at an Institution meeting on details of maps of St. Petersburg. There was a reciprocal succession of British visits to his works and in 1837–1838 William Fairbairn was consulted about his mills. Earlier, Matthew Murray Jr. had come to St. Petersburg on behalf of his father to install steam engines and in 1824–1825 a massive 400 ton hydraulic chain testing machine at Kolpino.

Wilson knew five successive Russian rulers personally and was held in high regard. He was also popular among his workforce. He did not finally retire on full state pension until 1860, by which time he held the rank of Engineer General in the Admiralty Engineering and Building Department. He died at Alexandrosk, aged nearly ninety, on 25 February 1866 and was buried in a family grave at Kolpino. A portrait hangs in the Institution of Civil Engineers, with his works in the background.

MIKE CHRIMES

[Membership records, ICE archives; A. Wilson (1818) Monument erected in 1782 at St. Petersburg to the memory of Peter the Great, and description of the proposed wooden bridge over the river Neva, OC 157, ICE archives; J. G. James collection, ICE archives; C. L. G. Eck (1841) *Traite de l'Application du Fer*, 54–55, plate 41; Obituary (1870) *Min Proc ICE*, **30**, 461–465; A. L. Punin (1981) *Arkhitektunie Pamyatniki, Petersburg*, 64–65; J. G. James (1983) *The Application of Iiron to Bridges and other Structures in Russia to about 1850*]

WILSON, Daniel (1790–1849), chemist and engineer, was born in Glasgow in 1790, the son of John Wilson and Margaret MacInness; he had a sister, Ann (Davidson). In 1810 he moved to Dublin as a manufacturing chemist, possibly contributing to the *Annals of Philosophy*. While there, he took out three patents relating to chemical processes, Nos. 4029 (1816), 4095 (1817) and 4106 (1817). His patent for refining sugar (No. 4095), which he further developed in 1818 (No. 4220), was adapted by Severn King and Company. This factory was destroyed by fire and the Insurance Company's refusal to pay the loss became a cause celebre of the time.

Wilson had evidently acquired some knowledge of the gas industry as his patent No. 4106 dealt with gas lighting. He was introduced by Major Taylor of Dublin to Aaron Manby (q.v.) to the Horseley Ironworks, possibly with a view to their supplying his equipment to the gas industry. Horseley began supplying equipment to the London-based Gas Light and Coke Company around 1817 and Wilson moved to London to act as Manby's agent around that time; in 1819 he is reported as crossing to Paris with Manby with a view to supplying the embryo French gas industry. He subsequently became involved with Manby in the establishment of the 'Compagnie d'Carriage par de gaz hydrogen' (1822), Charenton ironworks, and, subsequently, Le Creusot ironworks.

From 1816 the French were showing a growing interest in gas lighting and in 1822 Manby and Wilson took out a patent and in 1825 set up a gasworks at Ternes, on the outskirts of Paris, with a third partner, Jean Henry, who was also involved at Charenton. Of the 750,000 francs share capital, 500,000 was provided by their bankers and the remainder allocated to the partners, partly as a free allocation for their technical expertise. Henry had withdrawn, by 1827 but for Manby and Wilson it proved a tremendous investment; by 1830 the Company had absorbed two other gas concerns in Paris. With Henry's departure, Wilson became increasingly prominent, Manby moving to Belgium. At the time of his death Wilson's shares were valued at 1 million francs.

Manby and Wilson began the development of the Charenton ironworks in 1822, importing many English workers and setting a new standard for iron manufacture in France. The enterprise was soon overtaken by the ambition of the partners who, in January 1826, took over Le Creusot works formerly operated by the Wilkinsons (q.v.). Production was transferred to Le Creusot from 1827 with a view to closing Charenton but as a result of the 1830 Revolution and the associated financial crisis, Manby and Wilson's interests at Charenton and Le Creusot were sold off at a loss. Their company was declared bankrupt in 1833 and sold to the Schneiders in 1836. Another joint enterprise seems to have met with similar lack of success.

It would appear that in 1819 either Wilson or Manby persuaded (Admiral Sir) Charles Napier to support their attempt to introduce steamships on the river Seine, work being carried out at Charenton. The first iron boat, the *Aaron Manby*, was built at Horseley in 1821. Their company, the 'Societe des Transports acceleres par eau', had three (wooden) steamships on the Seine by 1822, when they distanced themselves from it and set up the 'Compagine des bateaux a vapeux en fer entre Paris et Havie', with the *Aaron Manby* as their first vessel. Ultimately, this company owned four iron vessels, two probably built at Charenton, which also began steam engine production. Nevertheless, it would appear that Wilson's talents were as a businessman as his Institution obituary, presumably written by Charles Manby who would have known him well, remarked that 'during the whole of his career he never acquired more than a superficial knowledge of mathematics, mechanics, or chemistry'.

Wilson was an early member of the Institution of Civil Engineers, joining in 1820 while resident at 6 Mount Place, Mile End; he provided its library with early works relating to the French engineering industry. He died suddenly of a heart attack on 2 September 1849 at his country estate, Chateau Econblay, Seine et Marne. In 1835 he had married Antoinette Casenave (c. 1807–1843) and left his estate of 1,419,353 francs to his surviving children, Margaret (b. 1836) and Daniel (b. 1840), his brother-in-law Antoine Casenave being appointed their guardian.

MIKE CHRIMES

[ICE membership records, ICE Archives; A. De Barante (1828) Rapport sur les etablissements du Creusot, *Min Proc ICE*, **IX**, 1849–1850, 102–103; S. Everard (1949) *History of the Gas Light and Coke Company*; W. H. Chaloner and W. O. Henderson (1953–1955) Aaron Manby, builder of the first iron steamship, *Trans Newc Soc*, **29**, 77–91; W. H. Chaloner and W. O. Henderson (1956–1957) The Manby's and the Industrial Revolution in France 1819–1884, *Trans Newc Soc*, **30**, 63–75; W. O. Henderson (1967) *The Industrial Revolution on the Continent*, 98–100; E. H. Wilson (c. 1980) A Daniel come to judgement, typescript, ICE]

Works

1822. Charenton Ironworks, Partner
1825. Ternes Gasworks, Paris, Partner
1826. Le Creusot Ironworks, development, Partner

WILSON, John (1770–1850), contractor, was the eldest son of a farmer from Inverneill in Argyll. He trained as a mason and may have worked on the nearby Crinan Canal in the 1790s. The name is a common one, indeed one of the earlier references to an 'ingineer' [*sic*] relates to the **John Wilson** who carried out work on the Tone Navigation in Somerset in 1699. Around 1800 confusion is possible as three John Wilsons were active in civil engineering. One **John Wilson**, who died 17 April 1840 at Kilmarnock, was surveyor of the Kilmarnock and Troon Railway (see: *JESSOP, William*), and carried out work on Ayr Harbour in the 1820s. The better known Cumbrian mason, also John Wilson (1773–1831 (see below)), was the partner of John Simpson (qq.v.). Both worked on the Caledonian Canal, this subject at the Clachnaharry (East) end of the Caledonian Canal as a mason, and it was there that he married Martha Brough, the sister-in-law of John Cargill (q.v.).

In 1813 Thomas Telford (q.v.) made his second visit to Sweden, accompanied by Wilson and Simpson's son, James. Wilson was given a two year contract to work on the Gotha Canal at a salary of £222 15s p.a., together with travelling expenses. He was based at Berg, the site of the sea lock into the Baltic, and acquitted himself well, being presented with a gold snuff box for his work there before returning to Scotland in the autumn of 1815. Wilson undoubtedly hoped to find work in Britain so that he could be with his family but none was available and, reluctantly, he returned to Sweden in May 1816, but on a higher salary. He returned to Scotland in the winter of 1817 and refused to return to Sweden, despite Von Platen's best efforts.

Wilson next became involved with Edinburgh's waterworks, specifically the Glencorse (compensation) reservoir where he was initially appointed resident engineer in January 1820 at a salary of £120 p.a., plus a house and expenses. As superintendent and keeper for Edinburgh Waterworks, he remained there for twenty-three years.

In 1843 work was being undertaken to repair the Caledonian Canal and George May, engineer for the canal, who had presumably known Wilson as a child, recommended him as Inspector of Works for the repairs at Gairlochy and in 1844 he moved to Gairlochy House.

Wilson died on 7 May 1850 at Merivil Cottage, Midcalder, to where he had returned, and was buried under the memorial to his eldest son at Glencorse. He had seven sons and five daughters, of whom three sons died at birth. Two of his sons, Edward (1820–1877) and Allan (1820–1897), became distinguished civil engineers.

MIKE CHRIMES

[Telford mss, ICE Library; J. Colston (1890) *Edinburgh and District Water Supply*; A. D. Cameron (1972) *The Caledonian Canal*; J. E. Russell (1982) The Wilsons: a line of engineers, *Ind Arch Rev*, **6**(3), 224–234; L. Stromback (1993) *Balzar von Platen, Thomas Telford och Gota Kanal*]

Works

1813–1817. Goth Canal, Berg Locks, superintendent of works
1819–1824. Glencose dam and reservoir, Edinburgh Waterworks, resident engineer, *c*. £32,000
1843–1847. Caledonian Canal, Gairlochy repairs, resident engineer, *c*. £21,000

WILSON, John (*c*. 1772–1831), contractor, was one of Thomas Telford's trusted workmen, originating at Dalston, Cumbria. He assisted John Simpson (q.v.) on his masonry contracts on the Ellesmere Canal at Pontcysyllte and then went with him to work on the Caledonian Canal from 1804 where he acted as Simpson's agent, and possibly later partner, on the masonry work at the west end of the Caledonian Canal at Corpach. Simpson and Wilson also carried out contracts on the Highland Roads.

At the end of March 1820, Messrs. John Straphen and John Hall (qq.v.), who carried out several contracts on the Holyhead Road, gave up their contract for the masonry of the Menai Bridge and John Wilson took over and successfully completed it. His success meant that in December 1826 Telford had no hesitation in recommending him as contractor on the Birmingham and Liverpool Junction Canal.

It was while working on the contracts of this canal that Wilson died at Market Drayton on 14 January 1831, aged fifty eight, and his two sons completed the work. Rickman remarked a short time later, in a letter to his wife of 12 September 1832, on 'Mr. Wilson, to whose practical remarks Mr. Telford is fond of acknowledging frequent obligation'. Among Wilson's pupils was David Lennox, the distinguished Australian bridge builder.

MIKE CHRIMES

[Telford mss, ICE Library; Rickman correspondence, Gibb collection, ICE Library; Notebooks on work performed at Menai Bridge, ICE archives; *Caledonian Canal Commissioners Annual Reports* (1804–); *Highland Road and Bridges, Annual Reports* (1809–1812); W. A. Provis (1828) *An Historical and Descriptive Account of the Suspension Bridge constructed over the Menai Strait*; C. Hadfield (1969) *Canals of the West Midlands*, 1965; A. D. Cameron (1972) *The Caledonian Canal*; L. Stromback (1993) *Baltzar von Platen, Thomas Telford och Gota Kanal*]

Works

(See: *SIMPSON, John*)

1820–1826. Menai Bridge, contractor
1827–1831. Birmingham–Liverpool Junction Canal, contractor, £198,000, tender

1830–1831. Birmingham–Liverpool Junction Canal, Church Eaton–Autherley section, contractor

WILSON, Thomas (1751–1820), schoolmaster and engineer, was baptised in Sunderland on 17 February 1751, the son of Thomas Wilson and his wife, Dorothy Woodmass; he was the younger son and the fifth child of a family of eight. As a young man he maintained a small school in Sunderland and in 1776 he married Frances Bruce (1738–1811), also of Sunderland.

Wilson's engineering career seemingly began in 1792 when he became associated with Rowland Burdon (q.v.) in connection with the building of the Wearmouth Bridge. Following the passing of the Act of Parliament for the bridge's construction in that year, Wilson became responsible for the supervision of its building. The subject of a patent taken out by Burdon, the 236 ft. span iron bridge—two and a half times as long as the earlier Coalbrookdale structure—was completed in 1796. The ribs of the arch had been positioned by supporting them on scaffolding, floated into position, for which method of construction Wilson has been given full credit; he had earlier put forward a proposal for supporting them by means of cables 'passed ten times across the river with proper length to secure the same at each end'. Wilson perpetuated his involvement with a small copper plate embedded in the bridge's masonry; it read: 'This was laid by Mr. Thos. Wilson, Engineer, A.D. 1794'.

In 1800 Wilson entered in a competition a design for a bridge over the river Thames. The spans were to have been 220, 240 and 220 ft. and bore strong similarities to the Wearmouth bridge. This project was shelved the following year, probably a result of the submission by Thomas Telford (q.v.) and James Douglass of the better-known design comprising a single span of 600 ft. although it, too, did not proceed. Also in 1800 he submitted a drawing and an estimate for widening the masonry bridge at Newcastle from 21½ ft. to 33½ ft. by forming additional cast-iron arches to both faces of the bridge, basing them on the existing piers and so obviating the need for extensive river works; his plans were not accepted and the widening was carried out in masonry by David Stephenson (q.v.).

At the same time, Wilson was responsible for an 82 ft. span bridge erected at Spanish Town, Jamaica. Like the Wearmouth bridge, it was cast in Rotherham where it was erected, dismantled, sent to Hull and then shipped to the West Indies, there to be erected in 1801. A more simple design than the Wearmouth bridge, it comprised four arched ribs with two longitudinal members rather than three.

The year 1802 saw Wilson involved in the building of a 40 ft. span estate bridge at Stratfield Saye and also, with Burdon, concerned with an application for a patent relating to improvements in Burdon's earlier method of arch construction. The modifications comprised principally the omission of the wrought-iron straps used at Sunderland and the substitution of cast-iron dowels secured by wrought-iron wedges. Flat perforated bars replaced the former tubular spacers used to connect the ribs transversely.

Wilson used the new method in the construction of the 180 ft. span bridge at Staines, opened in 1803. Problems soon became apparent and cracks in the ironwork were blamed on the abutments, supposed to have moved as a result of the thrust of the arching. Both John Rennie (q.v.) and Charles Hutton (q.v.) confirmed the inadequacies of the abutments and the former recommended that their rebuilding was necessary. The bridge was replaced by a hybrid structure in 1807 and was eventually replaced again in 1832.

Wilson had also experienced an earlier failure when in 1803 he had become responsible, through Burdon's influence, for a new bridge at Yarm to replace the fifteenth century masonry structure. The new bridge was to be an exact replica of the Staines bridge—problems there had not then become apparent and construction began in September 1803. Modifications to the design were made during construction, among them the use of vertical struts in the spandrels in place of circles, but completion was effected in September 1805 although the southern approach was unfinished. On 13 January 1806 the ironwork collapsed, failure again blamed on inadequate abutments. All remnants of the bridge were removed and the original masonry bridge was widened.

In March 1805 Wilson submitted a design for an iron bridge at Dublin but his proposals, perhaps because of events at Staines, were not accepted and the bridge was rebuilt in masonry. He was, at the same time, putting forward proposals for a bridge at Boston in conjunction with Rennie, an association which was to create some acrimony in conflicting views on design. The bridge, over the river Witham, was of 80 ft. span and although completed at the end of 1806 was not opened until six months later, a result of cracks having become of some concern. The problems led to a long dispute regarding responsibility and would seem not to have been completely resolved. Rennie was asked for his views on ensuring the safety of the structure and he suggested that the circle of the arch be plated; after this rectification the bridge survived until 1913 and its steel replacement is still supported on Rennie's abutments.

Wilson's last bridge, built jointly with Henry Provis (q.v.), was that at Newport Pagnell where a 60 ft. structure was erected to replace an earlier bridge. It was built to Wilson's revised patent and incorporates a series of circles in its spandrels, a feature first used at Wearmouth. It was opened in September 1810.

Wilson, later described as town surveyor, was said to have superintended the re-roofing of Sunderland church at the beginning of the century; at the time of his wife's death his occupation was given as 'gentleman'. He had lived during the later years of his life in the specially-built Bridge House at the south end of the Wearmouth bridge and it was there that he died, his burial taking place on 20 April 1820 when 'his remains were followed to the grave by a large body of Freemasons, attired in deep mourning'.

J. G. JAMES and R. W. RENNISON

[Documents regarding Tyne Bridge widening (1801) 612/345, Tyne and Wear Archives; *Durham County Advertiser*, 22 Apr. 1820; J. W. Summers (1858) *The History and Antiquities of Sunderland*, 174; J. G. James (1978–1979), Thomas Wilson's Cast-Iron Bridges 1800–1810, *Trans Newc Soc*, **50**, 55–72; J. G. James (1986) *The Cast Iron Bridge at Sunderland (1796)*; *Clayton and Gibson Collection*, DCG5/49–104, Durham County RO; Skempton (1987)]

Publications, etc.

1800. *A Design for widening Newcastle Bridge with Iron Arches*
1800. *Mr. Wilson's Estimate or Specification of a Cast Iron Bridge over the Thames*
1801. *Observations by Mr. T. Wilson* [on the design of a cast-iron bridge of three arches over the Thames], 13 April
1802. Patent No. 2635: *Certain New Methods of Uniting, Combining and Connecting the Metallick Patent Blocks of the said Rowland Burdon for the Construction of Arches* (with Rowland Burdon), 23 July

Works

1792–1796. Wearmouth bridge, Sunderland, 236 ft. span, £41,500
1800–1801. Spanish Town, Jamaica, 82 ft. span bridge, £4,000
1802. Stratfield Saye bridge, 40 ft. span
1802–1803. Staines bridge, 180 ft. span, c. £5,000
1803–1806. Yarm bridge, 180 ft. span, £5,560 (ironwork)
1800–1807. Boston bridge, 72 ft. span, £4,155 (ironwork);
1808–1810. Newport Pagnell bridge, 60 ft. span, £1,744 (ironwork)
n.d. Greatford bridge, Lincs, demolished c. 1960

WING family (fl. 1745–1852) members included three generations of land agents and surveyors who were involved in the drainage of the fens in Cambridgeshire, Norfolk and Lincolnshire.

John Wing (1723–1780) was baptised on 3 November 1723 at Pickworth, Rutland, the eldest of the five children of Tycho Wing (1696–1750), land surveyor and mathematician, and Eleanor Peach. The first known map by John Wing is dated 1745 and he practised with his father in 1747 and 1748. In the following year, he published a map of the Bedford Level, the Lincolnshire Fens and Wisbech North-side, in which he showed the drainage of the land. Wing then went on to be employed as steward for the 4th and 5th Dukes of Bedford at Thorney, Cambridgeshire, (1750–1760 and 1764–1780) and for the Dukes' Huntingdonshire estates at Thornhaugh (1750–1761). Regarded as an excellent administrator, he was sent to let the Bedford lands in Hampshire from 1761 to 1764. While at Thorney, Wing introduced windmills to pump water into the rivers and supervised drainage work; he was in charge of works on the North Level of the fens from 1762 to 1764 and in 1769 commissioned from James Brindley, John Golborne, John Smeaton and Thomas Yeoman (qq.v.) surveys of the river Nene from Peterborough to the sea, with the aim of improving drainage and navigation. He was then involved in deciding on the line of the cut to be constructed, as set out by Golborne, Richard Dunthorne and Langley Edwards (qq.v.), and in 1773 was appointed receiver to the works; the cut, 'Kinderley's Cut', was completed in 1774. In 1777, Wing collaborated with Golborne again when they and Golborne's nephew, James, surveyed the harbour at King's Lynn.

Wing married Anne Sisson (d. 1800) on 3 June 1751 at Cottesmore, Rutland, and they had two sons, John and William. The father made his will at Glinton, Northamptonshire, on 24 June 1780 and was buried there on 17 October. His will was proved on 3 November and he left freehold and copyhold estates in Cambridgeshire, Leicestershire, Northamptonshire and Rutland.

John Wing Jr. (1752–1812), the elder son of John and Anne, was born in March 1752. He married Katherine Elger of Peterborough on 30 August 1779 at Ingoldsby, Lincolnshire, and then succeeded his father as steward to the Duke of Bedford at Thorney. He married again on 19 August 1784 at Ingoldsby, this time to Jane Ansell (d. 1824, aged seventy); they had six children. John continued his father's concern with fen drainage and in 1788 he debated with George Maxwell (q.v.) about the revenues and problems of the North Level of the fens. Wing published the results of his inquiry and argued in favour of temporary taxes to repair banks, support the works of the commissioners and to pay debts; Maxwell opposed the introduction of any new tax. In 1811 Wing's management of the Thorney estate was commended by William Gooch, who lamented that he did not superintend drainage and embanking over the whole level of the fens; his vigilant and never-ceasing care over the heightening and strengthening of the barrier bank on the north of Whittlesea Wash was later praised by Samuel Wells. Wing was also steward of the Thornhaugh estates and managed the Duke's land in north Bedfordshire from 1798 to 1803. He died on 3 April 1812 after attacks of asthma and apoplexy and was buried at Thorney Abbey on 10 April.

Tycho Wing (1794–1851), the youngest child of John and Jane, was born on 23 November 1794 at Thorney Abbey and was baptised there on 13 January 1795. He was admitted a pensioner at Pembroke College, Cambridge, on 28 March 1812. On his father's death, his brother, John, succeeded to the stewardship of the Duke of Bedford's estates at Thorney and Thornhaugh until Tycho was deemed competent to take over. This occurred in April 1817 and, like his father and grandfather, Tycho Wing was quickly thought very able and a good manager of the tenantry. Known as 'King of the Fens', his work with drainage was such that in 1833 the Duke of Bedford praised him: 'Wing has immortalized himself'.

Wing's concerns with fen drainage extended beyond the boundaries of the Duke of Beford's estates and he became a clerk to the Wisbech Canal company. In 1820, he considered John Rennie's (q.v.) plan for the drainage of the North Level and Lincolnshire fens and its practical application. His works to improve of the outfall of the river Nene, under an Act of Parliament of 1827, brought him into contact with Thomas Telford (q.v.). He was honoured when the lower part of the cut was named 'Tycho Wing's Channel' and he was presented with a silver epergne at a public dinner on the completion of the works in 1834. The Nene outfall and drainage of the North Level were discussed by him in a publication of 1837 and again in 1849. Wing married Adelaide Basevi (d. 1885, aged eighty-nine) at St. James's Church, Westminster, on 28 March 1828; they had eight children. He died on 26 December 1851 and was buried at Thorney Abbey on 2 January 1852.

SARAH BENDALL

[Russell Estate Papers (introductory material and other, including especially R5/4200, R76/61 (b), R76/92), Bedfordshire RO; Parish register of Glinton, Northamptonshire, Northamptonshire RO; Parish registers of Thorney, Cambridgeshire, Cambridgeshire RO (Cambridge branch); Records deposited by the North Level Internal Drainage Board, especially R76/61, R76/92, R77/38, Cambridgeshire RO (Cambridge branch); Will of John Wing (I), PROB 11/1071, PRO; *In the shadow of the Wings; [Bedfordshire] County Record Office Exhibition* (n.d.); J. Golborne (1791) *The Report of James Golborne … in the Improvement of the Outfall of the River Ouse …*; S. Wells (1830–1828) *The History of the Drainage of the Great Level of the Fens, called Bedford Level; with the Constitution and Laws of the Bedford Level Corporation*; *History, Gazetteer and Directory of Cambridgeshire* (1851) 658–709; E. Green (1886) *Pedigree of the Family of Wing of North Luffenham, etc.*; Skempton; S. Bendall (1992) *Maps, Land and Society: a History, with a Carto-Bibliography of Cambridgeshire Estate Maps c. 1600–1836*; Bendall]

Publications
John Wing (I):

1749. *Actual Survey of the North Level … also of Crowland, Great Porsand, and Part of South Holland … and of Wisbeach North-side*

John Wing (II):

1788. *Inquiry into the State of the Revenues of the North Level … and the Causes of the Distresses …*

during the last Twenty-Five Years. With an Inquiry into the Probable Means of Preventing the Like in Future

Tycho Wing:

1820. *Considerations on the Principles of Mr. Rennie's Plan for the Drainage of the North Level, South Holland etc. With a view to their Practical Application*
1837. *Memoir of the Nene Outfall and the North Level Drainage* [In] Life of Thomas Telford. 1838.
1849. *Considerations of Sir John Rennie's Plan for Improving the Navigation of the River Nene from the Sea to Peterborough*

WING, John (1756–1826), builder and architect, no relation of the above, was the eldest son of the leading Leicestershire mason and architect, John Wing. In the early 1780s he moved to Bedford following his father's acquisition of a quarry at Totternoe, Beds. He became the leading architect and builder in the county, acting as surveyor of bridges, and designing and building 'The Bridge' in Bedford, 1811–1813.

MIKE CHRIMES

[Colvin]

WINGROVE, Benjamin (fl. 1814–1839), was a road surveyor who tangled with J. L. McAdam (q.v.) and eventually emerged second best from the encounter. His father was a brewer and coal merchant of Bath who was also an amateur in roadmaking. Wingrove himself first published an article on roads in a local newspaper in 1808, drawing attention to R. L. Edgworth's (q.v.) ideas on the subject. He became a trustee of the Bath turnpike roads in 1810, and about 1814 also surveyor of his local parish roads. The Bath Turnpike Trust was of medium size, being responsible for 49 miles of road, which were subject to a high degree of wear and tear. The trustees however were numerous and opinionated, so that there was little effective management. Wingrove had published an article in 1815 in praise of the Bristol roads, so late in 1816 he and others invited McAdam, recently appointed General Surveyor there, to come to Bath to confer with their Trust. McAdam brought his son William from Scotland to help in surveying the roads and their report was ready for a meeting of the Bath trustees on 1st February. Amongst its recommendations was the appointment of a General Surveyor, which McAdam clearly expected would be offered to him, but Wingrove came forward as a candidate and at a meeting on 15 March was elected by fifty-five votes to fifty-three, with one abstention. He was instructed to work to McAdam's system, but he was already beginning to doubt its desirability. In his first report he stated his objections but agreed to give it a trial on the Bath eastern road; to ensure fairness he employed some of McAdam's men to do the work. He claimed later that the trustees agreed that the resulting road

was not durable and that he received their approval to proceed on his own system, which included stronger foundations, better stone and greater convexity. He also claimed that he was the first to introduce proper drainage and that McAdam subsequently copied him.

While it was accepted that the standard of his roads was generally good, he was less efficient than McAdam in keeping his accounts. By 1823 his opponents at the Trust had picked on this aspect in order to attack him. He was unable to produce vouchers for some transactions; it appeared too that payments had been made for stone whose quantities had not been properly checked. His son Antony was appointed Surveyor jointly with him in March, but at a general meeting in May, at which 32 trustees were present, a motion was carried which subjected Mr. Wingrove 'to our animadversion, but upon his explanations, we are inclined to attribute those irregularities to the great labour of his office, partly to domestic affliction, and principally to his trusting to others what he himself ought to have performed, but certainly not to any dishonourable or interested motive; and from the great improvements to our roads under his management, we do not feel inclined to say any more than merely to express our regret'.

Wingrove gave evidence to the Parliamentary Committee which was considering McAdam's third application for public payment for his services. It seems that he was principally interested in seeking an award for himself but had neglected to present it in the proper form, so that it was ruled inadmissible. In order to maintain the good opinion of the committee, which he still hoped might allow him to present his petition, his criticisms of McAdam were somewhat subdued, but McAdam's friends at Bath continued their attack. Early in 1826 they passed a resolution which called for an entire district to be maintained by McAdam's system to determine which of the two could provide sound roads at least cost. Two weeks later Wingrove and his son resigned and in March McAdam obtained the position which had been denied to him nine years earlier. His salary was £600 p.a., compared with the £350 which Wingrove had had.

Wingrove was appointed Surveyor of the Taunton Trust, which had about 100 miles of roads, in August 1819, and was able to reduce the statute labour on them by two-thirds and the tolls by one-third in the following four years. He had left office before 1836, when he was employed only by three trusts in Wiltshire.

He applied unsuccessfully for the County Surveyorship of Somerset in 1818, on the resignation of Robert Anstice (q.v.). At the time of the enquiry into his accounts in 1823 he had private contracts for repairing the roads over and at the ends of some County bridges, for which the County rather than the turnpike was responsible.

Wingrove is thought to have died in 1839.

P. S. M. CROSS-RUDKIN

[*Philosophical Magazine* (1818); Minutes of Evidence to Parliamentary Committee on McAdam's Petition (1823); Minutes of Evidence to Select Committee on Turnpike Trusts and Tolls (1836); Sir Henry Parnell, Bt. (1838) *A Treatise on Roads*; inf. ex. D. J. Greenfield]

Publications

1821. *Remarks on Bill ... for regulating Turnpike Roads, in which are Introduced Strictures on the Opinions of Mr. M'Adam, Report to the Taunton Trust*

Works

1816–1826. Bath Turnpike Trust, General Surveyor
1819–. Taunton Turnpike Trust, Surveyor
18??. Chippenham Turnpike Trust, Surveyor
18??. Bradford (-on-Avon) Turnpike Trust, Surveyor
18??. Trowbridge Turnpike Trust, Surveyor

WINSTANLEY, Henry (*c.* 1644–1703), engineer and engraver, was baptised at Saffron Walden on 31 March 1644, the eldest son of Henry Winstanley (d. 1680). In 1665 he was a 'porter' in the service of James Howard, third earl of Suffolk, employed on Suffolk's buildings at Audley End and when, early in 1666, Suffolk sold the place to Charles II, Winstanley was transferred to the King's service and became clerk of the works there and at Newmarket. Between about 1676 and 1688 he engraved and published a set of twenty-four plans and views of Audley End. Winstanley obtained a certain notoriety from the whimsical mechanisms with which he embellished his house at Littlebury in Essex; he was also the inventor and proprietor of a place of entertainment known as the Water Theatre at the 'lower end of Piccadilly'. One Edmund Winstanley was the owner of the Dungeness lighthouse in 1636.

When William of Orange decided to make Plymouth his main naval arsenal Trinity House agreed in 1694 to let Henry Whitfield build a lighthouse on Eddystone Rock. The rock was a reef of red granite 9 miles SSW of Rame Head; the rocks are 600 yd. long underwater but show only as three jagged ridges, the centre being 200 yd. long except at high tide. Whitfield appears to have made an agreement with Winstanley to carry out the actual construction. Winstanley had a personal interest in the challenge as one of the merchant vessels he had had an interest in, *The Constant*, had capsized on the rock.

Work was begun in the summer of 1696 with the boring of 12 holes into the hard rock to act as foundations. The solid base, 12 ft. high and 14 ft. in diameter was, after two years work, increased to a diameter of 16 ft. Towards the end of June 1697, during the war with France, Winstanley was carried off by French privateers but Louis XIV saw that he was returned immediately, observing that 'he was not at war with mankind'. During the years 1696–1698 Winstanley succeeded in fixing a solid foundation on which was raised a

Henry Winstanley

lighthouse polygonal in shape with the upper half an elegant extravaganza of ironwork. The candles were lit on 14 November 1698 but during the following winter it was found that the lantern, 60 ft. from the rock, was frequently swept by the sea. The solid pillar foundation was braced and clamped until in 1700 its diameter had increased to 24 ft., braced by copper or iron, and the light was raised 40 ft. An engraving of the completed building is given by John Smeaton (q.v.) as 'drawn orthographically' from a perspective view made by Winstanley himself.

Winstanley was so confident of his structure that he wished only to be on the rock when the worst storm could be produced. On 26 November 1703 he was in Plymouth when there was every sign of a gale blowing and so he took a boat to the rock in the Great Storm of 1703 and nothing more was ever seen of Winstanley or his lighthouse. His widow received a grant of £200 in 1707 and by a warrant of 25 June 1708 a pension of £100 a year. Smeaton suggested that an insufficient knowledge of cement was one cause of the failure of the lighthouse.

While working at Eddystone, Winstanley designed a pier at St. Agnes, erected 1699–1705; it, too, was destroyed by the sea.

A brief display concerning Winstanley's life, including his self portrait, can be seen in the Saffron Walden Museum.

A. B. GEORGE

[Smeaton drawings, Royal Society; J. Smeaton (1793) *A Narrative of the Building and Description of the Construction of the Edystone Lighthouse*; DNB; M. M. Oppenheim (1968) *The Maritime History of Devon*, 96–98; A. C. Todd and P. Laws (1972) *Industrial Archaeology of Cornwall*; Anon. (1977) Smeaton's Tower, *Maritime History*, **V**(2), 136–137]

Publications

1699. *Edystone Lighthouse: a Narrative of the Building*, engraving

Works

1696–1700. Eddystone lighthouse
1699–1703. St. Agnes Pier
169?. Spire, lantern and church clock, St. Mary the Virgin, Saffron Walden

WOOD, Nicholas, FRS (1795–1865), colliery viewer and civil engineer, was the eldest child of a family of three boys and a girl. Born at Sowermires, near Ryton, on 24 April 1795, he was the son of Nicholas Wood and his wife Ann, the daughter of Robert Laws. Wood Sr. was a tenant farmer who, in addition, acted as agent for a member of the coalowning Simpson family at Bradley.

Wood's education began under his uncle and continued at the village school of Crawcrook. As a result of the influence of Sir Thomas Liddell, later Lord Ravensworth, he came to be employed at Killingworth colliery in 1811, Liddell being a principal partner there. It was at Killingworth, under Ralph Dodd, that Wood was initiated into the profession of colliery viewer and became associated with George Stephenson (q.v.).

Following the Felling colliery disaster of 1812, in which ninety men and boys lost their lives, Stephenson began experimenting with the design of a safety lamp, work in which Wood was also involved. The lamp was first demonstrated on 5 December 1815 at a meeting of the Literary and Philosophical Society in Newcastle, Wood taking an active part in the meeting and providing 'proofs and explanations ... of the merits and details of the invention' for which he had made the working drawings. Wood also defended Stephenson in the controversy which resulted from the claims of Sir Humphry Davy in respect of his contemporaneous invention of a similar lamp.

Continuing an interest which had begun earlier, in October 1818 Wood carried out experiments regarding horse and locomotive traction and resistance in relation to rail transport, together with tests on the durability of rails and fuel consumption. His work was undertaken in conjunction with Stephenson, who had by this time apprenticed his son, Robert, to Wood. He also came to be a party in the construction of Stephenson's first railway venture, the Hetton Colliery railway. Built to carry coal to the river Wear at Sunderland, the 8 mile long line was opened in 1822 and was notable for the fact that it employed locomotives as well as stationary engines

Wood's railway interests continued in his association with Stephenson regarding the Stockton and Darlington Railway. In 1821 he accompanied Stephenson on his initial inspection of the line and was present at the subsequent meeting with Edward Pease at Darlington. The following year he became involved in controversial correspondence with Benjamin Thompson (q.v.) on the use of stationary or locomotive engines. The appointment of Stephenson as engineer in 1823 was followed by the successful completion of the line in 1825. In 1827 Wood continued his railway experiments with the use of steel springs on a locomotive at Killingworth and also made a trial of a wrought iron tyre at Bedlington.

In 1827 Wood married Maria, daughter of Collingwood Forster Lindsay of Alnwick. When, in the same year, a railway from Liverpool to Manchester was being planned, Wood was called upon to give evidence before committees of both Houses of Parliament, perhaps involved as a result of the publication in 1825 of his *Practical Treatise on Railroads*, a pioneering classic on the subject. Wood's knowledge of locomotive practice also led to his appointment in 1829 as one of the three judges at the Rainhill Trials, held to establish the type and make of locomotive to be adopted by the company; the competition was won comfortably by the Stephensons' *Rocket*.

Wood published revised editions of his book in 1831, 1832 and 1838, chronicling the expanding use of railways. The first edition had noted that in 1825 only two public lines had been completed but by 1831 Wood was able to write that it had been proved that locomotives were 'capable of effecting a rapidity of transit greater than any other practical mode of travelling' and he had no doubts as to future development. He was elected a member of the Institution of Civil Engineers in 1829, one of his sponsors being Thomas Telford (q.v.).

Wood became a director of the Newcastle and Carlisle Railway, its engineer Francis Giles (q.v.). By 1832 differences led to the departure of Giles and the formation of a committee of management comprising George Johnson (q.v.), Benjamin Thompson (q.v.), and Wood to superintend the completion of the eastern section of the line, with John Blackmore (q.v.) as engineer. This railway came to be associated with two others, the Blaydon, Gateshead and Hebburn and the Brandling Junction. Wood became a director of the former on its establishment in 1834 and was appointed engineer to the latter in 1836, following the initial survey undertaken in conjunction with the Stephensons. He became a director of the Newcastle and Berwick Railway in 1845.

Wood was one of the leading colliery viewers in north east England and, in addition, became a leading mineowner. His first such speculation was with Stephenson and Michael Longridge at Bedlington and it was followed by a part-interest at Hetton; he moved from Killingworth to Hetton Hall in 1844 to assume the management of the colliery company there. He became a partner in John Bowes and Partners, formed in 1838 and renamed in 1847, which owned some 12 collieries, including Killingworth; it also embraced the making of coke and iron and the building of ships. Wood was also the owner, either fully or in part, of at least ten further collieries, including one in Germany, and during his career he became a partner in the glass manufacturer, Robert Walker Swinburne & Co.; he was also a ship-owner.

Wood's influence in the coalmining industry, especially in the North East, was significant. He gave extended evidence to the 1835 Select Committee on Accidents in Mines and on many other occasions. He was also instrumental in the establishment of the North of England Institute of Mining Engineers, the precursor of the Institution of Mining Engineers. Speaking at its inaugural meeting on 21 August 1852, as first President, he commented that 'there is no finality in science' and that the new body should strive for progress in its chosen subject. The Institute would be glad to accept as members 'any literary, scientific, or practical members of other Institutions, professions, or occupations, whose labours, talents, or professional experience' could in any way assist in reducing mining accidents or assist the mining industry. To help reduce the number of accidents in mines he envisaged the formation of an archive of mining records as one of the steps towards progress in this field. He often contributed to the discussions of the Institute although he was described as 'deficient' as a public speaker; he was praised, however, for his perseverance and investigative abilities and for his 'equanimity of temper'.

In 1858 Wood became a member of the Institution of Mechanical Engineers and in 1864 was elected a Fellow of the Royal Society when he was noted as being 'distinguished for his acquaintance with the sciences of Geology and Civil Engineering (and as being) eminent as an Engineer and Colliery-viewer'. He did not long enjoy this honour as, after a lengthy illness, he died in London on 19 December 1865 and was buried four days later at Hetton, Co. Durham. His wife had predeceased him and his fortune of some £400,000 was left to his four sons and three daughters; three of his sons continued his mining interests. Wood was described as being 'of commanding height, portly form, and had a ruddy, good-humoured countenance, which bore no trace of the hard work he got through'; he is commemorated by an imposing statue in the Wood Memorial Hall of the Mining Institute in Newcastle.

R. W. RENNISON

[*Report from the Select Committee on Accidents in Mines*, 1835; *Report of the Committee of Investigation of the Affairs of the Brandling Junction Railway*, 1843; *Transactions, North of England Institute of Mining Engineers*, 1852–1853, 1; 1859–1860, 8; 1866–1867, 15; *Engineer*, 12 January 1866; *Min Proc ICE*, **31**, 1870–1871, 236–238; R. Welford (1895) *Men of Mark Twixt Tyne and Tweed*; G. B. Hodgson (1903) *Borough of South Shields*; W. W. Tomlinson (1914) *North Eastern Railway*]

Publications

1825. *A Practical Treatise on Railroads* (revised editions 1831, 1832, 1838)
1834–1841. *Reports on the Newcastle and Carlisle Railway* (with Benjamin Thompson and George Johnson)
1838. *To the Directors of the Great Western Railway*, Killingworth, 10 Dec.
1840. *Northern Union Railway … from Gateshead to Ferryhill … 29 February* (with T. E. Harrison), Q/D/P.104, Durham CRO
1843. *Report on a Central Line of Railway into Scotland*, 19 Nov. (with George Johnson)

Works

1833–1841. Newcastle and Carlisle Railway, Committee of Management, 60 miles in length
1836–1839. Brandling Junction Railway, Engineer, 17 miles in extent.

WOOD, Ralph (1664–1730), stonemason, was the son of Francis and Elizabeth Wood of Bishop Middleham, County Durham. He was the eldest of a family of nine and was baptised on 4 September 1664. In 1686 he married Alice Whitfield (b. 1667) of Pittington and nine children resulted from the union, born between 1687 and 1703. Nothing is known of his early career but after the death of his father in 1705 Wood moved to Stockton-on-Tees, possibly to work on the construction of the new church there, completed in 1712.

The first recorded work undertaken by him was on the river Tees where in 1715 he became responsible for 'repairing and keeping in repair Yarm bridge'. In 1718 he undertook to build a bridge over the Waskerley Beck, its construction overseen by the Durham County Justices in the absence of an appointed bridge surveyor.

Wood's major work was the construction in 1725–1726 of a masonry bridge carrying an early wagonway over the Causey Burn, near Tanfield. With a span of 102 ft. it carried a double track and the line ran to the river Tyne at Dunston. Built by the coalowning 'Grand Allies' and some 80 ft. high, the bridge remained for more than 20 years the largest masonry span in Great Britain. Wood subsequently became responsible for the repair and maintenance of several County bridges and would seem to have acted as bridge surveyor, although not appointed as such. His activities included work at Durham and at Langley.

Wood died in Stockton-on-Tees on 22 October 1730 and was commemorated, with another Ralph Wood (1676–1743), by a stone inscribed:

We that have made tombs for others,
Now here we lie;
Once we were two flourishing woods,
But now we die.

In 1738 a further Ralph Wood (1712?–1747), probably the son of Ralph Wood (1676–1743) and also a stonemason, was appointed bridge surveyor to Durham County at a salary of £20 p.a., the post having been created only in 1732 when Thomas Shirley became the first to hold it.

R. W. RENNISON

[*Durham Quarter Sessions: Orders* (1700–1750) Q/S/OB/8–10, DRO; *George Bowes Coal Accounts with Liddell and others* (1725–1726) D/St/B1/6/77, DRO; J. Brand (1789) *History and Antiquities of Newcastle*; E. McKenzie and M. Ross (1834) *Historical, Topographical and Descriptive View of the County Palatinate of Durham*; M. J. T. Lewis (1970) *Early Wooden Railways*]

Works
1718. Bridge over Waskerley Beck, construction
1725–1726. Causey Arch, construction, cost *c.* £2,000

WOODHOUSE, John (*c.* 1776–1820), engineer, is the best-known of the Woodhouse family, active in the civil engineering world of the early nineteenth century. He was the son of Jonathan and Frances Woodhouse of Bedworth, his father practising as a mining engineer, installing pumping engines, before settling at a colliery at Bedworth. In 1796 he tendered to install two steam engines on the Grand Junction canal. John had a number of brothers and sisters, Ann (b. *c.* 1779), Frances, Charlotte, and Jonathan (q.v.). He followed his father's trade and was resident in Chilvers Coton, Nuneaton, where he had installed a steam mill in 1796. In April 1797 Woodhouse was asked by the Ashby Canal Company, in association with John Deakin, to carry out a mineral survey of the wolds through which the canal passed.

Woodhouse successfully tendered in June 1802, with his brother, Jonathan, for the Blisworth tunnel on the Grand Junction Canal, with George and Anthony Tissington of Ashby as partners; at that time John was described as civil engineer of Chilvers Coton. The rate was £15 13s per yard, with a bonus of £1,000 if the work was completed within two years and three months. Initially progress was good, despite William Jessop and James Barnes (qq.v.) altering the original plans, and by June 1803 642½ yd. were complete; that December 145 yd. were completed, an excellent month's work. There were, however, concerns over money advanced to the contractors, and it would appear that the contractors' financial controls may have been lax. When the company refused to honour a bill of exchange of the contractors for £1,930 in early February 1804 it became clear that the contractors had lost heavily, perhaps £13,000, partly due to circumstances outside their control. In this situation the contract was dissolved by the company and the contractors books were closed on 21 March 1804, with 1,618 yd. completed (of 3,075 yd.). The quality of work executed was high and although Jonathan attempted to take over the contract the

company decided to continue to work by small contracts let to small workforces.

In May 1804 John Woodhouse was made superintendent of work and following Barnes's retirement, in June 1805, as company engineer, in September Woodhouse was appointed engineer to the Northern district of the canal. In 1809 he surveyed a line for the Grand Union Canal from the Grand Junction at Norton to the Leicestershire and Northamptonshire Union canal, Northants. Succeeded by William Thompson, who remained there until his retirement in 1835, Woodhouse was responsible as contractor for the construction of side ponds—apparently poorly built—to conserve water at flights of locks such as those at Hanwell. He also installed the Tringford pumping station in 1817–1818.

By this time Woodhouse had been active elsewhere. Construction of the Worcester and Birmingham Canal had begun in the 1790s with Thomas Cartwright (q.v.) as engineer and in July 1809 Woodhouse succeeded him, his salary being confirmed at 500 guineas p.a. plus expenses in January 1811. By that time construction had reached Tardebigge old wharf and, looking to completion of the canal with a reduced number of locks, the company agreed to Woodhouse's proposal to erect a canal lift at Tardebigge. This was based on a system patented by Woodhouse in February 1806 while resident at Heyford, Northants, working on the Grand Junction. It was installed by 24 June 1808. Testing began in 1810, in preparation for the opening of the canal beyond Tardebigge tunnel, but the tests were unsatisfactory and in January 1811 it was decided to build conventional locks. Woodhouse still had supporters for his scheme and Jessop and John Rennie (q.v.) were asked to report. Jessop was in favour of the lift but Rennie felt that although it was working it was 'too delicate in its parts and requires much more attention and careful management than can possibly be expected to be given when in general use on a canal …'. In March 1811 it was decided to adopt locks and the trial lift itself was removed in 1815 prior to the canal's opening on 4 December. William Crosley (q.v.) had, in mid-1811, replaced Woodhouse as engineer, to enable Woodhouse to contract for the construction of the section between London Barn hill and Body Brook. Woodhouse ran into trouble for not carrying work out to specification and lack of progress; in 1815 the matter was referred to arbitration.

Woodhouse's other main project was on the Gloucester and Berkeley ship canal where work resumed in 1818. He succeeded John Upton (q.v.) as engineer at a salary of £500 p.a. and apparently took out a lease on a mill at Cambridge, Glos., where there was a feeder to the Canal. In 1819 he drew up a scheme, never executed, for a canal between Cambridge and Dursley, and in 1820 was dismissed, on Telford's recommendation, for purchasing from a son masonry of dubious quality for use in the sea wall. Woodhouse protested his innocence, blaming the contractor for using the masonry in the wrong place; nevertheless he was replaced as engineer by Thomas Fletcher (q.v.).

Woodhouse's subsequent career is unknown and has been overshadowed by that of his and his brother's sons. His work on the Gloucester–Berkeley line brought the family into contact with William Mackenzie (q.v.) and a close family friendship ensued. Woodhouse had married Dorothy Jackson at Bedworth Parish Church, Warwickshire, on 29 November 1791; of their children seven sons reached adulthood. The eldest, **Thomas Jackson** (1793–1855), and **George** (1801–1868) became civil engineers, the latter acting as an agent for William Mackenzie (q.v.) and Thomas Brassey, whereas Thomas Jackson, after trying a life at sea, worked with his father on the Grand Junction and Worcester and Birmingham canals, as well as working on the Gloucester and Berkeley ship canal. In 1825 he surveyed a proposed railway line from Birmingham to Bristol with Josias Jessop (q.v.) and in the following year became resident engineer on the Cromford and High Peak Railway. This was the beginning of a distinguished career as a civil engineer which eventually, on behalf of Mackenzie and Brassey, took him abroad on their projects.

MIKE CHRIMES

[Proceedings, Worcester and Birmingham Canal, RAIL 886, PRO; Rennie papers, NLS; Rennie reports, vols. 5–7, 1808–1812, ICE Library; W. Warcup (1818) Description of Woodhouse's lock, OC2, ICE; Frank Smith archive, ICE Library; Telford mss, ICE Library; Telford collection, Ironbridge Gorge Museum; Mackenzie collection, ICE Library; Membership records, ICE; Q/RUm/135, Glos. RO; *Repertory of Arts and Manufacturers*, 2nd series, **8**, 1806, 405; E. Smith (1810) A description of the patent perpendicular left, erected in the Worcester and Birmingham canal, at Tardebrigge, near Bromsgrove; obituary of T. J. Woodhouse, *Min Proc ICE*, **16**, 1856–1857, 150–154; Death notice of G. Woodhouse, *Engineer*, 23 October 1868; C. T. G. Boucher (1962–1963) The pumping station at Hawkesbury Junction, *Trans Newc Soc*, **35**, 59–68; Hadfield, *passim*; A. H. Faulkner (1972) *The Grand Junction Canal*; C. Owen (1984) *The Leicestershire and South Derbyshire Coalfield*; G. N. Fletcher (1989) Thomas Telford and the Gloucester and Berkeley canal, *IAR*, **xi**, 2; M. M. Chrimes *et al.* (1994) *Mackenzie: Giant of the Railways*]

Works

1802–1804. Grand Junction Canal, Blisworth tunnel, contractor, 1,618 yd.
1805–1809(?). Grand Junction Canal, Northern Division, resident engineer
1808–1811. Worcester and Birmingham Canal, patent lift, contractor/engineer
1809–1811. Worcester and Birmingham Canal, engineer

1811–1815. Worcester and Birmingham Canal, Tardebigge–Body Brook, contractor, £60,000
1815–1818(?). Grand Junction Canal, contractor
1819–1820. Gloucester and Berkeley Canal, engineer

WOODHOUSE, Jonathan (fl. 1795–1822), mechanical and mining engineer, of Overseal Cottage, Ashby-de-la-Zouch, was the brother of John Woodhouse (q.v.); their father, Jonathan Sr. was also a mining engineer in the Warwickshire area. Jonathan is best-known for his work with steam engines, particularly in the East Midlands, where he was involved with the estates of Lord Moira and erected a furnace for him at Ashby Wolds. According to his father, writing in January 1796, he was at that time installing engines for Earl Fitzwilliam in Yorkshire, as well as working for Wilkes, Boulbee, Fenton and Sir George Beaumont. He clearly was involved in railway or tramway construction as in 1803 he obtained a patent for the use of concave plates or rails on roads for the use of wagons etc, an idea similar to that employed by Benjamin Wyatt (q.v.) on the Penrhyn railway, although intended on this occasion for use on normal roads.

In 1802, probably as a result of his mining background, Woodhouse tendered for canal tunnel contracts. His offer to complete the Stanedge Tunnel on the Huddersfield Canal fell through in 1803 but his tender, with his brother, John, and others as partners, for the Blisworth Tunnel on the Grand Junction Canal was successful. By 1804, however, they were losing heavily and were taken off the contract; although Jonathan tried to resume the work, the Company would have none of it and it would appear that he returned to mining engineering. In 1804 he carried out a mineral survey on the Ashby Canal.

One of the Woodhouses had erected a steam engine at Fritchley cotton mill, Crich, in 1803 and Jonathan installed another (56 in.) engine at Measham cotton mill in 1807. In 1812 he negotiated a preferential rate for the carriage of Lord Moira's coal on the Ashby Canal. He may have worked 1817–1818 at Tringford pumping station, on the Grand Junction Canal. He supplied a Newcomen pumping engine at Hawkesbury Junction on the Coventry Canal in 1821–1822. Believed to have dated from the early eighteenth century, it was used at Oakthorpe Colliery, Measham, and apparently purchased and modified by Woodhouse before being sold to the Coventry Canal Company for £269 15s 4d.

Woodhouse must have died shortly afterwards as his son, John Thomas (1809–1878), was described as being orphaned at an early age and brought up by his guardian, Edward Mammatt, agent to Lord Moira. He was trained by John Twigg, mining engineer, of Chesterfield, and practised largely as a mining engineer in the East Midlands, rising to considerable eminence; he was born and is buried at Overseal. Little is known of his brother, William Henry, described as a civil engineer, who died in London in 1864.

MIKE CHRIMES

[F. Smith archive, ICE; Membership records, ICE; Specification ... for a new method of forming a cast iron rail (1803) *RAM*, 2nd series, 15–18, plate II; F. Nixon (1957–1958) The early steam engine in Derbyshire, *Trans Newc Soc*, **30**, 1–28; C. T. G. Boucher (1962–1963) The pumping station at Hawkesbury Junction, *Trans Newc Soc*, **35**, 59–68; R. B. Schofield (1981) The construction of the Huddersfield narrow canal, 1794–1811, *Trans Newc Soc*, **53**, 26; C. Owen (1984) *The Leicestershire and South Derbyshire Coalfield*]

Works

1802–1804. Blisworth tunnel, Grand Junction Canal, 1,618 yd., contractor

WOOLER, John (d. 1783) was a civil engineer working principally in north east England. His origins have not been ascertained and the first of his known engineering interests is his evidence given in 1749 to a Parliamentary Committee regarding an extension of the powers of the port commissioners of Whitby. The west pier there had been built under an Act of Parliament of 1734 and Wooler reported that £13,409 had been expended between 1735 and 1749. A further contact with the town was a communication to the Royal Society made jointly with Captain William Chapman in 1758 concerning a Jurassic fossil crocodile found in the harbour.

In 1762 Wooler, acting on behalf of the Board of Ordnance, inspected Cliffords Fort at the mouth of the Tyne and also reported on the fortification of Whitehaven harbour. He put forward suggestions regarding the siting of batteries, the provision of a barracks to house 100 men, a magazine and a building for storing ordnance in times of peace. He was paid 25 guineas for his work at Whitehaven but it did not then proceed although his proposals were revived in 1778 when the port's Trustees petitioned the Crown to erect a fort in accordance with Wooler's recommendations.

In 1763 he was called upon by the River Wear Commissioners to report on problems with quay encroachments, a task for which he received the sum of 21 guineas. He became involved in repairs and rebuilding work at the port of Bridlington in 1764 and the next year, from Whitby, reported jointly with John Smeaton (q.v.) on obstructions in the river Esk. In 1766 he again gave evidence on behalf of the Whitby port authorities regarding a new Act of Parliament, it then being noted that he had been employed by the port's Trustees for some years. He stated that although £12,884 had been spent in the past 15 years a further £11,000 were needed to repair both piers.

Wooler reported on the possible consequences of the installation of a water wheel in one of the arches of London bridge in 1767, while in the same year he was again retained by the River Wear Commissioners to report on Joseph Robson's

(q.v.) plan for a 1,100 ft. long mole on the South Rocks at the river's entrance, a proposal with which he generally agreed, although differing as to its cost; for this report he received 50 guineas. For planning the market place at South Shields in the same year he was paid 5 guineas by the Durham Dean and Chapter.

A disastrous flood on the rivers Tyne and Wear on 17 November 1771 brought Wooler to be involved in work on the two rivers where, on the Tyne, all bridges—except Corbridge—had been destroyed. With Smeaton, he submitted a report to Newcastle Corporation in January 1772 and, immediately following, he carried out a joint investigation with Robert Mylne (q.v.), taking soundings in the river and making proposals for a temporary bridge. Their report suggested that a new permanent bridge could be built a short distance up-river. A second report was submitted by him on the suggested new bridge in March of the same year and in it he advocated rebuilding on the original line. The temporary timber bridge was completed in 1772 and the rebuilding of the permanent structure, on the site of the former bridge, began in 1774 and was completed in 1781. Wooler acted on behalf of Newcastle Corporation, the owners of the northern six arches of the bridge, while Mylne acted for the Bishop of Durham, owner of the southern three arches.

Wooler was also asked to report on the Prebends bridge at Durham where an earlier structure crossing the river Wear had also been swept away. A design for a three-span masonry structure had been prepared by George Nicholson (q.v.), architect to the cathedral authorities, and Wooler's scrutiny of the design in 1774–1775 brought him £5. Later, he was invited, with Nicholson, to examine and report upon Durham cathedral, then recommending extensive repairs to the stonework and suggesting that pinnacles be added to the cathedral's towers; his recommendations were later adopted. The destruction of the Newcastle bridge had resulted in the need for a new magazine—earlier situated on the bridge—to provide safe storage for gunpowder in Newcastle and Wooler was asked in 1776 to prepare plans for a replacement magazine at Walker, to the east of the town. At the same time, he became responsible for rebuilding one of the pinnacles of St. Nicholas's church in Newcastle, together with providing 'proper conductors to preserve (the spire) from thunder and lightning'.

Up-river from Newcastle, Wooler was called upon to design a new bridge at Chollerford and a drawing was prepared by him. John Fryer (q.v.) was also involved in this work and he prepared two designs for a bridge of five spans, the form subsequently completed in 1775 although with slightly different details. It is uncertain which of the two men was responsible for the actual design.

The last of the bridges on the Tyne where Wooler was involved was that at Hexham. A bridge, completed only in 1770, had been destroyed by the 1771 flood and Wooler was requested to prepare designs for a replacement. Initially, he accepted a design drawn up by Fryer but it would seem that modifications were made to it. A start to building was made in 1775 by which time Wooler had secured the services of Jonathan Pickernell Sr. (q.v.) to supervise construction. Work had not proceeded beyond the first pier when ground conditions led Wooler to instigate more extensive site investigations. Such were the strata found, a gravel bed underlain by running sand, that he was forced to recommend, 'very sagaciously and prophetically', that the site should no longer be considered unless the entire river bed were to be paved with masonry laid over a timber raft which, in turn, should be piled at its edges. For his work at Chollerford and Hexham Wooler was paid £42. Wooler's association with Smeaton continued during the work at Hexham, Wooler writing some thirty letters to Smeaton while at Hexham.

In 1752, Wooler had been involved with bridge works at Hull and in 1773 he was approached by the Customs authorities in relation to the construction of a wet dock there, estimating its cost at £55,000 to £60,000. Five years later, again with Smeaton, he reported on a dock wall which had failed there and the following year both men were appointed to certify the work's completion. Between these two events, Wooler had inspected Wells harbour with a view to preventing the harbour deteriorating, his recommendations seemingly rejected on account of their expense.

Little is known of Wooler's personal life but in both 1762 and 1773 he was noted as acting for the Board of Ordnance and Smeaton referred to him as a military engineer. It is possible that he married Eleanor Oliver in Hull in 1772. He died in October 1783 in Woolwich, the Newcastle Courant then noting him as 'an eminent Engineer in the employment of Government; and whose services have frequently been had in this country; and particularly in building the Newcastle part of Tyne Bridge'.

R. W. RENNISON

[*Journal of the House of Commons* (1749); *Whitehaven Town and Harbour Commission Minutes* (1709–1783), 191–192. Cumbria RO, Whitehaven DH 4/1; *Plan and Prospect of the Town and Harbour (of Whitby)*, n.d. (*c*. 1750); *Sunderland Harbour: River Wear Commissioners: Minute Book* (1763 and 1767) Tyne & Wear Archives; *Newcastle Corporation: Chamberlain's Accounts; Committee Book* (1772–1778) Tyne & Wear Archives; *Dean and Chapter, Durham, Audit Book*, B, v. Special Collections, University of Durham; *Calendar of Common Council Book, Newcastle, 1766–1785*, Tyne and Wear Archives, 589/15; *Northumberland Quarter Sessions: Orders*, **11**, 1777, 344; *Proceedings on the Part of Henry Errington ... Relative to Hexham Bridge* (1777–1789) M17/57, Northumberland RO; *Newcastle Courant*, 1 November 1783; J. Smeaton, *Memorial*, **3**, 299–301; J.

Smeaton, *Reports*, **3**, 32; E. McKenzie (1827) *A Description and Historical Account ... of Newcastle*, 211–212; J. Sykes (1866) *Local Records*; M. W. Baldwin (1973–1974) The Engineering History of Hull's Earliest Docks, *Trans Newc Soc*, **46**, 1–12; Colvin (1978), 916; Skempton (1987); S. M. Linsley (1994) Tyne Crossings at Hexham up to 1795, *Archaeologia Aeliana*, 5 Series, **22**, 235–53]

Reports and drawings

1762. *Report to Sir James Lowther ... with Plan of the Port or Harbour of Whitehaven showing the disposition of the Ordnance*, D/Lons/W.Plan 150, Cumbria RO, Carlisle

1767. *The Report of John Wooler, Esq., Engineer, concerning a Pier proposed to be erected on the South Rocks at Sunderland in the County of Durham*, 042.4, 7, Lit. & Phil. Society, Newcastle

1772. *A Report relative to Tyne Bridge* (with J. Smeaton), Newcastle Central Library, Austhorpe

1772. *To Sir Walter Blackett, Bart., Mayor, the Aldermen and Common Council of Newcastle upon Tyne*, Newcastle Central Library, Newcastle

Works

c. 1750–1766. Whitby pier extension and harbour repairs

1772–1781. Rebuilding of Tyne Bridge, Newcastle, £26,000 (north end)

1772–1775. Possible rebuilding of Chollerford bridge

1777. Repairs to tower of St. Nicholas's Church, Newcastle

WOOLF, Arthur (1766–1837), Cornish steam engineer, was born in October 1766 of a Redruth family; his father, also Arthur Woolf, was employed as an engineman at Dolcoath mine in the early 1770s. Young Arthur was apprenticed to a carpenter and joiner and having served his time set out for London as a first rate journeyman. However, by 1795 he was erecting a second hand engine at a colliery at Newbottle, County Durham, for Hornblower and Maberly and this experience led him, upon his return to London the following year, to erect a steam engine at the brewery of Messrs. Meux and Reid and to become their resident engineer, a post he held until 1808. (Hornblower had been erecting engines in Cornwall before Watt's first engine in 1777.) While at the brewery he introduced several innovations and in experimenting with expansive steam conceived the idea of the compound engine with which his name is associated. During this time he kept in touch with Richard Trevithick (q.v.), supplying him with a steam gauge for his first road locomotive at Camborne, visiting him frequently when he was in London with his first carriage and persuading him, along with Davies Gilbert (q.v.) M.P., whom Thomas Telford (q.v.) consulted regarding the Menai Suspension Bridge, to witness a trial of Woolf's engine at the Meux's brewery. Although returning a 'duty' much better than that of engines then in use, Trevithick thought it could have nearly doubled its output with a better boiler.

After the brewery, Woolf entered into a partnership with Humphry Edwards constructing and improving smaller engines and by 1811 a standard form of a double-cylinder engine had been evolved. Following advertisements in Cornish newspapers, the Williams', Cornish mineowners, sent their engineer William Sims (q.v.) to London to inspect one of Woolf's engines and among others to be attracted were Trevithick and Henry Harvey, of the Hayle Foundry, whose tests found the engines to be far superior to those of Boulton and Watt. Notwithstanding the superiority of their engines, there was insufficient demand and in May 1811 the partnership ended, Edwards going to France and Woolf returning to Cornwall where Andrew Vivian, manager of several mines at this time, offered him the post of engineer at Wheal Abraham.

Now in his mid-forties Woolf was to enter the most productive period of his life and was to become engineer for some thirty mines. He supplied winding engines, including one for Wheal Abraham, where he also erected the largest double cylinder mine pumping engine for its time. He introduced the use of steam power for the stamping of ore and in May 1817, with Vivian and Thomas Lean, gave evidence before the Select Committee formed to investigate the cause of boiler explosions. His designs required greater accuracy of casting and working than was conventional for his day and he was often handicapped by waiting for large castings from South Wales. This led to his involvement with Harvey, with whom he became actively associated in the development of the Hayle foundry. It was during his time there that Francis Trevithick, son of Richard and author of the book on the life of his father, became his pupil. Woolf preferred cast iron boilers to his own patent design but following his 1817 visit to London he undertook the construction of a large wrought iron boiler of Trevithicks's design, later to be known as the Cornish boiler. This required the special design of a cylindrical punch and was the precursor of the boiler making department of Messrs. Harvey & Co.

Although Woolf is best known for his mechanical engineering abilities in the use of steam, he was a Member of the Institution of Civil Engineers. His civil engineering work is not so well documented but it is known that he undertook a survey of Hayle harbour and provided the drawings for the Penryn Swing Bridge, supplied by Harveys in 1829; it was replaced in 1935.

Woolf died on 26 October 1837 in Guernsey, where his last engine, made by Harveys, was erected in 1835 and it would seem appropriate to quote from his obituary notice in the proceedings of the Institution: 'His name is associated with the improvements in the drainage of the Cornish mines; and whatever share posterity may assign to his individual genius in these improvements, his name is recorded in the page of history among those who have dedicated their talents and the opportunities of a long life to the advancement of practical science'.

R. P. TRUSCOTT

[Membership records, ICE; Cornwall RO; The Courtney Library of the Royal Institution of Cornwall; *West Briton*; *Royal Cornwall Gazette*; Evidence (1817) *Select Committee on Steam Boats etc.*, HC 422, 39–40; Obituary (1838) *Min Proc ICE*, **I**; F. Trevithick (1872) *Life of Richard Trevithick*; T. R. Harris, *Arthur Woolf, The Cornish Engineer*; H. W. Dickinson and A. Titley (1934) *Richard Trevithick, The Engineer and The Man*; E. Vale *The Harveys of Hayle*; D. B. Barton (1966) *The Cornish Beam Engine*]

WREN, Sir Christopher, FRS (1632–1723), architect and scientist, was born on 20 October 1632, at East Knoyle, Wiltshire, the son of Christopher and Mary Wren. His father was rector there.

While Wren is now best known as an architect, he began this profession only after he had achieved eminence in mathematics, science and astronomy and been one of the founding fathers of the Royal Society. As an architect he designed churches and several buildings for the crown, including extensions to Hampton Court Palace and the Royal Hospital at Greenwich. Through his work as Surveyor of the King's Works following the Great Fire of London in 1666, he was to have considerable influence over the rebuilding of London, work which also developed his abilities as what we would nowadays call an engineer and project manager.

Wren's mother died when he was very young and he was brought up by an older sister and was rather frail for much of his childhood. Wren was introduced to mathematics at the age of eleven by his uncle, the mathematician William Holder. From the age of nine he attended Westminster School, London, where he excelled at Latin, natural science and mathematics. After leaving Westminster at fourteen he spent some time as the assistant to Dr. (later Sir) Charles Scarburgh demonstrating and making anatomical preparations for his lectures on anatomy. In 1649, at the age of seventeen, he entered at Wadham College, Oxford as a fellow-commoner, where he soon joined a society of philosophical inquirers. He graduated in 1651 and received his Master of Arts degree in 1653. He was appointed Professor of Geometry at Gresham College, London in 1657, at the age of twenty-five, and Professor of Astronomy at Oxford University four years later. Wren married Faith Coghill in 1669 who died of smallpox shortly after bearing his son, Christopher in 1675. He was married again in 1677 to Jane Fitzwilliam who bore him a daughter and son, and died in 1680.

Wren came to prominence as a mathematician when he solved a problem that had defeated many before him—calculating the length of an arc of the cycloid (and this before calculus had been invented). Like other eminent mathematicians of the day, he also turned his attentions to Kepler's Laws describing the motions of the planets, to the force of gravity, and the inverse square law. In 1669, and by that time fully engaged with the rebuilding of London after the Great Fire of 1666, he also proved that a hyperboloid of revolution can be generated with straight lines. Newton, who was seldom forthcoming with praise, later rated Wren as the equal of John Wallis and Cristian Huygens, the leading mathematicians of the day.

In London Wren began hosting a series of weekly or fortnightly meetings at Gresham College attended by a number of his colleagues from Oxford—including the chemist and physicist Robert Boyle (1627–1691), mathematician John Wallis (1616–1703) and Robert Hooke (q.v.). At these meetings they discussed all sorts of mathematical and scientific matters. Robert Hooke was the most practical man among them and he made instruments, apparatuses and experiments to demonstrate and test all manner of ideas, sometimes, it has to be said, in what appears to us to be a rather haphazard way. After the Restoration of Charles II to the throne in 1660, this group formed themselves into a society which attracted the King's attention and it was granted its Royal Charter in 1662 as *The Royal Society of London for the Promotion of Natural Knowledge*. Wren was prominent in this group and, although he soon moved back to Oxford as the Savile Professor of Astronomy, he continued to attend meetings regularly for many years and was the Society's president from 1680 to 1682.

Wren clearly impressed Charles with his practical and drawing skills as well as his intellect—he had, for example, presented to the king illustrations of insects seen down a microscope and a large model of the moon based on telescope observations—for, in 1661, the King invited Wren to take charge of building new fortifications at Tangier, a commission he turned down on the grounds of poor health.

Although Wren's work as an architect began with the Chapel at Pembroke College, Cambridge in 1663 it was famously boosted in the years following the Great Fire of London in 1666. The King appointed three Surveyors—Wren, Hugh May and Roger Pratt—to work alongside three appointed by the City of London—Robert Hooke, Edward German and Peter Mills. May and Pratt were established architects, while German and Mills were experienced surveyors. While Hooke was practical, inventive and a good project manager, it was clearly Wren who had the vision and grand view of the entire work. As an architect, Wren designed and supervised the construction of some fifty-three churches in the city, of which the largest and most famous is St. Paul's Cathedral. This legacy established his reputation as Britain's most talented and eminent architect until the twentieth century.

However, Wren's architectural accomplishments have overshadowed his skill in the field we now call structural engineering. The timber roof trusses of Wren's Sheldonian Theatre, one of the first buildings he designed, were not only large for their time (70 ft.—a span similar to Westminster Hall roof, but rising less than half the height) but they also had to support the considerable weight of books stored

there by the Clarenden Press. And while the 100 ft. (33 m) span of the masonry dome over St. Paul's was not the largest of its kind, but is undoubtedly one of the most ingenious and its span-to-thickness ratio ($s{:}t$) (approx. 37) is the largest of all the classical masonry domes. (The Pantheon, Rome, spans 43 m, approx. $s{:}t$ ratio 11; Brunelleschi's dome at Florence spans 42 m, approx. $s{:}t$ ratio 21; St. Peters, Rome spans 41.5 m, approx. $s{:}t$ ratio 30). In addition to these practical achievements, Wren also applied his intellectual skills to contemplating and explaining how masonry arches and domes carry their loads.

Wren had no training in building but, when plans for the Sheldonian were proposed, he would have been able to bring to the discussions an understanding of forces and of three-dimensional geometry far exceeding that of any architect or builder, an understanding he had developed in his astronomical studies concerning the motions of the planets. His design, as executed, was original in its details, but is also clearly adapted from other large trusses, especially some from Italy, which then led the world in such construction. In fact, Wren was not new to the problem of spanning large distances. His geometry teacher some years before had been John Wallis who had devised in 1652 an ingenious 'geometrical flat floor' capable of spanning about 50 ft. using timbers no longer than about 13 ft. Wren knew this structural idea and, furthermore, that Wallis had calculated its load capacity (though not its stiffness, which would have been too low for practical use). When a model of this floor was demonstrated in front of the King in 1660, shortly after the Restoration, he requested it take its place among his own treasures. Wren's achievement in the Sheldonian, however, was considerable. Unlike Wallis, he managed to combine his intellectual understanding of the problem with a feel for practicality and was thus able to gain sufficient confidence in his idea to go ahead and have it constructed at full scale. In his library at Trinity College, Wren was faced with the very problem to which Wallis's floor was intended to be a solution when he had to replace the original floor structure that had failed. Here he devised and used what are probably the earliest examples of trussed floor beams. The main timber beam is stiffened by the addition of inclined timber struts to create what is effectively a shallow timber arch—rather in the manner of the top chord of a roof truss of the type Wren devised for the Sheldonian.

There is wonderful evidence of Wren's practical understanding of structures in his several reports on large buildings which he undertook as Assistant to the Surveyor General, for instance his structural assessment of the ruin of the roof in the old St. Paul's (in 1667), and of the cathedral at Salisbury (in 1669). At the age of eighty-one, at the request of the Bishop of Rochester, he undertook a full report of the survey of Westminster Abbey he had conducted some years earlier together with his thoughts about its future repair and maintenance. This comprised a review of the historical development of masonry and arch building, as well as observations about how many of them had deteriorated, due not only to a poor choice of materials but also 'original faults in their first design'. Wren's particular concern was the manner in which walls were buttressed in order to resist the outward thrust of arches, vaults and domes, and he commented favourably on earlier work at Westminster when an architect had used iron bars to tie some arches until the tower would later be built above and provide the required stability by dead weight alone. The damage that Wren identified needing repair arose from the fact that this tower had never been built, and the iron ties had been removed, allowing the arches to spread. Wren shows his excellent understanding of stability in proposing iron ties to restrain the columns which supported the arches, before then continuing to build the tower, as originally intended, in order to provide the dead weight needed to fully stabilise the whole structure. In fact the latter recommendation, perhaps fortunately given the size of tower proposed, was never undertaken.

The most remarkable example of his structural understanding is, of course, displayed in St. Paul's cathedral where he demonstrates a remarkable sense of poise in how he conveys the weight of the 1,000-ton stone lantern down through the dome, piers, arches, vaults and walls to the foundations. He was no less careful to ensure the integrity of the masonry by using large blocks, well-interlocked rather than small ones. Especially ingenious is his brilliant adaptation of *inverted* arches from the foundations at his Trinity College Library (1776) to a load-bearing structure above ground. He had borrowed the idea from Hooke who had used inverted arches in the foundations of his Montagu House in 1775. Wren now took them aloft to spread the loads from piers at the base of the dome through inverted arches to distribute them safely into the great arches over the nave and transept, thus avoiding problems caused by applying point loads to an arch.

Wren already knew his dome was feasible—larger spans had been covered at the Pantheon in Rome and at Florence. However, in the structural conception and lightness of the dome, Wren's structural achievement at St. Paul's stands alongside that of Brunelleschi at Florence 200 years earlier. Wren's ingenuity derives largely from how he must have thought about the problem and the manner in which he set about the design, for at no time previously could anyone have brought Wren's understanding of forces and equilibrium to bear on how a building stands up. He was an eminent astronomer and mathematician and his work on the motions of the planets involved the very latest thinking about the nature of forces, especially the action of forces in three-dimensional space. It was also Wren's close friend Robert Hooke who first proposed (before Newton) a law of universal gravitation and the inverse square law of attraction between the sun and planets. Hooke also came upon the idea of using a hanging chain as the means of defining the shape (inverted) of a stable arch. In their work on the planets both were familiar with using mathematical models to understand phenomena many millions of times larger than man's own scale. Indeed, by comparison, St. Paul's was small and much more immediate.

Wren was the first designer of buildings who had such a solid background in statics. Indeed he anticipated by more than two centuries the idea that underpinned twentieth century engineering education, namely that one can derive an understanding of structures without having been a builder, by studying the mathematics and science of buildings. Wren is also probably the first engineer or architect to write about buildings in terms of forces and equilibrium and in ways that are almost the same as one would today.

Making use of Hooke's hanging chain analogy, Wren derived the most effective and economical shape for the brick structure to carry the massive weight of the lantern above an elongated, gently-paraboloidal cone, hidden from general view in the void between the two non-load-bearing hemispherical domes visible from outside and inside the cathedral. This was the probably first time that a design methodology based on statics had been used for a major, new structure. As Brunelleschi did before him, Wren also had to find a way of carrying the hoop stresses in the base of the cone. Three wrought-iron tension rings were installed by Jean Tijou, the French iron master who was currently working on the decorative wrought iron gates at Hampton Court, where Wren was also working at the time. By these means Wren was able to carry aloft the most massive of all stone lanterns using less material than similar domes and to rest his dome on a tall vertical drum, apparently in defiance of the outward thrust that domes exert.

In an age when building was almost exclusively a world of brick, stone and timber, Wren was not averse to experimenting with new materials. The tall railings erected around St. Paul's in 1710 are reputedly the first architectural use of cast iron. And, not only did he use wrought-iron chains in his dome, he also used straight wrought-iron ties on a more modest scale to support the bookcases at Trinity College Library. These are anchored in the masonry walls and incline inwards and downwards, hidden within the back-to-back bookcases which they support, to reduce the load they impose on the library floor at first storey level. He used the same idea in repairs at the Bodlean Library in Oxford. At Hampton Court, where, of course, Jean Tijou was working, an even bolder structural use of wrought iron was discovered recently when repairing severe fire damage. Within a partition wall, a wrought iron hanger supports a floor beam directly from the timber roof truss above.

Wren was probably also the first to use iron columns in a building. When expanding the capacity of the House of Commons in 1692 Wren introduced galleries. Supported mainly by brackets, he was forced, either side of the entry to the chamber, to use 'two iron pillars and capitalls in iron of Tijou's work' as he describes them in a memorandum. In 1693 Jean Tijou illustrated an iron capital, solid column and base in his book *A Newe Book of Drawing*. Though no dimensions are given, the scale of the ornate wrought iron decoration of the capital would suggest the columns are about 100 mm diameter. Although no surviving example is known, it was most likely the column was of wrought iron, given Tijou's trade. These two prototypes of 1692 were copied in 1707 when the galleries were widened and columns were added to support them. Judging from their size shown in an engraving of 1742, these would also seem to be of metal, though again no mention is made of whether the columns were of wrought or cast iron. Technically they could have been either, but wrought iron must be the more likely, especially given the Tijou pedigree. The new capitals, however, were wooden copies of the Tijou originals, carved by Grinling Gibbons. A drawing made in 1834 after the fire at the House of Commons shows some of the columns having 'wilted' which is clear evidence that at least some of the 1692 or 1707 columns were of wrought iron.

Wren never ceased bringing his understanding of physics into his thinking about building. Towards the end of his life he prepared several (unfinished) papers on architecture and masonry construction which were published only after his death, in the *Parentalia*. In one of these he demonstrates that he was on the point of making the greatest leap of all in proposing to break with the traditions of Vitruvius, Alberti and others who dealt only with proportions in architecture.

It seems very unaccountable, that the generality of our late architects dwell so much upon [the] ornamental, and so slightly pass over the geometrical (i.e. what we now call 'statics'), which is the most essential part of architecture. For instance, can an arch stand without butment sufficient? If the butment be more than enough, 'tis an idle expence of materials; if too little, it will fail; and so for any vaulting; and yet no author hath given a true and universal rule for this.

Later he points even more clearly to the way forward when he comes to consider how one should proportion an abutment to support an arch.

The design ... must be regulated by the art of staticks, or invention of the centers of gravity, and the duly poising all parts to equiponderate; without which, a fine design will fail and prove abortive. Hence I conclude, that all designs must, in the first place, be brought to this test, or rejected.

Thereafter, however, Wren appears to consider only vertical loads, due to gravity, and to ignore the part played in causing overturning by the lateral thrust of an arch, vault or dome, a fact some writers have claimed is a mistake by Wren. It is difficult to reconcile this apparent error with the countless occasions where he demonstrates he is fully aware of these non-vertical forces. It is possible he is using the word 'gravity' more as we would now use the word 'force'. More significant is the fact that this was the first time that anyone had incorporated the ideas of forces and loads in a discussion of how to design buildings. Unfortunately the fragment Wren did write serves only to heighten the regret that he did not complete his task.

Wren was knighted in 1673 and had been made the Surveyor General to the Crown in 1669 in which capacity he undertook a great many architectural and surveying commissions in addition to his continuing work on St. Paul's which was finally completed in 1710. Wren died in London on 25 February 1723 at the age of ninety-one. Christopher, his surviving son by his first wife, who became MP for Windsor, compiled the *Parentalia or Memoirs of the Family of the Wrens* which was published by his son, Stephen. While this has been a valuable source of information about Wren, from sources very close to the man himself, much recent research has shown it often gave a rather too favourable picture of Wren and does seem, for instance, to have been partly responsible for the partial eclipsing of some of his contemporaries, most notably, Robert Hooke.

W. A. ADDIS

[S. Wren (1750) *Parentalia or Memoirs of the Family of Wrens*; J. Elmes (1823) *Memoirs of the Life and Works of Sir Christopher Wren*; H. M. Fiebler (1923) Sir Christopher Wren's carpentry: a note on the library at Trinity College Cambridge, *Journal of the Royal Institute of British Architects*, **30**(12), 388–391; Wren Society (1924–1943) *Wren Society*, 20 vols.; S. B. Hamilton (1933–1934) The Place of Sir Christopher Wren in the history of structural engineering, *Trans Newcomen Soc*, **14**, 27–42; T. F. Reddaway (1940) *The Rebuilding of London after the Great Fire*; J. Harris (1961) Cast Iron Columns, 1706, *Architectural Review*, July, 60–61; H. Colvin *et al.* (1976) *History of the Kings Works*; L. Soo (1998) *Wren's 'Tracts' on Architecture and other writings*; J. W. P. Campbell (1999) *Sir Christopher Wren, the Royal Society and the Development of Structural Carpentry, 1660–1710*, Ph.D. thesis, University of Cambridge; Colvin (3)]

Works

Works include (see Colvin for full listing):

1664–1669. Sheldonian Theatre
1669–1674. The Customs House
1675. Royal Observatory, Greenwich
1675–1710. St. Paul's Cathedral, £750,000
1676–1684. Trinity College Library
1682–1691. Chelsea Hospital
1689–1702. Hampton Court Palace
1696–1702. Greenwich Hospital

WRIGHT, Edward (1561–1615), mathematical practitioner, was baptised at Garveston, Norfolk, on 8 October 1561, the younger son of Henry Wright and his wife, Margaret. He graduated from Gonville & Caius College, Cambridge, in 1581 and was elected to a Fellowship there in 1587. On leave of absence he accompanied the Earl of Cumberland on a voyage to the Azores in 1589. Returning to Cambridge he began working on what was to be an outstanding contribution, the correct mathematical basis of Mercator's projection. This enabled maps and sea-charts of vastly improved accuracy to be produced and was published, with some other matter, in his *Certaine Errors in Navigation ... Detected and Corrected* (1599; 2nd edn., 1610).

By 1594 he had moved to London. He married in the following year and soon afterwards was appointed Mathematical Lecturer to the City of London. He also gave private tuition. Before 1600 he was designing mathematical instruments of many types. Some were later made for Prince Henry to whom Wright gave mathematical instruction.

As a leading London practitioner Wright was chosen by Hugh Myddleton (q.v.) to carry out the demanding task of surveying the New River. This was to be an open channel bringing water for the City from springs at Chadwell and Amwell, near Ware in Hertfordshire, to Clerkenwell in Islington. It was a distance of 20 miles as the crow flies but nearly 40 miles by the winding course to be followed by the channel, its average gradient only 5 in. per mile (1:12,500).

In the New River account books the earliest record of expenditure, apart from legal charges, is for horse hire for the first survey on 16 March 1609. On 8 May there is an entry 'To Mr. Wright for his three severall surveis to Amwell and back againe to Islington' £20 3s. A week later Myddleton pays a further £40 to Wright 'before hande to be my Arts man' and from 9 September he is paid a weekly allowance of £2.

Work began on the first week of May with Edward Colthurst (q.v.) as Overseer or resident engineer. Thus Wright spent about five weeks on the three surveys and thereafter was employed in setting out the channel, the £40 being an advance for twenty weeks work followed by regular weekly payments. These continued until February 1610. Work then came practically to a standstill owing to financial problems and opposition from some local landowners, with some 10 miles of the channel completed. The difficulties having been resolved, work resumed in November 1611 but now with Edward Pond (d. 1626) as surveyor; he remained in the job until final completion of the channel in September 1613. The reason for this change in personnel is not known.

For taking levels the most accurate instrument available to Wright was probably a water-level 3 or 4 ft. in length of the type depicted by Blith (1652) and used by Jonas Moore (q.v.) in the Fens

in the 1650s. According to a contemporary, Mark Ridley, Wright was the first to fix 'perspective glasses', or finders, to his instruments. This would have assisted him, but should not be confused with the telescopic sights introduced in the late seventeenth century. By traversing the route back and forth from Amwell to Islington he could check his closing error and by repeating the exercise three times the errors would be minimised.

Prince Henry had intended to make Wright his librarian but he died in November 1612, aged eighteen, and Wright's only source of income in his last years seems to have been his Lectureship with a salary of £50 p.a. He died in London in November 1615 but his translation from the Latin of Napier's treatise on logarithms was brought out posthumously in 1616 by his life-long friend, Henry Briggs.

A. W. SKEMPTON

[W. Blith (1652) *The English Improver Improved*; E. G. R. Taylor (1954) *The Mathematical Practitioners of Tudor and Stuart England*; G. C. Berry (1956) Sir Hugh Myddleton and the New River, *Trans Hon Soc Cymmrodorion*, 17–46; P. J. Wallis (1976) Edward Wright, *Dictionary of Scientific Biography*, **4**, 513–515; F. Willmoth (1993) *Sir Jonas Moore*; N. A. F. Smith (1999) Edward Wright and his perspective glass: a surveying puzzle of the early seventeenth century, *Trans Newc Soc*, **70**(1), 109–122]

[W]RIGLEY, Joshua (fl. 1772–1810), millwright, was responsible for the design and installation of machinery and steam engines for the textile mills of the Manchester area in the late eighteenth century. He is known to have been responsible for more engines than Boulton and Watt at this time.

Wrigley first appears in a Manchester directory of 1772 as a pump maker and bell hanger. His pump making had evidently developed into the use of Savery-type steam engines to pump water for overshot waterwheels to drive textile machinery. According to John Farey (q.v.), John Smeaton (q.v.) carried out tests on the performance of two of his engines, erected in 1774, which 'operated by suction only'. Apparently there were two in Joseph Thackery's mill at Chorlton-on-Medlock, where the river Medlock had insufficient flow to drive the waterwheels. Wrigley operated in partnership with Rev. John Derbyshire and Joseph Young and in 1791 he had orders in hand for thirteen engines.

Wrigley must have been helped by the relatively low cost and simplicity of operation of the Savery-type steam engine although it appears that James Watt (q.v.) had a low opinion of his work. John Marshall of Leeds relied heavily on Wrigley when he set up his Holbeck mill in 1780–1781 and was still using him in 1791 when he ordered another steam engine and waterwheel from him, although he turned to Boulton and Watt the following year. Other engines known to have been installed by Wrigley were for Thomas Jourdain (1784), Paris Mill (Liverpool), Cark Cotton Company (1784), McConnel and Kennedy's Salvin's Mill (*c.* 1796) and Horridge (1799).

Wrigley was based at 75 Long Millgate and although best-known for his Savery-style engines he can be presumed to have also designed engines of the Newcomen-type. He owned Knott Mill for grinding log wood, etc., apparently an unsuccessful venture, and he left less than £40 at his death in 1810.

MIKE CHRIMES

[J. Farey (1825) *A Treatise on the Steam Engine*, 119–125; A. E. Musson and E. Robinson (1969) *Science and Technology in the Industrial Revolution, passim*; R. L. Hills (1989) *Power from Steam*, 43–45; M. Williams and D. A. Farrie (1992) *Cotton Mills in Greater Manchester*, 50–51]

WYATT, Benjamin (1745–1818), land agent, was born on 14 January 1745, the fifth son of Benjamin Wyatt, architect and builder, of Weeford, Staffordshire. In his early career he made drawings for the family firm and carried out surveying work in the Staffordshire area, including a new road through the Forest of Needwood, and improvements to the Sudbury–Ashbourne Road in 1766. In 1772 he married Sarah Ford, the daughter of the brewer William Ford, and thus became related by marriage to Dr. Samuel Johnson whom he knew. Whilst his brothers James and Samuel (q.v.) moved to London he continued to live at Blackbrook, the family home, and which he inherited, and eventually sold in 1798. While there he acted as agent to Sir Robert Lawley, the largest local landowner.

In 1785, on the recommendation of Samuel who was working at Penrhyn Castle, he moved to North Wales as chief agent to the Penrhyn Estates, a position he held for upwards of thirty years. Aside from architectural work in the area Wyatt was involved in several civil engineering projects on behalf of Lord Penrhyn, who had begun large scale quarrying operations at Penrhyn in 1782. The scale of the operation necessitated improvements in communications in the area and in 1785 the road to the coast was improved sufficiently to take wagons and was later incorporated into the Holyhead Road. The port facilities were steadily improved by Wyatt and a quay of 1,000 ft. was built. In 1794 15,000 tons of slate were being exported from Port Penrhyn; this had increased to 41,000 tons by 1844.

To deal with this traffic it was decided to build a tramroad. Built in 1800–1801, it had unusual edge rails with a concave surface designed by Wyatt. The line, divided into ten stages, enabled ten horses to do the work which would formerly have required four hundred. It was found that the concave form wore very quickly with oval wheels and after some years Wyatt modified the system.

Wyatt lived at Lime Grove, near Bangor, in a house designed by his brother, Samuel. Despite his relative geographical isolation he was well-known in engineering circles and in 1817 was consulted by Thomas Telford (q.v.) about improvements to the Holyhead Road. He died on 5 January 1818 and was buried in Llandegai churchyard. One of his six sons, James (1795–1882), succeeded him as agent to Lord Penrhyn.

MIKE CHRIMES

[T. Telford (1817) *Evidence, Select Committee on the Roads from London to Holyhead*, 1st, HC 313, 10–13; T. Tredgold (1822) *A Practical Treatise on Railroads and Carriages*; Parry (1848) *Railway Companion from Chester to Holyhead*; J. M. Robinson (1979) *The Wyatts: An Architectural Dynasty*, 136–140; Colvin (3)]

Published works
?. Account of the rail roads on the late Lord Penrhyn's Estate near Bangor, *Repertory of Arts and Manufactures*, 2nd series, **3**, 285–287, plate XI
1811. Account of the Penrhyn railway, *Rep Arts Man*, 2nd series, **19**, 15–16, plate II

Works
?. Port Penrhyn, quay, 1,000 ft.
1800–1801. Penrhyn railway, 6¼ miles

WYATT, Samuel (1737–1807), architect, builder and engineer was born on 8 September 1737 at Weeford, Staffordshire, the third son of Benjamin Wyatt (1709–1772), a farmer and timber merchant who developed a business as builder and architect in the mid-1750s. Benjamin's eldest son, William (1734–1780), worked as his principal assistant. One of their important commissions was the Soho Manufactory built in 1765–1766 for Matthew Boulton. Samuel and his brother, Joseph (1739–1788), were trained to help in the business, as carpenter and mason respectively, while the youngest son, James (1746–1813), studied in Italy and became a fashionable architect of the late 18th century.

Samuel Wyatt's first independent employment came in 1760 as master carpenter at Kedleston Hall, Derbyshire, where he soon became Clerk of Works under Robert Adam (q.v.). In 1765 he married Ann Sherwin, daughter of the land agent of his employer, Lord Scarsdale. He stayed at Kedleston until 1769, then built the large assembly rooms in London known as the Parthenon (1769–1771), designed by his brother, James, and carried out several relatively small works on his own before settling permanently in London in 1774, leasing a large house and timber yard in Berwick Street. There he established a successful business as builder and architect. His first big jobs were the main carpentry contract for Somerset House in 1776, which lasted until completion in 1801, and the design of Doddington Hall, Cheshire, also in 1776.

In 1792 Humphry Repton referred to 'the elegant simplicity of my ingenious friend Mr. S. Wyatt's design' and in these words he aptly defines

Samuel Wyatt

the nature of Wyatt's genius. As architect of the Commissioner's House, Portsmouth Dockyard (1784–1786), and the very fine Trinity House, London (1793–1796), and indeed at Tatton Hall, Cheshire (1785–1791) to which Repton referred, Wyatt is seen as master of a distinctive neo-classical style. Other examples—all listed in Colvin—include Heathfield House (1789–1790) for James Watt. Thomas Telford (q.v.) began his professional career as building superintendent on the Commissioner's House.

The 'ingenious Mr. S. Wyatt' in Repton's phrase denotes the second aspect of his work: as a skilled builder and engineer. In this capacity he designed and built the Albion Mill (1783–1786) on the south bank of the Thames at the foot of Blackfriars Bridge. Outstandingly the most advanced industrial structure of its day, it was the first to be planned from the start to incorporate the newly-invented Boulton & Watt rotative steam engines and the first so far known to be founded on a structural raft. Composed as a series of inverted barrel vaults covering an area 160 ft. long by 120 ft. wide, the raft distributed the building weight as evenly as possible, and with minimum pressure, to the underlying Thames alluvium. Above brick walls and stone pillars in the cellar the mill rose six storeys in a timber frame of posts, beams and floors exceptional in height and the large area supported without internal walls. One of Boulton & Watt's assistants erected the engines. The young John Rennie (q.v.) erected the mill-work.

The mill was completed and the first engine came into operation in March 1786, the opening being attended by a crowd of spectators including Henry Cavendish, Josiah Wedgwood and Sir Joseph Banks, President of the Royal Society. The second engine was working by December 1788 but the third, and last, had not yet been erected when the interior of the mill suffered total destruction by fire in the early morning of 2 March 1791.

Wyatt was devastated by the catastrophe and had the miserable task of finding what could be done with the remains of the building. He drew up plans for converting the shell into a warehouse, but encountered bureaucratic opposition. In 1795 he tried again, this time proposing that the internal structure be fire proofed by iron plating the floors and posts, possibly an adaptation of the 'fire plates' devised by David Hartley (q.v.) and used with some success for floors in domestic buildings. But this scheme also foundered. Moreover, advances in multi-storey 'fire proof' construction after the Albion Mill fire moved mainly on a different line: the use of brick arch floors supported by iron beams and columns developed for textile mills by William Strutt and Charles Bage (qq.v.) between 1792 and 1804.

If in this respect Wyatt's ingenuity seems inadequate, it is amply displayed in his design of twelve 'moveable hospitals' ordered by the Treasury for 'his Majesty's distant possessions'. Timber framed, they were 84 ft. long and 22 ft. wide, single storey. He demonstrated this early example of large-scale prefabrication to the King in April 1788, when one of these military hospitals was taken down and reassembled within an hour. That particular one was intended for the West Indies; another is almost certainly the moveable hospital which arrived in Sydney in 1790.

A new phase of Wyatt's career opened in 1790–1791 with the design and construction of Dungeness Lighthouse for a private client, the Earl of Exeter, for whom he had already designed several farm buildings. 86 ft. high, with the lantern in cast iron, it cost over £4,000. This probably helped in his successful application in February 1792 for the recently vacant post of Surveyor to Trinity House Corporation. As his first job for them he designed Longships Lighthouse off Land's End. Built between 1792 and 1795 it was a sturdy granite three-storey tower 52 ft. high, tapered in profile, with dovetailed masonry following the pattern adopted by John Smeaton (q.v.) in the Eddystone Lighthouse. Then came his commission to design and build the new Trinity House at Tower Hill. Excavations for the foundations began in June 1793 and building work was substantially completed by October 1796, though interior decoration continued for another year. Payments to Wyatt 'on account of the building' came to about £25,000, the total cost being over £30,000.

Wyatt held this post for the rest of his life. It was through the influence of Trinity House that he received an appointment to the Surveyorship of Ramsgate Harbour in March 1794. The two great piers of the harbour, the sluicing basin and a dry dock had been completed by 1792, latterly under the direction of Smeaton. Now required were a lighthouse on the head of the west pier and several buildings on the strip of land between the town and the harbour. The piers required constant maintenance; the magnitude of this task is indicated by the permanent employment of fifteen masons and forty-nine labourers under the master mason, Henry Cull, followed in 1801 by George Louch as 'deputy engineer'.

Wyatt, paid £200 p.a. plus travelling expenses and fees for drawings, visited the site on average three or four times a year. He produced designs for the lighthouse, built in 1794–1795; it had Argand lamps and was the first, or one of the first, to emit flashes at regular, specified intervals. The storehouse, Harbour Master's house and a gate lodge followed from 1795 until 1797, and the Pier House for the harbour Trustees' meetings between 1800 and 1802.

Meanwhile in 1796 plans were being considered for what were to become the huge London Docks at Wapping and the even larger West Indies Docks on the Isle of Dogs. Wyatt submitted a scheme for the latter, but this and other proposals, notably by William Jessop and Ralph Walker (qq.v.) were laid aside for the time being until more detailed studies in 1799; in these Wyatt played no part.

However, in 1799 he became involved in another engineering project: the tunnel under the Thames between Gravesend and Tilbury proposed by Ralph Dodd (q.v.). Wyatt made the site investigation borings in February, and when work began on the shaft at Gravesend he and Mr. Ludlam, a mining engineer, were in charge of the operations. Such large influx of water was encountered from the chalk at a depth of 40 ft. that a steam engine had to be installed, replacing the horse gin previously employed to operate the pumps. They struggled on but in October 1802, with the shaft at a depth of 85 ft., the engine house caught fire. Funds had now run out and the project was abandoned, some £15,000 having been spent.

In April 1800 a Select Committee of the House of Commons Improvements in the Port of London decided to hold an open competition for designs of a new London Bridge with 65 ft. headroom above high water, allowing ships of 200 tons burden to pass upstream as far as Blackfriars Bridge. Several engineers responded. Wyatt submitted a design illustrated by a 'very elegant painted Model' of a bridge entirely in cast iron except for the granite piers and chalk infill above the arches for the roadway paving. No details of the structural ironwork are given, but they were presumably based on a patent taken out by him in June 1800. This shows cast iron arched ribs formed by segmental voussoir elements, the upper and lower members of which are of hollow

oval section. The ribs are connected transversely by perforated iron plates running across the whole width of the bridge and arched plates between the ribs carry the infill up to road level.

In its report of July 1800, the Committee favoured a 3-arch design in iron by Thomas Wilson (q.v.), engineer of the famous Sunderland Bridge of 1796. However, a few months later Telford submitted a revolutionary proposal for bridging the river in a single iron arch of 600 ft. span. A lengthy enquiry followed, published in June 1801 with inconclusive results. In the same year Wyatt's assistant, John Harvey, exhibited drawings of a cast iron London Bridge at the Royal Academy.

By that time works were under way at the West India and the London Docks. The need for accommodation for ships above London Bridge receded and the scheme for a high level bridge, with its technical and planning problems, was abandoned.

Until 1996 Wyatt's patent remained little more than a curiosity in the history of iron bridges. But a long-forgotten bridge was then 'discovered' spanning a small lake in the grounds of Culford Hall, Suffolk, a house remodelled 1790–1796 for the Marquis Cornwallis by Wyatt and enlarged by him in 1804. The bridge spans 60 ft. between granite abutments. It is 20 ft. wide with six arch ribs each composed of five segments butted together with lead packing. As in the patent, the upper and lower members of the elements are tubular, oval in section, and transverse perforated plates tie the ribs together. Solid iron plates form the spandrels up to the gently sloping roadway. There can be no doubt the bridge is a Wyatt design. Further, it can be dated from a note by John Buddle (q.v.) in his *Memoranda Book* referring to castings for 'the Marquis Cornwallis's Cast Iron Bridge by Wm. Hawkes & Sons, Gateshead, 1804': weight 80 tons plus 2 tons of lead, cost £1,457. The total cost including the abutments, transport of materials and erection would have been at least double that amount. Culford bridge has the largest span of the eight surviving cast iron arch road bridges in the world built in what J. G. James has defined as the trial period, 1790–1810, preceding the master works of the next decade by Telford and others. Wyatt's bridge is now listed Grade I.

In his 1800 patent Wyatt also describes a system of fire proof construction with groined vault floors formed of cast iron plates supported on iron columns. This is the earliest proposal for complete multi-storey iron framing in which the columns take the whole vertical load; the structural function of the external walls is only to provide lateral stability. He incorporated it in a third scheme for rebuilding Albion Mill in 1802. Like the earlier rebuilding plans, nothing came of this, but the system is said to have been adopted in Kew Palace, built at enormous expense from 1801 until 1811 to designs by James Wyatt and demolished 1828. No later use is known.

Wyatt's last works include Flamborough Head Lighthouse for Trinity House, 85 ft. high and built in 1806 by a local builder, and plans for a new dry dock at Ramsgate which were not carried out.

Wyatt seems to have found the company of engineers and industrialists more congenial than that of his fellow architects. He was elected to the (Smeatonian) Society of Civil Engineers as early as 1781, joined the Society of Arts in 1788 and, though not a member, dined with the Lunar Society of Birmingham from time to time. In 1780 his elder brother, William, died leaving a four year old daughter, Louisa. Samuel and his wife, who were childless, adopted her and brought her up. In 1788 the family moved to Albion Mill House, on the corner of Blackfriars Bridge. Saved by the party wall, they escaped injury in the 1791 fire. But life there must have been unpleasant in the immediate aftermath and at Matthew Boulton's invitation, Ann and Louisa spent two months at Soho House, Handsworth, with his daughter, Nancy, keeping house after the death of her mother.

Meanwhile the Berwick Street premises were retained as an office. His pupils lived there, among whom was his nephew, Jeffrey Wyatt (later Sir Jeffrey Wyatville) son of Joseph Wyatt. They formed his drawing office under John Harvey. He also kept a permanent staff of craftsmen under the foreman, William Oldroyd, and was probably employing a total of twenty to thirty people.

In 1792 Wyatt had been appointed Clerk of Works to the Royal Hospital at Chelsea, an agreeable position with light duties, a salary of £200 p.a. and accommodation. He and Ann moved there in 1798, Louisa by this time having married the Rev. Thomas Cobb of Lydd; later as a widow she married Admiral Robert Lambert. Finally, Wyatt gave up Berwick Street in 1803 and transferred the building business to a new yard on the site of Albion Mill, while moving his architectural office to Surrey Street, off the Strand. At the same time Harvey left to set up an independent practice and Noah Siddons took his place.

Wyatt died quite suddenly on 8 February 1807 and was buried in the graveyard of the Royal Hospital. His will, leaving what was probably a comfortable fortune to 'my dear and affectionate wife Ann Wyatt', was proved ten days later. Several of his works left unfinished were completed by his nephew and former pupil, Lewis William Wyatt, son of his brother, Benjamin (q.v.) land agent to Lord Penrhyn.

A portrait of Wyatt by Lemuel Abbot (reproduced in Robinson 1979) is in a museum in Michigan, U.S.A. A group portrait by Dupont in Trinity House shows Wyatt submitting his design in 1793.

A. W. SKEMPTON

[Trinity House Corporation Minute Books and Cash Books (1790–1810) Trinity House; Report from the Select Committee (1800) on the Improvement of the Port of London, House of Commons; John Buddle (1804) Memoranda Book, Northumberland RO; H. W. Hart (1964) Ralph Dodd and the Gravesend and Tilbury tunnel scheme, *J Transport Hist*, **6**, 249–857; A. W. Skempton (1971) The Albion Mill foundation, *Geotechnique*, **21**, 203–210; A. W. Skempton (1971) Samuel Wyatt and the Albion Mill, *Architect Hist*, **14**, 53–73; J. M. Robinson (1973) Samuel Wyatt at Ramsgate, *Architect Hist*, **16**, 54–59; J. M. Crook and M. H. Port (1973) The royal palaces at Richmond and Kew, *Kings Works*, **6**, 354–359; J. M. Robinson (1979) *The Wyatts, an Architectural Dynasty*; J. G. James (1988) The evolution of early iron arched bridge designs, *Trans Newcomen Soc*, **59**, 153–186; J. G. Coad (1989) *The Royal Dockyards 1690–1850*; Colvin; David Perrett (1997) Culford's 'lost' bridge, *Indust Archaeol News*, **100**, 6; B. L. Hurst (1999) An iron heritage, *Structural Engineer*, **77**, 17–25]

Publications

1796. Explanation of the advantages to be attained by adoption of the docks in the Isle of Dogs [with] plan of the proposed docks at the Isle of Dogs, *Report from the Committee on the Port of London*, House of Commons

1800. Patent No. 2410. A method of making and constructing bridges, warehouses, and other buildings, without the use of wood, *Repertory of Arts and Manufacturers*, **14**, 1801, 145–149

Works

Works include:

1783–1786. Albion Mill, Southwark, £12,000
1784–1786. Commissioner's House, Portsmouth, £15,000
1790–1791. Dungeness Lighthouse, £4,000
1793–1796. Trinity House, London, £25,000
1794–1802. Lighthouse and buildings, Ramsgate Harbour
1804. Culford iron bridge
1806. Flamborough Lighthouse, £2,500

WYLD, James, Sr. (1790–1836), and **James, Jr.** (1812–1887), were both involved in the production and sale of maps.

James Wyld Sr. became Geographer Royal and served in the Quartermaster General's Office for fourteen years. As part of his duties he was responsible for supplying plans of the Peninsular War and, while doing this, introducing the art of lithography to England. He became one of the leading European map makers and a member of many learned societies in Britain and overseas. The quality of his maps no doubt brought him to the notice of civil engineers and in 1825 he was elected an Associate of the Institution of Civil Engineers.

He remained a member until his death, from overwork, on 14 October 1836.

James Wyld Jr., his son, followed his father's interests and acquired William Faden's (q.v.) business. His maps and atlases were well-known and popular and he owned three shops in London. From the 1830s he produced a large number of railway maps as well as printing Herapath's *Railway Magazine*, the first significant railway journal. Unfortunately, in the railway mania he produced a large number of plans for what proved to be unsuccessful companies and found it impossible to reclaim his costs. Other works of engineering interest included the sections of London Strata which he produced for R. W. Mylne. He was involved in politics, served as a Liberal MP for Bodmin, and was a great supporter of technical education.

He died at his house in South Kensington on 17 April 1887 leaving a son and daughter.

MIKE CHRIMES

[Membership records, ICE archives; *Min Proc ICE*, **1**, 1837, 7; *Gentleman's Magazine*, **2**, 1836, 656; DNB; *Journal of the Royal Geographical Society*, **21**, 69]

Y

YARNOLD, William (1668–*c*. 1720), entrepreneur, was baptised at Grimley, near Worcester, on 10 December 1668, the third of five sons of John and Elizabeth Yarnold.

Yarnold, with his brother John (b. 1644) and others, made an agreement with the Mayor of Oxford in 1694, enabling him to supply the town with water. The scheme envisaged the use of a water wheel located within the third arch of South, or Folly, Bridge with a forcing engine to transfer the water to a cistern, 'artificially and ornamentally erected upon columns or pillars of the heighte of ten feete from the pavement'. The partners were required to provide two sets of flood gates at the bridge but it was a condition that they should not impede navigation or drainage; they were granted a virtual monopoly of the water supply to the town.

Continuing his water interests, Yarnold arranged a lease with the town authorities of Newcastle in 1697 for the provision of a supply of water there. To safeguard his entitlement, he obtained an Act of Parliament the following year, after which three 'inquisitions' were held to assess damages; successful conclusion led to the commencement of work. The scheme comprised the tapping of springs on Gateshead Fell—some four miles south of Newcastle—and the transfer of water in elm or fir pipes into the town. An existing natural pond was utilised for storage and two earth-embankment reservoirs were built in Gateshead, from which water was carried across the Tyne bridge in lead pipes. Yarnold was also given consent to build or reconstruct three cisterns in Newcastle for the retention of water prior to its distribution to the houses of the town's wealthier inhabitants.

It is possible that Yarnold was involved, too, in the water supply to Windsor, where his brother, a 'plumber of Worcester', obtained a five hundred year lease from the Corporation and constructed works at Windsor bridge. Work began in 1697 and was completed in 1701, water from the river being pumped to a cistern 20 ft. square. It is perhaps worthy of notice that John Yarnold was granted a patent in 1698 for the construction of an engine 'useful for draining mines ... and for raising water for the supply of towns'.

In 1701, Yarnold—with Robert Walton—became involved in the formation of the Ravensbourne Waterworks, established in south-east London to supply Deptford and Greenwich and he petitioned for powers to take water from the river Ravensbourne. At the time of the petition some £600 had already been spent on a 'water house near the said river and fixing a forcing engine therein and laying pipes underground from thence' to East Greenwich. Letters Patent granted Yarnold and his associates a lease of five hundred years, obliging them, among other things, to provide a supply to Greenwich Hospital at rates to be approved by the Treasury. The pumping plant installed by Yarnold was reconditioned by John Smeaton (q.v.) in 1779.

R. W. RENNISON

[Patent No. 355, 19 July 1698: *Draining mines ...* [and] *raising water for supplying towns*; H. W. Dickinson (1954) *Water Supply of Greater London*, 52–54; H. E. Salter (1960) *Properties of the Corporation of Oxford*; G. C. Cullingham (1968) Windsor and Eton waterworks, *Industrial Archaeology*, **5**, 1; R. W. Rennison (1977) The supply of water to Newcastle upon Tyne and Gateshead, 1680–1837, *Archaeologia Aeliana*, 5 Series, **5**, 179–196]

Works
1694–1697(?). Water supply to Oxford
1697–1701. Water supply to Newcastle
1697–1701. Water supply to Windsor
1701–. Formation of Ravensbourne Waterworks, Kent

YARRANTON, Andrew (1619–1684), ironmaster, agriculturist and river engineer, was born at Larport in Astley, north Worcestershire, the son of Walter Yarranton (died 1631) and his wife Sarah, coming from a yeoman family long settled there, and was christened there on 29 August 1619. He was apprenticed to a linen draper in Worcester *c.* 1632 but left and 'lived a countrey-life for some years'. It is not known how he learnt his engineering skills. He served in the Parliamentary army during the Civil War, becoming a captain by 1645. In 1648 he discovered and interrupted a meeting at Boscobel of Royalists, who were intending to seize Dawley Castle, and arrested the leaders, including Colonel Dud Dudley, the iron innovator, as a result of which Dudley only barely escaped being executed; for this service Yarranton was awarded £200 by Parliament. From 1651 to 1653 he served as a Commissioner for Sequestrations for Worcestershire.

In 1651 he with his fellow Captain Godfrey Ellis of Gloucester and other soldiers used their arrears of pay to buy extensive coppices in the manor of Bewdley, part of the Wyre Forest. Subsequently he also acquired rights to Alton Woods in Rock and also in woods in Bayton, both nearby. He and his partners also bought 'Roman' iron cinders from Pitchcroft in Worcester and probably Forest of Dean iron ore and used them in a blast furnace they set up at Astley. This business was continued for some years, but the furnace was apparently in the other hands by 1662. However Yarranton still had a share in a furnace, at 'Sudley' (?Sudeley, near Winchcombe), which he was trying to sell in 1673.

In 1655 with Captain Wall he agreed, for £750 and a beneficial lease of salt vats there, to make the river Salwarpe navigable to connect Droitwich with the Severn so as to avoid the expense of carrying by land Droitwich salt to Worcester and coal to Droitwich. 'But times being unsettled, and Yarranton and Wall not rich, the scheme ... was never carried into execution'. Improvements were started on the Salwarpe some years after George, Earl of Bristol, Thomas, Lord Windsor (later Earl of Plymouth), and Thomas Smythe (a barrister) obtained an Act of Parliament in 1662 for making navigable the rivers Salwarpe and Stour. Yarranton suggested in December 1666 that 'Salwarpe ought not yet to be taken in hand', but work was begun by 1669; however according to Yarranton, when five of the six locks needed were 'compleated the plan was found not to answer and the sixth lock was never built'.

At the Restoration Yarranton was imprisoned by Lord Windsor—as Lord Lieutenant for Worcestershire—'for refusing his lordship's authority'. He was free in November 1661 when compromised by the discovery of some letters concerning an intended Presbyterian rising. On 16 November an order issued to Sir John Packington for his arrest, but in May 1662 'the escape of Andrew Yarranton, a person dangerous to the government from the custody' was reported. After 'meetings with several disaffected persons' he went to London where he was re-arrested. In his own version, *A Full Discovery of the First Presbyterian Sham Plot*, published in 1681 when he was again under political suspicion, he declared that the letters were forgeries planted by Sir John and that, after his wife discovered their origin in April 1662, he provoked an uproar in Worcester. When the Lord Lieutenant and his deputies called on him for an explanation of this they could not deny his allegations and crept away by the back stairs leaving one of their number who released him. He went up to London 'and prevails with the lord of Bristol to acquaint 'the King of the great wrong he had received', was arrested there, but was quickly released by the King's order. Less than six months later he was again arrested for having 'spoken Treasonable Words against the King' but was acquitted by the jury at Worcester assizes, because the prosecution witnesses were not credible.

Andrew Yarranton was among the first to appreciate the value of clover, one of the so-called 'artificial grasses', which were an important feature of the agricultural improvements of the seventeenth and eighteenth centuries. He wrote two pamphlets on the subject, *Yarranton's Improvement by Clover* (1662?) of which no surviving copy is known and *The Improvement improved by a Second Edition of the Great Improvement of Land by Clover* (1663), naming persons from whom seed could be obtained. He claimed by this means to have doubled the value of much of the land in the west Midlands.

Yarranton's main achievements as an engineer concern river navigation. The Navigation Act of 1662 included a clause for continuing the navigation of the river (Warwickshire) Avon. Lord Windsor acquired the navigation rights of that river from the Duke of York's trustees in 1664, and retained the river below Evesham himself. He employed Yarranton 'to set the river in order and to repair Pershore Sluice' and then granted him a proportion of the tonnage toll from the river for keeping it in repair. In 1665 Lord Windsor and Yarranton formed a syndicate, which spent £1,800 to restore the navigation of the upper Avon up to Stratford, Yarranton carrying out the necessary work, probably completed by the end of 1665. He assigned his share in the upper Avon in 1679 to his son Robert (q.v.), from whom it passed in 1681 to Mary Viccaris (perhaps a relative) and the following year to Andrew (another son), who was in 1684 seeking to recover his share of the profits from Francis Little, the manager; however Little stated that he had been instructed by some of the other partners not to pay him because the older Andrew had 'cozened them out of about £200'. This navigation was a profitable enterprise, the dowager countess of Plymouth claiming a decade or so later that she drew £400 p.a. from the Lower Avon and her third of the Upper Avon. Andrew Yarranton (the son) claimed his two-fifteenths of the Upper Avon yielded £30 p.a., though the manager reported much lower figures, probably because of the cost of repairs. This was a successful improvement and long remained in use.

The river Stour, a smaller river, was much encumbered by mills, mainly forges and slitting mills that served the Black Country iron manufacturing industry. The prime objective in making it navigable was to enable coal mined in Amblecote and Pensnett Chase (near Stourbridge) to be sold along the river Severn, but in this it was always liable to suffer from significant competition from the coalmasters of the Broseley and Madeley area in Shropshire, whose coal could be brought to larger barges via quite short railways. Nevertheless a great deal of effort was expended on the works on the river. It is probable Andrew Yarranton was a prime mover of this scheme from the start, since he witnessed the initial agreement in March 1661, providing for certain coalmasters mining in Pensnett Chase to sell 10,000 tons (of 25 cwt) or more of coal per year at the pithead at 3s per ton. The buyers then contracted to sell 15,000 tons to the navigation proprietors on the riverbank at 4s 6d per ton and a footrail (i.e. railway) was to be built from the mine to the river. They then obtained an Act of Parliament to enable the rivers Stour and Salwarpe to be made navigable. Andrew Yarranton joined the footrail partnership in July 1661, but in December 1664 disposed of much of his interest.

The navigation partnership agreement between Bristol, Windsor, and Smythe provided for Smythe to make the river navigable up to Stourbridge within two years of the Act of Parliament, but it was still not navigable in 1665. Yarranton was apparently employed by them to carry out the works, which proved more expensive than expected, forcing Smythe to sell part of his share to raise the excess. Eventually in November 1666 Yarranton was able to report that a boat of 7 tons had reached Stourmouth. Negotiations took place over the following winter for Yarranton to manage the business, but Windsor declined to advance money 'to finish the river ... and make a trade upon' it, unless the river was partitioned among the partners. Instead Windsor nominated a manager, but he declined to advance the necessary money (as was intended) lest he should die in debt.

In order to make the river navigable it was necessary to make a cut with a lock so that boats could pass each mill, something the millers did not like, as they considered that it reduced the amount of water available to drive their mills. The locks seem to have been pound locks, since there are references to locksful of water and to making the barges fit to carry 10 tons 'by enlarging them as well as the locks in length'. Additionally 'turnpikes', probably flash locks, were set up in other places. At Halfcot (in Kinver) a long cut was made from Wedge's (now Bells) mill and in 1664 a mill was built at the end of this to use such water as was not required for the navigation. This Trench Brook, as it was later known, continued in use as a mill leat until the closure of the mill in modern times. A similar strategy was adopted in respect of a shallow ford at Wolverley, where Joshua Newborough and Philip Foley were persuaded in 1669 to site a new slitting mill and forge, which was also intended to produce tinplate. The straight cut, still used as one of the river's channels past the Wyre Mill (as it is now known), was made at this time. By this means others could be persuaded to bear the cost of works that were also useful for the navigation.

Tinplate was at this period made in Saxony and imported from Hamburg. In March 1667 an agreement was made between the Stour navigation proprietors, several local ironmasters, Andrew Yarranton and Ambrose Crowley for establishing production in England, each sponsor putting up ten pounds. To this end Yarranton and Crowley 'a

competent fireman' travelled to Saxony to observe the process. On their return they carried out experiments at Kings Meadow (later Royal) Forge at Stourbridge and also at Wilden (below Kidderminster), where the slitting mill was used to roll plates, a process that was probably their own invention. The initial results were not wholly satisfactory and Yarranton perhaps made a second visit to Germany. Further experiments were carried out in January 1668, Crowley making the plates and Yarranton and his wife tinning them. The partners wanted a patent, but this was apparently unobtainable because Edward Chamberlaine and Dud Dudley already had one, although they did not know how to exploit it. Perhaps as a result of hostility going back to the Civil War, terms could not be reached between the patentees and the tinplate entrepreneurs, and as a result most of the partners decided not to go ahead. However Joshua Newborough and Philip Foley did build their mill at Wolverley and apparently made tinplate for a time. However in 1672 Chamberlaine succeeded in renewing his patent and tinplate manufacture at Wolverley was abandoned. Yarranton examined possible sites in Ireland for tinplate, and in 1675 his eldest son Andrew was to be sent 'into Germany to bring over the men intended', presumably intended to make tinplate. Possibly the works were to be at Clomolin in County Wexford, the subject of a report (perhaps by Yarranton), mentioning 'German plates'. However nothing came of any of this until the next century when the son of the slitter at Wolverley was connected with the successful production of tinplate at Pontypool.

In late 1667, after his return from Germany, the Stour Navigation was apparently in use and coal was being sold at Worcester, but money was short and it is not clear how much the navigation was used. In February 1670 Yarranton offered to make the river 'fully and sufficiently navigable' within two years at his own expense on the basis that he should have a third share, the proprietors contributing £400. The navigation was to have been completed by midsummer 1672, but in summer 1673 work was still in progress and Yarranton raised £500 by the sale of wood from his Alton Woods for his son Robert to lay out on the navigation.

After this Yarranton also undertook a number of other surveys. In July 1674 he surveyed the river Dee, drawing up a plan to reopen the river to shipping, at an estimated cost of £5,000. A slightly different scheme was carried out in the 1730s by Nathaniel Kinderley (q.v.) costing £52,000. That November Yarranton was in Ireland surveying the Enniscorthy Ironworks and the river Shane, near the intended tinplate works there, and later he went down to Hampshire to survey the Hampshire Avon to 'find whether that river might be made navigable [and] a safe harbour ... made at Christchurch'. That river is said to have been made navigable, though probably by others, for boats of up to 25 tons and a haven begun, but it fell into disrepair and was little used after 1705. While at Christchurch he identified a deposit of ironstone and suggested the use of this and wood from the New Forest to produce iron there.

While in Dublin in October 1674 Yarranton bought some coal deposits from Ananias Henzey, a member of one of the Stourbridge glassmaking families. To exploit this he proposed a 'footrail under ground', evidently also intended as a sough for dewatering the mines. It seems it was not made, but Yarranton was probably engaged in mining coal, until in November 1678, he handed transferred his interest in Henzey's coals to 'his son Osland'. Around this time negotiations were in progress for Sir Clement Clerke and Alderman John Forth, who had acquired certain local ironworks, to advance £2,000 and employ Yarranton to complete the Stour Navigation, for which they would receive Yarranton's share in the river together with a half share in the underground footrail. However the associated ironworks business was the subject of litigation due to Clerke having mortgaged his share of the ironworks partnership and nothing was achieved.

When that litigation was settled in 1676, Robert Yarranton and William Farnolls were appointed as contractors, but it was suggested the underground footrail should not now be made: 'Those works of darkness are best laid aside with the author', also: 'Go not on to spend your money on gimcracks that are unnecessary and uncertain'. However again nothing was apparently then done. In January 1678 Andrew Yarranton gave up his claim to an interest in the river in exchange for a life annuity of £30 out of the profits, as part of arrangements for Robert Yarranton and Farnolls to make a further attempt to complete the navigation.

Yarranton is today largely remembered as a result of his book *England's Improvement by Land and Sea ...* issued in 1677 with a continuation in 1681. This proposed a considerable number of improvements including the introduction of flax cultivation and linen manufacture. He also proposed connecting the Severn and Thames by making the river Cherwell navigable to Banbury and Warwickshire Stour to Shipston, greatly shortening the distance where land carriage was necessary between the two river systems. He also suggested making the Dee navigable to Bangor Bridge. Linked to the former scheme was one for establishing public warehouses for grain and other goods including one at Stratford. He claimed to have surveyed the Thames, Severn, and Humber, the former in summer 1676 with his son Robert. His book was ridiculed in a pamphlet, *The Coffee House Dialogue*, which suggested Yarranton wanted to turn London's streets into navigable rivers and this was followed by other pamphlets in reply.

In the period between the appearance of the two parts of *England's Improvement*, he carried out surveys on the river Mole up to Boxwood and of the river Chelm up to Chelmsford, as well as of the harbour at Newhaven, Sussex. In 1679 he

offered to advise on coal works near Ruperra in Monmouthshire, and improving the river Rhymney there, but nothing came of this. He also, probably on his own initiative suggested improvements to dockyards and the fortification of Tangier, his object almost certainly being to obtain employment for himself and his sons.

Yarranton's main residence throughout his life was at Astley, where he was born, but evidently visited London regularly. In some respects Yarranton's schemes were ahead of his time and anticipated until the era of stillwater canals a century later. On the other hand it seems evident that he grossly underestimated the difficulty of several of them particularly the river Stour where it is probable that considerable sums would have been required each year just to keep the works in order. His financial resources were probably always fairly modest and the loss of the money he invested in the river Stour, probably left him with relatively little. He was married and had two sons, Andrew and Robert, and at least three daughters. By religion he was probably a Presbyterian and was suspected in 1681 (possibly due to mistaken identity) of involvement in the Popish Plot, which was the occasion for his publication of the *First Presbyterian Sham Plot*. His death in 1684 was a violent one; according to John Aubrey, 'Captain Yarrington dyed at London in March last. The cause of death was a beating and throwne into a tub of water'.

P. W. KING

[Aqualate collection, Stour navigation papers, D(W) 1788/P37/B8, P43/B10, P59/B3, P61/B5–7, Staffs. RO; Chancery proceedings, C5/198/15 C8/222/72 C22/216/11&16, PRO; E12/VI/KE/24, f.1, Herefs. RO; E12/VI/KC/33-47B; E12/VI/C/6; E12/VI/KT/1-13; E12/IV/30, 26 June 1681, Herefs. RO; E12/S, Halfcot/1-2 and Kinver VIII (by kind permission of Mr. A. T. Foley), Herefs. RO; Diary of George Skippe, LC mss. W.920 oversize, Hereford library; E320/V20-V21 and SP24/15–16, various (on microfilm 'unpublished state papers of the interregnum', Harvester Press), PRO; Prattinton collection, Worcestershire Parishes 13, 59, 290, Society of Antiquaries; *Calendar of State Papers Domestic 1648–1649*, 206; *1660–1661*, 355–356; *1661–1662*, 138, 143, 148, 149, 383, 385 and 417; *1680–1681*, 93, 412, 424; DNB (New); T. R Nash (1782) *Collections for a History of Worcestershire*, i, 306; S. Smiles (1863) *Industrial Biography*, 60–76; E. A. B. Barnard (1929) The navigation of the Avon from Stratford to Evesham in 1664, *Evesham Journal*, 20 March–29 April 1929, 11, 13 col. 1–2; T. C. Cantill and M. Wight (1929) Yarranton's works at Astley, *Transactions of the Worcestershire Archaeological Society*, NS, **7**, 92–115; J. M. Palmer and M. I. Berrill (1958) Andrew Yarranton & the navigation works at Astley, *RCHS J*, **4**, 41–6 (comment 77–78); T. W. M. Johnson (1952–1953) Captain Andrew Yarranton and Herefordshire, *Transactions of the Woolhope Naturalists' Field Club*, 39–42; G. H. C.

Burley (1961) Andrew Yarranton: a seventeenth century Worcestershire Worthy, *Transactions of the Worcestershire Archaeological Society*, NS, **38**, 25–36; C. Hadfield (1968) The Avon navigation, in C Hadfield and John Norris (eds.), *Waterways to Stratford*, 2nd edn., 15–22; M. M. Hallett and G. R. Morton (1968) Yarranton's furnace at Sharpley Pool, *Worcs J Iron Steel Institute*, **206**, 689–692; J. H. Parker Oxspring (1979) Andrew Yarranton, 'Worcestershire Worthy' 1619–1684: his life and work with special reference to ... river Stour, TS, 2 vols., Worcester City Library, class WQ B/YAR; P. J. Brown (1982) The early industrial complex at Astley, Worcestershire, *Post-Mediaeval Archaeology*, **16**, 1–19; P. J. Brown (1988) Andrew Yarranton and the British tinplate industry, *Historical Metallurgy*, **22**, 42–48; P. W. King (1988) Wolverley Lower Mill and the beginnings of the tinplate industry, *Historical Metallurgy*, **22**, 104–113; P. J. Brown (1992) The military career of Andrew Yarranton, *Transactions of the Worcestershire Archaeological Society*, 3rd series, **13**, 193–202]

Publications

In the following an asterisk (*) denotes that the authorship is uncertain, and a dagger (†)denotes that no surviving copy is known.

1661. *Reasons for making Navigable the Rivers Stower and Salwerp ... in the Counties of Worcester and Stafford*, D(W) 1788/P59/B3, Staffs. RO*
1661. *Reasons wherefor the Making of the Rivers of Stower and Salwerp navigable ... will be of Advantage especially to the Town of Bridgenorth Wenlock Wellington and Towns adjoining the ... Severn*, 816.m.8(50), BL*
[1661?. Answered by anon., *An Answer as well to a Paper entitled Reasons wherefore the Making of the Rivers Stower and Salwerp will be of Great Advantage ... as also to a Paper intituled An Answer to some Partial Pretences called Reasons dispersed by some Shropshire Coal-masters*, 816.m.8(51), BL]
1662. A *New Map of the Town of Dunkirke New Harbour and Castle on the Sea*, n.d. (? before 1662), C.20.f.6(25), BL
1662. *Yarranton's Improvement by clover* (cited in Yarranton (1663) and (1677), 194)†
1663. *The Improvement Improved by a Second Edition of the Great Improvement of Land by Clover*
1677. *England's Improvement by Land and Sea: How to beat the Dutch without Fighting, to Pay Debts without Money, to Set at Work all the Poor of England with the Growth of our own Land, to Prevent unnecessary Suits at Law wit the Benefit of a Voluntary Register ... with Advantage of making the Great Rivers Navigable ...* (containing surveys of the rivers Dee (Cheshire), Slaney (Wexford), Avon (Hampshire) and Thames, amongst others)
c. 1677. Printed map (without title) showing how the great rivers of England may be made navigable, Add. MSS 4473,f.30, BL

1679?. *The Coffee House Dialogue Examined and Refuted ...*
1679?. *England's Improvement Justified and the Author thereof Captain Y. Vindicated from the Scandals of a Paper called the Coffee House Dialogue with some Animadversions upon the Popish Designs therein contained*
[1679. Replies by Anon., *The Coffee House Dialogue between Captain Y. and a Young Barrister with some Reflections on the Bill against the Duke of York London* and *A Continuation of the Coffee House Dialogue ...*]
1681. *England's Improvement by Land and Sea. Part ii* (containing a survey of Newhaven Harbour and details of various other schemes)
1681. *A Full Discovery of the First Presbyterian Sham Plot or a Letter from One in London to a Person in the Country*

Reports, etc., unpublished:

A Short Account of the River Avon and Mannor of Settlement with Interests upon the same and Mannor how they were Settled, D(W) 1788/P61/B7, Staffs. RO
Reasons why the River Stower is the best Place in Europe for Perfecting Tynne Manufacture, D(W) 1788/P61/B7, Staffs. RO
What Doctor Gorge hath to Parte with in Ireland as to Woods and Workes (at Clomolin, Co. Wexford), E12/VI/C/6, Herefs. RO*

Works

1662–1666. Stour Navigation, engineer, £4,500
1664. Avon Navigation (Lower), restoration, undertaker?, including Pershore sluice reconstruction, £40
1664. Avon Navigation (Lower), maintenance, undertaker
1664. Avon Navigation (Upper), restoration, engineer, c. £1,800
1665. Footrail at Amblecote, engineer?
c. 1668. Salwarpe Navigation, engineer, £658 or £720
1670–1673. Stour Navigation, restoration and improvement, undertaker, c. £1,200 (£400 and one-third share)
1674–1678. Amblecote Colliery, proprietor?

YARRANTON, Robert (*c.* 1650–1681), river engineer, was the second son of Andrew Yarranton (q.v.) of Astley, Worcestershire and his wife. He and his brother, Andrew, were apprenticed to Nicholas Baker, a Worcester mercer, who was appointed to manage the navigation of the river Avon above Evesham 'to breed them up in such trade as should be had upon the river'. Their father and Baker were partners with Lord Windsor in the Upper Avon navigation and their father was employed in restoring the Lower Avon navigation.

In summer 1673 Robert Yarranton worked on improvements to the river Stour in north Worcestershire, probably under the supervision of his father who had already been engaged on them for some ten years without achieving any permanent result. In July 1674 Sir Clement Clerke and Alderman John Forth agreed to advance a further £2,000 for paying off the accumulated debts of this enterprise and completing the Stour navigation, which was supposed to be done by Michaelmas 1676. This was associated with a substantial nearby ironworks business which they had bought, but that business ran into problems due to Sir Clement having mortgaged his share of that partnership and in consequence nothing was achieved except that debts of £242 were paid. After the resultant litigation between Forth and Clerke was settled by a sale of the ironworks business, Clerke, with George Skippe of Ledbury in Herefordshire, took over Forth's share of the navigation. In December 1676 Clerke and Skippe undertook to complete the navigation by Michaelmas 1678; apparently little was done, probably because they could not raise the money.

In January 1678 they employed William Farnolls of London and Yarranton to make the river navigable by midsummer 1679, this time for vessels of at least 12 tons. £1,758 was to be paid to Farnolls and Yarranton by instalments as work proceeded. The works progressed slowly, but Farnolls reported in January 1679, 'Upon Christmas Eve we brought down two barges with coales to Kidderminster one laden with 14 tons and the other with 15 tons, but can pass no more till supplied with money to discharge debts'. The difficulty was that Sir Clement's financial resources had been exhausted. An estimate, possibly of this date, indicates a further £750 was needed to reach the Severn and this evidently was not forthcoming, there was still £400 'remaining in Sir Clement's hands'.

A year later there was a rumour that Colonel Archer (of Umberslade, Warwickshire) was prepared to advance £1,000 to complete 'our Stour', being repaid in coal at the mouth of the Warwickshire Stour, but the rumour was either false or concerned an unsuccessful negotiation. A few days later George Skippe wrote that he relinquished his 'concern in the river to Sir Clement (and) had better given £500 then have medled with this and some other concerns with Sir Clement'. The following June some one surveyed the footrail at the head of the river and found one of Henzey's colliers had removed the rails and claiming Robert Yarranton had authorised this in satisfaction of a debt. This Henzey owned coal mines near the footrail earlier worked by Yarranton's father.

Yarranton joined his father in his survey of the river Thames in summer 1677 proposing improvements, which his father's *England's Improvement* (1677) suggested he should undertake; he also surveyed the river Cherwell in connection with his father's scheme for making the Cherwell and Warwickshire Stour navigable towards connecting the Thames and Severn. The rumour about Colonel Archer may have resulted from some discussions about this. In March 1679 Andrew

Yarranton assigned to Robert his share of the Upper Avon Navigation, which Robert in turn in April 1681 assigned to Mary Viccaris of Astley, probably a relative. Robert died that November having outlived Robert Yarranton junior, presumably his son, by six months.

P. W. KING

[Aqualate collection, Stour navigation papers, D(W) 1788/P37/B8, P59/B3, P61/B5–7, Staffs. RO; Diary of George Skippe, LC mss. W.920 oversize, Hereford library; Chancery proceedings, C5/198/15, PRO; Foley collection, E12/VI/KC/33–47B, Heref. RO (by kind permission of Mr. A. T. Foley); J. H. Parker Oxspring, *Andrew Yarranton, 'Worcestershire Worthy' 1619–84: his Life and Work with Special Reference to ... River Stour*, TS, 2 vols., Worcester City Library, class WQB/YAR; A. Yarranton (1677 and 1681) *England's Improvement by Land and Sea ...*, 64, 189]

Works

1673. Stour Navigation, improvement, undertaker, £500

1678–1680. Stour Navigation, restoration and improvement, undertaker, £1,350

YATES, William, Sr. (fl. *c.* 1750–1820), and **William, Jr.** (1781–1865), ironmasters, probably originated in the Rotherham area. William Yates Sr. was the cousin of Samuel Walker (II) and Joshua Walker (I) (qq.v.), ironmasters of Rotherham, and established a reputation as a foreman for the casting and erection of the iron bridges with which Walkers became involved from 1790 onwards. There are continuing references to the Yates's ability, Thomas Paine (q.v.) lamenting the father's unavailability for the erection of his model bridge in 1790. Yates generally worked on site for the erection of the bridges of Thomas Wilson (q.v.), advising on repairs to Boston bridge in 1814 and earning John Rennie's (q.v.) absolute confidence at Southwark, where he modified the details.

Yates Sr. was a stout and stalwart figure and his eldest son, William, 'a fine handsome man standing 6 ft. 3 in. high', joined with Samuel Walker (III) in 1817 in purchasing Gospel Oak ironworks, Staffordshire. In 1822 they transferred cannon making machinery from the Holmes works to their new manufactory and by early 1824 they were employing about 2,000 workers, including miners, with seven hundred men employed in the ironworks itself. In 1823 another son of Yates Sr., James, went into partnership with the Sandford family on the break up of the Walkers' concern and took over Clays New Foundry, renaming it the 'Phoenix Works, Rotherham', but carrying out no structural work of any significance.

In 1829 William Yates Jr. was elected a Corresponding Member of the Institution of Civil Engineers. In the late 1820s he was involved with the supply of the ironwork of a number of suspension bridges. In 1832, because of this experience, he was consulted by Count Szechenyi about suspension bridge design, referring Szechenyi to William Tierney Clark (q.v.) Szechenyi was promoting a bridge across the Danube to link Buda and Pest; Yates Jr. subsequently worked on its construction. The name of William Yates Jr. was erased from the roll of the Institution of Civil Engineers in 1844.

MIKE CHRIMES

[Rennie reports, ICE archives; Rennie papers, NLS; *Reports of Select Committee on Artizans and Machinery* (1824) 116–123; *Mechanics Magazine*, **8**, 1827, 335; *Civil Engineer and Architects Journal*, 1842, 397; J. Guest (1879) *Historic Notices of Rotherham*, 485–504; G. Rennie (1818) Account of experiences on the strength of materials, *Phil Trans*, **108**, 118–136; G. Rennie (1842) On expansion of arches, *Trans ICE*, **3**, 201–218; J. G. James (1979) The cast iron bridges of Thomas Wilson 1800–1810, *Trans Newc Soc*; J. G. James (1986) The cast iron bridge at Sunderland, *Newcastle-upon-Tyne Polytechnic, Occasional Papers in the History of Science and Technology*, **5**; J. G. James (1988) Some steps in the evolution of early iron arched bridge designs, *Trans Newc Soc*, **59**, 153–187]

Works

(For works before 1820 see: *WALKER family*)

1826. Hammersmith suspension bridge
1828–1829. Montrose suspension bridge
1829–1831. Scotswood suspension bridge
1839–1849. Budapest (Szechenyi) suspension bridge

YEOMAN, Thomas, FRS (1709/10–1781), millwright, land surveyor and civil engineer, was born in 1709/10. Neither the date nor the place of his birth are known but various indications, including a freehold property in Trudoxhill, suggest that he originated from Somerset, probably from one of the small villages to the south west of Frome. The first thirty years of his life are completely obscure and nothing is known of his education or whether he was apprenticed into any trade.

Yeoman can first be identified with certainty in 1742 when he was appointed by Edward Cave, the proprietor of the *Gentleman's Magazine*, to manage his cotton-mill in Northampton; most probably he came from the Abbey Mills at Bromley-by-Bow, where he had been employed as a wheelwright. Cave's cotton-mill, although ultimately a failure, was a significant milestone in the history of the textile factory, employing an early form of spinning machine, patented in 1738 by Lewis Paul. Yeoman was associated with the venture, seemingly as both mechanic and manager, until its eventual demise in *c.* 1756.

Moving to Northampton, Yeoman soon established himself in the town and, in addition to his position at the cotton-mill, promoted his own business as a millwright from premises in St. Mary's Street. In December 1743 he advertised his

services, including the construction of various agricultural devices, weighing machines, hydraulic engines and ventilators. From 1744–1746 he corresponded with the Rev. Stephen Hales, mostly relating to the ventilators invented by Hales which Yeoman undertook to build and install; he was responsible for erecting these machines in numerous hospitals and prisons, including those at Northampton, and later in the Drury Lane Theatre and the Houses of Parliament.

The first inkling of his future career as a civil engineer came in October 1744, when it was announced in the *Northampton Mercury* that 'Mr Yeoman, Engineer, is to survey the River Nene from Thrapston to Northampton and estimate the cost of completing the navigation'. Yeoman's suggested involvement in an earlier navigation scheme for the river Chelmer in 1735 cannot be substantiated in the records and it is not clear how he first came to be involved in surveying. Throughout the 1740s Yeoman's own business in Northampton seems to have prospered. He was admitted a freeman of the town, on payment of the traditional fee of 20 marks, in January 1745 and by 1746 had moved to Gold Street. From here he advertised an even wider range of products and services, including a trade in scientific instruments, and began to lecture on subjects of contemporary natural philosophy.

Yeoman's brief career as an itinerant philosophical lecturer probably came about through the Northampton Philosophical Society, founded in November 1743. Although it seems to have flourished for little more than a decade, it proved to be quite influential not only as a philosophical society, but also as a force for civic development within the town. Yeoman was one of the first members and was invited to formulate the rules; he was subsequently elected president on more than one occasion. It was in Yeoman's own house that the society held its weekly meetings and in 1748 he advertised 'an Experiment room' permanently maintained there. From this secure base he ventured farther afield and, in the winter months of 1746–1747, toured Coventry and Birmingham, styling himself a 'Member of the Philosophical Society at Northampton' and delivered a course of lectures on electricity and magnetism.

Starting with the river Nene navigation, surveying increasingly occupied Yeoman's time. He produced a new plan of the river in 1754, and completed a number of estate surveys for local landowners. He also surveyed turnpike routes such as that between St. Albans and Market Harborough (1750). It was in connection with the Nene navigation that Yeoman made the first of a number of appearances at Parliamentary Committees, when in January 1756 he testified to the benefit expected from the projected navigation by the towns of Higham Ferrers and Wellingborough. Later that same year he gave evidence on the proposed road from Towcester to Weston

Gate, Oxfordshire, and produced the plan that accompanied the Act the following year.

Yeoman continued living in Northampton, moving in 1755 to Bridge Street. He was a prominent member of the College Lane Baptist Church and a governor of the County Infirmary from 1746 until 1752. In 1742, a son, Samuel, was born but in 1746 his wife, Sarah, died. He remarried the following year, his new wife being Anne Remington, the sister of Joshua Remington, a Northampton grocer, and a further two children were born, Thomas in 1748 and Ann in 1752. The full extent of Yeoman's immediate family is unclear. Certainly he must already have had at least three children by the time he moved to Northampton; another son, James, was apprenticed in 1752 but in his will of 1779 Yeoman made bequests only to four daughters.

By about 1756, coinciding with the final demise of the cotton-mill, Yeoman's career was expanding in several directions and later that year he moved to London, first to Little Peter Street, Westminster. The business generated by the Hales ventilator continued to employ him. He introduced a number of his own improvements, adapted the machine to various agricultural applications and, in May 1751, installed ventilators in four British merchant vessels in Rotterdam; it was the only known occasion that he travelled abroad. The efficacy of this ventilation system was attested to by various ship owners and it evidently did something to alleviate the miseries endured in the transatlantic slave trade. He later applied the same system to British naval establishments and, in 1756, was appointed by the Admiralty to supervise the ventilation at the naval dockyards of Chatham, Sheerness, Portsmouth and Plymouth, and the naval hospital at Gosport. His preferment within the Admiralty was probably due to the influence of Lord Halifax, then First Lord, the leading Whig grandee in Northampton. By the time of his election to the Royal Society in 1764, Yeoman was described as 'Inspector of Ventilation in His Majesty's Fleet'.

Nevertheless, for the final twenty five years of Yeoman's career, it was civil engineering, particularly river and canal projects, that were to occupy most of his time, chiefly in the role of surveyor or consultant engineer. In 1755 he was appointed engineer for the Stroudwater Canal, although it was to be nearly twenty years before this work began in earnest. In 1757, together with Langley Edwards (q.v.), Yeoman was appointed Surveyor to the Ivel Navigation and both men gave evidence to the Parliamentary Committee in March of that year.

Meanwhile, the plans for the river Nene were proceeding and by April 1758 the Commissioners had received two estimates, from John Case and Ferdinando Stratford (q.v.), to complete the work within a budget of £14,500. On Yeoman's advice, Case was chosen, but both tenders were subsequently withdrawn and in June a further three bids were entertained. Again Yeoman's advice

was sought and the contract was finally awarded to John Smith Jr (q v) of Yorkshire. It was not until Yeoman had politely remonstrated with the Commissioners that his fee of 20 guineas was eventually paid. Once the work was nearing completion, Yeoman was again approached and in August 1760 was appointed to survey the works upstream of Thrapston for a further fee of 20 guineas. Yet again, in October 1761, he surveyed the final stretch from Hardwater lock, Wellingborough, to Northampton, at the same fee. The navigation had already been opened at Northampton on 7 August 1761, with considerable civic pomp, and Yeoman's report required only the correction of minor deficiencies.

In 1760, Yeoman was elected to the Society of Arts, founded six years previously by William Shipley, like Yeoman, a former resident of Northampton and member of the Philosophical Society. Yeoman remained an active supporter of the society for the rest of his life; he was a long-standing Chairman of the Committee of Mechanics and among the many new members whom he proposed were a number of distinguished surveyors and engineers. In 1764, he was elected a Fellow of the Royal Society, but seems to have participated very little in its proceedings. Unlike his more illustrious contemporary, John Smeaton (q.v.), Yeoman contributed nothing to the Royal Society's *Philosophical Transactions* and, besides the reports of his engineering surveys, published nothing. Evidence from his correspondence confirms that by 1746 he had already made substantial progress on a book of instruction for millwrights, but this was never published.

For the greater part of the 1760s, Yeoman was engaged in projects in Hertfordshire and Essex, specifically the navigation of the rivers Stort, Chelmer and Lee. He first surveyed the Stort from Bishop's Stortford to Hoddesdon in December 1758 and testified to the Parliamentary Committee that it could be made navigable at a cost of £10,000. He began work on the Lee Navigation, with Joseph Nickalls (q.v.), in 1762. He surveyed the river Chelmer from Chelmsford to Maldon in the same year and again, in 1765, estimated the cost of the project at £13,000 and produced a published map of the planned navigation. The following year, 1766, he was formally appointed assistant to Smeaton on the Lee Navigation, from Hertford to the Thames, a project that was considerably complicated by the mills and waterworks on the tidal reaches of the river at West Ham. Yeoman would, however, have been familiar with the situation from his previous experience in the early 1740s at one of these mills, the Abbey Mills. Smeaton's report of 1766 included an appendix by Yeoman on the Limehouse Cut, from Bromley to Limehouse, which he estimated at £5,310. The work was finally completed with the opening of the Cut in 1770.

In 1767 he also reported on the Level of Ancholme and, together with Smeaton, Robert Mylne and James Brindley (qq.v.), on the Thames water-works at London Bridge. By this time, surveying work was taking him to navigation projects throughout the country. In 1768, together with Brindley and Golborne (q.v.), he became drawn into the controversy surrounding the proposed route of the Forth-Clyde canal and in 1769 produced one of his most substantial published reports, on the outfall of the river Nene at Wisbech. In 1769 he spoke at the celebrations to mark the opening of the Stort Navigation at Bishop's Stortford, and in 1771 reported to Parliamentary Committees on the proposed navigation of the river Dee and the rebuilding of Chelmsford Gaol. Again, in 1771, he surveyed the river Medway and estimated the cost of extending the navigation.

By 1769, Yeoman had clearly established himself as a professional surveyor and civil engineer and it is notable that in Thomas Mortimer's *Universal Director* of 1763, both Smeaton and Yeoman were described as 'Surveyor & Civil Engineer', one of the earliest documented uses of the term. Presumably Yeoman's material wealth increased accordingly for, by 1769, he had moved house again and was living in Castle Street, Leicester Fields. He continued a devout Baptist as a member of the Grafton Street Church and in 1766 was appointed church secretary, a post that he held until 1774. This was to prove a turbulent period in the church's history involving the dismissal of its minister and a secession of some of the members. Yeoman played a leading role in guiding the church through these difficult times, and the members looked to him to act as guarantor of a mortgage of £100 on the church buildings.

On 15 March 1771, a Society of Civil Engineers was instituted at the King's Head in Holborn, London, with Yeoman presiding over the inaugural meeting; the minutes of the early meetings are in his own hand. Seven members attended the first meeting and precedence was probably given to Yeoman as the most senior in years, if not in his accomplishments. Although this first society never amounted to much more than a convivial dining group, it clearly signified that the profession of civil engineering had reached a milestone in its development and that several individuals were much in demand, particularly during Parliamentary sittings. Now towards the end of his career, Yeoman's work was almost exclusively confined to consulting. In 1774, he appeared on three occasions before Parliamentary Committees to testify on the Navigation of the Aire and Calder, the drainage of Deeping Fen and the state of the river Welland in Spalding, and finally, in a lengthy account, on the navigation of the Lee.

Yeoman examined Stonar Cut on the river Stour from Fordwich Bridge to the sea in 1775 and reported that he found great improvement since his earlier survey of 1765. He reiterated his earlier findings and concluded that proposed improvements to the drainage could be completed at an estimated £1,210 and would not be prejudicial to the Port of Sandwich.

One of the final major projects of his latter years was the Stroudwater canal, a long-delayed enterprise initiated by the wealthy local clothier, John Dallaway. The commissioners had first engaged Yeoman in 1755, when he reported on the scheme, together with John Willets, and estimated the cost. His renewed association appears to have begun in about 1774 under newly-appointed undertakers. Yeoman proposed a radically altered plan, completely avoiding the river Frome, some eight and a quarter miles in length from Framilode to Wallbridge. He considered the proposed route one of the most favourable he had ever surveyed, requiring no major construction of aqueducts, tunnels, cuttings or embankments and he estimated the total cost to be £16,766. Yeoman remained as consultant engineer to the Stroudwater company, with Thomas Dadford and John Priddey (qq.v.) in the role of resident engineers, and it was in December 1775 that he made his final appearance at Parliament to give evidence on the progress of the scheme. He reported that about one mile had already been cut but foresaw problems arising from the millers affected by the route; he also testified that the route would not be navigable for 70 ton trows from Bristol and that these would need unloading into 3 or 4 lighters. The navigation was finally opened at Wallbridge on 24 July 1779, by which time the costs had apparently soared to about £31,000.

In 1771, the Commissioners for the Navigation of the Thames appointed Nickalls to the position of General Surveyor for the completion of the third phase of the navigation from Maidenhead to beyond Reading. Much of the work was completed by the time that Yeoman, in October 1775, was consulted to resolve a problem with the channel at Maidenhead bridge. In August 1776, however, Nickalls was dismissed and Yeoman was consulted again, on this occasion to survey the pound lock at Marlow. He reported in November and was directed to proceed with repairs to the lock. The Commissioners appointed Yeoman as Surveyor in February 1777, in what was to be his final commission, and he completed the last two locks of the 3rd District, at Mapledurham (May 1777) and Caversham (July 1778).

One of Yeoman's final recorded deeds was to second the (unsuccessful) proposal of the surveyor, Joseph Hodskinson (q.v.), to a Fellowship of the Royal Society in March 1780. An extremely weak signature testifies to his declining health and on 28 April 1780 the Society of Civil Engineers commented in their minutes on his absence, 'our worthy President being ill and not capable of attending'. He never again appeared at their meetings and died, aged seventy-one on 23 January 1781. He was buried seven days later alongside many other distinguished dissenters in Bunhill Fields Cemetery, London. After her death on 6 November 1793, his wife, Anne, was buried in the same grave; the grave no longer survives but the inscription on the headstone was preserved in the nineteenth century.

Thomas Yeoman's career spanned the early formative years of the profession of civil engineering. Like many of his contemporaries, his skills were adapted and developed from related disciplines, in his case mechanical engineering (millwright) and surveying. Yet his talents also included the more refined accomplishments of the instrument maker and the philosophical lecturer. To some extent his legacy has been eclipsed by more illustrious contemporaries such as Brindley, Smeaton and John Grundy Jr. (q.v.) yet his achievements as a prolific consulting engineer and his role in the foundation of the Society of Civil Engineers mark him as one of the most important and influential practitioners of the mid-eighteenth century.

DAVID L. BATES

[E. Robinson (1962) The profession of civil engineer in the eighteenth century: a portrait of Thomas Yeoman, *Annals of Science*, **18**, 195–215; A. W. Skempton and E. Wright (1971) Early members of the Smeatonian Society of Civil Engineers, *Trans Newcomen Soc*, **44**, 23–42; J. Boyes and R. Russell (1977) *Canals of Eastern England*; M. Handford (1979) *The Stroudwater Canal*; A. W. Skempton (1983) Engineering on the Thames Navigation, *Trans Newcomen Soc*, **55**, 153–176; A. W. Skempton (1984) Engineering on the English river navigations in 1760, in M Baldwin and A Burton (eds.), *Canals: A New Look*; D. L. Bates (1993) All manner of natural knowledge: the Northampton Philosophical Society, *Northamptonshire Past & Present*, **8**, 363–377; D. L. Bates (1996) Cotton-spinning in Northampton: Edward Cave's mill 1742–1761, *Northamptonshire Past & Present*, **9**, 237–225; J. Harrison (1997) The ingenious Mr Yeoman and some associates: a practical man's contribution to the Society's formative years, *R Soc Arts J*, **145**, 53–68]

Publications

Publications include:

1754. *Plan of the River Nene, from Thrapston to Northampton with the Mills and Locks necessary for the Navigation*
1756. *Plan of the road from Towcester … to Weston Gate in the County of Oxford*
1762. *Plan of the River Chelmer from Chelmsford to Maldon* (revised plan, 1765)
1766. *Report of the State of the Level of Ancholme*
1766. *Estimate for Making a Navigable Cut from Limehouse-Hole to near the Four-Mills at Bromley*
1768. *Answers to Questions relative to a Navigable Communication between the Friths of Forth and Clyde and Observations thereupon*
1769. *Report concerning the Drainage of the North Level to the Fens and the Outfall of the Wisbeach River*
1771. *Survey and Estimate for extending the navigation [of the River Medway] from Tunbridge to Edenbridge and Six Miles above it*

1775. *Report of the State of the River Stour … and the Means of Improving the Drainage of the Marsh Land … and also of Preserving the Port of Sandwich*

Works

1757–1758. River Ivel navigation (with Langley Edwards), £6,000
1758–1761. River Nene navigation, consultant, £16,000
1766–1769. River Stort navigation, *c*. £18,000
1767–1771. River Lee navigation and Limehouse Cut, *c*. £25,000
1775–1779. Stroudwater Canal, consultant, £31,000
1776–1778. Stonar Cut
1776–1778. River Thames, two locks

YOUNG, John (fl. 1815–1830), Resident Engineer, had his first responsible appointment at Whitehaven Harbour, under John Rennie (q.v.), but only preliminary works were carried out before the Trustees decided that their funds were inadequate and a halt was called. He transferred to Portnessock in Galloway, for which Rennie had provided a design in 1813, but the works as built appear to have been varied. Portnessock was considered seriously as an alternative to Portpatrick for the Scottish end of the packet service to Ireland, but when the decision was given in favour of the latter, an 1820 proposal for a breakwater similar to that at Plymouth was abandoned. When (Sir) John Rennie (q.v.) was looking in 1827 for someone to direct the works at Portrush Harbour by direct labour, having failed to find a suitable contractor, he recommended Young as a 'person of considerable experience' and he was appointed in February 1828.

Sir John Rennie, in his autobiography, records that Young supervised the paving of the river Nene Outfall in 1837 and subsequently became a merchant and shipowner at Wisbech, where by his talents, energy and industry he realized an ample fortune, was elected Mayor several times, and became Member of Parliament for the county of Cambridge. He seems inexplicably to have confused John with Richard Young (1809–1871), who was Keeper of the North Level Sluice from 1830 to 1849, Superintendent of the Nene Outfall Works from 1833, Mayor of Wisbech for five consecutive terms from 1858 and Liberal MP from 1865 to 1868.

P. S. M. CROSS-RUDKIN

[Rennie reports, ICE; Rennie collection, NLS; *Autobiography of Sir John Rennie, FRS* (1875); B. Scott-Hindson (1994) *Whitehaven Harbour*]

Works

18??. Leith Docks, Assistant Engineer
1815–1818. Whitehaven Harbour, Resident Engineer; £200 p.a.
1818–1827. Portnessock (now Port Logan) Harbour
1828–. Portrush Harbour, Resident Engineer; £200 p.a.

YOUNG, Thomas, FRS (1773–1829), natural philosopher, physician and linguist was born in Milverton, Somerset, on 13 June 1773, the eldest son of Thomas Young and his wife, Sara, daughter of Robert Davis. His father was a banker and both his parents were Quakers in which sect Young was brought up. He was a precocious child, learning to read by the age of two. He attended two boarding schools between 1780 and 1786 where his ability to learn languages became marked. For the next few years he studied privately. In 1793 he entered St. Bartholomew's Hospital, London, to study medicine, and continued this study at Edinburgh from 1794—it is from this period that he began to distance himself from Quakerism—and Göttingen (1795–1796). From 1797 to 1803 he was attached to Emmanuel College, Cambridge, where he turned his attention to scientific matters. In 1797 an uncle left him £10,000 and a London house into which he moved in 1800, and in 1804 married Eliza Maxwell. Although he practised medicine and was Professor of Natural Philosophy at the newly founded Royal Institution from 1801 to 1803, he was not particularly successful at either occupation, though the latter did lead to his publishing *A Course of Lectures on Natural Philosophy and the Mechanical Arts* in two volumes in 1807. In this text Young appears to have been the first person to propose using the term 'energy' instead of the then normal term 'vis viva', what we now call kinetic energy. Also in this text Young defined the 'modulus of elasticity' although in a somewhat different form from that used today. Both these terms were not widely used until they were adopted by William Thomson and his colleagues fifty or so years later. One of the reasons for this neglect is Young's fairly impenetrable prose style in the 'Course of Lectures'.

Although the 'modulus' has proved Young's most enduring contribution to civil engineering, the range of his writings meant that his influence among early nineteenth century engineers was much greater. His great knowledge of scientific literature meant that he was able to publicise the work of foreign writers like Coulomb and Eytelwein to a British audience. Although his own use of language has been criticised as difficult to follow his work was widely referred to by Thomas Tredgold (q.v.) in his textbooks, introducing Young to a wider audience. Young himself wrote articles on carpentry and bridges for the *Encyclopaedia Britannica,* containing an interesting commentary on the 600 ft. cast iron arch proposal Thomas Telford (q.v.) for London Bridge. The bibliography appended to his *Course of Natural Philosophy* is possibly the most comprehensive guide to the scientific and technical literature of the preceding centuries.

In 1811 he became physician to St. George's Hospital. Elected a Fellow of the Royal Society in 1794, he served as its Foreign Secretary from 1804 until his death. Between 1816 and 1821 he was Secretary of the Royal Commission on Weights and Measures. From 1818 he was Secretary of the Board of Longitude and Superintendent of the Nautical Almanac, a position he retained following the abolition of the Board in 1828. After the abolition he was named one of the Admiralty's scientific advisers. In 1824 he was appointed inspector of calculations and physician to the Palladium Insurance Company.

It is for his optical work that Young in chiefly remembered. His earliest work was in physiological optics in which he sought from 1791 to explain how the eye functioned. He argued that the lens of an eye changed shape to focus light as necessary. In 1801 he suggested that the retina responded to three 'principle colours' only which combined to form all the other colours. In the later hands of James Clerk Maxwell and Hermann Helmholtz, this view came to be the standard theory of colour sensation.

It was a short step from physiological optics to considering the nature of light. Young's interest in this was reinforced by some work he had done in the mid-1790s on the transmission of sound which he came to believe was analogous to light. In 1800 he brought his ideas together in a paper where he argued for the transmission of light waves through an aether. Over the next two years Young developed his ideas on the nature of light discovering in the process, in May 1801, the principle of interference of light. Young's inability to sustain over time a consistent view of the nature of light and of the aether, contributed to little attention being paid to his theoretical work, though the principle of interference was viewed widely as an important discovery. The wave theory of light was not established until the work of Augustin Fresnel in the 1810s.

After his work on optics, Young returned to the study of languages and in particular from 1813 started to attempt to decipher Egyptian hieroglyphics, and translated the demotic script of the Rosetta stone. However, he published little at the time due, it would seem, to his official duties, and later (1823, 1830) publications were partial attempts to claim priority over Champollion.

That Young's work in virtually all the areas he worked in did not receive, for one reason or another, major contemporary recognition meant that throughout the nineteenth century his work was used as a place to locate precursor ideas in a given subject. This was especially so when dealing with overseas work. For instance, the historical work of William Whewell in the 1830s gave an unwarranted importance to Young's views over those of Fresnel's. It is only in recent times that Young's work generally has been properly contextualised.

Young died in London on 10 May 1829. After his death his brother, Robert, presented his extensive scientific library to the Institution of Civil Engineers and the Royal Institution.

FRANK A. J. L. JAMES

[H. Gurney (1831) *Memoir of the Life of Thomas Young*; G. Peacock (1855*) Life of Thomas Young*; Annual report (1840–1844) *Min Proc ICE*, 1–4; I. Todhunter and K. Pearson (1886) *A History of the Theory of Elasticity*, **1**, 80–86; A. Wood (1954) *Thomas Young, Natural Philosopher, 1773–1829*; A. W. Skempton (1979) Telford and the design for a new London Bridge, in A. Penfold (ed.), *Thomas Telford, Engineer*, 62–83]

Publications

1802. *A Syllabus of a Course of Lectures*

1807. A summary of the most useful parts of hydraulics …, *R Inst J*, **1**

1807. *A Course of Lectures on Natural Philosophy and the Mechanical Arts*, 2 vols

1808. Hydraulic investigations, *Phil Trans*, **98**

[1817]. 'Bridge' and 'Carpentry', articles in (supplement to) *Encyclopaedia Britannica*

1826. A summary of the most useful parts of hydraulics, chiefly extracted and abridged from Eytelwein's 'Handbuch der Mechanik', in T. Tredgold (ed.), *Tracts on Hydraulics*

Appendix I
Wages, costs, salaries and inflation

Menai Suspension Bridge was built in 1818–1826 to the design and under the direction of Telford at a construction cost of £180,000 with William Provis as resident engineer at a salary of £300 per annum. There were two assistant engineers and a clerk, and at that time masons and carpenters were paid typically 24s (£1.20) per week. If these and other historic costs and salaries are to have any meaning to the modern reader they need to be converted to equivalent present-day values.

The first step is to allow for the effect of monetary inflation. The purchasing power of the pound (PPP) in the 1820s was about 40 times that of the pound in 1995. Therefore, in what economists call 'real' terms, Provis's salary is £12,000, the cost of the bridge is £7.2 million and masons' wages about £50 per week.

But this is only part of the story. All the evidence shows that modern wages, costs and salaries exceed the 'real' values, in most cases by a considerable amount. Hence it is necessary to consider the modern value to be the product of two factors:

actual × PPP = 'real'

and

modern = 'real' × F

where F is, in general, greater than 1.0.

The salary today of a senior resident engineer on a bridge of comparable scale and importance would be around £36,000 or three times the 'real' salary of Provis, and building wages in 1995 were £300 per week, or six times the 'real' wages in the 1820s. These figures of $F = 3$ and 6, respectively, illustrate the increases in salaries and wages relative to the cost of living over the past 170 years.

For construction costs, the price of steel per ton in 1995 was roughly the same as the 'real' cost of wrought iron in the 1820s ($F = 1.0$), but the cost of masonry has risen since then by almost the same factor ($F = 6.0$) as wages, while the introduction of mechanical plant has kept the unit cost for earthwork to about twice the 'real' cost for excavation and embanking in the eighteenth and early nineteenth centuries.

With such widely differing factors the relevant value of F on a particular project depends on the relative proportions of the various components of the total cost. For the Menai Bridge, wrought iron contributed 40%, approach embankments about 5%, and the remaining 55% was mostly due to the masonry towers and arches (see the frontispiece of this book), together with the labour-intensive operations of fixing and painting the ironwork. Thus the weighted factor representing an increase on 'real' cost is appropriately

$F = 0.55 × 6.0 + 0.40 × 1.0 + 0.05 × 2.0 = 3.8$

giving the equivalent modern cost as

$3.8 × $ 'real' $= 3.8 × $ £7.2 million $=$ £27 million

The data on which those and other calculations are based are as follows.

Building wages and inflation

Wages for masons and carpenters from mediaeval times to 1954 are given by Phelps Brown and Hopkins (1955) and for construction workers from 1958 by Mitchell (1980) and the Office of National Statistics (1998). Typical values are summarised in Table A1.1 together with figures for the purchasing power of the pound derived from the price of consumables (Phelps Brown and Hopkins, 1956) and the Retail Price Index from 1914 (conveniently summarised in *Whitaker's Almanack*, 1998).

Also in Table A1.1 are the 'real' wages and the factor F by which the 1995 wages exceed the 'real'. During the seventeenth and early eighteenth centuries wages roughly doubled and 'real' wages increased by about 40%. From 1725 to 1830 wages again doubled but purchasing power halved, so 'real' wages remained approximately consistent (apart from a drop in the Napoleonic wars) at £50 per week. Since 1830 wages have increased by a multiple of 250, with 'real' wages doubling in Victorian times and trebling in the twentieth century, despite the pound falling to about one-fiftieth of its 1900 value.

Corresponding figures for the factor F decrease from about 8 in the early seventeenth century to 7 in the latter part of that century, remain essentially constant at around 6 from 1725 to 1830 (again except for the Napoleonic wars, with F about 8) and by definition approach 1.0 towards the end of the twentieth century. A detailed graph of 'real' wages from 1694 is given by MacFarlane and Mortimer-Lee (1994).

Construction costs

Masonry

The conversion of historical to modern costs can be made with some confidence for masonry structures since in one case the overall figures are available (Paxton, 1998). Laigh Milton viaduct on the Kilmarnock & Troon Railway was built in 1809–1810, with three 40-ft. span arches in sandstone, at an expenditure of about £3,600, and rebuilt in 1995–1996, as nearly as possible in its original form, for a little over £1 million: a multiple of 280 on the actual cost and a factor F of 7.9 on the 'real' cost (£120,000 with PPP = 35). These figures are almost identical with those for building wages in 1810, as seen in Table A1.1. With shallow foundations, a high proportion of the total cost can be attributed to the masonry.

Supporting evidence is provided by the price of Portland stone masonry, for which Smeaton in 1763 quotes 3s 6d per cu. ft. in London and 3s in 1771 at Dover Harbour. With PPP around 80 the 'real' cost is £13. However, the 1995 cost is about £70 per cu. ft., giving $F = 5.4$ compared with $F = 6.0$ for wages in the 1760s.

Brickwork

For the walls of Holderness Sluice, near Hull, Smeaton in 1764 estimated the brickwork at £6 per rod (308 cu. ft.), made up of £3.40 for 4500 bricks at 15s per thousand, £1.20 for mortar, and £1.40 for workmanship or about 25% of the total. Rennie in 1803 gives the cost of brickwork in the quay walls of London Docks as £15 per rod in grey stock bricks priced at 38s per thousand or £8.60 per rod. With the appropriate values of PPP the 'real' costs are: bricks £60 and £66 per thousand, respectively, and brickwork £480 and £525 per rod or £1.55 and £1.70 per cu. ft. In 1995 the price of machine-made 'common' bricks was £128 per thousand and brickwork in walls more than 3 ft. thick £4.50 per cu. ft. (including £2.10 per cu. ft. for labour). The values of F are therefore 128/63 = 2.0 for bricks and 4.5/1.63 = 2.8 for brickwork. The higher value of F for brickwork, compared with that for bricks, is due to labour costs contributing 47% of the total in 1995 in contrast to 25% in 1764.

Timber

For Riga fir in the floor of Holderness sluice, mostly in 12 × 12in. timber, Smeaton took 14d. cu. ft. for timber and 21d. cu. ft. for the finished work, allowing a small amount for wastage and workmanship as

Table A1.1. Wages and inflation

Date	Building wages per week, a	Wages multiple, b	Purchasing power of the pound (PPP), c	'Real' wages per week, $a \times c$ (£)	Factor, F = 1995/'real'
1600	6s	1,000	120	36	8.3
1660	7s	670	95	43	7.0
1700	10s	600	90	45	6.7
1725	11s	550	90	50	6.0
1760	12s	500	85	51	5.9
1790	15s	400	65	49	6.1
1800	18s	330	40	36	8.3
1810	22s	270	35	38	7.8
1820	24s	250	40	48	6.2
1830	24s	250	45	54	5.6
1900	£2	150	55	110	2.7
1960	£14	21	12	170	1.8
1995	£300	1	1	300	1.0

Table A1.2. Values of F *for unit costs*

Wages, F	Earthwork	Masonry	Brickwork	Timberwork	Iron
8	2.4	8.0	3.4	5.2	1.14
7	2.2	7.0	3.1	4.6	1.12
6	2.0	6.0	2.8	4.0	1.10
1	1.0	1.0	1.0	1.0	1.0

25% of the total. With PPP = 80 the 'real' costs are £4.70 cu. ft. for timber and £7.00 for timberwork. The modern equivalent is Douglas Fir which in 1995, in 12 × 12 in. timber, cost £14 per cu. ft., giving $F = 3.0$ for timber. By analogy with brickwork, F for timberwork can be taken as approximately one-third more or 4.

Ironwork
Between 1800 and 1812 the price of cast iron rails varied from £10.50 to £12 per ton, averaging £11.50 with a 'real' cost of £400 per ton. Wrought iron rails on the Liverpool & Manchester Railway in 1827 cost £12.75 per ton or £520 'real'. In 1995 steel rails cost £530 per ton. Steel has the advantage of greater strength, but on a weight-for-weight comparison the values of F for cast and wrought iron rails are 1.3 and 1.0, respectively. The average 'real' cost of cast iron pillars in 1795 and 1813 was £600 per ton, compared with £690 for steel columns in 1995. Telford in 1817 estimated £25 per ton for wrought iron chains in a suspension bridge, and this was probably the unit cost of the chains in Menai Bridge. With PPP = 40 the 'real' cost of £1,000 per ton compares with £1,020 for built-up steel girders and £1,200 for steel trusses in 1995. Erection costs added 14% and this is also the cost for fixing and painting the chains for Menai Bridge. Overall, then, a factor $F = 1.1$ is a reasonable value for ironwork.

Earthwork
Between 1760 and 1820 the unit costs for excavating earth and depositing the material nearby vary from 3d to 12d per cu. yd., depending on depth for excavation and date. The 'real' costs, independent of date, range from £1 per cu. yd. at 7ft. depth to £2 at 24 ft. and the corresponding 1995 figures are £2.40 and £3.20, respectively, giving values of F from 2.4 to 1.6. For most purposes it is sufficient to take $F = 2.0$.

The foregoing 'real' unit costs correspond closely to those adopted by Robert Stephenson in estimates for the London and Birmingham Railway in 1834 (taking PPP = 45): excavation for foundations £3 per cu. yd., brickwork £1.70 per cu. ft., stonework £8.50 per cu. ft., timberwork £7.50 per cu. ft., and cast iron £550 per ton.

Values of F corresponding to various wage factors are given in Table A1.2.

Fen drainage
The average costs of the main components of several fen drainage schemes between 1760 and 1820 are, as percentages of the total construction cost: earthworks 75%, sluices 15% and timber bridges 10%. Holderness Sluice is a typical eighteenth century outfall sluice. Its components are: masonry 40%, brickwork 18%, carpentry (floor and doors) 37% and ironwork 5%. With values of F for wages of 6.0, the overall figure is $F = 4.4$. Therefore for a typical drainage scheme:

$$F = 0.75 \times 2.0 + 0.15 \times 4.4 + 0.10 \times 4.0 = 2.6$$

For the Great Level Drainage of 1650–1656, with wages $F = 7.0$, the factors for earthworks and timber become 2.2 and 4.6, respectively (see Table A1.2) but earthwork probably constituted a higher proportion of the total, say 80%, and whereas the outfall sluice was stone-built, two other sluices were in timber. Taking all this into account the value of F is about 2.7. The actual cost was roughly £250,000, the 'real' cost (PPP = 95) about £24 million and the equivalent modern cost is £65 million.

Canals
From Smeaton's estimates for the Forth & Clyde Canal in 1767, with masonry locks and aqueducts and timber bridges, the percentages are: earthworks 50%, locks 32%, aqueducts 8% and bridges 10%, giving $F = 3.8$. There were 1.2 locks per mile, a characteristic figure for canals. If the locks and aqueducts had been in brickwork the percentages would be: earthwork 60%, locks 20%, aqueducts 5% and bridges 15%, giving $F = 2.5$.

River navigation
The timber-framed locks on the Wey Navigation (1651–1653) were appropriately half the size of those built in the 1780s on the Thames, the cost of which is known. Assuming cost is proportional to size, and

allowing for inflation, the ten locks on the Wey (on 9 miles of cut in a total length of 15 miles) cost about 20% of the total. To this must be added, say, 5% for weirs and floodgates. Therefore (with wages $F = 7.0$)

$$F = 0.75 \times 2.2 + 0.25 \times 4.6 = 2.8$$

The twenty locks on the Kennet Navigation (1719–1723), timber-framed except for three in brick-work, were slightly smaller than the Thames locks, and only two new weirs were requited. The locks (on 11 miles of cut in a total length of 18 miles) cost about 40% of the total, giving

$$F = 0.60 \times 2.0 + 0.40 \times 4.0 = 2.8$$

Docks

Estimates by Rennie for the London Docks in 1803, when work was well under way, show the following percentages of cost: excavations 32%, quay walls 38%, locks 28% and (iron) bridges 2%. These can be redistributed as: excavation 32%, stonework 22%, brickwork 27%, timber 17% and iron 2%. In 1803 the wages factor was about 8. With the corresponding factors for brickwork, etc., given in Table A1.2, the resulting value is $F = 4.3$.

In Jessop's Bristol docks or 'Floating Harbour' (1804–1810) with its 2 mile New Cut for the River Avon there is a much higher proportion of earthwork (51%), but this is balanced by a greater amount of stone-work (33%) which, with timber 11% and iron 5%, gives $F = 4.5$; not very different from the above figure for the more typical London Docks.

Bridges

Exact figures are available for the main components of Labelye's Westminster Bridge (1738–1750). Expressed as percentages of the total construction cost they are: dredging for foundations 5%, masonry 80%, and carpentry for caissons, piles and centering 15%. Thus

$$F = 0.05 \times 2.0 + 0.80 \times 6.0 + 0.15 \times 4.0 = 5.5$$

Table A1.3. Construction cost

Date	Work	Actual cost (£ thousand)	'Real' cost (£ million)	1995 cost (£ million)	Engineer
1583–1586	Dover Harbour	12	1.8	6	Digges
1609–1614	New River	18	2.2	7	Myddleton
1650–1656	Great Level Drainage III	250	24	65	Vermuyden
1651–1653	Wey Navigation	15	1.5	4	Weston
1691–1698	Portsmouth Docks	16	1.4	5	Dummer
1699–1704	Aire & Calder Navigation	25	2.2	9	Hadley
1716–1720	Dagenham Breach	40	3.6	9	Perry
1719–1723	Kennet Navigation	45	3.6	10	Hore
1729	Putney Bridge (timber)	12	1.1	4	Phillips
1731–1742	Newry Canal	52	47	12	Steers
1738–1750	Westminster Bridge	198	18	100	Labelye
1756–1759	Eddystone Lighthouse	16	1.4	8	Smeaton
1760–1769	Blackfriars Bridge	166	13	72	Mylne
1766–1777	Trent & Mersey Canal	300	23	58	Brindley
1768–1777	Forth and Clyde Canal I	164	12	45	Smeaton
1772–1780	Weighton Drainage	40	2.9	8	Grundy
1785–1788	King's Dock, Liverpool	25	1.6	5	Berry
1790–1801	Leeds & Liverpool Canal II	330	17	65	Whitworth
1799–1810	Beverley & Barmston Drainage	115	4.0	10	Chapman
1800–1806	West India Docks	515	18	80	Jessop
1803–1821	Highland Roads and Bridges	420	15	40	Telford
1804–1810	Bristol Harbour	470	17	76	Jessop
1811–1817	Waterloo Bridge	480	18	100	Rennie
1812–1816	Vauxhall Bridge (iron)	86	3.4	13	Walker
1818–1823	Eau Brink Cut	250	10	26	Rennie
1818–1826	Menai Bridge	185	7.2	27	Telford
1826–1830	Liverpool & Manchester Railway	600	27	73	Stephenson

Table A1.4. Salaries and construction costs 1790–1830

Item	Resident engineers			Chief engineers
	Group 1	Group 2	Group 3	
Salary (per annum)				
Actual (£)	100–200	200–300	400–600	630–1,000
Average actual (£)	160	260	500	900
'Real' (£ thousand)	4–8	9–13	16–24	28–42
Average 'real' (£ thousand)	6.5	11	20	38
Average 1995 (£ thousand)	25	34	52	92
F = 1995/'real'	3.8	3.1	2.6	2.4
Cost				
Actual (£ thousand)	40–130	80–420	254–860	130–600
Average actual (£ thousand)	80	190	430	300
'Real' (£ million)	1.5–5.2	3.5–15	10–30	5–27
Average 'real' (£ million)	3.0	7.7	17	13
'Real'/year (£ million)	0.2–0.4	0.6–1.0	1.7–3.0	1.1–5.4
Average 1995 (£ million)	10	29	64	43
1995/year (£ million)	1.3	3.1	8.6	8.0

Average figures from Smeaton's estimates for Perth Bridge (1763) and Banff Bridge (1772) are: excavations for foundations 12%, timber for cofferdams and centering 23% and masonry 65%, giving F = 5.1.

For Walker's Vauxhall Bridge (1812–1816) the cast iron for the nine 78 ft. span arches cost 32% of the total, the remainder being chiefly for masonry of the piers and abutments with, say, 5% for excavation and 15% for timber. Thus F = 3.9. The castings for Telford's single 150 ft. span arch of Craigellachie Bridge (1816–1845) cost 25% of the total. Probably all the rest would be for the masonry abutments, with perhaps 10% for the centering. Therefore:

$$F = 0.65 \times 6.0 + 0.10 \times 4.0 + 0.25 \times 1.1 = 4.6$$

And, as given in the introduction in this article, F = 3.8 for Menai Suspension Bridge.

Liverpool & Manchester Railway
Costs of the main components as percentages of the total construction expenditure 1826–1830 are: earthworks 56%, bridges 21% (mostly in brick with stone facing and some in masonry; say, brickwork 10% and stone 11%), stone blocks and formation 8%, wrought iron rails 13%, and fencing 2%. So F = 2.7. The actual construction cost was £600,000. With PPP = 45 the 'real' cost is £27 million and hence the equivalent modern cost is about £73 million.

Actual, 'real', and 1995 costs are given in Table A1.3 for selected works ranging in date from 1580 to 1830. The ten largest works listed, whether in the seventeenth century (1), the eighteenth century (4) or early nineteenth century (5), cost between £45 million and £100 million in modern terms, with annual expenditure mostly between £5 million and £15 million.

Excluding masonry structures, the average multiples of 1995 costs to actual costs for various periods are approximately: 310 (1600–1690), 280 (1690–1760), 220 (1760–1790) and 140 (1790–1830). For masonry the multiples are roughly double these figures.

Salaries and fees
Whereas building wages are subject only to minor variations at any given time, engineers' salaries vary widely; for instance, in the period around 1820, from £150 per annum for a resident engineer on a small project, to £500 for a senior resident on a major work, and double that amount for the chief engineer responsible for planning, design and direction.

Data have been collected on the salaries of nineteen resident engineers on works of known cost done in the period 1790–1830. Allowing for inflation, their 'real' salaries fall into three distinct groups, averaging £6,500, £11,000 and £20,000, and correlate broadly with the scale of the project and in particular with the rate of expenditure (see Table A1.4).

Engineers at the head of the profession (such as Jessop, Rennie, Telford and Chapman) form another well-defined group. Whether as consultants or chief engineers on a project, they usually charged fees on a daily basis and in fourteen examples between 1796 and 1826 their fees, with one exception, ranged from 4 to 7 guineas per day (average £6): in 'real' terms from £190 to £290 and averaging £240 per day. This is to be compared with consulting engineers in the mid-1990s normally charging daily fees of £500 to £640, a factor of about 2.4 times the 'real' and roughly 100 times the actual fees in the early nineteenth century. An income from this source, based for example on 150 days per year, would be £900 actual, £36,000 'real', and £86,000 for 1995.

On some projects chief engineers were paid a salary, often for part-time but in a few cases for practically full-time service. The average salary for four such engineers in the period 1790–1830, is £38,000 'real' (see Table A1.4). This is comparable to the income from consulting fees, mentioned above. The same factor of 2.4 may therefore be applied, giving an average equivalent modern salary of £92,000. The individual figures range from £70,000 to £100,000.

For the modern equivalents of resident engineers' salaries an element of choice is involved, and there is no exact correlation between the three groups in the 1790–1830 period and the various categories listed in the ICE Salary Survey. However, it is clear that a typical basic salary for engineers in 1995 at the lower end of the scale of those carrying responsibility (i.e. not in training or working under supervision) is around £25,000, ranging from about £20,000 to £30,000. If these figures apply to group 1, the average 'real' salary of £6,500 corresponds to £25,000, giving a factor of 3.8.

For group 2 the most relevant comparison with the mid-1990s is a senior engineer on a big bridge, who would have a basic salary of about £36,000. Compared with the 'real' £12,000 of William Provis this

Table A1.5. Values of F for wages and salaries (per annum)

Item	Wages	Salaries					
'Real' (£ thousand)	2.5	4	6	8	12	20	40
1995 (£ thousand)	15	20	24	28	36	52	98
F	6.0	5.0	4.0	3.5	3.0	2.6	2.4

Table A1.6. Engineers' salaries

Engineer	Date	Work	Actual (£)	'Real' (£ thousand)	1995 (£ thousand)
Resident engineers					
Edward Rubie	1725	Rye Habour	65	6	24
Joseph Nickalls	1759	Calder & Hebble Navigation	100	8.5	29
Hugh Henshall	1766	Trent & Mersey Canal	150	12	36
Robert Mackell	1768	Forth & Clyde Canal	315	24	60
Samuel Allam	1772	Weighton Drainage	55	4.5	21
John Gwyn	1775	Aberdeen Harbour	120	9	30
Benjamin Outram	1789	Cromford Canal	200	13	38
Archibald Millar	1793	Lancaster Canal	400	22	56
James Murray	1801	London Docks	500	18	48
John Mitchell	1809	Highland Roads	250	9	30
John Gibb	1810	Aberdeen Harbour	300	10.5	33
Thomas Townshend	1818	Eau Brink Cut	400	16	44
William Provis	1820	Menai Bridge	300	12	36
Peter Logan	1826	St. Katharine's Docks	500	21.5	55
James Leslie	1828	Leith Harbour	176	8	28
Long-term appointments					
William Vincent	1734	Scarborough Harbour	70	6.5	25
John Grundy Sr.	1742	Deeping Fen	100	9	30
Henry Berry	1771	Liverpool Docks	100	7.5	27
Matthias Scott	1774	Hatfield Chase	70	5.5	23
Samuel Bull	1777	Birmingham Canal	130	9.5	31
James Golborne	1778	Bedford Level	200	15	42
James Green	1808	Devon, County Surveyor	300	10.5	33
Stephen Leach	1810	Thames Navigation	300	10.5	33
John Dyson	1824	Bedford Level	250	10	32

Table A1.7. Chief engineers' salaries

Engineer	Date	Work	Salary per annum Actual (£)	'Real' (£ thousand)	1995 (£ thousand)
John Perry	1724	Rye Harbour	130[a]	12	36
William Jessop	1789	Cromford Canal	300[b]	19.5	50
Robert Whitworth	1790	Leeds & Liverpool Canal	630[c]	42	100
John Rennie	1801	London Docks	500[d]	18	50
James Walker	1810	Commercial Docks	800[c]	28	70
Marc Brunel	1824	Thames Tunnel	1,000[c]	40	95
George Stephenson	1826	Liverpool & Manchester Railway	800[a]	34	80
George Stephenson	1827	Liverpool & Manchester Railway	1,000[c]	42	100

Service required: [a]9 months per annum; [b]4 months per annum; [c]practically full-time; [d]90 days per annum.

Table A1.8. Consulting fees

Engineer	Date	Fees per day Actual (guineas)	'Real' (£)	1995 (£)
William Lellam	1732	1	95	260
Charles Labelye	1745	£1.25	110	300
John Grundy	1760	1	90	250
John Smeaton	1771	4.5[a]	350	800
Robert Mylne	1794	3	210	500
William Jessop	1796	5	240	570
John Rennie	1800	5	185	450
Thomas Telford	1810	5	185	450
John Rennie	1811	7	260	620
William Chapman	1824	6	250	600
Thomas Telford	1824	7	290	700

[a] 5 guineas per day, including personal expenses.

gives a factor of 3. For group 3 the senior resident engineer on a major project in 1995 might receive a 'package' of £60,000, including benefits, or say £52,000 basic salary: a factor of 2.6 on the average 'real' salary of £20,000 of engineers in this group. These data when put together form a consistent pattern, as summarised in Table A1.5.

Granted that salaries, like wages, kept pace with inflation from 1725 to 1830, the factors should apply throughout that period. They will be somewhat higher in the 17th century. Examples of salaries and fees are given in Tables A1.6, A1.7 and A1.8. The following points may be noted (1995 equivalents in brackets).

- Resident engineers' salaries in the period 1760–1790 did not exceed £150 (£38,000) even on the largest projects, apart from Robert Mackell's on the Forth & Clyde Canal (£60,000). But his responsibilities were exceptional, and he acted as engineer for four years at the same salary after Smeaton retired as engineer-in-chief in 1773.
- From 1790 to 1830 resident engineers on major works typically received £400 to £600 (£45,000 to £60,000). An apparent exception is John Mitchell on the Highland Roads & Bridges (£30,000). However, the rate of expenditure was moderate, representative of group 2 in this respect.
- Salaries of engineers on long-term appointments rarely fall outside the limits (£25,000 to £35,000) and show only a slight tendency to increase with time. They had security of tenure, often serving ten or twenty years in the same post, and in some cases received a salary increase to compensate for inflation.
- Before 1760 the standard consulting fee was 1 guinea per day (about £250) plus expenses. In the 1760s Grundy, Mylne and Smeaton raised their fees to at least double that amount. Allowing for inflation this level was maintained, e.g. 5 guineas (£470 in 1795–1810), and further increased to 6 or 7 guineas (over £600) before 1830.

Table A1.9. Civil engineering, average annual incomes 1790–1830

Role	Average annual income (£)
Chief engineer	900
Resident engineer, group 3	500
Resident engineer, group 2	260
Resident engineer, group 1	160
Foremen	65
Masons and carpenters	50
Labourers	35

Table A1.10. Average annual incomes 1801–1803 (Colquhuon, 1806)

Occupation	Average annual income (£)	Occupation	Average annual income (£)
Peers	8,000	Eminent merchants, bankers	2,600
Baronets	3,000	Lesser merchants	800
Knights and Esquires	1,500	Shopkeepers	150
Gentlemen living on income	700	Clerks	75
Persons in higher civil offices	800	Innkeepers	100
Persons in lower offices	200	Manufacturers	800
Eminent clergymen	500	Shipbuilders	700
Lesser clergymen	120	Artisans	55
Persons of the law	350	Labourers in mines and canals	40
Persons educating youth in universities and chief schools	600	Freeholders (agricultural)	200
Liberal arts and sciences	260	Farmers	120
		Labourers in husbandry	30

- In 1767 Grundy earned a total of £591 (£106,000). This is almost identical to Whitworth's £630 (£100,000) practically full-time salary in 1790 on the Leeds & Liverpool Canal and to George Stephenson's £1,000 (£100,000) in 1827 on the Liverpool & Manchester Railway. Jessop, Rennie and Telford had higher total incomes. As engineer-in-chief of the London & Birmingham Railway, a work of unprecedented magnitude (£350 million when completed in 1838), Robert Stephenson's salary in 1833 was £1,500 (£160,000), raised two years later to £2,000.

The converted salaries show approximately how much an engineer working in the eighteenth and early nineteenth century would be earning today if paid at present rates for an equivalent level of responsibility. What they do not, and cannot, do is to reflect the relative standard of living and social standing for the engineers in their own time. For this an indication may be obtained by comparing their actual salaries (Table A1.9) with those of other professions and occupations available in a list drawn up Patrick Colquhoun for the period 1801–1803 (Lindert, 1982) extracts from which are given in Table A1.10.

A. W. SKEMPTON

References

CES MM3 (1995) Price Database

Institution of Civil Engineers (1995) Salary Survey

Laxton (1996) Civil Engineering Price Book

Lindert, P. M. (1982) Revising England's social tables 1685–1812. Explorations in Economic History, 19, 385–408

MacFarlane, H., and P. Mortimer-Lee (1994) Inflation over 300 years. Bank of England Quarterly Bulletin, 34, 156–162

Mitchell, B. R. (1988) British Historical Statistics

Office of National Statistics (1998) Annual Abstracts of Statistics

Paxton, R. A. (1998) Conservation of Laigh Milton Viaduct, Ayrshire. Proc Inst Civ Eng, 126, 73–85

Phelps Brown, E. H., and S. V. Hopkins (1955) Seven centuries of building wages. Economica, 22, 195–206

Phelps Brown, E. H., and S. V. Hopkins (1956) Seven centuries of the prices of consumables, compared with builder's wage-rates. *Economica*, **23**, 296–314
Spon (1996) *Civil Engineering Price Book*
Whitaker's Almanack (1998). Cost of living and inflation rates, p. 604

Appendix II
Chronological table

Date	Work	Cost (£)	Engineer
1500–1502	Boston Sluice	1,100	Matthew Hake
1520–1527	Bridge of Dee, Aberdeen	10,000	Alexander Galloway, Thomas Franche
1535–1540	Dover Harbour piers	15,000	John Thompson
1541–1551	Dover Harbour, King's Pier	36,000	Richard Cavendish *et al.*
1547–1550	Boulogne Mole		John Rogers
1563–1566	Plumstead Marshes, drainage	5,000	Jacopo Aconcio
1564–1567	Exeter Canal	3,000	John Trew
1566–1575	Yarmouth Harbour piers	7,500	Joas Johnson
1580–1584	London Bridge waterworks I		Peter Morris
1583–1586	Dover Harbour, Pent	12,000	Thomas Digges
1590–1591	Plymouth Leat	200	Robert Lampen
1605–1610	Popham's Eau		John Hunt
1608–1621	Perth Bridge I		John Mylne
1609–1614	New River	18,500	Hugh Myddleton
1611–1624	Berwick Bridge	18,000	James Burrell
1618–1621	Great Ouse Navigation I		Arnold Spencer
1622–1624	Dagenham Breach (east)	4,000	Cornelius Vermuyden
1626–1631	Hatfield Chase Drainage I	58,000	Cornelius Vermuyden
1632–1635	Dutch River	33,000	John Liens
1631–1636	Great Level Drainage I	97,000	Andrewes Burrell *et al.*
1636–1639	Lindsey Level Drainage	45,000	John Liens
1636–1639	Avon Navigation	15,000	William Sandys
1640–1641	Great Level Drainage II	23,000	Cornelius Vermuyden
1650–1656	Great Level Drainage III	250,000	Cornelius Vermuyden
1651–1653	Wey Navigation	15,000	Richard Weston
1659–1661	Blackwall Wet Dock	4,700	Henry Johnson
1660–1662	Dover Harbour, Cross Wall		Bernard De Gomme
1662–1666	Stour Navigation	5,000	Andrew Yarranton
1663–1683	Tangier Mole	298,000	Hugh Cholmley, Henry Sheres
1670–1676	Deeping Fen Drainage II		Samuel Fortrey
1671–1674	Fleet Canal	51,000	Christopher Wren, Robert Hooke, Thomas Fitch
1676–1677	Exeter Canal extension	5,000	Richard Hurd
1680–1689	Great Ouse Navigation II	2,500	Henry Ashley
1682–1684	Denver Sluice (west) II	5,000	Ralph Peirson, Thomas Fitch
1689–1691	Worcester waterworks		John Hadley
1691–1693	Plymouth Docks	11,000	Edmund Dummer
1691–1698	Portsmouth Docks	16,000	Edmund Dummer
1692–1693	Chester waterworks		John Hadley
1692–1693	Derby waterworks		George Sorocold
1694–1695	Leeds waterworks		George Sorocold
1694–1696	Oxford waterworks		William Yarnold
1697–1700	Howland Dock	12,000	John Wells, William Ogbourne
1697–1701	Newcastle waterworks		William Yarnold
1699–1700	Denver Sluice (east)	3,000	Richard Russell

Date	Work	Cost (£)	Engineer
1699–1701	Exeter Canal, enlargement	6,000	Daniel Dunnell
1699–1704	Aire and Calder Navigation I	25,000	John Hadley, James Mitchell
1701–1702	London Bridge waterworks II		George Sorocold
1705–1706	Bridgnorth waterworks	1,000	George Sorocold
1706–1709	Eddystone Lighthouse II		John Rudyerd
1710–1712	Sea Mills Dock, Bristol	9,500	John Padmore
1710–1721	Liverpool Old Dock	16,000	Thomas Steers
1716–1720	Dagenham Breach (west)	40,000	John Perry
1719–1723	Kennet Navigation	40,000	John Hore
1720–1723	Yorkshire Derwent Navigation	4,000	Joshua Mitchell
1722–1724	Blenheim 'Canal'		John Armstrong
1723–1730	Sunderland Harbour pier	20,000	William Lellam
1724–1736	Mersey and Irwell Navigation		Thomas Steers
1725–1726	Causey Arch	2,000	Ralph Wood
1725–1727	Bristol Avon Navigation	12,000	John Hore
1725–1737	Roads in Scotland	23,000	George Wade
1725–1743	Rye New Harbour I	30,000	John Perry, Edward Rubie
1726–1751	Don Navigation	40,000	William Palmer, John Smith
1727–1728	Gunthorpe Sluice	3,500	Humphry Smith
1728–1730	Smith's Leam	7,000	Humphry Smith
1729	Fulham Bridge (timber)	12,000	Thomas Phillips
1729–1731	Combe Down Railway	10,000	John Padmore
1730–1731	Serpentine Lake	4,800	James Horne
1730–1733	Deeping Fen Drainage II	4,000	John Perry
1730–1735	Weaver Navigation	18,000	Thomas Robinson
1731–1735	Newhaven Harbour	4,000	John Reynolds
1731–1742	Newry Canal	52,000	Richard Castle, Thomas Steers
1733–1735	Aberfeldy Bridge	4,000	William Adam
1733–1736	Littlehampton Harbour	5,000	John Reynolds
1733–1743	Dee Canal	46,000	Nathaniel Kinderley, John Reynolds
1734–1746	Scarborough Harbour pier	12,000	William Vincent
1735–1765	Whitby Harbour I	26,000	John Wooler
1738–1742	Deeping Fen Drainage III	15,000	Humphry Smith, John Grundy Sr.
1738–1750	Westminster Bridge	198,000	Charles Labelye
1739–1753	Salthouse Dock, Liverpool	21,000	Thomas Steers
1740–1744	Bridport Harbour		John Reynolds
1746–1748	Grimsthorpe Great Water	1,000	John Grundy
1746–1750	Denver Sluice III	5,000	William Cole, John Leaford
1750–1754	Ramsgate Harbour I	45,000	Thomas Preston, William Etheridge
1753–1755	Essex Bridge, Dublin	21,000	George Semple
1753–1757	Naburn Lock and Weir	5,500	John Smith
1755–1757	Petworth Lake	1,200	Lancelot Brown
1755–1761	Sankey Canal	19,000	Henry Berry
1755–1762	Sunderland Harbour improvement	9,000	Joseph Robson
1756	Pontypridd Bridge	1,200	William Edwards
1756–1759	Eddystone Lighthouse III	16,000	John Smeaton
1756–1763	London Bridge improvement	66,000	Robert Taylor, George Dance
1756–1772	Grand Canal, Ireland I	74,000	Thomas Omer, John Trail
1758–1761	Nene Navigation	16,000	Thomas Yeoman, John Smith Jr.
1759–1763	Bridgewater Canal I	7,500	James Brindley
1760–1769	Blackfriars Bridge	166,000	Robert Mylne
1760–1770	Calder and Hebble Navigation	75,000	John Smeaton
1761–1772	Ramsgate Harbour II	100,000	Thomas Preston
1762–1766	Colstream Bridge	6,000	John Smeaton
1762–1773	Bridgewater Canal II	70,000	James Brindley
1763–1767	Witham Drainage	48,000	Langley Edwards
1764–1769	Stockton Bridge	8,000	Joseph Robson
1765–1766	Ferrybridge Bridge I	4,000	John Gott
1765–1769	Black Sluice Drainage	34,000	Langley Edwards
1765–1770	Blenheim Great Lake		Lancelot Brown
1765–1770	Louth Navigation	27,000	John Grundy
1765–1770	Holderness Drainage I	24,000	John Grundy

Date	Work	Cost (£)	Engineer
1765–1775	North Bridge, Edinburgh	16,000	William Mylne
1766–1769	Stort Navigation	18,000	Thomas Yeoman
1766–1771	Perth Bridge II	23,000	John Smeaton
1766–1777	Trent and Mersey Canal	300,000	James Brindley, Hugh Henshall
1766–1777	Potteric Carr Drainage	9,000	John Smeaton, Thomas Tofield
1767–1768	London Bridge waterworks III		John Smeaton
1767–1769	Adlingfleet Drainage	7,000	John Smeaton, John Grundy
1767–1770	Driffield Navigation	13,000	John Grundy
1767–1770	St. Ives Harbour pier	10,000	John Smeaton
1767–1771	Georges Dock, Liverpool	21,000	Henry Berry
1767–1771	Lea Navigation	25,000	John Smeaton, Thomas Yeoman
1767–1773	Ripon and Ure Navigation	16,000	John Smeaton
1768–1771	Linton Lock and Dam	8,000	John Smith Jr.
1768–1771	Coventry Canal I	50,000	James Brindley, Edmund Lingard
1768–1771	Droitwich Canal	23,000	James Brindley
1768–1772	Birmingham Canal I	112,000	James Brindley
1768–1772	Broomielaw Bridge	8,000	William Mylne
1768–1776	Atcham Bridge	9,000	John Gwynn
1768–1777	Forth and Clyde Canal I	164,000	John Smeaton, Robert Mackell
1769–1772	Laneham Drainage	15,000	John Grundy
1769–1773	Rye New Harbour II	19,000	John Smeaton
1769–1774	English Bridge, Shrewsbury	14,000	John Gwynn
1769–1778	Oxford Canal I	200,000	James Brindley, Samuel Simcock
1770–1773	Monkland Canal I	10,000	James Watt
1770–1775	Clyde improvement	13,000	John Golborne
1770–1777	Leeds and Liverpool Canal I	230,000	John Longbothom
1770–1777	Chesterfield Canal	150,000	James Brindley, Hugh Henshall
1770–1781	Severn Bridge, Worcester		John Gwynn
1771–1772	Battersea Bridge (timber)	16,000	Henry Holland, John Phillips
1771–1776	Thames Navigation I	26,000	Joseph Nickalls
1772–1766	Maidenhead Bridge	15,000	Robert Taylor
1772–1779	Chester Canal	71,000	Samuel Weston, Thomas Morris, Josiah Clowes
1772–1779	Banff Bridge	9,000	John Smeaton
1772–1778	Weighton Drainage	40,000	John Grundy, Samuel Allam
1773–1776	Kinderley's Cut	9,000	Richard Dunthorne, William Elstobb
1773–1791	Grand Canal, Ireland II	375,000	John Trail, Charles Tarrant, Richard Evans
1774–1777	Richmond Bridge	15,000	James Paine
1774–1780	Thames Navigation (City) I	23,000	Robert Whitworth
1774–1781	Tyne Bridge Newcastle	35,000	Robert Mylne, John Wooler
1775–1778	Hull Dock	73,000	John Grundy, Henry Berry
1775–1778	Selby Canal	20,000	William Jessop
1775–1779	Aire and Calder Navigation II	30,000	John Smeaton, William Jessop
1775–1779	Stroudwater Canal	31,000	Thomas Yeoman
1775–1780	Aberdeen Harbour I	16,000	John Smeaton
1775–1781	Peterhead Harbour I	6,000	John Gwyn
1776–1779	Stourbridge Canal	31,000	Thomas Dadford
1776–1780	Hatfield Chase Drainage II	26,000	John Smeaton, Samuel Foster
1776–1781	Ramsgate Harbour, Basin		John Smeaton, Thomas Preston
1777–1779	Erewash Canal	21,000	John Varley
1777–1781	Iron Bridge, Coalbrookdale	6,000	Abraham Darby
1780–1785	Chertsey Bridge	15,000	James Paine
1781–1714	Whitby Harbour II	55,000	Jonathan Pickernell
1783–1786	Albion Mill	12,000	Samuel Wyatt
1783–1789	Thames and Severn Canal	220,000	Josiah Clowes
1784–1788	Skerton Bridge, Lancaster	17,000	Thomas Harrison
1784–1789	Kew Bridge	16,000	James Paine
1785–1788	Kings Dock, Liverpool	25,000	Henry Berry
1785–1790	Coventry Canal II	40,000	Thomas Sheasby, Thomas Dadford
1785–1791	Forth and Clyde Canal II	140,000	Robert Whitworth
1785–1792	Dudley Canal tunnel	50,000	Thomas Dadford, Josiah Clowes
1785–1793	Witham Fens Drainage	14,000	John Hudson

Date	Work	Cost (£)	Engineer
1786–1798	Oxford Canal II	56,000	James Barnes
1786–1792	Thames Navigation II	29,000	John Clarke, John Treacher
1787–1789	Kildare Canal	10,000	William Chapman
1787–1705	Holderness Drainage II	16,000	William Jessop, Anthony Bower
1788–1790	Birmingham Canal II	25,000	Samuel Bull
1788–1792	Ramsgate Harbour IV	52,000	John Smeaton
1788–1793	Basingstoke Canal	150,000	William Jessop, Henry Eastburn
1789–1790	Brunswick Dock, Blackwall	25,000	John Powsey
1789–1791	Foyle Bridge, Londonderry	17,000	Lemuel Cox
1789–1794	Cromford Canal	79,000	William Jessop
1789–1800	Grand Canal, Ireland III	174,000	William Jessop, William Chapman
1789–1797	Royal Canal, Ireland	1,400,000	Richard Evans, John Rennie
1790–1793	Circular Line Canal, Dublin	52,000	William Jessop
1790–1794	Glamorganshire Canal	103,000	Thomas Dadford, Thomas Sheasby
1790–1795	Barrow Navigation	58,000	William Chapman
1790–1801	Leeds and Liverpool Canal II	330,000	Robert Whitworth, Samuel Fletcher
1791–	Neath Canal	40,000	Thomas Dadford, Thomas Sheasby
1791–1796	Leominster Canal	93,000	Thomas Dadford Jr.
1791–1797	Leicester Navigation	80,000	William Jessop, Christopher Staveley
1791–1815	Worcester and Birmingham Canal	610,000	Thomas Cartwright, John Woodhouse, William Crosley Jr.
1792–1795	Queens Dock, Liverpool	35,000	Thomas Morris Jr.
1792–1796	Ringsend Docks, Dublin	113,000	William Jessop
1792–1799	Monmouthshire Canal and Railway	220,000	Thomas Dadford Jr.
1793–1795	Belper West Mill	7,000	William Strutt
1793–1796	Sunderland Bridge (iron)	19,000	Rowland Burdon, Thomas Wilson
1793–1796	Derby Canal and Railway	100,000	Benjamin Outram
1793–1797	Chelmer Navigation	50,000	John Rennie
1793–1797	Shrewsbury Canal	70,000	Josiah Clowes, Thomas Telford
1793–1798	Hereford and Gloucester Canal I	106,000	Josiah Clowes, Robert Whitworth
1793–1798	Dudley Canal II	140,000	Josiah Clowes, William Underhill
1793–1802	Barnsley Canal	95,000	William Jessop
1793–1802	Stainforth and Keadby Canal	57,000	John Thompson, David Servant
1793–1803	Grantham Canal	118,000	William Jessop
1793–1803	Lancaster Canal I (including Lune Aqueduct)	490,000 (48,000)	John Rennie
1793–1804	Dearne and Dove Canal	100,000	John Thompson, Robert Whitworth Jr.
1793–1805	Grand Junction Canal	1,500,000	William Jessop, James Barnes
1794–1798	Swansea Canal	55,000	Charles Reynolds, Thomas Sheasby
1794–1799	Gloucester and Berkeley Canal I	112,000	Robert Mylne
1794–1801	Newport Bridge	16,000	DavidEdwards, Thomas Edwards
1794–1802	Rochdale Canal	580,000	William Jessop, William Crosley Jr.
1794–1804	Ashby Canal and Railway	166,000	Robert Whitworth, Benjamin Outram
1794–1804	Peak Forest Canal	177,000	Benjamin Outram, Thomas Brown
1794–1805	Ellesmere Canal (including Pontcysyllte)	460,000 (47,000)	William Jessop, Thomas Telford
1794–1805	Somerset Coal Canal	180,000	John Sutcliffe, William Bennet
1794–1809	Crinan Canal	165,000	John Rennie
1794–1810	Kennet and Avon Canal	860,000	John Rennie
1794–1811	Huddersfield Canal	400,000	Benjamin Outram, John Routh
1795–1796	Buildwas Bridge (iron)	6,500	Thomas Telford
1795–1810	Wiltshire and Berkshire Canal	255,000	Robert Whitworth, William Whitworth
1796–1797	Shrewsbury Mill		Charles Bage
1797–1801	Grimsby Dock	60,000	John Rennie
1797–1801	Isle of Axholme Drainage	20,000	William Jessop
1797–1803	Ferrybridge Bridge II	30,000	John Carr
1797–1805	Aberdeenshire Canal	44,000	John Rennie
1797–1812	Brecknock and Abergaveny Canal	100,000	Thomas Dadford Jr., Thomas Cartwright, William Crosley Jr.
1798–1799	Bewdley Bridge	9,000	Thomas Telford
1798–1800	Rudyard Dam	8,000	John Rennie
1799–1802	Penydarren Railway		George Overton

Date	Work	Cost (£)	Engineer
1799–1810	Beverley and Barmston Drainage	115,000	William Chapman
1800–1804	Kelso Bridge	13,000	John Rennie
1800–1806	West India Docks, London	515,000	William Jessop
1800–1808	Muston and Yedingham Drainage	42,000	William Chapman
1801–1803	Surrey Iron Railway	27,000	William Jessop
1801–1804	Grand Canal, Ireland IV	135,000	John Killaly
1801–1810	Axe and Brue Drainage	58,000	William Jessop, Robert Anstice
1801–1806	Leith East Dock	105,000	John Rennie
1801–1831	Scarborough Harbour	65,000	William Chapman
1802–1805	London Docks	350,000	John Rennie
1802–1805	Union Bridge, Aberdeen	13,000	Thomas Fletcher
1802–1805	Sirhowy Railway	64,000	John Hodgkinson
1802–1810	East, West and Wildmore Fens Drainage	430,000	John Rennie
1803–1805	Croydon and Merstham Railway	44,000	William Jessop
1803–1806	East India Docks, London	250,000	John Rennie, Ralph Walker
1803–1807	Boston Bridge (iron)	22,000	John Rennie, Thomas Wilson
1803–1809	Humber Dock, Hull	230,000	John Rennie, William Chapman
1803–1821	Highland Roads and Bridges	430,000	Thomas Telford
1803–1823	Caledonian Canal	855,000	Thomas Telford, William Jessop
1804	Culford iron bridge		Samuel Wyatt
1804–1810	Bristol Floating Harbour	470,000	William Jessop
1804–1817	Sunderland Harbour improvements		Matthew Shout
1806–1811	Glasgow and Paisley Canal	130,000	Thomas Telford
1805–1808	Castleford Bridge	14,000	Bernard Hartley
1806–1824	Highland Harbours	100,000	Thomas Telford
1807–1811	Bell Rock Lighthouse	61,000	John Rennie, Robert Stevenson
1808–1811	Kilmarnock and Troon Railway	44,000	William Jessop
1808–1811	Traeth Mawr Embankment	60,000	John Williams
1809–1810	Newport Pagnell Bridge (iron)		Thomas Wilson, Henry Provis
1809–1818	Howth Harbour	350,000	John Rennie
1810–1813	Commercial Docks, London	130,000	James Walker
1810–1814	Grand Union Canal	290,000	Benjamin Bevan
1810–1815	Thames Navigation (City) II	90,000	Stephen Leach
1810–1815	Aberdeen Harbour II	100,000	Thomas Telford
1810–1817	Leith West Dock	70,000	John Rennie
1810–1828	Holyhead Harbour	123,000	John Rennie, Thomas Telford
1811–1817	Waterloo Bridge	480,000	John Rennie
1812–1816	Eden Bridge, Carlisle	70,000	Robert Smirke
1812–1816	Vauxhall Bridge, London	86,000	James Walker
1812–1818	Hay Railway	51,000	John Hodgkinson
1812–1820	Regent's Canal	700,000	James Morgan
1812–1850	Plymouth Breakwater	1,500,000	John Rennie, Joseph Whidbey, William Stuart
1813–1816	Richmond Bridge, Dublin	26,000	James Savage
1813–1816	Wey and Arun Canal	103,000	Josias Jessop
1813–1825	Lancaster Canal II	110,000	Thomas Fletcher, William Crosley Jr.
1813–1830	Sheerness Dockyard	1,600,000	John Rennie, George Rennie, John Rennie Jr.
1814–1819	Southwark Bridge	660,000	John Rennie
1815–1816	Chepstow Bridge	22,000	John U. Rastrick
1815–1821	Princes Dock, Liverpool	530,000	John Foster
1815–1825	Dundee Harbour	120,000	Thomas Telford
1815–1830	Holyhead Road, Welsh section (including Menai Bridge)	360,000 (180,000)	Thomas Telford
1816–1819	Sheffield Canal	76,000	William Chapman
1817–1819	Mansfield and Pinxton Railway	32,000	Josias Jessop
1817–1821	Shoreham Harbour	58,000	William Chapman
1817–1821	Glasgow–Carlisle Road	50,000	Thomas Telford
1817–1822	Union Canal	460,000	Hugh Baird
1817–1836	Kingstown Harbour	700,000	John Rennie, John Rennie Jr.
1818–1820	Union Bridge	8,000	Samuel Brown

Date	Work	Cost (£)	Engineer
1818–1823	Eau Brink Cut	250,000	John Rennie
1818–1827	Gloucester and Berkeley Canal II	330,000	Thomas Telford
1819–1822	Edinburgh Waterworks	145,000	James Jardine
1819–1823	Bude Canal	118,000	James Green
1819–1824	Thames and Medway Canal tunnel		Tierney Clark
1819–1830	Holyhead Road, English section	130,000	Thomas Telford
1820–1823	Carlisle Canal	80,000	William Chapman
1821–1825	Brecon Forest Railway	40,000	Joseph Jones
1821–1836	Donaghadee Harbour	145,000	John Rennie Jr.
1821–1837	Portpatrick Harbour	150,000	John Rennie Jr.
1822–1823	Brighton Chain Pier	30,000	Samuel Brown
1822–1825	Stockton and Darlington Railway	140,000	George Stephenson
1822–1826	Goole Canal and Docks	300,000	George Leather
1822–1829	Chester–Bangor Road (including Conway Bridge)	68,000 (51,000)	Thomas Telford
1822–1830	Roads in western Ireland	167,000	Alexander Nimmo
1824–1827	Mythe Bridge, Tewkesbury	35,000	Thomas Telford
1824–1827	Laira Bridge, Plymouth	27,000	James M. Rendel
1824–1827	Harecastle Tunnel	112,000	Thomas Telford
1824–1827	Hammersmith Bridge	50,000	Tierney Clark
1824–1831	London Bridge	425,000	George Rennie, John Rennie Jr.
1824–1835	Wellesley Bridge and Docks, Limerick	89,000	Alexander Nimmo, Charles B. Vignoles
1825–1827	Knypersley Dam	16,000	Thomas Telford
1825–1827	Serpentine Bridge	44,000	George and John Rennie Jr.
1825–1828	Kingston Bridge	35,000	Edward Lapidge
1825–1828	Eastern Dock, London Docks	300,000	David Alexander, H. R. Palmer
1825–1828	Thames Tunnel I	138,000	Marc Brunel
1825–1829	Greenock water supply	51,000	Robert Thom
1825–1830	Birmingham Canal III	240,000	Thomas Telford
1825–1830	Exeter Canal IV	110,000	James Green
1825–1830	Canterbury and Whitstable Railway	68,000	George Stephenson
1825–1831	Cromford and High Peak Railway	160,000	Josias Jessop, Thomas Woodhouse
1826–1829	Eau Brink Cut, widening	33,000	Thomas Telford, John Rennie Jr.
1826–1828	Over Bridge, Gloucester	43,000	Thomas Telford
1826–1830	St. Katharine's Docks, London	250,000	Thomas Telford
1826–1830	Liverpool and Manchester Railway	600,000	George Stephenson
1826–1831	Macclesfield Canal	320,000	William Crosley Jr.
1826–1831	Woolwich Dockyard	340,000	John Rennie Jr.
1827–1831	Leith Harbour	50,000	William Chapman
1827–1835	Birmingham and Liverpool Junction Canal	800,000	Thomas Telford, William Cubitt
1827–1829	Junction Dock, Hull	95,000	James Walker
1827–1830	Nene Outfall channel	200,000	Thomas Telford, John Rennie Jr.
1827–1833	Norwich and Lowestoft Navigation	150,000	William Cubitt
1827–1833	Grosvenor Bridge, Chester	50,000	Thomas Harrison, George Rennie
1828–1834	Oxford Canal III	170,000	Charles B. Vignoles
1828–1835	Seaham Harbour	165,000	William Chapman, John Buddle
1829–1831	Dean Bridge, Edinburgh	19,000	Thomas Telford
1829–1832	Hutcheson Bridge, Glasgow	23,000	Robert Stevenson
1829–1832	High Bridge, Staffordshire (iron)	11,000	Joseph Potter
1829–1832	Aberdeen Harbour III	60,000	Thomas Telford
1829–1832	Marlow Suspension Bridge	22,000	Tierney Clark
1829–1834	Staines Bridge	30,000	George Rennie
1830–1831	Westminster Bridewell, raft foundations	30,000	Robert Abraham
1830–1834	North Level Drainage	150,000	Thomas Telford

Index of places